Valdosta Technical Institute Library
PO Box 928/4089 Val Tech Road
Valdosta, Georgia 31603-0928

MW00974797

to accompany

Technical Mathematics

and

Technical Mathematics

WITH CALCULUS

John C. Peterson
with
Alan Herweyer

Chattanooga State Technical College

Valdosta Technical Institute Library
PO Box 928/4089 Val Tech Road
Valdosta, Georgia 31603-0928

Delmar
Publishers Inc.™

NOTICE TO THE READER

Publisher does not warrant or guarantee any of the products described herein or perform any independent analysis in connection with any of the product information contained herein. Publisher does not assume, and expressly disclaims, any obligation to obtain and include information other than that provided to it by the manufacturer.

The reader is expressly warned to consider and adopt all safety precautions that might be indicated by the activities described herein and to avoid all potential hazards. By following the instructions contained herein, the reader willingly assumes all risks in connection with such instructions.

The publisher makes no representations or warranties of any kind, including but not limited to, the warranties of fitness for particular purpose or merchantability, nor are any such representations implied with respect to the material set forth herein, and the publisher takes no responsibility with respect to such material. The publisher shall not be liable for any special, consequential, or exemplary damages resulting, in whole or in part, from the readers' use of, or reliance upon, this material.

For information, address Delmar Publishers Inc.
3 Columbia Circle, Box 15-015
Albany, New York 12212

Copyright © 1994 by Delmar Publishers Inc.

The trademark ITP is used under license

All rights reserved. No part of this work covered by the copyright hereon may be reproduced or used in any form or by any means—graphic, electronic, or mechanical, including photocopying, recording, taping, or information storage and retrieval systems—without written permission of the publisher.

10 9 8 7 6 5 4 3 2 1

Printed in the United States of America
Published simultaneously in Canada by Nelson Canada
a Division of the Thomson Corporation

Library of Congress Cataloging in Publication Data
Peterson, John C. (John Charles), 1939–
Technical mathematics with calculus/John C. Peterson.
p. cm.
ISBN: 0-8273-4582-8
1. Mathematics. I. Title.
QA37.2.P49 1994 92–32757
510—dc20 CIP

Contents

TOPIC	WASHINGTON	CALTER	DAVIS	LYNG	NINESTEIN	PETERSON
The Real Number System	**Chapter 1, & Appendix B pp. 1–27**	**Chapter 1 pp. 1–43**	**Appendix A**	**Chapter 1 pp. 1–31**	**Chapter 1 pp. 1–40**	**Chapter 1 pp. 1–50**
Sets of Numbers	Section 1.1 pp. 2–6	Section 1.1 pp. 2–6	Appendix A-1	Section 1.1 pp. 2–10	Section 1.1 pp. 1–8	Section 1.1 pp. 2–6
Basic Operations On and Laws of Real Numbers	Section 1.2 Section 1.3 pp. 6–12	Section 1.2, 1.3, 1.4 pp. 7–17	Appendix A-2, A-3	Section 1.2 pp. 10–14	Section 1.2, 1.4 pp. 9–16, 24–28	Section 1.2, 1.3 pp. 7–23
Laws of Exponents	Section 1.5 pp. 14–20	Section 1.5 pp. 18–22 Section 2.3 pp. 52–56	Section 4.1 pp 128–133 Appendix A-4	Section 1.3 pp. 14–23	Section 2.4 pp. 60–67	Section 1.4 pp. 23–28
Significant Digits and Rounding off	Section 1.4 pp. 12–14 Appendix B	Section 1.1, 1.2, 1.3, 1.4, 1.5 pp. 2–22		Section 1.1 pp. 2–10	Section 1.3 pp. 16–24	Section 1.5 pp. 29–35
Scientific Notation and Roots	Section 1.6 Section 1.7 pp. 20–27	Section 1.5 pp. 18–22 Section 1.8 pp. 27–34	Appendix A-5, A-6	Section 1.4 p. 19 and pp. 23–31	Section 2.3 pp. 54–59	Section 1.6, 1.7
Algebraic Concepts and Operations	**Chapter 1 pp. 27–49**	**Chapter 2 & 3, pp. 44–75 pp. 76–97**	**Appendix A**	**Chapter 1 pp. 31–48**	**Chapter 2 pp. 41–94**	**Chapter 2 pp. 51–87**
Addition and Subtraction	Section 1.8 pp. 27–31	Section 2.1, 2.2 pp. 45–52	Appendix A-7	Section 1.6 pp. 33–36	Section 2.1 pp. 41–46	Section 2.1 pp. 52–56
Multiplication	Section 1.9 pp. 31–34	Section 2.4 pp. 58–66	Appendix A-8	Section 1.1 pp. 36–39	Section 2.2 pp. 47–54	Section 2.2 pp. 56–61
Division	Section 1.10 pp. 34–37	Section 2.5 pp. 67–74	Appendix A-9	Section 1.8 pp. 39–42	Section 2.2 pp. 47–54	Section 2.3 pp. 62–68
Solving Equations	Section 1.11, 1.12; pp. 38–44	Section 3.1 pp. 77–82	Section 1.1 pp. 2–8	Section 1.9 pp. 43–48	Section 2.5 pp. 67–74	Section 2.4 pp. 69–75
Applications of Equations	Section 1.13 pp. 44–49	Section 3.2, 3.3, 3.4, 3.5 pp. 82–96	Section 1.5 pp. 28–36		Section 2.7 pp. 81–88	Section 2.5 pp. 76–85

TOPIC	WASHINGTON	CALTER	DAVIS	LYNG	NINESTEIN	PETERSON
Geometry	**Appendix C**	**Chapter 5** **pp. 129–153**	**Appendix B**			**Chapter 3** **pp. 88–131**
Lines and Angles	Appendix C pp. A-18	Section 5.1 pp. 130–134	Appendix B			Section 3.1 pp. 89–97
Triangles	Appendix C pp. A-19	Section 5.2 pp. 134–143	Appendix B			Section 3.2 pp. 97–104
Other Polygons	Appendix C pp. A-20	Section 5.3 pp. 144–145	Appendix B			Section 3.3 pp. 104–110
Circles	Appendix C pp. A-20	Section 5.4 pp. 145–149	Appendix B			Section 3.4 pp. 110–116
Geometric Solids	Appendix C pp. A-21	Section 5.5 pp. 149–152	Appendix B			Section 3.5 pp. 117–126
Functions and Graphs	**Chapter 2 &** **Appendix E** **pp. 54–83**	**Chapter 4** **pp. 98–128**	**Chapter 2** **pp. 41–83**	**Chapter 3** **pp. 111–138**	**Chapter 3** **pp. 95–136**	**Chapter 4** **pp. 132–174**
Relations and Functions	Section 2.1 Section 2.2 pp. 54–64	Section 4.1, 4.2 pp. 99–114	Section 2.1 pp. 42–46	Section 3.5 pp. 130–135	Section 3.1 pp. 95–101	Section 4.1 pp. 133–143
Operations on Functions; Composite Functions		Section 4.2 pp. 105–114			Section 3.4 pp. 116–125	Section 4.2 pp. 143–148
Rectangular Coordinates	Section 2.3 pp. 64–6	Section 4.3 pp. 114–116	Section 2.2 pp. 47–50	Section 3.1 pp. 112–116	Section 3.2 pp. 101–110	Section 4.3 pp. 148–152
Graphs	Section 2.4 Section 2.5 pp. 66–77	Section 4.4 pp. 117–125 Section 13.5 pp. 357–363	Section 2.3 PP. 51–57	Section 3.2,3.3, 3.4 pp. 116–130	Section 3.3 pp. 110–116	Section 4.4 pp. 152–159
Graphing Calculators and Computer-aided Graphing	Appendix E pp. A-33 – A-40	Section 4.5 pp. 125–127			Section 1.5 pp. 28–34	Section 4.5 pp. 159–163
Using Graphs to Solve equations	Section 2.6 pp. 77–80	Section 4.5 pp. 125–127	Section 6.1 pp. 202–207	Section 4.1,4.2 pp. 140–145	Section 17.4 pp. 716–723	Section 4.6 pp. 163–172

TOPIC	WASHINGTON	CALTER	DAVIS	LYNG	NINESTEIN	PETERSON
Systems of Linear Equations and Determinants	**Chapter 4** **pp. 109–148**	**Chapter 9, 10** **pp. 240–265** **pp. 266–287**	**Chapter 6** **pp. 201–260**	**Chapter 4** **pp. 139–172**	**Chapter 17 & 18** **pp. 689–700** **pp. 729–764**	**Chapter 5** **pp. 175–215**
Linear Equations	Section 4.1 pp. 109–112	Section 9.1, 9.2, pp. 241–254	Section 6.1 pp. 202–207	Section 16.1, 16.2 p. 618–627	Section 9.1 pp. 338–345	Section 5.1 pp. 175–180
Graphical and Algebraic Methods for Solving Two Linear Equations in Two Variables	Section 4.3, 4.4, 4.5 pp. 116–126	Section 9.1, 9.2, 9.4 pp. 241–254 pp. 259–264	Section 6.1,6.2 pp. 201–217	Section 4.3,4.4 pp. 145–155 Section 11.1, 11.2 pp. 442–452	Section 17.1, 17.2 pp. 689–707	Section 5.2 pp. 180–188
Algebraic Methods for Solving Three Linear Equations in Three Variables	Section 4.6 pp. 133–138	Section 9.1, 9.2, 9.4 pp. 241–254 pp. 259–264	Section 6.3 pp. 217–225	Section 4.6 pp. 171–162	Section 17.3 pp. 707–715	Section 5.3 pp. 188–195
Determinants	Section 4.5 pp. 127–133 Section 4.7 pp. 138–145 Section 15.1 15.2,15.3 pp. 423–435	Section 10.1, 10.2, 10.3 pp. 267–287	Section 6.5 pp. 232–237	Section 4.5 pp. 156–161 Section 13.5, 13.6 pp. 538–548	Section 18.1 pp. 729–738	Section 5.4 pp. 195–205
Using Cramer's Rule to Solve Systems of Linear Equations	Section 4.7 pp. 138–145	Section 9.2 pp. 249–254 Section 10.2 pp. 249–254	Section 6.6 pp. 237–244	Section 4.5 pp. 158–160	Section 18.2 pp. 737–744	Section 5.5 pp. 205–212
Ratio, Proportion, and Variation	**Chapter 17** **pp. 486–500**	**Chapter 18** **pp. 475–503**	**Chapter 1** **pp. 1–39**	**Chapter 1** **pp. 48–68**		**Chapter 6** **pp. 216–237**
Ratio and Proportion	Section 17.1 pp. 486–490	Section 18.1 pp. 476–480	Section 1.5 pp. 28–36	Section 1.10 pp. 48–55	Section 2.6 pp. 74–81	Section 6.1 pp. 216–222
Similar Geometric Shapes	Section 3.2 pp. 87–92	Section 18.3 pp. 484–492	Section 13.2 pp. 454–461	Section 6.1 pp. 207–210	Section 2.6 pp. 74–81	Section 6.2 pp. 222–226
Direct and Inverse Variation	Section 17.2 pp. 490–497	Section 18.2, 18.4 pp. 480–484 pp. 492–496	Section 1.4 pp. 22–28	Section 1.11 pp. 56–61	Section 3.5 pp. 125–129	Section 6.3 pp. 226–231

TOPIC	WASHINGTON	CALTER	DAVIS	LYNG	NINESTEIN	PETERSON
Joint and Combined Variation	Section 17.2 pp. 490–497	Section 18.5 pp. 497–501	Section 1.4 pp. 22–28	Section 1.11 pp. 56–61	Section 3.5 pp. 125–129	Section 6.4 p.231–235
Factoring and Algebraic Fractions	**Chapter 5 pp. 149–184**	**Chapter 7 & 8 pp. 182–203 pp. 204–239**	**Chapter 3 pp. 86–126**	**Chapter 2 pp. 69 –110**	**Chapter 5 pp. 191–235**	**Chapter 7 pp. 238–274**
Special Products	Section 5.1 pp. 149–153	Section 7.6 pp. 199–202		Section 2.1 pp. 72–74	Section 5.2 pp. 198–203	Section 7.1 pp. 238–245
Factoring	Section 5.2 pp. 153–157	Section 7.1 pp. 183–189	Section 3.1 pp. 86–91	Section 2.2 pp. 74–84	Section 5.2 pp. 198–203	Section 7.2 pp. 245–249
Factoring Trinomials	Section 5.3 pp. 157–163	Section 7.2, 7.3, 7.4, 7.5 pp. 189–199	Section 3.2 pp. 92–97	Section 2.2 pp. 74–85	Section 5.1,5.3 pp. 191–198 pp. 203–209	*Section 7.3 pp. 249–256*
Fractions	Section 5.4 pp. 163–167	Section 8.1 pp. 205–210	Section 3.3 pp. 97–101	Section 2.3 pp. 85–91	Section 5.4 pp. 209–216	Section 7.4 pp. 256–259
Multiplication and Division of Fractions	Section 5.5 pp. 167–171	Section 8.2, 8.4 pp. 210–214 pp. 221–224	Section 3.4 pp. 102–108	Section 2.4 pp. 91–94	Section 5.4 pp. 209–216	Section 7.5 pp. 260–264
Addition and Subtraction of Fractions	Section 5.6 pp. 171–177	Section 8.3 pp. 215–221	Section 3.5 pp. 108–115	Section 2.5 pp. 95–101	Section 5.5 pp. 217–235	Section 7.6 pp. 264–273
Fractional and Quadratic Equations	**Chapter 6 pp. 185–204**	**Chapter 8 & 13 pp. 205–239 pp. 339–370**	**Chapter 5 pp. 161–200**	**Chapter 2 pp. 101–110 Chapter 5 pp. 174–200**	**Chapter 5 & 6 pp. 191–235 pp. 236–268**	**Chapter 8 pp. 275–301**
Fractional Equations	Section 5.7 pp. 177–181	Section 8.5, 8.6, 8.7; pp. 224–238	Section 3.6 pp. 115–121	Section 2.6 pp. 101–105	Section 5.6 pp. 223–230	Section 8.1 pp. 275–280
Quadratic Equations and Factoring	Section 6.1 pp. 185–190	Section 13.1 pp. 340–347	Section 5.1 pp. 162–168	Section 5.2 pp. 177–181	Section 5.1, 6.3 pp. 191–198 pp. 247–253	Section 8.2 pp. 281–288
Completing the Square	Section 6.2 pp. 190–193	Section 13.2 pp. 347–348	Section 5.2 pp. 168–173	Section 5.1, 5.3 pp. 174–170 pp. 181–186	Section 6.1 pp. 236–240	Section 8.3 pp. 289–293
The Quadratic Formula	Section 6.3 pp. 193–197	Section 13.3, 13.4 pp. 348–352	Section 5.3 pp. 174–182	Section 5.4 pp. 186–191	Section 6.2 pp. 240–247	Section 8.4 pp. 293–299

TOPIC	WASHINGTON	CALTER	DAVIS	LYNG	NINESTEIN	PETERSON
Trigonometric Functions	**Chapter 3 pp. 84–108 Chapter 7 pp. 205–227**	**Chapter 6 pp. 154–181**	**Chapter 9 & 10 pp. 318–348 pp. 349–378**	**Chapter 6 & 7 pp. 201–252 pp. 253–328**	**Chapter 4 pp. 137–190**	**Chapter 9 pp. 302–340**
Angles, Angle Measure and Trigonometric Functions (Radian Measure)	Section 3.1 pp. 84–92 Section 7.3 pp. 215–220	Section 6.1, 6.2 pp. 155–166	Section 9.1 9.2 pp. 320–329 Section 10.3, 10.4 pp. 363–373	Section 6.1, 6.2 pp. 202–220 Section 7.1 pp. 254–262	Section 4.1, 4.6 pp. 137–144 pp. 171–176	Section 9.1 pp. 302–308
Values of the Trigonometric Functions	Section 3.3 pp. 92–96	Section 6.2 pp. 158–166	Section 9.3 pp. 329–336	Section 6.3 pp. 220–229	Section 4.2 pp. 145–152	Section 9.2 pp. 308–315
The Right Triangle	Section 3.4,3.5 pp. 97–104	Section 6.3 pp. 166–171	Section 9.4 pp. 336–344	Section 6.3 pp. 220–229	Section 4.1,4.2 pp. 137–152	Section 9.3 pp. 315–322
Trigonometric Functions of Any Angle	Section 7.1, 7.2 pp. 205–215	Section 14.1 pp. 372–380	Section 10.1, 10.2 pp. 350–362	Section 7.2 pp. 262–273	Section 4.4,4.5 pp. 160–171	Section 9.4 pp. 322–327
Inverse Trigonometric Functions	Section 7.2 pp. 208–215 Section 19.6 pp. 547–554	Section 6.2 & 17.6 pp. 158–166 pp. 741–473	Section 13.5 pp. 473–481	Section 7.7 pp. 308–318	Section 11.6 pp. 473–481	Section 9.5 pp. 327–331
Applications of Trigonometry	Section 7.4 pp. 221–226	Section 6.4 pp. 171–174	Section 9.4 pp. 336–344		Section 4.3 pp. 152–160 Section 4.6,4.7 pp. 171–181	Section 9.6 pp. 331–335
Vectors and Trigonometric Functions	**Chapter 8 pp. 230–260**	**Chapter 6 & 14 pp. 154–181 pp. 371–397**	**Chapter 11 pp. 379–412**	**Chapter 6 pp. 229–246 Chapter 9 pp. 377–412**	**Chapter 10 pp. 395–439**	**Chapter 10 pp. 341–375**
Introduction to Vectors	Section 8.1, 8.2 pp. 230–238	Section 6.5 pp. 174–177	Section 11.1 pp. 380–386	Section 9.1 pp. 378–387	Section 10.1 pp. 395–402	Section 10.1 pp. 341–350
Adding and Subtracting Vectors	Section 8.3 pp. 238–242	Section 14.5 pp. 392–396	Section 11.1 pp. 380–386	Section 9.1 pp. 378–387	Section 10.2 pp. 402–411	Section 10.2 pp. 350–358
Applications of Vectors	Section 8.4 pp. 242–247	Section 6.6 pp. 177–180	Section 11.2 pp. 387–393	Section 9.1 pp. 378–387	Section 10.2 pp. 402–411	Section 10.3 pp. 358–362

TOPIC	WASHINGTON	CALTER	DAVIS	LYNG	NINESTEIN	PETERSON
Oblique Triangles: Law of Sines	Section 8.5 pp. 247–253	Section 14.2, 14.4 pp. 380–384 pp. 388–392	Section 11.3 pp. 394–401	Section 6.4 pp. 229–238	Section 10.3,10.4 pp. 411–424	Section 10.4 pp. 362–369
Oblique Triangles: Law of Cosines	Section 8.6 pp. 253–258	Section 14.3, 14.4 pp. 384–392 **Chapter 16** **pp. 411–446**	Section 11.4 pp. 402–407	Section 6.5 pp. 238–246	Section 10.5 pp. 425–433	Section 10.5 pp. 369–372
Graphs of Trigonometric Functions	**Chapter 9** **pp. 261–287**	Section 16.1, 16.2 pp. 399–429	**Chapter 12** **pp. 413–446**	**Chapter 7** **pp. 273–328**	**Chapter 11** **pp. 440–486**	**Chapter 11** **pp. 376–412**
Sine and Cosine Curves: Amplitude and Period	Section 9.1,9.2 pp. 261–268	Section 16.1, 16.2 pp. 399–429	Section 12.1,12.2 pp. 414–422	Section 7.3 pp. 273–285	Section 11.1 pp. 440–448	Section 11.1 pp. 376–382
Sine and Cosine Curves: Displacement or Phase Shift	Section 9.3 pp. 268–272	Section 16.1, 16.2 pp. 399–429	Section 12.3 pp. 422–427	Section 7.4 pp. 286–294	Section 11.2 pp. 448–453	Section 11.2 pp. 382–385
Composite Sine and Cosine Curves	Section 9.6 pp. 280–285	Section 16.1, 16.2 pp. 399–429	Section 12.6 pp. 437–442	Section 7.6 pp. 301–308	Section 11.4 pp. 460–466	Section 11.3 pp. 385–391
Graphs of Other Trigonometric Functions	Section 9.4 pp. 272–276	Section 16.1, 16.2 pp. 399–429	Section 12.4 pp. 428–431	Section 7.5 pp. 294–301	Section 11.3 pp. 454–460	Section 11.4 pp. 391–395
Applications of Trigonometric Graphs	Section 9.5 pp. 276–280	Section 16.3 pp. 429–438	Section 12.5 pp. 432–437		Section 11.5 pp. 467–473	Section 11.5 pp. 359–399
Parametric Equations		Section 16.3 pp. 429–438	Section 12.6 pp. 437–442		Section 11.4 pp. 460–466	Section 11.6 pp. 400–404
Polar Coordinates	Section 20.9, 20.10 pp. 601–608	Section 16.4 pp. 438–443				Section 11.7 pp. 404–410
Exponents and Radicals	**Chapter 10** **pp. 288–314**	**Chapter 12** **pp. 317–338**	**Chapter 4** **pp. 127–160**		**Chapter 13** **pp. 529–567**	**Chapter 12** **pp. 413–437**
Fractional Exponents	Section 10.1, 10.2 pp. 288–298	Section 12.1, 12.2 pp. 318–329	Section 4.2 pp. 134–138	Section 1.5 pp. 31–39	Section 13.1 pp. 529–536	Section 12.1 pp. 413–418

TOPIC	WASHINGTON	CALTER	DAVIS	LYNG	NINESTEIN	PETERSON
Laws of Radicals	Section 10.3 pp. 298–302	Section 12.3 pp. 329–334	Section 4.3 pp. 138–144		Section 13.2,13.3 13.4; pp. 535–553	Section 12.2 pp. 418–423
Basic Operations with Radicals	Section 10.4,10.5, 10.6; pp. 302–311	Section 12.3 pp. 329–334	Section 4.4 pp. 145–150		Section 13.2,13.3 13.4; pp. 535–553	Section 12.3 pp. 424–430
Equations with Radicals	Section 13.4 pp. 395–398	Section 12.4 pp. 334–337	Section 4.5 pp. 150–155	Section 11.3 pp. 452–456	Section 13.5 pp. 554–560	**Section 12.4** **pp. 430–434**
Exponential & Logarithmic Functions	**Chapter 12** **pp. 349–381**	**Chapter 19** **pp. 504–548**	**Chapter 8** **pp. 289–318**	**Chapter 10** **pp. 413–440**	**Chapter 15** **pp. 609–654**	**Chapter 13** **pp. 439–470**
Exponential Functions/Graphs of Exponential Functions	Section 12.1, 12.2 pp. 349–356	Section 19.2 pp. 505–509	Section 8.1 pp. 290–296	Section 10.1 pp. 414–419	Section 15.1 pp. 609–617	Section 13.1 pp. 438–443
The Exponential Function e^x	Section 12.5 pp. 365–368	Section 19.2 pp. 509–515	Section 8.1 pp. 290–296	Section 10.4 pp. 428–431		Section 13.2 pp. 443–449
Logarithmic Functions/Graphs of Logarithmic Functions	Section 12.1, 12.2 pp. 349–356	Section 19.3 pp. 515–520	Section 8.2 pp. 297–302	Section 10.2 pp. 419–423	Section 15.2,15.4 15.5 pp. 617–622 pp. 627–636	Section 13.3 pp. 449–453
Properties of Logarithms	Section 12.3, 12.4 pp. 356–365	Section 19.4 pp. 521–529	Section 8.3 pp. 302–308	Section 10.3 pp. 423–428	Section 15.3 pp. 622–626	Section 13.4 pp. 453–457
Exponential and Logarithmic Equations	Section 12.6 pp. 368–373	Section 19.5, 19.6 pp. 529–540	Section 8.4 pp. 308–315	Section 10.5 pp. 431–436	Section 15.6 pp. 637–643	Section 13.5 pp. 457–462
Graphs using Semilogarithmic and Logarithmic Paper	Section 12.7 pp. 373–378	Section 19.7 pp. 540–547			Section 15.7 pp. 643–649	Section 13.6 pp. 463–467
Complex Numbers	**Chapter 11** **pp. 315–348**	**Chapter 20** **pp. 549–588**	**Chapter 14** **pp. 487–524**	**Chapter 9** **pp. 377–412**	**Chapter 14** **pp. 568–608**	**Chapter 14** **pp. 471–510**
Imaginary and Complex Numbers	Section 11.1 pp. 315–319	Section 20.1 pp. 550–556	Section 14.1 pp. 488–492	Section 9.2 pp. 388–395	Section 14.1 pp. 568–575	Section 14.1 pp. 471–477
Operations with Complex Numbers	Section 11.2 pp. 320–322	Section 20.1 pp. 550–556	Section 14.2 pp. 492–497	Section 9.2 pp. 388–395	Section 14.1 pp. 568–575	Section 14.2 pp. 477–486

TOPIC	WASHINGTON	CALTER	DAVIS	LYNG	NINESTEIN	PETERSON
Graphing Complex Numbers; Polar Form of a Complex Number	Section 11.3, 11.4 pp. 322–329	Section 16.5, 20.2, 20.3 pp. 443–444 pp. 556–563	Section 14.3 pp. 497–503	Section 9.3 pp. 395–403	Section 14.2, 14.3 pp. 575–588	Section 14.3 pp. 486–491
Exponential Form of a Complex Number	Section 11.5 pp. 329–332	Section 20.4 pp. 563–566	Section 14.3 pp. 497–503	Section 9.4 pp. 403–409	Section 14.3 pp. 582–588	Section 14.4 pp. 492–496
Operations in Polar Form: DeMoivre's Formula	Section 11.6 pp. 332–338	Section 20.2 pp. 556–563	Section 14.4 pp. 504–510	Section 9.4 pp. 403–409	Section 14.3,14.4 pp. 582–596	Section 14.5 pp. 497–503
Complex Numbers in AC Circuits	Section 11.7 pp. 338–345	Section 20.6 pp. 569–573	Section 14.5 pp. 510–519		Section 14.5 pp. 596–603	Section 14.6 pp. 503–508
An Introduction to Plane Analytic Geometry	**Chapter 20 pp. 558–612**	**Chapter 25 & 26 pp. 673–732**	**Chapter 15 pp. 525–574**	**Chapter 16 pp. 617–678**	**Chapter 9 pp. 337–394**	**Chapter 15 pp. 511–559**
Basic Definitions and Straight Lines	Section 20.1, 20.2 pp. 559–571	Section 25.1, 25.2, 25.3 pp. 674–692	Section 2.4 pp. 57–64 Section 15.1 pp. 526–534	Section 16.1,16.2 pp. 618–627	Section 9.1 pp. 338–345	Section 15.1 pp. 511–519
The Circle	Section 20.3 pp. 571–575	Section 26.1 pp. 695–701	Section 15.2, 15.6; pp. 534–539 pp. 562–569	Section 16.3 pp. 627–632	Section 9.4 pp. 358–364	Section 15.2 pp. 519–525
The Parabola	Section 20.4 pp. 576–580	Section 26.2 pp. 702–710	Section 15.3,15.6 pp. 540–547 pp. 562–569	Section 16.4 pp. 632–642	Section 6.3 pp. 247–252 Section 9.2,9.3 pp. 345–358	Section 15.3 pp. 525–532
The Ellipse	Section 20.5 pp. 580–586	Section 26.3 pp. 710–720	Section 15.4, 15.6 pp. 547–554 pp. 562–569	Section 16.5 pp. 643–652	Section 9.5 pp. 364–371	Section 15.4 pp. 532–538
The Hyperbola	Section 20.6 pp. 586–593	Section 26.4 pp. 720–731	Section 15.5,15.6 pp. 554–569	Section 16.5 pp. 653–663	Section 9.6 pp. 371–379	Section 15.5 pp. 538–543
Translations of Axes	Section 20.7 pp. 593–597	Section 26.1 pp. 695–701				Section 15.6 pp. 543–548

TOPIC	WASHINGTON	CALTER	DAVIS	LYNG	NINESTEIN	PETERSON
Rotations of Axes: the General Second–Degree Equation	Section 20.8 pp. 597–601 Supplementary Topics pp. 965–969		Section 15.6 pp. 562–569			Section 15.7 pp. 548–553
Conic Sections in Polar Coordinates				Section 16.7 pp. 664–672	Section 9.7, 9.8 pp. 379–389	Section 15.8 p.553–558
Systems of Equations and Inequalities	**Chapter 13** **pp. 382–400** **Chapter 16** **pp. 460–486**	**Chapter 22** **pp. 589–606**		**Chapter 12** **pp. 461–498**	**Chapter 8** **pp. 299–236**	**Chapter 16** **pp. 560–588**
Solutions of Nonlinear Systems of Equations	Section 13.1, 13.2, 13.3 pp. 383–395	Section 13.8 pp. 367–369	Section 6.4 pp. 225–231		Section 17.4 pp. 716–722	Section 16.1 pp. 560–566
Properties of Inequalities; Linear Inequalities/ Inequalities Using Absolute Value	Section 16.1, 16.2 pp. 460–469 Section 16.4 pp. 475–478	Section 22.1 pp. 590–594	Section 1.3 pp. 14–22	Section 12.1,12.3 pp. 462–468 pp. 475–482	Section 8.1,8.2, 8.4 pp. 299–304 pp. 316–323	Section 16.2 pp. 566–572
Nonlinear Inequalities	Section 16.3 pp. 469–475	Section 22.2 pp. 594–599	Section 5.4 pp. 182–187	Section 5.5 pp. 192–196 Section 12.2 pp. 468–475	Section 8.3 pp. 309–316	Section 16.3 pp. 572–577
Inequalities in Two Variables	Section 16.5 pp. 478–483	Section 22.3 pp. 599–605	Section 2.5 pp. 65–72	Section 12.4 pp. 482–487	Section 8.5 pp. 324–329	Section 16.4 pp. 577–580
Systems of Inequalities; Linear Programming	Section 16.5 pp. 480–483	Section 22.3 pp. 599–605	Section 2.5 pp. 65–72	Section 12.5 pp. 487–495	Chapter 8 Appli-cation pp. 335–336	Section 16.5 pp. 580–586
Matrices	**Chapter 15** **pp. 423–459**	**chapter 11** **pp. 288–316**		**Chapter 13** **pp. 499–554**	**Chapter 19** **pp. 765–799**	**Chapter 17** **pp. 589–616**
Matrices and Determinants	Section 15.3 pp. 435–440	Section 11.1, 11.2 pp. 289–303		Section 13.1 pp. 500–507	Section 19.1 pp. 765–773	Section 17.1 pp. 589–598
Multiplications of Matrices	Section 15.4 pp. 440–445	Section 11.2 pp. 292–303		Section 13.3 pp. 517–529	Section 19.1 pp. 765–773	Section 17.2 pp. 598–604

TOPIC	WASHINGTON	CALTER	DAVIS	LYNG	NINESTEIN	PETERSON
Inverse of Matrices	Section 15.5 pp. 446–450	Section 11.4 pp. 308–314		Section 13.4 pp. 529–538	Section 19.3 pp. 780–785	Section 17.3 pp. 604–609
Matrices and Linear Equations	Section 15.6 pp. 450–454	Section 11.3, 11.4 pp. 303–314		Section 13.2 pp. 507–517	Section 19.2,19.4 pp. 773–780 pp. 786–790	Section 17.4 pp. 609–614
Higher Degree Equations	**Chapter 14 pp. 401–422**		**Chapter 7 pp. 261–288**	**Chapter 14 pp. 555–582**	**Chapter 7 pp. 269–298**	**Chapter 18 pp. 617–647**
The Remainder and Factor Theorems (Synthetic Division)	Section 14.1, 14.2 pp. 401–408	Section 13.3 pp. 363–366	Section 7.1 7.2 pp. 262–271	Section 14.1, 14.2 14.3 pp. 556–566	Section 7.1,7.2 pp. 269–280	Section 18.1 pp. 617–624
Roots of an Equation	Section 14.3 pp. 409–413	Section 4.5 pp. 125–127	Section 7.3 pp. 271–274	Section 14.3, 14.4 pp. 566–580	Section 7.3 pp. 280–286	Section 18.2 pp. 624–628
Rational Roots	Section 14.4 pp. 413–420		Section 7.4 pp. 274–279	Section 14.3,14.4 pp. 566–580	Section 7.3 pp. 280–286	Section 18.3 pp. 628–636
Irrational Roots	Section 14.4 pp. 413–420		Section 7.5 pp. 279–284		Section 7.4 pp. 286–292	Section 18.4 pp. 636–640
Rational Functions	Section 5.7 pp. 177–181					Section 18.5 pp. 640–646
Sequences, Series, and the Binomial Formula	**Chapter 18 pp. 501–521**	**Chapter 23 pp. 607–636**	**Chapter 17 pp. 609–632**	**Chapter 15 pp. 583–616**	**Chapter 16 pp. 655–688**	**Chapter 19 pp. 648–679**
Sequences/ Formula		Section 23.1 pp. 608–613	Section 17.1 pp. 610–615	Section 15.1 pp. 584–588	Section 16.1 pp. 655–661	Section 19.1 pp. 648–652
Arithmetic and Geometric Sequences	Section 18.1 & 18.2 pp. 501–510	Section 23.2, 23.3 pp. 613–624	Section 17.1, 17.2 pp. 610–620	Section 15.3,15.4 pp. 592–601	Section 16.1,16.2 pp. 661–667	Section 19.2 pp. 652–658
Series				Section 15.2 pp. 589–592	Section 16.1 pp. 655–661	Section 19.3 pp. 658–665
Infinite Geometric Series	Section 18.3 pp. 510–513	Section 23.4 pp. 624–627	Section 17.3 pp. 620–623	Section 15.5 pp. 601–604	Section 16.3 pp. 667–672	Section 19.4 pp. 665–670
The Binomial Theorem	Section 18.4 pp. 514–518	Section 23.5 pp. 627–635	Section 17.4 pp. 624–628	Section 15.6 pp. 605–612	Section 16.4,16.5 pp. 672–683	Section 19.5 pp. 670–767

TOPIC	WASHINGTON	CALTER	DAVIS	LYNG	NINESTEIN	PETERSON
Trigonometric Formulas, Identities, and Equations	**Chapter 19** **pp. 522–557**	**Chapter 17** **pp. 447–474**	**Chapter 13** **pp. 447–486**	**Chapter 8** **pp. 329–376**	**Chapter 12** **pp. 489–528**	**Chapter 20** **pp. 680–706**
Basic Identities	Section 19.1 pp. 522–530	Section 17.1 pp. 448–453	Section 13.1 pp. 448–454	Section 8.1 pp. 330–336	Section 12.1 pp. 489–496	Section 20.1 pp. 680–685
The Sum and Difference Identities	Section 19.2 pp. 530–535	Section 17.2 pp. 453–460	Section 13.2 pp. 454–461	Section 8.1,8.2, 8.3,8.4 pp. 336–363	Section 12.2 pp. 496–504	Section 20.2 pp. 685–691
The Double– and Half–Angle Identities	Section 19.3 & 19.4 pp. 535–543	Section 17.3, 17.4 pp. 453–465	Section 13.3 pp. 461–468	Section 8.1,8.2 8.3,8.4 pp. 336–363	Section 12.3 pp. 504–510	Section 20.3 pp. 691–698
Trigonometric Equations	Section 19.5 pp. 543–547	Section 17.5 pp. 465–471	Section 13.4 pp. 468–473	Section 8.5 pp. 363–370	Section 12.4,12.5 pp. 510–523	Section 20.4 pp. 699–704
Statistics and Empirical Methods	**Chapter 21** **pp. 613–641**	**Chapter 24** **pp. 637–672**	**Chapter 16** **pp. 57–608**	**Chapter 17** **pp. 679–736**	**Chapter 20** **pp. 800–829**	**Chapter 21** **pp. 707–738**
Probability		Section 24.4 pp. 654–665		Section 17.1 pp. 680–690		Section 21.1 pp. 707–712
Measures of Central Tendency	Section 21.1 21.2 pp. 613–623	Section 24.1,24.2, 24.3 pp. 638–654	Section 16.1, 16.2 pp. 576–586	Section 17.2 pp. 690–699	Section 20.1 pp. 800–805	Section 21.2 pp. 712–720
Measures of Dispersion	Section 21.3 pp. 623–628	Section 24.3 pp. 648–654	Section 16.3 pp. 587–592	Section 17.3 pp. 699–711	Section 20.2 pp. 805–809	Section 21.3 pp. 720–723
Fitting a Line to Data/Normal Curves	Section 21.4 pp. 628–634		Section 16.4 pp. 592–602	Section 17.4 pp. 711–720	Section 20.3,20.4 pp. 809–822	Section 21.4 pp. 724–731
Fitting Nonlinear Curves to Data	Section 21.5 pp. 634–638		Section 16.4 pp. 592–602	Section 17.5 pp. 720–729	Section 20.3, 20.4 pp. 809–822	Section 21.5 pp. 731–735
An Introduction to Calculus	**Chapter 22** **pp. 642–692**		**Chapter 18** **pp. 633–694**			**Chapter 22** **pp. 739–771**
The Tangent Question	Section 22.2 pp. 652–656		Section 18.2 pp. 645–658			Section 22.1 pp. 739–743
The Area Question	Section 24.4 pp. 745–750					Section 22.2 pp. 743–748

TOPIC	WASHINGTON	CALTER	DAVIS	LYNG	NINESTEIN	PETERSON
Limits: An Intuitive Approach	Section 22.1 pp. 643–652 Section 22.4 pp. 660–665	Section 27.1 pp. 734–740	Section 18.1 pp. 634–644			Section 22.3 pp. 748–754
One–Sided Limits						Section 22.4 pp. 754–758
Algebraic Techniques for Finding Limits	Section 22.2 22.3, 22.4 pp. 652–665		Section 18.2 pp. 645–658			Section 22.5 pp. 758–764
Continuity	Section 22.1 pp. 643–652		Section 18.1 pp. 634–644			Section 22.6 pp. 764–769
The Derivative	**Chapter 22 pp. 642–692**	**Chapter 27 pp. 733–768**	**Chapter 18 pp. 633–694**			**Chapter 23 pp. 772–809**
The Tangent Question and the Derivative	Section 22.2 22.3 pp. 652–660	Section 27.2 pp. 741–747	Section 18.2 pp. 645–658			Section 23.1 pp. 772–779
Derivatives of Polynomials	Section 22.5 pp. 665–670	Section 27.3 pp. 747–752	Section 18.3 pp. 658–670			Section 23.2 pp. 780–785
Derivatives of Products and Quotients	Section 22.6 pp. 670–675	Section 27.5 pp. 757–762	Section 18.4 pp. 670–680			Section 23.3 pp. 785–792
Derivatives of Composite Functions	Section 22.7 pp. 675–681	*Section 27.4 pp. 753–575*	Section 18.4 pp. 670–680			Section 23.4 pp. 792–799
Implicit Differentiation	Section 22.8 pp. 682–685	Section 27.6 pp. 762–766	Section 18.5 pp. 680–688			Section 23.5 pp. 799–804
Higher Order Derivatives	Section 22.9 pp. 685–689	Section 27.7 pp. 766–767	Section 18.2 pp. 645–658			Section 23.6 pp. 805–807
Applications of Derivatives	**Chapter 23 pp. 693–731**	**Chapter 28 & 29 pp. 769–827**	**Chapter 18 & 19 pp. 632–693 pp. 695–746**			**Chapter 24 pp. 810–849**
Rates of Change	Section 23.4 pp. 706–709	Section 29.1 pp. 792–797				Section 24.1 pp. 810–815

TOPIC	WASHINGTON	CALTER	DAVIS	LYNG	NINESTEIN	PETERSON
Extrema and the First Derivative Test	Section 23.5 pp. 710–717	Section 28.1, 28.2 pp. 770–778	Section 18.3 pp. 658–670			Section 24.2 pp. 815–821
Concavity and the Second Derivative Test	Section 23.5, 23.6 pp. 710–721	Section 28.3, 28.5 pp. 779–780 pp. 783–789	Section 18.3 pp. 658–670			Section 24.3 pp. 821–826
Applied Extrema Problems	Section 23.7 pp. 721–728	Section 29.4 pp. 812–820				Section 24.4 pp. 826–832
Related Rates	Section 23.4 pp. 706–710	Section 29.2, 29.3 pp. 798–812	Section 18.5 pp. 680–688			Section 24.5 pp. 832–839
Newton's Method	Section 23.2 pp. 698–701	Section 28.4 pp. 780–783				Section 24.6 pp. 839–840
Differentials	Section 24.1 pp. 732–736	Section 29.5 pp. 820–825	Section 19.1 pp. 696–700			Section 24.7 pp. 840–844
Antiderivatives	Section 24.2 pp. 737–739	Section 31.1 pp. 852–860	Section 19.2 pp. 701–709			**Section 24.8 pp. 844–847**
Integration	**Chapter 24 pp. 739–761**	**Chapter 32 pp. 882–902**	**Chapter 19 pp. 695–746**			**Chapter 25 pp. 850–882**
The Area Question and the Integral	Section 24.4, 25.2 pp. 745–776 pp. 771–776	Section 32.1, 32.2 pp. 883–892	Section 19.3 pp. 710–717			Section 25.1 pp. 850–854
The Fundamental Theorem of Calculus		Section 32.3 pp. 892–903	Section 19.4 pp. 718–733			Section 25.2 pp. 855–862
The Indefinite Integral	Section 24.3 pp. 739–745	Section 31.1 pp. 852–860	Section 19.2 pp. 701–709			Section 25.3 pp. 862–868
The Area Between Two Curves	Section 24.4 pp. 745–750	Section 33.5 pp. 923–932	Section 19.3 pp. 710–717			Section 25.4 pp. 867–874
Numerical Integration (The Trapezoid Rule and Simpsons Rule)	Section 24.6,24.7 pp. 754–761	Section 34.6 pp. 969–975	Section 19.5 pp. 734–743			Section 25.5 pp. 874–880
Applications of Integration	**Chapter 25 pp. 764–804**	**Chapter 33 pp. 905–946**				**Chapter 26 pp. 883–935**

TOPIC	WASHINGTON	CALTER	DAVIS	LYNG	NINESTEIN	PETERSON
Average Values and Other Antiderivative Applications	Section 25.1 pp. 764–771 Section 25.6 pp. 95–801 Section 27.5 pp. 854–859	Section 33.4 pp. 921–923				Section 26.1 pp. 883–891
Volumes of Revolution: Disk and Washer Methods	Section 25.3 pp. 776–782	Section 33.1 pp. 906–915				Section 26.2 pp. 891–899
Volumes of Revolution: Shell Method	Section 25.3 pp. 776–782	Section 33.1 pp. 906–915				Section 26.3 pp. 899–904
Arc Length and Surface Area		Section 33.2, 33.3 pp. 915–921				Section 26.4 pp. 904–908
Centroids	Section 25.4 pp. 783–790	Section 33.5 pp. 923–932	Section 19.4 pp. 718–733			Section 26.5 pp. 908–918
Moments of Inertia	Section 25.5 pp. 790–795	Section 33.8 pp. 939–945				Section 26.6 pp. 918–924
Work and Fluid Pressure	Section 25.6 pp. 795–800	Section 33.6–33.7 pp. 932–939				Section 26.7 pp. 924–933
Derivatives of Transcendental Functions	**Chapter 26 pp. 805–839**	**Chapter 30 pp. 828–850**	**Chapter 20 pp. 747–790**			**Chapter 27 pp. 936–972**
Derivatives of Sine and Cosine Functions	Section 26.1 pp. 805–810	Section 30.1 pp. 829–833	Section 20.1 pp. 748–759			Section 27.1 pp. 936–941
Derivatives of the Other Trigonometric Functions	Section 26.2 pp. 811–814	Section 30.2 pp. 833–837	Section 20.2 pp. 759–767			Section 27.2 pp. 942–945
Derivatives of Inverse Trigonometric Functions	Section 26.3 pp. 814–818	Section 30.3 pp. 837–839	Section 20.3 pp. 767–775			Section 27.3 pp. 946–950

TOPIC	WASHINGTON	CALTER	DAVIS	LYNG	NINESTEIN	PETERSON
Applications	Section 26.4 pp. 818–822	Section 30.1, 30.2, 30.3 pp. 829–839				Section 27.4 pp. 950–955
Derivatives of Logarithmic Functions	Section 26.5 pp. 823–826	Section 30.4 pp. 839–845	Section 20.4 pp. 776–781			Section 27.5 pp. 955–959
Derivatives of Exponential Functions	Section 26.6 pp. 827–831	Section 30.5 pp. 845–849	Section 20.5 pp. 782–786			Section 27.6 pp. 959–967
Applications	Section 26.7 pp. 831–834	Section 30.4, 30.5 pp. 839–849				Section 27.7 pp. 967–970
Techniques of Integration	**Chapter 27** **pp. 840–876**	**Chapter 34** **pp. 947–976**	**Chapter 21** **pp. 791–842**			**Chapter 28** **pp. 973–1019**
The General Power Formula	Section 27.1 pp. 840–843	Section 31.2 pp. 860–863	Section 21.1 pp. 792–796			Section 28.1 pp. 973–978
Basic Logarithmic and Exponential Integrals	Section 27.2, 27.3 pp. 844–850		Section 21.2 pp. 796–804			Section 28.2 pp. 978–984
Basic Trigonometric and Hyperbolic Integrals	Section 27.4 pp. 850–854	Section 31.4 pp. 865–872	Section 21.3 pp. 804–812			Section 28.3 pp. 894–990
More Trigonometric Integrals	Section 27.5 pp. 854–859		Section 21.4 pp. 812–820			Section 28.4 pp. 991–997
Integrals Related to Inverse Trigonometric and Inverse Hyperbolic Functions	Section 27.6 pp. 859–863		Section 21.5 pp. 820–829			Section 28.5 pp. 997–1002
Trigonometric Substitution	Section 27.8 pp. 867–870	Section 34.4 pp. 962–966				Section 28.6 pp. 1002–1007
Integration by Parts	Section 27.7 pp. 863–867	Section 34.1 pp. 948–952	Section 21.6 pp. 829–834			Section 28.7 pp. 1007–1015

TOPIC	WASHINGTON	CALTER	DAVIS	LYNG	NINESTEIN	PETERSON
Using Integration Tables	Section 27.9 pp. 870–873	Section 31.4 pp. 865–872	Section 21.7 pp. 834–836			Section 28.8 pp. 1015–1018
Parametric Equations and Polar Coordinates						**Chapter 29 pp. 1020–1043**
Derivatives of Parametric Equations	Section 23.3 pp. 701–705					Section 29.1 pp. 1020–1025
Differentiation in Polar Coordinates						Section 29.2 pp. 1025–1029
Arc Length and Surface Area Revisited						Section 29.3 pp. 1029–1034
Intersection of Graphs of Polar Coordinates						Section 29.4 pp. 1034–1037
Area in Polar Coordinates						Section 29.5 pp. 1037–1042
Partial Derivatives and Multiple Integrals						**Chapter 30 pp. 1044–1088**
Functions in Two Variables	Supplement 3 pp. 970–972					Section 30.1 pp. 1044–1048
Surfaces in Three Dimensions	Supplement 4 pp. 973–979					Section 30.2 pp. 1049–1053
Partial Derivatives	Supplement 5 pp. 979–984					Section 30.3 pp. 1053–1059
Some Applications of Partial Derivatives	Supplement 7 pp. 989–993					Section 30.4 pp. 1059–1065
Multiple Integrals	Supplement 6 pp. 984–989					Section 30.5 pp. 1065–1072
Cylindrical and Spherical Coordinates						Section 30.6 pp. 1072–1080

CHAPTER

1

The Real Number System

1.1 SOME SETS OF NUMBERS

1. 15 is a natural number, whole number, integer, rational number, and a real number
2. $-\frac{2}{3}$ is a rational number and a real number
3. $\frac{-\sqrt{7}}{8}$ is an irrational number and a real number
4. 0 is a whole number, integer, rational, and real number
5. $|15| = 15$
6. $\left|-\frac{2}{3}\right| = \frac{2}{3}$
7. $\left|-\frac{\sqrt{7}}{8}\right| = \frac{\sqrt{7}}{8}$
8. $|0| = 0$
9. $\frac{4}{7}$

   −6 −4 −2 0 2 4 6

10. 2.5

    −6 −4 −2 0 2 4 6

11. $\frac{-8}{3}$

    −6 −4 −2 0 2 4 6

12. $\frac{-\pi}{3}$

    −6 −4 −2 0 2 4 6

13. $2 < 3; 2 - 3 = -1$ negative
14. $5 > 3; 5 - 3 = 2$ positive
15. $-4 < 7; -4 - 7 = -11$ negative
16. $9 > -7; 9 - -7 = 16$ positive
17. $-3 > -8; -3 - (-8) = 5$ positive
18. $-15 < -7; -15 - -7 = -8$ negative
19. $-\frac{2}{3} < -\frac{1}{2}; -\frac{2}{3} - \left(-\frac{1}{2}\right) = -\frac{4}{6} - \left(-\frac{3}{6}\right) = -\frac{1}{6}$ negative
20. $0.7 > 0.5; 0.7 - 0.5 = 0.2$ positive
21. The reciprocal of -5 is $\frac{1}{-5} = -\frac{1}{5}$
22. The reciprocal of $\frac{1}{2}$ is $\frac{1}{1/2} = 1 \times \frac{2}{1} = 2$
23. The reciprocal of $\frac{17}{3}$ is $\frac{1}{17/3} = 1 \times \frac{3}{17} = \frac{3}{17}$
24. The reciprocal of $\frac{-2}{\pi}$ is $\frac{1}{-2/\pi} = 1 \times -\frac{\pi}{2} = -\frac{\pi}{2}$
25. $-5, -|4|, -\frac{2}{3}, \frac{-1}{3}, \frac{16}{3}, |-8|$
26. $-\sqrt{7}, -\left|\frac{7}{4}\right|, -\sqrt{2}, \sqrt{7} - \sqrt{2}, \frac{7}{5}, \pi, \frac{22}{7}$
27. (a) $\frac{1}{32}$ in., (b) $\frac{1}{8}$ in., (c) $\frac{1}{16} = 0.0625$ in.
28. $\frac{3}{4}$ h = 0.75 h, $1\frac{1}{2}$ h = 1.5 h, $2\frac{1}{4}$ h = 2.25 h
29. $|-19.4 - -16.8| = |-2.6| = 2.6$ so 2.6 V

1.2 BASIC LAWS OF REAL NUMBERS

1. Commutative law for addition (CA)
2. Associative law for multiplication (AM)
3. Commutative law for multiplication (CM)
4. Distributive law (DL)
5. Associative law for addition (AA)
6. Distributive law
7. Identity element for multiplication
8. Multiplicative inverses
9. Additive inverses
10. Commutative law for addition
11. Associative law for multiplication
12. Identity element for addition
13. Additive inverse of 91 is -91
14. 8
15. $-\sqrt{2}$
16. $-\frac{1}{3}$
17. Multiplicative inverse of $\frac{1}{2}$ is 2
18. $-\frac{1}{5}$
19. $\frac{2}{\sqrt{2}} = \frac{2}{\sqrt{2}}\frac{\sqrt{2}}{\sqrt{2}} = \frac{2\sqrt{2}}{2} = \sqrt{2}$
20. $-\frac{7}{3}$
21. $16 - 8 \div 4 = 16 - 2 = 14$
22. $16 \div 8 + 2 = 2 + 2 = 4$
23. $$\begin{aligned} 24 + 3 - 10 \div 5 \times 8 + 2 &= 24 + 3 - 2 \times 8 + 2 \\ &= 24 + 3 - 16 + 2 \\ &= 27 - 16 + 2 \\ &= 11 + 2 \\ &= 13 \end{aligned}$$
24. $13 \times 7 - 26 \div 5 + 5 = 91 - 5.2 + 5 = 85.8 + 5 = 90.8$
25. $(7 - 2) - (3 + 8 - 7) = 5 - (11 - 7) = 5 - 4 = 1$
26. $7-2-3+8-7 = 5-3+8-7 = 2+8-7 = 10-7 = 3$
27. $7 \times 3 + 5 \times 2 = 21 + 10 = 31$
28. $6 \times 4 - 3 \times 5 = 24 - 15 = 9$
29. $$\begin{aligned} &\{-[5 - (8 - 4) + (3 - 7)] - (4 - 2)\} \\ &= -[5 - 4 + -4] - 2 \\ &= -[1 + -4] - 2 \\ &= -[-3] - 2 \\ &= 3 - 2 = 1 \end{aligned}$$
30. $$\begin{aligned} (14 + 3(8 - 6) + 4(9 - 5)) &= 14 + 3 \times 2 + 4 \times 4 \\ &= 14 + 6 + 16 \\ &= 20 + 16 = 36 \end{aligned}$$

1.3 BASIC OPERATIONS WITH REAL NUMBERS

1. $27 + (+23)$. Same sign $27 + 23 = 50$; both positive so, $+50$
2. $8+(-19)$. Different signs $|-19|-|8| = 19-8 = 11$; larger negative so, -11
3. $27+(-13)$. Different signs $|27|-|-13| = 27-13 = 14$; larger positive so, 14
4. $-9 + (-8)$. Same sign $|-9| + |-8| = 9 + 8 = 17$; both negative so, -17
5. $7 - 16 = 7 + (-16)$. Different signs $|-16| - |7| = 16 - 7 = 9$; larger negative so, -9
6. $29 - (-8) = 29 + (+8) = 37$

7. $-8-16=-8+(-16)$. Same sign $|-8|+|-16|=8+16=24$; both negative so, -24

8. $-25-(-13)=-25+(+13)$. Different signs $|-25|-|13|=12$; larger negative so, -12

9. $-37-(-49)=-37+(+49)$. Different signs $|49|-|37|=49-37=12$; larger positive so, $+12$

10. $(-2)(6)$; $|-2|\cdot|6|=2\cdot 6=12$. Different signs so, -12

11. $(-3)(-5)$; $|-3|\cdot|-5|=3\cdot 5=15$. Same sign so $+15$

12. $(7)(-8)$; $|7|\cdot|-8|=7\cdot 8=56$. Different signs so -56

13. $-38\div 4$; $|-38|\div|4|=38\div 4=\frac{38}{4}=\frac{19}{2}$. Different signs so $-\frac{19}{2}$

14. $-45\div(-9)$; $\frac{|-45|}{|-9|}=\frac{45}{9}=5$. Same sign so $+5$

15. $\frac{3}{4}+\frac{-5}{8}$; $\frac{3}{4}=\frac{6}{8}$; $\frac{6}{8}+\frac{-5}{8}=\frac{6+(-5)}{8}=\frac{1}{8}$

16. $-1\frac{3}{4}+\frac{-2}{3}=-\frac{7}{4}+\frac{-2}{3}$. Same sign $\frac{7}{4}+\frac{2}{3}=\frac{21}{12}+\frac{8}{12}=\frac{29}{12}$. Both negative so $-\frac{29}{12}=-2\frac{5}{12}$

17. $\frac{-9}{5}+\frac{7}{3}$ common denominator is 15; $\frac{-9}{5}+\frac{7}{3}=\frac{-27}{15}+\frac{35}{15}=\frac{-27+35}{15}=\frac{8}{15}$

18. $\frac{-2}{3}+\frac{5}{6}$ common denominator is 6; $\frac{-2}{3}+\frac{5}{6}=\frac{-4}{6}+\frac{5}{6}=\frac{-4+5}{6}=\frac{1}{6}$

19. $\frac{2}{5}+\frac{-1}{4}=\frac{8}{20}+\frac{-5}{20}=\frac{8+(-5)}{20}=\frac{3}{20}$

20. $\frac{-4}{5}+\frac{-5}{6}=\frac{-24}{30}+\frac{-25}{30}=\frac{-24+(-25)}{30}=\frac{-49}{30}=-1\frac{19}{30}$

21. $\frac{3}{8}-\frac{-1}{4}=\frac{3}{8}-\frac{-2}{8}=\frac{3-(-2)}{8}=\frac{5}{8}$

22. $1\frac{1}{3}-\frac{-5}{6}=\frac{4}{3}-\frac{-5}{6}=\frac{8}{6}-\frac{-5}{6}=\frac{8-(-5)}{6}=\frac{13}{6}=2\frac{1}{6}$

23. $\frac{-9}{10}-\frac{2}{3}=\frac{-27}{30}-\frac{20}{30}=\frac{-27-20}{30}=\frac{-47}{30}=-1\frac{17}{30}$

24. $\frac{-5}{16}-\frac{-3}{8}=\frac{-5}{16}-\frac{-6}{16}=\frac{-5-(-6)}{16}=\frac{1}{16}$

25. $\frac{5}{32}-\frac{1}{8}=\frac{5}{32}-\frac{4}{32}=\frac{5-4}{32}=\frac{1}{32}$

26. $-\frac{7}{3}-\frac{-6}{7}=\frac{-49}{21}-\frac{-18}{21}=\frac{-49-(-18)}{21}=\frac{-31}{21}=-1\frac{10}{21}$

27. $\frac{-2}{3}\times\frac{4}{5}=\frac{-2\times 4}{3\times 5}=\frac{-8}{15}=-\frac{8}{15}$

28. $\frac{3}{4}\times\frac{-5}{8}=\frac{3\times(-5)}{4\times 8}=-\frac{15}{32}$

29. $\frac{-1}{8}\times\frac{-3}{4}=\frac{-1\times(-3)}{8\times 4}=\frac{3}{32}$

30. $\frac{9}{16}\times\frac{1}{2}=\frac{9\times 1}{16\times 2}=\frac{9}{32}$

31. $-\frac{4}{3}\times\frac{5}{2}=\frac{-4\times 5}{3\times 2}=\frac{-20}{6}=-\frac{10}{3}$ or $\frac{-2\times 2\times 5}{3\times 2}=-3\frac{1}{3}$

32. $\frac{-9}{5}\times\frac{-3}{8}=\frac{-9\times(-3)}{5\times 8}=\frac{27}{40}$

33. $-\frac{3}{4}\div\frac{-5}{8}=-\frac{3}{4}\times\frac{8}{-5}=\frac{-3\times 4\times 2}{4\times(-5)}=\frac{-6}{-5}=\frac{6}{5}=1\frac{1}{5}$

34. $1\frac{3}{4}\div\frac{-2}{3}=\frac{7}{4}\div\frac{-2}{3}=\frac{7}{4}\times\frac{3}{-2}=\frac{7\times 3}{4\times(-2)}=\frac{21}{-8}=-\frac{21}{8}=-2\frac{5}{8}$

35. $-\frac{3}{5}\div 4=\frac{-3}{5}\times\frac{1}{4}=\frac{-3\times 1}{5\times 4}=\frac{-3}{20}=-\frac{3}{20}$

36. $-\frac{3}{8}\div\frac{1}{4}=\frac{-3}{8}\times\frac{4}{1}=\frac{-3\times 4}{2\times 4}=-\frac{3}{2}=-1\frac{1}{2}$

37. $-\frac{7}{8}\div\left(-\frac{5}{7}\right)=-\frac{7}{5}\times\frac{-7}{5}=\frac{-7\times(-7)}{5\times 5}=\frac{49}{25}=1\frac{24}{25}$

38. $\frac{2}{3}\div\left(-\frac{7}{3}\right)=\frac{2}{3}\times\frac{-3}{7}=\frac{2\times(-3)}{3\times 7}=\frac{-2}{7}=-\frac{2}{7}$

39. $\left(-2+\frac{2}{3}\right)\times\frac{-1}{2}-\left[\frac{3}{2}\div(-3)\right]+\frac{7}{3}=\frac{-4}{3}\times\frac{-1}{2}-\left(\frac{-1}{2}\right)+\frac{7}{3}=\frac{-4\times(-1)}{3\times 2}-\left(\frac{-1}{2}\right)+\frac{7}{3}=\frac{2}{3}-\left(\frac{-1}{2}\right)+\frac{7}{3}=\frac{4}{6}-\frac{-3}{6}+\frac{7}{3}=\frac{7}{6}+\frac{7}{3}=\frac{7}{6}+\frac{14}{6}=\frac{21}{6}=\frac{7}{2}=3\frac{1}{2}$

40. $\frac{4}{3}\times\frac{-7}{8}\times\frac{3}{-5}\div\frac{4}{5}+\frac{-5}{8}-\frac{7}{8}=\frac{4\times(-7)}{3\times 8}\times\frac{3}{-5}\div\frac{4}{5}+\frac{-5}{8}-\frac{7}{8}=\frac{-7\times 3}{6\times(-5)}\div\frac{4}{5}+\frac{-5}{8}-\frac{7}{8}=\frac{7}{10}\div\frac{4}{5}+\frac{-5}{8}-\frac{7}{8}=\frac{7}{10}\times\frac{5}{4}+\frac{-5}{8}-\frac{7}{8}=\frac{7\times\pi}{2\times 5\times 4}+\frac{-5}{8}-\frac{7}{8}=\frac{7}{8}+\frac{-5}{8}-\frac{7}{8}=\frac{2}{8}-\frac{7}{8}=-\frac{5}{8}$

41. Add: $17\frac{1}{2}+15\frac{7}{8}+29\frac{3}{4}+15\frac{3}{8}=17\frac{4}{8}+15\frac{7}{8}+29\frac{6}{8}+15\frac{3}{8}=76\frac{20}{8}=78\frac{4}{8}=78\frac{1}{2}=6'6\frac{1}{2}''$

42. Add $\frac{5}{32}+\frac{3}{16}=\frac{5}{32}+\frac{6}{32}=\frac{11}{32}$

43. We add the voltages of the batteries: $30+15+(-12)+24=57$. So, the total voltage is 57 V.

44. Subtract $1\frac{5}{8}-\frac{1}{5}=\frac{13}{8}-\frac{1}{5}=\frac{65}{40}-\frac{8}{40}=\frac{57}{40}=1\frac{17}{40}''$

45. (a) $4\frac{5}{8}+3\frac{1}{5}=\frac{37}{8}+\frac{16}{5}=\frac{185}{40}+\frac{128}{40}=\frac{313}{40}=7\frac{33}{40}$ miles;

(b) We need to add the distance between the second and third checkpoints to the answer for (a). $7\frac{33}{40} + 3\frac{3}{8} = \frac{313}{40} + \frac{27}{8} = \frac{313}{40} + \frac{135}{40} = \frac{448}{40} = 11\frac{8}{40} = 11\frac{1}{5}$ miles;

(c) Using the answer from (b), we subtract that answer from the total length of the race. $14\frac{1}{2} - 11\frac{1}{5} = \frac{29}{2} - \frac{56}{5} = \frac{145}{10} - \frac{112}{10} = \frac{33}{10} = 3\frac{3}{10}$ miles

46. We divide $427\frac{1}{5}$ by $13\frac{3}{4}$. $427\frac{1}{5} \div 13\frac{3}{4} = \frac{2136}{5} \div \frac{55}{4} = \frac{2136}{5} \times \frac{4}{55} = \frac{8544}{275} = 31\frac{19}{275} \approx 31.07$ mi/gal

47. We want $37 \times 2'3\frac{1}{2}''$. There are $12''$ in 1 foot, so there are $2 \times 12'' = 24''$ in 2 feet. Thus, $2'3\frac{1}{2}'' = 24'' + 3\frac{1}{2}'' = 27\frac{1}{2}''$. Now, $37 \times 27\frac{1}{2} = \frac{37}{1} \times \frac{55}{2} = \frac{2035}{2} = 1017\frac{1}{2}'' = 84'9\frac{1}{2}''$

48. Each strip requires $1\frac{5}{16} + \frac{1}{8} = 1\frac{7}{16}$ in. of wood for its width. To determine the total number of strips we can get from the sheet of plywood, we will divide 8 ft by $1\frac{7}{16}$ in. Converting 8 feet to $8 \times 12 = 96$ in., we want the answer to $96 \div 1\frac{7}{16} = 96 \div \frac{23}{16} = 96 \times \frac{16}{23} = \frac{1536}{23} \approx 66.78$. You can get a total of 66 strips. (Note that the scrap is a strip $1\frac{1}{8}''$ wide.)

49. We subtract the two smaller lengths from the total length. $7\frac{5}{8} - 3\frac{3}{8} - 3\frac{3}{8} = \frac{61}{8} - \frac{27}{8} - \frac{27}{8} = \frac{34}{8} - \frac{27}{8} = \frac{7}{8}$. The missing length is $\frac{7}{8}''$.

1.4 LAWS OF EXPONENTS

1. $5^3 = 5 \times 5 \times 5 = 125$
2. $3.8^0 = 1$ by Rule 6
3. $\left(\frac{2}{3}\right)^{-1} = \frac{2^{-1}}{3^{-1}} = \frac{1/2}{1/3} = \frac{1}{2} \times \frac{3}{1} = \frac{3}{2}$
4. $\left(\frac{3}{5}\right)^{-2} = \frac{3^{-2}}{5^{-2}} = \frac{5^2}{3^2} = \frac{25}{9}$
5. $(-4)^2 = (-4)(-4) = 16$
6. $(-5)^4 = (-5)(-5)(-5)(-5) = 625$
7. $\frac{7}{7^3} = \frac{7^1}{7^3} = 7^{1-3} = 7^{-2} = \frac{1}{7^2} = \frac{1}{49}$
8. $-3^2 = -(3^2) = -(3 \cdot 3) = -9$
9. $3^2 \cdot 3^4 = 3^{2+4} = 3^6$
10. $d^8 d^5 = d^{8+5} = d^{13}$
11. $2^4 \cdot 2^3 \cdot 2^5 = 2^{4+3+5} = 2^{12}$
12. $f^3 f^4 f^1 = f^{3+4+1} = f^8$
13. $\frac{2^5}{2^3} = 2^{5-3} = 2^2$
14. $\frac{5^{14}}{5^3} = 5^{14-3} = 5^{11}$
15. $(2^3)^2 = 2^{3\times2} = 2^6$
16. $(5^7)^3 = 5^{7\times3} = 5^{21}$
17. $(x^4)^5 = x^{4\times5} = x^{20}$
18. $(xy^3)^4 = (x^1y^3)^4 = x^{1\times4}y^{3\times4} = x^4y^{12}$
19. $(a^{-2}b)^{-3} = a^{-2\times-3}b^{-3} = a^6b^{-3} = \frac{a^6}{b^3}$
20. $\left(\frac{2}{3}\right)^4 = \frac{2^4}{3^4}$; Rule 4a
21. $\left(\frac{x}{4}\right)^3 = \frac{x^3}{4^3}$
22. $\left(\frac{a}{b^3}\right)^5 = \frac{a^5}{b^{3\times5}} = \frac{a^5}{b^{15}}$
23. $\left(\frac{a^2b}{c^3}\right)^4 = \frac{a^{2\times4}b^4}{c^{3\times4}} = \frac{a^8b^4}{c^{12}}$
24. $4^{-3} = \frac{1}{4^3}$; Rule 7
25. $x^{-7} = \frac{1}{x^7}$

26. $\dfrac{1}{p^{-5}} = p^5$

27. $\left(\dfrac{1}{5}\right)^3 = \dfrac{1}{5^3}$

28. $\frac{x^4}{x^2} = x^{4-2} = x^2$

29. $\dfrac{7^3}{7^8} = 7^{3-8} = 7^{-5} = \dfrac{1}{7^5}$

30. $\dfrac{5^2}{5^{10}} = 5^{2-10} = 5^{-8} = \dfrac{1}{5^8}$

31. $\dfrac{a^2y^3}{a^5y^7} = a^{2-5}y^{3-7} = a^{-3}y^{-4} = \dfrac{1}{a^3}\dfrac{1}{y^4} = \dfrac{1}{a^3y^4}$

32. $\dfrac{x^4yb^2}{xy^3b^5} = x^{4-1}y^{1-3}b^{2-5} = x^3y^{-2}b^{-3} = \dfrac{x^3}{y^2b^3}$

33. $\dfrac{a^2p^5y^3}{a^6p^5y} = a^{2-6}p^{5-5}y^{3-1} = a^{-4}p^0y^2 = \dfrac{y^2}{a^4}$

34. $\dfrac{p^3q^4r^2}{p^4r^2} = p^{3-4}q^4r^{2-2} = p^{-1}q^4r^0 = \dfrac{q^4}{p}$

35. $(pr^2)^{-1} = \frac{1}{pr^2}$

36. $\left(\dfrac{2x^2}{y}\right)^{-1} = \dfrac{y}{2x^2}$

37. $\left(\dfrac{4y^3}{5^2}\right)^{-1} = \dfrac{5^2}{4y^3}$

38. $\left(\dfrac{4x^3}{y^2}\right)^{-2} = \dfrac{4^{-2}x^{-6}}{y^{-4}} = \dfrac{y^4}{4^2x^6}$

or $\left(\dfrac{4x^3}{y^2}\right)^{-2} = \left(\dfrac{y^2}{4x^3}\right)^2 = \dfrac{y^4}{16x^6}$

39. $\left(\dfrac{2b^2}{y^5}\right)^{-3} = \dfrac{2^{-3}b^{-6}}{y^{-15}} = \dfrac{y^{15}}{2^3b^6}$

or $\left(\dfrac{2b^2}{y^5}\right)^{-3} = \left(\dfrac{y^5}{2b^2}\right)^3 = \dfrac{y^{15}}{8b^6}$

40. $(-8pr^2)^{-3} = (-8)^{-3}p^{-3}r^{-6} = \dfrac{1}{-8^3p^3r^6} = \dfrac{-1}{8^3p^3r^6}$

41. $(-b^4)^6 = (-b^4)(-b^4)(-b^4)(-b^4)(-b^4)(-b^4) = b^{24}$

42. $ap^2(-a^2p^3)^2 = ap^2(a^4p^6) = a^{1+4}p^{2+6} = a^5p^8$

43. First we simplify $\left((2\pi fC)^{-1}\right)^2$ as $(2\pi fC)^{-2}$. Next, we compute $(2\pi fC)^{-2}$. Substituting the given values of f and C and using the $\boxed{\pi}$ key on the calculator, we obtain $(2\pi fC)^{-2} = 0.030159289^{-2} = 1099.405205$. To this, add 40^2, with the result of 2699.405205. Then $\sqrt{2699.405205} = 51.95580049$ gives the answer of $51.96\,\Omega$ when rounded off to 2 decimal places.

1.5 SIGNIFICANT DIGITS AND ROUNDING OFF

1. Exact
2. Approximate
3. Approximate
4. Exact; exact
5. Approximate
6. Approximate
7. 3
8. 3
9. 3
10. 3
11. 1
12. 3
13. 4
14. 6
15. (a) 6.05, (b) 6.05
16. (a) 6.324, (b) both
17. (a) 5.01, (b) 0.027
18. (a) 19,020, (b) 19,020
19. (a) $27{,}0\tilde{0}0$, (b) $27{,}0\tilde{0}0$
20. (a) both, (b) 0.003
21. (a) 86, (b) 0.2
22. (a) 305.00, (b) both
23. (a) 140.070, (b) 140.070
24. (a) 4, (b) 4.4, (c) 4.36

25. (a) 10, (b) 14, (c) 14.4

26. (a) 4, (b) 4.1, (c) 4.07

27. (a) 7, (b) 7.0, (c) 7.04

28. (a) 0.006, (b) 0.0062, (c) 0.00616

29. (a) 400, (b) $4\tilde{0}0$, (c) 403

30. (a) 0.04, (b) 0.037, (c) 0.0373

31. (a) 300, (b) 310, (c) 305

32. (a) 4, (b) 4.4, (c) 4.36

33. (a) 10, (b) 14, (c) 14.4

34. (a) 4, (b) 4.1, (c) 4.06

35. (a) 7, (b) 7.0, (c) 7.04

36. (a) 0.006, (b) 0.0062, (c) 0.00616

37. (a) 400, (b) $4\tilde{0}0$, (c) 403

38. (a) 0.04, (b) 0.037, (c) 0.0372

39. (a) 300, (b) $3\tilde{0}0$, (c) 305

40. (a) 30, (b) 25.3, (c) 25.335

41. (a) 90, (b) 89.9, (c) 89.899

42. (a) 130, (b) 125.4, (c) 125.376

43. (a) 240, (b) 237.3, (c) 237.302

44. (a) 100, (b) 97.0, (c) 96.999

45. (a) 440, (b) 438.0, (c) 437.998

46. (a) 10, (b) 12.3, (c) 12.341

47. (a) 80, (b) 78.7, (c) 78.671

48. absolute error = 0.005 m, relative error = 0.0004; percent error = 0.04%

49. Length: Absolute error is measured value minus true value, so $23.72 - 24 = -0.28$ mm. Relative error is absolute error divided by true value, so $\frac{-0.28}{24} = -0.0117$. Percent error is relative error × 100, so $-0.0117 \times 100 = -1.17\%$
Width: Absolute error; $8.35 - 8 = 0.35$ mm. Relative error; $\frac{0.35}{8} = 0.0438$. Percent error; $0.0438 \times 100 = 4.38\%$
Thickness: Absolute error; $2.98 - 3 = -0.02$ mm. Relative error; $\frac{-0.02}{3} = -0.0067$. Percent error; $-0.0067 \times 100 = -0.67\%$

50. 24.68

51. 99.37

52. 410.8

53. 1618

54. 82

55. 1020

56. 0.0085

57. 17

58. There are 2 front lamps, so they draw a total of $2 \times 0.417 = 0.834$ A. The 2 tail lamps draw a total current of $2 \times 0.457 = 0.914$ A. Thus the total for the 5 lamps is $0.834 + 0.914 + 0.736 = 2.484$. Rounded off to 3 significant digits, this is 2.48 A.

59. To find the total length, we need to multiply $37 \times 132 \times 0.072$. The product is 351.648, which is 351.65 when rounded to 5 significant digits.

60. (a) We want the product of 379 and 11 feet $10\frac{1}{8}$ inches. Convert 11 ft $10\frac{1}{8}$ in. to $11 \times 12 + 10\frac{1}{8}$ in. $= 142\frac{1}{8}$ in.= 142.125 in. Now when we multiply this by 379, we obtain 53865.375 in. This is correct to 3 decimal places, so the answer to (a) is 53,865.375 in. = 4,488 ft $9\frac{3}{8}$ in.

    (b) Rounded off to 3 significant digits, the answer is 53,900 in. = 4,491 ft 8 in.

61. The cutter removes 18×0.086 mm = 1.548 mm with each revolution. The total of 597.3 revolutions will remove $597.3 \times 1.548 = 924.6204$ mm. Rounded off to 4 significant digits, this is 924.6 mm.

1.6 SCIENTIFIC NOTATION

1. 4.2×10^4
2. 3.7×10^8
3. 3.8×10^{-4}
4. 7.5×10^{-6}
5. $9.807\ 00 \times 10^9$
6. 8.700×10^7
7. 9.70×10^{-5}
8. 4.00×10^{-1}
9. 4.3×10^0
10. 2.07×10^0
11. 4500
12. 370000
13. 40500000
14. 305000000
15. 0.000063
16. 0.0000000187
17. 72
18. 0.96
19. $(7.6 \times 10^5) \times (2.04 \times 10^{10}) = 15.504 \times 10^{15} = 1.5504 \times 10^{16}$
20. $(4.32 \times 10^7) \times (8.5 \times 10^8) = 36.72 \times 10^{15} = 3.672 \times 10^{16}$
21. $(3.5 \times 10^{-5}) \times (7.6 \times 10^{-7}) = 26.6 \times 10^{-12} = 2.66 \times 10^{-11}$
22. $(4.2 \times 10^{-4}) \times (7.5 \times 10^{-5}) = 31.5 \times 10^{-9} = 3.15 \times 10^{-8}$
23. $(8.4 \times 10^8) \times (3.5 \times 10^{-4}) = 29.4 \times 10^4 = 2.94 \times 10^5$
24. $(4.2 \times 10^{-6}) \times (2.3 \times 10^4) = 9.66 \times 10^{-2}$
25. $(7.04 \times 10^4) \times (3.2 \times 10^{-6}) = 22.528 \times 10^{-2} = 2.2528 \times 10^{-1}$
26. $(3.02 \times 10^{-4}) \times (4.37 \times 10^9) = 13.1974 \times 10^5 = 1.31974 \times 10^6$
27. $(2.88 \times 10^{10}) \div (2.4 \times 10^5) = 1.2 \times 10^5$
28. $(5.55 \times 10^{10}) \div (3.7 \times 10^5) = 1.5 \times 10^5$
29. $(3.75 \times 10^5) \div (1.5 \times 10^8) = 2.5 \times 10^{-3}$
30. $(7.98 \times 10^4) \div (8.4 \times 10^8) = 0.95 \times 10^{-4} = 9.5 \times 10^{-5}$
31. $(3.2 \times 10^{-3}) \div (1.6 \times 10^{-7}) = 2 \times 10^4$
32. $(4.8 \times 10^{-4}) \div (3 \times 10^{-7}) = 1.6 \times 10^3$
33. $(3.6 \times 10^{-7}) \div (2 \times 10^{-4}) = 1.8 \times 10^{-3}$
34. $(9.8 \times 10^{-9}) \div (1.4 \times 10^{-5}) = 7 \times 10^{-4}$
35. $(8.76 \times 10^6)(2.46 \times 10^8)(6.4 \times 10^9) = 137.91744 \times 10^{6+8+9} = 137.91744 \times 10^{23} = 1.3791744 \times 10^{25} \approx 1.38 \times 10^{25}$
36. $(4.36 \times 10^6)(6.25 \times 10^8)(3.87 \times 10^{10}) = 105.4575 \times 10^{24} = 1.054575 \times 10^{26} \approx 1.05 \times 10^{26}$
37. $(2.5 \times 10^5)(6.3 \times 10^8) \div (6.3 \times 10^{-9}) = 2.5 \times 10^{5+8-(-9)} = 2.5 \times 10^{22}$
38. $(2.52 \times 10^4)(8 \times 10^6) \div (3.97 \times 10^9) = 5.0780856 \times 10^{4+6-9} = 5.0780856 \times 10^1 \approx 5.08 \times 10^1$
39. $(5.2 \times 10^{-6})(4.8 \times 10^8) \div (6.4 \times 10^{-12}) = 3.9 \times 10^{-6+8-(-12)} = 3.9 \times 10^{14}$
40. $(9.6 \times 10^7)(8.1 \times 10^5) \div (2.43 \times 10^{14}) = 32 \times 10^{7+5-14} = 3.2 \times 10^{-1}$
41. 675 (answers may vary)
42. 2 200 (answers may vary)
43. $(3 \times 10^{11})(3.7 \times 10^{-6}) = 11.1 \times 10^5 = 1.11 \times 10^6$ mm
44. $(1.5 \times 10^8) \div (3 \times 10^5) = 0.5 \times 10^3 = 5 \times 10^2 = 500$ seconds; $500 \div 60 = 8.33$ minutes
45. $(1.6606 \times 10^{-27})(1.2 \times 10^1)(1.4 \times 10^7) = 2.789808 \times 10^{-27+1+7} = 2.789808 \times 10^{-19}$ kg
46. $(1.6606 \times 10^{-27})(5.5 \times 10^1)(2.3 \times 10^8) = 21.00659 \times 10^{-18} = 2.100659 \times 10^{-17}$ kg
47. One neutron; approximately $(1.6750 \times 10^{-27}) \div (9.1095 \times 10^{-31}) = 1.84 \times 10^3$ times heavier
48. $X_L = 2\pi fL = 2\pi(1 \times 10^7)(1.5 \times 10^{-2}) \approx 9.425 \times 10^5 = 942\ 500\ \Omega$

1.7 ROOTS

1. 5
2. 6
3. 12
4. 11
5. 2
6. −4
7. −3
8. 3
9. 2
10. −3
11. $\frac{2}{3}$
12. $\frac{-3}{10}$
13. 0.2
14. 0.5
15. −0.1
16. 0.5
17. 3
18. 5
19. 8.32
20. 7.91
21. $\sqrt[3]{5}\sqrt[3]{25} = \sqrt[3]{5 \cdot 25} = \sqrt[3]{5^3} = 5$
22. $\sqrt[5]{-3}\sqrt[5]{81} = \sqrt[5]{-3}\,\sqrt[5]{(-3)^4} = \sqrt[5]{(-3)^5} = -3$
23. $\sqrt[4]{8}\sqrt[4]{9} = \sqrt[4]{8 \cdot 9} = \sqrt[4]{72}$
24. $\sqrt[6]{12}\sqrt[6]{48} = \sqrt[6]{12 \cdot 48} = \sqrt[6]{576} = \sqrt[6]{64}\sqrt[6]{9} = 2\sqrt[6]{3^2} = 2\sqrt[3]{3}$
25. $\dfrac{\sqrt{75}}{\sqrt{3}} = \sqrt{\dfrac{75}{3}} = \sqrt{25} = 5$
26. $\dfrac{\sqrt{112}}{\sqrt{7}} = \sqrt{\dfrac{112}{7}} = \sqrt{16} = 4$
27. $\dfrac{\sqrt[3]{5}}{\sqrt[3]{40}} = \sqrt[3]{\dfrac{5}{40}} = \sqrt[3]{\dfrac{1}{8}} = \dfrac{1}{2}$
28. $\dfrac{\sqrt[3]{11}}{\sqrt[3]{297}} = \sqrt[3]{\dfrac{11}{297}} = \sqrt[3]{\dfrac{1}{27}} = \dfrac{1}{3}$
29. $\sqrt[3]{2^3} + \sqrt[4]{5^4} = 2 + 5 = 7$
30. $\sqrt{3^2} - \sqrt[3]{2^3} = 3 - 2 = 1$
31. $\sqrt[3]{\left(\dfrac{2}{3}\right)^3} - \sqrt[4]{\left(\dfrac{1}{3}\right)^4} = \dfrac{2}{3} - \dfrac{1}{3} = \dfrac{1}{3}$
32. $\sqrt[5]{\left(\dfrac{3}{4}\right)^5} + \sqrt[3]{\left(\dfrac{5}{4}\right)^3} = \dfrac{3}{4} + \dfrac{5}{4} = \dfrac{8}{4} = 2$
33. $\sqrt{5^{2/3}} = \left(\sqrt{5^{1/3}}\right)^2 = 5^{1/3}$
34. $\sqrt{7^{2/5}} = \left(\sqrt{7^{1/5}}\right)^2 = 7^{1/5}$
35. $\sqrt[3]{16^{3/4}} = \left(\sqrt[3]{16^{1/4}}\right)^3 = 16^{1/4} = \sqrt[4]{16} = 2$
36. $\sqrt[4]{27^{4/3}} = 27^{1/3} = 3$
37. $16^{3/4} = \sqrt[4]{16^3} = 2^3 = 8$
38. $(-27)^{2/3} = (-3)^2 = 9$
39. $(-8)^{2/3} = (-2)^2 = 4$
40. $25^{3/2} = \sqrt{25}^{\,3} = 5^3 = 125$
41. $8^{-2/3} = \left(\sqrt[3]{8}\right)^{-2} = 2^{-2} = \frac{1}{2^2} = \frac{1}{4}$
42. $9^{-3/2} = \sqrt{9}^{\,-3} = 3^{-3} = \frac{1}{3^3} = \frac{1}{27}$
43. $(-27)^{4/3} = (\sqrt[3]{-27})^4 = (-3)^4 = 81$
44. $(-32)^{3/5} = \sqrt[5]{-32}^{\,3} = -2^3 = -8$
45. $\sqrt{\dfrac{81}{(8)(0.01)}} = \dfrac{\sqrt{81}}{\sqrt{8}\sqrt{0.01}} = \dfrac{9}{2\sqrt{2}(0.1)} = \dfrac{9}{0.2\sqrt{2}}$
46. $\sqrt[3]{\dfrac{(27)(0.008)^2}{0.027}} = \dfrac{\sqrt[3]{27}\left(\sqrt[3]{0.008}\right)^2}{\sqrt[3]{0.027}} = \dfrac{3(0.2)^2}{0.3} = \dfrac{3(0.04)}{0.3} = 0.4$
47. $\sqrt[3]{\dfrac{(0.125)^3\sqrt{144}}{3/2}} = \dfrac{\sqrt[3]{(0.125)^3}\sqrt[3]{\sqrt{144}}}{\sqrt[3]{3/2}} = \dfrac{\sqrt[3]{(0.125)^3}\sqrt[3]{12}}{\frac{\sqrt[3]{3}}{\sqrt[3]{2}}} = 0.125\sqrt[3]{\dfrac{12 \cdot 2}{3}} = 0.125\sqrt[3]{8} = 0.125 \times 2 = 0.25$

48. $\sqrt{\dfrac{64}{(0.25)(0.16)}} = \dfrac{8}{(0.5)(0.4)} = \dfrac{8}{0.2} = 40$

49. $\lambda = \dfrac{3 \times 10^8}{1.435 \times 10^8} = 2.091$ m

50. about 4.601 days

51. $v_L = \sqrt{\frac{K}{\rho}} = \sqrt{\frac{2.1\times10^9}{1000}} = \sqrt{\frac{2.1\times10^9}{10^3}} = \sqrt{2.1 \times 10^{9-3}} = \sqrt{2.1 \times 10^6} \approx 1449.1377$ m/s

52. 2.473 m/s

CHAPTER 1 REVIEW

1. (a) Integers, rational numbers, real numbers, (b) Rational numbers, real numbers, (c) Irrational numbers, real numbers

2. (a) $\sqrt{42}$, (b) 16, (c) $\frac{5}{8}$

3. (a) $\frac{3}{2}$, (b) $\frac{-1}{8}$, (c) -5

4. (a) 5, (b) -17, (c) $-4\sqrt{2}$

5. (a) Commutative law for addition, (b) Distributive law, (c) Additive identity, (d) Multiplicative inverse

6. 64

7. -44

8. 53

9. $37 - (-16) = 37 + 16 = 98$

10. -32

11. $\frac{2}{3} + \frac{-5}{6} = \frac{4}{6} + \frac{-5}{6} = \frac{4-5}{6} = -\frac{1}{6}$

12. $\frac{4}{5} - \frac{5}{6} = \frac{24}{30} - \frac{25}{30} = -\frac{1}{30}$

13. -3

14. -3

15. $-\frac{2}{5}$

16. $3\frac{3}{4} \times -4\frac{1}{3} = \frac{15}{4} \times \frac{-13}{3} = \frac{5}{4} \times \frac{-13}{1} = -\frac{65}{4}$

17. $\frac{2}{3} \div \frac{1}{4} = \frac{2}{3} \times \frac{4}{1} = \frac{8}{3}$

18. $\frac{1}{5} \div \frac{-2}{15} = \frac{1}{5} \times \frac{15}{-2} = \frac{1}{1} \times \frac{3}{-2} = -\frac{3}{2}$

19. $2\frac{1}{2} \div 3\frac{1}{4} = \frac{5}{2} \div \frac{13}{4} = \frac{5}{2} \times \frac{4}{13} = \frac{5}{1} \times \frac{2}{13} = \frac{10}{13}$

20. $-4\frac{1}{3} \div -3\frac{1}{6} = -\frac{13}{3} \div -\frac{19}{6} = -\frac{13}{3} \times -\frac{6}{19} = \frac{26}{19}$

21. $2^5 = 2 \times 2 \times 2 \times 2 \times 2 = 32$

22. $(-3)^4 = (-3)(-3)(-3)(-3) = 81$

23. $(-4)^3 = (-4)(-4)(-4) = -64$

24. $8^{1/3} = \sqrt[3]{8} = 2$

25. $4^{1/2} = \sqrt{4} = 2$

26. $(-64)^{1/3} = \sqrt[3]{-64} = -4$

27. $2^5 \cdot 2^3 = 2^{5+3} = 2^8$

28. $3^5 \cdot 3^4 = 3^{5+4} = 3^9$

29. $2^{-3} \cdot 2^5 = 2^{-3+5} = 2^2$

30. $16^{-4} \cdot 16^{-3} = 16^{-7} = \frac{1}{16^7}$

31. $(4^3)^5 = 4^{3\cdot5} = 4^{15}$

32. $(2^{-3})^4 = 2^{(-3)4} = 2^{-12} = \frac{1}{2^{12}}$

33. $(4^{1/3})^3 = 4^1 = 4$

34. $(5^{1/4})^{2/3} = 5^{1/4\cdot2/3} = 5^{1/6}$

35. $\dfrac{a^2b^3}{ab^4} = \dfrac{a^{2-1}}{b^{4-3}} = \dfrac{a}{b}$

36. $\dfrac{x^2y^3z}{x^3yz} = \dfrac{y^{3-1}}{x^{3-2}} = \dfrac{y^2}{x}$

37. $(ax^2)^{-2} = a^{-2}x^{-4} = \dfrac{1}{a^2x^4}$ or $(ax^2)^{-2} = \dfrac{1}{(ax^2)^2} = \dfrac{1}{a^2x^4}$

38. $\frac{(by^{-2})^2}{(cy^{-4})^{1/4}} = \frac{b^2y^{-4}}{c^{1/4}y^{-1}} = \frac{b^2}{c^{1/4}y^3}$

39. (a) 2.37, (b) 2.37

40. (a) 2.02, (b) 0.002

41. (a) both, (b) 0.7

42. (a) both, (b) 0.0021

43. (a) 7.4, (b) 7.35, (c) 7.4, (d) 7.35

44. (a) 18, (b) 18.3, (c) 18.3, (d) 18.29

45. (a) 2.1, (b) 2.05, (c) 2.1, (d) 2.05

46. (a) 4.0, (b) 4.03, (c) 4.0, (d), 4.03

47. 3.71×10^{11}

48. 2.54×10^{15}

49. 2.4×10^{-11}

50. 4.91×10^{-20}

51. 12

52. −4

53. 5

54. $\frac{6}{11}$

CHAPTER 1 TEST

1. (a) $\left|\frac{4}{3}\right| = \frac{4}{3}$; (b) $\left|\frac{1}{2} - \frac{5}{8}\right| = \left|\frac{4}{8} - \frac{5}{8}\right| = \left|-\frac{1}{8}\right| = \frac{1}{8}$; (c) $|-6| = 6$

2. $-\frac{7}{3}$

3. $\frac{5}{8}$

4. 111

5. −112

6. −70

7. $\frac{5}{3} + 5\frac{2}{3} = \frac{5}{3} + \frac{17}{3} = \frac{22}{3}$ or $7\frac{1}{3}$

8. $\frac{7}{4} - (-\frac{3}{5}) = \frac{35}{20} + \frac{12}{20} = \frac{47}{20}$

9. $-4\frac{1}{2} \times 2\frac{1}{3} = -\frac{9}{2} \times \frac{7}{3} = -\frac{21}{2}$

10. $-\frac{5}{7} \div \frac{15}{28} = -\frac{5}{7} \times \frac{28}{15} = -\frac{4}{3}$

11. $-2\frac{1}{3} \div -4\frac{5}{6} = -\frac{7}{3} \div -\frac{29}{6} = -\frac{7}{3} \times -\frac{6}{29} = \frac{14}{29}$

12. $(-5)^3 = (-5)(-5)(-5) = -125$

13. $4^{3/2} = \left(\sqrt{4}\right)^3 = 2^3 = 8$

14. $2^6 \cdot 2^{-4} = 2^{6-4} = 2^2 = 4$

15. $\left(5^{1/4}\right)^8 = 5^{(1/4)8} = 5^2 = 25$

16. $3^{5/2} \div 3^{-3/2} = 3^{5/2-(-3/2)} = 3^{5/2+3/2} = 3^4 = 81$

17. $\frac{a^3b^5}{a^4b^2} = \frac{b^{5-2}}{a^{4-3}} = \frac{b^3}{a}$

18. 4.516

19. 0.000 51

20. 4.75×10^{13}

21. $\sqrt{\frac{49}{25}} = \frac{\sqrt{49}}{\sqrt{25}} = \frac{7}{5}$

CHAPTER

2

Algebraic Concepts and Operations

2.1 ADDITION AND SUBTRACTION

1. $4x + 7x = (4 + 7)x = 11x$
2. $5y - 2y = (5 - 2)y = 3y$
3. $3z - z = 3z - 1z = (3 - 1)z = 2z$
4. $7w + 4w - w = (7 + 4 - 1)w = 10w$
5. $8x + 9x^2 - 2x = 8x - 2x + 9x^2 = (8 - 2)x + 9x^2 = 6x + 9x^2$ or $9x^2 + 6x$
6. $11y - 7y + 6y^2 = (11 - 7)y + 6y^2 = 4y + 6y^2$ or $6y^2 + 4y$
7. $10w + w^2 - 8w^2 = 10w + (1 - 8)w^2 = 10w - 7w^2$ or $-7w^2 + 10w$
8. $y^2 - 6y^2 + 4y = (1 - 6)y^2 + 4y = -5y^2 + 4y$
9. $ax^2 + a^2x + ax^2 = ax^2 + ax^2 + a^2x = 2ax^2 + a^2x$
10. $by - by^2 + by = by + by - by^2 = 2by - by^2$
11. $7xy^2 - 5x^2y + 4xy^2 = 7xy^2 + 4xy^2 - 5x^2y = 11xy^2 - 5x^2y$
12. $12wz - 8w^2z + 6w^2z = 12wz + (-8 + 6)w^2z = 12wz - 2w^2z$
13. $(a + 6b) - (a - 6b) = a + 6b - a + 6b = a - a + 6b + 6b = (1 - 1)a + (6 + 6)b = 0a + 12b = 12b$
14. $(x - 7y) - (7y - x) = x - 7y - 7y + x = x + x - 7y - 7y = 2x - 14y$
15. $\left(2a^2 + 3b\right) + (2b + 4a) = 2a^2 + 3b + 2b + 4a = 2a^2 + (3 + 2)b + 4a = 2a^2 + 5b + 4a$
16. $\left(7c^2 - 8d\right) + (6d - 8c) = 7c^2 - 8d + 6d - 8c = 7c^2 + (-8 + 6)d - 8c = 7c^2 - 2d - 8c$
17. $\left(4x^2 + 3x\right) - \left(2x^2 - 3x\right) = 4x^2 + 3x - 2x^2 + 3x = (4 - 2)x^2 + (3 + 3)x = 2x^2 + 6x$
18. $\left(3y^2 - 4x\right) - (4y + 2x) = 3y^2 - 4x - 4y - 2x = 3y^2 + (-4 - 2)x - 4y = 3y^2 - 6x - 4y$ or $3y^2 - 4y - 6x$
19. $2\left(6y^2 + 7x\right) = 2 \cdot 6y^2 + 2 \cdot 7x = 12y^2 + 14x$
20. $5(3a + 4b) = 5 \cdot 3a + 5 \cdot 4b = 15a + 20b$
21. $-3(4b - 2c) = -3 \cdot 4b + (-3)(-2c) = -12b + 6c$
22. $-2(-6b + 3a) = (-2)(-6b) + (-2)(3a) = 12b - 6a$
23. $4(a + b) + 3(b + a) = 4a + 4b + 3b + 3a = 7a + 7b$
24. $2(c + d) + 8(d + c) = 2c + 2d + 8d + 8c = 10c + 10d$
25. $3(x + y) - 2(x + y) = 3x + 3y - 2x - 2y = (3 - 2)x + (3 - 2)y = x + y$
26. $3\left(x^2 - y\right) - 2\left(y + x^2\right) = 3x^2 - 3y - 2y - 2x^2 = x^2 - 5y$
27. $2(a + b + c) + 3(a + b - c) = 2a + 2b + 2c + 3a + 3b - 3c = 5a + 5b - c$
28. $4(x - y + z) + 2(x + y - z) = 4x - 4y + 4z + 2x + 2y - 2z = 6x - 2y + 2z$

29. $3[2(x+y)] = 3[2x+2y] = 6x+6y$

30. $4[3(x-y)] = 4[3x-3y] = 12x-12y$

31. $3(a+b)+4(a+b)-2(a+b) = 3a+3b+4a+4b-2a-2b = 5a+5b$ or $(3+4-2)(a+b) = 5(a+b) = 5a+5b$

32. $2(x+y)-3(x+y)-4(x+y) = (2-3-4)(x+y) = -5(x+y) = -5x-5y$ or $2x+2y-3x-3y-4x-4y = -5x-5y$

33. $2(a+b+c)+3(a-b+c)+(a-b-c) = 2a+2b+2c+3a-3b+3c+a-b-c = 6a-2b+4c$

34. $3(x-y+z)-2(x+y-z)+4(-x-y+z) = 3x-3y+3z-2x-2y+2z-4x-4y+4z = -3x-9y+9z$

35. $2(x+3y)-3(x-2y)+5(2x-y) = 2x+6y-3x+6y+10x-5y = 9x+7y$

36. $3(y-2a)-(a+3y)+4(a-3y) = 3y-6a-a-3y+4a-12y = -12y-3a$

37. $3(x+y-z)-2(3x+2y-z)-3(x-y+4z) = 3x+3y-3z-6x-4y+2z-3x+3y-12z = -6x+2y-13z$

38. $5(x-y+z)-(y-x+z)+2(x-2y+z) = 5x-5y+5z-y+x-z+2x-4y+2z = 8x-10y+6z$

39. $(x+y)-3(x-z)+4(y+4z)-2(x+y-3z) = x+y-3x+3z+4y+16z-2x-2y+6z = -4x+3y+25z$

40. $(a+b)-2(b-c)+4(c+2d)-5(d+2c-3b-a) = a+b-2b+2c+4c+8d-5d-10c+15b+5a = 6a+14b-4c+3d$

41. $x+[3x+2(x+y)] = x+[3x+2x+2y] = x+[5x+2y] = x+5x+2y = 6x+2y$

42. $x+[5y+3(y-x)] = x+[5y+3y-3x] = x+[8y-3x] = -2x+8y$

43. $y-[2z-3(y+z)+y] = y-[2z-3y-3z+y] = y-[-z-2y] = y+z+2y = 3y+z$

44. $2w-[4z-5(z+w)+2w] = 2w-[4z-5z-5w+2w] = 2w-[-z-3w] = 2w+z+3w = 5w+z$

45. $[2x+3(x+y)-2(x-y)+y]-2x = [2x+3x+3y-2x+2y+y]-2x = [3x+6y]-2x = x+6y$

46. $[4a-8(a+b)+2(a-b)+3a]-3b = [4a-8a-8b+2a-2b+3a]-3b = [a-10b]-3b = a-13b$

47. $-\{-[2a-(3b+a)]\} = -\{-[2a-3b-a]\} = -\{-[a-3b]\} = -\{-a+3b\} = a-3b$

48. $-\{-3[4x-(5x-4y)]\} = -\{-3[4x-5x+4y]\} = -\{-3[-x+4y]\} = -\{3x-12y\} = -3x+12y$

49. $5a-2\{4[a+2(4a+b)-b]+a\}-a = 5a-2\{4[a+8a+2b-b]+a\}-a = 5a-2\{4[9a+b]+a\}-a = 5a-2\{36a+4b+a\}-a = 5a-2\{37a+4b\}-a = 5a-74a-8b-a = -70a-8b$

50. $7x-3\{-[x+2(x-y)-y]+2x\}-3y = 7x-3\{-[x+2x-2y-y]+2x\}-3y = 7x-3\{-[3x-3y]+2x\}-3y = 7x-3\{-3x+3y+2x\}-3y = 7x-3\{-x+3y\}-3y = 7x+3x-9y-3y = 10x-12y$

51. $p+\frac{1}{2}p+\frac{2}{3}p = \frac{6}{6}p+\frac{3}{6}p+\frac{4}{6}p = \frac{6+3+4}{6}p = \frac{13}{6}p$

52. $x+2x+3x+3x = (1+2+3+3)x = 9x$

2.2 MULTIPLICATION

1. $(a^2x)(ax^2) = a^{2+1}\cdot x^{1+2} = a^3x^3$

2. $(by^2)(b^2y) = b^{1+2}y^{2+1} = b^3y^3$

3. $(3ax)(2ax^2) = 3\cdot 2\cdot a^{1+1}x^{1+2} = 6a^2x^3$

4. $(5by)(3b^2y) = 5\cdot 3b^{1+2}y^{1+1} = 15b^3y^2$

5. $(2xw^2z)(-3x^2w) = 2(-3)x^{1+2}w^{2+1}z = -6x^3w^3z$

6. $(-4ya^2b)(6y^2b) = (-4)6\cdot y^{1+2}a^2b^{1+1} = -24y^3a^2b^2$

7. $(3x)(4ax)(-2x^2b) = 3\cdot 4(-2)x^{1+1+2}ab = -24x^4ab$

8. $(4y)(3y^2b)(-5by^2) = 4\cdot 3(-5)y^{1+2+2}b^{1+1} = -60b^2y^5$

9. $2(5y-6)=2\cdot 5y+2(-6)=10y-12$

10. $4(3x-5)=4(3x)+4(-5)=12x-20$

11. $-5(4w-7)=-5(4w)+(-5)(-7)=-20w+35$ or $35-20w$

12. $-3(8+5p)=-24-15p$

13. $3x(7y+4)=3x(7y)+3x(4)=21xy+12x$

14. $6x(8y-7)=6x(8y)+6x(-7)=48xy-42x$

15. $-5t(-3+t)=(-5t)(-3)-5t(t)=15t-5t^2$

16. $-3n(2n-5)=-3n(2n)-3n(-5)=-6n^2+15n$ or $15n-6n^2$

17. $\frac{1}{2}a(4a-2)=\frac{1}{2}a(4a)+\frac{1}{2}a(-2)=2a^2-a$

18. $\frac{1}{3}x(-21x-15)=\frac{1}{3}x(-21x)+\frac{1}{3}x(-15)=-7x^2-5x$

19. $2x\left(3x^2-x+4\right)=2x\left(3x^2\right)+2x(-x)+2x(4)=6x^3-2x^2+8x$

20. $3y\left(4y^2-5y-7\right)=3y\left(4y^2\right)+3y(-5y)+3y(-7)=12y^3-15y^2-21y$

21. $4y^2\left(-5y^2+2y-5+3y^{-1}-6y^2\right)=4y^2\left(-5y^{-2}\right)+4y^2(2y)+4y^2(-5)+4y^2\left(3y^{-1}\right)+4y^2\left(-6y^{-2}\right)=-20y^4+8y^3-20y^2+12y-24$

22. $5p^2\left(-4p^3-3p+2+p^{-1}-7p^{-2}\right)=5p^2\left(-4p^3\right)+5p^2(-3p)+5p^2(2)+5p^2\left(p^{-1}\right)+5p^2\left(-7p^{-2}\right)=-20p^5-15p^3+10p^2+5p-35$

23. $(a+b)(a+c)=(a+b)a+(a+b)c=a^2+ab+ac+bc$ or by FOIL $a^2+ac+ab+bc$

24. $(s+t)(s+2t)=s^2+2st+st+2t^2=s^2+3st+2t^2$

25. $(x+5)\left(x^2+6\right)=x^3+6x+5x^2+30$ or $x^3+5x^2+6x+30$

26. $(y+3)\left(y^2+7\right)=y^3+7y+3y^2+21$ or $y^3+3y^2+7y+21$

27. $(2x+y)(3x-y)=6x^2-2xy+3xy-y^2=6x^2+xy-y^2$

28. $(4a+b)(8a-b)=32a^2-4ab+8ab-b^2=32a^2+4ab-b^2$

29. $(2a-b)(3a-2b)=6a^2-4ab-3ab+2b^2=6a^2-7ab+2b^2$

30. $(4p+q)(3p-2q)=12p^2-8pq+3pq-2q^2=12p^2-5pq-2q^2$

31. $(b-1)(2b+5)=2b^2+5b-2b-5=2b^2+3b-5$

32. $(4x-1)(3x-2)=12x^2-8x-3x+2=12x^2-11x+2$

33. $\left(7a^2b+3c\right)\left(8a^2b-3c\right)=56a^4b^2-21a^2bc+24a^2bc-9c^2=56a^4b^2+3a^2bc-9c^2$

34. $\left(6p^2r+2t\right)\left(5p^2r+4t\right)=30p^4r^2+24p^2rt+10p^2rt+8t^2=30p^4r^2+34p^2rt+8t^2$

35. $(x+4)(x-4)=x^2-16$ (Difference of Squares)

36. $(a+8)(a-8)=a^2-64$ (Difference of Squares)

37. $(p-6)(p+6)=p^2-36$

38. $(b-10)(b+10)=b^2-100$

39. $(ax+2)(ax-2)=a^2x^2-4$

40. $(xy-3)(xy+3)=x^2y^2-9$

41. $\left(2r^2+3x\right)\left(2r^2-3x\right)=4r^4-9x^2$

42. $\left(4p^3-7d\right)\left(4p^3+7d\right)=16p^6-49d^2$

43. $\left(5a^2x^3-4d\right)\left(5a^2x^3+4d\right)=25a^4x^6-16d^2$

44. $\left(3p^2st-\frac{11}{3}w^3\right)\left(3p^2st+\frac{11}{3}w^3\right)=9p^4s^2t^2-\frac{121}{9}w^6$

45. $\left(\frac{2}{3}pa^2f+\frac{3}{4}tb^3\right)\left(-\frac{2}{3}pa^2f+\frac{3}{4}tb^3\right)=\left(\frac{3}{4}tb^3+\frac{2}{3}pa^2f\right)\left(\frac{3}{4}tb^3-\frac{2}{3}pa^2f\right)=\frac{9}{16}t^2b^6-\frac{4}{9}p^2a^4f^2$

46. $\left(\frac{\sqrt{3}}{2}+\frac{7}{5}t^2u\right)\left(-\frac{\sqrt{3}}{2}+\frac{7}{5}t^2u\right)=\left(\frac{7}{5}t^2u+\frac{\sqrt{3}}{2}\right)\left(\frac{7}{5}t^2u-\frac{\sqrt{3}}{2}\right)=\frac{49}{25}t^4u^2-\frac{3}{4}$

47. $(x+y)^2=x^2+2xy+y^2$ (Square of a Binomial)

48. $(p+r)^2=p^2+2pr+r^2$ (Square of a Binomial)

49. $(x-5)^2=x^2-10x+25$

50. $(b-7)^2 = b^2 - 14b + 49$

51. $(a+3)^2 = a^2 + 6a + 9$

52. $(w+5)^2 = w^2 + 10w + 25$

53. $(2a+b)^2 = 4a^2 + 4ab + b^2$

54. $(3c+d)^2 = 9c^2 + 6cd + d^2$

55. $(3x-2y)^2 = 9x^2 - 12xy + 4y^2$

56. $(5a-6f)^2 = 25a^2 - 60af + 36f^2$

57. $4x(x+4)(3x-2) = 4x\left(3x^2 - 2x + 12x - 8\right) = 4x\left(3x^2 + 10x - 8\right) = 12x^3 + 40x^2 - 32x$

58. $5y(y-6)(2y+3) = 5y\left(2y^2 - 9y - 18\right) = 10y^3 - 45y^2 - 90y$

59. $(x+y-z)(x-y+z) = (x+y-z)(x) + (x+y-z)(-y) + (x+y-z)(z) = x^2 + xy - xz - xy - y^2 + yz + xz + yz - z^2 = x^2 - y^2 - z^2 + 2yz$ or $[x+(y-z)][x-(y-z)] = x^2 - (y-z)^2 = x^2 - y^2 - z^2 + 2yz$

60. $(a+b+c)(a-b-c) = [a+(b+c)][a-(b+c)] = a^2 - (b+c)^2 = a^2 - \left(b^2 + 2bc + c^2\right) = a^2 - b^2 - c^2 - 2bc$

61. $2[n(n+2)+n] = 2[n^2 + 2n + n] = 2[n^2 + 3n] = 2n^2 + 6n$

62. $(2R-x)^2 - x^2 - R^2 = \left(4R^2 - 4Rx + x^2\right) - x^2 - R^2 = 3R^2 - 4Rx$

2.3 DIVISION

1. x^7 by $x^3 = \frac{x^7}{x^3} = x^{7-3} = x^4$

2. y^8 by $y^6 = \frac{y^8}{y^6} = y^{8-6} = y^2$

3. $2x^6$ by $x^4 = \frac{2x^6}{x^4} = 2x^2$

4. $3w^4$ by $w^2 = \frac{3w^4}{w^2} = 3w^2$

5. $12y^5$ by $4y^3 = \frac{12y^5}{4y^3} = \frac{12}{4} \cdot \frac{y^5}{y^3} = 3y^2$

6. $15a^7$ by $3a^4 = \frac{15a^7}{3a^4} = 5a^3$

7. $-45ab^2$ by $15ab = \frac{-45ab^2}{15ab} = -3b$

8. $-55xy^3$ by $-11xy = \frac{-55xy^3}{-11xy} = 5y^2$

9. $33xy^2z$ by $3xyz = \frac{33xy^2z}{3xyz} = 11y$

10. $65x^2yz$ by $5xyz = \frac{65x^2yz}{5xyz} = 13x$

11. $96a^2xy^3$ by $-16axy^2 = \frac{96a^2xy^3}{-16axy^2} = -6ay$

12. $105b^3yw^2$ by $-15b^2yw = \frac{105b^3yw^2}{-15b^2yw} = -7bw$

13. $144c^3d^2f$ by $8cf = \frac{144c^3d^2f}{8cf} = 18c^2d^2$

14. $162x^2yz^3$ by $9x^2z^2 = \frac{162x^2yz^3}{9x^2z^2} = 18yz$

15. $9np^3$ by $-15n^3p^2 = \frac{9np^3}{-15n^3p^2} = \frac{3p}{-5n^2}$ or $\frac{-3p}{5n^2} = -\frac{3}{5}\frac{p}{n^2}$

16. $15rs^2t$ by $-27r^2st^3 = \frac{15rs^2t}{-27r^2st^3} = \frac{-5s}{9rt^2}$

17. $8abcdx^2y$ by $14adxy^2 = \frac{8abcdx^2y}{14adxy^2} = \frac{4bcx}{7y}$

18. $9efg^2hr$ by $24e^2fh^3r = \frac{9efg^2hr}{24e^2fh^3r} = \frac{3g^2}{8eh^2}$

19. $2a^3 + a^2$ by $a = \frac{2a^3+a^2}{a} = \frac{2a^3}{a} + \frac{a^2}{a} = 2a^2 + a$

20. $4x^4 - x^3$ by $x^2 = \frac{4x^4-x^3}{x^2} = \frac{4x^4}{x^2} - \frac{x^3}{x^2} = 4x^2 - x$

21. $36b^4 - 18b^2$ by $9b = \frac{36b^4-18b^2}{9b} = \frac{36b^4}{9b} - \frac{18b^2}{9b} = 4b^3 - 2b$

22. $49y^5 + 35y^3$ by $7y^2 = \frac{49y^5+35y^3}{7y^2} = \frac{49y^5}{7y^2} + \frac{35y^3}{7y^2} = 7y^3 + 5y$

23. $42x^2 + 28x$ by $7 = \frac{42x^2+28x}{7} = \frac{42x^2}{7} + \frac{28x}{7} = 6x^2 + 4x$

24. $56z^6 - 48z^3$ by $8 = \frac{56z^6-48z^3}{8} = \frac{56z^6}{8} - \frac{48z^3}{8} = 7z^6 - 6z^3$

25. $34x^5 - 51x^2$ by $17x^2 = \frac{34x^5-51x^2}{17x^2} = \frac{34x^5}{17x^2} - \frac{51x^2}{17x^2} = 2x^3 - 3$

26. $105w^6 + 63w^4$ by $21w^2 = \frac{105w^6+63w^4}{21w^2} = 5w^4 + 3w^2$

27. $24x^6 - 8x^4$ by $-4x^3 = \frac{24x^6-8x^4}{-4x^3} = \frac{24x^6}{-4x^3} - \frac{8x^4}{-4x^3} = -6x^3 + 2x$

28. $42y^7 - 24y^5$ by $6y^4 = \frac{42y^7-24y^5}{6y^4} = \frac{42y^7}{6y^4} - \frac{24y^5}{6y^4} = 7y^3 - 4y$

29. $5x^2y+5xy^2$ by $xy = \frac{5x^2y+5xy^2}{xy} = \frac{5x^2y}{xy} + \frac{5xy^2}{xy} = 5x+5y$

30. $7a^2b - 7ab^2$ by $ab = \frac{7a^2b-7ab^2}{ab} = \frac{7a^2b}{ab} - \frac{7ab^2}{ab} = 7a - 7b$

31. $10x^2y+15xy^2$ by $5xy = \frac{10x^2y+15xy^2}{5xy} = \frac{10x^2y}{5xy} + \frac{15xy^2}{5xy} = 2x + 3y$

32. $25p^2q - 15pq^2$ by $5pq = \frac{25p^2q-15pq^2}{5pq} = \frac{25p^2q}{5pq} - \frac{15pq^2}{5pq} = 5p - 3q$

33. $ap^2q - 2pq$ by $pq = \frac{ap^2q-2pq}{pq} = \frac{ap^2q}{pq} - \frac{2pq}{pq} = ap - 2$

34. bx^2w+3xw by $xw = \frac{bx^2w+3xw}{xw} = \frac{bx^2w}{xw} + \frac{3xw}{xw} = bx+3$

35. $a^2bc + abc$ by $abc = \frac{a^2bc+abc}{abc} = \frac{a^2bc}{abc} + \frac{abc}{abc} = a + 1$

36. $x^3yz - xyz$ by $xyz = \frac{x^3yz}{xyz} - \frac{xyz}{xyz} = x^2 - 1$

37. $9x^2y^2z-3xyz^2$ by $-3xyz = \frac{9x^2y^2z}{-3xyz} - \frac{3xyz^2}{-3xyz} = -3xy+z$

38. $12a^2b^2c + 4abc^2$ by $-4abc = \frac{12a^2b^2c}{-4abc} + \frac{4abc^2}{-4abc} = -3ab - c$

39. $b^3x^2 + b^3$ by $-b = \frac{b^3x^2}{-b} + \frac{b^3}{-b} = -b^2x^2 - b^2$

40. $c^5y^3 - cy^2$ by $-c = \frac{c^5y^3}{-c} - \frac{cy^2}{-c} = -c^4y^3 + y^2$

41. $x^2y + xy - xy^2$ by $xy = \frac{x^2y}{xy} + \frac{xy}{xy} - \frac{xy^2}{xy} = x + 1 - y$

42. $ab^2 - ab + a^2b$ by $ab = \frac{ab^2}{ab} - \frac{ab}{ab} + \frac{a^2b}{ab} = b - 1 + a$

43. $18x^3y^2z-24x^2y^3z$ by $-12x^2yz = \frac{18x^3y^2z}{-12x^2yz} - \frac{24x^2y^3z}{-12x^2yz} = -\frac{3}{2}xy + 2y^2$

44. $36a^4b^2c - 27a^2b^4c^2$ by $-27a^2bc = \frac{36a^4b^2c}{-27a^2bc} - \frac{27a^2b^4c^2}{-27a^2bc} = -\frac{4}{3}a^2b + b^3c$

45.
$$\begin{array}{r} x+4 \\ x+3\overline{)x^2+7x+12} \\ \underline{x^2+3x} \\ 4x+12 \\ \underline{4x+12} \\ 0 \end{array}$$

46.
$$\begin{array}{r} x-3 \\ x+4\overline{)x^2+x-12} \\ \underline{x^2+4x} \\ -3x-12 \\ \underline{-3x-12} \\ 0 \end{array}$$

47.
$$\begin{array}{r} x-1 \\ x-2\overline{)x^2-3x+2} \\ \underline{x^2-2x} \\ -x+2 \\ \underline{-x+2} \\ 0 \end{array}$$

48.
$$\begin{array}{r} x+3 \\ x-5\overline{)x^2-2x-15} \\ \underline{x^2-5x} \\ 3x-15 \\ \underline{3x-15} \\ 0 \end{array}$$

49.
$$\begin{array}{r} x-1 \\ x+2\overline{)x^2+x-2} \\ \underline{x^2+2x} \\ -x-2 \\ \underline{-x-2} \\ 0 \end{array}$$

50.
$$\begin{array}{r} x-2 \\ x+3\overline{)x^2+x-6} \\ \underline{x^2+3x} \\ -2x-6 \\ \underline{-2x-6} \\ 0 \end{array}$$

51.
$$\begin{array}{r} 2a+1 \\ 3a+7\overline{)6a^2+17a+7} \\ \underline{6a^2+14a} \\ 3a+7 \\ \underline{3a+7} \\ 0 \end{array}$$

52.
$$\begin{array}{r} b+3 \\ 4b-2\overline{)4b^2+10b-6} \\ \underline{4b^2-2b} \\ 12b-6 \\ \underline{12b-6} \\ 0 \end{array}$$

53.
$$\begin{array}{r} 4y+2 \\ 2y-3\overline{)8y^2-8y-6} \\ \underline{8y^2-12y} \\ 4y-6 \\ \underline{4y-6} \\ 0 \end{array}$$

54.
$$\begin{array}{r} 3t-2 \\ 4t+3\,\overline{\smash{\big)}\,12t^2+t-6} \\ \underline{12t^2+9t} \\ -8t-6 \\ \underline{-8t-6} \\ 0 \end{array}$$

55.
$$\begin{array}{r} x-2 \\ x^2+2x-1\,\overline{\smash{\big)}\,x^3+0x^2-5x+2} \\ \underline{x^3+2x^2-\ x} \\ -2x^2-4x+2 \\ \underline{-2x^2-4x+2} \\ 0 \end{array}$$

56.
$$\begin{array}{r} d \\ d^2+d+2\,\overline{\smash{\big)}\,d^3+d^2-3d+2} \\ \underline{d^3+d^2+2d} \\ -5d+2 \end{array}$$

57.
$$\begin{array}{r} 2a-1 \\ 3a^2-2a+4\,\overline{\smash{\big)}\,6a^3-7a^2+10a-4} \\ \underline{6a^3-4a^2+\ 8a} \\ -3a^2+\ 2a-4 \\ \underline{-3a^2+\ 2a-4} \\ 0 \end{array}$$

58.
$$\begin{array}{r} 3y-\ 3 \\ 3y^2+3y-4\,\overline{\smash{\big)}\,9y^3+0y^2-16y+\ 8} \\ \underline{9y^3+9y^2-12y} \\ -9y^2-\ 4y+\ 8 \\ \underline{-9y^2-\ 9y+12} \\ 5y-\ 4 \end{array}$$

59.
$$\begin{array}{r} 2x^2+\ x-1 \\ 2x-1\,\overline{\smash{\big)}\,4x^3+0x^2-3x+4} \\ \underline{4x^3-2x^2} \\ 2x^2-3x+4 \\ \underline{2x^2-\ x} \\ -2x+4 \\ \underline{-2x+1} \\ 3 \end{array}$$

60.
$$\begin{array}{r} \tfrac{7}{3}p^2-\tfrac{14}{9}p+\ \tfrac{46}{27} \\ 3p+2\,\overline{\smash{\big)}\,7p^3+\ 0p^2+\ 2p-\ 5} \\ \underline{7p^3+\tfrac{14}{3}p^2} \\ -\tfrac{14}{3}p^2+\ 2p-\ 5 \\ \underline{-\tfrac{14}{3}p^2-\tfrac{28}{9}p} \\ \tfrac{46}{9}p-\ 5 \\ \underline{\tfrac{46}{9}p+\ \tfrac{92}{27}} \\ -\tfrac{227}{27} \end{array}$$

61.
$$\begin{array}{r} r^2-3r \\ r+2\,\overline{\smash{\big)}\,r^3-\ r^2-6r+5} \\ \underline{r^3+2r^2} \\ -3r^2-6r+5 \\ \underline{-3r^2-6r} \\ 5 \end{array}$$

62.
$$\begin{array}{r} 2c^2+\ c+3 \\ c-2\,\overline{\smash{\big)}\,2c^3-3c^2+\ c-4} \\ \underline{2c^3-4c^2} \\ c^2+\ c-4 \\ \underline{c^2-2c} \\ 3c-4 \\ \underline{3c-6} \\ 2 \end{array}$$

63.
$$\begin{array}{r} x^3+3x^2+\ 9x+27 \\ x-3\,\overline{\smash{\big)}\,x^4\qquad\qquad\qquad -81} \\ \underline{x^4-3x^3\qquad\qquad\qquad} \\ 3x^3\qquad\qquad -81 \\ \underline{3x^3-9x^2\qquad\qquad} \\ 9x^2\qquad -81 \\ \underline{9x^2-27x\qquad} \\ 27x-81 \\ \underline{27x-81} \\ 0 \end{array}$$

64.

$$
\begin{array}{r}
y^3-3y^2+9y-27 \\
y+3\overline{)\,y^4 \qquad\qquad\quad -81} \\
\underline{y^4+3y^3 \qquad\qquad\quad} \\
-3y^3 \qquad\quad -81 \\
\underline{-3y^3-9y^2 \qquad\quad} \\
9y^2 \qquad -81 \\
\underline{9y^2+27y \qquad} \\
-27y-81 \\
\underline{-27y-81} \\
0
\end{array}
$$

65.

$$
\begin{array}{r}
4x^3-3x^2-x+6 \\
3x^2+x-2\overline{)\,12x^5-5x^4-14x^3+23x^2+8x-12} \\
\underline{12x^5+4x^4-8x^3 \qquad\qquad\qquad} \\
-9x^4-6x^3+23x^2 \qquad\qquad \\
\underline{-9x^4-3x^3+6x^2 \qquad\qquad} \\
-3x^3+17x^2+8x \qquad \\
\underline{-3x^3-x^2+2x \qquad} \\
18x^2+6x-12 \\
\underline{18x^2+6x-12} \\
0
\end{array}
$$

66.

$$
\begin{array}{r}
3a^4-2a^3 \qquad\qquad\quad +1 \\
7a^3+2a-7\overline{)\,21a^7-14a^6+6a^5-25a^4+21a^3+0a^2+10a-35} \\
\underline{21a^7 \qquad\quad +6a^5-21a^4 \qquad\qquad\qquad\qquad\qquad} \\
-14a^6+0a^5-4a^4+21a^3+0a^2+10a-35 \\
\underline{-14a^6 \qquad\quad -4a^4+14a^3 \qquad\qquad\qquad\quad} \\
7a^3+0a^2+10a-35 \\
\underline{7a^3 \qquad\quad +2a-7} \\
8a-28
\end{array}
$$

67.

$$
\begin{array}{r}
x+y \\
x-y\overline{)\,x^2 \qquad -y^2} \\
\underline{x^2-xy \qquad} \\
xy-y^2 \\
\underline{xy-y^2} \\
0
\end{array}
$$

68.

$$
\begin{array}{r}
a-b \\
a+b\overline{)\,a^2 \qquad -b^2} \\
\underline{a^2+ab \qquad} \\
-ab-b^2 \\
\underline{-ab-b^2} \\
0
\end{array}
$$

69.

$$
\begin{array}{r}
w^2+wz+z^2 \\
w-z\overline{)\,w^3 \qquad\qquad -z^3} \\
\underline{w^3-w^2z \qquad\qquad} \\
w^2z \qquad -z^3 \\
\underline{w^2z-wz^2 \qquad} \\
wz^2-z^3 \\
\underline{wz^2-z^3} \\
0
\end{array}
$$

70.

$$
\begin{array}{r}
x^2-xy+y^2 \\
x+y\overline{)\,x^3 \qquad\qquad +y^3} \\
\underline{x^3+x^2y \qquad\qquad} \\
-x^2y \qquad -y^3 \\
\underline{-x^2y-xy^2 \qquad} \\
xy^2+y^3 \\
\underline{xy^2+y^3} \\
0
\end{array}
$$

71.

$$
\begin{array}{r}
x^2+y^2 \\
x+y\overline{)\,x^3+x^2y+xy^2+y^3} \\
\underline{x^3+x^2y \qquad\qquad} \\
0+xy^2+y^3 \\
\underline{xy^2+y^3} \\
0
\end{array}
$$

72.

$$
\begin{array}{r}
a^2+b^2 \\
a-b\overline{)\,a^3-a^2b+ab^2-b^3} \\
\underline{a^3-a^2b \qquad\qquad} \\
0+ab^2-b^3 \\
\underline{ab^2-b^3} \\
0
\end{array}
$$

73.
$$\begin{array}{r} c^2d^2+2cd+4 \\ cd-2\,\overline{)\,c^3d^3 \qquad\qquad -8} \\ \underline{c^3d^3-2c^2d^2} \\ 2c^2d^2 \qquad -8 \\ \underline{2c^2d^2-4cd} \\ 4cd-8 \\ \underline{4cd-8} \\ 0 \end{array}$$

74.
$$\begin{array}{r} e^2f^2-3ef+9 \\ ef+3\,\overline{)\,e^3f^3 \qquad\qquad +27} \\ \underline{e^3f^3+3e^2f^2} \\ -3e^2f^2 \qquad +27 \\ \underline{-3e^2f^2-9ef} \\ 9ef+27 \\ \underline{9ef+27} \\ 0 \end{array}$$

75.
$$\begin{array}{r} x-y \\ x-y\,\overline{)\,x^2-2xy+y^2} \\ \underline{x^2-xy} \\ -xy+y^2 \\ \underline{-xy+y^2} \\ 0 \end{array}$$

76.
$$\begin{array}{r} a+3b \\ a+3b\,\overline{)\,a^2+6ab+9b^2} \\ \underline{a^2+3ab} \\ 3ab+9b^2 \\ \underline{3ab+9b^2} \\ 0 \end{array}$$

77.
$$\begin{array}{r} p^2r-2p+3r^2 \\ 5p-r\,\overline{)\,5p^3r-p^2r^2-10p^2+2pr+15pr^2-3r^3} \\ \underline{5p^3r-p^2r^2} \\ 0-10p^2+2pr+15pr^2-3r^3 \\ \underline{-10p^2+2pr} \\ 0+15pr^2-3r^3 \\ \underline{0+15pr^2-3r^3} \\ 0 \end{array}$$

78.
$$\begin{array}{r} 4x^2-4xy+y^2 \\ 2x-y\,\overline{)\,8x^3-12x^2y+6xy^2-y^3} \\ \underline{8x^3-4x^2y} \\ -8x^2y+6xy^2-y^3 \\ \underline{-8x^2y+4xy^2} \\ 2xy^2-y^3 \\ \underline{2xy^2-y^3} \\ 0 \end{array}$$

79.
$$\begin{array}{r} a+d+4 \\ a-3d-1\,\overline{)\,a^2-2ad-3d^2+3a-13d-8} \\ \underline{a^2-3ad \qquad -a} \\ ad-3d^2+4a-13d-8 \\ \underline{ad-3d^2 \qquad -d} \\ 4a-12d-8 \\ \underline{4a-12d-4} \\ -4 \end{array}$$

80.
$$\begin{array}{r} x-2y+3 \\ 2x+3y-5\,\overline{)\,2x^2-xy-6y^2+x+19y-15} \\ \underline{2x^2+3xy \qquad -5x} \\ -4xy-6y^2+6x+19y-15 \\ \underline{-4xy-6y^2 \qquad +10y} \\ 6x+9y-15 \\ \underline{6x+9y-15} \\ 0 \end{array}$$

81.
$$\begin{array}{r} af \\ a-f\,\overline{)\,a^2f-af^2} \\ \underline{a^2f-af^2} \\ 0 \end{array}$$

82.
$$\begin{array}{r} d^2-2m^2 \\ d-m\,\overline{)\,d^3-2dm^2+2m^3-d^2m} \\ \underline{d^3 \qquad\qquad -d^2m} \\ -2dm^2+2m^3 \\ \underline{-2dm^2+2m^3} \\ 0 \end{array}$$

83.
$$\begin{array}{r}
a+b-c\\
a-b+c\,\overline{)\,a^2-b^2+2bc-c^2}\\
\underline{a^2-ab+ac\qquad\qquad}\\
ab-ac-b^2+2bc-c^2\\
\underline{ab-b^2+bc\qquad}\\
-ac+bc-c^2\\
\underline{-ac+bc-c^2}\\
0
\end{array}$$

84.
$$\begin{array}{r}
e-f+h\\
e+f+h\,\overline{)\,e^2+2eh-f^2+h^2}\\
\underline{e^2+ef+eh\qquad\qquad}\\
-ef+eh-f^2+h^2\\
\underline{-ef-fh-f^2\qquad}\\
eh+fh+h^2\\
\underline{eh+fh+h^2}\\
0
\end{array}$$

85.
$$\begin{array}{r}
a^2-a+1\\
a^2+a+2\,\overline{)\,a^4+0a^3+2a^2-a+2}\\
\underline{a^4+a^3+2a^2\qquad\qquad}\\
-a^3+0a^2-a+2\\
\underline{-a^3-a^2-2a\qquad}\\
a^2+a+2\\
\underline{a^2+a+2}\\
0
\end{array}$$

86.
$$\begin{array}{r}
x^3-1\\
x^3-x+1\,\overline{)\,x^6-x^4+2x^2-1}\\
\underline{x^6-x^4+x^3\qquad\qquad}\\
-x^3+2x^2-1\\
\underline{-x^3+x-1}\\
2x^2-x
\end{array}$$

87. Reciprocal is $\frac{R_2R_3+R_1R_3+R_1R_2}{R_1R_2R_3} = \frac{R_2R_3}{R_1R_2R_3} + \frac{R_1R_3}{R_1R_2R_3} + \frac{R_1R_2}{R_1R_2R_3} = \frac{1}{R_1} + \frac{1}{R_2} + \frac{1}{R_3}$

2.4 SOLVING EQUATIONS

1. $x - 7 = 32$; $(x - 7) + 7 = 32 + 7$; $x = 39$
2. $y - 8 = 41$; $(y - 8) + 8 = 41 + 8$; $y = 49$
3. $a + 13 = 25$; $(a + 13) - 13 = 25 - 13$; $a = 12$
4. $b + 21 = 34$; $b + 21 - 21 = 34 - 21$; $b = 13$
5. $25 + c = 10$; $25 + c - 25 = 10 - 25$; $c = -15$
6. $28 + d = 12$; $28 + d - 28 = 12 - 28$; $d = -16$
7. $4.3 + w = 8.7$; $4.3 + w - 4.3 = 8.7 - 4.3$; $w = 4.4$
8. $5.1 + z = 9.1$; $5.1 + z - 5.1 = 9.1 - 5.1$; $z = 4$
9. $4x = 18$; $\frac{4x}{4} = \frac{18}{4}$; $x = \frac{9}{2}$ or $4\frac{1}{2}$
10. $5y = 12$; $\frac{5y}{5} = \frac{12}{5}$; $y = \frac{12}{5}$ or $2\frac{2}{5}$ or 2.4
11. $-3w = 24$; $\frac{-3w}{-3} = \frac{24}{-3}$; $w = -8$
12. $6z = -42 = \frac{6z}{6} = \frac{-42}{6}$; $w = -7$
13. $12a = 18$; $\frac{12a}{12} = \frac{18}{12}$; $a = \frac{3}{2}$
14. $15b = -25$; $\frac{15b}{15} = \frac{-25}{15}$; $b = -\frac{5}{3}$
15. $21c = -14$; $\frac{21c}{21} = \frac{-14}{21}$; $c = -\frac{2}{3}$
16. $24d = 16$; $\frac{24d}{24} = \frac{16}{24}$; $d = \frac{2}{3}$
17. $\frac{p}{3} = 5$; $3 \cdot \frac{p}{3} = 5 \cdot 3$; $p = 15$
18. $\frac{r}{5} = 4$; $5 \cdot \frac{r}{5} = 5 \cdot 4$; $r = 20$
19. $\frac{t}{4} = -6$; $4 \cdot \frac{t}{4} = -6 \cdot 4$; $t = -24$
20. $\frac{s}{-3} = -5$; $-3\frac{s}{-3} = -5(-3)$; $s = 15$
21. $4a + 3 = 11$; $4a + 3 - 3 = 11 - 3$; $4a = 8$; $\frac{4a}{4} = \frac{8}{4}$; $a = 2$
22. $3b + 4 = 16$; $3b + 4 - 4 = 16 - 4$; $3b = 12$; $\frac{3b}{3} = \frac{12}{3}$; $b = 4$
23. $7 - 8d = 39$; $7 - 8d - 7 = 39 - 7$; $-8d = 32$; $\frac{-8d}{-8} = \frac{32}{-8}$; $d = -4$
24. $9 - 7c = 44$; $9 - 7c - 9 = 44 - 9$; $-7c = 35$; $\frac{-7c}{-7} = \frac{35}{-7}$; $c = -5$

25. $4x - 3 = -37$; $4x - 3 + 3 = -37 + 3$; $4x = -34$; $\frac{4x}{4} = \frac{-34}{4}$; $x = -\frac{17}{2}$

26. $5y - 4 = -41$; $5y - 4 + 4 = -41 + 4$; $5y = -37$; $\frac{5y}{5} = \frac{-37}{5}$; $y = \frac{-37}{5}$

27. $2.3w + 4.1 = 13.3$; $2.3w + 4.1 - 4.1 = 13.3 - 4.1$; $2.3w = 9.2$; $\frac{2.3w}{2.3} = \frac{9.2}{2.3}$; $w = 4$

28. $3.5z + 5.2 = 22.7$; $3.5z + 5.2 - 5.2 = 22.7 - 5.2$; $3.5z = 17.5$; $\frac{3.5z}{3.5} = \frac{17.5}{3.5}$; $z = 5$

29. $2x + 5x = 28$; $7x = 28$; $\frac{7x}{7} = \frac{28}{7}$; $x = 4$

30. $3y + 8y = 121$; $11y = 121$; $\frac{11y}{11} = \frac{121}{11}$; $y = 11$

31. $3x = 7 - 10$; $3x = -3$; $\frac{3x}{3} = \frac{-3}{3}$; $x = -1$

32. $4x = 16 - 24$; $4x = -8$; $\frac{4x}{4} = \frac{-8}{4}$; $x = -2$

33. $3a+2(a + 5) = 45$; $3a+2a+10 = 45$; $5a+10 = 45$; $5a = 35$; $a = 7$

34. $4b+3(7 + b) = 56$; $4b+21+3b = 56$; $7b+21 = 56$; $7b = 35$; $b = 5$

35. $4(6 + c) - 5 = 21$; $24 + 4c - 5 = 21$; $4c + 19 = 21$; $4c = 2$; $c = \frac{1}{2}$

36. $5(7 + d) + 4 = 31$; $35 + 5d + 4 = 31$; $5d + 39 = 31$; $5d = -8$; $d = \frac{-8}{5}$

37. $2(p - 4) + 3p = 16$; $2p - 8 + 3p = 16$; $5p - 8 = 16$; $5p = 24$; $p = \frac{24}{5}$

38. $7(n - 5) + 4n = 16$; $7n - 35 + 4n = 16$; $11n - 35 = 16$; $11n = 51$; $n = \frac{51}{11}$

39. $3x = 2x + 5$; $3x - 2x = 5$; $x = 5$

40. $4y = 3y + 7$; $4y - 3y = 7$; $y = 7$

41. $4w = 6w + 12$; $4w - 6w = 12$; $-2w = 12$; $w = -6$

42. $7z = 10z + 42$; $7z - 10z = 42$; $-3z = 42$; $z = -14$

43. $9a = 54 + 3a$; $9a - 3a = 54$; $6a = 54$; $a = 9$

44. $8b = 55 + 3b$; $8b - 3b = 55$; $5b = 55$; $b = 11$

45. $\frac{5x}{2} = \frac{4x}{3} - 7$; $6\left(\frac{5x}{2}\right) = 6\left(\frac{4x}{3} - 7\right)$; $3 \cdot 5x = \frac{6 \cdot 4x}{3} - 6 \cdot 7$; $15x = 8x - 42$; $15x - 8x = -42$; $7x = -42$; $x = -6$

46. $\frac{3y}{7} = \frac{2y}{3} + 4$; $21\left(\frac{3y}{7}\right) = 21\left(\frac{2y}{3} + 4\right)$; $3 \cdot 3y = 7 \cdot 2y + 21 \cdot 4$; $9y = 14y + 84$; $-5y = 84$; $y = \frac{-84}{5}$

47. $\frac{6p}{5} = \frac{3p}{2} + 4$; $10\frac{6p}{5} = 10\left(\frac{3p}{2} + 4\right)$; $2 \cdot 6p = 5 \cdot 3p + 40$; $12p = 15p + 40$; $-3p = 40$; $p = -\frac{40}{3}$

48. $\frac{5z}{3} = \frac{4z}{5} - 3$; $15 \cdot \frac{5z}{3} = 15\left(\frac{4z}{5} - 3\right)$; $5 \cdot 5z = 3 \cdot 4z - 15 \cdot 3$; $25z = 12z - 45$; $13z = -45$; $z = -\frac{45}{13}$

49. $8n - 4 = 5n + 14$; $8n - 5n = 14 + 4$; $3n = 18$; $n = 6$

50. $9p - 5 = 6p + 37$; $9p - 6p = 37 + 5$; $3p = 42$; $p = 14$

51. $7r + 3 = 11r - 21$; $7r - 11r = -21 - 3$; $-4r = -24$; $r = 6$

52. $8s + 7 = 15s - 56$; $8s - 15s = -56 - 7$; $-7s = -63s$; $s = 9$

53. $9t + 6 = 3t - 5$; $9t - 3t = -5 - 6$; $6t = -11$; $t = -\frac{11}{6}$

54. $11u - 4 = 6u + 5$; $11u - 6u = 5 + 4$; $5u = 9$; $u = \frac{9}{5}$

55. $\frac{6x-3}{2} = \frac{7x+2}{3}$; $6\left(\frac{6x-3}{2}\right) = 6\left(\frac{7x+2}{3}\right)$; $3(6x - 3) = 2(7x + 2)$; $18x - 9 = 14x + 4$; $18x - 14x = 4 + 9$; $4x = 13$; $x = \frac{13}{4}$

56. $\frac{4r-3}{3} = \frac{5r+2}{2}$; $6\left(\frac{4r-3}{3}\right) = 6\left(\frac{5r+2}{2}\right)$; $2(4r - 3) = 3(5r + 2)$; $8r - 6 = 15r + 6$; $8r - 15r = 6 + 6$; $-7r = 12$; $r = -\frac{12}{7}$

57. $\frac{3t+4}{4} = \frac{2t-5}{2}$; $8\left(\frac{3t+4}{4}\right) = 8\left(\frac{2t-5}{2}\right)$; $2(3t + 4) = 4(2t - 5)$; $6t + 8 = 8t - 20$; $6t - 8t = -20 - 8$; $-2t = -28$; $t = 14$

58. $\frac{6a-5}{3} = \frac{7a+5}{6}$; $6\left(\frac{6a-5}{3}\right) = 6\left(\frac{7a+5}{6}\right)$; $2(6a - 5) = 1(7a + 5)$; $12a - 10 = 7a + 5$; $12a - 7a = 5 + 10$; $5a = 15$; $a = 3$

59. $3(x + 5) = 2x - 3$; $3x + 15 = 2x - 3$; $3x - 2x = -3 - 15$; $x = -18$

60. $2(y - 3) = 4 + 3y$; $2y - 6 = 4 + 3y$; $2y - 3y = 4 + 6$; $-y = 10$; $(-1)(-y) = (-1)\,10$; $y = -10$

61. $5(w - 7) = 2w + 4$; $5w - 35 = 2w + 4$; $5w - 2w = 4 + 35$; $3w = 39$; $w = 13$

62. $6(z+5) = 2z - 9$; $6z + 30 = 2z - 9$; $6z - 2z = -9 - 30$; $4z = -39$; $z = -\frac{39}{4}$

63. $\frac{x}{2} + \frac{x}{3} - \frac{x}{4} = 2$; $12\left(\frac{x}{2} + \frac{x}{3} - \frac{x}{4}\right) = 12 \cdot 2$; $12\frac{x}{2} + \frac{12x}{3} - \frac{12x}{4} = 24$; $6x + 4x - 3x = 24$; $7x = 24$; $x = \frac{24}{7}$

64. $\frac{p}{2} - \frac{p}{3} - \frac{p}{4} = 3$; $12\frac{p}{2} - 12\frac{p}{3} - 12\frac{p}{4} = 12 \cdot 3$; $6p - 4p - 3p = 36$; $-p = 36$; $p = -36$

65. $\frac{4(a-3)}{5} = \frac{3(a+2)}{4}$; $20\left(\frac{4(a-3)}{5}\right) = 20\left(\frac{3(a+2)}{4}\right)$; $4(4(a-3)) = 5(3(a+2))$; $16a - 48 = 15a + 30$; $16a - 15a = 30 + 48$; $a = 78$

66. $\frac{5(b+4)}{3} = \frac{4(b-5)}{5}$; $15\left(\frac{5(b+4)}{3}\right) = 15\left(\frac{4(b-5)}{5}\right)$; $5(5(b+4)) = 3(4(b-5))$; $25b + 100 = 12b - 60$; $25b - 12b = -60 - 100$; $13b = -160$; $b = -\frac{160}{13}$

67. Solve $ax + b = 3ax$ for x; $ax - 3ax = -b$; $x(a - 3a) = -b$; $x(-2a) = -b$; $x = \frac{-b}{-2a} = \frac{b}{2a}$

68. Solve $2by = 6 + 4by$ for y; $2by - 4by = 6$; $-2by = 6$; $y = \frac{6}{-2b} = -\frac{3}{b}$

69. Solve $ax - 3a + x = 5a$ for a; $ax + x = 8a$; $ax - 8a = -x$; $a(x-8) = -x$; $a = \frac{-x}{x-8} = -\frac{x}{x-8}$ or $\frac{x}{8-x}$

70. Solve $2(by - c) = 3\left(\frac{y}{2} - c\right)$ for y; $2by - 2c = \frac{3y}{2} - 3c$; $4by - 4c = 3y - 6c$; $4by - 3y = -2c$; $y(4b - 3) = -2c$; $y = \frac{-2c}{4b-3}$

71. $\frac{3}{x} + \frac{4}{x} = 3$; $x\left(\frac{3}{x} + \frac{4}{x}\right) = x \cdot 3$; $3 + 4 = 3x$; $7 = 3x$; $3x = 7$; $x = \frac{7}{3}$

72. $\frac{5}{y} - \frac{3}{y} = 6$; $y\left(\frac{5}{y} - \frac{3}{y}\right) = 6y$; $5 - 3 = 6y$; $2 = 6y$; $6y = 2$; $y = \frac{2}{6} = \frac{1}{3}$

73. $\frac{3}{4p} + \frac{1}{p} = \frac{7}{4}$; $4p\left(\frac{3}{4p} + \frac{1}{p}\right) = 4p \cdot \frac{7}{4}$; $3 + 4 = 7p$; $7 = 7p$; $p = 1$

74. $\frac{6}{5q} - \frac{2}{q} = \frac{6}{5}$; $5q\left(\frac{6}{5q} - \frac{2}{q}\right) = 5q\left(\frac{6}{5}\right)$; $6 - 10 = 6q$; $-4 = 6q$; $q = -\frac{4}{6} = -\frac{2}{3}$

75. $\frac{1}{x+1} - \frac{2}{x-1} = 0$; $(x+1)(x-1)\left[\frac{1}{x+1} - \frac{2}{x-1}\right] = 0(x+1)(x-1)$; $(x-1) - 2(x+1) = 0$; $x - 1 - 2x - 2 = 0$; $-x - 3 = 0$; $-x = 3$; $x = -3$

76. $\frac{3}{x+2} - \frac{4}{x-2} = 0$; $(x+2)(x-2)\left[\frac{3}{x+2} - \frac{4}{x-2}\right] = 0$; $3(x-2) - 4(x+2) = 0$; $3x - 6 - 4x - 8 = 0$; $-x - 14 = 0$; $-x = 14$; $x = -14$

77. $\frac{3}{2x} = \frac{1}{x+5}$; $(x+5)2x \cdot \frac{3}{2x} = (x+5)2x\left(\frac{1}{x+5}\right)$; $3(x+5) = 2x$; $3x + 15 = 2x$; $3x - 2x = -15$; $x = -15$

78. $\frac{4}{3x} = \frac{2}{x+1}$; $4(x+1) = 2 \cdot 3x$; $4x + 4 = 6x$; $4 = 2x$; $2x = 4$; $x = 2$

79. $\frac{2x+1}{2x-1} = \frac{x-1}{x-3}$; $(x-3)(2x+1) = (x-1)(2x-1)$; $2x^2 - 5x - 3 = 2x^2 - 3x + 1$; $2x^2 - 2x^2 - 5x + 3x = 1 + 3$; $-2x = 4$; $x = -2$

80. $\frac{2x+3}{2x+5} = \frac{5x+4}{5x+2}$; $(2x+3)(5x+2) = (5x+4)(2x+5)$; $10x^2 + 19x + 6 = 10x^2 + 33x + 20$; $19x - 33x = 20 - 6$; $-14x = 14$; $x = -1$

81. $F - 32 = \frac{9}{5}C$; $\frac{9}{5}C = F - 32$; $\frac{5}{9} \cdot \frac{9}{5}C = \frac{5}{9}(F - 32)$; $C = \frac{5}{9}(F - 32)$

82. (a) $v - v_0 = at$; $at = v - v_0$; $t = \frac{v - v_0}{a}$ (b) $t = \frac{97-12}{9.8} = \frac{85}{9.8} \approx 8.7$ seconds

2.5 APPLICATIONS OF EQUATIONS

1. $\frac{79+85+74+x}{4} = 80$; $\frac{238+x}{4} = 80$; $238 + x = 320$; $x = 82$

2. $\frac{65+72+x}{3} = 75$; $\frac{137+x}{3} = 75$; $137 + x = 225$; $x = 88$

3. $\frac{85+82+x}{3} = 75$; $167 + x = 225$; $x = 58$; minimum of 60

4. $\frac{69+73+68+t}{4} = 72$; $210 + t = 288$; $t = 78°$

5. $.80w = 920$; $w = \frac{920}{.8} = \$1150$; $1150 - 920 = \$230$

6. $c + .15c = 920$; $1.15c = 920$; $c = \frac{920}{1.15} = \$800$

7. $0.30c = 1839$; $c = \frac{1839}{0.30}$; $c = \$6130$

8. $p + .03p + .04(p + .03p) = 227.63$; $1.03p + .04(1.03p) = 227.63$; $1.03p + .0412p = 227.63$; $1.0712p = 227.63$; $p = \frac{227.63}{1.0712}$; $p = \$212.50$

9. a = amount at 7.5%; $(4500 - a)$ = amount at 6%; $0.075 \times a + .06(4500 - a) = 303$; $0.075a + 270 - .06a = 303$; $0.015a = 33$; $a = \frac{33}{.015} = \$2200$ at 7.5%; $4500 - 2200 = \$2300$ at 6%

10.
$$\begin{aligned} a &= \text{amount at } 8.2\% \\ 6200 - a &= \text{amount at } 7.25\% \\ 0.082 \times a + .0725(6200 - a) &= 482.75 \\ 0.082a + 449.5 - .0725a &= 482.75 \\ .0095a &= 33.25 \\ a &= \frac{33.25}{0.0095} = 3500 \text{ at } 8.2\% \\ 6200 - 3500 &= 2700 \text{ at } 7.25\% \end{aligned}$$

11. $40 \times 8.50 + x(1.5)8.50 = 429.25$; $340 + 12.75x = 429.25$; $12.75x = 89.25$; $x = 80.25 \div 12.75 = 7\text{h}$

12.
$$\begin{aligned} w &= \text{weekend} \\ 14 - w &= \text{regular overtime} \\ 40 \times 8.20 + (14 - w)(1.5)8.20 & \\ + 2(w)8.20 &= 524.80 \\ 328 + 172.20 - 12.30w + 16.40w &= 524.80 \\ 500.2 + 4.1w &= 524.80 \\ 4.1w &= 24.60 \\ w &= 6 \text{ h} \end{aligned}$$

13. $d = rt$; $d = 38 \text{ mph} \times 7 \text{ h} = 266 \text{ mi}$

14. $d = rt$; $475 = 38t$; $\frac{475}{38} = t$; $t = 12.5$ h

15. t = days of ships $380 - 12 \times (t + 1) = 80t$
$$\begin{aligned} 380 - 12t - 12 &= 80t \\ 368 &= 80t + 12t \\ 368 &= 92t \\ \frac{368}{92} &= t; t = 4 \text{ days} \\ 80 \times 4 &= 320 \text{ km} \end{aligned}$$

16. $60(t) = 80\left(t - \frac{1}{2}\right)$; $60t = 80t - 40$; $-20t = -40$; $t = 2$ h; $1:00 + 2 = 3:00$ p.m.; $60 \times 2 = 120$ mi

17. h = h together $\frac{1}{6} + \frac{1}{4} = \frac{1}{h}$; $12h = LCD$
$$\begin{aligned} 12h \cdot \frac{1}{6} + 12h \cdot \frac{1}{4} &= 12h \cdot \frac{1}{h} \\ 2h + 3h &= 12 \\ 5h &= 12 \\ h &= \frac{12}{5} \\ &= 2\tfrac{2}{5} \end{aligned}$$
2 hours 24 minutes

18. $\frac{1}{45} + \frac{1}{e} = \frac{1}{30}$; $90e\frac{1}{45} + 90e\frac{1}{e} = 90e \cdot \frac{1}{30}$; $2e + 90 = 3e$; $90 = e$; 90 days

19. $\frac{1}{4} + \frac{1}{2} = \frac{1}{h}$; $4h \cdot \frac{1}{4} + 4h \cdot \frac{1}{2} = 4h \cdot \frac{1}{h}$; $h + 2h = 4$; $3h = 4$; $h = \frac{4}{3} = 1$ h 20 min

20. $\frac{70}{12} + \frac{70}{4} = \frac{70}{h}$; $12h \cdot \frac{70}{12} + 12h \cdot \frac{70}{4} = 12h \cdot \frac{70}{h}$; $70h + 210h = 840$; $280h = 840$; $h = 3$ min

21. $50 \times 0.86 = (50 + w) \times 0.40$; $43 = 20 + 0.40$; $23 = .4w$; $\frac{23}{.4} = w$; 57.5 mL $= w$

22. x = amount of 8% alcohol; $1\,000\,000 - x$ = amount of 14% alcohol; $0.08x + 0.14(1\,000\,000 - x) = 0.09 \times 1\,000\,000$; $140\,000 - 0.06x = 90\,000$; $-0.06x = -50\,000$; $x = \frac{50\,000}{0.06} = 833\,333$ of 8%; $1\,000\,000 - 833\,333 = 166\,667$; The tank should be filled with 166 667 L of 14% alcohol and 833 333 L of 8% alcohol.

23. x = amount of 35%; $750 - x$ = amount of 75%; $0.35x + 0.75(750 - x) = 0.60 \times 750$; $0.35x + 562.5 - 0.75x = 450$; $-.4x = -112.5$; $x = 281.25$ of 35% copper; $750 - 281.25 = 468.75$ of 75% copper.

24. $0.20x + 0.30 \times 50 = 0.27(50 + x)$
$$\begin{aligned} 0.2x + 15 &= 13.5 + 0.27x \\ 15 - 13.5 &= .27x - .2x \\ 1.5 &= .07x \\ x &= \frac{1.5}{.07} = 21.428571 \text{ lb} \end{aligned}$$

25. $850 - 500 = 350$ lb; $500x = 350(20 - x)$; $500x = 7000 - 350x$; $(500 + 350)x = 7000$; $850x = 7000$; $x = 8.24$ ft from 500 lb end

26. $140\,000 - 60\,000 = 80\,000$ N on real wheels

$$\begin{aligned} x &= \text{distance from front} \\ 60\,000x &= 80\,000\,(4.2 - x) \\ 60\,000x &= 336\,000 - 80\,000x \\ 60\,000x + 80\,000x &= 336\,000 \\ 140\,000x &= 336\,000 \\ x &= 2.4 \text{ m from front axle} \end{aligned}$$

27. x = distance from left; $25x = 15\,(12 - x)$; $25x = 180 - 15x$; $40x = 180$; $x = 180 \div 40$; $x = 4.5$ ft from the left end

28. Let d be the location of the center of gravity in cm from the right end. The torque to the left of the center of gravity is $48 + 8(18 - d)$. To the right of the center of gravity, it is $8d$. Thus, $48 + 8(18 - d) = 8d$ or $192 = 16d$ and $d = 12$. The center of gravity is 12 cm from the right end.

29. Let d be the location of the center of gravity in cm from the right end. The torque to the left of the center of gravity is $4 + 2(8 - d)$. To the right of the center of gravity, it is $2d$. Thus, $4 + 2(8 - d) = 2d$ or $20 = 4d$ and $d = 5$. The center of gravity is 5 in. from the right end.

30. Let x be the thickness in cm of the right side. The torque to the left of the center of gravity is $72 + 48x$. To the right of the center of gravity, it is $84x$. Thus, $72 + 48x = 84x$ or $72 = 36x$ and $x = 2$. The thickness of the right side is 2 cm.

CHAPTER 2 REVIEW

1. $8y - 5y = (8 - 5)\,y = 3y$

2. $4z + 15z = (4 + 15)\,z = 19z$

3. $7x - 4x + 2x - 8 = (7 - 4 + 2)\,x - 8 = 5x - 8$

4. $-9a + 4a - 3a + 2 = (-9 + 4 - 3)\,a + 2 = -8a + 2$

5. $\left(2x^2 + 3x + 4\right) + \left(5x^2 - 3x + 7\right) = (2 + 5)\,x^2 + (3 - 3)\,x + (4 + 7) = 7x^2 + 11$

6. $\left(3y^2 - 4y - 3\right) + \left(5y - 3y^2 + 6\right) = (3 - 3)\,y^2 + (-4 + 5)\,y + (-3 + 6) = y + 3$

7. $2\,(8x + 4) = 2 \cdot 8x + 2 \cdot 4 = 16x + 8$

8. $-3\,(4a - 2) = -3 \cdot 4a - 3\,(-2) = -12a + 6$

9. $-3\,(x - 1) = -3x + 1$ or $1 - 3x$

10. $-\,(2z + 5) = -2z - 5$

11. $\left(4x^2 + 3x\right) - \left(2x - 5x^2 + 2\right) = 4x^2 + 3x - 2x + 5x^2 - 2 = 9x^2 + x - 2$

12. $\left(7y^2 + 6y\right) - \left(6y - 7y^2\right) + 2y - 5 = 7y^2 + 6y - 6y + 7y^2 + 2y - 5 = 14y^2 + 2y - 5$

13. $2\,(a + b) - 3\,(a - b) + 4\,(a + b) = 2a + 2b - 3a + 3b + 4a + 4b = 3a + 9b$

14. $6\,(c - d) - 4\,(d - c) + 2\,(c + d) = 6c - 6d - 4d + 4c + 2c + 2d = 12c - 8d$

15. $\left(ax^2\right)\left(a^2x\right) = \left(aa^2\right)\left(x^2x\right) = a^{1+2}x^{2+1} = a^3x^3$

16. $\left(cy^3\right)(dy) = cdy^{3+1} = cdy^4$

17. $\left(9ax^2\right)(3x) = (9 \cdot 3)\left(ax^{2+1}\right) = 27ax^3$

18. $\left(6cy^2z\right)\left(2cz^3\right) = 6 \cdot 2c^{1+1}y^2z^{1+3} = 12c^2y^2z^4$

19. $4\,(5x - 6) = 4 \cdot 5x - 4 \cdot 6 = 20x - 24$

20. $3\,(12y - 5) = 3 \cdot 12y - 3 \cdot 5 = 36y - 15$

21. $2x\,(4x - 5) = 2x \cdot 4x - 2x \cdot 5 = 8x^2 - 10x$

22. $3a\left(6a + a^2\right) = 3a \cdot 6a + 3a \cdot a^2 = 18a^2 + 3a^3$

23. $(a + 4)\,(a - 4) = a^2 - 16$ (Difference of Squares)

24. $(x - 9)\,(x + 9) = x^2 - 81$ (Difference of Squares)

25. $(2a - b)\,(3a - b) = 6a^2 - 2ab - 3ab + b^2 = 6a^2 - 5ab + b^2$ FOIL

26. $(4x + 1)\,(3x - 7) = 12x^2 - 28x + 3x - 7 = 12x^2 - 25x - 7$ FOIL

27. $\left(3x^2 + 2\right)\left(2x^2 - 3\right) = 6x^4 - 9x^2 + 4x^2 - 6 = 6x^4 - 5x^2 - 6$ FOIL

28. $(4a^3 + 2)(6a - 3) = 24a^4 - 12a^3 + 12a - 6$ FOIL

29. $(x + 2)^2 = x^2 + 2 \cdot x \cdot 2 + 2^2 = x^2 + 4x + 4$ Binomial Squared

30. $(3 - y)^2 = 3^2 - 2 \cdot 3 \cdot y + y^2 = 9 - 6y + y^2$ Binomial Squared

31. $5x(3x - 4)(2x + 1) = 5x[6x^2 + 3x - 8x - 4] = 5x[6x^2 - 5x - 4] = 30x^3 - 25x^2 - 20x$

32. $6a(4a + 3)(a - 2) = 6a[4a^2 - 8a + 3a - 6] = 6a[4a^2 - 5a - 6] = 24a^3 - 30a^2 - 36a$

33. $a^5 \div a^2 = \frac{a^5}{a^2} = a^{5-2} = a^3$

34. $x^7 \div x^3 = \frac{x^7}{x^3} = x^{7-3} = x^4$

35. $8a^2 \div 2a = \frac{8a^2}{2a} = \frac{8}{2} \cdot a^{2-1} = 4a$

36. $27b^3 \div 3b = \frac{27b^3}{3b} = \frac{27}{3} \cdot b^{3-1} = 9b^2$

37. $45a^2x^3 \div -5ax = \frac{45a^2x^3}{-5ax} = -9ax^2$

38. $52b^4c^2 \div -4b^2 = \frac{52b^4c^2}{-4b^2} = -13b^2c^2$

39. $(36x^2 - 16x) \div 2x = \frac{36x^2-16x}{2x} = \frac{36x^2}{2x} - \frac{16x}{2x} = 18x - 8$

40. $(39b^3 + 52b^5) \div 13b^2 = \frac{39b^3+52b^5}{13b^2} = \frac{39b^3}{13b^2} + \frac{52b^5}{13b^2} = 3b + 4b^3$

41.
$$\begin{array}{r} x-4 \\ x+3\overline{)x^2-x-12} \\ \underline{x^2+3x} \\ -4x-12 \\ \underline{-4x-12} \\ 0 \end{array}$$

42.
$$\begin{array}{r} x+5 \\ x-6\overline{)x^2-x-30} \\ \underline{x^2-6x} \\ 5x-30 \\ \underline{5x-30} \\ 0 \end{array}$$

43.
$$\begin{array}{r} x^2+3x+9 \\ x-3\overline{)x^3 \qquad -27} \\ \underline{x^3-3x^2} \\ 3x^2 \qquad -27 \\ \underline{3x^2-9x} \\ 9x-27 \\ \underline{9x-27} \\ 0 \end{array}$$

44.
$$\begin{array}{r} 4a^2-8a+16 \\ 2a+4\overline{)8a^3 \qquad +64} \\ \underline{8a^3+16a^2} \\ -16a^2 \qquad +64 \\ \underline{-16a^2-32a} \\ 32a+64 \\ \underline{32a+64} \\ 0 \end{array}$$

45.
$$\begin{array}{r} x-y \\ x+y\overline{)x^2 \qquad -y^2} \\ \underline{x^2+xy} \\ -xy-y^2 \\ \underline{-xy-y^2} \\ 0 \end{array}$$

46.
$$\begin{array}{r} y-a \\ y+a\overline{)y^2 \qquad -a^2} \\ \underline{y^2+ya} \\ -ya-a^2 \\ \underline{-ya-a^2} \\ 0 \end{array}$$

47.
$$\begin{array}{r} x-2y \\ x^2-y^2\overline{)x^3-2x^2y-xy^2+2y^3} \\ \underline{x^3 \qquad -xy^2} \\ -2x^2y \qquad +2y^3 \\ \underline{-2x^2y \qquad +2y^3} \\ 0 \end{array}$$

48.

$$\begin{array}{r l}
 & a^2 + 4ab + b^2 \\
a - b\,\big) & a^3 + 3a^2b + 2ba - 3ab^2 - b^3 \\
 & a^3 - a^2b \\
\hline
 & 4a^2b + 2ba - 3ab^2 - b^3 \\
 & 4a^2b \qquad - 4ab^2 \\
\hline
 & 2ba + ab^2 - b^3 \\
 & ab^2 - b^3 \\
\hline
 & 2ba
\end{array}$$

49. $x + 9 - 9 = 47 - 9; x = 38$

50. $y - 19 + 19 = -32 + 19; y = -13$

51. $\frac{2x}{2} = \frac{15}{2}; x = \frac{15}{2}$

52. $\frac{-3y}{-3} = \frac{14}{-3}; y = -\frac{14}{3}$

53. $4 \cdot \frac{x}{4} = 4 \cdot 9; x = 36$

54. $\frac{y}{3} = -7; 3\frac{y}{3} = 3(-7); y = -21$

55. $4x - 3 = 17; 4x - 3 + 3 = 17 + 3; 4x = 20; \frac{4x}{4} = \frac{20}{4}; x = 5$

56. $7 + 8y = 23; 7 + 8y - 7 = 23 - 7; 8y = 16; \frac{8y}{8} = \frac{16}{8}; y = 2$

57. $3.4a - 7.1 = 8.2; 3.4a - 7.1 + 7.1 = 8.2 + 7.1; 3.4a = 15.3; a = \frac{15.3}{3.4} = 4.5$

58. $6.2b + 19.1 = 59.4; 6.2b = 59.4 - 19.1; 6.2b = 40.3; b = \frac{40.3}{6.2} = 6.5$

59. $4x + 3 = 2x; 4x - 2x = -3; 2x = -3; x = -\frac{3}{2}$

60. $7a - 2 = 2a; 7a - 2a = 2; 5a = 2; a = \frac{2}{5}$

61. $4b + 2 = 3b - 5; 4b - 3b = -5 - 2; b = -7$

62. $7c + 9 = 12c - 4; 7c - 12c = -4 - 9; -5c = -13; c = \frac{13}{5}$

63. $\frac{4(x-3)}{3} = \frac{5(x+4)}{2}; 6\left(\frac{4(x-3)}{3}\right) = 6\left(\frac{5(x+4)}{2}\right); 2 \cdot 4(x-3) = 3 \cdot 5(x+4); 8x - 24 = 15x + 60; -7x = 84; x = -12$

64. $\frac{3(y-7)}{5} = \frac{5(y+4)}{2}; 10\left(\frac{3(y-7)}{5}\right) = 10\left(\frac{5(y+4)}{2}\right); 2 \cdot 3(y-7) = 5 \cdot 5(y+4); 6y - 42 = 25y + 100; -19y = 142; y = -\frac{142}{19}$

65. $\frac{2}{a} - \frac{3}{a} = 5; a\left(\frac{2}{a} - \frac{3}{a}\right) = a \cdot 5; 2 - 3 = 5a; -1 = 5a; a = -\frac{1}{5}$

66. $\frac{4}{x} + \frac{5}{x} = \frac{1}{8}; 8x\left(\frac{4}{x} + \frac{5}{x}\right) = 8x \cdot \frac{1}{8}; 32 + 40 = x; 72 = x$

67. $\frac{3}{2a} = \frac{3}{a+2}; 2a(a+2)\frac{3}{2a} = 2a(a+2) \cdot \frac{3}{a+2}; (a+2)3 = 2a \cdot 3; 3a + 6 = 6a; 6 = 3a; a = 2$

68. $\frac{9}{4b} = \frac{12}{b+4}; (b+4)9 = 4b(12); 9b + 36 = 48b; 36 = 39b; b = \frac{36}{39} = \frac{12}{13}$

69. $\frac{68+70+74+x}{4} = 72; \frac{212+x}{4} = 72; 212 + x = 72 \times 4; 212 + x = 288; x = 288 - 212 = 76$

70. $p + .065p = \$342.93; 1.065p = 342.93; p = 342.93 \div 1.065 = \$322$

71. $d = rt; \frac{2718}{755} = \frac{755t}{755}; t = 3.6$ hours or 3 hrs 36 min

72. t = satellite time; $t - 1\frac{3}{4}$ = shuttle time; $330t = 430(t - 1.75); 330t = 430t - 752.5; -100t = -752.5; t = 7.525$ hours or at $7{:}31\frac{1}{2}$ min $= 7{:}31{:}30$

73. 40 kg $\times$ 50% $+15 = 20 + 15 = 35$ kg lead; $\frac{35}{40+15} = \frac{35}{55} = 63.64\%$ lead

74. $460 - 320 = 140$N; $320x = 140(8 - x); 320x = 1120 - 140x; (320 + 140)x = 1120; 460x = 1120; x = 2.43$m from 320 force

75. Use the formula $F = ma$. $F_1 = 1538.6$ N; $F_2 = 1107.4$ N; $m_1 = \frac{F_1}{a} = \frac{1538.6}{9.8} = 157$ kg; $m_2 = \frac{F_2}{a} = \frac{1107.4}{9.8} = 113$ kg

CHAPTER 2 TEST

1. $5x^2 + (2-5)x = 5x^2 - 3x$

2. $(4a^3 - 2b) - (3b + a^3) = 4a^3 - 2b - 3b - a^3 = (4a^3 - a^3) - (2b + 3b) = 3a^3 - 5b$

3. $4x + [3(x + y - 2) - 5(x - y)] = 4x + [3x + 3y - 6 - 5x + 5y] = (4x + 3x - 5x) + (3y + 5y) - 6 = 2x + 8y - 6$

4. $(4xy^3z)\left(\frac{1}{2}xy^{-2}z^2\right) = 4\left(\frac{1}{2}\right)x^{1+1}y^{3-2}z^{1+2} = 2x^2yz^3$

5. $(2b-3)(2b+3) = (2b)^2 - 3^2 = 4b^2 - 9$

6. $(x^3 + 3x) \div x;\ \frac{x^3+3x}{x} = \frac{x^3}{x} + \frac{3x}{x} = x^{3-1} + 3x^{1-1} = x^2 + 3x^0 = x^2 + 3$

7. $(6x^5 + 4x^3 - 1) \div 2x^2;\ \frac{6x^5+4x^3-1}{2x^2} = \frac{6x^5}{2x^2} + \frac{4x^3}{2x^2} - \frac{1}{2x^2} = 3x^3 + 2x - \frac{1}{2x^2}$

8.
```
            3x²− 8x+17
    x + 2) 3x³−2x²+  x− 3
           3x³+6x²
           −8x²+  x− 3
           −8x²−16x
                17x− 3
                17x+34
                   −37
```

9. $\frac{y}{3y+2} - \frac{4}{y-1} = \frac{y(y-1)}{(3y+2)(y-1)} - \frac{4(3y+2)}{(3y+2)(y-1)} = \frac{y(y-1)-4(3y+2)}{(3y+2)(y-1)} = \frac{y^2-y-12y-8}{(3y+2)(y-1)} = \frac{y^2-13y-8}{(3y+2)(y-1)}$

10. $5x - 8 = 3x;\ 5x - 3x = 8;\ 2x = 8;\ x = \frac{8}{2};\ x = 4$

11. $\frac{7x+3}{2} - \frac{9x-12}{4} = 8;\ 4\left(\frac{7x+3}{2}\right) - 4\left(\frac{9x-12}{4}\right) = 4 \cdot 8;\ 14x + 6 - 9x + 12 = 32;\ 5x + 18 = 32;\ 5x = 14;\ x = \frac{14}{5}$

12. $p + .07p = 74.85;\ 1.07p = 74.85;\ p = 74.85 \div 1.07;\ p = \$69.95$

13. Let n be the amount of original solution to be removed. Then $9 - n$ is the amount left. You want to end with 60% of 9 qt or 5.4 qt of antifreeze. The antifreeze left from the original solution is 50% of $9-n$, so $50\%(9-n)+n = 5.4;\ 0.5(9-n)+n = 5.4;\ 4.5 - 0.5n + n = 5.4;\ 0.5n = 0.9;\ n = 1.8$ qt

CHAPTER

3

Geometry

3.1 LINES AND ANGLES

1. Using the proportion $\frac{D}{180^\circ} = \frac{R}{\pi}$, we substitute 15° for D and solve for R. Thus, $\frac{15^\circ}{180^\circ} = \frac{R}{\pi}$ or $R = \frac{15^\circ}{180^\circ}\pi = \frac{1}{12}\pi = \frac{\pi}{12} \approx 0.2617994$.

2. Substituting 75° for D in the proportion $\frac{D}{180^\circ} = \frac{R}{\pi}$, we obtain $\frac{75^\circ}{180^\circ} = \frac{R}{\pi}$ or $R = \frac{75^\circ}{180^\circ}\pi = \frac{5}{12}\pi = \frac{5\pi}{12} \approx 1.3089969$.

3. Multiplying 210° by $\frac{\pi}{180^\circ}$, we obtain $\frac{210^\circ}{180^\circ}\pi = \frac{7}{6}\pi = \frac{7\pi}{6} \approx 3.6651914$.

4. We first convert 10°45′ to decimal degrees. Now $45' = \frac{45}{60}^\circ = 0.75^\circ$, so $10^\circ 45' = 10.75^\circ$. Multiplying 10.75° by $\frac{\pi}{180^\circ}$, we obtain $\frac{10.75^\circ}{180^\circ}\pi = \frac{215}{3600} = \frac{43}{720}\pi = \frac{43\pi}{720} \approx 0.1876229$. You could also have multiplied 10.75° by 0.01745 with the result $10.75^\circ \times 0.01745 = 0.1875875$. Notice that the two methods give different decimal approximations because 0.01745 is rounded off before it is multiplied.

5. Multiplying 85.4° by $\frac{\pi}{180^\circ}$ produces $\frac{85.4^\circ}{180^\circ}\pi = \frac{854}{1800}\pi = \frac{427}{900}\pi \approx 1.4905112$. Multiplying $85.4 \times 0.01745 = 1.49023$.

6. As in Exercises 1–5, multiplying 48.6° by $\frac{\pi}{180^\circ}$ produces $\frac{48.6^\circ}{180^\circ}\pi = \frac{27}{100}\pi \approx 0.84823$. Multiplying 48.6° by 0.01745 gives 0.84807.

7. Multiplying 163.5° by $\frac{\pi}{180^\circ}$ we obtain $\frac{163.5^\circ}{180^\circ}\pi = \frac{109}{120}\pi \approx 2.8536133$. Multiplying $163.5 \times 0.01745 = 2.853075$.

8. Changing 242°35′ to decimal degrees results in $242^\circ 35' = 242\frac{35}{60}^\circ \approx 242.583^\circ$. Multiplying this by $\frac{\pi}{180^\circ}$ produces $\frac{2911}{2160}\pi \approx 4.2338779$ and multiplying 242.583 by 0.01745 yields 4.23307345.

9. Multiplying by $\frac{180^\circ}{\pi}$ produces $\frac{4\pi}{3} \cdot \frac{180^\circ}{\pi} = 240^\circ$.

10. Multiplying by $\frac{180^\circ}{\pi}$ results in $\frac{\pi}{6} \cdot \frac{180^\circ}{\pi} = 30^\circ$.

11. Multiplying by $\frac{180^\circ}{\pi}$ yields $(1.3\pi)\left(\frac{180^\circ}{\pi}\right) = 1.3 \times 180^\circ = 234^\circ$.

12. $2.15 \times \frac{180^\circ}{\pi} = \frac{387^\circ}{\pi} \approx 123.18593$ or multiplying by 57.296° results in 123.1864°.

13. Multiplying by 57.296° results in 14.324°.

14. Multiplying 1.1 by 57.296° yields 63.0256°.

15. (a) To find the supplement of a 35° angle in degrees, subtract 35° from 180° with the result $180^\circ - 35^\circ = 145^\circ$. (b) Converting 145° to radians, multiply by $\frac{\pi}{180^\circ}$ and obtain $\frac{145}{180}\pi = \frac{29}{36}\pi \approx 2.5307274$.

16. (a) Two angles are complementary if the sum of their measures is 90°. The complement of a 65° angle is $90^\circ - 65^\circ = 25^\circ$. (b) Converting 25° to radians, we obtain $25^\circ \times \frac{\pi}{180^\circ} = \frac{25\pi}{180} = \frac{5\pi}{36} \approx 0.4363323$.

17. Angle A and the given angle are supplementary, so $\angle A = 180^\circ - 33^\circ = 147^\circ$. Angle B and the given angle are vertical angles and so they are congruent. Thus, $\angle B = 33^\circ$.

18. Angles A and the given angle are complementary, so $\angle A = 90° - 42° = 48°$. Angle B is a vertical angle with the indicated right angle, so $\angle B = 90°$.

19. Angle A and the given angle are alternate exterior angles and so congruent. Then, $\angle A = \frac{3\pi}{8}$. Angle B and $\angle A$ are supplementary, so $\angle B = \pi - \frac{3\pi}{8} = \frac{5\pi}{8}$.

20. Angle A and the given angle are corresponding angles, so $\angle A = \frac{3\pi}{4}$. Angle B and $\angle A$ are supplementary, so $\angle B = \pi - \angle A = \pi - \frac{3\pi}{4} = \frac{\pi}{4}$.

21. Angle B and the given angle are opposite interior angles and so they are congruent. Thus, $\angle B = 70°30'$. Angle A and $\angle B$ are supplementary, so $\angle A = 180° - \angle B = 180° - 70°30' = 109°30'$.

22. Angle A and the given angle are complementary so $\angle A = \frac{\pi}{2} - \frac{5\pi}{16} = \frac{3\pi}{16}$. Angle B and the given angle are alternate interior angles and so they are congruent. Thus, $\angle B = \frac{5\pi}{16}$.

23. Since corresponding segments of transversals that intersect parallel lines are proportional, we have $\frac{AB}{BC} = \frac{DE}{EF}$. Substituting the given information in this proportion we obtain $\frac{AB}{3} = \frac{8}{5}$. Solving this proportion we obtain $AB = \frac{3 \cdot 8}{5} = \frac{24}{5} = 4.8$.

24. As in Exercise 23, we have the proportion $\frac{AB}{BC} = \frac{DE}{EF}$. Substituting the given data, we obtain $\frac{AB}{22.65} = \frac{8.64}{15.1}$ and so, $AB = \frac{(22.65)(8.64)}{15.1} = 12.96$.

25. The three angles form a straight angle. We have $\angle A + 85° + \angle B = 180°$ and so $\angle A + \angle B = 95°$. Since $\angle A \cong \angle B$, we have $\angle A = \frac{95°}{2} = 47.5°$.

26. Here we have three parallel lines (the cables of the bridge) and two transversals (the girder and the deck), so corresponding segments are proportional. Thus, $\frac{AB}{4.5} = \frac{3.6}{4.8}$ and $AB = \frac{(4.5)(3.6)}{4.8} = 3.375$ m.

27. Here $\omega = 2\pi f$ and, since $f = 60$, $\omega = 2\pi(60) = 120\pi$ rad/s.

28. (a) Since $\theta = 285°$ and $t = 0.6$ s, we have $\omega = \frac{285°}{0.6} = 475°/\text{s}$. (b) Since $\theta = \frac{11\pi}{16}$ and $t = 0.9$ s, we have $\omega = \frac{11\pi/16}{0.9} = \frac{110\pi}{144} = \frac{55\pi}{72}$ rad/s.

3.2 TRIANGLES

1. Since $x + 35° + 75° = 180°$, we have $x = 180° - (35° + 75°) = 180° - 110° = 70°$.

2. Since $x + 122°20' + 15°30' = 180°$, we know that $x = 180° - (122°20' + 15°30') = 180° - 137°50' = 179°60' - 137°50' = 42°10'$.

3. This is an isosceles triangle. Base angles (the angles opposite the congruent sides) of an isosceles triangle are congruent. So, x and the 55° angle are congruent and $x = 55°$. To find y, we add to obtain $55° + y + x = 180°$. Solving for y produces $y = 180° - 2(55°) = 180° - 110° = 70°$.

4. This is a right triangle, so $x + 2x = 90°$ and so $x = 30°$; $2x = 60°$.

5. This is an equilateral triangle, and so it is also an equiangular triangle and each angle is $\frac{180°}{3} = 60°$. Thus, $x = y = 60°$.

6. As in Exercise 3, base angles of isosceles triangles are congruent. Labeling the triangle as shown, we have $\angle A$ and $\angle ABD$ the base angles of $\triangle ABD$ and $\angle C$ and $\angle CBD$ the base angles of $\triangle CBD$. Thus, from $\triangle ABC$, we have $x + x + y + y = 180°$ or $2x + 2y = 180°$. Now, from $\triangle BCD$, we have $2x + 116° = 180°$, so $x = 32°$. Since $2x + 2y = 180°$ then $x + y = 90°$ and $y = 90° - x = 90° - 32° = 58°$.

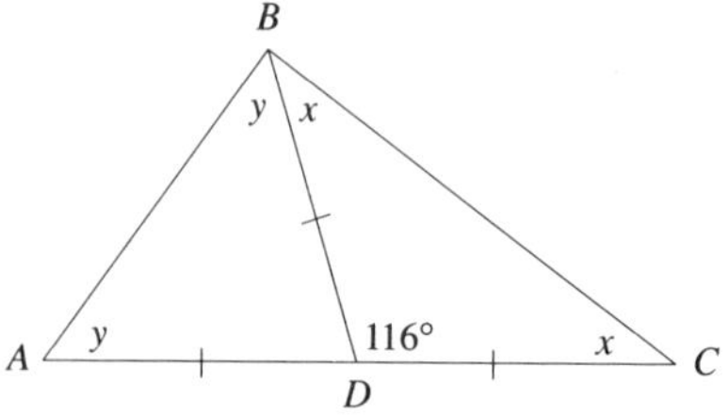

7. Labeling the triangle as shown, we see that $\angle ABC$ is a vertical angle of 53.13°. Since $\triangle ABC$ is

a right triangle, $x + 53.13° = 90°$ and we have $x = 90° - 53.13° = 36.87°$.

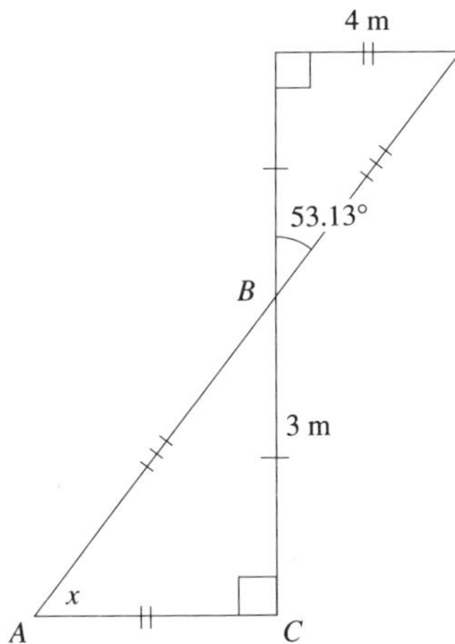

8. Label the figure as shown. From the given information, we know that $\angle ABE \cong \angle CBD$ because they are vertical angles, and $\angle D \cong \angle E$ because they are right angles. That means that $\triangle ABE$ and $\triangle CBD$ are similar triangles and so corresponding sides are proportional, or $\frac{AB}{CB} = \frac{AE}{CD} = \frac{BE}{BD}$. Substituting the given data, we have $\frac{AB}{y} = \frac{x}{8} = \frac{2}{5}$. Solving the right-hand proportion, $\frac{x}{8} = \frac{2}{5}$, we obtain $x = \frac{16}{5} = 3.2$. Using the Pythagorean Theorem on $\triangle BCD$, we see that $5^2 + 8^2 = y^2$ and so $89 = y^2$ and $y = \sqrt{89}$.

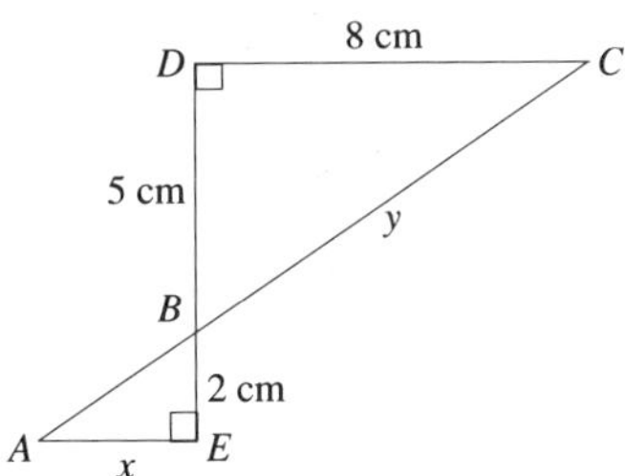

9. The perimeter $P = 6.5 + 7.2 + 9.7 = 23.4$ units. The area, $A = \frac{1}{2}bh$. Here $b = 7.2$ and $h = 6.5$, so $A = \frac{1}{2}bh = \frac{1}{2}(7.2)(6.5) = 23.4$ sq. units (or you may write this as 23.4 units^2).

10. $P = 15+41+52 = 108$ units. $A = \frac{1}{2}bh$ with $b = 52$ and $h = 9$; $A = \frac{1}{2}(52)(9) = 234$ units^2.

11. $A = \frac{1}{2}bh = \frac{1}{2}(28)(9) = 126$ units^2. Notice that the dotted line $\overline{BD}$ is not part of the indicated triangle. To determine the perimeter, we need to find the length of $\overline{AC}$. We will do this using the Pythagorean Theorem and $\triangle ADC$. That means we need the length of $\overline{DC} = DB + 28$. Using the Pythagorean Theorem, we determine $BD = \sqrt{15^2 - 9^2} = \sqrt{144} = 12$, so $DC = 40$ and $AC = \sqrt{9^2 + 40^2} = \sqrt{1681} = 41$. Using this information, $P = AB + BC + CA = 15 + 28 + 41 = 84$ units.

12. Using the Pythagorean Theorem on $\triangle ABD$ we see that $BD = \sqrt{25^2 - 20^2} = \sqrt{225} = 15$ and from $\triangle CAD$ we find $DC = \sqrt{101^2 - 20^2} = \sqrt{9801} = 99$. Thus, $BC = BD + DC = 15 + 99 = 114$ units. $P = 25 + 101 + 114 = 240$ units and $A = \frac{1}{2}(20)(114) = 1140$ units^2.

13. (a) The amount of fencing is the same as the perimeter, or $23.2 + 47.6 + 62.5 = 133.3$ m. (b) The area will provide the amount of sod that is needed. Using Hero's formula, we first determine the semiperimeter, $s = \frac{23.2+47.6+62.5}{2} = 66.65$. So, by Hero's formula, we find the area is $A = \sqrt{66.65(66.65 - 23.2)(66.65 - 47.6)(66.65 - 62.5)} = \sqrt{66.65(43.45)(19.05)(4.15)} = \sqrt{228{,}945.9742} \approx 478.5$ m^2.

14. Using proportions we have

$$\frac{\text{height of building}}{\text{shadow of building}} = \frac{\text{height of stick}}{\text{shadow of stick}}$$

or $\frac{h}{123'6''} = \frac{3'}{2'3''}$. Changing $123'6''$ to $123.5'$ and changing $2'3''$ to $2.25'$ and substituting in the proportion, we obtain $\frac{h}{123.5'} = \frac{3'}{2.25'}$. Solving for h, produces $h = \frac{(123.5)(3)}{2.25} = \frac{370.5}{2.25} = 164\frac{2}{3}' = 164'8''$.

15. Assuming that the building is perpendicular to the ground, we have a right triangle with the ladder forming the hypotenuse. So, $5^2 + s^2 = 15^2$ or $s^2 = 15^2 - 5^2 = 200$ and $s = \sqrt{200}$ m $= 10\sqrt{2}$ m $\approx$ 14.121 m.

16. The area of the triangle is $\frac{1}{2}$ the area of the rectangle, so $A = \frac{1}{2}(23)(16) = 184$ cm^2.

17. The diagonal, d, is the hypotenuse of the right triangle formed by the two streets. So, $d = \sqrt{45^2 + 62^2} = \sqrt{5869} \approx 76.6$ ft.

18. The length of the rafter, indicated by r on the drawing, is $x + 0.5$ m. The roof forms an isosceles triangle, so a perpendicular from the ridge to the base bisects the base. Thus, we have a right triangle with sides 3 and 4. Using the Pythagorean Theorem, we determine that $x = 5$ m and so, $r = 5.5$ m.

19. (a) Let s represent the length of the top of the second beam. The figure below shows two similar triangles $\triangle ABC$ and $\triangle ADE$. Similar triangles have proportional sides and altitudes, so $\frac{s}{2400} = \frac{5700}{2600}$ and we obtain $s = 5261.54$ mm (5262 mm to 4 significant digits). Let t represent the length of the top of the third beam. Reasoning as above, we have $\frac{t}{2400} = \frac{8800}{2600}$ or $t = 8123$ mm. (b) Reasoning as in (a), we have the proportion $\frac{b}{2400} = \frac{10\,000}{2600}$ or $b = 9231$ mm.

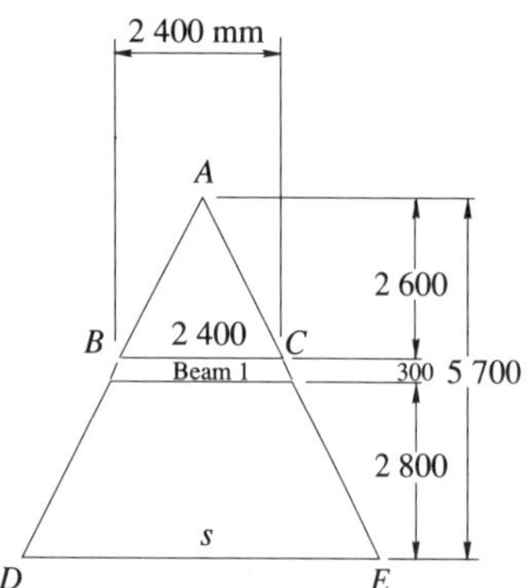

20. What we have are three right triangles each with one side 175 m and the other sides the length of the cable from the antenna base. We need to determine the length of the hypotenuse of each triangle. For the first, the length is $\sqrt{175^2 + 50^2} = \sqrt{33125} \approx 182$. For the second, the length is $\sqrt{175^2 + 75^2} = \sqrt{36250} \approx 190.4$. The length of the third cable is $\sqrt{175^2 + 100^2} = \sqrt{40625} \approx 201.6$.

21. Since the total length of the conduit is 30 m, the length of the "bent" section is $30 - 8 - 10 = 12$ m. This bent section is the hypotenuse of a right triangle with one leg 5 m. Using the Pythagorean Theorem, we see that $x^2 + 5^2 = 12^2$ and so $x = \sqrt{119}$ m ≈ 10.9 m.

22. The amount needed for the straight conduit is the length of the hypotenuse of a right triangle with one leg of length $x = \sqrt{119}$ and the other leg $8 + 5 + 10 = 23$. Using the Pythagorean Theorem, we obtain $\sqrt{119 + 23^2} = \sqrt{648} \approx 25.5$ m. Subtracting this from the length of the conduit that was actually used, produces $30 - 25.5 = 4.5$ m.

23. (a) The amount of pipe used is $97 + 62 + 53 = 212$ m. (b) We need to find the length of the hypotenuse AB of a right triangle. One side of the triangle is $65 - 28 = 37$ m. The length of the other side is made up of three horizontal distances, two for the pipe sections on the sides of the ravine and the 62 m section across the bottom. For the left-hand section, the horizontal distance is $\sqrt{97^2 + 65^2} = \sqrt{5184} = 72$ m. For the right-hand section, the horizontal distance is $\sqrt{53^2 - 28^2} = \sqrt{2025} = 45$ m. So, the length of the base of the right triangle is $72 + 62 + 45 = 179$ m and the length of the hypotenuse is $\sqrt{179^2 + 37^2} = \sqrt{33410} \approx 182.8$ m.

24. The three triangles are similar, so $\frac{AB}{BF} = \frac{BC}{EB}$. Substituting the given information in this proportion, we have $\frac{24}{0.5} = \frac{BC}{0.5}$. Solving the proportion for BC produces $BC = 24$ ft. Similarly, we have the proportion $\frac{EB}{DG} = \frac{BC}{CD}$ or $\frac{0.5}{0.2} = \frac{24}{CD}$. Solving for CD results in $CD = \frac{24(0.2)}{0.5} = 9.6$ ft.

3.3 OTHER POLYGONS

1. The figure is a square with $s = 15$ cm. $P = 4s = 4(15) = 60$ cm. $A = s^2 = 15^2 = 225$ cm^2.

2. The figure is a rectangle. We will let $a = 25$ and $b = 9$. $P = 2(a + b) = 2(25 + 9) = 2(34) = 68$ in. $A = ab = (25)(9) = 225$ in^2.

3. This figure is a parallelogram with $a = 15\frac{1}{4}$, $b = 33$, and $h = 11\frac{1}{2}$. $P = 2(a + b) = 2\left(15\frac{1}{4} + 33\right) = 2\left(48\frac{1}{4}\right) = 96.5$ in. $A = bh = 33\left(11\frac{1}{2}\right) = 379\frac{1}{2} = 379.5$ in^2.

4. Here the figure is a trapezoid. Since the left side is perpendicular to the base, it will act as the altitude. For this trapezoid, we have $b_1 = 23.4$, $b_2 = 24.9$, and $a = h = 11.2$. We must determine the length of c. If we drop a perpendicular from the top right vertex, we obtain a right triangle with legs 11.2 and $24.9 - 23.4 = 1.5$. Then $c = \sqrt{11.2^2 + 1.5^2} = 11.3$. Using these values, produces $P = a + b_1 + c + b_2 = 11.2 + 23.4 + 11.3 + 24.9 = 70.8$ mm. $A = \frac{1}{2}(b_1 + b_2)h = \frac{1}{2}(23.4 + 24.9)\,11.2 = \frac{1}{2}(48.3)\,11.2 = 270.48$ mm^2.

5. The figure for this exercise is a regular hexagon with $s = 11\frac{1}{2}$. Thus, $P = 6s = 6\left(11\frac{1}{2}\right) = 69$ in. Using the second formula for the area of regular hexagon that is given in Figure 3.27, we have $A = \frac{3\sqrt{3}}{2}s^2 = \frac{3\sqrt{3}}{2}\left(11\frac{1}{2}\right)^2 = \frac{3\sqrt{3}}{2}132.25 = 198.375\sqrt{3} \approx 343.6$ in^2.

6. This figure is a rhombus with $s = 12$ in and $h = 6\frac{1}{2}$ in. $P = 4s = 4(12) = 48$ in. $A = sh = 12\left(6\frac{1}{2}\right) = 78$ in^2.

7. Here we have a kite with $a = 10$ cm, $b = 12$ cm, $d_1 = 16$ cm, and $d_2 = 21$ cm. Thus, $P = 2(a + b) = 2(10 + 12) = 44$ cm. $A = \frac{1}{2}d_1d_2 = \frac{1}{2}(16)(21) = 168$ cm^2.

8. This is a trapezoid with the bases being the two vertical lines. Thus, $a = h = 20.8$ mm, $b_1 = 213$ mm, and $b_2 = 318$ mm. As in Exercise 4, we use the Pythagorean Theorem to determine that $c = \sqrt{20.8^2 + 105^2} = \sqrt{11457.64} \approx 107$ mm. Using all this, we determine that $P = a + b_1 + c + b_2 = 20.8 + 213 + 107 + 318 = 658.8$ mm. $A = \frac{1}{2}(b_1 + b_2)h = \frac{1}{2}(213 + 318)(20.8) = 5522.4$ mm^2.

9. The figure is a parallelogram with $a = 15.7$ mm, $b = 21.2$ mm, and $h = 13.2$ mm. $P = 2(a + b) = 2(15.7 + 21.2) = 73.8$ mm. $A = bh = (21.2)(13.2) = 279.84$ mm^2.

10. Here the figure is a trapezoid with $a = 12\frac{1}{2}$ in, $b_1 = 14\frac{3}{4}$ in, $c = 13$ in, $b_2 = 23\frac{1}{4}$ in, and $h = 12$ in. Using the given formulas we determine that $P = a + b_1 + c + b_2 = 12\frac{1}{2} + 14\frac{3}{4} + 13 + 23\frac{1}{4} = 63.5$ in. $A = \frac{1}{2}(b_1 + b_2)h = \frac{1}{2}\left(14\frac{3}{4} + 23\frac{1}{4}\right)(12) = \frac{1}{2}(38)(12) = 228$ in^2.

11. Draw a line across one side of the "L" to make this into two rectangles. If the line is drawn as shown in the figure, one rectangle will be $26' \times 14'$ with an area of 364 ft^2 and the other rectangle will be $8' \times 10'$ with an area of 80 ft^2. Adding the two areas together, we obtain a total area of 444 ft^2.

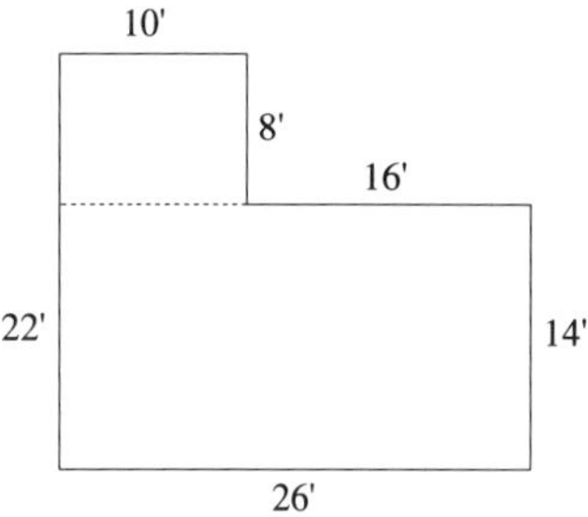

12. If there is no mortar and no space between the bricks, then this is almost the same as asking how many $8''$ bricks will divide into the perimeter of the patio. The patio has a perimeter of $22' + 10' + 8' + 16' + 14' + 26' = 96'$. Changing $96'$ to inches and dividing by 8, we obtain the number of bricks that will be needed $96 \times 12 \div 8 = 144$. But, if we use 144 bricks, there will be four corners with cut-outs. Each cut-out takes a half-brick, so the total is $144 + 4\left(\frac{1}{2}\right) = 146$ bricks.

13. Following the hint, we think of this as a rectangle that measures 100 mm by 300 mm for an area of 30 000 mm^2. Each trapezoid has bases $b_1 = 240$ mm and $b_2 = 260$ mm with $h = 40$ mm. Thus, each trapezoid has an area of $A = \frac{1}{2}(b_1 + b_2)h = \frac{1}{2}(240 + 260)(40) = 10\,000$ mm^2. Thus, the area of the I-beam is $A = 30\,000 - 2(10\,000) = 30\,000 - 20\,000 = 10\,000$ mm^2.

14. As in Exercise 13, we will think of this as a rectangle with two trapezoids removed. The area of the rectangle is $50' \times 45' = 2250$ ft^2. Each trapezoid has bases $30'$ and $35'$ and height $12'$ for an area $\frac{1}{2}(65)(12) = 390$ ft^2. Thus, the area of the highway support is $2250 - 2(390) = 1470$ ft^2.

15. Consulting Figure 3.27, we can see that the given dimension is $2h$ and so $h = \frac{7}{16}''$ and the area is $A = 2\sqrt{3}h^2 = 2\sqrt{3}\left(\frac{7}{16}\right)^2 = \frac{98}{256}\sqrt{3} \approx 0.663$ in^2. There are several ways we could determine s, but perhaps the easiest is to use the area formula $A = 3sh$. Since we know that $A = \frac{98}{256}\sqrt{3}$ and $h = \frac{7}{16}$, we find $s = A/3h = \left(\frac{98}{256}\sqrt{3}\right)/3\left(\frac{7}{16}\right) = \frac{7}{8\sqrt{3}} \approx 0.505$. Since the perimeter is $P = 6s = 6\frac{7}{8\sqrt{3}} = \frac{21}{4\sqrt{3}} \approx 3.031$ in.

16. As in Exercise 15, $h = 7.5$ mm and so $A = 2\sqrt{3}(7.5)^2 = 112.5\sqrt{3} \approx 194.856$ mm^2. Again, $s = A/3h = \left(112.5\sqrt{3}\right)/3(7.5) = 5\sqrt{3}$ and so $P = 6s = 6(5\sqrt{3}) = 30\sqrt{3} \approx 51.96$ mm.

17. (a) The cross-sectional area of this trapezoid is $A = \frac{1}{2}(85 + 40)\,12 = 750$ ft^2. (b) For the number of linear feet of concrete, we need the length along the bottom and the two sides. One side is $13'$ and the bottom is $40'$. We need to determine the length of the right-hand side. That side is the hypotenuse of a right triangle with sides 12 and 40 and so is $\sqrt{1744} \approx 41.76$ ft. Thus, the linear feet is $13 + 40 + 41.76 = 94.76$ ft.

18. The area of this floor is $20'6'' \times 15'3'' = 20.5' \times 15.25' = 312.625'$. Dividing by 9, we determine that there are 34.736 yd^2. Thus, the cost is $\$6.75 \times 34.736 = \$234.47$.

19. As in Exercise 11, we can draw a line to divide this into two rectangles. The area of one rectangle will be $5'' \times \frac{1}{4}'' = \frac{5}{4}$ in^2 and the area of the other rectangle is $3\frac{3}{4}'' \times \frac{1}{4}'' = \frac{15}{16}$ in^2. Adding these two areas together, we get the total area of $\frac{5}{4} + \frac{15}{16} = \frac{35}{16} = 2\frac{3}{16}$ in^2 =2.1875 in^2 or 2.19 in^2.

20. You can put $12' \div 9'' = 144'' \div 9'' = 16$ tiles along one wall and $17'3'' \div 9'' = 207'' \div 9'' = 23$ tiles along the other side. This produces a total of $16 \times 23 = 368$ tiles.

21. Drawing a line across the top of the base of the "T" we divide the figure into two rectangles. The vertical stem of the "T" measures 18 mm by 391 mm and has an area of 7038 mm^2. The top of the "T" is a rectangle that measures 29 mm by 400 mm with an area of 11 600 mm^2. The total area of the "T" is $7038 + 11\,600 = 18\,638$ mm^2

22. Draw a line across the house at the top of the vertical walls. The side is now divided into a rectangle which measures $38'6'' \times 12'$ with a trapezoid on top. The area of the rectangle is 462 ft^2. The area of the trapezoid is $A = \frac{1}{2}\left(38'6'' + 24'0''\right)\left(6'6''\right) = \frac{1}{2}\left(62.5'\right)\left(6.5'\right) = 203.125$ ft^2. Thus, the total area is $462 + 203.125 = 665.125$ ft^2.

3.4 CIRCLES

1. $A = \pi r^2 = \pi(4)^2 = 16\pi$ cm^2; $C = 2\pi r = 2\pi(4) = 8\pi$ cm

2. Here $d = 16$ in, so $r = 8$ in. $A = \pi r^2 = \pi(8)^2 = 64\pi$ in.^{2}; $C = 2\pi r = 2\pi(8) = 16\pi$ in.

3. $A = \pi r^2 = \pi(5)^2 = 25\pi$ in^2; $C = 2\pi r = 2\pi(5) = 10\pi$ in.

4. Since $d = 23$ mm, $r = 11.5$ mm. Thus, $A = \pi(11.5)^2 = 132.25\pi$ mm^2 and $C = 2\pi(11.5) = 23\pi$ mm

5. $A = \pi(14.2)^2 = 201.64\pi$ mm^2; $C = 2\pi(14.2) = 28.4\pi$ mm

6. $A = \pi\left(13\frac{1}{4}\right)^2 = \pi\left(\frac{53}{4}\right)^2 = \frac{2809}{16}\pi = 175.5625$ in^2; $C = 2\pi\left(13\frac{1}{4}\right) = 26.5\pi$ in.

7. Here $d = 24.20$ mm, so $r = 12.10$ mm and $A = \pi(12.10)^2 = 146.41\pi$ mm^2; $C = 2\pi(12.10) = 24.20\pi$ mm

8. Here $d = 23\frac{1}{2}$ in, so $r = 11\frac{3}{4}$ in. $A = \pi\left(11\frac{3}{4}\right)^2 = (11.75)^2\pi = 138.0625\pi$ in.^{2}; $C = 2\pi\left(11\frac{3}{4}\right) = 23.5\pi$ in.

9. (a) This table top has a diameter, d, of 48 in and so, $r = 24$ in. The area, A, is $A = \pi(24)^2 = 576\pi$ in^2. (b) The amount of metal edging is the same as the circumference. $C = \pi d = \pi(48) = 48\pi$ in ≈ 151 in.

10. Each turn of the wire is the same as the circumference of the coil. So, the total circumference of 42 turns is $42C = 42(\pi d) = 42\pi(0.5) = 21\pi$ m $\approx$ 66 m.

11. The diameter of this drum is 764 mm and so its circumference is $764\pi \approx 2400$ mm. If the rivets are 75 mm apart, then there are $2400 \div 75 = 32$ rivets.

12. Because the pulleys have the same radii, the belt can be considered as the length of two opposite sides of a rectangle and the circumference of two semicircles. Thus, the length, as shown in the figure, is $2a + 2\pi r$. The length a is the distance between the centers of the pulleys, so $a = 11'9\frac{1}{4}''$ and $r = 3'2''$. The length is $2\left(11'9\frac{1}{4}''\right) + 2\pi(3'2'') = 23'6\frac{1}{2}'' + 6'4''\pi \approx 23.542 + 19.897 = 43.439$ ft.

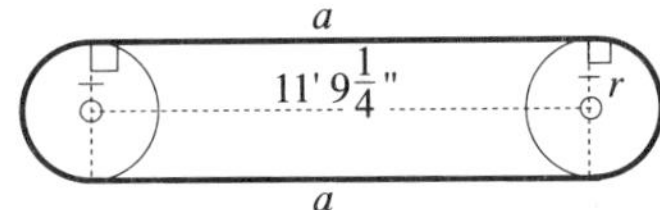

13. Since $\theta = 110°$, we need to change it to radians. $\theta = 110° = \frac{11}{18}\pi$. (a) The arc length $s = r\theta = 85\left(\frac{11}{18}\pi\right) = 51\frac{17}{18}\pi \approx 163.2$ mm. (b) For the area, $A = \frac{1}{2}r^2\theta = \frac{1}{2}\left(85^2\right)\left(\frac{11}{18}\pi\right) \approx 6\,935.5$ mm^2.

14. This is a semicircle on top of a rectangle. (a) The molding will be placed around the two vertical sides, the bottom of the rectangle, and the semicircle. This is $4'11'' + 4'2'' + 4'11'' + 2'1''\pi = 14' + 2'1''\pi \approx 20.545' \approx 20'6.5''$. (b) The area of the rectangle is $4'2'' \times 4'11'' = 20\frac{35}{72} \approx 20.486$ ft^2. The area of the semicircle is $\frac{1}{2}\pi\left(2'1''\right)^2 \approx 6.818$ ft^2. So, the total area is 27.304 ft^2.

15. (a) The amount of molding is the length of $\overset{\frown}{AB}$ + the length of $\overset{\frown}{BC}$ + the length of segment $\overline{AC}$. Since $60° = \frac{\pi}{3}$, $m\,\overset{\frown}{AB} = 28''\left(\frac{\pi}{3}\right) = m\,\overset{\frown}{BC}$. So, the length of molding needed is $28''\left(\frac{\pi}{3}\right) + 28''\left(\frac{\pi}{3}\right) + 28'' \approx 86.643$ in. (b) The area of this window is the area of the sector bounded by $\overset{\frown}{BC}$ and the sides $\overline{AB}$ and $\overline{AC}$ plus the area bounded by $\overline{AB}$ and $\overset{\frown}{AB}$. To find the area of this last figure, we will find the area of the sector bordered by $\overset{\frown}{AB}$, $\overline{BC}$ and $\overline{AC}$ and subtract the area of $\triangle ABC$. The area of each sector is $\frac{1}{2}\left(28''\right)^2\left(\frac{\pi}{3}\right) \approx 410.50$ in^2. Using Hero's formula, the area of $\triangle ABC$ is $\sqrt{42 \times 14^3} = 196\sqrt{3} \approx 339.48$. Subtracting this from the area of a sector produces $410.50 - 339.48 = 71.02$ and adding this to the area of a sector gives the desired answer, $410.50 + 71.02 = 481.52$ in^2.

16. (a) As in Exercise 15a, we need to find the length of the two arcs and add that to the length of three sides of the rectangle. The length of each arc is 2400mm $\left(\frac{\pi}{3}\right)$. So, the amount of molding needed is $3000 + 2400 + 3000 + 2(2400)\left(\frac{\pi}{3}\right)$ mm $= 8400 + 1600\pi \approx 13{,}426.5$ mm.

(b) The area of the glass needed is the area of the rectangle plus the area of the curved shape above the rectangle. As in Exercise 15b, the area of the curved shape is approximately $(2400\text{ mm})^2\left(\frac{\pi}{3}\right) - 2\,494\,153.16$ mm^2 $\approx 6\,031\,857.90 - 2\,494\,153.16 = 3\,537\,704.74$ mm^2. The area of the rectangular base is $(2400)(3000) = 7\,200\,000$ mm^2, so the total area is $3\,537\,704.74$ mm^2 + $7\,200\,000$ mm^2 = $10\,737\,704.74 \approx 10\,737\,705$ mm^2. Since 1 m^2 = 1000 mm^2 = $1\,000\,000$ mm^2, then $10\,737\,705$ mm^2 = $10\,737\,705 \div 1\,000\,000$ m^2 ≈ 10.738 m^2.

17. (a) This is the area of a rectangle measuring $42'' \times 36''$ and two semicircles of radius $18''$. $A = 42 \times 36 + \pi(18)^2 = 1512 + 324\pi \approx 2529.876$ in^2. (b) The edging needed is $2(42'') + 2\pi(18'') = 84'' + 36\pi'' \approx 197.1$ in.

18. The area is the difference between the area of the circles. So, $A = 40^2\pi - 30^2\pi = 1600\pi - 900\pi = 700\pi \approx 2199$ mm^2.

19. The area of a wire with a diameter of 8.42 mm is $4.21^2\pi = 17.7241\pi$. A wire with twice this area has an area of 35.4482π. If $\pi r^2 = 35.4482\pi$, then $r^2 = 35.4482$ and $r \approx 5.954$, so the diameter of the new wire is 11.91 mm.

20. The dotted line from the satellite to the Earth is tangent to the circle representing Earth, so it is perpendicular to the radii shown in the Figure. A line from the satellite to the Earth's center is the hypotenuse of two congruent right triangles and bisects the 120° angle. Thus, each triangle is a 30° − 60° right triangle. The length of the hypotenuse of a 30° − 60° right triangle is double the length of the shortest side, so the hypotenuse is 7920 mi. Subtracting the radius of the Earth, we see that the satellite is 3960 mi above the Earth.

3.5 GEOMETRIC SOLIDS

1. This is a cylinder with $r = 6''$ and $h = 18''$. $L = 2\pi rh = 2\pi(6'')(18'') = 216\pi \text{ in}^2 \approx 678.6 \text{ in}^2$. $T = 2\pi r(r+h) = 2\pi(6'')(6''+18'') = 2\pi(6'')(24'') = 288\pi \text{ in}^2 \approx 904.8 \text{ in}^2$. $V = \pi r^2 h = \pi(6'')^2(18'') = 648\pi \text{ in}^3 \approx 2{,}035.8 \text{ in}^3$.

2. This figure is a rectangular prism with $h = 6$ cm, $l = 14$ cm, and $w = 8$ cm. $L = 2h(l+w) = 2(6 \text{ cm})(14 \text{ cm} + 8 \text{ cm}) = 2(6 \text{ cm})(22 \text{ cm}) = 264 \text{ cm}^2$. $T = 2(lw + lh + hw) = 2[(14 \text{ cm})(8 \text{ cm}) + (14 \text{ cm})(6 \text{ cm}) + (6 \text{ cm})(8 \text{ cm})] = 2(112 \text{ cm}^2 + 84 \text{ cm}^2 + 48 \text{ cm}^2) = 488 \text{ cm}^2$. $V = lwh = (14 \text{ cm})(8 \text{ cm})(6 \text{ cm}) = 672 \text{ cm}^3$.

3. This is a right circular cone with $r = 8$ mm, $h = 15$ mm, and $s = 17$ mm. $L = \pi rs = \pi(8 \text{ mm})(17 \text{ mm}) = 136\pi \text{ mm}^2 \approx 427.3 \text{ mm}^2$. $T = \pi r(r+s) = \pi(8 \text{ mm})(8 \text{ mm} + 17 \text{ mm}) = 200\pi \text{ mm}^2 \approx 628.3 \text{ mm}^2$. $V = \frac{1}{3}\pi r^2 h = \frac{1}{3}\pi(8 \text{ mm})^2(15 \text{ mm}) = 320\pi \text{ mm}^3 \approx 1\,005.3 \text{ mm}^3$.

4. This is a square pyramid with $s = 17$ in and $h = 15$ in. Because it is a square, $p = 4(16 \text{ in}) = 64$ in and $B = (16 \text{ in})^2 = 256 \text{ in}^2$. $L = \frac{1}{2}ps = \frac{1}{2}(64 \text{ in})(17 \text{ in}) = 544 \text{ in}^2$. $T = \frac{1}{2}ps + B = L + B = 544 \text{ in}^2 + 256 \text{ in}^2 = 800 \text{ in}^2$. $V = \frac{1}{3}Bh = \frac{1}{3}(256 \text{ in}^2)(15 \text{ in}) = 1{,}280 \text{ in}^3$.

5. This triangular prism has $h = 21$ in and $p = 10 \text{ in} + 9 \text{ in} + 16 \text{ in} = 35$ in. Using Hero's formula, we see that $B \approx 40.91 \text{ in}^2$. $L = ph = (35 \text{ in})(21 \text{ in}) = 735 \text{ in}^2$; $T = L + 2B = 735 \text{ in}^2 + 2(40.91 \text{ in}^2) \approx 816.8 \text{ in}^2$; $V = Bh = (40.91 \text{ in}^2)(21 \text{ in}) \approx 859.1 \text{ in}^3$.

6. Here the figure is a frustum of a cone with $s = 10$ cm, $r_1 = 8$ cm, $r_2 = 16$ cm, and $h = 6$ cm. $L = \pi s(r_1 + r_2) = \pi(10 \text{ cm})(8 \text{ cm} + 16 \text{ cm}) = 240\pi \text{ cm}^2 \approx 754 \text{ cm}^2$; $T = L + \pi r_1^2 + \pi r_2^2 = 240\pi \text{ cm}^2 + \pi(8 \text{ cm})^2 + \pi(16 \text{ cm})^2 = 240\pi + 64\pi + 256\pi \text{ cm}^2 = 560\pi \text{ cm}^2 \approx 1759.3 \text{ cm}^2$. $V = \frac{1}{3}\pi h\left(r_1^2 + r_2^2 + r_1 r_2\right) = \frac{1}{3}\pi(6)\left(8^2 + 16^2 + (8)(16)\right) = 2\pi\,(64 + 256 + 128) = (640 + 256)\,\pi = 896\pi \approx 2\,814.9 \text{ cm}^3$.

7. This is a sphere with radius 8 cm. There is no lateral surface area. $T = 4\pi r^2 = 4\pi(8 \text{ cm})^2 = 256\pi \text{ cm}^2 \approx 804.25 \text{ cm}^2$. $V = \frac{4}{3}\pi r^3 = \frac{4}{3}\pi(8 \text{ cm})^3 = \frac{2048}{3}\pi \text{ cm}^3 = 682.7\pi \text{ cm}^3 \approx 2\,144.7 \text{ cm}^3$.

8. Here the figure is the frustum of a pyramid with square bases. We have $s = 50$ cm and $h = 26.46$ cm. We can determine that the perimeter of the top base is $p_1 = 80$ cm and its area is $B_1 = 400 \text{ cm}^2$. Similarly, the perimeter of the bottom base is $p_2 = 240$ cm and its area is $B_2 = 3\,600 \text{ cm}^2$. $L = \frac{s}{2}(p_1 + p_2) = \frac{50}{2}(80 + 240) = 25(320) = 8\,000 \text{ cm}^2$. $T = L + B_1 + B_2 = 8\,000 + 400 + 3600 = 12\,000 \text{ cm}^2$. $V = \frac{h}{3}\left(B_1 + B_2 + \sqrt{B_1 B_2}\right) = \frac{26.46}{3}\left(400 + 3600 + \sqrt{(400)(3600)}\right) = (8.82)(4000 + 1200) = 45\,864 \text{ cm}^3$.

9. The cross-section is a triangle on top of a rectangle. The area of the rectangle is 24 ft^2 (Note: $24' \times 12'' = 24' \times 1' = 24 \text{ ft}^2$) and the area of the triangle is $\frac{1}{2}(24')(6'') = \frac{1}{2}(24')(\frac{1}{2}') = 6 \text{ ft}^2$. Thus, the total area of the cross-section is $24 + 6 = 30 \text{ ft}^2$. Multiplying this area by the length of the road gives a volume of $30 \times 5280 = 158{,}400 \text{ ft}^3$. Thus, there are $158{,}400 \div 27 \approx 5866.7 \text{ yd}^3$.

10. We can think of the I-beam as a prism with the "I" the base. (a) The volume is the product of the area of the base and the altitude. The area of the base is 92 cm^2, so the volume is $92 \text{ cm}^2 \times 1000$ cm

$= 92\,000$ cm^3. (b) The painted portion amounts to the lateral surface area of the beam. The perimeter of the "I" is 96 cm, so the lateral surface area is 96 cm $\times 10$ m $= 96\,000$ cm^2. (c) The mass is the product of the volume ($92\,000$ cm^3) and the density (0.008 kg/cm^3) or 736 kg.

11. (a) The volume is $30(10)(12) = 3,600$ ft^3. (b) $T = 2\,[(30)(10) + (30)(12) + (10)(12)] = 2(300 + 360 + 120) = 1,560$ ft^2

12. $V = \pi r^2 h = \pi(48)^2 140 = 322,560\pi$ ft$^3 \approx 1,013,352.1$ ft^3; $T = 2\pi r(r + h) = 2\pi(48)(48 + 140) = 96\pi(188) = 18,048\pi \approx 56,699.5$ ft^2.

13. Here $r = 33$ mm. (a) $V = \pi(33)^2(95) = 103,455\pi \approx 325,013.5$ mm^3. (b) Because the label overlaps 5 mm, we need to add 5 mm to the circumference in order to determine the area of the paper needed for the label. $L = (2\pi r + 5)h = [2\pi(33) + 5]95 = (66\pi + 5)95 = 6270\pi + 475 \approx 20\,172.79$ mm^2.

14. The volume is the area of the cross section multiplied by the length of the pipe. The cross section has an area of $\pi(57^2 - 42^2) = 1\,485\pi$ mm^2, so the volume is $(1485\pi$ mm$^2)(2\,000$ mm$) = 2\,970\,000\pi$ mm$^3 \approx 9\,330\,530.2$ mm^3.

15. $V = \frac{4}{3}\pi r^3 = \frac{4\,000}{3}\pi$ m$^3 \approx 1\,333.3\pi$ m$^3 \approx 4\,188.8$ m^3

16. $L = \frac{s}{2}(p_1 + p_2)$. Here the perimeter of the top base is $p_1 = 4(380$ mm$) = 1\,520$ mm and the perimeter of the bottom base is $p_2 = 4(900$ mm$) = 3\,600$ mm. Since the slant height is 730 mm, the lateral surface area is $L = \frac{730}{2}(1\,520 + 3\,600) = (365)(5\,120) = 1\,868\,800$ mm$^2 \approx 1.8688$ m^2.

17. In Exercise Set 3.3, Exercise #14, we determined that the cross-sectional area of this highway support is 1470 ft^2. Since it is $2'6'' = 2.5$ ft thick, the total volume of the support is $(1470$ ft$^2)(2.5$ ft$) = 3675$ ft$^3 = \frac{3675}{27}$ yd$^3 \approx 136.1$ yd^3.

18. In Exercise Set 3.3, Exercise #17(a), we determined that the cross-sectional area of this river bed is 750 ft^2. Thus, the volume of a $\frac{1}{4}$-mile section is the product $(750$ ft$^2)(1320$ ft$) = 990,000$ ft^3.

19. When the piece of copper is joined along the straight edges, a cone is formed as shown in this figure below. To find the volume of this cone, we need the radius of the base and the height of the cone. We can determine the radius of the base from the circumference of the base. That circumference is the same as the arc length of the original sheet of copper. So, $C = 2\pi r = 85(110°)\left(\frac{\pi}{180°}\right)$ and $r = \frac{935}{36} \approx 25.97$. Using the Pythagorean Theorem, we get $h = \sqrt{85^2 - \left(\frac{935}{36}\right)^2} \approx 80.94$. So, $V = \frac{1}{3}\pi r^2 h \approx 18\,198.4\pi \approx 57\,172$ mm^3.

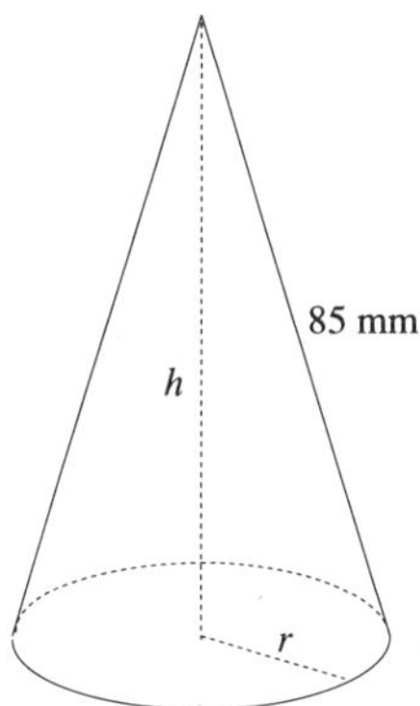

20. We need to find the lateral area of one frustum. By erecting a perpendicular to the bottom from its outer rim, we get a right triangle like the one shown. Its hypotenuse is the slant height of the frustum. So, $s = \sqrt{260^2 + 55^2} = \sqrt{70\,625} \approx 265.8$ mm. The lateral area is $L = \pi s(r_1 + r_2) \approx 265.8\pi(245 + 190) = 115\,623\pi$ mm. Thus, 500 tubs would require $500L \approx 57\,811\,500\pi$ mm$^2 \approx 181\,620\,180$ mm^2.

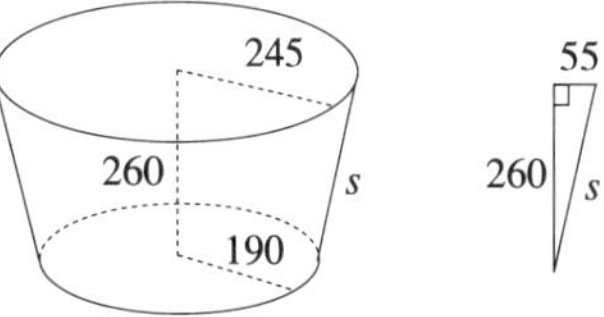

21. $L = \pi 12(1.5 + 4) = 66\pi$ in$^2 \approx 207.3$ in^2

22. This funnel is formed from two frustums of cones. The top frustum has a lateral area of $7.5\pi(5+1.5) = 48.75\pi$ cm^2. For the bottom frustum, the slant height, $s = \sqrt{13^2 + 1^2} = \sqrt{170}$. So, the lateral area is $\sqrt{170}\pi(1.5 + 0.5) = 2\sqrt{170}\pi \approx 26.08\pi$ cm^2. So, the total metal needed is $48.75\pi + 26.08\pi = 74.83\pi \approx 235.1$ cm^2.

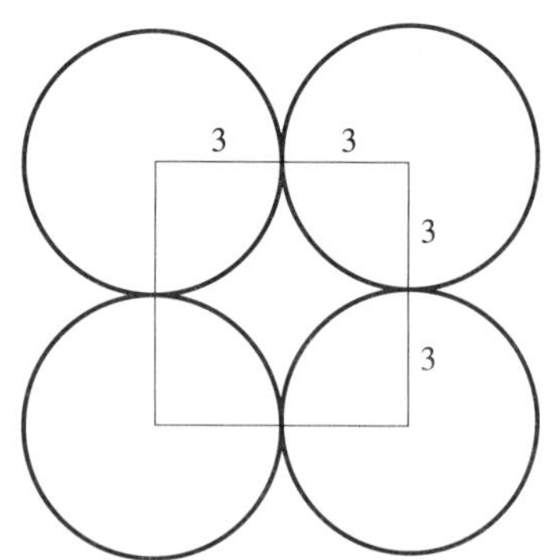

23. Draw the square formed by the centers of the silos. Each side of the square is 6 m, so the square has an area of 36 m^2. What remains outside the square are 4 sectors, each of which is $\frac{3}{4}$ of the original circle (silo). The area of each sector is $\frac{1}{2}r^2\theta = \frac{1}{2}(3^2)\left(\frac{3\pi}{2}\right) = \frac{27\pi}{4}$. So, the shaded area in Figure 3.34 is $4\left(\frac{27\pi}{4}\right) + 36 = 27\pi + 36$ m^2. The volume of grain is this area multiplied by the height of the silo, 10 m, or $270\pi + 360$ m$^3 \approx 1208.23$ m^3 of grain.

24. (a) The bore is 7 cm, so the radius of the cylinder is 3.5 cm and its height is 8 cm. Thus, the volume is $V = \pi r^2 h = \pi(3.5^2)8 = 98\pi \approx 307.9$ cm^3. (b) A 6-cylinder engine would have a total displacement of $6 \times 98\pi = 588\pi \approx 1847$ cm^3.

25. Here $h = 4$ ft and $r = 6$ ft, so $V = \frac{1}{3}\pi r^2 h = \frac{1}{3}\left(6^2\right)4 = 48\pi \approx 150.8$ ft^3

CHAPTER 3 REVIEW

1. Using the proportion $\frac{D}{180°} = \frac{R}{\pi}$, we substitute 27° for D and solve for R. Thus, $\frac{27°}{180°} = \frac{R}{\pi}$ or $R = \frac{27°}{180°}\pi = \frac{3}{20}\pi \approx 0.4712$.

2. Using the proportion $\frac{D}{180°} = \frac{R}{\pi}$, we substitute 212° for D and solve for R. Thus, $\frac{212°}{180°} = \frac{R}{\pi}$ or $R = \frac{212°}{180°}\pi = \frac{53}{45}\pi \approx 3.700$.

3. Multiplying by $\frac{180°}{\pi}$ produces 198°.

4. Multiplying by $\frac{180°}{\pi}$ produces $\frac{135°}{\pi} \approx 42.97°$.

5. The supplement of a 137° angle is $180° - 137° = 43°$.

6. The complement of a $\frac{\pi}{6}$ angle is $\frac{\pi}{2} - \frac{\pi}{6} = \frac{2\pi}{6} = \frac{\pi}{3}$.

7. Because $\angle A$ and the given angle are supplementary, $\angle A = 180° - 28° = 152°$. Angle A and $\angle B$ are corresponding angles formed by parallel lines cut (or intersected) by a transversal and so they are congruent. Thus, $\angle B = 152°$.

8. Angle B and the given angle are alternate exterior angles formed by parallel lines cut by a transversal and so they are congruent. Thus, $\angle B = \frac{7\pi}{16}$. Angle A and $\angle B$ are supplementary, so $\angle A = \pi - \angle B = \pi - \frac{7\pi}{16} = \frac{9\pi}{16}$.

9. Because corresponding segments are proportional, we have $\frac{2.7}{6.3} = \frac{AB}{7.7}$. Solving for AB produces $AB = \frac{(2.7)(7.7)}{6.3} = \frac{20.79}{6.3} = 3.3$

10. Because the sum of the angles of a triangle is 180°, we have $x = 180° - (32° + 21°) = 180° - 53° = 127°$.

11. This is an isosceles triangle, so the base angles are congruent and $x = 180° - 2(75°) = 180° - 150° = 30°$.

12. Using the Pythagorean Theorem, the missing side is the hypotenuse, so its length is $\sqrt{120^2 + 209^2} = \sqrt{58081} = 241$.

13. In this right triangle, the hypotenuse is 37 so, using the Pythagorean Theorem, the length of the missing side is $\sqrt{37^2 - 16^2} = \sqrt{1113} \approx 33.36$.

14. The perimeter of this triangle is $P = 18 + 5 + 16 = 39$ units. Using Hero's formula, we determine that the semiperimeter is $s = \frac{p}{2} = \frac{39}{2} = 19.5$ units and so the area is $A = \sqrt{19.5(19.5 - 18)(19.5 - 5)(19.5 - 16)} = \sqrt{19.5(1.5)(14.5)(3.5)} = \sqrt{1484.4375} \approx 38.53$ units^2.

15. This rectangle has an area of $A = (24.5)(12) = 294$ units^2. The perimeter is $P = 2(l + w) = 2(24.5 + 12) = 2(36.5) = 73$ units.

16. This parallelogram has a perimeter of $P = 2(l+w) = 2(23 + 14) = 2(37) = 74$ units. The area is $A = bh = 23(10) = 230$ units^2.

17. The circumference of a circle is $C = 2\pi r = 2\pi(9) = 18\pi \approx 56.55$ units. The area is $A = \pi r^2 = \pi 9^2 = 81\pi \approx 254.5$ units^2.

18. For this circle, we are given the diameter of 21 units and so the radius is $r = 10.5$ units. $C = \pi d = 21\pi \approx 65.97$ units. The area is $A = \pi r^2 = \pi(10.5)^2 = 110.25\pi \approx 346.36$ units^2.

19. Using the Pythagorean Theorem, we find that the length of the hypotenuse is $\sqrt{56^2 + 33^2} = \sqrt{4225} = 65$. Thus, $P = 33 + 56 + 65 = 154$ units and, since this is a right triangle, $A = \frac{1}{2}(33)(56) = 924$ units^2.

20. Converting 110° to radians, we get $110° = \frac{11}{18}\pi$, so the arc length is $r\theta = 7\left(\frac{11}{18}\pi\right) = \frac{77}{18}\pi \approx 4.277\pi \approx 13.44$ units. Adding the lengths of the two straight sides (radii) we get a perimeter of $13.44 + 7 + 7 = 27.44$ units. The area is $A = \frac{1}{2}r^2\theta = \frac{1}{2}7^2\left(\frac{11}{18}\pi\right) = \frac{539}{36}\pi \approx 14.972\pi \approx 47.04$ units^2.

21. The perimeter of this trapezoid is $P = 15 + 12 + 10.6 + 29.5 = 67.1$ units. The area is $\frac{1}{2}(b_1 + b_2)h = \frac{1}{2}(12 + 29.5)(9) = \frac{1}{2}(41.5)(9) = 186.75$ units^2.

22. This is a rectangular prism. We will let $l = 15.5$, $w = 8.4$, and $h = 35$. $L = 2h(l + w) = 2(35)(15.5 + 8.4) = 2(35)(23.9) = 1673$ units^2. $T = 2(lw + lh + hw) = 2[(15.5)(8.4) + (15.5)(35) + (35)(8.4)] = 2[130.2 + 542.5 + 294] = 2(966.7) = 1933.4$ units^2. $V = lwh = (15.5)(8.4)(35) = 4557$ units^3.

23. This is a right circular cylinder with $r = 12.1$ and $h = 9.6$. $L = 2\pi rh = 2\pi(12.1)(9.6) = 232.32\pi \approx 729.85$ units^2. $T = 2\pi r(r + h) = 2\pi(12.1)(12.1 + 9.6) = 2\pi(12.1)(21.7) = 525.14\pi \approx 1649.78$ units^2. $V = \pi r^2 h = \pi(12.1)^2(9.6) = \pi(146.41)(9.6) = 1405.54\pi \approx 4415.62$ units^3.

24. This figure is a right triangular prism with $p = 19 + 12 + 21 = 52$ units, $s = \frac{p}{2} = 26$ units, and $h = 17$ units. Using Hero's formula, we determine $B = \sqrt{26(26 - 21)(26 - 19)(26 - 12)} = \sqrt{26(5)(7)(14)} + \sqrt{12740} \approx 112.87$ units^2. $L = ph = (52)(17) = 884$ units^2. $T = ph + 2B = 884 + 2(112.87) = 1109.74$ units^2. $V = Bh \approx (112.87)(17) = 1918.79$ units^3.

25. This is a square pyramid with $p = 4(40) = 160$ units, $B = 40^2 = 1600$ units^2, $s = 29$ units, and $h = 21$ units. $L = \frac{1}{2}ps = \frac{1}{2}(160)(29) = 2320$ units^2. $T = \frac{1}{2}ps + B = L + B = 2320 + 1600$ units$^2 = 3920$ units^2. $V = \frac{1}{3}Bh = \frac{1}{3}(1600)(21) = 11,200$ units^3.

26. This figure is a right circular cone with $r = 8$ units and $h = 15$ units. Using the Pythagorean Theorem, we determine that $s = \sqrt{15^2 + 8^2} = \sqrt{225 + 64} = \sqrt{289} = 17$ units. $L = \pi rs = \pi(8)(17) = 136\pi \approx 427.26$ units^2. $T = \pi r(r + s) = \pi(8)(8 + 17) = 200\pi \approx 628.32$ units^2. $V = \frac{1}{3}\pi r^2 h = \frac{1}{3}\pi 8^2(15) = 320\pi \approx 1005.31$ units^3.

27. This is a frustum of a cone with $r_1 = 7$ units, $r_2 = 19$ units, $h = 16$ units, and $s = 20$ units. $L = \pi s(r_1 + r_2) = \pi(20)(7 + 19) = 520\pi \approx 1633.6$ units^2. $T = \pi[r_1(r_1 + s) + r_2(r_2 + s)] = \pi[(7)(7 + 20) + (19)(19 + 20)] = \pi(189 + 741) = 930\pi \approx 2921.68$ units^2. $V = \frac{1}{3}\pi h\left(r_1^2 + r_2^2 + r_1 r_2\right) = \frac{1}{3}\pi(16)\left(7^2 + 19^2 + (7)(19)\right) = \frac{1}{3}\pi(16(49 + 361 + 133) = \frac{1}{3}\pi(16)(543) = 2,896\pi \approx 9098.05$ units^3.

28. This is a frustum of a square pyramid with $p_1 = 4(24) = 96$ units, $p_2 = 4(14) = 56$ units, $s = 13$ units, and $h = 12$ units. Since the bases are squares we can determine that $B_1 = 24^2 = 576$ units^2 and that $B_2 = 14^2 = 196$ units^2. $L = \frac{1}{2}s(p_1 + p_2) =$

$\frac{1}{2}(13)(96+56) = 988$ units^2. $T = \frac{1}{2}s(p_1+p_2)+B_1+B_2 = L+B_1+B_2 = 988+576+196 = 1760$ units^2. $V = \frac{1}{3}h\left(B_1+B_2+\sqrt{B_1B_2}\right) = \frac{1}{3}(12)(576+196+\sqrt{(576)(196)}) = \frac{1}{3}(12)(576+196+336) = 4432$ units^3.

29. This sphere has a radius of 9 units. Because it is a sphere, it has no lateral area. $T = 4\pi r^2 = 4\pi 9^2 = 324\pi \approx 1017.88$ units^2. $V = \frac{4}{3}\pi r^3 = \frac{4}{3}\pi 9^3 = 972\pi \approx 3053.63$ units^3.

CHAPTER 3 TEST

1. $35\left(\frac{\pi}{180}\right) = \frac{35\pi}{180} = \frac{7\pi}{36} \approx 0.611$

2. $\left(\frac{7\pi}{15}\right)\left(\frac{180}{\pi}\right) = \frac{7\pi 180}{15\pi} = 84°$

3. $180° - 76° = 104°$

4. 133°

5. $\frac{5}{7} = \frac{9}{AB}$. Solving for $\overline{AB}$, we get $\overline{AB} = \frac{63}{5} = 12.6$ units

6. Using the Pythagorean Theorem, with $c = 3.9$ and $b = 3.6$, we have $a^2 + 3.6^2 = 3.9^2$ or $a^2 = 3.9^2 - 3.6^2 = 15.21 - 12.96 = 2.25$. Since $\sqrt{2.25} = 1.5$, $a = 1.5$ cm.

7. This rectangle has $l = 36.4$ cm and $w = 14.3$ cm, so $P = 2(l+w) = 2(36.4+14.3) = 2(50.7)101.4$ cm.

8. This trapezoid has been divided into a rectangle and two triangles. The length of the bottom is $4 + 12 + 4 = 20$ in and so the perimeter is $P = 12 + 5 + 20 + 5 = 42$ in.

9. This is a right trapezoid, so $h = 10.4$ cm and the area is $A = \frac{1}{2}h(b_1+b_2) = \frac{1}{2}(10.4)(21.6+32.4) = 280.8$ cm^2

10. This circle has a radius of $r = 9$ in and so its area is $A = \pi r^2 = \pi 9^2 = 81\pi \approx 254.47$ in^2

11. $\frac{14}{11.6} = \frac{x}{120}$. Solving for x produces $\frac{(14)(120)}{11.6} \approx 144.8$.

12. The circumference of the tire is $C = 2\pi r = \pi d = \pi(62)$ cm. To travel 25 m, the tire will have to make $\frac{25\text{ m}}{\pi(62)\text{ cm}}$ revolutions. Converting 25 m to cm, we obtain 2 500 cm. Substituting this in the previous statement, we get that the number of revolutions is $\frac{2\,500\text{ cm}}{\pi(62)\text{ cm}} \approx 12.84$

13. $V = lwh = (162)(52)(231) = 1\,945\,944$ mm^2 $= 1\,945\,944$ mm^3 $= 1\,945.944$ cm^3

14. $V = \frac{4}{3}\pi r^3$. We are given the diameter as 35′, so $r = \frac{35}{2} = 17.5$ ft. Thus $V = \frac{4}{3}\pi(17.5)^3 \approx 7145.83333\pi \approx 22,449.29750$ feet.

15. The label covers the lateral surface area of the cylinder plus an overlap. The lateral surface area is $L = 2\pi rh = \pi dh$ and the amount needed for the overlap is $0.8h$. So, the area is $\pi(6.5)(9.5) + (0.8)(9.5) = 61.75\pi + 7.6 \approx 201.59$ cm^2.

CHAPTER

4

Functions and Graphs

4.1 FUNCTIONS

1. This is a function because each value of x produces just one value for y.

2. This is a function because each value of x produces just one value for y.

3. This is not a function of x because the same value of x can give two different values for y. For example, the ordered pairs $(3, 2)$ and $(3, -2)$ both satisfy the relation.

4. This is a function because each value of x produces just one value for y.

5. Notice that $\sqrt{x}$ is defined only when $x \geq 0$. Since $\sqrt{x}$ produces a unique number for each x, this is a function for $x \geq 0$.

6. This is a function because each value of x produces just one value for y.

7. Since the square of any real number is nonnegative, this equation will be defined only for those values of x where $2x - 7 \geq 0$ or when $x \geq \frac{7}{2}$. However, this is not a function since $(4, 1)$ and $(4, -1)$ both satisfy the equation.

8. This is not a function of x because the same value of x can give two different values for y. For example, the ordered pairs $(10, 5)$ and $(10, -15)$ both satisfy the relation.

9. Yes, because for each value of x there is just one value of y.

10. Yes, because for each value of x there is just one value of y.

11. No, because when $x = 4$ there are two values of y: -2 and 2, or when $x = 1$, y is 1 or -1.

x	-40	-30	-20	-10	0	10	20	30	40
y	-40	-22	-4	14	32	50	68	86	104

12. Yes, because for each value of x there is just one value of y.

13. Domain: $\{-3, -2, -1, 0, 1, 2\}$;
Range: $\{-7, -5, -3, -1, 1, 3\}$

14. Domain: $\{-3, -2, -1, 0, 1, 2, 3\}$;
Range: $\{-6, -5, -2, 3\}$

x	-3	-2	-1	0	1	2	3
y	-25	-6	1	2	3	10	25

15. Domain: $\{-3, -2, -1, 0, 1, 2, 3\}$;
Range: $\{-25, -6, 1, 2, 3, 10, 25\}$

16. Domain and range are both all real numbers.

17. Domain is all real numbers except 5. Range is all real numbers except 1.

18. Domain is all non-negative real numbers, $x \geq 0$. Range is all real numbers greater than or equal to -2, that is, $y \geq -2$.

19. Domain is all real numbers. Range is all non-negative real numbers since $x^2 \geq 0$.

20. Domain is all real numbers except 2 and -3. Range is all positive real numbers and negative numbers less that -2.4; that is, $y \leq -2.4$ or $y > 0$.

21. $f(x) = 3x - 2$, so $f(0) = 3(0) - 2 = 0 - 2 = -2$.

22. $f(2) = 3(2) - 2 = 6 - 2 = 4$.

23. $f(-3) = 3(-3) - 2 = -9 - 2 = -11$.

24. $f(b) = 3(b) - 2 = 3b - 2$.

25. $f(b + 3) = 3(b + 3) - 2 = 3b + 9 - 2 = 3b + 7$.

26. Since $g(x) = x^2 - 5x$, then $g(0) = 0^2 - 5(0) = 0 - 0 = 0$.

27. Since $g(x) = x^2 - 5x$, then $g(-3) = (-3)^2 - 5(-3) = 9 - (-15) = 9 + 15 = 24$.

28. $g(2) = 2^2 - 5(2) = 4 - 10 = -6$.

29. $g(-x) = (-x)^2 - 5(-x) = x^2 - (-5x) = x^2 + 5x$.

30. $g(x - 5) = (x - 5)^2 - 5(x - 5) = (x^2 - 10x + 25) - 5x + 25 = x^2 - 15x + 50$.

31. We have $F(x) = \frac{2-x}{x^2+2}$, and so $F(0) = \frac{2-0}{0^2+2} = \frac{2}{0+2} = \frac{2}{2} = 1$.

32. $F(2) = \frac{2-2}{2^2+2} = \frac{0}{4+2} = \frac{0}{6} = 0$.

33. $F(1.5) = \frac{2-1.5}{(1.5)^2+2} = \frac{0.5}{2.25+2} = \frac{0.5}{4.25} = \frac{2}{17}$

34. $F(2x) = \frac{2-2x}{(2x)^2+2} = \frac{2-2x}{4x^2+2} = \frac{2(1-x)}{2(2x^2+1)} = \frac{1-x}{2x^2+1}$.

35. $\dfrac{F(2x)}{F(x)} = \dfrac{(1-x)/(2x^2+1)}{(2-x)/(x^2+2)} = \dfrac{1-x}{2x^2+1} \cdot \dfrac{x^2+2}{2-x} = \dfrac{(1-x)(x^2+2)}{(2x^2+1)(2-x)}$

36. $C(32) = \frac{5}{9}(32 - 32) = \frac{5}{9}(0) = 0$.

37. $C(98.6) = \frac{5}{9}(98.6 - 32) = \frac{5}{9}(66.6) = 37$.

38. $C(-40) = \frac{5}{9}(-40 - 32) = \frac{5}{9}(-72) = -40$.

39. $C(72) = \frac{5}{9}(72 - 32) = \frac{5}{9}(40) = \frac{200}{9} \approx 22.222222$.

40. explicit

41. implicit

42. implicit

43. explicit

44. Since $f(x, y) = 3x - 2y + 4$, then $f(1, 0) = 3(1) - 2(0) + 4 = 3 - 0 + 4 = 7$.

45. $f(0, 1) = 3(0) - 2(1) + 4 = 0 - 2 + 4 = 2$.

46. $f(-1, 2) = 3(-1) - 2(2) + 4 = -3 - 4 + 4 = -3$.

47. $f(2, -3) = 3(2) - 2(-3) + 4 = 6 + 6 + 4 = 16$.

48. Here $g(x, y) = x^2 - y^2 + 2xy$, and so $g(-2, 0) = (-2)^2 - 0^2 + 2(-2)(0) = 4 - 0 + 0 = 4$.

49. $g(0, -2) = 0^2 - (-2)^2 + 2(0)(-2) = 0 - 4 + 0 = -4$.

50. $g(5, 4) = 5^2 - 4^2 + 2(5)(4) = 25 - 16 + 40 = 49$.

51. $g(-3, 5) = (-3)^2 - 5^2 + 2(-3)(5) = 9 - 25 - 30 = -46$.

52. Since $h(x, y, z) = 3x - 4y - 2z + xyz$, then $h(-1, 2, 3) = 3(-1) - 4(2) - 2(3) + (-1)(2)(3) = -3 - 8 - 6 - 6 = -23$.

53. $h(2, -1, -3) = 3(2) - 4(-1) - 2(-3) + (2)(-1)(-3) = 6 + 4 + 6 + 6 = 22$.

54. $h(3, 2, -1) = 3(3) - 4(2) - 2(-1) + (3)(2)(-1) = 9 - 8 + 2 - 6 = -3$.

55. Solving $y = x - 5$ for x, we obtain $x = y + 5$. Exchanging the x- and y-variables, produces the inverse, $y = x + 5$.

56. Solving $2y = x + 7$ for x, produces $x = 2y - 7$. Exchanging the x- and y-variables, produces the inverse, $y = 2x - 7$.

57. Solving $y = 2x - 8$ for x, we obtain $x = \frac{y+8}{2}$, and so the inverse is $y = \frac{x+8}{2} = \frac{1}{2}(x + 8)$.

58. Solving $6y = 2x - 4$ for x, yields $x = \frac{6y+4}{2} = 3y + 2$, and so the inverse is $y = 3x + 2$.

59. Solving $9y = 3x + 5$ for x, we get $x = \frac{9y-5}{3}$, and so the inverse is $y = \frac{9x-5}{3} = \frac{1}{3}(9x - 5)$.

60. $y^2 = x - 5$ is not a function of x. The inverse of this relation is found as follows: Solving $y^2 = x - 5$ for x, we obtain $x = y^2 + 5$ and so $y = x^2 + 5$ is the inverse.

61. Solving $y = x^3 + 7$ for x, produces $x^3 = y - 7$ or $x = \sqrt[3]{y - 7}$ and the inverse is $y = \sqrt[3]{x - 7}$.

62. If $r = 20$, $V = 16{,}000\pi$ ft^3, $S = 2400\pi$ ft^2; if $r = 30$, $V = 36{,}000\pi$ ft^3, $S = 4200\pi$ ft^2

63. (a) $s(1) = 4.91(1^2) = 4.91 \times 1 = 4.91$ m; (b) $s(2) = 4.91(2^2) = 4.91 \times 4 = 19.64$ m; (c) $s(5) = 4.91(5^2) = 4.91 \times 25 = 122.75$ m.

64. Here $R(7) = 20 + 5(7 - 4) = 20 + 5(3) = 20 + 15 = 35$, so the cost of renting this chain saw is \$35.

4.2 OPERATIONS ON FUNCTIONS; COMPOSITE FUNCTIONS

1. $(f + g)(x) = (x^2 - 1) + (3x + 5) = x^2 - 1 + 3x + 5 = x^2 + 3x + 4$.

2. $(f + g)(4) = 4^2 + 3(4) + 4 = 16 + 12 + 4 = 32$.

3. $(f - g)(x) = (x^2 - 1) - (3x + 5) = x^2 - 1 - 3x - 5 = x^2 - 3x - 6$.

4. $(f - g)(4) = 4^2 - 3(4) - 6 = 16 - 12 - 6 = -2$.

5. $(g - f)(x) = (3x + 5) - (x^2 - 1) = 3x + 5 - x^2 + 1 = 3x - x^2 + 6$.

6. $(g - f)(5) = 3(5) - 5^2 + 6 = 15 - 25 + 6 = -4$.

7. $(f \cdot g)(x) = (x^2 - 1)(3x + 5) = 3x^3 + 5x^2 - 3x - 5$.

8. $(f \cdot g)(5) = 3(5^3) + 5(5^2) - 3(5) - 5 = 3(125) + 5(25) - 15 - 5 = 375 + 125 - 15 - 5 = 480$.

9. $(f/g)(x) = \left(\frac{f}{g}\right)(x) = \frac{f(x)}{g(x)} = \frac{x^2 - 1}{3x + 5}$

10. $(f/g)(-2) = \left(\frac{f}{g}\right)(-2) = \frac{f(-2)}{g(-2)} = \frac{(-2)^2 - 1}{3(-2) + 5} = \frac{4 - 1}{-6 + 5} = \frac{3}{-1} = -3$

11. $(g/f)(x) = \left(\frac{g}{f}\right)(x) = \frac{g(x)}{f(x)} = \frac{3x + 5}{x^2 - 1}$

12. $(g/f)(-2) = \left(\frac{g}{f}\right)(-2) = \frac{g(-2)}{f(-2)} = \frac{3(-2) + 5}{(-2)^2 - 1} = \frac{-6 + 5}{4 - 1} = \frac{-1}{3} = -\frac{1}{3}$

13. The domains of both f and g are all real numbers

14. The domain of f/g is all real numbers except $-\frac{5}{3}$, because when $x = -\frac{5}{3}$ the denominator of f/g is 0. The domain of g/f is all real numbers except -1 and $+1$, because the denominator of g/f, $x^2 - 1$, is 0 when $x = -1$ or $x = 1$.

15. $(f \circ g)(x) = f(g(x)) = f(3x + 5) = (3x + 5)^2 - 1 = 9x^2 + 30x + 24$.

16. One way to work this is: $(f \circ g)(-3) = f(g(-3)) = f(3(-3) + 5) = f(-4) = (-4)^2 - 1 = 16 - 1 = 15$. Another way is to use the value of $(f \circ g)(x) = 9x^2 + 30x + 24$ from the previous exercise. Then $(f \circ g)(-3) = 9(-3)^2 + 30(-3) + 24 = 9(9) - 90 + 24 = 81 - 90 + 24 = 15$.

17. $(g \circ f)(x) = g(f(x)) = g\left(x^2 - 1\right) = 3(x^2 - 1) + 5 = 3x^2 - 3 + 5 = 3x^2 + 2$.

18. Using the result of the previous exercise, $(g \circ f)(-3) = 3(-3)^2 + 2 = 3(9) + 2 = 27 + 2 = 29$.

19. $(f + g)(x) = f(x) + g(x) = (3x - 1) + (3x^2 + x) = 3x^2 + 4x - 1$

20. $(f + g)(4) = 3(4)^2 + 4(4) - 1 = 3 \cdot 16 + 16 - 1 = 48 + 16 - 1 = 63$

21. $(f - g)(x) = f(x) - g(x) = (3x - 1) - (3x^2 + x) = 3x - 1 - 3x^2 - x = -3x^2 + 2x - 1$

22. $(f - g)(-2) = -3(-2)^2 + 2(-2) - 1 = -3(4) + (-4) - 1 = -12 - 4 - 1 = -17$

23. $(g-f)(x) = g(x) - f(x) = (3x^2 + x) - (3x - 1) = 3x^2 + x - 3x + 1 = 3x^2 - 2x + 1$

24. $(g-f)(-2) = 3(-2)^2 - 2(-2) + 1 = 3(4) + 4 + 1 = 12 + 4 + 1 = 17$ Note: $(f-g)(-2) = -[(g-f)(-2)]$

25. $(f \cdot g)(x) = (3x-1)(3x^2+x) = 9x^3 + 3x^2 - 3x^2 - x = 9x^3 - x$

26. $(f \cdot g)(-1) = 9(-1)^3 - (-1) = -9 + 1 = -8$

27. $(f/g)(x) = \frac{f(x)}{g(x)} = \frac{3x-1}{3x^2+x}$

28. $(f/g)(-1) = \frac{3(-1)-1}{3(-1)^2+(-1)} = \frac{-3-1}{3-1} = \frac{-4}{2} = -2$

29. $(g/f)(x) = \frac{g(x)}{f(x)} = \frac{3x^2+x}{3x-1}$

30. $(g/f)(-1) = \frac{3(-1)^2+(-1)}{3(-1)-1} = \frac{3-1}{-3-1} = \frac{2}{-4} = -\frac{1}{2}$; note $(g/f)(-1) = \frac{1}{(f/g)(-1)}$

31. $(f \circ g)(x) = f(g(x)) = f(3x^2 + x) = 3(3x^2 + x) - 1 = 9x^2 + 3x - 1$

32. $(f \circ g)(5) = f(g(5)) = f(3 \cdot 5^2 + 5) = f(75 + 5) = f(80) = 3 \cdot 80 - 1 = 239$ or using #31 $(f \circ g)(5) = 9(5)^2 + 3(5) - 1 = 9 \cdot 25 + 15 - 1 = 225 + 15 - 1 = 239$

33. $(g \circ f)(x) = g(f(x)) = g(3x-1) = 3(3x-1)^2 + (3x-1) = 3(9x^2 - 6x + 1) + 3x - 1 = 27x^2 - 18x + 3 + 3x - 1 = 27x^2 - 15x + 2$

34. $(g \circ f)(5) = 27(5)^2 - 15(5) + 2 = 27 \cdot 25 - 15 \cdot 5 + 2 = 675 - 75 + 2 = 602$

35. The domains of both f and g are all real numbers.

36. The domain of f/g is all real numbers except 0 and $-\frac{1}{3}$. The domain of g/f is all real numbers except $\frac{1}{3}$.

37. (a) $C(n) = F(n) + V(n) = 7500 + 15n$ or $15n + 7500$;

(b) $C(100) = 15(100) + 7500 = 1500 + 7500 = \$9{,}000$;

(c) $C(1000) = 15(1000) + 7500 = 15000 + 7500 = \$22,500$

38. (a) $C(n) = S(n) + F(n) + V(n) = \frac{2{,}750{,}000}{n} + 12,500 + 8n$ dollars;

(b) $C(100) = \frac{2{,}750{,}000}{100} + 12,500 + 8(100) = 27,500 + 12,500 + 800 = \$40,800$;

(c) $C(1000) = \frac{2{,}750{,}000}{1000} + 12,500 + 8(1000) = 2750 + 12,500 + 8000 = \$23,250$

39. (a) $P(n) = R(n) - C(n) = 90n - (30n + 275) = 60n - 275$;

(b) $P(50) = 60(50) - 275 = 3000 - 275 = \$2{,}725$

40. (a) $R(p) = f(p) \times p = (750 - 4.2p)p = 750p - 4.2p^2$;

(b) $f(100) = 750 - 4.2(100) = 750 - 420 = 330$;

(c) $R(100) = 750(100) - 4.2(100)^2 = 75000 - 42000 = 33000$ or $330 \times 100 = \$33,000$;

(d) $f(80) = 750 - 4.20(80) = 750 - 336 = 414$;

(e) $R(80) = 750(80) - 4.2(80)^2 = 60,000 - 4.2(6400) = 60,000 - 26880 = \$33{,}120$ or $414 \times 80 = \$33{,}120$

41. $(P \circ n)(a) = P(7a+4) = 300(7a+4)^2 - 50(7a+4) = 300(49a^2 + 56a + 16) - 50(7a + 4) = 14700a^2 + 16800a + 4800 - 350a - 200 = 14700a^2 + 16450a + 4600$

42. (a) The volume V of a sphere of radius r is $V(r) = \frac{4}{3}\pi r^3$. We are given $r(t) = 3t$, where t represents the number of seconds since the weather balloon was begun to be inflated. Then $V(t) = (V \circ r)(t) = \frac{4}{3}\pi(3t)^3 = 36\pi t^3$ cm^3. (b) $V(10) = 36\pi(10^3) = 36,000\pi$ cm^3.

4.3 RECTANGULAR COORDINATES

1.

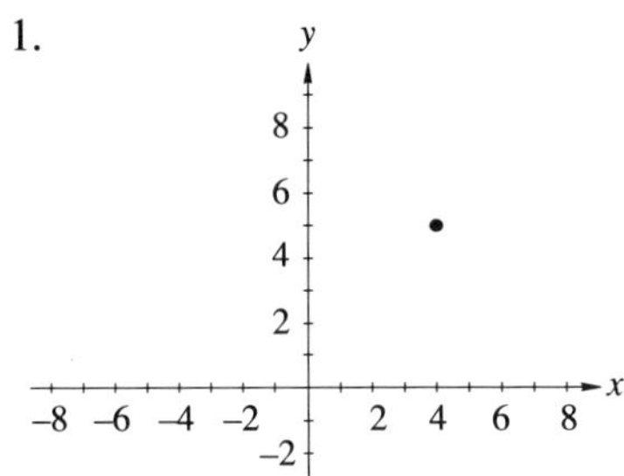

2.

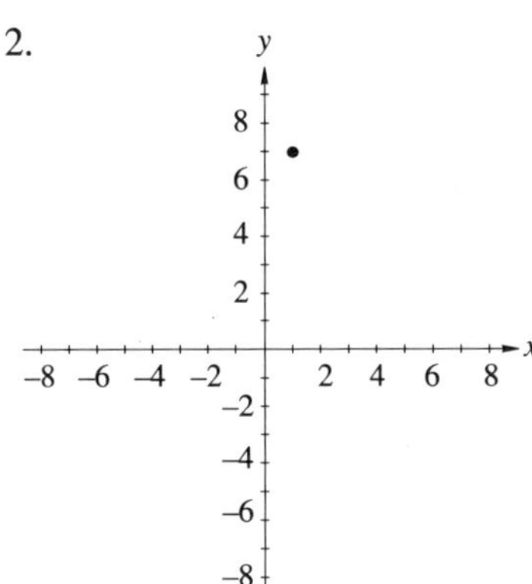

3.

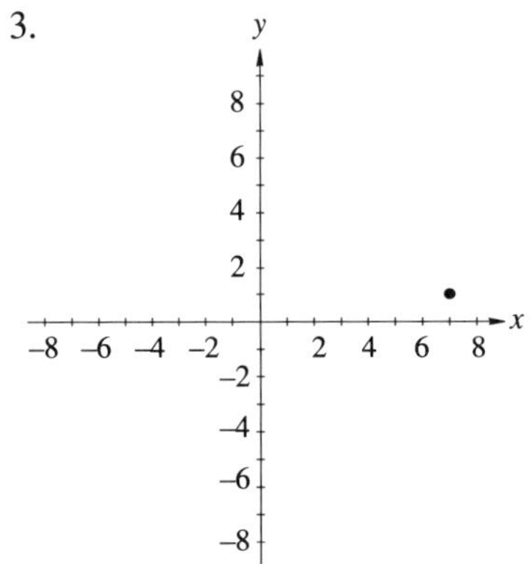

4.

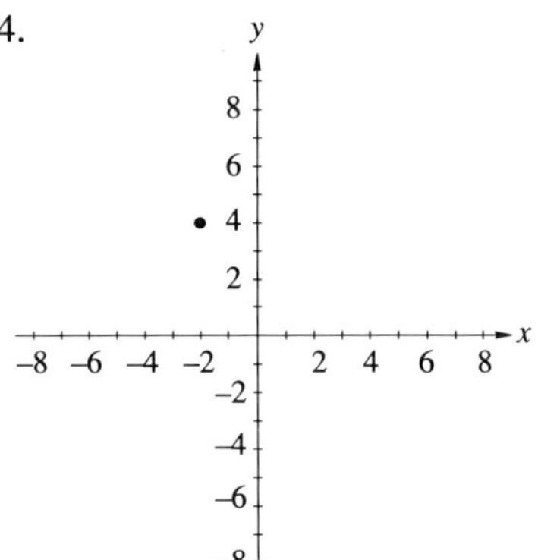

5.

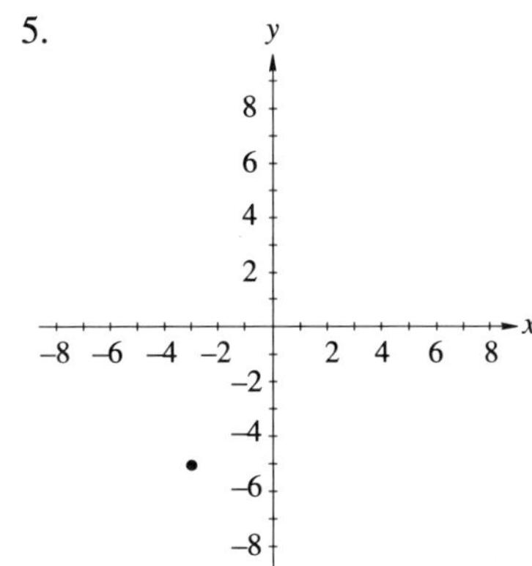

6.

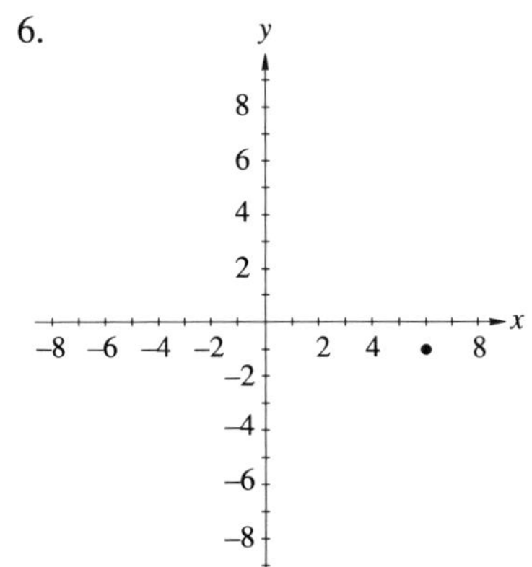

7.

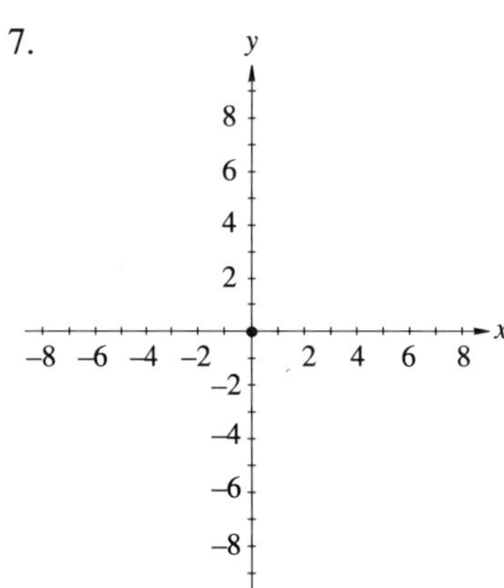

8.

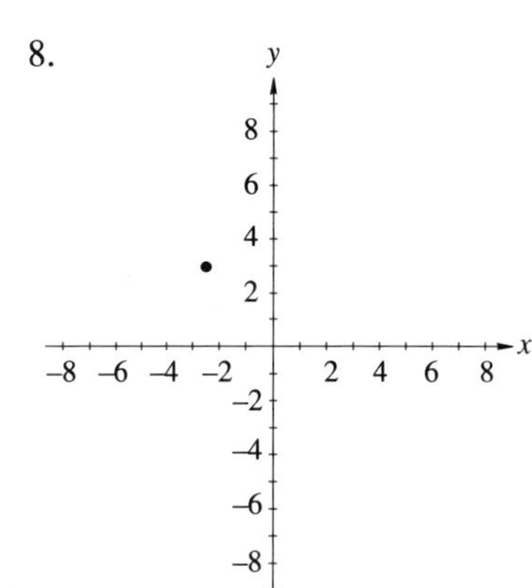

9.

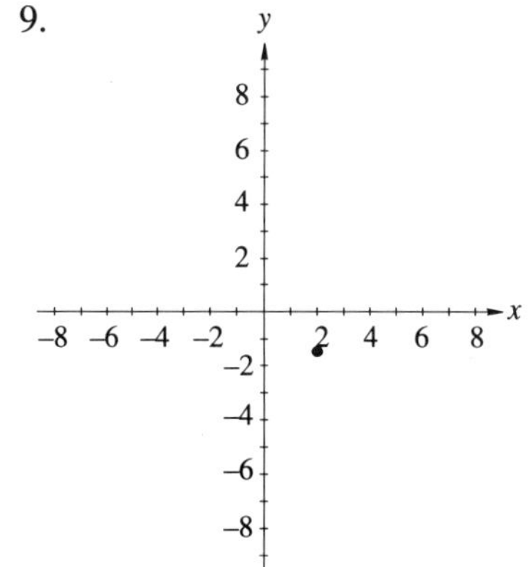

10.

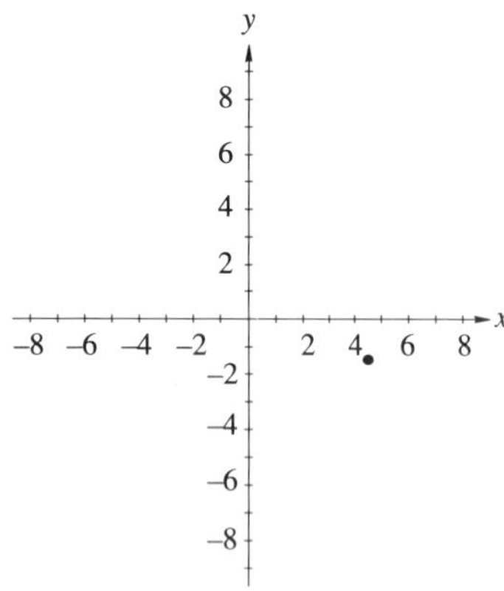

11. The fourth vertex will have the same x-coordinate as B, -1, and the same y-coordinate as C, -4. So its coordinates are $(-1, -4)$

12.

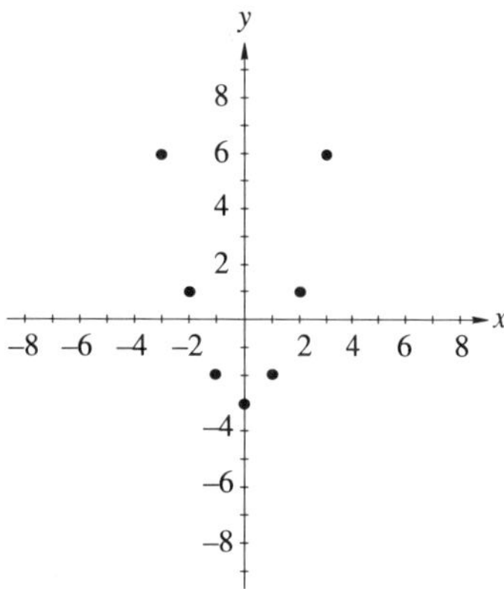

13. They are on a horizontal line and have the same y-coordinate.

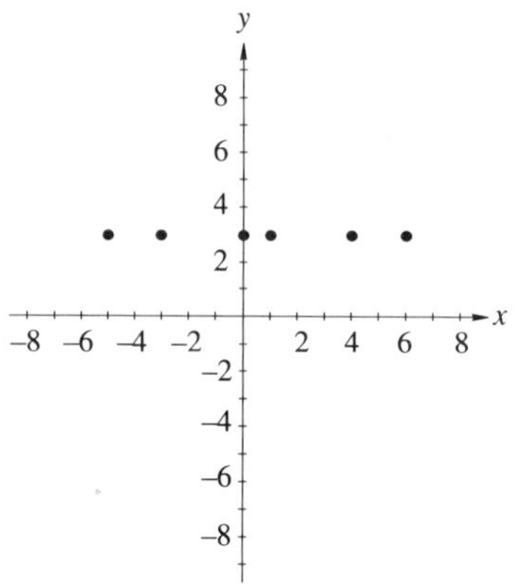

14. They are all on the same vertical line and have the same x-coordinate.

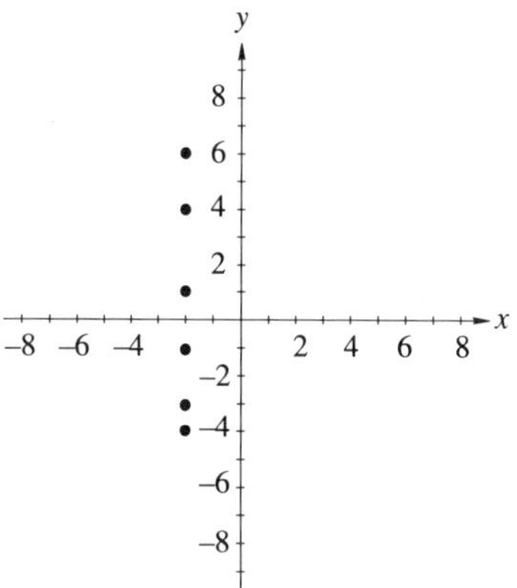

15. They all lie on the same straight line.

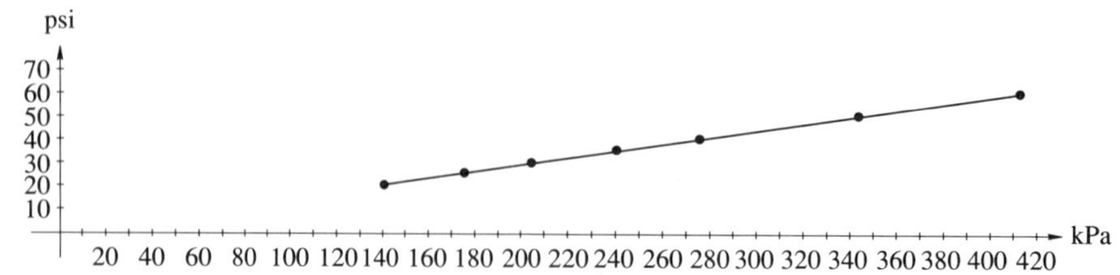

16. As in problem #14, these points will all lie on a vertical line through $x = 0$. This is the y-axis.

17. As in problem #13, these points all lie in a horizontal line through $y = -2$.

18. These points must all lie to the right of the vertical line $x = -3$.

19. These points must be to the right of the y-axis (see problem 16 and 18) and below the x-axis. This is the fourth quadrant.

20. To the right of the vertical line through the point $(1, 0)$ and below the horizontal line through the point $(0, -2)$.

21. (a)

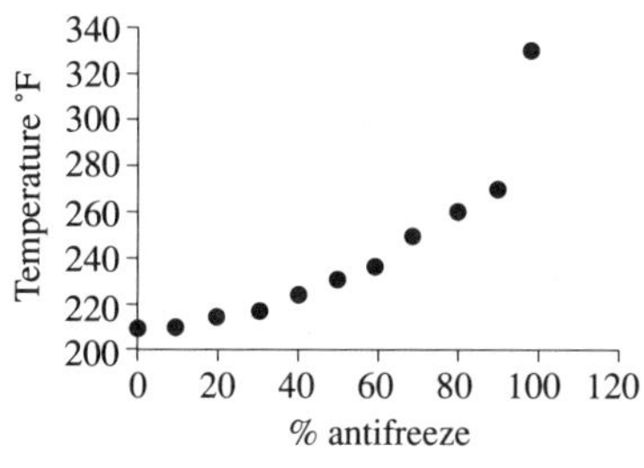

(b) 228°F; (c) 236°F

22. (a)

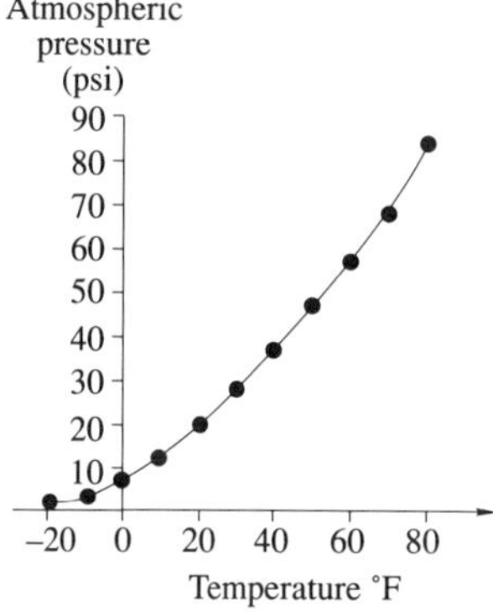

(b) 41.9 psi; (c) 99.6psi

4.4 GRAPHS

1. (a)

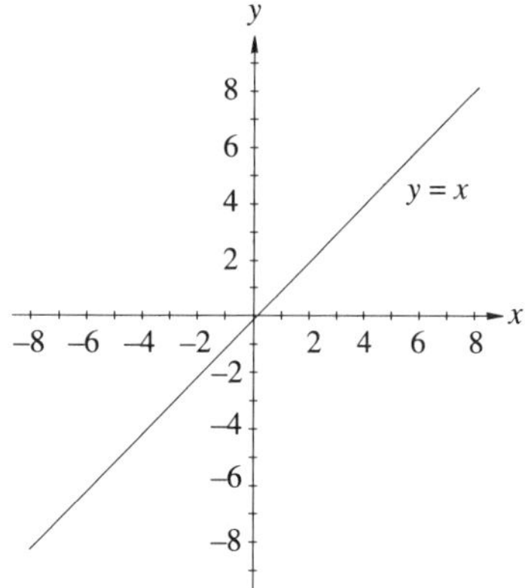

(b) x-intercept is 0; y-intercept is 0; (c) Let $(x_1, y_1) = (0, 0)$ and $(x_2, y_2) = (2, 2)$ thus $m = \frac{2-0}{2-0} = \frac{2}{2} = 1$

2. (a)

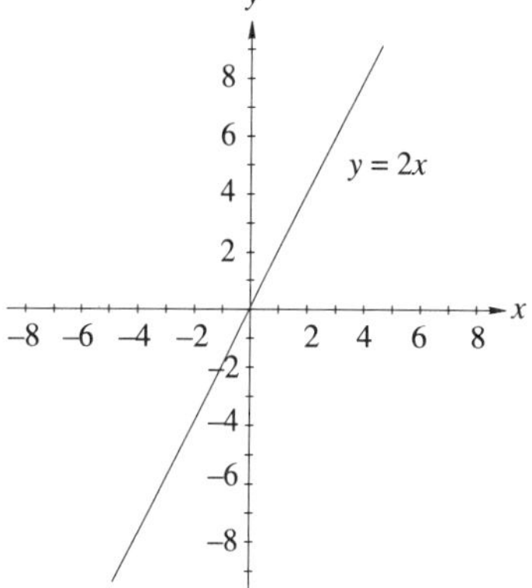

(b) x-intercept is 0; y-intercept is 0; (c) Let $(x_1, y_1) = (0, 0)$ and $(x_2, y_2) = (1, 2)$, then $m = \frac{2-0}{1-0} = \frac{2}{1} = 2$

3. (a)

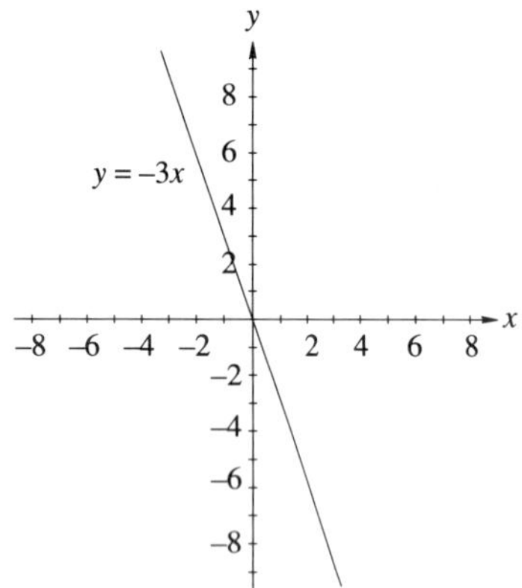

(b) x-intercept is 0; y-intercept is 0; (c) Let $(x_1, y_1) = (0, 0)$ and $(x_2, y_2) = (1, -3)$ then $m = \frac{-3-0}{1-0} = \frac{-3}{1} = -3$

4. (a)

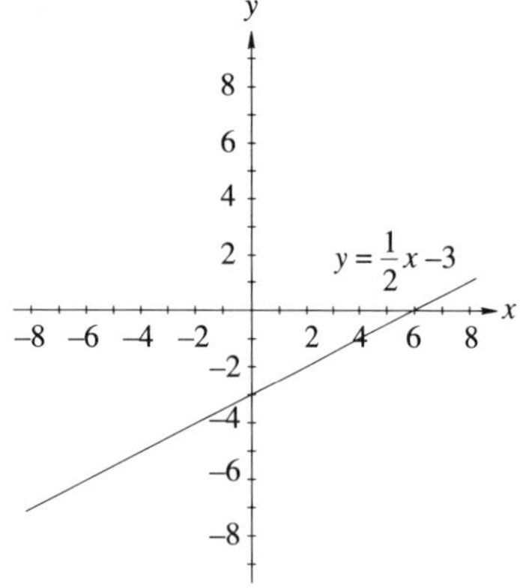

(b) x-intercept is 6; y-intercept is -3; (c) Let $(x_1, y_1) = (0, -3)$ and $(x_2, y_2) = (6, 0)$ then $m = \frac{0-(-3)}{6-0} = \frac{3}{6} = \frac{1}{2}$

5. (a)

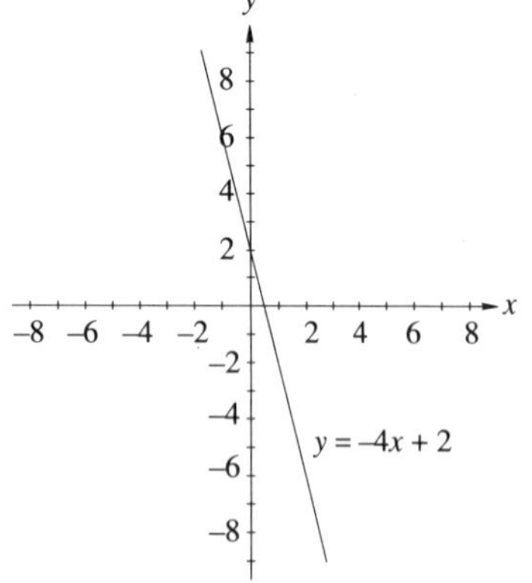

(b) x-intercept is $\frac{1}{2}$; y-intercept is 2; (c) $m = \frac{0-2}{\frac{1}{2}-0} = \frac{-2}{\frac{1}{2}} = -4$

6. (a)

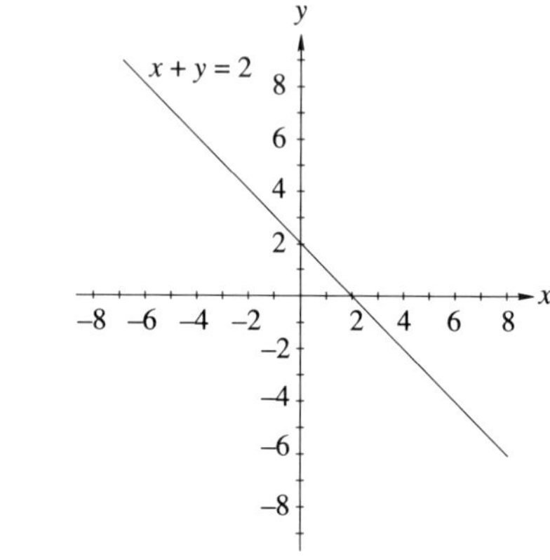

(b) x-intercept is 2; y-intercept is 2; (c) $m = \frac{0-2}{2-0} = -1$

7. (a)

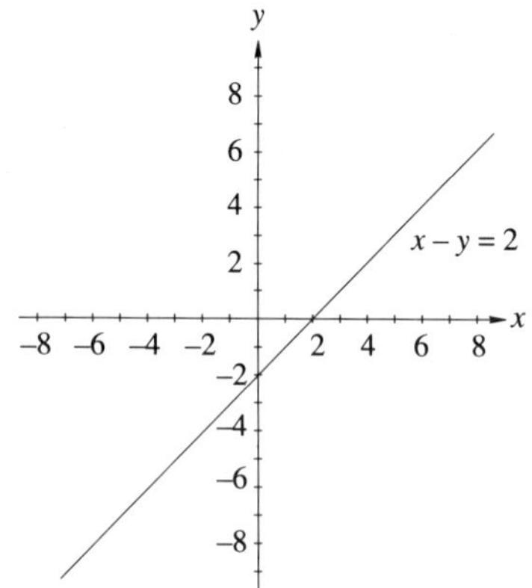

(b) x-intercept is 2; y-intercept is -2; (c) $m = \frac{0-(-2)}{2-0} = \frac{2}{2} = 1$

8. (a)

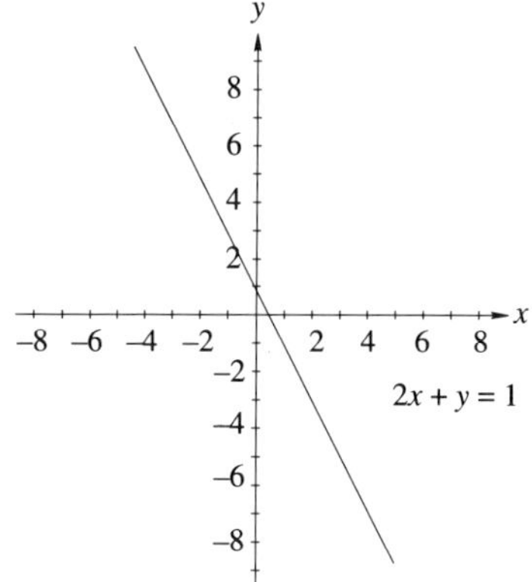

(b) x-intercept is $\frac{1}{2}$; y-intercept is 1; (c) $m = \frac{0-1}{\frac{1}{2}-0} = \frac{-1}{\frac{1}{2}} = -2$

9. (a)

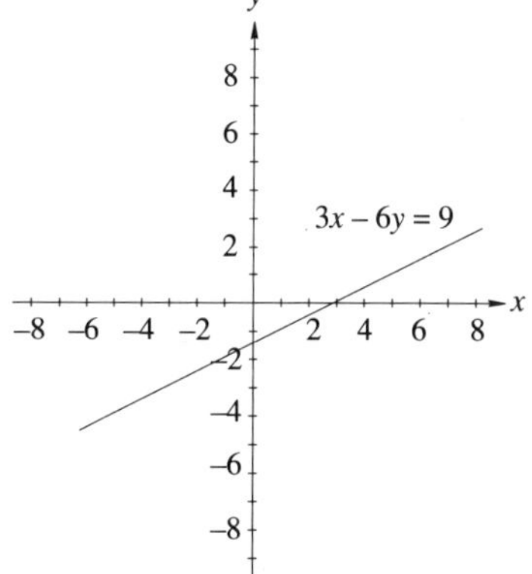

(b) x-intercept is 3; y-intercept is $-\frac{3}{2}$; (c) $m = \frac{0-(-\frac{3}{2})}{3-0} = \frac{\frac{3}{2}}{3} = \frac{1}{2}$

10. (a)

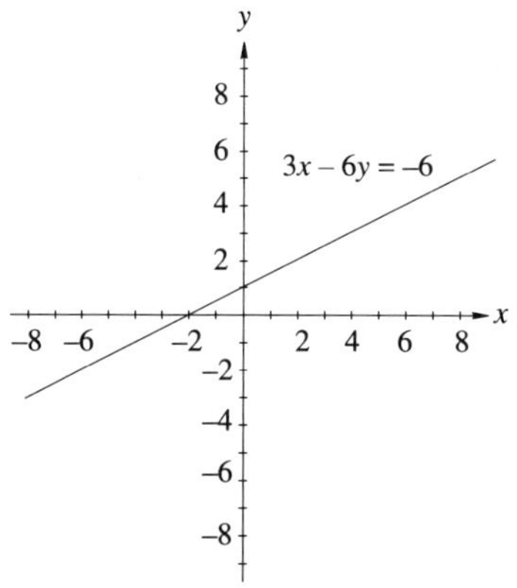

(b) x-intercept is -2; y-intercept is 1; (c) $m = \frac{0-1}{-2-0} = \frac{-1}{-2} = \frac{1}{2}$

11.

x	-3	-2	-1	0	1	2	3
y	9	4	1	0	1	4	9

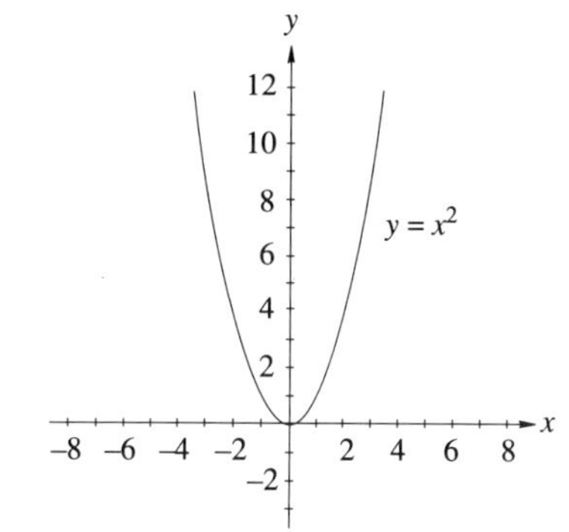

12.

x	-3	-2	-1	0	1	2	3
y	12	7	4	3	4	7	12

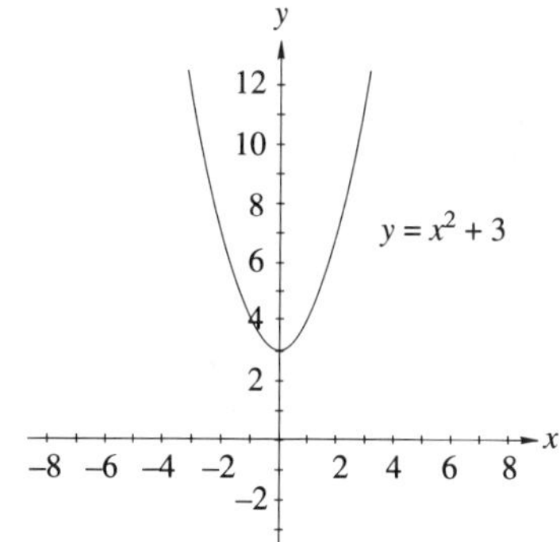

13.

x	-3	-2	-1	0	1	2	3
y	7	2	-1	-2	-1	2	7

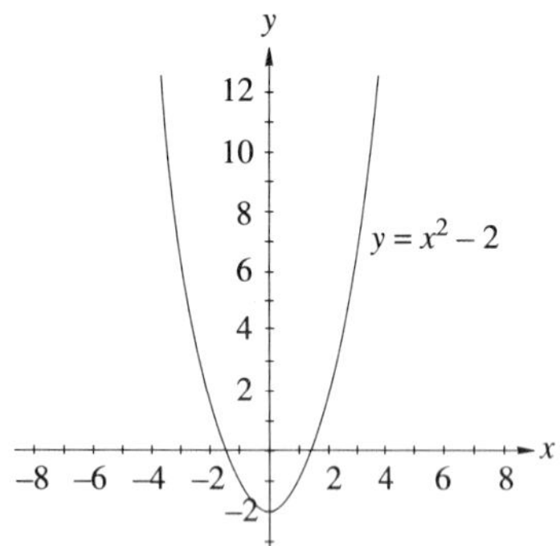

14.

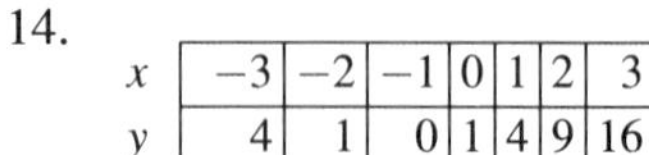

x	−3	−2	−1	0	1	2	3
y	4	1	0	1	4	9	16

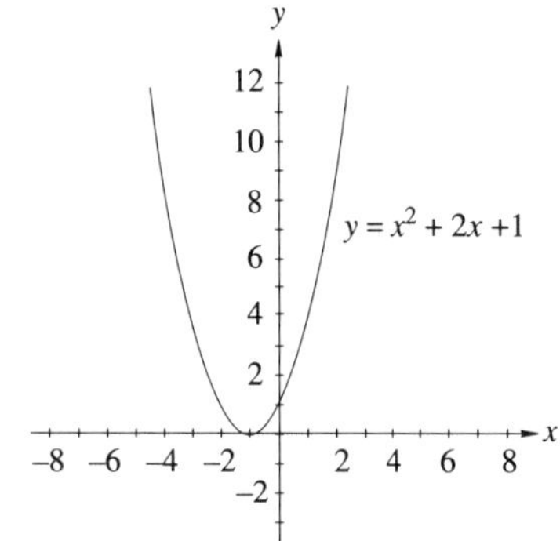

15.

x	−3	−2	−1	0	1	2	3	4	5	6
y	36	25	16	9	4	1	0	1	4	9

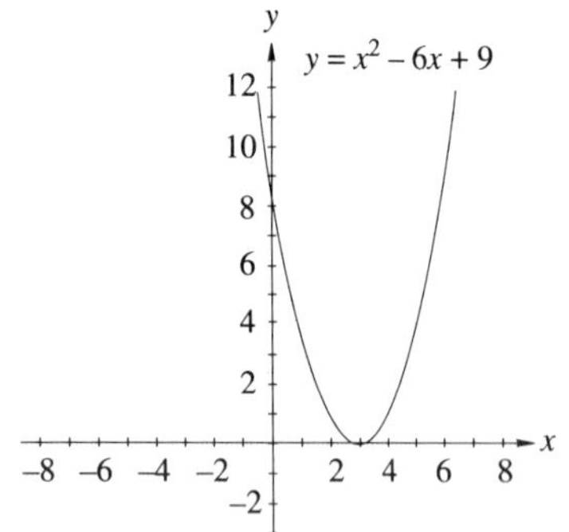

16.

x	−3	−2	−1	0	1	2	3
y	−7	−2	1	2	1	−2	−7

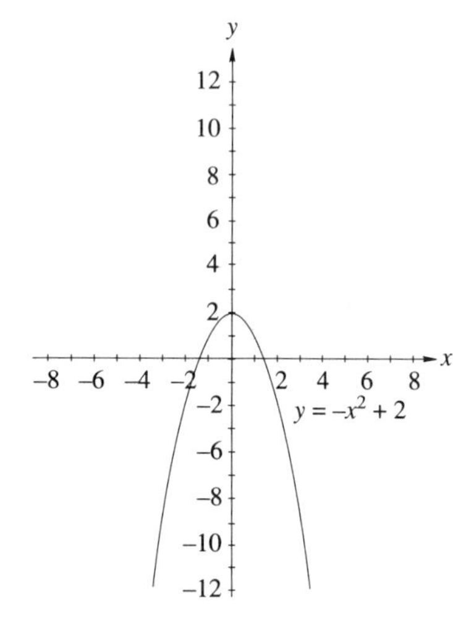

17.

x	−3	−2	−1	$-\frac{1}{2}$	$-\frac{1}{4}$	0	$\frac{1}{4}$	$\frac{1}{2}$	1	2	3
y	$-\frac{1}{3}$	$-\frac{1}{2}$	−1	−2	−4	Undefined	4	2	1	$\frac{1}{2}$	$\frac{1}{3}$

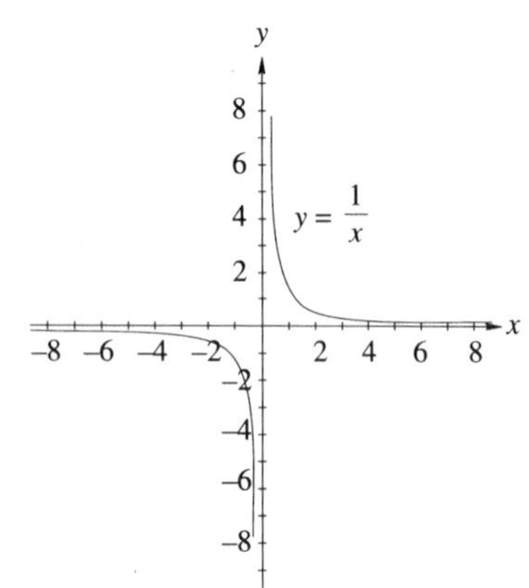

18.

x	-6	-5	-4	$-3\frac{1}{2}$	$-3\frac{1}{4}$	-3	$-2\frac{3}{4}$	$-2\frac{1}{2}$	-2	-1	0	1
y	$-\frac{1}{3}$	$-\frac{1}{2}$	-1	-2	-4	Undefined	4	2	1	$\frac{1}{2}$	$\frac{1}{3}$	$\frac{1}{4}$

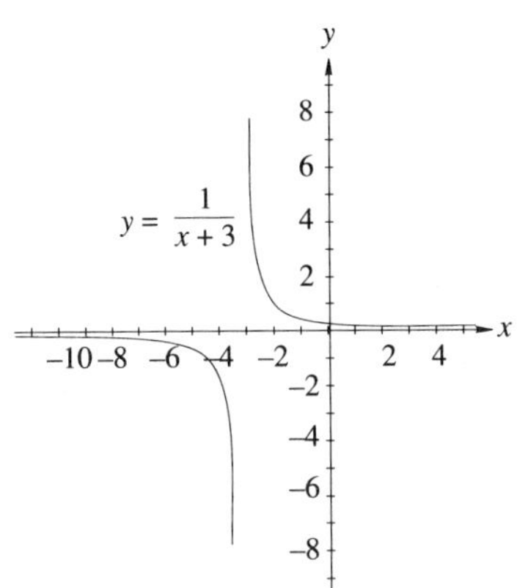

19.

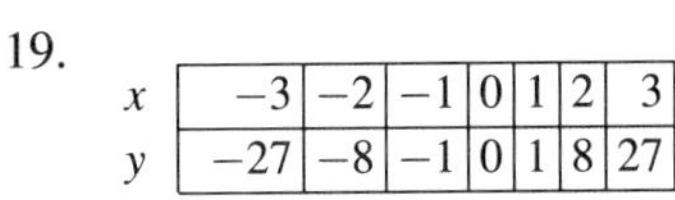

x	-3	-2	-1	0	1	2	3
y	-27	-8	-1	0	1	8	27

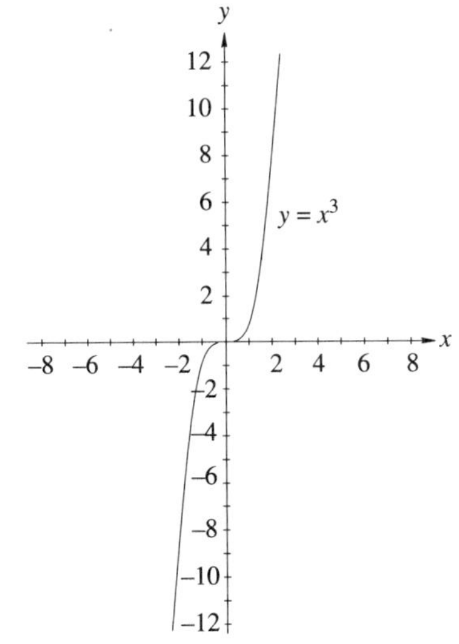

20.

x	-27	-8	-1	0	1	8	27
y	-3	-2	-1	0	1	2	3

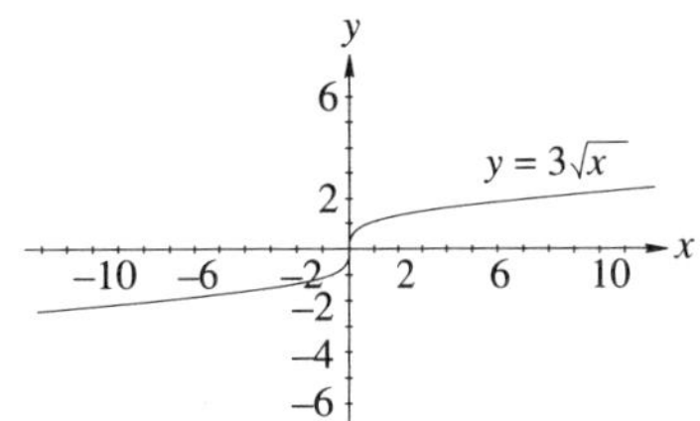

21.

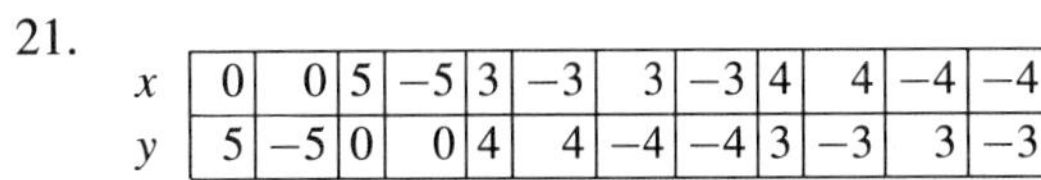

x	0	0	5	-5	3	-3	3	-3	4	4	-4	-4
y	5	-5	0	0	4	4	-4	-4	3	-3	3	-3

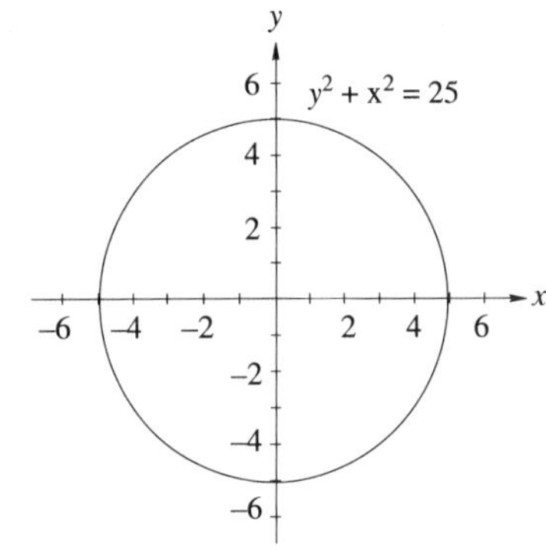

22.

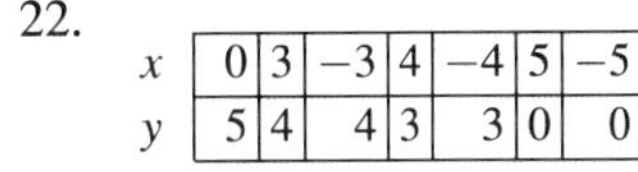

x	0	3	-3	4	-4	5	-5
y	5	4	4	3	3	0	0

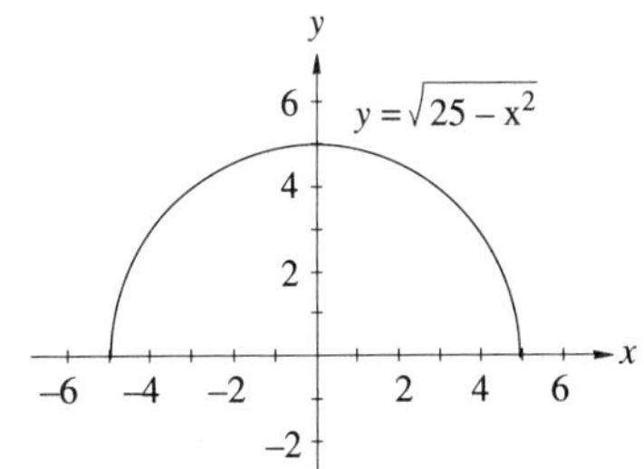

23.

x	-5	-4	-3	-2	-1	0	1	2
y	3	2	1	0	1	2	3	4

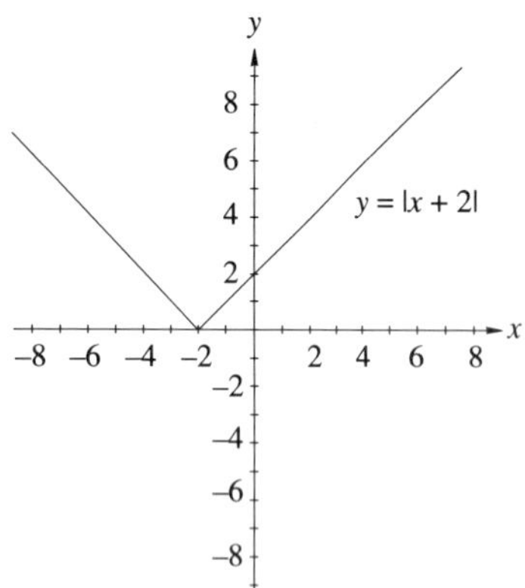

24.

x	0	1	−1	2	−2
y	±3	±2.6	±2.6	0	0

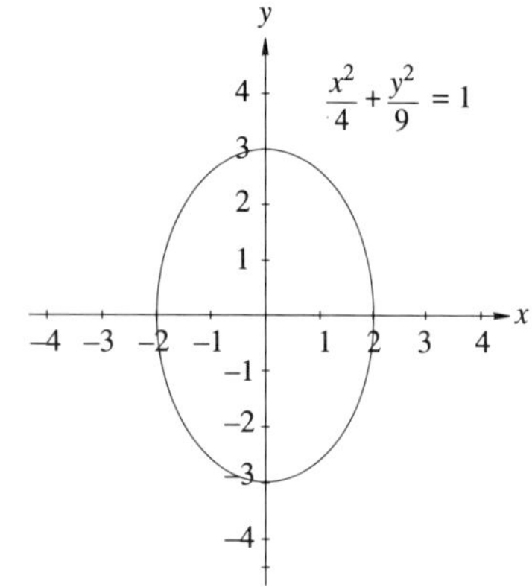

25.

x	0	1	−1	2	−2	3	−3	4	−4
y	Undefined			0	0	±3.35	±3.35	±5.2	±5.2

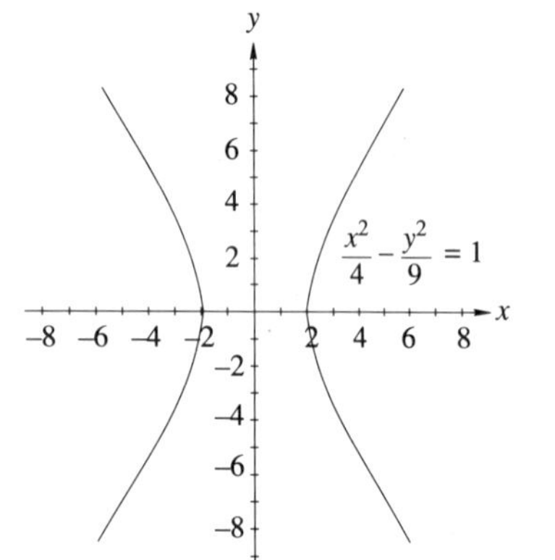

26.

x	−3	−2	−1	0	$\frac{1}{4}$	$\frac{1}{2}$	$\frac{3}{4}$	1	2	3
y	−36	−12	−2	0	−0.05	$-\frac{1}{8}$	−0.14	0	4	18

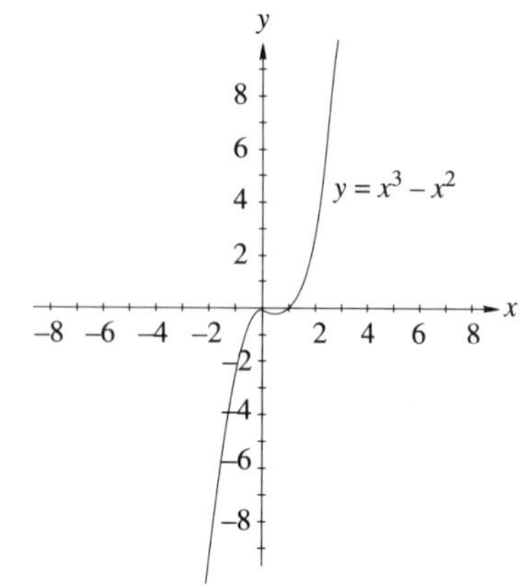

27. Function

28. Not a function

29. Function

30. Not a function

31. (a)

k	0	1	2	3	4	5	6	7	8	9
$p(k)$	25.0	8.33	5.00	3.57	2.78	2.27	1.92	1.67	1.47	1.32

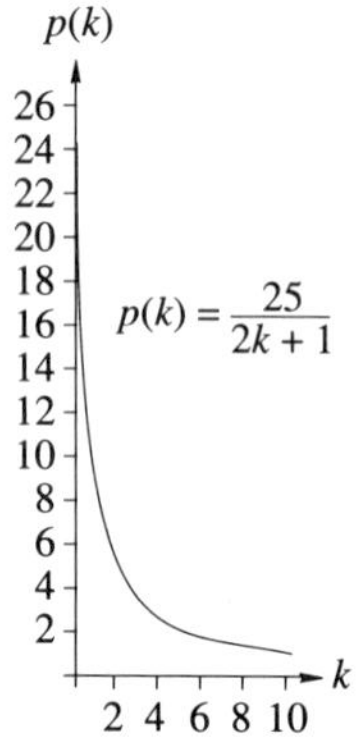

(b) yes

32.

k	0	1	2	3	4	5	6	7	8	9
$p(k)$	25.00	17.68	14.43	12.50	11.18	10.21	9.45	8.84	8.33	7.91

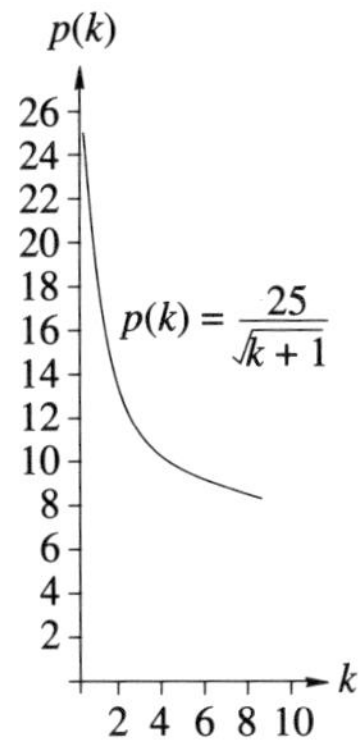

33. (a) $0 < p < \$4.65$; (b)

p	0.25	0.50	0.75	1.00	1.25	1.50	1.75	2.00
$d(p)$	10.56	11	11.31	11.50	11.56	11.50	11.31	11

;

(c)

$d(p) = -p^2 + 2.5p + 10$

(d) \$1.25

34. (a)

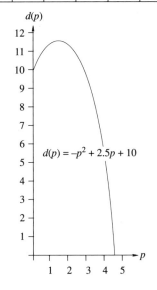

t	1	2	3	4	5	6	7	8
m	2.4	3.0	3.8	5.0	6.7	9.0	12.3	17.0
t	9	10	11	12	13	14	15	
m	23.6	33.0	46.3	65.0	91.5	129.0	182.0	

;

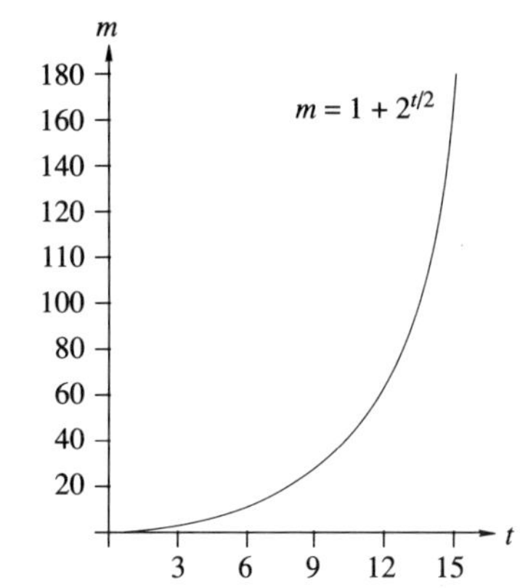

(b) About 13.3 from graph. More exactly it's 13.2587.

4.5 GRAPHING CALCULATORS AND COMPUTER-AIDED GRAPHING (OPTIONAL)

Note: All graphs were generated on a *TI-81* calculator. In each figure, the tick marks are 1 unit apart.

1. Graph $y = x$.

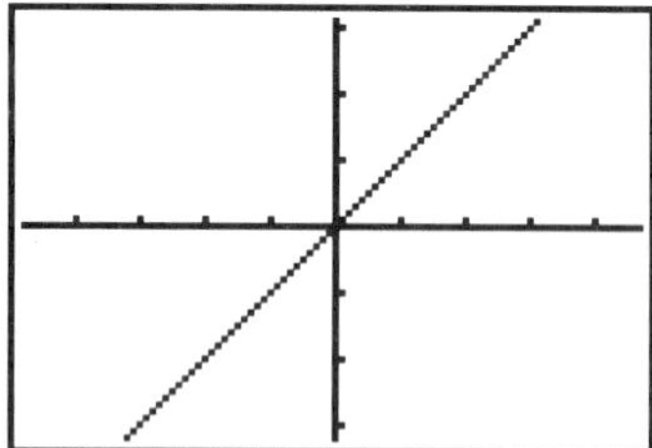

2. $y = 2x$

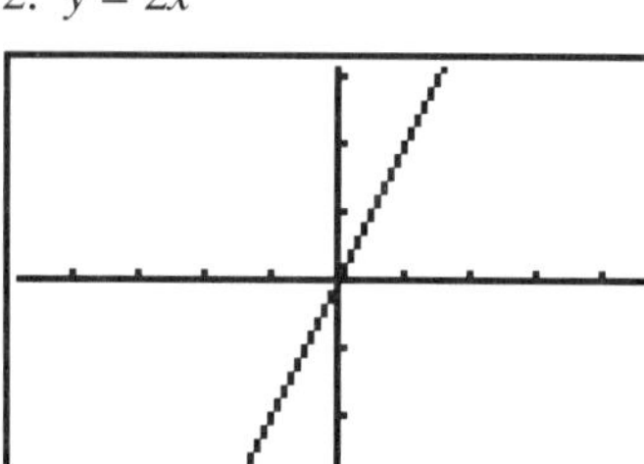

3. $y = -3x$

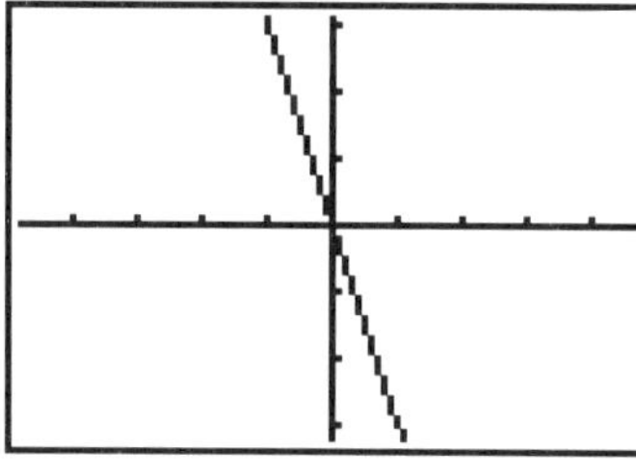

4. $y = \frac{1}{2}x - 3$

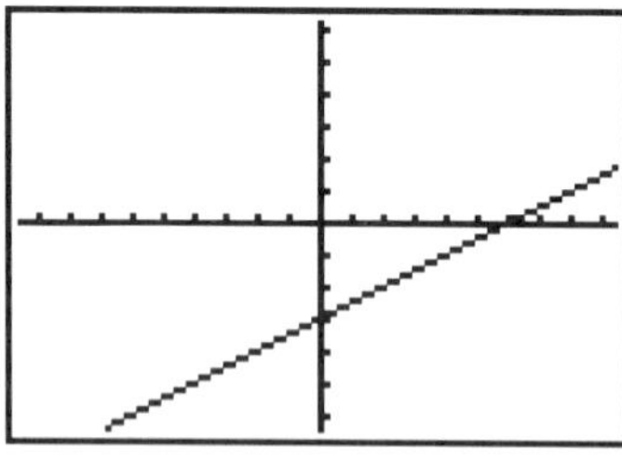

5. $y = -4x + 2$

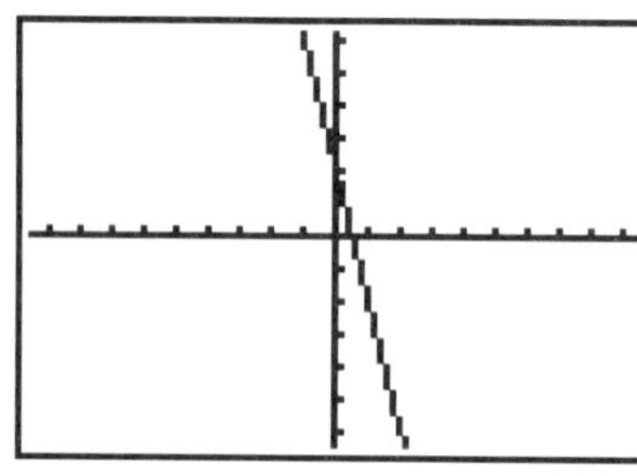

6. $x+y = 2$ graph either $y = 2-x$ or $y = -x+2$

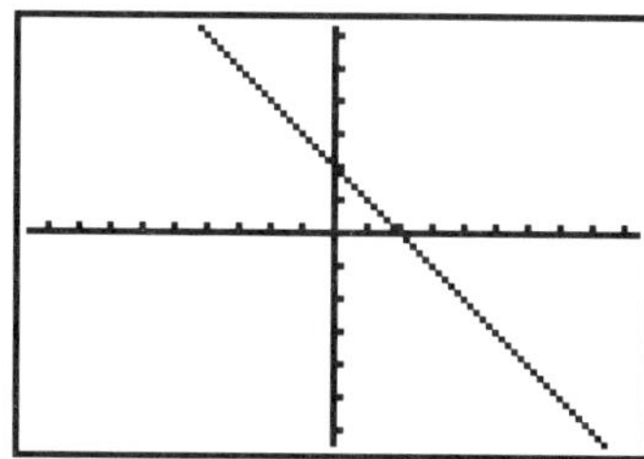

7. $x - y = 2$ graph as $y = x - 2$

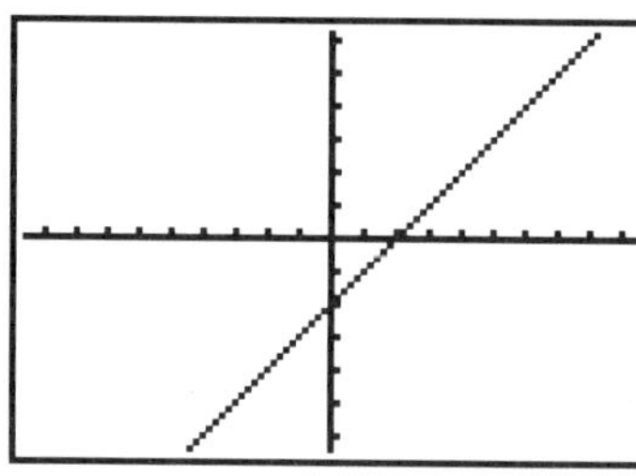

8. $2x + y = 1$ graph as $y = -2x + 1$

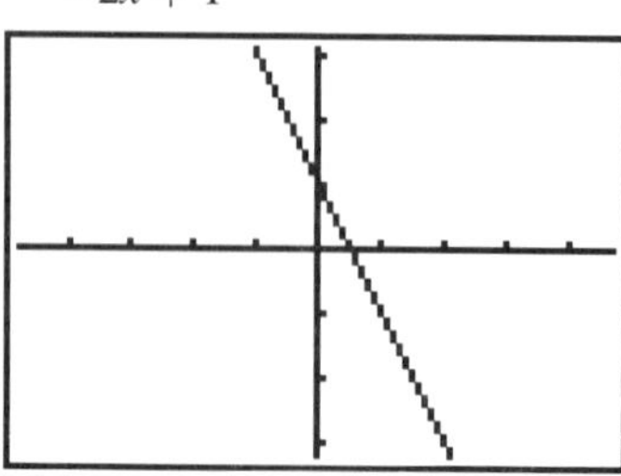

9. $3x - 6y = 9$ graph as $y = \frac{9-3x}{-6}$ or $y = \frac{1}{2}x - \frac{3}{2}$

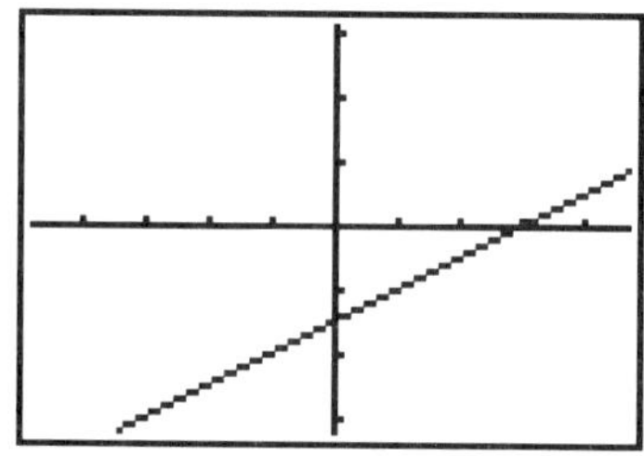

10. $3x - 6y = -6$ graph as $y = \frac{-6-3x}{-6}$ or $y = \frac{1}{2}x + 1$

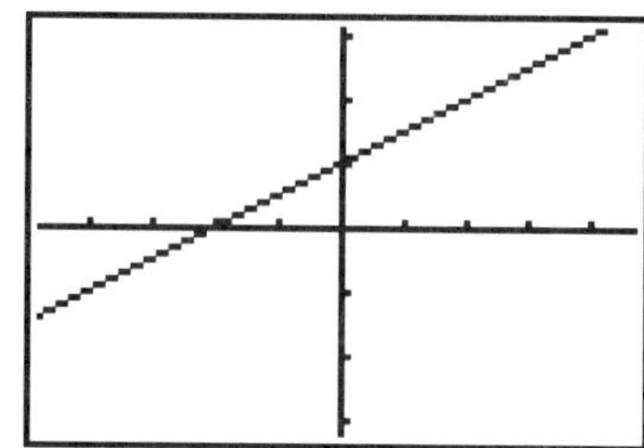

11. $y = x^2$

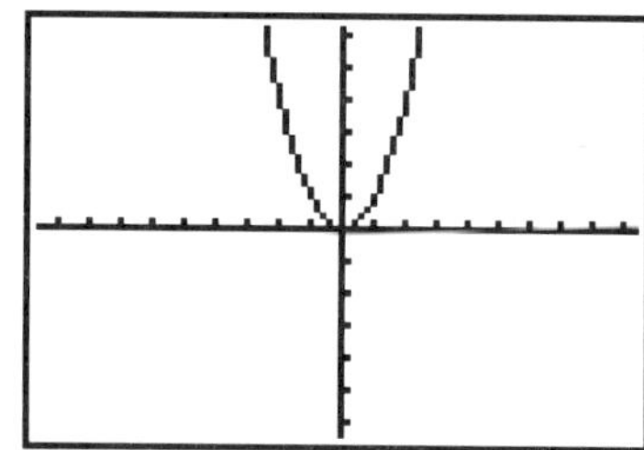

12. $y = x^2 + 3$

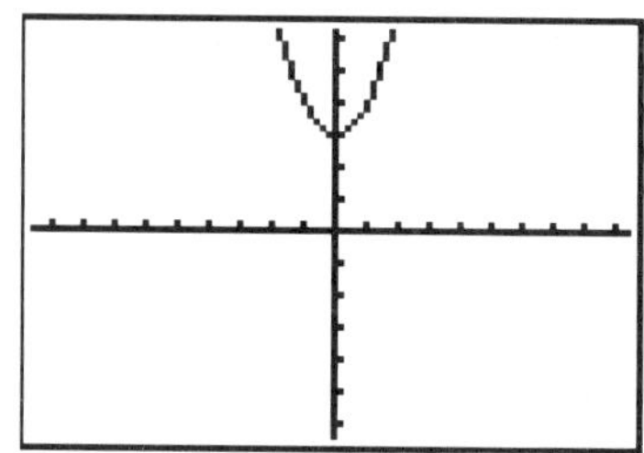

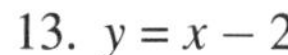

13. $y = x - 2$

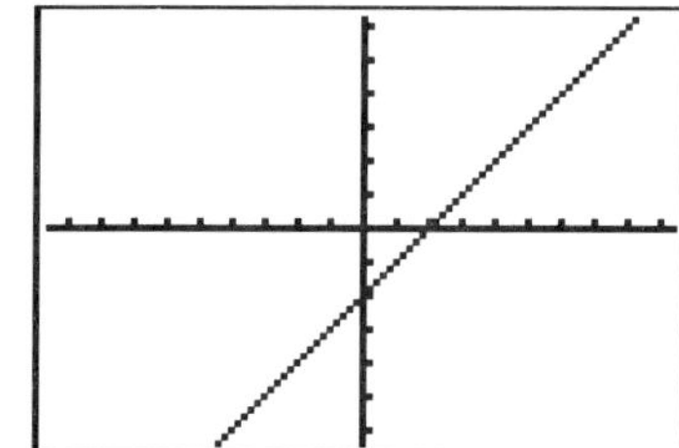

14. $y = x^2 + 2x + 1$

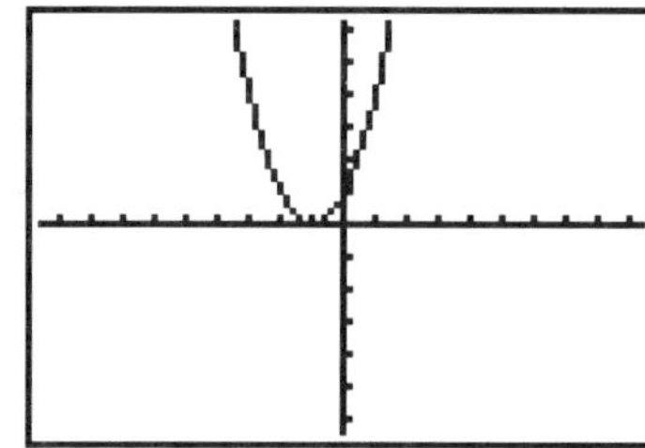

15. $y = x^2 - 6x + 9$

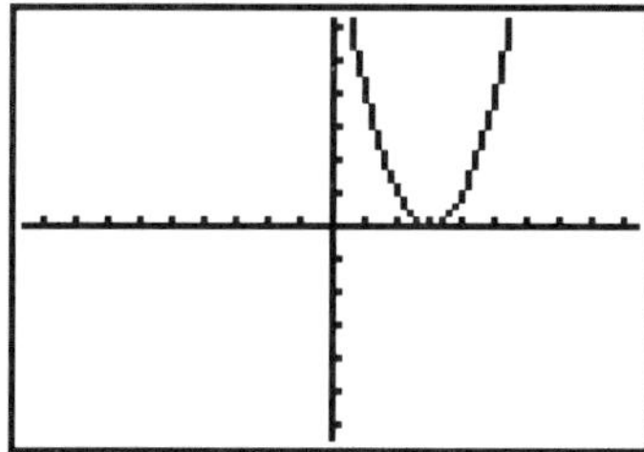

16. $y = -x^2 + 2$

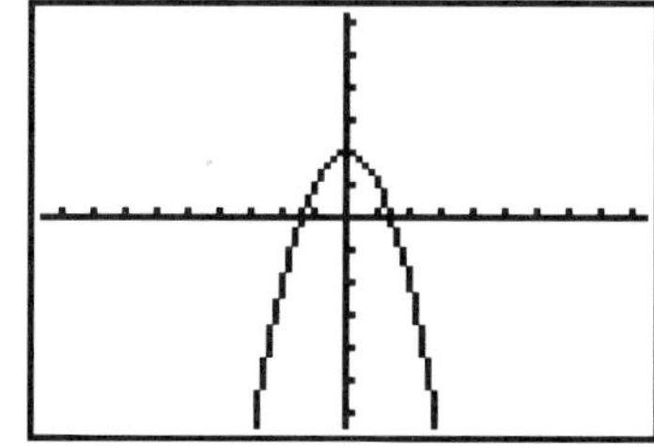

17. $y = \frac{1}{x}$

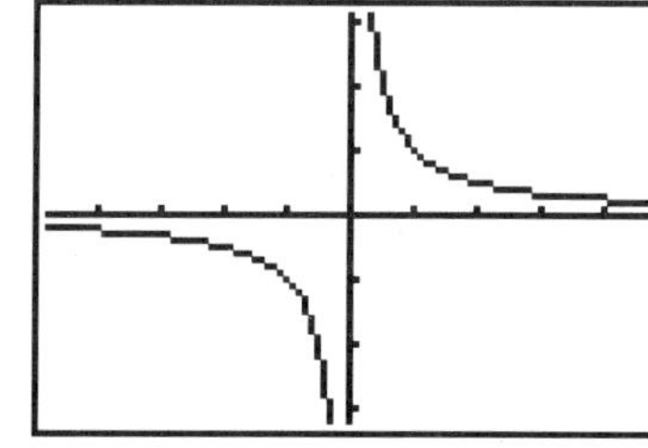

18. $y = \frac{1}{x+3}$

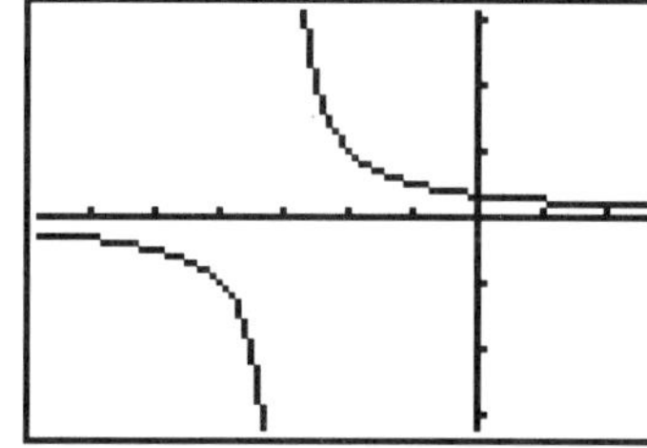

19. $y = x^3$

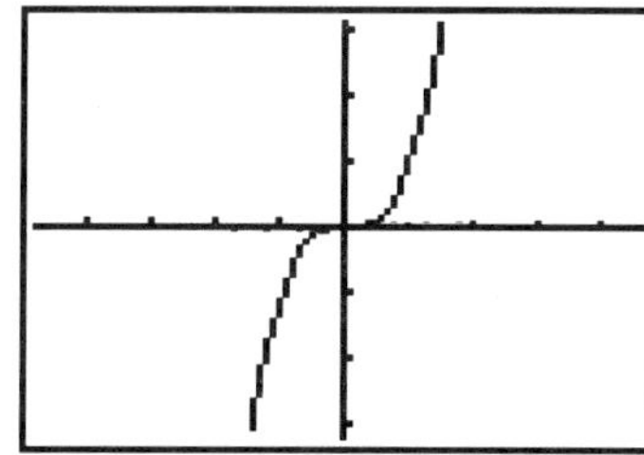

20. $y = \sqrt[3]{x}$

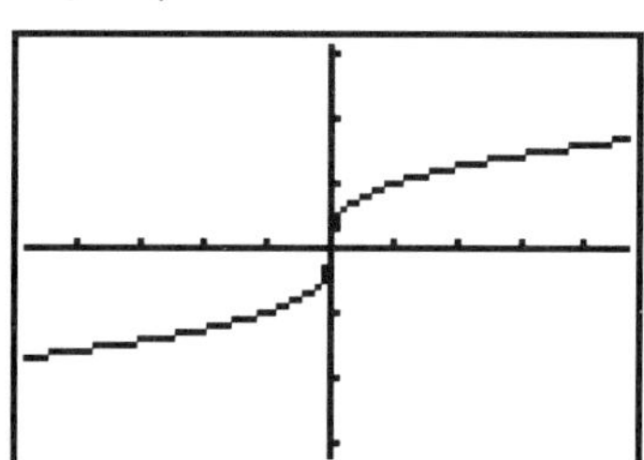

21. $y^2 + x^2 = 25$ or $y^2 = 25 - x^2$ graph both $y = \sqrt{25 - x^2}$ and $y = -\sqrt{25 - x^2}$.

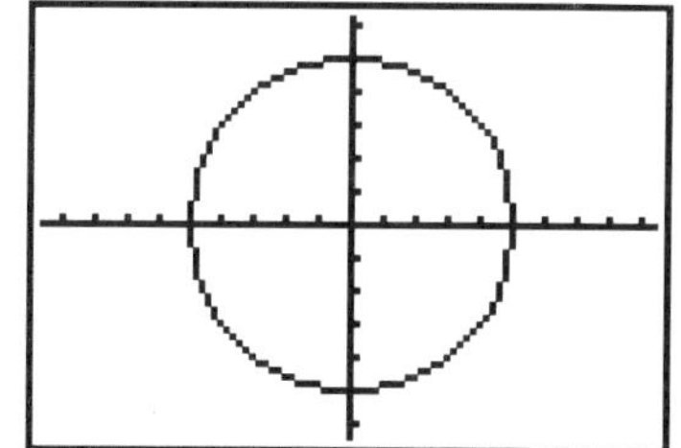

22. $y = \sqrt{25 - x^2}$

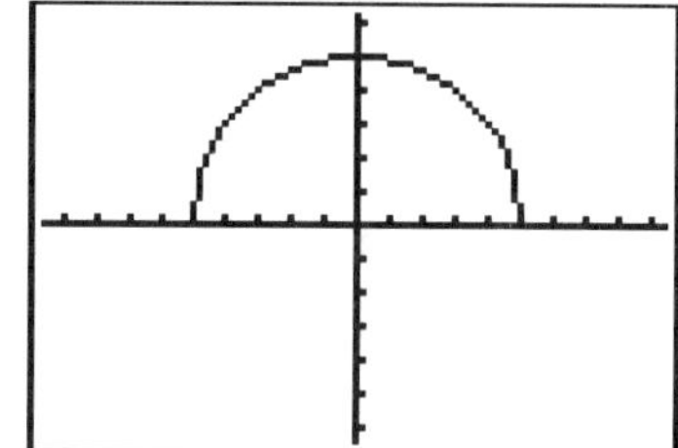

23. $y = |x + 2|$

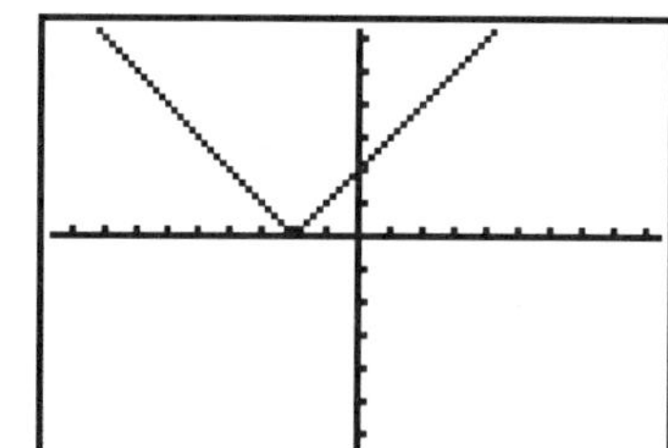

24. $\frac{x^2}{4} + \frac{y^2}{9} = 1$ or $9x^2 + 4y^2 = 36$ which becomes $4y^2 = 36 - 9x^2$ or $2y = \pm\sqrt{36 - 9x^2}$ graph as $y = \frac{\pm\sqrt{36-9x^2}}{2}$; make sure you graph both.

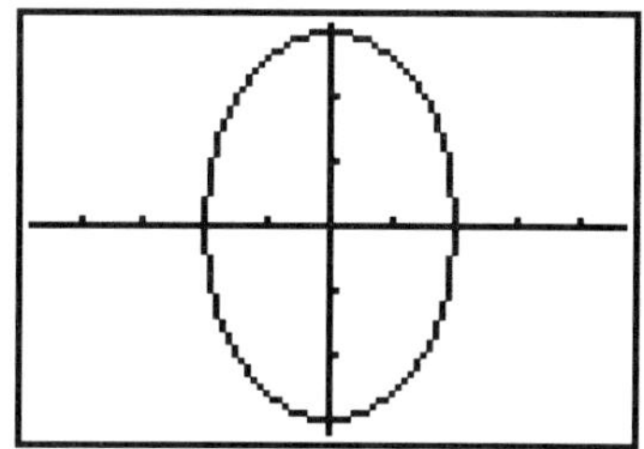

25. Here $\frac{x^2}{4} - \frac{y^x}{9} = 1$ can be written as $9x^2 - 4y^2 = 36$ or $-4y^2 = 36 - 9x^2$ or $4y^2 = 9x^2 - 36$ and so $2y = \pm\sqrt{9x^2 - 36}$ graph as $y = \frac{\pm\sqrt{9x^2-36}}{2}$; make sure you graph both.

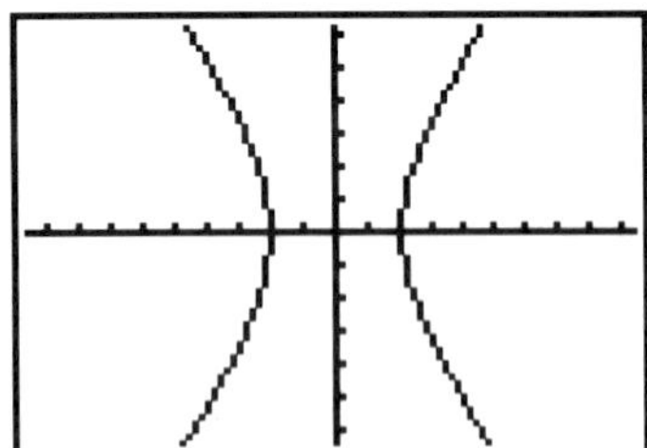

26. $y = x^3 - x^2$

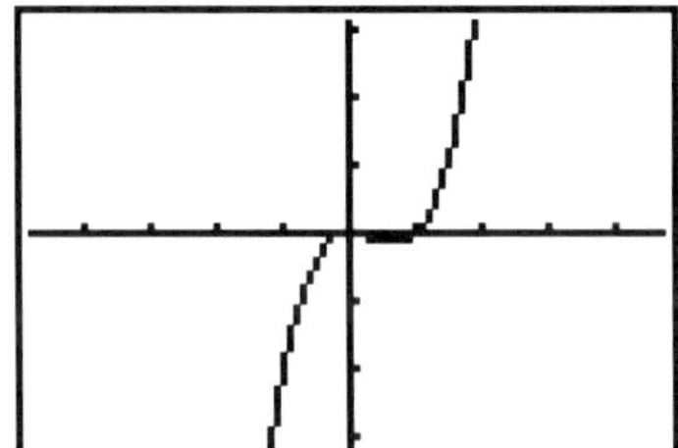

4.6 USING GRAPHS TO SOLVE EQUATIONS

1. The root is $x = -\frac{5}{2}$

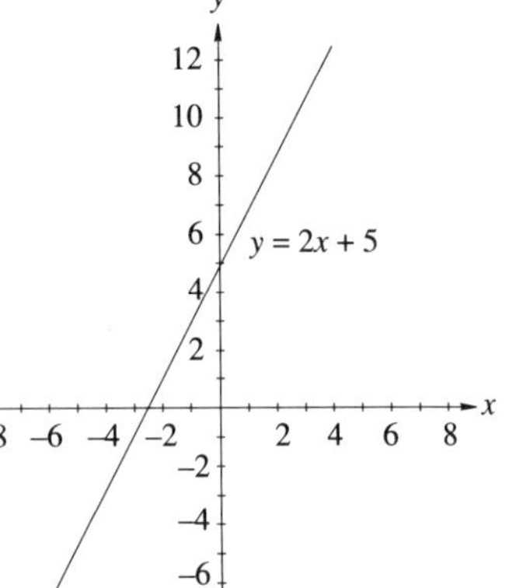

2. The root is $x = \frac{9}{5}$

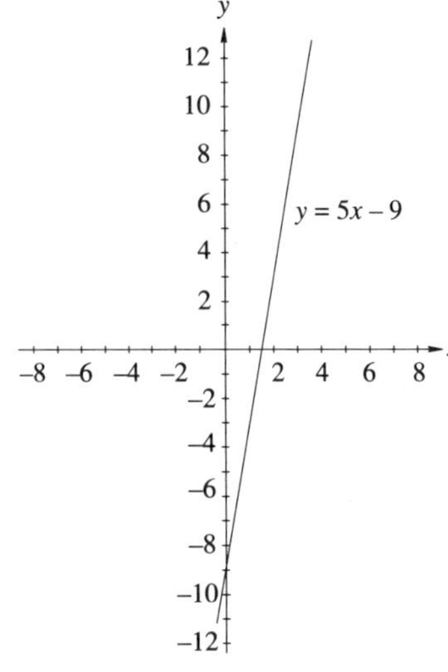

3. The roots are $x = -3$ and $x = 3$

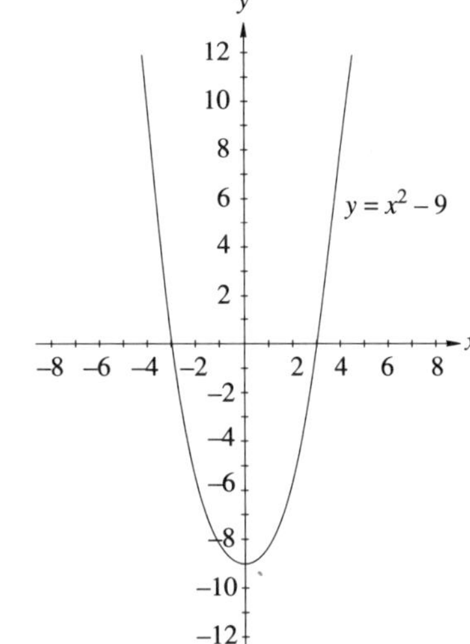

4. The roots are $x \approx -1.5811$ and $x \approx 1.5811$

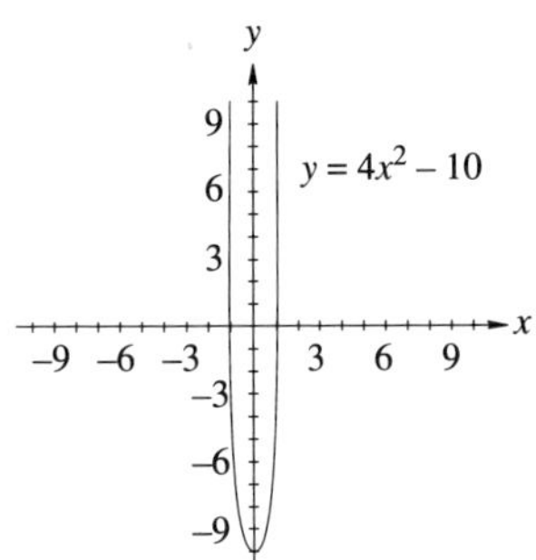

5. The roots are $x = 0$ and $x = 5$

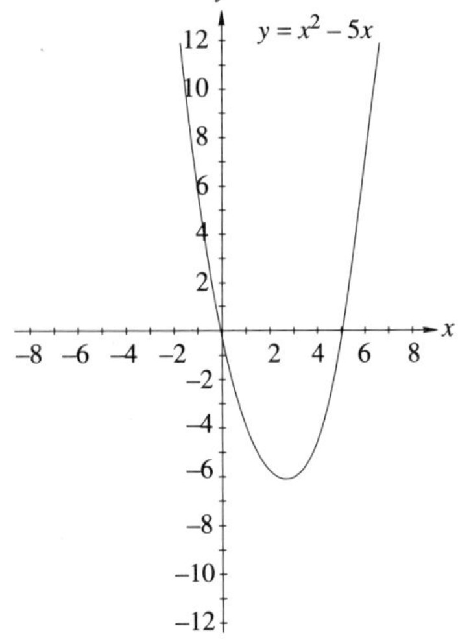

6. The roots are $x = 1$ and $x = 3$

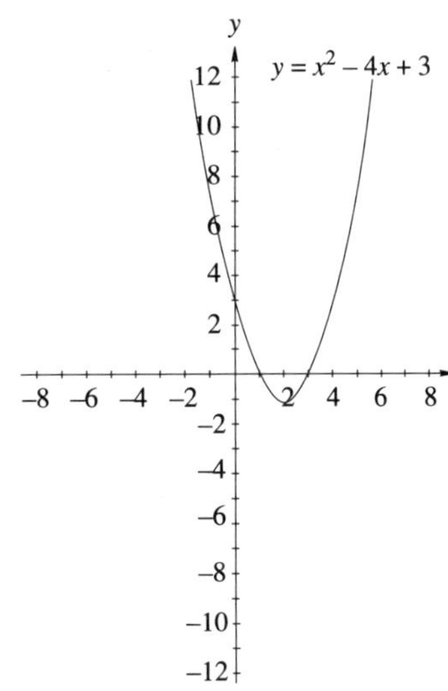

7. The roots are $x \approx -5.5414$ and $x \approx 0.5414$

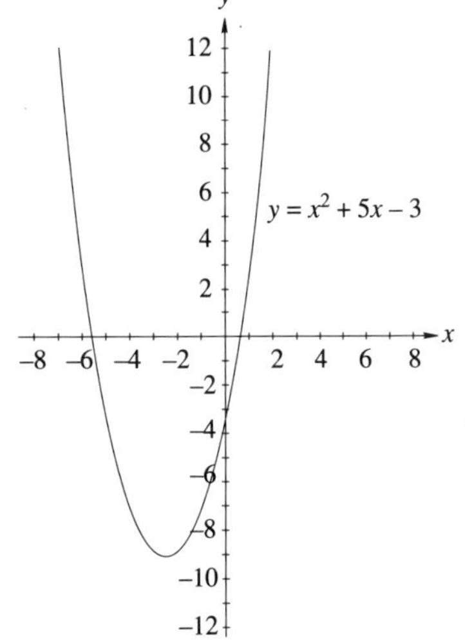

8. The roots are $x \approx -2.1350$ and $x \approx 0.4684$

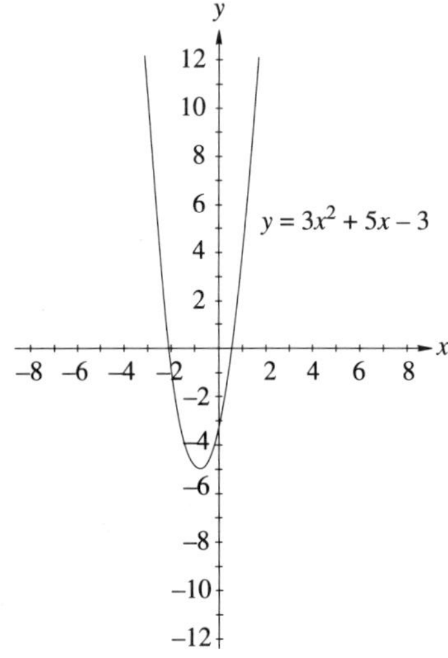

9. The roots are $x = -\frac{9}{4}$ and $x = \frac{6}{5}$

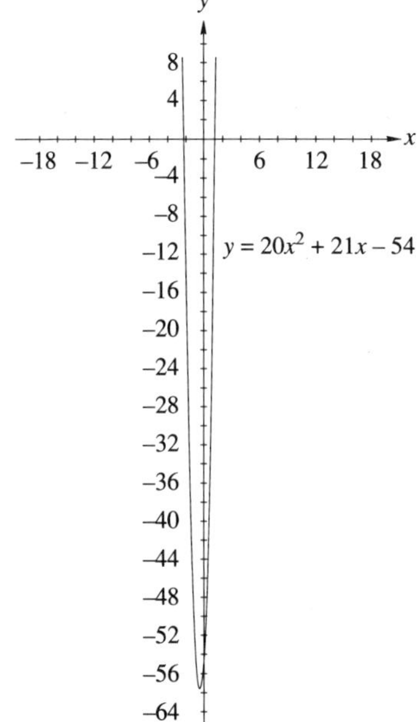

10. The roots are $x \approx -3.3635$, $x \approx -0.4916$, $x \approx 0.4929$, and $x \approx 7.3622$

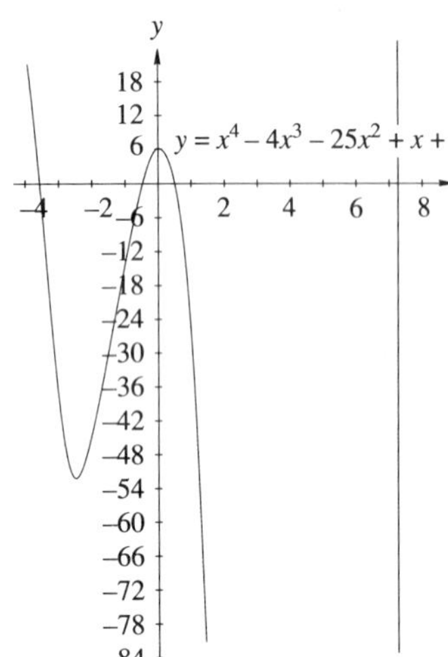

11. Has an inverse function

12. Has an inverse function

13. Does not have an inverse function

14. Does not have an inverse function

15.

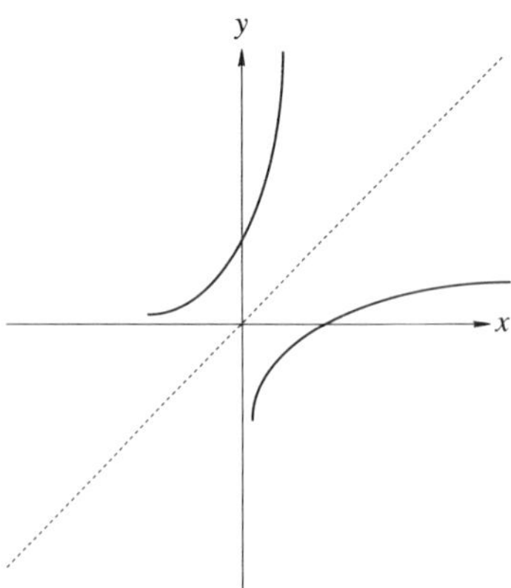

16.

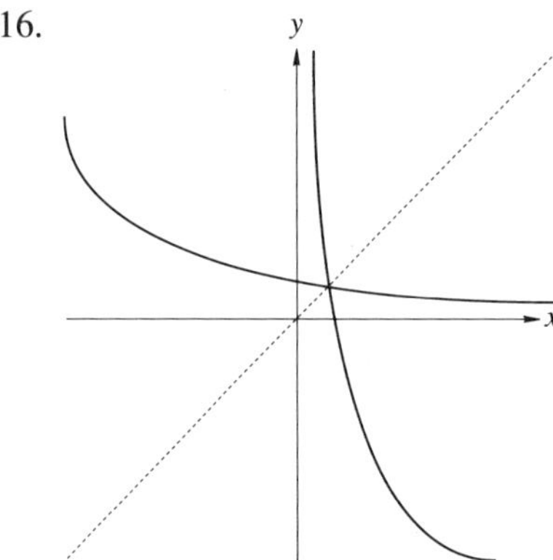

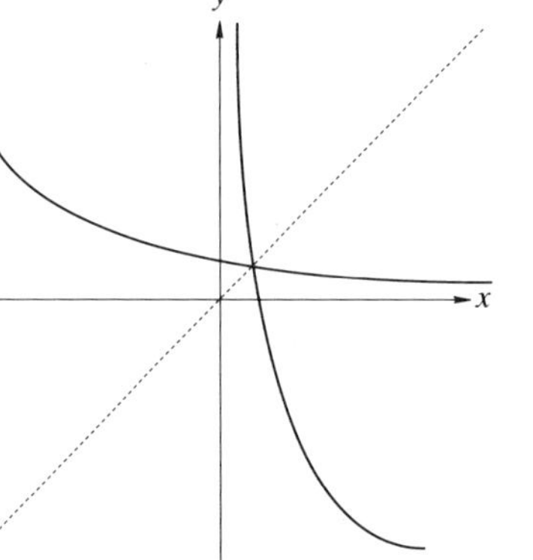

17.

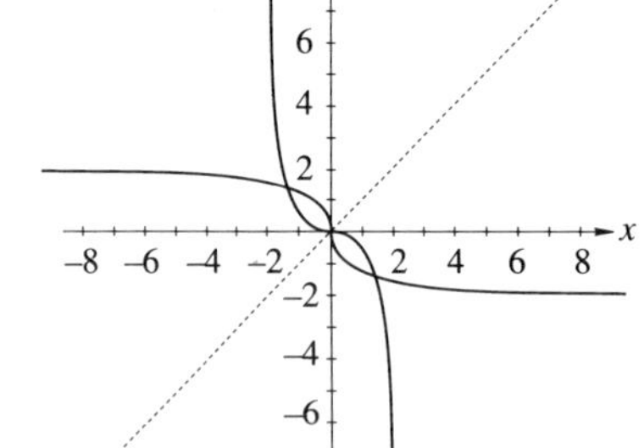

18.

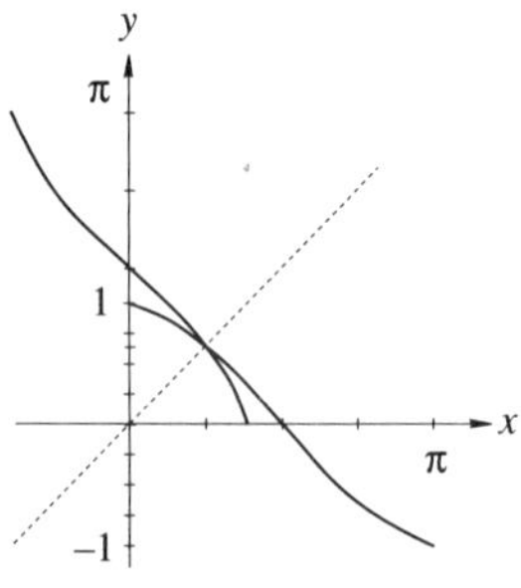

19. (a) 30%; (b) 60%; (c) 73%

20. (a) $P = 5\sqrt{n} - 10$; $P + 10 = 5\sqrt{n}$; $\frac{P+10}{5} = \sqrt{n}$; $n = \left(\frac{P+10}{5}\right)^2$ or $P^{-1}(n) = \left(\frac{n+10}{5}\right)^2$; (b) $n = \left(\frac{5+10}{5}\right)^2 = 3^2 = 9$;

(c)

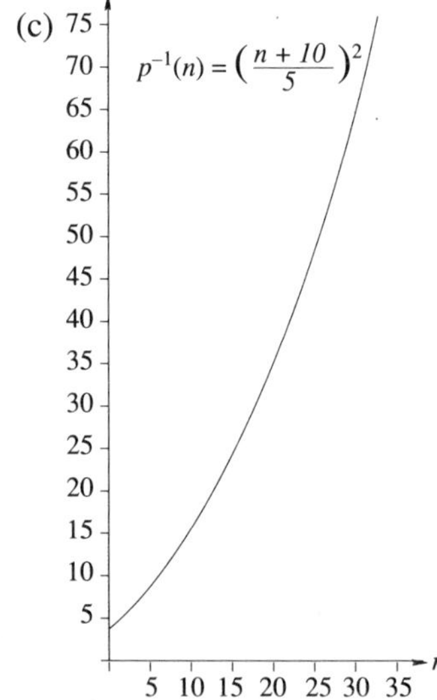

21. No

22. (a) $C(30) = \frac{6.4(30)}{100-30} = \frac{192}{70} \approx 2.74$ mill or 2,740,000;

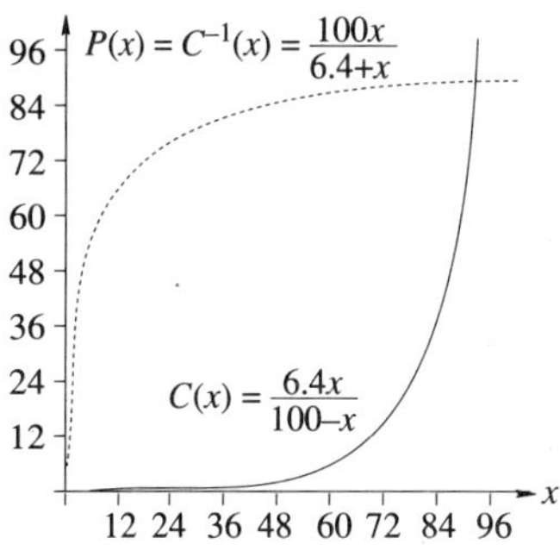

(b) $C(60) = \frac{6.4(60)}{100-60} = \frac{384}{40} = 9.6$ million or 9,600,000;

(c) $C = \frac{6.4x}{100-x}$; $C(100-x) = 6.4x$; $100C - Cx = 6.4x$; $100C = 6.4x + Cx$; $100C = x(6.4 + C)$; $x = \frac{100C}{6.4+C}$; or $P(x) = \frac{100x}{6.4+x}$ where x is the amount of money (in millions) that can be spent and $P(x)$ is the percent of pollutant that can be removed;

(d) $P(12) = \frac{100(12)}{6.4+12} = \frac{1200}{18.4} = 65.217 \approx 65.2\%$

CHAPTER 4 REVIEW

1.

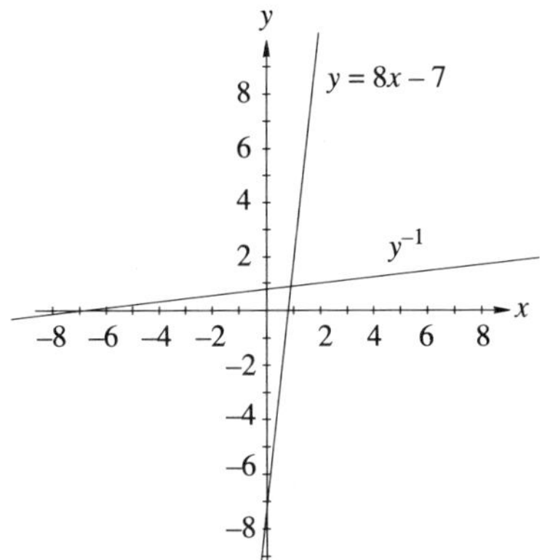

(b) Domain: all real numbers; Range: all real numbers; set $y = 0$ so $0 = 8x - 7$; $7 = 8x$; or $x = \frac{7}{8}$; x-intercept $\frac{7}{8}$; set $x = 0$; $y = 8(0) - 7 = -7$; y-intercept -7; (c) function; (d) has an inverse function

2.

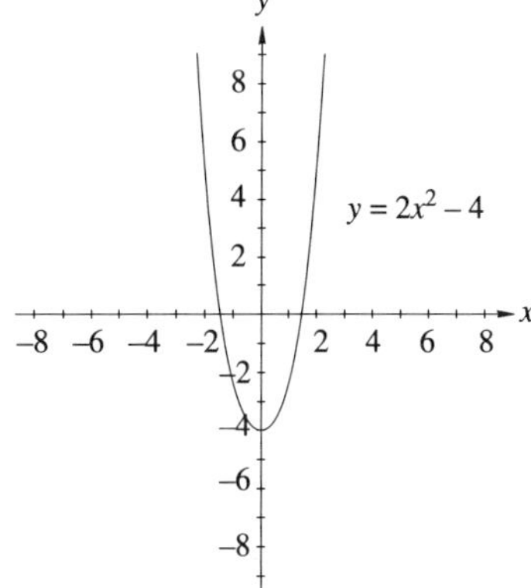

(b) Domain: all real numbers; Range: all real numbers greater than or equal to -4; set $y = 0$; $0 = 2x^2 - 4$; $2x^2 = 4$; $x^2 = 2$; $x = \pm\sqrt{2}$ x-intercept $\pm\sqrt{2}$; set $x = 0$; $y = 2(0)^2 - 4 = 0 - 4 = -4$; y-intercept -4; (c) function; (d) does not have an inverse function

3.

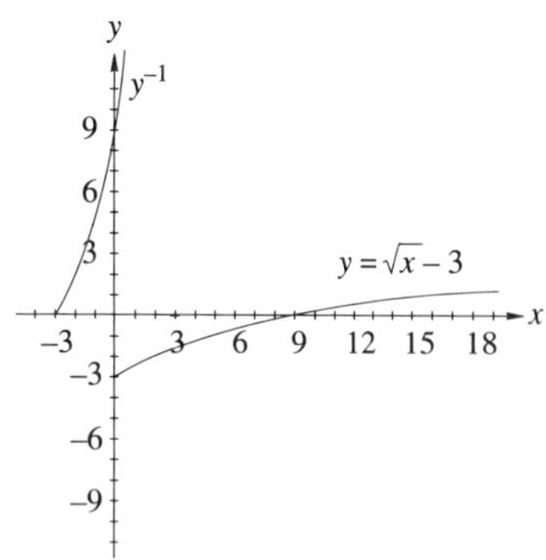

(b) Domain: all non-negative real numbers, $x \geq 0$; Range: all real numbers greater than or equal to -3. To find the x-intercept, set $y = 0$. Then, $0 = \sqrt{x} - 3$ or $\sqrt{x} = 3$ and so, $x = 3^2 = 9$ and the x-intercept $= 9$. To find the y-intercept, set $x = 0$. Then $y = \sqrt{0} - 3 = -3$ and hence, the y-intercept $= -3$; (c) function; (d) has an inverse function

4.

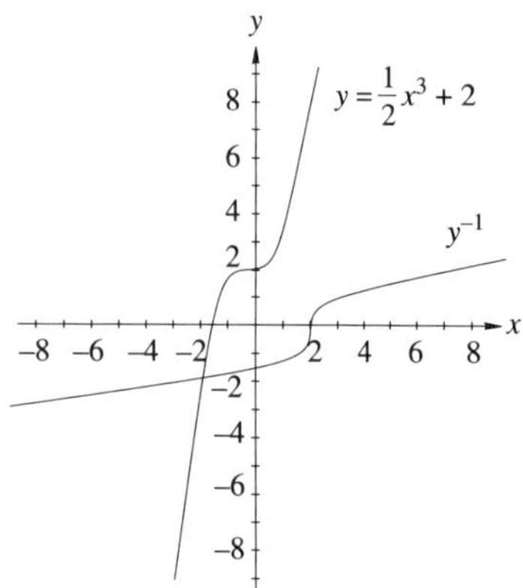

(b) Domain: all real numbers; Range: all real numbers; $0 = \frac{1}{2}x^3 + 2$; $\frac{1}{2}x^3 = -2$; $x^3 = -4$; $x = \sqrt[3]{-4}$; x-intercept $\sqrt[3]{-4}$; $y = \frac{1}{2}(0)^3 + 2 = 2$; y-intercept 2; (c) function; (d) has an inverse function

5.

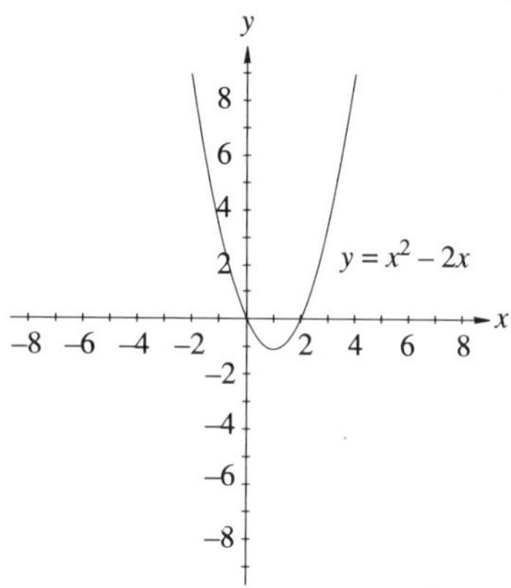

(b) Domain: all real numbers; Range: all real numbers greater than or equal to -1; $0 = x^2 - 2x$; $x(x-2) = 0$ so $x = 0$ or $x = 2$; x-intercepts $0, 2$; $y = 0^2 - 2(0) = 0$; y-intercept 0; (c) function; (d) does not have an inverse function

6.

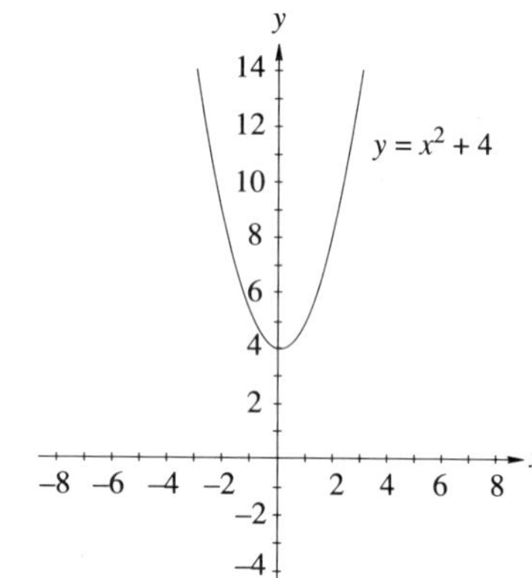

(b) Domain: all real numbers; Range: all real numbers 4 or larger; no x-intercept; y-intercept 4; (c) function; (d) does not have an inverse function

7. $f(0) = 4(0) - 12 = 0 - 12 = -12$

8. $f(-2) = 4(-2) - 12 = -8 - 12 = -20$

9. $f(3) = 4(3) - 12 = 12 - 12 = 0$

10. $f(a) = 4a - 12$

11. $f(a-2) = 4(a-2) - 12 = 4a - 8 - 12 = 4a - 20$

12. $f(x+h) = 4(x+h) - 12 = 4x + 4h - 12$

13. $g(0) = \frac{0^2-9}{0^2+9} = \frac{-9}{9} = -1$

14. $g(3) = \frac{3^2-9}{3^2+9} = \frac{9-9}{9+9} = \frac{0}{18} = 0$

15. $g(-3) = \frac{(-3)^2-9}{(-3)^2+9} = \frac{9-9}{9+9} = \frac{0}{18} = 0$

16. $g(-2) = \frac{(-2)^2-9}{(-2)^2+9} = \frac{4-9}{4+9} = \frac{-5}{13} \approx -0.3846$

17. $g(4) = \frac{4^2-9}{4^2+9} = \frac{16-9}{16+9} = \frac{7}{25} = 0.28$

18. $g(-5) = \frac{(-5)^2-9}{(-5)^2+9} = \frac{25-9}{25+9} = \frac{16}{34} = \frac{8}{17} \approx 0.4706$

19.

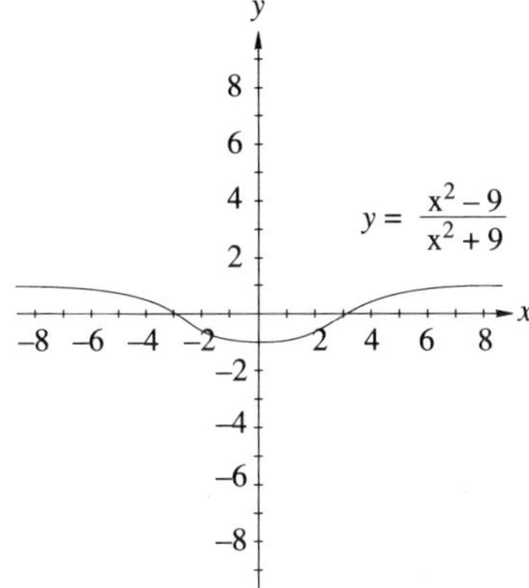

the zeros are -3 and 3

20. $(f+g)(x) = f(x) + g(x) = 4x - 12 + \frac{x^2-9}{x^2+9}$

21. $(f+g)(3) = f(3) + g(3) = (4 \cdot 3 - 12) + \frac{3^2-9}{3^2+9} = 0 + 0 = 0$

22. $(f-g)(x) = f(x) - g(x) = 4x - 12 - \frac{x^2-9}{x^2+9}$

23. $(f-g)(-2) = f(-2) - g(-2) = 4(-2) - 12 - \frac{-5}{13} = -20 + \frac{5}{13} = \frac{-260}{13} + \frac{5}{13} = \frac{-255}{13}$

24. $(f \cdot g)(x) = f(x) \cdot g(x) = (4x-12)\left(\frac{x^2-9}{x^2+9}\right)$

25. $(f \cdot g)(0) = f(0) \cdot g(0) = (-12)(-1) = 12$

26. $(f/g)(x) = \frac{4x-12}{\frac{x^2-9}{x^2+9}} = (4x-12)\left(\frac{x^2+9}{x^2-9}\right) = \frac{4(x-3)(x^2+9)}{(x-3)(x+3)} = \frac{4(x^2+9)}{x+3}$

27. $(f/g)(5) = \frac{4(5^2+9)}{5+3} = \frac{4(25+9)}{8} = \frac{1(34)}{2} = 17$

28. $(g/f)(x) = \frac{1}{(f/g)(x)} = \frac{x+3}{4(x^2+9)}$

29. $(g/f)(2) = \frac{2+3}{4(2^2+9)} = \frac{5}{4(4+9)} = \frac{5}{4(13)} = \frac{5}{52}$

30. $(f \circ g)(x) = f(g(x)) = f\left(\frac{x^2-9}{x^2+9}\right) = 4\left(\frac{x^2-9}{x^2+9}\right) - 12$

31. $(f \circ g)(4) = 4\left(\frac{4^2-9}{4^2+9}\right) - 12 = 4\left(\frac{16-9}{16+9}\right) - 12 = 4\left(\frac{7}{25}\right) - 12 = \frac{28}{25} - \frac{300}{25} = \frac{-272}{25}$

32. $(g \circ f)(x) = g(f(x)) = g(4x-12) = \frac{(4x-12)^2-9}{(4x-12)^2+9} = \frac{(16x^2-96x+144)-9}{(16x^2-96x+144)+9} = \frac{16x^2-96x+135}{16x^2-96x+153}$

33. $(g \circ f)(3) = \frac{16(3)^2-96(3)+135}{16(3)^2-96(3)+153} = \frac{144-288+135}{144-288+153} = \frac{-9}{9} = -1$

34. $4x + 7y = 0$; $7y = -4x$; $y = \frac{-4}{7}x$; roots are $0 = -\frac{4}{7}x$ or $x = 0$

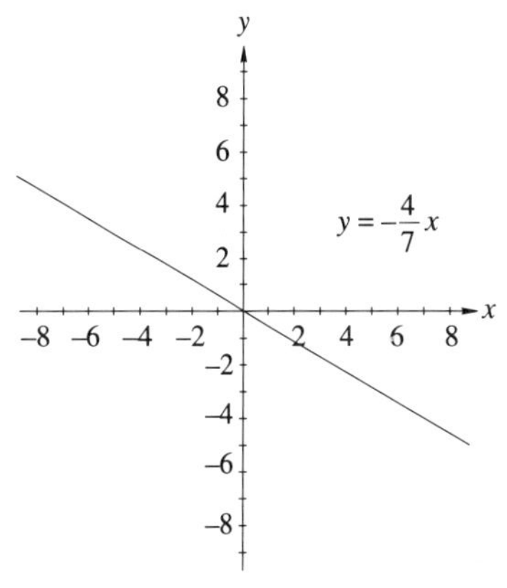

35. $x^2 - 20 = y$ or $y = x^2 - 20$; roots are $0 = x^2 - 20$; $x^2 = 20$; $x = \pm\sqrt{20} = \pm 2\sqrt{5} \approx \pm 4.4721$

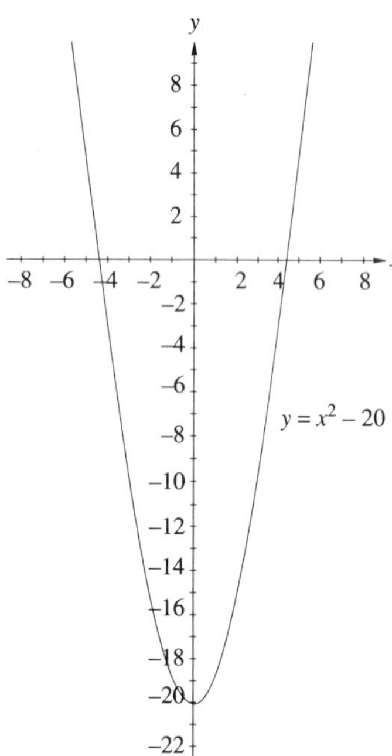

36. $2x^2 + 10x + 4 = 0$ or $2(x^2 + 5x + 2) = 0$ roots are about -0.44 and -4.50

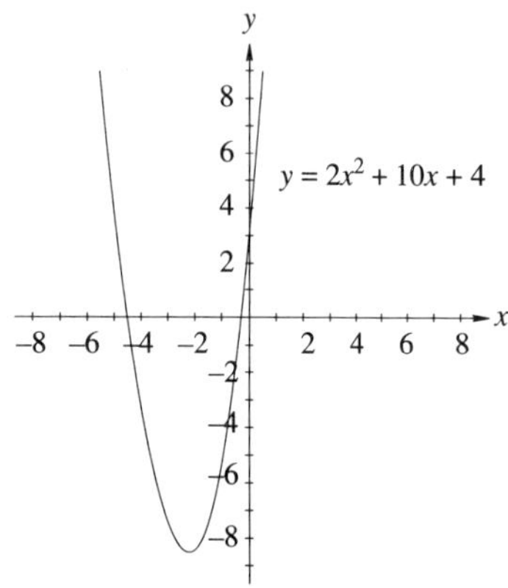

37. $y = 8x^3 - 20x^2 - 34x + 21$; roots: $x = -1.5, x = 0.5$, and $x = 3.5$

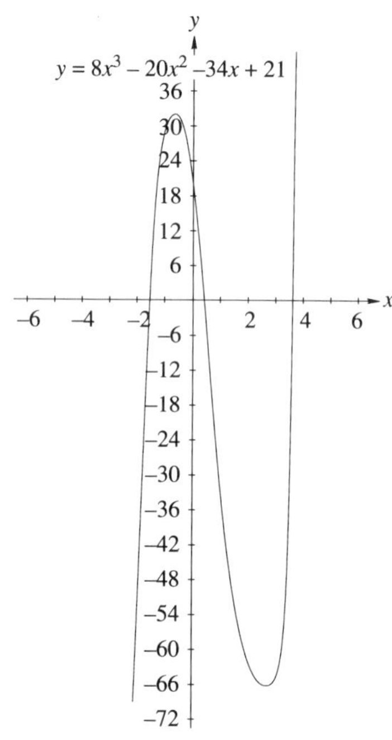

38. (a) $P(101) = 35 - \frac{101}{20} = 35 - 5.05 = \$29.95$; $P(350) = 35 - \frac{350}{20} = 35 - 17.50 = \$17.50$; $P(400) = 35 - \frac{400}{20} = 35 - 20 = \$15.00$;

(b) $R = D \times P = n\left(35 - \frac{n}{20}\right) = 35n - \frac{n^2}{20}$;

(c) $R = 101(29.95) = 3024.95$ or $R(n) = 35(101) - \frac{(101)^2}{20} = 3535 - 510.05 = 3024.95$; $R(350) = 35(350) - \frac{350^2}{20} = 6125$; $R(400) = 35(400) - \frac{400^2}{20} = 6000$

39. (a) $P(n) = \left(30n - \frac{n^2}{20}\right) - (550 + 10n) = 20n - \frac{n^2}{20} - 550$;

(b) $P(30) = 20(30) - \frac{30^2}{20} - 550 = \$5$; $P(100) = 20(100) - \frac{100^2}{20} - 550 = \$950$; $P(150) = \$1325$; $P(300) = \$950$; $P(400) = -\$550$

40.

t	0	1	2	3	4	5
$h(t)$	0	3.08	6.48	10.68	15.68	21.00
t	6	7	8	9	10	
$h(t)$	25.68	28.28	26.88	19.08	2.00	

$h(t) = -0.02t^4 + 0.2t^3 - 0.3t^2 + 3.2t$

41. (a)

v	0	10	20	30	40	50	60	70
$s(v)$	0	14	36	66	104	150	204	266

(b) about 69.8 mph;

(c)

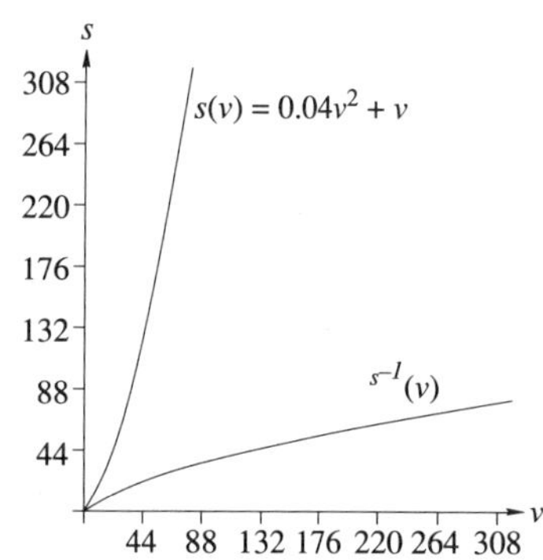

CHAPTER 4 TEST

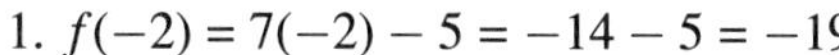

1. $f(-2) = 7(-2) - 5 = -14 - 5 = -19$

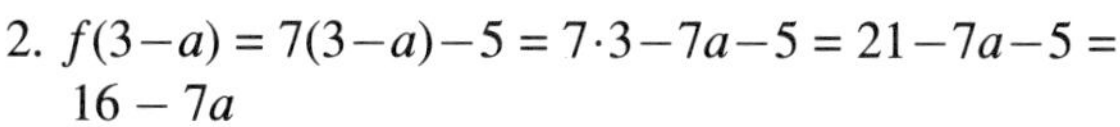

2. $f(3-a) = 7(3-a)-5 = 7\cdot 3-7a-5 = 21-7a-5 = 16 - 7a$

3. $g(0) = \frac{0^2-2\cdot 0-15}{0+3} = \frac{-15}{3} = -5$

4. $g(5) = \frac{5^2-2\cdot 5-15}{5+3} = \frac{25-10-15}{8} = \frac{0}{8} = 0$

5.

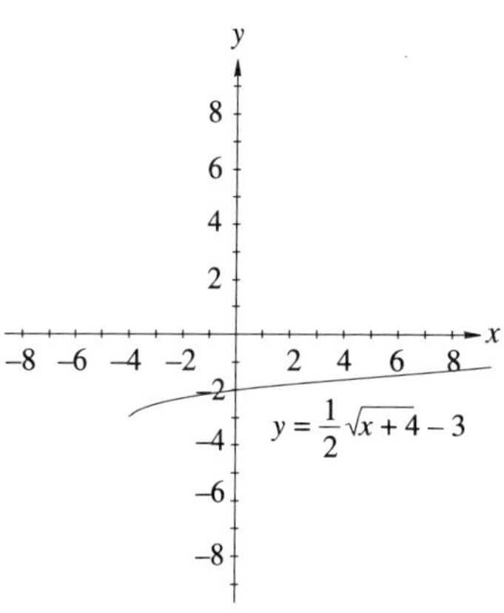

(b) This function is defined whenever $\sqrt{x+4} \geq 0$ or when $x \geq -4$;

(c) The low point of the graph is when $x = -4$. Here the y-value is -3. So, the graph can assume all values of $y \geq -3$;

(d) The x-intercept is when $y = 0$. Solving $0 = \frac{1}{2}\sqrt{x+4} - 3$, we get $3 = \frac{1}{2}\sqrt{x+4}$ or $6 = \sqrt{x+4}$. Squaring both sides we get $36 = x + 4$ or $x = 32$;

(e) The y-intercept is when $x = 0$. Solving $y = \frac{1}{2}\sqrt{0+4} - 3 = \frac{1}{2}\sqrt{4} - 3 = \frac{1}{2}2 - 3 = 1 - 3 = -2$.

6. $3x-15+\frac{x-5}{x+5} = \frac{(3x-15)(x+5)}{x+5} + \frac{x-5}{x+5} = \frac{3x^2-75}{x+5} + \frac{x-5}{x+5} = \frac{3x^2+x-80}{x+5}$

7. $3x-15-\frac{x-5}{x+5} = \frac{(3x-15)(x+5)}{x+5} - \frac{x-5}{x+5} = \frac{3x^2-75}{x+5} - \frac{x-5}{x+5} = \frac{3x^2-x-70}{x+5}$

8. $(3x-15)\left(\frac{x-5}{x+5}\right) = \frac{(3x-15)(x-5)}{x+5} = \frac{3(x-5)(x-5)}{x+5} = \frac{3(x-5)^2}{x+5}$

9. $(3x-15) \div \frac{x-5}{x+5} = (3x-15)\left(\frac{x+5}{x-5}\right) = \frac{3(x-5)(x+5)}{x-5} = 3(x+5) = 3x+15$

10. $3\left(\frac{x-5}{x+5}\right) - 15 = \frac{3x-15}{x+5} - \frac{15x+75}{x+5} = \frac{3x-15-15x-75}{x+5} = \frac{-12x-90}{x+5} = -6\left(\frac{2x+15}{x+5}\right)$

11. $\frac{(3x-15)-5}{(3x-15)+5} = \frac{3x-20}{3x-10}$

CHAPTER

5

Systems of Linear Equations and Determinants

5.1 LINEAR EQUATIONS

1. $(2,5), (3,8), m = \frac{y_2-y_1}{x_2-x_1} = \frac{8-5}{3-2} = \frac{3}{1} = 3$
2. $(4,7), (-2,1), m = \frac{1-7}{-2-4} = \frac{-6}{-6} = 1$
3. $(1,8), (5,3), m = \frac{3-8}{5-1} = \frac{-5}{4} = -\frac{5}{4}$
4. $(0,4), (5,0), m = \frac{0-4}{5-0} = \frac{-4}{5} = -\frac{4}{5}$
5. $(9,3), (2,-7), m = \frac{-7-3}{2-9} = \frac{-10}{-7} = \frac{10}{7}$
6. $m = \frac{-5-1}{2--6} = \frac{-6}{8} = -\frac{3}{4}$
7. $m = 4$, point: $(5,3)$. Using $y - y_1 = m(x - x_1)$ we get $y - 3 = 4(x - 5)$
8. $m = -3$, point: $(-6,1)$, $y - 1 = -3(x - -6)$ or $y - 1 = -3(x + 6)$
9. $m = \frac{2}{3}$, point $(1,-5)$, $y - (-5) = \frac{2}{3}(x - 1)$ or $y + 5 = \frac{2}{3}(x - 1)$
10. $m = \frac{3}{2}$, point $(0,5)$, $y - 5 = \frac{3}{2}(x - 0)$ or $y - 5 = \frac{3}{2}x$
11. $m = -\frac{5}{3}$, point: $(2,0)$, $y - 0 = -\frac{5}{3}(x - 2)$ or $y = -\frac{5}{3}(x - 2)$
12. $m = -\frac{3}{4}$, point: $(-3,-1)$, $y - -1 = -\frac{3}{4}(x - -3)$ or $y + 1 = -\frac{3}{4}(x + 3)$
13. points: $(1,5)$ and $(-3,2)$. First $m = \frac{2-5}{-3-1} = \frac{-3}{-4} = \frac{3}{4}$; then using $(-3,2)$ we get $y - 2 = \frac{3}{4}(x+3)$; using $(1,5)$ we get $y - 5 = \frac{3}{4}(x - 1)$
14. points: $(-5,6)$ and $(1,-6)$, $m = \frac{-6-6}{1+5} = \frac{-12}{6} = -2$, $y + 6 = -2(x - 1)$ or $y - 6 = -2(x + 5)$
15. $m = 2, b = 4$. Using $y = mx + b$ we get $y = 2x + 4$
16. $m = -3, b = 5, y = -3x + 5$
17. $m = 5, b = -3, y = 5x - 3$
18. $m = -4, b = -2, y = -4x - 2$
19. $y - 3x = 6$; adding $3x$ to both sides we get $y = 3x + 6$ so $m = 3$, $b = 6$, x-intercept $= -2$

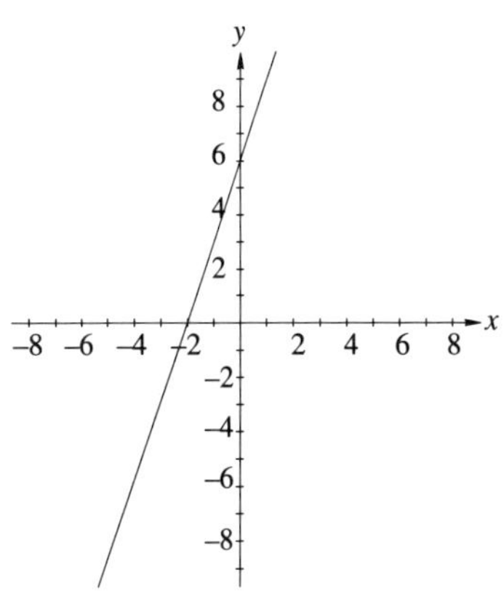

20. $2x - y = 5$. Subtracting $2x$ yields $-y = -2x + 5$, multiplying by -1 yields $y = 2x - 5$ so $m = 2$; $b = -5$; x-intercept $= \frac{5}{2}$

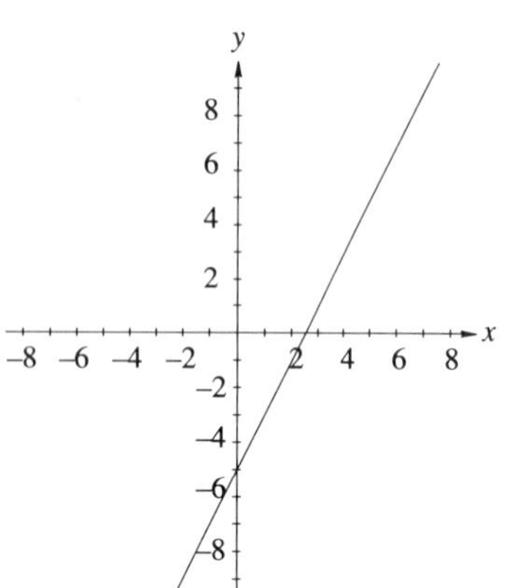

21. $2y - 5x = 8 \Rightarrow 2y = 5x + 8 \Rightarrow y = \frac{5}{2}x + 4$; $m = \frac{5}{2}$; $b = 4$; x-intercept $= -\frac{8}{5}$

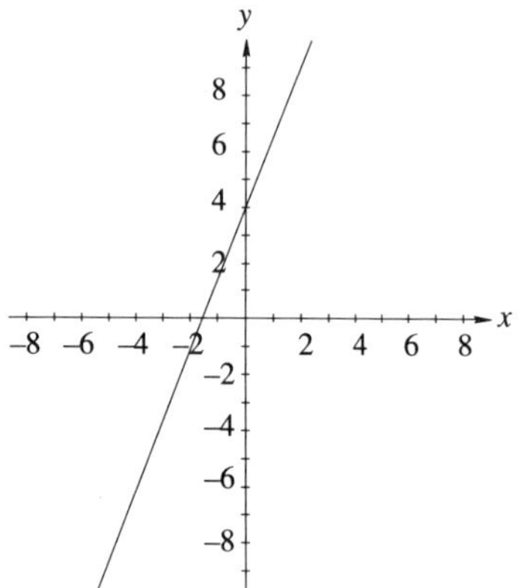

22. $2x - 3y = 9 \Rightarrow -3y = -2x + 9 \Rightarrow y = \frac{2}{3}x - 3$; $m = \frac{2}{3}$; $b = -3$; x-intercept $= \frac{9}{2}$

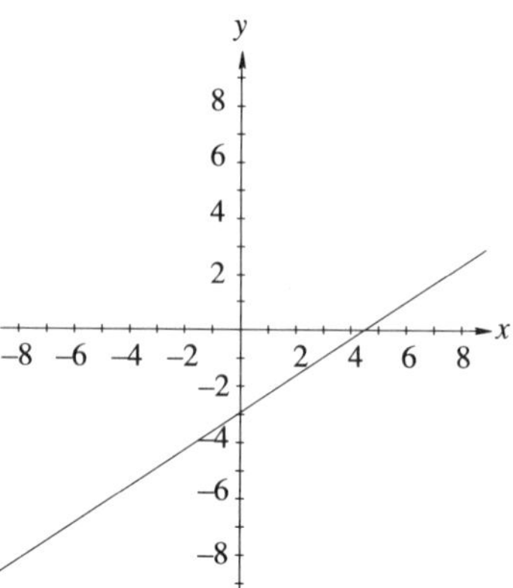

23. $5x - 2y - 10 = 0 \Rightarrow -2y = -5x + 10 \Rightarrow y = \frac{5}{2}x - 5$; $m = \frac{5}{2}$; $b = -5$; x-intercept $= 2$

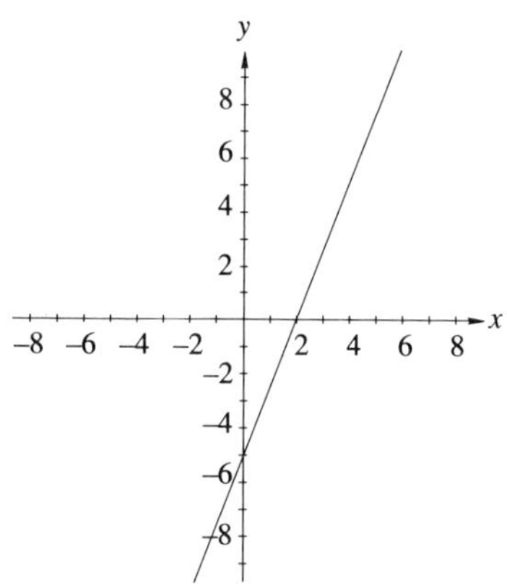

24. $4y - 3x - 4 = 0$; $4y = 3x + 4$; $y = \frac{3}{4}x + 1$; $m = \frac{3}{4}$; $b = 1$; x-intercept $= -\frac{4}{3}$

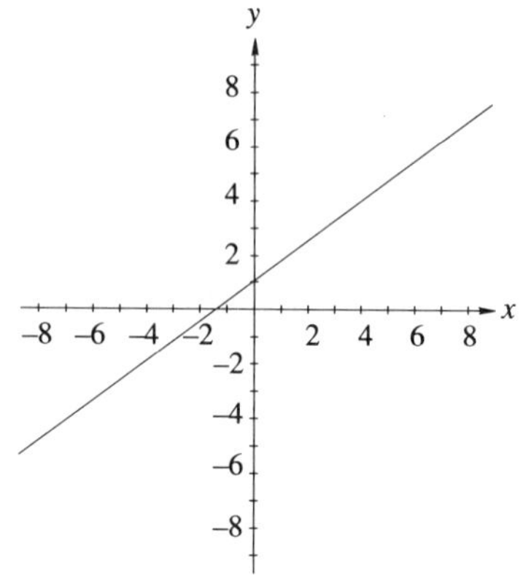

25. $x = -3y + 7$; $3y = -x + 7$; $y = -\frac{1}{3}x + \frac{7}{3}$; $m = -\frac{1}{3}$; $b = \frac{7}{3}$; x-intercept = 7

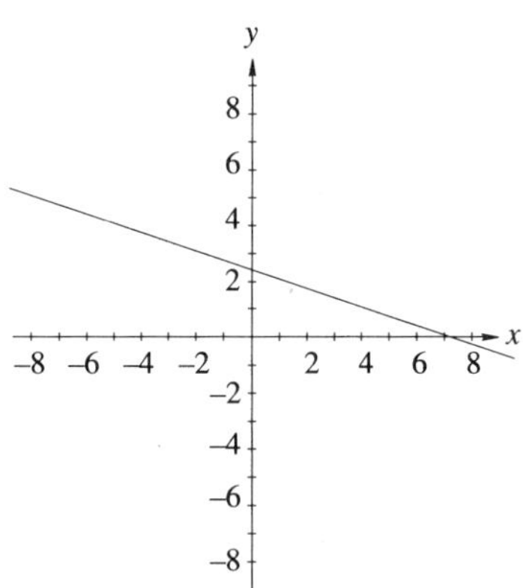

26. $3x = 5y - 6$, $5y = 3x + 6$; $y = \frac{3}{5}x + \frac{6}{5}$; $m = \frac{3}{5}$, $b = \frac{6}{5}$, x-intercept$= -2$

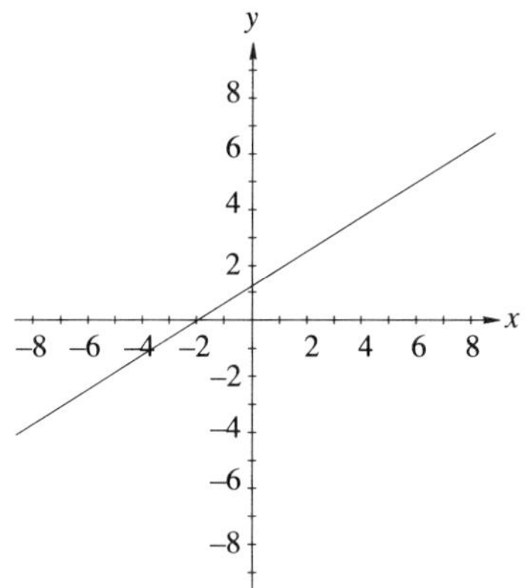

27. $s = md = \left(\frac{1780\pi}{60}\right) d$ so $m = \frac{1780\pi}{60} = \frac{178}{6}\pi = \frac{89\pi}{3} \approx 93.2$

28. (a) $m = \frac{0 - -40}{32 - -40} = \frac{40}{72} = \frac{5}{9}$

(b) equation is $y - 0 = \frac{5}{9}(x - 32) \Rightarrow y = \frac{5}{9}x - \frac{160}{9}$. Hence, the y-intercept is $-\frac{160}{9}$.

(c) $y = \frac{5}{9}(x - 32)$ or $y + 40 = \frac{5}{9}(x + 40)$ which has the slope-intercept form $y = \frac{5}{9}x - \frac{160}{9}$.

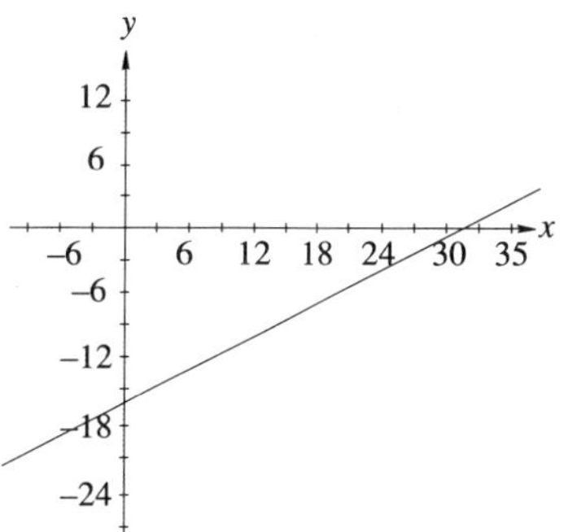

29. (a) $F = kx = k(L - L_0)$ or $F = kL - kL_0$; (b) $F = 4.5(L - 6)$ or $F = 4.5L - 27$; (c)

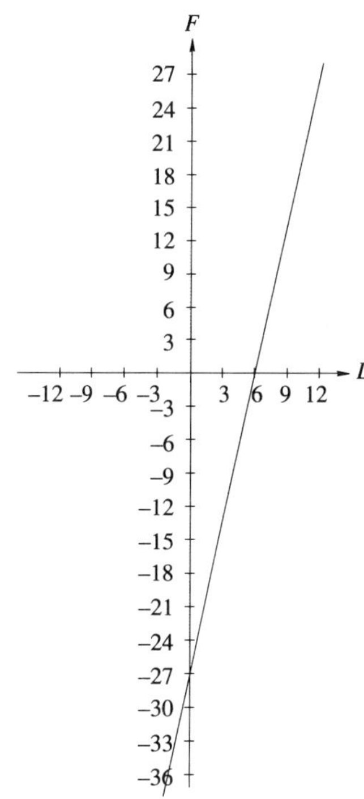

30. Using $(4, 250)$ and $(9, 562.5)$ as two points the slope $= \frac{562.5 - 250}{9 - 4} = \frac{312.5}{5} = 62.5$ hence $P - 250 = 62.5(h - 4)$; $P - 250 = 62.5h - 250$ or $P = 62.5h$

5.2 GRAPHICAL AND ALGEBRAIC METHODS FOR SOLVING TWO LINEAR EQUATIONS IN TWO VARIABLES

1.

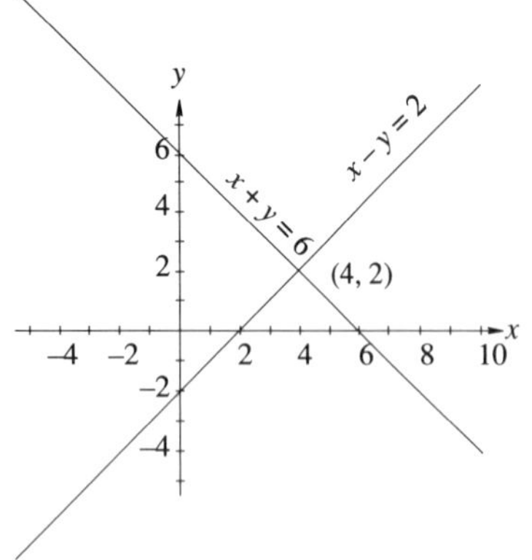

(4, 2)

2.

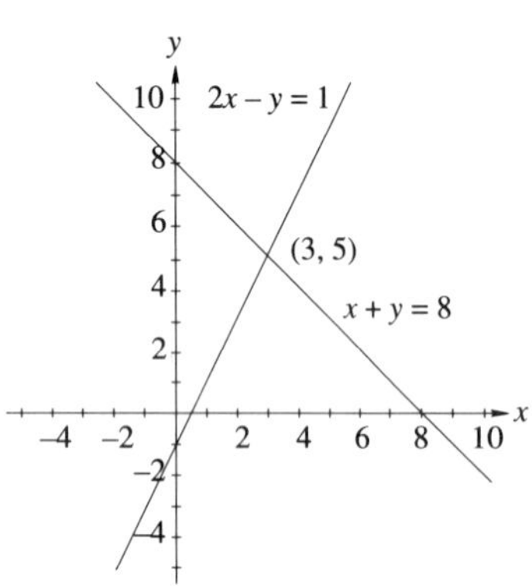

(3, 5)

3.

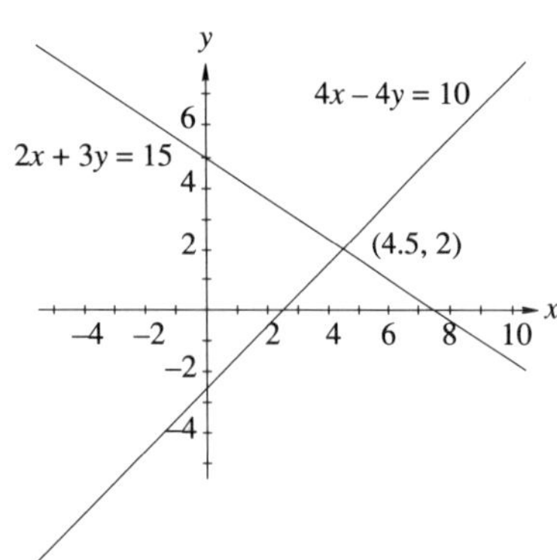

(4.5, 2)

4.

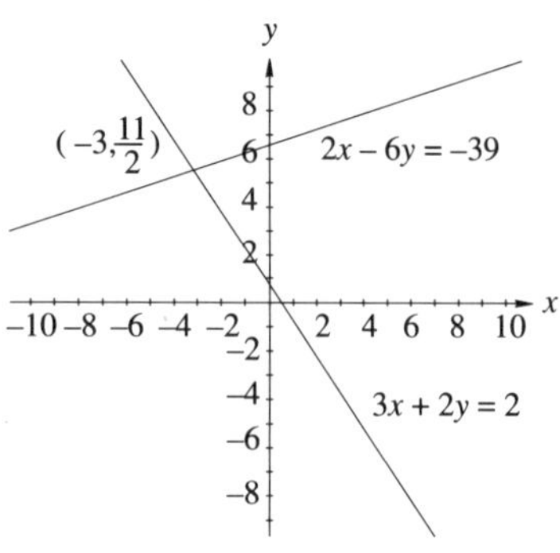

$\left(-3, \frac{11}{2}\right) = (-3, 5.5)$

5.

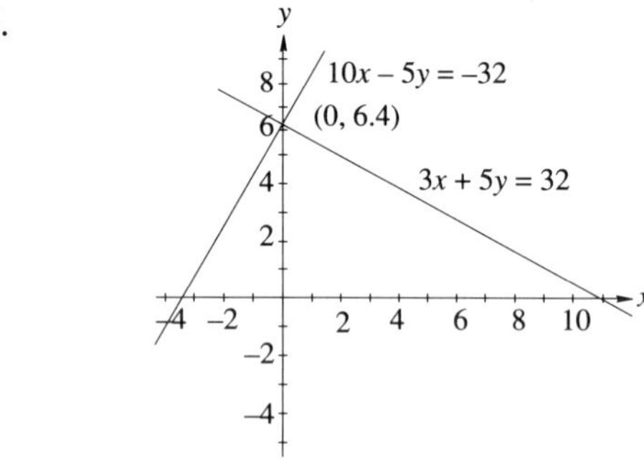

$\left(0, \frac{32}{5}\right) = (0, 6.4)$

6.

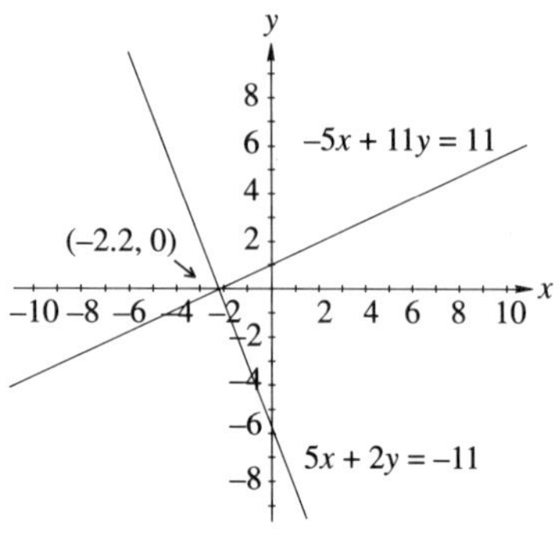

(−2.2, 0)

7.

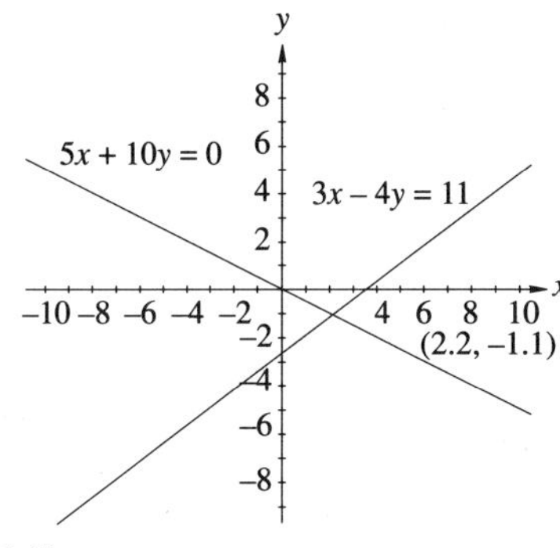

$(2.2, -1.1)$

8.

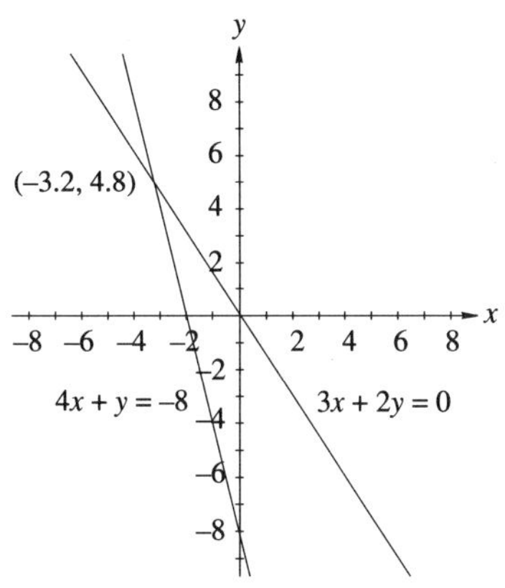

$(-3.2, 4.8)$

9. $\begin{cases} y = 3x - 4 & (1) \\ x + y = 8 & (2) \end{cases}$
Substituting $(3x - 4)$ for y in equation (2) yields $x + (3x - 4) = 8$. Solving $4x - 4 = 8$; $4x = 12$, $x = 3$ Substituting $x = 3$ in (1) $\Rightarrow y = 3(3) - 4 = 9 - 4 = 5$; $(3, 5)$

10. $\begin{cases} x = -2y + 12 & (1) \\ x + y = 5 & (2) \end{cases}$
Substituting $-2y + 12$ for x in (2) $\Rightarrow -2y + 12 + y = 5$ Solving $-y + 12 = 5$; $-y = -7$; $y = 7$. Substituting 7 for y in (1) $\Rightarrow x = -2(7) + 12 = -14 + 12 = -2$; $(-2, 7)$

11. $\begin{cases} y = -2x - 2 & (1) \\ 3x + 2y = 0 & (2) \end{cases}$
Substituting $-2x - 2$ for y in (2) $\Rightarrow 3x + 2(-2x - 2) = 0$; $3x - 4x - 4 = 0$; $-x = 4$ or $x = -4$. Back substituting in (1) $y = -2(-4) - 2 = +8 - 2 = 6$; $(-4, 6)$

12. $\begin{cases} x = 7 + 2y & (1) \\ 3x + 4y = 1 & (2) \end{cases}$
Substituting $7 + 2y$ in (2) $\Rightarrow 3(7 + 2y) + 4y = 1$; $21 + 6y + 4y = 1$; $21 + 10y = 1$; $10y = -20$; $y = -2$; $x = 7 + 2(-2)$; $x = 7 - 4 = 3$. Solution $(3, -2)$

13. $\begin{cases} 2x + 5y = 6 & (1) \\ x - y = 10 & (2) \end{cases}$
Solving (2) for $x \Rightarrow x = 10 + y$. Substituting in (1) $\Rightarrow 2(10 + y) + 5y = 6$; $20 + 2y + 5y = 6$; $7y = -14$; $y = -2$. $x = 10 + (-2) = 8$. Solution $(8, -2)$

14. $\begin{cases} 3x - 2y = 5 & (1) \\ -7x + 4y = -7 & (2) \end{cases}$
Solving (1) for y yields $-2y = 5 - 3x$ or $y = \frac{3x-5}{2}$ (3) Substituting in (2) yields $-7x + 4(\frac{3x-5}{2}) = -7$; $-7x + 6x - 10 = -7$; $-x = 3$ or $x = -3$. Substituting in (3) yields $y = \frac{3(-3)-5}{2} = \frac{-14}{2} = -7$; $(-3, -7)$

15. $\begin{cases} 2x + 3y = 3 & (1) \\ 6x + 4y = 15 & (2) \end{cases}$
Solving for x in (1) $\Rightarrow 2x = 3 - 3y$ or $x = \frac{3-3y}{2}$ (3). Substituting this in (2) yields $6\left(\frac{3-3y}{2}\right) + 4y = 15$; $3(3 - 3y) + 4y = 15$; $9 - 9y + 4y = 15$; $-5y = 6$ or $y = -\frac{6}{5}$ or -1.2. Substituting this for y in (3) yields $x = \frac{3-3(-1.2)}{2} = \frac{3+3.6}{2} = \frac{6.6}{2} = 3.3$. Solution $(3.3, -1.2)$

16. $\begin{cases} 2x + 2y = -3 & (1) \\ 4x + 9y = 5 & (2) \end{cases}$
Solving (1) for x we get $2x = -2y - 3$ or $x = \frac{-2y-3}{2}$. Substituting in (2) yields $4\left(\frac{-2y-3}{2}\right) + 9y = 5$; $-4y - 6 + 9y = 5$; $5y = 11$; $y = \frac{11}{5} = 2.2$; $x = \frac{-2(2.2)-3}{2} = \frac{-4.4-3}{2} = \frac{-7.4}{2} = -3.7$; $(-3.7, 2.2)$

17. $\begin{cases} x + y = 9 & (1) \\ x - y = 5 & (2) \end{cases}$
Adding (1) + (2) $\Rightarrow 2x = 14$; $x = 7$; Substituting in (1) yields $7 + y = 9$; $y = 2$; $(7, 2)$

18. $\begin{cases} 2x + 3y = 5 & (1) \\ -2x + 5y = 3 & (2) \end{cases}$
Adding (1) + (2) yields $8y = 8$ or $y = 1$. Substituting 1 for y in (1) yields $2x + 3 = 5$ or $2x = 2$ and so $x = 1$. The solution is $(1, 1)$.

19. $\begin{cases} -x + 3y = 5 & (1) \\ 2x + 7y = 3 & (2) \end{cases}$
Multiply (1) by 2 $\Rightarrow -2x+6y = 10$ (3). Adding (3) + (2) yields $13y = 13$ or $y = 1$. Back substituting in (1) yields $-x + 3 = 5$ or $-x = 2$ and so we see that $x = -2$. The solution is $(-2, 1)$.

20. $\begin{cases} 3x - 2y = 8 & (1) \\ 5x + y = 9 & (2) \end{cases}$
Multiply (2) by 2 $\Rightarrow 10x+2y = 18$ (3). Adding (1) + (3) gives $13x = 26$ or $x = 2$. Substituting 2 for x in (1) yields $3(2) - 2y = 8$ or $6 - 2y = 8$ and so $-2y = 2$; $y = -1$. The solution is $(2, -1)$

21. $\begin{cases} 3x - 2y = -15 & (1) \\ 5x + 6y = 3 & (2) \end{cases}$
Multiply (1) by 3 $\Rightarrow 9x - 6y = -45$ (3). Add (3) + (2): $14x = -42 \Rightarrow x = -3$. Substituting $x = -3$ in (1) we get $3(-3) - 2y = -15 \Rightarrow -9 - 2y = -15$; $-2y = -6$, so $y = 3$ and the solution is $(-3, 3)$.

22. $\begin{cases} 2x - 3y = 11 & (1) \\ 6x - 5y = 13 & (2) \end{cases}$
Multiply (1) by 3 $\Rightarrow 6x - 9y = 33(3)$; Subtract (3) − (2) $\Rightarrow -4y = 20 \Rightarrow y = -5$. Thus, $2x - 3(-5) = 11$ or $2x + 15 = 11$; $2x = -4$ and so $x = -2$. The solution is $(-2, -5)$.

23. $\begin{cases} 3x - 5y = 37 & (1) \\ 5x - 3y = 27 & (2) \end{cases}$
Multiply (1) by 5 $\Rightarrow 15x - 25y = 185$ (3). Multiply (2) by 3 $\Rightarrow 15x - 9y = 81$ (4). Subtract (3) − (4) $\Rightarrow -16y = 104$ and so $y = -6.5$. Substituting in (1) $\Rightarrow 3x - 5(-6.5) = 37$ or $3x + 32.5 = 37 \Rightarrow 3x = 4.5 \Rightarrow x = 1.5$. The solution is $(1.5, -6.5)$.

24. $\begin{cases} x + \frac{1}{2}y = 7 & (1) \\ 4x - 2y = 5 & (2) \end{cases}$
Multiply (1) by 4 $\Rightarrow 4x + 2y = 28$ (3). Adding (3) + (2) $\Rightarrow 8x = 33$ and so $x = \frac{33}{8} = 4\frac{1}{8} = 4.125$. Substituting $x = 4.125$ in (1), we get $4.125 + 0.5y = 7$ or $0.5y = 2.875$ and so we see that $y = 5.75$. The solution is $(4.125, 5.75)$.

25. $\begin{cases} 2x + 3y = 5 & (1) \\ x - 2y = 6 & (2) \end{cases}$
Solving (2) for x, we get $x = 2y + 6$. Substituting in (1) produces $2(2y+6)+3y = 5$ or $4y+12+3y = 5$; $7y = -7$; $y = -1$; $x = 2(-1) + 6 = -2 + 6 = 4$; $(4, -1)$

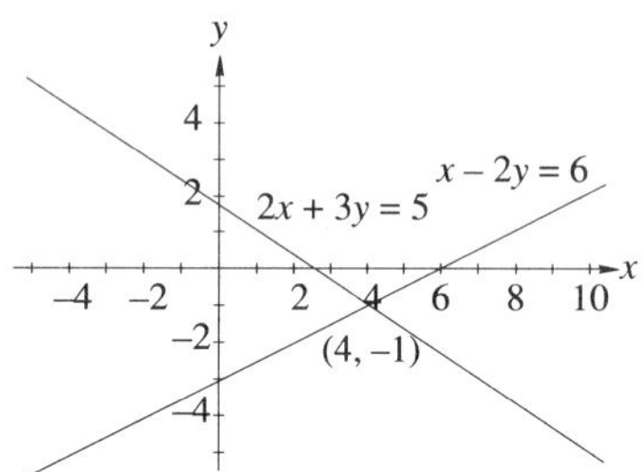

26. $\begin{cases} 2x - 3y = -14 & (1) \\ 3x + 2y = 44 & (2) \end{cases}$
Multiplying (1) by 2 we get $4x - 6y = -28$ (3). Multiplying (2) by 3 produces $9x + 6y = 132$ (4). Adding (3) and (4) we get $13x = 104$ and so $x = 8$. Substituting in (1) gives $2(8) - 3y = -14$ or $16 - 3y = -14 \Rightarrow -3y = -30$, and so we see that $y = 10$. The solution is $(8, 10)$.

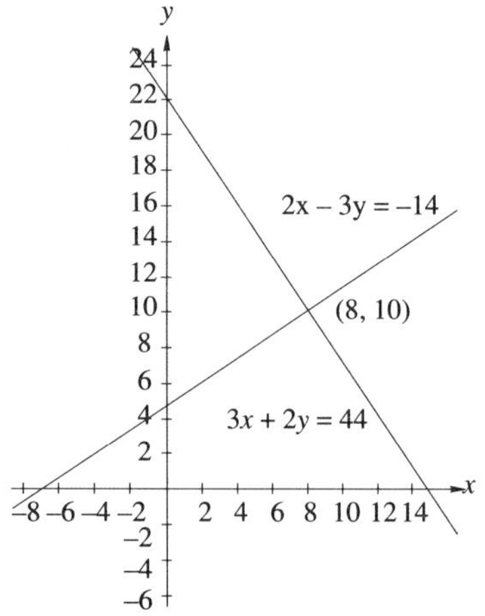

27. $\begin{cases} 8x + 3y = 13 & (1) \\ 3x + 2y = 11 & (2) \end{cases}$
(1) × 2 $\Rightarrow 16x + 6y = 26$ (3). (2) × 3 $\Rightarrow 9x + 6y =$

33 (4). (3) − (4) ⇒ $7x = -7$ or $x = -1$. Substituting in (1) we get $8(-1) + 3y = 13 \Rightarrow -8 + 3y = 13$ and so $3y = 21$ or $y = 7$. The solution is $(-1, 7)$.

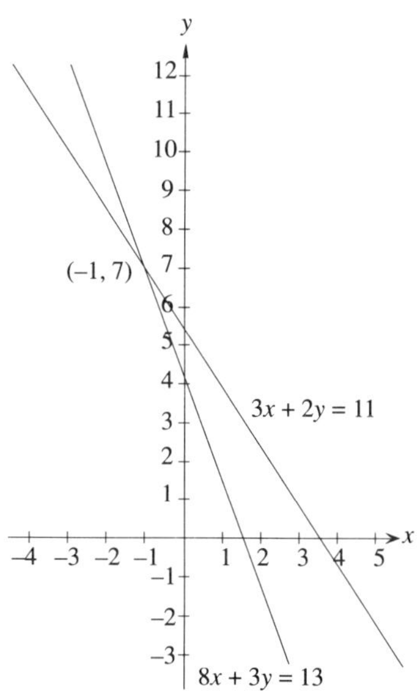

28. $\begin{cases} 6x + 12y = 7 & (1) \\ 8x - 15y = -1 & (2) \end{cases}$

(1)×4 ⇒ $24x+48y = 28$ (3). (2)×3 ⇒ $24x-45y = -3$ (4). Subtracting: (3)−(4) ⇒ $93y = 31$ and so $y = \frac{1}{3}$; Substituting in (1) yields $6x + 12(\frac{1}{3}) = 7$ or $6x + 4 = 7$ which simplifies to $6x = 3$, and so $x = \frac{1}{2}$. The solution is $\left(\frac{1}{2}, \frac{1}{3}\right)$.

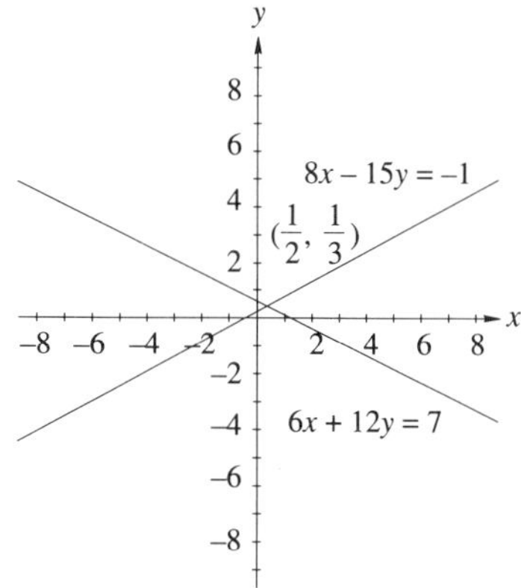

29. $\begin{cases} 10x - 9y = 18 & (1) \\ 6x + 2y = 1 & (2) \end{cases}$

(1)×2 ⇒ $20x-18y = 36$ (3); (2)×9 ⇒ $54x+18y = 9$ (4). Adding: (3)+(4) ⇒ $74x = 45$ and so $x = \frac{45}{74} \approx 0.608$. Substituting in (1), we get $10\left(\frac{45}{74}\right) - 9y = 18$ or $\frac{450}{74} - 9y = 18 \Rightarrow -9y = 18 - \frac{450}{74} = \frac{441}{37}$, and so $y = -\frac{1}{9}\left(\frac{441}{37}\right) = -\frac{49}{37} \approx -1.324$. The solution is $\left(\frac{45}{74}, -\frac{49}{37}\right) \approx (0.608, -1.324)$.

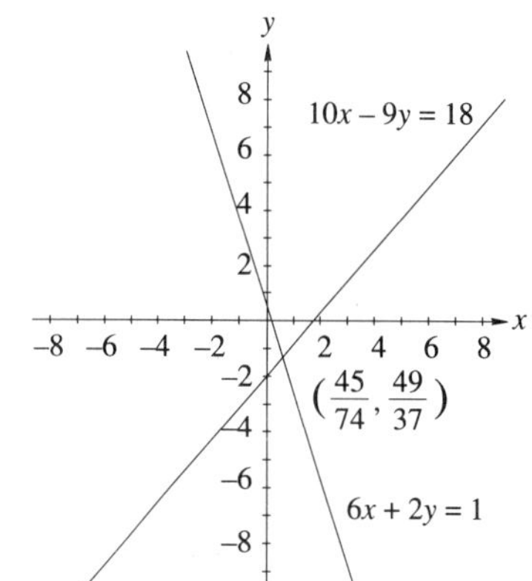

30. $\begin{cases} 4x - 5y = 7 & (1) \\ -8x + 10y = -30 & (2) \end{cases}$

(1) × 2 ⇒ $8x - 10y = 14$ (3); (3) + (2) ⇒ $0 = -16$. Since $0 \neq -16$ there is no solution. The system is inconsistent.

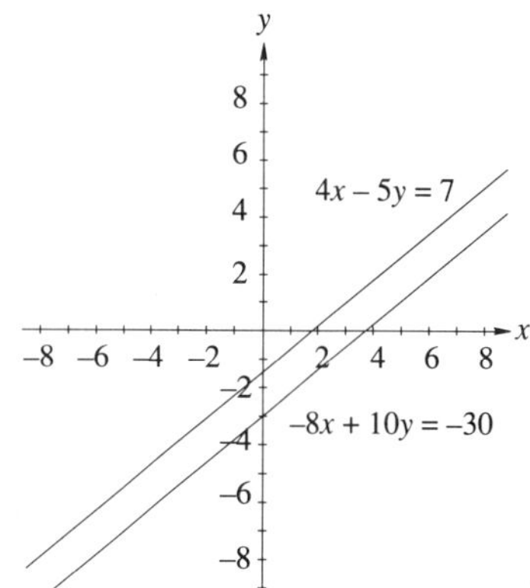

31. $\begin{cases} x - 9y = 0 & (1) \\ \frac{x}{3} = 2y + \frac{1}{3} & (2) \end{cases}$

(2) × 3 ⇒ $x = 6y+1$ or $x-6y = 1$ (3). Subtracting: (1) − (3) ⇒ $-3y = -1$ or $y = \frac{1}{3}$. Substituting in (1), we obtain $x - 9\left(\frac{1}{3}\right) = 0$ or $x - 3 = 0$ and so $x = 3$. The solution is $\left(3, \frac{1}{3}\right)$.

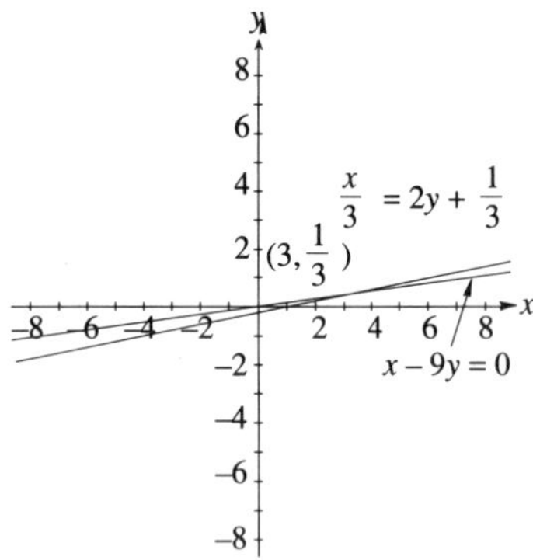

32. $\begin{cases} 5x + 3y = 7 & (1) \\ \frac{3}{2}x - \frac{3}{4}y = 9\frac{1}{4} & (2) \end{cases}$
$(2) \times 4 \Rightarrow 6x - 3y = 37$ (3). Adding: $(3) + (1) \Rightarrow 11x = 44$ or $x = 4$. Substituting for x in (1), we get $5(4) + 3y = 7$ or $20 + 3y = 7 \Rightarrow 3y = -13$, and so $y = -\frac{13}{3}$ or $-4\frac{1}{3}$. The solution is $\left(4, -4\frac{1}{3}\right)$.

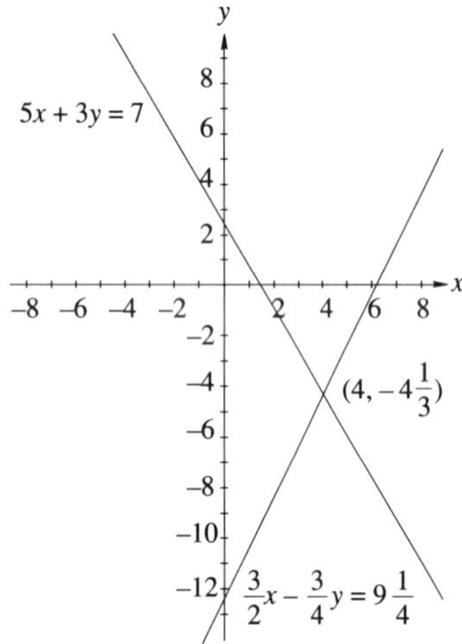

33. Substituting $w + 8$ for L in the first equation yields $2(w+8)+2w = 36$; $2w+16+2w = 36 \Rightarrow 4w+16 = 36$ or $4w = 20$ and so $w = 5$. $L = w+8$ or $5+8 = 13$. Length is 13 km and the width is 5 km.

34. Since the field is a rectangle, the perimeter is $2L + 2w = 72$. We are given $L = w + 9$. By substitution $2(w + 9) + 2w = 72$ or $4w + 18 = 72$ and so $4w = 54$ or $w = 13.5$. Now, $L = w+9 = 13.5+9 = 22.5$. Thus, $L = 22.5$ mi and $w = 13.5$ mi.

35. $2L + 2w = 45$ and $L = 3w$ are two equations that fit this situation. By substitution $2(3w) + 2w = 45 \Rightarrow 6w + 2w = 45$ or $8w = 45$ and so $w = 5.625$. Substituting, we get $L = 3(5.625) = 16.875$. Thus, the length is 16.875 km and the width is 5.625 km.

36. Let x = length of third side and y = length of the two equal sides. Equations are $y = 1\frac{1}{2}x$ and $x + y + y = 96$. By substitution $x + 1\frac{1}{2}x + 1\frac{1}{2}x = 96 \Rightarrow 4x = 96$ and so $x = 24$. By substitution, $y = 1\frac{1}{2}(24) = 36$. Thus, the three sides measure 36 yd, 36 yd, and 24 yd.

37. Multiplying (2) by $100 \Rightarrow 5x + 13y = 80{,}000$ and multiplying (1) by $5 \Rightarrow 5x + 5y = 50{,}000$. Subtracting yields $8y = 30{,}000$ and so $y = 3750$. Substituting in (1), we see that $x + 3750 = 10{,}000$ and so $x = 6{,}250$. As a result of these calculations, we have determined that we need 6,250 L of 5% gasohol and 3 750 L of 13% gasohol.

38. As in Exercise 37, $x + y = 20{,}000$; $0.04x + 0.12y = 0.09(20{,}000)$. Multiplying equation $(2) \times 100 \Rightarrow 4x + 12y = 180{,}000$ and multiplying equation $(1) \times 4 \Rightarrow 4x + 4y = 80\ 000$. Subtracting gives $8y = 100\ 000$ or $y = 12\ 500$. Since $x + y = 20\ 000$ we have $x + 12\ 500 = 20\,000$ and so $x = 7\,500$. Thus, we will need 7 500 L of 4% gasohol and 12 500 L of 12% gasohol.

39. Equations that fit this situation are $8x = 5y$ (1) and $12x = 5(y + 3.2)$ (2). Solving (1) for x produces $x = \frac{5}{8}y$. Substituting into (2) yields $12\left(\frac{5}{8}y\right) = 5(y + 3.2)$; $\frac{15}{2}y = 5y + 16$; $2.5y = 16$ or $y = 6.4$. $x = \frac{5}{8} \times 6.4$ or $x = 4$. Thus, the unknown mass is 6.4 kg and the 8-kg force is at a distance of 4 m.

40. Let x be the amount replaced. Then $(12-x)(0.25)+x(1.00) = 12(.45)$; $3 - 0.25x + x = 5.4$; $0.75x + 3 = 5.4$; $0.75x = 2.4$; $x = 2.4 \div 0.75$ or 3.2. You will need to replace 3.2 L of antifreeze.

41. $8 = I_1(3) + I_2(5)$; $5 = 5(I_2) - 6(I_1 - I_2)$ The second equation simplifies to $5 = 5I_2 - 6I_1 + 6I_2$ or $5 = 11I_2 - 6I_1$. Taking the first equation times 2 yields $16 = 6I_1 + 10I_2$. Adding these two gives $21 = 21I_2$ or $I_2 = 1$. Replacing I_2 with 1 in the first equation you get $8 = 3I_1 + 5$ or $3 = 3I_1$ or $I_1 = 1$. Since $I_1 = I_2 + I_3$; $I_3 = I_1 - I_2$ or $I_3 = 1 - 1 = 0$.

42. Using the set up of 41 you get $10 = 2I_1 + 4I_2$ and $15 = 4I_2 - 8(I_1 - I_2)$ or $15 = 12I_2 - 8I_1$. Multiplying the first equation by 4 gives $40 = 8I_1 + 16I_2$. Adding the last two equations gives $55 = 28I_2$ or $I_2 = \frac{55}{28} \approx 1.96$. Substituting in the first $10 = 2I_1 + 4\left(\frac{55}{28}\right)$ or $10 = 2I_1 + \frac{55}{7}$ and so $2I_1 = \frac{15}{7}$ and $I_1 = \frac{15}{14} \approx 1.07$. $I_1 = I_2 + I_3$ so $\frac{15}{14} = \frac{55}{28} + I_3$ and so $I_3 = \frac{15}{14} - \frac{55}{28} = -\frac{25}{28} \approx -0.89$; $I_1 = \frac{15}{14} \approx 1.07$ A, $I_2 = \frac{55}{28} \approx 1.96$ A; $I_3 = -\frac{25}{28} \approx -0.89$ A. Note that I_3 is in the reverse direction of the diagram.

43. Using $I_2 = I_1 + I_3$ is equivalent to $I_2 - I_1 = I_3$. Hence $E_2 = R_3I_3 + R_2I_2$ is equivalent to $E_2 = R_3(I_2 - I_1) + R_2I_2$. Substituting the given values yields $6 = 8I_1 + 4I_2$ and $10 = 7(I_2 - I_1) + 4I_2$ or $10 = 11I_2 - 7I_1$. Taking the first equation times $11 \Rightarrow 66 = 88I_1 + 44I_2$. The second times 4 gives $40 = 44I_2 - 28I_1$. Subtracting gives $26 = 116I_1$ or $I_1 = \frac{26}{116} = 0.224$ A. Thus, $6 = 8(0.224) + 4I_2 \Rightarrow 6 = 1.792 + 4I_2 \Rightarrow 4.208 = 4I_2$. Thus, $I_2 = 1.052$ A; $I_3 = 1.052 - 0.224 = .828$ A.

44. See *Computer Programs* in main text.

5.3 ALGEBRAIC METHODS FOR SOLVING THREE LINEAR EQUATIONS IN THREE VARIABLES

1. $\begin{cases} 2x + y + z = 7 & (1) \\ x - y + 2z = 11 & (2) \\ 5x + y - 2z = 1 & (3) \end{cases}$
Solving (2) for x yields $x = 11 + y - 2z$ (4). Substituting in (3) yields $5(11 + y - 2z) + y - 2z = 1$; $55 + 5y - 10z + y - 2z = 1$; or $6y - 12z = -54$ (5). Solving (5) for y we obtain $y = 2z - 9$ (6). Substituting (4) into (1) we get $2(11+y-2z)+y+z = 7$ or $22+2y-4z+y+z = 7$ or $3y-3z = -15$ (7). Substituting using (6) into (7) we get $3(2z-9)-3z = -15$; $6z - 27 - 3z = -15$; $3z = 12$; $z = 4$. Using (6) we get $y = 2 \cdot 4 - 9 = 8 - 9 = -1$. Using (4) we get $x = 11 + (-1) - 2(4) = 11 - 1 - 8 = 11 - 9 = 2$. The solution is $x = 2$; $y = -1$; $z = 4$.

2. $\begin{cases} x + y + 2z = 0 & (1) \\ 2x - y + z = 6 & (2) \\ 4x + 2y + 2z = 0 & (3) \end{cases}$
Solving equation (1) for y yields $y = -x - 2z$ (4). Substituting this into equation (2), we get $2x-(-x-2z)+z = 6$, which simplifies to $3x+3z = 6$ or $x + z = 2$ (5). We again use (4) to substitute into equation (3) to get $4x + 2(-x - 2z) + 2z = 0$ which simplifies to $2x - 2z = 0$ or $x - z = 0$ (6). Solving equation (6) for x we obtain $x = z$ which we substitute into (5) to obtain $z + z = 2$ or $2z = 2$ and so $z = 1$. Back substitution into equation (6) yields $x = 1$ and further back substitution into equation (4) yields $y = -1 - 2 = -3$. The solution is $(1, -3, 1)$.

3. $\begin{cases} 2x - y - z = -8 & (1) \\ x + y - z = -9 & (2) \\ x - y + 2z = 7 & (3) \end{cases}$
Solving equation (1) for y we obtain $y = 8 + 2x - z$ (4). Substituting this into (2) yields $x + (8 + 2x - z) - z = -9$ which simplifies to $3x - 2z = -17$ (5). Again, using equation (4), we substitute into equation (3), which yields $x - (8 + 2x - z) + 2z = 7$. This simplifies to $-x + 3z = 15$ (6). Solving equation (6) for x, we have $x = 3z - 15$ (7). Using equation (7) and substituting into equation (5) yields $3(3z - 15) - 2z = -17$ or $9z - 45 - 2z = -17$ or $7z = 28$. Hence, $z = 4$. Back substituting into equation (7) we get $x = 3(4) - 15 = 12 - 15 = -3$. Now, back substitution into equation (4) yields $y = 8 + 2(-3) - 4 = 8 - 6 - 4 = -2$. The solution is $(-3, -2, 4)$.

4. $\begin{cases} x + y + 5z = -10 & (1) \\ x - y - 5z = 11 & (2) \\ -x + y - 5z = 13 & (3) \end{cases}$
Since all three equations have $5z$ in them, let's solve equation (1) for $5z$. This gives $5z = -10 - x - y$ (4). Now, substituting for $5z$ in equation (2), we get $x-y-(-10-x-y) = 11$ or $x-y+10+x+y = 11$, which simplifies to $2x = 1$ or $x = \frac{1}{2}$. Using equation (4) again, we substitute for $5z$ in equation (3) with the result $-x+y+10+x+y = 13$. This simplifies to $2y = 3$ or $y = \frac{3}{2}$. Back substituting into equation (1) we get $\frac{1}{2} + \frac{3}{2} + 5z = -10$ or $2 + 5z = -10$ or

$5z = -12$ and so, $z = -\frac{12}{5} = -2.4$. The solution is $\left(\frac{1}{2}, \frac{3}{2}, -\frac{12}{5}\right) = (0.5, 1.5, -2.4)$.

5. $\begin{cases} 2x + y + z = 7 & (1) \\ x - y + 2z = 11 & (2) \\ 5x + y - 2z = 1 & (3) \end{cases}$
Adding (1)+(2) yields $3x + 3z = 18$ (4). Adding (2)+(3) yields $6x = 12$ or $x = 2$ (5). Substituting 2 for x in (4) yields $3 \cdot 2 + 3z = 18$; $6 + 3z = 18$; $3z = 12$, $z = 4$. Substituting $z = 4$, $x = 2$ in (1) yields $2(2) + y + 4 = 7$ or $4 + y + 4 = 7$; $y = -1$. The solution is $x = 2$, $y = -1$, $z = 4$.

6. $\begin{cases} x + y + 2z = 0 & (1) \\ 2x - y + z = 6 & (2) \\ 4x + 2y + 2z = 0 & (3) \end{cases}$
$(1) + (2) \Rightarrow 3x + 3z = 6$ or $x + z = 2$ (4). $2 \times (2) \Rightarrow 4x - 2y + 2z = 12$ (5). $(5) + (3) \Rightarrow 8x + 4z = 12$ (6). $4 \times (4) \Rightarrow 4x + 4z = 8$ (7). $(6) - (7) \Rightarrow 4x = 4$ or $x = 1$. Substituting into (4) yields $1 + z = 2$ or $z = 1$. Substituting into (1) yields $1 + y + 2(1) = 0$ or $y = -3$. The solution is $x = 1$, $y = -3$, $z = 1$.

7. $\begin{cases} 2x - y - z = -8 & (1) \\ x + y - z = -9 & (2) \\ x - y + 2z = 7 & (3) \end{cases}$
$(1) + (2) \Rightarrow 3x - 2z = -17$ (4). $(2) + (3) \Rightarrow 2x + z = -2$ (5). $2 \times (5) \Rightarrow 4x + 2z = -4$ (6). $(4) + (6) \Rightarrow 7x = -21$ or $x = -3$. Substituting into (5) yields $2(-3) + z = -2$ or $-6 + z = -2$ or $z = 4$. Substituting into (2) yields $(-3) + y - 4 = -9$ or $-7 + y = -9$ or $y = -2$. The solution is $x = -3$, $y = -2$, $z = 4$.

8. $\begin{cases} x + y + 5z = -10 & (1) \\ x - y - 5z = 11 & (2) \\ -x + y - 5z = 13 & (3) \end{cases}$
$(1) + (2) \Rightarrow 2x = 1$ or $x = 0.5$. $(2) + (3) \Rightarrow -10z = 24$ or $z = -2.4$ Substituting into (1) yields $0.5 + y + 5(-2.4) = -10$ or $0.5 + y - 12 = -10$ or $y - 11.5 = -10$ so $y = 1.5$. The solution is $x = 0.5$, $y = 1.5$, $z = -2.4$.

9. $\begin{cases} x + y + z = 2 & (1) \\ 8x - 2y + 4z = -3 & (2) \\ 6x - 4y - 3z = 3 & (3) \end{cases}$
$2 \times (1) \Rightarrow 2x + 2y + 2z = 4$ (4). $(4) + (2) \Rightarrow 10x + 6z = 1$ (5). $4 \times (1) \Rightarrow 4x + 4y + 4z = 8$ (6). $(6)+(3) \Rightarrow 10x+z = 11$ (7). $(5)-(7) \Rightarrow 5z = -10$ or $z = -2$. Substituting in (7) yields $10x - 2 = 11$ or $10x = 13$ or $x = 1.3$. Substituting in (1) yields $1.3 + y - 2 = 2$; $y - 0.7 = 2$ so $y = 2.7$. The solution is $x = 1.3$, $y = 2.7$, $z = -2$.

10. $\begin{cases} x + y - z = 7 & (1) \\ 8x + 4y + 2z = 21 & (2) \\ 4x + 3y + 6z = 2 & (3) \end{cases}$
$2 \times (1) \Rightarrow 2x + 2y - 2z = 14$ (4). Adding: $(4)+(2) \Rightarrow 10x + 6y = 35$ (5). $6 \times (1) \Rightarrow 6x + 6y - 6z = 42$ (6). Adding: $(6)+(3) \Rightarrow 10x + 9y = 44$ (7). Subtracting: $(7)-(5) \Rightarrow 3y = 9$ or $y = 3$. Substituting in (5) yields $10x + 6(3) = 35$; $10x + 18 = 35$; $10x = 17$ and so $x = 1.7$. Substituting in (1) we get $1.7 + 3 - z = 7$ or $4.7 - z = 7$ and so $z = -2.3$. The solution is $x = 1.7$, $y = 3$, $z = -2.3$.

11. $\begin{cases} 3x - y - 2z = 11 & (1) \\ -x + 3y + 2z = -1 & (2) \\ 2x - 2y - 4z = 17 & (3) \end{cases}$
$2 \times (2) \Rightarrow -2x + 6y + 4z = -2$ (4). Adding: $(4) + (3) \Rightarrow 4y = 15$ or $y = 3.75$ (5), and adding $(1)+(2) \Rightarrow 2x + 2y = 10$ or $x + y = 5$ (6). Substituting 3.75 for y into (6) yields $x + 3.75 = 5$; $x = 1.25$ Substituting into (1) yields $3(1.25) - (3.75) - 2z = 11$; $3.75 - 3.75 - 2z = 11$ or $-2z = 11$ and so $z = -5.5$. The solution is $x = 1.25$, $y = 3.75$, $z = -5.5$.

12. $\begin{cases} x - 2y + z = -4 & (1) \\ 2x + y + 3z = 5 & (2) \\ 6x + 3y + 12z = 6 & (3) \end{cases}$
$3 \times (2) \Rightarrow 6x + 3y + 9z = 15$ (4). Subtracting: $(4) - (3) \Rightarrow -3z = 9$ or $z = -3$ (5). $2 \times (2) \Rightarrow 4x + 2y + 6z = 10$ (6). $(1) + (6) \Rightarrow 5x + 7z = 6$ (7). Substituting -3 for z in (7) yields $5x + 7(-3) = 6$ or $5x - 21 = 6$ or $5x = 27$ and we get $x = 5.4$. Substituting in (1) yields $5.4 - 2y + (-3) = -4$; $2.4 - 2y = -4$; $-2y = -6.4$ and so $y = 3.2$. The solution is $x = 5.4$, $y = 3.2$, $z = -3$.

13. $\begin{cases} 2x + 3y + 3z = 9 & (1) \\ 5x - 2y + 8z = 6 & (2) \\ 4x - y + 5z = -1 & (3) \end{cases}$
To eliminate ys: $(3) \times 2 \Rightarrow 8x - 2y + 10z = -2$ (4).

(4) − (2) ⇒ $3x + 2z = -8$ (5). (3) × 3 ⇒ $12x - 3y + 15z = -3$ (6). (1) + (6) ⇒ $14x + 18z = 6$ (7). 9 × (5) ⇒ $27x + 18z = -72$ (8). (8) − (7) ⇒ $13x = -78$ or $x = -6$. Substituting in (5) yields $3(-6) + 2z = -8$; $-18 + 2z = -8$; $2z = 10$; $z = 5$. Substituting in (1) yields $2(-6) + 3y + 3(5) = 9$; $-12 + 3y + 15 = 9$; $3y + 3 = 9$; $3y = 6$; $y = 2$. The solution is $x = -6$, $y = 2$, $z = 5$.

14. $\begin{cases} x + 2y + 3z = 4 & (1) \\ 2x - 3y - 4z = -1 & (2) \\ 3x - 4y + 5z = 6 & (3) \end{cases}$
Eliminating xs: 2 × (1) ⇒ $2x + 4y + 6z = 8$ (4). (4) − (2) ⇒ $7y + 10z = 9$ (5). 3 × (1) ⇒ $3x + 6y + 9z = 12$ (6). (6) − (3) ⇒ $10y + 4z = 6$ or $5y + 2z = 3$ (7). Next, eliminate the ys: 5 × (5) − 7 × (7) ⇒ $36z = 24$; $z = \frac{24}{36} = \frac{2}{3}$. Substituting in (7) ⇒ $5y + 2\left(\frac{2}{3}\right) = 3$ or $5y + \frac{4}{3} = 3$ and $y = \frac{1}{3}$. Substituting in $x + 2\left(\frac{1}{3}\right) + 3\left(\frac{2}{3}\right) = 4$, we get $x = \frac{4}{3}$. Thus, we get the solution $x = 1\frac{1}{3}$, $y = \frac{1}{3}$, $z = \frac{2}{3}$.

15. $\begin{cases} I_1 - I_2 + I_3 = 0 & (1) \\ 6I_1 + 6I_2 = 18 \Rightarrow I_1 + I_2 = 3 & (2) \\ 6I_2 + I_3 = 14 & (3) \end{cases}$
(3)−(1) ⇒ $-I_1 + 7I_2 = 14$ (4). Adding (2)+(4) ⇒ $8I_2 = 17$; $I_2 = \frac{17}{8} = 2.125$. Substituting in (2) yields $I_1 + 2.125 = 3$ or $I_1 = 0.875$. Substituting in (3) yields $6(2.125) + I_3 = 14$; $12.75 + I_3 = 14$ or $I_3 = 1.25$. The values of the currents are $I_1 = 0.875$ A, $I_2 = 2.125$ A, $I_3 = 1.25$ A.

16. $\begin{cases} I_1 + I_2 - I_3 = 0 & (1) \\ 3I_1 - 5I_2 - 10 = 0 & (2) \\ 5I_2 + 6I_3 - 5 = 0 & (3) \end{cases}$
3 × (1) ⇒ $3I_1 + 3I_2 - 3I_3 = 0$ (4). Subtracting: (4) − (2) ⇒ $8I_2 - 3I_3 + 10 = 0$ (5). 2 × (5) ⇒ $16I_2 - 6I_3 + 20 = 0$ (6). Adding: (6) + (3) ⇒ $21I_2 + 15 = 0$; $21I_2 = -15$; $I_2 = \frac{-5}{7}$. Substituting in (2) yields $3I_1 - 5\left(\frac{-5}{7}\right) - 10 = 0$ or $3I_1 + \frac{25}{7} - \frac{70}{7} = 0$; $3I_1 = \frac{45}{7}$; $I_1 = \frac{15}{7} = 2\frac{1}{7}$. Substituting in (1), we get $2\frac{1}{7} - \frac{5}{7} = I_3$ and so $I_3 = 1\frac{3}{7}$. The solutions are $I_1 = 2\frac{1}{7}$ A, $I_2 = -\frac{5}{7}$ A, and $I_3 = 1\frac{3}{7}$ A.

17. Substituting the coordinates of $P(5, 1)$ in the standard equation for a circle, we get $5^2 + 1^2 + 5a + b + c = 0 \Rightarrow 5a + b + c = -26$ (1). Next we substitute $x = -2$ and $y = -6$, the coordinates of Q, in the standard equation of a circle, with the result: $(-2)^2 + (-6)^2 - 2a - 6b + c = 0 \Rightarrow -2a - 6b + c = -40$ or $2a + 6b - c = 40$ (2). Finally, substituting the coordinates of $R(-1, -7)$ in the standard equation of a circle, yields $(-1)^2 + (-7)^2 - a - 7b + c = 0$ or $-a - 7b + c = -50$ or $a + 7b - c = 50$ (3). We now proceed to solve this system of three equations. We begin by adding (1) + (2) ⇒ $7a + 7b = 14$ or $a + b = 2$ (4) and adding (1)+(3) ⇒ $6a + 8b = 24$ or $3a + 4b = 12$ (5). Then, 3 × (4) ⇒ $4a + 4b = 8$ (6) and subtracting (6) − (5), we find that $a = -4$. Substituting -4 for a in (4) ⇒ $-4 + b = 2$; $b = 6$. Substituting into (1) yields $5(-4) + 6 + c = -26$ or $-20 + 6 + c = -26$; $c - 14 = -26$ and so $c = -12$. Thus we see that $a = -4$, $b = 6$, and $c = -12$ and so equation of the circle through the points P, Q and R is $x^2 + y^2 - 4x + 6y - 12 = 0$.

18. As in exercise #17, we will substitute the coordinates of S, T, and U into the standard equation for a circle. Substituting the coordinates of S, we get $4^2 + 16^2 + 4a + 16b + c = 0$ or $16 + 256 + 4a + 16b + c = 0 \Rightarrow 4a + 16b + c = -272$ (1). Next, substituting the coordinates of T produces $(-6)^2 + (-8)^2 - 6a - 8b + c = 0$ or $36 + 64 - 6a - 8b + c = 0 \Rightarrow 6a + 8b - c = 100$ (2). Finally, the coordinates of U are substituted in the standard equation for a circle, with the result $11^2 + (-1)^2 + 11a - b + c = 0 \Rightarrow 11a - b + c = -122$ (3). We now solve this system of three equations. Adding (1) + (2) ⇒ $10a + 24b = -172$ or $5a + 12b = -86$ (4) and adding (2) + (3) ⇒ $17a + 7b = -22$ (5). 7 × (4) ⇒ $35a + 84b = -602$ (6). 12 × (5) ⇒ $204a + 84b = -264$ (7). (7) − (6) ⇒ $169a = 338$ or $a = 2$. Substituting in (4) yields $5(2) + 12b = -86$; $12b = -96$; $b = -8$. Substituting in (3) produces $11(2) + 8 + c = -122$; $30 + c = -122$; $c = -152$; $a = 2$, $b = -8$, $c = -152$ and the equation of the circle is $x^2 + y^2 + 2x - 8y - 152 = 0$.

19. Organizing the information as in Example 5.18, we get $\begin{cases} 7L + 6M + 8S = 64 & (1) \\ 6L + 3M + 1S = 33 & (2) \\ 4L + 2M + 2S = 26 & (3) \end{cases}$

(2) × 3 ⇒ $12L + 6M + 2S = 66$ (4). Subtracting (4) − (3) ⇒ $8L + 4M = 40$ (5). Multiplying 4 × (3) ⇒ $16L + 8M + 8S = 104$ (6) and then subtracting (6) − (1) ⇒ $9L + 2M = 40$ (7). 2 × (7) ⇒ $18L + 4M = 80$ (8). (8) − (5) ⇒ $10L = 40$ or $L = 4$. Substituting into (5) yields $8(4) + 4M = 40$; $32 + 4M = 40$; $4M = 8$ and $M = 2$. Substituting into (2) yields $6(4)+3(2)+S = 33$; $24+6+S = 33$ or $30 + S = 33$ and $S = 3$. So, the company needs 4 large, 2 medium and 3 small trucks.

20.

	Light	Medium	Heavy
Crude oil A	10%	20%	70%
Crude oil B	30%	40%	30%
Crude oil C	43%	44%	13%

Organizing the information as in Example 5.17.

$$\begin{cases} 0.10A + 0.30B + 0.43C = 0.24(450{,}000) \\ \quad = 108{,}000 & (1) \\ 0.20A + 0.40B + 0.44C = 0.32(450{,}000) \\ \quad = 144{,}000 & (2) \\ 0.70A + 0.30B + 0.13C = 0.44(450{,}000) \\ \quad = 198{,}000 & (3) \end{cases}$$

2 × (1) ⇒ $0.20A + 0.60B + 0.86C = 216{,}000$ (4). (4) − (2) ⇒ $0.2B + 0.42C = 72{,}000$ (5). 7 × (1) ⇒ $0.70A + 2.10B + 3.01C = 756{,}000$ (6). (6) − (3) ⇒ $1.80B + 2.88C = 558{,}000$ (7). 9×(5) ⇒ $1.80B+3.78C = 648{,}000$ (8). (8)−(7) ⇒ $0.90C = 90{,}000 \Rightarrow C = 100{,}000$. Substituting in (5) ⇒ $0.2B + 42{,}000 = 72{,}000$; $0.2B = 30{,}000$; $B = 150{,}000$. Substituting in (1) ⇒ $0.1A + 0.30 \times (150{,}000) + 43{,}000 = 108{,}000$; $0.1A + 45{,}000 + 43{,}000 = 108{,}000$; $0.1A = 20{,}000 \Rightarrow A = 200{,}000$. Thus, we see that 200 000 t of A, 150 000 t of B, and 100 000 t of C should be mixed. Note $A + B + C = 450\,000$.

5.4 DETERMINANTS

1. $\begin{vmatrix} 2 & 3 \\ 4 & -1 \end{vmatrix} = 2(-1) - 4(3) = -2 - 12 = -14$

2. $\begin{vmatrix} 4 & -5 \\ 6 & 2 \end{vmatrix} = 4(2) - (6)(-5) = 8 + 30 = 38$

3. $\begin{vmatrix} 5 & -1 \\ 8 & 1 \end{vmatrix} = 5(1) - 8(-1) = 5 + 8 = 13$

4. $\begin{vmatrix} 9 & -1 \\ -2 & 3 \end{vmatrix} = 9(3) - (-2)(-1) = 27 - 2 = 25$

5. $\begin{vmatrix} 4 & 7 \\ -3 & 1 \end{vmatrix} = 4(1) - (-3)7 = 4 + 21 = 25$

6. $\begin{vmatrix} -1 & -4 \\ 0 & 1 \end{vmatrix} = (-1)(1) - 0(-4) = -1 - 0 = -1$

7. $\begin{vmatrix} -9 & \frac{1}{2} \\ 2 & 1 \end{vmatrix} = (-9)(1) - (2)\frac{1}{2} = -9 - 1 = -10$

8. $\begin{vmatrix} 6 & -3 \\ \frac{1}{3} & -\frac{2}{3} \end{vmatrix} = 6(-\frac{2}{3}) - \frac{1}{3}(-3) = -4 + 1 = -3$

9. (a) 2, (b) $\begin{vmatrix} 3 & -4 \\ -2 & 1 \end{vmatrix} = -5$, (c) Since $1 + 2 = 3$ is odd, the cofactor is the negative of the minor or $-(-5) = 5$

10. (a) 3, (b) $\begin{vmatrix} 2 & -1 \\ 1 & 1 \end{vmatrix} = 3$, (c) Since $2 + 1 = 3$ is odd, the cofactor is the negative of the minor, or -3

11. (a) 1, (b) minor = $\begin{vmatrix} 4 & -1 \\ 3 & -4 \end{vmatrix} = -16 + 3 = -13$, (c) $3 + 2$ is odd so the cofactor is $-(-13) = 13$.

12. (a) −4, (b) minor = $\begin{vmatrix} 4 & 2 \\ -2 & 1 \end{vmatrix} = 4+4 = 8$; (c) $2+3$ odd so cofactor is $-(-8) = 8$

13. (a) −2, (b) minor = $\begin{vmatrix} 2 & -1 \\ 7 & -4 \end{vmatrix} = -8 + 7 = -1$, (c) $3 + 1$ even so cofactor = minor = -1

14. (a) −1, (b) minor = $\begin{vmatrix} 3 & 7 \\ -2 & 1 \end{vmatrix} = 3 + 14 = 17$, (c) $1 + 3$ even so cofactor = 17 also.

15. $-3\begin{vmatrix} 2 & -1 \\ 1 & 1 \end{vmatrix} + 7\begin{vmatrix} 4 & -1 \\ -2 & 1 \end{vmatrix} - (-4)\begin{vmatrix} 4 & 2 \\ -2 & 1 \end{vmatrix} =$ $-3(3)+7(2)+4(8) = -9+14+32 = 37$

16. $-1\begin{vmatrix} 3 & 7 \\ -2 & 1 \end{vmatrix} - (-4)\begin{vmatrix} 4 & 2 \\ -2 & 1 \end{vmatrix} + 1\begin{vmatrix} 4 & 2 \\ 3 & 7 \end{vmatrix} =$ $-1(17)+4(8)+1(22) = -17+32+22 = 37$

17. 0, rows 1 and 3 are identical

18. 0, row 2 is a multiple of row 1

19. 0, because row 3 is a multiple of row 1 (or because column 2 is a multiple of column 1).

20. $\begin{vmatrix} 2 & 4 & 3 \\ 0 & 1 & 19 \\ 0 & 0 & -3 \end{vmatrix} = 2(1)(-3) = -6$

21. $\begin{vmatrix} 5 & 2 & 3 \\ hfill4 & -5 & -6 \\ -2 & 5 & -9 \end{vmatrix} \begin{matrix} 5 & 2 \\ hfill4 & -5 \\ hfill-2 & 5 \end{matrix} = 5(-5)(-9) +$ $(2)(-6)(-2) + 3(4)5 - (-2)(-5)3 - 5(-6)5 - (-9)4(2) = 225+24+60-30+150+72 = 501$

22. First add $-1(R3)$ to $R1$ to get $\begin{vmatrix} 3 & 0 & 0 \\ -2 & 0 & 6 \\ 6 & 3 & 3 \end{vmatrix}$ and then, interchange $C2+C3$ to get $-\begin{vmatrix} 3 & 0 & 0 \\ -2 & 6 & 0 \\ 6 & 3 & 3 \end{vmatrix} =$ $-(3)(6)(3) = -54$

23. $\begin{vmatrix} 4 & 3 & 9 \\ -4 & -6 & 16 \\ 2 & 3 & 2 \end{vmatrix} \overset{R1+R2}{=} \begin{vmatrix} 4 & 3 & 9 \\ 0 & -3 & 25 \\ 2 & 3 & 2 \end{vmatrix} \overset{-2R3+R1}{=}$ $\begin{vmatrix} 0 & -3 & 5 \\ 0 & -3 & 25 \\ 2 & 3 & 2 \end{vmatrix} \overset{-R2+R1}{=} \begin{vmatrix} 0 & 0 & -20 \\ 0 & -3 & 25 \\ 2 & 3 & 2 \end{vmatrix} =$ $-\begin{vmatrix} 2 & 3 & 2 \\ 0 & -3 & 25 \\ 0 & 0 & -20 \end{vmatrix} = -2(-3)(-20) = -120$

24. $\begin{vmatrix} 9 & 18 & -1 \\ 0 & -7 & 5 \\ 0 & 0 & 2 \end{vmatrix} = 9(-7)(2) = -126$

25. 252.86583

26. $\begin{vmatrix} \sqrt{3} & \frac{1}{2} & -5 \\ \frac{1}{4} & \sqrt{2} & 4 \\ \frac{2}{5} & 7 & -1 \end{vmatrix} = -55.94248$. Use $\sqrt{3} = 1.732$ and $\sqrt{2} = 1.414$.

27. -0.057525

28. $\begin{vmatrix} \sqrt{7} & 3 & \sqrt{5} \\ -1.4 & \sqrt{6} & 2 \\ \sqrt{8} & 3 & -6 \end{vmatrix} = -87.873415$. Use $\sqrt{6} = 2.4495$, $\sqrt{7} = 2.6458$, $\sqrt{5} = 2.2361$, and $\sqrt{8} = 2.8284$.

29. $\begin{vmatrix} r & s & t \\ a & b & c \\ x & y & z \end{vmatrix} = -(-5) = 5$

30. $\begin{vmatrix} a & b & c \\ r & s & t \\ x & y & z \end{vmatrix} = -\begin{vmatrix} r & s & t \\ a & b & c \\ x & y & z \end{vmatrix} = \begin{vmatrix} s & r & t \\ b & a & c \\ y & x & z \end{vmatrix} =$ $-\begin{vmatrix} s & t & r \\ b & c & a \\ y & z & x \end{vmatrix} = -(-5) = 5$

31. $\begin{vmatrix} a & b & c \\ 3r & 3s & 3t \\ -2x & -2y & -2z \end{vmatrix} = 3(-2)(-5) = 30$

32. $\begin{vmatrix} a & b & c \\ r & s & t \\ x & y & z \end{vmatrix} = \begin{vmatrix} a & b & c \\ r & s & t \\ x-r & y-s & z-t \end{vmatrix} =$ $\begin{vmatrix} a & b & c \\ r-a & s-b & c-t \\ x-r & y-s & z-t \end{vmatrix} = -5$

33. $\begin{vmatrix} 1 & 2 & 3 & 4 \\ 5 & 0 & 7 & 0 \\ 9 & 10 & 11 & 12 \\ 13 & 14 & 15 & 16 \end{vmatrix} \begin{matrix} -3(R1)+R3 \\ -4(R1)+R4 \end{matrix} = \begin{vmatrix} 1 & 2 & 3 & 4 \\ 5 & 0 & 7 & 0 \\ 6 & 4 & 2 & 0 \\ 9 & 6 & 3 & 0 \end{vmatrix} =$ 0, since $R4 = 1.5 \times R3$.

34. $\begin{vmatrix} 2 & 0 & 3 & -1 \\ 4 & 0 & -2 & 5 \\ 9 & -5 & 0 & -2 \\ 0 & 3 & 1 & -6 \end{vmatrix} = -(-5)\begin{vmatrix} 2 & 3 & -1 \\ 4 & -2 & 5 \\ 0 & 1 & -6 \end{vmatrix} +$ $3\begin{vmatrix} 2 & 3 & -1 \\ 4 & -2 & 5 \\ 9 & 0 & -2 \end{vmatrix} = 5(82) + 3(149) = 410 +$ $447 = 857$ since $\begin{vmatrix} 2 & 3 & -1 \\ 4 & -2 & 5 \\ 0 & 1 & -6 \end{vmatrix} = 2\begin{vmatrix} -2 & 5 \\ 1 & -6 \end{vmatrix} -$ $4\begin{vmatrix} 3 & -1 \\ 1 & -6 \end{vmatrix} = 2(12-5) - 4(-18+1) = 2(7) -$

$4(-17) = 14 + 68 = 82$. $\begin{vmatrix} 2 & 3 & -1 \\ 4 & -2 & 5 \\ 9 & 0 & -2 \end{vmatrix} = -3\begin{vmatrix} 4 & 5 \\ 9 & -2 \end{vmatrix} - 2\begin{vmatrix} 2 & -1 \\ 9 & -2 \end{vmatrix} = -3(-8-45)-2(-4+9) = -3(-53) - 2(5) = 159 - 10 = 149$.

35. See *Computer Programs* in main text.

36. $\left|\begin{array}{ccc|cc} 5 & 2 & 3 & 5 & 2 \\ 4 & -5 & -6 & 4 & -5 \\ -2 & 5 & -9 & -2 & 5 \end{array}\right. = 5(-5)(-9) + 2(-6)(-2) + 3(4)(5) - (-2)(-5)(3) - 5(-6)(5) - (-9)(4)(2) = 225+24+60-30+150+72 = 501$

$\left|\begin{array}{ccc|cc} 9 & 3 & 3 & 9 & 3 \\ -2 & 0 & 6 & -2 & 0 \\ 6 & 3 & 3 & 6 & 3 \end{array}\right. = 9(0)(3) + 3(6)(6) + 3(-2)(3) - 6(0)(3) - (3)(6)(9) - 3(-2)(3) = 0 + 108 - 18 - 0 - 162 + 18 = -54$

$\left|\begin{array}{ccc|cc} 4 & 3 & 9 & 4 & 3 \\ -4 & -6 & 16 & -4 & -6 \\ 2 & 3 & 2 & 2 & 3 \end{array}\right. = 4(-6)(2)+3(16)(2)+9(-4)(3)-2(-6)(9)-3(16)(4)-2(-4)(3) = -48+96-108+108-192+24 = -120$

$\left|\begin{array}{ccc|cc} 9 & 18 & -1 & 9 & 18 \\ 0 & -7 & 5 & 0 & -7 \\ 0 & 0 & 2 & 0 & 0 \end{array}\right. = 9(-7)(2) + 18(5)(0) + (-1)(0)(0) - 0(-7)(-1) - 0(5)(9) - 2(0)(18) = -126$

37. See *Computer Programs* in main text.

5.5 USING CRAMER'S RULE TO SOLVE SYSTEMS OF LINEAR EQUATIONS

1. $D = \begin{vmatrix} 2 & 1 \\ 3 & -2 \end{vmatrix} = -4 - 3 = -7, D_x = \begin{vmatrix} 7 & 1 \\ -7 & -2 \end{vmatrix} = -14+7 = -7, D_y = \begin{vmatrix} 2 & 7 \\ 3 & -7 \end{vmatrix} = -14-21 = -35$, $x = \frac{-7}{-7} = 1, y = \frac{-35}{-7} = 5$

2. $D = \begin{vmatrix} 3 & 1 \\ 2 & -3 \end{vmatrix} = -9-2 = -11, D_x = \begin{vmatrix} 3 & 1 \\ 13 & -3 \end{vmatrix} = -9 - 13 = -22, D_y = \begin{vmatrix} 3 & 3 \\ 2 & 13 \end{vmatrix} = 39 - 6 = 33$, $x = \frac{-22}{-11} = 2, y = \frac{33}{-11} = -3$

3. $D = \begin{vmatrix} 4 & 3 \\ 3 & -2 \end{vmatrix} = -8 - 9 = -17, D_x = \begin{vmatrix} 4 & 3 \\ -14 & -2 \end{vmatrix} = -8 + 42 = 34, D_y = \begin{vmatrix} 4 & 4 \\ 3 & -14 \end{vmatrix} = -56 - 12 = -68, x = \frac{34}{-17} = -2, y = \frac{-68}{-17} = 4$

4. $D = \begin{vmatrix} \frac{1}{2} & -\frac{2}{3} \\ \frac{1}{3} & 2 \end{vmatrix} = 1 + \frac{2}{9} = \frac{11}{9}, D_x = \begin{vmatrix} \frac{3}{4} & -\frac{2}{3} \\ \frac{5}{6} & 2 \end{vmatrix} = \frac{3}{2} + \frac{5}{9} = \frac{37}{18}, D_y = \begin{vmatrix} \frac{1}{2} & \frac{3}{4} \\ \frac{1}{3} & \frac{5}{6} \end{vmatrix} = \frac{5}{12} - \frac{3}{12} = \frac{2}{12} = \frac{1}{6}$, $x = \frac{37/18}{11/9} = \frac{37}{18} \times \frac{9}{11} = \frac{37}{22} \approx 1.6818, y = \frac{1/6}{11/9} = \frac{1}{6} \cdot \frac{9}{11} = \frac{3}{22} \approx 0.13636$

5. $D = \begin{vmatrix} 1.2 & 3.7 \\ 4.3 & -5.2 \end{vmatrix} = -22.15, D_x = \begin{vmatrix} 9.1 & 3.7 \\ 8.3 & -5.2 \end{vmatrix} = -78.03, D_y = \begin{vmatrix} 1.2 & 9.1 \\ 4.3 & 8.3 \end{vmatrix} = -29.17, x = \frac{-78.03}{-22.15} \approx 3.5228, y = \frac{-29.17}{-22.15} = 1.3169$

6. $D = \begin{vmatrix} 0.02 & -1.22 \\ -0.14 & 0.32 \end{vmatrix} = -0.1644, D_x = \begin{vmatrix} 3.74 & -1.22 \\ -1.32 & 0.32 \end{vmatrix} = -0.4136, x = \frac{-4.136}{-.1644} \approx +2.5158, D_y = \begin{vmatrix} .02 & 3.74 \\ -.14 & -1.32 \end{vmatrix} = 0.4972, y = \frac{.4972}{-.1644} = -3.0243$

7. $D = \begin{vmatrix} 2.5 & 3.8 \\ .5 & .76 \end{vmatrix} = 0$, inconsistent system, no solution.

8. $D = \begin{vmatrix} 4.3 & -2.7 \\ 3.5 & 4.2 \end{vmatrix} = 27.51, D_x = \begin{vmatrix} -4 & -2.7 \\ -1 & 4.2 \end{vmatrix} = -19.5, D_y = \begin{vmatrix} 4.3 & -4 \\ 3.5 & -1 \end{vmatrix} = 9.7, x = \frac{-19.5}{27.51} \approx -.7088, y = \frac{9.7}{27.51} \approx 0.3526$

9. $D = \begin{vmatrix} -5.3 & 2.1 \\ 6.2 & -3.1 \end{vmatrix} = 3.41, D_x = \begin{vmatrix} 4.6 & 2.1 \\ 6 & -3.1 \end{vmatrix} =$

-26.86, $D_y = \begin{vmatrix} -5.3 & 4.6 \\ 6.2 & 6 \end{vmatrix} = -60.32$, $x = \frac{-26.86}{3.41} \approx -7.8768$, $y = \frac{-60.32}{3.41} \approx -17.689$

10. $D = \begin{vmatrix} 1.3 & -0.8 \\ 1.7 & -0.7 \end{vmatrix} = .45$, $D_x = \begin{vmatrix} 2.9 & -0.8 \\ 0.4 & -0.7 \end{vmatrix} = -1.71$, $D_y = \begin{vmatrix} 1.3 & 2.9 \\ 1.7 & 0.4 \end{vmatrix} = -4.41$, $x = \frac{-1.71}{.45} = -3.8$, $y = \frac{-4.41}{.45} = -9.8$

11. $D = \begin{vmatrix} 6 & 2.5 \\ 13.8 & 5.75 \end{vmatrix} = 0$, inconsistent system, no solution.

12. $D = \begin{vmatrix} 3.5 & -6.5 \\ 5.5 & 3.3 \end{vmatrix} = 47.3$, $D_x = \begin{vmatrix} 22.45 & -6.5 \\ -1.21 & 3.3 \end{vmatrix} = 66.22$, $D_y = \begin{vmatrix} 3.5 & 22.45 \\ 5.5 & -1.21 \end{vmatrix} = -127.71$, $x = \frac{66.22}{47.3} = 1.4$, $y = \frac{-127.71}{47.3} = -2.7$

13. $D = \begin{vmatrix} 4 & 3 \\ 3 & -2 \end{vmatrix} = -8 - 9 = -17$, $D_x = \begin{vmatrix} 2 & 3 \\ -24 & -2 \end{vmatrix} = -4 + 72 = 68$, $D_y = \begin{vmatrix} 4 & 2 \\ 3 & -24 \end{vmatrix} = -96 - 6 = -102$, $x = \frac{68}{-17} = -4$, $y = \frac{-102}{-17} = 6$

14. $D = \begin{vmatrix} 1 & 3 & -1 \\ 5 & -7 & 1 \\ 2 & -1 & -2 \end{vmatrix} = 42$, $D_x = \begin{vmatrix} 7 & 3 & -1 \\ 3 & -7 & 1 \\ 0 & -1 & -2 \end{vmatrix} = 126$, $D_y = \begin{vmatrix} 1 & 7 & -1 \\ 5 & 3 & 1 \\ 2 & 0 & -2 \end{vmatrix} = 84$, $D_z = \begin{vmatrix} 1 & 3 & 7 \\ 5 & -7 & 3 \\ 2 & -1 & 0 \end{vmatrix} = 84$, $x = \frac{126}{42} = 3$, $y = \frac{84}{42} = 2$, $z = \frac{84}{42} = 2$, $(3, 2, 2)$

15. $D = \begin{vmatrix} 1 & -.5 & 1.5 \\ 1 & .5 & -3.75 \\ -3 & -2.5 & 4 \end{vmatrix} = -12.5$, $D_x = \begin{vmatrix} 2 & -.5 & 1.5 \\ .25 & .5 & -3.75 \\ -.75 & -2.5 & 4 \end{vmatrix} = -16.03125$, $D_y = \begin{vmatrix} 1 & 2 & 1.5 \\ 1 & .25 & -3.75 \\ -3 & -.75 & 4 \end{vmatrix} = 12.6875$, $D_z = \begin{vmatrix} 1 & -.5 & 2 \\ 1 & .5 & .25 \\ -3 & -2.5 & -.75 \end{vmatrix} = -1.75$, $x = \frac{-16.03125}{-12.5} = 1.2825$, $y = \frac{12.6875}{-12.5} = -1.015$, $z = \frac{-1.75}{-12.5} = 0.14$

16. $D = \begin{vmatrix} 1 & 1 & -1 \\ .5 & -3.5 & 2.4 \\ 1.5 & 4.5 & -3.6 \end{vmatrix} = -0.3$, $D_x = \begin{vmatrix} -4.7 & 1 & -1 \\ 17.4 & -3.5 & 2.4 \\ -21.6 & 4.5 & -3.6 \end{vmatrix} = -.36$, $x = \frac{-.36}{-.3} = 1.2$, $D_y = \begin{vmatrix} 1 & -4.7 & -1 \\ .5 & 17.4 & 2.4 \\ 1.5 & -21.6 & -3.6 \end{vmatrix} = .72$, $y = \frac{.72}{-.3} = -2.4$, $D_z = \begin{vmatrix} 1 & 1 & -4.7 \\ .5 & -3.5 & 17.4 \\ 1.5 & 4.5 & -21.6 \end{vmatrix} = -1.05$, $z = \frac{-1.05}{-.3} = 3.5$. The solution is $x = 1.2$, $y = -2.4$, and $z = 3.5$.

17. See *Computer Programs* in main text.

18. $D = \begin{vmatrix} 1 & 1 & 1 \\ 8 & 0 & -10 \\ 6 & -3 & 0 \end{vmatrix} = -114$, $D_{I_1} = \begin{vmatrix} 0 & 1 & 1 \\ 8 & 0 & -10 \\ 12 & -3 & 0 \end{vmatrix} = -144$, $D_{I_2} = \begin{vmatrix} 1 & 0 & 1 \\ 8 & 8 & -10 \\ 6 & 12 & 0 \end{vmatrix} = 168$, $D_{I_3} = \begin{vmatrix} 1 & 1 & 0 \\ 8 & 0 & 8 \\ 6 & -3 & 12 \end{vmatrix} = -24$, $I_1 = \frac{-144}{-114} = \frac{24}{19}$, $I_2 = \frac{168}{-114} = -\frac{28}{19}$, $I_3 = \frac{-24}{-144} = \frac{4}{19}$.

19. First, we make the substitutions for t and s to obtain the following system of three equations. $\begin{cases} s_0 + 2v_0 + 2a = 212 \\ crs_0 + 5v_0 + \frac{25}{2}a = 156.5 \\ crs_0 + 7v_0 + \frac{49}{2}a = 70.5 \end{cases}$. $D = \begin{vmatrix} 1 & 2 & 2 \\ 1 & 5 & 12.5 \\ 1 & 7 & 24.5 \end{vmatrix} = 15$, $D_{s_0} = \begin{vmatrix} 212 & 2 & 2 \\ 156.5 & 5 & 12.5 \\ 70.5 & 7 & 24.5 \end{vmatrix} = 3000$, $D_{v_0} = \begin{vmatrix} 1 & 212 & 2 \\ 1 & 156.5 & 12.5 \\ 1 & 70.5 & 24.5 \end{vmatrix} = 237$, $D_a = \begin{vmatrix} 1 & 2 & 212 \\ 1 & 5 & 156.5 \\ 1 & 7 & 70.5 \end{vmatrix} = -147$, $s_0 = \frac{3000}{15} = 200$, $v_0 = \frac{237}{15} = 15.8$, $a = \frac{-147}{15} = -9.8$. Thus, $s = 200 + 15.8t - 4.9t^2$.

20. $D = \begin{vmatrix} 1 & 1 & 1 \\ 1 & -2 & 0 \\ -4 & 0 & 1 \end{vmatrix} = -11$, $D_A = \begin{vmatrix} 100 & 1 & 1 \\ 0 & -2 & 0 \\ 0 & 0 & 1 \end{vmatrix} = -200$, $D_B = \begin{vmatrix} 1 & 100 & 1 \\ 1 & 0 & 0 \\ -4 & 0 & 1 \end{vmatrix} = -100\begin{vmatrix} 1 & 0 \\ -4 & 1 \end{vmatrix} = -100$, $D_C = \begin{vmatrix} 1 & 1 & 100 \\ 1 & -2 & 0 \\ -4 & 0 & 0 \end{vmatrix} = 100\begin{vmatrix} 1 & -2 \\ -4 & 0 \end{vmatrix} = -800$, $A = \frac{-200}{-11} \approx 18.18$, $B = \frac{100}{11} \approx 9.09$, $C = \frac{800}{11} \approx 72.73$. Thus, the alloy consists of 18.18% metal A, 9.09% metal B, and 72.73% metal C.

21. Let A, B, and C represent the three distances. Then we get the equations $B = 90 + A$ and $C = 130 + B$, which leads to the system of three equations: $\begin{cases} -A + B = 90 \\ -B + C = 130 \\ A + B + C = 1900 \end{cases}$. $D = \begin{vmatrix} -1 & 1 & 0 \\ 0 & -1 & 1 \\ 1 & 1 & 1 \end{vmatrix} = 3$, $D_A = \begin{vmatrix} 90 & 1 & 0 \\ 130 & -1 & 1 \\ 1900 & 1 & 1 \end{vmatrix} = 1590$, $D_B = \begin{vmatrix} -1 & 90 & 0 \\ 0 & 130 & 1 \\ 1 & 1900 & 1 \end{vmatrix} = 1860$, $D_C = \begin{vmatrix} -1 & 1 & 90 \\ 0 & -1 & 130 \\ 1 & 1 & 1900 \end{vmatrix} = 2250$. $A = \frac{1590}{3} = 530$, $B = \frac{1860}{3} = 620$, $C = \frac{2250}{3} = 750$. Hence, The car traveled 530 km on fuel A, 620 km on fuel B, and 750 km on fuel C.

22. $D = \begin{vmatrix} 0 & \frac{6}{7} & -\frac{2}{3} \\ -1 & -\frac{3}{7} & \frac{1}{3} \\ 0 & \frac{2}{7} & \frac{2}{3} \end{vmatrix} = \frac{1}{21}\begin{vmatrix} 0 & 6 & -2 \\ -1 & -3 & 1 \\ 0 & 2 & 2 \end{vmatrix} = \frac{1}{21}(16) = \frac{16}{21}$, $D_{F_A} = \begin{vmatrix} 2{,}000 & \frac{6}{7} & -\frac{2}{3} \\ 0 & -\frac{3}{7} & \frac{1}{3} \\ 1{,}200 & \frac{2}{7} & \frac{2}{3} \end{vmatrix} = \frac{1}{21}\begin{vmatrix} 2{,}000 & 6 & -2 \\ 0 & -3 & 1 \\ 1{,}200 & 2 & 2 \end{vmatrix} = \frac{1}{21}(-16{,}000) = -\frac{16{,}000}{21}$, $D_{F_B} = \begin{vmatrix} 0 & 2{,}000 & -\frac{2}{3} \\ -1 & 0 & \frac{1}{3} \\ 0 & 1{,}200 & \frac{2}{3} \end{vmatrix} = \frac{1}{3}\begin{vmatrix} 0 & 2{,}000 & -2 \\ -1 & 0 & 1 \\ 0 & 1{,}200 & 2 \end{vmatrix} = \frac{1}{3}(6{,}400) = \frac{6{,}400}{3}$, $D_{F_C} = \begin{vmatrix} 0 & \frac{6}{7} & 2{,}000 \\ -1 & -\frac{3}{7} & 0 \\ 0 & \frac{2}{7} & 1{,}200 \end{vmatrix} = \frac{1}{7}\begin{vmatrix} 0 & 6 & 2{,}000 \\ -1 & -3 & 0 \\ 0 & 2 & 1{,}200 \end{vmatrix} = \frac{1}{7}(3{,}200) = \frac{3{,}200}{7}$. Thus, $F_A = \frac{-16{,}000/21}{16/21} = \frac{-16{,}000}{16} = -1{,}000$, $F_B = \frac{6{,}400/3}{16/21} = 2{,}800$, $F_C = \frac{3{,}200/7}{16/21} = 600$. The three forces are $F_A = -1{,}000$ N, $F_B = 2{,}800$ N, $F_C = 600$ N.

CHAPTER 5 REVIEW

1. $\begin{vmatrix} 2 & 5 \\ -3 & 6 \end{vmatrix} = 12 + 15 = 27$

2. $\begin{vmatrix} 4 & -7 \\ 3 & -6 \end{vmatrix} = -24 + 21 = -3$

3. $\begin{vmatrix} 5 & 9 \\ 8 & 1 \end{vmatrix} = 5 - 72 = -67$

4. $\begin{vmatrix} 3 & -12 \\ 0 & 5 \end{vmatrix} = 15 - 0 = 15$

5. $\begin{vmatrix} 6 & -2 \\ -4 & 3 \end{vmatrix} = 6 \times 3 - (-4)(-2) = 18 - 8 = 10$

6. $\begin{vmatrix} 8 & 9 \\ -1 & -2 \end{vmatrix} = -16 + 9 = -7$

7. $\begin{vmatrix} 9 & 2 & 1 \\ 3 & -4 & 6 \\ 7 & 2 & 1 \end{vmatrix} = 9(-4)1 + 2 \times 6 \times 7 + 1 \times 3 \times 2 - 7(-4)1 - 2(6)9 - 1(3)(2) = -36 + 84 + 6 + 28 - 108 - 6 = -32$

8. $\begin{vmatrix} 9 & -2 & 3 \\ 4 & -5 & 2 \\ 3 & 2 & 1 \end{vmatrix} = 9(-5)1 + (-2)2(3) + 3(4)2 - 3(-5)3 - 2(2)(9) - 1(4)(-2) = -45 - 12 + 24 + 45 - 36 + 8 = -16$

9. (a) $m = \frac{1-3}{5-(-2)} = \frac{-2}{7}$, (b) $y - 1 = \frac{-2}{7}(x - 5)$ or $y - 3 = \frac{-2}{7}(x + 2)$

10. (a) $m = \frac{-2-(-4)}{3-8} = \frac{2}{-5} = \frac{-2}{5}$. (b) $y + 2 = \frac{-2}{5}(x - 3)$ or $y + 4 = \frac{-2}{5}(x - 8)$

11. (a) $m = \frac{4-(-1)}{-2-9} = \frac{5}{-11} = \frac{-5}{11}$, (b) $y - 4 = \frac{-5}{11}(x + 2)$ or $y + 1 = \frac{-5}{11}(x - 9)$

12. (a) $m = \frac{-1-(-3)}{2-(-4)} = \frac{2}{6} = \frac{1}{3}$, (b) $y + 1 = \frac{1}{3}(x - 2)$ or $y + 3 = \frac{1}{3}(x + 4)$

13. $4x - 2y = 6$, $-2y = -4x + 6$, $y = 2x - 3$

14. $9x + 3y = 5$, $3y = -9x + 5$, $y = -3x + \frac{5}{3}$

15. $4x = 5y - 8$, $5y = 4x + 8$, $y = \frac{4}{5}x + \frac{8}{5}$

16. $7x + 3y - 8 = 0$, $3y = -7x + 8$, $y = \frac{-7}{3}x + \frac{8}{3}$

17.

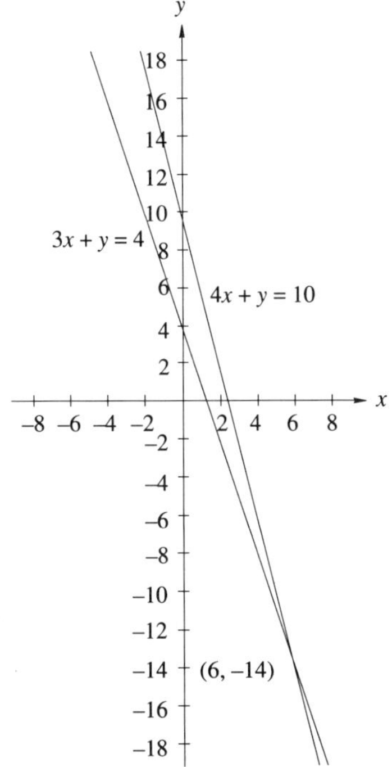

$x = 6, y = -14$

18.

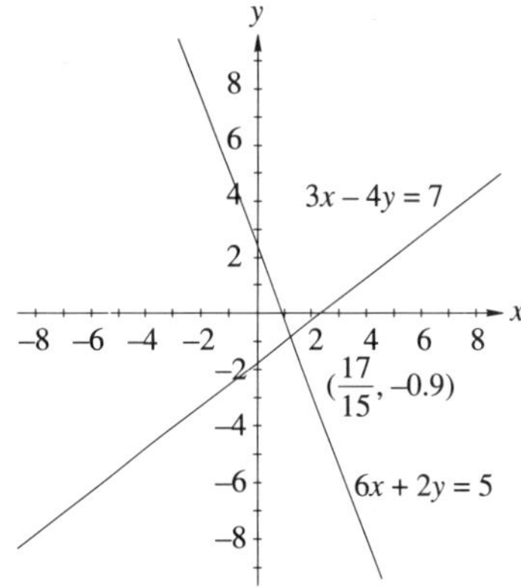

$x = \frac{17}{15} \approx 1.13, y = -0.9$

19. $\begin{cases} 3x + y = 5 \\ 4x - y = 16 \end{cases}$, $y = 5 - 3x$; $4x - (5 - 3x) = 16$; $7x - 5 = 16$; $7x = 21$; $x = 3$, $y = 5 - 3(3) = 5 - 9 = -4$; $(3, -4)$

20. $\begin{cases} 3x - 2y = 4 \\ x + 8y = 3 \end{cases}$, solving (2) for x, we get $x = 3 - 8y$ (3). Substituting in (1) yields $3(3 - 8y) - 2y = 4$ or $9 - 24y - 2y = 4 \Rightarrow -26y = -5$ and so $y = \frac{5}{26} = 0.1923$. Substituting this value of y in (3) produces $x = 3 - 8(0.1923) = 1.4615$. The solution is $(1.4615, 0.1923)$

21. $\begin{cases} 4x - y = 7 \\ 3x + 2y = 5 \end{cases}$. Solving (1) for y, we get $y = 4x - 7$ (3). Substituting this in (2) yields $3x + 2(4x - 7) = 5 \Rightarrow 3x + 8x - 14 = 5$ or $11x = 19$ and so, $x = \frac{19}{11} = 1.7273$. Substituting this value of x in (3), we see that $y = 4(\frac{19}{11}) - 7 = \frac{76}{11} - \frac{77}{11} = \frac{-1}{11} = -0.0909$. Thus, the solution is $\left(\frac{19}{11}, \frac{-1}{11}\right) = (1.7273, -0.0909)$.

22. $\begin{cases} 6x + y - 5 = 0 \\ 4y - 3x = -7 \end{cases}$. $y = 5 - 6x$; $4(5 - 6x) - 3x = -7$; $20 - 24x - 3x = -7$; $-27x = -27$, $x = 1$; $y = 5 - 6(1) = 5 - 6 = -1$

23. $\begin{cases} x - y = -2 \\ x + y = 8 \end{cases}$. Adding gives $2x = 6$; $x = 3$; $3 - y = -2$; $-y = -5$; $y = 5$

24. $\begin{cases} 6x + 5y = 7 \\ 3x - 7y = 13 \end{cases}$. (2) $\times$ 2 $\Rightarrow 6x - 14y = 26$ (3); (1) $-$ (3) $\Rightarrow 19y = -19$; $y = -1$; $6x + 5(-1) = 7$;

$6x = 12; x = 2$

25. $\begin{cases} x + \frac{1}{2}y = 2 \\ 3x - y = 1 \end{cases}$. $(1) \times 2 \Rightarrow 2x + y = 4$ (3); $(3) + (2) \Rightarrow 5x = 5; x = 1; 3(1) - y = 1; 3 - y = 1; -y = -2; y = 2$

26. $\begin{cases} \frac{1}{6}x + \frac{1}{4}y = \frac{1}{3} \\ \frac{1}{4}x - \frac{1}{2}y = 1 \end{cases}$. $(1) \times 12 \Rightarrow 2x + 3y = 4$ (3); $(2) \times 8 \Rightarrow 2x - 4y = 8$ (4); $(3) - (4) \Rightarrow 7y = -4$; $y = -\frac{4}{7}$; $2x + 3\left(-\frac{4}{7}\right) = 4$; $2x - \frac{12}{7} = 4$; $2x = \frac{40}{7}$; $x = \frac{20}{7}$

27. $\begin{matrix} 3x + 2y + 3z = -7 & (1) \\ 5x - 3y + 2z = -4 & (2) \\ 7x + 4y + 5z = 2 & (3) \end{matrix}$. Begin by multiplying $(1) \times 3 \Rightarrow 9x + 6y + 9z = -21$ (4). Next, multiply $(2) \times 2 \Rightarrow 10x - 6y + 4z = -8$ (5). Add: $(4) + (5) \Rightarrow 19x + 13z = -29$ (6). Now, multiply $(1) \times 2 \Rightarrow 6x + 4y + 6z = -14$ (7), and subtract $(7) - (3) \Rightarrow -x + z = -16$ (8). If you now multiply $19 \times (8) \Rightarrow -19x + 19z = -304$ (9), and add $(9) + (6) \Rightarrow 32z = -333$. As a result, we obtain $z = -\frac{333}{32} = -10.40625$. From (8), we obtain $x = z + 16$, and substituting for z yields $x = 16 - 10.40625 = 5.59375$. Substituting these values for x and z in (1) produces $3(5.59375) + 2y + 3(-10.40625) = -7 \Rightarrow 2y = -7 - 3(5.59375) + 3(10.40625)$, and so $2y = 7.4375$ or $y = 3.71875$. Thus, the solution is $x = 5.59375$, $y = 3.71875$, $z = -10.40625$.

28. $\begin{cases} x + y + z = 2.7 & (1) \\ 6x + 7y - 5z = -8.8 & (2) \\ 10x + 16y - 3z = -6.4 & (3) \end{cases}$. First multiply $(1) \times 6 \Rightarrow 6x + 6y + 6z = 16.2$ (4), and subtract: $(4) - (2) \Rightarrow -y + 11z = 25$ (5). Next, multiply $(1) \times 10 \Rightarrow 10x + 10y + 10z = 27$ (6), and subtract: $(3) - (6) \Rightarrow 6y - 13z = -33.4$ (7). Finally, multiply $(5) \times 6 \Rightarrow -6y + 66z = 150$ (8), and add $(7) + (8) \Rightarrow 53z = 116.6$, and so $z = 2.2$. From(5), $y = 11z - 25 = 11(2.2) - 25 = -0.8$. From (1), $x = 2.7 - y - z = 2.7 - (-.8) - 2.2 = 1.3$, and the solution is $(1.3, -0.8, 2.2)$

29. $D = \begin{vmatrix} 3 & -2 \\ 5 & 2 \end{vmatrix} = 6 + 10 = 16$; $D_x = \begin{vmatrix} 4 & -2 \\ 12 & 2 \end{vmatrix} = 8 + 24 = 32$; $D_y = \begin{vmatrix} 3 & 4 \\ 5 & 12 \end{vmatrix} = 36 - 20 = 16$; $x = \frac{32}{16} = 2$; $y = \frac{16}{16} = 1$

30. $D = \begin{vmatrix} 4 & 3 \\ 2 & -5 \end{vmatrix} = -20 - 6 = -26$; $D_x = \begin{vmatrix} 27 & 3 \\ -19 & -5 \end{vmatrix} = -78$; $D_y = \begin{vmatrix} 4 & 27 \\ 2 & -19 \end{vmatrix} = -130$; $x = \frac{-78}{-26} = 3$; $y = \frac{-130}{-26} = 5$

31. $D = \begin{vmatrix} 4 & -3 \\ 11 & 4 \end{vmatrix} = 16 + 33 = 49$; $D_x = \begin{vmatrix} 9 & -3 \\ 7 & 4 \end{vmatrix} = 57$; $x = \frac{57}{49}$; $D_y = \begin{vmatrix} 4 & 9 \\ 11 & 7 \end{vmatrix} = 28 - 99 = -71$; $y = \frac{-71}{49}$

32. $D = \begin{vmatrix} 2 & -5 \\ 4 & 6 \end{vmatrix} = 12 + 20 = 32$; $D_x = \begin{vmatrix} -8 & -5 \\ 17 & 6 \end{vmatrix} = 37$; $x = \frac{37}{32}$; $D_y = \begin{vmatrix} 2 & -8 \\ 4 & 17 \end{vmatrix} = 34 + 32 = 66$; $y = \frac{66}{32} = \frac{33}{16}$

33. $D = \begin{vmatrix} 5 & 3 & -2 \\ 3 & -4 & 3 \\ 1 & 6 & -4 \end{vmatrix} = -9$; $D_x = \begin{vmatrix} 5 & 3 & -2 \\ 13 & -4 & 3 \\ -8 & 6 & -4 \end{vmatrix} = -18$; $D_y = \begin{vmatrix} 5 & 5 & -2 \\ 3 & 13 & 3 \\ 1 & -8 & -4 \end{vmatrix} = 9$; $D_z = \begin{vmatrix} 5 & 3 & 5 \\ 3 & -4 & 13 \\ 1 & 6 & -8 \end{vmatrix} = -9$; $x = \frac{-18}{-9} = 2$; $y = \frac{9}{-9} = -1$; $z = \frac{-9}{-9} = 1$. $(2, -1, 1)$

34. $D = \begin{vmatrix} 5 & -2 & 3 \\ 6 & -3 & 4 \\ -4 & 4 & -9 \end{vmatrix} = 15$; $D_x = \begin{vmatrix} 6 & -2 & 3 \\ 10 & -3 & 4 \\ 4 & 4 & -9 \end{vmatrix} = 10$; $D_y = \begin{vmatrix} 5 & 6 & 3 \\ 6 & 10 & 4 \\ -4 & 4 & -9 \end{vmatrix} = -110$; $D_z = \begin{vmatrix} 5 & -2 & 6 \\ 6 & -3 & 10 \\ -4 & 4 & 4 \end{vmatrix} = -60$; $x = \frac{10}{15} = \frac{2}{3}$; $y = \frac{-110}{15} = -7.\overline{3}$; $z = \frac{-60}{15} = -4$. $\left(\frac{2}{3}, -7\frac{1}{3}, -4\right)$

35. Let x be the amount of Colombian Supreme and y the amount of Mocha Java coffee. Then we obtain two equations (1) $x + y = 50$ and (2)

$4.99x + 5.99y = 50 \times 5.39$ or $499x + 599y = 26950$. Multiplying (1) by 499 gives $499x + 499y = 24950$. Subtracting yields $100y = 2000$ or $y = 20$. This leaves 30 for x. Solution: 30 pounds of Colombian Supreme and 20 pounds of Mocha Java coffee.

36. Organizing the information in chart form we have

computer type	Chip A	Chip A
PC x	4	11
BC y	9	6

$4x + 9y = 670$; $11x + 6y = 1055$. Using Cramer's Rule $D = \begin{vmatrix} 4 & 9 \\ 11 & 6 \end{vmatrix} = 24 - 99 = -75$, $D_x = \begin{vmatrix} 670 & 9 \\ 1055 & 6 \end{vmatrix} = -5475$, $D_y = \begin{vmatrix} 4 & 670 \\ 11 & 1055 \end{vmatrix} = -3150$, $x = \frac{-5475}{-75} = 73$, $y = \frac{-3150}{-75} = 42$. They should produce 73 PC and 42 BC computers.

37. Let x be the number of 1 room offices and y be the number of 2 room offices and z be the number of 3 room offices; then $\begin{cases} x + y + z = 66 \\ x + 2y + 3z = 146 \\ 300x + 520y + 730z = 37{,}160 \end{cases}$.

Using Cramer's Rule $D = \begin{vmatrix} 1 & 1 & 1 \\ 1 & 2 & 3 \\ 300 & 520 & 730 \end{vmatrix} = -10$, $D_x = \begin{vmatrix} 66 & 1 & 1 \\ 146 & 2 & 3 \\ 37{,}160 & 520 & 730 \end{vmatrix} = -100$, $D_y = \begin{vmatrix} 1 & 66 & 1 \\ 1 & 146 & 3 \\ 300 & 37{,}160 & 730 \end{vmatrix} = -320$, $D_z = \begin{vmatrix} 1 & 1 & 66 \\ 1 & 2 & 146 \\ 300 & 520 & 37{,}160 \end{vmatrix} = -240$, $x = \frac{-100}{-10} = 10$, $y = \frac{-320}{-10} = 32$, $z = \frac{-240}{-10} = 24$. There are 10 small, 32 medium, and 24 large offices.

CHAPTER 5 TEST

1. (a) $m = \frac{6-2}{5--4} = \frac{4}{9}$. (b) $y - 2 = \frac{4}{9}(x + 4)$ or $y - 6 = \frac{4}{9}(x - 5)$; $y - 2 = \frac{4}{9}x + \frac{16}{9}$ and so $y = \frac{4}{9}x + \frac{34}{9}$

2. (a) $-5y = -6x + 12$; $y = \frac{6}{5}x - \frac{12}{5}$, (b) slope: $m = \frac{6}{5}$, (c) y-intercept is $-\frac{12}{5}$, (d)

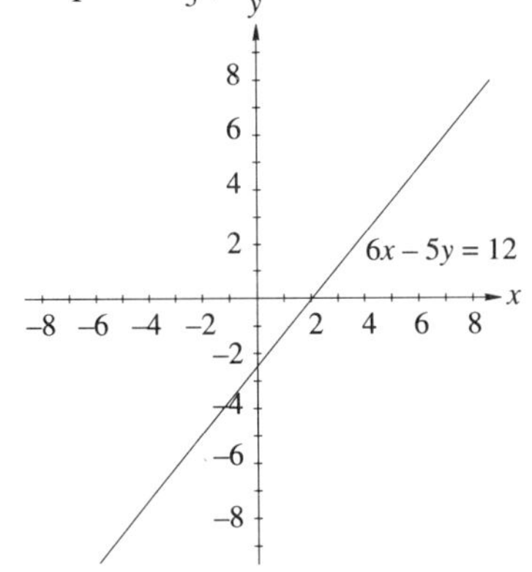

3. $\begin{cases} 2x + 3y = 1 & (1) \\ -3x + 6y = 16 & (2) \end{cases}$.
$(1) \times 2 \Rightarrow 4x + 6y = 2$ (3), (3) − (2) $\Rightarrow 7x = -14$, and we see that $x = -2$. Substitution in (1) yields $2(-2) + 3y = 1$ or $-4 + 3y = 1 \Rightarrow 3y = 5$ and so, $y = \frac{5}{3}$. Thus, the solution is $\left(-2, \frac{5}{3}\right)$.

4. $\begin{vmatrix} 4 & -2 \\ 5 & 6 \end{vmatrix} = 4(6) - 5(-2) = 24 + 10 = 34$

5. $\begin{vmatrix} 4 & 2 & -1 \\ 3 & 0 & 4 \\ -3 & 1 & 1 \end{vmatrix}$. Using the 2nd column and minors, $-2\begin{vmatrix} 3 & 4 \\ -3 & 1 \end{vmatrix} - 1\begin{vmatrix} 4 & -1 \\ 3 & 4 \end{vmatrix} = -2(15) - 1(19) = -30 - 19 = -49$

6. $\begin{cases} 2x + 3y = 5 \\ x - 4y = -14 \end{cases}$

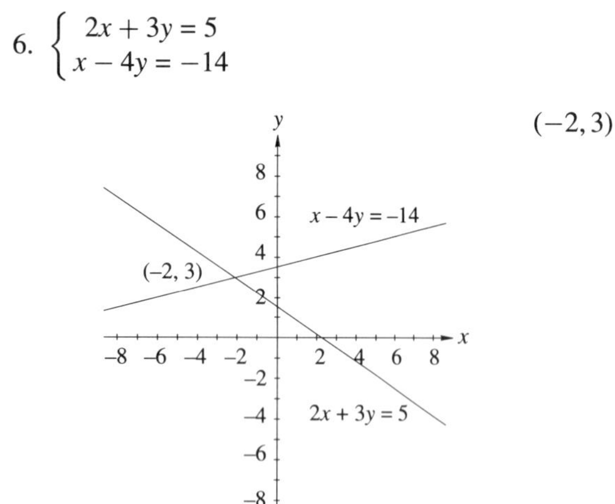

$(-2, 3)$

7. $\begin{cases} 4x + 3y = 9 & (1) \\ 2x + y = 2 & (2) \end{cases}$.
From (2), we get $y = 2 - 2x$ (3). Substituting in (1) yields $4x + 3(2 - 2x) = 9 \Rightarrow 4x + 6 - 6x = 9$, or $-2x + 6 = 9$ and so, $x = -\frac{3}{2}$. Substituting in (3) produces $y = 2 - 2(-\frac{3}{2}) = 2 + 3 = 5$. Hence, the solution is $\left(-\frac{3}{2}, 5\right)$.

8. We have $D = \begin{vmatrix} 2 & -4 \\ -6 & 8 \end{vmatrix} = 16 - 24 = -8$, $D_x = \begin{vmatrix} 2 & -4 \\ -9 & 8 \end{vmatrix} = 16 - 36 = -20$, and $D_y = \begin{vmatrix} 2 & 2 \\ -6 & -9 \end{vmatrix} = -18 + 12 = -6$. Using Cramer's rule, we see that $x = \frac{-20}{-8} = \frac{5}{2}$ and $y = \frac{-6}{-8} = \frac{3}{4}$. The solution is $\left(\frac{5}{2}, \frac{3}{4}\right)$.

9. x = cost of 1st, y = cost of 2nd, z = cost of 3rd
$\begin{cases} x + y + z = 60 & (1) \\ x = y + z & (2) \\ y = 2z + 3 & (3) \end{cases}$
or (2) can be written as $x - y - z = 0$ (4). Adding (1)+(4) we get $2x = 60$ or $x = 30$. Substituting this value for x in (2), we get $y + z = 30$ (5). Substituting $2z + 3$ for y in (5) yields $2z + 3 + z = 30$ or $3z = 27$, and so $z = 9$. Finally, substituting for z in (3), we see that $y = 2(9) + 3 = 18 + 3 = 21$. So, the cost of the first part is \$30, the cost of the second part \$21, and the cost of the third part is \$9.

CHAPTER

6

Ratio, Proportion, and Variation

6.1 RATIO AND PROPORTION

1. \$1.38 for 16 bolts is 1.38 : 16
2. 725 revolutions per minute is 725 : 1
3. 86 L per km is 86 : 1
4. 236 miles in 4 hours is 236 : 4
5. $\frac{9}{2} = \frac{4.5}{1}$ or 4.5 : 1
6. $\frac{7}{5} = \frac{1.4}{1}$ or 1.4 : 1
7. $\frac{23}{7} \approx \frac{3.2857}{1}$ or 3.2857 : 1
8. $\frac{37}{4} = \frac{9.25}{1}$ or 9.25 : 1
9. $2 \times 15 = 30$; $3 \times 10 = 30$ so equal
10. $1 \times 12 = 12, 3 \times 4 = 12$ so equal
11. $145 \times 5 = 725$; $25 \times 29 = 725$ so equal
12. $3 \times 33 = 99$; $8 \times 12 = 96$ so not equal
13. $8c = 7 \times 32$; $c = \frac{7 \times 32}{8}$; $c = 7 \times 4 = 28$
14. $9b = 3 \times 24$; $b = \frac{3 \times 24}{9} = 8$
15. $124d = 62 \times 158$; $d = \frac{62 \times 158}{124} = 79$
16. $20a = 8 \times 3.5$; $a = \frac{8 \times 3.5}{20} = 1.4$
17. $0.15a = 4 \times 0.16$; $a = \frac{4 \times 0.16}{0.15} \approx 4.267$
18. $10.5x = 7.5 \times 6.3$; $x = \frac{7.5 \times 6.3}{10.5} = 4.5$
19. $8b = 20 \times 5.6$; $b = \frac{20 \times 5.6}{8} = 14$
20. $2.4d = 10.8 \times 1.6$; $d = \frac{10.8 \times 1.6}{2.4} = 7.2$
21. $9x = 4 \times 4$; $x = \frac{4 \times 4}{9} \approx 1.78$
22. $xx = 3 \times 12$; $x^2 = 36$; $x = \pm\sqrt{36} = \pm 6$
23. $\frac{2}{x} = \frac{5}{14} = \frac{9}{z}$; $5x = 2 \times 14$; $x = 5.6$; $5z = 9 \times 14$; $z = \frac{9 \times 14}{5} = 25.2$
24. $\frac{a}{20} = \frac{6}{y} = \frac{9}{45}$; $\frac{a}{20} = \frac{9}{45}$; $45a = 9 \times 20$; $a = \frac{9 \times 20}{45} = 4$; $9y = 6 \times 45$; $y = \frac{6 \times 45}{9} = 30$
25. 54 : 13 or $\approx$ 4.1538 : 1
26. $\frac{15}{12} = \frac{5}{4}$ or 5 : 4
27. A steering wheel that makes $4\frac{2}{3}$ complete turns turns a total of $4\frac{2}{3} \times 360° = \frac{14}{3} \times 360° = 1680°$. Thus, the steering ratio is $\frac{1680}{60} = \frac{28}{1}$ or 28 : 1
28. 19 : 1 or just 19
29. $\frac{5 \times 10^{-4}}{300} \approx 1.67 \times 10^{-6}$ F$= 1.67\ \mu$F
30. $\frac{20 \text{ mm}}{80 \text{ mm}} = \frac{1}{4}$ or 1 : 4
31. $\frac{8.7}{1} = \frac{96}{x}$; $x = \frac{96}{8.7} \approx 11.03$ or 11.03 cm^3
32. $\frac{90 \text{ mL}}{20 \text{ L}} = \frac{x \text{ mL}}{54 \text{ L}}$; $x = \frac{90 \times 54}{20}$ mL$= 243$ mL

33. $\frac{80\text{ cal}}{1\text{ g}} = \frac{x\text{ cal}}{785\text{ kg}} = \frac{x\text{ cal}}{785{,}000\text{ g}}$; $x = 80 \times 785000$; $x = 62{,}800{,}000$ calories

34. $\frac{120}{V} = \frac{100}{750}$; $V = \frac{120 \times 750}{100} = 900$ V

35. $\frac{25.4\text{ mm}}{1\text{ in.}} = \frac{88.9\text{ mm}}{x\text{ in.}}$; $x = \frac{88.9}{25.4} = 3.5$; 3.5 in.

36. $\frac{9.78039}{32.0878} = \frac{9.83217}{x}$; $x = \frac{9.83217 \times 32.0878}{9.78039}$; $x = 32.257681\text{ ft/s}^2$

6.2 SIMILAR GEOMETRIC SHAPES

1. $\frac{a}{2} = \frac{3}{4} = \frac{b}{5}$; $4a = 3 \times 2$; $a = 6 \div 4 = 1.5$; $4b = 3 \times 5$; $b = 15 \div 4 = 3.75$

2. $\frac{a}{13} = \frac{2}{b} = \frac{9}{15}$; $15a = 9 \times 13$; $15a = 117$; $a = 7.8$; $9b = 2 \times 15$; $9b = 30$; $b = 3.33$ or $3\frac{1}{3}$

3. $\frac{a}{33} = \frac{24.3}{c} = \frac{22}{b} = \frac{34}{66}$; $66a = 33 \times 34$; $a = 17$; $34b = 22 \times 66$; $b \approx 42.71$; $34c = 24.3 \times 66$; $c \approx 47.17$

4. $\frac{a}{6} = \frac{b}{11} = \frac{c}{5} = \frac{d}{8} = \frac{4}{7}$; $7a = 4 \times 6$; $a = \frac{24}{7} \approx 3.43$; $7b = 4 \times 11$; $b = \frac{44}{7} \approx 6.29$; $7c = 5 \times 4$; $c = \frac{20}{7} \approx 2.86$; $7d = 4 \times 8$; $d = \frac{32}{7} \approx 4.57$

5. $\frac{a}{3} = \frac{b}{4} = \frac{c}{5} = \frac{d}{5} = \frac{1.7}{2}$; $2a = 3 \times 1.7$; $a = 2.55$; $2b = 4 \times 1.7$; $b = 3.4$; $2c = 5 \times 1.7$; $c = 4.25$; $d = c = 4.25$

6. $\frac{a}{5} = \frac{b}{3} = \frac{c}{4} = \frac{16}{9}$; $9a = 5 \times 16$; $a = \frac{80}{9} = 8\frac{8}{9} \approx 8.89$; $9b = 3 \times 16$; $b = \frac{48}{9} = 5\frac{3}{9} = 5\frac{1}{3}$; $9c = 4 \times 16$; $c = \frac{64}{9} = 7\frac{1}{9}$

7. $\frac{A}{20} = \frac{1^2}{(\frac{1}{8})^2}$, $\frac{1}{64}A = 20$, $A = 1280\text{ ft}^2$; $1280\text{ ft}^2 = 144 \times 1280\text{ in.}^2 = 184{,}320\text{ in.}^2$

8. Discharge is proportional to cross sectional area; $\frac{2000}{3000} = \frac{75^2}{x^2}$, $x^2 = \frac{75^2 \times 3000}{2000} = 8437.5$; $x = \sqrt{8437.5} \approx 91.86$

9. For a fixed length, the mass is proportional to the cross sectional area; $\frac{19^2}{38^2} = \frac{x}{22}$, $x = \frac{19^2 \times 22}{38^2} = 5.5$ kg

10. Cost proportional to length or perimeter; $\frac{652}{3 \times 652} = \frac{982^2}{x^2} = \frac{1}{3}$, $x^2 = 3 \times 982^2$; $x^2 = 2{,}892{,}972$, $x = \sqrt{2{,}892{,}972}$, $x = \$1{,}700.87$

11. Amount of paint is proportional to area; $\frac{700}{x} = \frac{20^2}{35^2}$; $x = \frac{35^2 \times 700}{20^2} = 2\ 143.75$ L

12. $\frac{20}{x} = \frac{12^3}{30^3}$; $x = \frac{20 \times 30^3}{12^3} = 312.5$ kL

13. $\frac{3930}{x} = \frac{15^3}{5^3} = \frac{3375}{125}$, $x = \frac{125 \times 3930}{3375} \approx 145.56\text{ mm}^3$

14. $\frac{4}{5} = \frac{2.5}{h}$; $4h = 5 \times 2.5$; $h = 3.125''$, $V_1 = \pi r^2 h = \pi \cdot 2^2 \times 5 \approx 62.83\text{ in.}^3$; $V_2 = \pi(1.25)^2 \cdot 3.125 \approx 15.34\text{ in.}^3$

15. $\frac{10^2}{2^2} = \frac{1256.64}{A}$; $A = \frac{4 \times 1256.64}{100} \approx 50.27\text{ cm}^2$, $\frac{10^3}{2^3} = \frac{4188.79}{V}$; $V = \frac{8 \times 4188.79}{1000} \approx 33.51\text{ cm}^3$

6.3 DIRECT AND INVERSE VARIATION

1. $R = kl$ or $\frac{R}{l} = k$

2. $P = kh$ or $\frac{P}{h} = k$

3. $A = kd^2$ or $\frac{A}{d^2} = k$

4. $E = kf$ or $\frac{E}{f} = k$

5. $IN = k$ or $I = \frac{k}{N}$

6. $Ed^2 = k$ or $E = \frac{k}{d^2}$

7. $k = \frac{r}{t} = \frac{4}{6} = \frac{2}{3}$

8. $ab = k$; $k = 8 \times 3 = 24$

9. $d = kr^2$; $k = \frac{d}{r^2} = \frac{8}{4^2} = \frac{8}{16} = \frac{1}{2}$

10. $ps^3 = k$; $k = 8 \times 2^3 = 8 \times 8 = 64$

11. $\frac{a_1}{d_1} = \frac{a_2}{d_2} \Rightarrow \frac{4}{8} = \frac{a}{18}$; $a = 4 \times 18 \div 8 = 9$

12. $r_1 t_1 = r_2 t_2 \Rightarrow r \times 9 = 6 \times 3, r = \frac{18}{9} = 2$

13. Note 1 min equals 60 s; $\frac{v_1}{t_1} = \frac{v_2}{t_2} \Rightarrow \frac{v}{60} = \frac{60}{20} \Rightarrow v = \frac{60 \times 60}{20} = 180$ m/s

14. $p_1 q_1 = p_2 q_2 \Rightarrow p \times 40 = 30 \times 10 \Rightarrow p = \frac{30 \times 10}{40} = \frac{30}{4} = 7.5$ mm^3

15. $\frac{n_1}{d_1^2} = \frac{n_2}{d_2^2} \Rightarrow \frac{n}{2^2} = \frac{10890}{10^2} \Rightarrow n = \frac{4 \times 10890}{100} = 435.6$ neutrons

16. $m_1 r_1^3 = m_2 r_2^3 \Rightarrow m5^3 = 25 \times 30^3$ or $125m = 25 \times 30^3$, and so $m = \frac{25 \times 27000}{125} = 5\,400$ kPa

17. (a) $V = kT \Rightarrow k = V/T = \frac{25}{293.18} \approx 0.08527 =$ constant; (b) $V = 0.08527 \times (85 + 273.18) \approx 30.542$ m^3; $\Delta V = 30.542 - 25 = 5.542$ m^3 = change in volume

18. $d = kt$; $k = \frac{d}{t} = \frac{475}{1} = 475$ mph; $1235 = 475t \Rightarrow t = \frac{1235}{475} = 2.6$ hr; or $\frac{d_1}{t_1} = \frac{d_2}{t_2} \Rightarrow \frac{475}{1} = \frac{1235}{t} \Rightarrow 475t = 1235$; $\Rightarrow t = \frac{1235}{475} = 2.6$ hr or 2 hr 36 min

19. $R = k\ell$ or $\frac{R_1}{l_1} = \frac{R_2}{l_2}$; $\frac{32}{1500} = \frac{R}{5000}$, $1500R = 32 \times 5000$; $R = \frac{32 \times 5000}{1500} = 106.67\ \Omega$

20. $d = kt^2$ or $\frac{d_1}{t_1^2} = \frac{d_2}{t_2^2}$;

(a) $k = \frac{d}{t^2} = \frac{29.4}{3^2} = \frac{29.4}{9} \approx 3.267$; $d = 3.267t^2$;

(b) $d = 3.267 \times 8^2 = 3.267 \times 64 \approx 209.09$ m;

(c) $150 = 3.267t^2 \Rightarrow t^2 = \frac{150}{3.267} \approx 45.91$ which means that $t = \sqrt{45.91} \approx 6.78$ s.

21. $P_1 V_1 = P_2 V_2$; $40 \times 200 = P \times 50 \Rightarrow P = \frac{40 \times 200}{50} = 40 \times 4 = 160$ psi

22. $Ed^2 = k$ or $E_1 d_1^2 = E_2 d_2^2$; $600 \times 2^2 = E \times 6^2$; $E = \frac{600 \times 4}{36} \approx 66.67$ lux

23. $In = k$, I = current, n =number of turns or $I_1 n_1 = I_2 n_2$; $I \times 200 = 0.3 \times 50$; $I = \frac{0.3 \times 50}{200} - 0.075$ A

24. $V = kn$ or $\frac{v_1}{n_1} = \frac{v_2}{n_2}$; $\frac{120}{300} = \frac{V}{40}$; $V = \frac{120 \times 40}{300} = 16$ V

25. $wr^2 = k$ or $w_1 r_1^2 = w_2 r_2^2$; $180 \times 4000^2 = w \times (4000 + 500)^2$; $w = \frac{180 \times 4000^2}{4500^2} \approx 142.22$ lb

6.4 JOINT AND COMBINED VARIATION

1. $K = kmv^2$

2. $V = klwh$

3. $f = \frac{kv}{l}$

4. $r = \frac{ks}{d}$

5. $E = \frac{kII}{d^2}$

6. $L = \frac{kN^2 A}{l}$

7. $a = kbc \Rightarrow 20 = k4(2)$; $k = \frac{20}{4 \cdot 2} = \frac{20}{8} = 2.5$

8. $p = krst$; $15 = k \cdot 3 \cdot 10 \cdot 25$; $k = \frac{15}{3 \cdot 10 \cdot 25} = \frac{1}{50} = 0.02$

9. $u = \frac{kv}{w}$; $20 = \frac{k \cdot 5}{2}$; $k = \frac{20 \times 2}{5} = 8$

10. $x = \frac{kyw^3}{z^2}$; $175 = \frac{k \cdot 3 \cdot 5^3}{4^2}$; $k = \frac{175 \times 16}{3 \times 125} \approx 7.47$

11. $r = kst$; $10 = k \cdot 6 \cdot 5$; $k = \frac{10}{30} = \frac{1}{3}$; $r = \frac{1}{3} \cdot 3 \cdot 4 = 4$

12. $f = \frac{kr}{\ell}$; $125 = \frac{k \cdot 10}{5}$; $k = \frac{125 \times 5}{10} = \frac{125}{2} = 62.5$; $f = \frac{125}{2} \cdot \frac{20}{10} = 125$

13. $a = kxy^2$; $9 = k \cdot 3 \cdot 5^2$; $k = \frac{9}{3 \cdot 5^2} = \frac{3}{25} = 0.12$; $a = 0.12 \cdot 6 \times 15^2 = 162$

14. $10 = 0.12 \cdot 9 \cdot y^2$; $y^2 = \frac{10}{0.12 \times 9} \approx 9.2593$; $y \approx \pm\sqrt{9.2593} \approx \pm 3.04$

15. $p = \frac{kr\sqrt{t}}{w}$; $9 = \frac{k 18\sqrt{9}}{6}$; $k = \frac{9 \times 6}{18 \times 3} = 1$; $p = \frac{36\sqrt{4}}{2} = 36$

16. $18 = \frac{9\sqrt{t}}{4}$; $\sqrt{t} = \frac{18 \cdot 4}{9} = 8$; $t = 8^2 = 64$

17. (a) $A = kbh$; $12 = k \cdot 3 \cdot 8$; $k = \frac{12}{24} = \frac{1}{2}$; (b) $A = \frac{1}{2} \cdot 10 \cdot 5 = 25$

18. $W = kIVt$; $2.4 = k \cdot 15 \cdot 240 \cdot \frac{45}{60}$; $k = \frac{2.4 \times 60}{15 \times 240 \times 45} \approx 8.89 \times 10^{-4}$; $W = 8.89 \times 1.25 \times 120 \times 8 \times 10^{-4}$; $W \approx 1.07$ kWh

19. $w = \frac{kV^2}{R}$; $2 = \frac{k6^2}{18}$; $k = \frac{2 \times 18}{36} = 1$; $V^2 = wR = 50 \times 10 = 500$; $V = \sqrt{500} \approx 22.36$ V

20. $C = \frac{kA}{d}$; $A = s^2 = 4^2$; $0.3 = \frac{k4^2}{0.5}$; $k = \frac{0.3 \times 0.5}{16}$; 0.009375; $C = \frac{.009375 \times 16}{0.1} = 1.5\ \mu$F

21. $R = \frac{k\ell}{A}$; $0.56 = \frac{k80}{2.5}$; $k = \frac{0.56 \times 2.5}{80} = 0.0175$; $3 = \frac{0.0175\ell}{0.1}$; $\ell = \frac{3 \times 0.1}{0.0175} \approx 17.14$ m

22. $\Delta R = k\Delta C$; $2.25 - 2.09 = k(25 - 10)$; $0.16 = k15$; $k = \frac{0.16}{15} \approx 0.01067$; $R - 2.25 = 0.01067(50 - 25)$; $R - 2.25 \approx 0.267$; $R \approx 0.267 + 2.25$; $R \approx 2.517\ \Omega$

23. $H = \frac{kA(\Delta T)}{d}$; $50 = \frac{k12(72-32)}{3}$; $k = 0.3125$; $H = \frac{0.3125(72-10) \times 12}{3} = 77.5$ Btu/hr

24. $F = \frac{k\sqrt{T}}{\ell}$; $256 = \frac{k\sqrt{25}}{500}$; $k = 25\ 600$; $F = \frac{25\ 600\sqrt{30}}{450} \approx 311.59$ Hz

25. $P = \frac{kT}{V}$; $15 = \frac{k(70+460)}{5}$; $k = \frac{15 \times 5}{530} \approx 0.1415$; $75 = \frac{0.1415(200+460)}{V}$; $V = \frac{0.1415 \times 660}{75} = 1.2452$ or $1.25\ \text{ft}^3$

CHAPTER 6 REVIEW

1. $3 : x = 4 : 6$; $4x = 3 \times 6$; $4x = 18$; $x = \frac{18}{4} = \frac{9}{2} = 4.5$

2. $x : 5 = 3 : 15$; $15x = 5 \times 3$; $x = 1$

3. $\frac{7}{9} = \frac{21}{d}$; $7d = 21 \cdot 9$; $d = \frac{21 \cdot 9}{7} = 27$

4. $\frac{14}{6} = \frac{c}{27}$; $6c = 14 \times 27$; $c = \frac{14 \times 27}{6} = 63$

5. $4 : 8 = 19 : x$; $4x = 8 \times 19$; $x = \frac{8 \times 19}{4} = 38$

6. $x : 12 = 15 : 32$; $32x = 12 \times 15$; $x = \frac{12 \times 15}{32} = 5.625$

7. $7 : 24.5 : x = 8 : y : 42$; $\frac{7}{8} = \frac{24.5}{y} = \frac{x}{42}$; $8x = 7 \times 42$; $x = 36.75$; $7y = 8 \times 24.5$; $y = 28$

8. $\frac{12.5}{x} = \frac{y}{47} = \frac{8}{5}$; $8x = 5 \times 12.5$; $x = 7.8125$; $5y = 8 \times 47$; $y = 75.2$

9. $TR = \frac{N_p}{N_s}$; $25 = \frac{4000}{N_s}$; $N_s = \frac{4000}{25} = 160$

10. $\Delta S = kI$; $12.50 = k6.5$; $k \approx 1.923$; $\Delta S = 1.923 \times 9.6 \approx \$18.46$

11. $\frac{S_1}{4} = \frac{S_2}{5} = \frac{180}{8}$; $S_1 = \frac{180 \times 4}{8} = 90$ mm; $S_2 = \frac{180 \times 5}{8} = 112.5$ mm

12. $TMA = \frac{90}{5} = 18$

13. $E = \frac{\text{AMA}}{\text{TMA}} = \frac{16}{18} = \frac{8}{9}$ or ≈ 0.889 or 88.9%

14. If rate of flow is constant, then $V_1A_1 = V_2A_2$; $A \propto d^2 \Rightarrow V_1d_1^2 = V_2d_2^2$; (a) $3 \cdot 1^2 = (.5)^2 \cdot V_2$; $V_2 = \frac{3}{.25} = 12$ m/s; (b) $30d^2 = 3$; $d^2 = \frac{1}{10}$; $d = \sqrt{.1} \approx .316$ cm $= 3.16$ mm diameter

15. $5 \times .5^2 = V \times 2.5^2$; $V = \frac{5 \times 0.25}{6.25} = 0.2$ m/s

16. $W = kT^4$; $k = \frac{W}{T^4} = \frac{6.5 \times 10^7}{(5800)^4} = 5.7438 \times 10^{-8}$

17. $F = \frac{kC_1C_2}{d^2}$; $k = \frac{Fd^2}{C_1C_2} = \frac{7.192 \times 10^{-4}(.05)^2}{4 \times 10^{-9} \times 5 \times 10^{-8}} = 8.99 \times 10^9$

18. $M = \frac{kI_1}{d}$; $4 \times 10^{-6} = \frac{k100}{5}$; $k = \frac{4 \times 10^{-6} \times 5}{100}$; $k = 2 \times 10^{-7}$; $M = \frac{2 \times 10^{-7} \times 150}{15} = 2 \times 10^{-6}$ T

19. $2.42 = \frac{3 \times 10^8}{s}$; $s = \frac{3 \times 10^8}{2.42} \approx 1.24 \times 10^8$ m/s

CHAPTER 6 TEST

1. $4:9 = x:51$; $9x = 4 \times 51$; $x = \frac{4\times 51}{9} = \frac{68}{3} = 22\frac{2}{3} \approx 22.67$
2. $8:x = 68:93.5$; $68x = 8 \times 93.5$; $x = 11$
3. $\frac{a}{28} = \frac{12}{20}$; $20a = 12 \times 28$; $a = 16.8$
4. $\frac{24}{42} = \frac{78}{d}$; $24d = 78 \times 42$; $d = 136.5$
5. $6:8:x = y:14:24.5$; $\frac{6}{y} = \frac{8}{14} = \frac{x}{24.5}$; $8y = 6 \times 14$; $y = 10.5$; $14x = 8 \times 24.5$; $x = 14$
6. $\frac{a}{9} = \frac{b}{30} = \frac{75}{45}$; $45a = 75 \times 9$; $a = 15$; $45b = 75 \times 30$; $b = 50$
7. $E = kd^2$
8. $h = \frac{kr}{t^3}$
9. $\frac{a}{5} = \frac{b}{12} = \frac{71.5}{13}$; $13a = 5 \times 71.5$; $a = 27.5$; $13b = 12 \times 71.5$; $b = 66$
10. $3x+2x = \frac{540}{2}$; $5x = 270$; $x = 54$; $\ell = 3\cdot 54 = 162\,\text{cm}$; $w = 2 \cdot 54 = 108$ cm
11. $A = ks^2$; $72 = k(12)^2$ and so $k = 0.5$. Thus, $A = (0.5)(18^2) = 162\ \text{m}^2$ or $A_1S_2^2 = A_2S_1^2$; $A \cdot 12^2 = 72 \cdot 18^2$; $A = 162\ \text{m}^2$
12. $d = \frac{k}{ph}$ and so $45 = \frac{k}{(80)(8)}$. Thus, $k = 28{,}800$ and $d = \frac{28{,}800}{(60)(10)} = 48$

CHAPTER

7

Factoring and Algebraic Fractions

7.1 SPECIAL PRODUCTS

1. $3(p+q)=3p+3q$
2. $7(2+x)=14+7x$
3. $3x(5-y)=15x-3xy$
4. $2a(4+a^2)=8a+2a^3$
5. $(p+q)(p-q)=p^2-q^2$
6. $(3a+b)(3a-b)=(3a)^2-b^2=9a^2-b^2$
7. $(2x-6p)(2x+6p)=(2x)^2-(6p)^2=4x^2-36p^2$
8. $\left(\frac{y}{2}+\frac{2p}{3}\right)\left(\frac{y}{2}-\frac{2p}{3}\right)=\left(\frac{y}{2}\right)^2-\left(\frac{2p}{3}\right)^2=\frac{y^2}{4}-\frac{4p^2}{9}$
9. $(r+w)^2=r^2+2rw+w^2$
10. $(q-f)^2=q^2-2qf+f^2$
11. $(2x+y)^2=(2x)^2+2\cdot 2x\cdot y+y^2=4x^2+4xy+y^2$
12. $\left(\frac{1}{2}a+b\right)^2=\left(\frac{1}{2}a\right)^2+2\cdot\frac{1}{2}ab+b^2=\frac{1}{4}a^2+ab+b^2$
13. $\left(\frac{2}{3}x+4b\right)^2=\left(\frac{2}{3}x\right)^2+2\cdot\frac{2}{3}x(4b)+(4b)^2=\frac{4}{9}x^2+\frac{16}{3}xb+16b^2$
14. $(a-\frac{1}{2}y)^2=a^2-2\cdot a\cdot\frac{1}{2}y+(\frac{1}{2}y)^2=a^2-ay+\frac{1}{4}y^2$
15. $(2p-\frac{3}{4}r)^2=(2p)^2-2\cdot 2p\cdot\frac{3}{4}r+(\frac{3}{4}r)^2=4p^2-3pr+\frac{9}{16}r^2$
16. $(\frac{2}{3}r-5t)^2=(\frac{2}{3}r)^2-2\cdot\frac{2}{3}r\cdot 5t+(5t)^2=\frac{4}{9}r^2-\frac{20}{3}rt+25t^2$
17. $(a+2)(a+3)=a^2+(2+3)a+2\cdot 3=a^2+5a+6$
18. $(x+5)(x+7)=x^2+(5+7)x+5\cdot 7=x^2+12x+35$
19. $(x-5)(x+2)=x^2+(-5+2)x+(-5)(2)=x^2-3x-10$
20. $(a+9)(x-12)=ax-12a+9x-108$ or $ax+9x-12a-108$
21. $(2a+b)(3a+b)=2\cdot 3a^2+(2+3)ab+b^2=6a^2+5ab+b^2$
22. $(4x+2)(3x+5)=4\cdot 3x^2+[4\cdot 5+2\cdot 3]x+2\cdot 5=12x^2+(20+6)x+10=12x^2+26x+10$
23. $(3x+4)(2x-5)=3\cdot 2x^2+[3(-5)+4\cdot 2]x+4(-5)=6x^2+[-15+8]x-20=6x^2-7x-20$
24. $(3m-n)(4m+2n)=3\cdot 4m^2+[3\cdot 2+(-1)4]mn+(-1)2n^2=12m^2+2mn-2n^2$
25. $(a+b)^3=a^3+3a^2b+3ab^2+b^3$
26. $(r-s)^3=r^3-3r^2s+3rs^2-s^3$
27. $(x+4)^3=x^3+3\cdot x^2\cdot 4+3x\cdot 4^2+4^3=x^3+12x^2+48x+64$
28. $(2a+b)^3=(2a)^3+3(2a)^2b+3(2a)b^2+b^3=8a^3+12a^2b+6ab^2+b^3$
29. $(2a-b)^3=(2a)^3-3(2a)^2b+3(2a)b^2-b^3=8a^3-12a^2b+6ab^2-b^3$

PROPERTY OF VALDOSTA STATE PRISt

· 019464

30. $(3x+2y)^3 = (3x)^3+3(3x)^2(2y)+3(3x)(2y)^2+(2y)^3 = 27x^3 + 54x^2y + 36xy^2 + 8y^3$

31. $(3x-2y)^3 = (3x)^3-3(3x)^2(2y)+3(3x)(2y)^2-(2y)^3 = 27x^3 - 54x^2y + 36xy^2 - 8y^3$

32. $(4r - s)^3 = (4r)^3 - 3(4r)^2s + 3(4r)s^2 - s^3 = 64r^3 - 48r^2s + 12rs^2 - s^3$

33. $(m + n)(m^2 - mn + n^2) = m^3 + n^3$

34. $(a + 2)(a^2 - 2a + 4) = a^3 + 8$

35. $(r - t)(r^2 + rt + t^2) = r^3 - t^3$

36. $(h - 3)(h^2 + 3h + 9) = h^3 - 27$

37. $(2x + b)(4x^2 - 2xb + b^2) = (2x + b)\left[(2x)^2 - 2x \cdot b + b^2\right] = (2x)^3 + b^3 = 8x^3 + b^3$

38. $\left(3a + \frac{2}{3}c\right)\left(9a^2 - 2ac + \frac{4}{9}c^2\right) = \left(3a + \frac{2}{3}c\right)\left[(3a)^2 - 3a \cdot \frac{2}{3}c + \left(\frac{2}{3}c\right)^2\right] = (3a)^3 + \left(\frac{2}{3}c\right)^3 = 27a^3 + \frac{8}{27}c^3$

39. $(3a - d)(9a^2 + 3da + d^2) = (3a - d)\left[(3a)^2 + 3ad + d^2\right]; = (3a)^3 - d^3 = 27a^3 - d^3$

40. $\left(\frac{2e}{5} - \frac{5r}{4}\right)\left(\frac{4e^2}{25} + \frac{er}{2} + \frac{25r^2}{16}\right) = \left(\frac{2e}{5} - \frac{5r}{4}\right)\left[\left(\frac{2e}{5}\right)^2 + \frac{2e}{5} \cdot \frac{5r}{4} + \left(\frac{5r}{4}\right)^2\right] = \left(\frac{2e}{5}\right)^3 - \left(\frac{5r}{4}\right)^3 = \frac{8e^3}{125} - \frac{125r^3}{64}$

41. $3(a + 2)^2 = 3(a^2 + 4a + 4) = 3a^2 + 12a + 12$

42. $5(2a - 4)^2 = 5(4a^2 - 16a + 16) = 20a^2 - 80a + 80$

43. $\frac{5r}{t}\left(t + \frac{r}{5}\right)^2 = \frac{5r}{t}\left(t^2 + \frac{2rt}{5} + \frac{r^2}{25}\right) = \frac{5r}{t} \cdot t^2 + \frac{5r}{t} \cdot \frac{2rt}{5} + \frac{5r}{t} \cdot \frac{r^2}{25} = 5rt + 2r^2 + \frac{r^3}{5t}$

44. $(x^2+4)(x+2)(x-2) = (x^2+4)(x^2-4) = x^4 - 4^2 = x^4 - 16$

45. $(x^2 - 6)(x^2 + 6) = x^4 - 6^2 = x^4 - 36$

46. $\left(\frac{2a}{b} + \frac{b}{2}\right)^2 = \left(\frac{2a}{b}\right)^2 + 2\frac{2a}{b}\frac{b}{2} + \left(\frac{b}{2}\right)^2 = \frac{4a^2}{b^2} + 2a + \frac{b^2}{4}$

47. $\left(\frac{3x}{y} - \frac{y}{x}\right)\left(\frac{3x}{y} + \frac{y}{x}\right)x^2y^2 = \left(\left(\frac{3x}{y}\right)^2 - \left(\frac{y}{x}\right)^2\right)x^2y^2 = \left(\frac{9x^2}{y^2} - \frac{y^2}{x^2}\right)x^2y^2 = 9x^4 - y^4$

48. $(x + 1)(x + 2)(x + 3) = \left[x^2 + 3x + 2\right](x + 3) = x^3 + 3x^2 + 3x^2 + 9x + 2x + 6 = x^3 + 6x^2 + 11x + 6$

49. $(x-y)^2-(y-x)^2 = (x^2-2xy+y^2)-(y^2-2yx+x^2) = x^2 - 2xy + y^2 - y^2 + 2xy - x^2 = 0$

50. $(x+y)^2-(y-x)^2 = (x^2+2xy+y^2)-(y^2-2xy+x^2) = 4xy$

51. $(x+3)(x-3)^2 = (x+3)(x^2 - 6x + 9) = x^3 - 6x^2 + 9x + 3x^2 - 18x + 27 = x^3 - 3x^2 - 9x + 27$

52. $(y-4)(y+4)^2 = (y-4)(y^2 + 8y + 16) = y^3 + 8y^2 + 16y - 4y^2 - 32y - 64 = y^3 + 4y^2 - 16y - 64$

53. $r(r-t)^2-t(t-r)^2 = r(r^2-2rt+t^2)-t(t^2-2rt+r^2) = r^3-2r^2t+rt^2-t^3+2rt^2-r^2t = r^3-3r^2t+3rt^2-t^3$

54. $5(x + 4)(x - 4)(x^2 + 16) = 5(x^2 - 16)(x^2 + 16) = 5\left(x^4 - 256\right) = 5x^4 - 1280$

55. $(5 + 3x)(25 - 15x + 9x^2) = (5 + 3x)\left[5^2 - 5 \cdot 3x + (3x)^2\right] = 5^3 + (3x)^3 = 125 + 27x^3$

56. $\left(2 - \sqrt{x}\right)\left(2 + \sqrt{x}\right)(4 + x) = \left(2^2 - \sqrt{x}^2\right)(4 + x) = (4 - x)(4 + x) = 16 - x^2$

57. $[(x + y) - (w + z)]^2 = (x + y)^2 - 2(x + y)(w + z) + (w + z)^2$

58. $[(r+s)+(t+u)]^2 = (r+s)^2+2(r+s)(t+u)+(t+u)^2$

59. $(x + y - z)(x + y + z) = [(x + y) - z][(x + y) + z] = (x + y)^2 - z^2 = x^2 + 2xy + y^2 - z^2$

60. $(a+b+2)(a-b+2) = [(a+2)+b][(a+2)-b] = (a + 2)^2 - b^2 = a^2 + 4a + 4 - b^2$

61. $z^2 = R^2 + [x_L^2 - 2x_Lx_C + x_C^2] = R^2 + x_L^2 - 2x_Lx_C + x_C^2$

62. $KE = \frac{1}{2}m(3t + 1)^2 = \frac{1}{2}m(9t^2 + 6t + 1) = \frac{9}{2}mt^2 + 3mt + \frac{1}{2}m$

63. $a_c = \frac{(2t^2-t)^2}{r} = \frac{4t^4-4t^3+t^2}{r}$

64. $W = \frac{1}{2}m(v_1^2 - v_2^2)$

7.2 FACTORING

1. $6x + 6 = 6(x + 1)$
2. $12x + 12 = 12(x + 1)$
3. $12a - 6 = 6(2a - 1)$
4. $15d - 5 = 5(3d - 1)$
5. $4x - 2y + 8 = 2(2x - y + 4)$
6. $6a + 9b - 3c = 3(2a + 3b - c)$
7. $5x^2 + 10x + 15 = 5(x^2 + 2x + 3)$
8. $16a^2 + 8b - 24 = 8(2a^2 + b - 3)$
9. $10x^2 - 15 = 5(2x^2 - 3)$
10. $14x^4 + 21 = 7(2x^4 + 3)$
11. $4x^2 + 6x = 2x(2x + 3)$
12. $8a - 4a^2 = 4a(2 - a)$
13. $7b^2y + 28b = 7b(by + 4)$
14. $9ax^2 + 27bx = 9x(ax + 3b)$
15. $3ax + 6ax^2 - 2ax = ax(3 + 6x - 2) = ax(1 + 6x)$
16. $6by - 12b^2y + 7by^2 = by(6 - 12b + 7y)$
17. $4ap^2 + 6a^2pq + 8apq^2 = 2ap(2p + 3aq + 4q^2)$
18. $12p^2r^2 - 8p^3r + 24pr^2 = 4pr(3pr - 2p^2 + 6r)$
19. $a^2 - b^2 = (a + b)(a - b)$
20. $p^2 - r^2 = (p + r)(p - r)$
21. $x^2 - 4 = x^2 - 2^2 = (x + 2)(x - 2)$
22. $a^2 - 16 = a^2 - 4^2 = (a + 4)(a - 4)$
23. $y^2 - 81 = y^2 - 9^2 = (y + 9)(y - 9)$
24. $m^2 - 49 = (m + 7)(m - 7)$
25. $4x^2 - 9 = (2x)^2 - 3^2 = (2x + 3)(2x - 3)$
26. $49y^2 - 64 = (7y)^2 - 8^2 = (7y + 8)(7y - 8)$
27. $9a^4 - b^2 = (3a^2)^2 - b^2 = (3a^2 + b)(3a^2 - b)$
28. $16t^6 - a^2 = (4t^3)^2 - a^2 = (4t^3 + a)(4t^3 - a)$
29. $25a^2 - 49b^2 = (5a)^2 - (7b)^2 = (5a + 7b)(5a - 7b)$
30. $121r^2 - 81t^2 = (11r)^2 - (9t)^2 = (11r + 9t)(11r - 9t)$
31. $144 - 25b^4 = (12)^2 - (5b^2)^2 = (12 + 5b^2)(12 - 5b^2)$
32. $81 - 49r^4 = 9^2 - (7r^2)^2 = (9 + 7r^2)(9 - 7r^2)$
33. $5a^2 - 125 = 5(a^2 - 25) = 5(a + 5)(a - 5)$
34. $7x^2 - 63 = 7(x^2 - 9) = 7(x + 3)(x - 3)$
35. $28a^2 - 63b^4 = 7(4a^2 - 9b^4) = 7[(2a)^2 - (3b^2)^2] = 7(2a + 3b^2)(2a - 3b^2)$
36. $81x^2 - 36t^6 = 9(9x^2 - 4t^6) = 9[(3x)^2 - (2t^3)^2] = 9(3x + 2t^3)(3x - 2t^3)$
37. $a^4 - 81 = (a^2 + 9)(a^2 - 9) = (a^2 + 9)(a + 3)(a - 3)$
38. $b^4 - 256 = (b^2 + 16)(b^2 - 16) = (b^2 + 16)(b + 4)(b - 4)$
39. $16x^4 - 256y^4 = 16(x^4 - 16y^4) = 16(x^2 + 4y^2)(x^2 - 4y^2) = 16(x^2 + 4y^2)(x + 2y)(x - 2y)$
40. $25a^5 - 400ab^8 = 25a(a^4 - 16b^8) = 25a(a^2 + 4b^4)(a^2 - 4b^4) = 25a(a^2 + 4b^4)(a + 2b^2)(a - 2b^2)$
41. $\pi r \left(r + \sqrt{h^2 + r^2}\right)$
42. $m(c\Delta t + L_f)$
43. $\frac{1}{2}d(v_2^2 - v_1^2) = \frac{1}{2}d(v_2 + v_1)(v_2 - v_1)$
44. $\pi R^2 - \pi r^2 = \pi(R^2 - r^2) = \pi(R + r)(R - r)$
45. $\frac{(\omega_f - \omega_0)(\omega_f + \omega_0)}{2\theta}$
46. $e\sigma T^4 - e\sigma T_0^4 = e\sigma(T^4 - T_0^4) = e\sigma(T^2 + T_0^2) \times (T + T_0)(T - T_0) = e\sigma(T - T_0)(T + T_0)(T^2 + T_0^2)$

7.3 FACTORING TRINOMIALS

1. x^2+9x-8; $b^2-4ac=113$; $\sqrt{113}\approx 10.6$; factors, but does not factor using rational numbers
2. x^2+7x-8; $b^2-4ac=81$; $\sqrt{81}=9$; factors
3. $3x^2-10x-8$; $b^2-4ac=196$; $\sqrt{196}=14$; factors
4. $2x^2+16x+14$; $b^2-4ac=144$; $\sqrt{144}=12$; factors
5. $5x^2+23x+18$; $b^2-4ac=169$; $\sqrt{169}=13$; factors
6. $7x^2-5x+16$; $b^2-4ac=-423$; $\sqrt{-423}$ is not a real number; so, $7x^2-5x+16$ does not factor using real numbers
7. $x^2+7x+10=(x+2)(x+5)$
8. $x^2+8x+15=(x+3)(x+5)$
9. $x^2-12x+27=(x-3)(x-9)$
10. $x^2-14x+33=(x-3)(x-11)$
11. $x^2-27x+50=(x-2)(x-25)$
12. $x^2+19x+48=(x+3)(x+16)$
13. $x^2-x-2=(x-2)(x+1)$
14. $x^2-4x-5=(x-5)(x+1)$
15. $x^2-3x-10=(x-5)(x+2)$
16. $p^2-16p+64=p^2-2\cdot 8p+8^2=(p-8)^2$
17. $r^2+10r+25=r^2+2\cdot 5r+5^2=(r+5)^2$
18. $v^2-14v+49=v^2-2\cdot 7v+7^2=(v-7)^2$
19. $a^2+22a+121=a^2+2\cdot 11a+11^2=(a+11)^2$
20. $e^2+26e+169=e^2+2\cdot 13e+13^2=(e+13)^2$
21. $f^2-30f+225=f^2-2\cdot 15f+15^2=(f-15)^2$
22. $3x^2+4x+1=(3x+1)(x+1)$
23. $6y^2-7y+1=(6y-1)(y-1)$
24. $3p^2+5p+2=(3p+2)(p+1)$
25. $7\times 2=14$; $7+2=9$; $7t^2+9t+2=7t^2+7t+2t+2$; $=7t(t+1)+2(t+1)=(7t+2)(t+1)$
26. $5(-3)=-15$; $15\times(-1)=-15$; $15+(-1)=14$; $5a^2+14a-3=5a^2+15a-1a-3=$ $5a(a+3)-1(a+3)=(5a-1)(a+3)$
27. $7(-5)=-35$; $-35\times 1=-35$; $-35+1=-34$; $7b^2-34b-5=7b^2-35b+b-5=$ $7b(b-5)+1(b-5)=(7b+1)(b-5)$
28. $2(-6)=-12$; $4\times(-3)=-12$; $4+(-3)=1$; $2y^2+y-6=2y^2+4y-3y-6=$ $2y(y+2)-3(y+2)=(2y-3)(y+2)$
29. $4(-5)=-20$; $20\times(-1)=-20$; $20+(-1)=19$; $4e^2+19e-5=4e^2+20e-e-5=$ $4e(e+5)-1(e+5)=(4e-1)(e+5)$
30. $6\times 3=18$; $(-18)(-1)=18$; $(-18)+(-1)=-19$; $6m^2-19m+3=6m^2-18m-m+3=$ $6m(m-3)-1(m-3)=(6m-1)(m-3)$
31. $3\times 8=24$; $4\times 6=24$; $4+6=10$; $3u^2+10u+8=3u^2+6u+4u+8=$ $3u(u+2)+4(u+2)=(3u+4)(u+2)$
32. $7(-2)=-14$; $14(-1)=-14$; $14+(-1)=13$; $7r^2+13r-2=7r^2+14r-r-2=$ $7r(r+2)-1(r+2)=(7r-1)(r+2)$
33. $9(-6)=-54$; $(-27)(2)=-54$; $(-27)+2=-25$; $9t^2-25t-6=9t^2-27t+2t-6=$ $9t(t-3)+2(t-3)=(9t+2)(t-3)$
34. $4\times 3=12$; $2\times 6=12$; $2+6=8$; $4x^2+8x+3=4x^2+6x+2x+3=$ $2x(2x+3)+1(2x+3)=(2x+1)(2x+3)$
35. $6(-5)=-30$; $15(-2)=-30$; $15+(-2)=13$; $6x^2+13x-5=6x^2+15x-2x-5=$ $3x(2x+5)-1(2x+5)=(3x-1)(2x+5)$
36. First factor out a 2, with the result $8y^2-8y-6=2(4y^2-4y-3)$. Then, $4(-3)=-12$; $-6\times 2=-12$; $-6+2=-4$; $8y^2-8y-6=2(4y^2-4y-3)=2[4y^2-6y+2y-3]=$ $2[2y(2y-3)+1(2y-3)]=2(2y+1)(2y-3)$

37. $15(-15) = -225$; $-25 \times 9 = -225$; $-25 + 9 = -16$; $15a^2 - 16a - 15 = 15a^2 - 25a + 9a - 15 = 5a(3a - 5) + 3(3a - 5) = (5a + 3)(3a - 5)$

38. $15(-15) - 225$; $25(-9) = -225$; $25 - 9 = 16$; $15d^2 + 16d - 15 = 15d^2 + 25d - 9d - 15 = 5d(3d + 5) - 3(3d + 5) = (5d - 3)(3d + 5)$

39. $15 \times 15 = 225$; $25 \times 9 = 225$; $25 + 9 = 34$; $15e^2 + 34e + 15 = 15e^2 + 25e + 9e + 15 = 5e(3e + 5) + 3(3e + 5) = (5e + 3)(3e + 5)$

40. $14 \times 10 = 140$; $(-35)(-4) = 140$; $-35 - 4 = -39$; $14a^2 - 39a + 10 = 14a^2 - 35a - 4a + 10 = 7a(2a - 5) - 2(2a - 5) = (7a - 2)(2a - 5)$

41. $10 \times 6 = 60$; $(-15) \times (-4) = 60$; $-15 + (-4) = -19$; $10x^2 - 19x + 6 = 10x^2 - 15x - 4x + 6 = 5x(2x - 3) - 2(2x - 3) = (5x - 2)(2x - 3)$

42. $3x^2 + 18x + 27 = 3(x^2 + 6x + 9) = 3(x + 3)^2$

43. $3r^2 - 18r - 21 = 3(r^2 - 6r - 7) = 3(r - 7)(r + 1)$

44. First factor out a 5, with the result $15x^2 + 50x + 35 = 5(3x^2 + 10x + 7)$. Then, $3 \times 7 = 21$; $3 + 7 = 10$; $15x^2 + 50x + 35 = 5(3x^2 + 10x + 7) = 5[3x^2 + 3x + 7x + 7] = 5[3x(x + 1) + 7(x + 1)] = 5(3x + 7)(x + 1)$

45. $49t^4 - 105t^3 + 14t^2 = 7t^2(7t^2 - 15t + 2) = 7t^2(7t - 1)(t - 2)$

46. $-2 \times -7 = 14$; $-2 + (-7) = -9$; $2y^4 - 9y^2 + 7 = 2y^4 - 2y^2 - 7y^2 + 7 = 2y^2(y^2 - 1) - 7(y^2 - 1) = (2y^2 - 7)(y^2 - 1) = (2y^2 - 7)(y - 1)(y + 1)$

47. $6(-10) = -60$; $(-15)4 = -60$; $-15 + 4 = -11$; $6x^2 - 11xy - 10y^2 = 6x^2 - 15xy + 4xy - 10y^2 = 3x(2x - 5y) + 2y(2x - 5y) = (3x + 2y)(2x - 5y)$

48. $4p^2 + 20pq + 25q^2 = (2p)^2 + 2(2p)(5q) + (5q)^2 = (2p + 5q)^2$

49. $8 \times (-9) = -72$; $-18(4) = -72$; $-18 + 4 = -14$; $8a^2 - 14ab - 9b^2 = 8a^2 - 18ab + 4ab - 9b^2 = 2a(4a - 9b) + b(4a - 9b) = (2a + b)(4a - 9b)$

50. $6d^9 + 15d^5e^2 + 6de^4 = 3d(2d^8 + 5d^4e^2 + 2e^4) = 3d(2d^4 + e^2)(d^4 + 2e^2)$

51. $a^3 - b^3 = (a - b)(a^2 + ab + b^2)$

52. $y^3 - 8 = y^3 - 2^3 = (y - 2)(y^2 + 2y + 4)$

53. $8x^3 - 27 = (2x)^3 - 3^3 = (2x - 3) \times \left[(2x)^2 + 2x \cdot 3 + 3^2\right] = (2x - 3)(4x^2 + 6x + 9)$

54. $64p^3 + 125t^6 = (4p)^3 + (5t^2)^3 = (4p + 5t^2)\left[(4p)^2 - 4p \cdot 5t^2 + (5t^2)^2\right] = (4p + 5t^2)(16p^2 - 20pt^2 + 25t^4)$

55. $i = 0.7(t^2 - 3t - 4) = 0.7(t - 4)(t + 1)$

56. $V = 180x - 58x^2 + 4x^3 = 2x(90 - 29x + 2x^2) = 2x(10 - x)(9 - 2x)$

57. $0.0001n^2(n - 2000) - 3(n - 2000) = (0.0001n^2 - 3)(n - 2000) = 0.0001(n^2 - 30{,}000)(n - 2000)$

58. $-16t^2 + 48t + 448 = -16(t^2 - 3t - 28) = -16(t - 7)(t + 4)$

59. (a) $(6 + 2x)(10 + 2x) - 60 = 36$; $60 + 32x + 4x^2 - 60 - 36 = 0$; $4x^2 + 32x - 36 = 0$. (b) $4(x^2 + 8x - 9) = 0$; $4(x + 9)(x - 1) = 0$

7.4 FRACTIONS

1. $\frac{7}{8}$ (by 5); $\frac{7}{8} \cdot \frac{5}{5} = \frac{35}{40}$; $7 \times 40 = 280$; $8 \times 35 = 280$

2. $\frac{-5}{9}$ (by -4); $\frac{-5}{9} \cdot \frac{-4}{-4} = \frac{20}{-36}$; $(-5)(-36) = 180$; $9(20) = 180$

3. $\frac{x}{y}$ (by a); $\frac{x}{y} \cdot \frac{a}{a} = \frac{ax}{ay}$. The cross-product is $axy = axy$.

4. $\frac{r}{t}$ (by z); $\frac{rz}{tz}$. The cross-product is $rtz = rtz$.

5. $\frac{x^2y}{a}$ (by $3ax$); $\frac{3ax^3y}{3a^2x}$. The cross-product is $3a^2x^3y = 3a^2x^3y$.

6. $\frac{a^3b}{ca}$ (by $3ab$); $\frac{3a^4b^2}{3ca^2b}$. The cross-product is $3a^5b^2c = 3a^5b^2c$.

7. $\frac{4}{x-y}$(by $x+y$); $\frac{4}{(x-y)} \cdot \frac{(x+y)}{(x+y)} = \frac{4(x+y)}{x^2-y^2} = \frac{4x+4y}{x^2-y^2}$. The cross-product is $4(x^2-y^2) = (4x+4y)(x-y) = 4x^2-4y^2$.

8. $\frac{a+b}{4}$(by $a-4$); $\frac{(a+b)}{4}\frac{(a-4)}{(a-4)} = \frac{a^2-4a+ab-4b}{4a-16}$. The cross-product is $(a+b)(4a-16) = 4a^2-16a+4ab-16b$.

9. $\frac{a+b}{a-b}$(by $a+b$); $\frac{(a+b)}{(a-b)} \cdot \frac{(a+b)}{(a+b)} = \frac{a^2+2ab+b^2}{a^2-b^2}$. The cross-product is $(a+b)(a^2-b^2) = a^3+a^2b-ab^2-b^3 = (a-b)(a^2+2ab+b^2)$.

10. $\frac{x-2}{x+3}$(by $x-3$); $\frac{(x-2)}{(x+3)} \cdot \frac{(x-3)}{(x-3)} = \frac{x^2-5x+6}{x^2-9}$. The cross-product is $(x-2)(x^2-9) = x^3-2x^2-9x+18 = (x+3)(x^2-5x+6)$.

11. $\frac{38}{24}$(by 2) $= \frac{19}{12}$; $12 \times 38 = 456$; $24 \times 19 = 456$

12. $\frac{51}{119}$(by 17) $= \frac{3}{7}$; $7 \times 51 = 357$; $119 \times 3 = 357$

13. $\frac{3x^2}{12x}$ (by $3x$) $= \frac{x}{4}$; $4 \cdot 3x^2 = 12x^2$; $12x \cdot x = 12x^2$

14. $\frac{15a^3x^2}{3a^4x}$ (by $3a^3x$) $= \frac{5x}{a}$; $a \cdot 15a^3x^2 = 15a^4x^2$; $3a^4x \cdot 5x = 15a^4x^2$

15. $\frac{4(x+2)}{(x+2)(x-3)}$ (by $x+2$) $= \frac{4}{x-3}$; $4(x+2)(x-3) = 4(x+2)(x-3)$.

16. $\frac{7(x-3)(x+5)}{14(x+5)(x-1)}$ [by $7(x+5)$] $= \frac{x-3}{2(x-1)}$; $2(x-1)(7)(x-3)(x+5) = 14(x-1)(x-3)(x+5)$

17. $\frac{x^2-16}{x^2+8x+16}$ (by $x+4$) $= \frac{x^2-16}{x^2+8x+10} = \frac{(x+4)(x-4)}{(x+4)^2} = \frac{x-4}{x+4}$; $(x+4)(x^2-16) = (x+4)^2(x-4)$

18. $\frac{(x-a)(x-b)(x-c)}{(x-c)(x-b)(x-d)}$ [by $(x-c)(x-b)$] $= \frac{x-a}{x-d}$; $(x-a)(x-c)(x-b)(x-d) = (x-d)(x-a)(x-b)(x-c)$

19. $\frac{4x^2}{12x} = \frac{4x \cdot x}{4x \cdot 3} = \frac{x}{3}$

20. $\frac{9y}{3y^2} = \frac{3y \cdot 3}{3y \cdot y} = \frac{3}{y}$

21. $\frac{x^2+3x}{x^3+5x} = \frac{x(x+3)}{x(x^2+5)} = \frac{x+3}{x^2+5}$

22. $\frac{y^2-4y}{2y+y^3} = \frac{y(y-4)}{y(2+y^2)} = \frac{y-4}{2+y^2}$

23. $\frac{6m^2-3m^3}{9m+18m^3} = \frac{3m^2(2-m)}{9m(1+2m^2)} = \frac{m(2-m)}{3(1+2m^2)}$ or $\frac{2m-m^2}{3+6m^2}$

24. $\frac{4r^2+12r^3}{8r+12r^2} = \frac{4r^2(1+3r)}{4r(2+3r)} = \frac{r(1+3r)}{2+3r} = \frac{r+3r^2}{2+3r}$

25. $\frac{x^2+3x}{x^2-9} = \frac{x(x+3)}{(x+3)(x-3)} = \frac{x}{x-3}$

26. $\frac{a^2-9a}{a^2-81} = \frac{a(a-9)}{(a+9)(a-9)} = \frac{a}{a+9}$

27. $\frac{2b^2-10b}{3b^2-75} = \frac{2b(b-5)}{3(b+5)(b-5)} = \frac{2b}{3(b+5)}$ or $\frac{2b}{3b+15}$

28. $\frac{4e^2-196}{14e-2e^2} = \frac{4(e^2-49)}{2e(7-e)} = \frac{4(e+7)(e-7)}{2e(7-e)} = \frac{-2 \cdot 2(e+7)(7-e)}{2e(7-e)} = \frac{-2(e+7)}{e}$

29. $\frac{z^2-9}{z^2-6z+9} = \frac{(z+3)(z-3)}{(z-3)^2} = \frac{z+3}{z-3}$

30. $\dfrac{x^2-16}{x^2+8x+16} = \dfrac{(x+4)(x-4)}{(x+4)^2} = \dfrac{x-4}{x+4}$

31. $\dfrac{x^2+4x+3}{x^2+7x+12} = \dfrac{(x+1)(x+3)}{(x+4)(x+3)} = \dfrac{x+1}{x+4}$

32. $\dfrac{a^2-5a+6}{a^2+5a-14} = \dfrac{(a-2)(a-3)}{(a+7)(a-2)} = \dfrac{a-3}{a+7}$

33. $\dfrac{2x^2+9x+4}{x^2+9x+20} = \dfrac{(2x+1)(x+4)}{(x+5)(x+4)} = \dfrac{2x+1}{x+5}$

34. $\dfrac{15m^2-22m-5}{3m^2+4m-15} = \dfrac{(5m+1)(3m-5)}{(3m-5)(m+3)} = \dfrac{5m+1}{m+3}$

35. $\dfrac{12y^3+12y^2+3y}{6y^2-3y-3} = \dfrac{3y(4y^2+4y+1)}{3(2y^2-y-1)} = \dfrac{y(2y+1)^2}{(2y+1)(y-1)} = \dfrac{y(2y+1)}{y-1}$

36. $\dfrac{45x^2-60x+20}{6x^2+5x-6} = \dfrac{5(9x^2-12x+4)}{(2x+3)(3x-2)} = \dfrac{5(3x-2)^2}{(2x+3)(3x-2)} = \dfrac{5(3x-2)}{2x+3}$

37. $\dfrac{x^3y^6-y^3x^6}{2x^3y^4-2x^4y^3} = \dfrac{x^3y^3(y^3-x^3)}{2x^3y^3(y-x)} = \dfrac{(y-x)(y^2+yx+x^2)}{2(y-x)} = \dfrac{y^2+xy+x^2}{2}$

38. $\dfrac{x^3-y^3}{y^2-x^2} = \dfrac{(x-y)(x^2+xy+y^2)}{(y-x)(y+x)} = -\dfrac{x^2+xy+y^2}{(x+y)}$

39. $\dfrac{x^2-y^2}{x+y} = \dfrac{(x+y)(x-y)}{(x+y)} = \dfrac{x-y}{1} = x-y$

40. $\dfrac{y-x}{x^2-y^2} = \dfrac{y-x}{(x+y)(x-y)} = \dfrac{-1}{x+y}$

7.5 MULTIPLICATION AND DIVISION OF FRACTIONS

1. $\dfrac{2}{x}\cdot\dfrac{5}{y} = \dfrac{2\times 5}{x\times y} = \dfrac{10}{xy}$

2. $\dfrac{4}{y}\cdot\dfrac{x}{3} = \dfrac{4\cdot x}{y\cdot 3} = \dfrac{4x}{3y}$

3. $\dfrac{4x^2}{5}\cdot\dfrac{3}{y^3} = \dfrac{4x^2\cdot 3}{5y^3} = \dfrac{12x^2}{5y^3}$

4. $\dfrac{7x}{6}\cdot\dfrac{5y}{2t} = \dfrac{7x\cdot 5y}{6\cdot 2t} = \dfrac{35xy}{12t}$

5. $\dfrac{3}{x}\div\dfrac{7}{y} = \dfrac{3}{x}\times\dfrac{y}{7} = \dfrac{3y}{7x}$

6. $\dfrac{a}{3}\div\dfrac{b}{4} = \dfrac{a}{3}\times\dfrac{4}{b} = \dfrac{4a}{3b}$

7. $\dfrac{2x^2}{3}\div\dfrac{7y}{4x} = \dfrac{2x^2}{3}\times\dfrac{4x}{7y} = \dfrac{8x^3}{21y}$

8. $\dfrac{9x}{2y}\div\dfrac{4y}{3a} = \dfrac{9x}{2y}\times\dfrac{3a}{4y} = \dfrac{27ax}{8y^2}$

9. $\dfrac{2x}{3y}\cdot\dfrac{5}{4x^2} = \dfrac{2\cdot x\cdot 5}{3\cdot y\cdot 2\cdot 2\cdot x\cdot x} = \dfrac{5}{3y2x} = \dfrac{5}{6xy}$

10. $\dfrac{3xy}{7}\cdot\dfrac{14x}{5y^2} = \dfrac{3\cdot x\cdot y\cdot 7\cdot 2\cdot x}{7\cdot 5\cdot y\cdot y} = \dfrac{3\cdot x\cdot 2\cdot x}{5\cdot y} = \dfrac{6x^2}{5y}$

11. $\dfrac{3a^2b}{5d}\cdot\dfrac{25ad^2}{6b^2} = \dfrac{3a^2b\cdot 5\cdot 5\cdot add}{5d\cdot 3\cdot 2\cdot bb} = \dfrac{5a^3d}{2b}$

12. $\dfrac{x^2y^2t}{abc}\cdot\dfrac{b^2c}{y^3t} = \dfrac{x^2y^2tbbc}{abcy^2yt} = \dfrac{x^2b}{ay}$

13. $\dfrac{3y}{5x}\div\dfrac{15x^2}{8xy} = \dfrac{3y}{5x}\times\dfrac{8xy}{15x^2} = \dfrac{3y8xy}{5x\cdot 3\cdot 5xx} = \dfrac{8y^2}{25x^2}$

14. $\dfrac{4y^2}{7x}\div\dfrac{8y^3}{21x} = \dfrac{4y^2}{7x}\times\dfrac{21x}{8y^3} = \dfrac{4y^2\cdot 3\cdot 7x}{7x\cdot 2\cdot 4y^2\cdot y} = \dfrac{3}{2y}$

15. $\dfrac{3x^2y}{7p}\div\dfrac{15x^2p}{7y^2} = \dfrac{3x^2y}{7p}\times\dfrac{7y^2}{15x^2p} = \dfrac{3x^2y\cdot 7y^2}{7p\cdot 3\cdot 5x^2p} = \dfrac{y^3}{5p^2}$

16. $\dfrac{9xyz}{7a}\div\dfrac{3ayz}{14z} = \dfrac{9xyz}{7a}\times\dfrac{14z}{3ayz} = \dfrac{3\cdot 3xyz\cdot 2\cdot 7z}{7a\cdot 3ayz} = \dfrac{6xz}{a^2}$

17. $\frac{4y+16}{5}\cdot\frac{15y}{3y+12}=\frac{4(y+4)\cdot 3\cdot 5y}{5\cdot 3(y+4)}=\frac{4y}{1}=4y$

18. $\frac{x^2+3x}{6a}\cdot\frac{a^2}{x^2-9}=\frac{x(x+3)a^2}{6a(x+3)(x-3)}=\frac{ax}{6(x-3)}$ or $\frac{ax}{6x-18}$

19. $\frac{a^2-b^2}{a+3b}\cdot\frac{5a+15b}{a+b}=\frac{(a+b)(a-b)5(a+3b)}{(a+3b)(a+b)}=$ $5(a-b)$ or $5a-5b$

20. $(x+y)\frac{x^2+2x}{x^2-y^2}=\frac{(x+y)(x)(x+2)}{(x+y)(x-y)}=\frac{x(x+2)}{x-y}$ or $\frac{x^2+2x}{x-y}$

21. $\frac{x^2-100}{10}\div\frac{2x+10}{15}=\frac{x^2-100}{10}\times\frac{15}{2x+10}=$ $\frac{(x+10)(x-10)\cdot 3\cdot 5}{2\cdot 5\cdot 2(x+5)}=\frac{3(x^2-100)}{4(x+5)}$

22. $\frac{5a^2}{x^2-49}\div\frac{25ax-25a}{x^2+7x}=\frac{5a^2}{x^2-49}\times\frac{x^2+7x}{25ax-25a}=$ $\frac{5\cdot aax(x+7)}{(x+7)(x-7)25a(x-1)}=\frac{ax}{5(x-7)(x-1)}$

23. $\frac{4x^2-1}{9x-3x^2}\div\frac{2x+1}{x^2-9}=\frac{4x^2-1}{9x-3x^2}\times\frac{x^2-9}{2x+1}=$ $\frac{(2x+1)(2x-1)(x+3)(x-3)}{3x(3-x)(2x+1)}=$ $-\frac{(2x-1)(x+3)}{3x}$

24. $\frac{x+y}{3x-3y}\div\frac{(x+y)^2}{x^2-y^2}=\frac{x+y}{3(x-y)}\times\frac{x^2-y^2}{(x+y)^2}=$ $\frac{(x+y)(x+y)(x-y)}{3(x-y)(x+y)^2}=\frac{1}{3}$

25. $\frac{a^2-8a}{a-8}\cdot\frac{a+2}{a}=\frac{a(a-8)(a+2)}{(a-8)a}=a+2$

26. $\frac{49-x^2}{x+y}\cdot\frac{x}{7-x}=\frac{(7+x)(7-x)\cdot x}{(x+y)(7-x)}=\frac{x(7+x)}{x+y}$

27. $\frac{2a-b}{4a}\cdot\frac{2a-b}{4a^2-4ab+b^2}=\frac{(2a-b)^2}{4a(2a-b)^2}=\frac{1}{4a}$

28. $\frac{x^4-81}{(x-3)^2}\cdot\frac{x-3}{4-x^2}=\frac{(x^2+9)(x+3)(x-3)}{(x-3)^2}\cdot$ $\frac{(x-3)}{(2+x)(2-x)}=\frac{(x^2+9)(x+3)}{4-x^2}$

29. $\frac{y^2}{x^2-1}\div\frac{y^2}{x-1}=\frac{y^2}{(x+1)(x-1)}\times\frac{x-1}{y^2}=\frac{1}{x+1}$

30. $\frac{m^2-49}{m^2-5m-14}\div\frac{m+7}{2m^2-13m-7}=$ $\frac{(m+7)(m-7)}{(m-7)(m+2)}\times\frac{(2m+1)(m-7)}{(m+7)}=$ $\frac{(2m+1)(m-7)}{m+2}$

31. $\frac{2y^2-y}{4y^2-4y+1}\div\frac{y^2}{8y-4}=\frac{y(2y-1)}{(2y-1)^2}\cdot\frac{4(2y-1)}{y^2}=$ $\frac{4}{y}$

32. $\frac{a-1}{a^2-1}\div\frac{(a-1)^2}{a^2-1}=\frac{a-1}{(a^2-1)}\cdot\frac{(a^2-1)}{(a-1)^2}=\frac{1}{a-1}$

33. $\frac{x^2-3x+2}{x^2+5x+6}\cdot\frac{x+3}{3x-6}=$ $\frac{(x-2)(x-1)}{(x+2)(x+3)}\cdot\frac{x+3}{3(x-2)}=\frac{x-1}{3(x+2)}$

34. $\frac{2x+2}{x^2+2x-8}\cdot\frac{x^2-4}{x^2+4x+4}=$ $\frac{2(x+1)}{(x+4)(x-2)}\cdot\frac{(x+2)(x-2)}{(x+2)^2}=\frac{2(x+1)}{(x+4)(x+2)}$

35. $\frac{x^2+xy-6y^2}{x^2+6xy+8y^2}\cdot\frac{x^2-9xy+20y^2}{x^2-4xy-21y^2}=$ $\frac{(x+3y)(x-2y)}{(x+4y)(x+2y)}\cdot\frac{(x-4y)(x-5y)}{(x-7y)(x+3y)}=$ $\frac{(x-2y)(x-4y)(x-5y)}{(x+4y)(x+2y)(x-7y)}$

36. $\frac{y^2+14xy+49x^2}{y^2-7xy-30x^2}\cdot\frac{y^2-100x^2}{y^3+7xy^2}=$ $\frac{(y+7x)^2}{(y-10x)(y+3x)}\cdot\frac{(y+10x)(y-10x)}{y^2(y+7x)}=$ $\frac{(y+7x)(y+10x)}{y^2(y+3x)}$

37. $\frac{9x^2-25}{x^2+6x+9}\div\frac{3x+5}{x+3}=$ $\frac{(3x+5)(3x-5)}{(x+3)^2}\times\frac{x+3}{3x+5}=\frac{3x-5}{x+3}$

38. $\dfrac{x^2-16}{x^2-6x+8} \div \dfrac{x^3+4x^2}{x^2-9x+14} =$
$\dfrac{(x+4)(x-4)}{(x-4)(x-2)} \cdot \dfrac{(x-7)(x-2)}{x^2(x+4)} = \dfrac{x-7}{x^2}$

39. $\dfrac{x^2+4xy+4y^2}{x^2-4y^2} \div \dfrac{x^2+xy-2y^2}{x^2-xy-2y^2} =$
$\dfrac{(x+2y)^2}{(x+2y)(x-2y)} \times \dfrac{(x-2y)(x+y)}{(x+2y)(x-y)} = \dfrac{x+y}{x-y}$

40. $\dfrac{p^3-27q^3}{3p^2+9pq+27q^2} \div \dfrac{9q^2-p^2}{6p+18q} =$
$\dfrac{(p-3q)(p^2+3pq+9q^2)}{3(p^2+3pq+9q^2)} \cdot \dfrac{6(p+3q)}{(3q+p)(3q-p)} = -2$

7.6 ADDITION AND SUBTRACTION OF FRACTIONS

1. $\dfrac{2}{7}+\dfrac{5}{7}=\dfrac{2+5}{7}=\dfrac{7}{7}=1$

2. $\dfrac{4}{5}+\dfrac{-11}{5}=\dfrac{4-11}{5}=\dfrac{-7}{5}$

3. $\dfrac{7}{3}-\dfrac{5}{3}=\dfrac{7-5}{3}=\dfrac{2}{3}$

4. $\dfrac{-2}{9}-\dfrac{8}{9}=\dfrac{-2-8}{9}=\dfrac{-10}{9}$

5. $\dfrac{1}{2}+\dfrac{1}{3}=\dfrac{3}{6}+\dfrac{2}{6}=\dfrac{3+2}{6}=\dfrac{5}{6}$

6. $\dfrac{3}{4}+\dfrac{-2}{3}=\dfrac{9}{12}+\dfrac{-8}{12}=\dfrac{9-8}{12}=\dfrac{1}{12}$

7. $\dfrac{4}{5}-\dfrac{2}{3}=\dfrac{12}{15}-\dfrac{10}{15}=\dfrac{12-10}{15}=\dfrac{2}{15}$

8. $-\dfrac{5}{7}-\dfrac{3}{5}=\dfrac{-25}{35}-\dfrac{21}{35}=\dfrac{-25-21}{35}=\dfrac{-46}{35}$

9. $\dfrac{1}{x}+\dfrac{5}{x}=\dfrac{1+5}{x}=\dfrac{6}{x}$

10. $\dfrac{2}{y}+\dfrac{-5}{y}=\dfrac{2-5}{y}=\dfrac{-3}{y}$

11. $\dfrac{4}{a}-\dfrac{3}{a}=\dfrac{4-3}{a}=\dfrac{1}{a}$

12. $\dfrac{-5}{p}-\dfrac{-7}{p}=\dfrac{-5+7}{p}=\dfrac{2}{p}$

13. $\dfrac{2x}{y}+\dfrac{3x}{y}=\dfrac{2x+3x}{y}=\dfrac{5x}{y}$

14. $\dfrac{4p}{q}-\dfrac{6p}{q}=\dfrac{4p-6p}{q}=\dfrac{-2p}{q}$

15. $\dfrac{3r}{2t}+\dfrac{-r}{2t}-\dfrac{5r}{2t}=\dfrac{3r-r-5r}{2t}=\dfrac{-3r}{2t}$

16. $\dfrac{3x}{2y}-\dfrac{5x}{2y}+\dfrac{x}{2y}=\dfrac{3x-5x+x}{2y}=\dfrac{-x}{2y}$

17. $\dfrac{3}{x+2}+\dfrac{x}{x+2}=\dfrac{3+x}{x+2}$

18. $\dfrac{5}{y-3}+\dfrac{y}{y-3}=\dfrac{5+y}{y-3}$

19. $\dfrac{t}{t+1}-\dfrac{2}{t+1}=\dfrac{t-2}{t+1}$

20. $\dfrac{a}{b-3}-\dfrac{4}{3-b}=\dfrac{a}{b-3}-\dfrac{4}{3-b}\dfrac{(-1)}{(-1)}=$
$\dfrac{a}{b-3}-\dfrac{-4}{b-3}=\dfrac{a+4}{b-3}$

21. $\dfrac{y-3}{x+2}+\dfrac{3+y}{x+2}=\dfrac{(y-3)+(3+y)}{x+2}=\dfrac{2y}{x+2}$

22. $\dfrac{x+4}{x-2}+\dfrac{x-5}{x-2}=\dfrac{(x+4)+(x-5)}{x-2}=\dfrac{2x-1}{x-2}$

23. $\dfrac{x+2}{a+b}-\dfrac{x-5}{a+b}=\dfrac{(x+2)-(x-5)}{a+b}=\dfrac{7}{a+b}$

24. $\dfrac{x+4}{y-5}-\dfrac{2-x}{y-5}=\dfrac{(x+4)-(2-x)}{y-5}=\dfrac{2x+2}{y-5}$

25. $\dfrac{2}{x}+\dfrac{3}{y}=\dfrac{2}{x}\cdot\dfrac{y}{y}+\dfrac{x}{x}\cdot\dfrac{3}{y}=\dfrac{2y}{xy}+\dfrac{3x}{xy}=\dfrac{2y+3x}{xy}$

26. $\dfrac{x}{y}+\dfrac{5}{x}=\dfrac{x}{y}\cdot\dfrac{x}{x}+\dfrac{5}{x}\cdot\dfrac{y}{y}=\dfrac{x^2}{xy}+\dfrac{5y}{xy}=\dfrac{x^2+5y}{xy}$

27. $\frac{a}{b} - \frac{4}{d} = \frac{a}{b} \cdot \frac{d}{d} - \frac{4}{d} \cdot \frac{b}{b} = \frac{ad}{bd} - \frac{4b}{bd} = \frac{ad - 4b}{bd}$

28. $\frac{2x}{y} - \frac{3y}{x} = \frac{2x}{y} \cdot \frac{x}{x} - \frac{3y}{x} \cdot \frac{y}{y} = \frac{2x^2}{xy} - \frac{3y^2}{xy} = \frac{2x^2 - 3y^2}{xy}$

29. $\frac{3}{x(x+1)} + \frac{4}{x^2 - 1} = \frac{3}{x(x+1)} + \frac{4}{(x+1)(x-1)} =$
$\frac{3(x-1)}{x(x+1)(x-1)} + \frac{4x}{x(x+1)(x-1)} =$
$\frac{3x - 3 + 4x}{x(x+1)(x-1)} = \frac{7x - 3}{x(x+1)(x-1)}$

30. $\frac{5}{y(x+1)} + \frac{x}{y(x+2)} =$
$\frac{5(x+2)}{y(x+1)(x+2)} + \frac{x(x+1)}{y(x+2)(x+1)} =$
$\frac{5x + 10 + x^2 + x}{x(x+1)(x+2)} = \frac{x^2 + 6x + 10}{x(x+1)(x+2)}$

31. $\frac{2}{x^2 - 1} - \frac{4}{(x+1)^2} = \frac{2}{(x+1)(x-1)} - \frac{4}{(x+1)^2} =$
$\frac{2(x+1)}{(x+1)^2(x-1)} - \frac{4(x-1)}{(x+1)^2(x-1)} =$
$\frac{(2x+2) - (4x-4)}{(x+1)^2(x-1)} = \frac{-2x + 6}{(x+1)^2(x-1)}$

32. $\frac{6}{y-2} - \frac{3}{y+2} = \frac{6}{y-2} \cdot \frac{(y+2)}{(y+2)} - \frac{3}{(y+2)} \cdot$
$\frac{(y-2)}{(y-2)} = \frac{(6y + 12) - (3y - 6)}{(y+2)(y-2)} = \frac{3y + 18}{(y+2)(y-2)}$

33. $\frac{x}{x^2 - 11x + 30} + \frac{2}{x^2 - 36} = \frac{x}{(x-6)(x-5)} +$
$\frac{2}{(x-6)(x+6)} = \frac{x(x+6) + 2(x-5)}{(x-6)(x+6)(x-5)} =$
$\frac{x^2 + 6x + 2x - 10}{(x-6)(x+6)(x-5)} = \frac{x^2 + 8x - 10}{(x-6)(x+6)(x-5)}$

34. $\frac{a}{(a-3)(a-6)} + \frac{a}{(a+3)(a-3)} =$
$\frac{a(a+3)}{(a+3)(a-3)(a-6)} + \frac{a(a-6)}{(a+3)(a-3)(a-6)} =$
$\frac{(a^2 + 3a) + (a^2 - 6a)}{(a+3)(a-3)(a-6)} = \frac{2a^2 - 3a}{(a+3)(a-3)(a-6)}$

35. $\frac{2}{x^2 - x - 6} - \frac{5}{x^2 - 4} =$
$\frac{2}{(x-3)(x+2)} - \frac{5}{(x+2)(x-2)} =$
$\frac{2(x-2)}{(x-3)(x+2)(x-2)} - \frac{5(x-3)}{(x-3)(x-2)(x+2)} =$
$\frac{(2x-4) - (5x-15)}{(x-3)(x+2)(x-2)} = \frac{-3x + 11}{(x-3)(x+2)(x-2)}$

36. $\frac{b}{b^2 - 10b + 21} - \frac{b}{b^2 - 9} =$
$\frac{b}{(b-7)(b-3)} - \frac{b}{(b+3)(b-3)} =$
$\frac{b(b+3)}{(b-7)(b-3)(b+3)} - \frac{b(b-7)}{(b+3)(b-3)(b-7)} =$
$\frac{(b^2 + 3b) - (b^2 - 7b)}{(b+3)(b-3)(b-7)} = \frac{10b}{(b+3)(b-3)(b-7)}$

37. $\frac{x-1}{3x^2 - 13x + 4} + \frac{3x+1}{4x - x^2} =$
$\frac{x-1}{(3x-1)(x-4)} + \frac{3x+1}{x(4-x)} =$
$\frac{x(x-1)}{x(3x-1)(x-4)} + \frac{(-1)(3x+1)(3x-1)}{x(x-4)(3x-1)} =$
$\frac{(x^2 - x) - (9x^2 - 1)}{x(x-4)(3x-1)} = \frac{-8x^2 - x + 1}{x(x-4)(3x-1)}$

38. $\frac{x-3}{x^2 + 3x + 2} + \frac{2x-5}{x^2 + x - 2} =$
$\frac{x-3}{(x+2)(x+1)} + \frac{2x-5}{(x+2)(x-1)} =$
$\frac{(x-3)(x-1)}{(x+2)(x+1)(x-1)} + \frac{(2x-5)(x+1)}{(x+2)(x+1)(x-1)} =$
$\frac{(x^2 - 4x + 3) + (2x^2 - 3x - 5)}{(x+2)(x+1)(x-1)} =$
$\frac{3x^2 - 7x - 2}{(x+2)(x+1)(x-1)}$

39. $\frac{x-3}{x^2 - 1} + \frac{2x-7}{x^2 + 5x + 4} =$
$\frac{x-3}{(x+1)(x-1)} + \frac{2x-7}{(x+1)(x+4)} =$
$\frac{(x-3)(x+4)}{(x+1)(x-1)(x+4)} + \frac{(2x-7)(x-1)}{(x+1)(x-1)(x+4)} =$
$\frac{(x^2 + x - 12) + (2x^2 - 9x + 7)}{(x+1)(x-1)(x+4)} =$
$\frac{3x^2 - 8x - 5}{(x+1)(x-1)(x+4)}$

40. $\frac{x+4}{x^2-9} - \frac{x-3}{x^2+6x+9} = \frac{x+4}{(x+3)(x-3)} - \frac{x-3}{(x+3)^2} = \frac{(x+4)(x+3)}{(x+3)^2(x-3)} - \frac{(x-3)(x-3)}{(x+3)^2(x-3)} = \frac{(x^2+7x+12)-(x^2-6x+9)}{(x+3)^2(x-3)} = \frac{13x+3}{(x+3)^2(x-3)}$

41. $\frac{y+3}{y^2-y-2} - \frac{2y-1}{y^2+2y-8} = \frac{y+3}{(y-2)(y+1)} - \frac{2y-1}{(y+4)(y-2)} = \frac{(y+3)(y+4)}{(y-2)(y+1)(y+4)} - \frac{(2y-1)(y+1)}{(y-2)(y+1)(y+4)} = \frac{(y^2+7y+12)-(2y^2+y-1)}{(y-2)(y+1)(y+4)} = \frac{-y^2+6y+13}{(y-2)(y+1)(y+4)}$

42. $\frac{1}{(a-b)(a-c)} + \frac{1}{(b-a)(b-c)} - \frac{1}{(b-c)(a-c)} = \frac{(b-c)-1(a-c)-(a-b)}{(a-b)(a-c)(b-c)} = \frac{-2a+2b}{(a-b)(a-c)(b-c)} = \frac{-2(a-b)}{(a-b)(a-c)(b-c)} = \frac{-2}{(a-c)(b-c)}$

43. $\frac{x}{(x^2+3)(x-1)} + \frac{3x^2}{(x-1)^2(x+2)} - \frac{x+2}{x^2+3} = \frac{x(x-1)(x+2)+3x^2(x^2+3)-(x+2)(x+2)(x-1)^2}{(x^2+3)(x-1)^2(x+2)} = \frac{(x^3+x^2-2x)+(3x^4+9x^2)-(x^4+2x^3-3x^2-4x+4)}{(x^2+3)(x-1)^2(x+2)} = \frac{2x^4-x^3+13x^2+2x-4}{(x^2+3)(x-1)^2(x+2)}$

44. $\frac{2x-1}{(x+2)(x+3)} - \frac{x-2}{(x+3)(x+1)} + \frac{x-4}{(x+2)(x+1)} = \frac{(2x-1)(x+1)-(x-2)(x+2)+(x-4)(x+3)}{(x+2)(x+3)(x+1)} = \frac{(2x^2+x-1)-(x^2-4)+(x^2-x-12)}{(x+2)(x+3)(x+1)} = \frac{2x^2+x-1-x^2+4+x^2-x-12}{(x+2)(x+3)(x+1)} = \frac{2x^2-9}{(x+2)(x+3)(x+1)}$

45. $\frac{1+\frac{2}{x}}{1-\frac{3}{x}} = \frac{\left(1+\frac{2}{x}\right)x}{\left(1-\frac{3}{x}\right)x} = \frac{x+2}{x-3}$

46. $\frac{x+\frac{1}{x}}{2-\frac{1}{x}} = \frac{\left(x+\frac{1}{x}\right)x}{\left(2-\frac{1}{x}\right)x} = \frac{x^2+1}{2x-1}$

47. $\frac{x-1}{1+\frac{1}{x}} = \frac{(x-1)x}{\left(1+\frac{1}{x}\right)x} = \frac{x^2-x}{x+1} = \frac{x(x-1)}{x+1}$

48. $\frac{x^2-25}{\frac{1}{x}-\frac{1}{5}} = \frac{\left(x^2-25\right)5x}{\left(\frac{1}{x}-\frac{1}{5}\right)5x} = \frac{\left(x^2-25\right)5x}{5-x} = \frac{(x+5)(x-5)5x}{5-x} = -5x(x+5)$

49. $\frac{\frac{x}{x+y}-\frac{y}{x-y}}{\frac{x}{x+y}+\frac{y}{x-y}} = \frac{\left(\frac{x}{x+y}-\frac{y}{x-y}\right)}{\frac{x}{x+y}+\frac{y}{x-y}} \cdot \frac{(x+y)(x-y)}{(x+y)(x-y)} = \frac{x(x-y)-y(x+y)}{x(x-y)+y(x+y)} = \frac{x^2-xy-xy-y^2}{x^2-xy+xy+y^2} = \frac{x^2-2xy-y^2}{x^2+y^2}$

50. $\frac{\left(x+3-\frac{16}{x+3}\right)}{\left(x-6+\frac{20}{x+6}\right)} \cdot \frac{(x+3)(x+6)}{(x+3)(x+6)} = \frac{\left(x^2+6x+9-16\right)(x+6)}{\left(x^2-36+20\right)(x+3)} = \frac{\left(x^2+6x-7\right)(x+6)}{\left(x^2-16\right)(x+3)}$

51. $\frac{1+\frac{3}{x}}{1+\frac{2}{x}} = \frac{\frac{x+3}{x}}{\frac{x+2}{x}} = \frac{x+3}{x} \times \frac{x}{x+2} = \frac{x+3}{x+2}$

52. $\frac{y+\frac{1}{y}}{3+\frac{2}{y}} = \frac{\frac{y^2+1}{y}}{\frac{3y+2}{y}} = \frac{y^2+1}{y} \times \frac{y}{3y+2} = \frac{y^2+1}{3y+2}$

53. $\frac{t-1}{t+\frac{1}{t}} = \frac{t-1}{\frac{t^2+1}{t}} = \frac{t-1}{1} \times \frac{t}{t^2+1} = \frac{t^2-t}{t^2+1}$

54. $\frac{x^2-36}{\frac{1}{6}-\frac{1}{x}} = \frac{x^2-36}{\frac{x-6}{6x}} = \frac{x^2-36}{1} \times \frac{6x}{x-6} = (x+6)(6x)$

55. $\dfrac{\frac{x}{x-y}-\frac{y}{x+y}}{\frac{1}{x-y}+\frac{1}{x+y}} = \dfrac{\frac{x^2+xy}{x^2-y^2}-\frac{xy-y^2}{x^2-y^2}}{\frac{x+y+x-y}{x^2-y^2}} = \dfrac{\frac{x^2+y^2}{x^2-y^2}}{\frac{2x}{x^2-y^2}} =$
$\dfrac{x^2+y^2}{x^2-y^2} \times \dfrac{x^2-y^2}{2x} = \dfrac{x^2+y^2}{2x}$

56. $\dfrac{t-5+\frac{25}{t-5}}{t+3+\frac{10}{t-3}} = \dfrac{\frac{t^2-10t+25+25}{t-5}}{\frac{t^2-9+10}{t-3}} =$
$\dfrac{t^2-10t+50}{t-5} \times \dfrac{t-3}{t^2+1} = \dfrac{(t^2-10t+50)(t-3)}{(t-5)(t^2+1)}$

57. $\dfrac{1}{R_1}+\dfrac{1}{R_2} = \dfrac{R_2}{R_1R_2}+\dfrac{R_1}{R_1R_2} = \dfrac{R_1+R_2}{R_1R_2}$

58. $\dfrac{1}{p}+\dfrac{1}{q} = \dfrac{q}{pq}+\dfrac{p}{pq} = \dfrac{q+p}{pq} = \dfrac{p+q}{pq}$

59. $\dfrac{1}{C_1}+\dfrac{1}{C_2}+\dfrac{1}{C_3} = \dfrac{C_2C_3}{C_1C_2C_3}+\dfrac{C_1C_3}{C_1C_2C_3}+\dfrac{C_1C_2}{C_1C_2C_3} =$
$\dfrac{C_1C_2+C_1C_3+C_2C_3}{C_1C_2C_3}$

60. $\dfrac{d_1+d_2}{\frac{d_1}{v_1}+\frac{d_2}{v_2}} \cdot \dfrac{v_1v_2}{v_1v_2} = \dfrac{v_1v_2(d_1+d_2)}{d_1v_2+d_2v_1}$

CHAPTER 7 REVIEW

1. $5x(x-y) = 5x^2 - 5xy$
2. $(3+x)^2 = 9+6x+x^2$
3. $(x-2y)^3 = x^3 - 3x^2(2y) + 3x(2y)^2 - (2y)^3 =$
   $x^3 - 6x^2y + 12xy^2 - 8y^3$
4. $(x+y)(x-6) = x^2 - 6x + xy - 6y$ using FOIL
5. $(2x+3)(x-6) = 2x^2 - 12x + 3x - 18 =$
   $2x^2 - 9x - 18$
6. $(x+7)(x-7) = x^2 - 7^2 = x^2 - 49$
7. $(x^2-5)(x^2+5) = (x^2)^2 - 5^2 = x^4 - 25$
8. $(7x-1)(x+5) = 7x^2+35x-x-5 = 7x^2+34x-5$
9. $(2+x)^3 = 2^3+3\cdot 2^2x+3\cdot 2x^2+x^3 = 8+12x+6x^2+x^3$
10. $(x-7)^2 = x^2 - 2\cdot x\cdot 7 + 7^2 = x^2 - 14x + 49$
11. $9+9y = 9(1+y)$
12. $x^2 - 4 = (x+2)(x-2)$
13. $7x^2 - 63 = 7(x^2-9) = 7(x+3)(x-3)$
14. $x^2 - 12x + 36 = x^2 - 2\cdot x\cdot 6 + 6^2 = (x-6)^2$
15. $x^2 - 11x + 30 = (x-5)(x-6)$ or $(x-6)(x-5)$
16. $x^2 + 15x + 36 = (x+12)(x+3)$
17. $x^2 + 6x - 16 = (x+8)(x-2)$
18. $x^2 - 4x - 45 = (x+5)(x-9)$
19. $2(-9) = -18$; $-6\cdot 3 = -18$; $-6+3 = -3$;
    $2x^2 - 3x - 9 = 2x^2 - 6x + 3x - 9 =$
    $2x(x-3) + 3(x-3) = (2x+3)(x-3)$
20. Begin by factoring out $2x$, getting
    $8x^3 + 6x^2 - 20x = 2x(4x^2+3x-10)$. Then, for the $4x^2+3x-10$ expression, we have
    $4(-10) = -40$; $8(-5) = -40$; $8+(-5) = 3$;
    $2x(4x^2+3x-10) = 2x[4x^2+8x-5x-10] =$
    $2x[4x(x+2) - 5(x+2)] = 2x(4x-5)(x+2)$
21. $\dfrac{2x}{6y} = \dfrac{x}{3y}$
22. $\dfrac{7x^2y}{9xy^2} = \dfrac{7x}{9y}$
23. $\dfrac{x^2-9}{(x+3)^2} = \dfrac{(x+3)(x-3)}{(x+3)^2} = \dfrac{x-3}{x+3}$
24. $\dfrac{x^2-4x-45}{x^2-81} = \dfrac{(x-9)(x+5)}{(x-9)(x+9)} = \dfrac{x+5}{x+9}$
25. $\dfrac{x^3+y^3}{x^2+2xy+y^2} = \dfrac{(x+y)(x^2-xy+y^2)}{(x+y)^2} =$
    $\dfrac{x^2-xy+y^2}{x+y}$
26. $\dfrac{x^3-16x}{x^2+2x-8} = \dfrac{x(x+4)(x-4)}{(x+4)(x-2)} = \dfrac{x(x-4)}{x-2}$

27. $\frac{x^2}{y} \cdot \frac{3y^2}{7x} = \frac{3x^2y^2}{7xy} = \frac{3xy}{7}$

28. $\frac{x^2-9}{x+4} \cdot \frac{x^3-16x}{x-3} =$
$\frac{(x+3)(x-3)(x)(x+4)(x-4)}{(x+4)(x-3)} =$
$x(x+3)(x-4)$

29. $\frac{4x}{3y} \div \frac{2x^2}{6y} = \frac{4x}{3y} \cdot \frac{6y}{2x^2} = \frac{4x \cdot 3 \cdot 2y}{3y \cdot 2x^2} = \frac{4}{x}$

30. $\frac{x^2-25}{x^2-4x} \div \frac{2x^2+2x-40}{x^3-x} =$
$\frac{x^2-25}{x^2-4x} \cdot \frac{x^3-x}{2x^2+2x-40} =$
$\frac{(x+5)(x-5)x(x+1)(x-1)}{x(x-4)2(x+5)(x-4)} =$
$\frac{(x-5)(x+1)(x-1)}{2(x-4)^2}$

31. $\frac{4x}{y} + \frac{3x}{y} = \frac{4x+3x}{y} = \frac{7x}{y}$

32. $\frac{4}{x-y} + \frac{6}{x+y} = \frac{4(x+y)}{x^2-y^2} + \frac{6(x-y)}{x^2-y^2} =$
$\frac{4x+4y+6x-6y}{x^2-y^2} = \frac{10x-2y}{x^2-y^2}$

33. $\frac{3(x-3)}{(x+2)(x-5)^2} + \frac{4(x-1)}{(x+2)^2(x-5)} =$
$\frac{3(x-3)(x+2)+4(x-1)(x-5)}{(x+2)^2(x-5)^2} =$
$\frac{3x^2-3x-18+4x^2-24x+20}{(x+2)^2(x-5)^2} =$
$\frac{7x^2-27x+2}{(x+2)^2(x-5)^2}$

34. $\frac{8a}{b} - \frac{3}{b} = \frac{8a-3}{b}$

35. $\frac{x}{y+x} - \frac{x}{y-x} = \frac{x(y-x)-x(y+x)}{(y+x)(y-x)} =$
$\frac{xy-x^2-xy-x^2}{y^2-x^2} = \frac{-2x^2}{y^2-x^2}$ or $\frac{2x^2}{x^2-y^2}$ or
$\frac{2x^2}{(x+y)(x-y)}$

36. $\frac{2(x+3)}{(x+1)^2(x+2)} - \frac{3(x-1)}{(x+1)(x+2)^2} =$
$\frac{2(x+3)(x+2)-3(x-1)(x+1)}{(x+1)^2(x+2)^2} =$
$\frac{2x^2+10x+12-3x^2+3}{(x+1)^2(x+2)^2} = \frac{-x^2+10x+15}{(x+1)^2(x+2)^2}$

37. $\frac{x^2-5x-6}{(x+6)(x+2)} + \frac{x^2+7x+6}{(x+2)(x-6)} =$
$\frac{(x^2-5x-6)(x-6)}{(x+6)(x+2)(x-6)} + \frac{(x^2+7x+6)(x+6)}{(x+2)(x-6)(x+6)} =$
$\frac{x^3-11x^2+24x+36+x^3+13x^2+48x+36}{(x+2)(x-6)(x+6)} =$
$\frac{2x^3+2x^2+72x+72}{(x+2)(x-6)(x+6)} =$
$\frac{2(x^3+x^2+36x+36)}{(x+2)(x-6)(x+6)} =$
$\frac{2[x^2(x+1)+36(x+1)]}{(x+2)(x-6)(x+6)} =$
$\frac{2(x^2+36)(x+1)}{(x+2)(x-6)(x+6)}$

38. $\frac{2x-1}{4x^2-12x+5} - \frac{x+1}{4x^2-4x-15} =$
$\frac{2x-1}{(2x-5)(2x-1)} - \frac{x+1}{(2x-5)(2x+3)} =$
$\frac{(2x-1)(2x+3)}{(2x-5)(2x-1)(2x+3)} -$
$\frac{(x+1)(2x-1)}{(2x-5)(2x+3)(2x-1)} =$
$\frac{(4x^2+4x-3)-(2x^2+x-1)}{(2x-5)(2x+3)(2x-1)} =$
$\frac{2x^2+3x-2}{(2x-5)(2x+3)(2x-1)} =$
$\frac{(2x-1)(x+2)}{(2x-5)(2x+3)(2x-1)} = \frac{x+2}{(2x-5)(2x+3)}$

39. $\frac{x^2-5x-6}{x^2+8x+12} \div \frac{x^2+7x+6}{x^2-4x-12} =$
$\frac{(x+1)(x-6)}{(x+6)(x+2)} \cdot \frac{(x-6)(x+1)}{(x+6)(x+1)} = \frac{(x-6)^2}{(x+6)^2}$

40. $\dfrac{x^2+x-2}{7a^2x^2-14a^2x+7a^2}\cdot\dfrac{14ax-28a}{1-2x+x^2}=$
$\dfrac{(x+2)(x-1)}{7a^2(x-1)^2}\cdot\dfrac{14a(x-2)}{(x-1)^2}=\dfrac{2(x+2)(x-2)}{a(x-1)^3}=$
$\dfrac{2\left(x^2-4\right)}{a(x-1)^3}$

41. $\dfrac{\frac{1}{x}-\frac{1}{y}}{\frac{1}{x}+\frac{1}{y}}=\dfrac{\frac{y-x}{xy}}{\frac{y+x}{xy}}\cdot\dfrac{xy}{xy}=\dfrac{y-x}{y+x}$

42. $\dfrac{\frac{1}{x}+\frac{1}{y}}{x+y}=\dfrac{\frac{y+x}{xy}}{x+y}\cdot\dfrac{xy}{xy}=\dfrac{y+x}{(x+y)(xy)}=\dfrac{1}{xy}$

43. $\dfrac{\frac{1}{x}-\frac{1}{y}}{\frac{x-y}{xy}}=\dfrac{\frac{y-x}{xy}}{\frac{x-y}{xy}}\cdot\dfrac{xy}{xy}=\dfrac{y-x}{x-y}=-1$

44. $\dfrac{1-\frac{1}{x}}{x-2+\frac{1}{x}}=\dfrac{\frac{x-1}{x}}{\frac{x^2-2x+1}{x}}\cdot\dfrac{x}{x}=\dfrac{x-1}{x^2-2x+1}=$
$\dfrac{x-1}{(x-1)^2}=\dfrac{1}{x-1}$

45. $\dfrac{\frac{x}{1+x}-\frac{1-x}{x}}{\frac{x}{1+x}+\frac{1-x}{x}}=\dfrac{\frac{x^2-(1-x)(1+x)}{x(1+x)}}{\frac{x^2+(1+x)(1-x)}{x(1+x)}}\cdot$
$\dfrac{x(1+x)}{x(1+x)}=\dfrac{x^2-\left(1-x^2\right)}{x^2+1-x^2}=\dfrac{2x^2-1}{1}=2x^2-1$

46. $\dfrac{x-\frac{xy}{x-y}}{\frac{x^2}{x^2-y^2}-1}=\dfrac{\frac{x(x-y)-xy}{x-y}}{\frac{x^2-(x^2-y^2)}{x^2-y^2}}\cdot\dfrac{(x-y)(x+y)}{(x-y)(x+y)}=$
$\dfrac{x(x+y)(x-y)-xy(x+y)}{x^2-\left(x^2-y^2\right)}=$
$\dfrac{x^3-xy^2-x^2y-xy^2}{y^2}=\dfrac{x^3-x^2y-2xy^2}{y^2}=$
$\dfrac{x\left(x^2-xy-2y^2\right)}{y^2}=\dfrac{x(x+y)(x-2y)}{y^2}$

CHAPTER 7 TEST

1. $(x+5)(x-3)=x^2-3x+5x-15=x^2+2x-15$

2. $(2x-3)(2x+3)=(2x)^2-3^2=4x^2-9$

3. $2x^2-128=2\left(x^2-64\right)=2(x+8)(x-8)$

4. $x^2-12x+32=(x-8)(x-4)$

5. $\dfrac{x^2-25}{x^2+6x+5}=\dfrac{(x+5)(x-5)}{(x+5)(x+1)}=\dfrac{x-5}{x+1}$

6. $\dfrac{3x}{x+2}\cdot\dfrac{x-1}{x+2}=\dfrac{3x(x-1)}{(x+2)(x+2)}=\dfrac{3x(x-1)}{(x+2)^2}=$
$\dfrac{3x^2-3x}{x^2+4x+4}$

7. $\dfrac{2x+6}{x-2}\div\dfrac{3x+9}{x^2-4}=\dfrac{2x+6}{x-2}\cdot\dfrac{x^2-4}{3x+9}=$
$\dfrac{2(x+3)(x+2)(x-2)}{(x-2)3(x+3)}=\dfrac{2(x+2)}{3}$

8. $\dfrac{6}{x-5}+\dfrac{x^2-2x}{x-5}=\dfrac{x^2-2x+6}{x-5}$

9. $\dfrac{2x}{x+3}-\dfrac{x+4}{x-2}=$
$\dfrac{2x(x-2)}{(x+3)(x-2)}-\dfrac{(x+4)(x+3)}{(x+3)(x-2)}=$
$\dfrac{\left(2x^2-4x\right)-\left(x^2+7x+12\right)}{(x+3)(x-2)}=\dfrac{x^2-11x-12}{(x+3)(x-2)}$

10. $\dfrac{x-\frac{1}{x}}{x-\frac{2}{x+1}}=\dfrac{\frac{x^2-1}{x}}{\frac{x^2+x-2}{x+1}}=\dfrac{\left(x^2-1\right)}{x}\cdot\dfrac{(x+1)}{x^2+x-2}=$
$\dfrac{(x+1)(x-1)(x+1)}{x(x+2)(x-1)}=\dfrac{(x+1)^2}{x(x+2)}$

11. $\dfrac{1}{R_1}+\dfrac{1}{R_2}+\dfrac{1}{R_3}=\dfrac{R_2R_3}{R_1R_2R_3}+\dfrac{R_1R_3}{R_1R_2R_3}+\dfrac{R_1R_2}{R_1R_2R_3}=$
$\dfrac{R_1R_2+R_1R_3+R_2R_3}{R_1R_2R_3}$

CHAPTER

8

Fractional and Quadratic Equations

8.1 FRACTIONAL EQUATIONS

1. $\frac{x}{2} + \frac{x}{3} = \frac{1}{4}$; LCD = 12; $12\left(\frac{x}{2} + \frac{x}{3}\right) = 12 \cdot \frac{1}{4}$; $12 \cdot \frac{x}{2} + 12 \cdot \frac{x}{3} = 12 \cdot \frac{1}{4}$; $6x + 4x = 3$; $10x = 3$; $x = \frac{3}{10}$ or 0.3

2. $\frac{x}{3} - \frac{x}{4} = \frac{1}{2}$; LCD = 12; $12\left(\frac{x}{3} - \frac{x}{4}\right) = 12 \cdot \frac{1}{2}$; $12 \cdot \frac{x}{3} - 12 \cdot \frac{x}{4} = 12 \cdot \frac{1}{2}$; $4x - 3x = 6$; $x = 6$

3. $\frac{y}{2} + 3 = \frac{4y}{5}$; LCD = 10; $10\left(\frac{y}{2} + 3\right) = 10\frac{4y}{5}$; $10 \cdot \frac{y}{2} + 10 \cdot 3 = 10 \cdot \frac{4y}{5}$; $5y + 30 = 2 \cdot 4y$; $5y + 30 = 8y$; $30 = 3y$; $10 = y$

4. $\frac{y}{5} - 5\frac{1}{2} = \frac{3y}{4}$; LCD = 20; $20\left(\frac{y}{5} - \frac{11}{2}\right) = 20\frac{3y}{4}$; $20\frac{y}{5} - 20 \cdot \frac{11}{2} = 20\frac{3y}{4}$; $4y - 10 \cdot 11 = 5 \cdot 3y$; $4y - 110 = 15y$; $-110 = 11y$; $y = -10$

5. $\frac{x-1}{2} + \frac{x+1}{3} = \frac{x-1}{4}$; LCD = 12; $12 \cdot \frac{x-1}{2} + 12 \cdot \frac{x+1}{3} = 12 \cdot \frac{x-1}{4}$; $6(x-1) + 4(x+1) = 3(x-1)$; $6x - 6 + 4x + 4 = 3x - 3$; $10x - 2 = 3x - 3$; $7x = -1$; $x = -\frac{1}{7}$

6. $\frac{x+2}{3} - \frac{x+4}{2} = \frac{x-1}{6}$; LCD = 6; $6 \cdot \frac{x+2}{3} - 6 \cdot \frac{x+4}{2} = 6 \cdot \frac{x-1}{6}$; $2(x+2) - 3(x+4) = x - 1$; $2x + 4 - 3x - 12 = x - 1$; $-x - 8 = x - 1$; $-7 = 2x$; $x = -\frac{7}{2}$

7. $\frac{1}{x} + \frac{2}{x} = \frac{1}{3}$; LCD = $3x$; $x \neq 0$; $3x \cdot \frac{1}{x} + 3x \cdot \frac{2}{x} = 3x \cdot \frac{1}{3}$; $3 + 6 = x$; $9 = x$

8. $\frac{3}{x} - \frac{4}{x} = \frac{2}{5}$; LCD = $5x$; $x \neq 0$; $5x \cdot \frac{3}{x} - 5x \cdot \frac{4}{x} = 5x \cdot \frac{2}{5}$; $15 - 20 = 2x$; $-5 = 2x$; $x = -\frac{5}{2}$

9. $\frac{7}{w-4} = \frac{1}{2w+5}$; LCD = $(w-4)(2w+5)$; $w \neq 4$; $w \neq -\frac{5}{2}$; $(w-4)(2w+5) \cdot \frac{7}{w-4} = (w-4)(2w+5) \cdot \frac{1}{2w+5}$; $(2w+5)7 = w - 4$; $14w + 35 = w - 4$; $13w = -39$; $w = -3$

10. $\frac{5}{y+1} = \frac{3}{y-3}$; LCD = $(y+1)(y-3)$; $y \neq -1$; $y \neq 3$; $(y+1)(y-3) \cdot \frac{5}{y+1} = (y+1)(y-3) \cdot \frac{3}{y-3}$; $(y-3)5 = (y+1)3$; $5y - 15 = 3y + 3$; $2y = +18$; $y = 9$

11. $\frac{2}{2x-1} = \frac{5}{x+5}$; LCD = $(2x-1)(x+5)$; $x \neq \frac{1}{2}$; $x \neq -5$; $(2x-1)(x+5) \cdot \frac{2}{2x-1} =$

$(2x-1)(x+5)\cdot\frac{5}{x+5}$; $(x+5)2 = (2x-1)5$; $2x+10 = 10x-5$; $15 = 8x$; $x = \frac{15}{8}$

12. $\frac{3}{4x+2} = \frac{1}{x+2}$; LCD $= (4x+2)(x+2)$; $x \neq -\frac{1}{2}$; $x \neq -2$; $(4x+2)(x+2)\cdot\frac{3}{4x+2} = (4x+2)(x+2)\cdot\frac{1}{x+2}$; $(x+2)3 = 4x+2$; $3x+6 = 4x+2$; $4 = x$

13. $\frac{4x}{x-3} - 1 = \frac{3x}{x+2}$; LCD $= (x-3)(x+2)$; $x \neq 3$; $x \neq -2$; $4x(x+2) - (x-3)(x+2) = 3x(x-3)$; $\left(4x^2+8x\right) - \left(x^2-x-6\right) = 3x^2-9x$; $3x^2+9x+6 = 3x^2-9x$; $18x = -6$; $x = -\frac{1}{3}$

14. $7 - \frac{3x}{x+2} = \frac{4x}{x+2}$; LCD $= x+2$; $x \neq -2$; $7(x+2) - (x+2)\cdot\frac{3x}{x+2} = (x+2)\frac{4x}{x+2}$; $7x + 14 - 3x = 4x$; $4x+14 = 4x$; $0 = 14$; no solution

15. $\frac{4}{x+2} - \frac{3}{x-1} = \frac{5}{(x-1)(x+2)}$; LCD $= (x+2)(x-1)$; $x \neq -2$; $x \neq 1$; $(x+2)(x-1)\cdot\frac{4}{x+2} - (x+2)(x-1)\cdot\frac{3}{x-1} = (x+2)(x-1)\cdot\frac{5}{(x+2)(x-1)}$; $4(x-1) - 3(x+2) = 5$; $4x-4-3x-6 = 5$; $x-10 = 5$; $x = 15$

16. $\frac{3}{x-3} + \frac{2}{2-x} = \frac{5}{(x-3)(x-2)}$; LCD $= (x-3)(x-2)$; $x \neq 3$; $x \neq 2$; $(x-3)(x-2)\cdot\frac{3}{x-3} + (x-3)(x-2)\cdot\frac{2}{2-x} = (x-3)(x-2)\frac{5}{(x-3)(x-2)}$; $3(x-2) - 2(x-3) = 5$; $3x-6-2x+6 = 5$; $x = 5$

17. $\frac{x+1}{x+2} + \frac{x+3}{x-2} = \frac{2x^2+3x-5}{x^2-4}$; LCD $= (x+2)(x-2)$ or x^2-4; $x \neq 2$; $x \neq -2$; $(x+2)(x-2)\cdot\frac{x+1}{x+2} + (x+2)(x-2)\cdot\frac{x+3}{x-2} = \left(x^2-4\right)\cdot\frac{2x^2+3x-5}{x^2-4}$; $(x-2)(x+1) + (x+2)(x+3) = 2x^2+3x-5$; $x^2-x-2+x^2+5x+6 = 2x^2+3x-5$; $2x^2+4x+4 = 2x^2+3x-5$; $x = -9$

18. $\frac{x+2}{x+3} - \frac{x+5}{x-3} = \frac{2x-1}{x^2-9}$; LCD $= (x+3)(x-3)$ or x^2-9; $x \neq 3$; $x \neq -3$; $(x-3)(x+2) - (x+3)(x+5) = 2x-1$; $x^2-x-6-x^2-8x-15 = 2x-1$; $-9x-21 = 2x-1$; $-11x = 20$; $x = -\frac{20}{11}$

19. $\frac{3}{a+1} + \frac{a+1}{a-1} = \frac{a^2}{a^2-1}$; LCD $= (a+1)(a-1)$ or a^2-1; $a \neq 1$; $a \neq -1$; $3(a-1) + (a+1)(a+1) = a^2$; $3a-3+a^2+2a+1 = a^2$; $5a-2 = 0$; $a = \frac{2}{5}$

20. $\frac{5}{x-4} - \frac{x+2}{x+4} = \frac{x^2}{16-x^2}$; LCD $= (x-4)(x+4)$ or x^2-16; $x \neq 4$; $x \neq -4$; $5(x+4) - (x+2)(x-4) = \left(x^2-16\right)\frac{x^2}{16-x^2}$; $5x+20-x^2+2x+8 = -x^2$; $7x+28 = 0$; $x = -4$; no solution, since -4 makes the denominator 0.

21. $\frac{2}{x-1} + \frac{5}{x+1} = \frac{4}{x^2-1}$; LCD $= (x-1)(x+1)$ or x^2-1; $x \neq 1$; $x \neq -1$; $2(x+1) + 5(x-1) = 4$; $2x+2+5x-5 = 4$; $7x-3 = 4$; $x = 1$; no solution since x cannot equal 1.

22. $\frac{3x+4}{x+2} - \frac{3x-5}{x-4} = \frac{12}{x^2-2x-8}$; LCD $= (x+2)(x-4)$ or x^2-2x-8; $x \neq -2$; $x \neq 4$; $(3x+4)(x-4) - (3x-5)(x+2) = 12$; $\left(3x^2-8x-16\right) - \left(3x^2+x-10\right) = 12$; $-9x-6 = 12$; $-9x = 18$; $x = -2$; no solution, since x cannot equal -2.

23. $\frac{5x-2}{x-3} + \frac{4-5x}{x+4} = \frac{10}{x^2+x-12}$; LCD $= (x-3)(x+4)$ or x^2+x-12; $x \neq 3$; $x \neq -4$; $(5x-2)(x+4) + (4-5x)(x-3) = 10$; $\left(5x^2+18x-8\right) + \left(-5x^2+19x-12\right) = 10$; $37x-20 = 10$; $37x = 30$; $x = \frac{30}{37}$

24. $\frac{2x}{x-1} - \frac{3}{x+2} = \frac{4x}{x^2+x-2} + 2$; LCD $= (x-1)(x+2) = x^2+x-2$; $x \neq 1$; $x \neq -2$; $2x(x+2) - 3(x-1) = 4x + 2(x^2+x-2)$; $2x^2 + 4x - 3x + 3 = 4x + 2x^2 + 2x - 4$; $-5x = -7$; $x = \frac{7}{5}$

25. $\frac{5}{x} + \frac{3}{x+1} = \frac{x}{x+1} - \frac{x+1}{x}$; LCD $= x(x+1)$; $x \neq 0$; $x \neq -1$; $5(x+1) + 3x = x^2 - (x+1)(x+1)$; $5x+5+3x = x^2 - (x^2+2x+1)$; $8x+5 = -2x-1$; $10x = -6$; $x = -\frac{6}{10} = -\frac{3}{5}$

26. $\frac{y}{y+2} + \frac{5}{y-1} = \frac{3}{y+2} + \frac{y}{y-1}$; LCD $= (y+2)(y-1)$; $y \neq -2$; $y \neq 1$; $y(y-1) + 5(y+2) = 3(y-1) + y(y+2)$; $y^2 - y + 5y + 10 = 3y - 3 + y^2 + 2y$; $4y + 10 = 5y - 3$; $13 = y$

27. $\frac{2t-4}{2t+4} = \frac{t+2}{t+4}$; LCD $= (2t+4)(t+4)$; $t \neq -2$; $t \neq -4$; $(2t-4)(t+4) = (t+2)(2t+4)$; $2t^2 + 4t - 16 = 2t^2 + 8t + 8$; $4t - 16 = 8t + 8$; $-4t = 24$; $t = -6$

28. $\frac{3x+5}{x-5} = \frac{3x-1}{x+3}$; LCD $= (x-5)(x+3)$; $x \neq 5$; $x \neq -3$; $(3x+5)(x+3) = (3x-1)(x-5)$; $3x^2 + 14x + 15 = 3x^2 - 16x + 5$; $30x = -10$; $x = -\frac{1}{3}$

29. $\frac{3x+1}{x-1} - \frac{x-2}{x+3} = \frac{2x-3}{x+3} + \frac{4}{x-1}$; LCD $= (x-1)(x+3)$; $x \neq 1$; $x \neq -3$; $(3x+1)(x+3) - (x-2)(x-1) = (2x-3)(x-1) + 4(x+3)$; $3x^2 + 10x + 3 - (x^2 - 3x + 2) = 2x^2 - 5x + 3 + 4x + 12$; $2x^2 + 13x + 1 = 2x^2 - x + 15$; $14x = 14$; $x = 1$; no solution

30. $\frac{7x+2}{x+2} + \frac{3x-1}{x+3} = \frac{6x+1}{x+3} + \frac{4x-3}{x+2}$; LCD $= (x+2)(x+3)$; $x \neq -2$; $x \neq -3$; $(7x+2)(x+3) + (3x-1)(x+2) = (6x+1)(x+2) + (4x-3)(x+3)$; $7x^2 + 23x + 6 + 3x^2 + 5x - 2 = 6x^2 + 13x + 2 + 4x^2 + 9x - 9$; $10x^2 + 28x + 4 = 10x^2 + 22x - 7$; $6x = -11$; $x = -\frac{11}{6}$

31. $\frac{1}{r} + \frac{1}{s} = \frac{1}{t}$; LCD $= rst$; $rst \cdot \frac{1}{r} + rst \cdot \frac{1}{s} = rst \cdot \frac{1}{t}$; $st + rt = rs$; $rs - st = rt$; $s(r-t) = rt$; $s = \frac{rt}{r-t}$

32. $\frac{P_1V_1}{T_1} = \frac{P_2V_2}{T_2}$; LCD $= T_1T_2$; $P_1V_1T_2 = P_2V_2T_1$; $T_1 = \frac{P_1V_1T_2}{P_2V_2}$

33. $\frac{1}{R} = \frac{1}{R_1} + \frac{1}{R_2} + \frac{1}{R_3}$; LCD $= RR_1R_2R_3$; $R_1R_2R_3 = RR_2R_3 + RR_1R_3 + RR_1R_2$; $R(R_2R_3 + R_1R_3 + R_1R_2) = R_1R_2R_3$; $R = \frac{R_1R_2R_3}{R_2R_3 + R_1R_3 + R_1R_2}$

34. $P = \frac{E^2}{R+r} - \frac{E^2}{(R+r)^2}$; LCD $= (R+r)^2$; $P(R+r)^2 = E^2(R+r) - E^2$; $E^2(R+r-1) = P(R+r)^2$; $E^2 = \frac{P(R+r)^2}{R+r-1}$

35. $V = 2\pi rh + 2\pi r^2$; $2\pi rh = V - 2\pi r^2$; $h = \frac{V - 2\pi r^2}{2\pi r}$

36. $\frac{5}{9}(F - 32) = C$; $(F - 32) = \frac{9}{5}C$; $F = \frac{9}{5}C + 32$

37. $\frac{1}{f} = (n-1)\left(\frac{1}{R_1} + \frac{1}{R_2}\right)$; LCD $= fR_1R_2$; $R_1R_2 = (n-1)(fR_2 + fR_1)$; $R_1R_2 - (n-1)(fR_2) = (n-1)fR_1$; $R_2(R_1 - fn + f) = (n-1)fR_1$; $R_2 = \frac{(n-1)fR_1}{R_1 - nf + f}$

38. $\frac{P_1}{g} + \frac{V_1^2}{2g} + h_1 = \frac{P_2}{dg} + \frac{V_2^2}{2g} + h_2$; LCD $= 2gd$; $2gd \cdot \frac{P_1}{g} + 2gd \cdot \frac{V_1^2}{2g} + 2gdh_1 = 2dg \cdot \frac{P_2}{dg} + 2dg \cdot \frac{V_2^2}{2g} + 2dgh_2$; $2dP_1 + dV_1^2 + 2gdh_1 = 2P_2 + dV_2^2 + 2dgh_2$; $2gdh_1 - 2dgh_2 = 2P_2 + dV_2^2 - 2dP_1 - dV_1^2$; $g \cdot 2d \cdot (h_1 - h_2) = 2P_2 + dV_2^2 - 2dP_1 - dV_1^2$; $g = \frac{2P_2 + dV_2^2 - 2dP_1 - dV_1^2}{2d(h_1 - h_2)}$ or $g = \frac{2dP_1 + dV_1^2 - 2P_2 - dV_2^2}{2d(h_2 - h_1)}$

39. h = hours B is open; $\frac{h+1}{6}+\frac{h}{4}=1$; LCD = 12; $12\cdot\frac{h+1}{6}+12\cdot\frac{h}{4}=12$; $2h+2+3h=12$; $5h=10$; $h=2$; $h+1=3$. It takes a total of 3 hours.

40. t = time together; $\frac{t}{5}+\frac{t}{8}=1$; $8t+5t=40$; $13t=40$; $t=\frac{40}{13}\approx 3.08\ \mu s$

41. $\frac{h+4}{18}+\frac{h}{12}=1$; LCD = 36; $2(h+4)+3h=36$; $5h+8=36$; $5h=28$; $h=\frac{28}{5}=5.6$ hr together; Total time is $5.6+4=9.6$ hr

42. r = speed of airplane in still air; $r+20$ = speed of airplane with tail wind; $r-20$ = speed of airplane with head wind; distance = rate × time so time = $\frac{\text{distance}}{\text{rate}}$; $\frac{500}{r-20}=\frac{650}{r+20}$; $500(r+20)=650(r-20)$; $500r+10{,}000=650r-13000$; $23{,}000=150r$; $r=\frac{23000}{150}\approx 153.3$ km/h

8.2 QUADRATIC EQUATIONS AND FACTORING

1. Factoring $x^2-9=0$ we get $(x+3)(x-3)=0$. By the zero product rule we have $x+3=0$ and $x=-3$ or $x-3=0$ and so $x=3$.

2. Factoring $x^2-100=0$ produces $(x+10)(x-10)=0$. By the zero product rule we have $x+10=0$ and so $x=-10$ or $x-10=0$ and we get $x=10$.

3. $x^2+x-6=0$; $(x+3)(x-2)=0$; $x+3=0\Rightarrow x=-3$ or $x-2=0\Rightarrow x=2$

4. $x^2-6x-7=0$; $(x-7)(x+1)=0$; $x-7=0\Rightarrow x=7$ or $x+1=0\Rightarrow x=-1$

5. $x^2-11x-12=0$; $(x-12)(x+1)=0$; $x-12=0\Rightarrow x=12$ or $x+1=0\Rightarrow x=-1$

6. $x^2-5x+4=0$; $(x-4)(x-1)=0$; $x-4=0\Rightarrow x=4$ or $x-1=0\Rightarrow x=1$

7. $x^2+2x-8=0$; $(x+4)(x-2)=0$; $x+4=0\Rightarrow x=-4$ or $x-2=0\Rightarrow x=2$

8. $x^2+2x-15=0$; $(x+5)(x-3)=0$; $x+5=0\Rightarrow x=-5$ or $x-3=0\Rightarrow x=3$

9. $x^2-5x=0$; $x(x-5)=0$; $x=0$ or $x-5=0\Rightarrow x=5$

10. $x^2+10x=0$; $x(x+10)=0$; $x=0$ or $x+10=0\Rightarrow x=-10$

11. $x^2+12=7x$; $x^2-7x+12=0$; $(x-3)(x-4)=0$; $x-3=0\Rightarrow x=3$; $x-4=0\Rightarrow x=4$

12. $x^2=7x-10$; $x^2-7x+10=0$; $(x-5)(x-2)=0$; $x-5=0\Rightarrow x=5$; $x-2=0\Rightarrow x=2$

13. $2x^2-3x-14=0$; $(2x-7)(x+2)=0$; $2x-7=0\Rightarrow x=\frac{7}{2}$; $x+2=0\Rightarrow x=-2$

14. $2x^2+x-15=0$; $(2x-5)(x+3)=0$; $2x-5=0\Rightarrow x=\frac{5}{2}$; $x+3=0\Rightarrow x=-3$

15. $2x^2+12=11x$; $2x^2-11x+12=0$; $(2x-3)(x-4)=0$; $2x-3=0\Rightarrow x=\frac{3}{2}$; $x-4=0\Rightarrow x=4$

16. $2x^2+18=15x$; $2x^2-15x+18=0$; $(2x-3)(x-6)=0$; $2x-3=0\Rightarrow x=\frac{3}{2}$; $x-6=0\Rightarrow x=6$

17. $3x^2-8x-3=0$; $(3x+1)(x-3)=0$; $3x+1=0\Rightarrow x=-\frac{1}{3}$; $x-3=0\Rightarrow x=3$

18. $3x^2-4x-4=0$; $(3x+2)(x-2)=0$; $3x+2=0\Rightarrow x=-\frac{2}{3}$; $x-2=0\Rightarrow x=2$

19. $4x^2-24x+35=0$; $(2x-5)(2x-7)=0$; $2x-5=0\Rightarrow x=\frac{5}{2}$; $2x-7=0\Rightarrow x=\frac{7}{2}$

20. $6x^2-13x+6=0$; $(3x-2)(2x-3)=0$; $3x-2=0\Rightarrow x=\frac{2}{3}$; $2x-3=0\Rightarrow x=\frac{3}{2}$

21. $6x^2+11x-35=0$; $(3x-5)(2x+7)=0$; $3x-5=0\Rightarrow x=\frac{5}{3}$; $2x+7=0\Rightarrow x=-\frac{7}{2}$

22. $10x^2+9x-9=0$; $(5x-3)(2x+3)=0$; $5x-3=0\Rightarrow x=\frac{3}{5}$; $2x+3=0\Rightarrow x=-\frac{3}{2}$

23. $10x^2 - 17x + 3 = 0$; $(5x - 1)(2x - 3) = 0$; $5x - 1 = 0 \Rightarrow x = \frac{1}{5}$; $2x - 3 = 0 \Rightarrow x = \frac{3}{2}$

24. $14x^2 - 29x - 15 = 0$; $(7x + 3)(2x - 5) = 0$; $7x + 3 = 0 \Rightarrow x = -\frac{3}{7}$; $2x - 5 = 0 \Rightarrow x = \frac{5}{2}$

25. $6x^2 = 31x + 60$; $6x^2 - 31x - 60 = 0$; $(3x - 20)(2x + 3) = 0$; $3x - 20 = 0 \Rightarrow x = \frac{20}{3}$; $2x + 3 = 0 \Rightarrow x = -\frac{3}{2}$

26. $15x^2 - 23x + 4 = 0$; $(5x - 1)(3x - 4) = 0$; $5x - 1 = 0 \Rightarrow x = \frac{1}{5}$; $3x - 4 = 0 \Rightarrow x = \frac{4}{3}$

27. $(x - 1)^2 = 4$; $x - 1 = \pm\sqrt{4}$; $x - 1 = \pm 2$; $x - 1 = 2 \Rightarrow x = 3$; $x - 1 = -2 \Rightarrow x = -1$

28. $(x + 2)^2 = 9$; $x + 2 = \pm\sqrt{9}$; $x + 2 = \pm 3$; $x + 2 = 3 \Rightarrow x = 1$; $x + 2 = -3 \Rightarrow x = -5$

29. $(5x - 2)^2 = 16$; $5x - 2 = \pm\sqrt{16}$; $5x - 2 = \pm 4$; $5x - 2 = 4 \Rightarrow x = \frac{6}{5}$; $5x - 2 = -4 \Rightarrow x = -\frac{2}{5}$

30. $(3x + 2)^2 = 64$; $3x + 2 = \pm 8$; $3x + 2 = 8 \Rightarrow x = 2$; $3x + 2 = -8 \Rightarrow x = -\frac{10}{3}$

31. $\frac{x}{x + 1} = \frac{x + 2}{3x}$; $3x \cdot x = (x + 2)(x + 1)$; $3x^2 = x^2 + 3x + 2$; $2x^2 - 3x - 2 = 0$; $(2x + 1)(x - 2) = 0$; $x = -\frac{1}{2}$ or 2

32. $(x + 2)^3 - x^3 = 56$; $x^3 + 3x^2 \cdot 2 + 3x2^2 + 2^3 - x^3 = 56$; $6x^2 + 12x + 8 = 56$; $6x^2 + 12x - 48 = 0$; $x^2 + 2x - 8 = 0$; $(x - 2)(x + 4) = 0$; $x = 2$ or $x = -4$

33. $\frac{1}{x - 3} + \frac{1}{x + 4} = \frac{1}{12}$; $12(x + 4) + 12(x - 3) = (x - 3)(x + 4)$; $12x + 48 + 12x - 36 = x^2 + x - 12$; $24x + 12 = x^2 + x - 12$; $0 = x^2 - 23x - 24$; $(x - 24)(x + 1) = 0$; $x = 24$ or $x = -1$

34. $\frac{1}{x - 5} + \frac{1}{x + 3} = \frac{1}{3}$; $3(x + 3) + 3(x - 5) = (x - 5)(x + 3)$; $3x + 9 + 3x - 15 = x^2 - 2x - 15$; $6x + 9 = x^2 - 2x$; $x^2 - 8x - 9 = 0$; $(x - 9)(x + 1) = 0$; $x = 9$ or $x = -1$

35. $-16t^2 + 64t + 192 = 192$; $-16t^2 + 64t = 0$; $-16t(t - 4) = 0$; $t = 0$ or $t = 4$; we want $t = 4$ s

36. $-16t^2 + 64t + 192 = 0$; $-16\left(t^2 - 4t - 12\right) = 0$; $-16(t - 6)(t + 2) = 0$; $t = 6$ or $t = -2$; we want $t = 6.0$ s.

37. If we let ℓ represent the length in cm and w its width in cm, then we are given $\ell = w + 5$. The area of a rectangle is $\ell w = (w + 5)w = 104$. Multiplying, we get $w^2 + 5w - 104 = 0$, which factors as $(w + 13)(w - 8) = 0$. Thus, by the zero product rule, we have $w = -13$ or $w = 8$. Choose $w = 8$ cm and then $\ell = w + 5 = 8 + 5 = 13$ cm.

38. $4.9t^2 = 411$; $t^2 = \frac{411}{4.9}$; $t = \sqrt{\frac{411}{4.9}} \approx 9.158$ s

39. Pythagorean Theorem yields $(x - 7)^2 + x^2 = 13^2$; $x^2 - 14x + 49 + x^2 = 169$; $2x^2 - 14x - 120 = 0$; $x^2 - 7x - 60 = 0$; $(x - 12)(x + 5) = 0$; $x = 12$ or $x = -5$; cannot be negative so $x = 12$; $x - 7 = 5$. The rafters are 5 m and 12 m.

40. x = time for B alone; $x - 6$ = time for A alone; $\frac{4}{x} + \frac{4}{x - 6} = 1 \Rightarrow 4(x - 6) + 4x = x(x - 6) \Rightarrow 8x - 24 = x^2 - 6x$, which can be written as $x^2 - 14x + 24 = 0$. This last equation factors as $(x - 12)(x - 2) = 0$. From this we see that B is 12 hr and A is $12 - 6 = 6$ hr

8.3 COMPLETING THE SQUARE

1. $x^2 + 6x + 8 = 0$; $x^2 + 6x + 9 = -8 + 9$; $(x + 3)^2 = 1$; $x + 3 = \pm 1$; $x = -3 + 1 = -2$ or $x = -3 - 1 = -4$

2. $x^2 - 7x - 8 = 0$; $x^2 - 7x + \frac{49}{4} = 8 + \frac{49}{4}$; $\left(x - \frac{7}{2}\right)^2 = \frac{81}{4}$; $x - \frac{7}{2} = \pm\sqrt{\frac{81}{4}}$; $x - \frac{7}{2} = \pm\frac{9}{2}$; $x = \frac{7}{2} + \frac{9}{2} = 8$ or $x = \frac{7}{2} - \frac{9}{2} = -1$

3. $x^2 - 10x = 11$; $x^2 - 10x + 25 = 11 + 25$; $(x - 5)^2 = 36$, $x - 5 = \pm 6$ and so, $x = 5 + 6 = 11$ or $x = 5 - 6 = -1$

4. $2x^2 - 3 = 14$; $x^2 - \frac{3}{2}x = 7$; $x^2 - \frac{3}{2}x + \frac{9}{16} = 7 + \frac{9}{16}$;

$\left(x-\frac{3}{4}\right)^2=\frac{121}{16}$; $x-\frac{3}{4}=\pm\frac{11}{4}$; $x=\frac{3}{4}+\frac{11}{4}=\frac{7}{2}$; $x=\frac{3}{4}-\frac{11}{4}=-2$

5. $x^2+6x+3=0$; $x^2+6x+9=-3+9$; $(x+3)^2=6$; $x+3=\pm\sqrt{6}$ and so $x=-3\pm\sqrt{6}$

6. $x^2+8x+10=0$; $x^2+8x+16=-10+16$; $(x+4)^2=6$; $x+4=\pm\sqrt{6}$; $x=-4\pm\sqrt{6}$

7. $x^2-5x+5=0$; $x^2-5x+\frac{25}{4}=-5+\frac{25}{4}$; $\left(x-\frac{5}{2}\right)^2=\frac{5}{4}$; $x-\dfrac{5}{2}=\pm\dfrac{\sqrt{5}}{2}$; $x=\dfrac{5}{2}\pm\dfrac{\sqrt{5}}{2}$ or $x=\dfrac{5\pm\sqrt{5}}{2}$

8. $x^2-7x+11=0$; $x^2-7x+\dfrac{49}{4}=-11+\dfrac{49}{4}$; $\left(x-\frac{7}{2}\right)^2=\frac{5}{4}$; $x-\dfrac{7}{2}=\pm\dfrac{\sqrt{5}}{2}$; $x=\dfrac{7\pm\sqrt{5}}{2}$

9. $2x^2-6x-10=0$; $x^2-3x-5=0$; $x^2-3x+\dfrac{9}{4}=5+\dfrac{9}{4}$; $\left(x-\frac{3}{2}\right)^2=\frac{29}{4}$; $x-\dfrac{3}{2}=\pm\dfrac{\sqrt{29}}{2}$; $x=\dfrac{3\pm\sqrt{29}}{2}$

10. $3x^2+12x-18=0$; $x^2+4x-6=0$; $x^2+4x+4=6+4$; $(x+2)^2=10$; $x+2=\pm\sqrt{10}$; $x=-2\pm\sqrt{10}$

11. $4x^2-12x-18=0$; $x^2-3x-\dfrac{9}{2}=0$; $x^2-3x+\dfrac{9}{4}=\dfrac{9}{2}+\dfrac{9}{4}$; $\left(x-\frac{3}{2}\right)^2=\frac{27}{4}$; $x-\dfrac{3}{2}=\dfrac{\pm\sqrt{27}}{2}$; $x=\dfrac{3\pm\sqrt{27}}{2}=\dfrac{3\pm3\sqrt{3}}{2}$

12. $3x^2-9x=33$; $x^2-3x=11$; $x^2-3x+\dfrac{9}{4}=11+\dfrac{9}{4}$; $\left(x-\frac{3}{2}\right)^2=\frac{53}{4}$; $x-\dfrac{3}{2}=\dfrac{\pm\sqrt{53}}{2}$; $x=\dfrac{3\pm\sqrt{53}}{2}$

13. $x^2+2kx+c=0$; $x^2+2kx=-c$; $x^2+2kx+k^2=-c+k^2$; $(x+k)^2=-c+k^2$; $x+k=\pm\sqrt{k^2-c}$; $x=-k\pm\sqrt{k^2-c}$

14. $px^2+2qx+r=0$; $x^2+\dfrac{2q}{p}x=-\dfrac{r}{p}$; $x^2+\dfrac{2q}{p}x+\left(\dfrac{q}{p}\right)^2=\dfrac{-r}{p}+\left(\dfrac{q}{p}\right)^2=\dfrac{q^2-rp}{p^2}$; $\left(x+\frac{q}{p}\right)^2=\frac{q^2-rp}{p^2}$; $x+\dfrac{q}{p}=\dfrac{\pm\sqrt{q^2-rp}}{p}$; $x=\dfrac{-q\pm\sqrt{q^2-rp}}{p}$

15. $196-16t^2=0$; $16t^2=196$; $t^2=12.25$; $t=\pm\sqrt{12.25}$ cannot be negative so $\sqrt{12.25}=3.5$ s

16. $q(2520-3q)=140{,}400$; ; $-3q^2+2520q=140{,}400$; $q^2-840q=-46{,}800$; $q^2-840q+176{,}400=-46{,}800+176{,}400=129{,}600$; $(q-420)^2=360^2$; $q=420\pm360$; $q=60$ or 780 objects

17. $q(1560-4q)=29{,}600$; $-4q^2+1560q=29{,}600$; $q^2-390q=-7400$; $q^2-390q+(195)^2=-7400+195^2$; $(q-195)^2=30625$; $q-195=\pm175$; $q=195+175=370$ objects; or $q=195-175=20$ objects

18. $p=2(\ell+w)=1500$ or $\ell+w=750$ and so $\ell=750-w$. $A=\ell\cdot w=(750-w)w=137{,}600$; $-w^2+750w=137{,}600$; $w^2-750w=-137{,}600$; $w^2-750w+375^2=-137{,}600+375^2$; $(w-375)^2=3025$; $w-375=\pm55$; $w=375+55=430\Rightarrow\ell=750-430=320$; or $w=375-55=320$ and $l=430$. Thus, we see that the dimensions are 430 ft by 320 ft

19. $P=4w+3\ell=2700$; $3\ell=2700-4w$; $\ell=900-\frac{4}{3}w$; $A=2\ell w=270{,}000$; $\ell w=135{,}000$; $\left(900-\frac{4}{3}w\right)w=135{,}000$; $900w-\frac{4}{3}w^2=135{,}000$; $-\frac{4}{3}w^2+900w=135{,}000$; $w^2-675w=-101{,}250$; $w^2-675w+\left(\frac{675}{2}\right)^2=-101{,}250+\left(\frac{675}{2}\right)^2$; $\left(w-\frac{675}{2}\right)^2=12656.25$; $w-337.5=\pm\sqrt{12656.25}$; $w-337.5=\pm112.5$; $w=337.5+112.5=450\Rightarrow\ell=300$; or $w=337.5-112.5=225\Rightarrow\ell=600$; 450 ft by 300 ft or 225 ft by 600 ft

8.4 THE QUADRATIC FORMULA

1. $x^2 + 3x - 4 = 0$; $x = \dfrac{-3 \pm \sqrt{(3)^2 - 4(1)(-4)}}{2 \cdot 1} = \dfrac{-3 \pm \sqrt{9+16}}{2} = \dfrac{-3 \pm \sqrt{25}}{2} = \dfrac{-3 \pm 5}{2}$; $\dfrac{-3+5}{2} = 1$; $\dfrac{-3-5}{2} = -4$; $\{1, -4\}$

2. $x^2 - 8x - 33 = 0$; $x = \dfrac{-(-8) \pm \sqrt{(-8)^2 - 4(-33)}}{2 \cdot 1} = \dfrac{8 \pm \sqrt{64+132}}{2} = \dfrac{8 \pm \sqrt{196}}{2} = \dfrac{8 \pm 14}{2}$; $\dfrac{8+14}{2} = \dfrac{22}{2} = 11$; $\dfrac{8-14}{2} = -\dfrac{6}{2} = -3$; $\{-3, 11\}$

3. $3x^2 - 5x - 2 = 0$; $x = \dfrac{-(-5) \pm \sqrt{(-5)^2 - 4(3)(-2)}}{2 \cdot 3} = \dfrac{5 \pm \sqrt{25+24}}{6} = \dfrac{5 \pm \sqrt{49}}{6} = \dfrac{5 \pm 7}{6}$; $\dfrac{5+7}{6} = \dfrac{12}{6} = 2$; $\dfrac{5-7}{6} = \dfrac{-2}{6} = -\dfrac{1}{3}$; $\{-\frac{1}{3}, 2\}$

4. $7x^2 + 5x - 2 = 0$; $x = \dfrac{-5 \pm \sqrt{5^2 - 4 \cdot 7(-2)}}{2 \cdot 7} = \dfrac{-5 \pm \sqrt{25+56}}{14} = \dfrac{-5 \pm \sqrt{81}}{14} = \dfrac{-5 \pm 9}{14}$; $\frac{-5+9}{14} = \frac{4}{14} = \frac{2}{7}$; $\dfrac{-5-9}{14} = \dfrac{-14}{14} = -1$; $\{-1, \frac{2}{7}\}$

5. $7x^2 + 6x - 1 = 0$; $x = \dfrac{-6 \pm \sqrt{6^2 - 4 \cdot 7(-1)}}{14} = \dfrac{-6 \pm \sqrt{36+28}}{14} = \dfrac{-6 \pm \sqrt{64}}{14} = \dfrac{-6 \pm 8}{14}$; $\dfrac{-6+8}{14} = \dfrac{2}{14} = \dfrac{1}{7}$; $\dfrac{-6-8}{14} = \dfrac{-14}{14} = -1$; $\{-1, \frac{1}{7}\}$

6. $2x^2 - 3x - 20 = 0$; $x = \dfrac{-(-3) \pm \sqrt{(-3)^2 - 4 \cdot 2(-20)}}{2 \cdot 2} = \dfrac{3 \pm \sqrt{9+160}}{4} = \dfrac{3 \pm \sqrt{169}}{4} = \dfrac{3 \pm 13}{4}$; $\dfrac{3+13}{4} = \dfrac{16}{4} = 4$; $\dfrac{3-13}{4} = \dfrac{-10}{4} = \dfrac{-5}{2}$; $\{4, -\frac{5}{2}\}$

7. $2x^2 - 5x - 7 = 0$; $x = \dfrac{-(-5) \pm \sqrt{(-5)^2 - 4 \cdot 2(-7)}}{2 \cdot 2} = \dfrac{5 \pm \sqrt{25+56}}{4} = \dfrac{5 \pm \sqrt{81}}{4} = \dfrac{5 \pm 9}{4}$; $\dfrac{5+9}{4} = \dfrac{14}{4} = \dfrac{7}{2}$; $\dfrac{5-9}{4} = \dfrac{-4}{4} = -1$; $\{-1, \frac{7}{2}\}$

8. $3x^2 + 4x - 7 = 0$; $x = \dfrac{-4 \pm \sqrt{4^2 - 4(3)(-7)}}{2 \cdot 3} = \dfrac{-4 \pm \sqrt{16+84}}{6} = \dfrac{-4 \pm \sqrt{100}}{6} = \dfrac{-4 \pm 10}{6}$; $\dfrac{-4+10}{6} = \dfrac{6}{6} = 1$; $\dfrac{-4-10}{6} = \dfrac{-14}{6} = -\dfrac{7}{3}$; $\{1, -\frac{7}{3}\}$

9. $3x^2 + 2x - 8 = 0$; $x = \dfrac{-2 \pm \sqrt{2^2 - 4(3)(-8)}}{2 \cdot 3} = \dfrac{-2 \pm \sqrt{4+96}}{6} = \dfrac{-2 \pm \sqrt{100}}{6} = \dfrac{-2 \pm 10}{6}$; $\dfrac{-2+10}{6} = \dfrac{8}{6} = \dfrac{4}{3}$; $\dfrac{-2-10}{6} = \dfrac{-12}{6} = -2$; $\{-2, \frac{4}{3}\}$

10. $9x^2 - 6x + 1 = 0$; $x = \dfrac{(-6) \pm \sqrt{-(-6)^2 - 4 \cdot 9(1)}}{2 \cdot 9} = \dfrac{6 \pm \sqrt{36-36}}{18} = \dfrac{6}{18} = \dfrac{1}{3}$ double root

11. $9x^2 + 12x + 4 = 0$; $x = \dfrac{-12 \pm \sqrt{12^2 - 4 \cdot 9 \cdot 4}}{2 \cdot 9} = \dfrac{-12 \pm \sqrt{144-144}}{18} = \dfrac{-12}{18} = \dfrac{-2}{3}$ double root

12. $3x^2 + 3x - 7 = 0$; $x = \dfrac{-3 \pm \sqrt{3^2 - 4(3)(-7)}}{2 \cdot 3} = \dfrac{-3 \pm \sqrt{9+84}}{6} = \dfrac{-3 \pm \sqrt{93}}{6}$

13. $2x^2 - 3x - 1 = 0$; $x = \dfrac{-(-3) \pm \sqrt{(-3)^2 - 4 \cdot 2 \cdot (-1)}}{2 \cdot 2} = \dfrac{3 \pm \sqrt{9+8}}{4} = \dfrac{3 \pm \sqrt{17}}{4}$

14. $2x^2 - 5x + 1 = 0$; $x = \dfrac{-(-5) \pm \sqrt{(-5)^2 - 4 \cdot 2 \cdot 1}}{2 \cdot 2} = \dfrac{5 \pm \sqrt{25-8}}{4} = \dfrac{5 \pm \sqrt{17}}{4}$

15. $x^2 + 5x + 2 = 0$; $x = \dfrac{-5 \pm \sqrt{5^2 - 4 \cdot 2 \cdot 1}}{2 \cdot 1} = \dfrac{-5 \pm \sqrt{25 - 8}}{2} = \dfrac{-5 \pm \sqrt{17}}{2}$

16. $3x^2 - 6x - 2 = 0$; $x = \dfrac{-(-6) \pm \sqrt{(-6)^2 - 4 \cdot 3 \cdot (-2)}}{2 \cdot 3} = \dfrac{6 \pm \sqrt{36 + 24}}{6} = \dfrac{6 \pm \sqrt{60}}{6} = \dfrac{6 \pm 2\sqrt{15}}{6} = \dfrac{2\left(3 \pm \sqrt{15}\right)}{2 \cdot 3} = \dfrac{3 \pm \sqrt{15}}{3}$

17. $2x^2 + 6x - 3 = 0$; $x = \dfrac{-6 \pm \sqrt{6^2 - 4(2)(-3)}}{2 \cdot 2} = \dfrac{-6 \pm \sqrt{36 + 24}}{4} = \dfrac{-6 \pm \sqrt{60}}{4} = \dfrac{-6 \pm 2\sqrt{15}}{4} = \dfrac{2\left(-3 \pm \sqrt{15}\right)}{2 \cdot 2} = \dfrac{-3 \pm \sqrt{15}}{2}$

18. $5x^2 + 2x - 1 = 0$; $x = \dfrac{-2 \pm \sqrt{2^2 - 4(5)(-1)}}{2 \cdot 5} = \dfrac{-2 \pm \sqrt{4 + 20}}{10} = \dfrac{-2 \pm \sqrt{24}}{10} = \dfrac{-2 \pm 2\sqrt{6}}{10} = \dfrac{2(-1 \pm \sqrt{6})}{2 \cdot 5} = \dfrac{-1 \pm \sqrt{6}}{5}$

19. $x^2 - 2x - 7 = 0$; $x = \dfrac{2 \pm \sqrt{(-2)^2 - 4 \cdot 1(-7)}}{2} = \dfrac{2 \pm \sqrt{4 + 28}}{2} = \dfrac{2 \pm \sqrt{32}}{2} = \dfrac{2 \pm 4\sqrt{2}}{2} = 1 \pm 2\sqrt{2}$

20. $x^2 + 3 = 0$; $x = \dfrac{0 \pm \sqrt{0 - 4(3)}}{2} = \dfrac{\pm\sqrt{-12}}{2}$ no real roots since the discriminant is -12

21. $2x^2 - 3 = 0$; $x = \dfrac{0 \pm \sqrt{0 - 4(2)(-3)}}{2 \cdot 2} = \dfrac{\pm\sqrt{24}}{4} = \dfrac{\pm 2\sqrt{6}}{4} = \dfrac{\pm\sqrt{6}}{2}$

22. $2x^2 = 5$; $2x^2 - 5 = 0$; $x = \dfrac{0 \pm \sqrt{0^2 - 4 \cdot 2(-5)}}{2 \cdot 2} = \dfrac{\pm\sqrt{40}}{4} = \dfrac{\pm 2\sqrt{10}}{4} = \dfrac{\pm\sqrt{10}}{2}$

23. $3x^2 + 4 = 0$; $x = \dfrac{0 \pm \sqrt{0^2 - 4(3)(4)}}{2 \cdot 3} = \dfrac{\pm\sqrt{-48}}{6}$ no real roots because the discriminant is -48

24. $3x^2 + 1 = 5x$; $3x^2 - 5x + 1 = 0$; $x = \dfrac{5 \pm \sqrt{(-5)^2 - 4 \cdot 3 \cdot 1}}{2 \cdot 3} = \dfrac{5 \pm \sqrt{25 - 12}}{6} = \dfrac{5 \pm \sqrt{13}}{6}$

25. $\dfrac{2}{3}x^2 - \dfrac{1}{9}x + 3 = 0$; $x = \dfrac{\frac{1}{9} \pm \sqrt{\left(-\frac{1}{9}\right)^2 - 4 \cdot \frac{2}{3} \cdot 3}}{2 \cdot \frac{2}{3}} = \dfrac{\frac{1}{9} \pm \sqrt{\frac{1}{81} - 8}}{\frac{4}{3}} = \dfrac{\frac{1}{9} \pm \sqrt{\frac{-647}{81}}}{\frac{4}{3}}$ no real roots because the discriminant is $\dfrac{-647}{81}$

26. $\dfrac{1}{2}x^2 - 2x + \dfrac{1}{3} = 0$; $x = \dfrac{2 \pm \sqrt{(-2)^2 - 4 \cdot \left(\frac{1}{2}\right)\left(\frac{1}{3}\right)}}{2 \cdot \frac{1}{2}} = \dfrac{2 \pm \sqrt{4 - \frac{2}{3}}}{1} = 2 \pm \sqrt{\dfrac{10}{3}}$; or $2 \pm \frac{\sqrt{30}}{3}$ or $\dfrac{6 \pm \sqrt{30}}{3}$

27. $0.01x^2 + 0.2x = 0.6$; $0.01x^2 + 0.2x - 0.6 = 0$; $x = \dfrac{-0.2 \pm \sqrt{.2^2 - 4 \cdot (0.01)(-0.6)}}{2 \cdot (0.01)} = \dfrac{-0.2 \pm \sqrt{0.064}}{0.02}$ or see the alternate solution.

Alternate Solution:

$0.01x^2 + 0.2x - 0.6 = 0$; $x^2 + 20x - 60 = 0$; $x = \dfrac{-20 \pm \sqrt{20^2 - 4(-60)}}{2} = \dfrac{-20 \pm \sqrt{400 + 240}}{2} = \dfrac{-20 \pm \sqrt{640}}{2} = \dfrac{-20 \pm 2\sqrt{160}}{2} = -10 \pm \sqrt{160} = -10 \pm 4\sqrt{10}$

28. $0.16x^2 = 0.8x - 1$; $0.16x^2 - 0.8x + 1 = 0$; $x = \dfrac{0.8 \pm \sqrt{(-0.8)^2 - 4(0.16)(1)}}{2(0.16)} = \dfrac{0.8 \pm \sqrt{0}}{0.32} = 2.5$ double root

29. $\dfrac{1}{4}x^2 + 3 = \dfrac{5}{2}x$; $\dfrac{1}{4}x^2 - \dfrac{5}{2}x + 3 = 0$ or $x^2 - 10x + 12 = 0$; $x = \dfrac{10 \pm \sqrt{(-10)^2 - 4 \cdot 12}}{2} = \dfrac{10 \pm \sqrt{52}}{2} = \dfrac{10 \pm 2\sqrt{13}}{2} = 5 \pm \sqrt{13}$

30. $\frac{3}{2}x^2 + 2x = \frac{7}{2}$; $\frac{3}{2}x^2 + 2x - \frac{7}{2} = 0$; $3x^2 + 4x - 7 = 0$; $x = \frac{-4 \pm \sqrt{4^2 - 4(3)(-7)}}{2 \cdot 3} = \frac{-4 \pm \sqrt{16 + 84}}{6} = \frac{-4 \pm \sqrt{100}}{6} = \frac{-4 \pm 10}{6}$; $\frac{-4 + 10}{6} = 1$; $\frac{-4 - 10}{6} = \frac{-7}{3}$

31. $1.2x^2 = 2x - 0.5 = 1.2x^2 - 2x + 0.5 = 0$; $12x^2 - 20x + 5 = 0$; $x = \frac{20 \pm \sqrt{(-20)^2 - 4(12)5}}{2 \cdot 12} = \frac{20 \pm \sqrt{400 - 240}}{24} = \frac{20 \pm \sqrt{160}}{24} = \frac{20 \pm 4\sqrt{10}}{24} = \frac{5 \pm \sqrt{10}}{6}$ or $x = \frac{2 \pm \sqrt{(-2)^2 - 4(1.2)(0.5)}}{2 \cdot 1.2} = \frac{2 \pm \sqrt{4 - 2.4}}{2.4} = \frac{2 \pm \sqrt{1.6}}{2.4}$

32. $1.4x^2 + 0.2x = 2.3$ or $1.4x^2 + 0.2x - 2.3 = 0$; Multiplying by 10 produces $14x^2 + 2x - 23 = 0$; and so $x = \frac{-2 \pm \sqrt{2^2 - 4(14)(-23)}}{2(14)} = \frac{-2 \pm \sqrt{4 + 1288}}{28} = \frac{-2 \pm \sqrt{1292}}{28} = \frac{-2 \pm 2\sqrt{323}}{28} = \frac{-1 \pm \sqrt{323}}{14}$; or $x = \frac{-0.2 \pm \sqrt{(0.2)^2 - 4 \cdot (1.4)(-2.3)}}{2(1.4)} = \frac{-0.2 \pm \sqrt{.04 + 12.88}}{28} = \frac{-0.2 \pm \sqrt{12.92}}{2.8}$

33. $3x^2 + \sqrt{3}x - 7 = 0$; $x = \frac{-\sqrt{3} \pm \sqrt{\sqrt{3}^2 - 4(3)(-7)}}{2 \cdot 3} = \frac{-\sqrt{3} \pm \sqrt{3 + 84}}{6} = \frac{-\sqrt{3} \pm \sqrt{87}}{6}$

34. $2x^2 - \sqrt{89}x + 5 = 0$ Here $a = 2$, $b = -\sqrt{89}$, and $c = 5$. $x = \frac{\sqrt{89} \pm \sqrt{\left(\sqrt{89}\right)^2 - 4(2)(5)}}{2 \cdot 2} = \frac{\sqrt{89} \pm \sqrt{89 - 40}}{4} = \frac{\sqrt{89} \pm 7}{4}$

35. $\frac{x - 3}{7} = 2x^2$; $x - 3 = 14x^2$; $14x^2 - x + 3 = 0$; $x = \frac{1 \pm \sqrt{(-1)^2 - 4 \cdot 14 \cdot 3}}{2} = \frac{1 \pm \sqrt{-167}}{2}$. Since the discriminant is negative, there are no real roots.

36. $\frac{x - 5}{3} = 5x^2$; Multiplying by 3 produces $x - 5 = 15x^2$ and so, we get $15x^2 - x + 5 = 0$. $x = \frac{1 \pm \sqrt{1 - 4 \cdot 15 \cdot 5}}{2(15)} = \frac{1 \pm \sqrt{-299}}{30}$. Since the discriminant, -299, is negative, there are no real roots.

37. $\frac{2}{x - 1} + 3 = \frac{-2}{x + 1}$; LCD $= (x - 1)(x + 1)$. Multiplying by the LCD produces $2(x + 1) + 3(x - 1)(x + 1) = -2(x - 1)$; $2x + 2 + 3x^2 - 3 = -2x + 2$; $3x^2 + 4x - 3 = 0$; $\frac{-4 \pm \sqrt{4^2 - 4(3)(-3)}}{2(3)}$; $\frac{-4 \pm \sqrt{16 + 36}}{6}$; $\frac{-4 \pm \sqrt{52}}{6} = \frac{-4 \pm 2\sqrt{13}}{6} = \frac{-2 \pm \sqrt{13}}{3}$

38. $\frac{3x}{x + 2} + 2x = \frac{2x^2 - 1}{x + 1}$; LCD $(x + 2)(x + 1) = x^2 + 3x + 2$. Multiplying by the LCD produces $3x(x + 1) + 2x\left(x^2 + 3x + 2\right) = \left(2x^2 - 1\right)(x + 2)$; $3x^2 + 3x + 2x^3 + 6x^2 + 4x = 2x^3 + 4x^2 - x - 2$; $9x^2 + 7x = 4x^2 - x - 2$; $5x^2 + 8x + 2 = 0$; $x = \frac{-8 \pm \sqrt{8^2 - 4 \cdot 5 \cdot 2}}{2 \cdot 5} = \frac{-8 \pm \sqrt{64 - 40}}{10} = \frac{-8 \pm \sqrt{24}}{10} = \frac{-8 \pm 2\sqrt{6}}{10} = \frac{-4 \pm \sqrt{6}}{5}$

39. $-4.9t^2 + 411 = 0$; $\frac{0 \pm \sqrt{0 - 4(-4.9)(411)}}{2(-4.9)} = \frac{\pm\sqrt{8055.6}}{-9.8} \approx 9.16$ s; only positive answer works

40. $\frac{20 \pm \sqrt{(-20)^2 - 4(-4.9)(411)}}{2(-4.9)} = \frac{20 \pm \sqrt{400 + 8055.6}}{-9.8} = \frac{20 \pm \sqrt{8455.6}}{-9.8} \approx \frac{20 \pm 91.95}{-9.8}$. So, $x \approx \frac{20 - 91.95}{-9.8} \approx 7.34$ or $x \approx \frac{20 + 91.95}{-9.8} = -11.42$. Only the positive answer makes sense, and so the ball hits the ground about 7.34 s after it is thrown.

41. $\frac{-20 \pm \sqrt{20^2 - 4(-4.9)(411)}}{2(-4.9)} = \frac{-20 \pm 91.95}{-9.8} = \frac{-20 - 91.95}{-9.8} \approx +11.42$; $\frac{-20+91.95}{-9.8} \approx -7.34$; only positive answer fits the problem situation 11.42 s

42. $4 \cdot (x-8)^2 = 100$; $(x-8)^2 = 25$; $x - 8 = \pm 5$; $x = 8 + 5 = 13$; only 13 cm fits problem since $x = 8 - 5 = 3$ gives negative box dimensions, so the desired dimensions are 13 cm $\times$13 cm.

43. $V = \ell \cdot w \cdot h = (1.5w - 6)(w - 6) \cdot 3 = 578$; $\left(1.5w^2 - 15w + 36\right)3 = 578$; $4.5w^2 - 45w + 108 = 578$; $4.5w^2 - 45w - 470 = 0$; $x = \frac{45 \pm \sqrt{(-45)^2 - 4(4.5)(-470)}}{2 \cdot 4.5} = \frac{45 \pm \sqrt{10{,}485}}{9} \approx \frac{45 + 102.4}{9} = \frac{147.4}{9} \approx 16.38$; $w = 16.38$ cm; $\ell = 1.5(16.38) = 24.57$ cm.

44. $20x + 15x - x^2 = 81.25$; $-x^2 + 35x - 81.25 = 0$; $x^2 - 35x + 81.25 = 0$; $x = \frac{35 \pm \sqrt{(-35)^2 - 4(81.25)}}{2} = \frac{35 \pm \sqrt{900}}{2} = \frac{35 \pm 30}{2}$; $\frac{35 - 30}{2} = 2.5$; $\frac{35 + 30}{2} = 32.5$; x must be less than 15 so answer is 2.5 cm

45. $x(12 - 2x) = 16.875$; $12x - 2x^2 = 16.875$; $-2x^2 + 12x - 16.875 = 0$; $2x^2 - 12x + 16.875 = 0$; $x = \frac{12 \pm \sqrt{(-12)^2 - 4 \cdot (2)(16.875)}}{2 \cdot 2} = \frac{12 \pm 3}{4}$; $\frac{12+3}{4} = \frac{15}{4} = 3\frac{3}{4}$; $\frac{12 - 3}{4} = \frac{9}{4} = 2\frac{1}{4}$; $12 - 2 \cdot 3\frac{3}{4} = 4.5$; $12 - 2 \cdot 2\frac{1}{4} = 7.5$; 2.25 in. by 7.5 in.; or 3.75 in. by 4.5 in.

46. $A = 2\pi r^2 + 2\pi rh$; $2\pi r^2 + 2\pi r \cdot 10 = 245$; $r^2 + 10r = \frac{245}{2\pi} \approx 39$; $r^2 + 10r - 39 = 0$; $r = \frac{-10 \pm \sqrt{10^2 - 4 \cdot 1 \cdot (-39)}}{2} = \frac{-10 \pm \sqrt{100 + 156}}{2}$; $\frac{-10+16}{2} = \frac{6}{2} = 3$ cm. Only positive answers fit the problem.

47. x = amount of increase; new price is $30 + x$; number of customers is $1000 - 5x$; revenue = $(1000 - 5x)(30 + x)$; $(1000 - 5x)(30 + x) = 45{,}000$; $30000 + 850x - 5x^2 = 45{,}000$; $-5x^2 + 850x - 15{,}000 = 0$; $x^2 - 170x + 3{,}000 = 0$; $x = \frac{170 \pm \sqrt{(-170)^2 - 4(3{,}000)}}{2} = \frac{170 \pm 130}{2}$. Thus, one value produces $x = \frac{170 - 130}{2} = \frac{40}{2} = 20$. The other value yields $x = \frac{170 + 130}{2} = 150$; a \$20 increase will lose the fewest customers.

48. $V^2 = V_R^2 + (V_L - V_c)^2$; $5.8^2 = 5^2 + (V_L - 10)^2$; $33.64 = 25 + (V_L - 10)^2$; $8.64 = (V_L - 10)^2$; $V_L - 10 = \pm\sqrt{8.64} \approx \pm 2.94$; $V_L \approx 10 + 2.94 = 12.94$ V or $10 - 2.94 = 7.06$ V

49. $610^2 = 300^2 + (X_L - 531)^2$; $372100 = 90000 + (X_L - 531)^2$; $282100 = (X_L - 531)^2$; $x_L - 531 = \pm\sqrt{282100} \approx \pm 531.1$; $x_L \approx 531 \pm 531.1$. Since x_L cannot be negative, we have $x_L = 1062.13\ \Omega$

50. $\frac{1}{2}w\ell x - \frac{1}{2}wx^2 = 0$; $\ell x - x^2 = 0$; $x(\ell - x) = 0$; $x = 0$ or $x = \ell$

51. See *Computer Programs* in main text.

CHAPTER 8 REVIEW

1. $\frac{x}{3} + \frac{x}{2} = 5$; LCD = 6; $6 \cdot \frac{x}{3} + 6 \cdot \frac{x}{2} = 6 \cdot 5 \Rightarrow 2x + 3x = 30 \Rightarrow 5x = 30$; $x = 6$

2. $\frac{2}{x} - \frac{3}{x} = \frac{1}{5}$; LCD = $5x, x \neq 0$; $5 \cdot 2 - 5 \cdot 3 = x$; $10 - 15 = x$; $-5 = x$

3. $\frac{x-2}{4} - \frac{x+2}{5} = \frac{x}{2}$; LCD 20; $20\left(\frac{x-2}{4}\right) - 20\left(\frac{x+2}{5}\right) = 20 \cdot \frac{x}{2}$; $(5x - 10) - (4x + 8) = 10x$; $5x - 10 - 4x - 8 = 10x$; $x - 18 = 10x$; $-18 = 9x$; $x = -2$

4. $\frac{2}{x-1} + \frac{3}{x-2} = \frac{4}{x^2 - 3x + 2}$; LCD $(x-1)(x-2) = x^2 - 3x + 2$; $x \neq 1$; $x \neq 2$;

$2(x-2)+3(x-1)=4$; $2x-4+3x-3=4$; $5x=11$; $x=\dfrac{11}{5}$

5. $\dfrac{3x}{x-1}-\dfrac{3x+2}{x}=3$; LCD $(x-1)(x)=x^2-x$; $3x(x)-(3x+2)(x-1)=3\left(x^2-x\right)$; $3x^2-3x^2+x+2=3x^2-3x$; $0=3x^2-4x-2$; $\dfrac{4\pm\sqrt{(-4)^2-4\cdot(3)(-2)}}{2\cdot 3}=\dfrac{4\pm\sqrt{40}}{6}=\dfrac{4\pm 2\sqrt{10}}{2\cdot 3}=\dfrac{2\pm\sqrt{10}}{3}$

6. $\dfrac{4x}{2x-3}+\dfrac{1}{x-1}=\dfrac{2x+1}{x-1}$; We see that the LCD $=(2x-3)(x-1)$ and that $x\neq\dfrac{3}{2}$ and $x\neq 1$. Multiplying by the LCD, we get $4x(x-1)+(2x-3)=(2x+1)(2x-3)$ or $4x^2-4x+2x-3=4x^2-4x-3$. Collecting terms, we get $2x=0$ and so, $x=0$.

7. $A=2lw+2(l+w)h$; $A-2lw=2(l+w)h$; $\dfrac{A-2lw}{2(l+w)}=h$

8. $A=2lw+2(l+w)h$; $A=2lw+2lh+2wh$; $2lw+2lh=A-2wh$; $l(2w+2h)=A-2wh$; $l=\dfrac{A-2wh}{2w+2h}$

9. $\dfrac{1}{f}=\dfrac{1}{p}+\dfrac{1}{q}$; LCD $=fpq$; $fpq\cdot\frac{1}{f}=fpq\cdot\frac{1}{p}+fpq\cdot\frac{1}{q}$; $pq=fq+fp$; $pq=f(q+p)$; $f=\dfrac{pq}{p+q}$

10. $X=wL-\dfrac{1}{wc}$; $Xwc=wLwC-1$; $Xwc=w^2Lc-1$; $w^2Lc-Xwc=1$; $c\left(w^2L-Xw\right)=1$; $c=\dfrac{1}{w^2L-Xw}$ or $c=\dfrac{1}{w(wL-X)}$

11. $x^2-8x+7=0$; $(x-7)(x-1)=0$; $x-7=0\Rightarrow x=7$; $x-1=0\Rightarrow x=1$

12. $x^2+4x+3=0$; $(x+3)(x+1)=0$; $x+3=0\Rightarrow x=-3$; $x+1=0\Rightarrow x=-1$

13. $x^2-11x+10=0$; $(x-10)(x-1)=0$; $x-10=0\Rightarrow x=10$; $x-1=0\Rightarrow x=1$

14. $x^2+15x+14=0$; $(x+14)(x+1)=0$; $x+14=0\Rightarrow x=-14$; $x+1=0\Rightarrow x=-1$

15. $x^2+15x+56=0$; $(x+8)(x+7)=0$; $x+7=0\Rightarrow x=-7$; $x+8=0\Rightarrow x=-8$

16. $x^2-8x+12=0$; $(x-6)(x-2)=0$; $x-6=0\Rightarrow x=6$; $x-2=0\Rightarrow x=2$

17. $2x^2-5x+3=0$; $(2x-3)(x-1)=0$; $2x-3=0\Rightarrow x=\frac{3}{2}$; $x-1=0\Rightarrow x=1$

18. $3x^2+10x+7=0$; $(3x+7)(x+1)=0$; $3x+7=0\Rightarrow x=-\frac{7}{3}$; $x+1=0\Rightarrow x=-1$

19. $6x^2+7x-10=0$; $(6x-5)(x+2)=0$; $6x-5=0\Rightarrow x=\frac{5}{6}$; $x+2=0\Rightarrow x=-2$

20. $8x^2+22x+9=0$; $8\times 9=72$; $18\times 4=72$; $18+4=22$; $8x^2+18x+4x+9=0$; $2x(4x+9)+1(4x+9)$; $(2x+1)(4x+9)=0$; $2x+1=0\Rightarrow x=-\frac{1}{2}$; $4x+9=0\Rightarrow x=-\frac{9}{4}$

21. $4x^2-9=0$; $(2x+3)(2x-3)=0$; $2x+3=0\Rightarrow x=-\frac{3}{2}$; $2x-3=0\Rightarrow x=\frac{3}{2}$

22. $9x^2-16=0$; $(3x+4)(3x-4)=0$; $3x+4=0\Rightarrow x=-\frac{4}{3}$; $3x-4=0\Rightarrow x=\frac{4}{3}$

23. $x^2+4x-5=0$; $x^2+4x+4=5+4$; $(x+2)^2=9$; $x+2=\pm 3$; $x=-2+3=1$; $x=-2-3=-5$

24. $x^2-7x+6=0$; $x^2-7x+\dfrac{49}{4}=-6+\dfrac{49}{4}$; $\left(x-\frac{7}{2}\right)^2=\frac{25}{4}$; $x-\dfrac{7}{2}=\pm\dfrac{5}{2}$; $x=\dfrac{7}{2}+\dfrac{5}{2}=\dfrac{12}{2}=6$; $x=\dfrac{7}{2}-\dfrac{5}{2}=\dfrac{2}{2}=1$

25. $x^2+19x=11$; $x^2+19x+\left(\dfrac{19}{2}\right)^2=11+\left(\dfrac{19}{2}\right)^2$; $\left(x+\frac{19}{2}\right)^2=\frac{44}{4}+\frac{361}{4}=\frac{405}{4}$; $x+\dfrac{19}{2}=\dfrac{\pm\sqrt{405}}{2}=\dfrac{\pm 9\sqrt{5}}{2}$; $x=\dfrac{-19\pm 9\sqrt{5}}{2}$

26. $2x^2 + 7x = 15$, upon dividing by 2 becomes $x^2+\frac{7}{2}x = \frac{15}{2}$ and after completing the square yields $x^2 + \frac{7}{2}x + \left(\frac{7}{4}\right)^2 = \frac{15}{2} + \left(\frac{7}{4}\right)^2 = \frac{120}{16} + \frac{49}{16}$ or $\left(x + \frac{7}{4}\right)^2 = \frac{169}{16}$, and after taking the square roots of both sides produces $x+\frac{7}{4} = \pm\frac{13}{4}$ or $x = -\frac{7}{4}+\frac{13}{4} = \frac{6}{4} = \frac{3}{2}$ and $x = \frac{-7}{4} - \frac{13}{4} = \frac{-20}{4} = -5$.

27. $3x^2 + 5x - 14 = 0$ or $3x^2 + 5x = 14$, and, after dividing by 3 yields $x^2 + \frac{5}{3}x = \frac{14}{3}$. Completing the square produces $x^2 + \frac{5}{3}x + \left(\frac{5}{6}\right)^2 = \frac{14}{3} + \left(\frac{5}{6}\right)^2$ or $\left(x + \frac{5}{6}\right)^2 = \frac{168}{36} + \frac{25}{36} = \frac{193}{36}$. Taking the square roots of both sides produces $x + \frac{5}{6} = \pm\frac{\sqrt{193}}{6}$ and so $x = \frac{-5 \pm \sqrt{193}}{6}$.

28. $4x^2+2x-5 = 0$ or $4x^2+2x = 5$, and, after dividing by 4 yields $x^2 + \frac{1}{2}x = \frac{5}{4}$. Completing the square produces $x^2+\frac{1}{2}x+\frac{1}{16} = \frac{5}{4}+\frac{1}{16}$ or $\left(x + \frac{1}{4}\right)^2 = \frac{21}{16}$. Taking the square roots of both sides produces $x = -\frac{1}{4} \pm \frac{\sqrt{21}}{4}$ and so $x = \frac{-1 \pm \sqrt{21}}{4}$.

29. $x^2 - 8x + 5 = 0$; $x = \frac{8 \pm \sqrt{(-8)^2 - 4(5)}}{2} = \frac{8 \pm \sqrt{64-20}}{2} = \frac{8 \pm \sqrt{44}}{2} = \frac{8 \pm 2\sqrt{11}}{2} = 4 \pm \sqrt{11}$

30. $x^2 + 7x - 6 = 0$; $x = \frac{-7 \pm \sqrt{7^2 - 4(-6)}}{2} = \frac{-7 \pm \sqrt{49+24}}{2} = \frac{-7 \pm \sqrt{73}}{2}$

31. $2x^3 + 3x - 5 = 0$; $x = \frac{-3 \pm \sqrt{3^2 - 4(2)(-5)}}{2 \cdot 2} = \frac{-3 \pm \sqrt{9+40}}{4} = \frac{-3 \pm 7}{4}$; $\frac{-3+7}{4} = 1$; $\frac{-3-7}{4} = \frac{-5}{2}$

32. $2x^2 - 7x + 4 = 0$; $x = \frac{7 \pm \sqrt{(-7)^2 - 4(2)(4)}}{2 \cdot 2} = \frac{7 \pm \sqrt{49-32}}{4} = \frac{7 \pm \sqrt{17}}{4}$

33. $3x^2 + 2x - 4 = 0$; $x = \frac{-2 \pm \sqrt{2^2 - 4(3)(-4)}}{2 \cdot 3} = \frac{-2 \pm \sqrt{4+48}}{6} = \frac{-2 \pm \sqrt{52}}{6} = \frac{-2 \pm 2\sqrt{13}}{2 \cdot 3} = \frac{-1 \pm \sqrt{13}}{3}$

34. $4x^2 - 3x = 1$; $4x^2 - 3x - 1 = 0$; $x = \frac{3 \pm \sqrt{(-3)^2 - 4(4)(-1)}}{2 \cdot 4} = \frac{3 \pm \sqrt{9+16}}{8} = \frac{3 \pm 5}{8}$; so $x = \frac{3+5}{8} = 1$ or, $x = \frac{3-5}{8} = -\frac{1}{4}$

35. $5x^2 + 2 = 8x$; $5x^2 - 8x + 2 = 0$; $x = \frac{8 \pm \sqrt{(-8)^2 - 4(5)(2)}}{2 \cdot 5} = \frac{8 \pm \sqrt{64-40}}{2 \cdot 5} = \frac{8 \pm \sqrt{24}}{2 \cdot 5} = \frac{8 \pm 2\sqrt{6}}{2 \cdot 5} = \frac{2(4 \pm \sqrt{6})}{2 \cdot 5} = \frac{4 \pm \sqrt{6}}{5}$

36. $6x^2 + 2x = 3$; $6x^2 + 2x - 3 = 0$; $x = \frac{-2 \pm \sqrt{2^2 - 4(6)(-3)}}{2 \cdot 6} = \frac{-2 \pm \sqrt{4+72}}{12} = \frac{-2 \pm \sqrt{76}}{12} = \frac{-2 \pm 2\sqrt{19}}{2 \cdot 6} = \frac{-1 \pm \sqrt{19}}{6}$

37. $3x^2 - 8x + 10 = 0$; $x = \frac{8 \pm \sqrt{(-8)^2 - 4(3)(10)}}{2 \cdot 3} = \frac{8 \pm \sqrt{64-120}}{6} = \frac{8 \pm \sqrt{-56}}{6}$; no answer because the discriminant is negative

38. $8x^2 = 4x + 3$; $8x^2 - 4x - 3 = 0$; $x = \frac{4 \pm \sqrt{(-4)^2 - 4(8)(-3)}}{2 \cdot 8} = \frac{4 \pm \sqrt{16+96}}{16} = \frac{4 \pm \sqrt{112}}{16} = \frac{4 \pm 4\sqrt{7}}{16} = \frac{1 \pm \sqrt{7}}{4}$

39. $\frac{x}{x-1} + \frac{2}{x+1} = 3$; $(x+1)x + 2(x-1) = 3(x+1)(x-1)$; $x^2 + x + 2x - 2 = 3x^2 - 3$;

$0 = 2x^2 - 3x - 1; x = \frac{3 \pm \sqrt{(-3)^2 - 4(2)(-1)}}{2 \cdot 2} = \frac{3 \pm \sqrt{9+8}}{4} = \frac{3 \pm \sqrt{17}}{4}$

40. $\frac{2}{x} - \frac{3}{x+2} = 4$; $2(x+2) - 3(x) = 4(x)(x+2)$; $2x + 4 - 3x = 4x^2 + 8x$; $0 = 4x^2 + 9x - 4$; $x = \frac{-9 \pm \sqrt{9^2 - 4(4)(-4)}}{2 \cdot 4} = \frac{-9 \pm \sqrt{81+64}}{8} = \frac{-9 \pm \sqrt{145}}{8}$

41. $12 = (i_1 + 0.4)^2 \cdot 50$; $(i_1 + 0.4)^2 = \frac{12}{50} = .24$; $(i_1 + 0.4) = \pm\sqrt{.24} \approx \pm.4899$; $i_1 \approx -0.4 + .4899$; $i_1 \approx 0.0899$ A; Only positive answers fit the problem

42. $\frac{h}{7} + \frac{h}{8} = 1$; $8h + 7h = 56$; $15h = 56$; $h = \frac{56}{15} \approx 3.73$ hr

43. $144 - 16t^2 = 0$; $9 - t^2 = 0$; $t^2 = 9$; $t = \pm\sqrt{9} = \pm 3$; only 3 fits the problem, so it takes 3 s for the stone to reach the bottom

44. $a^2 + (a-6)^2 = 27^2$; $a^2 + a^2 - 12a + 36 = 729$; $2a^2 - 12a - 693 = 0$; $a = \frac{12 \pm \sqrt{(-12)^2 - 4(2)(-693)}}{2 \cdot 2} = \frac{12 \pm \sqrt{5688}}{4} \approx \frac{12 + 75.42}{4} \approx 21.86$. Then $a - 6 = 21.86 - 6 = 15.86$ and the lengths are 15.86 mm and 21.86 mm

CHAPTER 8 TEST

1. $\frac{x}{8} + \frac{3x}{4} = \frac{7}{2}$; LCD = 8; $8 \cdot \frac{x}{8} + 8 \cdot \frac{3x}{4} = 8 \cdot \frac{7}{2}$; $x + 6x = 28$; $7x = 28$; $x = 4$

2. $\frac{12}{x^2 - 9} = \frac{x}{x-3} - \frac{x}{x+3}$; LCD $= (x-3)(x+3) = x^2 - 9$; $x \neq 3$; $x \neq -3$; $12 = x(x+3) - x(x-3)$; $12 = x^2 + 3x - x^2 + 3x$; $12 = 6x$; $x = 2$

3. $x^2 - 4x - 5 = 0 \Rightarrow (x-5)(x+1) = 0$; $x - 5 = 0 \Rightarrow x = 5$; $x + 1 = 0 \Rightarrow x = -1$

4. $3x^2 + x - 2 = 0$; Using the quadratic formula produces $x = \frac{-1 \pm \sqrt{1^2 - 4(3)(-2)}}{2 \cdot 3} = \frac{-1 \pm \sqrt{1+24}}{6} = \frac{-1 \pm 5}{6}$. Thus, we see that $x = \frac{-1-5}{6} = -1$ or $x = \frac{-1+5}{6} = \frac{2}{3}$. However, the given equation factors, and so this could have been solved by factoring where $(x+1)(3x-2) = 0$. By the zero product rule, $x + 1 = 0 \Rightarrow x = -1$ or $3x - 2 = 0 \Rightarrow x = \frac{2}{3}$.

5. $16x^2 + 9 = 24x$; $16x^2 - 24x + 9 = 0$; $(4x-3)^2 = 0$; $4x - 3 = 0 \Rightarrow x = \frac{3}{4}$ double root

6. $2x^2 + 3x + 3$; $x = \frac{-3 \pm \sqrt{3^2 - 4(2)(3)}}{2 \cdot 2} = \frac{-3 \pm \sqrt{-15}}{4}$; Discriminant is -15 so no solution

7. $x^2 + 3x + 1 = 0$; $\frac{-3 \pm \sqrt{3^2 - 4}}{2} = \frac{-3 \pm \sqrt{5}}{2}$

8. $x^2 - 8x + 16 = -5 + 16$; $(x-4)^2 = 11$; $x - 4 = \pm\sqrt{11}$; $x = 4 \pm \sqrt{11}$

9. $l = w + 3$; $(w+3)w = 46.75$; $w^2 + 3w - 46.75 = 0$; $w = \frac{-3 \pm \sqrt{9 + 4(46.75)}}{2} = \frac{-3 \pm \sqrt{196}}{2}$. Since a negative value does not make sense in this situation, we have $w = \frac{-3+14}{2} = \frac{11}{2} = 5.5$. Thus, $w = 5.5$ m and $l = 8.5$ m.

10. $s = vt + \frac{at^2}{2}$; $s - \frac{at^2}{2} = vt$ or $vt = \frac{2s - at^2}{2}$ and so $v = \frac{2s - at^2}{2t}$

CHAPTER

9

Trigonometric Functions

9.1 ANGLES, ANGLE MEASURE, AND TRIGONOMETRIC FUNCTIONS

1. $90° = 90° \times \frac{\pi}{180°} = \frac{\pi}{2}$ or 1.5708 or $90° \times 0.01745 = 1.5705$

2. $45° = 45° \cdot \frac{\pi}{180°} = \frac{\pi}{4}$ or 0.7854 or $45° \times 0.0175 = 0.78525$

3. $80° = 80° \times \frac{\pi}{180°} = \frac{4\pi}{9}$ or 1.3963 or $80° \times 0.01745 = 1.396$

4. $-15° = -15° \times \frac{\pi}{180°} = \frac{-\pi}{12} = -0.2618$ or $-15° \times 0.01745 = -0.26175$

5. $155° = 155° \cdot \frac{\pi}{180°} = \frac{31\pi}{36} = 2.7053$ or $155° \times 0.01745 = 2.70475$

6. $-235° = -235° \times \frac{\pi}{180°} = -4.1015$ or $-235° \times 0.01745 = -4.10075$

7. $215° = -215° \times \frac{\pi}{180°} = 3.7525$ or $215° \times 0.01745 = 3.75175$

8. $180° = \frac{180°\pi}{180°} = \pi \approx 3.1416$ or $180 \times 0.01745 = 3.141$

9. $2 \text{ rad} = 2 \cdot \frac{180°}{\pi} = \frac{360°}{\pi} \approx 114.592°$ or $2 \times 57.296° = 114.592°$

10. $3 = 3 \cdot \frac{180°}{\pi} \approx 171.887°$ or $3 \times 57.296° = 171.888°$

11. $1.5 = 1.5 \times \frac{180°}{\pi} \approx 85.944°$ or $1.5 \times 57.296 = 85.944°$

12. $\pi = \pi \cdot \frac{180°}{\pi} = 180°$

13. $\frac{\pi}{3} = \frac{\pi}{3} \cdot \frac{180°}{\pi} = 60°$

14. $\frac{5\pi}{6} = \frac{5\pi}{6} \times \frac{180°}{\pi} = 150°$

15. $-\frac{\pi}{4} = -\frac{\pi}{4} \cdot \frac{180°}{\pi} = -45°$

16. $-1.3 = -1.3 \times \frac{180°}{\pi} \approx -74.485°$ or $-1.3 \times 57.296° = -74.4848°$

17. (c) $150° = 150° + 360° = 510°$; $150° - 360° = -210°$

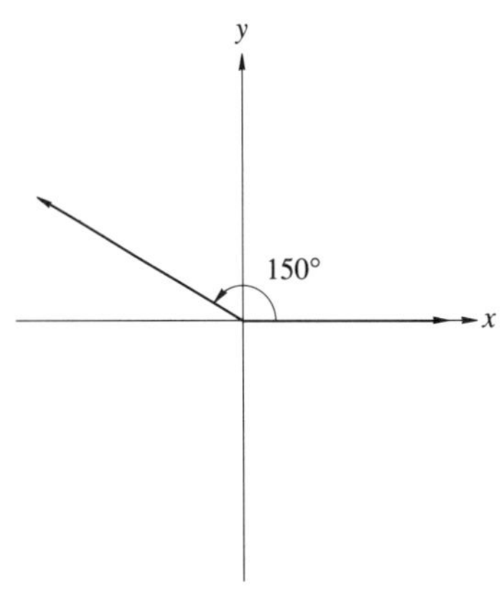

18. (c) $315° = 315° + 360° = 675°$; $315° - 360° = -45°$

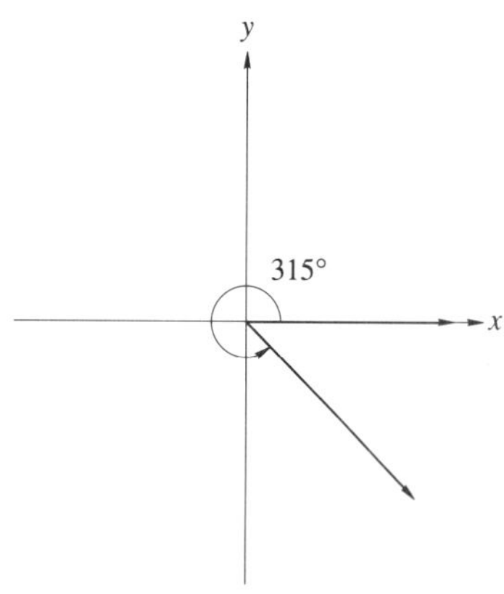

19. (c) $-135° = -135° + 360° = 225°$; $-135° - 360° = -495°$

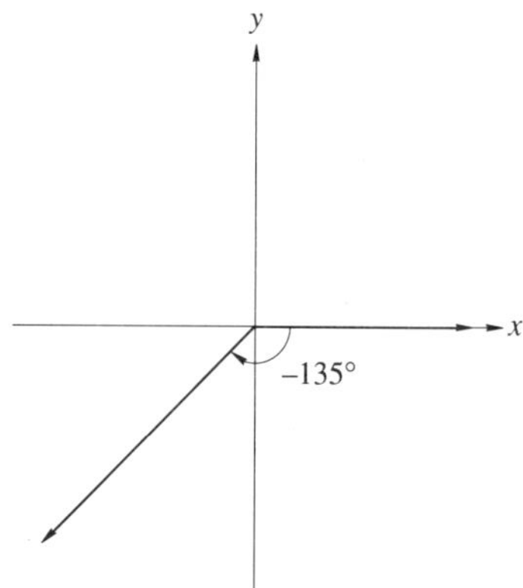

20. (c) $-30° = -30° + 360° = 330°$; $-30° - 360° = -390°$

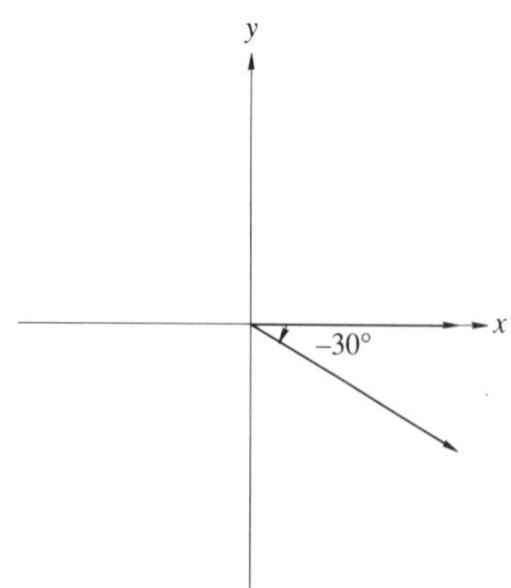

21. $r = \sqrt{4^2 + 3^2} = \sqrt{25} = 5$; $\sin\theta = \frac{3}{5}$; $\cos\theta = \frac{4}{5}$; $\tan\theta = \frac{3}{4}$; $\csc\theta = \frac{5}{3}$; $\sec\theta = \frac{5}{4}$; $\cot\theta = \frac{4}{3}$

22. $r = \sqrt{(-6)^2 + (-8)^2} = \sqrt{36 + 64} = \sqrt{100} = 10$; $\sin\theta = -\frac{4}{5}$; $\cos\theta = \frac{-3}{5}$; $\tan\theta = \frac{4}{3}$; $\csc\theta = \frac{-5}{4}$; $\sec\theta = \frac{-5}{3}$; $\cot\theta = \frac{3}{4}$

23. $r = \sqrt{8^2 + (-15)^2} = 17$; $\sin\theta = \frac{-15}{17}$; $\cos\theta = \frac{8}{17}$; $\tan\theta = \frac{-15}{8}$; $\csc\theta = \frac{-17}{15}$; $\sec\theta = \frac{17}{8}$; $\cot\theta = \frac{-8}{15}$

24. $r = \sqrt{(-20)^2 + (21)^2} = 29$; $\sin\theta = \frac{21}{29}$; $\cos\theta = \frac{-20}{29}$; $\tan\theta = \frac{-21}{20}$; $\csc\theta = \frac{29}{21}$; $\sec\theta = \frac{-29}{20}$; $\cot\theta = \frac{-20}{21}$

25. $r = \sqrt{1^2 + 2^2} = \sqrt{5}$; $\sin\theta = \frac{2}{\sqrt{5}}$; $\cos\theta = \frac{1}{\sqrt{5}}$; $\tan\theta = 2$; $\csc\theta = \frac{\sqrt{5}}{2}$; $\sec\theta = \sqrt{5}$; $\cot\theta = \frac{1}{2}$

26. $r = \sqrt{(-2)^2 + 4^2} = \sqrt{20} = 2\sqrt{5}$; $\sin\theta = \frac{2}{\sqrt{5}}$; $\cos\theta = \frac{-1}{\sqrt{5}}$; $\tan\theta = -2$; $\csc\theta = \frac{\sqrt{5}}{2}$; $\sec\theta = -\sqrt{5}$; $\cot\theta = -\frac{1}{2}$

27. $r = \sqrt{10^2 + (-8)^2} = \sqrt{164} = 2\sqrt{41}$; $\sin\theta = -\frac{4}{\sqrt{41}}$; $\cos\theta = \frac{5}{\sqrt{41}}$; $\tan\theta = \frac{-4}{5}$; $\csc\theta = -\frac{\sqrt{41}}{4}$; $\sec\theta = \frac{\sqrt{41}}{5}$; $\cot\theta = \frac{-5}{4}$

28. $r = \sqrt{(-5)^2 + (-2)^2} = \sqrt{25 + 4} = \sqrt{29}$; $\sin\theta = \frac{-2}{\sqrt{29}}$; $\cos\theta = -\frac{5}{\sqrt{29}}$; $\tan\theta = \frac{2}{5}$; $\csc\theta = -\frac{\sqrt{29}}{2}$; $\sec\theta = -\frac{\sqrt{29}}{5}$; $\cot\theta = \frac{5}{2}$

29. $r = \sqrt{\left(\sqrt{11}\right)^2 + 5^2} = \sqrt{11 + 25} = \sqrt{36} = 6$, $\sin\theta = \frac{5}{6}$; $\cos\theta = \frac{\sqrt{11}}{6}$; $\tan\theta = \frac{5}{\sqrt{11}}$; $\csc\theta = \frac{6}{5}$; $\sec\theta = \frac{6}{\sqrt{11}}$; $\cot\theta = \frac{\sqrt{11}}{5}$

30. $r = \sqrt{(-6)^2 + \left(\sqrt{13}\right)^2} = \sqrt{36 + 13} = \sqrt{49} = 7$, $\sin\theta = \frac{\sqrt{13}}{7}$; $\cos\theta = \frac{-6}{7}$; $\tan\theta = -\frac{\sqrt{13}}{6}$; $\csc\theta = \frac{7}{\sqrt{13}}$; $\sec\theta = \frac{-7}{6}$; $\cot\theta = \frac{-6}{\sqrt{13}}$

31. $r = 3$, $\sin\theta = 0$; $\cos\theta = 1$; $\tan\theta = 0$; $\csc\theta =$ Does not exist; $\sec\theta = 1$; $\cot\theta =$ Does not exist.

32. $r = \sqrt{0^2 + (-4)^2} = \sqrt{16} = 4$; $\sin\theta = -1$; $\cos\theta = 0$; $\tan\theta$: Does not exist; $\csc\theta = -1$; $\sec\theta$: Does not exist; $\cot\theta = 0$

33. $r = 5$; $\sin\theta = 0$; $\cos\theta = -1$; $\tan\theta = 0$; $\csc\theta =$ Does not exist; $\sec = -1$; $\cot\theta =$ Does not exist.

34. $r = 6$; $\sin\theta = 1$; $\cos\theta = 0$; $\tan\theta$: Does not exist; $\csc\theta = 1$; $\sec\theta$: Does not exist; $\cot\theta = 0$

35. $x = 6$, $r = 10$, $y > 0$, $y^2 = r^2 - x^2 = 10^2 - 6^2 = 100 - 36 = 64$, $y = \pm\sqrt{64}$ so $y = 8$; $\sin\theta = \frac{8}{10} = \frac{4}{5}$; $\cos\theta = \frac{x}{r} = \frac{6}{10} = \frac{3}{5}$; $\tan\theta = \frac{4}{3}$; $\csc\theta = \frac{5}{4}$; $\sec\theta = \frac{5}{3}$; $\cot\theta = \frac{3}{4}$

36. $y = -9$, $r = 15$, $x > 0$, $x^2 = r^2 - y^2$, $x^2 = 15^2 - (-9)^2 = 225 - 81 = 144$, $x = \pm\sqrt{144} = \pm 12$ so $x = 12$; $\sin\theta = \frac{-3}{5}$; $\cos\theta = \frac{4}{5}$; $\tan\theta = \frac{-3}{4}$; $\csc\theta = \frac{-5}{3}$; $\sec\theta = \frac{5}{4}$; $\cot\theta = \frac{-4}{3}$

37. $x = -20$, $r = 29$, $y > 0$, $y^2 = r^2 - x^2 = 29^2 - (-20)^2 = 841 - 400 = 441$, $y = \pm\sqrt{441} = \pm 21$ so $y = 21$, $\sin\theta = \frac{21}{29}$; $\cos\theta = \frac{-20}{29}$; $\tan\theta = \frac{-21}{20}$; $\csc\theta = \frac{29}{21}$; $\sec\theta = \frac{-29}{20}$; $\cot\theta = \frac{-20}{21}$

38. $y = 5$, $r = 13$, $x < 0$, $x^2 = r^2 - y^2 = 13^2 - 5^2 = 169 - 25 = 144$, $x = \pm\sqrt{144} = \pm 12$ so $x = -12$, $\sin\theta = \frac{5}{13}$; $\cos\theta = \frac{-12}{13}$; $\tan\theta = \frac{-5}{12}$; $\csc\theta = \frac{13}{5}$; $\sec\theta = \frac{-13}{12}$; $\cot\theta = \frac{-12}{5}$

39. $x = -7$, $r = 8$, $y < 0$, $y^2 = 8^2 - (-7)^2 = 64 - 49 = 15$, $y = \pm\sqrt{15}$ so $y = -\sqrt{15}$ $\sin\theta = -\frac{\sqrt{15}}{8}$; $\cos\theta = \frac{-7}{8}$; $\tan\theta = \frac{\sqrt{15}}{7}$; $\csc\theta = -\frac{8}{\sqrt{15}}$; $\sec\theta = \frac{-8}{7}$; $\cot\theta = \frac{7}{\sqrt{15}}$

40. $y = 5$, $r = 30$, $x < 0$, $x^2 = 30^2 - 5^2 = 900 - 25 = 875$, $x = \pm\sqrt{875} = \pm 5\sqrt{35}$ so $x = -5\sqrt{35}$, $\sin\theta = \frac{y}{r} = \frac{5}{30} = \frac{1}{6}$; $\cos\theta = \frac{x}{r} = \frac{-5\sqrt{35}}{30} = \frac{-\sqrt{35}}{6}$; $\tan\theta = \frac{y}{x} = \frac{5}{-5\sqrt{35}} = -\frac{1}{\sqrt{35}}$, $\csc\theta = \frac{r}{y} = \frac{6}{1} = 6$; $\sec\theta = \frac{r}{x} = \frac{30}{-5\sqrt{35}} = \frac{-6}{\sqrt{35}}$; $\cot\theta = \frac{x}{y} = \frac{-5\sqrt{35}}{5} = -\sqrt{35}$

9.2 VALUES OF THE TRIGONOMETRIC FUNCTIONS

1. $a = 5$, $c = 13$, $b = \sqrt{c^2 - a^2} = \sqrt{13^2 - 5^2} = \sqrt{169 - 25} = \sqrt{144} = 12$; $\sin\theta = \frac{5}{13}$; $\cos\theta = \frac{12}{13}$; $\tan\theta = \frac{5}{12}$; $\csc\theta = \frac{13}{5}$; $\sec\theta = \frac{13}{12}$; $\cot\theta = \frac{12}{5}$

2. $a = 7, b = 8, c = \sqrt{7^2 + 8^2} = \sqrt{49 + 64} = \sqrt{113}$; $\sin\theta = \frac{7}{\sqrt{113}}$; $\cos\theta = \frac{8}{\sqrt{113}}$; $\tan\theta = \frac{7}{8}$; $\csc\theta = \frac{\sqrt{113}}{7}$; $\sec\theta = \frac{\sqrt{113}}{8}$; $\cot\theta = \frac{8}{7}$

3. $a = 1.2$, $c = 2$, $b = \sqrt{2^2 - 1.2^2} = \sqrt{4 - 1.44} = \sqrt{2.56} = 1.6$, $\sin\theta = \frac{1.2}{2} = \frac{12}{20} = \frac{3}{5}$, $\cos\theta = \frac{1.6}{2} = \frac{16}{20} = \frac{4}{5}$, $\tan\theta = \frac{3}{4}$; $\csc\theta = \frac{5}{3}$; $\sec\theta = \frac{5}{4}$; $\cot\theta = \frac{4}{3}$

4. $a = 2.1$, $b = 2.8$, $c = \sqrt{2.1^2 + 2.8^2} = \sqrt{4.41 + 7.84} = \sqrt{12.25} = 3.5$; $\sin\theta = \frac{3}{5}$; $\cos\theta = \frac{4}{5}$; $\tan\theta = \frac{3}{5}$; $\csc\theta = \frac{5}{3}$; $\sec\theta = \frac{5}{4}$; $\cot\theta = \frac{4}{3}$

5. $a = 1.4$, $b = 2.3$, $c = \sqrt{1.4^2 + 2.3^2} = \sqrt{1.96 + 5.29} = \sqrt{7.25}$; $\sin\theta = \frac{1.4}{\sqrt{7.25}}$; $\cos\theta = \frac{2.3}{\sqrt{7.25}}$; $\tan\theta = \frac{14}{23}$; $\csc\theta = \frac{\sqrt{7.25}}{1.4}$; $\sec\theta = \frac{\sqrt{7.25}}{2.3}$; $\cot\theta = \frac{23}{14}$

6. $c = 3.5$, $b = 2.1$; $a = \sqrt{3.5^2 - 2.1^2} = \sqrt{12.25 - 4.41} = \sqrt{7.84} = 2.8$; $\sin\theta = \frac{4}{5}$; $\cos\theta = \frac{3}{5}$; $\tan\theta = \frac{4}{3}$; $\csc\theta = \frac{5}{4}$; $\sec\theta = \frac{5}{3}$; $\cot\theta = \frac{3}{4}$

7. Since $\cos\theta = \frac{8}{17}$, the side adjacent is 8 and the hypotenuse is 17, and so the side opposite is 15. $\sin\theta = \frac{\text{opp}}{\text{hyp}} = \frac{15}{17}$; $\tan\theta = \frac{\text{opp}}{\text{adj}} = \frac{15}{8}$, $\csc\theta = \frac{\text{hyp}}{\text{opp}} = \frac{17}{15}$, $\sec\theta = \frac{\text{hyp}}{\text{adj}} = \frac{17}{8}$, $\cot\theta = \frac{\text{adj}}{\text{opp}} = \frac{8}{15}$

8. Since $\tan\theta = \frac{21}{20}$, the side opposite is 21 and the side adjacent is 20, so the hypotenuse is 29. $\sin\theta = \frac{21}{29}$; $\cos\theta = \frac{20}{29}$; $\tan\theta = \frac{21}{20}$; $\csc\theta = \frac{29}{21}$; $\sec\theta = \frac{29}{20}$; $\cot\theta = \frac{20}{21}$

9. Since $\sin\theta = \frac{3}{5}$, the side opposite is 3 and the hypotenuse is 5, so side adjacent is 4. $\sin\theta = \frac{3}{5}$; $\cos\theta = \frac{4}{5}$; $\tan\theta = \frac{3}{4}$; $\csc\theta = \frac{5}{3}$; $\sec\theta = \frac{5}{4}$; $\cot\theta = \frac{4}{3}$

10. Since $\sec\theta = \frac{10}{6}$, the hypotenuse is 10 and the side adjacent is 6, so the side opposite is 8. $\sin\theta = \frac{8}{10} = \frac{4}{5}$; $\cos\theta = \frac{6}{10} = \frac{3}{5}$; $\tan\theta = \frac{8}{6} = \frac{4}{3}$; $\csc\theta = \frac{10}{8} = \frac{5}{4}$; $\sec\theta = \frac{10}{6} = \frac{5}{3}$; $\cot\theta = \frac{6}{8} = \frac{3}{4}$

11. Here $\csc\theta = \frac{29}{20}$, so the hypotenuse is 29 and the side opposite is 20. Thus, the side adjacent is 21. $\sin\theta = \frac{20}{29}$; $\cos\theta = \frac{21}{29}$; $\tan\theta = \frac{20}{21}$; $\csc\theta = \frac{29}{20}$; $\sec\theta = \frac{29}{21}$; $\cot\theta = \frac{21}{20}$

12. Since $\csc\theta = \frac{15}{12}$, the hypothenuse is 15 and the side opposite is 12, and so the side adjacent is 9. $\sin\theta = \frac{12}{15} = \frac{4}{5}$; $\cos\theta = \frac{9}{15} = \frac{3}{5}$; $\tan\theta = \frac{12}{9} = \frac{4}{3}$; $\csc\theta = \frac{15}{12} = \frac{5}{4}$; $\sec\theta = \frac{15}{9} = \frac{5}{3}$; $\cot\theta = \frac{9}{12} = \frac{3}{4}$

13. $\sin\theta = 0.866$, $\cos\theta = 0.5$, $\tan\theta = \frac{0.866}{0.5} = 1.732$, $\csc\theta = \frac{1}{0.866} \approx 1.155$, $\sec\theta = 2$, $\cot\theta = \frac{0.5}{0.866} \approx 0.577$

14. $\sin\theta = 0.975$, $\cos\theta = 0.222$, $\tan\theta = \frac{0.975}{0.222} \approx 4.392$, $\csc\theta = \frac{1}{0.975} \approx 1.026$, $\sec\theta = \frac{1}{0.222} \approx 4.505$, $\cot\theta = \frac{0.222}{0.975} \approx 0.228$

15. $\sin\theta = 0.085$, $\sec\theta = 1.004$, $\cos\theta = \frac{1}{1.004} \approx 0.996$, $\tan\theta \approx \frac{0.085}{0.996} \approx 0.085$, $\csc\theta = \frac{1}{0.085} \approx 11.765$, $\cot\theta \approx \frac{0.996}{0.085} \approx 11.718$

16. $\csc\theta = 4.872$, $\cos\theta = 0.979$, $\sin\theta = \frac{1}{\csc\theta} = \frac{1}{4.872} \approx 0.205$, $\tan\theta \approx \frac{0.205}{0.979} \approx 0.209$; $\cot\theta \approx \frac{0.979}{0.205} \approx 4.776$; $\sec\theta = \frac{1}{\cos\theta} = \frac{1}{0.979} \approx 1.021$

17. $\sin 18.6° = 0.3189593$

18. $\cos 38.4° = 0.7836935$

19. $\tan 18.3° = 0.3307184$

20. $\sin 20°15' = \sin 20.25° = 0.3461171$

21. $\tan 76°32' = \tan 76\frac{32}{60}° = 4.1760011$

22. $\sec 24°14' = \frac{1}{\cos(24+\frac{14}{60})} = 1.0966337$

23. $\cot 82.6° = \frac{1}{\tan 82.6°} = 0.1298773$

24. $\csc 19°50' = \frac{1}{\sin(19+\frac{50}{60})} = 2.9473725$

25. $\sin 0.25$ rad $= 0.247404$

26. $\cos 0.4$ rad $= 0.921061$

27. $\tan 0.63$ rad $= 0.7291147$

28. $\sec 1.35$ rad $= \frac{1}{\cos 1.35} = 4.5660706$

29. $\cot 1.43$ rad $= \frac{1}{\tan 1.43} = 0.1417341$

30. $\sin 1.21$ rad $= 0.935616$

31. $\csc 0.21$ rad $= \frac{1}{\sin 0.21} = 4.7970857$

32. $\tan 1.555$ rad ≈ 63.300591

33. $R = \frac{15^2}{9.8}\sin(2\cdot 40) \approx 22.61$ m

34. If the angle is 45°, then $R = \frac{15^2}{9.8}\sin(2\times 45°) \approx 22.96$ and the football would travel $22.96 - 22.61 = 0.35$ m farther than if it were thrown at an angle of 40°. If the angle is 48°, then $R = \frac{15^2}{9.8}\sin(2\cdot 48°) \approx 22.83$ and the football would travel $22.83 - 22.61 = 0.22$ m farther than at an angle of 40°.

35. $V_R = 5.8\cos 32° \approx 4.92$ V

36. See *Computer Programs* in main text.

37. See *Computer Programs* in main text.

9.3 THE RIGHT TRIANGLE

1. $\sin A = 0.732$, $A = 47.05°$

2. $\cos B = 0.285$, $B = 73.44°$

3. $\tan C = 4.671$, $C = 77.92°$

4. $\sin D = 0.049$, $D = 2.81°$

5. $\cos E = 0.839$, $E = 32.97°$

6. $\tan F = 0.539$, $F = 28.32°$

7. $\sec G = 3.421$, $G = 73.00°$

8. $\csc H = 1.924$, $H = 31.32°$

9. $\cot I = 0.539$, $I = 61.68°$

10. $\sec J = 1.734, J = 54.78^\circ$

11. $\csc K = 4.761, K = 12.12^\circ$

12. $\cot L = 4.021, L = 13.97^\circ$

13. $B = 90 - 16.5 = 73.5$; $\sin A = \frac{a}{c}$, $c = \frac{a}{\sin A} = \frac{7.3}{\sin 16.5} \approx 25.7$, $\tan A = \frac{a}{b}$, $b = \frac{a}{\tan A} = \frac{7.3}{\tan 16.5} \approx 24.6$. In summary, we have

$$\begin{array}{ll} a = \ 7.3 & A = 16.5^\circ \\ b \approx 24.6 & B = 73.5^\circ \\ c \approx 25.7 & C = \ 90^\circ. \end{array}$$

14. $A = 90^\circ - B = 90^\circ - 53^\circ = 37^\circ$; $\tan B = \frac{b}{a}$ and so $a = \frac{b}{\tan B} = \frac{9.1}{\tan 53^\circ} \approx 6.9$; $\sin B = \frac{b}{c}$ and so, $c = \frac{b}{\sin B} = \frac{9.1}{\sin 53^\circ} \approx 11.4$. In summary, we have

$$\begin{array}{ll} a \approx \ 6.9 & A = 37^\circ \\ b = \ 9.1 & B = 53^\circ \\ c \approx 11.4 & C = 90^\circ. \end{array}$$

15. $B = 90 - 72.6 = 17.4$, $\sin A = \frac{a}{c}$, $a = c \sin A = 20 \sin 72.6 \approx 19.1$, $\cos A = \frac{b}{c}$, $b = 20 \cos 72.6 \approx 6.0$. In summary, we have

$$\begin{array}{ll} a \approx 19.1 & A = 72.6^\circ \\ b \approx \ 6.0 & B = 17.4^\circ \\ c = 20.0 & C = \ 90^\circ. \end{array}$$

16. $\cos B = \frac{a}{c}$ and so $c = \frac{19.4}{\cos 12.7} \approx 19.9$. $A = 90^\circ - 12.7^\circ = 77.3^\circ$; $\tan B = \frac{b}{a}$ which means that $b = 19.4 \tan 12.7 \approx 4.4$. In summary, we have

$$\begin{array}{ll} a = 19.4 & A = 77.3^\circ \\ b \approx \ 4.4 & B = 12.7^\circ \\ c \approx 19.9 & C = \ 90^\circ. \end{array}$$

17. $B = 90^\circ - 43^\circ = 47^\circ$ and $\cos A = \frac{b}{c}$ which means that $c = \frac{34.6}{\cos 43} \approx 47.3$. Finally, $\tan A = \frac{a}{b}$, and so $a = b \tan A = 34.6 \tan 43 = 32.3$. Summarizing, we have

$$\begin{array}{ll} a \approx 32.3 & A = 43^\circ \\ b = 34.6 & B = 47^\circ \\ c \approx 47.3 & C = 90^\circ. \end{array}$$

18. We first find that $A = 90^\circ - 67^\circ = 23^\circ$. Next, we see that $\sin B = \frac{b}{c}$ and so $b = 32.4 \sin 67^\circ \approx 29.8$. Finally, we have $\cos B = \frac{a}{c}$, which means that $a = 32.4 \cos 67 \approx 12.7$. Summarizing, we have the following results:

$$\begin{array}{ll} a \approx 12.7 & A = 23^\circ \\ b \approx 29.8 & B = 67^\circ \\ c = 32.4 & C = 90^\circ. \end{array}$$

19. We begin by determining that $B = \frac{\pi}{2} - 0.92 \approx 0.65$ rad. Next, we determine that $\sin A = \frac{a}{c}$ and so $c = \frac{a}{\sin A} = \frac{6.5}{\sin 0.92} \approx 8.2$. Finally, we find that $\tan A = \frac{a}{b}$ and so $b = \frac{6.5}{\tan 0.92} \approx 4.9$. In summary, we have the following results:

$$\begin{array}{ll} a = 6.5 & A = 0.92\,\text{rad} \\ b \approx 4.9 & B \approx 0.65\,\text{rad} \\ c \approx 8.2 & C = \ \frac{\pi}{2}\,\text{rad} \end{array}$$

20. Using the given information, we see that $A = C - B = \frac{\pi}{2} - 1.13 \approx 0.44$ rad. Then $\tan B = \frac{b}{a}$ and so $a = \frac{24}{\tan 1.13} \approx 11.3$. Finally, $\sin B = \frac{b}{c}$ leads to $c = \frac{24}{\sin 1.13} \approx 26.5$. In summary, we have the following results:

$$\begin{array}{ll} a \approx 11.3 & A \approx 0.44\,\text{rad} \\ b = 24.0 & B = 1.13\,\text{rad} \\ c \approx 26.5 & C = \ \frac{\pi}{2}\,\text{rad} \end{array}$$

21. We first determine that $B = \frac{\pi}{2} - 0.15 \approx 1.42$ rad. Next, we find that $\sin A = \frac{a}{c}$ and so $a = c \sin A = 18 \sin 0.15 \approx 2.7$. Finally, we have $\cos A = \frac{b}{c}$ which produces $b = 18 \cos 0.15 \approx 17.8$. Summarizing, we have:

$$\begin{array}{ll} a \approx \ 2.7 & A = 0.15\,\text{rad} \\ b \approx 17.8 & B \approx 1.42\,\text{rad} \\ c = 18.0 & C = \ \frac{\pi}{2}\,\text{rad} \end{array}$$

22. We begin by finding that $A = \frac{\pi}{2} - 0.23 \approx 1.34$ rad. Then, $\cos B = \frac{a}{c}$ and so $c = \frac{a}{\cos B} = \frac{9.7}{\cos .23} \approx 10.0$. Third, we find that $\tan B = \frac{b}{a}$ and as a result, $b = a \tan B = 9.7 \tan 0.23 \approx 2.3$. In summary, we have the following results:

$$\begin{array}{ll} a = 9.7 & A \approx 1.34 \text{ rad} \\ b \approx 2.3 & B = 0.23 \text{ rad} \\ c \approx 10.0 & C = \frac{\pi}{2} \text{ rad} \end{array}$$

23. Using the given information, we find that $B \approx 1.57 - 1.41 \approx 0.16$ rad. Next, we see that $\tan A = \frac{a}{b}$ produces $a = 40 \tan 1.41 \approx 246.6$. Finally, we have $\cos A = \frac{b}{c}$, and so $c = \frac{40}{\cos 1.41} \approx 249.8$. In summary, we have the following results:

$$\begin{array}{ll} a \approx 246.6 & A = 1.41 \text{ rad} \\ b = 40.0 & B \approx 0.16 \text{ rad} \\ c \approx 249.8 & C = \frac{\pi}{2} \text{ rad} \end{array}$$

24. We begin by determining that $B \approx 1.57 - 1.15 \approx 0.42$ rad. Using the given information, we can use $\sin A = \frac{a}{c}$ to yield $a = 18 \sin 1.15 \approx 16.4$. Finally, we use $\cos A = \frac{b}{c}$ to find $b = 18 \cos 1.15 \approx 7.4$. Summarizing, we see that:

$$\begin{array}{ll} a \approx 16.4 & A = 1.15 \text{ rad} \\ b \approx 7.4 & B \approx 0.42 \text{ rad} \\ c = 18.0 & C = \frac{\pi}{2} \text{ rad} \end{array}$$

25. $a = 9$, $b = 15$, $c = \sqrt{9^2 + 15^2} \approx 17.5$, $\tan A = \frac{a}{b} = \frac{9}{15}$, $A = \tan^{-1} \frac{9}{15} = 31.0°$ or 0.54 rad, $B \approx 90° - 31° = 59°$ or $B \approx 1.57 - 0.54 = 1.03$ rad

26. $a = 19.3$, $c = 24.4$; $b = \sqrt{24.4^2 - 19.3^2} \approx 14.9$, $\sin A = \frac{a}{c} = \frac{19.3}{24.4}$, $A \approx 52.3°$ or 0.91 rad; $B \approx 90° - 52.3° = 37.7°$ or $B \approx 1.57 - 0.91 = 0.66$ rad

27. $b = 9.3$, $c = 18$, $a = \sqrt{18^2 - 9.3^2} \approx 15.4$, $\cos A = \frac{b}{c} = \frac{9.3}{18}$, $A \approx 58.9°$ or 1.03 rad, $B \approx 90° - 58.9° = 31.1°$ or $1.57 - 1.03 = 0.54$ rad

28. $a = 14$, $b = 9.3$, $c = \sqrt{14^2 + 9.3^2} \approx 16.8$, $\tan A = \frac{14}{9.3}$, $A \approx 56.4°$ or 0.98 rad, $B \approx 90° - 56.4° = 33.6°$ or $B \approx 1.57 - 0.98 = 0.59$ rad

29. $a = 20$, $c = 30$, $b = \sqrt{30^2 - 20^2} \approx 22.4$, $\sin A = \frac{a}{c} = \frac{20}{30}$, $A \approx 41.8°$ or 0.73 rad, $B \approx 90° - 41.8° = 48.2°$ or $B \approx 1.57 - 0.73 = 0.84$ rad

30. $b = 15$, $c = 25$, $a = \sqrt{25^2 - 15^2} = 20$, $\cos A \frac{b}{c} = \frac{15}{25}$, $A \approx 53.1°$ or 0.93 rad, $B = 90° - 53.1° = 36.9°$ or $B \approx 1.57 - 0.93 = 0.64$ rad

31. $\tan \phi = \frac{12.3}{19.7}$, $\phi = \tan^{-1} \frac{12.3}{19.7} \approx 31.9792$ or about $32°$

32. $\tan 23.2° = \frac{222}{x}$, $x = \frac{222}{\tan 23.2°} \approx 518$ ft

33. $\sin 37.6° = \frac{700}{d}$, $d = \frac{700}{\sin 37.6°} \approx 1147$ m

34. $\tan 57.62 = \frac{d}{95}$, $d = 95 \tan 57.62 \approx 149.8$ m

35. $F_x = 10 \cos 30° \approx 8.7$ N, $F_y = 10 \sin 30° = 5$ N

36. $c = \sqrt{20^2 + 12^2} \approx 23.3$ lb, $\tan \phi = \frac{20}{12}$, $\phi = \tan^{-1} \frac{20}{12} \approx 59.0°$

37. $\tan \phi = \frac{(80.67)^2}{32 \cdot 1200} \approx 0.1695$, $\phi = \tan^{-1}(0.1695) \approx 9.62°$

38. $\tan \phi = \frac{24.44^2}{9.8 \times 500} \approx 0.1219$, $\phi = \tan^{-1} 0.1219 \approx 6.95°$

39. $\cos 67° = \frac{x}{300}$, $x = 3.00 \cos 67° \approx 117.2$ m, $\sin 67° = \frac{d}{300}$ and so $d = 300 \sin 67° \approx 276.2$ m. So 117.2 m from intersection, 276.2 m from service station.

9.4 TRIGONOMETRIC FUNCTIONS OF ANY ANGLE

1. $87°$ is in Quadrant I, so $\theta_{ref} = 87°$

2. $200°$ is in Quadrant III, so $\theta_{ref} = 200° - 180° = 20°$

3. $137°$ is in Quadrant II, so $\theta_{ref} = 180° - 137° = 43°$

4. $298°$ is in Quadrant IV, so $\theta_{ref} = 360° - 298° = 62°$

5. $\frac{9\pi}{8}$ rad is in Quadrant III, so $\theta_{ref} = \frac{9\pi}{8} - \pi = \frac{9\pi}{8} - \frac{8\pi}{8} = \frac{\pi}{8}$

6. 2.1 rad is in Quadrant II, so $\theta_{ref} = \pi - 2.1 \approx 3.14 - 2.1 = 1.04$
7. 4.5 rad is in Quadrant III, so $\theta_{ref} = 4.5 - \pi \approx 4.5 - 3.14 = 1.36$
8. 5.85 rad is in Quadrant IV, so $\theta_{ref} = 2\pi - 5.85 \approx 6.28 - 5.85 = 0.43$
9. 518°; 518° − 360° = 158°
10. 1673°; 1673° − 4(360°) = 1673° − 1440° = 233°
11. −871°; 3 × 360° − 871° = 1080° − 871° = 209°
12. −137°; 360° − 137° = 223°
13. 7.3 rad; $7.3 - 2\pi \approx 7.3 - 6.28 = 1.02$
14. $\frac{16\pi}{4}$ rad; $\frac{16\pi}{3} - 4\pi = \frac{16\pi}{3} - \frac{12\pi}{3} = \frac{4\pi}{3}$
15. −2.17 rad; $-2.17 + 2\pi \approx -2.17 + 6.28 = 4.11$
16. −8.43 rad; $-8.43 + 4\pi \approx -8.43 + 12.56 = 4.13$
17. 165°; Quadrant II
18. 285°; Quadrant IV
19. −47°; Quadrant IV
20. 312°; Quadrant IV
21. 250°; Quadrant III
22. 197°; Quadrant III
23. 98°; Quadrant II
24. $\sin\theta$ is positive in Quadrants I and II.
25. $\tan\theta$ is negative in Quadrants II and IV.
26. $\cos\theta$ is positive in Quadrants I and IV.
27. $\sec\theta$ is negative in Quadrants II and III.
28. $\cot\theta$ is negative in Quadrants II and IV.
29. $\csc\theta$ is negative and $\cos\theta$ is positive in Quadrant IV.
30. $\tan\theta$ and $\sec\theta$ are negative in Quadrant II.
31. sin 105° is in Quadrant II, so the expression is positive.
32. sec 237° is in Quadrant III, so the expression is negative.
33. tan 372° is in Quadrant I, so the expression is positive.
34. cos(−53°) is in Quadrant IV, so the expression is positive.
35. cos 1.93 rad is in Quadrant II, so the expression is negative.
36. cot 4.63 rad is in Quadrant III, so the expression is positive.
37. sin 215° is in Quadrant III, so the expression is negative.
38. csc 5.42 rad IVQ = negative
39. sin 137° = 0.6820
40. cos 263° = −0.1219
41. tan 293° = −2.3559
42. sin 312° = −0.7431
43. tan 164.2° = −0.2830
44. csc 197.3° = −3.3628
45. sin 2.4 rad = 0.6755
46. cos 1.93 rad = −0.3515
47. tan 6.1 rad = −0.1853
48. sec 4.32 rad = −2.6151
49. tan 1.37 rad = 4.9131
50. sin 3.2 rad = −0.0584
51. sin 415.5° = 0.8241
52. tan 512.1° = −0.5295
53. cot −87.4° = −0.0454
54. cos 372.1° = 0.9778
55. csc −432.4° = −1.0491

56. $\cos 357.3° = 0.9989$

57. $\sin 6.5 \text{ rad} = 0.2151$

58. $\sec -4.3 \text{ rad} = -2.4950$

59. $\tan 8.35 \text{ rad} = -1.8479$

60. $\sin 9.42 \text{ rad} = 0.0048$

61. $\cos -9.43 \text{ rad} = 0.9090$

62. $\sec 9.34 \text{ rad} = -1.0036$

9.5 INVERSE TRIGONOMETRIC FUNCTIONS

1. $\sin\theta = \frac{1}{2}$ in Quadrants I and II.

2. $\tan\theta = \frac{3}{4}$ in Quadrants I and III.

3. $\cot\theta = -2$ in Quadrants II and IV.

4. $\cos\theta = -0.25$ in Quadrants II and III.

5. $\sec\theta = 4.3$ in Quadrants I and IV.

6. $\cos\theta = 0.8$ in Quadrants I and IV.

7. $\csc\theta = -6.1$ in Quadrants III and IV.

8. $\sin\theta = -\frac{\sqrt{3}}{2}$ in Quadrants III and IV.

9. $\sin\theta = \frac{1}{2}$; $\theta_r = \sin^{-1} 0.5 = 30.0°$; $\theta_{\text{I}} = 30.0°$, $\theta_{\text{II}} = 180° - 30° = 150°$

10. $\tan\theta = \frac{3}{4}$; $\theta_r = \tan^{-1} 0.75 = 36.9°$; $\theta_{\text{I}} = 36.9°$; $\theta_{\text{III}} = 180° + \theta_r = 180° + 36.9° = 216.9°$

11. $\cot\theta = -2$; $\theta_r = \cot^{-1} 2 = \tan^{-1}\frac{1}{2} = 26.6°$; $\theta_{\text{II}} = 180° - \theta_r = 153.4°$; $\theta_{\text{IV}} = 360° - \theta_r = 333.4°$

12. $\cos\theta = -0.25$; $\theta_r = \cos^{-1} 0.25 = 75.5°$; $\theta_{\text{II}} = 180° - \theta_r = 104.5°$; $\theta_{\text{III}} = 180° + \theta_r = 255.5°$

13. $\sec\theta = 4.3$; $\theta_r = \sec^{-1} 4.3 = \cos^{-1}\frac{1}{4.3} = 76.6°$; $\theta_{\text{I}} = \theta_r = 76.6°$; $\theta_{\text{IV}} = 360 - \theta_r = 283.4°$

14. $\cos\theta = 0.8$; $\theta_r = \cos^{-1} 0.8 = 36.9°$; $\theta_{\text{I}} = \theta_r = 36.9°$; $\theta_{\text{IV}} = 360 - \theta_r = 323.1°$

15. $\csc\theta = -6.1$; $\theta_r = \csc^{-1} 6.1 = \sin^{-1}\frac{1}{6.1} = 9.4°$; $\theta_{\text{III}} = 180° + \theta_r = 189.4°$; $\theta_{\text{IV}} = 360° - \theta_r = 350.6°$

16. $\sin\theta = -\frac{\sqrt{3}}{2}$; $\theta_r = \sin^{-1}\frac{\sqrt{3}}{2} = 60°$ exactly; $\theta_{\text{III}} = 180° + \theta_r = 240°$; $\theta_{\text{IV}} = 360° - \theta_r = 300°$

17. $\sin\theta = 0.75$; $\theta_r = \sin^{-1} 0.75 = 0.85$; $\theta_{\text{I}} = \theta_r = 0.85$; $\theta_{\text{II}} = \pi - \theta_r = 2.29$

18. $\tan\theta = 1.6$; $\theta_r = \tan^{-1} 1.6 = 1.01$; $\theta_{\text{I}} = \theta_r = 1.01$; $\theta_{\text{III}} = \pi + \theta_r = 4.15$

19. $\cot\theta = -0.4$; $\theta_r = \cot^{-1} 0.4 = \tan^{-1}\frac{1}{0.4} = 1.19$; $\theta_{\text{II}} = \pi - \theta_r = 1.95$; $\theta_{\text{IV}} = 2\pi - \theta_r = 5.09$

20. $\cos\theta = 0.25$; $\theta_r = \cos^{-1} 0.25 = 1.32$; $\theta_{\text{I}} = \theta_r = 1.32$; $\theta_{\text{IV}} = 2\pi - \theta_r = 4.96$

21. $\csc\theta = 4.3$; $\theta_r = \csc^{-1} 4.3 = \sin^{-1}\frac{1}{4.3} = 0.23$; $\theta_{\text{I}} = \theta_r = 0.23$; $\theta_{\text{II}} = \pi - \theta_r = 2.91$

22. $\sin\theta = -0.08$; $\theta_r = \sin^{-1} 0.08 = 0.08$; $\theta_{\text{III}} = \pi + \theta_r = 3.22$; $\theta_{\text{IV}} = 2\pi - \theta_r = 6.20$

23. $\sec\theta = 2.7$; $\theta_r = \sec^{-1} 2.7 = \cos^{-1}\frac{1}{2.7} = 1.19$; $\theta_{\text{I}} = \theta_r = 1.19$; $\theta_{\text{IV}} = 2\pi - \theta_r = 5.09$

24. $\cos\theta = -0.95$; $\theta_r = \cos^{-1} 0.95 = 0.32$; $\theta_{\text{II}} = \pi - \theta_r = 2.82$; $\theta_{\text{III}} = \pi + \theta_r = 3.46$

25. $\arcsin 0.84 = 57.1°$

26. $\arccos(-0.21) = 102.1°$

27. $\arctan 4.21 = 76.6°$

28. $\text{arccot}(-0.25) = \arctan\left(\frac{1}{-0.25}\right) = -76.0°$ but range of arccot is $0°$ to $180°$ so $180° - 76° = 104°$ is the correct answer.

29. $\sin^{-1} 0.32 = 18.7°$

30. $\cos^{-1} 0.47 = 62.0°$

31. $\tan^{-1}(-0.64) = -32.6°$

32. $\csc^{-1}(-3.61) = -16.1°$

33. $\arccos(-0.33) = 1.91$

34. arctan 1.55 = 1.00

35. arccos 0.29 = 1.28

36. arcsec(−3.15) = arccos $\left(\frac{1}{-3.15}\right)$ = 1.89

37. $\sin^{-1} 0.95 = 1.25$

38. $\cos^{-1}(-0.67) = 2.31$

39. $\tan^{-1} 0.25 = 0.24$

40. $\cot^{-1}(-0.75) = \tan^{-1} \frac{1}{-0.75} = -0.93$ but range of $\cot^{-1}$ is 0 to π so correct answer is $\pi - 0.93 = 2.21$

41. See *Computer Programs* in main text.

42. See *Computer Programs* in main text.

43. See *Computer Programs* in main text.

44. See *Computer Programs* in main text.

9.6 APPLICATIONS OF TRIGONOMETRY

1. $47° = 47 \times \frac{\pi}{180}$ rad ≈ 0.8203 rad; $s = r\theta = 1050.250 \times 0.8203 \approx 861.520$ m

2. $v = r\omega = 8 \cdot 5 = 40$ in./s

3. 30 rpm $= 30 \cdot 2\pi$ rad/min ≈ 188.5 rad/min. $v = r\omega = 188.5 \times 12 = 2262$ in./min= 2262 in/min $\left(\frac{1 \text{ min}}{60 \text{ s}}\right) = 37.7$ in/s

4. $v = \frac{4}{2} = 2$ m/s; $v = r\omega$; $\omega = \frac{v}{r} = \frac{2 \text{ m/s}}{0.18 \text{ m}} \approx 11.11$ rad/s

5. $s = r\theta = 15 \times \frac{5\pi}{8} \approx 29.45$ cm

6. $85° = \frac{85\pi}{180} \approx 1.48353$ rad; $s = r\theta = 235 \times 1.48353 \approx 348.63$ mm

7. $A = \frac{1}{2}r^2\theta = \frac{1}{2}(15)^2\left(\frac{5\pi}{8}\right) \approx 220.89 \text{ cm}^2$

8. $A = \frac{1}{2}r^2\theta = \frac{1}{2}(235)^2 \times 1.483\,53 \approx 40\,963.97 \text{ mm}^2$

9. $\theta = 2 \times 0.07 = 0.14$; $s = r\theta = 1.2 \times 0.14 = 0.168$ m

10. $\omega = 1$ rev/24 h $= 0.2617994$ rad/h $= 15°$/h; $v = r\omega = (22{,}300 + 3963) \times (0.2617994) \approx 6876$ mph

11. (a) $\omega = 45\frac{\text{rev}}{\text{min}} \times \frac{2\pi}{\text{rev}} \approx 282.74\frac{\text{rad}}{\text{min}}$ or $282.74\frac{\text{rad}}{\text{min}}\frac{1 \text{ min}}{60 \text{ s}} \approx 4.712$ rad/s; $r = 175.26/2 = 87.63$ mm; $v = r\omega = 87.63 \times 282.74 \approx 24\,776.5$ mm/min or $v = 87.63 \times 4.712 \approx 412.9$ mm/s; (b) 24 776.5 mm

12. $\alpha = \frac{\omega_f - \omega_0}{t} = \frac{0 - 4.2}{11.92} \approx -0.352 \text{ rad/s}^2$

13. (a) We use the given information to determine that $\cos D = (\sin 35°30')(\sin 40°40') + (\cos 35°30')(\cos 40°40')(\cos[-10°10'])$ and so $\cos D \approx 0.98624$ and so the angular distance between the cities is $D \approx 9.515\,478°$. (b) To find the linear distance s between the cities, we use $s = r\theta$ with $\theta = 9.515\,478°\left(\frac{\pi}{180°}\right) \approx 0.1660764$ rad. Using $r = 3960$ mi, we get $s = 3960(0.1660764) \approx 657.66$ mi.

14. $r = \frac{d}{2} = \frac{1}{2}$ ft; $v = r\omega = \frac{1}{2}$ ft $\cdot 1200\frac{\text{rev}}{\text{min}} \cdot \frac{2\pi}{\text{rev}} = 3769.9$ ft/min.

15. $\frac{8}{\frac{1}{4}} \times \frac{\pi}{3} \times 2.5 = 83.77$ Since you cannot have a fraction of a byte, the answer is 83

16. $C = 2\pi r = 2\pi \cdot 2.5 = 15.708$; $l = 15.708 \div 16 = 0.9818$ in.

17. $850\frac{\text{rev}}{\text{min}} \cdot \frac{360°}{1 \text{ rev}} \cdot \frac{1 \text{ min}}{60 \text{ s}} = 5100°/\text{s}$

18. $r = 7.5$ in.; $v = r\omega = 7.5 \text{ in.} \times 850\frac{\text{rev}}{\text{min}} \cdot \frac{2\pi}{\text{rev}} \cdot \frac{1 \text{ min}}{60 \text{ s}} \approx 667.59$ in./s

19. $v = r\omega = 4.2 \text{ m} \cdot \frac{20 \text{ rev}}{\text{min}} \cdot \frac{2\pi}{\text{rev}} \cdot \frac{1 \text{ min}}{60 \text{ s}} \approx 8.80$ m/s

20. $\sin 23° = \frac{F_x}{F}$; $F_x = 2500 \sin 23° \approx 977$ N; $F_y = F \cos 23° \approx 2301$ N

21. $\theta = 66.5° \times \frac{\pi}{180°} = 1.160\,644$ rad, $s = r\theta = 6370 \times 1.1606 = 7393.3$ km

CHAPTER 9 REVIEW

1. $60° \cdot \frac{\pi}{180} = \frac{\pi}{3}$ or $60° \times 0.01745 = 1.047$
2. $198° \times 0.01745 = 3.4551$ or $198 \times \frac{\pi}{180} \approx 3.456$
3. $325 \times 0.01745 = 5.67125$ or $325 \times \frac{\pi}{180} \approx 5.672$
4. $180° \times \frac{\pi}{180} = \pi \approx 3.1416$
5. $-115° \times 0.01745 = -2.00675$; $-115° \times \frac{\pi}{180} \approx -2.007$
6. $435° \times 0.01745 = 7.59075$; $435 \cdot \frac{\pi}{180} \approx 7.592$
7. $\frac{3\pi}{4}$ rad $= \frac{3\pi}{4} \cdot \frac{180°}{\pi} = 135°$
8. 1.10 rad $= 1.10 \times \frac{180°}{\pi} \approx 63.025$ or $1.10 \times 57.296 \approx 63.026$
9. 2.15 rad $= 2.15 \times 57.296 \approx 123.19°$ or $2.15 \times \frac{180°}{\pi} \approx 123.19°$
10. $\frac{7\pi}{3}$ rad $= \frac{7\pi}{3} \cdot \frac{180°}{\pi} = 420°$
11. -4.31 rad $= -4.31 \times 57.296 \approx -246.95°$ or $-4.31 \times \frac{180°}{\pi} \approx -246.94°$
12. 5.92 rad $= 5.92 \times 57.296 \approx 339.19°$ or $5.92 \times \frac{180°}{\pi} \approx 339.19°$
13. (a) I, (b) 60°, (c) $60° + 360 = 420°$; $60 - 360 = -300°$
14. (a) III, (b) $\theta_r = 198 - 180 = 18°$, (c) $198 + 360 = 558°$; $198 - 360 = -162°$
15. (a) IV, (b) $\theta_r = 360 - 325 = 35°$, (c) $325 + 360 = 685°$, $325 - 360 = 35°$
16. (a) Quadrantal angle between II and III, (b) 0°, (c) $180 + 360 = 540°$; $180 - 360° = -180°$
17. (a) III, (b) $\theta_r = 245 - 180 = 65°$, (c) $-115 + 360 = 245°$; $-115 - 360° = -475°$
18. (a) I, (b) 75°, (c) $435 + 360 = 795°$; $435 - 360 = 75°$; $75° - 360° = -285°$
19. (a) II, (b) $\pi - \frac{3\pi}{4} = \frac{\pi}{4}$, (c) $\frac{3\pi}{4} + 2\pi = \frac{11\pi}{4}$, $\frac{3\pi}{4} - 2\pi = \frac{-5\pi}{4}$
20. (a) I, (b) 1.10, (c) $1.10 + 2\pi = 1.10 + 6.28 = 7.38$; $1.10 - 6.28 = -5.18$
21. (a) II, (b) $\pi - 2.15 = 3.14 - 2.15 = 0.99$, (c) $2.15 + 2\pi = 2.15 + 6.28 = 8.43$; $2.15 - 2\pi = 2.15 - 6.28 = -4.13$
22. (a) I, (b) $\frac{7\pi}{3} - 2\pi = \frac{\pi}{3}$, (c) $\frac{7\pi}{3} + 2\pi = \frac{13\pi}{3}$, $\frac{7\pi}{3} - 2\pi = \frac{\pi}{3}$, or $\frac{7\pi}{3} - 4\pi = -\frac{5\pi}{3}$
23. (a) II, (b) $4.31 - \pi = 1.17$, (c) $-4.31 + 6.28 = 1.97$; $-4.31 - 6.28 = -10.59$
24. (a) IV, (b) $2\pi - 5.92 = 6.28 - 5.92 = 0.36$, (c) $5.92 - 2\pi = -0.36$; $5.92 + 6.28 = 12.20$
25. $r = \sqrt{3^2 + (-4)^2} = 5$; $\sin\theta = \frac{y}{r} = \frac{-4}{5}$; $\cos\theta = \frac{x}{r} = \frac{3}{5}$; $\tan\theta = \frac{y}{x} = \frac{-4}{3}$; $\csc\theta = \frac{r}{y} = \frac{-5}{4}$; $\sec\theta = \frac{r}{x} = \frac{5}{3}$; $\cot\theta = \frac{x}{y} = \frac{-3}{4}$
26. $r = \sqrt{5^2 + 12^2} = 13$; $\sin\theta = \frac{y}{r} = \frac{12}{13}$; $\cos\theta = \frac{x}{r} = \frac{5}{13}$; $\tan\theta = \frac{y}{x} = \frac{12}{5}$; $\csc\theta = \frac{r}{y} = \frac{13}{12}$; $\sec\theta = \frac{r}{x} = \frac{13}{5}$; $\cot\theta = \frac{x}{y} = \frac{5}{12}$
27. $r = \sqrt{(-20)^2 + 21^2} = \sqrt{841} = 29$; $\sin\theta = 21/29$, $\cos\theta = -20/29$, $\tan\theta = -21/20$, $\csc\theta = 29/21$, $\sec\theta = -29/20$, $\cot\theta = -20/21$
28. $r = \sqrt{(-4)^2 + (-7)^2} = \sqrt{16 + 49} = \sqrt{65}$; $\sin\theta = -7/\sqrt{65}$, $\cos\theta = -4/\sqrt{65}$, $\tan\theta = 7/4$, $\csc\theta = -\sqrt{65}/7$, $\sec\theta = -\sqrt{65}/4$, $\cot\theta = 4/7$
29. $r = \sqrt{7^2 + 1^2} = \sqrt{50} = 5\sqrt{2}$; $\sin\theta = \frac{1}{5\sqrt{2}}$; $\cos\theta = \frac{7}{5\sqrt{2}}$; $\tan\theta = \frac{1}{7}$; $\csc\theta = 5\sqrt{2}$; $\sec\theta = \frac{5\sqrt{2}}{7}$; $\cot\theta = 7$
30. $r = \sqrt{144 + 64} = \sqrt{208} = 4\sqrt{13}$; $\sin\theta = \frac{8}{4\sqrt{13}} = \frac{2}{\sqrt{13}}$; $\cos\theta = \frac{-12}{4\sqrt{13}} = \frac{-3}{\sqrt{13}}$; $\tan\theta = \frac{-8}{12} = \frac{-2}{3}$; $\csc\theta = \frac{\sqrt{13}}{2}$; $\sec\theta = -\frac{\sqrt{13}}{3}$; $\cot\theta = -\frac{3}{2}$
31. $b = \sqrt{17^2 - 8^2} = 15$; $\sin\theta = \frac{a}{c} = \frac{8}{17}$; $\cos\theta = \frac{b}{c} = \frac{15}{17}$; $\tan\theta = \frac{a}{b} = \frac{8}{15}$; $\csc\theta = \frac{c}{a} = \frac{17}{8}$; $\sec\theta = \frac{c}{b} = \frac{17}{15}$; $\cot\theta = \frac{b}{a} = \frac{15}{8}$

32. $c = \sqrt{5^2 + 12^2} = 13$; $\sin\theta = \frac{a}{c} = \frac{5}{13}$; $\cos\theta = \frac{b}{c} = \frac{12}{13}$; $\tan\theta = \frac{a}{b} = \frac{5}{12}$; $\csc\theta = \frac{c}{a} = \frac{13}{5}$; $\sec = \frac{c}{b} = \frac{13}{12}$; $\cot\theta = \frac{b}{a} = \frac{12}{5}$

33. $a = \sqrt{\left(\frac{40}{3}\right)^2 - 8^2} = \sqrt{\frac{1600}{9} - \frac{576}{9}} = \sqrt{\frac{1024}{9}} = \frac{32}{3}$; $\sin\theta = \frac{32}{40} = \frac{4}{5}$; $\cos\theta = \frac{8}{\frac{40}{3}} = \frac{24}{40} = \frac{3}{5}$; $\tan\theta = \frac{\frac{32}{3}}{8} = \frac{4}{3}$; $\csc\theta = \frac{5}{4}$; $\sec\theta = \frac{5}{3}$; $\cot = \frac{3}{4}$

34. $b = \sqrt{7.5^2 - 6^2} = 4.5$; $\sin\theta = \frac{6}{7.5} = .8$; $\cos\theta = \frac{4.5}{7.5} = 0.6$; $\tan\theta = \frac{6}{4.5} = 1.33$; $\csc\theta = \frac{7.5}{6} = 1.25$; $\sec\theta = \frac{7.5}{4.5} = 1.67$; $\cot\theta = \frac{4.5}{6} = 0.75$

35. $a = \sqrt{18.2^2 - 7^2} = 16.8$; $\sin\theta = \frac{16.8}{18.2} = \frac{168}{182} = \frac{84}{91}$; $\cos\theta = \frac{7}{18.2} = \frac{70}{182} = \frac{35}{91}$; $\tan\theta = \frac{16.8}{7} = \frac{168}{70} = \frac{84}{35}$; $\csc\theta = \frac{91}{84}$; $\sec\theta = \frac{91}{35}$; $\cot\theta = \frac{35}{84}$

36. $c = \sqrt{42^2 + 44.1^2} = 60.9$; $\sin\theta = \frac{42}{60.9} = \frac{420}{609}$; $\cos\theta = \frac{44.1}{60.9} = \frac{441}{609}$; $\tan\theta = \frac{42}{44.1} = \frac{420}{441} = \frac{20}{21}$; $\csc\theta = \frac{609}{420}$; $\sec\theta = \frac{609}{441}$; $\cot\theta = \frac{21}{20}$

37. $a = 12.8$; $c = 27.2$; $b = \sqrt{27.2^2 - 12.8^2} = 24$; $\cos\theta = \frac{24}{27.2}$; $\tan\theta = \frac{12.8}{24}$; $\csc\theta = \frac{27.2}{12.8}$; $\sec\theta = \frac{27.2}{24}$; $\cot\theta = \frac{24}{12.8}$

38. $a = 16$; $b = 16.8$; $c = \sqrt{16^2 + 16.8^2} = 23.2$; $\sin\theta = \frac{16}{23.2}$; $\cos\theta = \frac{16.8}{23.2}$; $\tan\theta = \frac{16}{16.8}$; $\csc\theta = \frac{23.2}{16}$; $\sec\theta = \frac{23.2}{16.8}$; $\cot\theta = \frac{16.8}{16}$

39. $c = 4$; $b = 2.5$; $a = \sqrt{4^2 - 2.5^2} \approx 3.12$; $\sin\theta = \frac{3.12}{4}$; $\cos\theta = \frac{2.5}{4}$; $\tan\theta = \frac{3.12}{2.5}$; $\csc\theta = \frac{4}{3.12}$; $\sec\theta = \frac{4}{2.5}$; $\cot\theta = \frac{2.5}{3.12}$

40. $\cos\theta = \frac{\sin\theta}{\tan\theta} = \frac{0.532}{0.628} \approx 0.847$; $\csc\theta = \frac{1}{0.532} \approx 1.880$; $\sec\theta = \frac{1}{0.847} \approx 1.181$; $\cot\theta = \frac{1}{0.628} \approx 1.592$

41. $\tan\theta = \frac{0.5}{0.866} \approx 0.577$; $\csc\theta = \frac{1}{0.5} = 2$; $\sec\theta = \frac{1}{0.866} \approx 1.155$; $\cot\theta = \frac{0.866}{0.5} = 1.732$

42. $\sin\theta = \frac{1}{1.364} \approx 0.733$; $\tan\theta \approx \frac{0.733}{0.680} \approx 1.078$; $\sec\theta = \frac{1}{0.680} \approx 1.471$; $\cot\theta \approx \frac{0.680}{0.733} \approx 0.928$

43. $\sin 45° = 0.7071$

44. $\cos 82.5° = 0.1305$

45. $\tan 213.5° = 0.6619$

46. $\sec(-81°) = 6.3925$

47. $\cos 2.3 \text{ rad} = -0.6663$

48. $\sin 4.75 \text{ rad} = -0.9993$

49. $\tan(-3.2) \text{ rad} = -0.0585$

50. $\csc 0.21 \text{ rad} = 4.7971$

51. $\sin\theta = 0.5$; $\theta_r = \sin^{-1} 0.5 = 30°$; $\theta_I = \theta_r = 30°$; $\theta_{II} = 180 - 30° = 150°$

52. $\tan\theta = 2.5$; $\theta_r = \tan^{-1} 2.5 = 68.2$; $\theta_I = 68.2$; $\theta_{III} = 180 + 68.2 = 248.2$

53. $\cos\theta = -0.75$; $\theta_r = \cos^{-1} 0.75 = 41.4$; $\theta_{II} = 180 - 41.4 = 138.6$; $\theta_{III} = 180 + 41.4 = 221.4$

54. $\csc\theta = 3.0$; $\theta_r = \sin^{-1} \frac{1}{3} = 19.5$; $\theta_I = 19.5$; $\theta_{II} = 180 - 19.5 = 160.5$

55. $\cos\theta = -0.5$; $\theta_r = \cos^{-1} 0.5 = 1.047 \approx 1.05$; $\theta_{II} = \pi - 1.05 = 3.14 - 1.05 = 2.09$; $\theta_{III} = \pi + 1.05 = 3.14 + 1.05 = 4.19$

56. $\sin\theta = 0.717$; $\theta_r = \sin^{-1} 0.717 = 0.80$; $\theta_I = 0.80$; $\theta_{II} = \pi - 0.80 = 3.14 - 0.80 = 2.34$

57. $\theta_r = \tan^{-1} 0.95 = 0.76$; $\theta_{II} = \pi - 0.76 = 3.14 - 0.76 = 2.38$; $\theta_{IV} = 2\pi - 0.76 = 6.28 - 0.76 = 5.52$

58. $\theta_r = \cos^{-1} \frac{1}{2.25} = 1.11$; $\theta_I = 1.11$; $\theta_{IV} = 2\pi - 1.11 = 6.28 - 1.11 = 5.17$

59. $\arcsin 0.866 = 60.0°$

60. $\arccos 0.5 = 60.0°$

61. $\arctan(-1) = -45.0°$

62. $\cos^{-1}(-0.707) = 135.0°$

63. $\sin^{-1} 0.385 = 22.6°$

64. $\cot^{-1}(3.5) = \tan^{-1} \frac{1}{3.5} = 15.9°$

65. $1.33 = \frac{\sin 30°}{\sin r}$, $\sin r = \frac{\sin 30°}{1.33} \approx 0.3759$; $r \approx \sin^{-1} 0.3759 \approx 22.1°$

66. $n = \frac{\sin 0.90}{\sin 0.55} \approx 1.50$

67. $I = 12.6 \sin \frac{\pi}{5} \approx 7.41$ A

68. $\sin\theta = \frac{4.732}{7.3} \approx 0.6482$; $\theta \approx \sin^{-1} 0.6482 \approx 40.4°$

69. $\tan\theta = \frac{44}{50}$; $\theta = \tan^{-1}\frac{44}{50} \approx 41.3°$; $V = \sqrt{V_L^2 + V_R^2} = \sqrt{44^2 + 50^2} \approx 66.6$ V

70. $\frac{1}{2}(57 - 46) = 5.5$; $\tan 3° = \frac{5.5}{h}$; $h = \frac{5.5}{\tan 3°} \approx$ 104.9 mm

71. We use the equation $s = r\theta$ with $\theta = 37° \times \frac{\pi}{180} \approx 0.6458$. Thus, we obtain $s \approx 900 \times 0.6458 \approx 581.2$ ft.

72. $300\dfrac{\text{rev}}{\text{min}} \cdot \dfrac{2\pi \text{ rad}}{\text{rev}} \cdot \dfrac{1 \text{ min}}{60 \text{ sec}} = 10\pi$ rad/s ≈ 31.416 rad/s; $v = r\omega = \dfrac{5.25 \text{ in.}}{2} \times 31.416 = 82.467$ in./s

73. $\omega = \frac{1 \text{ rev}}{365 \text{ days}} \cdot \frac{2\pi}{1 \text{ rev}} \cdot \frac{1 \text{ day}}{24 \text{ hr}} \approx 7.17 \times 10^{-4} \frac{\text{rad}}{\text{hr}}$; $v = r\omega = 9.3 \times 10^7 \times 7.17 \times 10^{-4} \approx 66{,}700$ mph

74. $\tan 68° = \frac{h}{15}$; $h = 15 \tan 68° \approx 37.1$ ft

75. Let y = length of common side, then $\cos 36° = \frac{y}{110.5}$; $y = 110.5 \cos 36° = 89.4$; $\cos x = \frac{82.3}{89.4}$; $x = \cos^{-1}\frac{82.3}{89.4}$; $x = 23.0°$

76. $\tan\theta = \frac{56}{43}$; $\theta = \tan^{-1}\frac{56}{43} \approx 52.5°$

77. $\tan 53.7° = \frac{h}{150}$; $h = 150 \tan 53.7° \approx 204.2$ ft

78. Draw a segment parallel to the axis from the top of the smaller circle. This makes a right triangle with legs of 128.3 m and $(50.2 - 26.1) = 24.1$ m. Then $\tan\phi = \frac{24.1}{128.3}$; $\phi \approx 10.64$; $\theta = 2\phi \approx 21.3°$

79. $F_x = 3500 \sin 34.3° \approx 1972.3$ lb;
$F_y = 3500 \cos 34.3 \approx 2891.3$ lb

80. Let h be height of the tower and x be the distance from the base of the tower to the closer anchor. Then $\tan 23.7° = \frac{h}{x+35}$ and $\tan 36.4° = \frac{h}{x}$; $\tan 23.7° \approx 0.439$; $\tan 36.4 \approx 0.737$, so $0.439(x + 35) = h$ and $0.737x = h$; by substitution $0.439(x+35) = 0.737x$; $0.439x + 15.365 = 0.737x$; $15.365 = 0.298x$; $x = \frac{15.365}{0.298} \approx 51.56$; $h = x \tan 36.4$; $h = 51.56 \tan 36.4 \approx$ 38.0 m

CHAPTER 9 TEST

1. $50° \times \frac{\pi}{180°} = \frac{5\pi}{18}$ or $50 \times 0.01745 = 0.8725$

2. $\frac{4\pi}{3} = \frac{4\pi}{3} \cdot \frac{180°}{\pi} = \frac{4}{3} \cdot 180° = 4(60°) = 240°$

3. (a) III; (b) $237° - 180° = 57°$

4. $r = \sqrt{(-5)^2 + 12^2} = 13$; $\sin\theta = \frac{12}{13}$; $\cos\theta = \frac{-5}{13}$; $\tan\theta = \frac{-12}{5}$; $\csc\theta = \frac{13}{12}$; $\sec\theta = \frac{-13}{5}$; $\cot\theta = \frac{-5}{12}$

5. $b = \sqrt{c^2 - a^2} = \sqrt{23.4^2 - 9^2} = \sqrt{466.56} = 21.6$; (a) $\sin\theta = \frac{9}{23.4} \approx 0.38462$; (b) $\tan\theta = \frac{9}{21.6} \approx 0.41667$

6. Here the side opposite $\theta = 30$ and the side adjacent to θ is 6.75. So, the hypotenuse $r = \sqrt{30^2 + 6.75^2} = 30.75$; (a) $\tan\theta = \dfrac{1}{\cot\theta} = \dfrac{30}{6.75} \approx 4.44444$; (b) $\sin\theta = \frac{30}{30.75} \approx 0.97561$

7. Using a calculator, we get (a) 0.79864; (b) 2.47509; (c) −1.66164; (d) −0.49026 [In (d) make sure that your calculator is in radian mode.]

8. (a) $\theta_r = \tan^{-1} 1.2 \approx 50.2°$; $\theta_{III} = 180° + \theta_r = 230.2°$; (b) $\theta_r = \sin^{-1} 0.72 = 46.1$; $\theta_{III} = 180° + \theta_r = 226.1°$; $\theta_{IV} = 360° - \theta_r = 313.9°$

9. $\tan 53° = \frac{h}{135}$; $h = 135 \tan 53° \approx 179.15$ ft

10. One revolution is 2π radians, so it rotates $3600 \times 2\pi = 7200\pi$ radians per minute. There are 60 seconds in one minute, so it rotates $7200\pi \div 60 = 120\pi$ radians per second.

CHAPTER

10

Vectors and Trigonometric Functions

10.1 INTRODUCTION TO VECTORS

1.

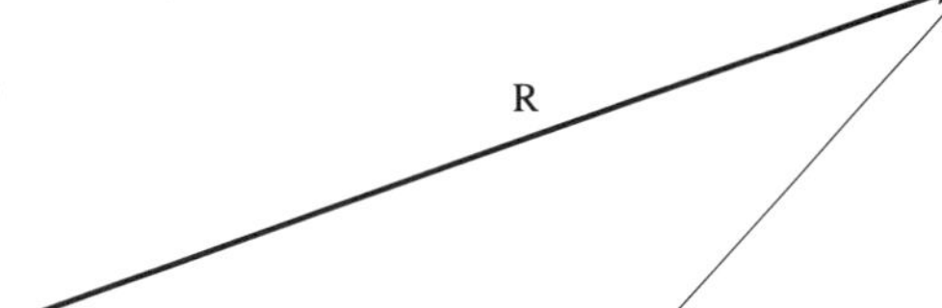

2.

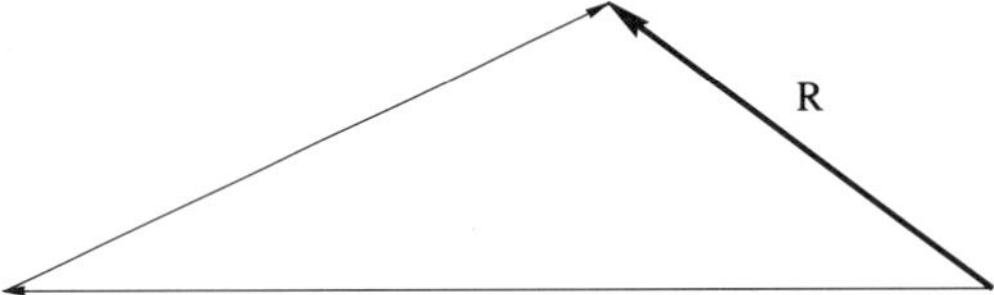

3.

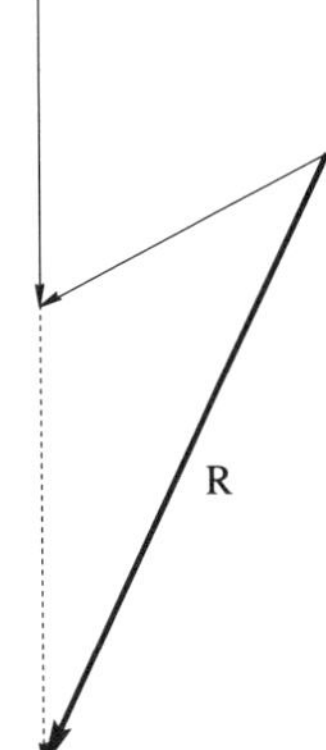

4.

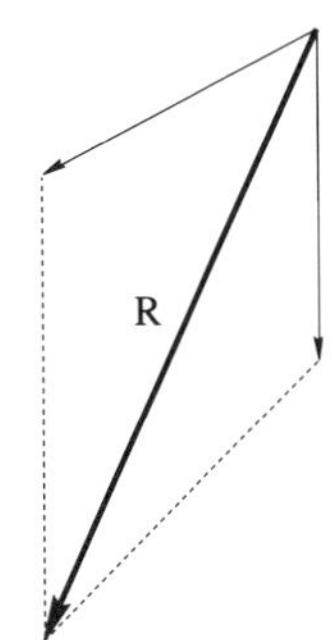

5.

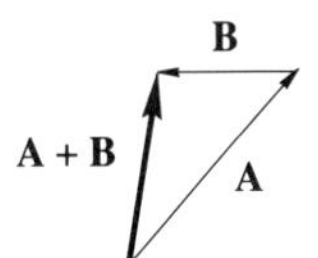

6.

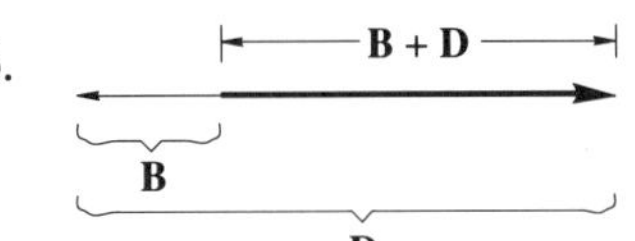

7.

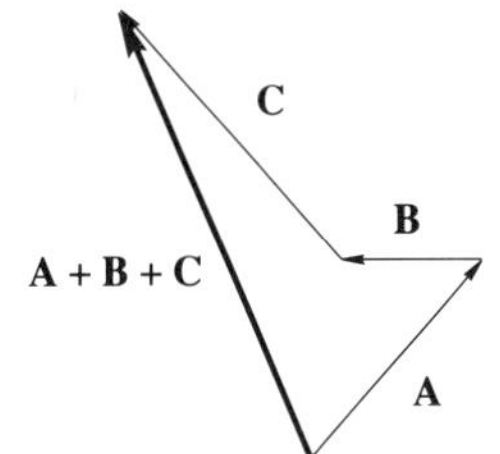

8.

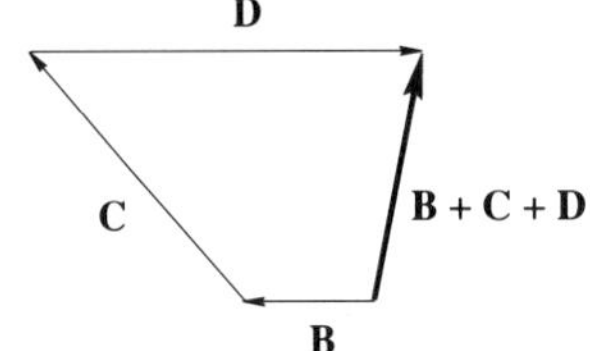

9.

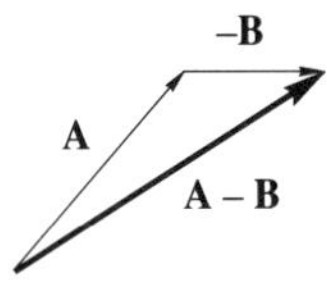

10.

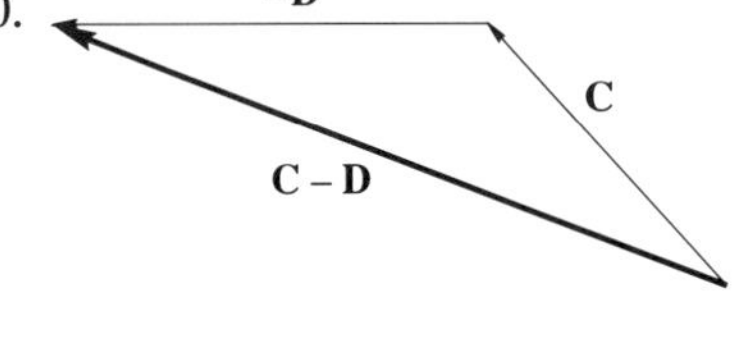

11.

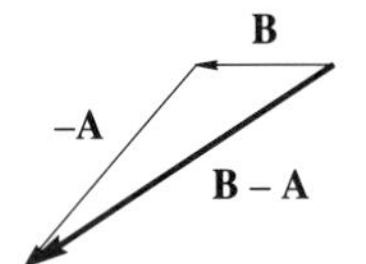

12. 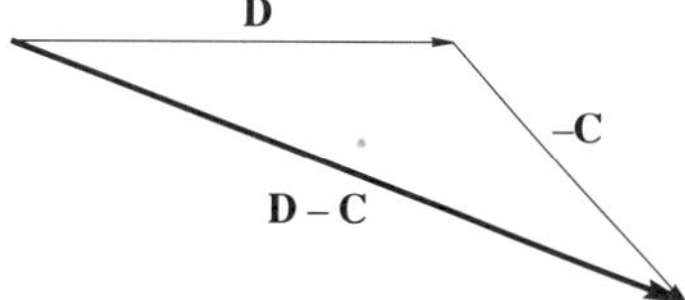

13. 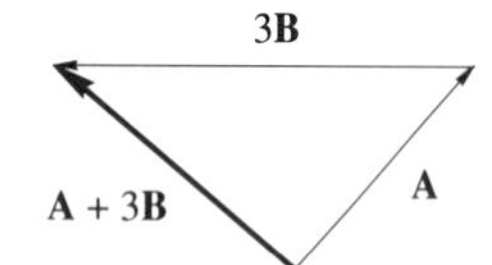

14. 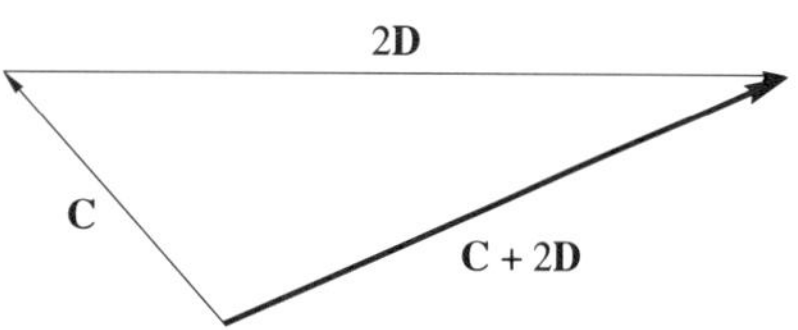

15. 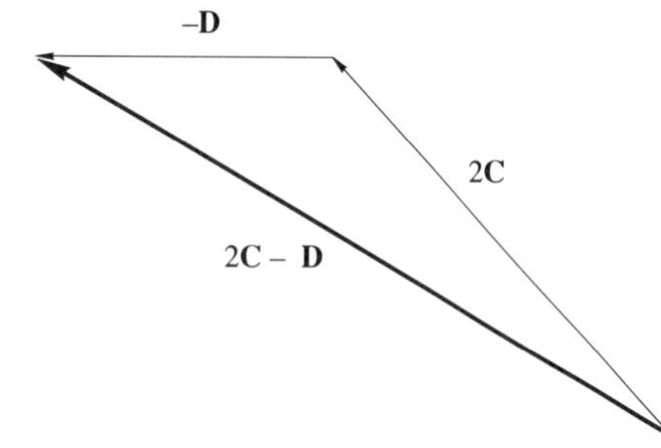

16. 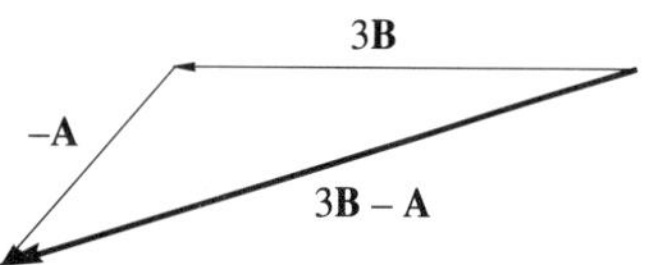

17. 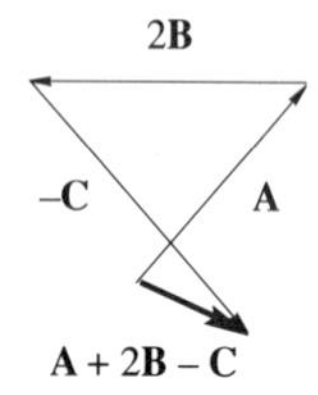

18. 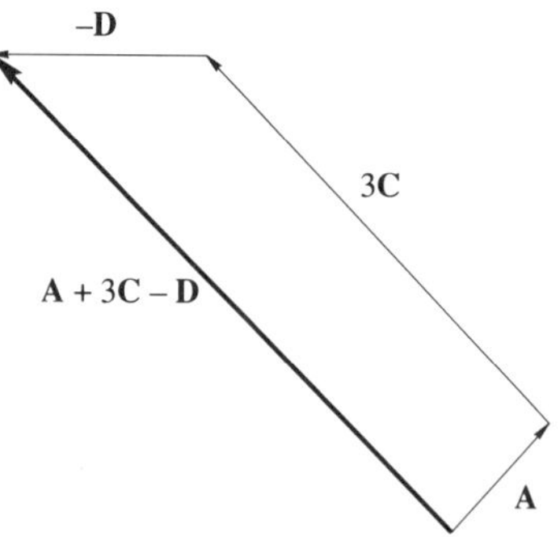

19. 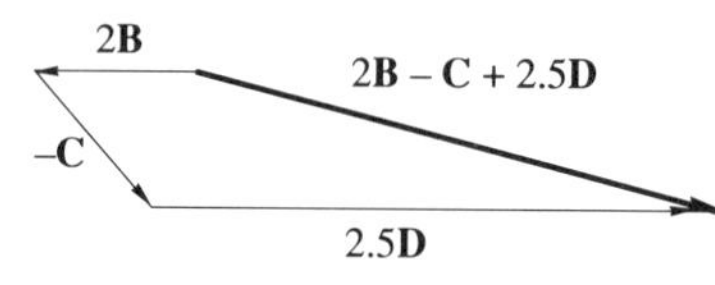

20. 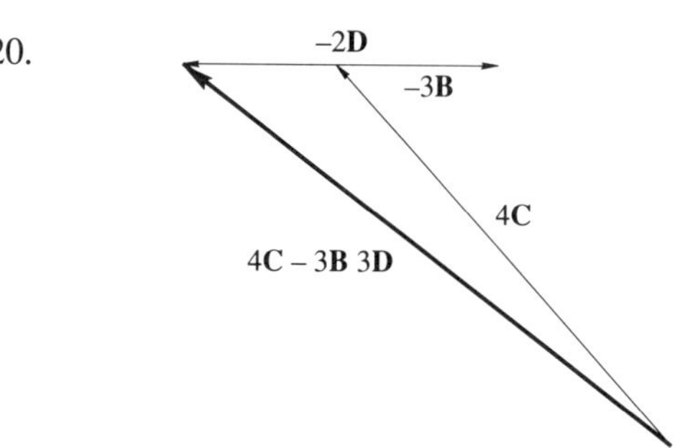

21. $\mathbf{P_x} = 20 \cos 75° = 5.176$; $\mathbf{P_y} = 20 \sin 75° = 19.319$

22. $\mathbf{P}_x = 16 \cos 212° = -13.569$; $\mathbf{P_y} = 16 \sin 212° = -8.479$

23. $\mathbf{P_x} = 18.4 \cos 4.97 \text{ rad} = 4.688$; $\mathbf{P_y} = 18.4 \sin 4.97 = -17.793$

24. $\mathbf{P_x} = 23.7 \cos 2.22 = -14.328$; $\mathbf{P_y} = 23.7 \sin 2.22 = 18.879$

25. $\mathbf{V_x} = 9.75 \cos 16° = 9.372$; $\mathbf{V_y} = 9.75 \sin 16° = 2.687$

26. $\mathbf{P_x} = 24.6 \cos 317° = 17.991$; $\mathbf{P_y} = 24.6 \sin 317° = -16.777$

27. $A = \sqrt{(-9)^2 + 12^2} = 15$; θ is in Quadrant II; $\theta_{Ref} = \tan^{-1} \frac{12}{9} = 53.13$; $\theta = 180° - 53.13° = 126.87°$

28. $B = \sqrt{10^2 + (-24)^2} = 26$; θ is in Quadrant IV; $\theta_{Ref} = \tan^{-1} \frac{24}{10} = 67.38°$; $\theta = 360 - 67.38 = 292.62°$

29. $C = \sqrt{8^2 + 15^2} = 17$; θ is in Quadrant I; $\theta_{Ref} = \tan^{-1} \frac{15}{8} = 61.93°$; $\theta = 61.93°$

30. $D = \sqrt{(-14)^2 + (-20)^2} = 24.413$; θ is in Quadrant III $\theta_{Ref} = \tan^{-1} \frac{20}{14} = 55.01°$; $\theta = 180° + 55.01° = 235.01°$

31. $v = \sqrt{5^2 + 12^2} = 13$ km/hr

32. $\mathbf{V_x} = 765.0$ $\mathbf{V_y} = 73$, $V = \sqrt{756^2 + 73^2} = 759.5$ km/hr; $\theta = \tan^{-1} \frac{73}{756} = 5.52°$ north of east

33. $\mathbf{V_x} = 976 \cos 72.4° = 295$ N; $\mathbf{V_y} = 976 \sin 72.4° = 930$ N

34. $F = \sqrt{125.0^2 + 26.50^2} = 127.8$ lb; $\theta = \tan^{-1} \frac{26.5}{125} = 11.97°$ with the large force or 11.97° from vertical.

35. Parallel to ramp $153 \times \sin 12° = 31.8$ lb. Perpendicular to ramp $153 \cos 12° = 149.7$ lb.

36. $V = \sqrt{V_R^2 + (V_L - V_C)^2} = \sqrt{12^2 + (5 - 10)^2} = 13$ V; $\phi = \tan^{-1} \frac{-5}{12} = -22.6°$ or 22.6° from $\mathbf{V}_R$ toward $(\mathbf{V}_L - \mathbf{V}_C)$.

37. $V = \sqrt{15^2 + (8 - 17)^2} = 17.5$ V; $\phi = \tan^{-1} \frac{-9}{15} \approx -30.96°$ or 30.96° from $\mathbf{V}_R$ towards $(\mathbf{V}_L - \mathbf{V}_C)$.

10.2 ADDING AND SUBTRACTING VECTORS

1. $R = |20.0 - 32.5| = 12.5$; θ_R = angle of larger so 180°

2. $R = |14.3 + 7.2| = 21.5$; $\theta_R = 90°$

3. $R = |121.7 - 86.9| = 34.8$; $\theta_R = 270°$ (angle of larger magnitude)

4. $R = |63.1 + 43.5| = 106.6$; $\theta_R = 180°$

5. $R = \sqrt{55^2 + 48^2} = 73.$; $\theta = \tan^{-1} \frac{55}{48} = 48.9$; since it's in Quadrant II $\theta_R = 180° - 48.9° = 131.1°$. (Note **C** is a vertical vector, **D** is a horizontal vector. $\theta_R = \tan^{-1} \frac{\text{vertical}}{\text{horizontal}}$)

6. $R = \sqrt{65^2 + 72^2} = 97$; $\theta = \tan^{-1} \frac{65}{72} = 42.08°$ Quadrant III so $180° + 42.08 = 222.08° = \theta_R$

7. $R = \sqrt{81.4^2 + 37.6^2} = 89.7$; $\theta_R = \tan^{-1} \frac{37.6}{81.4} = 24.79°$

8. $R = \sqrt{63.4^2 + 9.4^2} = 64.1$; $\theta = \tan^{-1} \frac{63.4}{9.4} = 81.57$ in Quadrant IV so $\theta_R = 360 - 81.57 = 278.43$

9. $A = \sqrt{33^2 + 56^2} = 65$; $\theta_A = \tan^{-1} \frac{56}{33} = 59.49°$

10. $B = \sqrt{231^2 + 520^2} = 569$; $\theta_B = \tan^{-1} \frac{520}{231} = 66.05°$

11. $C = \sqrt{11.7^2 + 4.4^2} = 12.5$; $\theta_C = \tan^{-1} \frac{4.4}{11.7} = 20.61°$

12. $D = \sqrt{31.9^2 + 36.0^2} = 48.1$; $\theta_D = \tan^{-1} \frac{36.0}{31.9} = 48.46°$

13. $E = \sqrt{6.3^2 + 1.6^2} = 6.5$; $\theta_E = \tan^{-1} \frac{1.6}{6.3} = 14.25°$

14. $F = \sqrt{5.1^2 + 14.0^2} = 14.9$; $\theta_F = \tan^{-1} \frac{14.0}{5.1} = 69.98°$

15. $G = \sqrt{8.4^2 + 12.6^2} = 15.1$; $\theta_G = \tan^{-1} \frac{12.6}{8.4} = 56.31°$

16. $H = \sqrt{15.3^2 + 9.2^2} = 17.9$; $\theta_H = \tan^{-1} \frac{9.2}{15.3} = 31.02°$

17. From these values for R_x and R_y we can determine $R = \sqrt{R_x^2 + R_y^2} = 12.33$ and $\theta_R = \tan^{-1} \dfrac{R_y}{R_x} = 32.03°$.

vector	horizontal component		vertical component
A	$\mathbf{A_x} = 4 \cos 60° =$	2.000	$\mathbf{A_y} = 4 \sin 60° = 3.464$
B	$\mathbf{B_x} = 9 \cos 20° =$	8.457	$\mathbf{B_y} = 9 \sin 20° = 3.078$
R	$\mathbf{R_x}$	10.457	$\mathbf{R_y}$ 6.542

18. $R = \sqrt{R_x^2 + R_y^2} = 25.55$; $\theta \tan^{-1} \dfrac{R_y}{R_x} = 53.81°$

vector	horizontal component		vertical component	
C	$\mathbf{C_x} = 12 \cos 75° =$	3.106	$\mathbf{C_y} = 12 \sin 75° =$	11.591
D	$\mathbf{D_x} = 15 \cos 37° =$	11.980	$\mathbf{D_y} = 15 \sin 37° =$	9.027
R	$\mathbf{R_x}$	15.086	$\mathbf{R_y}$	20.618

19. $R = \sqrt{R_x^2 + R_y^2} = 53.00$;
$\theta = \tan^{-1} \dfrac{R_y}{R_x} = 178.11°$

vector	horizontal component		vertical component	
C	$\mathbf{C_x} = 28 \cos 120° =$	−14.000	$\mathbf{C_y} = 28 \sin 120° =$	24.249
D	$\mathbf{D_x} = 45 \cos 210° =$	−38.971	$\mathbf{D_y} = 45 \sin 210° =$	−22.500
R	$\mathbf{R_x}$	−52.971	$\mathbf{R_y}$	1.749

20. $R = \sqrt{R_x^2 + R_y^2} = 97.00$;
$\theta = \tan^{-1} \dfrac{R_y}{R_x} = -30.92$; since this is Quadrant IV, we have
$\theta = 360 - 30.92 = 329.08°$

vector	horizontal component		vertical component	
E	$\mathbf{E_x} = 72 \cos 287° =$	21.051	$\mathbf{E_y} = 72 \sin 287° =$	−68.854
F	$\mathbf{F_x} = 65 \cos 17° =$	62.160	$\mathbf{F_y} = 65 \sin 17° =$	19.004
R	$\mathbf{R_x}$	83.211	$\mathbf{R_y}$	−49.850

21. $R = \sqrt{R_x^2 + R_y^2} = 88.50$;
$\theta = \tan^{-1} \dfrac{R_y}{R_x} = 223.83°$

vector	horizontal component		vertical component	
A	$\mathbf{A_x} = 31.2 \cos 197.5° =$	−29.756	$\mathbf{A_y} = 31.2 \sin 197.5° =$	−9.382
B	$\mathbf{B_x} = 62.1 \cos 236.7° =$	−34.094	$\mathbf{B_y} = 62.1 \sin 236.7° =$	−51.904
R	$\mathbf{R_x}$	−63.850	$\mathbf{R_y}$	−61.286

22. $R = \sqrt{R_x^2 + R_y^2} = 57.42$;
$\theta = \tan^{-1} \dfrac{R_y}{R_x} = 247.28°$

vector	horizontal component		vertical component	
C	$\mathbf{C_x} = 53.1 \cos 324.3° =$	43.122	$\mathbf{C_y} = 53.1 \sin 324.3° =$	−30.986
D	$\mathbf{D_x} = 68.9 \cos 198.6° =$	−65.301	$\mathbf{D_y} = 68.9 \sin 198.6° =$	−21.976
R	$\mathbf{R_x}$	−22.179	$\mathbf{R_y}$	−52.962

23. $R = \sqrt{R_x^2 + R_y^2} = 22.44$;
$\theta = \tan^{-1} \dfrac{R_y}{R_x} = 348.90°$

vector	horizontal component		vertical component	
E	$\mathbf{E_x} = 12.52 \cos 46.4° =$	8.634	$\mathbf{E_y} = 12.52 \sin 46.4° =$	9.067
F	$\mathbf{F_x} = 18.93 \cos 315° =$	13.386	$\mathbf{F_y} = 18.93 \sin 315° =$	−13.386
R	$\mathbf{R_x}$	22.020	$\mathbf{R_y}$	−4.319

24. $R = \sqrt{R_x^2 + R_y^2} = 116.86$;
$\theta = \tan^{-1} \dfrac{R_y}{R_x} = 65.61°$

vector	horizontal component		vertical component	
G	$\mathbf{G_x} = 76.2 \cos 15.7° =$	73.357	$\mathbf{G_y} = 76.2 \sin 15.7° =$	20.620
H	$\mathbf{H_x} = 89.4 \cos 106.3° =$	−25.092	$\mathbf{H_y} = 89.4 \sin 106.3° =$	85.807
R	$\mathbf{R_x}$	48.265	$\mathbf{R_y}$	106.427

25. $R = \sqrt{R_x + R_y} = 13.04$;
$\theta = \tan^{-1} \dfrac{R_y}{R_x} = 150.30°$

vector	horizontal component		vertical component	
A	$\mathbf{A_x} = 9.84 \cos 215.5° =$	−8.011	$\mathbf{A_y} = 9.84 \sin 215.5° =$	−5.714
B	$\mathbf{B_x} = 12.62 \cos 105.25° =$	−3.319	$\mathbf{B_y} = 12.62 \sin 105.25° =$	12.176
R	$\mathbf{R_x}$	−11.330	$\mathbf{R_y}$	6.462

26. $R = \sqrt{R_x^2 + R_y^2} = 104.99$;

$\theta = \tan^{-1}\frac{R_y}{R_x} = 240.29°$

vector	horizontal component	vertical component
C	$\mathbf{C_x} = 79.63\cos 262.75° = -10.049$	$\mathbf{C_y} = 79.63\sin 262.75° = -78.993$
D	$\mathbf{D_x} = 43.72\cos 196.2° = -41.984$	$\mathbf{D_y} = 43.72\sin 196.2° = -12.197$
R	$\mathbf{R_x}$ -52.033	$\mathbf{R_y}$ -91.190

27. $R = \sqrt{R_x^2 + R_y^2} = 63.58$;

$\theta = \tan^{-1}\frac{R_y}{R_x} = 4.62$

vector	horizontal component	vertical component
E	$\mathbf{E_x} = 42.0\cos 3.4 = -40.606$	$\mathbf{E_y} = 42.0\sin 3.4 = -10.733$
F	$\mathbf{F_x} = 63.2\cos 5.3 = 35.036$	$\mathbf{F_y} = 63.2\sin 5.3 = -52.599$
R	$\mathbf{R_x}$ -5.570	$\mathbf{R_y}$ -63.332

28. $R = \sqrt{R_x^2 + R_y^2} = 58.91$;

$\theta = \tan^{-1}\frac{R_y}{R_x} = 1.23$

vector	horizontal component	vertical component
G	$\mathbf{G_x} = 37.5\cos 0.25 = 36.334$	$\mathbf{G_y} = 37.5\sin 0.25 = 9.278$
H	$\mathbf{H_x} = 49.3\cos 1.92 = -16.868$	$\mathbf{H_y} = 49.3\sin 1.92 = 46.325$
R	$\mathbf{R_x}$ 19.466	$\mathbf{R_y}$ 55.603

29. $R = \sqrt{R_x^2 + R_y^2} = 13.13$;

$\theta = \tan^{-1}\frac{R_y}{R_x} = 64.41$

vector	horizontal component	vertical component
A	$\mathbf{A_x} = 15\cos 25° = 13.59$	$\mathbf{A_y} = 15\sin 25° = 6.34$
B	$\mathbf{B_x} = 20\cos 120° = -10.00$	$\mathbf{B_y} = 20\sin 120° = 17.32$
C	$\mathbf{C_x} = 12\cos 280° = 2.08$	$\mathbf{C_y} = 12\sin 280° = -11.82$
R	$\mathbf{R_x}$ 5.67	$\mathbf{R_y}$ 11.84

30. $R = \sqrt{R_x^2 + R_y^2} = 236.56$;

$\theta = \tan^{-1}\frac{R_y}{R_x} = 187.51$

vector	horizontal component	vertical component
A	$\mathbf{A_x} = 320\cos 105° = -82.82$	$\mathbf{A_y} = 320\sin 105° = 309.10$
B	$\mathbf{B_x} = 480\cos 215° = -393.19$	$\mathbf{B_y} = 480\sin 215° = -275.32$
C	$\mathbf{C_x} = 250\cos 345° = 241.48$	$\mathbf{C_y} = 250\sin 345° = -64.70$
R	$\mathbf{R_x}$ -234.53	$\mathbf{R_y}$ -30.92

31. The components for the airplane are $A_x = 480\cos 27° = 427.68$, $A_y = 480\sin 27° = 217.92$; while the components for the wind are $W_x = 45\cos -55° = 25.81$, $W_y = 45\sin -55° = -36.86$. Using these, we can determine that the components for the resultant are $R_x = 427.68 + 25.81 = 453.49$, $R_y = 217.92 - 36.86 = 181.06$ and the length of the component produces a ground speed of $R = \sqrt{R_x^2 + R_y^2} = 488.3$ mph. To determine the course and drift angle, we first find that $\theta = \tan^{-1}\frac{R_y}{R_x} = 21.8°$. Hence, the course is $90 - 21.8 = 68.2°$ and the drift angle is $68.2 - 63 = 5.2°$.

32. The components for the airplane are $A_x = 320\cos -82° = 44.54$ and $A_y = 320\sin -82° = -316.89$; while the components for the wind are $W_x = 72\cos 133° = -49.10$ and $W_y = 72\sin 133° = 52.66$. Using these, we can determine that the components for the resultant are $R_x = 44.54 - 49.10 = -4.56$ and $R_y = -316.89 + 52.66 = -264.23$ and so the length of the component produces a ground speed of $R = \sqrt{R_x^2 + R_y^2} = 264.3$ mph. To determine the course and drift angle, we first find that $\theta = \tan^{-1}\frac{R_y}{R_x} = -91°$. Hence, the course is $\approx 181°$ and the drift angle is $\approx 9°$.

33. See *Computer Programs* in main text.

34. See *Computer Programs* in main text.

10.3 APPLICATIONS OF VECTORS

1. Let the 70 lb force be at 0° and the 50 lb force at 35°. Then
$R = \sqrt{R_x^2 + R_y^2} = 114.60$ and
$\theta_R = \tan^{-1}\frac{R_y}{R_x} = 14.49°$ from the 70 lb force.

vector	horizontal component		vertical component	
A	$\mathbf{A_x} = 70\cos 0° =$	70.000	$\mathbf{A_y} = 70\sin 0° =$	0.000
B	$\mathbf{B_x} = 50\cos 35° =$	40.958	$\mathbf{B_y} = 50\sin 35° =$	28.679
R	$\mathbf{R_x}$	110.958	$\mathbf{R_y}$	28.679

2. Horizontal component is $35\cos 25° = 31.72$ lb. Vertical component is $35\sin 25° = 14.79$ lb.

3. a) Downward force is $25\sin 55° = 20.48$ lb. b) Horizontal force is $25\cos 55° = 14.34$ lb. c) for a 40° angle, Downward force $= 25\sin 40° = 16.07$ lb, Horizontal force is $25\cos 40° = 19.15$ lb.

4. The component forces perpendicular to the axis of the ship must be equal but opposite. Hence $1500\sin 37° = F\sin 36°$ or $F = \dfrac{1500\sin 37°}{\sin 36°} =$ 1535.80 N

5. The forward force is $F_f = 1700\cos 18° = 1616.80$ N; and the sideward force is $F_s = 1700\sin 18° =$ 525.33 N.

6. The ship's components are $S_x = 15\cos 135° = -10.61$ and $S_y = 15\sin 135° = 10.61$ and the components for the flow of the river are $R_x = 7$ and $R_y = 0$. Adding these, we get the components for the ship's velocity relative to the Earth's surface as $V_x = -3.61$ and $V_y = 10.61$. The resultant of the ship's components gives the ship's velocity as $V = \sqrt{V_x^2 + V_y^2} = 11.21$ km/h. Next we determine that $\theta = \tan^{-1}\frac{V_y}{V_x} = 108.79°$ and so, the angle is 18.79° west of north, which is a compass direction of 341.21°.

7. Since the heading and wind are perpendicular, the ground speed is $\sqrt{370^2 + 40^2} = 372.16$ mi/h. The angle east of south is $\theta = \tan^{-1}\frac{40}{370} = 6.17°$. This is a compass direction of $180° - 6.17° = 173.83°$.

8. $A_v = 420\cos 115° = -177.50$, $A_h = 420\sin 115° = 380.65$, $W_v = 62\cos 32° = 52.58$, $W_h = 62\sin 32° = 32.85$, $G_v = -124.92$, $G_h = 413.50$, $G = \sqrt{G_v^2 + G_h^2} = 431.96$ mi/hr; $\theta = \tan^{-1}\frac{G_h}{G_v} =$ 106.81°

9. Parallel component is $1200\sin 22° = 449.53$ kg. Perpendicular component is $1200\cos 22° =$ 1112.62 kg.

10. $F = 30\sin 15° = 7.76$ lb

11. The angle the guy wire makes with the ground is $\theta = \tan^{-1}\frac{35}{27} = 52.352°$. Horizontal tension $T_x = 195\cos 52.352° = 119.11$ lb. Vertical-downward $T_y = 195\sin 52.352° = 154.40$ lb.

12. Since I_C and I_R are perpendicular the magnitude of $I_T = \sqrt{I_C^2 + I_R^2} = \sqrt{2.4^2 + 1.6^2} = 2.88$ A. The phase angle $\theta = \tan^{-1}\frac{2.4}{1.6} = 56.31°$.

13. $I_C + I_R = I$; $I_C = I - I_R$; $I_C{}^2 = 9.6^2 - 7.5^2$, $I_C = 5.99$ or 6 A; $\phi = \tan^{-1}\frac{I_C}{I_R} = 38.66°$

14. $I = \sqrt{I_L^2 + I_R^2} = \sqrt{6.2^2 + 8.4^2} = 10.44$ A; $\theta = \tan^{-1}\frac{I_L}{I_R} = \tan^{-1}\frac{6.2}{8.4} = 36.43°$; Since θ is close-wise, $\theta = -36.43°$

15. $I_L = \sqrt{I^2 - I_R{}^2} = \sqrt{12.4^2 - 6.3^2} = 10.68$; $\theta = -\tan^{-1}\frac{I_L}{I_R} = -\tan^{-1}\frac{10.68}{6.3} = -59.46°$

16.

vector	horizontal component		vertical component	
$\mathbf{X}_C$	$\mathbf{X_{Cx}} =$	0	$\mathbf{X}_{Cy} =$	−50
$\mathbf{X}_L$	$\mathbf{X_{Lx}} =$	0	$\mathbf{X_{Ly}} =$	90
R	$\mathbf{R_x} =$	12	$\mathbf{R_y} =$	0
Z	$\mathbf{Z_x} =$	12	$\mathbf{Z_y} =$	40

$Z = \sqrt{Z_x{}^2 + Z_y{}^2} = \sqrt{12^2 + 40^2} = 41.76$; $\theta = \tan^{-1}\frac{40}{12} = 73.30^0$

17.

vector	horizontal component	vertical component
$\mathbf{X}_C$	$\mathbf{X}_{Cx} = 0$	$\mathbf{X}_{Cy} = -60$
$\mathbf{X}_L$	$\mathbf{X}_{Lx} = 0$	$\mathbf{X}_{Ly} = 40$
$\mathbf{R}$	$\mathbf{R}_x = 24$	$\mathbf{R}_y = 0$
$\mathbf{Z}$	$\mathbf{Z}_x = 24$	$\mathbf{Z}_y = -20$

$Z = \sqrt{Z_x^2 + Z_y^2} = \sqrt{24^2 + (-20)^2} = 31.24$; $\theta = \tan^{-1} \frac{-20}{24} = -39.81°$

18.

vector	horizontal component	vertical component
$\mathbf{X}_L$	$\mathbf{X}_{Lx} = 0$	$\mathbf{X}_{Ly} = 38$
$\mathbf{X}_C$	$\mathbf{X}_{Cx} = 0$	$\mathbf{X}_{Cy} = -265$
$\mathbf{R}$	$\mathbf{R}_x = 75$	$\mathbf{R}_y = 0$
$\mathbf{Z}$	$\mathbf{Z}_x = 75$	$\mathbf{Z}_y = -227$

$Z = \sqrt{75^2 + -227^2} = 239.07$; $\theta = \tan^{-1} \frac{-227}{75} = -71.72°$

19. $I = \sqrt{I_x^2 + I_y^2} = 55.84$;
$\theta = \tan^{-1} \frac{I_y}{I_x} = \tan^{-1} \frac{6.342}{55.478} = 6.52°$

vector	horizontal component	vertical component
$\mathbf{I}_A$	$\mathbf{I}_{Ax} = 20\cos 35° = 16.383$	$\mathbf{I}_{Ay} = 20\sin 35° = 11.472$
$\mathbf{I}_B$	$\mathbf{I}_{Bx} = 15\cos -20° = 14.095$	$\mathbf{I}_{By} = 15\sin -20 = -5.130$
$\mathbf{I}_C$	$\mathbf{I}_{Cx} = 25.000$	$\mathbf{I}_{Cy} = 0.000$
$\mathbf{I}$	$\mathbf{I}_x = 55.478$	$\mathbf{I}_y = 6.342$

20. $V = \sqrt{V_N^2 + V_E^2} = \sqrt{7.222^2 + 3.941^2} =$ 8.23 km/hr;
$\theta = \tan^{-1} \frac{V_E}{V_N} = \tan^{-1} \frac{3.941}{7.222} = 28.62°$

vector	horizontal component	vertical component
$\mathbf{S}$	$\mathbf{S}_N = 12\cos 10° = 11.818$	$\mathbf{S}_E = 12\sin 10° = 2.084$
$\mathbf{W}$	$\mathbf{W}_N = 2\cos 270° = 0.000$	$\mathbf{W}_E = 2\sin 270° = -2.000$
$\mathbf{C}$	$\mathbf{C}_N = 6\cos 140° = -4.596$	$\mathbf{C}_E = 6\sin 140° = 3.857$
$\mathbf{V}$	$\mathbf{V}_N = 7.222$	$\mathbf{V}_E = 3.941$

21. $V_x = 120$ m/s; $V_y = -9.8 \cdot 4$ m/s $= -39.2$ m/s; $V = \sqrt{{V_x}^2 + {V_y}^2} = 126.24$ m/s; $\theta = \tan^{-1} \frac{V_y}{V_x} = \tan^{-1} \frac{-39.2}{120} = -18.09°$

22. $4.9t^2 = 5000$; $t^2 = \frac{5000}{4.9}$; $t = \sqrt{\frac{5000}{4.9}}$; $t = 31.94$ sec. $V_y = -9.8(31.94) = -313.012$ m/s; $V = \sqrt{{V_x}^2 + {V_y}^2} = \sqrt{120^2 + 313.012^2} = 335.23$ m/s; $\theta = \tan^{-1} \frac{V_y}{V_x} = \tan^{-1} \frac{-313.012}{120} = -69.02°$

10.4 OBLIQUE TRIANGLES: LAW OF SINES

1. $A = 19.4°$; $B = 85.3°$; $c = 22.1$; $C = 180° - A - B = 180° - 19.4° - 85.3° = 75.3°$; $\frac{a}{\sin A} = \frac{c}{\sin C}$; $\frac{a}{\sin 19.4} = \frac{22.1}{\sin 75.3°}$; $a = \frac{(22.1)\sin 19.4}{\sin 75.3} = 7.59$; $\frac{b}{\sin B} = \frac{c}{\sin C}$; $\frac{b}{\sin 85.3} = \frac{22.1}{\sin 75.3}$; $b = \frac{(22.1)\sin 85.3}{\sin 75.3} = 22.77$. Summary: $C = 75.3°$, $a = 7.59$, $b = 22.77$

2. $A = 180° - (62.4° + 43.9°) = 73.7°$, $\frac{a}{\sin A} = \frac{b}{\sin B}$; $\frac{12.4}{\sin 73.7°} = \frac{b}{\sin 62.4°}$; $b = \frac{(12.4)\sin 62.4°}{\sin 73.7°} \approx 11.45$; $\frac{12.4}{\sin 73.7°} = \frac{c}{\sin 43.9°}$; $c = \frac{12.4 \sin 43.9°}{\sin 73.7°} \approx 8.96$. Summary: $A = 73.7°$, $b = 11.45$, $c = 8.96$

3. $a = 14.2$; $b = 15.3$; $B = 97°$: $\frac{a}{\sin A} = \frac{b}{\sin B}$; $\sin A = \frac{a \sin B}{b} = \frac{14.2 \sin 97°}{15.3}$; $\sin A = .9211866$; $A = 67.1°$; Since B is obtuse, A must be acute. $C = 180° - A - B = 180° - 67.1° - 97° = 15.9°$;

$\frac{c}{\sin C} = \frac{b}{\sin B}$; $c = \frac{15.3 \sin 15.9}{\sin 97°} = 4.22$; $A = 67.1°$; $C = 15.9°$; $c = 4.22$

4. $A = 27.42°$, $a = 27.3$, $b = 35.49$; $\frac{b}{\sin B} = \frac{a}{\sin A}$; $\sin B = \frac{b \sin A}{a} = \frac{35.49 \sin 27.42°}{27.3}$; $\sin B = 0.59866$; $B_1 = 36.77°$; $B_2 = 180° - B_1 = 143.23°$; $C_1 = 180° - A - B_1 = 180° - 27.42° - 36.77° = 115.81°$; $C_2 = 180° - A - B_2 = 180° - 27.42° - 143.23° = 9.35°$; $\frac{c_1}{\sin C_1} = \frac{a}{\sin A}$; $c_1 = \frac{27.3 \sin 115.81°}{\sin 27.42°} = 53.37$; $\frac{c_2}{\sin C_2} = \frac{a}{\sin A}$; $c_2 = \frac{27.3 \sin 9.35°}{\sin 27.42°} = 9.63$; $B_1 = 36.77°$; $C_1 = 115.81°$; $c_1 = 53.37$; $B_2 = 143.23°$; $C_2 = 9.35°$; $c_2 = 9.63$

5. $A = 86.32°$, $a = 19.19$, $c = 18.42$; $\frac{c}{\sin C} = \frac{a}{\sin A}$; $\sin C = \frac{18.42 \sin 86.32}{19.19}$; $\sin C = 0.9578958$, $C_1 = 73.31°$.; $C_2 = 180° - C_1 = 106.69°$ but $C_2 + A > 180°$ is not possible. $B = 180° - A - C = 180° - 86.32° - 73.31° = 20.37°$; $\frac{b}{\sin B} = \frac{a}{\sin A}$; $b = \frac{19.19 \sin 20.37}{\sin 86.32} = 6.69$; $C = 73.31°$; $B = 20.37°$; $b = 6.69$

6. $B = 75.46°$, $b = 19.4$, $C = 44.95°$; $\frac{c}{\sin C} = \frac{b}{\sin B}$; $c = \frac{b \sin C}{\sin B} = \frac{19.4 \sin 44.95°}{\sin 75.46°} \approx 14.16$; $A = 180° - (75.46° + 44.95°) = 59.59°$; $\frac{a}{\sin A} = \frac{b}{\sin B}$; $a = \frac{19.4 \sin 59.59°}{\sin 75.46°} \approx 17.28$. Summary: $c = 14.16, A = 59.59°, a = 17.28$

7. $B = 39.4°$, $b = 19.4$, $c = 35.2$; $\frac{c}{\sin C} = \frac{b}{\sin B}$; $\sin C = \frac{c \sin B}{b} = \frac{35.2 \sin 39.4}{19.4}$; $\sin C = 1.15$. Since the range of the sine function is $[-1, 1]$, there is no solution, and so this is not a triangle.

8. $A = 84.3°, b = 9.7, C = 12.7°$; $B = 180° - A - C = 180° - 84.3° - 12.7° = 83°$; $\frac{a}{\sin A} = \frac{b}{\sin B}$; $a = \frac{b \sin A}{\sin B} = \frac{9.7 \sin 84.3}{\sin 83} = 9.72$; $\frac{c}{\sin C} = \frac{b}{\sin B}$; $c = \frac{b \sin C}{\sin B} = \frac{9.7 \sin 12.7}{\sin 83} = 2.15$. Summary: $B = 83°, a = 9.72, c = 2.15$

9. $A = 45°$, $a = 16.3$, $b = 19.4$; $\frac{b}{\sin B} = \frac{a}{\sin A}$; $\sin B = \frac{b \sin A}{a} = \frac{19.4 \sin 45°}{16.3}$; $\sin B = 0.841587$; $B_1 = 57.31°$; $B_2 = 180° - B_1 = 122.69°$. $C_1 = 180° - A - B_1 = 180° - 45° - 57.31° = 77.69°$. $C_2 = 180° - A - B_2 = 180° - 45° - 122.69° = 12.31°$. $\frac{c_1}{\sin C_1} = \frac{a}{\sin A}$; $c_1 = \frac{16.3 \sin 77.69}{\sin 45} = 22.52$. $\frac{c_2}{\sin C_2} = \frac{a}{\sin A}$; $c_2 = \frac{16.3 \sin 12.31}{\sin 45} = 4.91$; $B_1 = 57.31°$; $C_1 = 77.69°$; $c_1 = 22.52$; $B_2 = 122.69°$; $C_2 = 12.31°$; $c_2 = 4.91$.

10. $a = 10.4$, $c = 5.2$, $C = 30°$: $\frac{a}{\sin A} = \frac{c}{\sin C}$; $\sin A = \frac{a \sin C}{c} = \frac{10.4 \sin 30}{5.2} = 1$; $A = 90°$, $B = 180° - A - C = 180° - 90° - 30° = 60°$; $\frac{b}{\sin B} = \frac{c}{\sin C}$; $b = \frac{5.2 \sin 60°}{\sin 30°} = 9.01$. Summary: $A = 90°, B = 60°, b = 9.01$.

11. $a = 42.3, B = 14.3°, C = 16.9°$: $A = 180° - B - C = 180° - 14.3° - 16.9° = 148.8°$; $\frac{b}{\sin B} = \frac{a}{\sin A}$; $b = \frac{42.3 \sin 14.3}{\sin 148.8} = 20.17$; $\frac{c}{\sin C} = \frac{a}{\sin A}$; $c = \frac{42.3 \sin 16.9}{\sin 148.8} = 23.74$. Summary: $A = 148.8°$, $b = 20.17, c = 23.74$

12. $A = 105.4°, B = 68.2°, c = 4.91$; $C = 180° - A - B = 180° - 105.4° - 68.2° = 6.4°$; Using $\frac{a}{\sin A} = \frac{c}{\sin C}$, we get $a = \frac{4.91 \sin 105.4°}{\sin 6.4°} = 42.47$. Next we use $\frac{b}{\sin B} = \frac{c}{\sin C}$, to get $b = \frac{4.91 \sin 68.2}{\sin 6.4} = 40.90$. Summary: $C = 6.4°, a = 42.47, b = 40.90$.

13. $b = 19.4$, $c = 12.5$, $C = 35.6°$; $\frac{b}{\sin B} = \frac{c}{\sin C}$; $\sin B = \frac{b \sin C}{c} = \frac{19.4 \sin 35.6}{12.5}$; $\sin B =$

0.9034549; $B_1 = 64.62°$; $B_2 = 180° - B_1 = 115.38°$; $A_1 = 180° - B_1 - C = 180° - 64.62° - 35.6° = 79.78°$; $A_2 = 180° - B_2 - C = 180° - 115.38° - 35.6° = 29.02°$; $a_1 = \dfrac{c \sin A_1}{\sin C} = \dfrac{12.5 \sin 79.78}{\sin 35.6°} = 21.13$; $a_2 = \dfrac{c \sin A_2}{\sin C} = \dfrac{12.5 \sin 29.02}{\sin 35.6°} = 10.42$. Summary: $B_1 = 64.62$, $A_1 = 79.78$, $a_1 = 21.13$, or $B_2 = 115.38$, $A_2 = 29.02$, $a_2 = 10.42$.

14. $a = 121.4$, $A = 19.7°$, $c = 63.4$; $\sin C = \dfrac{c \sin A}{a} = \dfrac{63.4 \sin 19.7°}{121.4} = 0.176045$; $C = 10.1°$. The other answer will not work. $B = 180° - A - C = 180° - 19.7° - 10.1° = 150.2°$; $b = \dfrac{a \sin B}{\sin A} = \dfrac{121.4 \sin 150.2°}{\sin 19.7°} = 179.0$.

15. $a = 19.7$, $b = 8.5$, $B = 78.4°$; $\sin A = \dfrac{a \sin B}{b} = \dfrac{19.7 \sin 78.4}{8.5} = 2.27$. Since 2.27 is not in the range of the sine function, there is no solution.

16. $b = 9.12$, $B = 1.3$ rad, $C = 0.67$ rad; $A = \pi - B - C = 3.14 - 1.3 - .67 = 1.17$ rad $a = \dfrac{b \sin A}{\sin B} = \dfrac{9.12 \sin 1.17}{\sin 1.3} = 8.71$; $c = \dfrac{b \sin C}{\sin B} = \dfrac{9.12 \sin 0.67}{\sin 1.3} = 5.88$: $A = 1.17$ rad, $a = 8.71$, $c = 5.88$

17. $b = 8.5$, $c = 19.7$, $C = 1.37$ rad; $\sin B = \frac{b \sin C}{c} = \frac{8.5 \sin 1.37}{19.7} = .4228$; $B_1 = 0.44$ rad; $B_2 = 3.14 - 0.44 = 2.7$; $A_1 = \pi - B_1 - C = 3.14 - .44 - 1.37 = 1.33$; $A_2 = \pi - B_2 - C = 3.14 - 2.7 - 1.37 = -0.93$ cannot be so only $B_1 + A_1$ are answers. $a = \dfrac{c \sin A}{\sin C} = \dfrac{19.7 \sin 1.33}{\sin 1.37} = 19.52$: $B = 0.44$, $A = 1.33$, $a = 19.52$.

18. $b = 19.7$, $c = 36.4$, $C = 0.45$ rad; $\sin B = \dfrac{b \sin C}{c} = \dfrac{19.7 \sin 0.45}{36.4} = 0.2354$. $B = 0.24$ rad only answer $A = \pi - B - C = 3.14 - 0.24 - 0.45 = 2.45$ rad. $a = \dfrac{c \sin A}{\sin C} = \dfrac{36.4 \sin 2.45}{\sin .45} = 53.37$: $B = 0.24$, $A = 2.45$, $a = 53.37$.

19. $A = 0.47$ rad, $b = 195.4$, $C = 1.32$ rad; $B = \pi - A - C = 3.14 - 0.47 - 1.32 = 1.35$. $a = \dfrac{b \sin A}{\sin B} = \dfrac{195.4 \sin 0.47}{\sin 1.35} = 90.7$; $c = \dfrac{b \sin C}{\sin B} = \dfrac{195.4 \sin 1.32}{\sin 1.35} = 194.0$: $B = 1.35$, $a = 90.7$, $c = 194.0$.

20. $a = 29.34$, $A = 1.23$ rad, $C = 1.67$ rad; $B = \pi - A - C = 3.14 - 1.23 - 1.67 = 0.24$ rad. $b = \dfrac{a \sin B}{\sin A} = \dfrac{29.34 \sin 0.24}{\sin 1.23} = 7.40$; $c = \dfrac{a \sin C}{\sin A} = \dfrac{29.34 \sin 1.67}{\sin 1.23} = 30.98$: $B = 0.24$, $b = 7.40$, $c = 30.98$.

21. The length between towers A and B is $AB = c = 360$ m. The angle at tower $B = 67.4°$ and at tower $A = 49.3°$. Hence, the angle at tower $C = 180° - A - B = 63.3°$. The distance between towers A and C is $AC = b = \dfrac{c \sin B}{\sin C} = \dfrac{360 \sin 67.4}{\sin 63.3} = 372.0$ m and the distance between towers B and C is $BC = a = \dfrac{c \sin A}{\sin C} = \dfrac{360 \sin 49.3}{\sin 63.3} = 305.5$ m.

22. $AC = b = 250$ m, $BC = a = 275$ m, $A = 43.62°$: $\sin B = \dfrac{b \sin A}{a} = \dfrac{250 \sin 43.62}{275} = 0.627157$; $B = 38.84°$; $C = 180° - A - B = 97.54°$. $AB = c = \dfrac{a \sin C}{\sin A} = \dfrac{275 \sin 97.54}{\sin 43.62} = 395.2$ m

23. Let C designate the present location of the airplane. Since the plane has a heading of 313°, then $\angle A$ of $\triangle ABC$ is 43°. We are given that the heading from B is 27° and so $B = 90° - 27° = 63°$. Using these two angles, we find that $C = 180° - 43° - 63° = 74°$. Now, using the Sine law, if the distance the plane has flown is b, then we have $\dfrac{b}{\sin B} = \dfrac{c}{\sin C}$. We were given that $c = 37$ mi, and so we have $b = \dfrac{c \sin B}{\sin C} = \dfrac{37 \sin 63°}{\sin 74°} \approx 34.3$ mi.

24. A football field has two 40-yard lines. The angle given is to the farther of the two lines from the end of the stadium whose height we want to determine.

Let A be the top of the given end of the stadium, B a point on the 50-yard line, and C a point on the given 40-yard line, so that A, B, and C are colinear and the line $\overleftrightarrow{BC}$ is perpendicular to the 40- (and 50-) yard lines. The distance, d, from the top of the stadium to the 40-yard line is AC. We can determine the following angle measures: $\angle ACB = 10.45°$, $\angle CAB = 0.91°$, and $\angle CBA = 168.64°$. From the Law of Sines, we see that

$$d = \frac{10 \sin 168.64°}{\sin 0.91°} = 124.02.$$

If h is the height of the stadium, then $\sin 10.45° = \frac{h}{d}$ and so $h = d \sin 10.45° = 22.49$ yards. Thus, the height of this one end of the stadium is about 22.5 yards.

25. $c = 180° - 47° - 67° = 66°$; $CB = a = \frac{400 \sin 47°}{\sin 66°} = 320.22$; $\sin 67° = \frac{h}{a}$; $h = 320.22 \sin 67° = 294.77$ m

10.5 OBLIQUE TRIANGLES: LAW OF COSINES

1. $a = 9.3$, $b = 16.3$, $C = 42.3°$ SAS so Cosine Law. First we find c by using the cosine law: $c^2 = a^2 + b^2 - 2ab \cos C = 9.3^2 + 16.3^2 - 2(9.3)(16.3) \cos 42.3°$, $c^2 = 127.9386$, $c = 11.3$. Next find the angle opposite the short side, namely A using Law of Sines.
$\frac{a}{\sin A} = \frac{c}{\sin C}$; $\sin A = \frac{a \sin C}{c} = \frac{9.3 \sin 42.3}{11.3}$;
$\sin A = 0.553895, A = 33.6°$ $B = 180° - A - C = 180° - 42.3° - 33.6° = 104.1°$. (Note: If you use the Law of Sines to find B, you may not recognize that it is obtuse. That's why you should find the angle opposite the shorter side first since it must be acute.)
$c = 11.31, A = 33.6°, B = 104.1°$.

2. $A = 16.25°$, $b = 29.43$, $c = 36.52$; $a^2 = b^2 + c^2 - 2bc \cos A = 29.43^2 + 36.52^2 - 2(29.43)(36.52) \cos 16.25° = 136.1436$; $a = 11.67$.

$$\sin B = \frac{b \sin A}{a} = \frac{29.43 \sin 16.25}{11.67} = .705687;$$

$B = 44.89$; $C = 180° - A - B = 180° - 16.25° - 44.89° = 118.86°$. (Note: If you used Law of Sines to find C, you may not recognize that it is obtuse. That is why you should find the angle opposite the shorter side first since it must be acute.)
$a = 11.67, B = 44.80°, C = 118.86°$.

3. $a = 47.85$, $B = 113.7°$, $c = 32.79$; $b^2 = a^2 + c^2 - 2ac \cos B = 47.85^2 + 32.79^2 - 2(47.85)(32.79) \cos 113.7° = 4626.1199$; hence, $b = 68.02$ and so, $\sin A = \frac{a \sin B}{b} = \frac{47.85 \sin 113.7°}{68.02} = 0.64414$. Thus, $A = 40.1°$.
$C = 180° - A - B = 180° - 40.1° - 113.7° = 26.2°$.
$b = 68.02, A = 40.1°, C = 26.2°$:
(Note: Since B is obtuse, A and C must be acute).

4. $a = 19.52$, $b = 63.42$, $c = 56.53$: Use the alternate form of the Law of Cosines to find one of the angles. Choose B since it is the largest and if it's obtuse, the Law of Cosines will yield this answer directly.
$\cos B = \frac{a^2 + c^2 - b^2}{2ac} = \frac{19.52^2 + 56.53^2 - 63.42^2}{2(19.52)(56.53)}$;
$\cos B \approx -0.2018$; $B = 101.6°$. $\sin A = \frac{a \sin B}{b} = \frac{19.52 \sin 101.6°}{63.42} = 0.3015$; $A = 17.5°$;
$C = 180° - A - C = 180° - 17.5° - 101.6° = 60.9°$:
$A = 17.5°, B = 101.6°, C = 60.9°$

5. $a = 29.43$, $b = 16.37$, $c = 38.62$: Use the alternate form of Law of Cosines to find one of the angles. Choose C since it's the largest and if it's obtuse, the Law of Cosines will yield this answer directly.
$\cos C = \frac{a^2 + b^2 - c^2}{2ab} = \frac{29.43^2 + 16.37^2 - 38.62^2}{2(29.43)(16.37)}$;
$\cos C \approx -0.37093$; $C \approx 111.8°$; $\sin A = \frac{a \sin C}{c} = \frac{29.43 \sin 111.8°}{38.62} \approx 0.7075$; $A = 45.0°$;

$B = 180° - A - C = 180° - 45.0° - 111.8° = 23.2°$: $A = 45.0°$, $B = 23.2°$, $C = 111.8°$.

6. $A = 121.37°$, $b = 112.37$, $c = 93.42$; $a^2 = b^2 + c^2 - 2bc\cos A = 112.37^2 + 93.42^2 - 2(112.37)(93.42)\cos 121.37° = 32283.636$; $a = 179.68$. $\sin C = \dfrac{c\sin A}{a} = \dfrac{93.42\sin 121.37}{179.68} = 0.443924$; $C = 26.35°$. $B = 180° - A - C = 180° - 121.37° - 26.35° = 32.28°$. $a = 179.68$, $B = 32.28°$, $C = 26.35°$.

7. $a = 63.92$, $B = 92.44°$, $c = 78.41$; $b^2 = a^2 + c^2 - 2ac\cos B = 63.92^2 + 78.41^2 - 2(63.92)(78.41)\cos 92.44° = 10660.645$; $b = 103.25$. $\sin A = \dfrac{a\sin B}{b} = \dfrac{63.92\sin 92.44}{103.25} = .6185186$; $A = 38.21$. $C = 180° - A - B = 180° - 38.21° - 92.44° = 49.35°$: $b = 103.25$, $A = 38.21°$, $C = 49.35°$

8. $a = 19.53$, $b = 7.66$, $C = 32.56°$; $c^2 = a^2 + b^2 - 2ab\cos C = 19.53^2 + 7.66^2 - 2(19.53)(7.66)\cos 32.56° = 187.9226$; $c = 13.71$. $\sin B = \dfrac{b\sin C}{c} = \dfrac{7.66\sin 32.56}{13.71} = .300691$; $B = 17.50°$. $A = 180° - B - C = 180° - 17.50° - 32.56° = 129.94°$: $c = 13.71$, $A = 129.94°$, $B = 17.50°$

9. $a = 4.527$, $b = 6.239$, $c = 8.635$: Find C first, $\cos C = \dfrac{a^2 + b^2 - c^2}{2ab} = \dfrac{4.527^2 + 6.239^2 - 8.635^2}{2(4.527)(6.239)}$; $\cos C = -0.268099$; $C = 105.55°$ $\sin A = \dfrac{a\sin C}{c} = \dfrac{4.527\sin 105.55°}{8.635} = 0.505072$; $A = 30.34$. $B = 180° - A - C = 180° - 30.34° - 105.55° = 44.11°$: $A = 30.34°$, $B = 44.11°$, $C = 105.55°$.

10. $A = 7.53°$, $b = 37.645$, $c = 42.635$; SAS, $a^2 = b^2 + c^2 - 2bc\cos A = 52.58182$; $a = 7.251$. $\sin B = \frac{b\sin A}{a} = 0.6803476$; $B = 42.87°$. $c = 180° - A - B = 129.6°$ $a = 7.251$, $B = 42.87°$, $C = 129.6°$.

11. $a = 8.5$, $b = 15.8$, $C = 0.82$ rad; SAS, $c^2 = a^2 + b^2 - 2ab\cos C = 138.6454$; $c = 11.77$; $\sin A = \frac{a\sin C}{c} = .5280$; $A = 0.56$ rad. $B = \pi - A - C = 3.14 - 0.82 - 0.56 = 1.76$ rad: $c = 11.77$, $A = 0.56$ rad, $B = 1.76$ rad.

12. $A = 0.31$, $b = 15.8$, $c = 38.47$; SAS, $a^2 = b^2 + c^2 - 2bc\cos A = 571.8747$; $a = 23.91$; $\sin B = \frac{b\sin A}{a} = 0.20159$; $B = 0.203$ rad. $C = \pi - A - B = 3.14 - 0.31 - 0.20 = 2.63$ rad: $a = 23.91$, $B = 0.20$ rad, $C = 2.63$ rad.

13. $a = 52.65$, $B = 1.98$, $c = 35.8$; $b^2 = a^2 + c^2 - 2ac\cos B = 5553.5624$; $b = 74.52$. $\sin C = \frac{c\sin B}{b} = .440744$, $C = 0.46$; $A = \pi - B - C = 3.14 - 1.98 - 0.46 = 0.70$: $b = 74.52$, $A = 0.70$, $C = 0.46$.

14. $a = 43.5$, $b = 63.4$, $c = 37.3$; SSS, $\cos B = \frac{a^2+c^2-b^2}{2ac} = -.2268$, $B = 1.80$ rad. $\sin A = \frac{a\sin B}{b} = 0.668176$; $A = 0.73$ rad. $C = \pi - A - B = 3.14 - 0.73 - 1.80 = 0.61$. $A = 0.73$, $B = 1.80$, $C = 0.61$

15. $a = 36.27$, $b = 24.55$, $c = 44.26$; $\cos C = \frac{a^2+b^2-c^2}{2ab} = -0.02287$; $C = 1.59$ rad. $\sin A = \frac{a\sin C}{c} = 0.8193$; $A = 0.96$. $B = \pi - A - C = 0.59$: $A = 0.96$, $B = 0.59$, $C = 1.59$

16. $A = 2.41$, $b = 153.21$, $c = 87.49$; SAS; $a^2 = b^2 + c^2 - 2bc\cos A = 51076.452$; $a = 226.00$. $\sin B = \frac{b\sin A}{a} = 0.45289$; $B = 0.47$. $C = \pi - A - B = 0.26$. $a = 226$, $B = 0.47$, $C = 0.26$

17. $a = 54.8$, $B = 1.625$, $c = 38.33$; SAS; $b^2 = a^2 + c^2 - 2ac\cos B = 4699.8253$; $b = 68.56$. $\sin A = \frac{a\sin B}{b} = 0.7981$; $A = 0.924$. $C = \pi - A - B = 3.142 - 0.924 - 1.625 = 0.593$: $b = 68.56$, $A = 0.924$, $C = 0.593$

18. $a = 7.621$, $b = 3.429$, $C = 0.183$; SAS, $c^2 = a^2 + b^2 - 2ab\cos C = 18.4456$; $c = 4.295$. $\sin B = \frac{b\sin C}{c} = 0.1452876$; $B = 0.146$. $A = \pi - B - C = 3.142 - 0.146 - 0.183 = 2.813$. $c = 4.295$, $A = 2.813$, $B = 0.146$

19. $a = 2.317$, $b = 1.713$, $c = 1.525$; SSS, $\cos A = (b^2 + c^2 - a^2)/2bc = -0.020766$, $A = 1.59$; $\sin B = \frac{b\sin A}{a} = .7392$, $B = 0.83$; $C = \pi - A - B = 3.14 - 1.59 - 0.83 = 0.72$: $A = 1.59$, $B = 0.83$, $C = 0.72$

20. $A = 0.09$, $b = 40.75$, $c = 50.25$; SAS, $a^2 = b^2 + c^2 - 2bc\cos A = 106.825$; $a = 10.34$. $\sin B = \frac{b\sin A}{a} = .35421188$; $B = 0.36$. $C = \pi - A - B = 3.14 - 0.09 - 0.36 = 2.69$: $a = 10.34$, $B = 0.36$, $C = 2.69$

21. From the diagram we know that $A = 90 + 34 = 124.7°$, $\overline{AC} = b = 6335$, $\overline{AB} = c = 12325$, $\overline{BC} = a$; $a^2 = b^2 + c^2 - 2bc\cos A = 6335^2 + 12325^2 - 2(6335)(12325)\cos 124.7° = 280\,935\,259.5$ and so $a = 16\,761$. Distance above the Earth's surface is $16\,761 - 6\,335 = 10\,426$ km.

22. $a = 21 \times 3 = 63$ km. $b = 21 \times 4 = 84$ km. $C = 180° - 37 = 143°$. $c^2 = a^2 + b^2 - 2ab\cos C = 63^2 + 84^2 - 2 \cdot 63 \cdot 84\cos 143$; $c^2 = 19477.76$; $c = 139.6$ km. Bearing $= B = \sin^{-1}\left(\frac{84\sin 143}{139.6}\right) = 21.2°$
Distance from port is 139.6 km bearing is 21.2° East of north.

23. The angle between the antenna and the hill is $90 + 12.37° = 102.37°$
If the length of the wire is ℓ then $\ell^2 = 40^2 + 75^2 - 2 \cdot 40 \cdot 75\cos 102.37°$; $\ell^2 = 8510.34$; $\ell = 92.25$ ft.

24. $R^2 = 50^2 + 35^2 - 2 \cdot 50 \cdot 35\cos 147.85° = 6688.3$, $R = 81.78$ lb

25. $R^2 = 12^2 + 23^2 - 2 \cdot 12 \cdot 23\cos 121.27° = 959.53$; $R = \sqrt{959.53} = 30.98$ or 31 N. $\theta = \sin^{-1}\left(\frac{23\sin 121.27}{31}\right) = 39.36°$

26. $\cos\phi = (36.4^2 + 15.5^2 - 30.1^2)/(2 \times 36.4 \times 15.5)$; $\cos\phi = 0.58419$; $\phi = 54.25°$. $\theta = 180° - \phi = 180° - 54.25° = 125.75°$.

CHAPTER 10 REVIEW

1.

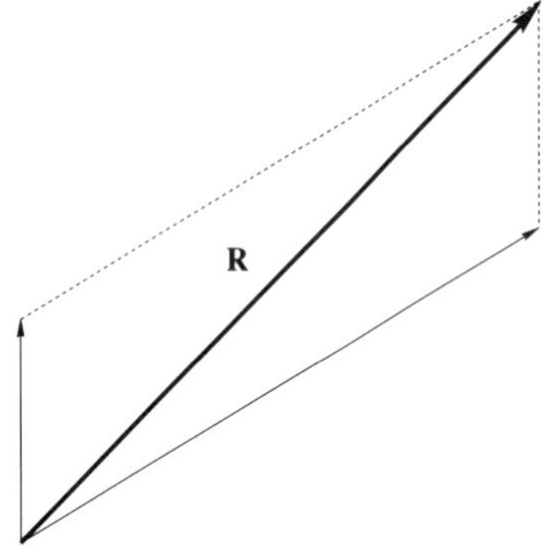

2.

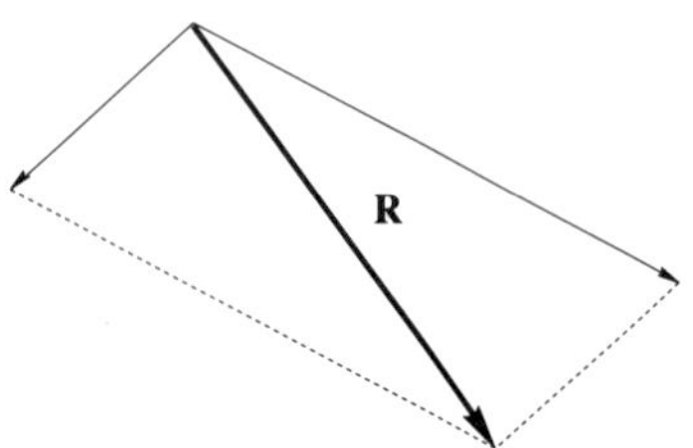

3.

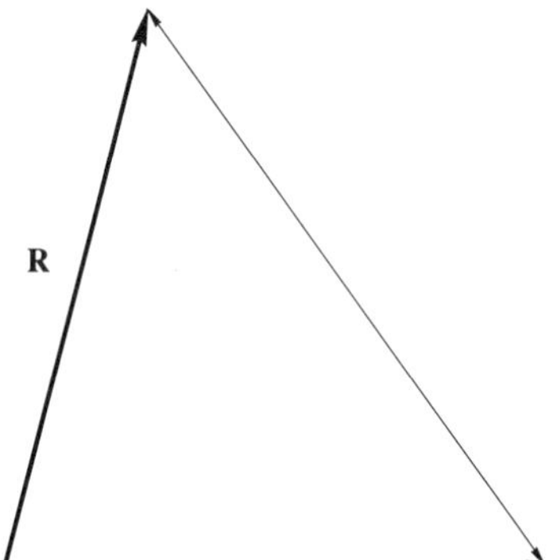

4.

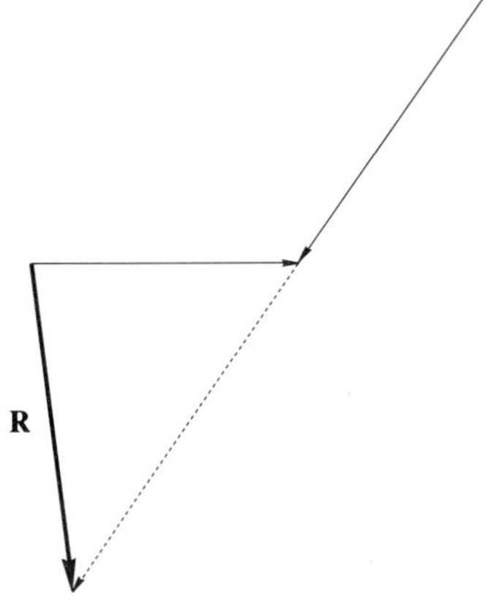

5. $P_x = 35\cos 67° = 13.68$; $P_y = 35\sin 67° = 32.22$

6. $P_x = 19.7\cos 237° = -10.73$; $P_y = 19.7\sin 237° = -16.52$

7. $P_x = 23.4\cos 172.4° = -23.19$; $P_y = 23.4\sin 172.4° = 3.09$

8. $P_x = 14.5\cos 338° = 13.44$; $P_y = 14.5\sin 338° = -5.43$

9. $A = \sqrt{A_x^2 + A_y^2} = \sqrt{16^2 + (-8)^2} = 17.89$ $\theta = \tan^{-1}\left(\frac{A_y}{A_x}\right) = \tan^{-1}\frac{-8}{16} = \tan^{-1} -\frac{1}{2} = -26.57°$ or $333.43°$

10. $B = \sqrt{B_x^2 + B_y^2} = \sqrt{(-27)^2 + 32^2} = 41.87$; $\theta_R = \tan^{-1}\frac{32}{27} = 49.84$ Second Quadrant so $180° - 49.84° = 130.16°$

11. $A_x = 38\cos 15° = 36.71$; $A_y = 38\sin 15° = 9.84$

12. $B_x = 43.5\cos 127° = -26.18$; $A_y = 43.5\sin 127° = 34.74$

13. $C_x = 19.4\cos 1.25 = 6.12$;
$C_y = 19.4\sin 1.25 = 18.41$

14. $D_x = 62.7\cos 5.37 = 38.32$; $D_y = 62.7\sin 5.37 = -49.62$

15. $R = \sqrt{R_x^2 + R_y^2} = 50.41$; $\theta = \tan^{-1}\frac{R_y}{R_x} = 20.69°$

vector	horizontal component	vertical component
A	$\mathbf{A_x} = 19\cos 32° = 16.11$	$\mathbf{A_y} = 19\sin 32° = 10.07$
B	$\mathbf{B_x} = 32\cos 14° = 31.05$	$\mathbf{B_y} = 32\sin 14° = 7.74$
R	$\mathbf{R_x}$ 47.16	$\mathbf{R_y}$ 17.81

16. $R = \sqrt{R_x^2 + R_y^2} = 36.96$;
$\theta_R = \tan^{-1}\frac{5.88}{36.49} = 9.15°$ Since θ is in 4th Quadrant $\theta = 360 - 9.15 = 350.85°$

vector	horizontal component	vertical component
C	$\mathbf{C_x} = 24\cos 57° = 13.07$	$\mathbf{C_y} = 24\sin 57° = 20.13$
D	$\mathbf{D_x} = 35\cos 312° = 23.42$	$\mathbf{D_y} = 35\sin 312° = -26.01$
R	$\mathbf{R_x}$ 36.49	$\mathbf{R_y}$ −5.88

17. $R = \sqrt{R_x^2 + R_y^2} = 72.77$;
$\theta_R = \tan^{-1}\frac{.92}{72.76} = .01264$. θ in 2nd Quadrant so $\theta = \pi - \theta_R = 3.13$ rad

vector	horizontal component	vertical component
E	$\mathbf{E_x} = 52.6\cos 2.53 = -43.07$	$\mathbf{E_y} = 52.6\sin 2.53 = 30.20$
F	$\mathbf{F_x} = 41.7\cos 3.92 = -29.69$	$\mathbf{F_y} = 41.7\sin 3.92 = -29.28$
R	$\mathbf{R_x}$ −72.76	$\mathbf{R_y}$ 0.92

18. $R = \sqrt{R_x^2 + R_y^2} = 57.67$;
$\theta_R = \tan^{-1}\frac{57.57}{3.41} = 1.51$; θ is in the 3rd Quadrant so $\theta = \pi + 1.51 = 4.65$

vector	horizontal component	vertical component
G	$\mathbf{G_x} = 43.7\cos 4.73 = 0.77$	$\mathbf{G_y} = 43.7\sin 4.73 = -43.69$
H	$\mathbf{H_x} = 14.5\cos 4.42 = -4.18$	$\mathbf{H_y} = 14.5\sin 4.42 = -13.88$
R	$\mathbf{R_x}$ −3.41	$\mathbf{R_y}$ −57.57

19. $a = 14$, $b = 32$, $c = 27$: SSS so use Law of Cosines to find the largest angle, namely B $\cos B = \frac{a^2 + c^2 - b^2}{2ac} = -0.1310$; $B = 97.5$. $\sin A = \frac{a \sin B}{b} = .4338$; $A = 25.7$. $C = 180° - A - B = 56.8°$: $A = 25.7°$, $B = 97.5°$, $C = 56.8°$

20. $a = 43$, $b = 52$, $B = 86.4°$: SSA so Law of Sines. $\sin A = \frac{a \sin B}{b} = .8253$; $A = 55.6$. $C = 180° - A - B = 38.0°$. $c = \frac{b \sin C}{\sin B} = 32.08$.

21. $b = 87.4$, $B = 19.57°$, $c = 65.3$: SSA, Sine Law. $\sin C = \frac{c \sin B}{b} = 0.25026$; $C = 14.49°$. $A = 180° - B - C = 145.94°$. $a = \frac{b \sin A}{\sin B} = 146.14$. $A = 145.94°$, $a = 146.14$, $C = 14.49°$

22. $A = 121.3°$, $b = 42.5$, $c = 63.7$: SAS, Cosine Law. $a^2 = b^2 + c^2 - 2bc \cos A = 8676.8762$; $a = 93.15$. $\sin B = \frac{b \sin A}{a} = 0.3898497$; $B = 22.9°$; $C = 180° - A - B = 35.8°$: $a = 93.15$, $B = 22.9°$, $C = 35.8°$

23. $a = 127.35$, $A = 0.12$, $b = 132.6$: SSA, Sine Law. $\sin B = \frac{b \sin A}{a} = 0.1246$; $B = 0.12$ $C = \pi - A - B = 3.14 - .12 - .12 = 2.90$. $c = \frac{a \sin C}{\sin A} = 254.51$. $B = 0.12$, $C = 2.90$, $c = 254.51$

24. $b = 84.3$, $c = 95.4$, $C = 0.85$: SSA so Sine Law. $\sin B = \frac{b \sin C}{c} = 0.6639$; $B = 0.73$. $A = \pi - B - C = 1.56$. $a = \frac{c \sin A}{\sin C} = 127.0$: $B = 0.73$, $A = 1.56$, $a = 127.0$.

25. $a = 67.9$, $b = 54.2$, $C = 2.21$: SAS so Cosine Law. $c^2 = a^2 + b^2 - 2ab \cos C = 11938.92$; $c = 109.3$. $\sin B = \frac{b \sin C}{c} = 0.39798$; $B = 0.41$: $A = \pi - B - C = 0.52$: $c = 109.3$, $A = 0.52$, $B = 0.41$.

26. $a = 53.1$, $b = 63.2$, $c = 74.3$: SSS so use the Cosine Law. $\cos C = (a^2 + b^2 - c^2)/2ab = 0.1927$; $C = 1.38$. $\sin A = \frac{a \sin C}{c} = 0.7017$; $A = 0.78$. $B = \pi - A - C = 0.98$: $A = 0.78$, $B = 0.98$, $C = 1.38$

27. $R = \sqrt{R_x^2 + R_y^2} = 2831.17$; $\theta = \tan^{-1} \frac{R_y}{R_x} = 83.97°$. Resultant force is 2831 kg directed at an angle of 83.97° to the axis of the vehicle.

vector	horizontal component	vertical component
A	$\mathbf{A_x} = 1650 \cos 68° = 618.10$	$\mathbf{A_y} = 1650 \sin 68° = 1529.85$
B	$\mathbf{B_x} = -1325 \cos 76° = -320.55$	$\mathbf{B_y} = 1325 \sin 76° = 1285.64$
R	$\mathbf{R_x}$ 297.55	$\mathbf{R_y}$ 2815.49

28. Call the alternate route a. Then by the Cosine Law we see that $a^2 = 3621^2 + 2342^2 - 2(3621)(2342) \cos 72.4°$; $a^2 = 13468181$; $a = 3669.9$ ft. Over the hill $3621 + 2342 = 5963$; $2.3 \times 3669.9 = 8440.77$ Over the hill is less expensive.

29. Parallel component is $126.5 \sin 31.7° = 66.47$lb. Perpendicular component $126.5 \cos 31.7° = 107.63$lb.

30. $X_C + X_L = -72 + 52 = -20\Omega$ $\theta = \tan^{-1}\left(\frac{-20}{35}\right) = -29.74°$

31. $AB = c$, $AC = b = 73$, $A = 123.4$, $C = 42.1$. $B = 180° - A - C = 14.5$. $c = \frac{b \sin C}{\sin B} = 195.47$ m. The distance is 195.47 m.

32. By Cosine Law $d^2 = 1235^2 + 962^2 - 2(1235)(962) \cos 52.57°$; $d^2 = 1006471$; $d = 1003.23$ m

CHAPTER 10 TEST

1. $V_x = 47 \cos 117° = -21.34$; $V_y = 47 \sin 117° = 41.88$

2. $A = \sqrt{A_x^2 + A_y} = \sqrt{12.91^2 + (-14.36)^2} = 19.31$; $\theta = \tan^{-1} \frac{-14.36}{12.91} = -48.04°$ or $311.96°$

3.

vector	horizontal component	vertical component
A	$\mathbf{A_x} = 25 \cos 64° = 10.96$	$\mathbf{A_y} = 25 \sin 64° = 22.47$
B	$\mathbf{B_x} = 40 \cos 112° = -14.98$	$\mathbf{B_y} = 40 \sin 112° = 37.09$
R	$\mathbf{R_x}$ -4.02	$\mathbf{R_y}$ 59.56

$R = \sqrt{R_x^2 + R_y^2} = 59.70$; $\theta = \tan^{-1} \frac{59.56}{-4.02} = 93.86°$

4. $b = \frac{a \sin B}{\sin A} = \frac{9.42 \sin 67.5}{\sin 35.6} = 14.95$.

5. $a^2 = b^2 + c^2 - 2bc \cos A = 4.95^2 + 6.24^2 - 2(4.95)(6.24) \cos 13.4$; $a^2 = 87.9743$; $a = 9.38$.

6. $A = 24°$, $b = 36.5$, and $C = 97°$: $B = 180° - A - C = 59°$. $a = \frac{b \sin A}{\sin B} = \frac{36.5 \sin 24°}{\sin 59°} = 17.32$. $c = \frac{b \sin C}{\sin B} = \frac{36.5 \sin 97°}{\sin 59°} = 42.26$. $B = 59°$, $a = 17.32$, $c = 42.26$.

7. $c^2 = 30^2 + 36^2 - 2 \cdot 30 \cdot 36 \cos 97° = 2459.237782$; $c = 49.59$

CHAPTER

11

Graphs of Trigonometric Functions

11.1 SINE AND COSINE CURVES: AMPLITUDE AND PERIOD

1. Period = $\frac{2\pi}{2} = \pi$, Amplitude $= |3| = 3$, Frequency = $\frac{1}{\pi}$

2. Period = $\frac{2\pi}{6} = \frac{\pi}{3}$, Amplitude $= |5| = 5$, Frequency $\frac{1}{\frac{\pi}{3}} = \frac{3}{\pi}$

3. Period $= 2\pi$, Amplitude $= |2| = 2$, Frequency $\frac{1}{2\pi}$

4. Period = $\frac{2\pi}{3}$, Amplitude = 7, Frequency $\frac{1}{\frac{2\pi}{3}} = \frac{3}{2\pi}$

5. Period = $\frac{2\pi}{2\pi} = 1$, Amplitude = 8, Frequency 1

6. Period = $\frac{2\pi}{\pi} = 2$, Amplitude = 7, Frequency $\frac{1}{2}$

7. Period = $\frac{2\pi}{4} = \frac{\pi}{2}$, Amplitude $= \frac{1}{2}$, Frequency $\frac{2}{\pi}$

8. Period = $\frac{2\pi}{\frac{1}{3}} = 6\pi$, Amplitude = 5, Frequency $\frac{1}{6\pi}$

9. Period = $\frac{2\pi}{\frac{1}{2}} = 4\pi$, Amplitude $= \frac{1}{3}$, Frequency $\frac{1}{4\pi}$

10. Period = $\frac{2\pi}{3}$, Amplitude $= \frac{1}{5}$, Frequency $\frac{3}{2\pi}$

11. Period = $\frac{2\pi}{\frac{1}{4}} = 8\pi$, Amplitude $= |-3| = 3$, Frequency $\frac{1}{8\pi}$

12. Period = $\frac{2\pi}{\frac{1}{3}} = 6\pi$, Amplitude $= |-5| = 5$, Frequency $\frac{1}{6\pi}$

13. Period $= 2\pi$, Amplitude $= \left|-\frac{1}{2}\right| = \frac{1}{2}$, Frequency $\frac{1}{2\pi}$

14. Period $= \frac{2\pi}{3}$, Amplitude $= \left|-\frac{1}{2}\right| = \frac{1}{2}$, Frequency $\frac{3}{2\pi}$

15.

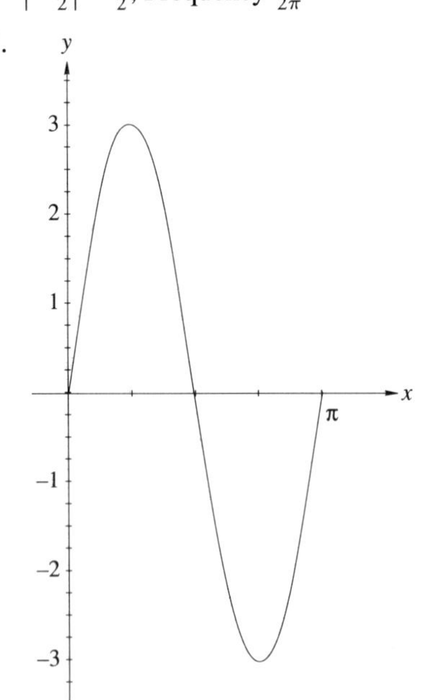

16.

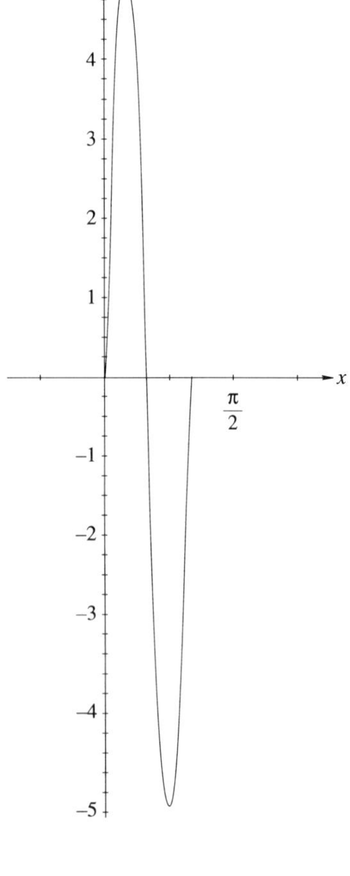

17.

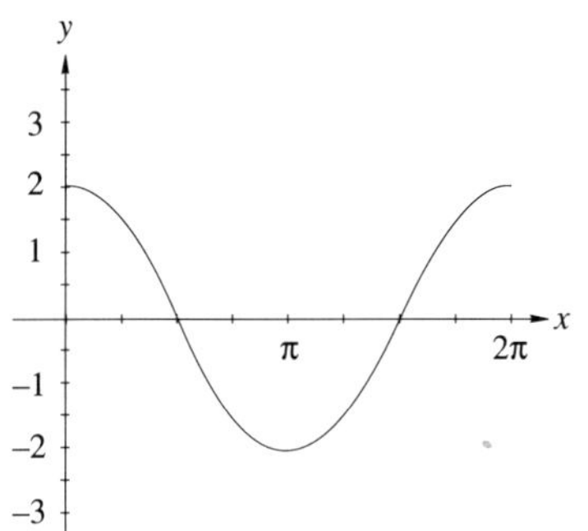

18.

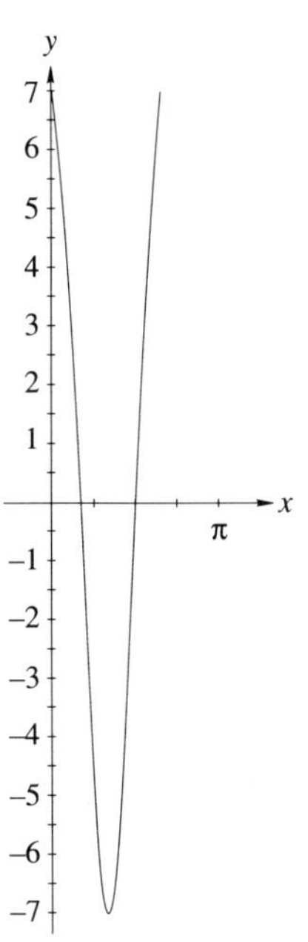

19.

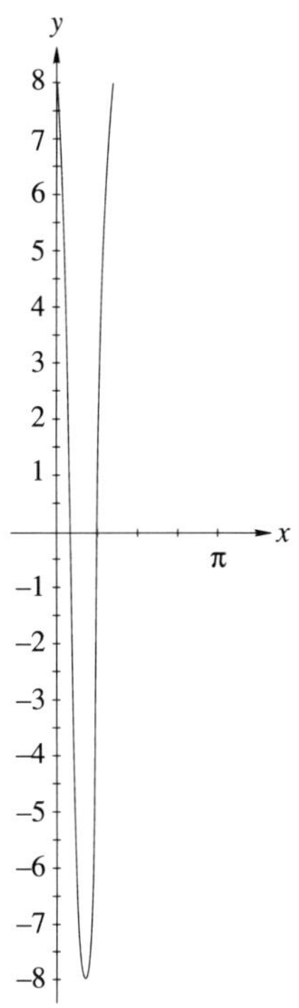

20.

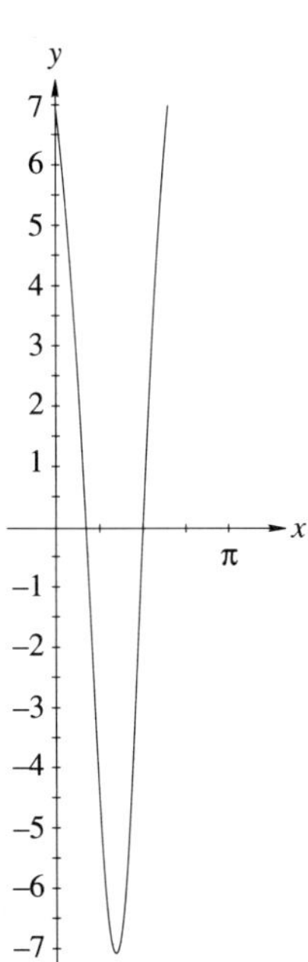

21.

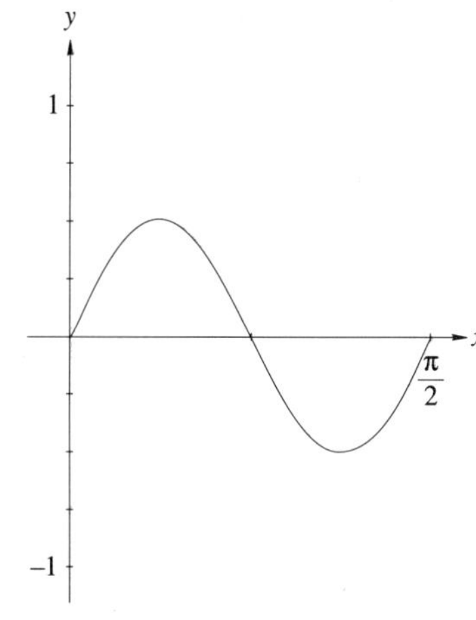

22.

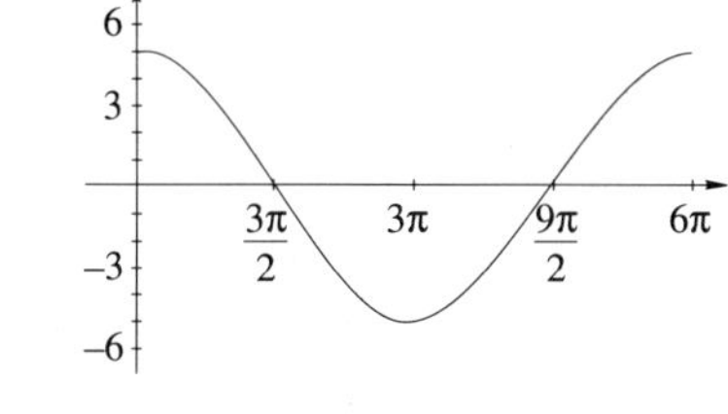

23.

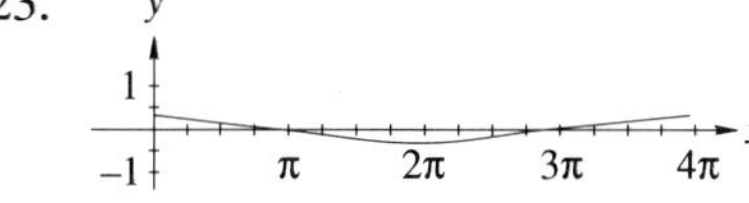

24.

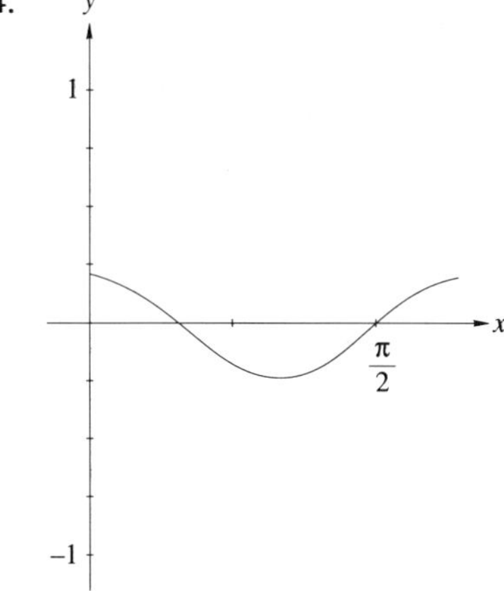

25.

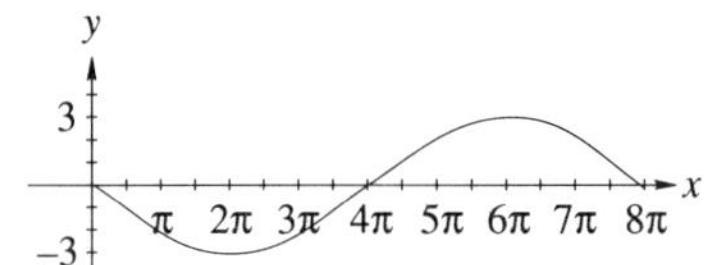

26.

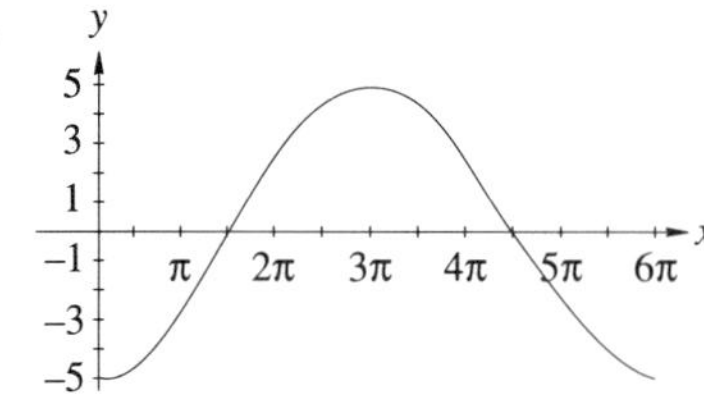

27.

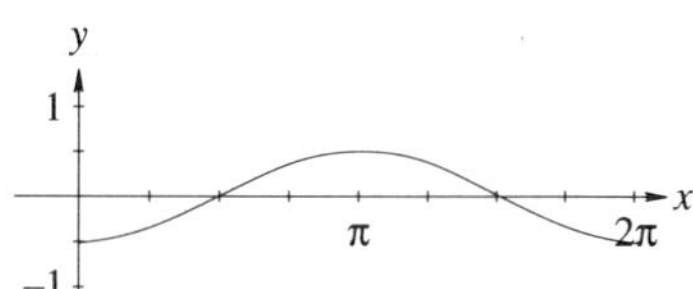

28.

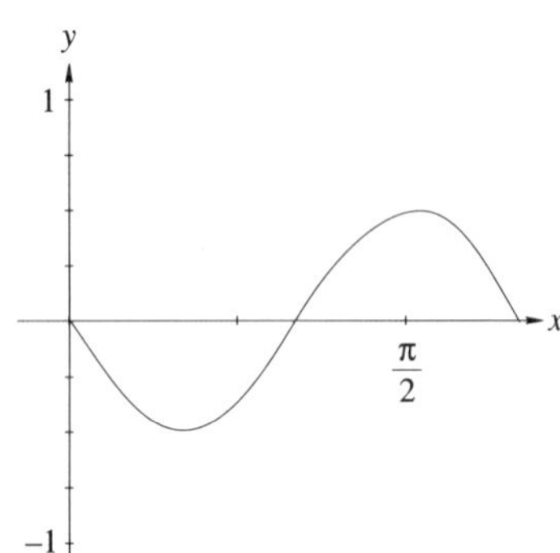

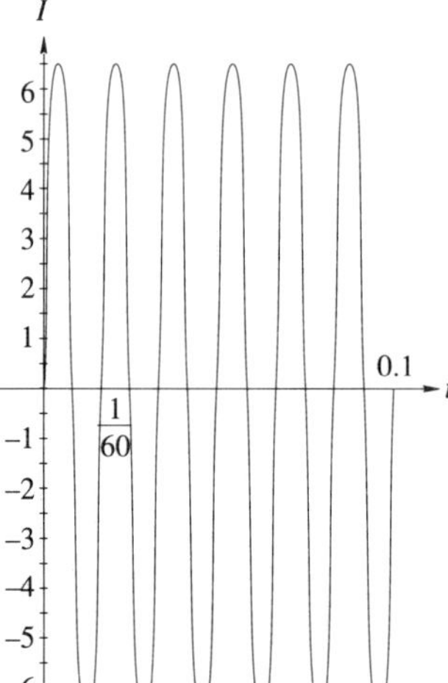

29. Period = $\frac{2\pi}{120\pi} = \frac{1}{60}$ s,
Amplitude = 6.5 A,
Frequency 60 Hz

11.2 SINE AND COSINE CURVES: DISPLACEMENT OR PHASE SHIFT

1. Amplitude 2, Period = 2π, Phase Shift = $-\frac{\pi}{4}$ or $\frac{\pi}{4}$ left.

2. Amplitude 1.5, Period = 2π, Phase Shift = $\frac{\pi}{3}$ or $\frac{\pi}{3}$ right.

3. Amplitude 2.5, Period = $\frac{2\pi}{3}$, Phase shift = $\frac{\pi}{3}$ or $\frac{\pi}{3}$ right.

4. Amplitude 3, Period = π, Phase Shift = $-\frac{\frac{\pi}{2}}{2} = -\frac{\pi}{4}$ or $\frac{\pi}{4}$ left.

5. Amplitude 6, Period = $\frac{2\pi}{1.5} = \frac{360}{1.5} = 240°$ or $\frac{4\pi}{3}$, Phase shift = $-\frac{180°}{1.5} = -120°$ or $-\frac{2\pi}{3}$.

6. Amplitude 8, Period = $\frac{360}{3.5} = 102.86°$ or $\frac{2\pi}{3.5} = \frac{4\pi}{7}$, Phase shift = $\frac{90}{3.5} = 25.71°$ or $\frac{\frac{\pi}{2}}{3.5} = \frac{\pi}{7}$.

7. Amplitude $|-2| = 2$, Period = $\frac{360}{3} = 120°$ or $\frac{2\pi}{3}$, Phase shift = $\frac{90}{3} = 30°$ or $\frac{\pi}{6}$

8. Amplitude $|-4| = 4$, Period = $\frac{360}{5} = 72°$ or $\frac{2\pi}{5}$, Phase shift = $-\frac{600}{5} = -120°$ or $-\frac{2\pi}{3}$

9. Amplitude 4.5, Period = $\frac{2\pi}{6} = \frac{\pi}{3}$, Phase shift $= \frac{8}{6} = \frac{4}{3}$

10. Amplitude 9, Period = $\frac{2\pi}{5}$, Phase shift = $-\frac{3}{5}$

11. Amplitude 0.5, Period = $\frac{2\pi}{\pi} = 2$, Phase shift $= \frac{-\frac{\pi}{8}}{\pi} = -\frac{1}{8}$

12. Amplitude $= |-0.25| = 0.25$, Period $= \frac{2\pi}{\frac{1}{\pi}} = 2\pi^2$, Phase Shift $= \frac{\frac{1}{4}}{\frac{1}{\pi}} = \frac{\pi}{4}$

13. Amplitude $= |-0.75| = 0.75$, Period $= \frac{2\pi}{\pi} = 2$, Phase Shift $= -\frac{\frac{\pi^2}{3}}{\pi} = -\frac{\pi}{3}$

14. Amplitude $= \pi$, Period $= \frac{2\pi}{\frac{1}{\pi}} = 2\pi^2$, Phase Shift $= -\frac{\frac{2}{3}}{\frac{1}{\pi}} = -\frac{2\pi}{3}$

15.

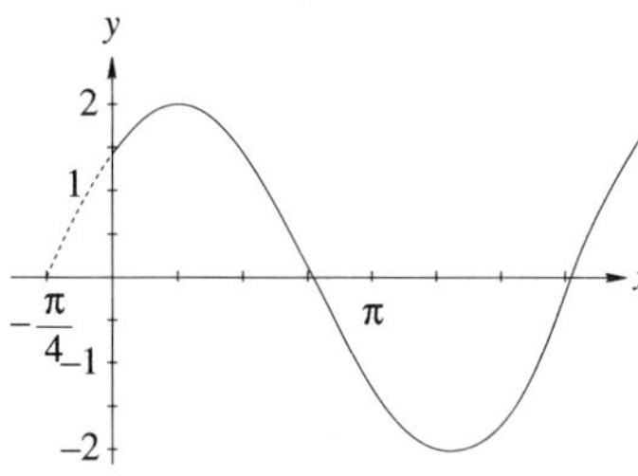

16.

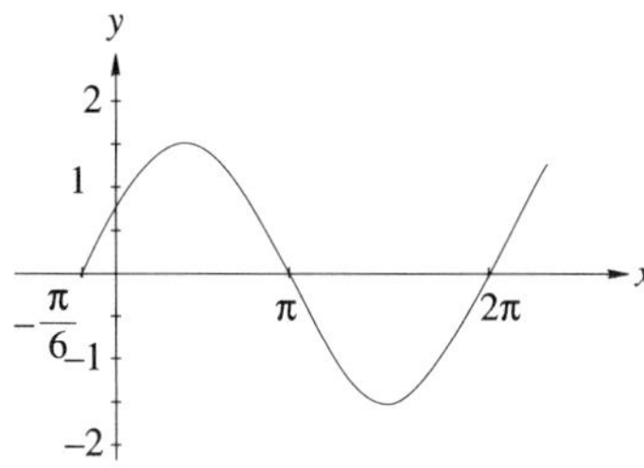

17.

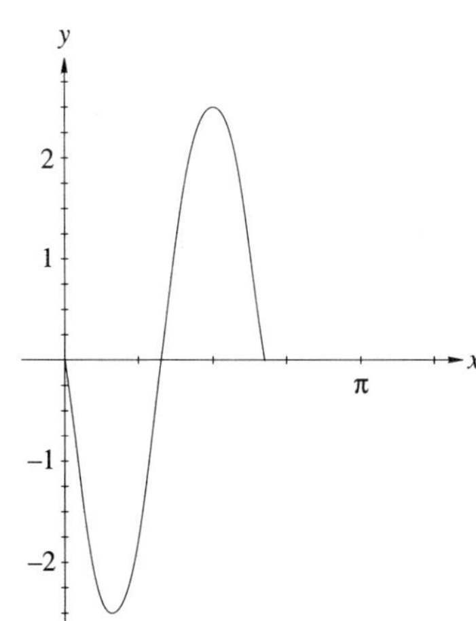

18.

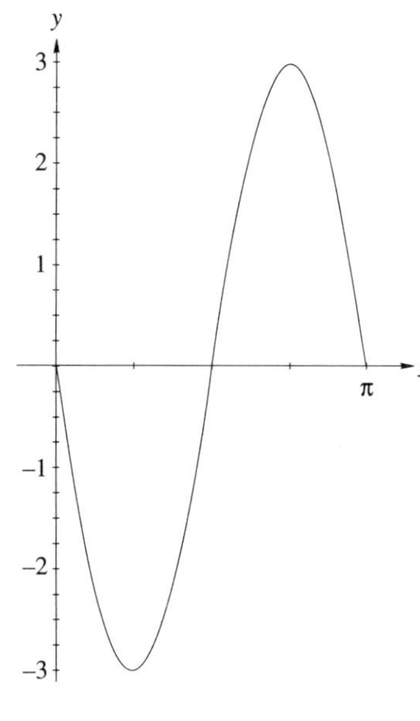

19.

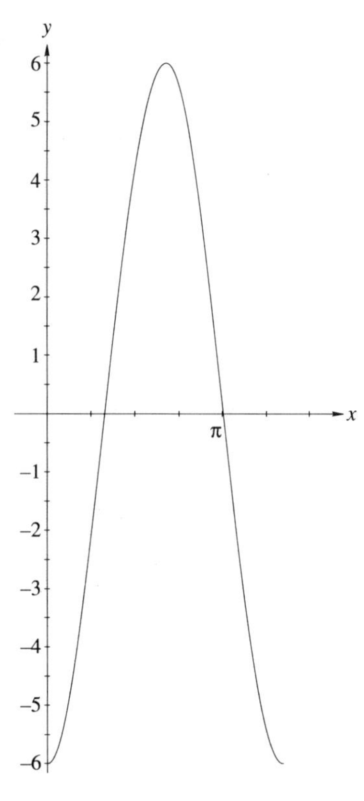

20.

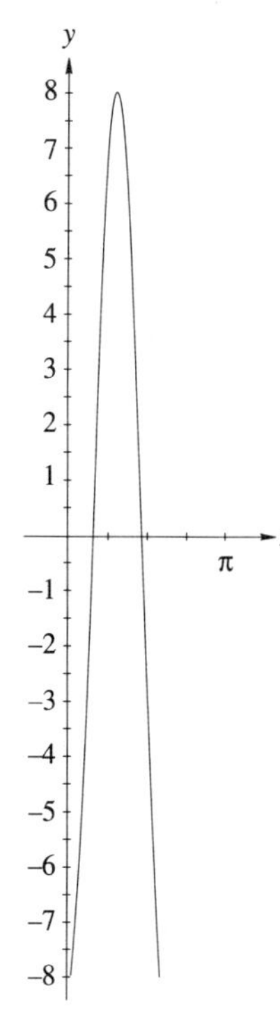

21.

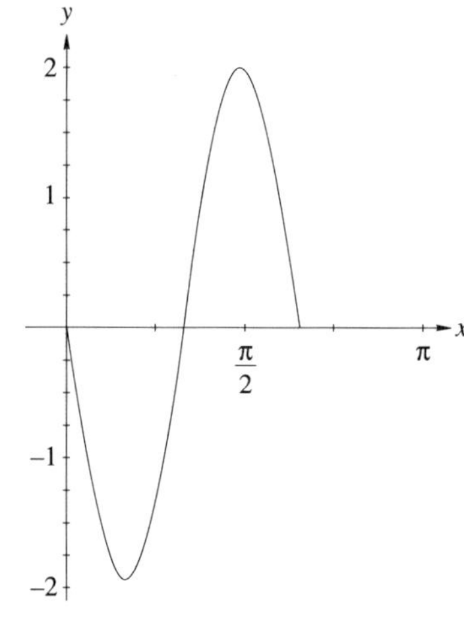

22.

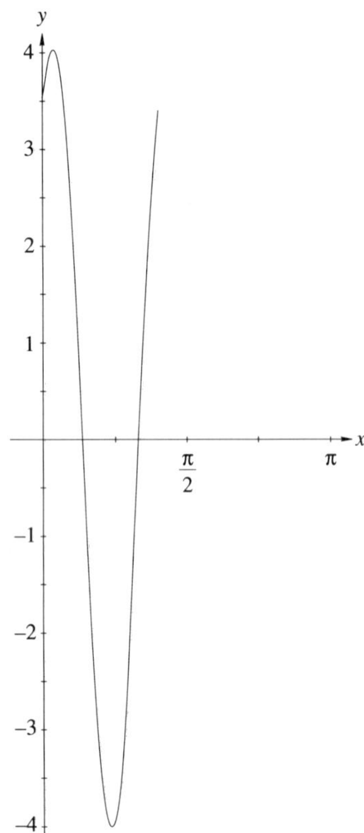
y
4
3
2
1
−1
−2
−3
−4
x
$\frac{\pi}{2}$
π

23.

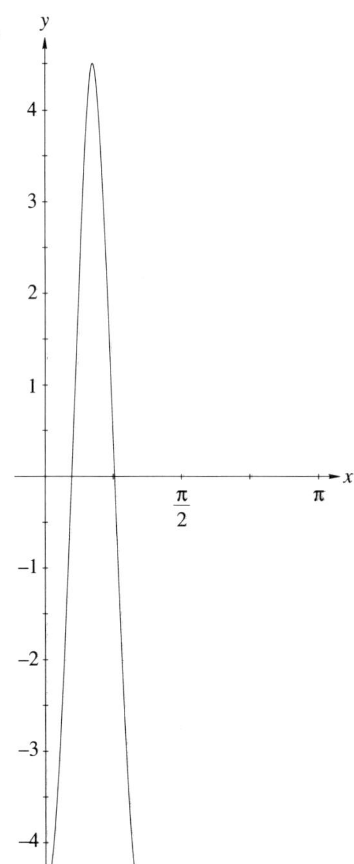
y
4
3
2
1
−1
−2
−3
−4
x
$\frac{\pi}{2}$
π

24.

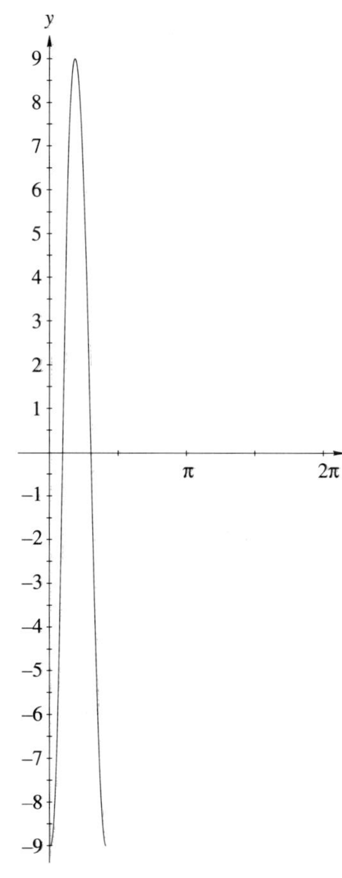
y
9
8
7
6
5
4
3
2
1
−1
−2
−3
−4
−5
−6
−7
−8
−9
x
π
2π

25.

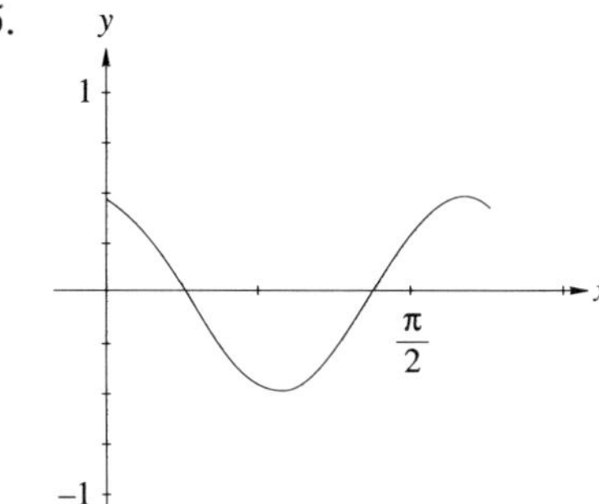
y
1
−1
x
$\frac{\pi}{2}$

26.

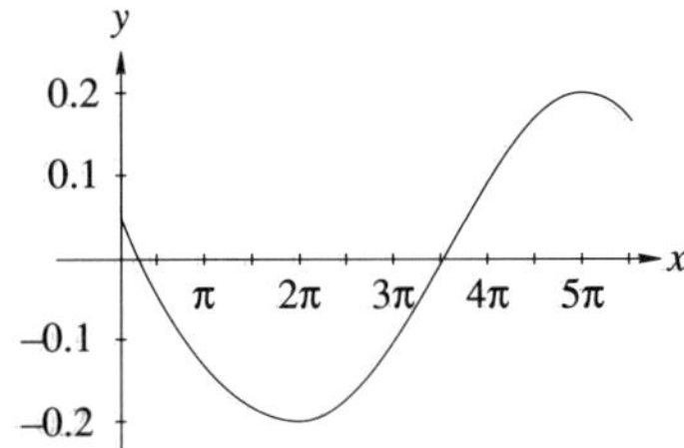

27.

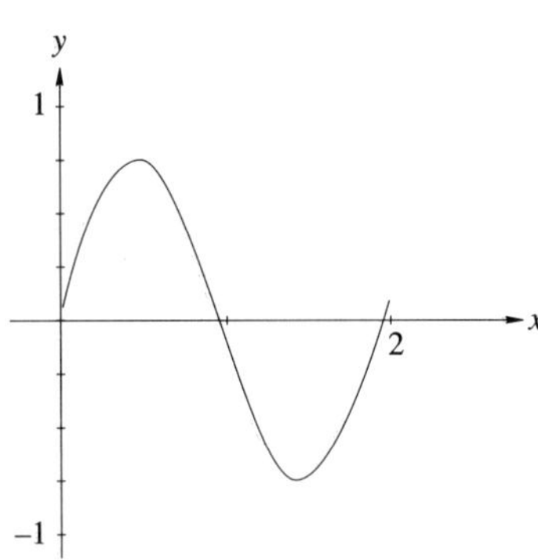

28.

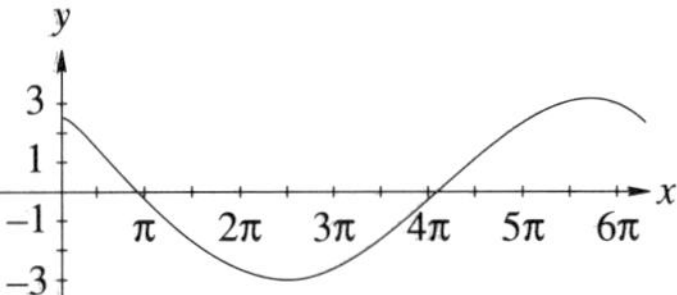

29. Amplitude 65 A, Period $= \frac{2\pi}{120\pi} = \frac{1}{60}$ s, Phase shift $= \dfrac{-\frac{\pi}{2}}{120\pi} = -\dfrac{1}{240}$ s

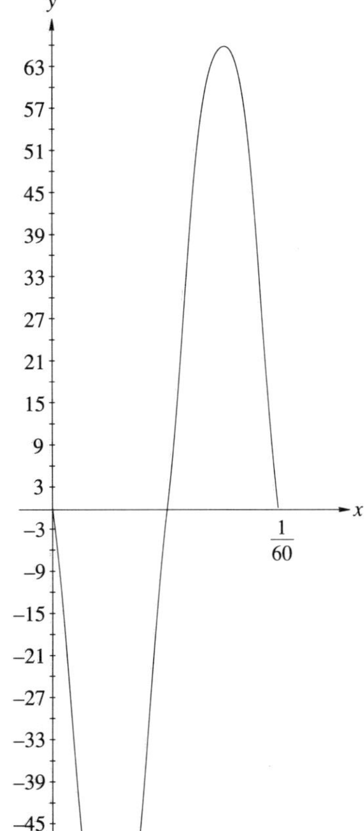

30. Rewrite the given equation as $y = 3\sin\left(\frac{\pi}{6}t - \frac{5\pi}{6}\right)$. Then, we get the following: Amplitude = 3 m, Period $= \dfrac{2\pi}{\frac{\pi}{6}} = 12$ s, Phase shift $= \dfrac{\frac{5\pi}{6}}{\frac{\pi}{6}} = 5$ s.

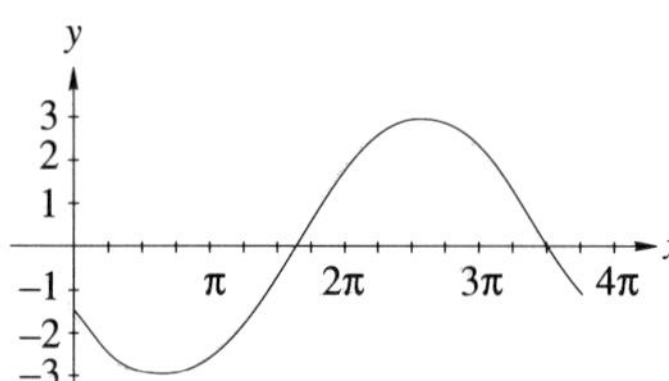

11.3 COMPOSITE SINE AND COSINE CURVES

1.

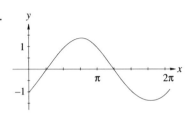

2.

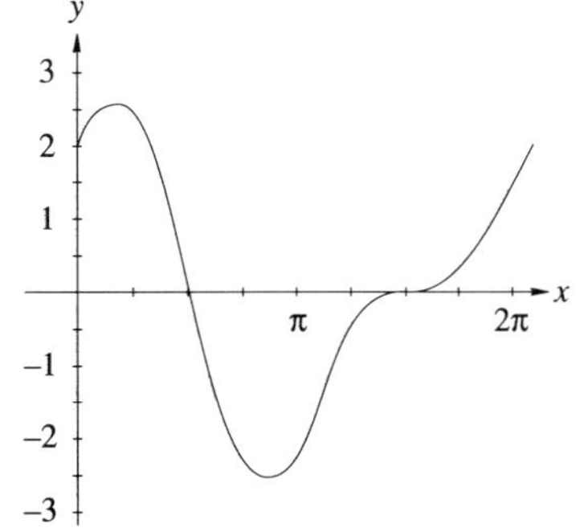

3.

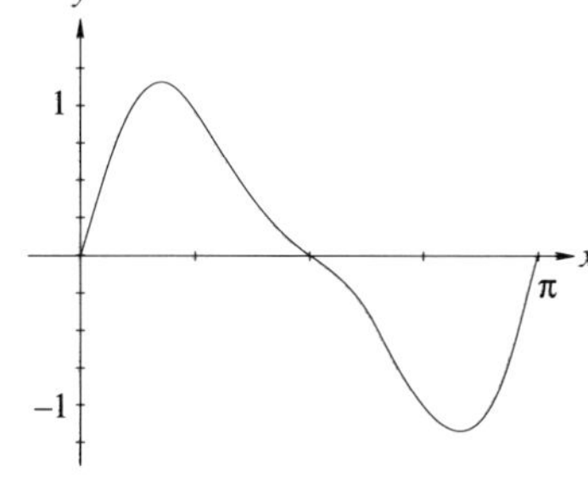

4.

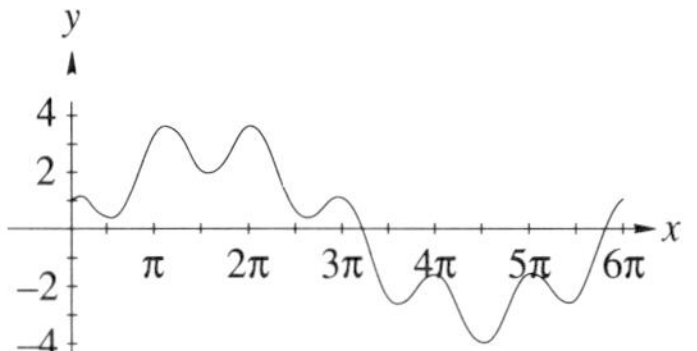

5.

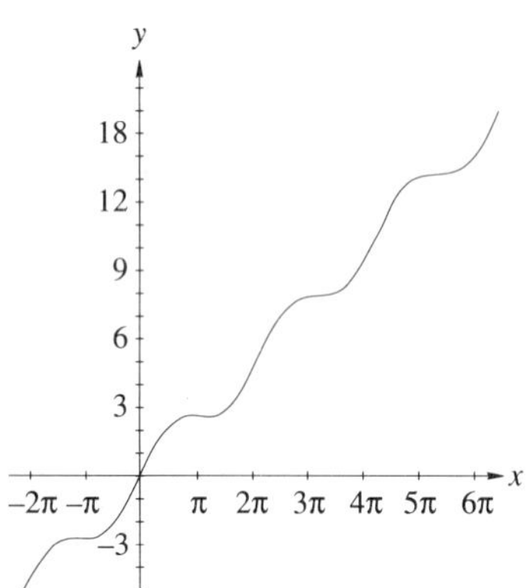

6.

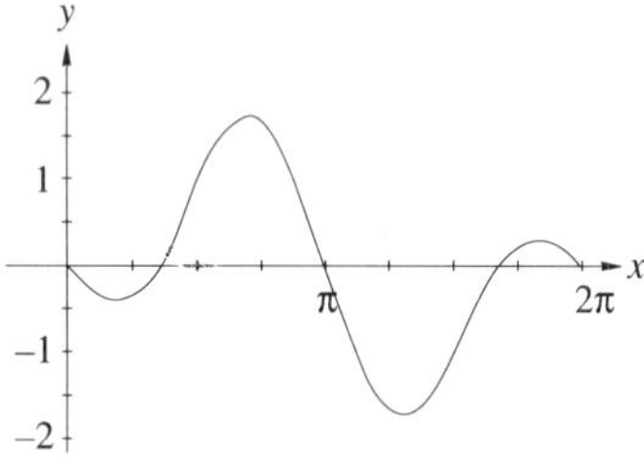

7.

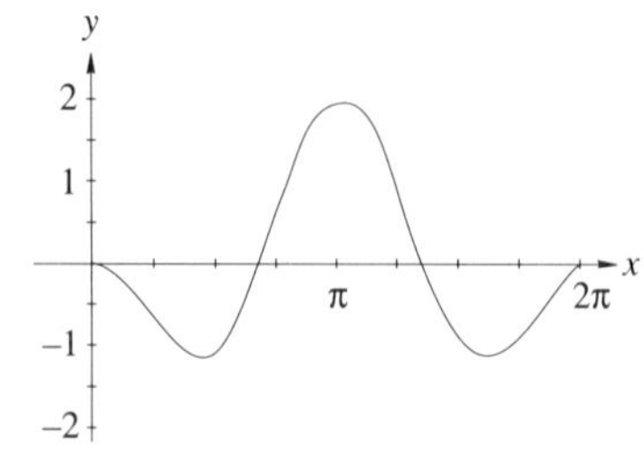

8.

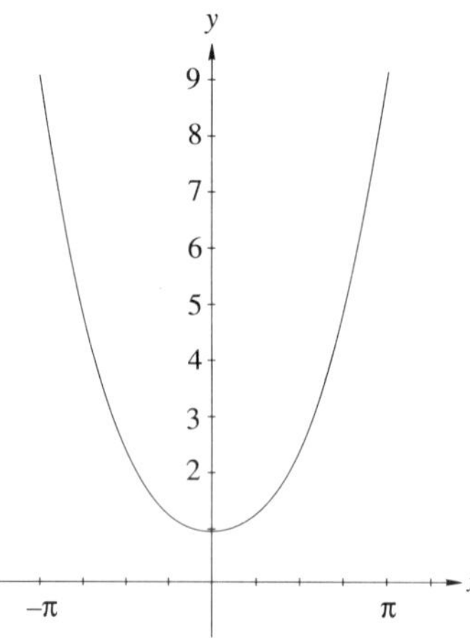

9.

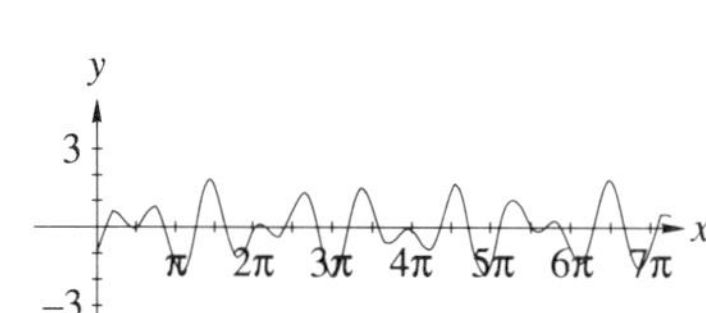

10.

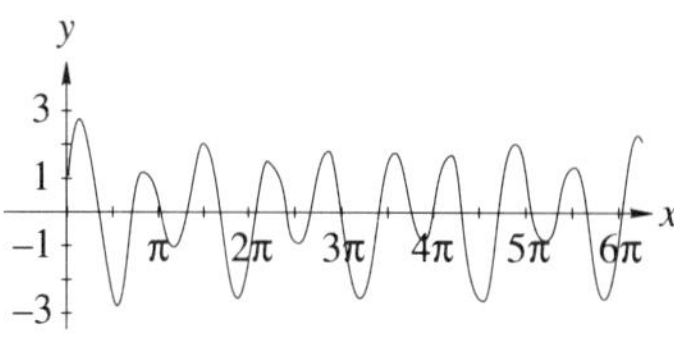

11.

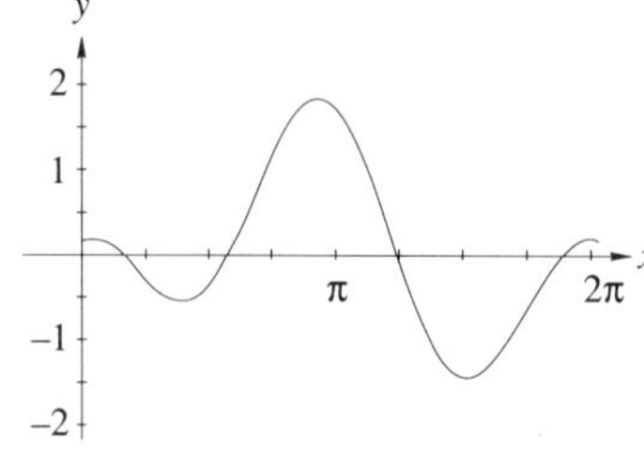

12.

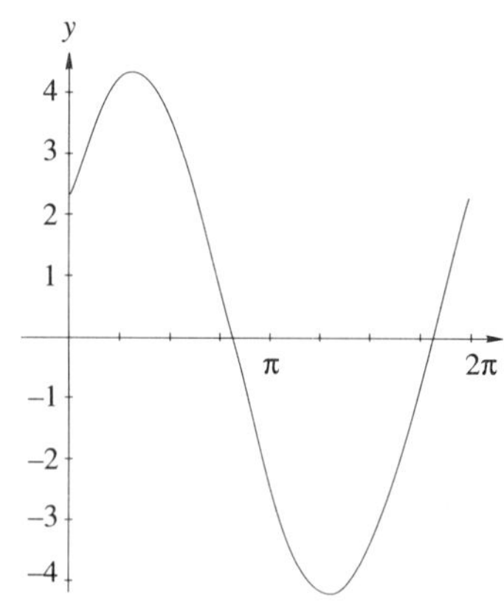

13.

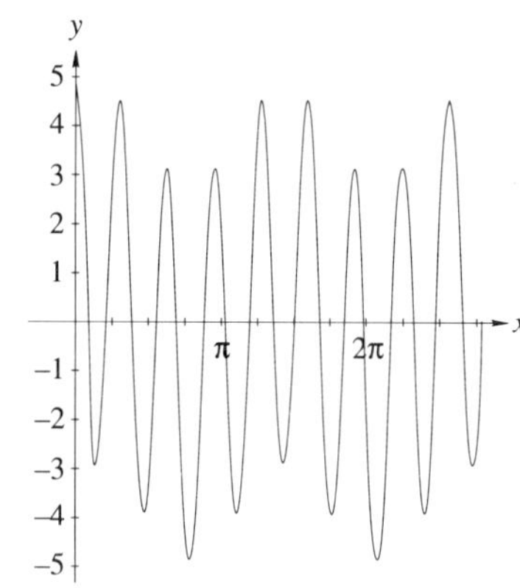

14.

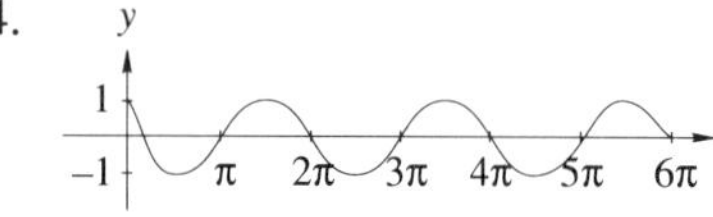

15.

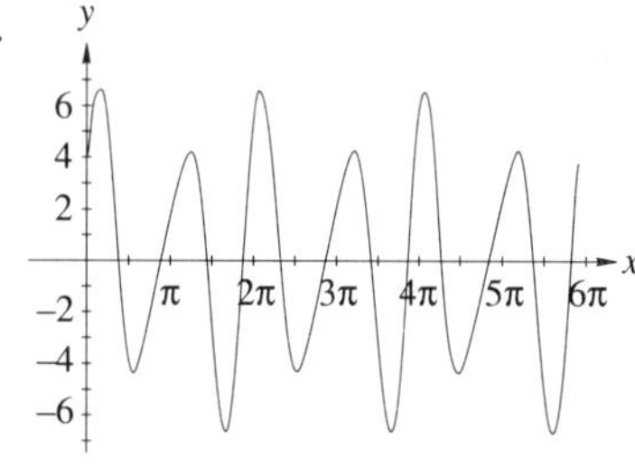

16.

17.

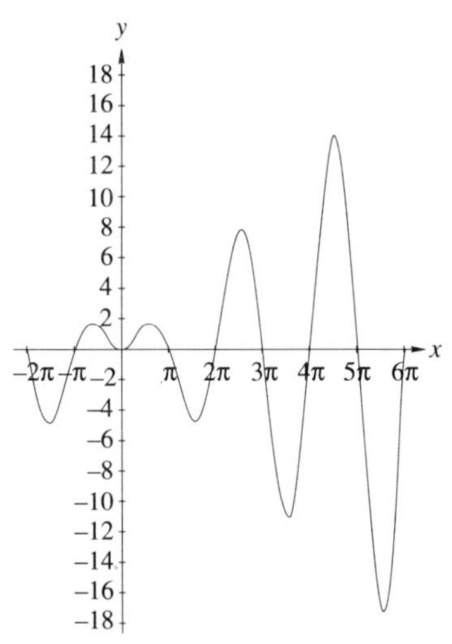

18.

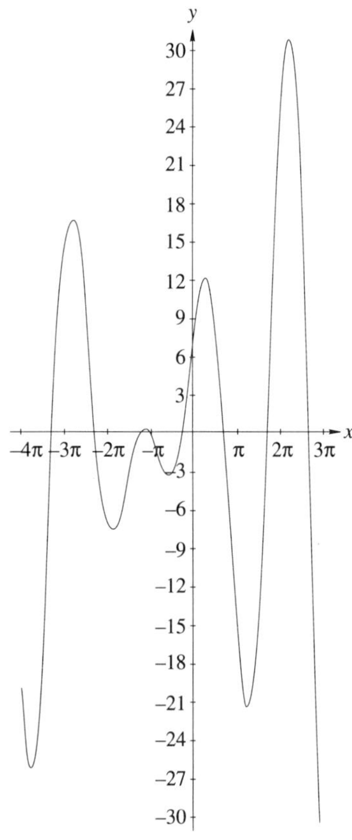

11.4 GRAPHS OF THE OTHER TRIGONOMETRIC FUNCTIONS

1.

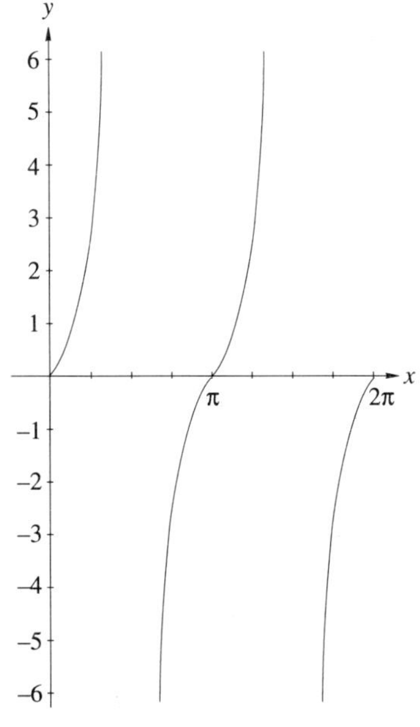

2.

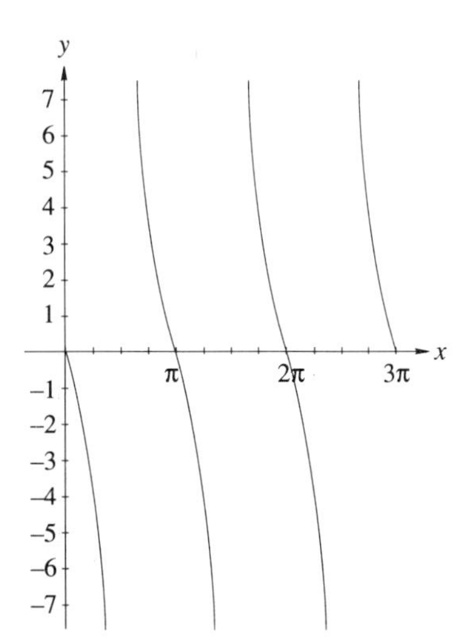

3.

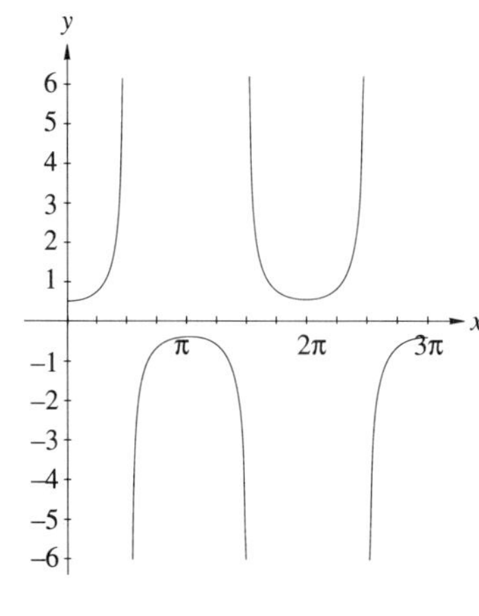

4.

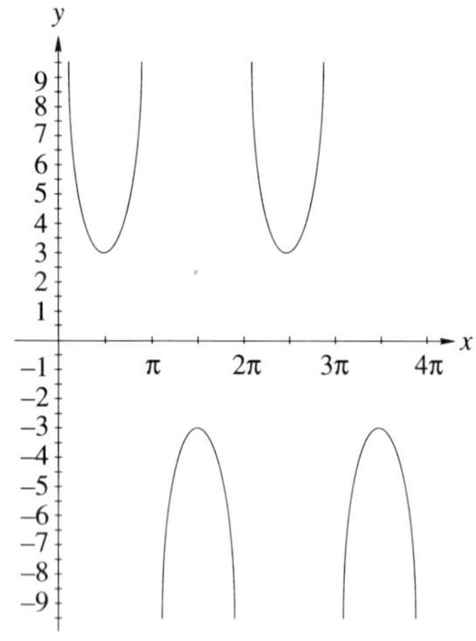

5.

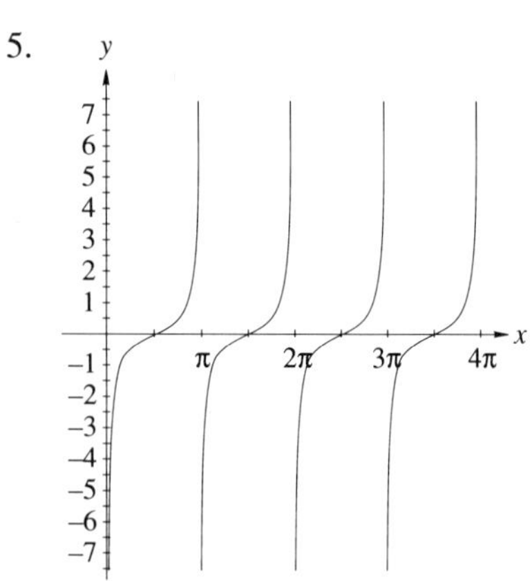

6.

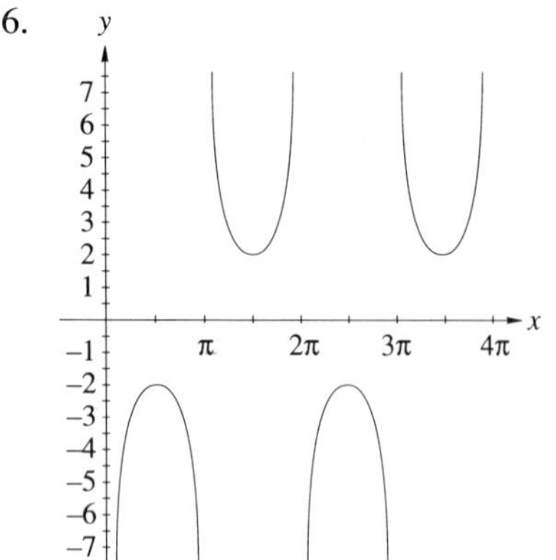

7.

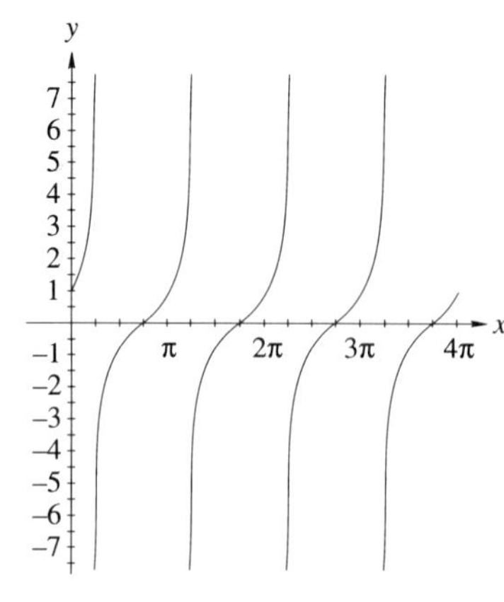

8.

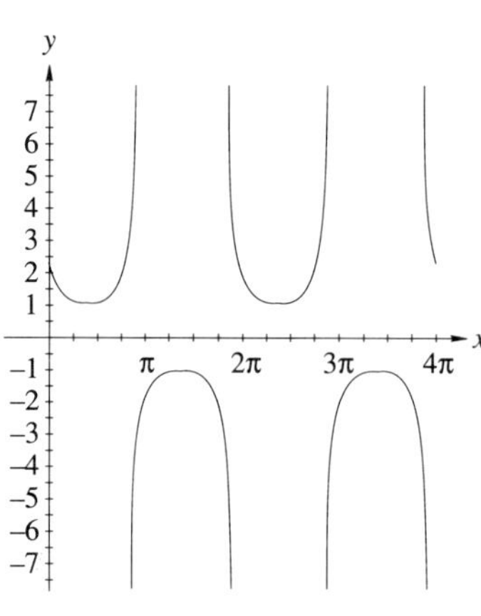

9.

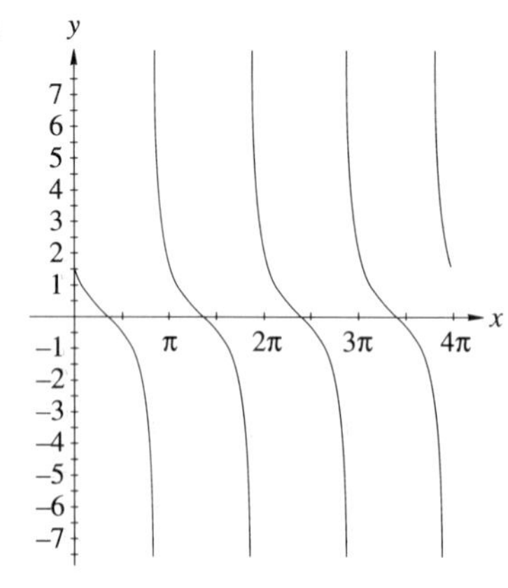

10.

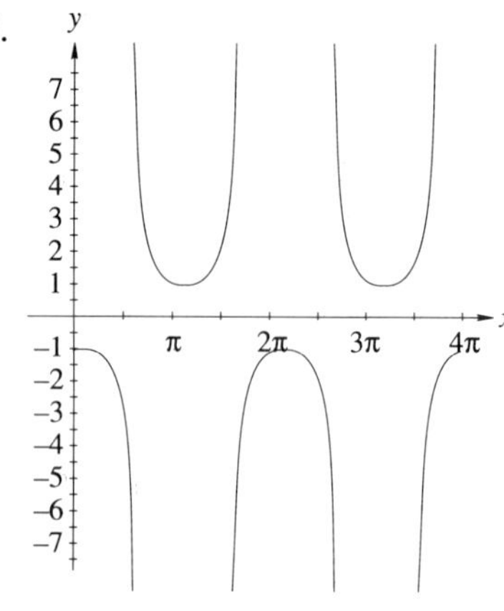

11.

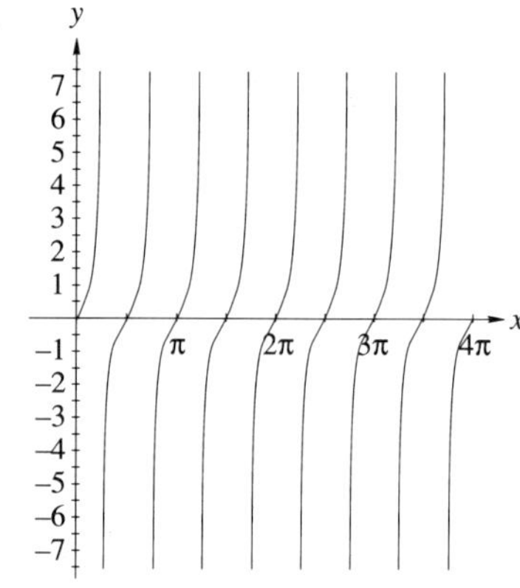

12.

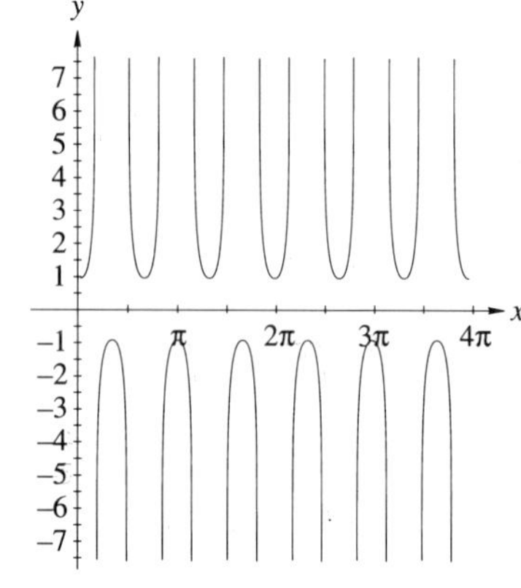

13.

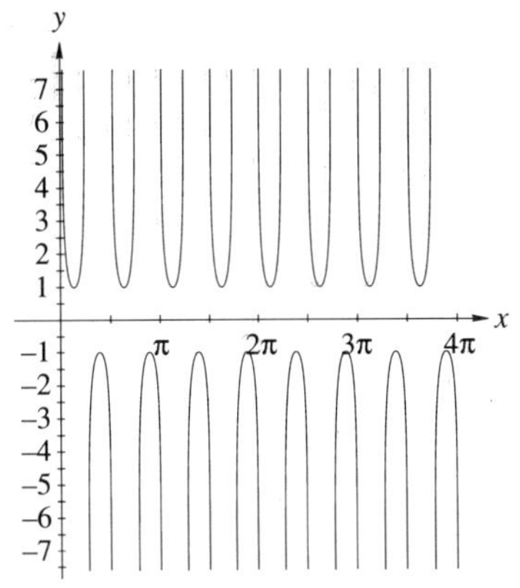

14.

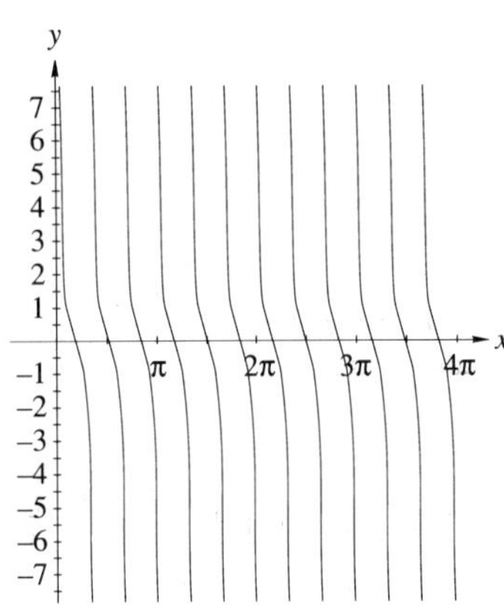

15.

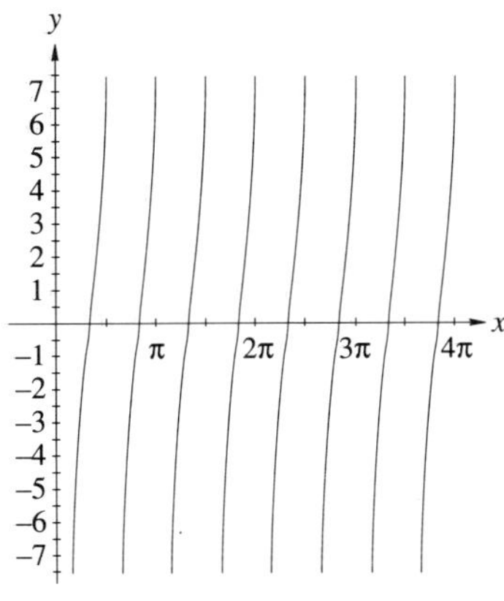

16.

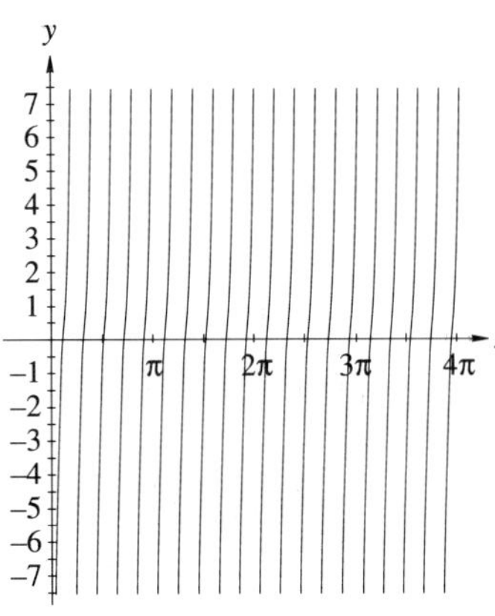

17.

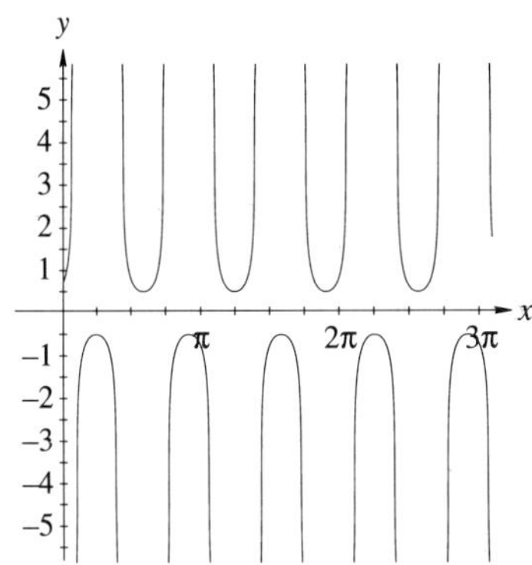

18.

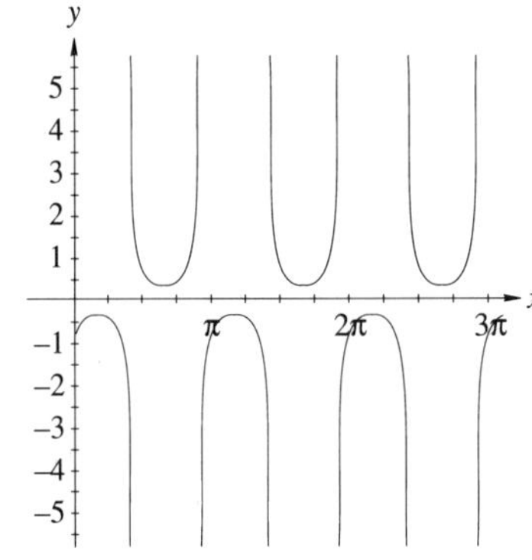

11.5 APPLICATIONS OF TRIGONOMETRIC GRAPHS

1. The amplitude is 10 so $A = 10$. Since $f = 4$, $\omega = 2\pi f = 8\pi$. $y = A \sin \omega t = 10 \sin 8\pi t$.

2. Using the answer for Exercise 1 and inserting a phase shift ϕ we have $y = 10 \sin(8\pi t - \phi)$. Substituting 8 for y and 0.1 for t solve for ϕ. $8 = 10 \sin(8\pi(0.1) - \phi)$; $\sin(0.8\pi - \phi) = 0.8$; $(0.8\pi - \phi) = \sin^{-1} 0.8 = 0.927295 \approx 0.927$; $-\phi = 0.927 - 0.8\pi = -1.586$ or $\phi = 1.586$. This gives the equation $y = 10 \sin(8\pi t - 1.586)$

3. Amplitude $A = 0.8$ m, $\omega = \frac{\pi}{3}$ rad/s. Since $y = 0$ when $t = 0$ there is no phase shift. Substituting into the form $y = A \sin \omega t$ we get $y = 0.8 \sin \frac{\pi}{3} t$.

4. $\omega = 2\pi f$ or $f = \frac{\omega}{2\pi} = \frac{\pi/3}{2\pi} = \frac{1}{6}$ Hz.

5. $a = -g \sin \theta = -32 \sin 5° = -2.79$ ft/s^2

6. $a = -g \sin \theta = -9.8 \sin(0.05) = -0.49$ m/s^2

7. $y = 8.5 \cos 2.8t$ (a) Amplitude is 8.5 cm, (b) period $= \frac{2\pi}{2.8} = \frac{\pi}{1.4} \approx 2.24$ s, (c) $f = \frac{2.8}{2\pi} = \frac{1.4}{\pi} \approx 0.45$ Hz.

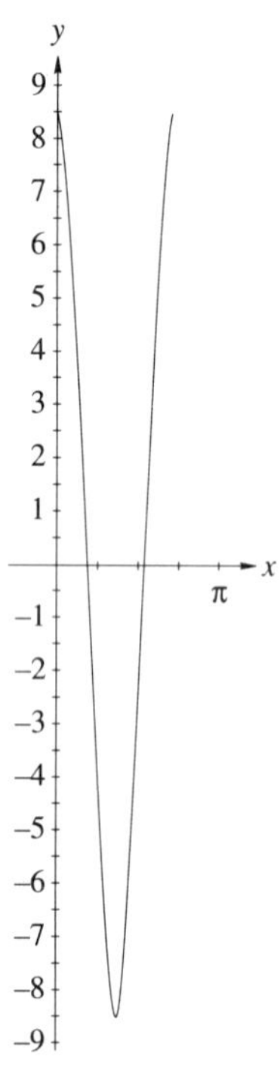

8. (a) $f = 3000$ rpm $= 50$ rps
(b) Using $y = A\cos 2\pi ft$ we get $y = \frac{10.5}{2}\cos 2\pi 50t$ or $y = 5.25\cos 100\pi t$. (c) $y = 5.25\cos\frac{4500}{60}\cdot 2\pi t = 5.25\cos 150\pi t$

9. $I = 10\sin 120\pi t$: amplitude 10 A, period $= \frac{2\pi}{120\pi} = \frac{1}{60}$ s; $f = \frac{120\pi}{2\pi} = 60$ Hz , $\omega = 2\pi f = 120\pi$ rad/s

10. Given that $I_{max} = 6.8$ and $f = 80$ Hz, then $I = I_{max}\sin 2\pi ft$, and so $I = 6.8\sin 160\pi t$

11. If $\phi = \frac{\pi}{3}$ then $I = 6.8\sin(160\pi t + \frac{\pi}{3})$

12. $V = I_{max}\sin(2\pi ft) = 220\sin(80\pi t)$

13. $V = 220\sin(80\pi t - \frac{\pi}{3})$

14. Since cosine and sine have the same shape and cosine is out of phase from the sine by $\frac{\pi}{2}$ we get $V = 220\cos\left(80\pi t + \frac{\pi}{2}\right)$

15. Since cosine is the same as sine except for a phase shift of $\frac{\pi}{2}$ we have $V = 220\cos(80\pi t - \frac{\pi}{3} + \frac{\pi}{2}) = 220\cos\left(80\pi t + \frac{\pi}{6}\right)$

16. $V = V_{max}\sin 2\pi ft = 250{,}000\sin(2\pi 10^{19}t)$

17. $y = 10^{-12}\sin 2\pi 10^{23}t$, frequency is 10^{23} Hz, amplitude is 10^{-12}

18. $I = I_{max}\sin(2\pi ft + \phi) = 3.6\sin(2\pi\cdot 60t + \frac{\pi}{2}) = 3.6\sin(120\pi t + \frac{\pi}{2})$

19. $I = I_{max}\sin(2\pi ft + \phi) = 1.2\sin(800\pi t + 37°)$ or $I = 1.2\sin(800\pi t + 0.6458)$

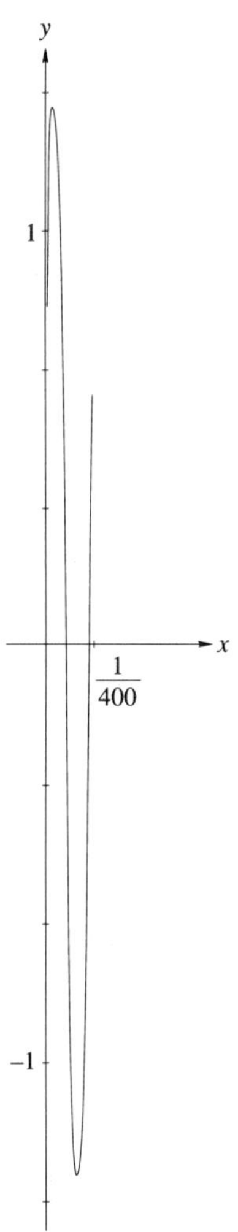

20. $\phi = \tan^{-1}\left(\frac{V_L - V_C}{V_R}\right) = \tan^{-1}\left(\frac{80 - 130}{120}\right) \approx \tan^{-1}(-0.4167)$; $\phi = -22.62° = -0.3948$ rad

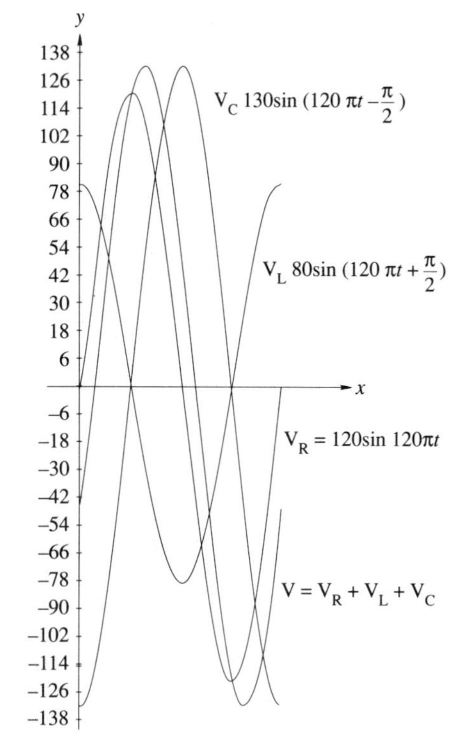

11.6 PARAMETRIC EQUATIONS

1.

t	-3	-2	-1	0	1	2	3
x	-3	-2	-1	0	1	2	3
y	-9	-6	-3	0	3	6	9

Given $x = t$ and $y = 3t$, substituting x for t in the second equation yields $y = 3x$.

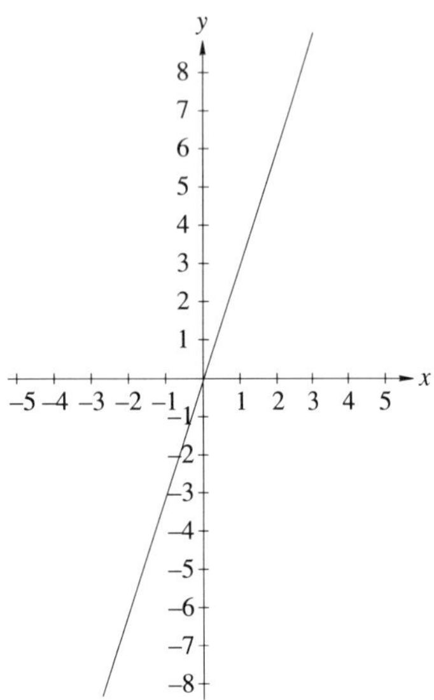

2.

t	-3	-2	-1	0	1	2	3
x	-6	-4	-2	0	2	4	6
y	-11	-7	-3	1	5	9	13

Here $x = 2t$ and $y = 4t + 1$. The first equation gives $t = \frac{x}{2}$. Substituting into the second equation yields $y = 4(\frac{x}{2}) + 1 = 2x + 1$ which simplifies to $y = 2x + 1$.

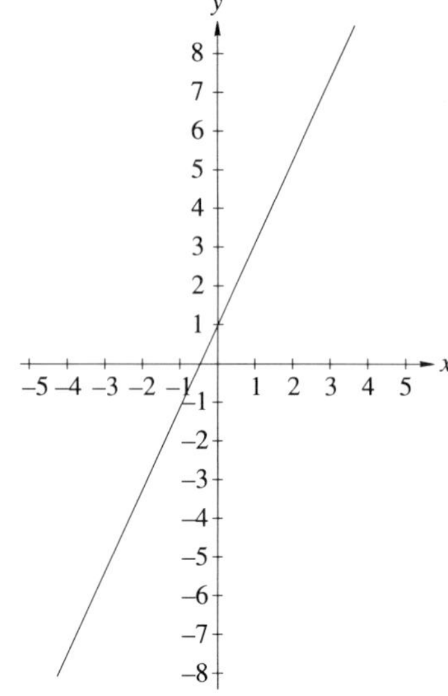

3.

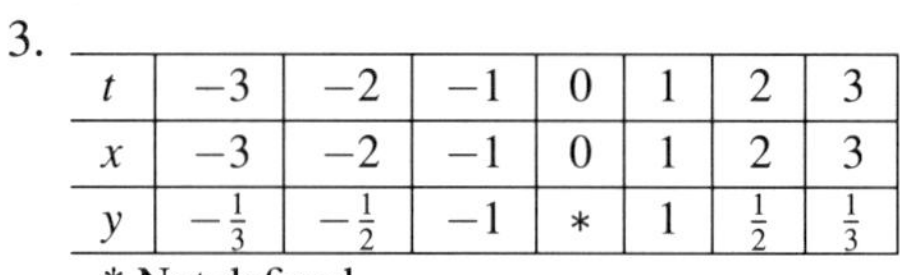

t	-3	-2	-1	0	1	2	3
x	-3	-2	-1	0	1	2	3
y	$-\frac{1}{3}$	$-\frac{1}{2}$	-1	*	1	$\frac{1}{2}$	$\frac{1}{3}$

* Not defined

Here $x = t$ and $y = \frac{1}{t}$. Direct substitution yields $y = \frac{1}{x}$.

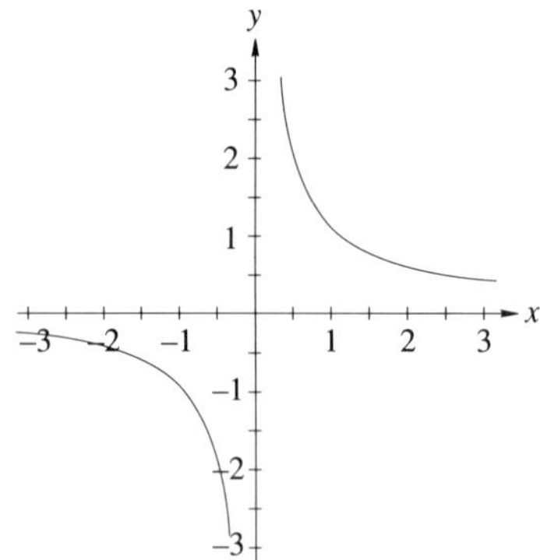

4.

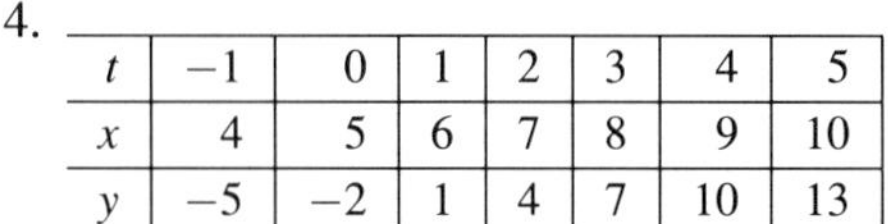

t	-1	0	1	2	3	4	5
x	4	5	6	7	8	9	10
y	-5	-2	1	4	7	10	13

Here $x = t + 5$ and $y = 3t - 2$. The first equation gives $t = x - 5$. Substitution into the second equation yields $y = 3(x-5) - 2 = 3x - 15 - 2 = 3x - 17$ which simplifies to $y = 3x - 17$.

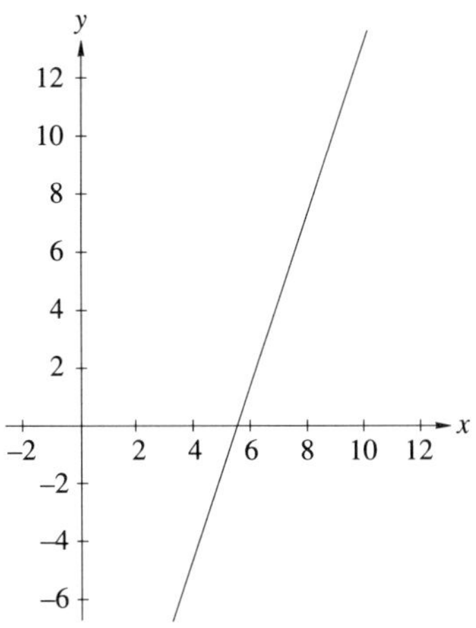

5.

t	−5	−4	−3	−2	−1	0	1	2	3	4	5
x	8	7	6	5	4	3	2	1	0	−1	−2
y	16	7	0	−5	−8	−9	−8	−5	0	7	16

Here $x = 3 - t$ and $y = t^2 - 9$. The first equation gives $t = 3 - x$. Substituting into the second equation yields $y = (3-x)^2 - 9 = 9 - 6x + x^2 - 9 = x^2 - 6x$ or $y = x^2 - 6x$.

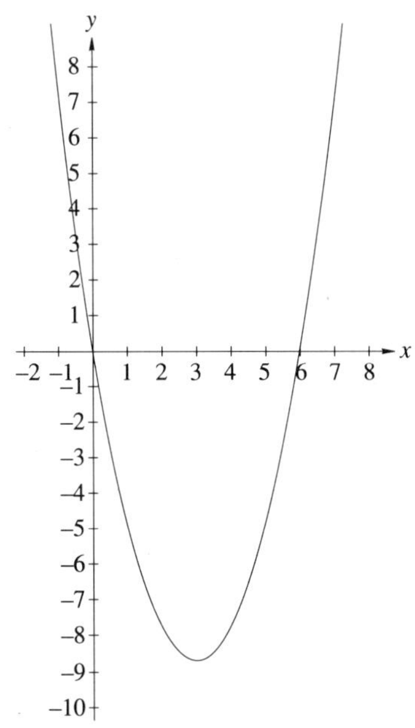

6.

t	−5	−4	−3	−2	−1	0	1	2	3	4	5
x	−3	−2	−1	0	1	2	3	4	5	6	7
y	30	20	12	6	2	0	0	2	6	12	20

Here $x = t + 2$ and $y = t^2 - t$. The first equation gives $t = x - 2$. Substitution into the second equation yields $y = (x-2)^2 - (x-2) = x^2 - 4x + 4 - x + 2 = x^2 - 5x + 6$ and so $y = x^2 - 5x + 6$.

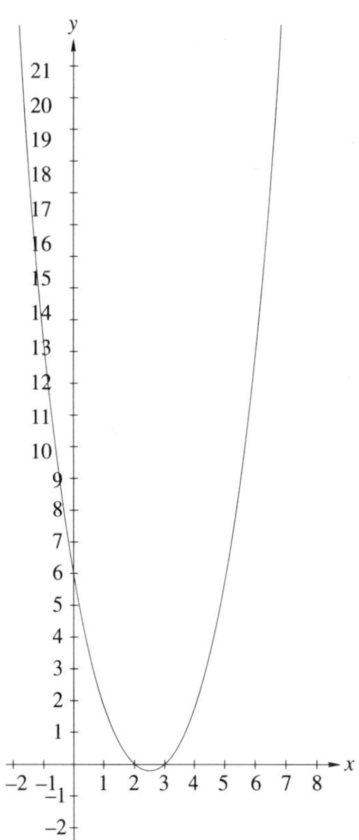

7.

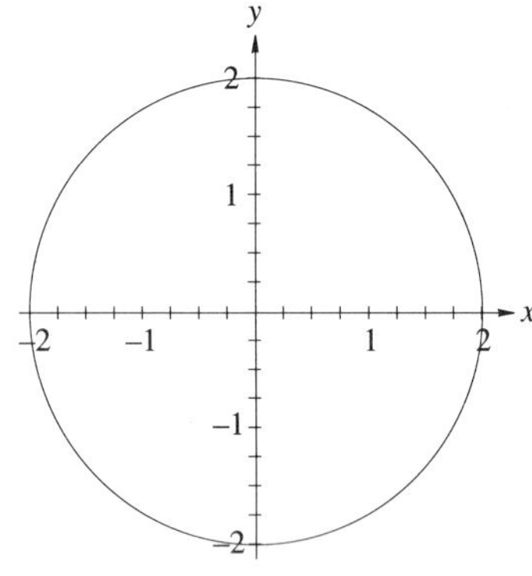

t	0	$\frac{\pi}{4}$	$\frac{\pi}{2}$	$\frac{3\pi}{4}$	π	$\frac{5\pi}{4}$	$\frac{3\pi}{2}$	$\frac{7\pi}{4}$
x	0	$\sqrt{2}$	2	$\sqrt{2}$	0	$-\sqrt{2}$	-2	$-\sqrt{2}$
y	2	$\sqrt{2}$	0	$-\sqrt{2}$	-2	$-\sqrt{2}$	0	$\sqrt{2}$

8.

t	0	$\frac{\pi}{4}$	$\frac{\pi}{2}$	$\frac{3\pi}{4}$	π	$\frac{5\pi}{4}$	$\frac{3\pi}{2}$	$\frac{7\pi}{4}$
x	0	$\sqrt{2}$	5	$\sqrt{2}$	0	$-\sqrt{2}$	-5	$-\sqrt{2}$
y	2	$\frac{5\sqrt{2}}{2}$	0	$-\frac{5\sqrt{2}}{2}$	-2	$-\frac{5\sqrt{2}}{2}$	0	$\frac{5\sqrt{2}}{2}$

9.

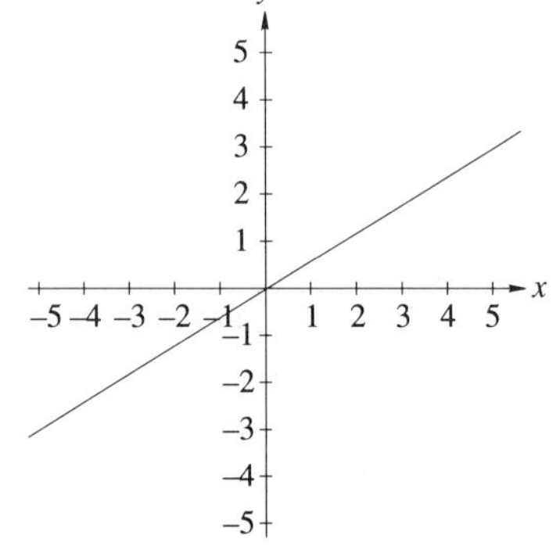

t	0	$\frac{\pi}{4}$	$\frac{\pi}{2}$	$\frac{3\pi}{4}$	π	$\frac{5\pi}{4}$	$\frac{3\pi}{2}$	$\frac{7\pi}{4}$	2π
x	2	$\frac{5\sqrt{2}}{2}$	5	$\frac{5\sqrt{2}}{2}$	0	$-\frac{5\sqrt{2}}{2}$	-5	$-\frac{5\sqrt{2}}{2}$	0
y	0	$\frac{3\sqrt{2}}{2}$	3	$\frac{3\sqrt{2}}{2}$	0	$-\frac{3\sqrt{2}}{2}$	-3	$-\frac{3\sqrt{2}}{2}$	0

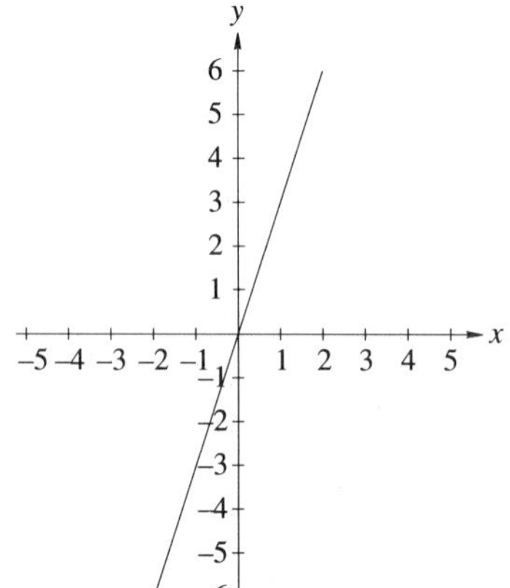

10.

t	0	$\frac{\pi}{4}$	$\frac{\pi}{2}$	$\frac{3\pi}{4}$	π	$\frac{5\pi}{4}$	$\frac{3\pi}{2}$	$\frac{7\pi}{4}$	2π
x	2	$\sqrt{2}$	0	$-\sqrt{2}$	-2	$-\sqrt{2}$	0	$\sqrt{2}$	2
y	6	$3\sqrt{2}$	0	$-3\sqrt{2}$	-6	$-3\sqrt{2}$	0	$3\sqrt{2}$	6

11.

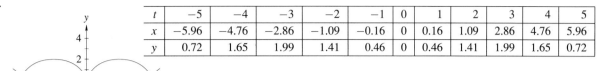

t	-5	-4	-3	-2	-1	0	1	2	3	4	5
x	-5.96	-4.76	-2.86	-1.09	-0.16	0	0.16	1.09	2.86	4.76	5.96
y	0.72	1.65	1.99	1.41	0.46	0	0.46	1.41	1.99	1.65	0.72

12.

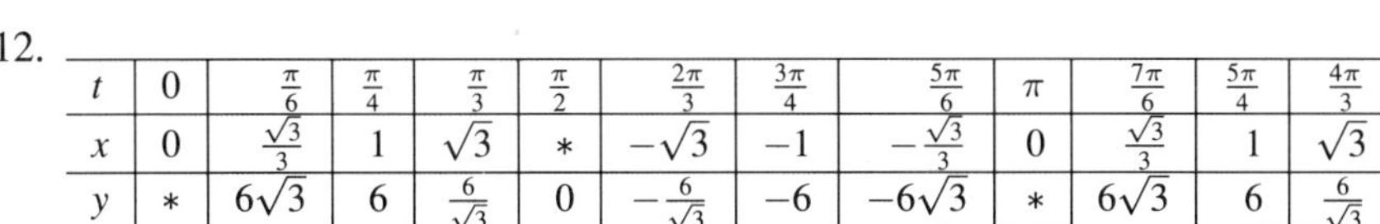

t	0	$\frac{\pi}{6}$	$\frac{\pi}{4}$	$\frac{\pi}{3}$	$\frac{\pi}{2}$	$\frac{2\pi}{3}$	$\frac{3\pi}{4}$	$\frac{5\pi}{6}$	π	$\frac{7\pi}{6}$	$\frac{5\pi}{4}$	$\frac{4\pi}{3}$
x	0	$\frac{\sqrt{3}}{3}$	1	$\sqrt{3}$	*	$-\sqrt{3}$	-1	$-\frac{\sqrt{3}}{3}$	0	$\frac{\sqrt{3}}{3}$	1	$\sqrt{3}$
y	*	$6\sqrt{3}$	6	$\frac{6}{\sqrt{3}}$	0	$-\frac{6}{\sqrt{3}}$	-6	$-6\sqrt{3}$	*	$6\sqrt{3}$	6	$\frac{6}{\sqrt{3}}$

* Not defined

13.

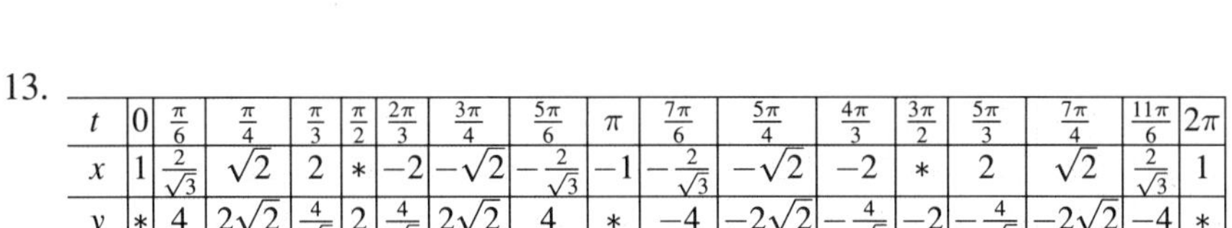

t	0	$\frac{\pi}{6}$	$\frac{\pi}{4}$	$\frac{\pi}{3}$	$\frac{\pi}{2}$	$\frac{2\pi}{3}$	$\frac{3\pi}{4}$	$\frac{5\pi}{6}$	π	$\frac{7\pi}{6}$	$\frac{5\pi}{4}$	$\frac{4\pi}{3}$	$\frac{3\pi}{2}$	$\frac{5\pi}{3}$	$\frac{7\pi}{4}$	$\frac{11\pi}{6}$	2π
x	1	$\frac{2}{\sqrt{3}}$	$\sqrt{2}$	2	*	-2	$-\sqrt{2}$	$-\frac{2}{\sqrt{3}}$	-1	$-\frac{2}{\sqrt{3}}$	$-\sqrt{2}$	-2	*	2	$\sqrt{2}$	$\frac{2}{\sqrt{3}}$	1
y	*	4	$2\sqrt{2}$	$\frac{4}{\sqrt{3}}$	2	$\frac{4}{\sqrt{3}}$	$2\sqrt{2}$	4	*	-4	$-2\sqrt{2}$	$-\frac{4}{\sqrt{3}}$	-2	$-\frac{4}{\sqrt{3}}$	$-2\sqrt{2}$	-4	*

* Not defined

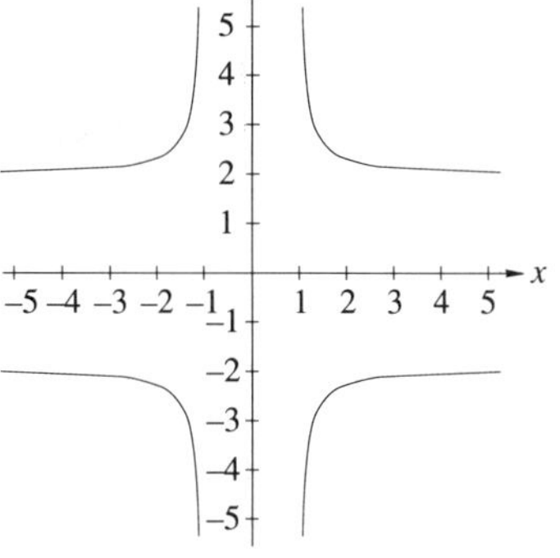

14.

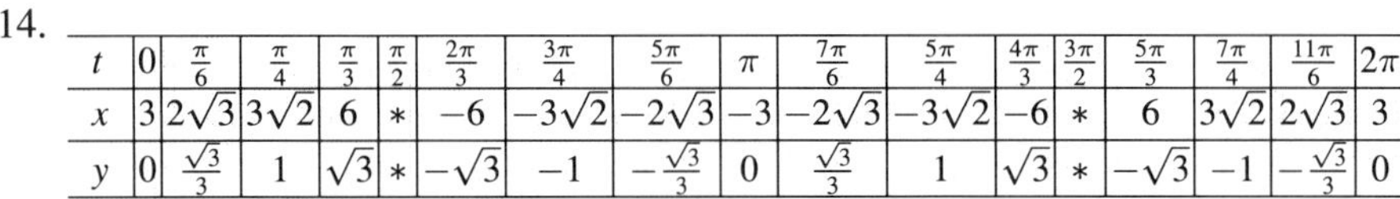

t	0	$\frac{\pi}{6}$	$\frac{\pi}{4}$	$\frac{\pi}{3}$	$\frac{\pi}{2}$	$\frac{2\pi}{3}$	$\frac{3\pi}{4}$	$\frac{5\pi}{6}$	π	$\frac{7\pi}{6}$	$\frac{5\pi}{4}$	$\frac{4\pi}{3}$	$\frac{3\pi}{2}$	$\frac{5\pi}{3}$	$\frac{7\pi}{4}$	$\frac{11\pi}{6}$	2π
x	3	$2\sqrt{3}$	$3\sqrt{2}$	6	*	-6	$-3\sqrt{2}$	$-2\sqrt{3}$	-3	$-2\sqrt{3}$	$-3\sqrt{2}$	-6	*	6	$3\sqrt{2}$	$2\sqrt{3}$	3
y	0	$\frac{\sqrt{3}}{3}$	1	$\sqrt{3}$	*	$-\sqrt{3}$	-1	$-\frac{\sqrt{3}}{3}$	0	$\frac{\sqrt{3}}{3}$	1	$\sqrt{3}$	*	$-\sqrt{3}$	-1	$-\frac{\sqrt{3}}{3}$	0

* Not defined

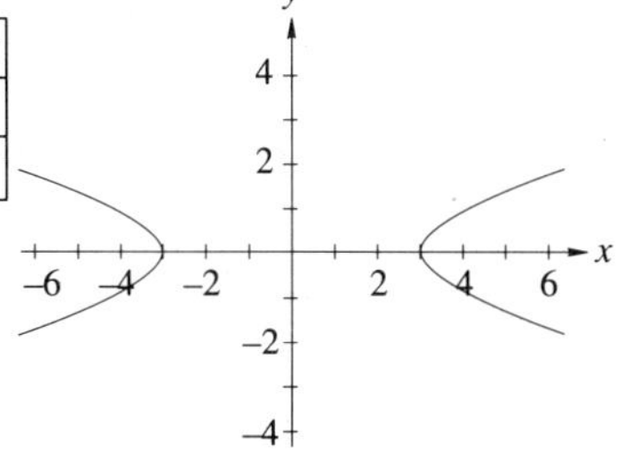

15. (a) 1 : 1;

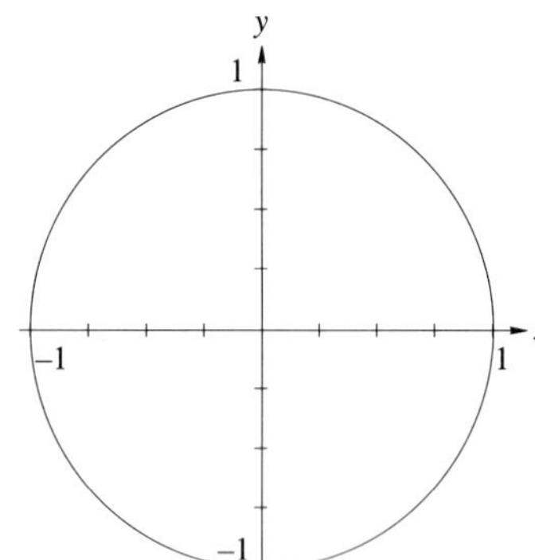

16. (a) 1 : 2;

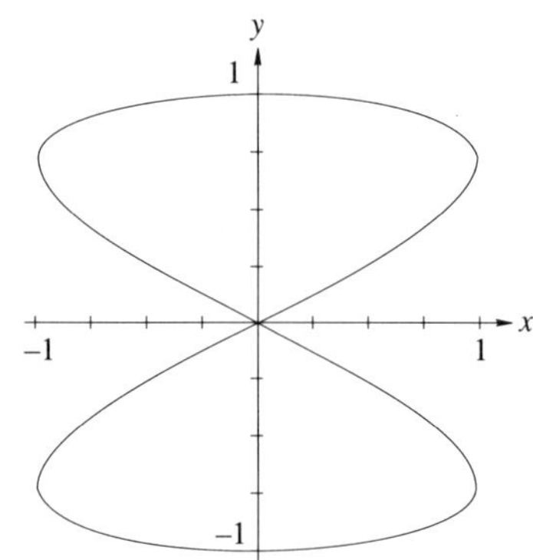

17. (a) 2 : 1;

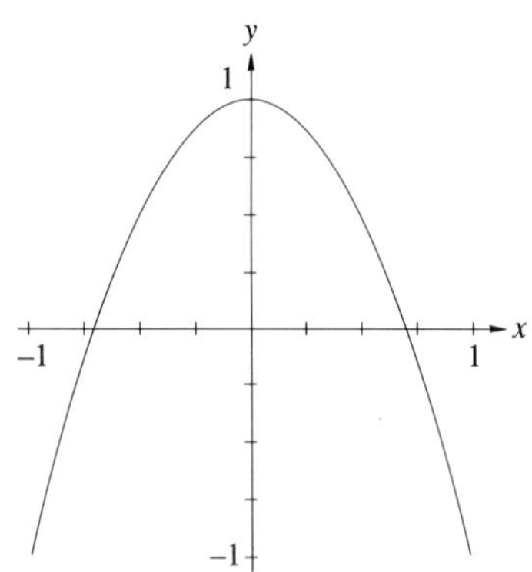

18. (a) 1 : 3;

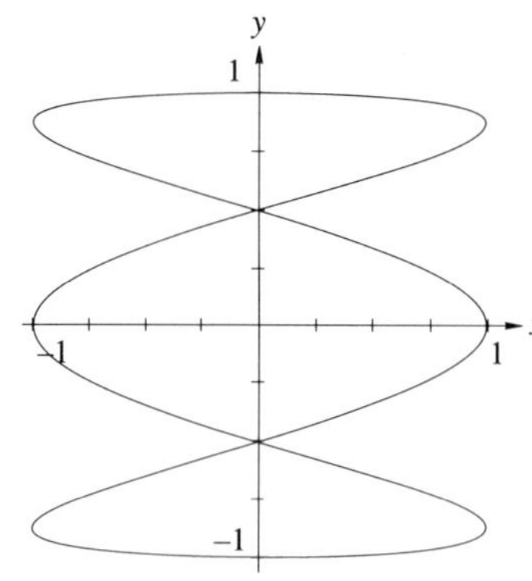

19. (a) 1 : 4;

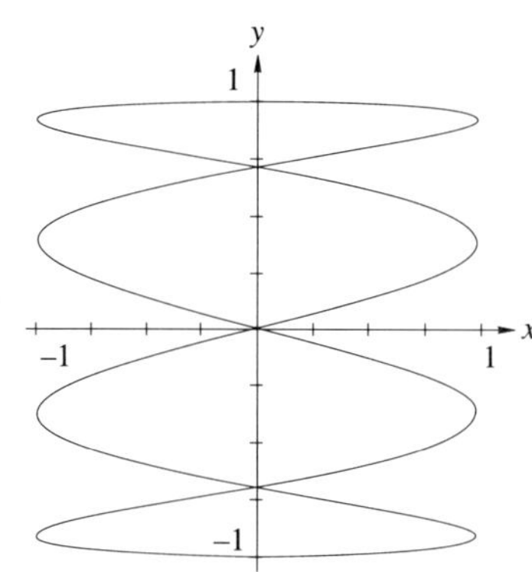

20. (a) 1 : 2;

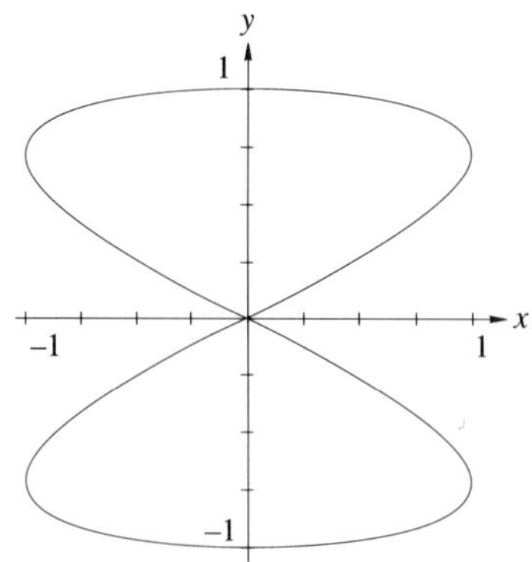

21. (a) 2 : 3 or 1 : 1.5;

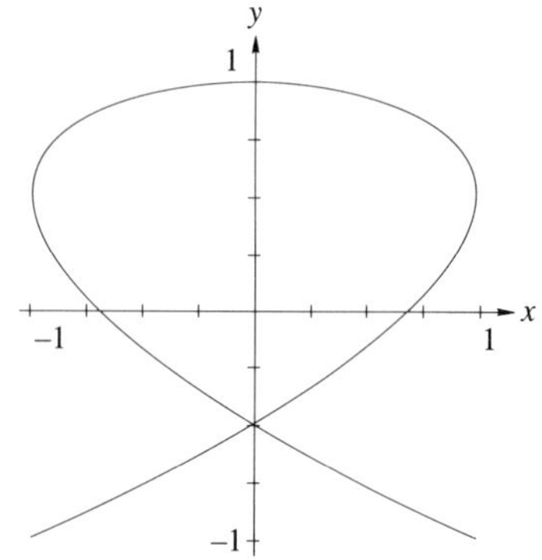

22. (a) 3 : 4;

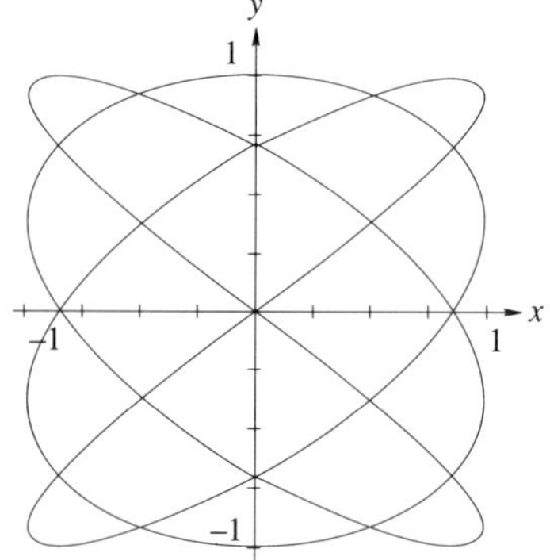

23. (a) 2 : 5 or 1 : 2.5;

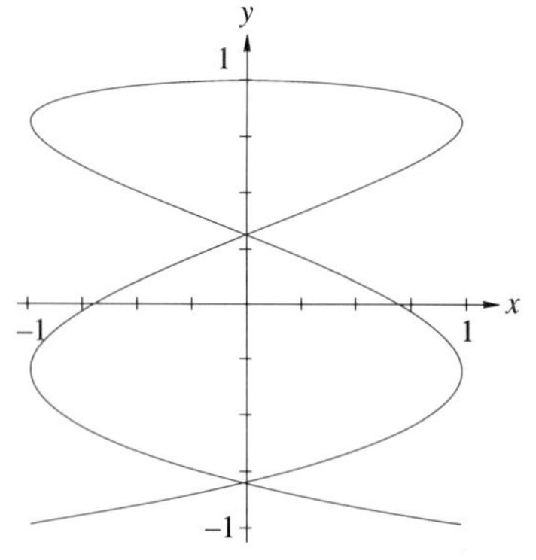

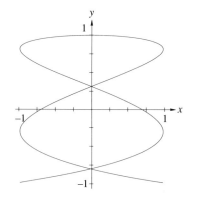

24. (a) 3 : 5;

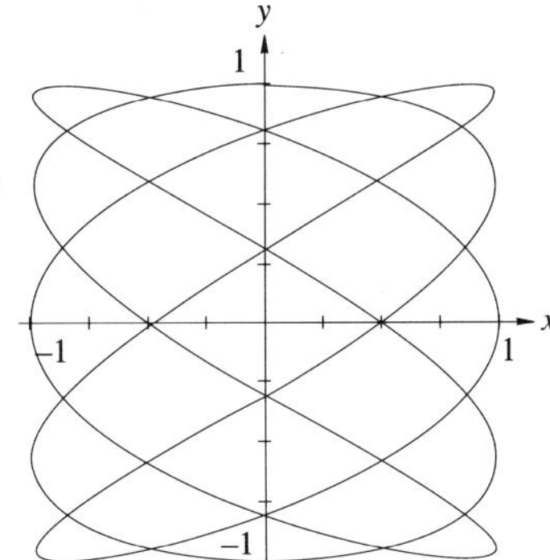

25.

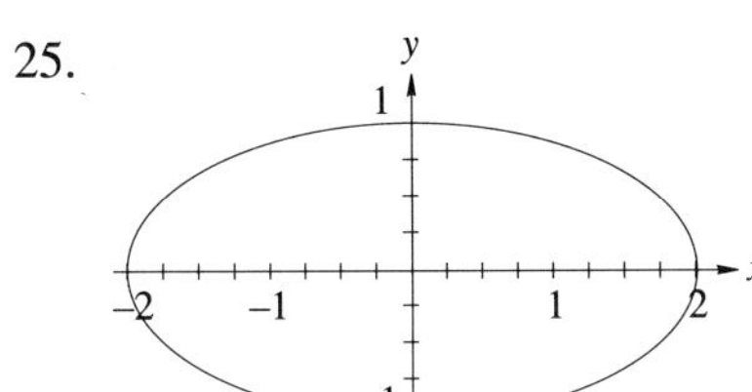

26.

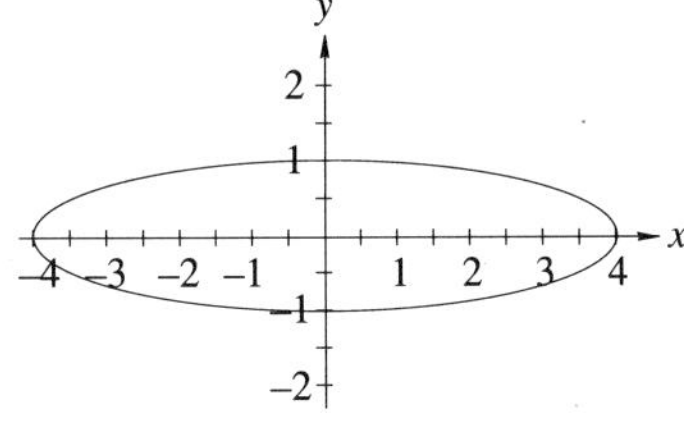

27.

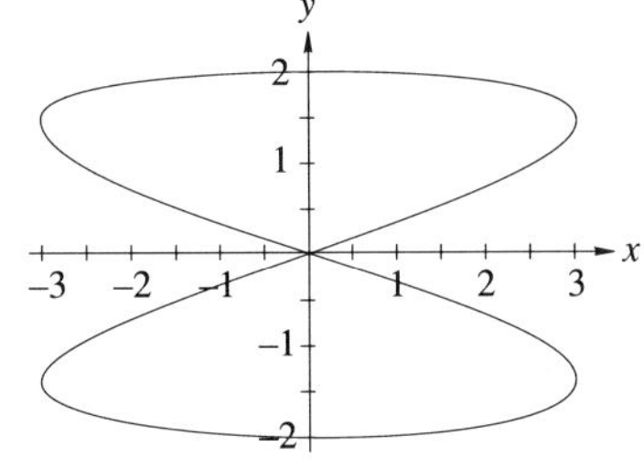

28.

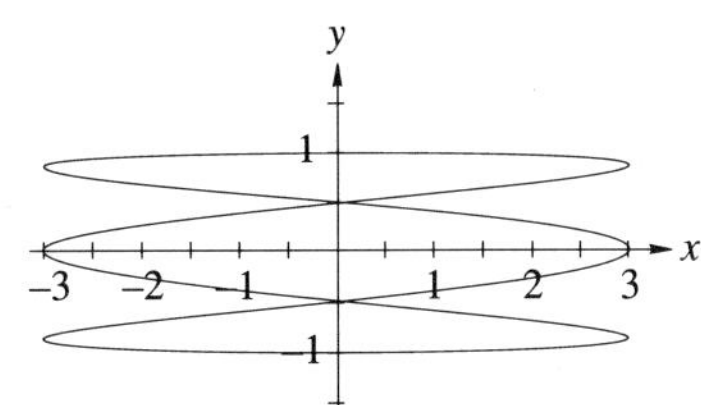

29.

30.

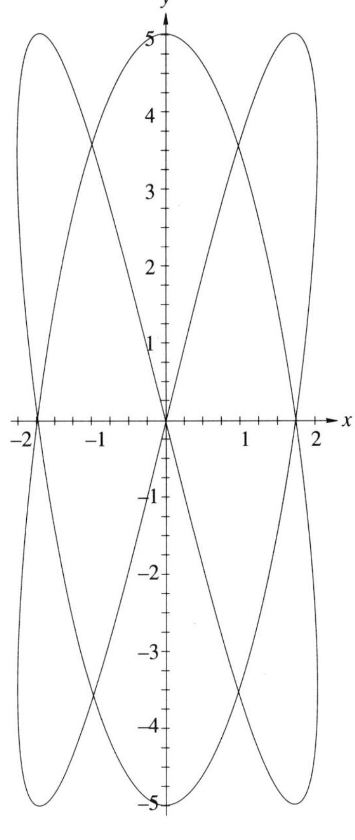

11.7 POLAR COORDINATES

1.

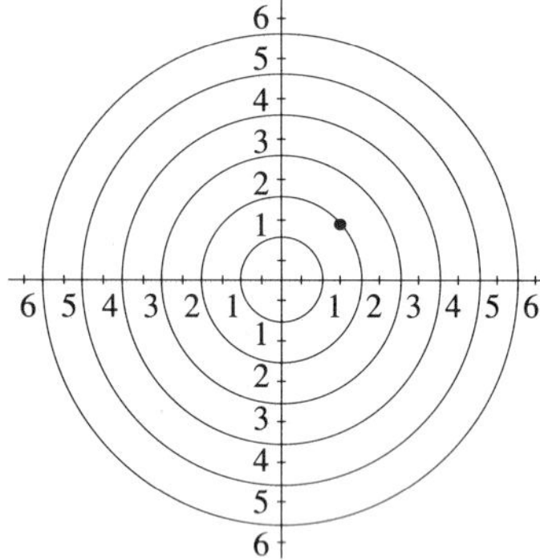

4.

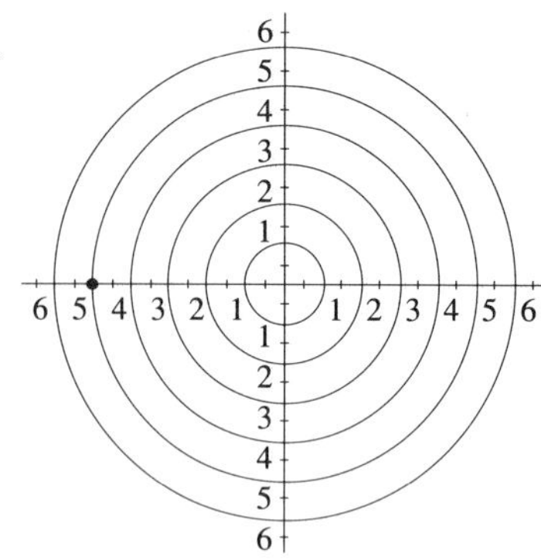

7.

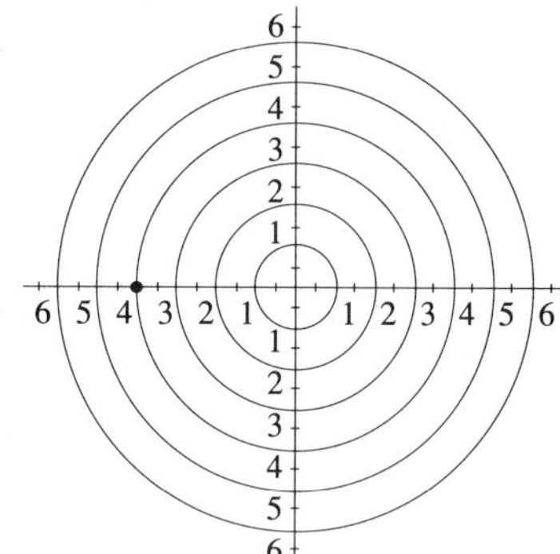

2.

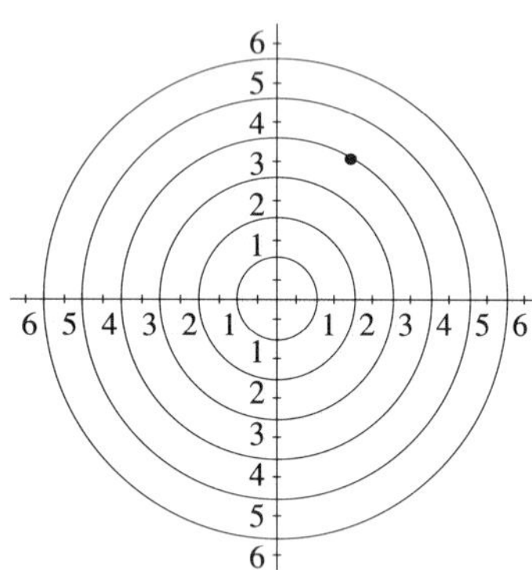

5.

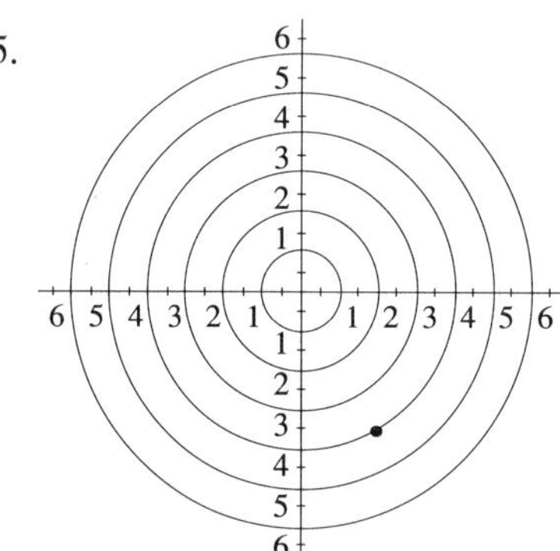

8.

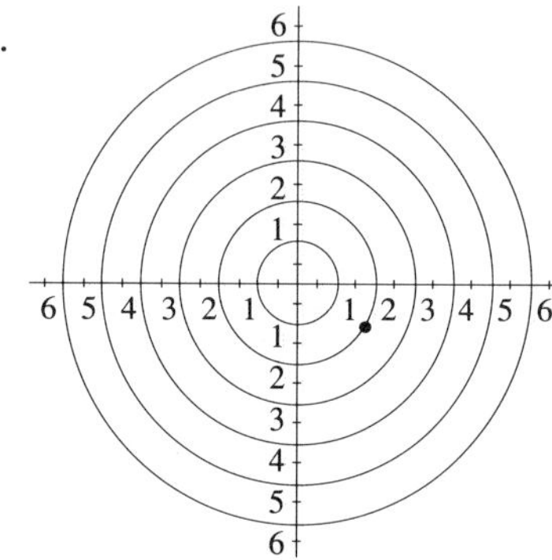

3.

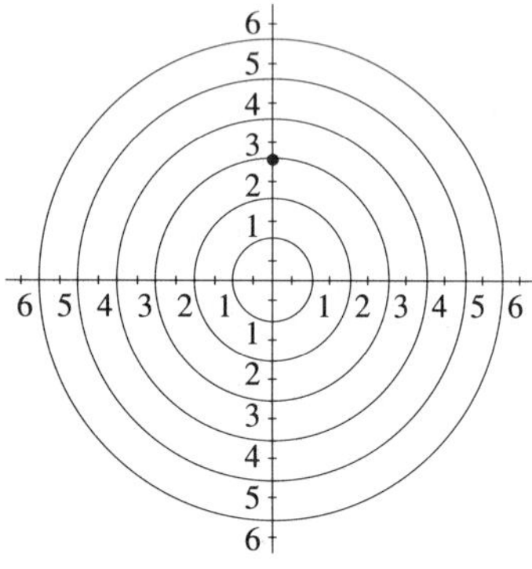

6.

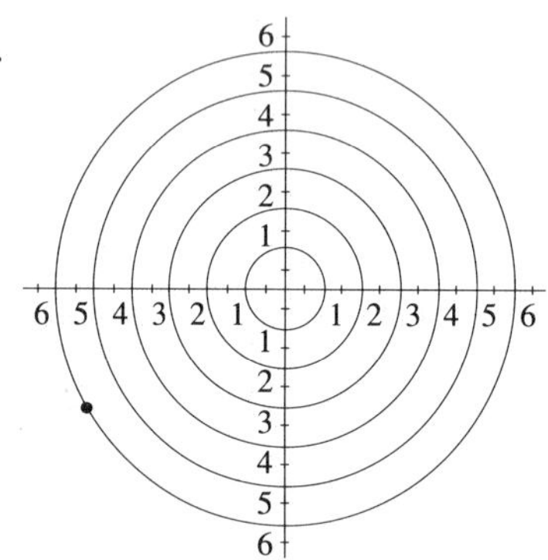

9.

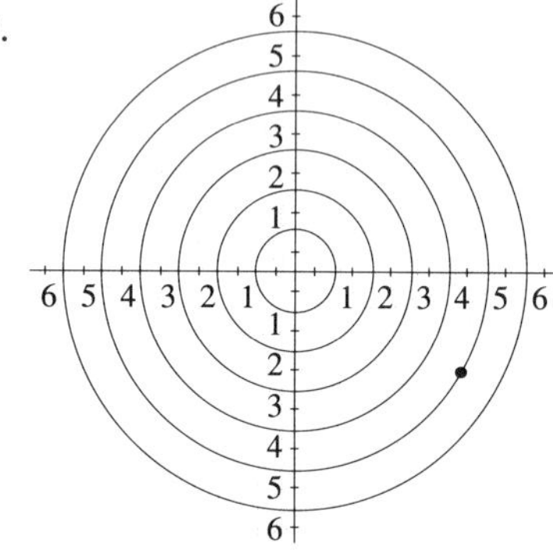

10. 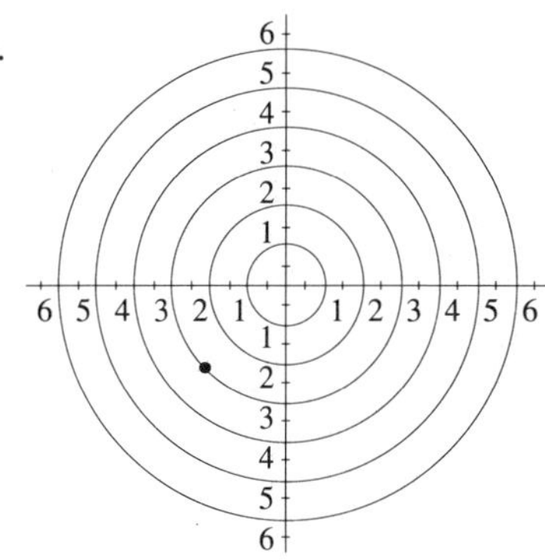

11. 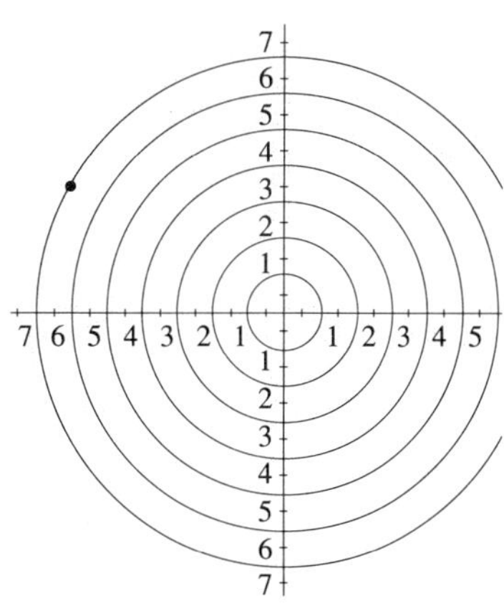

12. 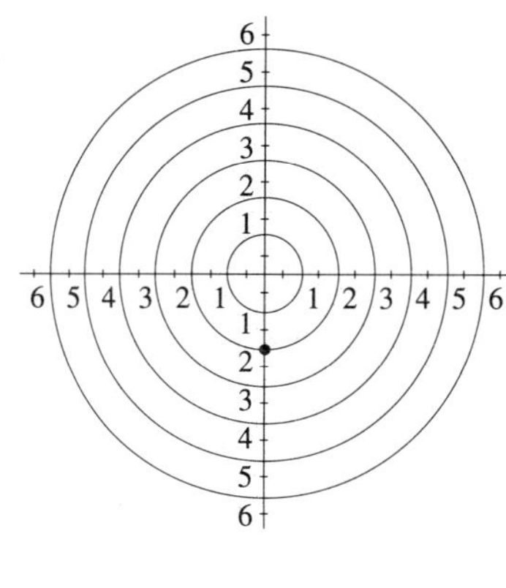

13. Using $x = r\cos\theta$, we find $x = 4\cos\frac{\pi}{3} = 4 \cdot \frac{1}{2} = 2$. Using $y = r\sin\theta$ yields $y = 4\sin\frac{\pi}{3} = 4 \cdot \frac{\sqrt{3}}{2} = 2\sqrt{3}$. Thus, the polar coordinates $(4, \frac{\pi}{3})$ have the rectangular coordinates $(2, 2\sqrt{3})$ or $(2, 3.46)$.

14. Using $x = r\cos\theta$, we find $x = 5\cos 75° = 1.294$, and using $y = r\sin\theta$ yields $y = 5\sin 75° = 4.830$. Thus, the polar coordinates $(5, 75°)$ have the rectangular coordinates $(1.29, 4.83)$.

15. Here $x = r\cos\theta$ yields $x = 2\cos 135° = 2 \cdot \frac{-\sqrt{2}}{2} = -\sqrt{2}$, and $y = r\sin\theta$ produces $y = 2\sin 135° = 2 \cdot \frac{\sqrt{2}}{2} = \sqrt{2}$. Thus, the polar coordinates $(2, 135°)$ have the rectangular coordinates $(-\sqrt{2}, \sqrt{2})$ or $(-1.414, 1.414)$.

16. Here $x = r\cos\theta$ yields $x = 3\cos\frac{3\pi}{2} = 3 \cdot 0 = 0$, and $y = r\sin\theta$ gives $y = 3\sin\frac{3\pi}{2} = 3 \cdot -1 = -3$. Thus, the polar coordinates $(3, \frac{3\pi}{2})$ have the rectangular coordinates $(0, -3)$.

17. Here $x = -6\cos 20° = -5.64$ and $y = -6\sin 20° = -2.05$ with the result $(-5.64, -2.05)$.

18. Here $x = -3\cos\frac{5\pi}{3} = -3 \cdot \frac{1}{2} = -\frac{3}{2}$ or -1.5 and $y = -3\sin\frac{5\pi}{3} = -3 \cdot \frac{-\sqrt{3}}{2} = \frac{3\sqrt{3}}{2}$ or 2.60 and we get $(-1.5, 2.60)$.

19. Here $x = -2\cos 4.3 = 0.80$ and $y = -2\sin 4.3 = 1.83$ with the result $(0.80, 1.83)$.

20. Here $x = -5\cos 255° = 1.29$ and $y = -5\sin 255° = 4.83$, hence $(1.29, 4.83)$.

21. $x = 3\cos -170° = -2.95$, $y = 3\sin -170° = -0.52$; $(-2.95, -0.52)$

22. $x = 4\cos\frac{-\pi}{8} = 3.70$, $y = 4\sin\frac{-\pi}{8} = -1.53$; $(3.70, -1.53)$

23. $x = -6\cos(-2.5) = 4.81$, $y = -6\sin(-2.5) = 3.59$; $(4.81, 3.59)$

24. $x = -3\cos -195° = 2.90$, $y = -3\sin -195° = -0.78$; $(2.90, -0.78)$

25. Here $r = \sqrt{x^2 + y^2} = \sqrt{4^2 + 4^2} = 4\sqrt{2} \approx 5.66$ and $\theta = \tan^{-1}\frac{4}{4} = \tan^{-1} 1 = 45°$ or $\frac{\pi}{4}$, so the rectangular coordinates $(4, 4)$ have a polar coordinate $(4\sqrt{2}, 45°)$ or $(4\sqrt{2}, \frac{\pi}{4})$ or $(5.66, 45°)$.

26. Here $r = \sqrt{x^2 + y^2} = \sqrt{3^2 + 6^2} = \sqrt{45} = 3\sqrt{5}$ or 6.71 and $\theta = \tan^{-1}\frac{6}{3} = \tan^{-1} 2 = 63.43°$ or 1.11, so the rectangular coordinates $(3, 6)$ have a polar coordinate $(6.71, 1.11)$ or $(6.71, 63.43°)$.

27. Here $r = \sqrt{4^2 + 3^2} = 5$ and $\theta = \tan^{-1}\frac{3}{4} = 36.87° = 0.64$ rad, so the rectangular coordinates $(4, 3)$ have a polar coordinate $(5, 36.87°)$ or $(5, 0.64)$.

28. Here $r = \sqrt{5^2 + 12^2} = 13$ and $\theta = \tan^{-1}\frac{12}{5} = 67.38°$ or 1.18 rad, so the rectangular coordinates $(5, 12)$ have a polar coordinate $(13, 67.38°)$ or $(13, 1.18)$.

29. Here $r = \sqrt{(-20)^2 + 21^2} = 29$, and $\theta_R = \tan^{-1}\frac{21}{20} = 46.40°$ or 0.81 rad. Since θ is in Quadrant II, $\theta = 180 - \theta_r = 133.60°$ or, $\theta = \pi - 0.81 = 2.33$, so the rectangular coordinates $(-20, 21)$ have a polar coordinate $(29, 133.60°)$ or $(29, 2.33)$.

30. Here $r = \sqrt{(-12)^2 + 5^2} = 13$ and $\theta_R = \tan^{-1}(\frac{5}{12}) = 22.62°$ or 0.39 rad. Since θ is in Quadrant II, we have $180 - 22.62 = 157.38°$ and so $\theta = \pi - 0.39 = 2.75$. Hence, the rectangular coordinates $(-12, 5)$ have a polar coordinate $(13, 157.38°)$ or $(13, 2.75)$.

31. $r = \sqrt{(-3)^2 + 4^2} = 5$, $\theta_R = \tan^{-1}\frac{4}{3} = 53.13° = 0.73$ rad, θ in Quadrant II so $180 - 53.13 = 126.87°$ or, $\theta = \pi - 0.93 = 2.21$; $(5, 126.87°)$ or $(5, 2.21)$

32. $r = \sqrt{9^2 + (-5)^2} = 10.30$, $\theta_R = \tan^{-1}\frac{5}{9} = 29.05° = 0.507$, θ in Quadrant IV so $360 - 29.05 = 330.95°$ or, $\theta = 2\pi - 0.507 = 5.776$; $(10.30, 330.95°)$ or $(10.30, 5.776)$

33. $r = \sqrt{(-7)^2 + (-10)^2} = 12.21$, $\theta_R = \tan^{-1}\frac{10}{7} = 55.01°$ or 0.96 rad. θ in Quadrant III so $180 + 55.01 = 235.01°$ or $\theta = \pi + .96 = 4.102$; $(12.21, 235.01°)$ or $(12.21, 4.102)$

34. $r = \sqrt{(-8)^2 + (-3)^2} = 8.54$, $\theta_R = \tan^{-1}\frac{3}{8} = 20.56°$ or 0.359. θ in Quadrant III so $180 + 20.56 = 200.56°$ or, $\theta = \pi + 0.359 = 3.501$; $(8.54, 200.56°)$ or $(8.54, 3.501)$

35. $r = \sqrt{2^2 + 9^2} = 9.22$, $\theta_R = \tan^{-1}\left(\frac{9}{2}\right) = 77.47°$ or 1.352; $(9.22, 77.47°)$ or $(9.22, 1.352)$

36. $r = \sqrt{(-6)^2 + 1^2} = \sqrt{37} = 6.08$, $\theta_R = \tan^{-1}\left(\frac{1}{6}\right) = 9.46°$ or 0.165 rad. θ is in Quadrant II so $\theta = 180 - 9.46 = 170.54°$ or $\theta = \pi - 0.165 = 2.977$; $(6.08, 170.54°)$ or $(6.08, 2.977)$

37.

39.

41.

38.

40.

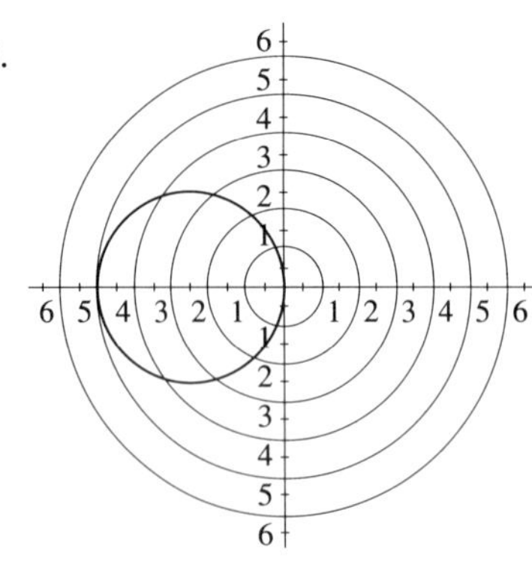

42.

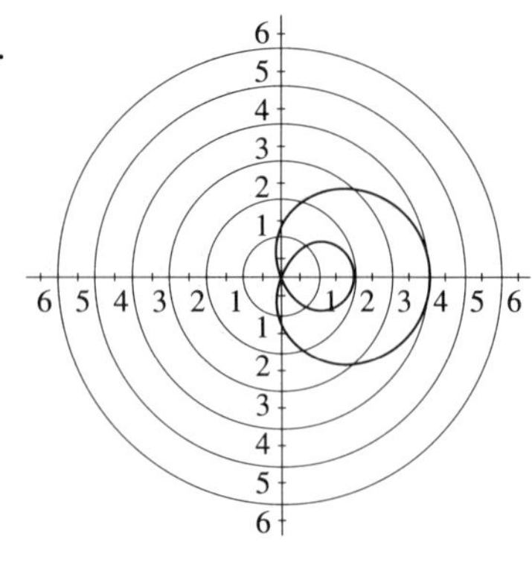

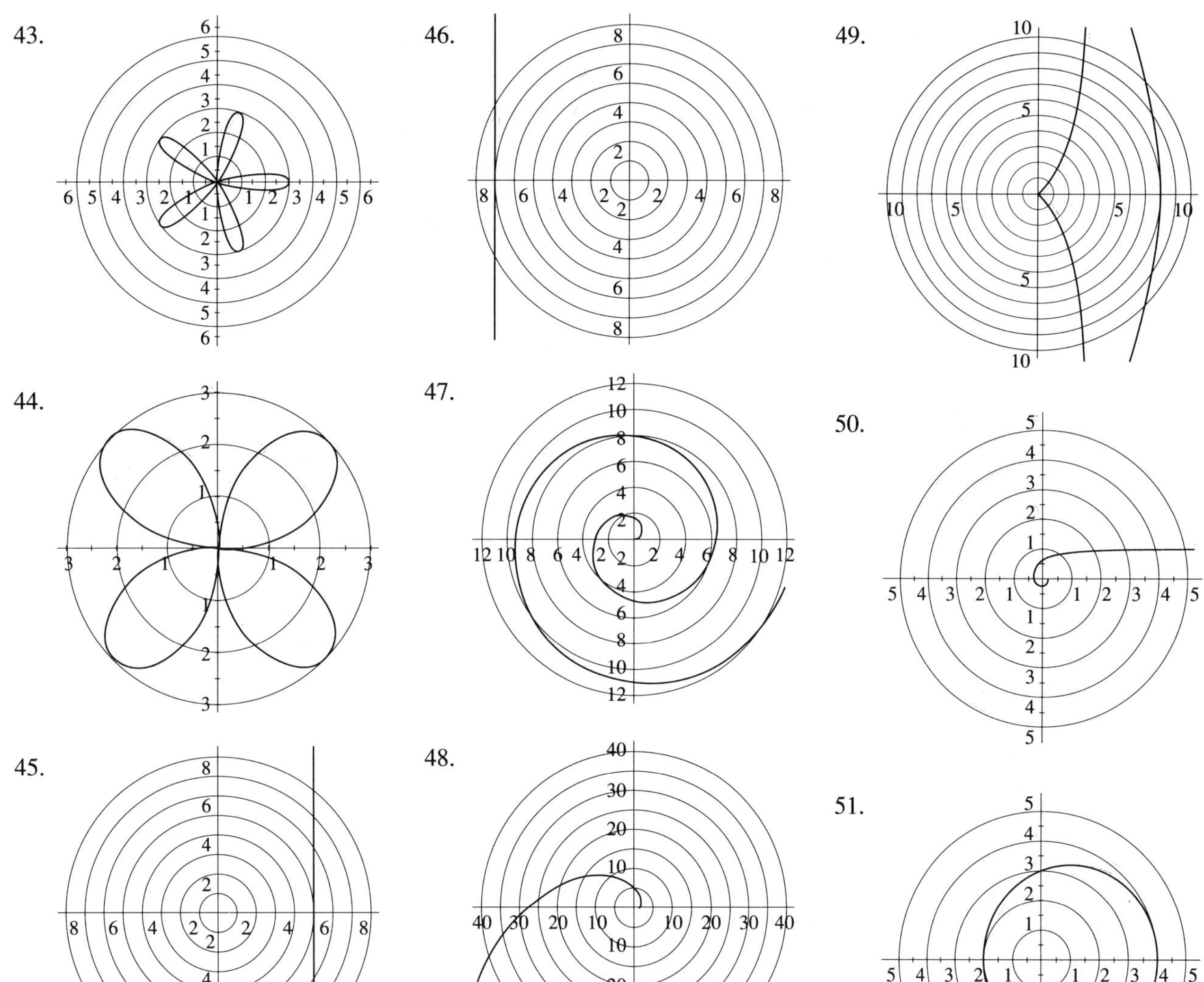
43.
46.
49.
44.
47.
50.
45.
48.
51.

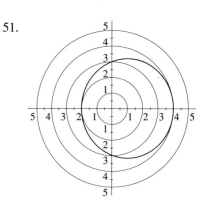

52.

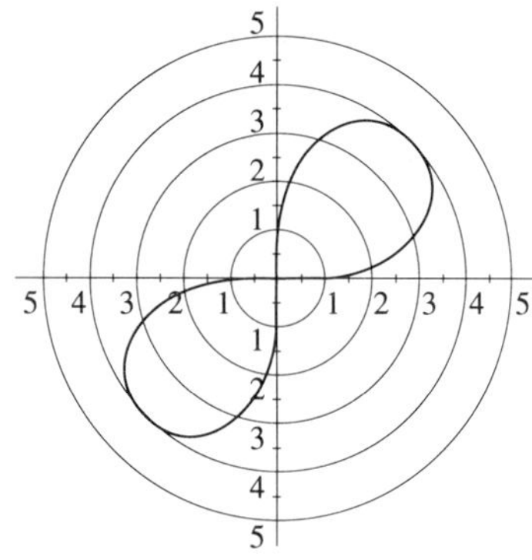

53.

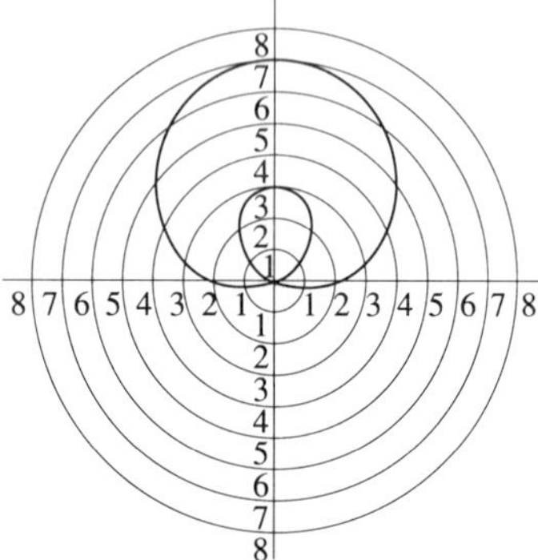

54.

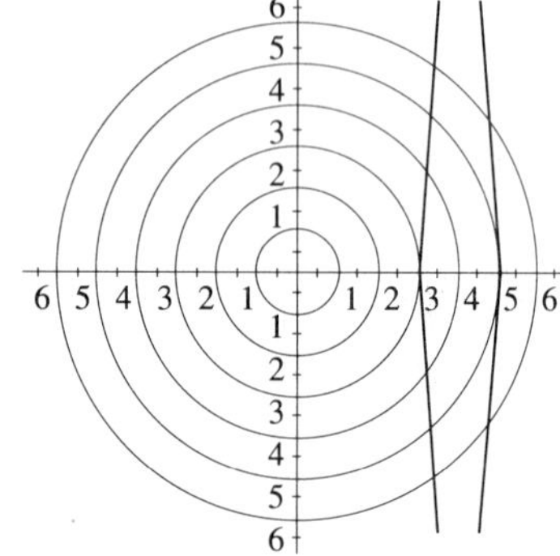

55.

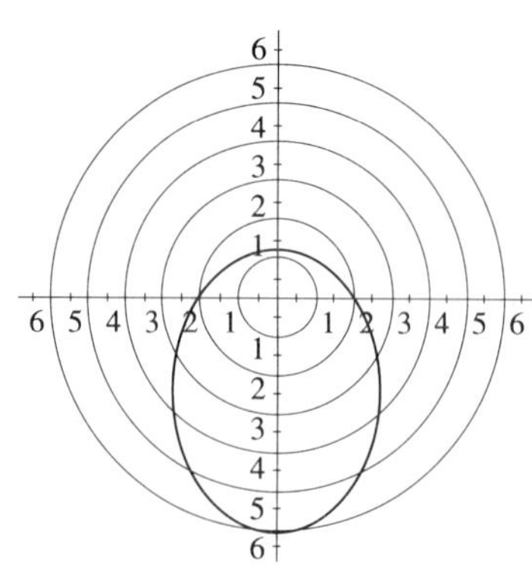

56.

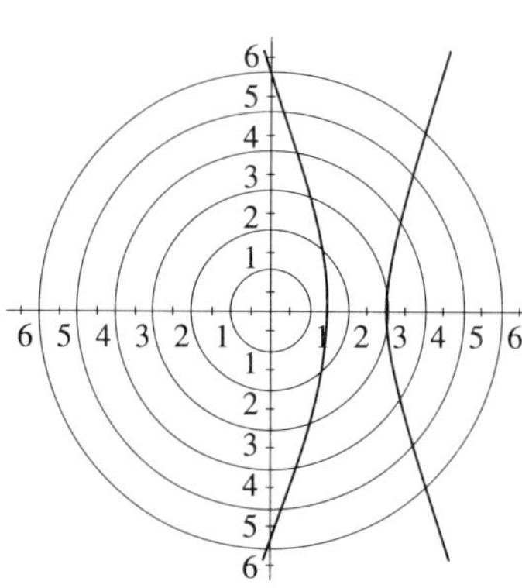

57.

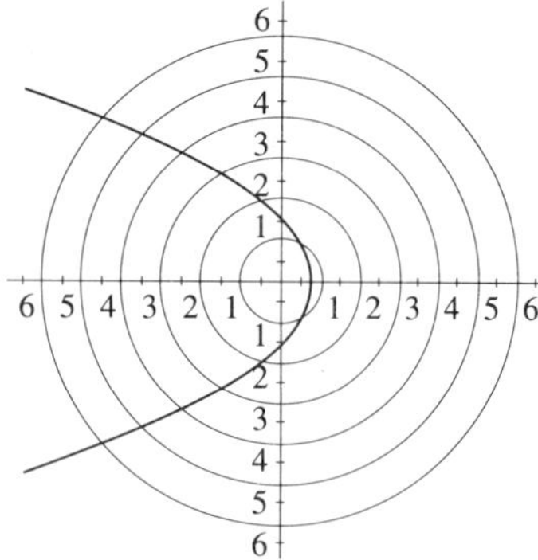

58.

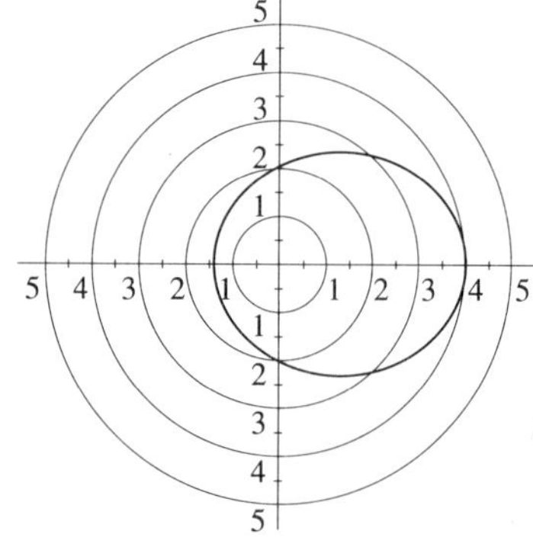

CHAPTER 11 REVIEW

1. Period = $\frac{2\pi}{4} = \frac{\pi}{2}$, Amplitude = $|8| = 8$, frequency $= \frac{2}{\pi}$, displacement or phase shift = 0.

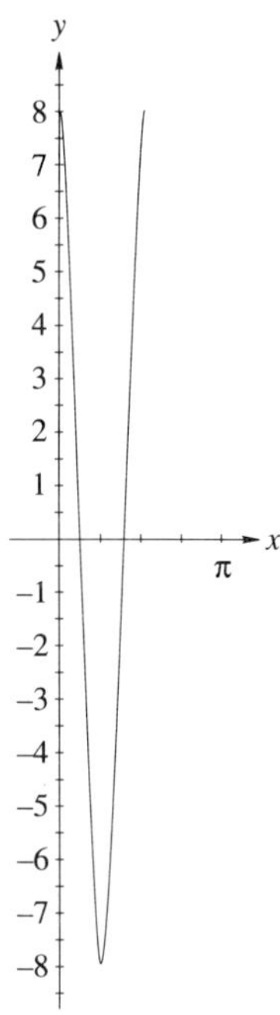

2. Period = $\frac{2\pi}{2} = \pi$, Amplitude = 3, frequency = $\frac{1}{\pi}$, displacement or phase shift = 0.

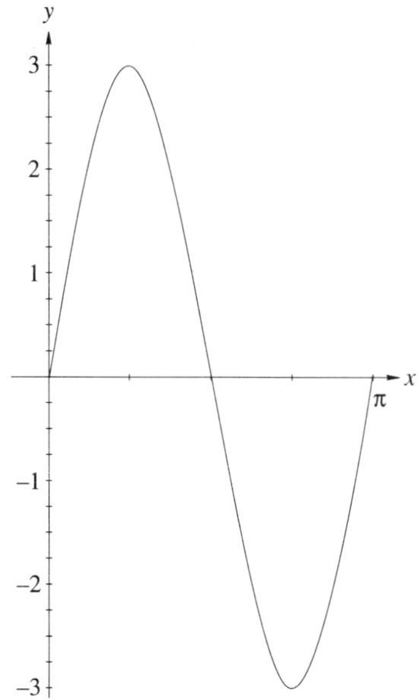

3. Period = $\frac{\pi}{3}$, Amplitude = ∞, frequency = $\frac{3}{\pi}$, displacement or phase shift = 0.

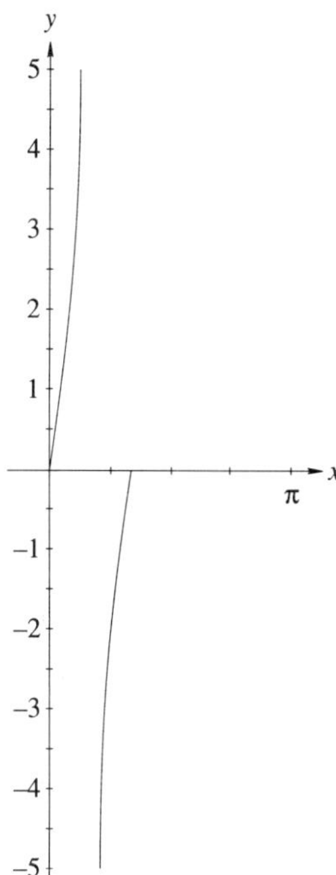

4. Period = $\frac{2\pi}{2} = \pi$, Amplitude = 3, frequency = π, displacement = $-\frac{\frac{\pi}{2}}{2} = -\frac{\pi}{4}$

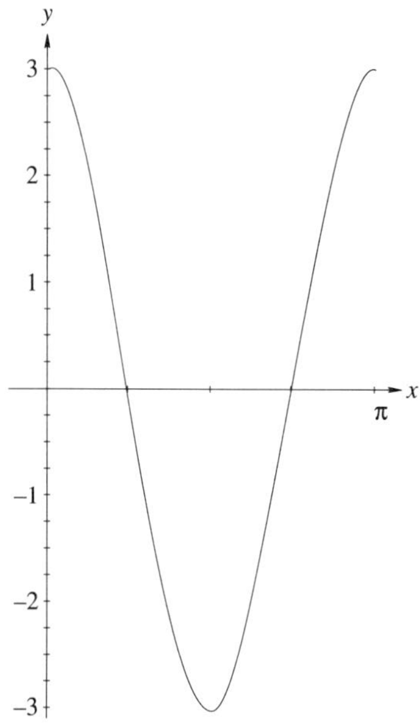

5. Period = $\frac{2\pi}{3}$, Amplitude = $\frac{1}{2}$, frequency = $\frac{3}{2\pi}$, displacement = $\frac{\frac{\pi}{3}}{3} = \frac{\pi}{9}$

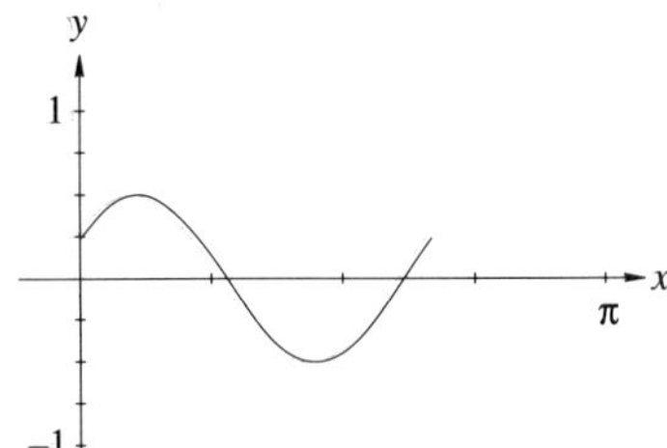

6. Period = $\frac{2\pi}{2} = \pi$, Amplitude = ∞, frequency = $\frac{1}{\pi}$, displacement = $\frac{\frac{\pi}{6}}{2} = \frac{\pi}{12}$

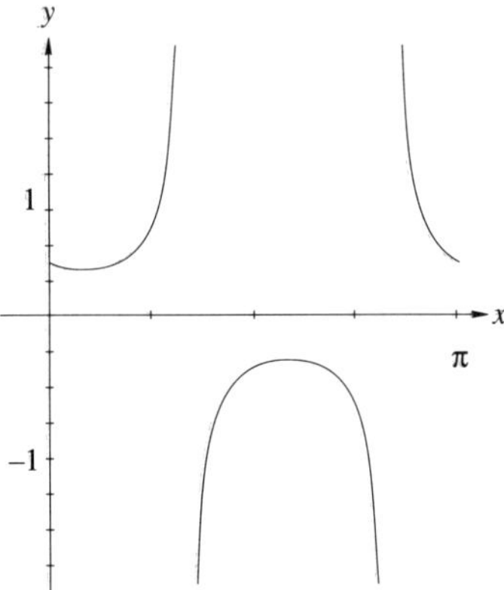

7. Period = π, Amplitude = ∞, frequency = $\frac{1}{\pi}$, displacement = $-\frac{\pi}{4}$

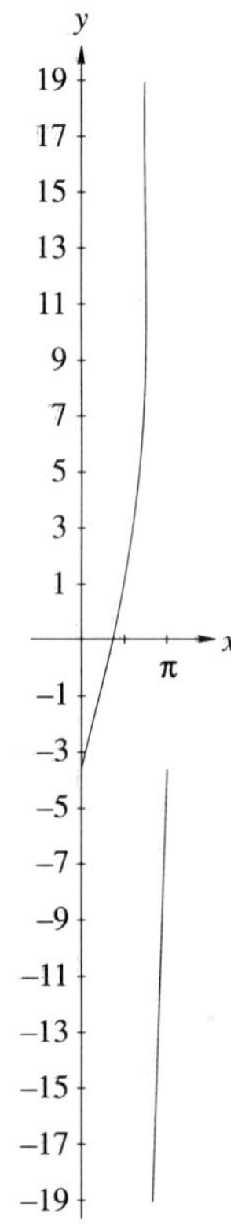

8. Period = $\frac{2\pi}{\frac{1}{3}} = 6\pi$, Amplitude = ∞, frequency = $\frac{1}{6\pi}$, displacement = $\frac{\frac{\pi}{5}}{\frac{1}{3}} = \frac{3\pi}{5}$

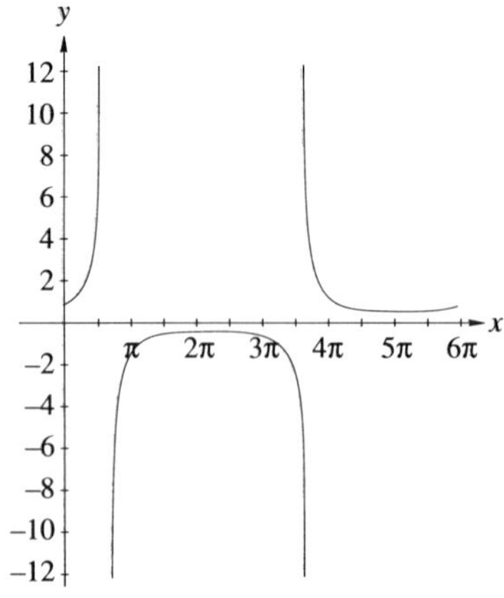

9. Period = $\frac{2\pi}{\frac{2}{3}} = 3\pi$, Amplitude = 2, frequency = $\frac{1}{3\pi}$, displacement = $-\frac{\frac{\pi}{6}}{\frac{2}{3}} = -\frac{\pi}{4}$

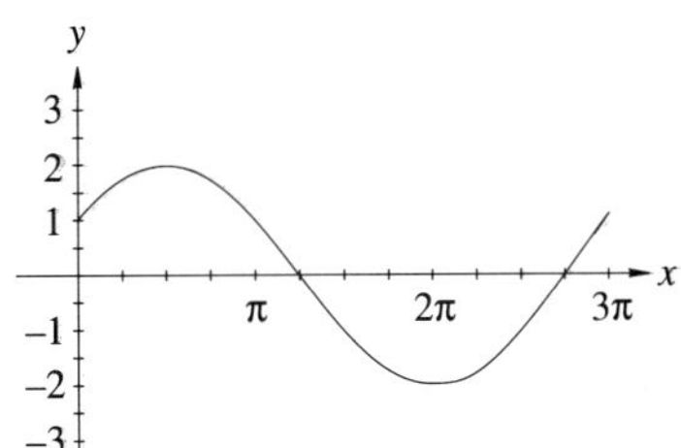

10. Period = $\frac{2\pi}{\frac{1}{2}} = 4\pi$, Amplitude = $\frac{1}{4}$, frequency = $\frac{1}{4\pi}$, displacement = $\frac{\frac{2\pi}{3}}{\frac{1}{2}} = \frac{4\pi}{3}$

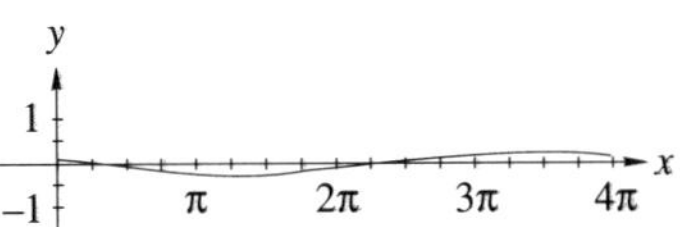

11.

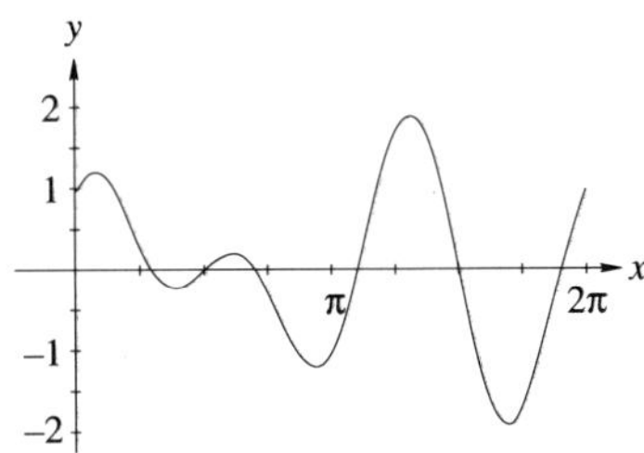

12.

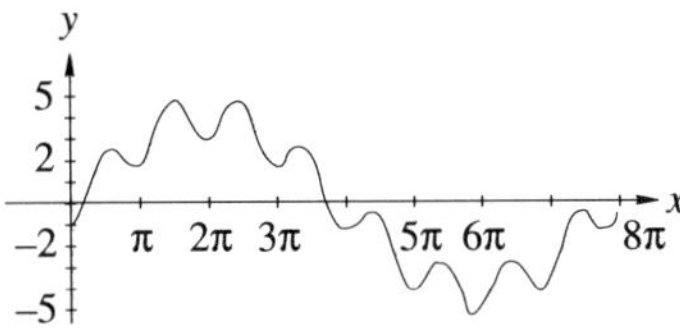

13.

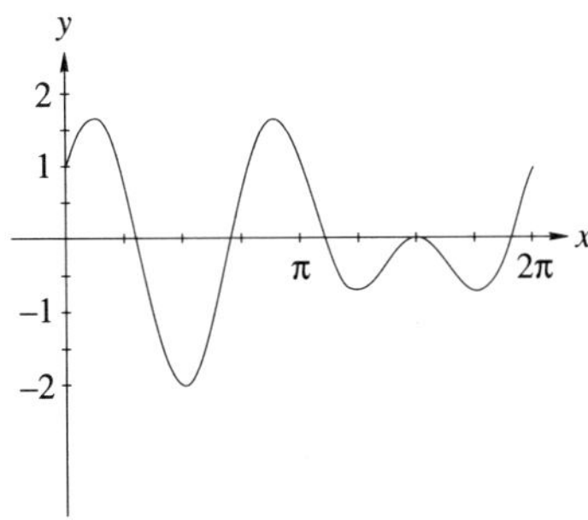

14.

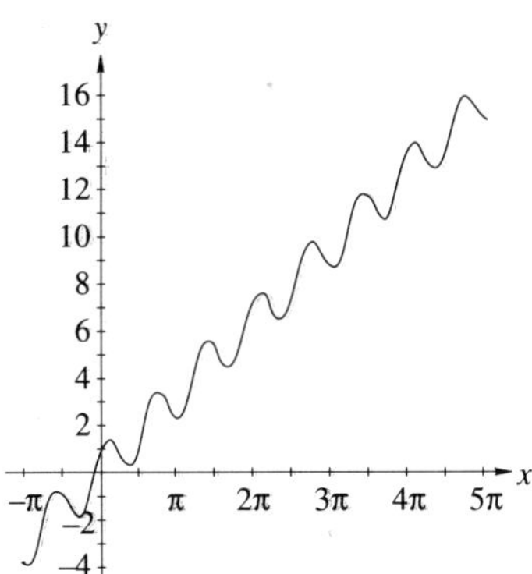

15.

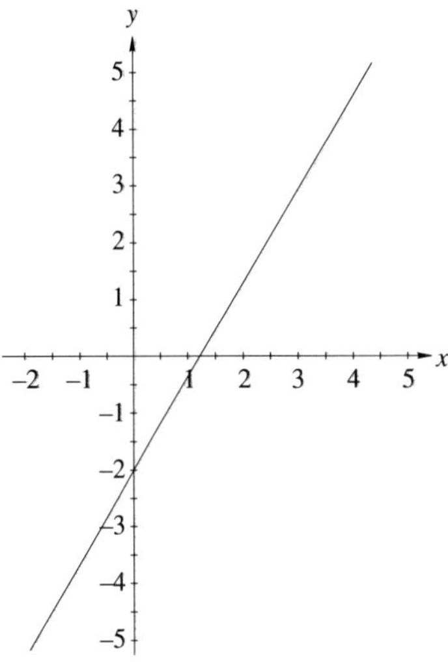

16.

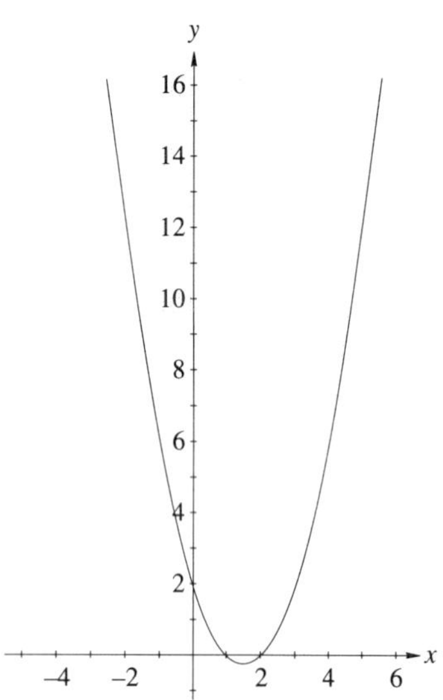

17.

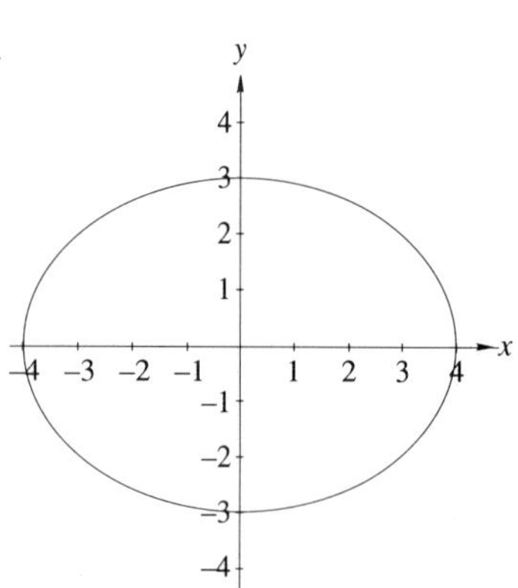

18.

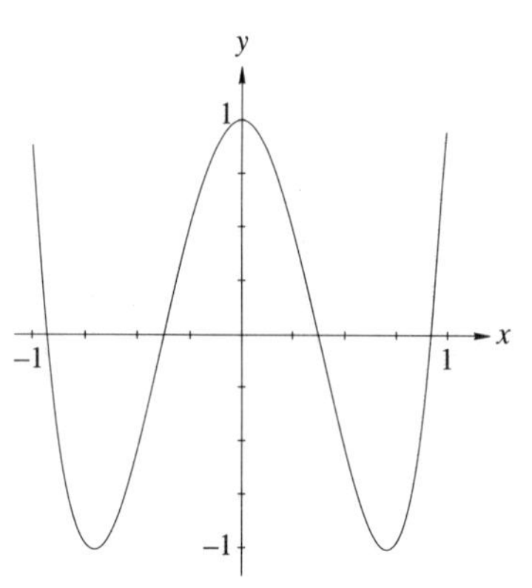

19.

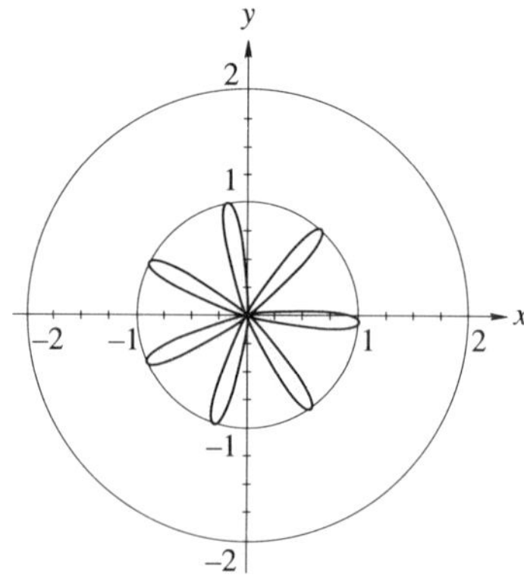

20.

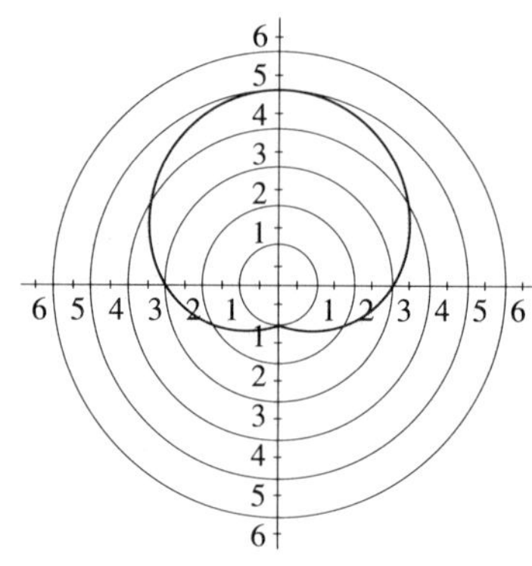

21. $y = A \sin 2\pi ft = 85 \sin 2\pi 5t$, $y = 85 \sin 10\pi t$

22. (a) $A = 1.7$ m, (b) Period $= \frac{2\pi}{3.4} = 1.85$ s, (c) $f = \frac{3.4}{2\pi} = 0.54$ Hz

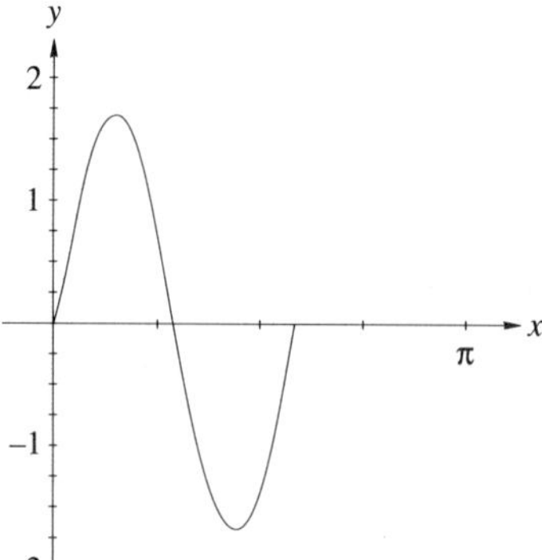

23. $V = V_{max} \sin 2\pi ft$, $I = I_{max} \sin(2\pi ft + \phi) = 4.8 \sin(2\pi 60t + \frac{\pi}{2})$, $I = 4.8 \sin(120\pi t + \frac{\pi}{2})$

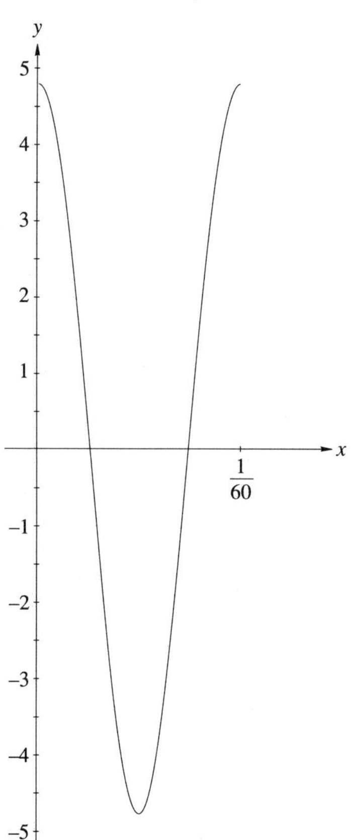

24.

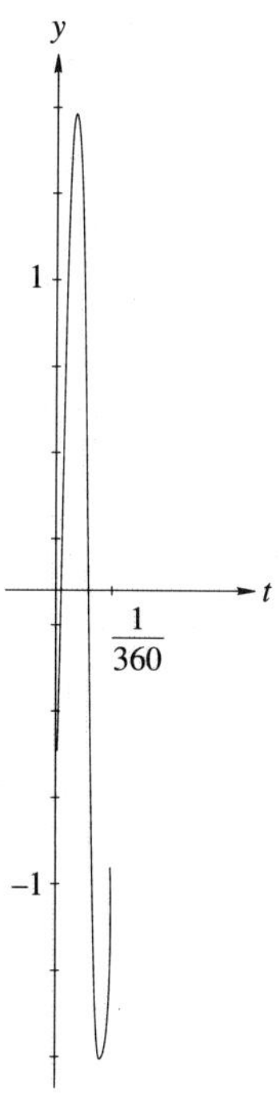

CHAPTER 11 TEST

1. (a) Period = $\frac{2\pi}{5}$, amplitude = $|-3|$ = 3, frequency = $\frac{5}{2\pi}$ displacement = 0

   (b) Period = $\frac{2\pi}{3}$, Amplitude = 2.4, frequency = $\frac{3}{2\pi}$, displacement = $\frac{\pi}{12}$

   (c) Period = $\frac{\pi}{2}$, Amplitude = ∞, frequency = $\frac{2}{\pi}$, displacement = $-\frac{\pi}{10}$

2. 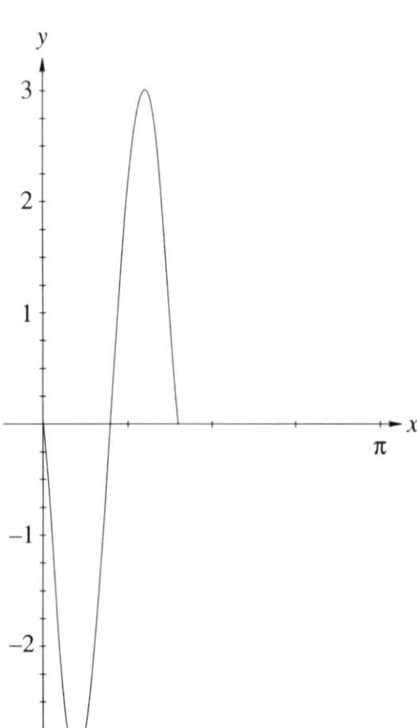

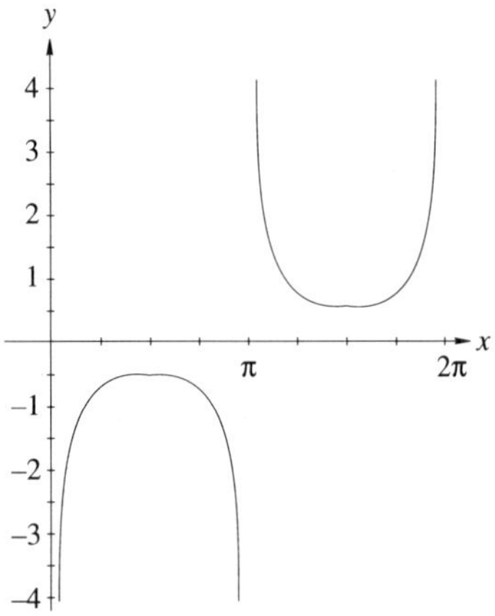

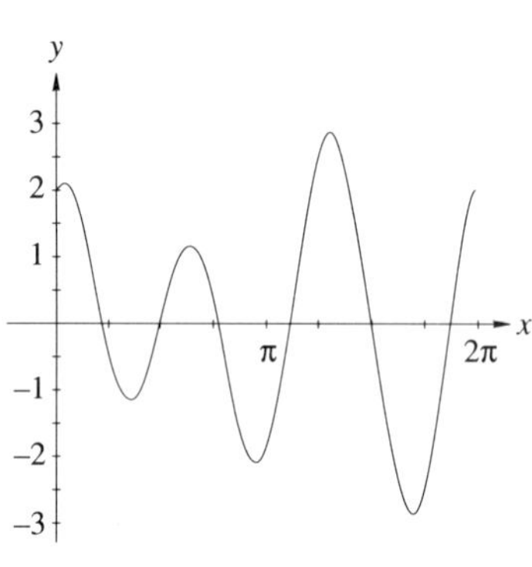

3. 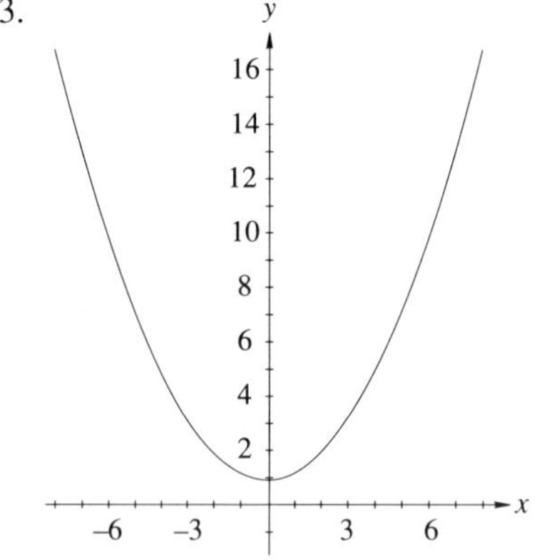

4. 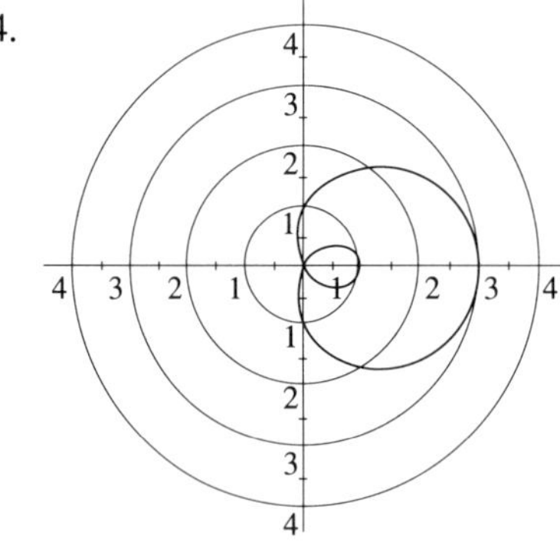

5. Using $x = r\cos\theta$, we find $x = 2\cos 55° = 1.147$, and using $y = r\sin\theta$ yields $y = 2\sin 55° = 1.638$. Thus, the polar coordinates $(2, 55°)$ have the rectangular coordinates $(1.147, 1.638)$.

6. Here $r = \sqrt{5^2 + (-12)^2} = 13$ and $\theta_R = \tan^{-1}\frac{+12}{5} = 67.38°$ or 1.176 rad. Since θ is in Quadrant IV, we see that $\theta = 360 - 67.38 = 292.52°$, or $\theta = 2\pi - 1.175 = 5.107$. Thus, the rectangular coordinates $(5, -12)$ have polar coordinates $(13, 292.62°)$ or $(13, 5.107)$.

7. $y = I_{\max}\sin(2\pi f t) = 5.7\sin(2\pi \cdot 60 \cdot t)$, $y = 5.7\sin 120\pi t$

CHAPTER

12

Exponents and Radicals

12.1 FRACTIONAL EXPONENTS

1. $25^{1/2} = \sqrt{25} = 5$
2. $27^{1/3} = \sqrt[3]{27} = 3$
3. $64^{1/3} = \sqrt[3]{64} = 4$
4. $125^{1/3} = \sqrt[3]{125} = 5$
5. $25^{-1/2} = \frac{1}{25^{1/2}} = \frac{1}{\sqrt{25}} = \frac{1}{5}$
6. $81^{-1/4} = \frac{1}{81^{1/4}} = \frac{1}{\sqrt[4]{81}} = \frac{1}{3}$
7. $32^{-1/5} = \frac{1}{\sqrt[5]{32}} = \frac{1}{2}$
8. $64^{-1/3} = \frac{1}{\sqrt[3]{64}} = \frac{1}{4}$
9. $27^{2/3} = \sqrt[3]{27}^{\,2} = 3^2 = 9$
10. $81^{3/4} = \sqrt[4]{81}^{\,3} = 3^3 = 27$
11. $125^{2/3} = \sqrt[3]{125}^{\,2} = 5^2 = 25$
12. $32^{3/5} = \sqrt[5]{32}^{\,3} = 2^3 = 8$
13. $16^{-3/4} = \frac{1}{\sqrt[4]{16}^{\,3}} = \frac{1}{2^3} = \frac{1}{8}$
14. $(-8)^{2/3} = \sqrt[3]{-8}^{\,2} = (-2)^2 = 4$
15. $(-8)^{-1/3} = \frac{1}{\sqrt[3]{-8}} = -\frac{1}{2}$
16. $(-27)^{-2/3} = \frac{1}{\sqrt[3]{-27}^{\,2}} = \frac{1}{(-3)^2} = \frac{1}{9}$
17. $\left(\frac{1}{8}\right)^{1/3} = \frac{1}{\sqrt[3]{8}} = \frac{1}{2}$
18. $\left(\frac{1}{25}\right)^{3/2} = \sqrt{\frac{1}{25}}^{\,3} = \left(\frac{1}{5}\right)^3 = \frac{1}{125}$
19. $\left(\frac{1}{16}\right)^{-5/4} = \sqrt[4]{16}^{\,5} = 2^5 = 32$
20. $\left(\frac{-1}{27}\right)^{4/3} = \sqrt[3]{\left(\frac{-1}{27}\right)}^{\,4} = \left(-\frac{1}{3}\right)^4 = \frac{1}{81}$
21. $3^2 \cdot 3^5 = 3^{2+5} = 3^7$
22. $5^9 5^8 = 5^{9+8} = 5^{17}$
23. $7^6 7^{-2} = 7^{6-2} = 7^4$
24. $11^9 11^{-6} = 11^{9-6} = 11^3$
25. $x^4 x^6 = x^{4+6} = x^{10}$
26. $y^7 y^9 = y^{7+9} = y^{16}$
27. $y^6 y^{-4} = y^{6-4} = y^2$
28. $x^8 x^{-2} = x^{8-2} = x^6$
29. $(9^5)^2 = 9^{5\times 2} = 9^{10}$
30. $(11^8)^{-5} = 11^{8(-5)} = 11^{-40} = \frac{1}{11^{40}}$
31. $(x^7)^3 = x^{7\times 3} = x^{21}$
32. $(P^9)^{-5} = P^{9\times(-5)} = P^{-45} = \frac{1}{P^{45}}$
33. $(xy)^5 = x^5 y^5$
34. $(yt)^3 = y^3 t^3$
35. $(ab)^{-5} = \frac{1}{(ab)^5} = \frac{1}{a^5 b^5}$
36. $(xyz)^{-9} = \frac{1}{x^9 y^9 z^9}$
37. $\frac{x^{10}}{x^2} = x^{10-2} = x^8$
38. $\frac{P^9}{P^3} = P^{9-3} = P^6$
39. $\frac{x^2}{x^8} = \frac{1}{x^{8-2}} = \frac{1}{x^6}$
40. $\frac{a^3}{a^{12}} = \frac{1}{a^{12-3}} = \frac{1}{a^9}$
41. $x^{1/2} x^{3/2} = x^{1/2+3/2} = x^2$
42. $a^{1/3} a^{4/3} = a^{1/3+4/3} = a^{5/3}$
43. $r^{3/4} r = r^{3/4+1} = r^{7/4}$
44. $a^2 a^{2/3} = a^{2+2/3} = a^{8/3}$
45. $a^{1/2} a^{1/3} = a^{1/2+1/3} = a^{3/6+2/6} = a^{5/6}$
46. $b^{2/3} b^{1/4} = b^{2/3+1/4} = b^{8/12+3/12} = b^{11/12}$
47. $d^{2/3} d^{-1/4} = d^{2/3-1/4} = d^{8/12-3/12} = d^{5/12}$

48. $x^{3/5}y^{-2/3} = \frac{x^{3/5}}{y^{2/3}}$

49. $\frac{a^2b^5}{a^5b^2} = \frac{b^{5-2}}{a^{5-2}} = \frac{b^3}{a^3}$

50. $\frac{x^3y^2}{x^7y} = \frac{y^{2-1}}{x^{7-3}} = \frac{y}{x^4}$

51. $\frac{r^5s^2t}{tr^3s^5} = r^{5-3}s^{2-5}t^{1-1} =$
$r^2s^{-3}t^0 = \frac{r^2}{s^3}$

52. $\frac{a^2bc^3}{(abc)^3} = \frac{a^2bc^3}{a^3b^3c^3} = \frac{1}{ab^2}$

53. $\frac{(xy^2z)^4}{x^4(yz^2)^2} = \frac{x^4y^8z^4}{x^4y^2z^4} = y^6$

54. $\frac{m^6n^7}{(m^2n)^3} = \frac{m^6n^7}{m^6n^3} = n^4$

55. $\left(\frac{a}{b^2}\right)^3\left(\frac{a}{b^3}\right)^2 = \frac{a^3}{b^6}\cdot\frac{a^2}{b^6} = \frac{a^5}{b^{12}}$

56. $\left(\frac{x^2}{y}\right)^4\left(\frac{x}{y^2}\right)^2 = \frac{x^8}{y^4}\cdot\frac{x^2}{y^4} = \frac{x^{10}}{y^8}$

57. $\frac{(xy^2b^3)^{1/2}}{(x^{1/4}b^4y)^2} = \frac{x^{1/2}yb^{3/2}}{x^{1/2}b^8y^2} = \frac{1}{b^{8-3/2}y} =$
$\frac{1}{b^{13/2}y}$

58. $\frac{(x^{1/3}y^3)^3}{(y^{10}x^5)^{1/5}} = \frac{xy^9}{y^2x} = y^7$

59. $\left(\frac{2x}{p^2}\right)^{-2}\left(\frac{p}{4}\right)^{-1} = \frac{p^4}{4x^2}\cdot\frac{4}{p} =$
$\frac{p^3}{x^2}$

60. $\left(\frac{5a^2}{6b}\right)^{-2}\left(\frac{6}{a}\right)^{-4} = \frac{6^2b^2}{5^2a^4}\cdot\frac{a^4}{6^4} =$
$\frac{b^2}{5^26^2} = \frac{b^2}{25\cdot 36} = \frac{b^2}{900}$

61. $8.3^{2/3} \approx 4.0993852$

62. $7.3^{2/5} \approx 2.21477261$

63. $92.47^{5/7} \approx 25.368006$

64. $81.94^{3/5} \approx 14.06363465$

65. $(-81.52)^{2/7} =$
$((-81.42)^2)^{1/7} \approx 3.5162154$

66. $(-78.64)^{1/3} \approx$
-4.284312769

67. $(432.61)^{1/4} \approx 4.5606226$

68. $(-537.15)^{2/3} =$
$((-537.15)^2)^{1/3} \approx$
66.07903926

69. $9.8(4.75)^2 \approx 221.1125$ or
221 m

70. $\frac{4}{3}\pi r^3 = \frac{4}{3}\pi(19.25)^3 =$
$9511.1042\pi = 29880.01498$
or 29,880 in.^{3}

71. $P = 5^5\left(\frac{15}{0.5}\right)^{5/3} =$
905 146.3 N/m^2

$T = 315\left(\frac{15}{0.5}\right)^{5/3} =$
91 238.747 K

72. $q(2000) = 75\cdot 2^{-2000/1600} =$
31.53361557 or about 32 mg

12.2 LAWS OF RADICALS

1. $\sqrt[3]{16} = \sqrt[3]{2^4} = \sqrt[3]{2^3}\sqrt[3]{2^1} = 2\sqrt[3]{2}$

2. $\sqrt[3]{81} = \sqrt[3]{3^3\cdot 3^1} = 3\sqrt[3]{3}$

3. $\sqrt{45} = \sqrt{9}\sqrt{5} = 3\sqrt{5}$

4. $\sqrt[3]{40} = \sqrt[3]{8\cdot 5} = 2\sqrt[3]{5}$

5. $\sqrt[3]{y^{12}} = y^{12/3} = y^4$

6. $\sqrt[4]{p^8} = p^{8/4} = p^2$

7. $\sqrt[5]{a^7} = \sqrt[5]{a^5a^2} = a\sqrt[5]{a^2}$

8. $\sqrt[7]{b^{10}} = \sqrt[7]{b^7b^3} = b\sqrt[7]{b^3}$

9. $\sqrt{x^2y^7} = \sqrt{x^2}\sqrt{y^6}\sqrt{y} = xy^3\sqrt{y}$

10. $\sqrt[3]{x^5y^3} = \sqrt[3]{x^3}\sqrt[3]{x^2}\sqrt[3]{y^3} = xy\sqrt[3]{x^2}$

11. $\sqrt[4]{a^5b^3} = \sqrt[4]{a^4}\sqrt[4]{ab^3} = a\sqrt[4]{ab^3}$

12. $\sqrt[5]{p^{12}y^8} = \sqrt[5]{p^{10}}\sqrt[5]{y^5}\sqrt[5]{p^2y^3} = p^2y\sqrt[5]{p^2y^3}$

13. $\sqrt[3]{8x^4} = \sqrt[3]{2^3x^3x} = 2x\sqrt[3]{x}$

14. $\sqrt[4]{81y^9} = \sqrt[4]{3^4y^8y} = 3y^2\sqrt[4]{y}$

15. $\sqrt{27x^3y} = \sqrt{3^2x^23xy} = 3x\sqrt{3xy}$

16. $\sqrt[3]{32a^5b^2} = \sqrt[3]{2^3a^32^2a^2b^2} = 2a\sqrt[3]{4a^2b^2}$

17. $\sqrt[3]{-8} = \sqrt[3]{(-2)^3} = -2$

18. $\sqrt[5]{-243} = \sqrt[5]{(-3)^5} = -3$

19. $\sqrt[3]{a^2b^4}\sqrt[3]{ab^5} = \sqrt[3]{a^3b^9} = ab^3$

20. $\sqrt[5]{x^3y^2z^4}\sqrt[5]{x^2y^8z} = \sqrt[5]{x^5y^{10}z^5} = xy^2z$

21. $\sqrt[4]{p^3q^2r^6}\sqrt[4]{pq^6r} = \sqrt[4]{p^4q^8r^7} = pq^2r\sqrt[4]{r^3}$

22. $\sqrt[6]{m^3n^2e^7}\sqrt[6]{m^2n^4e^5} = \sqrt[6]{m^5n^6e^{12}} = ne^2\sqrt[6]{m^5}$

23. $\sqrt[3]{\frac{8x^3}{27}} = \frac{\sqrt[3]{2^3x^3}}{\sqrt[3]{3^3}} = \frac{2x}{3}$

24. $\sqrt[4]{\dfrac{81y^8}{16}} = \dfrac{\sqrt[4]{3^4y^8}}{\sqrt[4]{2^4}} = \dfrac{3y^2}{2}$

25. $\sqrt[5]{\dfrac{x^5y^{10}}{z^5}} = \dfrac{\sqrt[5]{x^5y^{10}}}{\sqrt[5]{z^5}} = \dfrac{xy^2}{z}$

26. $\sqrt[3]{\frac{a^3b^9}{c^6}} = \frac{\sqrt[3]{a^3b^9}}{\sqrt[3]{c^6}} = \frac{ab^3}{c^2}$

27. $\sqrt[3]{\dfrac{16x^3y^2}{z^6}} = \dfrac{\sqrt[3]{2^3 \cdot 2x^3y^2}}{\sqrt[3]{z^6}} = \dfrac{2x\sqrt[3]{2y^2}}{z^2}$

28. $\sqrt{\dfrac{125a^5b^2}{c^4}} = \dfrac{\sqrt{5^2 \cdot 5a^4 \cdot a \cdot b^2}}{\sqrt{c^4}} = \dfrac{5a^2b\sqrt{5a}}{c^2}$

29. $\sqrt{\frac{64x^3y^4}{9z^4p^2}} = \frac{\sqrt{8^2x^2xy^4}}{\sqrt{3^2z^4p^2}} = \frac{8xy^2\sqrt{x}}{3z^2p}$

30. $\sqrt[3]{\dfrac{8a^5b^3}{27r^6s^9}} = \dfrac{\sqrt[3]{2^3a^3a^2b^3}}{\sqrt[3]{3^3r^6s^9}} = \dfrac{2ab\sqrt[3]{a^2}}{3r^2s^3}$

31. $\sqrt{\frac{16}{3}} = \frac{\sqrt{16}}{\sqrt{3}} \cdot \frac{\sqrt{3}}{\sqrt{3}} = \frac{4\sqrt{3}}{3}$

32. $\sqrt{\dfrac{4}{5}} = \dfrac{\sqrt{4}}{\sqrt{5}}\dfrac{\sqrt{5}}{\sqrt{5}} = \dfrac{2\sqrt{5}}{5}$

33. $\sqrt[3]{\dfrac{27}{4}} = \dfrac{\sqrt[3]{3^3}}{\sqrt[3]{2^2}} \cdot \dfrac{\sqrt[3]{2}}{\sqrt[3]{2}} = \dfrac{3\sqrt[3]{2}}{2}$

34. $\sqrt[3]{\dfrac{16}{25}} = \dfrac{\sqrt[3]{2^4}}{\sqrt[3]{5^2}} \cdot \dfrac{\sqrt[3]{5}}{\sqrt[3]{5}} = \dfrac{2\sqrt[3]{10}}{5}$

35. $\sqrt{\dfrac{25}{2x}} = \dfrac{\sqrt{25}}{\sqrt{2x}} \cdot \dfrac{\sqrt{2x}}{\sqrt{2x}} = \dfrac{5\sqrt{2x}}{2x}$

36. $\sqrt[3]{\dfrac{8}{5y^2}} = \dfrac{\sqrt[3]{2^3}}{\sqrt[3]{5y^2}} \cdot \dfrac{\sqrt[3]{5^2y}}{\sqrt[3]{5^2y}} = \dfrac{2\sqrt[3]{25y}}{5y}$

37. $\sqrt[4]{\dfrac{81}{32z^2}} = \dfrac{\sqrt[4]{3^4}}{\sqrt[4]{2^5z^2}} \cdot \dfrac{\sqrt[4]{2^3z^2}}{\sqrt[4]{2^3z^2}} = \dfrac{3\sqrt[4]{8z^2}}{4z}$

38. $\sqrt[3]{\dfrac{2}{25r^2}} = \dfrac{\sqrt[3]{2}}{\sqrt[3]{5^2r^2}} \cdot \dfrac{\sqrt[3]{5r}}{\sqrt[3]{5r}} = \dfrac{\sqrt[3]{10r}}{5r}$

39. $\sqrt[3]{\dfrac{16x^2y}{x^5}} = \sqrt[3]{\dfrac{16y}{x^3}} = \dfrac{\sqrt[3]{(2^3)2y}}{\sqrt[3]{x^3}} = \dfrac{2\sqrt[3]{2y}}{x}$

40. $\sqrt[4]{\dfrac{25a^3b^5}{a^7b}} = \sqrt[4]{\dfrac{5^2b^4}{a^4}} = \dfrac{b}{a}\sqrt[4]{5^2} = \dfrac{b}{a}5^{2/4} = \dfrac{b5^{1/2}}{a} =$
$\dfrac{b\sqrt{5}}{a}$

41. $\sqrt[3]{\dfrac{8x^3yz}{27b^2z^4}} = \sqrt[3]{\dfrac{8x^3y}{27b^2z^3}} = \dfrac{\sqrt[3]{2^3x^3y}}{\sqrt[3]{3^3b^2z^3}} \cdot \dfrac{\sqrt[3]{b}}{\sqrt[3]{b}} = \dfrac{2x\sqrt[3]{by}}{3bz}$

42. $\sqrt[4]{\dfrac{25a^2b^3}{16c^3b^6}} = \sqrt[4]{\dfrac{5^2a^2}{2^4c^3b^3}} = \dfrac{\sqrt[4]{5^2a^2}}{\sqrt[4]{2^4c^3b^3}} \cdot \dfrac{\sqrt[4]{cb}}{\sqrt[4]{cb}} =$
$\dfrac{\sqrt[4]{25a^2bc}}{2bc}$

43. $\sqrt{4 \times 10^4} = \sqrt{4} \times \sqrt{10^4} = 2 \times 10^2 = 200$

44. $\sqrt{9 \times 10^6} = \sqrt{9} \times \sqrt{10^6} = 3 \times 10^3 = 3000$

45. $\sqrt{25 \times 10^3} = \sqrt{25}\sqrt{10^1 10^2} = 5\sqrt{10} \times 10 = 50\sqrt{10}$

46. $\sqrt{16 \times 10^7} = \sqrt{16}\sqrt{10(10^6)} = 4\sqrt{10} \times 10^3 =$
$4000\sqrt{10}$

47. $\sqrt{4 \times 10^7} = \sqrt{4}\sqrt{10(10^6)} = 2\sqrt{10} \times 10^3 =$
$2000\sqrt{10}$

48. $\sqrt{9 \times 10^9} = \sqrt{9}\sqrt{10}\sqrt{10^8} = 3\sqrt{10} \times 10^4 =$
$30{,}000\sqrt{10}$

49. $\sqrt[3]{1.25 \times 10^{10}} = \sqrt[3]{125 \times 10^8} = \sqrt[3]{125}\sqrt[3]{10^2}\sqrt[3]{10^6} =$
$5\sqrt[3]{100} \times 10^2 = 500\sqrt[3]{100}$ or
$\sqrt[3]{1.25 \times 10^{10}} = \sqrt[3]{1.25 \times 10 \times 10^9} = \sqrt[3]{12.5} \times 10^3$

50. $\sqrt[5]{3.2 \times 10^{14}} = \sqrt[5]{32 \times 10^{13}} = \sqrt[5]{2^5 \cdot 10^{10} \times 10^3} =$
$2 \times 100 \times \sqrt[5]{1000} = 200 \times \sqrt[5]{1000}$ or
$\sqrt[5]{3.2 \times 10^{14}} = \sqrt[5]{3.2 \times 10^4 \times 10^{10}} =$
$\sqrt[5]{3.2 \times 10^4} \times 10^2$

51. $\sqrt{\dfrac{x}{y} + \dfrac{y}{x}} = \sqrt{\dfrac{x^2}{xy} + \dfrac{y^2}{xy}} = \sqrt{\dfrac{x^2 + y^2}{xy}} =$
$\dfrac{\sqrt{x^2 + y^2}}{\sqrt{xy}} \cdot \dfrac{\sqrt{xy}}{\sqrt{xy}} = \dfrac{\sqrt{xy(x^2 + y^2)}}{xy} = \dfrac{\sqrt{x^3y + xy^3}}{xy}$

52. $\sqrt{\dfrac{a}{b} - \dfrac{b}{a}} = \sqrt{\dfrac{a^2}{ab} - \dfrac{b^2}{ab}} = \sqrt{\dfrac{a^2 - b^2}{ab}} \cdot \dfrac{\sqrt{ab}}{\sqrt{ab}} =$
$\dfrac{\sqrt{ab(a^2 - b^2)}}{ab} = \dfrac{\sqrt{a^3b - ab^3}}{ab}$

53. $\sqrt{a^2 + 2ab + b^2} = \sqrt{(a + b)^2} = |a + b|$

54. $\sqrt{x^2 - 2xy + y^2} = \sqrt{(x-y)^2} = |x - y|$

55. $\sqrt{\frac{1}{a^2} + \frac{1}{b}} = \sqrt{\frac{b}{a^2b} + \frac{a^2}{a^2b}} = \sqrt{\frac{b+a^2}{a^2b}} =$
$\frac{\sqrt{b+a^2}}{\sqrt{a^2b}} \cdot \frac{\sqrt{b}}{\sqrt{b}} = \frac{\sqrt{b^2+a^2b}}{ab}$

56. $\sqrt{\frac{x}{y^2} + \frac{y}{x^2}} = \sqrt{\frac{x^3}{x^2y^2} + \frac{y^3}{x^2y^2}} = \sqrt{\frac{x^3+y^3}{x^2y^2}} =$
$\frac{\sqrt{x^3+y^3}}{\sqrt{x^2y^2}} = \frac{\sqrt{x^3+y^3}}{xy}$

57. $f = \frac{1}{2L}\sqrt{\frac{T}{\mu}} = \frac{1}{2L} \cdot \frac{\sqrt{T}}{\sqrt{\mu}} \cdot \frac{\sqrt{\mu}}{\sqrt{\mu}} = \frac{1}{2L} \cdot \frac{\sqrt{T\mu}}{\mu} = \frac{1}{2L\mu} \cdot \sqrt{T\mu}$

58. When T is quadrupled the frequency is doubled.

59. $z = \frac{1}{\sqrt{\frac{1}{x^2} + \frac{1}{R^2}}} = \frac{1}{\sqrt{\frac{R^2+x^2}{x^2R^2}}} = \sqrt{\frac{x^2R^2}{R^2+x^2}} =$
$\frac{xR\sqrt{R^2+x^2}}{R^2+x^2}$

60. $\sqrt[3]{\frac{M}{2\rho N}} \cdot \frac{\sqrt[3]{2^2\rho^2N^2}}{\sqrt[3]{2^2\rho^2N^2}} = \frac{\sqrt[3]{4M\rho^2N^2}}{2\rho N}$

12.3 BASIC OPERATIONS WITH RADICALS

1. $2\sqrt{3} + 5\sqrt{3} = 7\sqrt{3}$

2. $5\sqrt{6} - 3\sqrt{6} = 2\sqrt{6}$

3. $\sqrt[3]{9} + 4\sqrt[3]{9} = 5\sqrt[3]{9}$

4. $\sqrt[4]{8} + 3\sqrt[4]{8} = 4\sqrt[4]{8}$

5. $2\sqrt{3} + 4\sqrt{2} + 6\sqrt{3} = 8\sqrt{3} + 4\sqrt{2}$

6. $5\sqrt{3} - 6\sqrt{5} - 9\sqrt{3} = -4\sqrt{3} - 6\sqrt{5}$

7. $\sqrt{5} + \sqrt{20} = \sqrt{5} + 2\sqrt{5} = 3\sqrt{5}$

8. $\sqrt{8} + \sqrt{2} = 2\sqrt{2} + \sqrt{2} = 3\sqrt{2}$

9. $\sqrt{7} - \sqrt{28} = \sqrt{7} - 2\sqrt{7} = -\sqrt{7}$

10. $\sqrt{8} - \sqrt{32} = 2\sqrt{2} - 4\sqrt{2} = -2\sqrt{2}$

11. $\sqrt{60} - \sqrt{\frac{5}{3}} = 2\sqrt{15} - \sqrt{\frac{5}{3}}\sqrt{\frac{3}{3}} = 2\sqrt{15} - \frac{\sqrt{15}}{3} =$
$\frac{6\sqrt{15}}{3} - \frac{\sqrt{15}}{3} = \frac{5\sqrt{15}}{3}$

12. $\sqrt{84} + \sqrt{\frac{3}{7}} = 2\sqrt{21} + \frac{\sqrt{21}}{7} = \left(\frac{14}{7} + \frac{1}{7}\right)\sqrt{21} =$
$\frac{15}{7}\sqrt{21}$

13. $\sqrt{\frac{1}{2}} - \sqrt{\frac{9}{2}} = \frac{\sqrt{2}}{2} - \frac{3\sqrt{2}}{2} = \frac{-2\sqrt{2}}{2} = -\sqrt{2}$

14. $\sqrt{\frac{4}{3}} - \sqrt{\frac{25}{3}} = \frac{2\sqrt{3}}{3} - \frac{5\sqrt{3}}{3} = \frac{-3\sqrt{3}}{3} = -\sqrt{3}$

15. $\sqrt{x^3y} + 2x\sqrt{xy} = x\sqrt{xy} + 2x\sqrt{xy} = 3x\sqrt{xy}$

16. $\sqrt{a^5b^3} - 3ab\sqrt{a^3b} = a^2b\sqrt{ab} - 3a^2b\sqrt{ab} =$
$-2a^2b\sqrt{ab}$

17. $\sqrt[3]{24p^2q^4} + \sqrt[3]{3p^8q} = \sqrt[3]{8 \cdot 3p^2q^3q} + \sqrt[3]{3p^6p^2q} =$
$2q\sqrt[3]{3p^2q} + p^2\sqrt[3]{3p^2q} = (2q+p^2)\sqrt[3]{3p^2q}$

18. $\sqrt[4]{16a^2b} - \sqrt[4]{81a^6b} = 2\sqrt[4]{a^2b} - 3a\sqrt[4]{a^2b} =$
$(2-3a)\sqrt[4]{a^2b}$

19. $\sqrt{\frac{x}{y^3}} - \sqrt{\frac{y}{x^3}} = \sqrt{\frac{xy}{y^4}} - \sqrt{\frac{xy}{x^4}} = \left(\frac{1}{y^2} - \frac{1}{x^2}\right)\sqrt{xy}$
or $\frac{x^2-y^2}{x^2y^2}\sqrt{xy}$

20. $\sqrt{\frac{a^3}{b^3}} + \sqrt{\frac{b}{a^5}} = \frac{a}{b}\sqrt{\frac{a}{b}} + \frac{1}{a^2}\sqrt{\frac{b}{a}} =$
$\frac{a}{b^2}\sqrt{ab} + \frac{1}{a^3}\sqrt{ab} = \left(\frac{a}{b^2} + \frac{1}{a^3}\right)\sqrt{ab}$ or
$\frac{a^4+b^2}{a^3b^2}\sqrt{ab}$

21. $a\sqrt{\frac{b}{3a}} + b\sqrt{\frac{a}{3b}} = \frac{a\sqrt{3ab}}{3a} + \frac{b\sqrt{3ab}}{3b} =$
$\frac{\sqrt{3ab}}{3} + \frac{\sqrt{3ab}}{3} = \frac{2\sqrt{3ab}}{3}$

22. $x\sqrt{\frac{y}{5x}} - y\sqrt{\frac{x}{5y}} = \frac{x\sqrt{5xy}}{5x} - \frac{y\sqrt{5xy}}{5y} =$
$\frac{\sqrt{5xy}}{5} - \frac{\sqrt{5xy}}{5} = 0$

23. $\sqrt{5}\sqrt{8} = \sqrt{40} = 2\sqrt{10}$

24. $\sqrt[5]{-7}\sqrt[5]{11} = \sqrt[5]{-77}$

25. $\sqrt{3x}\sqrt{5x} = \sqrt{15x^2} = x\sqrt{15}$

26. $\sqrt[3]{7x^2}\sqrt[2]{3x} = \sqrt[2]{21x^3} = x\sqrt[3]{21}$

27. $\left(\sqrt{4x}\right)^3 = \left(2\sqrt{x}\right)^3 = 2^3\sqrt{x^3} = 8\sqrt{x^3} = 8x\sqrt{x}$

28. $\left(\sqrt[3]{2x^2y}\right)^4 = \sqrt[3]{2^4x^8y^4} = 2x^2y\sqrt[3]{2x^2y}$

29. $\sqrt{\frac{5}{8}} \cdot \sqrt{\frac{9}{10}} = \sqrt{\frac{5 \cdot 9}{8 \cdot 10}} = \sqrt{\frac{9}{16}} = \frac{3}{4}$

30. $\sqrt{\frac{7}{6}}\sqrt{\frac{12}{3}} = \sqrt{\frac{7 \cdot 12}{6 \cdot 3}} = \frac{\sqrt{7 \cdot 2}}{\sqrt{3}} \cdot \frac{\sqrt{3}}{\sqrt{3}} = \frac{\sqrt{42}}{3}$

31. $\sqrt{2}(\sqrt{x} + \sqrt{2}) = \sqrt{2x} + 2$

32. $\sqrt{3}(\sqrt{12} - \sqrt{y}) = \sqrt{36} - \sqrt{3y} = 6 - \sqrt{3y}$

33. $(\sqrt{x} + \sqrt{y})^2 = \sqrt{x}^2 + 2\sqrt{xy} + \sqrt{y}^2 = x + 2\sqrt{xy} + y$

34. $\left(\sqrt{a} + 3\sqrt{b}\right)^2 = \sqrt{a}^2 + 6\sqrt{ab} + (3\sqrt{b})^2 =$
$a + 6\sqrt{ab} + 9b$

35. $\sqrt[3]{\frac{5}{2}} \cdot \sqrt[3]{\frac{2}{7}} = \sqrt[3]{\frac{5 \cdot 2}{2 \cdot 7}} = \sqrt[3]{\frac{5}{7}} \cdot \sqrt[3]{\frac{7^2}{7^2}} = \frac{\sqrt[3]{245}}{7}$

36. $\sqrt{\frac{ab}{5c}}\sqrt{\frac{abc}{5}} = \sqrt{\frac{a^2b^2c}{5^2c}} = \sqrt{\frac{a^2b^2}{5^2}} = \frac{ab}{5}$

37. $\left(\sqrt{a} + \sqrt{b}\right)\left(\sqrt{a} - \sqrt{b}\right) = \sqrt{a}^2 - \sqrt{b}^2 = |a| - |b|$

38. $\left(\sqrt{x} - 2\sqrt{y}\right)\left(\sqrt{x} + 5\sqrt{y}\right) = x + 3\sqrt{xy} - 10y$

39. $\sqrt[3]{x}\sqrt{x} = x^{1/3}x^{1/2} = x^{2/6}x^{3/6} = x^{5/6} = \sqrt[6]{x^5}$

40. $\sqrt{5x}\sqrt[3]{2x^2} = (5x)^{1/2}(2x^2)^{1/3} = (5x)^{3/6}(2x^2)^{2/6} =$
$(5^3x^3)^{1/6}(2^2x^4)^{1/6} = \sqrt[6]{5^3x^3}\sqrt[6]{4x^4} = x\sqrt[6]{5^3x4} =$
$x\sqrt[6]{500x}$

41. $\frac{\sqrt{32}}{\sqrt{2}} = \sqrt{\frac{32}{2}} = \sqrt{16} = 4$

42. $\frac{\sqrt[3]{a^2}}{\sqrt[3]{4a}} = \sqrt[3]{\frac{a^2}{4a}} = \sqrt[3]{\frac{a}{4}} \cdot \sqrt[3]{\frac{2}{2}} = \frac{\sqrt[3]{2a}}{\sqrt[3]{8}} = \frac{\sqrt[3]{2a}}{2}$

43. $\frac{\sqrt[3]{4b^2}}{\sqrt[3]{16b}} = \sqrt[3]{\frac{4b^2}{16b}} = \sqrt[3]{\frac{b}{4}}\sqrt[3]{\frac{2}{2}} = \frac{\sqrt[3]{2b}}{2}$

44. $\frac{\sqrt{5a^3b}}{\sqrt{15ab^3}} = \sqrt{\frac{5a^3b}{15ab^3}} = \sqrt{\frac{a^2}{3b^2}}\sqrt{\frac{3}{3}} = \frac{a\sqrt{3}}{3b}$

45. $\frac{1}{x + \sqrt{5}} \cdot \frac{x - \sqrt{5}}{x - \sqrt{5}} = \frac{x - \sqrt{5}}{x^2 - 5}$

46. $\frac{1}{x + \sqrt{3}} \cdot \frac{x - \sqrt{3}}{x - \sqrt{3}} = \frac{x - \sqrt{3}}{x^2 - 3}$

47. $\frac{\sqrt{5} - \sqrt{3}}{\sqrt{5} + \sqrt{3}} \cdot \frac{\sqrt{5} - \sqrt{3}}{\sqrt{5} - \sqrt{3}} = \frac{5 - 2\sqrt{15} + 3}{5 - 3} =$
$\frac{8 - 2\sqrt{15}}{2} = \frac{2(4 - \sqrt{15})}{2} = 4 - \sqrt{15}$

48. $\frac{\sqrt{5} + \sqrt{7}}{\sqrt{7} - \sqrt{5}} \cdot \frac{\sqrt{7} + \sqrt{5}}{\sqrt{7} + \sqrt{5}} = \frac{5 + 2\sqrt{35} + 7}{7 - 5} =$
$\frac{12 + 2\sqrt{35}}{2} = \frac{2(6 + \sqrt{35})}{2} = 6 + \sqrt{35}$

49. $\frac{\sqrt{x+1}}{\sqrt{x-1}} + \frac{\sqrt{x-1}}{\sqrt{x+1}} = \frac{\sqrt{x+1}\sqrt{x-1}}{\sqrt{x-1}\sqrt{x-1}} +$
$\frac{\sqrt{x-1}}{\sqrt{x+1}}\frac{\sqrt{x+1}}{\sqrt{x+1}} = \frac{\sqrt{x^2-1}}{x-1} + \frac{\sqrt{x^2-1}}{x+1} =$
$\frac{(x+1)\sqrt{x^2-1}}{x^2-1} + \frac{(x-1)\sqrt{x^2-1}}{x^2-1} = \frac{2x\sqrt{x^2-1}}{x^2-1},$
$x \neq 1, -1$

50. $\frac{\sqrt{x+3}}{\sqrt{x-3}} - \frac{\sqrt{x-3}}{\sqrt{x+3}} = \frac{\sqrt{x+3}\sqrt{x-3}}{\sqrt{x-3}\sqrt{x-3}} -$
$\frac{\sqrt{x-3}\sqrt{x+3}}{\sqrt{x+3}\sqrt{x+3}} = \frac{\sqrt{x^2-9}}{x-3} - \frac{\sqrt{x^2-9}}{x+3} =$
$\frac{(x+3)\sqrt{x^2-9}}{x^2-9} - \frac{(x-3)\sqrt{x^2-9}}{x^2-9} = \frac{6\sqrt{x^2-9}}{x^2-9},$
$x \neq 3, -3$

51. $\frac{\sqrt{x+y}}{\sqrt{x-y} - \sqrt{x}} \cdot \frac{\sqrt{x-y} + \sqrt{x}}{\sqrt{x-y} + \sqrt{x}} =$
$\frac{\sqrt{x^2 - y^2} + \sqrt{x^2 + xy}}{x - y - x} =$
$-\frac{\sqrt{x^2 - y^2} + \sqrt{x^2 + xy}}{y}, y \neq 0$

52. $\dfrac{\sqrt{1+y}}{\sqrt{1-y}+\sqrt{y}} \cdot \dfrac{\sqrt{1-y}-\sqrt{y}}{\sqrt{1-y}-\sqrt{y}} = \dfrac{\sqrt{1-y^2}-\sqrt{y+y^2}}{(1-y)-y} = \dfrac{\sqrt{1-y^2}-\sqrt{y+y^2}}{1-2y}$, $y \neq \frac{1}{2}$

53. $\dfrac{-b+\sqrt{b^2-4ac}}{2a} + \dfrac{-b-\sqrt{b^2-4ac}}{2a} = \dfrac{-2b}{2a} = \dfrac{-b}{a}$

54. $v = \sqrt{\dfrac{\pi}{4\lambda d}} + \sqrt{\dfrac{4\pi}{\lambda d}} = \dfrac{1}{2}\sqrt{\dfrac{\pi}{\lambda d}} + \dfrac{2}{1}\sqrt{\dfrac{\pi}{\lambda d}} = \dfrac{5}{2}\sqrt{\dfrac{\pi}{\lambda d}} = \dfrac{5}{2\lambda d}\sqrt{\pi\lambda d}$

55. (a) $R = \dfrac{R_1R_2}{R_1+R_2} = \dfrac{x^{3/2}\sqrt{x}}{x^{3/2}+\sqrt{x}} = \dfrac{x^{3/2}x^{1/2}}{x^{3/2}+x^{1/2}} = \dfrac{x^2}{\sqrt{x^3}+\sqrt{x}} \cdot \dfrac{\sqrt{x^3}-\sqrt{x}}{\sqrt{x^3}-\sqrt{x}} = \dfrac{x^2x^{3/2}-x^2x^{1/2}}{x^3-x} = \dfrac{x^{7/2}-x^{5/2}}{x^3-x} = \dfrac{x^{5/2}(x-1)}{x(x^2-1)} = \dfrac{x \cdot x^{3/2}(x-1)}{x(x+1)(x-1)} = \dfrac{x^{3/2}}{x+1}$ (b) if $x = 20$; $\dfrac{20^{3/2}}{20+1} = \dfrac{89.44279191}{21} = 4.2591771 \approx 4.259\Omega$

56. $\dfrac{\sqrt{d_1}-\sqrt{d_2}}{\sqrt{d_1}+\sqrt{d_2}} \cdot \dfrac{\sqrt{d_1}-\sqrt{d_2}}{\sqrt{d_1}-\sqrt{d_2}} = \dfrac{d_1-2\sqrt{d_1d_2}+d_2}{d_1-d_2}$

57. $\left(\dfrac{-b+\sqrt{b^2-4ac}}{2a}\right)\left(\dfrac{-b-\sqrt{b^2-4ac}}{2a}\right) = \dfrac{b^2-\left(\sqrt{b^2-4ac}\right)^2}{4a^2} = \dfrac{b^2-(b^2-4ac)}{4a^2} = \dfrac{4ac}{4a^2} = \dfrac{c}{a}$

12.4 EQUATIONS WITH RADICALS

1. $\sqrt{x+3} = 5$; $\left(\sqrt{x+3}\right)^2 = 5^2$; $x+3 = 25$; $x = 22$. Check: $\sqrt{22+3} = \sqrt{25} = 5$. Solution is 22.

2. $\sqrt{x-16} = 5$; $\left(\sqrt{x-16}\right)^2 = 5^2$; $x-16 = 25$; $x = 41$. Check: $\sqrt{41-16} = \sqrt{25} = 5$. Solution is 41.

3. $\sqrt{2x+4}-7 = 0$; $\sqrt{2x+4} = 7$; $\left(\sqrt{2x+4}\right)^2 = 7^2$; $2x+4 = 49$; $2x = 45$; $x = \frac{45}{2}$. Check: $\sqrt{2\cdot\frac{45}{2}+4}-7 = \sqrt{45+4}-7 = \sqrt{49}-7 = 0$. Solution is $\frac{45}{2}$ or 22.5.

4. $\sqrt{3y-9}-9-18 = 0$; $\sqrt{3y-9} = 18$; $\left(\sqrt{3y-9}\right)^2 = 18^2$; $3y-9 = 324$; $3y = 333$; $y = 111$. Check: $\sqrt{3\cdot 111-9}-18 = \sqrt{333-9}-18 = \sqrt{324}-18 = 18-18 = 0$. Solution is 111.

5. $\sqrt{x^2+24} = x-4$; $\left(\sqrt{x^2+24}\right)^2 = (x-4)^2$; $x^2+24 = x^2-8x+16$; $24 = -8x+16$; $8 = -8x$; $x = -1$. Check: $\sqrt{(-1)^2+24} = \sqrt{1+24} = \sqrt{25} = 5$; $-1-4 = -5$ but $5 \neq -5$ so no solution.

6. $\sqrt{x^2-75} = x+5$; $\left(\sqrt{x^2-75}\right)^2 = (x+5)^2$; $x^2-75 = x^2+10x+25$; $-100 = 10x$; $x = -10$. Check: $\sqrt{(-10)^2-75} = \sqrt{100-75} = \sqrt{25} = 5$; $-10+5 = -5$ but $5 \neq -5$ so no solution.

7. $(y+12)^{1/3} = 3$; $((y+12)^{1/3})^3 = 3^3$; $y+12 = 27$; $y = 15$. Check: $(15+12)^{1/3} = 27^{1/3} = 3$. Solution is 15.

8. $(r+9)^{1/4} = 7$; $\left[(r+9)^{1/4}\right]^4 = 7^4$; $r+9 = 2401$; $r = 2392$. Check: $(2392+9)^{1/4} = (2401)^{1/4} = 7$. Solution is 2392.

9. $\left(\frac{x}{2}+1\right)^{1/2} = 3$; $\left[\left(\frac{x}{2}+1\right)^{1/2}\right]^2 = 3^2$; $\frac{x}{2}+1 = 9$; $\frac{x}{2} = 8$; $x = 16$. Check: $\left(\frac{16}{2}+1\right)^{1/2} = (8+1)^{1/2} = 9^{1/2} = 3$. Solution is 16.

10. $\left(\frac{y}{3}+4\right)^{1/3} = \frac{5}{2}$; $\left[\left(\frac{y}{3}+4\right)^{1/3}\right]^3 = \left(\frac{5}{2}\right)^3$; $\frac{y}{3}+4 = \frac{125}{8}$; $\frac{y}{3} = \frac{125}{8}-\frac{32}{8} = \frac{93}{8}$; $y = \frac{279}{8}$. Check:

$\left(\frac{\frac{279}{8}}{3}+4\right)^{1/3} = \left(\frac{93}{8}+4\right)^{1/3} = \left(\frac{93}{8}+\frac{32}{8}\right)^{1/3} =$
$\left(\frac{125}{8}\right)^{1/3} = \frac{5}{2}$. Solution is $\frac{279}{8}$ or $34\frac{7}{8}$.

11. $(x-1)^{3/2} = 27$; $\left((x-1)^{3/2}\right)^2 = 27^2$;
$(x-1)^3 = 729$; $(x-1) = \sqrt[3]{729} = 9$; $x = 10$.
Check: $(10-1)^{3/2} = 9^{3/2} = 3^3 = 27$. Solution is 10.

12. $(x+1)^{3/4} = 8$; $((x+1)^{3/4})^{4/3} = 8^{4/3}$; $x+1 = 16$ so $x = 15$. Check:
$(15+1)^{3/4} = 16^{3/4} = \sqrt[4]{16^3} = 2^3 = 8$. Solution is 15.

13. $\sqrt{x+1} = \sqrt{2x-1}$; $\sqrt{x+1}^2 = \sqrt{2x-1}^2$;
$x+1 = 2x-1$; $-x = -2$; $x = 2$. Check:
$\sqrt{2+1} = \sqrt{3}$; $\sqrt{2\cdot 2-1} = \sqrt{4-1} = \sqrt{3}$.
Solution is 2.

14. $\sqrt{x^2-4} = \sqrt{2x-1}$; $\sqrt{x^2-4}^2 = \sqrt{2x-1}^2$;
$x^2-4 = 2x-1$ or $x^2-2x-3 = 0$;
$(x-3)(x+1) = 0$; $x = 3$ or $x = -1$. Check:
$\sqrt{3^2-4} = \sqrt{2\cdot 3-1}$; $\sqrt{9-4} = \sqrt{6-1} = \sqrt{5}$;
$\sqrt{(-1)^2-4} = \sqrt{2(-1)-1}$; $\sqrt{-3} = \sqrt{-3}$ not real. Solution is 3 only.

15. $\sqrt{x+1}+\sqrt{x-5} = 7$; $\sqrt{x+1} = 7-\sqrt{x-5}$;
$\sqrt{x+1}^2 = (7-\sqrt{x-5})^2$;
$x+1 = 49-14\sqrt{x-5}+(x-5)$;
$x+1 = x-14\sqrt{x-5}+44$; $-43 = -14\sqrt{x-5}$;
$\frac{43}{14} = \sqrt{x-5}$; $\left(\frac{43}{14}\right)^2 = \sqrt{x-5}^2$; $\frac{1849}{196} = x-5$;
$x = \frac{2829}{196}$. Check: $\sqrt{\frac{2829}{196}+1}+\sqrt{\frac{2829}{196}-5} =$
$\sqrt{\frac{3025}{196}}+\sqrt{\frac{1849}{196}} = \frac{55}{14}+\frac{43}{14} = \frac{98}{14} = 7$. Solution is $\frac{2829}{196} \approx 14.43$.

16. $\sqrt{x-3}-\sqrt{x+2} = 19$; $\sqrt{x-3} = \sqrt{x+2}+19$;
$\sqrt{x-3}^2 = (\sqrt{x+2}+19)^2$;
$x-3 = (x+2)+38\sqrt{x+2}+361$;
$-366 = 38\sqrt{x+2}$. No solution because the right-hand side must be positive.

17. $\sqrt{x-1}+\sqrt{x+5} = 2$; $\sqrt{x-1} = 2-\sqrt{x+5}$;
$\sqrt{x-1}^2 = (2-\sqrt{x+5})^2$;
$x-1 = 4-4\sqrt{x+5}+(x+5)$; $-10 = -4\sqrt{x+5}$;
$\frac{5}{2} = \sqrt{x+5}$; $\left(\frac{5}{2}\right)^2 = \sqrt{x+5}^2$; $\frac{25}{4} = x+5$;
$x = \frac{25}{4}-\frac{20}{4} = \frac{5}{4}$. Check:
$\sqrt{\frac{5}{4}-1}+\sqrt{\frac{5}{4}+5} = \sqrt{\frac{1}{4}}+\sqrt{\frac{25}{4}} = \frac{1}{2}+\frac{5}{2} = \frac{6}{2} = 3$.
Does not check so there is no solution.

18. $\sqrt{x+3}-\sqrt{2x-4} = 3$; $\sqrt{x+3} = \sqrt{2x-4}+3$;
$\left(\sqrt{x+3}\right)^2 = (\sqrt{2x-4}+3)^2$;
$x+3 = (2x-4)+6\sqrt{2x-4}+9$;
$-x-2 = 6\sqrt{2x-4}$; $(-x-2)^2 = (6\sqrt{2x-4})^2$;
$x^2+4x+4 = 36(2x-4)$; $x^2+4x+4 = 72x-144$;
$x^2-68x+148 = 0$; by the quadratic formula $x \approx 2.25$ or $x \approx 65.75$. Neither checks so there is no solution. (Note: you can conclude there is no solution when $-x-2 = 6\sqrt{2x-4}$ since x must be > 2 for the radicand $2x-4$ to be positive. This yields a negative expression on the left and a positive on the right which cannot be equal.)

19. $\sqrt{\frac{1}{x}} = \sqrt{\frac{4}{3x-1}}$; square both sides to get
$\frac{1}{x} = \frac{4}{3x-1}$; $3x-1 = 4x$; $-1 = x$. This does not check as it gives negative radicands so no solution.

20. $\sqrt{\frac{2}{x-1}} = \sqrt{\frac{5}{x+1}}$; square both sides to get
$\frac{2}{x-1} = \frac{5}{x+1}$; $2(x+1) = 5(x-1)$;
$2x+2 = 5x-5$; $7 = 3x$ or $x = \frac{7}{3}$. Check:
$\sqrt{\frac{2}{\frac{7}{3}-1}} = \sqrt{\frac{2}{\frac{4}{3}}} = \sqrt{\frac{6}{4}} = \frac{\sqrt{6}}{2}$;
$\sqrt{\frac{5}{\frac{7}{3}+1}} = \sqrt{\frac{5}{\frac{10}{3}}} = \sqrt{\frac{3}{2}} = \frac{\sqrt{6}}{2}$ so the solution is $\frac{7}{3}$

21. Here $v = \sqrt{\mu g_s R}$. Squaring both sides gives
$v^2 = \mu_s g R$ and so $R = \frac{v^2}{\mu_s g}$.

22. Here $v = \sqrt{v_0^2-2gh}$. Squaring both sides yields
$v^2 = v_0^2-2gh$ or $2gh = v_0^2-v^2$, and so $h = \frac{v_0^2-v^2}{2g}$.

23. We are given $\frac{d_A}{d_B} - \frac{\sqrt{I_A}}{\sqrt{I_B}}$. Substituting the given data produces $\frac{11}{20} = \frac{\sqrt{3.5I_B-15}}{\sqrt{I_B}}$. Now, square

both sides to get $\frac{121}{400} = \frac{3.5I_B - 15}{I_B}$ and cross-multiply, with the result $121I_B = 1400I_B - 6000$ or $-1279I_B = -6000$ and so, $I_B = \frac{6000}{1279} \approx 4.691$. Since $I_A = 3.5I_B - 15$, we have $I_A = 3.5 \times 4.691 - 15 = 1.419$. The final result is $I_A = 1.419$, $I_B = 4.691$.

24. (a) $\frac{\sqrt{I_x}}{r_x} = \frac{\sqrt{I_y}}{r_y}$. Square both sides to get $\frac{I_x}{r_x^2} = \frac{I_y}{r_y^2}$ or $I_x = \frac{I_y r_x^2}{r_y^2} = \left(\frac{r_x}{r_y}\right)^2 I_y$.

(b) Substituting the given data yields $I_x = \left(\frac{2r_y + 3}{r_y}\right)^2 \cdot 30 = \left(2 + \frac{3}{r_y}\right)^2 \cdot 30 = \left(4 + \frac{12}{r_y} + \frac{9}{r_y^2}\right) 30 = 120 + \frac{360}{r_y} + \frac{270}{r_y^2}$, if $r_y \neq 0$.

CHAPTER 12 REVIEW

1. $49^{1/2} = \sqrt{49} = 7$
2. $81^{1/4} = \sqrt[4]{81} = 3$
3. $25^{3/2} = \sqrt{25}^3 = 5^3 = 125$
4. $64^{2/3} = \sqrt[3]{64^2} = 4^2 = 16$
5. $9^{-3/2} = \frac{1}{9^{3/2}} = \frac{1}{\sqrt{9}^3} = \frac{1}{3^3} = \frac{1}{27}$
6. $125^{-4/3} = \frac{1}{125^{4/3}} = \frac{1}{\sqrt[3]{125^4}} = \frac{1}{5^4} = \frac{1}{625}$
7. $5^4 5^9 = 5^{13}$
8. $13^6 13^{-9} = 13^{6-9} = 13^{-3} = \frac{1}{13^3}$
9. $(2^4)^8 = 2^{32}$
10. $(5^3)^6 = 5^{18}$
11. $(x^2y^3)^4 = x^8y^{12}$
12. $(y^6x)^4 = y^{24}x^4$
13. $\left(\frac{a^3}{ya^4}\right)^{-5} = \left(\frac{1}{ya}\right)^{-5} = (ya)^5 = y^5a^5$
14. $\left(\frac{a^5}{ab^4}\right)^3 = \left(\frac{a^4}{b^4}\right)^3 = \frac{a^{12}}{b^{12}}$ or $\left(\frac{a}{b}\right)^{12}$
15. $x^{1/2}x^{1/3} = x^{3/6}x^{2/6} = x^{5/6}$
16. $y^{2/3}y^{3/4} = y^{8/12}y^{9/12} = y^{17/12}$
17. $\frac{(xy^2a)^5}{x^4(ya^3)^2} = \frac{x^5y^{10}a^5}{x^4y^2a^6} = \frac{xy^8}{a}$
18. $\frac{(ab^3c^2)^4}{(a^2bc)^3} = \frac{a^4b^{12}c^8}{a^6b^3c^3} = \frac{b^9c^5}{a^2}$
19. $\sqrt[3]{-27} = -3$
20. $\sqrt[4]{81^3} = \sqrt[4]{81}^3 = 3^3 = 27$
21. $\sqrt{a^2b^5} = \sqrt{a^2}\sqrt{b^4}\sqrt{b} = ab^2\sqrt{b}$
22. $\sqrt[4]{a^3b^7} = \sqrt[4]{b^4}\sqrt[4]{a^3b^3} = b\sqrt[4]{a^3b^3}$
23. $\sqrt[3]{\frac{-8x^3y^6}{z}} = \frac{-2xy^2}{\sqrt[3]{z}} = \frac{-2xy^2}{\sqrt[3]{z}}\frac{\sqrt[3]{z^2}}{\sqrt[3]{z^2}} = \frac{-2xy^2\sqrt[3]{z^2}}{z}$ or $\frac{-2xy^2}{z}\sqrt[3]{z^2}$
24. $\sqrt[4]{\frac{32a^6b^8}{cd^2}} = \sqrt[4]{\frac{2^5a^6b^8}{cd^2}} \cdot \sqrt[4]{\frac{c^3d^2}{c^3d^2}} = \frac{2ab^2\sqrt[4]{2a^2c^3d^2}}{cd} = \frac{2ab^2}{cd}\sqrt[4]{2a^2c^3d^2}$
25. $\sqrt{\frac{a}{b^2} - \frac{b}{a^2}} = \sqrt{\frac{a^3}{a^2b^2} - \frac{b^3}{a^2b^2}} = \sqrt{\frac{a^3-b^3}{a^2b^2}} = \frac{\sqrt{a^3-b^3}}{ab}$
26. $\sqrt{a^2 + 8a + 16} = \sqrt{(a+4)^2} = a + 4$
27. $2\sqrt{5} + 6\sqrt{5} = 8\sqrt{5}$
28. $3\sqrt{7} - 6\sqrt{7} = -3\sqrt{7}$
29. $3\sqrt{6} + 5\sqrt{6} - 2\sqrt{3} = 8\sqrt{6} - 2\sqrt{3}$
30. $\sqrt{6} + \sqrt{24} = \sqrt{6} + 2\sqrt{6} = 3\sqrt{6}$

31. $\sqrt[4]{16a^3b} + \sqrt[4]{81a^7b} = 2\sqrt[4]{a^3b} + 3a\sqrt[4]{ab^3}$;
$(2 + 3a)\sqrt[4]{a^3b}$

32. $\sqrt[4]{32x^5} - x^2\sqrt[4]{2x} = 2x\sqrt[4]{2x} - x^2\sqrt[4]{2x}$;
$= (2x - x^2)\sqrt[4]{2x}$

33. $a\sqrt{\frac{a}{7b}} - b\sqrt{\frac{b}{7a}} = \frac{a}{7b}\sqrt{7ab} - \frac{b}{7a}\sqrt{7ab} =$
$\frac{a^2}{7ab}\sqrt{7ab} - \frac{b^2}{7ab}\sqrt{7ab} = \frac{(a^2 - b^2)\sqrt{7ab}}{7ab}$

34. $\sqrt{\frac{x-2}{x+2}} + \sqrt{\frac{x+2}{x-2}} = \frac{\sqrt{x-2}\sqrt{x+2}}{x+2} +$
$\frac{\sqrt{x+2}\sqrt{x-2}}{x-2} = \frac{\sqrt{x^2-4}}{x+2} + \frac{\sqrt{x^2-4}}{x-2} =$
$\frac{(x-2)\sqrt{x^2-4}}{(x-2)(x+2)} + \frac{(x+2)\sqrt{x^2-4}}{(x+2)(x-2)} =$
$\frac{x\sqrt{x^2-4} - 2\sqrt{x^2-4}}{x^2-4} + \frac{x\sqrt{x^2-4} + 2\sqrt{x^2-4}}{x^2-4} =$
$\frac{2x\sqrt{x^2-4}}{x^2-4}$

35. $\sqrt{3}\sqrt{5} = \sqrt{15}$

36. $\sqrt{7}\sqrt{6} = \sqrt{42}$

37. $\sqrt{2x}\sqrt{7x} = \sqrt{14x^2} = x\sqrt{14}$

38. $\sqrt[3]{5y}\sqrt[3]{25y} = \sqrt[3]{125y^2} = 5\sqrt[3]{y^2}$

39. $(\sqrt{a} + \sqrt{2b})(\sqrt{a} - \sqrt{2b}) = \sqrt{a}^2 - \sqrt{2b}^2 = a - 2b$

40. $(\sqrt{x} - 3\sqrt{y})(\sqrt{x} + 3\sqrt{y}) = \sqrt{x}^2 - (3\sqrt{y})^2 = x - 9y$

41. $\frac{\sqrt[3]{2a^2}}{\sqrt[3]{16a}} = \sqrt[3]{\frac{2a^2}{16a}} = \sqrt[3]{\frac{a}{8}} = \frac{\sqrt[3]{a}}{2}$

42. $\frac{\sqrt[3]{5a^2b}}{\sqrt[3]{25ab^2}} = \sqrt[3]{\frac{5a^2b}{25ab^2}} = \sqrt[3]{\frac{a}{5b}} \cdot \sqrt[3]{\frac{5^2b^2}{5^2b^2}} = \frac{\sqrt[3]{25ab^2}}{5b}$

43. $\sqrt{x-1} = 9$ so $x - 1 = 81$ and $x = 82$. Check: $\sqrt{82-1} = \sqrt{81} = 9$. Solution is 82.

44. $\sqrt{2x-7} - 5 = 0$ or $\sqrt{2x-7} = 5$. Squaring both sides gives $2x - 7 = 25$ or $2x = 32$, and so $x = 16$. Check: $\sqrt{2 \cdot 16 - 7} - 5$;
$= \sqrt{32-7} - 5 = \sqrt{25} - 5 = 5 - 5 = 0$. Solution is 16.

45. Here $\sqrt{x^2-7} = 2x - 4$. Squaring yields $\sqrt{x^2-7}^2 = (2x-4)^2$ or $x^2 - 7 = 4x^2 - 16x + 16$ and so $0 = 3x^2 - 16x + 25$. The discriminant, $b^2 - 4ac = -44$, is negative, so there is no solution.

46. Multiplying $\frac{3x}{4} - 1 = 25x^2 - 30x + 9$ by 4 gives $3x - 4 = 100x^2 - 120x + 36$. or $100x^2 - 123x + 40 = 0$. The discriminant, $b^2 - 4ac = -871$, so there is no solution.

47. Here $\sqrt{x+1} + \sqrt{x-5} = 8$ can be written as $\sqrt{x+1} = 8 - \sqrt{x-5}$. Squaring both sides produces $\sqrt{x+1}^2 = (8 - \sqrt{x-5})^2$ or $x + 1 = 64 - 16\sqrt{x-5} + (x-5)$ which simplifies to $-58 = -16\sqrt{x-5}$ or $\frac{58}{16} = \sqrt{x-5}$ or $3.625 = \sqrt{x-5}$. Squaring again yields $3.625^2 = x - 5$ or $13.140625 = x - 5$ and so $x = 18.140625$. This answer does check so $x \approx 18.14$.

48. Here $\sqrt{x+3} + \sqrt{x-2} = 20$ can be written as $\sqrt{x+3} = 20 - \sqrt{x-2}$. Squaring both sides gives $\sqrt{x+3}^2 = (20 - \sqrt{x-2})^2$ or $x + 3 = 400 - 40\sqrt{x-2} + (x-2)$ and so $-395 = -40\sqrt{x-2}$ or $\sqrt{x-2} = 9.875$. Squaring again gives $x - 2 = 9.875^2$ or $x - 2 = 97.51562$ and so, $x = 99.51562$. This answer does check so the solution is about 99.52.

49. Square both sides to get $\frac{1}{x} = \frac{3x}{4} - 1 = \frac{3x-4}{4}$. Multiplying by $4x$, we get $4 = 3x^2 - 4x$ or $3x^2 - 4x - 4 = 0$. This factors as $(3x+2)(x-2) = 0$ and so $x = -\frac{2}{3}$ or $x = 2$. We see that $-\frac{2}{3}$ yields a negative radicand so does not check. Thus, the only solution is 2.

50. Here $\sqrt{\frac{3}{x-1}} = \sqrt{\frac{4}{x+1}}$. Square both sides to get $\frac{3}{x-1} = \frac{4}{x+1}$ and so $3(x+1) = 4(x-1)$ or $3x + 3 = 4x - 4$ which simplifies to $x = 7$. Check: $\sqrt{\frac{3}{7-1}} = \sqrt{\frac{3}{6}} = \sqrt{\frac{1}{2}}$; $\sqrt{\frac{4}{7+1}} = \sqrt{\frac{4}{8}} = \frac{1}{2}$. Solution is 7.

51. $q(t) = q_0 2^{-t/12.5}$ (a) $q(10) = 50 \cdot 2^{-10/12.5} = 28.71745887$; or about 28.72 mg; (b) $q(20) = 50 \cdot 2^{-20/12.5} = 16.49384888$ or 16.49 mg.

52. (a) $v = \sqrt{\frac{2KE}{m}} = \frac{\sqrt{2KEm}}{m}$; (b) $v = \sqrt{\frac{2KE}{m}}$; $v^2 = \frac{2KE}{m}$; $m = \frac{2KE}{v^2}$

53. By the Pythagorean theorem $v^2 = v_x^2 + v_y^2$ or $v^2 = (3t+2)^2 + (4t-1)^2 = 9t^2 + 12t + 4 + 16t^2 - 8t + 1$; $v^2 = 25t^2 + 4t + 5$; $v = \sqrt{25t^2 + 4t + 5}$

54. From the diagram and the given information we have that $AB = 12, AC = x$ and let $BC = y$. Then $y - x = 5$ and $x^2 + 12^2 = y^2$; $y = 5 + x$ by substitution $x^2 + 144 = (5+x)^2$; $x^2 + 144 = 25 + 10x + x^2$; $10x = 119$; $x = 11.9$ m; $y = 5 + 11.16 = 16.9$ m. $AC = 11.9$ m, $BC = 16.9$ m

CHAPTER 12 TEST

1. $8^5 8^{-7} = 8^{-2} = \frac{1}{8^2}$

2. $(3^5)^4 = 3^{20}$

3. $\left(\frac{x^2}{xy^3}\right)^5 = \left(\frac{x}{y^3}\right)^5 = \frac{x^5}{y^{15}}$

4. $x^{1/4}x^{4/3} = x^{3/12}x^{16/12} = x^{19/12}$

5. $\sqrt[3]{-64} = \sqrt[3]{(-4)^3} = -4$

6. $\sqrt[4]{81} = \sqrt[4]{3^4} = 3$

7. $\sqrt{x^4y^3} = \sqrt{x^4}\sqrt{y^2}\sqrt{y} = x^2y\sqrt{y}$

8. $\sqrt[3]{\frac{-27x^6y}{z^2}} = \frac{\sqrt[3]{-27x^6yz}}{\sqrt[3]{z^3}} = \frac{-3x^2}{z}\sqrt[3]{yz}$

9. $\sqrt{x^2 + 6x + 9} = \sqrt{(x+3)^2} = x + 3$

10. $3\sqrt{6} + 5\sqrt{24} = 3\sqrt{6} + 10\sqrt{6} = 13\sqrt{6}$

11. $\sqrt{2x}\sqrt{6x} = \sqrt{12x^2} = 2\sqrt{3}x$

12. $(3\sqrt{x} + \sqrt{5y})(3\sqrt{x} - \sqrt{5y}) = (3\sqrt{x})^2 - \sqrt{5y}^2 = 9x - 5y$

13. $\frac{\sqrt[3]{4x^2y}}{\sqrt[3]{2x^2y^2}} = \sqrt[3]{\frac{4x^2y}{2x^2y^2}} = \sqrt[3]{\frac{2}{y}}\sqrt[3]{\frac{y^2}{y^2}} = \frac{\sqrt[3]{2y^2}}{y}$

14. $\sqrt{x+3} = 6$. Square both sides to get the result $x + 3 = 36$ or $x = 33$. Check: $\sqrt{33+3} = \sqrt{36} = 6$. Solution is 33.

15. $\sqrt{2x+1} = 5$. Squaring both sides produces the result $2x + 1 = 25$ or $2x = 24$ and so $x = 12$. Check: $\sqrt{2 \cdot 12 + 1} = \sqrt{24+1} = \sqrt{25} = 5$. Solution is 12

16. $\sqrt{2x^2 - 7} = x + 3$. Squaring both sides gives the result $\sqrt{2x^2-7}^2 = (x+3)^2$ or $2x^2 - 7 = x^2 + 6x + 9$ and so $x^2 - 6x - 16 = 0$. This factors as $(x-8)(x+2) = 0$, so $x = 8$ or $x = -2$. Check: $\sqrt{2 \cdot 8^2 - 7} = \sqrt{2 \cdot 64 - 7} = \sqrt{128 - 7} = \sqrt{121} = 11$; $8 + 3 = 11$; 8 works; $\sqrt{2(-2)^2 - 7} = \sqrt{2(4) - 7} = \sqrt{8 - 7} = 1$; $-2 + 3 = 1$ also, so -2 works. Solutions are 8 and -2.

17. $\sqrt{\frac{5}{x+3}} = \sqrt{\frac{4}{x-3}}$. Square both sides to get $\frac{5}{x+3} = \frac{4}{x-3}$ and so $5(x-3) = 4(x+3)$ or $5x - 15 = 4x + 12$ which simplifies to $x = 27$.
Check: $\sqrt{\frac{5}{27+3}} = \sqrt{\frac{5}{30}} = \sqrt{\frac{1}{6}}$; $\sqrt{\frac{4}{27-3}} = \sqrt{\frac{4}{24}} = \sqrt{\frac{1}{6}}$. Solution is 27.

18. (a) $d = (1.12 \times 10^{-2})\sqrt[3]{F\ell}$ or $\sqrt[3]{F\ell} = \frac{d}{1.12 \times 10^{-2}} = 8.92857 \times 10^1 d$. Cubing both sides produces $F\ell = (8.92857 \times 10^1 d)^3 = 7.12 \times 10^5 d^3$ and so $F = \frac{7.12 \times 10^5 d^3}{\ell}$, (b) $F = \frac{7.12 \times 10^5 (1.2)^3}{5} = 246{,}067.2$ kg

CHAPTER

13

Exponential and Logarithmic Functions

13.1 EXPONENTIAL FUNCTIONS

1. $3^{\sqrt{2}} = 4.7288$

2. $4^{\pi} = 77.8802$

3. $\pi^3 = 31.0063$

4. $\left(\sqrt{3}\right)^{\pi} = 5.6164$

5. $\left(\sqrt{3}\right)^{\sqrt{5}} = 3.4154$

6. $\left(\sqrt{4}\right)^{4/3} = 2^{4/3} = 2.5198$

7.

x	−3	−2	−1	0	1	2	3	4	5
$f(x)$	$\frac{1}{64}$	$\frac{1}{16}$	$\frac{1}{4}$	1	4	16	64	256	1024

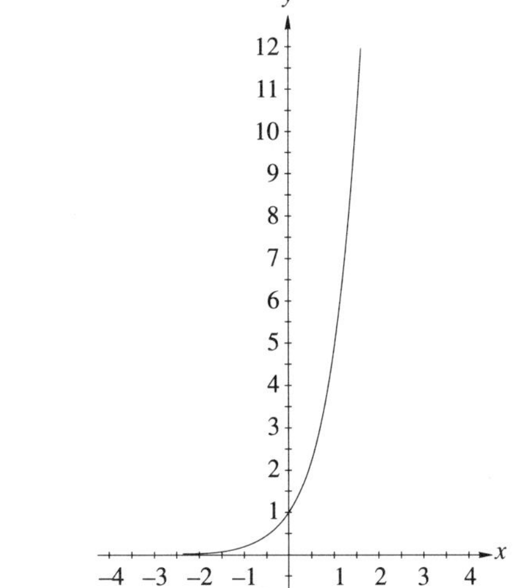

8.

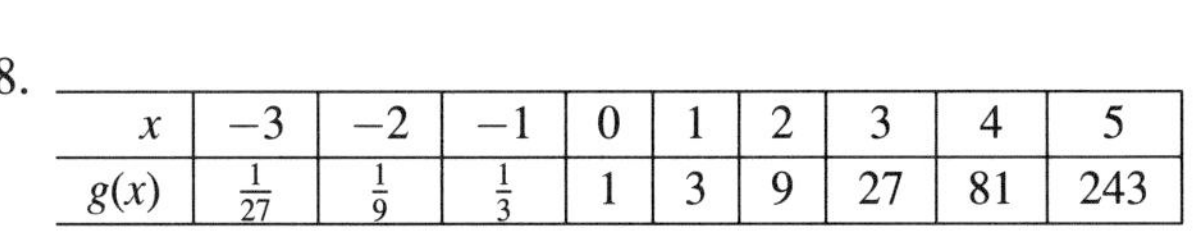

x	−3	−2	−1	0	1	2	3	4	5
$g(x)$	$\frac{1}{27}$	$\frac{1}{9}$	$\frac{1}{3}$	1	3	9	27	81	243

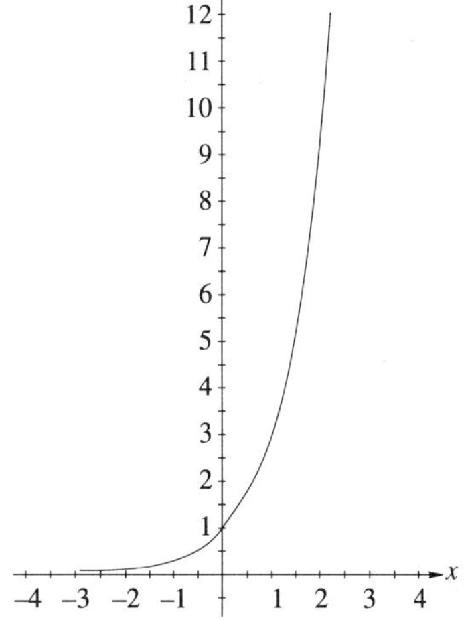

9.

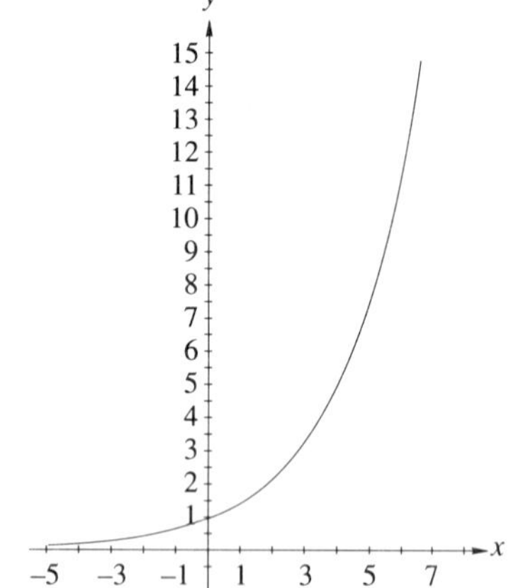

x	-3	-2	-1	0	1	2	3	4	5
$h(x)$	$\frac{8}{27}$	$\frac{4}{9}$	$\frac{2}{3}$	1	1.5	2.25	3.375	5.0625	7.59375

10.

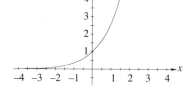

x	-3	-2	-1	0	1	2	3	4	5
$k(x)$	0.064	0.16	0.4	1	2.5	6.25	15.625	39.0625	97.65625

11.

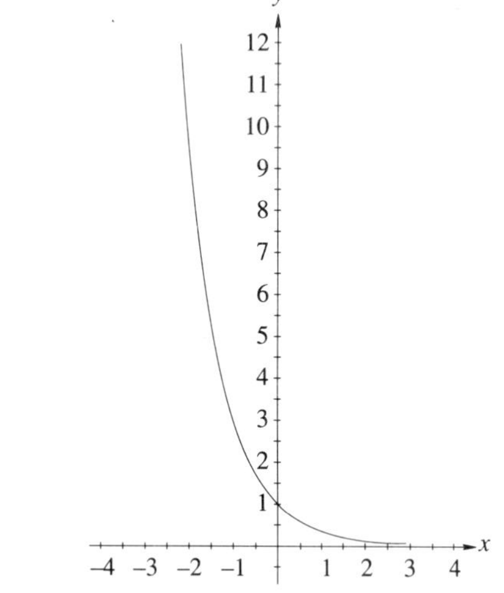

x	-3	-2	-1	0	1	2	3	4	5
$f(x)$	27	9	3	1	$\frac{1}{3}$	$\frac{1}{9}$	$\frac{1}{27}$	$\frac{1}{81}$	$\frac{1}{243}$

12.

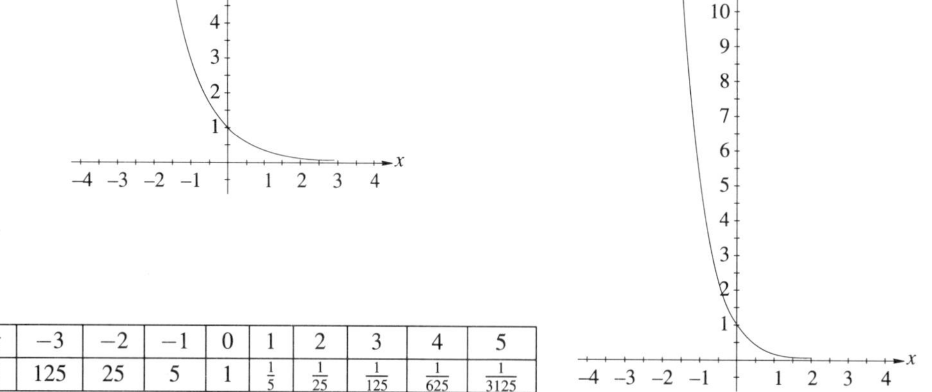

x	-3	-2	-1	0	1	2	3	4	5
$g(x)$	125	25	5	1	$\frac{1}{5}$	$\frac{1}{25}$	$\frac{1}{125}$	$\frac{1}{625}$	$\frac{1}{3125}$

13.

x	−3	−2	−1	0	1	2	3	4	5
$h(x)$	13.824	5.760	2.4	1	0.417	0.174	0.072	0.030	0.0126

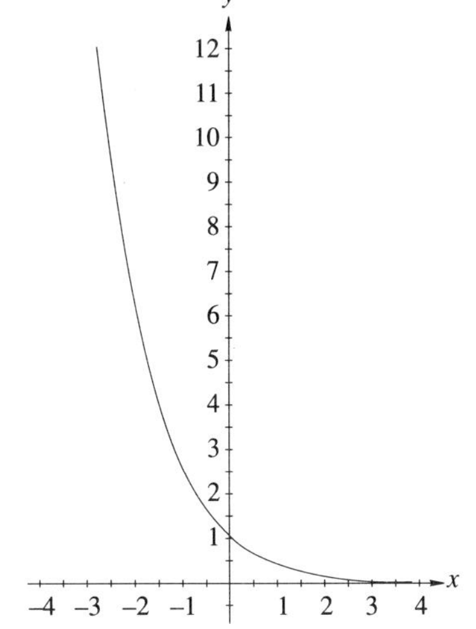

14.

x	−3	−2	−1	0	1	2	3	4	5
$k(x)$	0.192	$\frac{1}{3}$	0.577	1	$\sqrt{3}$	3	$3\sqrt{3}$	9	$9\sqrt{3}$

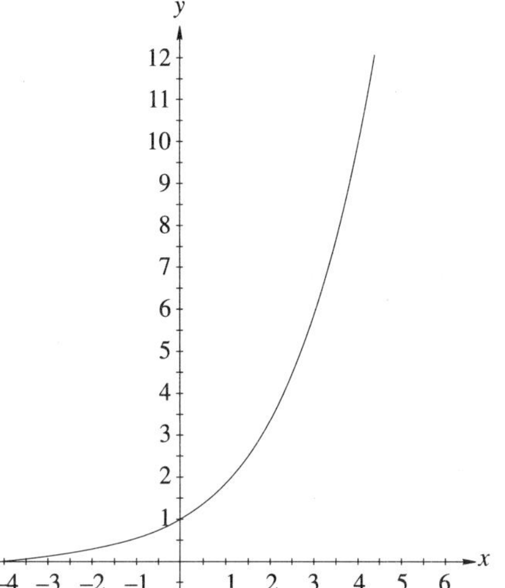

15.

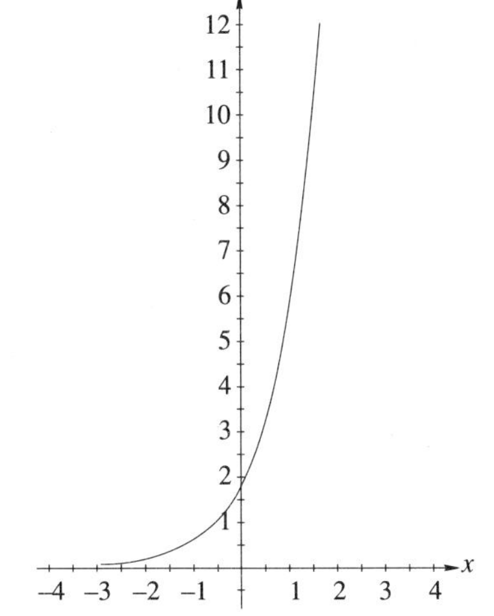

x	−3	−2	−1	0	1	2	3	4
$f(x)$	0.064	0.192	0.577	1.732	5.196	15.588	46.765	140.296

16.

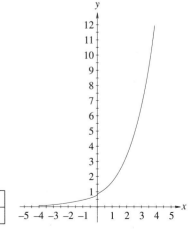

x	−3	−2	−1	0	1	2	3	4	5
$g(x)$	0.105	0.210	0.420	0.841	1.682	3.364	6.727	13.454	26.908

17. (a) $S = 1000(1 + 0.06)^5 = \$1338.23$, (b) $S = 1000(1 + \frac{0.06}{2})^{2\times5} = 1000(1.03)^{10} = \$1343.92$, (c) $S = 1000(1 + \frac{0.06}{4})^{4\times5} = 1000(1.015)^{20} = \$1346.86$, (d) $S = 1000(1 + \frac{0.06}{12})^{12\times5} = 1000(1.005)^{60} = \$1348.85$

18. $S = P(1 + \frac{r}{k})^{kt}$, (a) $S = 2000(1 + 0.08)^{10} = \$4317.85$, (b) $S = 2000(1 + \frac{0.08}{2})^{2\times10} = \$4382.25$, (c) $S = 2000(1 + \frac{0.08}{4})^{4\times10} = \$4416.08$, (d) $S = 2000(1 + \frac{0.08}{12})^{12\times10} = \$4439.28$

19. First Bank: $S = 1000(1 + \frac{0.06}{2})^{2\times5} = 1343.92$. Second Bank: $S = 1000(1+\frac{0.06}{12})^{12\times5} = 1348.85$. Since $1348.85 - 1343.92 = \$4.93$, you will earn \$4.93 more at the second bank.

20. First Bank: $S = 2000(1 + \frac{.08}{1})^{1\times10} = 4317.85$, Second Bank: $S = 2000(1 + \frac{.08}{4})^{4\times10} = 4416.08$, $4416.08 - 4317.85 = \$98.23$

13.2 THE EXPONENTIAL FUNCTION e^x

1. $e^3 = 20.085\,537$
2. $e^7 = 1096.633\,158$
3. $e^{4.65} = 104.584\,986$
4. $e^{5.375} = 215.939\,872$
5. $e^{-4} = 0.018\,316$
6. $e^{-9} = 1.234\,098 \times 10^{-4}$ or $0.000\,123\,4$
7. $e^{-2.75} = 0.063\,928$
8. $e^{-0.25} = 0.778\,801$

9.

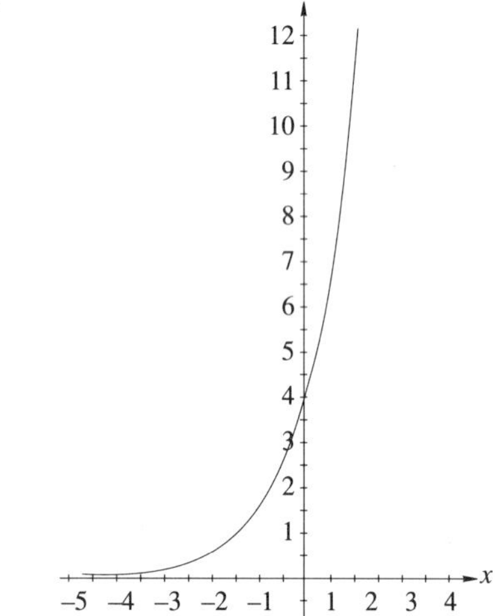

x	−2	−1	0	1	2	3	4
$f(x)$	0.5413	1.4715	4	10.8731	29.5562	80.3421	218.3926

10.

x	−1	0	1	2	3	4
$g(x)$	0.0236	3.5	519.446	77 092.630	11 441 560.8	1.6981×10^9

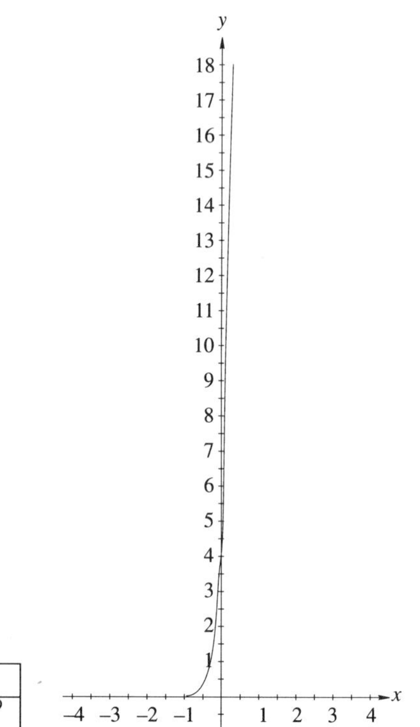

11.

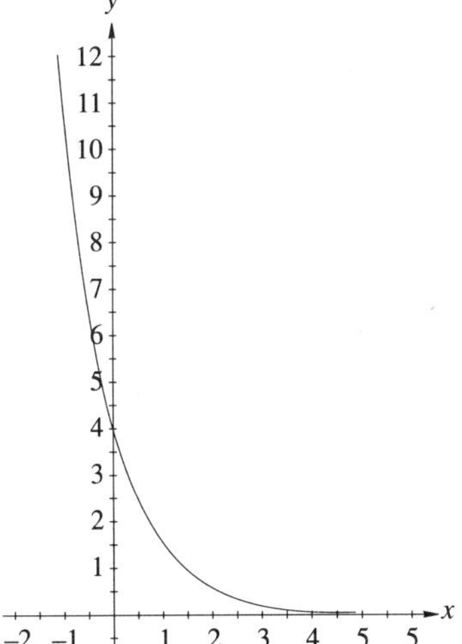

x	-4	-3	-2	-1	0	1	2
$h(x)$	218.3926	80.3421	29.5562	10.8731	4	1.4715	0.5413

12.

x	-4	-3	-2	-1	0	1	2
$k(x)$	2.252×10^{11}	5.581×10^{8}	1.383×10^{6}	3429.14	8.5	0.0211	5.223×10^{-5}

13. $Q = 125e^{-0.375t}$ (a) when $t = 0$, we get $e^0 = 1$ so Q is 125 mg (b) when $t = 1$, we have $Q = 125e^{-0.375} = 85.91$ mg (c) when $t = 16$, we obtain $Q = 125e^{-0.375\times 16} = 0.3098$ mg or 0.31 mg

14. A half life of 1600 years yields $\frac{1}{2} = e^{1600K}$. The initial amount is 100. $2000 \div 1600 = 1.25$. $y = 100e^{2000K} = 100(e^{1600K})^{1.25} = 100(\frac{1}{2})^{1.25} = 42.04$.

15. Initial amount is 5000. $y = ce^{kt}$. $15{,}000 = 5000e^{20kt}$; $3 = e^{20kt}$.

    (a) when $t = 10$, $y = 5000e^{k10} = 5000(e^{20k})^{1/2} = 5000 \cdot 3^{1/2} = 8{,}660$.

    (b) when $t = 30$, $y = 5000e^{30k} = 5000(e^{20k})^{1.5} = 5000 \times 3^{1.5} = 25{,}981$.

    (c) when $t = 3$ days $= 3 \times 24$ hr $= 72$ hr. $72 \div 20 = 3.6$, $y = 5000e^{72k} = 5000(e^{20k})^{3.6} = 5000 \cdot 3^{3.6} = 260{,}980$.

16. $S = Pe^{rt} = 5000e^{.05\times 10} = \$8243.61$.

17. Method 1. Use the compound interest formula $S = P(1 + r)^t$

    (a) $S = 200{,}000(1.07)^5 = 280{,}510$.

    (b) $S = 200{,}000(1.07)^{10} = 393{,}430$.

    Method 2. Use the growth formula $y = ce^{kt}$ with $c = 200{,}000$ and $e^k = 1.07$.

    (a) $y = 200{,}000e^{5k} = 200{,}000(e^k)^5 = 200{,}000(1.07)^5 = 280{,}510$.

    (b) $y = 200{,}000e^{10k} = 200{,}000(e^k)^{10} = 200{,}000(1.07)^{10} = 393{,}430$.

18. Use $S = Pe^{rt}$, with $S = 1000$, $r = 0.045$, and $t = 7 \times 24$. This gives $1000 = Pe^{rt} = Pe^{0.045\times 7\times 24} = Pe^{7.56}$, and so $P = \frac{1000}{e^{7.56}} = 0.521$ units.

19. Use $T = T_m + (T_0 - T_m)e^{-kt}$, with $T = 150°$, $T_0 = 180°$, $T_m = 60°$, and $t = 20$ min. This gives $150 = 60 + (180 - 60)e^{-k\cdot 20}$, or $90 = (120)e^{-k20}$, and so $e^{-k20} = 0.75$. When $t = 1$ hr $= 60$ min, we have $60 = 20 \times 3$. $T = 60 + (180 - 60)e^{-k\cdot 60} = 60 + 120(e^{-k\cdot 20})^3$ or $T = 60 + 120(0.75)^3 = 60 + 50.625 \approx 110.6°$F.

20. Use $T = T_m + (T_0 - T_m)e^{-kt}$, with $T = 90°$, $T_m = 30°$, $T_0 = 150°$, and $t = 15$. This gives $90 = 30 + (150 - 30)e^{-k15}$, or $60 = 120e^{-k\cdot 15}$ which leads to $\frac{1}{2} = e^{-k\cdot 15}$. When $t = 30$, you get $T = 30 + (150 - 30)e^{-k30} = 30 + 120(e^{-k\cdot 15})^2 = 30 + 120\left(\frac{1}{2}\right)^2 = 30 + 30 = 60°$C.

21. Using the formula for Newton's law of cooling $T = T_m + (T_0 - T_m)e^{-kt}$, with $T = 70$ when $t = 25$, $T_0 = 87°$, and $T_m = 37°$. This gives $70 = 37 + (87 - 37)e^{-k25}$, or $33 = 50e^{-k25}$ or $e^{-25k} = 0.66$ which is $(e^{-k})^{25} = 0.66$, and so $e^{-k} = 0.66^{1/25} = 0.9835$. Using this value of $e^{-k} = 0.9835$ with $T = 45$ you get $45 = 37 + (87 - 37)e^{-kt}$ or $8 = 50e^{-kt}$ or $e^{-kt} = 0.16$, and so $0.9835^t = 0.16$. Thus, $t = \frac{\log 0.16}{\log 0.9835} = 110.26$, and we have $t = 110.26$ or about 1 hr and 50 min.

22. $I = I_0(1 - e^{-t/T})$; $I_0 = \frac{V}{R}$, $T = \frac{L}{R}$; $V = 120$, $R = 40$, $L = 30$; $I_0 = \frac{120}{40} = 3$, $T = \frac{3.0}{40} = \frac{3}{40}$ (a) When $t = 0.01$ s; $-t/T = \frac{-0.01}{3/40} = -0.1333$; $I = 3(1 - e^{-.1333}) = 0.374$ A; (b) When $t = 0.1$ s; $-t/T = \frac{-0.1}{3/40} = -1.333$; $I = 3(1 - e^{-1.333}) = 2.209$ A (c) When $t = 1.0$ s; $-t/T = -13.33$; $I = 3(1 - e^{-13.33}) = 3.000$ A

23. Use $Q = Q_0e^{-t/T}$, where $Q_0 = CV$, and $T = RC$. We have $C = 130\ \mu\text{F} = 130 \times 10^{-6}$ F, $V = 120$ V, and $R = 4500\ \Omega$. So, $Q_0 = \left(130 \times 10^{-6}\right)120 = 1.560 \times 10^{-2}$, and $T = 4500 \times 130 \times 10^{-6} = 0.585$. Hence $Q = 1.56 \times 10^{-2} \cdot e^{\frac{-0.5}{0.585}} = 0.0066$ C.

24. A half life of 12.5 years means $e^{12.5k} = \frac{1}{2}$ so $y = 100e^{40k} = 100\left(e^{12.5k}\right)^{3.2} = 100(\frac{1}{2})^{3.2} = 10.88$ g. (Note $40 \div 12.5 = 3.2$.)

13.3 LOGARITHMIC FUNCTIONS

1. $\log_6 216 = 3$; $6^3 = 216$
2. $\log_9 6561 = 4$; $9^4 = 6561$
3. $\log_4 16 = 2$; $4^2 = 16$
4. $\log_7 16{,}807 = 5$; $7^5 = 16{,}807$
5. $\log_{1/7} \frac{1}{49} = 2$; $(\frac{1}{7})^2 = \frac{1}{49}$
6. $\log_{1/2} \frac{1}{64} = 6$; $(\frac{1}{2})^6 = \frac{1}{64}$
7. $\log_2 \frac{1}{32} = -5$; $2^{-5} = \frac{1}{32}$
8. $\log_3 \frac{1}{243} = -5$; $3^{-5} = \frac{1}{243}$
9. $\log_9 2187 = \frac{7}{2}$; $9^{\frac{7}{2}} = 2187$
10. $\log_8 2048 = \frac{11}{3}$; $8^{\frac{11}{3}} = 2048$
11. $5^4 = 625$; $\log_5 625 = 4$
12. $3^5 = 243$; $\log_3 243 = 5$
13. $2^7 = 128$; $\log_2 128 = 7$
14. $4^3 = 64$; $\log_4 64 = 3$
15. $7^3 = 343$; $\log_7 343 = 3$
16. $11^2 = 121$; $\log_{11} 121 = 2$
17. $5^{-3} = \frac{1}{125}$; $\log_5 \frac{1}{125} = -3$
18. $3^{-5} = \frac{1}{243}$; $\log_3 \frac{1}{243} = -5$
19. $4^{7/2} = 128$; $\log_4 128 = \frac{7}{2}$
20. $125^{5/3} = 3125$; $\log_{125} 3125 = \frac{5}{3}$
21. $\ln 5 = 1.609\,437\,912$
22. $\ln 19 = 2.944\,438\,979$
23. $\ln 4.751 = 1.558\,355\,122$
24. $\ln 35.62 = 3.572\,907\,278$
25. $\log 4 = 0.602\,059\,991\,3$
26. $\log 23 = 1.361\,727\,836$
27. $\log 12.67 = 1.102\,776\,615$
28. $\log 78.143 = 1.892\,890\,08$
29. $\log_5 8 = \frac{\ln 8}{\ln 5} = 1.292\,029\,674$
30. $\log_3 20 = \frac{\ln 20}{\ln 3} = 2.726\,833\,028$
31. $\log_{12} 16.4 = \frac{\ln 16.4}{\ln 12} = 1.125\,708\,821$
32. $\log_8 691.45 = \frac{\ln 691.45}{\ln 8} = 3.144\,493\,707$

33.

x	0.1	0.5	1	2	4	6	8	10
$\ln x$	−2.30	−0.69	0	0.69	1.39	1.79	2.08	2.30

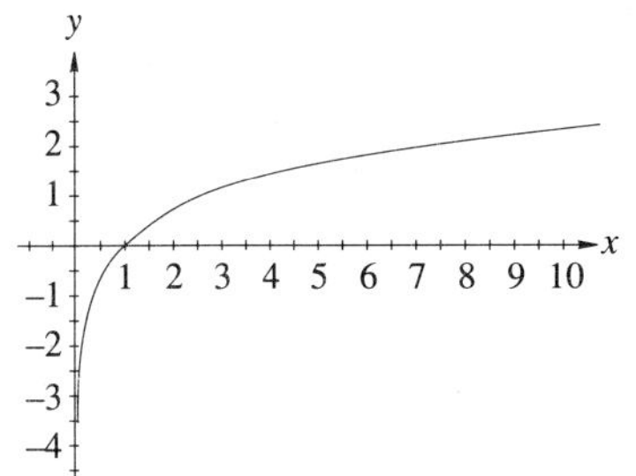

34.

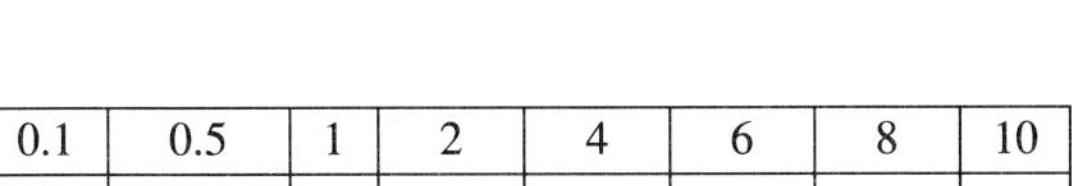

x	0.1	0.5	1	2	4	6	8	10
$\log x$	−1	−0.30	0	0.30	0.60	0.78	0.90	1

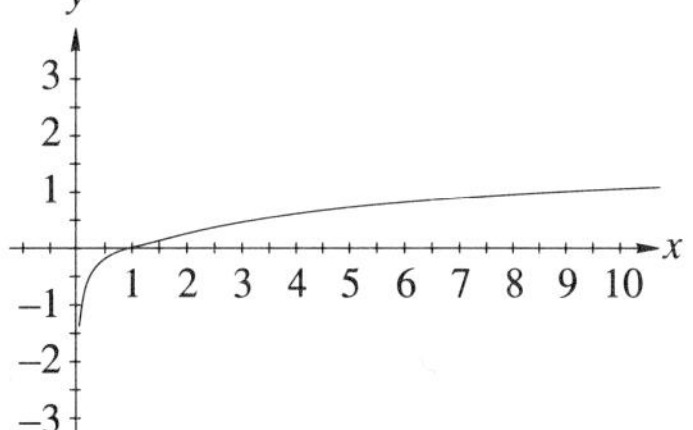

35.

x	0.1	0.5	1	2	4	6	8	10
$\log_2 x$	−3.32	−1	0	1	2	2.58	3	3.32

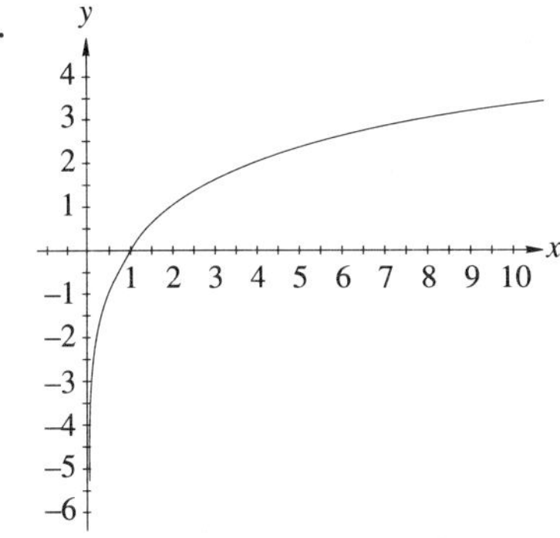

36.

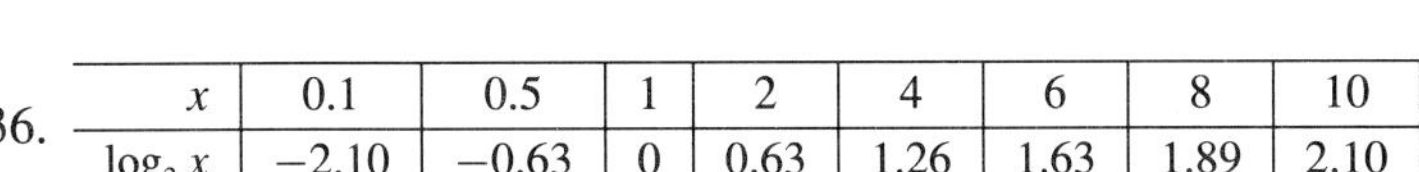

x	0.1	0.5	1	2	4	6	8	10
$\log_3 x$	−2.10	−0.63	0	0.63	1.26	1.63	1.89	2.10

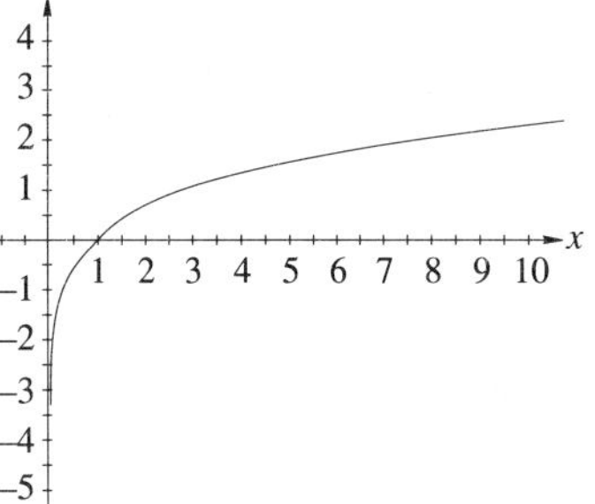

37.

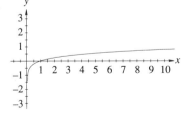

x	0.1	0.5	1	2	4	6	8	10
$\log_{12} x$	−0.93	−0.28	0	0.28	0.56	0.72	0.84	0.93

38.

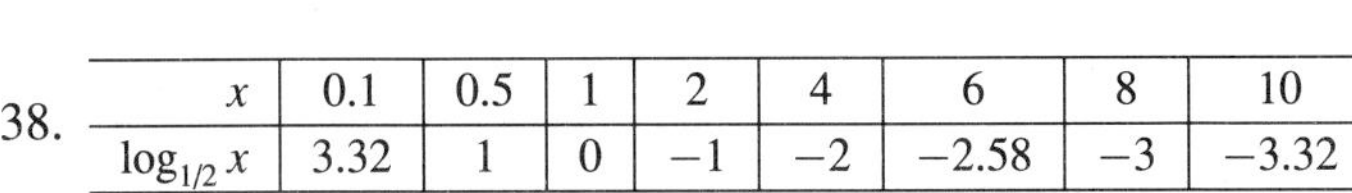

x	0.1	0.5	1	2	4	6	8	10
$\log_{1/2} x$	3.32	1	0	−1	−2	−2.58	−3	−3.32

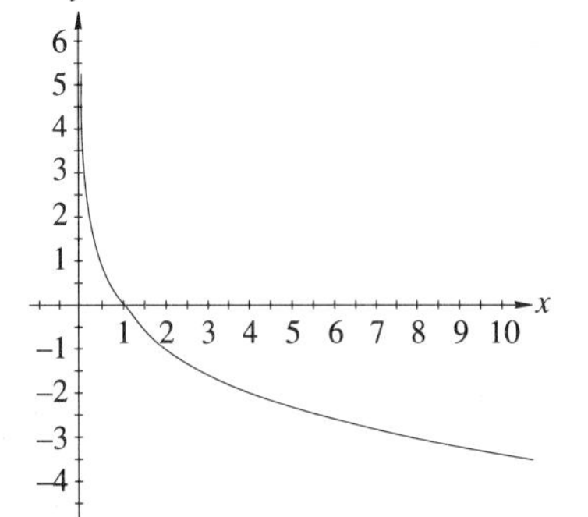

39.

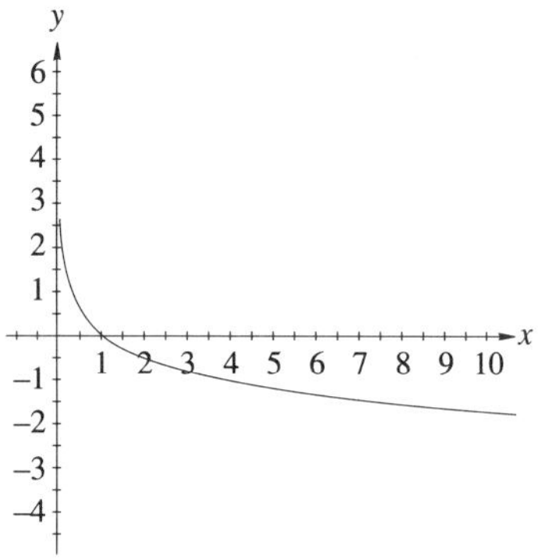

x	0.1	0.5	1	2	4	6	8	10
$\log_{1/4} x$	1.66	0.5	0	−0.5	−1	−1.29	−1.5	−1.66

40. $R = \log I = \log_{10} I$; $10^R = I$

41. (a) $\beta = 10 \log \frac{I}{I_o}$; $\frac{\beta}{10} = \log \frac{I}{I_0}$, $10^{\beta/10} = \frac{I}{I_0}$; $I = I_0 10^{\beta/10}$

(b) $\beta = 10 \log \left(\frac{10^{-7}}{10^{-12}} \right) = 10 \log 10^5 = 10 \cdot 5 =$ 50 dB

(c) $\beta = 10 \log \left(\frac{10^{-3}}{10^{-12}} \right) = 10 \log 10^9 = 10 \cdot 9 =$ 90 dB

13.4 PROPERTIES OF LOGARITHMS

1. $\log \frac{2}{3} = \log 2 - \log 3$

2. $\log \frac{5}{4} = \log 5 - \log 4$

3. $\log 14 = \log(2 \cdot 7) = \log 2 + \log 7$

4. $\log 55 = \log(5 \cdot 11) = \log 5 + \log 11$

5. $\log 12 = \log(2 \cdot 2 \cdot 3) = \log 2 + \log 2 + \log 3$

6. $\log 28 = \log(2 \cdot 2 \cdot 7) = \log 2 + \log 2 + \log 7$

7. $\log \frac{150}{7} = \log \left(\frac{2 \cdot 3 \cdot 5 \cdot 5}{7} \right) = \log 2 + \log 3 + \log 5 + \log 5 - \log 7$

8. $\log \left(\frac{588}{5x} \right) = \log 588 - \log 5x = \log(2 \cdot 2 \cdot 3 \cdot 7 \cdot 7) - \log(5 \cdot x) = \log 2 + \log 2 + \log 3 + \log 7 + \log 7 - \log 5 - \log x$

9. $\log 2x = \log 2 + \log x$

10. $\log \frac{1}{5x} = \log 1 - \log(5x) = 0 - \log 5 - \log x$; $= -\log 5 - \log x$

11. $\log \frac{2ax}{3y} = \log 2ax - \log 3y = \log 2 + \log a + \log x - \log 3 - \log y = \log 2 + \log a + \log x - (\log 3 + \log y) = \log 2ax - \log 3y$

12. $\log \frac{4bc}{3xy} = \log 4bc - \log 3xy = \log 4 + \log b + \log c - \log 3 - \log x - \log y = \log 4 + \log b + \log c - (\log 3 + \log x + \log y)$

13. $\log 2 + \log 11 = \log(2 \cdot 11) = \log 22$

14. $\log 3 + \log 13 = \log(3 \cdot 13) = \log 39$

15. $\log 11 - \log 3 = \log \frac{11}{3}$

16. $\log 17 - \log 23 = \log \frac{17}{23}$

17. $\log 2 + \log 2 + \log 3 = \log(2 \cdot 2 \cdot 3) = \log 12$

18. $\log 3 + \log 3 + \log 5 = \log(3 \cdot 3 \cdot 5) = \log 45$

19. $\log 4 + \log x - \log y = \log \frac{4x}{y}$

20. $\log 5 + \log 7 - \log 11 = \log \frac{5 \cdot 7}{11} = \log \frac{35}{11}$

21. $5 \log 2 + 3 \log 5 = \log 2^5 + \log 5^3 = \log(2^5 \cdot 5^3) = \log 4{,}000$

22. $7 \log 3 - 2 \log 8 = \log 3^7 - \log 8^2 = \log \frac{3^7}{8^2} = \log \frac{2187}{64}$

23. $\log \frac{2}{3} + \log \frac{6}{7} = \log(\frac{2}{3} \cdot \frac{6}{7}) = \log \frac{4}{7}$

24. $\log \frac{3}{24} + \log \frac{4}{7} - \log \frac{1}{3} = \log \left(\frac{3}{24} \cdot \frac{4}{7} \cdot 3 \right) = \log \frac{3}{14}$

25. $\log 8 = \log 2^3 = 3\log 2 = 3(0.3010) = 0.9030$

26. $\log 9 = \log 3^2 = 2\log 3 = 2(0.4771) = 0.9542$

27. $\log 12 = \log 2^2 \cdot 3 = 2\log 2 + \log 3 = 2(0.3010) + (0.4771) = 0.6020 + 0.4771 = 1.0791$

28. $\log 36 = \log 2^2 \cdot 3^2 = 2\log 2 + 2\log 3 = 2(0.3010) + 2(0.4771) = 0.6020 + .9542 = 1.5562$

29. $\log 15 = \log 3 \cdot 5 = \log 3 + \log 5 = 0.4771 + 0.6990 = 1.1761$

30. $\log 30 = \log(2 \cdot 3 \cdot 5) = \log 2 + \log 3 + \log 5 = 0.3010 + 0.4771 + 0.6990; = 1.4771$

31. $\log \frac{75}{2} = \log(3 \cdot 5 \cdot 5/2) = \log 3 + 2\log 5 - \log 2 = 0.4771 + 2(0.6990) - (0.3010) = 1.5741$

32. $\log \frac{45}{8} = \log(5 \cdot 3^2/2^3) = \log 5 + 2\log 3 - 3\log 2; = 0.6990 + 2(0.4771) - 3(0.3010) = 0.7502$

33. $\log 200 = \log 2 \cdot 10^2 = \log 2 + 2\log 10 = 0.3010 + 2 \cdot 1 = 2.3010$ (Note: $\log 10 = 1$)

34. $\log 150 = \log 10 \cdot 3 \cdot 5 = \log 10 + \log 3 + \log 5 = 1 + 0.4771 + 0.6990 = 2.1761$

35. $\log 5000 = \log 10^3 \cdot 5 = 3\log 10 + \log 5 = 3 \cdot 1 + 0.6990 = 3.6990$

36. $\log 4500 = \log 10^2 \cdot 5 \cdot 3^2 = 2\log 10 + \log 5 + 2\log 3 = 2 \cdot 1 + 0.6990 + 2(0.4771) = 3.6532$

37. $\log_b 16 = \log_b 2^4 = 4\log_b 2 = 4(0.3869) = 1.5476$

38. $\log_b 15 = \log_b(3 \cdot 5) = \log_b 3 + \log_b 5 = 0.6131 + 0.8982 = 1.5113$

39. $\log_b \frac{5}{3} = \log_b 5 - \log_b 3 = 0.8082 - 0.6131 = 0.2851$

40. $\log_b \frac{24}{5} = \log_b(2^3 \cdot 3/5) = 3\log_b 2 + \log_b 3 - \log_b 5 = 3(0.3869) + 0.6131 - 0.8082 = 0.8756$

41. $\log 2 = \frac{\ln 2}{\ln b} = 0.3869$; $\ln b = \ln 2/0.3869 = 1.7915$; Since e^x and $\ln x$ are inverse functions; $b = e^{1.7915} = 5.99$ or $b = 6$; (Note: $\log_b 6 = \log_b 2 + \log_b 3 = 0.3869 + 0.6131 = 1.0000$)

13.5 EXPONENTIAL AND LOGARITHMIC EQUATIONS

1. $5^x = 29$; $\log 5^x = \log 29$; $x\log 5 = \log 29$; $x = \frac{\log 29}{\log 5} = 2.0922$

2. $6^y = 32$; $\log 6^y = \log 32$; $y\log 6 = \log 32$; $y = \frac{\log 32}{\log 6} = 1.9343$

3. $4^{6x} = 119$; $\log 4^{6x} = \log 119$; $6x\log 4 = \log 119$; $x = \frac{\log 119}{6\log 4} = 0.5746$

4. $3^{-7x} = 4$; $\log 3^{-7x} = \log 4$; $-7x\log 3 = \log 4$; $x = \frac{\log 4}{-7\log 3} = -0.1803$

5. $3^{y+5} = 16$; $\ln 3^{y+5} = \ln 16$; $(y+5)\ln 3 = \ln 16$; $y + 5 = \frac{\ln 16}{\ln 3}$; $y = \frac{\ln 16}{\ln 3} - 5 = -2.4763$

6. $2^{y-7} = 67$; $\ln 2^{y-7} = \ln 67$; $(y-7)\ln 2 = \ln 67$; $y - 7 = \frac{\ln 67}{\ln 2}$; $y = \frac{\ln 67}{\ln 2} + 7 = 13.066$

7. $e^{2x-3} = 10$; $\ln e^{2x-3} = \ln 10$; $2x - 3 = \ln 10$; $2x = \ln 10 + 3$; $x = \frac{\ln 10+3}{2} = 2.6513$

8. $4^{3y+5} = 30$; $\ln 4^{3y+5} = \ln 30$; $(3y+5)\ln 4 = \ln 30$; $3y + 5 = \frac{\ln 30}{\ln 4}$; $3y = \frac{\ln 30}{\ln 4} - 5$; $y = \frac{(\ln 30/\ln 4)-5}{3} = -0.8489$

9. $3^{4x+1} = 9^{3x-5}$; $3^{4x+1} = (3^2)^{3x-5}$; $3^{4x+1} = 3^{6x-10}$; $\log_3 3^{4x+1} = \log_3 3^{6x-10}$; $4x + 1 = 6x - 10$; $-2x = -11$; $x = \frac{11}{2} = 5.5$

10. $2^{3-2x} = 8^{1+2x}$; $2^{3-2x} = (2^3)^{1+2x}$; $2^{3-2x} = 2^{3+6x}$; Taking $\log_2$ on both sides gives $3 - 2x = 3 + 6x$; $-2x = 6x$; $8x = 0$ or $x = 0$.

11. $e^{3x-1} = 5e^{2x}$; $\ln e^{3x-1} = \ln(5e^{2x})$; $3x - 1 = \ln 5 + 2x$; $x = \ln 5 + 1 = 2.6094$

12. $3^{2-5x} = 8(9^{x-1})$; $3^{2-5x} = 8 \cdot 3^{2x-2}$; $\log_3 3^{2-5x} = \log_3(8 \cdot 3^{2x-2})$; $2 - 5x = \log_3 8 + 2x - 2$; $-7x = \log_3 8 - 4$; $x = \frac{4-\log_3 8}{7} = \frac{4-(\ln 8/\ln 3)}{7} = 0.3010$

13. $\log x = 2.3$; rewrite as $x = 10^{2.3} \approx 199.53$

14. $\ln x = 5.4$; rewrite as $x = e^{5.4} = 221.406$

15. $\log(x-5)=17$; $x-5=10^{17}$; $x=10^{17}+5$

16. $\ln(y+3)=19$; $y+3=e^{19}$; $y=e^{19}-3$

17. $\ln 2x+\ln x=9$; $\ln 2x^2=9$; $2x^2=e^9$; $x^2=\frac{e^9}{2}$; $x=\sqrt{\frac{1}{2}e^9}\approx 63.6517$

18. $2\ln x=\ln 4$; $\ln x^2=\ln 4$; $x^2=4$; $x=\pm 2$. Positive 2 is ok but -2 is not in the domain of ln so the only answer is 2.

19. $2\ln 3x+\ln 2=\ln x$; $\ln 9x^2+\ln 2=\ln x$; $\ln 18x^2=\ln x$; $18x^2=x$; $18x=e^0$; $18x=1$; $x=\frac{1}{18}\approx 0.056$

20. $2\ln(x+3)=\ln x+4$; $\ln(x+3)^2-\ln x=4$; $\ln\left(\frac{(x+3)^2}{x}\right)=4$; $\frac{(x+3)^2}{x}=e^4=54.59815$; $(x+3)^2=54.59815x$; $x^2+6x+9=54.59815x$; $x^2+(6-54.59815)x+9=0$. Using the quadratic formula gives $x=48.4122$ or $x=0.1859$

21. $\log(x-1)=2$; $x-1=10^2=100$; $x=101$

22. $\log(x-4)=\log x-4$; $\log(x-4)-\log x=-4$; $\log\frac{x-4}{x}=-4$; $\frac{x-4}{x}=10^{-4}=0.0001$; $x-4=.0001x$; $0.9999x=4$; $x=\frac{4}{0.9999}=4.0004$

23. $2\log(x-4)=2\log x-4$; $\log(x-4)^2-\log x^2=-4$; $\log\left(\frac{x-4}{x}\right)^2=-4$; $\left(\frac{x-4}{x}\right)^2=10^{-4}=0.0001$; $\frac{x-4}{x}=\sqrt{0.0001}=.01$; $x-4=.01x$; $0.99x=4$; $x=\frac{4}{0.99}=4.040404$

24. $\ln x=1+3\ln x$; $\ln x-3\ln x=1$; $\ln x-\ln x^3=1$; $\ln\left(\frac{x}{x^3}\right)=1$; $\ln\left(\frac{1}{x^2}\right)=1$; $\frac{1}{x^2}=e^1$; $x^2=\frac{1}{e}$; $x=\sqrt{\frac{1}{e}}=e^{-0.5}=0.6065$

25. $t=\frac{\ln 2}{r}=\frac{\ln 2}{0.06}=11.55$ years

26. $t=\frac{\ln 2}{0.08}=8.66$ years

27. $t=\frac{\ln 2}{0.005}=138.63$ years

28. $t=\frac{\ln 2}{0.03}=23.10$ years

29. $12.5=\frac{\ln 2}{r}$; $r=\frac{\ln 2}{12.5}=0.055$ or 5.5%

30. $15=\frac{\ln 2}{r}$; $r=\frac{\ln 2}{15}=0.046=4.6\%$ per hour

31. $8.5=\frac{\ln 2}{r}$; $r=\frac{\ln 2}{8.5}=0.0815$ or 8.15%

32. $m=200-2^{t/2}$; $100=200-2^{t/2}$; $-100=-2^{t/2}$; $2^{t/2}=100$; $\ln 2^{t/2}=\ln 100$; $t/2\ln 2=\ln 100$; $t/2=\frac{\ln 100}{\ln 2}$; $t=\frac{2\ln 100}{\ln 2}=13.29$ min

33. $-\log(2.5\times 10^{-6})=5.6$

34. $5.3=-\log\left[H^+\right]$; $\log H^+=-5.3$; $H^+=10^{-5.3}=5.01\times 10^{-6}$ moles per liter.

35. $H=(30T+8000)\ln\frac{P_0}{P}=(30(-5)+8000)\ln\frac{76}{43.29}=(7850)\ln\left(\frac{76}{43.29}\right)=4418.07$ m

36. $\beta=10\log\frac{I}{I_0}$; $105=10\log\frac{I}{10^{-12}}$; $10.5=\log\frac{I}{10^{-12}}$; $10.5=\log I-\log 10^{-12}$; $10.5=\log I+12$; $-1.5=\log I$; $I=10^{-1.5}$ W/m^2 ≈ 0.0316 W/m^2

37. $Q=Q_0e^{-t/T}$; $T=RC=500\times 100\times 10^{-6}=0.05$; $0.05=0.40e^{-t/0.05}$; $0.125=e^{-t/0.05}$; $\ln 0.125=-t/0.05$; $t=-0.05\ln 0.125=0.104$ s

38. $T=T_0+ce^{-kt}$; $c=T_I-T_0=1200-20=1180$; $40=20+(1180)e^{-.08t}$; $20=1180e^{-.08t}$; $0.016949=e^{-0.08t}$; $\ln 0.016949=-0.08t$; $t=\frac{\ln 0.016949}{-0.08}=50.97$ hours

39. $I=I_0(1-e^{-Rt/L})L=0.2$, $R=4$; $I_0=1.5$; $I=1.4$; $1.4=1.5(1-e^{-4t/0.2})$; $0.9\overline{3}=1-e^{-20t}$; $-0.06\overline{6}=-e^{-20t}$; $e^{-20t}=0.0\overline{6}$; $-20t=\ln 0.0\overline{6}$; $t=\frac{\ln 0.0\overline{6}}{-20}=0.1354$ seconds

13.6 GRAPHS USING SEMILOGARITHMIC AND LOGARITHMIC PAPER

1.

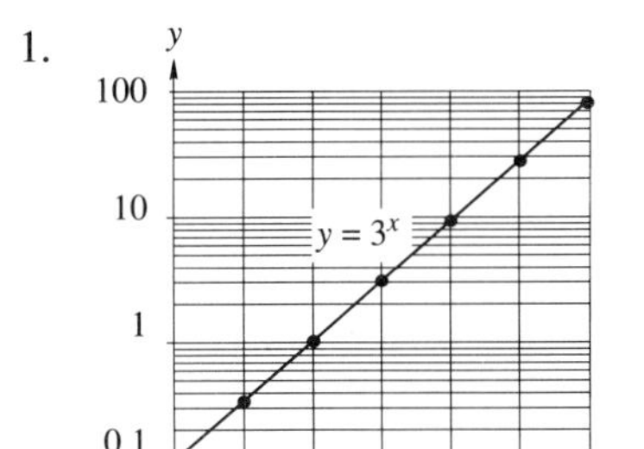

x	−3	−2	−1	0	1	2	3	4
y	0.037	0.11	0.33	1	3	9	27	81

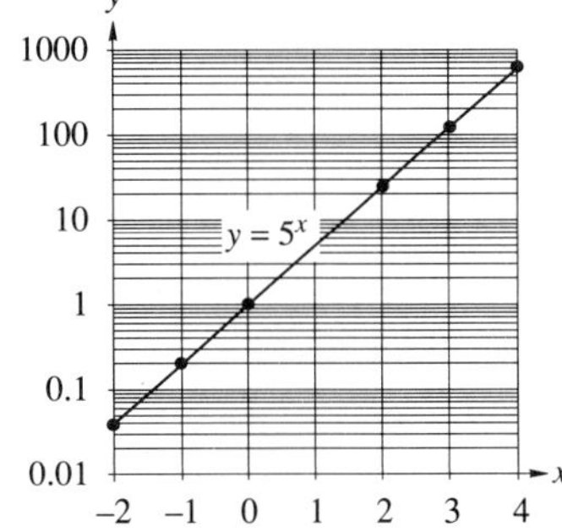

2.

x	−3	−2	−1	0	1	2	3	4
y	−.008	−.04	−.2	1	5	25	125	625

3.

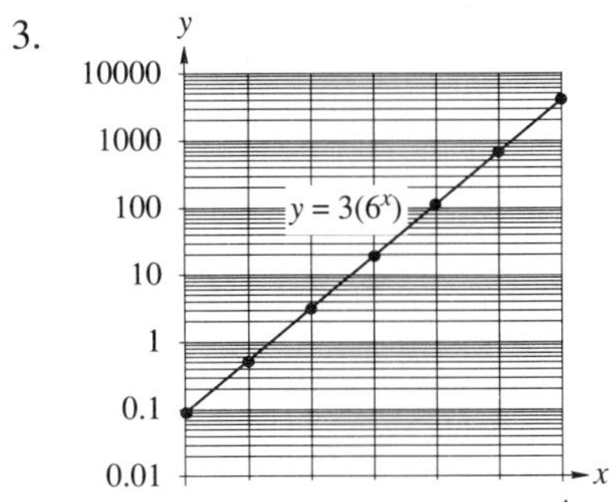

x	−3	−2	−1	0	1	2	3	4
y	0.014	0.083	0.5	3	18	108	648	3888

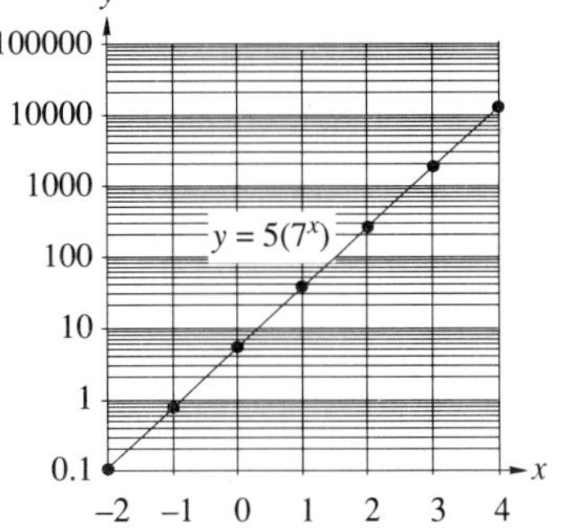

4.

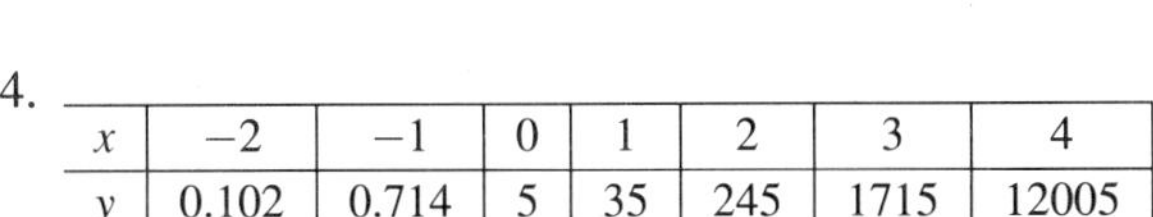

x	−2	−1	0	1	2	3	4
y	0.102	0.714	5	35	245	1715	12005

5.

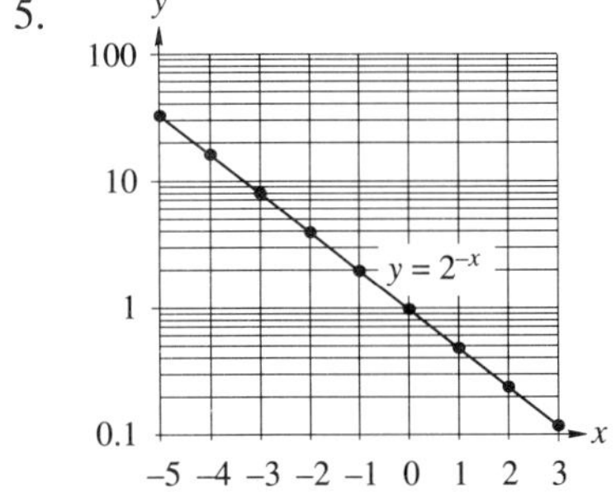

x	−5	−4	−3	−2	−1	0	1	2	3
y	32	16	8	4	2	1	1/2	1/4	1/8

6.

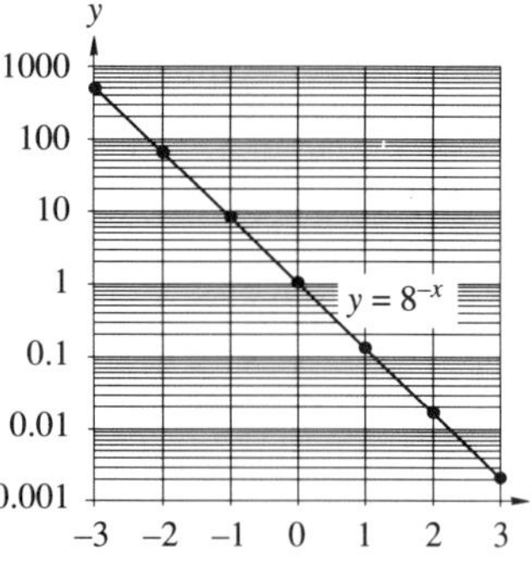

x	−3	−2	1	0	1	2	3
y	512	64	8	1	1/8	1/64	1/512

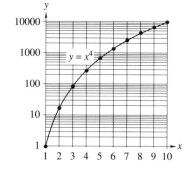

7.

x	1	2	3	4	5	6	7	8	9	10
y	1	16	81	256	625	1296	2401	4096	6561	10000

8.

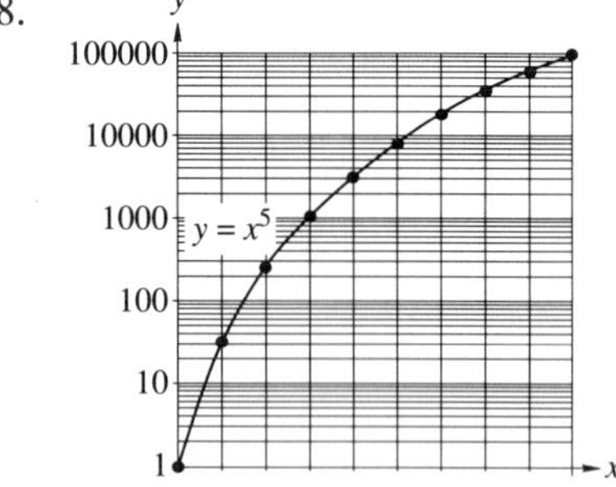

x	1	2	3	4	5	6	7	8	9
y	1	32	243	1024	3125	7776	16807	32768	59049

9.

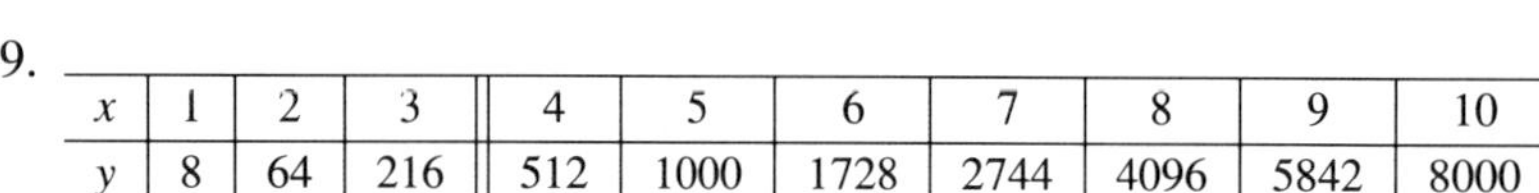

x	1	2	3	4	5	6	7	8	9	10
y	8	64	216	512	1000	1728	2744	4096	5842	8000

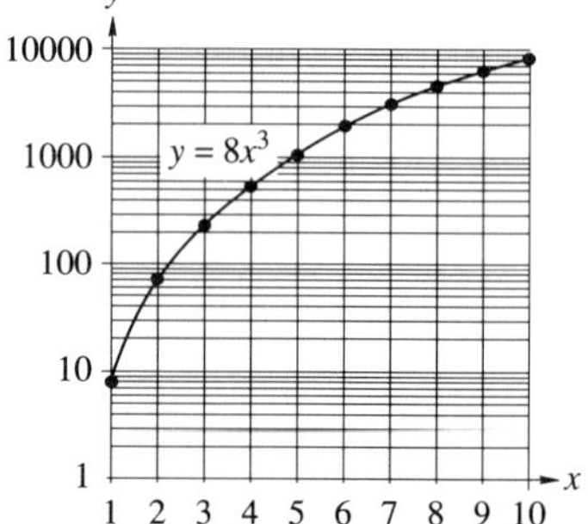

10.

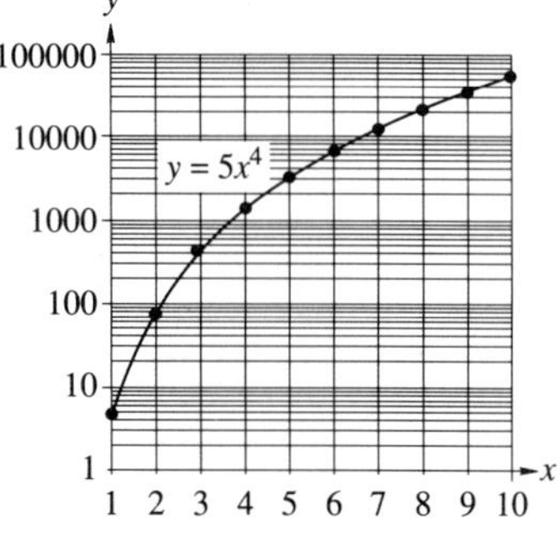

x	1	2	3	4	5	6	7	8	9
y	5	80	405	1280	3125	6480	12005	20480	32805

11.

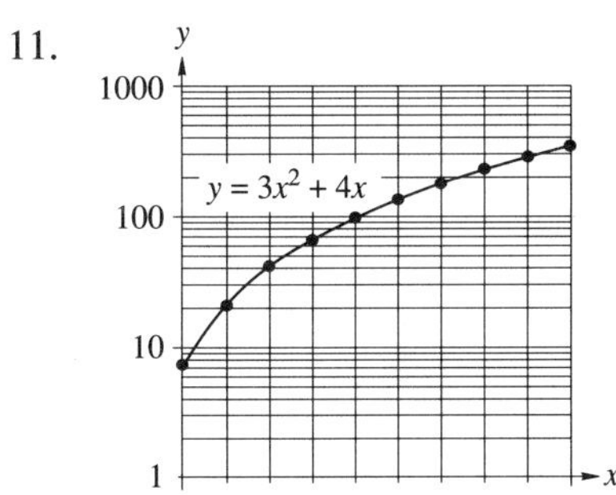

x	1	2	3	4	5	6	7	8
y	7	20	39	64	95	132	175	224

12.

x	1	2	3	4	5	6	7	8	9	10
y	11	74	237	548	1055	1806	2849	4232	6003	8210

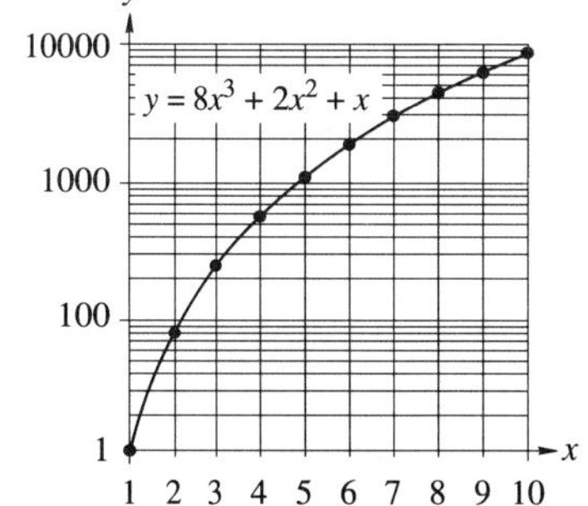

13.

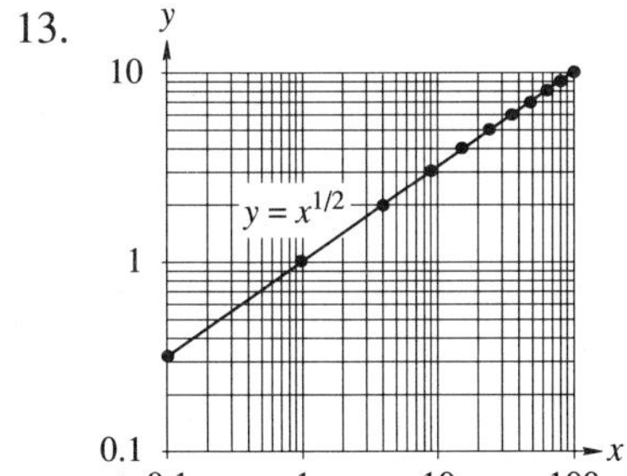

x	.1	.25	1	4	9	16	25	36	49	64	81
y	.316	.5	1	2	3	4	5	6	7	8	9

14.

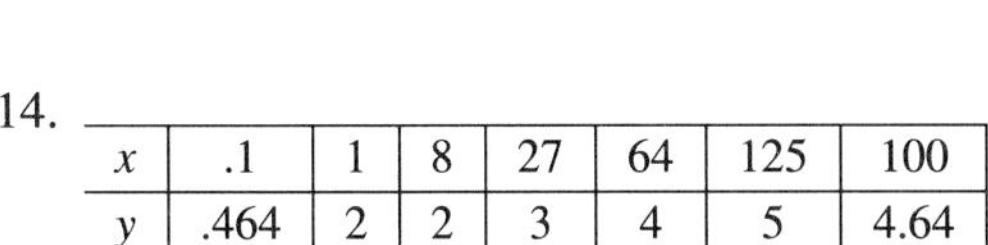

x	.1	1	8	27	64	125	100
y	.464	2	2	3	4	5	4.64

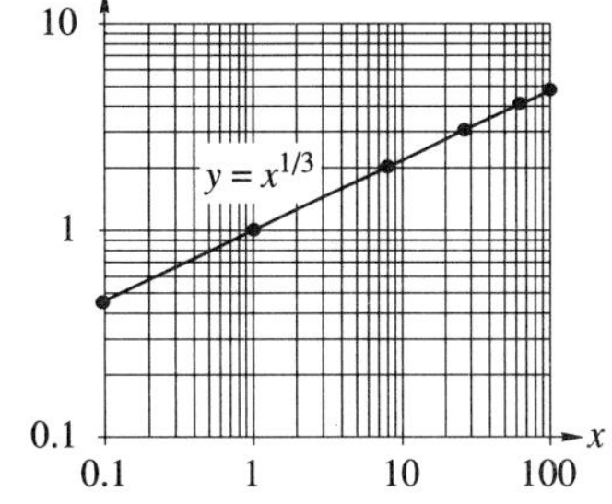

15.

x	.1	.5	1	2	3	4	5	6	7	8
y	.002	.25	2	16	54	128	250	432	686	1024

16.

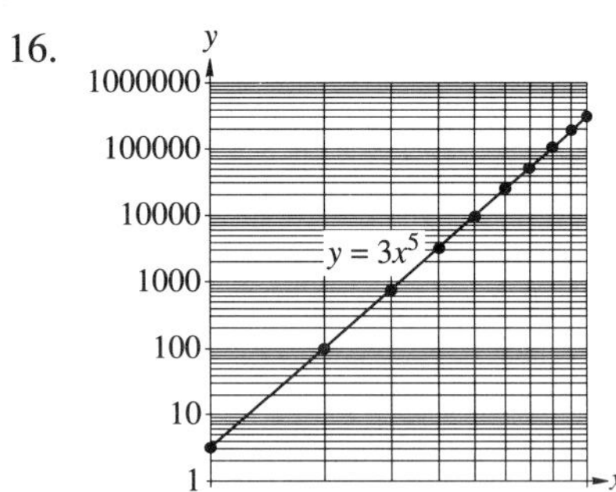

x	1	2	3	4	5	6	7	8	9
y	3	96	729	3072	9357	23328	50421	98304	177147

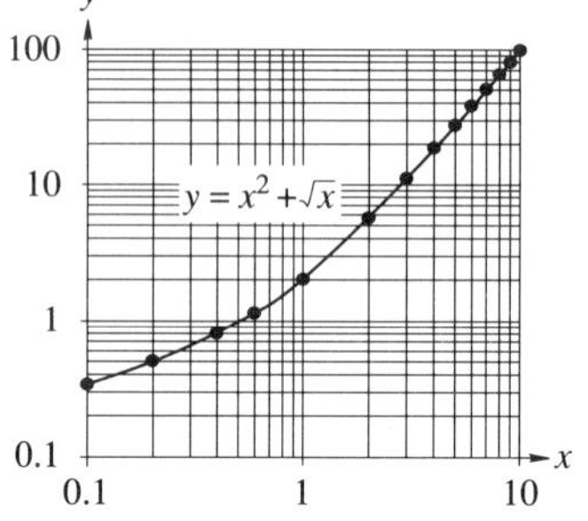

17.

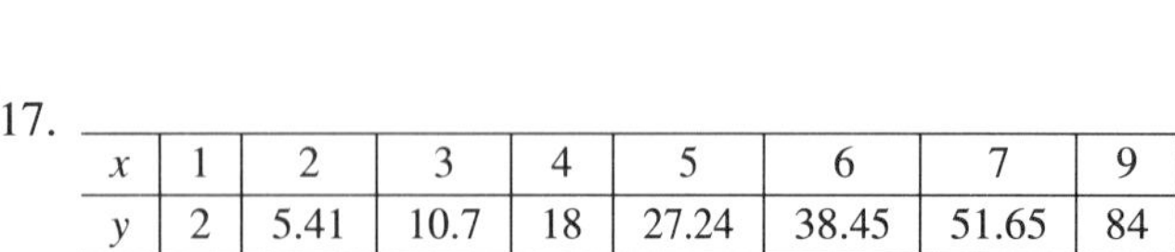

x	1	2	3	4	5	6	7	9
y	2	5.41	10.7	18	27.24	38.45	51.65	84

18.

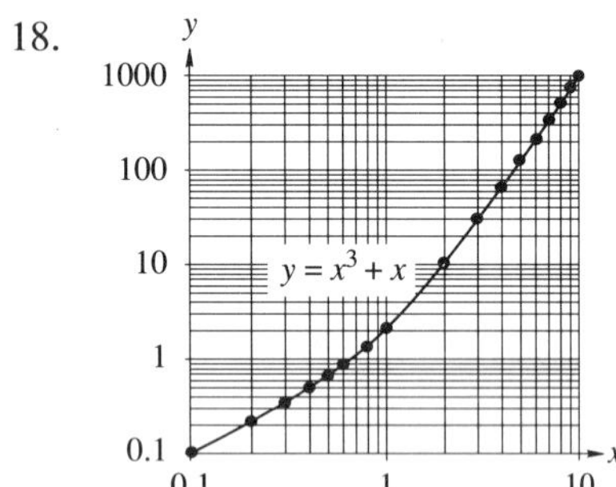

x	.1	1	2	3	4	5	6	8	10
y	.101	2	10	30	68	130	222	520	1010

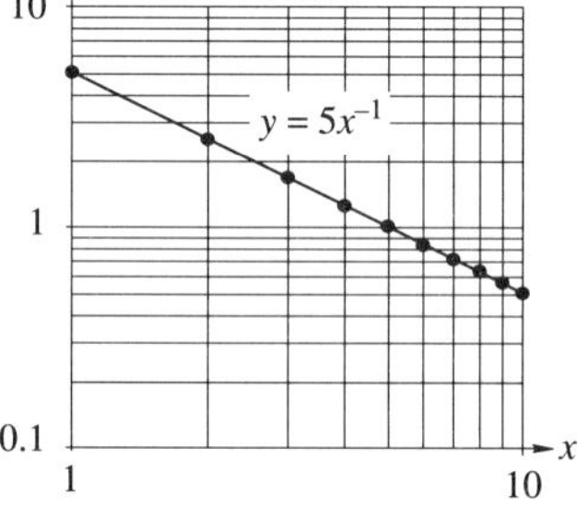

19.

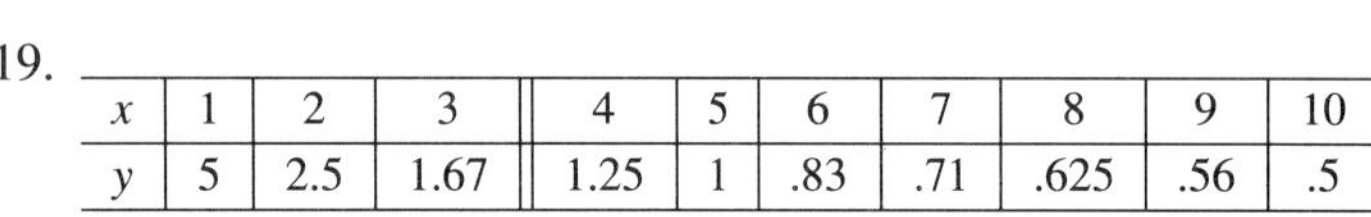

x	1	2	3	4	5	6	7	8	9	10
y	5	2.5	1.67	1.25	1	.83	.71	.625	.56	.5

20.

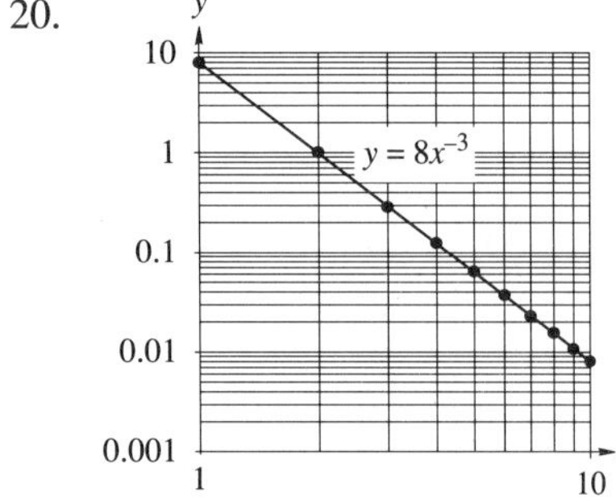

x	1	2	3	4	5	6	7	8	9	10
y	8	1	.3	.125	.064	.037	.023	.016	.011	.008

21.

x	1	2	3	4	5	7	9
y	2	5.66	10.39	16	22.36	37.04	54

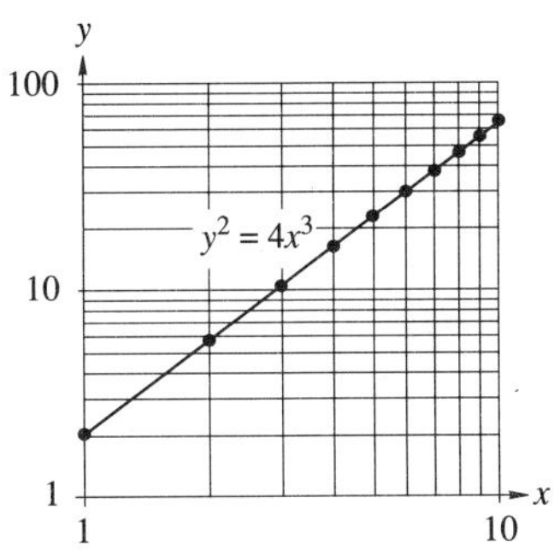

22. Rewrite $y^4 = 6x^5$ as $y = (6x^5)^{1/4}$ and use this second equation to complete the following table:

x	1	2	3	4	6	8	10
y	1.57	3.72	6.18	8.85	14.70	21.06	27.83

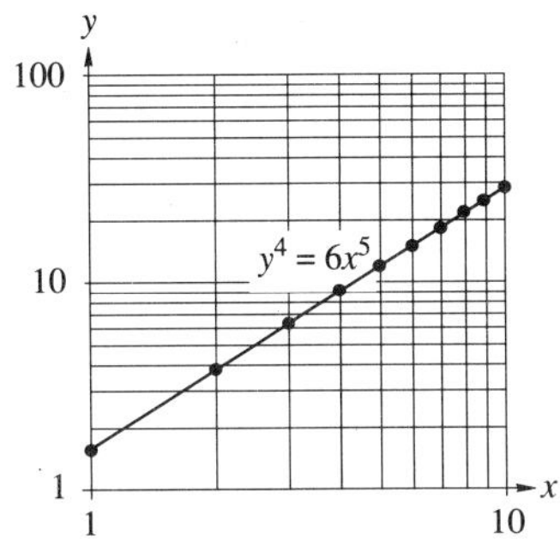

23. Rewrite $x^2y^3 = 8$, first as $y^3 = 8x^{-2}$, and then, solve for y with the result that $y = (8x^{-2})^{1/3}$ and use this third equation to complete the following table:

x	1	2	3	4	6	8	10
y	1	1.26	0.96	0.79	0.61	0.5	0.43

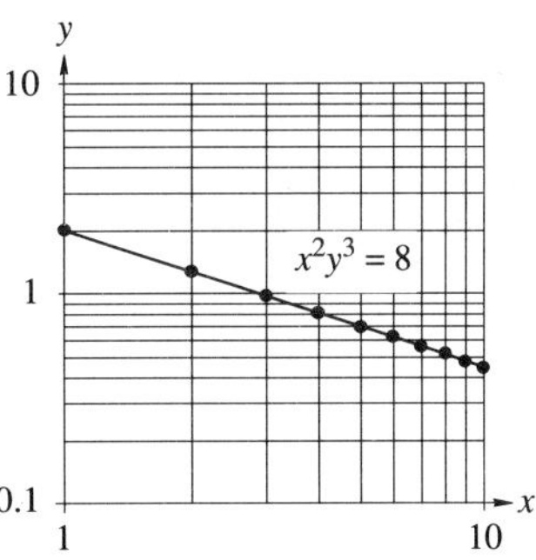

24. Rewrite $x^3y = 100$ as $y = 100x^{-3}$ and use this second equation to complete the following table:

x	1	2	3	4	6	8	10
y	100	12.5	3.7	1.56	0.46	0.195	0.1

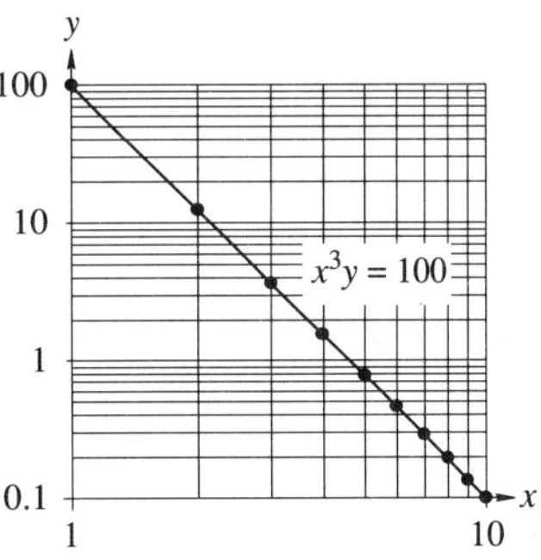

25. $P = e^{(-1.25\times10^{-4})h}$:

h	0	10	100	1000	5000	10000
P	1	0.999	0.987	0.882	0.535	0.287

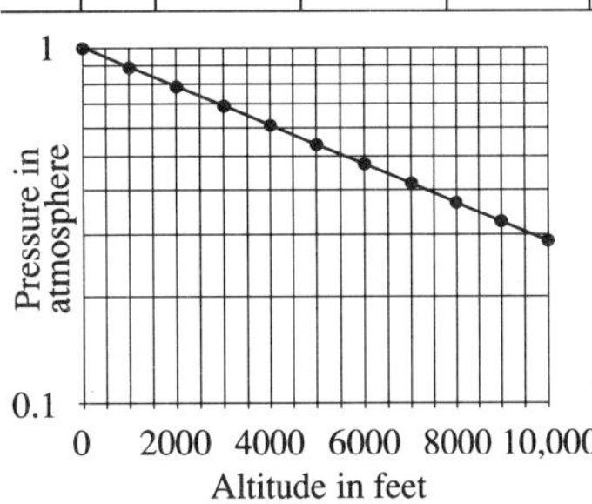

26. $P = 100e^{-0.3h}$:

h	0	2	4	6	8	10
P	100	54.9	30.1	16.5	9.07	4.98

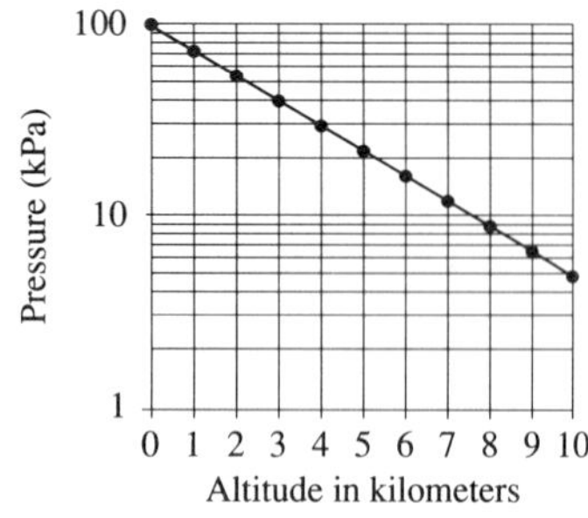

27. Rewrite $RD^2 = 0.9$ as $D^2 = 0.9R^{-1}$ and then, solve for D with the result $D = (0.9R^{-1})^{1/2}$. Now, use this last equation to complete the following table:

R	0.1	0.9	10	100	90
D	3	1	0.3	0.095	0.1

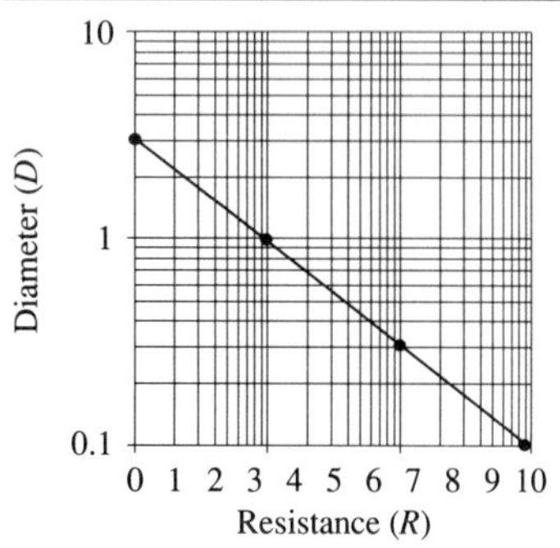

28. Rewrite $PV = 5$ as $P = 5V^{-1}$ and complete the table:

P	0.1	0.5	1	5	10
V	50	10	5	1	0.5

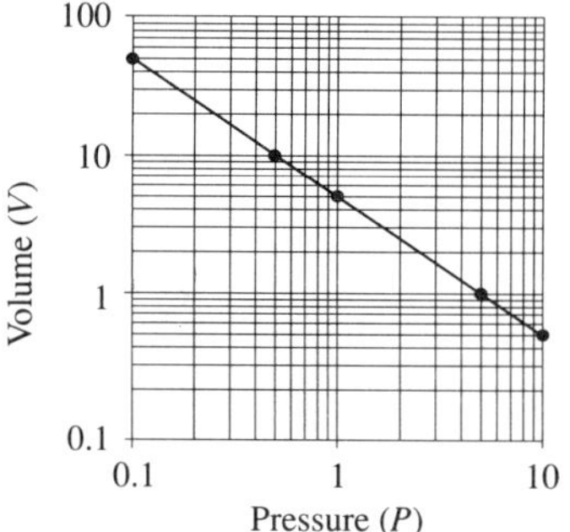

29.

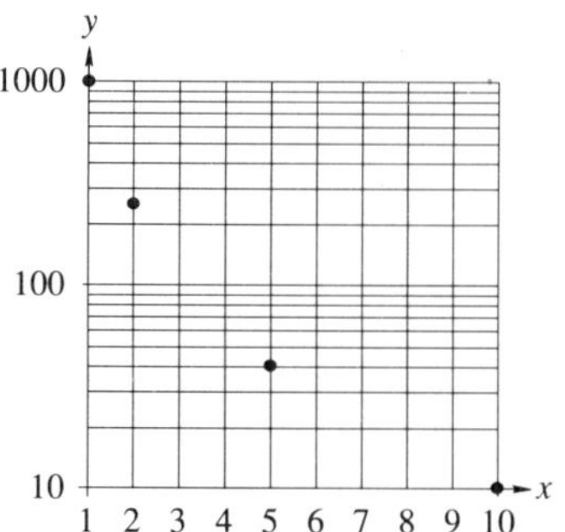

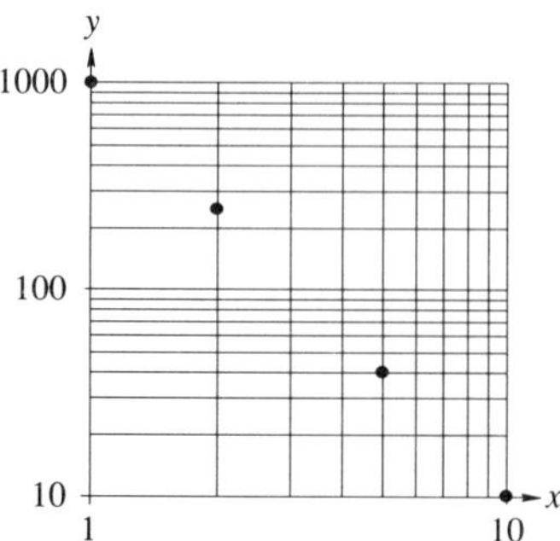

From the graph on the log-log paper we can conclude that the equation is of the form $y = ax^m$ when $x = 1y = 1000$, so we can conclude that $a = 1000$. To find m take

$$m = \frac{\log 250 - \log 1000}{\log 2 - \log 1} = \frac{2.39794 - 3}{0.30103 - 0} = -2$$

hence $y = 1000x^{-2}$

30. 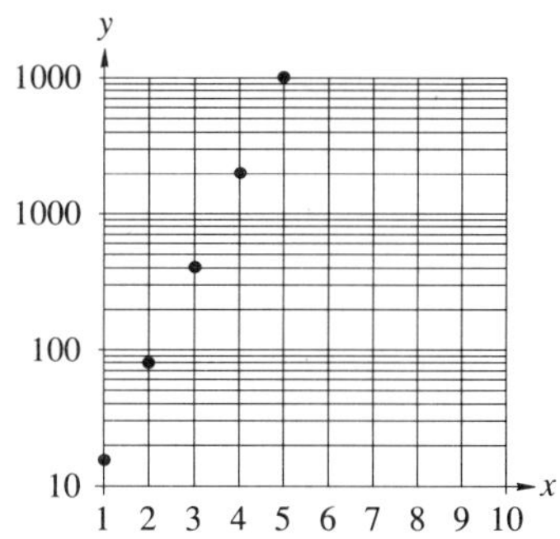

Since the graph is a straight line on semi-log paper, the equation is of the form $y = ab^x$ when $x = 1$, $y = 16$ so $16 = ab^1$ or $ab = 16$ when $x = 2$, $y = 80$ so $80 = ab^2$ or $ab^2 = 80$; $\frac{ab^2}{ab} = \frac{80}{16}$; $b = 5$. Since $ab = 16$, $a = \frac{16}{5} = 3.2$. Hence the equation is $y = 3.2(5^x)$.

CHAPTER 13 REVIEW

1. $e^5 = 148.41316$

2. $e^{-7} = 9.118819 \times 10^{-4} = 0.00091188$

3. $e^{4.67} = 106.69774$

4. $e^{-3.91} = 0.0200405$

5. $\log 8 = 0.90309$

6. $\log 196.5 = 2.2933626$

7. $\ln 81.3 = 4.398146$

8. $\ln 325.6 = 5.7856696$

9. 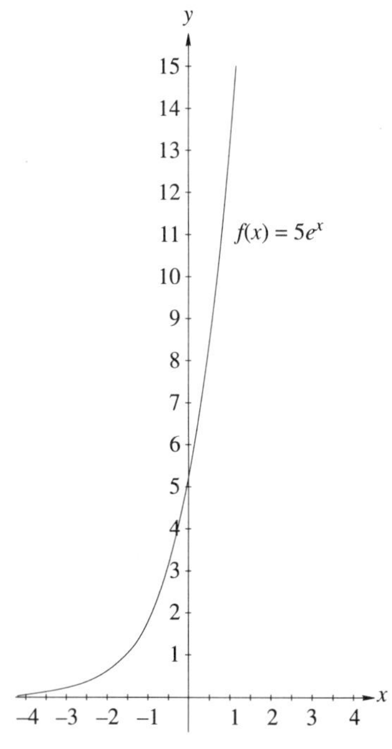

x	−2	−1	0	1	2	3	4
$5e^x$	0.677	1.839	5	13.59	36.95	100.43	272.99

10.

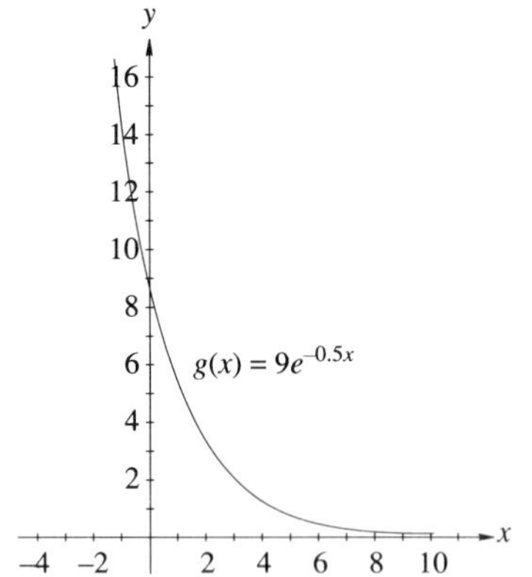

x	-4	-3	-2	-1	0	1	2
$9e^{-0.5x}$	66.50	40.34	24.46	14.84	9	5.46	3.31

11.

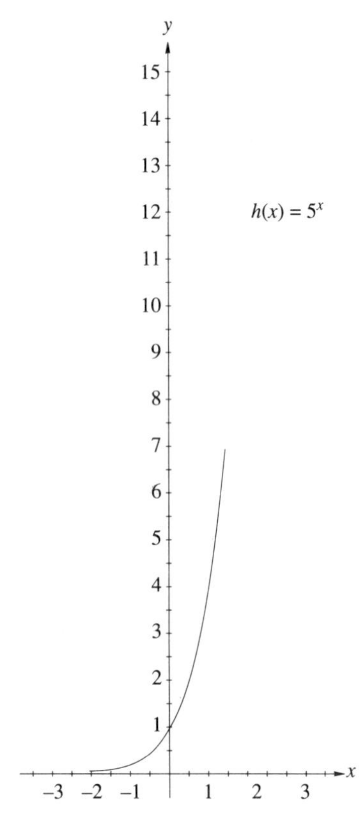

x	-3	-2	-1	0	1	2
5^x	0.008	0.04	0.2	1	5	25

12.

x	-4	-3	-2	-1	0	1	2
2.1^{-x}	19.45	9.26	4.41	2.1	1	0.48	0.23

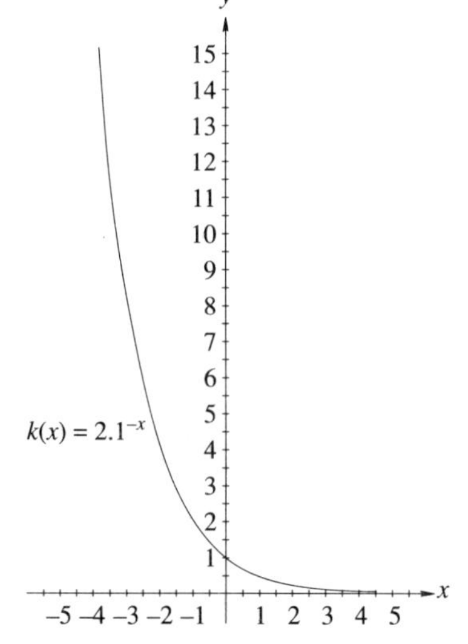

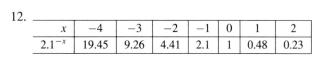

13.

x	0	1	2	3	4	5	6	7	8	9	10
$\log x$	not defined	0	0.301	0.477	0.602	0.699	0.778	0.845	0.903	0.954	1

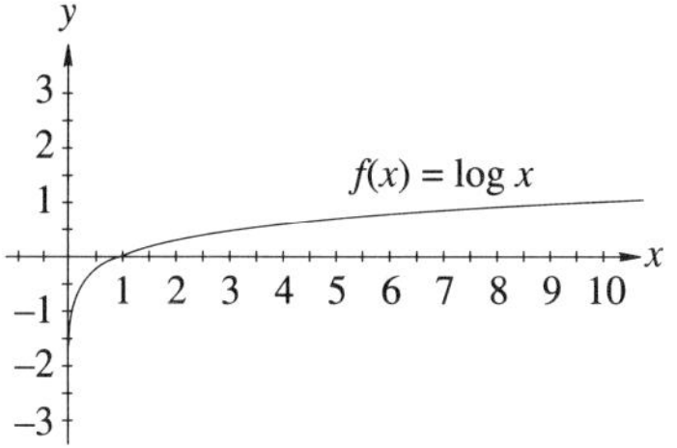

14.

x	0	1	2	3	4	5	6	7	8	9	10
$3 \log 2x$	not defined	0.9031	1.806	2.334	2.709	3	3.238	3.438	3.612	3.766	3.903

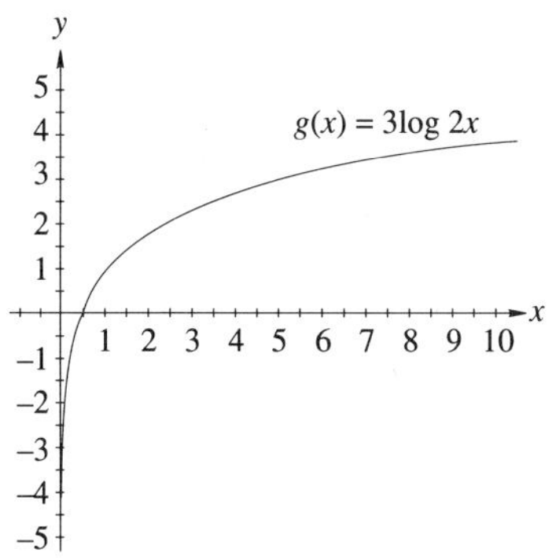

15.

x	$-\frac{1}{2}$	0	1	2	3	4	5	6	7	8	9	10
$\ln(2x+1)$	not defined	0	1.099	1.609	1.946	2.197	2.398	2.565	2.708	2.833	2.944	3.045

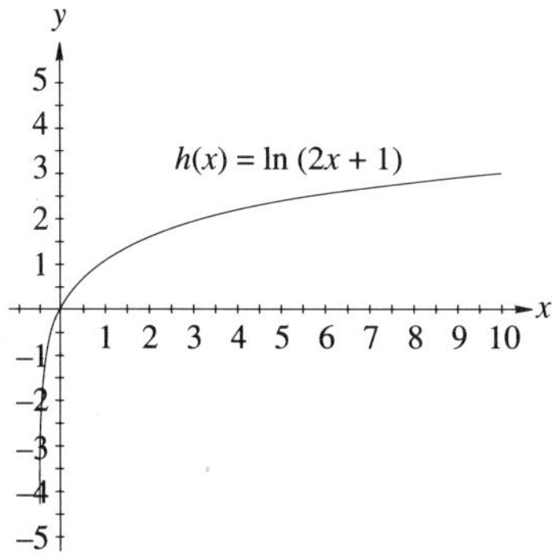

16.

x	0	1	2	3	4	5	6	7	8	9	10
$2 + 3\ln x$	not defined	2	4.079	5.296	6.159	6.828	7.375	7.838	8.238	8.592	8.908

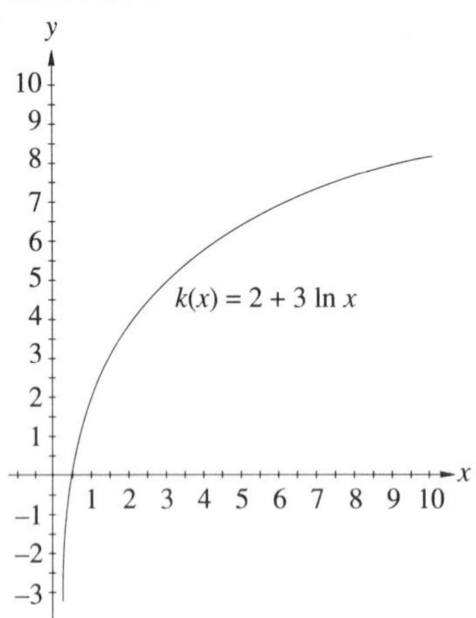

17. $\log \frac{3}{4} = \log 3 - \log 4$

18. $\log 77 = \log(7 \cdot 11) = \log 7 + \log 11$

19. $\log 5x = \log 5 + \log x$

20. $\ln \frac{7x}{3} = \ln 7 + \ln x - \ln 3$

21. $\ln(4x)^3 = 3\ln 4x = 3[\ln 4 + \ln x]$

22. $\ln 4x^3 = \ln 4 + 3\ln x$

23. $\log \sqrt{48} = \log 48^{1/2} = \frac{1}{2}\log 48 = \frac{1}{2}[\log 6 + \log 8]$. Answers may vary.

24. $\log \frac{\sqrt[3]{2x^5}}{4x} = \log(2x^5)^{1/3} - \log 4x = \frac{1}{3}[\log 2 + 5\log x] - \log 4 - \log x$. Answers may vary.

25. $\log 5 + \log 9 = \log(5 \cdot 9) = \log 45$

26. $\log 19 - \log 11 = \log \frac{19}{11}$

27. $4\log x + \log x = \log x^4 + \log x = \log x^5$

28. $\log 3a + \log b - \log a = \log\left(\frac{3ab}{a}\right) = \log 3b$

29. $7\log x + 2\log x - \log x = 8\log x = \log x^8$

30. $\ln \frac{2}{3} + \ln \frac{15}{8} = \ln\left(\frac{2}{3} \cdot \frac{15}{8}\right) = \ln\left(\frac{5}{4}\right)$

31. $4^x = 28$; $\log 4^x = \log 28$; $x\log 4 = \log 28$; $x = \frac{\log 28}{\log 4} = 2.404$

32. $5^{3x} = 20$; $\ln 5^{3x} = \ln 250$; $3x\ln 5 = \ln 250$; $x = \frac{\ln 250}{3\ln 5} = 1.1436$

33. $e^{5x} = 11$; $\ln e^{5x} = \ln 11$; $5x = \ln 11$; $x = \frac{\ln 11}{5} = 0.4796$

34. $e^{5x-1} = 2e^x$; $\ln^{5x-1} = \ln(2e^x) = \ln 2 + \ln e^x$; $5x - 1 = \ln 2 + x$; $4x = \ln 2 + 1$; $x = \frac{\ln 2 + 1}{4} = 0.4233$

35. $\ln x = 9.1$; $e^{\ln x} = e^{9.1}$; $x = e^{9.1} = 8955.29$

36. $4\log x = 49$; $\log x = \frac{49}{4}$; $x = 10^{49/4} = 1.778 \times 10^{12}$

37. $\log(2x + 1) = 50$; $2x + 1 = 10^{50}$; $x = \frac{10^{50} - 1}{2}$ or $0.5 \times 10^{50} - 0.5$ (Note: to 50 significant digits this is $0.5 \times 10^{50} = 5 \times 10^{49}$.)

38. $2\ln 4x = 100 + \ln 2x$; $\ln 16x^2 - \ln 2x = 100$; $\ln\left(\frac{16x^2}{2x}\right) = 100$; $\ln 8x = 100$; $8x = e^{100}$; $x = \frac{e^{100}}{8} = 3.36 \times 10^{42}$

39.

x	1	2	3	4	5	6	7	8	9
y	8	16	32	64	128	256	512	1024	2048

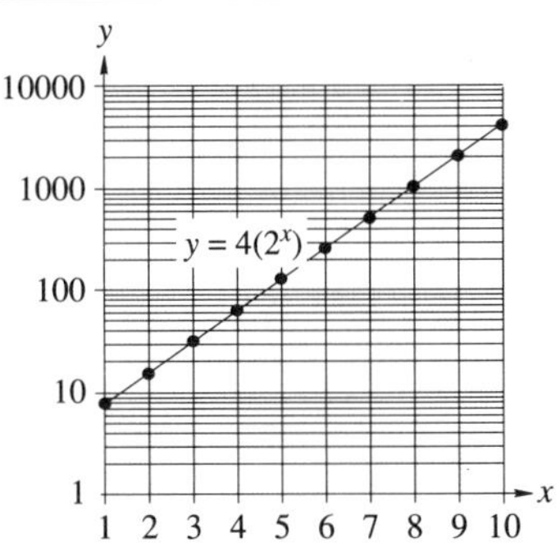

40.

x	1	2	4	6	8	10
y	8	128	2048	10,368	32,768	80,000

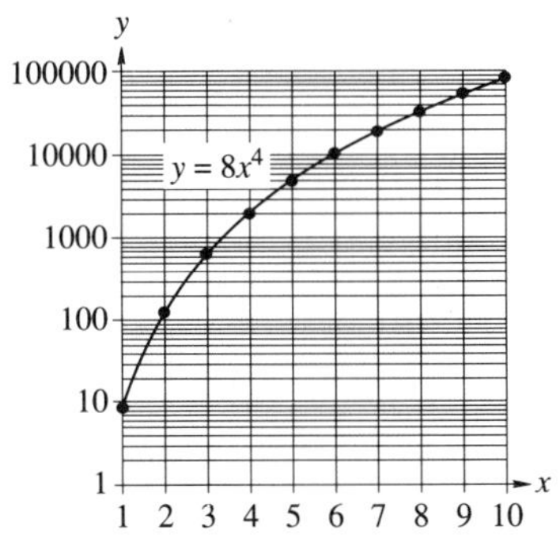

41.

x	1	2	3	4	5	6
y	32	128	512	2048	8192	32,768

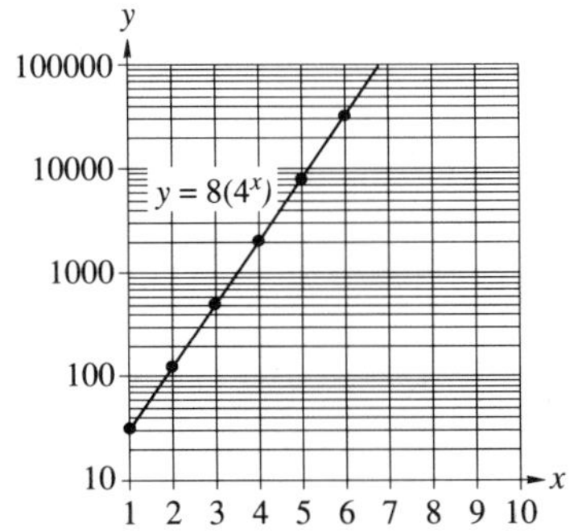

42.

x	1	2	3	4	6	8
y	3	12	48	192	3072	49,152

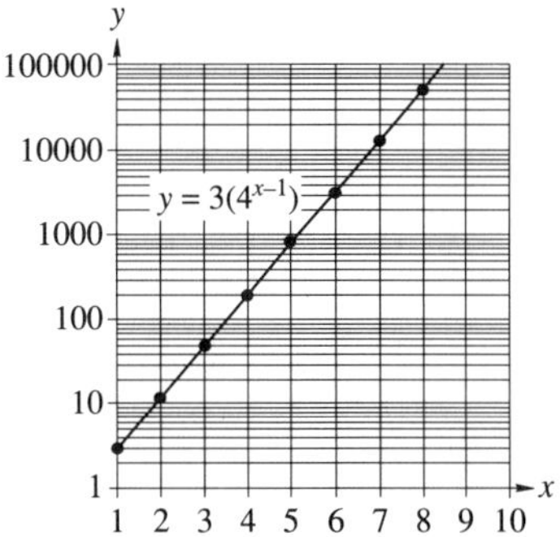

43.

x	1	10	16	81	256	625	2401	10000
y	1	1.78	2	3	4	5	7	10

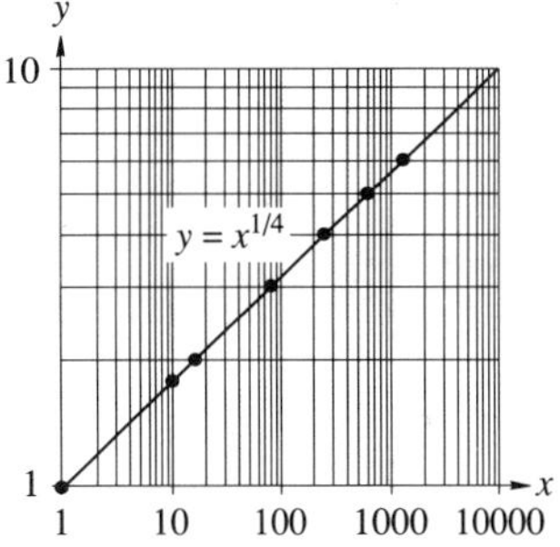

44.

x	1	2	3	4	6	8	10
y	5	40	135	320	1080	2560	5000

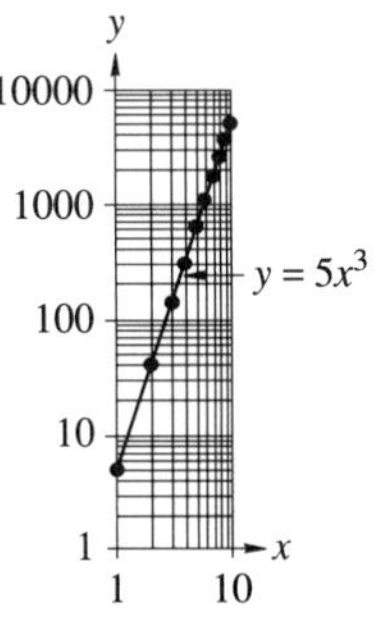

45.

x	0.001	0.01	0.1	1	10
y	211	37.6	6.69	1.18	0.211

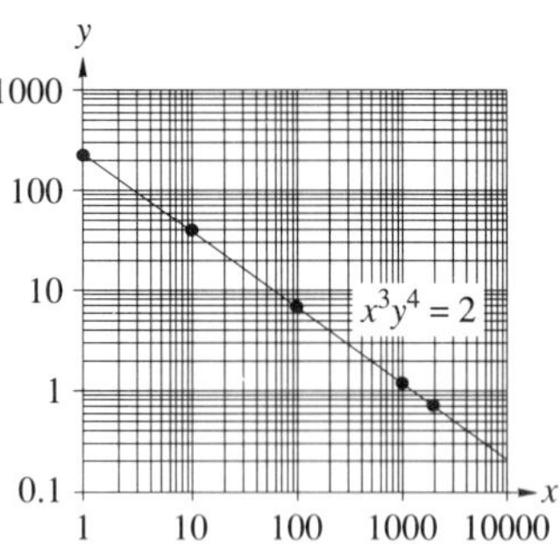

46. $y^3 = 10x^2$ or $y = (10x^2)^{1/3}$.

x	0.1	.1	10	100	1000
y	0.46	2.15	10	46.4	215.4

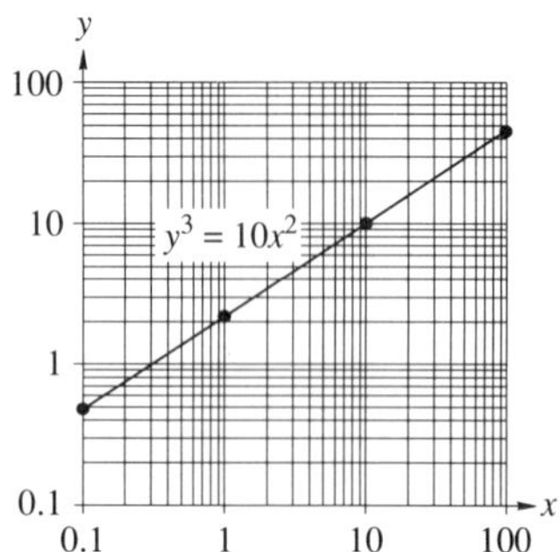

47. First Bank $S = 1000\left(1 + \frac{.075}{2}\right)^{2\times 10}$ 2088.15, Second Bank $S = 1000\left(1 + \frac{.075}{12}\right)^{2\times 10} = 2112.06$, Third Bank $S = 1000 \times e^{.075\times 10} = 2117.00$, Double time $= \frac{\ln 2}{r} = \frac{\ln 2}{.075} = 9.24$ years

48. $e^{1600k} = \frac{1}{2}$; After 100 years $y = 100e^{100k} = 100(e^{1600k})^{1/16} = 100(0.5)^{1/16} = 95.76$ mg. After 1000 yr, $y = 100e^{1000k} = 100(e^{1600k})^{\frac{10}{16}} = 100(.5)^{5/8} = 64.84$ mg.

49. $y = ce^{kt}$; $25{,}000 = 4000e^{24k}$; $e^{24k} = \frac{25}{4}$; $24k = \ln\frac{25}{4}$; $k = 0.0763571$. (a) $y = 4000e^{.07635761\times 12} = 10000$. (b) $y = 4000e^{0.7635761\times 48} = 156{,}250$.

50. $R = \log I$ or San Francisco of 1906 $I = 10^{8.3}$; (a) San Francisco of 1979 $I = 10^6$; $\frac{10^{8.3}}{10^6} = 10^{2.3} = 199.53$ times (b) San Francisco of 1989 $I = 10^{7.1}$; $\frac{10^{8.3}}{10^{7.1}} = 10^{1.2} = 15.85$ times

51. Since the graph is a straight line on semi log paper the equation must be of the form $y = ab^x$. Hence $15.76 = ab^2$ and $620.84 = ab^4$. This gives $\frac{ab^4}{ab^2} = \frac{620.84}{15.76}$ or $b^2 = 39.3934$ or $b = 6.2764$. Back substituting to find a, $15.76 = a(6.2764)^2$ $a = \frac{15.76}{6.2764^2} = 0.4$. Answer: $y = 0.4(6.2764)^x$

52. Since the graph is a straight line on log-log paper, the equation must be of the form $y = ax^m$. $m = \frac{\log 2.1875 - \log 8.75}{\log 4 - \log 2} = -2$. Back substituting to find a, $8.75 = a2^{-2}$ or $2^2 \cdot 8.75 = 35$. Answer $y = 35x^{-2}$

53. $M = 6 - 2.5\log\frac{I}{I_0}$ (a) $M - 6 = -2.5\log\frac{I}{I_0}$; $\log\frac{I}{I_0} = \frac{6-M}{2.5}$; $\frac{I}{I_0} = 10^{\frac{6-M}{2.5}}$; $I = I_0 \cdot 10^{\frac{6-M}{2.5}}$ (b) $\frac{I_0 10^{\frac{6+12.5}{2.5}}}{I_0 10^{\frac{6-2}{2.5}}} = \frac{10^{18.5/2.5}}{10^{4/2.5}} = 10^{5.8} = 630957.34$

54. Using $Q = \frac{MQ_0}{Q_0 + (m - Q_0)e^{-kmt}}$ we have $800 = \frac{2000\cdot 400}{400+(2000-400)e^{-k\cdot 2000\cdot\frac{1}{2}}}$ or $800 = \frac{800000}{400 + (1600)e^{-1000k}}$ and so $400 + 1600e^{-1000k} = \frac{800000}{800}$ or $400 + 1600e^{-1000k} = 1000$. This yields $1600e^{-1000k} = 600$ or $e^{1000k} = \frac{600}{1600} = \frac{3}{8} = 0.375$. Taking the natural logarithm of both sides produces $-1000k = \ln 0.375 = -0.9808$ and so $k = \frac{0.9808}{1000} = 0.0009808 = 9.808 \times 10^{-4}$.

CHAPTER 13 TEST

1. $e^{5.3} = 200.33681$

2. $\log 715.3 = 2.854488225$

3. $\ln 72.35 = 4.281515453$

4. $\log \frac{9}{15} = \log 9 - \log 15$ or $\log \frac{9}{15} = \log \frac{3}{5} = \log 3 - \log 5$

5. $\log 65x = \log 65 + \log x = \log 5 + \log 13 + \log x$

6. $\ln(7x)^{-2} = -2\ln 7x = -2[\ln 7 + \ln x] = -2\ln 7 - 2\ln x$

7. $\log \frac{\sqrt[5]{5x^3}}{9x} = \log \sqrt[5]{5x^3} - \log 9x$; $\frac{1}{5}\log 5x^3 - \log 9x = \frac{1}{5}\left[\log 5 + 3\log x\right] - \log 9 - \log x$

8. $\log 11 + \log 3 = \log(11 \cdot 3) = \log 33$

9. $\log 17 - 2\log x = \log 17 - \log x^2 = \log \frac{17}{x^2}$

10. $\ln 4 - \ln 5 + \ln 15 - \ln 8 = \ln \frac{4 \cdot 15}{5 \cdot 8} = \ln \frac{3}{2}$

11. $\ln x = 17.2$; $x = e^{17.2} = 29502925.92$

12. $3\log x = 55$; $\log x = \frac{55}{3}$, $x = 10^{55/3} = 2.154434 \times 10^{18}$

13. $3^x = 35$; $\ln 3^x = \ln 35$; $x\ln 3 = \ln 35$; $x = \frac{\ln 35}{\ln 3} = 3.23621727$

14. $e^{4x+1} = 2e^x$. Taking the natural logarithm of both sides gives $\ln e^{4x+1} = \ln 2e^x = \ln 2 + \ln e^x$ or $4x + 1 = \ln 2 + x$ and so $3x = \ln 2 - 1$, which means that $x = \frac{\ln 2 - 1}{3} \approx -0.10228$.

15. $\log(2x^2+4x) = 5 + \log 2x$; $\log(2x^2+4x) - \log 2x = 5$; $\log\left(\frac{2x^2+4x}{2x}\right) = 5$; $\log(x+2) = 5$; $x + 2 = 10^5$; $x = 10^5 - 2 = 99{,}998$

16.

x	−3	−2	−1	0	1	2	3	4
$f(x)$	0.037	0.111	0.333	1	3	9	27	81

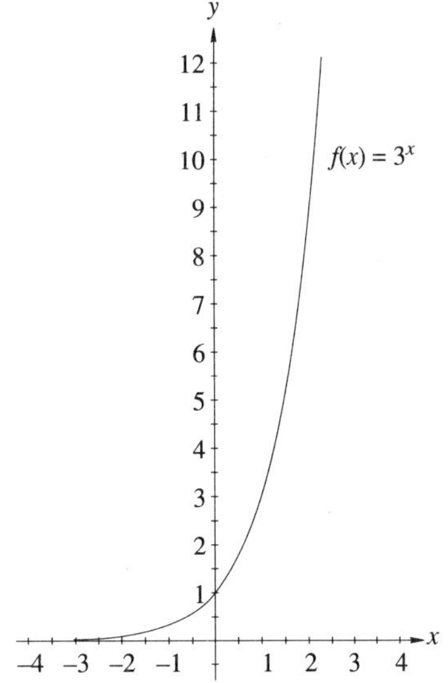

17.

x	1	2	3	4	5	6	7
y	1.2	3.6	10.8	32.4	97.2	291.6	874.8

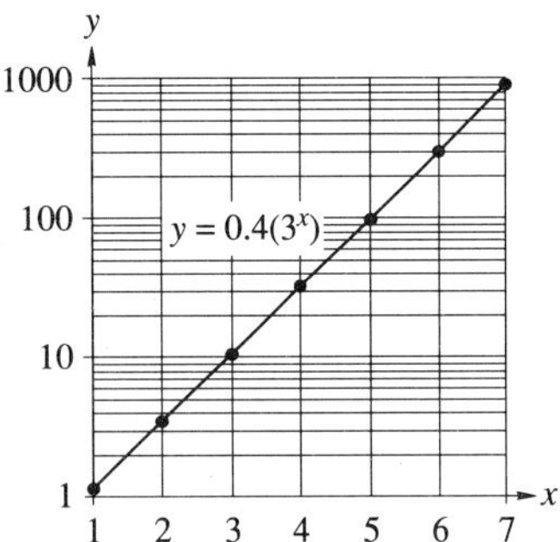

18.

x	0.1	1	2	10	32	243	1024
y	1.26	2	2.30	3.17	4	6	8

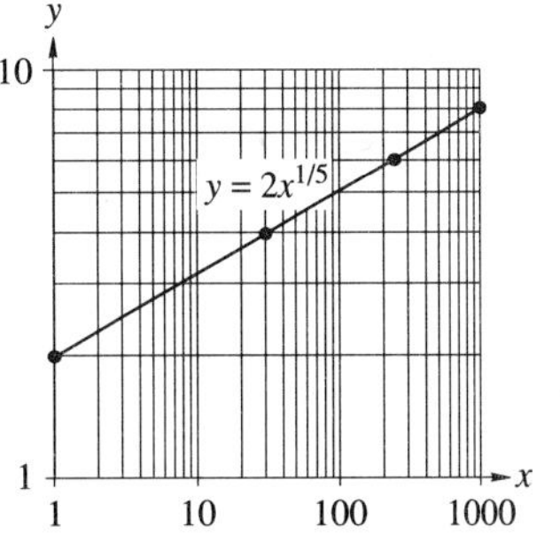

CHAPTER

14

Complex Numbers

14.1 IMAGINARY AND COMPLEX NUMBERS

1. $\sqrt{-25} = \sqrt{25}\sqrt{-1} = 5j$
2. $\sqrt{-81} = \sqrt{81}\sqrt{-1} = 9j$
3. $\sqrt{-0.04} = \sqrt{0.04}\sqrt{-1} = 0.2j$
4. $\sqrt{-1.44} = \sqrt{1.44}\sqrt{-1} = 1.2j$
5. $\sqrt{-75} = \sqrt{25}\sqrt{3}\sqrt{-1} = 5\sqrt{3}j$ or $5j\sqrt{3}$
6. $-\sqrt{-72} = -\sqrt{36}\sqrt{2}\sqrt{-1} = -6j\sqrt{2}$
7. $-3\sqrt{-20} = -3\sqrt{4}\sqrt{5}\sqrt{-1} = -3\cdot 2\cdot\sqrt{5}j = -6j\sqrt{5}$
8. $5\sqrt{-30} = 5\sqrt{30}\sqrt{-1} = 5j\sqrt{30}$
9. $\sqrt{\frac{-9}{16}} = \sqrt{\frac{9}{16}}\sqrt{-1} = \frac{3}{4}j$
10. $\sqrt{\frac{-25}{36}} = \sqrt{\frac{25}{36}}\sqrt{-1} = \frac{5}{6}j$
11. $-4\sqrt{\frac{-9}{16}} = -4\sqrt{\frac{9}{16}}\sqrt{-1} = -4\cdot\frac{3}{4}j = -3j$
12. $-3\sqrt{-\frac{10}{81}} = \frac{-3\sqrt{10}\sqrt{-1}}{\sqrt{81}} = \frac{-3\sqrt{10}j}{9} = \frac{-\sqrt{10}}{3}j = \frac{-j\sqrt{10}}{3}$
13. $\left(\sqrt{-11}\right)^2 = (\sqrt{11}j)^2 = \sqrt{11}^2 j^2 = 11(-1) = -11$
14. $\left(\sqrt{-7}\right)^2 = (\sqrt{7}j)^2 = \sqrt{7}^2 j^2 = 7(-1) = -7$
15. $\left(3\sqrt{-2}\right)^2 = 3^2\sqrt{2}^2 j^2 = 9\cdot 2(-1) = -18$
16. $(-2\sqrt{-3})^2 = (-2)^2\sqrt{3}^2 j^2 = 4\cdot 3(-1) = -12$
17. $\sqrt{-4}\sqrt{-25} = 2j\cdot 5j = 10j^2 = -10$
18. $\sqrt{-9}\sqrt{-16} = 3j\cdot 4j = 12j^2 = -12$
19. $(\sqrt{-49})(2\sqrt{-9}) = 7j\cdot 2\cdot 3j = 42j^2 = -42$
20. $(3\sqrt{-16})(\sqrt{-36}) = 3\cdot 4j\cdot 6j = 72j^2 = -72$
21. $(\sqrt{-5})(-\sqrt{5}) = \sqrt{5}j(-\sqrt{5}) = -5j$
22. $(-\sqrt{-7})(\sqrt{7}) = -\sqrt{7}j\cdot\sqrt{7} = -7j$
23. $(\sqrt{-0.5})(\sqrt{0.5}) = \sqrt{0.5}j\cdot\sqrt{0.5} = 0.5j$
24. $(\sqrt{0.8})(\sqrt{-0.2}) = \sqrt{0.8}\sqrt{0.2}j = \sqrt{0.16}j = 0.4j$
25. $(-2\sqrt{2.7})(\sqrt{-3}) = (-2\sqrt{2.7})(\sqrt{3}j) = -2\sqrt{8.1} = -2\sqrt{9}\sqrt{0.9}j = -2\cdot 3\sqrt{0.9}j = -6j\sqrt{0.9}$
26. $(-4\sqrt{1.6})(2\sqrt{-0.4}) = (-4\sqrt{1.6})(2j\sqrt{0.4}) = -8\sqrt{0.64}j = -8(0.8)j = -6.4j$
27. $(\sqrt{-5})(\sqrt{-6})(\sqrt{-2}) = \sqrt{5}j\sqrt{6}j\sqrt{2}j = \sqrt{60}j^3 = 2\sqrt{15}(-j) = -2\sqrt{15}j = -2j\sqrt{15}$ (Note: $j^3 = -j$.)
28. $(\sqrt{-3})(\sqrt{-9})(\sqrt{-15}) = \sqrt{3}\sqrt{9}\sqrt{15}j^3 = 9\sqrt{5}j^3 = -9j\sqrt{5}$ (Note: $j^3 = -j$.)
29. $(-\sqrt{-7})^2(\sqrt{-2})^2 j^3 = (-\sqrt{7}j)^2(\sqrt{2}j)^2 j^3 = 7j^2\cdot 2j^2 j^3 = 14j^3 = -14j$ (Note: $j^3 = -j$.)
30. $(\sqrt{-3})^2(\sqrt{-5})^2 j^2(\sqrt{-2}) = (\sqrt{3}j)^2(\sqrt{5}j)^2 j^2\sqrt{2}j = 3j^2 5j^2 j^2\sqrt{2}j = 15\sqrt{2}j^7 = -15j\sqrt{2}$ (Note: $j^7 = -j$.)
31. $x + yj = 7 - 2j$ so $x = 7$, $y = -2$

32. $x + yj = -9 + 2j$ so $x = -9, y = 2$

33. $x + 5 + yj = 15 - 3j$ so $x + 5 = 15$ or $x = 10$ and $y = -3$

34. $6j - x + yj = 4 + 2j$ so $-x = 4$ or $x = -4$ and $6 + y = 2$ or $y = -4$

35. $x - 5j + 2 = 4 - 3j + yj$ so $x + 2 = 4$ or $x = 2$ and $-5 = -3 + y$ or $y = -2$

36. $2x - 4j = 6j + 4 - yj$ so $2x = 4$ or $x = 2$ and $-4 = 6 - y$ or $y = 10$

37. $2(4 + 5j) = 8 + 10j$

38. $3(2 - 4j) = 6 - 12j$

39. $-5(2 + j) = -10 - 5j$

40. $-3(4 + 7j) = -12 - 21j$

41. $j(3 - 2j) = 3j - 2j^2 = 3j + 2 = 2 + 3j$

42. $j(5 + 4j) = 5j + 4j^2 = 5j - 4 = -4 + 5j$

43. $2j(4 + 3j) = 8j + 6j^2 = 8j - 6 = -6 + 8j$

44. $-3j(2 - 5j) = -6j + 15j^2 = -15 - 6j$

45. $\frac{1}{2}(6 - 8j) = 3 - 4j$

46. $\frac{2}{3}(6 + 9j) = 4 + 6j$

47. $\frac{5-10j}{5} = \frac{5}{5} - \frac{10}{5}j = 1 - 2j$

48. $\frac{6+12j}{3} = \frac{6}{3} + \frac{12}{3}j = 2 + 4j$

49. $\frac{6+\sqrt{-18}}{3} = \frac{6+3j\sqrt{2}}{3} = \frac{6}{3} + \frac{3j\sqrt{2}}{3} = 2 + j\sqrt{2}$

50. $\frac{7-\sqrt{-98}}{7} = \frac{7-\sqrt{98}j}{7} = \frac{7-7\sqrt{2}j}{7} = \frac{7}{7} - \frac{7j\sqrt{2}}{7} = 1 - j\sqrt{2}$

51. $\frac{8-\sqrt{-24}}{4} = \frac{8-\sqrt{24}j}{4} = \frac{8-2\sqrt{6}j}{4} = \frac{8}{4} - \frac{2j\sqrt{6}}{4} = 2 - \frac{1}{2}j\sqrt{6}$

52. $\frac{9+\sqrt{-27}}{6} = \frac{9+3j\sqrt{3}}{6} = \frac{9}{6} + \frac{3j\sqrt{3}}{6} = \frac{3}{2} + \frac{1}{2}j\sqrt{3}$

53. The conjugate of $7 + 2j$ is $7 - 2j$

54. The conjugate of $9 + \frac{1}{2}j$ is $9 - \frac{1}{2}j$

55. The conjugate of $6 - 5j$ is $6 + 5j$

56. The conjugate of $\frac{1}{2} - 9j$ is $\frac{1}{2} + 9j$

57. The conjugate of 19 is 19

58. The conjugate of $7j$ is $-7j$

59. The conjugate of $-8j$ is $8j$

60. The conjugate of -11 is -11

61. $x^2 + x + 2.5 = 0$. Here $a = 1$, $b = 1$, and $c = 2.5$. Thus $x = \frac{-1 \pm \sqrt{1^2 - 4 \cdot 1 \cdot 2.5}}{2 \cdot 1} = \frac{-1 \pm \sqrt{1 - 10}}{2} = \frac{-1 \pm \sqrt{-9}}{2} = \frac{-1 \pm 3j}{2}$. Hence $x = -\frac{1}{2} + \frac{3}{2}j$ and $x = -\frac{1}{2} - \frac{3}{2}j$.

62. $x^2 + 2x + 5 = 0$. Here $a = 1$, $b = 2$, and $c = 5$. Thus $x = \frac{-2 \pm \sqrt{2^2 - 4 \cdot 1 \cdot 5}}{2 \cdot 1} = \frac{-2 \pm \sqrt{4 - 20}}{2} = \frac{-2 \pm \sqrt{-16}}{2} = \frac{-2 \pm 4j}{2}$. Hence $x = -1 + 2j$ and $x = -1 - 2j$.

63. $x^2 + 9 = 0$. Here $a = 1$, $b = 0$, and $c = 9$. Thus $x = \frac{-0 \pm \sqrt{0^2 - 4 \cdot 9}}{2} = \frac{\pm\sqrt{-36}}{2} = \frac{\pm 6j}{2}$. Hence $x = 3j$ and $x = -3j$.

64. $x^2 + 25 = 0$. Here $a = 1$, $b = 0$, and $c = 25$. Thus $\frac{-0 \pm \sqrt{-4 \cdot 25}}{2} = \frac{\pm\sqrt{-100}}{2} = \frac{\pm 10j}{2}$. Hence $x = 5j$ and $x = -5j$.

65. $2x^2 + 3x + 7 = 0$. Here $a = 2$, $b = 3$, and $c = 7$. Thus $x = \frac{-3 \pm \sqrt{3^2 - 4 \cdot 2 \cdot 7}}{2 \cdot 2} = \frac{-3 \pm \sqrt{9 - 56}}{4} = \frac{-3 \pm \sqrt{-47}}{4} = \frac{-3 \pm j\sqrt{47}}{4}$. Hence $x = \frac{-3}{4} + \frac{\sqrt{47}}{4}j$ and $x = \frac{-3}{4} - \frac{\sqrt{47}}{4}j$.

66. $2x^2 + 7x + 9 = 0$. Here $a = 2$, $b = 7$, and $c = 9$. Thus $x = \frac{-7 \pm \sqrt{7^2 - 4 \cdot 2 \cdot 9}}{2 \cdot 2} = \frac{-7 \pm \sqrt{49 - 72}}{4} = \frac{-7 \pm \sqrt{-23}}{4} = \frac{-7 \pm j\sqrt{23}}{4}$. Hence $x = \frac{-7}{4} + \frac{\sqrt{23}}{4}j$ and $x = \frac{-7}{4} - \frac{\sqrt{23}}{4}j$.

67. $5x^2 + 2x + 5 = 0$. Here $a = 5$, $b = 2$, and $c = 5$. Thus $x = \frac{-2 \pm \sqrt{2^2 - 4 \cdot 5 \cdot 5}}{2 \cdot 5} = \frac{-2 \pm \sqrt{4 - 100}}{10} = \frac{-2 \pm \sqrt{-96}}{10} = \frac{-2 \pm 4j\sqrt{6}}{10}$. Hence $x = \frac{-1}{5} + \frac{2\sqrt{6}}{5}j$ and $x = \frac{-1}{5} - \frac{2\sqrt{6}}{5}j$.

68. $3x^2 + 2x + 10 = 0$. Here $a = 3$, $b = 2$, and $c = 10$. Thus $x = \frac{-2 \pm \sqrt{2^2 - 4 \cdot 3 \cdot 10}}{2 \cdot 3} = \frac{-2 \pm \sqrt{4 - 120}}{6} = \frac{-2 \pm \sqrt{-116}}{6} = \frac{-2 \pm \sqrt{116}j}{6} = \frac{-2}{6} \pm \frac{2j\sqrt{29}}{6}$. Hence $x = \frac{-1}{3} - \frac{\sqrt{29}}{3}j$ and $x = -\frac{1}{3} - \frac{\sqrt{29}}{3}j$.

14.2 OPERATIONS WITH COMPLEX NUMBERS

1. $(5 + 2j) + (-6 + 5j) = -1 + 7j$
2. $(9 - 7j) + (6 - 8j) = 15 - 15j$
3. $(11 - 4j) + (-6 + 2j) = 5 - 2j$
4. $(21 + 3j) + (-7 - 6j) = 14 - 3j$
5. $(4+\sqrt{-9})+(3-\sqrt{-16}) = (4+3j)+(3-4j) = 7-j$
6. $(-11 + \sqrt{-4}) + (9 + \sqrt{-36}) = (-11+2j)+(9+6j) = -2 + 8j$
7. $(2 + \sqrt{-9}) + (8j - \sqrt{5}) = (2+3j)+(8j-\sqrt{5}) = 2 - \sqrt{5} + 11j$
8. $(3+\sqrt{-8})+(3-\sqrt{8}) = (3+2j\sqrt{2})+(3-2\sqrt{2}) = 6 - 2\sqrt{2} + 2j\sqrt{2}$
9. $(14 + 3j) - (6 + j) = 8 + 2j$
10. $(-8 + 3j) - (4 - 3j) = -12 + 6j$
11. $(9 - \sqrt{-4}) - (\sqrt{-16} + 6) = (9 - 2j) - (4j + 6) = 3 - 6j$
12. $(\sqrt{-25} - 3) - (3 - \sqrt{-25}) = (5j - 3) - (3 - 5j) = -6 + 10j$
13. $(4 + 2j) + j + (3 - 5j) = 7 - 2j$
14. $(2 - 3j) - j - (6 + \sqrt{-81}) = (2 - 4j) - (6 + 9j) = -4 - 13j$
15. $(2 + j)3j = 6j + 3j^2 = -3 + 6j$
16. $(5 - 3j)2j = 10j - 6j^2 = 6 + 10j$
17. $(9 + 2j)(-5j) = -45j - 10j^2 = 10 - 45j$
18. $(11 - 4j)(-3j) = -33j + 12j^2 = -12 - 33j$
19. $(2 + j)(5 + 3j) = 10 + 6j + 5j + 3j^2 = 7 + 11j$
20. $(3 - 2j)(4 + 5j) = 12 + 15j - 8j - 10j^2 = 22 + 7j$
21. $(6 - 2j)(5 + 3j) = 30 + 18j - 10j - 6j^2 = 36 + 8j$
22. $(4 - 2j)(7 - 3j) = 28 - 12j - 14j + 6j^2 = 22 - 26j$
23. $(2\sqrt{-9} + 3)(5\sqrt{-16} - 2) = (6j + 3)(20j - 2) = 120j^2 - 12j + 60j - 6 = -126 + 48j$
24. $(6\sqrt{-25}-4)(-3-2\sqrt{-49}) = (30j-4)(-3-14j) = -90j - 420j^2 + 12 + 56j = 432 - 34j$
25. $(\sqrt{-3})^4 = (\sqrt{3}j)^4 = \sqrt{3}^4 j^4 = 9(1) = 9$; (Note: $j^4 = 1$)
26. $(\sqrt{-9})^3 = (3j)^3 = 27j^3 = -27j$; (Note: $j^3 = 1$)
27. $(1 + 2j)^2 = 1 + 4j + 4j^2 = -3 + 4j$
28. $(3 + 4j)^2 = 9 + 24j + 16j^2 = -7 + 24j$
29. $(7 - j)^2 = 49 - 14j + j^2 = 48 - 14j$
30. $(4 - 3j)^2 = 16 - 24j + 9j^2 = 7 - 24j$
31. $(5 + 2j)(5 - 2j) = 5^2 + 2^2 = 25 + 4 = 29$ (Note: $(a + bj)(a - bj) = a^2 + b^2$)
32. $(7 + 3j)(7 - 3j) = 7^2 + 3^2 = 49 + 9 = 58$ (Note: $(a + bj)(a - bj) = a^2 + b^2$)
33. $\frac{6 - 4j}{1 + j} \cdot \frac{1 - 1j}{1 - j} = \frac{6 - 6j - 4j + 4j^2}{1 + 1} = \frac{2 - 10j}{2} = 1 - 5j$

34. $\frac{4-8j}{2-2j}\cdot\frac{2+2j}{2+2j}=\frac{8+8j-16j-16j^2}{4+4}=\frac{24-8j}{8}=3-j$

35. $\frac{6-3j}{1+2j}\cdot\frac{1-2j}{1-2j}=\frac{6-12j-3j+6j^2}{1+4}=\frac{-15j}{5}=-3j$

36. $\frac{5-10j}{1-2j}\cdot\frac{1+2j}{1+2j}=\frac{5+10j-10j-20j^2}{1+4}=\frac{25}{5}=5$

37. $\frac{4+2j}{1-2j}\cdot\frac{1+2j}{1+2j}=\frac{4+8j+2j+4j^2}{1+4}=\frac{10j}{5}=2j$

38. $\frac{9+5j}{3+j}\cdot\frac{3-j}{3-j}=\frac{27-9j+15j-5j^2}{9+1}=\frac{32+6j}{10}=3.2+0.6j$

39. $\frac{2j}{5+j}\cdot\frac{5-j}{5-j}=\frac{10j-2j^2}{25+1}=\frac{2+10j}{26}=\frac{1}{13}+\frac{5}{13}j$

40. $\frac{5j}{6-j}\cdot\frac{6+j}{6+j}=\frac{30j+5j^2}{36+1}=\frac{-5+30j}{37}=-\frac{5}{37}+\frac{30}{37}j$

41. $\frac{\sqrt{3}-\sqrt{-6}}{\sqrt{-3}}=\frac{\sqrt{3}-\sqrt{6}j}{\sqrt{3}j}\cdot\frac{-\sqrt{3}j}{-\sqrt{3}j}=\frac{-3j+3\sqrt{2}j^2}{3}=\frac{-3\sqrt{2}-3j}{3}=-\sqrt{2}-j$

42. $\frac{\sqrt{5}+\sqrt{-10}}{\sqrt{-5}}=\frac{\sqrt{5}+\sqrt{10}j}{\sqrt{5}j}\cdot\frac{-\sqrt{5}j}{-\sqrt{5}j}=\frac{-5j-5\sqrt{2}j^2}{5}=\frac{5\sqrt{2}-5j}{5}=\sqrt{2}-j$

43. $\frac{(5+2j)(3-j)}{4+j}=\frac{15-5j+6j-2j^2}{4+j}=\frac{17+j}{4+j}\cdot\frac{4-j}{4-j}=\frac{68-17j+4j-j^2}{16+1}=\frac{69-13j}{17}=\frac{69}{17}-\frac{13}{17}j\approx 4.059-0.765j$

44. $\frac{(6-j)(2+3j)}{-1+3j}=\frac{12+18j-2j-3j^2}{-1+3j}=\frac{15+16j}{-1+3j}\cdot\frac{-1-3j}{-1-3j}=\frac{-15-45j-16j-48j^2}{1+9}=\frac{33-61j}{10}=3.3-6.1j$

45. $(1+j)^4=[(1+j)(1+j)]^2=(1+2j+j^2)^2=(2j)^2=4j^2=-4$

46. $(1-j)^4=\left[(1-j)^2\right]^2=(1-2j+j^2)^2=(-2j)^2=4j^2=-4$

47. $\left(\frac{1}{2}+\frac{\sqrt{3}}{2}j\right)^2=\frac{1}{4}+\frac{\sqrt{3}}{2}j+\frac{3}{4}j^2=-\frac{1}{2}+\frac{\sqrt{3}}{2}j=-0.5+\frac{j\sqrt{3}}{2}$

48. $\frac{4+j}{(3-2j)+(4-3j)}=\frac{4+j}{7-5j}\cdot\frac{7+5j}{7+5j}=\frac{28+20j+7j+5j^2}{49+25}=\frac{23+27j}{74}=\frac{23}{74}+\frac{27}{74}j\approx 0.3108+0.3649j$

49. $(a+bj)+(a-bj)=(a+a)+(b-b)j=2a+0j=2a$ which is a real number

50. $(a+bj)-(a-bj)=(a-a)+(b+b)j=2bj$ which is a a pure imaginary number

51. $(a+bj)(a-bj)=a^2-(bj)^2=a^2-b^2j^2=a^2+b^2$ which is a a real number

52. $9.32-6.12j)+(7.24+4.31j)=16.56-1.81j$ V

53. $(6.21-1.37j)+(4.32-2.84j)=10.53-4.21j$ V

54. $(19.2-3.5j)-(12.4+1.3j)=6.8-4.8j$ V

55. $(7.42+1.15j)-(2.34-1.73j)=5.08+2.88j$ V

56. $(0.25+0.20j)+(0.15-0.25j)=(0.40-0.05j)\ \Omega$

57. $(9.13-4.27j)-(3.29-5.43j)=5.84+1.16j\ \Omega$

58. $Z=\frac{6(3j)}{6+3j}=\frac{18j}{6+3j}\cdot\frac{6-3j}{6-3j}=\frac{54+108j}{36+9}=\frac{54+108j}{45}=\frac{9(6+12j)}{9\cdot 5}=\frac{6}{5}+\frac{12}{5}j$ or $1.2+2.4j\ \Omega$

59. $Z=\frac{(20+10j)(10-20j)}{(20+10j)+(10-20j)}=\frac{200-400j+100j-200j^2}{30-10j}=\frac{15000-5000j}{1000}=15-5j\ \Omega$

60. $V=IZ$; $V=(12.3+4.6j)(16.4-9.0j)=201.72-110.7j+75.44j-41.4j^2=243.12-35.26j$ V

61. $I = \frac{V}{Z} = \frac{5.2+3j}{4-2j} \cdot \frac{4+2j}{4+2j} = \frac{20.8+10.4j+12j+6j^2}{16+4} = \frac{14.8+22.4j}{20} = 0.74 + 1.12j$ A

62. $Z = \frac{V}{I} = \frac{10.6-6.0j}{4+j} \cdot \frac{4-j}{4-j} = \frac{42.4-10.6j-24j+6j^2}{16+1} = \frac{36.4-34.6j}{17} = 2.1412 - 2.0353j\ \Omega$

63. See *Computer Programs* in main text.

64. See *Computer Programs* in main text.

65. See *Computer Programs* in main text.

14.3 GRAPHING COMPLEX NUMBERS; POLAR FORM OF A COMPLEX NUMBER

1. $4(\cos 30° + j\sin 30°) = 2\sqrt{3} + 2j$; $a = 4\cos 30° = 2\sqrt{3} \approx 3.4641$; $b = 4\sin 30° = 2$

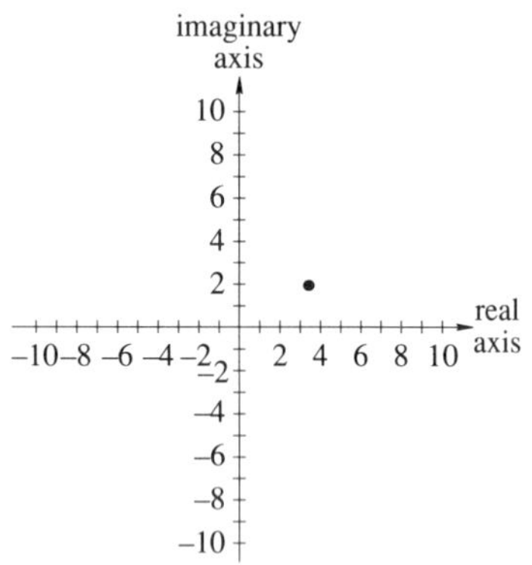

2. $10(\cos 45° + j\sin 45°) = 5\sqrt{2} + 5\sqrt{2}j \approx 7.0711 + 7.0711j$

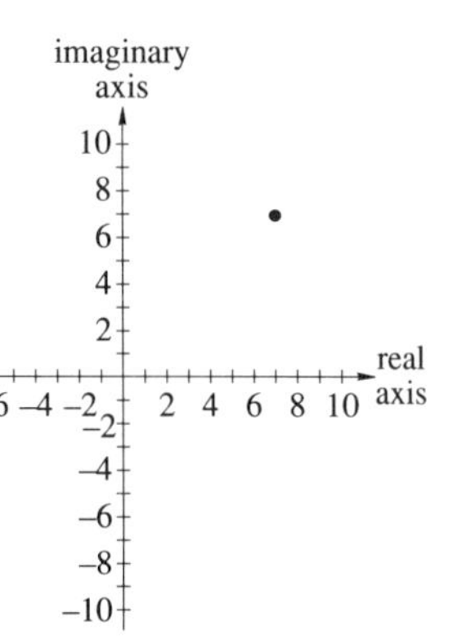

3. $5(\cos 135° + j\sin 135°) = \frac{-5\sqrt{2}}{2} + \frac{5j\sqrt{2}}{2} \approx -3.5355 + 3.5355j$

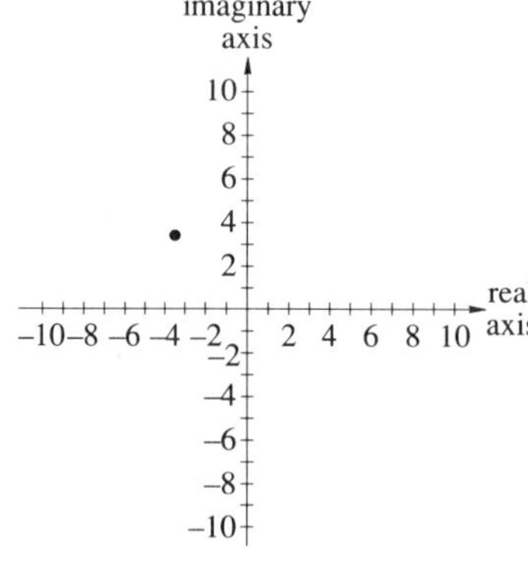

4. $8(\cos 305° + j\sin 305°) \approx 4.5886 - 6.5532j$

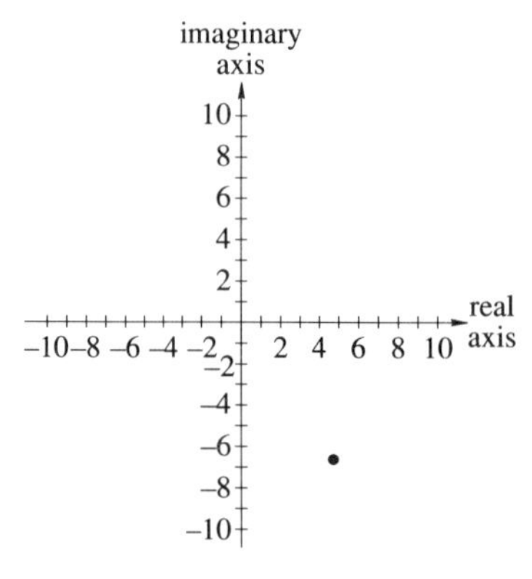

5. $7\text{ cis }260° = 7\cos 260° + 7j\sin 260° = -1.2155 - 6.8937j$

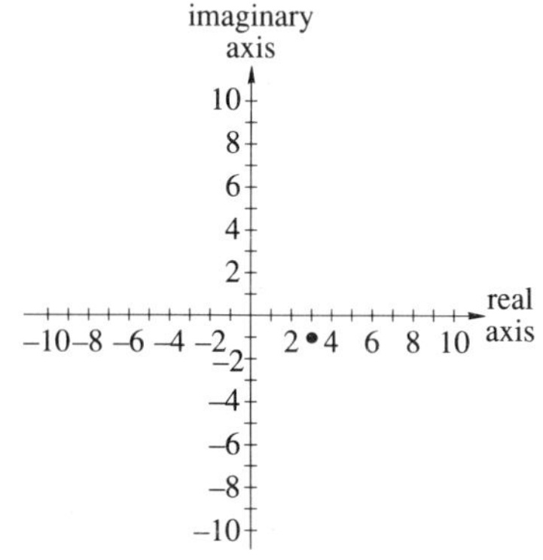

6. $3\text{ cis }340° = 3\cos 340° + 3j\sin 340° = 2.8191 - 1.0261j$

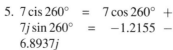

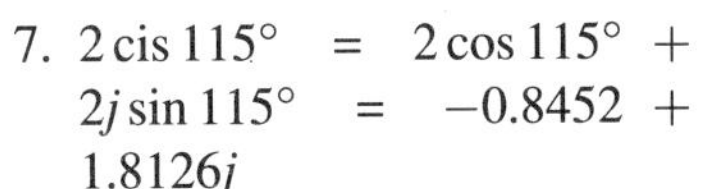

7. $2 \operatorname{cis} 115^\circ = 2\cos 115^\circ + 2j\sin 115^\circ = -0.8452 + 1.8126j$

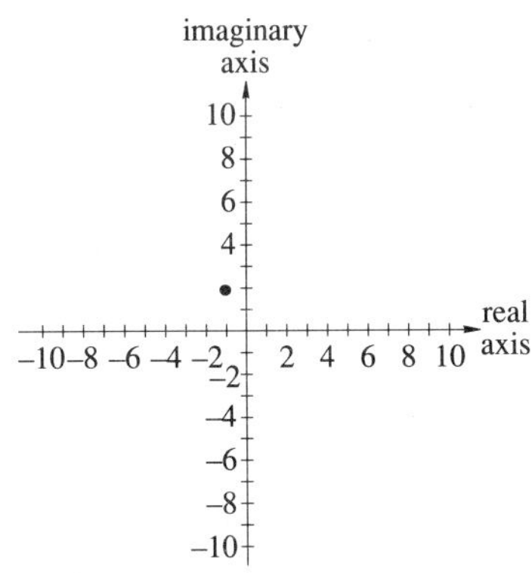

8. $5 \operatorname{cis} 285^\circ = 5\cos 285^\circ + 5j\sin 285^\circ = 1.2941 - 4.8296j$

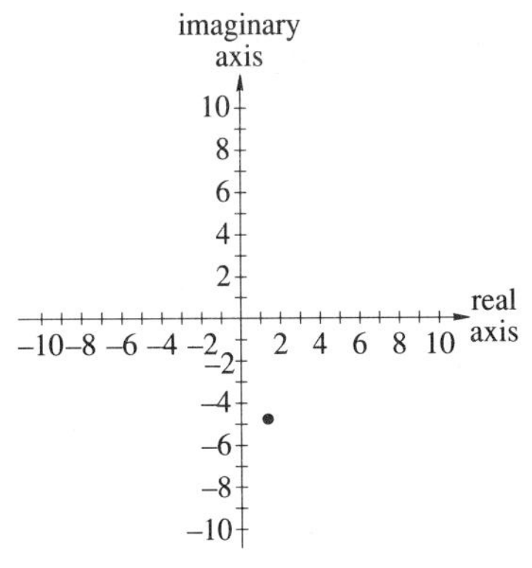

9. $3\underline{/25^\circ} = 3\cos 25^\circ + 3j\sin 25^\circ = 2.7189 + 1.2679j$

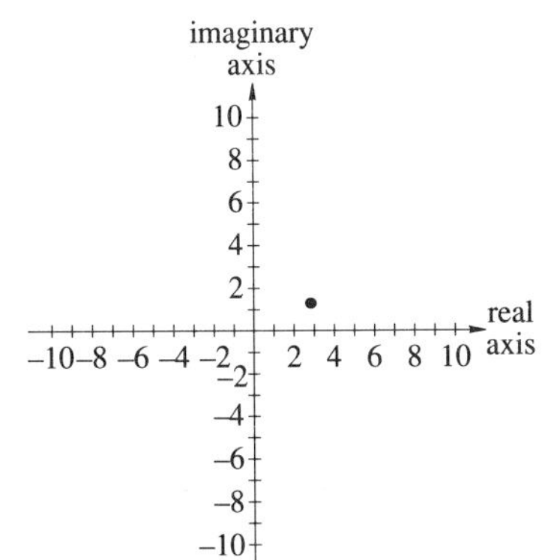

10. $4\underline{/240^\circ} = 4\cos 240^\circ + 4j\sin 240^\circ = -2 - 3.4641j$

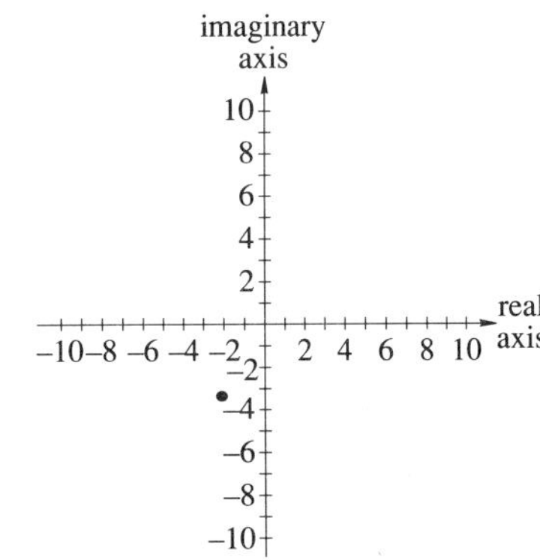

11. $5\underline{/340^\circ} = 5\cos 340^\circ + 5j\sin 340^\circ = 4.6985 - 1.7101j$

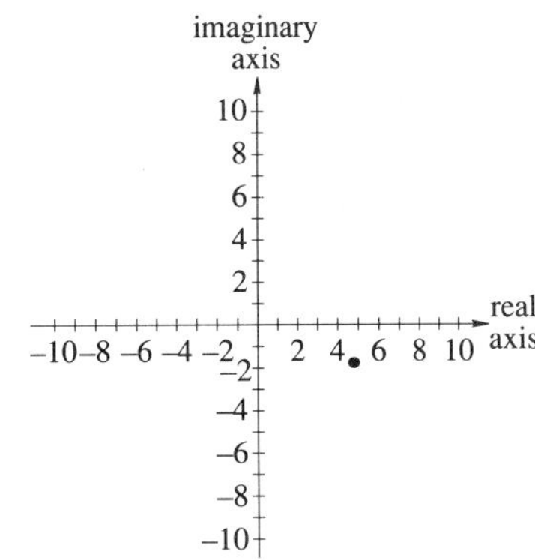

12. $6\underline{/90^\circ} = 6\cos 90^\circ + 6j\sin 90^\circ = 6j$

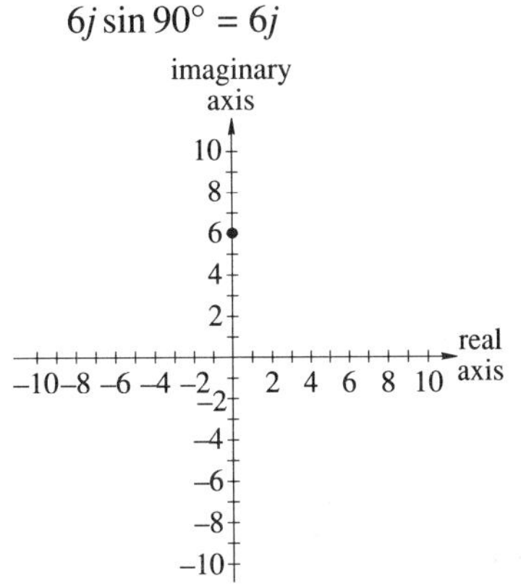

13. $4.5\underline{/245^\circ} = 4.5\cos 245^\circ + 4.5j\sin 245^\circ = -1.9018 - 4.0784j$

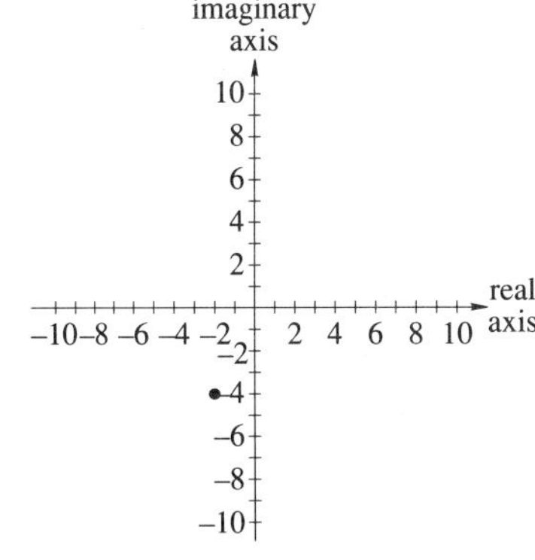

14. $6.8\underline{/10^\circ} = 6.8\cos 10^\circ + 6.8j\sin 10^\circ = 6.6967 + 1.1808j$

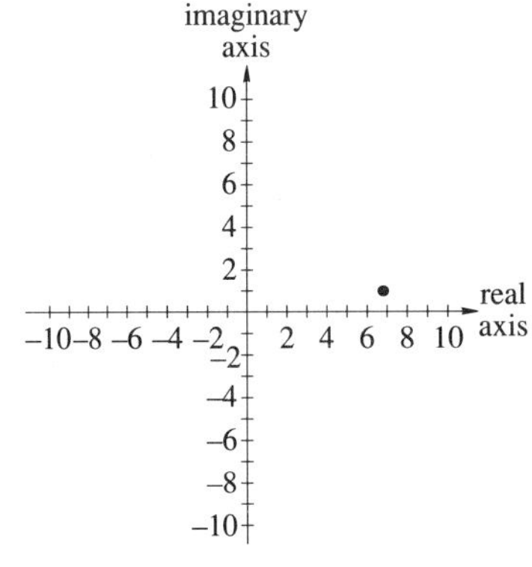

15. $2.5\underline{/180^\circ} = 2.5\cos 180^\circ + 2.5j\sin 180^\circ = -2.5$

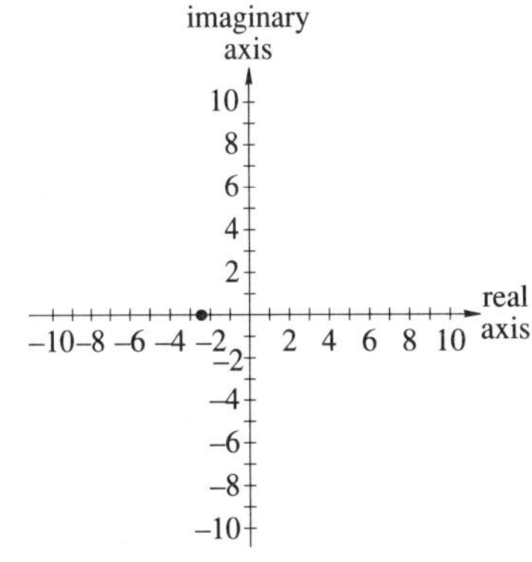

16. $5.9\,\underline{/270^\circ} = 5.9\cos 270^\circ + 5.9j\sin 270^\circ = -5.9j$

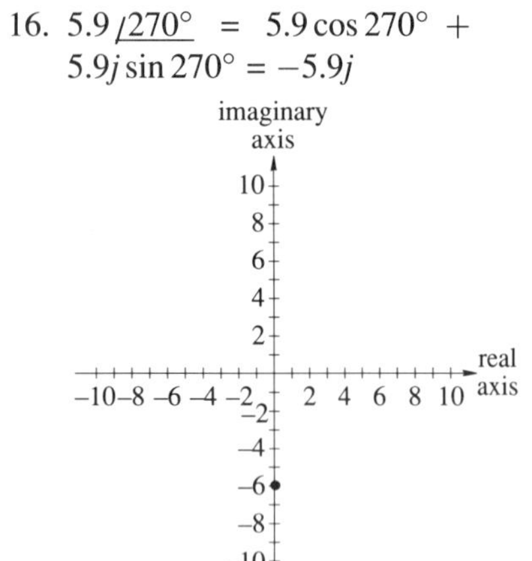

17. $5 + 5j$; $r = \sqrt{5^2 + 5^2} = \sqrt{50} = 5\sqrt{2} \approx 7.071$; $\theta = \tan^{-1}\frac{5}{5} = 45^\circ$; $7.071\,\underline{/45^\circ}$ or $7.071 \text{ cis } 45^\circ$

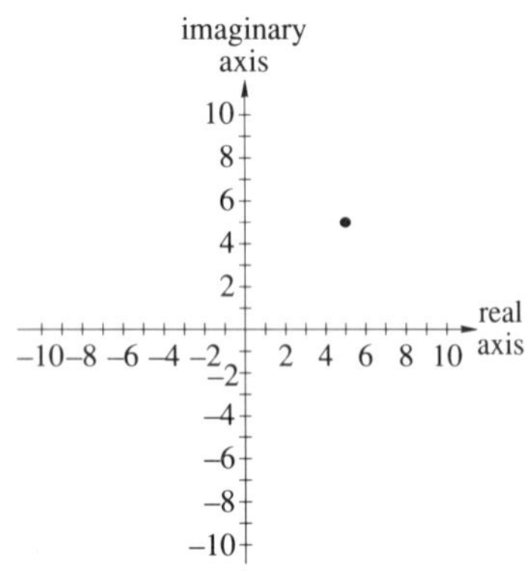

18. $6 + 3j$; $r = \sqrt{6^2 + 3^2} = 6.708$; $\theta = \tan^{-1}\frac{3}{6} = 26.6$; $6.708\,\underline{/26.6^\circ}$ or $6.708 \text{ cis } 26.6^\circ$

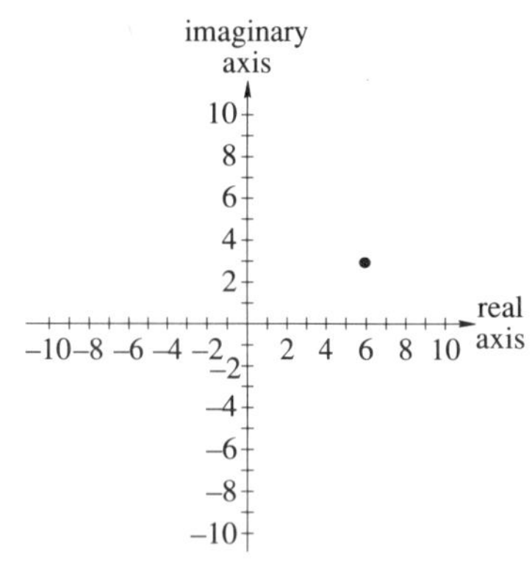

19. $4 - 8j$;
$r = \sqrt{4^2 + (-8)^2} = 8.944$; $\theta = \tan^{-1}\left(-\frac{8}{4}\right) = -63.4^\circ$ or 296.6°; $8.944\,\underline{/296.6^\circ}$ or $8.944 \text{ cis } 296.6^\circ$

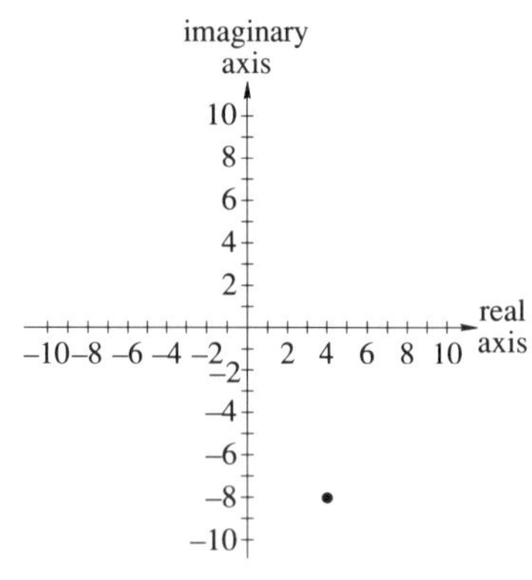

20. $8 - 2j$; $r = \sqrt{8^2 + (-2)^2} = \sqrt{68} = 8.246$; $\theta = \tan^{-1}\frac{-2}{8} = -14.0^\circ$ or 346°; $8.246\,\underline{/346^\circ}$ or $8.246 \text{ cis } 346^\circ$

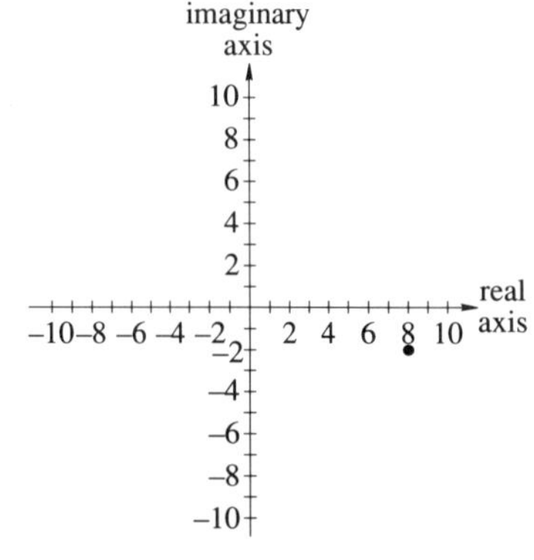

21. $-4 + 7j$; $r = \sqrt{(-4)^2 + 7^2} = 8.0623$; $\theta = \tan^{-1}\frac{-7}{4} = 119.7^\circ$; $8.0623\,\underline{/119.7^\circ}$

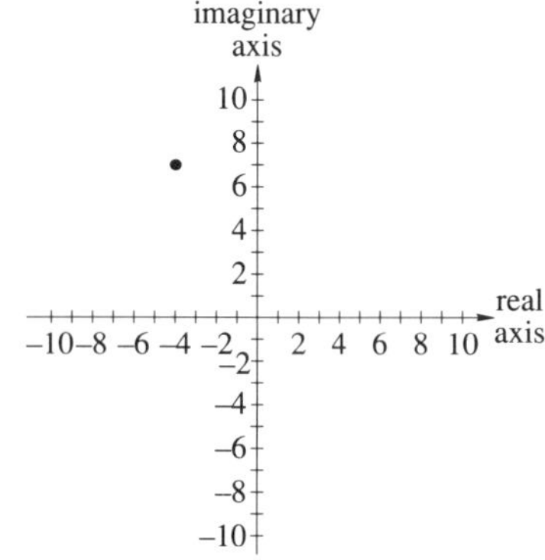

22. $-9 + 3j$; $r = \sqrt{(-9)^2 + 3^2} = 9.4868$; $\theta = \tan^{-1}\frac{3}{-9} = 161.6$; $9.4868\,\underline{/161.6^\circ}$

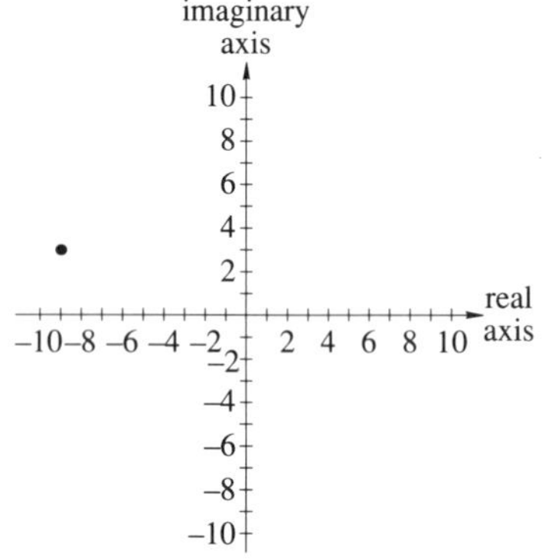

23. $-6 - 2j$;
$r = \sqrt{(-6)^2 + (-2)^2} = 6.3246$; $\theta = \tan^{-1}\frac{-2}{-6} = 198.4^\circ$; $6.3246\,\underline{/198.4^\circ}$

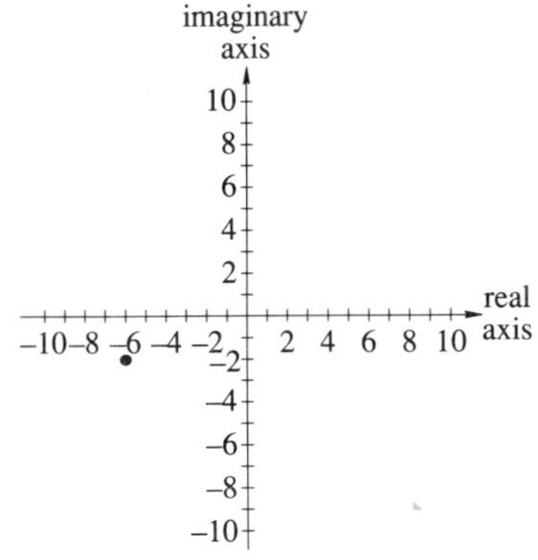

24. $-10-2j$; $r = \sqrt{(-10)^2+(-2)^2} = 10.1980$; $\theta = \tan^{-1}\frac{-2}{-10} = 191.3$; $10.1980\,\underline{/191.3^\circ}$

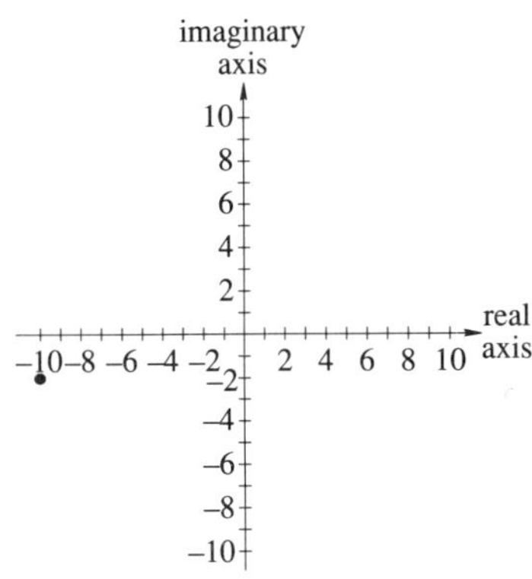

25. 6; $r = 6$; $\theta = 0$; $6\,\underline{/0^\circ}$

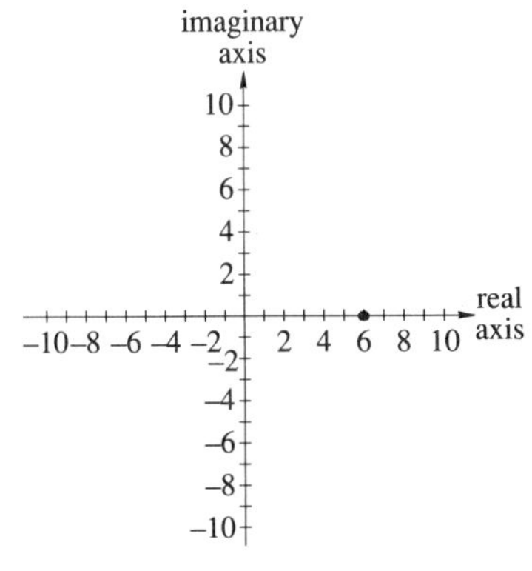

26. $1.2+7.3j$; $r = \sqrt{1.2^2+7.3^2} = 7.3980$; $\theta = \tan^{-1}\frac{7.3}{1.2} = 80.7$; $7.3980\,\underline{/80.7^\circ}$

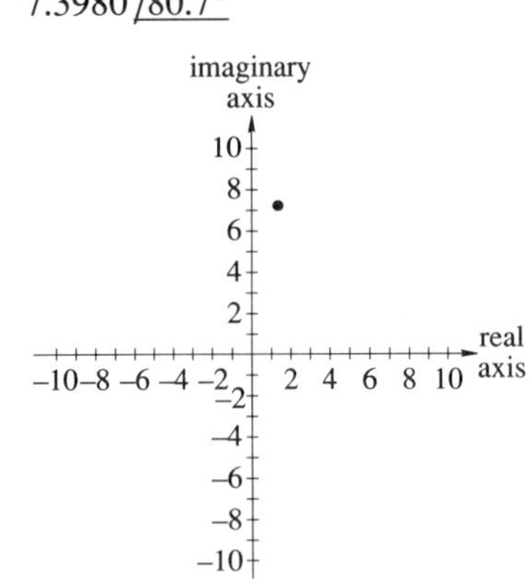

27. $4.2-6.3j$; $r = \sqrt{4.2^2+(-6.3)^2} = 7.5717$; $\theta = \tan^{-1}\frac{-6.3}{4.2} = -56.3$ or 303.7; $7.5717\,\underline{/303.7^\circ}$

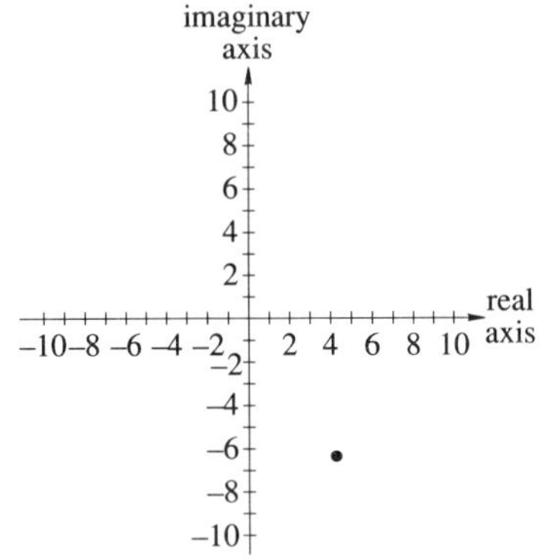

28. $9j$; $r = 9$; $\theta = 90^\circ$; $9\,\underline{/90^\circ}$

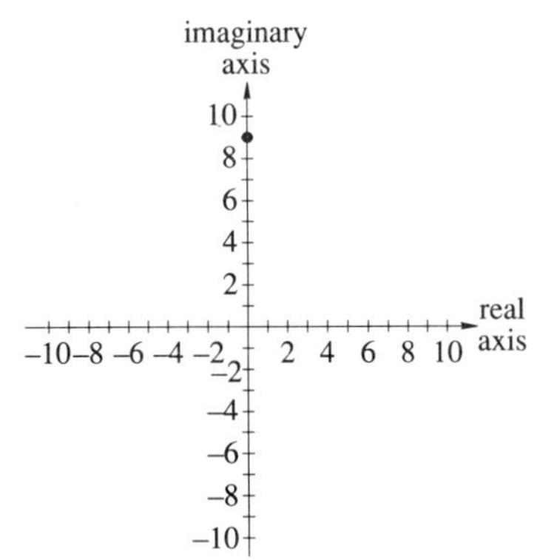

29. $-5.8+0.2j$; $r = \sqrt{(-5.8)^2+0.2^2} = 5.8034$; $\theta = \tan^{-1}\frac{0.2}{-5.8} = 178.0^\circ$; $5.8034\,\underline{/178.0^\circ}$

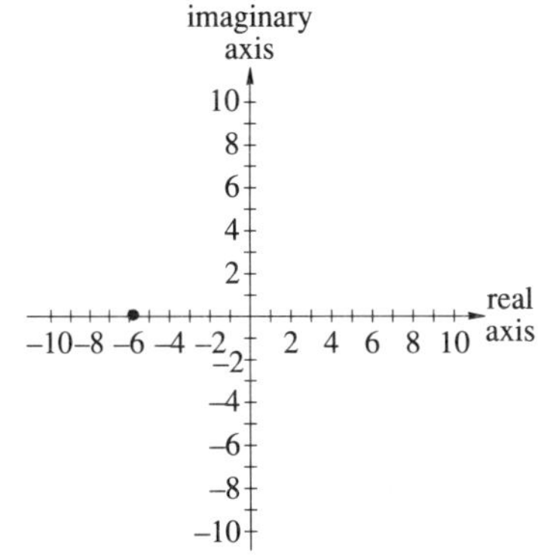

30. $-7j$; $r = 7$; $\theta = 270^\circ$; $7\,\underline{/270^\circ}$

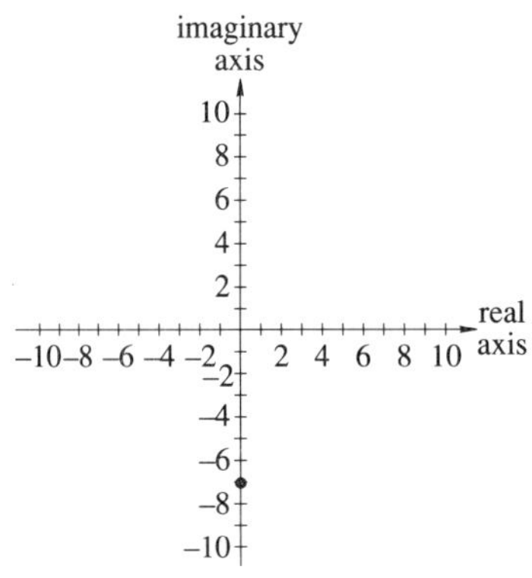

31. -2.7; $r = 2.7$; $\theta = 180^\circ$; $2.7\,\underline{/180^\circ}$

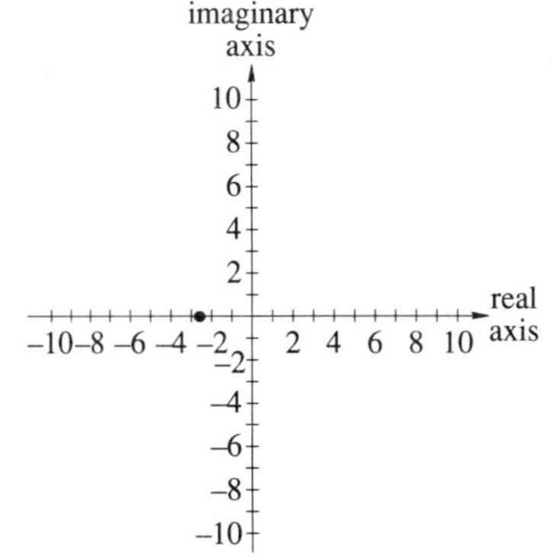

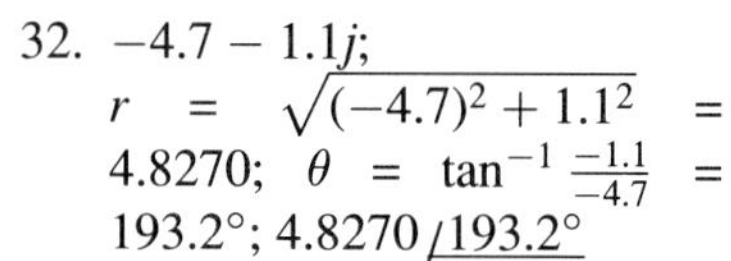

32. $-4.7 - 1.1j$; $r = \sqrt{(-4.7)^2 + 1.1^2} = 4.8270$; $\theta = \tan^{-1}\frac{-1.1}{-4.7} = 193.2°$; $4.8270\,\underline{/193.2°}$

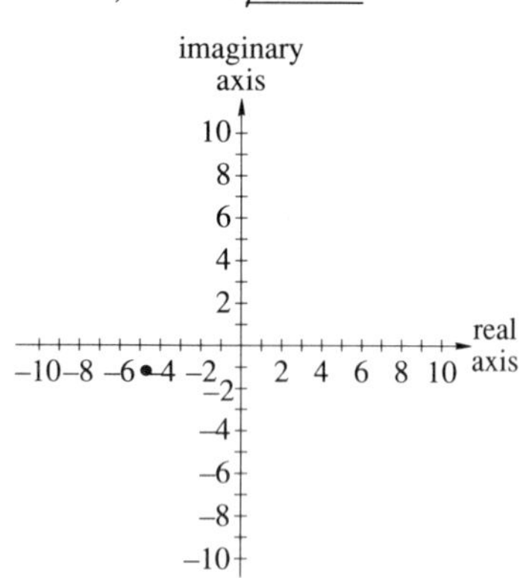

33.

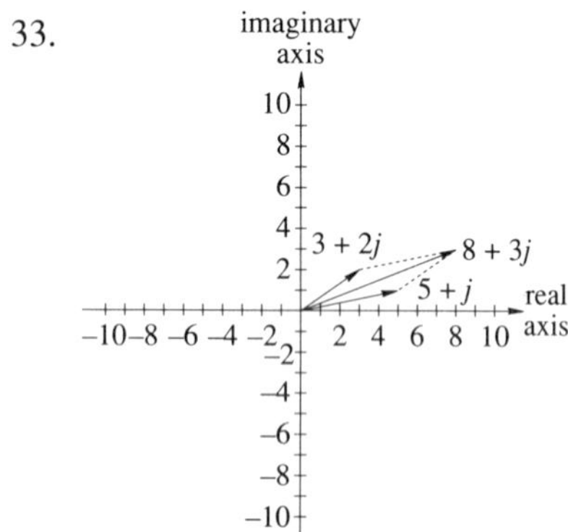

34.

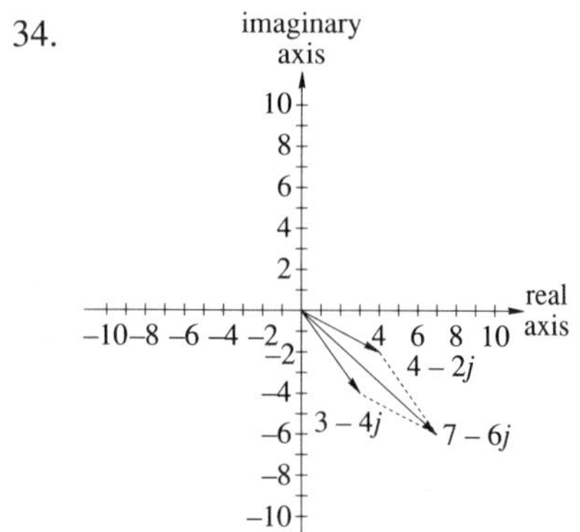

35.

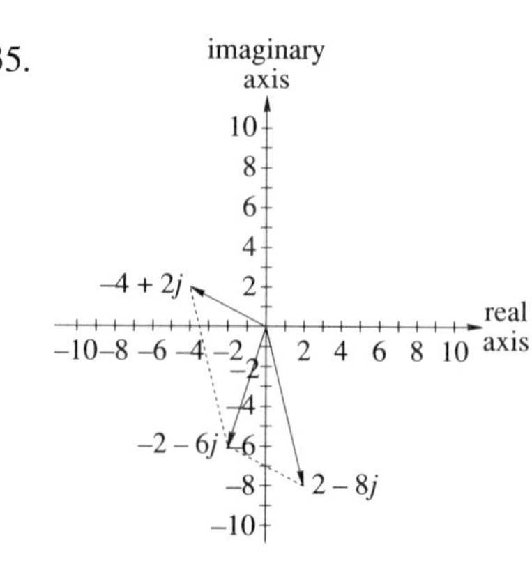

36.

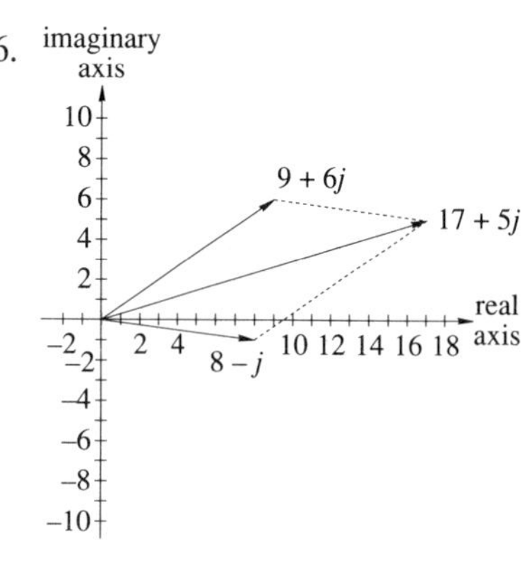

37.

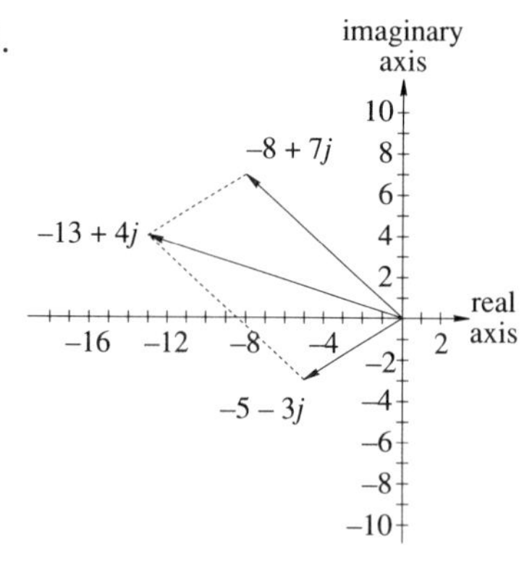

38.

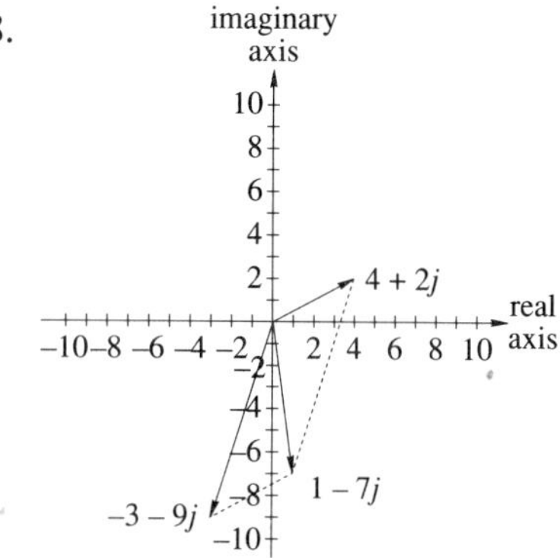

39.

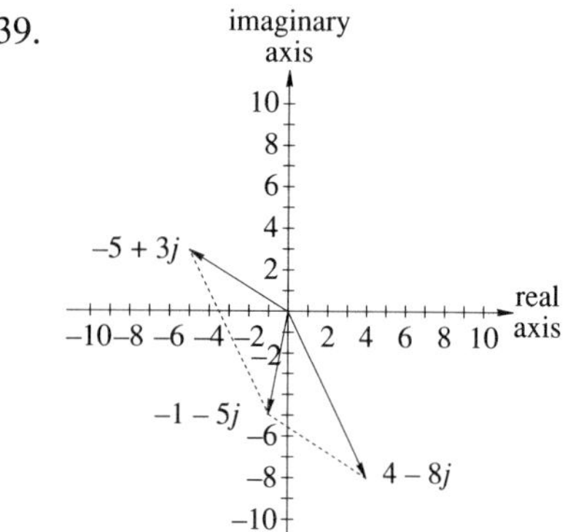

40.

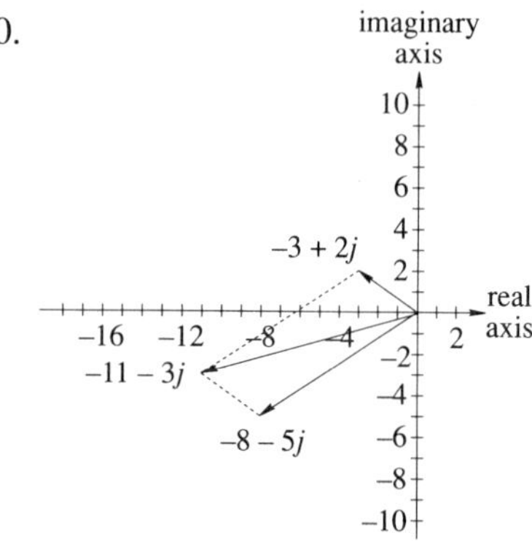

41. See *Computer Programs* in main text.

14.4 EXPONENTIAL FORM OF A COMPLEX NUMBER

1. $3(\cos\frac{3\pi}{2} + j\sin\frac{3\pi}{2})$; $r = 3$; $\theta = \frac{3\pi}{2}$; Answer: $3e^{3\pi/2j} \approx 3e^{4.7j}$
2. $7(\cos 1.4 + j\sin 1.4)$; $r = 7, \theta = 1.4$; Answer: $7e^{1.4j}$
3. $2(\cos 60° + j\sin 60°)$; $r = 2, \theta = 60° = \frac{\pi}{3}$; Answer: $2e^{\frac{\pi}{3}j} \approx 2e^{1.05j}$
4. $11(\cos 320° + j\sin 320°)$; $r = 11, \theta = 320° = 5.59$; Answer: $11e^{5.59j}$
5. $1.3(\cos 5.7 + j\sin 5.7) = 1.3e^{5.7j}$
6. $9.5(\cos 2.1 + j\sin 2.1) = 9.5e^{2.1j}$
7. $25° = 0.44$; $3.1(\cos 25° + j\sin 25°) = 3.1e^{0.44j}$
8. $195° = 3.40$; $10.5(\cos 195° + j\sin 195°) = 10.5e^{3.40j}$
9. $8 + 6j = 10\operatorname{cis} 0.64 = 10e^{0.64j}$
10. $12 - 5j = 13\operatorname{cis} 5.89 = 13e^{5.89j}$
11. $-9 + 12j = 15\operatorname{cis} 2.21 = 15e^{2.21j}$
12. $-8 - 2j = 8.25\operatorname{cis} 3.39 = 8.25e^{3.39j}$
13. $5e^{0.5j} = 5\operatorname{cis} 0.5 = 4.3879 + 2.3971j$
14. $8e^{1.9j} = 8\operatorname{cis} 1.9 = -2.5863 + 7.5704j$
15. $2.3e^{4.2j} = 2.3\operatorname{cis} 4.2 = -1.1276 - 2.0046j$
16. $4.5e^{\frac{7\pi}{6}j} = 4.5\operatorname{cis}\frac{7\pi}{6} = -3.8971 - 2.25j$
17. $2e^{3j} \cdot 6e^{2j} = (2 \cdot 6)e^{3j+2j} = 12e^{5j}$
18. $3e^{j} \cdot 4e^{2j} = (3 \cdot 4)e^{j+2j} = 12e^{3j}$
19. $e^{1.3j} \cdot 2.4e^{4.6j} = (1 \cdot 2.4)e^{(1.3+4.6)j} = 2.4e^{5.9j}$
20. $1.5e^{4.1j} \cdot 0.2e^{1.7j} = (1.5 \cdot 0.2)e^{(4.1+1.7)j} = 0.3e^{5.8j}$
21. $7e^{4.3j} \cdot 4e^{5.7j} = (7 \cdot 4)e^{(4.3+5.7)j} = 28e^{10j} = 28e^{3.72j}$ (Note: $10 - 2\pi \approx 3.72$)
22. $3.6e^{5.4j} \cdot 2.5e^{6.1j} = 9e^{11.5j} = 9e^{5.22j}$ (Note: $11.5 - 2\pi \approx 5.22$)
23. $8e^{3j} \div 2e^{j} = (8/2)e^{3j-j} = 4e^{2j}$
24. $28e^{5j} \div 7e^{2j} = (28/7)e^{5j-2j} = 4e^{3j}$
25. $17e^{4.3j} \div 4e^{2.8j} = 17/4e^{(4.3-2.8)j} = 4.25e^{1.5j}$
26. $8.5e^{3.4j} \div 2e^{5.3j} = (8.5/2)e^{(3.4-5.3)j} = 4.25e^{-1.9j} = 4.25e^{4.38j}$ (Note: $-1.9 + 2\pi \approx 4.38$)
27. $(3e^{2j})^4 = 3^4e^{2j\cdot 4} = 81e^{8j} \approx 81e^{1.72j}$ (Note: $8 - 2\pi \approx 1.72$.)
28. $(4e^{3j})^5 = 4^5e^{15j} = 1024e^{15j} \approx 1024e^{2.43j}$ (Note: $15 - 4\pi \approx 2.43$.)
29. $(2.5e^{1.5j})^4 = 2.5^4e^{1.5j\cdot 4} = 39.0625e^{6j}$
30. $(7.2e^{2.3j})^5 = 7.2^5e^{5\times 2.3j} = 19349e^{11.5j} = 19349e^{5.22j}$ or $7.2^5e^{5.22j}$
31. $(4e^{6j})^{1/2} = 4^{1/2}e^{\frac{6j}{2}} = 2e^{3j}$
32. $(16e^{6j})^{1/4} = 16^{1/4}e^{\frac{6j}{4}} = 2e^{1.5j}$
33. $(6.25e^{4.2j})^{1/2} = 6.25^{1/2}e^{\frac{4.2j}{2}} = 2.5e^{2.1j}$
34. $(1.728e^{2.1j})^{1/3} = 1.728^{1/3}e^{\frac{2.1j}{3}} = 1.2e^{0.7j}$
35. $V = 56.5 + 24.1j = 61.41\operatorname{cis} 0.4031 = 61.41e^{0.4031j}$ V
36. $I = 4.90 - 4.11j = 6.40\operatorname{cis} 5.59 = 6.40e^{5.59j}$ A; more exactly, $6.40e^{5.5852j}$ A
37. $Z = 135\underline{/-52.5°}\ \Omega = 135e^{5.37j}\ \Omega = 82.2 - 107.1j\ \Omega$
38. $X_L = 40.5\underline{/-\pi/2}\ \Omega = 40.5e^{-\pi/2j}\ \Omega = 40.5e^{-1.5708j}\ \Omega = 40.5e^{4.7124j}\ \Omega = -40.5j\ \Omega$
39. $V = IZ = (12.5e^{-0.7256j})(6.4e^{1.4285j}) = (12.5 \times 6.4)e^{(-0.7256+1.4285)j} = 80e^{0.7029j}$ V
40. $V_L = IX_L = (4.24e^{0.5627j})(28.5e^{-1.5708j}) = (4.24 \times 28.5)e^{(0.5627-1.5708)j} = 120.84e^{-1.0081j}$ V $\approx 120.84e^{5.2751j}$ V
41. $V = IZ \Rightarrow I = \frac{V}{Z} = \frac{115e^{-0.2125j}}{2.5e^{0.5792j}} = \frac{115}{2.5}e^{(-0.2145-0.5792)j} = 46e^{-0.7937j}$ A $\approx 46e^{5.49j}$ A
42. $R = \frac{V_R}{I} = \frac{35.1e^{1.3826j}}{0.78e^{1.3826j}} = 45e^{0j} = 45\ \Omega$

14.5 OPERATIONS IN POLAR FORM; DEMOIVRE'S FORMULA

1. $(3\,\text{cis}\,46°)\cdot(5\,\text{cis}\,23°) = 3\cdot 5\,\text{cis}(46° + 23°) = 15\,\text{cis}\,69° = 15(\cos 69° + j\sin 69°)$

2. $(4\,\text{cis}\,135°)\cdot(5\,\text{cis}\,63°) = 4\cdot 5\,\text{cis}(135° + 63°) = 20\,\text{cis}\,198° = 20(\cos 198° + j\sin 198°)$

3. $(2.5\,\text{cis}\,1.4°)\cdot(4\,\text{cis}\,2.67) = 2.5\cdot 4\,\text{cis}(1.43+2.67) = 10\,\text{cis}\,4.1 = 10(\cos 4.1° + j\sin 4.1°)$

4. $(6.4\,\text{cis}\,0.25°)\cdot(3.5\,\text{cis}\,1.1) = 6.4\times 3.5\,\text{cis}(0.25 + 1.1) = 22.4\,\text{cis}\,1.35 = 22.4(\cos 1.35° + j\sin 1.35°)$

5. $\frac{8\,\text{cis}\,85°}{2\,\text{cis}\,25°} = \frac{8}{2}\,\text{cis}(85° - 25°) = 4\,\text{cis}\,60° = 4(\cos 60° + j\sin 60°)$

6. $\frac{6\,\text{cis}\,273°}{3\,\text{cis}\,114°} = \frac{6}{3}\,\text{cis}(273° - 114°) = 2\,\text{cis}\,159° = 2(\cos 159° + j\sin 159°)$

7. $\frac{9\,\underline{/137°}}{2\,\underline{/26°}} = \frac{9}{2}\,\underline{/137° - 26°} = 4.5\,\underline{/111°}$

8. $\frac{18\,\underline{/3.52°}}{5\,\underline{/2.14°}} = \frac{18}{5}\,\underline{/3.52 - 2.14} = 3.6\,\underline{/1.38°}$

9. $(3\,\underline{/2.7})(4\,\underline{/5.3}) = 3\cdot 4\,\underline{/2.7+5.3} = 12\,\underline{/8} = 12\,\underline{/1.72}$ (Note: $8 - 2\pi \approx 1.72$.)

10. $[3\,\text{cis}\,20°]^4 = 3^4\,\text{cis}(4\cdot 20) = 81\,\text{cis}\,80°$

11. $[5\,\text{cis}\,84°]^6 = 5^6\,\text{cis}(6\cdot 84) = 5^6\,\text{cis}(504) = 15{,}625\,\text{cis}\,144°$

12. $[2.5\,\text{cis}\,118°]^3 = 2.5^3\,\text{cis}(3\cdot 118) = 15.625\,\text{cis}\,354°$

13. $[10.4\,\text{cis}\,3.42]^3 = 10.4^3\,\text{cis}(3\cdot 3.42) = 1124.864\,\text{cis}\,10.26 = 1124.864\,\text{cis}\,3.98$

14. $(2\,\underline{/1.38})^5 = 2^5\,\underline{/5\cdot 1.38} = 32\,\underline{/6.9} = 32\,\underline{/0.62}$

15. $(3.4\,\underline{/5.3})^4 = 3.4^4\,\underline{/4\cdot 5.3} = 133.6\,\underline{/21.2} \approx 133.6\,\underline{/2.35}$ (Note: $21.2 \approx 2.35 + 6\pi$.)

16. $(4.41\,\underline{/124°})^{1/2} = 4.41^{1/2}\,\underline{/1/2\cdot 124°} = 2.1\,\underline{/62°}$

17. $\left(4e^{2.1j}\right)\left(3e^{1.7j}\right) = 12e^{(2.1+1.7)j} = 12e^{3.8j}$

18. $6e^{1.5} \div 4e^{0.9j} = \frac{6}{4}e^{(1.5-0.9)j} = 1.5e^{0.6j}$

19. $(0.5e^{0.3j})^3 = 0.5^3e^{3\times 0.3j} = 0.125e^{0.9j}$

20. $(0.0625e^{4.2j})^{1/2} = 0.0625^{1/2}e^{\frac{4.2j}{2}} = 0.25e^{2.1j}$

21. Cube roots of 1. First the polar form of 1 is $1\,\text{cis}\,0° = \text{cis}\,0°$. $w_0 = \sqrt[3]{1}\,\text{cis}\left(\frac{0}{3} + \frac{360.0}{3}\right)° = 1\,\text{cis}\,0° = 1$; $w_1 = \sqrt[3]{1}\,\text{cis}\left(\frac{0}{3} + \frac{360\cdot 1}{3}\right)° = 1\,\text{cis}\,120° = -0.5 + 0.866j$; $w_2 = \sqrt[3]{1}\,\text{cis}\left(\frac{0}{3} + \frac{360\cdot 2}{3}\right)° = 1\,\text{cis}\,240° = -0.5 - 0.866j°$

22. Cube roots of j. First $j = 1\,\text{cis}\,90°$. $w_0 = \sqrt[3]{1}\,\text{cis}\left(\frac{90}{3} + \frac{360\cdot 0}{3}\right)° = 1\,\text{cis}\,30° = 0.866 + 0.5j$; $w_1 = \sqrt[3]{1}\,\text{cis}\left(\frac{90}{3} + \frac{360\cdot 1}{3}\right)° = 1\,\text{cis}\,150° = -0.866 + 0.5j$; $w_2 = \sqrt[3]{1}\,\text{cis}\left(30 + \frac{360\cdot 2}{3}\right)° = 1\,\text{cis}\,270° = -j$

23. We want the cube roots of $-8j$. Now, $-8j = 8\,\text{cis}\,270°$. $w_0 = \sqrt[3]{8}\,\text{cis}\left(\frac{270}{3} + \frac{360\cdot 0}{3}\right)° = 2\,\text{cis}\,90° = 2j$; $w_1 = \sqrt[3]{8}\,\text{cis}\left(90 + \frac{360\cdot 1}{3}\right)° = 2\,\text{cis}\,210° = -1.732 - j$; $w_2 = 2\,\text{cis}\left(90 + \frac{360\cdot 2}{3}\right)° = 2\,\text{cis}\,330° = 1.732 - j$

24. Fourth roots of -16; $-16 = 16\,\text{cis}\,180°$. $w_0 = \sqrt[4]{16}\,\text{cis}\left(\frac{180}{4} + \frac{360\cdot 0}{4}\right)° = 2\,\text{cis}\,45° = \sqrt{2} + \sqrt{2}j$; $= 1.414 + 1.414j$; $w_1 = 2\,\text{cis}\left(45 + \frac{360\cdot 1}{4}\right)° = 2\,\text{cis}\,135° = -1.414 + 1.414j$; $w_2 = 2\,\text{cis}\left(45 + \frac{360\cdot 2}{4}\right)° = 2\,\text{cis}(225°) = -1.414 - 1.414j$; $w_3 = 2\,\text{cis}\left(45 + \frac{360\cdot 3}{4}\right)° = 2\,\text{cis}(315°) = 1.414 - 1.414j$

25. Fourth roots of $-16j$. $-16j = 16\,\text{cis}\,270°$ $w_0 = \sqrt[4]{16}\,\text{cis}\left(\frac{270}{4} + \frac{360\cdot 0}{4}\right)° = 2\,\text{cis}\,67.5° = 0.7654 + 1.8478j$; $w_1 = 2\,\text{cis}\left(67.5 + \frac{360\cdot 1}{4}\right)° = 2\,\text{cis}\,157.5° = -1.8478 + 0.7654j$; $w_2 = 2\,\text{cis}\left(67.5 + \frac{360\cdot 2}{4}\right)° = 2\,\text{cis}\,247.5° = -0.7654 - 1.8478j$; $w_3 = 2\,\text{cis}\left(67.5 + \frac{360\cdot 3}{4}\right)° = 2\,\text{cis}\,337.5° = 1.8478 - 0.7654j$

26. Fourth roots of $1 - j$. $1 - j = \sqrt{2}\,\text{cis}\,315°$ $w_0 = \sqrt[4]{\sqrt{2}}\,\text{cis}\left(\frac{315}{4} + \frac{360\cdot 0}{4}\right)° = 1.0905\,\text{cis}\,78.75° = 0.2127 + 1.0696j$; $w_1 = \sqrt[8]{2}\,\text{cis}\left(\frac{315}{4} + \frac{360\cdot 1}{4}\right)° = 1.0905\,\text{cis}\,168.75$; $= -1.0696 + 0.2127j$; $w_2 = \sqrt[8]{2}\,\text{cis}\left(\frac{315}{4} + \frac{360\cdot 2}{4}\right)° = 1.0905\,\text{cis}\,258.75° =$

$-0.2127 - 1.0696j$; $w_3 = \sqrt[8]{2}\,\text{cis}\left(\frac{315}{4} + \frac{360 \cdot 3}{4}\right)^\circ = 1.0905\,\text{cis}\,348.75^\circ = 1.0696 - 0.2127j$

27. Fifth roots of $1 + j$. $1 + j = \sqrt{2}\,\text{cis}\,45^\circ$. $w_0 = \sqrt[5]{\sqrt{2}}\,\text{cis}\left(\frac{45}{5} + \frac{360 \cdot 0}{5}\right)^\circ = 1.0718\,\text{cis}\,9^\circ = 1.0586 + 0.1677j$; $w_1 = \sqrt[10]{2}\,\text{cis}\left(9 + \frac{360 \cdot 1}{5}\right)^\circ = 1.0718\,\text{cis}\,81^\circ = 0.1677 + 1.0586j$; $w_2 = \sqrt[10]{2}\,\text{cis}\left(9 + \frac{360 \cdot 2}{5}\right)^\circ = 1.0718\,\text{cis}\,153^\circ = -0.9550 + 0.4866j$; $w_3 = \sqrt[10]{2}\,\text{cis}\left(9 + \frac{360 \cdot 3}{5}\right)^\circ = 1.0718\,\text{cis}\,225^\circ = -0.7579 - 0.7579j$; $w_4 = \sqrt[10]{2}\,\text{cis}\left(9 + \frac{360 \cdot 4}{5}\right) = 1.0718\,\text{cis}\,297^\circ = 0.4866 - 0.9550j$

28. Sixth roots of $-1 - j = \sqrt{2}\,\text{cis}\,225^\circ$; $w_0 = \sqrt[6]{\sqrt{2}}\,\text{cis}\left(\frac{225}{6} + \frac{360 \cdot 0}{6}\right)^\circ = 1.0595\,\text{cis}\,37.5^\circ = 0.8405 + 0.6450j$; $w_1 = \sqrt[12]{2}\,\text{cis}\left(37.5 + \frac{360}{6}\right)^\circ = 1.0595\,\text{cis}\,97.5^\circ = -0.1383 + 1.0504j$; $w_2 = \sqrt[12]{2}\,\text{cis}\left(37.5 + \frac{360 \times 2}{6}\right)^\circ = 1.0595\,\text{cis}\,157.5^\circ = -0.9788 + 0.4054j$; $w_3 = \sqrt[12]{2}\,\text{cis}\left(37.5 + \frac{360 \times 3}{6}\right)^\circ = 1.0595\,\text{cis}\,217.5^\circ = -0.8405 - 0.6450j$; $w_4 = \sqrt[12]{2}\,\text{cis}\left(37.5 + \frac{360 \times 4}{6}\right)^\circ = 1.0595\,\text{cis}\,277.5^\circ = 0.1383 - 1.0504j$; $w_5 = \sqrt[12]{2}\,\text{cis}\,(37.5 + 60 \cdot 5)^\circ = 1.0595\,\text{cis}\,337.5^\circ = 0.9788 - 0.4054j$

29. $x^3 = -j$: the solutions are the three cube roots of $-j = 1\,\text{cis}\,270^\circ$. $w_0 = \sqrt[3]{1}\,\text{cis}(270/3)^\circ = 1\,\text{cis}\,90^\circ = j$; $w_1 = 1\,\text{cis}(90 + 360/3)^\circ = 1\,\text{cis}\,210^\circ = -0.8660 - 0.5j$; $w_2 = 1\,\text{cis}(90 + 120 \cdot 2)^\circ = 1\,\text{cis}\,330^\circ = 0.8660 - 0.5j$

30. $x^3 = 125j$: the solutions are the three cube roots of $125j = 125\,\text{cis}\,90^\circ$. $w_0 = \sqrt[3]{125}\,\text{cis}(90/3)^\circ = 5\,\text{cis}\,30^\circ = 4.3301 + 2.5j$; $w_1 = 5\,\text{cis}(30 + 360/3)^\circ = 5\,\text{cis}\,150^\circ = -4.3301 + 2.5j$; $w_2 = 5\,\text{cis}(30 + 120 \times 2)^\circ = 5\,\text{cis}\,270^\circ = -5j$

31. The solution of $x^6 - 64j = 0$ are the six sixth roots of $64j = 64\,\text{cis}\,90^\circ$. $w_0 = \sqrt[6]{64}\,\text{cis}(90/6)^\circ = 2\,\text{cis}\,15^\circ = 1.9319 + 0.5176j$; $w_1 = 2\,\text{cis}(15 + \frac{360}{6})^\circ = 2\,\text{cis}\,75^\circ = 0.5176 + 1.9319j$; $w_2 = 2\,\text{cis}(15 + 60 \times 2)^\circ = 2\,\text{cis}\,135^\circ = -1.4142 + 1.4142j$; $w_3 = 2\,\text{cis}(15 + 60 \times 3)^\circ = 2\,\text{cis}\,195^\circ = -1.9319 - 0.5176j$; $w_4 = 2\,\text{cis}(15 + 60 \times 4)^\circ = 2\,\text{cis}\,255^\circ = -0.5176 - 1.9319j$; $w_5 = 2\,\text{cis}(15 + 60 \times 5)^\circ = 2\,\text{cis}\,315^\circ = 1.412 - 1.4142j$

32. The solutions of $x^6 - 1 = j$ are the six sixth roots of $1 + j = \sqrt{2}\,\text{cis}\,45^\circ$. $w_0 = \sqrt[6]{\sqrt{2}}\,\text{cis}(45/6)^\circ = 1.0595\,\text{cis}\,7.5^\circ = 1.0504 + 0.1383j$; $w_1 = 1.0595\,\text{cis}(7.5 + \frac{360}{6})^\circ = 1.0595\,\text{cis}\,67.5^\circ = 0.4054 + 0.9788j$; $w_2 = 1.0595\,\text{cis}(7.5 + 60 \times 2)^\circ = 1.0595\,\text{cis}\,127.5^\circ = -0.6450 + 0.8405j$; $w_3 = 1.0595\,\text{cis}(7.5 + 60 \times 3)^\circ = 1.0595\,\text{cis}\,187.5^\circ = -1.0504 - 0.1383$; $w_4 = \sqrt[12]{2}\,\text{cis}(7.5 + 60 \times 4)^\circ = 1.0595\,\text{cis}\,247.5^\circ = -0.4054 - 0.9788j$; $w_5 = \sqrt[12]{2}\,\text{cis}(7.5 + 60 \times 5)^\circ = 1.0595\,\text{cis}\,307.5^\circ = 0.6450 - 0.8405j$

33. $V = IZ; I = 12\underline{/-23^\circ}$, $Z = 9\underline{/42^\circ}$; $V = (12\underline{/-23^\circ})(9\underline{/42^\circ}) = (12 \times 9)\underline{/-23^\circ + 42^\circ} = 108\underline{/19^\circ}$

34. $V = IZ$; $Z = \dfrac{V}{I} = \dfrac{20\underline{/30^\circ}}{5\underline{/40^\circ}} = \dfrac{20}{5}\underline{/30^\circ - 40^\circ} = 4\underline{/-10^\circ}$ or $4\underline{/350^\circ}$

35. See *Computer Programs* in main text.

36. See *Computer Programs* in main text.

37. See *Computer Programs* in main text.

14.6 COMPLEX NUMBERS IN AC CIRCUITS

1. (a) in Series
$Z = Z_1 + Z_2 = (2 + 3j) + (1 - 5j) = 3 - 2j$;

(b) in Parallel $Z = \dfrac{Z_1 Z_2}{Z_1 + Z_2} = \dfrac{(2 + 3j)(1 - 5j)}{3 - 2j} = \dfrac{2 - 7j - 15j^2}{3 - 2j} = \dfrac{17 - 7j}{3 - 2j} \cdot \dfrac{3 + 2j}{3 + 2j} = \dfrac{51 + 34j - 21j - 14j^2}{9 + 4} = \dfrac{65 + 13j}{13} = 5 + j$

2. (a) in Series
$Z = Z_1 + Z_2 = (4 - 7j) + (-3 + 4j) = 1 - 3j$;

(b) in Parallel $Z = \dfrac{Z_1 Z_2}{Z_1 + Z_2} = \dfrac{(4 - 7j)(-3 + 4j)}{1 - 3j} =$

$\frac{-12+16j+21j-28j^2}{1-3j} = \frac{16+37j}{1-3j} \cdot \frac{1+3j}{1+3j} = \frac{16+48j+37j+111j^2}{1+9} = \frac{-95+85j}{10} = -9.5+8.5j$

3. (a) in Series $Z = Z_1 + Z_2 = (1-j) + 3j = 1 + 2j$;

(b) in Parallel $Z = \frac{Z_1 Z_2}{Z_1 + Z_2} = \frac{(1-j)(3j)}{1+2j} = \frac{3+3j}{1+2j} \cdot \frac{1-2j}{1-2j} = \frac{3-6j+3j-6j^2}{1+4} = \frac{9-3j}{5} = 1.8 - 0.6j$

4. (a) in Series $Z = Z_1 + Z_2 = (2+j) + (4-3j) = 6 - 2j$; (b) in Parallel $Z = \frac{Z_1 Z_2}{Z_1 + Z_2} = \frac{(2+j)(4-3j)}{6-2j} = \frac{8-6j+4j-3j^2}{6-2j} = \frac{11-2j}{6-2j} \cdot \frac{6+2j}{6+2j} = \frac{66+22j-12j-4j^2}{36+4} = \frac{70+10j}{40} = \frac{7}{4} + \frac{1}{4}j = 1.75 + 0.25j$

5. $Z_1 = 2.19\underline{/18.4^\circ} = 2.0780 + 0.6913j$;
$Z_2 = 5.16\underline{/67.3^\circ} = 1.9913 + 4.7603j$;
(a) $Z = Z_1 + Z_2 = 4.0693 + 5.4516j = 6.803\underline{/53.26^\circ}$;

(b) $Z = \frac{Z_1 Z_2}{Z_1 + Z_2} = \frac{(2.19\underline{/18.4^\circ})(5.16\underline{/67.3^\circ})}{6.803\underline{/53.26^\circ}} = \frac{2.19 \cdot 5.16}{6.803}\underline{/18.4 + 67.3 - 53.26} = 1.6611\underline{/32.44^\circ}$

6. $Z_1 = 2\sqrt{3}\underline{/30^\circ} = 3 + \sqrt{3}j$;
$Z_2 = 2\underline{/120^\circ} = -1 + \sqrt{3}j$;
(a) $Z = Z_1 + Z_2 = 2 + 2\sqrt{3}j = 4\underline{/60^\circ}$;

(b) $Z = \frac{Z_1 Z_2}{Z_1 + Z_2} = \frac{(2\sqrt{3}\underline{/30^\circ})(2\underline{/120^\circ})}{4\underline{/60^\circ}} = \frac{2 \cdot 2\sqrt{3}}{4}\underline{/30^\circ + 120^\circ - 60^\circ} = \sqrt{3}\underline{/90^\circ} = \sqrt{3}j$ or $1.732j$

7. $Z_1 = 3\sqrt{5}\underline{/\pi/7} = 6.0439 + 2.9106j$;
$Z_2 = 1.5\underline{/0.45} = 1.3507 + 0.6524j$;
(a) $Z = Z_1 + Z_2 = 7.3946 + 3.5630j = 8.2082\underline{/0.449}$;

(b) $Z = \frac{Z_1 Z_2}{Z_1 + Z_2} = \frac{(3\sqrt{5}\underline{/\pi/7})(1.5\underline{/0.45})}{8.2082\underline{/0.449}} = \left(\frac{3\sqrt{5} \cdot 1.5}{8.2082}\right)\underline{/\pi/7 + 0.45 - 0.449} = 1.226\underline{/0.45}$

8. $Z_1 = 2.57\underline{/0.25} = 2.4901 + 0.6358j$;
$Z_2 = 1.63\underline{/1.38} = 0.3091 + 1.6004j$;
(a) $Z = Z_1 + Z_2 = 2.7992 + 2.2362j = 3.583\underline{/0.67}$;

(b) $Z = \frac{Z_1 Z_2}{Z_1 + Z_2} = \frac{(2.57\underline{/0.25})(1.63\underline{/1.38})}{3.583\underline{/0.674}} = \frac{2.57 \times 1.63}{3.583}\underline{/0.25 + 1.38 - 0.67} = 1.17\underline{/0.96}$

9. (a) $Z = Z_1 + Z_2 + Z_3 = (4+3j) + (3-2j) + (5+4j) = 12 + 5j$;

(b) $Z = \frac{Z_1 Z_2 Z_3}{Z_1 Z_2 + Z_1 Z_3 + Z_2 Z_3} = \frac{(4+3j)(3-2j)(5+4j)}{(4+3j)(3-2j)+(4+3j)(5+4j)+(3-2j)(5+4j)} = \frac{(12-8j+9j-6j^2)(5+4j)}{(12-8j+9j-6j^2)+(20+16j+15j+12j^2)+(15+12j-10j-8j^2)} = \frac{(18+j)(5+4j)}{(18+j)+(8+31j)+(23+2j)} = \frac{86+77j}{49+34j} \cdot \frac{49-34j}{49-34j} = \frac{6832+849j}{2401+1156} = 1.9207 + 0.2387j$

10. (a) $Z = Z_1 + Z_2 + Z_3 = (3-4j) + (1+5j) + (-2j) = 4 - j$;

(b) $Z = \frac{Z_1 Z_2 Z_3}{Z_1 Z_2 + Z_1 Z_3 + Z_2 Z_3} = \frac{(3-4j)(1+5j)(-2j)}{(3-4j)(1+5j)+(3-4j)(-2j)+(1+5j)(-2j)} = \frac{(23+11j)(-2j)}{(23+11j)+(-8-6j)+(10-2j)} = \frac{22-46j}{25+3j} \cdot \frac{25-3j}{25-3j} = \frac{412-1216j}{625+9} = 0.6498 - 1.9180j$

11. (a) $Z_1 = 1.64\underline{/38.2^\circ} = 1.2888 + 1.0142j$;
$Z_2 = 2.35\underline{/43.7^\circ} = 1.6990 + 1.6236j$;
$Z_3 = 4.67\underline{/-39.6^\circ} = 3.5983 - 2.9768j$; $Z = Z_1 + Z_2 + Z_3 = 6.5861 - 0.3390j = 6.595\underline{/-2.95^\circ}$;

(b) $Z_1 Z_2 = (1.64\underline{/38.2})(2.35\underline{/43.7^\circ}) = 3.854\underline{/81.9^\circ} = 0.5430 + 3.8156j$;

$Z_1 Z_3 = (1.64\underline{/38.2^\circ})(4.67\underline{/-39.6^\circ}) = 7.6588\underline{/-1.4^\circ} = 7.6565 - 0.1871j$;

$Z_2Z_3 = (2.35\,\underline{/43.7°})(4.67\,\underline{/-39.6°}) = 10.9745\,\underline{/4.1} = 10.9464 + 0.7846j$;
$Z_1Z_2 + Z_1Z_3 + Z_2Z_3 = 19.1459 + 4.4131j = 19.6479\,\underline{/12.98°}$;

$$Z = \frac{Z_1Z_2Z_3}{Z_1Z_2 + Z_1Z_3 + Z_2Z_3} = \frac{(1.64\,\underline{/38.2°})(2.35\,\underline{/43.7°})(4.67\,\underline{/-39.6°})}{19.648\,\underline{/12.98°}} = 0.916\,\underline{/29.32°}$$

12. (a) $Z_1 = 0.15\,\underline{/0.95} = 0.0873 + 0.1220j$;
$Z_2 = 2.17\,\underline{/1.39} = 0.3902 + 2.1346j$;
$Z_3 = 1.10\,\underline{/0.40} = 1.0132 + 0.4284j$;
$Z = Z_1+Z_2+Z_3 = 1.4907+2.6850j = 3.071\,\underline{/1.064}$;

(b) $Z_1Z_2 = (0.15\,\underline{/0.95})(2.17\,\underline{/1.39}) = 0.3255\,\underline{/2.34} = -0.2264 + 0.2339j$;

$Z_1Z_3 = (0.15\,\underline{/0.95})(1.10\,\underline{/0.40}) = 0.165\,\underline{/1.35} = 0.0361 + 0.1610j$;

$Z_2Z_3 = (2.17\,\underline{/1.39})(1.10\,\underline{/0.40}) = 2.387\,\underline{/1.79} = -0.5191 + 2.3299$; $Z_1Z_2 + Z_1Z_3 + Z_2Z_3 = -0.7094 + 2.7248j = 2.8156\,\underline{/1.825}$;

$$Z = \frac{Z_1Z_2Z_3}{Z_1Z_2 + Z_1Z_3 + Z_2Z_3} = \frac{(0.15\,\underline{/0.95})(2.17\,\underline{/1.39})(1.10\,\underline{/0.40})}{2.8156\,\underline{/1.825}} = 0.127\,\underline{/0.915}$$

13. $V = IZ = (4 - 3j)(8 - 15j) = 32 - 60j - 24j + 45j^2 = -13 - 84j$ V

14. $Z = \frac{V}{I} = \frac{5+5j}{4+3j} \cdot \frac{4-3j}{4-3j} = \frac{20-15j+20j-15j^2}{16+9} = \frac{35+5j}{25} = \frac{7}{5} + \frac{1}{5}j\ \Omega = 1.4 + 0.2j\ \Omega$

15. $V = IZ = (1 - j)(1 + j) = 2$ V

16. $I = \frac{V}{Z} = \frac{3+4j}{5-12j} \cdot \frac{5+12j}{5+12j} = \frac{15+36j+20j+48j^2}{25+144} = \frac{-33+56j}{169} = 0.1953 + 0.331j$ A

17. $Z = \frac{V}{I} = \frac{7\,\underline{/36.3°}}{2.5\,\underline{/12.6°}} = 2.8\,\underline{/23.7°}\ \Omega$

18. $I = \frac{V}{Z} = \frac{3\,\underline{/1.37}}{4\,\underline{/0.16}} = 0.75\,\underline{/1.21}$ A

19. $X_L = 2\pi fL = 2\pi 60 \cdot 0.2 = 75.40\ \Omega$;
$X_C = \frac{1}{2\pi fC} = \frac{1}{2\pi 60 \cdot 40 \times 10^{-6}} = 66.31\ \Omega$;
$X = X_L - X_C = 9.09\ \Omega$;
$R + Xj = 38 + 9.09j = 39.07\,\underline{/13.45°}$; $Z = 39.07\ \Omega$, $\phi = 13.45°$

20. $X_L = 2\pi fL = 2\pi 60 \cdot 0.15 = 56.55\ \Omega$;
$X_C = \frac{1}{2\pi fC} = \frac{1}{2\pi 60 \cdot 80 \times 10^{-6}} = 33.16\ \Omega$;
$X = X_L - X_C = 23.39\ \Omega$;
$R + Xj = 35 + 23.39j = 42.10\,\underline{/33.75°}$;
$Z = 42.10\ \Omega$, $\phi = 33.75°$

21. $X_L = 2\pi fL = 2\pi \times 60 \times 0.4 = 150.80\ \Omega$;
$X_C = \frac{1}{2\pi fC} = \frac{1}{2\pi 60 \cdot 60 \times 10^{-6}} = 44.21\ \Omega$;
$X = X_L - X_C = 106.59\ \Omega$;
$R + Xj = 20 + 106.59j = 108.45\,\underline{/79.37°}$;
$Z = 108.45\ \Omega$, $\phi = 79.37°$

22. $X_L = \omega L = 80 \times 0.3 = 24\ \Omega$;
$X_C = \frac{1}{\omega C} = \frac{1}{80 \times 250 \times 10^{-6}} = 50\ \Omega$;
$X = X_L - X_C = -26\ \Omega$;
$R + Xj = 12 - 26j = 28.64\,\underline{/-65.22°}$;
$Z = 28.64\ \Omega$, $\phi = -65.22°$

23. $X_L = \omega L = 50 \times 0.25 = 12.5\ \Omega$;
$X_C = \frac{1}{\omega C} = \frac{1}{50 \times 200 \times 10^{-6}} = 100\ \Omega$;
$X = X_L - X_C = -87.5\ \Omega$;
$R + Xj = 28 - 87.5j = 91.87\,\underline{/-72.26°}$;
$Z = 91.87\ \Omega$, $\phi = -72.26°$

24. $X_L = \omega L = 1000 \times 3 = 3000\ \Omega$;
$X_C = \frac{1}{\omega C} = \frac{1}{1000 \times 0.5 \times 10^{-6}} = 2000\ \Omega$;
$X = X_L - X_C = 1000\ \Omega$;
$R + Xj = 2000 + 1000j = 2236.07\,\underline{/26.57°}$;
$Z = 2236.07\ \Omega$, $\phi = 26.57°$

25. $X = X_L - X_C = 60 - 40 = 20\ \Omega$;
$R + Xj = 75 + 20j = 77.62\,\underline{/14.93°}$; $Z = 77.62\ \Omega$, $\phi = 14.93°$; $V = IZ = 77.62 \times 3.50 = 271.67$ V

26. $X = X_L - X_C = 30 - 60 = -30\ \Omega$;
$R + Xj = 40 - 30j = 50\,\underline{/-36.87°}$; $Z = 50\ \Omega$, $\phi = -36.87°$; $V = IZ = 7.50 \times 50 = 375$ V

27. $X = X_L - X_C = 6.0 - 5.0 = 1.0\ \Omega$;
$R + Xj = 3.0 + 1.0j = 3.162\,\underline{/18.43°}$; $Z = 3.162\ \Omega$, $\phi = 18.43°$; $V = IZ = 2.85 \times 3.162 = 9.01$ V

28. $X = X_L - X_C = 11.4 - 2.4 = 9\ \Omega$;
$R + Xj = 12.0 + 9.0j = 15\,\underline{/36.87°}$; $Z = 15\ \Omega$, $\phi = 36.87°$; $V = IZ = 0.60 \times 15 = 9$V

29. $I_2 = I_1+I_3 \Rightarrow I_3 = I_2-I_1 = (9-7j)-(7+2j) = 2-9j$

30. $\omega L = \frac{1}{\omega C}$; $100 \times 0.5 = \frac{1}{100 \times C}$;
$C = \frac{1}{100 \times 100 \times .5} = 2 \times 10^{-4} = 200\mu F$

31. $\frac{1}{2\pi fC} = 2\pi fL \Rightarrow 1 = 4\pi^2 f^2 LC; f^2 = \frac{1}{4\pi^2 LC}$;
$f = \frac{1}{2\pi\sqrt{LC}}; f = \frac{1}{2\pi\sqrt{2.5 \times 20 \times 10^{-6}}} = 22.51$ Hz

32. $Y = \frac{1}{Z} = \frac{1}{3-2j} = \frac{1}{3-2j} \cdot \frac{3+2j}{3+2j} = \frac{3+2j}{9+4} = \frac{3}{13} + \frac{2}{13}j =$
$0.2308 + 0.1538j$

33. $Z = \frac{1}{Y} = \frac{1}{4+3j} = \frac{1}{4+3j} \cdot \frac{4-3j}{4-3j} = \frac{4-3j}{25} = 0.16 - 0.12j$

34. $B = \frac{X}{Z^2}j = \frac{4}{(8+7j)^2}j = \frac{4}{15+112j}j$;
$\frac{4j}{15+112j} \cdot \frac{15-112j}{15-112j} = \frac{448+60j}{225+12544} = \frac{448+60j}{12769} =$
$0.0351 + 0.0047j$

35. See *Computer Programs* in main text.

36. See *Computer Programs* in main text.

CHAPTER 14 REVIEW

1. $\sqrt{-49} = \sqrt{49}\sqrt{-1} = 7j$

2. $\sqrt{-36} = \sqrt{36}\sqrt{-1} = 6j$

3. $\sqrt{-54} = \sqrt{9}\sqrt{6}\sqrt{-1} = 3\sqrt{6}j$ *or* $3j\sqrt{6}$

4. $(2j^3)^3 = 2^3 j^9 = 8j$ (Note: $j^8 = 1$)

5. $\sqrt{-2}\sqrt{-18} = (\sqrt{2}j)(3j\sqrt{2}) = 6j^2 = -6$

6. $\sqrt{-9}\sqrt{-27} = (3j)(3j\sqrt{3}) = 9j^2\sqrt{3} = -9\sqrt{3}$

7. $(2 - j) + (7 - 2j) = 9 - 3j$

8. $(9 + j)(4 + 7j) = 36 + 63j + 4j + 7j^2 = 29 + 67j$

9. $(5 + j) - (6 - 3j) = -1 + 4j$

10. $\frac{1}{11-j} = \frac{1}{11-j} \cdot \frac{11+j}{11+j} = \frac{11+j}{11^2+1^2} = \frac{11}{122} + \frac{j}{122} =$
$0.0902 + 0.0082j$

11. $(6+2j)(-5+3j) = -30+18j-10j+6j^2 = -36+8j$

12. $\frac{2-5j}{6+3j} \cdot \frac{6-3j}{6-3j} = \frac{12-6j-30j+15j^2}{36+9} = \frac{-3-36j}{45} = -0.0667 -$
$0.8j$

13. $(4 - 3j)(4 + 3j) = 4^2 + 3^2 = 16 + 9 = 25$

14. $\frac{-4}{\sqrt{3}+2j} \cdot \frac{\sqrt{3}-2j}{\sqrt{3}-2j} = \frac{-4\sqrt{3}+8j}{3+4} = \frac{-4\sqrt{3}}{7} + \frac{8}{7}j = -0.9897+$
$1.1429j$

15. $9 - 6j$; $r = \sqrt{9^2 + 6^2} = 10.82$; $\tan\theta = \frac{-6}{9}$;
$\theta = -33.69$ or 326.31; Answer: $10.82\underline{/326.31^\circ}$

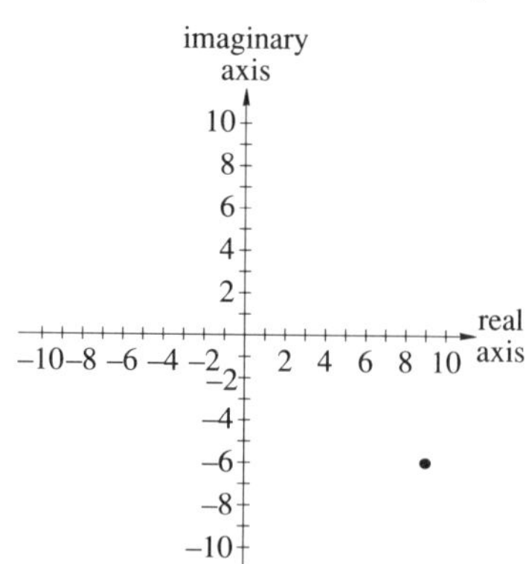

16. $-8 + 2j$; $r = \sqrt{8^2 + 2^2} = \sqrt{68} = 8.246$; $\tan\theta =$
$\frac{2}{-8} = 165.96^\circ$; Answer: $8.246\underline{/165.96^\circ}$

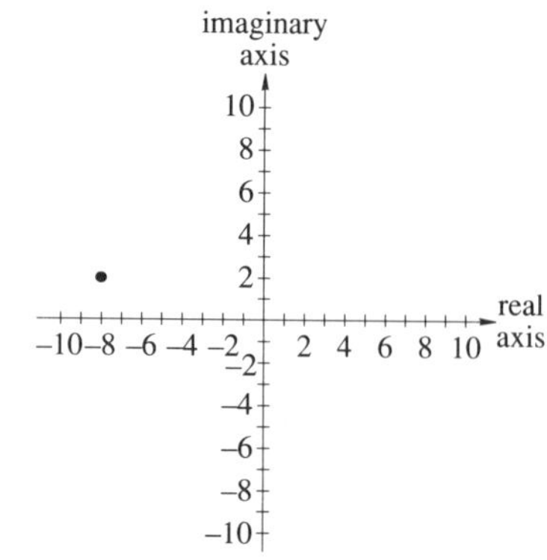

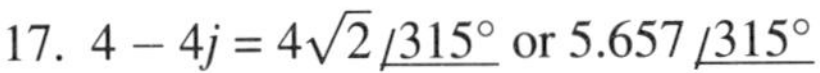

17. $4 - 4j = 4\sqrt{2}\,\underline{/315^\circ}$ or $5.657\,\underline{/315^\circ}$

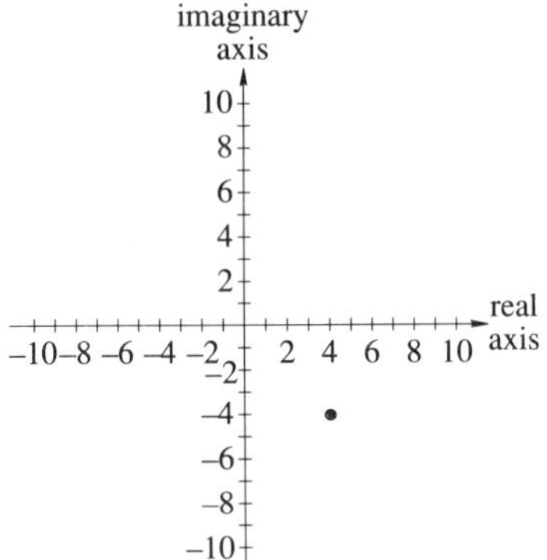

18. $4 \text{ cis } 60^\circ = 4\cos 60^\circ + 4j\sin 60^\circ = 2 + 3.464j$

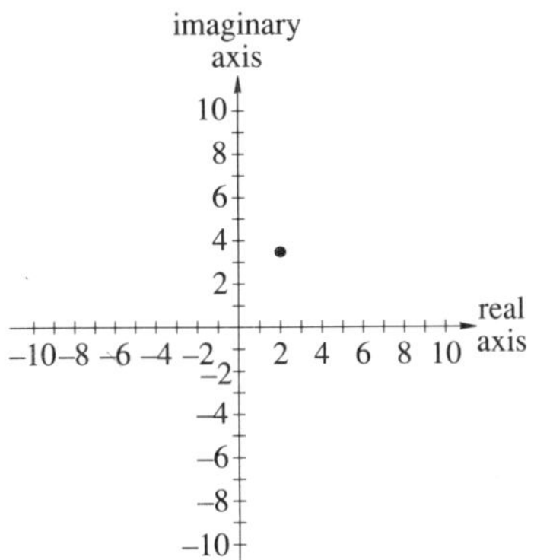

19. $6.5\,\underline{/2.3} = 6.5\cos 2.3 + 6.5j\sin 2.3 = -4.331 + 4.847j$

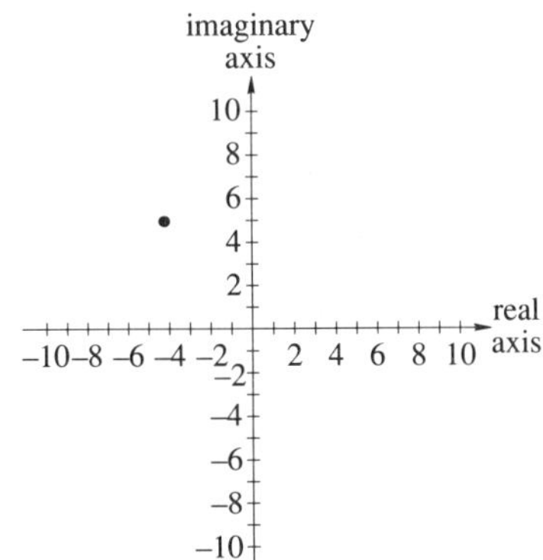

20. $10\,\underline{/20^\circ} = 10\cos 20^\circ + 10j\sin 20^\circ = 9.397 + 3.420j$

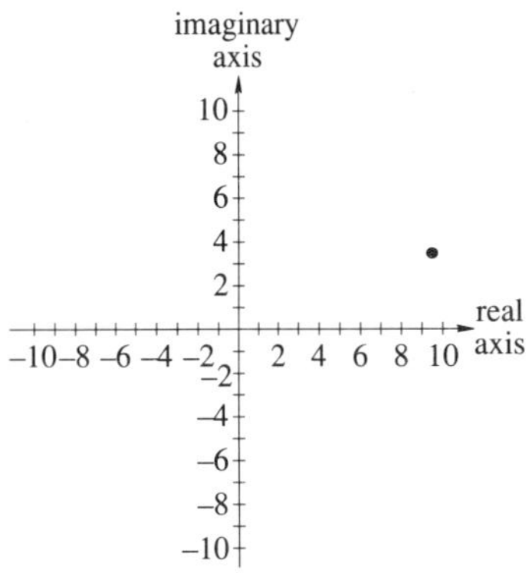

21. $(2 \text{ cis } 30^\circ)(5 \text{ cis } 150^\circ) = 10 \text{ cis } 180^\circ = -10$

22. $\frac{3 \text{ cis } 20^\circ}{6 \text{ cis } 80^\circ} = 0.5 \text{ cis}(-60^\circ) = 0.25 - 0.433j$

23. $(3 \text{ cis } \frac{5\pi}{4})^{14} = 3^{14} \text{ cis } \frac{14 \cdot 5\pi}{4} = 4{,}782{,}969 \text{ cis } \frac{35\pi}{2} = -4{,}782{,}969j$

24. $(324 \text{ cis } 225^\circ)^{1/5} = \sqrt[5]{324} \text{ cis } \frac{225}{5} = 3.178 \text{ cis } 45^\circ = 2.247 + 2.247j$

25. $(3\,\underline{/\pi/4})(9\,\underline{/2\pi/3}) = 3 \cdot 9\,\underline{/\pi/4 + 2\pi/3} = 27\,\underline{/11\pi/12} = -26.080 + 6.988j$

26. $(44\,\underline{/125^\circ}) \div (4\,\underline{/97^\circ}) = \frac{44}{4}\,\underline{/125 - 97} = 11\,\underline{/28} = 9.7124 + 5.1642j$

27. $(2\,\underline{/\pi/6})^{12} = 2^{12}\,\underline{/\frac{12\pi}{6}} = 4096\,\underline{/2\pi} = 4096$

28. $(2048\,\underline{/330^\circ})^{1/11} = \sqrt[11]{2048}\,\underline{/330/11} = 2\,\underline{/30} = 1.732 + j$

29. Cube roots of $-j = 1 \text{ cis } 270^\circ$; $w_0 = \sqrt[3]{1} \text{ cis } \frac{270}{3} = 1 \text{ cis } 90^\circ = j$; $w_1 = 1 \text{ cis}(90 + \frac{360}{3}) = 1 \text{ cis } 210^\circ = -0.866 - 0.5j$; $w_2 = 1 \text{ cis}(90 + 120 \times 2) = 1 \text{ cis } 330^\circ = 0.866 - 0.5j$

30. Fourth roots of $16 = 16 \text{ cis } 0^\circ$; $w_0 = \sqrt[4]{16} \text{ cis } \frac{0}{4} = 2$; $w_1 = 2 \text{ cis}(0 + \frac{360}{4}) = 2 \text{ cis } 90^\circ = 2j$; $w_2 = 2 \text{ cis } 180^\circ = -2$; $w_3 = 2 \text{ cis } 270^\circ = -2j$

31. Square roots of $16 \text{ cis } 120^\circ$; $w_0 = \sqrt[4]{16} \text{ cis } \frac{120}{2} = 4 \text{ cis } 60^\circ = 2 + 3.464j$; $w_1 = 4 \text{ cis}(60 + 180) - 4 \text{ cis } 240^\circ = -2 - 3.464j$

32. Cube roots of $27j = 27 \text{ cis } 90^\circ$; $w_0 = \sqrt[3]{27} \text{ cis } \frac{90}{3} = 3 \text{ cis } 30^\circ = 2.598 + 1.5j$; $w_1 = 3 \text{ cis}(30 + \frac{360}{3}) = 3 \text{ cis } 150^\circ = -2.598 + 1.5j$; $w_2 = 3 \text{ cis}(30 + 120 \times 2) = 3 \text{ cis } 270^\circ = -3j$

33. $(3+2j)(5-j) = (3.60555e^{0.588j})(5.099e^{-0.197j}) = 18.385e^{0.391j}$

34. $(4-7j)/(3+j) = (8.062e^{-1.052j})/(3.162e^{0.322j}) = 2.55e^{-1.374j}$

35. $(5+3j)^5 = (5.831e^{0.5404j})^5 = 5.831^5 e^{5\times 0.5404j} = 6740.6e^{2.702j}$

36. $(-7-2j)^{1/3} = (7.280e^{-2.863j})^{1/3} = \sqrt[3]{7.280}e^{\frac{-2.863}{3}j} = 1.938e^{-0.954j}$

37. $7.3 - 1.4j = 7.433\,\text{cis}(-10.86°)$; magnitude is 7.433, direction is $-10.86°$

38. $X = X_L - X_C = 7.0 - 4.5 = 2.5\ \Omega$; $R + Xj = 3 + 2.5j = 3.905\,\underline{/39.81°}$; $Z = 3.905\ \Omega$, $\phi = 39.81°$; $V = IZ = 1.5 \times 3.905 = 5.858$ V

39. $X_L = 2\pi fL = 2\pi(0.3)60 = 113.10\ \Omega$; $X_C = \frac{1}{2\pi fC} = \frac{1}{2\pi 60\times 5\times 10^{-6}} = 53.05\Omega$; $X = X_L - X_C = 60.05\ \Omega$; $R + Xj = 55 + 60.05j = 81.43\,\underline{/47.51°}$; $Z = 81.43\ \Omega$, $\phi = 47.51°$

40. (a) in series $Z = Z_1 + Z_2 = (3+5j) + (6-3j) = 9 + 2j$; (b) in parallel $Z = \frac{Z_1Z_2}{Z_1+Z_2} = \frac{(3+5j)(6-3j)}{9+2j} = \frac{18-9j+30j-15j^2}{9+2j} = \frac{33+21j}{9+2j}\cdot\frac{9-2j}{9-2j} = \frac{297-66j+189j-42j^2}{81+4} = \frac{339+123j}{85} = 3.988 + 1.447j$

CHAPTER 14 TEST

1. $\sqrt{-80} = \sqrt{80}\sqrt{-1} = \sqrt{16}\sqrt{5}j = 4j\sqrt{5}$

2. $7 - 2j$; $r = \sqrt{7^2 + 2^2} = \sqrt{53} \approx 7.28$; $\tan\theta = \frac{-2}{7} = -15.9$ or $344.1°$; Answer: $7.28\,\underline{/344.1°}$

3. $8\,\text{cis}\,150° = 8\cos 150° + 8j\sin 150° = -6.9282 + 4j$ or $-4\sqrt{3} + 4j$

4. $(5+2j) + (8-6j) = 13 - 4j$

5. $(-5+2j) - (8-6j) = -13 + 8j$

6. $(2+3j)(4-5j) = 8 - 10j + 12j - 15j^2 = 23 + 2j$

7. $\frac{6+5j}{-3-4j}\cdot\frac{-3+4j}{-3+4j} = \frac{-18+24j-15j+20j^2}{9+16} = \frac{-38+9j}{25} = -1.52 + 0.36j$

8. $(7\,\text{cis}\,75°)(2\,\text{cis}\,105°) = 14\,\text{cis}\,180° = -14$

9. $\frac{4\,\text{cis}\,115°}{3\,\text{cis}\,25°} = \frac{4}{3}\,\text{cis}\,90° = \frac{4}{3}j$ or $1.333j$

10. $\left(9\,\text{cis}\,\frac{2\pi}{3}\right)^{5/2} = (9)^{5/2}\,\text{cis}\,\frac{5\pi}{3} = 243\,\text{cis}\,\frac{5\pi}{3} = \frac{243}{2} - \frac{243}{2}j\sqrt{3}$

11. $(27\,\underline{/129°})^{1/3} = \sqrt[3]{27}\,\text{cis}\,\frac{129°}{3} = 3\,\text{cis}\,43° \approx 2.194 + 2.046j$

12. Since $j = 1\,\text{cis}\,90°$, then the four fourth roots of j are: $w_0 = \sqrt[4]{1}\,\text{cis}\,\frac{90°}{4} = 1\,\text{cis}\,22.5° = 0.924 + 0.383j$; $w_1 = 1\,\text{cis}(22.5 + 90)° = 1\,\text{cis}\,112.5° = -0.383 + 0.924j$; $w_2 = 1\,\text{cis}(22.5 + 90\times 2)° = 1\,\text{cis}\,202.5° = -0.924 - 0.383j$; $w_3 = 1\,\text{cis}(22.5 + 90\times 3) = 1\,\text{cis}\,292.5° = 0.383 - 0.924j$

13. (a) in series $Z = Z_1 + Z_2 = (4+2j) + (5-3j) = 9 - j$; (b) in parallel $Z = \dfrac{Z_1Z_2}{Z_1+Z_2} = \dfrac{(4+2j)(5-3j)}{9-j} = \dfrac{20-12j+10j-6j^2}{9-j} = \dfrac{26-2j}{9-j}\cdot\dfrac{9+j}{9+j} = \dfrac{234+26j-18j-2j^2}{81+1} = \dfrac{236+8j}{82} = \dfrac{118}{41} + \dfrac{4}{41}j = 2.878 + 0.098j$

CHAPTER

15

An Introduction to Plane Analytic Geometry

15.1 BASIC DEFINITIONS AND STRAIGHT LINES

1. $d = \sqrt{(2-7)^2+(4+9)^2} = \sqrt{(-5)^2+13^2} = \sqrt{25+169} = \sqrt{194} \approx 13.928$

2. $d = \sqrt{(-3-0)^2+(5-1)^2} = \sqrt{(-3)^2+4^2} = \sqrt{9+16} = \sqrt{25} = 5$

3. $d = \sqrt{(5+3)^2+(-6+5)^2} = \sqrt{8^2+(-1)^2} = \sqrt{64+1} = \sqrt{65} \approx 8.062$

4. $d = \sqrt{(-8-6)^2+(-4-10)^2} = \sqrt{(-14)^2+(-14)^2} = \sqrt{392} = 14\sqrt{2} \approx 19.799$

5. $d = \sqrt{(12-3)^3+(1+13)^2} = \sqrt{9^2+14^2} = \sqrt{81+196} = \sqrt{277} \approx 16.643$

6. $d = \sqrt{(11-4)^2+(-2+8)^2} = \sqrt{7^2+6^2} = \sqrt{49+36} = \sqrt{85} \approx 9.220$

7. $d = \sqrt{(-2-5)^2+(5-5)^2} = \sqrt{(-7)^2} = 7$

8. $d = \sqrt{(-4+4)^2+(-4-7)^2} = \sqrt{(-11)^2} = 11$

9. midpoint $= \left(\frac{2+7}{2}, \frac{4-9}{2}\right) = \left(\frac{9}{2}, \frac{-5}{2}\right)$

10. midpoint $= \left(\frac{-3+0}{2}, \frac{5+1}{2}\right) = (-3/2, 3)$

11. midpoint $= \left(\frac{5-3}{2}, \frac{-6-5}{2}\right) = \left(1, \frac{-11}{2}\right)$

12. midpoint $= \left(\frac{-8+6}{2}, \frac{-4+10}{2}\right) = (-1, 3)$

13. midpoint $= \left(\frac{12+3}{2}, \frac{1-13}{2}\right) = \left(\frac{15}{2}, -6\right)$

14. midpoint $= \left(\frac{11+4}{2}, \frac{-2-8}{2}\right) = \left(\frac{15}{2}, -5\right)$

15. midpoint $= \left(\frac{-2+5}{2}, \frac{5+5}{2}\right) = \left(\frac{3}{2}, -5\right)$

16. midpoint $= \left(\frac{-4-4}{2}, \frac{-4+7}{2}\right) = (-4, 3/2)$

17. slope $= \frac{-9-4}{7-2} = \frac{-13}{5} = -13/5 = -2.6$

18. slope $= \frac{1-5}{0+3} = \frac{-4}{3} = -4/3$

19. slope $= \frac{-5+6}{-3-5} = \frac{1}{-8} = -1/8$

20. slope $= \frac{10+4}{6+8} = \frac{14}{14} = 1$

21. slope $= \frac{-13-1}{3-12} = \frac{-14}{-9} = \frac{14}{9}$

22. slope $= \frac{-8+2}{4-11} = \frac{-6}{-7} = \frac{6}{7}$

23. slope $= \frac{5-5}{5+2} = \frac{0}{7} = 0$

24. slope $= \frac{7+4}{-4+4} = \frac{11}{0}$ is undefined

25. $y-4 = -2.6(x-2)$ or $y = -2.6x + 9.2$

26. $y-5 = -\frac{4}{3}(x+3)$ or $y-5 = -\frac{4}{3}x-4$ or $y = -\frac{4}{3}x+1$

27. $y+6 = -\frac{1}{8}(x-5)$ or $y+6 = -\frac{1}{8}x+\frac{5}{8}$ or $y = -\frac{1}{8}x-5\frac{3}{8}$

28. $y+4 = 1(x+8)$ or $y = x+4$

29. $y-1 = \frac{14}{9}(x-12)$ or $y-1 = \frac{14}{9}x-\frac{56}{3}$ or $y = \frac{14}{9}x-17\frac{2}{3}$

30. $y+2=\frac{6}{7}(x-11)$ or $y+2=\frac{6}{7}x-\frac{66}{7}$ or $y=\frac{6}{7}x-\frac{80}{7}$ or $y=\frac{6}{7}x-11\frac{3}{7}$

31. $y-5=0(x+2)$ or $y-5=0$ or $y=5$

32. Special case vertical line so $x = x$-intercept or $x=-4$

33. $m=\tan 45° = 1$

34. $m=\tan 175° \approx -0.0874887$

35. $m=\tan 0.15 \approx 0.1511352$

36. $m=\tan 1.65 \approx -12.599265$

37. $\alpha=\tan^{-1} 2.5 = 68.19859°$

38. $\alpha=\tan^{-1} 0.3 = 16.699244°$

39. $\alpha=180+\tan^{-1}(-0.50)=153.43495°$

40. $\alpha=180+\tan^{-1}(-1.475)=124.13594°$

41. $m=-\frac{1}{3}=-1/3$

42. $m=-\frac{1}{2/5}=-5/2$

43. $m=-\frac{1}{-1/2}=2$

44. $m=-\frac{1}{-5}=1/5$

45. $y+5=6(x-2)$ or $y+5=6x-12$ or $y=6x-17$

46. $y+2=-\frac{1}{2}(x-6)$ or $y+2=-\frac{1}{2}x+3$ or $y=-\frac{1}{2}x+1$

47. $m=\frac{-2-3}{4-2}=\frac{-5}{2}$; $y+2=-\frac{5}{2}(x-4)$ or $y+2=-\frac{5}{2}x+10$ or $y=\frac{-5}{2}x+8$

48. $m=\frac{-4-2}{-3+5}=\frac{-6}{2}=-3$; $y-2=-3(x+5)$ or $y-2=-3x-15$ or $y=-3x-13$

49. $m=\tan 60° = 1.732$; $y+4=1.732(x+2)$ or $y=1.732x-0.536$

50. $m=\tan 0.75 = 0.932$; $y-1=0.932(x-5)$ or $y=0.932x-3.66$

51. $m=\tan 20° = 0.364$; $y=0.364x+3$

52. $m=\tan 2.5=-0.747$; $y=-0.747x-2$

53. $2x-3y+4=0 \Rightarrow -3y=-2x+4 \Rightarrow y=\frac{2}{3}x-\frac{4}{3}$; so the slope is $\frac{2}{3}$.; $y-5=\frac{2}{3}(x-2)$ or $3y-15=2x-4$. Combining terms produces $-2x+3y=11$ or $-2x+3y-11=0$ or $2x-3y+11=0$

54. $3x+4y=12 \Rightarrow 4y=-3x+12 \Rightarrow y=-\frac{3}{4}x+3$; so the slope is $-\frac{3}{4}$. Thus, we obtain $y-2=-\frac{3}{4}(x+3)$ or $4y-8=-3x-9$ or $3x+4y=-1$ or $3x+4y+1=0$

55. $2x+5y=20 \Rightarrow 5y=-2x+20 \Rightarrow y=-\frac{2}{5}x+4$; the slope of this line is $-\frac{2}{5}$ and the slope of the perpendicular is $\frac{5}{2}$. Using the point-slope form, we get $y+1=\frac{5}{2}(x-4)$ or $2y+2=5x-20$ or $-5x+2y=-22$ or $5x-2y=22$

56. $8x-3y=24 \Rightarrow -3y=-8x+24 \Rightarrow y=\frac{8}{3}x-8$; $m=\frac{8}{3}$ so the slope of perpendicular is $-\frac{3}{8}$. Using the point-slope form, we get $y+6=-\frac{3}{8}(x+1)$ or $8y+48=-3x-3$ or $3x+8y=-51$

57. $3x+2y=12 \Rightarrow 2y=-3x+12 \Rightarrow y=-\frac{3}{2}x+6$; $m=-\frac{3}{2}$, y-intercept $=6$; when $y=0$, $x=4$ so x-intercept $=4$.

58. $5x-3y=15 \Rightarrow -3y=-5x+15 \Rightarrow y=\frac{5}{3}x-5$; $m=\frac{5}{3}$, y-intercept $=-5$; when $y=0$, $x=3$ and so, the x-intercept $=3$

59. $x-3y=9 \Rightarrow -3y=-x+9 \Rightarrow y=\frac{1}{3}x-3$; $m=\frac{1}{3}$, y-intercept $=-3$, when $y=0$, $x=9$ so x-intercept $=9$

60. $6x+y=9 \Rightarrow y=-6x+9$; $m=-6$, y-intercept $=9$. When $y=0$, we see that $x=\frac{3}{2}$ and so, the x-intercept $=\frac{3}{2}$.

61. $v=v_0+at$, $v_0=2.6$, $v=5.8$ when $t=8$; $5.8=2.6+a\cdot 8$; $3.2=8a$; $a=0.4$; $v=2.6+0.4t$

62. (a) $\alpha=\frac{\text{change in length}}{\text{change in temp.}}=\frac{1.000084-1.000000}{15-10}=\frac{0.000084}{5}=1.68\times 10^{-5}$; (b) $\alpha=\frac{.000336}{5}=6.72\times 10^{-5}$; (c) The amount of expansion is proportional to the original length of the solid. You need to divide by the original length of the solid, in part (b) you need to divide by 4.; (d) $\frac{\frac{72.005208-72.000024}{18-10}}{72.000024}=9.000\times 10^{-6}$

63. $E=IR+Ir$; $12=2.5\times 4+2.5\times r$; $12=10+2.5r$; $2=2.5r$; $r=\frac{2}{2.5}=0.8\ \Omega$

64. (a) $R=7+0.006T$ or $R=0.006T+7.000$, (b) $R=0.006\times 17+7.000=7.102\ \Omega$

15.2 THE CIRCLE

1. $C = (2,5)$, $r = 3$: The standard equation for the circle is $(x-2)^2+(y-5)^2 = 3^2$ or $(x-2)^2+(y-5)^2 = 9$. This gives $x^2 - 4x + 4 + y^2 - 10y + 25 = 9$ or $x^2 + y^2 - 4x - 10y + 20 = 0$ for the general equation.

2. $C = (4,1)$, $r = 8$: $(x-4)^2 + (y-1)^2 = 64$ is the standard form. $x^2 - 8x + 16 + y^2 - 2y + 1 = 64$ or $x^2 + y^2 - 8x - 2y - 47$ is the general form.

3. $C = (-2,0)$, $r = 4$: $(x+2)^2+y^2 = 16$ is the standard form. $x^2 + y^2 + 4x - 12 = 0$ is the general form

4. $C = (0,-7)$, $r = 2$: $x^2+(y+7)^2 = 4$ is the standard form. $x^2 + y^2 + 14y + 45 = 0$ is the general form.

5. $C = (-5,-1)$, $r = \frac{5}{2}$: $(x+5)^2 + (y+1)^2 = \frac{25}{4}$ is the standard form. $x^2+10x+25+y^2+2y+1 = \frac{25}{4}$ or $x^2 + y^2 + 10x + 2y + 19.75$ is the general form.

6. $C = (-4,-5)$, $r = \frac{5}{4}$: $(x+4)^2+(y+5)^2 = \frac{25}{16}$ is the standard form. Expanding the standard form and collecting terms, we get $x^2+8x+16+y^2+10y+25 = \frac{25}{16}$; $x^2+y^2+8x+10y+\frac{631}{16} = 0$ which is the general form.

7. $C = (2,-4)$, $r = 1$: $(x-2)^2 + (y+4)^2 = 1$ is the standard form. $x^2 - 4x + 4 + y^2 + 8y + 16 = 1$ or $x^2 + y^2 - 4x + 8y + 19 = 0$ is the general form

8. $C = (5,-2)$, $r = \sqrt{3}$: $(x-5)^2 + (y+2)^2 = 3$ is the standard form $x^2 - 10x + 25 + y^2 + 4y + 4 = 3$ or $x^2 + y^2 - 10x + 4y + 26 = 0$ is the general form.

9. $C = (3,4)$, $r = 3$

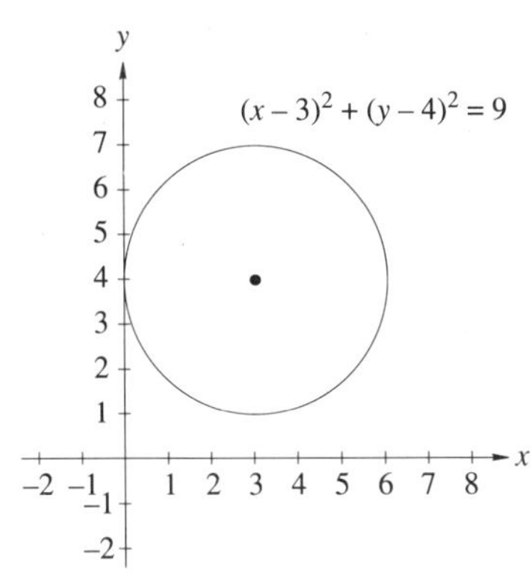

10. $C = (7,-5)$, $r = 5$

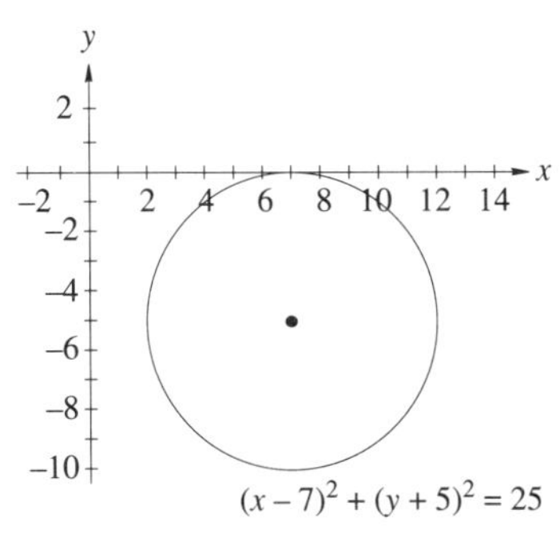

11. We first write this in the standard form for the equation of a circle, $(x-h)^2 + (y-k)^2 = r^2$ or $\left(x-\left(-\frac{1}{2}\right)\right)^2 + \left(y-\left(-\frac{13}{4}\right)\right)^2 = \left(\sqrt{7}\right)^2$. Thus, the center is $C = (h,k) = \left(-\frac{1}{2}, -\frac{13}{4}\right)$, and the radius is $r = \sqrt{7} \approx 2.646$.

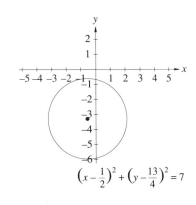

$\left(x-\frac{1}{2}\right)^2+\left(y-\frac{13}{4}\right)^2 = 7$

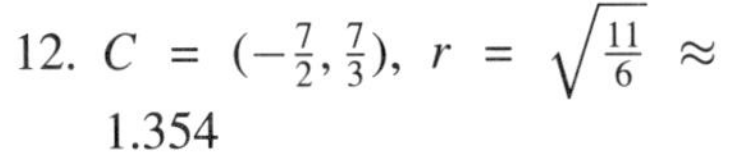
12. $C = \left(-\frac{7}{2}, \frac{7}{3}\right)$, $r = \sqrt{\frac{11}{6}} \approx 1.354$

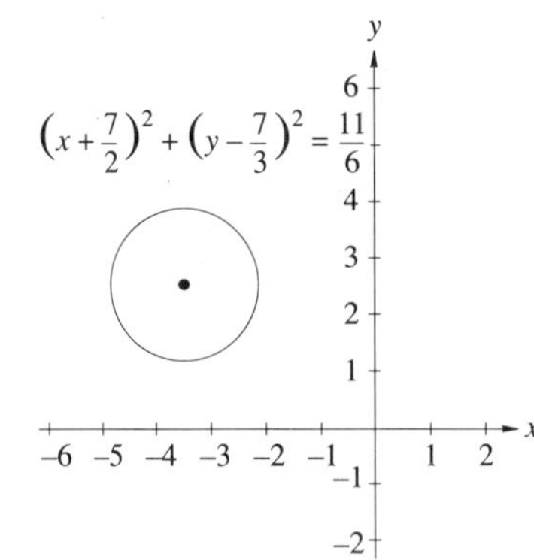

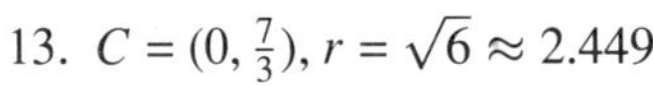
13. $C = \left(0, \frac{7}{3}\right)$, $r = \sqrt{6} \approx 2.449$

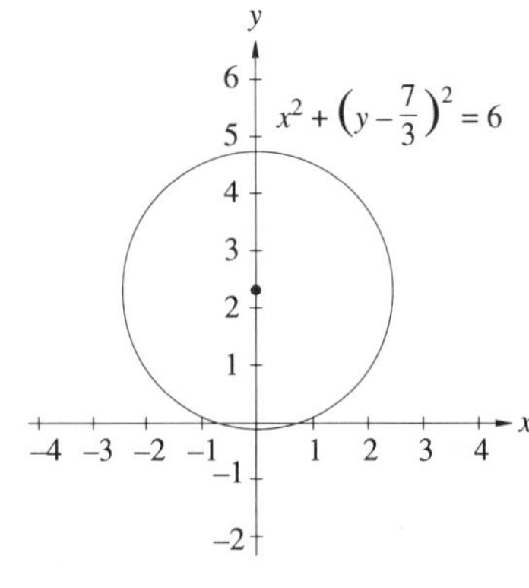

14. $C = (-3, 0)$, $r = \sqrt{1.21} = 1.1$

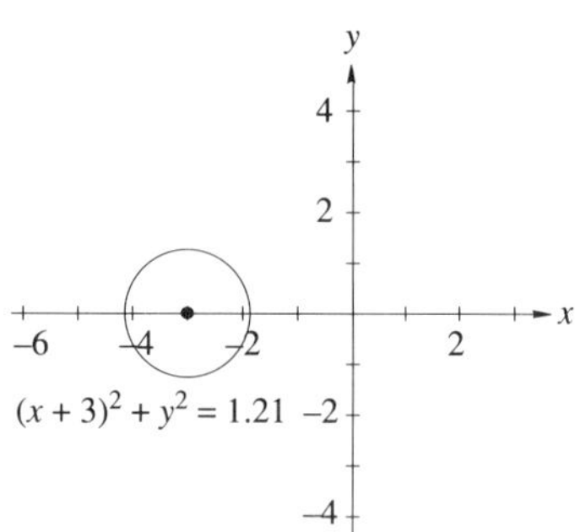

15. $x^2 + y^2 + 4x - 6y + 4 = 0$. Completing the square leads to $x^2 + 4x + 4 + y^2 - 6y + 9 = -4 + 4 + 9$ or $(x + 2)^2 + (y - 3)^2 = 9$. This is the standard equation for a circle with center $(-2, 3)$ and radius $r = 3$.

16. $x^2 + y^2 - 10x + 2y + 22$. Completing the square leads to $x^2 - 10x + 25 + y^2 + 2y + 1 = -22 + 25 + 1$ or $(x - 5)^2 + (y + 1)^2 = 4$. This is the standard equation for a circle with center $C = (5, -1)$ and radius $r = 2$.

17. $x^2 + y^2 + 10x - 6y - 47 = 0$; $x^2 + 10x + 25 + y^2 - 6y + 9 = 47 + 25 + 9$; $(x + 5)^2 + (y - 3)^3 = 81$; Circle: $C = (-5, 3)$, $r = 9$

18. $x^2 + y^2 + 2x - 12y - 27 = 0$; $x^2 + 2x + 1 + y^2 - 12y + 36 = 27 + 1 + 36$; $(x + 1)^2 + (y - 6)^2 = 64$; Circle: $C = (-1, 6)$, $r = 8$

19. $x^2+y^2-2x+2y+3 = 0$; $x^2-2x+1+y^2+2y+1 = -3 + 1 + 1$; $(x - 1)^2 + (y + 1)^2 = -1$; $r^2 = -1$ so it is not a circle.

20. $x^2+y^2+2x+2y-3 = 0$; $x^2+2x+1+y^2+2y+1 = 3 + 1 + 1$; $(x + 1)^2 + (y + 1)^2 = 5$; Circle: $C = (-1, -1)$, $r = \sqrt{5}$

21. $x^2 + y^2 + 6x - 16 = 0$; $x^2 + 6x + 9 + y^2 = 16 + 9$; $(x + 3)^2 + y^2 = 25$; Circle: $C = (-3, 0)$, $r = 5$

22. $x^2 + y^2 - 8y - 9 = 0$; $x^2 + y^2 - 8y + 16 = 9 + 16$; $(x^2) + (y - 4)^2 = 25$; Circle: $C = (0, 4)$, $r = 5$

23. $x^2+y^2+5x-9y = 9.5$; $x^2+5x+\frac{25}{4}+y^2-9y+\frac{81}{4} = \frac{19}{2}+\frac{25}{4}+\frac{81}{4}$; $(x+\frac{5}{2})^2+(y-\frac{9}{2})^2 = \frac{144}{4} = 36$; Circle: $C = (-\frac{5}{2}, \frac{9}{2})$, $r = 6$

24. $x^2+y^2-7x+3y+2.5 = 0$; $x^2-7x+\frac{49}{4}+y^2+3y+\frac{9}{4} = -\frac{5}{2} + \frac{49}{4} + \frac{9}{4}$; $(x - \frac{7}{2})^2 + (y + \frac{3}{2})^2 = 12$; Circle: $C = (\frac{7}{2}, -\frac{3}{2})$; $r = \sqrt{12} = 2\sqrt{3} \approx 3.464$

25. $9x^2+9y^2+18x-15y+27 = 0$; $9(x^2+2x)+9(y^2-\frac{5}{3}y) = -27$; $9(x^2 + 2x + 1) + 9(y^2 - \frac{5}{3}y + \frac{25}{36}) = -27+9\cdot 1+9\cdot\frac{25}{36}$; $9(x+1)^2+9(y-\frac{5}{6})^2 = -11.75$; $(x + 1)^2 + (y - \frac{5}{6})^2 = -1.306$; Not a circle since r^2 is negative.

26. $25x^2 + 25y^2 - 10x + 30y + 1 = 0$; $25(x^2 - \frac{2}{5}x) + 25(y^2 + \frac{6}{5}y) = -1$; $25(x^2 - \frac{2}{5}x + \frac{1}{25}) + 25(y^2 + \frac{6}{5}y + \frac{9}{25}) = -1+25\cdot\frac{1}{25}+25\cdot\frac{9}{25}$; $25(x-\frac{1}{5})^2+25(y+\frac{3}{5})^2 = 9$; $(x - \frac{1}{5})^2 + (y + \frac{3}{5})^2 = \frac{9}{25}$; Circle: $C = (\frac{1}{5}, -\frac{3}{5})$, $r = \frac{3}{5}$

27. $\sigma = \frac{mv}{\beta q} = \frac{(1.673\times 10^{-27})(1.186\times 10^{7})}{(4\times 10^{-4})(1.5\times 10^{-19})} = 330.70$. Choosing the center of the circle to be at the origin of the coordinate system we get $x^2 + y^2 = (330.70)^2$ or $x^2 + y^2 \approx 1.094 \times 10^5$.

28. $x^2 + y^2 = (1.495 \times 10^8)^2$ or $x^2 + y^2 = 2.235 \times 10^{16}$

29. $(x - 1.495 \times 10^8)^2 + y^2 = (3.844 \times 10^5)^2$ or $(x - 1.495 \times 10^8)^2 + y^2 = 1.478 \times 10^{11}$

30. First circle is centered at $(0, 0)$ with radius 6 so $x^2 + y^2 = 36$, Second circle center with x coordinate $6+5-1 = 10$ has center $C = (10, 0)$ and radius 5 so the equation is $(x - 10)^2 + y^2 = 25$.

31. If the center is 0.9 directly above the origin, the center is $(0, 0.9)$, the radius is 1.1/2 so $x^2 + (y - 0.9)^2 = 0.55^2$ or $x^2 + (y - 0.9)^2 = 0.3025$.

32. The x-coordinate of the center is $-(b + c)$, The y-coordinate is $-(a - c)$, the radius is c

$(x + (b + c))^2 + (y + (a - c))^2 = c^2$; $(x + 1.25)^2 + (y + 3)^2 = 0.25^2$

33. Center is at $(0, 5)$ and $r = \frac{26}{2} = 13$ (a) $x^2 + (y - 5)^2 = 13^2 = 169$ (b) Drawing a radius to where the floor would intersect the flywheel and the y-axis, we get a right triangle with hypothenuse 13 and one leg of 5. Using the Phythogorean theorem we find the other leg is 12, hence the chord is 2×12 or 24. Allowing for a 2 cm clearance on both sides, the opening must be $24 + 4 \times 2 = 28$ cm.

15.3 THE PARABOLA

1. $x^2 = 4y$, $p = 1$, vertical so $F = (0, 1)$, directrix: $y = -1$, opens upward

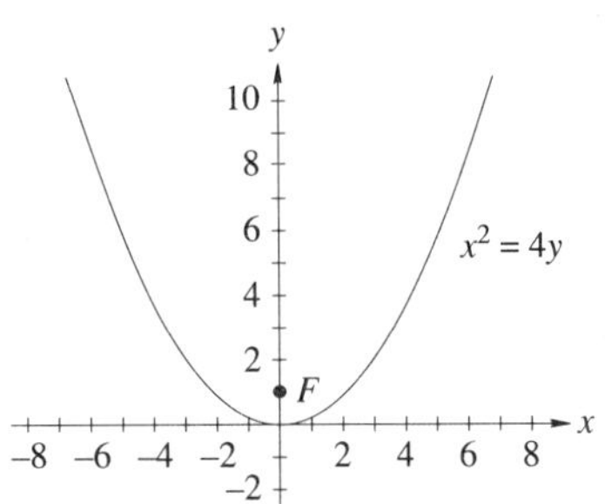

2. $x^2 = 12y$, $p = 3$, vertical so $F = (0, 3)$, directrix: $y = -3$, opens upward.

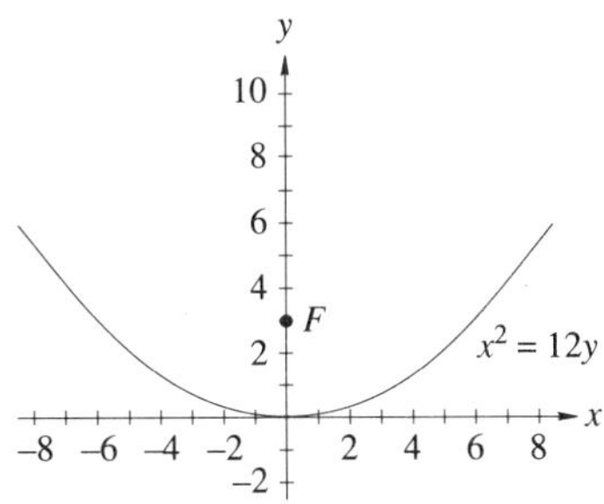

3. $y^2 = -4x$, $p = -1$, horizontal so $F = (-1, 0)$, directrix: $x = 1$, opens left.

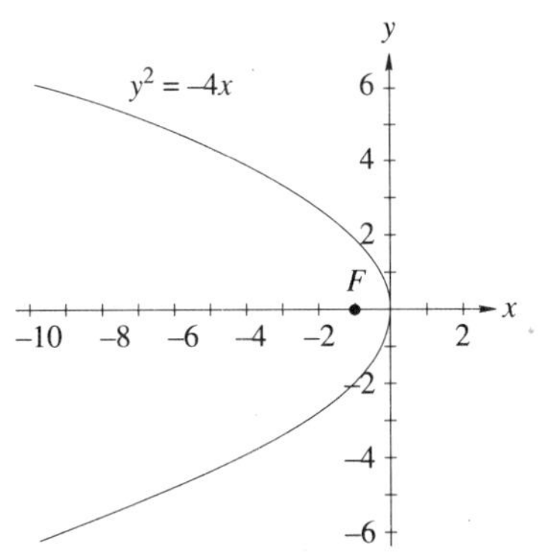

4. $y^2 = -20x$, $p = -5$, horizontal so $F = (-5, 0)$, directrix: $x = 5$, opens left.

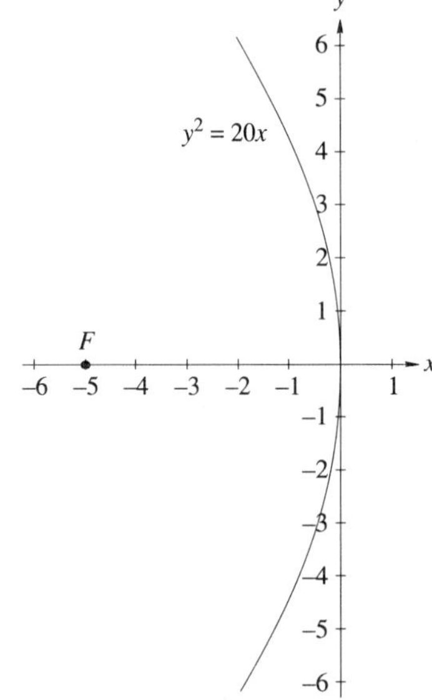

5. $x^2 = -8y$, $p = -2$, vertical so $F = (0, -2)$, directrix: $y = 2$, opens downward.

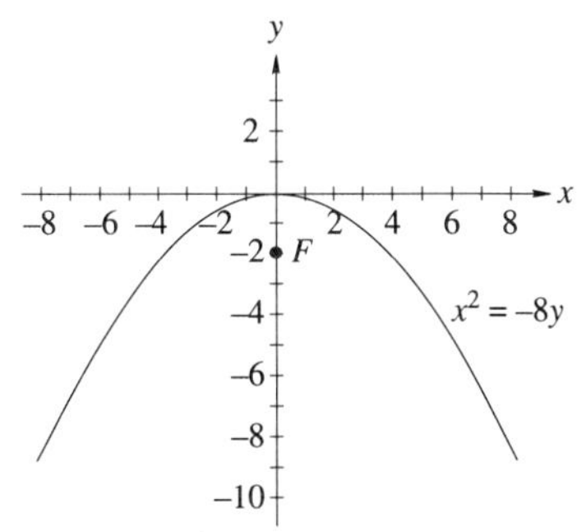

6. $x^2 = -16y$, $p = -4$, vertical so $F = (0, -4)$, directrix: $y = 4$, opens downwards.

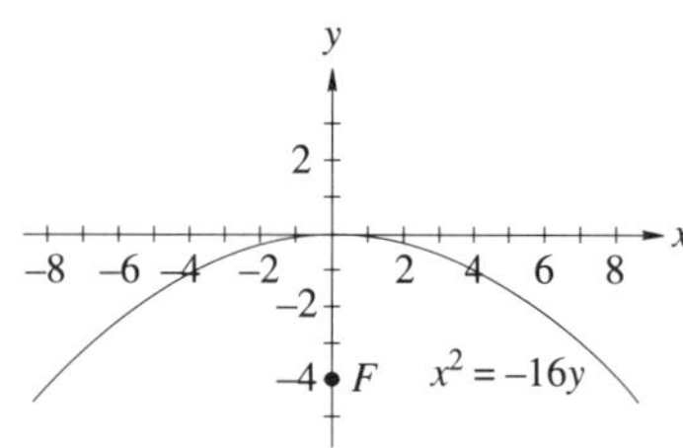

7. $y^2 = 10x$, $p = \frac{10}{4} = 2.5$, horizontal so $F = (2.5, 0)$, directrix: $x = -2.5$, opens right.

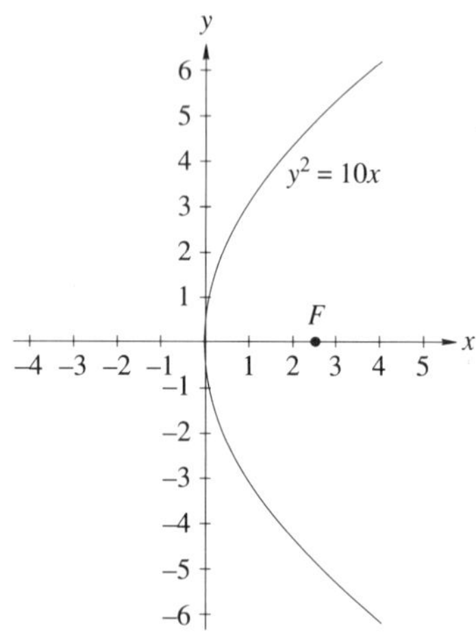

8. $y^2 = -14x$, $p = \frac{-14}{4} = -3.5$, horizontal so $F = (-3.5, 0)$, directrix: $x = 3.5$, opens left.

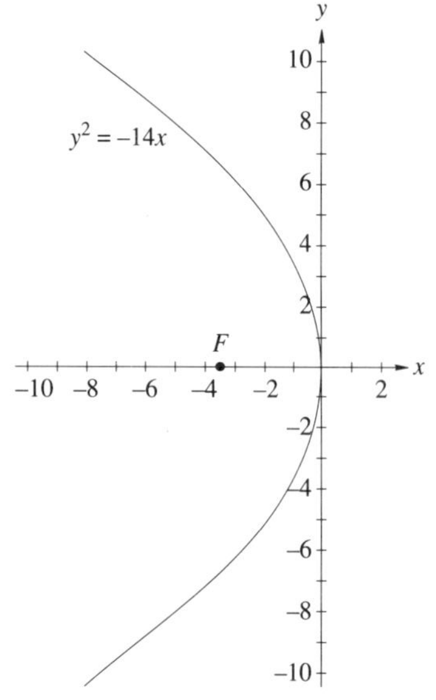

9. $x^2 = 2y$, $p = \frac{2}{4} = 0.5$, vertical so $F = (0, 0.5)$, directrix: $y = -0.5$, opens upward.

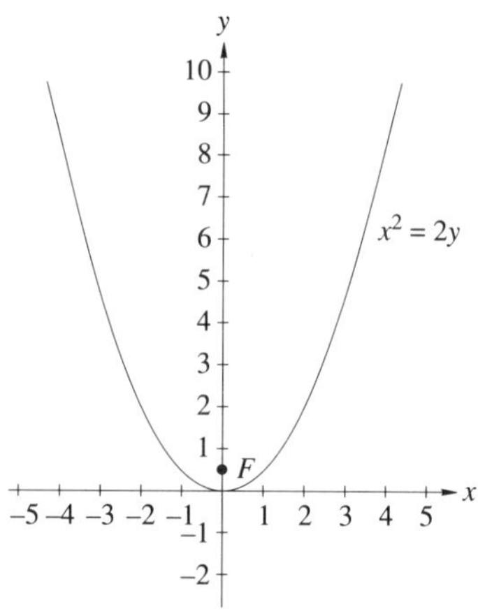

10. $x^2 = -17y$, $p = \frac{-17}{4} = -4.25$, vertical so $F = (0, -4.25)$, directrix: $y = 4.25$, opens downward.

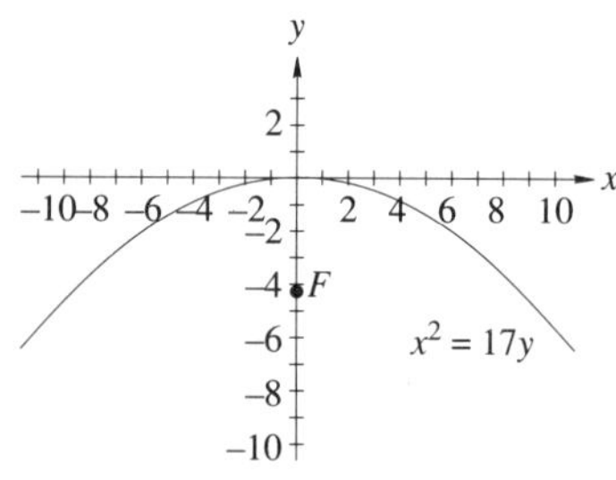

11. $y^2 = -21x$, $p = \frac{-21}{4} = -5.25$, horizontal so $F = (-\frac{21}{4}, 0)$, directrix: $x = \frac{21}{4}$, opens left.

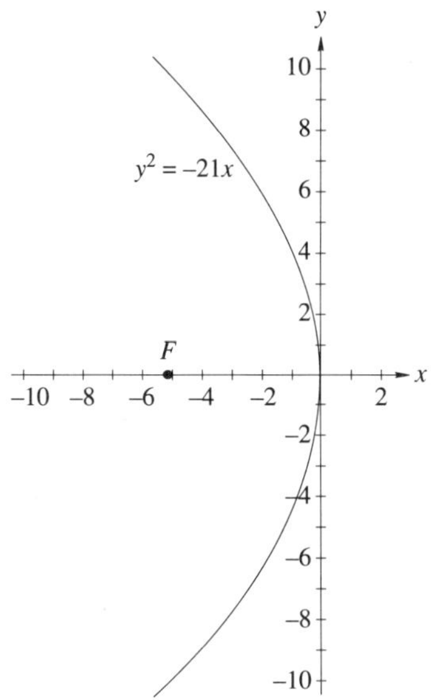

12. $y^2 = 5x$, $p = \frac{5}{4} = 1.25$, horizontal so $F = (1.25, 0)$, directrix: $x = -1.25$, opens right.

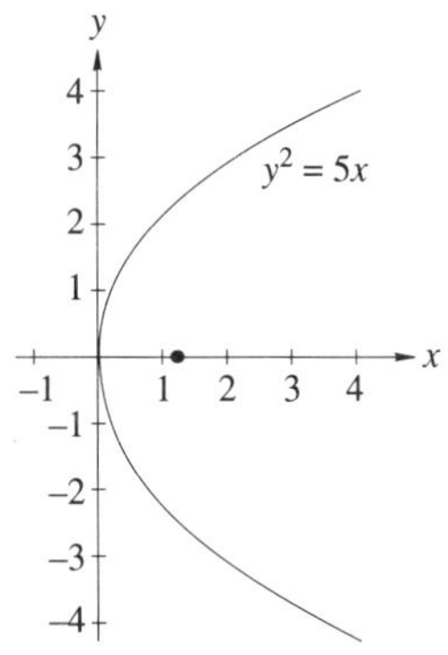

13. Focus $(0, 4)$, directrix: $y = -4$, $p = 4$ opens upward so $x^2 = 4 \cdot 4y$ or $x^2 = 16y$

14. Focus $(0, -5)$, directrix: $y = 5$, $p = -5$ opens downward so $x^2 = -5 \cdot 4y$ or $x^2 = -20y$.

15. Focus $(-6, 0)$, directrix: $x = 6$, $p = -6$ opens left so $y^2 = 4(-6)x$ or $y^2 = -24x$

16. Focus $(3, 0)$, directrix: $x = -3$, $p = 3$, opens right so $y^2 = 4 \cdot 3x$ or $y^2 = 12x$.

17. Focus $(2, 0)$, vertex $(0, 0)$, $p = 2$, opens right so $y^2 = 4 \cdot 2x$ or $y^2 = 8x$

18. Focus $(0, -7)$, vertex $(0, 0)$, $p = -7$, opens downward so $x^2 = 4(-7)y$ or $x^2 = -28y$

19. Vertex $(0, 0)$, directrix: $x = -\frac{3}{2}$, $p = \frac{3}{2}$, opens right so $y^2 = 4 \cdot \frac{3}{2}x$ or $y^2 = 6x$

20. Vertex $(0, 0)$, directrix $y = \frac{3}{4}$, $p = -\frac{3}{4}$, opens downward so $x^2 = 4(-\frac{3}{4})y$ or $x^2 = -3y$.

21. Choosing the vertex to be at the center of the bridge and the origin, we get $x^2 = 4py$ where $(180, 90)$ is a point on the parabola. Solving for p we get $(180)^2 = 4 \cdot 90p$ or $p = 90$. $x^2 = 360y$. Now setting $x = 100$ and solving for y, $100^2 = 360y$ or $y = \frac{10000}{360} = 27.78$ m.

22. If the diameter is 2 the radius is 1. Using $x^2 = 4py$ and the point $(1, 0.4)$ we can solve for p. $1 = 4(0.4)p$ or $p = \frac{1}{1.6} = 0.625$ The location is 0.625 m from the bottom or vertex.

23. $d = 5 \Rightarrow r = 2.5$ Using $x^2 = 4py$ and the point $(2.5, 1.5)$ we solve for p. $2.5^2 = 4(1.5)p$ or $p = \frac{6.25}{6}$ or 1.042 m from vertex.

24. $P = 10I^2$

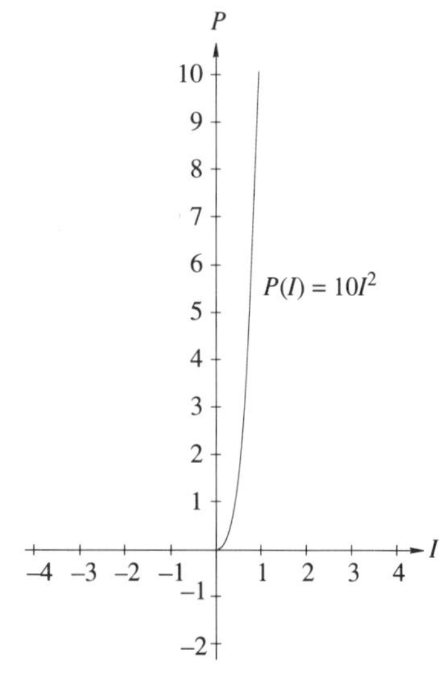

15.4 THE ELLIPSE

1. $F(4,0)$, $V(6,0)$, so $c = 4$, $a = 6$, $b^2 = a^2 - c^2$ or $b^2 = 6^2 - 4^2 = 20$ Major axis is horizontal so the equation is $\frac{x^2}{36} + \frac{y^2}{20} = 1$

2. $F(9,0)$, $V = (12,0)$, $c = 9$, $a = 12$, $b^2 = a^2 - c^2 = 12^2 - 9^2 = 63$ Major axis is horizontal so the equation is $\frac{x^2}{144} + \frac{y^2}{63} = 1$

3. $F(0,2)$, $V = (0,-4)$. Major axis is vertical so $b = 4$, $c = 2$, $a^2 = b^2 - c^2 = 4^2 - 2^2 = 12$. $\frac{x^2}{12} + \frac{y^2}{16} = 1$

4. $F(0,-3)$, $V(0,5)$. Major axis is vertical so $b = 5$, $c = 3$, $a^2 = b^2 - c^2 = 25 - 9 = 16$. $\frac{x^2}{16} + \frac{y^2}{25} = 1$

5. $F(3,0)$,$V(5,0)$. Major axis is horizontal so $a = 5$, $c = 3$, $b^2 = a^2 - c^2 = 25 - 9 = 16$; $\frac{x^2}{25} + \frac{y^2}{16} = 1$

6. $F(0,-2)$, $V(0,-3)$. Major axis vertical so $b = 3$, $c = 2$, $a^2 = b^2 - c^2 = 9 - 4 = 5$. $\frac{x^2}{5} + \frac{y^2}{9} = 1$

7. Minor axis length 6, vertex $(4,0)$. Major axis is horizontal so $a = 4$, $b = \frac{6}{2} = 3$;

$$\frac{x^2}{16} + \frac{y^2}{9} = 1$$

8. Minor axis length 10, $V(0,-6)$. Major axis is vertical so $b = 6$, $a = \frac{10}{2} = 5$, $\frac{x^2}{25} + \frac{y^2}{36} = 1$

9. $\frac{x^2}{4} + \frac{y^2}{9} = 1$, Major axis vertical $b = 3$, $a = 2$, $c^2 = b^2 - a^2 = 9 - 4 = 5$, $c = \sqrt{5}$, $V = (0,3)$, $V' = (0,-3)$, $F = (0,\sqrt{5})$, $F' = (0,-\sqrt{5})$, $M = (2,0)$, $M' = (-2,0)$

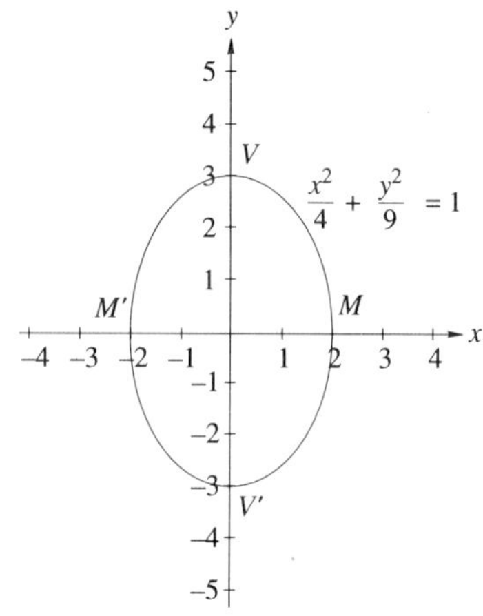

10. $\frac{x^2}{9} + \frac{y^2}{4} = 1$, Major axis is horizontal, $a = 3$, $b = 2$, $c^2 = 9 - 4 = 5$, $c = \sqrt{5}$. $V = (3,0)$, $V' = (-3,0)$, $F = (\sqrt{5},0)$, $F' = (-\sqrt{5},0)$, $M = (0,2)$, $M' = (0,-2)$

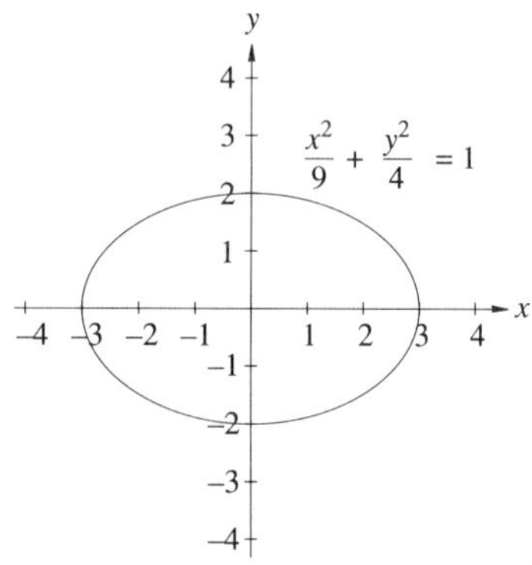

11. $\frac{x^2}{4} + \frac{y^2}{1} = 1$, $a = 2, b = 1$, $c^2 = a^2 - b^2 = 3$, $c = \sqrt{3}$, $V = (2,0)$, $V' = (-2,0)$, $F = (\sqrt{3},0)$, $F' = (-\sqrt{3},0)$, $M = (0,1)$, $M' = (0,-1)$

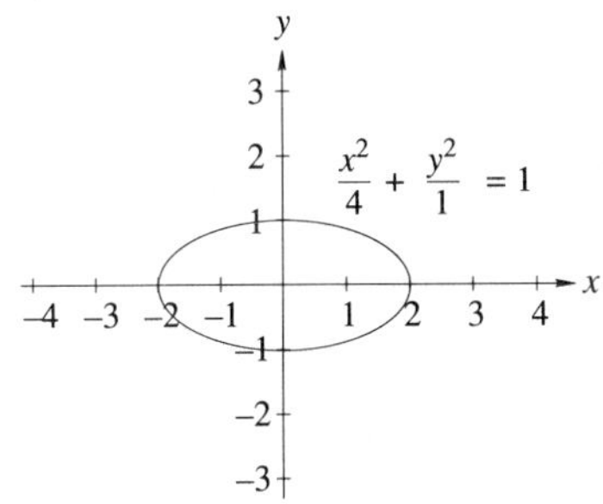

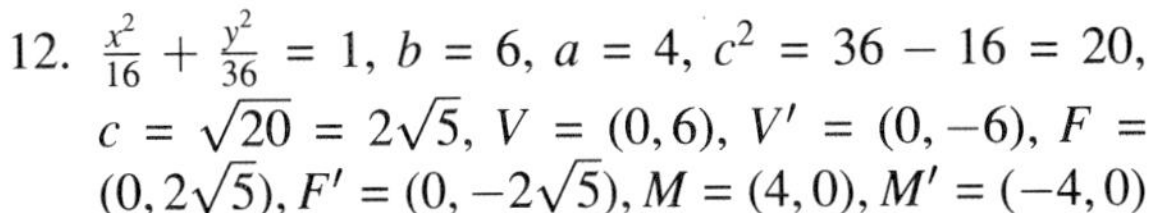

12. $\frac{x^2}{16} + \frac{y^2}{36} = 1$, $b = 6$, $a = 4$, $c^2 = 36 - 16 = 20$, $c = \sqrt{20} = 2\sqrt{5}$, $V = (0, 6)$, $V' = (0, -6)$, $F = (0, 2\sqrt{5})$, $F' = (0, -2\sqrt{5})$, $M = (4, 0)$, $M' = (-4, 0)$

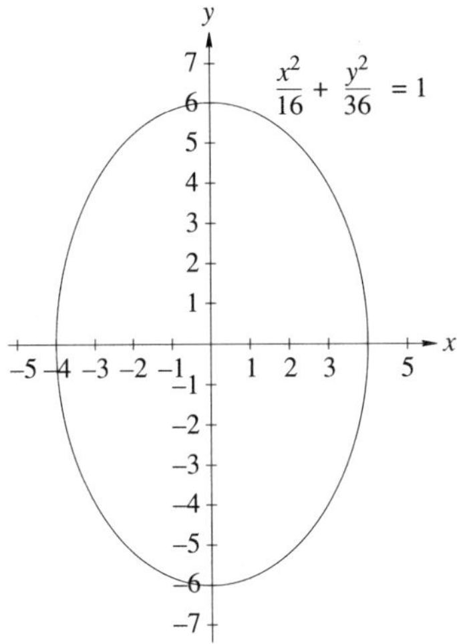

13. $4x^2 + y^2 = 4 \Rightarrow \frac{x^2}{1} + \frac{y^2}{4} = 1$, $a = 1$, $b = 2$, $c = \sqrt{3}$, $V = (0, \pm 2)$, $F = (-0, \pm\sqrt{3})$, $M = (\pm 1, 0)$

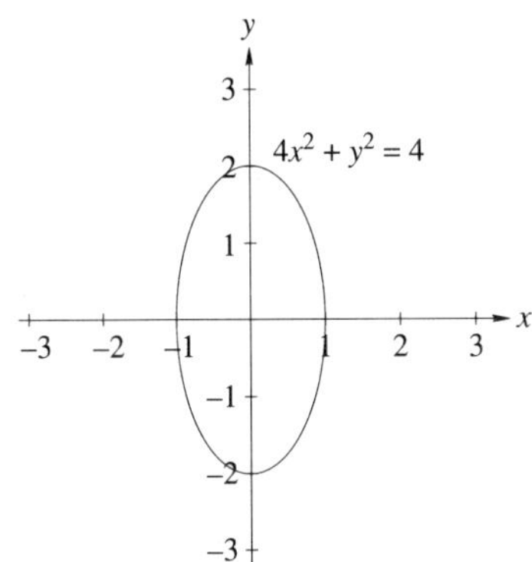

14. $9x^2 + 4y^2 = 36 \Rightarrow \frac{x^2}{4} + \frac{y^2}{9} = 1$, $a = 2$, $b = 3$, $c = \sqrt{5}$, $V = (0, \pm 3)$, $F = (0, \pm\sqrt{5})$, $M = (\pm 2, 0)$

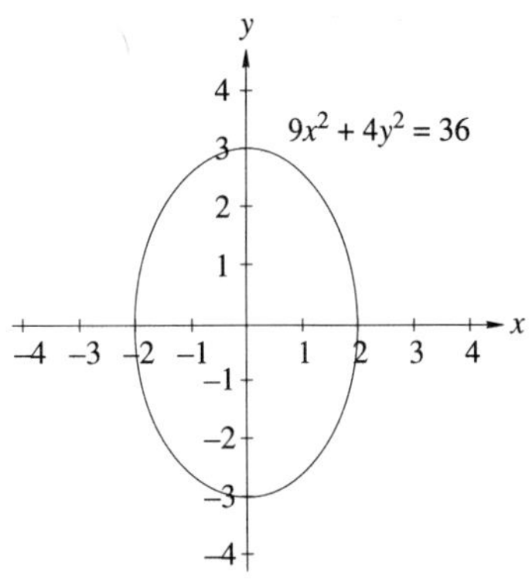

15. $25x^2 + 36y^2 = 900 \Rightarrow \frac{x^2}{36} + \frac{y^2}{25} = 1$, $a = 6$, $b = 5$, $c = \sqrt{11}$, $V = (\pm 6, 0)$, $F = (\pm\sqrt{11}, 0)$, $M = (0, \pm 5)$

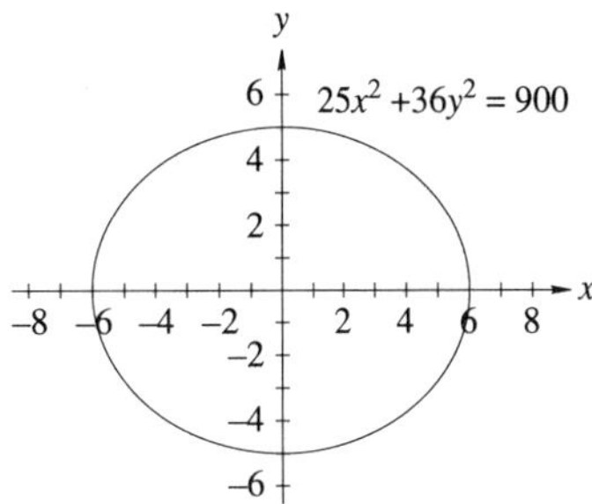

16. $9x^2 + y^2 = 18 \Rightarrow \frac{x^2}{2} + \frac{y^2}{18} = 1$, $a = \sqrt{2}$, $b = 3\sqrt{2}$, $c = 4$, $V = (0, \pm 3\sqrt{2})$, $F = (0, \pm 4)$, $M = (\pm\sqrt{2}, 0)$

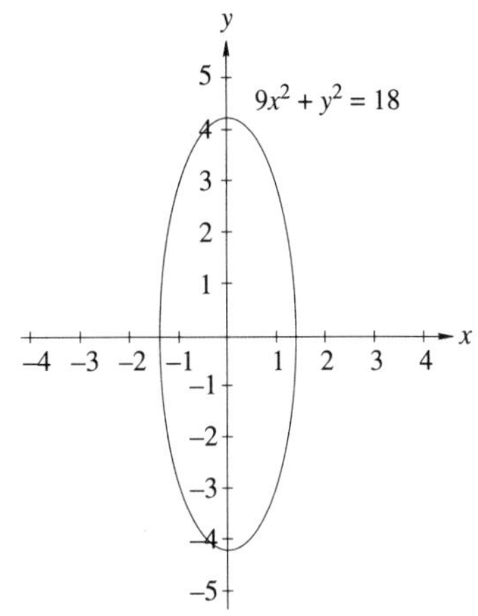

17. $a = 2.992\times 10^8/2 = 1.496\times 10^8, c = a/60 = 2.493\times 10^6, b^2 = a^2 - c^2$ or $b = \sqrt{a^2 - c^2} = 1.49579 \times 10^8$
Minor axis $= 2b = 2.9916 \times 10^8$ km

18. Major axis is 80 so $a = \frac{80}{2} = 40$ and $b = 24$. Substituting in the standard equation for an ellipse, we obtain $\frac{x^2}{40^2} + \frac{y^2}{24^2} = 1$ or $\frac{x^2}{1600} + \frac{y^2}{576} = 1$

19. Minor axis is the diameter of the pipe is 10 in. Major axis is $10\sqrt{2}$ or 14.14 in.

20. Distance between foci used in the two ellipses is $2a$ so $a = 3$. $c = a - 1 = 2, b^2 = 3^2 - 2^2 = 9 - 4 = 5$ so $b = \sqrt{5}$. Minor axis $= 2b = 2\sqrt{5}$ cm

21. $a = 3600/2 = 1800$, $b = 44/2 = 22$, $c = \sqrt{a^2 - b^2} = \sqrt{1800^2 - 22^2} = 1799.865551$, eccentricity $= \frac{c}{a} = \frac{1{,}799.865551}{1800} = 0.9999253$

15.5 THE HYPERBOLA

1. $F(6,0), V(4,0)$: Transverse axis is on the x-axis so $a = 4, c = 6$, and $b^2 = c^2 - a^2 = 36 - 16 = 20$. The equation is $\frac{x^2}{16} - \frac{y^2}{20} = 1$

2. $F(12,0), V = (9,0)$: Transverse axis is on the x-axis so $c = 12, a = 9$, and $b^2 = c^2 - a^2 = 12^2 - 9^2 = 63$. The equation is $\frac{x^2}{81} - \frac{y^2}{63} = 1$

3. $F(0,5), V(0,-3)$: Transverse axis is on y-axis so $c = 5, a = 3$, and $b^2 = 5^2 - 3^2 = 16$. The equation is $\frac{y^2}{9} - \frac{x^2}{16} = 1$

4. $F(0,-4), V(0,2)$: Transverse axis is on the y-axis and $c = 4, a = 2$, and $b^2 = 16-4 = 12$ The equation is $\frac{y^2}{4} - \frac{x^2}{12} = 1$

5. $F(5,0), V = (3,0) : c = 5, a = 3$, and $b^2 = 5^2 - 3^2 = 16$. Transverse axis is on the x-axis so the equation is $\frac{x^2}{9} - \frac{y^2}{16} = 1$.

6. $F(0,-3), V(0,-2), c = 3, a = 2$, and so $b^2 = 9 - 4 = 5$. The transverse axis is on the y-axis so the equation is $\frac{y^2}{4} - \frac{x^2}{5} = 1$.

7. $F(4,0)$, length of conjugate axis is 6 so $b = \frac{6}{2} = 3$, $c = 4, a^2 = c^2 - b^2 = 16 - 9 = 7$. The Transverse axis is on the x-axis so the equation is $\frac{x^2}{7} - \frac{y^2}{9} = 1$.

8. $F(0,-6)$, conjugate axis lenth is 10 so $b = \frac{10}{2} = 5, c = 6, a^2 = 6^2 - 5^2 = 11$. The transverse axis is on the y-axis so the equation is $\frac{y^2}{11} - \frac{x^2}{25} = 1$

9. $\frac{x^2}{4} - \frac{y^2}{9} = 1$, $a = 2$, $b = 3$, $c = \sqrt{4+9} = \sqrt{13}$. $V = (\pm 2,0), F(\pm\sqrt{13},0), M = (0,\pm 3)$

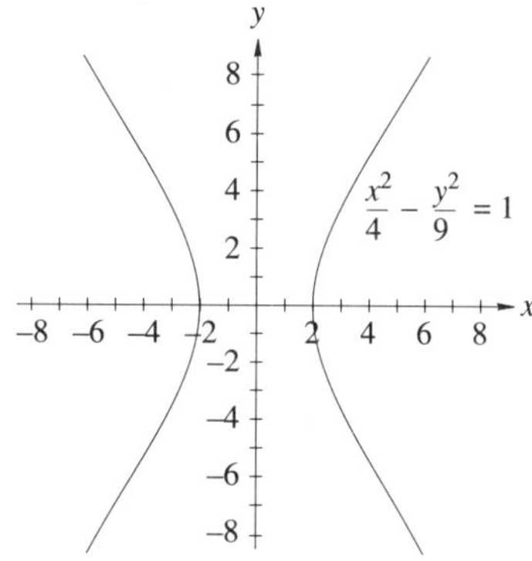

10. $\frac{x^2}{9} - \frac{y^2}{4} = 1$, $a = 3$, $b = 2$, $c = \sqrt{9+4} = \sqrt{13}$. The transverse axis is on the x-axis, so $V = (\pm 3,0)$, $F = (\pm\sqrt{13},0), M = (0,\pm 2)$

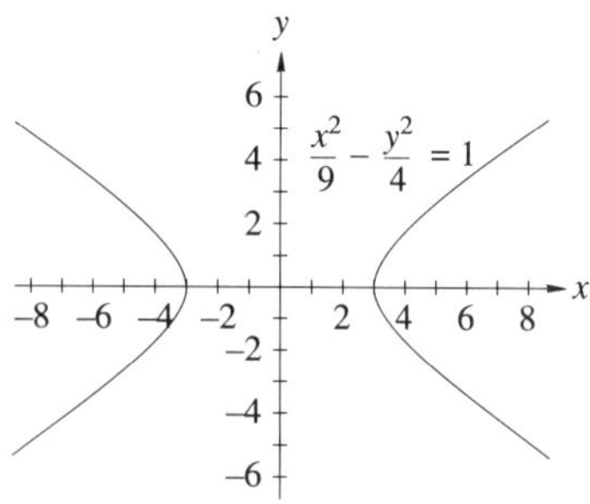

11. $\frac{y^2}{4} - \frac{x^2}{1} = 1$, $a = 2$, $b = 1$, $c = \sqrt{4+1} = \sqrt{5}$. The transverse axis is on the y-axis, so $V = (0,\pm 2)$, $F = (0,\pm\sqrt{5}), M = (\pm 1,0)$.

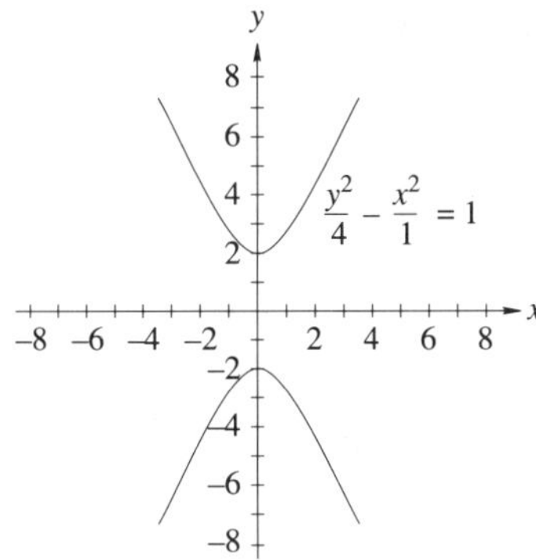

12. $\frac{y^2}{36} - \frac{x^2}{16} = 1$, $a = 6$, $b = 4$, $c = \sqrt{36 + 16} = \sqrt{52} = 2\sqrt{13}$. The transverse axis is on the y-axis, so $V = (0, \pm 6)$, $F = (0, \pm 2\sqrt{13})$, $M = (\pm 4, 0)$.

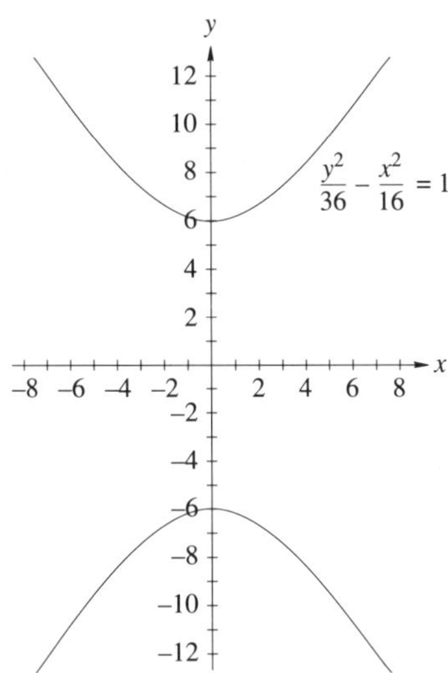

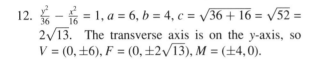

13. $4x^2 - y^2 = 4 \Rightarrow \frac{x^2}{1} - \frac{y^2}{4} = 1$, $a = 1$, $b = 2$, $c = \sqrt{1 + 4} = \sqrt{5}$. The transverse axis is on the x-axis, so $V = (\pm 1, 0)$, $F = (\pm\sqrt{5}, 0)$, $M = (0, \pm 2)$.

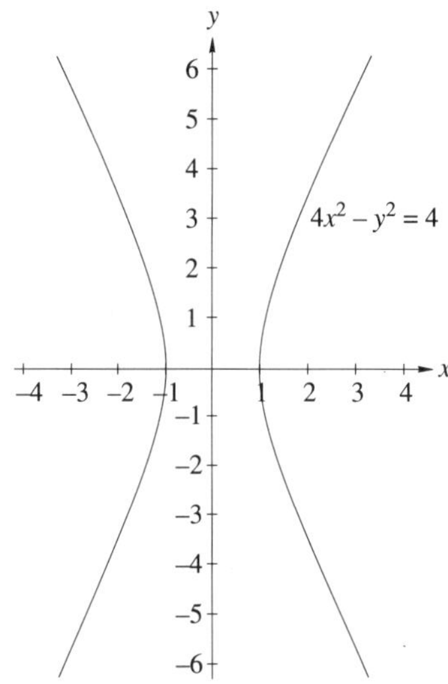

14. $9x^2 - 4y^2 = 36 \Rightarrow \frac{x^2}{4} - \frac{y^2}{9} = 1$, $a = 2$, $b = 3$, $c = \sqrt{4 + 9} = \sqrt{13}$. The transverse axis is on the x-axis, so $V = (\pm 2, 0)$, $F = (\pm\sqrt{13}, 0)$, $M = (0, \pm 3)$.

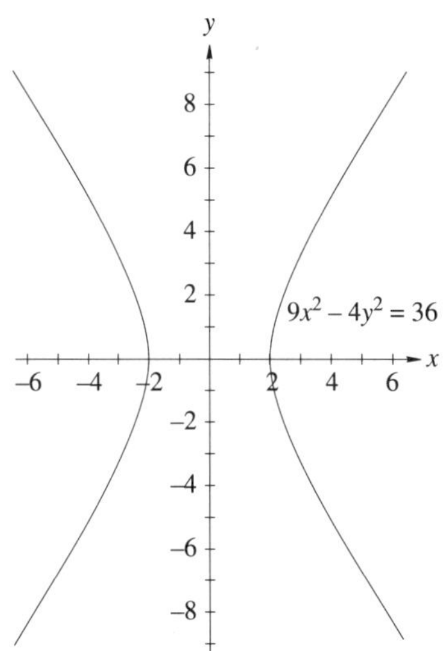

15. $36y^2 - 25x^2 = 900 \Rightarrow \frac{y^2}{25} - \frac{x^2}{36} = 1$, $a = 5$, $b = 6$, $c = \sqrt{25 + 36} = \sqrt{61}$. The transverse axis is on the y-axis so $V = (0, \pm 5)$, $F = (0, \pm\sqrt{61})$, $M = (\pm 6, 0)$

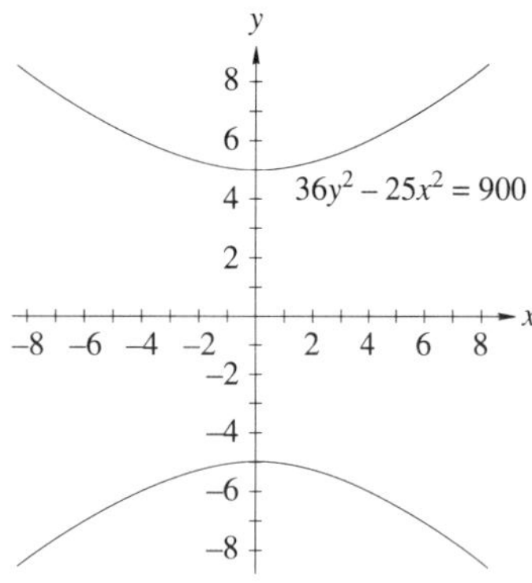

16. $y^2 - 9x^2 = 18 \Rightarrow \frac{y^2}{18} - \frac{x^2}{2} = 1$, $a = \sqrt{18} = 3\sqrt{2}$, $b = \sqrt{2}$, $c = \sqrt{18+2} = \sqrt{20} = 2\sqrt{5}$. The transverse axis is on the y-axis so $V = (0, \pm 3\sqrt{2})$, $F = (0, \pm 2\sqrt{5})$, $M = (\pm\sqrt{2}, 0)$.

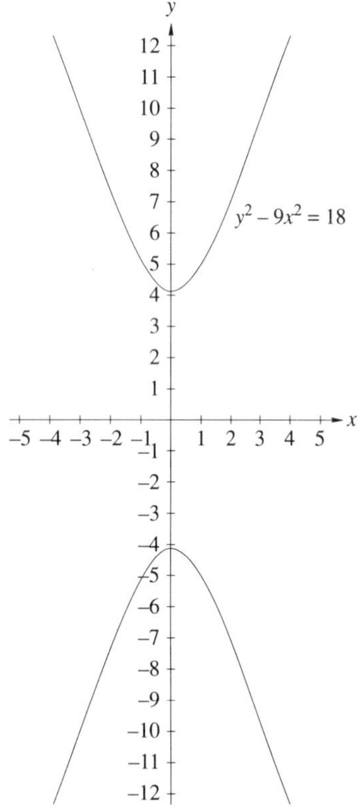

17. If the vertex is $(7, 0)$ then $a = 7$, $\frac{c}{a} = 1.5$ so $c = 1.5 \times 7 = 10.5$, $b^2 = c^2 - a^2 = 61.25$. The transverse axis is on the x-axis so the equation is $\dfrac{x^2}{49} - \dfrac{y^2}{61.25} = 1$.

18. (a)

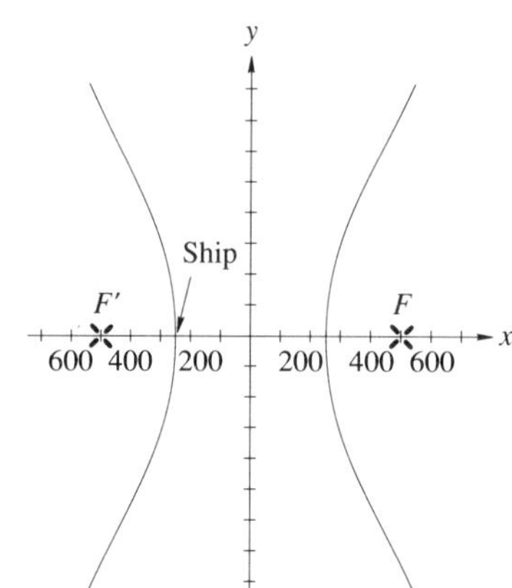

(b) $c = 1000/2 = 500$, $a = 500 - 250 = 250$, $b^2 = c^2 - a^2 = 187500$. $a^2 = 62500$, $\dfrac{x^2}{62{,}500} - \dfrac{y^2}{187{,}500} = 1$

19. If $a = b$ then $c = \sqrt{a^2 + b^2} = \sqrt{2a^2} = a\sqrt{2}$, $e = \dfrac{c}{a} = \dfrac{a\sqrt{2}}{a} = \sqrt{2}$

20. $a = 153$, $\dfrac{x^2}{153^2} - \dfrac{y^2}{b^2} = 1$, $\dfrac{238^2}{153^2} - \dfrac{(-616)^2}{b^2} = 1$, $\dfrac{616^2}{b^2} = \dfrac{238^2}{153^2} - 1$, $b^2 = \dfrac{616^2}{\frac{238^2}{153^2} - 1} = 267{,}269$; $b = 516.98$. The equation is $\dfrac{x^2}{153^2} - \dfrac{y^2}{516.98^2} = 1$

15.6 TRANSLATION OF AXES

1. $\dfrac{(x-4)^2}{9}+\dfrac{(y+3)^2}{4}=1$, $x'=x-4$, $y'=y+3$ so $(h,k)=(4,-3)$; $\dfrac{x'^2}{9}+\dfrac{y'^2}{4}=1$. This is an ellipse with $a=3$, $b=2$, and $c=\sqrt{9-4}=\sqrt{5}$. The major axis is horizontal.

	$x'y'$-system	xy-system
center	$(0',0')$	$(4,-3)$
vertices	$(\pm 3',0')$	$(7,-3)$, $(1,-3)$
foci	$(\pm\sqrt{5}',0')$	$(\sqrt{5}+4,-3)$, $(-\sqrt{5}+4,-3)$
endpoints of minor axis	$(0',\pm 2')$	$(4,-1)$, $(4,-5)$

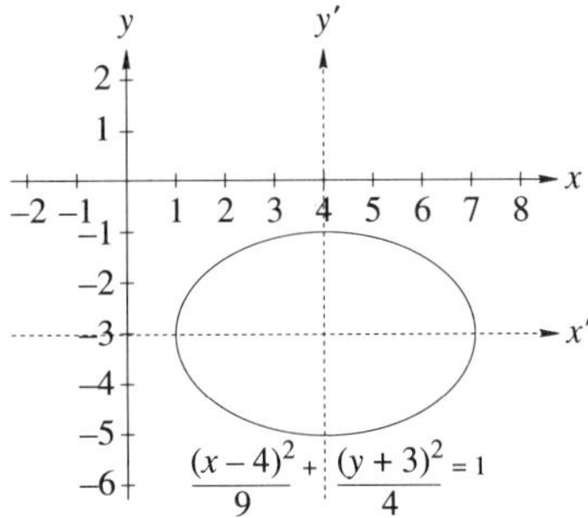

2. $\dfrac{(x+5)^2}{16}-\dfrac{(y+3)^2}{25}=1$. This is a hyperbola with $x'=x+5$ and $y'=y+3$ so $(h,k)=(-5,-3)$; $a=4$, $b=5$ so $c=\sqrt{16+25}=\sqrt{41}$. The transverse axis is horizontal.

	$x'y'$-system	xy-system
center	$(0',0')$	$(-5,-3)$
vertices	$(\pm 4',0')$	$(-9,-3)$, $(-1,-3)$
foci	$(\pm\sqrt{41}',0')$	$(\sqrt{41}-5,-3)$, $(-\sqrt{41}-5,-3)$
endpoints of conjugate axis	$(0',\pm 5')$	$(-5,2)$, $(-5,-8)$

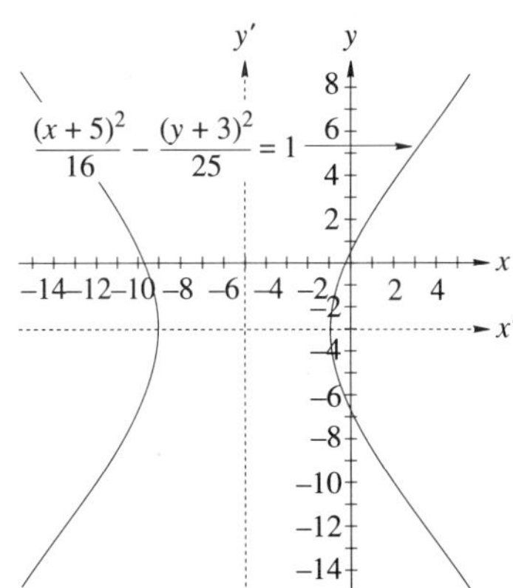

3. $\dfrac{(y-3)^2}{36}-(x+4)^2=1$, is a hyperbola with $y'=y-3$ and $x'=x+4$, so $(h,k)=(3,-4)$, $a=6$, $b=1$, $c=\sqrt{37}$. The transverse axis is horizontal.

	$x'y'$-system	xy-system
center	$(0',0')$	$(-4,3)$
vertices	$(0',\pm 6)$	$(-4,9)$, $(-4,-3)$
foci	$(0',\pm\sqrt{37}')$	$(-4,\sqrt{37}+3)$, $(-4,-\sqrt{37}+3)$
endpoints of conjugate axis	$(\pm 1',0')$	$(-3,3)$, $(-5,3)$

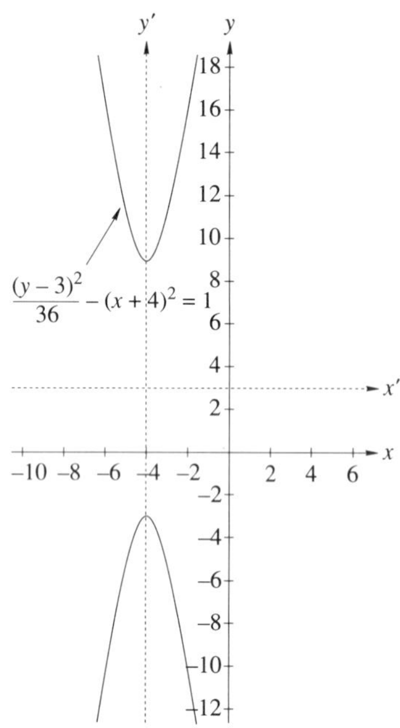

4. $(y+5)^2 = 12(x+1)$, $x' = x+1$, $y' = y+5$ so $(h,k) = (-1,-5)$; $p = 12/4 = 3$ parabola opens right

	$x'y'$-system	xy-system
vertex	$(0', 0')$	$(-1, -5)$
focus	$(3', 0')$	$(2, -5)$

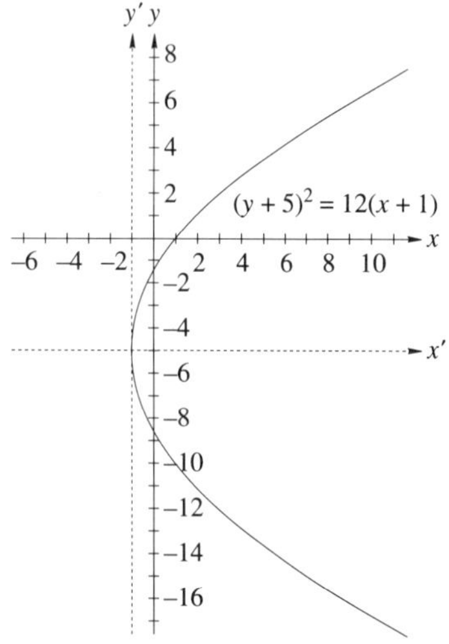

5. $100(x+5)^2 - 4y^2 = 400 \Rightarrow \dfrac{(x+5)^2}{4} - \dfrac{y^2}{100} = 1$;

This is a hyperbola with $x' = x+5$ and $y' = y$ so $(h,k) = (-5,0)$; $a = 2$, $b = 10$, $c = \sqrt{104} = 2\sqrt{26}$. The transverse axis is vertical.

	$x'y'$-system	xy-system
center	$(0', 0')$	$(-5, 0)$
vertices	$(\pm 2', 0')$	$(-3, 0)$, $(-7, 0)$
foci	$(\pm 2\sqrt{26}', 0')$	$(2\sqrt{26} - 5, 0)$, $(-2\sqrt{26} - 5, 0)$
endpoints of conjugate axis	$(0', \pm 10')$	$(-5, 10)$, $(-5, -10)$

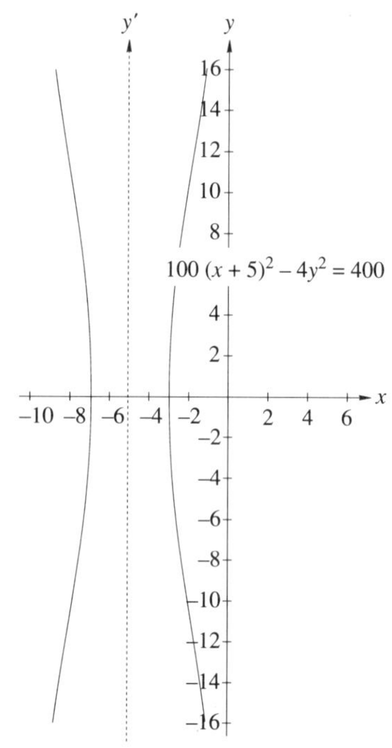

6. $(y+3)^2 = 8(x-2)$ is a parabla with $x' = x-2$ and $y' = y+3$. The vertex is $(h,k) = (2,-3)$ and $p = \frac{8}{4} = 2$, so the parabola opens right.

	$x'y'$-system	xy-system
vertex	$(0', 0')$	$(2, -3)$
focus	$(2', 0')$	$(4, -3)$

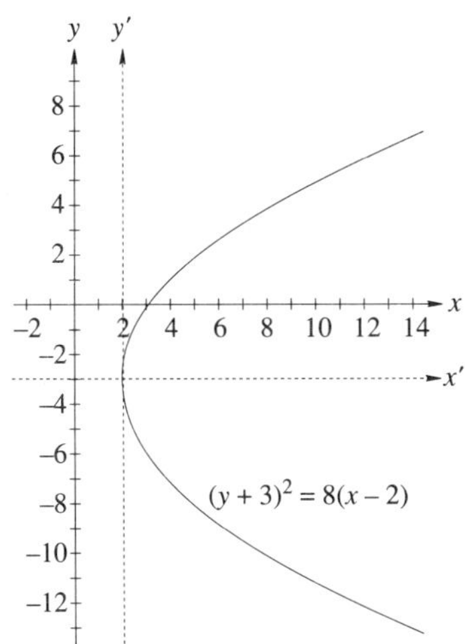

7. $16x^2 + 4y^2 + 64x - 12y + 57 = 0$ or $16(x^2 + 4x) + 4(y^2 - 3y) = -57$. Completing the square produces $16(x^2+4x+4)+4(y^2-3y+\frac{9}{4}) = -57+16\cdot 4+4\cdot\frac{9}{4}$ or $16(x+2)^2+4(y-\frac{3}{2})^2 = 16$. This has the standard form equation $\frac{(x+2)^2}{1} + \frac{(y-\frac{3}{2})}{4} = 1$ and so it is an ellipse with $x' = x+2$ and $y' = y - \frac{3}{2}$. Thus, $(h,k) = (-2, \frac{3}{2})$, $a = 1$, $b = 2$, and $c = \sqrt{4-1} = \sqrt{3}$. The major axis is vertical.

	$x'y'$-system	xy-system
center	$(0', 0')$	$(-2, \frac{3}{2})$
vertices	$(0', \pm 2')$	$(-2, \frac{7}{2})$, $(-2, -\frac{1}{2})$
foci	$(0', \pm\sqrt{3}')$	$(-2, \sqrt{3}+\frac{3}{2})$, $(-2, -\sqrt{3}+\frac{3}{2})$
endpoints of minor axis	$(\pm 1', 0')$	$(-1, \frac{3}{2})$, $(-3, \frac{3}{2})$

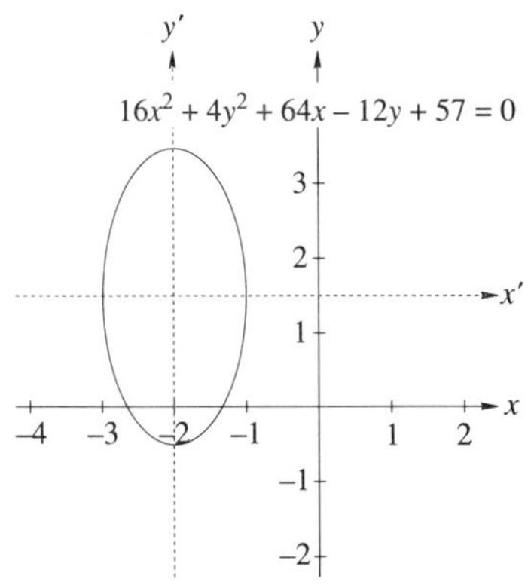

8. $x^2+y^2-2x+2y-2 = 0$; $(x^2-2x+1)+(y^2+2y+1) = 2+1+1$; $(x-1)^2 + (y+1)^2 = 4$; circle; center: $(1, -1)$, radius 2

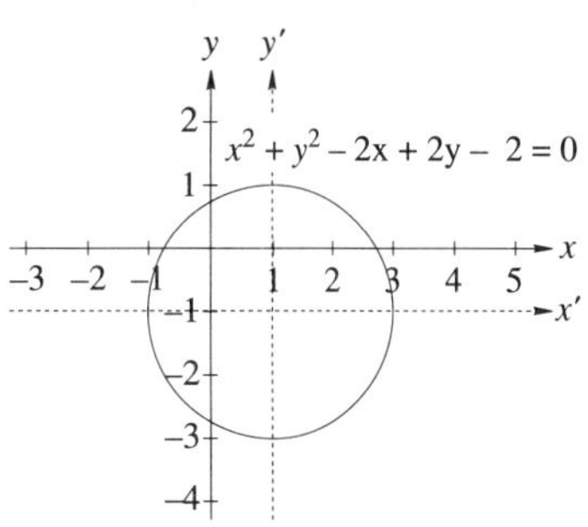

9. $25x^2 + 4y^2 - 250x - 16y + 541 = 0$ or $25(x^2 - 10x) + 4(y^2 - 4y) = -541$. Completing the square produces $25(x^2 - 10x + 25) + 4(y^2 - 4y + 4) = -541+25\cdot 25+4\cdot 4$ or $25(x-5)^2+4(y-2)^2 = 100$. This has the standard form $\frac{(x-5)^2}{4} + \frac{(y-2)^2}{25} = 1$ and so it is an ellipse with $x' = x-5$, $y' = y-2$. Here $(h,k) = (5,2)$, $a = 2$, $b = 5$, $c = \sqrt{21}$. The major axis is vertical.

	$x'y'$-system	xy-system
center	$(0', 0')$	$(5, 2)$
vertices	$(0', \pm 5')$	$(5, 7)$, $(5, -3)$
foci	$(0', \pm\sqrt{21}')$	$(5, 2+\sqrt{21})$, $(5, 2-\sqrt{21})$
endpoints of minor axis	$(\pm 2', 0')$	$(7, 2)$, $(3, 2)$

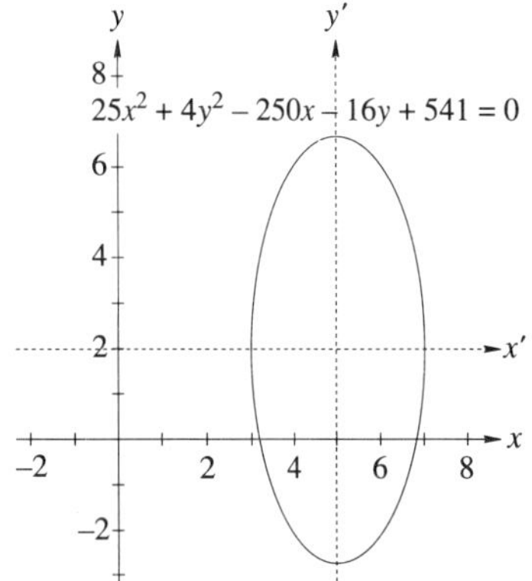

10. $100x^2 - 180x - 100y + 81 = 0$ or $100x^2 - 180x = 100y - 81$. Factoring, we get $100(x^2 - 1.8x) = 100y - 81$ and then completing the square yields $100(x^2 - 1.8x + 0.81) = 100y - 81 + 81$ or $100(x - 0.9)^2 = 100y$ which is equivalent to $(x - 0.9)^2 = y$. This is the standard form equation for a parabola with $(h, k) = (0.9, 0)$ and $p = \frac{1}{4} = 0.25$. The parabola opens upward.

	$x'y'$-system	xy-system
vertex	$(0', 0')$	$(0.9, 0)$
focus	$(0', 0.25')$	$(0.9, 0.25)$

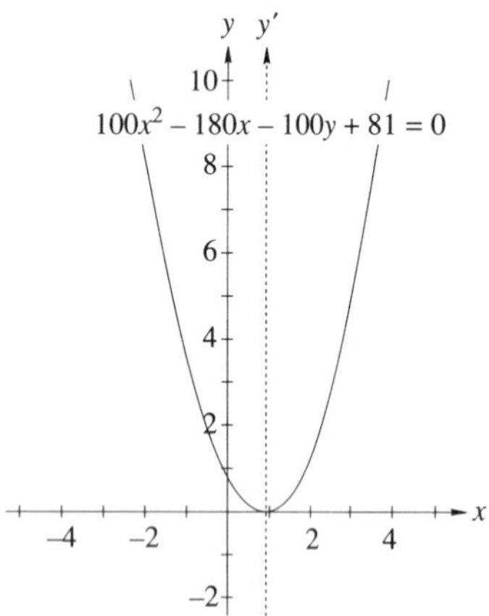

11. $2x^2 - y^2 - 16x + 4y + 24 = 0$ or $2(x^2 - 8x) - (y^2 - 4y) = -24$. Completing the square yields $2(x^2 - 8x + 16) - (y^2 - 4y + 4) = -24 + 2 \cdot 16 - 1 \cdot 4$ or $2(x - 4)^2 - (y - 2)^2 = 4$ which has the standard form equation $\dfrac{(x-4)^2}{2} - \dfrac{(y-2)^2}{4} = 1$. This is an hyperbola with $a = \sqrt{2}$, $b = 2$; $c = \sqrt{6}$, $x' = x - 4$, $y' = y - 2$, $(h, k) = (4, 2)$. The transverse axis is horizontal.

	$x'y'$-system	xy-system
center	$(0', 0')$	$(4, 2)$
vertices	$(\pm\sqrt{2}', 0')$	$(4 + \sqrt{2}, 2)$, $(4 - \sqrt{2}, 2)$
foci	$(\pm\sqrt{6}', 0')$	$(4 + \sqrt{6}, 2)$, $(4 - \sqrt{6}, 2)$
endpoints of conjugate axis	$(0', \pm 2')$	$(4, 0)$, $(4, 4)$

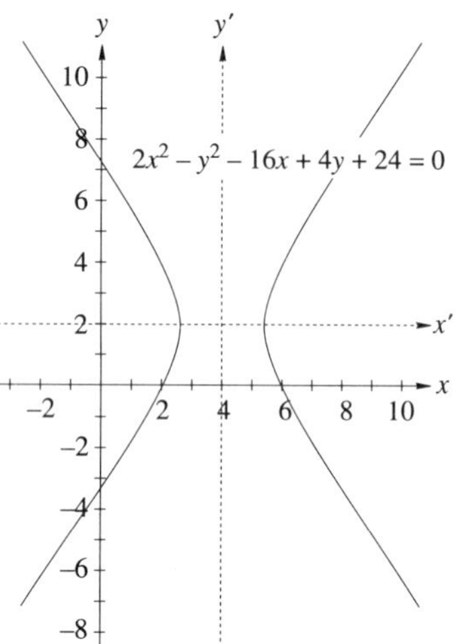

12. $9x^2 - 36x + 16y^2 - 32y - 524 = 0$ or $9(x^2 - 4x) + 16(y^2 - 2y) = 524$. Completing the square, we get $9(x^2 - 4x + 4) + 16(y^2 - 2y + 1) = 524 + 9 \cdot 4 + 16 \cdot 1$ or $9(x - 2)^2 + 16(y - 1)^2 = 576$. This has the standard form equation $\dfrac{(x-2)^2}{64} + \dfrac{(y-1)^2}{36} = 1$ and so we see that it is an ellipse, with $a = 8$, $b = 6$; $c = \sqrt{64 - 36} = \sqrt{28} = 2\sqrt{7}$, $x' = x - 2$, $y' = y - 1$; $(h, k) = (2, 1)$. The major axis is horizontal.

	$x'y'$-system	xy-system
center	$(0', 0')$	$(2, 1)$
vertices	$(\pm 8', 0')$	$(10, 1)$, $(-6, 1)$
foci	$(\pm 2\sqrt{7}', 0')$	$(2 + 2\sqrt{7}, 1)$, $(2 - 2\sqrt{7}, 1)$
endpoints of minor axis	$(0', \pm 6')$	$(2, 7)$, $(2, -5)$

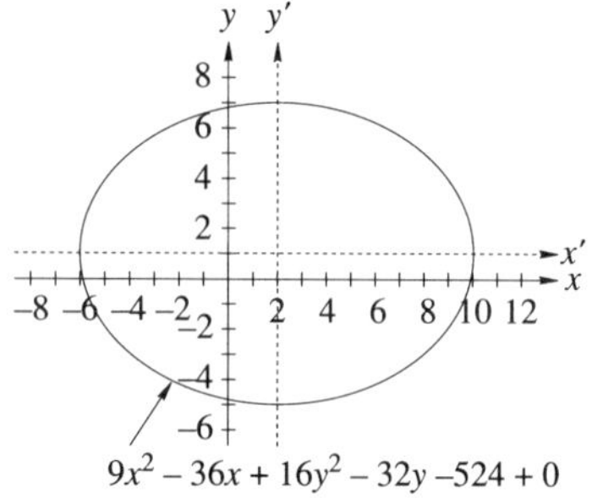

13. Parabola, Vertex $= (2, -3)$, $p = 8$, axis vertical $x' = x - 2$, $y' = y + 3$, $x'^2 = 4py'$ or $(x - 2)^2 = 32(y + 3)$

14. Hyperbola, vertex at $(2,1)$, foci at $(-6,1)$ and $(8,1)$. Foci are horizontally displaced so the tranverse axis is horizontal. the center is the midpoint between the foci or at $(1,1)$, $c = |8-1| = 7$, $a = |2-1| = 1$, $b = \sqrt{c^2 - a^2} = \sqrt{49-1} = \sqrt{48}$, $x' = x - 1$, $y' = y - 1$, $\frac{x'^2}{1} - \frac{y'^2}{48} = 1$ or $\frac{(x-1)^2}{1} - \frac{(y-1)^2}{48} = 1$

15. Ellipse, center at $(4,-3)$, focus at $(8,-3)$, vertex at $(10,-3)$, $a = |10-4| = 6$, $c = |8-4| = 4$, $b^2 = a^2 - c^2 = 36 - 16 = 20$. Major axis is horizontal. $x' = x - 4$, $y' = y + 3$, and so $\frac{x'^2}{36} + \frac{y'^2}{20} = 1$ or $\frac{(x-4)^2}{36} + \frac{(y+3)^2}{20} = 1$.

16. Ellipse, center at $(-2,0)$, focus at $(6,0)$ vertex at $(9,0)$ Major axis is horizontal, $a = |9 - -2| = 11$, $c = |6 - -2| = 8$, $b^2 = a^2 - c^2 = 11^2 - 8^2 = 121 - 64 = 57$, $x' = x + 2$, $y' = y$ and so $\frac{x'^2}{121} + \frac{y'^2}{57} = 1$ or $\frac{(x+2)^2}{121} + \frac{y^2}{57} = 1$.

17. Hyperbola, center at $(-3,2)$ focus at $(-3,7)$, transverse axis 6 units. Center and focus vertically displaces so transverse axis is vertical. $c = |7-2| = 5$, $a = \frac{6}{2} = 3$, $b^2 = c^2 - a^2 = 16$, $x' = x + 3$, $y' = y - 2$ and so $\frac{y'^2}{9} - \frac{x'^2}{16} = 1$ or $\frac{(y-2)^2}{9} -- \frac{(x+3)^2}{16} = 1$.

18. Ellipse, center at $(3,5)$, focus at $(3,8)$ minor axis 2 units. Major axis is vertical $c = |8-5| = 3$, $a = \frac{2}{2} = 1$, $b^2 = c^2 + a^2$ $b^2 = 9 + 1 = 10$; $x' = x - 3$, $y' = y - 5$ and so $\frac{x'^2}{1} + \frac{y'^2}{10} = 1$ or $\frac{(x-3)^2}{1} + \frac{(y-5)^2}{10} = 1$.

19. Parabola, vertex at $(-5,1)$ so $x' = x + 5$, $y' = y - 1$, $p = -4$, axis parallel to x axis $y'^2 = 4px'$ or $(y-1)^2 = -16(x+5)$

20. Parabola, vertex at $(-3,-6)$ so $x' = x+3$, $y' = y+6$, $p = -12$, axis parallel to y-axis $x'^2 = 4py'$ or $(x+3)^2 = -48(y+6)$

21. Ellipse, center at $(-4,1)$ so $x' = x + 4$, $y' = y - 1$, and the focus is at $(-4,9)$, so major axis is vertical and $c = |9-1| = 8$, minor axis is 12. Hence, $a = \frac{12}{2} = 6$, $b^2 = a^2 + c^2 = 64 + 36 = 100$. $\frac{x'^2}{36} + \frac{y'^2}{100} = 1$ or $\frac{(x+4)^2}{36} + \frac{(y-1)^2}{100} = 1$

22. Hyperbola, center at $(2,8)$ so $x' = x - 2$, $y' = y - 8$; vertex at $(6,8)$ so $a = |6-2| = 4$, transverse axis is horizontal, conjugate axis is 10 so $b = \frac{10}{2} = 5$. $\frac{x'^2}{16} - \frac{y'^2}{25} = 1$ or $\frac{(x-2)^2}{16} - \frac{(y-5)^2}{25} = 1$

23. $s = 29.4t - 4.9t^2$ or $4.9t^2 - 29.4t = -s$, $4.9(t^2 - 6t) = -s$, $4.9(t^2 - 6t + 9) = -s + 4.9 \times 9$, $4.9(t-3)^2 = -s + 44.1$ or $(t-3)^2 = -\frac{1}{4.9}(s - 44.1)$. This is a parabola with vertex at $(3, 44.1)$ it opens downward so the maximum height is 44.1 m.

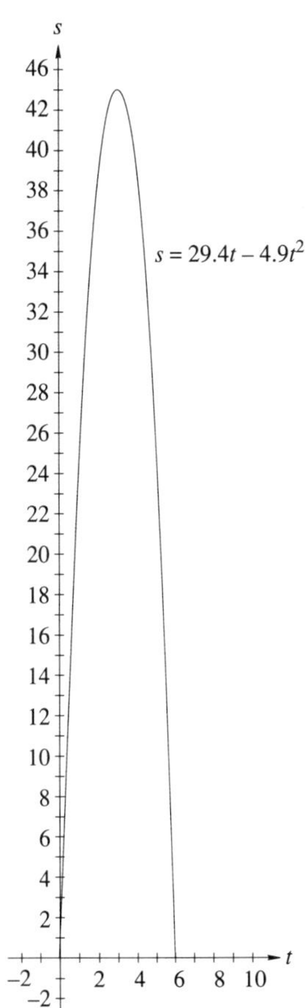

24. $2a = 90 + 140 + 2 \cdot 4000 = 8230$ so $a = 4115$ and $c = \frac{140-90}{2} = \frac{50}{2} = 25$. Hence $b^2 = a^2 - c^2 = 4115^2 -$

$25^2 = 16932600$ and so $\dfrac{x^2}{4115^2} + \dfrac{y^2}{16\,932\,600} = 1$ or $\dfrac{x^2}{4115^2} + \dfrac{y^2}{4114.924^2} = 1$.

25. Since the signal from A arrives $600\mu s$ after B the commmon difference on the hyperbola is $300 \times 600 = 180{,}000$ m $= 180$ km so $2a = 180$ hence $a = 90, b^2 = c^2 - a^2 = 250000 - 8100 = 241{,}900$ The equation is $\frac{x^2}{8100} - \frac{y^2}{241{,}900} = 1$. If the transverse axis passes through A and B and the conjugate axis passes through the midpoint of $\overline{AB}$, then A is at $(-500, 0)$ and B at $(500, 0)$. $AP + BP = 300 \times 8{,}000 = 2{,}400{,}000$ m $= 2{,}400$ km; $AP+BP = 2{,}400$; $AP - BP = 180$; $AB = 1000$; $2AP = 2580$; $AP = 1290$; $BP = 1110$; $\alpha = \cos^{-1}\left(\frac{1000^2+1290^2-1110^2}{2\cdot 1000\cdot 1290}\right) = 56.2866°$; $y = AP \sin\alpha = 1073.054$ km ; $x = AP\cos\alpha - 500 = 216$ km. The plane is at $(216, 1073.054)$.

26. Here $a = 18.09$ A.U.; $b = 4.56$ A.U. Combining these produces $c^2 = a^2 - b^2$, or $c = \sqrt{18.09^2 - 4.56^2} = 17.51$ A.U. The distance from a focus to the center is $c = 17.51$ A.U., so if the sun is at one focus and has coordinates $(0, 0)$, then the center is at $(17.51, 0)$ and we have the ellipse: $\dfrac{(x - 17.51)^2}{18.09^2} + \dfrac{y^2}{4.56^2} = 1$. The maximum distance is $a + c = 18.09 + 17.51 = 35.60$ A.U. The minimum distance is $a - c = 0.58$ A.U.

27. (a) $x^2 + 200y - 4.00 = 0$ or $x^2 = -200y + 4.00$ and so $x^2 = -200(y - 0.02)$

(b) In the no load position $y - 0.02 = 0$ or $y = 0.02$ m $= 2$ cm

15.7 ROTATION OF AXES; THE GENERAL SECOND-DEGREE EQUATION

1. (a) $xy = -9$. Here, $A = 0$, $B = 1$, and $C = 0$, and the discriminant is $B^2 - 4AC = 1 > 0$; so the curve is a hyperbola, (b) $\cot 2\theta = \frac{A-C}{B} = 0$ or $2\theta = 90°$ so the angle needed to rotate the coordinate axes to remove the xy-term is $\theta = 45°$, (c) The rotation formulas are $x = x''\cos\theta - y''\sin\theta$ and $y = y''\cos\theta + x''\sin\theta$ we get $x = 0.7071x'' - 0.7071y''$ and $y = 0.7071y'' + 0.7071x''$, (d) Substituting the equations from (c) into the original equation $xy = -9$ we get $(0.7071x'' - 0.7071y'')(0.7071y'' + 0.7071x'') = -9$ or $\frac{1}{2}x''^2 - \frac{1}{2}y''^2 = -9$ which is equivalent to $\dfrac{y''^2}{18} - \dfrac{x''^2}{18} = 1$.

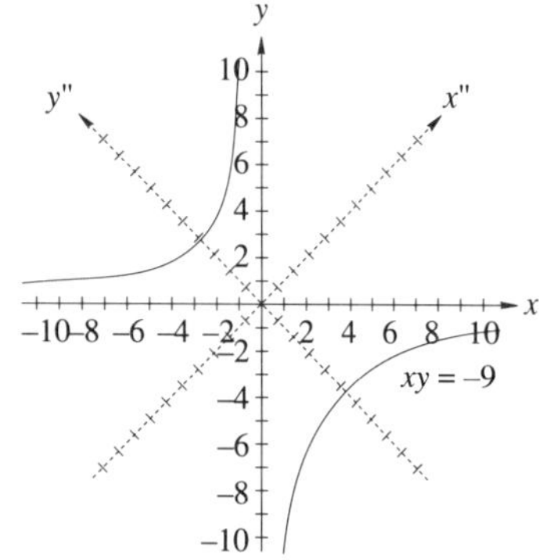

2. (a) $xy = 9$. Here $A = 0$, $B = 1$, and $C = 0$, so the discriminant is $B^2 = 4AC = 1 > 0$, so the curve is a hyperbola, (b) $\cot 2\theta = \frac{A-C}{B} = \frac{0}{1} = 0 \Rightarrow 2\theta = 90°$ so the angle needed to rotate the coordinate axes to remove the xy-term is $\theta = 45°$, (c) The rotation formulas are $x = x''\cos 45° - y''\sin 45°$ and $y = y''\cos 45° + x''\sin 45°$, (d) Substituting the equations from (c) into the original equation $xy = 9$ we get $\left(\frac{\sqrt{2}}{2}x'' - \frac{\sqrt{2}}{2}y''\right)\left(\frac{\sqrt{2}}{2}y'' + \frac{\sqrt{2}}{2}x''\right) = 9 \Rightarrow \frac{1}{2}x''^2 - \frac{1}{2}y''^2 = 9$ which is equivalent to $\dfrac{x''^2}{18} - \dfrac{y''^2}{18} = 1$.

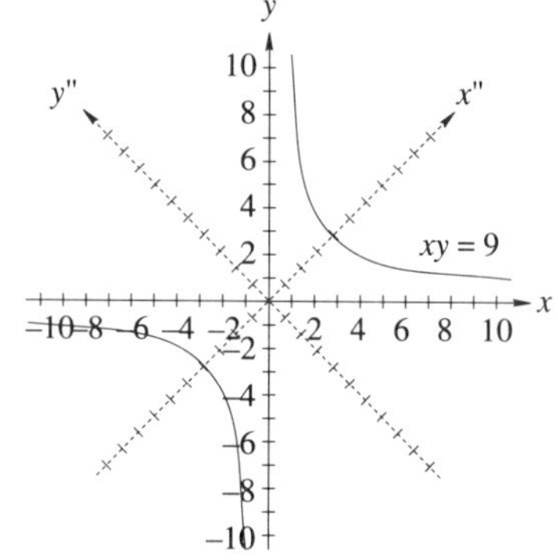

3. (a) $x^2 - 6xy + y^2 - 8 = 0$. Here, $A = 1$, $B = -6$, and $C = 1$. The discriminant is $B^2 - 4AC = 36 - 4 = 32 > 0$, so the curve is a hyperbola (b) $\cot 2\theta = \frac{A-C}{B} = \frac{0}{-6}$ so $2\theta = 90°$ and so the angle needed to rotate the coordinate axes to remove the xy-term is $\theta = 45°$, (c) The rotation formulas are $x = x'' \cos 45° - y'' \sin 45°$, $y = y'' \cos 45° + x'' \sin 45°$, (d) Substituting the equations from (c) into the original equation $x^2 - 6xy + y^2 - 8 = 0$ we get $\left(\frac{\sqrt{2}}{2}x'' - \frac{\sqrt{2}}{2}y''\right)^2 - 6\left(\frac{\sqrt{2}}{2}x'' - \frac{\sqrt{2}}{2}y''\right)\left(\frac{\sqrt{2}}{2}y'' + \frac{\sqrt{2}}{2}x''\right)+\left(\frac{\sqrt{2}}{2}y'' + \frac{\sqrt{2}}{2}x''\right)^2 - 8 = 0$; $\frac{1}{2}x''^2 - x''y'' + \frac{1}{2}y''^2 - 6(\frac{1}{2}x''^2 - \frac{1}{2}y''^2) + (\frac{1}{2}y''^2 + x''y'' + \frac{1}{2}x''^2) - 8 = 0$ or $-2x''^2 + 4y''^2 - 8 = 0$. This is equivalent to $\dfrac{y''^2}{2} - \dfrac{x''^2}{4} = 1$.

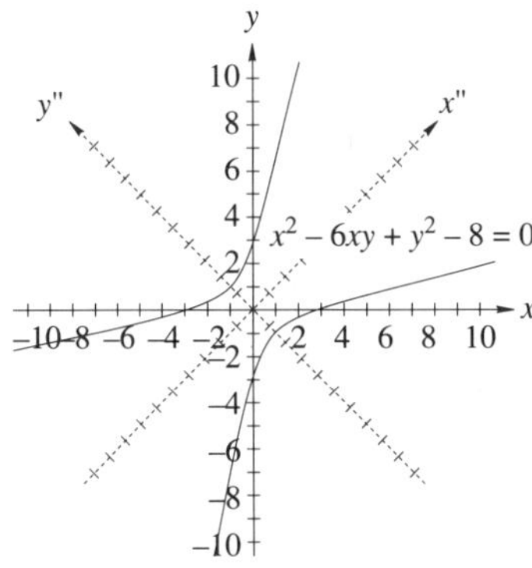

4. (a) $x^2 + 4xy - 2y^2 - 6 = 0$. Here $A = 1$, $B = 4$, and $C = -2$, so the discriminant is $B^2 - 4AC = 16 + 8 = 24 > 0$ and we see that this curve is a hyperbola, (b) $\cot 2\theta = \frac{A-C}{B} = \frac{1+2}{4} = \frac{3}{4}$, $2\theta = 53.13$ so the angle needed to rotate the coordinate axes to remove the xy-term is $\theta = 26.565°$, (c) The rotation formulas are $x = x'' \cos 26.565° - y'' \sin 26.565°$ or $x = 0.8944x'' - 0.4472y''$, and $y = y'' \cos 26.565° + x'' \sin 26.565°$ or $y = 0.8944y'' + 0.4472x''$ (d) Substituting the equations from (c) into the original equation $x^2 + 4xy - 2y^2 - 6 = 0$ we get $(0.8944x'' - 0.4472y'')^2 + 4(0.8944x'' - 0.4472y'')(0.8944y'' + 0.4472x'') - 2(0.8944y'' + 0.4472x'')^2 - 6 = 0$. Multiplying, we get $(0.8x''^2 - 0.8x''y'' + 0.2y'') + 4(0.4x''^2 + 0.6x''y'' - 0.4y'')^2 - 2(0.8y''^2 + 0.8x''y'' + 0.2x''^2) - 6 = 0$. Collecting like terms leads to $2x''^2 - 3y''^2 - 6 = 0$ which is equivalent to $\dfrac{x''^2}{3} - \dfrac{y''^2}{2} = 1$.

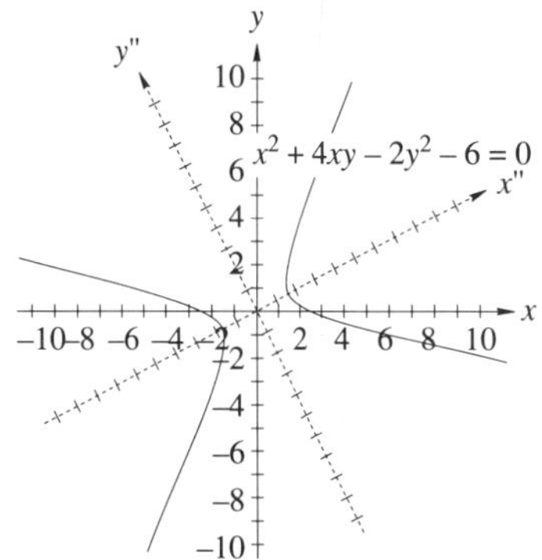

5. (a) $52x^2 - 72xy + 73y^2 - 100 = 0$. Here $A = 52$, $B = -72$, and $C = 73$, so the discriminant is $B^2 - 4AC = -10000 < 0$ so the curve is an ellipse, (b) $\cot 2\theta = \frac{A-C}{B} = \frac{52-73}{-72} = 0.29167$, $2\theta = 73.7398$ so the angle needed to rotate the coordinate axes to remove the xy-term is $\theta = 36.870$ with $\cos\theta = 0.8$ and $\sin\theta = 0.6$, (c) The rotation formulas are $x = 0.8x'' - 0.6y''$ and $y = 0.8y'' + 0.6x''$, (d) $52(0.8x'' - 0.6y'')^2 - 72(0.8x'' - 0.6y'')(0.8y'' + 0.6x'') + 73(0.8y'' + 0.6x'')^2 - 100 = 0$ which expands to $52(0.64x''^2 - 0.96x''y'' + 0.36y''^2) - 72(0.48x''^2 + 0.28x''y'' - 0.48y''^2) + 73(0.64y''^2 + 0.96x''y'' + 0.36x''^2) - 100 = 0$ or $25x''^2 + 100y''^2 = 100$ which is equivalent to $\dfrac{x''^2}{4} + y''^2 = 1$.

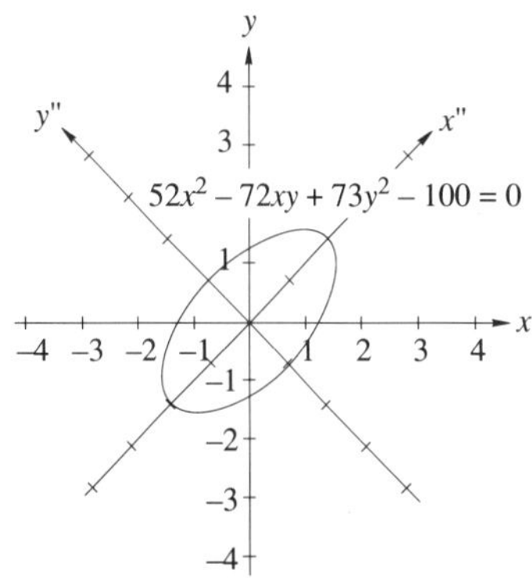

6. (a) $11x^2 + 10\sqrt{3}xy + y^2 - 4 = 0$. Here $A = 11$, $B = 10\sqrt{3}$, and $C = 1$, so the discriminant is

$B^2 - 4AC = 300 - 44 = 256 > 0$ so the curve is a hyperbola, (b) $\cot 2\theta = \frac{A-C}{B} = \frac{11-1}{10\sqrt{3}} = \frac{1}{\sqrt{3}}$, $2\theta = 60°$ so the angle needed to rotate the coordinate axes to remove the xy-term is $\theta = 30°$; so, $\cos\theta = \frac{\sqrt{3}}{2} \approx 0.866$ and $\sin\theta = 0.5$. (c) The rotation formulas are $x = \frac{\sqrt{3}}{2}x'' - \frac{1}{2}y''$ and $y = \frac{\sqrt{3}}{2}y'' + \frac{1}{2}x''$, (d) Substituting the equations from (c) into the original equation $11x^2 + 10\sqrt{3}xy + y^2 - 4 = 0$ we get $11\left(\frac{\sqrt{3}}{2}x'' - \frac{1}{2}y''\right)^2 + 10\sqrt{3}\left(\frac{\sqrt{3}}{2}x'' - \frac{1}{2}y''\right)\left(\frac{\sqrt{3}}{2}y'' + \frac{1}{2}x''\right) + \left(\frac{\sqrt{3}}{2}y''^2 + \frac{1}{2}x''\right)^2 - 4 = 0$; $11\left(\frac{3}{4}x''^2 - \frac{\sqrt{3}}{2}x''y'' + \frac{1}{4}y''^2\right) + 10\sqrt{3}\left(\frac{\sqrt{3}}{4}x''^2 + \frac{1}{2}x''y'' - \frac{\sqrt{3}}{4}y''\right) + \left(\frac{3}{4}y'' + \frac{\sqrt{3}}{2}x''y'' + \frac{1}{4}x''^2\right) - 4 = 0$. Collecting terms results in $\frac{33}{4}x''^2 - \frac{11\sqrt{3}}{2}x''y'' + \frac{11}{4}y''^2 + \frac{30}{4}x''^2 + \frac{10\sqrt{3}}{2}x''y'' - \frac{30}{4}y''^2 + \frac{3}{4}y''^2 + \frac{\sqrt{3}}{2}x''y'' + \frac{1}{4}x''^2 - 4 = 0$; $16x''^2 - 4y''^2 = 4$ which is equivalent to the standard form equation $4x''^2 - y''^2 = 1$ or $\frac{x''^2}{1/4} - \frac{y''^2}{1} = 1$.

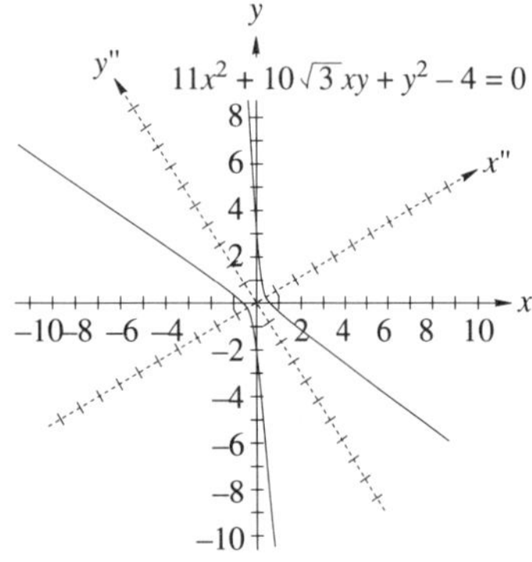

7. (a) $x^2 - 2xy + y^2 + x + y = 0$. The discriminant is $B^2 - 4AC = (-2)^2 - 4\cdot 1\cdot 1 = 4 - 4 = 0$ so the curve is a parabola, (b) $\cot 2\theta = \frac{A-C}{B} = \frac{1-1}{-2} = \frac{0}{-2} = 0$ so $2\theta = 90°$ and so $\theta = 45°$, (c) The rotation formulas are $x = \frac{\sqrt{2}}{2}x'' - \frac{\sqrt{2}}{2}y''$ and $y = \frac{\sqrt{2}}{2}y'' + \frac{\sqrt{2}}{2}x''$, (d) Substituting the equations from (c) into the original equation $x^2 - 2xy + y^2 + x + y = 0$ we get $(\frac{\sqrt{2}}{2}x'' - \frac{\sqrt{2}}{2}y'')^2 - 2(\frac{\sqrt{2}}{2}x'' - \frac{\sqrt{2}}{2}y'')(\frac{\sqrt{2}}{2}y'' + \frac{\sqrt{2}}{2}x'') + (\frac{\sqrt{2}}{2}y'' + \frac{\sqrt{2}}{2}x'')^2 + (\frac{\sqrt{2}}{2}x'' - \frac{\sqrt{2}}{2}y'') + (\frac{\sqrt{2}}{2}y'' + \frac{\sqrt{2}}{2}x'') = 0$; $(\frac{1}{2}x''^2 - x''y'' + \frac{1}{2}y''^2) - 2(\frac{1}{2}x''^2 - \frac{1}{2}y''^2) + (\frac{1}{2}y''^2 + x''y'' + \frac{1}{2}x''^2) + \sqrt{2}x'' = 0$; $2y''^2 + \sqrt{2}x'' = 0$ or $2y''^2 = -\sqrt{2}x''$ which is equivalent to $y''^2 = \frac{-\sqrt{2}}{2}x''$.

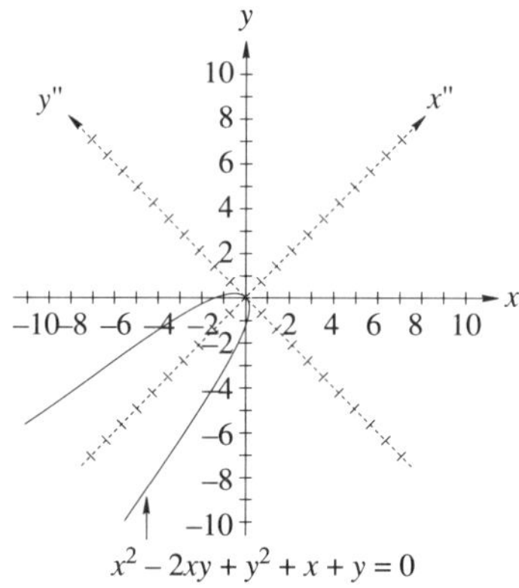

8. (a) $3x^2 + 2\sqrt{3}xy + y^2 - 2x + 2\sqrt{3}y = 0$. The discriminant is $B^2 - 4AC = 12 - 12 = 0$ so the curve is a parabola, (b) $\cot 2\theta = \frac{A-C}{B} = \frac{3-1}{2\sqrt{3}} = \frac{2}{2\sqrt{3}} = \frac{1}{\sqrt{3}}$ so $2\theta = 60°$ and so $\theta = 30°$, (c) The rotation formulas are $x = \frac{\sqrt{3}}{2}x'' - \frac{1}{2}y''$ and $y = \frac{\sqrt{3}}{2}y'' + \frac{1}{2}x''$, (d) Substituting the equations from (c) into the original equation $3x^2 + 2\sqrt{3}xy + y^2 - 2x + 2\sqrt{3}y = 0$ we get $3(\frac{\sqrt{3}}{2}x'' - \frac{1}{2}y'')^2 + 2\sqrt{3}(\frac{\sqrt{3}}{2}x'' - \frac{1}{2}y'')(\frac{\sqrt{3}}{2}y'' + \frac{1}{2}x'') + (\frac{\sqrt{3}}{2}y'' + \frac{1}{2}x'')^2 - 2(\frac{\sqrt{3}}{2}x'' - \frac{1}{2}y'') + 2\sqrt{3}(\frac{\sqrt{3}}{2}y'' + \frac{1}{2}x'') = 0$; $3(\frac{3}{4}x''^2 - \frac{\sqrt{3}}{2}x''y'' + \frac{1}{4}y''^2) + 2\sqrt{3}(\frac{\sqrt{3}}{4}x''^2 + \frac{1}{2}x''y'' - \frac{\sqrt{3}}{4}y''^2) + (\frac{3}{4}y''^2 + \frac{\sqrt{3}}{2}x''y'' + \frac{1}{4}x''^2) - 2(\frac{\sqrt{3}}{2}x'' - \frac{1}{2}y'') + 2\sqrt{3}(\frac{\sqrt{3}}{2}y'' + \frac{1}{2}x'') = 0$; $\frac{9}{4}x''^2 - \frac{3\sqrt{3}}{2}x''y'' + \frac{3}{4}y''^2 + \frac{6}{4}x''^2 + \frac{2\sqrt{3}}{2}x''y'' - \frac{6}{4}y''^2 + \frac{3}{4}y''^2 + \frac{\sqrt{3}}{2}x''y'' + \frac{1}{4}x''^2 - \sqrt{3}x'' + y'' + 3y'' + \sqrt{3}x'' = 0$; $4x''^2 + 4y'' = 0$ or $x''^2 = -y''$.

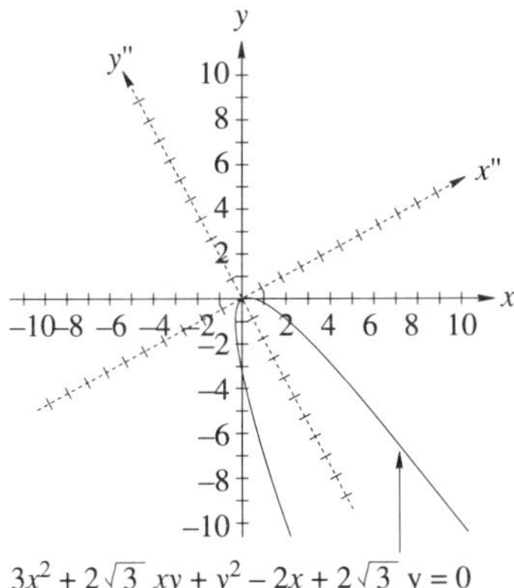

9. (a) $2x^2 + 12xy - 3y^2 - 42 = 0$. The discriminant is $B^2 - 4AC = 144 + 24 = 168 > 0$ so the curve is a hyperbola, (b) $\cot 2\theta = \frac{A-C}{B} = \frac{2+3}{12} = \frac{5}{12}$, $2\theta = 67.380$ so the angle needed to rotate the coordinate axes to remove the xy-term is $\theta = 33.690$; $\cos\theta = 0.8321$, $\sin\theta = 0.5547$, (c) The rotation formulas are $x = 0.8321x'' - 0.5547y''$ and $y = 0.8321y'' + 0.5547x''$, (d) Substituting the equations from (c) into the original equation $2x^2 + 12xy - 3y^2 - 42 = 0$ we get $2(0.8321x'' - 0.5547y'')^2 + 12(0.8321x'' - 0.5547y'')(0.8321y'' + 0.5547x'') - 3(0.8321y'' + 0.5547x'')^2 - 42 = 0$ or $2(0.6923x''^2 - 0.9231x''y'' + 0.3077y''^2) + 12(0.4615x''^2 + 0.3846x''y'' - 0.4615y''^2) - 3(0.6923y''^2 + 0.9231x''y'' + 0.3077x''^2) - 42 = 0$; $1.3846x''^2 - 1.8462x''y'' + 0.6154y''^2 + 5.538x''^2 + 4.6152x''y'' - 5.538y''^2 - 2.0769y''^2 - 2.7693x''y'' - .9231x''^2 - 42 = 0$; $6x''^2 - 7y''^2 = 42$ which is equivalent to $\frac{x''^2}{7} - \frac{y''^2}{6} = 1$.

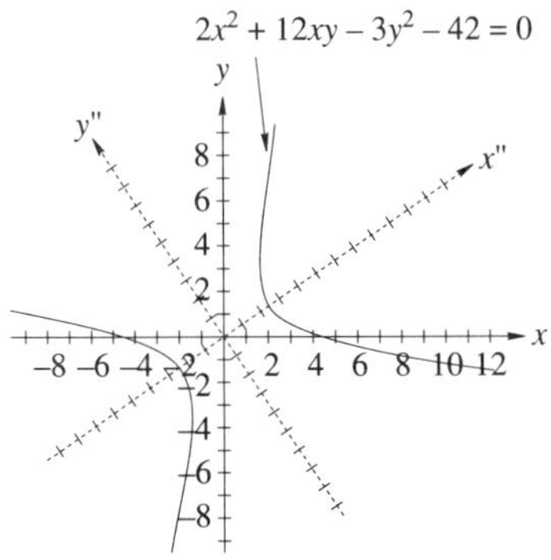

10. (a) $7x^2 - 20xy - 8y^2 + 52 = 0$. The discriminant is $B^2 - 4AC = 400 + 224 = 624 > 0$ so the curve is a hyperbola, (b) $\cot 2\theta = \frac{A-C}{B} = \frac{7+8}{-20} = \frac{15}{-20} = \frac{-3}{4}$, $2\theta = 126.8699$ so the angle needed to rotate the coordinate axes to remove the xy-term is $\theta = 63.435°$; $\cos\theta = 0.4472$, $\sin\theta = 0.8944$, (c) The rotation formulas are $x = 0.4472x'' - 0.8944y''$ and $y = 0.4472y'' + 0.8944x''$, (d) Substituting the equations from (c) into the original equation $7x^2 - 20xy - 8y^2 + 52 = 0$ produces $7(0.4472x'' - 0.8944y'')^2 - 20(0.4472x'' - 0.8944y'')(0.4472y'' + 0.8944x'') - 8(0.4472y'' + 0.8944x'')^2 + 52 = 0$; $7(0.2x''^2 - 0.8x''y'' + 0.8y''^2) - 20(0.4x''^2 - 0.6x''y'' - 0.4y''^2) - 8(0.2y''^2 + 0.8x''y'' + 0.8x''^2) + 52 = 0$ or $1.4x''^2 - 5.6x''y'' + 5.6y''^2 - 8x''^2 + 12x''y'' + 8y''^2 - 1.6y''^2 - 6.4x''y'' - 6.4x''^2 + 52 = 0$; $-13x''^2 + 12y''^2 + 52 = 0$ which is equivalent to $\frac{x''^2}{4} - \frac{y''^2}{4.33} = 1$.

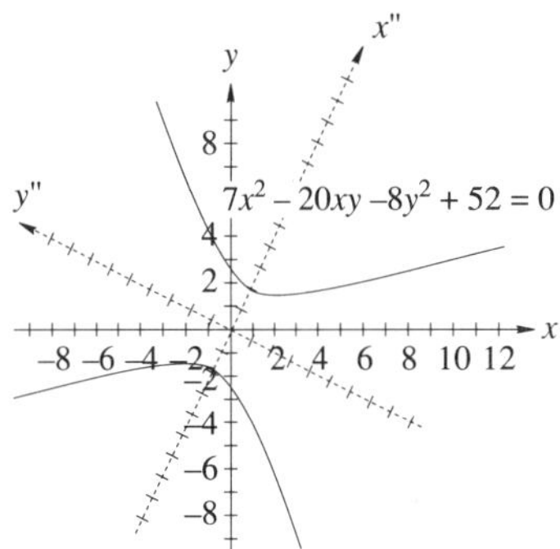

11. (a) $6x^2 - 5xy + 6y^2 - 26 = 0$. The discriminant is $B^2 - 4AC = 25 - 144 = -119$ so the curve is an ellipse, (b) $\cot 2\theta = \frac{A-C}{B} = \frac{6-6}{-5} = 0$ so $2\theta = 90°$ and we see that $\theta = 45°$, (c) The rotation formulas are $x = \frac{\sqrt{2}}{2}x'' - \frac{\sqrt{2}}{2}y''$ and $y = \frac{\sqrt{2}}{2}y'' + \frac{\sqrt{2}}{2}x''$, (d) Substituting the equations from (c) into the original equation $6x^2 - 5xy + 6y^2 - 26 = 0$ yields $6(\frac{\sqrt{2}}{2}x'' - \frac{\sqrt{2}}{2}y'')^2 - 5(\frac{\sqrt{2}}{2}x'' - \frac{\sqrt{2}}{2}y'')(\frac{\sqrt{2}}{2}y'' + \frac{\sqrt{2}}{2}x'') + 6(\frac{\sqrt{2}}{2}y'' + \frac{\sqrt{2}}{2}x'')^2 - 26 = 0$ or $6(\frac{1}{2}x''^2 - x''y'' + \frac{1}{2}y''^2) - 5(\frac{1}{2}x''^2 - \frac{1}{2}y''^2) + 6(\frac{1}{2}y''^2 + x''y'' + \frac{1}{2}x''^2) - 26 = 0$; $3x''^2 - 6x''y'' + 3y''^2 - \frac{5}{2}x''^2 + \frac{5}{2}y''^2 + 3y''^2 + 6x''y'' + 3x''^2 - 26 = 0$ which simplifies to $\frac{7}{2}x''^2 + \frac{17}{2}y''^2 = 26$ or $7x''^2 + 17y''^2 = 52$ or the standard form equation $\frac{x^2}{\frac{52}{7}} + \frac{y''^2}{\frac{52}{17}} = 1$. Note:

$a = \sqrt{\frac{52}{7}} \approx 2.726, b = \sqrt{\frac{52}{17}} \approx 1.749.$

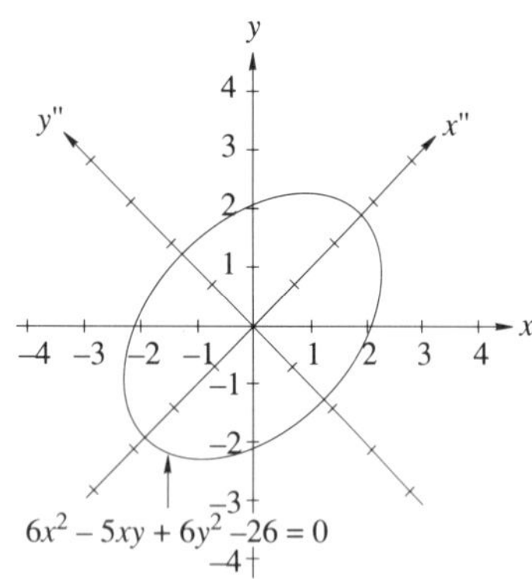

12. (a) $9x^2 - 6xy + y^2 - 12\sqrt{10}x - 36\sqrt{10}y = 0$. The discriminant is $B^2 - 4AC = 36 - 36 = 0$ so the curve is a parabola, (b) $\cot 2\theta = \frac{A-C}{B} = \frac{9-1}{-6} = \frac{8}{-6} = \frac{-4}{3}$, $2\theta = 143.1301$ and so the angle needed to rotate the coordinate axes to remove the xy-term is $\theta = 71.565$. Thus, we see that $\cos\theta = 0.3162$, $\cos^2\theta = 0.1$, $\sin\theta = 0.9487$, $\sin^2\theta = 0.9$, and $\sin\theta\cos\theta = 0.3$, (c) The rotation formulas are $x = 0.3162x'' - 0.9487y''$ and $y = 0.3162y'' + 0.9487x''$, (d) Substituting the equations from (c) into the original equation $9x^2 - 6xy + y^2 - 12\sqrt{10}x - 36\sqrt{10}y = 0$ yields $9(0.3162x'' - 0.9487y'')^2 - 6(0.3162x'' - 0.9487y'')(0.3162y'' + 0.9487x'') + (0.3162y'' + 0.9487x'')^2 - 12\sqrt{10}(0.3162x'' - 0.9487y'') - 36\sqrt{10}(0.3162y'' + 0.9487x'') = 0$; $9(0.1x''^2 - 0.6x''y'' + 0.9y''^2) - 6(0.3x''^2 - 0.8x''y'' - 0.3y''^2) + (0.1y''^2 + 0.6x''y'' + 0.9x''^2) - 12\sqrt{10}(0.3162x'' - 0.9487y'') - 36\sqrt{10}(0.3162y'' + 0.9487x'') = 0$ which simplifies to $10y''^2 - 120x'' = 0$ which is equivalent to $y''^2 = 12x''$.

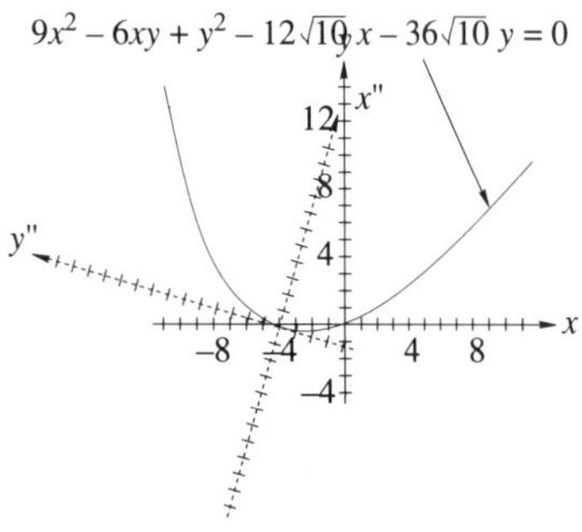

13. $k = PV = 15 \times 400 = 6000$, $PV = 6000$

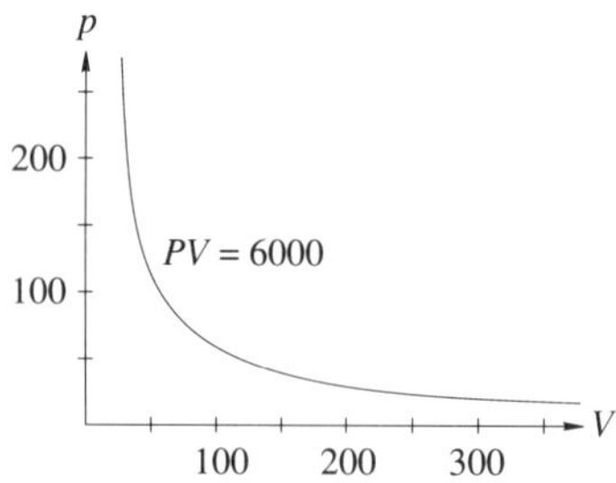

14. If w = contant 280, $X_c = \frac{1}{280C} or X_cC = \frac{1}{280}$

This is an equilateral hyperbola

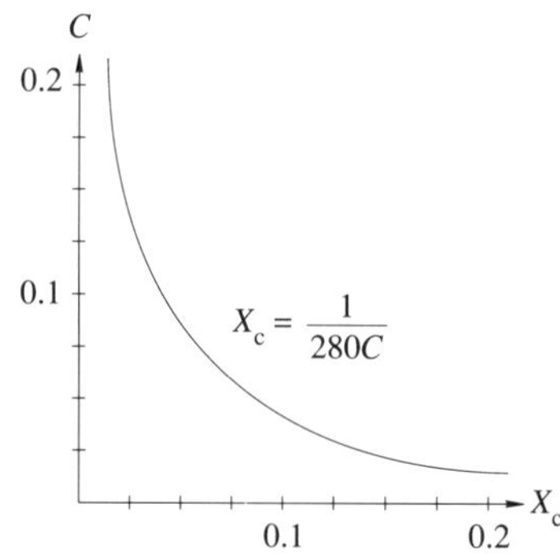

15. $(\overline{v_1} - \overline{v_2})^2 = 19.62h_L$ is a parabola; If $h_L < 5$ then $(\overline{v_1} - \overline{v_2}) < 9.9$

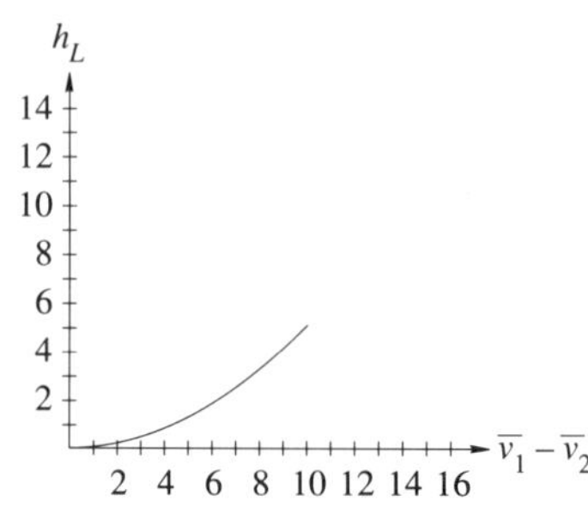

16. $(3x)^2 + y^2 = (x+10)^2$ or $9x^2 + y^2 = x^2 + 20x + 100$; $8x^2 - 20x + y^2 = 100$; $8(x^2 - 2.5x + 1.5625) + y^2 = 100 + 12.5$; $8(x - 1.25)^2 + y^2 = 112.5$; $\frac{(x-1.25)^2}{14.0625} + \frac{y^2}{112.5} = 1$; $a = 3.75$, $b = 10.61$ This is an ellipse with center at $(1.25, 0)$.

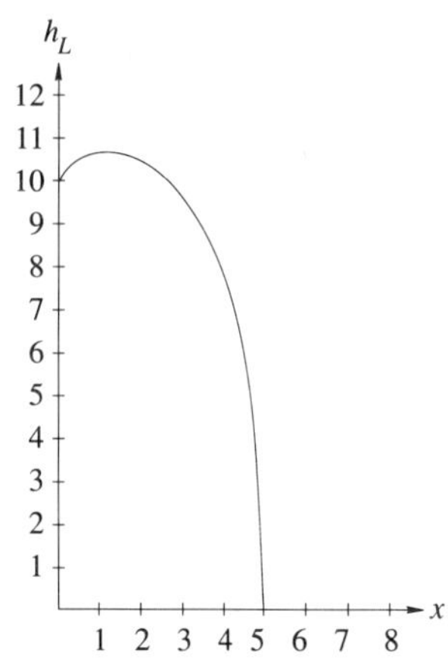

15.8 CONIC SECTIONS IN POLAR COORDINATES

1. $r = \dfrac{6}{1 + 3\cos\theta}$, $e = 3 > 1$, so this is a hyperbola.

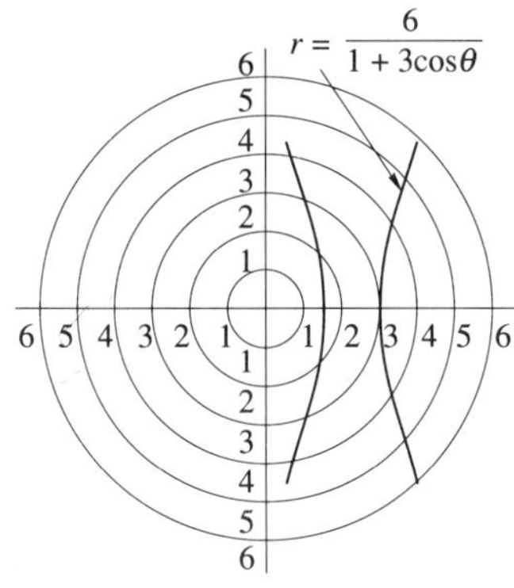

2. $r = \dfrac{8}{1 - 2\sin\theta}$, $e = 2$, so this is a hyperbola.

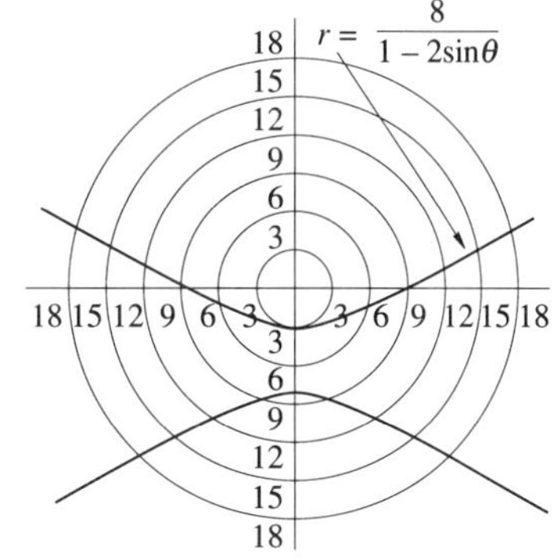

3. $r = \dfrac{12}{3 - \cos\theta} = \dfrac{4}{1 - \frac{1}{3}\cos\theta}$, $e = \frac{1}{3}$, so this is an ellipse.

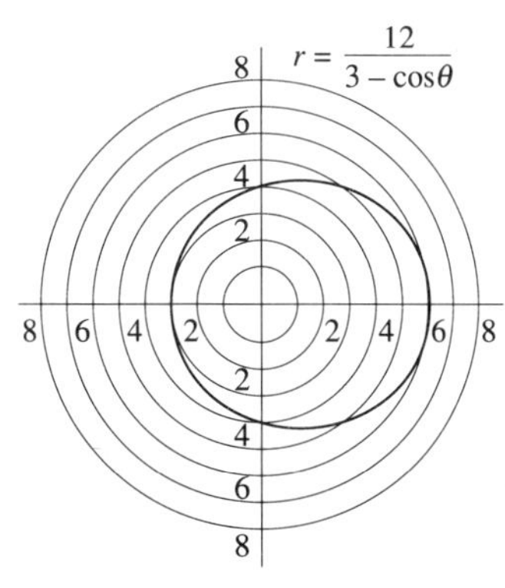

4. $r = \dfrac{15}{3 + 9\sin\theta} = \dfrac{5}{1 + 3\sin\theta}$, $e = 3$, so this is a hyperbola.

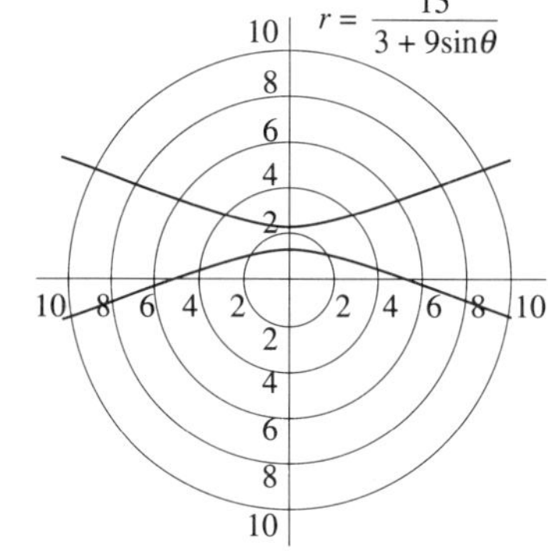

5. $r = \dfrac{12}{4 - 4\cos\theta} = \dfrac{3}{1 - 1\cos\theta}$, $e = 1$, so this is a parabola.

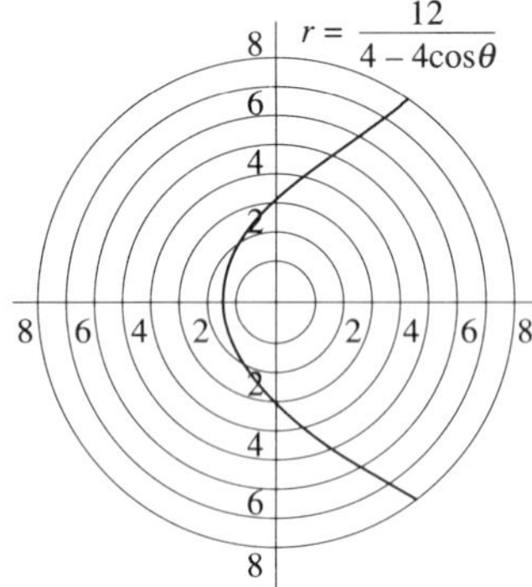

6. $r = \dfrac{8}{6 + 6\sin\theta} = \dfrac{\frac{4}{3}}{1 + \sin\theta}$, $e = 1$, so this is a parabola.

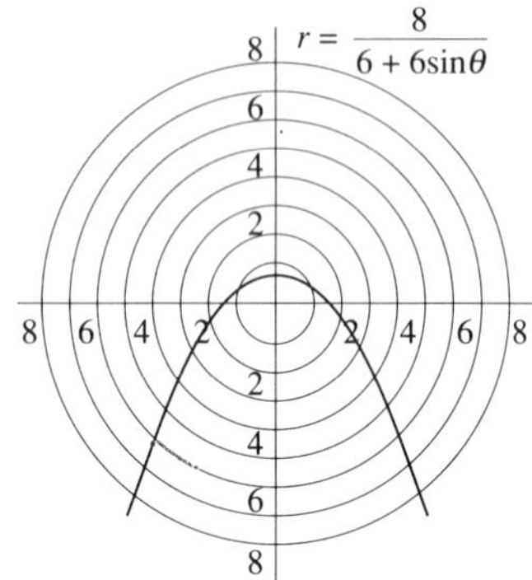

7. $r = \dfrac{12}{3 + 2\sin\theta} = \dfrac{4}{1 + \frac{2}{3}\cos\theta}$, $e = \frac{2}{3}$, so this is an ellipse.

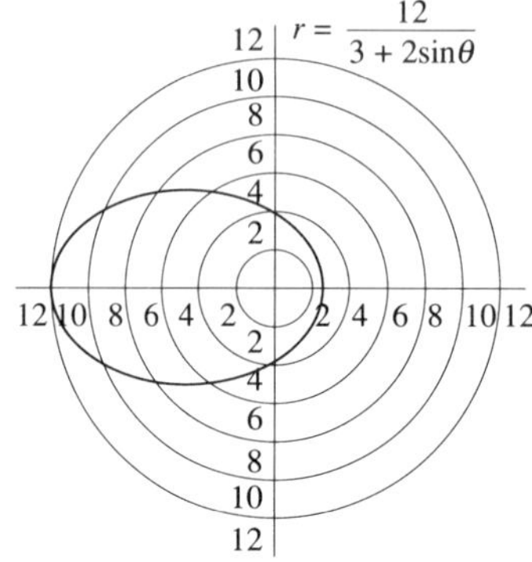

8. $r = \dfrac{12}{2 - 3\cos\theta} = \dfrac{6}{1 - \frac{3}{2}\cos\theta}$, $e = \frac{3}{2}$, so this is a hyperbola.

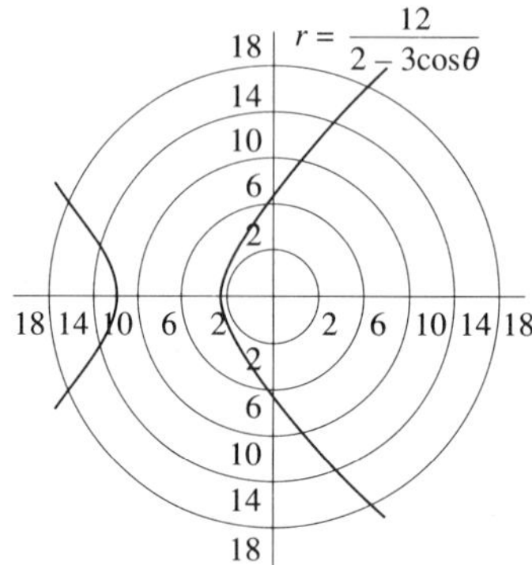

9. $r = \dfrac{12}{2 - 4\cos\left(\theta + \frac{\pi}{3}\right)} = \dfrac{6}{1 - 2\cos\left(\theta + \frac{\pi}{3}\right)}$, $e = 2$, so this is a hyperbola rotated $\frac{\pi}{3}$ counterclockwise.

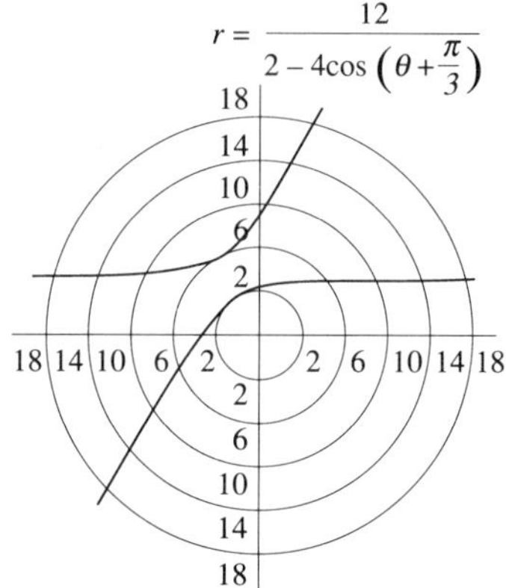

10. $r = \dfrac{6}{2 + \sin\left(\theta + \frac{\pi}{6}\right)} = \dfrac{3}{1 + \frac{1}{2}\sin\left(\pi + \frac{\pi}{6}\right)}$, $e = \frac{1}{2}$, so this is an ellipse rotated $\frac{\pi}{6}$ counterclockwise.

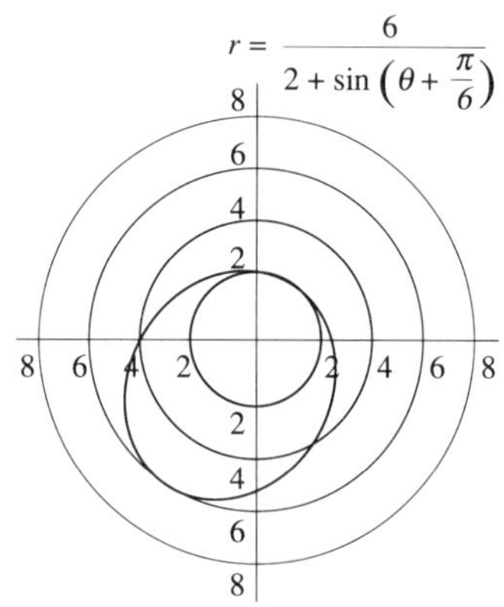

11. $r = \dfrac{pe}{1 - e\cos\theta} = \dfrac{4 \cdot \frac{3}{2}}{1 - \frac{3}{2}\cos\theta} = \dfrac{12}{2 - 3\cos\theta}$, so this is a hyperbola.

12. $r = \dfrac{2 \cdot \frac{3}{4}}{1 + \frac{3}{4}\cos\theta} = \dfrac{6}{4 + 3\cos\theta}$, so this is an ellipse.

13. $r = \dfrac{pe}{1 - e\sin\theta} = \dfrac{5}{1 - \sin\theta}$, so this is a parabola.

14. $r = \dfrac{3 \cdot 2}{1 + 2\sin\theta} = \dfrac{6}{1 + 2\sin\theta}$, so this is a hyperbola.

15. $r = \dfrac{1 \cdot \frac{2}{3}}{1 - \frac{2}{3}\cos\theta} = \dfrac{2}{3 - 2\cos\theta}$, so this is an ellipse.

16. $r = \dfrac{1 \cdot 5}{1 - \cos\theta} = \dfrac{5}{1 - \cos\theta}$, so this is a parabola.

17. Greatest distance is when $\theta = 0$, so $r = 4.335 \times 10^7$. The shortest distance is when $\theta = \pi$ so $r = 2.854 \times 10^7$.

18.

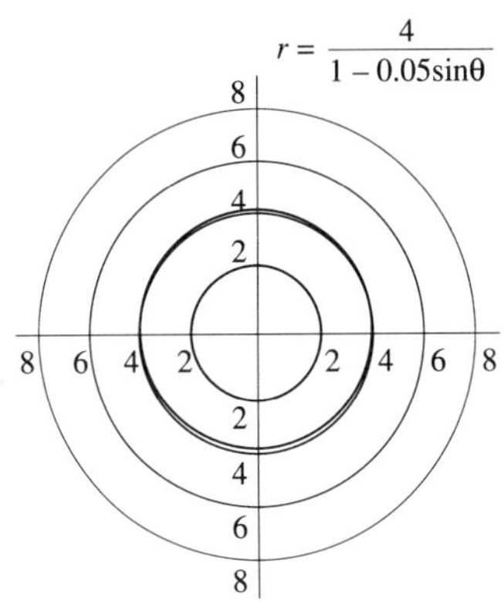

19.

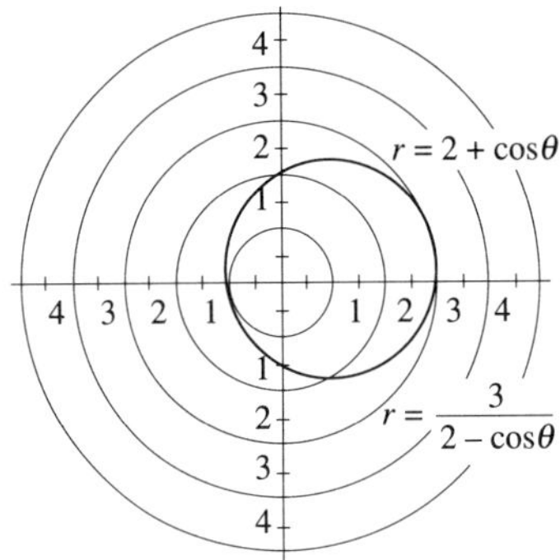

CHAPTER 15 REVIEW

1. (a) $d = \sqrt{(2+1)^2 + (5-9)^2} = \sqrt{3^2 + (-4)^2} = 5$
(b) midpoint $= \left(\frac{2-1}{2}, \frac{5+9}{2}\right) = \left(\frac{1}{2}, 7\right)$
(c) $m = \frac{9-5}{-1-2} = \frac{4}{-3} = -\frac{4}{3}$ (d) $y - 5 = -\frac{4}{3}(x-2)$ or $3y - 15 = -4x + 8$ or $4x + 3y = 23$

2. (a) $d = \sqrt{12^2 + 5^2} = 13$ (b) midpoint $= \left(\frac{8}{2}, \frac{-15}{2}\right) = \left(4, \frac{-15}{2}\right)$ (c) $m = \frac{-10+5}{10+2} = \frac{-5}{12}$
(d) $y + 5 = -\frac{5}{12}(x+2)$ or $12y + 60 = -5x - 10$ or $5x + 12y = -70$

3. (a) $d = \sqrt{2^2 + 10^2} = \sqrt{104} = 2\sqrt{26}$ (b) midpoint $= \left(\frac{4}{2}, \frac{2}{2}\right) = (2, 1)$ (c) $m = \frac{6+4}{3-1} = \frac{10}{2} = 5$
(d) $y + 4 = 5(x-1)$ or $y + 4 = 5x - 5$ or $5x - y = 9$ or $y = 5x - 9$ or $y - 5x = -9$

4. (a) $d = \sqrt{8^2 + 8^2} = 8\sqrt{2}$ (b) midpoint $= \left(\frac{2-6}{2}, \frac{-5+3}{2}\right) = \left(\frac{-4}{2}, \frac{-2}{2}\right) = (-2, -1)$
(c) $m = \frac{8}{-8} = -1$ (d) $y + 5 = -1(x-2)$ or $y + 5 = -x + 2$ or $x + y = -3$

5. #1 : $\frac{3}{4}$; #2 : $\frac{12}{5}$; #3 : $-\frac{1}{5}$; #4 : 1

6. #1: $y - 7 = \frac{3}{4}\left(x - \frac{1}{2}\right)$ or $4y - 28 = 3x - \frac{3}{2}$ or $8y - 56 = 6x - 3$ or $8y - 6x = 53$;
#2: $y + \frac{15}{2} = \frac{12}{5}(x-4)$ or $5y + \frac{75}{2} = 12x - 48$ or $10y + 75 = 24x - 96$ or $10y - 24x = -171$;
#3: $y - 1 = -\frac{1}{5}(x-2)$ or $5y - 5 = -x + 2$ or $x + 5y = 7$;
#4: $y + 1 = 1(x+2)$ or $y + 1 = x + 2$ or $y = x + 1$ or $y - x = 1$ or $x - y = -1$

7. The slope of $2y + 4x = 9$ is obtained by solving for y: This gives $y = -2x + \frac{9}{2}$ so the slope is -2.
$y - 5 = -2(x+3)$ or $y - 5 = -2x - 6$ or $y + 2x = -1$

8. $y + 7 = 4(x - 2)$ or $y + 7 = 4x - 8$ or
$y - 4x = -15$ or $4x - y = 15$ or $y = 4x - 15$

9. $x^2 + y^2 = 16$: circle with center at $(0, 0)$ and $r = 4$

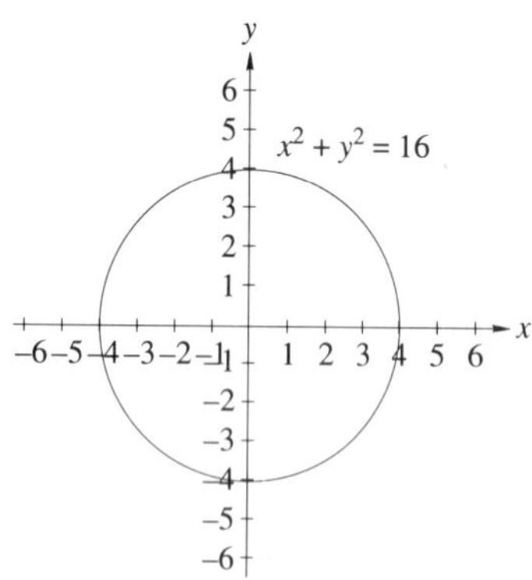

10. $x^2 - y^2 = 16 \Rightarrow \frac{x^2}{16} - \frac{y^2}{16} = 1$, hyperbola, center at $(0, 0)$, $a = 4$, $b = 4$, $c = 4\sqrt{2}$

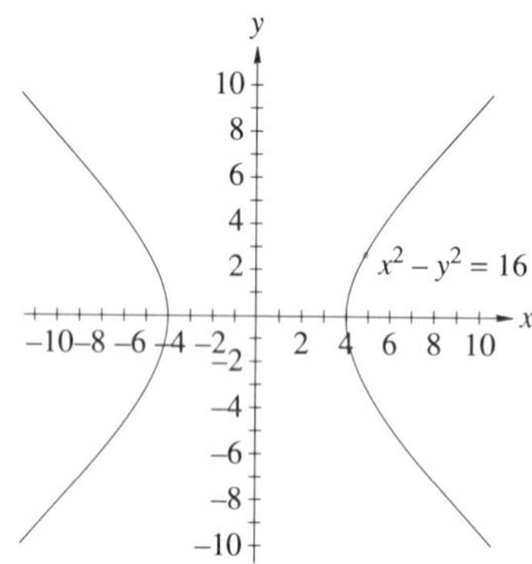

11. $x^2 + 4y^2 = 16 \Rightarrow \frac{x^2}{16} + \frac{y^2}{4} = 1$, ellipse, center at $(0, 0)$, $a = 4$, $b = 2$, $c = \sqrt{12} = 2\sqrt{3}$

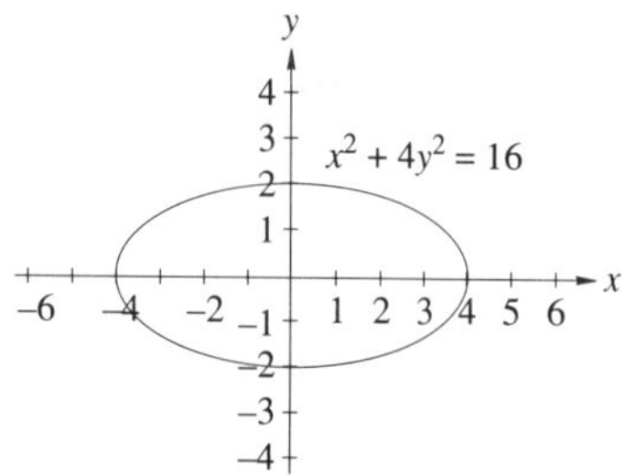

12. $y^2 = 16x$, parabola, vertex at $(0, 0)$ focus at $(4, 0)$, derictrix $x = -4$

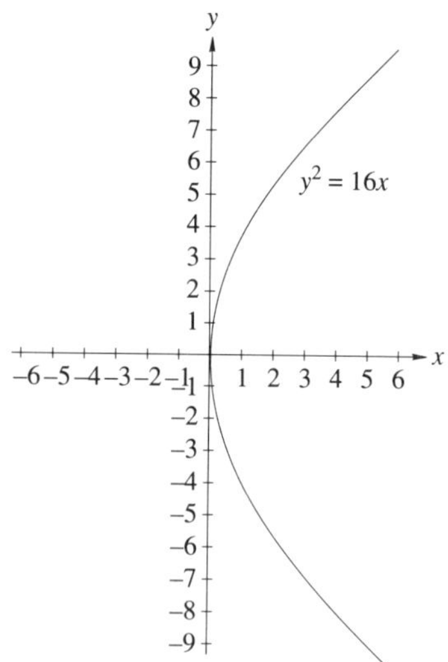

13. $(x - 2)^2 + (y + 4)^2 = 16$, circle center $(2, -4)$, radius 4

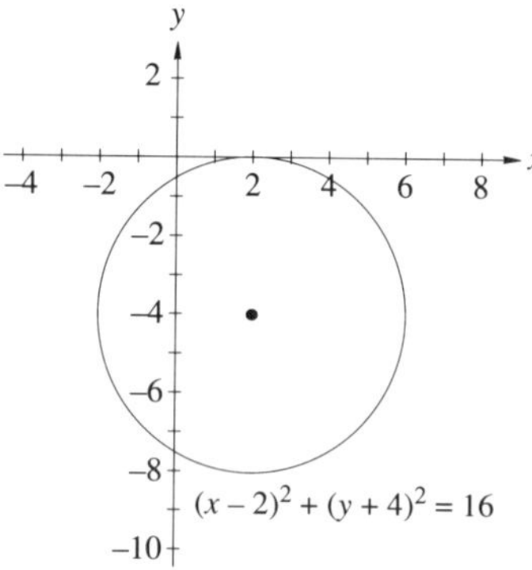

14. $(x - 2)^2 - (y + 4)^2 = 16 \Rightarrow \frac{(x-2)^2}{16} - \frac{(y+4)^2}{16} = 1$
hyperbola, center at $(2, -4)$, $a = 4$, $b = 4$, $c = 4\sqrt{2}$

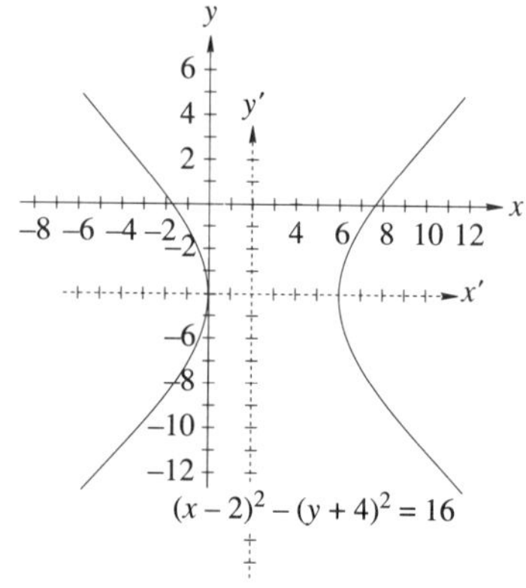

15. $(x-2)^2+4(y+4)^2=16 \Rightarrow \frac{(x-2)^2}{16}+\frac{(y+4)^2}{4}=1$
ellipse, center $(2,-4)$, $a=4$, $b=2$,
$c=\sqrt{12}=2\sqrt{3}$

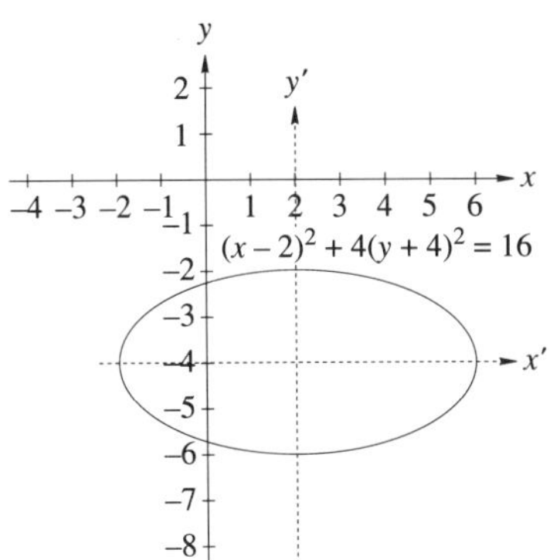

16. $(y+4)^2=16(x-2)$, parabola, vertex $(2,-4)$
opens right, focus at $(6,-4)$

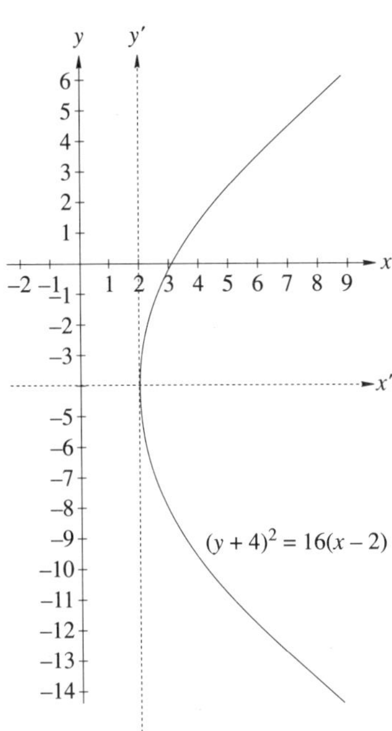

17. $x^2+y^2+6x-10y+18=0$,
$x^2+6x+9+y^2-10y+25=-18+9+25$,
$(x+3)^3+(y-5)^2=16$ circle, center $(-3,5)$
radius 4

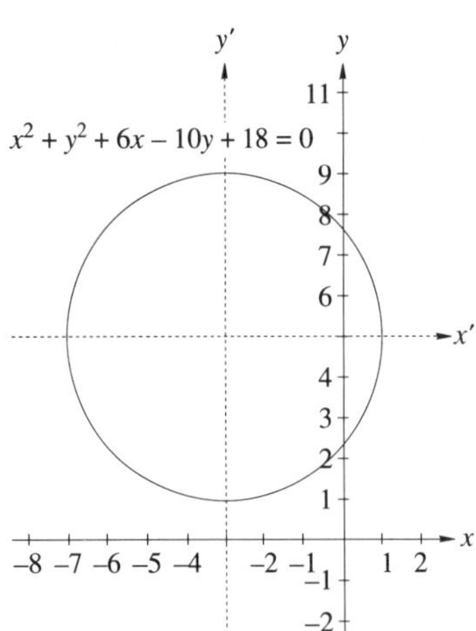

18. $x^2-4y^2+6x+40y-107=0$,
$x^2+6x-4(y^2-10y)=107$,
$(x^2+6x+9)-4(y^2-10y+25)=107+9-4\cdot 25$,
$(x+3)^2-4(y-5)^2=16$, $\frac{(x+3)^2}{16}-\frac{(y-5)^2}{4}=1$
Hyperbola, center $(-3,5)$, $a=4$, $b=2$,
$c=\sqrt{20}=2\sqrt{5}$

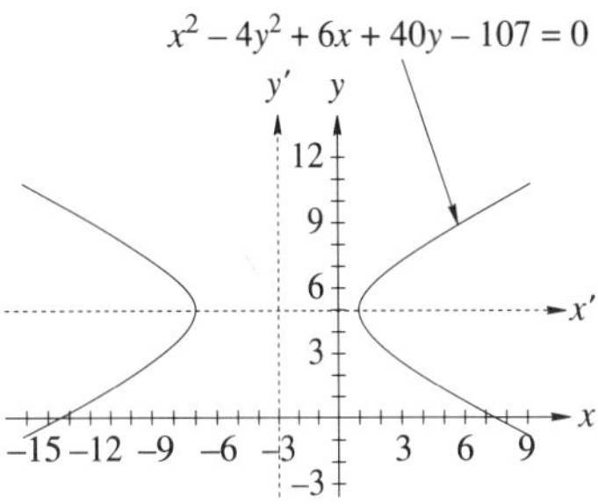

19. $x^2+4y^2+6x-40y+93=0$,
$(x^2+6x)+4(y^2-10y)=-93$,
$(x^2+6x+9)+4(y^2-10y+25)=-93+9+4\cdot 25$,
$(x+3)^2+4(y-5)^2=16$, $\frac{(x+3)^2}{16}+\frac{(y-5)^2}{4}=1$,
ellipse, center $(-3,5)$, $a=4$, $b=2$,
$c=\sqrt{12}=2\sqrt{3}$

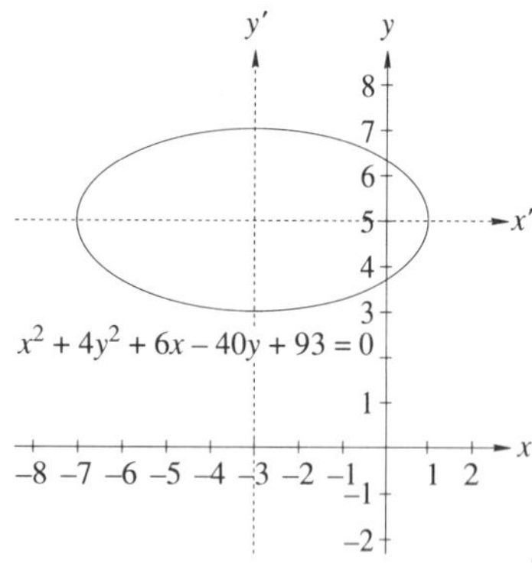

20. $x^2 + 6x - 4y + 29 = 0$, $x^2 + 6x = 4y - 29$, $(x^2 + 6x + 9) = 4y - 29 + 9$, $(x + 3)^2 = 4y - 20$, $(x + 3)^2 = 4(y - 5)$ parabola, vertex $(-3, 5)$, $p = 1$, opens upward, Focus $(-3, 6)$, directrix, $y = 4$

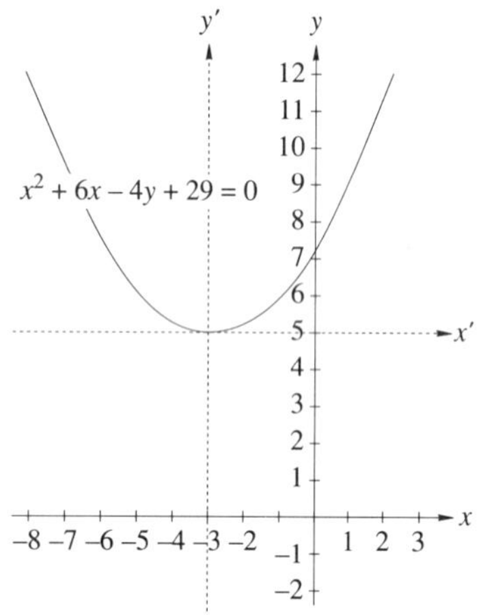

21. $2x^2 + 12xy - 3y^2 - 42 = 0$. The discriminant is $B^2 - 4AC = 144 + 24 = 168$ so the curve is a hyperbola. Thus, $\cot 2\theta = \frac{A-C}{B} = \frac{2+3}{12} = \frac{5}{12}$, and so $2\theta = 67.3801°$. Hence, the angle of rotation is $\theta = 33.69°$ and we obtain $\cos\theta = 0.83205$, $\cos^2\theta = 0.6923$, $\sin\theta = 0.5547$, $\sin^2\theta = 0.3077$, and $\sin\theta\cos\theta = 0.4615$. The rotation equations are $x = 0.83205x'' - 0.5547y''$ and $y = 0.83205y'' + 0.5547x''$. When these are substituted into the given equation, we obtain $2(0.83205x'' - 0.5547y'')^2 + 12(0.83205x'' - 0.5547y'') \times (0.83205y'' + 0.5547x'') - 3(0.83205y'' + 0.5547x'')^2 - 42 = 0$ or $2(0.6923x''^2 - 0.9230x''y'' + 0.3077y''^2) + 12(0.4615x''^2 + 0.3846x''y'' - 0.4615y''^2) - 3(0.6923y''^2 + 0.9230x''y'' + 0.3077x''^2) - 42 = 0$. This simplifies to $6x''^2 - 7y''^2 = 42$ or the standard form equation $\frac{x''^2}{7} - \frac{y''^2}{6} = 1$ with $a = \sqrt{7}$, $b = \sqrt{6}$, and $c = \sqrt{13}$.

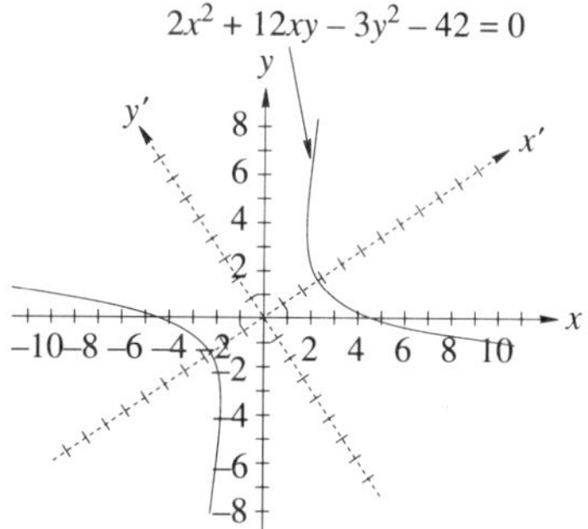

22. $5x^2 - 4xy + 8y^2 - 36 = 0$. The discriminant is $B^2 - 4AC = 16 - 160 = -144$ so the curve is an ellipse. Thus, $\cot 2\theta = \frac{A-C}{B} = \frac{5-8}{-4} = \frac{-3}{-4} = \frac{3}{4}$, so $2\theta = 53.1301$, and the angle of rotation is $\theta = 26.565$. As a result, we have $\cos\theta = 0.8944$, $\cos^2\theta = 0.8$, $\sin\theta = 0.4472$, $\sin^2\theta = 0.2$, and $\cos\theta\sin\theta = 0.4$. The rotation equations are $x = 0.8944x'' - 0.4472y''$ and $y = 0.8944y'' + 0.4472x''$. When these are substituted into the given equation, we obtain $5(0.8944x'' - 0.4472y'')^2 - 4(0.8944x'' - 0.4472y'')(0.8944y'' + 0.4472x'') + 8(0.8944y'' + 0.4472x'')^2 - 36 = 0$ or $5(0.8x''^2 - 0.8x''y'' + 0.2y''^2) - 4(0.4x''^2 + 0.6x''y'' - 0.4y''^2) + 8(0.8y''^2 + 0.8x''y'' + 0.2x''^2) - 36 = 0$. This simplifies to $4x''^2 + 9y''^2 = 36$ which has the standard form equation $\frac{x''^2}{9} + \frac{y''^2}{4} = 1$ with $a = 3$, $b = 2$, and $c = \sqrt{5}$.

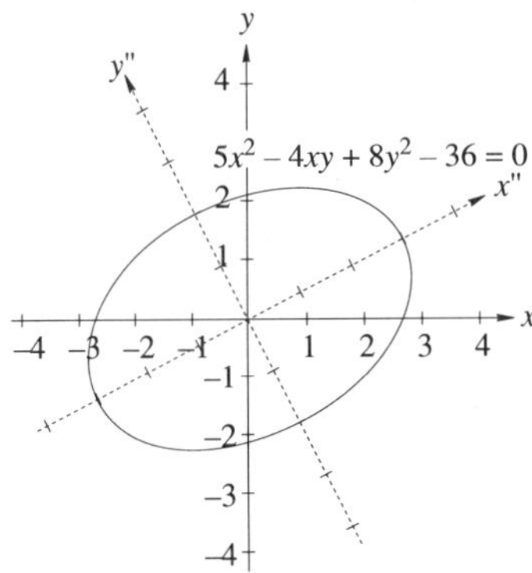

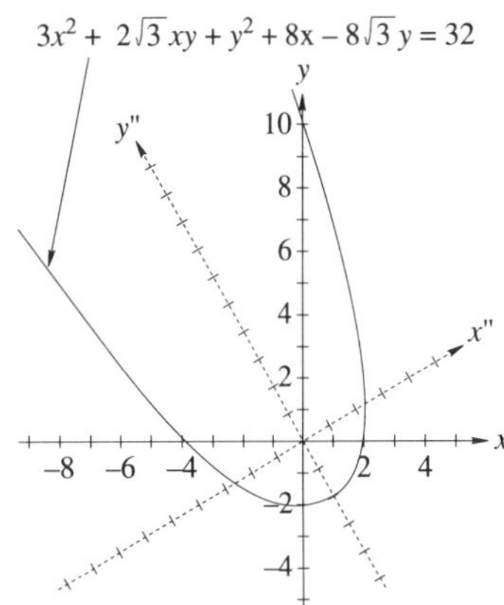

23. $3x^2 + 2\sqrt{3}xy + y^2 + 8x - 8\sqrt{3}y = 32$. The discriminant is $B^2 - 4AC = 12 - 12 = 0$ so the curve is a parabola. Thus, $\cot 2\theta = \frac{3-1}{2\sqrt{3}} = \frac{2}{2\sqrt{3}} = \frac{1}{\sqrt{3}}$ and so $2\theta = 60°$, and the angle of rotation is $\theta = 30°$. As a result, we have $\cos 30° = \frac{\sqrt{3}}{2}$, $\cos^2 30° = \frac{3}{4}$, $\sin 30° = \frac{1}{2}$, and $\sin^2 30° = \frac{1}{4}$. The rotation equations are $x = \frac{\sqrt{3}}{2}x'' - \frac{1}{2}y''$ and $y = \frac{\sqrt{3}}{2}y'' + \frac{1}{2}x''$. When these are substituted into the given equation, we obtain $3(\frac{\sqrt{3}}{2}x'' - \frac{1}{2}y'')^2 + 2\sqrt{3}(\frac{\sqrt{3}}{2}x'' - \frac{1}{2}y'')(\frac{\sqrt{3}}{2}y'' + \frac{1}{2}x'') + (\frac{\sqrt{3}}{2}y'' + \frac{1}{2}x'')^2 + 8(\frac{\sqrt{3}}{2}x'' - \frac{1}{2}y'') - 8\sqrt{3}(\frac{\sqrt{3}}{2}y'' + \frac{1}{2}x'') = 32$ or $3(\frac{3}{4}x''^2 - \frac{\sqrt{3}}{2}x''y'' + \frac{1}{4}y''^2) + 2\sqrt{3}(\frac{\sqrt{3}}{4}x''^2 - \frac{\sqrt{3}}{4}y''^2 + \frac{1}{2}x''y'') + (\frac{3}{4}y''^2 + \frac{\sqrt{3}}{2}x''y'' + \frac{1}{4}x''^2) + 4\sqrt{3}x'' - 4y'' - 12y'' - 4\sqrt{3}x'' = 32$. This simplifies to $4x''^2 = 16y'' + 32$ which has the standard form equation for a parabola $x''^2 = 4(y'' + 2)$ with $p = 1$. The parabola opens upward and has its vertex at $(0'', -2'')$

24. $2x^2 - 4xy - y^2 = 6$. The discriminant is $B^2 - 4AC = 16 + 8 = 24 > 0$ so the curve is a hyperbola. Thus, $\cot 2\theta = \frac{A-C}{B} = \frac{2+1}{-4} = -\frac{3}{4}$ and so $2\theta = 126.8699$ and the angle of rotation is $\theta = 63.4349$. As a result, we have $\cos\theta = 0.447$, $\cos^2\theta = 0.2$, $\sin\theta = 0.8944$, $\sin^2\theta = 0.8$, $\cos\theta\sin\theta = 0.4$. The rotation equations are $x = x''\cos\theta - y''\sin\theta = 0.447x'' - 0.8944y''$ and $y = y''\cos\theta + x''\sin\theta = 0.8944y'' + 0.447x''$. When these are substituted into the given equation, we obtain $2(x''\cos\theta - y''\sin\theta)^2 - 4(x''\cos\theta - y''\sin\theta)(y''\cos\theta + x''\sin\theta) - (y''\cos\theta + x''\sin\theta)^2 = 6$ or $2(0.2x''^2 - 0.8x''y'' + 0.8y''^2) - 4(0.4x''^2 - 0.6x''y'' - 0.4y''^2) - (0.2y''^2 + 0.8x''y'' + 0.8x''^2) = 6$. This simplifies to $-2x''^2 + 3y''^2 = 6$ or the standard form equation $\frac{y''^2}{2} - \frac{x''^2}{3} = 1$ with $a = \sqrt{2}$, $b = \sqrt{3}$, and $c = \sqrt{5}$.

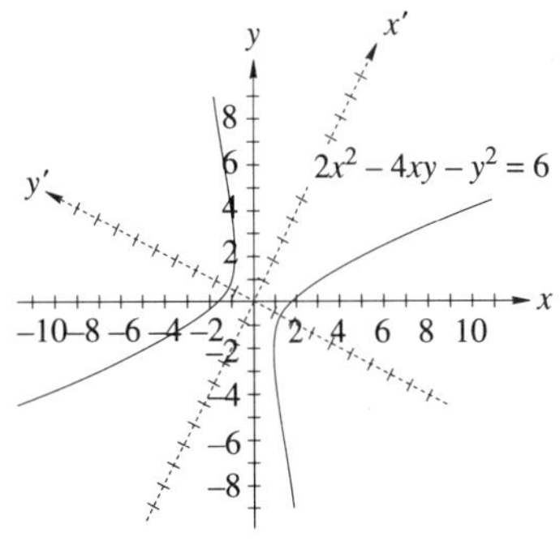

25. $r = \dfrac{16}{5 - 3\cos\theta} = \dfrac{\frac{16}{5}}{1 - \frac{3}{5}\cos\theta}$; $e = \frac{3}{5}$ so ellipse

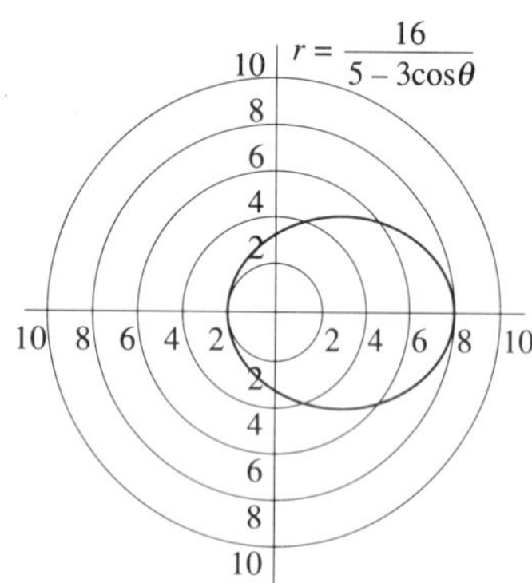

26. $r = \dfrac{9}{3 - 5\cos\theta} = \dfrac{3}{1 - \frac{5}{3}\cos\theta}$; $e = \frac{5}{3}$ so hyperbola

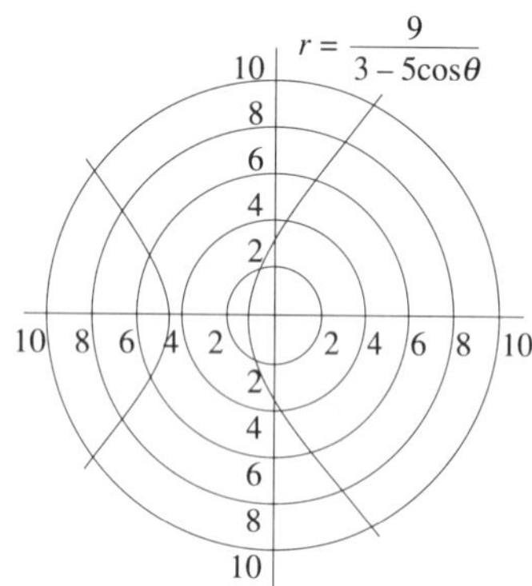

27. $r = \dfrac{9}{3 + 3\sin\theta} = \dfrac{3}{1 + \sin\theta}$; $e = 1$ so parabola

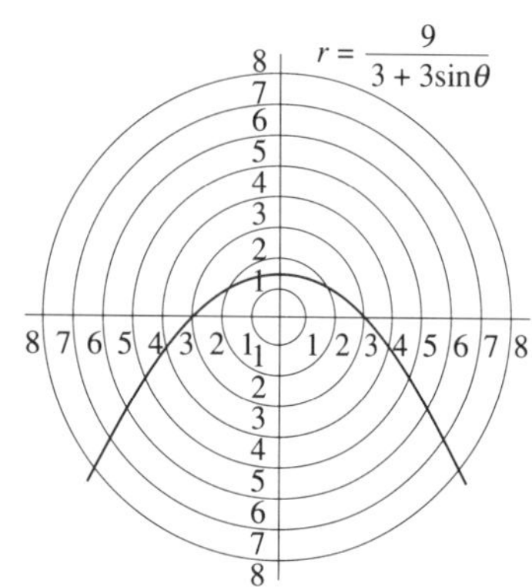

28. $r = \dfrac{2}{1 + \cos\theta}$; $e = 1$, so parabola

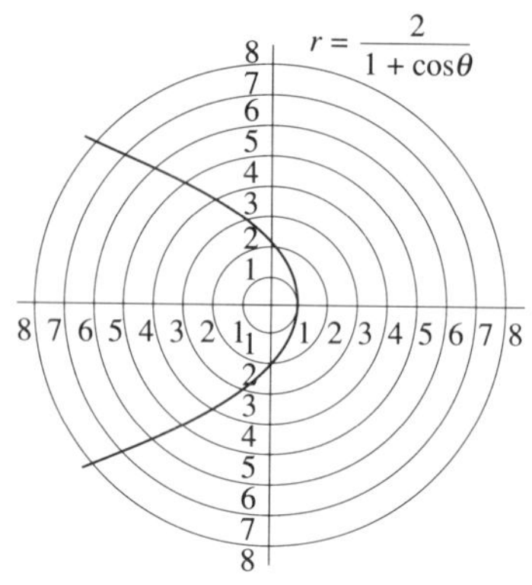

29. Since $2a = 100$, $a = 50$, this gives $\frac{x^2}{50^2} + \frac{y^2}{b^2} = 1$. The clearance constraint gives the point $(45, 6.1)$. Substituting and solving for b^2 results in $\dfrac{45^2}{50^2} + \dfrac{6.1^2}{b^2} = 1$ or $b^2 \cdot 45^2 + 6.1^2 \cdot 50^2 = 50^2 b^2$ or $50^2 b^2 - 45^2 b^2 = 50^2 \cdot 6.1^2$. As a result, $b^2 = \dfrac{50^2 \cdot 6.1^2}{50^2 - 45^2} = 195.84$ and we get the standard form equation $\dfrac{x^2}{2500} + \dfrac{y^2}{195.84} = 1$.

30. Taking the road to be the x-axis and the center of the road as the origin, the vertex of the parabola is at $(0, 10)$. This gives an equaton of the form $x^2 = 4p(y - 10)$ The top of the tower gives the point $(200, 120)$. Substituting and solving for $4p$ we obtain $4p = \dfrac{x^2}{y - 10} = \dfrac{200^2}{120 - 10} = \dfrac{40000}{110} = 363.64$. The equation is $x^2 = 363.64(y - 10)$

31. The quation is a hyperbola. The common difference is $AP - BP = 320 \times 500 = 160{,}000$ m $= 160$ km. Thus, $2a = 160$, $a = 80$, $2c = 400$, and so $c = 200$ and $b^2 = c^2 - a^2 = 33600$. The equation is $\dfrac{x^2}{6400} - \dfrac{y^2}{33600} = 1$. The y-coordiate for p is -50. Solving for x we get $\dfrac{x^2}{6400} = 1 + \dfrac{50^2}{33600} = 1.0744$

or $x^2 = 6876.19$ and so $x = 82.92$. The plane is at $(82.92, -50)$.

32. $x^2 + y^2 - 2x + 4y = 6361$, $x^2 - 2x + 1 + y^2 + 4y + 4 = 6361 + 1 + 4 = 6366$, $(x-1)^2 + (y+2)^2 = (79.7872)^2$. The radius is Earth's radius plus the number of units the satellite is orbiting above Earth. This is $79.7872 + 0.8 = 80.587$ km and so $r^2 = 6494.3$. Thus, we get the equation for a circle: $(x-1)^2 + (y+2)^2 = 6494.3$, $x^2 - 2x + 1 + y^2 + 4y + 4 = 6494.3$, $x^2 + y^2 - 2x + 4y - 6489.3 = 0$.

CHAPTER 15 TEST

1. Comparing the given equation with the basic equation $y^2 = 4px$, we see that $4p = -18$ and so $p = -\frac{18}{4} = -\frac{9}{2} = -4.5$. The focus of the parabola is at $(p, 0) = (-4.5, 0)$ and the directrix is the line $x = -p = 4.5$.

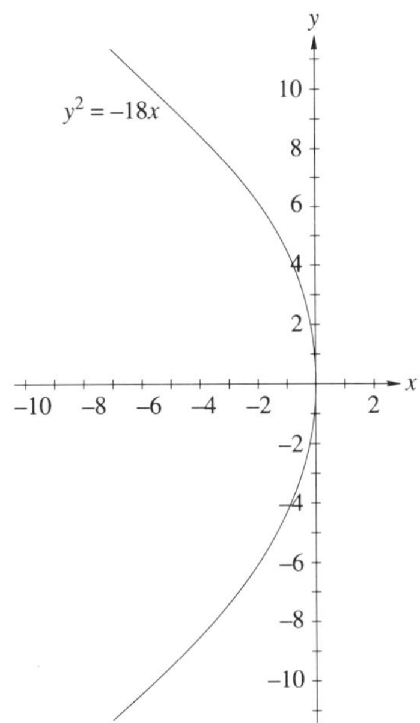

2. To write this in standard form, divide both sides by 36, obtaining $\frac{9x^2}{36} - \frac{4y^2}{36} = 1$ which simplifies to $\frac{x^2}{4} - \frac{y^2}{9} = 1$. From this we see that $a^2 = 4$ and $b^2 = 9$. Since $c^2 = a^2 + b^2 = 4 + 9 = 13$, we have $c = \sqrt{13}$. The foci are at $F' = (-c, 0) = (-\sqrt{13}, 0)$ and $F = (c, 0) = (\sqrt{13}, 0)$; the vertices are $V' = (-a, 0) = (-2, 0)$ and $V = (a, 0) = (2, 0)$; and the endpoints of the conjugate axis are $M' = (0, -b) = (0, -3)$ and $M = (0, b) = (0, 3)$.

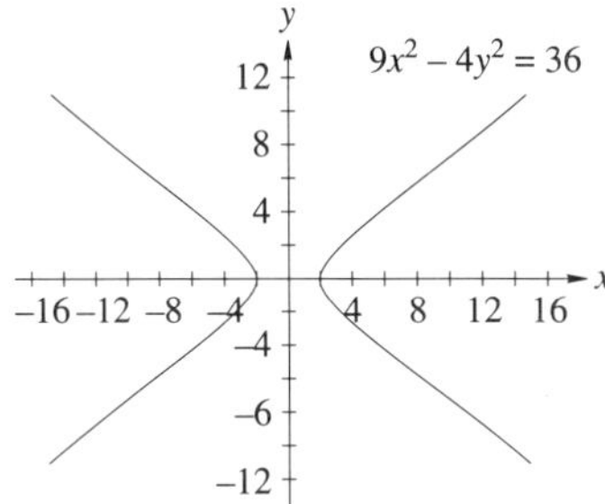

3. Using the distance formula for the distance between two points P_1 and P_2, we obtain $d(P_1, P_2) = \sqrt{(x_2 - x_1)^2 + (y_2 - y_1)^2} = \sqrt{(5 - (-2))^2 + (-6 - 4)^2} = \sqrt{7^2 + (-10)^2} = \sqrt{49 + 100} = \sqrt{149}$

4. The midpoint between these points is $\left(\frac{x_1+x_2}{2}, \frac{y_1+y_2}{2}\right) = \left(\frac{4+2}{2}, \frac{-5+(-8)}{2}\right) = \left(3, \frac{-13}{2}\right)$. The slope of the line through the two given points is $m = \frac{y_2-y_1}{x_2-x_1} = \frac{-8-(-5)}{2-4} = \frac{-3}{-2} = \frac{3}{2}$ and so the slope of a line perpendicular to this line is $-\frac{2}{3}$. Hence the desired line is $y - (\frac{-13}{2}) = -\frac{2}{3}(x - 3)$ or $y + \frac{13}{2} = -\frac{2}{3}(x - 3)$ or $6y + 4x + 27 = 0$.

5. A line with an angle of inclination 30° has a slope $m = \tan 30° = \frac{\sqrt{3}}{3}$. An x-intercept of 3 means it passes through $(3, 0)$. The equation has the point-slope form $y - 0 = \frac{\sqrt{3}}{3}(x - 3)$. Changing this to the slope-intercept form produces $y = \frac{\sqrt{3}}{3}x - \sqrt{3}$.

6. (a) Completing the square we obtain $(3x^2 - 6x) + (4y^2 + 16y) = -7$ or

$3(x^2 - 2x + 1) + 4(y^2 + 4y + 4) = -7 + 3 + 16$, which can be written as $3(x - 1)^2 + 4(y + 2)^2 = 12$. Dividing both sides by 12, we obtain a standard form equation for a horizontal ellipse $\frac{(x-1)^2}{4} + \frac{(y+2)^2}{3} = 1$.; (b) Ellipse; (c) center: $(1, -2)$, foci: $(0, -2)$, $(2, -2)$, vertices: $(-1, -2)$, $(3, -2)$, endpoints of minor axis: $(1, -2 - \sqrt{3})$, $(1, -2 + \sqrt{3})$

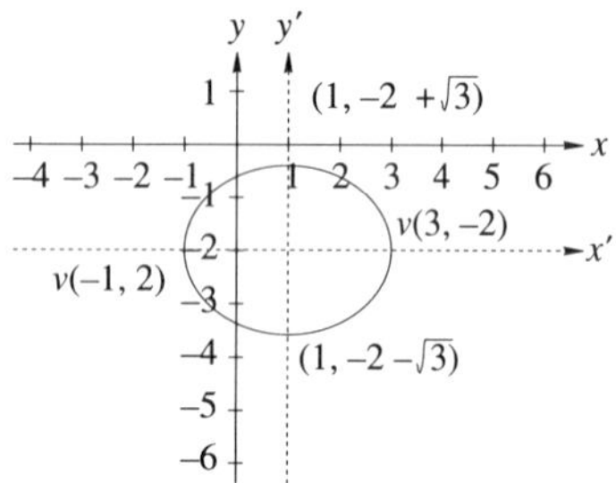

7. (a) The angle is determine by the equation $\cot 2\theta = \frac{A-C}{B}$. Here $A = 1$, $B = 1$, and $C = 1$, and so $\cot 2\theta = \frac{0}{1}$, which means that $\theta = \frac{\pi}{4}$. (b) Using $B^2 - 4AC = 1^2 - 4(1)(1) = -3$. Since this is negative, the conic is an ellipse.

8. The denominator has a constant term of 1 so $e = 4$. Since $e > 1$, the conic is a hyperbola

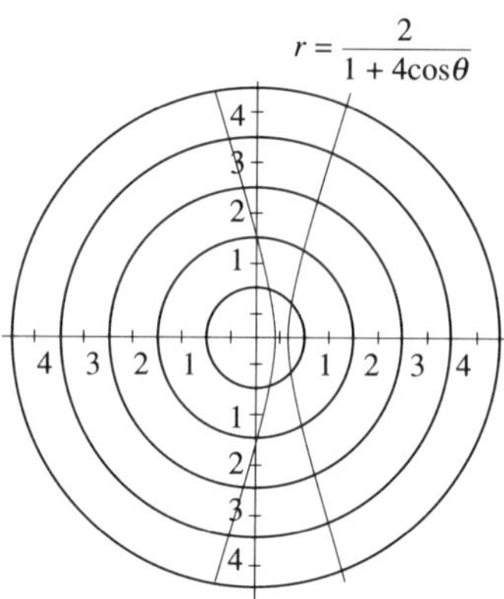

9. Consider this parabola to have its lowest point (vertex) at the origin and its axis the y-axis. Then the equation is of the form $x^2 = 4py$. Then the supports can be represented by the lines $x = -60$ and $x = 60$. The distance of 48 ft in from each support and 3 ft above the lowest point are the points $(-12, 3)$ and $(12, 3)$. Substituting the point $(12, 3)$ into the equation for the parabola we obtain $12^2 = 4p(3)$ and so $p = 12$. The equation is now $x^2 = 48y$. Substituting $x = 60$ in this equation and solving for y, we get $y = 75$.

CHAPTER

16

Systems of Equations and Inequalities

16.1 SOLUTIONS OF NONLINEAR SYSTEMS OF EQUATIONS

1. $\begin{cases} x - 2y = 5 & (1) \\ x^2 - 4y^2 = 45 & (2) \end{cases}$. Solve (1) for x: $x = 2y + 5$. Substitute this result into (2) to get $(2y+5)^2 - 4y^2 = 45$ or $4y^2 + 20y + 25 - 4y^2 = 45$ which simplifies to $20y = 20$ and so $y = 1$. Back substitutution produces $x = 2 \cdot 1 + 5 = 7$. The only solution is $(7, 1)$

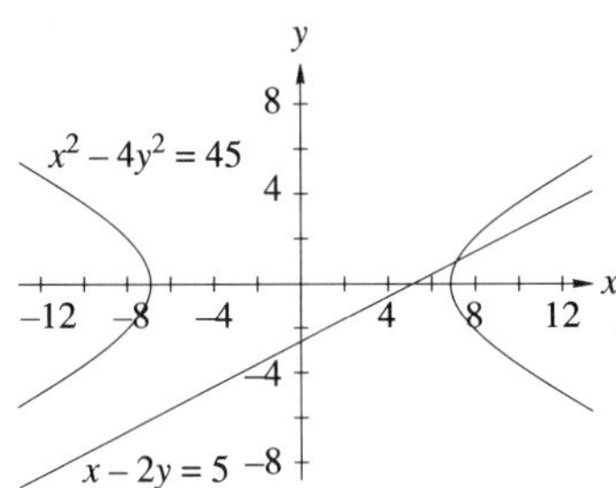

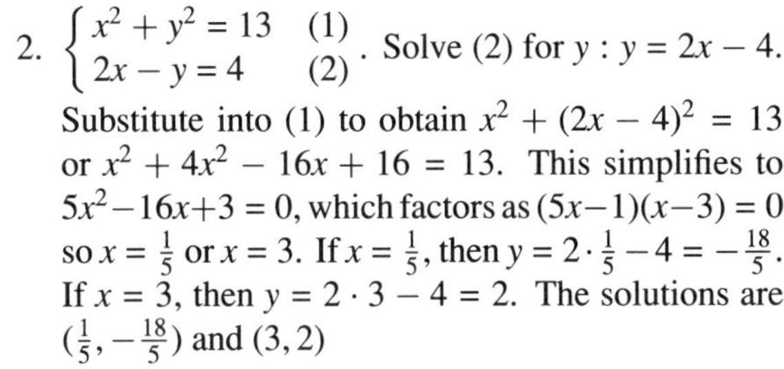

2. $\begin{cases} x^2 + y^2 = 13 & (1) \\ 2x - y = 4 & (2) \end{cases}$. Solve (2) for $y : y = 2x - 4$. Substitute into (1) to obtain $x^2 + (2x - 4)^2 = 13$ or $x^2 + 4x^2 - 16x + 16 = 13$. This simplifies to $5x^2 - 16x + 3 = 0$, which factors as $(5x-1)(x-3) = 0$ so $x = \frac{1}{5}$ or $x = 3$. If $x = \frac{1}{5}$, then $y = 2 \cdot \frac{1}{5} - 4 = -\frac{18}{5}$. If $x = 3$, then $y = 2 \cdot 3 - 4 = 2$. The solutions are $(\frac{1}{5}, -\frac{18}{5})$ and $(3, 2)$

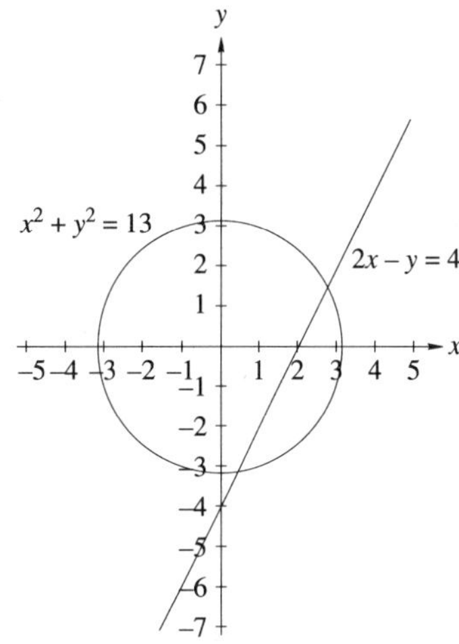

3. $\begin{cases} x^2 + 4y^2 = 32 & (1) \\ x + 2y = 0 & (2) \end{cases}$. Solving (2) for x: $x = -2y$ Substituting into (1) we get $(-2y)^2 + 4y^2 = 32$, $4y^2 + 4y^2 = 32$, $8y^2 = 32$, $y^2 = 4$; $y = \pm 2$. If $y = 2 \Rightarrow x = -2(2) = -4$; If $y = -2 \Rightarrow x = -2(-2) = 4$. Solutions $(-4, 2)$ and $(4, -2)$

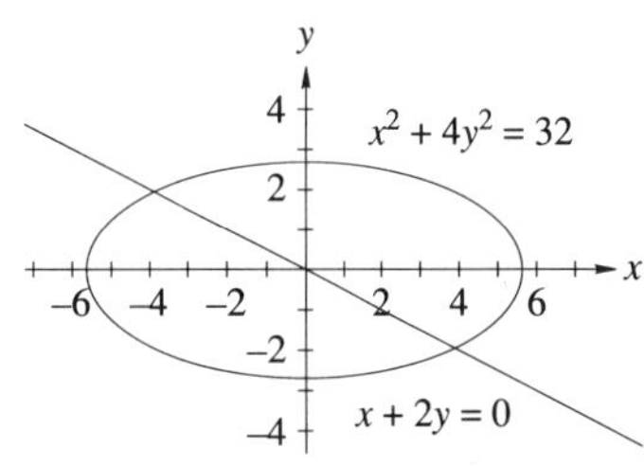

4. $\begin{cases} x - y = 4 & (1) \\ x^2 - y^2 = 32 & (2) \end{cases}$. Solving (1) for x we get $x = y+4$. Substituting into (2) we get $(y+4)^2 - y^2 = 32$, or $y^2 + 8y + 16 - y^2 = 32$, or $8y = 16$, and so $y = 2$. Back substituting, we have $x = 2 + 4 = 6$, and the solution is $(6, 2)$.

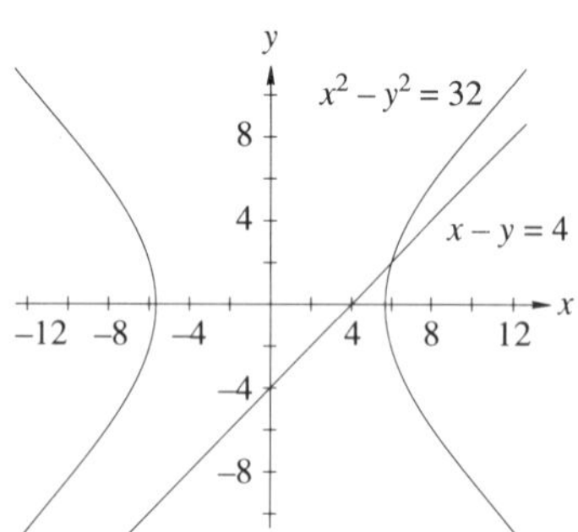

5. $\begin{cases} x^2 + y^2 = 4 & (1) \\ x^2 - 2y = 1 & (2) \end{cases}$. Subtracting (2) from (1) yields $y^2 + 2y = 3$ or $y^2 + 2y - 3 = 0$. This factors as $(y+3)(y-1) = 0$ and so $y = -3$ or $y = 1$. If $y = 1$, then, from (2), we obtain $x^2 - 2 = 1$, or $x^2 = 3$ and so $x = \pm\sqrt{3}$. If $y = -3$, then, from (2), we get $x^2 - 2 \cdot (-3) = 1$, or $x^2 + 6 = 1$, $x^2 = -5$, $x = \pm j\sqrt{5}$, which is not real. Solution: $(\sqrt{3}, 1)$, $(-\sqrt{3}, 1)$, $(j\sqrt{5}, -3)$, $(-j\sqrt{5}, -3)$

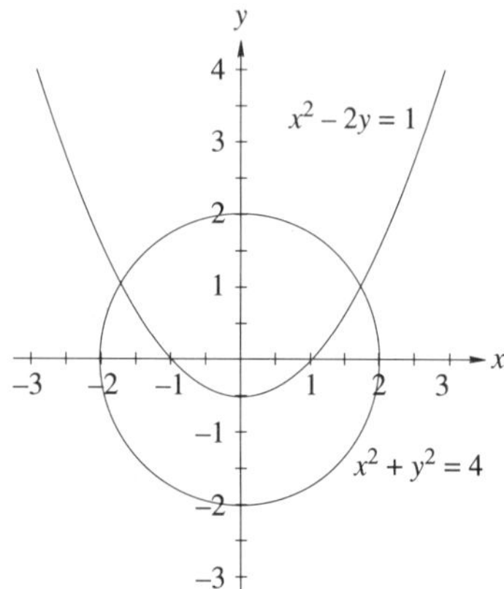

6. $\begin{cases} x^2 - y^2 = 4 & (1) \\ 2x - 3y = 10 & (2) \end{cases}$. Solving (2) for x: $x = \frac{3y+10}{2}$. Substituting into (1) we get $\left(\frac{3y+10}{2}\right)^2 - y^2 = 4$, or $\frac{9y^2+60y+100}{4} - y^2 = 4$. Multiplying by 4 produces $9y^2+60y+100-4y^2 = 16$, which simplifies as $5y^2 + 60y + 84 = 0$. By the quadratic formula $y = -1.618$ or $y = -10.382$. If $y = -1.618 \Rightarrow x = 2.573$. If $y = -10.382 \Rightarrow x = -10.573$. Thus, the solutions are $(2.573, -1.618)$ and $(-10.573, -10.382)$

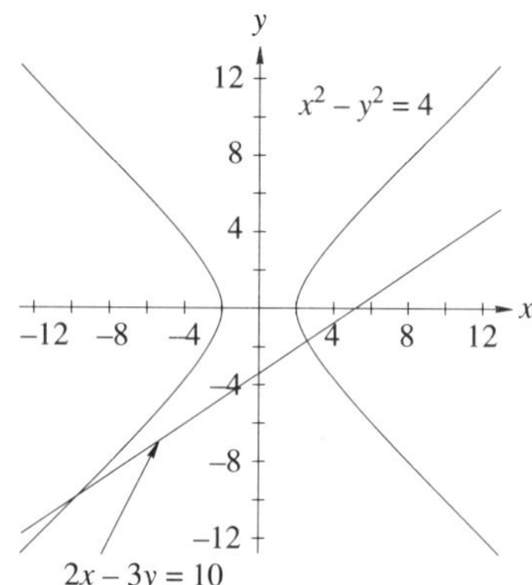

7. $\begin{cases} x^2 + y^2 = 7 & (1) \\ y^2 = 6x & (2) \end{cases}$. Substituting $6x$ for y^2 in (1) yields: $x^2 + 6x = 7, x^2 + 6x - 7 = 0, (x + 7)(x - 1) = 0, x = -7$ or $x = 1, x = -7$ yields $y^2 = 6(-7) = -42, y = \pm j\sqrt{42}$. $x = 1$ yields $y^2 = 6 \cdot 1 = 6, y = \pm\sqrt{6}$ Solution $(1, \sqrt{6})$, $(1, -\sqrt{6})$, $(-7, j\sqrt{42})$, $(-7, -j\sqrt{42})$. (Note: the last two solutions are not real numbers.)

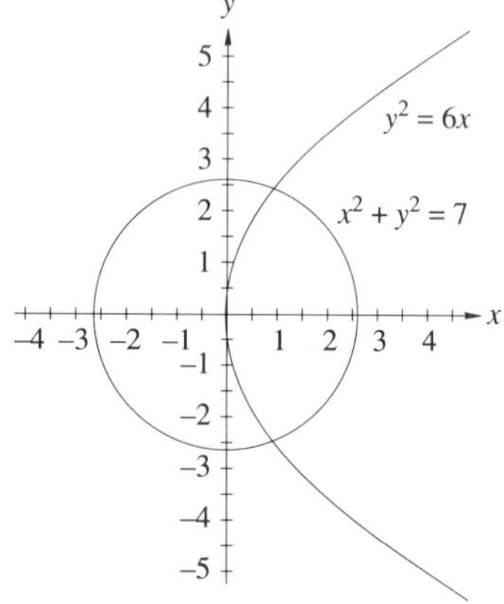

8. $\begin{cases} x^2 - y^2 + 6 = 0 & (1) \\ y^2 = 5x & (2) \end{cases}$. Substituting $5x$ for y^2 into (1) yields: $x^2 - 5x + 6 = 0, (x-2)(x-3) = 0, x = 2$ or $x = 3$, $x = 2$ gives $y^2 = 10$ or $y = \pm\sqrt{10}, x = 3$ gives $y^2 = 15$ or $y = \pm\sqrt{15}$. Solutions $(2, \sqrt{10})$, $(2, -\sqrt{10})$, $(3, \sqrt{15})$, $(3, -\sqrt{15})$

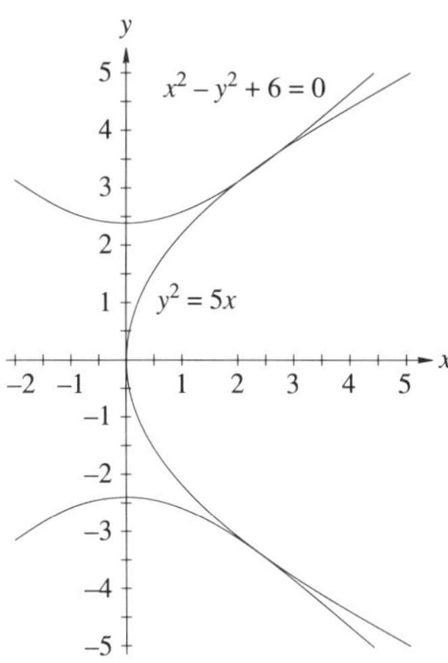

9. $\begin{cases} xy = 3 & (1) \\ 2x^2 - 3y^2 = 15 & (2) \end{cases}$. Solving (1) for x: $x = \frac{3}{y}$. Substituting into (2) $2 \cdot \left(\frac{3}{y}\right)^2 - 3y^2 = 15$, or $\frac{18}{y^2} - 3y^2 = 15$, or $18 - 3y^4 = 15y^2$, or $3y^4 + 15y^2 - 18 = 0$, which simplifies to $y^4 + 5y^2 - 6 = 0$, which factors as $(y^2 + 6)(y^2 - 1) = 0$. Thus, $y^2 = -6$ and $y = \pm j\sqrt{6} = \pm 3j\sqrt{\frac{2}{3}}$ or $y^2 = 1$ and $y = \pm 1$. Now, $y = 1 \Rightarrow x = 3$ which is the solution $(3, 1)$ and $y = -1 \Rightarrow x = -3$ and the solution $(-3, -1)$. Also, $y = j\sqrt{6} \Rightarrow x = \frac{3}{j\sqrt{6}} = -\frac{\sqrt{6}}{2}j = -j\sqrt{\frac{3}{2}}$ which produces the solution $\left(-\frac{\sqrt{6}}{2}j, j\sqrt{6}\right)$ or $\left(-j\sqrt{\frac{3}{2}}, 3j\sqrt{\frac{2}{3}}\right)$. Similarly, $y = -j\sqrt{6} \Rightarrow x = \frac{\sqrt{6}}{2}j = j\sqrt{\frac{3}{2}}$ and we have the solution $\left(\frac{\sqrt{6}}{2}j, -j\sqrt{6}\right)$ or $\left(j\sqrt{\frac{3}{2}}, -3j\sqrt{\frac{2}{3}}\right)$.

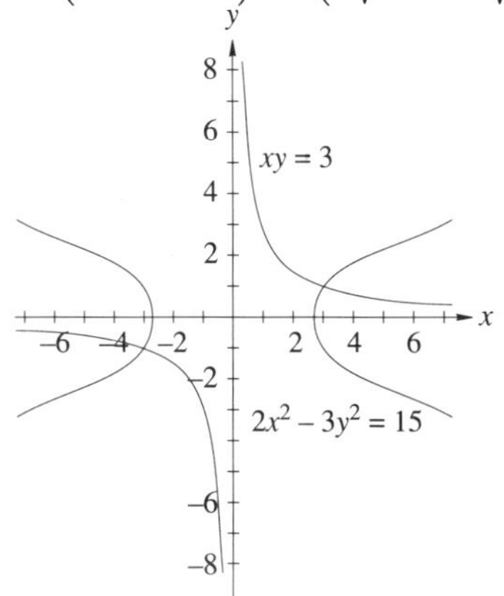

10. $\begin{cases} y = x^2 - 4 & (1) \\ x^2 + 3y^2 + 4y - 6 = 0 & (2) \end{cases}$. Substituting $x^2 - 4$ for y into (2) yields: $x^2 + 3(x^2 - 4)^2 + 4(x^2 - 4) - 6 = 0$, or $x^2 + 3(x^4 - 8x^2 + 16) + 4x^2 - 16 - 6 = 0$, or $x^2 + 3x^4 - 24x^2 + 48 + 4x^2 - 22 = 0$, which simplifies to $3x^4 - 19x^2 + 26 = 0$, which factors as $(3x^2 - 13)(x^2 - 2) = 0$. Thus, either $x^2 = \frac{13}{3}$ and $x = \frac{\pm\sqrt{39}}{3}$ or $x^2 - 2 = 0$ which means that $x^2 = 2$, and so $x = \pm\sqrt{2}$. Now, $x = \frac{\pm\sqrt{39}}{3} \Rightarrow y = \frac{39}{9} - 4 = \frac{13}{3} - 4 = \frac{1}{3}$, and $x = \pm\sqrt{2} \Rightarrow y = 2 - 4 = -2$. So the solutions are $\left(\frac{\sqrt{39}}{3}, \frac{1}{3}\right)$, $\left(-\frac{\sqrt{39}}{3}, \frac{1}{3}\right)$, $\left(\sqrt{2}, -2\right)$, $\left(-\sqrt{2}, -2\right)$.

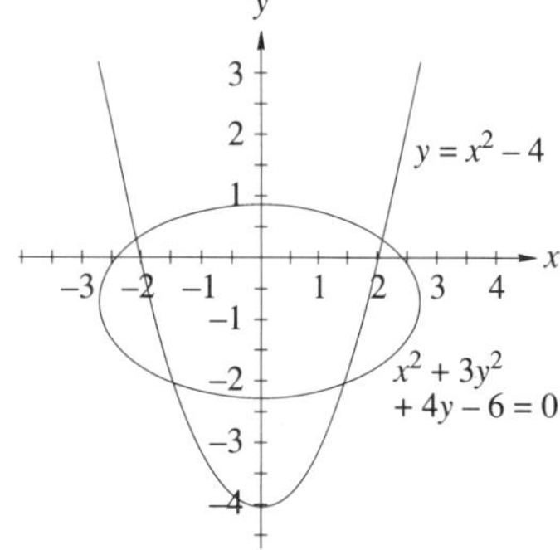

11. $t = \frac{d}{r}$, $t_1 = \frac{100}{r}$, $t_2 = \frac{20}{r-10}$, $t_1 + t_2 = \frac{100}{r} + \frac{20}{r-10} = 3$, $(r-10)100 + 20r = 3r(r-10)$, $100r - 1000 + 20r = 3r^2 - 30r$, $3r^2 - 150r + 1000 = 0$ Using the quadiatric formula, $r = 42.08$ mph for the first 100 miles, 32.08 mph for the last last 20 miles.

12. The perimeter is $2\ell + 2w = 160$ and the area is $\ell \cdot w = 1200$. From the equation for the perimeter, we get $\ell = 80 - w$ and substituting this in the equation for the area gives $(80 - w)w = 1200$, or $80w - w^2 = 1200$, or $w^2 - 80w + 1200 = 0$ which factors as $(w - 60)(w - 20) = 0$. Thus, $w = 60$ or $w = 20$ and we get that the length is 60 m and the width is 20 m.

13. When the resistors are connected is series, we get $R_1 + R_2 = 100$ and when they are connected in parallel $\frac{R_1 R_2}{R_1 + R_2} = 24$. Since $R_1 + R_2 = 100$, the second equation is equivalent to $R_1 R_2 = 2400$. Now, $R_1 = 100 - R_2$ and so $(100 - R_2)R_2 = 2400$, or $100R_2 - R_2^2 = 2400$, or $R_2^2 - 100R_2 + 2400 = 0$. This factors as $(R_2 - 60)(R_2 - 40) = 0$ which means

that $R_2 = 60$ or 40. The resistances are $40\ \Omega$ and $60\ \Omega$.

14. $R_1 + R_2 = 10$, $\frac{R_1R_2}{R_1+R_2} = 2.1$, $\frac{R_1R_2}{10} = 2.1$, $R_1R_2 = 21$, $R_1 = 10 - R_2$, $(10 - R_2)R_2 = 21$, $10R_2 - R_2^2 = 21$, $R_2^2 - 10R_2 + 21 = 0$, $(R_2 - 3)(R_2 - 7) = 0$, $R_2 = 3$ or 7. One is $3\ \Omega$, the other is $7\ \Omega$

15. $40v_1 + 60v_2 = 450 \Rightarrow 4v_1 + 6v_2 = 45$, $20v_1^2 + 30v_1^2 = 1087.5 \Rightarrow 2v_1^2 + 3v_2^2 = 108.75$. Solving the first equation for v_1 yields: $v_1 = \frac{45-6v_2}{4}$. Substituting into the second equation yields $2\left(\frac{45-6v_2}{4}\right)^2 + 3v_2^2 = 108.75$ or $2\left(\frac{2025-540v_2+36v_2^2}{16}\right) + 3v_2^2 = 108.75$, or $2025 - 540v_2 + 36v_2^2 + 24v_2^2 = 870$, which is equivalent to $60v_2^2 - 540v_2 + 1155 = 0$ or $4v_2^2 - 36v_2 + 77 = 0$. Factoring the last equation produces $(2v_2 - 7)(2v_2 - 11) = 0$, and so $v_2 = \frac{7}{2}$ or $\frac{11}{2}$. If $v_2 = \frac{7}{2} = 3.5$, then $v_1 = 6$. If $v_2 = 5.5$, then $v_1 = 3$. Either $v_1 = 6$ m/s and $v_2 = 3.5$ m/s or $v_1 = 3$ m/s and $v_2 = 5.5$ m/s.

16. $20x^2 + 40y^2 = 8000 \Rightarrow x^2 + 2y^2 = 400$ (1), and $30x^2 + 20y^2 = 6000 \Rightarrow 3x^2 + 2y^2 = 600$ (2). Subtracting (1) from (2) yields $2x^2 = 200$ or $x^2 = 100$ and so $x = \pm 10$. Since $2y^2 = 400 - x^2$, then if $x = \pm 10$, we get $2y^2 = 400 - 100 = 300$, or $y^2 = 150$, and so $y = \pm\sqrt{150} \approx \pm 12.25$. The points of intersection are approximately $(10, 12.25)$, $(-10, 12.25)$, $(10, -12.25)$ and $(-10, -12.25)$.

17. Since $F\ell = 200(0.8) = 160 \Rightarrow F = \frac{160}{\ell}$ and so $(F + 50)(\ell - 1) = 160 \Rightarrow \left(\frac{160}{\ell} + 50\right)(\ell - 1) = 160$, or $(160 + 50\ell)(\ell - 1) = 160\ell$, or $50\ell^2 + 110\ell - 160 = 160\ell$, which is equivalent to $50\ell^2 - 50\ell - 160 = 0$, and simplifies as $5\ell^2 - 5\ell - 16 = 0$. Using the quadratic formula $\ell = 2.36$ or -1.36. But, ℓ cannot be negative so 2.36 is the only answer. $F = \frac{160}{2.36} = 67.80$ gives $F = 67.80$ N, $\ell = 2.36$ m.

18. $y = 15t - 4.9t^2$, $y = 10t$ so $10t = 15t - 4.9t^2$, $4.9t^2 - 5t = 0$, $t(4.9t - 5) = 0$, so $t = 0$ or $t = 1.02$ If $t = 0$, $y = 0$; If $t = 1.02$, $y = 10.2$, so the solutions are $(0, 0)$ and $(1.02, 10.2)$

16.2 PROPERTIES OF INEQUALITIES; LINEAR INEQUALITIES

1. $x < 10 \Rightarrow x + 5 < 15$

2. $x < 10 \Rightarrow x - 7 < 3$

3. $x < 10 \Rightarrow 3x < 30$

4. $x < 10 \Rightarrow -4x > -40$

5. $x < 10 \Rightarrow -\frac{x}{2} > -5$

6. $x < 10 \Rightarrow \frac{x}{5} < 2$

7. $x < 10 \Rightarrow x^2 < 100$ (Note $x > 0$)

8. $x < 10 \Rightarrow \sqrt{x} < \sqrt{10}$

9. Dividing $3x > 6$ by 3 leads to $x > 2$.

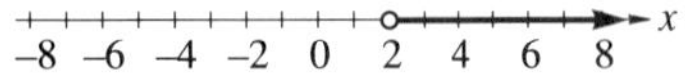

10. Dividing $4x < -8$ by 4 leads to $x < -2$.

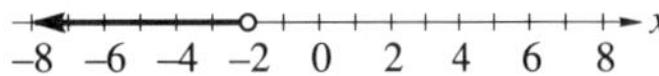

11. Subtracting 5 from both sides of the inequality $2x + 5 > -7$ produces $2x > -12$, which is equivalent to $x > -6$.

x
–8 –6 –4 –2 0 2 4 6 8

12. Adding 4 to both sides of $3x - 4 \leq 8$produces $3x \leq 12$, is equivalent to $x \leq 4$.

x
–8 –6 –4 –2 0 2 4 6 8

13. Adding 5 to both sides of the inequality $2x - 5 < x$ produces $2x < x + 5$ and subtracting x from both sides yields $x < 5$.

x
–8 –6 –4 –2 0 2 4 6 8

14. Adding $-x - 4$ to both sides of the inequality $3x + 4 \geq x$, we get $2x \geq -4$, and dividing by 2 produces $x \geq -2$.

x
–8 –6 –4 –2 0 2 4 6 8

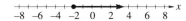

15. Adding $7 - 2x$ to both sides of the inequality $4x - 7 \geq 2x + 5$ gives $2x \geq 12$, and dividing by 2, produces $x \geq 6$.

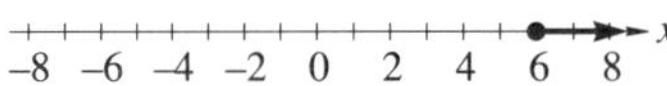

16. Adding $-7 - 5x$ to both sides of $7 - 3x \leq 3 + 5x$ gives $-8x \leq -4$. Then, dividing both sides by -8, we get $x \geq \frac{1}{2}$.

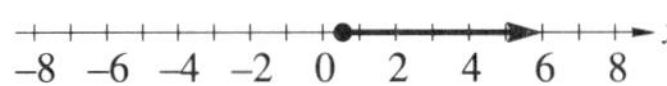

17. First we multiply the inequality $\frac{2}{3}x - 4 < \frac{1}{3}x + 2$ by 3, with the result $2x - 12 < x + 6$. Adding $12 - x$ to both sides, yields the result $x < 18$.

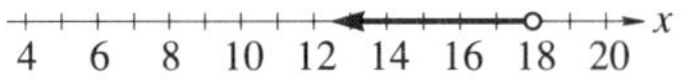

18. Multiplying $\frac{1}{5}x - \frac{2}{5} > \frac{3}{5}x + \frac{4}{5}$ by 5, produces $x - 2 > 3x + 4$. Adding $2 - 3x$ to both sides gives $-2x > 6$,; and dividing by -2 leads to $x < -3$.

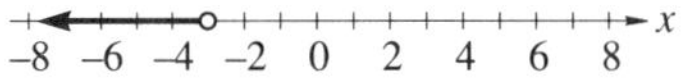

19. Multiplying $\frac{x-2}{4} \leq \frac{3}{8}$ by 8 produces $2(x-2) \leq 3$ or $2x - 4 \leq 3$. Adding 4 to both sides, gives $2x \leq 7$, and then dividing by 2, we get $x \leq \frac{7}{2}$ or $x \leq 3\frac{1}{2}$.

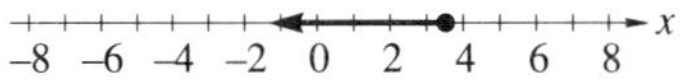

20. Multiplying $\frac{x+2}{3} \geq \frac{5}{6}$ by 3 gives $x + 2 \geq \frac{5}{2}$ and subtracting 2 produces $x \geq \frac{1}{2}$.

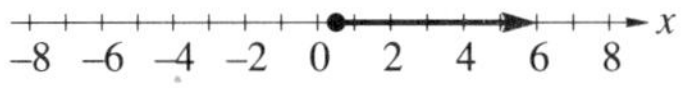

21. Multiplying $\frac{x-3}{4} \leq \frac{2x}{3}$ by 12 gives $3(x - 3) \leq 4 \cdot 2x$ which simplifies as $3x - 9 \leq 8x$, or $-5x \leq 9$, and so, after dividing by -5, we get $x \geq -\frac{9}{5}$.

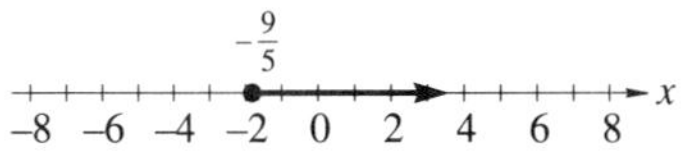

22. Multiplying the inequality $\frac{2x+5}{3} > \frac{3x-1}{2}$ by 6 produces $2(2x + 5) > 3(3x - 1)$ or $4x + 10 > 9x - 3$. This simplifies to $-5x > -13$ and, after dividing by -5, to $x < \frac{13}{5}$.

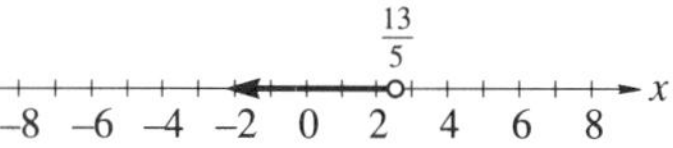

23. The absolute value inequality $|x + 1| < 5$ is equivalent to $-5 < x + 1 < 5$ which simplifies to $-6 < x < 4$.

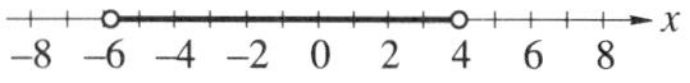

24. The absolute value inequality $|2x - 1| < 7$ is equivalent to $-7 < 2x - 1 < 7$, which simplifies to $-6 < 2x < 8$ or $-3 < x < 4$.

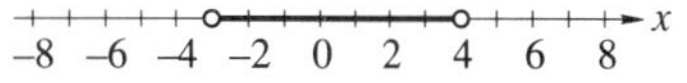

25. The inequality $|x+4| > 6$ is equivalent to $x+4 < -6$ or $x+4 > 6$. Simplifying each of these inequalities, we get $x < -10$ or $x > 2$.

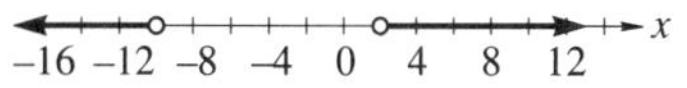

26. The inequality $|3x + 9| \geq 6$ is equivalent to $3x + 9 \leq -6$ or $3x + 9 \geq 6$. Adding -9 to each of these inequalities, produces $3x \leq -15$ or $3x \geq -3$ which simplify to $x \leq -5$ or $x \geq -1$.

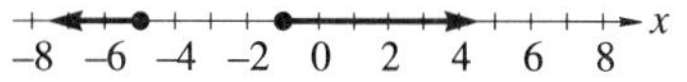

27. $|x + 5| < -3$. But, an absolute value can never be negative, so there are no real numbers that satisfy this inequality. This is a contradictory ineqality and so there is no solution.

28. The inequality $|3x + 2| < 12$ is equivalent to $-12 < 3x + 2 < 12$, or $-14 < 3x < 10$, and so $-\frac{14}{3} < x < \frac{10}{3}$.

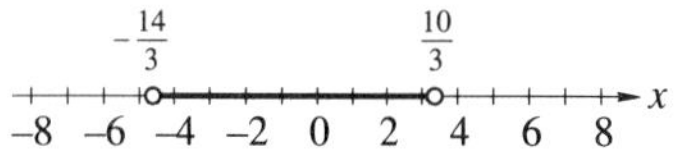

29. The inequality $-7 \leq 3x + 5 < 26$ is equivalent to $-12 \leq 3x < 21$, or $-4 \leq x < 7$

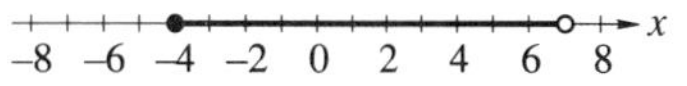

30. The inequality $-6 < 5 - 3x \leq 17$ is equivalent to $-11 < -3x \leq 12$, or, after dividing by -3, to $\frac{11}{3} > x \geq -4$ or $-4 \leq x < \frac{11}{3}$.

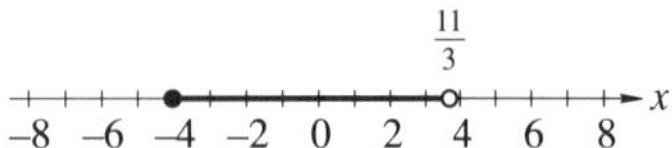

31. The inequality $3x + 1 < 5 < 2x - 3$ is equivalent to the two inequalities $3x + 1 < 5$ and $5 < 2x - 3$. Subtracting 1 from the first inequality, gives $3x < 4$ or $x < \frac{4}{3}$. Adding 3 to the second inequality, gives $8 < 2x$ which can be simplified as $4 < x$ or $x > 4$. Thus, the solution is $x < \frac{4}{3}$ and $x > 4$. This is a contradictory statement, there are no real numbers that satisfy the given inequality.

32. The inequality $4x - 6 \leq 11 < 9x + 1$ is equivalent to $4x - 6 \leq 11$ and $11 < 9x + 1$, $4x \leq 17$ and $10 < 9x$, $x \leq \frac{17}{4}$ and $\frac{10}{9} < x$. Thus, we get $\frac{10}{9} < x \leq \frac{17}{4}$.

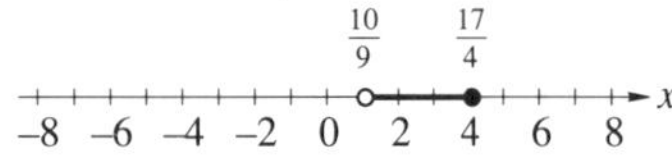

33. 6357 km $\leq r \leq$ 6378 km

34. 357 000 km $\leq d \leq$ 407 000 km

35. $0.60 - 0.05 \leq c \leq 0.60 + 0.05, 0.10° \leq c \leq 1.10°$

36. $2 \leq R - 12 \leq 4$ or 32 oz $\leq R - 12 \leq$ 64 oz

37. $200 < \frac{800R}{800+R} < 500{,}200(800 + R) < 800R < 500(800 + R)$, $160000 + 200R < 800R < 400000 + 500R$, so $160000 + 200R < 800R$ and $800R < 400000 + 500R$, $160000 < 600R$ and $300R < 400000$, $266.67 < R$ and $R < 1333.33$, $266.67\ \Omega < R < 1333.33\ \Omega$

38. $C = \frac{5}{9}(F - 32) \Rightarrow \frac{9}{5}C = F - 32$, $F = \frac{9}{5}C + 32$, $1800°C \Rightarrow \frac{9}{5} \cdot 1800 + 32 = 3272°F$, $2200°C \Rightarrow \frac{9}{5} \cdot 2200 + 32 = 3992°F$, so the temperature range is between 3272°F and 3992°F.

39. $R - C > 0$, $60x - (1{,}500 + 25x) > 0$, $35x - 1{,}500 > 0$, $35x > 1{,}500$, $x > 42.86$ or, since we cannot have a fraction, $x \geq 43$

40. $\ell \cdot h \cdot 600 =$ or $\ell \cdot 1.5 \cdot 600 = 900\ell$, $2400 < 900\ell < 4000$, 2.67 m $< \ell <$ 4.444 m.

16.3 NONLINEAR INEQUALITIES

1. $\{x | x < -1 \text{ or } x > 3\}$

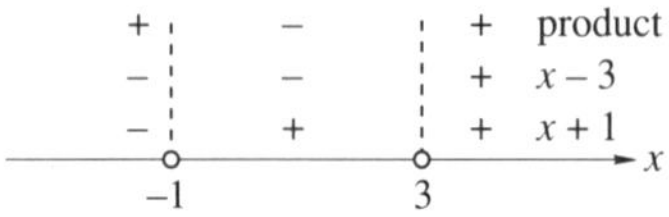

2. $\{x | x \leq -2 \text{ or } x \geq 4\}$

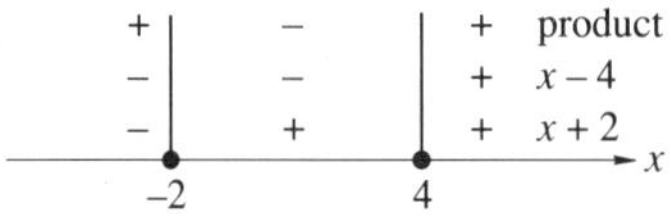

3. $\{x | -4 \leq x \leq 1\}$

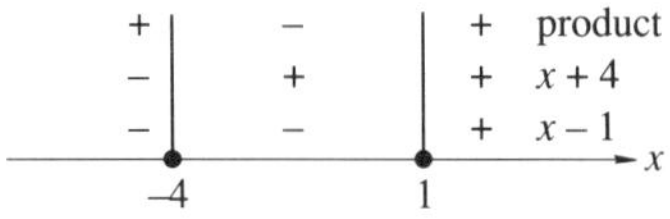

4. $\{x | -5 < x < 2\}$

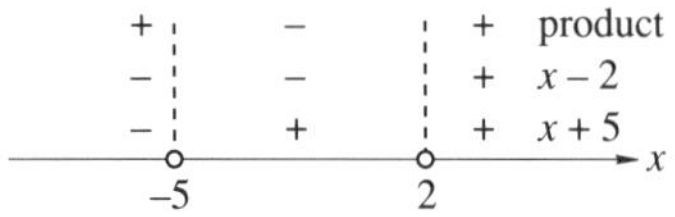

5. $x^2 - 1 < 0$ is equivalent to $(x + 1)(x - 1) < 0$. The zero values are -1 and 1, so the solution set is $\{x | -1 < x < 1\}$.

6. $x^2 \leq 16$ is equivalent to $x^2 - 16 \leq 0$ or $(x + 4)(x - 4) \leq 0$. The zero values are -4 and 4, so the solution set is $\{x | -4 \leq x \leq 4\}$.

7. $x^2 - 2x - 15 \leq 0$ is equivalent to $(x + 3)(x - 5) \leq 0$. The zero values are -3 and 5, so the solution set is $\{x | -3 \leq x \leq 5\}$.

8. $x^2 + x - 2 > 0$ is equivalent to $(x + 2)(x - 1) > 0$. The zero values are -2 and 1, so the solution set is $\{x | x < -2 \text{ or } x > 1\}$.

9. $x^2 - x - 2 \geq 0$ is equivalent to $(x + 1)(x - 2) \geq 0$. The zero values are -1 and 2, so the solution set is $\{x | x \leq -1 \text{ or } x \geq 2\}$.

10. $x^2 - 3x - 10 < 0$ is equivalent to $(x + 2)(x - 5) < 0$. The zero values are -2 and 5, so the solution set is $\{x | -2 < x < 5\}$.

11. $x^2 - 5x > -6$ is equivalent to $x^2 - 5x + 6 > 0$ which factors as $(x-2)(x-3) > 0$. The zero values are 2 and 3, so the solution set is $\{x|x < 2 \text{ or } x > 3\}$.

12. $x^2 + 7x \leq -12$ is equivalent to $x^2 + 7x + 12 \leq 0$ which factors as $(x+3)(x+4) \leq 0$. The zero values are -4 and -3, so the solution set is $\{x| -4 \leq x \leq -3\}$.

13. $2x^2 + 7x + 3 < 0$ is equivalent to $(2x+1)(x+3) < 0$. The zero values are -3 and $-\frac{1}{2}$, so the solution set is $\left\{x| -3 < x < -\frac{1}{2}\right\}$

14. $2x^2 - 5x + 2 \geq 0$ is equivalent to $(2x-1)(x-2) \geq 0$. The zero values are $\frac{1}{2}$ and 2, so the solution set is $\left\{x|x \leq \frac{1}{2} \text{ or } x \geq 2\right\}$.

15. $2x^2 - x < 1$ is equivalent to $2x^2 - x - 1 < 0$, or $(2x+1)(x-1) < 0$. The zero values are $-\frac{1}{2}$ and 1, so the solution set is $\left\{x| -\frac{1}{2} < x < 1\right\}$.

16. $2x^2 - x \leq 3$ is equivalent to $2x^2 - x - 3 \leq 0$ or $(2x-3)(x+1) \leq 0$. The zero values are -1 and $\frac{3}{2}$, so the solution set is $\left\{x| -1 \leq x \leq \frac{3}{2}\right\}$

17. $4x^2 + 2x > x^2 - 1$ is equivalent to $3x^2 + 2x + 1 > 0$. The discriminant of the equation $3x^2 + 2x + 1 = 0$ is $b^2 - 4ac = 4 - 4 \cdot 3 \cdot 1 = -8 < 0$, so this equation has no real solutions. When $x = 0$, we get $1 > 0$ which is true. Thus, the given inequality is an absolute inequality, all real numbers satisty this inequality.

18. $2x^2 + 5x < 2 - x^2$ is equivalent to $3x^2 + 5x - 2 < 0$ or $(3x-1)(x+2) < 0$. The zero values are $\frac{1}{3}$ and -2, so the solution set is $\left\{x| -2 < x < \frac{1}{3}\right\}$.

19. The zero values are -1, 2, and -3, so the solution is $x < -3$ or $-1 < x < 2$.

	$x<-3$	$-3<x<-1$	$-1<x<2$	$x>2$
product	−	+	−	+
$x+3$	−	+	+	+
$x-2$	−	−	−	+
$x+1$	−	−	+	+

Zero values on the x-axis: -3, -1, 2

20. $x \leq -2$ or $1 \leq x \leq 3$

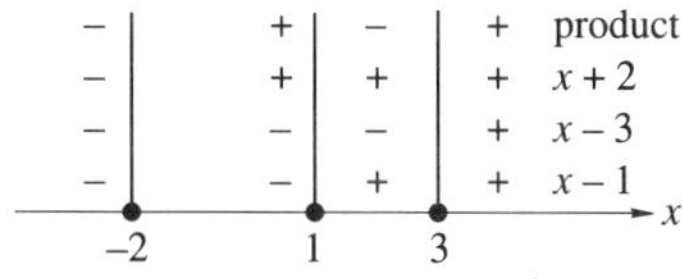

21. The zero values are -3, $\frac{5}{2}$, and -4, so the solution set is $\left\{x| -4 < x < -3 \text{ or } x > \frac{5}{2}\right\}$.

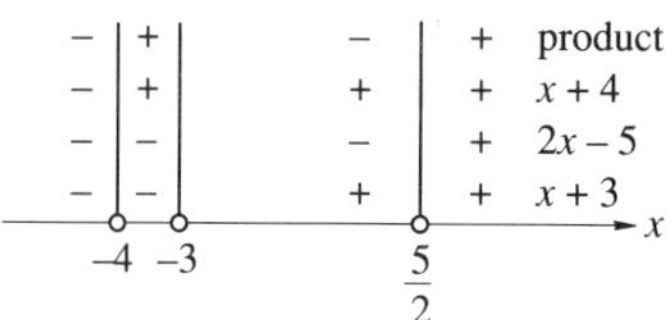

22. The zero values are -4, $\frac{1}{3}$, and 5, so the solution set is $\left\{x| -4 \leq x \leq \frac{1}{3} \text{ or } x \geq 5\right\}$.

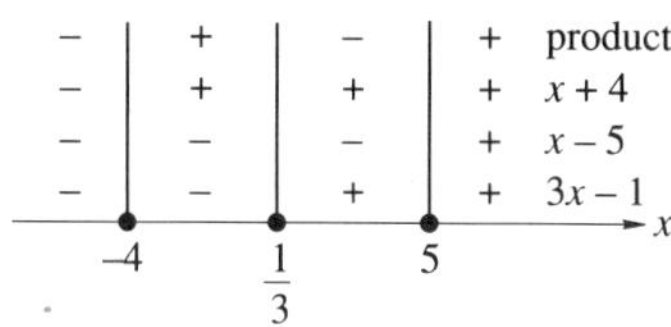

23. $(x-2)^2(x+4) < 0$. The zero values are 2 and -4, so the solution set is $\{x|x < -4\}$.

24. $(x+3)^2(x-5) \geq 0$. The zero values are -3 and 5, so the solution set is $\{x|x = -3 \text{ or } x > 5\}$.

25. $x^3 - 4x > 0$ is equivalent to $x(x+2)(x-2) > 0$. The zero values are 0, -2 and 2, so the solution set is $\{x| -2 < x < 0 \text{ or } x > 2\}$.

26. $x^4 - 9x^2 \leq 0$ is equivalent to $x^2(x^2 - 9) \leq 0$ which factors as $x^2(x+3)(x-3) \leq 0$. The zero values are 0, -3, and 3, so the solution set is $\{x| -3 \leq x \leq 3\}$.

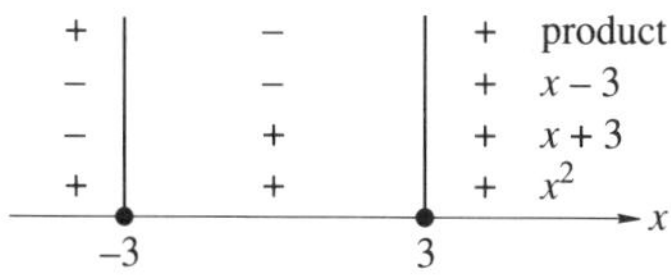

27. $x^2 + 2x + 3 \leq 0$. The discriminant of the equation $x^2 + 2x + 3 = 0$ is $2^2 - 4 \cdot 3 = -8 < 0$, so there are no real solutions to this equation. When $x = 0$, $3 \leq 0$ is false so the given inequality is a contradictory inequality and there are no real solutions.

28. $x^2+x+1>0$. The discriminant is $b^2-4ac=1-4=-3<0$, so the quadratic equation $x^2+x+1=0$ has no real solutions. When $x=0$, the given inequality gives $1>0$, which is always true. This is an absolute inequality, so all real numbers satisfy this inequality.

29. The zero values are 2, 5, and -1, so the solution set is $\{x|x<-1 \text{ or } 2<x<5\}$

	$x<-1$	$-1<x<2$	$2<x<5$	$x>5$
product	−	+	−	+
$x-5$	−	−	−	+
$x-2$	−	−	+	+
$x+1$	−	+	+	+

30. The zero values are -2, 4 and 2. Since the denominator is 0 when $x=2$, we know that 2 cannot be in the solution set. We see that the solution set is $\{x|-2\le x<2 \text{ or } x\ge 4\}$.

	$x<-2$	$-2<x<2$	$2<x<4$	$x>4$
product	−	+	−	+
$x-4$	−	−	−	+
$x-2$	−	−	+	+
$x+2$	−	+	+	+

31. $\dfrac{x}{(x+1)(x-2)}>0$. The zero values are 0, -1, and 2, so the solution set is $\{x|-1<x<0 \text{ or } x>2\}$.

32. $\dfrac{x(x-3)}{(x+1)(x-4)}\le 0$. The zero values are 0, 3, -1, and 4, so the solution set is $\{x|-1<x\le 0 \text{ or } 3\le x<4\}$.

33. $\dfrac{(3x+1)(x-3)}{2x-1}\le 0$. The zero values are $-\frac{1}{3}$, $\frac{1}{2}$, and 3, so the solution set is $\{x|x\le -\frac{1}{3} \text{ or } \frac{1}{2}<x\le 3\}$.

34. $\dfrac{(x-1)(x+6)}{(x+1)(x-3)}>0$. The zero values are -6, -1, 1, and 3, so the solution set is $\{x|x<-6 \text{ or } -1<x<1 \text{ or } x>3\}$.

35. The inequality $\dfrac{4}{x-1}<\dfrac{5}{x+1}$ is equivalent to $\dfrac{4}{x-1}-\dfrac{5}{x+1}<0$ or $\dfrac{4(x+1)-5(x-1)}{(x-1)(x+1)}<0$. This simplifies to $\dfrac{-x+9}{(x-1)(x+1)}<0$. The zero values are 9, -1, and 1, so the solution set is $\{x|-1<x<1 \text{ or } x>9\}$.

36. The inequality $\dfrac{-3}{x+2}\ge\dfrac{6}{x-3}$ is equivalent to $\dfrac{-3}{x+2}-\dfrac{6}{x-3}\ge 0$ or $\dfrac{-3(x-3)-6(x+2)}{(x+2)(x-3)}\ge 0$, which simplifies to $\dfrac{-9x-3}{(x+2)(x-3)}\ge 0$ which is equivalent to $\dfrac{3x+1}{(x+2)(x-3)}\le 0$. The zero values are $-\frac{1}{3}$, -2, and 3, so the solution set is $\{x|x<-2 \text{ or } -\frac{1}{3}\le x<3\}$.

	$x<-2$	$-2<x<-\frac{1}{3}$	$-\frac{1}{3}<x<3$	$x>3$
product	−	+	−	+
$x-3$	−	−	−	+
$x+2$	−	+	+	+
$3x+1$	−	−	+	+

37. $\dfrac{x^2+2x+3}{x-1}\ge 0$. The numerator x^2+2x+3 is always >0, so the original inequality is true when $x>1$.

38. $\dfrac{x^2+x+1}{x+2}<0$; x^2+x+1 is always >0 so the original inequality is true when $x<-2$

39. $|x^2+3x+2|\le 4$ is equivalent to $-4\le x^2+3x+2\le 4$. We first examine the right-hand inequality $x^2+3x+2\le 4$, which simplifies to $x^2+3x-2\le 0$. Using the quadratic formula, we see that the equation $x^2+3x-2=0$ is zero when $x=\frac{-3\pm\sqrt{9+8}}{2}=\frac{-3\pm\sqrt{17}}{2}$. Next, we examine the left-hand inequality and see that $-4\le x^2+3x+2$ is always true. So the solution set is $\left\{x\middle|\dfrac{-3-\sqrt{17}}{2}\le x\le\dfrac{-3+\sqrt{17}}{2}\right\}$.

40. $|x^2-3x+2|>5$ is equivalent to $x^2-3x+2<-5$ or $x^2-3x+2>5$. $x^2-3x+2<-5$ is never true, so the only answers are those that satisfy $x^2-3x+2>5$ or, equivalently, $x^2-3x-3>0$. By the quadratic formula, we see that $x^2-3x-3=0$

when $x = \frac{3 \pm \sqrt{9+12}}{2}$ or $x = \frac{3 \pm \sqrt{21}}{2}$. Thus, the solution set is $\left\{ x | x < \frac{3 - \sqrt{21}}{2} \text{ or } x > \frac{3 + \sqrt{21}}{2} \right\}$

41. Since $I = 75d^2$, the solution to $75 < I < 450$ is equivalent to $75 < 75d^2 < 450$ or $1 < d^2 < 6$. Since the context of the problem means that d is positive, and so we see that $1 \text{ m} < d < \sqrt{6} \text{ m}$.

42. We have the inequality $34.3t - 4.9t^2 > 49$ which is equivalent to $-4.9t^2 + 34.3t - 49 > 0$ which simplifies to $t^2 - 7t + 10 < 0$. This factors as $(t - 2)(t - 5) < 0$ and so $2 < t < 5$. Thus, the solution is $2 \text{ s} < t < 5 \text{ s}$.

43. $x^2 - 1.1x + 0.2 > 0.08$ is equiavlent to $x^2 - 1.1x + 0.12 > 0$. The equation $x^2 - 1.1x + 0.12 = 0$ has solutions $x = \dfrac{1.1 \pm \sqrt{1.21 - .48}}{2} = \dfrac{1.1 \pm \sqrt{0.73}}{3}$. So, the solution set to the given inequality is $x < 0.123$ or $x > 0.977$.

44. $L = \dfrac{90 \cdot 6d^2}{18} = 30d^2$. Since $L = 3000$, we need to solve the inequality $30d^2 > 3000$, which is equivalent to $d^2 > 100$ and so $d > \pm 10$. Since d is positive and < 12, we see that $10'' < d < 12''$.

16.4 INEQUALITIES IN TWO VARIABLES

1.

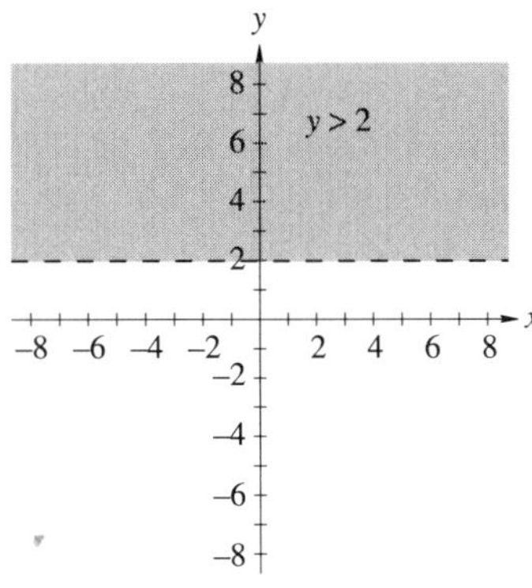

2.

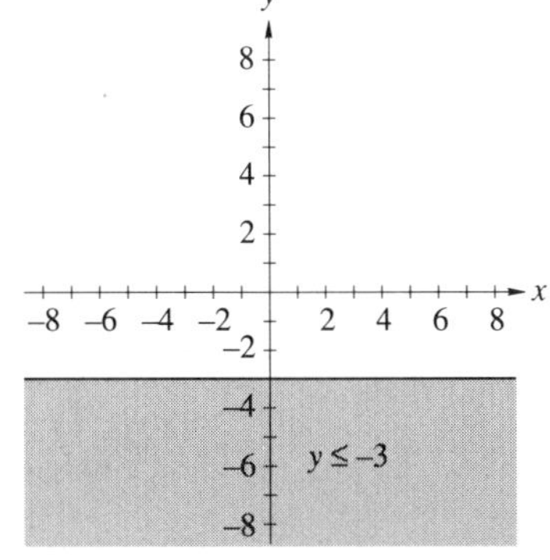

3.

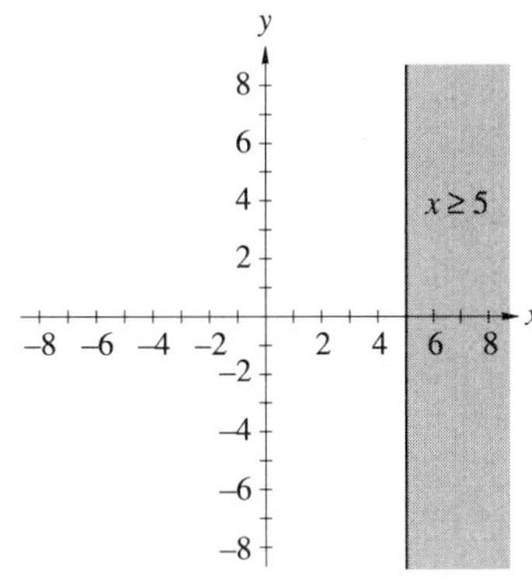

4.

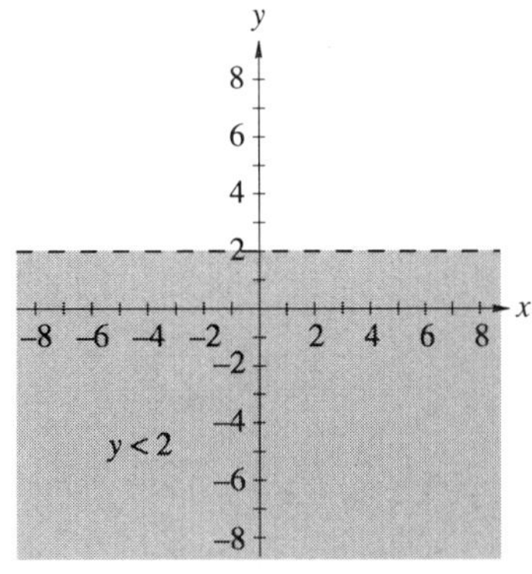

5.

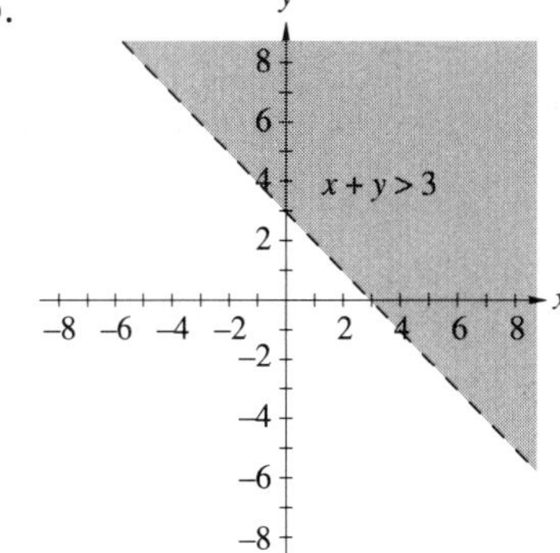

6.

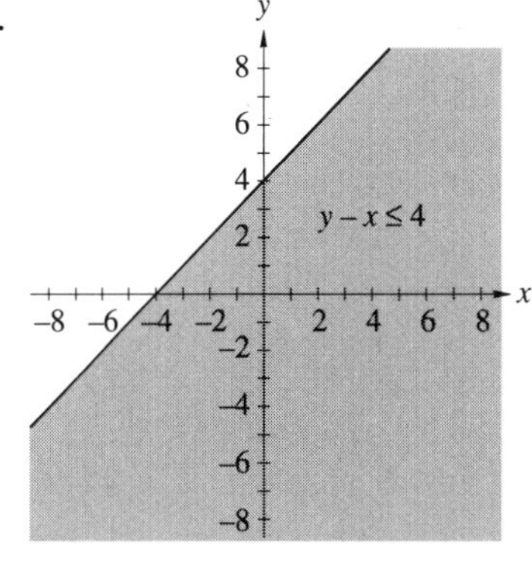

7.

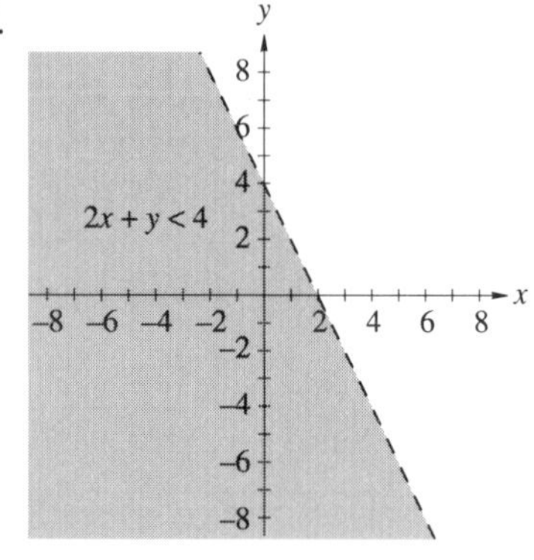

10.

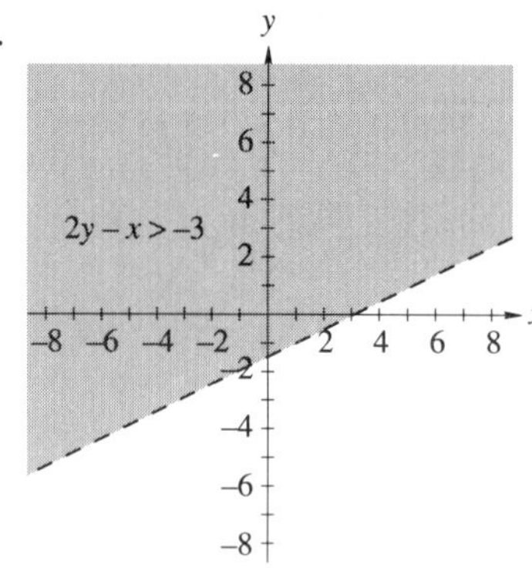

13.

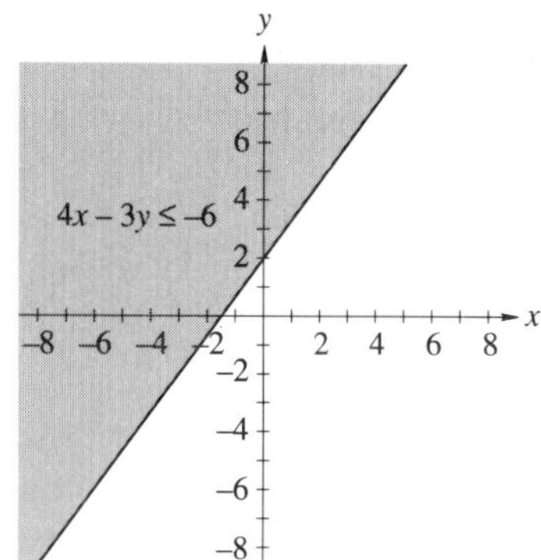

8.

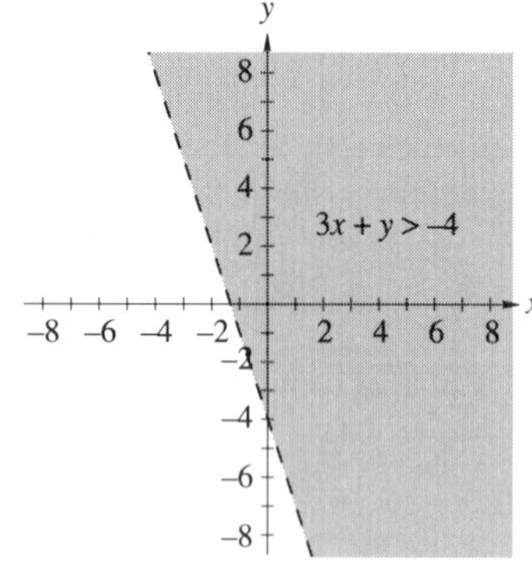

11.

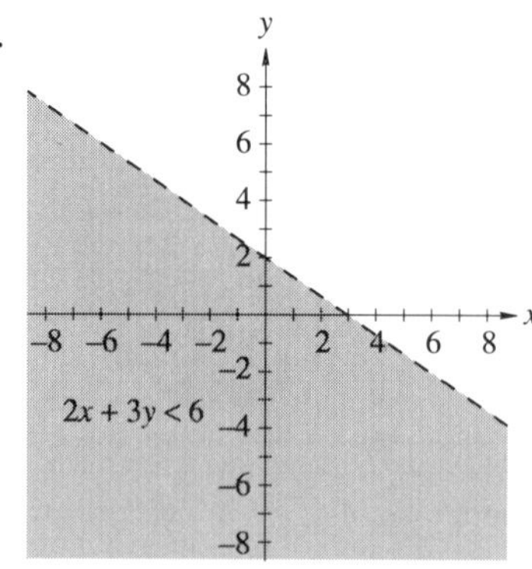

14.

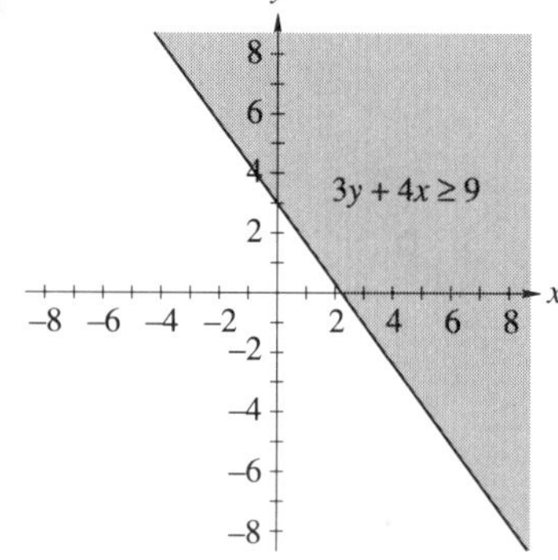

9.

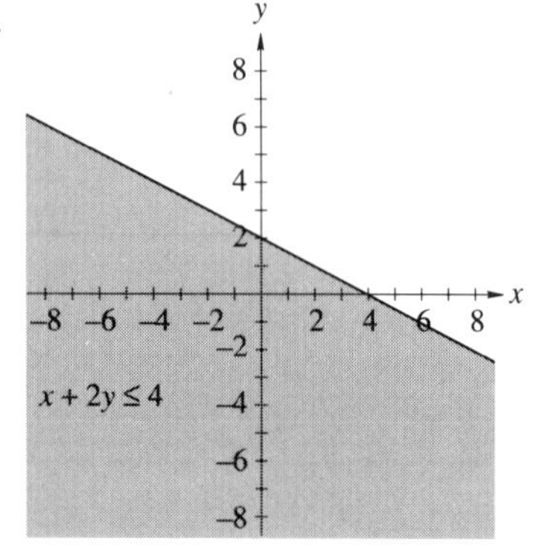

12.

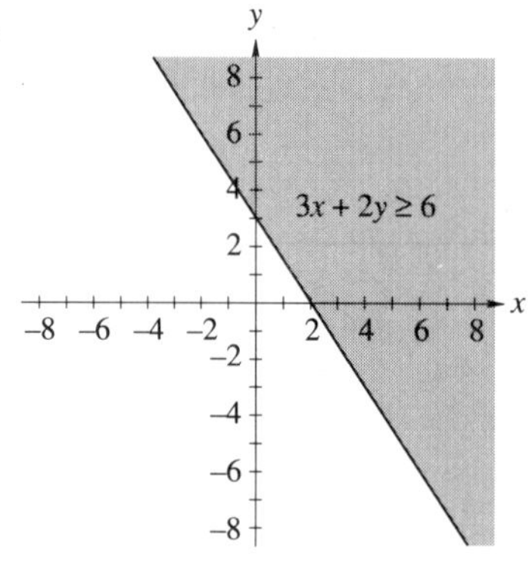

15.

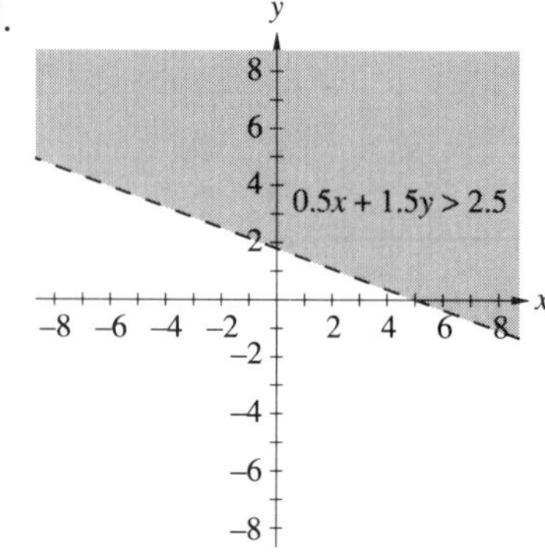

16.

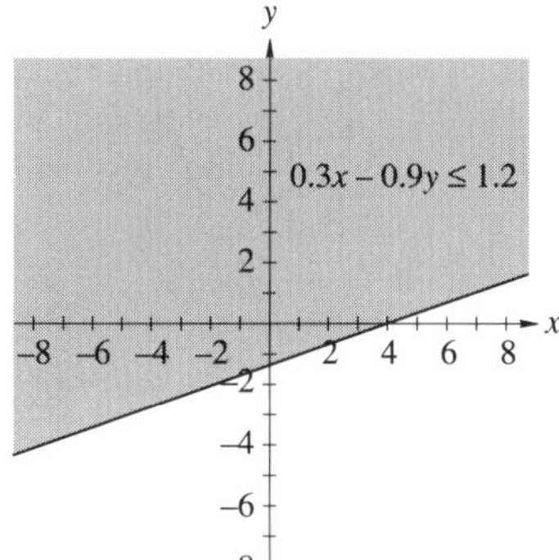

17.

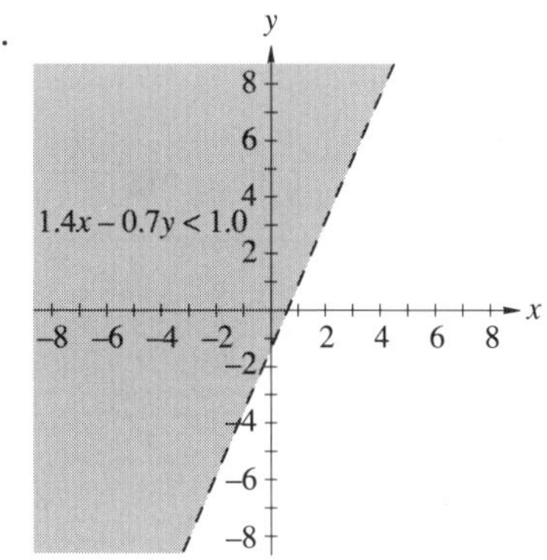

18.

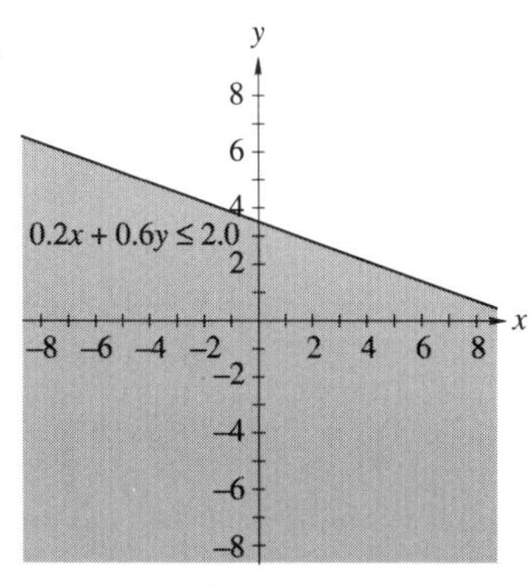

19.

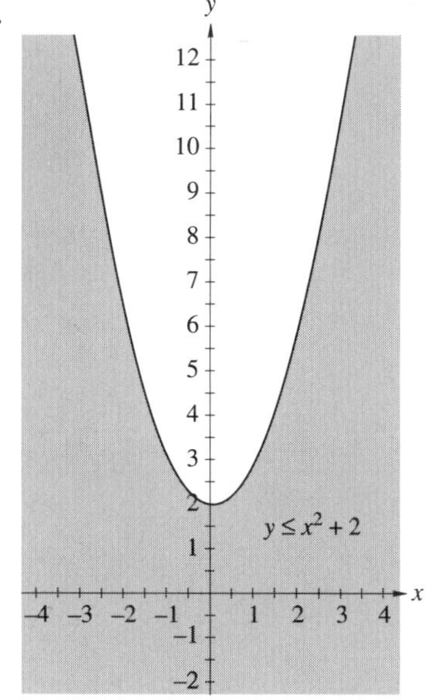

20.

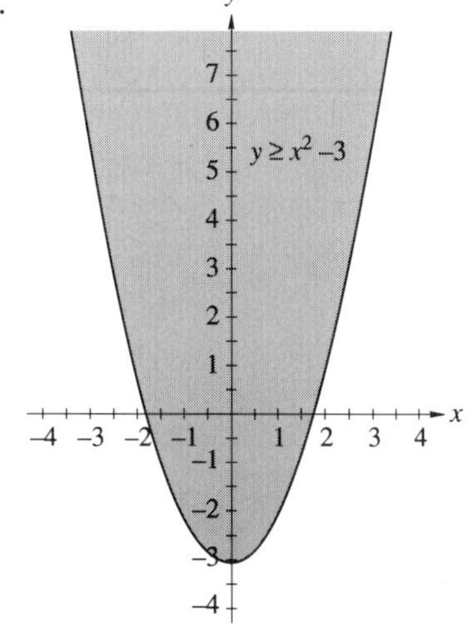

21.

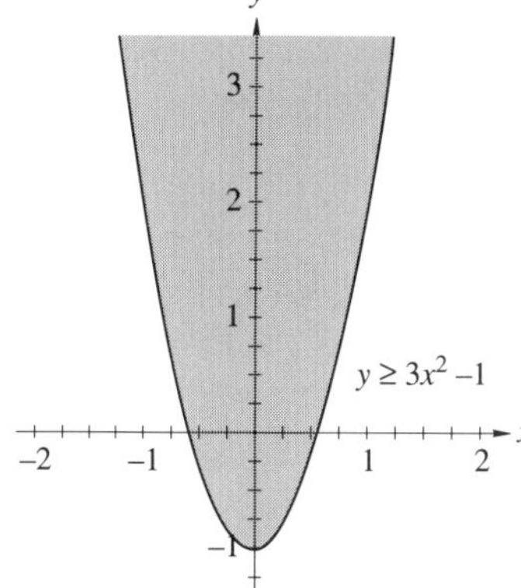

22.

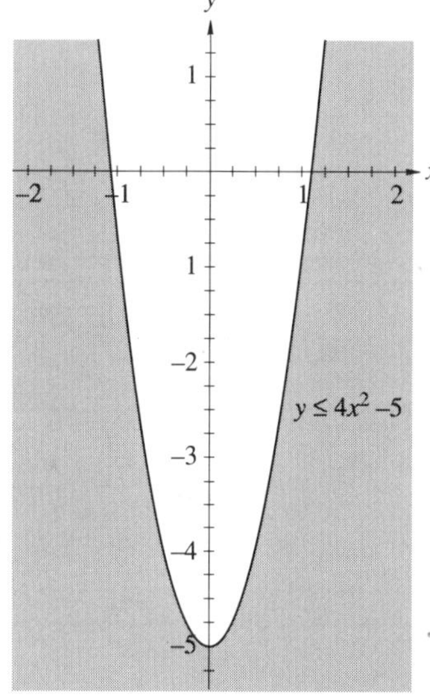

23.

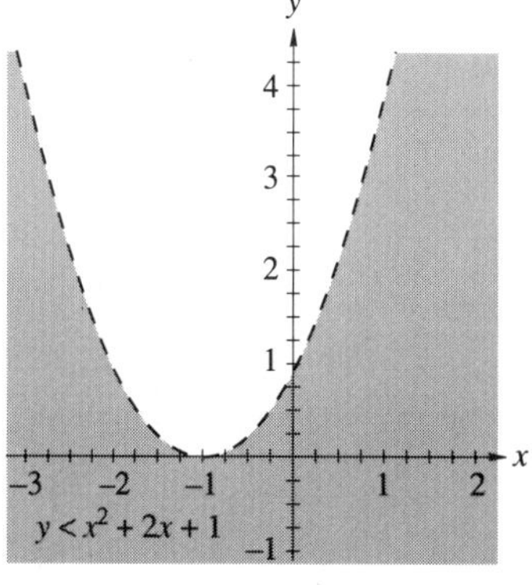

24.

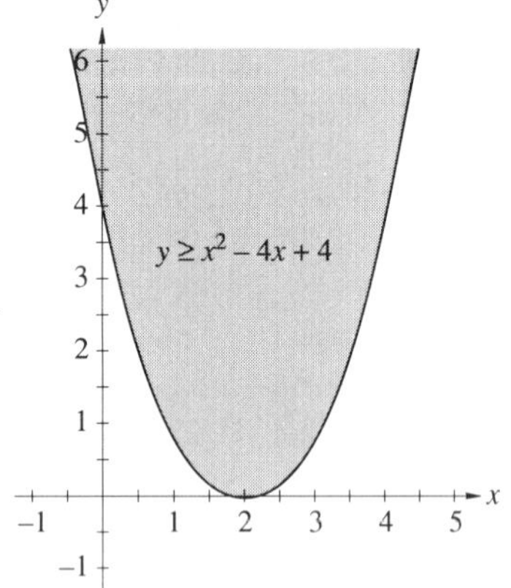

25.

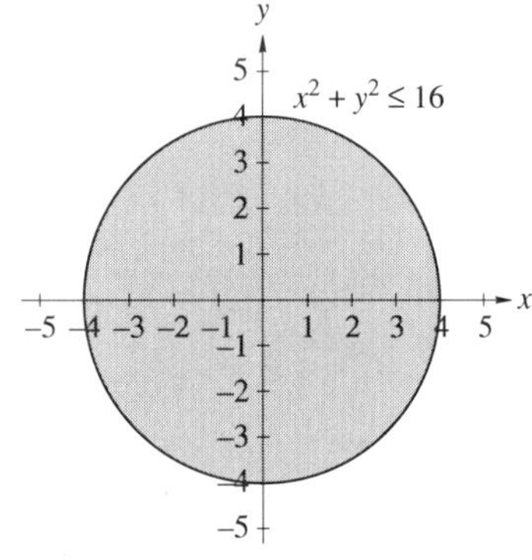

26.

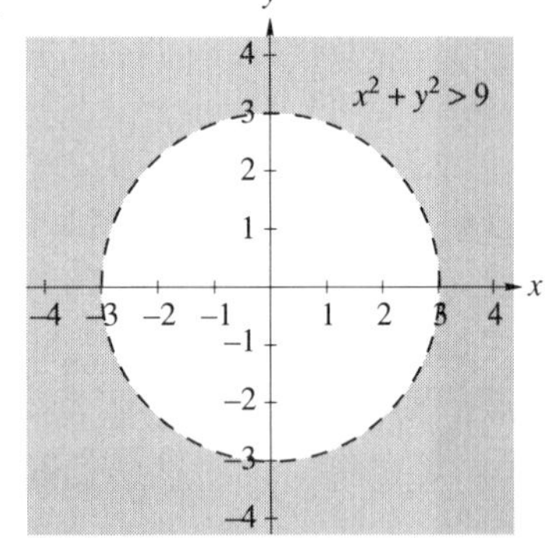

27.

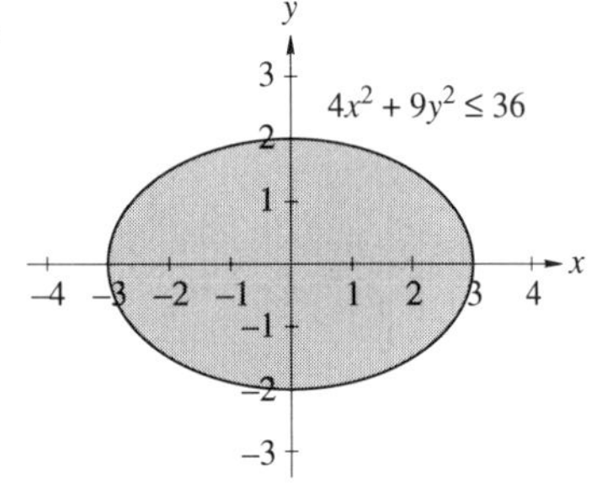

28.

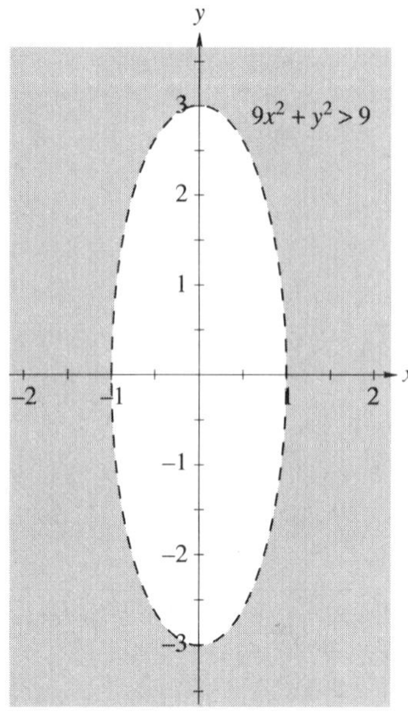

29. $|x + 1| < 4$, $-4 < x + 1 < 4$, $-5 < x < 3$

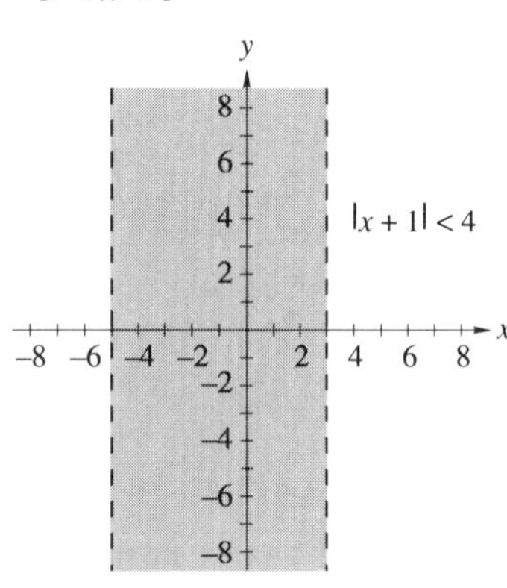

30. $|x - 5| \leq 3$, $-3 \leq x - 5 \leq 3$, $2 \leq x \leq 8$

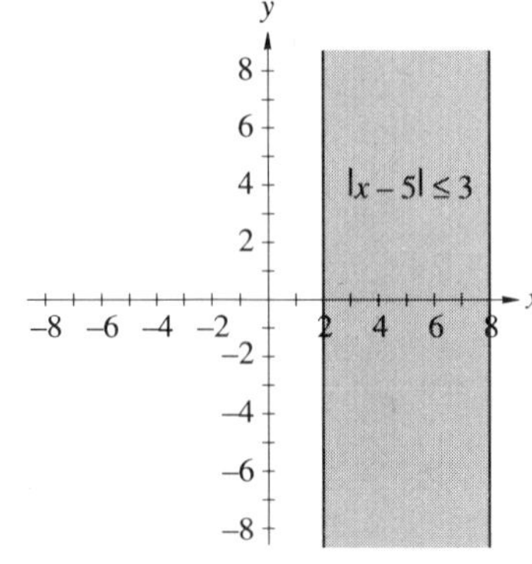

31.

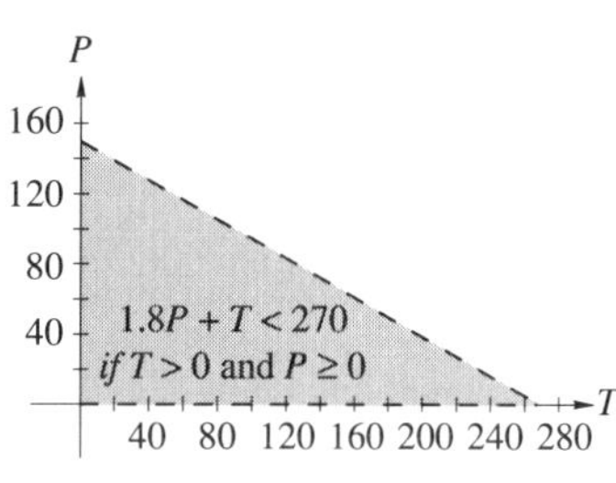

32.

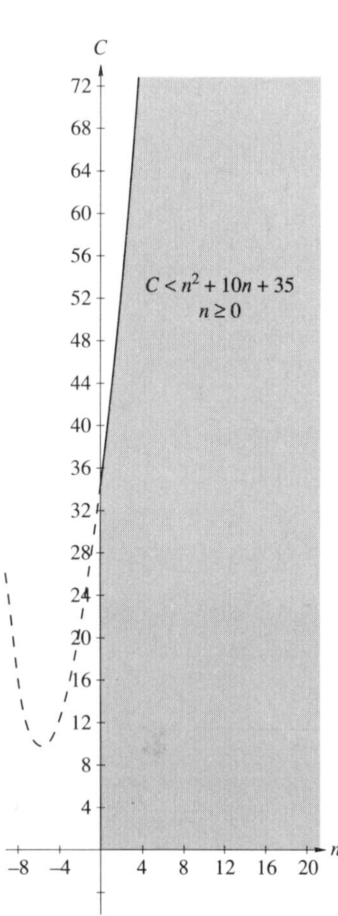

16.5 SYSTEMS OF INEQUALITIES: LINEAR PROGRAMMING

1.

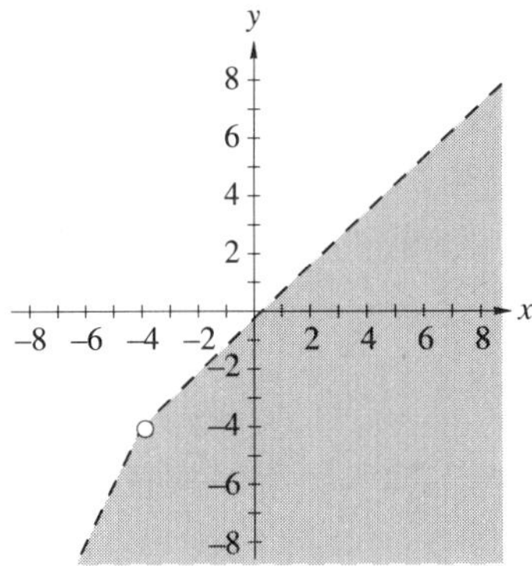

4.

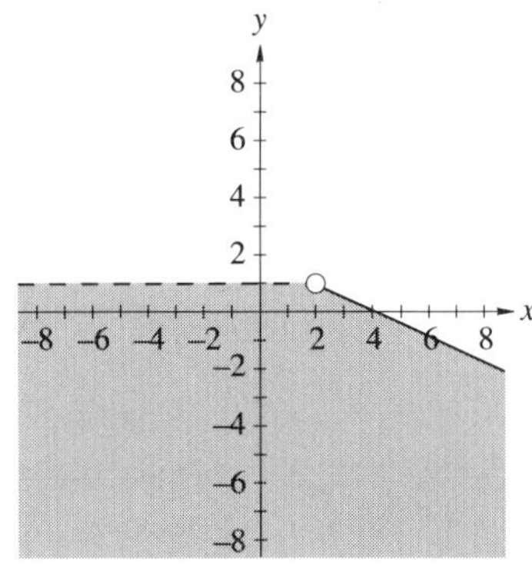

7.

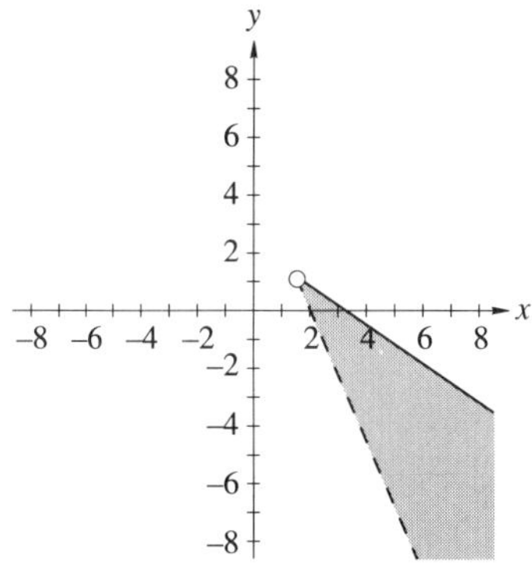

2.

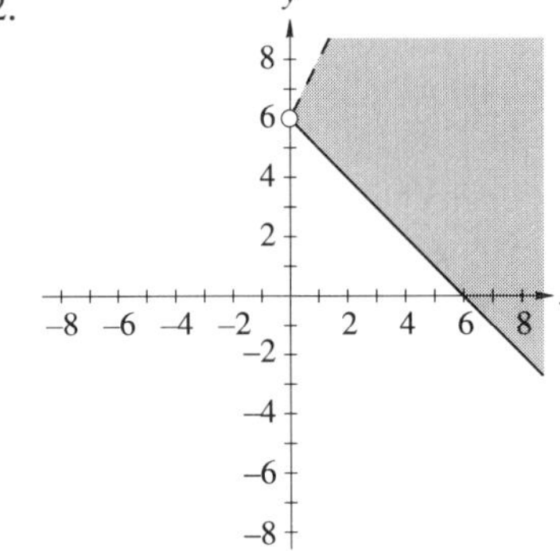

5.

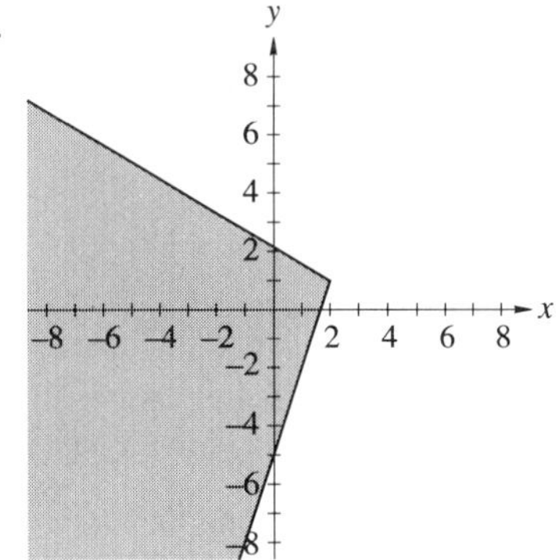

8.

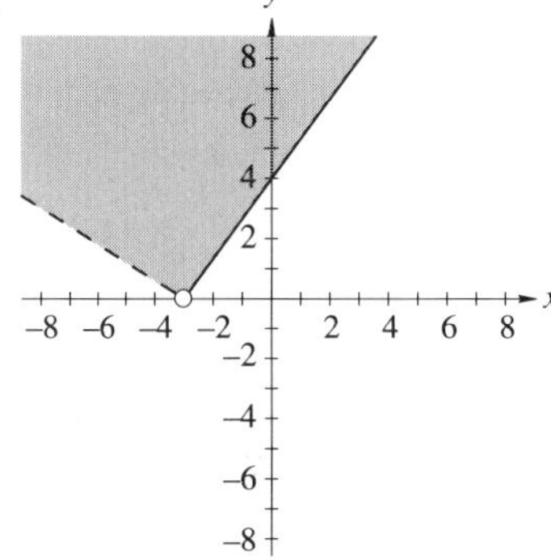

3.

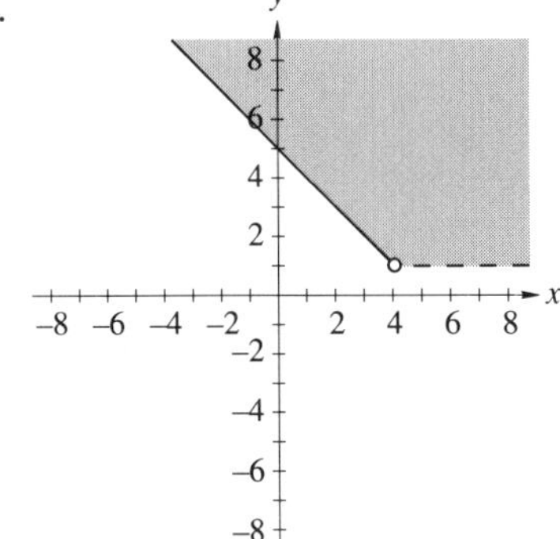

6.

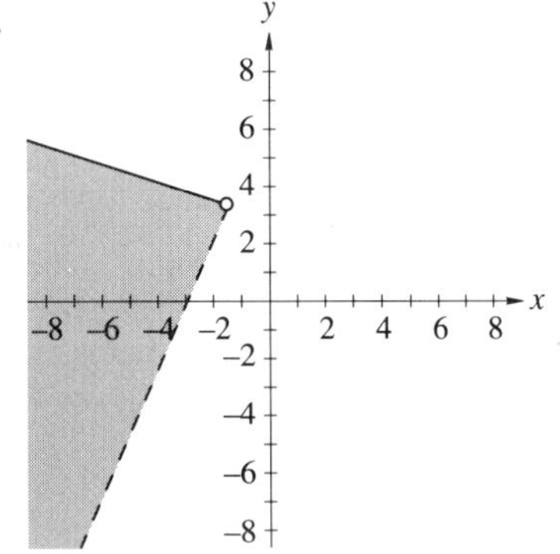

9.

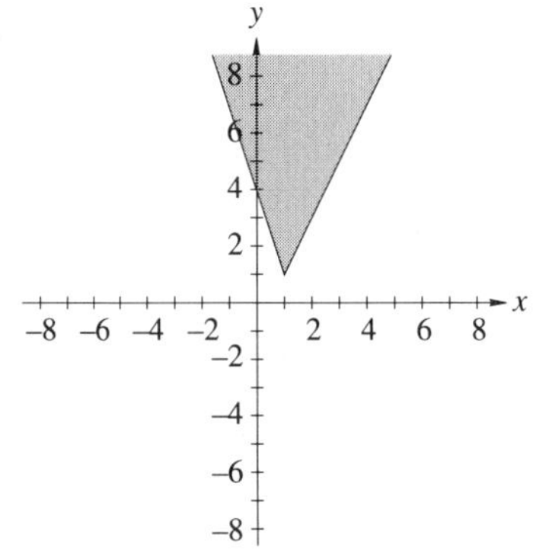

10.

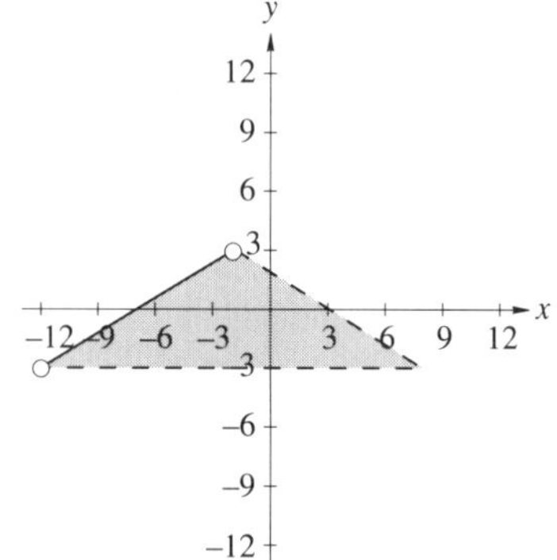

11.

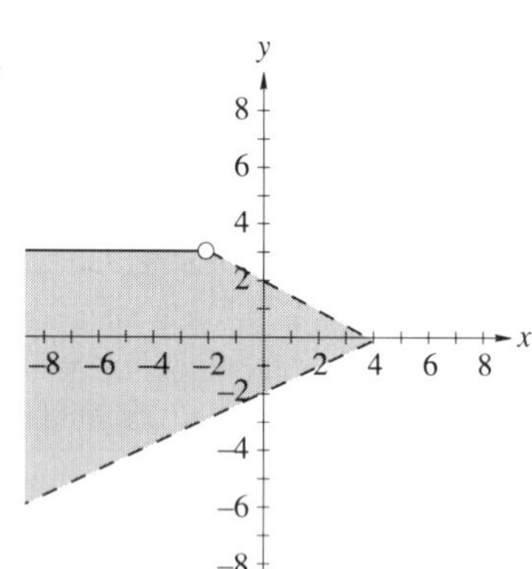

12.

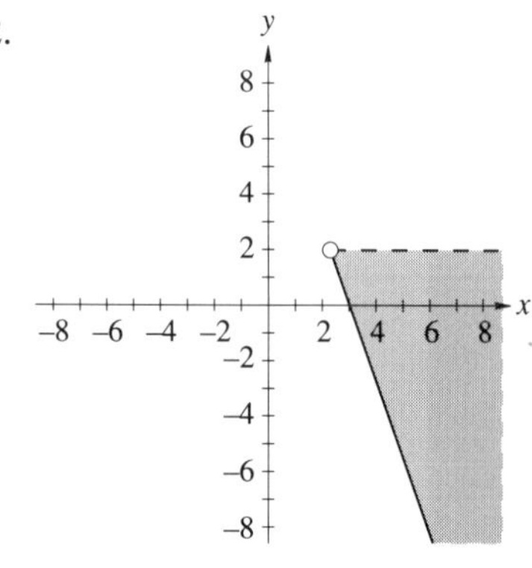

13.

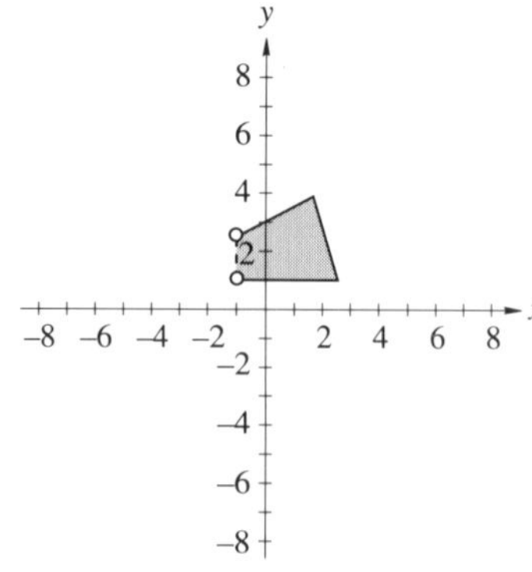

14.

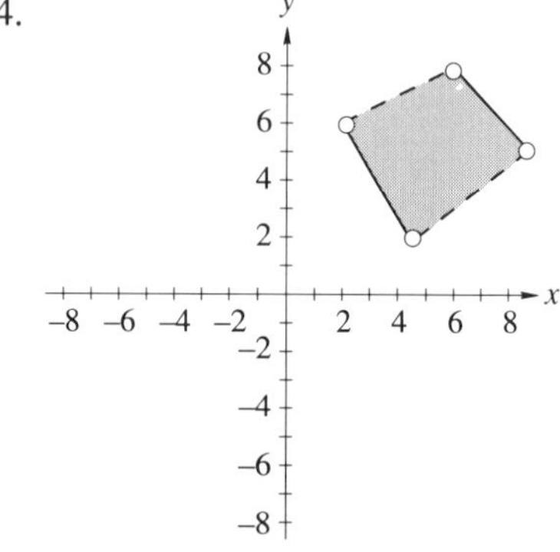

15. First graph the constraints. The vertex points of the feasible region are $(2,3)$, $(2,4.5)$ and $(4,3)$. (Fig. 16.79s). Evaluating $P = 2x + y$ at each of these points $(2,3)$: $2 \cdot 2 + 3 = 7$, $(2,4.5)$: $2 \cdot 2 + 4.5 = 8.5$, $(4,3): 2 \cdot 4 + 3 = 11$. Maximum of 11 at $(4,3)$.

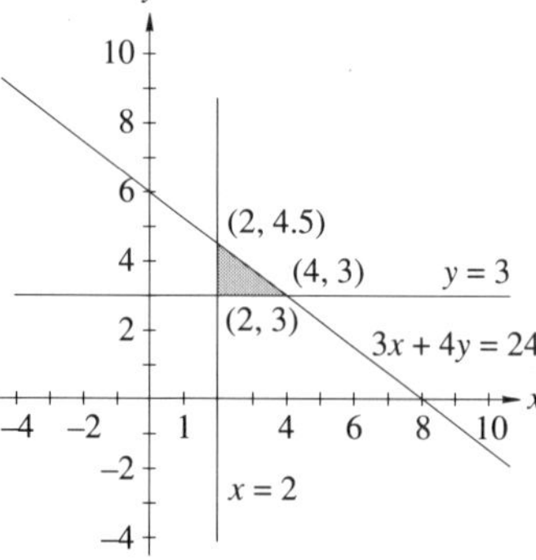

16. After graphing the constriants, the vertex points of the feasible region are at $(0,6.25)$, $(15,0)$ and $(3,4)$. $C(0,6.25) = 0 + 5 \times 6.25 = 31.25, C(15,0) = 9 \cdot 15 + 4 \cdot 0 = 135, C(4,3) = 9 \cdot 4 + 5 \cdot 3 = 51$. Minimum is 31.25 at $(0,6.25)$.

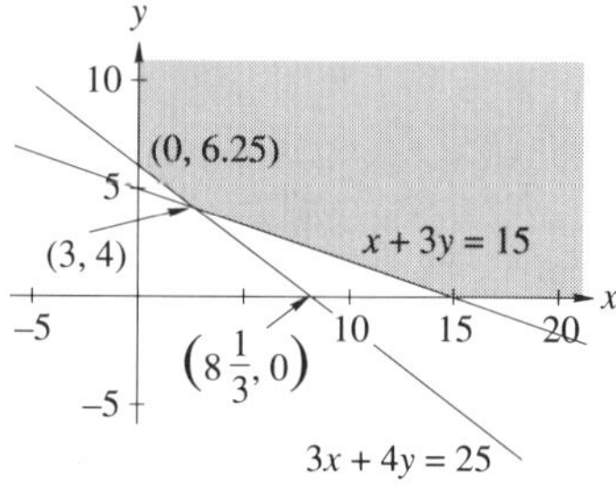

17. After graphing the constraints, the vertex points of the feasible region are $(0,0)$, $(0,23\frac{1}{3})$, $(26\frac{2}{3},0)$, and $(20,10)$. $F = 15x + 10y$, so $F(0,0) = 15 \cdot 0 + 10 \cdot 0 = 0$, $F(0,23\frac{1}{3}) = 10 \times 23\frac{1}{3} = 233.3, F(26\frac{2}{3},0) = 15 \times 26\frac{2}{3} = 400, F(20,10) = 15 \cdot 20 + 10 \cdot 10 = 400$ also. The maximum of 400 occurs at any point on the segment $3x + 2y = 80$ and $20 \leq x \leq 26\frac{2}{3}$.

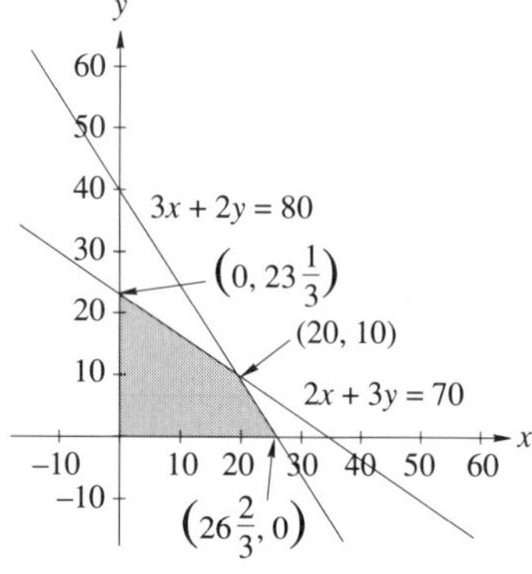

18. After graphing the constraints, the vertex points of the feasible region are at $(0, 2)$, $(0, 10)$, $(5, 7)$, and $(3\frac{1}{3}, 8\frac{2}{3})$. $F(0, 2) = 6 \cdot 0 + 9 \cdot 2 = 18$, $F(0, 10) = 6{\cdot}0+9{\cdot}10 = 90$, $F(5, 7) = 6{\cdot}5+9{\cdot}7 = 93$, $F\left(3\frac{1}{3}, 8\frac{2}{3}\right) = 6 \cdot 3\frac{1}{3} + 9 \cdot 8\frac{2}{3} = 20 + 78 = 98$. Minimum of 18 at $(0, 2)$

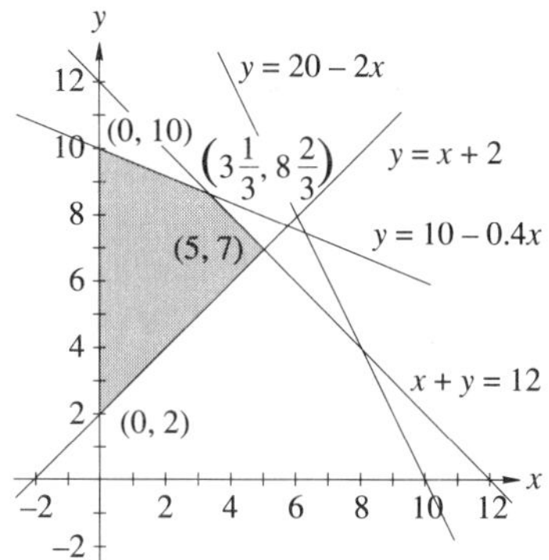

19. After graphing the constraints, the vertex points of the feasible region are at $(0, 3)$, $(0, 18)$, $(12, 12)$, and $(16.5, 3)$. Evaluating $P = 8x + 10y$ at each of these vertex points, we obtain $P(0, 3) = 30$, $P(0, 18) = 180$, $P(12, 12) = 216$, and $P(16.5, 3) = 162$. Maximum of 216 at $(12, 12)$.

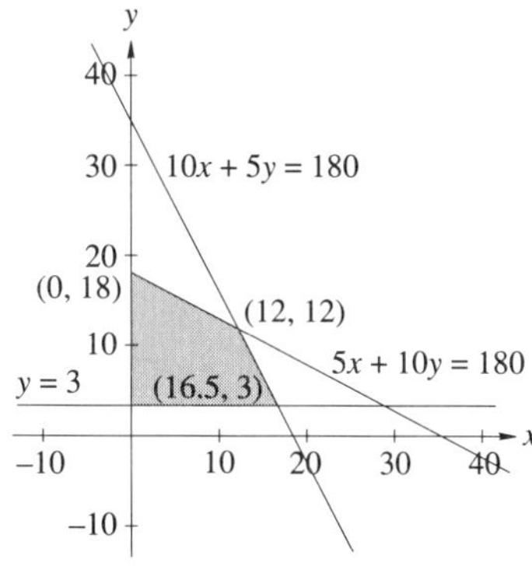

20. After graphing the constraints, the vertex points of the feasible region are at $(-2, -3)$, $(-5, 0)$, $(2, 7)$, and $(2, -3)$. Evaluating $M = 10x - 8y + 15$ at each of these vertex points, we obtain $M(-2, -3) = -20 + 24 + 15 = 19$, $M(-5, 0) = -50 + 15 = -35$, $M(2, 7) = 20 - 56 + 15 = -21$, and $M(2, -3) = 20 + 24 + 15 = 59$. Minimum of -35 at $(-5, 0)$.

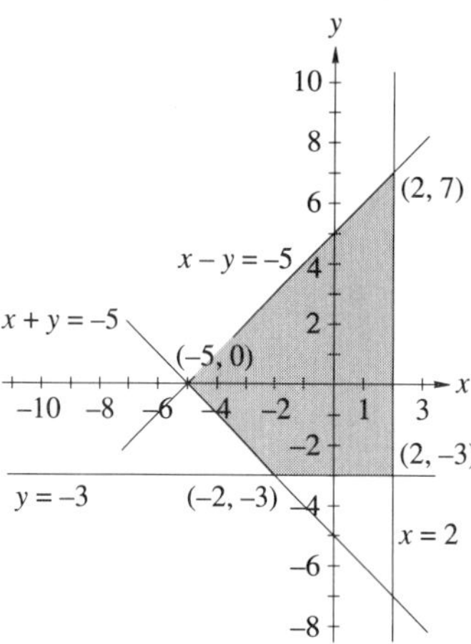

21. The constraint are as follows: $X \geq 0$, $Y \geq 0$, $\frac{1}{2}X + \frac{1}{4}Y \leq 100$, or $Y \leq 400 - 2X$, $\frac{3}{4}X + \frac{1}{6}Y \leq 80$ or $Y \leq 480 - \frac{9}{2}X$. Graphing the constraints gives vertex points at $(0,0)$, $(0{,}400)$, $\left(106\frac{2}{3}, 0\right)$, and $(32{,}336)$. Evaluating $P(x, y) = 10x + 8y$ at each of these vertex points, we get $P(0, 0) = 0$, $P(0{,}400) = 3200$, $P\left(106\frac{2}{3}, 0\right) = 1067$, and $P(32{,}336) = 3008$. Maximum profit of \$3200 when you produce 0 of X and 400 of Y. Adding people at A will increase profits.

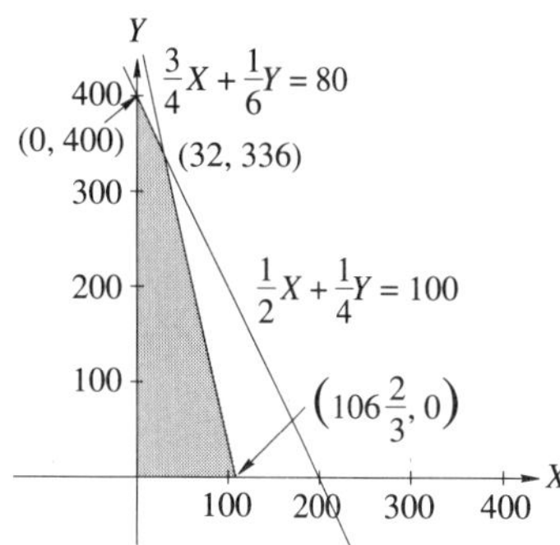

22. Labor constraint $A + B \leq 120$. Parts constraint $4A + 3B \leq 390$ along with the constraints $A \geq 0$ and $B \geq 0$. Graphing the constraints give vertex points of the feasible region at $(0, 0)$, $(0{,}120)$, $(30, 90)$, and $(97.5, 0)$. $P(A, B) = 7A + 5.5B$, $P(0, 0) = 0$, $P(0{,}120) = 660$, $P(30, 90) = 705$, $P(97.5, 0) = 682.5$. Thus, we see that 30 of A and 90 of B will produce a maximum profit of \$705.00.

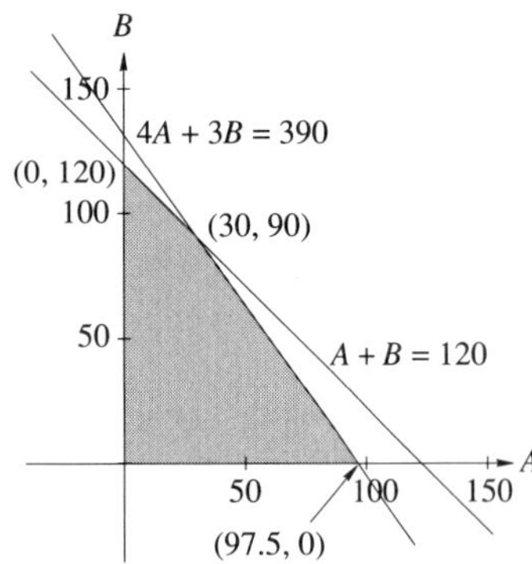

23. $P(A,B) = 8A + 9.50B$, $P(0,0) = 0$, $P(0,120) = 1140$, $P(30,90) = 1095$, $P(97.5,0) = 780$. The maximum profit will be when the company makes 120 of B and none of A for a profit of \$1,140.00.

24. Let P represent the number of PCs that the company makes and B the number of BCs that the company makes. The contraint on the AB chip means that $2P + 3B \le 200$, and the contraint on the EP chip means that $3P + 8B \le 450$. The nature of the problem means that $P \ge 0$ and $B \ge 0$. The feasible region is shown in the figure. Vertex points of the feasible region are at $(0,0)$, $(100,0)$, $(0,56.25)$, and $(35.71,42.86)$. The profit function is $P(P,B) = 145P + 230B$. Evaluating the profit function at each of the vertex points, we see that $P(0,0) = 0$, $P(100,0) = 14{,}500$, $P(35.71,42.86) = 15{,}035.75$, and $P(0,56.25) = 12{,}937.5$. So, making 35.71 of PC and 42.86 of BC will produce a maximum profit of \$15,035.75.

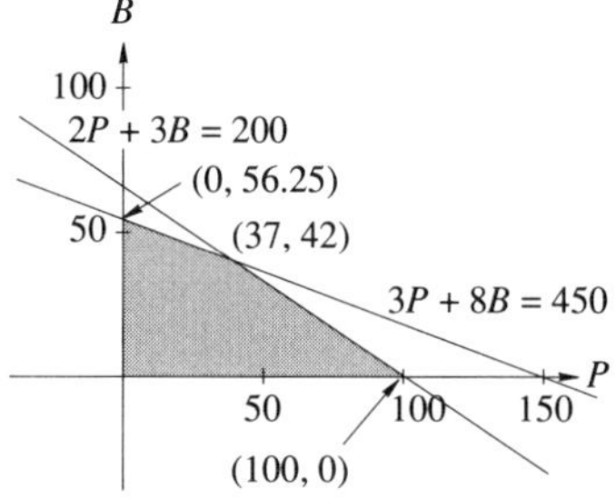

CHAPTER 16 REVIEW

1. $3x < -12$ or $x < -4$

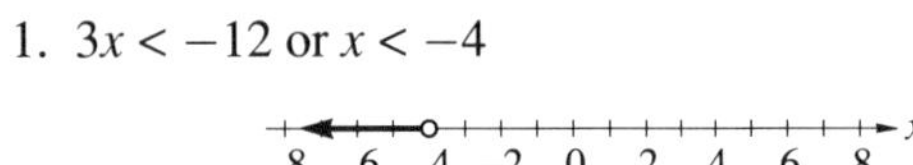

2. $4x + 5 < 6$ or $4x < 1$ and so $x < \frac{1}{4}$

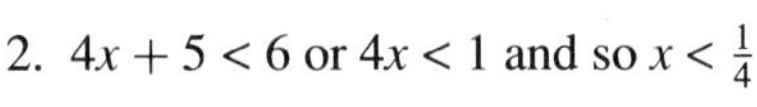

3. $2x - 7 \ge 15$ or $2x \ge 22$ which gives $x \ge 11$

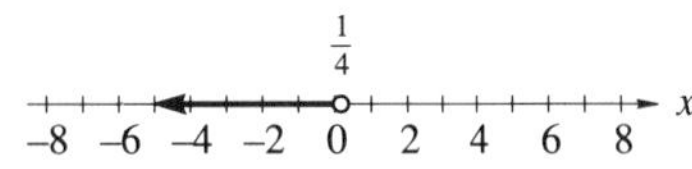

4. $7 - 5x < 2 + 3x$ or $-8x < -5$, and so $x > \frac{5}{8}$

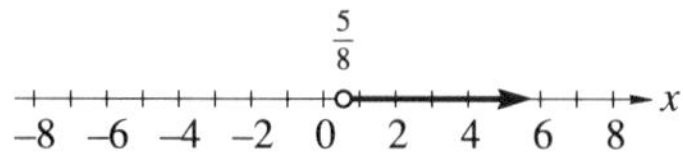

5. $\frac{2x+5}{4} < \frac{4x-1}{3}$ or $6x + 15 < 16x - 4$ which simplifies to $-10x < -19$, and so $x > \frac{19}{10}$.

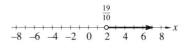

6. $|2x - 1| \le 5$ is equivalent to $-5 \le 2x - 1 \le 5$ or $-4 \le 2x \le 6$, and so $-2 \le x \le 3$.

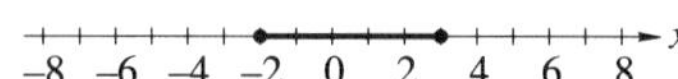

7. $|3x + 2| > 7$ is equivalent to
$3x + 2 < -7$ or $3x + 2 > 7$
$3x < -9$ or $3x > 5$
$x < -3$ or $x > \frac{5}{3}$

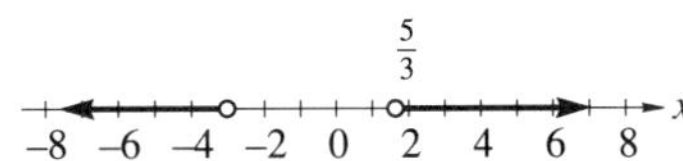

8. $2x + 3 < 13 \leq 3x - 9$ is equivalent to
$2x + 3 < 13$ and $13 \leq 3x - 9$
$2x < 10$ and $22 \leq 3x$
$x < 5$ and $x \geq \frac{22}{3}$
This is a contradictory inequality and so there is no solution.

9. $-2x + y < 5$ or $y < 2x + 5$

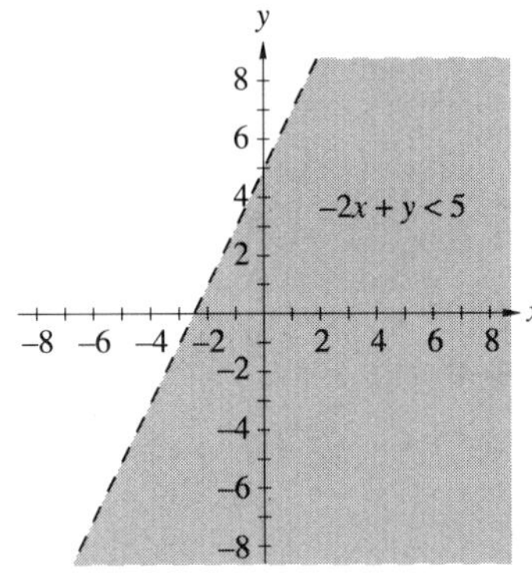

10. $2x + 3y \geq 6$ or $y \geq \frac{-2x}{3} + 2$

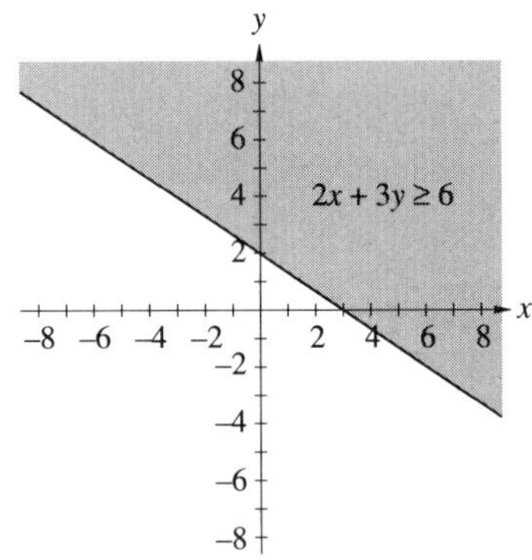

11. $4x - 3y < 3$ or $y > \frac{4}{3}x - 1$

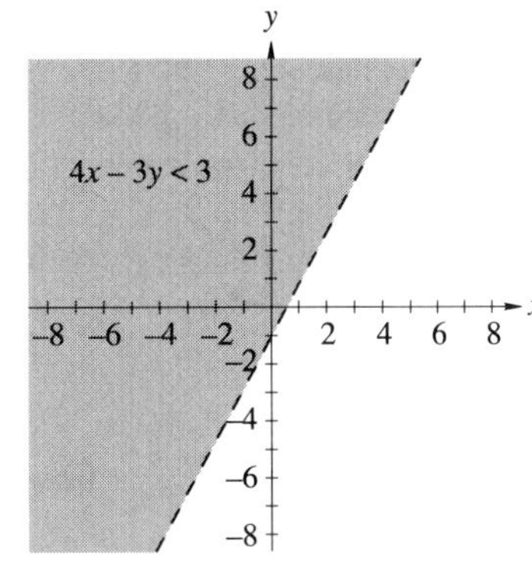

12. $y > 2x^2 - 5$

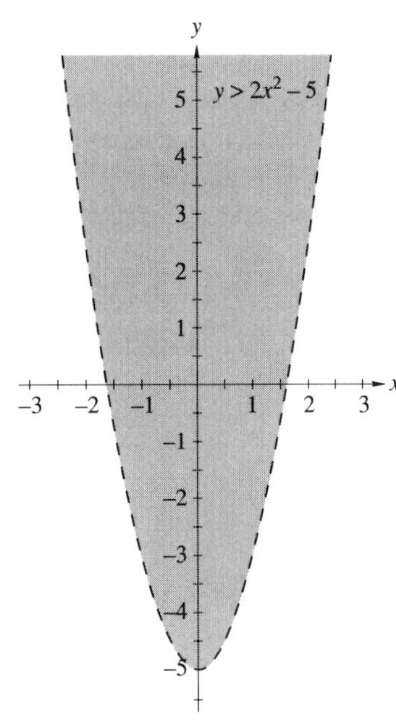

13. $y \leq x^2 - 6x + 9$

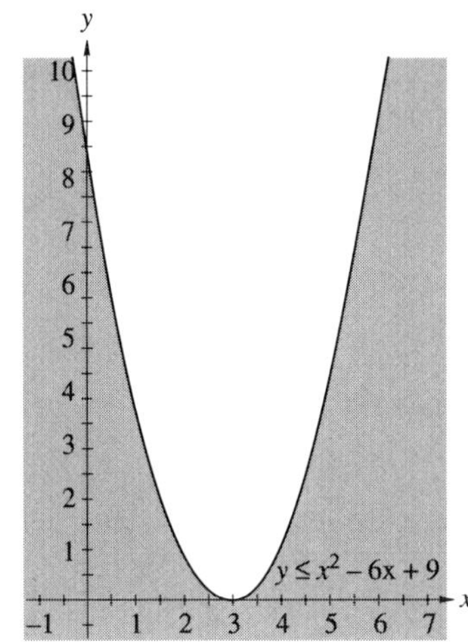

14.

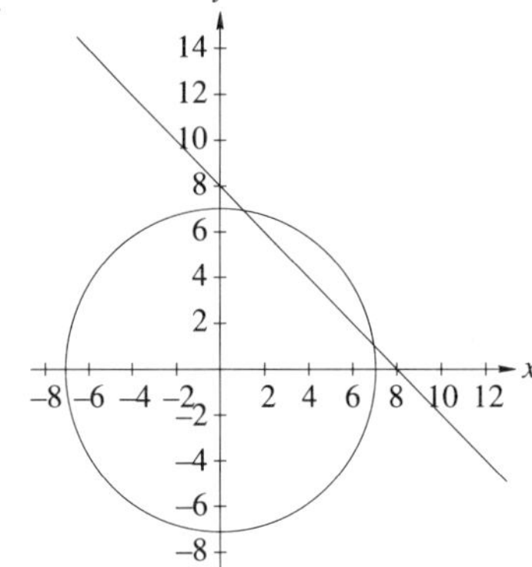

15.

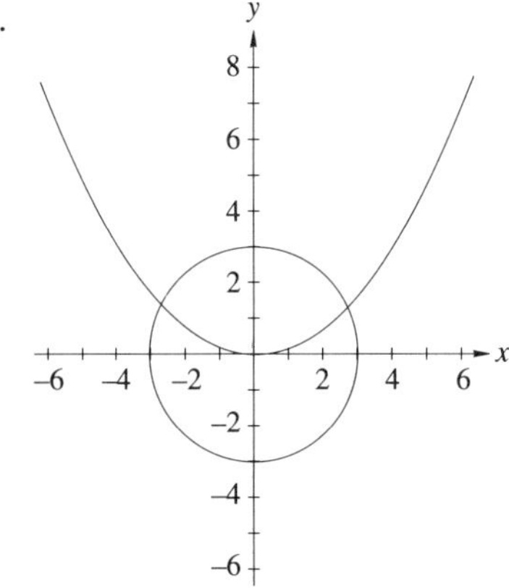

16.

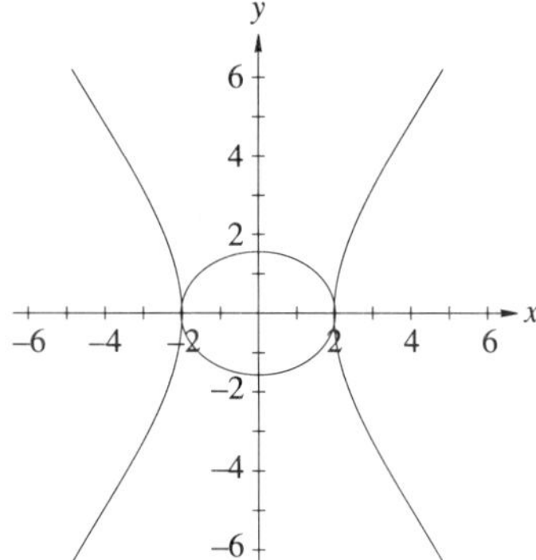

17.

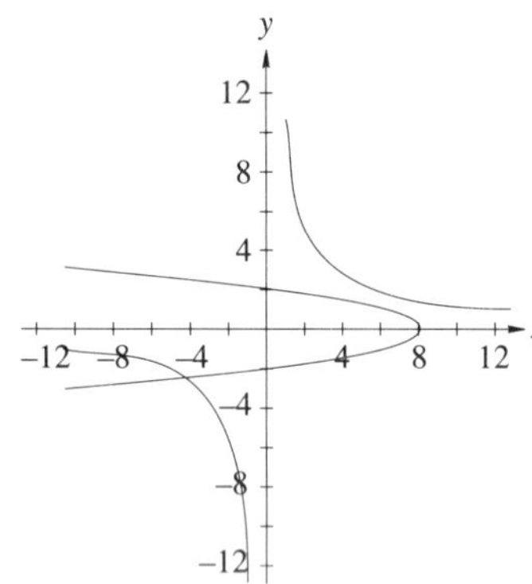

18.

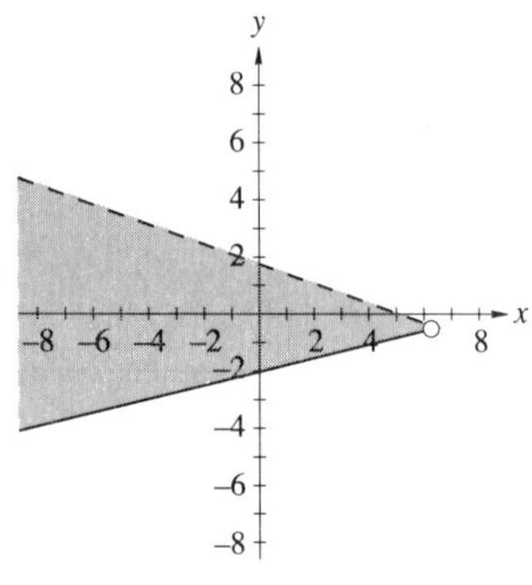

19.

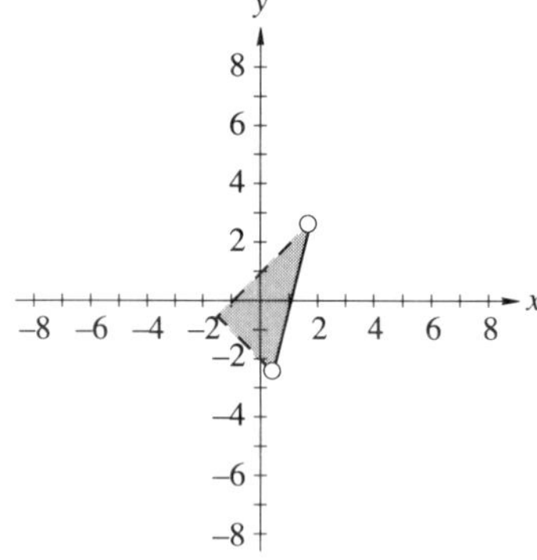

20.

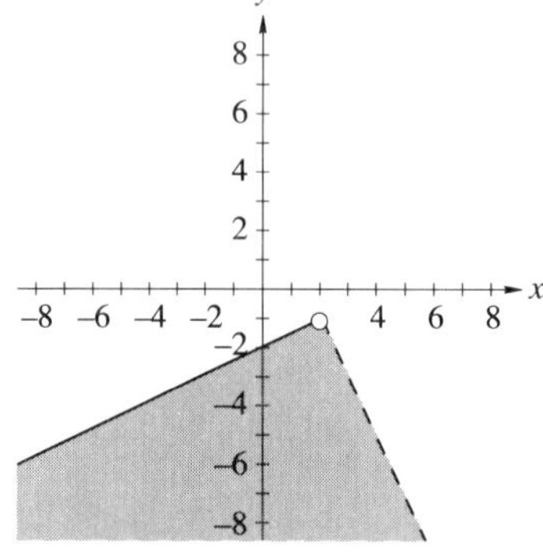

21.

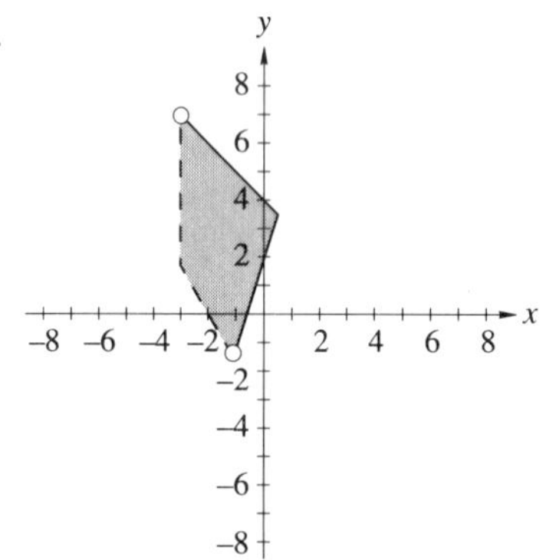

22. Graphing the constraints as shown, we see that the verticies of the feasible region are $(-1,6)$, $(-1,-3)$, $(3,1)$ and $(3,2)$. We have $P(x,y) = 3x + 5y$, $P(-1,6) = -3 + 30 = 27$, $P(-1,-3) = -3 - 15 = -18$, $P(3,1) = 9 + 5 = 14$, and $P(3,2) = 9 + 10 = 19$. We can see that there is a maximum of 27 at $(-1,6)$

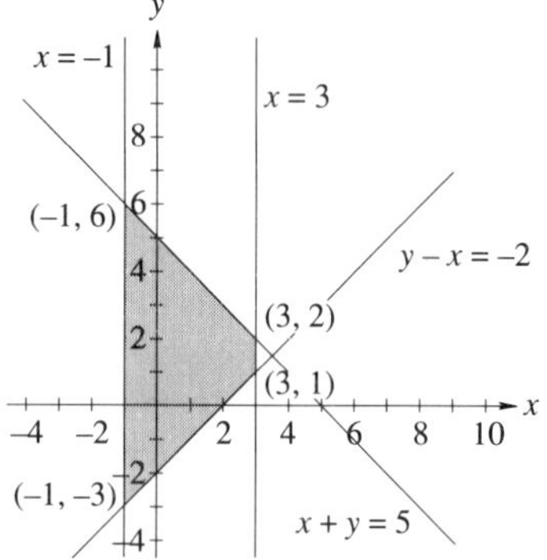

23. After graphing, we see that the vertices of the feasible region are $(\frac{-4}{7}, \frac{-23}{7})$, $(\frac{18}{5}, -\frac{6}{5})$, and $(\frac{7}{6}, \frac{11}{3})$. Checking $F(x,y) = 4x - 3y$ at these points, we see that $F(\frac{4}{7}, \frac{23}{7}) = 7\frac{4}{7}$, $F(\frac{7}{6}, \frac{11}{3}) = -6\frac{1}{3}$, and $F(\frac{18}{5}, -\frac{6}{5}) = 18$. So, the minimum of $-6\frac{1}{3}$ is at $\left(\frac{7}{6}, \frac{11}{3}\right)$ or $\left(1\frac{1}{6}, 3\frac{2}{3}\right)$.

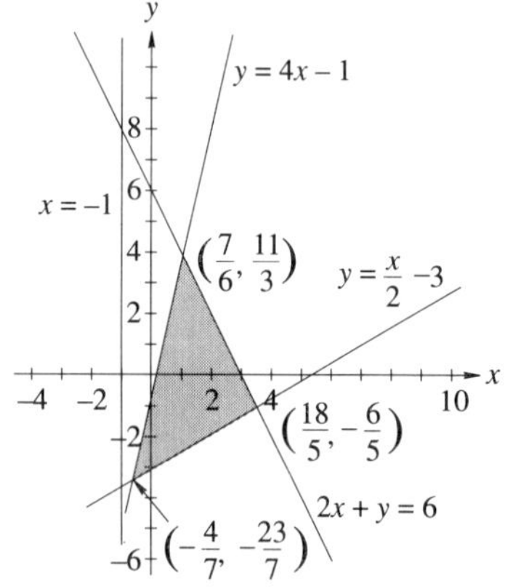

24. $A = lw.$, $7.0 \times 1.4 < A < 7.5 \times 1.6$, $9.8\text{ mm}^2 < A < 12\text{ mm}^2$

25. Since $R_1R_2 = 20$, we see that $\dfrac{R_1R_2}{R_1+R_2} \geq 4$ is equivalent to $\dfrac{20}{R_1+R_2} \geq 4$ or $20 \geq 4(R_1+R_2)$ and so $R_1+R_2 \leq 5\ \Omega$.

26.

	r	t	d
normal	r	8	$8r$
increased	$r+4$	7.5	$7.5(r+4)$

Thus, we have the equation $8r = 7.5r + 30$, or $0.5r = 30$, and so $r = 60$. Usual speed is 60 mph, distance is $8 \cdot 60 = 480$ mi.

27. We have the system: $12C + 18S \leq 220$, $15C + 10S \leq 180$, $C \geq 0$, $S \geq 0$, C and S must be integers. We graph C on the horizontal axis, and S on the vertical axis. Solving the equation derived from the second inequality, we get $S = 18 - 1.5C$. Substituting this value of S into the equation from the first inequality yields $12C+18(18-1.5C) = 220$ or $12C + 324 - 27C = 220$ which simplifies to $-15C = -104$. Thus, we find that these two lines intersect when $C = 6.9\overline{3}$. Possible vertex points are $(0,0)$, $(0,12)$, $(6,8)$, $(7,7)$ and $(12,0)$. Evaluating $P(C,S) = 1250C + 1620S$ at each of these vertex points, we find $P(0,0) = 0$, $P(0,12) = 19{,}440$, $P(6,8) = 20{,}460$, $P(7,7) = 20{,}090$, and $P(12,0) = 15{,}000$. Maximum profit of \$20,460 at $(6,8)$ or 6 of C and 8 of S.

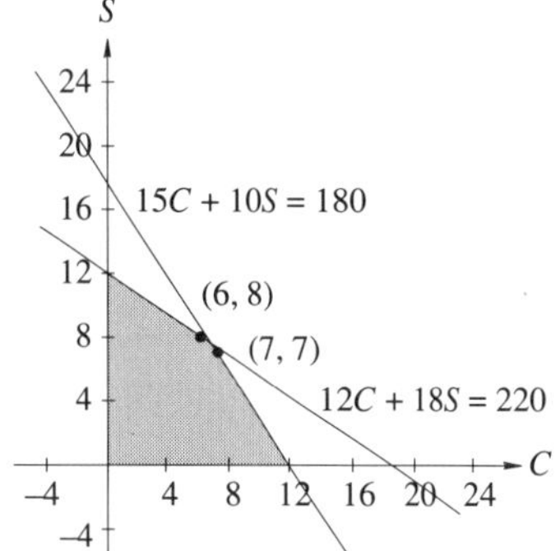

CHAPTER 16 TEST

1. $5x > 30$, $x > 6$

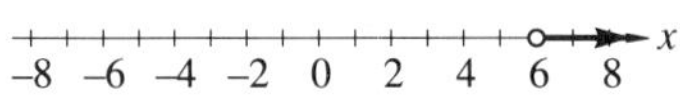

2. $|4x+5| < 17$, $-17 < 4x+5 < 17$, $-22 < 4x < 12$, $-\frac{11}{2} < x < 3$

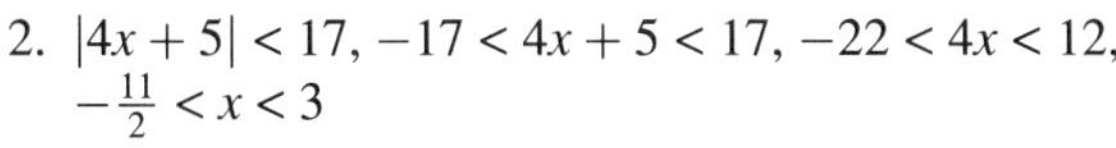

3. $\frac{4x-5}{2} \leq \frac{7x-2}{3}$, $12x - 15 \leq 14x - 4$, $-2x \leq 11$, $x \geq -\frac{11}{2}$

4. $(x-2)(x-5) > 0$ has two cases. In Case I, we have $x - 2 > 0$ or $x > 2$ and that $x - 5 > 0$ or $x > 5$. These both occur when $x > 5$. In Case II, we have $x - 2 < 0$ or $x < 2$ and $x - 5 < 0$ or $x < 5$. These both occur when $x < 2$. Thus the solution set is $\{x | x < 2 \text{ or } x > 5\}$

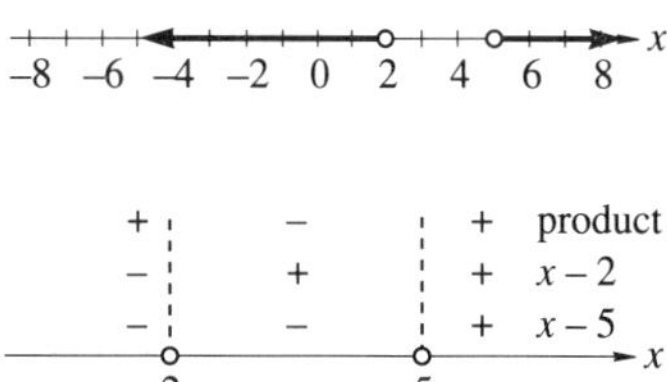

5. $5x^2+3x \leq 2x^2-x-1$, $3x^2+4x+1 \leq 0$, $(3x+1)(x+1) \leq 0$. The solution set is $\{x | -1 \leq x \leq -\frac{1}{3}\}$

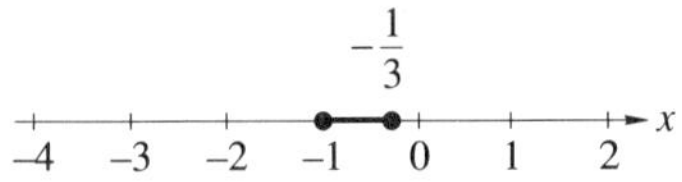

6. 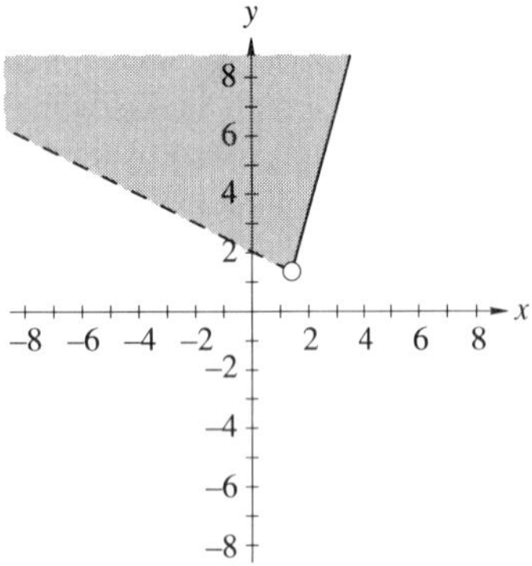

7. Iron constraint gives the inequality $12x + 5y \geq 80$, the zinc constraint gives the inequality $10x + 8y \geq 90$, while $x \geq 0$, and $y \geq 0$. The vertex points are on the axes are $(0, 16)$ and $(9, 0)$. The two equations $10x + 8y = 90$ and $12x + 5y = 80$ intersect at $\left(\frac{95}{23}, \frac{140}{23}\right) \approx (4.13, 6.09)$. The points $(5, 5)$ and $(4, 7)$ are the closest to this vertex. The cost is given by the equation $C(x, y) = 4x + 6y$. Evaluating the cost function at each of these four points we get $C(0, 15) = 90¢$, $C(4, 7) = 58¢$, $C(5, 5) = 50¢$, and $C(9, 0) = 36¢$. (a) The minimum cost is obtained by taking 9 of brand X and none of Y. (b) The minimum cost is 36¢.

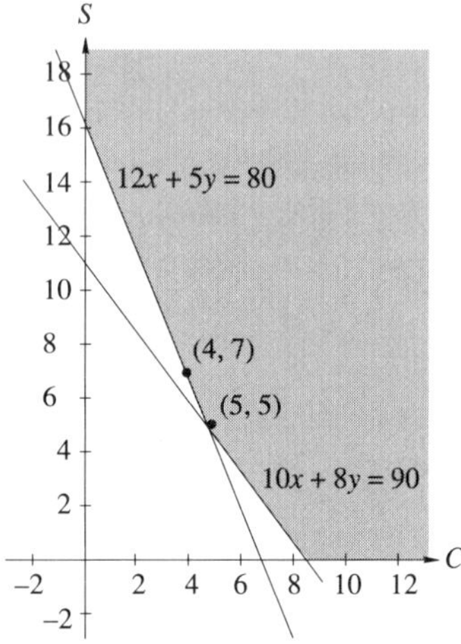

CHAPTER

17

Matrices

17.1 MATRICES

1. 2×3
2. 3×3
3. 3×6
4. 4×1
5. 4×2
6. 1×7
7. $a_{11} = 1$; $a_{24} = 13$; $a_{21} = 8$; $a_{32} = 6$
8. $b_{12} = 0$; $b_{21} = 4$; $b_{23} = 9$; $b_{32} = -8$
9. $x = 12$, $y = -6$, $z = 2$, $w = 5$
10. $x - 2 = 7$, so $x = 9$; $y + 5 = 3$, so $y = -2$; $3z = 3$, so $z = 1$; $w + 6 = 7$, so $w = 1$; and $2p = 10$, so $p = 5$.
11. $\begin{bmatrix} -3 & 6 & 6 & -8 \\ 12 & 9 & 17 & 0 \\ 9 & 0 & 5 & 10 \end{bmatrix}$
12. $\begin{bmatrix} 7 & -2 \\ -3 & 10 \\ 11 & 6 \end{bmatrix}$
13. $\begin{bmatrix} 2 & -1 & 4 \\ 5 & 9 & -2 \end{bmatrix}$
14. $\begin{bmatrix} 2 & 0 & 3 & 3 \\ 3 & 0 & 9 & 1 \\ 14 & 8 & 0 & -14 \\ 10 & 10 & 2 & 10 \\ 0 & 6 & 23 & 4 \end{bmatrix}$
15. $A + B = \begin{bmatrix} -1 & 7 & 4 \\ 8 & 1 & 15 \end{bmatrix}$
16. $B - C = \begin{bmatrix} -10 & -4 & -2 \\ -3 & 3 & -1 \end{bmatrix}$
17. $3A = \begin{bmatrix} 12 & 9 & 6 \\ 15 & 0 & 21 \end{bmatrix}$
18. $4B = \begin{bmatrix} -20 & 16 & 8 \\ 12 & 4 & 32 \end{bmatrix}$
19. $2B + C = \begin{bmatrix} -10 & 8 & 4 \\ 6 & 2 & 16 \end{bmatrix} + \begin{bmatrix} 5 & 8 & 4 \\ 6 & -2 & 9 \end{bmatrix} = \begin{bmatrix} -5 & 16 & 8 \\ 12 & 0 & 25 \end{bmatrix}$
20. $4A - 3C = \begin{bmatrix} 16 & 12 & 8 \\ 20 & 0 & 28 \end{bmatrix} - \begin{bmatrix} 15 & 24 & 12 \\ 18 & -6 & 27 \end{bmatrix} = \begin{bmatrix} 1 & -12 & -4 \\ 2 & 6 & 1 \end{bmatrix}$
21. $2B - 3A = \begin{bmatrix} -10 & 8 & 4 \\ 6 & 2 & 16 \end{bmatrix} - \begin{bmatrix} 12 & 9 & 6 \\ 15 & 0 & 21 \end{bmatrix} = \begin{bmatrix} -22 & -1 & -2 \\ -9 & 2 & -5 \end{bmatrix}$
22. $4C + 2B = \begin{bmatrix} 20 & 32 & 16 \\ 24 & -8 & 36 \end{bmatrix} + \begin{bmatrix} -10 & 8 & 4 \\ 6 & 2 & 16 \end{bmatrix} = \begin{bmatrix} 10 & 40 & 20 \\ 30 & -6 & 52 \end{bmatrix}$

23. $D - E = \begin{bmatrix} -3 \\ -7 \\ 3 \\ -5 \end{bmatrix}$

24. $E + F = \begin{bmatrix} 21 \\ 27 \\ -12 \\ 15 \end{bmatrix}$

25. $3E = \begin{bmatrix} 21 \\ 27 \\ -12 \\ 30 \end{bmatrix}$

26. $F - 2E = \begin{bmatrix} 14 \\ 18 \\ -8 \\ 5 \end{bmatrix} - \begin{bmatrix} 14 \\ 18 \\ -8 \\ 20 \end{bmatrix} = \begin{bmatrix} 0 \\ 0 \\ 0 \\ -15 \end{bmatrix}$

27. $3D + 2E = \begin{bmatrix} 12 \\ 6 \\ -3 \\ 15 \end{bmatrix} + \begin{bmatrix} 14 \\ 18 \\ -8 \\ 20 \end{bmatrix} = \begin{bmatrix} 26 \\ 24 \\ -11 \\ 35 \end{bmatrix}$

28. $2F - 7D = \begin{bmatrix} 28 \\ 36 \\ -16 \\ 10 \end{bmatrix} - \begin{bmatrix} 28 \\ 14 \\ -7 \\ 35 \end{bmatrix} = \begin{bmatrix} 0 \\ 22 \\ -9 \\ -25 \end{bmatrix}$

29. $7D - 2F = \begin{bmatrix} 28 \\ 14 \\ -7 \\ 35 \end{bmatrix} - \begin{bmatrix} 28 \\ 36 \\ -16 \\ 10 \end{bmatrix} = \begin{bmatrix} 0 \\ -22 \\ 9 \\ 25 \end{bmatrix}$

30. $5E - 2F = \begin{bmatrix} 35 \\ 45 \\ -20 \\ 50 \end{bmatrix} - \begin{bmatrix} 28 \\ 36 \\ -16 \\ 10 \end{bmatrix} = \begin{bmatrix} 7 \\ 9 \\ -4 \\ 40 \end{bmatrix}$

31. $\begin{bmatrix} 18 & 7 & 9 \\ 7 & 11 & 4 \end{bmatrix} + \begin{bmatrix} 9 & 5 & 6 \\ 11 & 4 & 12 \end{bmatrix} = \begin{bmatrix} 27 & 12 & 15 \\ 18 & 15 & 16 \end{bmatrix}$
27 chip A, 12 chip B, 15 EPROMS, 18 keyboards, 15 motherboards and 16 disk drives

32. The answer is the second matrix minus the first

Store	195/70SR-14	205/70SR-14	185/70SR-15	205/70SR-15	215/70SR-15
A	8	33	32	17	28
B	11	45	13	0	21
C	29	43	27	43	23

33. $\begin{bmatrix} 70 & 40 \\ 30 & 60 \end{bmatrix} + \begin{bmatrix} 65 & 42 \\ 35 & 58 \end{bmatrix} = \begin{bmatrix} 135 & 82 \\ 65 & 118 \end{bmatrix}$

34. The answer is the inventory matrix minus the sales matrix.

Chip type:	286	386	486	586
Warehouse A	930	1070	1590	635
Warehouse B	530	350	4076	625
Warehouse C	800	1974	3457	188

35. The answer is the last sales matrix times 1.1. Note: Answers are rounded to the nearest integer.

Chip type:	286	386	486	586
Warehouse A	297	2343	3531	292
Warehouse B	1232	4653	3436	83
Warehouse C	352	3439	3017	1113

36.

Model		S	M	L	XL
	Jogging	20	16	12	16
	Sweating	28	48	60	28
	Walking	16	32	48	56

37. See *Computer Programs* in main text.

38. See *Computer Programs* in main text.

39. See *Computer Programs* in main text.

17.2 MULTIPLICATION OF MATRICES

1. $\begin{bmatrix} 2 & 3 \\ 1 & -4 \end{bmatrix} \times \begin{bmatrix} 4 & -3 \\ 6 & 5 \end{bmatrix} = \begin{bmatrix} 8+18 & -6+15 \\ 4-24 & -3-20 \end{bmatrix} = \begin{bmatrix} 26 & 9 \\ -20 & -23 \end{bmatrix}$

2. $[1 \quad -2 \quad 4] \times \begin{bmatrix} 3 \\ 9 \\ -7 \end{bmatrix} = [3-18-28] = [-43]$

3. $[5 \quad 1 \quad 6 \quad 2] \times \begin{bmatrix} 2 \\ 9 \\ -8 \\ 3 \end{bmatrix} = [10+9-48+6] = [-23]$

4. $\begin{bmatrix} 2 & 3 & 1 \\ -4 & 2 & 0 \end{bmatrix} \times \begin{bmatrix} 4 & 1 \\ -5 & 2 \\ 3 & 0 \end{bmatrix} = \begin{bmatrix} 8-15+3 & 2+6+0 \\ -16-10+0 & -4+4+0 \end{bmatrix} = \begin{bmatrix} -4 & 8 \\ -26 & 0 \end{bmatrix}$

5. $\begin{bmatrix} 1 & 2 & 4 \\ 2 & 3 & 6 \end{bmatrix} \times \begin{bmatrix} 5 & -7 & 4 \\ 4 & 6 & 0 \\ -4 & 2 & 1 \end{bmatrix} = \begin{bmatrix} 5+8-16 & -7+12+8 & 4+0+4 \\ 10+12-24 & -14+18+12 & 8+0+6 \end{bmatrix} = \begin{bmatrix} -3 & 13 & 8 \\ -2 & 16 & 14 \end{bmatrix}$

6. $\begin{bmatrix} 4 & 7 \\ 2 & 3 \\ 6 & 2 \\ 9 & 1 \end{bmatrix} \times \begin{bmatrix} 3 & 7 & 6 & 1 & -5 \\ 4 & -3 & 2 & 0 & 10 \end{bmatrix} = \begin{bmatrix} 12+28 & 28-21 & 24+14 & 4+0 & -20+70 \\ 6+12 & 14-9 & 12+6 & 2+0 & -10+30 \\ 18+8 & 42-6 & 36+4 & 6+0 & -30+20 \\ 27+4 & 63-3 & 54+2 & 9+0 & -45+10 \end{bmatrix} = \begin{bmatrix} 40 & 7 & 38 & 4 & 50 \\ 18 & 5 & 18 & 2 & 20 \\ 26 & 36 & 40 & 6 & -10 \\ 31 & 60 & 56 & 9 & -35 \end{bmatrix}$

7. $\begin{bmatrix} 1 & 2 & 3 \\ 6 & 5 & 4 \\ -1 & 0 & 2 \end{bmatrix} \times \begin{bmatrix} 9 & 8 & 7 \\ -2 & 3 & -1 \\ 0 & 1 & 0 \end{bmatrix} = \begin{bmatrix} 9-4+0 & 8+6+3 & 7-2+0 \\ 54-10+0 & 48+15+4 & 42-5+0 \\ -9+0+0 & -8+0+2 & -7+0+0 \end{bmatrix} = \begin{bmatrix} 5 & 17 & 5 \\ 44 & 67 & 37 \\ -9 & -6 & -7 \end{bmatrix}$

8. $\begin{bmatrix} 9 & 8 & 7 \\ -2 & 3 & -1 \\ 0 & 1 & 0 \end{bmatrix} \times \begin{bmatrix} 1 & 2 & 3 \\ 6 & 5 & 4 \\ -1 & 0 & 2 \end{bmatrix} = \begin{bmatrix} 9+48-7 & 18+40+0 & 27+32+14 \\ -2+18+1 & -4+15+0 & -6+12-2 \\ 0+6+0 & 0+5+0 & 0+4+0 \end{bmatrix} = \begin{bmatrix} 50 & 58 & 73 \\ 17 & 11 & 4 \\ 6 & 5 & 4 \end{bmatrix}$

9. $\begin{bmatrix} 3 & 4 & -2 \\ 7 & 8 & 10 \end{bmatrix} \times \begin{bmatrix} 0 & 4 & 1 \\ 1 & 0 & 2 \\ 8 & -1 & 0 \end{bmatrix} = \begin{bmatrix} 0+4-16 & 12+0+2 & 3+8+0 \\ 0+8+80 & 28+0-10 & 7+16+0 \end{bmatrix} = \begin{bmatrix} -12 & 14 & 11 \\ 88 & 18 & 23 \end{bmatrix}$

10. $\begin{bmatrix} 1 & 2 & -1 \\ 2 & -1 & 3 \end{bmatrix} \times \begin{bmatrix} 0 & 1 & 0 & 2 \\ 1 & 0 & 2 & -1 \\ 3 & 4 & -1 & 0 \end{bmatrix} = \begin{bmatrix} 2-3 & 1-4 & 4+1 & 2-2 \\ -1+9 & 2+12 & -2-3 & 4+1 \end{bmatrix} = \begin{bmatrix} -1 & -3 & 5 & 0 \\ 8 & 14 & -5 & 5 \end{bmatrix}$

11. $AB = \begin{bmatrix} 1 & 2 \\ 3 & 4 \end{bmatrix} \times \begin{bmatrix} 3 & 6 \\ 2 & 5 \end{bmatrix} = \begin{bmatrix} 3+4 & 6+10 \\ 9+8 & 18+20 \end{bmatrix} = \begin{bmatrix} 7 & 16 \\ 17 & 38 \end{bmatrix}$

12. $BA = \begin{bmatrix} 3 & 6 \\ 2 & 5 \end{bmatrix} \times \begin{bmatrix} 1 & 2 \\ 3 & 4 \end{bmatrix} = \begin{bmatrix} 3+18 & 6+24 \\ 2+15 & 4+20 \end{bmatrix} = \begin{bmatrix} 21 & 30 \\ 17 & 24 \end{bmatrix}$

13. $AC = \begin{bmatrix} 1 & 2 \\ 3 & 4 \end{bmatrix} \begin{bmatrix} 4 & -2 \\ -6 & 2 \end{bmatrix} =$

$$\begin{bmatrix} 4-12 & -2+4 \\ 12-24 & -6+8 \end{bmatrix} = \begin{bmatrix} -8 & 2 \\ -12 & 2 \end{bmatrix}$$

14. $BC = \begin{bmatrix} 3 & 6 \\ 2 & 5 \end{bmatrix} \begin{bmatrix} 4 & -2 \\ -6 & 2 \end{bmatrix} =$
$\begin{bmatrix} 12-36 & -6+12 \\ 8-30 & -4+10 \end{bmatrix} = \begin{bmatrix} -24 & 6 \\ -22 & 6 \end{bmatrix}$

15. $A(BC) = A\begin{bmatrix} -24 & 6 \\ -22 & 6 \end{bmatrix} = \begin{bmatrix} 1 & 2 \\ 3 & 4 \end{bmatrix} \begin{bmatrix} -24 & 6 \\ -22 & 6 \end{bmatrix} =$
$\begin{bmatrix} -24-44 & 6+12 \\ -72-88 & 18+24 \end{bmatrix} = \begin{bmatrix} -68 & 18 \\ -160 & 42 \end{bmatrix}$

16. $(AB)C = \begin{bmatrix} 7 & 16 \\ 17 & 38 \end{bmatrix} \begin{bmatrix} 4 & -2 \\ -6 & 2 \end{bmatrix} =$
$\begin{bmatrix} 28-96 & -14+32 \\ 68-228 & -34+76 \end{bmatrix} = \begin{bmatrix} -68 & 18 \\ -160 & 42 \end{bmatrix}$

17. $A(B+C) = \begin{bmatrix} 1 & 2 \\ 3 & 4 \end{bmatrix} \left(\begin{bmatrix} 3 & 6 \\ 2 & 5 \end{bmatrix} + \begin{bmatrix} 4 & -2 \\ -6 & 2 \end{bmatrix} \right) =$
$\begin{bmatrix} 1 & 2 \\ 3 & 4 \end{bmatrix} \begin{bmatrix} 7 & 4 \\ -4 & 7 \end{bmatrix} = \begin{bmatrix} 7-8 & 4+14 \\ 21-16 & 12+28 \end{bmatrix} =$
$\begin{bmatrix} -1 & 18 \\ 5 & 40 \end{bmatrix}$

18. $(B+C)A = \begin{bmatrix} 7 & 4 \\ -4 & 7 \end{bmatrix} \begin{bmatrix} 1 & 2 \\ 3 & 4 \end{bmatrix} =$
$\begin{bmatrix} 7+12 & 14+16 \\ -4+21 & -8+28 \end{bmatrix} = \begin{bmatrix} 19 & 30 \\ 17 & 20 \end{bmatrix}$

19. $(2A)(3B)$: Scaler multiplication is commutative so $(2A)(3B) = 6(AB) = 6 \cdot \begin{bmatrix} 7 & 16 \\ 17 & 38 \end{bmatrix} = \begin{bmatrix} 42 & 96 \\ 102 & 228 \end{bmatrix}$

20. $\left(2A + \frac{1}{2}C\right)B = \left(2\begin{bmatrix} 1 & 2 \\ 3 & 4 \end{bmatrix} + \frac{1}{2}\begin{bmatrix} 4 & -2 \\ -6 & 2 \end{bmatrix}\right) \cdot$
$\begin{bmatrix} 3 & 6 \\ 2 & 5 \end{bmatrix} = \left(\begin{bmatrix} 2 & 4 \\ 6 & 8 \end{bmatrix} + \begin{bmatrix} 2 & -1 \\ -3 & 1 \end{bmatrix} \right) \begin{bmatrix} 3 & 6 \\ 2 & 5 \end{bmatrix} =$
$\begin{bmatrix} 4 & 3 \\ 3 & 9 \end{bmatrix} \cdot \begin{bmatrix} 3 & 6 \\ 2 & 5 \end{bmatrix} = \begin{bmatrix} 12+6 & 24+15 \\ 9+18 & 18+45 \end{bmatrix} =$
$\begin{bmatrix} 18 & 39 \\ 27 & 63 \end{bmatrix}$

21. $A = \begin{bmatrix} 2 & 10 \\ 3 & 15 \end{bmatrix}$, $B = \begin{bmatrix} -10 & 25 \\ 2 & -5 \end{bmatrix}$, $AB =$
$\begin{bmatrix} -20+20 & 50-50 \\ -30+30 & 75-75 \end{bmatrix} = \begin{bmatrix} 0 & 0 \\ 0 & 0 \end{bmatrix}$. We cannot *conclude* that, if $AB = 0$, then either $A = 0$ or $B = 0$. Hence, matrix multiplication does *not* have a zero product principal.

22. $\begin{bmatrix} 2 & x \\ y & 5 \end{bmatrix} \begin{bmatrix} 4 & 6 \\ 9 & -1 \end{bmatrix} = \begin{bmatrix} 8+9x & 12-x \\ 4y+45 & 6y-5 \end{bmatrix} =$
$\begin{bmatrix} 35 & 9 \\ 49 & 1 \end{bmatrix}$. So, $8 + 9x = 35$ and $12 - x = 9$. From the first, we obtain $9x = 27$ or $-x = -3$ and so $x = 3$. Also, $4y + 45 = 49$ and $6y - 5 = 1$. From the first, we get $4y = 4$ and so $y = 1$. The solutions are $x = 3$ and $y = 1$.

23. Here we have the product

$$\begin{bmatrix} 270 & 2130 & 3210 & 265 \\ 1120 & 4230 & 3124 & 75 \\ 320 & 3126 & 2743 & 1012 \end{bmatrix} \times \begin{bmatrix} 25 \\ 35 \\ 55 \\ 95 \end{bmatrix} = \begin{bmatrix} 283{,}025 \\ 354{,}995 \\ 364{,}415 \end{bmatrix}$$

From this wee see that Warehouse A generated sales of \$283,025, Warehouse B sales of \$354,995, and Warehouse C generated sales of \$364,415.

24. (a) The amount of material needed for each model house is determined by finding the product $PA = \begin{bmatrix} 30 & 20 & 15 \\ 10 & 10 & 15 \end{bmatrix} \cdot \begin{bmatrix} 10 & 2 & 2 & 3 \\ 15 & 3 & 4 & 4 \\ 25 & 5 & 6 & 3 \end{bmatrix} =$
$\begin{bmatrix} 975 & 195 & 230 & 215 \\ 625 & 125 & 150 & 115 \end{bmatrix}$; (b) The cost to build each size house is found from the product $AC =$
$\begin{bmatrix} 10 & 2 & 2 & 3 \\ 15 & 3 & 4 & 4 \\ 25 & 5 & 6 & 3 \end{bmatrix} \cdot \begin{bmatrix} 25 \\ 210 \\ 75 \\ 40 \end{bmatrix} = \begin{bmatrix} 940 \\ 1465 \\ 2245 \end{bmatrix}$; (c) The cost for each house model is the product of all three of the given matrices, $PAC = \begin{bmatrix} 30 & 20 & 15 \\ 10 & 10 & 15 \end{bmatrix} \cdot$
$\begin{bmatrix} 10 & 2 & 2 & 3 \\ 15 & 3 & 4 & 4 \\ 25 & 5 & 6 & 3 \end{bmatrix} \cdot \begin{bmatrix} 25 \\ 210 \\ 75 \\ 40 \end{bmatrix} = \begin{bmatrix} 91{,}175 \\ 57{,}725 \end{bmatrix}$; (d) To find the total cost for the subdivision you need to add the columns in the answer to (c) with the result that the total is \$148,900.

25. $A = \begin{bmatrix} 0 & 1 \\ 1 & 0 \end{bmatrix}$:

$A^2 = \begin{bmatrix} 0 & 1 \\ 1 & 0 \end{bmatrix} \times \begin{bmatrix} 0 & 1 \\ 1 & 0 \end{bmatrix} = \begin{bmatrix} 1 & 0 \\ 0 & 1 \end{bmatrix} = I,$

$B = \begin{bmatrix} 0 & -i \\ i & 0 \end{bmatrix}$: $B^2 = \begin{bmatrix} 0 & -i \\ i & 0 \end{bmatrix} \begin{bmatrix} 0 & -i \\ i & 0 \end{bmatrix} =$

$\begin{bmatrix} -i^2 & 0 \\ 0 & -i^2 \end{bmatrix} = \begin{bmatrix} 1 & 0 \\ 0 & 1 \end{bmatrix} = I$; $C = \begin{bmatrix} 1 & 0 \\ 0 & -1 \end{bmatrix}$:

$C^2 = \begin{bmatrix} 1 & 0 \\ 0 & -1 \end{bmatrix} \cdot \begin{bmatrix} 1 & 0 \\ 0 & -1 \end{bmatrix} = \begin{bmatrix} 1 & 0 \\ 0 & 1 \end{bmatrix} = I$

26. $AB = \begin{bmatrix} 0 & 1 \\ 1 & 0 \end{bmatrix} \begin{bmatrix} 0 & -i \\ i & 0 \end{bmatrix} = \begin{bmatrix} i & 0 \\ 0 & -i \end{bmatrix}$

$BA = \begin{bmatrix} 0 & -i \\ i & 0 \end{bmatrix} \begin{bmatrix} 0 & 1 \\ 1 & 0 \end{bmatrix} = \begin{bmatrix} -i & 0 \\ 0 & i \end{bmatrix} = -AB.$

$AC = \begin{bmatrix} 0 & 1 \\ 1 & 0 \end{bmatrix} \begin{bmatrix} 1 & 0 \\ 0 & -1 \end{bmatrix} = \begin{bmatrix} 0 & -1 \\ 1 & 0 \end{bmatrix}$

$CA = \begin{bmatrix} 1 & 0 \\ 0 & -1 \end{bmatrix} \begin{bmatrix} 0 & 1 \\ 1 & 0 \end{bmatrix} = \begin{bmatrix} 0 & 1 \\ -1 & 0 \end{bmatrix} = -AC,$

$BC = \begin{bmatrix} 0 & -i \\ i & 0 \end{bmatrix} \begin{bmatrix} 1 & 0 \\ 0 & -1 \end{bmatrix} = \begin{bmatrix} 0 & i \\ i & 0 \end{bmatrix},$

$CB = \begin{bmatrix} 1 & 0 \\ 0 & -1 \end{bmatrix} \times \begin{bmatrix} 0 & -i \\ i & 0 \end{bmatrix} = \begin{bmatrix} 0 & -i \\ -i & 0 \end{bmatrix} = -BC$

27. $AB - BA = \begin{bmatrix} i & 0 \\ 0 & -i \end{bmatrix} - \begin{bmatrix} -i & 0 \\ 0 & i \end{bmatrix} =$

$\begin{bmatrix} 2i & 0 \\ 0 & -2i \end{bmatrix} = 2i \begin{bmatrix} 1 & 0 \\ 0 & -1 \end{bmatrix} = 2iC$. $AC - CA =$

$\begin{bmatrix} 0 & -1 \\ 1 & 0 \end{bmatrix} - \begin{bmatrix} 0 & 1 \\ -1 & 0 \end{bmatrix} = \begin{bmatrix} 0 & -2 \\ 2 & 0 \end{bmatrix} = -2iB.$

$-2i \begin{bmatrix} 0 & -i \\ i & 0 \end{bmatrix} = \begin{bmatrix} 0 & 2i^2 \\ -2i^2 & 0 \end{bmatrix} = \begin{bmatrix} 0 & -2 \\ 2 & 0 \end{bmatrix}.$

$BC - CB = \begin{bmatrix} 0 & i \\ i & 0 \end{bmatrix} - \begin{bmatrix} 0 & -i \\ -i & 0 \end{bmatrix} = \begin{bmatrix} 0 & 2i \\ 2i & 0 \end{bmatrix} =$

$2i \begin{bmatrix} 0 & 1 \\ 1 & 0 \end{bmatrix} = 2iA$

28. $[x \quad y] \begin{bmatrix} 5 & -7 \\ 7 & 3 \end{bmatrix} \begin{bmatrix} x \\ y \end{bmatrix} =$

$[5x + 7y \quad -7x + 3y] \begin{bmatrix} x \\ y \end{bmatrix} =$

$5x^2 + 7xy - 7xy + 3y^2 = 30$. Collecting terms, we get $5x^2 + 3y^2 = 30$ or $\frac{x^2}{6} + \frac{y^2}{10} = 1$, which is the equation for an ellipse.

29. See *Computer Programs* in main text.

17.3 INVERSES OF MATRICES

1. $\begin{vmatrix} 6 & 1 \\ 5 & 1 \end{vmatrix} = 6 - 5 = 1$. Inverse is $\begin{bmatrix} 1 & -1 \\ -5 & 6 \end{bmatrix}$

2. $\begin{vmatrix} 5 & 1 \\ 9 & 2 \end{vmatrix} = 10 - 9 = 1$. Inverse is $\begin{bmatrix} 2 & -1 \\ -9 & 5 \end{bmatrix}$

3. $\begin{vmatrix} 10 & 4 \\ 8 & 3 \end{vmatrix} = 30 - 32 = -2$. Inverse is

$\frac{1}{-2} \begin{bmatrix} 3 & -4 \\ -8 & 10 \end{bmatrix} = \begin{bmatrix} \frac{-3}{2} & 2 \\ 4 & -5 \end{bmatrix} = \begin{bmatrix} -1.5 & 2 \\ 4 & -5 \end{bmatrix}$

4. $\begin{vmatrix} 9 & 6 \\ 4 & 3 \end{vmatrix} = 27 - 24 = 3$. Inverse is

$\frac{1}{3} \begin{bmatrix} 3 & -6 \\ -4 & 9 \end{bmatrix} = \begin{bmatrix} 1 & -2 \\ -\frac{4}{3} & 3 \end{bmatrix}$

5. $\begin{vmatrix} 8 & -6 \\ -6 & 4 \end{vmatrix} = 32 - 36 = -4$. Inverse is

$-\frac{1}{4} \begin{bmatrix} 4 & 6 \\ 6 & 8 \end{bmatrix} = \begin{bmatrix} -1 & -1.5 \\ -1.5 & -2 \end{bmatrix}$

6. $\begin{vmatrix} 12 & 9 \\ -9 & -7 \end{vmatrix} = -84 + 81 = -3$. Inverse is

$-\frac{1}{3} \begin{bmatrix} -7 & -9 \\ 9 & 12 \end{bmatrix} = \begin{bmatrix} \frac{7}{3} & 3 \\ -3 & -4 \end{bmatrix}$

7. $\begin{vmatrix} 15 & 10 \\ 4 & 3 \end{vmatrix} = 45 - 40 = 5$. Inverse is

$\frac{1}{5} \begin{bmatrix} 3 & -10 \\ -4 & 15 \end{bmatrix} = \begin{bmatrix} .6 & -2 \\ -0.8 & 3 \end{bmatrix}$

8. $\begin{vmatrix} 15 & 10 \\ 3 & 4 \end{vmatrix} = 60 - 30 = 30$. Inverse is

$\frac{1}{30} \begin{bmatrix} 4 & -10 \\ -3 & 15 \end{bmatrix} = \begin{bmatrix} \frac{2}{15} & -\frac{1}{3} \\ -\frac{1}{10} & \frac{1}{2} \end{bmatrix} =$

$$\begin{bmatrix} 0.133 & -0.333 \\ -0.1 & 0.5 \end{bmatrix}$$

9. $\left[\begin{array}{ccc|ccc} 0 & -3 & 0 & 1 & 0 & 0 \\ 1 & 0 & 0 & 0 & 1 & 0 \\ 0 & 0 & 4 & 0 & 0 & 1 \end{array}\right] \begin{matrix} -\frac{1}{3}R_1 \\ \Rightarrow \\ \frac{1}{4}R_3 \end{matrix}$
$\left[\begin{array}{ccc|ccc} 0 & 1 & 0 & -\frac{1}{3} & 0 & 0 \\ 1 & 0 & 0 & 0 & 1 & 0 \\ 0 & 0 & 1 & 0 & 0 & \frac{1}{4} \end{array}\right]$. Interchange R_1 and R_2 $\left[\begin{array}{ccc|ccc} 1 & 0 & 0 & 0 & 1 & 0 \\ 0 & 1 & 0 & -\frac{1}{3} & 0 & 0 \\ 0 & 0 & 1 & 0 & 0 & \frac{1}{4} \end{array}\right]$. Inverse is
$\begin{bmatrix} 0 & 1 & 0 \\ -\frac{1}{3} & 0 & 0 \\ 0 & 0 & \frac{1}{4} \end{bmatrix}$

10. $\left[\begin{array}{ccc|ccc} 1 & 0 & 0 & 1 & 0 & 0 \\ 0 & 4 & 0 & 0 & 1 & 0 \\ 0 & 0 & 2 & 0 & 0 & 1 \end{array}\right] \begin{matrix} \\ \frac{1}{4}R_2 \Rightarrow \\ \frac{1}{2}R_3 \end{matrix}$
$\left[\begin{array}{ccc|ccc} 1 & 0 & 0 & 1 & 0 & 0 \\ 0 & 1 & 0 & 0 & \frac{1}{4} & 0 \\ 0 & 0 & 1 & 0 & 0 & \frac{1}{2} \end{array}\right]$, so the inverse is
$\begin{bmatrix} 1 & 0 & 0 \\ 0 & \frac{1}{4} & 0 \\ 0 & 0 & \frac{1}{2} \end{bmatrix}$

11. $\begin{bmatrix} 1 & 2 & 6 \\ 0 & 0 & 2 \\ -3 & -6 & -9 \end{bmatrix}$ Since column 2 is twice column 1, this is a singular matrix and it has no inverse.

12. $\begin{bmatrix} 4 & 5 & 1 \\ 1 & 0 & 1 \\ 4 & 5 & 1 \end{bmatrix}$. Since $R_1 = R_3$ this is a singular matrix and it has no inverse.

13. $\left[\begin{array}{ccc|ccc} 8 & 7 & -1 & 1 & 0 & 0 \\ -5 & -5 & 1 & 0 & 1 & 0 \\ -4 & -4 & 1 & 0 & 0 & 1 \end{array}\right] \begin{matrix} 2R_3 + R_1 \\ \\ \\ \end{matrix}$

$\left[\begin{array}{ccc|ccc} 0 & -1 & 1 & 1 & 0 & 2 \\ -5 & -5 & 1 & 0 & 1 & 0 \\ -4 & -4 & 1 & 0 & 0 & 1 \end{array}\right] -1R_2$

$\left[\begin{array}{ccc|ccc} 0 & -1 & 1 & 1 & 0 & 2 \\ 5 & 5 & -1 & 0 & -1 & 0 \\ -4 & -4 & 1 & 0 & 0 & 1 \end{array}\right] R_3 + R_2$

$\left[\begin{array}{ccc|ccc} 0 & -1 & 1 & 1 & 0 & 2 \\ 1 & 1 & 0 & 0 & -1 & 1 \\ -4 & -4 & 1 & 0 & 0 & 1 \end{array}\right] 4R_2 + R_3$

$\left[\begin{array}{ccc|ccc} 0 & -1 & 1 & 1 & 0 & 2 \\ 1 & 1 & 0 & 0 & -1 & 1 \\ 0 & 0 & 1 & 0 & -4 & 5 \end{array}\right]$ Swap $R_1 + R_2$

$\left[\begin{array}{ccc|ccc} 1 & 1 & 0 & 0 & -1 & 1 \\ 0 & -1 & 1 & 1 & 0 & 2 \\ 0 & 0 & 1 & 0 & -4 & 5 \end{array}\right] \begin{matrix} R_2 + R_1 \\ \\ \\ \end{matrix}$

$\left[\begin{array}{ccc|ccc} 1 & 0 & 1 & 1 & -1 & 3 \\ 0 & -1 & 1 & 1 & 0 & 2 \\ 0 & 0 & 1 & 0 & -4 & 5 \end{array}\right] \begin{matrix} -R_3 + R_1 \\ -R_3 + R_2 \\ \\ \end{matrix}$

$\left[\begin{array}{ccc|ccc} 1 & 0 & 0 & 1 & 3 & -2 \\ 0 & -1 & 0 & 1 & 4 & -3 \\ 0 & 0 & 1 & 0 & -4 & 5 \end{array}\right] -1R_2$

$\Rightarrow \left[\begin{array}{ccc|ccc} 1 & 0 & 0 & 1 & 3 & -2 \\ 0 & 1 & 0 & -1 & -4 & 3 \\ 0 & 0 & 1 & 0 & -4 & 5 \end{array}\right]$. Inverse is
$\begin{bmatrix} 1 & 3 & -2 \\ -1 & -4 & 3 \\ 0 & -4 & 5 \end{bmatrix}$

14. $\left[\begin{array}{ccc|ccc} 1 & 2 & 3 & 1 & 0 & 0 \\ 2 & 5 & 7 & 0 & 1 & 0 \\ 1 & 1 & 1 & 0 & 0 & 1 \end{array}\right] \begin{matrix} \\ -2R_1 + R_2 \\ -R_1 + R_3 \end{matrix}$

$\left[\begin{array}{ccc|ccc} 1 & 2 & 3 & 1 & 0 & 0 \\ 0 & 1 & 1 & -2 & 1 & 0 \\ 0 & -1 & -2 & -1 & 0 & 1 \end{array}\right] \begin{matrix} 2R_3 + R_1 \\ \\ R_2 + R_3 \end{matrix}$

$\left[\begin{array}{ccc|ccc} 1 & 0 & -1 & -1 & 0 & 2 \\ 0 & 1 & 1 & -2 & 1 & 0 \\ 0 & 0 & -1 & -3 & 1 & 1 \end{array}\right] \begin{matrix} -R_3 + R_1 \\ R_3 + R_2 \\ -R_3 \end{matrix}$

$\left[\begin{array}{ccc|ccc} 1 & 0 & 0 & 2 & -1 & 1 \\ 0 & 1 & 0 & -5 & 2 & 1 \\ 0 & 0 & 1 & 3 & -1 & -1 \end{array}\right]$.

The inverse is $\begin{bmatrix} 2 & -1 & 1 \\ -5 & 2 & 1 \\ 3 & -1 & -1 \end{bmatrix}$

15. $\begin{bmatrix} 3 & -1 & 0 \\ -6 & 2 & 0 \\ 1 & 0 & 5 \end{bmatrix}$. Since $R_2 = -2R_1$ the matrix is singular, so there is no inverse.

16. $\left[\begin{array}{ccc|ccc} 1 & -1 & 1 & 1 & 0 & 0 \\ 7 & -8 & 5 & 0 & 1 & 0 \\ -4 & 5 & -3 & 0 & 0 & 1 \end{array}\right] \begin{matrix} \\ -7R_1 + R_2 \\ 4R_1 + R_3 \end{matrix}$

$\left[\begin{array}{ccc|ccc} 1 & -1 & 1 & 1 & 0 & 0 \\ 0 & -1 & -2 & -7 & 1 & 0 \\ 0 & 1 & 1 & 4 & 0 & 1 \end{array}\right] \begin{matrix} R_3 + R_1 \\ \\ R_2 + R_3 \end{matrix}$

$$\left[\begin{array}{ccc|ccc} 1 & 0 & 2 & 5 & 0 & 1 \\ 0 & -1 & -2 & -7 & 1 & 0 \\ 0 & 0 & -1 & -3 & 1 & 1 \end{array}\right] \begin{array}{l} 2R_3 + R_1 \\ 2R_3 - R_2 \\ -R_3 \end{array}$$

$$\left[\begin{array}{ccc|ccc} 1 & 0 & 0 & -1 & 2 & 3 \\ 0 & 1 & 0 & 1 & 1 & 2 \\ 0 & 0 & 1 & 3 & -1 & -1 \end{array}\right].$$ The inverse is

$$\begin{bmatrix} -1 & 2 & 3 \\ 1 & 1 & 2 \\ 3 & -1 & -1 \end{bmatrix}.$$

17. $$\left[\begin{array}{ccc|ccc} 2 & 0 & 0 & 1 & 0 & 0 \\ 2 & 2 & 0 & 0 & 1 & 0 \\ 2 & 2 & 2 & 0 & 0 & 1 \end{array}\right].$$

First multiply the entire matrix by $\frac{1}{2}$ to get

$$\left[\begin{array}{ccc|ccc} 1 & 0 & 0 & \frac{1}{2} & 0 & 0 \\ 1 & 1 & 0 & 0 & \frac{1}{2} & 0 \\ 1 & 1 & 1 & 0 & 0 & \frac{1}{2} \end{array}\right] \begin{array}{l} \\ -R_1 + R_2 \\ -R_1 + R_3 \end{array}$$

$$\left[\begin{array}{ccc|ccc} 1 & 0 & 0 & \frac{1}{2} & 0 & 0 \\ 0 & 1 & 0 & -\frac{1}{2} & \frac{1}{2} & 0 \\ 0 & 1 & 1 & -\frac{1}{2} & 0 & \frac{1}{2} \end{array}\right] \begin{array}{l} \\ \\ -R_2 + R_3 \end{array}$$

$$\left[\begin{array}{ccc|ccc} 1 & 0 & 0 & \frac{1}{2} & 0 & 0 \\ 0 & 1 & 0 & -\frac{1}{2} & \frac{1}{2} & 0 \\ 0 & 0 & 1 & 0 & -\frac{1}{2} & \frac{1}{2} \end{array}\right].$$

The inverse is $\begin{bmatrix} \frac{1}{2} & 0 & 0 \\ -\frac{1}{2} & \frac{1}{2} & 0 \\ 0 & -\frac{1}{2} & \frac{1}{2} \end{bmatrix}$ or

$$\begin{bmatrix} 0.5 & 0 & 0 \\ -0.5 & 0.5 & 0 \\ 0 & -0.5 & 0.5 \end{bmatrix}.$$

18. $$\left[\begin{array}{ccc|ccc} 3 & 1 & -1 & 1 & 0 & 0 \\ 1 & -2 & 0 & 0 & 1 & 0 \\ 0 & 3 & 1 & 0 & 0 & 1 \end{array}\right] \begin{array}{l} -3R_2 + R_1 \\ \\ \\ \end{array}$$

$$\left[\begin{array}{ccc|ccc} 0 & 7 & -1 & 1 & -3 & 0 \\ 1 & -2 & 0 & 0 & 1 & 0 \\ 0 & 3 & 1 & 0 & 0 & 1 \end{array}\right].$$ Swap R_1 and R_2

to get $$\left[\begin{array}{ccc|ccc} 1 & -2 & 0 & 0 & 1 & 0 \\ 0 & 7 & -1 & 1 & -3 & 0 \\ 0 & 3 & 1 & 0 & 0 & 1 \end{array}\right] \begin{array}{l} \\ R_3 + R_2 \\ \\ \end{array}$$

$$\left[\begin{array}{ccc|ccc} 1 & -2 & 0 & 0 & 1 & 0 \\ 0 & 10 & 0 & 1 & -3 & 1 \\ 0 & 3 & 1 & 0 & 0 & 1 \end{array}\right] \begin{array}{l} \frac{2}{10}R_2 + R_1 \\ \frac{1}{10}R_2 \rightarrow \\ -\frac{3}{10}R_2 + R_3 \end{array}$$

$$\left[\begin{array}{ccc|ccc} 1 & 0 & 0 & 0.2 & 0.4 & 0.2 \\ 0 & 1 & 0 & 0.1 & -0.3 & 0.1 \\ 0 & 0 & 1 & -0.3 & 0.9 & 0.7 \end{array}\right].$$

The inverse is $\begin{bmatrix} 0.2 & 0.4 & 0.2 \\ 0.1 & -0.3 & 0.1 \\ -0.3 & 0.9 & 0.7 \end{bmatrix}$.

19. $$\left[\begin{array}{ccc|ccc} 1 & -1 & 1 & 1 & 0 & 0 \\ 0 & 2 & -1 & 0 & 1 & 0 \\ 2 & 3 & 0 & 0 & 0 & 1 \end{array}\right] \begin{array}{l} \\ \\ -2R_1 + R_3 \end{array}$$

$$\left[\begin{array}{ccc|ccc} 1 & -1 & 1 & 1 & 0 & 0 \\ 0 & 2 & -1 & 0 & 1 & 0 \\ 0 & 5 & -2 & -2 & 0 & 1 \end{array}\right] \begin{array}{l} \frac{1}{2}R_2 + R_1 \\ \\ -\frac{5}{2}R_2 + R_3 \end{array}$$

$$\left[\begin{array}{ccc|ccc} 1 & 0 & \frac{1}{2} & 1 & \frac{1}{2} & 0 \\ 0 & 2 & -1 & 0 & 1 & 0 \\ 0 & 0 & \frac{1}{2} & -2 & -\frac{5}{2} & 1 \end{array}\right] \begin{array}{l} -R_3 + R_1 \\ 2R_3 + R_2 \\ \\ \end{array}$$

$$\left[\begin{array}{ccc|ccc} 1 & 0 & 0 & 3 & 3 & -1 \\ 0 & 2 & 0 & -4 & -4 & 2 \\ 0 & 0 & \frac{1}{2} & -2 & -\frac{5}{2} & 1 \end{array}\right] \begin{array}{l} \\ \frac{1}{2}R_2 \\ 2R_3 \end{array}$$

$$\left[\begin{array}{ccc|ccc} 1 & 0 & 0 & 3 & 3 & -1 \\ 0 & 1 & 0 & -2 & -2 & 1 \\ 0 & 0 & 1 & -4 & -5 & 2 \end{array}\right].$$

The inverse is $\begin{bmatrix} 3 & 3 & -1 \\ -2 & -2 & 1 \\ -4 & -5 & 2 \end{bmatrix}$.

20. $$\left[\begin{array}{ccc|ccc} 1 & 2 & -1 & 1 & 0 & 0 \\ 2 & -2 & 1 & 0 & 1 & 0 \\ 6 & 4 & 3 & 0 & 0 & 1 \end{array}\right] \begin{array}{l} \\ -2R_1 + R_2 \\ -6R_1 + R_3 \end{array}$$

$$\left[\begin{array}{ccc|ccc} 1 & 2 & -1 & 1 & 0 & 0 \\ 0 & -6 & 3 & -2 & 1 & 0 \\ 0 & -8 & 9 & -6 & 0 & 1 \end{array}\right] \begin{array}{l} \frac{1}{3}R_2 + R_1 \\ \\ \\ \end{array}$$

$$\left[\begin{array}{ccc|ccc} 1 & 0 & 0 & \frac{1}{3} & \frac{1}{3} & 0 \\ 0 & -6 & 3 & -2 & 1 & 0 \\ 0 & -8 & 9 & -6 & 0 & 1 \end{array}\right] \begin{array}{l} \\ \\ -4R_2 + 3R_3 \end{array}$$

$$\left[\begin{array}{ccc|ccc} 1 & 0 & 0 & \frac{1}{3} & \frac{1}{3} & 0 \\ 0 & -6 & 3 & -2 & 1 & 0 \\ 0 & 0 & 15 & -10 & -4 & 3 \end{array}\right] \begin{array}{l} \\ -\frac{1}{5}R_3 + R_2 \\ \\ \end{array}$$

$$\left[\begin{array}{ccc|ccc} 1 & 0 & 0 & \frac{1}{3} & \frac{1}{3} & 0 \\ 0 & -6 & 0 & 0 & \frac{9}{5} & -\frac{3}{5} \\ 0 & 0 & 15 & -10 & -4 & 3 \end{array}\right] \begin{array}{l} \\ -\frac{1}{6}R_2 \\ \frac{1}{15}R_3 \end{array}$$

$$\left[\begin{array}{ccc|ccc} 1 & 0 & 0 & \frac{1}{3} & \frac{1}{3} & 0 \\ 0 & 1 & 0 & 0 & -0.3 & 0.1 \\ 0 & 0 & 1 & -\frac{2}{3} & -\frac{4}{15} & 0.2 \end{array}\right].$$

The inverse is $\begin{bmatrix} \frac{1}{3} & \frac{1}{3} & 0 \\ 0 & -0.3 & 0.1 \\ -\frac{2}{3} & -\frac{4}{15} & 0.2 \end{bmatrix}$.

21. All problems should check.

22. See *Computer Programs* in main text.

17.4 MATRICES AND LINEAR EQUATIONS

1. The coefficient matrix of $A = \begin{bmatrix} 6 & 1 \\ 5 & 1 \end{bmatrix}$ has an inverse $A^{-1} = \begin{bmatrix} 1 & -1 \\ -5 & 6 \end{bmatrix}$. Thus, we see that $\begin{bmatrix} x \\ y \end{bmatrix} = \begin{bmatrix} 1 & -1 \\ -5 & 6 \end{bmatrix} \begin{bmatrix} -4 \\ -3 \end{bmatrix} = \begin{bmatrix} -1 \\ 2 \end{bmatrix}$, and so $x = -1$ and $y = 2$

2. The coefficient matrix $A = \begin{bmatrix} 10 & 4 \\ 8 & 3 \end{bmatrix}$ has the inverse $A^{-1} = \begin{bmatrix} -1.5 & 2 \\ 4 & -5 \end{bmatrix}$. Thus, $\begin{bmatrix} x \\ y \end{bmatrix} = \begin{bmatrix} -1.5 & 2 \\ 4 & -5 \end{bmatrix} \cdot \begin{bmatrix} 8 \\ 7 \end{bmatrix} = \begin{bmatrix} 2 \\ -3 \end{bmatrix}$, and so we see that $x = 2$ and $y = -3$

3. Here the coefficient matrix is $A = \begin{bmatrix} 8 & -6 \\ -6 & 4 \end{bmatrix}$, which has an inverse matrix $A^{-1} = \begin{bmatrix} -1 & -1.5 \\ -1.5 & -2 \end{bmatrix}$. Thus, we see that the solution is $\begin{bmatrix} x \\ y \end{bmatrix} = \begin{bmatrix} -1 & -1.5 \\ -1.5 & -2 \end{bmatrix} \begin{bmatrix} -27 \\ 19 \end{bmatrix} = \begin{bmatrix} -1.5 \\ 2.5 \end{bmatrix}$, and so $x = -1.5$ and $y = 2.5$

4. This coefficient matrix is $A = \begin{bmatrix} 15 & 10 \\ 4 & 3 \end{bmatrix}$ which has an inverse of $A^{-1} = \begin{bmatrix} 0.6 & -2 \\ -0.8 & 3 \end{bmatrix}$. The solution then, is found from $\begin{bmatrix} x \\ y \end{bmatrix} = \begin{bmatrix} 0.6 & -2 \\ -0.8 & 3 \end{bmatrix} \begin{bmatrix} -5 \\ 0 \end{bmatrix} = \begin{bmatrix} -3 \\ 4 \end{bmatrix}$, and we find that $x = -3$ and $y = 4$

5. This coefficent matrix $A = \begin{bmatrix} 1 & 2 & 6 \\ 0 & 0 & 2 \\ -3 & -6 & -9 \end{bmatrix}$, is a singular matrix, and so there is no solution.

6. The coefficient matrix is $A = \begin{bmatrix} 8 & 7 & -1 \\ -5 & -5 & 1 \\ -4 & -4 & 1 \end{bmatrix}$ which has an inverse $A^{-1} = \begin{bmatrix} 1 & 3 & -2 \\ -1 & -4 & 3 \\ 0 & -4 & 5 \end{bmatrix}$ and constant matrix $K = \begin{bmatrix} 9 \\ -1 \\ 0 \end{bmatrix}$. Thus, we have $\begin{bmatrix} x \\ y \\ z \end{bmatrix} = A^{-1}K = \begin{bmatrix} 6 \\ -5 \\ 4 \end{bmatrix}$; which gives the result that $x = 6$, $y = -5$, and $z = 4$.

7. The coefficient matrix is $A = \begin{bmatrix} 1 & 2 & 3 \\ 2 & 5 & 7 \\ 1 & 1 & 1 \end{bmatrix}$; which has the inverse $A^{-1} = \begin{bmatrix} 2 & -1 & 1 \\ -5 & 2 & 1 \\ 3 & -1 & -1 \end{bmatrix}$; and the constant matrix is $K = \begin{bmatrix} 4 \\ 7.5 \\ 1 \end{bmatrix}$. Thus we have $\begin{bmatrix} x \\ y \\ z \end{bmatrix} = A^{-1}K = \begin{bmatrix} 1.5 \\ -4 \\ 3.5 \end{bmatrix}$ with the result that $x = 1.5$, $y = -4$, and $z = 3.5$.

8. We have $A = \begin{bmatrix} 1 & -1 & 1 \\ 7 & -8 & 5 \\ -4 & 5 & -3 \end{bmatrix}$; and so $A^{-1} = \begin{bmatrix} -1 & 2 & 3 \\ 1 & 1 & 2 \\ 3 & -1 & -1 \end{bmatrix}$; and we have a constant matrix of $K = \begin{bmatrix} -6.6 \\ -43 \\ 26.9 \end{bmatrix}$. As a result, we have $\begin{bmatrix} x \\ y \\ z \end{bmatrix} = A^{-1}K = \begin{bmatrix} 1.3 \\ 4.2 \\ -3.7 \end{bmatrix}$, which gives the result that $x = 1.3$, $y = 4.2$, and $z = -3.7$.

9. The coefficient matrix $A = \begin{bmatrix} 2 & 0 & 0 \\ 2 & 2 & 0 \\ 2 & 2 & 2 \end{bmatrix}$ has inverse $A^{-1} = \begin{bmatrix} .5 & 0 & 0 \\ -.5 & .5 & 0 \\ 0 & -.5 & .5 \end{bmatrix}$ and the constant matrix is $K = \begin{bmatrix} 11 \\ 4 \\ -9 \end{bmatrix}$. Thus, we have

$\begin{bmatrix} x \\ y \\ z \end{bmatrix} = A^{-1}K = \begin{bmatrix} 5.5 \\ -3.5 \\ -6.5 \end{bmatrix}$, which produces the solution $x = 5.5$, $y = -3.5$, and $z = -6.5$.

10. The coefficient matrix $A = \begin{bmatrix} 1 & -1 & 1 \\ 0 & 2 & -1 \\ 2 & 3 & 0 \end{bmatrix}$ has inverse $A^{-1} = \begin{bmatrix} 3 & 3 & -1 \\ -2 & -2 & 1 \\ -4 & -5 & 2 \end{bmatrix}$. The constant matrix is $K = \begin{bmatrix} 22 \\ -23 \\ -11.2 \end{bmatrix}$ and so we use $\begin{bmatrix} x \\ y \\ z \end{bmatrix} = A^{-1}K = \begin{bmatrix} 8.2 \\ -9.2 \\ 4.6 \end{bmatrix}$ which leads to the solution that $x = 8.2$, $y = -9.2$, and $z = 4.6$.

11. $A = \begin{bmatrix} 2 & 1 \\ -3 & 2 \end{bmatrix}, K = \begin{bmatrix} 1 \\ 16 \end{bmatrix}, A^{-1} = \frac{1}{7}\begin{bmatrix} 2 & -1 \\ 3 & 2 \end{bmatrix}$,
$\begin{bmatrix} x \\ y \end{bmatrix} = A^{-1}K = \begin{bmatrix} -2 \\ 5 \end{bmatrix}, x = -2, y = 5.$

12. $A = \begin{bmatrix} 4 & 5 \\ 3 & -2 \end{bmatrix}, A^{-1} = \frac{1}{-23}\begin{bmatrix} -2 & -5 \\ -3 & 4 \end{bmatrix}$,
$K = \begin{bmatrix} 2 \\ 13 \end{bmatrix}, \begin{bmatrix} x \\ y \end{bmatrix} = A^{-1}K = \begin{bmatrix} 3 \\ -2 \end{bmatrix}, x = 3, y = -2.$

13. $A = \begin{bmatrix} 2 & 2 \\ 4 & 3 \end{bmatrix}, A^{-1} = -\frac{1}{2}\begin{bmatrix} 3 & -2 \\ -4 & 2 \end{bmatrix}, K = \begin{bmatrix} 4 \\ 1 \end{bmatrix}$,
$\begin{bmatrix} x \\ y \end{bmatrix} = A^{-1}K = \begin{bmatrix} -5 \\ 7 \end{bmatrix}, x = -5, y = 7.$

14. $A = \begin{bmatrix} 3 & 1 \\ 4 & 3 \end{bmatrix}, A^{-1} = \frac{1}{5}\begin{bmatrix} 3 & -1 \\ -4 & 3 \end{bmatrix}, K = \begin{bmatrix} -5 \\ 2 \end{bmatrix}$,
$\begin{bmatrix} x \\ y \end{bmatrix} = A^{-1}K = \begin{bmatrix} -3.4 \\ 5.2 \end{bmatrix}, x = -3.4, y = 5.2.$

15. $A = \begin{bmatrix} 1.5 & 2.5 \\ 3.2 & 2.6 \end{bmatrix}, A^{-1} = \frac{1}{-4.1}\begin{bmatrix} 2.6 & -2.5 \\ -3.2 & 1.5 \end{bmatrix}$,
$K = \begin{bmatrix} 0.3 \\ 7.2 \end{bmatrix}, \begin{bmatrix} x \\ y \end{bmatrix} = A^{-1}K = \begin{bmatrix} 4.2 \\ -2.4 \end{bmatrix}, x = 4.2,$
$y = -2.4.$

16. $A = \begin{bmatrix} 7 & 2 & 1 \\ 3 & -2 & 4 \\ 4 & 5 & -1 \end{bmatrix}, K = \begin{bmatrix} 2 \\ 13 \\ 1 \end{bmatrix}$,
$\begin{bmatrix} x \\ y \\ z \end{bmatrix} = A^{-1}K = \begin{bmatrix} -1 \\ 2 \\ 5 \end{bmatrix}, x = -1, y = 2, z = 3.$

17. $A = \begin{bmatrix} 5 & 2 & 3 \\ -3 & 2 & -8 \\ 4 & -2 & 9 \end{bmatrix}, K = \begin{bmatrix} 1 \\ 6 \\ -7 \end{bmatrix}$,
$\begin{bmatrix} x \\ y \\ z \end{bmatrix} = A^{-1}K = \begin{bmatrix} -2 \\ 4 \\ 1 \end{bmatrix}, x = -2, y = 4, z = 1.$

18. $A = \begin{bmatrix} 2 & 4 & 1 \\ 4 & 2 & 1 \\ 6 & 4 & 7 \end{bmatrix}, K = \begin{bmatrix} 10 \\ 8 \\ -2 \end{bmatrix}$,
$\begin{bmatrix} x \\ y \\ z \end{bmatrix} = A^{-1}K = \begin{bmatrix} 1.5 \\ 2.5 \\ -3 \end{bmatrix}, x = 1.5, y = 2.5, z = -3.$

19. $A = \begin{bmatrix} 1 & 2 & 4 \\ 3 & 1 & 4 \\ 2 & 9 & -2 \end{bmatrix}, K = \begin{bmatrix} 7 \\ -2 \\ 10 \end{bmatrix}$,
$\begin{bmatrix} x \\ y \\ z \end{bmatrix} = A^{-1}K = \begin{bmatrix} -3.4 \\ 2.2 \\ 1.5 \end{bmatrix}, x = -3.4, y = 2.2,$
$z = 1.5.$

20. Det $\begin{bmatrix} 2 & -1 & 1 \\ 4 & -2 & 1 \\ 6 & -3 & 5 \end{bmatrix} = 0$, so no solution.

21. $A = \begin{bmatrix} 1 & -1 & -1 \\ 16 & 0 & 4 \\ 0 & 12 & -4 \end{bmatrix}, K = \begin{bmatrix} 0 \\ 100, \\ 60 \end{bmatrix}$,
$\begin{bmatrix} x \\ y \\ z \end{bmatrix} = A^{-1}K = \begin{bmatrix} 6.053 \\ 5.263 \\ 0.789 \end{bmatrix}, I_A = 6.05, I_B = 5.26,$
$I_C = 0.79$ all in Amperes

22. $A = \begin{bmatrix} -1 & -1 & 1 \\ 20.5 & -15.4 & 0 \\ 0 & 15.4 & 10.6 \end{bmatrix}, K = \begin{bmatrix} 0 \\ 40 \\ 140 \end{bmatrix}$ and so
$\begin{bmatrix} I_1 \\ I_2 \\ I_3 \end{bmatrix} = \begin{bmatrix} 4.59 \\ 3.51 \\ 8.10 \end{bmatrix} = A^{-1}K$, and we obtain $I_1 = 4.59$,
$I_2 = 3.51$, $I_3 = 8.10$ all in amperes.

23. $A = \begin{bmatrix} 7.18 & -1 & 2.2 \\ -1 & 5.8 & 1.5 \\ 2.2 & 1.5 & 8.4 \end{bmatrix}, K = \begin{bmatrix} 10 \\ 15 \\ 20 \end{bmatrix}$,
$\begin{bmatrix} I_1 \\ I_2 \\ I_3 \end{bmatrix} = A^{-1}K = \begin{bmatrix} 1.22 \\ 2.37 \\ 1.64 \end{bmatrix}, I_1 = 1.22, I_2 = 2.37$ and

$I_3 = 1.64$ all in amperes.

24. Let x represent the number of RC-1 units that can be made and y the number of RC-2 units that can be made. The conditions leads to the system $\begin{cases} 8x + 9y = 1596 \\ 4x + 5y = 860 \end{cases}$, with $A = \begin{bmatrix} 8 & 9 \\ 4 & 5 \end{bmatrix}$ and $K = \begin{bmatrix} 1596 \\ 860 \end{bmatrix}$, so $A^{-1}K = \begin{bmatrix} 60 \\ 124 \end{bmatrix}$. They should make 60 of the RC-1 model and 124 units of the RC-2 model.

25. See *Computer Programs* in main text.

CHAPTER 17 REVIEW

1. $A + C = \begin{bmatrix} 4 & 3 & 2 & 5 \\ 6 & 7 & -1 & 4 \\ 9 & 10 & -8 & 3 \end{bmatrix} + \begin{bmatrix} 0 & 1 & 0 & 2 \\ 3 & 0 & 4 & 0 \\ 0 & -5 & 0 & -6 \end{bmatrix} = \begin{bmatrix} 4 & 4 & 2 & 7 \\ 9 & 7 & 3 & 4 \\ 9 & 5 & -8 & -3 \end{bmatrix}$

2. $B + C = \begin{bmatrix} 3 & -2 & 1 & 0 \\ 5 & -1 & 2 & 0 \\ 4 & 3 & -2 & 0 \end{bmatrix} + \begin{bmatrix} 0 & 1 & 0 & 2 \\ 3 & 0 & 4 & 0 \\ 0 & -5 & 0 & -6 \end{bmatrix} = \begin{bmatrix} 3 & -1 & 1 & 2 \\ 8 & -1 & 6 & 0 \\ 4 & -2 & -2 & -6 \end{bmatrix}$

3. $A - B = \begin{bmatrix} 4 & 3 & 2 & 5 \\ 6 & 7 & -1 & 4 \\ 9 & 10 & -8 & 3 \end{bmatrix} - \begin{bmatrix} 3 & -2 & 1 & 0 \\ 5 & -1 & 2 & 0 \\ 4 & 3 & -2 & 0 \end{bmatrix} = \begin{bmatrix} 1 & 5 & 1 & 5 \\ 1 & 8 & -3 & 4 \\ 5 & 7 & -6 & 3 \end{bmatrix}$

4. $C - B = \begin{bmatrix} 0 & 1 & 0 & 2 \\ 3 & 0 & 4 & 0 \\ 0 & -5 & 0 & -6 \end{bmatrix} - \begin{bmatrix} 3 & -2 & 1 & 0 \\ 5 & -1 & 2 & 0 \\ 4 & 3 & -2 & 0 \end{bmatrix} = \begin{bmatrix} -3 & 3 & -1 & 2 \\ -2 & 1 & 2 & 0 \\ -4 & -8 & 2 & -6 \end{bmatrix}$

5. $2A - 3C = \begin{bmatrix} 8 & 6 & 4 & 10 \\ 12 & 14 & -2 & 8 \\ 18 & 20 & -16 & 6 \end{bmatrix} - \begin{bmatrix} 0 & 3 & 0 & 6 \\ 9 & 0 & 12 & 0 \\ 0 & -15 & 0 & -18 \end{bmatrix} = \begin{bmatrix} 8 & 3 & 4 & 4 \\ 3 & 14 & -14 & 8 \\ 18 & 35 & -16 & 24 \end{bmatrix}$

6. $4A + B - 2C = \begin{bmatrix} 16 & 12 & 8 & 20 \\ 24 & 28 & -4 & 16 \\ 36 & 40 & -32 & 12 \end{bmatrix} + \begin{bmatrix} 3 & -2 & 1 & 0 \\ 5 & -1 & 2 & 0 \\ 4 & 3 & -2 & 0 \end{bmatrix} - \begin{bmatrix} 0 & 2 & 0 & 4 \\ 6 & 0 & 8 & 0 \\ 0 & -10 & 0 & -12 \end{bmatrix} = \begin{bmatrix} 19 & 8 & 9 & 16 \\ 23 & 27 & -10 & 16 \\ 40 & 53 & -34 & 24 \end{bmatrix}$

7. $\begin{bmatrix} 3 & 2 \\ 4 & 5 \\ 1 & 0 \end{bmatrix} \times \begin{bmatrix} 1 & 2 \\ 0 & 1 \end{bmatrix} = \begin{bmatrix} 3+0 & 6+2 \\ 4+0 & 8+5 \\ 1+0 & 2+0 \end{bmatrix} = \begin{bmatrix} 3 & 8 \\ 4 & 13 \\ 1 & 2 \end{bmatrix}$

8. $\begin{bmatrix} 1 & -4 \\ 5 & 1 \end{bmatrix} \begin{bmatrix} 3 & 4 & 1 \\ 2 & 5 & 0 \end{bmatrix} = \begin{bmatrix} 3-8 & 4-20 & 1+0 \\ 15+2 & 20+5 & 5+0 \end{bmatrix} = \begin{bmatrix} -5 & -16 & 1 \\ 17 & 25 & 5 \end{bmatrix}$

9. $\begin{bmatrix} 3 \\ 2 \\ -1 \end{bmatrix} [4 \quad 5 \quad 1] = \begin{bmatrix} 12 & 15 & 3 \\ 8 & 10 & 2 \\ -4 & -5 & -1 \end{bmatrix}$

10. $[4 \quad 5 \quad 1] \begin{bmatrix} -2 \\ 7 \\ -3 \end{bmatrix} = [-8 + 35 - 3] = [24]$

11. $\begin{bmatrix} 2 & 3 \\ -4 & -5 \end{bmatrix}^{-1} = \dfrac{1}{(-10+12)} \begin{bmatrix} -5 & -3 \\ 4 & 2 \end{bmatrix} = \begin{bmatrix} -\frac{5}{2} & -\frac{3}{2} \\ 2 & 1 \end{bmatrix}$

12. $\begin{bmatrix} 2 & -1 \\ 0 & 4 \end{bmatrix}^{-1} = \dfrac{1}{8} \begin{bmatrix} 4 & 1 \\ 0 & 2 \end{bmatrix} = \begin{bmatrix} \frac{1}{2} & \frac{1}{8} \\ 0 & \frac{1}{4} \end{bmatrix} = \begin{bmatrix} 0.5 & 0.125 \\ 0 & 0.25 \end{bmatrix}$

13. $\left[\begin{array}{ccc|ccc} -2 & 1 & 0 & 1 & 0 & 0 \\ 0 & 4 & 0 & 0 & 1 & 0 \\ 1 & 0 & 1 & 0 & 0 & 1 \end{array}\right]$ $2R_3 + R_1$

$\left[\begin{array}{ccc|ccc} 0 & 1 & 2 & 1 & 0 & 2 \\ 0 & 4 & 0 & 0 & 1 & 0 \\ 1 & 0 & 1 & 0 & 0 & 1 \end{array}\right]$ Swap R_1 and R_3

$\left[\begin{array}{ccc|ccc} 1 & 0 & 1 & 0 & 0 & 1 \\ 0 & 4 & 0 & 0 & 1 & 0 \\ 0 & 1 & 2 & 1 & 0 & 2 \end{array}\right]$ $-\frac{1}{4}R_2 + R_3$

$\left[\begin{array}{ccc|ccc} 1 & 0 & 1 & 0 & 0 & 1 \\ 0 & 4 & 0 & 0 & 1 & 0 \\ 0 & 0 & 2 & 1 & -\frac{1}{4} & 2 \end{array}\right]$ $-\frac{1}{2}R_3 + R_1$

$\left[\begin{array}{ccc|ccc} 1 & 0 & 0 & -\frac{1}{2} & \frac{1}{8} & 0 \\ 0 & 4 & 0 & 0 & 1 & 0 \\ 0 & 0 & 2 & 1 & -\frac{1}{4} & 2 \end{array}\right]$ $\frac{1}{4}R_2$, $\frac{1}{2}R_3$

$\left[\begin{array}{ccc|ccc} 1 & 0 & 0 & -\frac{1}{2} & \frac{1}{8} & 0 \\ 0 & 1 & 0 & 0 & \frac{1}{4} & 0 \\ 0 & 0 & 1 & \frac{1}{2} & -\frac{1}{8} & 1 \end{array}\right]$. The inverse is

$\begin{bmatrix} -\frac{1}{2} & \frac{1}{8} & 0 \\ 0 & \frac{1}{4} & 0 \\ \frac{1}{2} & -\frac{1}{8} & 1 \end{bmatrix}$ or $\begin{bmatrix} -0.5 & 0.125 & 0 \\ 0 & 0.25 & 0 \\ 0.5 & -0.125 & 1 \end{bmatrix}$.

14. $\left[\begin{array}{ccc|ccc} 2 & 3 & -1 & 1 & 0 & 0 \\ 1 & 2 & 1 & 0 & 1 & 0 \\ -1 & -1 & 3 & 0 & 0 & 1 \end{array}\right]$ $-R_1 + 2R_2$, $R_2 + R_3$

$\left[\begin{array}{ccc|ccc} 2 & 3 & -1 & 1 & 0 & 0 \\ 0 & 1 & 3 & -1 & 2 & 0 \\ 0 & 1 & 4 & 0 & 1 & 1 \end{array}\right]$ $-3R_2 + R_1$, $-R_2 + R_3$

$\left[\begin{array}{ccc|ccc} 2 & 0 & -10 & 4 & -6 & 0 \\ 0 & 1 & 3 & -1 & 2 & 0 \\ 0 & 0 & 1 & 1 & -1 & 1 \end{array}\right]$ $10R_3 + R_1$, $-3R_3 + R_2$

$\left[\begin{array}{ccc|ccc} 2 & 0 & 0 & 14 & -16 & 10 \\ 0 & 1 & 0 & -4 & 5 & -3 \\ 0 & 0 & 1 & 1 & -1 & 1 \end{array}\right]$ $\frac{1}{2}R_1$

$\left[\begin{array}{ccc|ccc} 1 & 0 & 0 & 7 & -8 & 5 \\ 0 & 1 & 0 & -4 & 5 & -3 \\ 0 & 0 & 1 & 1 & -1 & 1 \end{array}\right]$. The inverse is

$\begin{bmatrix} 7 & -8 & 5 \\ -4 & 5 & -3 \\ 1 & -1 & 1 \end{bmatrix}$

15. $A = \begin{bmatrix} 12 & 5 \\ 3 & 1 \end{bmatrix}$, $A^{-1} = -\frac{1}{3}\begin{bmatrix} 1 & -5 \\ -3 & 12 \end{bmatrix}$,
$K = \begin{bmatrix} -2 \\ 1.1 \end{bmatrix}$, $A^{-1}K = \begin{bmatrix} 2.5 \\ -6.4 \end{bmatrix}$, $x = 2.5$, $y = -6.4$

16. $A = \begin{bmatrix} 4 & 1 \\ -3 & 2 \end{bmatrix}$, $A^{-1} = \frac{1}{11}\begin{bmatrix} 2 & -1 \\ 3 & 4 \end{bmatrix}$, $K = \begin{bmatrix} -4 \\ 14 \end{bmatrix}$,
$A^{-1}K = \begin{bmatrix} -2 \\ 4 \end{bmatrix}$, $x = -2$, $y = 4$

17. $A = \begin{bmatrix} 2 & 3 & 5 \\ -2 & 3 & 5 \\ 5 & -3 & -2 \end{bmatrix}$, $K = \begin{bmatrix} 20 \\ 12 \\ 9 \end{bmatrix}$,
$A^{-1}K = \begin{bmatrix} 2 \\ -3 \\ 5 \end{bmatrix}$, $x = 2$, $y = -3$, $z = 5$

18. $A = \begin{bmatrix} 1 & 1 & 6 \\ -2 & 2 & 4 \\ 3 & 2 & 4 \end{bmatrix}$, $K = \begin{bmatrix} -3 \\ -2 \\ 14 \end{bmatrix}$,
$A^{-1}K = \begin{bmatrix} 3.2 \\ 6.4 \\ -2.1 \end{bmatrix}$, $x = 3.2$, $y = 6.4$, $z = -2.1$

19. $\begin{cases} -9a + T = 75.4 \\ -11a - T = -100.0 \end{cases}$ $A = \begin{bmatrix} -9 & 1 \\ -11 & -1 \end{bmatrix}$,
$K = \begin{bmatrix} 75.4 \\ -100.0 \end{bmatrix}$, $A^{-1}K = \begin{bmatrix} 1.23 \\ 86.47 \end{bmatrix}$, $a = 1.23$,
$T = 86.47$

20. $\begin{cases} 4P + 8B + 12T = 1872 \\ 2P + 3B + 5T = 771 \\ 7P + 6B = 11T = 1770 \end{cases}$ $A = \begin{bmatrix} 4 & 8 & 12 \\ 2 & 3 & 5 \\ 7 & 6 & 11 \end{bmatrix}$,
$K = \begin{bmatrix} 1872 \\ 771 \\ 1770 \end{bmatrix}$, $A^{-1}K = \begin{bmatrix} 45 \\ 72 \\ 93 \end{bmatrix}$, and so they can make 45 PCs, 72 BCs and 93 TCs.

21. The given systems can be expressed in matrices as follows: $\begin{bmatrix} \frac{1}{2} & \frac{\sqrt{3}}{2} \\ -\frac{\sqrt{3}}{2} & \frac{1}{2} \end{bmatrix}\begin{bmatrix} x \\ y \end{bmatrix} = \begin{bmatrix} x' \\ y' \end{bmatrix}$ and $\begin{bmatrix} -\frac{1}{2} & \frac{\sqrt{3}}{2} \\ -\frac{\sqrt{3}}{2} & -\frac{1}{2} \end{bmatrix}\begin{bmatrix} x' \\ y' \end{bmatrix} = \begin{bmatrix} x'' \\ y'' \end{bmatrix}$. Hence,
$\begin{bmatrix} x'' \\ y'' \end{bmatrix} = \begin{bmatrix} -\frac{1}{2} & \frac{\sqrt{3}}{2} \\ -\frac{\sqrt{3}}{2} & -\frac{1}{2} \end{bmatrix}\begin{bmatrix} \frac{1}{2} & \frac{\sqrt{3}}{2} \\ -\frac{\sqrt{3}}{2} & \frac{1}{2} \end{bmatrix}\begin{bmatrix} x \\ y \end{bmatrix} =$
$\begin{bmatrix} -1 & 0 \\ 0 & -1 \end{bmatrix}\begin{bmatrix} x \\ y \end{bmatrix} = \begin{bmatrix} -x \\ -y \end{bmatrix}$. Thus, we have shown that $x'' = -x$ and $y'' = -y$.

22. Since $x'' = -x$ and $y'' = -y$, we have that $x\cos\theta + y\sin\theta = -x$ or $\cos\theta = -1$ and $\sin\theta = 0$. This is a 180° rotation.

23. $M = \begin{bmatrix} 1 & -\frac{1}{f_2} \\ 0 & 1 \end{bmatrix} \cdot \begin{bmatrix} 1 & 0 \\ d & 1 \end{bmatrix} \cdot \begin{bmatrix} 1 & -\frac{1}{f_1} \\ 0 & 1 \end{bmatrix} =$
$\begin{bmatrix} 1-\frac{d}{f_2} & -\frac{1}{f_2} \\ d & 1 \end{bmatrix} \begin{bmatrix} 1 & -\frac{1}{f_1} \\ 0 & 1 \end{bmatrix} =$
$\begin{bmatrix} 1-\frac{d}{f_2} & -\frac{1}{f_1}+\frac{d}{f_1f_2}-\frac{1}{f_2} \\ d & -\frac{d}{f_1}+1 \end{bmatrix}$ or

$$M = \begin{bmatrix} 1-\frac{d}{f_2} & \frac{d}{f_1f_2}-\frac{1}{f_1}-\frac{1}{f_2} \\ d & 1-\frac{d}{f_1} \end{bmatrix}.$$

24. $-\frac{1}{f} = \frac{d}{f_1f_2} - \frac{1}{f_1} - \frac{1}{f_2}$ so $\frac{1}{f} = \frac{1}{f_1} + \frac{1}{f_2} - \frac{d}{f_1f_2}$

CHAPTER 17 TEST

1. (a) $A+B = \begin{bmatrix} 8 & 0 & -4 \\ 16 & -6 & 2 \end{bmatrix} + \begin{bmatrix} -1 & 5 & -3 \\ 3 & 0 & 4 \end{bmatrix} =$
$\begin{bmatrix} 7 & 5 & -7 \\ 19 & -6 & 6 \end{bmatrix}$
(b) $3A-2B = \begin{bmatrix} 24 & 0 & -12 \\ 48 & -18 & 6 \end{bmatrix} -$
$\begin{bmatrix} -2 & 10 & -6 \\ 6 & 0 & 8 \end{bmatrix} = \begin{bmatrix} 26 & -10 & -6 \\ 42 & -18 & -2 \end{bmatrix}$

2. $[1 \quad -2 \quad 3] \begin{bmatrix} -4 \\ -6 \\ 8 \end{bmatrix} = [-4+12+24] = [32]$

3. (a) $CD = \begin{bmatrix} 4 & 6 \\ -10 & 4 \end{bmatrix} \cdot \begin{bmatrix} \frac{1}{2} & 0 \\ -\frac{3}{2} & 4 \end{bmatrix} =$
$\begin{bmatrix} 2-9 & 0+24 \\ -5-6 & 0+16 \end{bmatrix} = \begin{bmatrix} -7 & 24 \\ -11 & 16 \end{bmatrix}.$
(b) $DC = \begin{bmatrix} \frac{1}{2} & 0 \\ -\frac{3}{2} & 4 \end{bmatrix} \begin{bmatrix} 4 & 6 \\ -10 & 4 \end{bmatrix} =$
$\begin{bmatrix} 2+0 & 3+0 \\ -6-40 & -9+16 \end{bmatrix} = \begin{bmatrix} 2 & 3 \\ -46 & 7 \end{bmatrix}.$

4. (a) $E = \begin{bmatrix} 2 & -3 \\ 7 & 9 \end{bmatrix}$,
$E^{-1} = \frac{1}{18+21} \begin{bmatrix} 9 & 3 \\ -7 & 2 \end{bmatrix} \approx \begin{bmatrix} 0.2308 & 0.0769 \\ -0.1795 & 0.0513 \end{bmatrix}$
(b) $EX = F$ so $X = E^{-1}F = \begin{bmatrix} -5 \\ 7 \end{bmatrix}.$

5. The coefficient matrix is $A = \begin{bmatrix} 1 & 3 & 1 \\ 2 & 5 & 1 \\ 1 & 2 & 3 \end{bmatrix}$ and the constant matrix is $K = \begin{bmatrix} -2 \\ -5 \\ 6 \end{bmatrix}$. If $X = \begin{bmatrix} x \\ y \\ z \end{bmatrix}$, we have the system $AX = K$, so $X = A^{-1}K$. You can determine that $A^{-1} = \begin{bmatrix} -\frac{13}{3} & \frac{7}{3} & \frac{2}{3} \\ \frac{5}{3} & -\frac{2}{3} & -\frac{1}{3} \\ \frac{1}{3} & -\frac{1}{3} & \frac{1}{3} \end{bmatrix} =$
$\frac{1}{3} \begin{bmatrix} -13 & 7 & 2 \\ 5 & -2 & -1 \\ 1 & -1 & 1 \end{bmatrix}$ and so, $A^{-1}K = \begin{bmatrix} 1 \\ -2 \\ 3 \end{bmatrix}.$
Thus, $x = 1, y = -2, z = 3$.

CHAPTER

18

Higher Degree Equations

18.1 THE REMAINDER AND FACTOR THEOREM

1. $P(2) = 3 \cdot 2^2 - 2 \cdot 2 + 1 = 12 - 4 + 1 = 9$
2. $P(-1) = 2(-1)^3 - 4(-1) + 5 = -2 + 4 + 5 = 7$
3. $P(-1) = (-1)^4 + (-1)^3 + (-1)^2 - (-1) + 1 = 1 - 1 + 1 + 1 + 1 = 3$
4. $P(-2) = (-2)^4 - 2(-2)^2 + (-2) = 16 - 8 - 2 = 6$
5. $P(3) = 5(3)^3 - 4(3) + 7 = 135 - 12 + 7 = 130$
6. $P(-2) = 7(-2)^4 - 5(-2)^2 + (-2) - 7 = 112 - 20 - 2 - 7 = 83$
7. $P(-1) = (-1)^5 - (-1)^4 + (-1)^3 + (-1)^2 - (-1) + 1 = -1 - 1 - 1 + 1 + 1 + 1 = 0$
8. $P(-3) = 3(-3)^5 + 4(-3)^2 - 3 = -729 + 36 - 3 = -696$
9. $R = P(1) = 1^3 + 2 \cdot 1^2 - 1 - 2 = 1 + 2 - 1 - 2 = 0$
10. $R = P(3) = 3^3 + 2 \cdot 3^2 - 12 \cdot 3 - 9 = 27 + 18 - 36 - 9 = 0$
11. $R = P(3) = 3^3 - 3 \cdot 3^2 + 2 \cdot 3 + 5 = 27 - 27 + 6 + 5 = 11$
12. $R = P(1) = 1^3 - 9 \cdot 1^2 + 23 \cdot 1 - 15 = 1 - 9 + 23 - 15 = 0$
13. $R = P(-2) = 4(-2)^4 + 13(-2)^3 - 13(-2)^2 - 40(-2) + 12 = 64 - 104 - 52 + 80 + 12 = 0$
14. $R = P(7) = 2 \cdot 7^4 - 2 \cdot 7^3 - 6 \cdot 7^2 - 14 \cdot 7 - 7 = 4802 - 686 - 294 - 98 - 7 = 3717$
15. $R = P(5) = 3 \cdot 5^4 - 12 \cdot 5^3 - 60 \cdot 5 + 4 = 1875 - 1500 - 300 + 4 = 79$
16. $R = P\left(\frac{1}{2}\right) = 4 \cdot \left(\frac{1}{2}\right)^3 - 4\left(\frac{1}{2}\right)^2 - 10\left(\frac{1}{2}\right) + 8 = \frac{1}{2} - 1 - 5 + 8 = 2\frac{1}{2}$ or 2.5
17. $P(3) = 3^3 + 2 \cdot 3^2 - 12 \cdot 3 - 9 = 27 + 18 - 36 - 9 = 0$; yes
18. $P(-1) = (-1)^4 - 9(-1)^3 + 18(-1)^2 - 3 = 1 + 9 + 18 - 3 = 25$; no
19. $P(-1) = 2(-1)^5 - 6(-1)^3 + (-1)^2 + 4(-1) - 1 = -2 + 6 + 1 - 4 - 1 = 0$; yes
20. $P(1) = 2(1)^5 - 6(1)^3 + 1^2 + 4 \cdot 1 - 1 = 2 - 6 + 1 + 4 - 1 = 0$; yes
21. $P(-3) = 3(-3)^5 + 3(-3)^4 - 14(-3)^3 + 4(-3)^2 - 24(-3) = -729 + 243 + 378 + 36 + 72 = 0$; yes
22. $P(2) = 3 \cdot 2^5 + 3 \cdot 2^4 - 14 \cdot 2^3 + 4 \cdot 2^2 - 24 \cdot 2 = 96 + 48 - 112 + 16 - 48 = 0$; yes
23. $P\left(\frac{5}{2}\right) = 6 \cdot \left(\frac{5}{2}\right)^4 - 15\left(\frac{5}{2}\right)^3 - 8\left(\frac{5}{2}\right)^2 + 20\left(\frac{5}{2}\right) = 234\frac{3}{8} - 234\frac{3}{8} - 50 + 50 = 0$; yes
24. $P\left(-\frac{3}{5}\right) = 20\left(-\frac{3}{5}\right)^4 + 12\left(-\frac{3}{5}\right)^3 + 10\left(-\frac{3}{5}\right) + 9 = 2.592 - 2.592 - 6 + 9 = 3$; no

25. $$\begin{array}{r|rrrrrr} 3 & 1 & 0 & -17 & 0 & 75 & 9 \\ & & 3 & 9 & -24 & -72 & 9 \\ \hline & 1 & 3 & -8 & -24 & 3 & 18 \end{array}$$

So, $Q(x) = x^4 + 3x^3 - 8x^2 - 24x + 3$, $R(x) = 18$

26. $$\begin{array}{r|rrrrrr} -2 & 2 & 0 & 0 & -1 & 8 & 44 \\ & & -4 & 8 & -16 & 34 & -84 \\ \hline & 2 & -4 & 8 & -17 & 42 & -40 \end{array}$$

So, $Q(x) = 2x^4 - 4x^3 + 8x^2 - 17x + 42$, $R(x) = -40$

27. $$\begin{array}{r|rrrr} -3 & 5 & 7 & 0 & 9 \\ & & -15 & 24 & -72 \\ \hline & 5 & -8 & 24 & -63 \end{array}$$

So, $Q(x) = 5x^2 - 8x + 24$; $R(x) = -63$

28. $$\begin{array}{r|rrrr} 2 & 1 & 3 & -2 & -4 \\ & & 2 & 10 & 16 \\ \hline & 1 & 5 & 8 & 12 \end{array}$$

So, $Q(x) = x^2 + 5x + 8$; $R(x) = 12$

29. $$\begin{array}{r|rrrrrr} \frac{1}{2} & 8 & 0 & -4 & 7 & -2 & 0 \\ & & 4 & 2 & -1 & 3 & \frac{1}{2} \\ \hline & 8 & 4 & -2 & 6 & 1 & \frac{1}{2} \end{array}$$

So, $Q(x) = 8x^4 + 4x^3 - 2x^2 + 6x + 1$ and $R(x) = \frac{1}{2}$

30. $$\begin{array}{r|rrrrrr} -\frac{1}{3} & 9 & 3 & -6 & -2 & 6 & 1 \\ & & -3 & 0 & 2 & 0 & -2 \\ \hline & 9 & 0 & -6 & 0 & 6 & -1 \end{array}$$

So, $Q(x) = 9x^4 - 6x^2 + 6$ and $R(x) = -1$

31. $$\begin{array}{r|rrrrr} \frac{3}{2} & 4 & -12 & 9 & -8 & 12 \\ & & 6 & -9 & 0 & -12 \\ \hline & 4 & -6 & 0 & -8 & 0 \end{array}$$

Since we used $\frac{2x-3}{2} = x - \frac{3}{2}$, we need to divide the depressed polynomial by 2 also. So, $Q(x) = \frac{4x^3 - 6x^2 - 8}{2} = 2x^3 - 3x^2 - 4$; $R(x) = 0$.

32. $$\begin{array}{r|rrrr} \frac{5}{4} & 4 & 7 & -3 & -15 \\ & & 5 & 15 & 15 \\ \hline & 4 & 12 & 12 & 0 \end{array}$$

Since we used $\frac{4x-5}{4} = x - \frac{5}{4}$, we need to divide the depressed polynomial by 4 also. So, $Q(x) = \frac{4x^2 + 12x + 12}{4} = x^2 + 3x + 3$; $R(x) = 0$

18.2 ROOTS OF AN EQUATION

1. Using synthetic division, we get

$$\begin{array}{r|rrrr} 1 & 5 & 0 & -8 & 3 \\ & & 5 & 5 & -3 \\ \hline & 5 & 5 & -3 & 0 \end{array}$$

We now use the quadratic equation on the depressed equation $5x^2 + 5x - 3 = 0$, with the results $x = \frac{-5 \pm \sqrt{5^2 - 4 \cdot 5 \cdot (-3)}}{2 \cdot 5} = \frac{-5 \pm \sqrt{25+60}}{10} = \frac{-5 \pm \sqrt{85}}{10}$. The roots are 1, $\frac{-5+\sqrt{85}}{10}$ and $\frac{-5-\sqrt{85}}{10}$.

2. Using synthetic division, we get

$$\begin{array}{r|rrrr} -4 & 2 & 5 & -11 & 4 \\ & & -8 & 12 & -4 \\ \hline & 2 & -3 & 1 & 0 \end{array}$$

The depressed equation $2x^2 - 3x + 1 = 0$ factors into $(2x - 1)(x - 1) = 0$, which has roots $\frac{1}{2}$ and 1. So, the roots of the given equation are -4, 1 and $\frac{1}{2}$.

3. Using synthetic division, we get

$$\begin{array}{r|rrrr} \frac{1}{3} & 9 & -3 & -81 & 27 \\ & & 3 & 0 & -27 \\ \hline & 9 & 0 & -81 & 0 \end{array}$$

Now, $9x^2 - 81 = 9(x^2 - 9) = 9(x + 3)(x - 3)$ and so the roots are $\frac{1}{3}$, 3, and -3.

4. With synthetic division, we get

$$\begin{array}{r|rrrr} \frac{2}{5} & 10 & -4 & -40 & 16 \\ & & 4 & 0 & -16 \\ \hline & 10 & 0 & -40 & 0 \end{array}$$

Now, $10x^2 - 40 = 10(x^2 - 4) = 10(x + 2)(x - 2)$. The roots are $\frac{2}{5}$, -2 and 2.

5. Synthetic division and the complex conjugate of r_1, which is $-j$, yields

$$\begin{array}{r|rrrrr} j & 1 & 0 & -3 & 0 & -4 \\ & & j & -1 & -4j & 4 \\ \hline -j & 1 & j & -4 & -4j & 0 \\ & & -j & 0 & 4j & \\ \hline & 1 & 0 & -4 & 0 & \end{array}$$

Since $x^2 - 4 = (x + 2)(x - 2)$, we see that the roots are 2, -2, j, and $-j$.

6. Synthetic division and the complex conjugate of r_1, which is $3j$, yields

$$\begin{array}{r|rrrrr} -3j & 3 & 0 & 6 & 0 & -189 \\ & & -9j & -27 & 63j & 189 \\ \hline 3j & 3 & -9j & -21 & 63j & 0 \\ & & 9j & 0 & -63j & \\ \hline & 3 & 0 & -21 & 0 & \end{array}$$

Since $3x^2 - 21 = 3(x^2 - 7) = 3(x + \sqrt{7})(x - \sqrt{7})$, the roots are $\sqrt{7}$, $-\sqrt{7}$, $3j$, and $-3j$.

7. Synthetic division and the complex conjugate of r_1, which is $-1 - 2j$, yields

$$\begin{array}{r|rrrrr} -1+2j & 1 & 2 & -4 & -18 & -45 \\ & & -1+2j & -5 & 9-18j & 45 \\ \hline -1-2j & 1 & 1+2j & -9 & -9-18j & 0 \\ & & -1-2j & 0 & 9+18j & \\ \hline & 1 & 0 & -9 & 0 & \end{array}$$

Since $x^2 - 9 = (x + 3)(x - 3)$, the roots are 3, -3, $-1 + 2j$, and $-1 - 2j$.

8. Synthetic division and the complex conjugate of r_1, which is $-1 + 2j$, yields

$$\begin{array}{r|rrrrr} -1-2j & 2 & 4 & 2 & -16 & -40 \\ & & -2-4j & -10 & 8+16j & 40 \\ \hline -1+2j & 2 & 2-4j & -8 & -8+16j & 0 \\ & & -2+4j & 0 & 8-16j & \\ \hline & 2 & 0 & -8 & 0 & \end{array}$$

Since $2x^2 - 8 = 2(x^2 - 4) = 2(x + 2)(x - 2)$, the roots are 2, -2, $-1 - 2j$, and $-1 + 2j$.

9. Repeated use of synthetic division yields

$$\begin{array}{r|rrrrr} 2 & 1 & -3 & -3 & 7 & 6 \\ & & 2 & -2 & -10 & -6 \\ \hline 3 & 1 & -1 & -5 & -3 & 0 \\ & & 3 & 6 & 3 & \\ \hline & 1 & 2 & 1 & 0 & \end{array}$$

Since $x^2 + 2x + 1 = (x + 1)^2$, the roots are -1, -1, 2, and 3.

10. Repeated use of synthetic division yields

$$\begin{array}{r|rrrrr} 2 & 1 & -3 & -12 & 52 & -48 \\ & & 2 & -2 & -28 & 48 \\ \hline -4 & 1 & -1 & -14 & 24 & 0 \\ & & -4 & 20 & -24 & \\ \hline & 1 & -5 & 6 & 0 & \end{array}$$

Since $x^2 - 5x + 6 = (x - 2)(x - 3)$, the roots are 2, -4, 2, and 3.

11. Repeated use of synthetic division yields

$$\begin{array}{r|rrrrr} -1 & 3 & 12 & 6 & -12 & -9 \\ & & -3 & -9 & 3 & 9 \\ \hline -1 & 3 & 9 & -3 & -9 & 0 \\ & & -3 & -6 & 9 & \\ \hline & 3 & 6 & -9 & 0 & \end{array}$$

Since $3x^2 + 6x - 9 = 3(x^2 + 2x - 3) = 3(x + 3)(x - 1)$, the roots are -1, -1, -3, and 1.

12. Repeated use of synthetic division yields

$$\begin{array}{r|rrrrr}
2 & 2 & 6 & -12 & -24 & 16 \\
 & & 4 & 20 & 16 & -16 \\ \hline
-2 & 2 & 10 & 8 & -8 & 0 \\
 & & -4 & -12 & 8 & \\ \hline
 & 2 & 6 & -4 & 0 &
\end{array}$$

Here $2x^2 + 6x - 4 = 2(x^2 + 3x - 2)$ and using the quadratic formula produces $x = \frac{-3 \pm \sqrt{3^2 - 4 \cdot 1(-2)}}{2} = \frac{-3 \pm \sqrt{9+8}}{2}$. The roots are 2, -2, $\frac{-3+\sqrt{17}}{2}$, $\frac{-3-\sqrt{17}}{2}$.

13. Repeated use of synthetic division yields

$$\begin{array}{r|rrrrr}
-\frac{1}{2} & 6 & 25 & 33 & 1 & -5 \\
 & & -3 & -11 & -11 & 5 \\ \hline
\frac{1}{3} & 6 & 22 & 22 & -10 & 0 \\
 & & 2 & 8 & 10 & \\ \hline
 & 6 & 24 & 30 & 0 &
\end{array}$$

Here $6x^2 + 24x + 30 = 6(x^2 + 4x + 5)$, so using the quadratic formula, we get $x = \frac{-4 \pm \sqrt{16-20}}{2} = -2 \pm j$. The roots are $-\frac{1}{2}$, $\frac{1}{3}$, $-2 + j$, and $-2 - j$.

14. Repeated use of synthetic division yields

$$\begin{array}{r|rrrrr}
\frac{1}{4} & 12 & -47 & 55 & 9 & -5 \\
 & & 3 & -11 & 11 & 5 \\ \hline
-\frac{1}{3} & 12 & -44 & 44 & 20 & 0 \\
 & & -4 & 16 & -20 & \\ \hline
 & 12 & -48 & 60 & 0 &
\end{array}$$

Here $12x^2 - 48x + 60 = 12(x^2 - 4x + 5)$, and so $x = \frac{4 \pm \sqrt{16-20}}{2} = 2 \pm j$. The roots are $\frac{1}{4}$, $-\frac{1}{3}$, $2 + j$, and $2 - j$.

15. Repeated use of synthetic division yields

$$\begin{array}{r|rrrrr}
\frac{2}{3} & 3 & -2 & 0 & -3 & 2 \\
 & & 2 & 0 & 0 & -2 \\ \hline
1 & 3 & 0 & 0 & -3 & 0 \\
 & & 3 & 3 & 3 & \\ \hline
 & 3 & 3 & 3 & 0 &
\end{array}$$

Here $3x^2 + 3x + 3 = 3(x^2 + x + 1)$, and by using the quadratic formula, we get $x = \frac{-1 \pm \sqrt{1-4}}{2} = -\frac{1}{2} \pm \frac{\sqrt{3}}{2}j$. The roots are 1, $\frac{2}{3}$, $-\frac{1}{2} + \frac{\sqrt{3}}{2}j$, and $-\frac{1}{2} - \frac{\sqrt{3}}{2}j$.

16. Repeated use of synthetic division yields

$$\begin{array}{r|rrrrr}
2 & 4 & 3 & 0 & -32 & -24 \\
 & & 8 & 22 & 44 & 24 \\ \hline
-\frac{3}{4} & 4 & 11 & 22 & 12 & 0 \\
 & & -3 & -6 & -12 & \\ \hline
 & 4 & 8 & 16 & 0 &
\end{array}$$

Here $4x^2 + 8x + 16 = 4(x^2 + 2x + 4)$ and use of the quadratic formula yields $x = \frac{-2 \pm \sqrt{4-16}}{2} = -1 \pm j\sqrt{3}$. The roots are 2, $-\frac{3}{4}$, $-1 + j\sqrt{3}$, and $-1 - j\sqrt{3}$.

17. Here we use of synthetic division three times, with the following results:

$$\begin{array}{r|rrrrrr}
1 & 1 & -1 & 1 & -7 & 10 & -4 \\
 & & 1 & 0 & 1 & -6 & 4 \\ \hline
1 & 1 & 0 & 1 & -6 & 4 & 0 \\
 & & 1 & 1 & 2 & -4 & \\ \hline
1 & 1 & 1 & 2 & -4 & 0 & \\
 & & 1 & 2 & 4 & & \\ \hline
 & 1 & 2 & 4 & 0 & &
\end{array}$$

Here $x^2 + 2x + 4$ has roots $x = \frac{-2 \pm \sqrt{4-16}}{2} = -1 \pm j\sqrt{3}$. The roots are 1, 1, 1, $-1 + j\sqrt{3}$, and $-1 - j\sqrt{3}$.

18. Here we use of synthetic division three times, with the following results:

$$\begin{array}{r|rrrrrr}
2 & 1 & -5 & 7 & -2 & 4 & -8 \\
 & & 2 & -6 & 2 & 0 & 8 \\ \hline
2 & 1 & -3 & 1 & 0 & 4 & 0 \\
 & & 2 & -2 & -2 & -4 & \\ \hline
2 & 1 & -1 & -1 & -2 & 0 & \\
 & & 2 & 2 & 2 & & \\ \hline
 & 1 & 1 & 1 & 0 & &
\end{array}$$

Here $x^2 + x + 1 = \frac{-1 \pm \sqrt{1-4}}{2} = \frac{-1 \pm j\sqrt{3}}{2}$ and the roots are 2, 2, 2, $-\frac{1}{2} + \frac{\sqrt{3}}{2}j$, and $-\frac{1}{2} - \frac{\sqrt{3}}{2}j$.

19. Here we use of synthetic division three times, with the following results:

3	3	-2	-24	1	28	-12
		9	21	-9	-24	12
-2	3	7	-3	-8	4	0
		-6	-2	10	-4	
$\frac{2}{3}$	3	1	-5	2	0	
		2	2	-2		
	3	3	-3	0		

Since $3x^2 + 3x - 3 = 3(x^2 + x - 1)$, we find that $x = \frac{-1\pm\sqrt{1+4}}{2} = \frac{-1\pm\sqrt{5}}{2}$. The roots are 3, -2, $\frac{2}{3}$, $\frac{-1+\sqrt{5}}{2}$, and $\frac{-1-\sqrt{5}}{2}$.

20. Repeated use of synthetic division yields

$-\frac{2}{3}$	3	-4	5	-18	-28	-8
		-2	4	-6	16	8
$2j$	3	-6	9	-24	-12	0
		$6j$	$-12-12j$	$24-6j$	12	
$-2j$	3	$-6+6j$	$-3-12j$	$-6j$	0	
		$-6j$	$12j$	$6j$		
	3	-6	-3	0		

Since $3x^2 - 6x - 3 = 3(x^2 - 2x - 1)$, we see that $x = \frac{2\pm\sqrt{4+4}}{2} = 1 \pm \sqrt{2}$. The roots are $-\frac{2}{3}$, $2j$, $-2j$, $1+\sqrt{2}$, and $1-\sqrt{2}$.

21. Use of synthetic division four times yields

j	1	-6	3	-60	-61	-54	-63
		j	$-1-6j$	$6+2j$	$-2-54j$	$54-63j$	63
$-j$	1	$-6+j$	$2-6j$	$-54+2j$	$-63-54j$	$-63j$	0
		$-j$	$6j$	$-2j$	$54j$	$63j$	
$-3j$	1	-6	2	-54	-63	0	
		$-3j$	$-9+18j$	$54+21j$	63		
$3j$	1	$-6-3j$	$-7+18j$	$21j$	0		
		$3j$	$-18j$	$-21j$			
	1	-6	-7	0			

Since $x^2 - 6x - 7 = (x+1)(x-7)$, the roots are j, $-j$, $3j$, $-3j$, -1, and 7.

22. Use of synthetic division four times yields

$2j$	6	-19	63	-152	216	-304	240
		$12j$	$-24-38j$	$76+78j$	$-156-152j$	$304+120j$	-240
$-2j$	6	$-19+12j$	$39-38j$	$-76+78j$	$60-152j$	$120j$	0
		$-12j$	$38j$	$-78j$	$152j$	$-120j$	
$2j$	6	-19	39	-76	60	0	
		$12j$	$-24-38j$	$76+30j$	-60		
$-2j$	6	$-19+12j$	$15-38j$	$30j$	0		
		$-12j$	$38j$	$-30j$			
	6	-19	15	0			

Here $6x^2 - 19x + 15 = (3x-5)(2x-3)$, and so the roots are $2j$, $2j$, $-2j$, $-2j$, $\frac{5}{3}$, and $\frac{3}{2}$.

18.3 RATIONAL ROOTS

1. $P(x) = x^3 - 3x - 2$ has one sign change, so there is one positive root. $P(-x) = -x^3 + 3x - 2$ has two sign changes so there are either 0 or 2 negative roots. $c = \pm1, \pm2$; $d = 1$, possible rational roots are ±1 and ±2 .

2. $P(x) = x^3 - 2x^2 - x + 2$ has two sign changes so 0 or 2 positive roots. $P(-x) = -x^3 - 2x^2 + x + 2$ has one sign change so one negative root. $c = \pm1, \pm2$; $d = 1$, so possible rational roots are ±1 and ±2

3. $P(x) = x^3 - 8x^2 - 17x + 6$ has two sign changes so 0 or 2 positive roots. $P(-x) = -x^3 - 8x^2 + 17x + 6$ has one sign change so 1 negative root. $c = \pm1$, ±2, ±3, ±6; $d = 1$ so possible rational roots are $\pm1, \pm2, \pm3, \pm6$

4. $P(x) = x^3 + 2x^2 - x - 2$ has one sign change so 1 positive root. $P(-x) = -x^3 + 2x^2 + x - 2$ has two sign changes so 0 or 2 negative roots. $c = \pm1, \pm2$; $d = 1$, so possible rational roots are ±1 and ±2.

5. $P(x) = x^3 - 2x^2 - 5x + 6$ has two sign change so 0 or 2 positive roots. $P(-x) = -x^3 - 2x^2 + 5x + 6$ has one sign change so 1 negative root. $c = \pm1$, ±2, ±3, ±6; $d = 1$ so possible rational roots are $\pm1, \pm2, \pm3, \pm6$.

6. $P(x) = 3x^3 - x^2 - 2x + 2$ has two sign changes so 0 or 2 postive roots. $P(-x) = -3x^3 - x^2 + 2x + 2$ has one sign change so one negative root. $c = \pm1$, ±2, $d = 1, 3$. Possible rational roots are ±1, ±2, $\pm\frac{1}{3}, \pm\frac{2}{3}$.

7. $P(x) = 2x^4 - x^3 - 5x^2 + x + 3$ has two sign changes so 0 or 2 positive roots. $P(-x) = 2x^4 + x^3 - 5x - x + 3$ has 2 sign changes so 0 or 2 negative roots. $c = \pm1$, ±3; $d = 1, 2$. Possible rational roots are ±1, ±3, $\pm\frac{1}{2}, \pm\frac{3}{2}$.

8. $P(x) = 2x^4 - 7x^3 - 10x^2 + 33x + 18 = 0$ has two sign changes so there are either 0 or 2 positive roots. $P(-x) = 2x^4 + 7x^3 - 10x^2 - 33x + 18$ has two sign changes so there are 0 or 2 negative roots. $c = \pm1$, ±2, ±3, ±6, ±9, ±18; $d = 1, 2$. Possible rational roots are $\pm1, \pm2, \pm3, \pm6, \pm9, \pm18, \pm\frac{1}{2}, \pm\frac{3}{2}, \pm\frac{9}{2}$.

9. $P(x) = 6x^4 - 7x^3 - 13x^2 + 4x + 4$ has two sign changes so 0 or 2 positive roots. $P(-x) = 6x^4 + 7x^3 - 13x^2 - 4x + 4$ has two sign changes so 0 or 2 negative roots. $c = \pm1, \pm2, \pm4$; $d = 1, 2, 3, 6$. Possible rational roots are ±1, ±2, ±4, $\pm\frac{1}{2}$, $\pm\frac{1}{3}$, $\pm\frac{2}{3}, \pm\frac{4}{3}, \pm\frac{1}{6}$.

10. $P(x) = 5x^6 - x^5 - 5x^4 + 6x^3 - x^2 - 5x + 1 = 0$ has 4 sign changes so $0, 2$ or 4 positive roots $P(-x) = 5x^6 + x^5 - 5x^4 - 6x^3 - x^2 + 5x + 1$ has 2 sign changes so 0 or 2 negative roots $c = \pm1$; $d = 1$, 5. Possible rational roots are $\pm1, \pm\frac{1}{5}$.

11. Possible rational roots are ±1, ±3, ±5, ±15. $P(-x) = -x^3 - 9x^2 - 23x - 15$ so no negative roots

1	1	−9	23	−15
		1	−8	15
	1	−8	15	0

Since $x^2 - 8x + 15 = (x - 3)(x - 5)$, we see that the roots are 1, 3, 5.

12. Possible rational roots are ±1, ±2, ±3, ±4, ±6, ±12 It has 0 or 2 positive roots and one negative. Using synthetic division, we see that

2	1	−3	−4	12
		2	−2	−12
	1	−1	−6	0

Since $x^2 - x - 6 = (x - 3)(x + 2)$ the roots are 2, 3, and -2.

13. Possible rational roots are ±1 and ±2. Using synthetic division, we obtain

−1	1	0	−3	−2
		−1	1	2
	1	−1	−2	0

Since $x^2 - x - 2 = (x - 2)(x + 1)$, the roots are -1, -1, and 2.

14. Possible rational roots are ±1, ±2, ±3, ±4, ±6, ±12.

2	1	−1	−8	12
		2	2	−12
	1	1	−6	0

Since $x^2 + x - 6 = (x-2)(x+3)$, the roots are 2, 2, and -3.

15. Possible roots are ±1, ±2, ±3, ±4, ±6, ±12, $\pm\frac{1}{2}$, $\pm\frac{3}{2}$, $\pm\frac{1}{4}$, and $\pm\frac{3}{4}$. Synthetic division produces

$-\frac{3}{4}$	4	−5	10	12
		−3	6	−12
	4	−8	16	0

We see that $4x^2 - 8x + 16 = 4(x^2 - 2x + 4)$ and so $x = \frac{+2\pm\sqrt{4-16}}{2} = 1 \pm j\sqrt{3}$ Roots are $-\frac{3}{4}$, $1 + j\sqrt{3}$, and $1 - j\sqrt{3}$.

16. Possible rational roots are ±1, ±2, ±3, ±6, $\pm\frac{1}{3}$, $\pm\frac{2}{3}$. Synthetic division produces

3	3	−4	−17	6
		9	15	−6
	3	5	−2	0

Since $3x^2 + 5x - 2 = (3x-1)(x+2)$, we see that the roots are 3, $\frac{1}{3}$, and -2.

17. Possible rational roots are ±1, ±2, ±3, ±4, ±6, ±12.

2	1	−4	7	−12	12
		2	−4	6	−12
2	1	−2	3	−6	0
		2	0	6	
	1	0	3	0	

Thus, $x^2 + 3 = 0$ and so $x^2 = -3$ and $x = \pm j\sqrt{3}$. The roots are 2, 2, $j\sqrt{3}$, and $-j\sqrt{3}$.

18. Possible rational roots are ±1, ±2, $\pm\frac{1}{2}$, $\pm\frac{1}{4}$. There are no positive roots.

$-\frac{1}{2}$	4	4	9	8	2
		−2	−1	−4	−2
$-\frac{1}{2}$	4	2	8	4	0
		−2	0	−4	
	4	0	8	0	

Here $4x^2 + 8 = 4(x^2 + 2)$. Solving $x^2 + 2 = 0$, we get $x^2 = -2$ and so $x = \pm j\sqrt{2}$. The roots are $-\frac{1}{2}$, $-\frac{1}{2}$, $j\sqrt{2}$, and $-j\sqrt{2}$.

19. Possible rational roots are ±1, ±2, ±3, ±6, $\pm\frac{1}{2}$, $\pm\frac{3}{2}$, $\pm\frac{1}{4}$, $\pm\frac{3}{4}$. There are either 0 or 2 positive roots, and 0 or 2 negative roots.

$-\frac{1}{2}$	4	−20	1	18	6
		−2	11	−6	−6
$-\frac{1}{2}$	4	−22	12	12	0
		−2	12	−12	
	4	−24	24	0	

Since $4x^2 - 24x + 24 = 4(x^2 - 6x + 6)$ we can use the quadratic formula to see that $x = \frac{6\pm\sqrt{36-24}}{2} = \frac{6\pm\sqrt{12}}{2} = 3 \pm \sqrt{3}$. Hence, the roots are $-\frac{1}{2}$, $-\frac{1}{2}$, $3 + \sqrt{3}$, $3 - \sqrt{3}$.

20. Possible rational roots are ±1, ±3, ±9, ±27. There are 1 or 3 positive roots and 0 or 2 negative roots. Using synthetic division three times, we get

1	1	0	−1	27	0	−27
		1	1	0	27	27
−1	1	1	0	27	27	0
		−1	0	0	−27	
−3	1	0	0	27	0	
		−3	9	−27		
	1	−3	9	0		

$\rightarrow$ no more negative roots

We have $x^2 - 3x + 9 = 0$ and so, $x = \frac{3\pm\sqrt{9-36}}{2} = \frac{3\pm 3j\sqrt{3}}{2}$. The roots are 1, -1, -3, $\frac{3}{2} + \frac{3j\sqrt{3}}{2}$, and $\frac{3}{2} - \frac{3j\sqrt{3}}{2}$.

21. Possible rational roots are ±1, ±2, ±5, ±10. There are no rational roots.

22. Possible rational roots are ±1, ±3, ±9, ±27, $\pm\frac{1}{2}$, $\pm\frac{3}{2}$, $\pm\frac{9}{2}$, $\pm\frac{27}{2}$, $\pm\frac{1}{4}$, $\pm\frac{3}{4}$, $\pm\frac{9}{4}$, $\pm\frac{27}{4}$, $\pm\frac{1}{8}$, $\pm\frac{3}{8}$, $\pm\frac{9}{8}$, $\pm\frac{27}{8}$. $P(-x) = -8x^3 - 36x^2 - 54x - 27$ so there are no negative roots. Using synthetic division we get

$\frac{3}{2}$	8	−36	54	−27
		12	−36	27
	8	−24	18	0

The depressed equation is $8x^2 - 24x + 18 = 0$. This factors as $2(2x-3)(2x-3) = 0$. From the synthetic division and the factoring, we see that that the roots are $\frac{3}{2}$, $\frac{3}{2}$ and $\frac{3}{2}$ again. [Thus, we say that $\frac{3}{2}$ is a triple root.]

23. $P(-x) = -2x^5 - 13x^4 - 26x^3 - 22x^2 - 24x - 9$ so no sign changes so no negative roots. Possible rational roots are 1, 3, 9, $\frac{1}{2}$, $\frac{3}{2}$, and $\frac{9}{2}$. Using synthetic division three times, we get

3	2	−13	26	−22	24	−9
		6	−21	15	−21	9
3	2	−7	5	−7	3	0
		6	−3	6	−3	
$\frac{1}{2}$	2	−1	2	−1	0	
		1	0	1		
	2	0	2	0		

We have $2x^2 + 2 = 0$ or $2x^2 = -2$ or $x^2 = -1$ and so, $x = \pm j$. The roots are $\frac{1}{2}$, 3, 3, j, and $-j$.

24. Possible rational roots are $\pm1, \pm\frac{1}{2}, \pm\frac{1}{4}, \pm\frac{1}{8}$

$\frac{1}{2}$	8	−4	6	−3	−2	1
		4	0	3	0	−1
	8	0	6	0	−2	0

Now, $8x^4 + 6x^2 - 2 = (8x^2 - 2)(x^2 + 1)$. We see that $8x^2 - 2 = 0$ or $x^2 = \frac{1}{4}$, which produces $x = \pm\frac{1}{2}$. We also see that $x^2 + 1 = 0$ or $x^2 = -1$, and so $x = \pm j$. The roots are $\frac{1}{2}$, $\frac{1}{2}$, $-\frac{1}{2}$, j, and $-j$.

25. Possible rational roots are ±1, ±2, ±3, ±5, ±6, ±10, ±15, ±30. Using synthetic division three times, we get

3	1	−5	6	11	−43	30
		3	−6	0	33	−30
1	1	−2	0	11	−10	0
		1	−1	−1	10	
−2	1	−1	−1	10	0	
		−2	6	−10		
	1	−3	5	0		

We have $x^2 - 3x + 5 = 0$ and so, $x = \frac{3\pm\sqrt{9-20}}{2} = \frac{3\pm j\sqrt{11}}{2}$. The roots are 1, 3, −2, $\frac{3}{2} + \frac{\sqrt{11}}{2}j$, and $\frac{3}{2} - \frac{\sqrt{11}}{2}j$.

26. Possible rational roots are ±1, ±2, ±4, ±5, ±8, ±20, ±40, $\pm\frac{1}{2}$, $\pm\frac{5}{2}$, ±10. There are either 0 or 2 positive roots and 1 or 3 negative roots.

$\frac{5}{2}$	2	−1	−6	−18	4	40
		5	10	10	−20	−40
	2	4	4	−8	−16	0

Thus, $\frac{5}{2}$ is the only rational root. The other roots are $\pm\sqrt{2}$ and $-1 \pm j\sqrt{3}$ but, at present, we have no way to use algebra to find them.

27. From the given information, we have the equation $(8-2x)(10-2x)x = 48$ where $0 < x < 4$. Expanding this equation produces $4x^3 - 36x^2 + 80x - 48 = 0$. Dividing both sides of this equation by 4 we get the equation $x^3 - 9x^2 + 20x - 12 = 0$. Possible rational roots are 1, 2, and 3. (Notice, that because $0 < x < 4$, we can ignore any possible negative roots.)
Using synthetic division, we get

1	1	−9	20	−12
		1	−8	12
	1	−8	12	0

The depressed equation factors: $x^2 - 8x + 12 = (x-2)(x-6) = 0$. Solutions to the equation are 1, 2 and 6. Solutions to the problem are 1 in. and 2 in.

28. If we let x = height, then the width = $x + 2.5$ and length = $x + 6.5$. Multiplying these we see that Volume = $x(x + 2.5)(x + 6.5) = 210$ or

$x^3 + 9x^2 + 16.25x - 210 = 0$. Multiplying by 4, we can eliminate the fractional parts, and work with the easier equation $4x^3 + 36x^2 + 65x - 840 = 0$. We know that x must be positive and by trial, we determine that 4 is too large. Since $\frac{7}{2}$ is a possible solution less then 4, we will use synthetic division to test and see if it is a root.
Using synthetic division, we get

$$\begin{array}{r|rrrr} \frac{7}{2} & 4 & 36 & 65 & -840 \\ & & 14 & 175 & 840 \\ \hline & 4 & 50 & 240 & 0 \end{array}$$

Thus, $\frac{7}{2} = 3.5$ is a solution. Since the original equation has only one sign change, there are no other positive solutions. Height = 3.5 cm, width = 6 cm, length = 10 cm.

29. The volume of a cylinder is $\pi r^2 h$ and the volume of a hemisphere is one-half the volume of a sphere or $\frac{1}{2}\frac{4}{3}\pi r^3 = \frac{2}{3}\pi r^3$. Since the total height of the silo is 34 ft, and the radius or the hemisphere is r ft, the height of the cylinder is $34 - r$ ft. Thus, the total volume of this grain silo is $V = \pi r^2 h + \frac{2}{3}\pi r^3 = \pi r^2(34 - r) + \frac{2}{3}\pi r^3 = 2511\pi$. Expanding, and dividing by π, we obtain $34r^2 - r^3 + \frac{2}{3}r^3 = 2511$. Collecting terms we get $-\frac{1}{3}r^3 + 34r^2 - 2511 = 0$ or $r^3 - 102r^2 + 7533 = 0$. Now, $7533 = 3^5 \cdot 31$ so factors include 1, 3, 9, 27, and 31. Synthetic division shows that 9 is a solution

$$\begin{array}{r|rrrr} 9 & 1 & -102 & 0 & 7533 \\ & & 9 & -837 & -7533 \\ \hline & 1 & -93 & -837 & 0 \end{array}$$

The radius is 9 ft.

30. As in the previous exercise, $V = \pi r^2 h + \frac{4}{3}\pi r^3 = 6\pi r^2 + \frac{4}{3}\pi r^3 = 18\pi$. Dividing this equation by π, we obtain $6r^2 + \frac{4}{3}r^3 = 18$ or $4r^3 + 18r^2 - 54 = 0$ Using synthetic division, we get

$$\begin{array}{r|rrrr} \frac{3}{2} & 4 & 18 & 0 & -54 \\ & & 6 & 36 & 54 \\ \hline & 4 & 24 & 36 & 0 \end{array}$$

There are no other positive solutions, so the radius is $\frac{3}{2}$ or 1.5 m.

31. Multiplying $\dfrac{1}{R} + \dfrac{1}{R+4} + \dfrac{1}{R+1} = 1$ by the common denominator $R(R+1)(R+4)$ yields

$$\begin{aligned} (R+1)(R+4) + R(R+1) + R(R+4) &= R(R+1)(R+4); \\ R^2 + 5R + 4 + R^2 + R + R^2 + 4R &= R^3 + 5R^2 + 4R; \\ 3R^2 + 10R + 4 &= R^3 + 5R^2 + 4R; \\ R^3 + 2R^2 - 6R - 4 &= 0 \end{aligned}$$

Using synthetic division, produces

$$\begin{array}{r|rrrr} 2 & 1 & 2 & -6 & -4 \\ & & 2 & 8 & 4 \\ \hline & 1 & 4 & 2 & 0 \end{array}$$

There are no other positive solutions, and so we have $R = 2$, $R + 4 = 6$ and $R + 1 = 5$.

32. Multiplying $\dfrac{1}{R} + \dfrac{1}{R+4} + \dfrac{1}{R+9} = \dfrac{1}{3}$ by the common denominator $3R(R+4)(R+9)$ produces

$$\begin{aligned} 3(R+4)(R+9) + 3(R)(R+9) + 3(R)(R+4) &= R(R+4)(R+9); \\ 3(R^2 + 13R + 36) + 3R^2 + 27R + 3R^2 + 12R &= R(R^2 + 13R + 36); \\ 9R^2 + 78R + 108 &= R^3 + 13R^2 + 36R; \\ R^3 + 4R^2 - 42R - 108 &= 0 \end{aligned}$$

Using synthetic division to divide by $x - 9$, we obtain

$$\begin{array}{r|rrrr} 9 & 1 & 4 & -42 & -108 \\ & & 9 & 117 & 675 \\ \hline & 1 & 13 & 75 & 576 \end{array}$$

From this we can see that 9 is not a factor but that it is an upper bound.

$$\begin{array}{r|rrrr} 6 & 1 & 4 & -42 & -108 \\ & & 6 & 60 & 108 \\ \hline & 1 & 10 & 18 & 0 \end{array}$$

This shows that 6 is a root. There are no other positive solutions, and we get $R = 6$, $R + 4 = 10$, and $R + 9 = 15$.

33. If the side of the square that is cut from each corner has length x, then the volume of the box is given by

$$\begin{aligned}(12 - 2x)(19 - 2x)x &= 210\\ (228 - 62x + 4x^2)x &= 210\\ 4x^3 - 62x^2 + 228x - 210 &= 0\\ 2x^3 - 31x^2 + 114x - 105 &= 0\end{aligned}$$

Now, x must be positive and the realistic domain is $0 < x < 6$. The possible rational solutions include $1, 5, 7, \frac{1}{2}, \frac{5}{2}$, and $\frac{7}{2}$.

$$\begin{array}{r|rrrr} \frac{7}{2} & 2 & -31 & 114 & -105\\ & & 7 & -84 & 105\\ \hline & 2 & -24 & 30 & 0\end{array}$$

Thus, $\frac{7}{2}$ is a root. The depressed polynomial $2x^2 - 24x + 30$ has roots at $x = \frac{24 \pm \sqrt{(-24)^2 - 4(2)(30)}}{2(2)} = \frac{24 \pm \sqrt{336}}{4} = \frac{24 \pm 4\sqrt{21}}{4} = 6 \pm \sqrt{21}$. There are no other real solutions. Since $6 + \sqrt{21} \approx 10.583$ is not in the realistic domain, it is not a solution. So, the only solutions are $x = \frac{7}{2} = 3.5$ cm or $6 - \sqrt{21} \approx 1.417$ cm.

34. $$\begin{aligned}M(d) &= 0.1d^4 - 2.2d^3 + 15.2d^2 - 32d\\ 10M(d) &= d^4 - 22d^3 + 152d^2 - 320d\\ &= d(d^3 - 22d^2 + 152d - 320)\end{aligned}$$

Thus, 0 is one root. Possible rational roots include $1, 2, 4, 5, 8, 10, 16, \ldots$.

$$\begin{array}{r|rrrr} 4 & 1 & -22 & 152 & -320\\ & & 4 & -72 & 320\\ \hline 8 & 1 & -18 & 80 & 0\\ & & 8 & -80 & \\ \hline 10 & 1 & -10 & 0 & \\ & & 10 & & \\ \hline & 1 & 0 & & \end{array}$$

Solutions are 0, 4, 8 and 10 meters.

18.4 IRRATIONAL ROOTS

1. $P(x) = x^3 + 5x - 3$, $P(0) = -3$, and $P(1) = 3$, so

a	$P(a)$	b	$P(b)$	$c = a - \frac{P(a)}{m}$	$P(c)$
0	−3	1	3	0.5	−0.375
0	−3	0.5	−0.375	0.5714	0.0437
0	−3	0.5714	0.0437	0.5632	−0.0052
0	−3	0.5632	−0.0052	0.5642	0.0006

Thus, $c \approx 0.56$

2. $P(x) = x^3 - 3x + 1$, $P(1) = -1$, and $P(2) = 3$

a	$P(a)$	b	$P(b)$	$c = a - \frac{P(a)}{m}$	$P(c)$
1	−1	2	3	1.25	−0.7969
1	−1	1.25	−0.7969	2.2308	5.4

The method fails because 1 is not close enough. We try $P(1.5) = -0.125$, with the following results:

a	$P(a)$	b	$P(b)$	$c = a - \frac{P(a)}{m}$	$P(c)$
1.5	−0.125	2	3	1.52	−0.04819
1.5	−0.125	1.52	−0.04819	1.5325	0.00186
1.5	−0.125	1.5325	0.00186	1.5321	−0.00007

Thus, $c \approx 1.53$.

3. Let $P(x) = x^4 + 2x^3 - 5x^2 + 1$, $P(0) = 1$ and $P(1) = -1$, with the following results:

a	$P(a)$	b	$P(b)$	$c = a - \frac{P(a)}{m}$	$P(c)$
0	1	1	−1	0.5	0.0625
0	1	0.5	0.0625	0.5333	−0.0379
0	1	0.5333	−0.0379	0.5139	0.0208
0	1	0.5139	0.0208	0.5248	−0.0121
0	1	0.5248	−0.0121	0.5185	0.0068
0	1	0.5185	0.0068	0.5221	−0.0039

Thus, $c \approx 0.52$.

4. Here we let $P(x) = x^4 + 2x^3 - 5x^2 + 1$. We see that $P(1) = -1$ and $P(2) = 13$ and get the following result:

a	$P(a)$	b	$P(b)$	$c = a - \frac{P(a)}{m}$	$P(c)$
1	−1	2	13	1.0714	−0.9621
1	−1	1.0714	−0.962	2.883	76.49

The method fails, so we try $a = 1.2$, with the following results:

a	$P(a)$	b	$P(b)$	$c = a - \frac{P(a)}{m}$	$P(c)$
1.2	−0.6704	2	13	1.239	−0.5139
1.2	−0.6704	1.239	−0.5139	1.368	0.2662
1.2	−0.6704	1.368	0.2662	1.3203	−0.0740
1.2	−0.6704	1.3203	−0.0740	1.3352	0.02549
1.2	−0.67045	1.3352	0.02549	1.3303	−0.0082
1.2	−0.67045	1.3303	−0.0082	1.3319	0.0027

Thus, $c \approx 1.33$.

5. Let $P(x) = x^3+x^2-7x+3$, $P(0) = 3$, and $P(1) = -2$. We get

a	$P(a)$	b	$P(b)$	$c = a - \frac{P(a)}{m}$	$P(c)$
0	3	1	−2	0.6	−0.624
0	3	0.6	−0.624	0.4967	−0.1076
0	3	0.4967	−0.1076	0.4795	−0.0163
0	3	0.4795	−0.0163	0.4769	−0.0024
0	3	0.4769	−0.0024	0.4765	−0.00036

Thus, $c \approx 0.48$.

6. Here $P(x) = x^3 + x^2 - 7x + 3$, $P(1) = -2$, and $P(2) = 1$. We get

a	$P(a)$	b	$P(b)$	$c = a - \frac{P(a)}{m}$	$P(c)$
1	−2	2	1	1.6667	−1.2593
1	−2	1.6667	−1.2593	2.8	13.192

The method fails, so we try $a = 1.8$, with the following results:

a	$P(a)$	b	$P(b)$	$c = a - \frac{P(a)}{m}$	$P(c)$
1.8	−0.528	2	1	1.8691	−0.0603
1.8	−0.528	1.8691	−0.0603	1.8780	0.00455
1.8	−0.528	1.8780	0.00455	1.8774	−0.00034

Thus, $c \approx 1.88$.

7. $P(x) = x^4 - x - 2$. Since $P(-1) = 0$, one root is at -1. Using synthetic division we get the depressed equation $D(x) = x^3 - x^2 + x - 2$. Since $D(1) = -1$ and $D(2) = 4$, there is a real root between 1 and 2.

a	$P(a)$	b	$P(b)$	$c = a - \frac{P(a)}{m}$	$P(c)$
1	−1	2	4	1.2	−0.512
1.2	−0.512	2	4	1.29	−0.2274
1.29	−0.2274	2	4	1.328	−0.0935
1.328	−0.0935	2	4	1.343	−0.03835
1.343	−0.03835	2	4	1.349	−0.0159
1.349	−0.0159	2	4	1.352	−0.0046

Thus, $c \approx 1.35$ and we have found an approximation of the second real root. The other two roots are complex numbers.

8. $x^5 - 2x^2 + 4 = 0$; Root is -1.10

9. Letting $P(x) = x^4 + x^3 - 2x^2 - 7x - 5$, we see that $P(-1) = 0$, so -1 is a root. Using synthetic divison, we obtain

$$\begin{array}{r|rrrrr} -1 & 1 & 1 & -2 & -7 & -5 \\ & & -1 & 0 & 2 & 5 \\ \hline & 1 & 0 & -2 & -5 & 0 \end{array}$$

This gives the depressed equation $D(x) = x^3-2x-5$. Since $D(2) = -1$ and $D(3) = 16$, we see that there is a root between 2 and 3.

a	$P_1(a)$	b	$P_1(b)$	$c = a - \dfrac{P_1(a)}{m}$	$P_1(c)$
2	−1	3	16	2.0588	−0.39105
2.0588	−0.39105	3	16	2.081	−0.1501
2.081	−0.1501	3	16	2.0895	−0.05622
2.0895	−0.05622	3	16	2.0927	−0.02064

Thus, $c \approx 2.09$ is a second root. The other two roots are complex numbers.

10. Here $P(x) = x^4 - 2x^3 - 3x + 4$ and, since $P(1) = 0$, 1 is a root. Synthetic divsion gives the depressed equaition $D(x) = x^3 - x^2 - x - 4$. Since $D(2) = -2$ and $D(3) = 11$, we know that there is a real root between −2 and 3. Using linear interpolation, we see that this root is approximately 2.24. There are no other real roots.

11. Here $P(x) = 2x^4 + 3x^3 - x^2 - 2x - 2$. $P(1) = 0$ and synthetic division gives the depressed equation $D(x) = 2x^3 + 5x^2 + 4x + 2$. Since $D(-2) = -2$ and $D(-1) = 1$, we see that there is a root between −2 and −1. Linear interpolation shows that this root is approximately −1.66.

12. Here $P(x) = 3x^4 + 3x^3 + x^2 + 4x + 3$. $P(-1) = 0$ and synthetic division gives the depressed equation $D(x) = 3x^3 + x + 3$. Since $D(-1) = -1$ and $D(0) = 3$, we see that there is a root between −1 and 0. Linear interpolation shows that this root is approximately −0.89.

13. Here $P(x) = 2x^5 - 5x^3 + 2x^2 + 4x - 1$. $P(-1) = 0$ and synthetic division gives the depressed equation $D(x) = 2x^4 - 2x^3 - 3x^2 + 5x - 1$. Since $D(-2) = 25$, $D(-1) = -5$, $D(0) = -1$, and $D(1) = 1$, we see that there is a root between −2 and −1 and another root between 0 and 1. Linear interpolation shows that these roots are approximately −1.43 and 0.24.

14. Here $P(x) = x^5 + 3x^4 - 5x^3 - 2x^2 + x + 2$. $P(1) = 0$ and synthetic division gives the depressed equation $D(x) = x^4 + 4x^3 - x^2 - 3x - 2$. Since $D(-5) = 113$, $D(-4) = -6$, $D(1) = -1$, and $D(2) = 36$, we see that there is a root between −5 and −4 and another root between 0 and 1. Linear interpolation shows that these roots are approximately −4.09 and 1.08.

15. If we let $P(x) = 8x^4 + 6x^3 - 15x^2 - 12x - 2$, we can see that there are rational roots at $-\frac{1}{2}$ and $-\frac{1}{4}$. Using synthetic division with these factors, we get the depressed equation $D(x) = 8x^2 - 16 = 8(x^2 - 2)$. This depressed equation has roots at $\pm\sqrt{2}$. Thus, this equation has the four real roots are $-\frac{1}{2}$, $-\frac{1}{4}$, −1.41, and 1.41.

16. If we let $P(x) = 9x^4 + 15x^3 - 20x^2 - 20x + 16 = 0$ we can see that there are rational roots at 1 and −2. Using synthetic division, we get the depressed equation $D(x) = 9x^2 + 6x - 8 = (3x + 4)(3x - 2)$ and so $x = -\frac{4}{3}$ and $x = \frac{2}{3}$. Thus, the roots are 1, −2, $-\frac{4}{3}$, and $\frac{2}{3}$.

17. Here we have $P(t) = 150t - 20t^2 + t^3$. We want to know the values of t when $P(t) = 400$. Thus, we want the solution to $P(t) = 150t - 20t^2 + t^3 - 400 = 0$. Since $P(6) = -4$ and $P(7) = 13$, we see that this will be between 6 and 7. Using linear interpolation, we obtain

a	$P(a)$	b	$P(b)$	$c = a - \dfrac{P(a)}{m}$	$P(c)$
6	−4	7	−13	6.23529	0.13759
6	−4	6.23529	0.13759	6.22747	0.002736
6	−4	6.22747	0.002736	6.22731	0.0000546
6	−4	6.22731	0.0000546	6.22731	0.000001

Thus, $c \approx 6.22731$ and so it will take about 6 years 83 days.

18. Let x = thickness of the tank in feet. Then, $h = 12 - 2x$ and $r = 4.5 - x$. Using the formula for the volume of a cylinder, we obtain

$$\begin{aligned}
V = \pi r^2 h = \pi(4.5 - x)^2(12 - 2x) &= 674 \\
(4.5 - x)^2(12 - 2x) &= 214.54 \\
(20.25 - 9x + x^2)(6 - x) &= 107.27 \\
-x^3 + 15x^2 - 74.25x + 121.5 &= 107.27 \\
x^3 - 15x^2 + 74.25x - 14.23 &= 0
\end{aligned}$$

Since $V = -14.23$ when $x = 0$ and $V = 46.02$ when $x = 1$, we have a solution between 0 and 1. Using linear interpolation, we get

a	$V(a)$	b	$V(b)$	$c = a - \frac{V(a)}{m}$	$V(c)$
0	−14.23	1	46.02	0.23618	2.4830
0	−14.23	0.23618	2.4830	0.2011	0.10276
0	−14.23	0.2011	0.10276	0.19965	0.00420
0	−14.23	0.19965	0.0042	0.19959	0.00017

Thus, $x \approx 0.1996 \approx 0.20$ ft = 2.4 in.

19. (a) Since $h = 2r$, we have the following

$$V = \pi(r + 15.5)^2(2r + 31) = 1\,000\,000\pi$$
$$(r^2 + 31r + 240.25)(2r + 31) = 1\,000\,000$$
$$2r^3 + 93r^2 + 1\,441.5r + 7\,447.75 = 1\,000\,000$$
$$2r^3 + 93r^2 + 1\,441.5r - 992\,552.25 = 0$$

Using linear interpolation, we get

a	$V(a)$	b	$V(b)$	$c = a - \frac{V(a)}{m}$	$V(c)$
63	−32526	64	4919	63.869	−54.17
63	−32526	63.869	−54.177	63.870	0.5983
63	−32526	63.870	0.5983	63.870	−0.0066

Thus, $r \approx 63.870$ cm. (b) $h = 2r$, $V = \pi(63.87)^2(2 \cdot 63.87) = 521\,100\pi$ cm^3 = 1 637 000 cm^3.

20. Letting $P(S) = S^3 - 6S^2 - 78S + 108$, we get

a	$P(a)$	b	$P(b)$	$c = a - \frac{P(a)}{m}$	$P(c)$
1	25	2	−64	1.2809	0.3472
1	25	1.2809	0.3472	1.2849	−0.00274
1	25	1.2849	−0.00274	1.2848	0.000022

Thus, $S \approx 1.2848$ psi.

21. R is an even function so all solutions come in pairs. Evaluing R at various points, we see that $R(0.3) = -0.675$; $R(0.5) = 1.536$; $R(0.8) = 0.864$; $R(0.9) = -0.509$; $R(1.4) = -1.278$; $R(1.5) = 0.700$; $R(1.7) = 1.0183$; $R(1.8) = -3.561$; $R(2.2) = -13.699$; and $R(2.3) = 51.874$. By the locator theorem, there are roots between 0.3 and 0.5, between 0.8 and 0.9, between 1.4 and 1.5, between 1.7 and 1.8 as well as between 2.2 and 2.3.

Linear interpolation for the one between 1.7 and 1.8 is as follows

a	$R_1(a)$	b	$R_1(b)$	$c = a - \frac{R_1(a)}{m}$	$R_1(c)$
1.7	1.0183	1.8	−3.5614	1.7222	0.3546
1.7	1.0183	1.7222	0.3546	1.7341	−0.0796
1.7	1.0183	1.7341	−0.0796	1.7316	0.01548
1.7	1.0183	1.7316	0.01548	1.7321	−0.0031
1.7	1.0183	1.7321	−0.0031	1.7320	0.00062
1.7	1.0183	1.7320	0.00062	1.7321	−0.0001
1.7	1.0183	1.7321	−0.00012	1.7321	0.000025

The approximate roots are ±0.3536, ±0.8654, ±1.4639, ±1.7321, and ±2.2323.

22. See *Computer Programs* in main text.

18.5 RATIONAL FUNCTIONS

1. Vertical asympotte: $x = -2$; Horizontal asymptote: $y = 0$

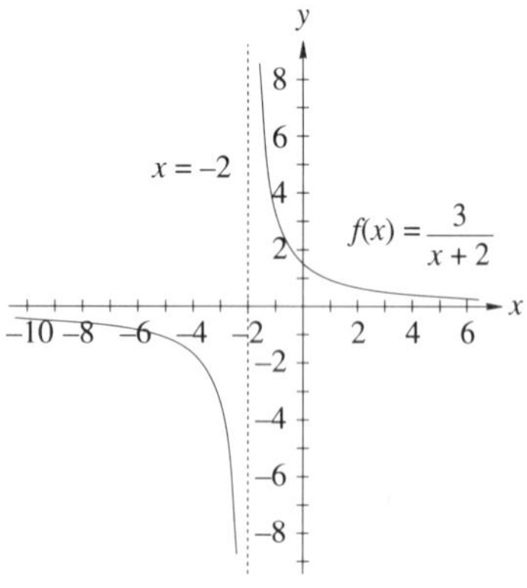

2. Vertical asymptote: $x = 3, x = -3$; Horizontal asymptote: $y = 0$

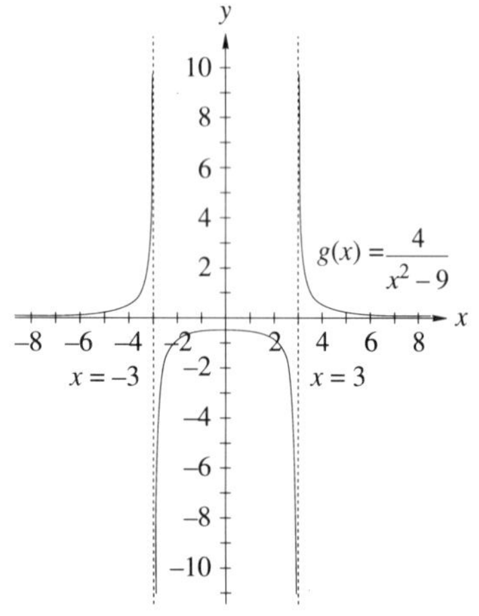

3. Vertical asymptote: $x = -4$; Horizontal asymptote: $y = 2$

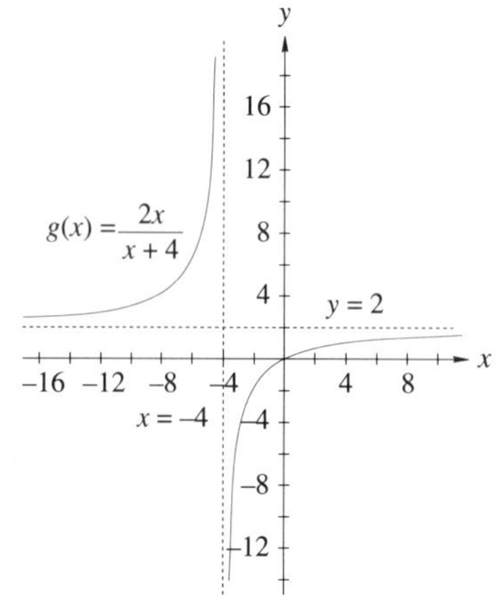

4. $\dfrac{5x}{x^2 + x - 6} = \dfrac{5x}{(x+3)(x-2)}$, Vertical asymptotes: $x = -3$ and $x = 2$; Horizontal asymptote: $y = 0$

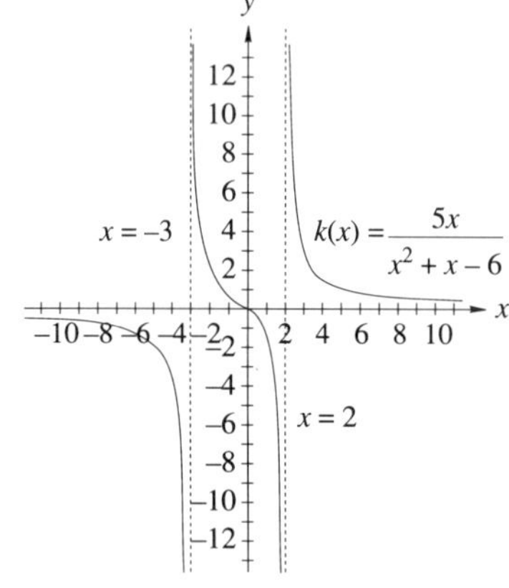

5. $\dfrac{3x^2}{x^2 + 6x + 8} = \dfrac{3x^2}{(x+4)(x+2)}$, Vertical asymptotes: $x = -2$ and $x = -4$; Horizontal asymptote: $y = 3$

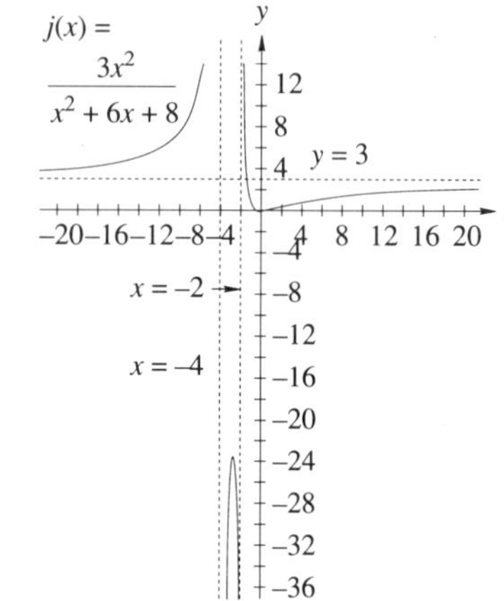

6. $\dfrac{x^2 - 4}{x^2 - 16} = \dfrac{(x+2)(x-2)}{(x+4)(x-4)}$; Vertical asymptotes: $x = -4$ and $x = 4$; Horizontal asymptote: $y = 1$

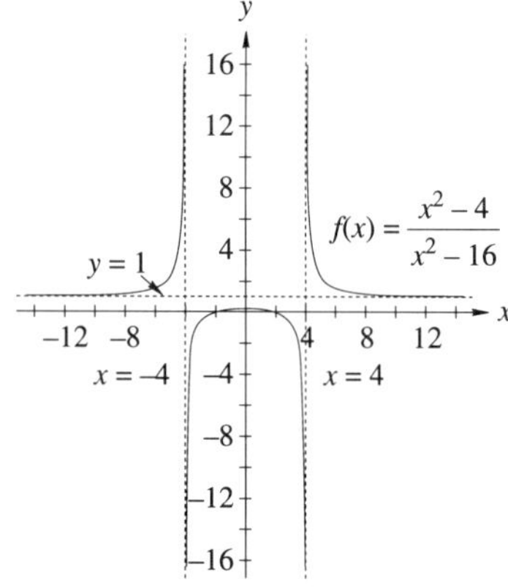

7. $\dfrac{x(x+1)}{(x+1)(x+2)}$; Vertical asymptote: $x = -2$; Horizontal asymptote: $y = 1$; (Note: hole at $x = -1$)

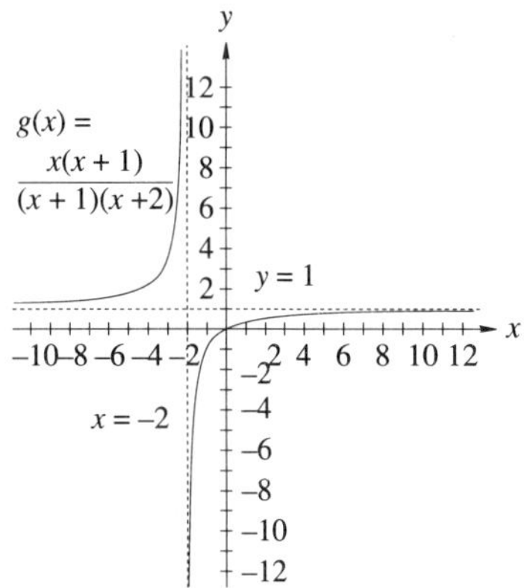

8. $\dfrac{x^2+1}{2x^2}$; Vertical asymptote: $x = 0$; Horizontal asymptote: $y = \frac{1}{2}$

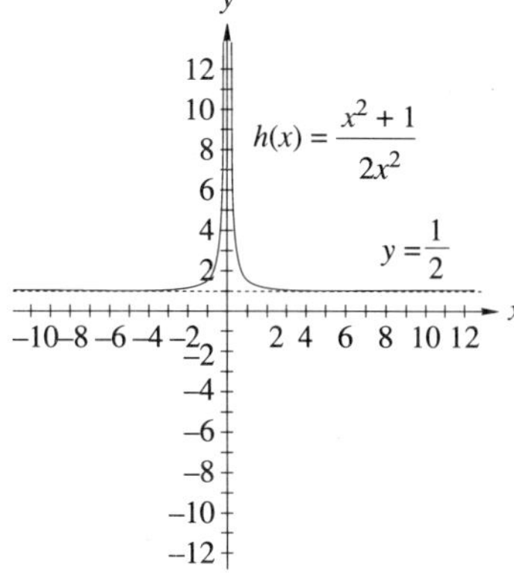

9. $\dfrac{4x^2 - 12x - 27}{x^2 - 6x + 8} = \dfrac{(2x-9)(2x+3)}{(x-2)(x-4)}$. Vertical asymptotes: $x = 2$ and $x = 4$; Horizontal asymptote: $y = 4$

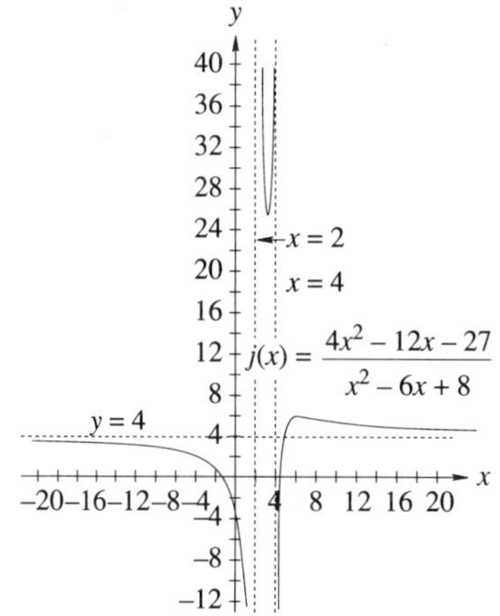

10. $\dfrac{4x^2 - 9}{2x^2 + 7x + 3} = \dfrac{(2x+3)(2x-3)}{(2x+1)(x+3)}$. Vertical asymptotes: $x = -\frac{1}{2}$ and $x = -3$; Horizontal asymptote: $y = \frac{4}{2} = 2$

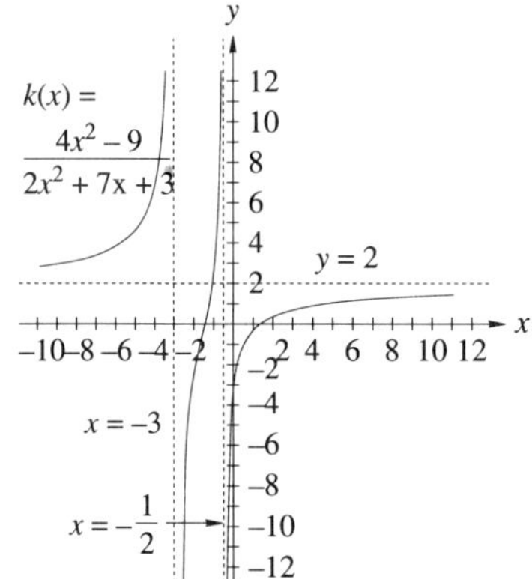

11. $\dfrac{(x+2)(x-3)}{(x+5)(x+4)} = 0 \Rightarrow (x+2)(x-3) = 0$; $x = -2$ *or* $x = 3$

12. $\dfrac{x^2 - 2x - 1}{x(x+1)} \Rightarrow x^2 - 2x - 1 = 0$; $\frac{2 \pm \sqrt{4+4}}{2} = 1 \pm \sqrt{2}$, $x = 1 + \sqrt{2}$ or $x = 1 - \sqrt{2}$

13. $\dfrac{x^2 - 1}{x^2 + 2x + 1} = \dfrac{(x+1)(x-1)}{(x+1)^2} = 0$; $x \neq -1$ so $x = 1$ is the only solution.

14. $\dfrac{3x - 4}{6x^2 - 5x - 4} = 0$; $\dfrac{3x - 4}{(3x-4)(2x+1)}$; Since x cannot equal $\frac{4}{3}$, there is no solution.

15. The given equation $\dfrac{4}{x-3} = \dfrac{2}{x+1}$. Subtracting, we get $\dfrac{4}{x-3} - \dfrac{2}{x+1} = 0$. This has a common denominator of $(x-3)(x+1)$. Multiplying the equation by this common denominator produces $\dfrac{4(x+1) - 2(x-3)}{(x-3)(x+1)} = 0$ which simplifies to $\dfrac{4x + 4 - 2x + 6}{(x-3)(x+1)} = 0$ or $\dfrac{2x + 10}{(x-3)(x+1)} = 0$. This has the solution $2x + 10 = 0$ or $x = -5$.

16. Subtracting, we get $\frac{5}{x+7} - \frac{2}{x-8} = 0$ and multiplying this equation by the common denominator $(x+7)(x-8)$ produces $\frac{5(x-8)-2(x+7)}{(x+7)(x-8)} = 0$. This can be simplified as $\frac{5x-40-2x-14}{(x+7)(x-8)} = 0$ and it has the solution $3x - 54 = 0$ or $3x = 54$ and so, we see that $x = 18$.

17. Subtracting $\frac{2}{3}$ from both sides of the equation produces $\frac{5}{2x} - \frac{3}{4x} - \frac{2}{3} = 0$. Finding a common denominator, we get $\frac{6 \cdot 5 - 3 \cdot 3 - 2 \cdot 4x}{12x} = 0$ or $\frac{30-9-8x}{12x} = 0$. Any solutions will be when $30 - 9 - 8x = 0$ or $21 - 8x = 0$ which means that $x = \frac{21}{8}$.

18. Subtracting 2 produces $\frac{2x-3}{x+5} - 2 = 0$ or $\frac{2x-3}{x+5} - \frac{2(x+5)}{x+5} = 0$ which is equivalent to $\frac{2x-3-2x-10}{x+5} = 0$. This simplifies to $\frac{-13}{x+5}$. Since the numerator can never be 0, there is no solution.

19. Finding that $(x+2)(x-2) = x^2 - 4$ is a common denominator for $\frac{3}{x-2} + \frac{5}{x+2} = \frac{20}{x^2-4}$, we get $\frac{3(x+2)}{x^2-4} + \frac{5(x-2)}{x^2-4} - \frac{20}{x^2-4} = 0$ or $\frac{3x+6+5x-10-20}{x^2-4} = 0$. The numerator simplifies to $8x - 24$ and is 0 only when $x = 3$.

20. $\frac{4}{x+3} - \frac{6}{x-3} = \frac{10}{x^2-9}$ is equivalent to $\frac{4(x-3)-6(x+3)-10}{x^2-9} = 0$. The numerator $4x - 12 - 6x - 18 - 10 = 0$, and hence the given equation has a solution, only when $x = -20$.

21. If we let $P(t) = 3t^4 + 2t^3 - 300t - 50$, we see that $P(4.4) = -75.20$ and $P(4.5) = 12.4375$. Using linear interpolation, we get

a	$P(a)$	b	$P(b)$	$c = a - \frac{P(a)}{m}$	$P(c)$
4.4	−75.20	4.5	12.4375	4.4858	−.469
4.4	−75.20	4.4858	−.469	4.4863	+0.01787
4.4	−75.20	4.4863	0.01787	4.4863	−0.00068

Thus, $t \approx 4.49$ hours.

CHAPTER 18 REVIEW

1. $P(2) = 9 \cdot 2^2 - 4 \cdot 2 + 2 = 36 - 8 + 2 = 30$, degree $= 2$

2. $P(3) = 2 \cdot 3^3 - 4 \cdot 3^2 + 7 = 54 - 36 + 7 = 25$, degree $= 3$

3. $P(-1) = 4(-1)^3 - 5(-1)^2 + 7(-1) + 3 = -4 - 5 - 7 + 3 = -13$, degree $= 3$

4. $P(-3) = 5(-3)^2 + 6(-3) - 3 = 45 - 18 - 3 = 24$, degree $= 2$

5.

$$\begin{array}{r|rrr} -3 & 4 & -6 & -3 \\ & & -12 & 54 \\ \hline & 4 & -18 & 51 \end{array}$$

So, $Q(x) = 4x - 18$; $R = 51$

6.

$$\begin{array}{r|rrrr} 1 & 6 & 0 & -2 & 1 \\ & & 6 & 6 & 4 \\ \hline & 6 & 6 & 4 & 5 \end{array}$$

So, $Q(x) = 6x^2 + 6x + 4$; $R = 5$

7.

$$\begin{array}{r|rrrrr} 1 & 9 & -5 & 2 & -7 & 1 \\ & & 9 & 4 & 6 & -1 \\ \hline & 9 & 4 & 6 & -1 & 0 \end{array}$$

So, $Q(x) = 9x^3 + 4x^2 + 6x - 1$; $R = 0$

8.
$$\begin{array}{r|rrrr} -2 & 7 & -3 & 0 & -12 \\ & & -14 & 34 & -68 \\ \hline & 7 & -17 & 34 & -80 \end{array}$$

So, $Q(x) = 7x^2 - 17x + 34$; $R = -80$

9. $x^3 - 3x^2 + 3x - 1 = 0$; possible rational roots are ± 1. Using synthetic divison, we find

$$\begin{array}{r|rrrr} 1 & 1 & -3 & 3 & -1 \\ & & 1 & -2 & 1 \\ \hline & 1 & -2 & 1 & 0 \end{array}$$

The quotient is $x^2 - 2x + 1 = (x - 1)(x - 1)$. Thus, the roots are 1, 1, and 1.

10. $x^4 - 5x^3 + 9x^2 - 7x + 2 = 0$; possible roots ± 1, ± 2. Using synthetic divison with 1, we find

$$\begin{array}{r|rrrrr} 1 & 1 & -5 & 9 & -7 & 2 \\ & & 1 & -4 & 5 & -2 \\ \hline 1 & 1 & -4 & 5 & -2 & 0 \\ & & 1 & -3 & 2 & \\ \hline & 1 & -3 & 2 & 0 & \end{array}$$

Since $x^2 - 3x + 2 = (x - 2)(x - 1)$. Thus, the roots are 1, 1, 1, and 2.

11. $x^4 - 6x^2 - 8x - 3 = 0$, Possible roots ± 1, and ± 3.

$$\begin{array}{r|rrrrr} -1 & 1 & 0 & -6 & -8 & -3 \\ & & -1 & 1 & 5 & 3 \\ \hline -1 & 1 & -1 & -5 & -3 & 0 \\ & & -1 & 2 & 3 & \\ \hline & 1 & -2 & -3 & 0 & \end{array}$$

Since $x^2 - 2x - 3 = (x + 1)(x - 3)$, the roots are $-1, -1, -1$, and 3.

12. $x^4 + 10x^3 + 25x^2 - 16 = 0$; Possible roots ± 1, ± 2, ± 4, ± 8, ± 16

$$\begin{array}{r|rrrrr} -1 & 1 & 10 & 25 & 0 & -16 \\ & & -1 & -9 & -16 & 16 \\ \hline -4 & 1 & 9 & 16 & -16 & 0 \\ & & -4 & -20 & 16 & \\ \hline & 1 & 5 & -4 & 0 & \end{array}$$

Using the quadratic formula on $x^2 + 5x - 4 = 0$, we get $x = \frac{-5\pm\sqrt{25+16}}{2} = \frac{-5\pm\sqrt{41}}{2}$. The roots are -1, -4, $\frac{-5+\sqrt{41}}{2}$, and $\frac{-5-\sqrt{41}}{2}$.

13. $x^4 + x^3 - 6x^2 + x + 1 = 0$ Only possible rational roots are ± 1. Neither works

a	$P(a)$	b	$P(b)$	$c = a - \frac{P(a)}{m}$	$P(c)$
−3.2	8.4496	−3	−2	−3.038	−0.2580
−3.2	8.4496	−3.038	−0.2580	−3.043	−0.3169
−3.2	8.4496	−3.043	−0.3169	−3.0436	−.00387
−3.2	8.4496	−3.0436	−0.00387	−3.0437	−.00047
−0.4	−0.3984	−0.3	0.1411	−0.3262	0.01221
−0.4	−0.3984	−0.3262	0.01221	−0.3283	0.00099
−0.4	−0.3984	−0.3283	0.00099	−0.3285	0.00008
−0.4	−3.984	−0.3285	0.00008	−0.3285	0.000006
0.5	0.1875	0.6	−0.2144	0.5467	0.00633
0.5	0.1875	0.5467	0.00633	0.5482	−0.00021
0.5	0.1875	0.5482	−0.00021	0.5482	0.000007
1.8	−0.3104	1.9	1.1311	1.8215	−0.03354
1.8	−0.3104	1.8215	−0.03354	1.8241	0.00121
1.8	−0.3104	1.8241	0.00121	1.8240	−0.00004

The approximate solutions are -3.044, -0.328, 0.548, and 1.824.

14. $2x^5 - 3x^4 - 18x^3 + 75x^2 - 104x + 48 = 0$; possible roots include ± 1, ± 2, ± 3, ± 4, ± 6, ± 8, $\pm\frac{1}{2}$, and $\pm\frac{3}{2}$.

$$\begin{array}{r|rrrrrr} 1 & 2 & -3 & -18 & 75 & -104 & 48 \\ & & 2 & -1 & -19 & 56 & -48 \\ \hline -4 & 2 & -1 & -19 & 56 & -48 & 0 \\ & & -8 & 36 & -68 & 48 & \\ \hline \frac{3}{2} & 2 & -9 & 17 & -12 & 0 & \\ & & 3 & -9 & 12 & & \\ \hline & 2 & -6 & 8 & 0 & & \end{array}$$

Here $2x^2 - 6x + 8 = 2(x^2 - 3x + 4)$. Using the quadratic formula, we get $x = \frac{3\pm\sqrt{9-16}}{2} = \frac{3\pm\sqrt{-7}}{2} = \frac{3}{2} \pm \frac{\sqrt{7}}{2}j$. The roots are 1, -4, $\frac{3}{2}$, $\frac{3}{2} + \frac{\sqrt{7}}{2}j$, and $\frac{3}{2} - \frac{\sqrt{7}}{2}j$.

15. $x^6 + 2x^5 + 3x^4 + 4x^3 + 3x^2 + 2x + 1 = 0$; possible solution is -1.

$$\begin{array}{r|rrrrrrr} -1 & 1 & 2 & 3 & 4 & 3 & 2 & 1 \\ & & -1 & -1 & -2 & -2 & -1 & -1 \\ \hline -1 & 1 & 1 & 2 & 2 & 1 & 1 & 0 \\ & & -1 & 0 & -2 & 0 & -1 & \\ \hline & 1 & 0 & 2 & 0 & 1 & 0 & \end{array}$$

Notice that $x^4 + 2x^2 + 1 = (x^2 + 1)^2$ and that $x^2 + 1 = 0 \Rightarrow x^2 = -1$ or $x = \pm j$. This happens twice. The solutions are $-1, -1, j, j, -j$, and $-j$.

16. $x^6 - 2x^5 - 4x^4 + 6x^3 - x^2 + 8x + 4 = 0$ Possible solutions are $\pm 1, \pm 2, \pm 4$

$$\begin{array}{r|rrrrrrr} -2 & 1 & -2 & -4 & 6 & -1 & 8 & 4 \\ & & -2 & 8 & -8 & 4 & -6 & -4 \\ \hline 2 & 1 & -4 & 4 & -2 & 3 & 2 & 0 \\ & & 2 & -4 & 0 & -4 & -2 & \\ \hline & 1 & -2 & 0 & -2 & -1 & 0 & \end{array}$$

Here $x^4 - 2x^3 - 2x - 1 = (x^2 + 1)(x^2 - 2x - 1)$ and that $x^2 + 1 = 0$ or $x^2 = -1$ and $x = \pm j$. Also, $x^2 - 2x - 1$ and $x = \frac{2 \pm \sqrt{4+4}}{2} = 1 \pm \sqrt{2}$. The roots are $-2, 2, 1 + \sqrt{2}, 1 - \sqrt{2}, j, -j$.

17. $\dfrac{5x}{x^2 + 3x - 4} = \dfrac{5x}{(x+4)(x-1)}$; Vertical asymptotes: $x = -4$ and $x = 1$, Horizontal asymptote: $y = 0$. A rational equation is 0 only when the numerator is 0, and so $5x = 0$ or $x = 0$, the solution is $x = 0$.

18. $\dfrac{7x^3}{x^3 + 1} = \dfrac{7x^3}{(x+1)(x^2 - x + 1)}$; Vertical asymptote: $x = -1$; Horizontal asymptote: $y = 7$. A rational equation is 0 only when the numerator is 0, and since $7x^3 = 0 \Rightarrow x = 0$, the solution is $x = 0$.

19. $\dfrac{4}{x-2} - \dfrac{5}{x+2} = \dfrac{4(x+2) - 5(x-2)}{(x-2)(x+2)} = \dfrac{4x + 8 - 5x + 10}{(x-2)(x+2)} = \dfrac{-x + 18}{(x-2)(x+2)}$ Vertical asymptote: $x = 2$, $x = -2$; Horizontal asymptote: $y = 0$. The solution is when $-x + 18 = 0$ or when $x = 18$.

20. $\dfrac{3}{x-1} + \dfrac{2}{x+1} - \dfrac{4}{x^2 - 1} = \dfrac{3(x+1) + 2(x-1) - 4}{(x-1)(x+1)} = \dfrac{3x + 3 + 2x - 2 - 4}{(x-1)(x+1)} = \dfrac{5x - 3}{(x-1)(x+1)}$. Vertical asymptotes: $x = 1$ and $x = -1$; Horizontal asymptote: $y = 0$. The solution is when $5x - 3 = 0$ or when $x = \frac{3}{5}$.

21. $P(n) = \dfrac{0.5 + 0.8(n-1)}{1.5 + 0.8(n-1)} = \dfrac{0.8n - 0.3}{0.8n + 0.7}$. Horizontal asymptote is $\frac{0.8}{0.8} = 1$ or 100%.

22. $N(t) = \dfrac{5t^2 + 6t + 15}{t^2}$ Horizontal asymptote is $y = \frac{5}{1} = 5$.

CHAPTER 18 TEST

1. $P(x) = 7x^5 - 2x^4 - 5$. Degree is 5

2.
$$\begin{aligned} P(x) &= 4x^3 - 5x^2 + 2x - 8 \\ P(-2) &= 4(-2)^3 - 5(-2)^2 + 2(-2) - 8 \\ &= -32 - 20 - 4 - 8 = -64 \end{aligned}$$

or

$$\begin{array}{r|rrrr} -2 & 4 & -5 & 2 & -8 \\ & & -8 & 26 & -56 \\ \hline & 4 & -13 & 28 & -64 \end{array}$$

and so, $R = -64$.

3.
$$\begin{array}{r|rrrrr} 6 & 1 & -8 & 3 & 44 & 75 \\ & & 6 & -12 & -54 & -60 \\ \hline & 1 & -2 & -9 & -10 & 15 \end{array}$$

and so, $Q(x) = x^3 - 2x^2 - 9x - 10$ and $R = 15$.

4. $c = \pm 1, \pm 2$, $d = 1, 3$ and so $\frac{c}{d} = \pm 1, \pm 2, \pm\frac{1}{3}, \pm\frac{2}{3}$.

5.
$$\begin{array}{r|rrrr} 1 & 3 & 5 & -4 & -4 \\ & & 3 & 8 & 4 \\ \hline & 3 & 8 & 4 & 0 \end{array}$$

and $3x^2 + 8x + 4 = (3x + 2)(x + 2)$ and the roots are $1, -2, -\frac{2}{3}$.

6. Possible roots include $\pm 1, \pm 2, \pm 3, \pm 4$, and ± 6.

$$\begin{array}{r|rrrrrrr} -1 & 1 & 2 & -4 & -10 & -41 & -72 & -36 \\ & & -1 & -1 & 5 & 5 & 36 & 36 \\ \hline -1 & 1 & 1 & -5 & -5 & -36 & -36 & 0 \\ & & -1 & 0 & 5 & 0 & 36 & \\ \hline 3 & 1 & 0 & -5 & 0 & -36 & 0 & \\ & & 3 & 9 & 12 & 36 & & \\ \hline -3 & 1 & 3 & 4 & 12 & 0 & & \\ & & -3 & 0 & -12 & & & \\ \hline & 1 & 0 & 4 & 0 & & & \end{array}$$

From the quotient we see that

$$\begin{aligned} x^2 + 4 &= 0 \\ x^2 &= -4 \\ x &= \pm 2j \end{aligned}$$

and so the roots are $-1, -1, -3, 3, 2j$, and $-2j$.

7. $\dfrac{3x^2 - 2x - 1}{x^2 + 5x + 6} = \dfrac{(3x + 1)(x - 1)}{(x + 2)(x + 3)}$. Vertical asymptote: $x = -2, x = -3$, Horizontal asymptote: $y = 3$ Solutions $x = -\frac{1}{3}, x = 1$.

CHAPTER

19

Sequences, Series, and the Binomial Formula

19.1 SEQUENCES

1. $a_1 = \frac{1}{1+1} = \frac{1}{2}$; $a_2 = \frac{1}{2+1} = \frac{1}{3}$; $a_3 = \frac{1}{3+1} = \frac{1}{4}$; $a_4 = \frac{1}{4+1} = \frac{1}{5}$; $a_5 = \frac{1}{5+1} = \frac{1}{6}$; $a_6 = \frac{1}{6+1} = \frac{1}{7}$

2. $b_1 = \frac{(-1)^1}{1} = -1$; $b_2 = \frac{(-1)^2}{2} = \frac{1}{2}$; $b_3 = \frac{(-1)^3}{3} = -\frac{1}{3}$; $b_4 = \frac{(-1)^4}{4} = \frac{1}{4}$; $b_5 = \frac{(-1)^5}{5} = -\frac{1}{5}$; $b_6 = \frac{(-1)^6}{6} = \frac{1}{6}$

3. $a_1 = (1+1)^2 = 4$; $a_2 = (2+1)^2 = 9$; $a_3 = (3+1)^2 = 16$; $a_4 = (4+1)^2 = 25$; $a_5 = (5+1)^2 = 36$; $a_6 = (6+1)^2 = 49$

4. $a_1 = \frac{1}{1(1+1)} = \frac{1}{2}$; $a_2 = \frac{1}{2(2+1)} = \frac{1}{6}$; $a_3 = \frac{1}{3(3+1)} = \frac{1}{12}$; $a_4 = \frac{1}{4(4+1)} = \frac{1}{20}$; $a_5 = \frac{1}{5(5+1)} = \frac{1}{30}$; $a_6 = \frac{1}{6(6+1)} = \frac{1}{42}$

5. $a_1 = (-1)^1 \cdot 1 = -1$; $a_2 = (-1)^2 \cdot 2 = 2$; $a_3 = (-1)^3 \cdot 3 = -3$; $a_4 = (-1)^4 \cdot 4 = 4$; $a_5 = (-1)^5 \cdot 5 = -5$; $a_6 = (-1)^6 \cdot 6 = 6$

6. $a_1 = \frac{1+(-1)^1}{1+4\cdot 1} = 0$; $a_2 = \frac{1+(-1)^2}{1+4\cdot 2} = \frac{2}{9}$; $a_3 = \frac{1+(-1)^3}{1+4\cdot 3} = 0$; $a_4 = \frac{1+(-1)^4}{1+4\cdot 4} = \frac{2}{17}$; $a_5 = \frac{1+(-1)^5}{1+4\cdot 5} = 0$; $a_6 = \frac{1+(-1)^6}{1+4\cdot 6} = \frac{2}{25}$

7. $a_1 = \frac{2\cdot 1-1}{2\cdot 1+1} = \frac{1}{3}$; $a_2 = \frac{2\cdot 2-1}{2\cdot 2+1} = \frac{3}{5}$; $a_3 = \frac{2\cdot 3-1}{2\cdot 3+1} = \frac{5}{7}$; $a_4 = \frac{2\cdot 4-1}{2\cdot 4+1} = \frac{7}{9}$; $a_5 = \frac{2\cdot 5-1}{2\cdot 5+1} = \frac{9}{11}$; $a_6 = \frac{2\cdot 6-1}{2\cdot 6+1} = \frac{11}{13}$

8. $a_1 = \left(-\frac{1}{2}\right)^1 = -\frac{1}{2}$; $a_2 = \left(-\frac{1}{2}\right)^2 = \frac{1}{4}$; $a_3 = \left(-\frac{1}{2}\right)^3 = -\frac{1}{8}$; $a_4 = \left(-\frac{1}{2}\right)^4 = \frac{1}{16}$; $a_5 = \left(-\frac{1}{2}\right)^5 = -\frac{1}{32}$; $a_6 = \left(-\frac{1}{2}\right)^6 = \frac{1}{64}$

9. $a_1 = \left(\frac{2}{3}\right)^0 = 1$; $a_2 = \left(\frac{2}{3}\right)^1 = \frac{2}{3}$; $a_3 = \left(\frac{2}{3}\right)^2 = \frac{4}{9}$; $a_4 = \left(\frac{2}{3}\right)^3 = \frac{8}{27}$; $a_5 = \left(\frac{2}{3}\right)^4 = \frac{16}{81}$; $a_6 = \left(\frac{2}{3}\right)^5 = \frac{32}{243}$

10. $a_1 = 1\cos\pi = -1$; $a_2 = 2\cos 2\pi = 2$; $a_3 = 3\cos 3\pi = -3$; $a_4 = 4\cos 4\pi = 4$; $a_5 = 5\cos 5\pi = -5$; $a_6 = 6\cos 6\pi = 6$

11. $a_1 = \left(\frac{1-1}{1+1}\right)^2 = 0$; $a_2 = \left(\frac{2-1}{2+1}\right)^2 = \frac{1}{9}$; $a_3 = \left(\frac{3-1}{3+1}\right)^2 = \frac{4}{16} = \frac{1}{4}$; $a_4 = \left(\frac{4-1}{4+1}\right)^2 = \frac{9}{25}$; $a_5 = \left(\frac{5-1}{5+1}\right)^2 = \frac{16}{36} = \frac{4}{9}$; $a_6 = \left(\frac{6-1}{6+1}\right)^2 = \frac{25}{49}$

12. $a_1 = \frac{1^2-1}{1^2+1} = 0$; $a_2 = \frac{2^2-1}{2^2+1} = \frac{3}{5}$; $a_3 = \frac{3^2-1}{3^2+1} = \frac{8}{10} = \frac{4}{5}$; $a_4 = \frac{4^2-1}{4^2+1} = \frac{15}{17}$; $a_5 = \frac{5^2-1}{5^2+1} = \frac{24}{26} = \frac{12}{13}$; $a_6 = \frac{6^2-1}{6^2+1} = \frac{35}{37}$

13. $a_1 = 1$; $a_2 = 2\cdot 1 = 2$; $a_3 = 3\cdot 2 = 6$; $a_4 = 4\cdot 6 = 24$; $a_5 = 5\cdot 24 = 120$; $a_6 = 6\cdot 120 = 720$

14. $a_1 = 3$; $a_2 = 3+2 = 5$; $a_3 = 5+3 = 8$; $a_4 = 8+4 = 12$; $a_5 = 12+5 = 17$; $a_6 = 17+6 = 23$

15. $a_1 = 5$; $a_2 = 5+3 = 8$; $a_3 = 8+3 = 11$; $a_4 = 11+3 = 14$; $a_5 = 14+3 = 17$; $a_6 = 17+3 = 20$

16. $a_1 = 2$; $a_2 = 2^2 = 4$; $a_3 = 4^3 = 64$; $a_4 = 64^4 = 16{,}777{,}216$; $a_5 = 16{,}777{,}216^5 = 1.329228 \times 10^{36}$; $a_6 = (1.329228 \times 10^{36})^6 = 5.51565 \times 10^{216}$

17. $a_1 = 2$; $a_2 = 2^1 = 2$; $a_3 = 2^2 = 4$; $a_4 = 4^3 = 64$; $a_5 = 64^4 = 16{,}777{,}216$; $a_6 = (16{,}777{,}216)^5 = 1.329228 \times 10^{36}$

18. $a_1 = 1$; $a_2 = \frac{-1}{2} \cdot 1 = -\frac{1}{2}$; $a_3 = \frac{-1}{3} \cdot \frac{-1}{2} = \frac{1}{6}$; $a_4 = \frac{-1}{4} \cdot \frac{1}{6} = \frac{-1}{24}$; $a_5 = \frac{-1}{5} \cdot \frac{-1}{24} = \frac{1}{120}$; $a_6 = \frac{-1}{6} \cdot \frac{1}{120} = \frac{-1}{720}$;

19. $a_1 = 1$; $a_2 = 2^1 = 2$; $a_3 = 3^2 = 9$; $a_4 = 4^9 = 262{,}144$; $a_5 = 5^{262{,}144}$; $a_6 = 6^{(5^{262{,}144})}$

20. $a_1 = \frac{1}{2}$; $a_2 = \left(\frac{1}{2}\right)^{-2} = 4$; $a_3 = 4^{-3} = \frac{1}{64}$; $a_4 = \left(\frac{1}{64}\right)^{-4} = 16{,}777{,}216$; $a_5 = (16{,}777{,}216)^{-5} = 7.52316385 \times 10^{-37}$; $a_6 = 5.5256524 \times 10^{216}$

21. $a_1 = 1$; $a_2 = \frac{1}{2}$; $a_3 = 1 \cdot \frac{1}{2} = \frac{1}{2}$; $a_4 = \frac{1}{2} \cdot \frac{1}{2} = \frac{1}{4}$; $a_5 = \frac{1}{2} \cdot \frac{1}{4} = \frac{1}{8}$; $a_6 = \frac{1}{4} \cdot \frac{1}{8} = \frac{1}{32}$

22. $a_1 = 1$; $a_2 = 1$; $a_3 = 1 + 1 = 2$; $a_4 = 1 + 2 = 3$; $a_5 = 2 + 3 = 5$; $a_6 = 3 + 5 = 8$

23. $a_1 = 0$; $a_2 = 2$; $a_3 = 2 - 0 = 2$; $a_4 = 2 - 2 = 0$; $a_5 = 0 - 2 = -2$; $a_6 = -2 - 0 = -2$

24. $a_1 = 1$; $a_2 = 2$; $a_3 = 1 \cdot 2 = 2$; $a_4 = 2 \cdot 2 = 4$; $a_5 = 2 \cdot 4 = 8$; $a_6 = 4 \cdot 8 = 32$

25. $a_1 = 25$ ft -3 in.; $a_{25} = 25$ ft $-3(25)$ in. $= 25$ ft -75 in.; Since 75 in. $= 6'3'' = 6.25$ ft, $a_{25} = 25 - 6.25 = 18.75$ ft

26. $a_1 = 320{,}000$; $a_2 = 240{,}000$; $a_3 = 240{,}000 + 12{,}000$; $= 252{,}000$; $a_4 = 252{,}000 + 12{,}000 = 264{,}000$; $a_n = 240{,}000 + (n - 2) \cdot 12{,}000$; $n \geq 2$; $a_{10} = 240{,}000 + (10 - 2)12{,}000 = 240{,}000 + 8 \cdot 12{,}000 = \$336{,}000$; $a_{20} = 240{,}000 + (20 - 2)12{,}000 = \$456{,}000$

27. $a_1 = 26000$, $a_2 = 26000 + 1100 = 27{,}100$; $a_n = 26000 + (n-1) \cdot 1100$; $a_5 = 26000 + (5-1)1100 = 26000 + 4400 = \$30{,}400$; $a_{10} = 26000 + 9(1100) = \$35{,}900$

28. $a_1 = 60$; $a_2 = 60 + 4$; $a_3 = 60 + 2 \cdot 4$; $a_{10} = 60 + 9 \cdot 4 = 60 + 36 = 96$ seats

19.2 ARITHMETIC AND GEOMETRIC SEQUENCES

1. Common difference, so arithmetic: $d = 9 - 1 = 8$; $a_9 = 1 + (9 - 1)8 = 1 + 64 = 65$

2. Common difference, so arithmetic: $d = -5 - 3 = -8$, $a_7 = 3 + 6(-8) = 3 - 48 = -45$

3. Common ratio, so geometric: $r = \frac{-4}{8} = -\frac{1}{2}$; $a_{10} = \left(-\frac{1}{2}\right)^9 \cdot 8 = -\frac{1}{512} \cdot 8 = -\frac{1}{64}$

4. Common Ratio, so geometric: $r = \frac{1}{4}$; $a_8 = \left(\frac{1}{4}\right)^7 \times 4 = \left(\frac{1}{4}\right)^6 = \frac{1}{4096}$

5. Geometric: $r = \frac{4}{-1} = -4$; $a_6 = (-4)^5 \cdot (-1) = 1024$

6. Arithmetic: $d = 4 - (-1) = 5$; $a_{12} = -1 + 11 \cdot 5 = 54$

7. Arithmetic: $d = -2 - 3 = -5$; $a_8 = 3 + 7(-5) = -32$

8. Geometric: $r = \frac{-3}{1} = -3$; $a_7 = (-3)^6 \cdot 1 = 729$

9. Neither a common difference nor a common ratio so it is neither arithmetic or geometric

10. Geometric: $r = \sqrt{2}$; $a_{12} = (\sqrt{2})^{11} \cdot 1 = 2^5 \cdot \sqrt{2} = 32\sqrt{2}$

11. Arithmetic: $d = 0.4$; $a_9 = 0.4 + 8(0.4) = 9(0.4) = 3.6$.

12. Neither: it could be $a_n = \sqrt{a_{n-1}}$ so $a_{11} = 4^{(1/2)^{11}} = 2^{(1/2)^{10}} = 2^{1/1024}$

13. Arithmetic: $d = -\frac{1}{6}$; $a_8 = \frac{2}{3} + 7\left(-\frac{1}{6}\right) = \frac{4}{6} - \frac{7}{6} = -\frac{3}{6} = -\frac{1}{2}$

14. Geometric: $r = \frac{1}{2}$; $a_8 = \left(\frac{1}{2}\right)^7 \cdot \frac{2}{3} = \frac{1}{64 \cdot 3} = \frac{1}{192}$

15. Geometric: $r = \frac{1}{10}$ or 0.1; $a_6 = (0.1)^5 \cdot 5 = 0.00005$

16. Geometric: $r = -\frac{1}{3}$; $a_7 = \left(-\frac{1}{3}\right)^6 \cdot 6 = \frac{2}{243}$

17. Arithmetic: $d = 2$; $a_9 = (1 + \sqrt{3}) + 8(2) = 17 + \sqrt{3}$

18. Geometric: $r = 10$; $a_{10} = 1.02 \times 10^9$

19. Neither. It looks like an arithmetic sequence with alternating signs, so try $[4.3 - (n-1)1.1](-1)^{n-1}$; $a_6 = 1.2$

20. Geometric: $r = -2$; $a_9 = (-2)^8 \cdot 3.4 = 870.4$

21. $d = a_6 - a_5 = 24 - 9 = 15$; $a_5 = a_1 + 4d$; $9 = a_1 + 4 \cdot 15$; $a_1 = 9 - 60 = -51$

22. $a_9 = a_7 + 2d$; $15 = 11 + 2d$; $2d = 4$; $d = 2$; $a_7 = a_1 + 6d$; $11 = a_1 + 12$; $a_1 = -1$

23. $r = \frac{a_5}{a_4} = \frac{3}{9} = \frac{1}{3}$; $a_4 = \left(\frac{1}{3}\right)^3 \cdot a_1$; $9 = \frac{1}{27} \cdot a_1$; $a_1 = 243$

24. $r = \frac{a_9}{a_3} = \frac{5}{\frac{1}{3}} = 15$; $a_8 = (15)^7 \cdot a_1$; $a_1 = \frac{1}{3} \div 15^7$; $a_1 = 1.9509221 \times 10^{-9}$

25. $81 = 5 + (n-1)4$; $76 = 4(n-1)$; $19 = (n-1)$; $n = 20$

26. $-31 = 20 + (n-1)(-3)$; $-51 = -3(n-1)$; $17 = n - 1$; $n = 18$

27. $\frac{1}{559{,}872} = 3\left(\frac{1}{6}\right)^{n-1}$; $\frac{1}{1{,}679{,}616} = \left(\frac{1}{6}\right)^{n-1}$; $6^{n-1} = 1{,}679{,}619$; $\ln 6^{n-1} = (n-1)\ln 6 = \ln 1{,}679{,}619$; $n - 1 = \frac{\ln 1{,}679{,}619}{\ln 6} = 8$; $n = 9$

28. $\frac{-1}{4{,}096} = (-4)\left(\frac{1}{2}\right)^{n-1}$; $\frac{1}{16{,}384} = \left(\frac{1}{2}\right)^{n-1}$; $2^{n-1} = 16{,}384$; $(n-1)\ln 2 = \ln 16{,}384$; $n - 1 = 14$; $n = 15$

29. $a_1 = -5$; $a_7 = 4$; $a_7 = a_1 + (n-1)d$; $4 = -5 + 6d$; $9 = 6d$; $d = \frac{3}{2}$; $a_2 = -5 + \frac{3}{2} = -3\frac{1}{2}$ or -3.5

30. $a_1 = \frac{1}{3}$; $a_8 = 729$; $a_8 = a_1 r^{n-1}$; $729 = \frac{1}{3}r^7$; $2187 = r^7$; $r = 3$; $a_3 = \frac{1}{3} \cdot 3^2 = 3$

31. $a_{61} = a_1 + 60d$; $8 = \frac{1}{2} + 60d$; $7\frac{1}{2} = 60d$; $d = \frac{7.5}{60} = 0.125 = \frac{1}{8}$ mi

32. $a_6 = a_1\left(\frac{7}{8}\right)^5$; $a_6 = \left(80 \cdot \frac{7}{8}\right) \cdot \left(\frac{7}{8}\right)^5 = 70 \cdot \left(\frac{7}{8}\right)^5 = 35.9036$ m

33. $a_{10} = a_1 + (10-1)d$; $a_{10} = 15 + 9(-0.3)$; $a_{10} = 15 - 2.7 = 12.3$ cm; $0 = 15 + (n-1)(-0.3)$; $-15 = -0.3(n-1)$; $n - 1 = 50$; $n = 51$

34. Geometric sequence; $r = 1 - 0.125 = 0.875$; $P = 100(1 - 0.125)^{8.8} = 100(0.875)^{8.8} = 30.88$ kPa

35. (a) Geometric sequence: Here $r = (1 - 0.162) = 0.838$ and $a_1 = 940$, so the concentration is reduced to 100 ppm when $100 = 940(0.838)^{n-1}$. Thus, $a_n = \frac{100}{940} = (0.838)^{n-1}$ and so, $(n-1)\ln(0.838) = \ln\frac{10}{94}$. Dividing, we get $n - 1 = \dfrac{\ln\left(\frac{10}{94}\right)}{\ln(0.838)} \approx 12.678$, and hence $n = 13.678$. Thus, the concentration will be reduced to 100 ppm about 13.678 mi from the first monitor or about 14.678 mi from the spill. (b) Here $a_n = 1.5 = 940(0.838)^{n-1}$ and so $n - 1 = \dfrac{\ln\left(\frac{1.5}{940}\right)}{\ln(0.838)} = 36.44$ which is $n = 37.44$ mi from the monitor so about 38.44 mi from the spill.

19.3 SERIES

1. $\sum_{k=1}^{20} 3\left(\frac{1}{2}\right)^k = 3\left(\frac{1}{2}\right)^1 + 3\left(\frac{1}{2}\right)^2 + 3\left(\frac{1}{2}\right)^3 + 3\left(\frac{1}{2}\right)^4 + \cdots = 3\left(\frac{1}{2}\right) + 3\left(\frac{1}{4}\right) + 3\left(\frac{1}{8}\right) + 3\left(\frac{1}{16}\right)\cdots = \frac{3}{2} + \frac{3}{4} + \frac{3}{8} + \frac{3}{16} + \cdots$

2. $\sum_{k=1}^{50}(4 + (n-1)3) = 4 + 7 + 10 + 13 + \cdots$

3. $\sum_{n=0}^{20}(-1)^n \frac{3^n}{n+1} = 1 - \frac{3}{2} + 3 - \frac{27}{4} + \cdots$

4. $\sum_{k=1}^{60}(-1)^k = -1 + 1 - 1 + 1 - \cdots$

5. $\sum_{k=1}^{5}(k-3) = -2 - 1 + 0 + 1 + 2 = 0$

6. $\sum_{k=0}^{4}(2k+1) = 1 + 3 + 5 + 7 + 9 = 25$

7. $\sum_{k=1}^{6} k^2 = 1 + 4 + 9 + 16 + 25 + 36 = 91$

8. $\sum_{k=2}^{8} \frac{k+2}{(k-1)k} = \frac{4}{2}+\frac{5}{6}+\frac{6}{12}+\frac{7}{20}+\frac{3}{30}+\frac{9}{42}+\frac{10}{56} = 4.342837$

9. $\sum \sum_{n=1}^{5} n^3 = 1 + 8 + 27 + 64 + 125 = 225$

10. $\sum_{n=1}^{8} \frac{(-1)^n(n^2+1)}{h} = -\frac{2}{7}+\frac{5}{2}-\frac{10}{3}+\frac{17}{4}-\frac{26}{5}+\frac{37}{6}-\frac{50}{7}+\frac{65}{8} = \frac{2827}{840} \approx 3.36547$

11. $\sum_{n=1}^{8} \frac{2^i}{3i+1} = \frac{8}{10}+\frac{16}{13}+\frac{32}{16}+\frac{64}{19}+\frac{128}{22}+\frac{256}{25} = \frac{1{,}593{,}342}{67{,}925} \approx 23.457$

12. $\sum_{i=0}^{7} \frac{(-1)^{i+1}(i+1)^2}{2i+1} = \frac{-1}{1}+\frac{4}{3}-\frac{9}{5}+\frac{16}{7}-\frac{25}{9}+\frac{36}{11}-\frac{49}{13}+\frac{64}{15} = \frac{244{,}788}{135{,}135} \approx 1.811$

13. Arithmetic; $d = 3$; $a_{10} = 3 + 9 \cdot 3 = 30$; $S_{10} = \frac{10(3+30)}{2} = 165$

14. Arithmetic: $d = -4$; $a_{12} = 5 - 11(4) = -39$; $S_{12} = \frac{12(5-39)}{2} = -204$

15. Geometric: $r = 2$; $S_8 = 1 \cdot \frac{1-2^8}{1-2} = \frac{-255}{-1} = 255$

16. Geometric: $r = 4$; $S_9 = \frac{1}{2} \cdot \frac{1-4^9}{1-4} = \frac{1}{2} \cdot \frac{-262143}{-3} = 43690.5$

17. Arithmetic: $d = 4$; $a_{12} = -6 + 11 \cdot 4 = 38$; $S_{12} = \frac{12(-6+38)}{2} = 192$

18. Geometric: $r = -\frac{1}{2}$; $S_{10} = 1 \cdot \frac{1-\left(\frac{-1}{2}\right)^{10}}{1+\frac{1}{2}} = \frac{2}{3}\left(1 - \left(\frac{1}{2}\right)^{10}\right) = \frac{341}{512} \approx 0.666016$

19. Arithmetic: $d = 0.25, a_{20} = 0.5 + 19(.25) = 5.25$; $S_{20} = \frac{20}{2}(0.5 + 5.25) = 57.5$

20. Geometric: $r = 0.4$; $S_{15} = 0.5\frac{(1-0.4^{15})}{1-0.4} \approx 0.8333324$

21. Geometric: $r = -\frac{1}{3}$; $S_{16} = \frac{3}{4}\frac{\left(1-\left(-\frac{1}{3}\right)^{16}\right)}{\left(1+\frac{1}{3}\right)} = \frac{9}{10}\left(1 - \left(\frac{1}{3}\right)^{16}\right) \approx 0.5625$

22. Geometric: $r = 0.1$; $S_6 = 0.444444$

23. Arithmetic: $d = 1.2$; $a_{14} = 0.4 + 13(1.2) = 16$; $S_{14} = \frac{14}{2}(0.4 + 16) = 114.8$

24. Geometric: $r = \frac{1}{2}$; $S_{10} = 16 \cdot \frac{1-\left(\frac{1}{2}\right)^{10}}{1-\frac{1}{2}} = 32(1 - 0.5^{10}) = 31.96875$

25. Geometric: $r = \sqrt{3}$; $S_{15} = 2\sqrt{3} \cdot \left(\frac{1-\sqrt{3}^{15}}{1-\sqrt{3}}\right) = 17{,}920.253$

26. Geometric: $r = 2\sqrt{5}$; $S_{12} = \sqrt{5} \cdot \left(\frac{1-(2\sqrt{5})^{12}}{1-2\sqrt{5}}\right) = 41{,}216{,}228.33$

27. Geometric: $r = 3$; $S_8 = \left(\frac{1-3^8}{1-3}\right) = 9840$

28. Geometric: $r = -5$; $S_{10} = 5\left(\frac{1-(-5)^{10}}{1+5}\right) = -8{,}138{,}020$

29. First time it hits the ground it has only fallen 80m. After this it rises and falls so $2 \times 80(.8)$. So this is a sequence $80 + 160(.8) + 160(0.8)^2 + 160(0.8)^3 = 160 + 160(0.8) + 160(0.8)^2 + 160(0.8)^3 - 80 = 160\left(\frac{1-0.8^4}{1-0.8}\right) - 80 = 392.32$m when it hits the fourth time. $160\left(\frac{1-0.8^{10}}{1-0.8}\right) - 80 = 634.10$ when it hits the tenth time.

30. Geometric with $a_1 = 50$; $r = 0.8$; $S_{10} = 50\left(\frac{1-0.8^{10}}{1-0.8}\right) = 223.156$ cm

31. In a 16 ft span there will be 12 studs 16″ apart. The longest is 8′ and the shortest is 8″ or $\frac{2}{3}$ ft. $S_{12} = \frac{12}{2}\left(8 + \frac{2}{3}\right) = 6\left(8\frac{2}{3}\right) = 52$ ft. This is only half of the roof. The whole roof will be twice that or 104 ft.

32. The rate of 8% compounded quarterly is 2% or 0.02 per quarter, 3 years is 12 quarters, so $A = 2000(1 + 0.02)^{12} = \$2536.48$.

33. The rate of 8% compounded monthly is $\frac{8}{12} = \frac{2}{3}\%$ per month $\approx 0.00\overline{6}$ per month, 3 years have 36 months $A = 2000(1 + 0.00\overline{6})^{36} = \$2540.47$. This is \$3.99 more than exercise #32.

34. $A = Pe^{rt} = 2000e^{(0.08\times3)} = 2542.50$

35. (a) Here the sequence is \$750,000 + \$320,000 + \$240,000 + \$252,000 + ⋯ + \$336,000. Here $a_1 =$ 240,000 is the cost of the 2nd floor, so $a_9 = 336{,}000$ and $S_9 = \frac{9(240{,}000 + 336{,}000)}{2} = 2{,}592{,}000$ is the cost of floors 2–10. So, the cost of a 10-story building is \$750,000 + \$320,000 + \$2,592,000 = \$3,662,000. (b) \$750,000 + \$320,000 + S_{19} = \$7,682,000. (S_{19} = cost of floors 2–20 = $\frac{19}{2}(240{,}000 + 456{,}000) = 6{,}612{,}000$.)

19.4 INFINITE GEOMETRIC SERIES

1. $r = \frac{-1}{2}$ so converges; $S = \frac{\frac{1}{2}}{1 - \frac{-1}{2}} = \frac{\frac{1}{2}}{\frac{3}{2}} = \frac{1}{3}$

2. $r = \frac{3}{4}$ so converges; $S = \frac{1}{1 - \frac{3}{4}} = \frac{1}{\frac{1}{4}} = 4$

3. $r = -\frac{2}{3}$ so converges; $S = \frac{1}{1 - \frac{-2}{3}} = \frac{1}{\frac{5}{3}} = \frac{3}{5}$ or 0.6

4. $r = 1.5$ so diverges

5. $r = \frac{1}{4}$ so converges; $S = \frac{3}{1 - \frac{1}{4}} = \frac{3}{\frac{3}{4}} = 4$

6. $r = -\frac{1}{4}$ so converges; $S = \frac{4}{1 - \frac{-1}{4}} = \frac{4}{\frac{5}{4}} = \frac{16}{5}$ or 3.2

7. $r = \frac{1}{10}$ so converges; $S = \frac{\frac{5}{4}}{1 - \frac{1}{10}} = \frac{\frac{5}{4}}{\frac{9}{10}} = \frac{5}{4} \cdot \frac{10}{9} = \frac{50}{36} = \frac{25}{18} = 1.3888\ldots$

8. $r = \frac{1}{4}$ so converges; $S = \frac{\frac{3}{4}}{1 - \frac{1}{4}} = \frac{\frac{3}{4}}{\frac{3}{4}} = 1$

9. $r = 0.1$ so converges; $S = \frac{0.03}{1-0.1} = \frac{0.03}{0.9} = \frac{3}{90} = \frac{1}{30} = 0.03333\ldots$

10. 0.01 so converges; $S = \frac{1.4}{1-0.01} = \frac{1.4}{0.99} = \frac{140}{99} = 1.41414\ldots$

11. $r = (-0.8)$ so converges; $S = \frac{1}{1+0.8} = \frac{1}{1.8} = \frac{10}{18} = \frac{5}{9} = 0.555\ldots$

12. $r = \frac{3}{4}$ so converges; $S = \frac{\frac{3}{4}}{1 - \frac{3}{4}} = \frac{\frac{3}{4}}{\frac{1}{4}} = 3$

13. $r = \frac{5}{4}$ so diverges

14. $r = -1.1$ So diverges

15. $r = 10$ so diverges

16. $2^{-n} = (2^{-1})^n = \frac{1}{2}^n$; $r = \frac{1}{2}$ so converges $S = \frac{\frac{1}{2}}{1 - \frac{1}{2}} = \frac{\frac{1}{2}}{\frac{1}{2}} = 1$

17. $r = \frac{2}{3}$ so converges; $S = \frac{\frac{2}{3}}{1 - \frac{2}{3}} = \frac{\frac{2}{3}}{\frac{1}{3}} = 2$

18. $r = -\frac{2}{3}$ so converges; $S = \frac{-4}{1 + \frac{2}{3}} = \frac{-4}{\frac{5}{3}} = -\frac{12}{5} = -2.4$

19. $r = \frac{1}{\sqrt{2}}$ so converges; $S = \frac{\frac{1}{\sqrt{2}}}{1 - \frac{1}{\sqrt{2}}} = \frac{\frac{1}{\sqrt{2}}}{\frac{\sqrt{2}-1}{\sqrt{2}}} = \frac{1}{\sqrt{2} - 1} = \sqrt{2} + 1 \approx 2.4142$

20. $r = \frac{\sqrt{3}}{3}$ So converges; $S = \frac{\frac{\sqrt{3}}{3}}{1 - \frac{\sqrt{3}}{3}} = \frac{\frac{\sqrt{3}}{3}}{\frac{3-\sqrt{3}}{3}} = \frac{\sqrt{3}}{3 - \sqrt{3}} = \frac{3\sqrt{3} + 3}{6} = \frac{\sqrt{3}}{2} + \frac{1}{2} \approx 1.366$

21. $r = \sqrt{5}$ so Diverges

22. Think of this as $\sum_{n=1}^{\infty} \left[\left(\sqrt{7}\right)^{-1}\right]^{n-1} = \sum_{n=1}^{\infty} \left(\frac{1}{\sqrt{7}}\right)^{n-1}$. Thus, $r = \frac{1}{\sqrt{7}}$ so the series converges; $S = \frac{1}{1 - \frac{1}{\sqrt{7}}} = \frac{1}{\frac{\sqrt{7}-1}{\sqrt{7}}} = \frac{\sqrt{7}}{\sqrt{7} - 1} \approx 1.6076$

23. $a_1 = 0.4; r = 0.1; S = \frac{0.4}{1-0.1} = \frac{0.4}{0.9} = \frac{4}{9}$

24. $a_1 = 0.7; r = 0.1; S = \frac{0.7}{1-0.1} = \frac{0.7}{0.9} = \frac{7}{9}$

25. $a_1 = 0.57; r = 0.01; S = \frac{0.57}{1-0.01} = \frac{0.57}{0.99} = \frac{57}{99} = \frac{19}{33}$

26. $a_1 = 0.84; r = 0.01; S = \frac{0.84}{1-0.01} = \frac{0.84}{0.99} = \frac{28}{33}$

27. $a_1 = 1.352; r = 0.0001; S = \frac{1.352}{1-0.0001} = \frac{1.352}{0.9999} = \frac{13520}{9999}$

28. $a_1 = 4.123; r = 0.0001; S = \frac{4.123}{1-0.0001} = \frac{4.123}{0.9999} = \frac{41230}{9999}$

29. $6.3021021\ldots = 6.3 + (0.0021 + 0.0000021 \cdots)$; $a_1 = 0.0021; r = 0.001; S = \frac{0.0021}{.999} = \frac{21}{9990} = \frac{7}{3330}$; $6.3 + \frac{7}{3330} = \frac{63}{10} + \frac{7}{3330} = \frac{20986}{3330} = \frac{10493}{1665}$

30. $2.1906906\ldots = 2.1 + (0.0906 + 0.0000906 + \cdots)$; $S = \frac{0.0906}{.999} = \frac{906}{9990} = \frac{302}{3330}$; $2.1 + \frac{302}{3330} = \frac{21}{10} + \frac{302}{3330} = \frac{6993+302}{3330} = \frac{7295}{3330} = \frac{1459}{666}$

31. After the first bounce it is a geometric series with $a_1 = 8; r = 0.8$ $S = \frac{8}{1-0.8} = \frac{8}{0.2} = 40$. The total distance is $5 + 40 = 45$ m.

32. Geometric series: $a_1 = 100; r = 0.9; S = \frac{100}{1-0.9} = 1000$ mm

33. Geometric series with $a_1 = 250; r = \frac{8}{10} = 0.8$ so $S = \frac{250}{1-0.8} = \frac{250}{0.2} = 1250$ rev.

34. Geometric series with $a_1 = 50$ and $r = 0.85$, hence $S = \frac{50}{1-0.85} = \frac{50}{0.15} = 333.33$ cm.

19.5 THE BINOMIAL THEOREM

1. $(a+1)^4 = 1a^4 + 4a^3 \cdot 1^1 + 6a^2 1^2 + 4a1^3 + 1^4 = a^4 + 4a^3 + 6a^2 + 4a + 1$

2. $(2x+b)^5 = (2x)^5 + 5(2x)^4 b + 10 \cdot (2x)^3 b^2 + 10(2x)^2 b^3 + 5(2x)b^4 + b^5 = 32x^5 + 80x^4 b + 80x^3 b^2 + 40x^2 b^3 + 10xb^4 + b^5$

3. $(3x-1)^4 = (3x)^4 + 4(3x)^3(-1) + 6(3x)^2(-1)^2 + 4(3x)(-1)^3 + (-1)^4 = 81x^4 - 108x^3 + 54x^2 - 12x + 1$

4. $(x-2y)^5 = x^5 + 5x^4(-2y)^1 + 10x^3(-2y)^2 + 10x^2(-2y)^3 + 5x(-2y)^4 + (-2y)^5 = x^5 - 10x^4 y + 40x^3 y^2 - 80x^2 y^3 + 80xy^4 - 32y^5$

5. $\left(\frac{x}{2}+d\right)^6 = \left(\frac{x}{2}\right)^6 + 6\left(\frac{x}{2}\right)^5 d + 15\left(\frac{x}{2}\right)^4 d^2 + 20\left(\frac{x}{2}\right)^3 d^3 + 15\left(\frac{x}{2}\right)^2 d^4 + 6\left(\frac{x}{2}\right) d^5 + d^6 = \frac{x^6}{64} + \frac{3x^5 d}{16} + \frac{15x^4 d^2}{16} + \frac{5x^3 d^3}{2} + \frac{15x^2 d^4}{4} + 3xd^5 + d^6$

6. $\left(xy - \frac{a}{3}\right)^6 = (xy)^6 + 6(xy)^5\left(-\frac{a}{3}\right)^1 + 15(xy)^4\left(-\frac{a}{3}\right)^2 + 20(xy)^3\left(-\frac{a}{3}\right)^3 + 15(xy)^2\left(-\frac{a}{3}\right)^4 + 6(xy)\left(-\frac{a}{3}\right)^5 + \left(-\frac{a}{3}\right)^6 = x^6y^6 - 2x^5y^5a + \frac{5}{3}x^4y^4a^2 - \frac{20}{27}x^3y^3a^3 + \frac{5}{27}x^2y^2a^4 - \frac{2}{81}xya^5 + \frac{1}{729}a^6$

7. $\left(\frac{a}{2} - \frac{4}{b}\right)^5 = \left(\frac{a}{2}\right)^5 + 5\left(\frac{a}{2}\right)^4\left(-\frac{4}{b}\right) + 10\left(\frac{a}{2}\right)^3\left(-\frac{4}{b}\right)^2 + 10\left(\frac{a}{2}\right)^2\left(-\frac{4}{b}\right)^3 + 5\left(\frac{a}{2}\right)\left(-\frac{4}{b}\right)^4 + \left(-\frac{4}{b}\right)^5 = \frac{a^5}{32} - \frac{5a^4}{4b} + 20\frac{a^3}{b^2} - 160\frac{a^2}{b^3} + 640\frac{a}{b^4} - \frac{1024}{b^5}$

8. $\left(\frac{x}{3} + \frac{2}{y}\right)^4 = \left(\frac{x}{3}\right)^4 + 4\left(\frac{x}{3}\right)^3\left(\frac{2}{y}\right) + 6\left(\frac{x}{3}\right)^2\left(\frac{2}{y}\right)^2 + 4\left(\frac{x}{3}\right)\left(\frac{2}{y}\right)^3 + \left(\frac{2}{y}\right)^4 = \frac{x^4}{81} + \frac{8x^3}{27y} + \frac{8x^2}{3y^2} + \frac{32x}{3y^3} + \frac{16}{y^4}$

9. $(a+b)^7 = \binom{7}{0}a^7 + \binom{7}{1}a^6b + \binom{7}{2}a^5b^2 + \binom{7}{3}a^4b^3 + \binom{7}{4}a^3b^4 + \binom{7}{5}a^2b^5 + \binom{7}{6}ab^6 + \binom{7}{7}b^7 = a^7 + 7a^6b + 21a^5b^2 + 35a^4b^3 + 35a^3b^4 + 21a^2b^5 + 7ab^6 + b^7$

10. $(a+b)^9 = \binom{9}{0}a^9 + \binom{9}{1}a^8b^1 + \binom{9}{2}a^7b^2 + \binom{9}{3}a^6b^3 + \binom{9}{4}a^5b^4 + \binom{9}{5}a^4b^5 + \binom{9}{6}a^3b^6 + \binom{9}{7}a^2b^7 + \binom{9}{8}ab^8 + b^9 = a^9 + 9a^8b + 36a^7b^2 + 84a^6b^3 + 126a^5b^4 + 126a^4b^5 + 84a^3b^6 + 36a^2b^7 + 9ab^8 + b^9$

11. $(t-a)^8 = \binom{8}{0}t^8 + \binom{8}{1}t^7(-a)^1 + \binom{8}{2}t^6(-a)^2 + \binom{8}{3}t^5(-a)^3 + \binom{8}{4}t^4(-a)^4 + \binom{8}{5}t^3(-a)^5 + \binom{8}{6}t^2(-a)^6 + \binom{8}{7}t(-a)^7 + \binom{8}{8}(-a)^8 = t^8 - 8t^7a + 28t^6a^2 - 56t^5a^3 + 70t^4a^4 - 56t^3a^5 + 28t^2a^6 - 8ta^7 + a^8$

12. $(x-2a)^7 = \binom{7}{0}x^7 + \binom{7}{1}x^6(-2a)^1 + \binom{7}{2}x^5(-2a)^2 + \binom{7}{3}x^4(-2a)^3 + \binom{7}{4}x^3(-2a)^4 + \binom{7}{5}x^2(-2a)^5 + \binom{7}{6}x(-2a)^6 + \binom{7}{7}(-2a)^7 = 1x^7 + 7x^6(-2a) + 21x^5(4a^2) + 35x^4(-8a^3) + 35x^3(-16a^4) + 21x^2(-32a^5) + 7x(64a^6) + 1(-128a^7) = x^7 - 14x^6a + 84x^5a^2 - 280x^4a^3 + 560x^3a^4 - 672x^2a^5 + 448xa^6 - 128a^7$

13. $(2a-1)^6 = \binom{6}{0}(2a)^6 + \binom{6}{1}(2a)^5(-1) + \binom{6}{2}(2a)^4(-1)^2 + \binom{6}{3}(2a)^3(-1)^3 + \binom{6}{4}(2a)^2(-1)^4 + \binom{6}{5}(2a)(-1)^5 + \binom{6}{6}(-1)^6 = 64a^6 - 192a^5 + 240a^4 - 160a^3 + 60a^2 - 12a + 1$

14. $(a^2-3)^9 = \binom{9}{0}(a^2)^9 + \binom{9}{1}(a^2)^8(-3) + \binom{9}{2}(a^2)^7(-3)^2 + \binom{9}{3}(a^2)^6(-3)^3 + \binom{9}{4}(a^2)^5(-3)^4 + \binom{9}{5}(a^2)^4(-3)^5 + \binom{9}{6}(a^2)^3(-3)^6 + \binom{9}{7}(a^2)^2(-3)^7 + \binom{9}{8}a^2(-3)^8 + \binom{9}{9}(-3)^9 = a^{18} - 27a^{16} + 324a^{14} - 2268a^{12} + 10206a^{10} - 30618a^8 + 61236a^6 - 78732a^4 + 59049a^2 - 19683$

15. $\left(x^2y+\frac{a}{2}\right)^7 = \binom{7}{0}(x^2y)^7 + \binom{7}{1}(x^2y)^6\left(\frac{a}{2}\right) + \binom{7}{2}(x^2y)^5\left(\frac{a}{2}\right)^2 + \binom{7}{3}(x^2y)^4\left(\frac{a}{2}\right)^3 + \binom{7}{4}(x^2y)^3\left(\frac{a}{2}\right)^4 + \binom{7}{5}(x^2y)^2\left(\frac{a}{2}\right)^5 + \binom{7}{6}(x^2y)\left(\frac{a}{2}\right)^6 + \binom{7}{7}\left(\frac{a}{2}\right)^7 = x^{14}y^7 + \frac{7}{2}x^{12}y^6a + \frac{21}{4}x^{10}y^5a^2 + \frac{35}{8}x^8y^4a^3 + \frac{35}{16}x^6y^3a^4 + \frac{21}{32}x^4y^2a^5 + \frac{7}{64}x^2ya^6 + \frac{a^7}{128}$

16. $\left(\frac{a}{3}-\frac{xy}{2}\right)^6 = \binom{6}{0}\left(\frac{a}{3}\right)^6 + \binom{6}{1}\left(\frac{a}{3}\right)^5\left(-\frac{xy}{2}\right) + \binom{6}{2}\left(\frac{a}{3}\right)^4\left(\frac{-xy}{2}\right)^2 + \binom{6}{3}\left(\frac{a}{3}\right)^3\left(-\frac{xy}{2}\right)^3 + \binom{6}{4}\left(\frac{a}{3}\right)^2\left(-\frac{xy}{2}\right)^4 + \binom{6}{5}\left(\frac{a}{3}\right)\left(-\frac{xy}{2}\right)^5 + \binom{6}{6}\left(-\frac{xy}{2}\right)^6 = \frac{a^6}{729} - \frac{a^5xy}{81} + \frac{5a^4x^2y^2}{108} - \frac{5a^3x^3y^3}{54} + \frac{5a^2x^4y^4}{48} - \frac{ax^5y^5}{16} + \frac{x^6y^6}{64}$

17. $(x+y)^{12} = \binom{12}{0}x^{12} + \binom{12}{1}x^{11}y + \binom{12}{2}x^{10}y^2 + \binom{12}{3}x^9y^3 + \cdots = x^{12} + 12x^{11}y + 66x^{10}y^2 + 220x^9y^3 + \cdots$

18. $(x-5)^{10} = \binom{10}{0}x^{10} + \binom{10}{1}x^9(-5) + \binom{10}{2}x^8(-5)^2 + \binom{10}{3}x^7(-5)^3 + \cdots = x^{10} - 50x^9 + 1125x^8 - 15000x^7 + \cdots$

19. $(1-a)^{-2} = 1 + (-2)(-a) + \frac{(-2)(-3)}{2!}(-a)^2 + \frac{(-2)(-3)(-4)}{3!}(-a)^3 + \cdots = 1 + 2a + 3a^2 + 4a^3 + \cdots$

20. $(1+x)^{\frac{1}{2}} = 1 + \frac{1}{2}x + \frac{\frac{1}{2}\cdot\left(-\frac{1}{2}\right)}{2!}x^2 + \frac{\frac{1}{2}\left(-\frac{1}{2}\right)\left(-\frac{3}{2}\right)}{3!}x^3 + \cdots = 1 + \frac{x}{2} - \frac{x^2}{8} + \frac{x^3}{16} + \cdots$

21. $(1+b)^{-\frac{1}{3}} = 1 + \left(-\frac{1}{3}\right)b + \frac{\left(-\frac{1}{3}\right)\left(-\frac{4}{3}\right)}{2!}b^2 + \frac{\left(-\frac{1}{3}\right)\left(-\frac{4}{3}\right)\left(-\frac{7}{3}\right)}{3!}b^3 + \cdots = 1 - \frac{b}{3} + \frac{2b^2}{9} - \frac{14b^3}{81} + \cdots$

22. $(1-y)^{-\frac{1}{4}} = 1 + \left(-\frac{1}{4}\right)(-y) + \frac{\left(-\frac{1}{4}\right)\left(-\frac{5}{4}\right)}{2!}(-y)^2 + \frac{\left(-\frac{1}{4}\right)\left(-\frac{5}{4}\right)\left(-\frac{9}{4}\right)}{3!}(-y)^3 = 1 + \frac{y}{4} + \frac{5y^2}{32} + \frac{15y^3}{128} + \cdots$

23. $(1.1)^4 = (1+0.1)^4 = 1 + 4(0.1) + \frac{4\cdot 3}{2!}(0.1)^2 + \frac{4\cdot 3\cdot 2}{3!}(0.1)^3 = 1 + 0.4 + 0.06 + 0.004 = 1.464$

24. $(0.85)^5 = (1-0.15)^5 = 1 + 5(-0.15) + \frac{5\cdot 4}{2!}(-0.15)^2 + \frac{5\cdot 4\cdot 3}{3!}(-0.15)^3 + \frac{5\cdot 4\cdot 3\cdot 2}{4!}(-0.15)^4 = 1 - 0.75 + 0.225 - 0.03375 + 0.00253 = 0.444$

25. $\sqrt{1.1} = (1+0.1)^{\frac{1}{2}} = 1 + \frac{1}{2}(0.1) + \frac{\frac{1}{2}\cdot\left(-\frac{1}{2}\right)}{2!}(0.1)^2 + \frac{\frac{1}{2}\left(-\frac{1}{2}\right)\left(-\frac{3}{2}\right)}{3!}(0.1)^3 = 1 + 0.05 - 0.00125 + 0.00006 = 1.049$

26. $\sqrt[3]{1.01} = (1+0.01)^{\frac{1}{3}} = 1 + \frac{1}{3}(0.01) + \frac{\frac{1}{3}\left(-\frac{2}{3}\right)}{2!}(0.01)^2 = 1 + 0.00333 - 0.00001 \approx 1.003$

27. $\sqrt[5]{1.04} = (1+0.04)^{\frac{1}{5}} = 1 + \frac{1}{5}(0.04) + \frac{\frac{1}{5}\left(-\frac{4}{5}\right)}{2!}(0.04)^2 = 1 + 0.008 - 0.000128 \approx 1.008$

28. $\sqrt[6]{0.98} = (1-0.02)^{\frac{1}{6}} = 1+\frac{1}{6}(-0.02)+\frac{\frac{1}{6}(-\frac{5}{6})}{2!}(-0.02)^2 = 1-0.0033-0.000028 \approx 0.997$

29. $\sqrt[3]{0.95} = (1-0.05)^{\frac{1}{3}} = 1+\frac{1}{3}(-0.05) = 1-0.017 \approx 0.983$

30. $\sqrt[4]{0.925} = (1-0.075)^{\frac{1}{4}} = 1+\frac{1}{4}(-0.075)+\frac{\frac{1}{4}(-\frac{3}{4})}{2!}(-0.075)^2 = 1-0.01875-0.000527 \approx 0.981$

31. Counting from 0, the sixth term is $\binom{15}{5}x^{10}y^5 = 3003x^{10}y^5$

32. Counting from 0, the eight term is $\binom{20}{7}a^{13}b^7 = 77{,}520a^{13}b^7$

33. Fifth term is $\binom{12}{4}(2x)^8(-y)^4 = 126{,}720x^8y^4$

34. Fourth term is $\binom{10}{3}(3x)^7(-2y)^3 = -2{,}099{,}520x^7y^3$

35. Term involving b^4 is the 5th term: $\binom{14}{4}a^{10}b^4 = 1001a^{10}b^4$

36. Term involving x^5 is the $(15-5+1) = 11$th term: $\binom{15}{10}x^5y^{10} = 3003x^5y^{10}$

37. (a) $\left(1-\frac{v^2}{c^2}\right)^{-\frac{1}{2}} = 1+\left(-\frac{1}{2}\right)\left(-\frac{v^2}{c^2}\right)+\frac{\left(-\frac{1}{2}\right)\left(-\frac{3}{2}\right)}{2!}\left(-\frac{v^2}{c^2}\right)^2 = 1+\frac{v^2}{2c^2}+\frac{3v^4}{8c^4}$ (b) $mc^2+\frac{mv^2}{2}+\frac{3mv^4}{8c^2}$

38. (a) $\sqrt{1-\frac{1}{4V^2}} = (1-\frac{1}{4V^2})^{\frac{1}{2}} = 1-\frac{1}{2}\cdot\frac{1}{4V^2} = 1-\frac{1}{8V^2}$,

(b) $1-\frac{v}{c} = 1-\left(1-\frac{1}{8V^2}\right) = \frac{1}{8V^2}$,

(c) $V = 100;\ \frac{1}{8V^2} = 1.25\times10^{-5}$,

(d) $V = 500;\ \frac{1}{8V^2} = 5\times10^{-7}$,

(e) $V = 125{,}000;\ \frac{1}{8V^2} = 8\times10^{-12}$,

(f) $V = 2.50\times10^5;\ \frac{1}{8V^2} = 2\times10^{-12}$

39.
$$\begin{aligned}(r^2)^{3/2}\left(1+\frac{\ell^2}{r^2}\right)^{3/2} &= r^3\left[1+\frac{3}{2}\cdot\frac{\ell^2}{r^2}+\frac{\frac{3}{2}\cdot\frac{1}{2}}{2!}\left(\frac{\ell^2}{r^2}\right)^2\right]\\ &= r^3\left[1+\frac{3\ell^2}{2r^2}+\frac{3\ell^4}{8r^4}\right]\\ &= r^3+\frac{3}{2}r\ell^2+\frac{3\ell^4}{8r}\end{aligned}$$

40.
$$\begin{aligned}\frac{(x+h)^4-x^4}{h} &= \frac{x^4+4x^3h+6x^2h^2+4xh^3+h^4-x^4}{h}\\ &= \frac{4x^3h+6x^2h^2+4xh^3+h^4}{h}\\ &= 4x^3+6x^2h+4xh^2+h^3\end{aligned}$$

CHAPTER 19 REVIEW

1. $\frac{1}{1+2} = \frac{1}{3};\ \frac{1}{2+2} = \frac{1}{4};\ \frac{1}{3+2} = \frac{1}{5};\ \frac{1}{4+2} = \frac{1}{6};\ \frac{1}{5+2} = \frac{1}{7};\ \frac{1}{6+2} = \frac{1}{8}$

2. $\frac{3}{2-1} = 3;\ \frac{3}{4-1} = 1;\ \frac{3}{6-1} = \frac{3}{5};\ \frac{3}{8-1} = \frac{3}{7};\ \frac{3}{10-1} = \frac{1}{3};\ \frac{3}{12-1} = \frac{3}{11}$

3. $\frac{-1}{3};\ \frac{1}{2\cdot5} = \frac{1}{10};\ \frac{-1}{3(7)} = \frac{-1}{21};\ \frac{1}{4(9)} = \frac{1}{36};\ \frac{-1}{5(11)} = \frac{-1}{55};\ \frac{1}{6(13)} = \frac{1}{78}$

4. $\frac{1+1}{3-1} = \frac{2}{2} = 1;\ \frac{5}{5} = 1;\ \frac{10}{8} = \frac{5}{4};\ \frac{17}{11};\ \frac{26}{14} = \frac{13}{7};\ \frac{37}{17}$

5. $a_1 = 1;\ a_2 = \frac{1}{2};\ a_3 = \frac{\frac{1}{2}}{3} = \frac{1}{6};\ a_4 = \frac{\frac{1}{6}}{4} = \frac{1}{24};\ a_5 = \frac{\frac{1}{24}}{5} = \frac{1}{120};\ a_6 = \frac{\frac{1}{120}}{6} = \frac{1}{720}$

6. $a_1 = 5; a_2 = 5 - 2 = 3; a_3 = 3 - 3 = 0; a_4 = 0 - 4 = -4; a_5 = -4 - 5 = -9; a_6 = -9 - 6 = -15$

7. $a_1 = 1; a_2 = 3; a_3 = 1 + 3 = 4; a_4 = 3 + 4 = 7; a_5 = 4 + 7 = 11; a_6 = 7 + 11 = 18$

8. $a_1 = 1; a_2 = 2; a_3 = 1 \cdot 2 = 2; a_4 = 2 \cdot 2 = 4; a_5 = 2 \cdot 4 = 8; a_6 = 4 \cdot 8 = 32$

9. Arithmetic; $d = -3, a_{10} = 10 + 9(-3) = -17$

10. Geometric: $r = \frac{1}{4}; a_8 = 8 \cdot \left(\frac{1}{4}\right)^7 = \frac{1}{2048}$

11. Arithmetic: $d = 4; a_7 = -1 + 6(4) = 23$

12. Neither; $7, 4, 0, -5, -11, -18, -26, -35, -45, = a_9$

13. Geometric: $r = -\frac{1}{6}; a_{10} = 3\left(-\frac{1}{6}\right)^9 = \frac{-1}{3{,}359{,}232} \approx 2.97687 \times 10^{-7} = 0.0000003$

14. Geometric: $r = 10; a_7 = 2.05 \times 10^6 = 2{,}050{,}000$

15. $d = 18 - 12 = 6; a_7 = a_1 + 6d; 12 = a_1 + 6 \cdot 6; a_1 = 12 - 36 = -24$

16. $r = \frac{\frac{9}{7}}{9} = \frac{1}{7}; a_8 = a_7 \cdot r = \frac{9}{7} \cdot \frac{1}{7} = \frac{9}{49}$

17. (a) $4 \cdot \frac{1}{3} + 4 \cdot \frac{1}{9} + 4 \cdot \frac{1}{27} + 4 \cdot \frac{1}{81} = \frac{4}{3} + \frac{4}{9} + \frac{4}{27} + \frac{4}{81}$
(b) $2 + \frac{3}{2} + \frac{4}{3} + \frac{5}{4}$

18. $2 + 3 + 4 + 5 = 14$

19. $-1 + 2 + 5 + 8 + 11 + 14 = 39$

20. $\frac{-1}{3} + \frac{2}{5} - \frac{3}{9} + \frac{4}{17} = \frac{-8}{255} = -0.03137$

21. Arithmetic: $d = 5; a_{10} = 4 + 9 \cdot 5 = 49; S_{10} = \frac{10}{2}(4 + 49) = 5(53) = 265$

22. Arithmetic: $d = -7; a_{12} = 2 + 11(-7) = -75; S_{12} = \frac{12}{2}(2 - 75) = 6(-73) = -438$

23. Geometric: $r = \frac{1}{3}; S_{14} = 1 \cdot \frac{1 - \frac{1}{3}^{14}}{1 - \frac{1}{3}} = 1.4999997$

24. Arithmetic: There are two possible ways to look at the sequence. One way is to let $a_1 = \sqrt{5} + 1$ and $d = \sqrt{5} + 1$. Then $a_8 = \sqrt{5} + 1 + (7)(\sqrt{5} + 1) = 8\sqrt{5} + 8$ and $S_8 = \frac{8}{2}(\sqrt{5} + 1 + 8\sqrt{5} + 8) = 4(9\sqrt{5} + 9) = 36\sqrt{5} + 36$. The other way to look at this sequence is to let $a_1 = \sqrt{5}$ and $d = 1 + \sqrt{5}$. Then, $a_8 = \sqrt{5} + 7(1 + \sqrt{5}) = 7 + 8\sqrt{5}$ and $S_8 = \frac{8}{2}(\sqrt{5} + 7 + 8\sqrt{5}) = 4(7 + 9\sqrt{5}) = 28 + 36\sqrt{5}$.

25. Converges: $r = \frac{1}{2}; S = \frac{\frac{1}{3}}{1 - \frac{1}{2}} = \frac{2}{3}$

26. Converges: $r = \frac{3}{4}; S = \frac{8}{1 - \frac{3}{4}} = 32$

27. Diverges: $r = 10$

28. Converges: $r = -\frac{1}{10}; S = \frac{1.5}{1 + 0.1} = \frac{1.5}{1.1} = \frac{15}{11} \approx 1.3636$

29. Converges: $r = \frac{1}{\sqrt{3}}; S = \frac{\frac{1}{\sqrt{3}}}{1 - \frac{1}{\sqrt{3}}} = \frac{\frac{1}{\sqrt{3}}}{\frac{\sqrt{3}-1}{\sqrt{3}}} = \frac{1}{\sqrt{3}-1} = \frac{\sqrt{3}+1}{2} \approx 1.3660$

30. Converges: $r = \frac{1}{\sqrt{2}}; a_1 = 1; S = \frac{1}{1 - \frac{1}{\sqrt{2}}} = \frac{1}{\frac{\sqrt{2}-1}{\sqrt{2}}} = \frac{\sqrt{2}}{\sqrt{2}-1} = \frac{2+\sqrt{2}}{1} \approx 3.4142$

31. $a_1 = 0.185; r = 0.001; S = \frac{0.185}{0.999} = \frac{185}{999} = \frac{5}{27}$

32. Start with $a_1 = 0.01; r = 0.1; S = \frac{0.01}{.9} = \frac{1}{90}; 0.6 + \frac{1}{90} = \frac{6}{10} + \frac{1}{90} = \frac{54}{90} + \frac{1}{90} = \frac{55}{90} = \frac{11}{18}$

33. $(a+2)^5 = a^5 + 5a^4 \cdot 2 + 10a^3 \cdot 2^2 + 10a^2 \cdot 2^3 + 5a^1 \cdot 2^4 + 2^5 = a^5 + 10a^4 + 40a^3 + 80a^2 + 80a + 32$

34. $(3x - y)^6 = (3x)^6 + 6(3x)^5(-y) + 15(3x)^4(-y)^2 + 20(3x)^3(-y)^3 + 15(3x)^2(-y)^4 + 6(3x)(-y)^5 + (-y)^6 = 729x^6 - 1458x^5y + 1215x^4y^2 - 540x^3y^3 + 135x^2y^4 - 18xy^5 + y^6$

35. $\left(\frac{x}{2} - 3y^2\right)^6 = \left(\frac{x}{2}\right)^6 + 6\left(\frac{x}{2}\right)^5(-3y^2) + 15\left(\frac{x}{2}\right)^4(-3y^2)^2 + 20\left(\frac{x}{2}\right)^3(-3y^2)^3 + 15\left(\frac{x}{2}\right)^2(-3y^2)^4 + 6\left(\frac{x}{2}\right)(-3y^2)^5 + (-3y^2)^6 = \frac{x^6}{64} - \frac{9}{16}x^5y^2 + \frac{135}{16}x^4y^4 - \frac{135}{2}x^3y^6 + \frac{1215}{4}x^2y^8 - 729xy^{10} + 729y^{12}$

36. $\left(\frac{2a^3}{5} + \frac{5b}{2}\right)^5 = \left(\frac{2a^3}{5}\right)^5 + 5\left(\frac{2a^3}{5}\right)^4\left(\frac{5b}{2}\right) + 10\left(\frac{2a^3}{5}\right)^3\left(\frac{5b}{2}\right)^2 + 10\left(\frac{2a^3}{5}\right)^2\left(\frac{5b}{2}\right)^3 + 5\left(\frac{2a^3}{5}\right)\left(\frac{5b}{2}\right)^4 + \left(\frac{5b}{2}\right)^5 = \frac{32a^{15}}{3125} + \frac{8a^{12}b}{25} + 4a^9b^2 + 25a^6b^3 + \frac{625a^3b^4}{8} + \frac{3125b^5}{32}$

37. $(2x + y)^{15} = (2x)^{15} + 15(2x)^{14}y + \binom{15}{2}(2x)^{13}y^2 + \binom{15}{3}(2x)^{12}y^3 + \cdots = 32{,}768x^{15} + 245{,}760x^{14}y + 860{,}160x^{13}y^2 + 1{,}863{,}680x^{12}y^3 + \cdots$

38. $(1 + ax^2)^{10} = 1^{10} + 10 \cdot 1^9(ax^2) + \binom{10}{2}1^8(ax^2)^2 + \binom{10}{3}1^7(ax^2)^3 + \cdots = 1 + 10ax^2 + 45a^2x^4 + 120a^3x^6 + \cdots$

39. $(1 - x)^{-5} = 1 - 5(-x) + \frac{(-5)(-6)}{2!}x^2 + \frac{(-5)(-6)(-7)}{3!}(-x)^3 + \cdots = 1 + 5x + 15x^2 + 35x^3 + \cdots$

40. $(1+b)^{-\frac{1}{4}} = 1 - \frac{1}{4}b + \frac{(-\frac{1}{4})(-\frac{5}{4})}{2!}b^2 + \frac{(-\frac{1}{4})(-\frac{5}{4})(-\frac{9}{4})}{3!}b^3 + \cdots = 1 - \frac{b}{4} + \frac{5b^2}{32} - \frac{15b^3}{128} + \cdots$

41. $\binom{20}{6}x^{14}(2y)^6 = 2{,}480{,}640x^{14}y^6$

42. $\binom{12}{4}a^8(-3b)^4 = 40{,}095a^8b^4$

43. $\sqrt[5]{1.02} = (1+0.02)^{\frac{1}{5}} = 1 + \frac{1}{5}(0.02) + \frac{\frac{1}{5}(-\frac{4}{5})}{2!}(0.02)^2 = 1 + 0.004 - 0.000032 \approx 1.004$

44. $\sqrt[7]{0.98} = (1 - 0.02)^{\frac{1}{7}} = 1 - \frac{1}{7}(0.02) \approx 1 - 0.003 = 0.997$

45. $\frac{1}{2}\%$ per month $= 0.005$; 10 years $= 120$ months; $A = 400(1.005)^{120} = \$727.76$

46. Geometric sequence: $a_1 = 20$; $r = \frac{3}{4}$; $a_6 = 20 \cdot (\frac{3}{4})^5 = 4.75$ cm

47. $S_6 = 20 \cdot \frac{1-(\frac{3}{4})^6}{1-\frac{3}{4}} = 65.76$ cm

48. $S = \frac{20}{1-\frac{3}{4}} = \frac{20}{\frac{1}{4}} = 80$cm

49. First convert 250 ft to inches. Since $250 \times 12 = 3000$, we see that 250 ft = 3000 in. $S_n = \frac{n}{2}(a_1 + a_n)$; $a_n = a_1 + (n-1)d$ so $a_n = 10 + (n-1)15 = -5 + 15n$. Substituting, we get $S_n = \frac{n}{2}(10 - 5 + 15n) = \frac{n}{2}(5 + 15n) = \frac{5n + 15n^2}{2} = 3000$. Hence, $15n^2 + 5n - 6000 = 0$. Dividing by 5, we have $3n^2 + n - 1200 = 0$. The quadratic formula gives answers of -20.167 and 19.834. Since n must be positive, the answer is 19.834 s

50. $a_{10} = 250{,}000 + 9(40{,}000) = \$610{,}000$. $S_{10} = \frac{10}{2}(250{,}000 + 610{,}000) = \$4{,}300{,}000$

CHAPTER 19 TEST

1. $\frac{5}{-1} = -5$; $\frac{5}{1} = 5$; $\frac{5}{3}$; $\frac{5}{5} = 1$; $\frac{5}{7}$; $\frac{5}{9}$

2. $a_1 = -2$; $a_2 = 3 + 2(-2) = -1$; $a_3 = 3 + 3(-1) = 0$; $a_4 = 3 + 4 \cdot 0 = 3$; $a_5 = 3 + 5 \times 3 = 18$; $a_6 = 3 + 6 \times 18 = 111$

3. Arithmetic: $d = -2.5$; $a_{10} = 15 + 9(-2.5) = -7.5$

4. $d = 14 - 10 = 4$; $a_5 = a_1 + 4d$; $10 = a_1 + 4 \cdot 4$; $a_1 = -6$

5. $\sum_{k=1}^{5}(2k - 1) = 1 + 3 + 5 + 7 + 9 = 25$

6. Converges: $r = -\frac{1}{2}$; $S = \frac{\frac{1}{5}}{1+\frac{1}{2}} = \frac{\frac{1}{5}}{\frac{3}{2}} = \frac{2}{15}$

7. $a_1 = 0.435$; $r = 0.001$; $S = \frac{0.435}{0.999} = \frac{435}{999} = \frac{145}{333}$

8. $(3x - 2y)^{12} = (3x)^{12} - 12(3x)^{11}(2y) + \binom{12}{2}(3x)^{10}(2y)^2 - \binom{12}{3}(3x)^9(2y)^3 + \cdots = 531{,}441x^{12} - 4{,}251{,}528x^{11}y + 15{,}588{,}936x^{10}y^2 - 34{,}642{,}080x^9y^3$

9. Arithmetic Series: $a_1 = 2000$; $d = 1500$; $a_{12} = 2000 + 11(1500) = 18{,}500$; $S_{12} = \frac{12}{2}(2000 + 18{,}500) = 123{,}000$ copies

CHAPTER

20 Trigonometric Formulas, Identities and Equations

20.1 BASIC IDENTITIES

1. $\sin^2\theta + \cos^2\theta = 1$; Dividing by $\sin^2\theta$ produces $\dfrac{\sin^2\theta}{\sin^2\theta} + \dfrac{\cos^2\theta}{\sin^2\theta} = \dfrac{1}{\sin^2\theta}$ or $1 + \cot^2\theta = \csc^2\theta$

2. $$\begin{array}{r|l} \multicolumn{2}{c}{\tan x \cot x = 1} \\ \hline \dfrac{\sin x}{\cos x} \cdot \dfrac{\cos x}{\sin x} & 1 \\ 1 & \end{array}$$

3. $$\begin{array}{r|l} \multicolumn{2}{c}{\sin\theta \sec\theta = \tan\theta} \\ \hline \sin\theta \dfrac{1}{\cos\theta} & \tan\theta \\ \dfrac{\sin\theta}{\cos\theta} & \\ \tan\theta & \end{array}$$

4. $$\begin{array}{r|l} \multicolumn{2}{c}{\cos\theta(\tan\theta + \sec\theta) = \sin\theta + 1} \\ \hline \cos\theta\tan\theta + \cos\theta\sec\theta & \sin\theta + 1 \\ \cos\theta\dfrac{\sin\theta}{\cos\theta} + \cos\theta \cdot \dfrac{1}{\cos\theta} & \\ \sin\theta + 1 & \end{array}$$

5. $$\begin{array}{r|l} \multicolumn{2}{c}{\dfrac{\sin\theta}{\cot\theta} = \sec\theta - \cos\theta} \\ \hline \dfrac{\sin\theta}{\frac{\cos\theta}{\sin\theta}} & \dfrac{1}{\cos\theta} - \cos\theta \\ \dfrac{\sin^2\theta}{\cos\theta} & \\ \dfrac{1 - \cos^2\theta}{\cos\theta} & \\ \dfrac{1}{\cos\theta} - \cos\theta & \end{array}$$

6. $$\begin{array}{r|l} \multicolumn{2}{c}{\tan x = \dfrac{\sec x}{\csc x}} \\ \hline \tan x & \dfrac{\frac{1}{\cos x}}{\frac{1}{\sin x}} \\ & \dfrac{1}{\cos x} \cdot \dfrac{\sin x}{1} \\ & \dfrac{\sin x}{\cos x} \\ & \tan x \end{array}$$

7.
$$\begin{array}{r|l}
\multicolumn{2}{r}{(1-\sin^2\theta)(1+\tan^2\theta) = 1} \\ \hline
1+\tan^2\theta-\sin^2\theta-\sin^2\theta\tan^2\theta & 1 \\
1+\dfrac{\sin^2\theta}{\cos^2\theta}-\sin^2\theta-\sin^2\theta\cdot\dfrac{\sin^2\theta}{\cos^2\theta} & \\
\dfrac{\cos^2\theta}{\cos^2\theta}+\dfrac{\sin^2\theta}{\cos^2\theta}-\dfrac{\sin^2\theta\cos^2\theta}{\cos^2\theta}-\dfrac{\sin^2\theta\sin^2\theta}{\cos^2\theta} & \\
\dfrac{\cos^2\theta+\sin^2\theta-\sin^2\theta(\cos^2\theta+\sin^2\theta)}{\cos^2\theta} & \\
\dfrac{1-\sin^2\theta}{\cos^2\theta} & \\
\dfrac{\cos^2\theta}{\cos^2\theta} & \\
1 &
\end{array}$$

8.
$$\begin{array}{r|l}
\multicolumn{2}{l}{\dfrac{\sin A}{\csc A}+\dfrac{\cos A}{\sec A} = 1} \\ \hline
\dfrac{\sin A}{\frac{1}{\sin A}}+\dfrac{\cos A}{\frac{1}{\cos A}} & 1 \\
\sin^2 A+\cos^2 A & \\
1 &
\end{array}$$

9.
$$\begin{array}{r|l}
\multicolumn{2}{l}{1-\dfrac{\sin A}{\csc A} = \cos^2 A} \\ \hline
1-\dfrac{\sin A}{1/\sin A} & \cos^2 A \\
1-\sin^2 A & \\
\cos^2 A &
\end{array}$$

10.
$$\begin{array}{r|l}
\multicolumn{2}{l}{(1+\tan\theta)(1-\tan\theta) = 2-\sec^2\theta} \\ \hline
1-\tan^2\theta & 2-\sec^2\theta \\
1-(\sec^2\theta-1) & \\
2-\sec^2\theta &
\end{array}$$

11.
$$\begin{array}{r|l}
\multicolumn{2}{l}{(1+\cos x)(1-\cos x) = \sin^2 x} \\ \hline
1-\cos^2 x & \sin^2 x \\
\sin^2 x &
\end{array}$$

12.
$$\begin{array}{r|l}
\multicolumn{2}{c}{\sec^4 x-\sec^2 x = \tan^4 x+\tan^2 x} \\ \hline
\sec^2 x(\sec^2 x-1) & \tan^4 x+\tan^2 x \\
\sec^2 x(\tan^2 x) & \\
(\tan^2 x+1)(\tan^2 x) & \\
\tan^4 x+\tan^2 x &
\end{array}$$

13.
$$\begin{array}{l|l}
\multicolumn{2}{l}{2\csc\theta = \dfrac{\sin\theta}{1+\cos\theta}+\dfrac{1+\cos\theta}{\sin\theta}} \\ \hline
2\csc\theta & \dfrac{\sin^2\theta}{\sin\theta(1+\cos\theta)}+\dfrac{(1+\cos\theta)(1+\cos\theta)}{\sin\theta(1+\cos\theta)} \\
& \dfrac{\sin^2\theta+1+2\cos\theta+\cos^2\theta}{\sin\theta(1+\cos\theta)} \\
& \dfrac{\sin^2\theta+\cos^2\theta+1+2\cos\theta}{\sin\theta(1+\cos\theta)} \\
& \dfrac{2+2\cos\theta}{\sin\theta(1+\cos\theta)} \\
& \dfrac{2(1+\cos\theta)}{\sin\theta(1+\cos\theta)} \\
& \dfrac{2}{\sin\theta} \\
& 2\csc\theta
\end{array}$$

14.
$$\begin{array}{l|l}
\multicolumn{2}{l}{\cos x = \sin x\cot x} \\ \hline
\cos x & \sin x\cdot\dfrac{\cos x}{\sin x} \\
& \cos x
\end{array}$$

15.
$$\begin{array}{r|l}
\multicolumn{2}{r}{(\sin\theta+\cos\theta)^2 = 1+2\sin\theta\cos\theta} \\ \hline
\sin^2\theta+2\sin\theta\cos\theta+\cos^2\theta & 1+2\sin\theta\cos\theta \\
\sin^2\theta+\cos^2\theta+2\sin\theta\cos\theta & \\
1+2\sin\theta\cos\theta &
\end{array}$$

16.
$$\begin{array}{r|l}
\multicolumn{2}{c}{\csc^2 x(1-\cos^2 x) = 1} \\ \hline
\csc^2 x-\csc^2 x\cos^2 x & 1 \\
\dfrac{1}{\sin^2 x}-\dfrac{\cos^2 x}{\sin^2 x} & \\
\dfrac{1-\cos^2 x}{\sin^2 x} & \\
\dfrac{\sin^2 x}{\sin^2 x} & \\
1 &
\end{array}$$

17. $\dfrac{\tan\theta + \cot\theta}{\tan\theta - \cot\theta} = \dfrac{\tan^2\theta + 1}{\tan^2\theta - 1}$

$\dfrac{\frac{\sin\theta}{\cos\theta} + \frac{\cos\theta}{\sin\theta}}{\frac{\sin\theta}{\cos\theta} - \frac{\cos\theta}{\sin\theta}}$	$\dfrac{\frac{\sin^2\theta}{\cos^2\theta} + \frac{\cos^2\theta}{\cos^2\theta}}{\frac{\sin^2\theta}{\cos^2\theta} - \frac{\cos^2\theta}{\cos^2\theta}}$
$\dfrac{\frac{\sin^2\theta+\cos^2\theta}{\sin\theta\cos\theta}}{\frac{\sin^2\theta-\cos^2\theta}{\sin\theta\cos\theta}}$	$\dfrac{\sin^2\theta + \cos^2\theta}{\sin^2\theta - \cos^2\theta}$
$\dfrac{\sin^2\theta + \cos^2\theta}{\sin^2\theta - \cos^2\theta}$	

18. $\dfrac{1 - \sin x}{\cos x} = \dfrac{\cos x}{1 + \sin x}$

$\dfrac{1 - \sin x}{\cos x} \cdot \dfrac{1 + \sin x}{1 + \sin x}$	$\dfrac{\cos x}{1 + \sin x}$
$\dfrac{1 - \sin^2 x}{\cos x(1 + \sin x)}$	
$\dfrac{\cos^2 x}{\cos x(1 + \sin x)}$	
$\dfrac{\cos x}{1 + \sin x}$	

19. $\dfrac{\sec\theta - \csc\theta}{\sec\theta + \csc\theta} = \dfrac{\tan\theta - 1}{\tan\theta + 1}$

$\dfrac{\frac{1}{\cos\theta} - \frac{1}{\sin\theta}}{\frac{1}{\cos\theta} + \frac{1}{\sin\theta}}$	$\dfrac{\tan\theta - 1}{\tan\theta + 1}$
$\dfrac{\left(\frac{1}{\cos\theta} - \frac{1}{\sin\theta}\right)\sin\theta}{\left(\frac{1}{\cos\theta} + \frac{1}{\sin\theta}\right)\sin\theta}$	
$\dfrac{\frac{\sin\theta}{\cos\theta} - \frac{\sin\theta}{\sin\theta}}{\frac{\sin\theta}{\cos\theta} + \frac{\sin\theta}{\sin\theta}}$	
$\dfrac{\tan\theta - 1}{\tan\theta + 1}$	

20. $(\sin A + \cos A)^2 + (\sin A - \cos A)^2 = 2$

$\sin^2 A + 2\sin A\cos A + \cos^2 A$	
$+ \sin^2 A - 2\sin A\cos A + \cos^2 A$	2
$\sin^2 A + \cos^2 A + \sin^2 A + \cos^2 A$	
$1 + 1$	
2	

21. $\tan^2 x\cos^2 x + \cot^2 x\sin^2 x = 1$

$\dfrac{\sin^2 x}{\cos^2 x} \cdot \dfrac{\cos^2 x}{1} + \dfrac{\cos^2 x}{\sin^2 x} \cdot \dfrac{\sin^2 x}{1}$	1
$\sin^2 x + \cos^2 x$	
1	

22. $\tan\theta + \dfrac{\cos\theta}{1 + \sin\theta} = \sec\theta$

$\dfrac{\sin\theta}{\cos\theta} + \dfrac{\cos\theta}{1 + \sin\theta}$	$\dfrac{1}{\cos\theta}$
$\dfrac{\sin\theta(1 + \sin\theta)}{\cos\theta(1 + \sin\theta)} + \dfrac{\cos^2\theta}{\cos\theta(1 + \sin\theta)}$	
$\dfrac{\sin\theta + \sin^2\theta + \cos^2\theta}{\cos\theta(1 + \sin\theta)}$	
$\dfrac{\sin\theta + 1}{\cos\theta(1 + \sin\theta)}$	
$\dfrac{1}{\cos\theta}$	

23. $\sec^4 x - \sec^2 x = \tan^2 x\sec^2 x$

$\sec^2 x(\sec^2 x - 1)$	$\sec^2 x\tan^2 x$
$\sec^2 x\tan^2 x$	

24. $\cos^2 A - \sin^2 A = 2\cos^2 A - 1$

$\cos^2 A - (1 - \cos^2 A)$	$2\cos^2 A - 1$
$\cos^2 A - 1 + \cos^2 A$	
$2\cos^2 A - 1$	

25. $\dfrac{\tan\theta - \sin\theta}{\sin^3\theta} = \dfrac{\sec\theta}{1 + \cos\theta}$

$\dfrac{\frac{\sin\theta}{\cos\theta} - \sin\theta}{\sin\theta \cdot \sin^2\theta}$	$\dfrac{\sec\theta}{1 + \cos\theta}$
$\dfrac{\sin\theta\left(\frac{1}{\cos\theta} - 1\right)}{\sin\theta(1 - \cos^2\theta)}$	
$\dfrac{\frac{1-\cos\theta}{\cos\theta}}{(1 + \cos\theta)(1 - \cos\theta)}$	
$\dfrac{\frac{1}{\cos\theta}}{1 + \cos\theta}$	
$\dfrac{\sec\theta}{1 + \cos\theta}$	

26. $\dfrac{\sin x-\cos x+1}{\sin x+\cos x-1}=\dfrac{\sin x+1}{\cos x}$

$\dfrac{\sin x-\cos x+1}{\sin x+\cos x-1}\cdot\dfrac{\sin x+\cos x+1}{\sin x+\cos x+1}$	$\dfrac{\sin x+1}{\cos x}$
$\dfrac{[(\sin x+1)-\cos x][(\sin x+1)+\cos x]}{[(\sin x+\cos x)-1][(\sin x+\cos x)+1]}$	
$\dfrac{(\sin x+1)^2-\cos^2 x}{(\sin x+\cos x)^2-1}$	
$\dfrac{\sin^2 x+2\sin x+1-\cos^2 x}{\sin^2 x+2\sin x\cos x+\cos^2 x-1}$	
$\dfrac{\sin^2 x+2\sin x+1-(1-\sin^2 x)}{2\sin x\cos x}$	
$\dfrac{2\sin^2 x+2\sin x}{2\sin x\cos x}$	
$\dfrac{2\sin x(\sin x+1)}{2\sin x\cos x}$	
$\dfrac{\sin x+1}{\cos x}$	

27. $\tan^2\theta\csc^2\theta\cot^2\theta\sin^2\theta=1$

$\dfrac{\sin^2\theta}{\cos^2\theta}\cdot\dfrac{1}{\sin^2\theta}\cdot\dfrac{\cos^2\theta}{\sin^2\theta}\cdot\dfrac{\sin^2\theta}{1}$	1
1	

28. $\tan x\sin x+\cos x=\sec x$

$\dfrac{\sin x}{\cos x}\cdot\dfrac{\sin x}{1}+\cos x$	$\dfrac{1}{\cos x}$
$\dfrac{\sin^2 x}{\cos x}+\dfrac{\cos^2 x}{\cos x}$	
$\dfrac{\sin^2 x+\cos^2 x}{\cos x}$	
$\dfrac{1}{\cos x}$	

29. $\dfrac{\sec A+\csc A}{\tan A+\cot A}=\sin A+\cos A$

$\dfrac{\left(\frac{1}{\cos A}+\frac{1}{\sin A}\right)\sin A\cos A}{\left(\frac{\sin A}{\cos A}+\frac{\cos A}{\sin A}\right)\sin A\cos A}$	$\sin A+\cos A$
$\dfrac{\sin A+\cos A}{\sin^2 A+\cos^2 A}$	
$\dfrac{\sin A+\cos A}{1}$	
$\sin A+\cos A$	

30. $\dfrac{\sin^3 x+\cos^3 x}{\sin x+\cos x}=1-\sin x\cos x$

$\dfrac{(\sin x+\cos x)\times(\sin^2 x-\sin x\cos x+\cos^2 x)}{\sin x+\cos x}$	$1-\sin x\cos x$
$\sin^2 x+\cos^2 x-\sin x\cos x$	
$1-\sin x\cos x$	

31. For $2\sin\theta\neq\sin 2\theta$ let $\theta=90°$. $2\sin 90°\neq\sin(2\cdot 90°)$, $2\times 1\neq\sin 180°$ and $2\neq 0$.

32. If $A=60°$, then $\frac{\tan 60°}{2}\neq\tan\left(\frac{60°}{2}\right)$ or $\frac{\sqrt{3}}{2}\neq\tan 30°$ which is equivalent to $\frac{\sqrt{3}}{2}\neq\frac{\sqrt{3}}{3}$

33. If $\theta=\pi$, then we have $\cos\pi^2\neq(\cos\pi)^2$ or $-0.90269\neq(-1)^2$, which is equivalent to $-0.90269\neq 1$.

34. If $x=60°$ and $y=30°$, then $\sin(60°-30°)\neq\sin 60°-\sin 30°$ or $\sin 30°\neq\frac{\sqrt{3}}{2}-\frac{1}{2}$, and so $\frac{1}{2}\neq\frac{\sqrt{3}-1}{2}$.

35. Here we let $x=120°$. Now, $\sin 120°=\frac{\sqrt{3}}{2}$ and $\dfrac{\tan 120°}{\sqrt{1+\tan^2 120°}}=\dfrac{-\sqrt{3}}{\sqrt{1+(-\sqrt{3})^2}}$, which is negative. Hence, we see that $\sin x\neq\dfrac{\tan x}{\sqrt{1+\tan^2 x}}$ when $x=120°$.

36. $\dfrac{(\sin x)(-\sin x)-(\cos x)(\cos x)}{\sin^2 x}=\dfrac{-\sin^2 x-\cos^2 x}{\sin^2 x}=\dfrac{-1}{\sin^2 x}=-\csc^2 x$

37. $-2\cot x\csc^2 x=-2\dfrac{\cos x}{\sin x}\cdot\csc^2 x=-2\cos x\cdot\dfrac{1}{\sin x}\cdot\csc^2 x=-2\cos x\csc x\csc^2 x=-2\cos x\csc^3 x$

38. $\sin^5 x=\sin^4 x\sin x=(\sin^2 x)^2\sin x=(1-\cos^2 x)^2\sin x=(1-2\cos^2 x+\cos^4 x)\sin x$

39. $\frac{(1.2\sin wt - 1.6\cos\omega t)^2 + (1.6\sin\omega t + 1.2\cos\omega t)^2}{2L}$ $= \frac{1.2^2\sin^2\omega t - 2(1.2)(1.6)\sin\omega t\cos\omega t + (1.6)^2\cos^2\omega t}{2L}$ $+ \frac{(1.6)^2\sin^2\omega t + 2(1.6)(1.2)\sin\omega t\cos\omega t + (1.2)^2\cos^2\omega t}{2L}$ $= \frac{(1.2)^2+(1.6)^2}{2L} = \frac{1.44+2.56}{2L} = \frac{4.00}{2L} = \frac{2.0}{L}$

20.2 THE SUM AND DIFFERENCE IDENTITIES

1. $\sin 15° = \sin(45° - 30°) = \sin 45°\cos 30° - \cos 45°\sin 30° = \frac{\sqrt{2}}{2}\cdot\frac{\sqrt{3}}{2} - \frac{\sqrt{2}}{2}\cdot\frac{1}{2} = \frac{\sqrt{6}-\sqrt{2}}{4}$

2. $\cos 75° = \cos(45° + 30°) = \cos 45°\cos 30° - \sin 45°\sin 30° = \frac{\sqrt{2}}{2}\cdot\frac{\sqrt{3}}{2} - \frac{\sqrt{2}}{2}\cdot\frac{1}{2} = \frac{\sqrt{6}-\sqrt{2}}{4}$

3. $\sin 120° = \sin(60° + 60°) = \sin 60°\cos 60° + \cos 60°\sin 60° = \frac{\sqrt{3}}{2}\cdot\frac{1}{2} + \frac{1}{2}\cdot\frac{\sqrt{3}}{2} = \frac{\sqrt{3}}{4} + \frac{\sqrt{3}}{4} = \frac{\sqrt{3}}{2}$

4. $\cos(-15°) = \cos(30° - 45°) = \cos 30°\cos 45° + \sin 30°\sin 45° = \frac{\sqrt{3}}{2}\cdot\frac{\sqrt{2}}{2} + \frac{1}{2}\cdot\frac{\sqrt{2}}{2} = \frac{\sqrt{6}+\sqrt{2}}{4}$

5. $\tan 15° = \tan(45° - 30°) = \frac{\tan 45° - \tan 30°}{1 + \tan 45°\tan 30°} = \frac{1 - \frac{\sqrt{3}}{3}}{1 + 1\cdot\frac{\sqrt{3}}{3}} = \frac{\frac{3-\sqrt{3}}{3}}{\frac{3+\sqrt{3}}{3}} = \frac{3-\sqrt{3}}{3+\sqrt{3}} = \frac{3-\sqrt{3}}{3+\sqrt{3}}\cdot\frac{3-\sqrt{3}}{3-\sqrt{3}} = \frac{9-6\sqrt{3}+3}{9-3} = \frac{12-6\sqrt{3}}{6} = 2-\sqrt{3}$

6. $\tan 135° = \tan(180° - 45°) = \frac{\tan 180° - \tan 45°}{1 + \tan 180°\tan 45°} = \frac{0-1}{1+0} = -1$

7. $\sin 150° = \sin(180° - 30°) = \sin 180°\cos 30° - \cos 180°\sin 30° = 0\cdot\frac{\sqrt{3}}{2} - (-1)\cdot\frac{1}{2} = \frac{1}{2}$

8. $\cos 105° = \cos(60° + 45°) = \cos 60°\cos 45° - \sin 60°\sin 45° = \frac{1}{2}\cdot\frac{\sqrt{2}}{2} - \frac{\sqrt{3}}{2}\cdot\frac{\sqrt{2}}{2} = \frac{\sqrt{2}-\sqrt{6}}{4}$

9. $\sin(x + 90°) = \sin x\cos 90° + \cos x\sin 90° = \sin x\cdot 0 + \cos x\cdot 1 = \cos x$

10. $\cos(x+\pi) = \cos x\cos\pi - \sin x\sin\pi = \cos x\cdot(-1) - \sin x\cdot 0 = -\cos x$

11. $\cos\left(x + \frac{\pi}{2}\right) = \cos x\cos\frac{\pi}{2} - \sin x\sin\frac{\pi}{2} = \cos x\cdot(0) - \sin x\cdot(1) = -\sin x$

12. $\sin\left(\frac{\pi}{2} - x\right) = \sin\frac{\pi}{2}\cos x - \cos\frac{\pi}{2}\sin x = 1\cos x - 0\sin x = \cos x$

13. $\cos(\pi - x) = \cos\pi\cos x + \sin\pi\sin x = -1\cos x + 0\sin x = -\cos x$

14. $\tan\left(x - \frac{\pi}{4}\right) = \frac{\tan x - \tan\frac{\pi}{4}}{1 + \tan x\tan\frac{\pi}{4}} = \frac{\tan x - 1}{1 + \tan x}$

15. $\sin(180° - x) = \sin 180°\cos x - \cos 180°\sin x = 0\cos x - (-1)\sin x = \sin x$

16. $\tan(180° + x) = \frac{\tan 180° + \tan x}{1 - \tan 180°\tan x} = \frac{0 + \tan x}{1 - 0\tan x} = \tan x$

Note: For exercises 17–24, if $\sin\alpha = \frac{3}{4}$, then $\cos\alpha = \sqrt{1 - \frac{9}{16}} = \frac{\sqrt{7}}{4}$ and $\tan\alpha = \frac{3}{\sqrt{7}} = \frac{3\sqrt{7}}{7}$. Similarly, if $\cos\beta = \frac{7}{8}$, then $\sin\beta = \sqrt{1 - \frac{49}{64}} = \frac{\sqrt{15}}{8}$ and $\tan\beta = \frac{\sqrt{15}}{7}$.

17. $\sin(\alpha+\beta) = \sin\alpha\cos\beta + \cos\alpha\sin\beta = \frac{3}{4}\cdot\frac{7}{8} + \frac{\sqrt{7}}{4}\cdot\frac{\sqrt{15}}{8} = \frac{21+\sqrt{105}}{32} \approx 0.97647$

18. $\cos(\alpha+\beta) = \cos\alpha\cos\beta - \sin\alpha\sin\beta = \frac{\sqrt{7}}{4}\cdot\frac{7}{8} - \frac{3}{4}\cdot\frac{\sqrt{15}}{8} = \frac{7\sqrt{7}-3\sqrt{15}}{32} \approx 0.21567$

19. $\tan(\alpha+\beta) = \frac{\sin(\alpha+\beta)}{\cos(\alpha+\beta)} = \frac{21+\sqrt{105}}{7\sqrt{7}-3\sqrt{15}} = 4.52768$ or $\tan(\alpha+\beta) = \frac{\tan\alpha + \tan\beta}{1 - \tan\alpha - \tan\beta} = \frac{\frac{3\sqrt{7}}{7} + \frac{\sqrt{15}}{7}}{1 - \frac{3\sqrt{7}}{7}\cdot\frac{\sqrt{15}}{7}} = \frac{\frac{3\sqrt{7}+\sqrt{15}}{7}}{\frac{49-3\sqrt{105}}{49}} = \frac{7(3\sqrt{7}+\sqrt{15})}{49-3\sqrt{105}} \approx 4.52768$

20. $\sin(\alpha-\beta)=\sin\alpha\cos\beta-\cos\alpha\sin\beta=\frac{3}{4}\cdot\frac{7}{8}-\frac{\sqrt{7}}{4}\cdot\frac{\sqrt{15}}{8}=\frac{21-\sqrt{105}}{32}\approx 0.33603$

21. $\cos(\alpha-\beta)=\cos\alpha\cos\beta+\sin\alpha\sin\beta=\frac{\sqrt{7}}{4}\cdot\frac{7}{8}+\frac{3}{4}\cdot\frac{\sqrt{15}}{8}=\frac{7\sqrt{7}+3\sqrt{15}}{32}\approx 0.94185$

22. $\tan(\alpha-\beta)=\frac{\sin(\alpha-\beta)}{\cos(\alpha-\beta)}=\frac{21-\sqrt{105}}{7\sqrt{7}+3\sqrt{15}}\approx 0.35678$ or $\tan(\alpha-\beta)=\frac{\tan\alpha-\tan\beta}{1+\tan\alpha\tan\beta}=\frac{\frac{3\sqrt{7}}{7}-\frac{\sqrt{15}}{7}}{1+\frac{3\sqrt{7}}{7}\cdot\frac{\sqrt{15}}{7}}=\frac{7(3\sqrt{7}-\sqrt{15})}{49+3\sqrt{105}}\approx 0.35678$

23. Since both $\sin(\alpha+\beta)$ and $\cos(\alpha+\beta)$ are positive $(\alpha-\beta)$ is in Quadrant I.

24. Since both $\sin(\alpha-\beta)$ and $\cos(\alpha-\beta)$ are positive, $(\alpha-\beta)$ is in Quadrant I.

25. $\sin(\alpha+\beta)=\sin\alpha\cos\beta+\cos\alpha\sin\beta=\frac{3}{4}\cdot\frac{-7}{8}+\frac{-\sqrt{7}}{4}\cdot\frac{-\sqrt{15}}{8}=\frac{-21+\sqrt{105}}{32}=\frac{\sqrt{105}-21}{32}\approx -0.33603$

26. $\cos(\alpha+\beta)=\cos\alpha\cos\beta-\sin\alpha\sin\beta=\frac{-\sqrt{7}}{4}\cdot\frac{-7}{8}-\frac{3}{4}\cdot\frac{-\sqrt{15}}{8}=\frac{7\sqrt{7}+3\sqrt{15}}{32}\approx 0.94185$

27. $\tan(\alpha+\beta)=\dfrac{\sin(\alpha+\beta)}{\cos(\alpha+\beta)}=\dfrac{\sqrt{105}-21}{7\sqrt{7}+3\sqrt{15}}\approx -0.35678$ or $\tan(\alpha+\beta)=\frac{\tan\alpha+\tan\beta}{1-\tan\alpha\tan\beta}=\frac{\frac{-3\sqrt{7}}{7}+\frac{\sqrt{15}}{7}}{1-\frac{-3\sqrt{7}}{7}\cdot\frac{\sqrt{15}}{7}}=\frac{7(\sqrt{15}-3\sqrt{7})}{49+3\sqrt{105}}\approx -0.35678$

28. $\sin(\alpha-\beta)=\sin\alpha\cos\beta-\cos\alpha\sin\beta=\frac{3}{4}\cdot\frac{-7}{8}-\frac{-\sqrt{7}}{4}\cdot\frac{-\sqrt{15}}{8}=\frac{-21-\sqrt{105}}{32}\approx -0.97647$

29. $\cos(\alpha-\beta)=\cos\alpha\cos\beta+\sin\alpha\sin\beta=\frac{-7}{8}\cdot\frac{-\sqrt{7}}{4}+\frac{3}{4}\cdot\frac{-\sqrt{15}}{8}=\frac{7\sqrt{7}-3\sqrt{15}}{32}\approx 0.21567$

30. $\tan(\alpha-\beta)=\frac{\sin(\alpha-\beta)}{\cos(\alpha-\beta)}=\frac{-(21+\sqrt{105})}{7\sqrt{7}-3\sqrt{15}}\approx -4.52768$ or $\tan(\alpha-\beta)=\frac{\tan\alpha-\tan\beta}{1+\tan\alpha\tan\beta}=\frac{\frac{-3\sqrt{7}}{7}-\frac{\sqrt{15}}{7}}{1+\frac{-3\sqrt{7}}{7}\cdot\frac{\sqrt{15}}{7}}=\frac{-7(3\sqrt{7}+\sqrt{15})}{49-3\sqrt{105}}\approx -4.52768$

31. Since $\sin(\alpha+\beta)$ is negative and $\cos(\alpha+\beta)$ is positive, $(\alpha+\beta)$ is in Quadrant IV.

32. Since $\sin(\alpha-\beta)$ is negative and $\cos(\alpha-\beta)$ is positive, $(\alpha-\beta)$ is in Quadrant IV.

33. $\sin 47^\circ\cos 13^\circ+\cos 47^\circ\sin 13^\circ=\sin(47^\circ+13^\circ)=\sin 60^\circ=\frac{\sqrt{3}}{2}$

34. $\sin 47^\circ\sin 13^\circ+\cos 47^\circ\cos 13^\circ=\cos(47^\circ-13^\circ)=\cos 34^\circ$

35. $\cos 32^\circ\cos 12^\circ-\sin 32^\circ\sin 12^\circ=\cos(32^\circ+12^\circ)=\cos 44^\circ$

36. $\frac{\tan 40^\circ+\tan 15^\circ}{1-\tan 40^\circ\tan 15^\circ}=\tan(40^\circ+15^\circ)=\tan 55^\circ$

37. $\cos(\alpha+\beta)\cos\beta+\sin(\alpha+\beta)\sin\beta=\cos(\alpha+\beta-\beta)=\cos\alpha$

38. $\sin(x-y)\cos y+\cos(x-y)\sin y=\sin(x-y+y)=\sin x$

39. $\cos(x+y)\cos(x-y)-\sin(x+y)\sin(x-y)=\cos(x+y+x-y)=\cos 2x=\cos x\cos x-\sin x\sin x=\cos^2 x-\sin^2 x=\cos 2x$

40. $\sin A\cos(-B)+\cos A\sin(-B)=\sin(A-B)$

41. $\sin(x+y)\sin(x-y)=\sin^2 x-\sin^2 y$

$(\sin x\cos y+\cos x\sin y)$ $\cdot(\sin x\cos y-\cos x\sin y)$	$\sin^2 x-\sin^2 y$
$\sin^2 x\cos^2 y-\cos^2 x\sin^2 y$	
$\sin^2 x(1-\sin^2 y)$ $-(1-\sin^2 x)\sin^2 y$	
$\sin^2 x-\sin^2 x\sin^2 y$ $-\sin^2 y+\sin^2 y\sin^2 x$	
$\sin^2 x-\sin^2 y$	

42. $(\sin A\cos B-\cos A\sin B)^2+(\cos A\cos B+\sin A\sin B)^2=1$

$[\sin(A-B)]^2+[\cos(A-B)]^2$	1
$\sin^2(A-B)+\cos^2(A-B)$	
1	

43. $\cos\theta=\sin(\theta+30^\circ)+\cos(\theta+60^\circ)$

$\cos\theta$	$\sin\theta\cos 30^\circ+\cos\theta\sin 30^\circ$ $+\cos\theta\cos 60^\circ-\sin\theta\sin 60^\circ$
	$\frac{\sqrt{3}}{2}\sin\theta+\frac{1}{2}\cos\theta+\frac{1}{2}\cos\theta-\frac{\sqrt{3}}{2}\sin\theta$
	$\cos\theta$

44. $$\frac{\sin(x+y)}{\cos(x-y)} = \frac{\tan x + \tan y}{1 + \tan x \tan y}$$

$\dfrac{\sin x \cos y + \cos x \sin y}{\cos x \cos y + \sin x \sin y}$	$\dfrac{\tan x + \tan y}{1 + \tan x \tan y}$
$\dfrac{\frac{\sin x \cos y}{\cos x \cos y} + \frac{\cos x \sin y}{\cos x \cos y}}{\frac{\cos x \cos y}{\cos x \cos y} + \frac{\sin x \sin y}{\cos x \cos y}}$	
$\dfrac{\tan x + \tan y}{1 + \tan x \tan y}$	

45. $$\tan x - \tan y = \frac{\sin(x-y)}{\cos x \cos y}$$

$\tan x - \tan y$	$\dfrac{\sin x \cos y - \cos x \sin y}{\cos x \cos y}$
	$\dfrac{\sin x \cos y}{\cos x \cos y} - \dfrac{\cos x \sin y}{\cos x \cos y}$
	$\tan x - \tan y$

46. $$\cos(A+B)\cos(A-B) = 1 - \sin^2 A - \sin^2 B$$

$(\cos A \cos B - \sin A \sin B)$ $\cdot (\cos A \cos B + \sin A \sin B)$	$1 - \sin^2 A - \sin^2 B$
$\cos^2 A \cos^2 B - \sin^2 A \sin^2 B$	
$(1 - \sin^2 A)(1 - \sin^2 B)$ $- \sin^2 A \sin^2 B$	
$1 - \sin^2 A - \sin^2 B$ $+ \sin^2 A \sin^2 B - \sin^2 A \sin^2 B$	
$1 - \sin^2 A - \sin^2 B$	

47. Both sides equal 0.8386705679

48. Both sides equal 0.2079116908

49. Both sides equal 14.10141995

50. Both sides equal 0.932039086

51. $\sin(55° + 37°) = 0.99939$;
$\sin 55° + \sin 37° = 1.420967$; not equal

52. $\cos(68° - 24°) = 0.7193398$;
$\cos 68° - \cos 24° = -0.538939$; not equal

53. $\cos(40° + 35°) = 0.258819$;
$\cos 40° + \cos 35° = 1.585196$; not equal

54. $\tan(76° - 37°) = 0.809784$;
$\tan 76° - \tan 37° = 3.257227$; not equal

55. $[r_1(\cos\theta_1 + j\sin\theta_1)][r_2(\cos\theta_2 + j\sin\theta_2)] =$
$r_1r_2[\cos\theta_1\cos\theta_2 + j\cos\theta_2\sin\theta_1 + j\sin\theta_2\cos\theta_1 +$
$j^2\sin\theta_1\sin\theta_2] = r_1r_2[(\cos\theta_1\cos\theta_2 -$
$\sin\theta_1\sin\theta_2) + j(\sin\theta_1\cos\theta_2 + \cos\theta_1\sin\theta_2)] =$
$r_1r_2[\cos(\theta_1+\theta_2) + j\sin(\theta_1+\theta_2)]$

56. $y_1 + y_2 = A_1\cos(\omega t + \pi) + A_2\cos(\omega t - \pi) =$
$A_1[\cos\omega t\cos\pi - \sin\omega t\sin\pi] +$
$A_2[\cos\omega t\cos\pi + \sin\omega t\sin\pi] =$
$A_1[-\cos\omega t] + A_2[-\cos\omega t] = -[A_1 + A_2]\cos\omega t$

57. $d = \frac{h}{\cos\theta_r}\cdot\sin(\theta_i - \theta_r) =$
$\frac{h}{\cos\theta_r}[\sin\theta_i\cos\theta_r - \cos\theta_i\sin\theta_r] =$
$h\left[\sin\theta_i\cos\theta_r\cdot\frac{1}{\cos\theta_r} - \cos\theta_i\sin\theta_r\cdot\frac{1}{\cos\theta_r}\right] =$
$h[\sin\theta_i - \cos\theta_i\tan\theta_r]$

58. $i(t) = I_p\sin(\theta - \omega t) =$
$14.8[\sin\theta\cos\omega t - \cos\theta\sin\omega t] =$
$14.8[\sin 45°\cos\omega t - \cos 45°\sin\omega t] =$
$14.8\left[\frac{\sqrt{2}}{2}\cos\omega t - \frac{\sqrt{2}}{2}\sin\omega t\right] =$
$7.4\sqrt{2}[\cos\omega t - \sin\omega t]$

20.3 THE DOUBLE- AND HALF-ANGLE IDENTITIES

1. $\cos 15° = \cos\frac{30°}{2} = \sqrt{\frac{1+\cos 30°}{2}} = \sqrt{\frac{1+\sqrt{3}/2}{2}} =$
$\sqrt{\frac{2+\sqrt{3}}{4}} \approx 0.96593$

2. $\sin 75° = \sin\frac{150°}{2} = \sqrt{\frac{1-\cos 150°}{2}} = \sqrt{\frac{1+\sqrt{3}/2}{2}} =$
$\sqrt{\frac{2+\sqrt{3}}{4}} \approx 0.96593$

3. $\sin 15° = \sin\frac{30°}{2} = \sqrt{\frac{1-\cos 30°}{2}} = \sqrt{\frac{1-\sqrt{3}/2}{2}} =$
$\sqrt{\frac{2-\sqrt{3}}{4}} \approx 0.25882$

4. $\cos 105° = \cos\frac{210°}{2} = -\sqrt{\frac{1+\cos 210°}{2}} =$
$-\sqrt{\frac{1-\sqrt{3}/2}{2}} = -\sqrt{\frac{2-\sqrt{3}}{4}} \approx -0.25882$

5. $\sin 105^\circ = \sin \frac{210^\circ}{2} = \sqrt{\frac{1-\cos 210^\circ}{2}} = \sqrt{\frac{1+\sqrt{3}/2}{2}} = \sqrt{\frac{2+\sqrt{3}}{4}} \approx 0.96593$

6. Using a double-angle formula and the value of $\cos 105^\circ$ from exercise 4, we obtain $\cos 210^\circ = 2(\cos 105^\circ)^2 - 1 = 2\left(\sqrt{\frac{2-\sqrt{3}}{4}}\right)^2 - 1 = \frac{2-\sqrt{3}}{2} - 1 = \frac{-\sqrt{3}}{2} \approx -0.86603.$

7. $\cos 7\frac{1}{2}^\circ = \cos \frac{15^\circ}{2} = \sqrt{\frac{1+\cos 15^\circ}{2}} = \sqrt{\frac{1+\sqrt{\frac{2+\sqrt{3}}{4}}}{2}} = \sqrt{\frac{2+\sqrt{2+\sqrt{3}}}{4}} \approx 0.99145$

8. $\tan 15^\circ = \tan \frac{30^\circ}{2} = \sqrt{\frac{1-\cos 30^\circ}{1+\cos 30^\circ}} = \sqrt{\frac{1-\sqrt{3}/2}{1+\sqrt{3}/2}} = \sqrt{\frac{2-\sqrt{3}}{2+\sqrt{3}}} \approx 0.26795$

9. $\tan 22\frac{1}{2}^\circ = \tan \frac{45^\circ}{2} = \sqrt{\frac{1-\cos 45^\circ}{1+\cos 45^\circ}} = \sqrt{\frac{1-\sqrt{2}/2}{1+\sqrt{2}/2}} = \sqrt{\frac{2-\sqrt{2}}{2+\sqrt{2}}} \approx 0.41421$

10. $\cos 67\frac{1}{2}^\circ = \cos \frac{135^\circ}{2} = \sqrt{\frac{1+\cos 135^\circ}{2}} = \sqrt{\frac{1+\frac{-\sqrt{2}}{2}}{2}} = \frac{\sqrt{2-\sqrt{2}}}{2} \approx 0.38268$

11. $\cos 75^\circ = \cos \frac{150^\circ}{2} = \sqrt{\frac{1+\cos 150^\circ}{2}} = \sqrt{\frac{1-\frac{\sqrt{3}}{2}}{2}} = \frac{\sqrt{2-\sqrt{3}}}{2} \approx 0.25882$

12. $\sin 37\frac{1}{2}^\circ = \sin \frac{75^\circ}{2} = \sqrt{\frac{1-\cos 75^\circ}{2}} = \sqrt{\frac{1-\frac{\sqrt{2-\sqrt{3}}}{2}}{2}} = \frac{\sqrt{2-\sqrt{2-\sqrt{3}}}}{2} \approx 0.60876$

13. $\sin 127\frac{1}{2}^\circ = \sin(105^\circ + 22.5^\circ)$. From exercise 5, we have $\sin 105^\circ = \frac{\sqrt{2+\sqrt{3}}}{2}$, and from exercise 4, $\cos 105^\circ = \frac{-\sqrt{2-\sqrt{3}}}{2}$. We determine that $\sin 22\frac{1}{2}^\circ = \sin \frac{45^\circ}{2} = \sqrt{\frac{1-\frac{\sqrt{2}}{2}}{2}} = \frac{\sqrt{2-\sqrt{2}}}{2}$ and that $\cos 22\frac{1}{2}^\circ = \cos \frac{45^\circ}{2} = \sqrt{\frac{1+\frac{\sqrt{2}}{2}}{2}} = \frac{\sqrt{2+\sqrt{2}}}{2}$. Using these values, we find the desired result as follows: $\sin(105^\circ + 22.5^\circ) = \sin 105^\circ \cos 22.5^\circ + \cos 105^\circ \sin 22.5^\circ = \left(\frac{\sqrt{2+\sqrt{3}}}{2}\right)\left(\frac{\sqrt{2+\sqrt{2}}}{2}\right) + \left(\frac{-\sqrt{2-\sqrt{3}}}{2}\right)\left(\frac{\sqrt{2-\sqrt{2}}}{2}\right)$. This "simplifies" as $\frac{1}{4}\left[\sqrt{4+2\sqrt{3}+2\sqrt{2}+\sqrt{6}} - \sqrt{4-2\sqrt{2}-2\sqrt{3}+\sqrt{6}}\right] \approx 0.79335$

14. $\tan -15^\circ = \tan \frac{-30^\circ}{2} = \frac{1-\cos(-30^\circ)}{\sin^- 30^\circ} = \frac{1-\frac{\sqrt{3}}{2}}{-\frac{1}{2}} = \sqrt{3} - 2 \approx -0.26795$

15. $\sin x = \frac{7}{25}$; $\cos x = -\sqrt{1-\left(\frac{7}{25}\right)^2} = -\frac{24}{25}$; $\sin 2x = 2\sin x \cos x = 2 \cdot \frac{7}{25} \cdot \frac{-24}{25} = \frac{-336}{625}$; $\cos 2x = \cos^2 x - \sin^2 x = \left(\frac{24}{25}\right)^2 - \left(\frac{7}{25}\right)^2 = \frac{527}{625}$; $\tan 2x = \frac{-336}{527}$; $\sin \frac{x}{2} = \sqrt{\frac{1-\cos x}{2}} = \sqrt{\frac{1+\frac{24}{25}}{2}} = \sqrt{\frac{49}{50}} = \frac{7\sqrt{2}}{10}$; $\cos \frac{x}{2} = \sqrt{\frac{1+\cos x}{2}} = \sqrt{\frac{1-\frac{24}{25}}{2}} = \sqrt{\frac{1}{50}} = \frac{\sqrt{2}}{10}$; $\tan \frac{x}{2} = \frac{7\sqrt{2}}{\sqrt{2}} = 7$

16. $\cos x = \frac{8}{17}$; $\sin x = -\sqrt{1-\left(\frac{8}{17}\right)^2} = -\frac{15}{17}$; $\sin 2x = 2\sin x \cos x = 2\frac{8}{27} \cdot \frac{-15}{17} = \frac{-240}{459}$; $\cos 2x = \cos^2 x - \sin^2 x = \left(\frac{64}{289}\right) - \left(\frac{225}{289}\right) = \frac{-161}{289}$; $\tan 2x = \frac{240}{161}$; $\sin \frac{x}{2} = \sqrt{\frac{1-\cos x}{2}} = \sqrt{\frac{1-\frac{8}{17}}{2}} = \sqrt{\frac{9}{34}} = \frac{3}{\sqrt{34}}$; $\cos \frac{x}{2} = -\sqrt{\frac{1+\cos x}{2}} = -\sqrt{\frac{1+\frac{8}{17}}{2}} = \frac{-5}{\sqrt{34}}$; $\tan \frac{x}{2} = -\frac{3}{5}$

17. $\sec x = \frac{29}{20}$; $\cos x = \frac{20}{29}$; $\sin x = \frac{21}{29}$; $\sin 2x = 2\left(\frac{21}{29}\right)\left(\frac{20}{29}\right) = \frac{840}{841}$; $\cos 2x = \cos^2 x - \sin^2 x = \frac{400-441}{841} = \frac{-41}{841}$; $\tan 2x = -\frac{840}{41}$; $\sin \frac{x}{2} = \sqrt{\frac{1-\frac{20}{29}}{2}} = \sqrt{\frac{9}{58}} = \frac{3}{\sqrt{58}}$; $\cos \frac{x}{2} = \sqrt{\frac{1+\frac{20}{29}}{2}} = \sqrt{\frac{49}{58}} = \frac{7}{\sqrt{58}}$; $\tan \frac{x}{2} = \frac{3}{7}$

18. $\csc x = \frac{-41}{9}$; $\sin x = -\frac{9}{41}$; $\cos x = -\frac{40}{41}$; $\sin 2x = 2\left(\frac{-9}{41}\right)\left(\frac{-40}{41}\right) = \frac{720}{1681}$; $\cos 2x = 1 - 2\sin^2 x = 1 - 2\left(\frac{9}{41}\right)^2 = \frac{1519}{1681}$; $\tan 2x = \frac{720}{1519}$; $\sin \frac{x}{2} = \sqrt{\frac{1+\frac{40}{41}}{2}} = \sqrt{\frac{81}{82}} = \frac{9}{\sqrt{82}}$; $\cos \frac{x}{2} = -\sqrt{\frac{1-\frac{40}{41}}{2}} = -\sqrt{\frac{1}{82}} = -\frac{1}{\sqrt{82}}$; $\tan \frac{x}{2} = -9$

19. $\tan x = \frac{35}{12}$; $\sqrt{u^2+v^2} = \sqrt{35^2+12^2} = 37$; $\sin x = -\frac{35}{37}$; $\cos x = -\frac{12}{37}$; $\sin 2x = 2\left(\frac{-35}{37}\right)\left(-\frac{12}{37}\right) = \frac{840}{1369}$;

$\cos 2x = 1 - 2\sin^2 x = 1 - 2\left(\frac{-35}{37}\right)^2 = \frac{-1081}{1369}$; $\tan 2x = -\frac{840}{1081}$; $\sin\frac{x}{2} = \sqrt{\frac{1+\frac{12}{37}}{2}} = \sqrt{\frac{49}{74}} = \frac{7}{\sqrt{74}}$; $\cos\frac{x}{2} = -\sqrt{\frac{1-\frac{12}{37}}{2}} = -\sqrt{\frac{25}{74}} = \frac{-5}{\sqrt{74}}$; $\tan\frac{x}{2} = -\frac{7}{5}$

20. $\cot x = \frac{-45}{28}$; $\sqrt{u^2+v^2} = 53$; $\sin x = \frac{28}{53}$; $\cos x = \frac{-45}{53}$; $\sin 2x = 2\left(\frac{28}{53}\right)\left(\frac{-45}{53}\right) = \frac{-2520}{2809}$; $\cos 2x = 2\cos^2 x - 1 = 2\left(\frac{-45}{53}\right)^2 - 1 = \frac{1241}{2809}$; $\tan 2x = \frac{-2520}{1241}$; $\sin\frac{x}{2} = \sqrt{\frac{1+\frac{45}{53}}{2}} = \sqrt{\frac{98}{106}} = \sqrt{\frac{49}{53}} = \frac{7}{\sqrt{53}}$; $\cos\frac{x}{2} = \sqrt{\frac{1-\frac{45}{53}}{2}} = \sqrt{\frac{8}{106}} = \sqrt{\frac{4}{53}} = \frac{2}{\sqrt{53}}$; $\tan\frac{x}{2} = \frac{7}{2}$

21. $\cos^2 x = \sin^2 x + \cos 2x$

$$\begin{array}{l|l}
\cos^2 x & \sin^2 x + 2\cos^2 x - 1 \\
 & \sin^2 x + \cos^2 x + \cos^2 x - 1 \\
 & 1 + \cos^2 x - 1 \\
 & \cos^2 x
\end{array}$$

22. $\cos^2 4x - \sin^2 4x = \cos 8x$

$$\begin{array}{l|l}
\cos^2 4x - \sin^2 4x & \cos(2\cdot 4x) \\
 & \cos^2 4x - \sin^2 4x
\end{array}$$

23. $\cos 4x = 1 - 8\sin^2 x\cos^2 x$

$$\begin{array}{r|l}
\cos(2\cdot 2x) & 1 - 8\sin^2 x\cos^2 x \\
1 - 2\sin^2 2x & \\
1 - 2(2\sin x\cos x)^2 & \\
1 - 2(4\sin^2 x\cos^2 x) & \\
1 - 8\sin^2 x\cos^2 x &
\end{array}$$

24. $\cos 3x = 4\cos^3 x - 3\cos x$

$$\begin{array}{r|l}
\cos(x + 2x) & 4\cos^3 x - 3\cos x \\
\cos x\cos 2x - \sin x\sin 2x & \\
\cos x(2\cos^2 x - 1) \qquad\qquad & \\
-\sin x(2\cos x\sin x) & \\
2\cos^3 x - \cos x - 2\sin^2 x\cos x & \\
2\cos^3 x - \cos x - 2(1-\cos^2 x)\cos x & \\
2\cos^3 x - \cos x - 2\cos x + 2\cos^3 x & \\
4\cos^3 x - 3\cos x &
\end{array}$$

25. $\dfrac{1+\tan^2\alpha}{1-\tan^2\alpha} = \sec 2\alpha$

$$\begin{array}{l|l}
\dfrac{1+\tan^2\alpha}{1-\tan^2\alpha} & \dfrac{1}{\cos 2\alpha} \\
 & \dfrac{1}{\cos^2\alpha - \sin^2\alpha} \\
 & \dfrac{\cos^2\alpha + \sin^2\alpha}{\cos^2\alpha - \sin^2\alpha} \\
 & \dfrac{\frac{\cos^2\alpha}{\cos^2\alpha} + \frac{\sin^2\alpha}{\cos^2\alpha}}{\frac{\cos^2\alpha}{\cos^2\alpha} - \frac{\sin^2\alpha}{\cos^2\alpha}} \\
 & \dfrac{1+\tan^2\alpha}{1-\tan^2\alpha}
\end{array}$$

26. $\sin 2x\cos 2x = \dfrac{1}{2}\sin 4x$

$$\begin{array}{l|l}
\sin 2x\cos 2x & \dfrac{1}{2}\sin(2\cdot 2x) \\
 & \dfrac{1}{2}\cdot 2\sin 2x\cos 2x \\
 & \sin 2x\cos 2x
\end{array}$$

27. $1 - 2\sin^2 3x = \cos 6x$

$$\begin{array}{l|l}
1 - 2\sin^2 3x & \cos(2\cdot 3x) \\
 & 1 - 2\sin^2 3x
\end{array}$$

28. $\dfrac{2\tan 3x}{1-\tan^2 3x} = \tan 6x$

$$\begin{array}{l|l}
\dfrac{2\tan 3x}{1-\tan^2 3x} & \tan(2\cdot 3x) \\
 & \dfrac{2\tan 3x}{1-\tan^2 3x}
\end{array}$$

29. $\sin^2 x\cos^2 x = \dfrac{1}{4}\sin^2 2x$

$$\begin{array}{l|l}
\sin^2 x\cos^2 x & \dfrac{1}{4}[2\sin x\cos x]^2 \\
 & \dfrac{1}{4}(4\sin^2 x\cos^2 x) \\
 & \sin^2 x\cos^2 x
\end{array}$$

30. $$1 + \tan\beta \tan\frac{\beta}{2} = \frac{1}{\cos\beta}$$

$1 + \tan\beta\left(\frac{\sin\beta}{1+\cos\beta}\right)$	$\frac{1}{\cos\beta}$
$1 + \frac{\sin\beta}{\cos\beta}\left(\frac{\sin\beta}{1+\cos\beta}\right)$	
$1 + \frac{\sin^2\beta}{\cos\beta(1+\cos\beta)}$	
$1 + \frac{1-\cos^2\beta}{\cos\beta(1+\cos\beta)}$	
$1 + \frac{(1+\cos\beta)(1-\cos\beta)}{\cos\beta(1+\cos\beta)}$	
$1 + \frac{1-\cos\beta}{\cos\beta}$	
$1 + \frac{1}{\cos\beta} - \frac{\cos\beta}{\cos\beta}$	
$\frac{1}{\cos\beta}$	

31. $$\begin{aligned}\tan\left(\frac{\alpha+\beta}{2}\right)\cot\left(\frac{\alpha-\beta}{2}\right) &= \left(\frac{1-\cos(\alpha+\beta)}{\sin(\alpha+\beta)}\right)\left(\frac{1+\cos(\alpha-\beta)}{\sin(\alpha-\beta)}\right)\\ &= \left(\frac{1-\cos\alpha\cos\beta+\sin\alpha\sin\beta}{\sin\alpha\cos\beta+\cos\alpha\sin\beta}\right)\left(\frac{1+\cos\alpha\cos\beta+\sin\alpha\sin\beta}{\sin\alpha\cos\beta-\cos\alpha\sin\beta}\right)\\ &= \frac{1+2\sin\alpha\sin\beta+\sin^2\alpha\sin^2\beta-\cos^2\alpha\cos^2\beta}{\sin^2\alpha\cos^2\beta-\cos^2\alpha\sin^2\beta}.\end{aligned}$$

In the denominator, let $\cos^2\beta = 1 - \sin^2\beta$ and $\cos^2\alpha = 1 - \sin^2\alpha$ to produce

$$\begin{aligned}&\frac{1+2\sin\alpha\sin\beta+\sin^2\alpha\sin^2\beta-(1-\sin^2\alpha)(1-\sin^2\beta)}{\sin^2\alpha(1-\sin^2\beta)-(1-\sin^2\alpha)\sin^2\beta}\\ &\quad= \frac{1+2\sin\alpha\sin\beta+\sin^2\alpha\sin^2\beta-1+\sin^2\alpha+\sin^2\beta-\sin^2\alpha\sin^2\beta}{\sin^2\alpha-\sin^2\alpha\sin^2\beta-\sin^2\beta+\sin^2\alpha\sin^2\beta}\\ &\quad= \frac{\sin^2\alpha+2\sin\alpha\sin\beta+\sin^2\beta}{\sin^2\alpha-\sin^2\beta} = \frac{(\sin\alpha+\sin\beta)^2}{\sin^2\alpha-\sin^2\beta}\end{aligned}$$

32. $\cot\theta = \dfrac{1}{2}\left(\cot\dfrac{\theta}{2} - \tan\dfrac{\theta}{2}\right)$

$\cot\theta$	$\dfrac{1}{2}\left(\dfrac{1}{\tan\frac{\theta}{2}} - \tan\dfrac{\theta}{2}\right)$
	$\dfrac{1}{2}\left[\dfrac{1+\cos\theta}{\sin\theta} - \dfrac{1-\cos\theta}{\sin\theta}\right]$
	$\dfrac{1}{2}\left[\dfrac{2\cos\theta}{\sin\theta}\right]$
	$\cot\theta$

33. $\cos 2\cdot 45° = \cos 90° = 0$; $2\cos 45° = 2\cdot\frac{\sqrt{2}}{2} = \sqrt{2}$; $0 \neq \sqrt{2}$

34. $\tan\frac{80°}{2} = \tan 40° \approx 0.8391$; $\frac{\tan 80°}{2} \approx 2.8356$; $0.8391 \neq 2.8356$

35. $\cot 2\cdot 150° = \cot 300° \approx -0.5774$; $2\cot 150° \approx -3.4641$; $-0.5774 \neq -3.4641$

36. $\sin\frac{210°}{2} = \sin 105° \approx 0.9659$; $\frac{\sin 210°}{2} = -0.25$; $0.9659 \neq -0.25$

37. $n = \dfrac{\sin\frac{\alpha+\phi}{2}}{\sin\frac{\alpha}{2}} = \dfrac{\sqrt{\frac{1-\cos(\alpha+\phi)}{2}}}{\sqrt{\frac{1-\cos\alpha}{2}}} = \sqrt{\dfrac{1-\cos(\alpha+\phi)}{1-\cos\alpha}} = \sqrt{\dfrac{1-\cos\alpha\cos\phi+\sin\alpha\sin\phi}{1-\cos\alpha}}$

38. Since $\cot\frac{\alpha}{2} = \dfrac{\cos\frac{\alpha}{2}}{\sin\frac{\alpha}{2}}$, we can write $n = \sqrt{\dfrac{1+\cos\phi}{2}} + \left(\cot\dfrac{\alpha}{2}\right)\sqrt{\dfrac{1-\cos\phi}{2}} = \sqrt{\dfrac{1+\cos\phi}{2}} + \left(\dfrac{\cos\frac{\alpha}{2}}{\sin\frac{\alpha}{2}}\right)\sqrt{\dfrac{1-\cos\phi}{2}}$. We know $\sqrt{\dfrac{1+\cos\phi}{2}} = \cos\dfrac{\phi}{2}$ and $\sqrt{\dfrac{1-\cos\phi}{2}} = \sin\dfrac{\phi}{2}$, and so we can rewrite the given expression as $n = \cos\dfrac{\phi}{2} + \left(\dfrac{\cos\frac{\alpha}{2}}{\sin\frac{\alpha}{2}}\right)\sin\dfrac{\phi}{2}$. Writing this with a common denominator of $\sin\frac{\alpha}{2}$, we get $n = \dfrac{\sin\frac{\alpha}{2}\cos\frac{\phi}{2} + \cos\frac{\alpha}{2}\sin\frac{\phi}{2}}{\sin\frac{\alpha}{2}} = \dfrac{\sin\frac{(\alpha+\phi)}{2}}{\sin\frac{\alpha}{2}}$.

39. $P = V_{max}I_{max}\cos\omega t\sin\omega t = V_{max}I_{max}\frac{1}{2}\cdot 2\cos\omega t\sin\omega t = \frac{V_{max}I_{max}}{2}\cdot\sin 2\omega t$

40. $A = \sqrt{e^{-2x}(1+\sin 2x)} = \sqrt{e^{-2x}}\sqrt{1+\sin 2x} = e^{-x}\sqrt{1+\sin 2x} = e^{-x}\sqrt{1+2\sin x\cos x} = e^{-x}\sqrt{\sin^2 x+\cos^2 x+2\sin x\cos x} = e^{-x}\sqrt{(\sin x+\cos x)^2} = e^{-x}(\sin x+\cos x)$

20.4 TRIGONOMETRIC EQUATIONS

1. $2\cos\theta = 0$; $\cos\theta = 0$; $\theta = 90°, 270°$

2. $2\sin\theta = -1$; $\sin\theta = -\frac{1}{2}$; $\theta = 210°, 330°$

3. $\sqrt{3}\tan x = 1$; $\tan x = \frac{1}{\sqrt{3}}$; $x = 30°$; $210°$

4. $\sqrt{3}\sec x = -2$; $\sec x = -\frac{2}{\sqrt{3}}$ or $\cos x = -\frac{\sqrt{3}}{2}$; $x = 150°, 210°$

5. $4\sin\theta = -3$; $\sin\theta = \frac{-3}{4}$; $\theta \approx 228.59°, 311.41°$

6. $2\cos x = 3$; $\cos x = \frac{3}{2}$; no solution

7. $4\tan\alpha = 5$; $\tan\alpha = \frac{5}{4}$; $\alpha = 51.34°, 231.34°$

8. $3\csc x = 1$; $\csc x = \frac{1}{3}$ or $\sin x = 3$; no solution

9. $\cos 2x = -1$; $2x = 180°$ or $540°$; $x = 90°$ or $270°$

10. $\sin 2x = \frac{1}{2}$; $2x = 30°, 150°, 390°, 510°$; $x = 15°, 75°, 195°, 255°$

11. $\tan\dfrac{\theta}{4} = 1$; $\dfrac{\theta}{4} = 45°$; $\theta = 180°$

12. $\cos\frac{\theta}{3} = -1$; $\frac{\theta}{3} = 180°$; $\theta = 540°$ is the smallest positive answer. No solutions between $0°$ and $360°$

13. $\sin^2\alpha = \sin\alpha$; $\sin^2\alpha - \sin\alpha = 0$; $\sin\alpha(\sin\alpha - 1) = 0$, so $\sin\alpha = 0$ or $\sin\alpha = 1$; $\alpha = 0°, 90°, 180°$

14. $\cos^2\beta = \frac{1}{2}\cos\beta$; $\cos^2\beta - \frac{1}{2}\cos\beta = 0$; $\cos\beta(\cos\beta - \frac{1}{2}) = 0$; $\cos\beta = 0$ or $\cos\beta = \frac{1}{2}$; $\beta = 60°, 90°, 270°, 300°$

15. $\sin x \cos x = 0$; $\sin x = 0$ $\cos x = 0$; $x = 0°, 90°, 180°, 270°$

16. $\frac{\sec\theta}{\csc\theta} = -1$; $\frac{\sin\theta}{\cos\theta} = -1$; $\tan\theta = -1$; $\theta = 135°, 315°$

17. $3\tan^2 x = 1$; $\tan^2 x = \frac{1}{3}$; $\tan x = \pm\sqrt{\frac{1}{3}}$; $x = 30°, 150°, 210°, 330°$

18. $\sec^2\theta = 2$; $\sec\theta = \pm\sqrt{2}$; $\cos\theta = \pm\sqrt{\frac{1}{2}}$; $\theta = 45°, 135°, 225°, 315°$

19. $4\sin\alpha\cos\alpha = 1$; $2(2\sin\alpha\cos\alpha) = 2(\sin 2\alpha) = 1$; $\sin 2\alpha = \frac{1}{2}$; $2\alpha = 30°, 150°, 390°, 510°$ and so, $\alpha = 15°, 75°, 195°, 255°$

20. $\sin^2\beta = \frac{1}{2}\sin\beta$; $\sin^2\beta - \frac{1}{2}\sin\beta = 0$; $\sin\beta(\sin\beta - \frac{1}{2}) = 0$; $\sin\beta = 0$ and $\beta = 0°, 180°$ or $\sin\beta = \frac{1}{2}$, and $\beta = 30°, 150°$

21. $\sin\theta - \cos\theta = 0$; $\sin\theta = \cos\theta$; $\frac{\sin\theta}{\cos\theta} = 1$ or $\tan\theta = 1$, and so $\theta = 45°, 225°$

22. $\tan\theta = \csc\theta$; $\frac{\sin\theta}{\cos\theta} = \frac{1}{\sin\theta}$; $\sin^2\theta = \cos\theta$; $1 - \cos^2\theta = \cos\theta$; $\cos^2\theta + \cos\theta - 1 = 0$; $\cos\theta = \frac{-1\pm\sqrt{1+4}}{2} = \frac{-1+\sqrt{5}}{2}$; $\cos\theta \approx 0.618$ or $\cos\theta \approx -1.618$; $\theta = 51.83°, 308.17°$. (Note: $\cos\theta = -1.618$ yields no solutions)

23. $\sin 6\theta + \sin 3\theta = 0$; $2\sin 3\theta\cos 3\theta + \sin 3\theta = 0$; $\sin 3\theta(2\cos 3\theta + 1) = 0$; $\sin 3\theta = 0$ or $\cos 3\theta = -\frac{1}{2}$. If $\sin 3\theta = 0$, then $3\theta = 0°, 180°, 360°, 540°, 720°, 900°$; and if $\cos 3\theta = -\frac{1}{2}$, then $3\theta = 120°, 240°$; $480°$; $600°$; $840°$; $960°$. Thus, we see that $\theta = 0°, 60°, 120°, 180°, 240°, 300°, 40°, 80°, 160°, 200°, 280°, 320°$

24. $4\tan^2 x = 3\sec^2 x$; $4\tan^2 x = 3(\tan^2 x + 1)$; $\tan^2 x = 3$; $\tan x = \pm\sqrt{3}$; $x = 60°, 120°, 240°, 300°$

25. $2\cos^2 x - 3\cos 2x = 1$; $2\cos^2 x - 3(2\cos^2 x - 1) = 1$; $-4\cos^2 x + 3 = 1$; $\cos^2 x = \frac{1}{2}$; $\cos x = \pm\frac{\sqrt{2}}{2}$; $x = 45°, 135°, 225°, 315°$

26. $\sin^2 4\alpha = \sin 4\alpha + 2$; $\sin^2 4\alpha - \sin 4\alpha - 2 = 0$; $(\sin 4\alpha - 2)(\sin 4\alpha + 1) = 0$; $\sin 4\alpha = 2$ or $\sin 4\alpha = -1$; $\sin 4\alpha = 2$ yields no solutions. $\sin 4\alpha = -1$ yields $4\alpha = 270°, 630°, 990°, 1350°$; $\alpha = 67.5°, 157.5°, 247.5°, 337.5°$

27. $\sec^2\theta + \tan\theta = 1$; $(\tan^2\theta + 1) + \tan\theta = 1$; $\tan^2\theta + \tan\theta = 0$; $\tan\theta(\tan\theta + 1) = 0$; $\tan\theta = 0$ or $\tan\theta = -1$; $\theta = 0°, 180°, 135°, 315°$

28. $\tan 2x + \sec 2x = 1$; $\frac{\sin 2x}{\cos 2x} + \frac{1}{\cos 2x} = \frac{\cos 2x}{\cos 2x}$; $\sin 2x + 1 = \cos 2x$; squaring both sides yields $\sin^2 2x + 2\sin 2x + 1 = \cos^2 2x = 1 - \sin^2 2x$; $2\sin^2 2x + 2\sin 2x = 0$; $\sin 2x(\sin 2x + 1) = 0$; and so $\sin 2x = 0$ or $\sin 2x = -1$. If $\sin 2x = 0$, then $2x = 0°, 180°, 360°, 540°$. If $\sin 2x = -1$, then $2x = 270°, 630°$ and $x = 0°, 90°, 180°, 270°, 135°, 315°$. Continue checking for extraneous solutions, the ones that are solutions are $x = 0°$ and $180°$

29. $\sin 2x = \cos x$; $2\sin x\cos x = \cos x$; $2\sin x\cos x - \cos x = 0$; $\cos x(2\sin x - 1) = 0$; $\cos x = 0$ or $\sin x = \frac{1}{2}$; $x = 30°, 90°, 150°, 270°$

30. $\csc^2\theta - \cot\theta = 1$; $\cot^2\theta + 1 - \cot\theta = 1$; $\cot^2\theta - \cot\theta = 0$; $\cot\theta(\cot\theta - 1) = 0$. so $\cot\theta = 0$ or $\cot\theta = 1$. Thus, $\theta = 45°, 90°, 225°, 270°$

31. $\frac{\sin 35°}{\sin\theta_r} = 1.61$; $\sin\theta_r = \frac{\sin 35°}{1.61} = 0.356258$; $\theta_r = 20.87°$

32. $\frac{\sin 27°}{\sin\theta_r} = 1.43$; $\sin\theta_4 = \frac{\sin 27°}{1.43} = 0.317476$; $\theta_r = 18.51°$

33. Since $E = 125\cos(\omega t - \phi)$, we have $60 = 125\cos(120\pi t - \frac{\pi}{2})$ so $\cos(120\pi t - \frac{\pi}{2}) = \frac{60}{125}$. Thus, $120\pi t - \frac{\pi}{2} \approx 1.070\,14$ and so $120\pi t = 2.640\,938$. Hence, $t = \frac{2.640\,938}{120\pi} \approx 0.007\,005$ s.

34. Substituting the given values for a, b and A into $A = a\sin\theta(b + a\cos\theta)$ we get $10 = 3\sin\theta(3.4 + 3\cos\theta)$ or $10 = 10.2\sin\theta + 9\sin\theta\cos\theta$. Since we know that $\sin\theta\cos\theta \approx 0.4675$, we have $10 \approx 10.2\sin\theta + 9(0.4675)$ or $10 \approx 10.2\sin\theta + 4.2075$. Thus, $10.2\sin\theta \approx 10 - 4.2075 = 5.7925$ or $\sin\theta \approx \frac{5.7925}{10.2} \approx 0.5679$ and so $\theta = \sin^{-1} 0.5679 \approx 34.6°$ or $145.4°$. But, $\sin(145.4°)\cos(145.4°) \approx -0.4662$ and so it does not satisfiy the stated conditions. Thus, $\theta \approx 34.6°$.

35. Since $d = \sin\omega t + \frac{1}{2}\sin 2\omega t$ we use a double-angle formula to get $0 = \sin\omega t + \frac{1}{2}\cdot 2\sin\omega t\cos\omega t$ which factors as $0 = \sin\omega t(1 + \cos\omega t)$. Thus, $\sin\omega t = 0$ or $\cos\omega t = -1$ and so, $\omega t = 0, \pi$

36. $\dfrac{a}{b} = \sqrt{\dfrac{\mu^2 - \sin^2\theta_i}{\cos^2\theta_i}}$; $\dfrac{14}{10} = \sqrt{\dfrac{1.333^2 - \sin^2\theta_i}{\cos^2\theta_i}}$; $1.4^2 = \dfrac{1.333^2 - \sin^2\theta_i}{\cos^2\theta_i}$; $1.96\cos^2\theta_i = 1.333^2 - \sin^2\theta_i = 1.7769 - (1 - \cos^2\theta_i)$; $0.96\cos^2\theta_i = 0.7769$; $\cos^2\theta_i = 0.80926$; $\cos\theta_i = 0.8996$; $\theta_i = 25.9°$

CHAPTER 20 REVIEW

1.

$\dfrac{\sin(x+y)}{\cos x\cos y}$	$= \tan x + \tan y$
$\dfrac{\sin x\cos y + \cos x\sin y}{\cos x\cos y}$	$\tan x + \tan y$
$\dfrac{\sin x\cos y}{\cos x\cos y} + \dfrac{\cos x\sin y}{\cos x\cos y}$	
$\dfrac{\sin x}{\cos x} + \dfrac{\sin y}{\cos y}$	
$\tan x + \tan y$	

2.

$(\sin x + \cos x)^2$	$= 1 + \sin 2x$
$\sin^2 x + 2\sin x\cos x + \cos^2 x$	$1 + \sin 2x$
$1 + 2\sin x\cos x$	

3.

$\dfrac{\sin 3x}{\sin x} - \dfrac{\cos 3x}{\cos x}$	$= 2$
$\dfrac{\sin(x+2x)}{\sin x} - \dfrac{\cos(x+2x)}{\cos x}$	2
$\dfrac{\sin x\cos 2x + \cos x\sin 2x}{\sin x} - \dfrac{\cos x\cos 2x - \sin x\sin 2x}{\cos x}$	
$\dfrac{\sin x(\cos^2 x - \sin^2 x) + \cos x(2\sin x\cos x)}{\sin x}$	
$- \dfrac{\cos x(\cos^2 x - \sin^2 x) - \sin x(2\sin x\cos x)}{\cos x}$	
$\cos^2 x - \sin^2 x + 2\cos^2 x - \cos^2 x + \sin^2 x + 2\sin^2 x$	
$2\cos^2 x + 2\sin^2 x$	
$2(\cos^2 x + \sin^2 x)$	
2	

4.

$\cos(\theta+\phi)\cos(\theta-\phi)$	$= \cos^2\phi - \sin^2\theta$
$(\cos\theta\cos\phi - \sin\theta\sin\phi)(\cos\theta\cos\phi + \sin\theta\sin\phi)$	$\cos^2\theta - \sin^2\theta$
$\cos^2\theta\cos^2\phi - \sin^2\theta\sin^2\phi$	
$(1 - \sin^2\theta)\cos^2\phi - \sin^2\theta(1 - \cos^2\phi)$	
$\cos^2\phi - \sin^2\theta\cos^2\phi - \sin^2\theta + \sin^2\theta\cos^2\phi$	
$\cos^2\phi - \sin^2\theta$	

5. $\sin(\alpha+\beta)\cos\beta - \cos(\alpha+\beta)\sin\beta = \sin\alpha$

$(\sin\alpha\cos\beta + \cos\alpha\sin\beta)\cos\beta$ $- (\cos\alpha\cos\beta - \sin\alpha\sin\beta)\sin\beta$	$\sin\alpha$
$\sin\alpha\cos^2\beta + \cos\alpha\sin\beta\cos\beta$ $- \cos\alpha\cos\beta\sin\beta + \sin\alpha\sin^2\beta$	
$\sin\alpha(\cos^2\beta + \sin^2\beta)$	
$\sin\alpha$	

6. $\tan 2x = \dfrac{2\cos x}{\csc x - 2\sin x}$

$\dfrac{2\tan x}{1-\tan^2 x}$	$\dfrac{2\cos x}{\csc x - 2\sin x}$
$\dfrac{2\frac{\sin x}{\cos x}}{1-\frac{\sin^2 x}{\cos^2 x}} \cdot \dfrac{\frac{\cos^2 x}{\sin x}}{\frac{\cos^2 x}{\sin x}}$	
$\dfrac{2\cos x}{\frac{\cos^2 x}{\sin x} - \frac{\sin^2 x}{\sin x}}$	
$\dfrac{2\cos x}{\frac{1}{\sin x}(\cos^2 x - \sin^2 x)}$	
$\dfrac{2\cos x}{\csc x(1-2\sin^2 x)}$	
$\dfrac{2\cos x}{\csc x - 2\sin x}$	

7. $\cos^4 x - \sin^4 x = \cos 2x$

$(\cos^2 x + \sin^2 x)(\cos^2 x - \sin^2 x)$	$\cos 2x$
$1 \cdot \cos 2x$	
$\cos 2x$	

8. $\dfrac{\sin 2x - \sin x}{\cos 2x + \cos x} = \tan\dfrac{x}{2}$

$\dfrac{2\sin x\cos x - \sin x}{2\cos^2 x - 1 + \cos x}$	$\dfrac{\sin x}{\cos x + 1}$
$\dfrac{\sin x(2\cos x - 1)}{(2\cos x - 1)(\cos x + 1)}$	
$\dfrac{\sin x}{\cos x + 1}$	

9. $\sin 3\theta = 2\sin\theta\cos 2\theta + \sin\theta$

$\sin(\theta + 2\theta)$	$2\sin\theta\cos 2\theta + \sin\theta$
$\sin\theta\cos 2\theta + \cos\theta\sin 2\theta$	
$\sin\theta(1-2\sin^2\theta) +$ $\cos\theta(2\sin\theta\cos\theta)$	
$\sin\theta - 2\sin^3\theta + 2\sin\theta\cos^2\theta$	
$\sin\theta + 2\sin\theta(\cos^2\theta - \sin^2\theta)$	
$2\sin\theta\cos 2\theta + \sin\theta$	

10. $\dfrac{\cos 3x - \cos 5x}{\sin 3x + \sin 5x} = \tan x$

$\dfrac{\cos(4x - x) - \cos(4x + x)}{\sin(4x - x) + \sin(4x + x)}$	$\tan x$
$\dfrac{(\cos 4x\cos x + \sin 4x\sin x) - (\cos 4x\cos x - \sin 4x\sin x)}{(\sin 4x\cos x - \cos 4x\sin x) + (\sin 4x\cos x + \cos 4x\sin x)}$	
$\dfrac{2\sin 4x\sin x}{2\sin 4x\cos x}$	
$\dfrac{\sin x}{\cos x}$	
$\tan x$	

11. $2\tan x = -\sqrt{3}$ and so $\tan x = -\frac{\sqrt{3}}{2}$ and $x \approx 139.11°$ or $319.11°$.

12. $3\sin x = -2$; $\sin x = \frac{-2}{3}$; $x = 221.81°, 318.19°$

13. $\cos 2x + \cos x = -1$; $2\cos^2 x - 1 + \cos x = -1$; $2\cos^2 x + \cos x = 0$; $\cos x(2\cos x + 1) = 0$; $\cos x = 0$ or $\cos x = \frac{-1}{2}$; $x = 90°, 120°, 240°, 270°$

14. $\cos x - \sin 2x - \cos 3x = 0$; $\cos x - 2\sin x\cos x - \cos x\cos 2x + \sin x\sin 2x = 0$; $\cos x - 2\sin x\cos x - \cos x(1 - 2\sin^2 x) + \sin x(2\sin x\cos x) = 0$; $\cos x - 2\sin x\cos x - \cos x + 2\sin^2 x\cos x + 2\sin^2 x\cos x = 0$; $4\sin^2 x\cos x - 2\sin x\cos x = 0$; $2\sin x\cos x(2\sin x - 1) = 0$. Hence $\sin x = 0$ or $\cos x = 0$ or $\sin x = \frac{1}{2}$ and so $x = 0°, 30°, 90°, 150°, 180°, 270°$

15. $\sin 4x - 2\sin 2x = 0$; $2\sin 2x\cos 2x - 2\sin 2x = 0$; $2\sin 2x(\cos 2x - 1) = 0$; $\sin 2x = 0$ or $\cos 2x = 1$; $2x = 0°, 180°, 360°, 540°$, $x = 0°, 90°, 180°, 270°$

16. $\sin(30° + x) - \cos(60° + x) = \frac{-\sqrt{3}}{2}$; $\sin 30°\cos x + \cos 30°\sin x - \cos 60°\cos x + \sin 60°\sin x = \frac{-\sqrt{3}}{2}$;

$\frac{1}{2}\cos x + \frac{\sqrt{3}}{2}\sin x - \frac{1}{2}\cos x + \frac{\sqrt{3}}{2}\sin x = \frac{-\sqrt{3}}{2}$; $\sqrt{3}\sin x = \frac{-\sqrt{3}}{2}$; $\sin x = -\frac{1}{2}$; $x = 210°, 330°$

17. $2\sin\theta = \sin 2\theta$; $2\sin\theta = 2\sin\theta\cos\theta$; $\sin\theta - \sin\theta\cos\theta = 0$; $\sin\theta(1-\cos\theta) = 0$, so $\sin\theta = 0$ or $\cos\theta = 1$ with the result that $\theta = 0$; $180°$.

18. $\sin^2\alpha + 5\cos^2\alpha = 3$; $\sin^2\alpha + \cos^2\alpha + 4\cos^2\alpha = 3$; $1 + 4\cos^2\alpha = 3$; $4\cos^2\alpha = 2$, or $\cos^2\alpha = \frac{1}{2}$. Thus, we have $\cos\alpha = \pm\frac{\sqrt{2}}{2}$, with the result that $\alpha = 45°$, $135°$, $225°$, $315°$.

19. $\sin^2 x = 1 + \sin x$; $\sin^2 x - \sin x - 1 = 0$; $\sin x = \frac{1\pm\sqrt{1+4}}{2} = \frac{1\pm\sqrt{5}}{2} \approx 1.618$ or -0.618. Since 1.618 is not in the range of the sine function, only -0.618 yields answers. They are $x = 218.17°$ and $321.83°$.

20. $\sin\theta - 2\csc\theta = -1$; $\sin\theta - \frac{2}{\sin\theta} = -1$; $\sin^2\theta - 2 = -\sin\theta$; $\sin^2\theta + \sin\theta - 2 = 0$; $(\sin\theta + 2)(\sin\theta - 1) = 0$; $\sin\theta = -2$ or $\sin\theta = 1$. But, $\sin\theta$ cannot be -2, so only $\sin\theta = 1$ yields an answer and it is $\theta = 90°$.

21. $\sin 2x = 2\sin x\cos x = 2\left(\frac{5}{13}\right)\left(\frac{-12}{13}\right) = \frac{-120}{169}$

22. $\cos\frac{x}{2} = \sqrt{\frac{1+\cos x}{2}} = \sqrt{\frac{1+\frac{-12}{13}}{2}} = \sqrt{\frac{1}{26}} \approx 0.196116$

23. $\sin\frac{x}{2} = \sqrt{\frac{1-\cos x}{2}} = \sqrt{\frac{1+\frac{12}{13}}{2}} = \sqrt{\frac{25}{26}} \approx 0.98058$

24. $\tan 2x = \frac{2\tan x}{1-\tan^2 x} = \frac{2\cdot\frac{-5}{12}}{1-\left(\frac{-5}{12}\right)^2} = \frac{\frac{-10}{12}}{\frac{119}{144}} = \frac{-120}{119}$

25. $\sin(x+y) = \sin x\cos y + \cos x\sin y = \frac{-5}{13}\cdot\frac{-15}{17} + \frac{-12}{13}\cdot\frac{8}{17} = \frac{75}{221} - \frac{96}{221} = \frac{-21}{221} \approx -0.09502$

26. $\cos(x-y) = \cos x\cos y + \sin x\sin y = \frac{-12}{13}\cdot\frac{-15}{17} + \frac{-5}{13}\cdot\frac{8}{17} = \frac{180}{221} - \frac{40}{221} = \frac{140}{221} \approx 0.63348$

27. $\cos(x+y) = \frac{-12}{13}\cdot\frac{-15}{17} - \frac{-5}{13}\cdot\frac{8}{17} = \frac{180}{221} + \frac{40}{221} = \frac{220}{221} \approx 0.99548$

28. $\tan(x+y) = \frac{\tan x + \tan y}{1-\tan x\tan y} = \frac{\frac{5}{12}+\frac{-8}{15}}{1-\frac{5}{12}\cdot\frac{-8}{15}} = \frac{\frac{-21}{180}}{\frac{220}{180}} = \frac{-21}{220} \approx -0.09545$ or $\tan(x+y) = \frac{\sin(x+y)}{\cos(x+y)} = \frac{\frac{-21}{221}}{\frac{220}{221}} = \frac{-21}{220} \approx -0.09545$

29. $\cos(y-x) = \cos y\cos x + \sin y\sin x = \frac{-15}{17}\cdot\frac{-12}{13} + \frac{8}{17}\cdot\frac{-5}{13} = \frac{140}{221} \approx 0.63348$ (Note #26 and #29 are the same answer.)

30. $\sin(x-y) = \sin x\cos y - \cos x\sin y = \frac{-5}{13}\cdot\frac{-15}{17} - \frac{-12}{13}\cdot\frac{8}{17} = \frac{75}{221} + \frac{96}{221} = \frac{171}{221} \approx 0.77376$

31. We begin with the given equation $a = 5.0(\sin\omega t + \cos 2\omega t)$ and substitute $a = 0$ and a double-angle identity for $\cos\omega t$ with the result that $0 = \sin\omega t + 1 - 2\sin^2\omega t$ which is equivalent to $2\sin^2\omega t - \sin\omega t - 1 = 0$. This last equation factors as $(2\sin\omega t + 1)(\sin\omega t - 1) = 0$. thus, by the zero product rule $\sin\omega t = -\frac{1}{2}$ or $\sin\omega t = 1$, and so $\omega t = 90°, 210°, 330°$.

32. $y = y_1 + y_2 = A_1\cos\omega_1 t + A_2\cos\omega_2 t$. If $\alpha = \frac{\omega_1+\omega_2}{2}$ and $\beta = \frac{\omega_2+\omega_1}{2}$, then $\alpha - \beta = \omega_1$ and $\alpha + \beta = \omega_2$. Substituting produces $A_1\cos\omega_1 t = A_1\cos(\alpha-\beta)t$ and $A_2\cos\omega_2 t = A_2\cos(\alpha+\beta)t$, and so $y_1 + y_2 = A_1\cos(\alpha-\beta)t + A_2\cos(\alpha+\beta)t$.

33. We want to shown that $x = 0.01e^{-6t}(\cos 8t + \sin 8t)$ and $x = \frac{\sqrt{2}}{100}e^{-6t}\cos(8t - \frac{\pi}{4})$ are identically equivalent. Now, using a difference formula, we see that the second equation can be rewritten as $\frac{\sqrt{2}}{100}e^{-6t}\cos(8t - \frac{\pi}{4}) = 0.01e^{-6t}\sqrt{2}\left[\cos 8t\cos\frac{\pi}{4} + \sin 8t\sin\frac{\pi}{4}\right] = 0.01e^{-6t}\sqrt{2}\left[\cos 8t\cdot\frac{1}{\sqrt{2}} + \sin 8t\cdot\frac{1}{\sqrt{2}}\right] = 0.01e^{-6t}(\cos 8t + \sin 8t)$, which is the first equation. Hence the two equations are identical.

CHAPTER 20 TEST

1. Since, $\sin\alpha = \frac{4}{5}$ and α is in Quadrant II, we have $\cos\alpha = -\frac{3}{5}$. Also, since $\cos\beta = -\frac{12}{13}$, and β is in Quadrant III, then $\sin\beta = -\frac{5}{13}$. Thus, $\sin(\alpha+\beta) = \frac{4}{5}\cdot\frac{-12}{13} + \frac{-3}{5}\cdot\frac{-5}{13} = \frac{-48}{65} + \frac{15}{65} = \frac{-33}{65}$.

2. Using the values of $\sin\alpha = \frac{4}{5}$, $\cos\alpha = -\frac{3}{5}$, $\cos\beta = -\frac{12}{13}$, and $\sin\beta = -\frac{5}{13}$ from Exercise #1, we see that $\cos(\alpha+\beta) = \frac{-3}{5}\cdot\frac{-12}{13} - \frac{4}{5}\cdot\frac{-5}{13} = \frac{36}{65} + \frac{20}{65} = \frac{56}{65}$.

3. Using the values of $\sin\alpha = \frac{4}{5}$, $\cos\alpha = -\frac{3}{5}$, $\cos\beta = -\frac{12}{13}$, and $\sin\beta = -\frac{5}{13}$ from Exercise #1, we see that $\sin(\alpha-\beta) = \frac{4}{5}\cdot\frac{-12}{13} - \frac{-3}{5}\cdot\frac{-5}{13} = \frac{-48}{65} - \frac{15}{65} = \frac{-63}{65}$.

4. Using the values of $\sin\alpha = \frac{4}{5}$, $\cos\alpha = -\frac{3}{5}$, $\cos\beta = -\frac{12}{13}$, and $\sin\beta = -\frac{5}{13}$ from Exercise #1, we see that $\cos(\alpha-\beta) = \frac{-3}{5}\cdot\frac{-12}{13} + \frac{4}{5}\cdot\frac{-5}{13} = \frac{36}{65} - \frac{20}{65} = \frac{16}{65}$.

5. Using the values of $\sin\alpha = \frac{4}{5}$, $\cos\alpha = -\frac{3}{5}$ from Exercise #1, we see that $\sin 2\alpha = 2\cdot\frac{4}{5}\cdot\frac{-3}{5} = \frac{-24}{25}$.

6. Using the values of $\cos\beta = -\frac{12}{13}$, and $\sin\beta = -\frac{5}{13}$ from Exercise #1, we see that $\cos 2\beta = 2\left(\frac{-12}{13}\right)^2 - 1 = \frac{288}{169} - \frac{169}{169} = \frac{119}{169}$.

7. Using the values of $\sin\alpha = \frac{4}{5}$, $\cos\alpha = -\frac{3}{5}$ from Exercise #1, we see that $\cos\frac{\alpha}{2} = \sqrt{\frac{1+\cos\alpha}{2}} = \sqrt{\frac{1-\frac{3}{5}}{2}} = \sqrt{\frac{8}{10}} = \sqrt{\frac{4}{5}} = \frac{2}{\sqrt{5}}$.

8. Using the values of $\cos\beta = -\frac{12}{13}$, and $\sin\beta = -\frac{5}{13}$ from Exercise #1, we see that $\cos\frac{\beta}{2} = -\sqrt{\frac{1+\cos\beta}{2}} = -\sqrt{\frac{1-\frac{12}{13}}{2}} = -\sqrt{\frac{1}{26}} = \frac{-1}{\sqrt{26}}$.

9. Using the double angle formula, $2\cos x\sin x = \sin 2x$, we see that can be rewritten as $8\cos 6x\sin 6x = 4\cdot 2\sin\left(\frac{12x}{2}\right)\cos\left(\frac{12x}{2}\right) = 4\sin 12x$.

10. $\tan x = \dfrac{\sec x}{\csc x}$; $\frac{\sec x}{\csc x} = \frac{\frac{1}{\cos x}}{\frac{1}{\sin x}} = \frac{\sin x}{\cos x} = \tan x$

11.
$$\frac{\sin x}{1-\cos x} + \frac{\sin x}{1+\cos x} = 2\csc x$$

$$\left(\frac{\sin x}{1-\cos x}\right)\left(\frac{1+\cos x}{1+\cos x}\right) + \left(\frac{\sin x}{1+\cos x}\right)\left(\frac{1-\cos x}{1-\cos x}\right) \qquad 2\csc x$$

$$\frac{\sin x(1+\cos x)}{1-\cos^2 x} + \frac{\sin x(1-\cos x)}{1-\cos^2 x}$$

$$\frac{\sin x(1+\cos x)}{\sin^2 x} + \frac{\sin x(1-\cos x)}{\sin^2 x}$$

$$\frac{1+\cos x}{\sin x} + \frac{1-\cos x}{\sin x}$$

$$\frac{2}{\sin x}$$

$$2\csc x$$

12. The equation $6\cos^2 x + \cos x = 2$ is equivalent to $6\cos^2 x + \cos x - 2 = 0$. This expression factors as $(3\cos x + 2)(2\cos x - 1) = 0$ and so by the zero -product rule, $3\cos x + 2 = 0$ or $2\cos x - 1 = 0$. If $3\cos x + 2 = 0$, then $\cos x = -\frac{2}{3}$ and $x \approx 131.81°$ or $228.19°$. If $2\cos x - 1 = 0$, then $\cos x = \frac{1}{2}$ and $x = 60°$ or $300°$. In radians the solutions are $\frac{\pi}{3}$, $\frac{5\pi}{3}$,2.3,and 3.98.

CHAPTER

21

Statistics and Empirical Methods

21.1 PROBABILITY

1. $\frac{13}{52} = \frac{1}{4}$
2. $\frac{4}{52} = \frac{1}{13}$
3. $\frac{4}{52} = \frac{1}{13}$
4. $\frac{8}{52} = \frac{2}{13}$
5. $\frac{12}{52} = \frac{3}{13}$
6. $\frac{26}{52} = \frac{1}{2}$
7. $\frac{1}{2} \cdot \frac{1}{2} = \frac{1}{4}$
8. $\frac{1}{2} \cdot \frac{3}{13} = \frac{3}{26}$
9. $\frac{1}{2} \cdot \frac{26}{51} = \frac{13}{51}$
10. $\frac{3}{13} \cdot \frac{4}{51} = \frac{4}{221}$
11. $\frac{1}{6}$
12. $\frac{2}{6} = \frac{1}{3}$
13. $\frac{3}{6} = \frac{1}{2}$
14. $\frac{3}{6} = \frac{1}{2}$
15. $\frac{1}{6} \cdot \frac{1}{6} = \frac{1}{36}$
16. $\frac{5}{6}$
17. $\frac{5}{6} \cdot \frac{5}{6} = \frac{25}{36}; P = 1 - \frac{25}{36} = \frac{11}{36}$
18. $\frac{1}{6} \cdot \frac{1}{6} \cdot \frac{1}{6} = \frac{1}{216}$
19.

Die 1 \ Die 2	1	2	3	4	5	6
1	(1, 1)	(1, 2)	(1, 3)	(1, 4)	(1, 5)	(1, 6)
2	(2, 1)	(2, 2)	(2, 3)	(2, 4)	(2, 5)	(2, 6)
3	(3, 1)	(3, 2)	(3, 3)	(3, 4)	(3, 5)	(3, 6)
4	(4, 1)	(4, 2)	(4, 3)	(4, 4)	(4, 5)	(4, 6)
5	(5, 1)	(5, 2)	(5, 3)	(5, 4)	(5, 5)	(5, 6)
6	(6, 1)	(6, 2)	(6, 3)	(6, 4)	(6, 5)	(6, 6)

20. $\frac{1}{36}$
21. $\frac{1}{36}$
22. $\frac{1}{36}$
23. 0
24. $\frac{2}{36} = \frac{1}{18}$
25. $\frac{6}{36} = \frac{1}{6}$
26. $\frac{6+2}{36} = \frac{8}{36} = \frac{2}{9}$
27. $\frac{1+2+3}{36} = \frac{6}{36} = \frac{1}{6}$
28. $\frac{1}{6} \cdot \frac{1}{6} = \frac{1}{36}$
29. $\frac{1}{36} \cdot \frac{1}{36} = \frac{1}{1{,}296}$
30. $\frac{6+5+4+3+2+1}{36} = \frac{21}{36} = \frac{7}{12}$
31. $\frac{6}{21} \cdot \frac{6}{21} = \frac{36}{441} = \frac{4}{49}$
32. $\frac{3}{21} \cdot \frac{4}{21} = \frac{4}{147}$
33. $\frac{2+4+6}{21} = \frac{12}{21} = \frac{4}{7}$
34. $\frac{6}{21} \cdot \frac{6}{21} = \frac{4}{49}$

35. (a) $.75 \times .75 = 0.5625 = \frac{9}{16}$, (b) $.25 \times .25 = 0.0625 = \frac{1}{16}$, (c) $1 - 0.5625 - 0.0625 = 0.375 = \frac{3}{8}$

36. $0.001 \times 0.005 = 0.000005$

37. (a) $0.7 \times 0.8 = 0.56$, (b) $0.3 \times 0.2 = 0.06$, (c) $1 - 0.56 - 0.06 = 0.38$

38. $\frac{25}{1000} = 0.025$

39. If one is defective, then the other is not defective. There are two ways this can happen, so the probability is $2(0.025)(1 - 0.025) = 2(0.025)(0.975) = 0.04875$.

40. The probability that it is not defective is $\frac{975}{1000} = 0.975$ The probability that four choosen at random are non-defective is $(0.975)^4 = 0.9037$.

21.2 MEASURES OF CENTRAL TENDENCY

1.

Number	2	3	4	6	7
Frequency	2	1	4	2	1

2.

Number	50	52	54	56	58
Frequency	1	2	4	1	2

3.

Number	73	74	77	80	82	83	84	87	89	92	94	96	100
Frequency	1	1	1	1	1	1	2	1	1	1	3	5	1

4.

Number	64	66	76	78	82	85	89	90	93	94	96	98	100
Frequency	3	1	1	2	1	1	2	1	3	1	4	2	1

5.

Interval	1 – 2	3 – 4	5 – 6	7 – 8
Frequency	2	5	2	1

6.

Interval	50 – 52	53 – 55	56 – 58
Frequency	3	4	3

7.

Interval	71 – 75	76 – 80	81 – 85	86 – 90	91 – 95	96 – 100
Frequency	2	2	4	2	4	6

8.

Interval	61 – 65	66 – 70	71 – 75	76 – 80	81 – 85	86 – 90	91 – 95	96 – 100
Frequency	3	1	0	3	2	3	4	7

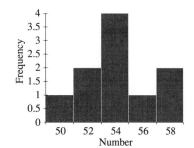

9.

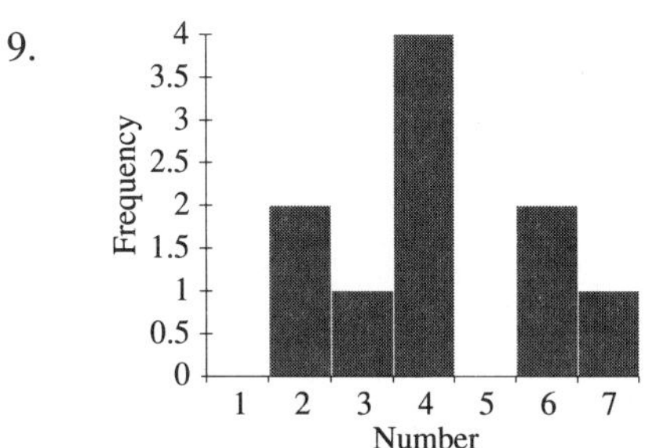

10.

Frequency
4
3.5
3
2.5
2
1.5
1
0.5
0
50 52 54 56 58
Number

11.

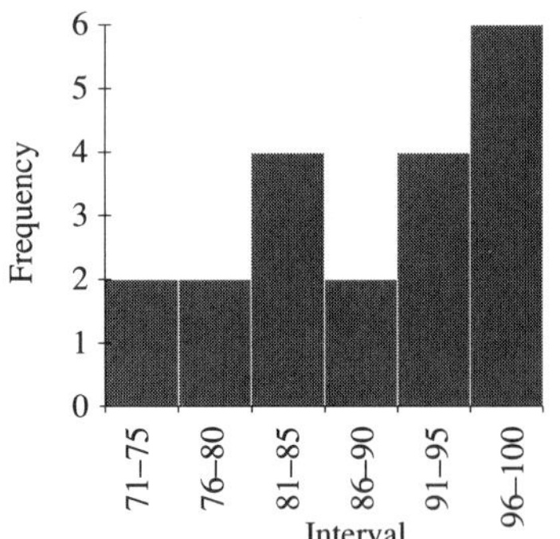

12.

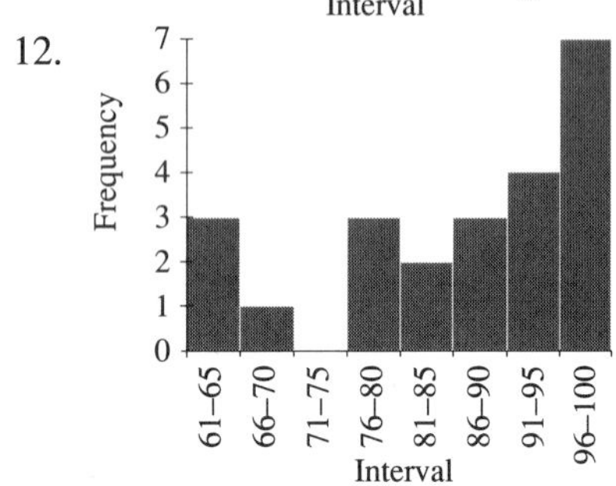

13.

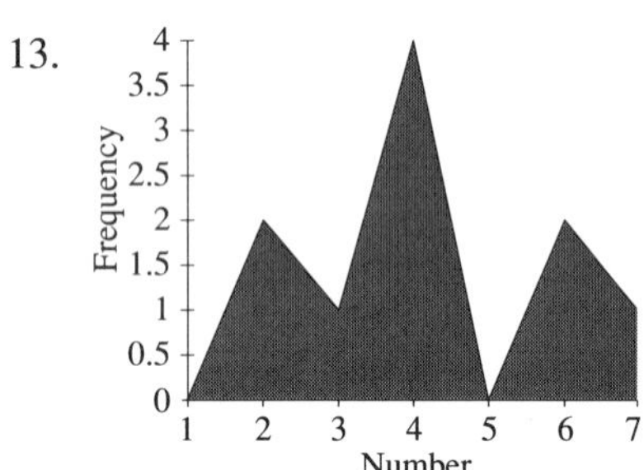

14.

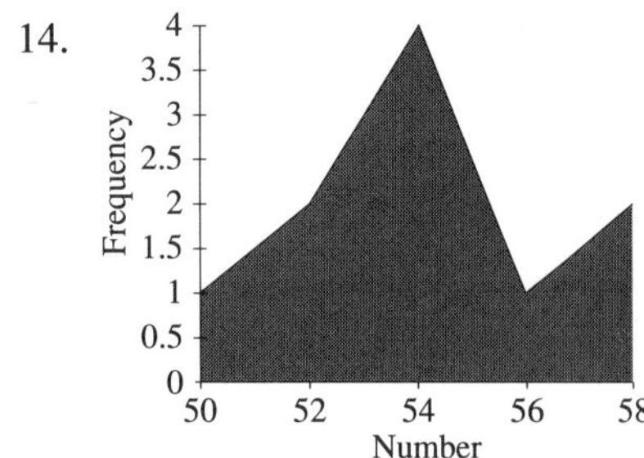

15.

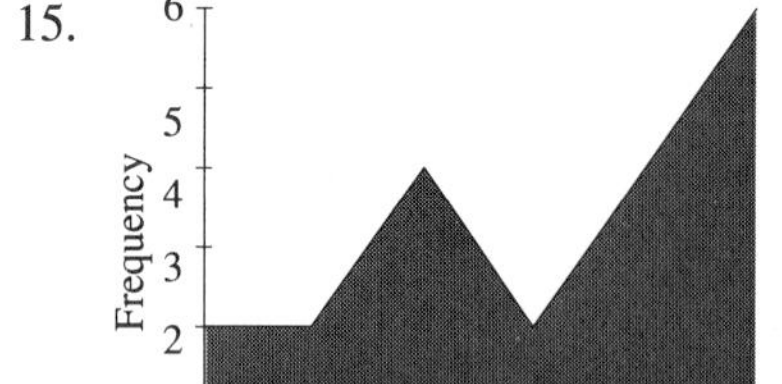

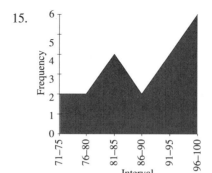

16.

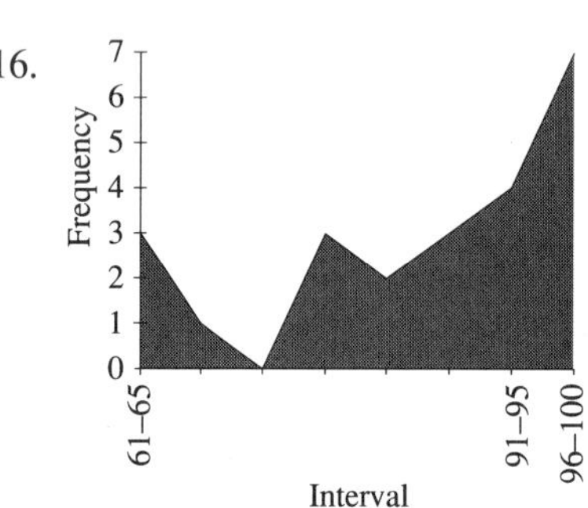

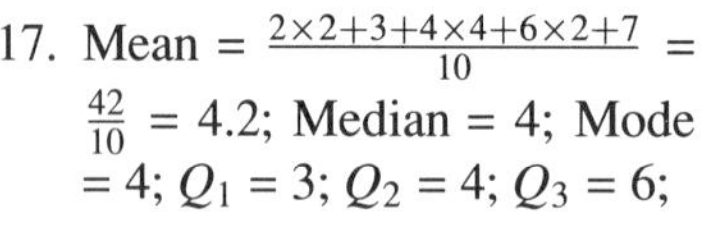

17. Mean $= \frac{2\times2+3+4\times4+6\times2+7}{10} = \frac{42}{10} = 4.2$; Median = 4; Mode = 4; $Q_1 = 3$; $Q_2 = 4$; $Q_3 = 6$;

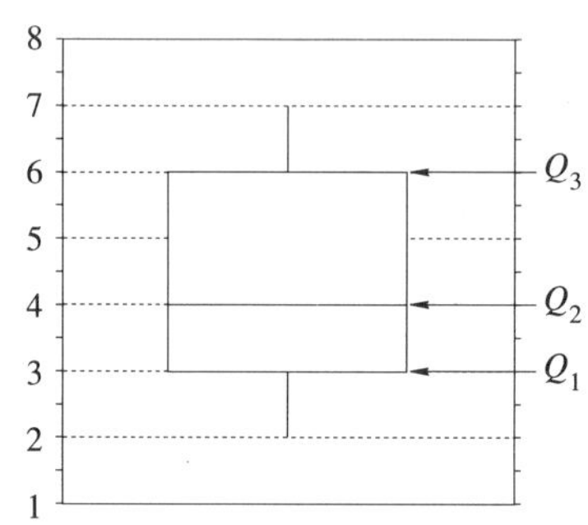

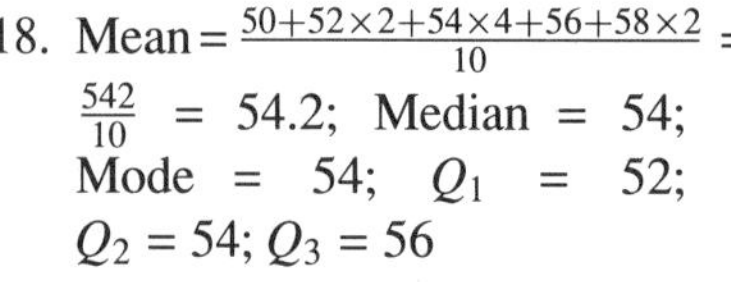

18. Mean $= \frac{50+52\times2+54\times4+56+58\times2}{10} = \frac{542}{10} = 54.2$; Median = 54; Mode = 54; $Q_1 = 52$; $Q_2 = 54$; $Q_3 = 56$

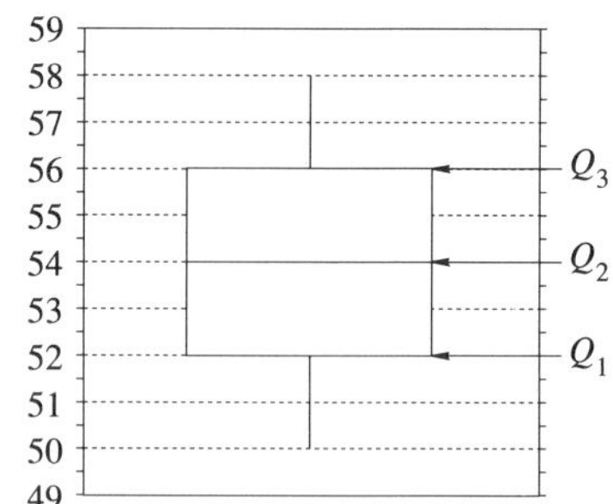

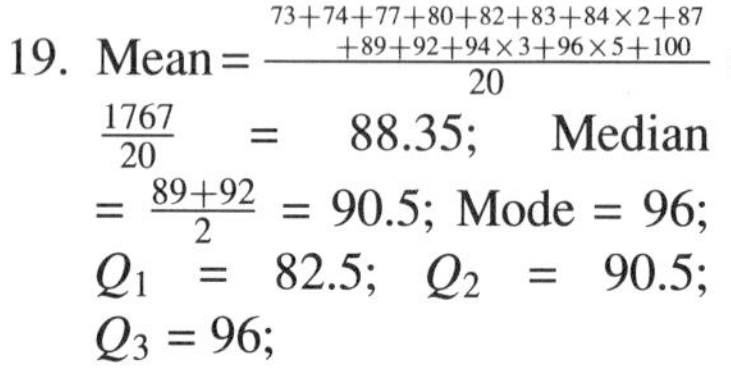

19. Mean $= \frac{73+74+77+80+82+83+84\times2+87+89+92+94\times3+96\times5+100}{20} = \frac{1767}{20} = 88.35$; Median $= \frac{89+92}{2} = 90.5$; Mode = 96; $Q_1 = 82.5$; $Q_2 = 90.5$; $Q_3 = 96$;

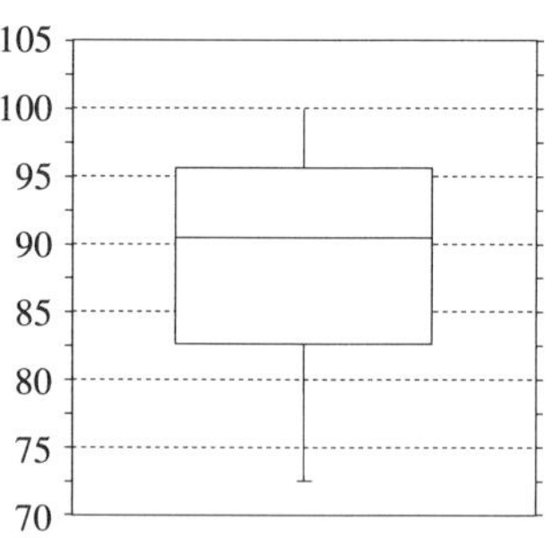

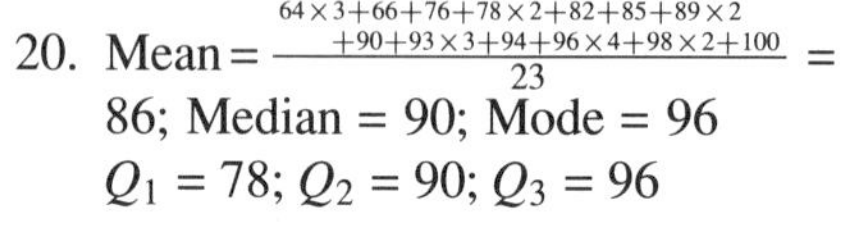

20. Mean $= \frac{64\times3+66+76+78\times2+82+85+89\times2+90+93\times3+94+96\times4+98\times2+100}{23} = 86$; Median = 90; Mode = 96 $Q_1 = 78$; $Q_2 = 90$; $Q_3 = 96$

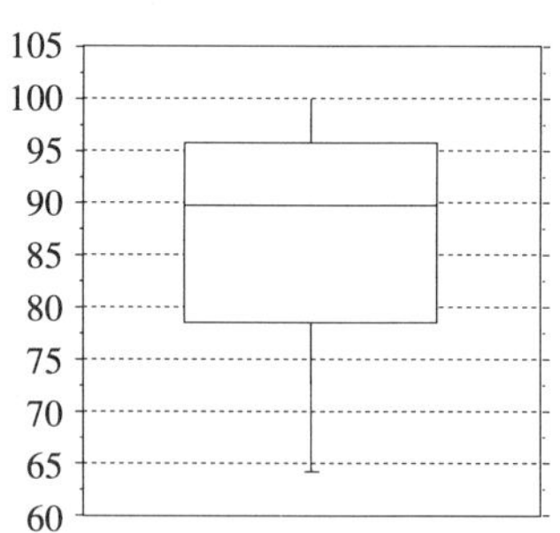

21.

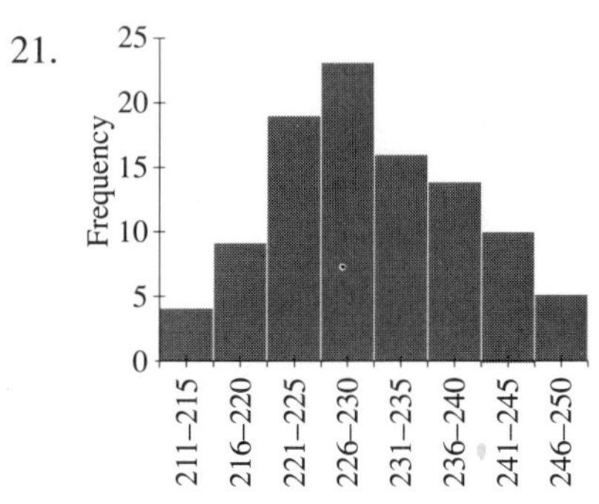

22.

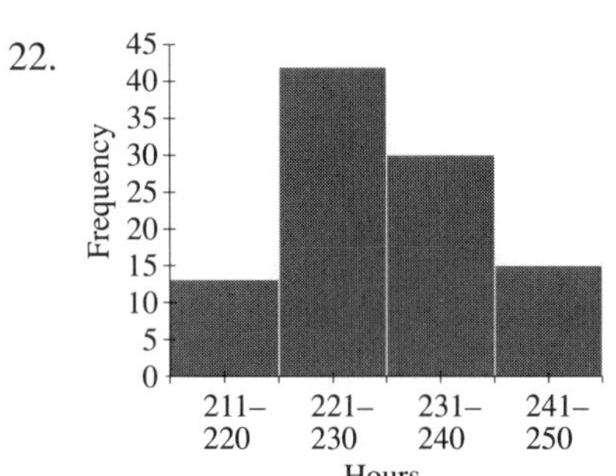

23.

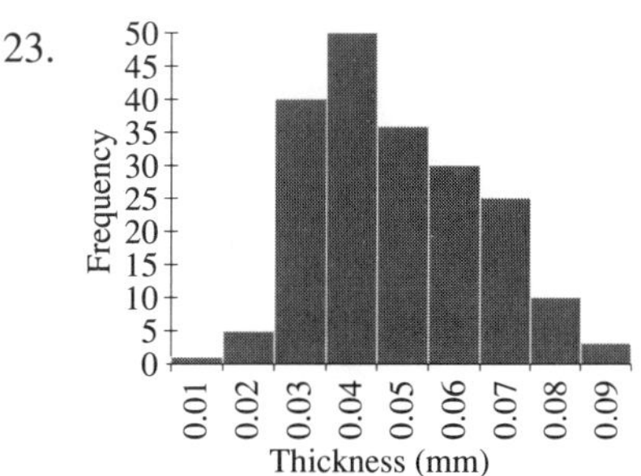

24.

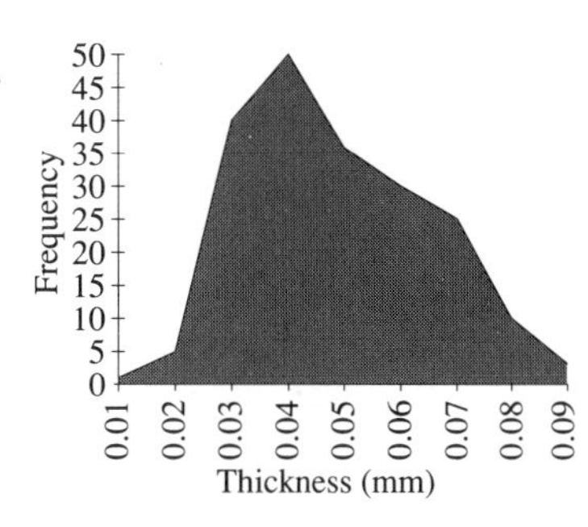

25. Frequency

Thickness (mm)	0.01–0.02	0.03–0.04	0.05–0.06	0.07–0.08	0.09–0.10

Thickness (mm)

26.

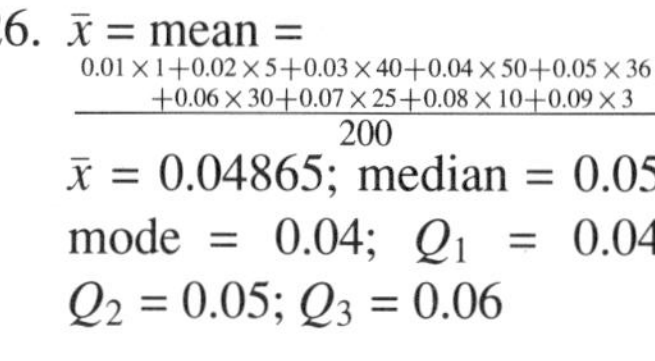

$\bar{x}$ = mean = $\dfrac{0.01\times1+0.02\times5+0.03\times40+0.04\times50+0.05\times36+0.06\times30+0.07\times25+0.08\times10+0.09\times3}{200}$; $\bar{x} = 0.04865$; median = 0.05; mode = 0.04; $Q_1 = 0.04$; $Q_2 = 0.05$; $Q_3 = 0.06$

27.

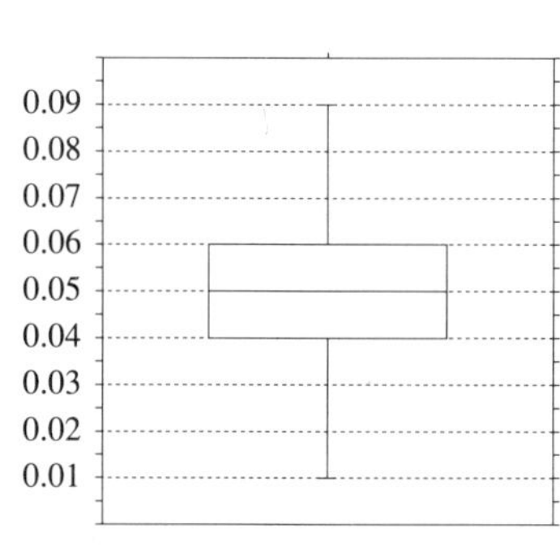

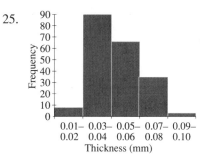

28.

mA	5.24	5.26	5.27	5.28	5.29	5.31	5.34	5.35	5.39	5.42	5.43	5.44	5.45	5.46	5.47
Frequency	1	2	2	1	2	3	4	3	2	2	1	1	1	3	2

29.

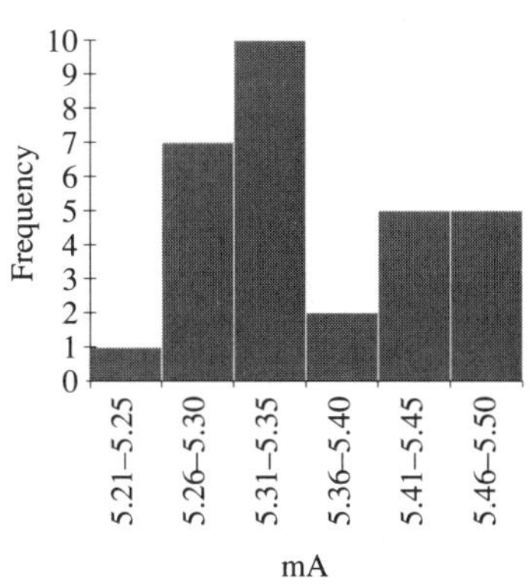

30. $\bar{x} = \frac{160.76}{30} = 5.3587$; Median = 5.345; mode = 5.34

31. $Q_1 = 5.29$; $Q_2 = 5.345$; $Q_3 = 5.43$

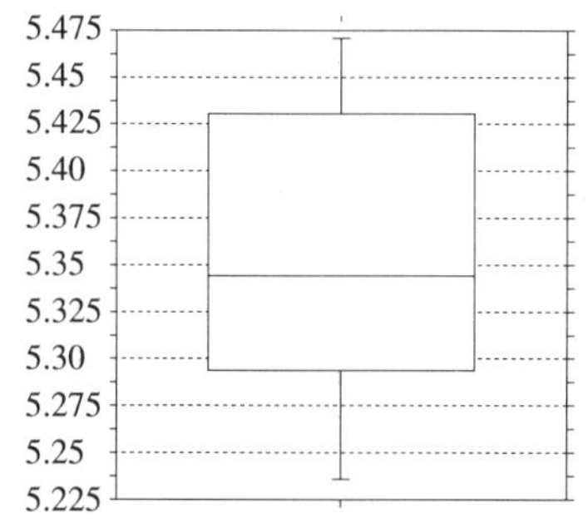

32. See *Computer Programs* in main text.

21.3 MEASURES OF DISPERSION

1. $v = 2.56$

2. $v = 5.96$

3. $v = 64.3275$

4. $v = 140.26087$

5. $\sigma_n = 1.6$

6. $\sigma_n = 2.44$

7. $\sigma = 8.02$

8. $\sigma = 11.84$

9. The mean is 4.2; $\bar{x} - \sigma = 2.6$; $\bar{x} + \sigma = 5.8$; (a) 5 of the 10 values are within one standard deviation. $\bar{x} - 2\sigma = 1$; $\bar{x} + 2\sigma = 7.4$; (b) All 10 values are within two standard deviations.

10. $\bar{x} = 54.2$; $\bar{x} - \sigma = 51.76$; $\bar{x} + \sigma = 56.64$; (a) there are 7 values within one standard deviation $\bar{x} - 2\sigma = 49.32$; $\bar{x} + 2\sigma = 59.08$;; (b) there are 10 values within two standard deviations.

11. The mean is 88.35; $\bar{x} - \sigma = 80.3$; $\bar{x} + \sigma = 96.4$. (a) There are 15 values within one standard deviation. $\bar{x} - 2\sigma = 72.3$; $\bar{x} + 2\sigma = 104.4$; (b) there are 20 values within two standard deviations.

12. $\bar{x} = 86$; $\bar{x} - \sigma = 74.16$; $\bar{x} + \sigma = 97.84$; (a) there are 16 values within one standard deviation. $\bar{x} - 2\sigma = 62.31$; $\bar{x} + 2\sigma = 109.68$; (b) there are 23 values within two standard deviations.

13. (a) The standard deviation is 0.0217. (b) The mean is 10.0214; $\bar{x} - \sigma = 9.9997$; $\bar{x} + \sigma = 10.0431$; The number of rods within one standard deviation is 390. (c) $\bar{x} - 2\sigma = 9.978$; $\bar{x} + 2\sigma = 10.0649$; The number of rods within two standard deviations is 480.

14. (a) The standard deviation is 8.84. (b) The mean is 230.25 so $\bar{x} - \sigma = 221.41$; $\bar{x} + \sigma = 239.09$; There are 72 within one standard deviation. (c) $\bar{x} - 2\sigma = 212.57$; $\bar{x} + 2\sigma = 247.93$. All 100 values are within two standard deviations.

15. (a) The standard deviation is 0.0164; (b) $\bar{x} = 0.04865$; $\bar{x} - \sigma = 0.03229$; $\bar{x} + \sigma = 0.06501$; 116 are within one standard deviation. (c) $\bar{x} - 2\sigma = 0.01593$; $\bar{x} + 2\sigma = 0.08137$; 196 are within two standard deviations.

16. (a) The standard deviation is 0.0727 (b) $\bar{x} = 5.3587$ so $\bar{x} - \sigma = 5.286$; $\bar{x} + \sigma = 5.4314$; there are 17 within one standard deviation. (c) $\bar{x} - 2\sigma = 5.2132$; $\bar{x} + 2\sigma = 5.5042$; all 30 are within two standard deviations.

17. See *Computer Programs* in main text.

18. See *Computer Programs* in main text.

21.4 FITTING A LINE TO DATA

1. $y = 0.833x - 1.333$, $r = 0.943$

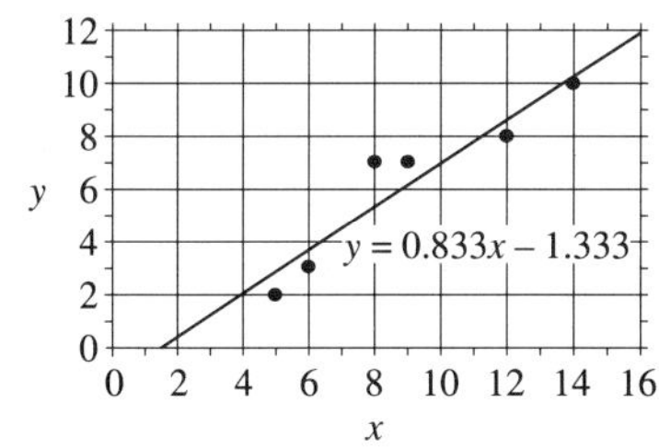

2. $y = 14.55x + 3.85$, $r = 0.994$

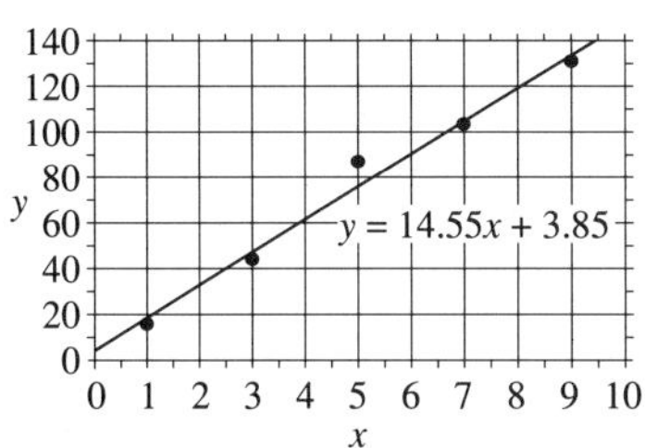

3. $y = 13.55x + 5.1$, $r = 0.986$

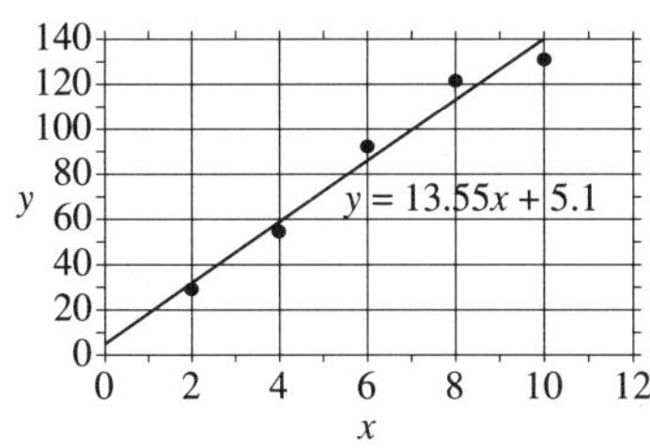

4. $y = 0.175x + 20.974$; $r = 0.150$

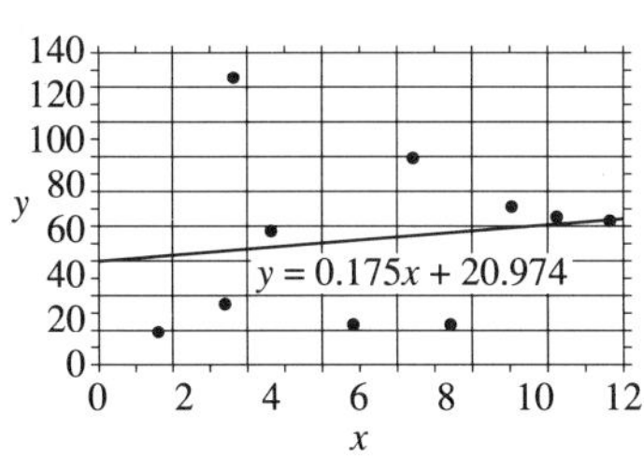

5. $y = 14.955x + 5.379$; $r = 0.996$

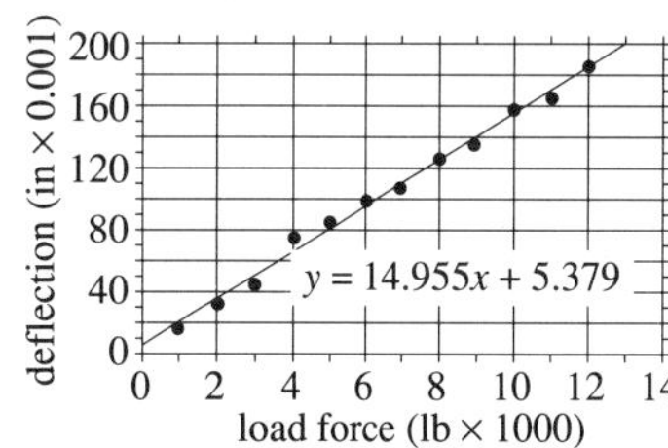

6. $y = 10.33x - 35$; $r = 0.692$

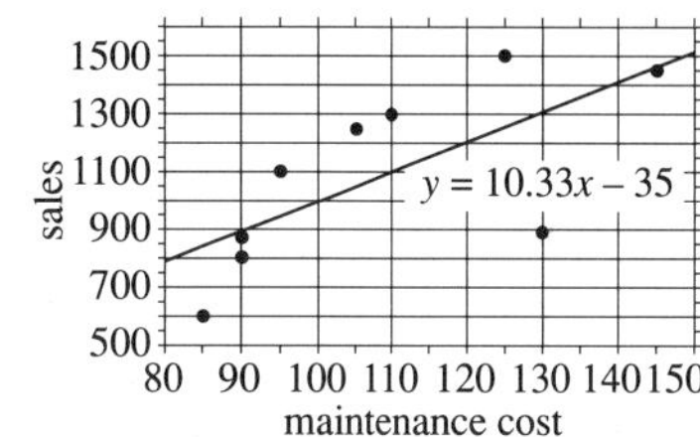

7. $y = 0.000425x + 0.072$; $r = 0.987$

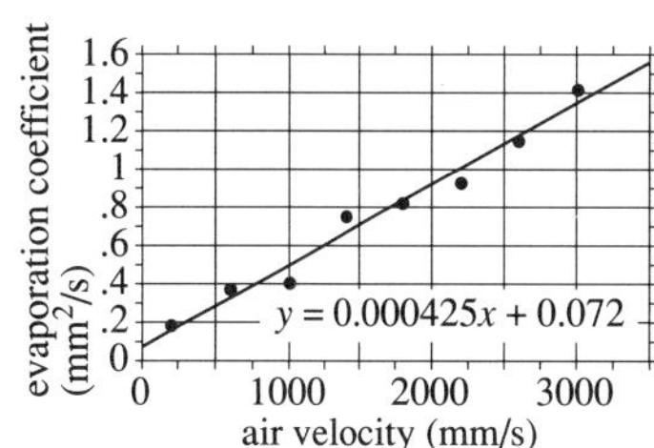

8. $y = -0.278x + 33.398$; $r = -0.784$

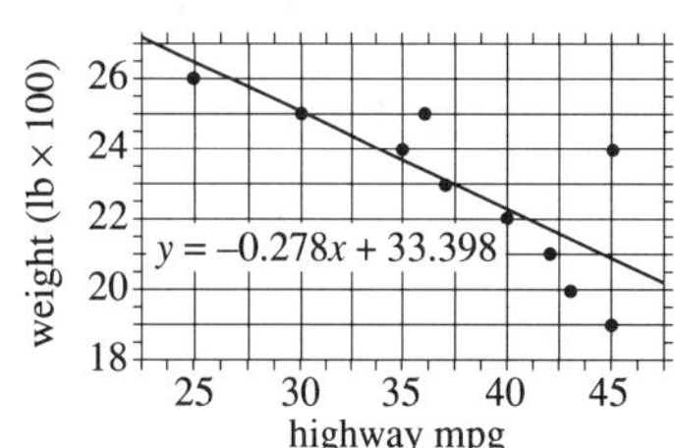

9. See *Computer Programs* in main text.

21.5 FITTING NONLINEAR CURVES TO DATA

1. (a)

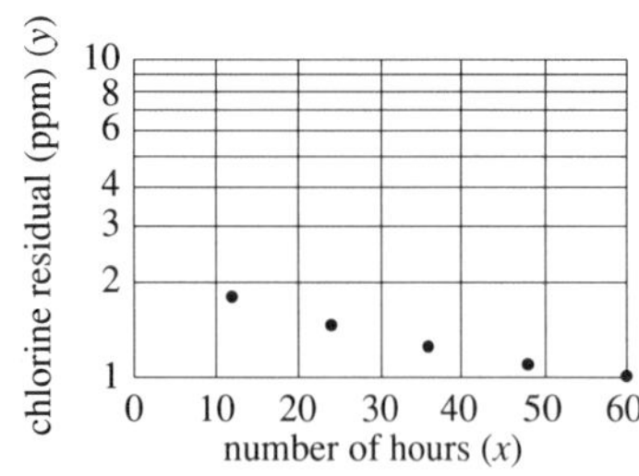

(b) $y = (2.002)(0.988)^x$ Calculator Function

2. (a)

Bacteria count (× 1000) (y)
1000
800
600
400
300
200
100
2 4 6 8 10 12 14 16 18
Days since inoculation (x)

(b) $y = 69.541 \times 1.149^x$ Calculator Function

3. (a) See the graph below.

(b)

x	5	10	4	10	7	8	8	5	10	5	12	6
y	30	51	26	52	40	43	45	31	52	30	59	36
x^2	25	100	16	100	49	64	64	25	100	25	144	36
x^3	125	1000	64	1000	343	512	512	125	1000	125	1728	216
x^4	625	10000	256	10000	2401	4096	4096	625	10000	625	20736	1296
xy	150	510	104	520	280	344	360	155	520	150	708	216
xy	750	5100	416	5200	1960	2752	2880	775	5200	750	8496	1296

$\sum x = 90$; $\sum x^2 = 748$; $\sum xy = 4017$; $\sum y = 495$; $\sum x^3 = 6750$; $\sum x^2 y = 35{,}575$; $\sum x^4 = 64{,}756$.
Hence:

$$495 = 12a_0 + 90a_1 + 748a_2$$
$$4017 = 90a_0 + 748a_1 + 6750a_2$$
$$35{,}575 = 748a_0 + 6750a_1 + 64{,}756a_2$$

Solving yields $a_0 = 4.084$; $a_1 = 5.848$, $a_2 = -0.107$. The equation is $y = 4.084 + 5.848x - 0.107x^2$

(c)

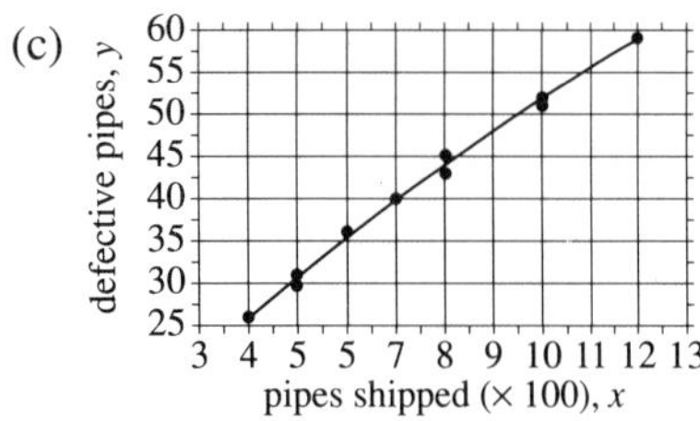

4. (a) See the graph below.

(b)

x	1	2	3	4	5	6	7	8	9	10	$\sum$ 55
y	1.28	1.47	1.00	0.85	0.72	0.63	0.89	0.91	1.05	1.08	9.88
x^2	1	4	9	16	25	36	49	64	81	100	385
x^3	1	8	27	64	125	216	343	512	729	1000	3025
x^4	1	16	81	256	625	1296	2401	4096	6561	10000	25333
xy	1.28	2.94	3.00	3.40	3.60	3.78	6.23	7.28	9.45	10.8	51.76
x^2y	1.28	5.88	9.00	13.6	18	22.68	43.61	58.24	85.05	108	365.34

Hence:

$$9.88 = 10a_0 + 55a_1 + 385a_2$$
$$51.76 = 55a_0 + 385a_1 + 3025a_2$$
$$365.34 = 385a_0 + 3025a_1 + 25333a_2$$

Solving this system of three equations in three variables, yields $a_0 = 1.716$, $a_1 = -0.309$, and $a_2 = 0.025$. Thus, we have $y = 1.716 - 0.309x + 0.025x^2$.

(c)

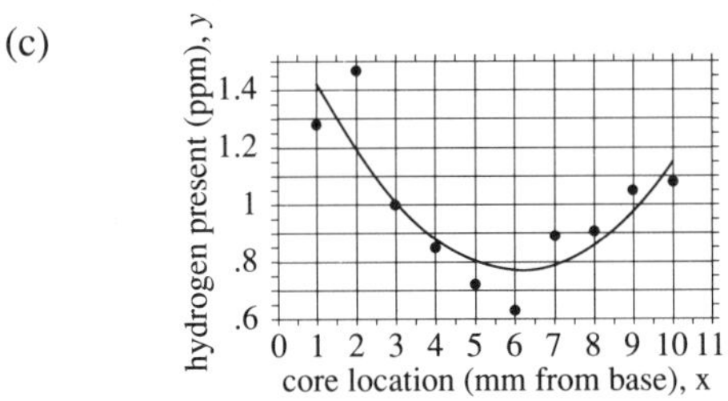

5. (a) See the graph below. (b) To find the least square curve for $y = ae^{-kx}$ we first take natural logarithms giving $\ln y = \ln a - kx$. To use the normal equations of the previous section we need to compute several sums. These are $\sum x = 36$; $\sum x^2 = 204$; $\sum \ln y = -4.48183$ and $\sum x \ln y = -25.3578$. Also $n = 8$.

$$\sum \ln y = nb - k\sum x$$
$$\sum x \ln y = b\sum x - k\sum x^2 \qquad \text{or}$$
$$-4.48183 = 8b - 36k$$
$$-25.3578 = 36b - 204k$$

Solving we get $b = -0.004208$ and $k = 0.12356$. Since $b = \ln a$; $a = e^{-0.004208} = 0.9958$. The final equation is $y = (0.9958)e^{-0.12356x}$. Note: Some calculators have built in functions to compute the least square curve of the form $y = ab^x$. This yields $a = 0.9958$ and $b = 0.883768$. To convert to $y = ae^{-kx}$ find $\ln b = \ln(0.883768) = -0.12356$ or $y = (0.9958)e^{-0.12356x}$.

(c)

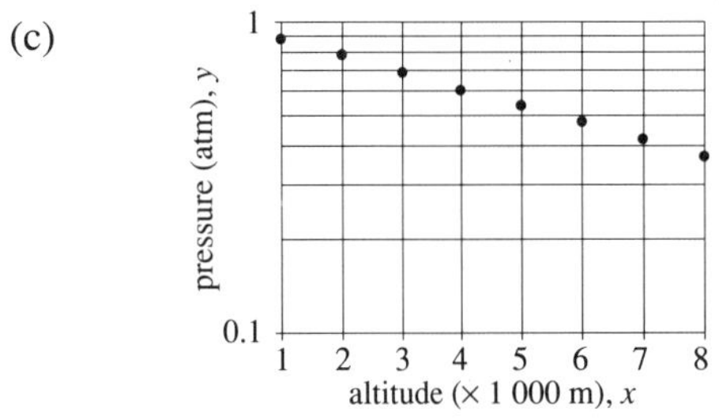

6. (a) See the graph below. (b) To use the least square curve we need the following sums $\sum y = 139$; $\sum \frac{1}{x^2} = 154.98$; $\sum \frac{y}{x^2} = 9726$ and $\sum \frac{1}{x^4} = 10{,}820.4$. The normal equations are $139 = 10b + 154.977m$; $9726.55 = 154.977b + 10{,}820.4m$. Solving yields $b = -0.039$ and $m = 0.899$. The final equation is $y = \frac{0.899}{x^2} - 0.039$.

(c)

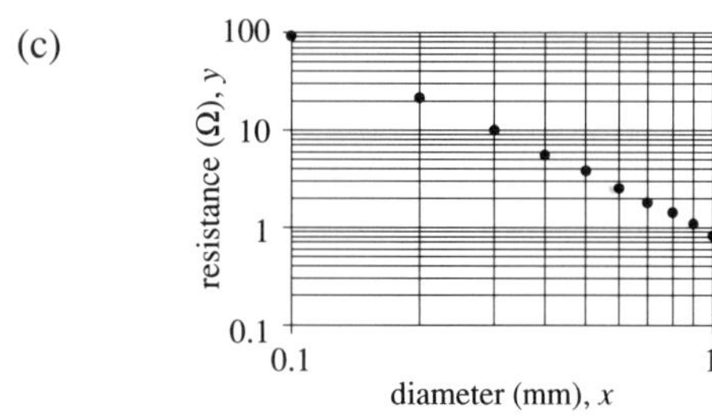

7. (a) See the graph below. (b) As in the previous examples we need sums for the following: $\sum \frac{1}{p}$; $\sum \frac{1}{p^2}$; $\sum y$; and $\sum \frac{y}{p}$. These are 2.0290; 1.5498; 15.2; and 8.1939, respectively. Solving the system $15.2 = 10b + 2.929m$; $8.1939 = 2.9292b + 1.5498m$ yields $b = -0.064$ and $m = 5.41$. The final equation is $y = \frac{5.41}{p} - 0.064$.

(c)

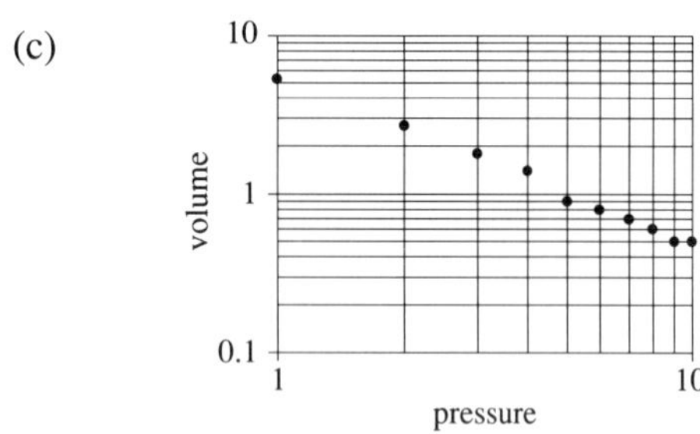

8. (c) $P = 3.46(1.172)^x$; (d) Shown by dashed curve in Figure.

(a)

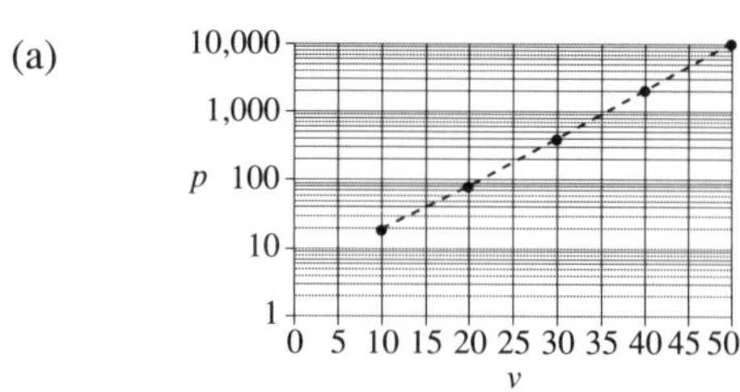

(b)

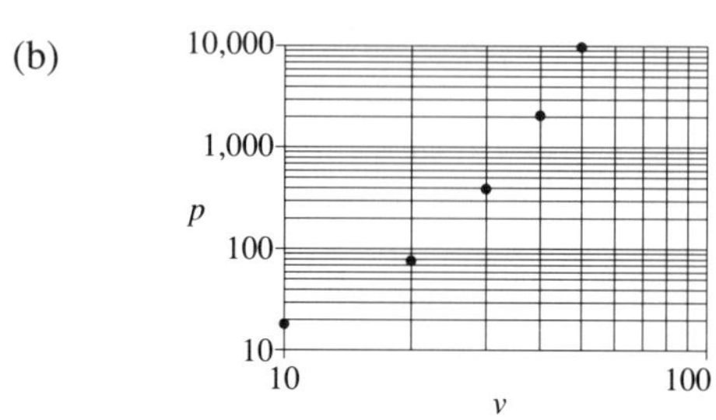

(c) From the graphs in a and b we can see that the best fit will be of the form $y = ab^x$. The equation is $y = 3.46(1.172)^x$.

(d) Shown by dashed curve in Figure 21.0.

CHAPTER 21 REVIEW

1. $\frac{1}{2}$
2. $\frac{2}{54} = \frac{1}{27}$
3. $\frac{10}{16} = \frac{5}{8}$
4. $\frac{48}{52} = \frac{12}{13}$
5. $\frac{2}{6} = \frac{1}{3}$
6. $\frac{1}{36} + \frac{2}{36} + \frac{3}{36} = \frac{6}{36} = \frac{1}{6}$
7. $\frac{1}{6} \times \frac{1}{6} = \frac{1}{36}$
8. $\frac{4}{52} \times \frac{3}{51} = \frac{1}{221}$
9. $\frac{46}{50} = \frac{23}{25}$ = probability that a drive is not defective $\left(\frac{23}{25}\right)^3 = \frac{12167}{15625}$
10. 297.8, 298.9, 299.6, 299.8, 300.1, 300.1, 300.3, 300.4, 300.5, 300.7, 301.2 The median is 300.1
11. $Q_1 = \frac{299.6+299.8}{2} = 299.7$; $Q_2 = 300.1$; $Q_3 = 300.45$

12.

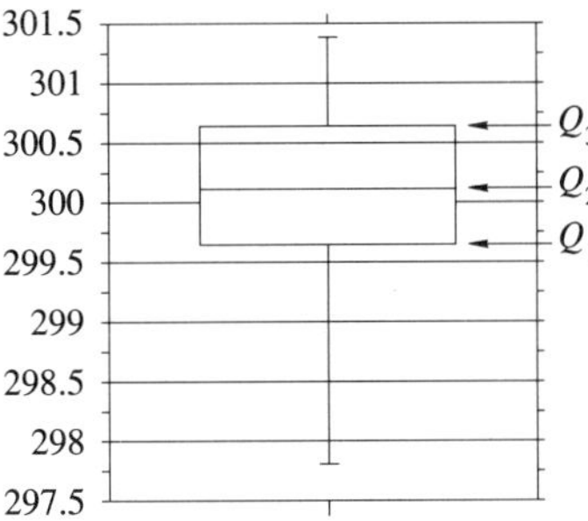

13. 300.1

14. $\bar{x} = 299.945$

15. 0.888

16. $\frac{8}{11} = 72.7\%$

17. $\frac{9}{11} = 81.8\%$

18. 23, 24, 26, 32, 33, 37, 37, 38, 39, 42, 43, 45, 48, 50,52; (a) 37.93, b) 38, (c) $Q_1 = 32.5$, $Q_2 = 38$, $Q_3 = 44$; (d) 8.79

19. (a) See the figure below. (b) $y = 43.94x + 18.94$, (d) $r = 0.996$

(c)

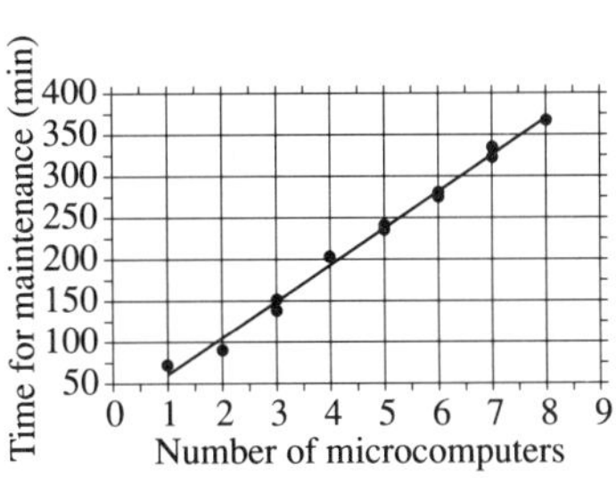

20. (a) See the figure below. (b) $P = 3983.96V^{-1.405}$ (Note: enter data with V first and P second),

(c)

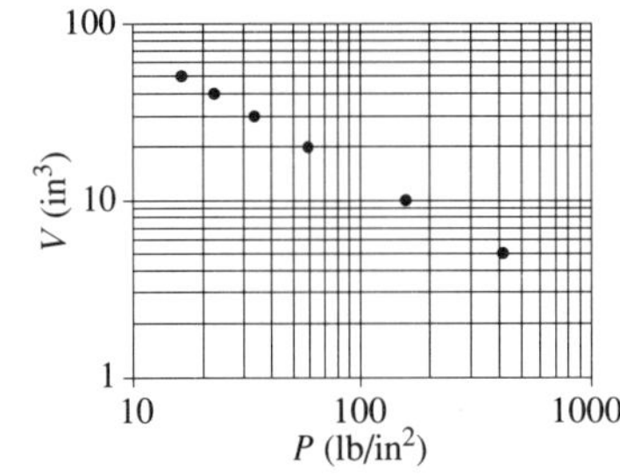

21. (a) See the figure below. (b) $n = -0.686p + 83.714$, $r = -0.9384$

(c) Shown by curve in the figure below.

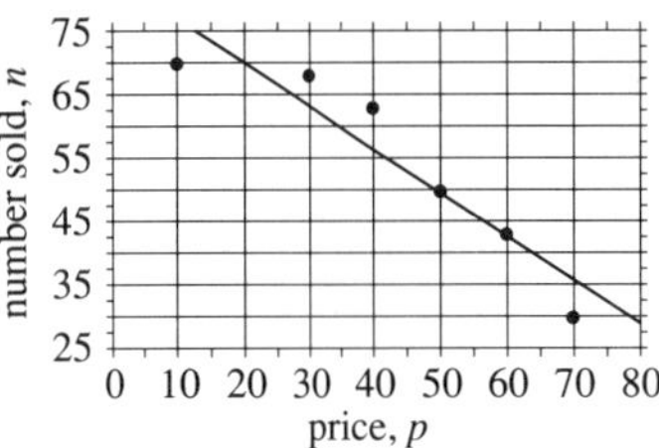

22. To find this equation we need the following sums $\sum\sqrt{x} = 31.775$; $\sum x = 110$; $\sum y = 111.5$; $\sum\sqrt{x}y = 346.772$. This gives the normal equations $111.5 = 11b + 31.775m$, $346.772 = 31.775b + 110m$. Solving this system yields $b = 6.221$, $m = 1.356$: Thus the equation is $y = 1.356\sqrt{x} + 6.221$.

CHAPTER 21 TEST

1. A deck of 52 cards has 2 red queens, so the probability of drawing a red queen from this deck is $\frac{2}{52} = \frac{1}{26}$.

2. The probability of no defect is $\frac{49}{50}$. To get 5 with no defect, the probability is $(\frac{49}{50})^5 \approx 0.904$.

3. 3598.2, 3598.6, 3598.7, 3598.9, 3599.2, 3599.2, 3599.6, 3599.6, 3599.9, 3600.1, 3600.1, 3600.2, 3600.4, 3600.4, 3600.4, 3600.4, 3600.8, 3601.2, 3601.4, 3601.6, median = 3600.1

4. $Q_1 = 3599.2, Q_2 = 3600.1, Q_3 = 3600.4$

5. 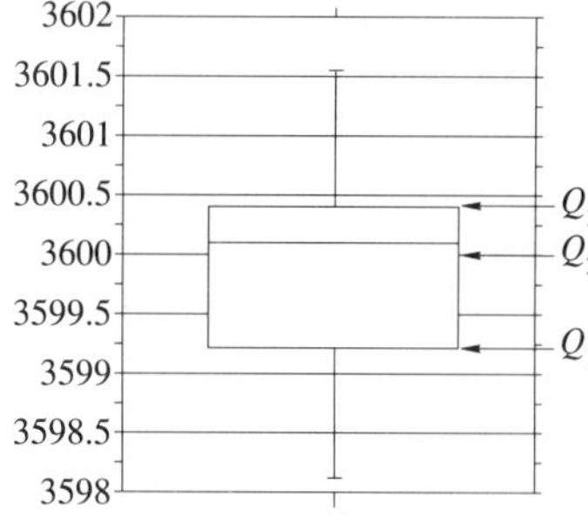

6. mode = 3600.4

7. mean = 3599.945

8. $\sigma = 0.9206$

9. $\frac{13}{20} = 65\%$

10. $\frac{20}{20} = 100\%$

11. (b) $y = -39.774x + 1063.76$, (c) See line in graph; (d) $r = -0.8997$

CHAPTER

22

An Introduction to Calculus

22.1 THE TANGENT QUESTION

1. $\frac{f(3)-f(1)}{3-1}=\frac{24-6}{3-1}=\frac{18}{2}=9$

2. $\frac{g(7)-g(4)}{7-4}=\frac{32-17}{7-4}=\frac{15}{3}=5$

3. $\frac{h(-1)-h(-3)}{-1--3}=\frac{-6--20}{-1+3}=\frac{14}{2}=7$

4. $\frac{j(0)-j(-2)}{0-(-2)}=\frac{-5-11}{2}=\frac{-16}{2}=-8$

5. $\frac{k(-1)-k(-2)}{-1--2}=\frac{-11--39}{-1+2}=\frac{28}{1}=28$

6. $\frac{m(1)-m(0)}{1-0}=\frac{-3--1}{1}=\frac{-2}{1}=-2$

7. $\frac{f(5)-f(2)}{5-2}=\frac{1-\frac{5}{2}}{5-2}=\frac{\frac{-3}{2}}{3}=-\frac{1}{2}$

8. $\frac{g(-3)-g(-6)}{-3--6}=\frac{-8--2}{-3+6}=\frac{-6}{3}=-2$

9. $\frac{h(0)-h(1)}{0-1}=\frac{0-1}{-1}=1$

10. $\frac{j(b)-j(-2)}{b--2}=\frac{b^2+3b+5-3}{b+2}=\frac{b^2+3b+2}{b+2}=\frac{(b+1)(b+2)}{b+2}=b+1$

11. $\frac{k(x_1)-k(1)}{x_1-1}=\frac{x_1^2+x_1-7--5}{x_1-1}=\frac{x_1^2+x_1-2}{x_1-1}=\frac{(x_1+2)(x_1-1)}{x_1-1}=x_1+2$

12. $\frac{m(x_1)-m(1)}{x_1-1}=\frac{x_1^3+4-5}{x_1-1}=\frac{x_1^3-1}{x_1-1}=\frac{(x_1-1)(x_1^2+x_1+1)}{x_1-1}=x_1^2+x_1+1$

13. $\frac{f(x_1+h)-f(x_1)}{(x_1+h)-x_1}=\frac{\left[(x_1+h)^2+1\right]-\left[x_1^2+1\right]}{h}=\frac{x_1^2+2x_1h+h^2+1-x_1^2-1}{h}=\frac{2x_1h+h^2}{h}=\frac{h(2x_1+h)}{h}=2x_1+h$

14. $\frac{g(x_1+h)-g(x_1)}{x_1+h-x_1}=\frac{\left[(x_1+h)^2+4(x_1+h)-1\right]-\left[x_1^2+4x_1-1\right]}{h}=\frac{x_1^2+2x_1h+h^2+4x_1+4h-1-x_1^2-4x_1+1}{h}=\frac{2x_1h+h^2+4h}{h}=\frac{h(2x_1+4+h)}{h}=2x_1+4+h$

15. $V(t)=5t^2+4t$; (a) $\frac{V(3)-V(2)}{3-2}=\frac{57-28}{1}=29$ L/min (b) $\frac{V(4)-V(3)}{4-3}=\frac{96-57}{1}=39$ L/min (c) $\frac{V(4)-V(2)}{4-2}=\frac{96-28}{2}=\frac{68}{2}=34$ L/min

16. The formula for area is $A=\pi(2t)^2=4\pi t^2$, thus $\overline{A}=\frac{A(4)-A(3)}{4-3}=\frac{64\pi-36\pi}{1}=28\pi\ \text{m}^2/\text{sec}$

17. (a) $\frac{f(6)-f(4)}{6-4} = \frac{103-43}{2} = \frac{60}{2} = 30$

(b) $\frac{f(5)-f(4)}{5-4} = \frac{70-43}{1} = 27$

(c) $\frac{f(4.5)-f(4)}{4.5-4} = \frac{55.75-43}{0.5} = \frac{12.75}{0.5} = 25.5$

(d) $\frac{f(4.25)-f(4)}{4.25-4} = \frac{49.1875-43}{0.25} = \frac{6.1875}{0.25} =$ 24.75

(e) $\frac{f(4.1)-f(4)}{4.1-4} = \frac{45.43-43}{0.1} = \frac{2.43}{0.1} = 24.3$

(f) $\frac{f(4.05)-f(4)}{4.05-4} = \frac{44.2075-43}{0.05} = \frac{1.2075}{0.05} =$ 24.15

18. (a) $\frac{f(2)-f(4)}{2-4} = \frac{7-43}{-2} = \frac{-36}{-2} = 18$

(b) $\frac{f(3)-f(4)}{3-4} = \frac{22-43}{-1} = \frac{-21}{-1} = 21$

(c) $\frac{f(3.5)-f(4)}{3.5-4} = \frac{31.75-43}{-0.5} = \frac{-11.25}{-0.5} = 22.5$

(d) $\frac{f(3.75)-f(4)}{3.75-4} = \frac{37.1875-43}{-0.25} = \frac{-5.8125}{-0.25} =$ 23.25

(e) $\frac{f(3.9)-f(4)}{3.9-4} = \frac{40.63-43}{-0.1} = \frac{-2.37}{-0.1} = 23.7$

(f) $\frac{f(3.95)-f(4)}{3.95-4} = \frac{41.8075-43}{-0.05} = \frac{-1.1925}{-0.05} =$ 23.85

19. See *Computer Programs* in main text.

22.2 THE AREA QUESTION

1. $\Delta x = \frac{b-a}{6} = \frac{7-1}{6} = 1$;

x^*	1.5	2.5	3.5	4.5	5.5	6.5
y^*	6.5	9.5	12.5	15.5	18.5	21.5

$\sum_{i=1}^{6} y_i^* \Delta x = 84$

2. $\Delta x = \frac{1-(-4)}{5} = 1$;

x^*	−3.5	−2.5	−1.5	−0.5	0.5
y^*	21	17	13	9	5

$\sum_{i=1}^{5} y_i^* \Delta x = 65$

3. $\Delta x = \frac{3-0}{6} = \frac{1}{2}$ or 0.5;

x^*	0.25	0.75	1.25	1.75	2.25	2.75
y^*	1.0625	1.5625	2.5625	4.0625	6.0625	8.5625

$\sum_{i=1}^{6} y_i^* \Delta x = 11.9375$

4. $\Delta x = \frac{3-1}{4} = \frac{1}{2}$ or 0.5;

x^*	1.25	1.75	2.25	2.75
y^*	2.6875	7.1875	13.1875	20.6875

$\sum_{i=1}^{4} y_i^* \Delta x = 43.75 \times 0.5 = 21.875$

5. $\Delta x = \dfrac{5-1}{8} = 0.5;$

x^*	1.25	1.75	2.25	2.75	3.25	3.75	4.25	4.75
y^*	5	12.5	22	33.5	47	62.5	80	99.5

$\sum_{i=1}^{8} y_i^* \Delta x = 362(0.5) = 181$

6. $\Delta x = \dfrac{2 - -1}{6} = 0.5;$

x^*	−0.75	−0.25	0.25	0.75	1.25	1.75
y^*	1.578125	1.984375	2.015625	2.421875	3.953125	7.359375

$\sum_{i=1}^{6} y_i^* \Delta x = 19.3125(0.5) = 9.65625$

7. $\Delta x = \dfrac{4 - -2}{8} = \dfrac{6}{8} = \dfrac{3}{4} = 0.75;$

x^*	−1.625	−0.875	−0.125	0.625	1.375	2.125	2.875	3.625
y^*	13.359375	15.234375	15.984375	15.609375	14.109375	11.484375	7.734375	2.859375

$\sum_{i=1}^{8} y_i^* \Delta x = 96.375(0.75) = 72.28125$

8. $\Delta x = \dfrac{2-0}{10} = \dfrac{1}{5} = 0.2;$

x^*	0.1	0.3	0.5	0.7	0.9	1.1	1.3	1.5	1.7	1.9
y^*	0.399	1.173	1.875	2.457	2.871	3.069	3.003	2.625	1.887	0.741

$\sum y_i^* \Delta x = 20.1(0.2) = 4.02$

9. $\Delta x = \dfrac{-1 - -5}{8} = \dfrac{1}{2} = 0.5;$

x^*	−4.75	−4.25	−3.75	−3.25	−2.75	−2.25	−1.75	−1.25
y^*	84.1875	69.6875	56.6875	45.1875	35.1875	26.6875	19.6875	14.1875

$\sum y_i^* \Delta x = 351.5(0.5) = 175.75$

10. $\Delta x = \dfrac{4-2}{8} = \dfrac{2}{8} = 0.25;$

x^*	2.125	2.375	2.625	2.875	3.125	3.375	3.625	3.875
y^*	17.3909	28.8167	44.4807	65.3206	92.3674	126.7463	169.6760	222.4690

$\sum y_i^* \Delta x = 767.26758(0.25) = 191.8169$ more accurately 191.8168945

11. $\Delta x = 1$, $\sum y_i^* = 4.05 + 3.85 + 3.9 + 3.75 + 3.45 + 3.65 + 4.0 + 4.2 = 30.85$; $\sum y_i^* \Delta x = 30.85$

12. $\Delta x = 0.25$, $\sum y_i^* = 16.4 + 16.605 + 16.575 + 16.4 + 16.335 + 16.27 = 98.585$; $\sum y_i^* \Delta x = 98.585(0.25) = 24.64625$

13. See *Computer Programs* in main text.

14. For Exercise 9 the answers are: (a) 175.928889, (b) 175.972222, (c) 175.993055, (d) 175.998264; and for Exercise 10 you should get: (a) 192.234094, (b) 192.335187, (c) 192.383798, (d) 192.395945

22.3 LIMITS: AN INTUITIVE APPROACH

1.

x	0.9	0.99	0.999	0.9999	1.0001	1.001	1.01	1.1
$f(x) = 3x$	2.7	2.97	2.997	2.9997	3.0003	3.003	3.03	3.3

$\lim_{x\to 1} 3x = 3$

2.

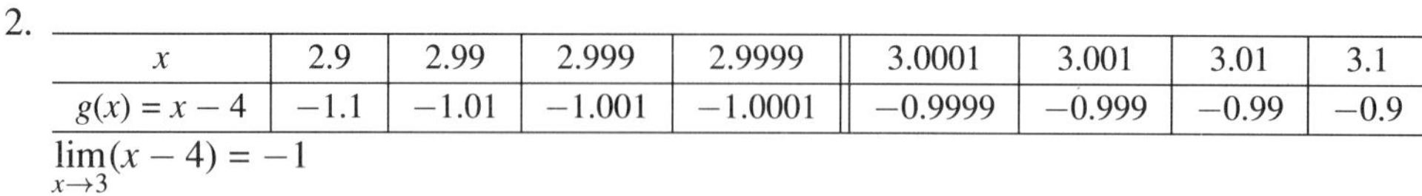

x	2.9	2.99	2.999	2.9999	3.0001	3.001	3.01	3.1
$g(x) = x - 4$	−1.1	−1.01	−1.001	−1.0001	−0.9999	−0.999	−0.99	−0.9

$\lim_{x\to 3}(x - 4) = -1$

3.

x	−1.1	−1.01	−1.001	−1.0001	−0.9999	−0.999	−0.99	−0.9
$h(x) = x^2 + 2$	3.21	3.0201	3.002	3.0002	2.9998	2.998	2.9801	2.81

$\lim_{x\to -1}(x^2 + 2) = 3$

4.

x	−2.1	−2.01	−2.001	−2.0001	−1.9999	−1.999	−1.99	−1.9
$k(x) = \frac{x^2-4}{x+2}$	−4.1	−4.01	−4.001	−4.0001	−3.9999	−3.999	−3.99	−3.9

$\lim_{x\to -2} \frac{x^2-4}{x+2} = -4$

5.

x	−0.1	−0.01	−0.001	−0.0001	0.0001	0.001	0.01	0.1
$f(x) = \frac{\tan x}{x}$	1.0033467	1.0000333	1.0000003	1	1	1.0000003	1.0000333	1.0033467

$\lim_{x\to 0} \frac{\tan x}{x} = 1$

6.

x	−0.1	−0.01	−0.001	−0.0001	0.0001	0.001	0.01	0.1
$g(x) = \dfrac{1 - \cos x}{x^2}$	0.49958	0.499996	0.5*	0.5*	0.5*	0.5*	0.499996	0.49958

*These answers may vary depending on the type of calculator or computer used.

$\lim_{x\to 0} \frac{1-\cos x}{x^2} = 0.5$

7.

x	0.9	0.99	0.999	0.9999	1.0001	1.001	1.01	1.1
$h(x) = \dfrac{x}{x-1}$	−9	−99	−999	−9999	10001	1001	101	11

$\lim_{x\to 1} \frac{x}{x-1}$ Does not exist

8.

x	−0.1	−0.01	−0.001	−0.0001	0.0001	0.001	0.01	0.1
$k(x) = \dfrac{1}{x^2}$	100	10000	10^6	10^8	10^8	10^6	10^4	100

$\lim_{x\to 0} \frac{1}{x^2}$ Does not exist

9. (a) -1, both sides of graph near $x = 0$ converge on -1. (b) Does not exist, graph goes toward -2 when $x < 2$ and goes toward 1 for $x > 2$; (c) 4, both sides of graph near $x = 3$ converge on 4.

10. (a) Does not exist, both sides of the graph keep getting larger as x nears -2. (b) Does not exist, graph goes toward 0 from the left of 0 and keeps getting smaller for $x > 0$; (c) 2, both sides of graph near $x = 3$ converge on 2.

11.

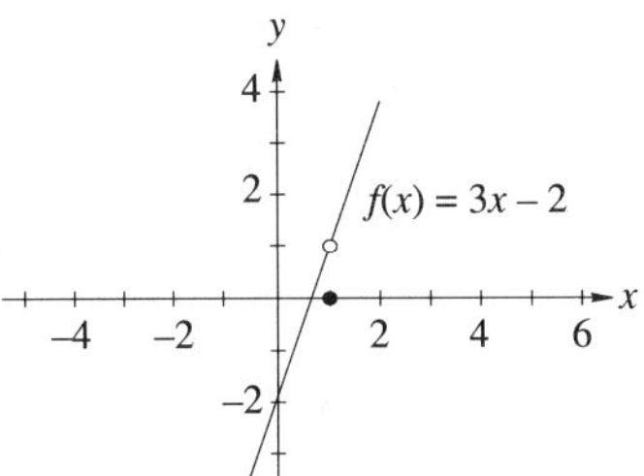

$\lim_{x\to 1} f(x) = 1$

12.

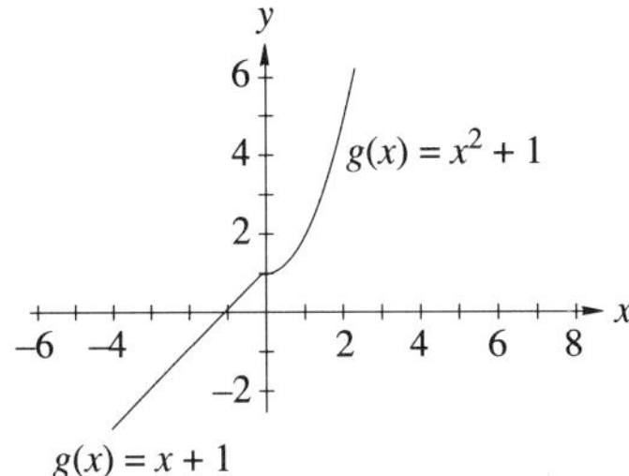

$\lim_{x\to 0} g(x) = 1$

13.

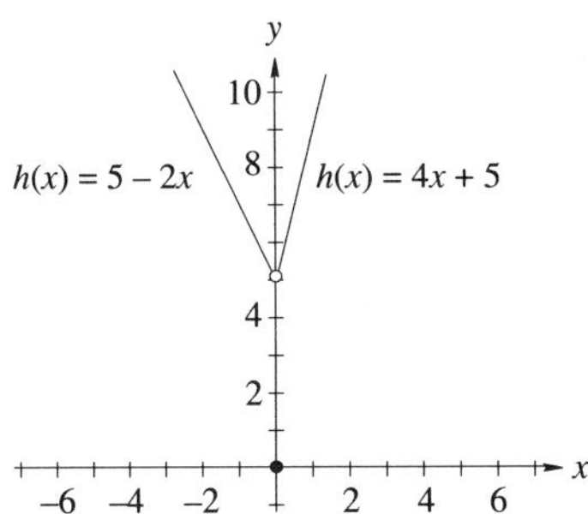

$\lim_{x\to 0} h(x) = 5$

14.

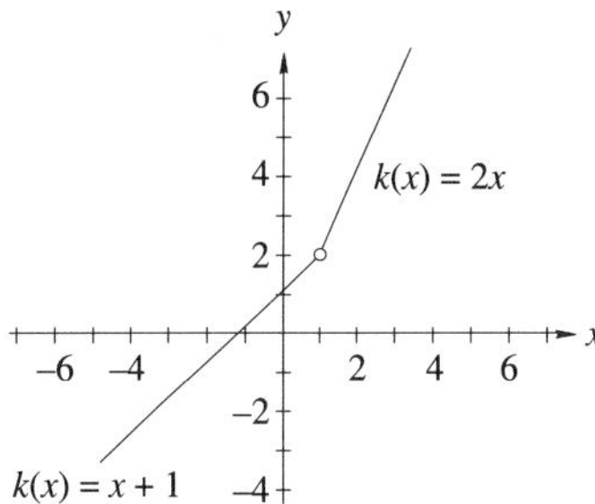

$\lim_{x\to 1} k(x) = 2$

15.

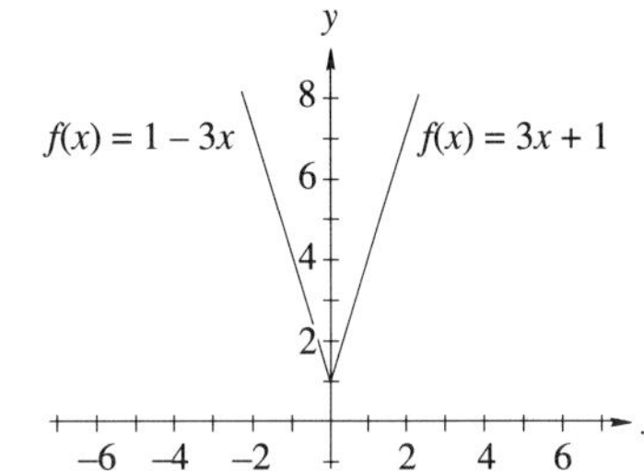

$\lim_{x\to 0} f(x) = 1$

16.

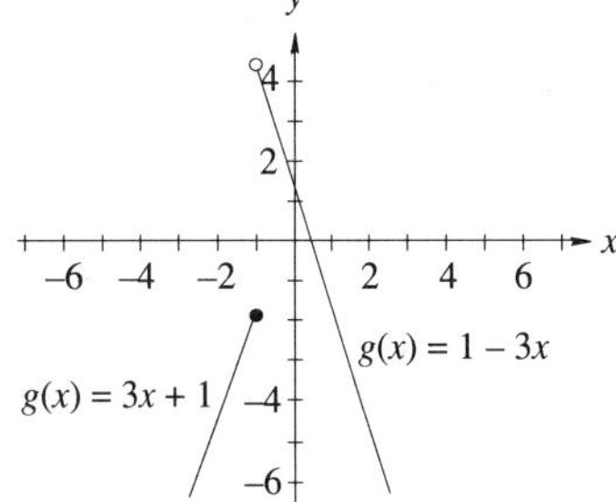

$\lim_{x\to -1} g(x)$ Does not exist

17.

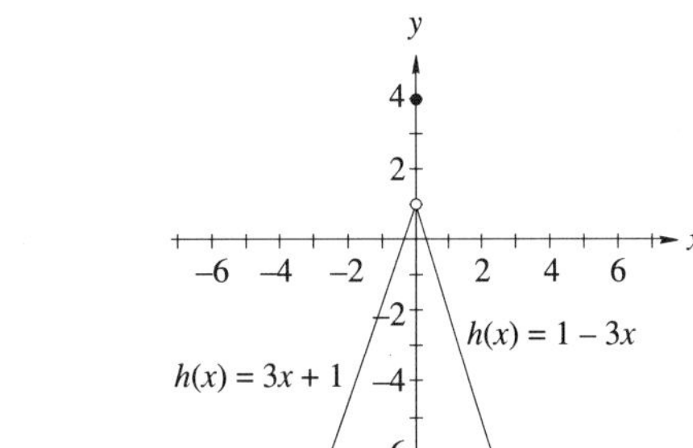

$\lim_{x\to 0} h(x) = 1$

18. $\lim_{x\to 0} k(x) = 1$

19. $\lim_{x\to 2}(-15) = -15$

20. $\lim_{x\to 0} x = 0$

21. $\lim_{x\to -4} x = -4$

22. $\lim_{x\to -7} 19 = 19$

23. $\lim_{x\to 8} x = 8$

24. $\lim_{x\to -5} (-9) = -9$

22.4 ONE-SIDED LIMITS

1. (a) $\lim_{x\to 2^-} f(x) = -2$; (b) $\lim_{x\to 2^+} f(x) = 1$; (c) $\lim_{x\to 3^-} f(x) = 4$; (d) $\lim_{x\to 3^+} f(x) = 4$

2. (a) $\lim_{x\to -1^-} g(x) = \infty$, so the limit does not exist; (b) $\lim_{x\to -1^+} g(x) = \infty$, so the limit does not exist; (c) $\lim_{x\to 0^-} g(x) = 0$; (d) $\lim_{x\to 0^+} g(x) = -\infty$, so the limit does not exist; (e) $\lim_{x\to\infty} g(x) = 2$; (f) $\lim_{x\to -\infty} g(x) = 0$

3. $\lim_{x\to 5^+} (x-2) = 3$;

x	6	5.5	5.1
$x-2$	4	3.5	3.1

4. $\lim_{x\to 2^-} (x^2-1) = 3$;

x	1	1.5	1.9	1.99
x^2-1	0	1.25	2.61	2.9601

5. $\lim_{x\to 4^+} \dfrac{x}{x-4} = \infty$;

x	5	4.5	4.1	4.01
$\frac{x}{x-4}$	5	9	41	401

6. $\lim_{x\to 3^-} \dfrac{2x}{x-3} = -\infty$;

x	2	2.5	2.9	2.99
$\frac{2x}{x-3}$	−4	−10	−58	−598

7. $\lim_{x\to -2^+} \dfrac{x^2-4}{x+2} = -4$;

x	−1	−1.5	−1.9	−1.99
$\frac{x^2-4}{x+2}$	−3	−3.5	−3.9	−3.99

8. $\lim_{x\to -5^-} \dfrac{x^2+25}{x+5} = -\infty$;

x	−6	−5.5	−5.1
$\frac{x^2+25}{x+5}$	−61	−110.5	−510.1

9. $\lim_{x\to 1^-} \dfrac{|x-1|}{x-1} = -1$;

x	0	0.5	0.9
$\frac{\lvert x-1\rvert}{x-1}$	−1	−1	−1

10. $\lim_{x\to 2^+} \dfrac{|x-2|}{x-2} = 1$;

x	3	2.5	2.1
$\frac{\lvert x-2\rvert}{x-2}$	1	1	1

11. $\lim_{x\to 2^+} \sqrt{x-2} = 0$;

x	3	2.5	2.1	2.01
$\sqrt{x-2}$	1	0.707	0.316	0.1

12. $\lim_{x\to -4^+} \sqrt{x+4} = 0$;

x	−3	−3.5	−3.9	−3.99
$\sqrt{x+4}$	1	0.707	0.316	0.1

13. $\lim_{x\to\infty} \dfrac{9}{3x} = 0$;

x	10	100	1000
$\frac{9}{3x}$	0.3	0.03	0.003

14. $\lim_{x\to -\infty} \dfrac{-8}{4x} = 0$;

x	−10	−100	−1000
$\frac{-8}{4x}$	0.2	0.02	0.002

15. $\lim_{x\to-\infty} \frac{1}{2x+1} = 0$;

x	-10	-100	-1000
$\frac{1}{2x+1}$	-0.0526	-0.00503	-0.0005

16. $\lim_{x\to\infty} \frac{-5}{9x-4} = 0$;

x	0	10	100
$\frac{-5}{9x-4}$	$\frac{5}{4}$	-0.0581	-0.00558

22.5 ALGEBRAIC TECHNIQUES FOR FINDING LIMITS

1. $\lim_{x\to5} 8 = 8$
2. $\lim_{x\to-4} x = -4$
3. $\lim_{x\to8}(x-3) = 8-3 = 5$
4. $\lim_{x\to-7}(x+4) = -7+4 = -3$
5. $\lim_{x\to-6}(x+7) = -6+7 = 1$
6. $\lim_{x\to6}(9-x) = 9-6 = 3$
7. $\lim_{x\to2} 3x = 3\cdot2 = 6$
8. $\lim_{t\to-3} 5t = 5(-3) = -15$
9. $\lim_{x\to6} \frac{2}{3}x+5 = \frac{2}{3}\cdot6+5 = 4+5 = 9$
10. $\lim_{x\to4}(\frac{3}{2}x-2) = \frac{3}{2}\cdot4-2 = 6-2 = 4$
11. $\lim_{p\to3} \frac{p^2+6}{p} = \frac{3^2+6}{3} = \frac{15}{3} = 5$
12. $\lim_{x\to-2} \frac{x^2-6}{x} = \frac{(-2)^2-6}{-2} = \frac{-2}{-2} = 1$
13. $\lim_{s\to5} \frac{3s^2+5}{2s} = \frac{3\cdot5^2+5}{2\cdot5} = \frac{80}{10} = 8$
14. $\lim_{x\to6} \frac{2x^2-3}{3-x} = \frac{2\cdot6^2-3}{3-6} = \frac{69}{-3} = -23$
15. $\lim_{x\to2}(x+1)^3 = (2+1)^3 = 3^3 = 27$
16. $\lim_{y\to5}(y-1)^3 = (5-1)^3 = 4^3 = 64$
17. $\lim_{x\to-1}(2x^2-1)^2 = (2(-1)^2-1)^2 = 1^2 = 1$
18. $\lim_{t\to4}(3t^2+2)^2 = (3\cdot4^2+2)^2 = 50^2 = 2500$
19. $\lim_{x\to3}\sqrt{x+3} = \sqrt{3+3} = \sqrt{6}$
20. $\lim_{x\to-6}\sqrt{6-7x} = \sqrt{6-7(-6)} = \sqrt{6+42} = \sqrt{48} = 4\sqrt{3}$
21. $\lim_{x\to2}\frac{x^2-2x}{x-2} = \lim_{x\to2}\frac{x(x-2)}{x-2} = \lim_{x\to2} x = 2$
22. $\lim_{w\to3}\frac{w^2-1}{w+1} = \frac{9-1}{3+1} = \frac{8}{4} = 2$
23. $\lim_{x\to-1}\frac{x^2-1}{x+1} = \lim_{x\to-1}\frac{(x+1)(x-1)}{x+1} = \lim_{x\to-1}(x-1) = -2$
24. $\lim_{y\to3}\frac{y^2-7y+12}{y-3} = \lim_{y\to3}\frac{(y-3)(y-4)}{y-3} = \lim_{y\to3}(y-4) = -1$
25. $\lim_{x\to-4}\frac{x^2+2x-8}{x^2+5x+4} = \lim_{x\to-4}\frac{(x+4)(x-2)}{(x+4)(x+1)} = \lim_{x\to-4}\frac{x-2}{x+1} = \frac{-6}{-3} = 2$
26. $\lim_{x\to\frac{1}{2}}\frac{2x^2+5x-3}{4x^2-2x} = \lim_{x\to\frac{1}{2}}\frac{(x+3)(2x-1)}{2x(2x-1)} = \lim_{x\to\frac{1}{2}}\frac{x+3}{2x} = 3\frac{1}{2}$ or $\frac{7}{2}$

27. $\lim_{t\to -7} \frac{2t^2+15t+7}{t^2+5t-14} = \lim_{t\to -7} \frac{(2t+1)(t+7)}{(t-2)(t+7)} = \lim_{t\to -7} \frac{2t+1}{t-2} = \frac{-14+1}{-7-2} = \frac{-13}{-9} = \frac{13}{9}$

28. $\lim_{x\to 0} \frac{(x+3)^2-9}{x} = \lim_{x\to 0} \frac{x^2+6x+9-9}{x} = \lim_{x\to 0}(x+6) = 6$

29. $\lim_{h\to 0} \frac{(4+h)^2-4^2}{h} = \lim_{h\to 0} \frac{16+8h+h^2-16}{h} = \lim_{h\to 0}(8+h) = 8$

30. $\lim_{x\to c} \frac{x^3-c^3}{x-c} = \lim_{x\to c} \frac{(x-c)(x^2+xc+c^2)}{x-c} = \lim_{x\to c}(x^2+xc+c^2) = c^2+c^2+c^2 = 3c^2$

31. $\lim_{x\to\infty} \frac{2x^2+4}{5x^2+3x+2} = \lim_{x\to\infty} \frac{2+\frac{4}{x^2}}{5+\frac{3}{x}+\frac{2}{x^2}} = \frac{2}{5}$

32. $\lim_{x\to\infty} \frac{6x^3+9x+1}{2x^3+4x+8} = \lim_{x\to\infty} \frac{6+\frac{9}{x^2}+\frac{1}{x^3}}{2+\frac{4}{x^2}+\frac{8}{x^3}} = \frac{6}{2} = 3$

33. $\lim_{x\to\infty} \frac{7x^4+5x^2}{x^5+2} = \lim_{x\to\infty} \frac{\frac{7}{x}+\frac{5}{x^3}}{1+\frac{2}{x^5}} = \frac{0}{1} = 0$

34. $\lim_{x\to\infty} \frac{4x^5+3x}{2x^4+1} = \lim_{x\to\infty} \frac{4x+\frac{3}{x^3}}{2+\frac{1}{x^4}} = \lim_{x\to\infty} 2x = \infty$ or does not exist.

22.6 CONTINUITY

1. $f(x) = \begin{cases} 3x+1 & \text{if } x \le 4 \\ x^2-3 & \text{if } x > 4 \end{cases}$ $c = 4 =$ Continuous
2. Not continuous, $\lim_{x\to 1^-} g(x) = 1$; $\lim_{x\to 1^+} g(x) = 2$
3. Not continuous, $\lim_{x\to 1} h(x) = 4 \neq 3 = h(1)$
4. Continuous
5. Not continuous, $k(2)$ is not defined
6. Not continuous, $\lim_{x\to 2} f(x) = 9 \neq f(2) = 0$
7. Not continuous, $\lim_{x\to -2^-} g(x) = 4$ and $\lim_{x\to -2^+} g(x) = 10$. Since $\lim_{x\to -2^-} g(x) \neq \lim_{x\to -2^+} g(x)$, the limit does not exist.
8. Not continuous, $\lim_{x\to -2^-} h(x) = 4 \neq -4 = \lim_{x\to -2^+} h(x)$
9. Not continuous, $\lim_{x\to 2} j(x) = 4 \neq 5 = j(2)$
10. Not continuous, $k(1)$ is not defined
11. None
12. 5
13. -1
14. None
15. $-3, 2$
16. $-2, -3$
17. None
18. 0
19. 2
20. 0, 5
21. $f(x) = \frac{x}{x+3}$; $(-\infty, -3), (-3, \infty)$
22. $g(x) = \frac{x}{x^2-1}$; $(-\infty, -1), (-1, 1), (1, \infty)$
23. $h(x) = \frac{x+2}{x^2-4}$; $(-\infty, -2), (-2, 2), (2, \infty)$
24. $j(x) = \frac{x^2+3x-4}{x+4}$; $(-\infty, -4), (-4, \infty)$
25. $k(x) = \sqrt{x+3}$; $[-3, \infty)$
26. $f(x) = \sqrt{5-x}$; $(-\infty, 5]$

27. $g(x) = \sqrt{x^2 - 9}$; $(-\infty, -3], [3, \infty)$

28. $h(x) = \sqrt{x^2 + 4x + 4}$; $(-\infty, \infty)$

29. $j(x) = \frac{1}{x}$; $(-\infty, 0), (0, \infty)$

30. $k(x) = \frac{1}{\sqrt{x}}$; $(0, \infty)$

CHAPTER 22 REVIEW

1. $\overline{m} = \frac{f(2) - f(0)}{2 - 0} = \frac{-3 - -7}{2} = \frac{4}{2} = 2$

2. $\overline{m} = \frac{g(-1) - g(-3)}{-1 - -3} = \frac{1 - -7}{-1 + 3} = \frac{8}{2} = 4$

3. $\overline{m} = \frac{h(-2) - h(-3)}{-2 - -3} = \frac{9 - 19}{-2 + 3} = \frac{-10}{1} = -10$

4. $\overline{m} = \frac{j\left(\frac{1}{2}\right) - j(0)}{\frac{1}{2} - 0} = \frac{\frac{8}{9/2} - 2}{\frac{1}{2}} = \frac{\frac{16}{9} - 2}{\frac{1}{2}} = \frac{\frac{-2}{9}}{\frac{1}{2}} = \frac{-4}{9}$

5. $\overline{m} = \frac{k(b) - k(-1)}{b - -1} = \frac{b^2 + 2b - (-1)}{b + 1} = \frac{b^2 + 2b + 1}{b + 1} = b + 1$

6. $\overline{m} = \frac{m(x_1) - m(6)}{x_1 - 6} = \frac{x_1^2 - 5x_1 - 6}{x_1 - 6} = \frac{(x_1 - 6)(x_1 + 1)}{x_1 - 6} = x_1 + 1$

7. $\Delta x = \frac{-2+5}{6} = \frac{3}{6} = 0.5$;

x^*	−4.75	−4.25	−3.75	−3.25	−2.75	−2.25
y^*	12	10	8	6	4	2

$\sum y_i^* \Delta x = 42(0.5) = 21$

8. $\Delta x = \frac{4-0}{8} = 0.5$;

x^*	0.25	0.75	1.25	1.75	2.25	2.75	3.25	3.75
y^*	5.125	6.125	8.125	11.125	15.125	20.125	26.125	33.125

$\sum y_i^* \Delta x = 125(0.5) = 62.5$

9. $\Delta x = \frac{4-2}{8} = \frac{2}{8} = 0.25$;

x^*	2.125	2.375	2.625	2.875	3.125	3.375	3.625	3.875
y^*	16.796875	20.671875	24.921875	29.546875	34.546875	39.921875	45.671875	51.796875

$\sum y_i^* \Delta x = 263.875(0.25) = 65.96875$

10. $\Delta x = \frac{1-(-1)}{8} = \frac{2}{8} = 0.25$;

x^*	−0.875	−0.625	−0.375	−0.125	0.125	0.375	0.625	0.875
y^*	3.46875	4.21875	4.71875	4.96875	4.96875	4.71875	4.21875	3.46875

$\sum y_i^* \Delta x = 34.75(0.25) = 8.6875$

11. $\lim_{x \to 1}(3x^2 + 2x - 5) = 3 + 2 - 5 = 0$

12. $\lim_{x \to -2} \frac{3x^2 - 12}{x + 2} = \lim_{x \to -2} \frac{3(x + 2)(x - 2)}{x + 2} = \lim_{x \to -2} 3(x - 2) = 3(-2 - 2) = -12$

13. $\lim_{x \to 0} \frac{3x^2 - 12}{x - 2} = \frac{-12}{-2} = 6$

14. $\lim_{x\to -3} \frac{x^2-9}{3x+9} = \lim_{x\to -3} \frac{(x+3)(x-3)}{3(x+3)} = \lim_{x\to -3} \frac{x-3}{3} = \frac{-6}{3} = -2$

15. $\lim_{x\to\infty} \frac{6x^2+3}{2x^2} = \lim_{x\to\infty} \frac{6+\frac{3}{x^2}}{2} = \frac{6}{2} = 3$

16. $\lim_{x\to\infty} \frac{9x^3+2x-1}{3x^2+x+1} = \lim_{x\to\infty} \frac{9x+\frac{2}{x}-\frac{1}{x^2}}{3+\frac{1}{x}+\frac{1}{x^2}} = \lim_{x\to\infty} \frac{9}{3}x = \infty$ or does not exist

17. $\lim_{x\to 5^+} \sqrt{x-5} = \sqrt{5-5} = \sqrt{0} = 0$

18. $\lim_{x\to 4^-} \frac{x+4}{x^2-16} = \lim_{x\to 4^-} \frac{1}{x-4} = -\infty$ or does not exist

19. $\lim_{x\to\infty} \frac{x+4}{2x^2+1} = \lim_{x\to\infty} \frac{\frac{1}{x}+\frac{4}{x^2}}{2+\frac{1}{x^2}} = \frac{0}{2} = 0$

20. Since both $\lim_{x\to 1^-} f(x) = 0$ and $\lim_{x\to 1^+} f(x) = 0$, we have $\lim_{x\to 1} f(x) = 0$.

21. At $x = 5$

22. At $x = 0$

23. At $x = -3$

24. At $x = -1$ and 3

25. x cannot equal -7 so $(-\infty,-7)$, $(-7,\infty)$

26. x cannot equal -3 or -2 so $(-\infty,-3)$, $(-3,-2)$, $(-2,\infty)$

27. x^2-25 must be greater than 0 so $x^2 > 25$; or $x < -5$ or $x > 5$; $(-\infty,-5)$, $(5,\infty)$

28. $3+x > 0$ and $3-x \geq 0$ so $x > -3$ and $x \leq 3$; $(-3,3]$

CHAPTER 22 TEST

1. $\bar{m} = \frac{f(1)-f(0)}{1-0} = \frac{-1-(-3)}{1-0} = 2$

2. Here $\Delta x = \frac{5-1}{4} = 1$, so the area is $\sum_{i=1}^{4} g\left(x_i^*\right)\Delta x = [g(1.5)+g(2.5)+g(3.5)+g(4.5)]\Delta x = (3.75 + 13.75 + 29.75 + 51.75)(1) = 99$.

3. $\lim_{x\to 2}(5x^2-7) = \left(\lim_{x\to 2} 5\right)\left(\lim_{x\to 2} x^2\right) - \lim_{x\to 2} 7 = 5(2^2) - 7 = 5\cdot 4 - 7 = 20 - 7 = 13$

4. $\lim_{x\to 2}\left(\frac{x^2-4}{3x-6}\right) = \lim_{x\to 2}\left(\frac{(x+2)(x-2)}{3(x-2)}\right) = \lim_{x\to 2}\left(\frac{x+2}{3}\right) = \frac{4}{3}$

5. $\lim_{x\to 1^+} \begin{cases} x^3-1 & \text{if } x<1 \\ 4 & \text{if } x=1 \\ 1+x & \text{if } x>1 \end{cases} = \lim_{x\to 1^+}(1+x) = 1+1 = 2$

6. $\lim_{x\to\infty} \frac{5x^3-4x+1}{7x^3-2} = \lim_{x\to\infty} \frac{\frac{5x^3-4x+1}{x^3}}{\frac{7x^3-2}{x^3}} = \lim_{x\to\infty} \frac{\frac{5x^3}{x^3}-\frac{4x}{x^3}+\frac{1}{x^3}}{\frac{7x^3}{x^3}-\frac{2}{x^3}} = \lim_{x\to\infty} \frac{5-\frac{4}{x^2}+\frac{1}{x^3}}{7-\frac{2}{x^3}} = \frac{5-0+0}{7-0} = \frac{5}{7}$

7. $\lim_{x\to 2^-} h(x) = \lim_{x\to 2^-}(x+7) = 9$; $\lim_{x\to 2^+} h(x) = \lim_{x\to 2^+} \frac{1}{x^2-9} = -\frac{1}{5}$. Since $\lim_{x\to 2^-} h(x) \neq \lim_{x\to 2^+} h(x)$, $x = 2$ is a point of discontinuity. Also, $\frac{1}{x^2-9} = \frac{1}{(x+3)(x-3)}$, so h is not defined for $x = 3$, and 3 is a point of discontinuity.

8. The only possible trouble points are where the denominator is 0. The denominator of $j(x) = \frac{x^2-9}{x^2-x-12} = \frac{x^2-9}{(x+3)(x-4)}$ is 0 when $x^2-x-12 = (x+3)(x-4) = 0$ or when $x = -3$ or $x = 4$. So, these are the only two points of discontinuity and j is continuous over the intervals $(-\infty,-3)$, $(-3,4)$, $(4,\infty)$.

CHAPTER

23

The Derivative

23.1 THE TANGENT QUESTION AND THE DERIVATIVE

1. Step 1: $f(x+h) = 3(x+h) + 2 = 3x + 3h + 2.$
   Step 2: $f(x+h) - f(x) =$
   $(3x + 3h + 2) - (3x + 2) = 3h.$
   Step 3: $\dfrac{3h}{h} = 3.$
   Step 4: $\lim\limits_{h\to 0} 3 = 3$: $f'(x) = 3$

2. Step 1: $g(x+h) = 5(x+h) - 7 = 5x + 5h - 7.$
   Step 2: $g(x+h) - g(x) =$
   $(5x + 5h - 7) - (5x - 7) = 5h.$
   Step 3: $\dfrac{5h}{h} = 5.$
   Step 4: $\lim\limits_{h\to 0} 5 = 5$: $g'(x) = 5$

3. Step 1: $f(x+h) = 4 - 3(x+h) = 4 - 3x - 3h.$
   Step 2: $f(x+h) - f(x) =$
   $(4 - 3x - 3h) - (4 - 3x) = -3h.$
   Step 3: $\dfrac{-3h}{h} = -3.$
   Step 4: $\lim\limits_{h\to 0} -3 = -3$: $f'(x) = -3$

4. Step 1: $g(t+h) = 6 - 4(t+h) = 6 - 4t - 4h.$
   Step 2: $g(t+h) - g(t) =$
   $(6 - 4t - 4h) - (6 - 4t) = -4h.$
   Step 3: $\dfrac{-4h}{h} = -4.$
   Step 4: $\lim\limits_{h\to 0} -4 = -4$: $g'(t) = -4$

5. Step 1: $3(x+h)^2 = 3(x^2 + 2xh + h^2) =$
   $3x^2 + 6xh + 3h^2.$
   Step 2: $(3x^2 + 6xh + 3h^2) - 3x^2 = 6xh + 3h^2.$
   Step 3: $\dfrac{6xh + 3h^2}{h} = 6x + 3h.$
   Step 4: $\lim\limits_{h\to 0}(6x + 3h) = 6x$: $y' = 6x$

6. Step 1: $-5(t+h)^2 = -5(t^2 + 2th + h^2) =$
   $-5t^2 - 10th - 5h^2.$
   Step 2: $(-5t^2 - 10th - 5h^2) - (-5t^2) =$
   $-10th - 5h^2.$
   Step 3: $\dfrac{-10th - 5h^2}{h} = -10t - 5h$
   Step 4: $\lim\limits_{h\to 0}(-10t - 5h) = -10t$: $y' = -10t$

7. Step 1: $2(x+h)^2 + 5 = 2(x^2 + 2xh + h^2) + 5 =$
   $2x^2 + 4xh + 2h^2 + 5.$
   Step 2: $(2x^2 + 4xh + 2h^2 + 5) - (2x^2 + 5) =$
   $4xh + 2h^2.$
   Step 3: $\dfrac{4xh + 2h^2}{h} = 4x + 2h.$
   Step 4: $\lim\limits_{h\to 0}(4x + 2h) = 4x$: $y' = 4x$

8. Step 1: $j(x+h) = 3(x+h)^2 - (x+h) + 2 =$
   $3(x^2 + 2xh + h^2) - (x+h) + 2 =$
   $3x^2 + 6xh + 3h^2 - x - h + 2.$
   Step 2: $(3x^2 + 6xh + 3h^2 - x - h + 2) - (3x^2 -$
   $x + 2) = 6xh + 3h^2 - h.$

Step 3: $\dfrac{6xh + 3h^2 - h}{h} = 6x - 1 + 3h.$

Step 4: $\lim_{h\to 0}(6x - 1 + 3h) = 6x - 1$: $j' = 6x - 1$

9. Step 1: $k(x + h) = 3 - 4(x + h)^2 = 3-4(x^2+2xh+h^2) = 3-4x^2-8xh-4h^2.$

Step 2: $(3 - 4x^2 - 8xh - 4h^2) - (3 - 4x^2) = -8xh - 4h^2$

Step 3: $\dfrac{-8xh - 4h^2}{h} = -8x - 4h$

Step 4: $\lim_{h\to 0}(-8x - 4h) = -8x$: $k'(x) = -8x$

10. Step 1: $g(x + h) = 2(x + h) - 7(x + h)^2 = 2x + 2h - 7x^2 - 14xh - 7h^2.$

Step 2: $(2x+2h-7x^2-14xh-7h^2)-(2x-7x^2) = 2h - 14xh - 7h^2.$

Step 3: $\dfrac{2h - 14xh - 7h^2}{h} = 2 - 14x - 7h.$

Step 4: $\lim_{h\to 0}(2 - 14x - 7h) = 2 - 14x$: $g'(x) = 2 - 14x$

11. Step 1: $-4(x+h)^3 = -4(x^3+3x^2h+3xh^2+h^3) = -4x^3 - 12x^2h - 12xh^2 - 4h^3.$

Step 2: $(-4x^3-12x^2h-12xh^2-4h^3)-(-4x^3) = -12x^2h - 12xh^2 - 4h^3.$

Step 3: $\dfrac{12x^2h - 12xh^2 - 4h^3}{h} = -12x^2 - 12xh - 4h^2.$

Step 4: $\lim_{h\to 0}(-12x^2 - 12xh - 4h^2) = -12x^2$: $y' = -12x^2$

12. Step 1: $(x + h)^2 - 2(x + h)^3 = x^2+2xh+h^2-2x^3-6x^2h-6xh^2-2h^3.$

Step 2: $(x^2+2xh+h^2-2x^3-6x^2h-6xh^2-2h^3)-(x^2-2x^3) = 2xh+h^2-6x^2h-6xh^2-2h^3.$

Step 3: $\dfrac{2xh + h^2 - 6x^2h - 6xh^2 - 2h^3}{h} = 2x + h - 6x^2 - 6xh - 2h^2.$

Step 4: $\lim_{h\to 0}(2x + h - 6x^2 - 6xh - 2h^2) = 2x - 6x^2$: $y' = 2x - 6x^2$

13. Step 1: $(t + h) - 3(t + h)^3 = t + h - 3t^3 - 9t^2h - 9th^2 - 3h^3.$

Step 2: $(t+h-3t^3-9t^2h-9th^2-3h^3)-(t-3t^3) = h - 9t^2h - 9th^2 - 3h^3.$

Step 3: $\dfrac{h - 9t^2h - 9th^2 - 3h^3}{h} = 1 - 9t^2 - 9th - 3h^2.$

Step 4: $\lim_{h\to 0}(1 - 9t^2 - 9th - 3h^2) = 1 - 9t^2$: $y' = 1 - 9t^2$

14. Step 1: $f(x + h) = \dfrac{2(x + h) + 1}{3} = \dfrac{2x + 2h + 1}{3}.$

Step 2: $\dfrac{2x + 2h + 1}{3} - \dfrac{2x + 1}{3} = \dfrac{(2x + 2h + 1) - (2x + 1)}{3} = \dfrac{2h}{3}.$

Step 3: $\dfrac{\frac{2h}{3}}{h} = \dfrac{2}{3}.$

Step 4: $\lim_{h\to 0}\dfrac{2}{3} = \dfrac{2}{3}$: $f'(x) = \dfrac{2}{3}$

15. Step 1: $f(x + h) = \dfrac{1}{x + h + 1}.$

Step 2: $\dfrac{1}{x + h + 1} - \dfrac{1}{x + 1} = \dfrac{(x + 1) - (x + h + 1)}{(x + 1)(x + h + 1)} = \dfrac{-h}{(x + 1)(x + h + 1)}.$

Step 3: $\dfrac{\frac{-h}{(x + 1)(x + h + 1)}}{h} = \dfrac{-1}{(x + 1)(x + h + 1)}.$

Step 4: $\lim_{h\to 0}\dfrac{-1}{(x + 1)(x + h + 1)} = \dfrac{-1}{(x + 1)^2}$: $f'(x) = \dfrac{-1}{(x + 1)^2}$

16. Step 1: $g(x + h) = \dfrac{5}{2(x + h) - 3} = \dfrac{5}{2x + 2h - 3}.$

Step 2: $\dfrac{5}{2x+2h-3} - \dfrac{5}{2x-3} =$
$\dfrac{5(2x-3) - 5(2x+2h-3)}{(2x+2h-3)(2x-3)} =$
$\dfrac{10x - 15 - 10x - 10h + 15}{(2x+2h-3)(2x-3)} =$
$\dfrac{-10h}{(2x+2h-3)(2x-3)}.$

Step 3: $\dfrac{\dfrac{-10h}{(2x+2h-3)(2x-3)}}{h} =$
$\dfrac{-10}{(2x+2h-3)(2x-3)}.$

Step 4: $\lim\limits_{h\to 0} \dfrac{-10}{(2x+2h-3)(2x-3)} =$
$\dfrac{-10}{(2x-3)^2}$: $g'(x) = \dfrac{-10}{(2x-3)^2}$

17. Step 1: $j(x+h) = \dfrac{4}{1-(x+h)^2}.$

Step 2: $\dfrac{4}{1-(x+h)^2} - \dfrac{4}{1-x^2} =$
$\dfrac{4(1-x^2) - 4[1-(x+h)^2]}{[1-(x+h)^2](1-x^2)} =$
$\dfrac{4 - 4x^2 - 4 + 4x^2 + 8xh + 4h^2}{[1-(x+h)^2](1-x^2)} =$
$\dfrac{8xh + 4h^2}{[1-(x+h)^2](1-x^2)}.$

Step 3: $\dfrac{\dfrac{8xh+4h^2}{[1-(x+h)^2](1-x^2)}}{h} =$
$\dfrac{8x+4h}{[1-(x+h)^2](1-x^2)}.$

Step 4: $\lim\limits_{h\to 0} \dfrac{8x+4h}{[1-(x+h)^2](1-x^2)} =$
$\dfrac{8x + 4\cdot 0}{[1-(x+0)^2](1-x^2)} =$
$\dfrac{8x}{(1-x^2)^2}$: $j'(x) = \dfrac{8x}{(1-x^2)^2}$

18. Step 1: $f(x+h) = \dfrac{9}{(x+h)^2 + 2(x+h)}.$

Step 2: $\dfrac{9}{(x+h)^2 + 2(x+h)} - \dfrac{9}{x^2+2x} =$
$\dfrac{9(x^2+2x) - 9[(x+h)^2 + 2(x+h)]}{[(x+h)^2 + 2(x+h)](x^2+2x)} =$
$\dfrac{9x^2+18x-9x^2-18xh-9h^2-18x-18h}{[(x+h)^2 + 2(x+h)](x^2+2x)} =$
$\dfrac{-18xh - 9h^2 - 18h}{[(x+h)^2 + 2(x+h)](x^2+2x)}.$

Step 3: $\dfrac{18x - 9h - 18}{[(x+h)^2 + 2(x+h)](x^2+2x)}.$

Step 4: $\lim\limits_{h\to 0} \dfrac{-18x - 9h - 18}{[(x+h)^2 + 2(x+h)](x^2+2x)} =$
$\dfrac{-18x - 18}{[x^2+2x]^2} = \dfrac{-18(x+1)}{[x^2+2x]^2}$
or $\dfrac{-9(2x+2)}{[x^2+2x]^2}$: $f'(x) = \dfrac{-9(2x+2)}{(x^2+2x)^2}$

19. Step 1: $16(t+h)^2 - 6(t+h) + 3 =$
$16t^2 + 32th + 16h^2 - 6t - 6h + 3.$

Step 2: $(16t^2 + 32th + 16h^2 - 6t - 6h + 3) -$
$(16t^2 - 6t + 3) = 32th + 16h^2 - 6h$

Step 3: $\dfrac{32th + 16h^2 - 6h}{h} = 32t + 16h - 6$

Step 4: $\lim\limits_{h\to 0}(32t + 16h - 6) = 32t - 6$: $s' = 32t - 6$

20. Step 1: $(t+h)^3 - 4(t+h)^2 + 5(t+h) - 2 =$
$t^3 + 3t^2h + 3th^2 + h^3 - 4(t^2 + 2th +$
$h^2) + 5(t+h) - 2 = t^3 + 3t^2h + 3th^2 +$
$h^3 - 4t^2 - 8th - 4h^2 + 5t + 5h - 2$

Step 2: $(t^3 + 3t^2h + 3th^2 + h^3 - 4t^2 - 8th -$
$4h^2 + 5t + 5h - 2) - (t^3 - 4t^2 + 5t - 2) =$
$3t^2h + 3th^2 + h^3 - 8th - 4h^2 + 5h$

Step 3: $\dfrac{3t^2h + 3th^2 + h^3 - 8th - 4h^2 + 5h}{h} =$
$3t^2 + 3th + h^2 - 8t - 4h + 5$

Step 4: $\lim\limits_{h\to 0}(3t^2 + 3th + h^2 - 8t - 4h + 5) =$
$3t^2 - 8t + 5$: $q' = 3t^2 - 8t + 5$

21. Step 1: $\sqrt{(x+h)^2 + 4}$

Step 2: $\sqrt{(x+h)^2 + 4} - \sqrt{x^2 + 4}$

Step 3: $\dfrac{\sqrt{(x+h)^2 + 4} - \sqrt{x^2+4}}{h} \cdot$
$\dfrac{\sqrt{(x+h)^2 + 4} + \sqrt{x^2+4}}{\sqrt{(x+h)^2 + 4} + \sqrt{x^2+4}} =$

$$\frac{((x+h)^2+4)-(x^2+4)}{h[\sqrt{(x+h)^2+4}+\sqrt{x^2+4}]} = \frac{(x^2+2xh+h^2+4)-(x^2+4)}{h[\sqrt{(x+h)^2+4}+\sqrt{x^2+4}]} = \frac{2xh+h^2}{h[\sqrt{(x+h)^2+4}+\sqrt{x^2+4}]} = \frac{2x+h}{\sqrt{(x+h)^2+4}+\sqrt{x^2+4}}$$

Step 4: $\lim_{h\to 0} \frac{2x+h}{\sqrt{(x+h)^2+4}+\sqrt{x^2+4}} = \frac{2x}{2\sqrt{x^2+4}} = \frac{x}{\sqrt{x^2+4}}$: $y' = \frac{x}{\sqrt{x^2+4}}$

22. Step 1: $\sqrt{5-(x+h)} = \sqrt{5-x-h}$

Step 2: $\sqrt{5-x-h}-\sqrt{5-x}$

Step 3: $\frac{\sqrt{5-x-h}-\sqrt{5-x}}{h} \cdot \frac{\sqrt{5-x-h}+\sqrt{5-x}}{\sqrt{5-x-h}+\sqrt{5-x}}$
$= \frac{(5-x-h)-(5-x)}{h[\sqrt{5-x-h}+\sqrt{5-x}]} = \frac{-h}{h[\sqrt{5-x-h}+\sqrt{5-x}]} = \frac{-1}{\sqrt{5-x-h}+\sqrt{5-x}}$

Step 4: $\lim_{h\to 0} \frac{-1}{\sqrt{5-x-h}+\sqrt{5-x}} = \frac{-1}{2\sqrt{5-x}}$: $y' = \frac{-1}{2\sqrt{5-x}}$

23. Step 1: $\sqrt{2(x+h)-(x+h)^2}$

Step 2: $\sqrt{2(x+h)-(x+h)^2}-\sqrt{2x-x^2}$

Step 3: $\frac{\sqrt{2(x+h)-(x+h)^2}-\sqrt{2x-x^2}}{h} \cdot \frac{\sqrt{2(x+h)-(x+h)^2}+\sqrt{2x-x^2}}{\sqrt{2(x+h)-(x+h)^2}+\sqrt{2x-x^2}} = \frac{[2(x+h)-(x+h)^2]-[2x-x^2]}{h[\sqrt{2(x+h)-(x+h)^2}+\sqrt{2x-x^2}]} = \frac{2h-2xh-h^2}{h[\sqrt{2(x+h)-(x+h)^2}+\sqrt{2x-x^2}]} = \frac{2-2x-h}{\sqrt{2(x+h)-(x+h)^2}+\sqrt{2x-x^2}}$

Step 4:

$$\lim_{h\to 0} \frac{2-2x-h}{\sqrt{2(x+h)-(x+h)^2}+\sqrt{2x-x^2}} = \frac{2-2x}{2\sqrt{2x-x^2}} = \frac{1-x}{\sqrt{2x-x^2}}$$

$y' = \frac{1-x}{\sqrt{2x-x^2}}$

24. Step 1: $\sqrt{5(x+h)-4(x+h)^3}$.

Step 2:

$\sqrt{5x+5h-4x^3-12x^2h-12xh^2-4h^3}-\sqrt{5x-4x^3} = \sqrt{E_1}-\sqrt{E_2}$

Step 3: $\frac{\sqrt{E_1}-\sqrt{E_2}}{h} \cdot \frac{\sqrt{E_1}+\sqrt{E_2}}{\sqrt{E_1}+\sqrt{E_2}} = \frac{E_1-E_2}{h\left[\sqrt{E_1}+\sqrt{E_2}\right]} = \frac{5x+5h-4x^3-12x^2h-12xh^2-4h^3-5x+4x^3}{h\left[\sqrt{E_1}+\sqrt{E_2}\right]} = \frac{5h-12x^2h-12xh^2-4h^3}{h\left[\sqrt{E_1}+\sqrt{E_2}\right]} = \frac{5-12x^2-12xh^2-4h^2}{\left[\sqrt{E_1}+\sqrt{E_2}\right]}$

Step 4:

$$\lim_{h\to 0} \frac{5-12x^2-12xh-4h^2}{\sqrt{5(x+h)-4(x+h)^3}+\sqrt{5x-4x^3}} = \frac{5-12x^2}{2\sqrt{5x-4x^3}}$$: $y' = \frac{5-12x^2}{2\sqrt{5x-4x^3}}$

25. $f(x) = 3x+7; f'(x) = 3; f'(2) = 3; f'(9) = 3$

26. $g(x) = x^2-2; g'(x) = 2x; g'(2) = 4; g'(9) = 18$

27. $s(t) = 16t^2+2t; s'(t) = 32t+2; s'(3) = 32\cdot 3+2 = 98; s'(2) = 32\cdot 2+2 = 66; s'(0) = 32\cdot 0+2 = 2$

28. $s(t) = 2t^2-1; s'(t) = 4t; s'(1) = 4; s'(5) = 20$

29. $q(t) = t^3-4t^2; q'(t) = 3t^2-8t; q'(0) = 0; q'(3) = 3\cdot 3^2-8\cdot 3 = 27-24 = 3$.

30. $j(x) = \frac{1}{\sqrt{x}}$; Since this function is unlike any in exercises 1–24, its derivative will be evaluated using the four-step process.

Step 1: $\frac{1}{\sqrt{x+h}}$

Step 2: $\dfrac{1}{\sqrt{x+h}} - \dfrac{1}{\sqrt{x}} = \dfrac{\sqrt{x} - \sqrt{x+h}}{\sqrt{x+h}\sqrt{x}}$

Step 3: $\dfrac{\sqrt{x} - \sqrt{x+h}}{h\left(\sqrt{x+h}\sqrt{x}\right)} \cdot \dfrac{\sqrt{x} + \sqrt{x+h}}{\sqrt{x} + \sqrt{x+h}} =$

$\dfrac{x - (x+h)}{h\left(\sqrt{x+h}\sqrt{x}\right)\left(\sqrt{x} + \sqrt{x+h}\right)} =$

$\dfrac{-1}{\sqrt{x+h}\sqrt{x}[\sqrt{x} + \sqrt{x+h}]}$

Step 4: $\lim\limits_{h\to 0} \dfrac{-1}{\sqrt{x+h}\sqrt{x}\left(\sqrt{x} + \sqrt{x+h}\right)} =$

$\dfrac{-1}{\sqrt{x}\sqrt{x}\left(\sqrt{x} + \sqrt{x}\right)} = \dfrac{-1}{2\sqrt{x}^3}$

$j'(x) = \dfrac{-1}{2\sqrt{x}^3}$

$j'(4) = \dfrac{-1}{2\sqrt{4}^3} = \dfrac{-1}{2 \cdot 2^3} = -\dfrac{1}{16}$

$j'(9) = \dfrac{-1}{2\sqrt{9}^3} = \dfrac{-1}{2 \cdot 3^3} = \dfrac{-1}{2 \cdot 27} = -\dfrac{1}{54}$

31. $k(x) = \dfrac{1}{x^2-4}$; $k'(x) = \dfrac{-2x}{(x^2-4)^2}$; $k'(5) = \dfrac{-10}{21^2} = \dfrac{-10}{441}$; $k'(-5) = \dfrac{-2(-5)}{((-5)^2-4)^2} = \dfrac{10}{(25-4)^2} = \dfrac{10}{21^2} = \dfrac{10}{441}$.

32. $f(x) = \dfrac{4}{x^2+1}$; $f'(x) = \dfrac{-8x}{(x^2+1)^2}$; $f'(0) = 0$; $f'(3) = \dfrac{-8 \cdot 3}{(3^2+1)^2} = \dfrac{-24}{10^2} = \dfrac{-24}{100} = \dfrac{-6}{25}$; $f'(-5) = \dfrac{-8(-5)}{(-5^2+1)^2} = \dfrac{40}{26^2} = \dfrac{40}{676} = \dfrac{10}{169}$

33. $j(t) = \sqrt{5t-t^2}$; $j'(t) = \dfrac{5-2t}{2\sqrt{5t-t^2}}$; $j'(1) = \dfrac{5-2}{2\sqrt{5-1}} = \dfrac{3}{2\sqrt{4}} = \dfrac{3}{4}$; $j'(4) = \dfrac{5-8}{2\sqrt{20-16}} = \dfrac{-3}{2\sqrt{4}} = \dfrac{-3}{4}$ or $-\frac{3}{4}$.

34. $g(x) = \dfrac{1}{\sqrt{x+4}}$; $g'(x) = \dfrac{-1}{2\sqrt{x+4}^3}$; $g'(0) = \dfrac{-1}{2\sqrt{4}^3} = \dfrac{-1}{16}$; $g'(5) = \dfrac{-1}{2\sqrt{5+4}^3} = \dfrac{-1}{2\sqrt{9}^3} = \dfrac{-1}{2 \cdot 27} = \dfrac{-1}{54}$; $g'(-2) = \dfrac{-1}{2\sqrt{-2+4}^3} = \dfrac{-1}{2\sqrt{2}^3} = \dfrac{-1}{4\sqrt{2}}$

35. $f(x) = \dfrac{5}{\sqrt{25-x^2}}$

$f'(x) = \dfrac{-5(-2x)}{2\sqrt{25-x^2}^3} = \dfrac{5x}{\sqrt{25-x^2}^3}$

$f'(0) = 0$; $f'(3) = \dfrac{15}{\sqrt{25-9}^3} = \dfrac{15}{\sqrt{16}^3}$

$= \dfrac{15}{4^3} = \dfrac{15}{64}$;

$f'(-4) = \dfrac{-20}{\sqrt{25-16}^3} = \dfrac{-20}{3^3} = \dfrac{-20}{27}$ or $-\dfrac{20}{27}$

The four-step process for finding $f'(x)$ is as follows:

Step 1: $f(x+h) = \dfrac{5}{\sqrt{25-(x+h)^2}}$

Step 2: $\dfrac{5}{\sqrt{25-(x+h)^2}} - \dfrac{5}{\sqrt{25-x^2}} =$

$\dfrac{5\sqrt{25-x^2} - 5\sqrt{25-(x+h)^2}}{\sqrt{25-(x+h)^2} \cdot \sqrt{25-x^2}} \times$

$\dfrac{5\sqrt{25-x^2} + 5\sqrt{25-(x+h)^2}}{5\sqrt{25-x^2} + 5\sqrt{25-(x+h)^2}} =$

$\dfrac{5(25-x^2-(25-x^2-2xh-h^2)}{\sqrt{E_1}\sqrt{E_2}(\sqrt{E_2}+\sqrt{E_1})} =$

$\dfrac{5(2xh+h^2)}{\sqrt{E_2}\sqrt{E_2}(\sqrt{E_2}+\sqrt{E_1})}$

Step 3: $\dfrac{10x+5h}{\sqrt{E_1}\sqrt{E_2}(\sqrt{E_2}+\sqrt{E_1})}$

Step 4:

$\lim\limits_{h\to 0} \dfrac{10x+5h}{\sqrt{25-(x+h)^2}\sqrt{25-x^2} \times (\sqrt{25-(x+h)^2} + \sqrt{25-x^2})} =$

$\dfrac{10x}{2\sqrt{25-x^2}^3} = \dfrac{5x}{\sqrt{25-x^2}^3}$

23.2 DERIVATIVES OF POLYNOMIALS

1. $f(x) = 19; f'(x) = 0$
2. $g(x) = -4; g'(x) = 0$
3. $h(x) = 7x - 5; h'(x) = 7$
4. $j(x) = -3x + 7; j'(x) = -3$
5. $k(x) = 9x^2; k'(x) = 9(2x) = 18x$
6. $m(x) = 15x^3; m'(x) = 15(3x^2) = 45x^2$
7. $f(x) = \frac{1}{3}x^{15}; f'(x) = \frac{1}{3}(15x^{14}) = 5x^{14}$
8. $g(x) = \frac{2}{5}x^{10}; g'(x) = \frac{2}{5}(10x^9) = 4x^9$
9. $h(x) = 5x^{-2}; h'(x) = 5(-2x^{-3}) = -10x^{-3}$
10. $j(x) = -3x^{-4}; j'(x) = -3(-4x^{-5}) = 12x^{-5}$
11. $k(x) = -\frac{2}{3}x^{3/2}; k'(x) = -\frac{2}{3}(\frac{3}{2}x^{1/2}) = -x^{1/2}$
12. $j(x) = \dfrac{4\sqrt{x}}{3} = \dfrac{4}{3}x^{1/2}; j'(x) = \frac{4}{3} \cdot \frac{1}{2}x^{-1/2} = \frac{2}{3}x^{-1/2}$ or $\dfrac{2}{3\sqrt{x}}$
13. $h(x) = 4\sqrt{3x} = 4\sqrt{3}x^{1/2}; h'(x) = 4\sqrt{3} \cdot \frac{1}{2}x^{-1/2} = 2\sqrt{3}x^{-1/2}$ or $\dfrac{2\sqrt{3}}{\sqrt{x}}$
14. $m(x) = \frac{3}{4}x^{-5/3}; m'(x) = \frac{3}{4}\left(-\frac{5}{3}x^{-8/3}\right) = \frac{-5}{4}x^{-8/3}$
15. $g(x) = \dfrac{3}{\sqrt[3]{x}} = 3x^{-1/3}; g'(x) = 3(-\frac{1}{3}x^{-4/3}) = -x^{-4/3}$ or $\dfrac{-1}{\sqrt[3]{x^4}} = \dfrac{-1}{x\sqrt[3]{x}}$
16. $f(x) = 4x^3 - 5x^2 + 2; f'(x) = 12x^2 - 10x$
17. $g(x) = 9x^2 + 3x - 4; g'(x) = 18x + 3$
18. $n(x) = 7x^7 - 5x^5; n'(x) = 49x^6 - 25x^4$
19. $h(x) = \frac{1}{3}x^3 + \frac{1}{2}x^2 - 5x + \frac{2}{3}x^{-2} + 5; h'(x) = x^2 + x - 5 - \frac{4}{3}x^{-3}$
20. $j(x) = \frac{3}{4}x^4 - \frac{2}{3}x^3 + \frac{5}{4}x^2 - \frac{1}{2}x + 2x^{-3} + x^{-4} + 7;$ $j'(x) = 3x^3 - 2x^2 + \frac{5}{2}x - \frac{1}{2} - 6x^{-4} - 4x^{-5}$
21. $f(x) = \frac{4}{5}x^5 - \frac{3}{2}x^3 - 7x^0 + \frac{1}{2}x^{-2} - \frac{2}{3}x^{-3}; f'(x) = 4x^4 - \frac{9}{2}x^2 - x^{-3} + 2x^{-4}$
22. $k(x) = \sqrt{2}x^5 - \sqrt{3}x^3 + 3x + \pi; k'(x) = 5\sqrt{2}x^4 - 3\sqrt{3}x^2 + 3$
23. $l(x) = 3\sqrt{3}x^4 + 3\sqrt{5x} + \sqrt{4x^4} = 3\sqrt{3}x^4 + 3\sqrt{5}x^{1/2} + 2x^2; l'(x) = 12\sqrt{3}x^3 + \frac{3\sqrt{5}}{2}x^{-1/2} + 4x = 12\sqrt{3}x^3 + \frac{3\sqrt{5}}{2\sqrt{x}} + 4x$
24. $m(x) = \sqrt{7}x^6 - 2x^3 + 5\sqrt{3}x^2 - \sqrt{7}; m'(x) = 6\sqrt{7}x^5 - 6x^2 + 10\sqrt{3}x$
25. $s(t) = 16t^2 - 32t + 5; s'(t) = 32t - 32$
26. $q(t) = 4.3t^3 - 2.7t + 3.0; q'(t) = 12.9t^2 - 2.7$
27. $\alpha(t) = 30 - 4.0t^2 + 2t^{1/2}; \alpha'(t) = -8.0t + t^{-1/2}$
28. $s(t) = \frac{5}{3}t^4 + 7\sqrt{2}t^3 - 8t^{2/3} + 7t^{-1/2}; s'(t) = \frac{20}{3}t^3 + 21\sqrt{2}t^2 - \frac{16}{3}t^{-1/3} - \frac{7}{2}t^{-3/2}$
29. First, rewrite $q(t) = \sqrt{6t} - 4t\sqrt{3t} + \sqrt[3]{t^2}$ as $q(t) = \sqrt{6}\,t^{1/2} - 4\sqrt{3}\,t^{3/2} + t^{2/3}$. Then, differentiating produces $q'(t) = \frac{\sqrt{6}}{2}t^{-1/2} - 6\sqrt{3}\,t^{1/2} + \frac{2}{3}t^{-1/3}$.
30. First, rewrite $f(x) = 3x\sqrt{x}$ as $f(x) = 3x^{3/2}$. Then, differentiating produces $f'(x) = \frac{9}{2}x^{1/2} = \frac{9}{2}\sqrt{x}$.
31. Rewrite $g(x) = 5x^2\sqrt[3]{x}$ as $g(x) = 5x^{2\frac{1}{3}}$ or $g(x) = 5x^{7/3}$. Then, differentiating gives $g'(x) = 5 \cdot \frac{7}{3}x^{4/3} = \frac{35}{3}x^{4/3}$ or $\frac{35}{3}x\sqrt[3]{x}$.
32. Rewrite $j(x) = \frac{3x^3}{\sqrt[3]{x}}$ as $j(x) = 3x^{3-\frac{1}{3}} = 3x^{8/3}$. Then, differentiating gives $j'(x) = 8x^{5/3}$.
33. Here, we rewrite $h(x) = \dfrac{4}{x^2} = 4x^{-2}$. Then, differentiation produces $h'(x) = -8x^{-3} = \dfrac{-8}{x^3}$

34. Rewrite $j(x) = 7x^2 + 2x\sqrt{x} + \frac{5}{x^3}$ as $j(x) = 7x^2 + 2x^{3/2} + 5x^{-3}$. Then, differentiating gives $j'(x) = 14x + 3x^{1/2} - 15x^{-4} = 14x + 3\sqrt{x} - \frac{15}{x^4}$.

35. Rewrite $K(x) = \frac{4\sqrt[3]{x}}{\sqrt{x}}$ as $K(x) = 4x^{1/3-1/2} = 4x^{-1/6}$. Then differentiation produces $K'(x) = -\frac{2}{3}x^{-7/6}$

36. Here $M(x) = \frac{3\sqrt{x^3}}{\sqrt{3x}} = \frac{3}{\sqrt{3}}x^{3/2-1/2} = \sqrt{3}x^{2/2} = \sqrt{3}x$, so $M'(x) = \sqrt{3}$.

37.
$$\begin{aligned} f(x) &= \frac{4\sqrt{3x} + 3x^2 - 7\sqrt{x^3} + 2x^{-1}}{\sqrt{x}} \\ &= \frac{4\sqrt{3x}}{\sqrt{x}} + \frac{3x^2}{\sqrt{x}} - \frac{7\sqrt{x^3}}{\sqrt{x}} + \frac{2x^{-1}}{\sqrt{x}} \\ &= 4\sqrt{3} + 3x^{3/2} - 7x + 2x^{-3/2} \\ f'(x) &= \frac{9}{2}x^{1/2} - 7 - 3x^{-5/2} \end{aligned}$$

38.
$$\begin{aligned} F(x) &= f(x) - g(x) \\ F'(x) &= \lim_{h\to 0} \frac{F(x+h) - F(x)}{h} \\ &= \lim_{h\to 0} \frac{[f(x+h) - g(x+h)] - [f(x) - g(x)]}{h} \\ &= \lim_{h\to 0} \frac{[f(x+h) - f(x)] - [g(x+h) - g(x)]}{h} \\ &= \lim_{h\to 0} \frac{f(x+h) - f(x)}{h} - \lim_{h\to 0} \frac{g(x+h) - g(x)}{h} \\ &= f'(x) - g'(x). \end{aligned}$$

Hence: $F'(x) = f'(x) - g'(x)$.

39. Here we have $f(x) = x^3 + 3x^2$, and so $f'(x) = 3x^2 + 6x$. Now, $f'(x) = 0$ means that $3x^2 + 6x = 0$ or $3x(x+2) = 0$. Thus, by the zero product principle, $x = 0$ or $x = -2$.

40. Here we have $\theta(t) = 3t^3 - t$, and so $\theta'(t) = 9t^2 - 1$. Since $\theta'(t) = 0$ is the same as $9t^2 - 1 = 0$, then $9t^2 = 1$, or $t^2 = \frac{1}{9}$, and so $t = \pm\frac{1}{3}$. Thus, $t = -\frac{1}{3}$ or $t = \frac{1}{3}$.

41. We are given $g(x) = 2x^3 + x^2 - 4x$ and determine that $g'(x) = 6x^2 + 2x - 4$. Now $g'(x) = 0$ is the same as $6x^2 + 2x - 4 = 0$, or $2(3x^2 + x - 2) = 0$. This factors as $2(3x-2)(x+1) = 0$ which, by the zero product principle, has solutions when $x = \frac{2}{3}$ or $x = -1$.

42. Since $\omega(t) = 4.0t^2 - 2.0t^{1/2}$, then $\omega'(t) = 8.0t - t^{-1/2}$. Now, $\omega'(t) = 0$ means that $8.0t - t^{-1/2} = 0$, or $8t = \frac{1}{\sqrt{t}}$, or $t^{3/2} = \frac{1}{8}$ and so $t = \left(\frac{1}{8}\right)^{\frac{2}{3}} = (1/2)^2 = \frac{1}{4}$.

43. Since $j(t) = 4t^3 + \frac{1}{t}$, then $j'(t) = 12t^2 - t^{-2} = 12t^2 - \frac{1}{t^2}$. Now, $j'(t) = 0$, means that $12t^2 - \frac{1}{t^2} = 0$ or $12t^2 = \frac{1}{t^2}$ and so $12t^4 = 1$, or $t^4 = \frac{1}{12}$. Hence, $t = \sqrt[4]{\frac{1}{12}} \approx 0.537285$.

23.3 DERIVATIVES OF PRODUCT AND QUOTIENTS

1. $f'(x) = (3x+1)(2) + (3)(2x-7) = 6x + 2 + 6x - 21 = 12x - 19$

2. $g'(x) = (6x-2)(-4) + (6)(5-4x) = -24x + 8 + 30 - 24x = -48x + 38$

3. $h'(x) = (2x^2 + x - 1)(3) + (4x+1)(3x-5) = 6x^2 + 3x - 3 + 12x^2 - 17x - 5 = 18x^2 - 14x - 8$

4. $k'(x) = (4x^3 - 1)(2x - 7) + (12x^2)(x^2 - 7x) = 8x^4 - 28x^3 - 2x + 7 + 12x^4 - 84x^3 = 20x^4 - 112x^3 - 2x + 7$

5.
$$\begin{aligned} j'(x) &= (4 - 3x)(12x + 6) + (-3)(6x^2 + 6x - 4) \\ &= 48x + 24 - 36x^2 - 18x - 18x^2 - 18x + 12 \\ &= -54x^2 + 12x + 36 \end{aligned}$$

6. $$\begin{aligned} s'(t) &= (6-4t)(6t-7)+(-4)(3t^2-7t) \\ &= 36t-42-24t^2+28t-12t^2+28t \\ &= -36t^2+92t-42 \end{aligned}$$

7. $$\begin{aligned} q(t) &= 4(t^2-3t)(t^2-3t) \\ q'(t) &= 4[(t^2-3t)(2t-3)+(2t-3)(t^2-3t)] \\ &= 4[2t^3-9t^2+9t+2t^3-9t^2+9t] \\ &= 16t^3-72t^2+72t \end{aligned}$$

(Note: we learn a better way to do this problem in the next unit).

8. $$\begin{aligned} r'(p) &= (4p-3p^2)(2)+(4-6p)(2p-4) \\ &= 8p-6p^2+8p-16-12p^2+24p \\ &= -18p^2+40p-16 \end{aligned}$$

9. $$\begin{aligned} f'(w) &= (3w^3-4w^2+2w-5)(2w+w^{-2}) \\ &\quad +(9w^2-8w+2)(w^2-w^{-1}) \\ &= 6w^4-8w^3+4w^2-10w+3w-4 \\ &\quad +2w^{-1}-5w^{-2}+9w^4-8w^3 \\ &\quad +2w^2-9w+8-2w^{-1} \\ &= 15w^4-16w^3+6w^2-16w+4-5w^{-2} \end{aligned}$$

10. $$\begin{aligned} g'(r) &= (2r^4-4r^2+2r)(2r+2) \\ &\quad +(8r^3-8r+2)(r^2+2r-1) \\ &= 4r^5+4r^4-8r^3-8r^2+4r^2 \\ &\quad +4r+8r^5+16r^4-8r^3 \\ &\quad -8r^3-16r^2+8r+2r^2+4r-2 \\ &= 12r^5+20r^4-24r^3-18r^2+16r-2 \end{aligned}$$

11. $$f'(x) = \frac{4(2x+3)-(4x-1)(2)}{(2x+3)^2} = \frac{8x+12-8x+2}{(2x+3)^2} = \frac{14}{(2x+3)^2}$$

12. $$\begin{aligned} h'(y) &= (y^3-1)[(y^2-1)\cdot 1+(2y)(y-1)] \\ &\quad +3y^2[(y^2-1)(y-1)] \\ &= (y^3-1)[y^2-1+2y^2-2y] \\ &\quad +3y^2[y^3-y^2-y+1] \\ &= (y^3-1)(3y^2-2y-1)+3y^2(y^3-y^2-y+1) \\ &= 3y^5-2y^4-y^3-3y^2+2y \\ &\quad +1+3y^5-3y^4-3y^3+3y^2 \\ &= 6y^5-5y^4-4y^3+2y+1 \end{aligned}$$

13. $$\begin{aligned} g'(x) &= \frac{(4x-1)(18x)-(9x^2+2)(4)}{(4x-1)^2} \\ &= \frac{72x^2-18x-36x^2-8}{(4x-1)^2} \\ &= \frac{36x^2-18x-8}{(4x-1)^2} \end{aligned}$$

14. $$h'(x) = 8x - \frac{x^2\cdot 0-4\cdot 2x}{x^4} = 8x+\frac{8}{x^3}$$

15. $$j'(s) = 12s - \frac{-12s}{36s^4} = 12s+\frac{1}{3s^3}$$

16. $$\begin{aligned} K'(s) &= \frac{(5s-2s^2)(-4)-(3-4s)(5-4s)}{(5s-2s^2)^2} \\ &= \frac{-20s+8s^2-15+32s-16s^2}{(5s-2s^2)^2} \\ &= \frac{-8s^2+12s-15}{(5s-2s^2)^2} \end{aligned}$$

17. $$f'(t) = 12t^2 - \frac{(t-2)2-2t}{(t-2)^2} = 12t^2+\frac{4}{(t-2)^2}$$

18. $$\begin{aligned} j'(x) &= \frac{(1-2x)2-(1+2x)(-2)}{(1-2x)^2} \\ &= \frac{2-4x+2+4x}{(1-2x)^2} \\ &= \frac{4}{(1-2x)^2} \end{aligned}$$

19. $H'(x) = \dfrac{(x-1)(3x^2) - (x^3-1)}{(x-1)^2}$

$= \dfrac{3x^3 - 3x^2 - x^3 + 1}{(x-1)^2}$

$= \dfrac{2x^3 - 3x^2 + 1}{(x-1)^2} = \dfrac{(2x+1)(x-1)(x-1)}{(x-1)^2}$

$= 2x + 1$

(Note: $H(x) = \dfrac{x^3-1}{x-1} = \dfrac{(x-1)(x^2+x+1)}{x-1} = x^2 + x + 1$ and so, $H'(x) = 2x + 1$.)

20. $k'(x) = \dfrac{3x(4x^3) - (x^4+4)3}{9x^2} = \dfrac{12x^4 - 3x^4 - 12}{9x^2}$

$= \dfrac{9x^4 - 12}{9x^2} = \dfrac{3x^4 - 4}{3x^2} = x^2 - \dfrac{4}{3x^2}$

21. $f'(\phi) = \dfrac{(3\phi^2 - 1)(2\phi) - (\phi^2)(6\phi)}{(3\phi^2-1)^2} = \dfrac{6\phi^3 - 2\phi - 6\phi^3}{(3\phi^2-1)^2}$

$= \dfrac{-2\phi}{(3\phi^2-1)^2}$

22. $L'(t) = \dfrac{(t^2 - 4t - 4)(2t+3) - (t^2+3t+2)(2t-4)}{(t^2-4t-4)^2}$

$= \dfrac{2t^3 - 5t^2 - 20t - 12 - 2t^3 - 2t^2 + 8t + 8}{(t^2-4t-4)^2}$

$= \dfrac{-7t^2 - 12t - 4}{(t^2-4t-4)^2} = \dfrac{-(7t^2+12t+4)}{(t^2-4t-4)^2}$

23. $h'(x) = \dfrac{(x^3 - 3x - 1)(9x^2 - 1) - (3x^3 - x + 1)(3x^2 - 3)}{(x^3-3x-1)^2}$

$= \dfrac{(9x^5 - 28x^3 - 9x^2 + 3x + 1) - (9x^5 - 12x^3 + 3x^2 + 3x - 3)}{(x^3-3x-1)^2}$

$= \dfrac{-16x^3 - 12x^2 + 4}{(x^3-3x-1)^2}$

24. $m'(x) = \dfrac{(x^2 - 5x + 6)(2x+5) - (x^2 + 5x + 6)(2x - 5)}{(x^2-5x+6)^2}$

$= \dfrac{(2x^3 - 5x^2 - 13x + 30) - (2x^3 + 5x^2 - 13x - 30)}{(x^2-5x+6)^2}$

$= \dfrac{-10x^2 + 60}{(x^2-5x+6)^2}$

25. $f'(t) = \dfrac{t^{1/3}(6t-1) - (3t^2 - t - 1)(\frac{1}{3}t^{-2/3})}{t^{2/3}} \cdot \dfrac{t^{2/3}}{t^{2/3}}$

$= \dfrac{t(6t-1) - (3t^2 - t - 1)\frac{1}{3}}{t^{4/3}}$

$= \dfrac{6t^2 - t - t^2 + \frac{1}{3}t + \frac{1}{3}}{t^{4/3}}$

$= \dfrac{5t^2 - \frac{2}{3}t + \frac{1}{3}}{t^{4/3}} = \dfrac{15t^2 - 2t + 1}{3t^{4/3}}$

26. $F'(x) = \dfrac{(4x^{1/2} - 2)(3x^{1/2} - 2x^{-1/2}) - (2x^{3/2} - 4x^{1/2})(2x^{-1/2})}{(4x^{1/2}-2)^2}$

$= \dfrac{12x - 6x^{1/2} - 8 + 4x^{-1/2} - 4x + 8}{(4x^{1/2}-2)^2}$

$= \dfrac{8x - 6x^{1/2} + 4x^{-1/2}}{(4x^{1/2}-2)^2}$

27. $g'(w) = 24w + \dfrac{(w+1) - (w-1)}{(w+1)^2} = 24w + \dfrac{2}{(w+1)^2}$

28. $j'(t) = \dfrac{(2t+1)(2t) - (t^2+1)2}{(2t+1)^2} + \dfrac{(2t+1) - (t-1)2}{(2t+1)^2}$

$= \dfrac{4t^2 + 2t - 2t^2 - 2 + 2t + 1 - 2t + 2}{(2t+1)^2}$

$= \dfrac{2t^2 + 2t + 1}{(2t+1)^2}$

29. $h'(x) = \dfrac{(x^2+1)0 - 1(2x)}{(x^2+1)^2} = \dfrac{-2x}{(x^2+1)^2}$

30. $k'(x) = \dfrac{-5(3x^2)}{(x^3-1)^2} = \dfrac{-15x^2}{(x^3-1)^2}$

31. $H'(x) = \dfrac{(2-x^2)(3x-x^3)(-3x^2) - (4-x^3)[(-2x)(3x-x^3)+(2-x^2)(3-3x^2)]}{[(2-x^2)(3x-x^3)]^2}$

$= \dfrac{(6x-5x^3+x^5)(-3x^2) - (4-x^3)[-6x^2+2x^4+6-9x^2+3x^4]}{[(2-x^2)(3x-x^3)]^2}$

$= \dfrac{-18x^3+15x^5-3x^7-(4-x^3)[5x^4-15x^2+6]}{(2-x^2)^2(3x-x^3)^2}$

$= \dfrac{-18x^3+15x^5-3x^7-20x^4+60x^2-24+5x^7-15x^5+6x^3}{(2-x^2)^2(3x-x^3)^2}$

$= \dfrac{2x^7-20x^4-12x^3+60x^2-24}{(2-x^2)^2(3x-x^3)^2}$

32. $L'(x) = \dfrac{(7x-1)[(2x+1)(6x-4)+(2)(3x^2-4x)] - (2x+1)(3x^2-4x)\cdot 7}{(7x-1)^2}$

$= \dfrac{(7x-1)[12x^2-2x-4+6x^2-8x] - [6x^3-5x^2-4x]\cdot 7}{(7x-1)^2}$

$= \dfrac{(7x-1)[18x^2-10x-4]-[6x^3-5x^2-4x]\cdot 7}{(7x-1)^2}$

$= \dfrac{126x^3-88x^2-18x+4-42x^3+35x^2+28x}{(7x-1)^2}$

$= \dfrac{84x^3-53x^2+10x+4}{(7x-1)^2}$

33. $n(s) = \dfrac{2s^3}{s^3-s^2-s+1}$

$n'(s) = \dfrac{(s^3-s^2-s+1)(6s^2)-2s^3(3s^2-2s-1)}{(s^3-s^2-s+1)^2}$

$= \dfrac{6s^5-6s^4-6s^3+6s^2-6s^5+4s^4+2s^3}{(s^3-s^2-s+1)^2}$

$= \dfrac{-2s^4-4s^3+6s^2}{(s^2-1)^2(s-1)^2}$

34. $m(z) = \dfrac{z^3-5z^2+7z-35}{z^2+3z};$

$m'(z) = \dfrac{(z^2+3z)(3z^2-10z+7) - (z^3-5z^2+7z-35)(2z+3)}{(z^2+3z)^2}$

$= \dfrac{3z^4-z^3-23z^2+21z-2z^4+7z^3+z^2+49z+105}{(z^2+3z)^2}$

$= \dfrac{z^4+6z^3-22z^2+70z+105}{(z^2+3z)^2}$

35. $y = \dfrac{(2x-3)(x^2-4x+1)}{3x^3+1} = \dfrac{2x^3-11x^2+14x-3}{3x^3+1}$

$y' = \dfrac{(3x^3+1)(6x^2-22x+14) - (2x^3-11x^2+14x-3)(9x^2)}{(3x^3+1)^2}$

$= \dfrac{18x^5-66x^4+42x^3+6x^2-22x+14-18x^5+99x^4-126x^3+27x^2}{(3x^3+1)^2}$

$= \dfrac{33x^4-84x^3+33x^2-22x+14}{(3x^3+1)^2}$

36. $y = \dfrac{5x}{5-x}(25-x^2) = \dfrac{5x(5+x)(5-x)}{5-x} = 25x+5x^2$

$y' = 25+10x$

37. $y = \dfrac{(t^2-1)(t-1)}{(2t+1)^2} = \dfrac{t^3-t^2-t+1}{4t^2+4t+1}$

$y' = \dfrac{(4t^2+4t+1)(3t^2-2t-1) - (t^3-t^2-t+1)(8t+4)}{(4t^2+4t+1)^2}$

$= \dfrac{12t^4+4t^3-9t^2-6t-1 - 8t^4+4t^3+12t^2-4t-4}{(4t^2+4t+1)^2}$

$= \dfrac{4t^4+8t^3+3t^2-10t-5}{(2t+1)^4}$

$= \dfrac{(2t+1)(2t^3+3t^2-5)}{(2t+1)^4}$

$= \dfrac{2t^3+3t^2-5}{(2t+1)^3}$

38. $y = \left(\dfrac{1-x}{x}\right)(1-x^2)$

$$= \frac{1 - x^2 - x + x^3}{x} = \frac{x^3 - x^2 - x + 1}{x}$$
$$= x^2 - x - 1 + \frac{1}{x}$$
$$y' = 2x - 1 + \frac{-1}{x^2} = 2x - 1 - \frac{1}{x^2}$$

39. $y' = (3x^2+2x-1)(3x^2-1)+(6x+2)(x^3-x+1)$. At $x = 1, y' = (3+2-1)(3-1)+(6+2)(1-1+1) = 4 \cdot 2 + 8 \cdot 1 = 8 + 8 = 16$. The slope at $(1,4)$ is 16.

40. $y' = \dfrac{(x^2+1)(3x^2) - x^3(2x)}{(x^2+1)^2}$. At $x = -1$, $y' = \dfrac{(1+1)(3) - (-1)(-2)}{(1+1)^2} = \dfrac{6-2}{4} = \dfrac{4}{4} = 1$. The slope at $\left(-1, -\frac{1}{2}\right)$ is 1.

41. $y' = \dfrac{-8}{(x+1)^2}$; at $x = 3$, $y' = \frac{-8}{4^2} = -\frac{1}{2}$. Slope of the tangent is $-\frac{1}{2}$. Equation of the tangent is $y - 2 = -\frac{1}{2}(x-3)$ or $2y + x = 7$. The slope of the normal is 2; the equation of the normal is $y - 2 = 2(x-3)$ or $y = 2x - 4$.

42. $y' = \dfrac{5x(4) - (4x+1) \cdot 5}{25x^2}$; at $x = 1$, $y' = \dfrac{20-25}{25} = -\dfrac{1}{5}$. The equation of the tangent line is $y - 1 = -\dfrac{1}{5}(x-1)$ or $5y + x = 6$. The equation of the normal line is $y - 1 = 5(x-1)$ or $y - 5x = -4$.

43. $y' = \dfrac{(2x+1)5 - (5x+4)2}{(2x+1)^2}$; at $x = 1$, $y' = \frac{15-18}{9} = -\frac{1}{3}$. The equation of the tangent is $y - 3 = -\frac{1}{3}(x-1)$ or $3y + x = 10$. The equation of the normal is $y - 3 = 3(x-1)$ or $y - 3x = 0$.

44. $y' = (x^2 - 3x - 6)(2) + (2x-3)(2x-6)$; at $x = -1$, $y' = (1+3-6)(2)+(-5)(-8) = -4+40 = 36$. The equation of the tangent line is $y - 16 = 36(x+1)$ or $y - 36x = 52$. The equation of the normal line is $y - 16 = -\frac{1}{36}(x+1)$ or $36y + x = 575$.

45. $y = \dfrac{9x+2}{x^3+4x^2}$. $y' = \dfrac{(x^3+4x^2)9 - (9x+2)(3x^2+8x)}{(x^3+4x^2)^2}$ at $x = -2$, $y' = \dfrac{(-8+16)9 - (-16)(12-16)}{(-8+16)^2} = \dfrac{72-64}{64} = \dfrac{8}{64} = \dfrac{1}{8}$. Tangent: $y + 2 = \frac{1}{8}(x+2)$ or $8y - x = -14$. Normal: $y + 2 = -8(x+2)$ or $y + 8x = -18$.

23.4 DERIVATIVES OF COMPOSITE FUNCTIONS

1. $f'(x) = 4(3x-6)^3(3) = 12(3x-6)^3$
2. $g'(x) = 3(7-2x)^2(-2) = -6(7-2x)^2$
3. $h'(x) = -5(5x-7)^{-6}(5) = -25(5x-7)^{-6}$
4. $k'(x) = -3(9x+5)^{-4}(9) = -27(9x+5)^{-4}$
5. $f'(x) = 4(x^2+3x)^3(2x+3) = 4(2x+3)(x^2+3x)^3$
6. $h'(x) = \frac{1}{2}(4x^2+7)^{-1/2}(8x) = 4x(4x^2+7)^{-1/2}$ or $\dfrac{4x}{\sqrt{4x^2+7}}$
7. $H(x) = (4x^2+7)^{-1/2}$, so $H'(x) = -\frac{1}{2}(4x^2+7)^{-3/2}(8x) = -4x(4x^2+7)^{-3/2}$ or $\dfrac{-4x}{\sqrt{4x^2+7}^{\,3}}$.
8. $g'(x) = 6(x^7-9)^5(7x^6) = 42x^6(x^7-9)^5$
9. $s'(t) = 3(t^4 - t^3 + 2)^2(4t^3 - 3t^2)$
10. $g'(t) = 10(3t^5 + 2t^3 - t)^9(15t^4 + 6t^2 - 1)$
11. $f'(u) = 3(u^3 + 2u^{-4})^2(3u^2 - 8u^{-5})$
12. $f'(u) = 2(4u^2 - 3u^{-1})(8u + 3u^{-2})$
13.
$$\begin{aligned} g'(x) &= 3[(x-2)(3x^2-x)]^2 \\ &\quad \times [(x-2)(6x-1) + 3x^2 - x] \\ &= 3[(x-2)(3x^2-x)]^2 \\ &\quad \times [6x^2 - 13x + 2 + 3x^2 - x] \\ &= 3[(x-2)(3x^2-x)]^2[9x^2 - 14x + 2] \end{aligned}$$

14. $j(t) = 4t^3(t^2 - 4)^{1/2}$

$$j'(t) = 4t^3\left[\frac{1}{2}(t^2 - 4)^{-1/2}(2t)\right] + (t^2 - 4)^{1/2}(12t^2)$$
$$= \frac{4t^4}{\sqrt{t^2 - 4}} + 12t^2\sqrt{t^2 - 4}$$

15. $h'(v) = (v^2 + 1)^2[3(2v - 5)^2(2)]$
$$+ (2v - 5)^3[2(v^2 + 1)2v]$$
$$= 6(v^2 + 1)^2(2v - 5)^2 + 4v(v^2 + 1)(2v - 5)^3$$

16.
$$g'(x) = 3\left[(x^2 - 4x)(2x^3 - 7)\right]^2\left[(x^2 - 4x)(6x^2); + (2x - 4)(2x^3 - 7)\right]$$
$$= 3\left[(x^2 - 4x)(2x^3 - 7)\right]^2\left[6x^4 - 24x^3 + 4x^4 - 8x^3 - 14x + 28\right]$$
$$= 3\left[(x^2 - 4x)(2x^3 - 7)\right]^2\left[10x^4 - 32x^3 - 14x + 28\right]$$

17.
$$h'(x) = 4\left[(x^3 - 6x^2)(5x - 6x^2 + x^3)\right]^3\left[(x^3 - 6x^2)(5 - 12x + 3x^2) + (3x^2 - 12x)(5x - 6x^2 + x^3)\right]$$
$$= 4\left[(x^3 - 6x^2)(5x - 6x^2 + x^3)\right]^3\left[3x^5 - 30x^4 + 77x^3 - 30x^2 + 3x^5 - 30x^4 + 87x^3 - 60x^2\right]$$
$$= 4\left[(x^3 - 6x^2)(5x - 6x^2 + x^3)\right]^3(6x^5 - 60x^4 + 164x^3 - 90x^2)$$

18.
$$j'(w) = (w^3 + 2w)^4(-2)(3w - 5)^{-3}(3) + 4(w^3 + 2w)^3(3w^2 + 2) \times (3w - 5)^{-2}$$
$$= -6(w^3 + 2w)^4(3w - 5)^{-3} + 4(w^3 + 2w)^3(3w^2 + 2)(3w - 5)^{-2}$$

19. $k'(x) = (3x^2 + 2)^3 \cdot \frac{1}{2}(x^3 - 7x)^{-1/2}(3x^2 - 7)$
$$+3(3x^2 + 2)^2(6x)(x^3 - 7x)^{1/2}$$
$$= \frac{(3x^2 + 2)^3(3x^2 - 7)}{2\sqrt{x^3 - 7x}} + 18x(3x^2 + 2)^2\sqrt{x^3 - 7x}$$
$$= \frac{(3x^2 + 2)^3(3x^2 - 7)}{2\sqrt{x^3 - 7x}} + \frac{36x(3x^2 + 2)^2(x^3 - 7x)}{2 \cdot \sqrt{x^3 - 7x}}$$
$$= \frac{(3x^2 + 2)^2[(3x^2 + 2)(3x^2 - 7) + 36x(x^3 - 7x)]}{2\sqrt{x^3 - 7x}}$$
$$= \frac{(3x^2 + 2)^2[9x^4 - 15x^2 - 14 + 36x^4 - 252x^2]}{2\sqrt{x^3 - 7x}}$$
$$= \frac{(3x^2 + 2)^2[45x^4 - 267x^2 - 14]}{2\sqrt{x^3 - 7x}}$$

20. $h'(x) = 4\left(\frac{3x + 4}{2x^2 - 1}\right)^3 \cdot \frac{(2x^2 - 1)(3) - (3x + 4)(4x)}{(2x^2 - 1)^2}$
$$= \frac{4(3x + 4)^3[6x^2 - 3 - 12x^2 - 16x]}{(2x^2 - 1)^5}$$
$$= \frac{4(3x + 4)^3(-6x^2 - 16x - 3)}{(2x^2 - 1)^5}$$
$$= \frac{-4(3x + 4)^3(6x^2 + 16x + 3)}{(2x^2 - 1)^5}$$

21.
$$g'(v) = \frac{(v^3-9)^2 \cdot 3(v^2-4v)^2(2v-4) - (v^2-4v)^3 2(v^3-9)(3v^2)}{(v^3-9)^4}$$
$$= \frac{3(v^3-9)(v^2-4v)^2(2v-4) - 6v^2(v^2-4v)^3}{(v^3-9)^3}$$
$$= \frac{(v^2-4v)^2[3(v^3-9)(2v-4) - 6v^2(v^2-4v)]}{(v^3-9)^3}$$
$$= \frac{(v^2-4v)^2[6v^4-12v^3-54v+108-6v^4+24v^3]}{(v^3-9)^3}$$
$$= \frac{(v^2-4v)^2[12v^3-54v+108]}{(v^3-9)^3}$$

22.
$$f'(u) = 5\left(\frac{4u^2+5}{6u^3-3u}\right)^4 \left(\frac{(6u^3-3u)(8u)-(4u^2+5)(18u^2-3)}{(6u^3-3u)^2}\right)$$
$$= \frac{5(4u^2+5)^4[48u^4-24u^2-72u^4-78u^2+15]}{(6u^3-3u)^6}$$
$$= \frac{5(4u^2+5)^4(-24u^4-102u^2+15)}{(6u^3-3u)^6}$$
$$= \frac{-5(4u^2+5)^4(24u^4+102u^2-15)}{(6u^3-3u)^6}$$

23. $f'(u) = 6u^5;\ u' = 5x^4$
$y' = 6u^5 \cdot 5x^4 = 6(x^5+4)^5 \cdot 5x^4 = 30x^4(x^5+4)^5$

24. $f'(u) = 3u^2;\ u' = (6x^2-5)$
$y' = 3u^2(6x^2-5) = 3(2x^3-5x)^2(6x^2-5)$

25. $f'(u) = \frac{1}{2}u^{-1/2};\ u' = 8x$
$y' = \frac{1}{2}u^{-1/2} \cdot 8x = 4x(4x^2-5)^{-1/2}$

26. $f'(u) = 20u^4;\ u' = \frac{1}{2}(3x^2+5x)^{-1/2}(6x+5)$
$y' = 20u^4(\frac{1}{2}(3x^2+5x)^{-1/2}(6x+5))$
$= 10\left(\sqrt{3x^2+5x}\right)^4 (3x^2+5x)^{-1/2}(6x+5)$
$= 10(3x^2+5x)^{3/2}(6x+5)$

27. $f'(u) = \frac{2}{3}u^{-1/3};\ u' = 21x^2-9$
$y' = \frac{2}{3}(7x^3-9x)^{-1/3}(21x^2-9)$
$= 2(7x^2-3)(7x^3-9x)^{-1/3}$

28. $f'(u) = 2(u+1);\ u' = -2x^{-2}$
$y' = 2(u+1)(-2x^{-2}) = 2\left(\frac{2}{x}+1\right)(-2x^{-2})$
$= \frac{-4}{x^2}\left(\frac{2}{x}+1\right)$

29. $f'(u) = 3(u^2+1)^2(2u);\ u' = -(x+3)^{-2}$

$$y' = 3(u^2+1)^2(2u)(-(x+3)^{-2})$$
$$= 3\left[\frac{1}{(x+3)^2}+1\right]^2\left[\frac{2}{x+3}\right]\left[\frac{-1}{(x+3)^2}\right]$$
$$= \frac{-6}{(x+3)^3}\left[\frac{1}{(x+3)^2}+1\right]^2$$

30. $f'(u) = 3u^2 - 4;\ u' = 4x^3$

$$y' = (3u^2-4)(4x^3) = [3(x^4+5)^2-4](4x^3)$$

31. $f'(u) = \frac{1}{3}(u^2-2u)^{-2/3}(2u-2);\ u' = 3x^2$

$$y' = \frac{1}{3}\left[(x^3+4)^2-2(x^3+4)\right]^{-2/3}$$
$$\times\left[2(x^3+4)-2\right]\cdot 3x^2$$
$$= 2x^2[(x^3+4)^2-2(x^3+4)]^{-2/3}(x^3+3)$$

32. $f'(u) = 6u^2 - 8;\ u' = 12x - 5$

$$y' = \left[6(6x^2-5x-1)^2-8\right](12x-5)$$

33. $g'(x) = 16x^3 - 3;\ x' = 6t - 4$

$$y'(t) = \left[16x^3-3\right](6t-4)$$
$$= \left[16(3t^2-4t)^3-3\right](6t-4)$$

34. $f'(u) = 15u^2 - 14u - 5;\ u' = 12t - 8$

$$y' = [15(6t^2-8t)^2-14(6t^2-8t)-5](12t-8)$$

35. $y' = 6(9x^2+4x)^5(18x+4)$

36. $y' = 5(2x^2-5x+1)^{2/3}(4x-5)$

37. $y' = 10(11x^5-2x+1)^9(55x^4-2)$

38. $y' = -32(x^3-5x+2)^{-5}(3x^2-5) = \dfrac{-32(3x^2-5)}{(x^3-5x+2)^5}$

39. $y = 7(9x^2-4)^{-8}$

$$y' = -56(9x^2-4)^{-9}(18x)$$
$$= \frac{-56(18x)}{(9x^2-4)^9}$$

40. $y = (2x^5-x^3)^{4/3}$

$$y' = \frac{4}{3}(2x^5-x^3)^{1/3}(10x^4-3x^2)$$
$$= \frac{4}{3}\cdot\sqrt[3]{2x^5-x^3}(10x^4-3x^2)$$

41. $y = (2x^2-5)^{3/4}$

$$y' = \frac{3}{4}(2x^2-5)^{-1/4}(4x)$$
$$= \frac{3x}{\sqrt[4]{2x^2-5}}$$

42. $y = (7x-4x^3)^{-2/5}$

$$y' = -\frac{2}{5}(7x-4x^3)^{-7/5}(7-12x^2)$$
$$= \frac{-2(7-12x^2)}{5\sqrt[5]{(7x-4x^3)^7}}$$

43. $\dfrac{dy}{dx} = \dfrac{dy}{du}\cdot\dfrac{du}{dv}\cdot\dfrac{dv}{dx};\ \dfrac{dy}{du} = 4u^3;\ \dfrac{du}{dv} = 6v^2;$

$$\frac{dv}{dx} = -8x^{-3} = \frac{-8}{x^3}$$
$$\frac{dy}{dx} = (4u^3)(6v^2)\left(\frac{-8}{x^3}\right) = (4(2v^3-1)^3)(6v^2)\left(\frac{-8}{x^3}\right)$$
$$= 4\left(2\cdot\left(\frac{4}{x^2}\right)^3-1\right)^3\left(6\left(\frac{4}{x^2}\right)^2\right)\left(\frac{-8}{x^3}\right)$$
$$= 4\left(\frac{128}{x^6}-1\right)^3\left(\frac{96}{x^4}\right)\left(\frac{-8}{x^3}\right)$$
$$= \frac{-3072}{x^7}\left(\frac{128}{x^6}-1\right)^3$$

44. $\frac{dy}{du} = 8u - 1$; $\frac{du}{dx} = 3x^2$; $\frac{dx}{dt} = 6$
$\frac{dy}{dt} = (8u-1)(3x^2)(6) = [(8)(x^3-8)-1](3x^2)(6)$
$= 18x^2(8x^3 - 65)$
$= 18(6t+4)^2[8(6t+4)^3 - 65]$

45. $\frac{dy}{du} = 6u$; $\frac{du}{dv} = -4v^{-2}$; $\frac{dv}{dx} = 5x^4$
$\frac{dy}{dx} = (6u)\left(\frac{-4}{v^2}\right)(5x^4) = 6\left(\frac{4}{v}\right)\left(\frac{-4}{v^2}\right)(5x^4)$
$= \frac{-96}{v^3}(5x^4) = \frac{-96}{(x^5)^3} \cdot 5x^4$
$= \frac{-480x^4}{x^{15}} = \frac{-480}{x^{11}}$

46. $y = (x-5)^{2/3}$; $y' = \frac{2}{3}(x-5)^{-1/3}$; at $x = 4$, $y' = \frac{2}{3}(4-5)^{-1/3} = -\frac{2}{3}$. Tangent equation is $y - 1 = -\frac{2}{3}(x-4)$; $3y - 3 = -2x + 8$; $2x + 3y = 11$

47. $y' = 3(4x^2 - 18)^2(8x)$; at $x = 2$, $y' = 3(4 \cdot 2^2 - 18)^2(8 \cdot 2) = 3(-2)^2(16) = 192$. Tangent equation is $y + 8 = 192(x-2)$ or $y = 192x - 392$ or $192x - y = 392$

48. Here $y = 4(25 - x^2)^{-1/2}$, so $y' = -2(25 - x^2)^{-3/2}(-2x)$. Thus, at $x = -3$, $y' = -2(25 - 9)^{-3/2}(-2(-3)) = -12(16)^{-3/2} = -\frac{12}{64} = -\frac{3}{16}$. The slope of the normal is $\frac{16}{3}$. The equation of the normal is $y - 1 = \frac{16}{3}(x+3)$; $3y - 3 = 16x + 48$ or $16x - 3y = -51$.

23.5 IMPLICIT DIFFERENTIATION

1. $\frac{d}{dx}(4x + 5y) = \frac{d}{dx}(0)$ gives $4 + 5\frac{dy}{dx} = 0$, and so $\frac{dy}{dx} = -\frac{4}{5}$.

2. $\frac{d}{dx}(6x - 7y) = \frac{d}{dx}(3)$ gives $6 - 7\frac{dy}{dx} = 0$, and so $\frac{dy}{dx} = \frac{6}{7}$.

3. $\frac{d}{dx}(x - y^2) = \frac{d}{dx}(4)$ yields $1 - 2y\frac{dy}{dx} = 0$. Thus, $\frac{dy}{dx} = \frac{1}{2y}$.

4. $\frac{d}{dx}(x^2 + y^2) = \frac{d}{dx}(9)$ produces $2x + 2y\frac{dy}{dx} = 0$. Thus, $\frac{dy}{dx} = -\frac{x}{y}$.

5. Here we see that $\frac{d}{dx}(x^2 - y^2) = \frac{d}{dx}(16)$ produces $2x - 2yy' = 0$. Solving for y', we get $2yy' = 2x$ or $y' = \frac{2x}{2y} = \frac{x}{y}$.

6. Here $\frac{d}{dx}(x^2 - 2y^2) = \frac{d}{dx}(2x)$ gives $2x - 4yy' = 2$. Solving for y', we get $y' = \frac{2-2x}{-4y}$ or $y' = \frac{x-1}{2y}$.

7. $\frac{d}{dx}(9x^2 + 16y^2) = \frac{d}{dx}(144)$ means that $18x + 32yy' = 0$. Solving for y', we get $y' = \frac{-18x}{32y} = \frac{-9x}{16y}$.

8. Here $\frac{d}{dx}(16x^2 - 9y^2) = \frac{d}{dx}(144)$ and so $32x - 18yy' = 0$. Solving for y', we get $y' = \frac{-32x}{-18y} = \frac{16x}{9y}$.

9. $\frac{d}{dx}(4x^2 - y^2) = \frac{d}{dx}4y$ produces $8x - 2yy' = 4y'$ or $8x = 4y' + 2yy' = y'(4 + 2y)$. Solving for y', we get $y' = \frac{8x}{4+2y} = \frac{4x}{2+y}$.

10. Here, we see that $\frac{d}{dx}(x^2y + xy^2) = \frac{d}{dx}x$ gives $2xy + x^2y' + y^2 + 2xyy' = 1$ or $x^2y' + 2xyy' = 1 - 2xy - y^2$. Solving for y', we get $y' = \frac{1 - 2xy - y^2}{x^2 + 2xy}$.

11. Here we have $\frac{d}{dx}(xy + 5xy^2) = \frac{d}{dx}x^2$ or $xy' + y + 5y^2 + 10xyy' = 2x$ which is the same as $xy' + 10xyy' = 2x - y - 5y^2$. Solving for y', we get $y' = \frac{2x - y - 5y^2}{x + 10xy}$.

12. $\frac{d}{dx}(x^2y - 3xy^2) = \frac{d}{dx}y^3$ produces $2xy + x^2y' - 3y^2 - 6xyy' = 3y^2y'$ or $2xy - 3y^2 = 3y^2y' - x^2y' + 6xyy'$. Solving for y', we get $y' = \frac{2xy - 3y^2}{3y^2 - x^2 + 6xy}$ or $\frac{3y^2 - 2xy}{x^2 - 6xy - 3y^2}$.

13. Here, $\frac{d}{dx}(x^2 + 4xy + y^2) = \frac{d}{dx}y$ gives $2x + 4y + 4xy' + 2yy' = y'$ or $4xy' + 2yy' - y' = -2x - 4y$, and so we get $y' = \frac{-2x - 4y}{4x + 2y - 1}$ or $\frac{2x + 4y}{1 - 2y - 4x}$.

14. $\frac{d}{dx}(x^3y - 3xy^2) = \frac{d}{dx}2x$ produces $3x^2y + x^3y' - 3y^2 - 6xyy' = 2$ or $x^3y' - 6xyy' = 2 + 3y^2 - 3x^2y$. Thus, $y' = \frac{2+3y^2-3x^2y}{x^3-6xy}$

15. Here, $\frac{d}{dx}(x + 5x^2 - 10y^2) = \frac{d}{dx}3y$ gives $1 + 10x - 20yy' = 3y'$ or $1 + 10x = 3y' + 20yy'$. Solving for y', we get $y' = \frac{1 + 10x}{3 + 20y}$.

16. Here $\frac{d}{dx}(x^2 - 2xy + y^2) = \frac{d}{dx}x$ yields $2x - 2y - 2xy' + 2yy' = 1$ or $2yy' - 2xy' = 1 + 2y - 2x$. Thus, $y' = \frac{1 + 2y - 2x}{2y - 2x}$.

17. $\frac{d}{dx}(4x^2 + y^3) = \frac{d}{dx}9$ yields $8x + 3y^2y' = 0$, and so $y' = \frac{-8x}{3y^2}$.

18. $\frac{d}{dx}(y^4 - 9x^2) = \frac{d}{dx}9$ gives $4y^3y' - 18x = 0$, and so $y' = \frac{18x}{4y^3} = \frac{9x}{2y^3}$.

19. Here $\frac{d}{dx}(6y^3 + x^3) = \frac{d}{dx}(xy)$ produces $18y^2y' + 3x^2 = xy' + y$ or $18y^2y' - xy' = y - 3x^2$. Solving for y', we get $y' = \frac{y - 3x^2}{18y^2 - x}$.

20. $\frac{d}{dx}(x^2y) = \frac{d}{dx}(7y)$ produces $2xy + x^2y' = 7y'$ or $x^2y' - 7y' = -2xy$. Solving for y', we get $y' = \frac{2xy}{7 - x^2}$.

21. $\frac{d}{dx}(x^2y) = \frac{d}{dx}(x + 1)$ yields $2xy + x^2y' = 1$ or $x^2y' = 1 - 2xy$. Solving for y', we get $y' = \frac{1 - 2xy}{x^2}$.

22. $\frac{d}{dx}(x^2 + y^{-1}) = \frac{d}{dx}(2x)$ yields $2x - y^{-2}y' = 2$. This simplifies to $\frac{y'}{y^2} = 2x - 2$ and so $y' = y^2(2x - 2)$.

23. $\frac{d}{dx}(x^{-1} + y^{-1}) = \frac{d}{dx}16$ produces $-x^{-2} - y^{-2}y' = 0$. Solving for y', we get $y' = \frac{x^{-2}}{-y^{-2}} = -\frac{y^2}{x^2}$.

24. $\frac{d}{dx}(x^{-2} - y^{-2}) = \frac{d}{dx}(xy)$, and so $-2x^{-3} + 2y^{-3}y' = y + xy'$ which can be rewritten as $2y^{-3}y' - xy' = y + 2x^{-3}$. Solving for y', we get $y' = \frac{y + 2x^{-3}}{2y^{-3} - x} = \frac{y + \frac{2}{x^3}}{\frac{2}{y^3} - x} = \frac{x^3y^4 + 2y^3}{2x^3 - x^4y^3}$.

25. $\frac{d}{dx}\left(\frac{3}{x+1} + x^2y\right) = \frac{d}{dx}5$ produces $-3(x + 1)^{-2} + 2xy + x^2y' = 0$. Solving for y', we get $y' = \frac{3(x + 1)^{-2} - 2xy}{x^2} = \frac{\frac{3}{(x+1)^2} - 2xy}{x^2} = \frac{3 - 2xy(x + 1)^2}{x^2(x + 1)^2}$.

26. Here $\frac{d}{dx}(y^2) = \frac{d}{dx}\left(\frac{x}{y + 1}\right)$ produces $2yy' = \frac{(y + 1) - xy'}{(y + 1)^2}$, or $y' \cdot 2y(y + 1)^2 = y + 1 - xy'$,

or $y'\left[2y(y+1)^2+x\right]=y+1$. Solving for y', we get $y'=\dfrac{y+1}{2y(y+1)^2+x}$.

27. $\dfrac{d}{dx}(xy+\dfrac{y}{x})=\dfrac{d}{dx}x$ gives $y+xy'+\dfrac{xy'-y}{x^2}=1$, or $yx^2+x^3y'+xy'-y=x^2$ which can be written as $x^3y'+xy'=x^2+y-yx^2$. Solving for y', we get $y'=\dfrac{x^2+y-yx^2}{x^3+x}$.

28. Here $\frac{d}{dx}(x^3-8x^2y^2+y)=\frac{d}{dx}9x$ produces $3x^2-8x^2(2yy')-16xy^2+y'=9$, or $y'-16x^2yy'=9+16xy^2-3x^2$. Thus, $y'=\dfrac{9+16xy^2-3x^2}{1-16x^2y}$.

29. Here, $\dfrac{d}{dx}(x^4-6x^2y^2+y^2)=\dfrac{d}{dx}(10x)$ yields $4x^3-6x^2(2yy')-12xy^2+2yy'=10$, or $2yy'-12x^2yy'=10+12xy^2-4x^3$, and so $y'=\dfrac{10+12xy^2-4x^3}{2y-12x^2y}=\dfrac{5+6xy^2-2x^3}{y-6x^2y}$.

30. Here, $\dfrac{d}{dx}(x+2y)^2=\dfrac{d}{dx}4x$ produces $2(x+2y)(1+2y')=4$, or $(x+2y)(1+2y')=2$, or $1+2y'=\dfrac{2}{x+2y}$. Thus, $2y'=\dfrac{2}{x+2y}-1$, and so $y'=\dfrac{1}{x+2y}-\dfrac{1}{2}=\dfrac{2-x-2y}{(x+2y)2}$.

31. $\dfrac{d}{dx}(x^2+y^2)^{1/2}=\dfrac{d}{dx}\left(\dfrac{2y}{x}\right)$ gives $\dfrac{1}{2}(x^2+y^2)^{-1/2}(2x+2yy')=\dfrac{x\cdot 2y'-2y}{x^2}$, or $x^2(2x+2yy')=2\sqrt{x^2+y^2}(2xy'-2y)$, or $2x^3+2x^2yy'=4x\sqrt{x^2+y^2}y'-4y\sqrt{x^2+y^2}$, and so $2x^2yy'-4x\sqrt{x^2+y^2}y'=-4y\sqrt{x^2+y^2}-2x^3$. Solving for y', we get $y'=\dfrac{4y\sqrt{x^2+y^2}+2x^3}{4x\sqrt{x^2+y^2}-2x^2y}=\dfrac{2y\sqrt{x^2+y^2}+x^3}{2x\sqrt{x^2+y^2}-x^2y}$.

32. Here, $\dfrac{d}{dx}(x+xy)^{1/2}=\dfrac{d}{dx}(x^2y^2)$ gives $\frac{1}{2}(x+xy)^{-1/2}(1+xy'+y)=2xy^2+2x^2yy'$, or $1+xy'+y=2\sqrt{x+xy}(2xy^2+2x^2yy')$ which is the same as $1+xy'+y=4xy^2\sqrt{x+xy}+4x^2yy'\sqrt{x+xy}$ or $xy'-4x^2yy'\sqrt{x+xy}=4xy^2\sqrt{x+xy}-y-1$. Solving for y', we get $y'=\dfrac{4xy^2\sqrt{x+xy}-y-1}{x-4x^2y\sqrt{x+xy}}$.

33. $\dfrac{d}{dx}(x^2+y^2)^3=\dfrac{d}{dx}y$ produces $3(x^2+y^2)^2(2x+2yy')=y'$, or $6x(x^2+y^2)^2+6yy'(x^2+y^2)^2=y'$, or $6x(x^2+y^2)^2=y'-6yy'(x^2+y^2)^2$. Solving for y', we get $y'=\dfrac{6x(x^2+y^2)^2}{1-6y(x^2+y^2)^2}$.

34. Here, $\dfrac{d}{dx}(x^2y+4)^2=\dfrac{d}{dx}(3x)$ gives $2(x^2y+4)(2xy+x^2y')=3$, or $(2x^2y+8)(2xy+x^2y')=3$, and so $4x^3y^2+2x^4yy'+16xy+8x^2y'=3$, or $2x^4yy'+8x^2y'=3-4x^3y^2-16xy$. Solving for y', we get $y'=\dfrac{3-4x^3y^2-16xy}{2x^4y+8x^2}=\dfrac{3-4xy(x^2y+4)}{2x^2(x^2y+4)}$.

35. $\dfrac{d}{dx}(x^2+y^2)=\dfrac{d}{dx}25$ gives $2x+2yy'=0$ and so $y'=\dfrac{-2x}{2y}=\dfrac{-x}{y}$. Thus, the slope of the tangent to the circle at the point $(3,-4)$ is $m=\frac{-3}{-4}=\frac{3}{4}$. Using the point-slope form for the equation of a line, we see that the tangent line is $y+4=\frac{3}{4}(x-3)$ or $4y+16=3x-9$ or $4y-3x=-25$.

36. $\dfrac{d}{dx}(x^2+y^2-2x-4y-20)=\dfrac{d}{dx}0$ produces $2x+2yy'-2-4y'=0$, or $2yy'-4y'=2-2x$, and so $y'=\dfrac{2-2x}{2y-4}=\dfrac{1-x}{y-2}$. Thus, the slope of the tangent to the circle at the point $(-2,6)$ is $m=\frac{1+2}{6-2}=\frac{3}{4}$ and the slope of the normal is $-\frac{4}{3}$. Hence, the equation of the normal is $y-6=-\frac{4}{3}(x+2)$ or $3y-18=-4x-8$ or $4x+3y=10$.

37. Here, $\dfrac{d}{dx}(9x^2+16y^2-100)=\dfrac{d}{dx}0$ gives $18x+32yy'=\frac{d}{dx}0$, and so $y'=\dfrac{-18x}{32y}=\dfrac{-9x}{16y}$. Thus, the slope of the tangent to the ellipse at the point

$(-2,2)$ is $m = \frac{-9(-2)}{16(2)} = \frac{+18}{32} = \frac{9}{16}$. Hence, the tangent line is $y-2 = \frac{9}{16}(x+2)$ or $16y-32 = 9x+18$ or $16y-9x = 50$ or $9x-16y+50 = 0$.

38. $\frac{d}{dx}y^2 = \frac{d}{dx}(-16x)$, or $2yy' = -16$, and so $y' = \frac{-16}{2y} = \frac{-8}{y}$. Thus, the slope of the tangent to the parabola at the point $(-1,4)$ is $m = \frac{-8}{4} = -2$ and the slope of the normal is $\frac{1}{2}$.

39. $\frac{d}{dx}(12x^2-16y^2-192) = \frac{d}{dx}0$ gives $24x-32yy' = 0$ and so $y' = \frac{24x}{32y} = \frac{3x}{4y}$. Thus, the slope of the tangent to the hyperbola at the point $(-8,-6)$ is $m = \frac{3(-8)}{4(-6)} = \frac{-24}{-24} = 1$ and the slope of the normal is -1.

23.6 HIGHER ORDER DERIVATIVES

1.
$$\begin{aligned} y &= 4x^3 - 6x^2 + 3x - 10 \\ y' &= 12x^2 - 12x + 3 \\ y'' &= 24x - 12 \end{aligned}$$

2.
$$\begin{aligned} y &= 9x^4 + x^3 - 10; \\ y' &= 36x^3 + 3x^2; \\ y'' &= 108x^2 + 6x; \\ y''' &= 216x + 6 \end{aligned}$$

3.
$$\begin{aligned} y &= 7x^5 - 3x^3 + x; \\ y' &= 35x^4 - 9x^2 + 1; \\ y'' &= 140x^3 - 18x; \\ y''' &= 420x^2 - 18; \\ y^{(4)} &= 840x \end{aligned}$$

4.
$$\begin{aligned} f(x) &= 10x - 1; \\ f'(x) &= 10; \\ f''(x) &= 0 \end{aligned}$$

5.
$$\begin{aligned} f(x) &= x + x^{-1}; \\ f'(x) &= 1 - x^{-2}; \\ f''(x) &= 2x^{-3} = \frac{2}{x^3} \end{aligned}$$

6.
$$\begin{aligned} f(t) &= t^3 - t^{-2}; \\ f'(t) &= 3t^2 + 2t^{-3}; \\ f''(t) &= 6t - 6t^{-4}; \\ f'''(t) &= 6 + 24t^{-5} \end{aligned}$$

7.
$$\begin{aligned} f(t) &= t^2 + \frac{1}{t+1}; \\ f'(t) &= 2t - \frac{1}{(t+1)^2}; \\ f''(t) &= 2 + \frac{2}{(t+1)^3} \end{aligned}$$

8.
$$\begin{aligned} h(x) &= (x^2+2)(x-1) = x^3 - x^2 + 2x - 2 \\ h'(x) &= 3x^2 - 2x + 2; \\ h''(x) &= 6x - 2 \end{aligned}$$

9.
$$\begin{aligned} g(u) &= \frac{u-1}{u^2}; \\ g'(u) &= \frac{u^2 - (u-1)2u}{u^4} = \frac{u^2 - 2u^2 + 2u}{u^4} \\ &= \frac{2u - u^2}{u^4} = \frac{2-u}{u^3} \\ g''(u) &= \frac{u^3(-1) - (2-u)3u^2}{u^6} = \frac{-u^3 - 6u^2 + 3u^3}{u^6} \\ &= \frac{2u^3 - 6u^2}{u^6} = \frac{2u-6}{u^4} \end{aligned}$$

$$g'''(u) = \frac{u^4(2) - (2u - 6)4u^3}{u^8} = \frac{2u - 8u + 24}{u^5} = \frac{-6u + 24}{u^5}$$

10. $$y = 3(x^2 + 1)^{1/3}; \quad \frac{dy}{dx} = (x^2 + 1)^{-2/3}(2x)$$

$$\frac{d^2y}{dx^2} = 2(x^2 + 1)^{-2/3} - \frac{2}{3}(2x)(x^2 + 1)^{-5/3}(2x)$$

$$= 2(x^2 + 1)^{-2/3} - \frac{8x^2}{3}(x^2 + 1)^{-5/3}$$

$$= \frac{2}{\sqrt[3]{x^2 + 1}^2} - \frac{8x^2}{3\left(\sqrt[3]{x^2 + 1}\right)^5}$$

$$= \frac{6(x^2 + 1) - 8x^2}{3\left(\sqrt[3]{x^2 + 1}\right)^5} = \frac{-2x^2 + 6}{3\left(\sqrt[3]{x^2 + 1}\right)^5}$$

11. $$y = (x + 1)^{1/2}$$

$$\frac{dy}{dx} = \frac{1}{2}(x + 1)^{-1/2}$$

$$\frac{d^2y}{dx^2} = -\frac{1}{4}(x + 1)^{-3/2} = \frac{-1}{4\sqrt{x + 1}^3}$$

12. $$y = (4x - x^2)^2;$$

$$\frac{dy}{dx} = 2(4x - x^2)(4 - 2x) = 4(2 - x)(4x - x^2);$$

$$\frac{d^2y}{dx^2} = 4(-1)(4x - x^2) + 4(2 - x)(4 - 2x)$$

$$= -4(4x - x^2) + 4(2 - x)(4 - 2x)$$

$$= -16x + 4x^2 + 8x^2 - 32x + 32$$

$$= 12x^2 - 48x + 32$$

Note: An easier method would be to first expand $y = (4x - x^2)^2$, to get $y = 16x^2 - 8x^3 + x^4$; $y' = 32x - 24x^2 + 4x^3$; $y'' = 12x^2 - 48x + 32$.

13. $$y = (2x + 1)^3 = 8x^3 + 12x^2 + 6x + 1$$

$$\frac{dy}{dx} = 24x^2 + 24x + 6; \quad \frac{d^2y}{dx^2} = 48x + 24$$

$$\frac{d^3y}{dx^3} = 48.$$

14. $$y = x^4 - 3x^2 + 2$$

$$\frac{dy}{dx} = 4x^3 - 6x;$$

$$\frac{d^2y}{dx^2} = 12x^2 - 6$$

$$\frac{d^3y}{dx^3} = 24x$$

15. $y = x^{-2/3}$; $\frac{dy}{dx} = -\frac{2}{3}x^{-5/3}$; $\frac{d^2y}{dx^2} = \frac{10}{9}x^{-8/3}$

16. Since $y = \dfrac{x^3}{\sqrt[3]{x}} = x^{8/3}$, we have $\dfrac{dy}{dx} = \dfrac{8}{3}x^{5/3}$. The second derivative is $\dfrac{d^2y}{dx^2} = \dfrac{40}{9}x^{2/3}$; and the third derivative is $\dfrac{d^3y}{dx^3} = \dfrac{80}{27}x^{-1/3} = \dfrac{80}{27\sqrt[3]{x}}$.

17. $$f(x) = x(x + 1)^{-1}; D_x f(x) = 1(x + 1)^{-1} - x(x + 1)^{-2}$$

$$D_x^2 f(x) = -(x + 1)^{-2} - (x + 1)^{-2} + 2x(x + 1)^{-3}$$

$$= -2(x + 1)^{-2} + 2x(x + 1)^{-3}$$

$$D_x^3 f(x) = 4(x + 1)^{-3} + 2(x + 1)^{-3} - 6x(x + 1)^{-4}$$

$$= 6(x + 1)^{-3} - 6x(x + 1)^{-4}$$

$$D_x^4 f(x) = -18(x + 1)^{-4} - 6(x + 1)^{-4} + 24x(x + 1)^{-5}$$

$$= -24(x + 1)^{-4} + 24x(x + 1)^{-5}$$

$$= -24(x + 1)(x + 1)^{-5} + 24x(x + 1)^{-5}$$

$$= -24x(x + 1)^{-5} - 24(x + 1)^{-5} + 24x(x + 1)^{-5}$$

$$= -24(x + 1)^{-5}$$

18. $$f(t) = (t + 1)(t + 2)^{-1};$$

$$D_t f(t) = (t + 1)(-1)(t + 2)^{-2} + (t + 2)^{-1}$$

$$= -(t + 1)(t + 2)^{-2} + (t + 2)^{-1};$$

$$D_t^2 f(t) = -(t + 2)^{-2} + 2(t + 1)(t + 2)^{-3} - (t + 2)^{-2}$$

$$= -2(t + 2)^{-2} + 2(t + 1)(t + 2)^{-3}$$

$$= -2(t + 2)(t + 2)^{-3} + 2(t + 1)(t + 2)^{-3}$$

$$= (-2t - 4)(t + 2)^{-3} + (2t + 2)(t + 2)^{-3}$$

$$= -2(t + 2)^{-3}$$

$$D_t^3 f(t) = 6(t+2)^{-4} = \frac{6}{(t+2)^4}$$

19. $$f(w) = (3w^2+1)^{-2};$$
$$D_w f(w) = -2(3w^2+1)^{-3}(6w)$$
$$= -12w(3w^2+1)^{-3}$$
$$D_w^2 f(w) = -12(3w^2+1)^{-3} + 36w(3w^2+1)^{-4}(6w)$$
$$= 216w^2(3w^2+1)^{-4} - 12(3w^2+1)^{-3}$$
$$= 216w^2(3w^2+1)^{-4} - 12(3w^2+1)(3w^2+1)^{-4}$$
$$= (216w^2 - 36w^2 - 12)(3w^2+1)^{-4}$$
$$= (180w^2 - 12)(3w^2+1)^{-4}$$

20. $$\frac{d}{dx}(x^4 - 3x^2 + 1) = 4x^3 - 6x;$$
$$\frac{d^2}{dx^2}(x^4 - 3x^2 + 1) = 12x^2 - 6;$$
$$\frac{d^3}{dx^3}(x^4 - 3x^2 + 1) = 24x;$$
$$\frac{d^4}{dx^4}(x^4 - 3x^2 + 1) = 24;$$
$$\frac{d^5}{dx^5}(x^4 - 3x^2 + 1) = 0$$

21. $$\frac{d}{dx}(x^7 - 4x^5 - x^3) = 7x^6 - 20x^4 - 3x^2;$$
$$\frac{d^2}{dx^2}(x^7 - 4x^5 - x^3) = 42x^5 - 80x^3 - 6x;$$
$$\frac{d^3}{dx^3}(x^7 - 4x^5 - x^3) = 210x^4 - 240x^2 - 6;$$
$$\frac{d^4}{dx^4}(x^7 - 4x^5 - x^3) = 840x^3 - 480x$$

22. $$\frac{d}{dx} = 8x^7 - 6x;$$
$$\frac{d^2}{dx^2} = 8 \cdot 7 \cdot x^6 - 6;$$
$$\frac{d^3}{dx^3} = 8 \cdot 7 \cdot 6x^5;\quad \frac{d^4}{dx^4} = 8 \cdot 7 \cdot 6 \cdot 5x^4;\ \ldots;$$
$$\frac{d^8}{dx^8} = 8! = 40{,}320$$

23. $$\frac{d}{dx}(x - x^{-2}) = 1 + 2x^{-3};$$
$$\frac{d^2}{dx^2}(x - x^{-2}) = -6x^{-4};$$
$$\frac{d^3}{dx^3}(x - x^{-2}) = 24x^{-5}$$

24. $$\frac{d}{dx}\left(\frac{x^2}{2x+3}\right)$$
$$= \frac{(2x+3)2x - x^2 \cdot 2}{(2x+3)^2} = \frac{4x^2 + 6x - 2x^2}{(2x+3)^2}$$
$$= \frac{2x^2 + 6x}{(2x+3)^2};$$
$$\frac{d^2}{dx^2}\left(\frac{x^2}{2x+3}\right)$$
$$= \frac{d}{dx}\left(\frac{2x^2 + 6x}{(2x+3)^2}\right)$$
$$= \frac{(2x+3)^2(4x+6) - (2x^2+6x)(2)(2x+3)2}{(2x+3)^4}$$
$$= \frac{(2x+3)(4x+6) - 4(2x^2+6x)}{(2x+3)^3}$$
$$= \frac{8x^2 + 24x + 18 - 8x^2 - 24x}{(2x+3)^3}$$
$$= \frac{18}{(2x+3)^3}$$

25. $$\frac{d}{dx}\left(\frac{x^3}{3x-1}\right)$$
$$= \frac{(3x-1)(3x^2) - x^3(3)}{(3x-1)^2} = \frac{9x^3 - 3x^2 - 3x^3}{(3x-1)^2}$$
$$= \frac{6x^3 - 3x^2}{(3x-1)^2};$$
$$\frac{d}{dx}\left(\frac{6x^3 - 3x^2}{(3x-1)^2}\right)$$
$$= \frac{(3x-1)^2(18x^2 - 6x) - (6x^3 - 3x^2)(2)(3x-1)(3)}{(3x-1)^4}$$

$$= \frac{(3x-1)(18x^2-6x) - 6(6x^3-3x^2)}{(3x-1)^3}$$

$$= \frac{54x^3 - 36x^2 + 6x - 36x^3 + 18x^2}{(3x-1)^3}$$

$$= \frac{18x^3 - 18x^2 + 6x}{(3x-1)^3} = \frac{6x(3x^2-3x+1)}{(3x-1)^3}$$

26. $$\frac{d}{dx}\left[x^3 + x^2 - 2x\right] = 3x^2 + 2x - 2$$

$$\frac{d}{dx}(3x^2 + 2x - 2) = 6x + 2$$

27. $$\frac{d}{dx}(x(x+1)^{-1/2})$$

$$= x(-\frac{1}{2})(x+1)^{-3/2} + (x+1)^{-1/2}$$

$$= -\frac{x}{2}(x+1)^{-3/2} + (x+1)^{-1/2}$$

$$\frac{d^2}{dx^2}\left[x(x+1)^{-1/2}\right]$$

$$= -\frac{x}{2}(-3/2)(x+1)^{-5/2} - \frac{1}{2}(x+1)^{-3/2}$$

$$- \frac{1}{2}(x+1)^{-3/2}$$

$$= \frac{3}{4}x(x+1)^{-5/2} - (x+1)^{-3/2}$$

$$= \frac{3}{4}x(x+1)^{-5/2} - (x+1)(x+1)^{-5/2}$$

$$= \frac{-x-4}{4(x+1)^{5/2}}$$

28. $2x + 2yy' = 0;\ y' = -\frac{x}{y};$

$$y'' = \frac{-y + xy'}{y^2} = \frac{-y + x\left(-\frac{x}{y}\right)}{y^2}$$

$$= \frac{\frac{-y^2}{y} - \frac{x^2}{y}}{y^2} = -\frac{x^2+y^2}{y^3} = \frac{-9}{y^3}$$

(Note: $x^2 + y^2 = 9$)

29. $2x - 8yy' = 1;$

$$y' = \frac{1-2x}{-8y} = \frac{2x-1}{8y}$$

$$y'' = \frac{8y(2) - (2x-1)8y'}{64y^2}$$

$$= \frac{2y - (2x-1)\left(\frac{2x-1}{8y}\right)}{8y^2}$$

$$= \frac{\frac{16y^2}{8y} - \frac{(2x-1)^2}{8y}}{8y^2} = \frac{16y^2 - (2x-1)^2}{64y^3}$$

$$= \frac{16y^2 - 4x^2 + 4x - 1}{64y^3}$$

$$= \frac{4(4y^2 - x^2) + 4x - 1}{64y^3}$$

$$= \frac{4(-x) + 4x - 1}{64y^3} = \frac{-1}{64y^3}$$

(Note: In the second-to-last step, we used the fact that since $x^2 - 4y^2 = x$, then $4y^2 - x^2 = -x$.)

30. $2x - 2yy' = 0;$

$$y' = \frac{x}{y}$$

$$y'' = \frac{y - xy'}{y^2} = \frac{y - x\cdot\frac{x}{y}}{y^2} = \frac{\frac{y^2}{y} - \frac{x^2}{y}}{y^2}$$

$$= \frac{y^2 - x^2}{y^3} = -\frac{(x^2-y^2)}{y^3} = \frac{-16}{y^3}$$

(Note: In the last step, remember that $x^2 - y^2 = 16$.)

31. Differentiating, we get $x(2yy') + y^2 + 4 = y'$. Then, solving for y' produces

$$2xyy' - y' = -y^2 - 4;$$

$$y' = \frac{y^2+4}{1-2xy}$$

$$\begin{aligned}
y'' &= \frac{(1-2xy)(2yy') - (y^2+4)[-2xy' - 2y]}{(1-2xy)^2} \\
&= \frac{(1-2xy)(2y \cdot \frac{y^2+4}{1-2xy}) + (y^2+4)(2x \cdot \frac{y^2+4}{1-2xy} + 2y)}{(1-2xy)^2} \\
&= \frac{(2y^3+8y) + (y^2+4)\left[\frac{2xy^2+8x}{1-2xy} + \frac{2y(1-2xy)}{1-2xy}\right]}{(1-2xy)^2} \\
&= \frac{(1-2xy)2y(y^2+4) + (y^2+4)(8x+2y-2xy^2)}{(1-2xy)^3} \\
&= \frac{(y^2+4)[2y - 4xy^2 + 8x + 2y - 2xy^2]}{(1-2xy)^3} \\
&= \frac{(y^2+4)(4y - 6xy^2 + 8x)}{(1-2xy)^3}
\end{aligned}$$

32. $x(2yy') + y^2 - 5 = 4y'$;

$$\begin{aligned}
2xyy' - 4y' &= 5 - y^2 \\
y' &= \frac{5-y^2}{2xy-4} \text{ or } \frac{y^2-5}{4-2xy} \\
y'' &= \frac{(4-2xy)(2yy') - (y^2-5)(-2xy' - 2y)}{(4-2xy)^2} \\
&= \frac{(4-2xy)\left(2y \cdot \frac{y^2-5}{4-2xy}\right) + (y^2-5)(2x \cdot \frac{y^2-5}{4-2xy} + 2y)}{(4-2xy)^2} \\
&= \frac{(y^2-5)(8y-4xy^2) + (y^2-5)[2xy^2 - 10x + 8y - 4xy^2]}{(4-2xy)^3} \\
&= \frac{(y^2-5)(16y - 6xy^2 - 10x)}{(4-2xy)^3}
\end{aligned}$$

33. $x^2y' + 2xy = 0$;

$$\begin{aligned}
y' &= \frac{-2xy}{x^2} = -\frac{2y}{x} \\
y'' &= \frac{x(-2y') + (2y \cdot 1)}{x^2} = \frac{-2x \cdot \left(\frac{-2y}{x}\right) + 2y}{x^2} \\
&= \frac{4y+2y}{x^2} = \frac{6y}{x^2}
\end{aligned}$$

34. $xy' + y + 2yy' = 4x$;

$$\begin{aligned}
xy' + 2yy' &= 4x - y \\
y' &= \frac{4x-y}{x+2y}
\end{aligned}$$

$$y'' = \frac{(x+2y)(4-y') - (4x-y)(1+2y')}{(x+2y)^2}$$

$$= \frac{(x+2y)\left(4 - \dfrac{4x-y}{x+2y}\right) - (4x-y)\left(1 + 2\cdot\dfrac{4x-y}{x+2y}\right)}{(x+2y)^2}$$

$$= \frac{(x+2y)\left[\dfrac{4x+8y-4x+y}{x+2y}\right] - (4x-y)\left[\dfrac{x+2y+8x-2y}{x+2y}\right]}{(x+2y)^2}$$

$$= \frac{(x+2y)9y - (4x-y)9x}{(x+2y)^3} = \frac{9xy + 18y^2 - 36x^2 + 9xy}{(x+2y)^3}$$

$$= \frac{18y^2 + 18xy - 36x^2}{(x+2y)^3} = \frac{18(y^2 + xy - 2x^2)}{(x+2y)^3}$$

CHAPTER 23 REVIEW

1. Step 1: $2(x+h)^2 - (x+h) = 2x^2 + 4xh + 2h^2 - x - h$

   Step 2: $\left[2x^2 + 4xh + 2h^2 - x - h\right] - \left[2x^2 - x\right] = 4xh + 2h^2 - h$

   Step 3: $\dfrac{4xh + 2h^2 - h}{h} = 4x - 1 + 2h$

   Step 4: $\lim\limits_{h\to 0}(4x - 1 + 2h) = 4x - 1 = f'(x)$

2. Step 1: $(x+h)^3 - 1 = x^3 + 3x^2h + 3xh^2 + h^3 - 1$

   Step 2: $\left[x^3 + 3x^2h + 3xh^2 + h^3 - 1\right] - \left[x^3 - 1\right] = 3x^2h + 3xh^2 + h^3$

   Step 3: $\dfrac{3x^2h + 3xh^2 + h^3}{h} = 3x^2 + 3xh + h^2$

   Step 4: $\lim\limits_{h\to 0}(3x^2 + 3xh + h^2) = 3x^2 = g'(x)$

3. Step 1: $f(x+h) = (x+h) - 2(x+h)^3 = x + h - 2x^3 - 6x^2h - 6xh^2 - 2h^3$

   Step 2: $\left[x + h - 2x^3 - 6x^2h - 6xh^2 - 2h^3\right] - (x - 2x^3) = h - 6x^2h - 6xh^2 - 2h^3$

   Step 3: $\dfrac{h - 6x^2h - 6xh^2 - 2h^3}{h} = 1 - 6x^2 - 6xh - 2h^2$

   Step 4: $\lim\limits_{h\to 0}(1 - 6x^2 - 6xh - 2h^2) = 1 - 6x^2$

   $f'(x) = 1 - 6x^2$

4. Step 1: $k(x+h) = \dfrac{4}{(x+h)^2}$

   Step 2: $\dfrac{4}{(x+h)^2} - \dfrac{4}{x^2} = \dfrac{4x^2 - 4(x+h)^2}{x^2(x+h)^2} = \dfrac{4x^2 - 4x^2 - 8xh - 4h^2}{x^2(x+h)^2} = \dfrac{-8xh - 4h^2}{x^2(x+h)^2}$

   Step 3: $\dfrac{-8xh - 4h^2}{x^2(x+h)^2} \cdot \dfrac{1}{h} = \dfrac{-8x - 4h}{x^2(x+h)^2}$

   Step 4: $\lim\limits_{h\to 0} \dfrac{-8x - 4h}{x^2(x+h)^2} = \dfrac{-8x - 0}{x^2(x+0)^2} = \dfrac{-8x}{x^4} = \dfrac{-8}{x^3}$

   $k'(x) = \frac{-8}{x^3}$

5. Step 1: $m(x+h) = \sqrt{x+h}$

   Step 2: $m(x+h) - m(x) = \sqrt{x+h} - \sqrt{x}$

   Step 3: $\dfrac{\sqrt{x+h} - \sqrt{x}}{h} \cdot \dfrac{\sqrt{x+h} + \sqrt{x}}{\sqrt{x+h} + \sqrt{x}} = \dfrac{x + h - x}{h(\sqrt{x+h} + \sqrt{x})} = \dfrac{h}{h(\sqrt{x+h} + \sqrt{x})} = \dfrac{1}{\sqrt{x+h} + \sqrt{x}}$

Step 4: $\lim_{h\to 0} \frac{1}{\sqrt{x+h}+\sqrt{x}} = \frac{1}{\sqrt{x}+\sqrt{x}} = \frac{1}{2\sqrt{x}}$

$m'(x) = \frac{1}{2\sqrt{x}}$

6. Step 1: $j(t+h) = \sqrt{t+h+4}$

Step 2: $\sqrt{t+h+4} - \sqrt{t+4}$

Step 3: $\frac{\sqrt{t+h+4} - \sqrt{t+4}}{h} \cdot \frac{\sqrt{t+h+4} + \sqrt{t+4}}{\sqrt{t+h+4} + \sqrt{t+4}} = \frac{(t+h+4)-(t+4)}{h(\sqrt{t+h+4} + \sqrt{t+4})} = \frac{1}{\sqrt{t+h+4} + \sqrt{t+4}}$

Step 4: $\lim_{h\to 0} \frac{1}{\sqrt{t+h+4} + \sqrt{t+4}} = \frac{1}{\sqrt{t+4} + \sqrt{t+4}} = \frac{1}{2\sqrt{t+4}}$

$j'(t) = \frac{1}{2\sqrt{t+4}}$

7. $f'(x) = 6x + 2$

8. $g'(x) = -20x^{-6} + 2$

9. $H'(x) = 15x^4 - 12x^{-4} + \frac{3}{2}x^{-1/2} = 15x^4 - \frac{12}{x^4} + \frac{3}{2\sqrt{x}}$

10. $y' = \frac{(x^2+1)(0) - 1(2x)}{(x^2+1)^2} = \frac{-2x}{(x^2+1)^2}$

11. $y' = \frac{(x+3)3 - 3x \cdot 1}{(x+3)^2} = \frac{3x+9-3x}{(x+3)^2} = \frac{9}{(x+3)^2}$

12. Method 1: Expand the given function as $y = (2x+1)(x^2+3) = 2x^3 + x^2 + 6x + 3$; and the differentiating produces $y' = 6x^2 + 2x + 6$. Method 2: Use the Product rule with the result: $y' = (2x+1)(2x) + 2(x^2+3) = 4x^2 + 2x + 2x^2 + 6 = 6x^2 + 2x + 6$.

13. $$\begin{aligned} y' &= (x-4)(3x^2-3) + (1)(x^3-3x) \\ &= 3x^3 - 12x^2 - 3x + 12 + x^3 - 3x \\ &= 4x^3 - 12x^2 - 6x + 12 \end{aligned}$$

14. $$\begin{aligned} F'(x) &= \frac{x \cdot \frac{1}{2}(x+1)^{-1/2} - \sqrt{x+1}}{x^2} \\ &= \frac{\frac{x}{2\sqrt{x+1}} - \sqrt{x+1}}{x^2} \\ &= \frac{\frac{x}{2\sqrt{x+1}} - \frac{2(x+1)}{2\sqrt{x+1}}}{x^2} = \frac{\frac{-x-2}{2\sqrt{x+1}}}{x^2} \\ &= \frac{-x-2}{2x^2\sqrt{x+1}} \end{aligned}$$

15. $$\begin{aligned} g'(x) &= \frac{\sqrt{x+1}(1) - (x+1) \cdot \frac{1}{2}(x+1)^{-1/2}}{\sqrt{x+1}^2} \\ &= \frac{\sqrt{x+1} - \frac{(x+1)}{2\sqrt{x+1}}}{x+1} = \frac{\frac{2(x+1)}{2\sqrt{x+1}} - \frac{(x+1)}{2\sqrt{x+1}}}{x+1} \\ &= \frac{\frac{x+1}{2\sqrt{x+1}}}{x+1} = \frac{1}{2\sqrt{x+1}} \end{aligned}$$

16. $$\begin{aligned} G'(x) &= \frac{\sqrt{x+1}(2x) - x^2 \cdot \frac{1}{2}(x+1)^{-1/2}}{x+1} \\ &= \frac{\frac{2(x+1)2x}{2\sqrt{x+1}} - \frac{x^2}{2\sqrt{x+1}}}{x+1} = \frac{4x^2+4x-x^2}{2\sqrt{x+1}(x+1)} \\ &= \frac{3x^2+4x}{2(x+1)^{3/2}} \end{aligned}$$

17. $$\begin{aligned} h'(x) &= (x^2+4)\left[(x-2)(3x^2) + (x^3-2)\right] \\ &\quad + (2x)(x-2)(x^3-2) \\ &= (x^2+4)\left[3x^3 - 6x^2 + x^3 - 2\right] \\ &\quad + (2x^2-4x)(x^3-2) \\ &= (x^2+4)(4x^3-6x^2-2) + (2x^2-4x)(x^3-2) \\ &= 4x^5 - 6x^4 + 16x^3 - 26x^2 - 8 \\ &\quad + 2x^5 - 4x^4 - 4x^2 + 8x \\ &= 6x^5 - 10x^4 + 16x^3 - 30x^2 + 8x - 8 \end{aligned}$$

18. $$\begin{aligned} f'(x) &= (3x^2-4x)(15x^2+2) \\ &\quad + (6x-4)(5x^3+2x-1) \\ &= 45x^4 - 60x^3 + 6x^2 - 8x \\ &\quad + 30x^4 - 20x^3 + 12x^2 - 14x + 4 \\ &= 75x^4 - 80x^3 + 18x^2 - 22x + 4 \end{aligned}$$

19. $$\begin{aligned} F'(x) &= \frac{1}{2}\left[(x^3+7)(x+5)\right]^{-1/2}\left[(x^3+7)+3x^2(x+5)\right] \\ &= \frac{4x^3+15x^2+7}{2\sqrt{(x^3+7)(x+5)}} \end{aligned}$$

20. $g'(x) = 3(x+5)^2$

21. $$\begin{aligned} y' &= (x^2+5x)^2(-2)(x^3+1)^{-3}(3x^2) \\ &\quad + 2(x^2+5x)(2x+5)(x^3+1)^{-2} \\ &= \frac{-6x^2(x^2+5x)^2}{(x^3+1)^3} + \frac{2(x^2+5x)(2x+5)}{(x^3+1)^2} \\ &= \frac{-6x^2(x^2+5x)^2 + 2(x^2+5x)(2x+5)(x^3+1)}{(x^3+1)^3} \\ &= \frac{(x^2+5x)\left[(-6x^4-30x^3)+2(2x^4+5x^3+2x+5)\right]}{(x^3+1)^3} \\ &= \frac{(x^2+5x)(-6x^4-30x^3+4x^4+10x^3+4x+10)}{(x^3+1)^3} \\ &= \frac{(x^2+5x)(-2x^4-20x^3+4x+10)}{(x^3+1)^3} \end{aligned}$$

22. $h'(x) = 4(x^2+7x-1)^3(2x+7)$

23. $$\begin{aligned} y' &= (x+1)^2 3(x-1)^2 + 2(x+1)(x-1)^3 \\ &= (x+1)(x-1)^2[3(x+1)+2(x-1)] \\ &= (x+1)(x-1)^2[5x+1] \end{aligned}$$

24. $y' = (x+4)^3\frac{1}{2}(x-1)^{-1/2} + 3(x+4)^2\sqrt{x-1}$

25. Here $y = 5(x^2+1)^{-3}$, and so $y' = -15(x^2+1)^{-4}(2x) = \dfrac{-30x}{(x^2+1)^4}$.

26. $$\begin{aligned} y' &= \frac{(3x+4)^2\cdot 4 - 4x\cdot 2(3x+4)\cdot 3}{(3x+4)^4} \\ &= \frac{4(3x+4)-24x}{(3x+4)^3} = \frac{-12x+16}{(3x+4)^3} \end{aligned}$$

27. Method 1: Using the Quotient rule produces $y' = \dfrac{(x^2-2)^{3/2}(2x) - x^2\cdot\frac{3}{2}(x^2-2)^{1/2}\cdot 2x}{\left(\sqrt{(x^2-2)^3}\right)^2} = \dfrac{(x^2-2)^{3/2}(2x) - x^2\cdot\frac{3}{2}(x^2-2)^{1/2}\cdot 2x}{(x^2-2)^3} = \dfrac{(x^2-2)2x-3x^3}{(x^2-2)^{5/2}} = \dfrac{2x^3-4x-3x^3}{(x^2-2)^{5/2}} = \dfrac{-x^3-4x}{(x^2-2)^{5/2}}$

Method 2: Rewriting as a product (using negative exponents), we obtain $y = x^2(x^2-2)^{-3/2}$, and then, using the Product rule produces $y' = x^2(-\frac{3}{2})(x^2-2)^{-5/2}(2x) + 2x(x^2-2)^{-3/2} = (x^2-2)^{-5/2}[-3x^3+2x(x^2-2)] = (x^2-2)^{-5/2}[-x^3-4x]$.

28. $$\begin{aligned} f'(t) &= \frac{(t^2-2)(12t^2)-(4t^3-3)(2t)}{(t^2-2)^2} \\ &= \frac{12t^4-24t^2-8t^4+6t}{(t^2-2)^2} = \frac{4t^4-24t^2+6t}{(t^2-2)^2} \end{aligned}$$

29. $g'(u) = -4(u^2+\frac{1}{u}-\frac{4}{u^2})^{-5}(2u-\frac{1}{u^2}+\frac{8}{u^3})$

30. $$\begin{aligned} R(t) &= (t^2-2t+1)^{1/2}(t^2+1)^{1/3} \\ R'(t) &= \frac{1}{3}(t^2+1)^{-2/3}\cdot(2t)(t^2-2t+1)^{1/2} \\ &\quad + \frac{1}{2}(t^2-2t+1)^{-1/2}(2t-2)(t^2+1)^{1/3} \\ &= \frac{2t}{3}(t^2-2t+1)^{1/2}(t^2+1)^{-2/3} \\ &\quad + (t-1)(t^2+1)^{1/3}(t^2-2t+1)^{-1/2} \end{aligned}$$

31. Here we have $\dfrac{dy}{du} = 3u^2-1$ and $\dfrac{du}{dx} = 2x$. Applying the Chain rule, we get

$$\frac{dy}{dx} = (3u^2-1)(2x) = [3(x^2+1)^2-1](2x)$$

$$= (3x^4 + 6x^2 + 3 - 1)(2x)$$
$$= (3x^4 + 6x^2 + 2)(2x)$$
$$= 6x^5 + 12x^3 + 4x$$

32. Here $\frac{dy}{du} = 3(u+4)^2$ and $\frac{du}{dt} = 2$, so if we apply the Chain rule, we obtain

$$\frac{dy}{dt} = 3(u+4)^2 \cdot 2 = 6(2t - 5 + 4)^2$$
$$= 6(2t - 1)^2 = 6(4t^2 - 4t + 1)$$
$$= 24t^2 - 24t + 6$$

33. $$\frac{d}{dx}(x^2 + 4xy + 4y^2) = \frac{d}{dx}16;$$
$$2x + 4xy' + 4y + 8yy' = 0$$
$$(4x + 8y)y' = -2x - 4y$$
$$y' = -\frac{2x + 4y}{4x + 8y} = -\frac{1}{2}$$

34. $$3x^2 - 2xy' - 2y = 2yy';$$
$$(-2x - 2y)y' = 2y - 3x^2;$$
$$y' = \frac{3x^2 - 2y}{2x + 2y}$$

35. $$2xy' + 2y - 2x = 2yy'x + y^2;$$
$$(2x - 2xy)y' = y^2 + 2x - 2y;$$
$$y' = \frac{y^2 + 2x - 2y}{2x - 2xy} \text{ or } \frac{2y - 2x - y^2}{2xy - 2x}$$

36. $$x^2y' + 2xy + y3x^2 + y'x^3 = 0;$$
$$(x^2 + x^3)y' = -2xy - 3x^2y;$$
$$y' = -\frac{2xy + 3x^2y}{x^2 + x^3}$$
$$= -\frac{2y + 3xy}{x + x^2}$$

37. $$x^2 3y^2 y' + 2xy^3 - \frac{y'}{y^2} = 1;$$
$$(3x^2y^2 - \frac{1}{y^2})y' = 1 - 2xy^3$$
$$(3x^2y^4 - 1)y' = y^2 - 2xy^5$$
$$y' = \frac{y^2 - 2xy^5}{3x^2y^4 - 1}$$

38. $$8x + x2yy' + y^2 - y'y^{-2} = 2$$
$$8xy^2 + 2xy^3y' + y^4 - y' = 2y^2$$
$$(2xy^3 - 1)y' = 2y^2 - 8xy^2 - y^4$$
$$y' = \frac{2y^2 - 8xy^2 - y^4}{2xy^3 - 1}$$

39. $$f(x) = x^7 + 2x^4 + 3x;$$
$$f'(x) = 7x^6 + 8x^3 + 3$$
$$f''(x) = 42x^5 + 24x^2$$

40. $$h(x) = x^{1/2} - x^{-1/2};$$
$$h'(x) = \frac{1}{2}x^{-1/2} + \frac{1}{2}x^{-3/2}$$
$$h''(x) = -\frac{1}{4}x^{-3/2} - \frac{3}{4}x^{-5/2};$$
$$h'''(x) = \frac{3}{8}x^{-5/2} + \frac{15}{8}x^{-7/2}$$

41. $$g(x) = 4x^3 - 5x + 2x^{-1};$$
$$g'(x) = 12x^2 - 5 - 2x^{-2}$$
$$g''(x) = 24x + 4x^{-3}$$
$$g'''(x) = 24 - 12x^{-4} \text{ or } 24 - \frac{12}{x^4}$$

42. $\frac{dy}{dx} = 20x^4 + 6x^{-4}$; $\frac{d^2y}{dx^2} = 80x^3 - 24x^{-5}$

43. $$\frac{dy}{dx} = 21x^2 - 21x^{-4};$$
$$\frac{d^2y}{dx^2} = 42x + 84x^{-5}$$
$$\frac{d^3y}{dx^3} = 42 - 420x^{-6}$$

44. $$\frac{dy}{dx} = -3x^{-4} - 4x^{-2};$$
$$\frac{d^2y}{dx^2} = 12x^{-5} + 8x^{-3}$$
$$\frac{d^3y}{dx^3} = -60x^{-6} - 24x^{-4}$$

45. $$D_x f(x) = 15x^2 + \frac{1}{2}x^{-1/2};$$
$$D_x^2 f(x) = 30x - \frac{1}{4}x^{-3/2}$$
$$D_x^3 f(x) = 30 + \frac{3}{8}x^{-5/2}$$

46. $$\frac{d}{dx}g(x) = 3(x^2+1)^2(2x) = 6x(x^2+1)^2$$
$$\frac{d^2}{dx^2}g(x) = 12x(x^2+1)(2x) + 6(x^2+1)^2$$
$$= 24x^4 + 24x^2 + 6x^4 + 12x^2 + 6$$
$$= 30x^4 + 36x^2 + 6$$
$$\frac{d^3}{dx^3}g(x) = 120x^3 + 72x$$

47. Using implicit differentiation, we get $x^2y' + 2xy = y' + 2$ or $(x^2-1)y' = 2 - 2xy$. Solving for y', we obtain $y' = \frac{2-2xy}{x^2-1}$. Differentiating again yields the following:

$$y'' = \frac{(x^2-1)(-2xy'-2y) - (2-2xy)2x}{(x^2-1)^2}$$
$$= \frac{(x^2-1)(-2x \cdot \frac{2-2xy}{x^2-1} - 2y) - (2-2xy)2x}{(x^2-1)^2}$$
$$= \frac{-2x(2-2xy) - (x^2-1)2y - 2x(2-2xy)}{(x^2-1)^2}$$
$$= \frac{-4x(2-2xy) - (x^2-1)2y}{(x^2-1)^2}$$
$$= \frac{-8x + 8x^2y - 2x^2y + 2y}{(x^2-1)^2} = \frac{6x^2y - 8x + 2y}{(x^2-1)^2}$$

48. $$2yy' = 2xyy' + y^2 + 3x^2$$
$$(2y - 2xy)y' = y^2 + 3x^2$$
$$y' = \frac{y^2+3x^2}{2y-2xy}$$

49. Using implicit differentiation, we get $xy' + y = 2x + 2yy'$ or $(x-2y)y' = 2x - y$. Solving for y', we obtain $y' = \frac{2x-y}{x-2y}$. Differentiating again yields the following:

$$y'' = \frac{(x-2y)(2-y') - (2x-y)(1-2y')}{(x-2y)^2}$$
$$= \frac{2x - xy' - 4y + 2yy' - 2x + 4xy' + y - 2yy'}{(x-2y)^2}$$
$$= \frac{3xy' - 3y}{(x-2y)^2} = \frac{3x\left(\frac{2x-y}{x-2y}\right) - 3y}{(x-2y)^2}$$
$$= \frac{3x(2x-y) - 3y(x-2y)}{(x-2y)^3}$$
$$= \frac{6x^2 - 3xy - 3xy + 6y^2}{(x-2y)^3} = \frac{6x^2 - 6xy + 6y^2}{(x-2y)^3}$$

50. $y' = 8x - 9$; at $x = 3$, $y' = 8 \cdot 3 - 9 = 24 - 9 = 15$. The slope of the tangent is 15 and the slope of the normal is $-\frac{1}{15}$.

51. $f'(x) = 21x^2 - 3$; at $x = -1, f'(-1) = 21 - 3 = 18$. The slope of the tangent is 18.

52. $8x + 18yy' = 0$; $y' = \frac{-8x}{18y} = \frac{-4x}{9y}$; at $(-1, 2)$, $y' = \frac{-4(-1)}{9 \cdot 2} = \frac{4}{18} = \frac{2}{9}$. The slope of the tangent is $\frac{2}{9}$.

53. Finding y' by implicit differentiation, we have

$$8x + 6yy' - 5y - 5xy' - 4 + 10y' = 0$$
$$(6y - 5x + 10)y' = 5y + 4 - 8x$$
$$y' = \frac{5y+4-8x}{6y-5x+10}$$

At $(2, -1)$, we see that

$$y' = \frac{5(-1) + 4 - 8(2)}{6(-1) - 5 \cdot 2 + 10}$$

$$= \frac{-5+4-16}{-6-10+10} = \frac{-17}{-6} = \frac{17}{6}.$$

The slope of the tangent is $\frac{17}{6}$ and the slope of the normal is $-\frac{6}{17}$.

54. $f'(x) = 3x^2 - 3$; $3x^2 - 3 = 0 \Rightarrow x^2 = 1 \Rightarrow x = \pm 1$

55. $$g'(x) = \frac{(x+1)2x - x^2 \cdot 1}{(x+1)^2} = \frac{2x^2+2x-x^2}{(x+1)^2} = \frac{x^2+2x}{(x+1)^2}$$

$g'(x) = 0$ when $x^2 + 2x = 0$ so $x(x+2) = 0 \Rightarrow x = 0$ or $x = -2$.

56. $y' = 2x$, at $(-1, 5)$, $y' = -2$. Equation of tangent is $y - 5 = -2(x+1)$ or $y = -2x + 3$ or $2x + y - 3 = 0$

57. Finding y' by implicit differentiation, we have

$$8x + 10yy' = 0;$$
$$y' = \frac{-4x}{5y}$$

At $(-2, 2)$, we see that $y' = \dfrac{-4(-2)}{5(2)} = \dfrac{4}{5}$.

Slope of the normal is $-\frac{5}{4}$. Equation of the normal is $y - 2 = -\frac{5}{4}(x+2)$ or $4y - 8 = -5x - 10$ or $5x + 4y + 2 = 0$

58. $y' = \dfrac{1}{2}(x+4)^{-1/2} = \dfrac{1}{2\sqrt{x+4}}$. This is never equal to zero so the answer is none.

59. $$f(x) = 3x^4 - 6x^2 + 2;\ f'(x) = 12x^3 - 12x$$
$$f''(x) = 36x^2 - 12$$

Now, $36x^2 - 12 = 0 \Rightarrow x^2 = \frac{1}{3}$ and so, $x = \pm\sqrt{\dfrac{1}{3}} = \dfrac{\pm\sqrt{3}}{3}$

60. $y' = 3x^2 - 18x$; $y'' = 6x - 18$. Now, $y'' = 0$ means that $6x - 18 = 0 \Rightarrow x - 3 = 0$ or $x = 3$.

CHAPTER 23 TEST

1. $f'(x) = 15x^2$

2. $g'(x) = -12x^{-5} + 10x$

3. Rewrite as $h(x) = (x^6 - 2)^{1/3}$ and, using the Chain Rule, we obtain $h'(x) = \frac{1}{3}(x^6-2)^{-2/3}(6x^5) = 2x^5(x^6-2)^{-2/3}$

4. Using the Product rule, we get $j'(x) = (2x+1)(6x-1) + (3x^2 - x)(2) = 12x^2 + 4x - 1 + 6x^2 - 2x = 18x^2 + 2x - 1$

5. Using the Quotient rule produces

$$k'(x) = \frac{(4x^2+1)^{1/2}(2) - (2x+1)\frac{1}{2}(4x^2+1)^{-1/2}(8x)}{[(4x^2+1)^{1/2}]^2}$$
$$= \frac{(4x^2+1)^{-1/2}[(4x^2+1)(2) - (2x+1)4x]}{4x^2+1}$$
$$= \frac{8x^2+2-4x(2x+1)}{(4x^2+1)^{3/2}} = \frac{2-4x}{(4x^2+1)^{3/2}}$$

6. $f'(x) = 20x^3 + 12x^{-4}$; $f''(x) = 60x^2 - 48x^{-5}$; $f'''(x) = 120x + 240x^{-6}$

7. Using implicit differentiation, we obtain $2y + 2xy' + 3y^2xy' + y^3 = 3x^2$. Rewriting, we get $(2x + 3y^2x)y' = 3x^2 - 2y - y^3$ and so $y' = \dfrac{3x^2 - 2y - y^3}{2x + 3y^2x}$.

8. Taking the derivative of g we obtain $g'(x) = 12x^2 - 4x$. When $x = 1$, $g'(1) = 8$, and so the slope of the tangent to the curve at $(1, 6)$ is 8. Using the point-slope form for the equation of a line, we see the equation of the tangent line is $y - 6 = 8(x - 1)$ or $y = 8x - 2$.

CHAPTER

24

Applications of Derivatives

24.1 RATES OF CHANGE

1. $s(t) = 3t^2 - 12t + 5$; $v(t) = s'(t) = 6t - 12$; $a(t) = s''(t) = 6$

t	0	1	2	3	4	5
$s(t)$	5	−4	−7	−4	5	20
$v(t)$	−12	−6	0	6	12	18
$a(t)$	6	6	6	6	6	6

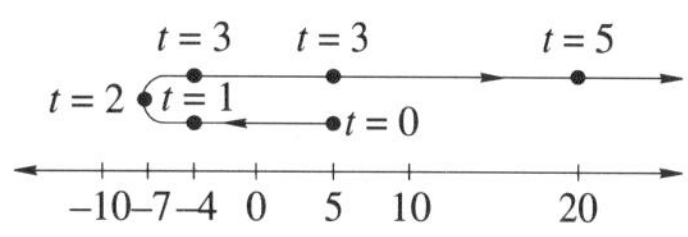

2. $s(t) = 3t^2 - 18t + 10$; $v(t) = s'(t) = 6t - 18$; $a(t) = s''(t) = 6$

t	0	1	2	3	4	5
$s(t)$	10	−5	−14	−17	−14	−5
$v(t)$	−18	−12	−6	0	6	12
$a(t)$	6	6	6	6	6	6

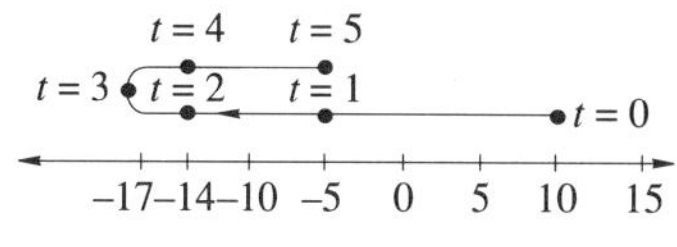

3. $s(t) = t^3 - 12t + 2$; $v(t) = s'(t) = 3t^2 - 12$; $a(t) = s''(t) = 6t$

t	−4	−3	−2	−1	0	1	2	3	4
$s(t)$	−14	11	18	13	2	−9	−14	−7	18
$v(t)$	36	15	0	−9	−12	−9	0	15	36
$a(t)$	−24	−18	−12	−6	0	6	12	18	24

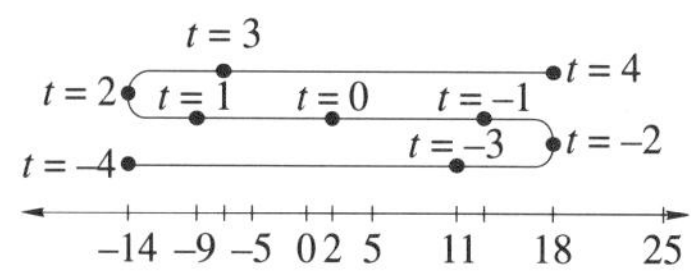

4. $s(t) = t^3 - 27t + 20$; $v(t) = 3t^2 - 27$; $a(t) = s''(t) = 6t$

t	−4	−3	−2	−1	0	1	2	3	4
$s(t)$	64	74	66	46	20	−6	−26	−34	−24
$v(t)$	21	0	−15	−24	−27	−24	−15	0	21
$a(t)$	−24	−18	−12	−6	0	6	12	18	24

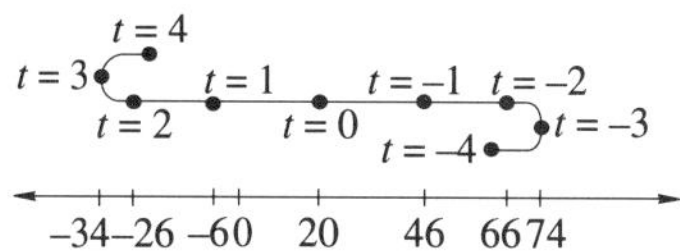

5. $s(t) = t + 4/t$; $v(t) = 1 - 4/t^2$; $a(t) = 8/t^3$

t	1	2	3	4
$s(t)$	5	4	4.3333	5
$v(t)$	−3	0	0.5555	0.75
$a(t)$	8	1	0.2963	0.125

t = 3, t = 4, t = 2, t = 1; 0, 4, 5, 6

6. $s(t) = t + 8/t$; $v(t) = 1 - 8/t^2$; $a(t) = 16/t^3$

t	1	2	3	4
$s(t)$	9	6	5.6667	6
$v(t)$	−7	−1	0.1111	0.5
$a(t)$	16	2	0.5926	0.25

t = 4, t = 3, t = 2, t = 1; 0, 4, 5, 6, 9

7. $s(t) = 2\sqrt{t} + \dfrac{1}{\sqrt{t}} = 2t^{1/2} + t^{-1/2}$; $v(t) = t^{-1/2} - \frac{1}{2}t^{-3/2}$;
$a(t) = -\frac{1}{2}t^{-3/2} + \frac{3}{4}t^{-5/2}$

t	1	2	3	4
$s(t)$	3	3.5355	4.0415	4.5
$v(t)$	0.5	0.5303	0.4811	0.4375
$a(t)$	0.25	−0.0442	−0.0481	−0.0391

t = 2, t = 1, t = 4, t = 3; 0, 3, 4, 5, 6

8. $s(t) = t^3 - 8$; $v(t) = 3t^2$; $a(t) = 6t$

t	0	1	2	3	4
$s(t)$	−8	−7	0	19	56
$v(t)$	0	3	12	27	48
$a(t)$	0	6	12	18	24

t = 1, t = 3, t = 0, t = 4, t = 2; −8 −7 0, 19, 56

9. (a) The object hits the ground when $s(t) = 0$. Solving $144t - 16t^2 = 0$ we get $t(144 - 16t) = 0$ so $t = 0$ or $t = 9$. The object is fired at $t = 0$ and hits the ground at $t = 9$ s. (b) $v(t) = 144 - 32t$; $v(9) = 144 - 32 \cdot 9 = -144$ ft/s; $a(t) = -32$; $a(9) = -32$ ft/s^2. (c) $s(0) = 0$ ft = height when fired.

10. (a) Solving for t when $250 + 256t - 16t^2 = 0$ the quadratic formula produces $t = -0.9233$ or $t = 16.9233$. Since t cannot be negative it hits the ground at 16.9233 s. (b) $v(t) = 256 - 32t$; $v(16.9233) = -285.55$ ft/s; $a(t) = -32$; $a(16.9233) = -32$ ft/s^2. (c) $s(0) = 250$ ft = height when fired.

11. (a) $s(t) = 29.4t - 4.9t^2 = 0 \Rightarrow t(29.4 - 4.9t) = 0$. By the zero product property, $t = 0$ or $t = \frac{29.4}{4.9} = 6$. Since t must be greater than 0, it hits the ground at 6 s. (b) $v(t) = 29.4 - 9.8t$; $v(6) = -29.4$ m/s; $a(t) = -9.8$; $a(6) = -9.8$ m/s^2. (c) $s(0) = 0$ m is the initial height.

12. (a) $s(t) = 180 + 98t - 4.9t^2 = 0$. Solving for t using the quadratic formula yields $t = -1.6934$ and $t = 21.6934$. Since $t > 0$ we get 21.6934 s. (b) $v(t) = 98 - 9.8t$ means that $v(21.6934) = -114.595$ m/s, and $a(t) = -9.8$ gives $a(21.6934) = -9.8$ m/s^2. (c) $s(0) = 180$ m is the initial height.

13. (a) The rate of change of $F(s)$ is $\dfrac{dF}{ds} = F'(s) = -100/s^3$ dynes/cm, (b) $F'(3) = -\frac{100}{27}$ dynes/cm

14. (a) $i = \dfrac{dq}{dt} = 10 - 4.0t$, (b) $q(0) = 25$ C, (c) $i(0) = 10$ A; $i(4) = 10 - 16 = -6$ A

15. (a) $\dfrac{dF}{dC} = \dfrac{9}{5}$, (b) $\dfrac{dC}{dF} = \dfrac{d}{dF}\left(\dfrac{5}{9}(F - 32)\right) = \dfrac{5}{9}$,
(c) $\dfrac{dC}{dF}$ at $C = 20°$ is $\dfrac{5}{9}$

16. (a) Since the area of a circle is related to (is a function of) its radius, we can think of $A = \pi r^2$ as $A(r) = \pi r^2$ and thus, the rate of change of the area with respect to the radius is $\frac{dA}{dr} = A'(r) = 2\pi r$ cm^2/cm, (b) at $r = 1.5$, $A'(1.5) = 2\pi(1.5) = 3\pi \approx 9.4248$ cm^2/cm; (c) at $r = 3$, we have $A'(3) = 2\pi \cdot 3 = 6\pi \approx 18.8496$ cm^2/cm.

17. (a) The rate of change in the quantity of water is $\frac{dQ}{dt} = Q'(t) = -50 + t$ gal/min, (b) $Q'(4) = -50+4 = -46$ gal/min, (c) $Q'(8) = -50+8 = -42$ gal/min.

18. (a) The rate of change of the distance with respect to time is called the velocity. Thus, $v(t) = s'(t) = 10t + 2$ mm/s; (b) The acceleration is the rate of change of the velocity, and so $a(t) = v'(t) = s''(t) = 10$ mm/s^2; (c) $s(3) = 5 \cdot 3^2 + 2 \cdot 3 = 45 + 6 = 51$ mm; $v(3) = 10 \cdot 3 + 2 = 32$ mm/s; $a(3) = 10$ mm/s^2; (d) $10t + 2 = 46$ or $10t = 44$, and so $t = 4.4$ s.

19. (a) The velocity, v, at any time t is given by $v(t) = s'(t) = 8.0 - 4.0t$. The initial velocity is when $t = 0$, and so the initial velocity is $v(0) = 8.0$ m/s. (b) When the ball stops rolling its velocity is 0. Thus, we want the solution to $v(t) = 0$ or $8.0 - 4.0t = 0$ or $t = 2$ s. When $t = 2$, the ball's position is $s(2) = 8.0 \cdot 2 - 2.0 \cdot 2^2 = 16 - 8 = 8$ m; (c) $s(t) = 0$ means that $8.0t - 2.0t^2 = 0$ and so $t(8 - 2t) = 0$. By the zero product property, this means that $t = 0$ or $t = 4$. It will take 4.0 s to return to its initial position.

20. The area of a circle is $A(r) = \pi r^2$ and its circumference $C = 2\pi r$. So, $r = \frac{C}{2\pi}$ and substitution produces the area of a circle as a function of its circumference $A(C) = \frac{C^2}{4\pi}$. The rate of change of the area with respect to its circumference is $A'(C)$ or $\frac{dA}{dC} = \frac{C}{2\pi}$.

21. (a) $v = -N\frac{d\phi}{dt} = -50\frac{d(2.4t^{1/2} - 0.4t^2)}{dt} = -50(1.2t^{-1/2} - 0.8t)$; $v(t) = -60t^{-1/2} + 40t$;
(b) $v(1.0) = \frac{-60}{\sqrt{1}} + 40 \cdot 1 = -60 + 40 = -20$ V and $v(9.0) = \frac{-60}{\sqrt{9}} + 40 \cdot 9 = \frac{-60}{3} + 360 = 340$ V.

22. (a) The formula for the volume of a cone is $V = \frac{1}{3}\pi r^2 h$. We are given $h = 2r$, so by substituting, we can write the volume as a function of the radius $V(r) = \frac{1}{3}\pi r^2 \cdot 2r = \frac{2}{3}\pi r^3$. Thus, $\frac{dV}{dr} = 2\pi r^2$; (b) $\frac{dV}{dr}(24.3) = 2\pi(24.3)^2 \approx 3710$ mm^2.

23. $A = \pi r^2$ and $r = 1.7t$ m so $A = 1.7^2\pi t^2 = 2.89\pi t^2$ m^2. Thus, $\frac{dA}{dt} = 5.78\pi t$ and when $t = 8$, the area is increasing at the rate of 46.24π m^2/s

24. (a) $i = \frac{V}{r} = \frac{t^2 - 1.2}{3.0t^{1/2}}$;
(b) In (a), we expressed i as a function of t, thus the derivative is

$$\begin{aligned}\frac{di}{dt} &= \frac{3.0t^{1/2}(2t) - (t^2 - 1.2)\frac{3}{2}t^{-1/2}}{9t}\\ &= \frac{6t^{3/2} - \frac{3}{2}t^{3/2} + 1.8t^{-1/2}}{9t}\\ &= \frac{\frac{9}{2}t^{3/2}}{9t} + \frac{1.8t^{-1/2}}{9t}\\ &= \frac{1}{2}t^{1/2} + 0.2t^{-3/2}\\ &= \frac{t^{1/2}}{2.0} + \frac{1}{5.0t^{3/2}}\end{aligned}$$

(c) At $t = 2.4$ s this is $\frac{\sqrt{2.4}}{2.0} + \frac{1}{5.0\sqrt{2.4}^3} \approx 0.83$ A/s.

25. (a) $P = \frac{k}{V}$ or $5.21 = \frac{k}{1.23}$ so $k = 6.4083 \approx 6.41$. Hence, $P = \frac{6.41}{V}$; (b) $\frac{dP}{dV} = -6.41V^{-2}$ or $\frac{-6.41}{V^2}$.

26. (a) $\omega(t) = \theta'(t) = 3.4 - 2.4t$ rad/s; (b) $\alpha(t) = \theta''(t) = -2.4$ rad/s^2; (c) $\omega(0) = 3.4$ rad/s; $\alpha(0) = -2.4$ rad/s^2

24.2 EXTREMA AND THE FIRST DERIVATIVE TEST

1. $f'(x) = 2x - 8$, so we have $2x - 8 = 0$, and $x = 4$ is the critical value. On the interval $(-\infty, 4)$, $f'(0) = -8$ so decreasing. On the interval $(4, \infty)$, $f'(5) = 10 - 8 = 2$, so f is increasing. No maximums; minimum at $x = 4, f(4) = 16 - 32 = -16$.

2. $g'(x) = 10-2x$, so we have $10-2x = 0$, and $x = 5$ is the critical value. On the interval $(-\infty, 5)$, $g'(0)$ is positive, so g is increasing. On the interval $(5, \infty)$, $g'(6)$ is negative, and so g is decreasing. Maximum at $x = 5$; $g(5) = 50 - 25 = 25$. No minimum.

3. $m'(x) = -2x$; $-2x = 0 \Rightarrow 0$ is the critical value. On the interval $(-\infty, 0)$, $m'(-1)$ is positive so increasing. On the interval $(0, \infty)$, $m'(1)$ is negative so decreasing. Maximum at $x = 0$; $m(0) = 16$. No minimum.

4. $h'(x) = 6x - 12$; $6x - 12 = 0 \Rightarrow x = 2$ is the critical value. On $(-\infty, 2)$, $h'(0)$ is negative so decreasing. On the interval $(2, \infty)$, $h'(3)$ is positive so increasing. No maximum. Minimum at $x = 2$; $h(2) = -2$.

5. $j'(x) = 16 - 8x$; $16 - 8x = 0 \Rightarrow x = 2$ is the critical value. On $(-\infty, 2)$, $j'(0)$ is positive so increasing. On the interval $(2, \infty)$, $j'(3)$ is negative so decreasing. Maximum at $x = 2$; $j(2) = 31$. Minimum: none.

6. $k'(x) = 2x - 4$; $2x - 4 = 0 \Rightarrow x = 2$ is the critical value. On the interval $(-\infty, 2)$, $k'(0)$ is negative so decreasing. On the interval $(2, \infty)$, $k'(3)$ is positive so increasing. Maximum: none. Minimum: $k(2) = 7$.

7. $f'(x) = 8x-16$; $x = 2$ is the critical value. On the interval $(-\infty, 2)$, $f'(0)$ is negative so decreasing. On the interval $(2, \infty)$, $f'(3)$ is positive so increasing. Maximum: none; Minimum: $f(2) = -15$.

8. $s'(t) = -2t+7$ and $-2t+7 = 0 \Rightarrow t = \frac{7}{2}$ is the critical value. On the interval $(-\infty, \frac{7}{2})$, $s'(0)$ is positive so increasing. On the interval $(\frac{7}{2}, \infty)$, $s'(4)$ is negative so decreasing. Maximum: $s(\frac{7}{2}) = \frac{65}{4} = 16.25$. Minimum: none.

9. Here $q'(t) = 3t^2-3$ and so, $3t^2-3 = 3(t+1)(t-1) = 0$. Thus, by the zero product property, -1 and 1 are the critical values. On $(-\infty, -1)$, $g'(-2)$ is positive so increasing. On the interval $(-1, 1)$, $q'(0)$ is negative so decreasing. On the interval $(1, \infty)$, $q'(2)$ is positive so increasing. Maximum: $q(-1) = 2$. Minimum: $q(1) = -2$.

10. $v'(t) = 21 + 18t - 3t^2$; $-3t^2 + 18t + 21 = -3(t^2 - 6t - 7) = -3(t - 7)(t + 1) \Rightarrow -1, 7$ are critical values. On the interval $(-\infty, -1)$, $v'(-2)$ is negative so decreasing. On the interval $(-1, 7)$, $v'(0)$ is positive so increasing. On the interval $(7, \infty)$, $v'(8)$ is negative so decreasing. Maximum: $v(7) = 245$; Minimum: $v(-1) = -11$.

11. The derivative is $y' = x^3 - 36x$. By factoring we can determine when the derivative is 0, thus $x^3-36x = x(x+6)(x-6) = 0$ and so, the critical values are at -6, 0, and 6. On the interval $(-\infty, -6)$, y' is negative so decreasing. On the interval $(-6, 0)$, y' is positive so increasing. On the interval $(0, 6)$, y' is negative so decreasing. On the interval $(6, \infty)$, y' is positive so increasing. Maximum: $y(0) = 16$. Minimum: $y(-6) = y(6) = -308$.

12. $t' = -25w + w^3$; $w^3 - 25w = w(w + 5)(w - 5) = 0$, and so the critical values are at -5, 0 and 5. On the interval $(-\infty, -5)$, t' is negative so decreasing. On $(-5, 0)$, t' is positive so increasing. On the interval $(0, 5)$, t' is negative so decreasing. On the interval $(5, \infty)$, t' is positive so increasing. Maximum at 0 of 3.2; Minimum at both ± 5 of -153.05.

13. $f'(z) = 3(z + 4)^2$, so $3(z + 4)^2 = 0$ when $x = -4$. Thus, -4 is only critical value. On the interval $(-\infty, -4)$, $f'(z)$ is positive so increasing. On $(-4, \infty)$, $f'(z)$ is positive so increasing. Hence $f(z)$ is always increasing and there are no extrema.

14. $g'(x) = 2(2x^2 - 8)(4x) = 8x(2x^2 - 8) = 0$, and so the critical values are at $x = -2, 0, 2$. On the interval $(-\infty, -2)$, $g'(x)$ is negative so g is decreasing. On the interval $(-2, 0)$, $g'(x)$ is positive so g is increasing. On the interval $(0, 2)$, $g'(x)$ is negative

so g decreasing. On the interval $(2, \infty)$, $g'(x)$ is positive, so g is increasing. Maximum: $g(0) = 64$; Minimum: $g(-2) = g(2) = 0$.

15. $g'(v) = 12v^3 - 12v^2 - 24v = 12v(v^2 - v - 2) = 12v(v-2)(v+1)$, so the critical values are at -1, 0, and 2. On the interval $(-\infty, -1)$, $g'(v)$ is negative so decreasing. On the interval $(-1, 0)$, $g'(v)$ is positive so increasing. On $(0, 2)$, $g'(v)$ is negative so decreasing. On the interval $(2, \infty)$, $g'(v)$ is positive so increasing. Maximum: $g(0) = 12$. Minima: $g(-1) = 7$, $g(2) = -20$.

16. Rewriting the given function as $y = 5x^{-1}$, we find that its derivative is $y' = -5x^{-2} = \frac{-5}{x^2}$. Since y is undefined at $x = 0$, 0 is *not* a critical value. On the interval $(-\infty, 0)$, y' is negative so y is decreasing. On the interval $(0, \infty)$, y' is negative so y is decreasing. There are no extrema.

17. $s'(t) = \frac{-6}{(t+1)^2}$, Since s is undefined at $t = -1$, -1 is *not* a critical value. On the interval $(-\infty, -1)$, $s'(t)$ is negative, so s decreasing over this interval. On the interval $(-1, \infty)$, $s'(t)$ is negative so s is decreasing over this interval. $s(t)$ is undefined at -1. There are no extrema.

18. We see that $x = 0$ is not in the domain of h (or h'). $h'(x) = 1 - \frac{1}{x^2} = \frac{x^2 - 1}{x^2}$. Hence, $x^2 - 1 = 0 \Rightarrow x = \pm 1$ are critical values. On the interval $(-\infty, -1)$, $h'(x)$ is positive so h increasing over this interval. On the interval $(-1, 0)$, $h'(x)$ is negative so h is decreasing over this interval. On the interval $(0, 1)$, $h'(x)$ is negative so h decreasing. On the interval $(1, \infty)$, $h'(x)$ is positive so h is increasing on this interval. $h(0)$ is undefined. Maximum: $h(-1) = -2$; Minimum: $h(1) = 2$.

19. $g(x)$ is not defined at $x = 0$. Thus, $g'(x) = 2x + \frac{2}{x^2} = \frac{2x^3 + 2}{x^2}$ which is 0 when $2x^3 + 2 = 0$ or $x = -1$. The critical value is -1. On the interval $(-\infty, -1)$, $g'(x)$ is negative so g is decreasing. On the interval $(-1, 0)$, $g'(x)$ is positive so g is increasing. On the interval $(0, \infty)$, $g'(x)$ is positive so g is increasing. Maximum: none; Minimum: $g(-1) = 3$.

20. $j(x)$ is not defined at -3. $j'(x) = 1 - \frac{1}{(x+3)^2} = \frac{(x+3)^2 - 1}{(x+3)^2} = \frac{x^2 + 6x + 8}{(x+3)^2}$. Thus, $j'(x) = 0$, when $x^2 + 6x + 8 = (x+4)(x+2) = 0$, and so critical values are $x = -4$ and $x = -2$. On the interval $(-\infty, -4)$, $j'(x)$ is positive so $j(x)$ is increasing. On the interval $(-4, -3)$, $j'(x)$ is negative so $j(x)$ is decreasing. On $(-3, -2)$, $j'(x)$ is negative, so $j(x)$ decreasing. On the interval $(-2, \infty)$, $j'(x)$ is positive so $j(x)$ increasing. Maximum: $j(-4) = -5$; Minimum: $j(-2) = -1$.

21. $k(x)$ is undefined at $x = 1$. $k'(x) = \frac{(1-x)2x + (x^2)}{(1-x)^2} = \frac{2x - x^2}{(1-x)^2}$. Critical values are at 0 and 2. On the interval $(-\infty, 0)$, $k'(x)$ is negative so $k(x)$ decreasing. On the interval $(0, 1)$, $k'(x)$ is positive, so $k(x)$ increasing. On the interval $(1, 2)$, $k'(x)$ is positive, so $k(x)$ increasing. On the interval $(2, \infty)$, $k'(x)$ is negative, so $k(x)$ is decreasing. Maximum: $k(2) = -4$; Minimum: $k(0) = 0$.

22. Here $m'(b) = \frac{2}{\sqrt{b}}$ and a critical number is $b = 0$. Domain of m is $[0, \infty)$. On the interval $(0, \infty)$, $m'(b)$ is positive so $m(b)$ is increasing. Maximum: none; Minimum, $m(0) = 0$.

23. The domain of f is all t when $1 - t^2 \geq 0$ or $[-1, 1]$. The derivative is

$$\begin{aligned} f'(t) &= \sqrt{1 - t^2} - t \cdot \frac{1}{2}(1 - t^2)^{-1/2}(2t) \\ &= \sqrt{1 - t^2} - \frac{t^2}{\sqrt{1 - t^2}} \\ &= \frac{1 - t^2 - t^2}{1\sqrt{1 - t^2}} \\ &= \frac{1 - 2t^2}{\sqrt{1 - t^2}} \end{aligned}$$

Critical values are when $1 - 2t^2 = 0$, or $t = \pm\frac{\sqrt{2}}{2}$ and when $1 - t^2 = 0$ or $t = \pm 1$. The domain of $f(t)$ is $[-1, 1]$. On the interval $(-1, -\frac{\sqrt{2}}{2})$, $f'(t)$

is negative, so $f(t)$ decreasing. On the interval $\left(-\frac{\sqrt{2}}{2}, \frac{\sqrt{2}}{2}\right)$, we see that $f'(t)$ is positive, so $f(t)$ increasing. On the interval $\left(\frac{\sqrt{2}}{2}, 1\right)$, $f'(t)$ is negative, so $f(x)$ decreasing. Maximum: $f(-1) = 0$ and $f\left(\frac{\sqrt{2}}{2}\right) = 0.5$; Minimum: $f(-\frac{\sqrt{2}}{2}) = -0.5$ and $f(1) = 0$.

24.
$$\begin{aligned} g'(s) &= 2(s+2)(s-1)^{2/3} + \frac{2}{3}(s-1)^{-1/3}(s+2)^2 \\ &= \frac{2(s+2)(s-1)^{2/3}}{1} + \frac{2(s+2)^2}{3(s-1)^{1/3}} \\ &= \frac{3 \cdot 2(s+2)(s-1) + 2(s+2)^2}{3(s-1)^{1/3}} \\ &= \frac{6(s^2+s-2) + 2(s^2+4s+4)}{3\sqrt[3]{s-1}} \\ &= \frac{8s^2+14s-4}{3\sqrt[3]{s-1}} \\ &= \frac{2(4s^2+7s-2)}{3\sqrt[3]{s-1}} = \frac{2(4s-1)(s+2)}{3\sqrt[3]{s-1}}. \end{aligned}$$

The critical values are $\frac{1}{4}$, -2, and 1. On the interval $(\infty, -2)$, we see that $g'(s)$ is negative so g is decreasing. On $(-2, \frac{1}{4})$, $g'(s)$ is positive so g is increasing. On the interval $\left(\frac{1}{4}, 1\right)$, $g'(s)$ is negative, so g is decreasing. On $(1, \infty)$, $g'(s)$ is positive so g is increasing. Maximum: $g(\frac{1}{4}) = 4.179$; Minimum: $g(-2) = 0$; $g(1) = 0$.

25.
$$\begin{aligned} y' &= (x-1)^{2/3} + \frac{2}{3}(x-1)^{-1/3} \cdot x \\ &= (x-1)^{2/3} + \frac{2x}{3(x-1)^{1/3}} \\ &= \frac{3(x-1)+2x}{3(x-1)^{1/3}} \\ &= \frac{5x-3}{3(x-1)^{1/3}} \end{aligned}$$

The critical values are 0.6 and 1. On the interval $(-\infty, 0.6)$, y' is positive so y is increasing. On the interval $(0.6, 1)$, y' is negative so y is decreasing. On the interval $(1, \infty)$, y' is positive so y is increasing. Maximum: $(0.6, 0.326)$; Minimum: $(1, 0)$.

26. The derivative is $f'(x) = 3x^{-2/3} - 4 = \frac{3}{x^{2/3}} - 4 = \frac{3 - 4x^{2/3}}{x^{2/3}}$. The derivative is 0 when the numerator is 0, or when $3 - 4x^{2/3} = 0$, and so $4x^{2/3} = 3$, or $2x^{1/3} = \pm\sqrt{3}$, or $x^{1/3} = \pm\frac{\sqrt{3}}{2}$, and so $x = \pm\left(\frac{\sqrt{3}}{2}\right)^3 = \frac{\pm 3\sqrt{3}}{8}$. The derivative is undefined when the denominator is 0, or when $x^{2/3} = 0$, or $x = 0$. Thus, the critical values are 0 and $\pm\frac{3\sqrt{3}}{8}$. On the interval $\left(-\infty, \frac{-3\sqrt{3}}{8}\right)$, $f'$is negative so f is decreasing. On the interval $\left(\frac{-3\sqrt{3}}{8}, 0\right)$, f' is positive, so f increasing. On the interval $\left(0, \frac{3\sqrt{3}}{8},\right)$, f' is positive, so f increasing. On the interval $\left(\frac{3\sqrt{3}}{8}, \infty\right)$, f' is negative, so f decreasing. Maximum: $f\left(\frac{3\sqrt{3}}{8}\right) = 5.19615$; Minimum: $f\left(\frac{-3\sqrt{3}}{8}\right) = -5.19615$

27. (a) $s(t) = 250 + 256t - 16t^2$; $s'(t) = 256 - 32t$; Critical value when $256 - 32t = 0$ or $t = 8$. On the interval $[0, 8)$, $s'(t)$ is positive so increasing. On the interval $(8, \infty)$, $s'(t)$ is negative, so s is decreasing. Maximum at 8 s (b) $s(8) = 1274$ ft.

28. $s(t) = 180 + 98t - 4.9t^2$; $s'(t) = 98 - 9.8t$; The critical value is 10 and this gives a maximum (a) 10 s (b) $s(10) = 670$ m.

29. $P(t) = 4.7(t - 2.0)^2$; $P'(t) = 9.4(t - 2.0)$ (a) The critical value is 2.0 s and this will give the lowest power $P(2) = 0.0$ W (b) This occurs at 2.0 s.

30. $i(t) = 4.8t - 1.2t^2$; $i'(t) = 4.8 - 2.4t$; the critical value is $t = 2$. (a) The current is maximum at $t = 2.0$ s. (b) $i(2) = 4.8$ A.

31. $R = \frac{9R_2}{9+R_2}$;

$$R' = \frac{(9+R_2)9 - 9R_2(1)}{(9+R_2)^2} = \frac{81 + 9R_2 - 9R_2}{(9+R_2)^2} = \frac{81}{(9+R_2)^2}.$$

(a) the only critical value is -9 but R_2 must be greater than or equal to 0. R' is always positive hence R is always increasing. There is no maximum and the minimum value ocurs at the minimum value for R_2 which is 0. (b) The minimum value for R is 0.

32. $x(t) = 8t^3 - t^4$; $x'(t) = 24t^2 - 4t^3$; $a(t) = x''(t) = 48t - 12t^2$; $a'(t) = 48 - 24t$. The critical value for $a(t)$ is 2 which yields a maximum $v(2) = x'(2) = 24 \cdot 2^2 - 4 \cdot 2^3 = 64$

24.3 CONCAVITY AND THE SECOND DERIVATIVE TEST

1. Since $f(x) = -3x^2 + 12x$, $f'(x) = -6x + 12 = 0$ when $x = 2$. $f''(x) = -6$ is always negative, so f is concave down over the interval $(-\infty, \infty)$. Maximum at $(2, 12)$.

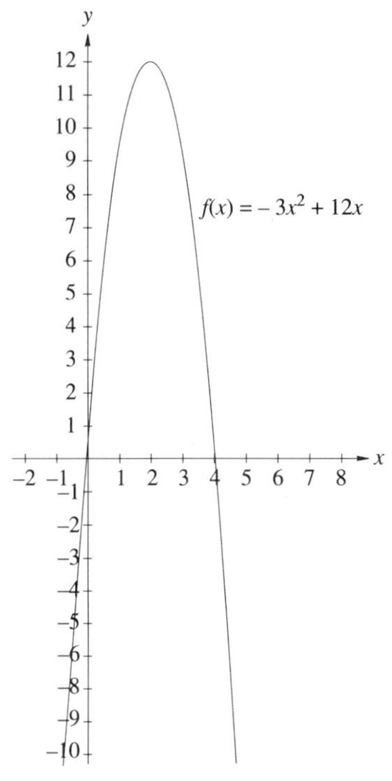

2. $g(x) = 3x^2 - 6x + 5$; $g'(x) = 6x - 6 = 0$ when $x = 1$. $g''(x) = 6$ always positive so concave up $(-\infty, \infty)$. Minimum at $(1, 2)$.

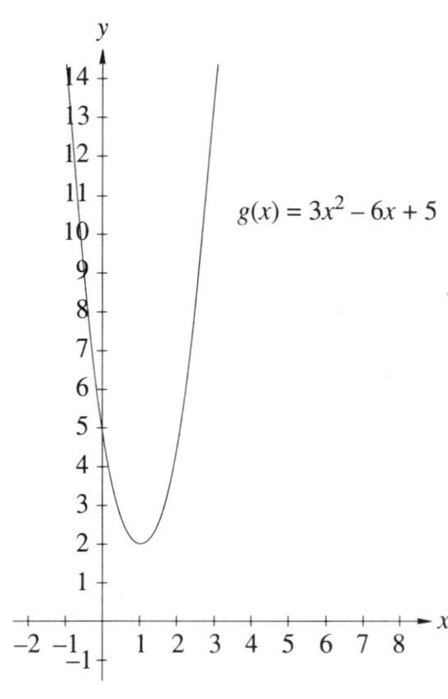

3. $g(x) = x^3 - 12x + 12$; $g'(x) = 3x^2 - 12 = 0$ when $x = \pm 2$; and $g''(x) = 6x = 0$ when $x = 0$. Concave up: $(0, \infty)$; Concave down: $(-\infty, 0)$. Maximum: $(-2, 28)$; Minimum: $(2, -4)$; Inflection Point: $(0, 12)$.

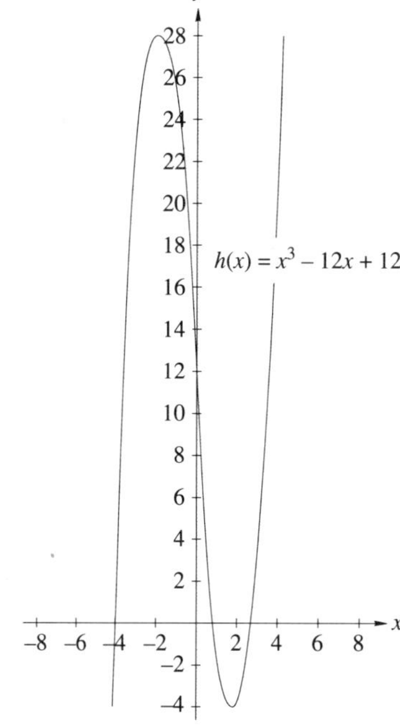

4. Since $f(x) = x^3 + 3x^2 - 5$, then $f'(x) = 3x^2 + 6x$ and $f''(x) = 6x + 6$. We have $f'(0) = 0, f'(-2) = 0$, and $f''(-1) = 0$. Hence, f is concave up on the interval $(-1, \infty)$; Concave down over $(-\infty, -1)$. Maximum: $(-2, -1)$; Minimum: $(0, -5)$; Inflection point: $(-1, -3)$.

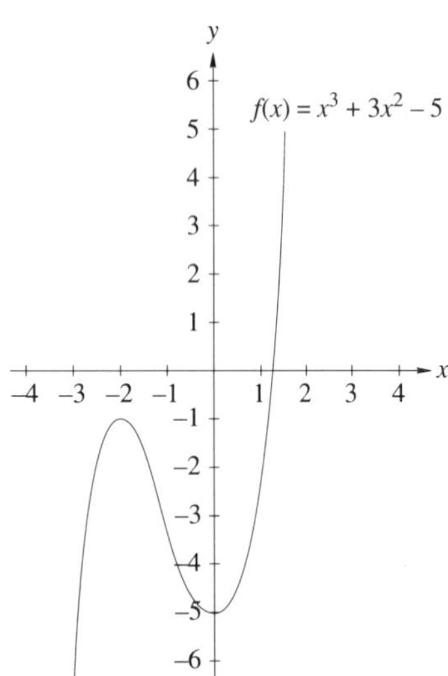

5. $j(x) = x^3 - 6x^2 + 12x - 4$; $j'(x) = 3x^2 - 12x + 12 = 3(x - 2)^2 = 0$ when $x = 2$; $j''(x) = 6x - 12 = 0$ when $x = 2$. Inflection point at $(2, 4)$. No extrema. Concave up: $(2, \infty)$; Concave down: $(-\infty, 2)$.

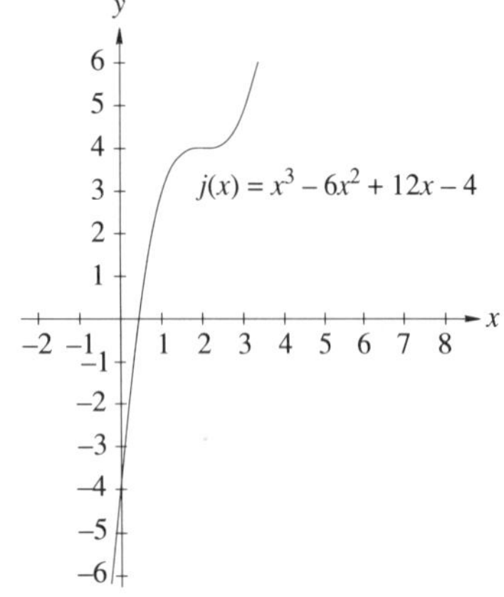

6. $k(x) = x^3 - 6x^2 - 20$; $k'(x) = 3x^2 - 12x = 3x(x-4) = 0$ when $x = 0$ or 4; $k''(x) = 6x - 12 = 0$ when $x = 2$. Concave up: $(2, \infty)$; Concave down: $(-\infty, 2)$ Inflection point: $(2, -36)$; Maximum: $(0, -20)$; Minimum: $(4, -52)$.

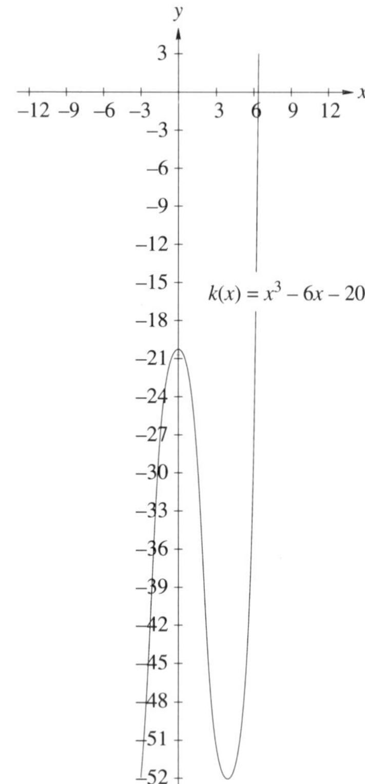

7. $h(x) = -2x^3 + 3x^2 - 12x + 5$; $h'(x) = -6x^2 + 6x - 12 = -6(x^2 - x + 2)$ never equals 0 since the discriminant $= 1 - 8 = -7$ is negative: hence no extrema. $h''(x) = -12x + 6 = 0$ when $x = \frac{1}{2}$. Concave up: $(-\infty, \frac{1}{2})$; Concave down: $(\frac{1}{2}, \infty)$; Inflection point: $(\frac{1}{2}, -\frac{1}{2})$.

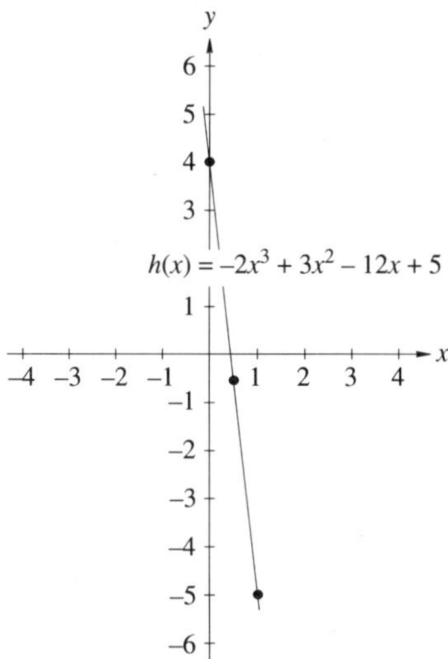

8. $m(x) = x^4 - 4x^3 + 15$; $m'(x) = 4x^3 - 12x^2 = 4x^2(x-3) = 0$ when $x = 0$ or $x = 3$; $m''(x) = 12x^2 - 24x = 12x(x-2) = 0$ when $x = 0$ or $x = 2$. Concave up: $(-\infty, 0)$ and $(2, \infty)$: Concave down: $(0, 2)$; Inflection points: $(0, 15)$ and $(2, -1)$; Minimum: $(3, -12)$, Maximum: none.

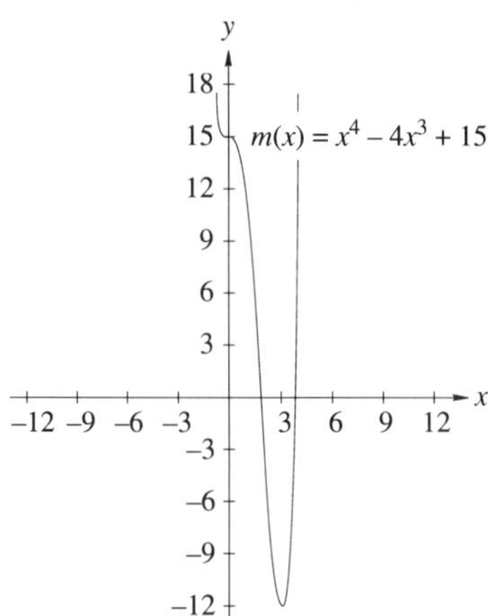

9. $j(x) = 4x^2 - \frac{4}{3}x^3 - x^4$; $j'(x) = 8x - 4x^2 - 4x^3 = -4x(x^2 + x - 2) = -4x(x+2)(x-1) = 0$ when $x = 0, -2$ and 1; $j''(x) = 8 - 8x - 12x^2 = -4(3x^2 + 2x - 2) = 0$ when $x = -1.22$ and 0.55. Concave up: $(-1.22, 0.55)$; Concave down $(-\infty, -1.22)$ and $(0.55, \infty)$; Inflection points: $(-1.22, 6.16)$ and $(0.55, 0.9)$; Maximum: $(-2, 10.67)$ and $(1, 1.67)$; Minimum: $(0, 0)$

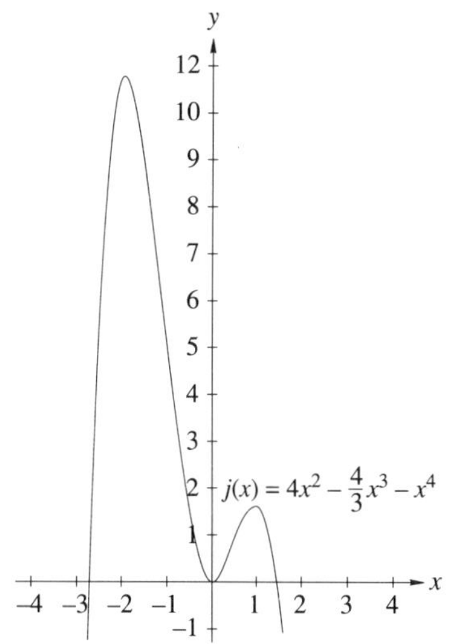

10. $f(x) = 3x^4 - 4x^3 + 8$; $f'(x) = 12x^3 - 12x^2 = 12x^2(x-1) = 0$ when $x = 0$ or $x = 1$; $f''(x) = 36x^2 - 24x = 12x(3x-2) = 0$ when $x = 0$ or $x = \frac{2}{3}$. Thus, f is concave up on the interval $(-\infty, 0)$ and $(\frac{2}{3}, \infty)$ and concave down over $(0, \frac{2}{3})$. Inflection points are $(0, 8)$ and $\left(\frac{2}{3}, 7\frac{11}{27}\right)$; Maximum: none; Minimum: $(1, 7)$.

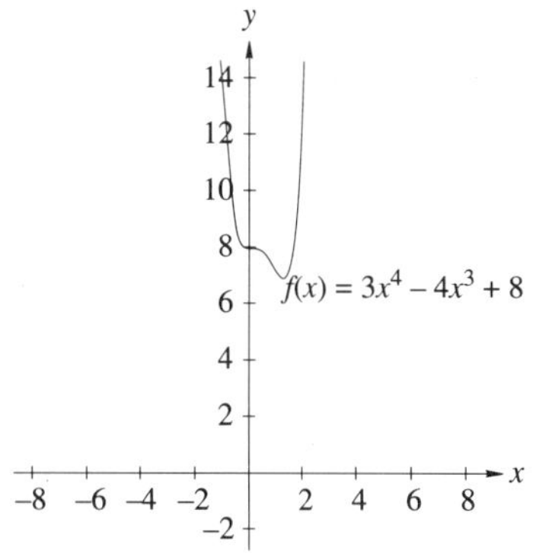

11. $h(x) = (1-x)^3$; $h'(x) = -3(1-x)^2 = 0$ when $x = 1$; $h''(x) = 6(1-x) = 0$ when $x = 1$; Concave down: $(1, \infty)$; Concave up: $(-\infty, 1)$; No extrema. Inflection point: $(1, 0)$.

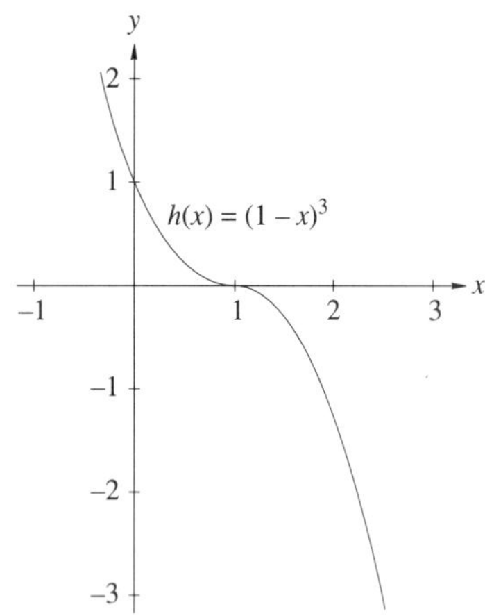

12. $g(x) = 3x^5 - 5x^3$; $g'(x) = 15x^4 - 15x^2 = 15x^2(x^2 - 1) = 0$ when $x = 0$ and $x = \pm 1$; $g''(x) = 60x^3 - 30x = 30x(2x^2 - 1) = 0$ when $x = 0$ and $x = \pm\frac{\sqrt{2}}{2}$; Concave up: $\left(-\frac{\sqrt{2}}{2}, 0\right)$ and $\left(\frac{\sqrt{2}}{2}, \infty\right)$; Concave down: $\left(-\infty, -\frac{\sqrt{2}}{2}\right)$ and $\left(0, \frac{\sqrt{2}}{2}\right)$; Inflection points: $\left(\frac{\sqrt{2}}{2}, 1.2374\right)$, $(0,0)$ and $\left(\frac{\sqrt{2}}{2}, -1.2374\right)$; Maximum: $(-1, 2)$; Minimum: $(1, -2)$;

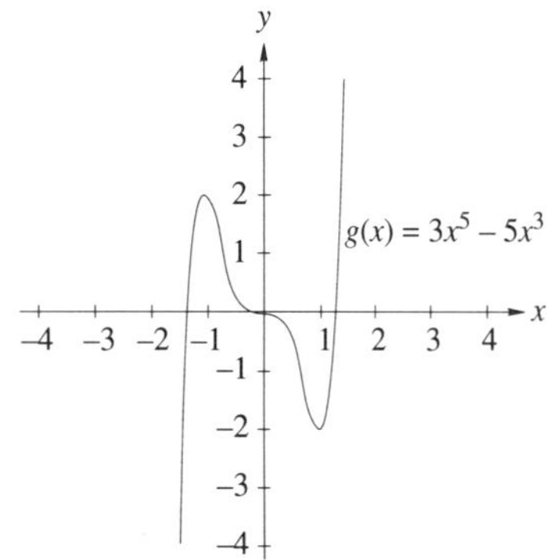

13. Since $h(x) = \sqrt[3]{x} + 2 = x^{1/3} + 2$, then $h'(x) = \frac{1}{3}x^{-2/3}$ and $h''(x) = -\frac{2}{9}x^{-5/3}$. We see that h is defined at $x = 0$, but h' is undefined at $x = 0$, so there is a critical value at $x = 0$; also $h''(x)$ is undefined at $x = 0$. Concave up: $(-\infty, 0)$; Concave down: $(0, \infty)$; Inflection point: $(0, 2)$; No extrema.

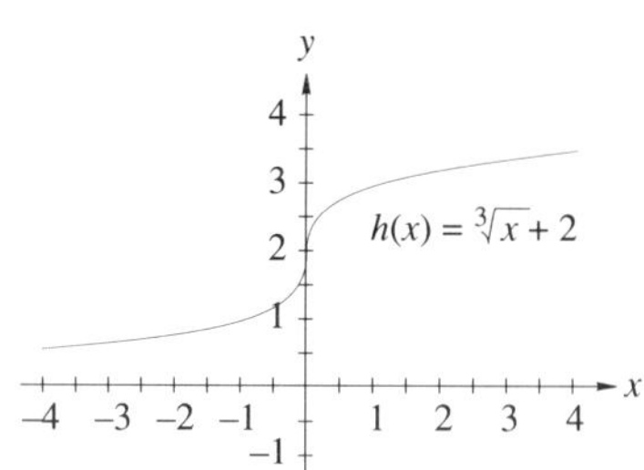

14. Here $j(x) = x - x^{2/3}$, so $j'(x) = 1 - \frac{2}{3}x^{-1/3} = 1 - \frac{2}{3\sqrt[3]{x}} = \frac{3\sqrt[3]{x} - 2}{3\sqrt[3]{x}}$. Critical values at $x = 0$ and $x = \frac{8}{27}$. $j''(x) = \frac{2}{9}x^{-4/3}$ which is undefined at $x = 0$. Concave up: $(-\infty, 0)$ and $(0, \infty)$; Increasing from $(-\infty, 0)$ and from $\left(\frac{8}{27}, \infty\right)$; decreasing from $\left(0, \frac{8}{27}\right)$. Maximum: $(0, 0)$, Minimum: $\left(\frac{8}{27}, \frac{-4}{27}\right)$

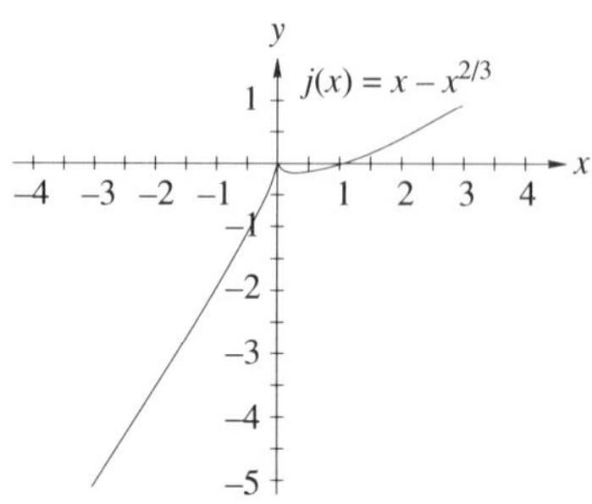

15. $k(x) = x + \frac{4}{x}$ has the derivative $k'(x) = 1 - \frac{4}{x^2} = \frac{x^2 - 4}{x^2}$. Now, $k'(x) = 0$ when $x = \pm 2$ and k' undefined at $x = 0$, so critical values are at $x = -2$ and $x = 2$. $k''(x) = \frac{8}{x^3}$ which is undefined at $x = 0$; Concave up: $(0, \infty)$; Concave down: $(-\infty, 0)$; Maximum: $(-2, -4)$; Minimum: $(2, 4)$; Vertical Asymptote at $x = 0$; No inflection point

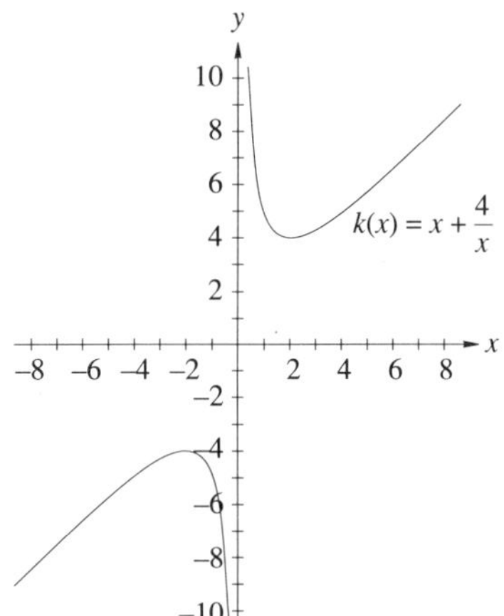

16. $m(x) = x - \dfrac{9}{x}$ has the derivative $m'(x) = 1 + \dfrac{9}{x^2}$ which is undefined at $x = 0$. But $x = 0$ is *not* a critical value, since 0 is not in the domain of m. The second derivative is $m''(x) = -\dfrac{18}{x^3}$, which is also undefined at $x = 0$; Concave up: $(-\infty, 0)$; Concave down: $(0, \infty)$. No extrema; Vertical Asymptote at $x = 0$; No inflection point

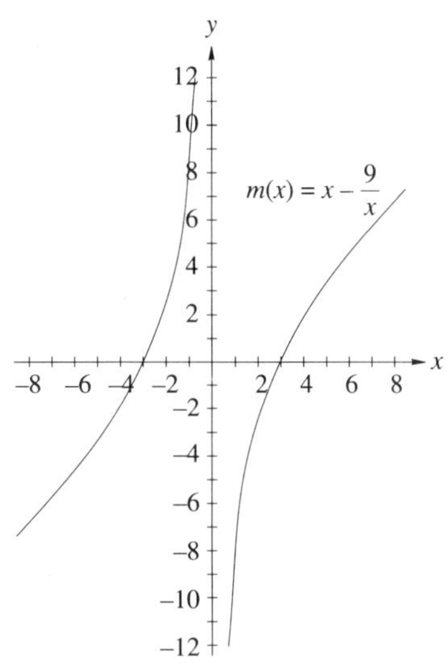

17. $n(x) = x^2 - \dfrac{2}{x}$ has the derivative $n'(x) = 2x + \dfrac{2}{x^2} = \dfrac{2x^3 + 2}{x^2}$. Critical value at -1. The second derivative is $n''(x) = 2 - \dfrac{4}{x^3} = \dfrac{2x^3 - 4}{x^3}$; this is undefined at 0 and is zero at $x = \sqrt[3]{2}$. Concave up $(-\infty, 0)$ and $\left(\sqrt[3]{2}, \infty\right)$, Concave down $\left(0, \sqrt[3]{2}\right)$. Maximum: none; Minimum: $(-1, 3)$; Vertical Asymptote: $x = 0$; Inflection point: $\left(\sqrt[3]{2}, 0\right)$.

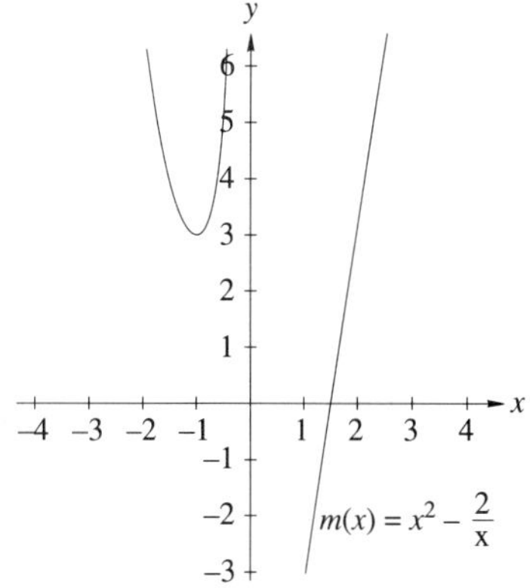

18. $f(x) = \dfrac{1}{x - 2} = (x - 2)^{-1}$; $f'(x) = -(x - 2)^{-2}$ which is undefined at $x = 2$. $f''(x) = 2(x - 2)^{-3}$ which is also undefined at $x = 2$. Concave up: $(2, \infty)$; Concave down: $(-\infty, 2)$ No extrema; Vertical Asymptote: $x = 2$; No inflection point.

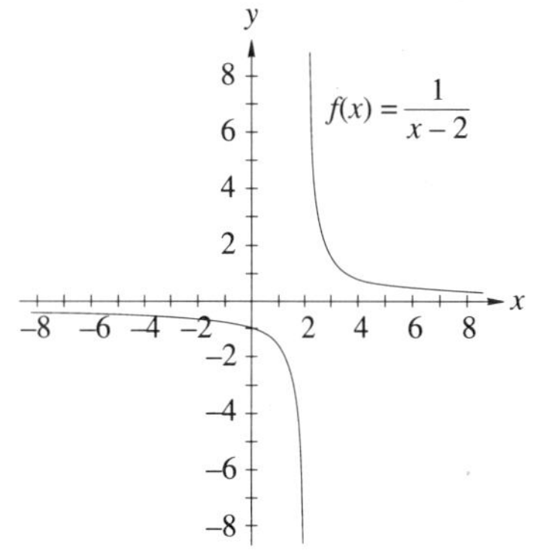

19. $g(x) = \dfrac{x^3}{3} - \dfrac{x^2}{2} - 6x$ has the derivative $g'(x) = x^2 - x - 6 = (x - 3)(x + 2)$, and so $g'(x) = 0$ when $x = -2$ or $x = 3$. The second derivative $g''(x) = 2x - 1 = 0$ when $x = \frac{1}{2}$. Concave up: $(\frac{1}{2}, \infty)$; Concave down: $(-\infty, \frac{1}{2})$. Maximum:

$(-2, 7\frac{1}{3})$; Minimum: $(3, -13\frac{1}{2})$; Inflection point: $(\frac{1}{2}, -3.0833)$.

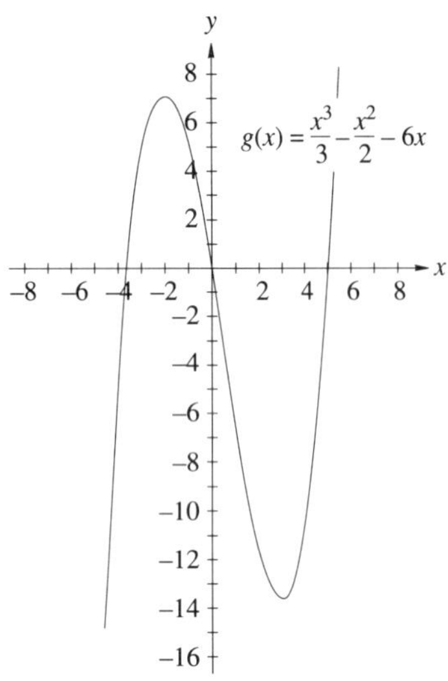

20. $h(x) = x^4 - 32x + 48$; $h'(x) = 4x^3 - 32 = 4(x^3 - 8) = 0$ when $x = 2$; $h''(x) = 12x^2 = 0$ when $x = 0$. Concave up $(-\infty, \infty)$. Minimum: $(2, 0)$; no inflection points

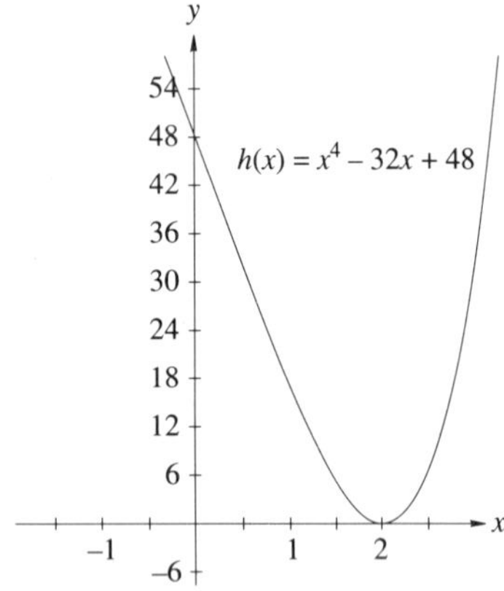

21. $j(x) = 2x^{1/2} - x$ has the derivative $j'(x) = x^{-1/2} - 1$, and $j'(x) = 0$ when $x = 1$ and j' is undefined at $x = 0$. The second derivative is $j''(x) = -\frac{1}{2}x^{-3/2}$ which is undefined at $x = 0$; Concave up: nowhere; Concave down: $(0, \infty)$. Maximum: $(1, 1)$; Minimum: $(0, 0)$; Inflection points: none.

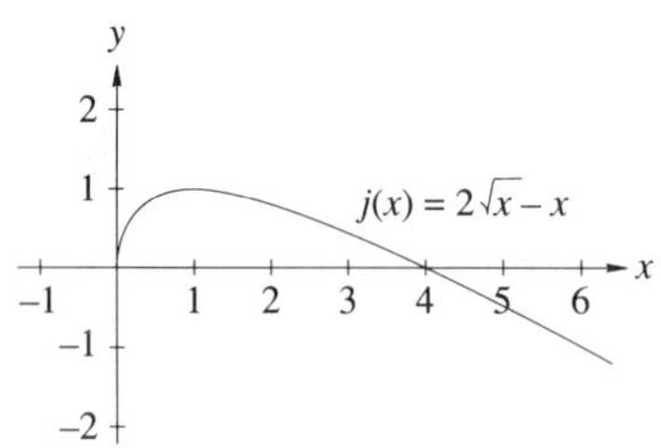

22. $k(x) = -(x^2 - 4)^{1/2}$, so $k'(x) = -\frac{1}{2}(x^2 - 4)^{-1/2}(2x) = \frac{-x}{\sqrt{x^2 - 4}}$. Since the domain of $k(x)$ is $(-\infty, 2)$ and $(2, \infty)$ the only critical values are -2 and 2. $k''(x) = -(x^2 - 4)^{-1/2} + (x^2 - 4)^{-3/2}x^2 = \frac{-(x^2 - 4) + x^2}{(x^2 - 4)^3} = \frac{4}{(x^2 - 4)^3}$. On the domain of $k(x)$, we see that k'' is positive, so Concave up on the intervals $(-\infty, -2)$ and $(2, \infty)$. We also see that $k'(x)$ is positive on $(-\infty, -2)$ so is increasing and yields a maximum at $(-2, 0)$. $k'(x)$ is negative on $(2, \infty)$ so decreasing and yields a maximum at $(2, 0)$ also. No inflection points.

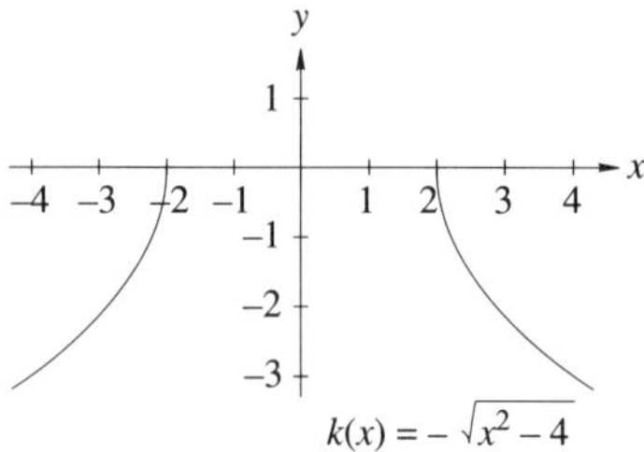

23. $g(x) = 2 + x^{2/3}$, so $g'(x) = \frac{2}{3}x^{-1/3}$. Since g is defined at $x = 0$ and g' is undefined at $x = 0$, 0 is a critical value. $g''(x) = \frac{-2}{9}x^{-4/3}$ is undefined at 0 but negative everywhere else, so concave down over $(-\infty, 0)$ and $(0, \infty)$ and $g'(x) > 0$ on $(0, \infty)$ and $g'(x) < 0$ on $(-\infty, 0)$, so Minimum value: $(0, 2)$. No inflection points.

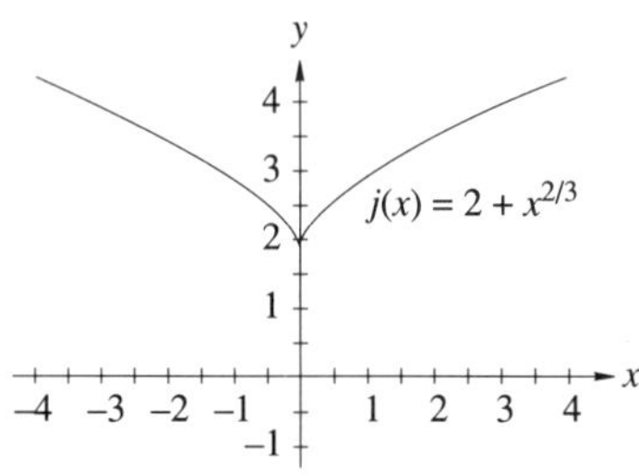

24. $h(x) = 2x - 3x^{2/3}$; $h'(x) = 2 - 2x^{-1/3} = \dfrac{2x^{1/3} - 2}{x^{1/3}}$. Critical values at 1 and 0. $h''(x) = \frac{2}{3}x^{-4/3}$, which is undefined at $x = 0$ and positive everywhere else. Hence, concave up on the invervals $(-\infty, 0)$ and $(0, \infty)$. Maximum: $(0, 0)$; Minimum: $(1, -1)$. No inflection points.

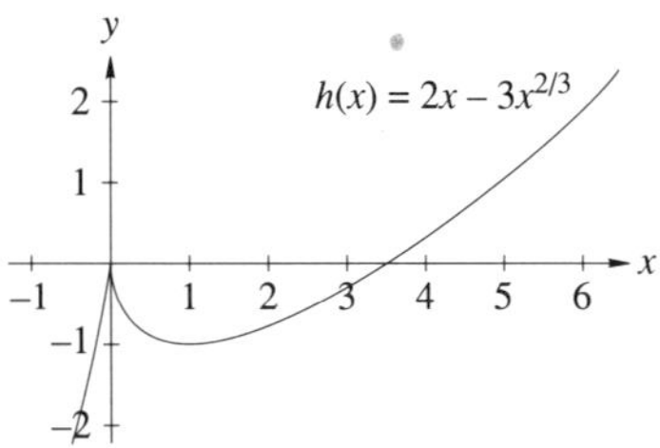

25.
$$
\begin{aligned}
f(x) &= x^{2/3}(x - 10) = x^{5/3} - 10x^{2/3} \\
f'(x) &= \frac{5}{3}x^{2/3} - \frac{20}{3}x^{-1/3} \\
&= \frac{5x^{2/3}}{3} - \frac{20}{3x^{1/3}} \\
&= \frac{5x - 20}{3x^{1/3}}
\end{aligned}
$$

Critical values are at 0 and 4. $f''(x) = \dfrac{10}{9}x^{-1/3} + \dfrac{20}{9}x^{-4/3} = \dfrac{10x + 20}{9x^{4/3}} = 0$ when $x = -2$, undefined at $x = 0$. Concave down $(-\infty, -2)$. Concave up $(-2, 0)$ and $(0, \infty)$. Maximum: $(0, 0)$; Minimum: $(4, -15.119)$. Inflection point: $(-2, -19.049)$.

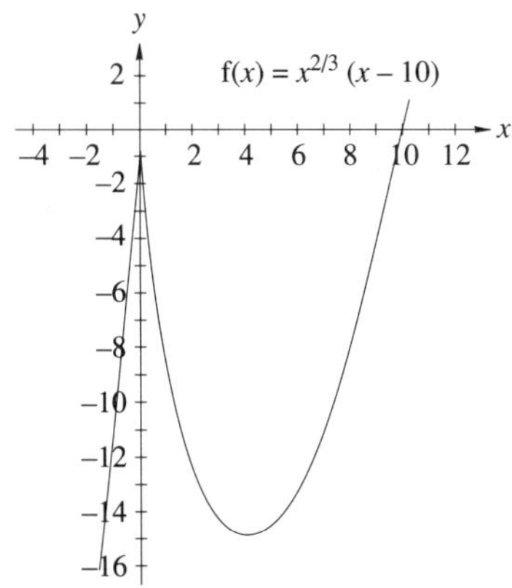

26. $g(x) = \dfrac{2}{x^2 - 4}$; Domain excludes $x = \pm 2$. Vertical Asymptotes at $x = 2, x = -2$. Now, $g'(x) = -2(x^2 - 4)^{-2}(2x) = \dfrac{-4x}{(x^2 - 4)^2}$. Critical value at 0. $g''(x) = 8x(x^2 - 4)^{-3}(2x) - 4(x^2 - 4)^{-2} = \dfrac{16x^2}{(x^2 - 4)^3} - \dfrac{4}{(x^2 - 4)^2} = \dfrac{16x^2 - 4x^2 + 16}{(x^2 - 4)^3} = \dfrac{12x^2 + 16}{(x^2 - 4)^3}$. We can see that g'' is undefined at ± 2. Concave up: $(-\infty, -2)$ and $(2, \infty)$; Concave down: $(-2, 2)$; Maximum: $(0, -\frac{1}{2})$. No inflection points.

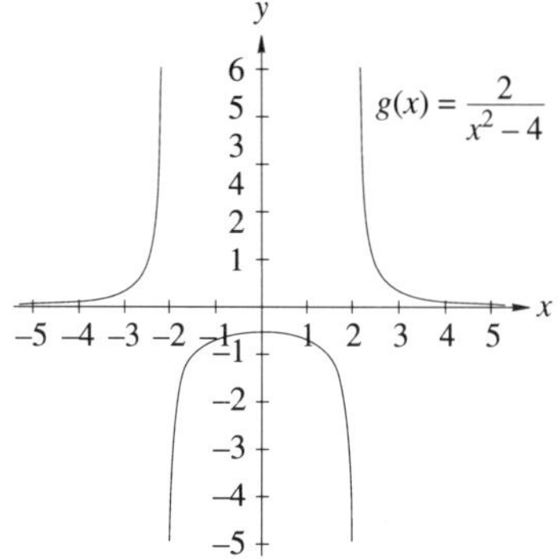

27.
$$
\begin{aligned}
h(x) &= x(x^2 + 1)^{-1} \\
h'(x) &= (x^2 + 1)^{-1} - 2x^2(x^2 + 1)^{-2} \\
&= \frac{1}{x^2 + 1} - \frac{2x^2}{(x^2 + 1)^2} = \frac{x^2 + 1 - 2x^2}{(x^2 + 1)^2} \\
&= \frac{1 - x^2}{(x^2 + 1)^2}
\end{aligned}
$$

Critical values at $x = \pm 1$.

$$
\begin{aligned}
h''(x) &= \frac{(x^2 + 1)^2(-2x) - (1 - x^2)(x^2 + 1)(2x)2}{(x^2 + 1)^4} \\
&= \frac{-2x^3 - 2x - 4x + 4x^3}{(x^2 + 1)^3} \\
&= \frac{2x^3 - 6x}{(x^2 + 1)^3}
\end{aligned}
$$

You can determine that $h''(x) = 0$ when $x = 0$ and $x = \pm\sqrt{3}$. Thus, h is concave up over the intervals $(-\sqrt{3}, 0)$ and $(\sqrt{3}, \infty)$ and h is concave down over the intervals $(-\infty, -\sqrt{3})$ and $(0, \sqrt{3})$.

Maximum: $(1, \frac{1}{2})$; Minimum: $(-1, -\frac{1}{2})$; Inflection point: $\left(-\sqrt{3}, \frac{-\sqrt{3}}{4}\right)$, $(0,0)$, and $\left(\sqrt{3}, \frac{\sqrt{3}}{4}\right)$.

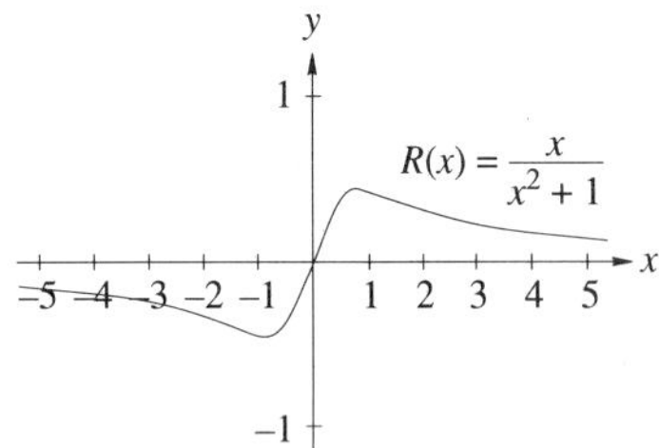

28. $j(x) = x^{2/3}(x-4)^{-1}$

$$j'(x) = \frac{2}{3}x^{-1/3}(x-4)^{-1} - (x-4)^{-2}x^{2/3}$$

$$= \frac{2}{3x^{1/3}(x-4)} - \frac{x^{2/3}}{(x-4)^2} = \frac{2(x-4)-3x}{3x^{1/3}(x-4)^2}$$

$$= \frac{-x-8}{3x^{1/3}(x-4)^2}.$$

Critical values at -8 and 0. j is undefined at $x = 4$, so there is a vertical asymptote at $x = 4$.

$$j''(x) = \frac{3x^{1/3}(x-4)^2(-1)+(x+8)\left[x^{-2/3}(x-4)^2+6x^{1/3}(x-4)\right]}{9x^{2/3}(x-4)^4}$$

$$= \frac{-3x^{1/3}(x-4)^2+(x+8)(x-4)\left[x^{1/3}-4x^{-2/3}+6x^{1/3}\right]}{9x^{2/3}(x-4)^4}$$

$$= \frac{-3x(x-4)+(x+8)(7x-4)}{9x^{4/3}(x-4)^3}$$

$$= \frac{-3x^2+12x+7x^2+52x-32}{9x^{4/3}(x-4)^3}$$

$$= \frac{4x^2+64x-32}{9x^{4/3}(x-4)^3} = \frac{4(x^2+16x-8)}{9x^{4/3}(x-4)^3}$$

j'' is undefined at 0 and 4 and zero at $x \approx -16.485$ and $x \approx 0.485$. Thus, j is concave up over the intervals $(-16.485, 0)$, $(0, 0.485)$, and $(4, \infty)$ and concave down over $(-\infty, -16.485)$ and $(0.485, 4)$. Maximum: $(0,0)$; Minimum: $\left(-8, -\frac{1}{3}\right)$; Inflection points: $(-16.485, -0.316)$ and $(0.485, -0.176)$.

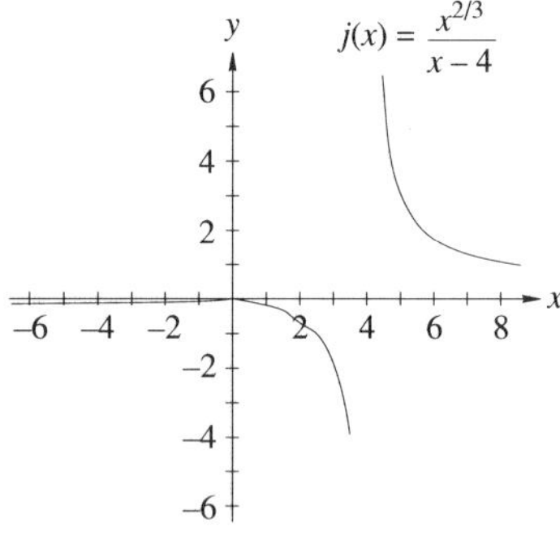

24.4 APPLIED EXTREMA PROBLEMS

1. $A = \ell \cdot w$; $\ell + 2w = 10{,}000 \Rightarrow \ell = 10{,}000 - 2w$. Substituting into the area formula yields $A = (10{,}000 - 2w)w = 10{,}000w - 2w^2$, so $A' = 10{,}000 - 4w = 0 \Rightarrow w = 2{,}500$. $A'' = -4 < 0$ so $w = 2{,}500$ yields a maximum area, and $\ell = 10{,}000 - 5000 = 5000$; $A = 5{,}000 \times 2{,}500 = 12{,}500{,}000 \text{ ft}^2$.

2. This time $2\ell + 2w = 10{,}000 \Rightarrow \ell + w = 5{,}000 \Rightarrow \ell = 5{,}000 - w$. Substituting into the area formula, we have $A = (5000 - w)w = 5000w - w^2$; $A' = 5000 - 2w = 0 \Rightarrow w = 2{,}500$. $\ell = 5000 - 2500 = 2{,}500$. $A = 2{,}500 \times 2{,}500 = 6{,}250{,}000 \text{ ft}^2$.

3. (a) Let x be the length of the side along the river, as well as the side opposite. Then the length of the other two sides is $6{,}000{,}000/x$. The cost $C(x) = 2x + x + 2(6{,}000{,}000)/x$, so $C(x) = 3x + \frac{12{,}000{,}000}{x}$. $C'(x) = 3 - \frac{12{,}000{,}000}{x^2} = 0 \Rightarrow 3x^2 = 12{,}000{,}000 \Rightarrow x^2 = 4{,}000{,}000 \Rightarrow x = 2{,}000$. The other sides are 3,000 ft. The dimensions are 2,000 ft $\times$3,000 ft. (b) The cost is $3(2000) + 2(3{,}000) = \$12{,}000$.

4. Let x be length of a side of square and $200 - 4x$ be the circumference of the circle. The radius of the circle is $\frac{100-2x}{\pi}$. The total area for the square and circle is $A = x^2 + \pi\left(\frac{100-2x}{\pi}\right)^2 = x^2 + \frac{1}{\pi}(100-2x)^2$ and

$A' = 2x + \frac{2}{\pi}(100 - 2x)(-2) = 2x - \frac{4}{\pi}(100 - 2x) = 2x - \frac{400}{\pi} + \frac{8}{\pi}x = \left(2 + \frac{8}{\pi}\right)x - \frac{400}{\pi}$. We see that $A'(x) = 0$ when $x = \frac{200}{\pi + 4} \approx 28.005$. When $x < 28.005$, $A'(x) < 0$ and when $x > 28.005$, $A'(x) > 0$ and so, when $x \approx 28.005$, the area is a minimum. Thus, the fencing should not be cut and it all should be used to form a circular fence.

5. After cutting out the squares, the resulting length and width of the box will both be $(24 - 2x)$. The height will be x; $V = \ell \cdot w \cdot h = (24 - 2x)^2 x = 4x^3 - 96x^2 + 576x$; $V' = 12x^2 - 192x + 576 = 12(x^2 - 16x + 48) = 12(x - 4)(x - 12)$. Now, $V' = 0$ when $x = 4$ or 12. $V'' = 24x - 192$; $V''(4) < 0$ so $V(4)$ is maximum. $V''(12) > 0$ so $V(12)$ is minimum. $V(4) = 4 \cdot 4^3 - 96 \cdot 4^2 + 576 \cdot 4 = 1024\text{ cm}^3$. The length of the side of a removed square should be 4 cm.

6. The bottom center of the building can be placed on the origin of a coordinate system. Then the formula for the edge of the building is $y = \sqrt{100 - x^2}$. If $2x$ is the length of the sign, then $\sqrt{100 - x^2}$ will be the height. Its area is given by the formula $A = 2x\sqrt{100 - x^2}$.

$$\begin{aligned} A' &= 2(100 - x^2)^{1/2} - (100 - x^2)^{-1/2} \cdot 2x^2 \\ &= 2(100 - x^2)^{1/2} - \frac{2x^2}{(100 - x^2)^{1/2}} \\ &= \frac{2(100 - x^2)}{(100 - x^2)^{1/2}} - \frac{2x^2}{(100 - x^2)^{1/2}} = \frac{200 - 4x^2}{(100 - x^2)^{1/2}} \end{aligned}$$

$200 - 4x^2 = 0$, or $4x^2 = 200$, or $x^2 = 50$, and so $x = \pm\sqrt{50} \approx \pm 7.071$. Since x cannot be negative, 7.071 is the answer we seek. Length of the sign is $2x = 14.14$ m. The height is $\sqrt{100 - x^2} = \sqrt{50} = 7.07$ m.

7. First solve $9x^2 + 16y^2 = 144$ for y. This yields $y = \sqrt{9 - \frac{9}{16}x^2}$. The area of the rectangle is $A = xy$, or after substituting for y, can be written as $A = x\sqrt{9 - \frac{9}{16}x^2}$. $A' = \left(9 - \frac{9}{16}x^2\right)^{1/2} + x \cdot \frac{1}{2}\left(9 - \frac{9}{16}x^2\right)^{-1/2}\left(\frac{-18}{16}x\right) = \frac{9 - \frac{9}{16}x^2 - \frac{9}{16}x^2}{\left(9 - \frac{9}{16}x^2\right)^{1/2}} = \frac{9 - \frac{9}{8}x^2}{\left(9 - \frac{9}{16}x^2\right)^{1/2}}$. Critical values are when $9 - \frac{9}{8}x^2 = 0$, or $x^2 = 8$, or $x = 2\sqrt{2}$, and when $9 - \frac{9}{16}x^2 = 0$ or $x = 4$. When $x = 4$, $y = 0$ so $A = 0$ which is the minimum. When $x = 2\sqrt{2}$, we see that $y = \sqrt{9 - \frac{9}{2}} = \sqrt{\frac{9}{2}} = \frac{3}{\sqrt{2}} = \frac{3\sqrt{2}}{2}$ and the area is $A = 2\sqrt{2} \cdot \frac{3\sqrt{2}}{2} = 6$, which is a maximum. The dimensions are $2\sqrt{2} \approx 2.83$ units horizontally and $\frac{3}{2}\sqrt{2} \approx 2.12$ units vertically.

8. Similar to exercise #5, the length will be $15 - 2x$, the width $8 - 2x$ and the height x. $V = \ell \cdot w \cdot h = (15 - 2x)(8 - 2x)x = 120x - 46x^2 + 4x^3$. $V' = 12x^2 - 92x + 120$ and solving yields $x = \frac{5}{3}$ or $x = 6$. Since $0 \le x \le 4$, the answer is $\frac{5}{3}$ in (Note: $V'' = 24x - 92 < 0$ when $x = \frac{5}{3} \Rightarrow$ Concave down $\Rightarrow$ is maximum.)

9. Let x be the length and width of the box and $y = \frac{32}{x^2}$ be the height. The material used is proportional to the area of the 4 sides plus the base. $A = x^2 + 4 \cdot x \cdot \frac{32}{x^2} = x^2 + \frac{128}{x}$, and so $A' = 2x - \frac{128}{x^2} = \frac{2x^3 - 128}{x^2}$. Solving $2x^3 - 128 = 0$ yields $x^3 = 64$ or $x = 4$. $A'' = 2 + \frac{256}{x^3}$ is positive at $x = 4$, so this yields a minimum. The height is $\frac{32}{16} = 2$. The dimensions are $4'' \times 4'' \times 2''$.

10. This time the area is $A = 2x^2 + \frac{128}{x}$, so $A' = 4x - \frac{128}{x^2} = \frac{4x^3 - 128}{x^2}$. Solving $4x^3 - 128 = 0 \Rightarrow x^3 = 32 \Rightarrow x = \sqrt[3]{32} \approx 3.175$. Height is $\frac{32}{\sqrt[3]{32^2}} = \sqrt[3]{32} \approx 3.175$. The box is a cube $\sqrt[3]{32} \approx 3.175$ in. on each side.

11. $A = 2\pi r^2 + 2\pi rh = 480 \Rightarrow h = \frac{480 - 2\pi r^2}{2\pi r} = \frac{240 - \pi r^2}{\pi r}$, so the volume is $V = \pi r^2 h = \pi r^2\left(\frac{240 - \pi r^2}{\pi r}\right) = 240r - \pi r^3$, which has the

derivative $V' = 240 - 3\pi r^2$. Setting $240 - 3\pi r^2 = 0$ and solving for r yields $\pi r^2 = 80$ or $r = \sqrt{\frac{80}{\pi}} \approx 5.05$. $h = 10.09$. Radius = 5.05 cm; height = 10.09 cm.

12. Here $s(t) = \sqrt{t - 9t^2}$ has the derivative $s'(t) = \frac{1}{2}(t - 9t^2)^{-1/2}(1 - 18t) = \frac{1 - 18t}{2\sqrt{t - 9t^2}}$. We see that $s'(t) = 0$, when $1 - 18t = 0 \Rightarrow t = \frac{1}{18}$. The other critical values are 0 and $\frac{1}{9}$ which yield a speed of 0. $s\left(\frac{1}{18}\right) = 0.16667$ m/min.

13. $A = \ell \cdot w = 580$, and solving for ℓ, we get $\ell = \frac{580}{w}$. The printing area of the page is $P = (\ell - 5)(w - 6) = \left(\frac{580}{w} - 5\right)(w - 6) = 580 - \frac{3480}{w} - 5w + 30 = 610 - 5w - \frac{3480}{w}$. In order to determine when the printed area of the page is the largest, we take the derivative, with the result $P' = -5 + \frac{3480}{w^2}$. Setting the derivative equal to 0, we have $-5 + \frac{3480}{w^2} = 0$ or $w^2 = \frac{3480}{5} = 696$ and so $w = \sqrt{696} = 26.38$ and $\ell = \frac{580}{w} = 21.98$. The largest printed area will have a length of about 21.98 cm and a width of about 26.38 cm.

14. $d = \sqrt{(x - 5)^2 + (y - 1)^2}$, so $d^2 = (x - 5)^2 + (y - 1)^2$. The distance is minimum when d^2 is minimum. $d^{2'} = 2(x - 5) + 2(y - 1)y'$; $y = 2x^2$; $y' = 4x$; $d^{2'} = 2(x - 5) + 2(2x^2 - 1)(4x) = 2x - 10 + 16x^3 - 8x = 16x^3 - 6x - 10 = 2(8x^3 - 3x - 5) = 2(x - 1)(8x^2 + 8x + 5)$. The only critical number is 1. $d^{2''} = 48x^2 - 6 > 0$ when $x = 1$ so this yields a minimum $y = 2 \cdot 1^2 = 2$. The point is $(1, 2)$.

15. $r = 15$. Placing the center of the log at the origin yields the equation $x^2 + y^2 = 225$ and so, $y = \sqrt{225 - x^2}$, $d = 2y$, and $w = 2x$. Thus, the strength is $S = wd^2 = 2x\left(2\sqrt{225 - x^2}\right)^2 = 8x(225 - x^2) = 1800x - 8x^3$. To find when the strength is the most, we take the derivative, getting $S' = 1800 - 24x^2$, then determine the critical values by finding when the derivative is 0. Thus, $1800 - 24x^2 = 0 \Rightarrow 24x^2 = 1800 \Rightarrow x^2 = 75$, and so $x = 5\sqrt{3}$. As a result, the width is $w = 2x = 10\sqrt{3} \approx 17.32$ in. and the depth is $d = 2y = 2\sqrt{225 - 75} = 2\sqrt{150} = 24.49$ in. (b) This time our equation is $x^2 + y^2 = r^2$ so $y = \sqrt{r^2 - x^2}$ and so the strength is $S = 2x(2\sqrt{r^2 - x^2})^2 = 8x\left(r^2 - x^2\right) = 8xr^2 - 8x^3$. Taking the derivative, we obtain $S' = 8r^2 - 24x^2$, and to find when the derivative is 0, we get $8r^2 - 24x^2 = 0 \Rightarrow 24x^2 = 8r^2 \Rightarrow x^2 = \frac{1}{3}r^2$. Thus, $x = \sqrt{\frac{r^2}{3}} = \frac{r}{\sqrt{3}} = \frac{r\sqrt{3}}{3}$. For the width we have $w = 2x = 2r\sqrt{3}/3$ and the depth is $d = 2y = 2\sqrt{r^2 - \frac{r^2}{3}} = 2\sqrt{\frac{2r^2}{3}} = \frac{2r\sqrt{6}}{3}$.

16. $A = (s(s - a)(s - b)(s - c))^{1/2}$

$$= \left[\frac{1}{2}(300 + c)\left(\frac{300 + c}{2} - 150\right) \times \left(\frac{300 + c}{2} - 150\right)\left(\frac{300 + c}{2} - c\right)\right]^{1/2}$$

$$= \left[\left(150 + \frac{c}{2}\right)\left(\frac{c}{2}\right)\left(\frac{c}{2}\right)\left(150 - \frac{c}{2}\right)\right]^{1/2}$$

$$= \left[\left(22500 - \frac{c^2}{4}\right)\left(\frac{c^2}{4}\right)\right]^{1/2} = \left[5625c^2 - \frac{c^4}{16}\right]^{1/2}$$

$$A' = \frac{1}{2}\frac{11250c - \frac{c^3}{4}}{\sqrt{5625c^2 - \frac{c^4}{16}}}$$

Critical values are when the numerator is 0 or the denominator is 0. Now, if $11250c - \frac{c^3}{4} = 0$, then $\frac{c}{4}(45000 - c^2) = 0$, and we have $c = 0$, $c = \pm\sqrt{45000} = \pm 212.13$. The denominator yields $5625 - \frac{c^2}{16} = 0$, or $c = \sqrt{90000}$ and $c = 300$. Since c cannot be negative, 0 and 300 yield an area of 0 which is a minimum, so 212.13 gives maximum. Use the Law of Cosines to determine θ. Then, $\cos\theta = \frac{150^2 + 150^2 - 212.13^2}{2 \cdot 150 \cdot 150} = 0$ and

$\theta = \cos^{-1} 0 = 90°$. The angle that gives the maximum capacity is 90°.

17. $$A = \ell \cdot w;\ \ell = 320 - 2w$$
$$A = (320 - 2w)w = 320w - 2w^2$$
$$A' = 320 - 4w = 0 \Rightarrow w = 80 \text{ mm}$$
$$\ell = 320 - 2 \cdot 80 = 320 - 160 = 160 \text{ mm}$$

The gutter is 80 mm deep and 160 mm wide.

18. (a) $$I = \frac{E}{\sqrt{R^2 + X^2}} = E(R^2 + X^2)^{-1/2}$$
$$I' = \frac{-1}{2}E(R^2 + X^2)^{-3/2}(2X) = \frac{-XE}{(R^2 + X^2)^{3/2}}$$

The only critical value is 0. Since $I' < 0$, I is decreasing and this yields a maximum.

(b) Maximum $= \dfrac{E}{\sqrt{R^2}} = \dfrac{E}{R}$.

19. $P = VI - RI^2 = 10I - 4I^2$ and so $P' = 10 - 8I$. Critical value is $\frac{5}{4} = 1.25$. $P'' = -8$ so critical value yields a maximum. Maximum power given by current of 1.25 A.

20. $$v(x) = 5x(a - x) = 5ax - 5x^2$$
$$v' = 5a - 10x = 0 \Rightarrow x = \frac{a}{2}$$
$$v'' = -10$$

so the reaction is a maximum at $x = \dfrac{a}{2}$.

21. $C(A) = bA + \dfrac{c}{A}$; $C' = b - \frac{c}{A^2} = 0$ when $A = \sqrt{c/b}$; $C'' = 2c/A^3 > 0$, so minimum cost at $A = \sqrt{c/b}$.

22. Let ℓ = height of rectangular part of window; $w = d$ = width. Then,

$$A = \ell \cdot w = 2 \text{ so } w = \frac{2}{\ell} = d$$
$$r = \frac{1}{2}w = \frac{1}{\ell}$$

Perimeter of window $= 2\ell + w + \pi r = 2\ell + \dfrac{2}{\ell} + \dfrac{\pi}{\ell} = 2\ell + \dfrac{2+\pi}{\ell}$. $P' = 2 - (2+\pi)/\ell^2 = 0$ and $\frac{2+\pi}{\ell^2} = 2$; $2\ell^2 = 2 + \pi$; $\ell^2 = \frac{2+\pi}{2}$; $\ell = \sqrt{\frac{2+\pi}{2}} \approx 1.6$; $d = \frac{2}{1.6} = 1.247 \approx 1.25$ m.

23. ℓ = height; $w = d$ = width $= 2r$. Thus,

$$P = 8 = 2\ell + w + \frac{\pi w}{2} = 2\ell + \frac{2+\pi}{2}w$$
$$2\ell = 8 - \frac{2+\pi}{2}w;\ \ell = 4 - \frac{2+\pi}{4}w$$
$$A = \ell \cdot w + \frac{1}{2}\pi\left(\frac{w}{2}\right)^2 = \left(4 - \frac{2+\pi}{4} \cdot w\right) w + \frac{\pi}{8}w^2$$
$$= 4w - \frac{2+\pi}{4}w^2 + \frac{\pi}{8}w^2 = 4w - \frac{4+\pi}{8}w^2$$
$$A' = 4 - \frac{4+\pi}{4}w = 0;\ (4+\pi)w = 16;$$
$$w = \frac{16}{4+\pi} = 2.24 \text{ m.}$$

The diameter is 2.24 m.

24. (a) Time $= \dfrac{800}{s}$. Total cost is
$$C = \left(0.35 + \frac{s}{300}\right) 800 + 12\left(\frac{800}{s}\right).$$
$$C = 800\left(0.35 + \frac{s}{300} + \frac{12}{s}\right).$$
$C' = 800\left(\dfrac{1}{300} - \dfrac{12}{s^2}\right) = 0$. This is 0 when $\dfrac{s^2}{12} = 300$, or $s^2 = 300 \times 12 = 3600$, and so $s = \sqrt{3600} = 60$ mph yields the least cost.
(b) $C = \left(0.35 + \dfrac{60}{300} + \dfrac{12}{60}\right)800 = (0.35 + 0.2 + 0.2)800 = (0.75)800 = \$600$ is the least cost.

25. (a) $$P = I^2R = (4.5 - t)^2(25.4)$$
$$P' = 2(4.5 - t)(25.4)(-1) = 0$$

when $t = 4.5$ s. P'' is always positive so this is a minimum. (b) $P = (4.5 - 4.5)^2(25.4) = 0$ W.

26. If $i(t)$ is the current, then

$$q = -4.36t^2 + 2.14t^3$$
$$i(t) = \frac{dq}{dt} = -8.72t + 6.42t^2$$
$$i'(t) = -8.72 + 12.84t$$

$i'(t) = 0$ when $t = 0.679$ s. Thus, $i(0.679) = -2.96$ A.

27. Let x = distance from less intense light and $8 - x$ = distance from more intense light. Light intensity $= \frac{1}{x^2} + \frac{4}{(8-x)^2}$. Taking the derivative, we get $L' = \frac{-2}{x^3} + \frac{8}{(8-x)^3}$. Critical values will be when $\frac{(8-x)^3}{8} = \frac{x^3}{2}$, or $(8-x)^3 = 4x^3$, or $512-192x+24x^2-x^3 = 4x^3$, or when $512 - 192x + 24x^2 - 5x^3 = 0$. This has one solution between 0 and 8. It is about 3.1 m. Thus, the desired answer is that the light is the least about 3.1 m from the less intense light.

28.
$$D = av^2 + \frac{b}{v^2}$$
$$D' = 2av - 2\frac{b}{v^3} = 0$$
$$2av = \frac{2b}{v^3}$$
$$v^4 = \frac{b}{a}$$
$$v = \sqrt[4]{b/a}$$

(Note: $D'' = 2a + 6\frac{b}{v^4} > 0$, so we have a minimum.)

29. Here, we have $Y = k(2x^4 - 5Lx^3 + 3L^2x^2)$, and so $Y' = k(8x^3 - 15Lx^2 + 6L^2x) = 0$ or $x(8x^2 - 15Lx + 6L^2) = 0$. This yields critical values when $x = 0$, $0.5785L$, and $1.2965L$. Now, $Y'' = k(24x^2 - 30Lx + 6L^2)$ and, since $L > 0$, we find that $Y''(0) > 0$, $Y''(0.5785L) < 0$; $Y''(1.2965L) > 0$. Thus, the maximum deflection occurs at $x = 0.5785L$.

30.
$$E = \frac{T(1 - T\mu)}{T + \mu} = \frac{T - T^2\mu}{T + \mu};$$

$$E' = \frac{(T + \mu)(1 - 2T\mu) - (T - T^2\mu)1}{(T + \mu)^2}$$
$$= \frac{T + \mu - 2T^2\mu - 2T\mu^2 - T + T^2\mu}{(T + \mu)^2}$$
$$= \frac{\mu - T^2\mu - 2T\mu^2}{(T + \mu)^2} = \frac{\mu(-T^2 - 2T\mu + 1)}{(T + \mu)^2}$$

Setting the numerator equal to 0, we have $-T^2 - 2T\mu + 1 = 0$; $T^2 + 2\mu T - 1 = 0$ and so $T = \frac{-2\mu + \sqrt{4\mu^2 + 4}}{2} = -\mu + \sqrt{\mu^2 + 1}$ is the critical value that yields the angle of greatest efficiency.

31. Let d represent the diameter (or width) and h the height of the vertical walls of the tunnel. Then the area of the opening is $A = hd + \frac{1}{2}\pi\left(\frac{d}{2}\right)^2 = hd + \frac{\pi d^2}{8}$, or $hd = A - \frac{\pi d^2}{8}$ and so $h = \frac{A}{d} - \frac{\pi d}{8}$ where A is the constant area. Now, the cost, C, is given by $C = 2h + d + \frac{1}{2}\pi d \cdot 4$. We begin by substituting the above value of h to get the cost as a function of just one variable, d. Thus, we have

$$C = 2h + d + 2\pi d = 2\left(\frac{A}{d} - \frac{\pi d}{8}\right) + (2\pi + 1)d$$
$$= \frac{2A}{d} - \frac{\pi d}{4} + (2\pi + 1)d$$

To determine when the cost is a minumum, we take the derivative, getting

$$C' = \frac{-2A}{d^2} - \frac{\pi}{4} + 2\pi + 1$$
$$= \frac{-2A}{d^2} + \frac{7\pi}{4} + 1$$

Setting this equal to 0, we get $\frac{2A}{d^2} = \frac{7\pi + 4}{4}$. Substituting for A, we have

$$\frac{2\left(hd + \frac{\pi d^2}{8}\right)}{d^2} = \frac{7\pi + 4}{4}$$

or $\frac{hd}{d^2} + \frac{\pi d^2}{8d^2} = \frac{7\pi + 4}{8}$ which produces $\frac{h}{d} = \frac{7\pi + 4}{8} - \frac{\pi}{8} = \frac{6\pi + 4}{8} = \frac{3\pi + 2}{4}$.

32. Let x be the distance down the shore from P. Then, by the Pythagorean theorem, the distance that must be rowed is $\sqrt{x^2+4^2}$. The distance that will be walked is $20-x$. The total time is then given by the equation $T = \frac{\sqrt{x^2+4^2}}{2.5} + \frac{20-x}{4.5}$.

To find the shortest time, we find T' and set it equal to 0. Thus,

$$T' = \frac{1}{2.5}\cdot\frac{1}{2}\frac{2x}{\sqrt{x^2+4^2}} - \frac{1}{4.5}$$
$$= \frac{x}{2.5\sqrt{x^2+4^2}} - \frac{1}{4.5}$$

Setting this equal to 0, we have $\frac{x}{2.5\sqrt{x^2+4^2}} = \frac{1}{4.5}$ or $4.5x = 2.5\sqrt{x^2+4^2}$ or $1.8x = \sqrt{x^2+4^2}$. Squaring both sides, we get $3.24x^2 = x^2+16$, or $2.24x^2 = 16$, or $x^2 \approx 7.14286$, and so $x \approx \pm 2.6726$. Since the distance must be positive, the answer is 2.673 km from P. Checking the critical value of $x = 2.673$ with the second derivative, we find T'' as

$$T'' = \frac{1}{2.5}\frac{\sqrt{x^2+4^2} - x\frac{1}{2}\frac{2x}{\sqrt{x^2+4^2}}}{x^2+4^2}$$
$$= \frac{1}{2.5}\frac{x^2+16-x^2}{(x^2+16)^{3/2}}$$
$$= \frac{16}{2.5}(x^2+16)^{-3/2}$$

This is always positive, so the critical value is a minimum.

24.5 RELATED RATES

1. As in example 24.19 we have variables t, r, S, and V with formulas $S = 4\pi r^2$ and $V = \frac{4}{3}\pi r^3$. $\frac{dS}{dt} = 8\pi r\frac{dr}{dt}$; $\frac{dV}{dt} = 4\pi r^2\frac{dr}{dt}$; and $\frac{dV}{dt} = 20$ L/min $= 20{,}000$ cm^3/min; $20{,}000 = 4\pi r^2\frac{dr}{dt} \Rightarrow \frac{dr}{dt} = \frac{5{,}000}{\pi r^2}$. This yields $\frac{dS}{dt} = 8\pi r\cdot\frac{5000}{\pi r^2} = \frac{40{,}000}{r}$. When $r = 300$ cm, then $\frac{dS}{dt} = \frac{40{,}000}{300} = 133\frac{1}{3}$ cm^2/min.

2. The variables are t, r, h, and V with $h = 2r$, $V = \frac{1}{3}\pi r^2h = \frac{1}{3}\pi r^2(2r) = \frac{2}{3}\pi r^3$. $\frac{dV}{dt} = 2\pi r^2\frac{dr}{dt}$; $\frac{dV}{dt}$ is given as 10 ft^3/s and h is 4 ft. When $h = 4$; $r = 2$ ft. Substituting we get $10 = 2\pi(2)^2\frac{dr}{dt}$ or $\frac{dr}{dt} = \frac{10}{8\pi} = \frac{5}{4\pi}$ ft/s ≈ 0.40 ft/s.

3. In this problem we are not concerned about the volume, only the radius and height. These are related by the equation $(24-h)^2 + r^2 = 24^2$ or $(24^2 - 48h + h^2) + r^2 = 24^2$ or $-48h + h^2 = -r^2$. Taking the derivatives with respect to time we get

$$-48\frac{dh}{dt} + 2h\frac{dh}{dt} = -2r\frac{dr}{dt}$$

and thus

$$\frac{dr}{dt} = \frac{24-h}{r}\frac{dh}{dt}$$

When $h = 8$, then $r^2 = 24^2 - (24-8)^2 = 24^2 - 16^2 = 576 - 256$, so $r^2 = 320$ and $r = \sqrt{320} = 8\sqrt{5}$ ft. We were given $\frac{dh}{dt} = 2$ ft/min. Substituting we get

$$\frac{dr}{dt} = \frac{24-8}{8\sqrt{5}}(2)$$
$$= \frac{32}{8\sqrt{5}} = \frac{4}{\sqrt{5}} = \frac{4\sqrt{5}}{5} \approx 1.789 \text{ ft/min}$$

4. $V = \pi r^2h = 10$ m^3 $= 10$. Taking derivatives with respect to time yields $2\pi rh\frac{dr}{dt} + \pi r^2\frac{dh}{dt} = 0$ or $2h\frac{dr}{dt} + r\frac{dh}{dt} = 0$. Solving for $\frac{dr}{dt}$, we obtain $2h\frac{dr}{dt} = -r\frac{dh}{dt}$, or $\frac{dr}{dt} = \frac{-r}{2h}\frac{dh}{dt}$. When

$r = 12$ m, then $h = \frac{10}{\pi 12^2} \approx 0.02210485$, so we get $\frac{dr}{dt} = \frac{-12}{2(0.0221)} \cdot (-1)$ mm/h $= 271.4$ mm/hr $= 0.2714$ m/h.

5. Let x be the distance the helicopter flies after passing over you and y the distance from you to the helicopter. x and y are related by the equation $x^2 + (0.75)^2 = y^2$. Taking derivatives with respect to time yields $2x\frac{dx}{dt} + 0 = 2y\frac{dy}{dt}$ or $\frac{dy}{dt} = \frac{x}{y}\frac{dx}{dt}$. After 1 minute, $x = 3$ km and $y = 3.0923$ km. Substituting yields $\frac{dy}{dt} = \frac{3}{3.0923} \cdot 3 = 2.9104$ km/min $= 174.6$ km/h.

6. In this problem the width of the water is a constant of 10m but the length and depth vary. They are related by the equation $\ell = 5h$ when $1 \le h \le 4$. The volume of the pool is $V = \frac{1}{2}\ell wh$, and so $V = \frac{1}{2}(5h) \cdot 10 \cdot h$ or $V = 25h^2$. Taking the derivative, we have $\frac{dV}{dt} = 50 \cdot h \cdot \frac{dh}{dt}$. Since we are given $\frac{dV}{dt} = 50$ L/min $= 50$L/min$\cdot \frac{1 \text{ m}^3}{1000 \text{ L}} = 0.05$ m^3/min, substituting, we get $0.05 = 50 \cdot 2\frac{dh}{dt} \Rightarrow \frac{dh}{dt} = \frac{0.05}{100} = 0.0005$ m/min.

7. Using the formula $V = \frac{1}{3}\pi h^2(3R - h)$ with $R = 5$ we get

$$V = \frac{1}{3}\pi h^2(15 - h) = 5\pi h^2 - \frac{1}{3}\pi h^3$$
$$\frac{dV}{dt} = 10\pi h\frac{dh}{dt} - \pi h^2\frac{dh}{dt} = (10\pi h - \pi h^2)\frac{dh}{dt}$$
$$\frac{dh}{dt} = \frac{dV/dt}{10\pi h - \pi h^2}$$

We are given $\frac{dV}{dt} = 10$ ft^3/min, so $\frac{dh}{dt} = \frac{10}{10\pi h - \pi h^2} = \frac{10}{(10h - h^2)\pi}$.

(a) When $h = 2$, we get $\frac{dh}{dt} = \frac{10}{(10 \cdot 2 - 2^2)\pi} = \frac{10}{(20 - 4)\pi} = \frac{10}{16\pi} = \frac{5}{8\pi} \approx 0.1989$ ft/min

(b) When $h = 3$, then $\frac{dh}{dt} = \frac{10}{(30 - 9)\pi} = \frac{10}{21\pi} \approx 0.1516$ ft/min

(c) When $h = 4$, we get $\frac{dh}{dt} = \frac{10}{(40 - 16)\pi} = \frac{10}{24\pi} = \frac{5}{12\pi} \approx 0.1326$ ft/min

8. As in Exercise #7, $\frac{dh}{dt} = \frac{dV/dt}{(10h - h^2)\pi}$, and since $\frac{dV}{dt} = -45$ ft^3/min, we have $\frac{dh}{dt} = \frac{-45}{(10h - h^2)\pi}$.

(a) When $h = 4$, we have $\frac{dh}{dt} = \frac{-45}{(40 - 16)\pi} = \frac{-45}{24\pi} \approx -0.5968$ ft/min

(b) When $h = 2$, then $\frac{dh}{dt} = \frac{-45}{(20 - 4)\pi} = \frac{-45}{16\pi} \approx -0.8952$ ft/min.

9. Let h be the height of the surface of the water. Since the trough is an equilateral triangle we know that the width of the surface and the height are related by the equation $w = \frac{2\sqrt{3}h}{3}$. The volume of the trough is $V = \frac{1}{2}wh \cdot 10 = \frac{10\sqrt{3}h^2}{3}$ and $\frac{dV}{dt} = 25.0$ L/min $= 25$ L/min$\cdot\frac{1 \text{ m}^3}{1000 \text{ L}} = 0.025$ m^3/min; $\frac{dV}{dt} = \frac{20\sqrt{3}}{3}h\frac{dh}{dt} \Rightarrow \frac{dh}{dt} = \frac{3dV/dt}{20\sqrt{3}h}$. When $h = 0.175$, we have $\frac{dh}{dt} = \frac{0.025}{\frac{20\sqrt{3}}{3}(0.175)} \approx 0.0124$ m/min $= 12.4$ mm/min.

10. $A = \frac{1}{2}bh$. Differentiating, with respect to t, we have $\frac{dA}{dt} = \frac{1}{2}b\frac{dh}{dt} + \frac{1}{2}h\frac{db}{dt}$. Now, substituting the given information yields $\frac{dA}{dt} = \frac{1}{2}(262.5 \cdot (-4.00) + (137.5) \cdot (5.00)) = -181.25$ cm^2/s.

11. We start with the formulas $V = \frac{4}{3}\pi r^3$ and $S = 4\pi r^2$. Differentiating V and setting it equal to 60, we get

$\frac{dV}{dt} = 4\pi r^2 \frac{dr}{dt} = 60 \Rightarrow \frac{dr}{dt} = \frac{60}{4\pi r^2} = \frac{15}{\pi r^2}$. Now, differentiating S, we find $\frac{dS}{dt} = 8\pi r \frac{dr}{dt} = 8\pi r \cdot \frac{15}{\pi r^2} = \frac{120}{r}$. When $r = 8$ cm, $\frac{dS}{dt} = \frac{120}{8} = 15\text{ cm}^2/\text{s}$.

12. $LA = \frac{1}{2} \cdot 2\pi r \cdot \ell$ where ℓ is the slant height. Thus, $\ell = \sqrt{r^2 + h^2}$ so the lateral area is $LA = \pi r\sqrt{r^2 + h^2}$. From the given information, we see that $\frac{r}{h} = \frac{2.43}{4.27} \Rightarrow r = 0.57h$. Substituting, we get $LA = \pi(0.57h)\sqrt{(0.57h)^2 + h^2} = \pi(0.57h)(1.15h)$; $LA = 0.656\pi h^2$. Differentiating and substituting the given values, produces $\frac{dLA}{dt} = 1.312\pi h \frac{dh}{dt} = 1.312(\pi)(4.27)(0.425) = 7.48\text{ m}^2/\text{h}$.

13. We have $V = \frac{4}{3}\pi r^3 + \pi r^2 h$ and $h = 4r$ so $V = \frac{4}{3}\pi r^3 + 4\pi r^3 = \frac{16}{3}\pi r^3$. Differentiating, we obtain $\frac{dV}{dt} = 16\pi r^2 \frac{dr}{dt} = 60 \Rightarrow \frac{dr}{dt} = \frac{60}{16\pi r^2} = \frac{15}{4\pi r^2}$. The formula for the surface area is $S = 4\pi r^2 + 2\pi rh = 4\pi r^2 + 2\pi r(4r) = 4\pi r^2 + 8\pi r^2 = 12\pi r^2$. Differentiating this, we get $\frac{dS}{dt} = 24\pi r \frac{dr}{dt} = \frac{24\pi r \cdot 15}{4\pi r^2} = \frac{90}{r} = \frac{90}{8} = 11.25\text{ cm}^2/\text{s}$.

14. $A = \pi r^2$; $\frac{dA}{dt} = 2\pi r \frac{dr}{dt} = 2\pi(20)(0.2) = 8\pi \approx 25.13\text{ mm}^2/\text{s}$.

15. $A = s^2$; $\frac{dA}{dt} = 2s\frac{ds}{dt} = 2(40) \cdot (0.3) = 24\text{ mm}^2/\text{s}$.

16. As in example 24.21 $PV^{1.4} = k$, so $k = 60 \times 100^{1.4} = 37857$. $V^{1.4} = kP^{-1}$; $1.4V^{0.4}\frac{dV}{dt} = -kP^{-2}\frac{dP}{dt}$. Solving for dV/dt, we obtain $\frac{dV}{dt} = \frac{-kP^{-2}}{1.4V^{0.4}}\frac{dP}{dt} = -\frac{37857}{1.4}(100)^{-0.4}(4)(60)^{-2} = -4.76\text{ cm}^2/\text{s}$.

17. Here $PV^{1.2} = 400$, so $P = 400V^{-1.2}$. $\frac{dP}{dt} = -1.2(400)V^{-2.2}\frac{dV}{dt} = \frac{-480}{V^{2.2}}\frac{dV}{dt}$. We are given that $\frac{dP}{dt} = 0.1P = 0.1(400V^{-1.2}) = 40V^{-1.2}$. Thus, $40V^{-1.2} = \frac{-480}{V^{2.2}}\frac{dV}{dt}$. Solving for dV/dt, we get $\frac{dV}{dt} = \frac{40V^{-1.2} \times V^{2.2}}{-480} = -0.0833V = -8.3\%$ of the volume per hour.

18. The height of the balloon, h, and the distance from the balloon to the observer, s, are related by the following equation $s^2 = h^2 + 500^2$. Taking the derivative with respect to t, we get $2s\frac{ds}{dt} = 2h\frac{dh}{dt}$ so $\frac{ds}{dt} = \frac{h}{s}\frac{dh}{dt}$. When $h = 2000$, $s = \sqrt{2000^2 + 500^2} = 2061.6$. Hence, $\frac{ds}{dt} = \frac{2000}{2061} \cdot 5 = 4.85$ ft/s.

19. Let a = altitude of the rocket and s be the distance from the radar station to the rocket. Then $a^2 + 2.430^2 = s^2$. Taking the derivatives, we obtain

$$2a\frac{da}{dt} = 2s\frac{ds}{dt} \Rightarrow \frac{da}{dt} = \frac{s}{a}\frac{ds}{dt}$$

When $a = 3.750$, $s = \sqrt{3.750^2 + 2.430^2} \approx 4.468$, and so $\frac{da}{dt} = \frac{4.468}{3.750}(325.0\text{ m/s}) = 387.2$ m/s or 1394 km/hr.

20. $PV = k$ so $k = 20 \cdot 50 = 1000$ in.·lb.; $V = kP^{-1}$; $\frac{dV}{dt} = -kP^{-2}\frac{dP}{dt} = \frac{-k}{P^2}\frac{dP}{dt} = \frac{-1000}{20^2} \cdot (1.5) = -3.75$ ft^3/min.

21. (a) The derivative of $v^2 = 1200 - 36.0s$ is $2v\frac{dv}{dt} = -36.0\frac{ds}{dt}$. Now, $\frac{dv}{dt} = a$ and $\frac{ds}{dt} = v$, so we get $2va = -36.0v$ or $a = -18.0\text{ m/s}^2$. (b) The acceleration is constant so when $s = 2.45$ m, then a is -18.0 m/s^2.

22. Differentiating $h^2 + x^2 = 10^2$ we obtain $2h\frac{dh}{dt} + 2x\frac{dx}{dt} = 0$ and so $\frac{dh}{dt} = \frac{-x}{h}\frac{dx}{dt} = \frac{-x}{h}(0.5\text{ m/s})$.

(a) When $x = 2$, $h = \sqrt{100 - 2^2} = \sqrt{96}$ and $\frac{dh}{dt} = \frac{-2}{\sqrt{96}}(0.5) = -0.1021$ m/s.

(b) When $x = 3$, $h = \sqrt{100 - 9} = \sqrt{91}$, and $\frac{dh}{dt} = \frac{-3}{\sqrt{91}}(0.5) = -0.1572$ m/s.

(c) When $x = 4$, $h = \sqrt{100-16} = \sqrt{84}$, and $\frac{dh}{dt} = \frac{-4}{\sqrt{84}}(0.5) = -0.2182$ m/s.

(d) When $x = 6$, $h = \sqrt{100-36} = \sqrt{64} = 8$, and $\frac{dh}{dt} = \frac{-6}{8}(0.5) = -0.375$ m/s.

23. (a) We let A be at the origin and B at $(8, 0)$. The location of the ship at 8 mi from A means that the ship is on the circle $x^2 + y^2 = 64$; and the fact that the ship is 6 mi from B, means that it is on the circle $(x-8)^2 + y^2 = 36$. Solving the first equation for $y^2 = 64 - x^2$ and substituting into the second gives $(x-8)^2 + (64 - x^2) = 36$ or $x^2 - 16x + 64 + 64 - x^2 = 36$, or $-16x = -92$, or $x = 5.75$ and so, $y = \pm\sqrt{64 - 5.75^2} = \pm 5.56$. The ship is 5.75 mi east of A and 5.56 mi north or south of A.
(b) Using the same setup as (a), the velocity vector from A has magnitude 14 and direction $\tan^{-1}\left(\frac{5.56}{5.75}\right) = 44.0°$. The velocity vector from B has magnitude 2 and direction $\tan^{-1}\left(\frac{5.56}{2.25}\right) = 112.0°$. Using component vectors to add, gives

$A_x = 14\cos 44 =$	10.07	$A_y = 14\sin 44 =$	9.73
$B_x = 2\cos 112° =$	−0.75	$B_y = 2\sin 112° =$	1.85
$R_x =$	9.32	$R_y =$	11.88

Thus, $|R| = \sqrt{9.32^2 + 11.88^2} = 15.1$ mph.
(c) $\theta = 51.9°$ north of east or $38.1°$ east of north.

24. (a) The first car's position t hours after the second car leaves is $y = 40 + 80t$. The second car's position is $x = -100t$. The distance between them is $s^2 = x^2 + y^2$.

$$2s\frac{ds}{dt} = 2x\frac{dx}{dt} + 2y\frac{dy}{dt} \text{ or } \frac{ds}{dt} = \frac{xx' + yy'}{s}$$

When $t = 1$, then $x = -100$ and $y = 120$. We are given $x' = -100$ and $y' = 80$. Thus, $s = \sqrt{100^2 + 120^2} = 156.2$; $\frac{ds}{dt} = \frac{(-100)(-100)+120\cdot 80}{156.2} = \frac{19600}{156.2} = 125.5$ km/h.
(b) When $t = 2$; $x = -200$; $x' = -100$; $y = 200$, $y' = 80$; $s = \sqrt{200^2 + 200^2} = 200\sqrt{2} = 282.8$, and so, $\frac{ds}{dt} = \frac{(-200)(-100) + (200)(80)}{282.8} = 127.3$ km/h.

25. (a) Let x be the distance from the person to the base of the light and y be the distance from the light to the top of the shadow. Then we have similar triangles so $\frac{y}{20} = \frac{y-x}{6}$. Simplifying: $6y = 20y - 20x$ or $-14y = -20x$ or $y = \frac{10}{7}x$, we get $\frac{dy}{dt} = \frac{10}{7}\,dx/dt = \frac{10}{7}(5) = \frac{50}{7} \approx 7.143$ ft/s.
(b) Same as (a) $\frac{50}{7} = 7.143$ ft/s. (c) Same as (a) 7.143 ft/s.

26. Let s be the distance from the truck to the pulley, x the horizontal distance, and y the height of the weight above the ground. Then, $s^2 = x^2 + 18^2$ and $(20-y)+s = 50$, so $s = 50-20+y = 30+y$ and $s' = y'$, which means that $2ss' = 2xx'$ and so, $s' = \frac{x}{s}x'$. When the weight is 10 ft above the ground, then $s = 30 + 10 = 40$ and $x = \sqrt{40^2 - 18^2} = 35.721$, so $s' = \frac{35.721}{40}(8) = 7.144$ ft/s. $y' = s' = 7.144$ ft/s.

27. Here $R = 6 + 0.008T^2$, so $\frac{dR}{dt} = 0.016T\frac{dt}{dt} = 0.016(40)(0.01) = 0.0064\ \Omega$/s.

28. The first car's position is $x = 80t$ and the second is $y = 40t$. The distance between the two cars is $d^2 = x^2 + y^2$ so $2dd' = 2xx' + 2yy'$ or

$$d' = \frac{xx' + yy'}{d}$$

(a) After 3 min or 0.05 h, $x = 30(0.05) = 1.5$ mi; $x' = 30$; $y = 40(0.05) = 2$ mi and $d = \sqrt{1.5^2 + 2^2} = 2.5$; $d' = \frac{1.5(30)+2(40)}{2.5} = 50$ mph.
(b) After 6 min or 0.10 h, $x = 3$ mi; $x' = 30$; $y = 4$ mi; $y' = 40$; $d = \sqrt{3^2 + 4^2} = 5$; $d' = \frac{3(30)+4(40)}{5} = 50$ mph

29. $Z = \frac{RX}{R+X}$. If R is constant 3 then $Z = \frac{3X}{3+X}$ and $Z' = \frac{(3+X)3X' - 3X(X')}{(3+X)^2} = \frac{9X'}{(3+X)^2}$, or $Z' = \frac{9(1.45)}{(3+1.05)^2} = 0.80\ \Omega$/min.

30. $Z^2 = R^2 + X^2 = R^2 + 12^2$, and differentiating gives $2ZZ' = 2RR'$ or $Z' = \frac{R}{Z}R'$. When $R = 6$ and $X = 12$, $Z = \sqrt{6^2 + 12^2} = 6\sqrt{5}$ so $Z' = \frac{6}{6\sqrt{5}}(3) = \frac{3\sqrt{5}}{5} \approx 1.3416\ \Omega$/s.

31. $R = \frac{1}{R_1} + \frac{1}{R_2} = R_1^{-1} + R_2^{-1}$; $R' = \frac{-R_1'}{R_1^2} + \frac{-R_2'}{R_2^2} = \frac{-0.5}{4^2} + \frac{-0.4}{5^2} = -0.04725\ \Omega/\text{s}$. Thus, R is decreasing at the rate of $0.04725\ \Omega/\text{s}$.

32. $P = RI^2 = 80.00I^2$, so $P' = 160.00II' = 160.00(2.5)(0.24) = 96.00\ \text{W/s}$.

33. $R = 35.0 + 0.0174T^2$, so $R' = 0.0348TT'$, and after substituting the given values, we have $R' = 0.0348(47.0)(-1.25) = -2.04\ \Omega/\text{min}$.

34. $v = 331\left(\frac{T}{273}\right)^{1/2}$ and $C + 273 = K$ so $C' = K'$. When $C = -30°$, we have $K = 243°$. $C' = -5.8 = K' = T'$; $v' = \frac{1}{2}\left(\frac{331}{273}\right)\left(\frac{T}{273}\right)^{-1/2} T' = \frac{1}{2}\left(\frac{331}{273}\right)\left(\frac{273}{243}\right)^{1/2}(-5.8) = -3.727\ \text{m/s}$.

35. (a) $f = f_s\frac{v_L}{v_L - v_s}$; $f_s = 200$; $v_L = 343$, $v_s = 29$; $f = 200\left(\frac{343}{343-29}\right) = 218.47$ Hz. (b) Here f_s and v_L are constants. We want to find f' and, in this case, we will write f as $f = f_s v_L(v_L - v_s)^{-1}$. Here v_s is the only variable, so $f' = f_s v_L(v_L - v_s)^{-2} = \frac{f_s v_L}{(v_L - v_s)^2}$. Substituting the given values, we obtain $f' = \frac{200(343)}{(343-29)^2} \approx 20.17$ Hz/s.

24.6 NEWTON'S METHOD

1. See *Computer Programs* in main text.

2. $x^3 + 5x - 8 = 0 : P(x) = x^3 + 5x - 8; P'(x) = 3x^2 + 5$ There is a solution near 1. Using Newton's method, we obtain the following results:

i	x_i	$P(x_i)$	$P'(x_i)$	$x_{i+1} = x_i - \frac{P(x_i)}{P'(x_i)}$	$P(x_{i+1})$
1	1	−2	8	1.25	0.2031
2	1.25	0.2031	9.6875	1.2290	0.0013
3	1.229	0.0013	9.5313	1.2289	0.0004
4	1.2289	0.0004	9.5306	1.22885	−0.000002

Hence, $x \approx 1.2289$

3. $x^3 - 4x + 2 = 0$; $P(x) = x^3 - 4x + 2$; $P'(x) = 3x^2 - 4$. There are three solutions; one each near −2; 0.5, and 1.5. Newton's method produces the following results for each solution.

i	x_i	$P(x_i)$	$P'(x_i)$	$x_{i+1} = x_i - \frac{P(x_i)}{P'(x_i)}$	$P(x_{i+1})$
1	−2	2	8	−2.25	−0.3906
2	−2.25	−0.3906	11.1875	−2.2151	−0.0084
3	−2.2151	−0.0084	10.7200	−2.2143	2.11×10^{-4}

Hence, one solution is $x \approx -2.2143$.

i	x_i	$P(x_i)$	$P'(x_i)$	$x_{i+1} = x_i - \frac{P(x_i)}{P'(x_i)}$	$P(x_{i+1})$
1	0.5	0.125	−3.25	0.5385	0.0022
2	0.5385	0.0022	−3.1301	0.5392	-3.48×10^{-5}

Another solution is $x \approx 0.5392$.

i	x_i	$P(x_i)$	$P'(x_i)$	$x_{i+1} = x_i - \frac{P(x_i)}{P'(x_i)}$	$P(x_{i+1})$
1	1.5	−0.625	2.75	1.7273	0.2443
2	1.7273	.2443	4.9507	1.6780	0.0127
3	1.6780	.0127	4.4471	1.6751	-1.36×10^{-4}

And the final solution is $x \approx 1.6751$.

4. Let $x^4 - x - 4 = P(x)$, so $P'(x) = 4x^3 - 1$. Solutions are near −1.2 and 1.4. Newton's method produces the following results for each solution.

i	x_i	$P(x_i)$	$P'(x_i)$	$x_{i+1} = x_i - \frac{P(x_i)}{P'(x_i)}$	$P(x_{i+1})$
1	−1.2	−0.7264	−7.912	−1.2918	0.0766
2	−1.2918	0.0766	−9.6228	−1.2838	1.74×10^{-4}

Hence, one solution is $x \approx -1.2838$.

i	x_i	$P(x_i)$	$P'(x_i)$	$x_{i+1} = x_i - \frac{P(x_i)}{P'(x_i)}$	$P(x_{i+1})$
1	1.4	−1.5584	9.9760	1.5562	0.3088
2	1.5562	0.3088	14.0750	1.5343	0.0074
3	1.5343	0.0074	13.4474	1.5337	-6.9×10^{-4}

And the other solution is $x \approx 1.5337$.

5. $P(x) = x^4 - 2x^3 - 3x + 2$ and $P'(x) = 4x^3 - 6x^2 - 3$. Solutions are near 0.6 and 2.3. Newton's method produces the following results for each solution.

i	x_i	$P(x_i)$	$P'(x_i)$	$x_{i+1} = x_i - \frac{P(x_i)}{P'(x_i)}$	$P(x_{i+1})$
1	0.6	−0.1024	−4.2960	0.5762	-9.76×10^{-4}
2	0.5762	-9.76×10^{-4}	−4.2268	0.5760	-1.3×10^{-4}

Hence, one solution is $x \approx 0.5760$.

i	x_i	$P(x_i)$	$P'(x_i)$	$x_{i+1} = x_i - \frac{P(x_i)}{P'(x_i)}$	$P(x_{i+1})$
1	2.3	−1.2499	13.9280	2.3897	0.1490
2	2.3897	0.1490	17.3231	2.3811	0.0015
3	2.3811	0.0015	16.9821	2.3810	-1.6×10^{-4}

And the other solution is $x \approx 2.3810$.

6. $x^4 = 125 \Rightarrow x^4 - 125 = 0$ so let $P(x) = x^4 - 125$, and then $P'(x) = 4x^3$. Since $P(x)$ is an even function, solutions will be in positive and negative pairs. One solution is near 3.3. Newton's method will show that $x \approx -3.3437$ and 3.3437.

7. $x^3 = 91 \Rightarrow x^3 - 91 = 0$; $P(x) = x^3 - 91$; $P'(x) = 3x^2$. A solution is near 4.5. Using Newton's method, you can show that $x \approx 4.4979$.

8. $x^5 = 111 \Rightarrow x^5 - 111 = 0$ so $P(x) = x^5 - 111$ and $P'(x) = 5x^4$. A solution is near 2.5. Using Newton's method, you can show that $x \approx 2.5649$.

24.7 DIFFERENTIALS

1. $dy = f'(x)\,dx = (4x^3 - 2x)\,dx$

2. $dy = g'(x)\,dx = (6x + 2)\,dx$

3. $dy = (15x^2 - 2x + 1)\,dx$

4. $dy = 3(x^2 - 4)^2(2x)\,dx = 6x(x^2 - 4)^2\,dx$

5. $dy = \frac{1}{3}(4 - 2x)^{-2/3}(-2)\,dx = \frac{-2}{3}(4 - 2x)^{-2/3}\,dx$

6. $dy = \dfrac{(x-4) - x}{(x-4)^2}\,dx = \dfrac{-4}{(x-4)^2}\,dx$ or $-4(x-4)^{-2}\,dx$

7. $\Delta y = ((3.2)^2 - 3.2) - (3^2 - 3) = 7.04 - 6 = 1.04$, $dy = (2x - 1)\,dx = 5(0.2) = 1.0$

8. $\Delta y = f(2.1) - f(2) = 11.361 - 10 = 1.361$, $dy = f'(2)\,dx = 13(0.1) = 1.3$

9. $\Delta y = f(5.15) - f(5) = 59.9675 - 56 = 3.9675$, $dy = f'(5)(0.15) = 26(0.15) = 3.9$

10. $\Delta y = g(4.2) - g(4.1) = 311.1696 - 282.5761 = 28.5935$, $dy = g'(4.1)(0.1) = 27.5684$

11. $\Delta y = h(3.2) - h(3) = 0.0305175781 - 0.037037037 = -0.00651946$, $dy = h'(3)(0.2) = -0.037037037(0.2) = -0.00740741$

12. $\Delta y = j(0.6) - j(0.5) = 0.375 - 0.333333 = 0.0416667$, $dy = j'(0.5)(0.1) = 0.444444(0.1) = 0.0444444$

13. (a) $V = \frac{1}{3}\pi r^2 h = \frac{1}{3}\pi(2)^2 \cdot 4 = \frac{16\pi}{3} = 16.755\text{ m}^3$ (b) $dV = \dfrac{2}{3}\pi r h dr = \dfrac{2}{3}(\pi \cdot 2 \cdot 4)(0.01) = \dfrac{0.16\pi}{3} = \dfrac{16}{300}\pi \approx 0.16755\text{ m}^3$ (c) Relative error $= \dfrac{dV}{V} = \dfrac{\frac{16\pi}{300}}{16\frac{\pi}{3}} = \dfrac{1}{100}$; $\frac{1}{100} = 0.01 = 1\%$.

14. (a) $A = \pi r^2$; $dA = 2\pi r dr = 2\pi(200)(1.5) = 600\pi\text{ mm}^2$ (b) $\dfrac{dr}{r} = \dfrac{1.5}{200} = 0.0075 = 0.75\%$ (c) $\dfrac{dA}{A} = \dfrac{600\pi}{\pi(200)^2} = \dfrac{6}{400} = 0.015 = 1.5\%$

15. (a) $A = s^2$; $dA = 2s\,ds = 2(24)(0.02) = 0.96\text{ in.}^2$ (b) $\dfrac{dA}{A} = \dfrac{0.96}{24^2} = 0.00167 = 0.167\%$

16. $V = s^3$; $dV = 3s^2\,ds = 3(1.452)^2(0.0005) = 0.00316\text{ m}^3 = 3162\text{ cm}^3$ (Note: 0.5 mm = 0.0005 m and $1\text{ m}^3 = (100\text{ cm})^3$.)

17. $S = 2\pi r^2$; $dS = 4\pi r\,dr = 4\pi(100)(0.01) = 4\pi\text{ m}^2$. (Note: 10 mm = 0.01 m.)

18. $\dfrac{dS}{S} = \dfrac{4\pi r\,dr}{2\pi r^2} = \dfrac{2\,dr}{r} = 0.01\% = 0.0001$, $dr = \dfrac{0.0001r}{2} = \dfrac{0.01}{2} = 0.005\text{ m} = 5\text{ mm}$.

19. (a) The volume of a sphere is $V = \frac{4}{3}\pi r^3$. Since we are given $d = 12.4$ m, we see that $r = 6.2$ m, and $V = \frac{4}{3}\pi(6.2)^3 \approx 317.8\pi$. Taking the derivative of the formula for the volume, we obtain $dV = 4\pi r^2\,dr = 4\pi(6.2)^2(\pm 0.05) = \pm 7.69\pi\text{ m}^3 \approx \pm 24.2\text{ m}^3$ (b) $\dfrac{dV}{V} = \dfrac{7.69\pi}{317.8\pi} = 0.024 = \pm 2.4\%$.

20. (a) $dV = 4\pi r^2\,dr = 4\pi(6.20)^2(0.005) = \pm 0.7688\pi \approx 2.42\text{ m}^3$ (b) $\dfrac{dV}{V} = \dfrac{0.7688\pi}{317.8\pi} = \pm 0.00242 = \pm 0.242\%$

21. $R = \sigma T^4$, so $dR = 4\sigma T^3\,dT$ and $\frac{dT}{T} = 0.02$; $\frac{dR}{R} = \frac{4\sigma T^3\,dT}{\sigma T^4} = 4\frac{dT}{T} = 4(0.02) = 0.08$.

22. Here $R = 250r^{-2}$ and $\frac{dr}{r} = \pm 0.4\%$. (a) $dR = -500r^{-3}\,dr = \frac{-500}{r^2}\cdot\frac{dr}{r} = \frac{-500}{r^2}\cdot(\pm 0.4\%) = \frac{\pm 200\%}{r^2} = \pm\frac{2}{r^2}\ \Omega$ (b) $\frac{dR}{R} = \frac{\pm\frac{2}{r^2}}{\frac{250}{r^2}} = \frac{\pm 2}{250} = \pm 0.008 = \pm 0.8\%$

23. $V = IR$; $30 = IR$; $I = 30R^{-1}$; $dI = -30R^{-2}\,dR = -30(10)^{-2}(0.1) = -0.03$ A

24. $R = \frac{20R_v}{20+R_v}$; $dR = \frac{(20+R_v)20\,(dR_v) - 20R_v(dR_v)}{(20+R_v)^2} = \frac{400dR_v}{(20+R_v)^2} = \frac{400(0.1)}{(20+10)^2} = \frac{40}{900} = 0.0444\ \Omega$

25. $R = 35.0 + 0.0174T^2$; $dR = 0.0348T \cdot dT = 0.0348(125)(\pm 0.5) = \pm 2.175\ \Omega$.

26. Since 1 dm^3 = 1 L, we will convert all units to decimeters. Here $V = \frac{4}{3}\pi r^3$, and so, $dV = 4\pi r^2\,dr$. Since 20 m = 200 dm and 0.5 mm = 0.005 dm, we have $dV = 4\pi(200)^2(0.005) \approx 2\ 531$ dm^3 = 2513 L.

27. $S = 4\pi r^2$; $\frac{dS}{S} = \frac{8\pi r dr}{4\pi r^2} = 2\frac{dr}{r} = 2(0.5\%) = 1.0\%$

28. (a) $V = \frac{4}{3}\pi r^3 = \frac{4}{3}\pi(8)^3 = 2144.7\ \text{mm}^3$, mass = (2144.7)(0.00794) g = 17.03 g, (b) $dV = 4\pi r^2 dr = 4\pi\cdot 8^2\cdot(.005\times 8) = 32.17$, so the mass is 32.17(0.00794) = 0.255 g. (Note: $\frac{dr}{r} = 0.5\% = 0.005$ so $dr = 0.005r = 0.005 \times 8$.)

24.8 ANTIDERIVATIVES

1. $f(x) = 7; F(x) = 7x + C$

2. $g(x) = 4x; G(x) = 4\cdot\frac{1}{2}x^2 + C = 2x^2 + C$

3. $h(x) = 2 - 3x^2; H(x) = 2x - 3\left(\frac{1}{3}\right)x^3 + C = 2x - x^3 + C$

4. $j(x) = x^2 - 3x + 5; J(x) = \frac{1}{3}x^3 - 3\cdot\frac{1}{2}x^2 + 5x + C = \frac{1}{3}x^3 - \frac{3}{2}x^2 + 5x + C$

5. $k(x) = 4x^3 - 3x^2 + 2x + 9; K(x) = 4\cdot\frac{1}{4}x^4 - 3\cdot\frac{1}{3}x^3 + 2\cdot\frac{1}{2}x^2 + 9x + C = x^4 - x^3 + x^2 + 9x + C$

6. $m(x) = \frac{4}{5}x^3 - 6x^2; M(x) = \frac{4}{5}\cdot\frac{1}{4}x^4 - 6\cdot\frac{1}{3}x^3 + C = \frac{1}{5}x^4 - 2x^3 + C$

7. $f(x) = x^{-4} + 2x^{-3} - x^{-2} + 5; F(x) = \frac{1}{-3}x^{-3} + 2\cdot\frac{1}{-2}x^{-2} - \frac{1}{-1}x^{-1} + 5x + C = \frac{-1}{3}x^{-3} - x^{-2} + x^{-1} + 5x + C$

8. $g(x) = \frac{1}{x^3} + 2 = x^{-3} + 2; G(x) = \frac{1}{-2}x^{-2} + 2x + C = -\frac{1}{2}x^{-2} + 2x + C$

9. $h(x) = \sqrt{x} + x - \frac{1}{x^2} = x^{1/2} + x - x^{-2}; H(x) = \frac{1}{3/2}x^{3/2} + \frac{1}{2}x^2 - \frac{1}{-1}x^{-1} + C = \frac{2}{3}x^{3/2} + \frac{1}{2}x^2 + x^{-1} + C$

10. $j(x) = 3x^2 - x^{-1/3} - 4x^{-2}; J(x) = 3\cdot\frac{1}{3}x^3 - \frac{1}{2/3}x^{2/3} - 4\cdot\frac{1}{-1}x^{-1} + C = x^3 - \frac{3}{2}x^{2/3} + 4x^{-1} + C$

11. $k(x) = 2x^{-1.4} + 3.5x^{-6} + 1.2x^{-1.3}; K(x) = 2\cdot\frac{1}{-.4}x^{-.4} + 3.5\cdot\frac{1}{-5}x^{-5} + 1.2\cdot\frac{1}{-0.3}x^{-.3} + C = -5x^{-0.4} - 0.7x^{-5} - 4x^{-0.3} + C$

12. $m(x) = 2x^{-1/2} + 3x^{5/2}; M(x) = 2\cdot\frac{1}{1/2}x^{1/2} + 3\cdot\frac{1}{7/2}x^{7/2} + C = 4x^{1/2} + \frac{6}{7}x^{7/2} + C$

13. $s(t) = t^2 + 2t; S(t) = \frac{1}{3}t^3 + t^2 + C$

14. $a(t) = \frac{2}{3}t - \frac{3}{4}t^2 + t^{3/2}; A(t) = v(t) = \frac{2}{3}\cdot\frac{1}{2}t^2 - \frac{3}{4}\cdot\frac{1}{3}t^3 + \frac{1}{5/2}t^{5/2} = \frac{1}{3}t^2 - \frac{1}{4}t^3 + \frac{2}{5}t^{5/2} + C$

15. $v(t) = 42t - 5; V(t) = \frac{42}{2}t^2 - 5t + C = 21t^2 - 5t + C$

16. $p(v) = \frac{1}{3}v^3 - 3v^{-4}; P(v) = \frac{1}{3}\cdot\frac{1}{4}v^4 - 3\cdot\frac{1}{-3}v^{-3} + C = \frac{1}{12}v^4 + v^{-3} + C$

17. $a(t) = -32$ ft/s^2; $v(t) = A(t) = -32 \cdot t + C_1$. When the ball is dropped at $t = 0$, we have $v(0) = 0$, so we can see that $C_1 = 0$. Hence, $v(t) = -32t$. Now, $s(t) = V(t) = -16t^2 + C_2$. When $t = 0$, we are given that $s(0) = 600$, so $C_2 = 600$ and we have $s(t) = -16t^2 + 600$.
(a) When the ball hits the ground, $s(t) = 0$. Solving $s(t) = 0$ for t, we have $16t^2 = 600$ or $t^2 = \frac{600}{16}$ and so the ball hits the ground after $t = \sqrt{\frac{600}{16}} \approx 6.12$ s
(b) $v(6.12) = -32(6.12) = -195.96$ ft/s

18. $a(t) = -9.8$ m/s^2; $v(t) = -9.8t + C$. Since at the time the ball is dropped ($t = 0$), the velocity is 0, we have $C = 0$, and so $v(t) = -9.8t$. Thus, we find that $s(t) = -4.9t^2 + C_2$. When $t = 0$, $s = 140$ m, and so $s(t) = -4.9t^2 + 140$. (a) $s(t) = 0$ when $4.9t^2 = 140$ or $t \approx 5.345$ s. (b) $v(5.345) = -52.38$ m/s

19. $a(t) = -32$; $v(t) = -32t + C$. Since the ball was thrown down at $t = 0$ at 25 ft/s, we have $C = -25$ and thus, $v(t) = -32t - 25$. (a) $v(t) = -185 = -32t - 25 \Rightarrow 32t = 160 \Rightarrow t = 5$ s. (b) $s(t) = -16t^2 - 25t + C$, $s(5) = 0 = -16 \cdot 5^2 - 25 \cdot 5 + C$, $C = 16 \cdot 5^2 + 25 \cdot 5 = 525$. The building is 525 feet high.

20. $a(t) = -9.8$ m/s^2; $v(t) = -9.8t + C$; when $t = 0$, $v = -12.1$ so $C = -12.1$. This gives $v(t) = -9.8t - 12.1$.
(a) $v(t) = -9.8t - 12.1 = -66 \Rightarrow 9.8t = 53.9$; $t = 5.5$ s
(b) $s(t) = -4.9t^2 - 12.1t + C$; $s(5.5) = -4.9(5.5)^2 - 12.1(5.5) + C = 0$, so $C = 4.9(5.5)^2 + 12.1(5.5) = 214.775$ m

21. $a(t) = -32$ and $v(t) = -32t + 160$. (a) Since $v(t) = -32t + 160 = -384$, then $-32t = -544$ and $t = 17$ s. (b) $s(t) = -16t^2 + 160t + C$, so $s(17) = -16 \cdot 17^2 + 160 \cdot 17 + C = 0$, thus $C = 16 \cdot 17^2 - 160 \cdot 17 = 1904$ ft.

22. $a(t) = -9.8$; $v(t) = -9.8t + 45$. (a) $-9.8t + 45 = -120$ or $-9.8t = -165$ and so $t = 16.837$ s. (b) $s(t) = -4.9t^2 + 45t + C$, and $s(16.837) = -4.9(16.837)^2 + 45(16.837) + C = 0$, $C = 4.9(16.837)^2 - 45(16.837) = 631.41$ m

23. $a(t) = 3.2$; $v(t) = 3.2t + 0$; $3.2t = 32 \Rightarrow t = 10$ s, $s(t) = 1.6t^2$, $s(10) = 1.6(10)^2 = 160$ m

24. 45 mph $= 45\frac{\text{mi}}{\text{hr}} \times 5280$ ft/mi $\times \frac{1 \text{ hr}}{3600 \text{ s}} = 66$ ft/s
(a) $a(t) = -20$ ft/s^2; $v(t) = -20t + 66 = 0 \Rightarrow t = \frac{66}{20} = 3.3$ s
(b) $s(t) = -10t^2 + 66t$; $s(3.3) = -10(3.3)^2 + 66 \cdot 3.3 = 108.9$ ft

25. $v(t) = a \cdot t + 50 \Rightarrow a = \frac{-50}{t}$, $s(t) = \frac{a}{2}t^2 + 50t + 500 = 0$, $\frac{-50}{2t}t^2 + 50t = -500$. This simplifies to $-25t + 50t = 500$, or $25t = 500$, or $t = 20$ s. Finally, $a = \frac{-50}{20} = -2.5$ m/s^2.

26. (a) $\alpha(t) = 4 - t$, $\omega(t) = 4t - \frac{1}{2}t^2 + C$, $\omega(0) = 10 = 4t - \frac{1}{2}t^2 + C$ *so* $C = 10$, $\omega(t) = 10 + 4t - \frac{1}{2}t^2$, $\theta(t) = 10t + 2t^2 - \frac{1}{6}t^3 + C$, $\theta(0) = 0$ *so* $C = 0$, $\theta(t) = 10t + 2t^2 - \frac{1}{6}t^3$
(b) The wheel begins to turn in the opposite direction when $\omega(t) = 0$. This is at -2 and 10 seconds. Since $t \geq 0$, $t = 10$ and $\theta(10) = 133.3$ rad.
(c) $\omega(t) = -14 = 10 + 4t - \frac{1}{2}t^2$, $0 = 24 + 4t - \frac{1}{2}t^2 \Rightarrow t^2 - 8t - 48 = 0$, $(t-12)(t+4) = 0$; $t = 12$ or $t = -4$. In this problem time cannot be negative, so $t = 12$ s.

27. (a) $\theta(t) = 12.4t - \frac{3.4}{2}t^2 + \frac{0.30}{3}t^3$ = antiderivative of $\omega(t) = 12.4t - 1.7t^2 + 0.10t^3$; $\theta(3) = 24.6$ rad $= \frac{24.6}{2\pi} = 3.9$ rev.
(b) $\alpha(t) = \omega'(t) = -3.4 + 0.6t = -1$, $0.6t = 2.4$, $t = 4$ s, $\theta(4) = 28.8$ rad $= \frac{28.8}{2\pi} = 4.6$ rev

28. $a(t) = kt$; $v(t) = \frac{k}{2}t^2 + C$. Since the ball starts from rest, let $C = 0$. $s(t) = \frac{k}{6}t^3$; $\frac{k}{6}4^3 = 24$; $k = \frac{24 \cdot 6}{4^3} = 2.25$, $s(t) = \frac{2.25}{6}t^3 = 0.375t^3$; $s(t) = 0.375t^3$

29. (a) $i = 4.4t - 2.1t^2$; $q = 2.2t^2 - 0.7t^3 + C = 2.2t^2 - 0.7t^3 + 5$
(b) $q(3.2) \approx 4.59 \approx 4.6$ C

30. $i(t) = 6\sqrt{t} = 6t^{1/2}$, $q(t) = 6 \cdot \frac{1}{3/2}t^{3/2} + C = 4t^{3/2} + C$. Now, finding C, we have $q(0.16) = 4(0.16)^{3/2} + C = 0.347$, $C = 0.347 - 4(0.16)^{3/2} = 0.091$. Hence, we get $q(t) = 4t^{3/2} + 0.091$, $q(0.25) = 4(0.25)^{3/2} + 0.091 = 0.591$ C

31. $v(t) = -N\phi'(t)$ (a) $-200\phi'(t) = 2t - 4t^{1/3}$, $\phi'(t) = -\frac{t}{100} + \frac{1}{50}t^{1/3}$, $\phi(t) = \frac{-t^2}{200} + \frac{3}{200}t^{4/3} + C$, $C = 0.02$; $\phi(t) = -0.005t^2 + 0.015t^{4/3} + 0.02$, or $-\frac{1}{200}(t^2 - 3t^{4/3} - 4)$
(b)$\phi(0.729) = -\frac{1}{200}(0.729^2 - 3(0.729)^{4/3} - 4) =$ 0.027 Wb

32. $m = 4x + 2 = y'$; $y =$ antiderivative of $4x + 2$, so $y = 2x^2 + 2x + C$; $5 = 2(-1)^2 + 2(-1) + C$, so $C = 5$ and we have $y = 2x^2 + 2x + 5$.

33. $m = \sqrt{x}$ so $y' = \dfrac{-1}{\sqrt{x}} = -x^{-1/2}$, $y = -\frac{1}{1/2}x^{1/2} = -2x^{1/2} + C$, and $-3 = -2(4)^{1/2} + C$ gives us that $C = -3 + 4 = 1$. Hence, $f(x) = -2x^{1/2} + 1 = -2\sqrt{x} + 1$.

CHAPTER 24 REVIEW

1. $f(x) = x^4 - x^2$; $f'(x) = 4x^3 - 2x = 2x(2x^2 - 1) = 0$ when $x = 0$ and $\pm\frac{\sqrt{2}}{2}$. These are the critical values. $f''(x) = 12x^2 - 2$; $f''(0) < 0$, so there is a maximum at $(0, 0)$. $f''\left(\frac{\sqrt{2}}{2}\right) = f''\left(-\frac{\sqrt{2}}{2}\right) > 0$, so there are minimums at $\left(\frac{\pm\sqrt{2}}{2}, -\frac{1}{4}\right)$. $f''(x) = 12x^2 - 2 = 0$ when $x = \pm\frac{1}{\sqrt{6}} = \frac{\pm\sqrt{6}}{6}$. Inflection points are at $\left(\frac{\pm\sqrt{6}}{6}, -0.1389\right)$. Concave up over the intervals $\left(-\infty, \frac{-\sqrt{6}}{6}\right)$ and $\left(\frac{\sqrt{6}}{6}, \infty\right)$. Concave down over $\left(\frac{-\sqrt{6}}{6}, \frac{\sqrt{6}}{6}\right)$.

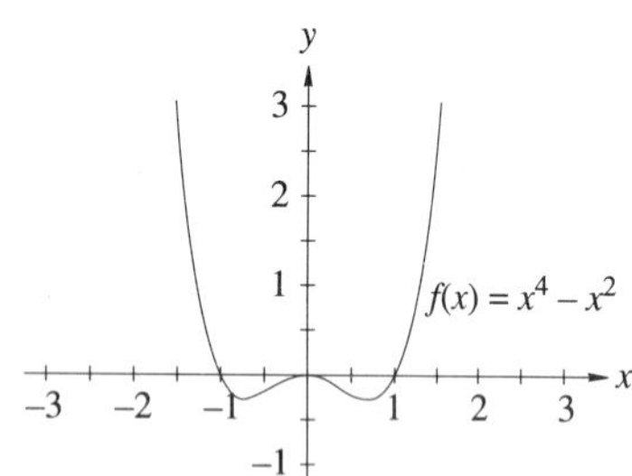

2. $g(x) = x^4 - 32x$; $g'(x) = 4x^3 - 32 = 4(x^3 - 8)$. Critical value is 2. $g''(x) = 12x^2$ is always ≥ 0 so there is a minimum at $(2, -48)$. There are no maxima. $g''(x) = 0$ when $x = 0$. Since $g''(x)$ is never negative there is no inflection point and $g(x)$ is concave up over the interval $(-\infty, \infty)$.

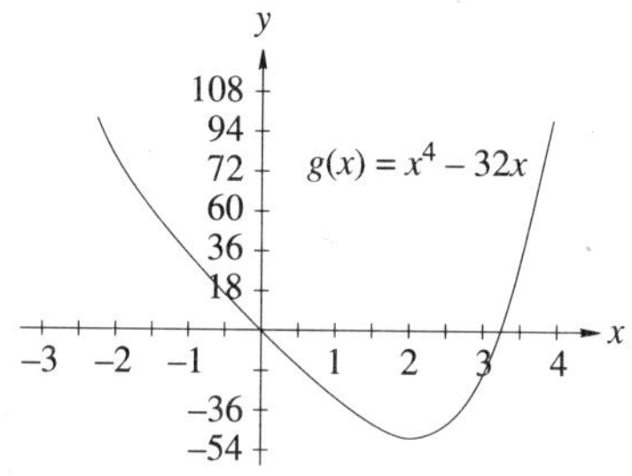

3. $h(x) = 18x^2 - x^4$, so $h'(x) = 36x - 4x^3 = 4x(9 - x^2)$. Critical values are at 0 and ± 3. $h''(x) = 36 - 12x^2 = 12(3 - x^2)$. Now, $h''(0) > 0$, so there is a minimum at $(0, 0)$; and $h''(\pm 3) < 0$, so there is a maximum at $(\pm 3, 81)$. Inflection points at $\left(\pm\frac{\sqrt{3}}{3}, 5.889\right)$. Concave up $\left(\frac{-\sqrt{3}}{3}, \frac{\sqrt{3}}{3}\right)$. Concave down $\left(-\infty, \frac{-\sqrt{3}}{3}\right)$, and $\left(\frac{\sqrt{3}}{3}, \infty\right)$.

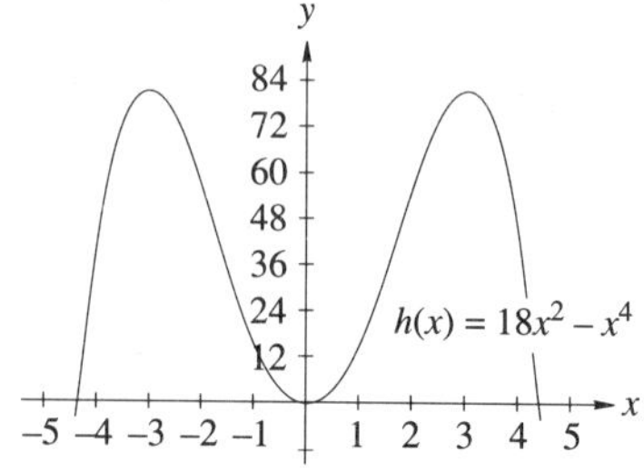

4. $j(x) = 3x^4 - 4x^3 + 1$; $j'(x) = 12x^3 - 12x^2 = 12x^2(x - 1)$. Critical values at $0, 1$. $j''(x) = 36x^2 - 24x$; $j''(0) = 0$ so the Second Deriva-

tive Test fails. $j''(1) > 0$ so minimum at $(1,0)$. Use the First Derivative Test to determine if the critical value 0 yields a maximum or minimum. On the interval $(-\infty,0)$; $j' < 0$. Likewise, on $(0,1)$, $j' < 0$ so also decreasing. No maxima. $j''(x) = 36x^2 - 24x = 12x(3x-2)$. Inflection points at $(0,1)$ and $\left(\frac{2}{3}, 0.4074\right)$. Concave up $(-\infty,0)$ and $(\frac{2}{3},\infty)$. Concave down $(0,\frac{2}{3})$.

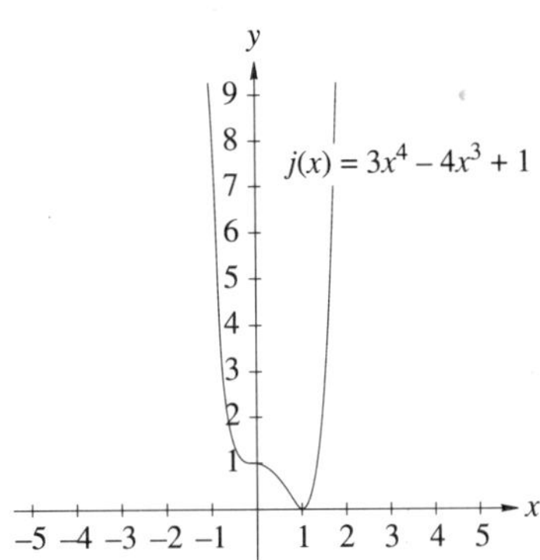

5. $k(x) = \sqrt[5]{x} = x^{1/5}$; $k'(x) = \frac{1}{5}x^{-4/5} = \frac{1}{5} \cdot \frac{1}{\sqrt[5]{x^4}}$. Critical value at 0, but since $k'(x)$ is always positive or undefined there are no extrema. $k''(x) = -\frac{4}{25}x^{-8/5}$. This is undefined at 0. On the int rval $(-\infty,0)$, $k'' > 0$ so Concave up. On the interval $(0,\infty)$, $k'' < 0$ so concave down. Inflection point: $(0,0)$.

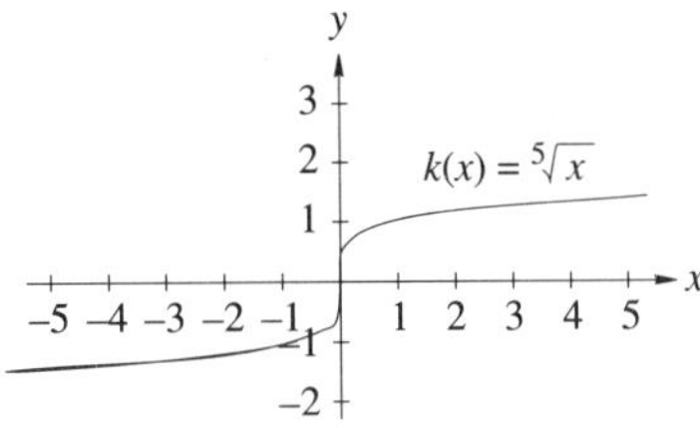

6. Here $f(x) = x\sqrt[3]{4-x}$, so $f'(x) = \sqrt[3]{4-x} + x \cdot \frac{1}{3}(4-x)^{-2/3}(-1) = \sqrt[3]{4-x} - \frac{x}{3\left(\sqrt[3]{4-x}\right)^2} = \frac{3(4-x)-x}{3\left(\sqrt[3]{4-x}\right)^2} = \frac{12-4x}{3\left(\sqrt[3]{4-x}\right)^2}$. Critical values are at 3 and 4. $f''(x) = \frac{3\sqrt[3]{4-x}^2(-4) - (12-4x)2(4-x)^{-1/3}(-1)}{9\left(\sqrt[3]{4-x}\right)^4} = \frac{3(4-x)(-4)+(12-4x)2}{9\left(\sqrt[3]{4-x}\right)^5} = \frac{-48+12x+24-8x}{9\left(\sqrt[3]{4-x}\right)^5} = \frac{4x-24}{9\left(\sqrt[3]{4-x}\right)^5}$. The second derivative is undefined at 4 and zero at 6. Concave down over $(-\infty,4)$ and $(6,\infty)$; concave up over (4 6). Inflection points at $(4,0)$ and $(6,-7.560)$. Maximum at $(3,3)$, no minimum.

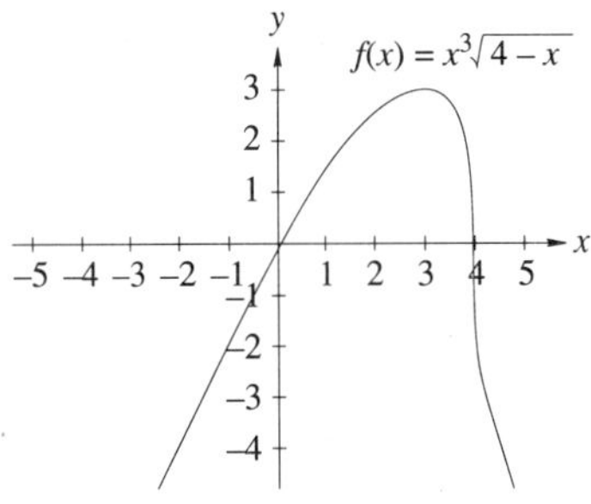

7. $g(x) = \frac{x^2+1}{x^2-4}$; $g'(x) = \frac{(x^2-4)2x - (x^2+1)(2x)}{(x^2-4)^2} = \frac{-10x}{(x^2-4)^2}$. Critical values are at at 0 and ± 2. Vertical asymptotes at $x = 2$ and $x = -2$, Horizontal asymptote at $y = 1$. $g''(x) = \frac{(x^2-4)^2(-10) + 10x(2)(x^2-4)(2x)}{(x^2-4)^4} = \frac{(x^2-4)(-10)+40x^2}{(x^2-4)^3} = \frac{30x^2+40}{(x^2-4)^3}$. $g''(x)$ is undefined at $x = \pm 2$. Concave up over the intervals $(-\infty,-2)$ and $(2,\infty)$. Conc^ve down over $(-2,2)$. Maximum at the point $\left(0,-\frac{1}{4}\right)$, no minimum. No inflection points.

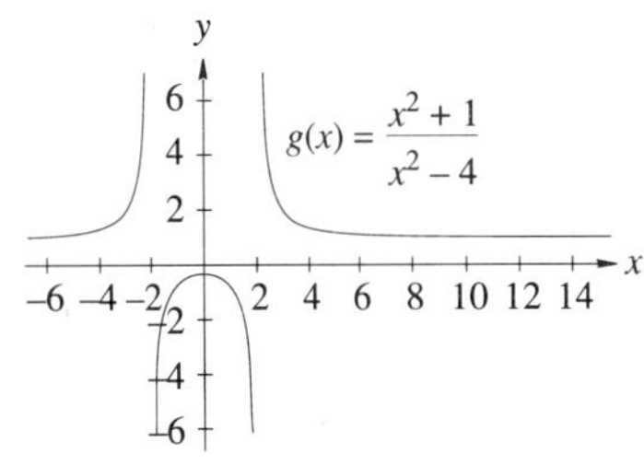

8. $h(x) = \frac{x-1}{x+2}$; $h'(x) = \frac{(x+2)-(x-1)}{(x+2)^2} = \frac{3}{(x+2)^2}$. Critical value $x = -2$; Vertical asymptote: $x = -2$, horizontal asymptote: $y = 1$. No extrema. $h''(x) = -6(x+2)^{-3}$ undefined at $x = -2$.

Concave up: $(-\infty, -2)$; Concave down $(-2, \infty)$; No inflection points.

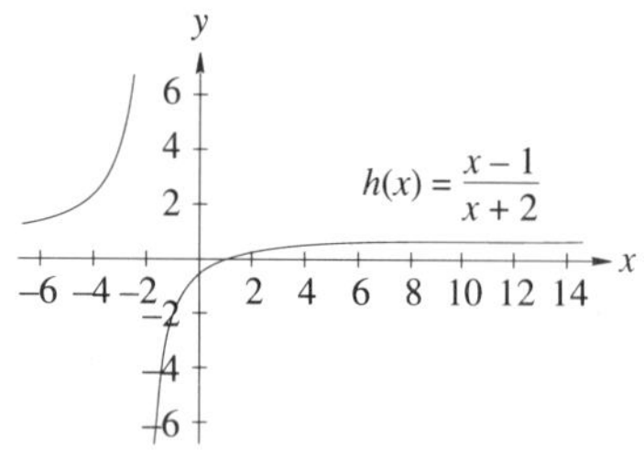

9. $j(x) = \dfrac{x}{x^2+x-2}$; $j'(x) = \dfrac{(x^2+x-2)-x(2x+1)}{(x^2+x-2)^2} = \dfrac{x^2+x-2-2x^2-x}{(x^2+x-2)^2} = \dfrac{-x^2-2}{(x^2+x-2)^2}$. Since $j(x)$ is undefined at $x=-2$ and 1, there are no critical values. No extrema. Vertical asymptotes: $x=-2$ and $x=1$; Horizontal asymptote: $y=0$.

$$j''(x) = \frac{(x^2+x-2)^2(-2x) + (x^2+2)(2)(x^2+x-2)(2x+1)}{(x^2+x-2)^4}$$
$$= \frac{(x^2+x-2)(-2x)+2(x^2+2)(2x+1)}{(x^2+x-2)^3}$$
$$= \frac{(-2x^3-2x^2+4x)+(4x^3+2x^2+8x+4)}{(x^2+x-2)^3}$$
$$= \frac{2x^3+12x+4}{(x^2+x-2)^3}$$

Using Newton's method to solve $2x^3+12x+4=0$ we get an inflection point when $x=-0.3275$. Concave up: $(-2, -0.3275)$ and $(1, \infty)$, Concave down: $(-\infty, -2)$ and $(-0.3275, 1)$.

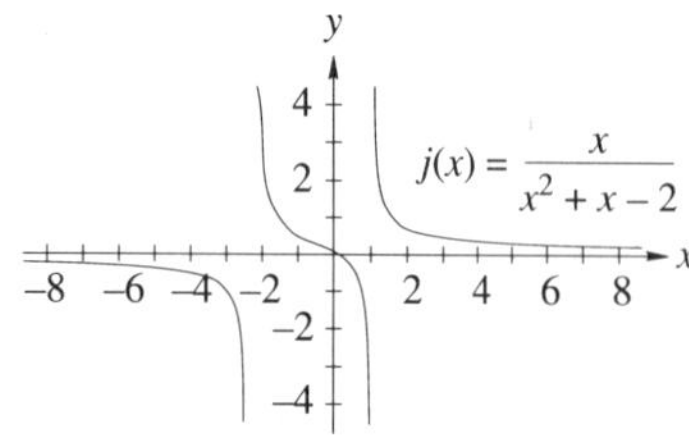

10. $k(x) = \dfrac{x^3}{x^2-9}$; $k'(x) = \dfrac{(x^2-9)3x^2-x^3(2x)}{(x^2-9)^2} = \dfrac{3x^4-27x^2-2x^4}{(x^2-9)^2} = \dfrac{x^4-27x^2}{(x^2-9)^2}$. Critical values at 0 and $\pm 3\sqrt{3}$. Vertical asymptotes at $x=\pm 3$. No horizontal asymptote but there is a slant asymptote $y=x$.

$$k''(x) = \frac{(x^2-9)^2(4x^3-54x)-(x^4-27x^2)(2)(x^2-9)(2x)}{(x^2-9)^4}$$
$$= \frac{(x^2-9)(4x^3-54x)-4x(x^4-27x^2)}{(x^2-9)^3}$$
$$= \frac{4x^5-90x^3+486x-4x^5+108x^3}{(x^2-9)^3}$$
$$= \frac{18x^3+486x}{(x^2-9)^3}$$

This is undefined at $x=\pm 3$ and zero at $x=0$. Inflection point: $(0,0)$; Maximum: $(-3\sqrt{3}, -7.794)$; Minimum: $(3\sqrt{3}, 7.794)$; Concave up: $(-3,0)$; $(3,\infty)$; Concave down: $(-\infty,-3)$; $(0,3)$

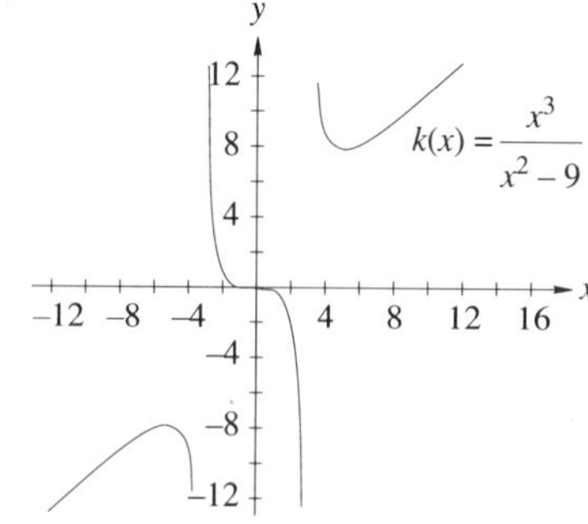

11. $s(t) = t^3-9t^2+t, v(t) = s'(t) = 3t^2-18t+1; a(t) = v'(t) = 6t-18, a'(t) = 6; v(t) = 0 = 3t^2-18t+1 = 0$ when $t = 0.056$ and 5.944. Distance: maximum at $(0.056, 0.028)$; minimum $(-4, -212)$ and $(4, -76)$; $a(t) = 0$ at $t = 3$. Velocity: minimum $(3, -26)$; maxima at $(-4, 121)$ and $(4, -23)$. Acceleration is always increasing. Minimum: $(-4, -42)$, Maximum: $(4, 6)$.

12. $s(t) = t + 4t^{-1}$; $v(t) = 1-4t^{-2}$; $a(t) = 8t^{-3}$; $v(t) = 0 = 1 - \dfrac{4}{t^2} \Rightarrow \dfrac{4}{t^2} = 1$; $t^2 = 4$ or $t = 2$; distance: Maximum: $(4, 5)$; Minimum: $(1, 5)$, $(2, 4)$; $a(t) = 0 = 8t^{-3}$ is never 0. $v(1) = -3$; $v(4) = 0.75$. Velocity: Maximum: $(4, 0.75)$; Minimum: $(1, -3)$. Acceleration is always decreasing on $[1, 4]$, so maximum $(1, 8)$; minimum $(4, 0.125)$.

13. $f(x) = 3x^2 - 4x$; $F(x) = 3 \cdot \frac{1}{3}x^3 - 4 \cdot \frac{1}{2}x^2 + C$; $F(x) = x^3 - 2x^2 + C$

14. $g(x) = 2x^{-3} + 5x^6$; $G(x) = 2 \cdot \frac{1}{-2}x^{-2} + 5 \cdot \frac{1}{7}x^7 + C$; $G(x) = -x^{-2} + \frac{5}{7}x^7 + C$

15. $h(x) = x^{1/2} + x^2 - 3x^{-2}$; $H(x) = \frac{1}{3/2}x^{3/2} + \frac{1}{3}x^3 - 3 \cdot \frac{1}{-1}x^{-1} + C$; $H(x) = \frac{2}{3}x^{3/2} + \frac{1}{3}x^3 + 3x^{-1} + C$

16. $s(t) = 4.9t^2 - 3.6t + 14$; $S(t) = 4.9 \cdot \frac{1}{3}t^3 - 3.6 \cdot \frac{1}{2}t^2 + 14t + C$; $S(t) = 1.63t^3 - 1.8t^2 + 14t + C$

17. (a) $s(t) = 288t - 16t^2$; $v(t) = s'(t) = 288 - 32t$ ft/s; $a(t) = v'(t) = -32$ ft/s^2 (b) $s(4) = 896$ ft; $v(4) = 160$ ft/s; $a(4) = -32$ ft/s^2 (c) Maximum height is when $v(t) = 0$; $288 - 32t = 0 \Rightarrow 32t = 288 \Rightarrow t = 9$; $s(9) = 1296$ ft. (d) at $t = 9$ s. (e) $s(t) = 0 = 288t - 16t^2 = t(288 - 16t)$; $t = 288/16 = 18$ s. (f) $v(18) = -288$ ft/s

18. (a) $s(t) = 15t^2 + 5$; $v(t) = s'(t) = 30t$ cm/s (b) $v(2) = 60$ cm/s (c) $v(t) = 30t = 105 \Rightarrow t = 3.5$ s.

19. The total cost is $C = (1 + 0.00058v^{3/2}) \times 1000 + 25\left(\frac{1000}{v}\right) = (1000 + 0.58v^{3/2}) + \left(\frac{25{,}000}{v}\right)$. To find the minimum cost, we find the derivative and set it equal to 0. Thus, $C' = \frac{3}{2}(0.58)v^{1/2} - \frac{25{,}000}{v^2} = 0$. Hence, $0.87v^{5/2} - 25{,}000 = 0$ or $v^{5/2} = \frac{25{,}000}{0.87} \approx 28{,}735$ and $v = (28{,}735)^{2/5} \approx 60.7$. Checking the second derivative, we have $C'' = 0.435v^{-1/2} + 50{,}000v^{-3}$, and we see that $C''(60.7) > 0$, so, by the second derivative test, $v \approx 60.7$ is a minimum. Thus, the trucker should drive at 60.7 mph in order to minimize the total cost of the trip.

20. $V = \pi r^2 h = 246\pi \Rightarrow h = \frac{246}{r^2}$. Area of top and bottom $= 2\pi r^2$. Area of side $= 2\pi rh = 2\pi r \cdot \frac{246}{r^2} = 492\frac{\pi}{r}$. Cost is $C(r) = 2\pi r^2 + 3 \cdot 492\frac{\pi}{r} = 2\pi r^2 + 1476\pi r^{-1}$ and has the derivative $C'(r) = 4\pi r - 1476\pi r^{-2} = 0 \Rightarrow r - \frac{369}{r^2} = 0$, or $r = \frac{369}{r^2}$, or $r^3 = 369$. Thus, $r = \sqrt[3]{369} \approx 7.17258$ ft; $h = \frac{246}{7.17258^2} = 4.78$ ft.

21. $A = \ell \cdot w$; $2\ell + 2w = 1200$ m; $\ell + w = 600$; $w = 600 - \ell$; $A = \ell(600 - \ell) = 600\ell - \ell^2$; $A' = 600 - 2\ell \Rightarrow \ell = 300$ m; $w = 600 - 300 = 300$ m; $A = (300 \times 300) = 90000$ m^2

22. Since the radius of the tank is 8 ft and the height is 11 ft, the tank is more than half full. The height left is $16 - 11 = 5$ ft and the volume remaining is given by the formula $V = \frac{1}{3}\pi h^2(3R - h)$ where h is the height left and $R = 8$ ft. $V = \frac{1}{3}\pi h^2(24) - \frac{1}{3}\pi h^3$; $\frac{dV}{dt} = (16\pi h - \pi h^2)dh/dt$ and $\frac{dV}{dt} = 20$ gal/min $= 20(0.13358)$ ft^3/min; $\frac{dh}{dt} = \frac{20(0.13358)}{[16\pi \cdot 5 - \pi \cdot 5^2]} = \frac{2.6716}{55\pi} = 0.01546$ ft/min.

23. $V = \frac{4}{3}\pi r^3$; $dV = 4\pi r^2 dr$; $dV = 4\pi(1.5)^2(0.01) = 0.2827$ m^3; $A = 4\pi r^2$; $dA = 8\pi r dr = 8\pi(1.5)(0.01) \approx 0.3770$ m^2

24. The largest capacity is when the cross-sectional area is the largest. $A = \ell \cdot w$; $\ell + 2w = 380$ mm; $\ell = 380 - 2w$; $A = (380 - 2w)w = 380w - 2w^2$; $A' = 380 - 4w$; maximum when $A' = 0$; $4w = 380$; $w = 95$; $\ell = 380 - 2 \cdot 95 = 190$. The gutter should measure 95 mm $\times$ 190 mm.

25. $v = 4.5T + 0.0003T^3$; $\Delta v \approx dv = (4.5 + 0.0009T^2)\,dT = (4.5 + 0.0009 \cdot 100^2)(1) = 13.5$ V

26. The distance between the ships is related by the equation $d^2 = x^2 + y^2$ so $2dd' = 2xx' + 2yy'$ or $d' = \frac{xx' + yy'}{d}$. After 3 hours, we have $x = 12 \cdot 3 = 36$, $y = 5 \cdot 3 = 15$, $d = \sqrt{36^2 + 15^2} = 39$, and $d' = \frac{36 \cdot 12 + 15 \cdot 5}{39} = 13$ km/h.

27. $V = \pi r^2 h = 20$ kL $= 20\,000$ L $= 20\,000$ dm^3 $= 20$ m^3; $h = \frac{20}{\pi r^2}$; $A = 2\pi r^2 + 2\pi rh = 2\pi r^2 + 2\pi r\left(\frac{20}{\pi r^2}\right)$. $\frac{dA}{dr} = 4\pi r - 40r^{-2}$; minimized when $\frac{dA}{dr} = 0$ or $40r^{-2} = 4\pi r$; $\frac{10}{\pi} = r^3$; $r = \sqrt[3]{\frac{10}{\pi}} \approx 1.4710$ m; $h = \frac{20}{\pi r^2} = 2.942$ m.

28. Distance in the desert is $\sqrt{8^2+x^2} = \sqrt{64+x^2}$. Distance on the road is $10 - x$. Total time $= \frac{\sqrt{64+x^2}}{15} + \frac{10-x}{55}$; $\frac{dt}{dx} = \frac{\frac{1}{2}(64+x^2)^{-1/2}(2x)}{15} - \frac{1}{55}$. Set $\frac{dt}{dx} = 0$ and solve $\frac{x}{15\sqrt{64+x^2}} = \frac{1}{55}$, or $55x = 15\sqrt{64+x^2}$, and so $\frac{11}{3}x = \sqrt{64+x^2}$, or $\frac{121}{9}x^2 = 64 + x^2$; $\frac{112}{9}x^2 = 64$; $112x^2 = 576$; $x^2 = 5.14287$ and we get $x = 2.2678$; 2.2678 mi from A

29. (a) $a = -32$ ft/s^2; $v = -32t$; $s(t) = -16t^2 + C$; $s(7) = 0 = -16{\cdot}7^2 + C$; $C = 16{\cdot}7^2 = 16{\cdot}49 = 784$ ft. The building is 784 ft tall.
(b) $v(7) = -32 \cdot 7 = -224$ ft/s; Speed $= |v| = |-224| = 224$ ft/s.

CHAPTER 24 TEST

1. $f(x) = x^4 - 8x^2; f'(x) = 4x^3 - 16x = 4x(x^2-4)$
(a) Critical vlaues are $0, \pm 2$. (b) $f''(x) = 12x^2 - 16$; $x = \pm\sqrt{\frac{16}{12}} = \pm\sqrt{4/3} = \frac{\pm 2\sqrt{3}}{3}$ are inflection points. Maximum: $(0,0)$; Minima: $(-2,-16), (2,16)$
(c) Concave up: $\left(-\infty, \frac{-2\sqrt{3}}{3}\right), \left(\frac{2\sqrt{3}}{3}, \infty\right)$
(d) Concave down $\left(\frac{-2\sqrt{3}}{3}, \frac{2\sqrt{3}}{3}\right)$

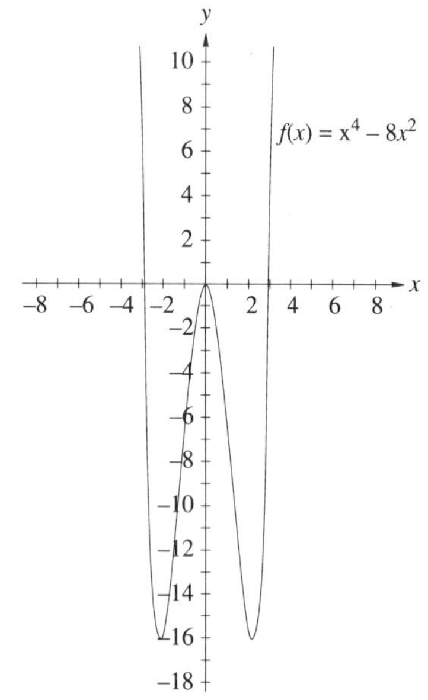

2. $g(x) = \frac{x^2}{9-x^2}$; $g'(x) = \frac{(9-x^2)(2x) - x^2(-2x)}{(9-x^2)^2} = \frac{18x - 2x^3 + 2x^3}{(9-x^2)^2} = \frac{18x}{(9-x^2)^2}$.
(a) Critical value: $x = 0$. Vertical asymptote at $x = 3, x = -3$ (b) $g''(x) = \frac{(9-x^2)^2(18) - 18x(2)(9-x^2)(-2x)}{(9-x^2)^4} = \frac{18(9-x^2) + 18 \cdot 4x^2}{(9-x^2)^3} = \frac{54x^2+162}{(9-x^2)^3}$. The second derivative is undefined at ± 3. Maximum: none; Minimum: $(0,0)$, (c) Concave up: $(-3,3)$, (d) Concave down: $(-\infty,-3), (3,\infty)$, (e) Vertical asymptotes: $x = \pm 3$. Horizontal asymptote: $y = -1$

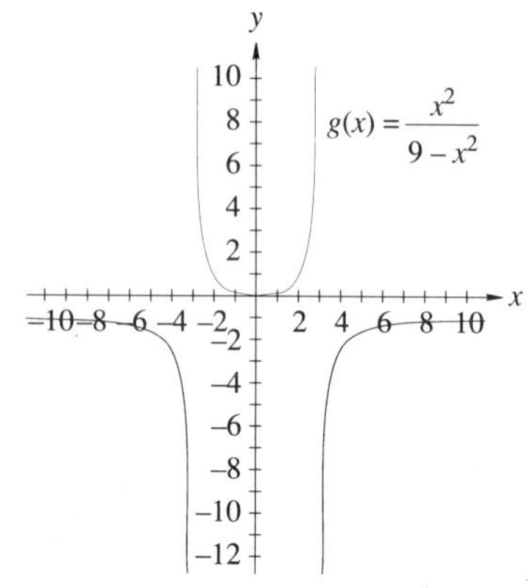

3. $x^2 - 7 = 0$; $P(x) = x^2 - 7$; $P'(x) = 2x$

i	x_i	$P(x_i)$	$P'(x_i)$	$x_{i+1} = x_i - \dfrac{P(x_i)}{P'(x_i)}$	$P(x_{i+1})$
1	3	2	6	2.6667	0.1113
2	2.6667	0.1113	5.3333	2.6458	2.58×10^{-4}
3	2.6458	2.58×10^{-4}	5.2916	2.64575	-6.9×10^{-6}

Hence, $x \approx 2.646$

4. $f(x) = 5x^4 - 7x$; $F(x) = 5 \cdot \frac{1}{5}x^5 - 7 \cdot \frac{1}{2}x^2 + C$; $F(x) = x^5 - \frac{7}{2}x^2 + C$

5. $s(t) = t^3 - 12t^2 + 5$ (a) $v(t) = s'(t) = 3t^2 - 24t$; $a(t) = v'(t) = 6t - 24$ (b) Extrema for $s(t)$ are when $v(t) = 0$ that is at 0 and 8; $s(-2) = -51$; $s(0) = 5$; $s(4) = -123$. Position: Maximum at $(0, 5)$; Minima at $(-2, -51)$ and $(4, -123)$: Extrema for velocity when $a(t) = 0$ or $t = 4$; $v(-2) = 60$; $v(4) = -48$ Velocity: Maximum at $(-2, 60)$, Minimum at $(4, -48)$. Acceleration is always increasing on the interval $[-2, 4]$, so Maximum: $(4, 0)$, Minimum: $(-2, -36)$

6. Let ℓ and w be the dimensions of the printed part. Then $\ell \cdot w = 24$ so $\ell = \frac{24}{w}$. The dimensions of the page are $\ell + 3$ and $w + 2$. Area of the page is $(\ell + 3)(w + 2) = \left(\frac{24}{w} + 3\right)(w + 2) = 24 + \frac{48}{w} + 3w + 6 = 3w + \frac{48}{w} + 30$; $A' = 3 - 48/w^2 = 0$; $48/w^2 = 3$; $16 = w^2$; $w = 4$. The width is 4 in. and the length is 6 in. for the printed portion. The page is 6 in. wide by 9 in. long.

CHAPTER

25

Integration

25.1 THE AREA QUESTION AND THE INTEGRAL

1. $\int_1^3 x\,dx$. The interval [1, 3] has length 2 so each subinterval is length $\frac{2}{n}$. Using the right end point of each subinterval, $x_i = 1 + \frac{2i}{n} = \frac{n+2i}{h}$. Since $f(x) = x; f(x_i) = x_i = \frac{n+2i}{n}$

$$\begin{aligned}\sum_{i=1}^{n} f(x_i)nx &= \sum_{i=1}^{n}\left(\frac{n+2i}{n}\right)\cdot\frac{2}{n}\\ &= \sum_{i=1}^{n}\left(\frac{2}{n}+\frac{4i}{n^2}\right)\\ &= \frac{2}{n}\sum_{i=1}^{n}1+\frac{4}{n^2}\sum_{i=1}^{n}i\\ &= \frac{2}{n}\cdot n+\frac{4}{n^2}\left(\frac{n^2+n}{2}\right)\\ &= 2+2+\frac{2}{n}=4+\frac{2}{n}\end{aligned}$$

$$\lim_{n\to\infty}\left(4+\frac{2}{n}\right)=4$$

2. $\int_0^2 2x\,dx; \Delta x = \frac{2}{n}; x_i = \frac{2i}{n}; f(x_i) = 2x_i = \frac{4i}{n};$

$$\begin{aligned}\sum_{i=1}^{n} f(x_i)\Delta x &= \sum_{i=1}^{n}\frac{4i}{n}\cdot\frac{2}{n}=\frac{8}{n^2}\sum_{i1}^{n}i=\frac{8}{n^2}\left(\frac{n^2+n}{2}\right)\\ &= 4+\frac{4}{n};\end{aligned}$$

$$\lim_{n\to\infty}\left(4+\frac{4}{n}\right)=4$$

3. $\int_0^4(3x+1)\,dx; \Delta x = \frac{4}{n}; x_i = \frac{4i}{n}; f(x_i) = 3\left(\frac{4i}{n}\right) + 1 = \frac{12i}{n} + 1;$

$$\begin{aligned}\sum_{i=1}^{n}\left(\frac{12i}{n}+1\right)\cdot\frac{4}{n} &= \frac{48}{n^2}\sum_{i=1}^{n}i+\frac{4}{n}\sum_{i=1}^{n}1\\ &= \frac{48}{n^2}\left(\frac{n^2+n}{2}\right)+\frac{4}{n}\cdot n\\ &= 24+\frac{18}{n}+4;\end{aligned}$$

$$\lim_{n\to\infty}\left(28+\frac{18}{n}\right)=28.$$

4. $\int_1^3(2x+2)\,dx;\ \Delta x = \frac{2}{n};\ x_i = 1+\frac{2i}{n};\ f(x_i) = 2(1+\frac{2i}{n})+2 = 4+\frac{4i}{n}$

$$\begin{aligned}\sum_{i=1}^{n} &= \left(4+\frac{4i}{n}\right)\frac{2}{n}=\sum_{i=1}^{n}\left(\frac{8}{n}+\frac{8i}{n^2}\right)\\ &= \frac{8}{n}\sum_{i=1}^{n}1+\frac{8}{n}\sum_{i=1}^{n}i\\ &= \frac{8}{n}n+\frac{8}{n^2}\left(\frac{n^2+n}{2}\right)=8+4+\frac{4}{n};\end{aligned}$$

$$\lim_{n\to\infty}\left(12+\frac{4}{n}\right)=12$$

5. $\int_1^3 (2x-2)\,dx;\ \Delta x = \frac{2}{n};\ x_i = 1+\frac{2i}{n};\ f(x_i) = 2\left(1+\frac{2i}{n}\right)-2=\frac{4i}{n};$

$$\sum_{i=1}^{n}\frac{4i}{n}\cdot\frac{2}{n}=\frac{8}{n^2}\sum_{i=1}^{n}i=\frac{8}{n^2}\left(\frac{n^2+n}{2}\right)=4+\frac{4}{n};$$

$$\lim_{n\to\infty}\left(4+\frac{4}{n}\right)=4$$

6. $\int_0^4 (1-3x)\,dx;\ \Delta x=\frac{4}{n};\ x_n=\frac{4i}{n};\ f(x_i)=1-3\cdot\frac{4i}{n}=1-\frac{12i}{n};$

$$\sum_{i=1}^{n}\left(1-\frac{12i}{n}\right)\frac{4}{n}=\sum_{i=1}^{n}\left(\frac{4}{n}-\frac{48i}{n^2}\right)$$

$$=\frac{4}{n}\sum_{i=1}^{n}1-\frac{48}{n^2}\sum_{i=1}^{n}i$$

$$=\frac{4}{n}\cdot n-\frac{48}{n^2}\cdot\left(\frac{n^2+n}{2}\right)$$

$$=4-24-\frac{24}{n}=-20-\frac{24}{n};$$

$$\lim_{n\to\infty}\left(-20-\frac{24}{n}\right)=-20$$

7. $\int_1^4 (3-2x)\,dx;\ \Delta x=\frac{3}{n};\ x_i=1+\frac{3i}{n};\ f(x_i)=3-2\left(1+\frac{3i}{n}\right)=3-2-\frac{6i}{n}=1-\frac{6i}{n};$

$$\sum_{i=1}^{n}\left(1-\frac{6i}{n}\right)\frac{3}{n}=\sum_{i=1}^{n}\left(\frac{3}{n}-\frac{18i}{n^2}\right)$$

$$=\frac{3}{n}\sum_{i=1}^{n}1-\frac{18}{n^2}\sum_{i=1}^{n}i$$

$$=\frac{3}{n}\cdot n-\frac{18}{n^2}\left(\frac{n^2+n}{2}\right)$$

$$=3-9-\frac{9}{n}=-6-\frac{9}{n};$$

$$\lim_{n\to\infty}\left(-6-\frac{9}{n}\right)=-6$$

8. $\int_0^4 (x^2)\,dx;\ \Delta x=\frac{4}{n};\ x_i=\frac{4i}{n};\ f(x_i)=\left(\frac{4i}{n}\right)^2=\frac{16i^2}{n^2};$

$$\sum_{i=1}^{n}\frac{16i^2}{n^2}\cdot\frac{4}{n}=\frac{64}{n^3}\sum_{i=1}^{n}i^2=\frac{64}{n^3}\left(\frac{n(n+1)(2n+1)}{6}\right)$$

$$=\frac{64}{n^3}\left(\frac{2n^3+3n^2+n}{6}\right)$$

$$=\frac{64}{3}+\frac{32}{n}+\frac{32}{3n^2}$$

$$\lim_{n\to\infty}\left(\frac{64}{3}+\frac{32}{n}+\frac{32}{3n^2}\right)=\frac{64}{3}$$

9. $\int_0^2 (x^2-1)\,dx;\ \Delta x=\frac{2}{n};\ x_i=\frac{2i}{n};\ f(x_i)=\left(\frac{2i}{n}\right)^2-1=\frac{4i^2}{n^2}-1;$

$$\sum_{i=1}^{n}\left(\frac{4i^2}{n^2}-1\right)\cdot\frac{2}{n}=\sum_{i=1}^{n}\frac{8i^2}{n^3}-\sum_{i=1}^{n}\frac{2}{n}$$

$$=\frac{8}{n^3}\sum_{i=1}^{n}i^2-\frac{2}{n}\sum_{i=1}^{n}1$$

$$=\frac{8}{n^3}\left(\frac{n^3}{3}+\frac{n^2}{2}+\frac{n}{6}\right)-\frac{2}{n}\cdot n$$

$$=\frac{8}{3}+\frac{4}{n}+\frac{4}{3n^2}-2$$

$$=\frac{2}{3}+\frac{4}{n}+\frac{4}{3n^2};$$

$$\lim_{n\to\infty}\left(\frac{2}{3}+\frac{4}{n}+\frac{4}{3n^2}\right)=\frac{2}{3}$$

10. $\int_0^4 (x^2+2)\,dx;\ \Delta x=\frac{4}{n};\ x_i=\frac{4i}{n};\ f(x_i)=\left(\frac{4i}{n}\right)^2+2=\frac{16i^2}{n^2}+2;$

$$\sum_{i=1}^{n}\left(\frac{16i^2}{n^2}+2\right)\frac{4}{n}=\frac{64}{n^3}\sum_{i=1}^{n}i^2+\frac{8}{n}\sum_{i=1}^{n}1$$

$$= \frac{64}{n^3}\left(\frac{n^3}{3} + \frac{n^2}{2} + \frac{n}{6}\right) + \frac{8}{n}\cdot n$$
$$= \frac{64}{3} + \frac{32}{n} + \frac{32}{3n^2} + 8$$
$$= \frac{88}{3} + \frac{32}{n} + \frac{32}{3n^2};$$

$$\lim_{n\to\infty}\left(\frac{88}{3} + \frac{32}{n} + \frac{32}{3n^2}\right) = \frac{88}{3}$$

11. $\int_0^2 (2x^2 + 1)\,dx$; $\Delta x = \frac{2}{n}$; $x_i = \frac{2i}{n}$; $f(x_i) = 2\cdot\left(\frac{2i}{n}\right)^2 + 1 = \frac{8i^2}{n} + 1$;

$$\sum_{i=1}^{n}\left(\frac{8i^2}{n^2} + 1\right)\cdot\frac{2}{n} = \sum_{i=1}^{n}\left(\frac{16i^2}{n^3} + \frac{2}{n}\right)$$
$$= \frac{16i}{n^3}\sum_{i=1}^{n} i^2 + \frac{2}{n}\sum_{i=1}^{n} 1$$
$$= \frac{16}{n^3}\left(\frac{n^3}{3} + \frac{n^3}{2} + \frac{n}{6}\right) + \frac{2}{n}(n)$$
$$= \frac{16}{3} + \frac{8}{n} + \frac{8}{3n^2} + 2$$
$$= \frac{22}{3} + \frac{8}{n} + \frac{8}{3n^2};$$

$$\lim_{n\to\infty}\left(\frac{22}{3} + \frac{8}{n} + \frac{8}{3n^2}\right) = \frac{22}{3}.$$

12. $\int_0^4 (2 - x^2)\,dx$; $\Delta x = \frac{4}{n}$; $x_i = \frac{4i}{n}$; $f(x_i) = 2 - \left(\frac{4i}{n}\right)^2 = 2 - \frac{16i^2}{n^2}$;

$$\sum_{i=1}^{n}\left(2 - \frac{16i^2}{n^2}\right)\frac{4}{n}$$
$$= \sum_{i=1}^{n}\left(\frac{8}{n} - \frac{64i^2}{n^3}\right)$$
$$= \frac{8}{n}\cdot n - \frac{64}{n^3}\left(\frac{n^3}{3} + \frac{n^2}{2} + \frac{n}{6}\right)$$
$$= 8 - \frac{64}{3} - \frac{32}{n} - \frac{32}{3n^2};$$

$$\lim_{n\to\infty}\left(8 - \frac{64}{3} - \frac{32}{n} - \frac{32}{3n^2}\right) = 8 - \frac{64}{3} = -\frac{40}{3}$$

13. $\int_1^3 x^3\,dx$; $\Delta x = \frac{2}{n}$; $x_i = 1 + \frac{2i}{n}$; $f(x_i) = \left(1 + \frac{2i}{n}\right)^3 = 1 + \frac{6i}{n} + \frac{12i^2}{n^2} + \frac{8i^2}{n^3}$;

$$\sum_{i=1}^{n}\left(1 + \frac{6i}{n} + \frac{12i^2}{n^2} + \frac{8i^3}{n^3}\right)\frac{2}{n}$$
$$= \sum_{i=1}^{n}\left(\frac{2}{n} + \frac{12i}{n^2} + \frac{24i^2}{n^3} + \frac{16i^3}{n^4}\right)$$
$$= \frac{2}{n}\cdot n + \frac{12}{n^2}\left(\frac{n^2+n}{2}\right) + \frac{24}{n^3}\left(\frac{n^3}{3} + \frac{n^2}{2} + \frac{n}{6}\right) + \frac{16}{n^4}\left(\frac{n^4}{4} + \frac{n^3}{2} + \frac{n^2}{4}\right)$$
$$= 2 + 6 + 8 + 4 + \frac{6}{n} + \frac{12}{n} + \frac{4}{n^2} + \frac{8}{n} + \frac{4}{n^2};$$

$$\lim_{n\to\infty}\left(20 + \frac{26}{n} + \frac{8}{n^2}\right) = 20$$

14. $\int_1^3 (1 - x^3)\,dx$; $\Delta x = \frac{2}{n}$; $x_i = 1 + \frac{2i}{n}$; $f(x_i) = 1 - \left[1 + \frac{2i}{n}\right]^3 = 1 - \left[1 + \frac{6i}{n} + \frac{12i^2}{n^2} + \frac{8i^3}{n^3}\right]$;

$$\sum_{i=1}^{n} -\left[\frac{6i}{n} + \frac{12i^2}{n^2} + \frac{8i^3}{n^3}\right]\frac{2}{n}$$
$$= -\left[\frac{12}{n^2}\cdot\left(\frac{n^2+n}{2}\right) + \frac{24}{n^3}\left(\frac{n^3}{3} + \frac{n^2}{2} + \frac{n}{6}\right) + \frac{16}{n^4}\left(\frac{n^4}{4} + \frac{n^3}{2} + \frac{n^2}{4}\right)\right]$$

$$\lim_{n\to\infty} -\left[\frac{12}{n^2}\cdot\left(\frac{n^2+n}{2}\right) + \frac{24}{n^3}\left(\frac{n^3}{3} + \frac{n^2}{2} + \frac{n}{6}\right) + \frac{16}{n^4}\left(\frac{n^4}{4} + \frac{n^3}{2} + \frac{n^2}{4}\right)\right] = -[6 + 8 + 4] = -18$$

15. $\int_0^4 (x^3+2)\,dx$; $\Delta x = \frac{4}{n}$; $x_i = \frac{4i}{n}$; $f(x_i) = \left(\frac{4i}{n}\right)^3 + 2 = \frac{64i^3}{n^3} + 2$;

$$\sum_{i=1}^{n}\left(\frac{64i^3}{n^3}+2\right)\frac{4}{n} = \sum_{i=1}^{n}\left(\frac{256i^3}{n^4}+\frac{8}{n}\right)$$
$$= \frac{256}{n^4}\left[\frac{n^4}{4}+\frac{n^3}{2}+\frac{n^2}{2}\right]+\frac{8}{n}\cdot n$$

$$\lim_{n\to\infty}\left[\frac{256}{n^4}\left[\frac{n^4}{4}+\frac{n^3}{2}+\frac{n^2}{2}\right]+\frac{8}{n}\cdot n\right] = 64+8 = 72$$

16. $\int_0^3 (2x^3-1)\,dx$; $\Delta x = \frac{3}{n}$; $x_i = \frac{3i}{n}$; $f(x_i) = 2\cdot\left(\frac{3i}{b}\right)^3 - 1 = \frac{54i^3}{n^3} - 1$;

$$\sum_{i=1}^{n}\left(\frac{54i^3}{n}-1\right)\frac{3}{n} = \sum_{i=1}^{n}\left(\frac{162i^3}{n^2}-\frac{3}{n}\right)$$
$$= \frac{162}{n^4}\left(\frac{n^4}{4}+\frac{n^3}{2}+\frac{n^2}{2}\right)-\frac{3}{n}\cdot n$$

$$\lim_{n\to\infty}\left[\frac{162}{n^4}\left(\frac{n^4}{4}+\frac{n^3}{2}+\frac{n^2}{2}\right)-\frac{3}{n}\cdot n\right] = 40.5-3 = 37.5 \text{ or } \frac{75}{2}$$

17. $\int_0^2 (x^2+x)\,dx$; $\Delta x = \frac{2}{n}$; $x_i = \frac{2i}{n}$; $f(x_i) = \left(\frac{2i}{n}\right)^2 + \frac{2i}{n} = \frac{4i^2}{n^2} + \frac{2i}{n}$;

$$\sum_{i=1}^{n}\left(\frac{4i^2}{n^2}+\frac{2i}{n}\right)\frac{2}{n} = \sum_{i=1}^{n}\left(\frac{8i^2}{n^3}+\frac{4i}{n^2}\right)$$
$$= \frac{8}{n^3}\left(\frac{n^3}{3}+\frac{n^2}{2}+\frac{n}{6}\right)+\frac{4}{n^2}\left(\frac{n^2}{2}+\frac{n}{2}\right)$$

$$\lim_{n\to\infty}\left[\frac{8}{n^3}\left(\frac{n^3}{3}+\frac{n^2}{2}+\frac{n}{6}\right)+\frac{4}{n^2}\left(\frac{n^2}{2}+\frac{n}{2}\right)\right] = \frac{8}{3}+2 = \frac{14}{3}$$

18. $\int_0^1 x^4\,dx$; $\Delta x = \frac{1}{n}$; $x_i = \frac{i}{n}$; $f(x_i) = \frac{i^4}{n^4}$;

$$\sum_{i=1}^{n}\frac{i^4}{n^4}\cdot\frac{1}{n}$$
$$= \frac{1}{n^5}\sum_{i=1}^{n} i^4$$
$$= \frac{1}{n^5}\left[\frac{n(n+1)(2n+1)(3n^2+3n-1)}{30}\right]$$
$$= \frac{1}{n^5}\left[\left(\frac{6n^5+15n^4+10n^3-n}{30}\right)\right]$$
$$= \frac{1}{n^5}\left[\left(\frac{n^5}{5}+\frac{15n^4}{2}+\frac{10n^3}{3}-\frac{n}{30}\right)\right]$$

$$\lim_{n\to\infty}\frac{1}{n^5}\left[\left(\frac{n^5}{5}+\frac{15n^4}{2}+\frac{10n^3}{3}-\frac{n}{30}\right)\right] = \frac{1}{5}$$

25.2 THE FUNDAMENTAL THEOREM OF CALCULUS

1. $\int_0^5 6\,dx = 6x\Big|_0^5 = 6\cdot 5 - 6\cdot 0 = 30$

2. $\int_1^2 (x-6)\,dx = \frac{1}{2}x^2 - 6x\Big|_1^2 = \left(\frac{1}{2}\cdot 2^2 - 6\cdot 2\right) - \left(\frac{1}{2}\cdot 1^2 - 6\cdot 1\right) = 2 - 12 - \frac{1}{2} + 6 = -\frac{9}{2}$

3. $\int_1^5 2x\,dx = x^2\Big|_1^5 = 5^2 - 1^2 = 25 - 1 = 24$

4. $\int_{-1}^2 (6-2x)\,dx = 6x - x^2\Big|_{-1}^2 = (6\cdot 2 - 2^2) - (6(-1) - (-1)^2) = 12 - 4 + 6 + 1 = 15$

5. $\int_{-1}^3 \frac{5}{3}x^3\,dx = \frac{5}{12}x^4\Big|_{-1}^3 = \frac{5}{12}\cdot 3^4 - \frac{5}{12}\cdot(-1)^4 =$

$\frac{5 \cdot 81}{12} - \frac{5}{12} = \frac{400}{12} = \frac{100}{3}$

6. $\int_0^4 4x^2\,dx = \frac{4}{3}x^3\Big|_0^5 = \frac{4}{3}\cdot 5^3 - \frac{4}{3}\cdot 0^3 = \frac{4}{3}\cdot 125 = \frac{500}{3}$

7. $\int_1^2 \frac{8}{5}x^3\,dx = \frac{2}{5}x^4\Big|_1^2 = \frac{2}{5}\cdot 2^4 - \frac{2}{5}\cdot 1^4 = \frac{32}{5} - \frac{2}{5} =$
$\frac{30}{5} = 6$

8. $\int_{-2}^2 6x\,dx = 3x^2\Big|_{-2}^2 = 3\cdot 2^2 - 3\cdot(-2)^2 = 12 - 12 = 0$

9. $\int_0^2 (4x+5)\,dx = 2x^2 + 5x\Big|_0^2 =$
$(2\cdot 2^2 + 5\cdot 2) - (2\cdot 0^2 + 5\cdot 0) = 8 + 10 = 18$

10. $\int_1^3 (-4)\,dx = (-4x)\Big|_1^3 = -4\cdot 3 - (-4\cdot 1) =$
$-12 + 4 = -8$

11. $\int_2^0 (4x+5)\,dx = 2x^2 + 5x\Big|_2^0 =$
$(2\cdot 0^2 + 5\cdot 0) - (2\cdot 2^2 + 5\cdot 2) = 0 - 18 = -18$

12. $\int_{-1}^2 (2t - t^2)\,dx = t^2 - \frac{1}{3}t^3\Big|_{-1}^2 = \left(2^2 - \frac{1}{3}\cdot 2^3\right) -$
$\left((-1)^2 - \frac{1}{3}(-1)^3\right) = \left(4 - \frac{8}{3}\right) - \left(1 + \frac{1}{3}\right) = 0$

13. $\int_1^3 (-4x)\,dx = -2x^2\Big|_1^3 = -2\cdot 3^2 - (-2\cdot 1^2) =$
$-2\cdot 9 - (-2\cdot 1) = -18 + 2 = -16$

14. $\int_2^5 (5w^2 + 2w)\,dw = \frac{5}{3}w^3 + w^2\Big|_2^5 =$
$\left(\frac{5}{3}\cdot 5^3 + 5^2\right) - \left(\frac{5}{3}\cdot 2^3 + 2^2\right) = \left(\frac{625}{3} + 25\right) -$
$\left(\frac{40}{3} + 4\right) = \frac{585}{3} + 21 = 195 + 21 = 216$

15. $\int_0^3 (t^2 + 2t)\,dt = \frac{1}{3}t^3 + t^2\Big|_0^3 = \frac{1}{3}\cdot 3^3 + 3^2 - 0 = 18$

16. $\int_{-2}^0 (5 - t^3)\,dt = 5t - \frac{t^4}{4}\Big|_{-2}^0 =$
$0 - \left(5(-2) - \frac{(-2)^4}{4}\right) = -(-10 - 4) = 14$

17. $\int_{-1}^3 (3y^2 - 2y)\,dy = y^3 - y^2\Big|_{-1}^3 =$
$(3^3 - 3^2) - ((-1)^3 - (-1)^2) = 18 + 2 = 20$

18. $\int_0^4 (s^2 - 4s + 16)\,ds = \frac{s^3}{3} - 2s^2 + 16s\Big|_0^4 = \frac{4^3}{3} -$
$2\cdot 4^2 + 16\cdot 4 - 0 = \frac{64}{3} - 32 + 64 = \frac{64}{3} + 32 = \frac{160}{3}$

19. $\int_1^4 (4 - 2w^2)\,dw = 4w - \frac{2}{3}w^3\Big|_1^4 =$
$\left(4\cdot 4 - \frac{2}{3}\cdot 4^3\right) - \left(4\cdot 1 - \frac{2}{3}\cdot 1^3\right) =$
$\left(16 - \frac{128}{3}\right) - \left(4 - \frac{2}{3}\right) = 12 - \frac{126}{3} =$
$12 - 42 = -30$

20. $\int_{-2}^0 -\,dt = -t\Big|_{-2}^0 = 0 - (-(-2)) = -2$

21. $\int_2^3 \left(x^2 + 2x + 1\right)\,dx = \frac{x^3}{3} + x^2 + x\Big|_2^3 =$
$\left(\frac{3^3}{3} + 3^2 + 3\right) - \left(\frac{2^3}{3} + 2^2 + 2\right) = 21 - \frac{26}{3} = \frac{37}{3}$

22. $\int_{-3}^0 (3y - y^3)\,dy = \frac{3}{2}y^2 - \frac{y^4}{4}\Big|_{-3}^0 =$
$0 - \left(\frac{3}{2}(-3)^2 - \frac{(-3)^4}{4}\right) = -\left(\frac{27}{2} - \frac{81}{4}\right) =$
$\frac{81}{4} - \frac{54}{4} = \frac{27}{4}$

23. $\int_3^2 (x^2 + 2x + 1)\,dx = \frac{x^3}{3} + x^2 + x\Big|_3^2 =$
$\left(\frac{2^3}{3} + 2^2 + 2\right) - \left(\frac{3^3}{3} + 3^2 + 3\right) = \frac{26}{3} - 21 =$
$-\frac{37}{3}$

24. $\int_1^4 \left(\sqrt{x} - \frac{1}{\sqrt{x}}\right) dx = \int_1^4 \left(x^{1/2} - x^{-1/2}\right) dx = \frac{2}{3}x^{3/2} - 2x^{1/2}\Big|_1^4 = \left(\frac{2}{3} \cdot 4^{3/2} - 2 \cdot 4^{1/2}\right) - \left(\frac{2}{3} \cdot 1^{3/2} - 2 \cdot 1^{1/2}\right) = \left(\frac{16}{3} - 4\right) - \left(\frac{2}{3} - 2\right) = \frac{14}{3} - 2 = \frac{8}{3}$

25. $\int_6^8 dx = x\Big|_6^8 = 8 - 6 = 2$

26. $\int_{1/2}^1 \frac{4}{x^2} dx = 4\int_{1/2}^1 x^{-2} dx = 4 \cdot -x^{-1}\Big|_{1/2}^1 = -4 \cdot 1 + 4 \cdot \frac{2}{1} = -4 + 8 = 4$

27. $\int_{-4}^{-2} (3y^2 - y - 1)\, dy = y^3 - \frac{1}{2}y^2 - y\Big|_{-4}^{-2} = \left((-2)^3 - \frac{1}{2}(-2)^2 - (-2)\right) - \left((-4)^3 - \frac{1}{2}(-4)^2 - (-4)\right) = (-8 - 2 + 2) - (-64 - 8 + 4) = -8 - (-68) = 60$

28. $\int_1^4 \left(\frac{3}{x^3} - \frac{2}{x^2}\right) dx = \int_1^4 (3x^{-3} - 2x^{-2})\, dx = -\frac{3}{2}x^{-2} + 2x^{-1}\Big|_1^4 = \left(\left(-\frac{3}{2}\right)4^{-2} + 2 \cdot 4^{-1}\right) - \left(-\frac{3}{2}1^{-2} + 2 \cdot 1^{-1}\right) = \left(\frac{-3}{32} + \frac{1}{2}\right) - \left(-\frac{3}{2} + 2\right) = -\frac{3}{32} + \frac{1}{2} + \frac{3}{2} - 2 = -\frac{3}{32}$

29. $\int_2^4 (x^2 - 2x + 1)\, dx = \frac{x^3}{3} - x^2 + x\Big|_2^4 = \left(\frac{4^3}{3} - 4^2 + 4\right) - \left(\frac{2^3}{3} - 2^2 + 2\right) = \left(\frac{64}{3} - 16 + 4\right) - \left(\frac{8}{3} - 4 + 2\right) = \frac{64}{3} - 16 + 4 - \frac{8}{3} + 4 - 2 = \frac{56}{3} - 10 = \frac{26}{3}$

30. $\int_0^4 \sqrt{x^3}\, dx = \int_0^4 x^{3/2}\, dx = \frac{2}{5}x^{5/2}\Big|_0^4 = \frac{2}{5}4^{5/2} - 0 = \frac{2}{5} \cdot 32 = \frac{64}{5}$

31. $\int_1^2 -3x^{-4}\, dx = x^{-3}\Big|_1^2 = \frac{1}{8} - 1 = -\frac{7}{8}$

32. $\int_0^1 3x^{5/6}\, dx = 3 \cdot \frac{6}{11}x^{11/6}\Big|_0^1 = \frac{18}{11}$

33. $\int_{-2}^{-1} \frac{t^{-2}}{3}\, dx = -\frac{1}{3}t^{-1}\Big|_{-2}^{-1} = \frac{1}{3} - \frac{1}{6} = \frac{1}{6}$

34. $\int_0^2 (x^4 - x^3)\, dx = \frac{1}{5}x^5 - \frac{1}{4}x^4\Big|_0^2 = \frac{1}{5} \cdot 2^5 - \frac{1}{4} \cdot 2^4 - 0 = \frac{32}{5} - \frac{16}{4} = \frac{32}{5} - 4 = \frac{12}{5}$

35. $\int_0^2 \sqrt{x}\, dx = \int_0^2 x^{1/2}\, dx = \frac{2}{3}x^{3/2}\Big|_0^2 = \frac{2}{3} \cdot 2^{3/2} - 0 = \frac{4}{3}\sqrt{2} \approx 1.8856$

36. $\int_1^3 x\sqrt[3]{x}\, dx = \int_1^3 x^{4/3}\, dx = \frac{3}{7}x^{7/3}\Big|_1^3 = \frac{3}{7} \cdot 3^{7/3} - \frac{3}{7} \approx 5.1344$

37. $\int_{1/2}^3 \left(\frac{1}{\sqrt{x}} - 2\right) dx = \int_{1/2}^3 \left(x^{-1/2} - 2\right) dx = 2x^{1/2} - 2x\Big|_{1/2}^3 = \left(2\sqrt{3} - 2 \cdot 3\right) - \left(2 \cdot \sqrt{1/2} - 1\right) = 2\sqrt{3} - 6 - \sqrt{2} + 1 = 2\sqrt{3} - \sqrt{2} - 5 \approx -2.9501$

38. To use the Fundamental Theorem of Calculus, the function must be continuous over a closed interval. This function, $f(x) = x^{-2}$ is not continuous at $x = 0$.

39. Since velocity is the derivative of displacement, the displacement is the integral of the velocity. $\int_0^4 (4.00t - 1.00t^2)\, dt = 2.00t^2 - \frac{1}{3}t^3\Big|_0^4 = 2 \cdot 4^2 - \frac{1}{3} \cdot 4^3 - 0 = 32 - \frac{64}{3} = 10.7 \text{ m}$

40. Angular displacement is the integral of angular velocity. Hence, we have $\int_1^3 (1.00t^3 - 27.0t)\,dt = \frac{1}{4}t^4 - \frac{27}{2}t^2\Big|_1^3 = \left(\frac{3^4}{4} - \frac{27\cdot 3^2}{2}\right) - \left(\frac{1}{4} - \frac{27}{2}\right) =$ -88.0 rad

41. (a) $a(t) = -9.8$ m/s^2; $v(t)$ = antiderivative of $a(t)$; $v(t) = -9.8t + C_1$; at $t = 0$; $v(t) = -10$ so $v(t) = -9.8t - 10$; $v(3) = -9.8\cdot 3 - 10 = -39.4$ m/s. The negative means its velocity is downward. Its speed is $|-39.4| = 39.4$ m/s

(b) Displacement is the integral of the velocity. Hence $\int_0^3 (-9.8t - 10)\,dt = -4.9t^2 - 10t\Big|_0^3 =$ $-4.9(3^2) - 10.3 - 0 = -74.1$. The negative indicates a downward movement. Distance is $|-74.1| = 74.1$ m.

(c) $4.9t^2 - 10t = 180$; $4.9t^2 + 10t - 180 = 0$. Using the quadratic formula, $t = -7.167$ or $t = 5.1258$. Since t must be positive, $t = 5.1258$ s.

42. $v(t) = \int a(t)\,dt = \int -32\,dt = -32t + C_1$. $s(t) = \int v(t)\,dt = -16t^2 + C_1t + C_2$. Since the sandbag is dropped from an elevation of 840 ft, $s(0) = 840$ and so, $C_2 = 840$, which means that $s(t) = -16t^2 + C_1t + 840$. We also have $s(8) = 0 = -16\cdot 8^2 + C_1\cdot 8 + 840 = -1024 + C_18 + 840$ and so, $8C_1 = 184$, or $C_1 = \frac{184}{8} = 23$ ft/s.

43. (a) Acceleration is change in velocity divided by time so $a = (10{,}500 - 500)/0.01 = 10{,}000/0.01 =$ 1 000 000 m/s^2.

(b) $v = \int a(t)\,dt = 1\,000\,000t + 500$; $s = \int_0^{0.01} v(t) = 500\,000t^2 + 500t\Big|_0^{0.01} = 50 + 5 =$ 55 m.

44. We have $i = 1.00t + 1.00\sqrt{t}$. Since charge is the integral of current, we have $q = \int_1^9 (t + t^{1/2})\,dt = \frac{t^2}{2} + \frac{2}{3}t^{3/2}\Big|_1^9 = \left(\frac{9^2}{2} + \frac{2}{3}\cdot 9^{3/2}\right) - \left(\frac{1}{2} + \frac{2}{3}\right) = \frac{81}{2} + \frac{54}{3} - \frac{1}{2} - \frac{2}{3} = \frac{80}{2} + \frac{52}{3} \approx 57.3$ C

25.3 THE INDEFINITE INTEGRAL

1. $\int 9\,dx = 9x + C$

2. $\int 3x\,dx = \frac{3}{2}x^2 + C$

3. $\int 6x^2\,dx = 6\cdot\frac{1}{3}x^3 + C = 2x^3 + C$

4. $\int \sqrt{x}\,dx = \int x^{1/2}\,dx = \frac{2}{3}x^{3/2} + C$

5. $\int x^2\sqrt{x}\,dx = \int x^{5/2}\,dx = \frac{2}{7}x^{7/2} + C$

6. $\int x^{-2/3}\,dx = 3x^{1/3} + C$

7. $\int (t^3 + 1)\,dt = \frac{1}{4}t^4 + t + C$

8. $\int (2 - 4s^3)\,ds = 2s - s^4 + C.$

9. $\int (y^2 + 4y - 3)\,dy = \frac{1}{3}y^3 + 2y^2 - 3y + C.$

10. $\int \left(4\sqrt{y} + \frac{1}{4\sqrt{y}}\right)dy =$ $4\int y^{1/2} + \frac{1}{4}\int y^{-1/2}\,dy =$ $4\cdot\frac{2}{3}y^{3/2} + \frac{1}{4}\cdot 2y^{1/2} + C = \frac{8}{3}y^{3/2} + \frac{1}{2}y^{1/2} + C.$

11. $\int \left(\sqrt[3]{x} - \frac{3}{\sqrt[3]{x^2}}\right)dx = \int (x^{1/3} - 3x^{-2/3})\,dx =$ $\frac{3}{4}x^{4/3} - 9x^{1/3} + C = \frac{3x\sqrt[3]{x}}{4} - 9\sqrt[3]{x} + C.$

12. $\int \left(x^{-3} - 3x^{-1/2} + \frac{1}{3}x^{1/2}\right) dx = -\frac{1}{2}x^{-2} - 3 \cdot 2x^{1/2} + \frac{1}{3} \cdot \frac{2}{3}x^{3/2} + C = \frac{-1}{2x^2} - 6\sqrt{x} + \frac{2}{9}x\sqrt{x} + C.$

13. $\int (x^2 + 3)2x\,dx$. Let $u = x^2 + 3$, then $du = 2x\,dx$. Then, substituting we get $\int u\,du = \frac{1}{2}u^2 + C = \frac{1}{2}(x^2 + 3)^2 + C.$

14. $\int (3x^2 - 5)6x\,dx$. Let $u = 3x^2 - 5$, then $du = 6x\,dx$. Substituting, we see that $\int u\,du = \frac{1}{2}u^2 + C = \frac{1}{2}(3x^2 - 5)^2 + C.$

15. $\int (4 - 2x^2)4x\,dx$. Let $u = 4 - 2x^2$, then $du = -4x\,dx$. Substituting, we can rewrite the given integral as $\int (4 - 2x^2)4x\,dx = -\int u\,du = -\frac{1}{2}u^2 + C = -\frac{1}{2}(4 - 2x^2)^2 + C.$

16. $\int (2x^3 + 1)^4 6x^2\,dx$. Let $u = 2x^3 + 1$, then $du = 6x^2\,dx$. Substituting, we have $\int u^4\,du = \frac{1}{5}u^5 + C = \frac{1}{5}(2x^3 + 1)^5 + C.$

17. $\int (3 - x^2)^3 2x\,dx$. Let $u = 3 - x^2$, then $du = -2x\,dx$ or $-du = 2x\,dx$. Substituting, we have $-\int u^3\,du = -\frac{1}{4}u^4 + C = -\frac{1}{4}(3 - x^2)^4 + C.$

18. $\int (x^2 - 3)^4 x\,dx$. Let $u = x^2 - 3$, then $du = 2x\,dx$ or $x\,dx = \frac{1}{2}\,du$. Substituting, produces $\frac{1}{2}\int u^4\,du = \frac{1}{2} \cdot \frac{1}{5}u^5 + C = \frac{1}{10}(x^2 - 3)^5 + C.$

19. $\int (x^2 + 4)^{1/2} x\,dx$. Let $u = x^2 + 4$, then $du = 2x\,dx$ or $x\,dx = \frac{1}{2}\,du$. Substituting, we obtain $\frac{1}{2}\int u^{1/2}\,du = \frac{1}{2} \cdot \frac{2}{3}u^{3/2} + C = \frac{1}{3}(x^2 + 4)^{3/2} + C.$

20. $\int (x^3 + 1)^5 x^2\,dx$. Let $u = x^3 + 1$, then $du = 3x^2\,dx$ or $x^2\,dx = \frac{1}{3}\,du$. Substituting, we see that $\frac{1}{3}\int u^5\,du = \frac{1}{3} \cdot \frac{1}{6}u^6 + C = \frac{1}{18}(x^3 + 1)^6 + C.$

21. $\int \frac{(\sqrt{x} - 1)^3}{\sqrt{x}}\,dx = \int (\sqrt{x} - 1)^3 \cdot x^{-1/2}\,dx$. Let $u = \sqrt{x} - 1$, then $du = \frac{1}{2}x^{-1/2}$ or $2\,du = x^{-1/2}\,dx$. Substituting, we have $2\int u^3\,du = 2 \cdot \frac{1}{4}u^4 + C = \frac{1}{2}(\sqrt{x} - 1)^4 + C.$

22. $\int (x^4 - 5)^7 x^3\,dx$. Let $u = x^4 - 5$, then $du = 4x^3\,dx$ or $x^3\,dx = \frac{1}{4}\,du$. Substituting, we have $\frac{1}{4}\int u^7\,du = \frac{1}{4} \cdot \frac{1}{8}u^8 + C = \frac{1}{32}(x^4 - 5)^8 + C.$

23. $\int (3x^3 + 1)^4 x^2\,dx$. Let $u = 3x^3 + 1$, then $du = 9x^2\,dx$ or $x^2\,dx = \frac{1}{9}\,du$. Substituting, we obtain $\frac{1}{9}\int u^4\,du = \frac{1}{9} \cdot \frac{1}{5}u^5 + C = \frac{1}{45}(3x^3 + 1)^5 + C.$

24. $\int (3 - 6x^2)^{3/2} 2x\,dx$. Let $u = 3 - 6x^2$, then $du = -12x\,dx$, or $2x\,dx = -\frac{1}{6}\,du$. Substituting, we see that $-\frac{1}{6}\int u^{3/2}\,du = -\frac{1}{6} \cdot \frac{2}{5}u^{5/2} + C = -\frac{1}{15}(3 - 6x^2)^{5/2} + C.$

25. $\int \frac{x\,dx}{\sqrt{x^2 + 3}} = \int (x^2 + 3)^{-1/2} x\,dx$. Let $u = x^2 + 3$, then $du = 2x\,dx$ or $x\,dx = \frac{1}{2}\,du$. Substituting produces $\frac{1}{2}\int u^{-1/2}\,du = \frac{1}{2} \cdot 2u^{1/2} + C = (x^2 + 3)^{1/2} + C = \sqrt{x^2 + 3} + C.$

26. $\int \frac{x\,dx}{\sqrt{4x^2-1}} = \int (4x^2-1)^{-3/2} x\,dx$. Let $u = 4x^2-1$, then $du = 8x\,dx$ and so $\frac{1}{8}\,du = x\,dx$. Substituting, we have $\frac{1}{8}\int u^{-3/2}\,du = \frac{1}{8}(-2)u^{-1/2} + C = -\frac{1}{4}(4x^2-1)^{-1/2} + C = \frac{-1}{4\sqrt{4x^2-1}} + C = -\frac{1}{4}(4x^2-1)^{-1/2} + C$.

27. $\int \frac{x\,dx}{(x^2+3)^3} = \int (x^2+3)^{-3} x\,dx$. Let $u = x^2+3$, then $du = 2x\,dx$, or $x\,du = \frac{1}{2}\,du$ Substitution yields $\frac{1}{2}\int u^{-3}\,du = \frac{1}{2}\cdot\frac{-1}{2}u^{-2}+C = \frac{-1}{4}(x^2+3)^{-2}+C = \frac{-1}{4(x^2+3)^2} + C = -\frac{1}{4}(x^2+3)^{-2} + C$.

28. $\int (x^2+3)^2\,dx$. We cannot use substitution, so we expand $(x^2+3)^2$ to get $\int (x^4+6x^2+9)\,dx = \frac{1}{5}x^5 + 2x^3 + 9x + C$.

29. $\int (3x^2-1)^2\,dx$. We cannot cannot use substitution so we expand to get $\int (9x^4-6x^2+1)\,dx = \frac{9}{5}x^5 - 2x^3 + x + C$.

30. $\int \frac{x^2-1}{x+1}\,dx = \int \frac{(x+1)(x-1)}{x+1}\,dx = \int (x-1)\,dx = \frac{1}{2}x^2 - x + C, x \neq -1$.

31. $\int \frac{x^2-x-6}{x+1}\,dx = \int \frac{(x-3)(x+2)}{x+2}\,dx = \int (x-3)\,dx = \frac{1}{2}x^2 - 3x + C, x \neq -2$.

32. $\int \frac{3x\,dx}{(x^2+4)^5}$. Let $u = x^2+4$, and then $du = 2x\,dx$, or $3x\,dx = \frac{3}{2}\,du$. Substitution yields $\frac{3}{2}\int u^{-5}\,du = \frac{3}{2}\cdot\frac{-1}{4}u^{-4} + C = -\frac{3}{8}(x^2+4)^{-4} + C$.

33. $\int \frac{5x\,dx}{(x^2-1)^{1/3}}$. Let $u = x^2-1$, then $du = 2x\,dx$ or $5x\,dx = \frac{5}{2}\,du$. Substitution produces $\frac{5}{2}\int u^{-1/3}\,du = \frac{5}{2}\cdot\frac{3}{2}u^{2/3} + C = \frac{15}{4}(x^2-1)^{2/3} + C$.

34. $\int (4-x)^{-1/2}\,dx$. Let $u = 4-x$, and then $du = -dx$. . Substituting we see that $-\int (u)^{-1/2}\,du = -\frac{2}{1}u^{1/2} + C = -2(4-x)^{1/2} + C$.

35. $\int (1+3x^2)^2 x\,dx$. Let $u = 1+3x^2$, and then $du = 6x\,dx$, or $x\,dx = \frac{1}{6}\,du$. Substitution produces $\frac{1}{6}\int u^2\,du = \frac{1}{6}\cdot\frac{1}{3}u^3 + C = \frac{1}{18}(1+3x^2)^3 + C$.

36. $\int 2x\sqrt{5x^2+3}\,dx$. Let $u = 5x^2+3$, then $du = 10x\,dx$, or $\frac{1}{5}\,du = 2x\,dx$. Substitution produces $\frac{1}{5}\int u^{1/2}\,du = \frac{1}{5}\cdot\frac{2}{3}u^{3/2} + C = \frac{2}{15}(5x^2+3)^{3/2} + C$.

37. $\int 3x^2(4x^3-5)^{2/3}\,dx$. Let $u = 4x^3-5$, then $du = 12x^2\,dx$, and we have $\frac{1}{4}\,du = 3x^2\,dx$. Thus, $\frac{1}{4}\int u^{2/3}\,du = \frac{1}{4}\cdot\frac{3}{5}u^{5/3} + C = \frac{3}{20}(4x^3-5)^{5/3} + C$.

38. $\int \frac{2x^2\,dx}{(5-3x^3)^{2/3}}$. Let $u = 5-3x^3$, and $du = -9x^2\,dx$; $2x^2\,dx = -\frac{2}{9}\,du$. Thus, $-\frac{2}{9}\int u^{-2/3}\,du = -\frac{2}{9}\cdot\frac{3}{1}u^{1/3} + C = -\frac{2}{3}(5-3x^3)^{1/3} + C$.

39. $\int (1+3x^2)^2\,dx$. We cannot cannot use substitution so we expand to get $\int (1+6x^2+9x^4)\,dx = x + 2x^3 + \frac{9}{5}x^5 + C$.

40. $\int (2x^2+4x)^3(4x+4)\,dx$. Let $u = 2x^2+4x$, and then $du = (4x+4)\,dx$. Hence, $\int u^3\,du = \frac{1}{4}u^4 + C = \frac{1}{4}(2x^2+4x)^4 + C$.

41. $\int \sqrt{4x^2+2x}(4x+1)\,dx$. Let $du = 4x^2+2x$, and then $du = (8x+2)\,dx$, or $\frac{1}{2}\,du = (4x+1)\,dx$. Thus, substituting results in $\frac{1}{2}\int u^{1/2}\,du = \frac{1}{2}\cdot\frac{2}{3}u^{3/2}+C = \frac{1}{3}(4x^2+2x)^{3/2}+C$.

42. $\int (x^2+x-5)^4(2x+1)\,dx$. Let $u = x^2+x-5$, then $du = (2x+1)\,dx$. Substituting, we obtain $\int u^4\,du = \frac{1}{5}u^5 + C = \frac{1}{5}(x^2+x-5)^5+C$.

43. $\int \frac{x-1}{(x-1)^3}\,dx = \int (x-1)^{-2}\,dx = -1(x-1)^{-1} + C = \frac{-1}{x-1}+C$

44. $\int \frac{2x+1}{(x^2+x-5)^3}\,dx$. Letting $u = x^2+x-5$, we obtain $du = (2x+1)\,dx$. Substitution yields $\int u^{-3}\,du = -\frac{1}{2}u^{-2}+C = -\frac{1}{2}\left(x^2+x-5\right)^{-2} + C = \frac{-1}{2(x^2+x-5)^2}+C$.

45. $\int (2x^4-3)^2 8x\,dx$. Let $u = 2x^4-3$ and $du = 8x^3$. We do not have x^3 in the original integrand, so substitution will not work. Expand the indicated binomial to get $\int (4x^8-12x^4+9)8x\,dx = \int (32x^9-96x^5+72x)\,dx = \frac{32}{10}x^{10} - \frac{96}{6}x^6 + \frac{72}{2}x^2 + C = \frac{16}{5}x^{10} - 16x^6 + 36x^2 + C$.

46. $\int \frac{x+3}{(x^2+6x)^{1/3}}\,dx$. Let $u = x^2+6x$, then $du = (2x+6)\,dx$, and we have $\frac{1}{2}\,du = (x+3)\,dx$. Thus, $\frac{1}{2}\int u^{-1/3}\,du = \frac{1}{2}\cdot\frac{3}{2}u^{2/3}+ = \frac{3}{4}(x^2+6x)^{2/3}+C$

47. $\int (x^3-3x)^{2/3}(x^2-1)\,dx$. Let $u = x^3-3x$, then $du = (3x^2-3)\,dx$, or $\frac{1}{3}\,du = (x^2-1)\,dx$; $\frac{1}{3}\int u^{2/3}\,du = \frac{1}{3}\cdot\frac{3}{5}u^{5/3}+C = \frac{1}{5}(x^3-3x)^{5/3}+C$

48. $\int (x^3-2)(x^4-8x)^{1/2}\,dx$. Let $u = x^4-8x$, then $du = (4x^3-8)\,dx$, or $\frac{1}{4}\,du = (x^3-2)\,dx$. Thus, $\int (x^3-2)(x^4-8x)^{1/2}\,dx = \frac{1}{4}\int u^{1/2}\,du = \frac{1}{4}\cdot\frac{2}{3}u^{3/2}+C = \frac{1}{6}(x^4-8x)^{3/2}+C$.

49. $\int (3x+1)^3\,dx$. Let $u = 3x+1$, then $du = 3\,dx$, or $\frac{1}{3}\,du = dx$. Hence, we obtain $\frac{1}{3}\int u^3\,du = \frac{1}{3}\cdot\frac{1}{4}u^4+C = \frac{1}{12}(3x+1)^4+C$.

50. $\int 7x(x^3-2)^2\,dx$. For substitution to work, we must have an x^2 term. We do not, so expand $\int 7x(x^6-4x^3+4)\,dx = \int (7x^7-28x^4+28x)\,dx = \frac{7}{8}x^8 - \frac{28}{5}x^5 + 14x^2 + C$.

51. $\int (2x^3-1)^4\sqrt{x^4-2x}\,dx$. Let $u = x^4-2x$, then $du = (4x^3-2)\,dx$, or $\frac{1}{2}\,du = (2x^3-1)\,dx$. Substitution results in $\frac{1}{2}\int u^{1/4}\,du = \frac{1}{2}\cdot\frac{1}{5}u^{5/4}+C = \frac{2}{5}(x^4-2x)^{5/4}+C$.

52. $\int \frac{x^2}{(4-x^3)^{3/2}}\,dx$. Setting $u = 4-x^3$, we have $du = -3x^2\,dx$, or $x^2\,dx = -\frac{1}{3}\,du$. Thus, $-\frac{1}{3}\int u^{-3/2}\,du =$

$-\frac{1}{3} \cdot \frac{-2}{1} u^{-1/2} + C = \frac{2}{3}(4 - x^3)^{-1/2} + C$ or $\frac{2}{3\sqrt{4 - x^3}} + C$.

53. $\int (t+7)^{1/2}\,dt = \frac{2}{3}(t+7)^{3/2} + C$

54. $\int \frac{x^4 - 1}{x - 1}\,dx = \int \frac{(x-1)(x+1)(x^2+1)}{x-1}\,dx = \int (x^3 + x^2 + x + 1)\,dx = \frac{x^4}{4} + \frac{x^3}{3} + \frac{x^2}{2} + x + C$, $x \neq 1$.

55. $\int \frac{y^3 - 8}{y - 2}\,dy = \int \frac{(y-2)(y^2+2y+4)}{y-2}\,dy = \int (y^2 + 2y + 4)\,dy = \frac{1}{3}y^3 + y^2 + 4y + C, y \neq 2$.

56. $\int (6x-7)^8\,dx$. Let $u = 6x - 7$, then $du = 6\,dx$; $\frac{1}{6}\,du = dx$. Hence, we obtain $\frac{1}{6}\int u^8\,du = \frac{1}{6} \cdot \frac{u^9}{9} + C = \frac{1}{54}(6x-7)^9 + C$.

57. $\int_0^2 (x^2 - 4)2x\,dx$. Let $u = x^2 - 4$, then $du = 2x\,dx$. For the limits of integration, we have $x = 0 \Rightarrow u = -4$ and $x = 2 \Rightarrow u = 0$. Thus, $\int_{-4}^0 u\,du = \left.\frac{u^2}{2}\right|_{-4}^0 = 0 - \frac{16}{2} = -8$.

58. $\int_0^4 x(4x^2+2)^3\,dx$. Let $u = 4x^2 + 2$, then $du = 8x\,dx$, or $\frac{1}{8}\,du = x\,dx$. For the limits of integration, we have $x = 0 \Rightarrow u = 2$ and $x = 4 \Rightarrow u = 66$. Thus, by substitution, we have $\frac{1}{8}\int_2^{66} u^3\,du = \frac{1}{8} \cdot \left.\frac{1}{4}u^4\right|_2^{66} = \frac{1}{32}\left(66^4 - 2^4\right) = \frac{18974720}{32} = 592960$.

59. $\int_1^2 6x(3x^2 - 7)\,dx$. Letting $u = 3x^2 - 7$, we obtain $du = 6x\,dx$ and $x = 1 \Rightarrow u = -4$, and $x = 2 \Rightarrow u = 5$. Substitution produces $\int_{-4}^5 u\,du = \left.\frac{1}{2}u^2\right|_{-4}^5 = \frac{25}{2} - \frac{16}{2} = \frac{9}{2}$.

60. $\int_1^3 2x^2(3x^3 - 1)^2\,dx$. Letting $u = 3x^3 - 1$, we see that $du = 9x^2\,dx$, or $\frac{2}{9}\,du = 2x^2\,dx$. For the limits of integration, we have $x = 1 \Rightarrow u = 2$ and $x = 3 \Rightarrow u = 80$. Thus, the integral becomes $\frac{2}{9}\int_2^{80} u^2\,du = \frac{2}{9} \cdot \left.\frac{u^3}{3}\right|_2^{80} = \frac{2}{27}(80^3 - 2^3) = 37{,}925\frac{1}{3}$.

61. $\int_0^2 2x(20 - 3x^2)\,dx$. Let $u = 20 - 3x^2$, then $du = -6x\,dx$, and so $-\frac{1}{3}\,du = 2x\,dx$. Here, we see that the limits of integration can be changed as $x = 0 \Rightarrow u = 20$ and $x = 2 \Rightarrow u = 8$. Substitution produces $-\frac{1}{3}\int_{20}^8 u\,du = -\frac{1}{3} \cdot \left.\frac{1}{2}u^2\right|_{20}^8 = -\frac{1}{6}(8^2 - 20^2) = 56$.

62. $\int_0^5 x(x^2 + 144)^{1/2}\,dx$. Let $u = x^2 + 144$, then $du = 2x\,dx$, and so $\frac{1}{2}\,du = x\,dx$. Here we see that $x = 0 \Rightarrow u = 144$ and $x = 5 \Rightarrow u = 169$. Using all these substutions, you get $\frac{1}{2}\int_{144}^{169} u^{1/2}\,du = \frac{1}{2} \cdot \left.\frac{2}{3}u^{3/2}\right|_{144}^{169} = \frac{1}{3}(169^{3/2} - 144^{3/2}) = \frac{1}{3}(13^3 - 12^3) = \frac{1}{3}(2197 - 1728) = \frac{469}{3} = 156\frac{1}{3}$.

63. $\int_0^3 (x^3 + 2)x^2\,dx$. Let $u = x^3 + 2$, then $du = 3x^2\,dx$, or $\frac{1}{3}\,du = x^2\,dx$. Here, we see that $x = 0 \Rightarrow u = 2$ and $x = 3 \Rightarrow u = 29$. Substitution yields $\frac{1}{3}\int_2^{29} u\,du = \frac{1}{3} \cdot \left.\frac{1}{2}u^2\right|_2^{29} = \frac{1}{6}(29^2 - 2^2) = 139.5$.

64. $\int_1^2 \frac{dx}{(x+4)^2}$. Let $u = x + 4$, then $du = dx$. Here $x = 1 \Rightarrow u = 5$; and $x = 2 \Rightarrow u = 6$. Substitution produces $\int_5^6 u^{-2}\,du = -u^{-1}\Big|_5^6 = -\frac{1}{6} + \frac{1}{5} = \frac{1}{30}$

65. $\int_1^4 (3x^2 - 2)^4 x\,dx$. Let $u = 3x^2 - 2$, then $du = 6x\,dx$, or $\frac{1}{6}du = x\,dx$ and we see that $x = 1 \Rightarrow u = 1$ and $x = 4 \Rightarrow 4 = 46$. Substitution produces $\frac{1}{6}\int_1^{46} u^4\,du = \frac{1}{6}\cdot\frac{1}{5}u^5\Big|_1^{46} = \frac{1}{30}(46^5 - 1^5) =$ 6,865,432.5

66. $\int_{-1}^0 (6x^2 - 1)^3 x\,dx$. Let $u = 6x^2 - 1$, then $du = 12x\,dx$, or $\frac{1}{12}du = x\,dx$ and we see that $x = -1 \Rightarrow u = 5$ and $x = 0 \Rightarrow u = -1$. Substitution produces $\frac{1}{12}\int_5^{-1} u^3\,du = \frac{1}{12}\cdot\frac{1}{4}u^4\Big|_5^{-1} = \frac{1}{48}((-1)^4 - 5^4) = \frac{1}{48}(-624) = -13$

25.4 THE AREA BETWEEN TWO CURVES

1.

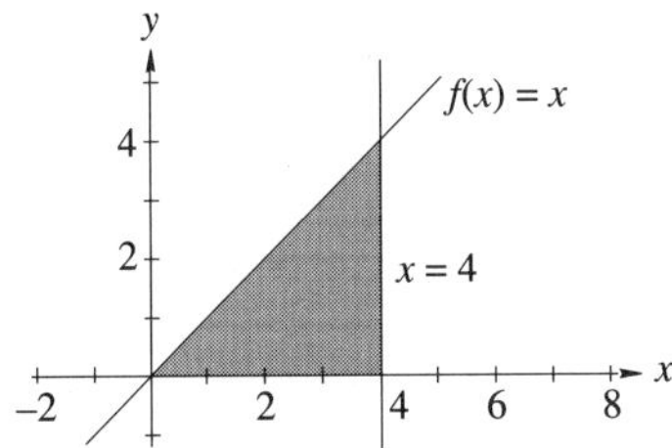

From the graph we can determine that all the area is above the x-axis so we only need to integrate from 0 to 4. $\int_0^4 x\,dx = \frac{x^2}{2}\Big|_0^4 = \frac{4^2}{2} - \frac{0^2}{2} = 8$

2.

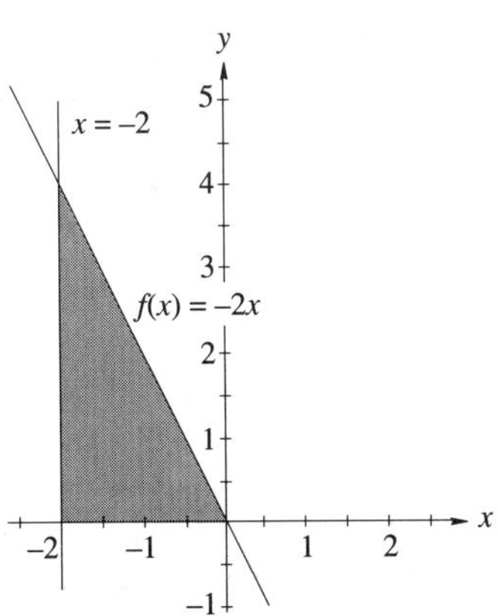

From the graph we can determine that all the area is above the x-axis so we need only integrate from -2 to 0. $\int_{-2}^0 -2x\,dx = -x^2\Big|_{-2}^0 = -0^2 + (-2)^2 = 4$

3.

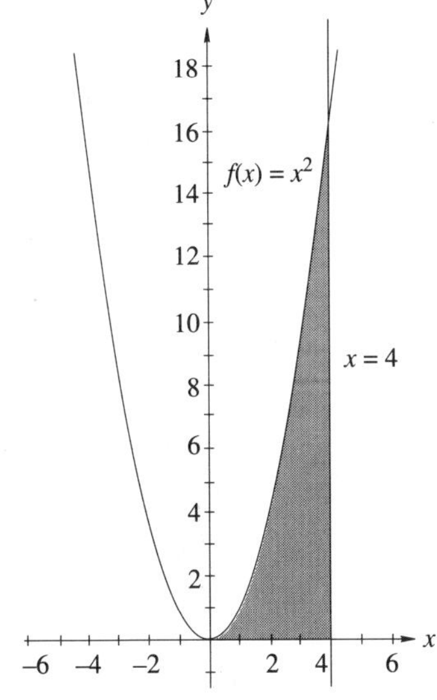

All the area is above the x-axis $\int_0^4 x^2 = \frac{x^3}{3}\Big|_0^4 = \frac{4^3}{3} - \frac{0^3}{3} = \frac{64}{3}$

4.

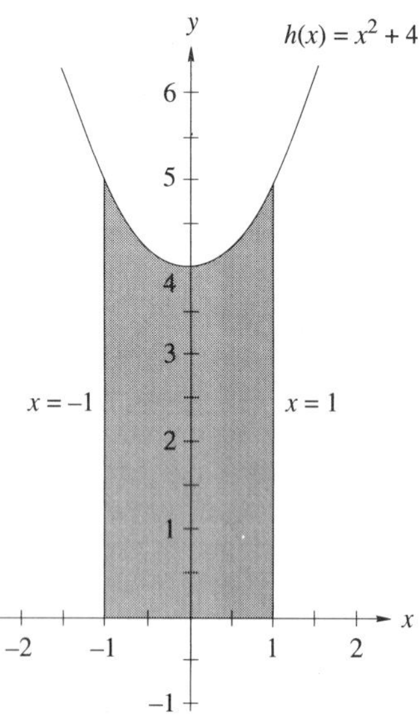

All the area above the x-axis $\int_{-1}^{1} (x^2 + 4)\,dx =$ $\frac{x^3}{3} + 4x\Big|_{-1}^{1} = \left(\frac{1}{3} + 4\right) - \left(-\frac{1}{3} - 4\right) = 8\frac{2}{3}$.

5.

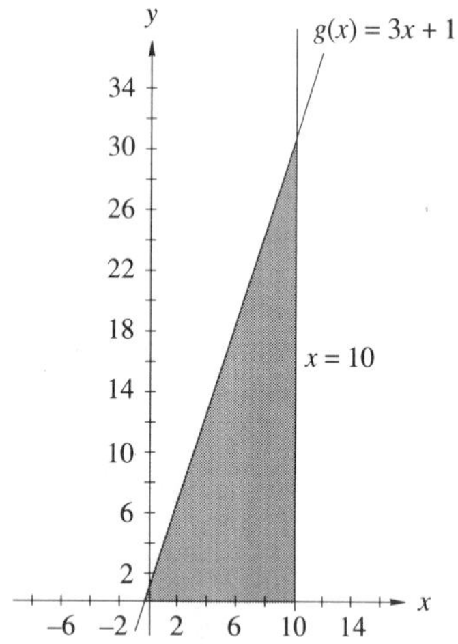

All the area is above the x-axis. We can determine that $3x + 1$ intersects the x-axis at $-\frac{1}{3}$. Hence, the desired area is

$$\int_{-1/3}^{10} (3x + 1)\,dx = \frac{3}{2}x^2 + x\Big|_{-1/3}^{10}$$
$$= \left(\frac{300}{2} + 10\right) - \left(\frac{1}{6} - \frac{1}{3}\right) = 160\frac{1}{6}$$

6.

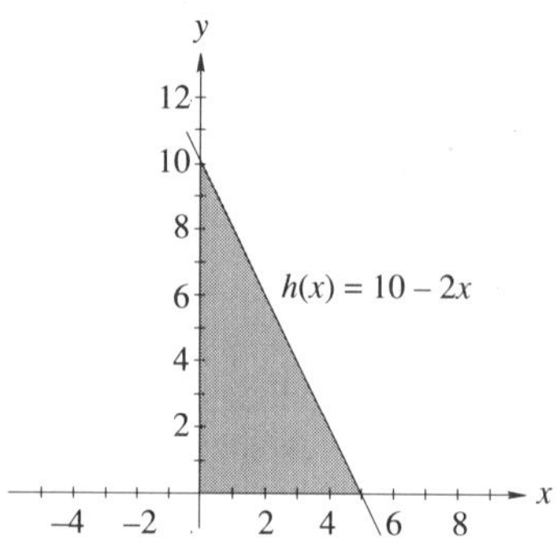

All the area is above the x-axis. We can see that $h(x)$ intersects the x-axis at 5. $\int_0^5 (10 - 2x)\,dx =$ $10x - x^2\big|_0^5 = 50 - 25 = 25$

7.

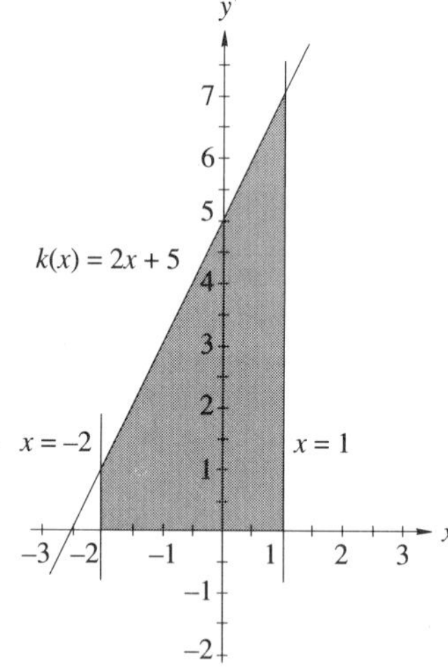

Between -2 and 1, $k(x)$ is above the x-axis. Thus, the desired area is $\int_{-2}^{1} (2x + 5)\,dx = x^2 + 5x\big|_{-2}^{1} =$ $(1 + 5) - (4 - 10) = 6 + 6 = 12$.

8.

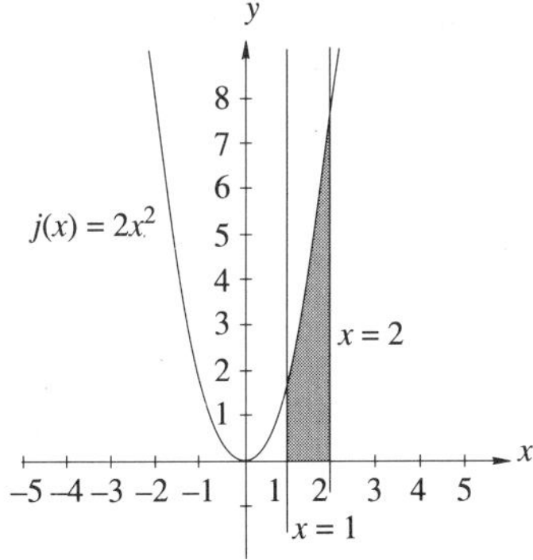

Between 1 and 2, j is above the x-axis. Hence, the desired area is $\int_1^2 2x^2\,dx = \left.\frac{2x^3}{3}\right|_1^2 = \frac{16}{3} - \frac{2}{3} = \frac{14}{3}$.

9.

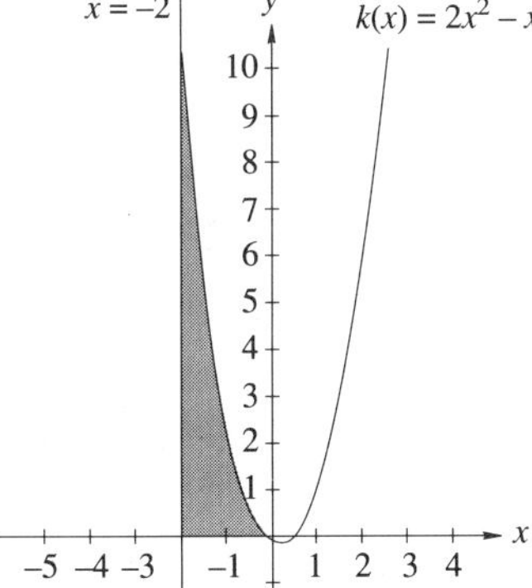

Between -2 and 0, we can see that k is above the x-axis. Thus, the desired area is $\int_{-2}^0 (2x^2 - x)\,dx = \left.\frac{2x^3}{3} - \frac{x^2}{2}\right|_{-2}^0 = 0 - \left(\frac{-16}{3} - \frac{4}{2}\right) = 7\frac{1}{3} = \frac{22}{3}$.

10.

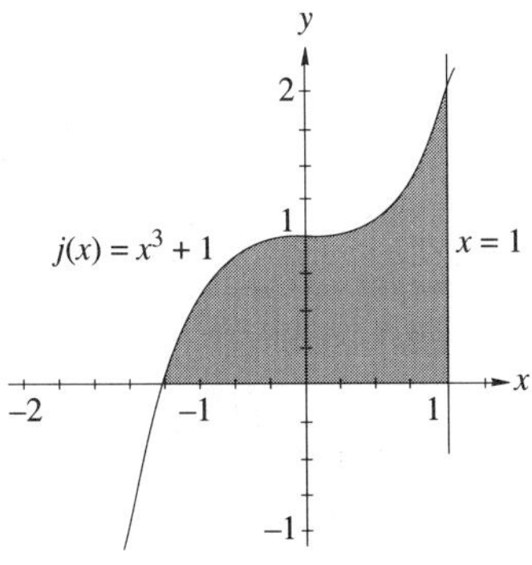

Between -1 and 1, we see that j is above the x-axis. As a result, the desired area is $\int_{-1}^1 (x^3 + 1)\,dx = \left.\frac{x^4}{4} + x\right|_{-1}^1 = \left(\frac{1}{4} + 1\right) - \left(\frac{1}{4} - 1\right) = 2.$

11.

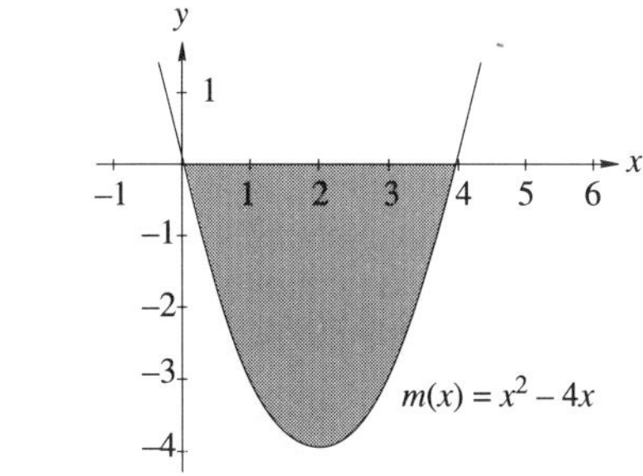

From the graph we can determine that m intersects the x-axis at 0 and 4 and is below the axis. Hence we take the absolute value of the integral.

$$\left|\int_0^4 (x^2 - 4x)\,dx\right| = \left|\left[\frac{x^3}{3} - 2x^2\right]_0^4\right| = \left|\frac{64}{3} - 32\right|$$

$$= \left|-10\frac{2}{3}\right| = 10\frac{2}{3} \text{ or } \frac{32}{3}$$

12.

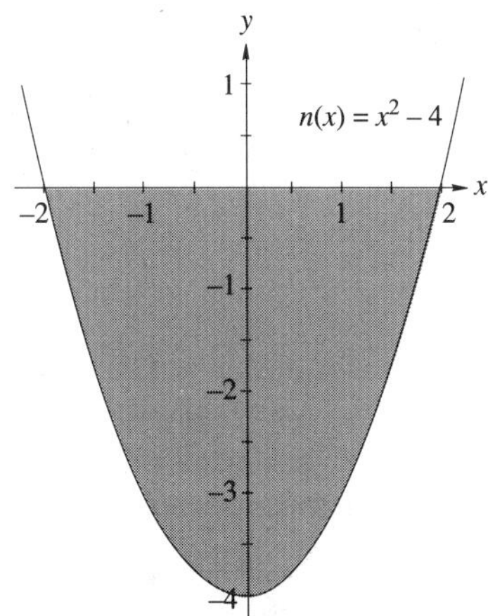

From the graph we determine that n intersects the x-axis at -2 and 2 and the area is below the axis between these points. We need the absolute value of the integral or its negative.

$$\left|\int_{-2}^{2}(x^2-4)\,dx\right|$$

$$= -\int_{-2}^{2}(x^2-4)\,dx = -\left(\frac{x^3}{3}-4x\right)\Big|_{-2}^{2}$$

$$= -\left[\left(\frac{8}{3}-8\right)-\left(\frac{-8}{3}+8\right)\right]$$

$$= -\left[\frac{16}{3}-16\right] = -\left[\frac{-32}{3}\right] = \frac{32}{3} = 10\frac{2}{3}$$

13.

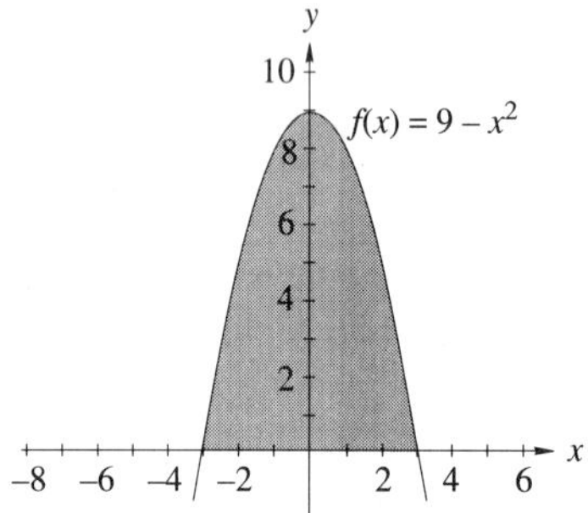

The area is above the x-axis and f intersects the x-axis at -3 and 3. Thus, we have $\int_{-3}^{3}(9-x^2)\,dx = 9x - \frac{x^3}{3}\Big|_{-3}^{3} = (27-9)-(-27+9) = 36$

14.

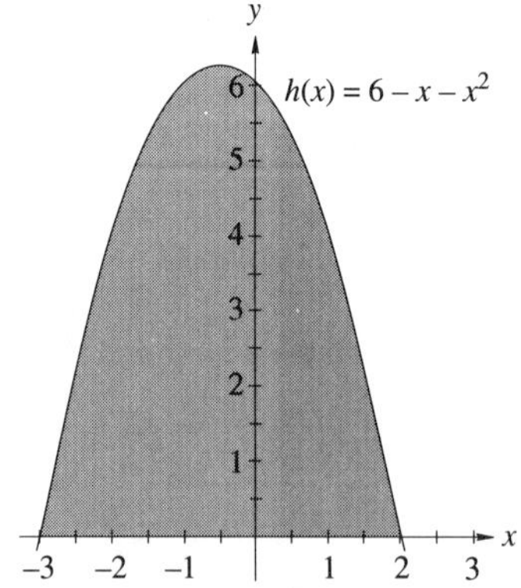

The area is above the x-axis and h intersects the axis at -3 and 2.

$$\int_{-3}^{2}(6-x-x^2)\,dx = 6x - \frac{x^2}{2} - \frac{x^3}{3}\Big|_{-3}^{2}$$

$$= \left(12-2-\frac{8}{3}\right) - \left(-18-\frac{9}{2}+9\right) = 20\frac{5}{6}$$

15.

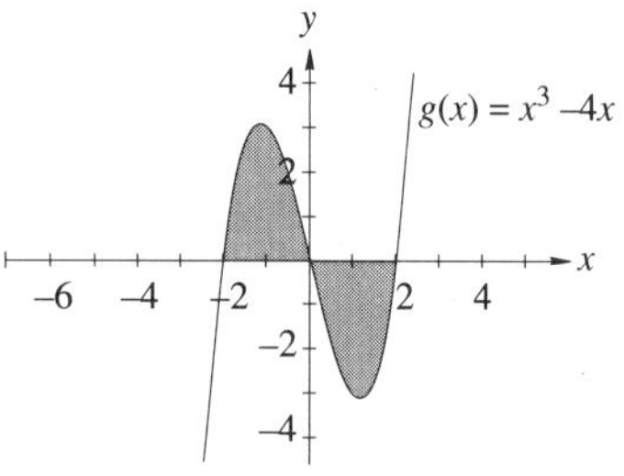

The graph is above the x-axis from -2 to 0 and below the x-axis from 0 to 2.

$$\int_{-2}^{0}(x^3-4x)\,dx - \int_{0}^{2}(x^3-4x)\,dx$$

$$= \left(\frac{x^4}{4}-2x^2\right)\Big|_{-2}^{0} - \left(\frac{x^4}{4}-2x^2\right)\Big|_{0}^{2}$$

$$= 0 - \left(\frac{16}{4}-8\right) - \left(\frac{16}{4}-8\right) + 0 = 8$$

16.

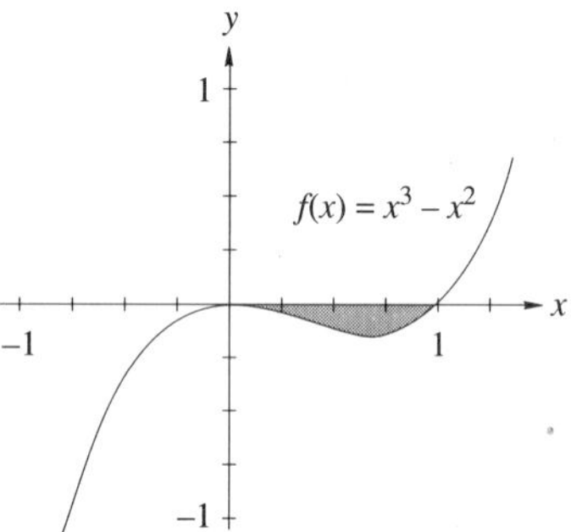

The graph is below the axis from 0 to 1, the two places it intersects the x-axis. $-\int_{0}^{1}(x^3-x^2)\,dx = -\left(\frac{x^4}{4}-\frac{x^3}{3}\right)\Big|_{0}^{1} = -\left(\frac{1}{4}-\frac{1}{3}\right) = \frac{1}{12}$.

17.

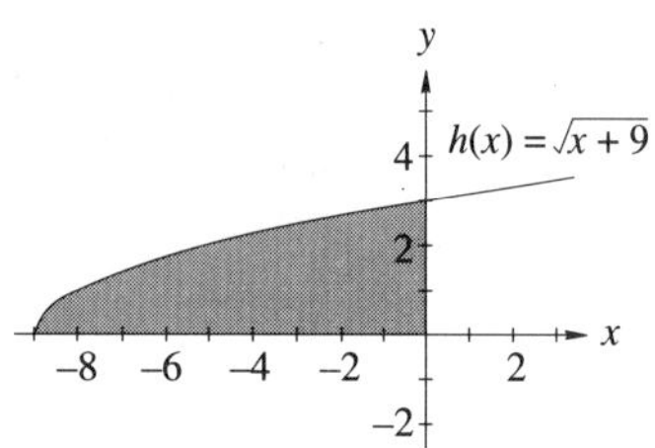

The graph is above the x-axis and intersects at -9. Thus, the desired area is $\int_{-9}^{0} \sqrt{x+9}\, dx = \frac{2}{3}(x+9)^{3/2}\Big|_{-9}^{0} = \frac{2}{3}\left[(0+9)^{3/2} - (-9+9)^{3/2}\right] = \frac{2}{3} \cdot 27 = 18.$

18.

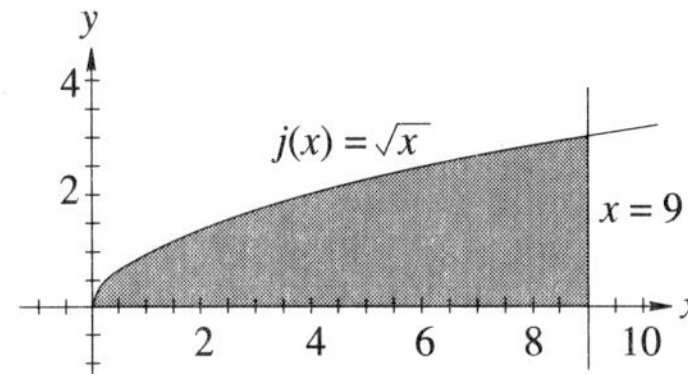

The graph is above the x-axis and intersects the axis at 0. Thus, the desired area is $\int_{0}^{9} \sqrt{x}\, dx = \frac{2}{3}x^{3/2}\Big|_{0}^{9} = \frac{2}{3} \cdot 9^{3/2} = 18.$

19.

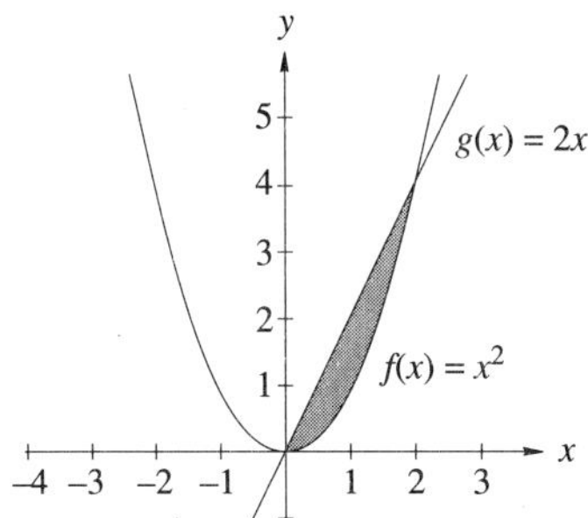

From the graph we determine that $g(x) \geq f(x)$ on $[0, 2]$. Hence, the desired area is $\int_{0}^{2} (2x - x^2)\, dx = x^2 - \frac{x^3}{3}\Big|_{0}^{2} = 4 - \frac{8}{3} = \frac{4}{3}.$

20.

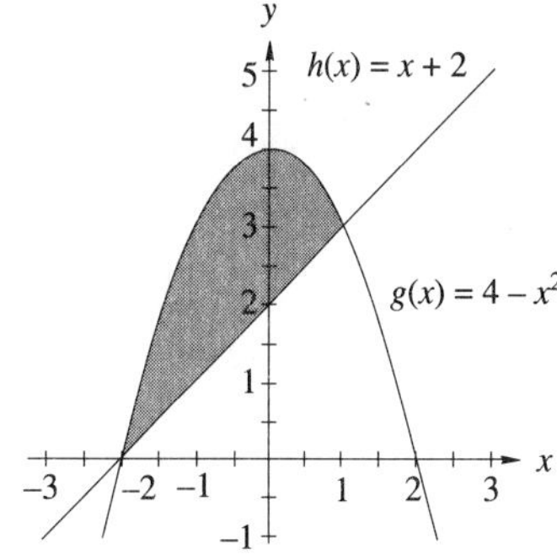

The graphs intersect at $x = -2$ and $x = 1$. Since $g(x) \geq h(x)$, the desired area is

$$\int_{-2}^{1} (g(x) - h(x))\, dx = \int_{-2}^{1} ((4 - x^2) - (x + 2))\, dx$$

$$= \int_{-2}^{1} (2 - x - x^2)\, dx = 2x - \frac{x^2}{2} - \frac{x^3}{3}\Big|_{-2}^{1}$$

$$= \left(2 - \frac{1}{2} - \frac{1}{3}\right) - \left(-4 - 2 + \frac{8}{3}\right)$$

$$= 2 - \frac{1}{2} - \frac{1}{3} + 4 + 2 - \frac{8}{4} = 4\frac{1}{2}$$

21.

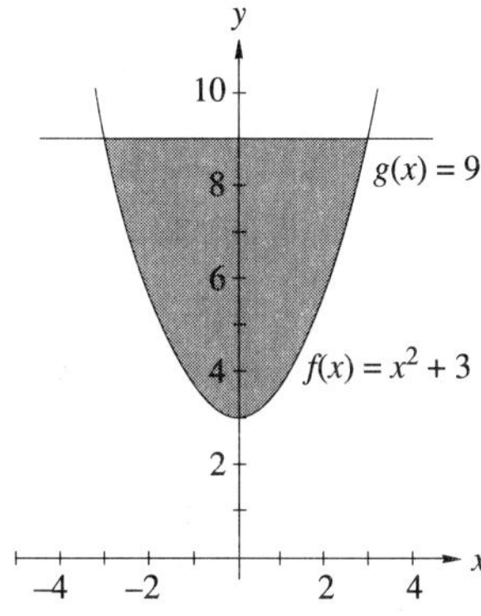

f and g intersect at $\left(\pm\sqrt{6}, 9\right)$. On the interval $\left[-\sqrt{6}, \sqrt{6}\right]$, we see that $g(x) \geq f(x)$. Thus, the desired area is

$$\int_{-\sqrt{6}}^{\sqrt{6}} (9 - (x^2 + 3))\, dx$$

$$= \int_{-\sqrt{6}}^{\sqrt{6}} (6 - x^2)\, dx = 6x - \frac{x^3}{3}\Big|_{-\sqrt{6}}^{\sqrt{6}}$$

$$= \left(6\sqrt{6} - \frac{\sqrt{6}^3}{3}\right) - \left(-6\sqrt{6} + \frac{\sqrt{6}^3}{3}\right)$$
$$= 12\sqrt{6} - 4\sqrt{6} = 8\sqrt{6} \approx 19.596$$

22.

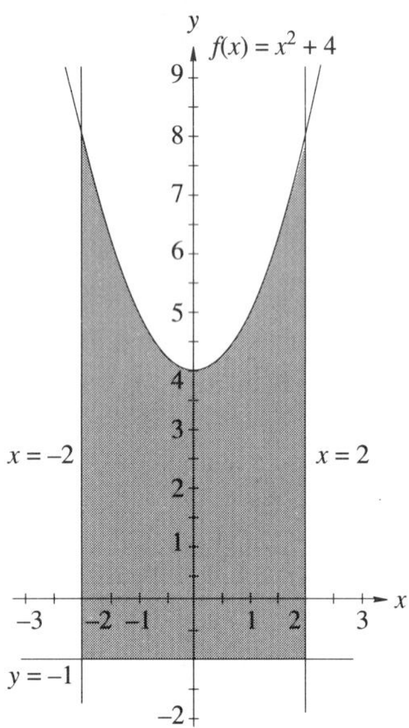

From the figure we determine that we need to integrate from -2 to 2 and that the needed integrand is $f(x) - (-1)$.

$$\int_{-2}^{2} (x^2 + 4 - (-1))\, dx = \int_{-2}^{2} (x^2 + 5)\, dx$$
$$= \frac{x^3}{3} + 5x\Big|_{-2}^{2} = \left(\frac{8}{3} + 10\right) - \left(\frac{-8}{3} - 10\right) = 25\frac{1}{3}$$

23.

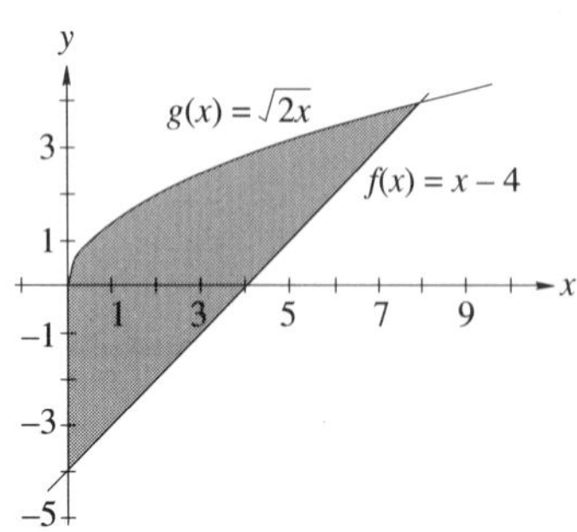

f and g intersect at $(8, 4)$. g is defined on $[0, \infty)$.

On $[0, 8]$, we see that $g(x) \geq f(x)$.

$$\int_0^8 (g(x) - f(x))\, dx = \int_0^8 (\sqrt{2x} - x + 4)\, dx$$
$$= \int_0^8 \sqrt{2x}\, dx + \int_0^8 (4 - x)\, dx$$

For the first integral let $u = 2x$, then $du = 2\,dx$, so $dx = \frac{1}{2}\, du$ and the limits of integration become $x = 0 \Rightarrow u = 0$ and $x = 8 \Rightarrow u = 16$. Thus, by substitution into the first integral, we get

$$\frac{1}{2}\int_0^{16} \sqrt{u}\, du + \int_0^8 (4 - x)\, dx$$
$$= \frac{1}{2}\frac{2}{3}u^{3/2}\Big|_0^{16} + \left(4x - \frac{x^2}{2}\right)\Big|_0^8$$
$$= \left(\frac{64}{3} - 0\right) + \left(32 - \frac{64}{2}\right) = 21\frac{1}{3} = 21.333$$

24.

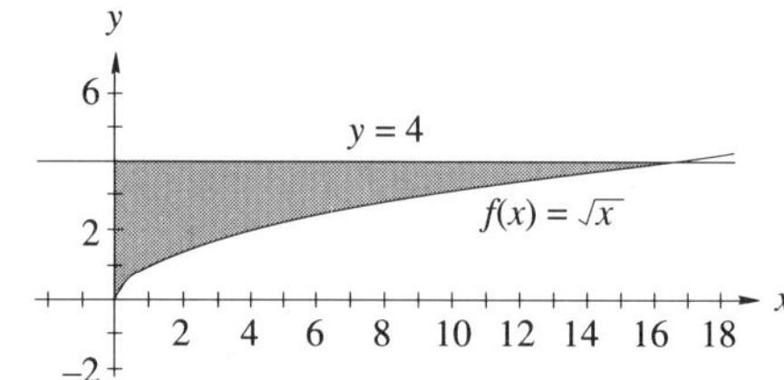

$$\int_0^{16} (4 - \sqrt{x})\, dx = 4x - \frac{2}{3}x^{3/2}\Big|_0^{16} = 64 - \frac{128}{3} = \frac{64}{3} = 21\frac{1}{3}$$

25.

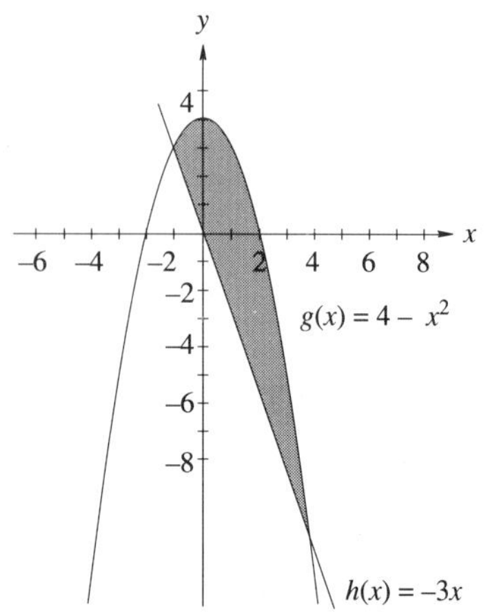

f and g intersect at $(-1, 3)$ and $(4, -12)$

$$\int_{-1}^{4} ((4 - x^2) - (-3x))\,dx$$

$$= \int_{-1}^{4} (4 + 3x - x^2)\,dx$$

$$= 4x + \frac{3x^2}{2} - \frac{x^3}{3}\Big|_{-1}^{4}$$

$$= \left(16 + 24 - \frac{64}{3}\right) - \left(-4 + \frac{3}{2} + \frac{1}{3}\right)$$

$$= 44 - \frac{3}{2} - \frac{65}{3} = 20.8333$$

26.

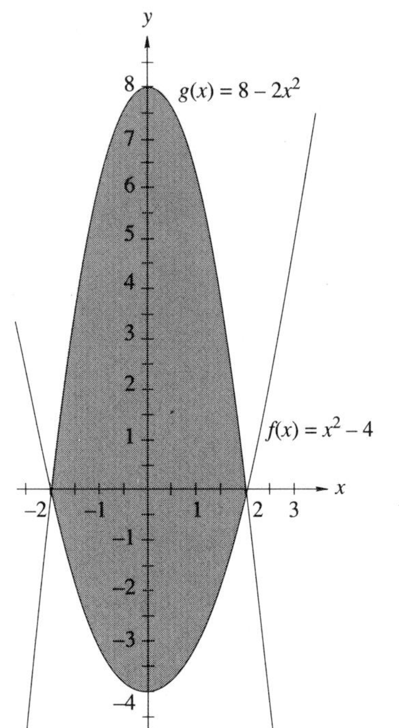

f and g intersect at $(\pm 2, 0)$. On $[-2, 2]$, $g(x) \geq f(x)$.

$$\int_{-2}^{2} \left[(9 - 2x^2) - (x^2 - 4)\right] dx$$

$$= \int_{-2}^{2} (12 - 3x^2)\,dx$$

$$= 12x - x^3\Big|_{-2}^{2} = (24 - 8) - (-24 + 8) = 32$$

27.

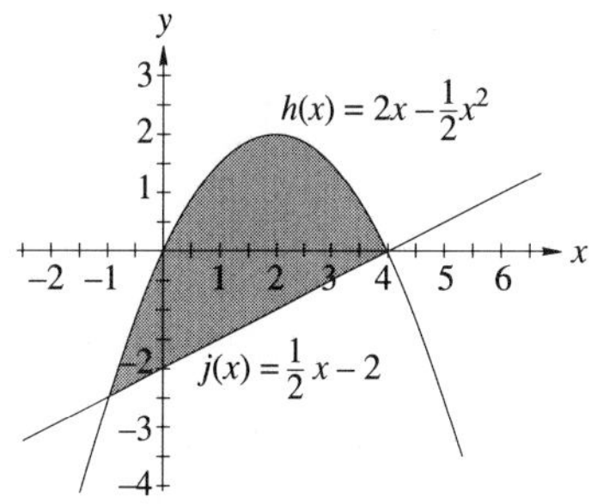

h and j intersect at $(-1, -2\frac{1}{3})$ and $(4, 0)$

$$\int_{-1}^{4} (h(x) - j(x))\,dx$$

$$= \int_{-1}^{4} \left(\frac{3}{2}x - \frac{1}{2}x^2 + 2\right) dx$$

$$= \frac{3}{4}x^2 - \frac{1}{6}x^3 + 2x\Big|_{-1}^{4}$$

$$= \left(12 - \frac{64}{6} + 8\right) - \left(\frac{3}{4} + \frac{1}{6} - 2\right)$$

$$= 10.417$$

28.

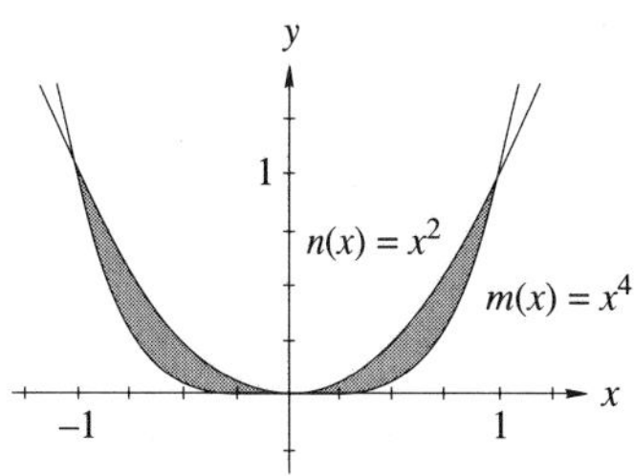

$$\int_{-1}^{1} (x^2 - x^4)\,dx = \frac{x^3}{3} - \frac{x^5}{5}\Big|_{-1}^{1} = \left(\frac{1}{3} - \frac{1}{5}\right) - \left(\frac{1}{5} - \frac{1}{3}\right) = \frac{4}{15}$$

29.

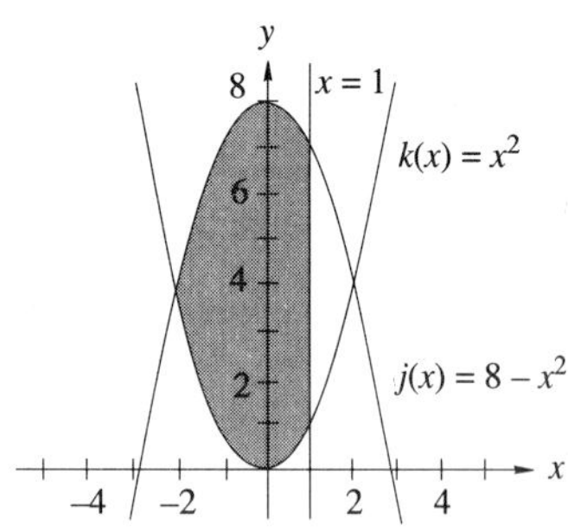

$$\int_{-2}^{1}((8-x^2)-(x^2))\,dx$$
$$= \int_{-2}^{1}(8-2x^2)\,dx$$
$$= 8x - \frac{2}{3}x^3\Big|_{-2}^{1}$$
$$= \left(8-\frac{2}{3}\right)-\left(-16+\frac{16}{3}\right)$$
$$= 24-\frac{18}{3} = 18$$

30.

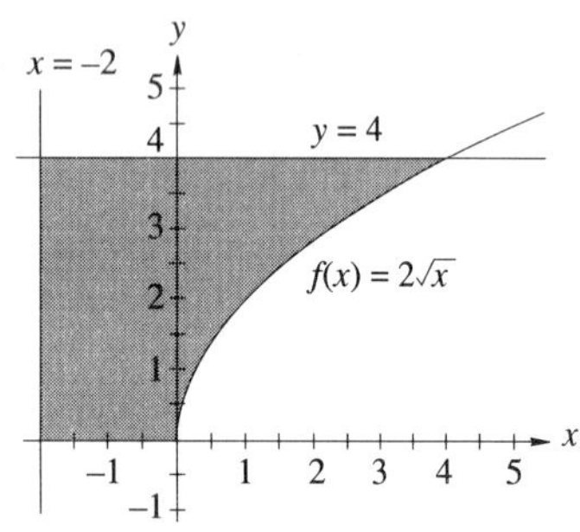

$$\int_{-2}^{0}4\,dx + \int_{0}^{4}(4-2\sqrt{x})\,dx$$
$$= 4x|_{-2}^{0} + \left(4x-\frac{4}{3}x^{3/2}\right)\Big|_{0}^{4}$$
$$= 8+\left(16+\frac{32}{3}\right) = 24-\frac{32}{3} = \frac{40}{3} = 13\frac{1}{3}$$

31.

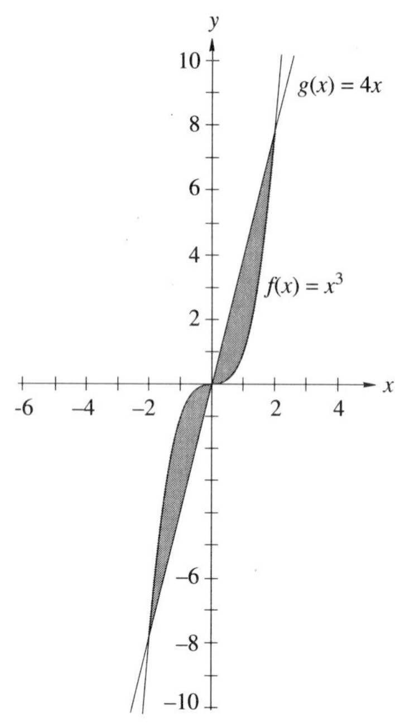

$$\int_{-2}^{0}(x^3-4x)\,dx + \int_{0}^{2}(4x-x^3)\,dx$$
$$= \left(\frac{x^4}{4}-2x^2\right)\Big|_{-2}^{0} + \left(2x^2-\frac{x^4}{4}\right)\Big|_{0}^{2}$$
$$= 0-\left(\frac{16}{4}-18\right)+\left(8-\frac{16}{4}\right)-0 = 8$$

32.

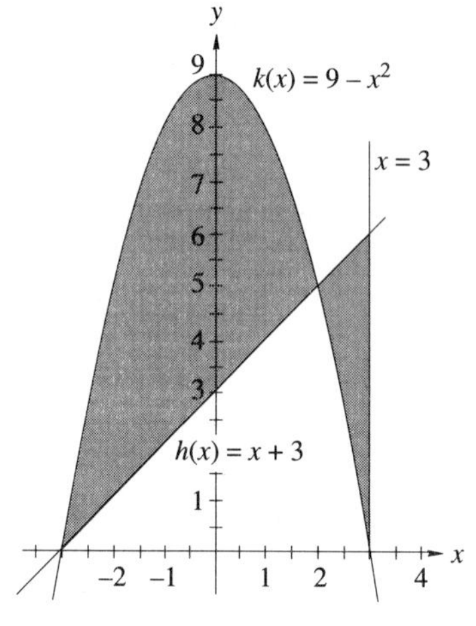

h intersects k at $(-3, 0)$ and $(2, 5)$

$$\int_{-3}^{2} \left[(9 - x^2) - (x + 3)\right] dx$$

$$+ \int_{2}^{3} \left[(x + 3) - (9 - x^2)\right] dx$$

$$= \int_{-3}^{2} (6 - x - x^2)\,dx + \int_{2}^{3} (x^2 + x - 6)\,dx$$

$$= \left(6x - \frac{x^2}{2} - \frac{x^3}{3}\right)\Bigg|_{-3}^{2} + \left(\frac{x^3}{3} + \frac{x^2}{2} - 6x\right)\Bigg|_{2}^{3}$$

$$= \left(12 - 2 - \frac{8}{3}\right) - \left(-18 - \frac{9}{2} + 9\right)$$

$$+ \left(9 + \frac{9}{2} - 18\right) - \left(\frac{8}{3} + 2 - 12\right)$$

$$= 12 - 2 - \frac{8}{3} + 18 + \frac{9}{2} - 9 + 9$$

$$+ \frac{9}{12} - 18 - \frac{8}{3} - 2 + 12$$

$$= 24 - 4 + 9 - \frac{16}{3} = 29 - \frac{16}{3} = 23\frac{2}{3}$$

33.

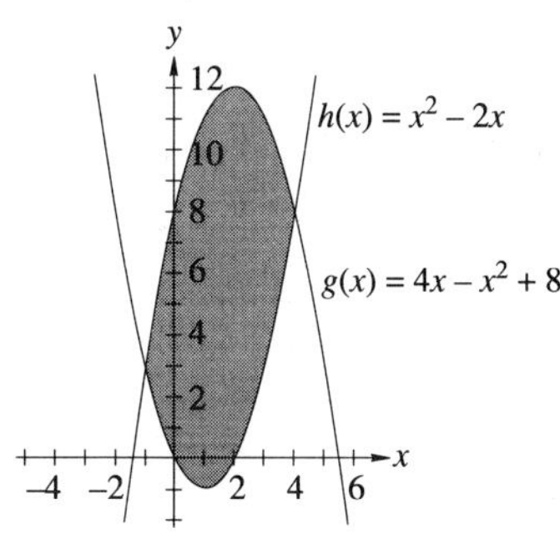

h and g intersect at $(-1, 3)$ and $(4, 8)$

$$\int_{-1}^{4} (g(x) - h(x))\,dx$$

$$= \int_{-1}^{4} (8 + 6x - 2x^2)\,dx$$

$$= 8x + 3x^2 - \frac{2}{3}x^3\Bigg|_{-1}^{4}$$

$$= \left(32 + 48 - \frac{128}{3}\right) - \left(-8 + 3 + \frac{2}{3}\right)$$

$$= 80 - \frac{128}{3} + 8 - 3 - \frac{2}{3}$$

$$= 85 - \frac{130}{3} = \frac{125}{3} = 41\frac{2}{3}$$

34.

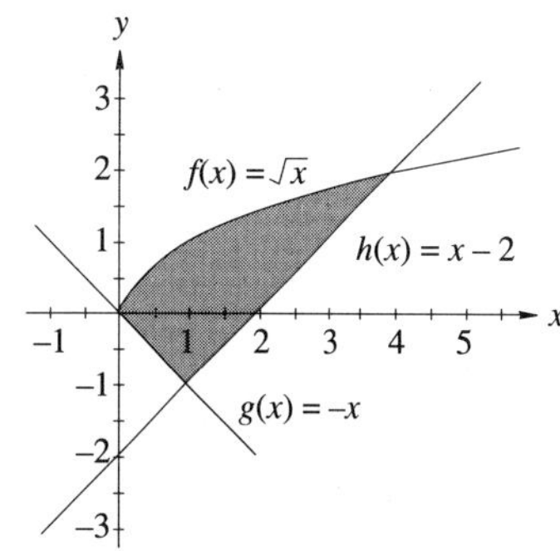

f is always the top graph. From 0 to 1, g is the bottom; from 1 to 4, h is the bottom.

$$\int_{0}^{1} (\sqrt{x} + x)\,dx + \int_{1}^{4} (\sqrt{x} - x + 2)\,dx$$

$$= \left(\frac{2}{3}x^{3/2} + \frac{x^2}{2}\right)\Bigg|_{0}^{1} + \left(\frac{2}{3}x^{3/2} - \frac{x^2}{2} + 2x\right)\Bigg|_{1}^{4}$$

$$= \left(\frac{2}{3} + \frac{1}{2}\right) - 0 + \left(\frac{16}{3} - \frac{16}{2} + 8\right)$$

$$- \left(\frac{2}{3} - \frac{1}{2} + 2\right)$$

$$= \frac{2}{3} + \frac{1}{2} + \frac{16}{3} - 8 + 8 - \frac{2}{3} + \frac{1}{2} - 2$$

$$= \frac{16}{3} - 1 = 4\frac{1}{3}$$

35.

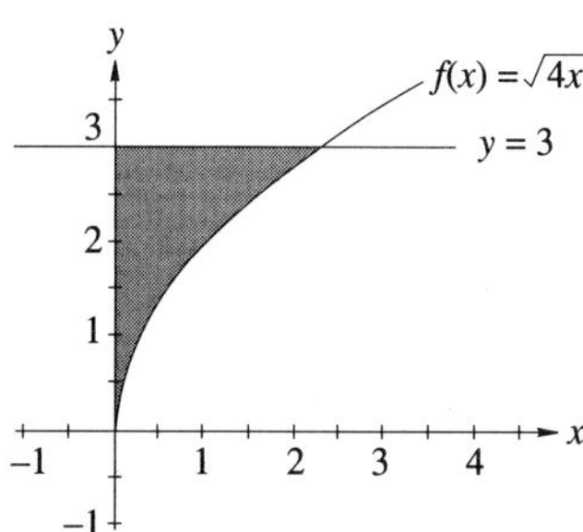

Solving for the point of intersection, we get $\sqrt{4x} = 2\sqrt{x} = 3$; $\sqrt{x} = \frac{3}{2}$, and $x = \frac{9}{4}$. Now integrating, we have

$$\int_0^{9/4} (3 - 2\sqrt{x})\,dx = 3x - \frac{4}{3}x^{3/2}\Big|_0^{9/4}$$
$$= \frac{27}{4} - \frac{4}{3}\cdot\frac{27}{8} = \frac{27}{4} - \frac{27}{6} = 2.25$$

36.

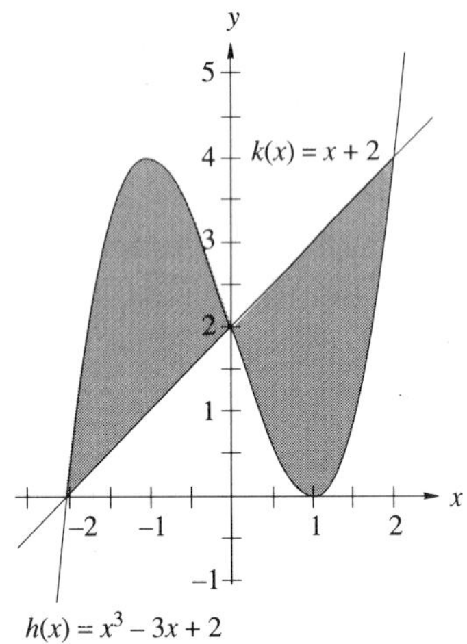

h and k intersect at $(-2, 0)$, $(0, 2)$, and $(2, 4)$. Thus, the desired area is

$$\int_{-2}^{0} (h(x) - k(x))\,dx + \int_0^2 (k(x) - h(x))\,dx$$
$$= \int_{-1}^{0} (x^3 - 4x)\,dx + \int_0^2 (4x - x^3)\,dx$$
$$= \left(\frac{x^4}{4} - 2x^2\right)\Big|_{-2}^{0} + \left(2x^2 - \frac{x^4}{4}\right)\Big|_0^2$$
$$= 0 - (-4 + 8) + (8 - 4) - 0 = 8$$

37.

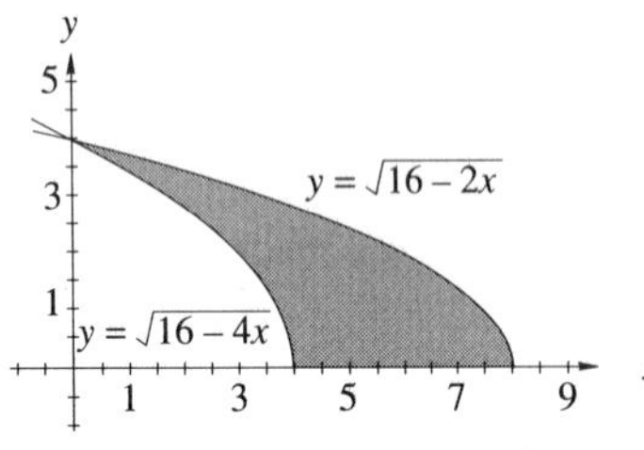

$$\int_0^4 (\sqrt{16-2x} - \sqrt{16-4x})\,dx + \int_4^8 \sqrt{16-2x}\,dx = \int_0^4 \sqrt{16-2x}\,dx - \int_0^4 \sqrt{16-4x}\,dx + \int_4^8 \sqrt{16-2x}\,dx.$$

For the first and third integrals let $u = 16 - 2x$, then $du = -2\,dx$ or $dx = -\frac{1}{2}\,du$. For the limits of integration, we obtain $x = 0 \Rightarrow u = 16$; $x = 4 \Rightarrow u = 8$ and $x = 8 \Rightarrow u = 0$.

$$-\frac{1}{2}\int_{16}^{8} u^{1/2}\,du = -\frac{1}{2}\cdot\frac{2}{3}u^{3/2}\Big|_{16}^{8} = -\frac{1}{3}\left(8^{3/2} - 64\right)$$
$$-\frac{1}{2}\int_{8}^{0} u^{1/2}\,dx = \frac{1}{2}\cdot\frac{2}{3}u^{3/2}\Big|_{8}^{0} = -\frac{1}{3}(0 - 8^{3/2})$$

The sum of these two integrals is $\frac{64}{3}$. For the second integral let $u = 16 - 4x$, and then $du = -4\,dx$. or $dx = -\frac{1}{4}\,du$. Here the limits of integration are $x = 0 \Rightarrow u = 16$ and $x = 4 \Rightarrow u = 0$.

$$-\frac{1}{4}\int_{16}^{0} u^{1/2}\,du = -\frac{1}{4}\cdot\frac{2}{3}u^{3/2}\Big|_{16}^{0}$$
$$= 0 + \frac{1}{6}\cdot 64 = \frac{64}{6}$$

For the final answer, we subtract this last answer from the earlier result, and obtain $\frac{64}{3} - \frac{64}{6} = \frac{64}{6} = \frac{32}{3} = 10\frac{2}{3} = 10.667$.

38.

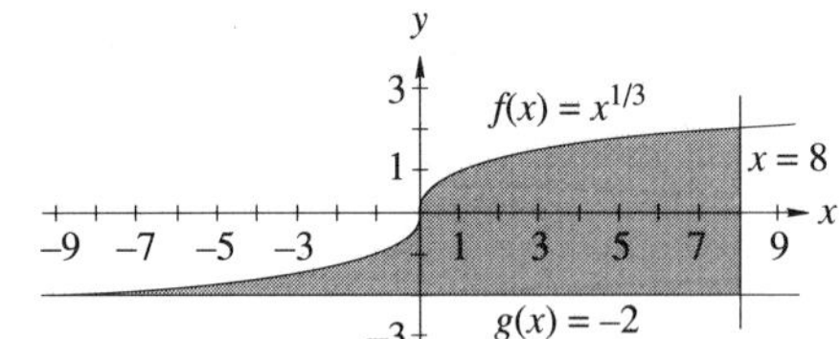

$\int_{-8}^{8} (x^{1/3} - (-2))\,dx = \frac{3}{4}x^{4/3} + 2x\Big|_{-8}^{8} = (12 + 16) - (12 - 16) = 32.$

39.

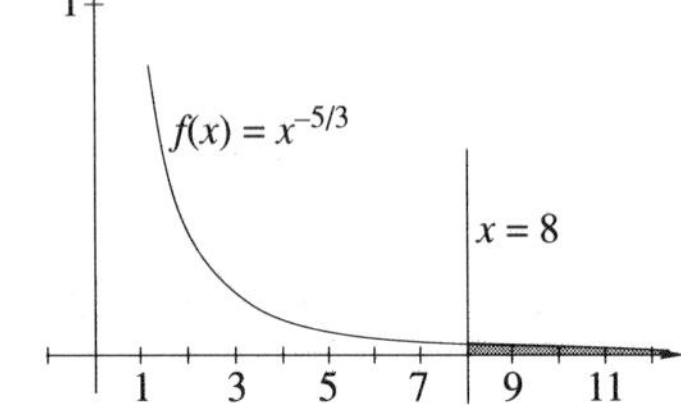

$$\lim_{b\to\infty}\int_8^b x^{-5/3}\,dx = \lim_{b\to\infty} -\frac{3}{2}x^{-2/3}\Big|_8^b$$
$$= \lim_{b\to\infty} -\frac{3}{2}b^{-3/2} + \frac{3}{2}(8)^{-2/3}$$
$$= \frac{3}{8} \text{ or } 0.375$$

40.

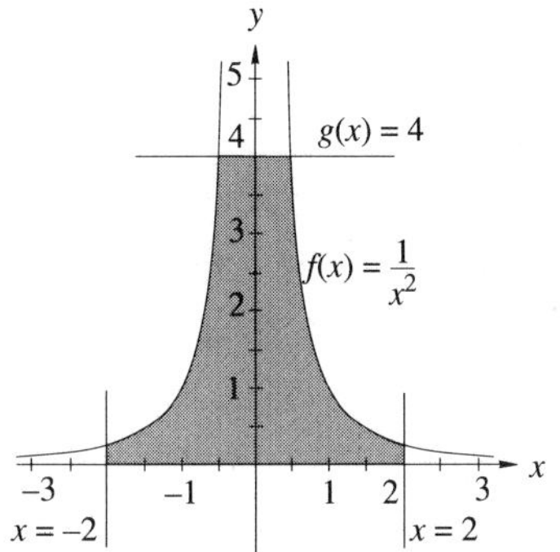

The graphs of f and g intersect when $\frac{1}{x^2} = 4$, or when $x^2 = \frac{1}{4} \Rightarrow x = \pm\frac{1}{2}$. Thus, the desired area is

$$\int_{-2}^{-1/2}\frac{1}{x^2}\,dx + \int_{-1/2}^{1/2}4\,dx + \int_{1/2}^{2}\frac{1}{x^2}\,dx$$
$$= -x^{-1}\Big|_{-2}^{-1/2} + 4x\Big|_{-1/2}^{1/2} + -x^{-1}\Big|_{1/2}^{2}$$
$$= 2 - \frac{1}{2} + 2 - (-2) - \frac{1}{2} + 2 = 7$$

41.

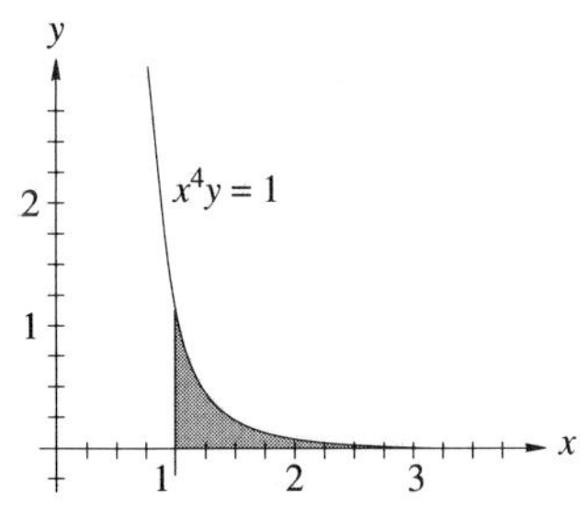

$x^4y = 1 \Rightarrow y = \frac{1}{x^4} = x^{-4}$; $\lim_{b\to\infty}\int_1^b x^{-4} =$

$$\lim_{b\to\infty} -\frac{1}{3}x^{-3}\Big|_1^b$$
$$\lim_{b\to\infty}\left(\frac{-1}{3b^3} + \frac{1}{3}\right) = \frac{1}{3}$$

42.

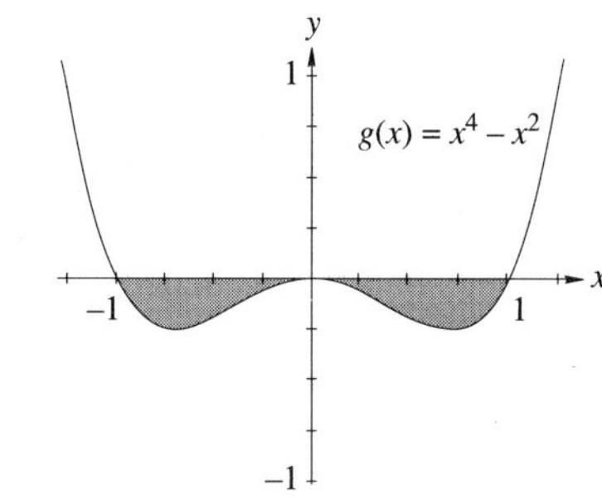

$$-\int_{-1}^{1}(x^4 - x^2)\,dx$$
$$= -\left(\frac{x^5}{5} - \frac{x^3}{3}\right)\Big|_{-1}^{1}$$
$$= -\left(\frac{-1}{5} - \frac{-1}{3}\right) + \left(\frac{-1}{5} - \frac{-1}{3}\right) = \frac{4}{15}$$

43. The area of the blue is $\int_0^1 (\sqrt{x} - x^3)\,dx = \frac{2}{3}x^{3/2} - \frac{x^4}{4}\Big|_0^1 = \frac{2}{3} - \frac{1}{4} = \frac{5}{12}$. The total area is 1 so the area of the white is $1 - \frac{5}{12} = \frac{7}{12}$. More white enamel is needed.

25.5 NUMERICAL INTEGRATION

1. $\int_1^2 x^4\,dx;\ n=6;\ \Delta x=\frac{1}{6}$

(a) $\frac{1}{2}\left[1^4+2\cdot\left(\frac{7}{6}\right)^4+2\left(\frac{8}{6}\right)^4+2\left(\frac{9}{6}\right)^4+2\left(\frac{10}{6}\right)^4+2\left(\frac{11}{6}\right)^4+2^4\right]\cdot\frac{1}{6}=\frac{1}{12}(75.17746914)=6.2647891$

(b) $\frac{1}{3}\left[1^4+4\cdot\left(\frac{7}{6}\right)^4+2\left(\frac{8}{6}\right)^4+4\left(\frac{9}{6}\right)^4+2\left(\frac{10}{6}\right)^4+4\left(\frac{11}{6}\right)^4+2^4\right]\cdot\frac{1}{6}=\frac{1}{18}(111.6018519)=6.20010286$

(c) $\int_1^2 x^4\,dx=\left.\frac{x^4}{5}\right|_1^2=\frac{32}{5}-\frac{1}{5}=\frac{31}{5}=6.2$

2. $\int_1^2 2x^5\,dx;\ n=8;\ \Delta x=\frac{1}{4}$

(a) $\frac{1}{2}[0+2\cdot 2\left(\frac{1}{4}\right)^5+2\cdot 2\left(\frac{1}{2}\right)^5+4\left(\frac{3}{4}\right)^5+4(1)^5+4\left(\frac{5}{4}\right)^5+4\left(\frac{3}{2}\right)^5+4\left(\frac{7}{4}\right)^5+2(2)^5]\cdot\frac{1}{4}=\frac{1}{8}(177.3125)=22.1640625$

(b) $\frac{1}{3}[0+4\cdot 2\left(\frac{1}{4}\right)^5+2\cdot 2\left(\frac{1}{2}\right)^5+4\cdot 2\cdot\left(\frac{3}{4}\right)^5+2\cdot 2\cdot 1^5+4\cdot 2\left(\frac{5}{4}\right)^5+2\cdot 2\left(\frac{3}{2}\right)^5+4\cdot 2\cdot\left(\frac{7}{4}\right)^5+2(2)^5]\cdot\frac{1}{4}=\frac{1}{12}(256.125)=21.34375$

(c) $\int_1^2 2x^5\,dx=\left.\frac{1}{3}x^6\right|_0^2=\frac{2^6}{3}=\frac{64}{3}=21\frac{1}{3}.$

3. $\int_0^1 \sqrt{x}\,dx;\ n=4;\ \Delta x=\frac{1}{4}$

(a) $\frac{1}{2}\left(0+2\sqrt{\frac{1}{4}}+2\sqrt{1/2}+2\sqrt{\frac{3}{4}}+1\right)\cdot\frac{1}{4}=\frac{1}{8}(5.14626437)=0.643283$

(b) $\frac{1}{3}\left(0+4\sqrt{\frac{1}{4}}+2\sqrt{1/2}4\sqrt{\frac{3}{4}}+1\right)\cdot\frac{1}{4}=\frac{1}{12}(7.878315178)=0.6565263$

(c) $\int_0^1 x^{1/2}\,dx=\left.\frac{2}{3}x^{3/2}\right|_0^1=\frac{2}{3}=0.6667$

4. $\int_1^2 x^{1/3}\,dx$

(a) 1.139241, (b) 1.139875, (c) $\int_1^2 x^{1/3}\,dx=\left.\frac{3}{4}x^{\frac{4}{3}}\right|_1^2=\frac{3}{4}\cdot 2^{4/3}-\frac{3}{4}=1.139882$

5. $\int_1^5 \frac{dx}{x},\ n=8$

(a) 1.628968, (b) 1.610847

6. $\int_1^4 \frac{dx}{x^2};\ n=6$

(a) 0.789219, (b) 0.754838, actual value $\int_1^4 x^{-2}\,dx=-x^{-1}\big|_1^4=-\frac{1}{4}+1=0.75$

7. $\int_1^5 \frac{dx}{1+x},\ n=8$

(a) 1.1032107, (b) 1.098726

8. $\int_0^2 x^{\pi}\,dx;\ n=4$

(a) 4.550118, (b) 4.262661 actual $\left.\frac{1}{\pi+1}x^{\pi+1}\right|_0^2=\frac{2^{\pi+1}}{\pi+1}=4.261635$

9. $\int_0^1 \frac{4}{1+x^2}\,dx,\ n=10$

(a) 3.139926; (b) 3.1415926 (Note: The actual value is π.)

10. $\int_0^2 \sqrt{1+x^3}\,dx,\ n=8$

(a) 6.521610, (b) 6.391210

11. $\int_0^2 \sqrt{1+x^3}\,dx;\ n=8$

(a) 3.251744, (b) 3.241238

12. $\int_0^{0.2} \sin x\,dx;\ n=4$

(a) 0.019929, (b) 0.019933

13. $\int_{-2}^{2} \sqrt{x+4}\,dx, n = 8$
(a) 7.909233, (b) 7.912321

14. $\int_{1}^{3} \ln x\,dx, n = 8$
(a) 1.292375, (b) 1.295798

15. $\int_{0}^{4} \frac{\sqrt{x}}{1+\sqrt{x}}\,dx; n = 8$
(a) 2.141030, (b) 2.170342

16. $\int_{0}^{1} \frac{e^x + e^{-x}}{2}\,dx; n = 4$
(a) 1.181316, (b) 1.175227

17. (a) $\Delta x = 1$; $\frac{1}{2}(4.2+2(3.9)+2(3.8)+2(4.0)+2(3.5)+2(3.4)+2(3.9)+2(4.1)+4.3) = \frac{1}{2}(61.7) = 30.85$
(b) $\frac{1}{3}(4.2+4(3.9)+2(3.8)+4(4.0)+2(3.5)+4(3.4)+2(3.9)+4(4.1)+4.3) = \frac{1}{3}(92.5) = 30.83333$

18. $\Delta x = 0.1$
(a) $\frac{1}{2}(4.32+2(4.57)+2(5.14)+2(5.78)+2(6.84)+2(6.62)+6.51)(0.1) = \frac{1}{2}(68.73)(0.1) = 3.4365$
(b) $\frac{1}{3}(4.32+4(4.57)+2(5.14)+4(5.78)+2(6.84)+4(6.62)+6.51)(0.1) = \frac{1}{3}(102.67)(0.1) = 3.4223$

19. (a) $\Delta x = 0.25$; $\frac{1}{2}(16.32+2(16.48)+2(16.73)+2(16.42)+2(16.38)+2(16.29)+16.25)(0.25) = \frac{1}{2}(197.17)(0.25) = 24.64625$
(b) $\frac{1}{3}(16.32+4(16.48)+2(16.73)+4(16.42)+2(16.38)+4(16.29)+16.25)(0.25) = \frac{1}{3}(295.55)(0.25) = 24.629167$

20. $\Delta x = 5$;
(a) $\frac{1}{2}(0+2\cdot21+2\cdot18+2\cdot17+2\cdot18+2\cdot14+0)\cdot5 = \frac{5}{2}(142) = 355\text{ m}^2$
(b) $\frac{1}{3}(0+4\cdot21+2\cdot18+4\cdot17+2\cdot18+4\cdot14+0)\cdot5 = \frac{5}{3}(280) = 466.7\text{ m}^2$

21. $\int_{1}^{2} 4.0e^{-10t} - 4.0e^{-20t}\,dt; \Delta t = \frac{1}{4}$
(a) $2.676 \times 10^{-5}C = 26.76\ \mu C$
(b) $20.34\ \mu C$

22. $\Delta x = 5$ milliseconds moving from 10 to 60.
(a) $\frac{1}{2}[0.3+2(0.7)+2(0.8)+2(0.7)+2(0.8)+2(0.7)+2(0.5)+2(0.4)+2(0.2)+2(0.2)+0.1](5) = \frac{5}{2}(10.8) = 27$
(b) $\frac{5}{3}[0.3+4(0.7)+2(0.8)+4(0.7)+2(0.8)+4(0.7)+2(0.5)+4(0.4)+2(0.2)+4(0.2)+0.1] = \frac{5}{3}(16.2) = 27$

23. See *Computer Programs* in main text.

CHAPTER 25 REVIEW

1. $\int x^5\,dx = \frac{1}{6}x^6 + C$

2. $\int (\sqrt[3]{x}+4)\,dx = \int (x^{1/3}+4))\,dx = \frac{3}{4}x^{4/3} + 4x + C$

3. $\int_{0}^{3} (x^2+4x-6)\,dx = \frac{1}{3}x^3 + 2x^2 - 6x\Big|_0^3 = 9 + 18 - 18 = 9$

4. $\int_{1}^{4} (x+4)^2\,dx = \int_{1}^{4} (x^2+8x+16)\,dx =$

$\frac{1}{3}x^3 + 4x^2 + 16x\Big|_1^4 = \left(\frac{64}{3} + 64 + 64\right) - \left(\frac{1}{3} + 4 + 16\right) = 129$

5. $\int_1^8 \frac{dt}{\sqrt[3]{t}} = \int_1^8 t^{-1/3}\,dt = \frac{3}{2}t^{2/3}\Big|_1^8 = 6 - \frac{3}{2} = \frac{9}{2}$

6. $\int_1^4 (\sqrt{x} + 4)\,dx = \frac{2}{3}x^{3/2} + 4x\Big|_1^4 = \left(\frac{16}{3} + 16\right) - \left(\frac{2}{3} + 4\right) = \frac{14}{3} + 12 = 16\frac{2}{3} = \frac{50}{3}$

7. $\int \frac{dt}{(t+5)^4} = \int (t+5)^{-4}\,dt = -\frac{1}{3}(t+5)^{-3} + C$

8. $\int \frac{x\,dx}{\sqrt[3]{x^2+4}} = \int (x^2+4)^{-1/3}x\,dx$. Let $u = x^2 + 4$, and then $du = 2x\,dx$, or $\frac{1}{2}\,du = x\,dx$. Substitution produces $\frac{1}{2}\int u^{-1/3}\,du = \frac{1}{2} \cdot \frac{3}{2}u^{2/3} + C = \frac{3}{4}(x^2+4)^{2/3} + C$.

9. $\int_1^8 x\sqrt[3]{x}\,dx = \int_1^8 x^{4/3}\,dx = \frac{3}{7}x^{7/3}\Big|_1^8 = \frac{3}{7} \cdot 2^7 - \frac{3}{7} = \frac{381}{7} \approx 54.429.$

10. $\int_1^4 \frac{4\,dx}{3\sqrt{x^3}} = \frac{4}{3}\int_1^4 x^{-3/2}\,dx = \frac{4}{3} \cdot (-2)x^{-1/2}\Big|_1^4 = -\frac{8}{3}\left(\frac{1}{2} - 1\right) = -\frac{8}{3} \cdot \frac{-1}{2} = \frac{4}{3}$

11. $\int (u^4 - u^{-4})\,du = \frac{1}{5}u^5 + \frac{1}{3}u^{-3} + C$

12. $\int \frac{x^4 - 1}{x^3}\,dx = \int (x - x^{-3})\,dx = \frac{x^2}{2} + \frac{1}{2}x^{-2} + C = \frac{x^2}{2} + \frac{1}{2x^2} + C$

13. $\int 2x\sqrt{x^2 - 5}\,dx$. Let $u = x^2 - 5$, then $du = 2x\,dx$. Thus, $\int u^{1/2}\,du = \frac{2}{3}u^{3/2} + C = \frac{2}{3}(x^2-5)^{3/2} + C$

14. $\int_0^4 x\sqrt{x^2+9}\,dx$. Let $u = x^2 + 9$, then $du = 2x\,dx$, or $\frac{1}{2}\,du = x\,dx$. For the limits of integration, we have $x = 0 \Rightarrow u = 9$ and $x = 4 \Rightarrow u = 25$. Thus, $\frac{1}{2}\int_9^{25} u^{1/2}\,du = \frac{1}{2} \cdot \frac{2}{3}u^{3/2}\Big|_9^{25} = \frac{1}{5}(125 - 27) = \frac{1}{3}(98) = \frac{98}{3} = 32\frac{2}{3}.$

15. $\int x\sqrt[3]{6 - x^2}\,dx$. Let $u = 6 - x^2$, then $du = -2x\,dx$, or $-\frac{1}{2}\,du = x\,dx$. Substitution yields $-\frac{1}{2}\int u^{1/3}\,du = -\frac{1}{2}\cdot\frac{3}{4}u^{4/3} + C = \frac{-3}{8}(6-x^2)^{4/3} + C.$

16. $\int 3x(4 - x^2)^3\,dx$. Let $u = 4 - x^2$, then $du = -2x\,dx$, or $3x\,dx = -\frac{3}{2}\,du$. Substituting, we have $-\frac{3}{2}\int u^3\,du = -\frac{3}{2} \cdot \frac{1}{4}u^4 + C = \frac{-3}{8}(4 - x^2)^4 + C.$

17. $\int_{-1}^2 (x^2+4)(x^3-3)\,dx = \int_{-1}^2 (x^5 + 4x^3 - 3x^2 - 12)\,dx = \frac{x^6}{6} + x^4 - x^3 - 12x\Big|_{-1}^2 = \left(\frac{64}{6} + 16 - 8 - 24\right) - \left(\frac{1}{6} + 1 + 1 + 12\right) = -\frac{39}{2}$

18. $\int_1^2 x(x+2)^2\,dx = \int_1^2 (x^3 + 4x^2 + 4x)\,dx = \frac{x^4}{4} + \frac{4}{3}x^3 + 2x^2\Big|_1^2 = \left(4 + \frac{32}{3} + 8\right) - \left(\frac{1}{4} + \frac{4}{3} + 2\right) = 19\frac{1}{12}$

19. $\int \frac{y^2\,dy}{(y^3-5)^{2/3}}$; Let $u = y^3 - 5$, then $du = 3y^2\,dy$, or $\frac{1}{3}\,du = y^3\,dy$. Substitution produces $\frac{1}{3}\int u^{-2/3}\,du = \frac{1}{3} \cdot \frac{3}{1}u^{1/3} + C = (y^3-5)^{1/3} + C.$

20. $\int \frac{5-6x+x^2}{1-x}\,dx = \int \frac{(5-x)(1-x)}{1-x}\,dx = \int (5-x)\,dx = 5x - \frac{x^2}{2} + C, x \neq 1$

21. $\int_0^1 (1-x^2)^3\,dx = \int_0^1 (1-3x^2+3x^4-x^6)\,dx = \left(x - x^3 + \frac{3}{5}x^5 - \frac{1}{7}x^7\right)\Big|_0^1 = 1 - 1 + \frac{3}{5} - \frac{1}{7} = \frac{21}{35} - \frac{5}{35} = \frac{16}{35}$

22. $\int \frac{x+2}{(x^2+4x)}\,dx$. Let $u = x^2 + 4x$, then $du = (2x+4)\,dx$, or $\frac{1}{2}du = (x+2)\,dx$. Thus, $\frac{1}{2}\int u^{-2}\,du = \frac{1}{2}\cdot(-1)u^{-1} + C = -\frac{1}{2}(x^2+4x)^{-1} + C$ or $\frac{-1}{2(x^2+4x)} + C$.

23. $\int \left(\sqrt{2x} - \frac{1}{\sqrt{2x}}\right) dx = \sqrt{2}\cdot\frac{2}{3}x^{3/2} - \frac{1}{\sqrt{2}}\cdot\frac{2}{1}x^{1/2} + C = \frac{2\sqrt{2}}{3}x^{3/2} - \sqrt{2}x^{1/2} + C = \frac{1}{3}(2x)^{3/2} - (2x)^{1/2} + C$

24. $\int (x^3-4x)(3x^2-4)\,dx$. Let $u = x^3 - 4x$, then $du = (3x^2-4)\,dx$. Thus, $\int u\,du = \frac{u^2}{2} + C = \frac{1}{2}(x^3-4x)^2 + C$.

25. $\int_1^4 \frac{9-t^3}{3t^2}\,dt = \int_1^4 \left(3t^{-2} - \frac{1}{3}t\right) dt = \left(-3t^{-1} - \frac{1}{3}\cdot\frac{1}{2}t^2\right)\Big|_1^4 = \left(\frac{-3}{4} - \frac{8}{3}\right) - \left(-3 - \frac{1}{2}\right) = -\frac{1}{4}$

26. $\int_0^1 \sqrt{x}\,(3\sqrt{x}-5)\,dx = \int_0^1 (3x - 5\sqrt{x})\,dx = \frac{3}{2}x^2 - 5\cdot\frac{2}{3}x^{3/2}\Big|_0^1 = \frac{3}{2} - \frac{10}{3} = -\frac{11}{6}$

27. $\int_{-1}^8 \frac{4\sqrt[3]{u}\,du}{(1+u^{4/3})3}$. Let $v = 1 + u^{4/3}$, then $dv = \frac{4}{3}u^{1/3}\,du$, or $3dv = 4u^{1/3}\,du$. The limits of integration become $u = -1 \Rightarrow v = 2$ and $u = 8 \Rightarrow v = 17$. Thus, we have $3\int_2^{17} v^{-3}dv = 3\cdot\frac{-1}{2}v^{-2}\Big|_2^{17} = \frac{-3}{2\cdot 17^2} + \frac{3}{2\cdot 2^2} \approx -0.00519 + 0.375 = 0.36981$.

28. $\int_1^2 \frac{3x^3-2x^{-3}}{x^2}\,dx = \int_1^2 (3x - 2x^{-5})\,dx = \left(\frac{3}{2}x^2 + \frac{1}{2}x^{-4}\right)\Big|_1^2 = \left(6 + \frac{1}{32}\right) - \left(\frac{3}{2} + \frac{1}{2}\right) = 6\frac{1}{32} - 2 = 4\frac{1}{32}$

29. $\int_1^\infty \frac{dx}{x^4} = \lim_{b\to\infty}\int_1^b x^{-4}\,dx = \lim_{b\to\infty}\left(\frac{-1}{3}x^{-3}\right)\Big|_1^b = \lim_{b\to\infty}\left(\frac{-1}{3b^3} + \frac{1}{3}\right) = \frac{1}{3}$

30. $\int_0^\infty \frac{4x\,dx}{(3x^2+1)^5}$; Let $u = 3x^2+1$, then $du = 6x\,dx$, or $4x\,dx = \frac{2}{3}du$. For the limits of integration, we obtain $x = 0 \Rightarrow u = 1$ and $x \to \infty \Rightarrow u \to \infty$. Thus, $\frac{2}{3}\int_1^\infty u^{-5}\,du = \frac{2}{3}\left(-\frac{1}{4}\right)u^{-4}\Big|_1^\infty = \lim_{b\to\infty}\left(\frac{-1}{6b^4} + \frac{1}{6}\right) = \frac{1}{6}$.

31. $f(x) = 4x - 6 = 0$; $4x = 6$; $x = \frac{3}{2}$. Thus, $-\int_{-3}^{3/2}(4x-6)\,dx + \int_{3/2}^4 (4x-6)\,dx = -(2x^2-6x)\Big|_{-3}^{3/2} + (2x^2-6x)\Big|_{3/2}^4 = -\left(\frac{9}{2} - 9\right) + (18+18) + (32-24) - \left(\frac{9}{2} - 9\right) = \frac{9}{2} + 36 + 8 + \frac{9}{2} = 53$.

32. $g(x) = 2 - x - x^2 = (2+x)(1-x)$. This crosses the x-axis at -2 and 1. $\int_{-2}^1 (2-x-x^2)\,dx = 2x - \frac{x^2}{2} - \frac{x^3}{3}\Big|_{-2}^1 =$

$\left(2-\frac{1}{2}-\frac{1}{3}\right)-\left(-4-2+\frac{8}{3}\right)=2-\frac{5}{6}+6-\frac{8}{3}=$
$8-\frac{21}{6}=8-\frac{7}{2}=\frac{9}{2}=4\frac{1}{2}$ or 4.5.

33.

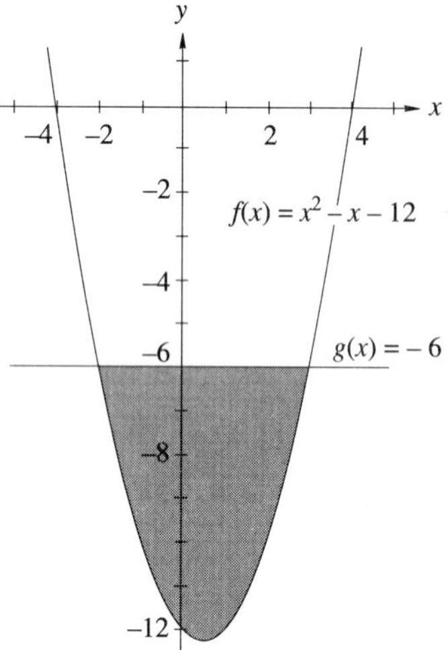

The graphs will intersect when $f(x) = g(x)$ or when $x^2-x-12 = -6$. This can be rewritten as $x^2-x-6 = 0$ which factors as $(x-3)(x+2) = 0$. The graph intersect at $x = -2$ and $x = 3$. Thus, the desired area is $\left|\int_{-2}^{3}(x^3-x-6)\,dx\right| = \left|\frac{x^3}{3}-\frac{x^2}{2}-6x\Big|_{-2}^{3}\right| = \left|\left(9-\frac{9}{2}-18\right)-\left(\frac{-8}{3}-2+12\right)\right| = \left|-\frac{125}{6}\right| = \frac{125}{6}$.

34. $h(x) = \sqrt{x+2} = 0$ at $x = -2$. Thus, $\int_{-2}^{7}\sqrt{x+2}\,dx = \frac{2}{3}(x+2)^{3/2}\Big|_{-2}^{7} = \frac{2\cdot 27}{3} - 0 = 18$.

35. $4x^3 = x \Rightarrow 4x^3 - x = 0 \Rightarrow x(4x^2-1) = 0 \Rightarrow x = 0, -\frac{1}{2}, \frac{1}{2}$.

$$\int_{-1/2}^{0}(4x^2-x)\,dx + \int_{0}^{1/2}(x-4x^3)\,dx$$
$$= x^4 - \frac{x^2}{2}\Big|_{-1/2}^{0} + \frac{x^2}{2} - x^4\Big|_{0}^{1/2}$$
$$= -\left(\frac{1}{16}-\frac{1}{8}\right)+\left(\frac{1}{8}-\frac{1}{16}\right) = \frac{1}{8}$$

36.

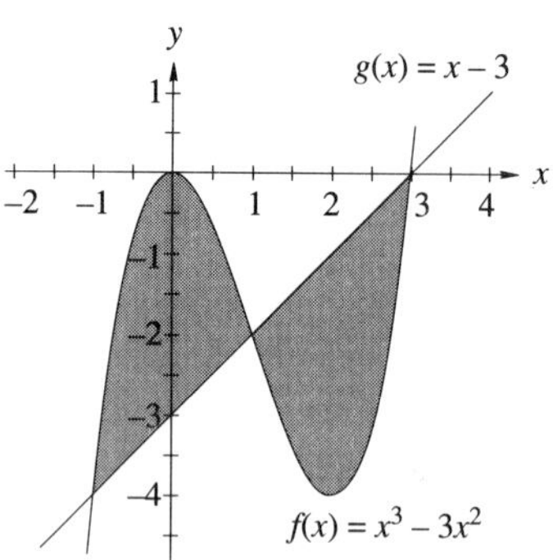

The graphs intersect when $f(x) = g(x)$ or $x^3 - 3x^2 = x - 3$. This can be rewritten as $x^2(x-3) = (x-3)$, or $x^2(x-3)-(x-3) = 0$, or $(x^2-1)(x-3) = 0$. Thus, we see that the graphs intersect at -1, 1 and 3. The desired area is

$$\int_{-1}^{1}(x^3-3x^2-x+3)\,dx + \int_{1}^{3}(x-3-x^3+3x^2)\,dx$$
$$= \left(\frac{x^4}{4}-x^3-\frac{x^2}{2}+3x\right)\Big|_{-1}^{1}$$
$$+ \left(\frac{x^2}{2}-3x-\frac{x^4}{4}+x^3\right)\Big|_{1}^{3}$$
$$= \left(\frac{1}{4}-1-\frac{1}{2}+3\right)-\left(\frac{1}{4}+1-\frac{1}{2}-3\right)$$
$$+ \left(\frac{9}{2}-9-\frac{81}{4}+27\right)-\left(\frac{1}{2}-3-\frac{1}{4}+1\right)$$
$$= \left(1\frac{3}{4}\right)-\left(-2\frac{1}{4}\right)+\left(2+\frac{1}{4}\right)-\left(-1\frac{3}{4}\right)$$
$$= 8$$

37. $y^2 + x = 0$; $y = 2x + 1$; $(2x+1)^2 + x = 0$; $4x^2+4x+1+x = 0$; $4x^2+5x+1 = 0$; $(4x+1)(x+1)$
The graphs intersect at $(-1,-2)$ and $(-\frac{1}{4}, \frac{1}{2})$

$$\int_{-1}^{-1/4}(2x+1+\sqrt{-x})\,dx$$
$$+ \int_{-1/4}^{0} 2\sqrt{-x}\,dx$$

$$= \left(x^2 + x - \frac{2}{3}(-x)^{3/2}\right)\Bigg|_{-1}^{-1/4} + \left(-2 \cdot \frac{2}{3}(-x)^{3/2}\right)\Bigg|_{-1/4}^{0}$$
$$= \left(\frac{1}{16} - \frac{1}{4} - \frac{1}{12}\right) - \left(1 - 1 - \frac{2}{4}\right) + 0 + \frac{1}{6}$$
$$= \frac{9}{16} = 0.5625$$

38. $2x^3 = x^4; x^4 - 2x^3 = 0; x^3(x-2) = 0$ they intersect at $(0, 0)$ and $(2, 16)$ $\int_0^2 (2x^3 - x^4)\,dx = \frac{1}{2}x^4 - \frac{x^5}{5}\Big|_0^2 = 8 - \frac{32}{5} = \frac{8}{5} = 1\frac{3}{5} = 1.6$

39. $\int_0^1 \sqrt{1-x^2}\,dx$; $n = 10$ (a) 0.776130, (b) 0.781752

40. $\int_1^4 \frac{1}{3x}\,dx$; $n = 6$ (a) 0.468452, (b) 0.462566

41. $\int_2^{2.5} \sqrt{x^3 - 1}\,dx$; $n = 10$ (a) 1.613695, (b) 1.613656

42. $\int_{-1}^{4} \sqrt{1 + x^5}\,dx$; $n = 10$ (a) 38.795198, (b) 38.464568

43. $\int_0^1 (e^x - e^{-x})\,dx$; $n = 4$(a) 1.091812, (b) 1.086185

44. (a) $\frac{1}{2}(4.25 + 2(5.72) + 2(5.13) + 2(3.19) + 2(2.10) + 2(0.15) + 2(1.65) + 2(3.10) + 3.70)(0.25) = \frac{1}{8}(50.03) = 6.25375$

(b) $\frac{1}{3}(4.25 + 4(5.72) + 2(5.13) + 4(3.19) + 2(2.10) + 4(0.15) + 2(1.65) + 4(3.10) + 3.70)(0.25) = \frac{1}{12}(74.35) = 6.195833$

45. (a) $\int_0^{15} \left(t + \frac{1}{\sqrt{1+t}}\right) dt = \frac{t^2}{2} + 2\sqrt{1+t}\Big|_0^{15} = \frac{225}{2} + 2\sqrt{16} - 2 = 112.5 + 8 - 2 = 118.5$ m (b) The acceleration is given by $a(t) = v'(t) = 1 - \frac{1}{2}(1+t)^{-3/2}$. Thus, the acceleration at $t = 15$ is $a(15) = 1 - \frac{1}{2}(16)^{-3/2} = 1 - \frac{1}{128} = \frac{127}{128} \approx 0.992$ m/s^2.

46. $\omega = \int_3^5 10t^3\,dt = \frac{10}{4}t^4\Big|_0^5 = \frac{5}{2}(625 - 81) = \frac{5}{2}(544) = 1360$ J

47. $2.0 - 1.0t^2 = -2\frac{di}{dt} \Rightarrow \frac{di}{dt} = \frac{1}{2}t^2 - 1$. Hence, $i = \int_1^6 \left(\frac{1}{2}t^2 - 1\right) dt = \left(\frac{1}{6}t^3 - t\right)\Big|_1^6 = (36 - 6) - \left(\frac{1}{6} - 1\right) = 30 + \frac{5}{6} = 30\frac{5}{6} \approx 30.8$ A.

48. (a) When the car stops its velocity is 0. Hence, we want to know when $v(t) = 88 - 4t = 0$, which is the same as $4t = 88$, and so $t = 22$ s (b) The distance the car travels is given by $s(t) = \int v(t) = 88t - 2t^2$, and so, the distance it travels in those 22 s is $s(22) = 88 \cdot 22 - 2 \cdot 22^2 = 968$ ft.

CHAPTER 25 TEST

1. $\int_0^1 2x\,dx = x^2\big|_0^1 = 1 - 0 = 1.$

2. $\int_1^4 (x+1)(x^2+1)\,dx = \int_1^4 (x^3 + x^2 + x + 1)\,dx = \left(\frac{x^4}{4} + \frac{x^3}{3} + \frac{x^2}{2} + x\right)\Big|_1^4 = \left(64 + \frac{64}{3} + 8 + 4\right) - \left(\frac{1}{4} + \frac{1}{3} + \frac{1}{2} + 1\right) = 95.25.$

3. $\int (x^2+3)^2\,dx = \int (x^4 + 6x^2 + 9)\,dx = \frac{1}{5}x^5 + 2x^3 + 9x + C.$

4. $\int 4(x^4+7)^5x^3\,dx$; Let $u = x^4+7$, then $du = 4x^3\,dx$. Thus, $\int u^5\,du = \frac{1}{6}u^6 + C = \frac{1}{6}(x^4+7)^6 + C$.

5. $\int \frac{x^2+2}{x^2}\,dx = \int (1+2x^{-2})\,dx = x - 2x^{-1} + C$.

6. $\int (x^2+8x)^3(x+4)\,dx$. Let $u = x^2+8x$, then $du = (2x+8)\,dx$, or $\frac{1}{2}\,du = (x+4)\,dx$. Substitution yields $\frac{1}{2}\int u^3\,du = \frac{1}{2}\cdot\frac{1}{4}u^4 + C = \frac{1}{8}(x^2+8x)^4 + C$.

7. $\int_1^2 \left(3x^2\sqrt{x^3+x} + \sqrt{x^3+x}\right)dx = \int_1^2 \sqrt{x^3+x}\,\times$ $(3x^2+1)\,dx$. Let $u = x^3+x$, then $du = (3x^2+1)\,dx$. When $x = 1 \Rightarrow u = 2$ and when $x = 2 \Rightarrow u = 10$. Thus, with these substitutions we have $\int_2^{10} \sqrt{u}\,du = \frac{2}{3}u^{3/2}\Big|_1^{10} = \frac{2}{3}(10^{3/2} - 2^{3/2}) \approx 19.196$.

8. $4x^3 - 4 = 4(x^3-1) = 4(x-1)(x^2+x+1)$. This crosses the x-axis at $x = 1$. $\int_{-1}^{1} (4-4x^3)\,dx +$ $\int_1^2 (4x^3-4)\,dx = \left(4x - x^4\right)\Big|_{-1}^{1} + \left(x^4 - 4x\right)\Big|_1^2 =$ $(4-1) - (-4-1) + (16-8) - (1-4) = 3+5+8+3 = 19$.

9. The graphs intersect at $(-3,7)$ and $(2,2)$. Thus, the desired area is $\int_{-3}^{2} \left[(4-x)-(x^2-2)\right]dx =$ $\int_{-3}^{2} (6-x-x^2)\,dx = \left(6x - \frac{x^2}{2} - \frac{x^3}{3}\right)\Big|_{-3}^{2} =$ $\left(12 - 2 - \frac{8}{3}\right) - \left(-18 - \frac{9}{2} + 9\right) = \frac{125}{6} \approx 20.833$.

10. $\int_2^4 \frac{1}{x}\,dx$; $n = 8$; $\Delta x = \frac{2}{8} = \frac{1}{4}$

The trapezoidal rule yields

$$\frac{1}{2}\left(\frac{1}{2} + \frac{2}{2.25} + \frac{2}{2.5} + \frac{2}{2.75} + \frac{2}{3} + \frac{2}{3.25} + \frac{2}{3.5} + \frac{2}{3.75} + \frac{1}{4}\right)\cdot\frac{1}{4} = \frac{1}{8}(5.55297) = 0.694122$$

Simpson's rule produces

$$\frac{1}{3}\left(\frac{1}{2} + \frac{4}{2.25} + \frac{2}{2.5} + \frac{4}{2.75} + \frac{2}{3} + \frac{4}{3.25} + \frac{2}{3.5} + \frac{4}{3.75} + \frac{1}{4}\right)\cdot\frac{1}{4} = \frac{1}{12}(8.31785) = 0.693154$$

CHAPTER

26

Applications of Integration

26.1 AVERAGE VALUES AND OTHER ANTIDERIVATIVE APPLICATIONS

1. $\bar{y} = \frac{1}{1}\int_0^1 x^2\,dx = \frac{1}{3}x^3\Big|_0^1 = \frac{1}{3}$

$f_{\text{rms}} = \sqrt{\frac{1}{1}\int_0^1 x^4\,dx} = \sqrt{\frac{1}{5}x^5\Big|_0^1} = \sqrt{\frac{1}{5}} = \frac{\sqrt{5}}{5} \approx 0.4472$

2. $\bar{y} = \frac{1}{4-2}\int_2^4 3x^2\,dx = \frac{1}{2}\left[x^3\right]_2^4 = \frac{1}{2}(64-8) = 28$

$g_{\text{rms}} = \sqrt{\frac{1}{2}\int_2^4 9x^4\,dx} = \sqrt{\frac{9}{2}\left(\frac{x^5}{5}\right)\Big|_2^4} = \sqrt{\frac{9}{2}\left(\frac{1024-32}{5}\right)} = \sqrt{892.8} = 29.8798$

3. $\bar{y} = \frac{1}{3-1}\int_1^3 (x^2-1)\,dx = \frac{1}{2}\left[\frac{x^3}{3}-x\right]_1^3 = \frac{1}{2}\left[(9-3)-\left(\frac{1}{3}-1\right)\right] = \frac{1}{2}\left[\frac{20}{3}\right] = \frac{10}{3}$

$h_{\text{rms}} = \sqrt{\frac{1}{2}\int_1^3 (x^2-1)^2\,dx} = \sqrt{\frac{1}{2}\int_1^3 (x^4-2x^2+1)\,dx} = \sqrt{\frac{1}{2}\left[\frac{x^5}{5}-\frac{2x^3}{3}+x\right]_1^3} = \sqrt{\frac{1}{2}\left[\left(\frac{243}{5}-\frac{54}{3}+3\right)-\left(\frac{1}{5}-\frac{2}{3}+1\right)\right]} = \sqrt{\frac{248}{15}} = \sqrt{16.53} \approx 4.0661$

4. $\bar{y} = \frac{1}{3+3}\int_{-3}^3 (9-x^2)\,dx = \frac{1}{6}\left[9x-\frac{x^3}{3}\right]_{-3}^3 = \frac{1}{6}\left[(27-9)-(-27+9)\right] = \frac{1}{6}(36) = 6.$

$j_{\text{rms}} = \sqrt{\frac{1}{6}\int_{-3}^3 (9-x^2)^2\,dx} = \sqrt{\frac{1}{6}\int_{-3}^3 (81-18x^2+x^4)\,dx} = \sqrt{\frac{1}{6}\left[81x-6x^3-\frac{x^5}{5}\right]_1^3} = \sqrt{43.2} \approx 6.5727$

5. $\bar{y} = \int_0^5 \frac{1}{5}\sqrt{x+4}\,dx = \frac{1}{5}\left[\frac{2}{3}(x+4)^{3/2}\right]_0^5 = \frac{2}{15}\left[(27-8)\right] = \frac{38}{15}.$

$f_{\text{rms}} = \sqrt{\frac{1}{5}\int_0^5 (x+4)\,dx} = \sqrt{\frac{1}{6}\left[\frac{x^2}{2}+4x\right]_0^5} = \sqrt{\frac{1}{5}\left[\frac{25}{2}+20\right]} = \sqrt{\frac{5}{2}+4} = \sqrt{\frac{13}{2}} = 2.5495$

6. $\bar{y} = \frac{1}{3-1}\int_1^3 (x^3-1)\,dx = \frac{1}{2}\left[\frac{x^4}{4}-x\right]_1^3 = \frac{1}{2}\left[\frac{81}{4}-3-\frac{1}{4}+1\right] = \frac{1}{2}[18] = 9.$

$k_{\text{rms}} = \sqrt{\frac{1}{2}\int_1^3 (x^6 - 2x^3 + 1)\,dx} =$
$\sqrt{\frac{1}{2}\left[\frac{x^7}{7} - \frac{x^4}{2} + x\right]_1^3} =$
$\sqrt{\frac{1}{2}\left[\frac{2187}{7} - \frac{81}{2} + 3 - \frac{1}{7} + \frac{1}{2} - 1\right]} = \sqrt{\frac{960}{7}} =$
$\sqrt{137.1428} \approx 11.7108$

7. $\bar{y} = \frac{1}{4}\int_0^4 (t^2 + 4)\,dt = \frac{1}{4}\left[\frac{t^3}{3} + 4t\right]_0^4 =$
$\frac{1}{4}\left[\frac{64}{3} + 16\right] = \frac{16}{3} + 4 = 9\frac{1}{3} = \frac{28}{3}$
$f_{\text{rms}} = \sqrt{\frac{1}{4}\int_0^4 (t^4 + 8t^2 + 16)\,dt} =$
$\sqrt{\frac{1}{4}\left[\frac{t^5}{5} + \frac{8t^3}{3} + 16t\right]_0^4} =$
$\sqrt{\frac{1}{4}\left[\frac{1024}{5} + \frac{512}{3} + 64\right]} = \sqrt{\frac{1648}{15}} =$
$\sqrt{109.8667} \approx 10.4817$

8. $\bar{y} = \frac{1}{4-2}\int_2^4 (16t^2 - 25)\,dt = \frac{1}{2}\left[\frac{16t^3}{3} - 25t\right]_2^4 =$
$\frac{1}{2}\left(\frac{1024}{3} - 100\right) - \left(\frac{128}{3} - 50\right) = \frac{1}{2}\left[248\frac{2}{3}\right] =$
124.3333
$g_{\text{rms}} = \sqrt{\frac{1}{2}\int_2^4 (256t^4 - 800t^2 + 625)\,dt} =$
$\sqrt{\frac{1}{2}\left[\frac{256t^5}{5} + \frac{800t^3}{3} + 625t\right]_0^4} = \sqrt{18553.53} =$
136.2114

9. $\bar{y} = \frac{1}{1}\int_1^2 (6x - x^2)\,dx = \left[3x^2 - \frac{x^3}{3}\right]_1^2 =$
$\left(12 - \frac{8}{3}\right) - \left(3 - \frac{1}{3}\right) = \frac{20}{3}$
$f_{\text{rms}} = \sqrt{\int_1^2 (36x^2 - 12x^3 + x^4)\,dx} =$
$\sqrt{\left[12x^3 - 3x^4 + \frac{x^5}{5}\right]_1^2} = \sqrt{45.2} \approx 6.7231$

10. $\bar{y} = \frac{1}{3}\int_0^3 (16t^2 - 8t + 1)\,dt =$
$\frac{1}{3}\left[\frac{16t^2}{3} - 4t^2 + t\right]_0^3 = \frac{1}{3}[144 - 36 + 3] = 37$
$h_{\text{rms}} =$
$\sqrt{\frac{1}{3}\int_0^3 (256t^4 - 256t^3 + 96t^2 - 16t + 1)\,dt} =$
$\sqrt{\frac{1}{3}\left[\frac{256t^5}{5} - 64t^4 + 32t^3 - 8t^2 + t\right]_0^3} =$
$\sqrt{2684.2} \approx 51.8093$

11. $\bar{y} = \frac{1}{2.5}\int_0^{2.5} (4.9t^2 - 2.8t + 4)\,dt =$
$\frac{1}{2.5}\left[\frac{4.9t^3}{3} - 1.4t^2 + 4t\right]_0^{2.5} = \frac{1}{2.5}(26.7708) =$
10.7083
$j_{\text{rms}} = \sqrt{\frac{1}{2.5}\int_0^{2.5} (4.9t^2 - 2.8t + 4)^2\,dt} =$
$\sqrt{\frac{1}{2.5}(415.9765)} = 12.8992$

12. $\bar{y} = \frac{1}{1.6}\int_0^{1.6} (4.9t^2 - 4.2t + 9)\,dt =$
$\frac{1}{1.6}(15.71413) = 9.8213$
$k_{\text{rms}} = \sqrt{\frac{1}{1.6}\int_0^{1.6} (4.9t^2 - 4.2t^3 + 9)^2\,dt} =$
$\sqrt{\frac{1}{1.6}(160.25496)} = 10.0080$

13. $\frac{1}{90}\int_0^{90} 0.0002(4991 + 366x - x^2)\,dx =$
$\frac{0.0002}{90}\left[4991x + 183x^2 - \frac{x^3}{3}\right]_0^{90} =$
$\frac{0.0002}{90}[1688490] = 3.7522$ cm for 90 days.
$\frac{0.0002}{365}\int_0^{365} (4991 + 366x - x^2)\,dx =$
$\frac{0.0002}{365}\left[4991x + 183x^2 - \frac{x^3}{3}\right]_0^{365} =$
$\frac{0.0002}{365}(9992848) = 5.4755$ cm for the year.

14. $\frac{1}{4}\int_0^4 (1 + x - \sqrt{x})\,dx = \frac{1}{4}\left[x + \frac{x^2}{2} - \frac{2}{3}x^{3/2}\right]\Big|_0^4 = \frac{1}{4}\left[4 + 8 - \frac{16}{3}\right] = 1\frac{2}{3} = 1.6667$ kg

15. $a(t) = -32$; $v(t) = -32t$; $s(t) = -16t^2 + 360$. Setting $s(t) = 0$, and solving for t, we obtain $-16t^2 + 360 = 0$, or $16t^2 = 360$, and so $t = 4.743416$. Thus, $\bar{s} = \frac{1}{4.74}\int_0^{4.74} (-16t^2 + 360)\,dt = \frac{1}{4.74}\left[-\frac{16}{3}t^3 + 360t\right]_0^{4.74} = 240$ ft. The average velocity is $\bar{v} = \frac{1}{4.74}\int_0^{4.74} (-32t)\,dt = \left[\frac{-16t^2}{4.74}\right]_0^{4.74} = -75.8947$ ft/s.

16. $V(t) = 200(100 - t^2)$. $\bar{V} = \frac{200}{10}\int_0^{10} (100 - t^2)\,dt = 20\left[100t - \frac{t^3}{3}\right]_0^{10} = 20\left(1000 - \frac{1000}{3}\right) = 13333.33$ L

17. (a) $\frac{1}{6}\int_0^6 (60 + 4t - t^{2/3})\,dt = \frac{1}{6}\left[60t + 2t^2 - \frac{3}{5}t^{5/3}\right]_0^6 = 60 + 2 \cdot 6 - \frac{3}{5}6^{2/3} = 70.018$ or 70°F

(b) $\frac{1}{6}\int_3^9 (60 + 4t - t^{2/3})\,dt = \frac{1}{6}\left[60t + 2t^2 - \frac{3}{5}t^{5/3}\right]_3^9 = \frac{1}{6}[484.3797] = 80.7$°F

18. $\frac{1}{30}\int_0^{30} (900 - 15\sqrt{120t})\,dt = \frac{1}{30}\left[900t - 10\sqrt{120}t^{3/2}\right]_0^{30} = \frac{1}{30}[27000 - 18000] = \frac{1}{30}[9{,}000] = 300$ cases

19. $\frac{1}{3}\int_0^3 (1.0t + \sqrt{t})\,dt = \frac{1}{3}\left[\frac{t^2}{2} + \frac{2}{3}t^{3/2}\right]_0^3 = \frac{3}{2} + \frac{2}{3} \cdot \sqrt{3} \approx 2.6547$ A.

20. $i_{\text{eff}} = \sqrt{\frac{1}{3}\int_0^3 \left(t + \sqrt{t}\right)^2 dt} = \sqrt{\frac{1}{3}\int_2^3 \left(t^2 + 2t^{3/2} + t\right) dt} = \sqrt{\frac{1}{3}\left[\frac{t^3}{3} + \frac{4}{5}t^{5/2} + \frac{t^2}{2}\right]_0^3} = \sqrt{3 + \frac{4}{5} \cdot 3^{3/2} + \frac{3}{2}} = \sqrt{8.6569} = 2.9423$ A.

21. $\frac{1}{4}\int_0^4 4t\sqrt{t^2 + 1} = \int_0^4 t\sqrt{t^2 + 1}$. Let $u = t^2 + 1$ and then, $du = 2t\,dt$. When $t = 0 \Rightarrow u = 1$, and when $t = 4 \Rightarrow u = 17$. Making these substitutions, we have $\frac{1}{2}\int_1^{17} \sqrt{u}\,du = \frac{1}{2} \cdot \frac{2}{3}u^{3/2}\Big|_1^{17} = \frac{1}{3}(69.092796) = 23.0309$ A.

22. $i_{\text{eff}} = \sqrt{\frac{1}{4}\int_0^4 \left[(4t)^2(t^2 + 1)\right] dt} = \sqrt{4\int_0^4 (t^4 + t^2)\,dt} = 2 \cdot \sqrt{\frac{t^5}{5} + \frac{t^3}{3}\Big|_0^4} = 2 \cdot \sqrt{\frac{1024}{5} + \frac{64}{3}} = 30.0755$ A

23. $P = (i_{\text{eff}})^2 \cdot R = (30.0755)^2 \cdot 30 = 27136$ W

24. $i_{\text{eff}} = \sqrt{\frac{1}{0.25}\int_0^{0.25} (t^3 - 2t^2)^2\,dt} = \sqrt{4\int_0^{0.25} (t^6 - 4t^5 + 4t^4)\,dt} = 2\sqrt{\frac{1}{7}t^7 - \frac{2}{3}t^6 + \frac{4}{5}t^5\Big|_0^{0.25}} = 2\sqrt{0.000\,627\,2}$. Thus, $(i_{\text{eff}})^2 = 4(0.000\,627\,2) \approx 0.002\,509$ and so, $P = i_{\text{eff}}^2 \cdot R = 0.002\,509 \cdot 5 = 0.012\,545 \approx 12.5$ mW.

25. We have $v = 4\sqrt{t} - 2t = 2(2\sqrt{t} - t)$.

(a) So, $v_{\text{eff}} = \sqrt{\frac{1}{6}\int_0^6 \left[2(2\sqrt{t} - t)\right]^2 dt} = \sqrt{\frac{2}{3}\int_0^6 (4t - 4t^{3/2} + t^2)\,dt} = \sqrt{\frac{2}{3}\left[2t^2 - \frac{8}{5}t^{5/2} + \frac{1}{3}t^3\right]_0^6} = \sqrt{\frac{2}{3}\left(72 - \frac{288}{5}\sqrt{6} + 72\right)} = \sqrt{96 - \frac{192}{5}\sqrt{6}} \approx \sqrt{1.939593877} \approx 1.3927$. So, the effective voltage is about 1.3927 V. (b) The average power is $P = i_{\text{eff}}^2 \cdot R = \left(\sqrt{1.939593877}\right)^2 (10) \approx 19.40$. Hence, the average power is about 19.40 W.

26. $V = 8.0\sqrt{t} - 16\sqrt{t^3} = 8\left(\sqrt{t} - 2\sqrt{t^3}\right) = 8\sqrt{t}(1 - 2t)$ and is greater than 0 from $t = 0$ to $t = \frac{1}{2}$ s. $(V_{\text{rms}})^2 = 2\int_0^{1/2}\left[8\left(\sqrt{t} - 2\sqrt{t^3}\right)\right]^2 dt = 2 \cdot 64\int_0^{1/2}\left(t - 4t^2 + 4t^3\right) dt = 128\left[\frac{t^2}{2} - \frac{4}{3}t^3 + t^4\right]_0^{1/2} = 128[0.02083] = 2.6667$. Hence, $P = (V_{\text{rms}})^2 R = 2.6667 \times 9.75 = 26.0$ W.

27. $q = \int_0^{0.1} 2t\,dt = t^2\Big|_0^{0.1} = 0.01$ C is the increase. The new charge is $0.01 + 0.01 = 0.02$ C.

28. $V_c = \frac{1}{C}\int i\,dt = \frac{1}{80 \times 10^{-6}}\int 0.04t^3\,dt = \frac{1}{80 \times 10^{-6}}(0.01t^4) + k$; $k = 100$, so $V_c = 125t^4 + 100$ $225 = 125t^5 + 100$ $125 = 125t$, $t^4 = 1$, so $t = 1$ s.

29. $V = 2t + 1$, $i = 0.03t$ W$=$ $\int P\,dt = \int Vi\,dt = \int (2t+1)(0.03t)\,dt = \int_0^{50}(0.06t^2 + 0.03t)\,dt = \left[0.02t^2 + \frac{0.03}{2}t^2\right]_0^{50} = 2537.5$ J

30. $V = \frac{1}{C}\int_0^{0.001} 0.2\,dt = \frac{1}{90 \times 10^{-6}} \cdot 0.0002 = 2.222$ V

31. $V = \frac{1}{C}\int 0.04t\,dt = \frac{1}{C}0.2t^2 + k$, so $V_c = \frac{0.2t^2}{7.5\times 10^{-6}}$. At $t = 0.005$, $V = \frac{0.2(0.005)^2}{7.5 \times 10^{-6}} = 0.6667$ V.

32. $q = \int i\,dt$; $q = \int_0^3 \sqrt[3]{1+5t}\,dt$. Let $u = 1 + 5t$, then $du = 5\,dt$. When $t = 0$, then $u = 1$ and when $t = 3$, then $u = 16$. Substituting, we get $\frac{1}{5}\int_1^{16} u^{1/3}\,du = \frac{1}{5}\cdot\frac{3}{4}u^{4/3}\Big|_1^{16} = \frac{3}{20}\left[16^{4/3} - 1^{4/3}\right] = 5.8976$ C.

26.2 VOLUMES OF REVOLUTION: DISK AND WASHER METHODS

1. $\pi\int_1^5 (4x)^2\,dx = \frac{16\pi}{3}r^3\Big|_1^5 = \frac{16\pi}{3}(124) = 661.33\pi \approx 2077.64$

2. $\pi\int_0^2 x^4\,dx = \frac{\pi}{5}x^5\Big|_0^2 = \frac{32\pi}{5} \approx 20.11$

3. $\pi\int_0^4 x^2\,dx = \frac{\pi}{3}x^3\Big|_0^4 = \frac{64\pi}{3} \approx 67.02$

4. $\pi\int_0^6 (6-x)^2\,dx = \pi\int_0^6 (36 - 12x + x^2)\,dx = \pi\left[36x - 6x^2 + \frac{x^3}{3}\right]_0^6 = 72\pi \approx 226.19$

5. $\pi\int_1^7 (x+1)\,dx = \pi\left[\frac{x^2}{2} + x\right]_1^7 = \pi\left[\frac{49}{2} + 7 - \frac{1}{2} - 1\right] = 30\pi \approx 94.25$

6. $\pi\int_1^2 (x^6 - 1)\,dx = \pi\left[\frac{x^7}{7} - x\right]_1^2 = \pi\left[\frac{128}{7} - 2 - \frac{1}{7} + 1\right] = \frac{120\pi}{7}$

7. $y = x^2 \Rightarrow x = \sqrt{y}$; $\pi\int_0^4 \sqrt{y}^2\,dy = \pi\frac{y^2}{2}\Big|_0^4 = 8\pi$

8. $x + y = 5 \Rightarrow x = 5 - y$; $\pi\int_0^5 (5-y)^2\,dy = \pi\int_0^5 (25 - 10y + y^2)\,dy = \pi\left[25y - 5y^2 + \frac{y^3}{3}\right]_0^5 = \frac{125\pi}{3}$

9. The two curves intersect at $(2,2)$. Thus, $V = \pi\int_0^2 ((4-x)^2 - x^2)\,dx = \pi\int_0^2\left(16 - 8x + x^2 - x^2\right)\,dx = \pi\left[16x - 4x^2\right]_0^2 = 16\pi$

10. $y = \sqrt[3]{x} \Rightarrow x = y^3$ and so $V = \pi\int_0^8 y^6\,dy = \pi\frac{y^7}{7}\Big|_0^8 = \frac{8^7\pi}{7} = \frac{2\,097\,152\pi}{7}$.

11. $y = \sqrt[3]{x} \Rightarrow x = y^3$ and so $V\pi\int_0^2 (8^2 - y^{3\cdot 2})\,dy = \pi\left[64y - \frac{y^7}{7}\right]_0^2 = \pi\left[128 - \frac{128}{7}\right] = \frac{768\pi}{7}$.

12. $y = \sqrt{x} \Rightarrow x = y^2$; $y = x^2 \Rightarrow x = \sqrt{y}$. The curves intersect at $(1, 1)$. Thus, $\pi \int_0^1 (y - y^4)\, dy = \pi \left[\frac{y^2}{2} - \frac{y^5}{5}\right]_0^1 = \pi \left[\frac{1}{2} - \frac{1}{5}\right] = \frac{3\pi}{10}$

13. They intersect at $(\pm 2, 0)$ with $8-2x^2$ on the outside. Thus, we have $\pi \int_{-2}^{2} ((8-2x^2)^2 - (4-x^2)^2)\, dx = \pi \int_{-2}^{2} \left[(64 - 32x^2 + 4x^4) - (16 - 8x^2 + x^4)\right] dx = \pi \int_{-2}^{2} (48 - 24x^2 + 3x^4)\, dx = \pi \left[48x - 8x^3 + \frac{3}{5}x^5\right]_{-2}^{2} = 102.4\pi$ or $\frac{512\pi}{5}$

14. $y = x^2 \Rightarrow x = \sqrt[2]{y}$; $y = x^3 \Rightarrow x = \sqrt[3]{y}$. Thus, we have $\pi \int_0^1 (\sqrt[3]{y}^2 - \sqrt[2]{y}^2)\, dy = \pi \left(\frac{3}{5}y^{5/3} - \frac{y^2}{2}\right)\Big|_0^1 = \pi \left(\frac{3}{5} - \frac{1}{2}\right) = \frac{\pi}{10}$

15. $\pi \int_0^1 (x^4 - x^6)\, dx = \pi \left[\frac{x^5}{5} - \frac{x^7}{7}\right]_0^1 = \frac{2\pi}{35}$

16. They intersect at $(0, 0)$ and $(4, 16)$. Thus, we have $\pi \int_0^4 ((4x)^2 - x^4)\, dx = \pi \left[\frac{16x^3}{3} - \frac{x^5}{5}\right]_0^4 = \frac{2048\pi}{15} = 428.93$

17. $x = \sqrt{y}$; $x = \frac{y}{4}$. Hence, we obtain $\pi \int_0^{16} \left(\sqrt{y}^2 - \left(\frac{y}{4}\right)^2\right) dy = \pi \left[\frac{y^2}{2} - \frac{y^3}{48}\right]_0^{16} = \pi \left(\frac{256}{2} - \frac{256}{3}\right) = \pi \left(\frac{256}{6}\right) = \frac{128\pi}{3}$

18. $\pi \int_0^2 \sqrt{8 - x^3}^2\, dx = \pi \left[8x - \frac{x^4}{4}\right]_0^2 = \pi \left(16 - \frac{16}{4}\right) = 12\pi$

19. $y = \sqrt{8 - x^2}$; $y^2 = 8 - x^2$; $x^2 = 8 - y^2$. So, $x = \sqrt{8 - y^2}$. Now, we obtain $\pi \int_0^{\sqrt{8}} (8 - y^2)\, dy = \pi \left[8y - \frac{y^3}{3}\right]_0^{\sqrt{8}} = \pi \left[8\sqrt{8} - \frac{8\sqrt{8}}{3}\right] = \pi \left[16\sqrt{2} - \frac{16\sqrt{2}}{3}\right] = \frac{32\sqrt{2}\pi}{3}$

20. Take the line $y = \frac{r}{h}x$. This line rotated about the x-axis forms a cone of height h and radius r. $V = \pi \int_0^h \frac{r^2}{h^2}x^2\, dx = \pi \left[\frac{r^2}{h^2}\frac{x^3}{3}\right]_0^h = \pi \cdot \frac{r^2}{h^2} \cdot \frac{h^3}{3} = \frac{1}{3}\pi r^2 h$

21. Use the equation for the top of a semicircle $y = \sqrt{r^2 - x^2}$ rotated about the x-axis.

$$V = \pi \int_{-r}^{r} \sqrt{r^2 - x^2}\, dx = \pi \int_{-r}^{r} (r^2 - x^2)\, dx$$

$$= \pi \left[r^2 x - \frac{x^3}{3}\right]_{-r}^{r} = \pi \left[r^3 - \frac{r^3}{3} + r^3 - \frac{r^3}{3}\right]$$

$$= \frac{4}{3}\pi r^3$$

22. $\frac{x^2}{a^2} + \frac{y^2}{b^2} = 1$; $\frac{y^2}{b^2} = 1 - \frac{x^2}{a^2} = \frac{a^2 - x^2}{a^2}$, so $y^2 = \frac{a^2b^2 - b^2x^2}{a^2}$, and we have $y = \sqrt{\frac{a^2b^2 - b^2x^2}{a^2}}$. The volume is $\pi \int_{-a}^{a} \left(\frac{a^2b^2}{a^2} - \frac{b^2}{a^2}x^2\right) dx = \pi \left[b^2x - \frac{b^2x^3}{3a^2}\right]_{-a}^{a} = \pi \left[b^2a - \frac{b^2a}{3} + b^2a - \frac{b^2a}{3}\right] = \frac{4}{3}ab^2\pi$ or $\frac{4b^2a\pi}{3}$. (Note: if $a = b = r$ we have the formula for the volume of a circle.)

23. (a) The equation must be of the form $y = mx^2$ assuming it passes through the origin. If must also contain the point $(1.5, 1)$ so $1 = m(1.5)^2$ $1 = m(2.25)$ or $m = \frac{1}{2.25} = \frac{4}{9}$ The equation is $y = \frac{4}{9}x^2$ or $x = \sqrt{\frac{9}{4}y} = \frac{3}{2}\sqrt{y}$.

(b) To find its volume rotate about the y-axis we evaluate $\pi \int_0^1 \sqrt{\frac{9}{4}y}^{\,2}\, dy =$

$\pi \int_0^1 \frac{9}{4} y\, dy = \pi \left[\frac{9}{4} \cdot \frac{y^2}{2}\right]_0^1 = \frac{9\pi}{8}$ m^2 = 1.125π m^2.

24. $\frac{dV}{dy} = V' = \frac{9\pi}{4} y$; $dV = \frac{9\pi}{4} y\, dy = \frac{9\pi}{4} \cdot 1 \cdot (0.004) =$ 0.009π, and $\Delta v = v'\, dy = \frac{9\pi}{4} y\, dy = \frac{9\pi}{4} \cdot 1(0.004) =$ $0.009\pi \approx 0.028\,27$ m^3.

25. $\pi \int_{-442}^{123} \sqrt{(147^2 + 0.16y^2)^2}\, dy = \pi \int_{-442}^{123} (147^2 + 0.16y^2)\, dy = \pi \left[21609y + \frac{0.16y^3}{3}\right]_{-442}^{123} =$ 16913711π ft^3 $\approx$ 53135993 ft^3.

26. $\pi \int_{-442}^{123} (146.5^2 + 0.16y^2)\, dy = \pi \left[(146.5)^5 \cdot y + \frac{10.16y^3}{3}\right]_{-442}^{123}$ $= 16{,}830{,}798\pi$ ft^3 or 52,875,552 ft^3. The volume is $(16{,}913{,}711 - 16{,}830{,}798)\pi$ ft^3 $= 82{,}913\pi$ or 260,479 ft^3

27. Use the equation for a circle $x^2 + y^2 = 18^2$ or $x = \sqrt{18^2 - y^2}$

(a) $\pi \int_{-18}^{-14} (18^2 - y^2)\, dy = \pi \left[18^2 y - \frac{y^3}{3}\right]_{-18}^{-14} =$ $266\frac{2}{3}\pi$ or 837.758 m^3

(b) $\pi \int_{-18}^{6} (18^2 - y^2)\, dy = \pi \left[18^2 y - \frac{y^3}{3}\right]_{-18}^{6} =$ $5760\pi \approx 18\,095.574$ m^3

26.3 VOLUMES OF REVOLUTION: SHELL METHOD

1. $2\pi \int_0^2 x(x^2 - 0)\, dx = 2\pi \int_0^2 x^3\, dx = 2\pi \left[\frac{x^4}{4}\right]_0^2 =$ 8π

2. $x = y^3 \Rightarrow y = \sqrt[3]{x} = x^{1/3}$. Hence we have $= 2\pi \int_0^8 x \cdot x^{1/3}\, dx = 2\pi \int_0^8 x^{4/3}\, dx = 2\pi \left[\frac{3}{7} x^{7/3}\right]_0^8 =$ $\frac{768\pi}{7}$

3. When $y = 8$ we see that $x = \pm 3$, we will rotate the portion in the first quadrant. Thus, $2\pi \int_1^3 x(8 - x^2 + 1)\, dx = 2\pi \int_1^3 (9x - x^3)\, dx =$ $2\pi \left[\frac{9x^2}{2} - \frac{x^4}{4}\right]_1^3 = 2\pi \left[\frac{81}{4} - \frac{17}{4}\right] = 32\pi$

4. $2\pi \int_0^2 x(x^3 + x)\, dx = 2\pi \int_0^2 (x^4 + x^2)\, dx =$ $2\pi \left[\frac{x^5}{5} + \frac{x^3}{3}\right]_0^2 = 2\pi \left[\frac{32}{5} + \frac{8}{3}\right] = \frac{272\pi}{15}$

5. $2\pi \int_0^5 x(25 - x^2)\, dx = 2\pi \int_0^5 (25x - x^3)\, dx =$ $2\pi \left[\frac{25x^2}{2} - \frac{x^4}{4}\right]_0^5 = \frac{625\pi}{2}$

6. $x^2 = 8 - x^2$, or $2x^2 = 8$, or $x^2 = 4$, so $x = \pm 2$. Now we can solve this using $2\pi \int_0^2 x(8 - x^2 - x^2)\, dx =$ $2\pi \int_0^2 (8x - 2x^3)\, dx = 2\pi \left[4x^2 - \frac{x^4}{2}\right]_0^2 = 16\pi$

7. $y = x^3 \Rightarrow x = \sqrt[3]{y} = y^{1/3}$. Hence, we have $2\pi \int_0^8 y(y^{1/3} - 0)\, dy = 2\pi \int_0^8 y^{4/3}\, dy =$ $2\pi \left[\frac{3}{7} y^{7/3}\right]_0^8 = \frac{768\pi}{7}$

8. $y = x^4 \Rightarrow x = y^{\frac{1}{4}}$; $y = x^3 \Rightarrow x = y^{1/3}$. They intersect at $(0, 0)$ and $(1, 1)$, so we have $2\pi \int_0^1 y(y^{1/4} - y^{1/3})\, dy = 2\pi \int_0^1 (y^{5/4} - y^{4/3})\, dy =$ $2\pi \left[\frac{4}{9} y^{9/4} - \frac{3}{7} y^{7/3}\right] = 2\pi \left(\frac{1}{63}\right) = \frac{2\pi}{63}$

9. $y = \sqrt[3]{x} \Rightarrow x = y^3$; $2\pi \int_0^2 y(y^3)\, dy = 2\pi \int_0^2 y^4\, dy =$ $2\pi \left[\frac{y^5}{5}\right]^2 = \frac{64\pi}{5}$

10. $x = y^{2/3}$; $2\pi \int_0^8 y \cdot y^{2/3}\, dy = 2\pi \left[\frac{3}{8} y^{8/3}\right]_0^8 = 192\pi$

11. $2\pi \int_0^1 x(x^{1/2} - x^{3/2})\, dx = 2\pi \int_0^1 (x^{3/2} - x^{5/2})\, dx =$ $2\pi \left[\frac{2}{5} x^{5/2} - \frac{2}{7} x^{7/2}\right]_0^1 = \frac{8\pi}{35}$

12. $x+2y=3 \Rightarrow x=3-2y$; $x=y$; $y=0$. These graphs intersect at $(1,1)$. $2\pi\int_0^1 y[(3-2y)-y]\,dy = 2\pi\int_0^1 (3y-3y^2)\,dy = 2\pi\left[\frac{3}{2}y^2-y^3\right]_0^1 = 2\pi(0.5) = \pi$.

13. $y=2x^3 \Rightarrow x=\sqrt[3]{\frac{y}{2}}$; $y^2=4x \Rightarrow x=\frac{1}{4}y^2$. These graphs intersect at $(0,0)$ and $(1,2)$.

$$\begin{aligned} 2\pi\int_0^2 y\left[\left(\frac{y}{2}\right)^{1/3}-\frac{y^2}{4}\right]dy &= 2\pi\int_0^2\left[\frac{1}{\sqrt[3]{2}}y^{4/3}-\frac{y^3}{4}\right]dy \\ &= 2\pi\left[\frac{1}{\sqrt[3]{2}}\frac{3}{7}y^{7/3}-\frac{y^4}{16}\right]_0^2 \\ &= 2\pi\left[\frac{12}{7}-1\right]=\frac{10\pi}{7} \end{aligned}$$

14. They intersect at $(0,0)$ and $(0,2)$

$$\begin{aligned} 2\pi\int_0^2 y(4y-y^3)\,dy = 2\pi\int_0^2(4y^2-y^4)\,dy &= 2\pi\left[\frac{4y^3}{3}-\frac{y^5}{5}\right]_0^2 \\ &= \left(\frac{64}{3}-\frac{64}{5}\right)\pi=\frac{128\pi}{15} \end{aligned}$$

15. They intersect at $(0,0)$ and $(2,4)$. Rotating about the line $y=6$ means we must integrate with respect to dy. $y=x^2 \Rightarrow x=\sqrt{y}$ and $y=2x \Rightarrow x=\frac{1}{2}y$

$$\begin{aligned} 2\pi\int_0^4(6-y)\left(y^{1/2}-\frac{y}{2}\right)dy &= 2\pi\int_0^4\left(6y^{1/2}-3y-y^{3/2}+\frac{y^2}{2}\right)dy \\ &= 2\pi\left[4y^{3/2}-\frac{3y^2}{2}-\frac{2}{5}y^{5/2}+\frac{y^3}{6}\right]_0^4 \\ &= 2\pi\left[32-24-\frac{64}{5}+\frac{32}{3}\right] \\ &= \frac{176}{15}\pi\approx 11.7333\pi \end{aligned}$$

16. $y=2-x$ and $y=x^2$ intersect at $(-2,4)$ and $(1,1)$. This region in bounded on the left by $y=2-x$ or $x=2-y$ and on the right by $y=x^2$ or $x=-\sqrt{y}$.

$$\begin{aligned} 2\pi\int_1^4 y\left[(2-y)-(-\sqrt{y})\right]dy &= 2\pi\int_1^4(2y-y^2+y\sqrt{y})\,dy \\ &= 2\pi\left[y^2-\frac{y^3}{3}+\frac{2}{5}y^{5/2}\right]_1^4 \\ &= 2\pi\left[16-\frac{64}{3}+\frac{64}{5}-1+\frac{1}{3}-\frac{2}{5}\right]=\frac{64\pi}{5} \end{aligned}$$

17.
$$\begin{aligned} 2\pi\int_{-2}^1(1-x)(2-x-x^2)\,dx &= 2\pi\int_{-2}^1(2-3x+x^3)\,dx \\ &= 2\pi\left[2x-\frac{3}{2}x^2+\frac{x^4}{4}\right]_{-2}^1 \\ &= 2\pi\left[2-\frac{3}{2}+\frac{1}{4}+4+6-4\right] \\ &= 2\pi\left(6\frac{3}{4}\right)=2\pi\left(\frac{27}{4}\right)=\frac{27\pi}{2} \end{aligned}$$

18. $$\begin{aligned} 2\pi\int_{-2}^2(4-x)(4-x^2)\,dx &= 2\pi\int_{-2}^2(16-4x-4x^2+x^3)\,dx \\ &= 2\pi\left[16x-2x^2-\frac{4}{3}x^3+\frac{x^4}{4}\right]_{-2}^2 \\ &= 2\pi\left[(32-8-\frac{32}{3}+4) - (-32-8+\frac{32}{3}+4)\right] \\ &= 2\pi\left[32-8-\frac{32}{3}+4+32+8-\frac{32}{3}-4\right] \\ &= 2\pi\left[\frac{128}{3}\right]=\frac{256\pi}{3} \end{aligned}$$

19. Start with the equation $x = \pm\sqrt{25-y^2}$ rotated about the x axis. We will take the shells starting with $y = 2$ to $y = 5$. The height of a shell is $2\sqrt{25-y^2}$. $2\pi\int_2^5 y(2\sqrt{25-y^2})\,dy$; Let $u = 25 - y^2$; $du = -2y\,dy$; $y = 2 \Rightarrow u = 21$; $y = 5 \Rightarrow u = 0$. Substituting, we have

$$2\pi\int_{21}^{0} -u^{1/2}\,du = 2\pi\left[-\frac{2}{3}u^{3/2}\right]_{21}^{0}$$
$$= 2\pi\left(\frac{2}{3}\cdot 21^{3/2}\right) = 28\pi\sqrt{21}$$
$$= 403.1$$

By the washer method:

$$\pi\int_{-\sqrt{21}}^{\sqrt{21}}((25-y^2)-4)\,dy$$
$$= \pi\int_{-\sqrt{21}}^{\sqrt{21}}(21-y^2)\,dy$$
$$= \pi\left[21y - \frac{y^3}{3}\right]_{-\sqrt{21}}^{\sqrt{21}}$$
$$= \pi(21\sqrt{21} - 7\sqrt{21})\times 2 = 28\pi\sqrt{21}$$

20. (a) First we obtain the value of y when $x = 7$, using $y = -\sqrt{100-x^2}$; $= \sqrt{10^2-7^2} = 100 - 49 = \sqrt{51}$. Now, integrating, we have

$$2\pi\int_0^{\sqrt{51}} x(\sqrt{100-x^2}-7)\,dx$$
$$= 2\pi\int_0^{\sqrt{51}}(x\sqrt{100-x^2}-7x)\,dx$$
$$= 2\pi\int_0^{\sqrt{51}} x\sqrt{100-x^2} - 2\pi\int_0^{\sqrt{51}} 7x\,dx$$

For the first integral, let $u = 100 - x^2$; $du = -2x\,dx$ so $-\frac{1}{2}\,du = x\,dx$. When $x = 0$, then $u = 100$ and when $x = \sqrt{51}$, then $u = 49.1$ Substituting, we obtain

$$2\pi\cdot\frac{-1}{2}\int_{100}^{49} u^{1/2}\,du = -\pi\cdot\frac{2}{3}u^{\frac{3}{2}}\Big|_{100}^{49}$$
$$= -\frac{2\pi}{3}\left[7^3 - 10^3\right] = 438\pi$$

The second integral is equal to $2\pi\left[\frac{7}{2}x^2\right]_0^{\sqrt{51}} = 7\cdot 51\pi = 357\pi$. Substituting these two answers, we have $438\pi - 357\pi = 81\pi$ m^3

(b) $81\pi\cdot 1000 = 254\,469$ kg

26.4 ARC LENGTH AND SURFACE AREA

1. $y = \frac{1}{3}(x^2+2)^{3/2}$; $y' = \frac{1}{2}(x^2+2)^{1/2}\cdot 2x = x(x^2+2)^{1/2}$

$$\int_0^3 \sqrt{1+\left[x(x^2+2)^{1/2}\right]^2}\,dx$$
$$= \int_0^3 \sqrt{1+x^2(x^2+2)}\,dx$$
$$= \int_0^3 \sqrt{1+x^4+2x^2}\,dx = \int_0^3\sqrt{x^4+2x^2+1}\,dx$$
$$= \int_0^3 (x^2+1) = \frac{x^3}{3}+x\Big|_0^3 = 9+3 = 12$$

2. $y = x^{3/2}$; $y' = \frac{3}{2}x^{1/2}$

$$\int_0^8 \sqrt{1+\left(\frac{3}{2}x^{1/2}\right)} = \int_0^8\sqrt{1+\frac{9x}{4}}\,dx$$
$$= \frac{4}{9}\int_0^8 \frac{9}{4}\sqrt{1+\frac{9}{4}x}\,dx$$
$$= \frac{4}{9}\cdot\frac{2}{3}\left(1+\frac{9}{4}x\right)^{3/2}\Bigg|_0^8$$
$$= \frac{8}{27}\left[19^{3/2}+1\right] = 24.8353$$

3. $9x^2 = 4y^3$, or $x^2 = \frac{4}{9}y^3$, or $x = \frac{2}{3}y^{3/2}$, and so $x' = y^{1/2}$

$$\int_0^3 \sqrt{1 + \left(y^{1/2}\right)^2}\, dy = \int_0^3 \sqrt{1 + y}\, dy$$
$$= \frac{2}{3}(1 + y)^{3/2}\Big|_0^3$$
$$= \frac{2}{3}\left[4^{3/2} - 1^{3/2}\right]$$
$$= \frac{2}{3}[8 - 1] = \frac{14}{3}$$

4. $y = \frac{x^4}{4} + \frac{1}{8}x^{-2}$ and $y' = x^3 - \frac{1}{4}x^{-3}$

$$\int_1^2 \sqrt{1 + \left(x^3 - \frac{1}{4x^3}\right)^2}\, dx$$
$$= \int_1^2 \sqrt{1 + x^6 - \frac{1}{2} + \frac{1}{16x^{65}}}\, dx$$
$$= \int_1^2 \sqrt{x^6 + \frac{1}{2} + \frac{1}{16x^6}} = \int_1^2 \sqrt{\left(x^3 + \frac{1}{4x^3}\right)}$$
$$= \int_1^2 \left(x^3 + \frac{1}{4}x^{-3}\right) dx = \left(\frac{x^4}{4} - \frac{1}{8}x^{-2}\right)\Big|_1^2$$
$$= 4 - \frac{1}{32} - \frac{1}{4} + \frac{1}{8} = \frac{123}{32} = 3.84375$$

5. $y = \frac{x^3}{6} + \frac{1}{2x}$, so $y' = \frac{x^2}{2} - \frac{1}{2x^2}$

$$\int_1^3 \sqrt{1 + \left[\frac{x^2}{2} - \frac{1}{2x^2}\right]^2}\, dx$$
$$= \int_1^3 \sqrt{1 + \left(\frac{x^4}{4} - \frac{1}{2} + \frac{1}{4x^4}\right)}\, dx$$
$$= \int_1^3 \sqrt{\frac{x^4}{4} + \frac{1}{2} + \frac{1}{4x^4}}\, dx$$
$$= \int_1^3 \left(\frac{x^2}{2} + \frac{1}{2x^2}\right) dx$$
$$= \frac{x^3}{6} - \frac{1}{2x}\Big|_1^3 = \frac{9}{2} - \frac{1}{6} - \frac{1}{6} + \frac{1}{2}$$
$$= 5 - \frac{1}{3} = 4\frac{2}{3} = \frac{14}{3}$$

6. $x = \frac{y^3}{3} + \frac{1}{4y}$, so $x' = y^2 - \frac{1}{4y^2}$

$$\int_1^3 \sqrt{1 + \left(y^2 - \frac{1}{4y^2}\right)^{-2}}\, dy$$
$$= \int_1^3 \sqrt{1 + y^4 - \frac{1}{2} + \frac{1}{16y^4}}\, dy$$
$$= \int_1^3 \sqrt{\left(y^4 + \frac{1}{2} + \frac{1}{16y^4}\right)}\, dy$$
$$= \int_1^3 \sqrt{\left(y^2 + \frac{1}{4y^2}\right)}\, dy$$
$$= \int_1^3 \left(y^2 + \frac{1}{4}y^{-2}\right) dy = \frac{y^3}{3} - \frac{y^{-1}}{4}\Big|_1^3$$
$$= \left(9 - \frac{1}{12} - \frac{1}{3} + \frac{1}{4}\right)$$
$$= \frac{108 - 1 - 4 + 3}{12} = \frac{106}{12} = \frac{53}{6}$$

7. $2\pi \int_0^2 4x\sqrt{1 + 4^2}\, dx = 8\pi\sqrt{17}\int_0^2 x\, dx = 8\pi\sqrt{17}\left[\frac{x^2}{2}\right]_0^2 = 16\pi\sqrt{17}$

8. $2\pi \int_0^8 \frac{12}{5}x\sqrt{1 + \left(\frac{12}{5}\right)^2}\, dx = \pi 2 \cdot \frac{12}{5} \cdot \frac{13}{5}\int_0^8 x\, dx = \frac{312\pi}{25}\frac{x^2}{2}\Big|_0^8 = \frac{9984\pi}{25} = 399.36\pi$

9. $y = x$; $y' = 1$; $2\pi \int_0^2 x\sqrt{1 + 1^2}\, dx = 2\sqrt{2}\pi\frac{x^2}{2}\Big|_0^2 = 4\pi\sqrt{2}$

10. $y^2 = 4x$, so $y = 2x^{1/2}$; $y' = x^{-1/2} = \frac{1}{\sqrt{x}}$; $\left(\frac{1}{\sqrt{x}}\right)^2 = \frac{1}{x}$

$$2\pi \int_0^8 2x^{1/2}\sqrt{1 + \frac{1}{x}}\, dx = 4\pi \int_0^8 \sqrt{x}\sqrt{\frac{x+1}{x}}\, dx$$

$$= 4\pi \int_0^8 \sqrt{x+1}\,dx$$
$$= \frac{2}{3} 4\pi (x+1)^{3/2} \Big|_0^8$$
$$= \frac{8\pi}{3}[27-1] = \frac{208\pi}{3}$$

11. $x = 4y \Rightarrow y = \frac{x}{4}$, $y = 1 \Rightarrow x = 4$, and $y = 3 \Rightarrow x = 12$. We have $y' = \frac{1}{4}$, and so, the surface area is

$$2\pi \int_4^{12} \frac{x}{4}\sqrt{1+\frac{1}{16}}\,dx = \frac{\pi\sqrt{17}}{8}\int_4^{12} x\,dx$$
$$= \frac{\pi\sqrt{17}}{8}\left[\frac{x^2}{2}\right]_4^{12}$$
$$= \frac{\pi\sqrt{17}}{8}[72-8]$$
$$= 8\pi\sqrt{17} \approx 103.62$$

12. Since $x = \sqrt{y} = y^{1/2}$, then $x' = \frac{1}{2}y^{-1/2} = \frac{1}{2\sqrt{y}}$ and $x'^2 = \frac{1}{4y}$. Thus, the surface area is

$$2\pi \int_0^6 \sqrt{y}\sqrt{1+\frac{1}{4y}}\,dy$$
$$= 2\pi \int_0^6 \sqrt{y}\sqrt{\frac{4y+1}{4y}}\,dy$$
$$= 2\pi \int_0^6 \frac{\sqrt{4y+1}}{2}\,dy = \pi \int_0^6 \sqrt{4y+1}\,dy$$
$$= \pi \cdot \frac{1}{4} \cdot \frac{2}{3}(4y+1)^{3/2}\Big|_0^6$$
$$= \frac{\pi}{6}\left(25^{3/2} - 1^{3/2}\right) = \frac{62\pi}{3}$$

13. Since $y = \frac{2}{75}x^{3/2}$, then $y' = \frac{1}{25}x^{1/2}$. Place the origin at the point where the cable meets the ground, then the length of the cable is given by $L = \int_0^{225}\sqrt{1+\left(\frac{1}{25}x^{1/2}\right)^2}\,dy = \int_0^{225}\sqrt{1+\left(\frac{1}{25}\right)^2 x}\,dy = \int_0^{225}\left(1+\frac{1}{625}x\right)^{1/2}dy = 625\cdot\frac{2}{3}\left(1+\frac{1}{625}x\right)^{3/2}\Big|_0^{225} \approx 660.841\ 2147 - 416.666\ 6667 = 244.174\ 548$ or approximately 244.17 m.

14. Here $y = (4-x)^3$, so $y' = -3(4-x)^2$. Thus, we find that the surface area is $S = 2\pi\int_2^4 (4-x)^3\sqrt{1+\left[-3(4-x)^2\right]^2}\,dx = 2\pi\int_2^4 (4-x)^3\sqrt{1+9(4-x)^4}\,dx$. Let $u = (4-x)^4$, and then $du = -4(4-x)^3\,dx$. When $x = 4$ we see that $u = 0$, and when $x = 2$, we find that $u = 16$. Substituting these values in S, we obtain $S = -\frac{\pi}{2}\int_{16}^0\sqrt{1+9u}\,du = -\frac{\pi}{2}\int_{16}^0(1+9u)^{1/2}\,du = -\frac{\pi}{2}\cdot\frac{2}{3}\cdot\frac{1}{9}(1+9u)^{3/2}\Big|_{16}^0 = -\frac{\pi}{27}(1-576) = \frac{575}{27}\pi \approx 66.904$ in.^{2}

26.5 CENTROIDS

1. $\bar{x} = \frac{4\cdot5+6(-3)}{4+6} = \frac{20-18}{10} = \frac{1}{5}$; $\bar{y} = \frac{4\cdot2+6\cdot7}{4+6} = \frac{8+42}{10} = 5$. The centroid is at $(\bar{x},\bar{y}) = (0.2, 5)$.

2. $\bar{x} = \frac{3(-1)+4(-2)+13(1)}{3+4+13} = \frac{-3-8+13}{20} = \frac{1}{10} = 0.1$; $\bar{y} = \frac{3(4)+4(-5)+13(2)}{3+4+13} = \frac{12-20+26}{20} = \frac{18}{20} = 0.9$. The centroid is at $(\bar{x},\bar{y}) = (0.1, 0.9)$.

3. $\bar{x} = \frac{2\cdot1+3(-1)+5(6)}{2+3+5} = \frac{2-3+30}{10} = 2.9$; $\bar{y} = \frac{2\cdot5+3(4)+5(-4)}{2+3+5} = \frac{2}{10} = 0.2$. The centroid is at $(2.9, 0.2)$.

4. $\bar{x} = \frac{1\cdot1+2\cdot2+3(-3)+4(-4)}{1+2+3+4} = \frac{1+4-9-16}{10} = \frac{-20}{10} = -2$; $\bar{y} = \frac{1\cdot1+2(-2)+3\cdot3+4(-4)}{10} = \frac{1-4+9-16}{10} = \frac{-10}{10} = -1$. The centroid is at $(-2,-1)$.

5. The left rectangle has center at $(-\frac{1}{2}, 2)$ and area 4. The right rectangle has center at $(2, 1)$ and area 8. $\bar{x} = \dfrac{4(-\frac{1}{2})+8(2)}{4+8} = \dfrac{-2+16}{12} = \dfrac{14}{12} = \dfrac{7}{6}$;

$\bar{y} = \frac{4 \cdot 2 + 8 \cdot 1}{12} = \frac{8+8}{12} = \frac{16}{12} = \frac{4}{3}$. The centroid is at $\left(\frac{7}{6}, \frac{4}{3}\right)$.

6. The upper rectangle has center at $(3, 1)$ and area 12. The lower rectangle has center at $\left(\frac{1}{2}, -\frac{1}{2}\right)$ and area 7. $\bar{x} = \frac{12(3) + 7\,(1/2)}{12+7} = \frac{36 + 3\frac{1}{2}}{19} = \frac{79}{38}$; $\bar{y} = \frac{12 - \frac{7}{2}}{19} = \frac{17}{38}$. The centroid is at $\left(\frac{79}{38}, \frac{17}{38}\right)$.

7. The upper left rectangle has center $(-3.5, 1)$ and area 6. Lower retangle has center $(-1, -1)$ and area 16. The upper right rectangle has center $(2.5, 1.5)$ and area 15. $\bar{x} = \frac{6(-3.5) + 16(-1) + 15(2.5)}{6 + 16 + 15} = \frac{0.5}{37} = \frac{1}{74}$; $\bar{y} = \frac{6 - 16 + 15 \cdot (1.5)}{37} = \frac{12.5}{37} = \frac{25}{74}$. The centroid is at $\left(\frac{1}{74}, \frac{25}{74}\right)$.

8. Divide the region into several rectangles giving areas and centers similar to the following: 4, $(-2, 3.5)$; 10, $(-3.0.5)$; 16, $(0, 0)$; 2, $(3 - 1.5)$; 8, $(4, 1)$; and 2, $(5.5, 3)$.

$$\bar{x} = \frac{4(-2)+10(-3)+16(0)+2(3)+8(4)+2(5.5)}{4 + 10 + 16 + 2 + 8 + 2} = \frac{11}{42}$$
$$\bar{y} = \frac{4(3.5)+10(0.5)+16(0)+2(-1.5)+8(1)+2(3)}{42} = \frac{30}{42} = \frac{5}{7}$$

The centroid is at $\left(\frac{11}{42}, \frac{5}{7}\right)$.

9. $y = 2x + 3[0, 3]$

$$M_y = \int_0^3 x(2x+3)\,dx = \int_0^3 (2x^2 + 3x)\,dx$$
$$= \frac{2x^3}{3} + \frac{3x^2}{2}\bigg|_0^3 = 18 + \frac{27}{2} = \frac{63}{2}$$
$$M_x = \frac{1}{2}\int_0^3 (2x+3)^2\,dx = \frac{1}{2}\int_0^3 (4x^2 + 12x + 9)\,dx$$
$$= \frac{1}{2}\left[\frac{4x^3}{3} + 6x^2 + 9x\right]_0^3 = \frac{1}{2}[36 + 54 + 27]$$
$$= \frac{117}{2}$$
$$m = \int_0^3 (2x+3)\,dx = x^2 + 3x\Big|_0^3 = 9 + 9 = 18$$

$$\bar{x} = \frac{63}{36} = \frac{7}{4} = 1.75$$
$$\bar{y} = \frac{117}{36} = \frac{13}{4} = 3.25$$

The centroid is $(1.75, 3.25)$.

10. $y = x^3\ [0, 2]$

$$M_y = \int_0^2 x \cdot x^3\,dx = \frac{x^5}{5}\bigg|_0^2 = \frac{32}{5}$$
$$M_x = \frac{1}{2}\int_0^2 x^6\,dx = \frac{1}{2} \cdot \frac{x^7}{7}\bigg|_0^2 = \frac{64}{7}$$
$$m = \int_0^2 x^3\,dx = \frac{x^4}{4}\bigg|_0^2 = 4$$
$$\bar{x} = \frac{32}{20} = \frac{8}{5}$$
$$\bar{y} = \frac{64}{7.4} = \frac{16}{7}.$$

The centroid is $\left(\frac{8}{5}, \frac{16}{7}\right)$.

11. $y = x^{1/3}\ [0, 8]$

$$M_y = \int_0^8 x \cdot x^{2/3}\,dx = \int_0^8 x^{4/3}\,dx$$
$$= \frac{3}{7}x^{7/3}\bigg|_0^8 = \frac{384}{7}$$
$$M_x = \frac{1}{2}\int_0^8 x^{2/3}\,dx = \frac{1}{2} \cdot \frac{3}{5}x^{5/3}\bigg|_0^8 = \frac{48}{5}$$
$$m = \int_0^8 x^{1/3}\,dx = \frac{3}{4} \cdot x^{4/3}\bigg|_0^8 = 12$$
$$\bar{x} = \frac{384}{7 \cdot 12} = \frac{32}{7}$$
$$\bar{y} = \frac{48}{5 \cdot 12} = \frac{4}{5}.$$

The centroid is $\left(\frac{32}{7}, \frac{4}{5}\right)$.

12. $y = x^4\ [-1, 2]$

$$M_y = \int_{-1}^{2} x \cdot x^4\,dx = \int_{-1}^{2} x^5\,dx$$
$$= \left.\frac{x^6}{6}\right|_{-1}^{2} = \frac{69 - 1}{6} = \frac{63}{6} = \frac{21}{2}$$
$$M_x = \frac{1}{2}\int_{-1}^{2} x^8\,dx = \frac{1}{2} \cdot \left.\frac{x^9}{9}\right|_{-1}^{2}$$
$$= \frac{513}{18} = \frac{171}{6} = \frac{57}{2}$$
$$m = \int_{-1}^{2} x^4\,dx = \left.\frac{x^5}{5}\right|_{-1}^{2} = \frac{32 + 1}{5} = \frac{33}{5};$$
$$\bar{x} = \frac{21}{2} \cdot \frac{5}{33} = \frac{35}{22}$$
$$\bar{y} = \frac{57}{2} \cdot \frac{5}{33} = \frac{285}{66} = \frac{95}{22}.$$

The centroid is $\left(\frac{35}{22}, \frac{95}{22}\right)$.

13. $y = \sqrt{x+4}$ on $[0, 5]$ $M_y = \int_0^5 x\sqrt{x+4}\,dx$; $u = x + 4$; $du = dx$; $x = u - 4$ $x = 0 \Rightarrow u = 4$ and $x = 5 \Rightarrow u = 9$. Substituting, we obtain

$$\int_4^9 (u - 4)\sqrt{u}\,du$$
$$= \int_4^9 \left(u^{3/2} - 4u^{1/2}\right)\,du = \left[\frac{2}{5}u^{5/2} - 4 \cdot \frac{2}{3}u^{3/2}\right]_4^9$$
$$= \frac{2 \cdot 243}{5} - 72 - \frac{64}{5} + \frac{64}{3}$$
$$= \frac{1458 - 1080 - 192 + 320}{15} = \frac{506}{15}$$
$$M_x = \frac{1}{2}\int_0^5 (x + 4)\,dx = \frac{1}{2}\left[\frac{x^2}{2} + 4x\right]_0^5$$
$$= \frac{1}{2}\left[\frac{25}{2} + 20\right] = \frac{65}{4}$$
$$m = \int_0^5 \sqrt{x+4}\,dx = \left.\frac{2}{3}(x + 4)^{3/2}\right|_0^5$$
$$= \frac{2}{3}[27 - 8] = \frac{38}{3}$$
$$\bar{x} = \frac{506}{15} \cdot \frac{3}{38} = \frac{253}{95}$$
$$\bar{y} = \frac{65}{4} \cdot \frac{3}{38} = \frac{195}{152}.$$

The centroid is $\left(\frac{253}{95}, \frac{195}{152}\right)$.

14. $y = x^2 + 16$; $[0, 4]$

$$M_y = \int_0^4 x(x^2 + 16)\,dx = \left[\frac{x^4}{4} + 8x^2\right]_0^4$$
$$= 64 + 128 = 192$$
$$M_x = \frac{1}{2}\int_0^4 (x^2 + 16)^2\,dx$$
$$= \frac{1}{2}\int_0^4 (x^4 + 32x^2 + 256)\,dx$$
$$= \frac{1}{2}\left[\frac{x^5}{5} + \frac{32x^3}{3} + 256x\right]_0^4$$
$$= \frac{1}{2}\left[\frac{1024}{5} + \frac{2048}{3} + 1024\right]$$
$$= \frac{512}{5} + \frac{1024}{3} + 512$$
$$= \frac{1536 + 5120 + 7680}{15} = \frac{14336}{15}$$
$$m = \int_0^4 (x^2 + 16)\,dx = \left[\frac{x^3}{3} + 16x\right]_0^4$$
$$= \frac{64}{3} + 64 = \frac{256}{3}$$
$$\bar{x} = 192 \cdot \frac{3}{256} = \frac{9}{4} = 2.25$$
$$\bar{y} = \frac{14336}{15} \cdot \frac{3}{256} = \frac{56}{5} = 11.2$$

The centroid is (2.25, 11.2).

15. $$M_y = \int_0^2 x(4x - x^2)\,dx = \int_0^2 (4x^2 - x^3)\,dx$$
$$= \left[\frac{4x^3}{3} - \frac{x^4}{4}\right]_0^2 = \frac{32}{3} - 4 = \frac{20}{3}$$
$$M_x = \frac{1}{2}\int_0^2 \left[(4x)^2 - (x^2)^2\right]\,dx$$

$$= \frac{1}{2}\int_0^2 (16x^2 - x^4)\,dx$$
$$= \frac{1}{2}\left[\frac{16x^3}{3} - \frac{x^5}{5}\right]_0^2 = \frac{1}{2}\left[\frac{128}{3} - \frac{32}{5}\right]$$
$$= \frac{64}{3} - \frac{16}{5} = \frac{272}{15}$$
$$m = \int_0^2 (4x - x^2)\,dx = \left[2x^2 - \frac{x^3}{3}\right]_0^2$$
$$= 8 - \frac{8}{3} = \frac{16}{3}$$
$$\bar{x} = \frac{20}{3}\cdot\frac{3}{16} = \frac{5}{4}$$
$$\bar{y} = \frac{272}{15}\cdot\frac{3}{16} = \frac{17}{5}.$$

The centroid is $\left(\frac{5}{4}, \frac{17}{5}\right)$.

16. $$M_y = \int_0^1 x(2x - x^3)\,dx = \int_0^1 (2x^2 - x^4)\,dx$$
$$= \left[\frac{2x^3}{3} - \frac{x^5}{5}\right]_0^1 = \frac{2}{3} - \frac{1}{5} = \frac{7}{15}$$
$$M_x = \frac{1}{2}\int_0^1 \left[(2x)^2 - (x^3)^2\right]\,dx$$
$$= \frac{1}{2}\int_0^1 (4x^4 - x^6)\,dx$$
$$= \left[\frac{2x^5}{5} - \frac{x^7}{14}\right]_0^1 = \frac{2}{5} - \frac{1}{14} = \frac{23}{70}$$
$$m = \int_0^1 (2x - x^3)\,dx = \left[x^2 - \frac{x^4}{4}\right]_0^1 = 1 - \frac{1}{4} = \frac{3}{4}$$
$$\bar{x} = \frac{7}{15}\cdot\frac{4}{3} = \frac{28}{45}$$
$$\bar{y} = \frac{23}{70}\cdot\frac{4}{3} = \frac{92}{210} = \frac{46}{105}.$$

The centroid is $\left(\frac{28}{45}, \frac{46}{105}\right)$.

17. The graphs intersect at $(-2, 4)$ and $(1, 1)$

$$M_y = \int_{-2}^1 x(2 - x - x^2)\,dx = \int_{-2}^1 (2x - x^2 - x^3)\,dx$$
$$= \left[x^2 - \frac{x^3}{3} - \frac{x^4}{4}\right]_{-2}^1$$
$$= 1 - \frac{1}{3} - \frac{1}{4} - 4 - \frac{8}{4} + 4 = -\frac{9}{4}$$
$$M_x = \frac{1}{2}\int_{-2}^1 ((2 - x)^2 - (x^2)^2)\,dx$$
$$= \frac{1}{2}\int_{-2}^1 (4 - 4x + x^2 - x^4)\,dx$$
$$= \frac{1}{2}\left[4x - 2x^2 + \frac{4^3}{3} - \frac{x^5}{5}\right]_{-2}^1$$
$$= \frac{1}{2}\left[4 - 2 + \frac{1}{3} - \frac{1}{5} + 8 + 8 + \frac{8}{4} - \frac{32}{5}\right]$$
$$= \frac{1}{2}\left[18 + 3 - \frac{33}{5}\right]$$
$$= \frac{36}{5}$$
$$m = \int_{-2}^1 (2 - x - x^2)\,dx = \left[2x - \frac{x^2}{2} - \frac{x^3}{3}\right]_{-2}^1$$
$$= 2 - \frac{1}{2} - \frac{1}{3} + 4 + 2 - \frac{8}{3} = 8 - \frac{1}{2} - 3$$
$$= 4\frac{1}{2} = \frac{9}{2}$$
$$\bar{x} = -\frac{9}{4}\cdot\frac{2}{9} = -\frac{1}{2}$$
$$\bar{y} = \frac{36}{5}\cdot\frac{2}{9} = \frac{8}{5}.$$

The centroid is $\left(-\frac{1}{2}, \frac{8}{5}\right)$.

18. The graphs intersect at $(0, 0)$ and $(1, 1)$

$$M_y = \int_0^1 x\left(x - x^{3/2}\right)dx = \int_0^1 \left(x^2 - x^{5/2}\right)dx$$
$$= \left[\frac{x^3}{3} - \frac{2}{7}x^{7/2}\right]_0^1 = \frac{1}{3} - \frac{2}{7} = \frac{1}{21}$$
$$M_x = \frac{1}{2}\int_0^1 \left(x^2 - x^3\right)dx = \frac{1}{2}\left[\frac{x^3}{3} - \frac{x^4}{4}\right]_0^1$$
$$= \frac{1}{2}\left(\frac{1}{3} - \frac{1}{4}\right) = \frac{1}{24}$$

$$m = \int_0^1 \left(x - x^{3/2}\right) dx = \left[\frac{x^2}{2} - \frac{2}{5}x^{5/2}\right]_0^1$$
$$= \frac{1}{2} - \frac{2}{5} = \frac{1}{10}$$
$$\bar{x} = \frac{1}{21} \cdot \frac{10}{1} = \frac{10}{21}$$
$$\bar{y} = \frac{1}{24} \cdot \frac{10}{1} = \frac{10}{24} = \frac{5}{12}.$$

The centroid is $\left(\frac{10}{21}, \frac{5}{12}\right)$

19. The graphs intersect at $(\pm 3, 9)$

$$M_y = \int_{-3}^{3} x(18 - x^2 - x^2)\, dx$$
$$= \int_{-3}^{3} (18x - 2x^3)\, dx = 0$$
$$M_x = \frac{1}{2}\int_{-3}^{3} ((18 - x^2)^2 - (x^2)^2)\, dx$$
$$= \frac{1}{2}\int_{-3}^{3} (324 - 36x^2 + x^4 - x^4)\, dx$$
$$= \frac{1}{2}\left[324x - 12x^3\right]_{-3}^{3}$$
$$= \frac{1}{2}[972 - 324 + 972 - 324] = 648$$
$$m = \int_{-3}^{3} (18 - x^2 - x^2)\, dx = \left[18x - \frac{2x^3}{3}\right]_{-3}^{3}$$
$$= (54 - 18 + 54 - 18) = 72$$
$$\bar{x} = 0$$
$$\bar{y} = \frac{648}{72} = 9$$

The centroid is $(0, 9)$.

20. The intersect at $(0, 0)$ and $(1, 1)$

$$M_y = \int_0^1 x(\sqrt{x} - x)\, dx = \int_0^1 \left(x^{3/2} - x^2\right) dx$$
$$= \left[\frac{2}{5}x^{5/2} - \frac{x^3}{3}\right]_0^1$$
$$= \frac{2}{5} - \frac{1}{3} = \frac{1}{15}$$
$$M_x = \frac{1}{2}\int_0^1 (x - x^2)\, dx = \frac{1}{2}\left[\frac{x^2}{2} - \frac{x^3}{3}\right]_0^1 = \frac{1}{12}$$
$$m = \int_0^1 (\sqrt{x} - x)\, dx = \left[\frac{2}{3}x^{3/2} - \frac{x^2}{2}\right]_0^1 = \frac{2}{3} - \frac{1}{2} = \frac{1}{6}$$
$$\bar{x} = \frac{1}{15} \cdot \frac{6}{1} = \frac{6}{15} = \frac{2}{5}$$
$$\bar{y} = \frac{1}{12} \cdot \frac{6}{1} = \frac{1}{2}.$$

The centroid is $\left(\frac{2}{5}, \frac{1}{2}\right)$.

21. They intersect at $(-4, -4)$ and $(3, 3)$

$$M_y = \int_{-4}^{3} x(12 - x^2 - x)\, dx = \int_{-4}^{3} (12x - x^3 - x^2)\, dx$$
$$= \left[6x^2 - \frac{x^4}{4} - \frac{x^3}{3}\right]_{-4}^{3}$$
$$= 54 - \frac{81}{4} - 9 - 96 + 64 - \frac{64}{3}$$
$$= 13 - \frac{81}{4} - \frac{64}{3} = -\frac{343}{12}$$
$$M_x = \frac{1}{2}\int_{-4}^{3} (12 - x^2)^2 - (x)^2\, dx$$
$$= \frac{1}{2}\int_{-4}^{3} (144 - 24x^2 + x^4 - x^3)\, dx$$
$$\times \frac{1}{2}\left[144x - \frac{25x^3}{3} + \frac{x^5}{5}\right]_{-4}^{3}$$
$$= \frac{1}{2}\left[432 - 225 + \frac{243}{5} + 576 - \frac{1600}{3} + \frac{1024}{5}\right]$$
$$= \frac{3773}{15}$$
$$m = \int_{-4}^{3} (12 - x^2 - x)\, dx = \left[12x - \frac{x^3}{3} - \frac{x^3}{2}\right]_{-4}^{3}$$
$$= 36 - 9 - \frac{9}{2} + 48 - \frac{64}{3} + 8 = 57\frac{1}{6} = \frac{343}{6}$$
$$\bar{x} = -\frac{343}{12} \times \frac{6}{343} = -\frac{1}{2} = -0.5$$
$$\bar{y} = \frac{3773}{15} \cdot \frac{6}{343} = 4.4$$

The centroid is $(-0.5, 4.4)$.

22. $y = x^2$ and $y = x^3$ intersect at $(0,0)$ and $(1,1)$.

$$M_y = \int_0^1 x(x^2 - x^3)\,dx = \left[\frac{x^4}{4} - \frac{x^5}{5}\right]_0^1 = \frac{1}{4} - \frac{1}{5} = \frac{1}{20}$$

$$M_x = \frac{1}{2}\int_0^1 (x^4 - x^6)\,dx = \frac{1}{2}\left[\frac{x^5}{5} - \frac{x^7}{7}\right]_0^1$$

$$= \frac{1}{2}\left(\frac{1}{5} - \frac{1}{7}\right) = \frac{2}{23}\cdot\frac{1}{2} = \frac{1}{35}$$

$$m = \int_0^1 (x^2 - x^3)\,dx = \left[\frac{x^3}{3} - \frac{x^4}{4}\right]_0^1 = \frac{1}{3} - \frac{1}{4} = \frac{1}{12}$$

$$\overline{x} = \frac{1}{20}\cdot\frac{12}{1} = \frac{3}{5}$$

$$\overline{y} = \frac{1}{35}\cdot\frac{12}{1} = \frac{12}{35}$$

The centroid is $\left(\frac{3}{5}, \frac{12}{35}\right)$.

23. $y = x^3$;

$$M_y = \pi\int_0^2 x(x^2)\,dx = \pi\frac{x^3}{8}\Big|_0^2 = 32\pi$$

$$m = \pi\int_0^2 (x^3)^2 = \pi\cdot\frac{x^7}{7}\Big|_0^2 = \frac{128\pi}{7}.$$

Thus, we find that $\overline{x} = \frac{32\pi}{\frac{128\pi}{7}} = 32\pi\left(\frac{7}{128\pi}\right) = \frac{7}{4}$ and $\overline{y} = 0$ since it's rotated about the x-axis. The centroid is $\left(\frac{7}{4}, 0\right)$.

24. $$y = x^3 \Rightarrow x = y^{1/3}$$

$$M_x = \pi\int_0^8 y(y^{2/3})\,dy = \pi\frac{3}{8}y^{8/3}\Big|_0^8 = 96\pi$$

$$m = \pi\int_0^8 y^{2/3}\,dy = \pi\cdot\frac{3}{5}y^{5/3}\Big|_0^8 = \frac{96\pi}{5}$$

Thus, we see that $\overline{x} = 0$ since it is rotated about the y-axis, and $\overline{y} = \frac{96\pi}{96\frac{\pi}{5}} = 96\pi\left(\frac{5}{96\pi}\right) = 5$. The centroid is $(0, 5)$.

25. $$y = x^4 \Rightarrow x = y^{1/4}$$

$$M_x = \pi\int_0^1 y\cdot\left(y^{1/4}\right)^2 dy = \pi\frac{2}{5}\cdot y^{5/2}\Big|_0^1 = \frac{2\pi}{5}$$

$$m = \pi\int_0^1 \left(y^{1/4}\right)^2 dy = \pi\cdot\frac{2}{3}y^{3/2}\Big|_0^1 = \frac{2\pi}{3}$$

$$\overline{x} = 0$$

$$\overline{y} = \frac{\frac{2\pi}{5}}{\frac{2\pi}{3}} = \frac{2\pi}{5}\cdot\frac{3}{2\pi} = \frac{3}{5}.$$

The centroid is $\left(0, \frac{3}{5}\right)$.

26. $$y = (x-1)^2 \Rightarrow x - 1 = \pm y^{1/2} \Rightarrow x = 1 \pm \sqrt{y}$$

$$M_x = \pi\int_0^1 y\left[(1+\sqrt{y})^2 - (1-\sqrt{y})^2\right]dy$$

$$= \pi\int_0^1 y\left[1 + 2\sqrt{y} + y - 1 + 2\sqrt{y} - y\right]dy$$

$$= \pi\int_0^1 4y^{3/2}\,dy$$

$$= \pi\left[4\cdot\frac{2}{5}y^{5/2}\right]_0^1 = \frac{8\pi}{5}$$

$$m = \pi\int_0^1 4y^{1/2}\,dy = \pi 4\cdot\frac{2}{3}y^{3/2}\Big|_0^1 = \frac{8\pi}{3}$$

$$\overline{x} = 0$$

$$\overline{y} = \frac{\frac{8\pi}{5}}{\frac{8\pi}{3}} = \frac{8\pi}{5}\cdot\frac{3}{8\pi} = \frac{3}{5}.$$

The centroid is $\left(0, \frac{3}{5}\right)$.

27. $x + y = 6 \Rightarrow y = 6 - x$; $x = 3$

$$M_y = \pi\int_0^3 x\left[(6-x)^2\right]dx$$

$$= \pi\int_0^3 (36x - 12x^2 + x^3)\,dx$$

$$= \pi\left[18x^2 - 4x^3 + \frac{x^4}{4}\right]_0^3 = 74.25\pi = \frac{297}{4}\pi$$

$$m = \pi\int_0^3 (6-x)^2\,dx = \pi\int_0^3 (36 - 12x + x^2)\,dx$$

$$= \pi\left(36x - 6x^2 + \frac{x^3}{3}\right)\Big|_0^3 = 63\pi$$

$$\bar{x} = \frac{297}{4} \cdot \frac{1}{63} = \frac{33}{28}$$
$$\bar{y} = 0$$

The centroid is $\left(\frac{33}{28}, 0\right)$.

28. $y = 8 - x$

$$M_y = \pi \int_4^8 x(8-x)^2\,dx$$
$$= \pi \int_4^8 (64x - 16x^2 + x^3)\,dx$$
$$= \pi\left(32x^2 - \frac{16}{3}x^3 + \frac{1}{4}x^4\right)\Big|_4^8 = 106\frac{2}{3}$$
$$m = \pi \int_4^8 (8-x)^2\,dx = -\frac{1}{3}(8-x)^3\Big|_4^8 = 21\frac{1}{3}$$
$$\bar{x} = 5$$
$$\bar{y} = 0$$

The centroid is $(5, 0)$.

29. $$M_x = \pi \int_0^4 y \cdot 4^2\,dy + \pi \int_4^8 y(8-y)^2\,dy$$
$$= \pi\left[128 + 106\frac{2}{3}\right] = 234\frac{2}{3}\pi$$
$$m = \pi\left[\int_0^4 16\,dy + \int_4^8 (8-y)^2\,dy\right]$$
$$= \pi\left[64 + 21\frac{1}{3}\right] = 85\frac{1}{3}\pi$$
$$\bar{x} = 0$$
$$\bar{y} = \frac{234\frac{2}{3}}{85\frac{1}{3}} = 2.75$$

The centroid is $(0, 2.75)$.

30. $y = \sqrt{16 - x^2}$

$$M_y = \pi \int_0^4 x\sqrt{16-x^2}^{\,2}\,dx = \pi \int_0^4 16x - x^2\,dx$$
$$= \pi\left[8x^2 - \frac{x^4}{4}\right]_0^4 = \pi(128 - 64) = 64\pi$$
$$m = \pi \int_0^4 (16 - x^2)\,dx = \pi\left[16x - \frac{x^3}{3}\right]_0^4 = \frac{128\pi}{3}$$
$$\bar{x} = \frac{64\pi}{\frac{128\pi}{3}} = \frac{3}{2} \text{ or } 1.5$$
$$\bar{y} = 0$$

The centroid is $(1.5, 0)$.

31. $x^2 - y^2 = 1$; $-y^2 = 1 - x^2$; $y^2 = x^2 - 1$; $y = \sqrt{x^2 - 1}$

$$M_y = \pi \int_1^3 x(x^2-1)\,dx = \pi\left[\frac{x^4}{4} - \frac{x^2}{2}\right]_1^3$$
$$= \pi\left[\frac{81}{4} - \frac{9}{2} - \frac{1}{4} + \frac{1}{2}\right] = \pi[20 - 4] = 16\pi$$
$$m = \pi \int_1^3 (x^2 - 1)\,dx = \pi\left[\frac{x^3}{3} - x\right]_1^3$$
$$= \pi\left(9 - 3 - \frac{1}{3} + 1\right) = \pi 6\frac{2}{3} \text{ or } \frac{20}{3}\pi$$
$$\bar{x} = \frac{16}{\frac{20}{3}} = \frac{4 \cdot 3}{5} = \frac{12}{5}.$$

The centroid is $\left(\frac{12}{5}, 0\right)$ or $(2.4, 0)$.

32. $x^2 - y^2 = 1 \Rightarrow x = \sqrt{y^2 + 1}$ and $x = 3 \Rightarrow y = \sqrt{8}$

$$M_x = \pi \int_0^{\sqrt{8}} y(3^2 - y^2 - 1)\,dy$$
$$= \pi \int_0^{\sqrt{8}} y(8 - y^2)\,dy$$
$$= \pi\left[4y^2 - \frac{y^4}{4}\right]_0^{\sqrt{8}} = \pi[32 - 16] = 16\pi$$
$$m = \pi \int_0^{\sqrt{8}} (9 - y^2 - 1)\,dy = \pi\left[8y - \frac{y^3}{3}\right]_0^{\sqrt{8}}$$
$$= \left[8\sqrt{8} - \frac{8\sqrt{8}}{3}\right] = \frac{\pi 16\sqrt{8}}{3} = \frac{32\pi\sqrt{2}}{3}$$
$$\bar{x} = 0$$
$$\bar{y} = \frac{16\pi}{\frac{32\pi\sqrt{2}}{3}} = \frac{3\sqrt{2}}{4} = 1.0607$$

The centroid is $(0, 1.0607)$.

33. $y = x^2$; $y = \sqrt{x}$

$$M_y = \pi \int_0^1 x\left[x - x^4\right] dx = \pi \left[\frac{x^3}{3} - \frac{x^6}{6}\right]_0^1 = \frac{\pi}{6}$$

$$m = \pi \int_0^1 x - x^4\, dx = \pi \left[\frac{x^2}{2} - \frac{x^5}{5}\right] = \frac{3\pi}{10}$$

$$\bar{x} = \frac{\frac{\pi}{6}}{\frac{3\pi}{10}} = \frac{10}{18} = \frac{5}{9}.$$

The centroid is $\left(\frac{5}{9}, 0\right)$.

34. These two curves intersect at $(0,0)$ and $(1,1)$. $y = x^4 \Rightarrow x = y^{1/4}$; $y = x^2 \Rightarrow x = y^{1/2}$.

$$M_x = \pi \int_0^1 y\left(y^{1/2} - y\right) dy = \pi \left[\frac{2}{5}y^{5/2} - \frac{y^3}{3}\right]_0^1$$

$$= \pi \left[\frac{2}{5} - \frac{1}{3}\right] = \frac{\pi}{15}$$

$$m = \pi \int_0^1 \left(y^{1/2} - y\right) dy = \pi \left[\frac{2}{3}y^{3/2} - \frac{y^2}{2}\right]_0^1$$

$$= \pi \left[\frac{2}{3} - \frac{1}{2}\right] = \frac{\pi}{6}$$

$$\bar{y} = \frac{\frac{1}{15}}{\frac{1}{6}} = \frac{6}{15} = \frac{2}{5}$$

The centroid is $\left(0, \frac{2}{5}\right)$.

35. (a) Make two rectangles by cutting off the top piece. Then the bottom has center at $(50, \frac{1}{2})$ and area 100 cm^2. The left piece has center at $\left(\frac{1}{2}, \frac{101}{2}\right)$ and area 99 cm^2.

$$\bar{x} = \frac{100 \times 50 + 99 \cdot \frac{1}{2}}{100 + 99} = \frac{5049.5}{199} \approx 25.374$$

$$\bar{y} = 100 \times \frac{1}{2} + 101\left(\frac{99}{2}\right) = \frac{5049.5}{199} \approx 25.374$$

The centroid of the frame is $(25.374, 25.374)$. (b) The bottom area is $h \cdot w$ with center $\left(\frac{h}{2}, \frac{w}{2}\right)$ and the left area is $(h - w)$, center $\left(\frac{w}{2}, \frac{h+w}{2}\right)$

$$\bar{x} = \frac{hw\left(\frac{h}{2}\right) + w(h-w) \cdot \frac{w}{2}}{hw + (h-w)w} = \frac{\frac{h^2w}{2} + \frac{hw^2}{2} - \frac{w^3}{2}}{2wh - w^2}$$

$$= \frac{h^2 + hw - w^2}{2(2h - w)}.$$

$$\bar{y} = \frac{\left(\frac{hw \cdot w}{2}\right) + (h-w)w\left(\frac{h+w}{2}\right)}{w(2h - w)}$$

$$= \frac{hw + h^2 - w^2}{2(2h - w)}$$

The centroid is $\left(\frac{h^2 + hw - w^2}{2(2h - w)}, \frac{h^2 + hw - w^2}{2(2h - w)}\right)$.

36. (a) Use the center of the bottom of the cone as the origin and the y-axis as the axis of the cone. Then the cone is formed by rotating the line $x = 7 - \frac{4.5}{28}y$ or $x = 7 - \frac{45}{280}y = 7 - \frac{9}{56}y$.

$$M_x = \pi \int_0^{28} y\left(7 - \frac{9}{56}y\right)^2 dy$$

$$= \pi \int y\left(49 - \frac{9}{4}y + \frac{81}{3136}y^2\right)$$

$$= \pi \int_0^{28} \left(49y - \frac{9}{4}y^2 + \frac{81}{3136}y^3\right) dy$$

$$= \pi \left[\frac{49y^2}{2} - \frac{3y^3}{4} + \frac{81y^4}{12544}\right]_0^{28} = 6713\pi$$

$$m = \pi \int_0^{28} \left(7 - \frac{9}{56}y\right)^2 dy$$

$$= \pi \int_0^{28} \left(49 - \frac{9}{4}y + \frac{81}{3136}y^2\right) dy$$

$$= \pi \left[49y - \frac{9y^2}{8} + \frac{27y^3}{3136}\right]_0^{28} = 679\pi$$

$\bar{y} = \frac{6713}{679} = 9.8866$ inches from the top of the base. The centroid is $(0, 9.8866)$.
(b) The bottom has volume is $18^2 = 324$ in.^{2} and centroid $\frac{1}{2}$ in. below the top of the base.

$$\bar{y} = \frac{9.8866 \times 679\pi - \frac{1}{2}(324)}{679\pi + 324}$$

$$= \frac{21089.52 - 162}{2457.14} = \frac{20{,}927.52}{2457.14} \approx 8.517$$

The centroid is 8.517 in. above the top of the base.

37. (a) $m = \int_0^{40} (5x+1)\,dx = \left[\frac{5x^2}{2} + x\right]_0^{40} = 4040$ g (b) $M_y = \int_0^{40} x(5x+1)\,dx = \left[\frac{5x^3}{3} + \frac{x^3}{2}\right]_0^{40} = 107{,}467$; $\bar{x} = \frac{107{,}467}{4040} = 26.6$ cm from the less dense end.

38. (a) $m = \int_0^{8} (x+5)\,dx = \left[\frac{x^2}{2} + 5x\right]_0^{8} = 32 + 40 =$ 72 kg
(b) $M_y = \int_0^{8} x(x+5)\,dx = \left[\frac{x^3}{3} + \frac{5x^2}{2}\right]_0^{8} = 330.67$; $\bar{x} = \frac{330.67}{72} = 4.59$ m from the one end.

26.6 MOMENTS OF INERTIA

1. $$I_y = (3\cdot 4^2) + 5(-3)^2 = 48 + 45 = 93$$
$$I_x = 3(0)^2 + 5(0)^2 = 0$$
$$r_y = \sqrt{\tfrac{93}{3}} = 3.41;\ r_0 = 0;\ {}_0r = \sqrt{r_x^2 + r_y^2} = 3.4095$$

2. $$I_y = 6\cdot 0^2 + 3\cdot 0^2 + 1\cdot 0^2 = 0$$
$$I_x = 6\cdot 2^2 + 3\cdot(-5)^2 + 1\cdot 4^2 = 115$$
$$r_y = 0;\ r_x = \sqrt{\tfrac{115}{10}} = 3.39;\ r_0 = \sqrt{r_x^2 + r_y^2} = 3.3912$$

3. $I_y = 4\cdot 2^2 + 3\cdot 1^2 = 19;\ r_y = \sqrt{\frac{19}{7}} = 1.6475$

$I_x = 4\cdot 1^2 + 3\cdot 4^2 = 52;\ r_x = \sqrt{\frac{52}{7}} = 2.7255;$ $r_0 = \sqrt{\frac{19+52}{7}} = 3.1847$

4. $$I_y = 3\cdot 2^2 + 4\cdot 2^2 = 3\cdot 4^2 = 76$$
$$I_x = 3\cdot 3^2 + 4\cdot 4^2 + 3\cdot 5^2 = 166$$
$$r_0 = \sqrt{\frac{76 + 166}{10}} = 4.9193$$

5. $$I_y = \rho\int_0^{\sqrt{2}} x^2\cdot(2 - x^2)\,dx = \rho\left[\frac{2x^3}{3} - \frac{x^5}{5}\right]_0^{\sqrt{2}}$$
$$= \rho\left[\frac{4\sqrt{2}}{3} - \frac{4\sqrt{2}}{5}\right] = \rho\cdot\frac{8\sqrt{2}}{15} \approx 0.7542\rho$$
$$m = \rho\int_0^{\sqrt{2}} (2 - x^2)\,dx = \rho\left[2x - \frac{x^3}{3}\right]_0^{\sqrt{2}}$$
$$= \rho\left(2\sqrt{2} - \frac{2\sqrt{2}}{3}\right)$$
$$= \rho\cdot\frac{4\sqrt{2}}{3};\ r_y = \sqrt{\frac{\frac{8}{15}}{4/3}} = \sqrt{\frac{2}{5}} = 0.6325$$

6. $$I_x = \rho\int_0^{2} y^2(\sqrt{y})\,dy = \rho\left[\frac{2}{y}y^{7/2}\right]_0^{3} = \frac{16\sqrt{2}}{7}\rho$$
$$= 3.2325\rho$$
$$m = \frac{4\sqrt{2}}{3}\rho$$
$$r_x = \sqrt{\frac{\frac{16\sqrt{2}}{7}}{\frac{4\sqrt{2}}{3}}} = \sqrt{\frac{12}{7}} = 1.3093$$

7. $$I_x = \rho\int_0^{5} y^2\cdot 3\,dy = \rho y^3\big|_0^5 = 125\rho$$
$$m = \rho\int_0^{5} 3\,dx = 15$$
$$r_x = \sqrt{\frac{125}{15}} = \sqrt{\frac{25}{3}} \approx 2.8868$$

8. $$I_y = \rho\int_0^{3} x^2\cdot 5\,dx = \rho\frac{5x^3}{3}\bigg|_0^3 = 45\rho$$
$$m = 15$$
$$r_y = \sqrt{\frac{45}{15}} = \sqrt{3} \approx 1.7321$$

9. $$I_x = \rho \int_0^1 y^2(y^{1/3} - y^{1/2})\,dy = \rho\left[\frac{3}{10}y^{10/3} - \frac{2}{7}y^{7/2}\right]_0^1 = \frac{\rho}{70} \approx 0.014286\rho$$
$$m = \rho\int_0^1 (x^2 - x^3)\,dx = \rho\left[\frac{x^3}{3} - \frac{x^4}{4}\right]_0^1 = \frac{1}{12}\rho$$
$$r_x = \sqrt{\frac{\frac{1}{70}}{\frac{1}{12}}} = \sqrt{\frac{12}{70}} \approx 0.4140$$

10. $$I_y = \rho\int_0^1 x^2(x^2 - x^3)\,dx = \rho\left[\frac{x^5}{5} - \frac{x^6}{6}\right]_0^1 = \frac{\rho}{30} \approx 0.0333\rho$$
$$m = \frac{\rho}{12};\ r_y = \sqrt{\frac{\frac{1}{30}}{\frac{1}{12}}} = \sqrt{\frac{12}{30}} = \sqrt{\frac{2}{5}} \approx 0.6325$$

11. $$I_y = 5\int_0^4 x^2(4x - x^2)\,dx = 5\left[x^4 - \frac{x^5}{5}\right]_0^4 = 5\left[256 - \frac{1024}{5}\right] = 256\text{ g}\cdot\text{cm}^2$$
$$m = 5\int_0^4 (4x - x^2)\,dx = 5\left[2x^2 - \frac{x^3}{3}\right]_0^4 = 5\left[32 - \frac{64}{3}\right] = 53.333;\ r_y = \sqrt{\frac{256}{53.333}} \approx 2.1909\text{ cm}$$

12. $$I_x = 4\int_0^4 y^2(y^{3/2})\,dy = 4\int_0^4 y^{7/2}\,dy = 4\left[\frac{2}{9}y^{9/2}\right]_0^4 = \frac{4096}{9} = 455.11\text{ g}\cdot\text{cm}^2$$
$$m = 4\int_0^8 x^{2/3}\,dx = 4\left[\frac{3}{5}x^{5/3}\right]_0^8 = \frac{32\cdot 3\cdot 4}{5} = 51.2\text{ g}$$
$$r_x = \sqrt{\frac{455.1111}{51.2}} \approx 2.9814\text{ cm}$$

13. $$I_y = 8\int_1^2 x^2\left(x^2 - \frac{1}{x^2}\right)dx = 8\left[\frac{x^5}{5} - x\right]_1^2 = 8\left[\frac{32}{5} - 2 - \frac{1}{5} + 1\right] = 8\left[\frac{26}{5}\right] = 41.6\text{ g}\cdot\text{cm}^2$$
$$m = 8\int_1^2\left(x^2 - \frac{1}{x^2}\right)dx = 8\left[\frac{x^3}{3} + \frac{1}{x}\right]_1^2 = 8\left[\frac{8}{3} + \frac{1}{2} - \frac{1}{3} - 1\right] = 14\frac{2}{3}\text{ g}$$
$$r_y = \sqrt{\frac{41.6}{14.6667}} \approx 1.6842\text{ cm}$$

14. $$I_x = 8\int_0^1 y^2\left(2 - y^{-1/2}\right)dy + 8\int_1^4 y^2\left(2 - y^{1/2}\right)dy$$
$$= 8\left[\frac{2y^3}{3} - \frac{2}{5}y^{5/2}\right]_0^1 + 8\left[\frac{2y^3}{3} - \frac{2}{7}y^{7/2}\right]_1^4$$
$$= 8\left[\frac{2}{3} - \frac{2}{5}\right] + 8\left[\frac{128}{3} - \frac{256}{7} - \frac{2}{3} + \frac{2}{7}\right]$$
$$= 8\left[\frac{128}{3} - \frac{2}{5} - \frac{254}{7}\right] = 47.8467\text{ g}\cdot\text{cm}^2$$
$$r_x = \sqrt{\frac{47.8476}{14.6666}} = 1.8062\text{ cm}$$

15. $$I_y = 2\pi\rho\int_0^{\sqrt{2}} x^3\left[2 - x^2\right]dx = 2\pi\rho\left[\frac{2x^4}{4} - \frac{x^6}{6}\right]_0^{\sqrt{2}} = 2\pi\rho\left[2 - \frac{8}{6}\right] = \frac{4\pi\rho}{3}$$
$$m = 2\pi\rho\int_0^{\sqrt{2}} x\left(2 - x^2\right)dx = 2\pi\rho\left[x^2 - \frac{x^4}{4}\right]_0^{\sqrt{2}} = 2\pi\rho$$
$$r_y = \sqrt{\frac{4/3}{2}} = \sqrt{2/3} \approx 0.8165$$

16. $$I_x = 2\pi\rho\int_0^2 y^3\left[y^{1/2}\right]dy = 2\pi\rho\left[\frac{2}{9}y^{9/2}\right]_0^2 = 2\pi\rho\cdot\frac{2}{9}\cdot 16\sqrt{2} = \frac{64\pi\rho\sqrt{2}}{9} \approx 31.5938\rho$$

$$m = 2\pi\rho \int_0^2 y^{3/2}\,dy = 2\pi\rho \left[\frac{2}{5}y^{5/2}\right]_0^2 = \frac{16\pi\sqrt{2}\rho}{5}$$

$$r_x = \sqrt{\frac{\frac{64\sqrt{2}}{2}}{\frac{16\sqrt{2}}{5}}} = \sqrt{\frac{20}{9}} = \frac{2}{3}\sqrt{5} \approx 1.4907$$

17. $$I_y = 2\pi\rho \int_0^4 x^3\left(4x - x^2\right)dx = 2\pi\rho\left[\frac{4x^5}{5} - \frac{x^6}{6}\right]_0^4$$

$$= 2\pi\rho\left(\frac{4096}{30}\right) \approx 857.86\rho \text{ or } 273.07\pi\rho$$

$$m = 2\pi\rho \int_0^4 x(4x - x^2)\,dx = 2\pi\rho\left[\frac{4x^3}{3} - \frac{x^4}{4}\right]_0^4$$

$$= \frac{128\pi\rho}{3}$$

$$\approx 134.04\rho \text{ or } 42.667\pi\rho$$

$$r_y = \sqrt{\frac{273.07}{42.667}} = 2.5298$$

18. $$I_x = 2\pi\rho \int_0^{16} y^3\left(y^{1/2} - \frac{1}{4}y\right)dy$$

$$= 2\pi\rho \int_0^{16}\left(y^{7/2} - \frac{1}{4}y^4\right)dy$$

$$= 2\pi\rho\left[\frac{2}{9}y^{9/2} - \frac{4y^5}{5}\right]_0^{16} \approx 11{,}650.84\pi\rho$$

$$m = 2\pi\rho \int_0^{16} y\left(y^{1/2} - \frac{1}{4}y\right)dy$$

$$= 2\pi\rho \int_0^{16}\left(y^{3/2} - \frac{1}{4}y^2\right)dy$$

$$= 2\pi\rho\left[\frac{2}{5}y^{5/2} - \frac{1}{12}y^3\right]_0^{16} \approx 136.53\pi\rho$$

$$r_x \approx \sqrt{\frac{11{,}650.84}{136.53}} = 9.24$$

19. $$I_x = 2\pi\rho \int_0^6 y^3 \cdot 4\,dy = 2\pi\rho y^4 \,\Big|_0^6 = 2592\pi\rho$$

$$m = 2\pi\rho \int_0^6 y \cdot 4\,dx = 2\pi\rho\left[2y^2\right]_0^6 = 144\pi\rho$$

$$r_x = \sqrt{\frac{2592}{144}} \approx 4.2426$$

20. $$I_y = 2\pi\rho \int_0^4 x^3 \cdot 6\,dx = 2\pi\rho\left[\frac{6x^4}{4}\right]_0^4 = 768\pi\rho$$

$$m = 2\pi\rho \int_0^4 x \cdot 6\,dx = 2\pi\rho\left[3x^2\right]_0^4 = 96\pi\rho$$

$$r_y = \sqrt{\frac{768}{96}} = \sqrt{8} \approx 2.8284$$

21. $$I_y = 2\pi \cdot 3 \int_1^4 x^3(4x - x^2(4 - x))\,dx$$

$$= 6\pi \int_1^4 x^3(5x - x^2 - 4)\,dx$$

$$= 6\pi \int_1^4 (5x^4 - x^5 - 4x^3)\,dx$$

$$= 6\pi\left[x^5 - \frac{x^6}{6} - x^4\right]_1^4$$

$$= 6\pi(85.5) = 513\pi \text{ g} \cdot \text{cm}$$

$$m = 6\pi \int_1^4 x(5x - x^2 - 4)\,dx$$

$$= 6\pi \int_1^4 (5x^2 - x^3 - 4x)\,dx$$

$$= 6\pi\left[\frac{5x^3}{3} - \frac{x^3}{4} - 2x^2\right]_1^4 = 67.5\pi \text{ g}$$

$$r_y = \sqrt{\frac{513}{67.5}} = 2.7568 \text{ cm}$$

22. $$y = \sqrt{4 - x} \Rightarrow y^2 = 4 - x \Rightarrow x = 4 - y^2$$

$$x + 2y = 4 \Rightarrow x = 4 - 2y$$

$$I_x = 2\pi \cdot 5 \int_0^2 y^3(4 - y^2 - (4 - 2y))\,dy$$

$$= 10\pi \int_0^2 y^3(2y - y^2)\,dy$$

$$= 10\pi \int_0^2 (2y^4 - y^5)\,dy$$

$$= 10\pi\left[\frac{2y^5}{5} - \frac{y^6}{6}\right]_0^2 = 21.333\pi \text{ g} \cdot \text{cm}^2$$

$$m = 10\pi \int_0^2 y(2y - y^2) = 10\pi \int_0^2 (2y^2 - y^3)\,dy$$
$$= 10\pi \left[\frac{2y^3}{3} - \frac{y^4}{4}\right]_0^2 = 13.333 \text{ g}$$
$$r_x = \sqrt{\frac{21.333}{13.333}} = \sqrt{1.6} = 1.2649 \text{ cm}$$

23. $I_y = 2\pi \cdot 5 \int_1^2 x^3(x^3 - x^{-1})\,dx$
$$= 10\pi \int_1^2 (x^6 - x^2)\,dx = 10\pi \left[\frac{x^7}{7} - \frac{x^3}{3}\right]_1^2$$
$$= 158.0952\pi \text{ g} \cdot \text{cm}^2$$
$$m = 10\pi \int_1^2 x(x^3 - x^{-1})\,dx = 10\pi \int_1^2 (x^4 - 1)\,dx$$
$$= 10\pi \left[\frac{x^5}{5} - x\right]_1^2 = 52\pi \text{ g}$$
$$r_y = \sqrt{\frac{158.0952}{52}} = 1.7436 \text{ cm}$$

24. $I_x = 10\pi \int_{1/2}^1 y^3(2 - y^{-1})\,dy + 10\pi \int_1^8 y^3(2 - y^{1/3})\,dy$
$$= 10\pi \int_{1/2}^1 (2y^3 - y^2)\,dy + 10\pi \int_1^8 (2y^3 - y^{10/3})\,dy$$
$$= 10\pi \left[\frac{y^4}{2} - \frac{y^3}{3}\right]_{1/2}^1 + 10\pi \left[\frac{y^4}{2} - \frac{3}{13}y^{\frac{13}{3}}\right]_1^8$$
$$= 1574.4631\pi \text{ g} \cdot \text{cm}^2$$
$$m = 10\pi \int_{1/2}^1 y(2 - y^{-1})\,dy + 10\pi \int_1^8 y(2 - y^{1/3})\,dy$$
$$= 10\pi \left[y^2 - y\right]_{1/2}^1 + 10\pi \left[y^2 - \frac{3}{7}y^{7/3}\right]_1^8$$
$$= 88.2143\pi \text{ g}$$
$$r_x = \sqrt{\frac{1574.4631}{88.2143}} = 4.2247 \text{ cm}$$

26.7 WORK AND FLUID PRESSURE

1. $F = kx$; $3 = k(1/3)^1$ yields $k = 9$. Hence, $F = 9x$; 10 in. $= \frac{5}{6}$ ft. Work $= \int_0^{5/6} 9x\,dx = \frac{9}{2}x^2\Big|_0^{5/6} = \frac{9}{2} \cdot \frac{25}{36} = \frac{25}{8} = 3.125$ ft·lb

2. $\int_1^2 9x\,dx = \frac{9}{2}x^2\Big|_1^2 = \frac{9}{2}(4 - 1) = \frac{27}{2} = 13.5$ ft·lb.

3. 20 cm =0.2 m; $6 = k(0.2)$; so $k = 30$ and we get $F = 30x$. Thus, work $= \int_0^1 30x = \frac{30x^2}{2}\Big|_0^1 = 15$ J

4. $\int_{0.1}^{0.2} 30x = 15x^2\Big|_{0.1}^{0.2} = 15(0.04 - 0.01) = 0.45$ J

5. $\int_0^{0.1} kx\,dx = \frac{k}{2}x^2\Big|_0^{0.1} = \frac{k(0.01)}{2} = 0.20$. Hence, $k = \frac{0.4}{0.01} = 40$; $F = kx = 40 \cdot (0.1) = 4$ N

6. $F = kx$; $5 = k\left(\frac{1}{6}\right)$, so $k = 30$. Work $= \int_{1/12}^{4/12} 30x\,dx = 15x^2\Big|_{1/12}^{1/3} = 15\left(\frac{1}{9} - \frac{1}{144}\right) = 15\left(\frac{15}{144}\right) = \frac{225}{144}$ ft·lb or 1.5625 ft·lb

7. $\int_0^b 9x\,dx = \frac{9}{2}x^2\Big|_0^b = \frac{9}{2}b^2$; $\frac{9}{2}b^2 = 105$; $b^2 = 23.333$; $b = 4.8305$ ft

8. The weight of a section is $2\,dy$; distance moved is $50 - y$. Work $= \int_0^{20} 2(50 - y)\,dy = 2\left[50y - \frac{y^2}{2}\right]_0^{20} = 2[1000 - 200] = 2(800) = 1600$ ft·lb.

9. $\int_0^{30} 2.5(30 - y)\,dy = 2.5\left[30y - \frac{y^2}{2}\right]_0^{30} = 2.5(450) = 1125$ J

10. The mass = 4.7 kg/m; weight = 4.7 × 9.8 = 46.06 N. A mass of 87 kg has a weight of

$87 \times 9.8 = 852.6$ N. The work in lifting the chain is $\int_0^{10} 46.06(15 - y)\,dy = 46.06\left[15 - \frac{y^2}{2}\right]_0^{10} =$ 4606 J. The work is lifting the load is $852.6 \times 10 =$ 8526 J. The total work is $4606 + 8526 = 13{,}132$ J or about 13 kJ.

11. For the first 12 feet; each section will have a weight of $0.75\,dy$. $\int_0^{12} 0.75(12 - y)\,dy = 0.75\left(12y - \frac{y^2}{2}\right)_0^{12} =$ $\frac{3}{4}(72) = 54$ ft·lb. The total weight of the chain is $12 \cdot \frac{3}{4} = 9$ lb. The work in lifting the last 8 ft is 72 ft·lb. The total work is $54 + 72 = 126$ ft·lb.

12. For the first 5 meters; each section has a weight of $0.4 \times 0.8\Delta y = 3.92\Delta y$ N. The work in raising one end 5 meters is $\int_0^5 3.92(5 - y)\,dy =$ $3.92\left(5y - \frac{y^2}{2}\right)\Big|_0^5 = 3.92 \times 12.5 = 49$ N·m = 49 J. The total weight of the chain is $5(3.92) = 19.6$ N. The work in lifting the chain the last 5 meter is $19.6\times 5 = 98$ J. The total work done is $49+98 = 147$ J.

13. When the tank is x meters above the ground it has taken $2x$ minutes to get there and has used $20 \cdot 2x = 40x$ L. Hence the volume of the tank is $(1000 - 40x)$ L. Water has a dersity of 1 kg/L so its mass is $(1000 - 40x)$kg. The total mass is $(1047 - 40x)$ kg so its weight is $9.8(1047 - 40x)$. $W = \int_0^{20} 9.8(1047 - 40x)\,dx =$ $9.8\left[1047x - 20x^2\right]_0^{20} = 9.8(12940) = 126{,}812$ J or 127 kJ.

14. 1 cm = 0.01 m; 3 mm = 0.003 m

$$W = \int_{0.003}^{0.01} 8.988 \times 10^9 \cdot 5 \times 10^{-7}$$

$$\times(-2) \times 10^{-7} \cdot \frac{1}{r^2}dr$$

$$= -8.988 \times 10^{-4}\left[\frac{-1}{r}\right]_{0.003}^{0.01}$$

$$= -8.988 \times 10^{-4}(233.33)$$

$$= -2.0972 \times 10^{-1} = -0.20972 \text{ J}$$

15. $W = \int_{0.05}^{10\times 10^{-9}} 8.988 \times 4 \times 10^{-19} \times 5 \times 10^{-8}\frac{1}{r^2}dr$

$$= 1.7976 \times 10^{-16}\left[\frac{-1}{r}\right]_{0.05}^{10^{-8}}$$

$$= 1.7976 \times 10^{-16} \times 10^8$$

$$= 1.7976 \times 10^{-8} \text{ J} = 179.76\ \mu\text{J}$$

16. $W = \int_{5.3\times 10^{-7}}^{10^{-3}} 8.988\times 1.6\times 10^{-18}\times(-1.6)\times 10^{-19}\cdot$ $\frac{1}{r^2}dr = 2.3009\times 10^{-27}\left[\frac{-1}{r}\right]_{5.3\times 10^{-7}}^{10^{-3}} = -2.3009\times$ $10^{-27}(1.8858 \times 10^6) = -4.3391 \times 10^{-21}$ J

17. $W = \int F(r)dr = \int_{0.01}^{1} \frac{k}{r^2} = -k\left[\frac{1}{r}\right]_{0.01}^{1} =$ $-k\,[1 - 100] = 99k$ J.

18. $W = \int_1^{20} \frac{k}{r^2}dr = \frac{-k}{r}\Big|_1^{20} = -k\left[\frac{1}{20} - 1\right] = \frac{19k}{20}$ ft·lb.

19. $W = \int_{4000}^{5000} \frac{k}{r^2} = -k\left[\frac{1}{r}\right]_{4000}^{5000} = k\left[5 \times 10^{-5}\right]$. Since $k = 16\times 10^9$, we get $W = 16\times 10^9\cdot 5\times 10^{-5} =$ 80,000 mi·lb $= 4.224 \times 10^9$ ft·lb.

20. $\int_0^8 (54.8)(16\pi)x\,dx = 876.8\pi\frac{x^2}{2}\Big|_0^8 = 28057.6\pi \approx$ 88145.6 ft·lb.

21. $\int_0^8 880(4\pi)x\,dx = 1.264\pi \times 10^5$ kg·m. Multiplying by the gravitational constant to get joules, we have $1.264\pi \times 10^5 \times 9.8 = 1.23872\pi \times 10^6$ J $= 3.89155 \times 10^6$ J

22. $\int_0^2 70 \cdot 9.8 \cdot 10^3 \cdot x\,dx = 6.86 \times 10^5\frac{x^2}{2}\Big|_0^2 =$ 1.372×10^6 J = 1.372 MJ. (Note: 1 m^3 of water has a mass of 10^3 kg and a weight of 9.8×10^3 N.)

23. $\int_0^{1.5} 6.86 \times 10^5(2-x)\,dx = 6.86 \times 10^5 \left[2x - \frac{x^2}{2}\right]_0^{1.5} =$ $6.86 \times 10^5 \times 1875 = 1.28625$ MJ $= 1{,}286{,}250$ J

24. $\int_0^{10} \rho(9\pi)(x)\,dx + 9\pi\rho 10 \int_0^5 1\,dx = 9\pi\rho\frac{x^2}{2}\bigg|_0^{10} +$ $450\pi\rho = 450\pi\rho + 450\pi\rho = 900\pi\rho = 900 \cdot 54.8\pi =$ $49{,}320\pi = 1.549 \times 10^5$ ft·lb.

25. Position the tank so that the bottom is at $(0,0)$. Then the radius will be $\frac{2}{3}y$ where y is the height. The volume will be $\pi\left(\frac{2}{3}y\right)^2 dy$; $\rho = 847\text{ kg/m}^3 =$ 8300.6 N/m^3. $W = 8300.6\pi \int_0^{3.6} \left(\frac{2}{3}y\right)^2 (18.6 - y)\,dy = 8300.6\pi \cdot \left[\frac{4}{9}(18.6)\frac{y^3}{3} - \frac{4}{9}\cdot\frac{y^4}{4}\right]_0^{3.6} =$ $8300.6\pi\,[109.9] = 912\ 236\pi$ J $= 2\ 866\ 000$ J $= 2.866$ MJ

26. $r = \sqrt{16-y^2} = r^2 = (16-y^2)$; distance raised is $(-y)$ $W = \int_{-4}^{0} (1000 \cdot 9.8)\pi(16-y^2)(-y)\,dy = 9800\pi\left[-8y^2 + \frac{y^4}{4}\right]_{-4}^{0} =$ $9800\pi\,[64] = 627200\pi$ J.

27. $W = 9800\pi \int_{-2}^{0} (16 - y^2)(-y)\,dy =$ $9800\pi\left[-8y^2 + \frac{y^4}{4}\right]_{-2}^{0} = 9800\pi\,[32-4] =$ $9800\pi\,[28] = 274400\pi$ J

28. Using the same set-up as Example 26.40 we get $9800\pi \int_1^4 (16-x^2)(14-x)\,dx$ where x is the distance from the top of the tank. This is equal to $9800\pi \int_1^4 (224 - 16x - 14x^2 + x^3)\,dx =$ $9800\pi\left[224x - 8x^2 - \frac{14x^3}{3} + \frac{x^4}{4}\right]_1^4 = 9\,800\pi(321.75) =$ $3\,153\,150\pi$ J.

29. (a) $P = \rho g h = 1000 \cdot 9.8 \cdot 2 = 19{,}600\text{ N/m}^2$
(b) $F = PA = 19{,}600 \times 10 \times 7 = 1372000$ N

30. (a) $P = \rho g h = 54.8 \cdot 10 = 548\text{ lb/ft}^2$; (b) $F = PA =$ $548 \cdot 9\pi = 4932\pi = 15494$ lb.

31. The area of the short side is 7 m $\times 2$ m. So, we get $= 9800 \int_0^2 h(y)\cdot L(y)\,dy = 9800 \int_0^2 (2-y)\cdot 7\,dy =$ $9800\left[14y - \frac{7y^2}{2}\right]_0^2 = 9800 \cdot 14 = 137200$ N.

32. Long side $9800 \int_0^2 (2 - y) \cdot 10 =$ $9800\left[20y - 5y^2\right]_0^2 = 9800 \cdot 20 = 196\,000$ N

33. $9800 \int_0^2 (2-y)\left(3 - \frac{3}{2}y\right)dy = 9800 \int_0^2 \left(6 - 6y + \frac{3}{2}y^2\right)dy =$
$9800\left[6y - 3y^2 + \frac{y^3}{2}\right]_0^2 = 9800\,[4] = 39\,200$ N

34. $\rho g \int_0^1 h(40)\,dh + \rho g \int_0^2 (1+h)(40-20h)\,dh$
$= 9800 \int_0^1 40h + 9800 \int_0^2 (40 + 20h - 20h^2)dh$
$= 9800\left[20h^2\right]_0^1 + 9800\left[40h + 10h^2 - \frac{20h^3}{3}\right]_0^2$
$= 9800\,[20] + 9800\,[66.67] = 9800\,[86.67]$
$= 849300$ N $= 849$ kN

35. $r = 1.6$ m; $\rho = 680\text{ kg/m}^3$; $\rho g = 680 \times 9.8 = 6664\text{ N/m}^3$. Thus, we have $\rho g \int_{-1.6}^{0} (1\cdot 6 + y)(2\sqrt{2.56-y^2})\,dy = \rho g \cdot 3.2 \int_{-1.6}^{0} \sqrt{2.56-y^2} - \rho g \int_{-1.6}^{0} -2y\sqrt{2.56-y^2}\,dy$. As in example 26.43. The left integeral will be $\frac{1}{4}$ the area of a circle so $\frac{1}{4}\cdot\pi(1.6)^2 = 2.0106$. The right integral is $\frac{2}{3}(2.56-y^2)^{3/2}\Big|_{-1.6}^{0} =$ $4.096 \cdot \frac{2}{3} = 2.7307$; $\rho g\,[3.2 \times 2.0106 - 2.7307] =$ $\rho g\,[3.7032] = 6664(3.7032) = 24\,578$ N.

36. $x^2 + y^2 = 16$. $2\rho\int_{-4}^{0}(4+y)\sqrt{16-y^2}\,dy = 2\rho\int_{-4}^{0}4\sqrt{16-y^2} - 2\rho\int_{-4}^{0}-y\sqrt{16-y^2}\,dy = 8\rho\int_{-4}^{0}\sqrt{16-y^2}\,dy - \rho\int_{-4}^{0}-2y\sqrt{16-y^2}\,dy$. The left integral is $\frac{1}{4}$ the area of a circle of radius 4 so 4π. The right integral is $\frac{2}{3}(16-y^2)^{3/2}\Big|_{-4}^{0} = \frac{128}{3}$. Putting this together $8\rho\cdot 4\pi - \rho\cdot\frac{128}{3} = 42.5(57.86) = 2{,}459.2$ lb.

CHAPTER 26 REVIEW

1. $$\bar{y} = \frac{1}{5}\int_1^6 4x\,dx = \frac{1}{5}2x^2\Big|_1^6 = \frac{1}{5}(72-2) = \frac{20}{5} = 14$$
$$f_{\text{rms}} = \sqrt{\frac{1}{5}\int_1^6 16x\,dx} = \sqrt{\frac{1}{5}\left[\frac{16x^3}{3}\right]_1^6}$$
$$= \sqrt{229.\overline{3}} = 15.1438$$

2. $$\bar{y} = \frac{1}{2}\int_0^2 x^3\,dx = \frac{1}{2}\frac{x^4}{4}\Big|_0^2 = \frac{16}{8} = 2$$
$$y_{\text{rms}} = \sqrt{\frac{1}{2}\int_0^2 x^6\,dx} = \sqrt{\frac{1}{2}\frac{x^7}{7}\Big|_0^7}$$
$$= \sqrt{9.142857} = 3.0237$$

3. $$\bar{y} = \frac{1}{2}\int_2^4 (x^2-4)\,dx = \frac{1}{2}\left[\frac{x^3}{3} - 4x\right]_2^4$$
$$= \frac{1}{2}\cdot\frac{32}{3} = \frac{16}{3}$$
$$h_{\text{rms}} = \sqrt{\frac{1}{2}\int_2^4 (x^4 - 8x^2 + 16)\,dx}$$
$$= \sqrt{\frac{1}{2}\left[\frac{x^5}{5} - \frac{8x^3}{3} + 16x\right]_2^4}$$
$$= \sqrt{40.5\overline{3}} = 6.3666$$

4. $$\bar{y} = \frac{1}{2.5}\int_0^{2.5}(4.9t^2 - 2.8t - 4)\,dt = \frac{6.7708\overline{3}}{2.5}$$
$$= 2.7083$$
$$k_{\text{rms}} = \sqrt{\frac{1}{2.5}\int_0^{2.5}(4.9t^2 - 2.8t - 4)^2\,dt}$$
$$= \sqrt{\frac{147.6432}{2.5}}$$
$$= \sqrt{59.0573} = 7.6849$$

5. $$V = \pi\int_1^5 (6x)^2\,dx = 36\pi\cdot\frac{x^3}{3}\Big|_1^5$$
$$= 12\pi\,[125 - 1] = 1488\pi$$
$$M_y = \pi\int_1^5 x(6x)^2 = 36\pi\int_1^5 x^3\,dx = 36\pi\left[\frac{x^4}{4}\right]_1^5$$
$$= 9\pi\,[625 - 1] = 5616\pi$$
$$\bar{x} = \frac{5676}{1488} = 3.7742$$

The centroid is $(3.7742, 0)$.

6. $y = -x \Rightarrow x = -y$. Hence,
$$V = \pi\int_1^4 (-y)^2\,dy = \pi\frac{y^3}{3}\Big|_1^4 = \frac{\pi}{3}(64-1) = 2\pi$$
$$M_x = \pi\int_1^4 \pi y^3\,dy = \pi\frac{y^4}{4}\Big|_1^4 = \pi\left(64 - \frac{1}{4}\right)$$
$$= 63.75\pi$$
$$\bar{y} = \frac{63.75}{21} = 3.0357$$

The centroid is $(0, 3.0357)$.

7. $$V = \pi\int_1^2 \left(x^4\right)^2 dx = \pi\frac{x^9}{9}\Big|_1^2 = \frac{\pi}{9}(512-1) = \frac{511\pi}{9}$$

$$M_y = \pi \int_1^2 x^9\,dx = \pi \frac{x^{10}}{10}\bigg|_1^2 = \frac{\pi}{10}(1024-1) = \frac{1023\pi}{10}$$

$$\bar{x} = \frac{1023}{10} \times \frac{9}{511} = 1.8018$$

The centroid is $(1.8018, 0)$.

8. $$V = \pi \int_0^1 \left(x^{2/3}\right)^2 dx = \pi \frac{3}{7} x^{7/3}\bigg|_0^1 = \frac{3\pi}{7}$$

$$M_y = \pi \int_0^1 x\left(x^{4/3}\right) dx = \pi \frac{3}{10} x^{10/3} = \frac{3\pi}{10}$$

$$\bar{x} = \frac{3}{10} \cdot \frac{7}{3} = \frac{7}{10} \text{ or } 0.7$$

The centroid is $(0.7, 0)$.

9. $$V = \pi \int_0^1 \left(y^{2/3} - y^6\right) dy = \pi \left[\frac{3}{5}y^{5/3} - \frac{y^7}{y}\right]_0^1 = \frac{16\pi}{35}$$

$$\approx 0.4571\pi$$

$$M_x = \pi \int_0^1 \left(y^{5/3} - y^7\right) dy = \pi \left[\frac{3}{8}y^{8/3} - \frac{y^8}{3}\right]_0^1 = \frac{\pi}{4}$$

$$\bar{y} = \frac{1}{4} \cdot \frac{35}{16} = \frac{35}{64} \approx 0.5469$$

The centroid is $(0, 0.5469)$.

10. $$V = \pi \int_0^1 \left(x^{2/3} - x^6\right) dx = \frac{16\pi}{35}$$

$$M_y = \pi \int_0^1 x\left(x^{2/3} - x^6\right) = \frac{\pi}{4}$$

$$\bar{x} = \frac{1}{4} \cdot \frac{35}{16} = \frac{35}{64} \approx 0.5469$$

The centroid is $(0.5469, 0)$.

11. Here $y = x^2 \Rightarrow x = y^{1/2}$, $y = 9 - x^2 \Rightarrow x = \sqrt{9-y}$, and $x^2 = 9 - x^2 \Rightarrow 2x^2 = 9$. Hence $x = \pm\frac{3}{\sqrt{2}}$ and $y = \frac{9}{2} = 4.5$.

$$V = \pi \int_0^{4.5} y\,dy + \pi \int_{4.5}^9 (9-y)\,dy$$

$$= \pi \frac{y^2}{2}\bigg|_0^{4.5} + \pi \left[9y - \frac{y^2}{2}\right]_{4.5}^9 = 20.25\pi$$

$$M_x = \pi \int_0^{4.5} y^2\,dy + \pi \int_{4.5}^9 (9y - y^2)\,dy$$

$$= \pi \left(\left[\frac{y^3}{3}\right]_0^{4.5} + \left[\frac{9y^2}{2} - \frac{y^3}{3}\right]_{4.5}^9\right) = 91.125\pi$$

$$\bar{y} = \frac{91.125}{20.25} = 4.5$$

The centroid is $(0, 4.5)$. (Note: this is the answer you would expect due to the symmetry of the curves.)

12. Curves intersect at $(0,0)$ and $(2,4)$. Here $x = y^{1/3}$ and $x = \frac{y}{4}$.

$$V = \pi \int_0^4 \left(5 - \frac{y}{4}\right)^2 - \left(5 - y^{1/3}\right)^2 dy$$

$$= \pi \int_0^4 \left[\left(25 - \frac{5}{2}y + \frac{y^2}{16}\right) - \left(25 - 10y^{1/3} + y^{2/3}\right)\right] dy$$

$$= \pi \int_0^4 \left(-\frac{5}{2}y + \frac{y^2}{16} + 10y^{1/3} - y^{2/3}\right) dy$$

$$= \pi \left[\frac{-5y^2}{4} + \frac{y^3}{48} + \frac{15}{2}y^{4/3} - \frac{3}{5}y^{5/3}\right]_0^4$$

$$= \pi\,[22.9077]$$

$$M_x = \pi \int_0^4 \left(-\frac{5}{2}y^2 + \frac{y^3}{16} + 10y^{4/3} - y^{5/3}\right) dy$$

$$= \pi \left[-\frac{5y^3}{6} + \frac{y^3}{64} + \frac{30}{7}y^{7/3} - \frac{3}{8}y^{8/3}\right]_0^4$$

$$= 44.3980\pi$$

$$\bar{y} = \frac{44.3980}{22.9077} = 1.93812$$

Since its rotated about $x = 5$, then $\bar{x} = 5$. The centroid is $(5, 1.0169)$.

13. $x = y^{3/2}$; $x' = \frac{3}{2}y^{1/2}$; $x'^2 = \frac{9}{4}y$

$$L = \int_0^4 \sqrt{1 + x'^2}\,dy = \int_0^4 \sqrt{1 + \frac{9}{4}y}\,dy$$

$$= \frac{4}{9}\int_0^4 \frac{9}{4}\sqrt{1+\frac{9}{4}y}\,dy$$

$$= \frac{4}{9}\cdot\frac{2}{3}\left(1+\frac{9}{4}y\right)^{3/2}\Bigg|_0^4 = \frac{8}{27}\left(10^{3/2}-1\right)$$

$$= 9.0734$$

14. $y = \left(8x^3\right)^{1/2};$

$$y' = \frac{1}{2}\left(8x^3\right)^{-1/2}\cdot 24x^2 = \frac{24x^{1/2}}{4\sqrt{2}}\cdot 24x^2$$

$$= \frac{6x^{3/2}}{\sqrt{2}} = 3\sqrt{2}x^{3/2}$$

$$L = \int_0^2 \sqrt{1+(3\sqrt{2}x^{1/2})^2}\,dx = \int_0^2 \sqrt{1+18x}\,dx$$

$$= \frac{1}{18}\cdot\frac{2}{3}(1+18x)^{3/2}\Bigg|_0^2 = 8.2986$$

15. (a) Here $y = \frac{x^4}{4} + \frac{1}{8x^2}$ and $y' = x^3 - \frac{1}{4x^3}$.

$$L = \int_1^3 \sqrt{1+\left(x^3-\frac{1}{4x^3}\right)^2}\,dx$$

$$= \int_1^3 \sqrt{1+\left(x^6-\frac{1}{2}+\frac{1}{16x^6}\right)}\,dx$$

$$= \int_1^3 \sqrt{x^6+\frac{1}{2}+\frac{1}{16x^6}}\,dx = \int_1^3 \left(x^3+\frac{1}{4x^3}\right)dx$$

$$= x - \frac{1}{4}\cdot\frac{1}{2}x^{-2}\Bigg|_1^3 = \frac{81}{4} - \frac{1}{72} - \frac{1}{4} + \frac{1}{8} = 20.111$$

(b)

$$S = 2\pi\int_1^3 \left(\frac{x^4}{4}+\frac{1}{8x^2}\right)\sqrt{1+\left(x^6-\frac{1}{2}+\frac{1}{16x^2}\right)}\,dx$$

$$= 2\pi\int_1^3 \left(\frac{x^4}{4}+\frac{1}{8x^3}\right)\left(x^3+\frac{1}{4x^3}\right)dx$$

$$= 2\pi\int_1^3 \left(\frac{x^7}{4}+\frac{x}{6}+\frac{1}{8}+\frac{1}{32x^6}\right)dx$$

$$= 2\pi\left[\frac{x^8}{32}+\frac{x^2}{32}+\frac{x}{8}-\frac{1}{32}\frac{1}{5}x^{-5}\right]_1^3$$

$$= 2\pi\,[205.5062] = 411.0124\pi$$

16. (a) $x = \frac{1}{6}y^6 + \frac{1}{16}y^{-4}$; $x' = y^5 - \frac{1}{4}y^{-5}$; $x'^2 = \left(y^{10} - \frac{1}{2} + \frac{1}{16}y^{-10}\right)$

$$L = \int_1^2 \sqrt{1+\left(y^{10}-\frac{1}{2}+\frac{1}{16}y^{-10}\right)}\,dy$$

$$= \int_1^2 \left(y^5+\frac{1}{4}y^{-5}\right)dy$$

$$= \left(\frac{y^6}{6}-\frac{1}{16}y^{-4}\right)\Bigg|_1^2 = 10.5586$$

$$S = 2\pi\int_1^2 \left(\frac{y^6}{6}+\frac{1}{16y^4}\right)\left(y^5+\frac{1}{4y^5}\right)dy$$

$$= 2\pi\int_1^2 \left(\frac{y^{11}}{6}+\frac{y}{24}+\frac{y}{16}+\frac{1}{64y^9}\right)dy$$

$$= 2\pi\left[\frac{y^{12}}{72}+\frac{y^2}{48}+\frac{y^2}{32}-\frac{1}{512y^8}\right]_1^2$$

$$= 2\pi\,[57.08332] = 114.0664\pi$$

17. $F = kx$; $5 = k(0.2)$; $k = 25$; $W = \int_0^{0.6} 25x\,dx = \frac{25x^2}{2}\Big|_0^{0.6} = 4.5\text{ N·m} = 4.5\text{ J}$

18. $V = \frac{1}{60\times10^{-6}}\int_0^{0.001} 0.40\,dt = \frac{1}{60\times10^{-6}}0.40t\Big|_0^{0.001} = \frac{1}{60\times10^{-6}}(0.004) = 6.6667\text{ V}$

19. $I_{avg} = \frac{1}{2}\int_0^2 (4t-t^3)\,dt = \frac{1}{2}\left[2t^2-\frac{t^4}{4}\right]_0^2 = \frac{1}{2}[8-4] = 2\text{ A}$

20. $a(t) = -32$; $v(t) = -32t$; $s(t) = -16t^2 + 555$; $s(t) = 0 \Rightarrow 16t^2 = 555 \Rightarrow t = 5.8896$.

$$\bar{s} = \frac{1}{5.8896}\int_0^{5.8896}\left(-16t^2+555\right)dt$$

$$= \frac{1}{5.8896}\left[\frac{-16t^3}{3}+555t\right]_0^{5.8896}$$

$$\bar{s} = 370\text{ ft}$$

$$\bar{v} = \frac{1}{5.8896}\int_0^{5.8896}(-32t)\,dt$$

$$= \frac{1}{5.8896}\left[-16t^2\right]_0^{5.8896} = -94.2338\text{ ft/s}$$

21. $$W = \int_0^{30} [1000 + 5(50 - x)]\,dx = \int_0^{30} (1250 - 5x)\,dx = \left[1250x - \frac{5x^2}{2}\right]_0^{30} = 35250 \text{ ft} \cdot \text{lb}$$

22. $h(y) = (100 - y)$; $L(y) = 200 + 2y$

$$P = 62.4 \int_0^{100} (100 - y)(200 + 2y)\,dy = 62.4 \int_0^{100} (20000 - 2y^2)\,dy = 62.4 \left[20000y - \frac{2}{3}y^3\right]_0^{100} = 62.4(1.333 \times 10^6) = 8.32 \times 10^7 \text{ lb} = 4.16 \times 10^4 \text{ tons}$$

23. The top of the tank is at $x = 0$ and the bottom at $x = 16$ $W = 880 \int_0^{16} 36\pi \cdot x\,dx = 31680\pi \frac{x^2}{2}\Big|_0^{16} =$ $4\,055\,040\pi$ kg · m $= 39\,739\,392\pi$ J

24. $$62.4\pi \int_0^{16} \left(6 - \frac{3x}{8}\right)^2 x\,dx = 62.4\pi \int_0^{10} \left(36 - \frac{9}{2}x + \frac{9}{64}x^2\right) x\,dx = 62.4\pi \int_0^{10} \left(36x - \frac{9}{2}x^2 + \frac{9}{64}x^3\right) dx = 62.4\pi \left[18x^2 - \frac{3}{2}x^3 + \frac{9}{256}x^4\right]_0^{16} = 62.4\pi(768) = 4.79232 \times 10^4\pi \text{ ft} \cdot \text{lb}$$

CHAPTER 26 TEST

1. $\bar{y} = \frac{1}{5-1} \int_1^5 (3x^2 + 1)\,dx = \frac{1}{4}(x^3 + x)\Big]_1^5 = \frac{1}{4}(130 - 2) = 32$

2. $g_{\text{rms}} = \sqrt{\frac{1}{3-1} \int_1^3 (x^3 - 1)^2\,dx} = \sqrt{\frac{1}{2} \int_1^3 (x^6 - 2x^3 + 1)\,dx} = \sqrt{\frac{1}{2} \left[\frac{1}{7}x^7 - \frac{1}{2}x^4 + x\right]_1^3} \approx \sqrt{137.1429} \approx 11.71$

3. $V = \pi \int_0^7 [f(x)]^2\,dx = \pi \int_0^7 \sqrt{9x + 1}^2\,dx = \pi \int_0^7 (9x + 1)\,dx = \pi \left[\frac{9}{2}x^2 + x\right]_0^7 = 227.5\pi$

4. We begin be determining the moment: $M_y = \pi \int_0^4 \rho x(x^2 + 1)^2\,dx = \rho\pi \int_0^4 (x^5 + 2x^3 + x)\,dx = \rho\pi \left[\frac{1}{6}x^6 + \frac{1}{2}x^4 + \frac{1}{2}x^2\right]_0^4 = \frac{2456}{3}\rho\pi$. Now we find the mass, $m = \rho V$, from $V == \pi \int_0^4 (x^2 + 1)^2\,dx = \pi \int_0^4 (x^4 + 2x^2 + 1)\,dx = \pi \left[\frac{1}{5}x^5 + \frac{2}{3}x^3 + x\right]_0^4 = \frac{7544}{30}\pi$. So, $m = \frac{7544}{30}\rho\pi$ and $\bar{x} = \frac{M_y}{m} = \left(\frac{2456}{3}\rho\pi\right) / \left(\frac{7544}{30}\rho\pi\right) = \frac{24560}{7544} \approx 3.26$. Thus, the centroid is at $(3.26, 0)$

5. Since $y' = \frac{3}{2}x^{1/2}$, the arc length $L = \int_1^6 \sqrt{1 + \left(\frac{3}{2}x^{1/2}\right)^2}\,dx = \int_1^6 \sqrt{1 + \frac{9}{4}x}\,dx = \frac{1}{2} \int_1^6 \sqrt{4 + 9x}\,dx$. Let $u = 4+9x$ and then $du = 9\,dx$, and if $x = 1$, $u = 13$ and if $x = 6$, $u = 58$, so $L = \frac{1}{18} \int_{13}^{58} u^{1/2}\,du = \frac{1}{27}u^{3/2}\Big]_{13}^{58} \approx 14.6238$.

6. Here $y' = \frac{1}{2}x^2$ and so, the surface area $S = \int_1^3 2\pi \left(\frac{1}{6}x^3\right) \sqrt{1 + \left(\frac{1}{2}x^2\right)^2}\,dx = \frac{\pi}{3} \int_1^3 x^3 \sqrt{1 + \frac{1}{4}x^4}\,dx = \frac{4\pi}{9} \left(1 + \frac{1}{4}x^4\right)^{3/2}\Big|_1^3 \approx 42.9156\pi \approx 134.82$

7. $W = \int_0^{10} 25\pi\rho(y + 4)\,dy = 25\pi\rho \int_0^{10} (y + 4)\,dy = 25\pi\rho \left[\frac{1}{2}y^2 + 4y\right]_0^{10} = 25\pi\rho(50 + 40) = 2250\pi\rho = 2250\,000\pi$ J

8. $F = \rho \int_5^8 y2(8 - y)\,dy = \rho \int_5^8 (16y - 2y^2)\,dy = \rho \left[8y^2 - \frac{2}{3}y^3\right]_5^8 \approx 54\rho$. Since the density of water is $\rho = 62.5$ lb, we have a force of $F = 54(62.5) = 3375$.

CHAPTER

27

Derivatives of Transcendental Functions

27.1 DERIVATIVES OF THE SINE AND COSINE FUNCTIONS

1. $y = \sin 3x$; $y' = 3\cos 3x$

2. $y = \cos 4x$; $y' = -4\sin 4x$

3. $y = 3\cos 2x$; $y' = 3(-\sin 2x)\cdot 2 = -6\sin 2x$

4. $y = 5\sin 6x$; $y' = 5(\cos 6x)\cdot 6 = 30\cos 6x$

5. $y = \sin(x^2+1)$; $y' = 2x\cos(x^2+1)$

6. $y = \cos(x^3-5)$; $y' = -\sin(x^3-5)\cdot 3x^2 = -3x^2\sin(x^3-5)$

7. $y = 4\sin^2 3x = 4(\sin 3x)^2$; $y' = 4\cdot 2(\sin 3x)(\cos 3x)(3) = 24\sin 3x\cos 3x$

8. $y = 5\cos^2 2x = 5(\cos 2x)^3$; $y' = 5\cdot 3(\cos 2x)^2(-\sin 2x)(2) = -30\cos^2 2x\sin 2x$

9. $y = \cos(3x^2-2)$; $y' = -6x\sin(3x^2-2)$

10. $y = \sin(2x^3+1)$; $y' = 6x^2\cos(2x^3+1)$

11. $y = \sin\sqrt{x} = \sin x^{1/2}$; $y' = \cos x^{1/2}\cdot\frac{1}{2}x^{-1/2} = \frac{\cos\sqrt{x}}{2\sqrt{x}}$

12. $y = \cos x^{3/2}$; $y' = -\sin x^{3/2}\cdot\frac{3}{2}x^{1/2} = -\frac{3}{2}x\sin x^{3/2}$

13. $y = \cos\sqrt{2x^3-4}$; $y' = -\sin\sqrt{2x^3-4}\times\left(\frac{1}{2}(2x^3-4)^{-1/2}(6x^2)\right) = \frac{-3x^2\sin\sqrt{2x^3-4}}{\sqrt{2x^3-4}}$

14. $y = \sin^2\sqrt{x+1}$; $y' = 2\sin^2\cos\sqrt{x+1}\cdot\frac{1}{2}(x+1)^{-1/2} = \frac{\sin\sqrt{x+1}\cos\sqrt{x+1}}{\sqrt{x+1}}$

15. $y = x^2+\sin^2 x$; $y' = 2x + 2\sin x\cos x$

16. $y = \cos x + \sin x$; $y' = -\sin x + \cos x = \cos x - \sin x$

17. $y = \sin x\cos x$; $y' = \sin x(-\sin x) + \cos x\cos x = \cos^2 x - \sin^2 x = \cos 2x$

18. $y = (\sin x - \cos x)^2$; $y' = 2(\sin x - \cos x)(\cos x + \sin x) = 2(\sin^2 x - \cos^2 x) = -2\cos 2x$

19. $y = \frac{2\cos x}{\sin 2x}$;

$$\begin{aligned} y' &= \frac{\sin 2x(-2\sin x) - 2\cos x\cos 2x\cdot 2}{\sin^2 2x} \\ &= \frac{-2\sin x\sin 2x - 4\cos x\cos 2x}{\sin^2 2x} \\ &= \frac{-2\sin x(2\sin x\cos x) - 4\cos x(2\cos^2 x - 1)}{(2\sin x\cos x)^2} \\ &= \frac{-4\cos x(\sin^2 x + 2\cos^2 x - 1)}{4\sin^2 x\cos^2 x} \\ &= \frac{-4\cos x(\sin^2 x + \cos^2 x + \cos^2 x - 1)}{4\sin^2 x\cos^2 x} \\ &= \frac{-4\cos x(\cos^2 x)}{4\sin^2 x\cos^2 x} = \frac{-\cos x}{\sin^2 x} \end{aligned}$$

20. $y = \frac{\sin 2x}{2x}$; $y' = \frac{2x\cos 2x\cdot 2 - \sin 2x\cdot 2}{4x^2} =$

$\dfrac{4x\cos 2x - 2\sin 2x}{4x^2} = \dfrac{2x\cos 2x - \sin 2x}{2x^2}$

21. $y = x^2 \sin x$; $y' = 2x\sin x + x^2\cos x$

22. $y = x^3\cos x$; $y' = 3x^2\cos x - x^3\sin x$

23. $y = \sqrt{x}\sin x$; $y' = \frac{1}{2}x^{-1/2}\sin x + \sqrt{x}\cos x$

24. $y = \dfrac{\sin^2 x}{x}$; $y' = \dfrac{3x\sin^2 x\cos x - \sin^3 x}{x^2} = \dfrac{\sin^2 x(3x\cos x - \sin x)}{x^2}$

25. $y = (\sin 2x)(\cos 3x)$; $y' = \sin 2x(-3\sin 3x) + 2\cos 2x\cos 3x = 2\cos 2x\cos 3x - 3\sin 2x\sin 3x$

26. $y = x^2\cos(3x^2 - 1)$; $y' = 2x\cos(3x^2 - 1) + x^2(-\sin(3x^2 - 1))(6x) = 2x\cos(3x^2 - 1) - 6x^3\sin(3x^2 - 1)$

27. $y = \sin^3(x^4)$; $y' = 3\sin^2(x^4)\cos(x^4)\cdot(4x^3) = 12x^3\sin^2(x^4)\cos(x^4)$

28. $y = \sqrt{\cos\sqrt{x}} = \left(\cos\sqrt{x}\right)^{\frac{1}{2}}$;

$y' = \frac{1}{2}\left(\cos\sqrt{x}\right)^{-1/2}(-\sin\sqrt{x})\left(\frac{1}{2}x^{-1/2}\right) = \dfrac{-\sin\sqrt{x}}{4\sqrt{x}\sqrt{\cos\sqrt{x}}}$

29. $y = \sin^2 x + \cos^2 x = 1$; $y' = 0$

30. $y = \sin^2 x - 2\cos^2 x$; $y' = 2\sin x\cos x + 4\cos x\sin x = 6\sin x\cos x = 3\sin 2x$

31. $f(x) = \sin x; f'(x) = \cos x; f''(x) = -\sin x$

32. $g(x) = \cos x$; $g'(x) = -\sin x$; $g''(x) = -\cos x$

33. $y = \sin x$; $y' = \cos x$; $y'' = -\sin x$; $y''' = -\cos x$

34. $y = \cos x$; $y' = -\sin x$; $y'' = -\cos x$; $y''' = \sin x$

35. $y = \sin x$; $y' = \cos x$; $y'' = -\sin x$; and $y''' = -\cos x$; $y^{(4)} = \sin x$, so $\dfrac{d^4}{dx^4}(\sin x) = \sin x$

36. $y = \cos x$; $y' = -\sin x$; $y'' = -\cos x$; $y''' = \sin x$, and $y^{(4)} = \cos x$, so $\dfrac{d^4}{dx^4}(\cos x) = \cos x$.

27.2 DERIVATIVES OF THE OTHER TRIGONOMETRIC FUNCTIONS

1. $y = \tan^2\sqrt{x}$; $y' = 2\tan\sqrt{x}\cdot\sec^2\sqrt{x}\cdot\frac{1}{2}x^{-1/2} = \dfrac{\tan\sqrt{x}\sec^2\sqrt{x}}{\sqrt{x}}$

2. $y = \tan 4x$; $y' = 4\sec^2 4x$

3. $y = \sec 5x$; $y' = 5\sec 5x\tan 5x$

4. $y = \cot(1 + 2x)$; $y' = -2\csc^2(1 + 2x)$

5. $y = \csc(2x - 1)$; $y' = -2\csc(2x - 1)\cot(2x - 1)$

6. $y = \tan\frac{x}{2}$; $y' = \frac{1}{2}\sec^2\frac{x}{2}$

7. $y = \sin x\tan x$; $y' = \sin x\sec^2 x + \tan x\cos x = \sin x\sec^2 x + \sin x = \sin x(\sec^2 x + 1)$

8. $y = \sin x\cot x$; $y' = -\sin x\csc^2 x + \cot x\cos x = \dfrac{-1}{\sin x} + \dfrac{\cos^2 x}{\sin x} = \dfrac{\cos^2 x - 1}{\sin x} = \dfrac{-\sin^2 x}{\sin x} = -\sin x$

9. $y = \tan^3 x$; $y' = 3\tan^2 x\cdot\sec^2 x$

10. $y = \cot^3 4x$; $y' = 3\cot^2 4x\cdot(-\csc^2 4x)\cdot 4x = -12x\cot^2 4x\csc^2 4x$

11. $y = \sec^4(x^2)$; $y' = 4\sec^3(x^2)(\sec(x^2)\tan(x^2))(2x) = 8x\sec^4(x^2)\tan(x^2)$

12. $y = \csc^3\sqrt{x^3} = \left[\csc\left(x^{3/2}\right)\right]^3$; $y' = 3\left[\csc^2\left(x^{3/2}\right)\right]\left[-\csc\left(x^{3/2}\right)\cot\left(x^{3/2}\right)\right]\left(\frac{3}{2}x^{1/2}\right) = -\frac{9}{2}x^{1/2}\csc^3\left(x^{3/2}\right)\cot\left(x^{3/2}\right)$

13. $y = \tan x \cot x$; $y' = \tan x(-\csc^2 x) + \cot x(\sec^2 x) = \sec^2 x \cot x - \csc^2 x \tan x$

14. $y = \sec x \csc x$; $y' = \sec x \tan x \csc x - \sec x \csc x \cot x = \sec x \cdot \frac{\sin x}{\cos x} \cdot \frac{1}{\sin x} - \frac{1}{\cos x} \cdot \frac{\cos x}{\sin} \csc x = \sec^2 x - \csc^2 x$

15. $y = \sin^2 x \cot x$; $y' = 2 \sin x \cos x \cot x + \sin^2 x(-\csc^2 x) = 2 \sin x \cos x \cdot \frac{\cos x}{\sin x} - \sin^2 x \cdot \frac{1}{\sin^2 x} = 2\cos^2 x - 1 = \cos 2x$

16. $y = \sin 5x^2$; $y' = \cos 5x^2(10x) = 10x \cos 5x^2$

17. $y = \tan \frac{1}{x}$; $y' = \sec^2 \frac{1}{x} \cdot (-x^{-2}) = -\frac{1}{x^2} \sec^2 \frac{1}{x}$

18.
$$y = \cot \sqrt{x^2+1};$$
$$y' = -\csc^2 \sqrt{x^2+1} \left(\frac{1}{2}(x^2+1)^{-1/2}\right) 2x$$
$$= \frac{-x}{\sqrt{x^2+1}} \csc \sqrt{x^2+1}$$

19. $y = \sqrt{1+\tan x^2} = \left(1+\tan x^2\right)^{1/2}$; $y' = \frac{1}{2}\left(1+\tan x^2\right)^{-1/2}\left(\sec^2 x^2\right)(2x) = \frac{x \sec^2 x^2}{\sqrt{1+\tan x^2}}$

20. $y = \sqrt{\tan x + \cot x} = (\tan x + \cot x)^{1/2}$; $y' = \frac{1}{2}(\tan x + \cot x)^{-1/2}\left(\sec^2 x - \csc^2 x\right) = \frac{\sec^2 x - \csc^2 x}{2\sqrt{\tan x + \cot x}}$

21.
$$y = \frac{\tan x}{1+\sec x};$$
$$y' = \frac{(1+\sec x)\sec^2 x - \tan x(\sec x \tan x)}{(1+\sec x)^2}$$
$$= \frac{\sec^2 x + \sec^3 x - \sec x \tan^2 x}{(1+\sec x)^2}$$
$$= \frac{\sec x\left[\sec x + \sec^2 x - \tan^2 x\right]}{(1+\sec x)^2}$$
$$= \frac{\sec x\,[\sec x + 1]}{(1+\sec x)^2} = \frac{\sec x}{1+\sec x}$$

22. $y(\csc x + \cot x)^3$; $y' = 3(\csc x + \cot x)^2(-\csc x \cot x - \csc^2 x) = -3\csc x(\csc x + \cot x)^3$

23. Here $y = \frac{\cot x}{1-\csc x}$, so $y' = \frac{(1-\csc x)(-\csc^2 x) - \cot x(\csc x \cot x)}{(1-\csc x)^2} = \frac{-\csc^2 x + \csc^3 x - \csc x \cot^2 x}{(1-\csc x)^2} = \frac{\csc x(\csc^2 x - \cot^2 x - \csc x)}{(1-\csc x)^2} = \frac{\csc x(1-\csc x)}{(1-\csc x)^2} = \frac{\csc x}{1-\csc x}$

24. Here $y = \sqrt{x} + \tan\sqrt{x}$, so $y' = \frac{1}{2\sqrt{x}} + \frac{\sec^2\sqrt{x}}{2\sqrt{x}} = \frac{1}{2\sqrt{x}}(1+\sec^2\sqrt{x})$

25. $y = (\csc x + 2\tan x)^3$; $y' = 3(\csc x + 2\tan x)^2(-\csc x \cot x + 2\sec^2 x) = 3(\csc x + 2\tan x)^2(2\sec^2 x - \csc x \cot x)$

26. $y = (1 + \cot^2 x)^{1/2}$; $y' = \frac{1}{2}(1 + \cot^2 x)^{-1/2}(2\cot x)(-\csc^2 x) = \frac{-\cot x \csc^2 x}{\sqrt{1+\cot^2 x}} = \frac{-\cot x \csc^2 x}{\csc x} = -\cot x \csc x$ or $y = \sqrt{1+\cot^2 x} = \csc x$; $y' = -\csc x \cot x$

27. $y = (\tan 2x)^{3/5}$; $y' = \frac{3}{5}(\tan 2x)^{-2/5}(\sec^2 2x)(2) = \frac{6}{5}\sec^2 2x\,(\tan 2x)^{-2/5}$

28. $y = x\csc x$; $y' = x(-\csc x \cot x) + \csc x = \csc x(1 - x\cot x)$

29. Here $y = \frac{\sec x}{1+\tan x}$, and so $y' = \frac{(1+\tan x)(\sec x \tan x) - \sec^3 x}{(1+\tan x)^2} = \frac{\sec x(\tan x + \tan^2 x - \sec^2 x)}{(1+\tan x)^2} = \frac{\sec x(\tan x - 1)}{(1+\tan x)^2}$

30. $y = \dfrac{\cos x}{1+\sec^2 x}$;

$$y' = \frac{(1+\sec^2 x)(-\sin x) - \cos x(2\sec^2 x \tan x)}{(1+\sec^2 x)^2}$$
$$= \frac{-\sin x - \sin x \sec^2 x - 2\sec^2 x \sin x}{(1+\sec^2)^2}$$
$$= \frac{-\sin x(1+3\sec^2 x)}{(1+\sec^2 x)^2}$$

31. $\dfrac{d}{dx}\cot u = \dfrac{d}{dx}\left(\dfrac{\cos u}{\sin u}\right) = \dfrac{\sin u \frac{d}{dx}\cos u - \cos u^2 \sin u}{\sin^2 u} = \dfrac{\sin u(-\sin u)\frac{du}{dx} - \cos u \cos u \frac{du}{dx}}{\sin^2 u} = \dfrac{(-\sin^2 u - \cos^2 u)\frac{du}{dx}}{\sin^2 u} = -\csc^2 u \dfrac{du}{dx}$

32. $\dfrac{d}{dx}\csc u = \dfrac{d}{du}\dfrac{1}{\sin x} = \dfrac{d}{dx}(\sin u)^{-1} = -1(\sin u)^{-2}\dfrac{d}{dx}\sin u = -1(\sin u)^{-2}\cos u\dfrac{du}{dx} = -\dfrac{1}{\sin u}\cdot\dfrac{\cos u}{\sin u}\dfrac{du}{dx} = -\csc u \cot u \dfrac{du}{dx}$

33. $y = \tan 2x$; $y' = 2\sec^2 2x = 2(\sec 2x)^2$; $y'' = 4(\sec 2x)(\sec 2x \tan 2x)2 = 8\sec^2 2x \tan 2x$

34. $y = \sec 3x$; $y' = 3\sec 3x \tan 3x$; $y'' = 3\sec 3x(\sec^2 3x \cdot 3) + 3\tan 3x(\sec 3x \tan 3x \cdot 3) = 9\sec 3x(\sec^2 3x + \tan^2 3x)$

35. $y = x\tan x$; $y' = x(\sec^2 x) + \tan x$; $y'' = x(2\sec x \cdot \sec x \tan x) + \sec^2 x + \sec^2 x = 2x\sec^2 x \tan x + 2\sec^2 x = 2\sec^2 x(x\tan x + 1)$

36. Here $y = \dfrac{\cos x}{x}$, and so $y' = \dfrac{x(-\sin x) - \cos x \cdot 1}{x^2} = \dfrac{-x\sin x - \cos x}{x^2}$; $y'' = \dfrac{x^2(-x\cos x - \sin x + \sin x) + 2x(x\sin x + \cos)}{x^4} = \dfrac{-x^2\cos x + 2x\sin x + 2\cos x}{x^3}$ or $\dfrac{2\cos x + 2x\sin x - x^2\cos x}{x^3}$

37. Since $y = x\sin y$, then $y' = x(\cos y)y' + \sin y$, or $y' - x(\cos y)y' = \sin y$, or $y'(1 - x\cos y) = \sin y$, and so $y' = \dfrac{\sin y}{1 - x\cos y}$.

38. Since $\sin xy + xy = 0$, then $\cos xy(xy'+y) + (xy'+y) = 0$, or $(xy'+y)[\cos xy + 1] = 0 \Rightarrow xy' + y = 0$, and so $y' = \dfrac{-y}{x}$.

39. Here $x + y = \sin(x+y)$, and differentiating, we obtain $1 + y' = \cos(x+y)[1+y'] = \cos(x+y) + y'\cos(x+y)$, or $y' - y'\cos(x+y) = \cos(x+y) - 1$, and so $y' = \dfrac{\cos(x+y) - 1}{1 - \cos(x+y)}$ or $\dfrac{1 - \cos(x+y)}{\cos(x+y) - 1}$.

40. Differentiating $y^2 = \tan 4xy$ we get $2yy' = \sec^2 4xy(4xy' + 4y)$, or $2yy' = 4xy'\sec^2 4xy + 4y\sec^2 4xy$, or $y'(2y - 4x\sec^2 4xy) = 4y\sec^2 4xy$, and so $y' = \dfrac{2y\sec^2 4xy}{y - 2x\sec^2 4xy}$.

27.3 DERIVATIVES OF INVERSE TRIGONOMETRIC FUNCTIONS

1. $y = \sin^{-1} 2x$, so $y' = \dfrac{2}{\sqrt{1-4x^2}}$.

2. $y = \cos^{-1} 4x$, so $y' = \dfrac{-4}{\sqrt{1-16x^2}}$.

3. $y = \tan^{-1}\dfrac{x}{2}$, so $y' = \dfrac{\frac{1}{2}}{1+\frac{x^2}{4}} = \dfrac{2}{4+x^2}$.

4. $y = \sin^{-1}\left(\dfrac{x}{3}\right)$, so $y' = \dfrac{\frac{1}{3}}{\sqrt{1-\frac{x^3}{9}}} = \dfrac{\frac{1}{3}}{\frac{\sqrt{9-x^2}}{3}} = \dfrac{1}{\sqrt{9-x^2}}$

5. $y = \cos^{-1}(1-x^2)$ and $y' = \dfrac{2x}{\sqrt{1-(1-x^2)^2}} = \dfrac{2x}{\sqrt{2x^2-x^4}}$

6. $y = \sin^{-1}\sqrt{x}$, and $y' = \dfrac{\frac{1}{2\sqrt{x}}}{\sqrt{1-\sqrt{x}^2}} = \dfrac{1}{2\sqrt{x}\sqrt{1-x}} = \dfrac{1}{2\sqrt{x-x^2}}$

7. $y = \sin^{-1}(1-2x)$, so $y' = \dfrac{-2}{\sqrt{1-(1-2x)^2}} = \dfrac{-2}{\sqrt{4x-4x^2}} = \dfrac{-1}{\sqrt{x-x^2}}$

8. $y = \tan^{-1}(4x-1)$; $y' = \dfrac{4}{1+(4x-1)^2} = \dfrac{4}{16x^2-8x+2} = \dfrac{2}{8x^2-4x+1}$ or $\dfrac{2}{1-4x+8x^2}$

9. $y = \cos^{-1}(x^3-x)$; $y' = \dfrac{1-3x^2}{\sqrt{1-(x^3-x)^2}} = \dfrac{1-3x^2}{\sqrt{1-x^2+6x^4-x^6}}$

10. $y = \sin^{-1}(x-3)^2$; $y' = \dfrac{2(x-3)}{\sqrt{1-(x-3)^4}}$

11. $y = \sec^{-1}(4x+2)$; $y' = \dfrac{4}{(4x+2)\sqrt{(4x+2)^2-1}} = \dfrac{2}{(2x+1)\sqrt{(4x+2)^2-1}}$

12. $y = \csc^{-1}(3x+1)$; $y' = \dfrac{-3}{(3x+1)\sqrt{(3x+1)^2-1}}$

13. $y = x\sin^{-1}x$; $y' = x \cdot \dfrac{1}{\sqrt{1-x^2}} + \sin^{-1}x$

14. $y = \sin^{-1}\sqrt{x+1}$; $y' = \dfrac{\frac{1}{2\sqrt{x+1}}}{\sqrt{1-\left(\sqrt{x+1}\right)^2}} = \dfrac{1}{2\sqrt{x+1}\sqrt{-x}}$

15. $y = x^2\cos^{-1}x$; $y' = 2x\cos^{-1}x - \dfrac{x^2}{\sqrt{1-x^2}}$

16. $y = \cos^{-1}\sqrt{1-x^2}$; $y' = \dfrac{-\frac{1}{2}(1-x^2)^{-1/2}(-2x)}{\sqrt{1-(1-x^2)}} = \dfrac{x(1-x^2)^{-1/2}}{\sqrt{x^2}} = \dfrac{1}{\sqrt{1-x^2}}$

17. Since $y = \tan^{-1}\left(\dfrac{2x-1}{2x}\right)$, then $y' = \dfrac{1}{1+\left(\dfrac{2x-1}{2x}\right)^2}\left(\dfrac{2x\cdot 2-(2x-1)2}{4x^2}\right) = \dfrac{1}{\dfrac{4x^2+4x^2-4x+1}{4x^2}}\left(\dfrac{4x-4x+2}{4x^2}\right) = \dfrac{2}{8x^2-4x+1}$

18. $y = \csc^{-1}\sqrt{x}$; $y' = \dfrac{-1}{\sqrt{x}\sqrt{x-1}} \cdot \dfrac{1}{2\sqrt{x}} = \dfrac{-1}{2x\sqrt{x-1}}$

19. $y = x\tan^{-1}(x+1)$; $y' = \tan^{-1}(x+1) + \dfrac{x}{1+(x+1)^2}$

20. $y = \left(1+\tan^{-1}x\right)^2$; $y' = 2\left(1+\tan^{-1}x\right)\left(\dfrac{1}{1+x^2}\right) = \dfrac{2(1+\tan^{-1}x)}{1+x^2}$

21. Here we are given $y = \sqrt{\sin^{-1}(1-x^2)}$, and so $y' = \dfrac{1}{2}\left(\sin^{-1}(1-x^2)\right)^{-1/2}\left(\dfrac{-2x}{\sqrt{1-(1-x^2)^2}}\right) = \dfrac{-x}{\sqrt{\sin^{-1}(1-x^2)}\sqrt{1-(1-x^2)^2}} = \dfrac{-x}{\sqrt{\sin^{-1}(1-x^2)}\sqrt{2x^2-x^4}} = \dfrac{-1}{\sqrt{\sin^{-1}(1-x^2)}\sqrt{2-x^2}}$.

22. $y = (1-x^2)^{1/2}\sin^{-1}x$; $y' = \dfrac{1}{2}\left(1-x^2\right)^{-1/2} \times (-2x)\sin^{-1}x + (1-x^2)^{1/2} \cdot \dfrac{1}{\sqrt{1-x^2}} = \dfrac{-\sin^{-1}x}{\sqrt{1-x^2}} + 1$ or $1 - \dfrac{\sin^{-1}x}{\sqrt{1-x^2}}$

23. $y = x\cot^{-1}(1+x^2)$; $y' = \cot^{-1}(1+x^2) + x\dfrac{-1}{1+(1+x^2)^2}\cdot 2x = \cot^{-1}(1+x^2) - \dfrac{2x^2}{2+2x^2+x^4}$

24. Here $y = \sin^{-1}\left(\dfrac{x-1}{x+1}\right)$, and so $y' = \dfrac{1}{\sqrt{1-\dfrac{(x-1)^2}{(x+1)^2}}}\cdot\dfrac{(x+1)-(x-1)}{(x+1)^2} = \dfrac{1}{\dfrac{\sqrt{(x+1)^2-(x-1)^2}}{x+1}}\cdot\dfrac{2}{(x+1)^2} = \dfrac{2}{(x+1)\sqrt{4x}} = \dfrac{1}{(x+1)\sqrt{x}}$

25. Since $y = \dfrac{\sin^{-1}x}{x}$, then $y' = \dfrac{x\left(\frac{1}{\sqrt{1-x^2}}\right) - \sin^{-1}x}{x^2} = \dfrac{1}{x\sqrt{1-x^2}} - \dfrac{\sin^{-1}x}{x^2} = \dfrac{1}{x}\left(\dfrac{1}{\sqrt{1-x^2}} - \dfrac{\sin^{-1}x}{x}\right)$.

26. $y = \dfrac{\arccos x^2}{x}$; $y' = \dfrac{x\cdot\dfrac{-1}{\sqrt{1-x^4}}\cdot 2x - \arccos x^2}{x^2} = \dfrac{-2}{\sqrt{1-x^4}} - \dfrac{\arccos x^2}{x^2}$

27. $y = (\arcsin x)^{1/3}$; $y' = \dfrac{1}{3}(\arcsin x)^{-2/3}\left(\dfrac{1}{\sqrt{1-x^2}}\right) = \dfrac{1}{3\sqrt{1-x^2}(\arcsin x)^{2/3}}$

28. $y = \left(\arctan x^2\right)^{1/3}$; $y' = \dfrac{1}{3}\left(\arctan x^2\right)^{-2/3}\times\left(\dfrac{1}{1+x^4}\right)2x = \dfrac{2x}{3(1+x^4)(\arctan x^2)^{2/3}}$

29. Let $y = \cos^{-1}u$ or $u = \cos y$. Hence $\dfrac{du}{dx} = \dfrac{d}{dx}\cos y = -\sin y\dfrac{dy}{dx}$ so $\dfrac{dy}{dx} = \dfrac{-1}{\sin y}\dfrac{du}{dx}$. Since $\sin^2 y + \cos^2 y = 1$ we have $\sin y = \sqrt{1-\cos^2 y} = \sqrt{1-u^2}$. and so $\dfrac{dy}{dx} = \dfrac{d}{dx}\cos^{-1}u = \dfrac{-1}{\sqrt{1-u^2}}\dfrac{du}{dx}$.

30. Let $y = \cot^{-1}u$ and then, $u = \cot y$. Thus, we have $\dfrac{du}{dx} = -\csc^2 y\dfrac{dy}{dx} \Rightarrow \dfrac{dy}{dx} = \dfrac{-1}{\csc^2 y}\dfrac{du}{dx}$. By the Pythagrean identities we have $\csc^2 = 1+\cot^2 y = 1+u^2$, and so $\dfrac{dy}{dx} = \dfrac{-1}{1+u^2}\dfrac{du}{dx}$ or $\dfrac{d}{dx}\cot^{-1}u = \dfrac{-1}{1+u^2}\dfrac{du}{dx}$.

27.4 APPLICATIONS

1. If $y = \sin x$, then $y' = \cos x$. Since $\cos x > 0$ on $\left(-\frac{\pi}{2}, \frac{\pi}{2}\right)$, $\sin x$ is increasing on $\left[-\frac{\pi}{2}, \frac{\pi}{2}\right]$.

2. If $y = \tan x, y' = \sec^2 x$. Since $\sec^2 x \geq 0$ everywhere it is defined, $\tan x$ is increasing for all values where it is defined.

3. If $y = \arccos x$, then $y' = \dfrac{-1}{\sqrt{1-x^2}}$ when $|x| < 1$. Hence $y' < 0$ which means that the $\arccos x$ is decreasing for $|x| \leq 1$.

4. If $y = \arctan x$, then $y' = \frac{1}{1+x^2}$ which is always positive. As a result, we see that $\arctan x$ is always increasing.

5. $f(x) = 2\sin x + \cos 2x$; $f'(x) = 2\cos x - 2\sin 2x$. The critical values occur when $f'(x) = 0$. Hence, we have $2\cos x - 2\sin 2x = 0$; $2\cos x - 4\sin x\cos x = 0$. Factoring, we obtain $2\cos x(1-2\sin x) = 0$. If $\cos x = 0 \Rightarrow x = \frac{\pi}{2}, \frac{3\pi}{2}$ and if $1 - 2\sin x = 0 \Rightarrow \sin x = \frac{1}{2} \Rightarrow x = \frac{\pi}{6}, \frac{5\pi}{6}$. Differentiating again, we get $f''(x) = -2\sin x - 4\cos 2x = 0$, or $-2(\sin x + 2(1-2\sin^2 x)) = 0$, or $-2(\sin x + 2 - 4\sin^2 x) = 0$, and so $2(4\sin^2 x - \sin x - 2) = 0$. The quadratic formula yields $\sin x = -0.59307$ or $\sin x = 0.84307$. Hence, $x = 3.7765$, $x = 5.6483$, $x = 1.0030$, or $x = 2.1386$. Maxima are at $\left(\frac{\pi}{6}, 1.5\right), \left(\frac{5\pi}{6}, 1.5\right)$, minima are at $\left(\frac{\pi}{2}, 1\right)$ and $\left(\frac{3\pi}{2}, -3\right)$. Inflection points are at $(1.0030, 1.2646)$, $(2.1386, 1.2646)$,

(3.7765, −0.8896) and (5.6483, −0.8896).

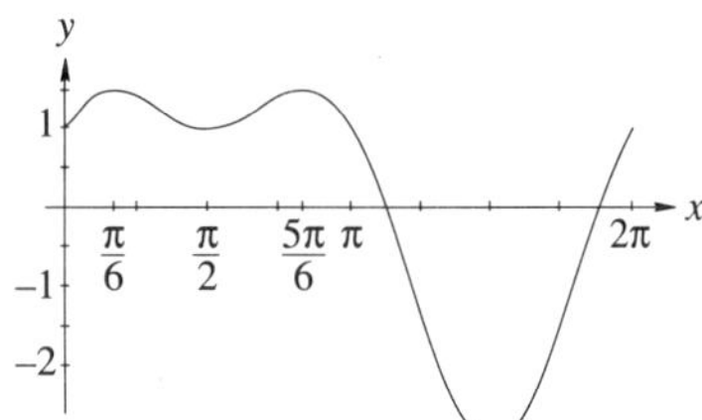

6. $g(x) = 2\cos x + \sin 2x$. To find the critical values, we take the derivative and set it equal to 0. $g'(x) = -2\sin x + 2\cos 2x = 0 \Rightarrow -2\sin x + 2(1 - 2\sin^2 x) = 0$ or $-4\sin^2 x - 2\sin x + 2 = 0$. Factoring, we have $(-4\sin x + 2)(\sin x + 1) = 0$. This gives $\sin x = -1$ or $\sin x = \frac{1}{2}$. Hence $x = \frac{3\pi}{2}$ or $x = \frac{\pi}{6}$ or $\frac{5\pi}{6}$. Taking the second derivative and setting it equal to 0 produces $g''(x) = -2\cos x - 4\sin 2x = -2\cos x - 4(2\sin x\cos x) = -2\cos x(1 + 4\sin x) = 0$, or $\cos x = 0 \Rightarrow x = \frac{\pi}{2}$ and so, $x = \frac{3\pi}{2}$ or $\sin x = -\frac{1}{4}$, and so, $x = 3.3943$ or $x = 6.0305$. As a result, the maximum is at $(\frac{\pi}{6}, 2.5981)$, the minimum at $(\frac{5\pi}{6}, -2.5981)$, and inflection points are at $(\frac{\pi}{2}, 0)$, $(3.3943, -1.4523)$, $(\frac{3\pi}{2}, 0)$, and $(6.0305, 1.4524)$.

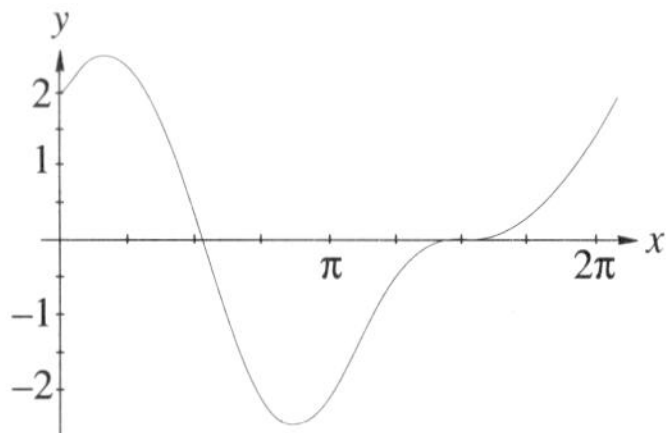

7. $h(x) = \cos x - \sin x$. To find the critical values, we take the derivative and set it equal to 0. Thus, $h'(x) = -\sin x - \cos x = 0$ and so, $-\sin x = \cos x$, or $\tan x = -1$, which means that $x = \frac{3\pi}{4}$ or $\frac{7\pi}{4}$. The second derivative is $h''(x) = -\cos x + \sin x$ and setting it equal to 0 produces $\sin x = \cos x$ or $\tan x = 1$. Thus, $x = \frac{\pi}{4}$ or $x = \frac{5\pi}{4}$. As a result, the maxima are $\left(\frac{7\pi}{4}, \sqrt{3}\right)$; the minimum is $\left(\frac{3\pi}{4}, -\sqrt{3}\right)$, and the inflection points are $\left(\frac{\pi}{4}, 0\right)$ and $\left(\frac{5\pi}{4}, 0\right)$.

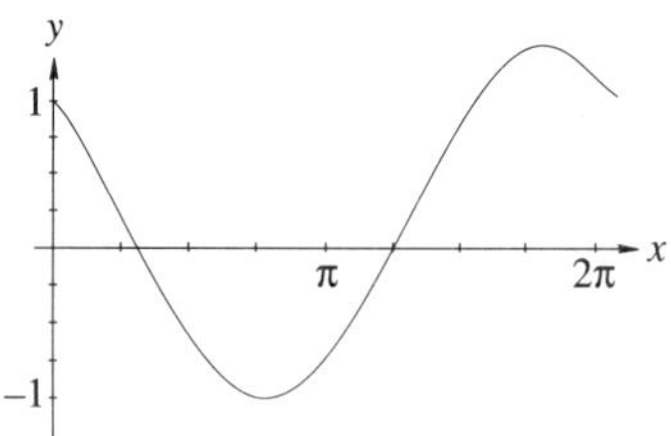

8. $j(x) = \cos^2 x - \sin x$. Taking the derivative and setting it equal to 0, we have $j'(x) = -2\cos x\sin x - \cos x = -\cos x(2\sin x + 1) = 0$. This yields $\cos x = 0$ which produces the critical values $x = \frac{\pi}{2}$ and $\frac{3\pi}{2}$ or $\sin x = -\frac{1}{2}$ gives the critical values $x = \frac{7\pi}{6}$ and $\frac{11\pi}{6}$. For inflection points take the second derivative $j''(x) = -\cos x(2\cos x) + \sin(2\sin x + 1) = -2\cos^2 x + 2\sin^2 x + \sin x = -2(1 - 2\sin^2 x) + 2\sin^2 x + \sin x = -2 + 6\sin^2 x + \sin x = 6\sin^2 x + \sin x - 2 = (3\sin x + 2)(2\sin x - 1)$. Setting the second derivative equal to 0 and solving, we obtain $\sin x = \frac{1}{2} \Rightarrow x = \frac{\pi}{6}$ or $\frac{5\pi}{6}$ or $\sin x = -\frac{2}{3} \Rightarrow x = 3.8713$ or $x = 5.5535$. From the above information, we obtain maxima at $\left(\frac{7\pi}{6}, 1.25\right)$ and $\left(\frac{11\pi}{6}, 1.25\right)$; minima at $\left(\frac{\pi}{2}, -1\right)$ and $\left(\frac{3\pi}{2}, 1\right)$; and inflection points at $\left(\frac{\pi}{6}, 0\right)$, $\left(\frac{5\pi}{6}, 0\right)$, $(3.8713, 1.2222)$, and $(5.5535, 1.2222)$.

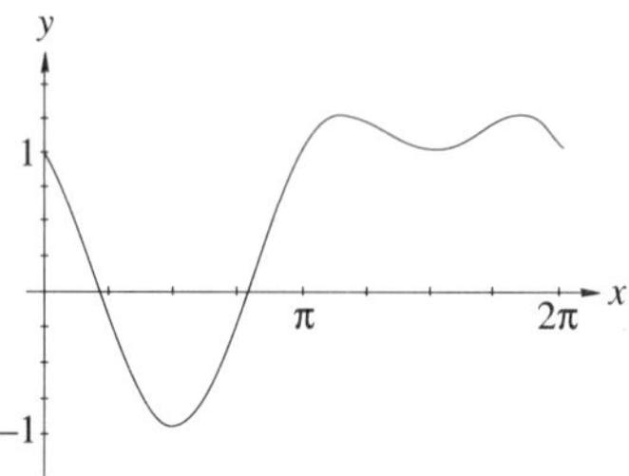

9. $k(x) = \cos^2 x + \sin x$. To find the critical values, we take the derivative and set it equal to 0. $k'(x) = -2\cos x\sin x + \cos x = \cos x(-2\sin x + 1) = 0$. This yields cos = 0 which means that there are critical values at $x = \frac{\pi}{2}$ and $x = \frac{3\pi}{2}$. We also get $\sin x = \frac{1}{2}$, and so there are additional critical values at $x = \frac{\pi}{6}$ and $x = \frac{5\pi}{6}$. For inflection points, we take the second derivatives and set them equal to 0: $k''(x) = \cos x(-2\cos x) - \sin x(-2\sin x + 1) = -2\cos^2 x + 2\sin^2 x - \sin x = -2(1 - 2\sin^2 x) + 2\sin^2 x - \sin x = 6\sin^2 - \sin x - 2 = (3\sin x - 2)(2\sin x + 1) = 0$. Hence, $\sin x = \frac{2}{3} \Rightarrow x = 0.7297$ or $x = 2.4119$, or

$\sin x = -\frac{1}{2} \Rightarrow x = \frac{7\pi}{6}$ or $\frac{11\pi}{6}$. Thus, we see that this function has the following maxima: $(\frac{\pi}{6}, 1.25)$ and $(\frac{5\pi}{6}, 1.25)$; minima: $(\frac{\pi}{2}, 1)$ and $(\frac{3\pi}{2}, -1)$; and inflection points: $(0.7297, 1.2222)$, $(2.4119, 1.2222)$, $(\frac{7\pi}{6}, 0)$ and $(\frac{11\pi}{6}, 0)$.

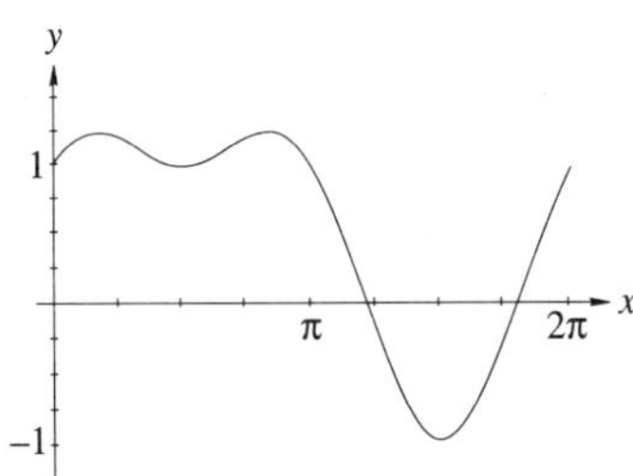

10. $m(x) = \sec x - \tan x$. For critical values $m'(x) = \sec x \tan x - \sec^2 x = \sec x(\tan x - \sec x) = 0$. Hence, either $\sec x = 0$ which yields no solutions or $\tan x = \sec x$, or $\dfrac{\sin x}{\cos x} = \dfrac{1}{\cos x} \Rightarrow \sin x = 1$ or $x = \frac{\pi}{2}$. This is not a solution since both sec and tan are undefined at $\frac{\pi}{2}$. Hence, there are not critical values and m has no maxima or minima. To determine inflection points we take the second derivative and set it equal to 0. $m''(x) = \sec x \tan^2 x + \sec^3 x - 2\sec^2 x \tan x = \sec x(\tan^2 x + \sec^2 x - 2\sec \tan x) = \sec x(\tan x - \sec x)^2 = 0$. Once again undefined at $\frac{\pi}{2}$ and $\frac{3\pi}{2}$. There is an apparent inflection point at $\frac{\pi}{2}, 0)$, but this is actually an undefined point and hence a hole. Thus, there are no inflection points. Concave up $\left(\frac{-\pi}{2}, \frac{\pi}{2}\right)$ Concave down $\left(\frac{-\pi}{2}, \frac{3\pi}{2}\right)$.

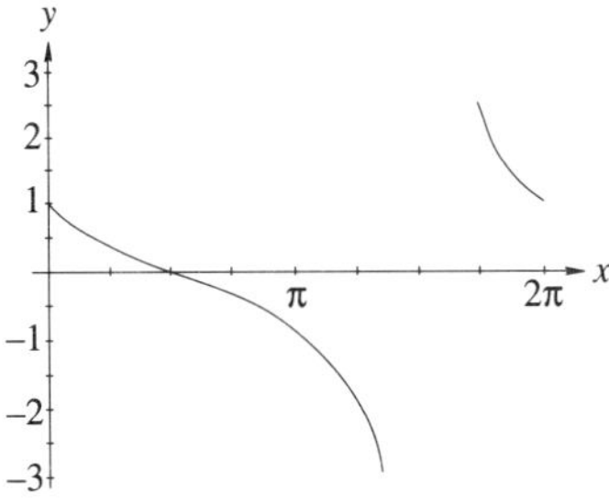

11. $y = \sin x$, at $x = \frac{\pi}{6}$, we get $y = \sin\frac{\pi}{6} = 0.5$. Thus, we want to find the line tangent at this point $\left(\frac{\pi}{6}, 0.5\right)$. To find the slope of the tangent line we take the derivative $y' = \cos x$ and evaluate it when $x = \frac{\pi}{6}$ to get $\cos\frac{\pi}{6} = \frac{\sqrt{3}}{2}$. Using the point-slope form for the equation of the line, we obtain the desired equation: $y - 0.5 = \frac{\sqrt{3}}{2}\left(x - \frac{\pi}{6}\right)$.

12. $y = \sin x$, at $x = \frac{\pi}{6}$, we get $y = \sin\frac{\pi}{6} = 0.5$. Thus, we want to find the line normal at the point $\left(\frac{\pi}{6}, 0.5\right)$. As in Exercise #11, the slope of the tangent line when $x = \frac{\pi}{6}$ is $\frac{\sqrt{3}}{2}$, and so the slope of the normal line is $-\frac{2}{\sqrt{3}}$. Using the point-slope form for the equation of the line, we obtain the desired equation: $y - 0.5 = \frac{-2}{\sqrt{3}}\left(x - \frac{\pi}{6}\right)$ or $y - 0.5 = \frac{-2\sqrt{3}}{3}\left(x - \frac{\pi}{6}\right)$.

13. $y = \tan^3 x$; at $x = \frac{\pi}{4}$, we get $y = 1$, so point is $\left(\frac{\pi}{4}, 1\right)$. To find the slope, we take the derivative $y' = 3\tan x(\sec^2 x)$ and at $\frac{\pi}{4}$, we get $y' = 3 \cdot 1^2(\sqrt{2})^2 = 6$. Hence, the equation of the tangent line is $y - 1 = 6\left(x - \frac{\pi}{4}\right)$.

14. As in Exercise #13, the point is $\left(\frac{\pi}{4}, 1\right)$, but the slope of the normal line is $-\frac{1}{6}$. Thus, the equation of the normal line is $y - 1 = -\frac{1}{6}\left(x - \frac{\pi}{4}\right)$.

15. The horizontal distance from the light to the plane is 8000 ft. The velocity is $\dfrac{dx}{dt} = 400 \text{ mi/h} = 400 \cdot \dfrac{\text{mile}}{\text{hr}} \cdot \dfrac{5280 \text{ ft}}{1 \text{ mi}} \cdot \dfrac{1 \text{ hr}}{3600 \text{ s}} = 586.7 \text{ ft/s}$; $\tan\theta = \dfrac{x}{6000}$. Taking derivatives we have $\sec^2\theta\dfrac{d\theta}{dt} = \dfrac{1}{6000}\dfrac{dx}{dt}$. Also, $\tan\theta = \dfrac{8000}{6000} = \dfrac{4}{3}$, and so $\theta = 0.9273$ rad. Hence, $\dfrac{d\theta}{dt} = \dfrac{dx/dt}{6000\sec^2\theta} = \dfrac{(dx/dt)\cos^2\theta}{6000} = \dfrac{(586.7)\cos^2(0.9273)}{6000} = 0.0352$ rad/s.

16. 300 m= 0.3 km; 400 m= 0.4 km. The ship is $\sqrt{400^2 - 300^2} = 264.6$ m $= 0.2646$ km from its closest point. $\tan\theta = \frac{x}{0.3}$. Taking derivatives, we have $\sec^2\theta\dfrac{d\theta}{dt} = \dfrac{1}{0.3}dx/dt$. We know $\cos\theta = \frac{3}{4}$, and so $\sec\theta = \frac{4}{3}$. Substituting, we obtain $\dfrac{d\theta}{dt} = \dfrac{10}{0.3\left(\frac{4}{3}\right)^2} = 18.75$ rad/hr $= 0.0052$ rad/s. (Note: $\dfrac{dx}{dt}$ = velocity = 10km/h.)

17. $\frac{d\theta}{dt} = 1$ rpm $= 2\pi$ rad/min $= 0.10472$ rad/s. $\tan\theta = \frac{x}{500}$. Taking derivatives we have $\sec^2\theta\frac{d\theta}{dt} = \frac{1}{500}\frac{dx}{dt}$ or $\frac{dx}{dt} = \sec^2\theta\cdot\frac{d\theta}{dt}\cdot 500 = \left(\sec^2\frac{\pi}{4}\right)(0.10472)(500) = 104.72$ m/s.

18. $P_{\text{app}} = P\sec\theta$; $P'_{textapp} = P\sec\theta\tan\theta\theta' = 15\sec\frac{\pi}{3}\tan\frac{\pi}{3}(0.6) = 31.1769$ W/min

19. $\frac{d\theta}{dt} - 30$ rpm $= 3.1416$ rad/s. We know that $\tan\theta = \frac{x}{8}$, and taking derivatives of both sides we have $\sec^2\theta\frac{d\theta}{dt} = \frac{1}{8}\frac{dx}{dt}$. Hence, $\frac{dx}{dt} = \left(8\sec^2 40°\right)(3.1416) = 42.8284$ mi/s.

20. (a) $x(t) = 2 + 0.75\cos 2\pi t$; $x(0) = 2.75$ m; $x(\frac{1}{2}) = 1.245$ m; $x(1) = 2.75$ m
 (b) $v(t) = x'(t) = -1.5\pi\sin 2\pi t$
 (c) $v(0) = 0$ m/s; $v(\frac{1}{2}) = 0$ m/s; $v(1) = 0$ m/s
 (d) $a(t) = v'(t) = -3.0\pi^2\cos 2\pi t$
 (e) Velocity is a maximum when $a(t) = 0$. Hence $-3.0\pi^2\cos 2\pi t = 0$; or $\cos 2\pi t = 0 \Rightarrow 2\pi t = \frac{\pi}{2} + k\pi \Rightarrow t = \frac{1}{4} + \frac{k}{2} : k = 0, 1, 2\ldots$; Using the second derivative test, we have $a'(t) = 6.0\pi^3\sin 2\pi t$. At $= \frac{1}{4} + \frac{k}{2}$, with k is even, we see that $a'(t)$ is positive, so $v(t)$ is minimum. When k is odd, $a'(t)$ is negative so $v(t)$ is maximum. Answer: $t = \frac{1}{4} + \frac{k}{2}$; $k = 1, 3, 5\ldots$ or $t = \frac{4n+3}{4}$, $n = 0, 1, 2, \ldots$.

21. (a) $\overline{Q} = 1.33(2)^{5/2}\tan\frac{\theta}{2} = 7.523\tan\frac{\theta}{2}$; $\overline{Q}' = \frac{1}{2}(7.5236)\sec^2\frac{\theta}{2}$ (a) $Q'(30) = 3.7618\sec^2 15° = 3.7518(1.0353)^2 = 4.0319$ m^3/s
 (b) $\overline{Q}(40) = 3.7168\sec^2 20° = 4.2602$
 (c) $\overline{Q}' = k\sec^2\frac{\theta}{2}$; critical values at $\frac{\theta}{2} = 90°$ or $\theta = 180°$. $\overline{Q}$ is increasing so maximum at $\theta = 180°$.

22. $\cos\theta = \frac{12}{s} = 12s^{-1}$. Taking derivatives we have $-\sin\theta\frac{d\theta}{dt} = -12s^{-2}\frac{ds}{dt}$. When $s = 20, \cos\theta = \frac{12}{20} \Rightarrow \theta = 0.9273$ rad. $\frac{d\theta}{dt} = \frac{12\times(0.4)}{\sin(0.9273)20^2} = 0.015$ rad/min smaller.

23. $I = \sin^2 3t$; $I' = 6\sin 3t\cos 3t$; $I'(1.5) = 6(\sin 4.5)(\cos 4.5) = 1.2364$ A/s

24. Using $\tan(\theta+\phi) = \frac{6.5}{x}$, with $\tan(\phi) = \frac{2.5}{x}$, then $\tan\theta = \tan[(\theta+\phi)-\phi] = \frac{\tan(\theta+\phi)-\tan\phi}{1+\tan(\theta+\phi)\tan\phi} = \frac{\frac{6.5}{x}-\frac{2.5}{x}}{1+\frac{6.5}{x}\cdot\frac{2.5}{x}} = \frac{\frac{9}{x}}{1+\frac{16.25}{x^2}} = \frac{9x}{x^2+16.25}$. Hence $\theta = \arctan\frac{9x}{x^2+162.5}$. Taking derivatives we have $\frac{d\theta}{dx} = \frac{1}{1+\left(\frac{9x}{x^2+16.25}\right)^2}\cdot\frac{d}{dx}\left(\frac{9x}{x^2+16.25}\right) = \frac{(x^2+16.25)^2}{x^4+32.5x^2+18x^2+264.0625}\cdot\frac{(x^2+16.25)9-9x(2x)}{(x^2+16.25)^2} = \frac{146.25-9x^2}{x^4+50.5x^2+264.0625}$. Now, $\frac{d\theta}{dx} = 0$ when $9x^2 = 146.25$ or $x = 4.0311$. By the first derivative test this yields a maximum. Answer: 4.0311 ft.

25. $x(t) = \frac{1}{4}\cos 2t$; $v(t) = -\frac{1}{2}\sin 2t$; $a(t) = v(t) = -\cos 2t$; $a(t) = 0$ when $2t = \frac{\pi}{2} + 2k\pi$ or $\frac{3\pi}{2} + 2k\pi$ By the first derivative test the maximums occur at $2t = \frac{3\pi}{2} + 2k\pi$ or $t = \frac{3\pi}{4} + k\pi$ or $t = \frac{3\pi+4k\pi}{4}$, $k = 0, 1, 2, 3, \ldots$. Thus, $v\left(\frac{3\pi}{4}\right) = -\frac{1}{2}\sin\left(\frac{3\pi}{2}\right) = -\frac{1}{2}(-1) = \frac{1}{2}$ m/s $= 0.5$ m/s.

26. $A = \frac{1}{2}(b_1+b_2)h$; $b_1 = 10$; $b_2 = 10 + 20\sin\theta$; $h = 10\cos\theta\left(0 \le \theta \le \frac{\pi}{2}\right)$. Hence, we ahve $A = \frac{1}{2}(10 + 10 + 20\sin\theta)(10\cos\theta) = 5\cos\theta(20+20\sin\theta) = 100\cos\theta(1+\sin\theta)$. Taking the derivative, we get $A' = 100\cos\theta(+\cos\theta) - 100\sin\theta(1+\sin\theta) = 100(\cos^2\theta - \sin\theta - \sin^2\theta)$ and setting $A' = 0$ we obtain $\cos^2\theta - \sin\theta - \sin^2\theta = 0$; or $1 - \sin^2\theta - \sin\theta - \sin^2\theta = 0$. Hence, $1 - \sin\theta - 2\sin^2\theta = 0$ and factoring, we have $(1-2\sin\theta)(1+\sin\theta) = 0$.

Thus, $\sin\pi = \frac{1}{2} \Rightarrow \theta = \frac{\pi}{6}$; or $\sin\theta = -1 \Rightarrow \theta = \pi$ which is out of range. Hence the only critical value is $\frac{\pi}{6}$. By the first derivative test, this is a maximum. Answer: $\theta = \frac{\pi}{6}$ or 30°.

27. $f(x) = \cos x + x - 2$; $f'(x) = -\sin x + 1$. Use first $x = 3$.

n	x_n	$f(x_n)$	x_{n+1}	$f(x_{n+1})$
1	3	0.0100	2.98835	6.716×10^{-5}
2	2.98835	6.716×10^{-5}	2.9883	3.105×10^{-9}

Thus, $x \approx 2.9883$ (more exactly 2.988268926).

28. $f(x) = \sin x - x^2$; $f'(x) = \cos x - 2x$. $x = 0$ is one answer. Using Newton's method to find the other, $x = 0.8767$ more exactly 0.8767262154.

29. $g(x) = \sin x + \frac{x^2}{4} - 2x$; $g'(x) = \cos x + \frac{3}{4}x^2 - 2$; $x = 0$ is one answer. By Newton's method $x = 2.71995214$. Since $g(x)$ is odd, $x = -2.71995214$ is also an answer.

30. $h(x) = \cos x + x^2 - 3$; $h'(x) = -\sin x + 2x$. Since $h(x)$ is even, roots come in pairs, so $x = \pm 1.795116642$.

27.5 DERIVATIVES OF LOGARITHMIC FUNCTIONS

1. Since $y = \log 5x$, its derivative is $y' = \frac{1}{5x} \cdot \log e \cdot 5 = \frac{1}{x} \cdot \frac{1}{\ln 10}$.

2. Here $y = \log_5 x^3$, and its derivative is $y' = \frac{3x^2}{x^3} \cdot \log_5 e \cdot = \frac{3}{x} \cdot \frac{1}{\ln 5}$.

3. $y = \ln(x^2 + 4x)$; $y' = \frac{2x+4}{x^2+4x}$

4. $y = \ln\sqrt{x^3 + 4x}$; $y' = \frac{1}{\sqrt{x^2+4x}} \cdot \frac{1}{2}\left(x^2+4x\right)^{-1/2}(2x+4) = \frac{x+2}{x^2+4x}$

5. $y = \sqrt{\ln(x^2+4x)}$; $y' = \frac{1}{2}(\ln(x^2+4x))^{-1/2} \cdot \frac{2x+4}{x^2+4x} = \frac{x+2}{\sqrt{\ln(x^2+4x)} \cdot (x^2+4x)}$

6. $y = \sqrt{\ln x^2 + 4x}$; $y' = \frac{1}{2}(\ln x^2 + 4x)^{-1/2} \cdot \left[\frac{2x}{x^2} + 4\right] = \left(\frac{1}{x} + 2\right)\left(\ln x^2 + 4x\right)^{-1/2}$

7. Here $y = \ln\frac{1}{x}$ and its derivative is $y' = \frac{1}{1/x} \cdot (-1x^{-2}) = -\frac{1}{x}$.

By an alternate method: $y = \ln x^{-1} = -1\ln x$, so $y' = -1\frac{1}{x} = \frac{-1}{x}$.

8. $y = \ln\cos x$; $y' = \frac{1}{\cos x}(-\sin x) = -\tan x$

9. $y = \ln\tan x$; $y' = \frac{1}{\tan x}(\sec^2 x) = \sec x \csc x$

10. $y = \log_4 \sqrt{x}$; $y' = \frac{1}{\sqrt{x}} \cdot \frac{1}{\ln 4} \cdot x^{-1/2} = \frac{1}{2x\ln 4}$

11. $y = \frac{\ln x}{x}$; $y' = \frac{x \cdot \frac{1}{x} - \ln x}{x^2} = \frac{1 - \ln x}{x^2}$

12. $y = \ln\left(\frac{x}{1+x}\right)$; $y' = \frac{1+x}{x} \cdot \left(\frac{1+x-x}{(1+x)^2}\right) = \frac{1}{x(1+x)}$

13. $y = \left[\ln\left(\frac{1+x}{1-x}\right)\right]^{-1/2}$;

$y' = \frac{1}{2}\left[\ln\left(\frac{1+x}{1-x}\right)\right]^{-1/2} \cdot \left(\frac{1-x}{1+x}\right) \cdot \left(\frac{(1-x)+(1+x)}{(1-x)^2}\right) = \frac{1}{2}\left[\ln\left(\frac{1+x}{1-x}\right)\right]^{-1/2}\left(\frac{1-x}{1+x}\right) \cdot \left(\frac{2}{(1-x)^2}\right) = \frac{1}{(1+x)(1-x)} \cdot \left[\ln\left(\frac{1+x}{1-x}\right)\right]^{-1/2} =$

$\frac{1}{1-x^2}\left[\ln\left(\frac{1+x}{1-x}\right)\right]^{-1/2}$

14. $y = \frac{1}{2}\ln\left(\frac{1+x^2}{1-x^2}\right)$; $y' = \frac{1}{2}\left(\frac{1-x^2}{1+x^2}\right)\cdot\left(\frac{(1-x^2)2x+2x(1+x^2)}{(1-x^2)^2}\right) = \frac{1}{2}\left(\frac{1}{1+x^2}\right)\times\left(\frac{4x}{1-x^2}\right) = \frac{2x}{1-x^4}$

15. $y = \ln\frac{4x^3}{\sqrt{x^2+4}}$; $y' = \frac{\sqrt{x^2+4}}{4x^3}\cdot\left(\frac{\sqrt{x^2+4}\cdot 12x^2 - 4x^3\cdot\frac{1}{2}(x^2+4)^{-1/2}\cdot 2x}{x^2+4}\right) = \frac{\sqrt{x^2+4}}{4x^3}\cdot\frac{(x^2+4)\cdot 12x^2-4x^4}{(x^2+4)^{3/2}} = \frac{1}{4x^3}\cdot\frac{4x^2\left[(x^2+4)\cdot 3-x^2\right]}{x^2+4} = \frac{2x^2+12}{x(x^2+4)}$

16. $y = \frac{\ln x}{x^3}$; $y' = \frac{x^3\cdot\frac{1}{x}-\ln x(3x^2)}{x^6} = \frac{x^2-3x^2\ln x}{x^6} = \frac{1-3\ln x}{x^4}$

17. $y = (\ln x)^2$; $y' = 2\ln x\cdot\frac{1}{x} = \frac{2\ln x}{x}$

18. $y = \sin(\ln x)$; $y' = \cos(\ln x)\cdot\frac{1}{x} = \frac{\cos(\ln x)}{x}$ or $\frac{1}{x}\cos(\ln x)$

19. $y = \ln(\ln x)$; $y' = \frac{1}{\ln x}\cdot\frac{1}{x} = \frac{1}{x\ln x}$

20. $y = \frac{x}{\ln x}$; $y' = \frac{\ln x - x\cdot\frac{1}{x}}{(\ln x)^2} = \frac{\ln x - 1}{(\ln x)^2}$

21. Since $y = \ln x$, then $y' = \frac{1}{x} = x^{-1}$, and $y'' = -1x^{-2} = \frac{-1}{x^2}$.

22. Here $y = x\ln x$, and $y' = 1+\ln x$, so $y'' = \frac{1}{x}$.

23. $y = \frac{1}{x}\ln x$; $y' = \frac{1}{x}\cdot\frac{1}{x}+\ln x\cdot(-1x^{-2}) = \frac{1}{x^2}-\frac{\ln x}{x^2} = \frac{1-\ln x}{x^2}$; $y'' = \frac{x^2\left(-\frac{1}{x}\right)-(1-\ln x)2x}{x^4} = \frac{-x-2x(1-\ln x)}{x^4} = \frac{-1-2+2\ln x}{x^3} = \frac{2\ln x-3}{x^3} = \frac{\ln x^2-3}{x^3}$

24. $y = \sqrt{\ln x} = (\ln x)^{1/2}$; $y' = \frac{1}{2}(\ln x)^{-1/2}\cdot\frac{1}{x} = \frac{1}{2x(\ln x)^{1/2}}$; $y'' = -1\left(2x(\ln x)^{1/2}\right)^{-2}\left(2(\ln x)^{1/2}+2x\cdot\frac{1}{x}\cdot\frac{1}{2}(\ln x)^{-1/2}\right) = \frac{-1}{4x^2\ln x}\cdot\frac{2(\ln x)^{1/2}+(\ln x)^{-1/2}}{1} = \frac{-1}{4x^2(\ln x)^{3/2}}\cdot\frac{2\ln +1}{1} = -\frac{2\ln x+1}{4x^2(\ln x)^{3/2}}$

25. $y = \ln\sqrt{\frac{\sin^2 x}{x^3}} = \frac{1}{2}\ln\frac{\sin^2 x}{x^3}$; $y' = \frac{1}{2}\cdot\frac{x^3}{\sin^2 x}\cdot\frac{2x^3\sin x\cos x-3x^2\sin^2 x}{x^6} = \frac{2x\cos x-3\sin x}{2\sin x\cdot x} = \cot x-\frac{3}{2x}$; $y'' = -\csc^2 x-\frac{3}{2}\cdot\frac{(-1)}{x^2} = \frac{3}{2x^2}-\csc^2 x$

26. $y = \ln\sqrt[3]{\frac{\sin^3 x}{x}} = \frac{1}{3}\ln\left(\frac{\sin^3 x}{x}\right)$; $y' = \frac{1}{3}\cdot\frac{x}{\sin^3 x}\cdot\frac{x\cdot 3\sin^2 x\cdot\cos x-\sin^3 x}{x^2} = \frac{1}{3}\cdot\frac{3x\cos x-\sin x}{x\sin x} = \cot x-\frac{1}{3x}$; $y'' = -\csc^2 x+\frac{1}{3x^2} = \frac{1}{3x^x}-\csc^2 x$

27. $y = \sqrt{\ln x^2} = \left(\ln x^2\right)^{1/2}$; $y' = \frac{1}{2}\left(\ln x^2\right)^{-1/2}\cdot\frac{1}{x^2}\cdot 2x = \frac{1}{x\left(\ln x^2\right)^{1/2}}$; $y'' = \frac{-1}{x^2\ln x^2}\left(x\cdot\frac{1}{2}\left(\ln x^2\right)^{-1/2}\cdot\frac{2}{x}+\left(\ln x^2\right)^{1/2}\right) = \frac{-1}{x^2\ln x^2}\left(\frac{1}{\left(\ln x^2\right)^{1/2}}+\left(\ln x^2\right)^{1/2}\right) = -\frac{1+\ln x^2}{x^2\left(\ln x^2\right)^{3/2}}$

28. Here $y = \ln(x+\cos x)$, and $y' = \frac{1}{x+\cos x}\cdot|1-\sin x| = \frac{1-\sin x}{x+\cos x}$, which means that $y'' = \frac{(x+\cos x)(-\cos x)-(1-\sin x)(1-\sin x)}{(x+\cos x)^2} =$

$\dfrac{-x\cos x - \cos^2 x - 1 + 2\sin x - \sin^2 x}{(x+\cos x)^2} = \dfrac{2\sin x - x\cos x - 2}{(x+\cos x)^2}$.

29. Since $y = x^2 \ln(\sin 2x)$, then $y' = 2x\ln(\sin 2x) + \dfrac{2x^2}{sin2x}\cos 2x = 2x\ln(\sin 2x) + 2x^2\cot 2x$, and so $y'' = 2\ln(\sin 2x) + 2x \cdot \dfrac{1}{\sin 2x} \cdot \cos 2x \cdot 2 - 2x^2\csc^2 2x \cdot 2 + 4x\cot 2x = 2\ln(\sin 2x) + 4x\cot 2x - 4x^2\csc^2 2x + 4x\cot 2x = 2\ln(\sin 2x) + 8x\cot 2x - 4x^2\csc^2 2x$.

30. $y = (\cos x)(\ln x)$; $y' = \cos x \cdot \dfrac{1}{x} + \ln x(-\sin x) = \dfrac{1}{x}\cos x - (\sin x)(\ln x)$; $y'' = \dfrac{-1}{x}\sin x - \dfrac{1}{x^2}\cos x - \sin x \cdot \dfrac{1}{x} - (\cos x)(\ln x) = -\dfrac{2}{x}\sin x - \dfrac{1}{x^2}\cos x - (\cos x)(\ln x) = -\dfrac{2x\sin x + \cos x + x^2(\cos x)(\ln x)}{x^2}$

27.6 DERIVATIVES OF EXPONENTIAL FUNCTIONS

1. $y = 4^x$; $y' = 4^x \cdot 1 \cdot \ln 4 = 4^x \ln 4$

2. $y = e^{x^2}$; $y' = e^{x^2} \cdot 2x \cdot = 2xe^{x^2}$

3. $y = 5^{\sqrt{x}}$; $y' = 5^{\sqrt{x}} \cdot \dfrac{1}{2\sqrt{x}} \cdot \ln 5 = \dfrac{5^{\sqrt{x}}}{2\sqrt{x}} \ln 5$

4. $y = 6^{x^2+x}$; $y' = 6^{x^2+x}(2x+1)\ln 6 = (2x+1)6^{x^2+x} \cdot \ln 6$

5. $y = 2^{\sin x}$; $y' = 2^{\sin x} \cdot \cos x \cdot \ln 2 = \cos x \cdot 2^{\sin x} \cdot \ln 2$

6. $y = e^{5x+3}$; $y' = e^{5x+3} = 5e^{5x+3}$

7. $y = e^{x^2+x}$; $y' = (2x+1)e^{x^2+x}$

8. $y = e^{2x}$; $y' = 2e^{2x}$

9. $y = 4^{x^4}$; $y' = 4^{x^4} \cdot 4x^3 \cdot \ln 4 = 4x^3 4^{x^3} \ln 4$

10. $y = x + e^x$; $y' = 1 + e^x$

11. $y = \dfrac{e^x}{x^2}$; $y' = \dfrac{x^2 e^x - e^x(2x)}{x^4} = \dfrac{e^x(x-2)}{x^3}$

12. $y = x^3 e^x$; $y' = 3x^2 e^x + x^3 e^x = x^2 e^x(3+x)$

13. $y = \dfrac{1+e^x}{x^2}$; $y' = \dfrac{x^2(e^x) - (1+e^x)2x}{x^4} = \dfrac{xe^x - 2 - 2e^x}{x^3} = \dfrac{e^x(x-2) - 2}{x^3}$

14. $y = \dfrac{1+e^x}{e^x}$; $y' = \dfrac{e^x(e^x) - (1+e^x)e^x}{(e^x)^2} = \dfrac{e^{2x} - e^x - e^{2x}}{e^{2x}} = \dfrac{-e^x}{e^{2x}} = \dfrac{-1}{e^x} = -e^{-x}$

15. $y = e^{\tan x}$; $y' = e^{\tan x} \cdot \sec^2 x = \sec^2 x \cdot e^{\tan x}$

16. $y = e^{\cos 3x}$; $y' = e^{\cos 3x}(-\sin 3x)(3) = -3\sin 3xe^{\cos 3x}$

17. $y = \sin e^{x^2}$; $y' = \cos e^{x^2} \cdot e^{x^2} \cdot 2x = 2xee^{x^2} \cos e^{x^2}$

18. $y = \cos 2^{x^x}$; $y' = -\sin 2^{x^x} \cdot 2^{x^2} \cdot 2x \cdot \ln 2 = -2x \cdot 2^{x^2} \cdot \sin 2^{x^2} \ln 2$

19. $y = 3^x(x^3-1)$; $y' = 3^x(3x^2) + (x^3-1)3^x \cdot \ln 3 = 3x^2 3^x + (x^3-1)3^x \ln 3$

20. $y = e^{x^2 - \ln 2}$; $y' = e^{x^2 - \ln x}\left(2x - \dfrac{1}{x}\right) = \left(2x - \dfrac{1}{x}\right)e^{x^2 - \ln x}$

21. $y = x^{\cos x}$. Using logarithmic differentation, we obtain $\ln y = \ln x^{\cos x}$ or $\ln y = \cos x \ln x$, or $\dfrac{1}{y}y' = \cos x \cdot \dfrac{1}{x} + (-\sin x)\ln x$, and so $y' = y\left[\dfrac{1}{x}\cos x - \sin x \cdot \ln x\right]$, or $y' = x^{\cos x} \cdot \left[\dfrac{\cos x}{x} - (\sin x)(\ln x)\right]$.

22. $y = x^{x^3}$; Use logarithmic differentiation. $\ln y = x^3 \ln x$; $\frac{1}{y}y' = x^3 \cdot \frac{1}{x} + 3x^2 \ln x$; $y' = y\left[x^2 + 3x^2 \ln x\right]$; $y' = x^{x^3}\left[x^2 + 3x^x \ln x\right] = x^{x^3} \cdot x^2\left[1 + 3\ln x\right] = x^{x^3+2}\left[1 + 3\ln x\right]$

23. $y = (\sin x)^x$; Use logarithmic differentiation. Then, $\ln y = x\ln(\sin x)$, or $\frac{1}{y}y' = x \cdot \frac{1}{\sin x} \cdot \cos x + \ln(\sin x)$, and so $y' = y\,[x\cot x + \ln(\sin x)]$ or $y' = (\sin x)^x\,[x\cot x + \ln(\sin x)]$.

24. $y = e^{\arctan x}$; $y' = e^{\arctan x} \cdot \frac{1}{1+x^2} = \frac{e^{\arctan x}}{1+x^2}$

25. $y = \ln \sin e^{3x}$; $y' = \frac{1}{\sin e^{3x}} \cos e^{3x} \cdot e^{3x} \cdot 3 = \frac{3e^{3x}\cos e^{3x}}{\sin e^{3x}} = 3e^{3x}\cot e^{3x}$

26. $y = e^{\sin x}\ln\sqrt{x}$; Use the product rule with the chain rule. $y' = e^{\sin x} \cdot \frac{d}{dx}\ln\sqrt{x} + \ln\sqrt{x}\frac{d}{dx}e^{\sin x} = e^{\sin x} \cdot \frac{1}{\sqrt{x}} \cdot \frac{1}{2\sqrt{x}} + \ln\sqrt{x} \cdot e^{\sin x} \cdot \cos x = e^{\sin x}\left[\frac{1}{2x} + \cos x \ln\sqrt{x}\right]$

27. $y = e^y + y + x$; this is equivalent to $0 = e^y + x$. By implicit differentiation we obtain $0 = e^y y' + 1$ or $e^y y' = -1$ so $y' = \frac{-1}{e^y}$ or $y' = -e^{-y}$

28. $ye^x - xe^y = 1$; $(ye^x + y'e^x) - (e^y + xe^y y') = 0$; $y'e^x - xe^y y' = e^y - ye^x$; $y' = \frac{e^y - ye^x}{e^x - xe^y}$

29. $\frac{d}{dx}(\cosh x) = \frac{d}{dx}\left(\frac{e^x + e^{-x}}{2}\right) = \frac{1}{2}\left(e^x + e^{-x}(-1)\right) = \frac{e^x - e^{-x}}{2} = \sinh x$

30. $\cosh^2 x - \sinh^2 x = \left(\frac{e^x + e^{-x}}{2}\right)^2 - \left(\frac{e^x - e^{-x}}{2}\right)^2 = \frac{e^{2x} + 2 + e^{-2x}}{4} - \frac{e^{2x} - 2 + e^{-2x}}{4} = \frac{2}{4} + \frac{2}{4} = 1$

31. $\cosh x + \sin hx = \left(\frac{e^x + e^{-x}}{2}\right) + \left(\frac{e^x - e^{-x}}{2}\right) = \frac{2e^x}{2} = e^x$

32. $\sinh 2x = \frac{e^{2x} - e^{-2x}}{2} = \frac{(e^x + e^{-x})(e^x - e^x)}{2} = 2 \cdot \left(\frac{e^x - e^{-x}}{2}\right)\left(\frac{e^x + e^{-x}}{2}\right) = 2\sinh x \cosh x$

33. $y = e^{\sinh 3x}$; $y' = e^{\sinh 3x} \cdot \cosh 3x \cdot 3 = 3(\cosh 3x)e^{\sinh 3x}$

34. $f(x) = \sinh(2x + 3)$; $f'(x) = \cosh(2x + 3) \cdot 2 = 2\cosh(2 + 3)$

35. $g(x) = \tanh\sqrt{x^3 + 4}$; $g'(x) = \operatorname{sech}^2\sqrt{x^3 + 4} \cdot \frac{1}{2}\left(x^3 + 4\right)^{-1/2} \cdot 3x^2 = \frac{3x^2}{2\sqrt{x^3 + 4}} \cdot \operatorname{sech}^2\sqrt{x^3 + 4}$

36. $h(x) = \sinh\left[\ln(x^2 + 3x)\right]$; $h'(x) = \cos h\left[\ln(x^2 + 3x\right] \cdot \frac{1}{x^2 + 3x} \cdot (2x + 3) = \frac{(2x + 3)\cosh\ln(x^2 + 3x)}{x^2 + 3x}$

37. $j(x) = \cosh^3(x^4 + \sin x)$; $j'(x) = 3\cosh^2(x^4 + \sin x) \cdot \sinh(x^4 + \sin x)(4x^3 + \cos x) = 3(4x^3 + \cos x)\cosh^2(x^4 + \sin x)\sinh(x^4 + \sin x)$

38. $k(x) = \tanh^2 x + \operatorname{sech}^2 x$; by one of the hyperbolic identities, $k(x) = 1$ so $k'(x) = 0$

39. Here $f(x) = \sinh^{-1} 7x$, so $f'(x) = \frac{1}{\sqrt{(7x)^2 + 1}} \cdot 7 = \frac{7}{\sqrt{49x^2 + 1}}$.

40. Since $g(x) = \cosh^{-1}\sqrt{x}$, we have $g'(x) = \frac{1}{\sqrt{x^2 - 1}} \cdot \frac{1}{2\sqrt{x}} = \frac{1}{2\sqrt{x - 1}\sqrt{x}}$.

41. $j(x) = x\sinh^{-1}\frac{1}{x}$. Using the product rule with the chain rule we obtain $j'(x) = \sinh^{-1}\frac{1}{x} + x \cdot \frac{1}{\sqrt{\left(\frac{1}{x}\right)^2 + 1}} \cdot \frac{-1}{x^2} = \sinh^{-1}\frac{1}{x} - \frac{1}{x\sqrt{\frac{1}{x^2} + 1}} =$

$\sinh^{-1}\frac{1}{x} - \frac{|x|}{x\sqrt{\frac{1+x^2}{x^2}}} = \sinh^{-1}\frac{1}{x} - \frac{|x|}{x\sqrt{1+x^2}}$. Recall that $\sqrt{x^2} = |x|$.

42. $k(x) = \tanh^{-1}(\sin x)$; $k'(x) = \frac{1}{1-\sin^2 x} \cdot \cos x = \frac{\cos x}{\cos^2 x} = \frac{1}{\cos x} = \sec x$

27.7 APPLICATIONS

1. $y = x\ln x$. The domain is $(0,\infty)$. $y' = x \cdot \frac{1}{x} + 1\ln x = 1+\ln x$. Setting $y' = 0$ we obtain the critical values $1+\ln x = 0$; $\ln x = -1$, $x = e^{-1} = 0.3679$; $y'' = \frac{1}{x}$; since $x > 0$, y'' is always positive so y is always concave up. This also means that the critical value yields a minimum. Summary: maxima: none, minima: $(e^{-1}, -e^{-1})$ or $(0.3679 - 0.3679)$; inflection points: none.

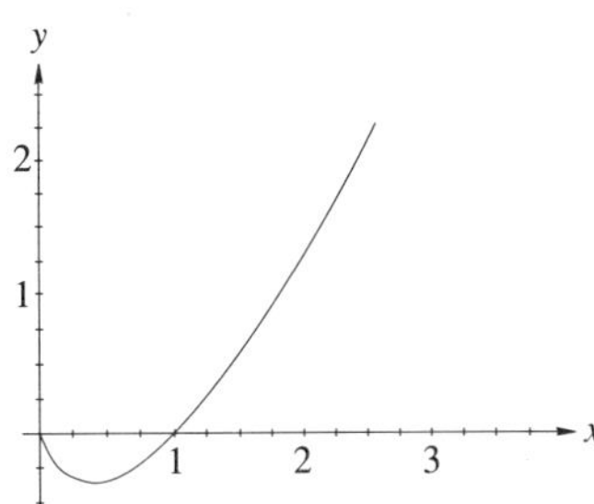

2. $y = \frac{\ln x}{x}$. The domain is $(0,\infty)$. We determine that $y' = \frac{1-\ln x}{x^2}$; and the denominative cannot be zero since the domain is $(0,\infty)$. The critical values occur when $1 - \ln x = 0$ or $x = e$. Now, $y'' = \frac{x^2\left(-\frac{1}{x}\right) - (1-\ln x)2x}{x^4} = \frac{-1-2+2\ln x}{x^3} = \frac{2\ln x - 3}{x^3}$. Inflection point when $2\ln x - 3 = 0$ or $x = e^{3/2}$. At $x = e$, we see that $y'' < 0$, so the critical value yields a maximum. Summary: Maxima: $(e, \frac{1}{e})$, minima: none; inflection point: $\left(e^{3/2}, \frac{3}{2e^{3/2}}\right)$.

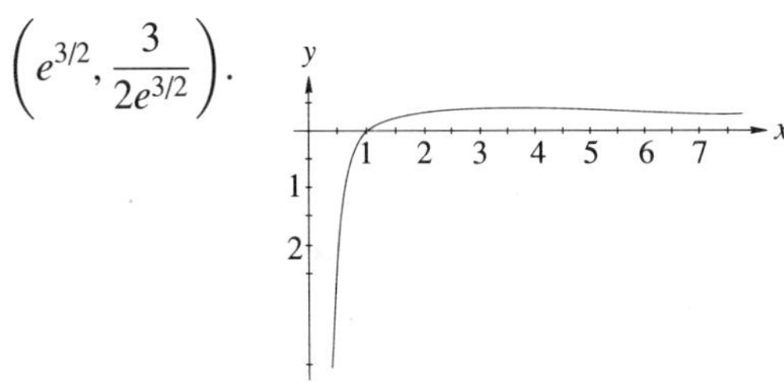

3. $y = xe^x$; $y' = xe^x + e^x = e^x(x+1)$. The critical values occur when $y' = 0$. Since e^x cannot be 0, the only one is when $x+1 = 0$ or $x = -1$. $y'' = e^x(1) + (x+1)e^x = 2e^x + xe^x = e^x(2+x)$. Inflection point is when $x = -2$; $y''(-1) > 0$ so at $x = -1$ there is a minimum. Summary: Maxima: none; minimum: $\left(-1, -e^{-1}\right)$; inflection point: $\left(-2, -2e^{-2}\right)$

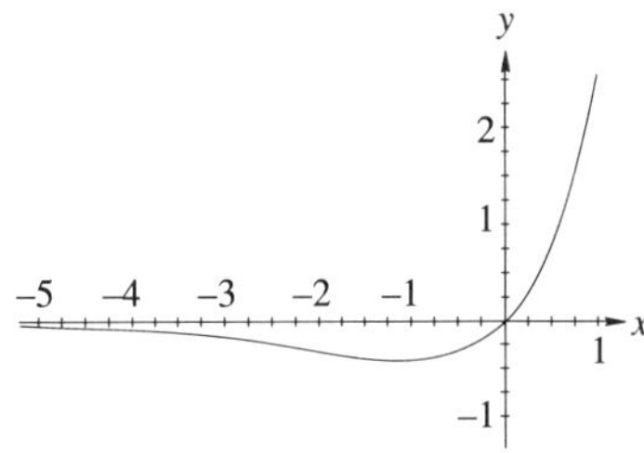

4. $y = x^2e^x$; $y' = x^2e^x + 2xe^x = e^x(x^2+2x)$. Critical values occur when $x^2+2x =$ or at $x = 0$ and $x = -2$. $y'' = e^x(2x+2) + (x^2+2x)e^2 = e^x[x^2+4x+2]$ Inflection points may occur when $x^2+4x+2 = 0$. The quadriatic formula yields $x = -2+\sqrt{2}$. When $x = -2$; $y'' < 0$ so y is maximum at $x = -2$. when $x = 0$; $y'' > 0$ so y is minimum at $x = 0$. Summary: Maxima: $(-2, 4e^{-2})$; minima: $(0,0)$; inflection points: $(-2-\sqrt{2}, 0.3835370)$ and $(-2+\sqrt{2}, 0.1910182)$.

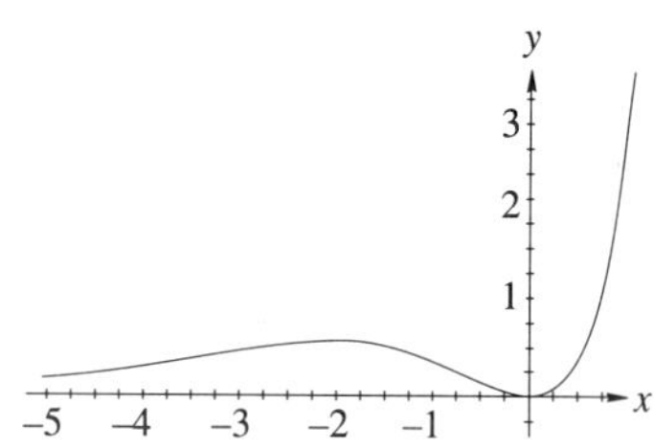

5. Here $y = \ln \dfrac{1}{x^2+1}$, and so $y' = \dfrac{x^2+1}{1} \cdot \dfrac{-2x}{(x^2+1)^2} = \dfrac{-2x}{x^2+1}$. Since the denominator cannot be 0, the only critical value is when $x = 0$. $y'' = \dfrac{(x^2+1)(-2)+2x(2x)}{(x^2+1)^2} = \dfrac{-2x^2-2+4x^2}{(x^2+1)^2} = \dfrac{2x^2-2}{(x^2+1)^2}$. Inflection points when $x^2 - 1 = 0$ or at $x = \pm 1$. At $x = 0$; $y'' < 0$ so y is a maximum. Summary: maxima; $(0,0)$; minima: none; inflection points: $(-1, \ln \frac{1}{2})$; $(1, \ln \frac{1}{2})$.

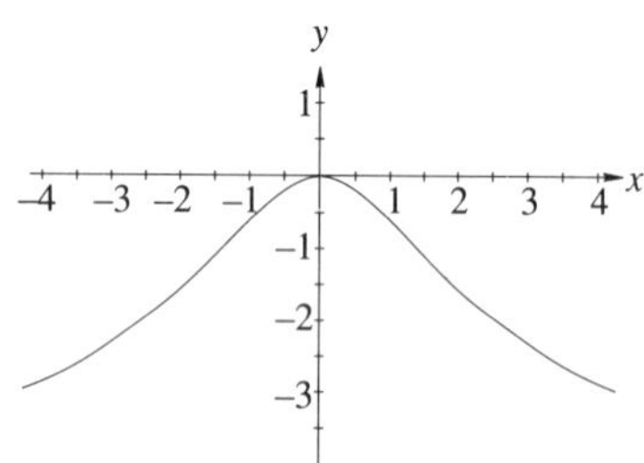

6. $y = \dfrac{4\ln^2 x}{x}$; the domain is $(0, \infty)$. $y' = \dfrac{x \cdot 8 \ln x \cdot \frac{1}{x} - 4\ln^2 x}{x^2} = \dfrac{4\ln x(2 - \ln x)}{x^2}$. Critical values occur when $\ln x = 0$ or $x = 1$ and when $2 - \ln x = 0$ or $x = e^2$. $y'' = \dfrac{x^2\left[\frac{8}{x} - \frac{8\ln x}{x}\right] - 2x \cdot 4\ln x(2-\ln x)}{x^4} = \dfrac{8 - 8\ln x - 8\ln x(2 - \ln x)}{x^3} = \dfrac{8 - 24\ln x + 8\ln^2 x}{x^3} = \dfrac{8}{x^3}(1 - 3\ln x + \ln^2 x)$. The inflection occur when $\ln^2 x - 3\ln x + 1 = 0$ or, by the quadratic formula, when $\ln x = 0.381966$ or $\ln x = 2.618034$. This yields $x = 1.46516$ or $x = 13.70875$. At $x = 1$; $y'' > 0$ so y is a minimum. At $x = e^2$, we see that $y'' < 0$, so y is a maximum. Summary: Maxima: $(e^2, 2.16536)$; minima $(1, 0)$; Inflection points: $(1.46516, 0.39831)$ and $(13.70875, 1.99992)$.

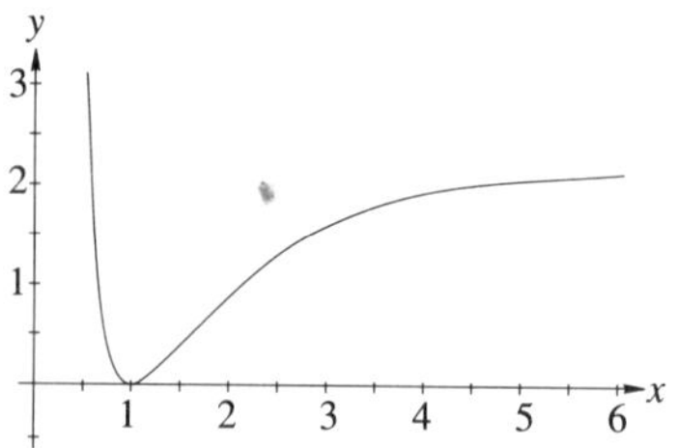

7. $y = \sinh x$; $y' = \cosh x = \dfrac{e^x + e^{-x}}{2}$, which is never 0. $y'' = \sinh x = \dfrac{e^x - e^{-x}}{2}$, which is 0 when $e^x = e^{-x}$ or when $x = 0$. Inflection point at $x = 0$. Summary: Maxima: none, minima: none, inflection point $(0, 0)$

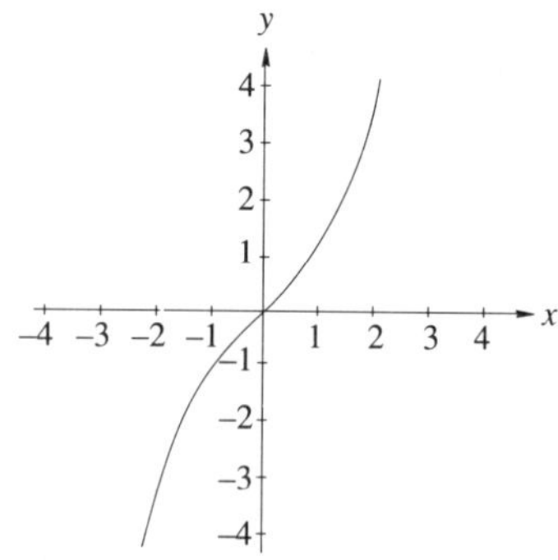

8. Since $y = \cosh x$, its derivative is $y' = \sinh x$; which is 0 when $x = 0$. The second derivative is $y'' = \cosh x$, which is never 0. At $x = 0$, $y'' > 0$, so minimum. Summary: maxima: none; minima: $(0, 1)$; inflection points: none.

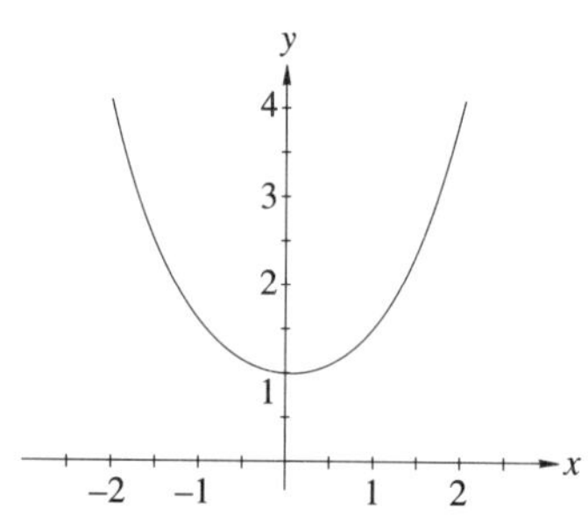

9. $y = e^{-x}\cos x$; $-2\pi \le x \le 2\pi$; $y' = -e^{-x}\sin x - \cos x e^{-x} = -e^{-x}(\sin x + \cos x)$. Critical values occur when $\sin x + \cos x = 0$ or $\sin x = -\cos x$ or $\tan x = -1$. This is at $-\frac{5\pi}{4}, -\frac{\pi}{4}, \frac{3\pi}{4}$, and $\frac{7\pi}{4}$. $y'' = -e^{-x}(\cos x - \sin x) + e^{x}(\sin x + \cos x) = e^{-x}(-\cos x + \sin x + \sin x + \cos x) = 2e^{-x}\sin x$. Inflection points occur when $\sin x = 0$ or $x = -2\pi, -\pi, 0, \pi, 2\pi$. At $x = -\frac{5\pi}{4}$, $y'' > 0$ so y is minima. At $x = -\frac{\pi}{4}$, $y'' < 0$ so y is maxima. At $x = \frac{3\pi}{4}$, $y'' > 0$ so y is minima. At $x = \frac{7\pi}{4}$, $y'' < 0$ so y is maxima. Summary: maxima $\left(\frac{-\pi}{4}, 1.5509\right)$, $\left(\frac{7\pi}{4}, 0.002896\right)$, minima at $\left(\frac{-5\pi}{4}, -35.8885\right)$ and $\left(\frac{3\pi}{4}, 0.06702\right)$; Inflection points at $(-2\pi, 535.49)$, $(-\pi, -23.1407)$, $(0, 1)$, $(\pi, -0.0432)$ and $(2\pi, 0.001867)$.

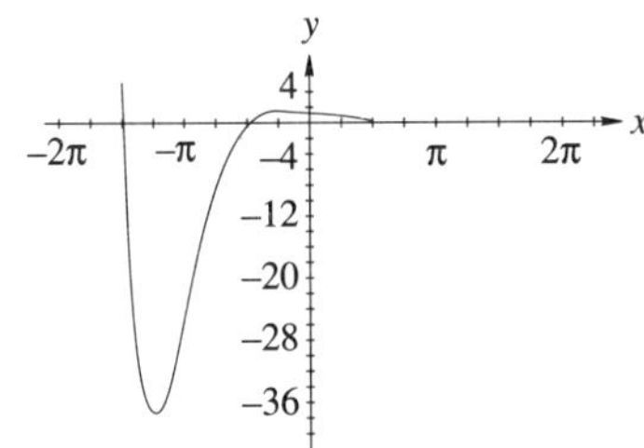

10. $y = \dfrac{\sin x}{\ln(x+2)}$; $0 \le x \le 4\pi$; $y' = \dfrac{\ln(x+2)\cos x - \frac{\sin x}{x+2}}{\ln^2(x+2)}$. This is 0 when $\ln(x+2)\cos x - \frac{\sin x}{x+2} = 0$. We can solve this using Newton's method. $f(x) = y' = \ln(x+2)\cos x - \dfrac{\sin x}{x+2}$; $f'(x) = y'' = -\ln(x+2)\sin x + \dfrac{1}{x+2}\cos x + \dfrac{(x+2)\cos x + \sin x}{(x+2)^2}$; $x = 1.3256, 4.6329, 7.8094$ or 10.9655. Using a computer algebraic system, we can determine the inflection points occur at 2.1469, 5.1439, 8.2376, and 11.3605. Summary: maxima: $(1.3256, 0.8073)$, $(7.8094, 0.4375)$; minima: $(4.6329, -0.5269)$, $(10.9655), -0.3901)$ inflection points: $(2.1469, 0.5896)$, $(5.1439, -0.4620)$, $(8.2376, 0.3987)$, $(11.3605, -0.3604)$

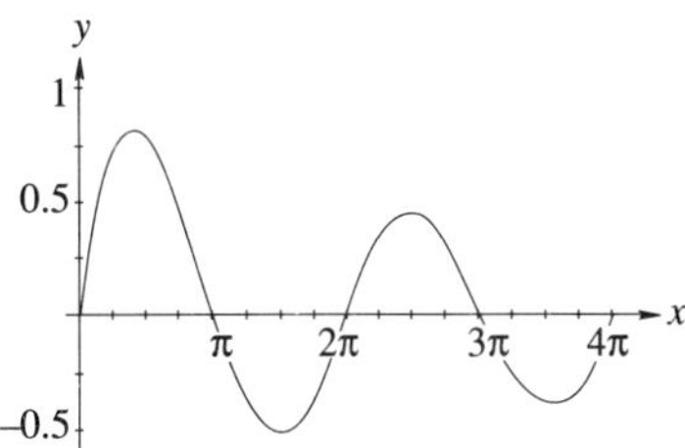

11. $y = x^3\ln x$; $y' = 3x^2\ln x + x^2$; at $x = 1$, $y' = 3\cdot 1^2 \cdot \ln 1 + 1^2 = 1^2 = 1$ so the slope is 1. $y - 0 = 1(x-1)$ or $y = x - 1$

12. If the slope in Exercise #1 is 1, then the normal is -1 so we get $y = -1(x-1)$ or $y = -x + 1$.

13. $y = x^2e^x$; $y' = 2xe^x + x^2e^x$; at $x = 1$; $y' = 2e + e = 3e$. The equation of the line is $y - e = 3e(x-1)$

14. The normal has slope $-\dfrac{1}{3e}$. Its equation is $y - e = -\dfrac{1}{3e}(x-1)$

15. (a) $s(t) = e^{-3t}$; $v(t) = s'(t) = -3e^{-3t}$, (b) $v'(t) = 9e^{-3t}$ which is never undefined or 0 so velocity is never maximum.

16. $s(t) = t^2 + 4\ln(t+1)$

 (a) $v(t) = s'(t) = 2t + \dfrac{4}{t+1}$

 (b) $a(t) = v'(t) = 2 - \dfrac{4}{(t+1)^2} = \dfrac{2(t+1)^2 - 4}{(t+1)^2}$

 (c) $a(t) = 0$ when $2(t+1)^2 - 4 = 0$ or $t + 1 = \pm\sqrt{2}$ or $t = -1 \pm \sqrt{2}$; $a(t) < 0$ when $-1 - \sqrt{2} < t < -1 + \sqrt{2}$. This means the v will be a maximum at $= -1 - \sqrt{2}$ and minimum at $t = -1 + \sqrt{2}$

 (d) This minimum is $v(-1+\sqrt{2}) = 2(-1+\sqrt{2}) + \dfrac{4}{\sqrt{2}} = -2 + 2\sqrt{2} + 2\sqrt{2} = 4\sqrt{2} - 2 \approx 3.6569$ units/s

17. $s(t) = \sin e^t$;

 (a) $v(t) = s'(t) = \cos e^t \cdot e^t$; $v(t) = e^t \cos e^t$; $a(t) = v'(t) = e^t\cos e^t - \sin e^t(e^{2t})$ or $a(t) = e^t[\cos e^t - e^t \sin e^t]$

(b) $v(t) = 0$ when $\cos e^t = 0$ or $e^t = \frac{\pi}{2}$; this gives $t = \ln\frac{\pi}{2} \approx 0.4516$ s; $a\left(\ln\frac{\pi}{2}\right) = \frac{\pi}{2}\left[0 - \frac{\pi}{2}\sin\frac{\pi}{2}\right] = -\left(\frac{\pi}{2}\right)^2$; $-\left(\frac{\pi}{2}\right)^2 \approx -2.4674$ units/s.

18. $x = 5000e^{0.5t}$; the growth rate is the derivative of x or $x' = 2500e^{0.5t}$ bacteria/h.

19. $\rho(x) = xe^{-x^{2/3}}$; This will have a maximum when $\rho'(x) = 0$. $\rho'(x) = x \cdot e^{-x^{2/3}} \cdot \left(-\frac{2}{3}x^{-1/3}\right) + e^{-x^{2/3}} = e^{-x^{2/3}}\left(1 - \frac{2}{3}x^{2/3}\right) = 0$ when $\frac{2}{3}x^{\frac{2}{3}} = 1$ or $x^{2/3} = \frac{3}{2}$ or $x = \left(\frac{3}{2}\right)^{3/2} \approx 1.8371$ m. We can use the first derivative test to varify that this is a maximum.

20. $P(x) = 10^4e^{-0.0012x}$. The rate of change is $P'(x)$; $P'(x) = -0.0012 \times 10^4e^{-0.0012x} = -12e^{-0.0012x}$ kg/m^3

21. $i = 1 - e^{-t^2/2L}$; $i = \frac{t}{L}e^{-t^2/2L}$; The critical value is when $t = 0$. When $t < 0$, $i' < 0$ and when $t > 0$, $i' > 0$ so this yields a maximum.

22. $Q(t) = e^{-6t}(4\cos 8t + 3\sin 8t) - 0.4\cos 10t$.

(a) $Q'(t) = e^{-6t}(-4\sin 8t \cdot 8 + 3\cos 8t \cdot 8) - 6e^{-6t}(4\cos 8t + 3\sin 8t) + 4\sin 10t = e^{-6t}(-50\sin 8t) + 4\sin 10t = 4\sin 10t - 50e^{-6t}\sin 8t$

(b) $Q'\left(\frac{\pi}{16}\right) = -11.69768$

23. $P(t) = 100e^{-0.015t}$; $P'(t) = -1.5e^{-0.015t}$; $P'(50) = -1.5e^{-0.015\cdot 50} = -0.7085$ W/day

24. $T = 10 + 15e^{-0.875t}$; $T' = -0.875 \cdot 15 \cdot e^{-0.875t} = -13.125e^{-0.875t}$; $T'(30) = -13.125e^{-0.875\cdot 30} = -5.2224 \times 15^{-11}$°C/min

25. $f(x) = x - \ln x - 2$; $f'(x) = 1 - \frac{1}{x}$. The roots are 0.1586; 3.1462.

26. $g(x0 = \ln x - \sin x$; $g'(x) = \frac{1}{x} - \cos x$. The root is 2.2191.

27. $h(x) = e^x \cdot \cos x$; $h'(x) = e^x\cos x - e^x\sin x = e^x(\cos x - \sin x)$. Newton's method yield 1.570796327 and several other answers. Reexamining $h(x)$ we can see that e^x is never 0 and $\cos x = 0$ when $x = \frac{\pi}{2} + k\pi$ where k is an integer.

28. $j(x) = e^x + 2x - 2$; $j'(x) = e^x + 2$. The root is 0.3149.

CHAPTER 27 REVIEW

1. $y = \sin 2x + \cos 3x$; $y' = \cos 2x \cdot 2 - \sin 3x \cdot 3 = 2\cos 2x - 3\sin 3x$

2. $y = \tan 3x^2$; $y' = \sec^2 3x^2 \cdot 6x = 6x\sec^2 3x^2$

3. $y = \tan^2 3x$; $y' = 2\tan 3x \cdot \sec^2 3x \cdot 3 = 6\tan 3x\sec^2 3x$

4. $y = \sqrt{\sin 2x} = (\sin 2x)^{1/2}$; $y' = \frac{1}{2}(\sin 2x)^{-1/2} \cdot (\cos 2x) \cdot 2 = \frac{\cos 2x}{\sqrt{\sin 2x}} = \frac{\cos 2x\sqrt{\sin 2x}}{\sin 2x} = \cot 2x\sqrt{\sin 2x}$

5. $y = x^2\sin x$. Using the product rule we obtain; $y' = x^2\cos x + 2x\sin x = x(x\cos x + 2\sin x)$

6. $y = \frac{\cos 3x}{3x}$. Using the quotient rule we obtain $y' = \frac{3x(-\sin 3x)3 - \cos 3x \cdot 3}{9x^2} = \frac{-3x\sin x - \cos 3x}{3x^2}$

7. $y = \sin^{-1}(3x - 2)$; $y' = \frac{1}{\sqrt{1 - (3x - 2)^2}} \cdot 3 = \frac{3}{\sqrt{12x - 9x^2 - 3}}$

8. $y = \cos^{-1}x^2$; $y' = \frac{-1}{\sqrt{1 - x^4}} \cdot 2x = \frac{-2x}{\sqrt{1 - x^4}}$

9. $y = \arctan 3x^2$; $y' = \frac{1}{1 + 9x^4} \cdot 6x = \frac{6x}{1 + 9x^4}$

10. $y = \arctan\left(\frac{1 + x}{1 - x}\right)$; $y' = \frac{1}{1 + \left(\frac{1+x}{1-x}\right)^2} \cdot \frac{(1 - x)1 + (1 + x)}{(1 - x)^2} = \frac{2}{(1 - x)^2 + (1 + x)^2} =$

$\frac{1}{1+x^2}$

11. Here $y = \arcsin\sqrt{1-x^2}$, and so $y' = \frac{1}{\sqrt{1-(1-x^2)}} \cdot \frac{1}{2}\left(1-x^2\right)^{-1/2} \cdot (-2x) = \frac{-x}{\sqrt{x^2}\sqrt{1-x^2}} = \frac{-x}{|x|\sqrt{1-x^2}}$. If $x > 0$ then $y' = \frac{-1}{\sqrt{1-x^2}}$, $x < 0$, $y' = \frac{1}{\sqrt{1-x^2}}$

12. Here $y = \arccos\sqrt{-3+4x-x^2}$, and so $y' = \frac{-1}{\sqrt{1-(-3+4x-x^2)}} \cdot \frac{1}{2}\left(-3+4x-x^2\right)^{-1/2}(4-2x) = \frac{-1}{\sqrt{x^2-4x+4}} \cdot \frac{2-x}{\sqrt{-3+4x-x^2}} = \frac{-1}{|x-2|} \cdot \frac{2-x}{\sqrt{-3+4x-x^2}}$. If $x-2 > 0$ then $y' = \frac{1}{-3+4x-x^2}$.

13. $y = \log_4(3x^2-5)$; $y' = \frac{1}{3x^2-5}(6x) \cdot \frac{1}{\ln 4} = \frac{6x}{3x^2-5} \cdot \frac{1}{\ln 4}$

14. $y = \ln(x+4)^3 = 3\ln(x+4)$; $y' = \frac{3}{x+4}$

15. $y = \ln^2(x^3+4)$; $y' = \frac{2\ln(x^3+4)}{x^3+4} \cdot 3x^2 = \frac{6x^2\ln(x^3+4)}{x^3+4}$

16. $y = \ln\cos x$; $y' = \frac{1}{\cos x} \cdot (-\sin x) = -\tan x$

17. $y = \ln\frac{x^3}{(4x-3)^2}$; $y' = \frac{(4x-3)^2}{x^3} \cdot \frac{(4x-3)^2 \cdot 3x^2 - x^3 \cdot 2 \cdot (4x-3) \cdot 4}{(4x-3)^4} = \frac{(4x-3)^2}{x^3} \cdot \frac{3x^2(4x-3)^2 - 8x^3(4x-3)}{(4x-3)^4} = \frac{3(4x-3)-8x}{x(4x-3)} = \frac{4x-9}{x(4x-3)}$

18. $y = \ln\arctan x$; $y' = \frac{1}{\arctan x} \cdot \frac{1}{1+x^2} = \frac{1}{(1+x^2)\arctan x}$

19. $y = e^{4x^2}$; $y' = 8xe^{4x^2}$

20. $y = e^{\ln x^2} = x^2$; $y' = 2x$

21. $y = e^{\sin x^2}$; $y' = 2x\cos x^2 e^{\sin x^2}$

22. $y = ex^{x^2} + \sin x$; $y' = xe^{x^2} \cdot 2x + e^{x^2} + \cos x = e^{x^2}(2x^2+1) + \cos x$

23. $y = e^{-x}\sin 5x$; $y' = e^{-x}\cos 5x \cdot 5 + (-1)e^{-x}\sin 5x = e^{-x}(5\cos 5x - \sin 5x)$

24. $y = e^{-x}\ln x$; $y' = e^{-x} \cdot \frac{1}{x} + e^{-x}(-1)\ln x = e^{-x}\left(\frac{1}{x} - \ln x\right)$

25. $y = x^2e^x$; $y' = 2xe^x + x^2e^2 = e^x(x^2+2x)$. The critical values occur when $x^2+2x = 0$ or $x = 0$ and $x = -2$. For the inflection points, we find $y'' = 2e^x + 2xe^x + x^2e^x + 2xe^x = e^x(x^2+4x+2)$. Inflection points occur when $x^2+4x+2 = 0$. The quadratic formula yields $x = -2 \pm \sqrt{2}$. When $x = -2$, then $y'' < 0$, and so y is a maximum. When $x = 0$, then $y'' > 0$, and so y is a minimum.
Summary: Maxima: $(-2, 0.5413)$; minima: $(0,0)$; inflection points: $(-2-\sqrt{2}, 0.3835)$ and $(-2+\sqrt{2}, 0.1910)$

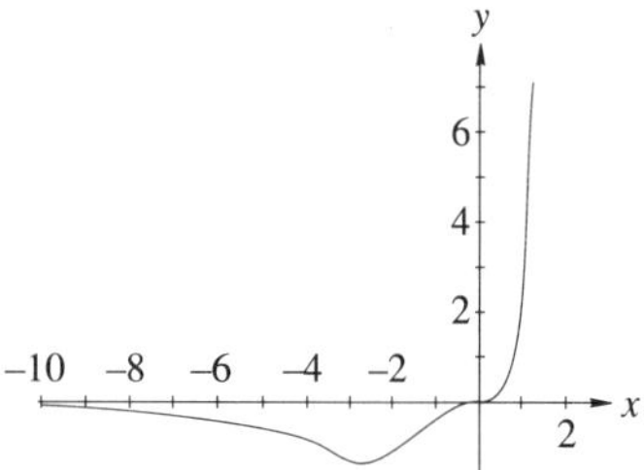

26. $y = 4e^{-9x^2}$. Finding y', we have $y' = (-18x)4e^{-9x^2} = -72xe^{-9x^2}$. The only critical value is 0. Next, we find $y'' = -72e^{-9x^2} + 72x(18x)e^{-9x^2} = -72e^{-9x^2} + 1296x^2e^{-9x^2}$. Setting this equal to 0 we find $1296x^2 - 72 = 0$, or $18x^2 = 1$,

or $x = \pm\frac{1}{\sqrt{18}} = \frac{\pm\sqrt{2}}{6}$. Summary: maxima: $(0, 4)$; minima: none; inflection points: $\left(\pm\frac{\sqrt{2}}{6}, 2.4261\right)$

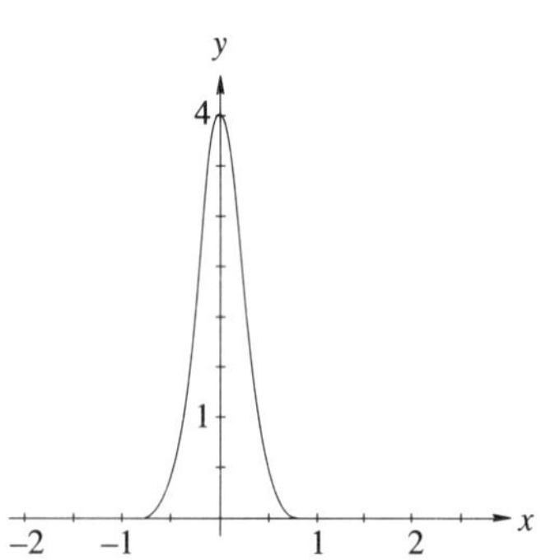

27. $y = e^{-x/2}\sin 2x$ on $[0, 2\pi]$; $y' = -\dfrac{1}{2}e^{-x/2}\sin 2x + 2\cos 2xe^{-x/2} = e^{-x/2}(2\cos 2x - \dfrac{1}{2}\sin 2x)$. Since $e^{-\frac{1}{2}x}$ never equals 0, the critical values occur when $2\cos 2x - \frac{1}{2}\sin 2x = 0$. Solving we get $2\cos 2x = \frac{1}{2}\sin 2x$ or $\tan 2x = 4$; so $2x = 1.325817$, 4.45741, 7.6090, and 10.75060, and as a result $x = -.6629, 2.2337, 3.8045, 5.3753$. Taking the second derivative, we obtain $y'' = -\frac{1}{2}e^{-x/2}(2\cos 2x - \frac{1}{2}\sin 2x) + e^{-x/2}(-4\sin 2x - \cos 2x) = e^{-x/2}(-\cos 2x + \frac{1}{4}\sin 2x - 4\sin 2x - \cos 2x) = e^{-x/2}(-2\cos 2x - \frac{15}{4}\sin 2x)$. Solving $2\cos 2x + \frac{15}{4}\sin 2x = 0$, we obtain $\frac{15}{4}\sin 2x = -2\cos 2x$, or $\dfrac{\sin 2x}{\cos 2x} = \dfrac{-2}{\frac{15}{4}}$, or $\tan 2x = -\dfrac{8}{14}$ which means that $2x = 2.6516, 5.7932, 8.9348$, and 12.0764, and as a result, that $x = 1.3258, 2.8966, 4.4675, 6.0382$.
Summary: maxima: $(0.6629, 0.6964)$, $(3.8045, 0.1448)$; minima: $(2.2337, -0.3175)$, $(5.3753, -0.0660)$; Inflection points: $(1.3258, 0.2425)$, $(2.8966, -0.1106)$, $(4.4674, 0.0504)$, and $(6.0382, -0.0230)$

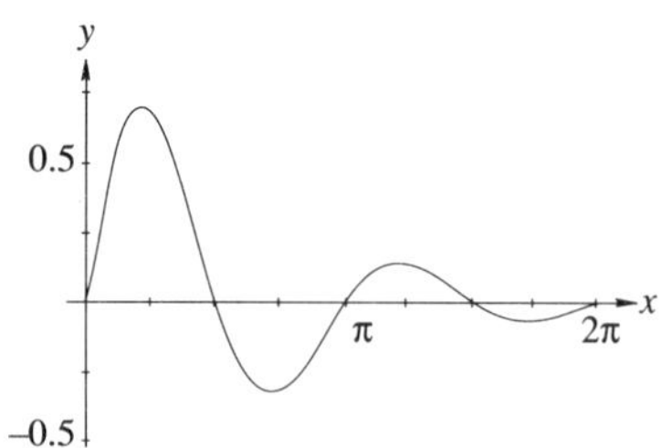

28. $y = e\sin 3x - e^{-1}\cos x$. Noting the e and e^{-1} are constants, we find $y' = e\cos x + e^{-1}\sin x$. Setting y' equal to 0, we have $e\cos x + e^{-1}\sin x = 0$, or $e^{-1}\sin x = -e\cos x$, or $\tan x = -e^2$. Hence, $x = 1.7053$ or 4.8469 are the critical values. Now, $y'' = -e\sin x + e^{-1}\cos x$. Setting this equal to 0 we have $\tan x = e^{-2}$ or $x = 0.1345$ or 3.2761. Thus, we have a maxima at $(1.7053, 2.7431)$, minima at $(4.8469, -2.7431)$, and inflection points at $(1.4363, 0)$, and $(3.2761, 0)$.

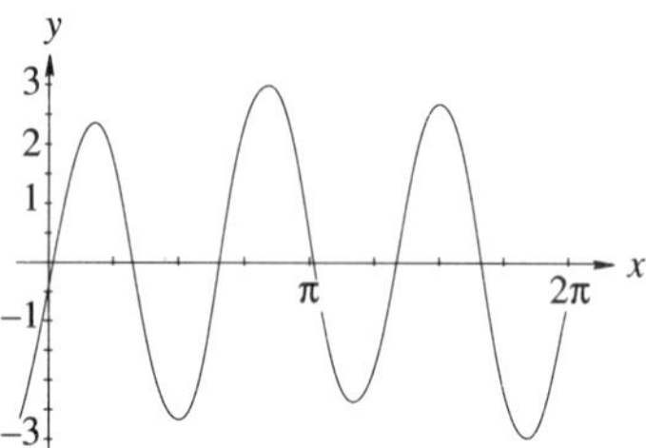

29. With $y = x^3\ln x$, we get $y' = 3x^2\ln x + x^2 = x^2(3\ln x + 1)$. Solving $3\ln x + 1 = 0$, we obtain $\ln x = -\frac{1}{3}$ or $x = e^{-1/3} \approx 0.7165$. For the second derivative, we have $y'' = 6x\ln x + 3x + 2x = 6x\ln x + 5x = x(6\ln x + 5)$. Solving $6\ln x + 5 = 0$, we find $\ln x = -\frac{5}{6}$ or $x = e^{-\frac{5}{6}} \approx 0.4346$. At $x = -0.7165$, y'' is positive so y has a minimum at this point.
Summary: Maxima: none; minima; $(0.7165, -0.1226)$; Inflection point $(0.4346, 0.0684)$

30. 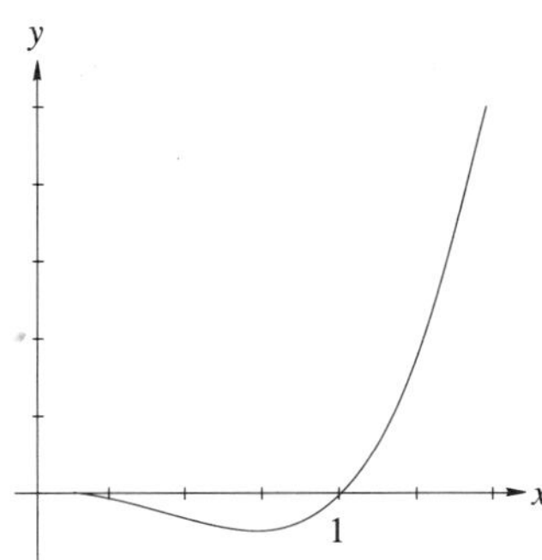

$y = 2\arccos\frac{x}{4}$; $y' = \dfrac{1}{2} \cdot \dfrac{-1}{\sqrt{1-\frac{x^2}{16}}} = \dfrac{1}{2} \cdot \dfrac{-1}{\sqrt{\frac{16-x^2}{16}}} = \dfrac{-2}{\sqrt{16-x^2}}$. This is undefined at -4 and 4 which are the ends of the domain. $y(-4) = 2\pi, y(4) = 0$, $y' = -2(16-x)^{-1/2}$; $y'' = (16-x^2)^{-3/2}(-2x)$. This is 0 when $x = 0$

Summary: maxima: $(-4, 2\pi)$; minima: $(4, 0)$; inflection point $(0, \pi)$

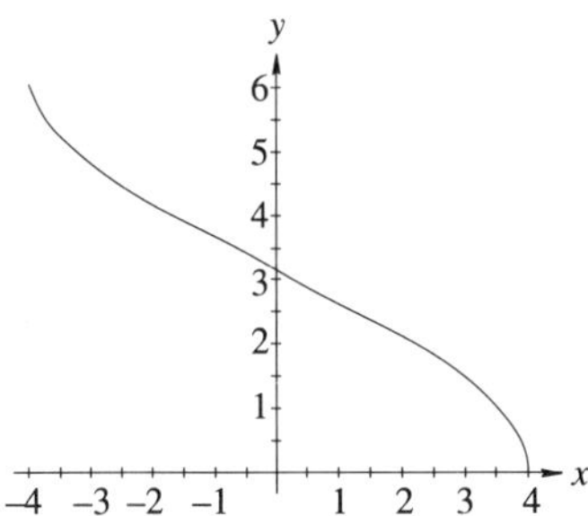

31. $y = \arctan x$; $y' = \dfrac{1}{1+x^2}$; $y'(1) = \dfrac{1}{1+1} = \dfrac{1}{2}$. (a) The slope of the tangent is $\frac{1}{2}$ and its equation is $y - \frac{\pi}{4} = \frac{1}{2}(x-1)$, (b) The slope of the normal is -2 and its equation is $y - \frac{\pi}{4} = -2(x-1)$

32. $y = \sin^3 x$. The first derivative is $y' = 3\sin^2 x \cos x$ and so $y'\left(\dfrac{\pi}{4}\right) = 3 \cdot \left(\dfrac{\sqrt{2}}{2}\right)^2 \cdot \dfrac{\sqrt{2}}{2} = \dfrac{3\sqrt{2}}{4}$.

(a) The slope of the tangent line is $\frac{3\sqrt{2}}{4}$ and its equation is $y - \frac{\sqrt{2}}{4} = \frac{3\sqrt{2}}{4}\left(x - \frac{\pi}{4}\right)$.

(b) The slope of the normal is $-\frac{4}{3\sqrt{2}} = \frac{-2\sqrt{2}}{3}$ and its equation is $y - \frac{\sqrt{2}}{4} = -\frac{2\sqrt{2}}{3}\left(x - \frac{\pi}{4}\right)$.

33. $y = \ln 3x$ has derivative $y' = \dfrac{1}{x}$ and, when $x = 1$, the derivative is $y'(1) = 1$. (a) The slope of the tangent at the point $(1, 0)$ is 1 and its equation is $y = x - 1$. (b) The slope of the normal line at the point $(1, 0)$ is -1 and its equation is $y = -x + 1$ or $y = 1 - x$.

34. $y = \dfrac{e^x}{x}$ has the derivative $y' = \dfrac{xe^x - e^x}{x^2}$ and, when $x = 1$, the derivative is $y'(1) = \dfrac{1e - e}{1^2} = 0$. (a) The slope of the tangent at $(1, e)$ is 0 and its equation is $y = e$. (b) The slope of the normal at $(1, e)$ is undefined, which means that the normal line is a vertical line. Its equation is $x = 1$.

35. $y = \ln(\sin x)$ has the derivative $y' = \dfrac{1}{\sin x} \cdot \cos x = \tan x$ and, when $x = \frac{\pi}{4}$, the derivative is $y'\left(\frac{\pi}{4}\right) = \tan\left(\frac{\pi}{4}\right) = 1$. (a) The slope of the tangent is 1 and its equation is $y + \ln\sqrt{2} = x - \frac{\pi}{4}$. (b) The slope of the normal is -1 and its equation is $y + \ln\sqrt{2} = \frac{\pi}{4} - x$.

36. $y = \sin(\ln x)$ has the derivative $y' = \cos(\ln x) \cdot \dfrac{1}{x}$ and, when $x = 1$, the derivative is $y'(1) = \cos(\ln 1) \cdot \frac{1}{1} = \cos 0 = 1$. (a) The slope of the tanget is 1 and its equation is $y = x - 1$. (b) The slope of the normal is -1 and its equation is $y = 1 - x$.

37. $T = 75e^{-0.05109t} + 20$ (a) $T(10) = 75e^{-0.5109} + 20 = 64.9967°\text{C}$, (b) $T' = (-0.05109)75e^{-0.05109t}$ and $T'(10) = (-0.05109)75e^{-0.5109} = -2.2989°\text{C}$

38. $s(t) = 10{,}000 - \frac{400}{3}\ln(\cosh 1.6t)$ (a) $v(t) = s'(t) = \dfrac{400}{3} \cdot \dfrac{1}{\cosh 1.6t} \cdot \sinh 1.6t \cdot 1.6 = \dfrac{-640}{3}\tanh 1.6t$ (b) $v(60) = \frac{-640}{3}\tan 1.6(60) = -213.33\text{ft/s}$

39. Using the same set-up as Example 27.20 we have $\tan(\theta + \phi) = \frac{48}{x}$ and $\tan\phi = \dfrac{8}{x}$. Writing $\tan\theta = \tan[(\theta + \phi) - \phi]$, we obtain $\tan\theta = \tan[(\theta+\phi) - \phi] = \dfrac{\tan(\theta+\phi) - \tan\phi}{1 + \tan(\theta+\phi)\tan\phi} =$

$\frac{\frac{48}{x}-\frac{8}{x}}{1+\frac{48}{x}\cdot\frac{8}{x}} = \frac{\frac{40}{x}}{1+\frac{384}{x^2}} = \frac{40x}{x^2+384}$. Hence, $\theta = \arctan\left(\frac{40x}{x^2+384}\right)$ and $\theta' = \frac{1}{1+\left(\frac{40x}{x^2+384}\right)^2} \cdot \frac{(x^2+384)40-40x(2x)}{(x^2+384)^2} = \frac{(x^2+384)^2}{(x^2+384)^2+(40x)^2} \cdot \frac{40x^2+15360-80x^2}{(x^2+384)^2} = \frac{15360-40x^2}{(x^2+384)^2+(40x)^2}$. The derivative is 0 when $15360-40x^2=0$ or $40x^2=15360$, or $x^2=384$, which is $x=19.5959$. By the first derivative test this is a maximum. Answer 19.5959 ft.

40. $v(t)=2$ m/s. Letting y represent the height of the balloon above the ground, we have $\tan\theta = \frac{y}{100}$. Taking the derivative of $\tan\theta = \frac{y}{100}$, we obtain $\sec^2\theta\frac{d\theta}{dt} = \frac{1}{100}\frac{dy}{dt}$ and so, $\frac{d\theta}{dt} = \frac{1}{dt}\cdot\frac{dy}{dt}\cdot\frac{1}{\sec^2\theta} = \frac{\cos^2\theta\frac{dy}{dt}}{100}$. Since the balloon is rising at the rate of 2 m/s, we have $\frac{dy}{dt}=2$ m/s. Thus, when the balloon is 30 m above the ground, $y=30$, and we see that we obtain $\cos\theta = \frac{100}{\sqrt{100^2+30^2}} = 0.957826$ and the rate of increase of the angle of inclination of the observer's line of sight when the balloon is 30 m high is $\frac{d\theta}{dt} = \frac{(0.957826)^2\cdot 2}{100} = 0.0183$ rad/s.

CHAPTER 27 TEST

1. $f'(x) = \cos 7x\frac{d}{dx}(7x) = 7\cos 7x$

2. $g'(x) = \sec^2(3x^2+2x)\left[\frac{d}{dx}(3x^2+2x)\right] = [\sec^2(3x^2+2x)](6x+2) = (6x+2)\sec^2(3x^2+2x)$

3. $h'(x) = e^{2x}\frac{d}{dx}(2x) = 2e^{2x}$

4. $j'(x) = \frac{1}{5x^2}\frac{d}{dx}(5x^2) = \frac{1}{5x^2}(10x) = \frac{10x}{5x^2} = \frac{2}{x}$

5. $k'(x) = \frac{1}{\sqrt{1-\left(e^{x^2}\right)^2}}\frac{d}{dx}\left(e^{x^2}\right) = \frac{1}{\sqrt{1-e^{2x^2}}}\left(e^{x^2}\right)\frac{d}{dx}\left(x^2\right) = \frac{2xe^{x^2}}{\sqrt{1-e^{2x^2}}}$

6. Using the quotient rule: $f'(x) = \frac{(1+e^{2x})4\cos 4x - \sin 4x(2e^{2x})}{\left(1+e^{2x}\right)^2}$

7. Using the product rule: $g'(x) = (7x^2+3x)\frac{d}{dx}\left(\tan^2 5x\right) + \left(\tan^2 5x\right)\frac{d}{dx}(7x^2+3x)$. Then, by using the chain rule we obtain $g'(x) = (7x^2+3x)(2\tan 5x)\left(\sec^2 5x\right)5 + \left(\tan^2 5x\right)(14x+3) = (\tan 5x)\left[10(7x^2+3x)\left(\sec^2 5x\right) + (14x+3)(\tan 5x)\right]$.

8. Using the product rule: $h'(x) = e^{\sin x}\frac{d}{dx}(\ln\sqrt{x}) + \left(\ln\sqrt{x}\right)\frac{d}{dx}\left(e^{\sin x}\right)$. We simplify this by using the chain rule: $h'(x) = e^{\sin x}\frac{1}{\sqrt{x}}\frac{d}{dx}\sqrt{x} + \left(\ln\sqrt{x}\right)e^{\sin x}\frac{d}{dx}(\sin x) = e^{\sin x}\frac{1}{\sqrt{x}}\frac{1}{2\sqrt{x}} + \left(\ln\sqrt{x}\right)e^{\sin x}(\cos x) = e^{\sin x}\left(\frac{1}{2x} + \cos x\ln\sqrt{x}\right)$.

9. The current, i, at a particular time is found by taking the derivative of the charge. So, $i(t) = q'(t) = -3e^{-t}\cos 2.5t - 7.5e^{-t}\sin 2.5t$. At $t=0.65$ sec, $i(0.65) \approx -3.82$ A.

10. The velocity at time t, $v(t)$, is the derivative of the position function. So, $v(t) = x'(t) = 1.6(\tan 5.0t)^{-1/2}(5\sec^2 5.0t) = 8\frac{\sec^2 5.0t}{\sqrt{\tan 5.0t}}$. When $t=1.5$, we see that $v(1.5) = 8\frac{\sec^2 7.5}{\sqrt{\tan 7.5}} \approx 40.4743$. The particle is at 40.4743 m.

CHAPTER

28

Techniques of Integration

28.1 THE GENERAL POWER FORMULA

1. $\int \sin^3 x \cos x\, dx$. Let $u = \sin x$ and $du = \cos x\, dx$, then $\displaystyle\int u^3\, du = \frac{u^4}{4} + C = \frac{1}{4}\sin^4 x + C$.

2. $\int \cos^4 x \sin x\, dx$. If you let $u = \cos x$ and $du = -\sin x\, dx$, then you get $\displaystyle -\int u^4\, du = -\frac{1}{5}u^5 + C = -\frac{1}{5}\cos^5 x + C$.

3. $\int \sqrt{\sin^3 x} \cos x\, dx$. Let $u = \sin x$ and $du = \cos x\, dx$, so we have $\displaystyle\int u^{3/2}\, du = \frac{2}{5}u^{5/2} + C = \frac{2}{5}\sin^{5/2} x + C$.

4. $\int \sin 2x \cos 2x\, dx$. Let $u = \sin 2x$ and $du = 2\cos 2x\, dx$, so we have $\cos 2x\, dx = \frac{1}{2}\, du$. Substituting, we have $\displaystyle\frac{1}{2}\int u\, du = \frac{1}{4}u^2 + C = \frac{1}{4}\sin^2 2x + C$.

5. $\int x \sec^2 x^2\, dx$. If you let $u = x^2$ and $du = 2x\, dx$, or $\frac{1}{2}\, du = x\, dx$, then you get $\frac{1}{2}\int \sec^2 u\, du = \frac{1}{2}\tan u + C = \frac{1}{2}\tan x^2 + C$.

6. $\int \cos^3 x \sin x\, dx$. If you let $u = \cos x$ and $du = -\sin x\, dx$, then you get $-\int u^3\, du = -\frac{1}{4}u^4 + C = -\frac{1}{4}\cos^4 x + C$.

7. $\int \sec^2 x \tan x\, dx$. If you let $u = \tan x$ and $du = \sec^2 x\, dx$, then you get $\displaystyle\int u\, du = \frac{1}{2}u^2 + C = \frac{1}{2}\tan^2 x + C$ or rewrite the original integral as $\int \sec^2 x \tan x\, dx = \int \sec x(\sec x \tan x\, dx)$ and let $u = \sec x$ and $du = \sec x \tan x\, dx$. Thus, $\int u\, du = \frac{1}{2}u^2 + C = \frac{1}{2}\sec^2 x + C$.

8. $\displaystyle\int \frac{e^{2x}}{(1+e^{2x})^{1/2}}\, dx$. Here $u = 1 + e^{2x}$ and $du = 2e^{2x}\, dx$, so we have $\frac{1}{2}\, du = e^{2x}\, dx$. Thus, we rewrite the given integral as $\frac{1}{2}\int (u)^{1/2}\, du = \frac{1}{2}\cdot 2u^{1/2} + C = u^{1/2} + C = (1+e^{2x})^{1/2} + C$.

9. $\displaystyle\int \frac{\arccos x}{\sqrt{1-x^2}}\, dx$. Let $u = \arccos x$ and then $\displaystyle du = \frac{-1}{\sqrt{1-x^2}}\, dx$, with the result $\displaystyle -\int u\, du = \frac{-1}{2}u^2 + C = \frac{-1}{2}(\arccos x)^2 + C$.

10. $\displaystyle\int \frac{\arctan x}{1+x^2}\, dx$. Let $u = \arctan x$ and then $\displaystyle du = \frac{1}{1+x^2}\, dx$, with the result $\displaystyle\int u\, du = \frac{1}{2}u^2 + C = \frac{1}{2}(\arctan x)^2 + C$.

11. $\displaystyle\int \frac{1}{x}\sqrt{\frac{\operatorname{arccsc} x}{x^2-1}}\, dx$. Let $u = \operatorname{arccsc} x$ and $\displaystyle du = \frac{-1}{x\sqrt{x^2-1}}\, dx$, with the result $-\int u^{1/2}\, du = -\frac{2}{3}u^{3/2} + C = -\frac{2}{3}(\operatorname{arccsc} x)^{3/2} + C$.

12. $\int \frac{\arctan 6x}{1+36x^2}\,dx$. Let $u = \arctan 6x$ and $du = 6 \cdot \frac{1}{1+36x^2}\,dx$, so $\frac{1}{6}\,du = \frac{1}{1+36x^2}\,dx$. Substitution yields $\frac{1}{6}\int u\,du = \frac{1}{6}\cdot\frac{1}{2}u^2 + C = \frac{1}{12}(\arctan 6x)^2 + C$.

13. $\int \frac{(\arcsin 2x)^3}{\sqrt{1-4x^2}}\,dx$. Let $u = \arcsin 2x$ and $du = 2\cdot\frac{1}{\sqrt{1-4x^2}}\,dx$, so $\frac{1}{2}\,du = \frac{dx}{\sqrt{1-4x^2}}$. Substitution yields $\frac{1}{2}\int u^3\,du = \frac{1}{2}\cdot\frac{1}{4}u^4 + C = \frac{1}{8}(\arcsin 2x)^4 + C$.

14. $\int \frac{(\ln x)^3}{x}\,dx$. Let $u = \ln x$ and $du = \frac{dx}{x}$, and then $\int u^3\,du = \frac{1}{4}u^4 + C = \frac{1}{4}(\ln x)^4 + C$.

15. $\int \frac{[\ln(x+4)]^4}{x+4}\,dx$. Let $u = \ln(x+4)$ and $du = \frac{dx}{x+4}$, then $\int u^4\,du = \frac{1}{5}u^5 + C = \frac{1}{5}[\ln(x+4)]^5 + C$.

16. $\int \frac{[5+2\ln x]^3}{x}\,dx$. If you let $u = 5 + 2\ln x$ and $du = \frac{2\,dx}{x}$, then $\frac{1}{2}\,du = \frac{dx}{x}$. Thus, we have $\frac{1}{2}\int u^3\,du = \frac{1}{2}\cdot\frac{1}{4}u^4 + C = \frac{1}{8}[5+2\ln x]^4 + C$.

17. $\int \frac{e^x\,dx}{(e^x+4)^2}$. Let $u = e^x + 4$, and $du = e^x\,dx$, then you have $\int u^{-2}\,du = -1u^{-1} + C = -\left(e^x+4\right)^{-1} + C$.

18. $\int \left(e^x + e^{-x}\right)^{1/3}\left(e^x - e^{-x}\right)\,dx$. Let $u = e^x + e^{-x}$ and $du = \left(e^x - e^{-x}\right)\,dx$. Then we have $\int u^{1/3}\,du = \frac{3}{4}u^{4/3} + C = \frac{3}{4}\left(e^x + e^{-x}\right)^{4/3} + C$.

19. $\int \left(e^{2x} - 1\right)^3 e^{2x}\,dx$. Let $u = e^{2x} - 1$ and $du = 2e^{2x}\,dx$ so $\frac{1}{2}\,du = e^{2x}\,dx$. Then, $\frac{1}{2}\int u^3\,du = \frac{1}{2}\cdot\frac{1}{4}u^4 + C = \frac{1}{8}\left(e^{2x}-1\right)^4 + C$.

20. $\int \left(3e^{3x}+1\right)^{2/3} e^{3x}\,dx$. Let $u = 3e^{3x} + 1$ and $du = 9e^{3x}\,dx$ so $\frac{1}{9}\,du = e^{3x}\,dx$. Then we have $\frac{1}{9}\int u^{2/3}\,du = \frac{1}{9}\cdot\frac{3}{5}u^{5/3} + C = \frac{1}{15}\left(3e^{3x}+1\right)^{5/3} + C$.

21. $\int_0^{\pi/4} \sin^3 x \cos x\,dx$. If you let $u = \sin x$ and $du = \cos x\,dx$, then you have $\int u^3\,du = \frac{1}{4}u^4$. Substituting for u produces $\frac{1}{4}\sin^4 x\Big|_0^{\pi/4} = \frac{1}{4}\left[\sin^4\frac{\pi}{4} - \sin^4 0\right] = \frac{1}{4}\left[\left(\frac{\sqrt{2}}{2}\right)^4 - 0^4\right] = \frac{1}{4}\cdot\frac{1}{4} = \frac{1}{16}$.

22. $\int_{\frac{\pi}{4}}^{\frac{\pi}{3}} \tan^3 x \sec^2 x\,dx$. Let $u = \tan x$ and $du = \sec^2 x\,dx$, which leads to $\int u^3\,du = \frac{1}{4}u^4$. Substituting for u produces $\frac{1}{4}\tan^4 x\Big|_{\pi/4}^{\pi/3} = \frac{1}{4}\left[\tan^4\frac{\pi}{3} - \tan^4\frac{\pi}{4}\right] = \frac{1}{4}\left[\sqrt{3}^4 - 1^4\right] = \frac{1}{4}[9-1] = 2$.

23. $\int_1^2 \frac{\left[\ln(x^2+1)\right]^3}{x^2+1}\,dx$. Let $u = \ln(x^2+1)$ and $du = \frac{2x}{x^2+1}\,dx$ and so $\frac{1}{2}\,du = \frac{x}{x^2+1}\,dx$. Thus, we get $\frac{1}{2}\int u^3\,du = \frac{1}{8}u^4$ or $\frac{1}{8}\ln^4(x^2+1)\Big|_1^2 = \frac{1}{8}\left[\ln^4 5 - \ln^4 2\right] \approx 0.8098$.

24. $\int_1^{\sqrt{3}} \frac{(\arctan x)^3}{1+x^2}\,dx$. Let $u = \arctan x$ and $du = \frac{dx}{1+x^2}$. Then, we get $\int u^3\,du = \frac{1}{4}u^4$ or $\frac{1}{4}(\arctan x)^4\Big|_1^{\sqrt{3}} = \frac{1}{4}\left[\left(\frac{\pi}{3}\right)^4 - \left(\frac{\pi}{4}\right)^4\right] \approx 0.2055$.

25. $\int_1^2 \frac{e^{2x}\,dx}{(e^{2x}-1)^2}$. Let $u = e^{2x} - 1$ and $du = 2e^{2x}\,dx$ so $\frac{1}{2}\,du = e^{2x}\,dx$. Then, $\frac{1}{2}\int u^{-2}\,du = -\frac{1}{2}u^{-1}$ or $-\frac{1}{2}\left(e^{2x}-1\right)^{-1}\Big|_1^2 = -\frac{1}{2}\left((e^4-1)^{-1} - (e^2-1)^{-1}\right) \approx 0.0689$.

26. $\int_0^2 8\left(e^{2x}-1\right)^3 e^{2x}\,dx$. Let $u = e^{2x} - 1$ and

$du = 2e^{2x}\,dx$, so $4\,du = 8e^{2x}\,dx$. This leads to $4\int u^3\,du = 4\cdot\frac{1}{4}u^4$. Substituting for u produces $\int_0^2 8\left(e^{2x}-1\right)^3 e^{2x}\,dx = \left(e^{2x}-1\right)^4\Big|_0^2 = \left[(e^4-1)^4-(e^0-1)^4\right] = \left(e^4-1\right)^4 = 8.25276\times 10^6$.

27. $\int_0^{\pi}\sin^2 x\cos x\,dx$. Let $u=\sin x$ and $du=\cos x\,dx$, and so $\int u^2\,du = \frac{1}{3}u^3$. The curve is above the x axis from 0 to $\frac{\pi}{2}$ and below from $\frac{\pi}{2}$ to π. Thus, the area is $\frac{1}{3}\sin^3 x\Big|_0^{\pi/2} - \frac{1}{3}\sin^3 x\Big|_{\pi/2}^{\pi} = \frac{1}{3}\left(\sin^3\frac{\pi}{2}-\sin^3 0-\sin^3\pi+\sin^3\frac{\pi}{2}\right) = \frac{1}{3}(1-0-0+1) = \frac{2}{3}$.

28. $\int (e^{3x}+1)e^{3x}\,dx$. Let $u=e^{3x}+1$ and $du = 3e^{3x}\,dx$. Then, $\frac{1}{3}\int u\,du = \frac{1}{6}u^2$. Substituting for u gives $\frac{1}{6}\left(e^{3x}+1\right)^2\Big|_0^1 = \frac{1}{6}\left[\left(e^3+1\right)^2-2^2\right] = \frac{1}{6}[444.599-4]\approx 73.43331$.

29. $\int_{0.7}^{3} e^{-t}(e^{-t}-2)^2\,dt$. Let $u=e^{-t}-2$ and $du = -e^{-t}\,dt$, and so $-\int u^2\,du = -\frac{1}{3}u^3$. Substituting for u we obtain $\int_{0.7}^{3} e^{-t}(e^{-t}-2)^2\,dt = -\frac{1}{3}\left[(e^{-t}-2)^3\right]_{0.7}^{3} = -\frac{1}{3}\left[(e^{-3}-2)^3-(e^{-0.7}-2)^3\right]\approx 1.3397$ C.

30. $\int_{0.25}^{2.50} e^{-t}(e^{-t}-5)^2\,dt$. Substituting $u = e^{-t}-5$ and $du = -e^{-t}\,dt$, produces $-\int u^2\,du = -\frac{1}{3}u^3$, which leads to $\int_{0.25}^{2.50} e^{-t}(e^{-t}-5)^2\,dt = -\frac{1}{3}\left[(e^{-t}-5)^3\right]_{0.25}^{2.50} = -\frac{1}{3}\left[(e^{-2.50}-5)^3-(e^{-0.25}-5)^3\right]\approx 14.5762$ C.

28.2 BASIC LOGARITHMIC AND EXPONENTIAL INTEGRALS

1. $\int \frac{dx}{4x+1}$. Let $u=4x+1$ and $du=4\,dx$, so $\frac{1}{4}\,du = dx$. Then, $\frac{1}{4}\int\frac{du}{u} = \frac{1}{4}\ln|u|+C = \frac{1}{4}\ln|4x+1|+C$.

2. $\int \frac{dx}{9x-5}$. Let $u=9x-5$ and $du=9\,dx$ and $\frac{1}{9}\,du = dx$. Then, $\frac{1}{9}\int\frac{du}{u} = \frac{1}{9}\ln|u|+C = \frac{1}{9}\ln|9x-5|+C$.

3. $\int e^{-3x}\,dx$. Let $u=-3x$ and $du=-3\,dx$, or $-\frac{1}{3}\,du = dx$, which leads to $-\frac{1}{3}\int e^u\,du = -\frac{1}{3}e^u + C = -\frac{1}{3}e^{-3x}+C$.

4. $\int e^{6x}\,dx$. Here let $u=6x$ and $du=6\,dx$ or $\frac{1}{6}\,du = dx$. This leads to $\frac{1}{6}\int e^u\,du = \frac{1}{6}e^u + C = \frac{1}{6}e^{6x}+C$.

5. $\int \frac{\sin x}{\cos x}\,dx$. Let $u=\cos x$ and $du = -\sin x\,dx$, then you have $-\int\frac{du}{u} = -\ln|u|+C = -\ln|\cos x|+C$.

6. $\int \frac{10x+3}{5x^2+3x-7}\,dx$. Here $u = 5x^2+3x-7$ and $du = (10x+3)\,dx$, and so $\int\frac{du}{u} = \ln|u|+C = \ln\left|5x^2+3x-7\right|+C$.

7. $\int \frac{2\sec^2 x}{\tan x}$. If you let $u=\tan x$ and $du=\sec^2 x\,dx$, then $2\,du = 2\sec^2 x\,dx$. Substituting we get $2\int\frac{du}{u} = 2\ln|u|+C = 2\ln|\tan x|+C$.

8. $\int\frac{2e^x}{e^x+4}\,dx = 2\int\frac{e^x}{e^x+4}\,dx$. If you let $u=e^x+4$, then $du = e^x\,dx$. Substituting we get $2\int\frac{du}{u} = 2\ln|u|+C = 2\ln|e^x+4|+C$. Since $e^x>0$ the absolute value signs are not needed. The final answer is $2\ln(e^x+4)+C$.

9. $\int 4^x\,dx = \frac{4^x}{\ln 4}+C$.

10. $\int (e^x - e^{-x})\,dx = \int e^x\,dx + \int -e^{-x}\,dx = e^x + e^{-x} + C$. (Note: for the second integral let $u = x$ then $du = -\,dx$).

11. $\int (e^x + e^{-x})\,dx = \int e^x\,dx - \int -e^{-x}\,dx = e^x - e^{-x} + C$. (Note: for the second integral let $u = -x$, then $du = -\,dx$)

12. $\int \sec^2 x e^{\tan x}\,dx$. Here you should let $u = \tan x$, and $du = \sec^2 x\,dx$. Then, you have $\int e^u\,du = e^u + C = e^{\tan x} + C$.

13. $\int \frac{e^{2x}}{1+e^{2x}}\,dx$. Let $u = 1 + e^{2x}$, so $du = e^{2x}2\,dx$ or $\frac{1}{2}\,du = e^{2x}\,dx$. Substituting we get $\frac{1}{2}\int \frac{du}{u} = \frac{1}{2}\ln|u| + C = \frac{1}{2}\ln(1+e^{2x}) + C = \ln\sqrt{1+e^{2x}} + C$.

14. $\int \frac{x}{1+x^2}\,dx$. Here $u = 1 + x^2$ and $du = 2x\,dx$, so $\frac{1}{2}\,du = x\,dx$. Substituting we get $\frac{1}{2}\int \frac{du}{u} = \frac{1}{2}\ln|u| + C = \frac{1}{2}\ln(1+x^2) + C$ or $\ln\sqrt{1+x^2} + C$.

15. $\int \frac{\ln(1/x)}{x}\,dx$. Let $u = \ln\frac{1}{x} = \ln(x)^{-1} = -\ln x$, and then $du = -\frac{1}{x}\,dx$. Substituting we get $-\int u\,du = -\frac{1}{2}u^2 + C = -\frac{1}{2}\left[\ln\left(\frac{1}{x}\right)\right]^2 + C$.

16. $\int \frac{e^{\sqrt{x}}}{8\sqrt{x}}\,dx$. Let $u = \sqrt{x} = x^{1/2}$; $du = \frac{1}{2}x^{-1/2}\,dx = \frac{1}{2\sqrt{x}}\,dx$ so $\frac{1}{4}\,du = \frac{1}{8\sqrt{x}}\,dx$. Substituting we get $\frac{1}{4}\int e^u\,du = \frac{1}{4}e^u + C = \frac{1}{4}e^{\sqrt{x}} + C$.

17. $\int \frac{e^{\sqrt[3]{x}}}{x^{2/3}}\,dx$. Let $u = \sqrt[3]{x} = x^{1/3}$; $du = \frac{1}{3}x^{-2/3}\,dx$ so $3\,du = \frac{1}{x^{2/3}}\,dx$. Substituting we get $3\int e^u\,du = 3e^u + C = 3e^{\sqrt[3]{x}} + C$.

18. $\int \frac{(\ln x)^{3/4}}{x}\,dx$. Let $u = \ln x$ and $du = \frac{1}{x}\,dx$. Substituting we get $\int u^{3/4}\,du = \frac{4}{7}u^{7/4} + C = \frac{4}{7}(\ln x)^{7/4} + C$.

19. $\int_0^4 \frac{x}{x^2+1}\,dx$. Let $u = x^2 + 1$ and $du = 2x\,dx$ or $\frac{1}{2}\,du = x\,dx$. Substituting we get $\frac{1}{2}\int \frac{du}{u} = \frac{1}{2}\ln|u| = \frac{1}{2}\ln(x^2+1)$. This leads to $\frac{1}{2}\ln(x^2+1)\Big|_0^4 = \frac{1}{2}[\ln(17) - \ln(1)] = \frac{1}{2}\ln 17 \approx 1.4166$.

20. $\int_0^1 \frac{dx}{e^{2x}} = \int_0^1 e^{-2x}\,dx$. Let $u = -2x$ and $du = -2\,dx$ or $-\frac{1}{2}\,du = dx$. Substituting we get $-\frac{1}{2}\int e^u\,du = -\frac{1}{2}e^u = -\frac{1}{2}e^{-2x}$. Evaluating from 0 to 1 we have $-\frac{1}{2}e^{-2x}\Big|_0^1 = -\frac{1}{2}\left[e^{-2} - e^0\right] = -\frac{1}{2}\left[e^{-2} - 1\right] = \frac{1}{2}\left[1 - e^{-2}\right] \approx 0.4323$.

21. $\int_1^3 \frac{1}{\ln e^x}\,dx$. Recall that $\ln e^x = x$ so $\int_1^3 \frac{1}{x}\,dx = \ln|x|\Big|_1^3 = \ln 3 - \ln 1 = \ln 3 \approx 1.0986$.

22. $\int_0^3 \frac{x^2}{x^3+1}\,dx$. Let $u = x^3 + 1$ and $du = 3x^2\,dx$ or $\frac{1}{3}\,du = x^2\,dx$. Thus, we have $\frac{1}{3}\int \frac{1}{u}\,du = \frac{1}{3}\ln|u|$. Evaluating, produces $\frac{1}{3}\ln(x^3+1)\Big|_0^3 = \frac{1}{3}(\ln 28 - \ln 1) = \frac{1}{3}\ln 28 \approx 1.1107$.

23. $\int_0^1 x^3 e^{x^4}\,dx$. Let $u = x^4$ and $du = 4x^3\,dx$ or $\frac{1}{4}\,du = x^3\,dx$. Substituting produces $\frac{1}{4}\int e^u\,du = \frac{1}{4}e^u = \frac{1}{4}e^{x^4}$ which leads to $\frac{1}{4}e^{x^4}\Big|_0^1 = \frac{1}{4}\left(e^1 - e^0\right) = \frac{1}{4}(e-1) \approx 0.4296$.

24. $\int_{\pi/4}^{\pi/3} \frac{dx}{(x^2+1)\arctan x}$. Let $u = \arctan x$, then $du = \frac{1}{x^2+1}\,dx$. Substituting we get $\int \frac{1}{u}\,du = \ln|u| = \ln|\arctan x|_{\pi/4}^{\pi/3} = \ln\arctan\frac{\pi}{3} - \ln\arctan\frac{\pi}{4} = 0.19417$.

25. $\int_2^4 \left(\frac{1}{5}\right)^{x/2} dx$. Let $u = \dfrac{x}{2}$ and $du = \frac{1}{2}\,dx$. Substituting, we get $2\displaystyle\int \left(\frac{1}{5}\right)^u du = 2 \cdot \frac{\left(\frac{1}{5}\right)^u}{\ln \frac{1}{5}} = \frac{2\left(\frac{1}{5}\right)^{x/2}}{\ln \frac{1}{5}}$. Evaluating, we have $\left.\dfrac{2\left(\frac{1}{5}\right)^{x/2}}{\ln\frac{1}{5}}\right|_2^4 = \dfrac{2}{\ln\frac{1}{5}}\left[\left(\frac{1}{5}\right)^2 - \frac{1}{5}\right] = 0.1998.$

26. $\displaystyle\int_{\ln 2}^{\ln 5} \frac{e^x}{e^x+1}\,dx$. Let $u = e^x + 1$ and $du = e^x\,dx$. Hence, we have $\displaystyle\int \frac{1}{u}\,du = \ln|u| = \ln(e^x+1)$. Evaluating, we get $\ln(e^x+1)\big|_{\ln 2}^{\ln 5} = \ln(e^{\ln 5}+1) - \ln(e^{\ln 2}+1) = \ln(5+1) - \ln(2+1) = \ln 6 - \ln 3 = \ln\frac{6}{3} = \ln 2 \approx 0.6931.$

27. Since $e^{x+1} > x^2$ on $[0,1]$, the area we want is $\int_0^1 (e^{x+1} - x^2)\,dx = \int_0^1 e^{x+1} - \int_0^1 x^2\,dx = e^{x+1}\big|_0^1 - \left.\dfrac{x^3}{3}\right|_0^1 = e^2 - e^1 - \dfrac{1^3}{3} = e^2 - e - \dfrac{1}{3} \approx 4.3374.$

28. This area is $\int_0^4 e^{-x}\,dx = -e^{-x}\big|_0^4 = -e^{-4} + e^0 = 1 - e^{-4} = 0.9817.$

29. $s(t) = \int (394e^{-0.025t} - 384)\,dt = -15760e^{-0.025t} - 384t + C$. At $t = 0$, the object is on the ground, so $s(0) = 0$ and we see that, $C = 15{,}760$. Thus, the position function is $s(t) = -15{,}760e^{-0.025t} - 384t + 15{,}760$;
(b) The object reaches its maximum height when $v(t) = 0$ or when $e^{-0.025t} = \frac{384}{394}$. Taking the natural logarithm of both sides we get $-0.025t = \ln\left(\frac{384}{394}\right) = \ln 384 - \ln 394$. So, $t = \frac{\ln 384 - \ln 394}{-0.025} \approx 1.028$;
(c) $s(1.028) = 5.1196$.

30. The average value is $\frac{1}{2}\int_0^2 e^{(-2/3)t}\,dt$. Let $u = -\frac{2}{3}t$, then $du = -\frac{2}{3}\,dt$ or $-\frac{3}{2}\,du = dt$. Then, $\frac{1}{2}\cdot\frac{-3}{2}\int e^u\,du = \frac{-3}{4}e^{(-2/3)t}$. Hence, $-\frac{3}{4}e^{-(2/3)t}\big|_0^2 = -\frac{3}{4}\left[e^{-4/3} - e^0\right] = \frac{3}{4}\left[1 - e^{-4/3}\right] = 0.5523.$

31. Using the disc method this volume is $\pi\int_0^2 (e^{-x})^2\,dx = \pi\int_0^2 e^{-2x}\,dx$. Let $u = -2x$, then $du = -2\,dx$ or $-\frac{1}{2}\,du = dx$. Thus, the integral becomes $\pi\cdot -\frac{1}{2}\int e^u\,du = \frac{-\pi}{2}e^u = -\frac{\pi}{2}e^{-2x}$. Evaluationg, we get $\left.\dfrac{-\pi}{2}e^{-2x}\right|_0^2 = \dfrac{-\pi}{2}(e^{-4} - e^0) = \dfrac{\pi}{2}(1 - e^{-4}) \approx 1.5420.$

32. $V = 200e^{-50t}$. The average voltage is $\frac{1}{0.02}\int_0^{0.02} 200e^{-50t}\,dt = \frac{200}{0.02}\cdot\frac{-1}{50}e^{-50t}\big|_0^{0.02} = -200e^{-50t}\big|_0^{0.02} = -200(e^{-1} - e^0) = 200(1 - e^{-1}) = 126.4241$ V.

33. Since $\dfrac{dV}{ds} = -0.45e^{-0.15s}$, we see that $V = \int -0.45e^{-0.15s}\,ds = \frac{-0.45}{-0.15}e^{-0.15s} + C = 3e^{-0.15s} + C$. Hence, $V(0) = 3 = 3e^0 + C = 3 + C$ so $C = 0$ and $V(s) = 3e^{-0.15s}$. Also, $3e^{-0.15s} = 1.5$ yields $e^{-0.15s} = \frac{1}{2}$. Taking the natural logarithm of both sides we get $-0.15s = \ln\frac{1}{2}$ or $s = \ln(0.5) \div (-0.15) = 4.6210$ mi.

34. Since $P = \frac{k}{V}$, we can solve for $k = PV = 500 \cdot 0.200 = 100$. Thus, the work done by the gas is given by $W = \displaystyle\int_{V_0}^{V_1} \frac{k}{V}\,dV = \int_{0.200}^{0.800} \frac{100}{V}\,dV = 100\ln|V|\Big|_{0.2}^{0.8} = 100(\ln 0.8 - \ln 0.2) \approx 138.6294$ N·m $= 138.6$ J.

35. First we find $k = PV = 20 \times 0.4 = 8$. Then, the work done by the gas is $W = \displaystyle\int_{0.400}^{1.500} \frac{8}{V}\,dN = 8(\ln 1.5 - \ln 0.4) \approx 10.574$ ft·lb.

28.3 BASIC TRIGONOMETRIC AND HYPERBOLIC INTEGRALS

1. $\int \sin\frac{x}{2}\,dx$. Let $u = \frac{x}{2}$ and $du = \frac{1}{2}\,dx$ gives $2\int \sin u\,du = -2\cos u + C = -2\cos\frac{x}{2} + C$.

2. $\int \cos 3x\,dx$. Let $u = 3x$, then $du = 3\,dx$ and $\frac{1}{3}\,du = dx$. This gives $\frac{1}{3}\int \cos u\,du = \frac{1}{3}\sin u + C = \frac{1}{3}\sin 3x + C$.

3. $\int \tan 5x\,dx$. Let $u = 5x$, then $du = 5\,dx$ and $\frac{1}{5}\,du = dx$. This gives $\frac{1}{5}\int \tan u\,du = -\frac{1}{5}\ln|\cos u| + C = -\frac{1}{5}\ln|\cos 5x| + C$ or $\frac{1}{5}\ln|\sec 5x| + C$.

4. $\int x\csc x^2\,dx$. Let $u = x^2$ then $du = 2x\,dx$ and $\frac{1}{2}\,du = x\,dx$. This gives $\frac{1}{2}\int \csc u\,du = \frac{1}{2}\ln|\csc u - \cot u| + C = \frac{1}{2}\ln\left|\csc x^2 - \cot x^2\right| + C$.

5. $\int \frac{\tan\sqrt{x}}{\sqrt{x}}\,dx$. Let $u = \sqrt{x}$ then $du = \frac{1}{2\sqrt{x}}\,dx$ and $2\,du = \frac{1}{\sqrt{x}}\,dx$. Substituting produces $2\int \tan u\,du = -2\ln|\cos u| + C = -2\ln\left|\cos\sqrt{x}\right| + C$ or $2\ln\left|\sec\sqrt{x}\right| + C$.

6. $\int \tan^2 6x\,dx = \int (\sec^2 6x - 1)\,dx = \frac{1}{6}\tan 6x - x + C$.

7. $\int \sin(4x-1)\,dx = -\frac{1}{4}\cos(4x-1) + C$

8. $\int \frac{\sin x + \cos x}{\cos x}\,dx = \int (\tan x + 1)\,dx = \ln|\sec x| + x + C$ or $x - \ln|\cos x| + C$

9. $\int \frac{\sin x}{\cos^2 x}\,dx$. Let $u = \cos x$ and $du = -\sin x\,dx$. Substituting, we obtain $-\int u^{-2}\,du = u^{-1} + C = (\cos x)^{-1} + C = \sec x + C$.

10. $\int \frac{1 + \sin x}{\cos x}\,dx = \int \left(\frac{1}{\cos x} + \tan x\right)dx = \int (\sec x + \tan x)\,dx = \ln|\sec x + \tan x| + \ln|\sec x| + C$.

11. $\int \frac{\sec 3x\tan 3x}{5 + 2\sec 3x}\,dx$. Let $u = 5 + 2\sec 3x$, and then $du = 6\sec 3x\tan 3x\,dx$ or $\frac{1}{6}\,du = \sec 3x\tan 3x\,dx$. This yields $\frac{1}{6}\int \frac{1}{u}\,du = \frac{1}{6}\ln|u| + C = \frac{1}{6}\ln|5 + 2\sec 3x| + C$.

12. $\int \cos^2 x\sin x\,dx$. Let $u = \cos x$ and $du = -\sin x\,dx$. Substituting yields $-\int u^2\,du = -\frac{1}{3}u^3 + C = -\frac{1}{3}\cos^3 x + C$.

13. $\int \sec^4 3x\tan 3x\,dx = \int \sec^3 3x\sec 3x\tan 3x\,dx$. If you let $u = \sec 3x$, then $du = 3\sec 3x\tan 3x\,dx$. This leads to $\frac{1}{3}\int u^3\,du = \frac{1}{12}u^4 + C = \frac{1}{12}\sec^4 3x + C$.

14. $\int \tan 3x\sec^2 3x\,dx = \int \sec 3x\tan 3x\sec 3x\,dx$. If you let $u = \sec 3x$, then $du = 3\sec 3x\tan 3x\,dx$. Hence, $\frac{1}{3}\int u\,du = \frac{1}{6}u^2 + C = \frac{1}{6}\sec^2 3x + C$. An alternate method is to let $u = \tan 3x$ and $du = 3\sec^2 3x\,dx$, and so $\frac{1}{6}\int u\,du = \frac{1}{6}u^2 + C = \frac{1}{6}\tan^2 3x + C$.

15. $\int \frac{1 + \cos 4x}{\sin^2 4x}\,dx = \int \left(\frac{1}{\sin^2 4x} + \frac{\cos 4x}{\sin^2 4x}\right)dx = \int \csc^2 4x\,dx + \int \frac{\cos 4x}{\sin^2 4x}\,dx$. Now, the first integral is $\int \csc^2 4x = \frac{-1}{4}\cot^2 4x + C$. For the second integral, let $u = \sin 4x$ then $du = 4\cos 4x\,dx$ or $\frac{1}{4}\,du = \cos 4x\,dx$. Substitution yields $\frac{1}{4}\int u^{-2}\,du = \frac{1}{4}\cdot -1u^{-1} + C = -\frac{1}{4}(\sin 4x)^{-1} + C = -\frac{1}{4}\csc 4x + C$ Putting these together we get $-\frac{1}{4}\cot^2 4x - \frac{1}{4}\csc 4x + C = -\frac{1}{4}\left[\cot^2 4x + \csc 4x\right] + C$ or $-\frac{1}{4}\left(\frac{\cos 4x + 1}{\sin 4x}\right) + C$.

16. $\int x\csc^2 x^2\,dx$. Let $u = x^2$ and $du = 2x\,dx$ or $\frac{1}{2}\,du = x\,dx$. Then, you get $\frac{1}{2}\int \csc^2 u\,du = \frac{-1}{2}\cot u + C = -\frac{1}{2}\cot x^2 + C$.

17. $\int \tan\frac{x}{4}\,dx = 4\ln\left|\sec\frac{x}{4}\right| + C$

18. $\int x\sec^2(x^2+1)\tan(x^2+1)\,dx$. Let $u = \sec(x^2+1)$ then $du = 2x\sec(x^2+1)\tan(x^2+1)\,dx$. Then, you get $\frac{1}{2}\int u\,du = \frac{1}{4}u^2 + C = \frac{1}{4}\sec^2(x^2+1) + C$ or let $u = \tan(x^2+1)$ which yields the answer $\frac{1}{4}\tan^2(x^2+1) + C$.

19. Using trig identities, we get $\int_{\pi/4}^{\pi/2} \frac{1+\cot^2 x}{\csc^2 x}\,dx = \int_{\pi/4}^{\pi/2}\left(\frac{1}{\csc^2 x} + \cos^2 x\right)dx = \int_{\pi/4}^{\pi/2} (\sin^2 x + \cos^2 x)\,dx = \int_{\pi/4}^{\pi/2} dx = x\Big|_{\pi/4}^{\pi/2} = \frac{\pi}{2} - \frac{\pi}{4} = \frac{\pi}{4}$.

20. $\int_0^{\pi/2} \frac{\cos x}{1+\sin x}\,dx$. Let $u = 1 + \sin x$ and $du = \cos x\,dx$. This leads to $\int \frac{1}{u}\,du = \ln|u| = \ln|1+\sin x|_0^{\pi/2} = \ln|2| - \ln|1| = \ln 2 \approx 0.6931$.

21. $\int_{\pi/4}^{\pi/2} \frac{\csc\sqrt{x}\cot\sqrt{x}}{\sqrt{x}}\,dx$. Let $u = \sqrt{x}$ and $du = \frac{1}{2\sqrt{x}}$. Then, $2\int \csc u \cot u\,du = -2\csc u$. Evaluating, we obtain $-2\csc\sqrt{x}\Big|_{\pi/4}^{\pi/2} = 0.4765$.

22. $\int_0^{\pi/8} \sec^5 2x \tan 2x\,dx = \int_0^{\pi/8} \sec^4 2x \sec 2x \tan 2x\,dx$. Let $u = \sec 2x$ and $du = 2\sec 2x \tan 2x$. Dividing by 2 and substituting gives $\frac{1}{2}\int u^4\,du = \frac{1}{10}u^5$ or $\frac{1}{10}\sec^5 2x\Big|_0^{\pi/8} = \frac{1}{10}\left(\sec^5\frac{\pi}{4} - \sec^5 0\right) = \frac{1}{10}\left(\sqrt{2}^5 - 1^5\right) = \frac{1}{10}\left(4\sqrt{2}-1\right) \approx 0.4657$.

23. Using a Pythagorean trigonometric substitution gives the result $\int_{\pi/6}^{\pi/2} \frac{\cos^2 x}{\sin x}\,dx = \int_{\pi/6}^{\pi/2} \frac{1-\sin^2 x}{\sin x}\,dx = \int_{\pi/6}^{\pi/2}\left(\frac{1}{\sin x} - \sin x\right)dx = \int_{\pi/6}^{\pi/2} (\csc x - \sin x)\,dx = \Big[\ln|\csc x - \cot x| + \cos x\Big]_{\pi/6}^{\pi/2} = \ln|1-0| + 0 - \ln\left|2-\sqrt{3}\right| - \frac{\sqrt{3}}{2} = 0.4509$.

24. $\int_0^{5\pi/12} \frac{\sec^2 x}{2\tan x + 4}\,dx$; Let $u = 2\tan x + 4$, and then $du = 2\sec^2 x\,dx$ or $\frac{1}{2}\,du = \sec^2 x\,dx$. Substituting yields $\frac{1}{2}\int \frac{1}{u}\,du = \frac{1}{2}\ln|u| = \frac{1}{2}\ln|2\tan x + 4|_0^{5\pi/12} = \frac{1}{2}\ln\left|2\tan\frac{5\pi}{12} + 4\right| - \frac{1}{2}\ln|4| = 0.5265$.

25. $\int x \sinh x^2\,dx$. Let $u = x^2$ and $du = 2x\,dx$ or $\frac{1}{2}\,du = x\,dx$. Substituting we get $\frac{1}{2}\int \sinh u\,du = \frac{1}{2}\cosh u + C = \frac{1}{2}\cosh x^2 + C$

26. $\int \frac{\sinh x}{1+\cosh x}\,dx$. Let $u = 1 + \cosh x$ and then $du = \sinh x\,dx$. This leads to $\int \frac{1}{u}\,du = \ln|u| + C = \ln(1+\cosh x) + C$. Absolute value signs are not needed as $1 + \cosh x$ is never negative.

27. $\int 3x\cosh x^2\sqrt{\sinh x^2}\,dx$. Let $u = \sinh x^2$, then $du = 2x\cosh x^2\,dx$ or $\frac{3}{2}\,du = 3x\cosh x^2\,dx$. Substituting we get $\frac{3}{2}\int \sqrt{u}\,du = \frac{3}{2}\cdot\frac{2}{3}u^{3/2} + C = (\sinh x^2)^{3/2} + C$

28. $\int \operatorname{sech}^2 5x\,dx = \frac{1}{5}\tanh 5x + C$

29. $\int \sinh^3 x \cdot \cosh^2 x\,dx$. Recall that $\sinh^3 x = \sinh x \sinh^2 x = \sinh x\left(\cosh^2 x - 1\right)$, and then we have $\int \sinh x(\cosh^2 x - 1)\cosh^2 x\,dx = \int \cosh^4 x \sinh x\,dx - \int \cosh^2 x \sinh x\,dx = \frac{1}{5}\cosh^5 x - \frac{1}{3}\cosh^3 x + C$.

30. $\int \tanh 3x \operatorname{sech} 3x\,dx = -\frac{1}{3}\operatorname{sech} 3x + C$.

31. We begin by rewriting $\cot u$ as $\frac{\cos u}{\sin u}$. Here, if we let $v = \sin u$, then $dv = \cos u\,du$, and the integral is $\int \cot u\,du = \int \frac{dv}{v} = \ln|v| + C$. Back substitution gives the desired result: $\int \cot u\,du = \ln|\sin u| + C$.

32. We will follow a procedure similar to the one we followed prior to Example 28.18. Begin by multiplying the integand by $\frac{\csc u - \cot u}{\csc u - \cot u}$. This produces $\int \csc u\,du = \int \csc u\left(\frac{\csc u - \cot u}{\csc u - \cot u}\right)du = \int \frac{\csc u(\csc u - \cot u)}{\csc u - \cot u}\,du = \int \frac{\csc^2 u - \csc u \cot u}{\csc u - \cot u}\,du$. Let $v = \csc u - \cot u$. Then, $dv = -\csc u \cot u + \csc^2 u)\,du$ and we can write the original integral as $\int \csc u\,du = \int \frac{dv}{v} = \ln|v| + C$. Back substitution produces the result we wanted to show, namely that $\int \csc u\,du = \ln|\csc u - \cot u| + C$.

33. For $0 \le x \le \frac{\pi}{2}$, we see that $\sin 2x \ge 0$, so the area is the integral $\int_0^{\pi/2} \sin 2x\,dx = \frac{-1}{2}\cos 2x\Big|_0^{\pi/2} = -\frac{1}{2}(\cos\pi - \cos 0) = \frac{-1}{2}(-1-1) = 1$.

34. For $0 \le x \le \frac{\pi}{4}$, we see that $\sec x \ge x$, so the area is $\displaystyle\int_0^{\pi/4}(\sec x - x)\,dx = \left(\ln|\sec x + \tan x| - \frac{x^2}{2}\right)\Bigg|_0^{\pi/4} = \ln(\sqrt{2}+1) - \frac{(\pi/4)^2}{2} - \ln(1+0) - 0 \approx 0.5729$.

35. The average value is $\displaystyle\frac{1}{\pi/4}\int_0^{\pi/4}\tan x\,dx = \frac{4}{\pi}\ln|\sec x|_0^{\pi/4} = \frac{4}{\pi}\left(\ln\sqrt{2} - \ln 1\right) = \frac{4\ln\sqrt{2}}{\pi} \approx 0.4413$.

36. Since $v(t) = 5\sin 2t$, then $s(t) = \int v(t)\,dt = \int 5\sin 2t\,dt = \frac{-5}{2}\cos 2t + C$. Since $s(0) = 4$, we can solve for C. $C = 4 + \frac{5}{2}\cos 0 = 4 + \frac{5}{2} = \frac{13}{2}$. Thus, we see that $s(t) = -\frac{5}{2}\cos 2t + \frac{13}{2}$.

37. $\bar{E} = \frac{1}{1-0}\int_0^1 5\sin 4t\,dt = -\frac{5}{4}\cos 4t\Big|_0^1 = -\frac{5}{4}(-0.6536 - 1) \approx 2.067$ V.

38. The time to complete one cycle is $\frac{2\pi}{377}$ sec, so the average rate at which heat is produced is $\displaystyle\bar{P} = \frac{1}{\frac{2\pi}{377} - 0}\int_0^{2\pi/377} 40(6.5\sin 377t)^2\,dt = \frac{377}{2\pi}\int_0^{2\pi/377} 1690\sin^2 377t\,dt$. Using the half-angle identity, $\sin\frac{\alpha}{2} = \pm\sqrt{\frac{1-\cos\alpha}{2}}$, we see that $\sin^2 377t = \dfrac{1-\cos 745t}{2}$. Thus, we get $\displaystyle\frac{377}{2\pi}\int_0^{2\pi/377} 845(1-\cos 754t)\,dt = \frac{(377)(845)}{2\pi}\int_0^{2\pi/377}(1-\cos 754t)\,dt = \frac{(377)(845)}{2\pi}(t - \sin 754t)\Bigg|_0^{2\pi/377} = \frac{(377)(845)}{2\pi} \times \left(\frac{2\pi}{377} - 0\right) = 845$ W.

39. Here $y = 12\cosh\frac{x}{12}$ and so $y' = \sinh\frac{x}{12}$. The length of the cable is the arc length $= \int_{-18}^{18}\sqrt{1+\sinh^2\frac{x}{12}}\,dx = \int_{-18}^{18}\cosh\frac{x}{12}\,dx = 12\sinh\frac{x}{12}\Big|_{-18}^{18} = 12\sinh\frac{3}{2} - 12\sinh\frac{-3}{2} = 51.1027$ m.

40. The center of the towers is 0 so the two towers are at -60 and 60. $y = 60\cosh\frac{x}{60}$; $y' = \sinh\frac{x}{60}$. The arc length is $\int_{-60}^{60}\sqrt{1+\sinh^2\frac{x}{60}}\,dx = \int_{-60}^{60}\cosh\frac{x}{60}\,dx = 60\sinh\frac{x}{60}\Big|_{-60}^{60} = 60\sinh 1 - 60\sinh(-1) = 141.02$ ft.

41. The center of the towers is at 0 and the towers are at -150 and 150. We are given $y = 80\cosh\frac{x}{80}$ and differentiating, we obtain $y' = \sinh\frac{x}{80}$. The arc length is $\int_{-150}^{150}\sqrt{1+\sinh^2\frac{x}{80}}\,dx = \int_{-150}^{150}\cosh\frac{x}{80}\,dx = 80\left[\sinh\frac{x}{80}\right]_{-150}^{150} = 509.397$ ft.

42. (a) The maximum will occur at the center or when $x = 0$. $(-\cosh x)' = -\sinh x = 0$ when $x = 0$. Hence, $y(0) = -127.7\cosh 0 + 757.7 = 630$ ft.

(b) $315 - (-315) = 630$ ft.

(c) Area $\displaystyle = \int_{-315}^{315}\left[-127.7\cosh\left(\frac{x}{127.7}\right) + 757.7\right]dx = -(127.7)^2\sinh\left(\frac{x}{127.7}\right) + 757.7x\Bigg|_{-315}^{315} = 286{,}574\ \text{ft}^2$.

(d) To find the arc length we first find y' as $y' = -\sinh\left(\frac{x}{127.7}\right)$. Substituting this in the arc length formula produces $\displaystyle\int_{-315}^{315}\sqrt{1+\left[-\sinh\left(\frac{x}{127.7}\right)\right]^2}\,dx = \int_{-315}^{315}\cosh\left(\frac{x}{127.7}\right)dx = 127.7\sinh\left(\frac{x}{127.7}\right)\Bigg|_{-315}^{315} = 1493.94$ ft.

28.4 MORE TRIGONOMETRIC INTEGRALS

1. $\int \sin^2 3x \cos 3x\,dx$. Let $u = \sin 3x$ and $du = 3\cos 3x\,dx$. Then, $\frac{1}{3}\int u^2\,du = \frac{1}{9}u^3 + C = \frac{1}{9}\sin^3 3x + C$

2. $\int \sin 2x \cos^2 2x\,dx$. Let $u = \cos 2x$ and $du = -2\sin 2x\,dx$. Then, $-\frac{1}{2}\int u^2\,du = -\frac{1}{6}u^3 + C = -\frac{1}{6}\cos^3 2x + C$

3. $\int \cos x \sin^3 x\,dx = \frac{1}{4}\sin^4 x + C$

4. $\int \sin^3 4x \cos^2 4x\,dx = \int \sin 4x(1 - \cos^2 4x)\cos^2 4x\,dx$. Let $u = \cos 4x$ and $du = -4\sin 4x\,dx$. Then, $-\frac{1}{4}\int (1-u^2)u^2\,du = -\frac{1}{4}\int u^2 - u^4\,du = -\frac{1}{4}\left(\frac{1}{3}u^3 - \frac{1}{5}u^5\right) + C = \frac{1}{20}\cos^5 4x - \frac{1}{12}\cos^3 4x + C$

5. $\int \cos 5x \sin^4 5x\,dx = \frac{1}{25}\sin^5 5x + C$

6. $\int \cos^2 y \sin^5 y\,dy = \int \cos^2 y(1 - \cos^2 y)^2 \sin y\,dy$. Let $u = \cos y$ and $du = -\sin y\,dy$, and we get $-\int u^2(1-u^2)^2\,du = -\int (u^2 - 2u^4 + u^6)\,du = -\frac{1}{3}u^3 + \frac{2}{5}u^5 - \frac{1}{7}u^7 + C = -\frac{1}{3}\cos^3 y + \frac{2}{5}\cos^5 y - \frac{1}{7}\cos^7 y + C$.

7.
$$\begin{aligned}
\int \sin^2 x \cos^4 x\,dx &= \int \left(\frac{1-\cos 2x}{2}\right)\left(\frac{1+\cos 2x}{2}\right)^2 dx = \frac{1}{8}\int (1-\cos 2x)\left(1 + 2\cos 2x + \cos^2 2x\right) dx \\
&= \frac{1}{8}\int (1 + \cos 2x - \cos^2 2x - \cos^3 2x)\,dx = \frac{1}{8}\int (1 + \cos 2x - \cos^3 2x) - \frac{1}{8}\int \cos^2 2x\,dx \\
&= \frac{1}{8}x - \frac{1}{16}\sin 2x - \frac{1}{8}\int \left(1 - \sin^2 2x\right)\cos 2x\,dx - \frac{1}{8}\int \frac{1+\cos 4x}{2}\,dx \\
&= \frac{1}{8}x - \frac{1}{16}\sin 2x + \frac{1}{16}\sin 2x + \frac{1}{48}\sin^3 2x - \frac{1}{16}x + \frac{1}{16}\cdot\frac{1}{4}\sin 4x + C \\
&= \frac{1}{16}x + \frac{1}{48}\sin^3 2x + \frac{1}{64}\sin 4x + C
\end{aligned}$$

8.
$$\begin{aligned}
\int \sin^8 x\,dx &= \int (\sin^2 x)^4\,dx = \int \left(\frac{1-\cos 2x}{2}\right)^4 dx = \frac{1}{16}\int (1-\cos 2x)^4\,dx \\
&= \frac{1}{16}\int (1 - 4\cos 2x + 6\cos^2 2x - 4\cos^3 2x + \cos^4 2x)\,dx \\
&= \frac{1}{16}\int 1 - 4\cos 2x + 3(1+\cos 4x) - 4(1-\sin^2 2x)\cos 2x + \left(\frac{1+\cos 4x}{2}\right)^2 dx \\
&= \frac{1}{16}\int (1 - 4\cos 2x + 3 + 3\cos 4x - 4\cos 2x + 4\cos 2x\sin^2 2x + \frac{1}{4} + \frac{1}{2}\cos 4x + \frac{1}{4}\cos^2 4x)\,dx \\
&= \frac{1}{16}\int \frac{17}{4} - 8\cos 2x + \frac{7}{2}\cos 4x + 4\cos 2x\sin^2 2x + \frac{1}{4}\left(\frac{1+\cos 8x}{2}\right) dx \\
&= \frac{1}{16}\int \left(\frac{35}{8} - 8\cos 2x + \frac{7}{2}\cos 4x + 4\cos 2x\sin^2 2x + \frac{1}{8}\cos 8x\right) dx \\
&= \frac{1}{16}\left(\frac{35}{8}x - 4\sin 2x + \frac{7}{8}\sin 4x + \frac{2}{3}\sin^3 2x + \frac{1}{64}\sin 8x\right) + C
\end{aligned}$$

9. $$\int \cos^6 3x\,dx = \int \left(\frac{1+\cos 6x}{2}\right)^3 dx = \frac{1}{8}\int (1+3\cos 6x + 3\cos^2 6x + \cos^3 6x)\,dx$$
$$= \frac{1}{8}\int \left[1+3\cos 6x + \frac{3}{2}(1+\cos 12x) + (1-\sin^2 6x)\cos 6x\right] dx$$
$$= \frac{1}{8}\int \left[1+3\cos 6x + \frac{3}{2} + \frac{3}{2}\cos 12x + \cos 6x - \sin^2 6x\cos 6x\right] dx$$
$$= \frac{1}{8}\int \left[\frac{5}{2} + 4\cos 6x + \frac{3}{2}\cos 12x - \sin^2 6x\cos 6x\right] dx$$
$$= \frac{1}{8}\left[\frac{5x}{2} + \frac{2}{3}\sin 6x + \frac{1}{8}\sin 12x - \frac{1}{18}\sin^3 6x\right] + C = \frac{5x}{16} + \frac{1}{12}\sin 6x + \frac{1}{64}\sin 12x - \frac{1}{144}\sin^3 6x + C$$

10. $\int \frac{\cos x}{\sin^3 x}\,dx$. Let $u = \sin x$ and $du = \cos x\,dx$, then $\int \frac{du}{u^3} = -\frac{1}{2}u^{-2} + C = \frac{-1}{2\sin^2 x} + C.$

11. $$\int \sin^2 2\theta \cos^4 2\theta\,d\theta = \int \left(\frac{1-\cos 4\theta}{2}\right)\left(\frac{1+\cos 2\theta}{2}\right)^2 d\theta = \frac{1}{8}\int \left(1+\cos 4\theta - \cos^2 4\theta - \cos^3 4\theta\right) d\theta$$
$$= \frac{1}{8}\int \left[1+\cos 4\theta - \frac{1}{2}(1+\cos 8\theta) - (1-\sin^2 4\theta)\cos 4\theta\right] d\theta$$
$$= \frac{1}{8}\int \left[1+\cos 4\theta - \frac{1}{2} - \frac{1}{2}\cos 8\theta - \cos 4\theta + \sin^2 4\theta\cos 4\theta\right] d\theta$$
$$= \frac{1}{8}\int \left[\frac{1}{2} - \frac{1}{2}\cos 8\theta + \sin^2 4\theta\cos 4\theta\right] d\theta = \frac{1}{8}\left[\frac{\theta}{2} - \frac{1}{16}\sin 8\theta + \frac{1}{12}\sin^3 4\theta\right] + C$$
$$= \frac{\theta}{16} - \frac{1}{128}\sin 8\theta + \frac{1}{96}\sin^3 4\theta + C$$

12. $$\int \sin^4 3\theta \cos^2 3\theta\,d\theta = \int \left(\frac{1-\cos 6\theta}{2}\right)^2\left(\frac{1+\cos 6\theta}{2}\right) d\theta = \frac{1}{8}\int \left(1-2\cos 6\theta + \cos^2 6\theta\right)(1+\cos 6\theta)\,d\theta$$
$$= \frac{1}{8}\int \left(1-\cos 6\theta - \cos^2 6\theta + \cos^3 6\theta\right) d\theta$$
$$= \frac{1}{8}\int \left[1-\cos 6\theta - \frac{1}{2}(1+\cos 12\theta) + (1-\sin^2 6\theta)\cos 6\theta\right] d\theta$$
$$= \frac{1}{8}\int \left[\frac{1}{2} - \cos 6\theta - \frac{1}{2}\cos 12\theta + \cos 6\theta - \sin^2 6\theta\cos 6\theta\right] d\theta$$
$$= \frac{1}{8}\int \left[1/2 - \frac{1}{2}\cos 12\theta - \sin^2 6\theta\cos 6\theta\right] d\theta = \frac{1}{8}\left[\frac{\theta}{2} - \frac{1}{24}\sin 12\theta - \frac{1}{18}\sin^3 6\theta\right] + C$$
$$= \frac{\theta}{16} - \frac{1}{192}\sin 12\theta - \frac{1}{144}\sin^3 6\theta + C$$

13. $\int \sec^2 x \tan^2 x\,dx$. Let $u = \tan x$ and then $du = \sec^2 x\,dx$. Hence, we get $\int u^2\,du = \frac{1}{3}u^3 + C = \frac{1}{3}\tan^3 x + C$.

14. $\int \sec^4 y \tan^3 y\,dy = \int \sec^2 y(\sec^2 y)\tan^3 y\,dy = \int \left[\sec^2 y(1+\tan^2 y)\tan^3 y\right] dy = \int \sec^2 y[\tan^3 y + \tan^5 y]\,dy$. Let $u = \tan y$ and then $du = \sec^2 y\,dy$, and you have $\int \left(u^3 + u^5\right) du = \frac{1}{4}u^4 + \frac{1}{6}u^6 + C = \frac{1}{4}\tan^4 y + \frac{1}{6}\tan^6 y + C$.

15. First factor $\int \csc^4 x \cot x\,dx = \int \csc^3 x(\csc x \cot x\,dx)$. Then let $u = \csc x$ and $du = -\csc x \cot x\,dx$. These produce $-\int u^3\,du = -\frac{1}{4}u^4 + C = -\frac{1}{4}\csc^4 x + C_1$ or $\int \csc^4 x \cot x\,dx = \int \csc^2 x(1+\cot^2 x)\cot x\,dx = \int \left(\cot^3 x + \cot x\right)\csc^2 x\,dx$. Let $u = \cot x$ and $du = -\csc^2 x\,dx$, and then you have $-\int (u^3 + u)\,du = -\frac{1}{4}u^4 - \frac{1}{2}u^2 + C = -\frac{1}{4}\cot^4 x - \frac{1}{2}\cot^2 x + C_2$ where $C_2 = C_1 - 0.25$.

16. $\displaystyle\int \frac{\sin^2\theta}{\cos^4\theta}\,d\theta = \int \sec^2\theta \tan^2\theta\,d\theta$. Let $u = \tan\theta$ and $du = \sec^2\theta\,d\theta$, then $\int u^2\,du = \frac{1}{3}u^3 + C = \frac{1}{3}\tan^3\theta + C$.

17. We first factor the integrand and use a Pythagorean identity: $\int \sin^{1/2}3\theta \cos^3 3\theta\,d\theta = \int \sin^{1/2}3\theta(1 - \sin^2 3\theta)\cos 3\theta\,d\theta$. Then multiply and integrate, with the result $\int \left(\sin^{1/2}3\theta - \sin^{5/2}3\theta\right)\cos 3\theta\,d\theta = \frac{1}{3}\cdot\frac{2}{3}\sin^{3/2}3\theta - \frac{1}{3}\cdot\frac{2}{7}\sin^{7/2}3\theta + C = \frac{2}{9}\sin^{3/2}3\theta - \frac{2}{21}\sin^{7/2}3\theta + C$.

18. The given integral $\int \csc^6 2x \cot^2 2x\,dx$ can be written as $\int \csc^4 2x \cot^2 2x \csc^2 2x\,dx$ or, using a Pythagorean identity, as $\int (1+\cot^2 2x)^2 \cot^2 2x \csc^2 2x\,dx$. This expands to $\int (1 + 2\cot^2 2x + \cot^4 2x)\cot^2 2x \csc^2 2x\,dx$ or $\int (\cot^2 2x + 2\cot^4 2x + \cot^6 2x)\csc^2 2x\,dx$. Let $u = \cot 2x$ and $du = -2\csc^2 2x\,dx$ and we rewrite the integral as $-\frac{1}{2}\int \left(u^2 + 2u^4 + u^6\right) du = -\left(\frac{1}{6}u^3 + \frac{1}{5}u^5 + \frac{1}{14}u^7\right) + C = -\left(\frac{1}{6}\cot^3 2x + \frac{1}{5}\cot^5 2x + \frac{1}{14}\cot^7 2x\right) + C$.

19. $\int \csc x \cot^3 x\,dx = \int \left(\csc^2 x - 1\right)\csc x \cot x\,dx = -\frac{1}{3}\csc^3 x + \csc x + C$.

20. $\int \sec t \tan^5 t\,dt = \int \sec t \tan t(\tan^4 t)\,dt = \int \sec t \tan t(\tan^2 t)^2\,dt = \int \sec t \tan t(\sec^2 - 1t)^2\,dt = \int \sec t \tan t(\sec^4 - 2\sec^2 t + 1)\,dt = \frac{1}{5}\sec^5 t - \frac{2}{3}\sec^3 t + \sec t + C$.

21. $\int \tan^3 x \sec^2 x\,dx = \frac{1}{4}\tan^4 x + C$.

22. $\int \cot 2x \csc^4 2x\,dx$. Let $u = \csc 2x$ and then $du = -2\csc 2x \cot 2x\,dx$. Thus, we then intgrate $-\frac{1}{2}\int u^3\,du = -\frac{1}{8}u^4 + C = -\frac{1}{8}\csc^4 2x + C_1$ or $\int \cot 2x \csc^4 2x = \int \cot 2x(\cot^2 2x + 1)\csc^2 2x\,dx = \int (\cot^2 2x + \cot 2x)\csc^2 2x\,dx = -\frac{1}{8}\cot^4 2x - \frac{1}{4}\cot^2 2x + C_2$ where $C_2 = C_1 - \frac{1}{8}$.

23. $\int \tan^6 x \sec^2 x\,dx = \frac{1}{7}\tan^7 x + C$

24. $\int \tan^5 x\,dx = \int (\sec^2 x - 1)^2 \tan x\,dx = \int (\sec^4 x - 2\sec^2 x + 1)\tan x\,dx = \int \sec^4 x \tan x\,dx + 2\int \sec^2 x \tan x\,dx + \int \tan x\,dx = \frac{1}{4}\sec^4 x - \sec^2 x + \ln|\sec x| + C$. An alternate method is $\int \tan^5 x\,dx = \int \tan^3 x(\sec^2 x - 1)\,dx$ which can be written as $\int \tan^3 x \sec^2 x - \int \tan x(\sec^2 - 1)\,dx = \frac{1}{4}\tan^4 x - \frac{1}{2}\tan^2 x + \ln|\sec x| + C$.

25. $\int \cot^6 x\,dx = \int \cot^4 x(\csc^2 x - 1)\,dx = \int \cot^4 x \csc^2 x\,dx - \int \cot^2 x(\csc^2 x - 1)\,dx = \int \cot^4 x \csc^2 x\,dx - \int \cot^2 x \csc^2 x\,dx + \int \cot^2 x\,dx = -\frac{1}{5}\cot^5 x + \frac{1}{3}\cot^3 x + \int (\csc^2 x - 1)\,dx = -\frac{1}{5}\cot^5 x + \frac{1}{3}\cot^3 x - \cot x - x + C$. Recall: $\cot^2 x = \csc^2 x - 1$.

26. $\int \sec^6 2\theta\,d\theta = \int \sec^2 2\theta(\tan^2 2\theta + 1)^2\,d\theta = \int \sec^2 2\theta(\tan^4 2\theta + 2\tan^2 2\theta + 1)\,d\theta = \frac{1}{10}\tan^5 2\theta + \frac{1}{3}\tan^3 2\theta + \frac{1}{2}\tan 2\theta + C$

27. $\int_0^{\pi/4} \tan^2 x\,dx = \int_0^{\pi/4}(\sec^2 x - 1)\,dx = \tan x - x\Big|_0^{\pi/4} = 1 - \frac{\pi}{4} - 0 = 1 - \frac{\pi}{4}$

28. $\tan^3 2t$ is not defined at $\frac{\pi}{4}$ and so, $\tan^3 2t$ is not continuous on the interval $[\pi/6, \pi/3]$. Hence, this integral does not exist.

29. $\displaystyle\int_0^{\pi/2} \sin^4 x\,dx = \int_0^{\pi/2}\left(\frac{1-\cos 2x}{2}\right)^2 dx$ by a half-angle identity. This expands as $\frac{1}{4}\int_0^{\pi/2}(1 - 2\cos 2x + \cos^2 2x)\,dx$ which becomes, by another half-angle identity,

$\frac{1}{4}\int_0^{\pi/2}\left(1-2\cos 2x+\frac{1+\cos 4x}{2}\right)dx$ or $\frac{1}{4}\int_0^{\pi/2}\left(\frac{3}{2}-2\cos 2x+\frac{\cos 4x}{2}\right)dx = \frac{1}{4}\left[\frac{3x}{2}-\sin 2x+\frac{1}{8}\sin 4x\right]_0^{\pi/2} = \frac{1}{4}\left[\frac{3\pi}{4}-\sin\pi+\frac{1}{8}\sin 2\pi\right] = \frac{1}{4}\left[\frac{3\pi}{4}-0+0\right] = \frac{3\pi}{16}$.

30. $\int_0^{\pi/6}\sec^3 2x\tan 2x\,dx = \frac{1}{2}\cdot\frac{1}{3}\sec^3 2x\Big|_0^{\pi/6} = \frac{1}{6}\left(\sec^3\frac{\pi}{3}-\sec^3 0\right) = \frac{1}{6}(8-1) = \frac{7}{6}$

31. Using the disc method, you get $\pi\int_0^{\pi/3}\cos^2 x\,dx = \pi\int_0^{\pi/3}\frac{1+\cos 2x}{2}\,dx = \frac{\pi}{2}\left[x+\frac{1}{2}\sin 2x\right]_0^{\pi/3} = \frac{\pi}{2}\left[\frac{\pi}{3}+\frac{\sqrt{3}}{4}\right] \approx 2.3251$.

32. $V = 80\sin 120\pi t$. One period is $\frac{2\pi}{120\pi} = \frac{1}{60}$. Thus, the root mean square of the voltage for one period is $\sqrt{\frac{1}{1/60}\int_0^{1/60}(80\sin 120\pi t)^2\,dt} = 80\sqrt{60\int_0^{1/60}\sin^2 120\pi t\,dt} = 80\sqrt{60\int_0^{1/60}\frac{1-\cos 240\pi t}{2}\,dt} = 80\sqrt{60\left[\frac{1}{2}t-\frac{\sin 240\pi t}{480\pi}\right]_0^{1/60}} = 80\sqrt{60\left(\frac{1}{120}\right)} = 80\sqrt{1/2} = 80\frac{\sqrt{2}}{2} = 40\sqrt{2} \approx 56.5685$ V.

28.5 INTEGRALS RELATED TO INVERSE TRIGONOMETRIC AND INVERSE HYPERBOLIC FUNCTIONS

1. $\int\frac{dx}{\sqrt{4-x^2}} = \arcsin\frac{x}{2}+C$

2. $\int\frac{dx}{\sqrt{1-9x^2}}$. Let $u = 3x$ and $du = 3\,dx$ so $\frac{1}{3}\,du = dx$. Substituting, we have $\frac{1}{3}\int\frac{du}{\sqrt{1-u^2}} = \frac{1}{3}\arcsin u+C = \frac{1}{3}\arcsin 3x+C$.

3. $\int\frac{dx}{\sqrt{4-9x^2}}$. Let $u = 3x$ and $du = 3\,dx$ so $\frac{1}{3}\,du = dx$. Substituting, we have $\frac{1}{3}\int\frac{du}{\sqrt{4-u^2}} = \frac{1}{3}\arcsin\frac{u}{2}+C = \frac{1}{3}\arcsin\frac{3x}{2}+C$.

4. $\int\frac{8\,dx}{1+16x^2}$. Let $u = 4x$ and then $du = 4\,dx$ so $2\,du = 8\,dx$. Thus, $2\int\frac{du}{1+u^2} = 2\cdot\arctan u+C = 2\arctan 4x+C$.

5. $\int\frac{-x\,dx}{\sqrt{9-x^2}}$. Let $u = 9-x^2$ and $du = -2x\,dx$ so $\frac{1}{2}\,du = -x\,dx$. Then, $\frac{1}{2}\int u^{-1/2}\,du = \frac{1}{2}\cdot\frac{2}{1}u^{1/2}+C = \sqrt{9-x^2}+C$.

6. $\int\frac{dx}{x\sqrt{x^2-16}} = \frac{1}{4}\operatorname{arcsec}\frac{x}{4}+C$.

7. $\int\frac{dx}{3x\sqrt{9x^2-4}}$. Let $u = 3x$, then $du = 3\,dx$, so $\frac{1}{3}\,du = dx$. Thus, we can rewrite the integral as $\frac{1}{3}\int\frac{du}{u\sqrt{u^2-4}} = \frac{1}{3}\cdot\frac{1}{2}\operatorname{arcsec}\frac{u}{2}+C = \frac{1}{6}\operatorname{arcsec}\frac{3x}{2}+C$.

8. $\int\frac{x\,dx}{\sqrt{1-25x^2}}$. Here $u = 1-25x^2$ and $du = -50x\,dx$ and the integral becomes $-\frac{1}{50}\int u^{-1/2}\,du = -\frac{1}{50}\cdot\frac{2}{1}u^{\frac{1}{2}}+C = -\frac{1}{25}\sqrt{1-25x^2}+C$.

9. $\int\frac{dx}{1+(3-x)^2}$. Here $u = 3-x$ and $du = -\,dx$ and so we get $-\int\frac{du}{1+u^2} = -\arctan u+C =$

$-\arctan(3-x)+C$.

10. $\int \frac{dx}{(2x+1)\sqrt{(2x+1)^2-4}}$. Let $u = 2x+1$ and $du = 2\,dx$, then $\frac{1}{2}\,du = dx$. $\frac{1}{2}\int \frac{du}{u\sqrt{u^2-4}} = \frac{1}{2}\cdot\frac{1}{2}\operatorname{arcsec}\frac{u}{2}+C = \frac{1}{4}\operatorname{arcsec}\frac{2x+1}{2}+C$

11. $\int \frac{4x-6}{4x^2+25}\,dx = \int \frac{4x}{4x^2+25}\,dx - \int \frac{6}{4x^2+25}\,dx$. In the first integral let $u = 4x^2+25$; $du = 8x\,dx$ and we get by substitution $2\int \frac{du}{u} = 2\ln|u| = 2\ln(4x^2+25)$ The second integral is $-6\int \frac{dx}{4x^2+25} = -6\cdot\frac{1}{10}\arctan\frac{2x}{5}$. Putting these together we have $2\ln(4x^2+25)-\frac{3}{5}\arctan\frac{2x}{5}+C$

12. $\int \frac{dx}{(x+3)\sqrt{(x+3)^2-1}} = \operatorname{arcsec}(x+3)+C$.

13. Completing the square, we obtain $\int \frac{dx}{x^2+6x+10} = \int \frac{dx}{(x+3)^2+1} = \arctan(x+3)+C$.

14. We begin by completing the square on the expression under the radical. $\int \frac{dx}{\sqrt{-4x^2+4x+15}} = \int \frac{dx}{\sqrt{-(4x^2-4x+1)+15+1}} = \int \frac{dx}{\sqrt{16-(2x-1)^2}}$. Now let $u = 2x-1$ and $du = 2\,dx$ and you get $\frac{1}{2}\int \frac{du}{\sqrt{16-u^2}} = \frac{1}{2}\cdot\arcsin\frac{u}{2}+C = \frac{1}{2}\arcsin\frac{2x-1}{2}+C$.

15. $\int \frac{x\,dx}{1+x^4}$; Let $u = x^2$ and then $du = 2x\,dx$. Substituting, we obtain $\frac{1}{2}\int \frac{du}{1+u^2} = \frac{1}{2}\cdot\arctan u + C = \frac{1}{2}\arctan x^2+C$.

16. $\int \frac{\sec^2 x\,dx}{4+\tan^2 x}$. Let $u = \tan x$ and $du = \sec^2 x\,dx$. Substituting, produces $\int \frac{du}{4+u^2} = \frac{1}{2}\arctan\frac{u}{2}+C = \frac{1}{2}\arctan\left(\frac{\tan x}{2}\right)+C$.

17. $\int \frac{\sin x\,dx}{\sqrt{1-\cos^2 x}}$. Let $u = \cos x$ and $du = -\sin x\,dx$. Substituting produces $-\int \frac{du}{\sqrt{1-u^2}} = -\arcsin u + C = -\arcsin(\cos x)+C = -\left(\frac{\pi}{2}-x\right)+C = x-\frac{\pi}{2}+C$.

18. $\int \frac{x-5}{\sqrt{x^2-10x+16}}\,dx$. Let $u = x^2-10x+16$; $du = (2x-10)\,dx$ or $\frac{1}{2}\,du = (x-5)\,dx$. Hence, we have $\frac{1}{2}\int \frac{du}{\sqrt{u}} = \frac{1}{2}\cdot\frac{2}{1}u^{1/2}+C = \sqrt{x^2-10x+16}+C$.

19. $\int_0^{\pi/4} \frac{\cos x\,dx}{1+\sin^2 x} = \arctan(\sin x)\big|_0^{\pi/4} = \arctan\left(\sin\frac{\pi}{4}\right) - \arctan(\sin 0) = \arctan\frac{\sqrt{2}}{2} - \arctan 0 = 0.6155-0 = 0.6155$.

20. Completing the square, we have $\int_3^4 \frac{dx}{\sqrt{-x^2+8x-15}} = \int_3^4 \frac{dx}{\sqrt{1-(x^2-8x-16)}} = \int_3^4 \frac{dx}{\sqrt{1-(x-4)^2}} = \arcsin(x-4)\big|_3^4 = \arcsin 0 - \arcsin(-1) = 0-\frac{-\pi}{2} = \frac{\pi}{2}$.

21. $\int \frac{dx}{\sqrt{x^2-25}} = \cosh^{-1}\frac{x}{5}+C$.

22. $\int \frac{dx}{\sqrt{(x-3)^2+16}} = \sinh^{-1}\left(\frac{x-3}{4}\right)+C$.

23. $\int \frac{dx}{\sqrt{25+9x^2}}$. Let $u = 3x$ and $du = 3\,dx$, and so $\frac{1}{3}\,du = dx$. Substituting produces $\int \frac{du}{\sqrt{25+u^2}} = \frac{1}{3}\sinh^{-1}\frac{u}{5}+C = \frac{1}{3}\sinh^{-1}\frac{3x}{5}+C$.

24. $\int \frac{e^{2x}}{25-e^{4x}}\,dx$. Let $u = e^{2x}$ and then $du = 2e^{2x}\,dx$. Substituting, we obtain $\frac{1}{2}\int \frac{du}{25-u^2} = \frac{1}{2}\cdot\frac{1}{5}\tanh^{-1}\frac{u}{5} + C = \frac{1}{10}\tanh^{-1}\frac{e^{2x}}{5} + C$.

25. $\int_0^1 \frac{1}{1+x^2}\,dx = \arctan x\big|_0^1 = \arctan 1 - \arctan 0 = \frac{\pi}{4} - 0 = \frac{\pi}{4}$.

26. $\int_{-1/4}^{1/4} \frac{1}{\sqrt{1-4x^2}}\,dx = \frac{1}{2}\arcsin 2x\Big|_{-1/4}^{1/4} = \frac{1}{2}\left[\arcsin\frac{1}{2} - \arcsin\left(-\frac{1}{2}\right)\right] = \frac{1}{2}\left[\frac{\pi}{6} - \frac{-\pi}{6}\right] = \frac{\pi}{6}$.

27. These two graphs intersect at (0, 1) and, from 0 to 1, e^x is always larger. Using the washer method we get

$$\begin{aligned}
&\pi\int_0^1 \left[(e^x)^2 - \left(\frac{1}{\sqrt{x^2-1}}\right)^2\right] dx \\
&= \pi\int_0^1 \left(e^{2x} - \frac{1}{x^2-1}\right) dx \\
&= \pi\int_0^1 e^{2x}\,dx - \pi\int_0^1 \frac{1}{x^2+1}\,dx \\
&= \frac{\pi}{2}e^{2x}\Big|_0^1 - \pi\arctan x\big|_0^1 \\
&= \frac{\pi}{2}\left(e^2-1\right) - \pi(\arctan 1 - \arctan 0) \\
&= \frac{\pi}{2}\left(e^2-1\right) - \pi\left(\frac{\pi}{4}-0\right) \\
&= \frac{\pi}{2}\left(e^2-1-\frac{\pi}{2}\right) \approx 7.5685
\end{aligned}$$

28. $M_y = \int_1^2 x\left(\frac{1}{1+x^4}\right) dx = \int_1^2 \frac{x\,dx}{1+x^4}$. Let $u = x^2$ and $du = 2x\,dx$ or $\frac{1}{2}\,du = x\,dx$. Substituting we get $\frac{1}{2}\int\frac{du}{1+u^2} = \frac{1}{2}\cdot\tan^{-1}u = \frac{1}{2}\tan^{-1}x^2\Big|_1^2 = \frac{1}{2}\left[\tan^{-1}4 - \tan^{-1}1\right] = \frac{1}{2}[1.3258 - 0.7854] = 0.2702$.

29.

$$\begin{aligned}
y &= \sqrt{4-x^2} \\
y' &= \frac{1}{2}(4-x)^{-1/2}\cdot(-2x) = -\frac{x}{\sqrt{4-x^2}} \\
y'^2 &= \frac{x^2}{4-x^2}
\end{aligned}$$

$$\begin{aligned}
\int_{-1}^1 \sqrt{1+(y')^2} &= \int_{-1}^1 \sqrt{1+\frac{x^2}{4-x_2}}\,dx \\
&= \int_{-1}^1 \sqrt{\frac{4-x^2+x^2}{4-x^2}}\,dx \\
&= \int_{-1}^1 \sqrt{\frac{4}{4-x^2}}\,dx = \int_{-1}^1 \frac{2\,dx}{\sqrt{4-x^2}} \\
&= 2\arcsin\frac{x}{2}\Big|_{-1}^1 \\
&= 2\left[\arcsin\frac{1}{2} - \arcsin\frac{-1}{2}\right] \\
&= 2\left[\frac{\pi}{6} - \frac{-\pi}{6}\right] = \frac{2\pi}{3}
\end{aligned}$$

30.

$$\begin{aligned}
m &= \int_0^1 \frac{4}{\sqrt{4-x^2}} \\
&= 4\sin^{-1}\frac{x}{2}\Big|_0^1 \\
&= 4\left(\frac{\pi}{6}-0\right) = \frac{2\pi}{3} \approx 2.0944 \\
M_x &= \frac{1}{2}\int_0^1 \left(\frac{4}{\sqrt{4-x^2}}\right)^2 dx \\
&= 8\int_0^1 \frac{1}{4-x^2}\,dx \\
&= 8\cdot\frac{1}{4}\ln\left|\frac{x+2}{x-2}\right|_0^1 \\
&= 2\left[\ln 3 - \ln -1\right] = 2\ln 3 \approx 2.1972 \\
M_y &= \int_0^1 x\left(\frac{4}{\sqrt{4-x^2}}\right) dx.
\end{aligned}$$

Let $u = 4-x^2$; $du = -2x\,dx$ so $-2\,du = 4x\,dx$. When $x - 0$, then $u = 4 - 0^2 = 4$, and when $x = 1$, then $u = 4 - 1^2 = 3$. Substituting we get

$$-2\int_0^1 \frac{du}{\sqrt{u}} = -2\int_4^3 u^{-1/2}\,du = -2\cdot 2u^{1/2}\Big|_4^3$$
$$= -4\left(\sqrt{3}-\sqrt{4}\right) = 1.0718$$
$$\bar{x} = \frac{M_y}{m} = \frac{1.0718}{2.0944} = 0.51174$$
$$\bar{y} = \frac{M_x}{m} = \frac{2.1972}{2.0944} = 1.0491.$$

31. $$f'(x) = -\frac{1}{x\sqrt{1-x^2}} + \frac{x}{\sqrt{1-x^2}}$$

$$f(x) = \int f'(x)\,dx = \int \frac{-1}{x\sqrt{1-x^2}}\,dx + \int \frac{x}{\sqrt{1-x^2}}\,dx$$
$$= -\operatorname{sech}^{-1} x + \int \frac{x\,dx}{\sqrt{1-x^2}}.$$

For the second integral let $u = 1 - x^2$, then $du = -2x\,dx$ or $-\frac{1}{2}\,du = x\,dx$. These substitutions make $\int \frac{x\,dx}{\sqrt{1-x^2}} = -\frac{1}{2}\int \frac{du}{\sqrt{u}} = -\frac{1}{2}\frac{2}{1}u^{1/2} = -\sqrt{u} = -\sqrt{1-x^2}$. Putting these together we get $f(x) = -\operatorname{sech}^{-1} x - \sqrt{1-x^2} + C$. Since $f(1) = 0$ we have $C = \operatorname{sech}^{-1} 1 + \sqrt{1-1^2} = \operatorname{sech}^{-1} 1$. Thus, the desired equation is $y = -\operatorname{sech}^{-1} x - \sqrt{1-x^2} + \operatorname{sech}^{-1} 1$.

28.6 TRIGONOMETRIC SUBSTITUTION

1. $\int \frac{x}{\sqrt{9-x^2}}\,dx$. This can be done with trigonometric substitutions but it is much easier to do the following. Let $u = 9 - x^2$; $du = -2x\,dx$ so $x\,dx = -\frac{1}{2}\,du$. Substituting yields $-\frac{1}{2}\int \frac{du}{\sqrt{u}} = -\frac{1}{2}\frac{2}{1}u^{1/2} + C = -\sqrt{9-x^2} + C$.

2. $\int \frac{dx}{\sqrt{9-x^2}} = \arcsin \frac{x}{3} + C.$

3. $\int \frac{x^2}{\sqrt{9-x^2}}\,dx$. Let $x = 3\sin\theta$ and $dx = 3\cos\theta\,d\theta$. Substituting produces

$$\int \frac{(3\sin\theta)^2 3\cos\theta\,d\theta}{\sqrt{9-9\sin^2\theta}} = \int 9\sin^2\theta\,d\theta$$
$$= 9\int \frac{1-\cos 2\theta}{2}$$
$$= \frac{9}{2}\int 1\,d\theta - \frac{9}{2}\int \cos 2\theta\,d\theta$$
$$= \frac{9}{2}\theta - \frac{9}{2}\cdot\frac{1}{2}\sin 2\theta + C.$$

Since $x = 3\sin\theta$, we have $\sin\theta = \frac{x}{3}$ and $\theta = \sin^{-1}\frac{x}{3}$. Hence $\sin 2\theta = 2\sin\theta\cos\theta$, which yields $\sin 2\theta = 2\cdot\frac{x}{3}\cdot\frac{\sqrt{9-x^2}}{3} = \frac{2}{9}x\sqrt{9-x^2}$. Thus, the final answer is $\frac{9}{2}\arcsin\frac{x}{3} - \frac{1}{2}x\sqrt{9-x^2} + C$.

4. $\int (9-x^2)^{3/2}\,dx$. Let $x = 3\sin\theta$ and $dx = 3\cos\theta\,d\theta$. This produces

$$\int (9-9\sin^2\theta)^{3/2} 3\cos\theta\,d\theta)$$
$$= \int (9\cos^2\theta)^{3/2} 3\cos\theta\,d\theta$$
$$= 81\int \cos^4\theta\,d\theta = 81\int \left(\frac{1+\cos 2\theta}{2}\right)^2 d\theta$$
$$= \frac{81}{4}\int (1 + 2\cos 2\theta + \cos^2 2\theta)\,d\theta$$
$$= \frac{81}{4}\int [1 + 2\cos 2\theta + \frac{1}{2}(1+\cos 4\theta)]\,d\theta$$
$$= \frac{81}{4}\int \left[\frac{3}{2} + 2\cos 2\theta + \frac{1}{2}\cos 4\theta\right] d\theta$$
$$= \frac{81}{4}\left[\frac{3}{2}\theta + \sin 2\theta + \frac{1}{8}\sin 4\theta\right] + C$$

Since $\theta = \sin^{-1}\frac{x}{3}$, then $\sin\theta = \frac{x}{3}$ and $\cos\theta =$

$\frac{\sqrt{9-x^2}}{3}$, and we have

$$\sin 2\theta = 2\sin\theta\cos\theta = 2\cdot\frac{x}{3}\cdot\frac{\sqrt{9-x^2}}{3} = \frac{2x\sqrt{9-x^2}}{9}$$
$$\sin 4\theta = 2\sin 2\theta\cos 2\theta = 2\sin\theta\cos\theta(1-2\sin^2\theta)$$
$$= 2\cdot\frac{x}{3}\cdot\frac{\sqrt{9-x^2}}{3}\left[1-2\left(\frac{x}{3}\right)^2\right]$$
$$= \frac{2x\sqrt{9-x^2}}{9}\left(1-\frac{2x^2}{9}\right).$$

Back substitution yields

$$\frac{81}{4}\left[\frac{3}{2}\sin^{-1}\frac{x}{3}+\frac{2x\sqrt{9-x^2}}{9} + \frac{1}{8}\left(\frac{2x\sqrt{9-x^2}}{9}\right)\left(1-\frac{2x^2}{9}\right)\right]$$
$$= \frac{243}{8}\sin^{-1}\frac{x}{3}+\left(\frac{81}{4}\right)\frac{2x\sqrt{9-x^2}}{9}\left(\frac{9}{8}-\frac{2x^2}{72}\right)$$
$$= \frac{243}{8}\sin^{-1}\frac{x}{3}+\left(\frac{9}{2}\right)x\sqrt{9-x^2}\left(\frac{9}{8}-\frac{2x^2}{72}\right)$$
$$= \frac{243}{8}\sin^{-1}\frac{x}{3}+x\sqrt{9-x^2}\left(\frac{81}{16}-\frac{x^2}{8}\right)$$

5. $\int x^3\sqrt{9-x^2}\,dx$. Again, this can be done using trig substitutions but the following method is easier.. Let $u = 9-x^2$ and $du = -2x\,dx$, and so $x^2 = 9-u$ and $-\frac{1}{2}\,du = x\,dx$. thus, $\int x^3\sqrt{9-x^2} = \int x^2\sqrt{9-x^2}x\,dx$. Now substitute

$$\int (9-u)\sqrt{u}\left(-\frac{1}{2}\,du\right)$$
$$= -\frac{1}{2}\int\left(9\sqrt{u}-u^{3/2}\right)du$$
$$= -\frac{1}{2}\cdot 9\cdot\frac{2}{3}u^{3/2}+\frac{1}{2}\frac{2}{5}u^{5/2}+C$$
$$= -3(9-x^2)^{3/2}+\frac{1}{5}(9-x^2)^{5/2}+C$$
$$= -3(9-x^2)^{3/2}+\frac{1}{5}(9-x^2)(9-x^2)^{3/2}+C$$
$$= -\frac{15}{5}(9-x^2)^{3/2}+\frac{1}{5}(9-x^2)(9-x^2)^{3/2}+C$$
$$= -\frac{1}{5}(9-x^2)^{3/2}(6+x^2)+C.$$

6. $\int\frac{x}{\sqrt{x^2-9}}\,dx$: Let $u = x^2-9$; $du = 2x\,dx$ or $x\,dx = \frac{1}{2}\,du$. Substituting yields $\frac{1}{2}\int\frac{du}{\sqrt{u}} = \frac{1}{2}\cdot\frac{2}{1}u^{1/2}+C = \sqrt{x^2-9}+C$.

7. $\int\frac{dx}{\sqrt{x^2-9}}$. Let $x = 3\sec\theta$ and then $dx = 3\sec\theta\tan\theta\,d\theta$. Substitution into the given integral yields $\int\frac{3\sec\theta\tan\theta}{\sqrt{9\sec^2\theta-9}}\,d\theta = \int\frac{3\sec\theta\tan\theta\,d\theta}{3\tan\theta} = \int\sec\theta\,d\theta = \ln|\sec\theta+\tan\theta|+C$. Since $\sec\theta = \frac{x}{3}$ and $\tan\theta = \frac{\sqrt{x^2-9}}{3}$, the answer is $\ln\left|\frac{x}{3}+\frac{\sqrt{x^2-9}}{3}\right| + C = \ln\frac{1}{3}\left|x+\sqrt{x^2-9}\right| + C = \ln\frac{1}{3} + \ln\left|x+\sqrt{x^2-9}\right|+C$ or $\ln\left|x+\sqrt{x^2-9}\right|+k$ where $k = C+\ln\frac{1}{3} = C-\ln 3$.

8. $\int\frac{x^3}{\sqrt{x^2-9}}\,dx$. Once again we use a substitution but do not need trigonometric functions. Let $u = x^2-9$; $du = 2x\,dx$ or $\frac{1}{2}\,du = x\,dx$ also $u = x^2-9$ yields $x^2 = u+9$

$$\int\frac{x^3}{\sqrt{x^2-9}}\,dx = \int\frac{x^2x\,dx}{\sqrt{x^2-9}} = \frac{1}{2}\int\frac{(u+9)\,du}{\sqrt{u}}$$
$$= \frac{1}{2}\int\left(\sqrt{u}+\frac{9}{\sqrt{u}}\right)du$$
$$= \frac{1}{2}\cdot\frac{2}{3}u^{3/2}+\frac{1}{2}\cdot 9\cdot\frac{2}{1}u^{1/2}+C$$
$$= \frac{1}{3}(x^2-9)^{3/2}+9(x^2-+9)^{1/2}+C$$
$$= \frac{1}{3}(x^2-9)\sqrt{x^2-9}+9\sqrt{x^2-9}+C$$
$$= \left(\frac{x^2}{3}+6\right)\sqrt{x^2-9}+C$$

9. $\int\left(x^2-9\right)^{1/2}dx = \int\sqrt{x^2-9}\,dx$. Let $x = 3\sec\theta$, then $dx = 3\sec\theta\tan\theta\,d\theta$. Substituting yields $\int\sqrt{9\sec^2\theta-9}\cdot 3\sec\theta\tan\theta\,d\theta =$

$\int 9\tan^2\theta\sec\theta\,d\theta = 9\int(\sec^2\theta-1)\sec\theta\,d\theta = 9\int\left(\sec^3\theta-\sec\theta\right)d\theta = 9\int\sec^2\theta\,d\theta - 9\int\sec\theta\,d\theta$. The first integral is example 20.50 and the second is a formula. They yield $9\left[\frac{1}{2}\sec\theta\tan\theta + \frac{1}{2}\ln|\sec\theta+\tan\theta| - \ln|\sec\theta+\tan\theta|\right] = \frac{9}{2}\sec\theta\tan\theta - \frac{9}{2}\ln|\sec\theta+\tan\theta|$. Since $\sec\theta = \frac{x}{3}$ and $\tan\theta = \dfrac{\sqrt{x^2-9}}{3}$, we get $\int(x^2-9)^{1/2}\,dx = \dfrac{9}{2}\cdot\dfrac{x}{3}\cdot\dfrac{\sqrt{x^2-9}}{3} - \dfrac{9}{2}\ln\left|\dfrac{x}{3}+\dfrac{\sqrt{x^2-9}}{3}\right| + K = \dfrac{1}{2}\left[x\sqrt{x^2-9} - 9\ln\left|x+\sqrt{x^2-9}\right|\right] + C$ where $C = K - \frac{9}{2}\ln 3$.

10. $\int x^3\sqrt{x^2-9}\,dx$; Let $u = x^2-9$; $du = 2x\,dx$ so $x\,dx = \frac{1}{2}\,du$ and $x^2 = u+9$. Substituting, we get $\frac{1}{2}\int(u+9)\sqrt{u}\,du = \frac{1}{2}\int\left(u^{3/2}+9u^{1/2}\right)du = \frac{1}{3}\frac{2}{5}u^{5/2} + \frac{9}{2}\frac{2}{3}u^{3/2} + C = \frac{1}{5}(x^2-9)^{5/2} + 3(x^2-9)^{3/2} = \left(\frac{6+x^2}{5}\right)(x^2-9)^{3/2} + C$.

11. $\displaystyle\int\frac{x}{\sqrt{x^2+9}}\,dx$. Letting $u = x^2+9$ and $du = 2x\,dx$, produces $\dfrac{1}{2}\displaystyle\int\frac{du}{\sqrt{u}} = \frac{1}{2}\cdot\frac{2}{1}u^{1/2} + C = \sqrt{x^2+9} + C$.

12. $\displaystyle\int\frac{dx}{\sqrt{x^2+9}}$. Let $x = 3\tan\theta$ and $dx = 3\sec^2\theta\,d\theta$. This gives $\displaystyle\int\frac{3\sec^2\theta\,d\theta}{\sqrt{9\tan^2\theta+9}} = \int\frac{3\sec^2\theta\,d\theta}{3\sec\theta} = \int\sec\theta\,d\theta = \ln|\sec\theta+\tan\theta| + C = \ln\left|\frac{\sqrt{x^2+9}}{3}+\frac{x}{3}\right| + C = \ln\left|\sqrt{x^2+9}+x\right| + K$ where $K = C - \ln 3$.

13. $\displaystyle\int\frac{dx}{x^2+9} = \frac{1}{3}\arctan\frac{x}{3} + C$.

14. $\displaystyle\int\frac{x^2}{x^2+9}\,dx$. Let $x = 3\tan\theta$ and $d\theta = 3\sec^2\theta\,d\theta$. Substituting this into the given integral produces $\displaystyle\int\frac{9\tan^2\theta}{9\tan^2\theta+9}\cdot 3\sec^2\theta\,d\theta = \int\frac{9\tan^2\theta}{9\sec^2\theta}3\sec^2\theta\,d\theta = \int 3\tan^2\theta\,d\theta = 3\int(\sec^2\theta-1)\,d\theta = 3\int\sec^2\theta - 3\int d\theta = 3\tan\theta - 3\theta + C = 3\frac{x}{3} - 3\tan^{-1}\frac{x}{3} + C = x - 3\tan^{-1}\frac{x}{3} + C$.

15. $\int x^3\sqrt{9+x^2}\,dx$. Let $u = 9+x^2$ and $du = 2x\,dx$ or $\frac{1}{2}\,du = x\,dx$. Thus, $x^2 = u-9$. . Substituting these in the given integral yields $\dfrac{1}{2}\displaystyle\int(u-9)\sqrt{u}\,du = \frac{1}{2}\int(u^{3/2}-9u^{1/2})\,du = \frac{1}{2}\cdot\frac{2}{5}u^{5/2} - \frac{9}{2}\cdot\frac{2}{3}u^{3/2} + C = \frac{1}{5}u^{5/2} - 3u^{3/2} + C = \frac{1}{5}uu^{3/2} - 3u^{3/2} = \frac{1}{5}(9+x^2)(9+x^2)^{3/2} - 3(9+x^2)^{3/2} = \left(\frac{x^2-6}{5}\right)(9+x^2)^{3/2}$.

16. $\displaystyle\int\frac{\sqrt{1-x^2}}{x^2}\,dx$. Let $x = \sin\theta$ and $dx = \cos\theta\,d\theta$. Substitution yields $\displaystyle\int\frac{\sqrt{1-\sin^2\theta}}{\sin^2\theta}\cdot\cos\theta\,d\theta = \int\frac{\cos^2\theta}{\sin^2\theta}\,d\theta = \int\cot^2\theta\,d\theta = \int(\csc^2\theta-1)\,d\theta = -\cot\theta - \theta + C$; $\theta = \sin^{-1}x$; $\cot\theta = \dfrac{\sqrt{1-x^2}}{x}$. The final answer is $-\dfrac{\sqrt{1-x^2}}{x} - \sin^{-1}x + C$.

17. $\displaystyle\int\frac{\sqrt{4-3x^2}}{x^4}\,dx$. Let $\sqrt{3}x = 2\sin\theta$, so $x = \frac{2}{\sqrt{3}}\sin\theta$ and $dx = \frac{2}{\sqrt{3}}\cos\theta\,d\theta$. Substitution produces

$$\int\frac{\sqrt{4-4\sin^2\theta}}{\left(\frac{2}{\sqrt{3}}\sin\theta\right)^4}\frac{2}{\sqrt{3}}\cos\theta\,d\theta$$

$$= \int\frac{2\cos\theta\frac{2}{\sqrt{3}}\cos\theta\,d\theta}{\frac{16}{9}\sin^4\theta}$$

$$= \frac{4}{\sqrt{3}}\cdot\frac{9}{16}\int\frac{\cos^2\theta}{\sin^4\theta}\,d\theta$$

$$= \frac{3\sqrt{3}}{4}\int\csc^2\theta\cot^2\theta\,d\theta$$

$$= \frac{-3\sqrt{3}}{4} \cdot \frac{1}{3} \cot^3 \theta + C$$
$$= \frac{-\sqrt{3}}{4} \left(\frac{\sqrt{4-3x^2}}{\sqrt{3}x} \right)^3 + C$$
$$= -\frac{1}{12} \frac{\left(\sqrt{4-3x^2}\right)^3}{x^3} + C$$
$$= -\frac{1}{12} \frac{(4-3x^2)^{3/2}}{x^3} + C$$

18. $\int \frac{\sqrt{4x^2-9}}{x^3}\,dx$. Let $2x = 3\sec\theta$, then $x = \frac{3}{4}\sec\theta$ and $dx = \frac{3}{2}\sec\theta\tan\theta\,d\theta$. Substituting we get

$$\int \frac{\sqrt{9\sec^2\theta - 9}}{\frac{27}{8}\sec^2\theta} \frac{3}{2} \sec\theta\tan\theta\,d\theta$$
$$= \int \frac{8 \cdot 3\tan\theta}{27\sec^2\theta} \frac{3}{2} \sec\theta\tan\theta\,d\theta$$
$$= \int \frac{8\tan^2\theta\,d\theta}{6\sec^2\theta} = \frac{4}{3} \int \frac{\sec^2\theta - 1}{\sec^2\theta}\,d\theta$$
$$= \frac{4}{3} \int \left(1 - \frac{1}{\sec^2\theta}\right) d\theta = \frac{4}{3} \int (1-\cos^2\theta)\,d\theta$$
$$= \frac{4}{3} \int \sin^2\theta\,d\theta = \frac{4}{3} \int \frac{1-\cos 2\theta}{2}\,d\theta$$
$$= \frac{4}{3} \cdot \frac{1}{2}\theta - \frac{4}{3} \cdot \frac{1}{4} \sin 2\theta + C$$
$$= \frac{2}{3}\theta - \frac{\sin 2\theta}{3} + C$$
$$= \frac{2}{3} \sec^{-1} \frac{2x}{3} - \frac{2\sin\theta\cos\theta}{3} + C$$
$$= \frac{2}{3} \sec^{-1} \frac{2x}{3} - \frac{2}{3} \frac{\sqrt{4x^2-9}}{2x} \cdot \frac{3}{2x} + C$$
$$= \frac{2}{3} \sec^{-1} \frac{2x}{3} - \frac{\sqrt{4x^2-9}}{2x^2} + C$$

19. $\int \frac{x^3}{\sqrt{9x^2+4}}\,dx$; $u = 3x$; $3x = 2\tan\theta$ and so $x = \frac{2}{3}\tan\theta$; and $dx = \frac{2}{3}\sec^2\theta\,d\theta$. Substituting these values in the given integral produces

$$\int \frac{\left(\frac{2}{3}\tan\theta\right)^3}{\sqrt{(2\tan\theta)^2+4}} \frac{2}{3} \sec^2\theta\,d\theta$$
$$= \int \frac{\frac{16}{81}\tan^3\theta\sec^2\theta}{2\sec\theta}\,d\theta$$
$$= \frac{8}{81} \int \tan^3\theta\sec\theta\,d\theta$$
$$= \frac{8}{81} \int \tan\theta(\sec^2\theta - 1)\sec\theta\,d\theta$$
$$= \frac{8}{81} \int (\tan sec^3\theta - \tan\theta\sec\theta)\,d\theta$$
$$= \frac{8}{81} \left(\frac{1}{3}\sec^3\theta - \sec\theta\right) + C$$
$$= \frac{8}{81} \left[\frac{1}{3} \left(\frac{\sqrt{9x^2+4}}{2}\right)^3 - \left(\frac{\sqrt{9x^2+4}}{2}\right) \right] + C$$
$$= \frac{1}{243} \sqrt{(9x^2+4)^3} - \frac{4}{81} \sqrt{9x^2+4} + C$$

20. $\int \frac{dx}{(3x^2+6)^{3/2}} = \frac{1}{3^{3/2}} \int \frac{dx}{\sqrt{x^2+2}^3}$. Let $x = \sqrt{2}\tan\theta$ and $dx = \sqrt{2}\sec^2\theta$. Substituting we get $-\frac{1}{3\sqrt{3}} \cdot \int \frac{\sqrt{2}\sec^2\theta}{\sqrt{(2\tan^2\theta+2)^3}}\,d\theta = \frac{1}{3\sqrt{3}} \int \frac{\sqrt{2}\sec^2\theta}{\left(\sqrt{2}\sec\theta\right)^3}\,d\theta = \frac{1}{3\sqrt{3}} \int \frac{1}{2\sec\theta}\,d\theta = \frac{1}{6\sqrt{3}} \int \cos\theta\,d\theta = \frac{1}{6\sqrt{3}} \sin\theta + C = \frac{1}{6\sqrt{3}} \cdot \frac{x}{\sqrt{x^2+2}} + C = \frac{x\sqrt{3}}{6\sqrt{3x^2+6}} + C.$

21. $\int \frac{\sqrt{x^2+1}}{x^2}\,dx$. Let $x = \tan\theta$ and $dx = \sec^2\theta$. Then

$$\int \frac{\sqrt{\tan^2\theta+1}}{\tan^2\theta} \sec^2\theta\,d\theta = \int \frac{\sec\theta}{\tan^2\theta} \sec^2\theta\,d\theta$$
$$= \int \frac{\sec^3\theta}{\tan^2\theta}\,d\theta = \int \frac{(1+\tan^2\theta)\sec\theta}{\tan^2\theta}\,d\theta$$
$$= \int \frac{\sec\theta}{\tan^2\theta} + \int \sec\theta\,d\theta$$
$$= \int \frac{\frac{1}{\cos\theta}}{\frac{\sin^2\theta}{\cos^2\theta}}\,d\theta + \int \sec\theta\,d\theta$$

$$= \int \frac{\cos\theta}{\sin^2\theta}\,d\theta + \int \sec\theta\,d\theta$$
$$= -(\sin\theta)^{-1} + \ln|\sec\theta + \tan\theta| + C$$
$$= -\frac{\sqrt{x^2+1}}{x} + \ln\left(\sqrt{x^2+1}+x\right) + C$$

22. $\int \frac{dx}{\sqrt{4-(x-1)^2}}\,dx = \arcsin\left(\frac{x-1}{2}\right) + C$

23. $\int \sqrt{4-(x-1)^2}\,dx$. Here we have $a = 2$ and $u = x - 1$. Let $x - 1 = 2\sin\theta$, then $x = 2\sin\theta + 1$ and $dx = 2\cos\theta\,d\theta$. Substitution yields $\int \sqrt{4-4\sin^2\theta}(2\cos\theta)\,d\theta = \int (2\cos\theta)(2\cos\theta)\,d\theta = 4\int \cos^2\theta\,d\theta = 4\int \frac{1+\cos 2\theta}{2}\,d\theta = 2\int (1+\cos 2\theta)\,d\theta = 2\theta + \sin 2\theta + C$. Thus $\theta = \sin^{-1}\frac{x-1}{2}$ and using the identity $\sin 2\theta = 2\sin\theta\cos\theta = 2\cdot\frac{x-1}{2}\cdot\frac{\sqrt{4-(x-1)^2}}{2}$ so the answer is $2\sin^{-1}\left(\frac{x-1}{2}\right) + \frac{(x-1)\sqrt{4-(x-1)^2}}{2} + C$

24. $\int \frac{dx}{\sqrt{4+(x-1)^2}}$. Let $x - 1 = 2\tan\theta$, then $dx = 2\sec^2\theta\,d\theta$. Substituting produces

$$\int \frac{2\sec^2\theta\,d\theta}{\sqrt{4+4\tan^2\theta}} = \int \frac{2\sec^2\theta\,d\theta}{2\sec\theta} = \int \sec\theta\,d\theta$$
$$= \ln|\sec\theta + \tan\theta| + C$$
$$= \ln\left|\frac{\sqrt{4+(x-1)^2}}{2} + \frac{x-1}{2}\right| + C$$

25. $\int \frac{dx}{\sqrt{(x-1)^2-4}}\,dx$. Let $x - 1 = 2\sec\theta$ and $dx = 2\sec\theta\tan\theta\,d\theta$. Then, you get

$$\int \frac{2\sec\theta\tan\theta\,d\theta}{\sqrt{(2\sec\theta)^2-4}}$$
$$= \int \frac{2\sec\theta\tan\theta\,d\theta}{2\tan\theta} = \int \sec\theta\,d\theta$$
$$= \ln|\sec\theta + \tan\theta| + C$$
$$= \ln\left|\frac{x-1}{2} + \frac{\sqrt{(x-1)^2-4}}{2}\right| + C$$
$$= \ln\frac{1}{2}\left|x - 1 + \sqrt{(x-1)^2-4}\right| + C$$
$$= \ln\left|x - 1 + \sqrt{(x-1)^2-4}\right| + K$$

where $K = C - \ln 2$.

26. Completing the square gives $x^2 - 2x + 5 = x^2 - 2x + 1 + 4 = (x-1)^2 + 4$. Thus, the given integral can be written as $\int \sqrt{x^2-2x+5}\,dx = \int \sqrt{(x-1)^2+4}\,dx$. If we let $x - 1 = 2\tan\theta$ then $dx = 2\sec^2\theta\,d\theta$. Substitution yields $\int \sqrt{(2\tan\theta)^2+4}(2\sec^2\theta)\,d\theta = \int (2\sec\theta)(2\sec^2\theta)\,d\theta = 4\int \sec^3\theta\,d\theta = 4\left[\frac{1}{2}\sec\theta\tan\theta + \frac{1}{2}\ln|\sec\theta+\tan\theta|\right] + C = 2\cdot\frac{\sqrt{x^2-2x+5}}{2}\cdot\frac{x-1}{2} + 2\ln\left|\frac{\sqrt{x^2-2x+5}}{2} + \frac{x-1}{2}\right| + C = \frac{1}{2}\sqrt{x^2-2x+5}(x-1) + 2\ln\left|\sqrt{x^2-2x+5}+x-1\right| + K$ where $K = C - \ln 2$.

27. Completing the square produces $x^2 - 2x - 3 = x^2 - 2x + 1 - 4 = (x-1)^2 - 4$. Thus, $\int \sqrt{x^2-2x-3}\,dx = \int \sqrt{(x-1)^2-4}\,dx$. Let $x - 1 = 2\sec\theta$, then $dx = 2\sec\theta\tan\theta\,d\theta$. Substitution gives

$$\int \sqrt{(2\sec\theta)^2-4}(2\sec\theta\tan\theta)\,d\theta$$
$$= \int 2\tan\theta\cdot 2\sec\theta\tan\theta\,d\theta$$
$$= 4\int \tan^2\theta\sec\theta\,d\theta$$
$$= 4\int (\sec^2\theta - 1)\sec\theta\,d\theta$$

$$= 4\int \sec^3\theta\,d\theta - 4\int \sec\theta\,d\theta$$
$$= 4\cdot\left[\frac{1}{2}\sec\theta\tan\theta + \frac{1}{2}\ln|\sec\theta+\tan|\right.$$
$$\left.-4\ln|\sec\theta+\tan\theta|\right] + C$$
$$= 4\left[\frac{1}{2}\sec\theta\tan\theta - \frac{1}{2}\ln|\sec\theta+\tan\theta|\right] + C$$
$$= 2\cdot\frac{x-1}{2}\cdot\frac{\sqrt{x^2-2x-3}}{2}$$
$$-2\ln\left|\frac{x-1}{2}+\frac{\sqrt{x^2-2x-3}}{2}\right| + C$$
$$= \frac{1}{2}(x-1)\sqrt{x^2-2x-3}$$
$$-2\ln\left|x-1+\sqrt{x^2-2x-3}\right| + K$$

where $K = C + \ln 2$.

28. $\int \frac{dx}{(x^2-2x+10)^{3/2}} = \int \frac{dx}{[(x^2-2x+1)+9]^{3/2}} = \int \frac{dx}{[(x-1)^2+9]^{3/2}}$. Here $u = x-1$ and $a = 3$. Let $x-1 = 3\tan\theta$ and $dx = 3\sec^2\theta\,d\theta$ Then, $\int \frac{3\sec^2\theta\,d\theta}{[(3\tan\theta)^2+9]^{3/2}} = \int \frac{3\sec^2\theta\,d\theta}{27\sec^2\theta} = \int \frac{d\theta}{9\sec\theta} = \frac{1}{9}\int \cos\theta = \frac{1}{9}\sin\theta + C = \frac{1}{9}\left(\frac{x-1}{\sqrt{x^2-2x+10}}\right) + C.$

29. $\int \frac{dx}{(x-3)\sqrt{x^2-6x+25}} = \int \frac{dx}{(x-3)\sqrt{(x-3)^2+16}}$. Here $u = x-3$ and $a = 4$. Let $x-3 = 4\tan\theta$, so $dx = 4\sec^2\theta\,d\theta$. Then, $\int \frac{4\sec^2\theta\,d\theta}{4\tan\theta\sqrt{(4\tan\theta)^2+16}} = \int \frac{4\sec^2\theta\,d\theta}{4\tan\theta 4\sec\theta} = \frac{1}{4}\int \frac{\sec\theta}{\tan\theta}d\theta = \frac{1}{4}\int \frac{1}{\sin\theta} = \frac{1}{4}\int \csc\theta\,d\theta = \frac{1}{4}\ln|\csc\theta - \cot\theta| + C = \frac{1}{4}\ln\left|\frac{\sqrt{x^2-6x+25}}{x-3} - \frac{4}{x-3}\right| + C = \frac{1}{4}\ln\left|\frac{\sqrt{x^2-6x+25}-4}{x-3}\right| + C.$

30. $\int \frac{x-3}{\sqrt{x^2-6x+25}}\,dx$. Let $u = x^2-6x+25$, and then $du = (2x-6)\,dx$ or $\frac{1}{2}\,du = (x-3)\,dx$. Substitution yields $\frac{1}{2}\int \frac{du}{\sqrt{u}} = \frac{1}{2}\cdot\frac{2}{1}u^{1/2} = u^{1/2} + C = \sqrt{x^2-6x+25} + C.$

31. The area under $\frac{x^3}{\sqrt{16-x^2}}$ from 0 to 3 is $\int_0^3 \frac{x^2\,dx}{\sqrt{16-x^2}}$. Let $x = 4\sin\theta$, then $dx = 4\cos\theta\,d\theta$. Substituting gives

$$\int \frac{(4\sin\theta)^3(4\cos\theta)\,d\theta}{\sqrt{16-(4\sin\theta)^2}}$$
$$= \int \frac{64\sin^3\theta 4\cos\theta}{4\cos\theta}$$
$$= 64\int \sin^3\theta\,d\theta = 64\int (1-\cos^2\theta)\sin\theta\,d\theta$$
$$= 64\left[-\cos\theta + \frac{1}{3}\cos^3\theta\right]$$
$$= 64\left[\frac{1}{3}\left(\frac{\sqrt{16-x^2}}{4}\right)^3 - \frac{\sqrt{16-x^2}}{4}\right]_0^3$$
$$= \frac{1}{3}\sqrt{7}^3 - 16\sqrt{7} - \frac{1}{3}\cdot 4^3 + 16\cdot 4$$
$$= \frac{1}{3}(7\sqrt{7} - 48\sqrt{7} - 64 + 192)$$
$$= \frac{1}{3}(128 - 41\sqrt{7}) \approx 6.5080654$$

32. $y = 10x - x^2$ is above the x-axis between 0 and 10 and $y' = 10 - 2x$. The arc length is $\int_0^{10}\sqrt{1+(10-2x)^2}\,dx = \int_0^{10}\sqrt{1+(2x-10)^2}\,dx$. Let $u = 2x - 10 = \tan\theta; x - 5 = \frac{1}{2}\tan\theta$ so $dx = \frac{1}{2}\sec^2\theta\,d\theta$. Thus, $\int \sqrt{1+\tan^2\theta}\cdot\frac{1}{2}\sec^2\theta\,d\theta = \int \sec\theta\cdot\frac{1}{2}\sec^2\theta\,d\theta = \frac{1}{2}\int \sec^3\theta\,d\theta$. As in Example 28.50, this is $\frac{1}{2}\left[\frac{1}{2}\sec\theta\tan\theta + \right.$

$\frac{1}{2}\ln|\sec\theta+\tan\theta| = \frac{1}{4}\left[\sqrt{1+(2x-10)^2}\cdot(2x-10)+\ln\left|\sqrt{1+(2x-10)^2}+2x-10\right]_0^{10} = \frac{1}{4}\left[10\sqrt{101}+10\sqrt{101}+\ln\left(\sqrt{101}+10\right)-\ln\left(\sqrt{101}-10\right)\right] = 5\sqrt{101}+\frac{1}{4}\ln\left(\frac{\sqrt{101}+10}{\sqrt{101}-10}\right) \approx 51.7485$.

33. The area described is $\int_0^4 \sqrt{9+x^2}\,dx$. Let $x = 3\tan\theta$ and $dx = 3\sec^2\theta\,d\theta$. Then, $\int\sqrt{9+9\tan^2\theta}(3\sec^2\theta)\,d\theta = \int(3\sec\theta)(3\sec^2\theta)\,d\theta = 9\int\sec^3\theta\,d\theta = 9\left[\frac{1}{2}\sec\theta\tan\theta+\frac{1}{2}\ln|\sec\theta+\tan\theta|\right] = \frac{9}{2}\left[\frac{\sqrt{9+x^2}}{3}\cdot\frac{x}{3}+\ln\left[\frac{\sqrt{9+x^2}}{3}+\frac{x}{3}\right]\right]_0^4$ or, we get, $\frac{9}{2}\left[\frac{5}{3}\cdot\frac{4}{3}+\ln\left|\frac{5}{3}+\frac{4}{3}\right|-\frac{3}{3}\cdot\frac{0}{3}-\ln\left|\frac{3}{3}\right|\right] = \frac{9}{2}\left[\frac{20}{9}+\ln|3|-0-0\right] = 5+\frac{9}{2}\ln 3$ or $5+\ln 3^{9/2} \approx 9.9438$.

34. Since $\rho(x) = \sqrt{108-3x^2}$, then the mass is $m = \int_0^6 \sqrt{108-3x^2}\,dx = \sqrt{3}\int_0^6\sqrt{36-x^2}\,dx$. Let $x = 6\sin\theta$ and $dx = 6\cos\theta\,d\theta$. Substituting these into the original integral gives $\sqrt{3}\int\sqrt{36-36\sin^2\theta}\cdot 6\cos\theta\,d\theta = \sqrt{3}\int(6\cos\theta)(6\cos\theta)\,d\theta = 36\sqrt{3}\int\cos^2\theta\,d\theta = 36\sqrt{3}\int\frac{1+\cos 2\theta}{2}\,d\theta = 18\sqrt{3}\left[\theta+\frac{1}{2}\sin 2\theta\right] = 18\sqrt{3}\left[\theta+\sin\theta\cos\theta\right] = 18\sqrt{3}\left[\sin^{-1}\frac{x}{6}+\frac{x}{6}\cdot\frac{\sqrt{36-x^2}}{6}\right]_0^6 = 18\sqrt{3}\left[\frac{\pi}{2}+0-0-0\right] = 9\sqrt{3}\pi \approx 48.9726$.

35. To find the desired area we first solve $\frac{x^2}{25}+\frac{y^2}{4}=1$ for y: $\frac{y^2}{4} = 1-\frac{x^2}{25}$ so $y^2 = 4-\frac{4x^2}{25}$ or $y^2 = \frac{100-4x^2}{25}$ and we get $y = \sqrt{\frac{100-4x^2}{25}} = \frac{2}{5}\sqrt{25-x^2}$. We also need $y' = \frac{2}{5}(25-x^2)^{-1/2}\cdot\frac{1}{2}\cdot 2x = \frac{2}{5}x(25-x^2)^{-1/2} = \frac{2x}{5\sqrt{25-x^2}}$. Thus, $y'^2 = \frac{4x^2}{25(25-x^2)}$ and so

$$S = \int_{-5}^{5} 2\pi y\sqrt{1+y'^2}\,dx$$

$$= \int_{-5}^{5} 2\pi\left(\frac{2}{5}\sqrt{25-x^2}\right)\sqrt{1+\frac{4x^2}{25(25-x^2)}}\,dx$$

$$= \int_{-5}^{5} 2\pi\cdot\frac{2}{5}\sqrt{25-x^2}\sqrt{\frac{625-21x^2}{25(25-x^2)}}\,dx$$

$$= \int_{-5}^{5} 2\pi\cdot\frac{2}{5}\sqrt{25-x^2}\cdot\frac{1}{5}\frac{\sqrt{625-21x^2}}{\sqrt{25-x^2}}\,dx$$

$$= \frac{4\pi}{25}\int_{-5}^{5}\sqrt{625-21x^2}\,dx.$$

If you let $\sqrt{21}x = 25\sin\theta$, then $dx = \frac{25}{\sqrt{21}}\cos\theta\,d\theta$. Substitution gives

$$\frac{4\pi}{25}\int\sqrt{625-625\sin^2\theta}\cdot\frac{25}{\sqrt{21}}\cos\theta\,d\theta$$

$$= \frac{4\pi}{25}\int 25\cos\theta\cdot\frac{25}{\sqrt{21}}\cos\theta\,d\theta$$

$$= \frac{100\pi}{25}\int\cos^2\theta\,d\theta = \frac{100\pi}{\sqrt{21}}\int\frac{1+\cos 2\theta}{2}\,d\theta$$

$$= \frac{50\pi}{\sqrt{21}}\left[\theta+\frac{1}{2}\sin 2\theta\right] = \frac{50\pi}{\sqrt{21}}\left[\theta+\sin\theta\cos\theta\right]$$

$$= \frac{50\pi}{\sqrt{21}}\left[\sin^{-1}\frac{\sqrt{21}x}{25}+\frac{\sqrt{21}x}{25}\cdot\frac{\sqrt{625-21x^2}}{25}\right]_{-5}^{5}$$

$$= \frac{50\pi}{\sqrt{21}}\left[\sin^{-1}\frac{\sqrt{21}}{5}+\frac{\sqrt{21}\cdot 5\sqrt{625-21\cdot 25}}{625}\right](2)$$

$$= \frac{100\pi}{\sqrt{21}}\left[\sin^{-1}\frac{\sqrt{21}}{5}+\frac{\sqrt{21}\cdot 50}{625}\right]$$

$$= \frac{100\pi\cdot 50}{625}+\frac{100\pi}{\sqrt{21}}\sin^{-1}\frac{\sqrt{21}}{5}$$

$$= 8\pi+\frac{100\pi}{\sqrt{21}}\sin^{-1}\frac{\sqrt{21}}{5} \approx 104.6073.$$

36. Charge is the integral of the current. As a result, we have $\int_0^1 t\sqrt{t+2}\,dt$. Let $u = t+2$ with $du = dt$ and $t = u-2$. Substituting we get $\int (u-2)\sqrt{u} = \int u^{3/2} - 2u^{1/2}\,du = \frac{2}{5}u^{5/2} - \frac{4}{3}u^{3/2} = \frac{2}{5}(t+2)^{5/2} - \frac{4}{3}(t+2)^{3/2}\Big|_0^1 = \frac{2}{5}(3)^{5/2} - \frac{4}{3}(3)^{3/2} - \frac{2}{5}(2)^{5/2} + \frac{4}{3}(2)^{3/2} \approx 0.8157$.

37. Let $u = t = \tan\theta$ and $du = dt = \sec^2\theta\,d\theta$. Then

$$\int \frac{\sqrt{t^2+1}}{9t^2}\,dt = \int \frac{\sqrt{\tan^2\theta+1}}{9\tan^2\theta}\sec^2\theta\,d\theta$$
$$= \frac{1}{9}\int \frac{\sec^3\theta}{\tan^2\theta}\,d\theta = \frac{1}{9}\int \csc^2\sec\theta\,d\theta$$
$$= \frac{1}{9}\int \left(1+\cot^2\theta\right)\sec\theta\,d\theta$$
$$= \frac{1}{9}\int \left(\sec\theta + \frac{\cos\theta}{\sin^2\theta}\right)d\theta$$
$$= \frac{1}{9}\left[\ln|\sec\theta+\tan\theta| - (\sin\theta)^{-1}\right]$$
$$= \frac{1}{9}\left[\ln\left(\sqrt{1+t^2}+t\right) - \frac{\sqrt{1+t^2}}{t}\right]$$

So from $t = 0.5$ to $t = 1$, we have $i = \frac{1}{9}\left[\ln\left(\sqrt{1+t^2}+t\right) - \frac{\sqrt{1+t^2}}{t}\right]_{0.5}^{1} \approx 0.1829$ A

28.7 INTEGRATION BY PARTS

1. $\int x\ln x\,dx$. Completing the table, we get

$u = \ln x$	$v = \frac{1}{2}x^2$
$du = \frac{1}{x}\,dx$	$dv = x\,dx$

Thus, using the table, we obtain $\int x\ln x\,dx = \frac{1}{2}x^2\ln x - \int \frac{1}{2}x^2\cdot\frac{1}{x}\,dx = \frac{1}{2}x^2\ln x - \frac{1}{2}\int x\,dx = \frac{1}{2}x^2\ln x - \frac{1}{4}x^2 + C$ or $\frac{1}{2}x^2\left(\ln x - \frac{1}{2}\right) + C$.

2. $\int \tan^{-1}x\,dx$

$u = \tan^{-1}x$	$v = x$
$du = \frac{1}{1+x^2}\,dx$	$dv = dx$

Thus, $\int \tan^{-1}x\,dx = x\tan^{-1}x - \int \frac{x\,dx}{1+x^2} = x\tan^{-1}x - \frac{1}{2}\ln(1+x^2) + C$.

3. $\int x\sin 2x\,dx$

$u = x$	$v = -\frac{1}{2}\cos 2x$
$du = dx$	$dv = \sin 2x\,dx$

Using the table, we see that $\int x\sin 2x\,dx = -\frac{1}{2}x\cos 2x - \int -1/2\cos 2x\,dx = -\frac{1}{2}x\cos 2x + \frac{1}{4}\sin 2x + C$.

4. $\int x^2\cos 3x\,dx$

$u = x^2$	$v = \frac{1}{3}\sin 3x$
$du = 2x\,dx$	$dv = \cos 3x\,dx$

Using the table, we see that $\int x^2\cos 3x\,dx = \frac{1}{3}x^2\sin 3x - \int \frac{2}{3}x\sin 3x$. We again use integration by parts to determine the second integral.

$u = \frac{2}{3}x$	$v = -\frac{1}{3}\cos 3x$
$du = \frac{2}{3}\,dx$	$dv = \sin 3x\,dx$

Thus, $\frac{1}{3}x^2\sin 3x - \int \frac{2}{3}x\sin 3x = \frac{1}{3}x^2\sin 3x + \frac{2}{9}x\cos 3x + \int -\frac{2}{9}\cos 3x\,dx = \frac{1}{3}x^2\sin 3x + \frac{2}{9}x\cos 3x - \frac{2}{27}\sin 3x + C$ or $\frac{1}{9}\left[2x\cos 3x + (3x^2 - \frac{2}{3})\sin 3x\right] + C$.

5. $\int x^2\ln x\,dx$

$u = \ln x$	$v = \frac{1}{3}x^3$
$du = \frac{1}{x}\,dx$	$du = x^2\,dx$

Thus, from the table, we see that $\int x^2\ln x\,dx = \frac{1}{3}x^3\ln x - \int \frac{1}{3}x^3\cdot\frac{1}{x}\,dx = \frac{1}{3}x^3\ln x - \frac{1}{9}x^3 + C$.

6. $\int x\sqrt{1+x}\,dx$. Let $u = 1+x$, then $du = dx$ and $x = u-1$. We can rewrite the given integral

as $\int (u-1)\sqrt{u}\,du = \int \left(u^{3/2} - u^{1/2}\right) du = \frac{2}{5}u^{5/2} - \frac{2}{3}u^{3/2} + C = \frac{2}{5}(1+x)^{\frac{5}{2}} - \frac{2}{3}(1+x)^{3/2} + C = \frac{2}{5}(1+x)(1+x)^{3/2} - \frac{2}{3}(1+x)^{3/2} + C = \frac{2}{15}(3x-2)(1+x)^{3/2} + C$.

7. $\int x \sin^{-1} x^2\,dx$.

$u = \sin^{-1} x^2$	$v = \frac{1}{2}x^2$
$du = \dfrac{2x}{\sqrt{1-x^2}}\,dx$	$dv = x\,dx$

Hence, $\displaystyle\int x \sin^{-1} x^2\,dx = \frac{1}{2}x^2 \sin^{-1} x^2 - \int \frac{x^3}{\sqrt{1-x^2}}\,dx$. For the second integral let $w = 1 - x^4$, $dw = -4x^3\,dx$ or $\frac{1}{4}dw = -x^3\,dx$. Substituting, we get $\displaystyle\frac{1}{2}x^2 \sin^{-1} x^2 - \int \frac{x^3}{\sqrt{1-x^2}}\,dx = \frac{1}{2}x^2 \sin^{-1} x^2 - \frac{1}{4}\int \frac{dw}{\sqrt{w}} = \frac{1}{2}x^2 \sin^{-1} x^2 - \frac{1}{4}\int w^{-1/2}\,dw = \frac{1}{2}x^2 \sin^{-1} x^2 + \frac{1}{2}w^{1/2} + C = \frac{1}{2}x^2 \sin^{-1} x^2 + \frac{1}{2}(1-x^4)^{1/2} + C$. The final answer is $\frac{1}{2}x^2 \sin^{-1} x^2 + \frac{1}{2}(1-x^4)^{1/2} + C = \frac{1}{2}\left[x^2 \sin^{-1} x^2 + \sqrt{1-x^4}\right] + C$.

8. $\int x^2 \sin x\,dx$

$u = x^2$	$v = -\cos x$
$du = 2x\,dx$	$dv = \sin x\,dx$

From the table, we see that $\int x^2 \sin x\,dx = -x^2 \cos x + \int 2x \cos x\,dx$. We now use a second table.

$u = 2x$	$v = \sin x$
$du = 2\,dx$	$dv = \cos x\,dx$

Hence, $-x^2 \cos x + \int 2x \cos x\,dx = -x^2 \cos x + 2x \sin x - \int 2 \sin x\,dx = -x^2 \cos x + 2x \sin x - 2\cos x + C = 2x \sin x - (x^2+2)\cos x + C$.

9. $\int x^3 e^{2x}\,dx$

$u = x^3$	$v = \frac{1}{2}e^{2x}$
$du = 3x^2\,dx$	$dv = e^{2x}\,dx$

So, $\int x^3 e^{2x}\,dx = \frac{1}{2}x^3 e^{2x} - \frac{3}{2}\int x^2 e^{2x}\,dx$. Using a second table produces

$u = x^2$	$v = \frac{1}{2}e^{2x}$
$du = 2x\,dx$	$dv = e^{2x}\,dx$

This, combined with the result from the first table, produces $\frac{1}{2}x^3 e^{2x} - \frac{3}{2}\int x^2 e^{2x}\,dx = \frac{1}{2}x^3 e^{2x} - \frac{3}{2}\left[\frac{1}{2}x^2 e^{2x} - \int x e^{2x}\,dx\right]$. A third table produces

$u = x$	$v = \frac{1}{2}e^{2x}$
$du = dx$	$dv = e^{2x}\,dx$

When combined with the previous result, this yields $\frac{1}{2}x^3 e^{2x} - \frac{3}{2}\left[\frac{1}{2}x^2 e^{2x} - \int x e^{2x}\,dx\right] = \frac{1}{2}x^3 e^{2x} - \frac{3}{4}x^2 e^{2x} + \frac{3}{4}x e^{2x} - \int \frac{1}{2}e^{2x}\,dx = \frac{1}{2}x^3 e^{2x} - \frac{3}{4}x^2 e^{2x} + \frac{3}{4}x e^{2x} - \frac{1}{4}e^{2x} + C = e^{2x}\left(\frac{1}{2}x^3 - \frac{3}{4}x^2 + \frac{3}{4}x - \frac{1}{4}\right) + C$

10. $\int x \tan^{-1} x\,dx$

$u = \tan^{-1} x$	$v = \frac{1}{2}x^2$
$du = \dfrac{1}{1+x^2}\,dx$	$dv = x\,dx$

$\displaystyle\int x \tan^{-1} x\,dx = \frac{1}{2}x^2 \tan^{-1} x - \int \frac{1}{2}x^2 \frac{1}{1+x^2}\,dx.$ Now in the second integral let $x = \tan\theta$ and $dx = \sec\theta\,d\theta$. Then, we have $\displaystyle -\int \frac{1}{2}x^2 \frac{1}{1+x^2}\,dx = -\frac{1}{2}\int \frac{\tan^2\theta}{1+\tan^2\theta} \cdot \sec^2\theta\,d\theta = -\frac{1}{2}\int \frac{\tan^2\theta}{\sec^2\theta} \cdot \sec^2\theta\,d\theta = -\frac{1}{2}\int \tan^2\theta\,d\theta = -\frac{1}{2}\int \left(\sec^2\theta - 1\right) d\theta = \frac{1}{2}\int \left(1 - \sec^2\theta\right) d\theta = \frac{1}{2}\theta - \frac{1}{2}\tan\theta + C = \frac{1}{2}\tan^{-1} x - \frac{1}{2}x + C$. Thus, the final answer is $\frac{1}{2}x^2 \tan^{-1} x + \frac{1}{2}\tan^{-1} x - \frac{1}{2}x + C = \frac{1}{2}\tan^{-1} x(x^2+1) - \frac{1}{2}x + C$.

11. $\int x\sqrt{4x+1}\,dx$. Let $u = 4x+1$ and $du = 4\,dx$. So, $\frac{u-1}{4} = x$ and $\frac{1}{4}du = dx$. Thus, the given integral is $\int \frac{u-1}{4} \cdot \sqrt{u} \cdot \frac{1}{4}\,dx = \frac{1}{16}\int u^{3/2} - u^{1/2}\,du = \frac{1}{16}\left[\frac{2}{5}u^{5/2} - \frac{2}{3}u^{3/2}\right] + C = \frac{1}{16}\left[\frac{2}{5}(4x+1)^{5/2} - \frac{2}{3}(4x+1)^{3/2}\right] + C = \frac{1}{16}\left[\frac{2}{5}(4x+1)(4x+1)^{3/2} - \frac{2}{3}(4x+1)^{3/2}\right] + C = \frac{1}{16}\left[\left(\frac{8x}{5} - \frac{4}{15}\right)(4x+1)^{3/2}\right] + C = \frac{1}{16}\left[\frac{24x-4}{15}(4x+1)^{3/2}\right] + C = \frac{1}{60}(6x-1)(4x+1)^{3/2} + C$.

12. $\int x^5 \ln x\, dx$

$u = \ln x$	$v = \frac{1}{6}x^6$
$du = \dfrac{1}{x}\, dx$	$dv = x^5\, dx$

Thus, we have $\int x^5 \ln x\, dx = \frac{1}{6}x^6 \ln x - \int \frac{1}{6}x^5\, dx = \frac{1}{6}x^6 \ln x - \frac{1}{36}x^6 + C = \frac{1}{6}x^6\left[\ln x - \frac{1}{6}\right] + C$.

13. $\int x^2 e^{x/4}\, dx$

$u = x^2$	$v = 4e^{x/4}$
$du = 2x\, dx$	$dv = e^{x/4}\, dx$

Thus, $\int x^2 e^{x/4}\, dx = 4x^2 e^{x/4} - \int 8xe^{x/4}$. Using integration by parts a second time, we get the following table:

$u = 8x$	$v = 4e^{x/4}$
$du = 8\, dx$	$dv = e^{x/4}\, dx$

Hence, we get $4x^2e^{x/4} - \int 8xe^{x/4} = 4x^2e^{x/4} - 32xe^{x/4} + \int 32e^{x/4}\, dx = 4x^2e^{x/4} - 32xe^{x/4} + 128e^{x/4} + C = e^{x/4}(4x^2 - 32x + 128) + C$.

14. $\int e^{4x} \sin 2x\, dx$

$u = e^{4x}$	$-\frac{1}{2}\cos 2x$
$du = 4e^{4x}$	$dv = \sin 2x\, dx$

This produces $\int e^{4x} \sin 2x\, dx = -\frac{1}{2}e^{4x}\cos 2x + 2\int e^{4x}\cos 2x$. But, we need to use integration by parts a second time. Here we get the table

$u = 2e^{4x}$	$v = \frac{1}{2}\sin 2x$
$du = 8e^{4x}$	$dv = \cos 2x\, dx$

Thus, we have $-\frac{1}{2}e^{4x}\cos 2x + 2\int e^{4x}\cos 2x = -\frac{1}{2}e^{4x}\cos 2x + e^{4x}\sin 2x - 4\int e^{4x}\sin 2x$. So, $5\int e^{4x}\sin 2x\, dx = e^{4x}\sin 2x - \frac{1}{2}e^{4x}\cos 2x + C$ or $\int e^{4x}\sin 2x\, dx = \frac{1}{10}e^{4x}\left[2\sin 2x - \cos 2x\right] + C$.

15. $\int e^x \cos x\, dx$

$u = e^x$	$v = \sin x$
$du = e^x\, dx$	$dv = \cos x\, dx$

This results in $\int e^x \cos x\, dx = e^x \sin x - \int e^x \sin x\, dx$. Using integration by parts again produces

$u = e^x$	$v = -\cos x$
$du = -e^x\, dx$	$dv = \sin x\, dx$

Combining this with the previous result yields $e^x \sin x - \int e^x \sin x\, dx = e^x \sin x + e^x \cos x - \int e^x \cos x\, dx$. Hence $2\int e^x \cos x\, dx = e^x \sin x + e^x \cos x$ or $\int e^x \cos x\, dx = \frac{1}{2}e^x(\sin x + \cos x) + C$.

16. $\int x^4 e^{2x}\, dx$

$u = x^4$	$v = \frac{1}{2}e^{2x}$
$du = 4x^3\, dx$	$dv = e^{2x}\, dx$

Thus, $\int x^4 e^{2x}\, dx = x^4 \cdot \frac{1}{2}e^{2x} - \int 2x^3 e^{2x}$. From Exercise #9, we know that

$$\int x^3 e^{2x}\, dx = e^{2x}\left(\frac{1}{2}x^3 - \frac{3}{4}x^2 + \frac{3}{4}x - \frac{3}{8}\right) + C$$

Using this result, we have

$$\begin{aligned}
&\int x^4 e^{2x}\, dx \\
&= x^4 \cdot \frac{1}{2}e^{2x} - 2\int x^3 e^{2x} \\
&= x^4 \cdot \frac{1}{2}e^{2x} - 2\left[e^{2x}\left(\frac{1}{2}x^3 - \frac{3}{4}x^2 + \frac{3}{4}x - \frac{3}{8}\right)\right] + C \\
&= e^{2x}\left(\frac{1}{2}x^4 - x^3 + \frac{3}{2}x^2 - \frac{3}{2}x + \frac{3}{4}\right) + C
\end{aligned}$$

17. Rewrite the given integral as $\int x^3 \cos x^2\, dx = \int \frac{x^2}{2} \cdot 2x \cos x^2\, dx$, and then use integration by parts.

$u = \dfrac{x^2}{2}$	$v = \sin x^2$
$du = x\, dx$	$dv = 2x \cos x^2\, dx$

So the integral equals $\frac{x^2}{2}\sin x^2 - \int x \sin x^2 = \frac{x^2}{2}\sin x^2 + \frac{1}{2}\cos x^2 + C$

18. $\int x^2 e^{-x}\, dx$

$u = x^2$	$v = -e^{-x}$
$du = 2x\, dx$	$dv = e^{-x}\, dx$

Hence $\int x^2 e^{-x}\, dx = -x^2 e^{-x} + \int 2xe^{-x}\, dx$. Integrating by parts again,

$u = 2x$	$v = -e^{-x}$
$du = 2\, dx$	$dv = e^{-x}\, dx$

Thus, $-x^2e^{-x} - 2xe^{-x} + \int 2e^{-x}\, dx = -x^2e^{-x} - 2xe^{-x} - 2e^{-x} + C = -e^{-x}(x^2 + 2x + 2) + C$.

19. $\int x^3 e^{-x}\, dx$

$u = x^3$	$v = -e^{-x}$
$du = 3x^2\, dx$	$dv = e^{-x}\, dx$

Thus, we obtain $\int x^3 e^{-x}\, dx = -x^3 e^{-x} + \int 3x^2 e^{-x}$.

Using integration by parts a second time, we have the following table.

$u = 3x^2$	$v = -e^{-x}$
$du = 6x\,dx$	$dv = e^{-x}\,dx$

Combined with the earlier result, this produces $-x^3e^{-x} + \int 3x^2e^{-x} = -x^3e^{-x} - 3x^2e^{-x} + \int 6xe^{-x}$. We now use integration by parts a third time.

$u = 6x$	$v = -e^{-x}$
$du = 6\,dx$	$dv = e^{-x}\,dx$

Hence, $-x^3e^{-x} - 3x^2e^{-x} + \int 6xe^{-x} = -x^3e^{-x} - 3x^2e^{-x} - 6xe^{-x} + \int 6e^{-x} = -x^3e^{-x} - 3x^2e^{-x} - 6xe^{-x} - 6e^{-x} + C = -e^{-x}(x^3 + 3x^2 + 6x + 6) + C$

20. $\displaystyle\int \sin^2 x\,dx = \int \frac{1-\cos 2x}{2}\,dx = \frac{1}{2}x - \frac{1}{4}\sin 2x + C$ or $= \frac{1}{2}x - \frac{1}{2}\sin x\cos x + C$.

21. $\int e^{-x}\cos x\,dx$

$u = e^{-x}$	$v = \sin x$
$du = -e^{-x}\,dx$	$dv = \cos x\,dx$

Thus, $\int e^{-x}\cos x\,dx = e^{-x}\sin x + \int \sin xe^{-x}\,dx$. Applying integration by parts again, we have the following table.

$u = e^{-x}$	$v = -\cos x$
$du = -e^{-x}\,dx$	$dv = \sin x\,dx$

Hence, $e^{-x}\sin x + \int \sin xe^{-x}\,dx = e^{-x}\sin x - e^{-x}\cos x - \int e^{-x}\cos x\,dx$ which produces $2\int e^{-x}\cos x\,dx = e^{-x}\sin x - e^{-x}\cos x$ and means that $\int e^{-x}\cos x = \frac{1}{2}e^{-x}(\sin x - \cos x) + C$.

22. $\int x4^x\,dx$

$u = x$	$v = \dfrac{4^x}{\ln 4}$
$du = dx$	$dv = 4^x\,dx$

Thus, $\displaystyle\int x4^x\,dx = x\cdot\frac{1}{\ln 4}4^x - \int \frac{4^x}{\ln 4}\,dx = \frac{x4^x}{\ln 4} - \frac{1}{\ln 4}\cdot\frac{1}{\ln 4}4^x + C = \frac{4^x}{\ln 4}\left(x - \frac{1}{\ln 4}\right) + C.$

23. $\int x\cos 4x\,dx$

$u = x$	$v = \frac{1}{4}\sin 4x$
$du = dx$	$dv = \cos 4x\,dx$

Hence, $\int x\cos 4x\,dx = \frac{x}{4}\sin 4x - \frac{1}{4}\int \sin 4x\,dx = \frac{x}{4}\sin 4x + \frac{1}{16}\cos 4x + C$.

24. $\int e^{4x}\cos 2x\,dx$

$u = e^{4x}$	$v = \frac{1}{2}\sin 2x$
$du = 4e^{4x}$	$dv = \cos 2x\,dx$

Which means $\int e^{4x}\cos 2x\,dx = \frac{1}{2}e^{4x}\sin 2x - \int 2e^{4x}\sin 2x\,dx$. Using integration by parts a second time, we get

$u = 2e^{4x}$	$v = -\frac{1}{2}\cos 2x$
$du = 8e^{4x}$	$dv = \sin 2x\,dx$

Combined with the earlier results, we get $\frac{1}{2}e^{4x}\sin 2x - \int 2e^{4x}\sin 2x\,dx = \frac{1}{2}e^{4x}\sin 2x + e^{4x}\cos 2x - 4\int e^{4x}\cos 2x\,dx$. This leads to $5\int e^{4x}\cos 2x\,dx = \frac{1}{2}e^{4x}\sin 2x + e^{4x}\cos 2x$ and so, the desired result is $\int e^{4x}\cos 2x\,dx = \frac{1}{10}e^{4x}(\sin 2x + 2\cos 2x) + C$.

25. Using the disc method, we see that $V = \pi\displaystyle\int_0^{\pi/2}\cos^2 x\,dx = \pi\int_0^{\pi/2}\frac{1+\cos 2x}{2} = \pi\left[\frac{1}{2}x + \frac{1}{4}\sin 2x\right]_0^{\pi/2} = \pi\left(\frac{\pi}{4} + 0 - 0 - 0\right) = \frac{\pi^2}{4}$.

26. To find this volume, the disc method will lead to an integral we haven't worked yet, hence we use the shell method, with $V = 2\pi\int_0^{\pi/2} x\cos x\,dx$. This requires integration by parts, with the following table:

$u = x$	$v = \sin x$
$du = dx$	$dv = \cos x\,dx$

Hence, $V = 2\pi\int_0^{\pi/2} x\cos x\,dx = 2\pi\left[x\sin x - \int \sin x\,dx\right] = 2\pi\left[x\sin x + \cos x\right]_0^{\pi/2} = 2\pi\left[\frac{\pi}{2} + 0 - 0 - 1\right] = \pi^2 - 2\pi = 3.5864$.

27. Charge is the integral of current. Hence we must find $\int 4t\sin 2t\,dt$. Use the following integration by parts table.

$u = 4t$	$v = -\frac{1}{2}\cos 2t$
$du = 4\,dt$	$dv = \sin 2t\,dt$

Thus, $\int 4t\sin 2t\,dt = -2t\cos 2t + \int 2\cos 2t\,dt = -2t\cos 2t + \sin 2t$. Evaluated from 0 to 4 we have $[-8\cos 8 + \sin 8] = 2.1534$ C.

28. The total mass is given by $m = \int_0^6 xe^{-x}\,dx$. Use the following integration by parts table.

$u = x$	$v = -e^{-x}$
$du = dx$	$dv = e^{-x}\,dx$

Hence, $m = \int_0^6 xe^{-x}\,dx = -xe^{-x} + \int e^{-x}\,dx = \left[-xe^{-x} - e^{-x}\right]_0^6 = -6x^{-6} - e^{-6} + 0 + e^0 = 0.9826$ kg.

29. $M_y = \int_{-1}^2 x \cdot x^2 e^x\,dx = \int_{-1}^2 x^3 e^x\,dx$. From Exercise #9, we have $\int x^3 e^{2x}\,dx = e^{2x}\left(\frac{1}{2}x^3 - \frac{3}{4}x^2 + \frac{3}{4}x - \frac{3}{8}\right) + C$. Hence, we see that

$$M_y = \int_{-1}^{2} x \cdot x^2 e^x\,dx = e^x\left[x^3 - 3x^2 + 6x - 6\right]_{-1}^2$$
$$= 14.7781 + 5.8861 = 20.6642$$

$M_x = \frac{1}{2}\int_{-1}^2 (x^2e^x)^2\,dx = \frac{1}{2}\int_{-1}^2 x^4e^{2x}\,dx$. From Exercise #16, we know

$$\int x^4e^{2x}\,dx = e^{2x}\left(\frac{1}{2}x^4 - x^3 + \frac{3}{2}x^2 - \frac{3}{2}x + \frac{3}{4}\right) + C$$

Thus,

$$M_x = \frac{1}{2}\int_{-1}^{2} (x^2e^x)^2\,dx = \frac{1}{2}\int_{-1}^{2} x^4e^{2x}\,dx$$
$$= \frac{1}{8}e^{2x}\left[2x^4 - 4x^2 + 6x^2 - 6x + 3\right]_{-1}^2$$
$$= \frac{15}{8}e^4 - \frac{21}{8}e^{-2} = 102.0163$$

The mass is $m = \int_{-1}^2 x^2e^x\,dx$. Once again, we need to use integration by parts.

$u = x^2$	$v = e^x$
$du = 2x\,dx$	$dv = e^x\,dx$

So, $\int x^2e^x\,dx = x^2e^x - \int 2xe^x\,dx$. A second application of integration by parts produces

$u = 2x$	$v = e^x$
$du = 2\,dx$	$dv = e^x\,dx$

With the result that $\int x^2e^x\,dx = x^2e^x - 2xe^x + \int 2e^x\,dx = x^2e^x - 2xe^x + 2e^x$. Thus, $m = \int_{-1}^2 x^2e^x\,dx = \left[x^2e^x - 2xe^x + 2e^x\right]_{-1}^2 = \left[e^x\left(x^2 - 2x + 2\right)\right]_{-1}^2 = 12.9378$.

Using the above result, we see that $\bar{x} = \frac{M_y}{m} = \frac{20.6642}{12.9387} = 1.5971$ and $\bar{y} = \frac{M_x}{m} = \frac{102.0163}{12.9387} = 7.8846$.

30. $$\sqrt{\frac{1}{1}\int_0^1 \left[\sqrt{\cos^{\ 1} x}\right]^2\,dx}$$
$$= \sqrt{\int_0^1 \cos^{-1} x\,dx}$$
$$= \sqrt{\left[x\cos^{-1}x - \sqrt{1 - x^2}\right]_0^1}$$
$$= \sqrt{\left[1 \cdot 0 - \sqrt{0} - 0 \cdot 1 + \sqrt{1}\right]}$$
$$= \sqrt{1} = 1$$

31. Work is the integral of force so we get $W = \int x^3 \cos x\,dx$. Using integration by parts, we have

$u = x^3$	$v = \sin x$
$du = 3x^2\,dx$	$dv = \cos x\,dx$

Thus, $\int x^3 \cos x\,dx = x^3 \sin x - \int 3x^2 \sin x\,dx$. Using integration by parts again yields

$u = -3x^2$	$v = -\cos x$
$du = -6x\,dx$	$dv = \sin x\,dx$

Hence, $\int x^3 \cos x\,dx = x^3 \sin x + 3x^2 \cos x - \int 6x \cos x$. We need integration by parts one more time.

$u = -6x$	$v = \sin x$
$du = -6\,dx$	$dv = \cos x\,dx$

Thus,

$$\int x^3 \cos x\,dx$$
$$= x^3 \sin x + 3x^2 \cos x - 6x \sin x + \int 6 \sin x\,dx$$
$$= x^3 \sin x + 3x^2 \cos x - 6x \sin x + 6 \cos x\Big|_0^{\pi/2}$$
$$= \left(\frac{\pi}{2}\right)^3 + 0 - 6\frac{\pi}{2} - 0 - 0 + 0 - 0 - 6$$
$$= \left(\frac{\pi}{2}\right)^3 - 3\pi + 6 \approx 0.4510$$

32. Since v is always positive we need only integrate $v(t)$ from 0 to 1. Expanding, we obtain $\int t^5\left(1-t^2\right)^{1/2}dt = \int t^3\cdot t^2(1-t^3)^{1/2}\,dt$. Let $u = 1-t^3$, then $t^3 = 1-u$ and $du = -3t^2\,dt$ or $-\frac{1}{3}\,du = t^2\,dt$. Substituting we have $\int (1-u)u^{\frac{1}{2}}\cdot\left(-\frac{1}{3}\right)du = -\frac{1}{3}\int\left(u^{1/2}-u^{3/2}\right)du = -\frac{1}{3}\left[\frac{2}{3}u^{3/2}-\frac{2}{5}u^{5/2}\right] = -\frac{2}{9}u^{3/2}+\frac{2}{15}u^{5/2} = \left[-\frac{2}{9}\left(1-t^3\right)+\frac{2}{15}\left(1-t^3\right)\right]_0^1 = \frac{2}{9}-\frac{2}{15} = \frac{10-6}{45} = \frac{4}{45}$.

33. As in example 28.52 set up the coordinate axis so that $(0,0)$ is at the center of the road between the towers. Since it is a parabola the equation of the main cable fits the form $y = 4px^2$ and contains the point (500,200). Thus, $200 = 4p\cdot(500)$ or $p = \frac{2}{10000}$. The equation of the cable is thus $y = \frac{1}{1250}x^2$. Differentiating we get $y' = \frac{1}{625}x$. The length of the cable is $\int_{-500}^{500}\sqrt{1+\left(\frac{x}{625}\right)^2}\,dx$. Evaluate this integral as follows: $\int\sqrt{1+\left(\frac{x}{625}\right)^2} = \frac{1}{625}\int\sqrt{625^2+x^2}$. Let $x = 625\tan\theta$ and $dx = 625\sec^2\theta\,d\theta$. This leads to $\frac{1}{625}\int\sqrt{625^2+25^2\tan^2\theta}\cdot 625\sec\theta\,d\theta = \frac{1}{625}\int 625\sec\theta\cdot 625\sec^2\theta\,d\theta = 625\int\sec^3\theta\,d\theta$. Using Example 28.50, we get $\frac{625}{2}\left[\sec\theta\tan\theta+\ln|\sec\theta+\tan\theta|\right] = \frac{625}{2}\left[\frac{\sqrt{625+x^2}}{625}+\frac{x}{625}+\ln\left|\frac{\sqrt{625+x^2}}{625}+\frac{x}{625}\right|\right]_{-500}^{500} = 625(1.02445+\ln(2.08062)) = 625(1.02445+0.73267)$. Thus, the length of the main cable is $625(1.75712) = 1098$ ft.

28.8 USING INTEGRATION TABLES

1. $\int(1+\tan 3x)^2\,dx = \int\left(1+2\tan 3x+\tan^2 3x\right)dx$ by formulas #12 and #58 we get $x+2\cdot\frac{1}{3}\ln|\sec 3x| + \frac{1}{3}(\tan 3x-3x)+C = x+\frac{2}{3}\ln|\sec 3x|+\frac{1}{3}\tan 3x - x + C = \frac{1}{3}\left(2\ln|\sec 3x|+\tan 3x\right)+C$

2. $\int\frac{\sqrt{16+25x^2}}{x}\,dx$. Let $u = 5x$, $du = 5\,dx$, so $x = \frac{u}{5}$ and $dx = \frac{du}{5}$. Let $a = 4$ and use formula #32. $\int\frac{\sqrt{4^2+u^2}}{\frac{4}{5}}\cdot\frac{1}{5}\,du = \int\frac{\sqrt{4^2+u^2}}{u}\,du = \sqrt{u^2+4^2} - 4\ln\left|\frac{4+\sqrt{u^2+4^2}}{u}\right| + C = \sqrt{26x^2+16}-4\ln\left|\frac{4+\sqrt{25x^2+16}}{5x}\right|+C.$

3. $\int\frac{dx}{x(4x-3)} = \int\frac{dx}{x(-3+4x)}$ Let $u = x$, $a = -3$, and $du = dx$. By formula #48 this is $\frac{1}{a}\ln\left|\frac{u}{a+bu}\right|+C = \frac{-1}{3}\ln\left|\frac{x}{4x-3}\right|+C.$

4. $\int\frac{5x}{3+7x}\,dx = 5\int\frac{x}{3+7x}\,dx$. Using formula #46 we obtain $5\left(\frac{1}{7^2}(3+7x-3\ln|3+7x|)\right)+C = \frac{5}{49}\left(3+7x-3\ln|3+7x|\right)+C.$

5. $\int\frac{x^2\,dx}{x^6\sqrt{16+x^6}}$. Let $u = x^3$ and $du = 3x^2\,dx$ or $\frac{1}{3}\,du = 3x^2\,dx$. Substituting yields $\frac{1}{3}\int\frac{du}{u^2\sqrt{4^2+u^2}}$ By formula #39 this is $\frac{1}{3}\left(\frac{-\sqrt{16+u^2}}{16u}\right)+C = \frac{1}{3}\left(\frac{-\sqrt{16+x^6}}{16x^3}\right)+C = -\frac{\sqrt{16+x^6}}{48x^3}+C.$

6. $\int(25-4x^2)^{3/2}\,dx$. Let $u = 2x$ and then $du = 2x$, or $\frac{1}{2}\,du = dx$. Substitution yields $\frac{1}{2}\int\left(25-u^2\right)^{3/2}$. By formula #28 we get $\frac{1}{2}\left(\frac{u}{8}(2u^2-5\cdot 5^2)\sqrt{25-u^2}+\frac{3\cdot 5^4}{8}\sin^{-1}\frac{u}{5}\right)+ C = \frac{1}{2}\left(\frac{2x}{8}(8x^2-125)\sqrt{25-4x^2}+\frac{1875}{8}\sin^{-1}\frac{2x}{5}\right)+ C = \frac{x}{8}(8x^2-125)\sqrt{25-4x^2}+\frac{1875}{16}\sin^{-1}\frac{2x}{5}+C.$

7. $\int \frac{x^2}{\sqrt{9-x^2}}\,dx$. By formula #25 this is $-\frac{x}{2}\sqrt{9-x^2}+\frac{9}{2}\sin^{-1}\frac{x}{3}+C$.

8. $\int x^2\sqrt{9x^2-49}\,dx$ Let $u = 3x$ and $du = 3\,dx$ or $x = \frac{1}{3}u$ and $x^2 = \frac{1}{9}u^2$ and $\frac{1}{3}\,du = dx$. Substitution yields $\int \frac{1}{9}u^2\sqrt{u^2-7^2}\cdot\frac{1}{3}\,du = \frac{1}{27}\int u^2\sqrt{u^2-7^2}\,du$. By formula #31, we have $\frac{1}{27}\left(\frac{u}{8}(2u^2-7^2)\sqrt{u^2-7^2}-\frac{7^4}{8}\ln\left|u+\sqrt{u^2-7^2}\right|\right)+C = \frac{1}{27}\left(\frac{3x}{8}(18^2-49)\sqrt{9x^2-49}-\frac{2401}{8}\ln\left|3x+\sqrt{9x^2-49}\right|\right)+C$.

9. $\int \cos^3 x\,dx$. By formula #65 this is $\frac{1}{3}\cos^2 x\sin x + \frac{2}{3}\int\cos x\,dx = \frac{1}{3}\cos^2 x\sin x + \frac{2}{3}\sin x + C = \frac{1}{3}(1-\sin^2 x)\sin x + \frac{2}{3}\sin x + C = \sin x - \frac{1}{3}\sin^3 x + C$. Alternate solution: $\int\cos^3 x\,dx = \int(1-\sin^2 x)\cos x\,dx = \int\cos x\,dx - \int\sin^2 x\cos x\,dx = \sin x - \frac{1}{3}\sin^3 x + C$.

10. $\int \sin 5x\sin 2x\,dx$. By formula #70 this is $\frac{\sin 3x}{6}-\frac{\sin 7x}{14}+C$

11. $\int \sin^6 3x\,dx$. Let $u = 3x$, then $du = 3\,dx$ or $\frac{1}{3}\,du = dx$. Substituting we get $\frac{1}{3}\int\sin^6 u\,du$. By formula #64 this is $\frac{1}{3}\left[-\frac{1}{6}\sin^5 u\cos u+\frac{5}{6}\int\sin^4 u\,du\right]$. Using formula #64 once more, you obtain $\frac{1}{3}\left[-\frac{1}{6}\sin^5 u\cos u + \frac{5}{6}\left(-\frac{1}{4}\sin^3 u\cos u + \frac{3}{4}\int\sin^2 u\,du\right)\right]$. By formula #56 we get $\frac{1}{3}\left[-\frac{1}{6}\sin^5 u\cos u + \frac{5}{6}\left(-\frac{1}{4}\sin^3 u\cos u + \frac{3}{4}\left\{\frac{1}{2}u - \frac{1}{2}\sin u\cos u\right\}\right)\right] + C = -\frac{1}{18}\sin^5 u\cos u - \frac{5}{72}\sin^3 u\cos u - \frac{5}{48}\sin u\cos u + \frac{5}{48}u + C = -\frac{1}{18}\sin^5 3x\cos 3x - \frac{5}{72}\sin^3 3x\cos 3x - \frac{5}{48}\sin 3x\cos 3x + \frac{5}{48}x + C$.

12. $\int e^{-6x}\sin 10x\,dx$: By formula #87 we have $\frac{e^{-6x}}{36+100}(-6\sin 10x - 10\cos 10x) + C = \frac{1}{136}e^{-6x}(-6\sin 10x - 10\cos 10x)+C$.

13. $\int e^{10x}\cos 6x\,dx$: By formula #88 we have this is $\frac{e^{10x}}{10^2+6^2}(10\cos 6x + 6\sin 6x) + C = \frac{1}{136}e^{10x}(10\cos 6x + 6\sin 6x)+C$.

14. $\int x^2\tan^{-1}x\,dx$. By formula #83 we get $\frac{1}{2+1}\left[x^{2+1}\tan^{-1}x - \int\frac{x^{2+1}}{1+x^2}\right] = \frac{1}{3}\left[x^3\tan^{-1}x - \int\frac{x^3}{1+x^2}\right]$. Now for the second integral let $u = 1+x$ with $du = 2x\,dx$ and $x^2 = u-1$. Substituting we have $\int\frac{x^3}{1+x^2}\,dx = \int\frac{u-1}{u}\,du = \int 1\,du - \int u^{-1}\,du = u - \ln u + C = 1 + x^2 - \ln(1+x^2)+C$. Putting it all together we get $\frac{1}{3}(x^3\tan^{-1}x - x^2 + \ln(1+x^2))+C$.

15. $\int x^7\ln x\,dx$. By formula #90 we obtain $x^8\left(\frac{1}{8}\ln x - \frac{1}{64}\right)+C$

16. $\int\frac{\sqrt{7-9x^2}}{x}\,dx$. Let $u = 3x$ and $du = 3\,dx$, then $x = \frac{1}{3}\,du$ and we have $\int\frac{\sqrt{7-u^2}}{u}\,du$. Using formula #23 and replacing u with $3x$ we get $\sqrt{7-9x^2}-\sqrt{7}\ln\left|\frac{\sqrt{7}+\sqrt{7-9x^2}}{3x}\right|+C$.

17. $\int\arcsin 4x\,dx$. Letting $u = 4x$ and $du = 4\,dx$ and then substituting we get $\frac{1}{4}\int\arcsin u\,du$. By formula #78, $\int\arcsin 4x\,dx = \frac{1}{4}\left[4x\sin^{-1}4x + \sqrt{1-(4x)^2}\right] + C = \frac{1}{4}\left[4x\sin^{-1}4x + \sqrt{1-16x^2}\right]+C$.

18. $\int\frac{dx}{(4-3x^2)^{3/2}}$. Let $u = \sqrt{3}x$, then $du\sqrt{3}\,dx$ and substituting produces $\frac{1}{\sqrt{3}}\int\frac{du}{(2^2-u^2)^{3/2}}$. By for-

mula #29 this is $\frac{1}{\sqrt{3}}\left[\frac{u}{4\sqrt{4-u^2}}\right]+C=\frac{1}{\sqrt{3}}\cdot\left[\frac{\sqrt{3}x}{4\sqrt{4-3x^2}}\right]+C=\frac{x}{4\sqrt{4-3x^2}}+C.$

19. $\int \frac{\sqrt{9+x^2}}{x}\,dx$. By formula #32 this is $\sqrt{9+x^2}-3\ln\left|\frac{3+\sqrt{9+x^2}}{x}\right|+C.$

20. $\int e^{\sin x}\sin x\cos x\,dx$. Let $u=\sin x$, then $du=\cos x\,dx$. Substituting we get $\int ue^u\,du$. By formula #84 this is $\frac{1}{1}(1u-1)e^u+C=(\sin x-1)e^{\sin x}+C.$

21. $\int x^3e^{2x}\,dx$. By formula #86 we get $\frac{1}{2}x^3e^{2x}-\frac{3}{2}\int x^2e^{2x}\,dx$. Repeatng this process, we get $\frac{1}{2}x^3e^{2x}-\frac{3}{2}\cdot\frac{1}{2}x^2e^{2x}+\frac{3}{2}\cdot\frac{2}{2}\int xe^{2x}\,dx$ or $\frac{1}{2}x^3e^{2x}-\frac{3}{4}x^2e^{2x}+\frac{3}{2}\int xe^{2x}\,dx$. Repeatng again produces $\frac{1}{2}x^3e^{2x}-\frac{3}{4}x^2e^{2x}+\frac{3}{4}xe^{2x}-\frac{3}{2}\frac{1}{2}\int e^{2x}\,dx=\frac{1}{2}x^3e^{2x}-\frac{3}{4}x^2e^{2x}+\frac{3}{4}xe^{2x}-\frac{3}{8}e^{2x}+C=\frac{1}{8}e^{2x}(4x^3-6x^2+6x-3)+C.$

22. $\int x^4\ln 2x\,dx$. Let $u=2x$ then $u^4=16x^4$ or $x^4=\frac{1}{16}u^4$ and $du=2\,dx$ or $dx=\frac{1}{2}\,du$. Substituting we get $\frac{1}{32}\int u^4\ln u\,du$. By formula #90 this is $\frac{1}{32}\left[u^5\left(\frac{\ln u}{5}-\frac{1}{5^2}\right)+C=\frac{1}{32}\left((2x)^5\cdot\left(\frac{\ln 2x}{5}-\frac{1}{25}\right)\right)+C=\frac{1}{5}x^5\left(\ln 2x-\frac{1}{5}\right)+C.\right.$

23. $\int x^3\sin 2x\,dx$. Let $u=2x$, then $u^3=8x^3$ or $x^3=\frac{1}{8}u^3$, and also, $du=2\,dx$ or $dx=\frac{1}{2}\,du$. Substitution produces $\frac{1}{16}\int u^3\sin u\,du$. By formula #75 this is $\frac{1}{16}\left[-u^3\cos u+3\int u^2\cos u\,du\right]$. Now using formula #76, on the integral in this result, we obtain
$\frac{1}{16}\left[-u^3\cos u+3(u^2\sin u-2\int u\sin u\,du)\right]$.
Finally, using formula #75 one more time produces the result
$\frac{1}{16}\left[-u^3\cos u+3u^2\sin u+6u\cos u-6\int\cos u\right]$
or
$\frac{1}{16}\left[-u^3\cos u+3u^2\sin u+6u\cos u-6\sin u\right]+C.$
Back substituting for u produces
$\int x^3\sin 2x\,dx=\frac{1}{16}\left[-8x^3\cos 2x+12x^2\sin 2x+12x\cos 2x-6\sin 2x\right]+C=$
$-\frac{1}{2}x^3\cos 2x+\frac{3}{4}x^2\sin 2x+\frac{3}{4}x\cos 2x-\frac{3}{8}\sin 2x+C.$

24. $\int\frac{\sqrt{\tan^2 2x-9}}{\cos^2 2x}\,dx$. Let $u=\tan 2x$, then $du=2\sec^2 2x\,dx$ or $\frac{1}{2}\,du=\sec^2 2x\,dx=\frac{1}{\cos^2 2x}\,dx$. Substituting we get $\frac{1}{2}\int\sqrt{u^2-3^2}\,du$. Using formula #30, we obtain
$\frac{1}{2}\left[\frac{u}{2}\sqrt{u^2-3^2}-\frac{3^2}{2}\ln\left|u+\sqrt{u^2-3^2}\right|\right]+C.$
Back substituting for u, we get
$\frac{1}{2}\left[\frac{\tan 2x}{2}\sqrt{\tan^2 2x-9}-\frac{9}{2}\ln\left|\tan 2x+\sqrt{\tan^2 2x-9}\right|\right]+C.$

CHAPTER 28 REVIEW

1. $\int xe^{3x}\,dx$. Using integration by parts with the table

$u=x$	$v=\frac{1}{3}e^{3x}$
$du=dx$	$dv=e^{3x}\,dx$

We have, $\int xe^{3x}\,dx=\frac{1}{3}xe^{3x}-\frac{1}{3}\int e^{3x}\,dx=\frac{1}{3}xe^{3x}-\frac{1}{9}e^{3x}+C=\frac{1}{9}e^{3x}(3x-1)+C.$

2. $\int\sin^3 x\cos^2x\,dx=\int\sin x(1-\cos^2 x)\cos^2x\,dx=\int\sin x(\cos^2 x-\cos^4)\,dx=-\frac{1}{3}\cos^3 x+\frac{1}{5}\cos^5 x+C.$

3. $\int\frac{x}{\sqrt{25-x^2}}\,dx$. Let $x=5\sin\theta$, then $dx=5\cos\theta\,d\theta$. Thus, substitution produces $\int\frac{5\sin\theta}{\sqrt{25-5\sin^2\theta}}\cdot 5\cos\theta\,d\theta=\int\frac{5\sin\theta\cdot 5\cos\theta\,d\theta}{5\cos\theta}=5\int\sin\theta\,d\theta=-5\cos\theta+C=5\cdot\frac{\sqrt{25-x2}}{5}+C=-\sqrt{25-x^2}+C.$
Alternate solution: Let $u=25-x^2$, then $du=-2x\,dx$ or $-\frac{1}{2}du=x\,dx$. Substituting, we

obtain

$$-\frac{1}{2}\int\frac{du}{\sqrt{u}} = -\frac{1}{2}\int u^{-1/2}\,du = -\frac{1}{2}\cdot\frac{2}{1}u^{1/2}+C$$
$$= -\sqrt{25-x^2}+C$$

4. $\int x^3\sqrt{25-x^2}\,dx$. Let $u = 25-x^2$ or $x^2 = 25-u$, and $du = -2x\,dx$ or $-\frac{1}{2}\,du = x\,dx$. Substituting we get $-\frac{1}{2}\int(25-u)\sqrt{u}\cdot du = -\frac{1}{2}\int 25\sqrt{u}\,du + \frac{1}{2}\int u^{3/2}\,du = -\frac{25}{2}\cdot\frac{2}{3}u^{3/2}+\frac{1}{2}\frac{2}{5}u^{5/2}+C = -\frac{25}{3}(25-x^2)^{3/2}+\frac{1}{5}(25-x^2)^{5/2}+C = -\frac{25}{3}(25-x^2)^{3/2}+\frac{1}{5}(25-x^2)(25-x^2)^{3/2}+C = \left(5-\frac{25}{3}-\frac{x^2}{5}\right)(25-x^2)^{3/2}+C = \left(-\frac{1}{5}x^2-\frac{10}{3}\right)(25-x^2)+C$.

5. $\int \sin^4 2x\cos 2x\,dx$. Let $u = \sin 2x$ then $du = 2\cos 2x\,dx$ or $\frac{1}{2}\,du = \cos 2x\,dx$. Substitution produces $\frac{1}{2}\int u^4\,du = \frac{1}{2}\cdot\frac{1}{5}u^5+C = \frac{1}{10}\sin^5 2x+C$.

6. $\int 2(e^x-e^{-x})\,dx = 2\int e^x\,dx - 2\int e^{-x}\,dx = 2e^x + 2e^{-x}+C$.

7. $\int\frac{dx}{9x+5}$. Let $u = 9x+5$ and $du = 9\,dx$ or $\frac{1}{9}\,du = dx$. Then $\frac{1}{9}\int\frac{du}{u} = \frac{1}{9}\ln|u| = \frac{1}{9}\ln|9x+5|+C$.

8. $\int\tan^2 8x\,dx = \int\frac{\sin^2 8x}{\cos^2 8x}\,dx = \int\frac{1-\cos^2 8x}{\cos^2 8x}\,dx = \int\frac{1}{\cos^2 8x}\,dx - \int 1\,dx = \frac{1}{8}\tan 8x - x + C$.

9. $\int\sin(7x+2)\,dx$. Let $u = 7x+2$ and $du = 7\,dx$ or $\frac{1}{7}\,du = dx$. Then, substitution produces $\frac{1}{7}\int\sin u\,du = \frac{1}{7}(-\cos u)+C = -\frac{1}{7}\cos(7x+2)+C$.

10. $\int\sin^3 2x\cos 2x\,dx$. Let $u = \sin 2x$ and then $du = 2\cos 2x\,dx$. Substitution produces $\frac{1}{2}\int u^3\,du = \frac{1}{2}\cdot\frac{1}{4}u^4+C = \frac{1}{8}\sin^4 2x+C$.

11. $\int\sin^5 3x\cos^2 3x\,dx = \int\sin 3x(\sin^4 3x)\cos^2 3x\,dx = \int\sin 3x(1-\cos^2 3x)^2\cos^2 3x\,dx$. This expands as $\int(\cos^2 3x - 2\cos^4 3x+\cos^6 3x)\sin 3x\,dx$. Let $u = \cos 3x$ and $du = -3\sin 3x\,dx$ or $-\frac{1}{3}\,du = \sin 3x\,dx$. Substituting we get $-\frac{1}{3}\int(u^2-2u^4+u^6)\,du = -\frac{1}{3}\left[\frac{1}{3}u^3-\frac{2}{5}u^5+\frac{1}{7}u^7\right]+C = -\frac{1}{9}\cos^3 3x + \frac{2}{15}\cos^5 3x - \frac{1}{21}\cos^7 3x + C$.

12. $\int\frac{dx}{x^2+4x+20} = \int\frac{dx}{(x+2)^2+16}$. Let $u = x+2$, $a = 4$, and $du = dx$. Substituting, we get $\int\frac{du}{u^2+4^2} = \frac{1}{4}\tan^{-1}\left(\frac{u}{4}\right)+C = \frac{1}{4}\tan^{-1}\left(\frac{x+2}{4}\right)+C$.

13. $\int\frac{dx}{(4x^2+49)^{3/2}}$. Let $2x = 7\tan\theta$, or $x = \frac{7}{2}\tan\theta$ and then $dx = \frac{7}{2}\sec^2\theta\,d\theta$. Substituting yields $\frac{7}{2}\int\frac{\sec^2\theta\,d\theta}{((7\tan\theta)^2+49)^{3/2}} = \frac{7}{2}\cdot\int\frac{\sec^2\theta\,d\theta}{7^3\sec^3\theta} = \frac{1}{98}\int\frac{1}{\sec\theta}\,d\theta = \frac{1}{98}\int\cos\theta\,d\theta = \frac{1}{98}\cdot\sin\theta+C = \frac{1}{98}\cdot\frac{2x}{\sqrt{4x^2+49}}+C = \frac{x}{49\sqrt{4x^2+49}}+C$.

14. $\int x^3\ln x\,dx$. Using integration by parts with the table

$u = \ln x$	$v = \frac{1}{4}x^4$
$du = \frac{1}{x}\,dx$	$dv = x^3\,dx$

We obtain $\int x^3\ln x\,dx = \frac{1}{4}x^4\ln x - \int\frac{1}{4}x^3\,dx = \frac{1}{4}x^4\ln x - \frac{1}{16}x^4+C = x^4\left(\frac{1}{4}\ln x-\frac{1}{16}\right)+C$.

15. $\int x^2e^{x^3}\,dx$. Let $u = x^3$ and $du = 3x^2\,dx$ or $\frac{1}{3}\,du = x^2\,dx$. Substituting, we get $\frac{1}{3}\int e^u\,du = \frac{1}{3}e^u+C = \frac{1}{3}e^{x^3}+C$.

16. $\int\frac{dx}{\sqrt{4x^2+49}}$. Let $2x = 7\tan\theta$ or $x = \frac{7}{2}\tan\theta$ and then $dx = \frac{7}{2}\sec^2\theta\,d\theta$. Substituting, we get $\frac{7}{2}\int\frac{\sec^2\theta\,d\theta}{\sqrt{49\tan^2\theta+49}} = \frac{1}{2}\int\frac{\sec^2\theta\,d\theta}{\sec\theta} = \frac{1}{2}\int\sec\theta\,d\theta = \frac{1}{2}\ln|\sec\theta+\tan\theta| + K = \frac{1}{2}\ln\left|\frac{\sqrt{4x^2+49}}{7}+\frac{2x}{7}\right| + K = \frac{1}{2}\ln\left|\sqrt{4x^2+49}+2x\right| - \frac{1}{2}\ln 7 + K = \frac{1}{2}\ln\left|\sqrt{4x^2+49}+2x\right|+C$ where $C = K-\frac{1}{2}\ln 7$.

17. $\int\frac{x\,dx}{\sqrt{4x^2+49}}$. Let $u = 4x^2+49$ and $du = 8x\,dx$ or $\frac{1}{8}\,du = x\,dx$. Substituting we get $\frac{1}{8}\int\frac{du}{\sqrt{u}} = \frac{1}{8}\int u^{-\frac{1}{2}}\,du = \frac{1}{8}\cdot\frac{2}{1}u^{1/2}+C = \frac{1}{4}\sqrt{4x^2+49}+C$.

18. $\int \cot 5x \csc^4 5x\,dx = \int \cot 5x \csc 5x \csc^3 5x\,dx$. Let $u = \csc 5x$ and $du = -5\cot 5x \csc 5x\,dx$. Substituting we get $-\frac{1}{5}\int u^3\,du = -\frac{1}{5}\cdot\frac{1}{4}u^4 + C = -\frac{1}{20}\csc^4 5x + C$ or $-\frac{1}{20}\cdot\sin^{-4} 5x + C$.

19. $\displaystyle\int \frac{\sec 4x \tan 4x}{9 + 2\sec 4x}\,dx$. Let $u = 9 + 2\sec 4x$ and $du = 8\sec 4x \tan 4x\,dx$. Substituting we get $\displaystyle\frac{1}{8}\int\frac{du}{u} = \frac{1}{8}\ln|u| + C = \frac{1}{8}\ln|9 + 2\sec 4x| + C$.

20. $\displaystyle\int \sin^6 \frac{3x}{2}\,dx$. We begin by rewriting this as a cubic expression, then using a half-angle trig identity, and finally expanding the rewritten expression.

$$\begin{aligned}\int \sin^6\frac{3x}{2}\,dx &= \int\left(\sin^2\frac{3x}{2}\right)^3 dx\\ &= \int\left(\frac{1-\cos 3x}{2}\right)^3 dx\\ &= \frac{1}{8}\int(1 - 3\cos 3x + 3\cos^2 3x - \cos^3 3x)\,dx\\ &= \frac{1}{8}x - \frac{1}{8}\sin 3x + \frac{3}{8}\int\cos^2 3x - \frac{1}{8}\int\cos^3 3x\,dx\end{aligned}$$

Now $= \frac{3}{8}\int\cos^2 3x\,dx = \frac{3}{8}\int\left(\frac{1+\cos 6x}{2}\right)dx = \frac{3}{16}\int(1+\cos 6x)\,dx = \frac{3}{16}x + \frac{3}{96}\sin 6x = \frac{3}{16}x + \frac{1}{32}\sin 6x$. And also, $-\frac{1}{8}\int\cos^3 3x\,dx = -\frac{1}{8}\int(1-\sin^2 3x)\cos 3x\,dx = -\frac{1}{8}\cdot\frac{1}{3}\sin 3x + \frac{1}{8}\cdot\frac{1}{9}\sin^3 3x + C$. Putting this all together we get $\frac{1}{8}x - \frac{1}{8}\sin 3x + \frac{3}{16}x + \frac{1}{32}\sin 6x - \frac{1}{24}\sin 3x + \frac{1}{72}\sin^3 3x + C = \frac{5}{16}x - \frac{5}{24}\sin 3x + \frac{1}{32}\sin 6x + \frac{1}{72}\sin^3 3x + C$.

21. $\displaystyle\int\frac{[\ln(2x+1)]^5}{2x+1}\,dx$. Let $u = \ln(2x+1)$ and then $du = \dfrac{2\,dx}{2x+1}$. Substituting yields $\frac{1}{2}\int u^5\,du = \frac{1}{12}u^6 + C = \frac{1}{12}[\ln(2x+1)]^6 + C$.

22. $\displaystyle\int\frac{\arctan 7x}{1+49x^2}\,dx$. Let $u = \arctan 7x$ and $du = \dfrac{7\,dx}{1+49x^2}$. Substitution produces $\frac{1}{7}\int u\,du = \frac{1}{7}\cdot\frac{1}{2}u^2 + C = \frac{1}{14}(\arctan 7x)^2 + C$.

23. $\displaystyle\int\frac{x^2}{x^3+4}\,dx$. Let $u = x^3 + 4$ and $du = 3x^2\,dx$ or $\frac{1}{3}\,du = x^2\,dx$. Now substituting we get $\displaystyle\frac{1}{3}\int\frac{du}{u} = \frac{1}{3}\ln|u| + C = \frac{1}{3}\ln\left|x^3+4\right| + C$.

24. $\int 4x^3 e^{x^4}\,dx$. Let $u = x^4$ and $du = 4x^3\,dx$. Substituting yields $\int e^u\,du = e^u + C = e^{x^4} + C$.

25. $\int \tan\frac{x}{5}\,dx = 5\ln\left|\sec\frac{x}{5}\right| + C$ or $-5\ln\left|\cos\frac{x}{5}\right| + C$.

26. $\int x\sin^3 x\,dx = \int x(1-\cos^2 x)\sin x\,dx = \int x\sin x\,dx - \int x\cos^2 x\sin x\,dx$. For the first integral we use integration by parts with

$u = x$	$v = -\cos x$
$du = dx$	$dv = \sin x\,dx$

Then, $\int x\sin x\,dx = -x\cos x + \int\cos x\,dx = -x\cos x + \sin x + C$. For the second integral we also use integration by parts.

$u = x$	$v = \frac{1}{3}\cos^3 x$
$du = dx$	$dv = -\cos^2 x\sin x\,dx$

Thus, $\displaystyle\int x(-\cos^2 x\sin x)\,dx = \frac{x}{3}\cos^3 x - \frac{1}{3}\int\cos^3 x\,dx = \frac{x}{3}\cos^3 x - \frac{1}{3}\int(1-\sin^2 x)\cos x\,dx = \frac{x}{3}\cos^3 x - \frac{1}{3}\sin x + \frac{1}{9}\sin^3 x + C$. Putting these together we have $\displaystyle\int x\sin^3 x\,dx = -x\cos x + \sin x - \left(\frac{x}{3}\cos^3 x - \frac{1}{3}\sin x + \frac{1}{9}\sin^3 x\right) + C = -x\cos x + \sin x - \frac{x}{3}\cos^3 x + \frac{1}{3}\sin x - \frac{1}{9}\sin^3 x + C = -x\cos x + \frac{x}{3}\cos^3 x + \frac{4}{3}\sin x - \frac{1}{9}\sin^3 x + C$.

27. $\int e^{\cos x}\sin x\,dx = -e^{\cos x} + C$

28. $\int x^4 e^{-x}\,dx$. We use integration by parts with

$u = x^4$	$v = -e^{-x}$
$du = 4x^3\,dx$	$dv = e^{-x}\,dx$

Hence, $\int x^4e^{-x}\,dx = x^4(-e^{-x}) + \int 4x^3e^{-x}\,dx$. Using integration by parts a second time produces

$u = 4x^3$	$v = -e^{-x}$
$du = 12x^2\,dx$	$dv = e^{-x}\,dx$

Thus, $x^4(-e^{-x})+\int 4x^3e^{-x}\,dx = -x^4e^{-x}-4x^3e^{-x}+\int 12x^2e^{-x}$. Using integration by parts a third time, we have

$u = 12x^2$	$v = -e^{-x}$
$du = 24x\,dx$	$dv = e^{-x}\,dx$

This produces $-x^4e^{-x} - 4x^3e^{-x} - 12x^2e^{-x} - 24xe^{-x} + \int 24xe^{-x}\,dx$. Finally, using integration by parts a fourth time, we obtain

$u = 24x$	$v = -e^{-x}$
$du = 24\,dx$	$dv = e^{-x}\,dx$

$-x^4e^{-x}-4x^3e^{-x}-12x^2e^{-x}-24xe^{-x}+\int 24e^{-x}\,dx$
$= -x^4e^{-x}-4x^3e^{-x}-12x^2e^{-x}-24xe^{-x}-24e^{-x}+C$
$= e^{-x}(-x^4-4x^3-12x^2-24x-24)+C$.

29. $\displaystyle\int \frac{\cos x}{\sin^2 x+9}dx$. Let $u = \sin x$ and $du = \cos x\,dx$. Substituting we get $\displaystyle\int \frac{du}{u^2+9} = \frac{1}{3}\tan^{-1}\frac{u}{3} + C = \frac{1}{3}\tan^{-1}\left(\frac{\sin x}{3}\right) + C$.

30. $\displaystyle\int \frac{e^x}{e^x+16}dx$. Let $u = e^x + 16$ and $du = e^x\,dx$. Substitution yields $\displaystyle\int \frac{du}{u} = \ln|u| + C = \ln|e^x+16| + C = \ln(e^x+16)+C$.

31. $\int x^3 \ln x\,dx$. Use integration by parts with

$u = \ln x$	$v = \frac{1}{4}x^4$
$du = \frac{1}{x}\,dx$	$dv = x^3\,dx$

Hence, $\int x^3\ln x\,dx = \frac{1}{4}x^4\ln x - \frac{1}{4}\int x^3\,dx = \frac{1}{4}x^4\ln x - \frac{1}{16}x^4 + C = x^4\left(\frac{1}{4}\ln x - \frac{1}{16}\right) + C$.

32. $\int e^{8x}\cos 2x\,dx$. Use integration by parts with $u = e^{8x}$ and $dv = \cos 2x\,dx$. Then $v = \frac{1}{2}\sin 2x$ and $du = 8e^{8x}\,dx$. This produces $\int e^{8x}\cos 2x\,dx = \frac{1}{2}e^{8x}\sin 2x - \int 4e^{8x}\sin 2x\,dx$. Use integration by parts again, with

$u = 2e^{8x}$	$v = -\cos 2x$
$du = 16e^{8x}\,dx$	$dv = 2\sin 2x\,dx$

Thus we have $\int e^{8x}\cos 2x\,dx = \frac{1}{2}e^{8x}\sin 2x - \int 4e^{8x}\sin 2x = \frac{1}{2}e^{8x}\sin 2x + 2e^{8x}\cos 2x - 16\int e^{8x}\cos 2x$. Adding $16\int e^{8x}\cos 2x$ to both sides of this last equation, we obtain $17\int e^{8x}\cos 2x\,dx = \frac{1}{2}e^{8x}\sin 2x + 2e^{8x}\cos 2x$, and, dividing by 17 produces the desired integral $\int e^{8x}\cos 2x\,dx = \frac{1}{34}e^{8x}\sin 2x + \frac{2}{17}e^{8x}\cos 2x + C = \frac{1}{34}e^{8x}(\sin 2x + 4\cos 2x) + C$.

33. $\int \sin^{1/3} 4x\cos^5 4x\,dx =$
$\int \sin^{1/3} 4x(\cos^2 4x)^2\cos 4x\,dx =$
$\int \sin^{1/3} 4x(1-\sin^2 4x)^2\cos 4x\,dx =$
$\int (\sin^{1/3} 4x - 2\sin^{7/3} 4x + \sin^{13/3} 4x)\cos 4x\,dx =$
$\frac{1}{4}(\frac{3}{4}\sin^{4/3} 4x - 2\cdot\frac{1}{10}\sin^{10/3} 4x + \frac{3}{16}\sin^{16/3} 4x) + C =$
$\frac{3}{320}\sin^{4/3} 4x(20 - 16\sin^2 4x + 5\sin^4 4x) + C$.

34. $\displaystyle\int \frac{\sin^3 x\,dx}{\cos^4 x} = \int \frac{\sin^3 x}{\cos^3 x}\cdot\frac{dx}{\cos x} =$
$\displaystyle\int \tan^3 x\sec x\,dx = \int \tan x(\sec^2 x - 1)\sec x\,dx =$
$\displaystyle\int \sec^2 x(\tan x\sec x\,dx) - \int \tan x\sec x\,dx =$
$\displaystyle\frac{1}{3}\sec^3 x - \sec x + C$.

35. $\displaystyle\int \frac{\cos x\,dx}{\sqrt{16-4\sin^2 x}}$. Let $u = 2\sin x$ and $du = 2\cos x\,dx$ or $\frac{1}{2}\,du = \cos x\,dx$. Substituting we get $\displaystyle\frac{1}{2}\int \frac{du}{\sqrt{4^2-u^2}} = \frac{1}{2}\sin\frac{u}{4} + C = \frac{1}{2}\sin^{-1}\left(\frac{2\sin x}{4}\right) + C = \frac{1}{2}\sin^{-1}\left(\frac{\sin x}{2}\right) + C$.

36. $\displaystyle\int \frac{e^{5x}}{4-e^{5x}}\,dx$. Let $u = 4 - e^{5x}$ and $du = -5e^{5x}\,dx$ or $-\frac{1}{5}\,du = e^{5x}\,dx$. Substituting we get $-\frac{1}{5}\int\frac{du}{u} = -\ln|u| + C = -\frac{1}{5}\ln\left|4-e^{5x}\right| + C$.

37. $\int x^5e^{x^2}\,dx$. Using integration by parts, we have the folowing table.

$u = x^4$	$v = \frac{1}{2}e^{x^2}$
$du = 4x^3\,dx$	$dv = xe^{x^2}\,dx$

This produces $\int x^5e^{x^2}\,dx = \frac{1}{2}e^{x^2}x^4 - 2\int x^3e^{x^2}\,dx$. Using integration by parts again with $u = -2x^2$ and $dv = xe^{x^2}\,dx$, we get $v = \frac{1}{2}e^{x^2}$ and $du = -4x\,dx$. Thus,

$\frac{1}{2}e^{x^2}x^4 - 2\int x^3 e^{x^2}\,dx = \frac{1}{2}e^{x^2}x^4 - e^{x^2}\cdot x^2 + 2\int xe^{x^2} =$
$\frac{1}{2}e^{x^2}x^4 - e^{x^2}x^2 + e^{x^2} + C =$
$e^{x^2}\left(\frac{x^4}{2} - x^2 - 1\right) + C = \frac{e^{x^2}}{2}(x^4 - 2x^2 + 2) + C.$

38. $\int \frac{e^{3x}}{(e^{3x}-1)^2}\,dx$. Let $u = e^{3x} - 1$ and $du = 3e^{3x}\,dx$ so $\frac{1}{3}\,du = e^{3x}\,dx$. Substituting we get
$$\frac{1}{3}\int \frac{du}{u^2} = -\frac{1}{3}u^{-1} + C = -\frac{1}{3}(e^{3x} - 1)^{-1} + C$$

39. $\int \frac{(\arctan 2x)^4}{1+4x^2}\,dx$. Let $u = \arctan 2x$, then $du = \frac{2}{1+4x^2}\,dx$ and substituting gives
$\frac{1}{2}\int u^4\,du = \frac{1}{10}u^5 + C = \frac{1}{10}(\arctan 2x)^5 + C.$

40. $\int \frac{\sec^2 5x\,dx}{2\tan 5x + 9}$. Let $u = 2\tan 5x + 9$, and then $du = 10\sec^2 5x\,dx$ or $\frac{1}{10}\,du = \sec^2 5x\,dx$. Substituting we get
$$\frac{1}{10}\int \frac{du}{u} = \frac{1}{10}\ln|u| + C = \frac{1}{10}\ln|2\tan 5x + 9| + C.$$

CHAPTER 28 TEST

1. Let $u = 9 - e^x$, then $du = -e^x\,dx$ and $\int \frac{e^x\,dx}{\sqrt{9-e^x}} = -\int u^{-1/2}\,du = -2u^{1/2} + C = -2\sqrt{9-e^x} + C$

2. $\int \tan^2 4x\cos^2 4x\,dx = \int \frac{\sin^2 4x}{\cos^2 4x}\cdot\cos^4 4x\,dx =$
$\int \sin^2 4x\cos^2 4x\,dx = \int \sin^2 4x(1 - \sin^2 4x)\,dx =$
$\int \sin^2 4x\,dx - \int \sin^4 4x\,dx = \int \frac{1-\cos 8x}{2}\,dx -$
$\int \left(\frac{1-\cos 8x}{2}\right)^2 dx = \frac{1}{2}x - \frac{1}{16}\sin 8x -$
$\frac{1}{4}\int (1 - 2\cos 8x + \cos^2 8x)\,dx = \frac{1}{2}x -$
$\frac{1}{10}\sin 8x - \frac{1}{4}x + \frac{1}{16}\sin 8x - \frac{1}{4}\int \left(\frac{1+\cos 16x}{2}\right) dx =$
$\frac{1}{4}x - \frac{1}{8}x - \frac{1}{128}\sin 16x + C = \frac{1}{8}x - \frac{1}{128}\sin 16x + C.$

3. $\int \frac{5\,dx}{x^2+1} = 5\tan^{-1}x + C$

4. Let $u = x^2 + 1$, then $du = 2x\,dx$ and so $\int \frac{4x\,dx}{(x^2+1)^3} = 2\int u^{-3}\,du = -u^{-2} + C = \frac{-1}{(x^2+1)^2} + C.$

5. Use integration by parts with $u = x$ and $dv = e^{4x}\,dx$. Then $du = dx$ and $v = \frac{1}{4}e^{4x}$ and so $\int xe^{4x}\,dx = uv - \int v\,du = \frac{1}{4}xe^{4x} - \int \frac{1}{4}e^{4x}\,dx = \frac{1}{4}xe^{4x} - \frac{1}{16}e^{4x} + C = \frac{1}{16}e^{4x}(4x - 1) + C.$

6. Let $u = \tan x$ and then $du = \sec^2 x\,dx = \frac{1}{\cos^2 x}\,dx$ and $\int \frac{e^{\tan x}}{\cos^2 x}\,dx = \int e^u\,du = e^u + C = e^{\tan x} + C.$

7. Using the disk method, we have $V = \pi\int_0^1 \left(\sqrt{x}\,e^x\right)^2 dx = \pi\int_0^1 xe^{2x}\,dx = \frac{\pi}{4}e^{2x}(2x-1)\Big]_0^1 = \frac{\pi}{4}(e^2 + 1).$

8. The arc length L, uses $(y')^2 = (2x)^2$ and so, $L = \int_0^2 \sqrt{1+(2x)^2}\,dx = \int_0^2 \sqrt{1+4x^2}\,dx$. From Formula 30 in Appendix C, we see that $L\int_0^2 \sqrt{1+4x^2}\,dx = \frac{2x}{2}\sqrt{1+4x^2} + \frac{1}{2}\ln\left|2x + \sqrt{1+4x^2}\right|\Big]_0^2 = 2\sqrt{17} + \frac{1}{2}\ln|4 + \sqrt{17}| \approx 9.2936.$

CHAPTER

29

Parametric Equations and Polar Coordinates

29.1 DERIVATIVES OF PARAMETRIC EQUATIONS

1. $x = t^2 + t$, $y = t + 1$. Thus, $\frac{dx}{dt} = 2t + 1$ and $\frac{dy}{dt} = 1$, which means that $\frac{dy}{dx} = \frac{1}{2t+1}$. Next, $\frac{dy}{dt}\left(\frac{d}{dt}\right) = \frac{d}{dt}(2t+1)^{-1} = \frac{-2}{(2t+1)^2}$ and we get $\frac{d^2y}{dx^2} = \frac{-2}{(2t+1)^3}$. Critical values occur when $\frac{dx}{dy} = 0$ or is undefined. The derivative $\frac{dy}{dx}$ is underfined at $t = -\frac{1}{2}$. The second derivative is also undefined. There are no extrema. Inflection point at $t = -\frac{1}{2} : \left(-\frac{1}{4}, \frac{1}{2}\right)$

2. Here $x = t + 3$ and $y = t^2 + t$ and so the first derivative is $\frac{dy}{dx} = \frac{dy/dt}{dx/dt} = \frac{2t+1}{1} = 2t+1$ and the second derivative is $\frac{d^2y}{dx^2} = \frac{\frac{d}{dt}\left(\frac{dy}{dx}\right)}{\frac{dx}{dt}} = \frac{2}{1} = 2$. Critical value at $t = -\frac{1}{2}$. Since the second derivative is positive, this is a minimum. There are no inflection points. Minimum at $t = -\frac{1}{2} : \left(\frac{5}{2}, -\frac{1}{4}\right)$

3. Here $x = t^2 - 6t + 12$ and $y = t + 4$, and so, $\frac{dy}{dx} = \frac{dy/dt}{dx/dt} = \frac{1}{2t-6}$. Critical value at $t = 3$. The second derivative is $\frac{d^2y}{dx^2} = \frac{\frac{d}{dt}\left(\frac{dy}{dx}\right)}{\frac{dx}{dt}} = \frac{\frac{-2}{(2t-6)^2}}{2t-6} = \frac{-2}{(2t-6)^3}$. Inflection point at $t = 3$: $(3, 7)$. No extrema.

4. Since $x = t^2 + 6t + 12$ and $y = t + 4$, the first derivative is $\frac{dy}{dx} = \frac{dy/dt}{dx/dt} = \frac{1}{2t+6}$. Critical value at $t = -3$. The second derivative is $\frac{d^2y}{dx^2} = \frac{\frac{d}{dt}\left(\frac{dy}{dx}\right)}{\frac{dx}{dt}} = \frac{\frac{-2}{(2t+6)^2}}{2t+6} = \frac{-2}{(2t+6)^3}$. Inflection point at $t = -3$: $(3, 1)$

5. Here $x = t^2 + t$ and $y = t^2 - t$ which means that the first derivative is $\frac{dy}{dx} = \frac{dy/dt}{dx/dt} = \frac{2t-1}{2t+1}$. Critical values are at $t = \frac{1}{2}, -\frac{1}{2}$. The second derivative is $\frac{d^2y}{dx^2} = \frac{\frac{d}{dt}\left(\frac{dy}{dx}\right)}{\frac{dx}{dt}} = \frac{\frac{(2t+1)2-(2t-1)2}{(2t+1)^2}}{2t+1} = \frac{4}{(2t+1)^3}$. Inflection point at $t = -\frac{1}{2} : \left(-\frac{1}{4}, \frac{3}{4}\right)$. Minimum at $t = \frac{1}{2}$: $\left(\frac{3}{4}, -\frac{1}{4}\right)$. It is a minimum since the second derivative is positive.

6. Since $x = 4t + 1$ and $y = 9t^2$, the first derivative is $\frac{dy}{dx} = \frac{dy/dt}{dx/dt} = \frac{18t}{4} = \frac{9}{2}t$. Critical value at $t = 0$. The second derivative is $\frac{d^2y}{dx^2} = \frac{\frac{d}{dt}\left(\frac{dy}{dx}\right)}{\frac{dx}{dt}} = \frac{\frac{18}{4}}{4} =$

$\frac{18}{16} = \frac{9}{8}$. there are no inflection points. Since the second derivative is positive the critical value yields a minimum. Minimum at $t = 0$: $(1, 0)$.

7. Here $x = 3+4\cos t$ and $y = -1+\cos t$ which means that the first derivative is $\frac{dy}{dx} = \frac{dy/dt}{dx/dt} = \frac{-\sin t}{-4\sin t} = \frac{1}{4}$. The second derivative is $\frac{d^2y}{dx^2} = 0$ since $\frac{dy}{dx}$ is a constant. This is segment where the point (x, y) moves between a minimum when t is an odd multiple of π giving $(-1, -2)$, and a maximum when t is an even multiple of π giving $(7, 0)$.

8. Here $x = 3 + 4\cos t$ and $y = 1 - \sin t$, which means that the first derivative is $\frac{dy}{dx} = \frac{dy/dt}{dx/dt} = \frac{-\cos t}{-4\sin t} = \frac{1}{4}\cot t$. The second derivative is $\frac{d^2y}{dx^2} = \frac{\frac{d}{dt}\left(\frac{dy}{dx}\right)}{\frac{dt}{dt}} = \frac{-\frac{1}{4}\csc^2 t}{-4\sin t} = \frac{1}{16}\csc^3 t$. The critical values occur when $\cos t = 0$, or when $t = \frac{\pi}{2} + 2k\pi$ yields a minimum $(3, 0)$, as $\frac{d^2y}{dx^2} > 0$. The values of $t = \frac{-\pi}{2} + 2k\pi$ yield a maximum at $(3, 2)$, since $\frac{d^2y}{dx^2} < 0$. Inflection points occure when $\sin t = 0$ or at multiples of π: $t = 0 + 2k\pi$ yields $(7, 1)$; $t = \pi + 2k\pi$ yields $(-1, 1)$. (Note: This curve is an ellipse with major axis between $(-1, 1)$ and $(7, 1)$ and minor axis between $(3, 0)$, and $(3, 2)$.)

9. Since $x = 2+\sin t$ and $y = -1+\cos t$, the first derivative is $\frac{dy}{dx} = \frac{dy/dt}{dx/dt} = \frac{-\sin t}{\cos t} = -\tan t$. The second derivative is $\frac{d^2y}{dx^2} = \frac{\frac{d}{dt}\left(\frac{dy}{dx}\right)}{\frac{dx}{dt}} = \frac{-\sec^2 t}{\cos t} = -\sec^3 t$. Critical values occur when $\sin t = 0$. At $t = 2k\pi$, the second derivative is negative so maximum $(2, 0)$. At $t = \pi + 2k\pi$, y'' is positive so we have a minimum $(2, -2)$. Inflection points are at $t = \frac{\pi}{2} + 2k\pi$ and $-\frac{\pi}{2} + 2k\pi$. These are $(3, -1)$ and $(1, -1)$. (Note: this is an circle with vertical diameter between $(2, 0)$ and $(2, -2)$ are horizontal diameter between $(3, -1)$ and $(1, -1)$).

10. Since $x = -2 + 4e^t$ and $y = 3 + 2e^{-t}$, the first derivative is $dy/dx = \frac{dy/dt}{dx/dt} = \frac{-2e^{-t}}{4e^t} = -\frac{1}{2}e^{-2t}$ or $\frac{-1}{2e^{2t}}$. The second derivative is $\frac{d^2y}{dx^2} = \frac{\frac{d}{dt}\left(\frac{dy}{dx}\right)}{\frac{dx}{dt}} = \frac{e^{-2t}}{4e^t} = \frac{1}{4e^{3t}}$. Since $e^t \neq 0$ there are no extrema or inflection points.

11. Here $x = 2t - 1$ and $y = 4t^2 - 2t$ and the first derivative is $\frac{dy}{dx} = \frac{dy/dt}{dx/dt} = \frac{8t-2}{2}$. At $t = 1$, the slope is $m = \frac{dy}{dx} = \frac{8-2}{2} = 3$ and the desired point is $(x, y) = (1, 2)$. Thus, the equation of the tangent is $y-2 = 3(x-1) \Rightarrow y = 3x-1$ and the equation of the normal line is $(y-2) = -\frac{1}{3}(x-1) \Rightarrow y = -\frac{1}{3}x+2\frac{1}{3}$ or $3y + x = 7$.

12. $x = t - 4$, $y = t^3 + 2t^2 - 5t - 2$; $t = 1$. The desired point is $(x, y) = (-3, -4)$. The derivative is $\frac{dy}{dx} = \frac{3t^2 + 4t - 5}{1}$ and when $t = 1$, the slope is $m = 2$. The requested equations of the lines are: Tangent: $y + 4 = 2(x + 3)$ or $y = 2x + 2$; Normal: $y + 4 = -\frac{1}{2}(x + 3)$ or $2y + x = -11$.

13. $x = t^3$, $y = t^2$, $t = -3$. Thus, $(x, y) = (-27, 9)$. The derivative is $\frac{dy}{dx} = \frac{2t}{3t^2} = \frac{2}{3t}$; at $t = -3$, $m = -\frac{2}{9}$. Tangent: $y - 9 = -\frac{2}{9}(x + 27)$ or $9y + 2x = 27$; Normal: $y - 9 = \frac{9}{2}(x + 27)$ or $2y - 9x = 261$.

14. $x = 2\cos t$, $y = 3\sin t$, $t = \frac{\pi}{4}$, and so $(x, y) = \left(\sqrt{2}, \frac{3\sqrt{2}}{2}\right)$. The derivative is $\frac{dy}{dx} = \frac{3\cos t}{-2\sin t} = -\frac{3}{2}\tan t$ and so, at $\frac{\pi}{4}$, the slope is $m = -\frac{3}{2}$. Tangent: $y - \frac{3\sqrt{2}}{2} = -\frac{3}{2}(x - \sqrt{2})$ or $2y + 3x = 6\sqrt{2}$; Normal: $y - \frac{3\sqrt{2}}{2} = \frac{2}{3}(x - \sqrt{2})$ or $3y - 2x = \frac{5\sqrt{2}}{2}$.

15. $x = 2+\cos t$, $y = 2\sin t$, $t = \frac{\pi}{2}$, $(x, y) = (2, 2)$. Thus, $\frac{dy}{dx} = \frac{2\cos t}{-\sin t} = -2\cot t$. At $t = \frac{\pi}{2}$, $m = 0$ which means that the desired lines are Tangent: $y = 2$ and Normal: $x = 2$.

16. $x = e^t + 1$, $y = e^t + e^{-t}$, $t = 1$, so $(x, y) = \left(e + 1, e + \frac{1}{e}\right)$ and the derivative is $\frac{dy}{dx} = \frac{e^t - e^{-t}}{e^t}$. At $t = 1$, we determine that $m = \frac{e - e^{-1}}{e} = \frac{e^2 - 1}{e^2}$. Tangent: $y - \frac{e^2 + 1}{e} = \frac{e^2 - 1}{e^2}(x - (e + 1))$. Multiplying by e^2, we obtain $e^2 y - e^3 - e = (e^2 - 1)x - e^3 - e^2 + e + 1$or $e^2 y - (e^2 - 1)x = -e^2 + 2e + 1$ or about $7.389y - 6.389x = -0.9525$. Normal: $y - \frac{e^2 + 1}{e} = \frac{e^2}{1 - e^2}[x - (e + 1)]$ or $(1 - e^2)y - \left(\frac{1 - e^4}{e}\right) = e^2 x - e^3 - e^2$, or $(1 - e^2)y - e^2 x = -e^2 + e^{-1} - 2e^3$, or $(e^2 - 1)y + e^2 x = e^2 - e^{-1} + 2e^3$. This is about $6.389y + 7.389x = 47.192$.

17. $x = t + 3$, $y = t^2 - 4t$. The derivative is $\frac{dy}{dx} = \frac{dy/dt}{dx/dt} = \frac{2t - 4}{2}$. Horizontal tangent when $2t - 4 = 0$ or $t = 2$: $(5, -4)$. Vertical tangents: none.

18. $x = t - 4$, $y = (t^2 + t)^2$. The derivative is $\frac{dy}{dt} = 2(t^2 + t)(2t + 1) = 2t(t + 1)(2t + 1)$. Horizontal tangents are at $t = 0$: $(-4, 0)$, $t = -1$: $(-5, 0)$, and $t = -\frac{1}{2}$: $\left(-4\frac{1}{2}, \frac{1}{16}\right)$. Since $\frac{dx}{dt} = 1$ is never 0 there are no vertical tangents.

19. $x = 3\cos t$, $y = 5\sin t$. To find any horizontal tangents, we have $\frac{dy}{dt} = 5\cos t$ and $5\cos t = 0$ when $t = \frac{\pi}{2} + 2k\pi$: $(0, 5)$ and $t = -\frac{\pi}{2} + 2k\pi$: $(0, -5)$. To find vertical tangents, we take $\frac{dx}{dt} = -3\sin t$; $-3\sin t = 0$ when $t = 2k\pi$: $(3, 0)$ and $t = \pi + 2k\pi$: $(-3, 0)$.

20. $x = t^2 + 1$, $y = \cos t$. To find horizontal tangents, we take $\frac{dy}{dt}$ and set it equal to 0, with the result that $\frac{dy}{dt} = -\sin t = 0$ when $t = 2k\pi$, $k \neq 0$; $(4\pi^2 + 1, 1)$, $(16\pi^2 + 1, 1)$, etc. or $t = \pi + 2k\pi$: $(\pi^2 + 1, -1)$, $(9\pi^2, -1)$, etc. To find the vertical tangents, we use $\frac{dx}{dt} = 2t = 0$ when $t = 0$: $\frac{dy/dt}{dx/dt} = \frac{-\sin t}{2t}$. Since $\lim_{t\to 0} \frac{-\sin t}{2t} = -\frac{1}{2}$ there are no vertical tangents.

21. $s_x(t) = 3780t\cos\frac{4\pi}{15}$; $s_y(t) = 3780t\sin\frac{4\pi}{15} - 16t^2$.

(a) $s_x(5) = 3780 \cdot 5 \cdot \cos\frac{4\pi}{15} \approx 12{,}646.57$ ft; $s_y(5) = 3780 \cdot 5 \cdot \sin\frac{4\pi}{15} - 16 \cdot 5^2 \approx 13{,}645.44$ ft.

(b) $v_x(5) = 3780\cos\frac{4\pi}{15} = 2529.31$ ft/sec, and $v_y(5) = 3780\sin\frac{4\pi}{15} - 32 \cdot 5 = 2649.09$ ft/sec

(c) $a_x(t) = 0$;
$a_y(t) = -32$ ft/sec^2

22. $s_x(t) = 1250t\cos\frac{3\pi}{11}$; $s_y(t) = 1250t\sin\frac{3\pi}{11} - 4.9t^2$

(a) $s_x(8) = 1250 \cdot 8 \cdot \cos\frac{3\pi}{11} = 6548.61$ m and $s_y(8) = 1250 \cdot 8 \cdot \sin\frac{3\pi}{11} - 4.9 \cdot 8^2 = 7243.90$ m

(b) $v_x(t) = \frac{ds_x(t)}{dt} = 1250\cos\frac{3\pi}{11}$, so $v_x(8) = 1250\cos\frac{3\pi}{11} \approx 818.58$ m/s, and $v_y(t) = \frac{ds_y(t)}{dt} = 1250\sin\frac{3\pi}{11} - 9.8t$, so $v_y(8) = 1250\sin\frac{3\pi}{11} - 9.8 \cdot 8 \approx 866.29$ m/s

(c) $a_x(t) = 0$, so $a_x(8) = 0$, and $a_y(t) = -9.8$, so $a_y(8) = -9.8$ m/s^2

29.2 DIFFERENTIATION IN POLAR COORDINATES

1. Since $r = 3\sin\theta$, then $r' = 3\cos\theta$ which leads to $\frac{dy}{dx} = \frac{r'\sin\theta + r\cos\theta}{r'\cos\theta - r\sin\theta} = \frac{3\cos\theta\sin\theta + 3\sin\theta\cos\theta}{3\cos\theta\cos\theta - 3\sin\theta\sin\theta} = \frac{2\cos\theta\sin\theta}{\cos^2\theta - \sin^2\theta} = \frac{\sin 2\theta}{\cos 2\theta} = \tan 2\theta$

2. Since $r = -2\cos\theta$, then $r' = 2\sin\theta$ and so $\frac{dy}{dx} = \frac{r'\sin\theta + r\cos\theta}{r'\cos\theta - r\sin\theta} = \frac{2\sin\theta\sin\theta - 2\cos\theta\cos\theta}{2\sin\theta\cos\theta + 2\cos\theta\sin\theta} = \frac{\sin^2\theta - \cos^2\theta}{2\sin\theta\cos\theta} = \frac{-\cos 2\theta}{\sin 2\theta} = -\cot 2\theta$

3. Since $r = 1 + \cos\theta$, then $r' = -\sin\theta$ and we have $\frac{dy}{dx} = \frac{r'\sin\theta + r\cos\theta}{r'\cos\theta - r\sin\theta} = \frac{-\sin\theta\sin\theta + (1+\cos\theta)\cos\theta}{-\sin\theta\cos\theta - (1+\cos\theta)\sin\theta} = \frac{-\sin^2\theta + \cos\theta + \cos^2\theta}{-2\sin\theta\cos\theta - \sin\theta} = \frac{\cos\theta + \cos 2\theta}{-\sin 2\theta - \sin\theta} = -\frac{\cos 2\theta + \cos\theta}{\sin 2\theta + \sin\theta}$.

4. Since $r = \cos 3\theta$, then $r' = -3\sin 3\theta$ and we have $\frac{dy}{dx} = \frac{r'\sin\theta + r\cos\theta}{r'\cos\theta - r\sin\theta} = \frac{-3\sin 3\theta\sin\theta + \cos 3\theta\cos\theta}{-3\sin 3\theta\cos\theta - \cos 3\theta\sin\theta} = \frac{3\sin 3\theta\sin\theta - \cos 3\theta\cos\theta}{3\sin 3\theta\cos\theta + \cos 3\theta\sin\theta}$.

5. We are given $r = 1 + \cos 3\theta$, and so $r' = -3\sin 3\theta$, which leads to $\frac{dy}{dx} = \frac{-3\sin 3\theta\sin\theta + (1+\cos 3\theta)\cos\theta}{-3\sin 3\theta\cos\theta - (1+\cos 3\theta)\sin\theta} = \frac{3\sin 3\theta\sin\theta - \cos 3\theta\cos\theta - \cos\theta}{3\sin 3\theta\cos\theta + \cos 3\theta\sin\theta + \sin\theta}$.

6. Since $r = \csc\theta$, we have $r' = -\csc\theta\cot\theta$, and so the desired derivative is $\frac{dy}{dx} = \frac{-\csc\theta\cot\theta\sin\theta + \csc\theta\cos\theta}{-\csc\theta\cot\theta\cos\theta - \csc\theta\sin\theta} = \frac{-\cot\theta + \cot\theta}{-\cot^2\theta - 1} = 0.$

7. Here $r = \tan\theta$, which means that $r' = \sec^2\theta$, and so the desired derivative is $\frac{dy}{dx} = \frac{\sec^2\theta\sin\theta + \tan\theta\cos\theta}{\sec^2\theta\cos\theta - \tan\theta\sin\theta} = \frac{\tan\theta\sec\theta + \tan\theta\cos\theta}{\sec\theta - \tan\theta\sin\theta}\cdot\frac{\cos\theta}{\cos\theta} = \frac{\tan\theta + \tan\theta\cos^2\theta}{1 - \sin^2\theta} = \tan\left(\frac{1+\cos^2\theta}{\cos^2\theta}\right) = \tan\theta(\sec^2\theta + 1)$.

8. $r = e^\theta, r' = e^\theta$; $\frac{dy}{dx} = \frac{e^\theta\sin\theta + e^\theta\cos\theta}{e^\theta\cos\theta - e^\theta\sin\theta} = \frac{\sin\theta + \cos\theta}{\cos\theta - \sin\theta}$

9. $r = \sin\theta, r' = \cos\theta$; $\frac{dy}{dx} = \frac{\cos\theta\sin\theta + \sin\theta\cos\theta}{\cos\theta\cos\theta - \sin\theta\sin\theta} = \frac{2\sin\theta\cos\theta}{\cos^2\theta - \sin^2\theta} = \frac{\sin 2\theta}{\cos 2\theta} = \tan 2\theta$. At the point $\left(1, \frac{\pi}{2}\right), m = \tan 2\cdot\frac{\pi}{2} = \tan\pi = 0$. Thus, $x = r\cos\theta = 1\cos\frac{\pi}{2} = 0$ and $y = r\sin\theta = 1\sin\frac{\pi}{2} = 1$. Tangent: $y = 1$; Normal: $x = 0$

10. $r = 2\cos\theta$, $\left(\sqrt{2}, \frac{\pi}{4}\right)$, $r' = -2\sin\theta$; $\frac{dy}{dx} = \frac{-2\sin\theta\sin\theta + 2\cos\theta\cos\theta}{-2\sin\theta\cos\theta - 2\cos\theta\sin\theta} = \frac{\sin^2\theta - \cos^2\theta}{2\sin\theta\cos\theta} = -\cot 2\theta$. At $\left(\sqrt{2}, \frac{\pi}{4}\right)$, we find $m = -\cot\left(2\cdot\frac{\pi}{4}\right) = -\cot\frac{\pi}{2} = 0$. Thus, $x = r\cos\theta = \sqrt{2}\cdot\cos\frac{\pi}{4} = \sqrt{2}\cdot\frac{\sqrt{2}}{2} = 1$, and $y = r\sin\theta = \sqrt{2}\cdot\sin\frac{\pi}{4} = 1$. Tangent: $y = 1$, Normal: $x = 1$

11. $r = 5\sin 3\theta$, $\left(\frac{5}{\sqrt{2}}, \frac{\pi}{12}\right)$, $r' = 15\cos 3\theta$; $\frac{dy}{dx} = \frac{15\cos 3\theta\sin\theta + 5\sin 3\theta\cos\theta}{15\cos 3\theta\cos\theta - 5\sin 3\theta\sin\theta}$. When $\theta = \frac{\pi}{12}$, then $m = \frac{dy}{dx} \approx 0.660254 \approx 0.6603$. Thus, $x = \frac{5}{\sqrt{2}}\cos\frac{\pi}{12} = 3.4151$ and $y = \frac{5}{\sqrt{2}}\sin\frac{\pi}{12} = 0.9151$. Tangent:

$y = -0.9151 = 0.6603(x - 3.4151)$. Normal: $y - 0.9151 = -1.5146(x - 3.4151)$

12. $r = 2 - 3\sin\theta$, $\left(\frac{1}{2}, \frac{5\pi}{6}\right)$; $r' = -3\cos\theta$; $\frac{dy}{dx} = \frac{-3\cos\theta\sin\theta + (2 - 3\sin\theta)\cos\theta}{-3\cos\theta\cos\theta - (2 - 3\sin\theta)\sin\theta}$. At $\theta = \frac{5\pi}{6}$, we find that $m = \frac{dy}{dx} = \frac{-3\left(-\frac{\sqrt{3}}{2}\right)\frac{1}{2} + \left(2 - 3\cdot\frac{1}{2}\right)\cdot\frac{-\sqrt{3}}{2}}{-3\cdot\frac{-\sqrt{3}}{2}\cdot\frac{-\sqrt{3}}{2} - \left(2 - 3\cdot\frac{1}{2}\right)\frac{1}{2}} = \frac{\frac{3\sqrt{2}}{4} + \frac{-\sqrt{3}}{4}}{\frac{-9}{4} - \frac{1}{4}} = \frac{2\sqrt{3}}{-10} = \frac{-\sqrt{3}}{5} \approx -0.3464$. Thus, we get $x = r\cos\theta = \frac{1}{2}\cdot\frac{-\sqrt{3}}{2} = \frac{-\sqrt{3}}{4}$ and $y = r\sin\theta = \frac{1}{2}\cdot\frac{1}{2} = \frac{1}{4}$. As a result, Tangent: $y - \frac{1}{4} = \frac{-\sqrt{3}}{5}\left(x + \frac{\sqrt{3}}{4}\right) = 4y + \frac{\sqrt{3}}{5}x = \frac{2}{5}$, or $20y + \sqrt{3}x = 2$ or about $20y + 1.732x = 2$, or $y = -0.25 = -0.3464(x + 1.4330)$. Normal: $y - \frac{1}{4} = \frac{5}{\sqrt{3}}\left(x + \frac{\sqrt{3}}{5}\right)$ or $4y - 1 = \frac{20}{\sqrt{3}}x + 5$ or $4y - \frac{20}{\sqrt{3}}x = 6$, or $y - 0.24 = 2.8858(x + 0.433)$, or $4y - 11.547x = 6$.

13. We are given $r = 6\sin^2\theta$ and $\left(4.5, \frac{2\pi}{3}\right)$, and so $r' = 12\sin\theta\cos\theta$ which leads to the derivative $\frac{dy}{dx} = \frac{12\sin\theta\cos\theta\sin\theta + 6\sin^2\theta\cos\theta}{12\sin\theta\cos\theta\cos\theta - 6\sin^2\theta\sin\theta} = \frac{3\sin^2\theta\cos\theta}{\sin\theta(2\cos^2\theta - \sin^2\theta)} = \frac{3\sin\theta\cos\theta}{2\cos^2\theta - \sin^2\theta}$. At $\theta = \frac{2\pi}{3}$ we get $m = \frac{3\cdot\frac{\sqrt{3}}{2}\cdot\frac{-1}{2}}{2\cdot\frac{1}{4} - \frac{3}{4}} = \frac{\frac{-3\sqrt{3}}{4}}{-\frac{1}{4}} = 3\sqrt{3} \approx 5.1962$. At $\left(4.5, \frac{2\pi}{3}\right)$, we find that $x = 4.5\cos\frac{2\pi}{3} = \frac{-9}{4}$ and $y = 4.5\sin\frac{2\pi}{3} = \frac{9\sqrt{3}}{4}$. Tangent: $y - \frac{9\sqrt{3}}{4} = 3\sqrt{3}\left(x + \frac{9}{4}\right)$ or $4y - 9\sqrt{3} = 12\sqrt{3}x + 27\sqrt{3}$ or $4y - 12\sqrt{3}x = 36\sqrt{3}$ or $y - 3\sqrt{3}x = 9\sqrt{3}$. Normal: $y - \frac{9\sqrt{3}}{4} = \frac{-1}{3\sqrt{3}}\left(x + \frac{9}{4}\right)$, or $4y - 9\sqrt{3} = \frac{-4}{3\sqrt{3}}x - \frac{9}{3\sqrt{3}}$, or $4y + \frac{4}{3\sqrt{3}}x = 8\sqrt{3}$, or $y + \frac{1}{3\sqrt{3}}x = 2\sqrt{3}$, or $9y + \sqrt{3}x = 18\sqrt{3}$.

14. $r = 2 + 3\sec\theta$; $\left(-4, \frac{2\pi}{3}\right)$, and so $r' = 3\sec\theta\tan\theta$, which leads to $\frac{dy}{dx} = \frac{3\sec\theta\tan\theta\sin\theta + (2 + 3\sec\theta)\cos\theta}{3\sec\theta\tan\theta\cos\theta - (2 + 3\sec\theta)\sin\theta} = \frac{3\tan^2\theta + 2\cos\theta + 3}{3\tan\theta - 2\sin\theta - 3\tan\theta} = \frac{3\tan^2\theta + 2\cos\theta + 3}{-2\sin\theta}$. So, $\theta = \frac{2\pi}{3}$, yields $m = \frac{3(-\sqrt{3})^2 + 2\left(\frac{-1}{2}\right) + 3}{-2\left(\frac{\sqrt{3}}{2}\right)} = \frac{9 - 1 + 3}{-\sqrt{3}} = \frac{11}{-\sqrt{3}} = \frac{-11}{3}\sqrt{3}$. As a result, we find that $x = -4\cos\frac{2\pi}{3} = -4\cdot\frac{-1}{2} = 2$ and $y = -4\sin\frac{2\pi}{3} = -4\frac{\sqrt{3}}{2} = -2\sqrt{3}$. Tangent: $y + 2\sqrt{3} = \frac{-11\sqrt{3}}{3}(x - 2)$, or , or $3y + 11\sqrt{3}x = 16\sqrt{3}$; $y + 6.3509x = 9.2376$. Normal: $y + 2\sqrt{3} = \frac{\sqrt{3}}{11}(x - 2)$, or $11y + 22\sqrt{3} = \sqrt{3}x - 2\sqrt{3}$, or $11y - \sqrt{3}x = -24\sqrt{3}$, or $y - 0.1574x = -3.7790$.

15. $r = e^{\theta}$; $\left(2.8497, \frac{\pi}{3}\right)$; By exercise #8, we see that $\frac{dy}{dx} = \frac{\sin\theta + \cos\theta}{\cos\theta - \sin\theta}$, and at $\frac{\pi}{3}$, we obtain $m = \frac{\frac{\sqrt{3}}{2} + \frac{1}{2}}{\frac{1}{2} - \frac{\sqrt{3}}{2}} = \frac{\sqrt{3} + 1}{1 - \sqrt{3}} = \frac{(\sqrt{3} + 1)(\sqrt{3} + 1)}{(1 - \sqrt{3})(1 + \sqrt{3})} = \frac{4 + 2\sqrt{3}}{-2} = -2 - \sqrt{3}$. Thus, $x = 2.8497\cdot\cos\frac{\pi}{3} = 1.4249$ and $y = 2.8497\cdot\sin\frac{\pi}{3} = 2.4679$. Tangent: $y - 2.4679 = -(2 + \sqrt{3})(x - 1.4249)$, or $y - 2.4697 = -3.7321x + 5.3178$, or $y + 3.7321x = 7.7875$. Normal: $y - 2.4679 = \frac{1}{2 + \sqrt{3}}(x - 1.4249)$, or $y - 2.4679 = 0.2679x - 0.3818$, or $y - 0.2679x = 2.0861$.

16. $r = \tan\theta$, $\left(-1, \frac{3\pi}{4}\right)$. By exercise #7, $\frac{dy}{dx} = (\sec^2\theta + 1)\tan\theta$. At $\theta = \frac{3\pi}{4}$, we obtain $m = (2 + 1)\cdot(-1) = -3$. Thus, $x = -1\cos\frac{3\pi}{4} = \frac{\sqrt{2}}{2}$ and $y = -1\sin\frac{3\pi}{4} = -\frac{\sqrt{2}}{2}$. Tangent: $y + \frac{\sqrt{2}}{2} = -3\left(x - \frac{\sqrt{2}}{2}\right)$, or $2y + \sqrt{2} = -6x + 3\sqrt{2}$, or $2y + 6x = 2\sqrt{2}$, or $y + 3x = \sqrt{2}$. Normal: $y + \frac{\sqrt{2}}{2} = \frac{1}{3}\left(x - \frac{\sqrt{2}}{2}\right)$, or $6y + 3\sqrt{2} = 2x - \sqrt{2}$, or $6y - 2x = -4\sqrt{2}$, or $3y - x = -2\sqrt{2}$.

17. Since $r = 3\cos 2\theta$ we have $r' = -6\sin 2\theta$, which gives the derivative

$\dfrac{dy}{dx} = \dfrac{-6\sin 2\theta \sin\theta + 3\cos 2\theta \cos\theta}{-6\sin 2\theta \cos\theta - 3\cos 2\theta \sin\theta} = \dfrac{2\sin 2\theta \sin\theta - \cos 2\theta \cos\theta}{2\sin 2\theta \cos\theta + \cos 2\theta \cos\theta}$. If we set the numerator equal to 0 we get $2\sin 2\theta \sin\theta - \cos 2\theta \cos\theta = 0$ which expands to $2(2\sin\theta\cos\theta)\sin\theta - (1 - 2\sin^2\theta)\cos\theta = 0$; $\cos\theta(4\sin^2\theta - 1 + 2\sin^2\theta) = 0$. Which simplifies to $\cos\theta(6\sin^2\theta - 1) = 0$. Hence either $\cos\theta = 0$ or $\sin^2\theta = \frac{1}{6}$ or $\sin\theta = \pm\frac{\sqrt{6}}{6}$. This gives critical values of $\theta = \frac{\pi}{2}, \frac{3\pi}{2}$ and $\theta = 0.4205$, 2.7211, 3.5621, and 5.8627.

Set the denominator to 0 and solve $2\sin 2\theta \cos\theta + \cos 2\theta \sin\theta = 0$. Using an expansion identity for $2\sin\alpha\cos\alpha$, we get $4\sin\theta\cos\theta\cos\theta + (2\cos^2\theta - 1)\sin\theta = 0$. Factoring, we get $\sin\theta\left[4\cos^2\theta + 2\cos^2\theta - 1\right] = 0$, so $\sin\theta = 0$ or $\cos^2\theta = \frac{1}{6}$ or $\cos\theta = \pm\frac{\sqrt{6}}{6}$. This gives critical values of 0, π, 1.1503, 1.9913, 4.2918, and 5.1329.

The graph of $r = 3\cos 2\theta$ is a four leaf rose. The first set of critical values gives the locations of horizontal tangents and the second set gives the locations of vertical tangents. Horizontal tangents at $\left(-3, \frac{\pi}{2}\right)$, $\left(-3, \frac{3\pi}{2}\right)$, $2, 0.4205)$, $(2, 2.7211)$, $(2, 3.5621)$, and $(2, 5.8627)$. Vertical tangents at $(3, 0)$, $(3, \pi)$, $(-2, 1.1503)$, $(-2, 1.9913)$, $(-2, 4.2918)$, and $(-2, 5.1329)$.

18. $r = 1 - \cos\theta$; $r' = \sin\theta$, and so $\dfrac{dy}{dx} = \dfrac{\sin\theta\sin\theta + (1-\cos\theta)\cos\theta}{\sin\theta\cos\theta - (1-\cos)\sin\theta}$.

The numerator is 0 when $\sin^2\theta + \cos\theta - \cos^2\theta = 0$ or $(1 - \cos^2\theta) + \cos\theta - \cos^2\theta = 0$ or $1 + \cos\theta - 2\cos^2\theta = 0$, which factors as $(1 + 2\cos\theta)(1 - \cos\theta) = 0$. So $\cos\theta = 1$ or $\cos\theta = -\frac{1}{2}$. This gives critical values of $\theta = 0, \frac{2\pi}{3}$, and $\frac{4\pi}{3}$. The denominator is 0 when $\sin\theta\cos\theta - (1 - \cos\theta)\sin\theta = 0$; $\sin\theta(\cos\theta - 1 + \cos\theta) = 0$; $\sin\theta = 0$ or $\cos\theta = \frac{1}{2}$. This gives critical values of $0, \pi, \frac{\pi}{3}$ and $\frac{5\pi}{3}$. This shape is a cardioid with horizontal tangents at $\theta = \frac{2\pi}{3}$ and $\frac{4\pi}{3}$ and vertical tangents at $\frac{\pi}{3}$, π, and $\frac{5\pi}{3}$. At $\theta = 0$, there is no tangent. Horizontal tangents at $\left(1.5, \frac{3\pi}{3}\right)$, $\left(1.5, \frac{4\pi}{3}\right)$. Vertical tangents at $(0.5, \frac{\pi}{3})$, $(2, \pi)$, and $\left(0.5, \frac{5\pi}{3}\right)$.

19. $r = 1 + 2\cos\theta$, $r' = -2\sin\theta$. Hence, we obtain

$$\begin{aligned}\frac{dy}{dx} &= \frac{-2\sin\theta\sin\theta + (1+2\cos\theta)\cos\theta}{-2\sin\theta\cos\theta - (1+2\cos\theta)\sin\theta} \\ &= \frac{-2\sin^2\theta + \cos^2\theta + 2\cos\theta}{-2\sin\theta\cos\theta - \sin\theta + 2\cos\theta\sin\theta} \\ &= \frac{-2(1-\cos^2\theta) + \cos\theta + 2\cos^2\theta}{-4\sin\theta\cos theta - \sin\theta} \\ &= \frac{4\cos^2\theta + \cos\theta - 2}{-\sin\theta(4\cos theta + 1)}\end{aligned}$$

The numerator equals 0 when $4\cos^2\theta + \cos\theta - 2 = 0$ or $\cos\theta = \left(-1 \pm \sqrt{33}\right)8$ or when $\theta \approx 0.9359$, 5.3476, 2.5738, or 3.7094. The denominator equals 0 when $\sin\theta = 0$ or $\cos\theta = -\frac{1}{4}$ which yields $\theta = 0$, π, 1.8234, or 4.4597. This is a limaçon. Horizontal tangents are at $(2.1861, 0.9359)$, $(-0.6861, 2.5738)$, $(-0.6861, 3.7094)$, and $(2.1861, 5.3476)$. Vertical tangents are at $(3, 0)$, $(0.5, 1.8234)$, $(-1, \pi)$, and $(0.5, 4.4597)$.

20. $r = \sin^2\theta$, $r' = 2\sin\theta\cos\theta$. Hence, we obtain

$$\begin{aligned}\frac{dy}{dx} &= \frac{2\sin\theta\cos\theta\sin\theta + \sin^2\theta\cos\theta}{2\sin\theta\cos\theta\cos\theta - \sin^2\theta\sin\theta} \\ &= \frac{2\sin^2\theta\cos\theta + \sin^2\theta\cos\theta}{2\sin\theta\cos^2\theta - \sin^3\theta} \\ &= \frac{3\sin^2\theta\cos\theta}{2\cos^2\theta - \sin^2\theta}\end{aligned}$$

The numerator equals 0 when $\sin\theta = 0$ or $\cos\theta = 0$. These will be when $\theta = 0$, $\frac{\pi}{2}$, π, and $\frac{3\pi}{2}$. The denominator equals 0 when $2\cos^2\theta - \sin^2\theta = 0$ or $2\cos^2\theta = \sin^2\theta$, or $2 = \tan^2\theta$. This will be when $\theta = 0.9553$, 2.1863, 4.0969, or 5.3279. Horizontal tangents accur at $(0, 0)$, $(1, \frac{\pi}{2})$, and $(1, \frac{3\pi}{2})$. Vertical tangents are at $(\frac{2}{3}, 0.9553)$, $(\frac{2}{3}, 2.1863)$, $(\frac{2}{3}, 4.0969)$, and $(\frac{2}{3}, 5.3279)$. (Note: $(0, \pi)$ is the same point as $(0, 0)$.)

29.3 ARC LENGTH AND SURFACE AREA REVISITED

1. $x = 4t^3$; $y = 3t^2$ so $\frac{dx}{dt} = 12t^2$ and $\frac{dy}{dt} = 6t$. The arc length is $L = \int_0^1 \sqrt{(12t^2)^2 + (6t)^2}\,dt = \int_0^1 \sqrt{144t^4 + 36t^2}\,dt = \int_0^1 6t\sqrt{4t^2+1}\,dt$. Let $u = 4t^2 + 1$ and $du = 8t\,dt$. Substituting, we get $\frac{6}{8}\int \sqrt{u}\,du = \frac{3}{4}\frac{2}{3}u^{3/2} = \frac{1}{2}(4t^2+1)^{3/2}\Big|_0^1 = \frac{1}{2}\left[5^{3/2} - 1^{3/2}\right] \approx \frac{1}{2}(10.1803) = 5.0902$.

2. $x = \sin t$; $\frac{dx}{dt} = \cos t$; $y = \cos t$; $\frac{dy}{dt} = -\sin t$ The arc length is $L = \int_0^\pi \sqrt{(\cos t)^2 + (-\sin t)^2}\,dt = \int_0^\pi \sqrt{\cos^2 t + \sin^2 t}\,dt = \int_0^\pi 1\,dt = t\Big|_0^\pi = \pi$.

3. $x = 3\cos t$; $\frac{dx}{dt} = -3\sin t$; $y = 3\sin t$; $\frac{dy}{dt} = 3\cos t$. The arc length is $L = \int_0^{2\pi} \sqrt{(-3\sin t)^2 + (3\cos t)^2}\,dt = \int_0^{2\pi} \sqrt{9\sin^2 t + 9\cos^2 t}\,dt = \int_0^{2\pi} 3\,dt = 3t\Big|_0^{2\pi} = 6\pi$. (Note: this curve is a circle of radius 3.)

4. $x = \cos^3 t$; $\frac{dx}{dt} = -3\cos^2 t\sin t$; $y = \sin^3 t$; $\frac{dy}{dt} = 3\sin^2 t\cos t$. The arc length is $L = \int_0^{\pi/2} \sqrt{(-3\cos^2 t\sin t)^2 + (3\sin^2 t\cos t)^2}\,dt$, or $L = \int_0^{\pi/2} \sqrt{9\cos^4 t\sin^2 t + 9\sin^4 \cos^2 t}\,dt$, or $L = \int_0^{\pi/2} 3\sqrt{\cos^2 t\sin^2 t(\cos^2 t + \sin^2 t)}\,dt$, or $L = \int_0^{\pi/2} 3\cos t\sin t\,dt$. Let $u = \sin t$; $du = \cos t\,dt$ so by sustituting we get $3\int u\,du = 3\cdot\frac{1}{2}u^2 = \left[\frac{3}{2}\sin^2 t\right]_0^{\pi/2} = \frac{3}{2}$.

5. $x = \cos t + t\sin t$; $\frac{dx}{dt} = -\sin t + t\cos t + \sin t = t\cos t$. $y = \sin t - t\cos t$; $\frac{dy}{dt} = \cos t + t\sin t - \cos t = t\sin t$. The arc length is $L = \int_0^\pi \sqrt{(t\cos t)^2 + (t\sin t)^2}\,dt = \int_0^\pi \sqrt{t^2\cos^2 t + t^2\sin^2 t}\,dt = \int_0^\pi t\,dt = \frac{t^2}{2}\Big|_0^\pi = \frac{1}{2}\pi^2$.

6. $r = \theta$; $r' = 1$; The arc length is $L = \displaystyle\int_0^{\pi/2} \sqrt{r^2 + r'^2}\,d\theta = \int_0^{\pi/2} \sqrt{\theta^2 + 1}\,d\theta$. By formula #30 in Appendix C, this is $\left[\frac{\theta}{2}\sqrt{\theta^2+1} + \frac{1}{2}\ln\left|\theta + \sqrt{\theta^2+1}\right|\right]_0^{\pi/2} = \frac{\pi}{4}\sqrt{\frac{\pi^2}{4} + 1} + \frac{1}{2}\ln\left|\frac{\pi}{2} + \sqrt{\frac{\pi^2}{4} + 1}\right| \approx 2.0792$.

7. Since $r = 1 + \cos\theta$ we have $r' = -\sin\theta$. Thus, the arc length is $L = \displaystyle\int_0^\pi \sqrt{(1+\cos\theta)^2 + (-\sin\theta)^2}\,d\theta = \int_0^\pi \sqrt{1 + 2\cos\theta + \cos^2\theta + \sin^2\theta}\,d\theta = \int_0^\pi \sqrt{2 + 2\cos\theta}\,d\theta = \sqrt{2}\int_0^\pi \sqrt{1+\cos\theta}\,d\theta$. Since $\cos^2\frac{\theta}{2} = \dfrac{1+\cos\theta}{2}$ we get $\sqrt{1+\cos\theta} = \sqrt{2}\cos\frac{\theta}{2}$. Now our integral is $\sqrt{2}\cos\displaystyle\int_2^\pi \sqrt{2}\cos\frac{\theta}{2}\,d\theta = 2\int_0^\pi \cos\frac{\theta}{2}\,d\theta$. Letting $u = \frac{\theta}{2}$ we have $du = \dfrac{1}{2}\,d\theta$ or $2\,du = d\theta$, and so, we get by substitution, $4\int \cos u\,du = 4\sin u = 4\sin\frac{\theta}{2}\Big|_0^\pi = 4\left(\sin\frac{\pi}{2} - \sin 0\right) = 4$.

8. Here $r = \cos^2\theta$ and so $r' = -2\cos\theta\sin\theta$. Thus, the arc length is given by

$$\begin{aligned} L &= \int_0^{\pi/2} \sqrt{\cos^4\theta + (-2\cos\theta\sin\theta)^2}\,d\theta \\ &= \int_0^{\pi/2} \sqrt{\cos^4\theta + 4\cos^2\theta\sin^2\theta}\,d\theta \\ &= \int_0^{\pi/2} \cos\theta\sqrt{\cos^4\theta + 4\sin^2\theta}\,d\theta \\ &= \int_0^{\pi/2} \cos\theta\sqrt{1 + 3\sin^2\theta}\,d\theta. \end{aligned}$$

Letting $u = \sqrt{3}\sin\theta$, we get $du = \sqrt{3}\cos\theta\,d\theta$. Substituting produces $L = \dfrac{1}{\sqrt{3}}\displaystyle\int \sqrt{1+u^2}\,du$. By formula #30 in Appendix C this is

$$\frac{1}{\sqrt{3}}\left[\frac{u}{2}\sqrt{1+u^2}+\frac{1}{2}\ln\left|u+\sqrt{1+u^2}\right|\right]$$

$$=\frac{1}{\sqrt{3}}\left[\frac{3\sin\theta}{2}\sqrt{1+3\sin^2\theta}\right.$$

$$\left.+\frac{1}{2}\ln\left|3\sin\theta+\sqrt{1+3\sin^2\theta}\right|\right]_0^{\pi/2}$$

$$=\frac{1}{\sqrt{3}}\left[\frac{\sqrt{3}}{2}\cdot\sqrt{4}+\frac{1}{2}\ln\left(\sqrt{3}+\sqrt{4}\right)\right]$$

$$=\frac{1}{\sqrt{3}}\left[\sqrt{3}+\frac{1}{2}\ln\left(2+\sqrt{3}\right)\right]$$

$$=1+\frac{\sqrt{3}}{6}\ln(2+\sqrt{3})\approx 1.3802.$$

9. Here $r=\sin^2\theta$, and so $r'=2\sin\theta\cos\theta=\sin 2\theta$. The arc length is

$$L=\int_0^{\pi/2}\sqrt{\sin^4\theta+(2\sin\theta\cos\theta)^2}\,d\theta$$

$$=\int_0^{\pi/2}\sqrt{\sin^4\theta+4\sin^2\theta\cos^2\theta}\,d\theta$$

$$=\int_0^{\pi/2}\sin\theta\sqrt{\sin^2\theta+4\cos^2\theta}\,d\theta$$

$$=\int_0^{\pi/2}\sin\theta\sqrt{1+3\cos^2\theta}\,d\theta.$$

Letting $u=\sqrt{3}\cos\theta$, then $du=-\sqrt{3}\sin\theta$ and substituting we get $-\frac{1}{\sqrt{3}}\int\sqrt{1+u^2}\,du$. By formula #30 in Appendix C we get $-\frac{1}{\sqrt{3}}\left[\frac{u}{2}\sqrt{1+u^2}+\frac{1}{2}\ln\left|u+\sqrt{1+u^2}\right|\right]=$ $-\frac{1}{\sqrt{3}}\left[\frac{\sqrt{3}\cos\theta}{2}\sqrt{1+3\cos^2\theta}+\right.$ $\left.\frac{1}{2}\ln\left|\sqrt{3}\cos\theta+\sqrt{1+3\cos^2\theta}\right|\right]_0^{\pi/2}=$ $-\frac{1}{\sqrt{3}}\left(-\sqrt{3}-\frac{1}{2}\ln(\sqrt{3}+2)\right)=$ $1+\frac{\sqrt{3}}{6}\ln(2+\sqrt{3})\approx 1.3802.$

10. Since $r=e^{\theta/2}$, then $r'=\frac{1}{2}e^{\theta/2}$. The arc length is $L=\int_0^4\sqrt{\left(e^{\theta/2}\right)^2+\left(\frac{1}{2}e^{\theta/2}\right)^2}\,d\theta=$ $\int_0^4\sqrt{e^\theta+\frac{1}{4}e^\theta}\,d\theta.$ Thus, we have $L=$ $\int_0^4\sqrt{\frac{5e^\theta}{4}}\,d\theta=\frac{\sqrt{5}}{2}\int_0^4\sqrt{e^\theta}\,d\theta=\frac{\sqrt{5}}{2}\int_0^4 e^{\theta/2}\,d\theta.$ Letting $u=\frac{\theta}{2}$, then $du=\frac{1}{2}\,d\theta$, so we get $\frac{\sqrt{5}}{2}\cdot$ $2\int e^u\,du=\sqrt{5}e^u=\sqrt{5}e^{\theta/2}\Big|_0^4=\sqrt{5}\left(e^2-e^0\right)=$ $\sqrt{5}\left(e^2-1\right)=14.2864.$

11. $x=t+4$, so $\frac{dx}{dt}=1$ and $y=t^3$, so $\frac{dy}{dt}=3t^2$. Thus, the area of the surface of revolution is $S=2\pi\int_0^2 t^3\sqrt{1+(3t^2)^2}\,dt=2\pi\int_0^2 t^3\sqrt{1+9t^4}\,dt$. Let $u=1+9t^4$, and then $du=36t^3\,dt$. Substituting we get $\frac{2\pi}{36}\int\sqrt{u}\,du=\frac{\pi}{18}\cdot\frac{2}{3}u^{3/2}=\frac{\pi}{27}(1+9t^4)^{3/2}\Big|_0^2=$ $\frac{\pi}{27}(145^{3/2}-1)\approx 203.0436.$

12. Here $x=t$, so $\frac{dx}{dt}=1$ and $y=4-t^2$, so $\frac{dy}{dt}=-2t$. Thus, the area of the surface of revolution is $S=2\pi\int_0^2 t\sqrt{1+(4t^2)}\,dt$. Let $u=$ $1+4t^2$, and then $du=8t\,dt$. Substituting we get $2\pi\cdot\frac{1}{8}\int\sqrt{u}\,du=\frac{\pi}{4}\frac{2}{3}u^{3/2}=\frac{\pi}{6}\left(1+4t^2\right)^{3/2}\Big|_0^2=$ $\frac{\pi}{6}\left(17^{3/2}-1\right)\approx 36.1769.$

13. Here $x=\cos t$, so $\frac{dx}{dt}=-\sin t$ and $y=\sin t$, so $\frac{dy}{dt}=\cos t$. Thus,

$$S=2\pi\int_0^{\pi/2}\sin t\sqrt{(-\sin t)^2+(\cos t)^2}\,dt$$

$$=2\pi\int_0^{\pi/2}\sin t\sqrt{\sin^2 t+\cos^2 t}\,dt$$

$$=2\pi\int_0^{\pi/2}\sin t\,dt$$

$$=2\pi(-\cos t)\Big|_0^{\pi/2}=2\pi(0+1)=2\pi.$$

14. $x = 1 + \sin t$; $\frac{dx}{dt} = \cos t$; $y = \cos t$; $\frac{dy}{dt} = -\sin t$. Thus,

$$S = 2\pi \int_0^{\pi/2} (1 + \sin t)\sqrt{\cos^2 t + \sin^2 t}\, dt$$
$$= 2\pi \int_0^{\pi/2} (1 + \sin t)\, dt = 2\pi\, [t - \cos t]\Big|_0^{\pi/2}$$
$$= 2\pi \left[\frac{\pi}{2} - 0 - 0 + 1\right] = 2\pi \left[\frac{\pi}{2} + 1\right]$$
$$= \pi^2 + 2\pi \approx 16.1528$$

15. $x = 1 + \sin t$, $\frac{dx}{dt} = \cos t$; $y = \cos t$; $\frac{dy}{dt} = -\sin t$. Thus, the area is $S = 2\pi \int_0^{\pi/2} \cos t\sqrt{\cos^2 t + \sin^2 t}\, dt = 2\pi \int_0^{\pi/2} \cos t\, dt = 2\pi[\sin t]_0^{\pi/2} = 2\pi$.

16. $r = \sin\theta$ and so $\frac{dr}{d\theta} = \cos\theta$. Thus,

$$S = 2\pi \int_0^{\pi/2} r\sin\theta\sqrt{\sin^2\theta + \cos^2\theta}\, d\theta$$
$$= 2\pi \int_0^{\pi/2} \sin\theta \sin\theta \cdot 1\, d\theta$$
$$= 2\pi \int_0^{\pi/2} \sin^2\theta\, d\theta = 2\pi \int_0^{\pi/2} \frac{1 - \cos 2\theta}{2}\, d\theta$$
$$= \pi \int_0^{\pi/2} (1 - \cos 2\theta)\, d\theta$$
$$= \pi\left(\theta - \frac{1}{2}\sin 2\theta\right)\Big|_0^{\pi/2} \pi\left(\frac{\pi}{2} - 0 - 0 - 0\right) = \frac{\pi^2}{2}.$$

17. $r = 1 + \cos\theta$; $\frac{dr}{d\theta} = -\sin\theta$. Thus, the area of the surface of revolution is

$$S = 2\pi \int_0^{\pi} r\sin\theta\sqrt{(1 + \cos t)^2 + (-\sin\theta)^2}\, d\theta$$
$$= 2\pi \int_0^{\pi} (1 + \cos\theta)\sin\theta \times \sqrt{1 + 2\cos\theta + \cos^2\theta + \sin^2\theta}\, d\theta$$
$$= 2\pi \int_0^{\pi} (1 + \cos\theta)\sin\theta\sqrt{2 + 2\cos\theta}\, d\theta$$
$$= 2\sqrt{2}\pi \int_0^{\pi} (1 + \cos\theta)^{3/2}\sin\theta\, d\theta$$

Let $u = 1 + \cos\theta$, then $du = -\sin\theta\, d\theta$, and we get $-2\sqrt{2}\pi \int u^{3/2}\, du = -2\sqrt{2}\pi \cdot \frac{2}{5}u^{5/2} = \frac{-4\sqrt{2}\pi}{5}(1 + \cos\theta)^{5/2}\Big|_0^{\pi} = \frac{-4\sqrt{2}\pi}{5}\left((1 - 1)^{5/2} - (1 - 1)^{5/2}\right) = \frac{-4\sqrt{2}\pi}{5}(2)^{5/2} = \frac{2^5}{5}\pi = \frac{32}{5}\pi \approx 20.1062$.

18. $r = e^{\theta/2}$; $\frac{dr}{d\theta} = \frac{1}{2}e^{\theta/2}$

$$S = 2\pi \int_0^{\pi} e^{\theta/2}\sin\theta\sqrt{\left(e^{\theta/2}\right)^2 + \left(\frac{1}{2}e^{\theta/2}\right)^2}\, d\theta$$
$$= 2\pi \int_0^{\pi} e^{\theta/2}\sin\theta\sqrt{e^{\theta} + \frac{1}{4}e^{\theta}}\, d\theta$$
$$= 2\pi \int_0^{\pi} e^{\theta/2}\sin\theta\left(\frac{\sqrt{5}}{2}e^{\theta/2}\right) d\theta$$
$$= \sqrt{5}\pi \int_0^{\pi} e^{\theta}\sin\theta\, d\theta.$$

By formula #87 of Appendix C, this is $\sqrt{5}\pi\left[\frac{e^{\theta}}{2}(\sin\theta - \cos\theta)\right]_0^{\pi} = \frac{1}{2}\sqrt{5}\pi\left[e^{\pi}(1) - e^{0}(-1)\right] = \frac{1}{2}\sqrt{5}\pi\,(e^{\pi} + 1) \approx 84.7919$.

29.4 INTERSECTION OF GRAPHS OF POLAR COORDINATES

1. $r = 3\theta, r = \frac{\pi}{2}$: By direct substitution we get $\frac{\pi}{2} = 3\theta$ or $\theta = \frac{\pi}{6}$. This gives the one point of intersection $\left(\frac{\pi}{2}, \frac{\pi}{6}\right)$

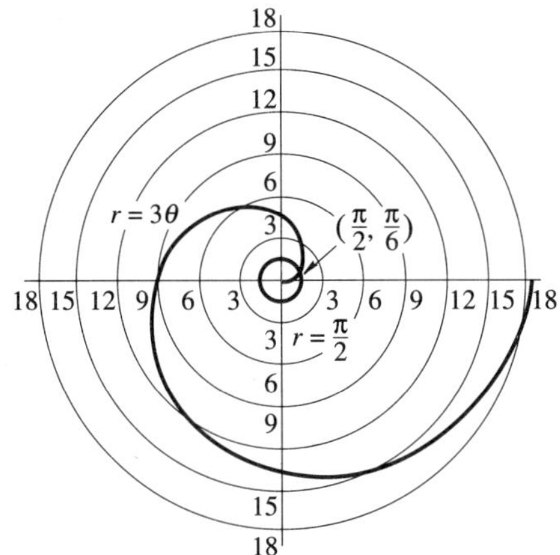

2. $r = \frac{\theta}{4}; r = \frac{\pi}{6}$: By direct substitution we obtain the equation $\frac{\pi}{6} = \frac{\theta}{4}$ or $\theta = \frac{2\pi}{3}$. This gives the one point of intersection $\left(\frac{\pi}{6}, \frac{2\pi}{3}\right)$.

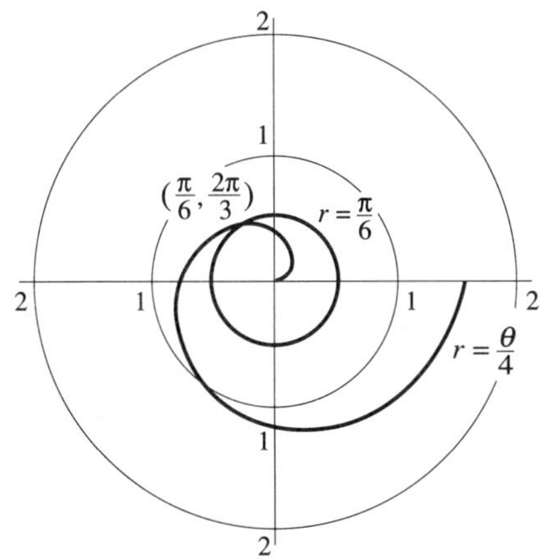

3. $r = \frac{1}{2}$; $r = \cos\theta$: By direct substitution we get $\cos\theta = \frac{1}{2}$ or $\theta = \frac{\pi}{3}$ and $\frac{5\pi}{3}$. This gives the two points of intersection $\left(\frac{1}{2}, \frac{\pi}{3}\right)$ and $\left(\frac{1}{2}, \frac{5\pi}{3}\right)$

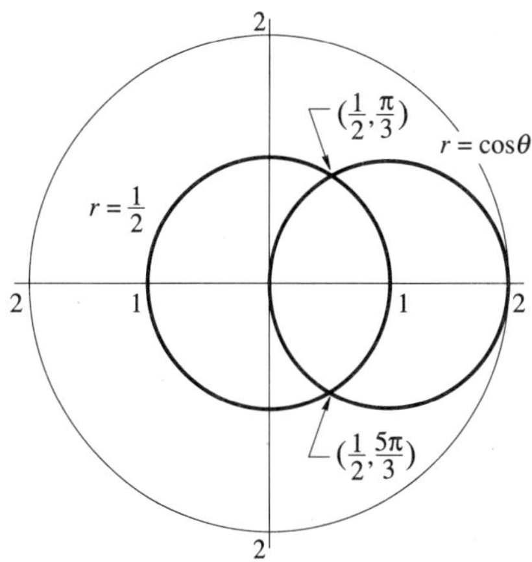

4. $r = 2 - 2\sin\theta, r = 2 - 2\cos\theta$: Direct substitution yields $2-2\sin\theta = 2-2\cos\theta$ or $\sin\theta = \cos\theta$. This is equivalent to $\tan\theta = 1$ or $\theta = \frac{\pi}{4}$ and $\frac{5\pi}{4}$. Hence the two points $\left(2 - \sqrt{2}, \frac{\pi}{4}\right)$ and $\left(2 + \sqrt{2}, \frac{5\pi}{4}\right)$. Looking at the graph we can see a third point, the pole. $(0, 0)$ also checks.

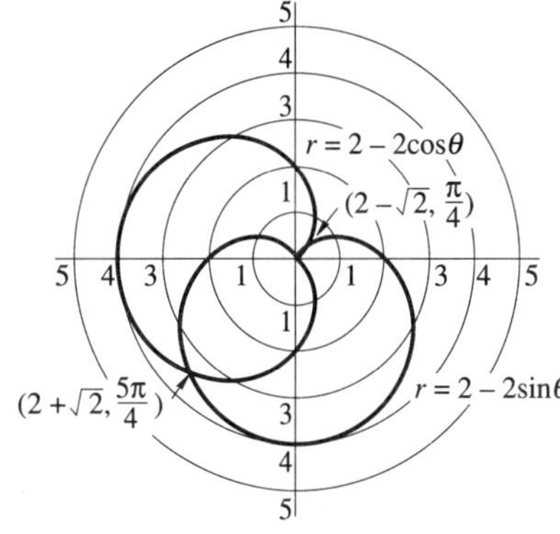

5. $r = 1 - \sin\theta$; $r = 1 + \cos\theta$: Direct substitution yields $1 - \sin\theta = 1 + \cos\theta$ or $-\sin\theta = \cos\theta$. This is equivalent to $\tan\theta = -1$. Hence we get $\theta = \frac{3\pi}{4}$ and $\frac{7\pi}{4}$ and the points $\left(1 - \frac{\sqrt{2}}{2}, \frac{3\pi}{4}\right)$ and $\left(1 + \frac{\sqrt{2}}{2}, \frac{7\pi}{4}\right)$. Looking at the graph we also wee that the pole $(0, 0)$ is a common point.

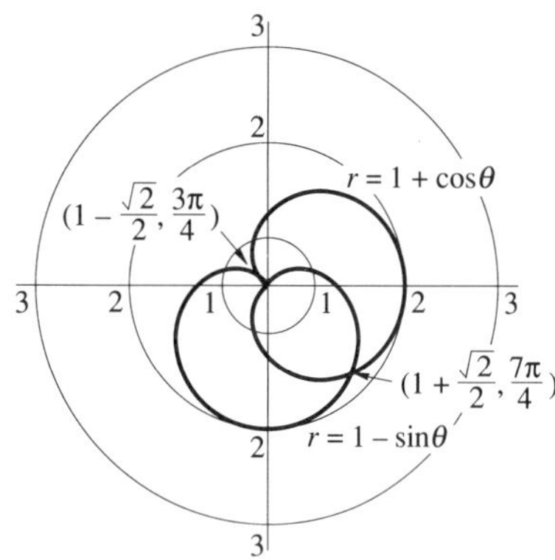

6. $r = -4 + 4\cos\theta$; $r = -4 + 4\sin\theta$. By direct substitution we get $-4 + 4\cos\theta = -4 + 4\sin\theta$ or $\cos\theta = \sin\theta$. This is true at $\theta = \frac{\pi}{4}$ and $\frac{5\pi}{4}$ giving the point $\left(-4 + 2\sqrt{2}, \frac{\pi}{4}\right)$ and $\left(-4 - 2\sqrt{2}, \frac{5\pi}{4}\right)$. Inspection of the graph and checking also gives the pole $(0, 0)$.

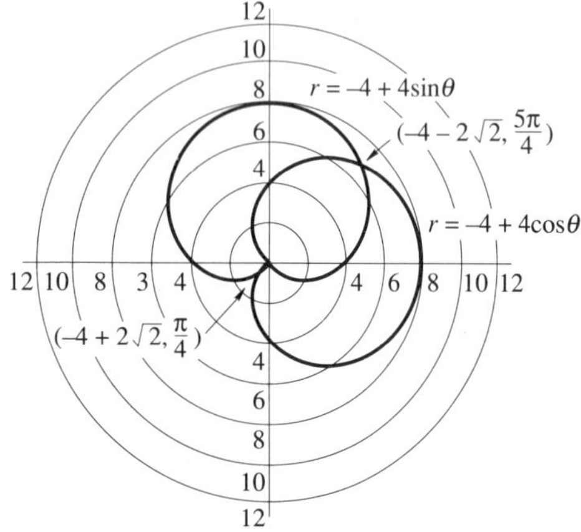

7. $r = \sqrt{3}$; $r = 2\cos\theta$. Direct substitution yields $2\cos\theta = \sqrt{3}$ or $\cos\theta = \frac{\sqrt{3}}{2}$. Hence we get $\theta = \frac{\pi}{6}, \frac{11\pi}{6}$ and the two points $\left(\sqrt{3}, \frac{\pi}{6}\right)$ and $\left(\sqrt{3}, \frac{11\pi}{6}\right)$.

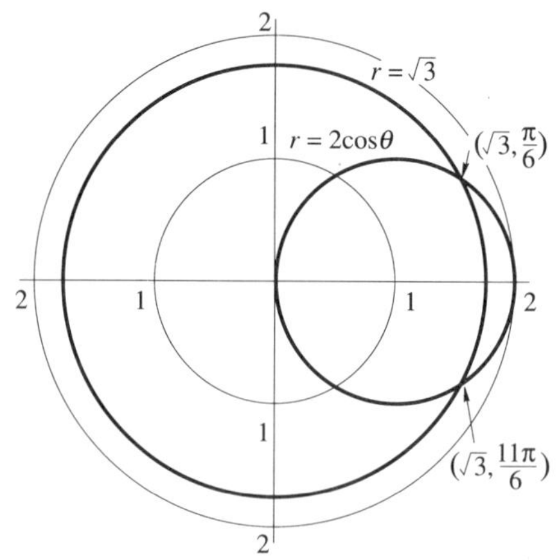

8. $r = \sin 2\theta$; $r = \sin\theta$. Direct substitution gives $\sin 2\theta = \sin\theta$. Using identities we get $2\sin\theta\cos\theta = \sin\theta$ or $2\sin\theta\cos\theta - \sin\theta = 0$. Factoring we get $\sin\theta(2\cos\theta - 1) = 0$. Hence $\sin\theta = 0$ and $\cos\theta = \frac{1}{2}$ or $\theta = 0, \pi, \frac{\pi}{3}$, or $\frac{5\pi}{3}$. This gives four points $(0, 0)$, $\left(\frac{\sqrt{3}}{2}, \frac{\pi}{3}\right)$, $(0, \pi)$, and $\left(\frac{-\sqrt{3}}{2}, \frac{5\pi}{3}\right)$. Since $(0, 0)$ and $(0, \pi)$ are the same we have only three points of intersection.

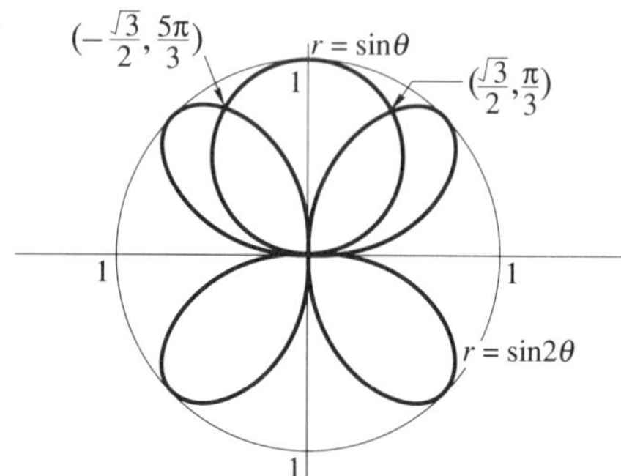

9. $r = \sin 2\theta$; $r = \sqrt{2}\sin\theta$: Direct substitution yields $\sin 2\theta = \sqrt{2}\sin\theta$ or using a trig identity $2\sin\theta\cos\theta = \sqrt{2}\sin\theta$. Hence $2\sin\theta\cos\theta - \sqrt{2}\sin\theta = 0$ or $\sin\theta(\cos\theta - \sqrt{2}) = 0$. Hence, $\sin\theta = 0 \Rightarrow \theta = 0$ or π and $\cos\theta = \frac{\sqrt{2}}{2} \Rightarrow \frac{\pi}{4}, \frac{7\pi}{4}$. Now, $\theta = 0$ and $\theta = \pi$ yield the same point $(0, 0)$, and so we have three solutions $(0, 0)$, $\left(1, \frac{\pi}{4}\right)$, and $\left(-1, \frac{7\pi}{4}\right)$.

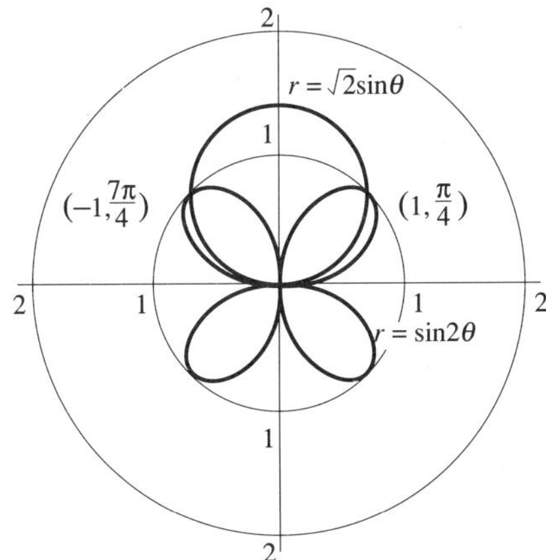

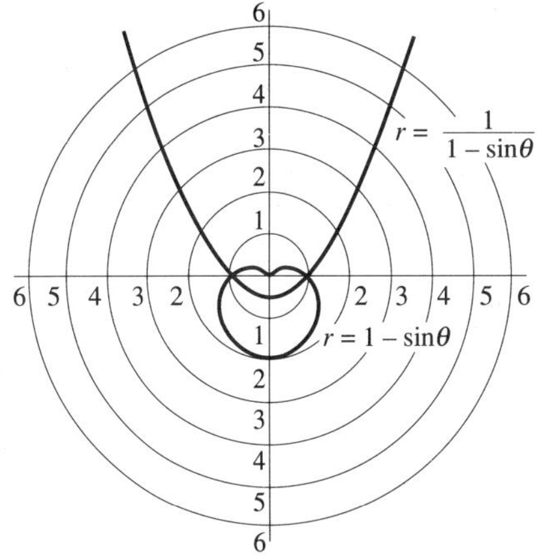

10. $r = 2 + 2\cos\theta$; $r = \dfrac{1}{1-\cos\theta}$. Direct substitution yields $2 + 2\cos\theta = \dfrac{1}{1-\cos\theta}$. Multiplying both sides by $1 - \cos\theta$ we get $2 - 2\cos^2\theta = 1$. Hence, $-2\cos^2\theta = -1$ or $\cos^2\theta = \frac{1}{2}$ and $\cos\theta = \pm\frac{\sqrt{2}}{2}$. We now have $\theta = \frac{\pi}{4}$, $\frac{3\pi}{4}$, $\frac{5\pi}{4}$, and $\frac{7\pi}{4}$ and the points $\left(2+\sqrt{2}, \frac{\pi}{4}\right)$, $\left(2-\sqrt{2}, \frac{3\pi}{4}\right)$, $\left(2-\sqrt{2}, \frac{5\pi}{4}\right)$, and $\left(2+\sqrt{2}, \frac{7\pi}{4}\right)$.

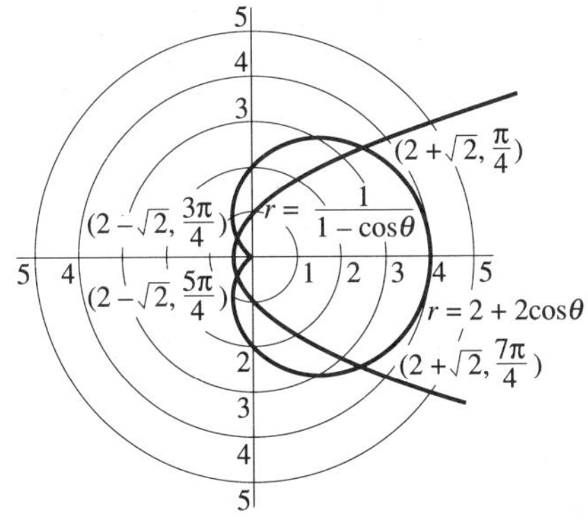

11. $r = 1 - \sin\theta$, $r = \dfrac{1}{1-\sin\theta}$. Direct substitution gives $1 - \sin\theta = \dfrac{1}{1-\sin\theta}$. Multiplying both sides by $1-\sin\theta$ we get $(1-\sin\theta)^2 = 1$ or $1-\sin\theta = \pm 1$; $-\sin\theta = -1 \pm 1$ or $\sin\theta = 1 \pm 1$; $\sin\theta = 0$ or 2. Since 2 is not in the range of the sine function, $\theta = 0$ or π. This gives the two points $(1, 0)$ and $(1, \pi)$.

12. $r = \sin 2\theta$; $r = \cos 2\theta$: Direct substitution yields $\sin 2\theta = \cos 2\theta$ or $\dfrac{\sin 2\theta}{\cos 2\theta} = \tan 2\theta = 1$. From 0 to 4π, $2\theta = \frac{\pi}{4}$, $\frac{5\pi}{4}$, $\frac{9\pi}{4}$, and $\frac{13\pi}{4}$ so θ is $\frac{\pi}{8}$, $\frac{5\pi}{8}$, $\frac{9\pi}{8}$, or $\frac{13\pi}{8}$. Hence the points $\left(\frac{\sqrt{2}}{2}, \frac{\pi}{8}\right)$, $\left(-\frac{\sqrt{2}}{2}, \frac{5\pi}{8}\right)$, $\left(\frac{\sqrt{2}}{2}, \frac{9\pi}{8}\right)$, and $\left(-\frac{\sqrt{2}}{2}, \frac{13\pi}{8}\right)$. Inspection of the graph we see four more points of intersection. By symmetry these are $\left(\frac{\sqrt{2}}{2}, \frac{3\pi}{8}\right)$, $\left(-\frac{\sqrt{2}}{2}, \frac{5\pi}{8}\right)$, $\left(\frac{\sqrt{2}}{2}, \frac{11\pi}{8}\right)$, and $\left(-\frac{\sqrt{2}}{2}, \frac{15\pi}{8}\right)$.

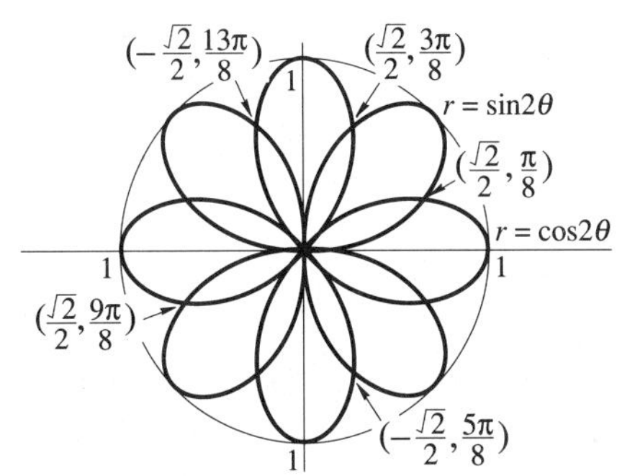

29.5 AREA IN POLAR COORDINATES

1. $r = 4\sin\theta$. This curve is complete when θ goes from 0 to π so the area is $\frac{1}{2}\int_0^{\pi}(4\sin\theta)\,d\theta =$ $\frac{1}{2}\int_0^{\pi}16\sin^2\theta\,d\theta = \frac{1}{2}\int_0^{\pi}16\frac{1-\cos 2\theta}{2}\,d\theta =$ $4\int_0^{\pi}(1-\cos 2\theta)\,d\theta = 4\left[\theta - \frac{1}{2}\sin\theta\right]_0^{\pi} =$ $4[\pi - 0 - 0 + 0] = 4\pi$

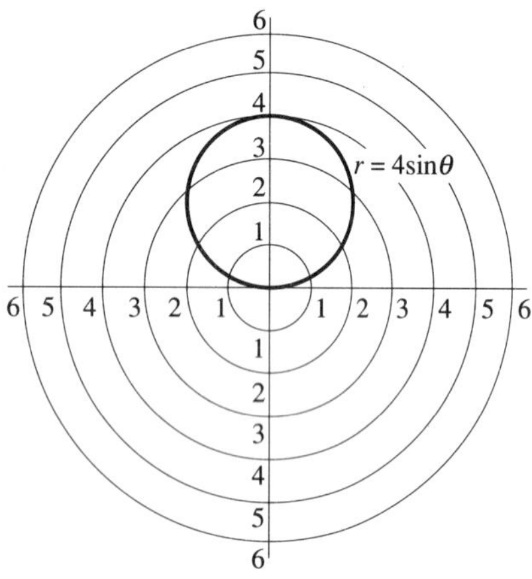

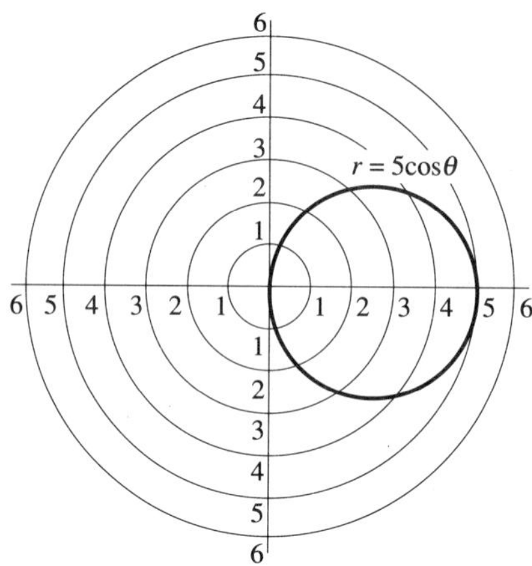

2. $r = 5\cos\theta$. This curve is completed from $\theta = 0$ to $\theta = \pi$, so the area is $\frac{1}{2}\int_0^{\pi}(5\cos\theta)^2\,d\theta =$ $\frac{1}{2}\int_0^{\pi}25\cos^2\theta\,d\theta = \frac{25}{2}\int_0^{\pi}\frac{1+\cos 2\theta}{2}\,d\theta =$ $\frac{25}{2}\left[\frac{1}{2}\theta + \frac{1}{4}\sin 2\theta\right]_0^{\pi} = \frac{25}{2}\left[\frac{1}{2}\pi + 0 - 0 - 0\right] =$ $\frac{25}{4}\pi$

3. $r = 1 - \cos\theta$. For this curve we need to integrate from 0 to 2π with the result $\frac{1}{2}\int_0^{2\pi}(1-\cos\theta)^2\,d\theta =$ $\frac{1}{2}\int_0^{2\pi}(1 - 2\cos\theta + \cos^2\theta)\,d\theta =$ $\frac{1}{2}\int_0^{2\pi}\left(1 - 2\cos\theta + \frac{1+\cos 2\theta}{2}\right)d\theta =$ $\frac{1}{2}\left[\frac{3}{2}\theta - 2\sin\theta + \frac{1}{4}\sin 2\theta\right]_0^{2\pi} =$ $\frac{1}{2}[3\pi - 0 + 0 - 0 + 0 - 0] = \frac{3}{2}\pi$

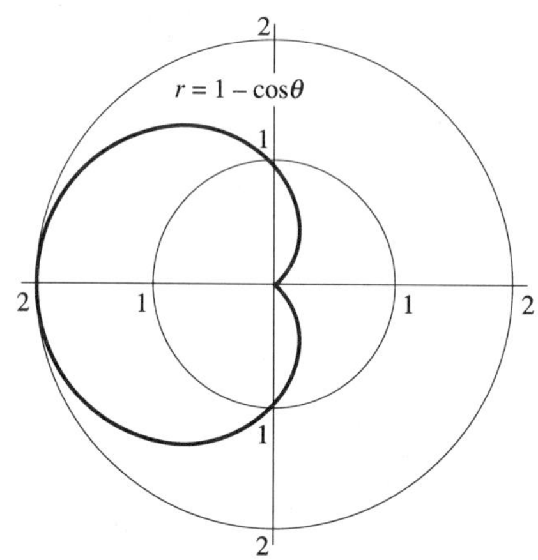

4. $r = 4 + 4\sin\theta$. Integrating from 0 to 2π getting

$$\begin{aligned}
&\frac{1}{2}\int_0^{4\pi}(4+4\sin\theta)^2\,d\theta \\
&= \frac{1}{2}\int_0^{2\pi}16(1+2\sin\theta+\sin^2\theta)\,d\theta \\
&= \frac{1}{2}\int_0^{2\pi}16(1+2\sin\theta+\frac{1-\cos 2\theta}{2})\,d\theta \\
&= \frac{1}{2}\int_0^{2\pi}(24+32\sin\theta-8\cos 2\theta)\,d\theta \\
&= \frac{1}{2}\left[24\theta-32\cos\theta-4\sin 2\theta\right]_0^{2\pi} \\
&= \frac{1}{2}\left[48\pi-32-0-0-(-32)+0\right] = 24\pi
\end{aligned}$$

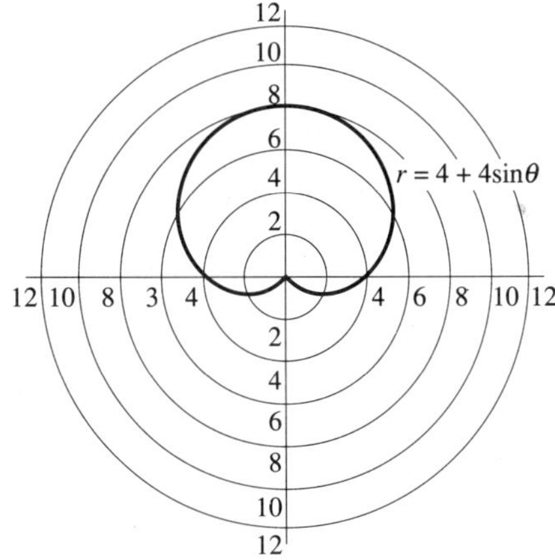

5. $r = \sin 2\theta$. This curve is a four leaf rose. By symmetry we can get the area of one leaf and multiply by 4. The result is $4\cdot\frac{1}{2}\int_0^{\pi/2}(\sin 2\theta)^2\,d\theta = 2\int_0^{\pi/2}\sin^2 2\theta\,d\theta = 2\int_0^{\pi/2}\frac{1-\cos 4\theta}{2}\,d\theta = \theta - \frac{1}{4}\sin 4\theta\Big|_0^{\pi/2} = \frac{\pi}{2} - 0 = \frac{\pi}{2}$

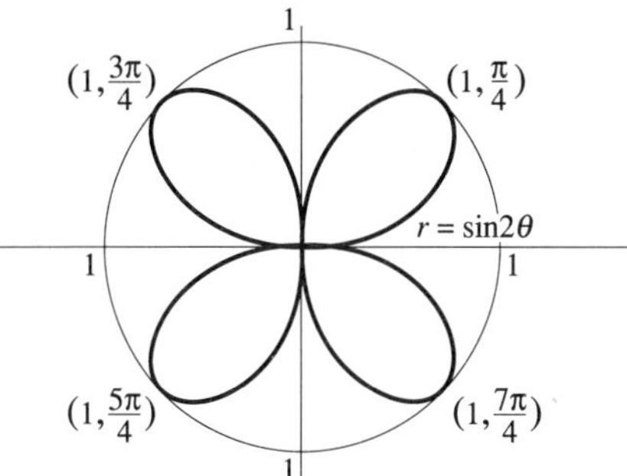

6. $r = \cos 3\theta$. This is a three leaf rose and we can get the area of the right leaf by integrating from $-\frac{\pi}{6}$ to $\frac{\pi}{6}$ and the whole area by multiplying this answer by 3. The result is $3\cdot\frac{1}{2}\int_{-\pi/6}^{\pi/6}(\cos 3\theta)^2\,d\theta = \frac{3}{2}\int_{-\pi/6}^{\pi/6}\cos^2 3\theta\cos\theta\,d\theta = \frac{3}{2}\int_{-\pi/6}^{\pi/6}\frac{1+\cos 6\theta}{2}\,d\theta = \frac{3}{2}\left[\frac{1}{2}\theta+\frac{1}{12}\sin 6\theta\right]_{-\pi/6}^{\pi/6} = \frac{3}{2}\cdot\left[\frac{\pi}{12}+0-\left(-\frac{\pi}{12}\right)-0\right] = \frac{3}{2}\cdot\frac{\pi}{6} = \frac{\pi}{4}$.

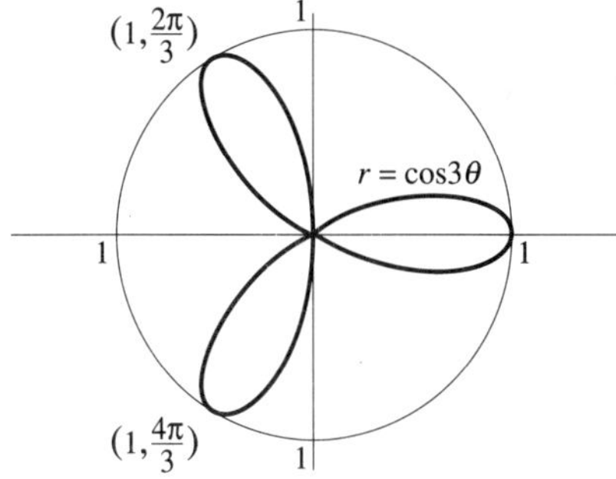

7. $r = 4 + \cos\theta$. Integrating from 0 to 2π, the area is

$$\begin{aligned}
&\frac{1}{2}\int_0^{2\pi}(4+\cos\theta)\,d\theta \\
&= \frac{1}{2}\int_0^{2\pi}(16+8\cos\theta+\cos^2\theta)\,d\theta \\
&= \frac{1}{2}\int_0^{2\pi}\left(16+8\cos\theta+\frac{1+\cos 2\theta}{2}\right)d\theta \\
&= \frac{1}{2}\left[16\theta+8\sin\theta+\frac{1}{2}\theta+\frac{1}{4}\sin 2\theta\right]_0^{2\pi} \\
&= \frac{1}{2}\left[32\pi+0+\pi+0\right] = \frac{33}{2}\pi
\end{aligned}$$

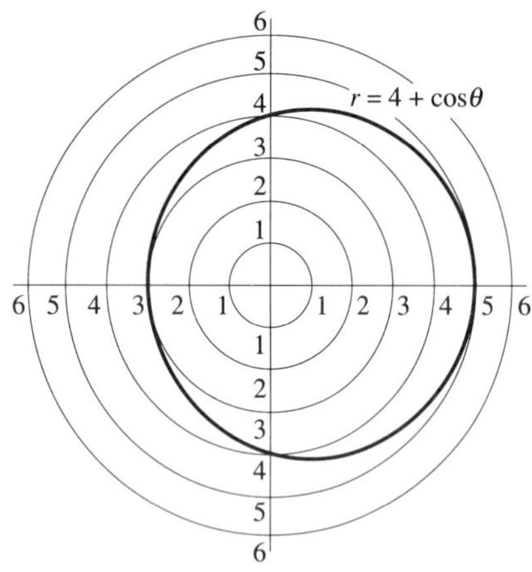

8. $r = 4 + 3\cos\theta$. By symmetry we can get the total area by doubling the area from $\theta = 0$ to $\theta = \pi$. $2 \cdot \frac{1}{2}\int_0^{\pi}(4 + 3\cos\theta)^2\,d\theta = \int_0^{\pi}(16 + 24\cos\theta + 9\cos^2\theta)\,d\theta = \int_0^{\pi}\left(16 + 24\cos\theta + 9\left(\frac{1+\cos 2\theta}{2}\right)\right)d\theta = \left[16\theta + 24\sin\theta + \frac{9}{2}\theta + \frac{9}{4}\sin 2\theta\right]_0^{\pi} = 16\pi + \frac{9}{2}\pi = \frac{41}{2}\pi$.

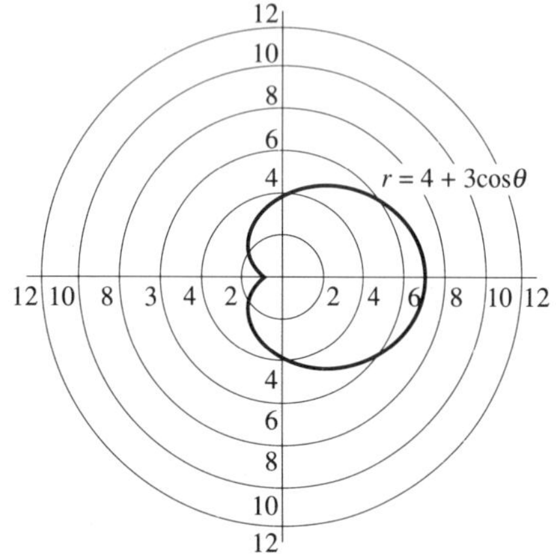

9. $r^2 = \sin\theta \Rightarrow r = \pm\sqrt{\sin\theta}$; $r = \sqrt{\sin\theta}$ gives the top half of the curve with $0 \le \theta \le \pi$. The bottom half is $r = -\sqrt{\sin\theta}$. By symmetry we can integrate $\sqrt{\sin\theta}$ from 0 to π and double the answer. $2 \cdot \frac{1}{2}\int_0^{\pi}\left(\sqrt{\sin\theta}\right)^2 d\theta = \int_0^{\pi}\sin\theta\,d\theta = -\cos\theta\big|_0^{\pi} = -[(-1) - 1] = 2$.

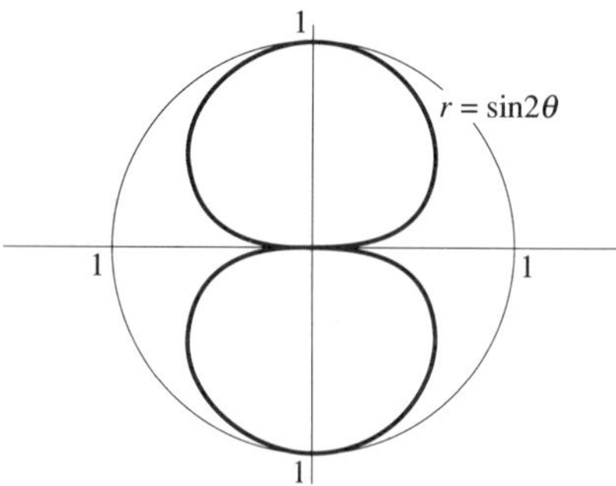

10. $r^2 = 4\cos 2\theta$ or $r = \pm\sqrt{4\cos 2\theta}$. This curve is undefined when $\cos 2\theta < 0$ or $\frac{\pi}{4} < \frac{3\pi}{4}$ and $\frac{5\pi}{4} < \theta < \frac{7\pi}{4}$. By symmetry we can integrate from 0 to $\frac{\pi}{4}$ and multiply the answer by 4. This produces $4 \cdot \frac{1}{2}\int_0^{\pi/4}\sqrt{4\cos 2\theta}^{\,2}\,d\theta = 2\int_0^{\pi/4} 4\cos 2\theta\,d\theta = 2 \cdot 2\sin 2\theta\big|_0^{\pi/4} = 4 \cdot 1 = 4$.

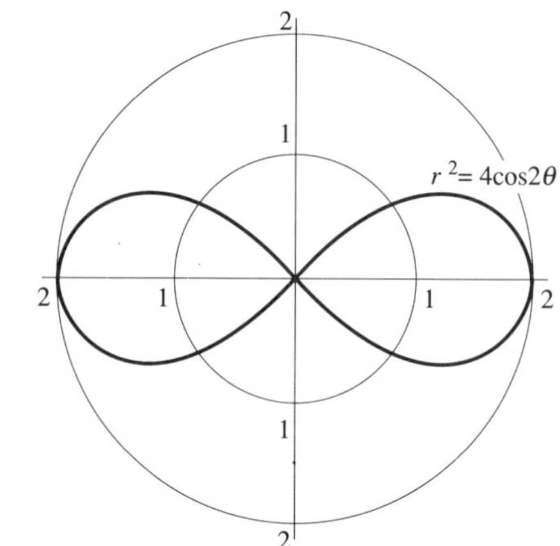

11. $\frac{1}{2}\int_0^{\pi/4}(3\cos\theta)^2\,d\theta = \frac{1}{2}\int_0^{\pi/4} 9\cos^2\theta\,d\theta = \frac{9}{2}\int_0^{\pi/4}\frac{1+\cos 2\theta}{2}\,d\theta = \frac{9}{2}\left[\frac{\theta}{2} + \frac{\sin 2\theta}{4}\right]_0^{\pi/4} = \frac{9}{3}\left[\frac{\pi}{8} + \frac{1}{4}\right] = \frac{9}{16}[\pi + 2] \approx 2.8921$.

12. $\frac{1}{2}\int_0^{\pi/3}(4\cos\theta)^2\,d\theta = \frac{1}{2}\int_0^{\pi/3}16\cos^2\theta\,d\theta = \frac{1}{2}\int_0^{\pi/3}8(1-\cos 2\theta)\,d\theta = \frac{1}{2}[8\theta - 4\sin 2\theta]_0^{\pi/3} = \frac{1}{2}\left[\frac{8\pi}{3} - 4\cdot\frac{\sqrt{3}}{2}\right] = \frac{4\pi}{3} - \sqrt{3} \approx 2.4567.$

13. $\frac{1}{2}\int_0^{\pi/4}(\cos 2\theta)^2\,d\theta = \frac{1}{2}\int_0^{\pi/4}\cos^2\theta\,d\theta = \frac{1}{2}\int_0^{\pi/4}\frac{1+\cos 4\theta}{2}\,d\theta = \frac{1}{2}\left[\frac{1}{2}\theta + \frac{1}{4}\sin 4\theta\right]_0^{\pi/4} = \frac{1}{2}\left[\frac{\pi}{8}\right] = \frac{\pi}{16} \approx 0.1963.$

14. $\frac{1}{2}\int_0^{\pi/8}(\sin 4\theta)^2\,d\theta = \frac{1}{2}\int_0^{\pi/8}\frac{1-\cos 8\theta}{2}\theta\,d\theta = \frac{1}{2}\left[\frac{1}{2}\theta - \frac{\sin 8\theta}{4}\right]_0^{\pi/8} = \frac{\pi}{32} \approx 0.0982.$

15. $\frac{1}{2}\int_0^{\pi/2}(e^{2\theta})^2\,d\theta = \frac{1}{2}\int_0^{\pi/2}e^{4\theta}\,d\theta = \frac{1}{2}\left[\frac{1}{4}e^{4\theta}\right]_0^{\pi/2} = \frac{1}{8}\left[e^{2\pi} - 1\right] \approx 66.8115$

16. $\frac{1}{2}\int_0^{\pi/6}(5\theta)^2\,d\theta = \frac{1}{2}\int_0^{\pi/6}25\theta^2\,d\theta = \frac{25}{2}\cdot\frac{1}{3}\theta^3\Big|_0^{\pi/6} = \frac{25}{6}\left(\frac{\pi}{6}\right)^3 = \frac{25\pi^3}{1296} \approx 0.5981$

17. To get the area A of the region enclosed by one loop of $r = 4\cos 2\theta$ we can integrate from $-\frac{\pi}{4}$ to $\frac{\pi}{4}$ with the result $A = \frac{1}{2}\int_{-\pi/4}^{\pi/4}(4\cos 2\theta)^2\,d\theta = 8\int_{-\pi/4}^{\pi/4}\frac{1+\cos 4\theta}{2}\,d\theta = 8\left[\frac{1}{2}\theta + \frac{\sin 4\theta}{8}\right]_{-\pi/4}^{\pi/4} = 8\left[\frac{\pi}{8} + \frac{\pi}{8}\right] = 2\pi.$

18. $r^2 = 4\cos 2\theta$; $r = \sqrt{4\cos 2\theta}$. The area of the region enclosed by one loop of this curve can be obtained by integrating from $-\frac{\pi}{4}$ to $\frac{\pi}{4}$. If A is the area, we obtain $\frac{1}{2}\int_{-\pi/4}^{\pi/4}\sqrt{4\cos 2\theta}^2\,d\theta = 2\int_{-\pi/4}^{\pi/4}\cos 2\theta\,d\theta = \sin 2\theta|_{-\pi/4}^{\pi/4} = 1 + 1 = 2.$

19. $r = 2\sin 3\theta$, and to get the area of the region enclosed by one loop if we integrate from 0 to $\frac{\pi}{3}$. Thus, if A is the area, we have $A = \frac{1}{2}\int_0^{\pi/3}(2\sin 3\theta)^2\,d\theta = \int_0^{\pi/3}2\sin 3\theta\,d\theta = \int_0^{\pi/3}(1-\cos 6\theta)\,d\theta = \left[\theta - \frac{\sin 6\theta}{6}\right]_0^{\pi/3} = \frac{\pi}{3}.$

20. $r = \sin 6\theta$. To get the area of the region enclosed by one loop we integrate from 0 to $\frac{\pi}{6}$. The result is that the area A is $A = \frac{1}{2}\int_0^{\pi/6}(\sin 6\theta)^2\,d\theta = \frac{1}{2}\int_0^{\pi/6}\frac{1-\cos 6\theta}{2}\,d\theta = \left[\frac{1}{4}\theta - \frac{\sin 6\theta}{24}\right]_0^{\pi/6} = \frac{\pi}{24}.$

21. Looking at the figure 29.23S we can determine that the graphs intersect at $(1, -\frac{\pi}{2})$ and $(1, \frac{\pi}{2})$ so we need to integrate as follows: $\frac{1}{2}\int_{-\pi/2}^{\pi/2}\left((1+\cos\theta)^2 - 1^2\right)d\theta = \frac{1}{2}\int_{-\pi/2}^{\pi/2}(1 + 2\cos\theta + \cos^2\theta - 1)\,d\theta = \frac{1}{2}\int_{-\pi/2}^{\pi/2}\left(2\cos\theta + \frac{1+\cos 2\theta}{2}\right)d\theta = \left[\sin\theta + \frac{1}{4}\theta + \frac{1}{8}\sin 2\theta\right]_{-\pi/2}^{\pi/2} = \left(1 + \frac{\pi}{8} + 0\right) - \left(-1 - \frac{\pi}{8} + 0\right) = 2 + \frac{\pi}{4} \approx 2.7854.$

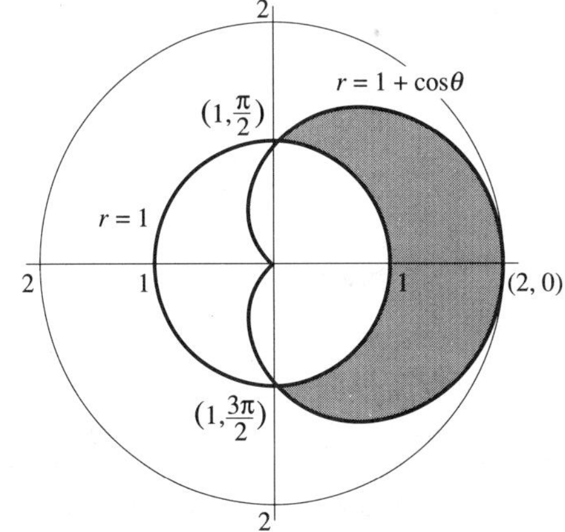

22. For this area we integrate from $\frac{\pi}{6}$ to $\frac{5\pi}{6}$. Thus, $A = \frac{1}{2}\int_{\pi/6}^{5\pi/6}\left((2\sin\theta)^2 - 1^2\right)d\theta =$ $\frac{1}{2}\int_{\pi/6}^{5\pi/6}(4\sin^2\theta - 1)\,d\theta =$ $\frac{1}{2}\int_{\pi/6}^{5\pi/6}(2(1-\cos 2\theta) - 1)\,d\theta =$ $\frac{1}{2}\int_{\pi/6}^{5\pi/6}(1-\cos 2\theta)\,d\theta = \left[\frac{1}{2}\theta - \frac{\sin 2\theta}{4}\right]_{\pi/6}^{5\pi/6} =$ $\frac{\pi}{12} + \frac{\sqrt{3}}{8} - \frac{\pi}{12} + \frac{\sqrt{3}}{8} = \frac{\pi}{3} + \frac{\sqrt{3}}{4} \approx 1.4802.$

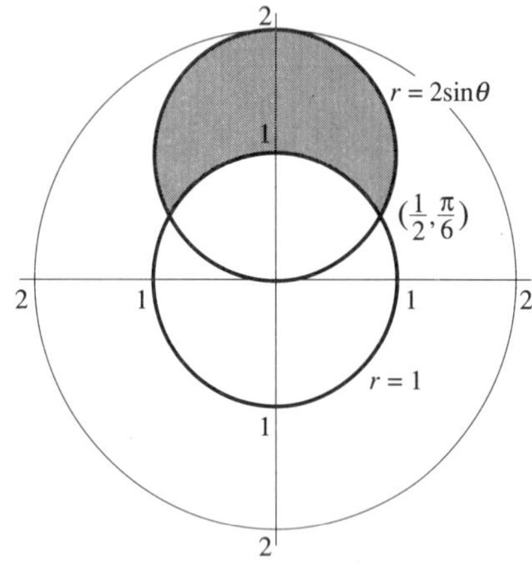

23. First we need to solve for θ. Using direct substitution we get $1 - \sin\theta = 2\cos\theta$. Squaring both sides we have $1 - 2\sin\theta + \sin^2\theta = 4\cos^2\theta$. Since $\cos^2\theta = 1 - \sin^2\theta$ we have $1 - 2\sin\theta + \sin^2\theta = 4 - 4\sin^2\theta$ or $0 = 3 + 2\sin\theta - 5\sin^2\theta$. This factors into $(3 + 5\sin\theta)(1 - \sin\theta) = 0$. Hence $\sin\theta = 1$ or $\sin\theta = -\frac{3}{5}$, and so $\theta = \frac{\pi}{2}, \frac{3\pi}{2}$, 3.78509, or 5.63968. Checking we see that $\frac{\pi}{2}$ and 5.63968 are the answers we need. Let $5.63968 = T$, Then,

$$\frac{1}{2}\int_{\pi/6}^{T}\left((1-\sin\theta)^2 - (2\cos\theta)^2\right)d\theta$$

$$= \frac{1}{2}\int_{\pi/2}^{T}(1 - 2\sin\theta + \sin^2\theta - 4\cos^2\theta)\,d\theta$$

$$= \frac{1}{2}\int_{\pi/2}^{T}\left(1 - 2\sin\theta + \sin^2\theta - 4 + 4\sin^2\theta\right)d\theta$$

$$= \frac{1}{2}\int_{\pi/2}^{T}\left[-3 - 2\sin\theta + 5\left(\frac{1-\cos 2\theta}{2}\right)\right]d\theta$$

$$= \frac{1}{2}\left[-3\theta + 2\cos\theta + \frac{5}{2}\theta - \frac{5\sin 2\theta}{4}\right]_{\pi/2}^{T}$$

$$= \left[-\frac{1}{4}\theta + \cos\theta - \frac{5\sin 2\theta}{8}\right]_{\pi/2}^{T}$$

$$\approx -0.009921 + 0.392699 = 0.382778$$

So, the desired area is about 0.3828.

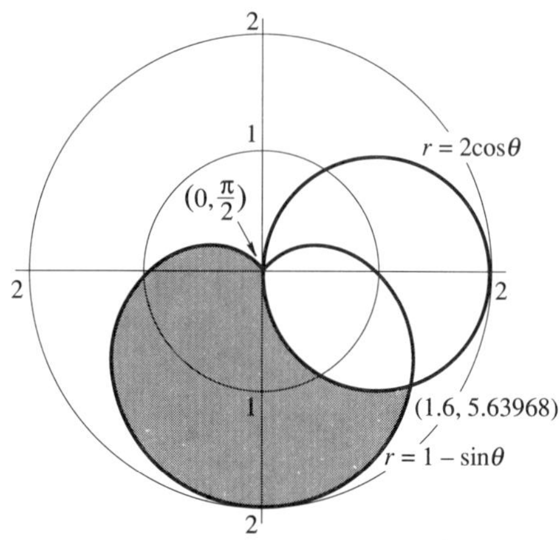

24. These two curves intersect at $\theta = \frac{\pi}{6}, \frac{5\pi}{6}, \frac{7\pi}{6}$, and $\frac{11\pi}{6}$. By symmetry we can integrate from $\frac{\pi}{6}$ to $\frac{5\pi}{6}$ and double the answer. This gives $2 \cdot \frac{1}{2}\int_{\pi/6}^{5\pi/6}\left[\sqrt{(8\sin\theta)}^2 - 2^2\right]d\theta =$ $\int_{\pi/6}^{5\pi/6}(8\sin\theta - 4)\,d\theta = [-8\cos\theta - 4\theta]_{\pi/6}^{5\pi/6} =$ $-8\left(-\frac{\sqrt{3}}{2}\right) - \frac{20\pi}{6} + 8\frac{\sqrt{3}}{2} + \frac{4\pi}{6} = 8\sqrt{3} - \frac{8\pi}{3} \approx$ 5.4788.

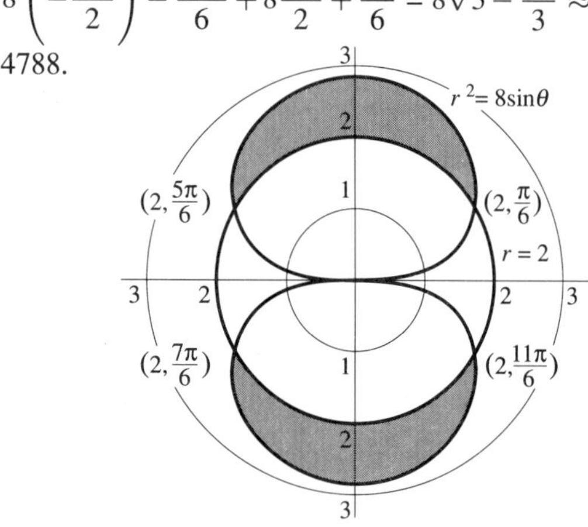

25. $\frac{1}{2}\int_0^{\pi/2}\left(\cos^2\theta-(1-\sin\theta)^2\right)d\theta=\frac{1}{2}\int_0^{\pi/2}(\cos^2\theta-1+2\sin\theta-\sin^2\theta)\,d\theta=\int_0^{\pi/2}(\cos 2\theta-1+2\sin\theta)\,d\theta=\frac{1}{2}\left[-\theta+\frac{\sin 2\theta}{2}-2\cos\theta\right]_0^{\pi/2}=\frac{1}{2}\left[-\frac{\pi}{2}+0-0-0-0+2\right]=-\frac{\pi}{4}+1\approx 0.2146.$

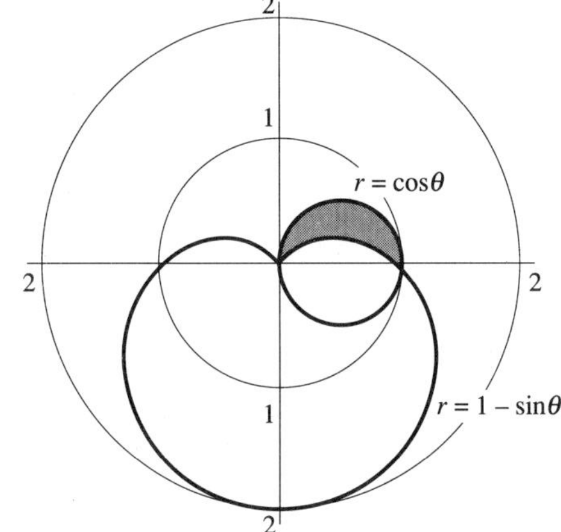

26. The area is $\frac{1}{2}\int_0^{\pi/3}\sin^2\theta\,d\theta+$

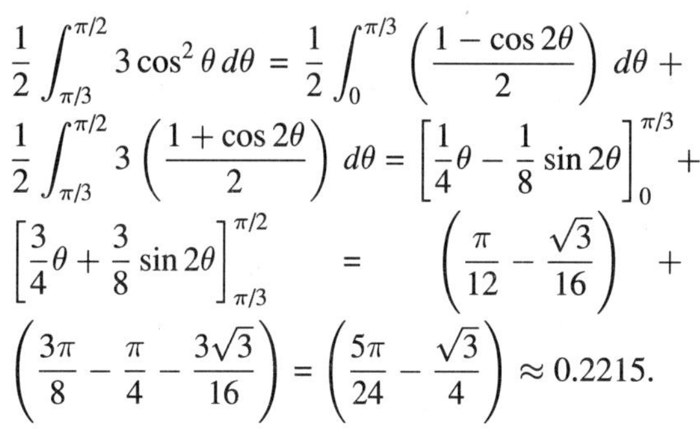

$\frac{1}{2}\int_{\pi/3}^{\pi/2}3\cos^2\theta\,d\theta=\frac{1}{2}\int_0^{\pi/3}\left(\frac{1-\cos 2\theta}{2}\right)d\theta+\frac{1}{2}\int_{\pi/3}^{\pi/2}3\left(\frac{1+\cos 2\theta}{2}\right)d\theta=\left[\frac{1}{4}\theta-\frac{1}{8}\sin 2\theta\right]_0^{\pi/3}+\left[\frac{3}{4}\theta+\frac{3}{8}\sin 2\theta\right]_{\pi/3}^{\pi/2}=\left(\frac{\pi}{12}-\frac{\sqrt{3}}{16}\right)+\left(\frac{3\pi}{8}-\frac{\pi}{4}-\frac{3\sqrt{3}}{16}\right)=\left(\frac{5\pi}{24}-\frac{\sqrt{3}}{4}\right)\approx 0.2215.$

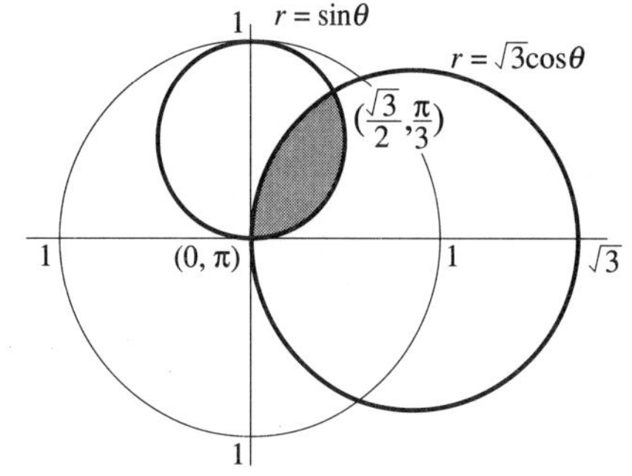

CHAPTER 29 REVIEW

1. $x=t^2,\ y=t^3+t.$ (a) $\frac{dy}{dx}=\frac{dy/dt}{dx/dt}=\frac{3t^2+1}{2t}$; $\frac{d^2y}{dx^2}=\frac{2t(6t)-(3t^2+1)2}{(2t)^3}=\frac{12t^2-6t^2-2}{(2t)^3}=\frac{6t^2-2}{2(4t^3)}=\frac{3t^2-1}{4t^3}$ (b) At $t=1$, $\frac{dy}{dx}=\frac{3+1}{2}=2$.

2. (a) $\frac{dy}{dx}=\frac{dy/dt}{dx/dt}=\frac{-2t^{-2}}{-1t^{-2}}=2$. Since $\frac{dy}{dx}$ is a constant, $\frac{d^2y}{dx^2}=0$, (b) At $t=1$, $\frac{dy}{dx}=2$.

3. (a) $\frac{dy}{dx}=\frac{\cos t}{2\cos t(-\sin t)}=-\frac{1}{2}\csc t$, and the second derivative is $\frac{d^2y}{dx^2}=\frac{\frac{1}{2}\csc t\cot t}{-2\cos t\sin t}=-\frac{1}{4}\csc^3 t$

(b) At $t=1$, $\frac{dy}{dx}=-\frac{1}{2}\csc 1\approx -0.5942$

4. (a) $\frac{dy}{dx}=\frac{-e^{-t}}{\frac{1}{2}t^{-1/2}}=-2\sqrt{t}e^{-t}$; $\frac{d^2y}{dx^2}=\frac{2\sqrt{t}e^{-t}-t^{-1/2}e^{-t}}{\frac{1}{2}t^{-1/2}}=4te^{-t}-2e^{-t}=(4t-2)e^{-t}$

(b) At $t=1$, $\frac{dy}{dx}=-2e^{-1}\approx -0.7358$

5. (a)

$$\frac{dy}{dx}=\frac{-2\sin\theta\sin\theta+(3+2\cos\theta)\cos\theta}{-2\sin\theta\cos\theta-(3+2\cos\theta)\sin\theta}$$
$$=\frac{-2\sin^2\theta+3\cos\theta+2\cos^2\theta}{-2\sin\theta\cos\theta-3\sin\theta-2\cos\theta\sin\theta}$$

$$= \frac{2(\cos^2\theta - \sin^2\theta) + 3\cos\theta}{-4\sin\theta\cos\theta - 3\sin\theta}$$
$$= \frac{2\cos 2\theta + 3\cos\theta}{-2\sin 2\theta - 3\sin\theta}$$
$$= -\frac{2\cos 2\theta + 3\cos\theta}{2\sin 2\theta + 3\sin\theta}$$

(b) If $\theta = \frac{\pi}{4}$, then

$$\frac{dy}{dx} = -\frac{2\cos\frac{\pi}{2} + 3\cos\frac{\pi}{4}}{2\sin\frac{\pi}{2} + 3\sin\frac{\pi}{4}}$$
$$= -\frac{2\cdot 0 + \frac{3\sqrt{2}}{2}}{2\cdot 1 + \frac{3\sqrt{2}}{2}} = -\frac{3\sqrt{2}}{4 + 3\sqrt{2}} \approx -0.5147.$$

6. (a)

$$\frac{dy}{dx} = \frac{-12\sin 2\theta\sin\theta + 6\cos 2\theta\cos\theta}{-12\sin 2\theta\cos\theta - 6\cos 2\theta\sin\theta}$$
$$= \frac{2\sin 2\theta\sin\theta - \cos 2\theta\cos\theta}{2\sin 2\theta\cos\theta + \cos 2\theta\sin\theta}$$
$$= \frac{2(2\sin\theta\cos\theta)\sin\theta - (\cos^2\theta - \sin^2\theta)\cos\theta}{2(\sin\theta\cos\theta)\cos\theta + (\cos^2\theta - \sin^2\theta)\sin\theta}$$
$$= \frac{\cos\theta(4\sin^2\theta - \cos^2\theta + \sin^2\theta}{\sin\theta(4\cos^2\theta + \cos^2\theta - \sin^2\theta}$$
$$= \frac{\cos\theta(5\sin^2\theta - \cos^2\theta)}{\sin\theta(5\cos^2\theta - \sin^2\theta)}$$

(b) At $\theta = \dfrac{\pi}{4}, \dfrac{dy}{dx}$ is $\dfrac{\cos\frac{\pi}{4}\left(5\sin^2\frac{\pi}{4} - \cos^2\frac{\pi}{4}\right)}{\sin\frac{\pi}{4}\left(5\cos^2\frac{\pi}{4} - \sin^2\frac{\pi}{4}\right)} = 1$

7. $r^2 = \sin\theta$; $r = (\sin\theta)^{1/2}$; $r' = \frac{1}{2}(\sin\theta)^{-1/2}\cos\theta$

(a)

$$\frac{dy}{dx} = \frac{\frac{1}{2}\sin^{-1/2}\theta\cos\theta\sin\theta + \sin^{1/2}\theta\cos\theta}{\frac{1}{2}\sin^{-1/2}\theta\cos\theta\cos\theta - \sin^{1/2}\theta\sin\theta}$$
$$= \frac{\cos\theta\sin\theta + 2\sin\theta\cos\theta}{\cos^2\theta - 2\sin^2\theta}$$
$$= \frac{3\cos\theta\sin\theta}{\cos^2\theta - 2\sin^2\theta}$$

(b) At $\theta = \frac{\pi}{4}$ we have $\dfrac{dy}{dx} = \dfrac{3\cdot\frac{\sqrt{2}}{2}\frac{\sqrt{2}}{2}}{\left(\frac{\sqrt{2}}{2}\right)^2 - 2\left(\frac{\sqrt{2}}{2}\right)^2} =$

$$\frac{\frac{3}{2}}{-\frac{1}{2}} = -3$$

8. (a) If $r = \frac{8}{3+\cos\theta} = 8(3+\cos\theta)^{-1}$, then $r' = -8(3+\cos\theta)^{-2}(-\sin\theta) = \dfrac{8\sin\theta}{(3+\cos\theta)^2}$. So,

$$\frac{dy}{dx} = \frac{\frac{8\sin\theta}{(3+\cos\theta)^2}\sin\theta + \frac{8}{3+\cos\theta}\cos\theta}{\frac{8\sin\theta}{(3+\cos\theta)^2}\cos\theta - \frac{8}{3+\cos\theta}\sin\theta}$$
$$= \frac{8\sin^2\theta + 8\cos\theta(3+\cos\theta)}{8\sin\theta\cos\theta - 8\sin\theta(3+\cos\theta)}$$
$$= \frac{\sin^2\theta + 3\cos\theta + \cos^2\theta}{\sin\theta\cos\theta - 3\sin\theta - \sin\theta\cos\theta}$$
$$= \frac{1 + 3\cos\theta}{-3\sin\theta}.$$

(b) At $\theta = \frac{\pi}{4}$, then $\dfrac{dy}{dx} = \dfrac{1 + 3\cdot\frac{\sqrt{2}}{2}}{-3\cdot\frac{\sqrt{2}}{2}} \approx -1.4714.$

9. $\dfrac{dx}{dt} = 2t$; $\dfrac{dy}{dt} = 3t^2$ and $L = \displaystyle\int_0^2 \sqrt{(2t)^2 + (3t^2)^2}\,dt = \int_0^2 \sqrt{4t^2 + 9t^4}\,dt = \int_0^2 t\sqrt{4 + 9t^2}\,dt$. Let $u = 4 + 9t^2$, and then $du = 18t\,dt$, so we get $\frac{1}{18}\int u^{1/2}\,du = \frac{1}{16}\cdot\frac{2}{5}u^{3/2} = \frac{1}{27}\left(4 + 9t^2\right)^{3/2}\Big|_0^2 = \frac{1}{27}\left(40^{3/2} - 4^{3/2}\right) \approx 9.0734.$

10. $\dfrac{dx}{dt} = 2t$; $\dfrac{dy}{dt} = 2$, Thus, $L = \int_0^3 \sqrt{(2t)^2 + 2^2}\,dt = \int_0^3 \sqrt{4t^2 + 4}\,dt = 2\int_0^3 \sqrt{t^2 + 1}\,dt = 2\left[\frac{t}{2}\sqrt{t^2+1} + \frac{1}{2}\ln\left|t + \sqrt{t^2+1}\right|\right]_0^3 = \left(3\sqrt{10} + \ln\left|3 + \sqrt{10}\right|\right) \approx 11.3053.$

11. $\dfrac{dx}{dt} = 4\cos t$; $\dfrac{dy}{dt} = (-4\sin t)$, and as a result, $L = \int_0^{\pi/2} \sqrt{(4\cos t)^2 + (-4\sin t)^2}\,dt = \int_0^{\pi/2} 4\,dt = 4t\Big|_0^{\pi/2} = 2\pi.$

12. $r = e^{2\theta}$; $r' = 2e^{2\theta}$, then $L = \displaystyle\int_0^{\pi} \sqrt{(e^{2\theta})^2 + (2e^{2\theta})}\,d\theta = \int_0^{\pi} \sqrt{e^{4\theta} + 4e^{4\theta}}\,d\theta =$

$\sqrt{5}\int_0^{\pi} e^{2\theta} = \frac{\sqrt{5}}{2}e^{2\theta}\Big|_0^{\pi} = \frac{\sqrt{5}}{2}(e^{2\pi} - 1) \approx$ 597.5798.

13. Since $r = \cos^2\left(\frac{\theta}{2}\right)$, we have $r' = 2\cos\frac{\theta}{2}\left(-\frac{1}{2}\sin\frac{\theta}{2}\right) = -\cos\frac{\theta}{2}\sin\frac{\theta}{2}$. Thus, $L = \int_0^{\pi}\sqrt{\cos^2\frac{\theta}{2} + \cos^2\frac{\theta}{2}\sin^2\frac{\theta}{2}}\,d\theta = \int_0^{\pi}\cos\frac{\theta}{2}\,d\theta = 2\sin\frac{\theta}{2}\Big|_0^{\pi} = 2\cdot\sin\frac{\pi}{2} = 2.$

14. $r = 1 - \cos\theta$ so $r' = \sin\theta$

$$L = \int_{-\pi}^{0}\sqrt{(1-\cos\theta)^2 + \sin^2\theta}\,d\theta$$
$$= \int_{-\pi}^{0}\sqrt{1 - 2\cos\theta + \cos^2\theta + \sin^2\theta}\,d\theta$$
$$= \int_{-\pi}^{0}\sqrt{2 - 2\cos\theta} = \int_{-\pi}^{0} 2\sqrt{\frac{1-\cos\theta}{2}}\,d\theta$$
$$= 2\int_{-\pi}^{0}\sin\frac{\theta}{2}\,d\theta = -4\cos\frac{\theta}{2}\Big|_{-\pi}^{0} = 4$$

15. $\frac{dx}{dt} = 2t;\ \frac{dy}{dt} = 1$

$$S = 2\pi\int_0^2 t\sqrt{(2t)^2 + 1^2}\,dt = 2\pi\int_0^2 t\sqrt{4t^2+1}\,dt.$$

Let $u = 4t^2 + 1$, and then $du = 8t\,dt$ or $\frac{1}{8}\,du = t\,dt$. Substituting we get $\frac{1}{4}\pi\int\sqrt{u}\,du = \frac{\pi}{4}\cdot\frac{2}{3}u^{3/2} = \frac{\pi}{6}(4t^2+1)^{3/2}\big|_0^2 = \frac{\pi}{6}(17^{3/2} - 1) \approx 36.1769.$

16. $\frac{dx}{dt} = 2t$ and $\frac{dy}{dt} = (1 - t^2).$

$$S = 2\pi\int_0^1\left(t - \frac{t^3}{3}\right)\sqrt{(2t)^2 + (1-t^2)^2}\,dt$$
$$= 2\pi\int_0^1\left(t - \frac{t^3}{3}\right)\sqrt{4t^2 + 1 - 2t^2 + t^4}\,dt$$
$$= 2\pi\int_0^1\left(t - \frac{t^3}{3}\right)\sqrt{t^4 + 2t^2 + 1}\,dt$$
$$= 2\pi\int_0^1\left(t - \frac{t^3}{3}\right)(t^2+1)\,dt$$
$$= 2\pi\int_0^1\left(t^3 - \frac{t^3}{3} - \frac{t^5}{3} + t\right)dt$$
$$= 2\pi\left[\frac{2}{3}\cdot\frac{1}{4}t^4 - \frac{1}{3}\frac{1}{6}t^6 + \frac{t^2}{2}\right]_0^1 = 2\pi\left(\frac{1}{6} - \frac{1}{18} + \frac{1}{2}\right)$$
$$= 2\pi\left(\frac{3 - 1 + 9}{18}\right) = \frac{11\pi}{9} \approx 3.8397$$

17. Here $\frac{dx}{dt} = 2t$ and $\frac{dy}{dt} = 3t^2$, so the surface area is $S = 2\pi\int_0^2 t^2\sqrt{(2t)^2 + (3+2)^2}\,dt = 2\pi\int_0^2 t^2\sqrt{4t^2 + 9t^4}\,dt = 2\pi\int_0^2 t^3\sqrt{4 + 9t^2}\,dt.$ Let $u = 4 + 9t^2$, then $t^2 = \frac{u-4}{9}$ and $du = 18t\,dt$ or $\frac{1}{18}du = t\,dt.$ Substituting, we get $2\pi\cdot\frac{1}{18}\int\left(\frac{u-4}{9}\right)u^{1/2}\,du = \frac{\pi}{81}\int\left(u^{3/2} - 4u^{1/2}\right)du = \frac{\pi}{81}\left(\frac{2}{5}u^{5/2} - 4\frac{2}{3}u^{3/2}\right) = \frac{\pi}{81}\left[\frac{2}{5}(4+9t^2)^{5/2} - \frac{8}{3}(u + 9t^2)^{3/2}\right]_0^2 = \frac{\pi}{81}\left[\frac{2}{5}40^{5/2} - \frac{8}{3}40^{3/2} - \frac{64}{5} + \frac{64}{3}\right] \approx 41.7485\pi \approx 131.1568.$

18. Since $x = e^t\sin t$, then $dx = (e^t\cos t + e^t\sin t)\,dt$ and since $y = e^t\cos$, then $dy = (e^t\cos t - e^t\sin t)\,dt$. Thus, the desired surface area is given by

$$S = 2\pi\int_0^{\pi/2} e^t\sin t\sqrt{\begin{matrix}(e^t\cos t + e^t\sin t)^2 + \\ (e^t\cos t - e^t\sin t)^2\end{matrix}}\,dt$$
$$= 2\pi\int_0^{\pi/2} e^t\sin t\sqrt{2e^{2t}\cos^2 t + 2e^{2t}\sin^2 t}\,dt$$
$$= 2\pi\int_0^{\pi/2} e^t\sin t\sqrt{2}e^t\,dt = 2\sqrt{2}\pi\int_0^{\pi/2} e^{2t}\sin t$$

By formula #87, Appendix C, this integral is $2\sqrt{2}\pi\left(\frac{e^{2t}}{2^2+1^2}(2\sin t - \cos t)\right)\Big|_0^{\pi/2} = 2\sqrt{2}\pi\left(\frac{e^{\pi}}{5}\cdot 2 + \frac{1}{5}\right) = \frac{2\sqrt{2}\pi}{5}(2e^{\pi} + 1) \approx$ 84.0263.

19. $r = 6\sin\theta$ and $r' = 6\cos\theta$; $y = 6\sin^2\theta$. Hence, we find that the desired surface area is $S = 2\pi\int_0^\pi \sqrt{36\sin^2\theta + 36\cos^2\theta}(6\sin^2 6)\,dt = 2\pi\int_0^\pi 36\sin^2\theta\,d\theta = 36\pi\int_0^\pi (1-\cos 2\theta)\,d\theta = 36\pi\left[\theta - \dfrac{\sin 2\theta}{2}\right]_0^\pi = 36\pi^2 \approx 355.3058.$

20. Here $r = 4 + 4\cos\theta$, $r' = -4\sin\theta$, and $y = 4\sin\theta + 4\sin\theta\cos\theta$. Hence, the desired surface area is

$$S = 2\pi\int_0^\pi (4\sin\theta + 4\sin\theta\cos\theta) \times \sqrt{(4+4\cos\theta)^2 + (4\sin\theta)^2}\,d\theta$$
$$= 2\pi\int_0^\pi (4\sin\theta + 4\sin\theta\cos\theta) \times 4\sqrt{1 + 2\cos\theta + \cos^2\theta + \sin^2\theta}\,d\theta$$
$$= 32\pi\int_0^\pi (\sin\theta + \sin\theta\cos\theta)\sqrt{2 + 2\cos\theta}\,d\theta$$
$$= 32\sqrt{2}\pi\int_0^\pi (1+\cos\theta)\sqrt{1+\cos\theta}\sin\theta\,d\theta$$
$$= 32\sqrt{2}\pi\int_0^\pi (1+\cos\theta)^{3/2}\sin\theta\,d\theta$$

Let $u = 1 + \cos\theta$ and $du = -\sin\theta$. Substituting, we get $-32\sqrt{2}\pi\left[\int u^{3/2}\,du\right] = -32\sqrt{2}\pi\left[\frac{2}{5}u^{5/2}\right] = -32\sqrt{2}\pi\left[\frac{2}{5}(1+\cos\theta)^{5/2}\right]_0^\pi = \frac{-64\sqrt{2}}{5}(0 - 2^{5/2}) = \frac{64\sqrt{2}\pi}{5}(4\sqrt{2}) = \frac{2^9\pi}{5} \approx 102.4\pi \approx 321.6991.$

21. By direct substitution, $4 = 4 + 4\sin\theta$; $0 = \sin\theta$ so $\theta = 0$ or π; $(4, 0)$, $(4, \pi)$.

22. By direct substitution, $3 = 6\sin\theta$; $\sin\theta = \frac{1}{2}$; $\theta = \frac{\pi}{6}, \frac{5\pi}{6}$; $\left(3, \frac{\pi}{6}\right)$, $\left(3, \frac{5\pi}{6}\right)$.

23. By direct substitution, $4 + 4\cos\theta = 2$; $4\cos\theta = -2$; $\cos\theta = -\frac{1}{2}$; $\theta = \frac{2\pi}{3}, \frac{4\pi}{3}$. $\left(2, \frac{2\pi}{3}\right)$, $\left(2, \frac{4\pi}{3}\right)$.

24. By direct substitution, $2\sin 2\theta = 1$; $\sin 2\theta = \frac{1}{2}$, and so $2\theta = \frac{\pi}{6}, \frac{5\pi}{6}, \frac{13\pi}{6}$, and $\frac{17\pi}{6}$. Thus, $\theta = \frac{\pi}{12}, \frac{5\pi}{12}, \frac{13\pi}{12}$, and $\frac{17\pi}{12}$. Graphing yields symmetric points at $\frac{-\pi}{12}, \frac{-5\pi}{12}, \frac{-13\pi}{12}$, and $\frac{-17\pi}{12}$. The resulting points are $\left(1, \frac{\pi}{12}\right)$, $\left(1, \frac{5\pi}{12}\right)$, etc.

25. The area is $A = \frac{1}{2}\int_0^{2\pi}(1+\cos\theta)^2\,d\theta = \frac{1}{2}\int_0^{2\pi}(1 + 2\cos\theta + \cos^2\theta)\,d\theta = \frac{1}{2}\int_0^{2\pi}\left(1 + 2\cos\theta + \frac{1+\cos\theta}{2}\right)d\theta = \frac{1}{2}\left[\frac{3}{2}\theta + 2\sin\theta + \frac{\sin 2\theta}{4}\right]_0^{3\pi} = \frac{3\pi}{2}.$

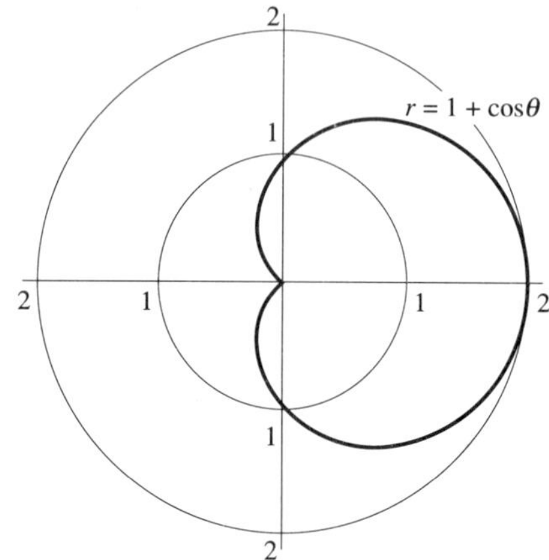

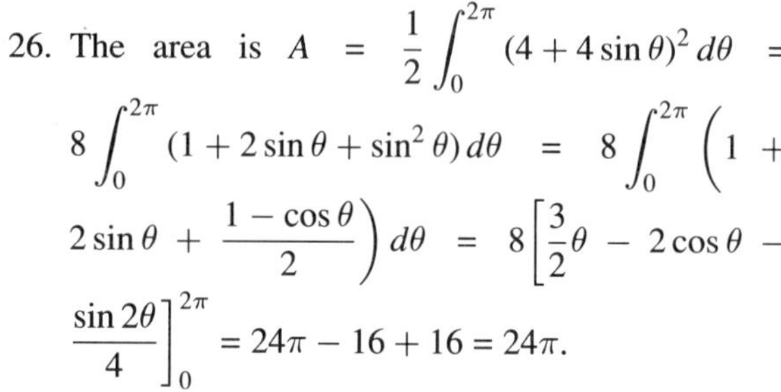

26. The area is $A = \frac{1}{2}\int_0^{2\pi}(4 + 4\sin\theta)^2\,d\theta = 8\int_0^{2\pi}(1 + 2\sin\theta + \sin^2\theta)\,d\theta = 8\int_0^{2\pi}\left(1 + 2\sin\theta + \frac{1-\cos\theta}{2}\right)d\theta = 8\left[\frac{3}{2}\theta - 2\cos\theta - \frac{\sin 2\theta}{4}\right]_0^{2\pi} = 24\pi - 16 + 16 = 24\pi.$

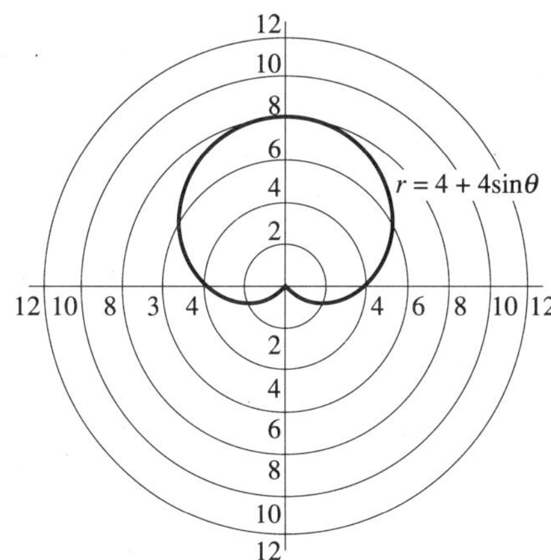

27. $A = \frac{1}{2}\int_0^{\pi} ((4 + 4\sin\theta)^2 - 4^2)\,d\theta =$
$\frac{1}{2}\int_0^{\pi} (16 + 32\sin\theta + 16\sin^2\theta - 16)\,d\theta =$
$\frac{1}{2}\int_0^{\pi} [32\sin\theta + 8(1 - \cos 2\theta)]\,d\theta =$
$\frac{1}{2}[8\theta - 32\cos\theta - 4\sin 2\theta]_0^{\pi} = 4\pi + 32 \approx$ 44.5664.

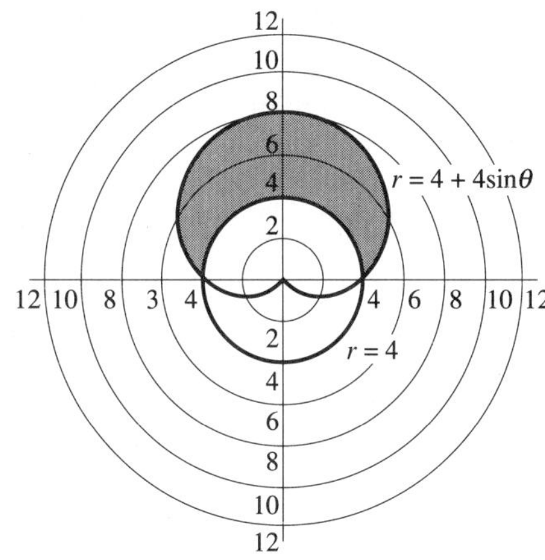

28. The area is $A = \frac{1}{2}\int_{\pi/6}^{5\pi/6} \left[(6\sin\theta)^2 - 3^2\right] d\theta =$
$\frac{1}{2}\int_{\pi/6}^{5\pi/6} (36\sin^2\theta - 9)\,d\theta =$
$\frac{1}{2}\int_{\pi/6}^{5\pi/6} [18(1 - \cos 2\theta) - 9]\,d\theta =$
$\frac{1}{2}[9\theta - 9\sin 2\theta]_{\pi/6}^{5\pi/6} =$
$\frac{9}{2}\left[\frac{5\pi}{6} + \frac{\sqrt{3}}{2} - \frac{\pi}{6} + \frac{\sqrt{3}}{2}\right] = \frac{9}{2}\left[\frac{2\pi}{3} + \sqrt{3}\right] =$
$3\pi + \frac{9\sqrt{3}}{2} \approx 17.2190.$

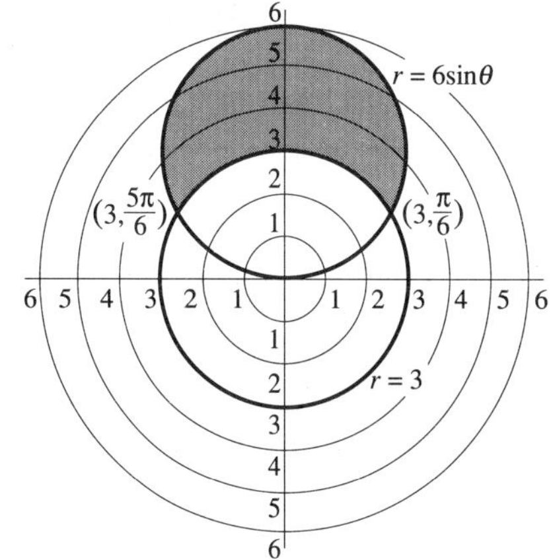

29. $\frac{1}{2}\int_{\pi/2}^{\pi} \left[(\sin\theta)^2 - (1 + \cos\theta)^2\right] d\theta =$
$\frac{1}{2}\int_{\pi/2}^{\pi} \left[\sin^2\theta - (1 + 2\cos\theta + \cos^2\theta)\right] d\theta =$
$\frac{1}{2}\int_{\pi/2}^{\pi} \left[\sin^2\theta - \cos^2\theta - 1 - 2\cos\theta\right] d\theta =$
$\frac{1}{2}\int_{\pi/2}^{\pi} [-\cos 2\theta - 1 - 2\cos\theta]\,d\theta$. Evaluating this integral, we obtain $\frac{1}{2}\left[-\frac{\sin 2\theta}{2} - 0 - 2\sin\theta\right]_{\pi/2}^{\pi} =$
$\frac{1}{2}\left[-\pi + \frac{\pi}{2} + 2\right] = 1 - \frac{\pi}{4} \approx 0.2146.$

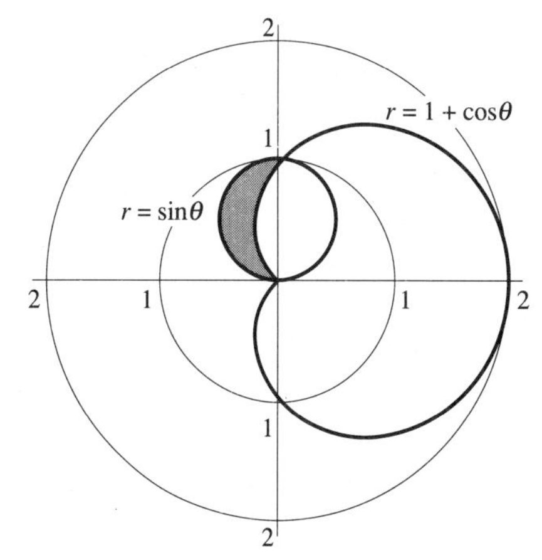

30. In problem #24 we solved for θ. We will use the first pair and multiply the answer by 4. $A = 4 \cdot \frac{1}{2}\int_{\pi/12}^{5\pi/12} [(2\sin 2\theta)^2 - 1^2]\,d\theta = 2\int_{\pi/12}^{5\pi/12} [4\sin^2 2\theta - 1]\,d\theta = 2\int_{\pi/12}^{5\pi/12} [2(1 - \cos 4\theta) - 1]\,d\theta = 2\left[\theta - \frac{\sin 4\theta}{2}\right]_{\pi/12}^{5\pi/12} = 2\left[\frac{5\pi}{12} + \frac{\sqrt{3}}{4} - \frac{\pi}{4} + \frac{\sqrt{3}}{4}\right] = \frac{2\pi}{3} + \sqrt{3} \approx 3.8264.$

CHAPTER 29 TEST

1. $\frac{dy}{dx} = \frac{dy/dt}{dx/dt} = \frac{2t+5}{3t^2}$; $\frac{d^2y}{dx^2} = \frac{\frac{d}{dt}\left(\frac{dy}{dx}\right)}{dx/dt} = \frac{[(3t^2)(2) - (2t+5)(6t)]/(3t^2)^2}{3t^2} = \frac{-6t^2 - 30t}{(3t^2)^3} = \frac{-2t - 10}{9t^5}$.

2. The first derivative is $\frac{dy}{dx} = \frac{dy/dt}{dx/dt} = \frac{-2\cos t \sin t}{\cos t} = -2\sin t$. When $t = \frac{\pi}{4}$, then $\frac{dy}{dx} = -\sqrt{2}$, so the slope is $-\sqrt{2}$. The tangent line is $y - \frac{1}{2} = -\sqrt{2}\left(x - \frac{\sqrt{2}}{2}\right)$ or $y = -\sqrt{2}x + \frac{3}{2}$.

3. We will use $\frac{dy}{dx} = \frac{r'\sin\theta + r\cos\theta}{r'\cos\theta - r\sin\theta}$. Differentiating $r = 4 - 2\sin\theta$ we obtain $r' = \frac{dr}{d\theta} = -2\cos\theta$ and so

$$\begin{aligned}\frac{dy}{dx} &= \frac{-2\cos\theta\sin\theta + (4 - 2\sin\theta)\cos\theta}{-2\cos\theta\cos\theta - (4 - 2\sin\theta)\sin\theta}\\ &= \frac{-4\sin\theta\cos\theta + 4\cos\theta}{-2\cos^2\theta - 4\sin\theta + 2\sin^2\theta}\\ &= \frac{4\cos\theta - 4\sin\theta\cos\theta}{-2(1 - \sin^2\theta) - 4\sin\theta + 2\sin^2\theta}\\ &= \frac{4\cos\theta(1 - \sin\theta)}{-2 + 2\sin^2\theta - 4\sin\theta + 2\sin^2\theta}\\ &= \frac{4\cos\theta(1 - \sin\theta)}{4\sin^2\theta - 4\sin\theta - 2} = \frac{2\cos\theta(1 - \sin\theta)}{2\sin^2\theta - 2\sin\theta - 1}\end{aligned}$$

4. We find that $r' = 24\cos 3\theta$, so $\frac{dy}{dx} = \frac{r'\sin\theta + r\cos\theta}{r'\cos\theta - r\sin\theta} = \frac{24\cos 3\theta\sin\theta + 8\sin 3\theta\cos\theta}{24\cos 3\theta\cos\theta - 8\sin 3\theta\sin\theta}$. When $\theta = \frac{\pi}{4}$, we see that $\frac{dy}{dx} = \frac{1}{2}$, so the slope of the normal line at this point is -2. Using the relationships that $x = r\cos\theta$ and $y = r\sin\theta$, we see that the polar coordinate $\left(4\sqrt{2}, \frac{\pi}{4}\right)$ has the rectangular coordinate $(4, 4)$ and so the equation of the normal line is $y - 4 = -2(x - 4)$ or $y = -2x + 12$.

5. $L = \int_0^2 \sqrt{(e^t)^2 + (2e^t)^2}\,dt = \int_0^2 \sqrt{e^{2t} + 4e^{2t}}\,dt = \int_0^2 \sqrt{5}e^t\,dt = \sqrt{5}e^t\Big|_0^2 = \sqrt{5}(e^2 - 1)$

6. Here $f(\theta) = 4\sin\theta$, $f'(\theta) = 4\cos\theta$, and $y = 4\sin^2\theta$. So, the surface area is

$$\begin{aligned}S &= 2\pi\int_0^{\pi/2} (4\sin^2\theta)\sqrt{(4\sin\theta)^2 + (4\cos\theta)^2}\,d\theta\\ &= 8\pi\int_0^{\pi/2} (\sin^2\theta)\sqrt{16\sin^\theta + 16\cos^2\theta}\,d\theta\\ &= 32\pi\int_0^{\pi/2} (\sin^2\theta)\,d\theta = 32\pi\left(\frac{1}{2} - \frac{1}{4}\sin 2\theta\right)\Big|_0^{\pi/2}\end{aligned}$$

(by Formula 56 in Appendix C). Evaluating this we obtain $32\pi\left(\frac{\pi}{4}\right) = 8\pi^2 \approx 78.9568$

7. Setting these equations equal to each other, we obtain $4\cos\theta = 1 - \cos\theta$ or $5\cos\theta = 1$ and so the points of intersection are when $\cos\theta = \frac{1}{5} = 0.2$ or when $\theta \approx 1.3694$ and -1.3694 and the points of intersection are $(0.2, 1.3694)$ and $(0.2, -1.3694) = (0.2, 4.9137)$

8. The points of intersection are $-\frac{\pi}{2}$ and $\frac{\pi}{2}$, so

$$A = \int_{-\pi/2}^{\pi/2} \frac{1}{2}\left(r_2^2 - r_1^2\right)\,d\theta$$

$$= \int_{-\pi/2}^{\pi/2} \frac{1}{2}\left[\left(1^2 - (1 - \cos\theta)^2\right)\right]\,d\theta$$

$$= \int_{-\pi/2}^{\pi/2} \frac{1}{2}\left[\left(1 - (1 - 2\cos\theta + \cos^2\theta)\right)\right]\,d\theta$$

$$= \frac{1}{2}\left[2\sin\theta - \frac{\sin 2\theta}{4} - \frac{\theta}{2}\right]_{-\pi/2}^{\pi/2} = 2 - \frac{\pi}{4}$$

CHAPTER

30

Partial Derivatives and Multiple Integrals

30.1 FUNCTIONS IN TWO VARIABLES

1. $f(x, y) = 3x + 4y - xy$: (a) $f(1, 0) = 3 \cdot 1 = 3$; (b) $f(0, 1) = 4 \cdot 1 = 4$; (c) $f(2, 1) = 3 \cdot 2 + 4 \cdot 1 - 2 \cdot 1 = 6 + 4 - 2 = 8$; (d) $f(x + h, y) = 3(x + h) + 4y - (x + h)y = 3x + 3h + 4y - xy - hy$; (e) $f(x, y + h) = 3x + 4(y + h) - x(y + h) = 3x + 4y + 4h - xy - xh$

2. $g(x, y) = x^2y + \sin x - \cos y$: (a) $g\left(\frac{\pi}{2}, \pi\right) = \left(\frac{\pi}{2}\right)^2 \pi + \sin\frac{\pi}{2} - \cos\pi = \frac{\pi^3}{4} + 2 \approx 9.7516$; (b) $g\left(\pi, \frac{\pi}{2}\right) = \pi^2 \cdot \frac{\pi}{2} + \sin\pi - \cos\frac{\pi}{2} = \frac{\pi^3}{2} \approx 15.5031$; (c) $g\left(\frac{\pi}{2}, \frac{\pi}{4}\right) = \left(\frac{\pi}{2}\right)^2 \frac{\pi}{4} + \sin\frac{\pi}{2} - \cos\frac{\pi}{4} = \frac{\pi^3}{16} + 1 - \frac{\sqrt{2}}{2} \approx 2.2308$; (d) $g(x+h, y) = (x+h)^2y + \sin(x+h) - \cos y = x^2y + 2xhy + h^2y + \sin(x+h) - \cos y$; (e) $g(x, y+h) = x^2(y+h) + \sin x - \cos(y+h) = x^2y + x^2h + \sin x - \cos(y+h)$

3. $j(x, y) = \sqrt{xy} - x + \frac{4}{y}$: (a) $j(-1, -2) = \sqrt{2} + 1 + \frac{4}{-2} = \sqrt{2} - 1 \approx 0.4142$; b) $j(-1, -4) = \sqrt{4} + 1 - 1 = 2$; (c) $j(4, 1) = \sqrt{4} - 4 + 4 = 2$; (d) $j(x + h, y) = \sqrt{(x + h)y} - (x + h) + \frac{4}{y}$; (e) $j(x, y + h) = \sqrt{x(y + h)} - x + \frac{4}{y+h}$

4. $k(x, y) = e^x + e^{xy} - y^2$: (a) $k(1, 2) = e^1 + e^2 - 4$; (b) $k(2, 1) = e^2 + e^2 - 1 = 2e^2 - 1$; (c) $k(2, 3) = e^2 + e^6 - 9$; (d) $k(x + h, y) = e^{x+h} + e^{(x+h)y} - y^2$; (e) $k(x, y + h) = e^x + e^{x(y+h)} - (y + h)^2$

5. $f(x, y) = \dfrac{2xy - x^2}{y - x}$: (a) $f(1, 0) = \frac{0-1}{-1} = 1$; (b) $f(0, 1) = \frac{0}{1} = 0$; (c) $f(2, 1) = \dfrac{2 \cdot 2 - 2^2}{1 - 2} = 0$; (d) $f(1, 2) = \frac{4-1}{1} = 3$; (e) the domain is all real ordered pairs (x, y) such that $x \neq y$.

6. $g(x, y) = \dfrac{\ln(x + y)}{y}$: (a) $g(1, 1) = \frac{\ln(2)}{1} = \ln 2$; (b) $g(2, 1) = \frac{\ln(2+1)}{1} = \ln 3$ (c) $g(3, 6) = \frac{\ln(3+6)}{6} = \frac{1}{6}\ln 9$; (d) $g(4, -3) = \frac{\ln(4-3)}{-3} = \frac{\ln 1}{-3} = 0$; (e) the domain is the set of all real ordered pairs (x, y) such that $x + y > 0$.

7. $V = \frac{1}{3}\pi r^2 h$

8. $V = Bh$, where B is the area of the base.

9. $A = 2\pi rh + 2\pi r^2$

10. $A = \pi rs + \pi r^2 = \pi r(s + r)$

11. The area of the storage tank is $A = 2\pi r^2 + 2\pi rh$. For the cost we have $C_{\text{bottom}} + C_{\text{top}} = 200(\pi r^2) + 200(\pi r^2) = 400\pi r^2$ and $C_{\text{side}} = 1000(2\pi rh) = 2000\pi rh$. Thus, the total cost is $C_{\text{bottom}} + C_{\text{top}} + C_{\text{side}} = 400\pi r^2 + 2000\pi rh = 400\pi r(r + 5h)$.

12. The area of the sewage tank is $A = \pi rs + \pi r^2$. For the cost of the base we have $C_{\text{base}} = 1500\pi r^2$ and for the side the cost is $C_{\text{side}} = 200\pi rs$. Thus, the total cost is $C_{\text{base}} + C_{\text{side}} = 1500\pi r^2 + 200\pi rs = 100\pi r(15r + 2s)$.

13. $v(R, r) = c(R^2 - r^2)$; $v(0.0075, 0.0045) = (0{,}0075^2 - 0.0045^2) = 3.6 \times 10^{-5} = 0.000\,036$ cm/min

14. $S(a,V,d) = \dfrac{aV}{0.51d^2}$ (a) $S(0.62, 2.4 \times 10^6, 100) = \dfrac{0.62 \times 2.4 \times 10^6}{0.51 \times 100^2} = 291.8$ mph, (b) $S(0.62, 2.4 \times 10^6, 200) = \dfrac{0.62 \times 2.4 \times 10^6}{0.51 \times 200^2} = 72.9$ mph.

15. (a) $T = 20°F$ and $v = 20$ mph so we use the following part of the formula: $WCI = 91.4 - \dfrac{(10.45 + 6.69\sqrt{v} - 0.447v)(91.4 - F)}{22} = -10.6$ or $-11°$F, (b) $T = 10°$, $v = 20$ mph, using the same formula as in part (a) we get the $WCI = -25°$F, (c) Since $v = 4$mph, $WCI = F = 10°$F, (d) As in parts (a) and (b) $WCI = -15°$F

16. $ATI(F, RH) = 0.885F - 78.7RH + 1.20F \cdot RH + 2.70$: (a) $ATI(95, 0.30) = 97.365 \approx 97°$F (b) $ATI(95, 0.6) = 107.955 \approx 108°$F (c) $ATI(95, 9.90) = 118.545 \approx 119°$F (d) $ATI(25, 0.75) = -11.7 \approx -12°$

30.2 SURFACES IN THREE DIMENSIONS

1. $x + 2y + 3z - 6 = 0$: A plane whose intercepts are $(6, 0, 0)$, $(0, 3, 0)$ and $(0, 0, 2)$.

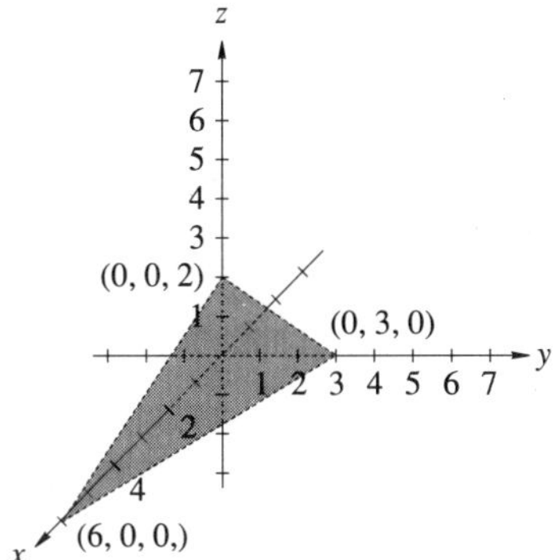

2. $3x + 2y - z - 6 = 0$: A plane whose intercepts are $(2, 0, 0)$, $(0, 3, 0)$ and $(0, 0, -6)$.

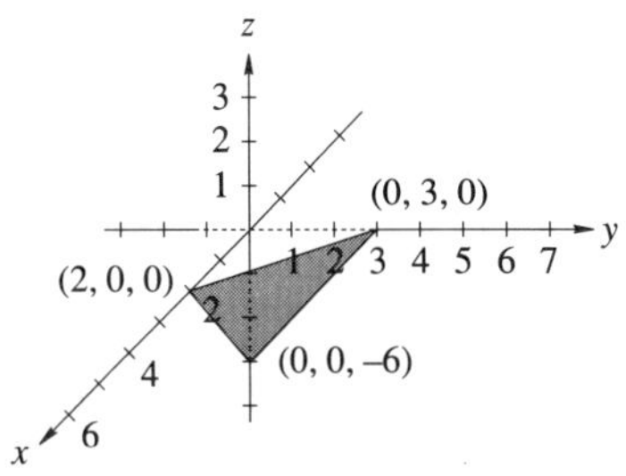

3. $x = 8y^2$: A parabolic cylinder perpendicular to the xy-plane.

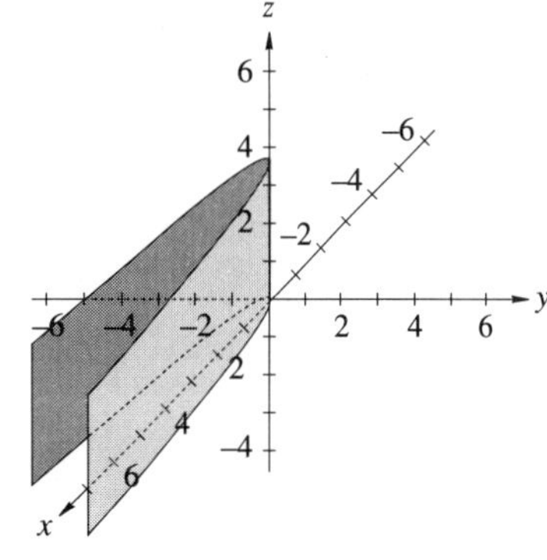

4. $x^2 + y^2 + z^2 = 9$: A sphere of radius 3.

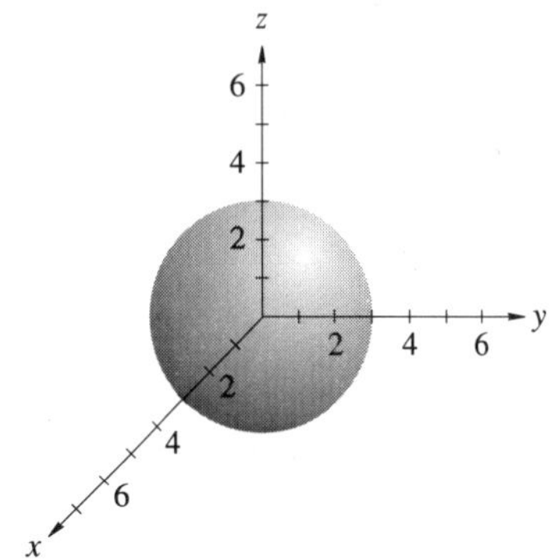

5. $x^2 + 2y^2 + z^2 = 4 \Rightarrow \frac{x^2}{4} + \frac{z^2}{4} = 1$: Ellipsoid with intercepts $(\pm 2, 0, 0)$, $(0, \pm\sqrt{2}, 0)$, and $(0, 0, \pm 2)$.

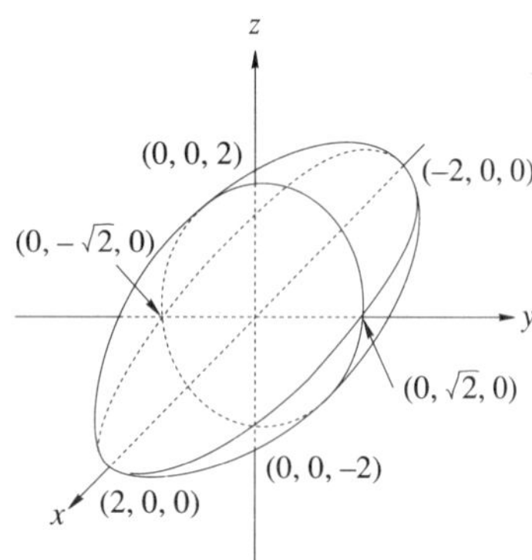

6. $4x^2 + 9y^2 = 36 \Rightarrow \frac{x^2}{9} + \frac{y^2}{4} = 1$: Elliptic cylinder perpendicular to the xy-plane with intercepts $(\pm 3, 0, 0)$ and $(0, \pm 2, 0)$.

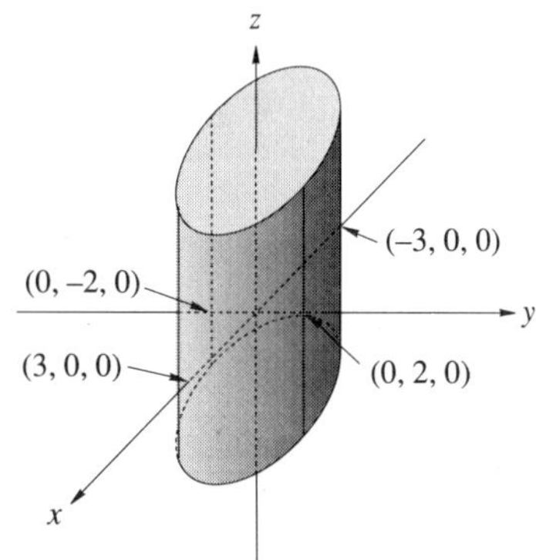

7. $2x^2 + y^2 + 2z^2 = 8 \Rightarrow \frac{x^2}{4} + \frac{y^2}{8} + \frac{z^2}{4} = 1$: Ellipsoid with intercepts $(\pm 2, 0, 0)$, $(0, \pm 2\sqrt{2}, 0)$, and $(0, 0, \pm 2)$.

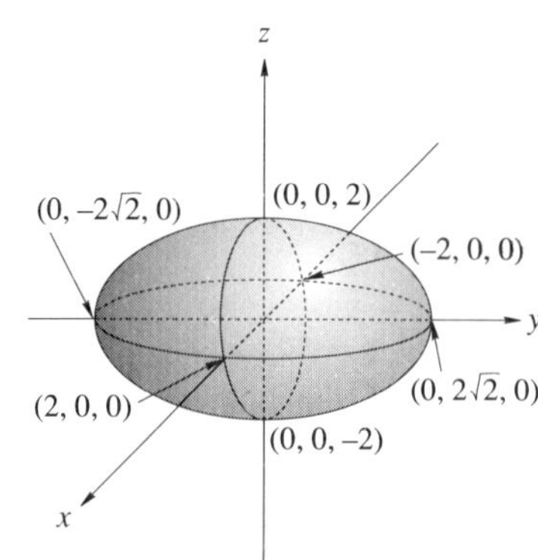

8. $4x^2 + y^2 - z^2 = 1$: Hyperboloid of one sheet with intercepts $(\pm\frac{1}{2}, 0, 0)$, $(0, \pm 1, 0)$.

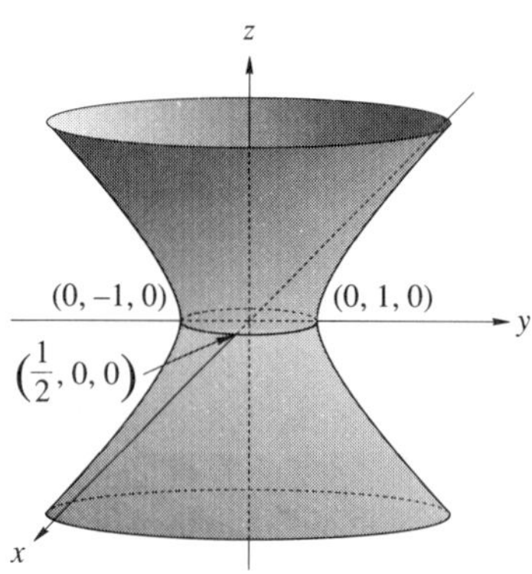

9. $(4x^2 + y^2 - z^2 = -1$: Hyperboloid of two sheets with intercepts $(0, 0, \pm 1)$.

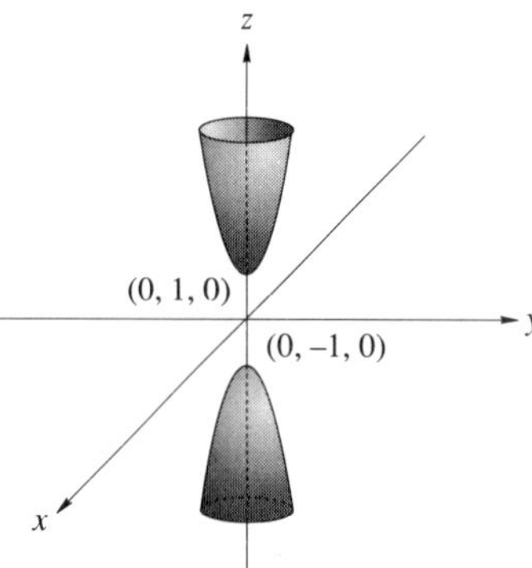

10. $4x^2 - 9y^2 = 1$: Hyperbolic cylinder with intercepts $(\pm\frac{1}{2}, 0, 0)$.

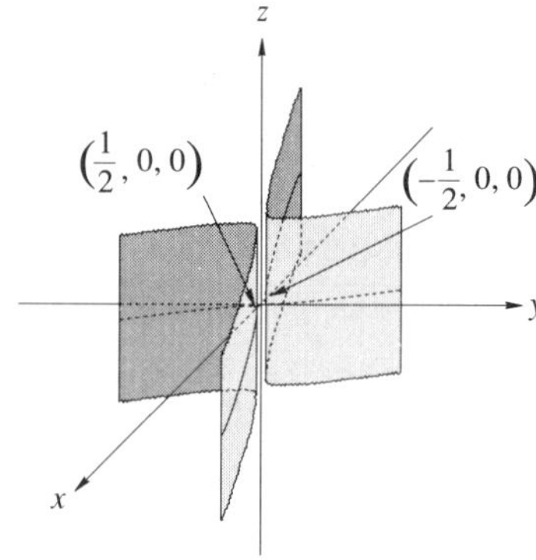

11. $3x - 4y - 6z = 12$: Plane with intercepts $(4, 0, 0)$, $(0, -3, 0)$, and $(0, 0, -2)$.

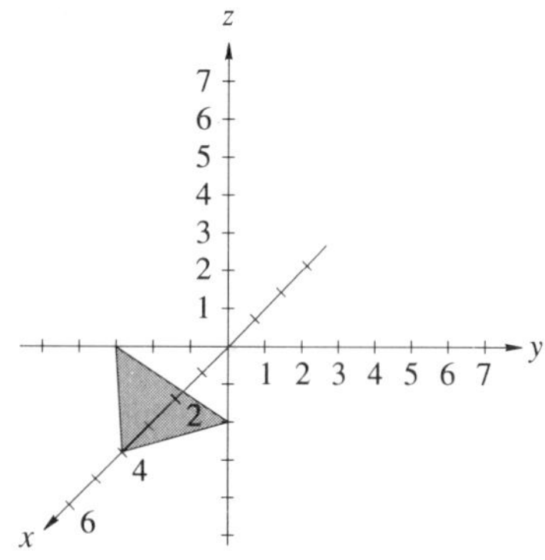

12. $xy = 4$: Hyperbolic cylinder whose asymptote planes are the xz-plane and the yz-plane.

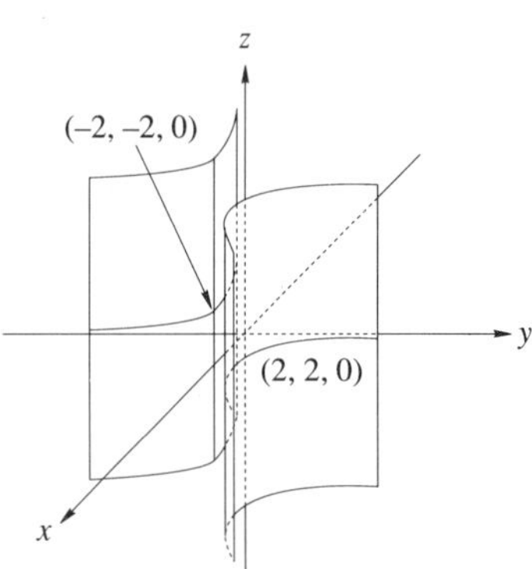

13. $z = x^2 + 4y^2 - 16$: This is an elliptic paraboliod with a z-translation of -16. The intercepts are $(\pm 4, 0, 0)$, $(0, \pm 2, 0)$, and $(0, 0, -16)$.

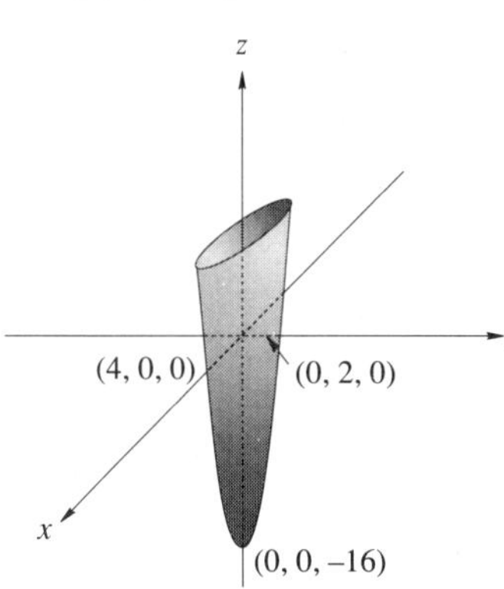

14. $2x^2 + y^2 - 2z^2 = -8 \Rightarrow \frac{x^2}{4} + \frac{y^2}{8} - \frac{z^2}{4} = -1$: Hyperboloid of two sheets with intercepts at $(0, 0, \pm 2)$.

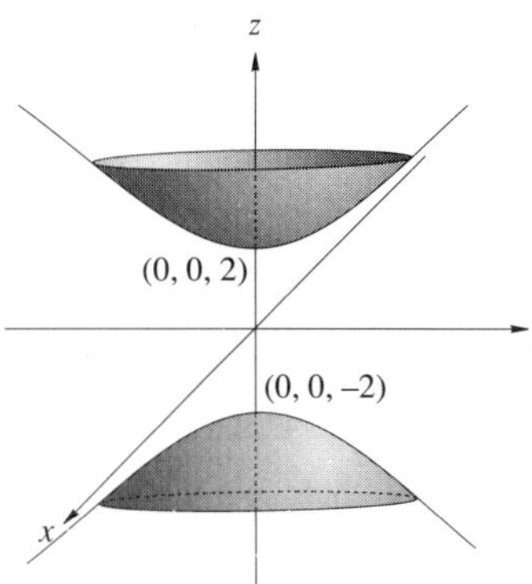

15. $36z = 4x^2 + 9y^2 \Rightarrow z = \frac{x^2}{9} + \frac{y^2}{4}$: Elliptic paraboloid with vertex $(0, 0, 0)$. Four other points are $(\pm 3, 0, 1)$ and $(0, \pm 2, 1)$.

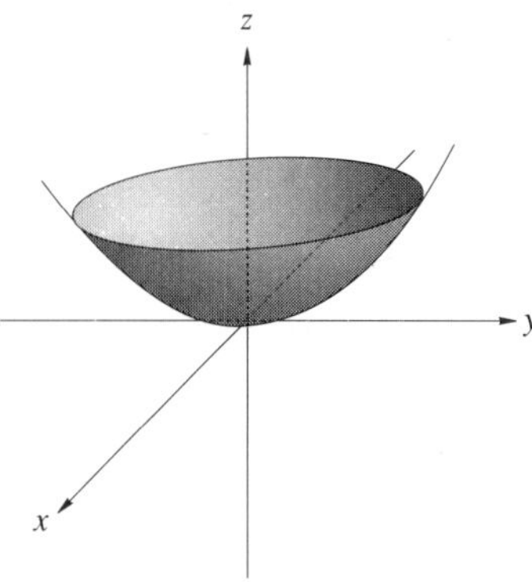

16. $36z = 4x^2 + 9y^2 \Rightarrow z = \frac{x^2}{9} + \frac{y^2}{4}$: Elliptic Cone with vertex at the origin. Eight other points are $(\pm 3, 0, \pm 1)$ and $(0, \pm 2, \pm 1)$.

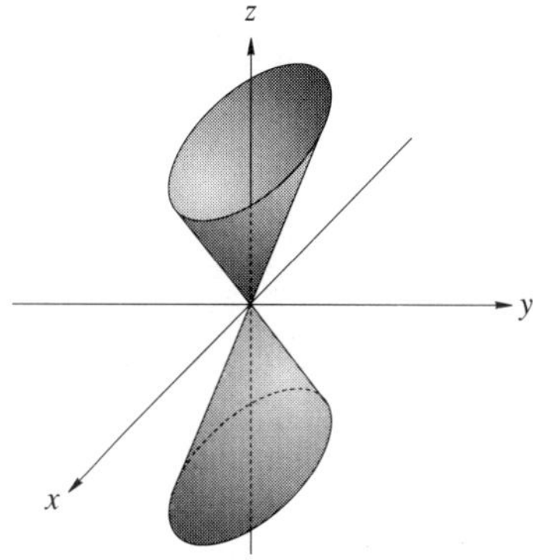

17. $36z = 4x^2 - 9y^2 \Rightarrow z = \frac{x^2}{9} - \frac{y^2}{4}$: Hyperbolic paraboloid with saddle point $(0, 0, 0)$. Four other points are $(\pm 3, 0, 1)$ and $(0, \pm 2, -1)$.

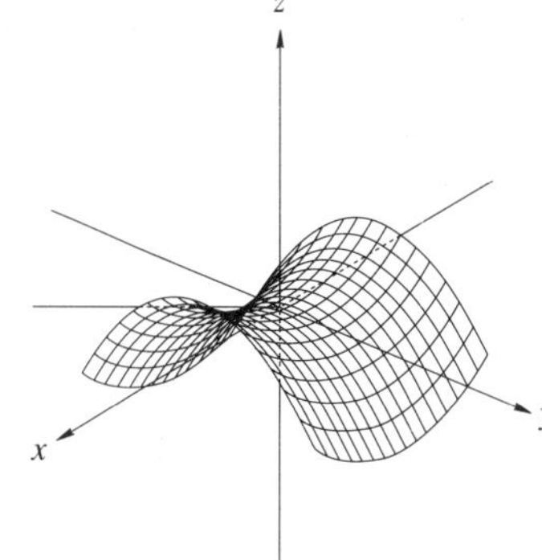

18. $36z^2 = 1 - 4x^2 - 9y^2 \Rightarrow 4x^2+9y^2+36z^2 = 1$: Ellipsoid with intercepts $(\pm\frac{1}{2}, 0, 0)$, $(0, \pm\frac{1}{3}, 0)$, and $(0, 0, \pm\frac{1}{6})$.

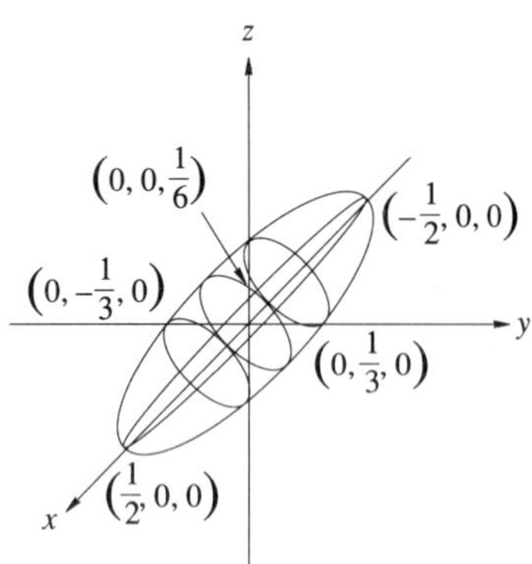

19. $2x = y^2 \Rightarrow x = \frac{1}{2}y^2$: Parabolic cylinder perpendicular to the xy-plane.

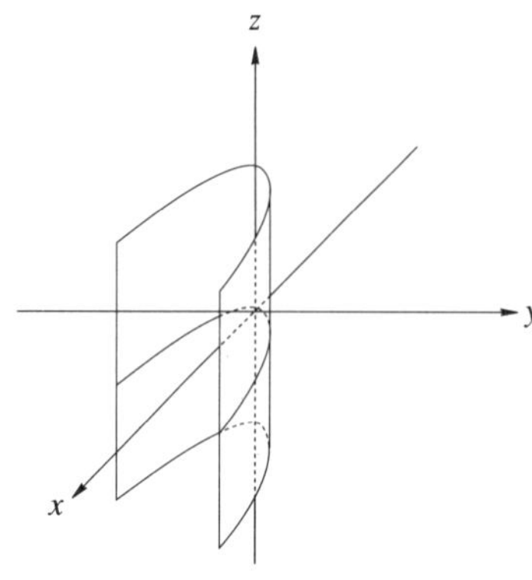

20. $3x + 2y - 3z = 6$: Plane with intercepts $(2, 0, 0)$, $(0, 3, 0)$, and $(0, 0, -2)$.

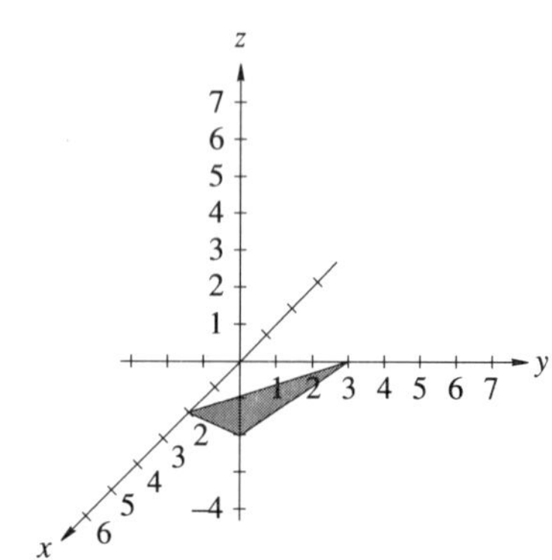

30.3 PARTIAL DERIVATIVES

1. $f(x, y) = x^2y + y^2x; f_x = 2xy + y^2; f_y = x^2 + 2xy.$

2. $f(x, y) = x^2y + 7y^2; f_x = 2xy; f_y = x^2 + 14y$

3. $f(x, y) = 3x^2 + 6xy^3; f_x = 6x + 6y^3; f_y = 18xy^2$

4. $f(x, y) = \dfrac{x}{y^2} = xy^{-2}; f_x = y^{-2} = \dfrac{1}{y^2}; f_y = -2xy^{-3} = \dfrac{-2x}{y^3}$

5. $f(x, y) = \dfrac{x^2 + y^2}{y} = \dfrac{x^2}{y} + y; f_x = \dfrac{2x}{y}; f_y = \dfrac{-x^2}{y^2} + 1 = \dfrac{y^2 - x^2}{y^2}$

6. $f(x, y) = \dfrac{x + y}{\sqrt{xy}}$. So, $f_x = \dfrac{\sqrt{xy} \cdot 1 - (x + y)(xy)^{-1/2}y}{xy} = \dfrac{xy - (x + y)y}{(xy)^{3/2}} = \dfrac{-y^2}{(xy)^{3/2}} = \dfrac{-y}{x\sqrt{xy}}$, and $f_y = \dfrac{\sqrt{xy} \cdot 1(x + y)(xy)^{-1/2} \cdot x}{(xy)} = \dfrac{xy - (x + y)x}{(xy)^{3/2}} = -\dfrac{-x^2}{(xy)^{3/2}} = \dfrac{-x}{y\sqrt{xy}}$.

7. $f(x, y) = e^y \cos x + e^x \sin y; f_x = -e^y \sin x + e^x \sin y; f_y = e^y \cos x + e^x \cos y$

8. $f(x, y) = x^2 \cos y + y^2 \cos x; f_x = 2x \cos y - y^2 \sin x; f_y = -x^2 \sin y + 2y \cos x = 2y \cos x - x^2 \sin y$

9. $f(x, y) = e^{2x+3y}; f_x = 2e^{2x+3y}; f_y = 3e^{2x+3y}$

10. $f(x, y) = e^{x^2+y^2}; f_x = 2xe^{x^2+y^2}; f_y = 2ye^{x^2+y^2}$

11. $f(x, y) = \sin(x^2y^3); f_x = 2xy^3 \cos(x^2y^3); f_y = 3x^2y^2 \cos(x^2y^3)$

12. $f(x, y) = \ln(x^2y^3); f_x = \dfrac{2xy^3}{x^2y^3} = \dfrac{2}{x}; f_y = \dfrac{3y^2x^2}{x^2y^3} = \dfrac{3}{y}$

13. $f(x, y) = \ln\sqrt{x^2 + y^2}$. So, $f_x = \dfrac{\frac{1}{2}(x^2 + y^3)^{-1/2} \cdot 2x}{\sqrt{x^2 + y^2}} = \dfrac{x}{x^2 + y^2}$ and $f_y = \dfrac{\frac{1}{2}(x^2 + y^2)^{-1/2}(2y)}{\sqrt{x^2 + y^2}} = \dfrac{y}{x^2 + y^2}$.

14. $f(x,y) = \tan -1\frac{y}{x}$; $f_x = \frac{-yx^{-2}}{1+\left(\frac{y}{x}\right)^2} = \frac{\frac{-y}{x^2}}{1+\frac{y^2}{x^2}} = \frac{-y}{x^2+y^2}$; $f_y = \frac{\frac{1}{x}}{1+\frac{y^2}{x^2}} = \frac{x}{x^2+y^2}$

15. $f(x,y) = \sin^2(3xy)$; $f_x = 2\sin(3xy)\cdot\cos(3xy)\cdot 3y = 6y\sin(3xy)\cos(3xy)$; $f_y = 2\sin 3xy\cos 3xy\cdot 3x = 6x\sin(3xy)\cos(3xy)$

16. $f(x,y) = e^{\sqrt{x^2+y^2}}$; $f_x = \frac{1}{2}\cdot(x^2+y^2)^{-1/2}(2x)e^{\sqrt{x^2+y^2}} = \frac{x}{\sqrt{x^2+y^2}}e^{\sqrt{x^2+y^2}}$; $f_y = \frac{y}{\sqrt{x^2+y^2}}e^{\sqrt{x^2+y^2}}$

17. $f(x,y) = x^3y^2 - xy^5$; $f_x = 3x^2y^2 - y^5$; $f_y = 2x^3y - 5xy^4$; $f_{xx} = 6xy^2$; $f_{yx} = 6x^2y - 5y^4$; $f_{yy} = 2x^3 - 20xy^3$; $f_{xy} = 6x^2y - 5y^4$

18. $f(x,y) = \sin xy^3$; $f_x = y^3\cos xy^3$; $f_y = 3y^2x\cos xy^3$; $f_{xx} = -y^6\sin xy^3$; $f_{yx} = 3y^2\cos xy^3 - 3xy^5\sin xy^3$; $f_{yy} = 6yx\cos xy^3 - 9x^2y^3\sin xy^3$; $f_{xy} = 3y^2\cos xy^3 - 3xy^5\sin xy^3$

19. $f(x,y) = 3e^{xy^3}$; $f_x = 3y^3e^{xy^3}$; $f_y = 9xy^2e^{xy^3}$; $f_{xx} = 3y^6e^{xy^3}$; $f_{yx} = f_{xy} = 9e^{xy^3}y^2(1+xy^3)$; $f_{yy} = 27x^2y^4e^{xy^3} + 18xye^{ey^3} = 9xye^{xy^3}(3xy^3+2)$

20. $f(x,y) = \sin(x+y^2)$; $f_x = \cos(x+y^2)$; $f_y = 2y\cos(x+y^2)$; $f_{xx} = -\sin(x+y^2)$; $f_{yx} = f_{xy} = -2y\sin(x+y^2)$; $f_{yy} = 2\cos(x+y^2) - 4y^2\sin(x+y^2)$

21. $f(x,y) = \frac{2x}{y^5}$; $f_x = \frac{2}{y^5}$; $f_y = \frac{-10x}{y^6}$; $f_{xx} = 0$; $f_{yx} = f_{xy} = \frac{-10}{y^6}$; $f_{yy} = \frac{60x}{y^7}$

22. $f(x,y) = e^x\tan y$; $f_x = e^x\tan y$; $f_y = e^x\sec^2 y$; $f_{xx} = e^x\tan y$; $f_{yx} = e^x\sec^2 x = f_{xy}$; $f_{yy} = 2e^x\sec^2 y\tan y$

23. $f(x,y) = \ln(x^3y^5)$; $f_x = \frac{3}{x}$; $f_y = \frac{5}{y}$; $f_{xx} = \frac{-3}{x^2}$; $f_{yx} = f_{xy} = 0$; $f_{yy} = \frac{-5}{y^2}$

24. $f(x,y) = \sqrt{x^2y+3y^2}$; $f_x = xy(x^2y+3y^2)^{-1/2} = \frac{xy}{\sqrt{x^2y+3y^2}}$; $f_y = \frac{1}{2}(x^2+6y)(x^2y+3y^2)^{-1/2}$;

$$f_{xx} = \frac{\sqrt{x^2y+3y^2}\cdot y - xy\cdot xy(x^2y+3y^2)^{-1/2}}{x^2y+3y^2}$$
$$= \frac{(x^2y+3y^2)y - x^2y^2}{(x^3y+3y^2)^{3/2}} = \frac{x^2y^2+3y^3-x^2y^2}{(x^2+3y^2)^{3/2}}$$
$$= \frac{3y^2}{(x^2+3y^2)^{3/2}}$$

$$f_{yx} = f_{xy} = \frac{\sqrt{x^2y+3y^2}x - \frac{1}{2}xy(x^2+6y)(x^2y+3y^2)^{-1/2}}{x^2y+3y^2}$$
$$= \frac{(x^2y+3y^2)x - \frac{1}{2}x^3y - 3xy^2}{(x^2y+3y^2)^{3/2}} = \frac{x^3y}{2(x^2y+3y^2)^{3/2}}$$

$$f_{yy} = \frac{2\sqrt{x^2y+3y^2}\cdot 6 - (x^2+6y)\cdot(x^26y)\cdot(x^2y+3y^2)^{-1/2}}{4(x^2y+3y^2)}$$
$$= \frac{12(x^2y+3y^2) - (x^4+12x^2y+36y^2)}{4(x^2y+3y^2)^{3/2}}$$
$$= \frac{12x^2y+36y^2-x^4-12x^2y-36y^2}{4(x^2+3y^2)^{3/2}}$$
$$= \frac{-x^4}{4(x^2+3y^2)^{3/2}}$$

25. $P = \frac{20\cdot R\cdot 295}{V}$; $P_V = \frac{-5900R}{V^2}$ at $V = 3$; $P_V = \frac{-5900R}{9}$ kg/m^2

26. $T = \frac{100}{\ln 2}\ln(x^2+y^2)$; $T_x = \frac{100\cdot 2x}{(x^2+y^2)\ln 2} = \frac{200x}{(x^2+y^2)\ln 2}$; T_x at $(1,0)$ is $\frac{200}{\ln 2} \approx 288.539$, and at $(0,1)$, $T_x = 0$.

27. $V = \frac{82.06T}{P}$; $V_P = \frac{-82.06T}{P^2}$ at $T = 300°$, $P = 5$ atm; $V = \frac{-82.06\times 300}{5^2} = -984.72$ cm^3/atm

28. $V_T = \frac{82.06}{P}$. At $P = 5$, then $V_T = \frac{82.06}{5} = 16.412$ cm^3/°K.

29. $z = 16 - x^2 - y^2$; $z_y = -2y$. At $(1,3,6)$, $z_y = -2\cdot 3 = -6$.

30. $PV = nRT$; $V = \frac{nRT}{P}$; $\frac{\partial V}{\partial T} = \frac{nR}{P}$; $T = \frac{PV}{nR}$; $\frac{\partial T}{\partial P} \cdot \frac{V}{nR}$; $P = \frac{nRT}{V}$; $\frac{\partial P}{\partial V} = \frac{-nRT}{V^2}$; $\left(\frac{\partial V}{\partial T}\right)\left(\frac{\partial T}{\partial P}\right)\left(\frac{\partial P}{\partial V}\right) = \frac{nR}{P} \cdot \frac{PV}{nR} \cdot \frac{-nRT}{V^2} = \frac{-nRT}{V} = \frac{-nRT}{\frac{nRT}{P}} = -P$

30.4 SOME APPLICATIONS OF PARTIAL DERIVATIVES

1. $z = x^2 + y^2$; $z_x = 2x$; $z_y = 2y$; $dz = 2xdx + 2ydy$

2. $z = 2x^2 + xy - y^2$; $z_x = 4x + y$; $z_y = x - 2y$; $dz = (4x + y)dx + (x - 2y)dy$

3. $z = 3x^2 + 4xy - 2y^3$; $z_x = 6x + 4y$; $z_y = 4x - 6y^2$; $dz = (6x + 4y)dx + (4x - 6y^2)dy$

4. $z = xye^{x+y}$; $z_x = (xy + y)e^{x+y}$; $z_y = (xy + x)e^{x+y}$; $dz = (1 + x)ye^{x+y}dx + (1 + y)xe^{x+y}dy$

5. $z = \arctan\left(\frac{y}{x}\right)$; $z_x = \frac{-\frac{y}{x^2}}{1+\frac{y^2}{x^2}} = \frac{-y}{x^2 + y^2}$; $z_y = \frac{\frac{1}{x}}{1+\frac{y^2}{x^2}} = \frac{x}{x^2 + y^2}$; $dz = \frac{-y}{x^2 + y^2}dx + \frac{x}{x^2 + y^2}dy$

6. $z = \ln(x^2 + y^2)$; $z_x = \frac{2x}{x^2 + y^2}$; $z_y = \frac{2y}{x^2 + y^2}$; $dz = \frac{2x}{x^2 + y^2}dx + \frac{2y}{x^2 + y^2}dy$

7. $z = x\tan yx$; $z_x = \tan yx + xy\sec^2 yx$; $z_y = x^2\sec^2 yx$; $dz = (\tan yx + xy\sec^2 yx)dx + x^2\sec^2 yxdy$

8. $z = \ln\left(\frac{x}{y}\right)$; $z_x = \frac{1/y}{x/y} = \frac{1}{x}$; $z_y = \frac{-x/y^2}{x/y} = -\frac{1}{y}$; $dz = \frac{1}{x}dx - \frac{1}{y}dy$

9. $z = x^2 + y^2$; $x = te^t$; $y = t^2e^t$; $z_x = 2x$; $z_y = 2y$; $\frac{dx}{dt} = (t+1)e^t$; $\frac{dy}{dt} = (t^2 + 2t)e^t$; $\frac{dz}{dt} = 2x(t+1)e^t + 2y(t^2+2t)e^t = 2(te^t)(t+1)e^t + 2(t^2e^t)(t^2+2t)e^t = (t+1)2te^{2t} + (t^2 + 2t)2t^2e^{2t} = 2te^{2t}\left[1 + t + 2t^2 + t^3\right]$

10. $z = \frac{1}{x^2 + y^2}$; $x = \cos 2t$; $y = \sin 2t$; $z_x = -1(x^2 + y^2)^{-2} \cdot 2x = \frac{-2x}{(x^2 + y^2)^2}$; $z_y = \frac{-2y}{(x^2 + y^2)^2}$; $\frac{dx}{dt} = -2\sin 2t$; $\frac{dy}{dt} = 2\cos 2t$; $\frac{dz}{dt} = \frac{-2x(-2\sin 2t)}{(x^2 + y^2)^2} + \frac{-2y(2\cos 2t)}{(x^2 + y^2)^2} = \frac{4x\sin 2t - 4y\cos 2t}{(x^2 + y^2)^2} = \frac{4(\cos 2t)\sin 2t - 4(\sin 2t)\cos 2t}{(x^2 + y^2)^2} = 0$

11. $z = e^u\sin v$; $u = \sqrt{t}$; $v = \pi t$; $z_u = e^u\sin v$; $z_v = e^u\cos v$; $\frac{du}{dt} = \frac{1}{2\sqrt{t}}$; $\frac{dv}{dt} = \pi$; $\frac{dz}{dt} = e^u\sin v \cdot \frac{1}{2\sqrt{t}} + e^u\cos v \cdot \pi = \frac{e^{\sqrt{t}}\sin \pi t}{2\sqrt{t}} + \pi e^{\sqrt{t}}\cos \pi t$

12. $z = x^3 - y$; $x = te^{-t}$; $y = \sin t$; $z_x = 3x^2$; $z_y = -1$; $\frac{dx}{dt} = (1 - t)e^{-t}$; $\frac{dy}{dt} = \cos t$; $\frac{dz}{dt} = 3x^2(1 - t)e^{-t} + (-1)\cos t = 3t^2e^{-2t}(1 - t)e^{-t} - \cos t = 3t^2e^{-3t}(1 - t) - \cos t$

13. $z = x^2 + 4y^2 + x + 8y - 1$; $z_x = 2x + 1$; $z_y = 8y + 8$; $z_x = 0 \Rightarrow x = -\frac{1}{2}$; $z_y = 0 \Rightarrow y = -1$; $z_{xx} = 2$; $z_{yx} = 0$; $z_{xy} = 0$; $z_{yy} = 8$; $\Delta = \begin{vmatrix} z_{xx} & z_{xy} \\ z_{yx} & z_{yy} \end{vmatrix} = \begin{vmatrix} 2 & 0 \\ 0 & 8 \end{vmatrix} = 16 > 0$. Since z_{xx} is $2 > 0$ we have a local minimum at $\left(-\frac{1}{2}, -1\right)$.

14. $z = x^2 + y^2 - 2x + 4y + 2$; $z_x = 2x - 2 = 0 \Rightarrow x = 1$; $z_y = 2y + 4 = 0 \Rightarrow y = -2$; $z_{xx} = 2$; $z_{xy} = z_{yx} = 0$; $z_{yy} = 2$; $\Delta = \begin{vmatrix} 2 & 0 \\ 0 & 2 \end{vmatrix} = 4 > 0$. Since $z_{yy} = 2$ we have a local minimum at $(1, -2)$.

15. $z = 20 + 12x - 12y - 3x^2 - 2y^2$; $z_x = 12 - 6x = 0 \Rightarrow x = 2$; $z_y = -12 - 4y = 0 \Rightarrow y = -3$; $z_{xx} = -6$; $z_{xy} = z_{yx} = 0$; $z_{yy} = -4$; $\Delta = \begin{vmatrix} -6 & 0 \\ 0 & -4 \end{vmatrix} = 24 > 0$. Since $z_{xx} = 2 > 0$ we have a local minimum at $(2, -3)$.

16. $z = xy + 3x - 2y + 4$; $z_x = y + 3 = 0 \Rightarrow y = -3$; $z_y = x - 2 = 0 \Rightarrow x = 2$; $z_{xx} = 0$; $z_{xy} = z_{yx} = 1$; $z_{yy} = 0$; $\Delta = \begin{vmatrix} 0 & 1 \\ 1 & 0 \end{vmatrix} = -1 < 0$, so there is a saddle point at $(2, -3)$.

17. $z = x^2 + 2xy - y^2$; $z_x = 2x + 2y$; $z_y = 2x - 2y$; $2x + 2y = 0$ and $2x - 2y = 0 \Rightarrow x = 0$ and $y = 0$; $z_{xx} = 2$; $z_{xy} = z_{yx} = 2$; $z_{yy} = -2$. $\Delta \begin{vmatrix} 2 & 2 \\ 2 & -2 \end{vmatrix} = -8 < 0$, so saddle point at $(0, 0)$.

18. $z = x^2 - 3xy - y^2$; $z_x = 2x - 3y$; $z_y = -3x - 2y$; $2x - 3y = 0$ and $-3xy - 2y = 0 \Rightarrow x$ and y both 0. $z_{xx} = 2$; $z_{xy} = z_{yx} = -3$; $z_{yy} = -2$ $\Delta = \begin{vmatrix} 2 & -3 \\ -3 & -2 \end{vmatrix} = -4 - 9 = -13 < 0$, so we have a saddle point at $(0, 0)$.

19. $z = x^3 + x^2y + y^2$; $z_x = 3x^2 + 2xy$; $z_y = x^2 + 2y$; $3x^2 + 2xy = 0$, and $x^2 + 2y = 0$. Solving for $2y$ in the second equaiton we get $2y = -x^2$. Substituting into the first equation we have $3x^2 - x^3 = 0$ or $x^2(3 - x) = 0$ so $x = 0$ or $x = 3$. Back substituting we get the ordered pairs $(0, 0)$ and $(3, -\frac{9}{2})$. $z_{xx} = 6x + 2y$; $z_{xy} = z_{yx} = 2x$; $z_{yy} = 2$. $\Delta = \begin{vmatrix} 6x + 2y & 2x \\ 2x & 2 \end{vmatrix} = 12x + 4y - 4x^2$. At $(0, 0)$ the test fails. At $(3, -\frac{9}{2})$, $\Delta = 36 - 18 - 81 < 0$, so there is a saddle point at $(3, -\frac{9}{2})$.

20. $z = x^3 + y^3 + 3xy$; $z_x = 3x^2 + 3y$; $z_y = 3y^2 + 3x$. Setting $3x^2 + 3y = 0$ and $3y^2 + 3x = 0$ we solve for y in the first equation getting $y = -x^2$. Substituting this into the second equation yields $3(-x^2)^2 + 3x = 0$ or $x^4 + x = 0$. Hence $x = 0$ or -1. Backing substituting we have $y = 0$ or $y = -1$ or the order pairs $(0, 0)$ and $(-1, -1)$. $z_{xx} = 6x$; $z_{xy} = z_{yx} = 3$; $z_{yy} = 6y$; $\Delta = \begin{vmatrix} 6x & 3 \\ 3 & 6y \end{vmatrix} = 36xy - 9$; at $(0, 0)\Delta = -9 < 0$ so we have a saddle point at $(0, 0)$. At $(-1, -1)$, $\Delta = 36 - 9 > 0$; $z_{xx} = 6(-1) = -6 < 0$ and we have a local maximum at $(-1, -1)$.

21. $A = bh$; $A_b = h$; $A_h = b$; $dA = hdb + bdh = 25(0.1) + 10(0.1) = 3.5$ cm^2

22. (a) The volume of a right circular cone is $V = \frac{1}{3}\pi r^2 h$. Differentiating with respect to r, we get $V_r = \frac{2}{3}\pi rh$, and with respect to h, produces $V_h = \frac{1}{3}\pi r^2$. Thus, $dV = V_r dr + V_h dh = \frac{2}{3}\pi rhdr + \frac{1}{3}\pi r^2 dh = \frac{2}{3}\pi \cdot 4 \cdot 12 \cdot \frac{1}{16} + \frac{1}{3}\pi(4)^2 \cdot \frac{1}{16} = 2\pi + \frac{\pi}{3} = \frac{7\pi}{3}$ in.^{2}
(b) The total surface area is $A = \pi r^2 + \pi r\ell$, where $\ell = \sqrt{h^2 + r^2} = \sqrt{12^2 + 4^2} = 4\sqrt{10}$. Differentiating with respect to r, yields $A_r = 2\pi r + \pi\sqrt{h^2 + r^2} + \frac{1}{2} \cdot \pi r(2r)(h^2 + r^2)^{-1/2} = 2\pi r + \pi\sqrt{h^2 + r^2} + \frac{\pi r^2}{\sqrt{h^2 + r^2}}$. Differentiating with respect to h, gives $A_h = \frac{1}{2}\pi r(2h)(h^2 + r^2)^{-1/2} = \frac{\pi rh}{\sqrt{h^2 + r^2}}$. Thus we have $dA = A_r dr + A_h dh = \frac{1}{16}\left(8\pi + 4\sqrt{10}\pi + \frac{16\pi}{8\sqrt{10}} + \frac{48\pi}{4\sqrt{10}}\right) = \frac{\pi}{16}\left(8 + 4\sqrt{10} + \frac{2\sqrt{10}}{5} + \frac{6\sqrt{10}}{5}\right) = \frac{\pi}{16}\left(8 + \frac{28\sqrt{10}}{5}\right) = \frac{\pi}{2} + \frac{7\sqrt{10}\pi}{20}$ in.^{2}

23. $R_{R_1} = \frac{(R_1 + R_2)R_2 + R_1R_2}{(R_1 + R_2)^2}$; $R_{R_2} = \frac{(R_1 + R_2)R_1 + R_1R_2}{(R_1 + R_2)^2} dR =$ $R_{R_1}dR_1 + R_{R_2}dR_2 = \frac{(500)400 + 40000}{(500)^2}1 + \frac{500 \cdot 100 + 40000}{500^2}4 = 0.96 + 1.44 = 2.4\ \Omega$. (Note $dR_1 = 0.01 \times 100 = 1\Omega$; $dR_2 = 0.01 \times 400 = 4\Omega$.)

24. $PV = kT$; $P = \frac{kT}{V} dP = P_V dV + P_T dT = \frac{-kT}{V^2} \cdot (dV) + \frac{k}{V}(dT)$. $T = 110°F = (110 - 32)\frac{5}{9} + 273°K = 316.3°K$. $T = 112°F = (112 - 32)\frac{5}{9} + 273°K = 317.4°K$. $dT = 1.1°K$; $dV = 70 - 66 = 4$ in.^{3}; $k = \frac{PV}{T} = \frac{0.5 \times 66}{316.3} = 0.0104$; $dP = \frac{-0.0104 \cdot 316.3}{66^2} \times 4 + \frac{0.0104}{66} \times 1.1 = -0.00285$ psi.

25. $PV = kT$ or $T = \frac{PV}{k}$. $\frac{dT}{dt} = \frac{\partial T}{\partial P}\frac{dP}{dt} + \frac{\partial T}{\partial V}\frac{dV}{dt} = \frac{V}{k}\cdot\frac{dP}{dt} + \frac{P}{k}\frac{dV}{dt} = \frac{1}{k}\left(V\frac{dP}{dt} + P\cdot\frac{dV}{dt}\right)$.

26. $\frac{dT}{dt} = \frac{1}{0.4}(66\cdot 0.05 - 0.5\times 1) = 7$ in· lb/min.

27. $W = \frac{Li^2}{2}$; $dw = \frac{\partial W}{\partial L}dL + \frac{\partial w}{\partial i}di = \frac{i^2}{2}(0.02) + Li(-0.1) = \frac{1^2}{2}(0.02) + 30(1)(-0.1) = 0.01 - 0.3 = -0.29$.

28. $A = xy$; $\frac{dA}{dt} = \frac{\partial A}{\partial x}\frac{dx}{dt} + \frac{\partial A}{\partial y}\frac{dy}{dt} = y\frac{dx}{dt} + x\frac{dy}{dt} = 7\cdot 2 + 6\cdot 3 = 14 + 18 = 32\text{ cm}^2/\text{min}$.

29. $V = \ell\cdot w\cdot h$; $h = \frac{500}{\ell w}$. $A = \ell w + 2\ell\cdot h + 2\ell h = \ell w + 2\cdot\frac{500}{w} + 2\cdot\frac{500}{\ell}$. $A_\ell = w - \frac{1000}{w^2}$; $A_w = \ell - \frac{1000}{\ell^2}$; $A_\ell = A_w = 0$ so $w - \frac{1000}{\ell^2} = 0$ and $\ell - \frac{1000}{w^2} = 0$. Hence $\ell = \frac{1000}{w^2}$ and substituting we get $w - \frac{1000}{\left(\frac{1000}{w^2}\right)^2} = 0$ or $w - \frac{w^4}{1000} = 0$. Now $w\left(1 - \frac{w}{1000}\right) = 0$ so $w = 0$ or $w^3 = 1000 \Rightarrow w = 10$. But, w cannot be 0, so $w = 10$ and $\ell = \frac{1000}{10^2} = 10$; $h = \frac{500}{10\times 10} = 5$. To conclude that this is a minimum we compute $A_{\ell\ell} = \frac{2000}{\ell^3}$; $A_{ww} = \frac{2000}{w^3}$. $A_{w\ell} = A_{\ell w} = 1$; $\Delta = \begin{vmatrix} \frac{2000}{10^3} & 1 \\ 1 & \frac{2000}{10^3}\end{vmatrix} = 4 - 1 = 3 > 0$. $A_{\ell\ell} = 2 > 0$ so these dementions yield a minimum; $h = 5$ cm, $\ell = 10$ cm, $w = 10$ cm.

30. In this case $A = 2(\ell w + \ell h + hw) = 2(\ell w + \frac{500}{w} + \frac{500}{\ell})$. $A_\ell = 2\left(w - \frac{500}{\ell^2}\right)$, $A_w = 2\left(\ell - \frac{500}{w^2}\right)$. Setting these both equal to zero and solving we have $w = \frac{500}{\ell^2}$ and substituting $\ell - \frac{500}{\left(\frac{500}{\ell^2}\right)^2} = 0$ or $\ell - \frac{\ell^4}{500} = 0$. Hence, $\ell = 0$ or $\ell = \sqrt[3]{500} = 5\sqrt[3]{4}$. Back substituting we get $h = w = \ell = 5\sqrt[3]{4} \approx 7.937$ cm. As in exercise #29, this yields a minimum.

31. (a) $V = \ell wh = 4$ or $h = \frac{4}{\ell w}$. Cost $= 1.25\ell w + 1.50(2\ell h + 2\ell w) = 1.25\ell w + 3\left(\frac{4}{w} + \frac{4}{\ell}\right) = 1.25\ell w + 12\left(\frac{1}{w} + \frac{1}{\ell}\right)$. $C_\ell = 1.25w - \frac{12}{\ell^2}$, $C_w = 1.25\ell - \frac{12}{w^2}$. Setting these both equal to zero we get $w = \frac{9.6}{\ell^2}$. Substituting, we have $1.25\ell - \frac{12}{\left(\frac{9.6}{\ell^2}\right)^2} = 0$ or $1.25\ell - 0.1302\ell^4 = 0$, or $\ell(1.25 - 0.1302\ell^3) = 0$, so $\ell = 0$ or $\ell = \sqrt[3]{\frac{1.25}{0.1302}} = 2.125$. Since ℓ cannot be 0, we have $\ell = 2.125$ ft, $w = 2.125$ ft and $h = \frac{4}{(2.125)^2} = 0.886$ ft. This will yield a minimum cost.
(b) $C = 1.25(2.125)^2 + 6(2.125\times 0.886) = \$16.94$

32. $T = x^2 + 2y^2 - x$; $T_x = 2x - 1$; $T_y = 4y$; $T_y = 0$; and $T_x = 0$ yields the point $(\frac{1}{2}, 0)$. Thus, $T_{xx} = 2$, $T_{xy} = T_{yx} = 0$, $T_{yy} = 4$; $\Delta = \begin{vmatrix} 2 & 0 \\ 0 & 4\end{vmatrix} = 8 > 0$; $T_{xx} = 2$ so minimum. $T(\frac{1}{2}, 0) = \frac{1}{2}^2 + 2\cdot 0 - \frac{1}{2} = -\frac{1}{4} = -0.25$.

30.5 MULTIPLE INTEGRALS

1. $\int_0^1\int_0^{x^2} xy\,dy\,dx = \int_0^1\left[\frac{1}{2}xy^2\right]_0^{x^2} dx = \int_0^1 \frac{1}{2}x^5\,dx = \frac{1}{12}x^6\Big]_0^1 = \frac{1}{12}$

2. $\int_{-1}^{1}\int_{-2}^{2}(2xy-3y^2)\,dy\,dx =$
$\int_{-1}^{1}\left[xy^2-y^3\right]_{-2}^{2}dx =$
$\int_{-2}^{1}[(4x-8)-(4x+8)]\,dx = \int_{-1}^{1}(-16)\,dx =$
$-16x|_{-1}^{1} = (-16)-(+16) = -32$

3. $\int_{0}^{\pi}\int_{-\pi/2}^{\pi/2}\sin x\cos y\,dy\,dx =$
$\int_{0}^{\pi}[-\sin x\sin y]_{-\pi/2}^{\pi/2}dx =$
$\int_{0}^{\pi}[-\sin x(1+1)]\,dx = -2\cos x|_0^{\pi} = 2+2 = 4$

4. $\int_{0}^{1}\int_{0}^{2}\sqrt{x+y}\,dy\,dx = \int_{0}^{1}\left[\frac{2}{3}(x+y)^{3/2}\right]_0^2 dx =$
$\frac{2}{3}\int_{0}^{1}\left[(x+2)^{3/2}-x^{3/2}\right]dx =$
$\frac{4}{15}\left[(x+2)^{5/2}-x^{5/2}\right]_0^1 = \frac{4}{15}\left[(3)^{5/2}-1-2^{5/2}\right] \approx$
2.3818

5. $\int_{0}^{\ln 2}\int_{0}^{\ln 5}e^{x+y}\,dy\,dx = \int_{0}^{\ln 2}\left[e^{x+y}\right]_0^{\ln 5}dx =$
$\int_{0}^{\ln 2}e^x(5-1) = 4e^x|_0^{\ln 2} = 4(2-1) = 4$

6. $\int_{0}^{1}\int_{y}^{y^2}(x+y)\,dx\,dy = \int_{0}^{1}\left[\frac{x^2}{2}+xy\right]_y^{y^2}dy =$
$\int_{0}^{1}\left(\frac{y^4}{2}+y^3-\frac{y^2}{2}-y^2\right)dy =$
$\left[\frac{y^5}{10}+\frac{y^4}{4}-\frac{y^3}{2}\right]_0^1 = \frac{1}{10}+\frac{1}{4}-\frac{1}{2} = \frac{2+5-10}{20} =$
$\frac{-3}{20}$

7. $\int_{0}^{1}\int_{x}^{\sqrt{x}}(x+y)\,dy\,dx = \int_{0}^{1}\left[xy+\frac{y^2}{2}\right]_x^{\sqrt{x}}dx =$
$\int_{0}^{1}\left(x^{3/2}-\frac{x}{2}-x^2-\frac{x^2}{2}\right)dx =$
$\left[\frac{2}{5}x^{5/2}+\frac{x^2}{4}-\frac{x^3}{2}\right]_0^1 = \frac{2}{5}+\frac{1}{4}-\frac{1}{2} =$
$\frac{8+5-10}{20} = \frac{3}{20}.$

8. $\int_{0}^{4}\int_{-\sqrt{2x}}^{\sqrt{2x}}(2x+y)\,dy\,dx =$
$\int_{0}^{4}\left(2xy+\frac{y^2}{2}\right)_{-\sqrt{2x}}^{\sqrt{2x}}dx =$
$\int_{0}^{4}\left(2\sqrt{2}x^{3/2}+x+2\sqrt{2}x^{3/2}-x\right)dx =$
$\int_{0}^{4}4\sqrt{2}x^{3/2}\,dx = \frac{8\sqrt{2}}{5}x^{5/2}\Big|_0^4 = \frac{8\sqrt{2}\cdot 32}{5} \approx$
72.4077

9. $\int_{0}^{1}\int_{x^3}^{x}(y-x)\,dy\,dx = \int_{0}^{1}\left[\frac{y^2}{2}-xy\right]_{x^3}^{x}dx =$
$\int_{0}^{1}\left(\frac{x^2}{2}-x^2-\frac{x^6}{2}+x^4\right)dx =$
$\left[-\frac{x^3}{6}-\frac{x^7}{14}+\frac{x^5}{5}\right]_0^1 = -\frac{1}{6}-\frac{1}{14}+\frac{1}{5} =$
$\frac{-35-15+42}{210} = \frac{-8}{210} = \frac{-4}{105}.$

10. $\int_{0}^{4}\int_{y}^{\sqrt{y}}(3x+2y)\,dx\,dy =$
$\int_{0}^{4}\left[\frac{3x^2}{2}+2xy\right]_y^{\sqrt{y}}dy =$
$\int_{0}^{4}\left(\frac{3y}{2}+2y^{3/2}-\frac{3y^2}{2}-2y^2\right)dy =$
$\int_{0}^{4}\left(\frac{3y}{2}+2y^{3/2}-\frac{7y^2}{2}\right)dy =$
$\left[\frac{3y^2}{4}+\frac{4}{5}y^{5/2}-\frac{7}{6}y^3\right]_0^4 = 3\cdot 4+\frac{4}{5}\cdot 4^{5/2}-\frac{7\cdot 64}{6} =$
$12+\frac{128}{5}-\frac{7\cdot 64}{6} = \frac{-1112}{30} = \frac{-556}{15} \approx -37.0667.$

11. $\int_{0}^{2}\int_{0}^{3}(xy+x-y)\,dy\,dx =$
$\int_{0}^{2}\left(\frac{xy^2}{2}+xy-\frac{y^2}{2}\right)_0^3 dx =$

$\int_0^2 \left(\frac{9}{2}x + 3x - \frac{9}{2}\right) dx = \int_0^2 \left(\frac{15}{2}x - \frac{9}{x}\right) dx = \left[\frac{15x^2}{4} - \frac{9x}{2}\right]_0^2 = 15 - 9 = 6.$

12. $\int_0^1 \int_0^x xy\,dy\,dx = \int_0^1 \left.\frac{xy^2}{2}\right|_0^x dx = \int_0^1 \frac{x^3}{2}\,dx = \left.\frac{x^4}{8}\right|_0^1 = \frac{1}{8}.$

13. $\int_0^1 \int_0^{1-x} (x^2y + xy^2)\,dy\,dx$

$= \int_0^1 \left[\frac{x^2y^2}{2} + \frac{xy^3}{3}\right]_0^{1-x} dy$

$= \int_0^1 \left(\frac{x^2(1-x)^2}{2} + \frac{x(1-x)^3}{3}\right) dx$

$= \int_0^1 \left(\frac{x^2 - 2x^3 + x^4}{2} + \frac{x - 3x^2 + 3x^2 - x^4}{3}\right) dx$

$= \int_0^1 \left(\frac{x}{3} - \frac{x^2}{2} + \frac{x^4}{6}\right) dx$

$= \left[\frac{x^2}{6} - \frac{x^3}{6} + \frac{x^5}{30}\right]_0^1 = \frac{1}{30}$

14. $\int_0^2 \int_0^{x^2} (x^2 - y^2)\,dy\,dx = \int_0^2 \left.\left(x^2y - \frac{y^3}{3}\right)\right|_0^{x^2} = \int_0^2 \left[x^4 - \frac{x^6}{3}\right] dx = \left[\frac{x^5}{5} - \frac{x^7}{21}\right]_0^2 = \frac{32}{5} - \frac{128}{21} = \frac{32}{105} \approx 0.3048.$

15. $\int_0^1 \int_{y^2}^y (xy + 1)\,dx\,dy = \int_0^1 \left[\frac{x^2y}{2} + x\right]_{y^2}^y dy = \int_0^1 \left(\frac{y^3}{2} + y - \frac{y^5}{2} - y^2\right) dy = \left[\frac{y^4}{8} + \frac{y^2}{2} - \frac{y^6}{12} - \frac{y^3}{3}\right]_0^1 = \frac{1}{8} + \frac{1}{2} - \frac{1}{12} - \frac{1}{3} = \frac{3 + 12 - 2 - 8}{24} = \frac{5}{24}.$

16. $\int_1^2 \int_y^{y^2} \frac{x}{y}\,dx\,dy = \int_1^2 \left.\frac{x^2}{2y}\right|_y^{y^2} dy = \int_1^2 \left(\frac{y^3}{2} - \frac{y}{2}\right) dy = \left[\frac{y^4}{8} - \frac{y^2}{4}\right]_1^2 = \frac{16}{8} - 1 - \frac{1}{8} + \frac{1}{4} = \frac{16 - 8 - 1 + 2}{8} = \frac{9}{8}.$

17. For this region we can integrate with respect to y first from x to 1 and then with respect to x from 0 to 1. $\int_0^1 \int_x^1 (x^2 - y^2)\,dy\,dx = \int_0^1 \left[x^2y - \frac{y^3}{3}\right]_x^1 dx = \int_0^1 (x^2 - \frac{1}{3} - x^3 + \frac{x^3}{3})\,dx = \int_0^1 (x^2 + \frac{2x^3}{3} - \frac{1}{3})\,dx = \left[\frac{x^3}{3} + \frac{x^4}{6} - \frac{x}{3}\right]_0^1 = \frac{1}{6}.$

18. $\int_0^2 \int_0^{x^2} x^2y\,dy\,dx = \int_0^2 \left.\frac{x^2y^2}{2}\right|_0^{x^2} dx = \int_0^2 \frac{x^6}{2}\,dx = \left.\frac{x^7}{14}\right|_0^2 = \frac{128}{14} = \frac{64}{7}.$

19. $\int_0^8 \int_0^{\sqrt[3]{y}} (x^3 - y^3)\,dx\,dy = \int_0^8 \left.\left(\frac{x^4}{4} - y^3x\right)\right|_0^{\sqrt[3]{y}} dy = \int_0^8 \left(\frac{y^{4/3}}{4} - y^{10/3}\right) dy = \left[\frac{3y^{7/3}}{28} - \frac{3y^{13/3}}{13}\right]_0^8 = \frac{3 \cdot 2^7}{28} - \frac{3 \cdot 2^{13}}{13} = -1876.7473.$

20. $\int_0^1 \int_{x^4}^x xy\,dy\,dx = \int_0^1 \left.\frac{xy^2}{2}\right|_{x^4}^x dx = \int_0^1 \left(\frac{x^3}{2} - \frac{x^9}{2}\right) dx = \left.\frac{x^4}{8} - \frac{x^{10}}{20}\right|_0^1 = \frac{1}{8} - \frac{1}{20} = \frac{5 - 2}{40} = \frac{3}{40}$

21. $\int_1^8 \int_{8/x}^{9-x} (x + y)\,dy\,dx$

$= \int_1^8 \left[xy + \frac{y^2}{2}\right]_{8/x}^{9-x} dx$

$$= \int_1^8 \left(x(9-x) + \frac{(9-x)^2}{2} - 8 - \frac{32}{x^2} \right) dx$$

$$= \int_1^8 \left(9x - x^2 + \frac{81}{2} - 9x + \frac{x^2}{2} - 8 - \frac{32}{x^2} \right) dx$$

$$= \int_1^8 \left(\frac{-x^2}{2} - \frac{32}{x^2} + \frac{65}{2} \right) dx$$

$$= \left. \frac{-x^3}{6} + \frac{32}{x} + \frac{65x}{2} \right|_1^8$$

$$= \frac{-8^3}{6} + \frac{32}{8} + \frac{65 \cdot 8}{2} + \frac{1}{6} - 32 - \frac{65}{2}$$

$$= \frac{343}{3} = 114\frac{1}{3}$$

22. $\int_1^4 \int_0^5 (x^2 + y^2)\, dy\, dx = \int_1^4 \left[x^2 + \frac{y^3}{3} \right]_0^5 dx =$
$\int_1^4 \left(5x^2 + \frac{125}{3} \right) dx = \left[\frac{5x^3}{3} + \frac{125x}{3} \right]_1^4 =$
$\left(\frac{5 \cdot 64}{3} + \frac{126 \cdot 4}{3} \right) - \left(\frac{5}{3} + \frac{125}{3} \right) = \frac{1279}{3} = 230$

23. $\int_0^1 \int_{x^3}^x (x+y)^2\, dy\, dx$

$$= \int_0^1 \int_{x^3}^x (x^2 + 2xy + y^2)\, dy\, dx$$

$$= \int_0^1 \left[x^2y + xy^2 + \frac{y^3}{3} \right]_{x^3}^x dx$$

$$= \int_0^1 \left[\left(x^3 + x^3 + \frac{x^3}{3} \right) - \left(x^5 + x^7 + \frac{x^9}{3} \right) \right]_{x^3}^x dx$$

$$= \int_0^1 \left[\frac{7}{3}x^3 - x^5 - x^7 - \frac{1}{3}x^9 \right] dx$$

$$= \left[\frac{7}{12}x^4 - \frac{1}{6}x^6 - \frac{1}{8}x^8 - \frac{1}{30}x^{10} \right]_0^1$$

$$= \frac{7}{12} - \frac{1}{6} - \frac{1}{8} - \frac{1}{30}$$

$$= \frac{60 + 10 - 20 - 15 - 4}{120} = \frac{31}{120} \approx 0.2583$$

24. $x = y^2$ and $x = y + 4$ intersect when $y + 4 = y^2$ or $y^2 - y - 4 = 0$. By the quadratic formula, $y = \frac{1 \pm \sqrt{17}}{2} \approx -1.5616$ and $= 2.5616$. Thus, the desired integral is
$\int_{-1.5616}^{2.5616} \int_{y^2}^{y+4} (xy)\, dx\, dy = \int_{-1.5616}^{2.5616} \left. \frac{x^2y}{2} \right|_{y^2}^{y+4} dy =$
$\frac{1}{2} \int_{-1.5616}^{2.5616} \left[(y+4)^2 y - y^5 \right] dy =$
$\frac{1}{2} \int_{-1.5616}^{2.5616} \left(y^3 + 8y^2 + 16y - y^5 \right) dy =$
$\frac{1}{2} \left[\frac{1}{4}y^4 + \frac{8}{3}y^3 + 8y^2 - \frac{1}{6}y^6 \right]_{-1.5616}^{2.5616} \approx$
$\frac{1}{2}(60.9931 - 8.4235) = 26.2848.$

25. The desired area is symmetric about the x-axis, so we can need only to integrate the y-values from 0 to 3, and double that answer. Thus, we have
$\int_{-3}^3 \int_0^{y^2-9} xy\, dx\, dy = 2 \int_0^3 \int_0^{y^2-9} xy\, dx\, dy =$
$2 \int_0^3 \left. \frac{x^2y}{2} \right|_0^{y^2-9} dy = \int_0^3 (y^2 - 9)^2 y\, dy =$
$\int_0^3 (y^5 - 18y^3 + 81y)\, dy =$
$\left[\frac{1}{6}y^6 - \frac{9}{2}y^4 + \frac{81}{2}y^2 \right]_0^3 = 121.5.$

26. The area is
$\int_0^1 \int_{x^2}^1 \sqrt{xy}\, dy\, dx = \int_0^1 \left. \frac{2}{3}\sqrt{x}y^{3/2} \right|_{x^2}^1 dx =$
$\frac{2}{3} \int_0^1 \left(x^{1/2} - x^{7/2} \right) dx = \frac{2}{3} \left[\frac{2}{3}x^{3/2} - \frac{2}{9}x^{9/2} \right]_0^1 =$
$\frac{2}{3} \left[\frac{2}{3} - \frac{2}{9} \right] = \frac{2}{3} \left[\frac{6-2}{9} \right] = \frac{2}{3} \cdot \frac{4}{9} = \frac{2}{27}$. (Note: $\sqrt{xy}$ is undefined if $xy < 0$ so the region can only be in the first or third quadrants.) The region described is in the second and first. We can only use the first. Hence $0 \le x \le 1$ and $x^2 \le y \le 1$).

27. Solving $x + 2y + 2 = 4$ for z we get $z = 4 - x - 2y$. The x-intercept is 4 and the trace in the xy plane is $y = 2 - \frac{1}{2}x$. Thus, the volume is

$$V = \int_0^4 \int_0^{2 - \frac{1}{2}x} (4 - x - 2y)\, dy\, dx$$

$$= \int_0^4 \left[4y - xy - y^2\right]_0^{2-\frac{1}{2}x} dx$$
$$= \int_0^4 \left[4\left(2 - \frac{1}{2}x\right) - x\left(2 - \frac{1}{2}x\right) - \left(2 - \frac{1}{2}x\right)^2\right] dx$$
$$= \int_0^4 \left(8 - 2x - 2x + \frac{1}{2}x^2 - 4 + 2x - \frac{1}{4}x^2\right) dx$$
$$= \int_0^4 \left(4 - 2x + \frac{1}{4}x^2\right) dx$$
$$= 4x - x^2 + \frac{1}{12}x^3\Big]_0^4$$
$$= 16 - 16 + \frac{1}{12}64 = \frac{16}{3} = 5\frac{1}{3}$$

28. Solving $x^2 + z^2 = 4$ for z we have $z = \pm\sqrt{4 - x^2}$. To get the volume we integrate from 0 to 2 and double the answer. Thus,
$$V = 2\int_0^2 \int_0^2 \sqrt{4 - x^2}\, dx\, dy =$$
$$2\int_0^2 \left(\frac{x}{4}\sqrt{4 - x^2} + 2\sin^{-1}\frac{x}{2}\right)_0^2 =$$
$$2\int_0^2 (2\sin^{-1} 1 - 2\sin^{-1} 0)\, dy = 2\int_0^2 2\pi =$$
$$4\pi y|_0^2 = 8\pi.$$

29. As in problem #28, $z = \pm\sqrt{2 - x^2}$. We can integrate from 0 to $\sqrt{2}$ and double the answer.
Thus, $V = 2\int_0^{\sqrt{2}} \int_0^x \sqrt{2 - x^2}\, dy\, dx =$
$2\int_0^{\sqrt{2}} \sqrt{2 - x^2} \cdot x\, dx$. Let $u = 2 - x^2$, and $du = -2x\, dx$. Substituting, we get
$$V = -\int_0^{\sqrt{2}} u^{1/2} du = -\frac{2}{3}u^{3/2} =$$
$$-\frac{2}{3}\sqrt{2 - x^2}^3\Big|_0^{\sqrt{2}} = \frac{2}{3}\sqrt{2}^3 = \frac{4\sqrt{2}}{3} \approx 1.8856.$$

30. $$\int_0^\infty \int_0^\infty e^{-(x+y)}\, dy\, dx = \int_0^\infty -e^{-(x+y)}\Big|_0^\infty dx =$$
$$\int_0^\infty e^{-x}\, dx = -e^{-x}\big|_0^\infty = e^0 = 1.$$

30.6 CYLINDRICAL AND SPHERICAL COORDINATES

1. Here $x = 2\cos\frac{\pi}{4} = \sqrt{2}$ and $y = 2\sin\frac{\pi}{4} = \sqrt{2}$. Thus, the cylindrical coordinate $(2, \pi/4, 2)$ has the equivalent rectangular coordinate $(\sqrt{2}, \sqrt{2}, 2)$.

2. Here $x = 3\cos\frac{2\pi}{3} = -\frac{3}{2}$ and $y = 3\sin\frac{2\pi}{3} = \frac{3\sqrt{3}}{2}$. Thus, the cylindrical coordinate $(3, 2\pi/3, -2)$ has the equivalent rectangular coordinate $\left(-\frac{3}{2}, \frac{3\sqrt{3}}{2}, -2\right)$ or $(-1.5, 2.598, -2)$.

3. $x = 2\cos 0 = 2$, $y = 2\sin 0 = 0$. Thus, the cylindrical coordinate $(2, 0, 4)$ has the equivalent rectangular coordinate $(2, 0, 4)$.

4. $x = y = 0$. Thus, the cylindrical coordinate $(0, 5\pi/4, -5)$ has the equivalent rectangular coordinate $(0, 0, -5)$.

5. Here $x = 5\cos\frac{5\pi}{4} = \frac{-5\sqrt{2}}{2}$ and $y = 5\sin\frac{5\pi}{4} = -\frac{5\sqrt{2}}{2}$, so the cylindrical coordinate $(5, 5\pi/4, 0)$ has the equivalent rectangular coordinate $\left(\frac{-5\sqrt{2}}{2}, \frac{-5\sqrt{2}}{2}, 0\right)$ or $(-3.5355, -3.5355, 0)$.

6. Here $x = 4\cos\frac{5\pi}{3} = 2$ and $y = 4\sin\frac{5\pi}{3} = -2\sqrt{3}$, which means that the cylindrical coordinate $(4, 5\pi/3, 7)$ has the equivalent rectangular coordinate $(2, -2\sqrt{3}, 7)$ or $(2, -3.4641, 7)$.

7. Here $r = \sqrt{x^2 + y^2} = \sqrt{2^2 + 2^2} = \sqrt{8} = 2\sqrt{2}$ and $\theta = \tan^{-1}\frac{y}{x} = \tan^{-1}\frac{2}{2} = \tan^{-1} 1 = \frac{\pi}{4}$. This means that the rectangular coordinates $(2, 2, 5)$ have cylindrical coordinates $\left(2\sqrt{2}, \frac{\pi}{4}, 5\right) = (2.8284, 0.7854, 5)$.

8. Here $r = \sqrt{3^2 + (-4)^2} = 5$ and $\theta = \tan^{-1}\frac{-4}{3} = 5.3559$. This means that the rectangular coordinates $(3, -4, -5)$ have cylindrical coordinates $(5, 5.3559, -5)$.

9. Here $r = \sqrt{(-4)^2 + 3^2} = 5$ and $\theta = \tan^{-1}\frac{-3}{4} = 2.4981$. This means that the rectangular coordinates $(-4, 3, 2)$ have cylindrical coordinates $(5, 2.4981, 2)$.

10. Here $r = \sqrt{(-1)^2 + (-\sqrt{3})^2} = \sqrt{1+3} = 2$ and $\theta = \tan^{-1}\frac{-\sqrt{3}}{-1} = 4.1888$. Thus, the rectangular coordinates $(-1, -\sqrt{3}, 4)$ have cylindrical coordinates $(2, 4.1888, 4)$ or $(2, \frac{4\pi}{3}, 4)$.

11. Here $r = \sqrt{12^2 + (-5)^2} = 13$ and $\theta = \tan^{-1}\frac{-5}{12} = 5.8884$. Thus, the rectangular coordinates $(12, -5, -3)$ have cylindrical coordinates $(13, 5.8884, -3)$

12. Here $r = \sqrt{6^2 + 8^2} = 10$ and $\theta = \tan^{-1}\frac{8}{6} = 0.9273$. Thus, the rectangular coordinates $(6, 8, -4)$ have cylindrical coordinates $(10, 0.9273, -4)$

13. Here $x = 4\sin\frac{\pi}{6}\cos\frac{\pi}{4} = 4 \cdot \frac{1}{2} \cdot \frac{\sqrt{2}}{2} = \sqrt{2}$ and $y = 4\sin\frac{\pi}{6}\sin\frac{\pi}{4} = 4 \cdot \frac{1}{2} \cdot \frac{\sqrt{2}}{2} = \sqrt{2}$ and $z = 4\cos\frac{\pi}{6} = 4 \cdot \frac{3}{2} = 2\sqrt{3}$. Thus, the spherical coordinate $(4, \frac{\pi}{4}, \frac{\pi}{6})$ has the equivalent rectangular coordinate $(\sqrt{2}, \sqrt{2}, 2\sqrt{3}) \approx (1.4142, 1.4142, 3.4641)$

14. Here $x = \sin\frac{\pi}{2}\cos\frac{\pi}{6} = 1 \cdot \frac{1}{2} \cdot \frac{\sqrt{3}}{2} = \frac{\sqrt{3}}{2}$ and $y = \sin\frac{\pi}{2}\sin\frac{\pi}{6} = \frac{1}{2}$ and $z = \cos\frac{\pi}{2} = 0$. Thus, the spherical coordinate $(1, \frac{\pi}{6}, \frac{\pi}{2})$ has the equivalent rectangular coordinate $\left(\frac{\sqrt{3}}{3}, \frac{1}{2}, 0\right) = (0.8660, 0.5, 0)$.

15. Here $x = 3\sin\frac{5\pi}{3}\cos\frac{\pi}{2} = 3 \cdot \frac{-\sqrt{3}}{2} \cdot 0 = 0$, $y = \sin\frac{5\pi}{3}\sin\frac{\pi}{2} = \frac{-3\sqrt{3}}{2}$, and $z = 3\cos\frac{5\pi}{3} = \frac{3}{2}$. Thus, the spherical coordinate $\left(3, \frac{\pi}{2}, \frac{5\pi}{3}\right)$ has the equivalent rectangular coordinate $\left(0, \frac{-3\sqrt{3}}{2}, \frac{3}{2}\right) = (0, -2.5981, 1.5)$.

16. Here $x = 2\sin\frac{3\pi}{4}\cos\frac{5\pi}{6} = 2 \cdot \frac{-\sqrt{2}}{2} \cdot \frac{-\sqrt{3}}{2} = \frac{-\sqrt{6}}{2}$, $y = 2\sin\frac{3\pi}{4}\sin\frac{5\pi}{6} = 2 \cdot \frac{\sqrt{2}}{2} \cdot \frac{1}{2} = \frac{\sqrt{2}}{2}$, and $z = 2\cos\frac{3\pi}{4} = -\sqrt{2}$. Thus, the spherical coordinate $\left(2, \frac{5\pi}{6}, \frac{3\pi}{4}\right)$ has the equivalent rectangular coordinate $\left(\frac{-\sqrt{6}}{2}, \frac{\sqrt{2}}{2}, -\sqrt{2}\right) = (-1.2247, 0.7071, -1.4142)$.

17. Here $x = 5\sin\frac{2\pi}{3}\cos\frac{7\pi}{6} = 5 \cdot \frac{-\sqrt{3}}{2} \cdot \frac{-\sqrt{3}}{2} = \frac{-15}{4} = -3.75$, $y = 5\sin\frac{2\pi}{3}\sin\frac{7\pi}{6} = 5 \cdot \frac{\sqrt{3}}{2} \cdot \frac{-1}{2} = \frac{-5\sqrt{3}}{4} \approx -2.1651$, and $z = 5\cos\frac{2\pi}{3} = 5 \cdot -\frac{1}{2} = -2.5$. Thus, the spherical coordinate $\left(5, \frac{7\pi}{6}, \frac{2\pi}{3}\right)$ has the equivalent rectangular coordinate $(-3.75, -2.1651, -2.5)$.

18. Here $x = 5\sin\frac{\pi}{4}\cos\frac{5\pi}{3} = 5 \cdot \frac{\sqrt{2}}{2} \cdot \frac{1}{2} = \frac{5\sqrt{2}}{4} = 1.7678$, $y = 5\sin\frac{\pi}{4}\sin\frac{5\pi}{3} = 5 \cdot \frac{\sqrt{2}}{2} \cdot \frac{-\sqrt{3}}{2} = \frac{-5\sqrt{6}}{4} \approx -3.0619$, and $z = 5\cos\frac{\pi}{4} = \frac{5\sqrt{2}}{2} = 3.5355$. Thus, the spherical coordinate $\left(5, \frac{5\pi}{3}, \frac{\pi}{4}\right)$ has the equivalent rectangular coordinate $(1.7678, -3.0619, 3.5355)$.

19. Here $\rho = \sqrt{4^2 + 3^2} = 5$, $\theta = \tan^{-1}\frac{3}{4} = 0.6435$, and $\phi = \cos^{-1}\frac{0}{5} = \frac{\pi}{2} \approx 1.5708$. Thus, the rectangular coordinate $(4, 3, 0)$ has the spherical coordinate $(5, 0.6435, \frac{\pi}{2})$.

20. Here $\rho = \sqrt{4^2 + \sqrt{5}^2 + (-2)^3} = \sqrt{16+5+4} = 5$, $\theta = \tan^{-1}\frac{\sqrt{5}}{4} = 0.5097$, and $\phi = \cos^{-1}\frac{-2}{5} \approx 1.9823$. Thus, the rectangular coordinate $(4, \sqrt{5}, -2)$ has the spherical coordinate $(5, 0.5097, 1.9823)$.

21. Here $\rho = \sqrt{2^2 + 1^2 + 2^2} = 3$, $\theta = \tan^{-1}\frac{1}{2} = 0.4636$, and $\phi = \cos^{-1}\frac{-2}{3} = 2.3005$. Thus, the rectangular coordinate $(2, 1, -2)$ has the spherical coordinate $(3, 0.4636, 2.3005)$.

22. Here $\rho = \sqrt{4^2 + (-\sqrt{3})^2 + 2^2} = \sqrt{16+3+4} = \sqrt{23} = 4.7958$, $\theta = \tan^{-1}\frac{-\sqrt{3}}{4} = 2\pi - 0.408637 = 5.874547$, and $\phi = \cos^{-1}\frac{2}{\sqrt{23}} = 1.1406$. Thus, the rectangular coordinate $(4, -\sqrt{3}, 2)$ has the spherical coordinate $(4.7958, 5.8745, 1.1406)$.

23. Here $\rho = \sqrt{1^2 + 1^2 + \sqrt{2}^2} = 2$, $\theta = \tan^{-1}\frac{1}{1} = \frac{\pi}{4} = 0.7854$, and $\phi = \cos^{-1}\frac{\sqrt{2}}{2} = \frac{\pi}{4}$. Thus, the rectangular coordinate $(1, 1, \sqrt{2})$ has the spherical coordinate $\left(2, \frac{\pi}{4}, \frac{\pi}{4}\right) = (2, 0.7854, 0.7854)$.

24. Here $\rho = \sqrt{3+3+3} = 3$, $\theta = \tan^{-1}\frac{\sqrt{3}}{-\sqrt{3}} = \frac{3\pi}{4} = 2.3562$, and $\phi = \cos^{-1}\frac{-\sqrt{3}}{3} = 2.1863$. Thus, the rectangular coordinate $(-\sqrt{3}, \sqrt{3}, -\sqrt{3})$ has that spherical coordinate $\left(3, \frac{3\pi}{4}, 2.1863\right)$

25. The graph is a line that makes an angle of $\frac{\pi}{4}$ with the z-axis and whose image on the xy-plane make an angle of $\frac{\pi}{4}$ with the x-axis.

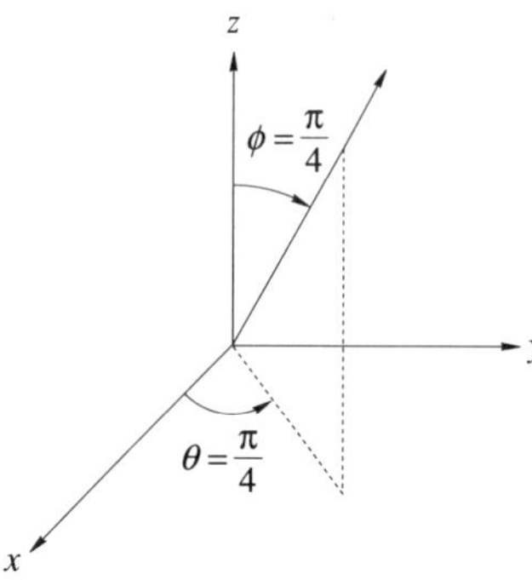

26. A plane through the z-axis and that makes an angle of $\frac{3\pi}{4}$ with the x-axis.

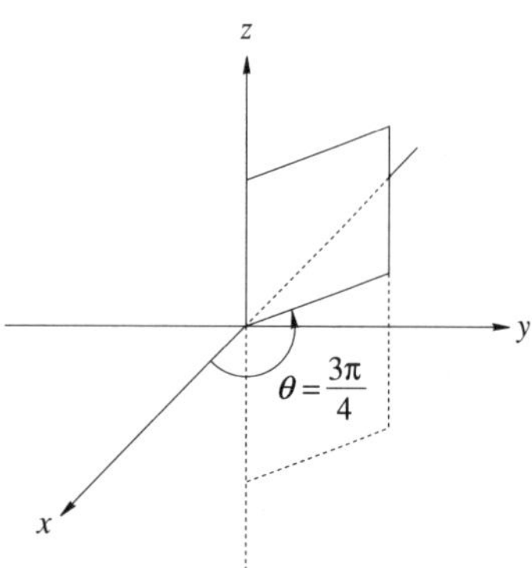

27. The region between the two spheres $\rho = 3$ and $\rho = 5$.

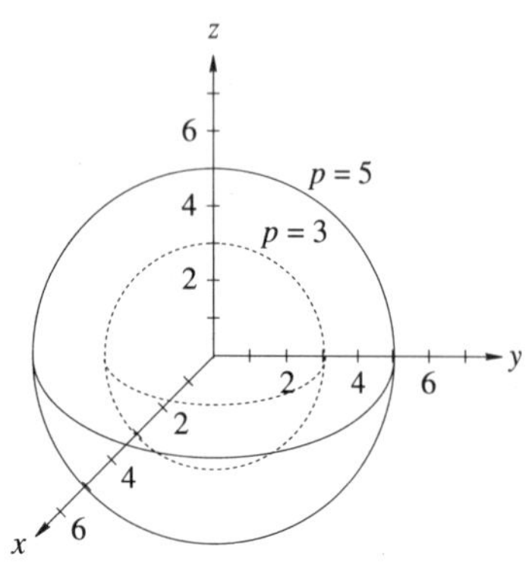

28. A circle of radius 5 in the plane form by $\theta = \pi/3$

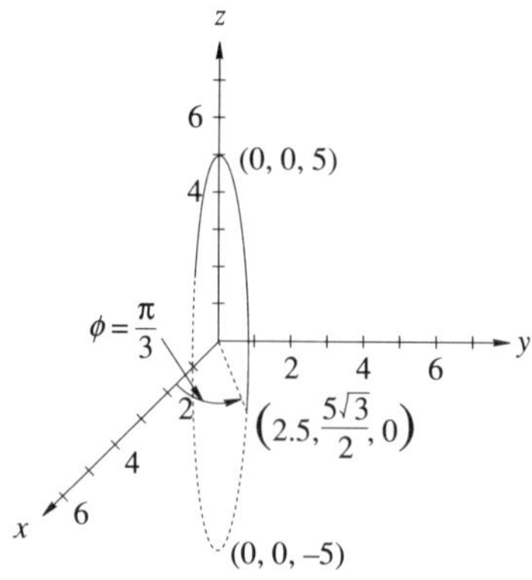

29. A cone shaped figure with spherical base.

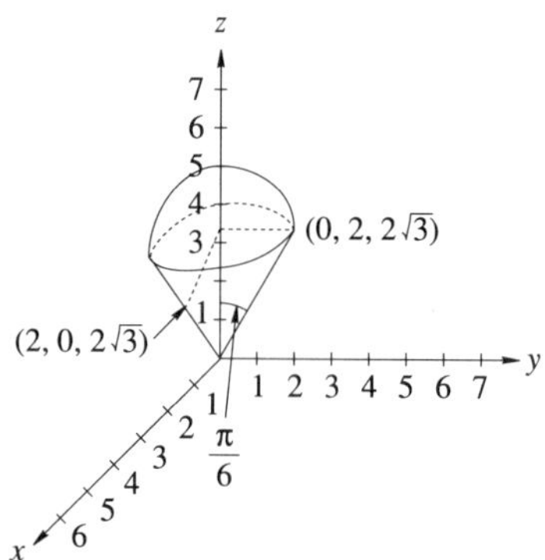

30. A wege that is a section of a sphere of radius 4 on top of a section of a cone

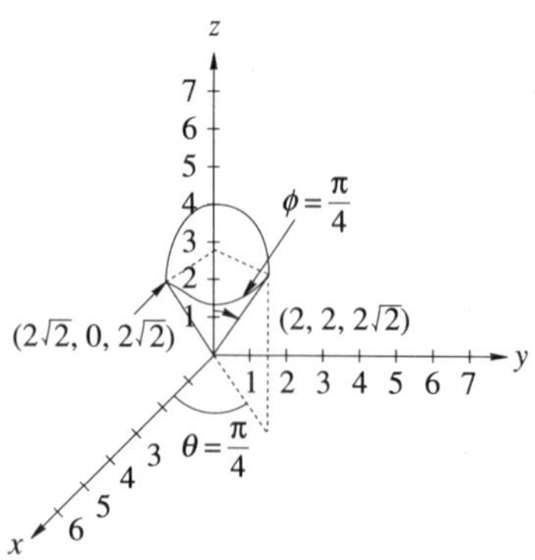

31. See *Computer Programs* in the main text.

30.7 MOMENTS AND CENTROIDS

1. $A = \int_0^3 \int_0^{2x+3} dy\,dx = \int_0^3 y\Big|_0^{2x+3} dx =$ $\int_0^3 (2x+3)\,dx = x^2 + 3x\Big|_0^3 = 18.$
$m = \rho A = 18\rho.$
$M_x = \int_0^3 \int_0^{2x+3} y\rho\,dy\,dx = \rho \int_0^3 \frac{y^2}{2}\Big|_0^{2x+3} dx =$ $\frac{1}{2}\rho \int_0^3 (2x+3)^2\,dx = \frac{1}{2}\rho \int_0^3 (4x^2+12x+9)\,dx =$ $\frac{1}{2}\rho \left[\frac{4}{3}x^3 + 6x^2 + 9x\right]_0^3 = 58.5\rho.$
$M_y = \int_0^3 \int_0^{2x+3} x\rho\,dy\,dx = \rho \int_0^3 x[y]_0^{2x+3} dx =$ $\rho \int_0^3 (2x^2+3x)\,dx = \rho\left[\frac{2}{3}x^3 + \frac{3}{2}x^2\right]_0^3 = 31.5\rho.$
$\bar{x} = \frac{M_y}{m} = \frac{31.5\rho}{18\rho} = 1.75;\ \bar{y} = \frac{M_y}{m} = \frac{58.5\rho}{18\rho} = 3.25$

2. $m = \rho \int_0^2 \int_0^{x^3} dy\,dx = \rho \int_0^2 x^3\,dx = \rho \frac{x^4}{4}\Big|_0^2 = 4\rho.$
$M_x = \rho \int_0^2 \int_0^{x^3} y\,dy\,dx = \rho \int_0^2 \frac{y^2}{2}\Big|_0^{x^3} dx =$
$\frac{\rho}{2} \int_0^2 x^6\,dx = \frac{\rho}{2} \cdot \frac{x^7}{7}\Big|_0^2 = \frac{64}{7}\rho \approx 9.142857\rho.$
$M_y = \rho \int_0^2 \int_0^{x^3} x\,dy\,dx = \rho \int_0^2 xy\Big|_0^{x^3} dx =$
$\int_0^2 x^4\,dx = \frac{\rho}{2} \cdot \frac{x^5}{5}\Big|_0^2 = \frac{32}{5}\rho = 6.4\rho;$
$\bar{x} = \frac{M_y}{m} = \frac{6.4}{4} = 1.6;\ \bar{y} = \frac{M_x}{m} = \frac{\frac{64}{7}}{4} = \frac{16}{7} \approx 2.2857$

3. $m = \rho \int_0^8 \int_0^{x^{1/3}} dy\,dx = \rho \int_0^8 x^{1/3}\,dx = \rho \frac{3}{4}x^{4/3}\Big|_0^8 =$ $12\rho.\ M_x = \rho \int_0^8 \int_0^{x^{1/3}} y\,dy\,dx = \rho \int_0^8 \frac{y^2}{2}\Big|_0^{x^{1/3}} dx =$ $\rho \int_0^8 \frac{1}{2}x^{2/3}\,dx = \rho\left[\frac{1}{2} \cdot \frac{3}{5}x^{5/3}\right]_0^8 = 9.6\rho.$
$M_y = \rho \int_0^8 \int_0^{x^{1/3}} x\,dy\,dx = \rho \int_0^8 x^{4/3}\,dx = \rho \frac{3}{7}x^{7/3}\Big|_0^8 =$ $54.8571.\ \bar{x} = \frac{M_y}{m} = \frac{32}{7} = 4.5714;\ \bar{y} = \frac{M_x}{m} = 0.8$

4. $m = \rho \int_{-1}^2 \int_0^{x^4} dy\,dx = \rho \int_{-1}^2 x^4\,dx = \rho \frac{x^5}{5}\Big|_{-1}^2 =$ $6.6\rho.\ M_x = \rho \int_{-1}^2 \int_0^{x^4} y\,dy\,dx = \rho \int_{-1}^2 \frac{y^2}{2}\Big|_0^{x^4} dx =$
$\rho \int_{-1}^2 \frac{x^8}{2}\,dx = \rho \frac{x^9}{18}\Big|_{-1}^2 = 28.5\rho.\ M_y =$
$\rho \int_{-1}^2 \int_0^{x^4} x\,dy\,dx = \rho \int_{-1}^2 x^5\,dx = \rho \frac{x^6}{6}\Big|_{-1}^2 = 10.5\rho.$
$\bar{x} = \frac{M_y}{m} = \frac{35}{22} = 1.5909;\ \bar{y} = \frac{M_x}{m} = \frac{95}{22} = 4.3182.$

5. $m = \rho \int_0^5 \int_0^{\sqrt{x+4}} dy\,dx = \rho \int_0^5 \sqrt{x+4}\,dx =$ $\rho \cdot \frac{2}{3}(x+4)^{3/2}\Big|_0^5 = \left(18 - \frac{16}{3}\right)\rho = \frac{38}{3}\rho = 12.6667\rho.$
$M_x = \rho \int_0^5 \int_0^{\sqrt{x+4}} y\,dy\,dx = \rho \int_0^5 \frac{y^2}{2}\Big|_0^{\sqrt{x+4}} dx =$
$\rho \int_0^5 \frac{x+4}{2}\,dx = \rho\left[\frac{x^2}{4} + 2x\right]_0^5 = \frac{65}{4}\rho = 16.25\rho.$
$M_y = \rho \int_0^5 \int_0^{\sqrt{x+4}} x\,dy\,dx = \rho \int_0^5 xy\Big|_0^{\sqrt{x+4}} dx =$
$\rho \int_0^5 x\sqrt{x+4}\,dx$. Let $u = x+4$ and $du = dx$. Then $x = u-4$, $x = 0 \Rightarrow u = 4$, and $x = 5 \Rightarrow u = 9$. Substituting, we obtain
$\rho \int_4^9 (u-u)\sqrt{4}du = \rho \int_4^9 \left(u^{3/2} - 4u^{1/2}\right) =$
$\rho\left[\frac{2}{5}u^{5/2} - \frac{8}{3}u^{3/2}\right]_4^9 = 33.7333\rho.\ \bar{x} = \frac{M_y}{m} = 2.6632,$
$\bar{y} = \frac{M_x}{m} = 1.2829.$

6. $m = \rho \int_0^4 \int_0^{\sqrt{x^2+16}} dy\,dx = \rho \int_0^4 \sqrt{x^2+16}\,dx =.$
By Formula #30, in Appendix C, we have
$m = \rho\left[\frac{x}{2}\sqrt{x^2+16} + 8\ln\left|x + \sqrt{x^2+16}\right|\right]_0^4 =$
$(29.4551 - 11.0904)\rho = 18.3647\rho.$
$M_x = \rho \int_0^4 \int_0^{\sqrt{x^2+16}} y\,dy\,dx =$
$\rho \int_0^4 \frac{y^2}{2}\Big|_0^{\sqrt{x^2+16}} dx = \rho \int_0^4 \frac{x^2+16}{2}\,dx =$
$\rho\left[\frac{x^3}{6} + 8x\right]_0^4 = 42.6667\rho.$
$M_y =$

$\rho \int_0^4 \int_0^{\sqrt{x^2+16}} x\,dy\,dx = \rho \int_0^4 (x\sqrt{x^2+16})\,dx.$
Let $u = x^2 + 16$, and then $du = 2x\,dx$. When $x = 0 \Rightarrow u = 16$, and when $x = 4 \Rightarrow u = 32$. Substitution yields $\frac{1}{2}\rho \int_{16}^{32} \sqrt{u}du = \frac{1}{2}\rho \frac{2}{3} u^{3/2}\Big|_{16}^{32} = \frac{1}{3}\rho\left[32^{3/2} - 16^{3/2}\right] = 39.0064\rho$. Thus, $\bar{x} = \frac{M_y}{m} = 2.1240$, $\bar{y} = \frac{M_x}{m} = 2.3233$.

7. $m = \rho \int_0^2 \int_{x^2}^{4x} dy\,dx = \rho \int_0^2 (4x - x^2)\,dx = \left(2x^2 - \frac{x^3}{3}\right)\Big|_0^2 \rho = \left(8 - \frac{8}{3}\right)\rho = \frac{16}{3}\rho = 5.3333\rho.$
$M_x = \rho \int_0^2 \int_{x^2}^{4x} y\,dy\,dx = \rho \int_0^2 \frac{y^2}{2}\Big|_{x^2}^{4x} dx = \rho \int_0^2 \left(8x^2 - \frac{x^4}{2}\right) dx = \rho\left[\frac{8}{3}x^3 - \frac{x^5}{10}\right]_0^2 = 18.1333\rho$. $M_y = \rho \int_0^2 \int_{x^2}^{4x} x\,dy\,dx = \rho \int_0^2 x(4x - x^2)\,dx = \rho \int_0^2 (4x^2 - x^3)\,dx = \rho\left[\frac{4}{3}x^3 - \frac{x^4}{4}\right]_0^2 = 6.6667 = 6\frac{2}{3}\rho$. $\bar{x} = \frac{M_y}{m} = 1.25$, $\bar{y} = \frac{M_x}{m} = 3.4$

8. $m = \rho \int_0^2 \int_{x^3}^{8x} dy\,dx = \rho \int_0^2 (8x - x^3)\,dx = \rho\left[4x^2 - \frac{x^4}{4}\right]_0^2 = 12\rho$. $M_x = \rho \int_0^2 \int_{x^3}^{8x} y\,dy\,dx = \rho \int_0^2 \frac{y^2}{2}\Big|_{x^3}^{8x} dx = \rho \int_0^2 \left((32x^2 - \frac{x^6}{2}\right) dx = \rho\left[\frac{32}{3}x^3 - \frac{x^7}{14}\right]_0^2 = 76.1905\rho$. $M_y = \rho \int_0^2 \int_{x^3}^{8x} x\,dy\,dx = \rho \int_0^2 (8x^2 - x^4)\,dx = \rho\left[\frac{8}{3}x^3 - \frac{x^5}{5}\right]_0^2 = 14.9333\rho.$
Thus, $\bar{x} = \frac{M_y}{m} = 1.2444$, $\bar{y} = \frac{M_x}{m} = 6.3492$.

9. $m = \rho = \int_0^1 \int_{x^{3/2}}^{x} dy\,dx = \rho \int_0^1 (x - x^{3/2})\,dx = \rho\left[\frac{x^2}{2} - \frac{2}{5}x^{5/2}\right]_0^1 = \rho\left(\frac{1}{2} - \frac{2}{5}\right) = \frac{1}{10}\rho$. Thus, we have $M_x = \rho \int_0^1 \int_{x^{3/2}}^{x} y\,dy\,dx = \rho \int_0^1 \frac{y^2}{2}\Big|_{x^{3/2}}^{x} dx = \rho \int_0^1 \left(\frac{x^2}{2} - \frac{x^3}{2}\right) dx = \rho\left[\frac{x^3}{6} - \frac{x^4}{8}\right]_0^1 = \frac{1}{24}$. We also have
$M_y = \rho = \int_0^1 \int_{x^{3/2}}^{x} x\,dy\,dx = \rho \int_0^1 \left(x^2 - x^{5/2}\right) dx = \rho[\frac{x^3}{3} - \frac{2}{7}x^{7/2}]_0^1 = \left(\frac{1}{3} - \frac{2}{7}\right)\rho = \frac{1}{21}\rho$. Hence,
$\bar{x} = \frac{M_y}{m} = \frac{1/21}{1/10} = \frac{10}{21} \approx 0.4762$, and
$\bar{y} = \frac{M_x}{m} = \frac{1/24}{1/10} = \frac{5}{12} \approx 0.4167$.

10. $m = \int_{-3}^{3} \int_{x^2}^{18-x^2} \rho\,dy\,dx = \rho \int_{-3}^{3} (18 - 2x^2)\,dx = \rho\left(18x - \frac{2}{3}x^3\right)\Big|_{-3}^{3} = 72\rho.$
$M_x = \rho \int_{-3}^{3} \int_{x^2}^{18-x^2} y\,dy\,dx = \rho \int_{-3}^{3} \frac{y^2}{2}\Big|_{x^2}^{18-x^2} dx = \frac{\rho}{2}\left(18 - x^2\right)^2 - (x^2)^2\,dx = \frac{\rho}{2}\int_{-3}^{3} 324 - 36x^2\,dx = \frac{\rho}{2}(324x - 12x^3)\Big|_{-3}^{3} = 648\rho$. $M_y = \rho \int_{-3}^{3} \int_{x^2}^{18-x^2} x\,dy\,dx = \rho \int_{-3}^{3} (18x - 2x^3)\,dx = 0.$
$\bar{x} = 0$, $\bar{y} = 9$

11. $y = x$ and $y = 12 - x^2$ intersect at $x = -4$ and $x = 3$. First, we determine that $m = \rho \int_{-4}^{3} \int_{x}^{12-x^2} dy\,dx = \rho \int_{-4}^{3} (12 - x - x^2)\,dx = \rho\left(12x - \frac{x^2}{2} - \frac{x^3}{3}\right)_{-4}^{3} = 57\frac{1}{6}\rho$. Then, we have
$M_x = \rho \int_{-4}^{3} \int_{x}^{12-x^2} y\,dy\,dx = \rho \int_{-4}^{3} \frac{y^2}{2}\Big|_{z}^{12-x^2} dx = \frac{\rho}{2}\int_{-4}^{3} (144 - 24x^2 + x^4 - x^2)\,dx =$

$\frac{\rho}{2}\left[144x - \frac{25x^3}{3} + \frac{x^5}{5}\right]_{-4}^{3} = \frac{(260.1 + 276.8)}{2}\rho =$ 251.5333ρ. We also have $M_y =$ $\rho\int_{-4}^{3}\int_{x}^{12-x^2} x\,dy\,dx = \rho\int_{-4}^{3}[12x - x^3 - x^2]\,dx =$ $\rho\left[6x^2 - \frac{x^3}{3} - \frac{x^4}{4}\right]_{-4}^{3} = -28.5833\rho$. Hence, $\bar{x} = \frac{-28.5833}{57.1667} = -0.5$ and $\bar{y} = \frac{251.5333}{57.1667} = 4.4$.

12. $m = \rho\int_0^1\int_{x^3}^{x^2} dy\,dx = \rho\int_0^1 (x^2 - x^3)\,dx =$ $\rho\left[\frac{x^3}{3} - \frac{x^4}{4}\right]_0^1 = \frac{\rho}{12}$.

$M_x = \rho\int_0^1\int_{x^3}^{x^2} y\,dy\,dx = \rho\int_0^1 \left.\frac{y^2}{2}\right|_{x^3}^{x^2} dx =$ $\frac{\rho}{2}\int_0^1 (x^4 - x^6)\,dx = \frac{\rho}{2}\left[\frac{x^5}{5} - \frac{x^9}{7}\right]_0^1 = \frac{\rho}{35}$.

$M_y = \rho\int_0^1\int_{x^3}^{x^2} x\,dy\,dx = \rho\int_0^1 \left(x^3 - x^4\right)\,dx =$ $\rho\left[\frac{x^4}{4} - \frac{x^5}{5}\right]_0^1 = \frac{\rho}{20}$. $\bar{x} = \frac{12}{20} = \frac{3}{5} = 0.6$. $\bar{y} = \frac{12}{35} \approx 0.3429$

13. Here $I_y = \rho\int_0^{\sqrt{2}}\int_{x^2}^{2} x^2\,dy\,dx = \rho\int_0^{\sqrt{2}} \left. x^2y\right|_{x^2}^{2} dx =$ $\rho\int_0^{\sqrt{2}} (2x^2 - x^4)\,dx = \rho\left[\frac{2x^3}{3} - \frac{x^5}{5}\right]_0^{\sqrt{2}} =$ $\left(\frac{4}{3}\sqrt{2} - \frac{4}{5}\sqrt{2}\right)\rho = \frac{8\sqrt{2}}{15}\rho = 0.7542\rho$. We also find that $m = \rho\int_0^{\sqrt{2}}\int_{x^2}^{2} dy\,dx =$ $\rho\int_0^{\sqrt{2}} (2 - x^2)\,dx = \rho\left[2x - \frac{x^3}{3}\right]_0^{\sqrt{2}} = \rho\frac{4\sqrt{2}}{3}$.

Hence, $r_y = \sqrt{\frac{\frac{8}{15}}{\frac{4}{3}}} = \sqrt{\frac{2}{5}} \approx 0.6325$.

14. $I_x = \rho\int_0^{\sqrt{2}}\int_{x^2}^{2} y^2\,dy\,dx = \rho\int_0^{\sqrt{2}} \left.\frac{y^3}{3}\right|_{x^2}^{2} dx =$ $\rho\int_0^{\sqrt{2}}\left[\frac{8}{3} - \frac{x^6}{3}\right]dx = \rho\left[\frac{8x}{3} - \frac{x^7}{21}\right]_0^{\sqrt{2}} =$ $\left(\frac{8\sqrt{2}}{3} - \frac{8\sqrt{2}}{21}\right)\rho = \frac{48\sqrt{2}}{21}\rho = \frac{16\sqrt{2}}{7}\rho$.

$r_x = \sqrt{\frac{\frac{16}{7}}{4/3}} = \sqrt{\frac{12}{7}} \approx 1.3093$.

15. $m = \rho\int_0^3\int_0^5 dy\,dx = \rho\int_0^3 5\,dx = 15$;

$I_x = \rho\int_0^3\int_0^5 y^2\,dy\,dx = \rho\int_0^3 \left.\frac{y^3}{3}\right|_0^5 dx =$ $\rho\int_0^3 \frac{125}{3}\,dx = 125\rho$; $r_x = \sqrt{\frac{125}{15}} = \sqrt{\frac{25}{3}} \approx 2.8868$.

16. $I_y = \rho\int_0^3\int_0^5 x^2\,dy\,dx = \rho\int_0^3 5x^2\,dx = \rho\left[\frac{5x^3}{3}\right]_0^3 =$ 45ρ; $r_y = \sqrt{\frac{45}{15}} = \sqrt{3} \approx 1.7321$.

17. $m = \rho\int_0^4\int_0^{4x-x^2} dy\,dx = \rho\int_0^4 4x - x^2\,dx =$ $\rho\left[2x^2 - \frac{x^3}{3}\right]_0^4 = \left(32 - \frac{64}{3}\right)\rho = \frac{32}{3}\rho$;

$I_x = \rho\int_0^4\int_0^{4x-x^2} y^2\,dy\,dx = \rho\int_0^4 \left.\frac{y^3}{3}\right|_0^{4x-x^2} dx =$ $\frac{\rho}{3}\int_0^4 (64x - 48x^4 + 12x^5 - x^6)\,dx =$ $\frac{\rho}{3}\left[16x^4 - \frac{48}{5}x^5 - 2x^6 - \frac{x^6}{5}\right]_0^4 \approx 39.010$ or $\frac{4096}{105}$; $r_x = \sqrt{\frac{39.010}{10.667}} = 1.9124$.

18. $I_y = \rho\int_0^4\int_0^{4x-x^2} x^2\,dy\,dx = \rho\int_0^4 (4x^3 - x^4)\,dx =$ $\rho\left[x^4 - \frac{x^5}{5}\right]_0^4 = 51.2\rho$; $r_y = \sqrt{\frac{51.2}{10.667}} = 2.1909$

19. $m = \rho\int_1^2\int_{1/x}^{x^2} dy\,dx = \rho\int_1^2 \left[x^2 - x^1\right]dx =$ $\rho\left[\frac{x^3}{3} - \ln x\right]_1^2 = 1.6402\rho$;

$I_y = \rho \int_1^2 \int_{1/x}^{x^2} x^2\, dy\, dx = \rho \int_1^2 (x^4 - x)\, dx =$
$\left[\frac{x^5}{5} - \frac{x^2}{2}\right]_1^2 \rho = \left(\frac{3^2}{5} - 2 - \frac{1}{5} + \frac{1}{2}\right) \rho = 4.7\rho;$
$r_y = \sqrt{\frac{4.7}{1.6402}} \approx 1.6928$

20. $I_x = \rho \int_1^2 \int_{1/x}^{x^2} y^2\, dy\, dx = \rho \int_1^2 \left.\frac{y^3}{3}\right|_{1/x}^{x^2} dx =$
$\rho \int_1^2 \left(\frac{x^6}{3} - \frac{1}{3x^3}\right) dx = \rho \left[\frac{x^7}{21} + \frac{1}{6x^2}\right]_1^2 =$
$5.9226\rho;\ r_x = \sqrt{\frac{5.9226}{1.6402}} \approx 1.90024.$

CHAPTER 30 REVIEW

1. A plane whose intercepts are (6, 0, 0), (0, 2, 0), and (0, 0, 3)

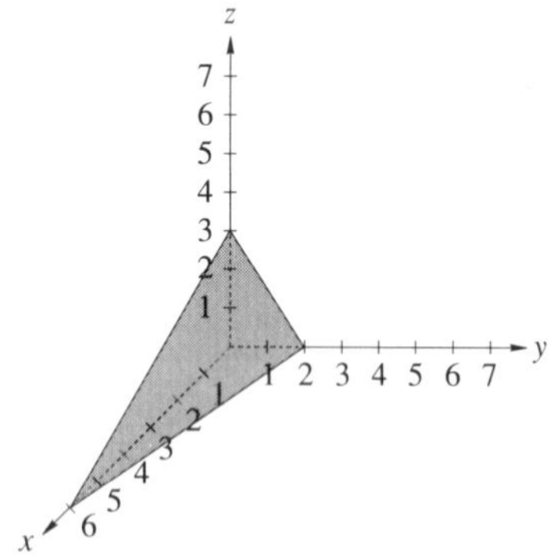

2. A plane whose intercepts are (3, 0, 0), (0, −4, 0), and (0, 0, −6)

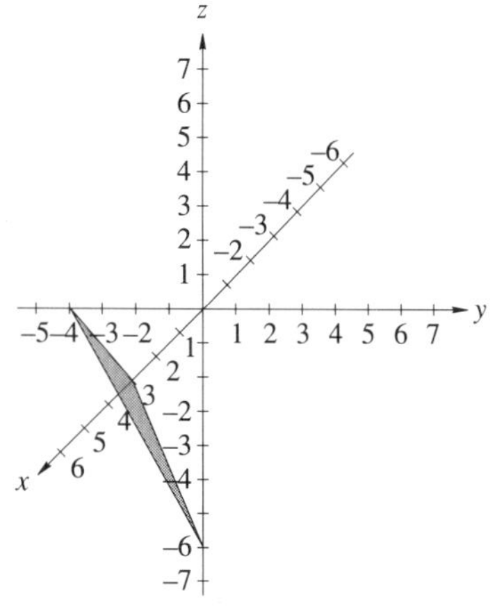

3. An elliptical cxylinder whose axis is sthe x-axis, intercepts are $(0, \pm 1, 0)$, $(0.0. \pm 2)$

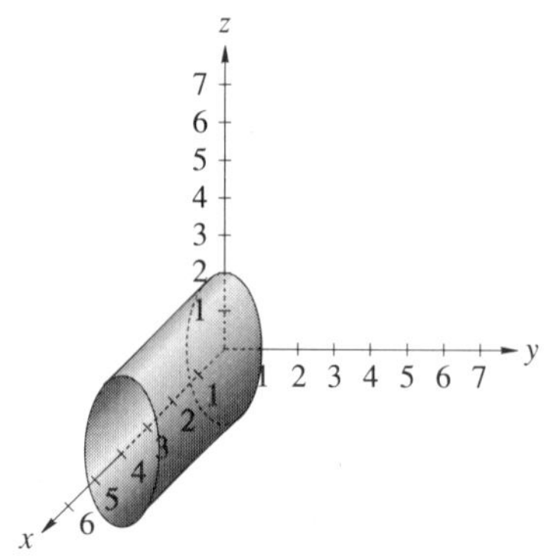

4. Sphere of radius 4.

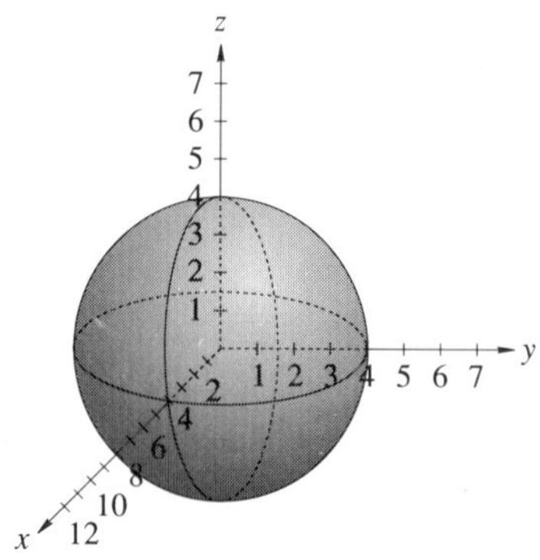

5. A parabolic cylinder whose axis is the y-axis and has the trace in the xy-plane of $y = 4x^2$

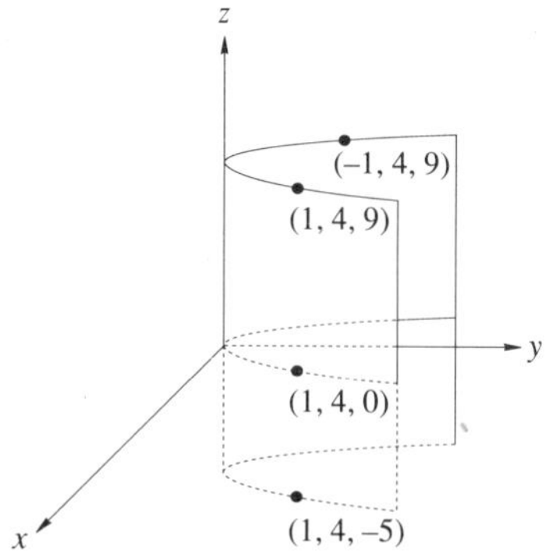

6. A hyperbolaic cylinder whose axis is the x-axis and has the trace in the xy-plane of $9x^2 - 4y^2 = 1$.

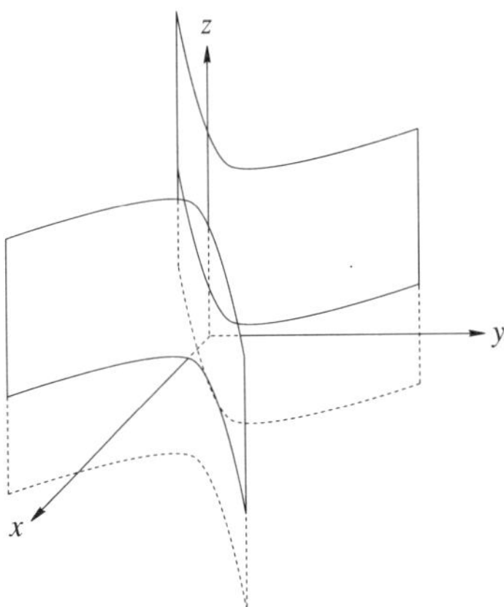

7. Hyperboloid of one-sheet. The trace in the xy-plane is the ellipse $\frac{x^2}{16} + \frac{y^2}{4} = 1$. The trace in the yz-plane is the hyperbola $\frac{y^2}{4} - \frac{z^2}{16} = 1$ and the trace in the xz-plane is the hyperbola $\frac{x^2}{16} - \frac{z^2}{16} = 1$.

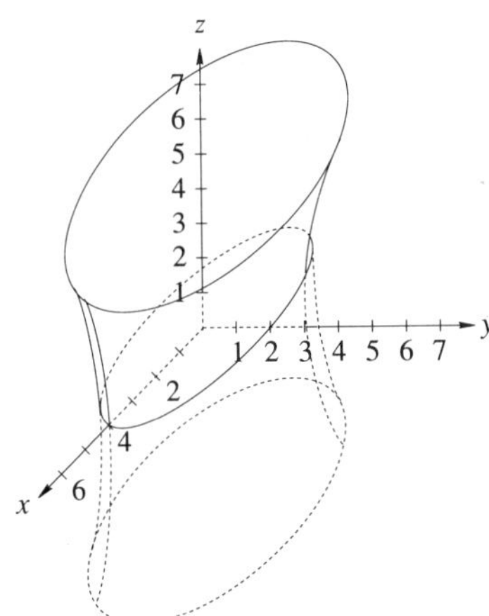

8. An elliptic cone

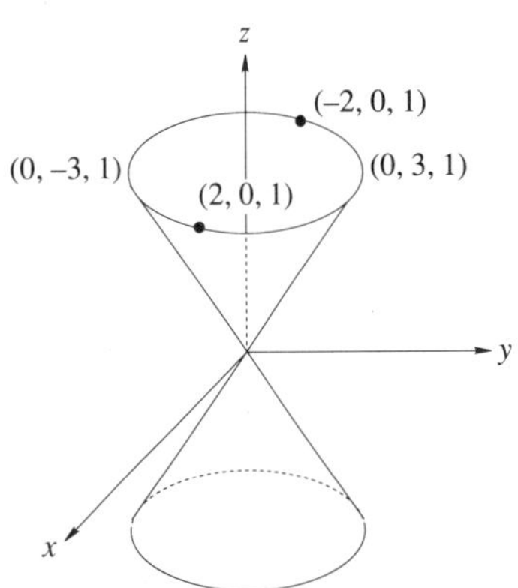

9. $z = 3x^2 + 6xy^3$; $\frac{\partial z}{\partial x} = 6x + 6y^3$; $\frac{\partial z}{\partial y} = 18xy^2$;
$\frac{\partial^2 z}{\partial x^2} = 6$; $\frac{\partial^2 z}{\partial y^2} = 36xy$; $\frac{\partial^2 z}{\partial x \partial y} = 18y^2$

10. $z = \frac{x^2 + y^2}{y} = x^2y^{-1} + y$; $\frac{\partial z}{\partial x} = \frac{2x}{y}$;
$\frac{\partial z}{\partial y} = \frac{-x^2}{y^2} + 1 = \frac{y^2 - x^2}{y^2}$; $\frac{\partial^2 z}{\partial x^2} = \frac{2}{y}$; $\frac{\partial^2 z}{\partial y^2} = \frac{2x^2}{y^3}$;
$\frac{\partial^2 z}{\partial x \partial y} = \frac{-2x}{y^2}$

11. $z = e^x \cos y - e^y \sin x$; $\frac{\partial z}{\partial x} = e^x \cos y - e^y \cos x$;
$\frac{\partial z}{\partial y} = -e^x \sin y - e^y \sin x$; $\frac{\partial^2 z}{\partial x^2} = e^x \cos y + e^y \sin x$;
$\frac{\partial^2 z}{\partial y^2} = -e^x \cos y - e^y \sin x$;
$\frac{\partial^2 z}{\partial x \partial y} = -e^x \sin y - e^y \cos x$

12. $z = \ln \sqrt{x^2 + y^3} + \sin^2(3xy)$;

$$\frac{\partial z}{\partial x} = \frac{2x}{2(x^2 + y^3)} + 2\sin(3xy) \cdot \cos 3xy \cdot 3y$$
$$= \frac{x}{x^2 + y^3} + 6y\sin(3xy)\cos(3xy)$$
$$\frac{\partial z}{\partial y} = \frac{3y^2}{2(x^2 + y^3)} + 6x \sin 3xy \cos 3xy$$

$$\frac{\partial^2 z}{\partial x^2} = \frac{x^2+y^2-2x^2}{(x^2+y^3)^2} + 18y^2\cos^2 3xy - 18y^2\sin^2 3xy$$
$$= \frac{y^3-x^2}{(x^2+y^3)^2} + 18y^2(\cos^2 3xy - \sin^2 3xy)$$
$$= \frac{y^3-x^2}{(x^2+y^3)^2} + 18y^2\cos 6xy$$
$$\frac{\partial^2 z}{\partial y^2} = \frac{2(x^2+y^3)6y - 3y^2\cdot 2\cdot 3y^2}{4(x^2+y^3)^2} + 18x^2(\cos^2 3xy - \sin^2 3xy)$$
$$= \frac{6x^2y-3y^4}{2(x^2+y^3)^2} + 18x\cos 6xy$$
$$\frac{\partial^2 z}{\partial x\partial y} = \frac{12xy^2}{4(x^2+y^3)^2} + 6\sin 3xy\cos 3xy + 18xy\cos^2 3xy - 18xy\sin^2 3xy$$
$$= \frac{3xy^2}{(x^2+y^3)^2} + 6\sin 3xy\cos 3xy + 18xy\cos 6xy$$

13. $\int_0^2\int_0^{x^3} xy\,dy\,dx = \int_0^2 \frac{xy^2}{2}\Big|_0^{x^3} dx = \int_0^2 \frac{x^2}{2}\,dx = \frac{x^8}{18}\Big|_0^2 = 16$

14. $$\int_0^{\pi/2}\int_0^{\pi} \sin x\cos y\,dy\,dx = \int_0^{\pi/2} \sin x[\cos\pi - \cos 0]\,dx = -2\int_0^{\pi/2}\sin x\,dy = -2\,[\cos x]_0^{\pi/2} = -2\left[\cos\left(\frac{\pi}{2}\right) - \cos 0\right] = 2$$

15. $\int_0^1\int_0^4 (x+y)^{1/2}\,dy\,dx = \int_0^1 \frac{2}{3}(x+y)^{3/2}\Big|_0^4 dx = \frac{2}{3}\int_0^1 \left[(x+4)^{3/2} - x^{3/2}\right]dx = \frac{2}{3}\left[\frac{2}{5}(x+4)^{5/2} - \frac{2}{5}x^{5/2}\right]_0^1 = \frac{4}{15}\left(5^{5/2} - 1 - 4^{5/2}\right) = \frac{4}{15}\left(5^{5/2} - 33\right) \approx 6.1071.$

16. $\int_0^4\int_0^9 xy\sqrt{x^2+y^2}\,dy\,dx$: Let $u = x^2+y^2$, and then $du = 2y\,dy$. When $y = 0 \Rightarrow u = x^2$, and when $y = 9 \Rightarrow u = x^2+9^2$. Substitution yields $\frac{1}{2}\int_0^4\int_{x^2}^{x^2+9^2} x(u)^{1/2}du = \frac{1}{2}\int_0^4 \frac{2}{3}xu^{3/2}\Big|_{x^2}^{x^2+9^2} dx = \frac{1}{2}\int_0^4 (x(x^2+9^2)^{3/2} - x^4)\,dx = \frac{1}{3}\left[\frac{1}{2}\cdot\frac{2}{5}(x^2+9^2)^{5/2} - \frac{x^5}{5}\right]_0^4 = \frac{1}{15}\left[(4^2+9^2)^{5/2} - 4^5 - 9^5\right] = 2172.9935.$

17. $\int_0^{\ln 4}\int_0^{\ln 10} e^{x+2y}\,dy\,dx = \int_0^{\ln 4} \frac{1}{2}e^{x+2y}\Big|_0^{\ln 10} dx = \int_0^{\ln 4}\left(\frac{1}{2}e^{x+2\ln 10} - \frac{1}{2}e^x\right)dx = \int_0^{\ln 4} 49\frac{1}{2}e^x\,dx = 49.5e^x\big|_0^{\ln 4} = 49.5(4-1) = 148.5.$

18. $\int_0^1\int_0^{x^2} x\,dy\,dx = \int_0^1 x^3\,dx = \frac{x^4}{4}\Big|_0^1 = \frac{1}{4}$

19. $\left(4, \frac{\pi}{6}, 2\right), x = 4\cos\frac{\pi}{6} = 2\sqrt{3} \approx 3.4641$, $y = 4\sin\frac{\pi}{6} = 2$, and $z = 2$; rectangular: $(2\sqrt{3}, 2, 2)$; $\rho = \sqrt{(2\sqrt{3})^2+2^2+2^2} = \sqrt{12+4+4} = \sqrt{20} = 2\sqrt{5} \approx 4.4721.$ $\theta = \tan^{-1}\frac{2}{2\sqrt{3}} = \frac{\pi}{6}, \phi = \cos^{-1}\frac{2}{2\sqrt{5}} = 1.1071$; spherical $\left(2\sqrt{5}, \frac{\pi}{6}, 1.1071\right)$

20. $x = 9\cos\frac{\pi}{6} = -\frac{9\sqrt{3}}{2} \approx -7.7942, y = 9\sin\frac{5\pi}{6} = \frac{9}{2} = 4.5$; $z = -3$; rectangular: $(-7.7942, 4.5, -3)$. $\rho = \sqrt{\left(\frac{9\sqrt{3}}{2}\right)^2 + \left(\frac{9}{2}\right)^2 + 3^2} = 9.4868$; $\theta = \tan^{-1}\left(-\frac{1}{\sqrt{3}}\right) = \frac{5\pi}{6}, \phi = \cos^{-1}\frac{-3}{9.4868} = 1.8925$; spherical: $\left(9.4868, \frac{5\pi}{6}, 1.8925\right)$.

21. $x = 4\sin\frac{\pi}{6}\cos\frac{3\pi}{4} = 2\cdot\cos\frac{3\pi}{4} = -\sqrt{2}$, $y = 4\sin\frac{\pi}{6}\sin\frac{3\pi}{4} = \sqrt{2}, z = 4\cos\frac{\pi}{6} = 2\sqrt{3}$; rectangular: $(-\sqrt{2}, \sqrt{2}, 2\sqrt{3})$. $r = \sqrt{(-\sqrt{2})^2 + \sqrt{2}^2} = 2, \theta = \tan^{-1}\frac{\sqrt{2}}{-\sqrt{2}} = \frac{3\pi}{4}$; cylindrical: $\left(2, \frac{3\pi}{4}, 2\sqrt{3}\right)$.

22. $x = 5\sin\frac{2\pi}{3}\cos\frac{7\pi}{6} = 5\cdot\frac{\sqrt{3}}{2}\cdot\frac{-\sqrt{3}}{2} = \frac{-15}{4} = -3.75$; $y = 5\sin\frac{2\pi}{3}\sin\frac{7\pi}{6} = 5\cdot\frac{\sqrt{3}}{2}\cdot\left(\frac{-1}{2}\right) = \frac{-5\sqrt{3}}{4} \approx$

-2.1651; $z = 5\cos\frac{2\pi}{3} = 5\left(\frac{-1}{2}\right) = -2.5$;
rectangular: $(-3.75, -2.1651, -2.5)$.
$r = \sqrt{(3.75)^2 + (-2.1651)^2} = 4.3301$, $\theta = \theta = \frac{7\pi}{6}$,
$z = z = -2.5$; cylindrical: $\left(4.3301, \frac{7\pi}{6}, -2.5\right)$.

23. $V = \ell \cdot \omega \cdot h = 300$ so $h = \frac{300}{\ell \cdot w}$.
$A = \ell \cdot \omega + 2\ell h + 2\omega h = \ell \cdot \omega + \frac{600}{\omega} + \frac{600}{\ell}$;
$A_\ell = \omega - \frac{600}{\ell^2} = 0$; $A_w = \ell - \frac{600}{\omega^2} = 0$. Solving the first equation for ω we have $\omega = \frac{600}{\ell^2}$. Substituting this into the second equation, we get
$\ell - \frac{600}{\left(\frac{600}{\ell^2}\right)^2} = \ell - \frac{\ell^4}{600} = 0$ or $\ell = \frac{\ell^4}{600} \Rightarrow \ell^3 = 600$
or $\ell = \sqrt[3]{600} \approx 8.434$ in. Now
$\omega = \frac{600}{\ell^2} = \frac{600}{\sqrt[3]{600^2}} = \sqrt[3]{600} \approx 8.434$ in.
$h = \frac{300}{\sqrt[3]{600}} = \frac{300\sqrt[3]{600}}{600} = \frac{\sqrt[3]{600}}{2} \approx 4.217$ in.

24. $R = \frac{R_1R_2}{R_1 + R_1}$;
$\frac{\partial R}{\partial R_1} = \frac{(R_1 + R_2)R_2 - R_1R_2}{(R_1 + R_2)^2} = \frac{R_2^2}{(R_1 + R_2)^2}$.
$\frac{\partial R}{\partial_1 R_2} = \frac{R_1^2}{(R_1 + R_2)^2}$, $dR_1 = 300 \times 0.01 = 3$,
$dR_2 = 600 \times 0.01 = 6$.
$dR = \frac{\partial R}{\partial R_1}dR_1 + \frac{\partial R}{\partial R_2}dR_2 = \frac{R_2^2}{(R_1 + R_2)^2}dR_1 + \frac{R_1^2}{(R_1 + R_2)^2}dR_2 = \frac{600^2}{(900)^2} \times 3 + \frac{300^2}{900^2} \times 6 = 2\ \Omega$.

25. $\frac{\partial R}{\partial V} = \frac{1}{I}$, $\frac{\partial R}{\partial I} = -VI^{-2}$, $dR = \frac{\partial R}{\partial V}dV + \frac{\partial R}{\partial I}dI = \frac{1}{I}dV - \frac{V}{I^2}dI = \frac{1}{2} \times 0.2 - \frac{116 \times 0.01}{2} = -0.19\ \Omega$.

26. $V = \ell \cdot \omega \cdot h = 25$, so $h = \frac{25}{\ell\omega}$.
$A = \ell\omega + 2\ell h + 2\omega h = \ell\omega + \frac{50}{\omega} + \frac{50}{\ell}$.
$A_\ell = \omega - \frac{50}{\ell^2}$; $A_\omega = \ell - \frac{50}{\omega^2}$; $A_\ell = A_\omega = 0$;
$\ell - \frac{50}{\omega^2} = 0 \Rightarrow \ell = \frac{50}{\omega^2}$; $\omega - \frac{50}{\left(\frac{50}{\omega^2}\right)^2} = \omega - \frac{\omega^4}{50} = 0$;
$\omega = \frac{\omega^4}{50}$; $50 = \omega^3$ or $\omega = \sqrt[3]{50} \approx 3.684$ m. In like manner,$\ell = \sqrt[3]{50} = 3.684$ m. Since $h = \frac{25}{\ell \cdot \omega}$, we get $h = \frac{25}{\sqrt[3]{50^2}} = \frac{25\sqrt[3]{50}}{50} = \frac{\sqrt[3]{50}}{2} \approx 1.842$ m.

27. $PV = nRT \Rightarrow V = \frac{nRT}{P}$;
$T = \frac{PV}{nR} = \frac{4 \times 1000}{8R} = \frac{500}{R}$;
$dV = \frac{\partial V}{\partial T}dT + \frac{\partial V}{\partial P}dP = \frac{nR}{P}dT - \frac{nRT}{P^2}dP = \frac{8R}{4}(0.5) - \frac{8R \cdot \frac{500}{R}}{4^2}(0.4) = R - 100\ \text{cm}^3/\text{min}$.

28. $V = \frac{1}{3}\pi r^2 h$; $dV = \frac{\partial V}{\partial r}dr = \frac{1}{3}\pi r^2 \cdot dh + \frac{2}{3}\pi rhdr = \frac{1}{3}\pi(180)^2 15 + \frac{2}{3}\pi(180)(270)10 = 486{,}000\pi$ cm^2/s.

29. $m = \int_1^5 \int_0^{6x} dy\,dx = \int_1^5 6x\,dx = 3x^2\big|_1^5 = 75 - 3 = 72$, $M_x = \int_1^5 \int_0^{6x} y\,dy\,dx = \int_1^5 \frac{y^2}{2}\Big|_0^{6x} dx = \int_1^5 18x^2\,dx = 6x^3\Big|_1^5 = 744$,
$M_y = \int_1^5 \int_0^{6x} x\,dy\,dx = \int_1^5 6x^2\,dx = 2x^2\big|_1^5 = 248$,
$\bar{x} = \frac{M_y}{m} = \frac{248}{72} = 3.4444$;
$\bar{y} = \frac{M_x}{m} = \frac{744}{72} \approx 10.3333$.

30. $m = \int_1^2 \int_0^{x^4} dy\,dx = \int_1^2 x^4\,dx = \frac{x^5}{5}\Big|_1^2 = \frac{31}{5}$,
$M_x = \int_1^2 \int_0^{x^4} y\,dy\,dx = \int_1^2 x^4\,dx = \frac{y^2}{2}\Big|_0^{x^4} dx = \int_1^2 \frac{x^8}{2}\,dx = \frac{x^9}{18}\Big|_{1_y}^2 = \frac{511}{18}$,
$M_y = \int_1^2 \int_0^{x^4} x\,dy\,dx = \int_1^2 x^5\,dx = \frac{x^6}{6}\Big|_1^2 = \frac{63}{6}$,
$\bar{x} = \frac{63/6}{31/5} = \frac{105}{62} \approx 1.6935$,
$\bar{y} = \frac{511/18}{31/5} = \frac{2555}{558} = 4.579$

31. $m = \int_0^1 \int_{x^3}^{\sqrt[3]{x}} dy\,dx = \int_0^1 \left(x^{1/3} - x^3\right) dx = \frac{3}{4}x^{4/3} - \frac{1}{4}x^4\Big|_0^1 = \frac{1}{2}$; $M_x = \int_0^1 \int_{x^3}^{\sqrt[3]{x}} y\,dy\,dx = \int_0^1 \frac{y^2}{2}\Big|_{x^3}^{\sqrt[3]{x}} dx = \frac{1}{2}\int_0^1 \left(x^{2/3} - x^6\right) dx = \frac{1}{2}\left[\frac{3}{5}x^{5/3} - \frac{1}{7}x^7\right]_0^1 = \frac{1}{2}\left(\frac{3}{5} - \frac{1}{7}\right) = \frac{8}{35}$;

$M_y = \int_0^1 \int_{x^3}^{\sqrt[3]{x}} x\,dy\,dx = \int_0^1 \left(x^{4/3} - x^4\right) dx = \left[\frac{4}{7}x^{7/4} - \frac{1}{5}x^5\right]_0^2 = \frac{8}{35}$; $\bar{x} = \bar{y} = \frac{8/35}{1/2} = \frac{16}{35} \approx 0.4571$

32. These two graphs intersect when $x^2 = 9 - x^2$ or $2x^2 = 9 \Rightarrow x = \pm\frac{3/\sqrt{2}}{2} = \pm\frac{3}{\sqrt{2}} \approx \pm 2.1213$.

$m = \int_{-3/\sqrt{2}}^{3/\sqrt{2}} \int_{x^2}^{9-x^2} dy\,dx = \int_{-3/\sqrt{2}}^{3/\sqrt{2}} (9 - 2x^2)\,dx = 9x - \frac{2}{3}x^3\Big|_{-3/\sqrt{2}}^{3/\sqrt{2}} = \frac{54}{\sqrt{2}} - \frac{9}{2\sqrt{2}} = 27\sqrt{2} - 9\sqrt{2} = 18\sqrt{2}$; $M_x = \int_{-3/\sqrt{2}}^{3/\sqrt{2}} \int_{x^2}^{9-x^2} y\,dy\,dx = \int_{-3/\sqrt{2}}^{3/\sqrt{2}} \frac{y^2}{2}\Big|_{x^2}^{9-x^2} dx = \frac{1}{2}\int_{-3/\sqrt{2}}^{3/\sqrt{2}} (81 - 18x^2)\,dx = \frac{1}{2}\left[81x - 6x^3\right]_{-3/\sqrt{2}}^{3/\sqrt{2}} = \frac{243}{\sqrt{2}} - \frac{81}{\sqrt{2}} = 81\sqrt{2}$.

$M_y = \int_{-3/\sqrt{2}}^{3/\sqrt{2}} \int_{x^2}^{9-x^2} x\,dy\,dx = \int_{-3/\sqrt{2}}^{3/\sqrt{2}} (9x - 2x^3)\,dx = 0$. This integral is 0 since $9x - 2x^3$ is an odd function and we are integrating from $-\frac{3}{\sqrt{2}}$ to $\frac{3}{\sqrt{2}}$. $\bar{x} = 0, \bar{y} = \frac{81\sqrt{2}}{18\sqrt{2}} = \frac{9}{2} = 4.5$. This answer is exactly what you would expect to get due to the symetry of the region.

33. $I_x = \rho \int_0^5 \int_0^{6x} y^2\,dy\,dx = \rho \int_0^5 \frac{y^3}{3}\Big|_0^{6x} dx = \rho \int_0^5 72x^3\,dx = \rho 18x^4\big|_0^5 = 11{,}250\rho$.

$m = \rho \int_0^5 dy\,dx = \rho \int_0^5 6x\,dx = \rho 3x^2\big|_0^5 = 75\rho$.

$r_x = \sqrt{\frac{I_x}{m}} = \sqrt{\frac{11{,}250\rho}{75\rho}} = \sqrt{150} \approx 12.2474$

34. $I_y = \rho \int_0^4 \int_0^{4-x} x^2\,dy\,dx = \rho \int_0^4 (4x^2 - x^3)\,dx = \rho\left[\frac{4}{3}x^3 - \frac{x^4}{4}\right]_0^4 = 21.3333\rho$ or $\frac{64}{3}\rho$.

$m = \rho \int_0^4 \int_0^{4-x} dy\,dx = \rho \int_0^4 (4 - x)\,dx = \rho\left[4x - \frac{x^2}{2}\right]_0^4 = 8\rho$.

$r_y = \sqrt{\frac{I_y}{m}} = \sqrt{\frac{\frac{64}{3}\rho}{8}} = \sqrt{\frac{8}{3}\rho} \approx 1.6330$.

35. $I_x = \rho \int_0^1 \int_0^{x^{2/3}} y^2\,dy\,dx = \rho \int_0^1 \frac{y^3}{3}\Big|_0^{x^{2/3}} dx = \rho \int_0^1 \frac{x^2}{3}\,dx = \rho \frac{x^3}{8}\Big|_0^1 = \frac{1}{9}\rho$. $m = \rho \int_0^1 \int_0^{x^{2/3}} dy\,dx = \rho \int_0^1 x^{2/3}\,dx = \rho\,\frac{3}{5}x^{5/3}\Big|_0^1 = \frac{3}{5}\rho$.

$r_x = \sqrt{\frac{1/9\rho}{3/5\rho}} = \sqrt{\frac{5}{27}} \approx 0.4303$

36. $I_y = \rho \int_0^1 \int_{x^3}^{\sqrt[3]{x}} y^2\,dy\,dx = \rho \int_0^1 \frac{y^3}{3}\Big|_{x^3}^{\sqrt[3]{x}} dx = \rho \int_0^1 \left(\frac{x}{3} - \frac{x^9}{3}\right) dx = \rho\left[\frac{x^2}{6} - \frac{x^{10}}{30}\right]_0^1 = \left(\frac{1}{6} - \frac{1}{30}\right)\rho = \frac{4}{30}\rho = \frac{2}{15}\rho$.

$m = \rho \int_0^1 \int_{x^3}^{\sqrt[3]{x}} dy\,dx = \rho \int_0^1 \left(x^{1/3} - x^3\right) dx = \rho\left(\frac{3}{4}x^{4/3} - \frac{x^4}{4}\right)\Big|_0^1 = \frac{1}{2}\rho$.

$r_y = \sqrt{\frac{\frac{2}{15}\rho}{\frac{1}{2}\rho}} = \sqrt{\frac{4}{15}} \approx 0.5164$

37. $\int_0^2 \int_0^x (x^2 - y)\,dy\,dx + \int_2^4 \int_0^{4-x} (x^2 - y)\,dy\,dx =$

$\int_0^2 \left(x^2 y - \frac{y^2}{2}\right)\Big|_0^x dx + \int_2^4 \left(x^2 y - y\frac{2}{2}\right)\Big|_0^{4-x} dx =$

$\int_0^2 \left(x^3 - \frac{x^2}{2}\right) dx +$

$\int_2^4 \left(4x^2 - x^3 - 8 + 4x - \frac{x^2}{2}\right) dx =$

$\left[\frac{x^4}{4} - \frac{x^3}{6}\right]_0^2 + \left[\frac{4}{3}x^3 - \frac{x^4}{4} - 8x + 2x^2 - \frac{x^3}{6}\right]_2^4 =$

$4 - \frac{8}{6} + \frac{256}{3} - 64 - 32 + 32 - \frac{64}{4} - \frac{32}{3} + 4 +$

$16 - 8 + \frac{8}{6} = 16.$

38. $\int_{-\sqrt{5}}^{\sqrt{5}} \int_{x^2+x-5}^{x} (x - 4y)\,dy\,dx$

$= \int_{-\sqrt{5}}^{\sqrt{5}} (xy - 2y^2)\Big|_{x^2+x-5}^{x} dx$

$= \int_{-\sqrt{5}}^{\sqrt{5}} [x^2 - 2x^2 - (x^3 + x^2 - 5x)$

$+ 2(x^2 + x - 5)^2]\,dx$

$= \int_{-\sqrt{5}}^{\sqrt{5}} [-x^3 - 2x^2 + 5x$

$+ 2(x^4 + 2x^3 - 9x^2 - 10x + 25)]\,dx$

$= \int_{-\sqrt{5}}^{\sqrt{5}} (2x^4 + 3x^3 - 20x^2 - 15x + 50)\,dx$

$= \left[\frac{2}{5}x^5 + \frac{3}{4}x^4 - \frac{20}{3}x^3 - \frac{15}{2}x^2 + 50x\right]_{-\sqrt{5}}^{\sqrt{5}}$

$= 2\left(\frac{2}{5} \cdot 25\sqrt{5} - \frac{20}{3} \cdot 5\sqrt{5} + 50\sqrt{5}\right)$

$= 2\left(10\sqrt{5} - \frac{100}{3}\sqrt{5} + 50\sqrt{5}\right) = \frac{160}{3}\sqrt{5}$

≈ 119.2570

(Note: to evaluate from $-\sqrt{5}$ to $\sqrt{5}$, we can double the odd terms and cancel the even.)

CHAPTER 30 TEST

1. A plane whose intercepts are $(4, 0, 0)$, $(0, -8, 0)$, and $(0, 0, 2)$

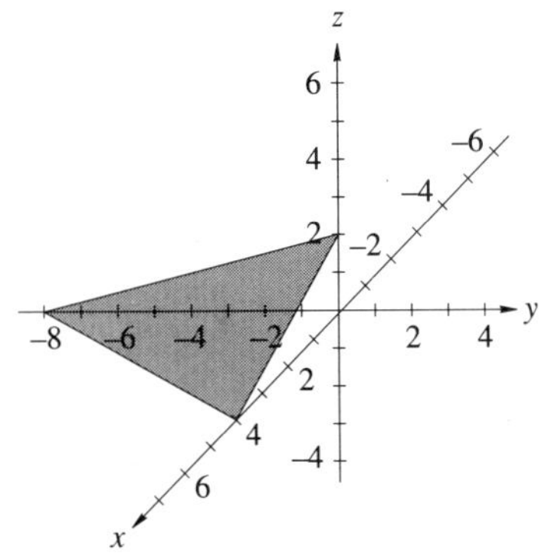

2. Sphere of radius 3

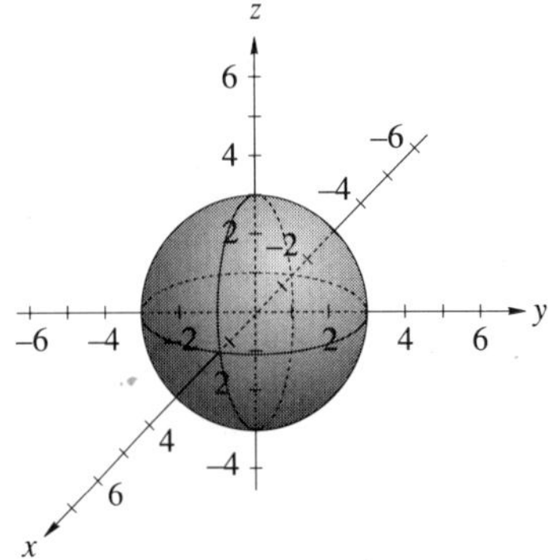

3. Hyperboloid of one-sheet. The trace in the xy-plane is the ellipse $\frac{x^2}{4}+\frac{y^2}{9}=1$. The trace in the yz-plane is the hyperbola $\frac{y^2}{9}-\frac{z^2}{4}=1$ and the trace in the xz-plane is the hyperbola $\frac{x^2}{4}-\frac{z^2}{4}=1$.

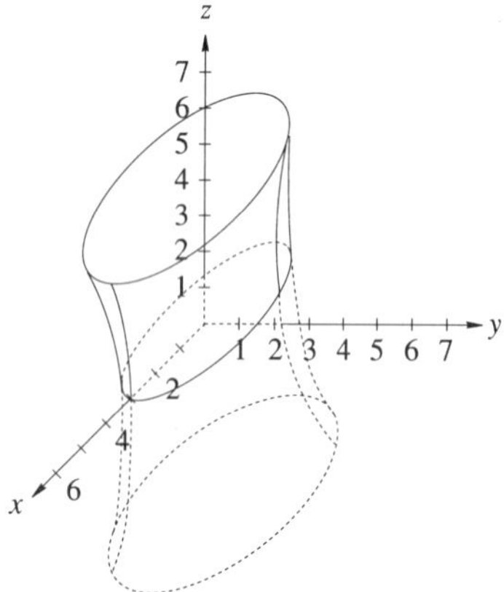

4. Elliptic paraboloid with intercepts $(\pm 2, 0, 0)$, $(0, \pm 3, 0)$, and $(0, 0, -36)$

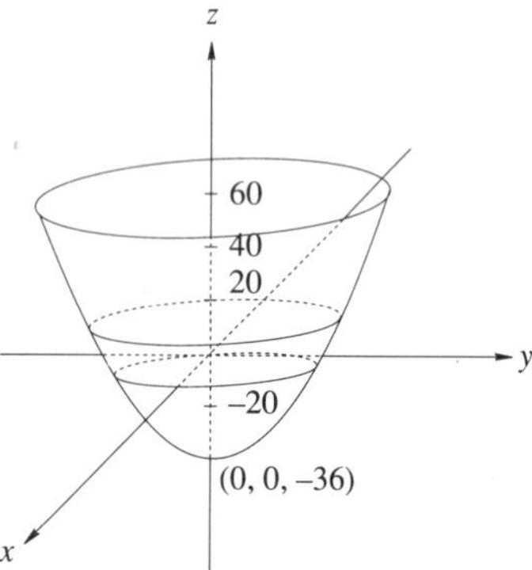

5. Treating y as a constant and differentiating z with respect to x, produces $\frac{\partial z}{\partial x}=12x^2-10xy^4$. Treating x as a constant and differentiating z with respect to y, produces $\frac{\partial z}{\partial y}=-20x^2y^3$

6. Rewriting z as $z=\left(x^2+4y^3\right)^{1/2}+\ln(x^2y)$, treating y as a constant and differentiating z with respect to x, produces $\frac{\partial z}{\partial x}=\frac{1}{2}\left(x^2+4y^3\right)^{-1/2}(2x)+\frac{2xy}{x^2y}$ which simplifies to $\frac{\partial z}{\partial x}=x\left(x^2+4y^3\right)^{-1/2}+\frac{2}{x}$. Treating y as a constant and differentiating $\frac{\partial z}{\partial x}$ with respect to x, produces $\frac{\partial^2 z}{\partial x^2}=x\left[-\frac{1}{2}\left(x^2+4y^3\right)^{-3/2}(2x)\right]+\left(x^2+4y^3\right)^{-1/2}-\frac{2}{x^2}=-x^2\left(x^2+4y^3\right)^{-3/2}+\left(x^2+4y^3\right)^{-1/2}-\frac{2}{x^2}$

7.
$$\begin{aligned}\int_0^4\int_0^{\pi/4} x\sin y\,dy\,dx &= \int_0^4 (-x\cos y)\Big|_{y=0}^{y=\pi/4} dx\\ &= \int_0^4 -x\left(\frac{\sqrt{2}}{2}-1\right) dx\\ &= \left(1-\frac{\sqrt{2}}{2}\right)\int_0^4 x\,dx\\ &= \left(1-\frac{\sqrt{2}}{2}\right)\frac{1}{2}x^2\Big|_0^4\\ &= 8-4\sqrt{2}\approx 2.3431\end{aligned}$$

8. $\int_0^2\int_0^{x^2} 3xy^2\,dy\,dx = \int_0^2 xy^3\big|_0^{x^2} dx = \int_0^2 x^7\,dx = \frac{1}{8}x^8\big|_0^2 = \frac{256}{8} = 32$

9. First we will convert these coordinates from cylindrical coordinates to rectangular coordinates. A point with the cylindrical coordinates (r, θ, z), has the rectangular coordinates (x, y, z), where $x = r\cos\theta, y = r\sin\theta, z = z$. We begin with $x = r\cos\theta = 3\cos\frac{5\pi}{6} = 3\left(-\frac{\sqrt{3}}{2}\right) = -\frac{3\sqrt{3}}{2}$. Next, $y = r\sin\theta = 3\sin\frac{5\pi}{6} = 3\left(\frac{1}{2}\right) = \frac{3}{2}$. Finally, $z = z = 8$. So, the cylindrical coordinates $\left(3, \frac{5\pi}{6}, 8\right)$ are equivalent to the rectangular coordinates $\left(-\frac{3\sqrt{3}}{2}, \frac{3}{2}, 8\right)$.
Next, we convert these rectangular coordinates to spherical coordinates. A point with the rectangular coordinates (x, y, z), has the spherical coordinates (ρ, θ, ϕ), where $\rho = \sqrt{x^2+y^2+z^2}$, $\tan\theta = \frac{y}{x}$, $\cos\phi = \frac{z}{\rho}$. First, we obtain

$\rho = \sqrt{x^2+y^2+z^2} = \sqrt{\left(-\frac{3\sqrt{3}}{2}\right)^2 + \left(\frac{3}{2}\right)^2 + 8^2} = \sqrt{\frac{27}{4}+\frac{9}{4}+64} = \sqrt{73}$. Next we determine $\tan\theta = \frac{y}{x} = \frac{3/2}{\frac{-3\sqrt{3}}{2}} = -\frac{1}{\sqrt{3}}$ and so $\theta = \frac{5\pi}{6}$. Finally, we determine $\cos\phi = \frac{z}{\rho} = \frac{8}{\sqrt{73}}$ and so $\phi = \cos^{-1}\frac{8}{\sqrt{73}} \approx 0.3588$. [The value of θ is the same in both the cylindrical and spherical coordinate systems, so we could have used the given value $\theta = \frac{5\pi}{6}$.]

10. First we will convert these coordinates from spherical coordinates to rectangular coordinates. We have $\rho = 4$, $\theta = \frac{7\pi}{6}$ and $\phi = \frac{3\pi}{4}$. Using the formulas in (3), we have $x = \rho\sin\phi\cos\theta = 4\sin\frac{3\pi}{4}\cos\frac{7\pi}{6} = 4\left(\frac{\sqrt{2}}{2}\right)\left(-\frac{\sqrt{3}}{2}\right) = -\sqrt{6}$; $y = \rho\sin\phi\sin\theta = 4\left(\frac{\sqrt{2}}{2}\right)\left(-\frac{1}{2}\right) = -\sqrt{2}$; and $z = \rho\cos\phi = 4\left(-\frac{\sqrt{2}}{2}\right) = -2\sqrt{2}$. So, the spherical coordinates of $\left(4, \frac{7\pi}{6}, \frac{3\pi}{4}\right)$ are, in rectangular coordinates, $\left(-\sqrt{6}, -\sqrt{2}, -2\sqrt{2}\right)$.
Next, we convert these rectangular coordinates to cylindrical coordinates. A point with the rectangular coordinates (x, y, z), has the cylindrical coordinates (r, θ, z), where $r = \sqrt{x^2+y^2}$, $\tan\theta = \frac{y}{x}$, $z = z$.
First, $r = \sqrt{x^2+y^2} = \sqrt{\left(-\sqrt{6}\right)^2 + \left(-\sqrt{2}\right)^2} = \sqrt{6+2} = \sqrt{8} = 2\sqrt{2}$. Next, $\tan\theta = \frac{y}{x} = \frac{-\sqrt{2}}{-\sqrt{6}} = \frac{1}{\sqrt{3}}$. Since the coordinates of x and y are in the third quadrant of the xy-plane, we have $\theta = \pi + \tan^{-1}\frac{1}{\sqrt{3}} = \frac{7\pi}{6}$. [Notice that since θ is the same in both the spherical and cylindrical coordinate systems, we could have used the given value of *theta*.] Finally, $z = z = -2\sqrt{2}$. So, the spherical coordinates of $\left(4, \frac{7\pi}{6}, \frac{3\pi}{4}\right)$ are, in cylindrical coordinates, $\left(2\sqrt{2}, \frac{7\pi}{6}, -2\sqrt{2}\right)$.

11. Let $z = 0$. Then the base of the region in the xy-plane is the ellipse $x^2 + 4y^2 = 4$ or $\frac{x^2}{4} + y^2 = 1$. Solving this for y, we see that $y = \pm\sqrt{\frac{4-x^2}{4}}$ and so the bounds for y are $-\sqrt{(4-x^2)/4} \le y \le \sqrt{(4-x^2)/4}$ and $-2 \le x \le 2$. Thus, the volume is $V = \int_{-2}^{2}\int_{-\sqrt{(4-x^2)/4}}^{\sqrt{(4-x^2)/4}} (4 - x^2 - 4y^2)\,dy\,dx = \int_{-2}^{2}\left[4y - x^2y - \frac{4}{3}y^3\right]_{-\sqrt{(4-x^2)/4}}^{\sqrt{(4-x^2)/4}} dx = \frac{2}{3}\int_{-2}^{2}\left(4-x^2\right)^{3/2} dx$. To integrate this, we use Formula #28, Appendix C, and obtain $\frac{2}{3}\left[\frac{x}{4}(x^2-10)\sqrt{4-x^2} + \frac{3}{2}\sin^{-1}\frac{x}{2}\right]_{-2}^{2} \approx 12.5664 = 4\pi$

12. The two functions intersect at the points $(-1, 4)$ and $(3, 4)$ as shown in Figure 30.0. So, we have $f(x) = 4$, $g(x) = x^2 - 2x + 1$, $a = -1$ and $b = 3$. The area of this region is $A = \int_{-1}^{3}\int_{x^2-2x+1}^{4} dy\,dx = \int_{-1}^{3} y\Big|_{x^2-2x+1}^{4} dx = \int_{-1}^{3}(3 + 2x - x^2)\,dx = 3x + x^2 - \frac{1}{3}x^3\Big]_{-1}^{3} = 9 - \left(-\frac{5}{3}\right) = \frac{32}{3}$. So, the mass is $\frac{32}{3}\rho$. Next, we find the first moments of this region. $M_y = \rho\int_{-1}^{3}\int_{x^2-2x+1}^{4} y\,dy\,dx = \rho\int_{-1}^{3}(3x + 2x^2 - x^3)\,dx = \rho\left[\frac{3}{2}x^2 + \frac{2}{3}x^3 - \frac{1}{4}x^4\right]_{-1}^{3} = \frac{32}{3}\rho$ and $M_x = \rho\int_{-1}^{3}\int_{x^2-2x+1}^{4} y\,dy\,dx = \rho\int_{-1}^{3}\int_{(x-1)^2}^{4} y\,dy\,dx = \rho\int_{-1}^{3}\frac{y^2}{2}\Big|_{(x-1)^2}^{4} dx = \rho\int_{-1}^{3}\frac{1}{2}\left[16 - (x-1)^4\right] dx = \frac{\rho}{2}\left[16x - \frac{1}{5}(x-1)^5\right]_{-1}^{3} = 25.6$. So, $\bar{x} = \frac{M_y}{m} = \frac{32}{3}\Big/\frac{32}{3} = 1$ and $\bar{y} = \frac{M_x}{m} = 25.6\Big/\frac{32}{3} = 2.4$. The centroid is $(\bar{x}.\bar{y}) = (1, 2.4)$

CHAPTER

31

Infinite Series

31.1 MACLAURIN SERIES

1. We first find the derivatives as $f(x) = \sin x$, $f'(x) = \cos x$, $f''(x) = -\sin x$, $f'''(x) = -\cos x$, $f^{(4)}(x) = \sin x$, $f^{(5)}(x) = \cos x$, $f^{(6)}(x) = -\sin$, and $f^{(7)}(x) = -\cos x$. Then, we evaluate f and its of its derivatives at $x = 0$, with the following results: $f(0) = 0$, $f'(0) = 1$, $f''(0) = 0$, $f'''(0) = -1$, $f^{(4)}(0) = 0$, $f^{(5)}(0) = 1$, $f^{(6)}(0) = 0$, and $f^{(7)} = -1$. Hence, $a_0 = 0$, $a_1 = 1$, $a_2 = 0$, $a_3 = -1$, $a_4 = 0$, $a_4 = 0$, $a_5 = 1$, $a_6 = 0$, and $a_7 = -1$. Using these results, we get $\sin x = x - \frac{x^3}{3!} + \frac{x^5}{5!} - \frac{x^7}{7!}$.
2. First we find the derivatives of $f(x) = \sinh x$, $f'(x) = \cosh x$, $f''(x) = \sinh(x)$, etc. Then, we evaluate f and its of its derivatives at $x = 0$, with the following results: $f(0) = 0$, $f'(0) = 1$, $f''(0) = 0$, $f'''(0) = 1$, etc. Using these results, we get $\sinh(x) = x + \frac{x^3}{3!} + \frac{x^5}{5!} + \frac{x^7}{7!}$.
3. Finding the derivatives of $\cosh(x) = f(x)$, we obtain $f'(x) = \sinh x$, $f''(x) = \cosh x$, etc. Evaluating f and its derivatives at $x = 0$ produces $f(0) = 1$, $f'(0) = 0$, $f''(0) = 1$, $f'''(0) = 0$, $f^{(4)}(0) = 1$, etc. Thus, we obtain the desired result $\cosh(x) = 1 + \frac{x^2}{2!} + \frac{x^4}{4!} + \frac{x^6}{6!}$.
4. Finding the derivatives of $f(x) = \ln(1 + x)$, we get $f'(x) = \frac{1}{1+x}$, $f''(x) = -1(1 + x)^{-2}$, $f'''(x) = 2(1 + x)^{-3}$, $f^{(4)}(x) = -6(1 + x)^{-4}$. Evaluating f and its derivatives at $x = 0$ produces $f(0) = 0$, $f'(0) = 1$, $f''(0) = -1$, $f'''(0) = 2$, $f^{(4)}(0) = -6$, etc. Hence, we obtain $\ln(1 + x) = x - \frac{x^2}{2!} + \frac{2x^3}{3!} - \frac{6x^4}{4!} = x - \frac{x^2}{2} + \frac{x^3}{3} - \frac{x^4}{4}$.
5. Here $f(x) = e^{3x}$, $f'(x) = 3e^{3x}$, $f''(x) = 9e^{3x}$, and $f'''(x) = 27e^{3x}$. Thus, $f(0) = 1$, $f'(0) = 3$, $f''(0) = 9$, and $f'''(0) = 27$. Using these results, we obtain the Maclaurin series $e^{3x} = 1 + 3x + \frac{9x^2}{2!} + \frac{27x^3}{3!} = 1 + 3x + \frac{9}{2}x^2 + \frac{9}{2}x^3$.

6. Here we have

$$\begin{aligned}
f(x) &= \sin^3 x, & f(0) &= 0\\
f'(x) &= 3\sin^2 x\cos x, & f'(0) &= 0\\
f''(x) &= 6\sin x\cos^2 x - 3\sin^3 x, & f''(0) &= 0\\
f'''(x) &= 6\cos^3 x - 12\sin^2 x\cos x - 9\sin^2 x\cos x & &\\
&= 6\cos^3 x - 21\sin^2 x\cos x, & f'''(0) &= 6\\
f^{(4)}(x) &= -18\cos^3 x\sin x - 42\sin x\cos^2 x + 21\sin^3 x & &\\
&= -60\cos^2 x\sin x + 21\sin^3 x, & f^{(4)}(0) &= 0\\
f^{(5)}(x) &= 120\cos x\sin^2 x - 60\cos^3 x + 63\sin^2 x\cos x & &\\
&= 183\cos x\sin^2 x - 60\cos^3 x, & f^{(5)}(0) &= -60\\
f^{(6)}(x) &= -183\sin^2 x + 366\cos^2 x\sin x + 180\cos^2 x\sin x & &\\
&= -183\sin^3 x + 546\cos^2 x\sin x, & f^{(6)}(0) &= 0\\
f^{(7)}(x) &= -549\sin^2 x\cos x - 1092\cos x\sin^2 x + 546\cos^3 x & &\\
&= -1641\cos x\sin^2 x + 546\cos^3 x, & f^{(7)}(0) &= 546\\
f^{(8)}(x) &= -4920\sin x\cos^2 x + 1641\sin^3 x, & f^{(8)}(0) &= 0\\
f^{(9)}(x) &= -4920\cos^3 x + 14763\sin^2 x\cos x, & f^{(9)}(0) & -4920
\end{aligned}$$

Hence, we have $\sin^3 x = \dfrac{6x^3}{3!} - \dfrac{60x^5}{5!} + \dfrac{546x^7}{7!} + \dfrac{-4920x^9}{9!} = x^3 - \dfrac{1}{2}x^5 + \dfrac{13}{120}x^7 - \dfrac{4920}{9!}x^9 = x^3 - \dfrac{1}{2}x^5 + \dfrac{13}{120}x^7 - \dfrac{41}{3024}x^9.$

7. Here we have

$$\begin{aligned}
f(x) &= \ln(1+x^2), & f(0) &= 0\\
f'(x) &= \frac{2x}{(x^2+1)}, & f'(0) &= 0\\
f''(x) &= \frac{-2(x^2-1)}{(x^2+1)^2}, & f''(0) &= 2\\
f'''(x) &= 4x(x^2-3)/(x^2+1)^3, & f'''(0) &= 0\\
f^{(4)}(x) &= \frac{-12(x^4-6x^2+1)}{(x^2+1)^4}, & f^{(4)}(0) &= -12\\
f^{(5)}(x) &= \frac{48x(x^4-10x^2+5)}{(x^2+1)^3}, & f^{(5)}(0) &= 0\\
f^{(6)}(x) &= \frac{-240(x^6-15x^4+15x^2-1)}{(x^2+1)^6}, & f^{(6)}(0) &= 240\\
f^{(7)}(x) &= \frac{1440(x^7-21x^5+35x^3-7x)}{(x^2+1)^7}, & f^{(7)}(0) &= 0\\
f^{(8)}(x) &= \frac{-10{,}080(x^8-28x^6+70x^4-28x^2+1)}{(x^2+1)^8}, & f^{(8)}(0) &= -10{,}080
\end{aligned}$$

Hence, we have $\ln(1+x^2) = \dfrac{2x^2}{2!} + \dfrac{-12x^4}{4!} + \dfrac{240x^6}{6!} + \dfrac{-10{,}080x^8}{8!} = x^2 - \dfrac{x^4}{2} + \dfrac{x^6}{3} - \dfrac{x^8}{4} = x^2 - \dfrac{1}{2}x^4 + \dfrac{1}{3}x^6 - \dfrac{1}{4}x^8.$

8. Here we have

$$\begin{aligned}
f(x) &= e^{-x}, & f(0) &= 1\\
f'(x) &= -e^{-x}, & f'(0) &= -1\\
f''(x) &= e^{-x}, & f''(0) &= 1\\
f'''(x) &= -e^{-x}, & f'''(0) &= -1
\end{aligned}$$

Thus, $e^{-x} = 1 - x + \frac{x^2}{2!} - \frac{x^3}{3!} + = 1 - x + \frac{1}{2}x^2 - \frac{1}{6}x^3$.

9. This time we have

$$\begin{aligned}
f(x) &= \cos x^2, & f(0) &= 1\\
f'(x) &= -2x\sin x^2, & f'(0) &= 0\\
f''(x) &= -4x^2\cos x^2 - 2\sin x^2, & f''(0) &= 0\\
f'''(x) &= 8x^3\sin x^2 - 12x\cos x^2, & f'''(0) &= 0\\
f^{(4)}(x) &= 16x^4\cos x^2 - 12\cos x^2 + 48x^2\sin x^2, & f^{(4)}(0) &= -12\\
f^{(5)}(x) &= 160x^3\cos x^2 - 32x^5\sin x^2 + 120x\sin x^2, & f^{(5)}(0) &= 0\\
f^{(6)}(x) &= -64x^6\cos x^2 + 720x^2\cos x^2 - 480x^4\sin x^2 & &\\
&\quad +120\sin x^2, & f^{(6)}(0) &= 0\\
f^{(7)}(x) &= -1{,}344x^5\cos x^2 + 1{,}680x\cos x^2 + 128x^7\sin x^2 & &\\
&\quad -3{,}360x^3\sin x^2, & f^{(7)}(0) &= 0\\
f^{(8)}(x) &= 256x^8\cos x^2 - 13{,}440x^4\cos x^2 + 1{,}680\cos x^2 & &\\
&\quad +3{,}584x^2\sin x^2 - 13{,}440x^2\sin x^2, & f^{(8)}(0) &= 1680\\
f^{(9)}(x) &= -512x^9\sin x^2 + 9{,}216x^7\cos x^2 + 48{,}384x^5\sin x^2 & &\\
&\quad -80{,}640x^3\cos x^2 - 30240x\sin x^2 & f^{(9)}(0) &= 0\\
f^{10}(x) &= -1{,}024x^{10}\cos x^2 - 23{,}040x^8\sin x^2 + 161{,}280x^6\cos x^2 & &\\
&\quad +403{,}200x^4\sin x^2 - 302{,}400x^2\cos x^2 - 30{,}240\sin x^2, & f^{(10)}(0) &= 0\\
f^{(11)}(x) &= 2{,}048x^{11}\sin x^2 - 56{,}320x^9\cos x^2 - 506{,}880x^7\sin x^2 & &\\
&\quad +1774080x^5\cos x^2 + 2{,}217{,}600x^3\sin x^2 - 665{,}280x\cos x^2, & f^{(11)}(0) &= 0\\
f^{(12)}(x) &= 4{,}096x^{12}\cos x^2 + 135{,}168x^{10}\sin x^2 - 1{,}520{,}640x^8\cos x^2 & &\\
&\quad -7{,}096{,}320x^6\sin x^2 + 13{,}305{,}600x^4\cos x^2 & &\\
&\quad +7{,}983{,}360x^2\sin x^2 - 665{,}280\cos x^2, & f^{(12)}(0) &= -665{,}280
\end{aligned}$$

Hence, $\cos x^2 = 1 - \frac{12x^4}{4!} + \frac{1680x^8}{8!} - \frac{665{,}280x^{12}}{12!} = 1 - \frac{1}{2}x^4 + \frac{1}{24}x^8 - \frac{1}{720}x^{12} = 1 - \frac{1}{2!}x^4 + \frac{1}{4!}x^8 - \frac{1}{6!}x^{12}$.

10. Here we have

$$\begin{aligned}
f(x) &= e^{-x^2}, & f(0) &= 1\\
f'(x) &= -2xe^{-x^2}, & f'(0) &= 0\\
f''(x) &= 4x^2e^{-x^2} - 2e^{-x^2}, & f''(0) &= -2\\
f'''(x) &= 12xe^{-x^2} - 8x^3e^{-x^2}, & f'''(0) &= 0\\
f^{(4)}(x) &= 16x^4e^{-x^2} - 48x^2e^{-x^2} + 12e^{-x^2}, & f^{(4)}(0) &= 12\\
f^{(5)}(x) &= -32x^5e^{-x^2} + 160x^3e^{-x^2} - 96xe^{-x^2} - 24e^{-x^2} & f^{(5)}(0) &= 0\\
f^{(6)}(x) &= 64x^6e^{-x^2} - 480x^4e^{-x^2} + 672x^2e^{-x^2} + 48xe^{-x^2} - 120e^{-x^2} & f^{(6)}(0) &= -120
\end{aligned}$$

Hence, we have $e^{-x^2} = 1 - \frac{2x^2}{2!} + \frac{12x^4}{4!} - \frac{120x^6}{6!} = 1 - x^2 + \frac{1}{2}x^4 - \frac{1}{6}x^6$.

11. Here we have

$$\begin{aligned}
f(x) &= e^x \sin x, & f(0) &= 0\\
f'(x) &= e^x \sin x + e^x \cos x, & f'(0) &= 1\\
f''(x) &= e^x \sin x + e^x \cos x + e^x \cos x - e^x \sin x & &\\
&= 2e^2 \cos x, & f''(0) &= 2\\
f'''(x) &= 2e^x \cos x - 2e^x \sin x, & f'''(0) &= 2\\
f^{(4)}(x) &= 2e^x \cos x - 2e^x \sin x - 2e^x \sin x - 2e^x \cos x & &\\
&= -4e^x \sin x, & f^{(4)}(0) &= 0\\
f^{(5)}(x) &= -4e^x \sin x - 4e^x \cos x, & f^{(5)}(0) &= -4
\end{aligned}$$

Thus, $e^x \sin x = x + \dfrac{2x^2}{2!} + \dfrac{2x^3}{3!} - \dfrac{4x^5}{5!} = x + x^2 + \dfrac{1}{3}x^3 - \dfrac{x^5}{30}.$

12. Here we have

$$\begin{aligned}
f(x) &= xe^x, & f(0) &= 0\\
f'(x) &= xe^x + e^x, & f'(0) &= 1\\
f''(x) &= xe^x + 2e^x, & f''(0) &= 2\\
f'''(x) &= xe^x + 3e^x, & f'''(0) &= 3\\
f^{(4)}(x) &= xe^x + 4e^x, & f^{(4)}(0) &= 4
\end{aligned}$$

Thus, $xe^x = x + \dfrac{2x^2}{2!} + \dfrac{3x^2}{3!} + \dfrac{4x^2}{4!} = x + x^2 + \dfrac{x^3}{2} + \dfrac{x^4}{6}.$

13. This time we have

$$\begin{aligned}
f(x) &= x^2 e^{-x^2}, & f(0) &= 0\\
f'(x) &= (2x - 2x^3)e^{-x^2}, & f'(0) &= 0\\
f''(x) &= (4x^4 - 10x^2 + 2)e^{-x^2}, & f''(0) &= 2\\
f'''(x) &= (-8x^5 + 36x^3 - 24x)e^{-x^2}, & f'''(0) &= 0\\
f^{(4)}(x) &= (16x^6 - 112x^4 + 156x^2 - 24)e^{-x^2}, & f^{(4)}(0) &= -24\\
f^{(5)}(x) &= (-32x^7 + 320x^5 - 760x^3 + 360x)e^{-x^2}, & f^{(5)}(x) &= 0\\
f^{(6)}(x) &= (64x^8 - 864x^6 + 3120x^4 - 3000x^2 + 360)e^{-x^2}, & f^{(6)}(0) &= 360\\
f^{(7)}(x) &= (-128x^9 + 2240x^7 - 11424x^5 + 18{,}480x^3 - 6720x)e^{-x^2}, & f^{(7)}(0) &= 0\\
f^{(8)}(x) &= (256x^{10} - 5632x^8 + 38{,}528x^6 - 94{,}080x^4 + 68{,}880x^2 - 6720)e^{-x^2}, & f^{(8)}(0) &= -6720
\end{aligned}$$

Hence, $x^2e^{-x^2} = \dfrac{2x^2}{2!} - \dfrac{24x^2}{4!} + \dfrac{360x^6}{6!} - \dfrac{6720x^8}{8!} = x^2 - x^4 + \dfrac{1}{2}x^6 - \dfrac{1}{6}x^8 = x^2\left(1 - x^2 + \dfrac{1}{2}x^4 - \dfrac{1}{6}x^6\right).$

14. For this problem, we have

$$\begin{aligned}
f(x) &= e^{\sin x}, & f(0) &= 1\\
f'(x) &= \cos x e^{\sin x}, & f'(0) &= 1\\
f''(x) &= \cos^2 x \cdot e^{\sin x} - \sin x e^{\sin x}, & f(0) &= 1\\
f'''(x) &= -2\cos x \sin x e^{\sin x} + \cos^3 x e^{\sin x} - \cos x e^{\sin x} - \sin x \cos x e^{\sin x}, & f'''(0) &= 0\\
f^{(4)}(x) &= 3\sin^2 x e^{\sin x} - 3\cos^2 x e^{\sin x} - 3\cos^2 x \sin x e^{\sin x}\\
&= 3\cos^2 x \sin x e^{\sin x} + \cos^4 x e^{\sin x} + \sin x e^{\sin x} - \cos^2 x e^{\sin x}\\
&= 3\sin^2 x e^{\sin x} - 4\cos^2 x e^{\sin x} - 6\cos^2 x \sin e^{\sin x} + \cos^4 x e^{\sin x}\\
&\quad + \sin x e^{\sin x}, & f^{(4)}(0) &= -3
\end{aligned}$$

Thus, $e^{\sin x} = 1 + x + \frac{x^2}{2!} - \frac{3x^4}{4!}$.

15. Here we have

$$\begin{aligned}
f(x) &= x\sin 3x, & f(0) &= 0\\
f'(x) &= 3x\cos 3x + \sin 3x, & f'(0) &= 0\\
f''(x) &= 6\cos 3x - 9x\sin 3x, & f''(0) &= 6\\
f'''(x) &= -27x\cos 3x - 27\sin 3x, & f'''(0) &= 0\\
f^{(4)}(x) &= 81x\sin 3x - 108\cos 3x, & f^{(4)}(0) &= -108\\
f^{(5)}(x) &= 243x\cos 3x + 405\sin 3x, & f^{(5)}(0) &= 0\\
f^{(6)}(x) &= 1458\cos 3x - 729x\sin 3x, & f^{(6)}(0) &= 1458\\
f^{(7)}(x) & -5103\sin 3x - 2187x\sin 3x & f^{(7)}(0) &= 0\\
f^{(8)}(x) &= -17{,}496\cos 3x - 6561x\cos 3x & f^{(8)}(0) &= -17{,}496
\end{aligned}$$

Hence, we have $x\sin 3x = \frac{6x^2}{2!} - \frac{108x^4}{4!} + \frac{1458x^6}{6!} - \frac{17{,}496x^8}{8!} = 3x^2 - \frac{9}{2}x^4 + \frac{81}{40}x^6 - \frac{243}{560}x^8$.

16. Here we have

$$\begin{aligned}
f(x) &= e^{-x}\cos x, & f(0) &= 1\\
f'(x) &= -e^{-x}\cos x - e^{-x}\sin x, & f'(0) &= -1\\
f''(x) &= e^{-x}\cos x + e^{-x}\sin x + e^{-x}\sin x - e^{-x}\cos x\\
&= 2e^{-x}\sin x, & f''(0) &= 0\\
f'''(x) &= -2e^{-x}\sin x + 2e^{-x}\cos x, & f'''(0) &= 2\\
f^{(4)}(x) &= 2e^{-x}\sin x - 2e^{-x}\cos x - 2e^{-x}\cos x - 2e^{-x}\sin x\\
&= -4e^{-x}\cos x, & f^{(4)}(0) &= -4
\end{aligned}$$

Combining these results, we see that $e^{-x}\cos x = 1 - x + \frac{2}{3!}x^3 - \frac{4}{4!}x^4 = 1 - x + \frac{1}{3}x^3 - \frac{1}{6}x^4$.

17. As in example 31.5, let $f(t) = A'(t)$. Hence $f(t) = \frac{-60}{t^2+30}$, then $f(0) = -2$; $f'(t) = \frac{120t}{(t^2+30)^2}$, then $f'(0) = 0$, and $f''(t) = \frac{120(t^2+30)^2 - 120t(2)(t^2+30)2t}{(t^2+30)^4} = \frac{120t^2 + 120\cdot 30 - 480t^2}{(t^2+30)^3}$, then $f''(0) = \frac{4}{30} = \frac{2}{15}$.

This gives the first two terms of the Maclauren series for $A'(t)$ as $A'(t) = -2 + \frac{2}{15}\frac{t^2}{2} = -2 + \frac{t^2}{15}$. Thus, $A(t) = \int A'(t)dt = \int \left(-2 + \frac{t^2}{15}\right)dt = -2t + \frac{t^3}{45} + C$. Since the initial wound was 12 cm², $C = 12$. Hence, $A(t) = 12 - 2t + \frac{t^3}{45}$, and so $A(2) = 12 - 2\cdot 2 + \frac{2^3}{45} = 12 - 4 + \frac{8}{45} = 8\frac{8}{45} \approx 8.178$ cm².

31.2 OPERATIONS WITH SERIES

1. Since $e^x = 1 + x + \frac{x^2}{2!} + \frac{x^3}{3!}$, we see that $e^{3x} = 1 + 3x + \frac{(3x)^2}{2!} + \frac{(3x)^3}{3!} = 1 + 3x + \frac{9x^2}{2} + \frac{9x^3}{2}$.

2. Since $e^x = 1 + x + \frac{x^2}{2!} + \frac{x^3}{3!}$, then $e^{-4x} = 1 + (-4x) + \frac{(-4x)^2}{2!} + \frac{(-4x)^3}{3!} = 1 - 4x + 8x^2 - \frac{32}{3}x^3$.

3. Since $\cos x = 1 - \frac{x^2}{2!} + \frac{x^4}{4!} - \frac{x^6}{6!}$, we can find $\cos\frac{x}{2} = 1 - \frac{x^2}{8} + \frac{x^4}{16\cdot 4!} - \frac{x^6}{2^6 6!} = 1 - \frac{x^2}{8} + \frac{x^4}{384} - \frac{x^6}{46{,}080}$

4. Since $\sin x = x - \frac{x^3}{3!} + \frac{x^5}{5!} - \frac{x^7}{7!}$, then, we determine that $\sin x^3 = x^3 - \frac{(x^3)^3}{3!} + \frac{(x^3)^5}{5!} - \frac{(x^3)^7}{7!} = x^3 - \frac{x^9}{3!} + \frac{x^{15}}{5!} - \frac{x^{21}}{7!}$.

5. Since $\cos x = 1 - \frac{x^2}{2!} + \frac{x^4}{4!} - \frac{x^6}{6!}$, and so $\cos x^3 = 1 - \frac{x^6}{2!} + \frac{x^{12}}{4!} - \frac{x^{18}}{6!}$.

6. $\sin 3x = 3x - \frac{(3x)^3}{3!} + \frac{(3x)^5}{5!} - \frac{(3x)^7}{7!} = 3x - \frac{9}{2}x^3 + \frac{3^5x^5}{5!} - \frac{3^7x^7}{7!}$

7. $\sin 2x^2 = 2x^2 - \frac{(2x^2)^3}{3!} + \frac{(2x^2)^5}{5!} - \frac{(2x^2)^7}{7!} = 2x^2 - \frac{4}{3}x^6 + \frac{2^5x^{10}}{5!} - \frac{2^7x^{14}}{7!} = 2x^2 - \frac{8x^6}{3!} + \frac{32x^{10}}{5!} - \frac{128x^{14}}{7!}$

8. $\ln(1 - x) = (-x) - \frac{(-x)^2}{2} + \frac{(-x)^3}{3} - \frac{(-x)^4}{4} = -x - \frac{x^2}{2} - \frac{x^3}{3} - \frac{x^4}{4}$

9. $\sin x = x - \frac{x^3}{3!} + \frac{x^5}{5!} - \frac{x^7}{7!} + \frac{x^9}{9!} + \cdots$, and hence $\frac{d}{dx}\sin x = 1 - \frac{x^2}{2!} + \frac{x^4}{4!} - \frac{x^6}{6!} + \frac{x^8}{8!} + \cdots$, which is $\cos x$.

10. $\cos x^2 = 1 - \frac{x^4}{2!} + \frac{x^8}{4!} - \frac{x^{10}}{5!} + \frac{x^{12}}{6!}$; $\frac{d}{dx}\cos x^2 = -\frac{4x^3}{2!} + \frac{8x^7}{4!} - \frac{10x^9}{5!} + \frac{12x^{11}}{6!} - \cdots = -2x^3 + \frac{2x^7}{3!} - \frac{2x^9}{4!} + \frac{2x^{11}}{5!} - \cdots = -2x\left(x^2 - \frac{x^6}{3!} + \frac{x^8}{4!} - \frac{x^{10}}{5!}\right) = -2x\sin x^2$

11. $e^{2x} = 1 + 2x + \frac{(2x)^2}{2!} + \frac{(2x)^3}{3!} + \frac{(2x)^4}{4!} = 1 + 2x + \frac{4x^2}{2!} + \frac{8x^3}{3!} + \frac{16x^4}{4!} + \cdots$. Hence, $\frac{d}{dx}e^{2x} = 2 + 4x + \frac{8x^2}{2!} + \frac{16x^3}{3!} + \cdots = 2\left(1 + 2x + \frac{4x^2}{2!} + \frac{8x^3}{3!} + \cdots\right) = 2\left(1 + 2x + \frac{(2x)^2}{2!} + \frac{(2x)^3}{3!} + \cdots\right) = 2e^{2x}$.

12. By problem #6, $\sin 3x = 3x - \frac{3^3x^3}{3!} + \frac{3^5x^5}{5!} - \frac{3^7x^7}{7!} + \cdots$. Hence, $\frac{d}{dx}\sin 3x = 3 - \frac{3^3x^2}{2!} + \frac{3^5x^4}{4!} - \frac{3^7x^6}{6!} = 3\left(1 + \frac{3^2x^2}{2!} + \frac{3^4x^4}{4!} - \frac{3^6x^6}{6!}\right) = 3\cos 3x$.

13. $\int_0^1 \sin x^2\,dx \approx \int_0^1 \left(x^2 - \frac{x^6}{3!} + \frac{x^{10}}{5!}\right)dx = \left[\frac{x^3}{3} - \frac{x^3}{7\cdot 3!} + \frac{x^{11}}{11\cdot 5!}\right]_0^1 = \frac{1}{3} - \frac{1}{42} + \frac{1}{1320} \approx 0.3103$.

14. Using the expansion for e^x, we have $e^x - 1 = x + \frac{x^2}{2!} + \frac{x^3}{3!} + \frac{x^4}{4!}$ and dividing by x produces $\frac{e^x - 1}{x} = 1 + \frac{x}{2!} + \frac{x^2}{3!} + \frac{x^3}{4!}$. Hence, we have $\int_0^{0.1} \frac{e^x - 1}{x}\,dx \approx \int_0^1 \left(1 + \frac{x}{2!} + \frac{x^2}{3!}\right) dx \approx \left[x + \frac{x^2}{4} + \frac{x^3}{18}\right]_0^{0.1} = 0.1 + \frac{0.1^2}{4} + \frac{(0.1)^3}{18} \approx 0.1026.$

15. $\int_0^{0.2} \sin\sqrt{x}\,dx = \int_0^{0.2} \left(x^{1/2} - \frac{x^{3/2}}{3!} + \frac{x^{5/2}}{5!}\right) dx = \left[\frac{2}{3}x^{3/2} - \frac{2x^{5/2}}{5 \cdot 31} + \frac{2x^{7/2}}{7 \cdot 5!}\right]_0^{0.2} = \frac{2}{3}(0.2)^{3/2} - \frac{1}{15}(0.2)^{5/2} + \frac{2}{7 \cdot 5!}(0.2)^{7/2} \approx 0.05844.$

16. $\int_0^1 \frac{\sin x}{x}\,dx \approx \int_0^1 \left(1 - \frac{x^2}{3!} + \frac{x^4}{5!}\right) dx = \left[x - \frac{x^3}{3 \cdot 3!} + \frac{x^5}{5 \cdot 5!}\right]_0^1 = 1 - \frac{1}{18} + \frac{1}{600} \approx 0.9461$

17. We know that $e^{-x} = 1 - x + \frac{x^2}{2!} - \frac{x^3}{3!} + \frac{x^4}{4!}$ and that $\cos x = 1 - \frac{x^2}{2!} + \frac{x^4}{4!} - \frac{x^6}{6!}$. Multiplying produces $e^{-x}\cos x = \left(1 - x + \frac{x^2}{2!} - \frac{x^3}{6} + \frac{x^4}{4!}\right)\left(1 - \frac{x^2}{2} + \frac{x^4}{4!} - \cdots\right) - 1 - x - \frac{x^2}{2} + \frac{x^2}{2} + \frac{x^3}{2} - \frac{x^3}{6} + \frac{x^4}{24} - \frac{x^4}{4} + \frac{x^4}{4!} = 1 - x + \frac{2x^3}{3!} - \frac{4x^4}{4!} + \cdots.$

18. Here we subtract the expansion for e^{-x} from that of e^x and multiply this answer by $\frac{1}{2}$. The result is

$$\frac{1}{2}\left(e^x - e^{-x}\right) = \frac{1}{2}\left[\left(1 + x + \frac{x^2}{2!} + \frac{x^3}{3!} + \frac{x^4}{4!}\right) - \left(1 - x = \frac{x^2}{2!} - \frac{x^3}{3!} + \frac{x^4}{4!} + \cdots\right)\right]$$
$$= \frac{1}{2}\left[2x + 2\frac{x^3}{3!} + 2\frac{x^5}{5!} + 2\frac{x^7}{7!} + \cdots\right]$$
$$= x + \frac{x^3}{3!} + \frac{x^5}{5!} + \frac{x^7}{7!} + \cdots$$

This is the same as Exercise #2 in Exercise Set 30.1.

19. $e^{-x^2} = 1 - x^2 + \frac{x^4}{2!} - \frac{x^6}{3!} + \frac{x^8}{4!} + \cdots$. Hence $x^2e^{-x^2} = x^2 - x^4 + \frac{x^6}{2!} - \frac{x^8}{3!} + \frac{x^{10}}{4!} - \cdots$

20. $\tan x = \frac{\sin x}{\cos x} = \frac{x - \frac{x^3}{3!} + \frac{x^5}{5!} - \frac{x^7}{7!}\cdots}{1 - \frac{x^2}{2!} + \frac{x^4}{4!} - \frac{x^6}{6!}\cdots}$. To compute this we must use a procedure like long division. We get the following:

$$\begin{array}{r|l}
 & x + \tfrac{1}{3}x^3 + \tfrac{2}{15}x^5 + \cdots \\
\hline
1 - \tfrac{1}{2}x^2 + \tfrac{1}{24}x^4 - \cdots & x - \tfrac{1}{6}x^3 + \tfrac{1}{120}x^5 - \cdots \\
 & x - \tfrac{1}{2}x^3 + \tfrac{1}{24}x^5 - \cdots \\
\hline
 & \tfrac{1}{3}x^3 - \tfrac{1}{30}x^5 + \cdots \\
 & \tfrac{1}{3}x^3 - \tfrac{1}{6}x^5 + \cdots \\
\hline
 & \tfrac{2}{15}x^5 + \cdots
\end{array}$$

Hence, $\tan x = x + \frac{x^3}{3} + \frac{2}{15}x^5 + \cdots$.

21. $5e^{0.8j} = 5(\cos 0.8 + j\sin 0.8) = 3.4835 + 3.5868j$

22. $5 - 12j$; $r = \sqrt{5^2 + 12^2} = 13$, and $\theta = \tan^{-1}\frac{-12}{5} = -1.176$. Thus, $5 - 12j = 13\,[\cos(-1.176) + j\sin(-1.176)]$

23. $6\,\text{cis}\,\frac{4\pi}{3} = 6e^{\frac{4\pi}{3}j} = 6e^{4\pi j/3}$

24. $(-3 + 4j)$; $r = 5$, and $\theta = \tan^{-1}\frac{4}{-3} = 2.2143$. Thus, $(-3 + 4j) = 5e^{2.2143j}$

25. $y = e^{x^2} = 1 + x^2 + \frac{x^4}{2} + \cdots$. Hence, we obtain $A = \int_0^1 \left(1 + x^2 + \frac{x^4}{2}\right) dx = \left[x + \frac{x^3}{3} + \frac{x^5}{10}\right]_0^1 = 1 + \frac{1}{3} + \frac{1}{10} = \frac{43}{30} = 1.4333.$

26. $y = \cos x^2 = 1 - \frac{x^4}{2!} + \frac{x^8}{4!} - \cdots$. As a result, we obtain $\int_0^{\pi/2} \left(1 - \frac{x^4}{2!} + \frac{x^8}{4!}\right) dx = \left[x - \frac{x^5}{10} + \frac{x^9}{9 \cdot 4}\right]_0^{\pi/2} =$

$\frac{\pi}{2} - \frac{\left(\frac{\pi}{2}\right)^5}{10} + \frac{\left(\frac{\pi}{2}\right)^9}{9 \cdot 4!} \approx 0.8840$. (The actual value is 0.8491.)

27. $x^2e^2 = x^2 + x^3 + \frac{x^4}{2}$. Integrating, we obtain $\int_0^{0.2} \left(x^2 + x^3 + \frac{x^4}{2}\right) dx = \left[\frac{x^3}{3} + \frac{x^4}{4} + \frac{x^5}{5}\right]_0^{0.2} = \frac{(0.2)^2}{3} + \frac{(0.2)^4}{4} + \frac{(0.2)^5}{10} = 0.0031$.

28. $e^{-x^2} = 1 - x^2 + \frac{x^4}{2} + \cdots$. Hence, we get $\int_0^1 \left(1 - x^2 + \frac{x^4}{2}\right) dx = \left[x - \frac{x^3}{3} + \frac{x^5}{10}\right]_0^1 = 1 - \frac{1}{3} + \frac{1}{10} = \frac{23}{30} = 0.7667$.

29. $\cos 0.1 = 1 - \frac{(0.1)^2}{2!} + \frac{(0.1)^4}{4!} - \frac{(0.1)^6}{6!} + \frac{(0.1)^8}{8!} \approx 0.995004$. This produces $15\cos(0.1) = 15 \times 0.9950 = 14.9251$.

30. $\sinh x + \cosh x = \left[\left(x + \frac{x^3}{3!} + \frac{x^5}{5!} + \cdots\right) + \left(1 + \frac{x^2}{2!} + \frac{x^4}{4!} + \cdots\right)\right] = 1 + x + \frac{x^2}{2!} + \frac{x^3}{3!} + \frac{x^4}{4!} + \frac{x^5}{5!} + \cdots = e^x$.

31. (a) $i = \sin t^2 = t^2 - \frac{t^6}{3!} + \frac{t^{10}}{5!} - \frac{t^{14}}{7!} + \cdots$

(b) Charge $= \int_0^{0.02} \left(t^2 - \frac{t^6}{3!} + \frac{t^{10}}{5!} - \frac{t^{14}}{7!}\right) dx = \left[\frac{t^3}{3} - \frac{t^7}{7 \cdot 3!} + \frac{t^{11}}{11 \cdot 5!} - \frac{t^{15}}{15 \cdot 7!}\right]_0^{0.02} = \frac{(0.02)^3}{3} - \frac{(0.02)^7}{7 \cdot 3!} + \frac{(0.02)^{11}}{11 \cdot 5!} = 2.7 \times 10^{-6} = 0.0000027$.

32. To find this volume, we can use the shell method. $V = 2\pi \int_0^{0.1} (xe^{-x})\, dx$. Since $xe^{-x} = x - x^2 + \frac{x^3}{2} - \cdots$, we get $V = 2\pi \int_0^{0.1} \left(x - x^2 + \frac{x^3}{2}\right) dx = 2\pi \left[\frac{x^2}{2} - \frac{x^3}{3} + \frac{x^4}{8}\right]_0^{0.1} = 2\pi \left[\frac{0.01}{2} - \frac{0.001}{3} + \frac{0.0001}{8}\right] = 0.009358\pi \approx 0.02940$.

31.3 NUMERICAL TECHNIQUES USING SERIES

1. $e^{-0.3} = 1 - 0.3 + \frac{0.3^2}{2!} - \frac{(0.3)^3}{3!} + \frac{(0.3)^4}{4!} = 1 - 0.3 + 0.045 - 0.0045 + 0.0003375$. Using the first four terms and rounding we get $e^{-0.3} = 0.7408$.

2. $\sin(0.1) = 0.1 - \frac{(0.1)^3}{3!} + \frac{(0.1)^5}{5!} = 0.1 - 0.000167 + 0.000000083$. Using only the first two terms and rounding we get $\sin(0.1) = 0.0998$

3. $\cos(0.1) = 1 - \frac{(0.1)^2}{2!} + \frac{(0.1)^4}{4!} - \frac{(0.1)^6}{6!} = 1 - 0.005 + 0.00000417$. Using the first two terms we get 0.9950

4. $e^{0.2} = 1 + 0.2 + \frac{(0.2)^2}{2!} + \frac{(0.2)^3}{3!} + \frac{(0.2)^4}{4!} = 1 + 0.2 + 0.02 + 0.00133 + 0.000067$. Using the first four terms and rounding we obtain 1.2213. If we use the first five terms we get 1.2214.

5. $5° = \frac{\pi}{36}$. Hence, $\sin 5° - \sin\frac{\pi}{36} = \frac{\pi}{36} - \frac{\left(\frac{\pi}{36}\right)^3}{3!} - \frac{\left(\frac{\pi}{36}\right)^5}{5!} = 0.087266 - 0.00011 = 0.0872$

6. $\ln(1.5) = \ln(1+0.5) = 0.5 - \frac{(0.5)^2}{2} + \frac{(0.5)^3}{3} - \frac{(0.5)^4}{4} + \frac{(0.5)^5}{5} + \cdots = 0.5 - 0.125 + 0.041667 - 0.015626 + 0.00625 - 0.002605 + 0.001116 - 0.000488 + 0.000217 - 0.000097 = 0.4054$

7. $\ln(0.97) = \ln(1 - 0.03) = -0.03 - \frac{(0.03)^2}{2} - \frac{(0.03)^3}{3} - \frac{(0.03)^4}{4} = -0.03 - 0.00045 = -0.0305$

8. $\sqrt{e} = e^{1/2} = 1 + \frac{1}{2} + \frac{(0.5)^2}{2!} + \frac{(0.5)^3}{3!} + \frac{(0.5)^4}{4!} = 1 + 0.5 + 0.125 + 0.020833 + 0.002604 + 0.00026 + 0.0000217 = 1.6487$

9. $\ln(0.5) = \ln(1 - 0.5) = -0.5 - \frac{(0.5)^2}{2} - \frac{(0.5)^3}{3} - \frac{(0.5)^4}{4} - \cdots = -0.5 - 0.125 - 0.041667 - 0.015625 - 0.00625 - 0.002604 - 0.001116 - 0.000488 - 0.000217 - 0.000097 = -0.6931$

10. $\cos 5^\circ = \cos\frac{\pi}{36} = 1 - \frac{\left(\frac{\pi}{36}\right)^2}{2!} + \frac{\left(\frac{\pi}{36}\right)^4}{4!} = 1 - 0.003808 + 0.0000024 = 0.9962$

11. First we must find a series for $\frac{1-\cos x}{x}$. Since $\cos x = 1 - \frac{x^2}{2!} + \frac{x^4}{4!}\cdots$, we have then $1-\cos x = \frac{x^2}{2!} - \frac{x^4}{4!} + \frac{x^6}{6!}$, and $\frac{1-\cos x}{x} = \frac{x}{2!} - \frac{x^3}{4!} + \frac{x^5}{6!}$. As a result, we get

$$\int_0^{0.5}\frac{1-\cos x}{x}\,dx = \int_0^{0.5}\left(\frac{x}{2!} - \frac{x^3}{4!} + \frac{x^5}{6!} + \cdots\right)dx$$
$$= \frac{x^2}{4} - \frac{x^4}{4\cdot 4!} + \frac{x^6}{6\cdot 6!} - \frac{x^8}{8\cdot 8!}\Big|_0^{0.5}$$
$$= \frac{(0.5)^2}{4} - \frac{(0.5)^4}{4\cdot 4!} + \frac{(0.5)^6}{6\cdot 6!} - \frac{(0.5)^8}{8\cdot 8!}$$
$$= 0.0526 - 0.000651 + 0.0000036$$
$$= 0.0619.$$

12. Since $\frac{x-\sin x}{x} = \int_0^1 \frac{x - \left(x - \frac{x^3}{3!} + \frac{x^5}{5!}\right)}{x}\,dx = \frac{\frac{x^3}{3!} - \frac{x^5}{5!} + \frac{x^7}{7!}}{x} = \frac{x^2}{3!} - \frac{x^4}{5!} + \frac{x^6}{7!}$, we have $\int_0^{0.5}\frac{x-\sin x}{x}\,dx = \int_0^{0.5}\left(\frac{x^2}{3!} - \frac{x^4}{5!} + \frac{x^6}{7!}\right)dx = \left[\frac{x^3}{3\cdot 3!} - \frac{x^5}{5\cdot 5!} + \frac{x^7}{7\cdot 7!}\right]_0^{0.5} = 0.006944 - 0.000052 = 0.00689$ or 0.0069. (Note: we only needed the first term for the desired accuracy.)

13. $\int_0^1 \frac{\sin x - x}{x^2}\,dx = \int_0^1 \frac{-x}{3!} + \frac{x^3}{5!} - \frac{x^5}{7!}\,dx = \left[\frac{-x^2}{12} + \frac{x^4}{4\cdot 5!} + \frac{x^6}{6\cdot 7!}\right]_0^1 = -\frac{1}{12} + \frac{1}{480} - 0.000033 = -0.0813.$

14. $$\int_0^{0.5}\sqrt{x}\cos\sqrt{x}\,dx$$
$$= \int_0^{0.5}\sqrt{x}\left(1 - \frac{x}{2!} + \frac{x^2}{4!} - \frac{x^3}{6!} + \cdots\right)dx$$
$$= \int_0^{0.5}\left(x^{1/2} - \frac{x^{3/2}}{2!} + \frac{x^{5/2}}{4!} - \frac{x^{7/2}}{6!}\right)dx$$
$$= \left[\frac{2}{3}x^{3/2} - \frac{1}{5}x^{5/2} + \frac{2x^{7/2}}{7\cdot 4!} - \frac{2x^{9/2}}{9\cdot 6!}\right]_0^{0.5}$$
$$= 0.235702 - 0.035355 + 0.001052 - 0.0000136$$
$$\approx 0.2014$$

15. $$\int_0^{0.2}\frac{e^x - 1}{x} = \int_0^{0.2}\left(1 + \frac{x}{2!} + \frac{x^2}{3!} + \frac{x^3}{4!} + \cdots\right)dx$$
$$= \left[x + \frac{x^2}{4} + \frac{x^3}{3\cdot 3!} + \frac{x^4}{4\cdot 4!}\right]_0^{0.2}$$
$$= 0.2 + 0.01 + 0.000444 + 0.0000167$$
$$= 0.2105$$

16. $$\int_{0.5}^1 \frac{\ln(1+x)}{x}\,dx$$
$$= \int_{0.5}^1\left(1 - \frac{x}{2} + \frac{x^2}{3} - \frac{x^3}{4} + \frac{x^4}{5}\right)dx$$
$$= \left[x - \frac{x^2}{4} + \frac{x^3}{9} + \frac{x^4}{16} + \frac{x^5}{25}\right]_{0.5}^1$$
$$= \left(1 - \frac{1}{4} + \frac{1}{9} - \frac{1}{16} + \frac{1}{25} - \frac{1}{36} + \frac{1}{49}\right)$$
$$- \left(0.5 - \frac{(0.5)^2}{4} + \frac{(0.5)^3}{9}\right) = 0.3828$$

17. (a) $\sin\theta = \theta - \frac{\theta^3}{3!} + \frac{\theta^4}{5!} - \cdots$, so $\sin\theta = \theta$ when $\frac{\theta^3}{3!}$ is negligable. (b) If we want accuracy to 0.0001, then set $\frac{\theta^3}{3!} < 0.0001$ or $\theta^3 < 0.0006$ or $\theta < \sqrt[3]{0.0006} \approx 0.0843$.

18. To solve this, we need to find $\frac{1}{C}\int_0^{0.2} i\,dt$. Since 1 μF 10^{-6}F, this integral is $\frac{1}{10^{-6}}\int_0^{0.2}\frac{0.1(t-\sin t)}{t}\,dt = 10^5\int\left(\frac{t^2}{3!} -\right.$

$\frac{t^4}{5!} + \frac{t^6}{7!}\Big) dt = 10^5 \left[\frac{t^3}{18} - \frac{t^5}{5\cdot 5!} + \frac{t^7}{7\cdot 7!}\right]_0^{0.2} =$ $10^5[4.\bar{4}\times 10^{-4} - 5.33\times 10^{-7}] = 10^5 \times 4.43911 \times 10^{-4} = 44.3911$ V.

19. The Maclaurin series for e^x is $1+x+\frac{x^2}{2!}+\frac{x^3}{3!}+\cdots$, so $e^{-0.05t^2} = 1 - 0.05t^2 + \frac{0.0025t^4}{2!} - \frac{0.000125t^6}{3!} + \cdots$. Thus,

$$q = \int_0^{0.5} \left(1 - e^{-0.05t^2}\right) dt$$
$$= \int_0^{0.5} \left(0.05t^2 - \frac{0.0025t^4}{2!} + \frac{0.000125t^6}{3!} + \cdots\right) dt$$
$$= \left[\frac{0.05}{3}t^3 - \frac{0.0025t^5}{5\cdot 2!} + \frac{0.000125t^7}{7\cdot 3!} + \cdots\right]_0^{0.5}$$
$$\approx 0.002083 - .000008 + \cdots \approx 0.0021.$$

The charge is about 0.0021 μC.

20. The Maclaurin series for e^x is $1+x+\frac{x^2}{2!}+\frac{x^3}{3!}+\cdots$, so $e^{-0.25t^2} = 1 - 0.25t^2 + \frac{0.0625t^4}{2!} - \frac{0.015625t^6}{3!} + \cdots$.

Thus,

$$q = \int_0^{1.5} \left(1 - e^{-0.25t^2}\right) dt$$
$$= \int_0^{1.5} \left(0.25t^2 - \frac{0.0625t^4}{2!} + \frac{0.015625t^6}{3!} + \cdots\right) dt$$
$$= \left[\frac{0.25}{3}t^3 - \frac{0.0625t^5}{5\cdot 2!} + \frac{0.015625t^7}{7\cdot 3!} + \cdots\right]_0^{1.5}$$
$$\approx 0.28125 - 0.0474609 + 0.006356 - 0.000695$$
$$\approx 0.23945$$

The charge is about 0.23945 μF.

21. (a) The Maclaurin series for e^x is $1 + x + \frac{x^2}{2!} + \frac{x^3}{3!} + \cdots$, so the desired Maclaurin series is $V = 0.75e^{0.75t} = 0.75\left(1 + 0.75t + \frac{(0.75)^2}{2!}t^2 + \frac{(0.75)^3}{3!}t^3 + \cdots\right)$; (b) When $t = 0.45$, this becomes $0.75(1 + 0.3375 + 0.05695 + 0.00641 + 0.00054\cdots) \approx 0.75(1.4014) \approx 1.05105$.

31.4 TAYLOR SERIES

1. $f(x) = e^x$, $f(2) = e^2$, $f'(x) = e^x$, $f'(2) = e^2$, etc. $e^x = e^2 + e^2(x-2) + e^2\frac{(x-2)^2}{2!} + e^2\frac{(x-2)^3}{3!} + \cdots =$ $e^2\left[x - 1 + \frac{(x-2)^2}{2!} + \frac{(x-2)^3}{3!} + \cdots\right]$

2. $g(x) = e^{2x}$, $g(-1) = e^{-2}$, $g'(x) = 2e^{2x}$, $g'(-1) = 2e^{-2}$, $g''(x) = 2^2e^{2x}$, $g''(-1) = 2^2e^{-2}$, ... Hence, $e^{2x} = e^{-2} + 2e^{-2}(x+1) + 2e^{-2}\frac{(x+1)^2}{2!} + \cdots =$ $e^{-2}\left[1 + 2(x+1) + \frac{2^2(x+1)^2}{2!} + \frac{2^3(x+1)^3}{3!} + \cdots\right]$.

3. $h(x) = \sin x$, $h\left(\frac{\pi}{4}\right) = \frac{\sqrt{2}}{2}$, $h'(x) = \cos x$, $h'\left(\frac{\pi}{4}\right) = \frac{\sqrt{2}}{2}$, $h''(x) = \sin x$, $h''\left(\frac{\pi}{4}\right) = -\frac{\sqrt{2}}{2}$, $h'''(x) = -\cos x$, $h'''\left(\frac{\pi}{2}\right) = -\frac{\sqrt{2}}{2}$. As a result, we get

$$\sin x = \frac{\sqrt{2}}{2} + \frac{\sqrt{2}}{2}\left(x - \frac{\pi x}{4}\right) - \frac{\sqrt{2}}{2}\frac{\left(x-\frac{\pi}{2}\right)^2}{2!} - \frac{\sqrt{2}}{2}\frac{\left(x-\frac{\pi}{4}\right)3}{3!} + \cdots$$
$$= \frac{\sqrt{2}}{2}\left[1 + \left(x - \frac{\pi}{4}\right) - \frac{\left(x-\frac{\pi}{4}\right)^2}{2!} - \frac{\left(x-\frac{\pi}{4}\right)^3}{3!} + \frac{\left(x-\frac{\pi}{4}\right)^4}{4!} + \cdots\right]$$

4. $j(x) = \cos x$, $j\left(\frac{\pi}{4}\right) = \frac{\sqrt{2}}{2}$, $j'(x) = -\sin x$, $j'\left(\frac{\pi}{4}\right) = -\frac{\sqrt{2}}{2}$, $j''(x) = -\cos x$, $j''\left(\frac{\pi}{4}\right) = -\frac{\sqrt{2}}{2}$,

$j'''(x) = \sin x, j'''\left(\frac{\pi}{4}\right) = \frac{\sqrt{2}}{2}$. Hence,

$$\cos x = \frac{\sqrt{2}}{2} - \frac{\sqrt{2}}{2}\left(x - \frac{\pi}{2}\right) - \frac{\sqrt{2}}{2} \cdot \frac{\left(x - \frac{\pi}{4}\right)^2}{2!} + \frac{\sqrt{2}}{2} \cdot \frac{\left(x - \frac{\pi}{4}\right)^3}{3!} + \cdots$$
$$= \frac{\sqrt{2}}{2}\left[1 - \left(x - \frac{\pi}{4}\right) - \frac{1}{2!}\left(x - \frac{\pi}{4}\right)^2 + \frac{1}{3!}\left(x - \frac{\pi}{4}\right)^3 + \frac{1}{4!}\left(x - \frac{\pi}{4}\right)^4 - \cdots\right]$$

5. $k(x) = \frac{1}{x}$, so $k(2) = \frac{1}{2}$, $k'(x) = -x^{-2}$, so $k'(2) = -\frac{1}{4} = \frac{-1}{2^2}$, $k''(x) = 2x^{-3}$, so $k''(2) = \frac{2}{2^3}$, and $k'''(x) = -6x^{-4}$, so $k'''(2) = \frac{-6}{2^4}$. Hence, we see that $\frac{1}{x} = \frac{1}{2} - \frac{(x-2)}{2^2} + \frac{2(x-2)^2}{2^3 2!} - \frac{6(x-2)^3}{2^4 3!} + \frac{4!(x-2)^4}{2^5 4!} - \cdots = \frac{1}{2}\left[1 - \frac{x-2}{2} + \frac{(x-2)^2}{2^2} - \frac{(x-2)^3}{2^3} + \frac{(x-2)^4}{2^4} - \cdots\right]$.

6. $m(x) = \sqrt{x}$, $m(4) = 2$, $m'(x) = \frac{1}{2}x^{-\frac{1}{2}}$, $m'(4) = \frac{1}{4} = \frac{1}{2^2}$, $m''(x) = -\frac{1}{4}x^{-\frac{3}{2}}$, $m''(4) = \frac{-1}{32}$, $m'''(x) = \frac{3}{8}x^{-\frac{5}{2}}$, $m'''(4) = \frac{3}{2^8}$. Hence, $\sqrt{x} = 2 + \frac{1}{4}(x-4) - \frac{1}{32}\frac{(x-4)^2}{2!} + \frac{3}{2^8}\frac{(x-4)^3}{3!} - \cdots = 2 + \frac{1}{2^2}(x-4) - \frac{1}{2^5}\frac{(x-4)^2}{2!} + \frac{3}{2^8}\frac{(x-4)^3}{3!} - \frac{15(x-4)^4}{2^{11}4!} + \cdots$.

7. $f(x) = \frac{1}{(x+1)^2} = (x+1)^{-2}$, $f(0) = 1$, $f'(x) = -2(x+1)^{-3}$, $f'(0) = -2$ $f''(x) = 6(x+1)^{-4}$, $f''(0) = 6, f'''(x) = -24(x+1)^{-5}, f'''(0) = -24$. As a result, we obtain $\frac{1}{(x+1)^2} = 1 - 2(x) + \frac{6(x)^2}{2!} - \frac{24(x)^3}{3!} + \cdots = 1 - 2(x) + 3(x)^2 - 4(x)^3 + 5(x)^4 - \cdots$.

8. $g(x) = e^{1+x}$, $g(1) = e^2$, $g'(x) = e^{1+x}$, $g'(1) = e^2$, ... Hence, we get $e^{1+x} = e^2 + e^2(x-1) + e^2\frac{(x-1)^2}{2!} + e^2\frac{(x-1)^2}{3!} + \cdots = e^2\left[1 + (x-1) + \frac{(x-1)^2}{2!} + \frac{(x-1)^3}{3!} + \cdots\right]$.

9. $h(x) = e^{-x}$, so $h(1) = e^{-1}$, and $h'(x) = -e^{-x}$, so $h'(1) = -e^{-1}$, and $h''(x) = e^{-x}$, so $h''(1) = e^{-1}\cdots$. Combining these results, we see that $e^{-x} = e^{-1}e^{-1}(x-1) + e^{-1}\frac{(x-1)^2}{2!} - e^{-1}\frac{1}{3!}(x-1)^3 \cdots = e^{-1}\left[1 - (x-1) + \frac{1}{2!}(x-1)^2 - \frac{1}{3!}(x-1)^3 + \cdots\right]$.

10. $j(x) = \ln(x), j(4) = \ln 4, j'(x) = \frac{1}{x}, j'(4) = \frac{1}{4}, j''(x) = -x^{-2}$, $j''(4) = \frac{-1}{4^2}$, $j'''(x) = 2x^{-3}$, $j'''(4) = \frac{2}{4^3}$. Hence, $\ln(x) = \ln 4 + \frac{1}{4}(x-4) - \frac{1}{4^2}\frac{(x-4)^2}{2!} + \frac{2}{4^3}\frac{(x-4)^3}{3!} - \cdots = \ln 4 + \frac{1}{4}(x-4) - \frac{1}{4^2}\frac{(x-4)^2}{2} + \frac{1}{4^3}\frac{(x-4)^3}{3} - \frac{1}{4^4}\frac{(x-4)^4}{4} + \cdots$

11. $k(x) = \sqrt[3]{x}, k(8) = 2, k'(x) = \frac{1}{3}x^{-2/3}, k'(8) = \frac{1}{3} \cdot \frac{1}{2^2}$, $k''(x) = \frac{-2}{9}x^{-5/3}$, $k''(8) = \frac{-2}{9} \cdot \frac{1}{2^5} = \frac{-1}{9 \cdot 2^4}$, $k'''(x) = \frac{10}{27}x^{-8/3}$. Hence, we obtain $\sqrt[3]{x} = 2 + \frac{1}{3} \cdot \frac{1}{2^2}(x-8) - \frac{1}{9 \cdot 2^4} \cdot \frac{(x-8)^2}{2!} + \frac{10}{27} \cdot \frac{1}{2^8} \cdot \frac{(x-8)^3}{3!} + \cdots$.

12. $m(x) = \tan x$, $m\left(\frac{\pi}{4}\right) = \tan\left(\frac{\pi}{4}\right) = 1$, $m'(x) = \sec^2 x$, $m'\left(\frac{\pi}{4}\right) = 2$, $m''(x) = 2\sec^2 x \tan x$, $m''\left(\frac{\pi}{4}\right) = 4$, $m'''(x) = 4 \sec x \tan x + 2\sec^4 x$, $m'''\left(\frac{\pi}{4}\right) = 8 + 8 = 16$. As a result, we obtain $\tan x = 1 + 2\left(x - \frac{\pi}{4}\right) + 4\frac{\left(x - \frac{\pi}{4}\right)^2}{2!} + 16\frac{\left(x - \frac{\pi}{4}\right)3}{3!} + \cdots$.

13. Using Exercise #1, with $x = 2.1$, and $c = 2$, we see that $x - c = 0.2$. As a result, we obtain

$e^{2.1} = e^2\left[1+(0.1)+\frac{(0.1)^2}{2!}+\frac{(0.1)^3}{3!}+\cdots\right] = e^2[1.1052] = 8.1662$

14. Here, we use Exercise #2, with $x = -0.4$, $c = -1$, and $x-c = x+1 = -0.6+ = 0.6$. As a result, we get $e^{-0.8} \approx e^{-2}\left[1+2(0.6)+\frac{2^2(0.6)^2}{2!}+\frac{2^3(0.6)^3}{3!}+\frac{2^4(0.6)^4}{4!}\right] = e^{-2}(3.2944) = 0.4458$.

15. In this exercise we use the result from Exercise #3. We convert 43° to $\frac{43\pi}{180}$ radians. We have $x = \frac{\pi}{4}$ and $c = \frac{43\pi}{180}$, with the result that $x - c = \frac{43\pi}{180} - \frac{\pi}{4} = -\frac{\pi}{90}$. Thus, $\sin 43° = \sin\frac{43\pi}{180} \approx \sin\left(-\frac{\pi}{90}\right) = \frac{\sqrt{2}}{2}\left[1+\left(-\frac{\pi}{90}\right)-\left(\frac{-\frac{\pi}{90}}{2!}\right)^2-\frac{\left(-\frac{\pi}{90}\right)^3}{3!}+\frac{\left(-\frac{\pi}{90}\right)^4}{4!}\right] = \frac{\sqrt{2}}{2}[0.9645] = 0.6820$.

16. $\cos 43° = \cos\left(-\frac{\pi}{90}\right) = \frac{\sqrt{2}}{2}\left[1-\left(-\frac{\pi}{90}\right)-\frac{1}{2!}\left(-\frac{\pi}{90}\right)^2+\frac{1}{3!}\left(-\frac{\pi}{90}\right)^3+\frac{1}{4!}\left(-\frac{\pi}{90}\right)^4\right] = \frac{\sqrt{2}}{2}[1.0343] = 0.7314$.

17. Using Exercise \$5, with $x = 1.8$, $c = 2$, and $x - c = -0.2$, we get $\frac{1}{1.8} \approx \frac{1}{2}\left[1-\frac{(-0.2)}{2}+\frac{(-0.2)^2}{2^2}-\frac{(-0.2)^3}{2^3}+\frac{(0.2)^4}{2^4}\right] = \frac{1}{2}[1+0.1+0.01+0.001+0.0001] = \frac{1}{2}[1.1111] = 0.5555$

18. $\sqrt{3.9} = \sqrt{4-0.1} \approx 2+\frac{1}{2^2}(-0.1)-\frac{1}{2^6}(-0.1)^2 = 2-0.025-0.000156 = 1.9748$

19. $\frac{1}{1.12^2} = \frac{1}{(0.12+1)^2} = 1-2(0.12)+3(0.12)^2-4(0.12)^3 = 1-0.24+0.0432-0.006912+0.0010368-0.000149 = 0.7972$

20. $e^{1.9} = e^{1+.9} \approx e^2\left[1+(-0.1)+\frac{(-0.1)^2}{2!}+\frac{(-0.1)^3}{3!}+\frac{(-0.1)^4}{4!}\right] = e^2\,[0.9048] = 6.6859$.

21. $e^{-0.8} \approx e^{-1}\left[1-(-0.2)+\frac{1}{2!}(-0.2)^2-\frac{1}{3!}(-0.2)^3\right] = e^{-1}\,[1.2214] = 0.4493$.

22. $\ln(4.1) = \ln(4+0.1) = \ln 4+\frac{1}{4}(0.1)-\frac{1}{16}\frac{(0.1)^2}{2} = \ln 4+0.025-0.0003125 = 1.4110$

23. $\sqrt[3]{7.8} = \sqrt[3]{8-0.2} = 2+\frac{1}{3}\cdot\frac{1}{2^2}(-0.2)-\frac{1}{9}\cdot\frac{1}{2^4}\frac{(-0.2)^2}{2} = 2-0.01667-0.000139 = 1.9832$

24. $\tan 43° = \tan\left(-\frac{\pi}{90}\right) \approx 1+2\left(-\frac{\pi}{90}\right)+2\left(-\frac{\pi}{90}\right)^2+\frac{16}{6}\left(-\frac{\pi}{90}\right)^3 = 1-0.069813+0.002437-0.0001134 = 0.9325$

25. From Example 31.0, we know that $\frac{\sin\pi t}{t} = \pi\left(\frac{t-2}{t}\right)-\frac{\pi^3}{3!}\frac{(t-2)^3}{t}+\cdots$. When we let $t = 2.1$, we get $\frac{\sin\pi t}{t} = \pi\left(\frac{0.1}{2.1}\right)-\frac{\pi^3}{3!}\frac{(0.1)^3}{2.1}+\cdots \approx 0.14960-(0.00246)+\cdots \approx 0.14714$. A calculator gives $\frac{\sin(2.1\pi)}{2.1} \approx 0.14715$.

26. Using a procedure similar to the one used in Example 31.0, we determine the following fundamental information needed for the Taylor's expansion of $\sin x$ around $t = 1$. $f(x) = \sin 2\pi x$, so $f(1) = \sin 2\pi = 0$, $f'(x) = 2\pi\cos 2\pi x$, so $f'(1) = 2\pi\cos 2\pi = 2\pi$, $f''(x) = -4\pi^2\sin 2\pi x$, so $f''(1) = -4\pi^2\sin 2\pi = 0$, $f'''(x) = -8\pi^3\cos 2\pi x$, so $f'''(1) = -8\pi^3\cos 2\pi = -8\pi^3$. Using these values we obtain

$$\begin{aligned}\sin 2\pi t &= 0+2\pi(t-1)-0\cdot\frac{1}{2!}(t-1)^2\\ &\quad -8\pi^3\frac{1}{3!}(t-1)^3+\cdots\\ &= 2\pi(t-1)-\frac{8\pi^3}{3!}(t-1)^3+\cdots\end{aligned}$$

Dividing by t, produces $\frac{\sin 2\pi t}{t} = 2\pi\left(\frac{t-1}{t}\right) - 8\frac{\pi^3}{3!}\frac{(t-1)^3}{t} + \cdots$ and when we let $t = 1.05$, we get $\dfrac{\sin 2\pi t}{t} = 2\pi\left(\dfrac{0.05}{1.05}\right) - 8\dfrac{\pi^3}{3!}\dfrac{(0.05)^3}{1.05} + \cdots \approx 0.2991993 - (0.0049216) + \cdots \approx 0.294278$. A calculator gives $\dfrac{\sin 2(1.05\pi)}{1.05} \approx 0.294302$.

31.5 FOURIER SERIES

1. $f(x) = \begin{cases} 2, & -\pi < x < 0 \\ -2, & 0 < x < \pi \end{cases}$, period 2π.

$2L = 2\pi$, so $L = \pi$, First, $a_0 = \dfrac{1}{2\pi}\displaystyle\int_{-\pi}^{\pi} f(x)\,dx = \dfrac{1}{2\pi}\int_{-\pi}^{0}(-2)\,dx + \dfrac{1}{2\pi}\int_{0}^{\pi} 2\,dx = \dfrac{1}{2\pi}[-2x]_{-\pi}^{0} + \dfrac{1}{2\pi}[2x]_{0}^{-\pi} = \dfrac{-2\pi}{2\pi} + \dfrac{2\pi}{2\pi} = 0$. We have $a_k = \dfrac{1}{\pi}\displaystyle\int_{-\pi}^{\pi} f(x)\cos\frac{k\pi x}{\pi}\,dx = \dfrac{1}{\pi}\int_{-\pi}^{0} 2\cos kx\,dx + \int_{0}^{\pi} -2\cos kx\,dx = \dfrac{2}{\pi}\left[\dfrac{\sin kx}{k}\right]_{-\pi}^{0} - \dfrac{2}{\pi}\left[\dfrac{\sin kx}{k}\right]_{0}^{\pi} = 0 - \dfrac{2}{k\pi}\sin(-k\pi) - \dfrac{2}{k\pi}\sin k\pi = 0$. We also have $b_k = \dfrac{1}{\pi}\displaystyle\int_{-\pi}^{\pi} f(x)\sin\frac{k\pi x}{\pi}\,dx = \dfrac{1}{\pi}\int_{-\pi}^{0} 2\sin kx\,dx + \dfrac{1}{\pi}\int_{0}^{\pi} = 2\sin kx\,dx = \dfrac{-2}{\pi}\left[\dfrac{\cos kx}{k}\right]_{-\pi}^{0} + \dfrac{2}{\pi}\left[\dfrac{\cos kx}{k}\right]_{0}^{\pi} = \dfrac{-2}{k\pi}[1 - \cos k\pi] + \dfrac{2}{k\pi}[\cos k\pi - 1] = \dfrac{-4}{k\pi}(1 - \cos k\pi) = \begin{cases} 0 & \text{If } k \text{ is even} \\ \frac{-8}{k\pi} & \text{if } k \text{ is odd} \end{cases}$. Hence $f(x) = \dfrac{-8}{\pi}\sin x + \dfrac{-8}{3\pi}\sin 3x + \dfrac{-8}{5\pi}\sin 5x + \cdots = \dfrac{-8}{\pi}\left(\sin x + \dfrac{1}{3}\sin 3x + \dfrac{1}{5}\sin 5x + \cdots\right) = \dfrac{-8}{\pi}\displaystyle\sum_{k=1}^{\infty}\dfrac{1}{2k-1}\sin(2k-1)x$.

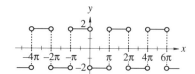

2. $g(x) = 1$, $-2 < x < 2$, period 4, $L = 2$. First, $a_0 = \dfrac{1}{4}\displaystyle\int_{-2}^{2} 1\,dx = \dfrac{1}{4}x\Big|_{-2}^{2} = \dfrac{1}{4}(2+2) = 1$ and $a_k = \dfrac{1}{2}\displaystyle\int_{-2}^{2}\cos\frac{k\pi x}{2}\,dx = \dfrac{1}{2}\dfrac{2}{k\pi}\sin\dfrac{k\pi x}{2}\Big|_{-2}^{2} = \dfrac{1}{k\pi}[\sin k\pi - \sin(-k\pi)] = 0$ for all k. We also have $b_k = \dfrac{1}{2}\displaystyle\int_{-2}^{2}\sin\frac{k\pi x}{2}\,dx = 0$ since sine is an odd function. Hence $g(x) = 1$.

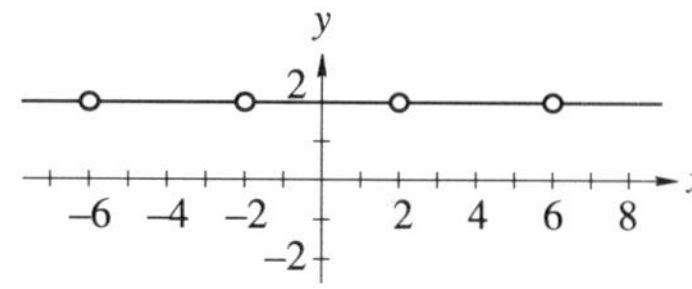

3. $h(x) = \begin{cases} 0, & -3 < x < 0 \\ 2, & 0 < x < 3 \end{cases}$, period 6 so $L = 3$. First, $a_0 = \dfrac{1}{6}\displaystyle\int_{-3}^{3} h(x)\,dx = \dfrac{1}{6}\int_{0}^{3} 2\,dx = 1$. We have $a_k = \dfrac{1}{3}\displaystyle\int_{-3}^{3} h(x)\cos\frac{k\pi x}{3}\,dx = \dfrac{1}{3}\int_{0}^{3} 2\cos\frac{k\pi x}{3}\,dx = \dfrac{2}{3}\dfrac{3}{k\pi}\left[\sin\dfrac{k\pi x}{3}\right]_{0}^{3} = \dfrac{2}{k\pi}\sin k\pi = 0$ for all k. We also have $b_k = \dfrac{1}{3}\displaystyle\int_{-3}^{3} h(x)\sin\frac{k\pi x}{3}\,dx = \dfrac{1}{3}\int_{0}^{3} 2\sin\frac{k\pi x}{3}\,dx = \dfrac{2}{3}\cdot\dfrac{3}{k\pi}\left[-\cos\dfrac{k\pi x}{3}\right]_{0}^{3} =$

$\frac{2}{k\pi}[-\cos k\pi + 1] = \begin{cases} 0 & \text{if } k \text{ is even} \\ \frac{4}{k\pi} & \text{if } k \text{ is odd} \end{cases}$. Hence,
$h(x) =$
$1 + \frac{4}{\pi}\sin\frac{\pi x}{3} + \frac{4}{3\pi}\sin\frac{3\pi x}{3} + \frac{4}{5\pi}\sin\frac{5\pi x}{3} + \cdots$.

4. $j(x) = \begin{cases} 0, & -1 < x < 0 \\ x, & 0 < x < 1 \end{cases}$, period $= 2$ so $L = 1$.

First, $a_0 = \frac{1}{2}\int_{-1}^{1} j(x)\,dx = \frac{1}{2}\int_0^1 x\,dx = \frac{1}{2}\frac{x^2}{2}\Big|_0^1 = \frac{1}{4}$.

We have $a_k = \frac{1}{1}\int_{-1}^{1} j(x)\cos k\pi x\,dx =$

$\int_0^1 x\cos k\pi x\,dx = \left(\frac{1}{k\pi}\right)^2\int_0^1 (k\pi x)\cos k\pi x\,dx =$

$\left(\frac{1}{k\pi}\right)^2\Big[\cos k\pi x + k\pi x\sin k\pi x\Big]_0^1 =$

$\left[\left(\frac{1}{k\pi}\right)^2\cos k\pi x + \frac{x}{k\pi}\sin k\pi x\right]_0^1 =$

$\left(\left(\frac{1}{k\pi}\right)^2\cos k\pi + 0\right) - \left(\frac{1}{k\pi}\cos 0 + 0\right)$

$= \begin{cases} 0 & \text{if } k \text{ is even} \\ \frac{-2}{(k\pi)^2} & \text{if } k \text{ is odd} \end{cases}$.

We also have

$b_k = \int_{-1}^{1} j(x)\sin k\pi x\,dx = \int_0^1 x\sin k\pi x\,dx =$

$\left[\frac{1}{(k\pi)^2}\sin k\pi x - \frac{x}{k\pi}\cos k\pi x\right]_0^1 = -\frac{1}{k\pi}\cos k\pi =$

$\begin{cases} -\frac{1}{k\pi}, & \text{if } k \text{ is even} \\ \frac{1}{k\pi}, & \text{if } k \text{ is odd} \end{cases}$. So,

$$j(x) = \frac{1}{4} - \frac{2}{\pi^2}\cos\pi x - \frac{2}{9\pi^2}\cos 3\pi x - \frac{2}{25\pi^2}\cos 5\pi x - \cdots + \frac{1}{\pi}\sin\pi x - \frac{1}{2\pi}\sin 2\pi x + \frac{1}{3\pi}\sin 3\pi x - \frac{1}{4\pi}\sin 4\pi x + \cdots$$

$$= \frac{1}{4} + \sum_{k=1}^{\infty}\left[\frac{-2}{(2k-1)^2\pi^2}\cos(2k-1)\pi x + (-1)^{k+1}\frac{1}{k\pi}\sin k\pi x\right].$$

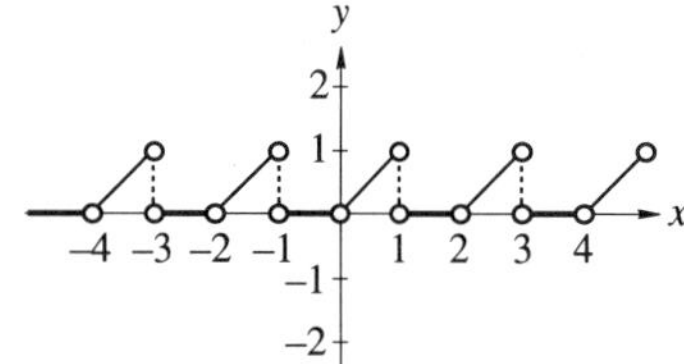

5. $f(x) = \begin{cases} -x, & -\pi < x < 0 \\ x, & 0 < x < \pi \end{cases}$, period $= 2\pi$ so $L = \pi$.

First, $a_0 = \frac{1}{2\pi}\int_{-\pi}^{\pi} f(x)\,dx =$

$\frac{1}{2\pi}\int_{-\pi}^{0} -x\,dx + \frac{1}{2\pi}\int_0^{\pi} x\,dx =$

$\frac{1}{2\pi}\frac{-x^2}{2}\Big|_{-\pi}^{0} + \frac{1}{2\pi}\frac{x^2}{2}\Big|_0^{\pi} = \frac{\pi^2}{4\pi} + \frac{\pi^2}{4\pi} = \frac{\pi}{2}$. Next,

$a_k = \frac{1}{\pi}\int_{-\pi}^{\pi} f(x)\cos kx\,dx =$

$\frac{1}{\pi}\int_{-\pi}^{0} -x\cos kx\,dx + \frac{1}{\pi}\int_0^{\pi} x\cos kx\,dx =$

$\frac{2}{\pi}\int_0^{\pi} x\cos kx\,dx = \frac{2}{\pi}\left[\frac{\cos kx}{k^2} + \frac{x\sin kx}{k}\right]_0^{\pi} =$

$\frac{2}{\pi k^2}[\cos k\pi - 1] = \begin{cases} 0 & \text{if } k \text{ is even} \\ \frac{-4}{k^2\pi} & \text{if } k \text{ is odd} \end{cases}$. Also, we

see that $b_k = \frac{1}{2\pi}\int_{-\pi}^{\pi} f(x)\sin kx = 0$ since

$f(x)\sin k\pi$ is an odd function. Hence,

$f(x) = \frac{\pi}{2} - \frac{4}{\pi}\cos x - \frac{4}{9\pi}\cos 3x - \frac{4}{25\pi}\cos 5\pi - \cdots =$

$\frac{\pi}{2} - \frac{4}{\pi}\sum_{k=1}^{\infty}\frac{1}{(2k-1)^2}\cos(2k-1)x.$

6. $g(x) = \begin{cases} 3, & 0 < x < 4 \\ -3, & 4 < x < 8 \end{cases}$, period = 8 so $L = 4$.
First,
$a_0 = \frac{1}{8}\int_0^8 g(x)\,dx = \frac{1}{8}\int_0^4 3\,dx + \frac{1}{8}\int_4^8 (-3)\,dx = 0.$
Next, $a_k = \dfrac{1}{4}\displaystyle\int_0^8 g(x)\cos\frac{k\pi x}{4}\,dx =$
$\dfrac{1}{4}\displaystyle\int_0^4 3\cos\frac{k\pi x}{4}\,dx - \frac{1}{4}\int_4^8 3\cos\frac{k\pi x}{4}\,dx =$
$\dfrac{3}{4}\cdot\dfrac{4}{k\pi}\sin\dfrac{k\pi x}{4}\Big|_0^4 - \dfrac{3}{4}\cdot\dfrac{4}{k\pi}\sin\dfrac{k\pi x}{4}\Big|_4^8 = 0.$ We also have
$b_k = \dfrac{1}{4}\displaystyle\int_0^4 3\sin\frac{k\pi x}{4}\,dx - \frac{1}{4}\int_4^8 3\sin\frac{k\pi x}{4}\,dx =$
$-\dfrac{3}{4}\cdot\dfrac{4}{k\pi}\cos\dfrac{k\pi x}{4}\Big|_0^4 + \dfrac{3}{4}\cdot\dfrac{4}{k\pi}\cos\dfrac{k\pi x}{4}\Big|_4^8 =$
$-\dfrac{3}{k\pi}[\cos k\pi - 1] + \dfrac{3}{k\pi}[1 - \cos k\pi] =$
$\dfrac{6}{k\pi}[1 - \cos k\pi] = \begin{cases} 0 & \text{if } k \text{ is even} \\ \frac{12}{k\pi} & \text{if } k \text{ is odd} \end{cases}$. Thus,
$g(x) = \dfrac{12}{\pi}\sin\dfrac{\pi x}{4} + \dfrac{12}{3\pi}\sin\dfrac{3\pi x}{4} + \dfrac{12}{5\pi}\sin\dfrac{5\pi x}{4}\cdots =$
$\dfrac{12}{\pi}\displaystyle\sum_{k=1}^{\infty}\frac{1}{2k-1}\sin\frac{(2k-1)\pi x}{4}$

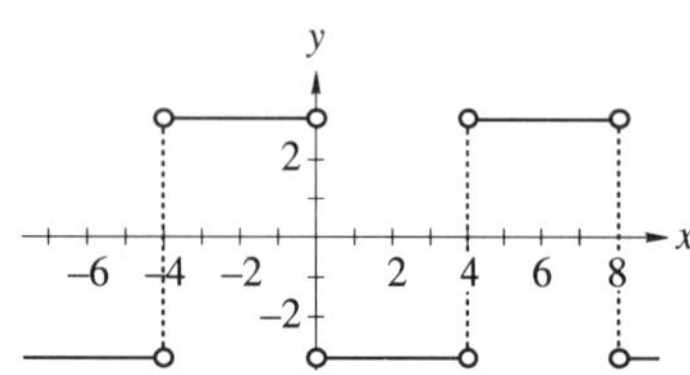

7. $h(x) = x$, $-\pi < x < \pi$, period 2π so $L = \pi$. First,
$a_0 = \dfrac{1}{2\pi}\displaystyle\int_{-\pi}^{\pi} x\,dx = 0.$ Next,
$a_k = \dfrac{1}{\pi}\displaystyle\int_{-\pi}^{\pi} x\cos kx\,dx = 0$, since $[x\cos kx]$ is an odd function. Also, $b_k = \dfrac{1}{\pi}\displaystyle\int_{-\pi}^{\pi} x\sin kx\,dx =$
$\dfrac{1}{\pi}\left[\dfrac{\sin kx}{k^2} - \dfrac{x\cos kx}{k}\right]_{-\pi}^{\pi} =$
$\dfrac{1}{\pi}\left[\left(-0 - \dfrac{\pi\cos k\pi}{k}\right) - \left(0 + \dfrac{\pi\cos k\pi}{k}\right)\right] =$
$\begin{cases} \dfrac{-2}{k} & \text{if } k \text{ even} \\ \dfrac{2}{k} & \text{if } k \text{ is odd} \end{cases}$. Hence,
$h(x) = 2\sin\pi - \dfrac{2}{2}\sin 2\pi + \dfrac{2}{3}\sin 3\pi - \cdots =$
$2\displaystyle\sum_{k=1}^{\infty}\frac{(-1)^{k+1}}{k}\sin kx.$

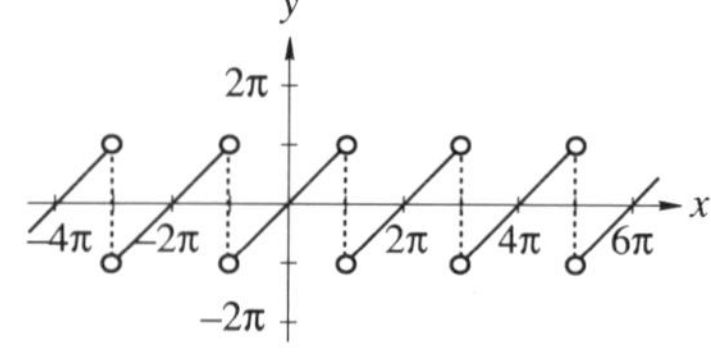

8. $j(x) = \begin{cases} 0, & -\dfrac{\pi}{2} < x < 0 \\ 2, & 0 < x < \dfrac{\pi}{2} \end{cases}$, period $= \pi$, so $L = \frac{\pi}{2}$.
First,
$a_0 = \dfrac{1}{\pi}\displaystyle\int_{-\pi/2}^{\pi/2} j(x)\,dx = \frac{1}{\pi}\int_0^{\pi/2} 2\,dx = \frac{1}{\pi}2x\Big|_0^{\pi/2} = 1.$
Next, we have $a_k = \dfrac{2}{\pi}\displaystyle\int_{-\pi/2}^{\pi/2} j(x)\cos\frac{k\pi x}{\pi/2}\,dx =$
$\dfrac{2}{\pi}\displaystyle\int_0^{\pi/2} 2\cos 2kx\,dx = \frac{4}{\pi}\cdot\frac{1}{2k}\Big[\sin 2kx\Big]_0^{\pi/2} =$
$\dfrac{2}{k\pi}(\sin\pi - \sin 0) = 0.$ Also,
$b_k = \dfrac{2}{\pi}\displaystyle\int_0^{\pi/2} 2\sin 2kx\,dx =$
$\dfrac{4}{\pi}\cdot\dfrac{1}{2k}\Big[-\cos 2kx\Big]_0^{\pi/2} = \dfrac{2}{k\pi}[-\cos k\pi + \cos 0] =$
$\dfrac{2}{k\pi}[1 - \cos k\pi] = \begin{cases} 0 & \text{if } k \text{ is even} \\ \dfrac{4}{k\pi} & \text{if } k \text{ is odd} \end{cases}$. Thus,
$j(x) = 1 + \dfrac{4}{\pi}\sin 2x + \dfrac{4}{3\pi}\sin 6x + \dfrac{4}{5\pi}\sin 10x + \cdots =$
$1 + \dfrac{4}{\pi}\displaystyle\sum_{k=1}^{\infty}\frac{1}{2k-1}\sin(4k-2)x.$

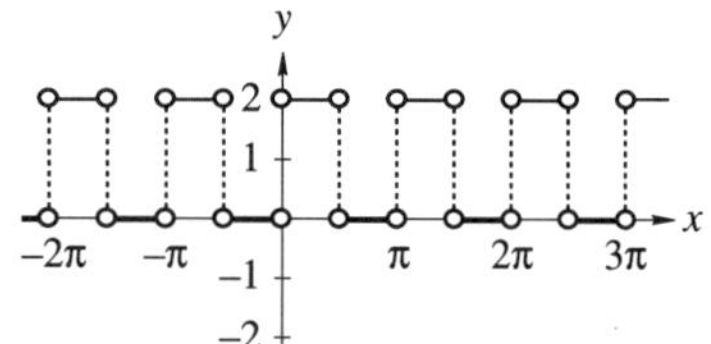

9. $f(x) = \begin{cases} 1, & 0 < x < \frac{2\pi}{3} \\ 0, & \frac{2\pi}{3} < x < \frac{4\pi}{3} \\ -1, & \frac{4\pi}{3} < x < 2\pi \end{cases}$, period $= 2\pi$, so $L = \pi$.

First, $a_0 = \frac{1}{2\pi}\int_0^{2\pi} f(x)\,dx = \frac{1}{2\pi}\left[\int_0^{2\pi/3} 1\,dx + \int_{4\pi/3}^{2\pi} -1\,dx\right] = 0$. Then, we have $a_k = \frac{1}{\pi}\int_0^{2\pi} f(x)\cos kx\,dx = \frac{1}{2}\left[\int_0^{2\pi/3}\cos kx\,dx - \int_{4\pi/3}^{2\pi}\cos kx\,dx\right] = \frac{1}{k\pi}\left[\sin kx\Big|_0^{2\pi/3} - \sin kx\Big|_{4\pi/3}^{2\pi}\right] = \frac{1}{\pi k}\left[\sin\frac{2k\pi}{3} - 0 - 0 + \sin\frac{4\pi k}{3}\right]$. This is always 0. Now, we have $b_k = \frac{1}{\pi}\int_0^{2\pi} f(x)\sin kx\,dx = \frac{1}{\pi}\int_0^{2\pi/3}\sin kx - \frac{1}{\pi}\int_{4\pi/3}^{2\pi}\sin kx\,dx = \frac{-1}{k\pi}\cos kx\Big|_0^{2\pi/3} + \frac{1}{k\pi}\cos kx\Big|_{4\pi/3}^{2\pi} = \frac{-1}{k\pi}\left[1 - \cos\frac{2k\pi}{3} + 1 - \cos\frac{4k\pi}{3}\right]$. Thus, $b_1 = \frac{3}{\pi}$, $b_2 = \frac{3}{2\pi}$, $b_3 = 0$, $b_4 = \frac{3}{4\pi}$, $b_5 = \frac{3}{5\pi}$, $b_6 = 0$, and we obtain

$f(x) = \frac{3}{\pi}\sin x + \frac{3}{2\pi}\sin 2x + 0\cdot\sin 3x + \frac{3}{4\pi}\sin 4x + \cdots = \frac{3}{\pi}\sum_{k=1}^{\infty}\left(\frac{\sin(3k-2)x}{3k-2} + \frac{\sin(3k-1)x}{3k-1}\right).$

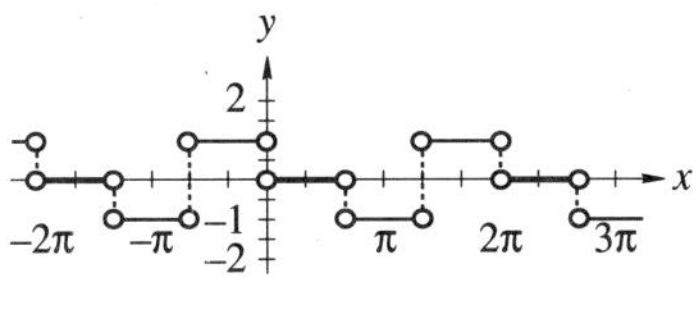

10. $g(x) = \begin{cases} x, & 0 < x \le \pi \\ \pi, & \pi < x < 2\pi \end{cases}$, period $= 2\pi$ so $L = \pi$.

Thus, $a_0 = \frac{1}{2\pi}\int_0^{2\pi} g(x)\,dx = \frac{1}{2\pi}\int_0^{\pi} x\,dx + \frac{1}{2\pi}\int_{\pi}^{2\pi}\pi\,dx = \frac{1}{2\pi}\cdot\frac{x^2}{2}\Big|_0^{\pi} + \frac{1}{2\pi}(\pi x)\Big|_{\pi}^{2\pi} = \frac{\pi}{4} + \pi - \frac{\pi}{2} = \frac{3\pi}{4}$.

Next we have $a_k = \frac{1}{\pi}\int_0^{2\pi} g(x)\cos kx\,dx = \frac{1}{\pi}\int_0^{\pi} x\cos kx\,dx + \frac{1}{\pi}\int_{\pi}^{2\pi}\pi\cos kx\,dx = \frac{1}{\pi}\left[\frac{\cos kx}{k^2} + \frac{x\sin kx}{k}\right]_0^{\pi} + \left[\frac{1}{k}\sin kx\right]_{\pi}^{2\pi} = \frac{1}{\pi k^2}[\cos\pi - 1] = \begin{cases} 0 & \text{if } k \text{ is even} \\ \frac{-2}{\pi k}, & \text{if } k \text{ is odd} \end{cases}$. We also have $b_k = \frac{1}{\pi}\int_0^{\pi} x\sin kx\,dx + \frac{1}{\pi}\int_{\pi}^{2\pi}\pi\sin kx\,dx = \frac{1}{\pi}\left[\frac{\sin kx}{k^2} - \frac{x\cos kx}{k}\right]_0^{\pi} + \frac{1}{k}(-\cos kx)\Big|_{\pi}^{2\pi} = \frac{1}{\pi}\left(-\frac{\pi\cos k}{k}\right)\frac{1}{k}(\cos k\pi - \cos 2\pi k) = -\frac{\cos k\pi}{k} + \frac{\cos k\pi}{k} - \frac{\cos 2\pi k}{k} = -\frac{1}{k}$. Hence,

$g(x) = \frac{3\pi}{4} - \frac{2}{\pi}\left(\cos x + \frac{1}{9}\cos 3x + \frac{1}{25}\cos 5x + \cdots\right) - \left(\sin x + \frac{1}{2}\sin 2x + \frac{1}{3}\sin 3x + \cdots\right) = \frac{3\pi}{4} - \frac{2}{\pi}\sum_{k=1}^{\infty}\frac{\cos(2k-1)x}{(2k-1)^2} - \sum_{k=1}^{\infty}\frac{\sin k\pi}{k}$

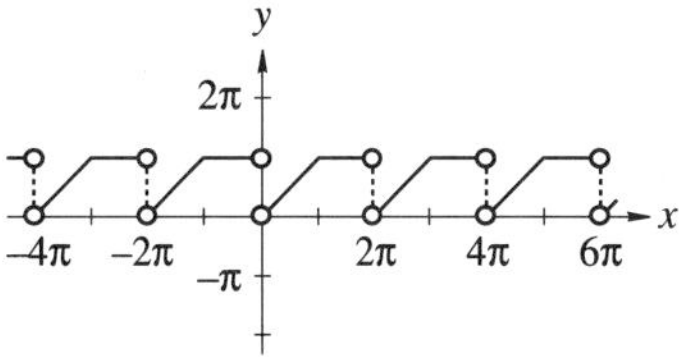

11. $h(x) = \begin{cases} 0, & \text{if } -\pi < x < 0 \\ \sin x, & \text{if } 0 < x < \pi \end{cases}$, period $= 2\pi$, so $L = \pi$. First, $a_0 = \frac{1}{2\pi}\int_{-\pi}^{\pi} h(x)\,dx = \frac{1}{2\pi}\int_0^{\pi}\sin x\,dx = \frac{1}{2\pi}[-\cos x]_0^{\pi} = \frac{1}{2\pi}[1+1] = \frac{1}{\pi}$.

For a_1 we have $a_1 = \frac{1}{\pi}\int_0^{\pi} \sin x \cos x\, dx =$
$\frac{1}{2\pi}\int_0^{\pi} 2\sin x \cos x\, dx = \frac{1}{2\pi}\int_0^{\pi} \sin 2x\, dx =$
$\left[-\frac{1}{4\pi}\cos 2x\right]_0^{\pi} = 0$. For a_k, $k > 1$, we have $a_k =$
$\frac{1}{\pi}\int_{-\pi}^{\pi} h(x)\cos kx\, dx = \frac{1}{\pi}\int_0^{\pi} \sin x \cos kx\, dx =$
$\frac{1}{\pi}\left[-\frac{\cos(1-k)x}{2(1-k)} - \frac{\cos(1+k)x}{2(1+k)}\right]_0^{\pi} =$
$-\frac{1}{\pi}\left[\frac{\cos(1-k)\pi}{2(k-1)} + \frac{\cos(1+k)x}{2(1+k)} - \frac{1}{2(1-k)} - \frac{1}{2(1+k)}\right]$. If k is odd,
$\cos(1-k)\pi = \cos(1+k)\pi = 1$, so $a_k = 0$. If k is even, $\cos(1-k)\pi = \cos(1+k)\pi = -1$, so $a_k =$
$\frac{1}{\pi}\left[\frac{1}{1-k} + \frac{1}{1+k}\right] = \frac{1}{\pi}\frac{2}{1-k^2} = -\frac{2}{(k^2-1)\pi}$.
In particular, $a_2 = -\frac{2}{(2^2-1)\pi} = -\frac{2}{3\pi}$,
$a_4 = -\frac{2}{(4^2-1)\pi} = -\frac{2}{15\pi}$, $a_6 = -\frac{2}{35\pi}$.

For b_1 we have
$b_1 = \frac{1}{\pi}\int_0^{\pi} \sin^2 x\, dx = \frac{1}{\pi}\left[\frac{x}{2} - \frac{1}{4}\sin 2x\right]_0^{\pi} = \frac{1}{2}$.
For $k > 1$, we have $b_k = \frac{1}{\pi}\int_0^{\pi} \sin x \sin kx\, dx =$
$\frac{1}{\pi}\left[\frac{\sin(1-k)x}{2(1-k)} - \frac{\sin(1+k)x}{2(1+k)}\right]_0^{\pi} = 0$. Hence,
$h(x) =$
$\frac{1}{\pi} + \frac{1}{2}\sin x - \frac{2}{\pi}\left[\frac{1}{3}\cos 2x + \frac{1}{15}\cos 4x + \cdots\right] =$
$\frac{1}{\pi} + \frac{1}{2}\sin x - \frac{2}{\pi}\sum_{k=1}^{\infty}\frac{\cos 2kx}{4k^2-1}$.

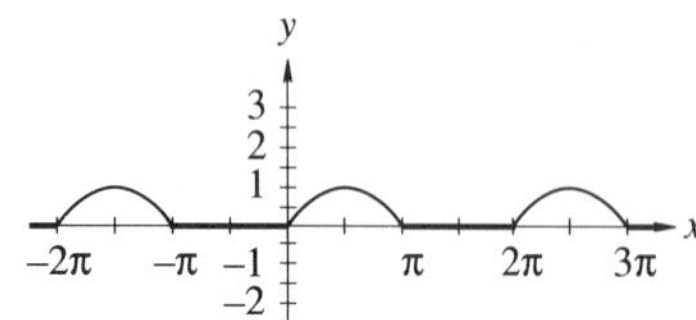

12. $f(x) = \begin{cases} \sin x, & 0 < x < \pi \\ -\sin x, & \pi < x < 2\pi \end{cases}$, period 2π, so
$L = \pi$. First $a_0 = \frac{1}{2\pi}\int_0^{2\pi} f(x)\, dx =$
$\frac{1}{2\pi}\int_0^{\pi} \sin x\, dx + \frac{1}{2\pi}\int_{\pi}^{2\pi} (-\sin x)\, dx =$
$\frac{1}{2\pi}\left[-\cos x\big|_0^{\pi} + \frac{1}{2\pi}\cos x\Big|_{\pi}^{2\pi}\right] =$
$\frac{1}{2\pi}(1+1+1+1) = \frac{2}{\pi}$.

For a_1 we have
$\frac{1}{\pi}\int_0^{\pi} \sin x \cos x\, dx - \frac{1}{\pi}\int_{\pi}^{2\pi} \sin x \cos x\, dx =$
$\frac{1}{2x}\int_0^{\pi} 2\sin x \cos x\, dx - \frac{1}{2\pi}\int_{\pi}^{2\pi} 2\sin x \cos x\, dx =$
$\frac{1}{2\pi}\int_0^{\pi} \sin 2x\, dx - \frac{1}{2\pi}\int_{\pi}^{2\pi} \sin 2x\, dx =$
$\frac{1}{4\pi}[-\cos 2x]_0^{\pi} - \frac{1}{4\pi}[-\cos 2x]_{\pi}^{2\pi} = 0$.

For $k > 1$, $a_k =$
$\frac{1}{\pi}\int_0^{\pi} \sin x \cos kx\, dx - \frac{1}{\pi}\int_{\pi}^{2\pi} \sin x \cos kx\, dx =$
$-\frac{1}{\pi}\left[\frac{\cos(1-k)x}{2(1-k)} + \frac{\cos(1+k)x}{2(1+k)}\right]_0^{\pi} +$
$\frac{1}{\pi}\left[\frac{\cos(1-k)x}{2(1-k)} + \frac{\cos(1+k)x}{2(1+k)}\right]_{\pi}^{2\pi} =$
$-\frac{1}{\pi}\left[\frac{\cos(1-k)\pi}{2(1-k)} + \frac{\cos(1+k)\pi}{2(1+k)} - \frac{1}{2(1-k)} - \frac{1}{2(1+k)}\right] + \frac{1}{\pi}\left[\frac{1}{2(1-k)} + \frac{1}{2(1+k)} - \frac{\cos(1-k)\pi}{2(1-k)} - \frac{\cos(1-k)\pi}{2(1+k)}\right]$. If k is odd,
$\cos(1-k)\pi = \cos(1+k)\pi 1$, so $a_k = 0$. If k is even, $\cos(1-k)\pi = \cos(1+k)\pi = -1$, so
$a_k = \frac{1}{\pi}\left[\frac{2}{1-k} + \frac{2}{1+k}\right] = \frac{4}{\pi(1-k^2)}$. In
particular, $a_2 = \frac{4}{\pi(1-2^2)} = -\frac{4}{3\pi}$,
$a_4 = \frac{4}{\pi(1-4^2)} = -\frac{4}{15\pi}$, $a_6 = -\frac{4}{35\pi}$.

In a similar manner we find that
$b_1 = \frac{1}{\pi}\int_0^{\pi} \sin^2 x\, dx - \frac{1}{\pi}\int_{\pi}^{2\pi} \sin^2 x\, dx =$

$\frac{1}{\pi}\int_0^{\pi}\sin^2 x\,dx - \frac{1}{\pi}\int_{-\pi}^{0}\sin^2 x\,dx$. Since $\sin^2 x$ is an even function, these two integrals are equal, so their difference is 0. The same argument holds for b_k when $k > 1$. Hence,
$f(x) = \frac{2}{\pi} - \frac{4}{3\pi}\cos 2x - \frac{4}{15\pi}\cos 4x - \cdots = \frac{2}{\pi}\left(1 - \sum_{k=1}^{\infty}\frac{2\cos 2kx}{4k^2-1}\right)$.

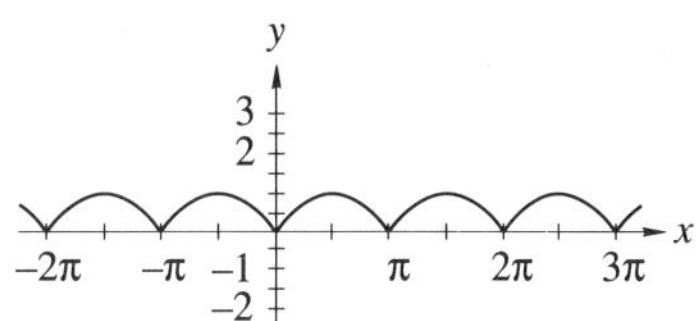

CHAPTER 31 REVIEW

1. Since $\cos x = 1 - \frac{x^2}{2!} + \frac{x^4}{4!} - \frac{x^6}{6!} + \cdots$ we have $1 - \cos x = \frac{x^6}{2!} - \frac{x^4}{4!} + \frac{x^6}{6!} - \cdots$ and $\frac{1-\cos x}{x}$ equals $\frac{x}{2!} - \frac{x^3}{4!} + \frac{x^5}{6!} - \frac{x^7}{8!} + \frac{x^9}{10!} - \frac{x^{11}}{12!} + \cdots$.

2. Since $e^x = 1 + x + \frac{x^2}{2!} + \frac{x^3}{3!} + \cdots$ we get $e^{2x} = 1 - 2x + \frac{4x^2}{2!} - \frac{8x^3}{3!} + \cdots$ and so $xe^{-2x} = x - 2x^2 + \frac{4x^3}{2!} - \frac{8x^4}{3!} + \frac{16x^5}{4!} - \frac{32x^6}{5!} + \cdots$

3. $\sin x\cos x = \left(x - \frac{x^3}{3!} + \frac{x^5}{5!} - \frac{x^7}{7!} + \cdots\right)\left(1 - \frac{x^2}{2!} + \frac{x^4}{4!} - \frac{x^6}{6!} + \cdots\right) = x - \frac{x^3}{2} - \frac{x^3}{3} + \frac{x^5}{4!} + \frac{x^5}{6} + \frac{x^5}{6!} - \frac{x^7}{3!4!} - \frac{x^7}{2\cdot 5!} - \frac{x^7}{6!} - \frac{x^7}{7!} + \cdots = x - \frac{5x^3}{3!} + \frac{16x^5}{5!} - \frac{64x^7}{7!} + \cdots$

4. $\ln(2 + x) = \ln(x + 1 + 1) = x + 1 - \frac{(x+1)^2}{2} + \frac{(x+1)^3}{3} - \frac{(x+1)^4}{4} + \cdots$

5. $f(x) = (1 + x)^{3/4}$, $f(0) = 1$; $f'(x) = \frac{3}{4}(1 + x)^{-1/4}$; $f'(0) = \frac{3}{4}$; $f''(x) = -\frac{3}{16}(1 + x)^{-5/4}$; $f''(0) = -\frac{3}{16}$; $f'''(x) = \frac{15}{16}(1 + x)^{-9/4}$; $f'''(0) = \frac{15}{64}$. Hence $(1 + x)^{3/4} = 1 + \frac{3}{4}(1+x) - \frac{3}{16}\frac{(1+x)^2}{2!} + \frac{15}{64}\frac{(1+x)^3}{3!} - \frac{135}{256}\frac{(1+x)^4}{4!} + \cdots$.

6. Since $e^x = 1 + x + \frac{x^2}{2!} + \frac{x^3}{3!} + \cdots$ we get $e^{-x^4} = 1 - x^4 + \frac{x^8}{2!} - \frac{x^{12}}{3!} + \frac{x^{16}}{4!} - \frac{x^{20}}{5!} + \cdots$.

7. Since $\sin x = x - \frac{x^8}{3!} + \frac{x^5}{5!} - \frac{x^7}{7!} + \cdots$ we obtain $\sin x^2 = x^2 - \frac{x^6}{3!} + \frac{x^{10}}{5!} - \frac{x^{14}}{7!} + \frac{x^{18}}{9!} - \cdots$.

8. $1 - e^x = -x - \frac{x^2}{2!} - \frac{x^3}{3!} - \frac{x^4}{4!} - \cdots$ and we get $\frac{1-e^x}{x} = -1 - \frac{x}{2!} - \frac{x^2}{3!} - \frac{x^3}{4!} - \frac{x^4}{5!} - \cdots$

9. First we obtain the Maclaurin series for $\frac{\sin x}{\sqrt{x}}$ which is $x^{1/2} - \frac{x^{5/2}}{3!} + \frac{x^{9/2}}{5!} - \cdots$. Now we integrate $\int_0^1\left(x^{1/2} - \frac{x^{5/2}}{3!} + \frac{x^{9/2}}{5!} - \frac{x^{13/2}}{7!}\right) = \left[\frac{2}{3}x^{3/2} - \frac{2x^{3/2}}{7\cdot 3!} + \frac{2}{11}\frac{x^{11/2}}{5!} - \frac{2x^{15/2}}{157!}\right]_0^1 = \frac{2}{3} - \frac{2}{42} + \frac{2}{1320} - \frac{2}{75600} =$ $0.66667 - 0.04762 + 0.001515 - 0.000026 = 0.6205$

10. $x^3e^{-x^3} = x^3 - x^6 + \frac{x^9}{2!} - \frac{x^{12}}{3!} + \frac{x^{15}}{4!}$. Hence, we obtain $\int_0^1 x^3e^{-x^3}\,dx = \int_0^1 (x^3 - x^6 + \frac{x^9}{2!} - \frac{x^{12}}{3!} + \cdots)\,dx =$ $\left[\frac{x^4}{4} - \frac{x^7}{7} + \frac{x^{10}}{20} - \frac{x^{13}}{13\cdot 13!} + \frac{x^{16}}{16\cdot 4!} - \frac{x^{19}}{19\cdot 5!} + \right.$

$\cdots \Big]_0^1 = \frac{1}{4} - \frac{1}{7} + \frac{1}{20} - \frac{1}{78} + \frac{1}{384} - \frac{1}{2280} + \frac{1}{15840} = 0.1466.$

11. $e^x = 1 + x + \frac{x^2}{2!} + \frac{x^3}{3!} + \frac{x^4}{4!} + \cdots$. Hence, $e^{0.25} = 1 - 0.25 + \frac{0.25^2}{2!} - \frac{0.25^3}{3!} + \cdots = 1 - 0.25 + 0.03125 - 0.002604 + 0.000163 - 0.000008 = 0.7788.$

12. $\cos x^4 = 1 - \frac{x^8}{2!} + \frac{x^{16}}{4!} - \cdots$. Hence, $\int_0^1 \cos x^4\,dx = \int_0^1 \left(1 - \frac{x^8}{2!} + \frac{x^{16}}{4!} - \frac{x^{24}}{6!} + \cdots\right) dx = \left(x - \frac{x^9}{18} + \frac{x^{17}}{17 \cdot 4!} - \frac{x^{25}}{25 \cdot 6!} + \cdots\right)\Big|_0^1 = 1 - \frac{1}{18} + \frac{1}{408} - \frac{1}{18000} = 0.9468.$

13. $f(x) = \cos x$, $\cos \frac{\pi}{6} = \frac{\sqrt{3}}{2}$, $f'(x) = -\sin x$, $f'\left(\frac{\pi}{6}\right) = -\frac{1}{2}$, $f''(x) = -\cos x$, $f''\left(\frac{\pi}{6}\right) = -\frac{\sqrt{3}}{2}$, $f'''(x) = \sin x$, $f'''\left(\frac{\pi}{6}\right) = \frac{1}{2}$. Hence, we obtain $\cos x = \frac{\sqrt{3}}{2} - \frac{1}{2}\left(x - \frac{\pi}{6}\right) - \frac{\sqrt{3}}{2} \cdot \frac{\left(x - \frac{\pi}{6}\right)^2}{2!} + \frac{1}{2}\frac{\left(x - \frac{\pi}{6}\right)}{3!} + \cdots$

14. $f(x) = \sinh x$, $f(\ln 2) = \frac{3}{4}$, $f'(x) = \cosh x$, $f'(\ln(2)) = \frac{5}{4}$; $f''(x) = f(x)$, $f'''(x) = f'(x)$, etc. Hence, $\sinh x = \frac{3}{4} + \frac{5}{4}(x - \ln 2) + \frac{3}{4} \cdot \frac{(x - \ln 2)^2}{2!} + \frac{5}{4}\frac{(x - \ln 2)^3}{3!} + \cdots$

15. $f(x) = x^{1/2}$, $f(9) = 3$; $f'(x) = \frac{1}{2}x^{-1/2}$; $f'(9) = \frac{1}{6}$, $f''(x) = -\frac{1}{4}x^{-3/2}$; $f''(9) = \frac{-1}{4 \cdot 27} = \frac{-1}{108}$; $f'''(x) = \frac{3}{8}x^{-5/2}$, $f'''(9) = \frac{3}{8 \cdot 3^5}$. Thus, we obtain $\sqrt{x} = 3 + \frac{1}{6}(x - 9) - \frac{1}{4 \cdot 3^3}\frac{(x-9)^2}{2!} + \frac{3}{8 \cdot 3^5}\frac{(x-9)^3}{3!} - \cdots$.

16. $f(x) = \ln x$, $f(1) = 0$, $f'(x) = \frac{1}{x}$, $f'(1) = 1$, $f''(x) = -x^{-2}$, $f''(1) = -1$, $f'''(x) = 2x^{-3}$; $f'''(1) = 2$, $f^{(4)}(x) = -6x^{-4}$; $f^{(4)} = -6$. Thus, we get $\ln x = (x - 1) - \frac{(x-1)^2}{2} + \frac{(x-1)^3}{3} - \frac{(x-1)^4}{4} + \cdots$.

17. $\cos 29^\circ = \cos(30 - 1)^2 = \cos\left(\frac{\pi}{6} - \frac{\pi}{180}\right) = \frac{\sqrt{3}}{2} - \frac{1}{2}\left(-\frac{\pi}{180}\right) - \frac{\sqrt{3}}{4}\left(\frac{-\pi}{180}\right)^2 + \frac{1}{2}\left(\frac{-\pi}{180}\right)^3 + \cdots = 0.866025 + 0.008727 - 0.00013 = 0.8746$

18. $\sinh(\ln 1.9) = \sinh(\ln(2 - 0.1)) = \sinh\left(\frac{\ln 2}{\ln 0.1}\right)$; $\sinh(\ln 1.9) = \frac{3}{4} + \frac{5}{4}(\ln 1.9 - \ln 1) + \frac{3}{4}\frac{(\ln 1.9 - \ln 2)^2}{2} + \cdots = 0.75 - 0.06412 + 0.00099 - 0.00003 = 0.6868.$

19. $\sqrt{9.1} = 3 + \frac{1}{6}(.1) - \frac{1(0.1)^2}{4 \cdot 3^2 2} = 3 + 0.016667 - 0.000046 = 3.0166$

20. $\ln(1.1) = 0.1 - \frac{0.1^2}{2} + \frac{0.1^3}{3} - \frac{0.1^4}{4} = 0.1 - 0.005 + 0.0003333 - 0.000025 = 0.09531$

21. $f(x) = \begin{cases} 1 & -\pi < x < 0 \\ 0 & 0 < x < \pi \end{cases}$, period $= 2\pi$ so $L = \pi$. First, $a_0 = \frac{1}{2\pi}\int_{-\pi}^{\pi} f(x)\,dx = \frac{1}{2\pi}\int_{-\pi}^{0} 1\,dx = \frac{1}{2\pi}x\Big|_{-\pi}^{0} = \frac{1}{2}$. Next, we have $a_k = \frac{1}{\pi}\int_{-\pi}^{\pi} f(x)\cos kx\,dx = \frac{1}{\pi}\int_{-\pi}^{0}\cos kx\,dx = \frac{1}{k\pi}\sin kx\Big|_{-\pi}^{0} = 0$. Also, $b_k = \frac{1}{\pi}\int_{-\pi}^{0}\sin kx\,dx = \frac{1}{k\pi}(-\cos kx)\Big|_{-\pi}^{0} = \frac{1}{k\pi}(-1 + \cos(-k\pi)) = \begin{cases} 0, & \text{if } k \text{ is even} \\ \frac{-2}{k\pi} & \text{if } k \text{ is odd} \end{cases}$. Thus, $f(x) = \frac{1}{2} - \frac{2}{\pi}\sin x + \frac{2}{3\pi}\sin 3x - \frac{2}{5\pi}\sin 5x - \cdots = \frac{1}{2} - \frac{2}{\pi}\sum_{k=1}^{\infty}\frac{\sin(2k-1)x}{2k-1}$.

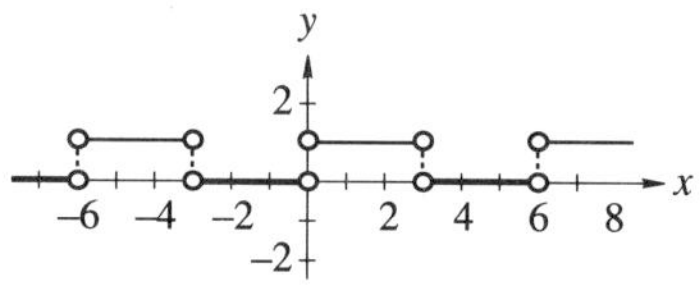

22. $g(x) = \begin{cases} 1 & -\pi < x < 0 \\ 2 & 0 < x < \pi \end{cases}$, period 2π so

$L = \pi$. First, $a_0 = \dfrac{1}{2\pi}\displaystyle\int_{-\pi}^{\pi} g(x)\,dx = \dfrac{1}{2\pi}\int_{-\pi}^{0} 1\,dx + \dfrac{1}{2\pi}\int_{0}^{\pi} 2\,dx = \dfrac{1}{2\pi}(0+\pi) + \dfrac{1}{2\pi}(2\pi) = \dfrac{1}{2} + 1 = \dfrac{3}{2}$. Next, $a_k = \dfrac{1}{\pi}\displaystyle\int_{-\pi}^{\pi} g(x)\cos kx\,dx = \dfrac{1}{\pi}\int_{-\pi}^{0} \cosh\,dx + \dfrac{1}{\pi}\int_{0}^{\pi} 2\cos kx\,dx = \dfrac{1}{k\pi}\sin kx\Big|_{-\pi}^{0} + \dfrac{2}{k\pi}\sin kx\Big|_{0}^{\pi} = 0$. We also have $b_k = \dfrac{1}{\pi}\displaystyle\int_{-\pi}^{0} \sin kx\,dx + \dfrac{1}{\pi}\int_{0}^{\pi} 2\sin kx\,dx = \dfrac{-1}{k\pi}(\cos kx)\Big|_{-\pi}^{0} + \dfrac{-2}{k\pi}\cos kx\Big|_{0}^{\pi} = \dfrac{-1}{k\pi}(1-\cos(-k\pi) - \dfrac{2}{k\pi}(\cos k\pi - 1) = \begin{cases} 0 & \text{if } k \text{ is even} \\ \frac{2}{k\pi} & \text{if } k \text{ is odd} \end{cases}$. This gives $g(x) = \frac{3}{2} + \frac{2}{\pi}\sin x + \frac{2}{3\pi}\sin 3x + \cdots = \dfrac{3}{2} + \dfrac{2}{\pi}\displaystyle\sum_{k=1}^{\infty}\dfrac{\sin(2k-1)x}{2k-1}$.

23. $h(x) = x^2, -\pi < x < \pi$, period 2π. First, $a_0 = \dfrac{1}{2\pi}\displaystyle\int_{-\pi}^{\pi} x^2\,dx = \dfrac{x^3}{6\pi}\Big|_{-\pi}^{\pi} = \dfrac{\pi^2}{3}$. Next, $a_k = \dfrac{1}{\pi}\displaystyle\int_{-\pi}^{\pi} x^2\cos kx\,dx = \dfrac{x^2}{\pi k}\sin kx - 2\int x\sin kx = \dfrac{x^2}{k\pi}\sin kx + \dfrac{2^x}{k^2\pi}\cos kx - \dfrac{2}{\pi}\int \cos kx = \left[\dfrac{x^2}{k\pi}\sin kx + \dfrac{2x}{k^2\pi}\cos kx - \dfrac{1}{k^3\pi}\sin kx\right]_{-\pi}^{\pi} = \begin{cases} -\dfrac{4}{k^2}, & k \text{ odd} \\ \dfrac{4}{k^2}, & k \text{ even} \end{cases}$. Also $b_k = 0$ since $x^2 \sin kx$ is an odd function. Hence, we obtain $h(x) = \dfrac{\pi^2}{3} - 4\cos x + \dfrac{4}{2^2}\cos 2x - \dfrac{4}{3^2}\cos 3x + \dfrac{4}{4^2}\cos 4x - \cdots = \dfrac{\pi^2}{3} + 4\displaystyle\sum_{k=1}^{\infty}\dfrac{(-1)^k}{k^2}\cos kx$.

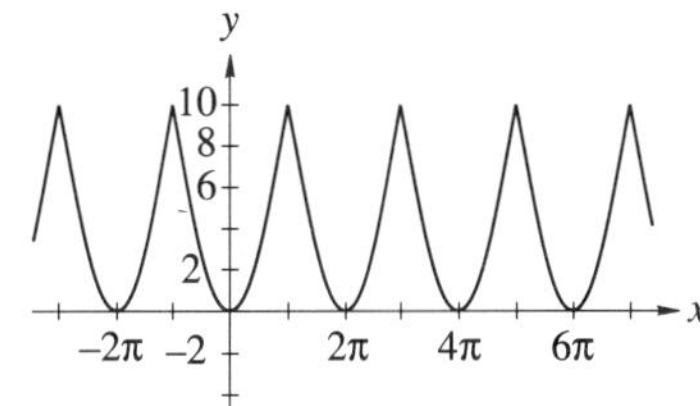

24. $j(x) = 4 \cdot x^2$, $-\pi < x < \pi$, period $= 2\pi$. First, $a_0 = \frac{1}{2\pi}\int_{-\pi}^{\pi}(4-x^2)\,dx = \frac{1}{2\pi}\left(4x - \frac{x^3}{3}\right)\Big|_{-\pi}^{\pi} = \frac{1}{\pi}\left(4\pi - \frac{\pi^3}{3}\right) = 4 - \frac{\pi^2}{3}$. (Note: the function is even).

$$\begin{aligned} a_k &= \frac{1}{\pi}\int_{-\pi}^{\pi}(4-x^2)\cos kx\,dx \\ &= \frac{1}{\pi}\int_{-\pi}^{\pi} 4\cos kx - \frac{1}{\pi}\int_{-\pi}^{\pi} x^2\cos kx\,dx \\ &= \Bigg[\frac{4}{\pi k}\sin kx - \frac{x^2}{k\pi}\sin kx \\ &\qquad + \frac{x}{k^2\pi}\cos kx - \frac{1}{k^3\pi}\cos kx\Bigg]_{-\pi}^{\pi} \\ &= \begin{cases} \dfrac{-4}{k^2} & k \text{ even} \\ \dfrac{4}{k^2} & k \text{ odd} \end{cases} \end{aligned}$$

$b_k = 0$ since $(4-x^2)\sin kx$ is an odd function. Hence $j(x) = 4 - \dfrac{\pi^2}{3} + 4\displaystyle\sum_{k=1}^{\infty}\dfrac{(-1)^{k+1}}{k^2}\cos kx$.

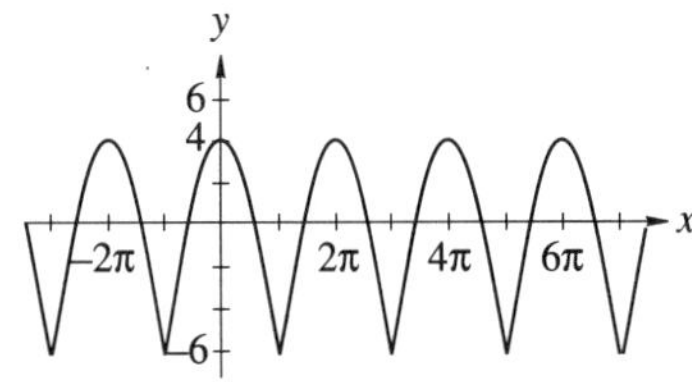

CHAPTER 31 TEST

1. Since $\sin x = x - \frac{x^3}{3!} + \frac{x^5}{5!} - \frac{x^7}{7!} + \cdots$, we get $\sin x^2 = x^2 - \frac{x^6}{3!} + \frac{x^{10}}{5!} - \frac{x^{14}}{7!} + \cdots$

2. Since $\cos x = 1 - \frac{x^2}{2!} + \frac{x^4}{4!} - \frac{x^6}{6!} + \cdots$, we obtain $\cos\sqrt{x} = 1 - \frac{x}{2!} + \frac{x^2}{4!} - \frac{x^3}{6!} + \cdots$

3. $h(x) = \frac{1}{x}$; $h(2) = \frac{1}{2}$; $h'(x) = -x^2$; $h'(2) = -\frac{1}{4}$; $h''(x) = 2x^{-3}$; $h''(2) = \frac{1}{4}$; $h'''(x) = -6x^{-4}$; $h'''(2) = \frac{-6}{2^4}$. Thus, we see that $h(x) = \frac{1}{x} = \frac{1}{2} - \frac{1}{4}(x-2) + \frac{1}{4}\cdot\frac{(x-2)^2}{2!} - \frac{6}{2^4}\cdot\frac{(x-2)^3}{3!} + \cdots = \frac{1}{2} - \frac{(x-2)}{4} + \frac{(x-2)^2}{8} - \frac{(x-2)^3}{16} + \cdots = \sum_{k=0}^{\infty}\frac{(-1)^k(x-2)^k}{2^{k+1}}$.

4. $\frac{1}{2.1} = \frac{1}{2} - \frac{0.1}{4} + \frac{(0.1)^2}{8} - \frac{(0.1)^3}{16} = 0.5 - 0.025 + 0.00125 - 0.0000625 = 0.47619$

5. $j(x) = x^{1/3}$; $j(-1) = -1$; $j'(x) = \frac{1}{3}x^{-2/3}$; $y'(-1) = \frac{1}{3}$; $y''(x) = -\frac{2}{9}(x)^{-5/3}$; $y''(-1) = \frac{2}{9}$; $y'''(x) = \frac{10}{27}x^{-8/3}$; $y'''(-1) = \frac{10}{27}$. Hence, $j(x) = -1 + \frac{1}{3}(x+1) + \frac{2}{9}\cdot\frac{(x+1)^2}{2!} + \frac{10}{27}\frac{(x+1)^3}{3!} = -1 + \frac{(x+1)}{3} + \frac{(x+1)^2}{9} + \frac{5(x+1)^3}{81}$.

6. $\sqrt[3]{0.9} = -1 + \frac{(0.1)}{3} + \frac{(0.1)^2}{9} + \frac{5(0.1)^3}{81} = -1 + 0.03333 + 0.001111 + 0.000062 = -0.96549$

7. $f(x) = x$, $0 < x < 2\pi$, period is 2π so $L = \pi$. First, $a_0 = \frac{1}{2\pi}\int_0^{2\pi} x\,dx = \frac{1}{2\pi}\frac{x^2}{2}\Big|_0^{2\pi} = \pi$. Next, we see that $a_k = \frac{1}{\pi}\int_0^{2\pi} x\cos kx\,dx = \frac{1}{k^2\pi}\int_0^{2\pi} kx\cos kx\,k\,dx = \frac{1}{k^2\pi}[\cos kx + kx\sin kx]_0^{2\pi} = 0$. We also determine that, $b_k = \frac{1}{\pi}\int_0^{2\pi} x\sin kx\,dx = \frac{1}{k^2\pi}\int_0^{2\pi} kx\sin kx\,k\,dx = \frac{1}{k^2\pi}[\sin kx - kx\cos kx]_0^{2\pi} = \frac{1}{k^2\pi}[-2k\pi] = \frac{-2}{k}$. Hence, we obtain $f(x) = \pi - 2\left(\sin x + \frac{\sin 2x}{2} + \frac{\sin 3x}{3} + \cdots\right) = \pi - 2\sum_{k=1}^{\infty}\frac{\sin kx}{k}$.

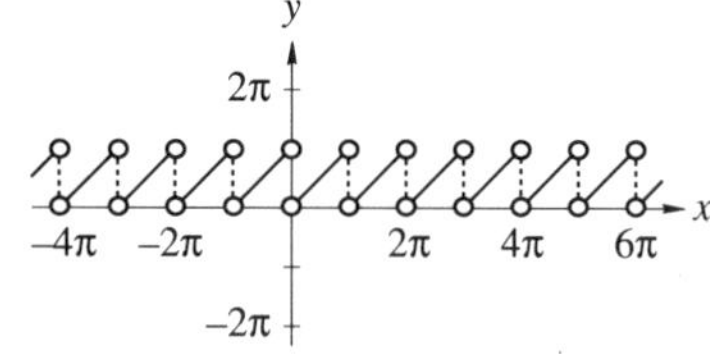

8. $g(x) = \begin{cases} -\cos x & -\pi < x < 0 \\ \cos x & 0 < x < \pi \end{cases}$, period is 2π so $L = \pi$. First $a_0 = \frac{1}{2\pi}\int_{-\pi}^{\pi} g(x)\,dx = \frac{1}{2\pi}\int_{-\pi}^{0} -\cos x\,dx + \frac{1}{2\pi}\int_0^{\pi}\cos x\,dx = \frac{1}{2\pi}(-\sin x)\Big|_{-\pi}^{0} + \frac{1}{2\pi}(\sin x)\Big|_0^{\pi} = 0$. Then, we have $a_k = \frac{1}{\pi}\int_{-\pi}^{\pi} g(x)\cos kx\,dx = \frac{1}{\pi}\int_{-\pi}^{0} -\cos x\cos kx\,dx + \frac{1}{\pi}\int_0^{\pi}\cos x\cos kx\,dx$. These integrals add up to 0 since the cosine function is even and the product of the cosine functions is even. Hence, $\int_{-\pi}^{0} = \int_0^{\pi}$ and, since one of the integrals is negative, their sum is 0. We also have, for $k > 1$, $b_k = \frac{1}{\pi}\int_{-\pi}^{0} -\cos x\sin kx\,dx +$

$\frac{1}{\pi}\int_0^{\pi}\cos x\sin kx\,dx$. By form 72 this is $\frac{1}{\pi}\left[\frac{\cos(k-1)x}{2(k-1)}+\frac{\cos(k+1)x}{2(k+1)}\right]_{-\pi}^{0}$ + $\frac{1}{\pi}\left[-\frac{\cos(k-1)x}{2(k-1)}-\frac{\cos(k+1)x}{k+1}\right]_0^{\pi}=\frac{1}{\pi}\left[\frac{-\cos(k-1)(-\pi)}{2(k-1)}-\frac{\cos(k+1)(-\pi)}{2(k+1)}+\frac{1}{2(k-1)}+\frac{1}{2(k+1)}-\frac{\cos(k-1)\pi}{2(k-1)}-\frac{\cos(k+1)\pi}{2(k+1)}+\frac{1}{2(k-1)}+\frac{1}{2(k+1)}\right]$. When k is odd, $\cos(k-1)(-\pi)=\cos(k+1)(-\pi)=\cos(k-1)\pi=\cos(k+1)\pi=1$, and so $b_k=0$. When k is even, $\cos(k-1)(-\pi)=\cos(k+1)(-\pi)=\cos(k-1)\pi=\cos(k+1)\pi=-1$, and so we get $b_k=\frac{1}{\pi}\left[\frac{4}{2(k-1)}+\frac{4}{2(k+1)}\right]=\frac{4k}{\pi(k^2-1)}$. In particular, $b_2=\frac{4\cdot 2}{\pi(2^2-1}=\frac{8}{3\pi}$, $b_4=\frac{4\cdot 4}{(4^2-1)\pi}=\frac{16}{15\pi}$, $b_6=\frac{24}{35\pi}$. Now, $k=1$ yields $b_1=\frac{1}{\pi}\int_{-\pi}^{0}(-\cos x\sin x\,dx)+\frac{1}{\pi}\int_0^{\pi}\cos x\sin x\,dx=0$. Putting it all together, we get $g(x)=\frac{8}{\pi}\left(\frac{\sin 2x}{3}+\frac{2\sin 2x}{15}+\frac{3\sin 2x}{35}+\cdots\right)=\frac{8}{\pi}\sum_{k=1}^{\infty}\frac{k\sin 2x}{4k^2-1}$.

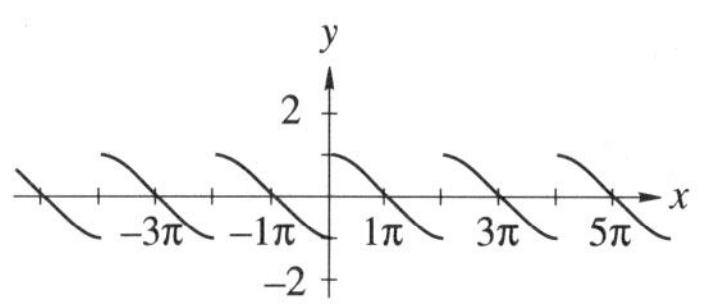

9. The Maclaurin series for $\cos x$ is $1-\frac{x^2}{2!}+\frac{x^4}{4!}+\cdots$, so $\cos x^3=1-\frac{(x^3)^2}{2!}+\frac{(x^3)^4}{4!}+\cdots=1-\frac{x^6}{2!}+\frac{x^{12}}{4!}+\cdots$. Thus, $\int_1^2\cos x^3\,dx=\int_1^2\left(1-\frac{x^6}{2!}+\frac{x^{12}}{4!}+\cdots\right)dx=\left[x-\frac{x^7}{7\cdot 2!}+\frac{x^{13}}{13\cdot 4!}+\cdots\right]_1^2\approx 19.1136-0.9318=18.1818$.

10. The Maclaurin series for e^x is $1+x+\frac{x^2}{2!}+\frac{x^3}{3!}+\cdots$, so $e^{-0.30t^2}=1-0.30t^2+\frac{0.09t^4}{2!}-\frac{0.027t^6}{3!}+\cdots$. Thus, $q=\int_0^{0.15}e^{-0.30t^2}\,dt=\int_0^{0.15}\left(1-0.30t^2+\frac{0.09t^4}{2!}-\frac{0.027t^6}{3!}+\cdots\right)dt=\left[t-0.10t^3+0.009t^5-\frac{0.027t^7}{7\cdot 3!}+\cdots\right]_0^{0.15}\approx 0.149663\ \mu\text{C}$

CHAPTER

32

First Order Differential Equations

32.1 SOLUTIONS OF DIFFERENTIAL EQUATIONS

1. $y = e^x$; $y' = e^x$, $y'' = e^x$. Hence $y'' - y = e^x - e^x = 0$

2. $y = e^{2x} + e^x$; $y' = 2e^{2x} + e^x$; $y'' = 4e^{2x} + e^x$. Hence $y'' - 4y' + 4y = (4e^{2x} + e^x) - 4(2e^{2x} + e^x) + 4(e^{2x} + e^x) = 4e^{2x} + e^x - 8e^{2x} - 4e^x + 4e^{2x} + 4e^x = (4 - 8 + 4)e^{2x} + (1 - 4 + 4)e^x = e^x$.

3. $y = 2x^3$, $y' = 6x^2$. Hence $xy' = x6x^2 = 6x^3 = 3 \cdot 2x^3 = 3y$

4. $y = x^3 + 3, y' = 3x^2$. Hence $3y - xy' = 3(x^3 + 3) - x(3x^2) = 3x^3 + 9 - 3x^3 = 9$

5. $y = 5\cos x$; $y' = -5\sin x$. Hence $y' + y\tan x = -5\sin x + 5\cos x\tan x = -5\sin x + 5\sin x = 0$

6. $y = e - 1, y' = e^x$. Hence $y' - y = e^x - (e^x - 1) = 1$

7. $y = x^2 + 3x$; $y' = 2x + 3$; $y'' = 2$. Hence $xy' - 2y + 3xy'' - 3x = x(2x + 3) - 2(x^3 + 3x) + 3x(2) - 3x = 2x^2 + 3x - 2x^2 - 6x + 6x - 3x = 0$

8. $y = 4x^2$, $y' = 8x$. Hence $xy' = x(8x) = 8x^2 = 2 \cdot 4x^2 = 2y$

9. $y = 2e^x + 3e^{-x} - 4x$, $y' = 2e^x - 3e^{-x} - 4$, $y'' = 2e^x + 3e^{-x}$. Hence $y'' - y = (2e^x + 3e^{-x}) - (2e^x + 3e^{-x} - 4x) = 4x$

10. $x^2 - y^2 + 2xy = 9$. Using implicit differeniation we get $2x - 2yy' + 2xy' + 2y = 0$ or $y'(2x - 2y) = -2x - 2y$ or $y' = \dfrac{2x + 2y}{2y - 2x} = \dfrac{y + x}{y - x}$

11. $y = 9x^2 + 6x - 54$; $y' = 18x + 6$; $y'' = 18, y''' = 0$

12. $xy^2 = 7x - 3$ so $y^2 = 7 - 3x^{-1}$, Using implicit differciation we get $2yy' = 3x^{-2}$ or $y' = \dfrac{3}{2yx^2} = \dfrac{3}{2}y^{-1}x^{-2}$, $y'' = -\frac{3}{2}y^{-2}y'x^{-1} - 3y^{-1}x^{-3} = -\frac{3}{2}y^{-2}\left(\frac{3}{2}y^{-1}x^{-2}\right)x^{-2} - 3y^{-1}x^{-3} = -\frac{9}{4}y^{-3}x^{-4} - 3y^{-1}x^{-3}$. Hence $x(y')^2 + 2yy' + xyy'' = x\left(\frac{3}{2}y^{-1}x^{-2}\right)^2 + 2y\left(\frac{3}{2}y^{-1}x^{-2}\right) + xy\left(-\frac{9}{4}y^{-3}x^{-4} - 3y^{-1}x^{-3}\right) = \frac{9}{4}y^{-2}x^{-3} + 3x^{-2} + \frac{9}{4}y^{-2}x^{-3} - 3x^{-2} = 0$

13. $y = x^2 + 4$; $y' = 2x$

14. $2y = \sin x$, $y' = \frac{1}{2}\cos x$, $y'' = -\frac{1}{2}\sin x$. Hence $y'' + 5y = -\frac{1}{2}\sin x + 5 \cdot \frac{1}{2}\sin x = 2\sin x$

15. $y = 5e^{-3x} + 2e^{2x}$, $y' = -15e^{-3x} + 4e^{2x}$, $y'' = 45e^{-3x} + 8e^{2x}$. Hence $y'' + y' - 6y = (45e^{-3x} + 8e^{2x}) + (-15e^{-2x} + 4e^{2x}) - 6(5e^{-3x} + 2e^{2x}) = (45 - 15 - 30)e^{-3x} + (8 + 4 - 12)e^{2x} = 0$

16. $y = 5\cos x$; $y' = -5\sin x$, $y'' = -5\cos x$. Hence $y'' + y = -5\cos x + 5\cos x = 0$.

17. $y' = \int y''\,dx = \int 6x^2\,dx = 2x^3 + C_1$. Since $y'(1) = -2, 2(1)^3 + C_1 = 2 + C_1 = 2 + C_1 = -2$ or $C_1 = -4$ so $y' = 2x^3 - 4$. Next, we see that $y = \displaystyle\int y'\,dx = \int (2x^3 - 4)\,dx = \frac{x^4}{2} - 4x + C_2$. Since $y(1) = 3, \frac{1}{2} - 4 + C_2 = 3$ or $C_2 = 6\frac{1}{2}$. Hence $y = \frac{1}{2}x^4 - 4x + 6\frac{1}{2}$

18. $y' = \int y''\,dx = \int \sin x\,dx = -\cos x + C_1$, since $y'(0) = 2, -\cos 0 + C_1 = 2 \Rightarrow C_1 = 3$. Thus, $y = \int y'\,dx = \int(-\cos x + 3)\,dx = -\sin x + 3x + C_2$. Since $y(\pi) = \pi, -\sin\pi + 3\pi + C_2 = \pi \Rightarrow C_2 = -2\pi$. $y = -\sin x + 3x - 2\pi$.

19. $y = \int \frac{dy}{dx}\,dx = \int \sec^2 x\,dx = \tan x + C$. Since $y\left(\frac{\pi}{4}\right) = -6, \tan\frac{\pi}{4} + C = -6 \Rightarrow C = -7$ So $y = \tan x - 7$.

20. $y' = \int y''\,dx = \int e^{-x}\,dx = -e^{-x} + C_1$, since $y'(0) = 0, -e^0 + C_1 = 0 \Rightarrow -1 + C_1 = 0$ or $C_1 = 1$. So, $y = \int y'\,dx = \int(-e^{-x} + 1)\,dx = e^{-x} + x + C_2$. Since $y(1) = -1, e^{-1} + 1 + C_2 = -1 \Rightarrow C_2 = -2 - e^{-1}$ so $y = e^{-x} + x - 2 - e^{-1}$.

21. As in Example 32.0, we have $a(t) = -30$, so $v(t) = -30t + C_1$. We were given $v(0) = 65$ mph $= 95\frac{1}{3}$ ft/s and so $v(t) = -30t + 95\frac{1}{3}$ ft/s. By integrating, we obtain $s(t) = -15t^2 + 95\frac{1}{3}t + C_2$. Since $s(0) = 0$, we see that $C_2 = 0$ and thus $s(t) = -15t^2 + 95\frac{1}{3}t$. (a) When the car stops $v(t) = -30t + 95\frac{1}{3} = 0$. Solving for t, we see that the car will stop in about $t = 3.178$ s. (b) Substituting $t = 3.178$ s in the equation for $s(t)$, we get $s(3.178) \approx 151.474$. The car takes about 151.474 ft to stop.

22. Here $a(t) = -28$, so $v(t) = -28t + C_1$. Since $v(0) = 70$ mph $= 102\frac{2}{3}$ ft/s, we have $v(t) = -28t + 102\frac{2}{3}$. Integrating, we get $s(t) = -14t^2 + 102\frac{2}{3}$. (a) Setting $v(t) = 0$ and solving for t, we see that it takes $t = 3\frac{2}{3}$ s to stop the car. (b) Substituting $t = 3\frac{2}{3}$ s in the equation for $s(t)$, we get $s\left(3\frac{2}{3}\right) = 188\frac{2}{9}$ ft. It takes the car $188\frac{2}{9}$ ft to stop.

23. (a) The cars will strike when $s(t) = 151$. Solving $s(t) = -15t^2 + 95\frac{1}{3}t = 151$ or $15t^2 - 95\frac{1}{3}t + 151 = 0$ for t we get $t = 3$ s. (b) When they strike $v(3) = -30(3) + 95\frac{1}{3} = 5\frac{1}{3}$. So, the first car is going $5\frac{1}{3}$ mph.

24. The cars hit when $s(t) = 184.5$. Solving $s(t) = -14t^2 + 102\frac{2}{3} = 184.5$, we get $t \approx 3.15$ s. (b) At $t = 3.15$, $v(3.15) = -28t + 102\frac{2}{3} = -28(3.15) + 102\frac{2}{3} \approx 14.47$. The car is moving at around 14.47 mph when the cars hit.

25. (a) Acceleration is gravity, $g = -32$ ft/s^2. So, $v(t) = -32t + C_1$. Since $v(0) = 64$, we have $C_1 = 64$ and $v(t) = -32t + 64$. Thus, the ball's position is $s(t) = -16t^2 + 64t + C_2$. At $t = 0$, we are given $s(0) = 192$, so $C_2 = 192$ and $s(t) = -16t^2 + 64t + 192 = -16(t^2 - 4t - 12) = -16(t-6)(t+2)$. The ball will strike the ground at $t = 6$ s. (b) At $t = 6$, $v(6) = -32(6) + 64 = -128$, so it is going 128 ft/s downward when it strikes the ground.

26. (a) Since $a(t) = -9.8$ m/s^2 we have $v(t) = -9.8t + C_1$. Since $v(0) = 29.4$, we have $v(t) = -9.8t + 29.4$. Integrating, we obtain $s(t) = -4.9t^2 + 29.4t + C_2$. We are given $s(0) = 34.3$ m, so $s(t) = -4.9t^2 + 34.3 = -4.9(t^2 - 6t - 7) = -16(t-7)(t+1)$. The ball will strike the ground at $t = 7$ s. (b) When it hits the ground, its velocity is $v(7) = -9.8(7) + 29.4 = -39.2$, so it is going 39.2 m/s downward when it strikes the ground.

27. (a) Let $p(t)$ be the number of mg of pain reliever still in her system t hours after taking the tablet. We are given $\frac{dp(t)}{dt} = -75$ as the rate of metabolism. Integrating, we get $p(t) = -75t + C$. At $t = 0$, there are 500 mg in the system, so $C = 500$ and $p(t) = -75t + 500$. (b) We want to find t so that $p(t) = -115t + 500 = 0$. Solving for t, we get $t = 6\frac{2}{3}$ h.

28. We begin with the result of Exercise 0 with $p(t) = -75t + 500$. At $t = 3$, she has 275 mg from the first tablet plus 500 mg from the second tablet. Now, solving $p(t) = -75 + C$ when $t = 3$ and $p(t) = 775$ we get $C = 1000$. Putting these results with those from Exercise 0 we find that $p(t) = \begin{cases} -75t + 500 & \text{if } 0 \le 3 \\ -75t + 1000 & \text{if } t \ge 3 \end{cases}$ where t represents the number of hours since the first pill was taken. Solving $p(t) = -75t + 1000$ for t we get $t = 13\frac{1}{3}$ h after the first pill was taken or $10\frac{1}{3}$ h after the second pill was taken before all the pain reliever is metabolized.

29. Let $c(t)$ be the number of mg of cortisone still in the patients system t hours after taking the 5 mg

dose. We are given $\frac{dc(t)}{dt} = -\frac{1}{2\sqrt{t}} = -\frac{1}{2}t^{-1/2}$ as the rate of metabolism. Integrating, produces $c(t) = -t^{1/2} + C$. At $t = 0$, there are 5 mg in the system, so $C = 5$ and $c(t) = -t^{1/2} + 5$. We want to find t so that $c(t) = -t^{1/2} + 5 = 0$. Solving for t, we get $\sqrt{t} = 5$ and so $t = 25$ h.

32.2 SEPARATION OF VARIABLES

1. $y' = \frac{dy}{dx} = \frac{2x}{y} \Rightarrow 2x\,dx - y\,dy = 0$. Integrating we get $x^2 - \frac{1}{2}y^2 = C_1$ or $\frac{1}{2}y^2 - x^2 = C_2$ or $y^2 - 2x^2 = C_3$.

2. $\frac{dy}{dx} = \frac{2y}{x} \Rightarrow 2y\,dx - x\,dy = 0$. The integrating factor $I(x,y)$ is $\frac{1}{xy}$. Multiplying and cancelling we get $\frac{2}{x}\,dx - \frac{1}{y}\,dy = 0$. Integration yields $2\ln|x| - \ln|y| = \ln C \Rightarrow \ln x^2 - \ln y = \ln C \Rightarrow \ln\left(\frac{x^2}{y}\right) = \ln C \Rightarrow \frac{x^2}{y} = C$. Hence $x^2 = Cy$ or $y = Cx^2$.

3. $5x\,dx + 3y^2\,dy = 0$. Integrating we get $\frac{5}{2}x^2 + y^3 = C$

4. $5y\,dx + 3y^2\,dy = 0$. Multiply first by $I(x,y) = \frac{1}{y}$. This yields $5\,dx + 3y\,dy = 0$. Now by integration $5x + \frac{3}{2}y^2 = C$.

5. $5y\,dx + 3\,dy = 0$. Multiply by $I(x,y) = \frac{1}{y}$ to get $5\,dx + \frac{3}{y} = 0$. Now by integrating we get $5x + 3\ln|y| = C$

6. $4y^5\,dx - 3x^2\,dy = 0$. Multiplying by $\frac{1}{x^2y^5}$ we get $\frac{4}{x^2}\,dx - \frac{3}{y^5}\,dy = 0$. Integration yields $-4x^{-1} + \frac{3}{4}y^{-4} = C$.

7. $4y^5\,dx - 3x^2y\,dy = 0$. Multiplying by $I(x,y) = \frac{1}{y^5x^2}$ we get $\frac{4}{x^2}\,dx - \frac{3}{y^4}\,dy = 0$. Now integration yields $-4x^{-1} + y^{-3} = C$.

8. $xyy' + \sqrt{1+y^2} = 0$ is the same as $\frac{dy}{dx} + \frac{\sqrt{1+y^2}}{xy} = 0$. Hence we get $\sqrt{1+y^2}\,dx + xy\,dx = 0$. $I(x,y) = \frac{1}{\sqrt{1+y^2}x}$ and multiplying by this integrating factor gives $\frac{1}{x}\,dx + \frac{y}{\sqrt{1+y^2}}\,dy = 0$. Integration now yields $\ln|x| + \frac{1}{2}\frac{2}{1}\left(1+y^2\right)^{1/2} + C$ or $\ln|x| + \sqrt{1+y^2} = C$.

9. $x^3\left(y^2+4\right) + yy' = 0$ is the same as $x^3(y^2+4)\,dx + y\,dy = 0$. Multiplying by $\frac{1}{y^2+4}$ yields $x^3\,dx + \frac{y}{y^2+4}\,dy = 0$. Now integration gives us $\frac{1}{4}x^4 + \frac{1}{2}\ln\left(y^2+4\right) = C_1$ or $\frac{1}{4}x^4 + \ln\sqrt{y^2+4} = C_1$ or $\frac{1}{2}x^4 + \ln(y^2+4) = C_2$.

10. $x^4(y^3-3)\,dx + y^2(x^5-2)\,dy = 0$ becomes $\frac{x^4}{x^5-2}\,dx + \frac{y^2}{y^3-3}\,dy = 0$. Integration yields $\frac{1}{5}\ln\left|x^5-2\right| + \frac{1}{3}\ln\left|y^3-3\right| = \ln|C|$. This is the same as $\left|x^5-2\right|^{1/5}\cdot\left|y^3-3\right|^{1/3} = C$

11. $ye^{x^2}\,dy = 2x(y^2+4)\,dx$. The integrating factor is $I(x,y) = \frac{1}{\left(y^2+4\right)3^{x^2}}$. Multiplying by $I(x,y)$ yields $\frac{y}{y^2+4}\,dy = \frac{2x}{e^{x^2}}\,dx$. Now integration yields $\frac{1}{2}\ln\left(y^2+4\right) = -e^{-x^2} + C$ or $\ln\sqrt{y^2+4} = -e^{-x^2} + C$ or $\ln(y^2+4) = -2e^{-x^2} + C_2$.

12. $e^{x^2}\,dy + x\sqrt{4-y^2}\,dx = 0$ is the same as $\frac{1}{\sqrt{4-y^2}}\,dy + \frac{x}{e^{x^2}}\,dx = 0$. Integration yields

$\arcsin \frac{y}{2} - \frac{1}{2}e^{-x^2} + C = 0.$

13. $2y + e^{-3x}y' = 0 \Rightarrow 2y\,dx + e^{-3x}\,dy = 0 \Rightarrow 2e^{3x}\,dx + \frac{1}{y}\,dy = 0$. Integration yields $\frac{2}{3}e^{3x} + \ln|y| = C$

14. $\sin^2 y\,dx + \cos^2 x\,dy = 0$. First multiply by $\frac{1}{\sin^2 y \cos^2 x}$ to obtain $\frac{1}{\cos^2 x}\,dx + \frac{1}{\sin^2 y}\,dy = 0$ or $\sec^2 x\,dx + \csc^2 y\,dy = 0$. This integrates to $\tan x - \cot y = C$

15. $(y^2 - 4)\cos x\,dx + 2y \sin x\,dy = 0$. Multiplying by the integrating factor yields $\frac{\cos x}{\sin x}\,dx + \frac{2y}{y^2 - 4}\,dy = 0$. Integration yields $\ln|\sin x| + \ln|y^2 - 4| = \ln C$. This simplifies to $|y^2 - 4|\,|\sin x| = C$ or $|(y^2 - 4)\sin x| = C$.

16. $(1 + x)^2 y' = 1 \Rightarrow (1 + x)^2 \frac{dy}{dx} = 1$ or $dy = (1 + x)^{-2}\,dx$. Integration yield $y = -(1+x)^{-1} + C$, or $y + \frac{1}{1 + x} = C$.

17. $y' = \frac{x^2 y - y}{y^2 + 3} = \frac{y(x^2 - 1)}{y^2 + 3} = \frac{dy}{dx}$. Hence $(x^2 - 1)\,dx = \frac{y^2 + 3}{y}\,dy$. Integratng we get $\frac{1}{3}x^3 - x = \frac{y^2}{2} + 3\ln|y| + C$ or $\frac{1}{2}y^2 + 3\ln|y| = \frac{1}{3}x^3 - x + C$.

18. $\frac{dy}{dx} = \frac{y}{x^2 + 6x + 9}$ is equivalent to $\frac{1}{y}dy = \frac{1}{(x + 3)^2}\,dx$. Integration yields $\ln|y| = -(x+3)^{-1} + C$ or $\ln|y| + \frac{1}{x + 3} = C$.

19. $\sec 3x\,dy + y^2 e^{\sin 3x}\,dx = 0$ is equivalent to $\frac{1}{y^2}\,dy + \cos 3x e^{\sin 3x}\,dx = 0$. Integration yields $-\frac{1}{y} + \frac{1}{3}e^{\sin 3x} = C$

20. $2xy + (1 + x^2)y' = 0$. The integrating factor is $\frac{1}{y(1 + x^2)}$. Multiplying yields $\frac{2x}{1 + x^2}\,dx + \frac{1}{y}\,dy = 0$. Integerating yields $\ln(1 + x^2) + \ln|y| = \ln C$. This simplifies to $(1 + x^2)\,|y| = C$

21. $y' + y^2 x^3 = 0$ becomes $\frac{1}{y^2}\,dy + x^3\,dx = 0$. Integration yields the general solution $-\frac{1}{y} + \frac{x^4}{4} = C$. Substituting -1 for y and 2 for x we get $1 + 4 = C$ so $C = 5$. The particular solution is $-\frac{1}{y} + \frac{1}{4}x^4 = 5$.

22. $x\,dy + y\,dx = 0$ becomes $\frac{1}{y}\,dy + \frac{1}{x}\,dx = 0$. Integration yields the general solution $\ln|y| + \ln|x| = \ln C$ or $xy = C$. Substituting 3 for y and 2 for x we get $C = 6$. The particular solution is $xy = 6$.

23. $\frac{dy}{dx} = \frac{3x + xy^2}{y + x^2 y} = \frac{x(3 + y^2)}{y(1 + x^2)}$. This is equivalent to $\frac{y}{3 + y^2}\,dy = \frac{x}{1 + x^2}\,dx$. Integration yields the general solution $\frac{1}{2}\ln(3 + y^2) = \frac{1}{2}\ln(1 + x^2) + \frac{1}{2}\ln C$ or $(3 + y^2)/(1 + x^2) = C$. Substituting 3 for y and 1 for x we get $\frac{12}{2} = 6 = C$. The particular solution is $\frac{3 + y^2}{1 + x^2} = 6$.

24. $\sin^2 y\,dx + \cos^2 x\,dy = 0$ is the same as $\sec^2 x\,dx + \csc^2 y\,dy = 0$. Integration yields the general solution $\tan x - \cot y = C$. Substituting $\frac{3\pi}{4}$ for y and $\frac{\pi}{4}$ for x we get $\tan\frac{\pi}{4} - \cot\frac{3\pi}{4} = 1 + 1 = 2 = C$. The particular solution is $\tan x - \cot y = 2$.

25. $y' = \frac{dy}{dx} = 3x^2 y - 2y = y(3x^2 - 2)$. Hence we get $\frac{1}{y}\,dy = (3x^2 - 2)\,dx$. Integration yields the general solution $\ln|y| = x^3 - 2x + C$. Substituting $y = x = 1$ we get $\ln 1 = 1 - 2 + C$ or $C = 1$. The particular solution is $\ln|y| = x^3 - 2x + 1$

26. $\sqrt{1 + 9x^2}\,dy = y^4 x\,dx$. Multiplying by the integrating factor $\frac{1}{y^4\sqrt{1 + 9x^2}}$ we get $\frac{1}{y^4}\,dy =$

$\frac{x}{\sqrt{1+9x^2}}\,dx$. Integrating to get the general solution produces $\frac{-1}{3y^3} = \frac{1}{9}(1+9x^2)^{1/2} + C$. This is equivalent to $\sqrt{1+9x^2} + \frac{3}{y^3} = C$. Substituting $y = -1$ and $x = 0$ we get $\sqrt{1} + (-3) = -2 = C$. Hence the particular solution is $\sqrt{1+9x^2} + \frac{3}{y^3} = -2$ or $\sqrt{1+9x^2} + \frac{3}{y^3} + 2 = 0$.

32.3 INTEGRATING FACTORS

1. No, $\frac{\partial(x^2-y)}{\partial y} = -1$, $\frac{\partial x}{\partial x} = 1$
2. No, $\frac{\partial(x^2+y^2)}{\partial y} = 2y$, $\frac{\partial xy}{\partial x} = y$
3. Yes, $\frac{\partial(x^2+y^2)}{\partial y} = 2y = \frac{\partial 2xy}{\partial x}$
4. No, $\frac{\partial(x+\sin y)}{\partial y} = \cos y$, $\frac{\partial(-\sin x)}{\partial x} = -\cos x$
5. Yes, $\frac{\partial(x+y\cos x)}{\partial y} = \cos x = \frac{\partial \sin x}{\partial x}$
6. Yes, $\frac{\partial(2ye^{2x})}{\partial y} = 2e^{2x} = \frac{\partial\left(e^{2x}-1\right)}{\partial x}$
7. Yes, $\frac{\partial(2xy+x)}{\partial y} = 2x = \frac{\partial(y+x^2)}{\partial x}$
8. No, $\frac{\partial(4x^3y^2)}{\partial y} = 8x^3y$, $\frac{\partial(3x^4y+5x^4)}{\partial x} = 12x^3y + 20x^3$
9. No, $\frac{\partial(x-y)}{\partial y} = -1$, $\frac{\partial(y+x)}{\partial x} = 1$
10. No, $\frac{\partial(y\sin x + x\cos y)}{\partial y} = \sin x - s\sin y$, $\frac{\partial(\frac{1}{2}y^2\cos x + \sin y)}{\partial x} = -\frac{1}{2}y^2\sin x$.
11. $y\,dx - x\,dy = 0$. By Table 32.1 we can use one of four Integrating factors. Choosing $-\frac{1}{xy}$ we get $-\frac{y\,dx - x\,dy}{xy} = 0$. Integration yields $\ln\frac{y}{x} = C_1$ or letting $C_2 = e^{C_1}$ we get $\frac{y}{x} = C_2$.
12. $y\,dx - x\,dy - x^3\,dx = 0$. This time choose $-\frac{1}{x^2}$ to be the integration factor. Multiplication yields $-\frac{y\,dx - x\,dy}{x^2} + x\,dx = 0$. Integrating we get $\frac{y}{x} + \frac{1}{2}x^2 = C$.
13. $y\,dx - x\,dy - 5y^4\,dy = 0$. Choose $I(x,y) = \frac{1}{y^2}$. Multiplying by $\frac{1}{y^2}$ we get $\frac{y\,dx - x\,dy}{y^2} - 5y^2\,dy = 0$. Integration yields $\frac{x}{y} - \frac{5}{3}y^3 = C$.
14. $y\,dx - x\,dy - x^2y\,dx = 0$. Choose $I(x,y) = -\frac{1}{xy}$ giving $-\frac{y\,dx - x\,dy}{xy} + x\,dx = 0$. Integrating we get $\ln\frac{y}{x} + \frac{1}{2}x^2 = C$.
15. $y\,dx + y^2\,dx = x\,dy - x^2\,dx$ is equivalent to $y\,dx - x\,dy + x^2\,dx + y^2\,dx = 0$. This time we choose $I(x,y) = \frac{-1}{x^2+y^2}$ and multiplying we get $-\frac{y\,dx - x\,dy}{x^2+y^2} - \frac{x^2+y^2}{x^2+y^2}\,dx = 0$. Now integration yields $\arctan\frac{y}{x} - x = C$.
16. $y\,dx - x\,dy - 3\,dx = 0$. Multiplying by $I(x,y) = -\frac{1}{x^2}$ we get $-\frac{y\,dx - x\,dy}{x^2} + \frac{3}{x^2}\,dx = 0$. Integration yields $\frac{y}{x} - \frac{3}{x} = C$ or $\frac{y-3}{x} = C$.

17. $y\,dx + x\,dy - 3x^3y\,dx = 0$. The only integrating factor is $\frac{1}{xy}$. Multiplying by this we get $\frac{y\,dx + x\,dy}{xy} - 3x^2\,dx = 0$. Integration yields $\ln(xy) - x^3 = C$.

18. $x\,dx + y\,dy - 3x^2\,dy - 3y^2\,dy = 0$. Using $I(x, y) = \frac{1}{x^2 + y^2}$ we get $\frac{x\,dx + y\,dy}{x^2 + y^2} - 3\,dy = 0$. Integrating yields $\frac{1}{2}\ln(x^2 + y^2) - 3y = C$.

19. $xy^2\,dx + x^2y\,dy = 0$. This does not fit a set form but using $I(x, y) = \frac{1}{x^2y^2}$ we get $\frac{1}{x}\,dx + \frac{1}{y}\,dy = 0$. Integrating we get $\ln x + \ln y = C_1$ or $\ln(xy) = C_1$ or $xy = C_2$ where $C_2 = e^{C_1}$.

20. $4y\,dx - 3x\,dy - x^{-3}y^2\,dy = 0$. The first two terms fit form 8 of Table 32.1. Using $I(x, y) = x^3y^{-4}$ we get $(4y\,dx - 3x\,dy)x^3y^{-4} - x^{-3}y^2(x^3y^{-4})\,dy = 0$ or $x^3y^{-3}\,dx - 3x^4y^{-4}\,dy - y^{-2}\,dy = 0$. Integration yields $x^4y^{-3} + y^{-1} = C$.

21. $2y\,dx + x\,dy - 3x\,dx = 0$. Using $I(x, y) = x$ we get $2xy\,dx + x^2\,dy - 3x^2\,dx = 0$. Integrating we have $x^2y - x^3 = C$.

22. $x^2\,dx + 2xy\,dy - y^2\,dx = 0$. The second two terms are somewhat similar to the terms of form 4 in Table 32.1. Multiplying by $\frac{1}{x^2}$ we get $dx + \frac{2xy\,dy - y^2\,dx}{x^2} = 0$. Integrating we get $x + \frac{y^2}{x} = C$.

23. $3x^2(x^2 + y^2)\,dx + y\,dx - x\,dy = 0$. Using $I(x, y) = \frac{-1}{x^2 + y^2}$ we get $-3x^2\,dx - \frac{y\,dx - x\,dy}{x^2 + y^2} = 0$. Integration yields $-x^3 + \arctan\frac{y}{x} = C$.

24. $(x^2 + y^2)\,dx = 4x\,dx + 4y\,dy = 4(x\,dx + y\,dy)$. Using $I(x, y) = \frac{1}{x^2 + y^2}$ we get $dx = 4\frac{x\,dx + y\,dy}{x^2 + y^2}$. Integration yields $x = 4 \cdot \frac{1}{2}\ln(x^2 + y^2) + C$ or $x - 2\ln(x^2 + y^2) = C$.

25. $\tan(x^2 + y^2)\,dy = x\,dx + y\,dy$. First we can multiply both sides by $\cot(x^2 + y^2)$ giving us $dy = \cot(x^2 + y^2)(x\,dx + y\,dy)$. Upon examination you will discover that $(x\,dx + y\,dy)$ is the derivative of $x^2 + y^2$. Hence we can take the integral of both sides giving us $y = \ln\left|\sin(x^2 + y^2)\right| + C$.

26. $8x\,dy + 4y\,dy + 9x^{-3}y\,dy = 0$. Let $I(x, y) = x^3y^7$ and multiplying yields $(8x^4y^7\,dy + 4x^3y^8\,dx) + 9y^8\,dy = 0$. Integration yields $x^4y^8 + y^9 = C$.

27. $2x\,dy + 2y\,dx = 3x^3y\,dx$. Multiplying by $\frac{1}{xy}$ we get $2\frac{x\,dy + y\,dx}{xy} = 3x^2\,dx$ so the general solution is $2\ln(xy) = x^3 + C$ or $\ln x^2y^2 = x^3 + C$ or $x^2y^2 = e^{x^3} + C$. Now letting $x = 1$ and $y = 3$ and solving for C we get $9 = e' + C$ or $C = 9 - e$. Hence the particular solution is $x^2y^2 = e^{x^3} + 9 - e$.

28. $y\,dx - x\,dy = x^2\,dx$. Multiplying by $-\frac{1}{x^2}$ we get $-\frac{y\,dx - x\,dy}{x^2} = -\,dx$. Integration yields the general solution $\frac{y}{x} = -x + C$. Now substituting $y = 4$ and $x = 2$ we solve for C. $\frac{4}{2} = -2 + C$ or $C = 4$. Hence the particular solution is $\frac{y}{x} = -x + 4$ or $\frac{y}{x} + x = 4$.

29. $(x^2 + y + y^2)\,dx = x\,dy$ is equivalent to $(x^2 + y^2)\,dx = x\,dy - y\,dx$. Now let $I(x, y) = \frac{-1}{x^2 + y^2}$ and multiplying yield $-\,dx = \frac{dyd - x\,dy}{x^2 + y^2}$. Integration yields $-x = -\arctan\frac{y}{x} + C$ or $\arctan\left(\frac{y}{x}\right) = x + C$. Now setting $y = \frac{\pi}{3}$ and $x = \frac{\pi}{3}$ we solve for C $\arctan 1 = \frac{\pi}{3} + C$ or $C = -\frac{\pi}{12}$. Hence the particular solution is $\arctan\frac{y}{x} = x - \frac{\pi}{12}$.

30. $2y\,dx + 3x\,dy = 0$. Let $I(x, y) = xy^2$ and multiply to get $2xy^3\,dx + 3x^2y^2\,dy = 0$. Integration yields $x^2y^3 = C$. Now setting $y = -2$ and $x = 2$ we get $C = 2^2(-2)^3 = -32$. The particular solution is $x^2y^3 = -32$.

32.4 LINEAR FIRST-ORDER DIFFERENTIAL EQUATIONS

1. $dy + 2xy\,dx = 6x\,dx$ is equivalent to $\frac{dy}{dx} + 2xy = 6x$. The indegrating factor is $I(x,y) = e^{\int 2x\,dx} = e^{x^2}$. Multiplying we get $\frac{dy}{dx}e^{x^2} + 2xye^{x^2} = 6xe^{x^2}$. Now taking the integral of both sides we get $ye^{x^2} = 3e^{x^2} + C$.

2. $y' + y = 2$ or $\frac{dy}{dx} + y = 2, I(x,y) = e^x$. Multiplying we get $\frac{dy}{dx}e^x + ye^x = 2e^x$. Integrating yields $ye^x = 2e^x + C$.

3. $y' - 2y = 4$ or $\frac{dy}{dx} - 2y = 4$. $I(x,y) = e^{-2x}$ and multiplying we get $\frac{dy}{dx}e^{-2x} - 2ye^{-2x} = 4e^{-2x}$. Integrating we get $ye^{-2x} = -2e^{-2x} + C$.

4. $\frac{dy}{dx} - 3y = e^{4x}$, so $I(x,y) = e^{-3x}$ and multiplying we get $\frac{dy}{dx}e^{-3x} - 3ye^{-3x} = e^x$. Now integration yields $ye^{-3x} = e^x + C$.

5. $\frac{dy}{dx} + \frac{y}{x} = x^2$. $I(x,y) = e^{\int x^{-1}\,dx} = e^{\ln x} = x$. Multiplying we get $\frac{dy}{dx}x + y = x^3$. Integration yield $xy = \frac{1}{4}x^4 + C$.

6. $y' - 6y = e^x$, so $I(x,y) = e^{-6x}$. Multiplying we get $\frac{dy}{dx}e^{-6x} - 6ye^{-6x} = e^{-5x}$. Now integration yields $ye^{-6x} = -\frac{1}{5}e^{-5x} + C$.

7. $y' - 6y = 12x$. $I(x,y) = e^{-6x}$ and multiplying we get $\frac{dy}{dx}e^{-6x} - 6ye^{-6x} = 12xe^{-6x}$. The left-hand side integrates to ye^{-6x}. The right-hand side we can integrate of parts with $u = 12x$, $du = 12\,dx$, $dv = e^{-6x}\,dx$, $v = \frac{-1}{6}e^{-6x}$. Hence $\int 12xe^{-6x}\,dx = 12x\left(\frac{-1}{6}e^{-6x}\right) - \int \frac{-1}{6}e^{-6x} \cdot 12\,dx = -2xe^{-6x} - \frac{1}{3}e^{-6x}$. Putting it all together we get $ye^{-6x} = -\frac{1}{3}e^{-6x}(-6x - 1) + C$.

8. $\frac{dy}{dx} + x^2y = x^2$. $I(x,y) = e^{x^3/3}$ and multiplying we get $\frac{dy}{dx}e^{x^3/3} + x^2ye^{x^3/3} = x^2e^{x^3/3}$. Integration yields $ye^{x^3/3} = e^{x^3/3} + C$.

9. $\frac{dy}{dx} = e^x - \frac{y}{x}$ is equivalent to $\frac{dy}{dx} + \frac{y}{x} = e^x$. $I(x,y) = e^{\int \frac{1}{x}\,dx} = e^{\ln x} = x$. Multiplying we get $\frac{dy}{dx}x + y = xe^x$. Integration yields $xy = e^x(x-1) + C$. (Use form 84 from Appendix C).

10. $\frac{dy}{dx} - 6xy = 0$, $I(x,y) = e^{-3x^2}$, hence we get $\frac{dy}{dx}e^{-3x^2} - 6xye^{-3x^2} = 0$. Integration yield $ye^{-3x^2} = C$.

11. $y' - \frac{1}{x^2}y = \frac{1}{x^2}$. $I(x,y) = e^{\int \frac{-1}{x^2}\,dx} = e^{1/x}$. Now multiplying by $e^{1/x}$ we get $\frac{dy}{dx}e^{1/x} - \frac{1}{x^2}ye^{1/x} = \frac{1}{x^2}e^{1/x}$. Integration yields $ye^{1/x} = -e^{1/x} + C$.

12. $y' - 4xy = xe^{x^2}$. $I(x,y) = e^{-2x^2}$ and multiplication yields $\frac{dy}{dx}e^{-2x^2} - 4xye^{-2x^2} = xe^{-x^2}$. Now integration yields $ye^{-2x^2} = -\frac{1}{2}e^{-x^2} + C$.

13. $y\,dy - 2y^2\,dx = y\sin 2x\,dx$. Dividing by $y\,dx$ we get $\frac{dy}{dx} - 2y = \sin 2x$. $I(x,y) = e^{-2x}$ and multiplying we get $\frac{dy}{dx}e^{-2x} - 2ye^{-2x} = e^{-2x}\sin 2x$. Integratng using form 87 we get $ye^{-2x} = \frac{e^{-2x}}{8}(-2\sin 2x + 2\cos 2x) + C$ or $ye^{-2x} = -\frac{1}{4}e^{-2x}(\sin 2x + \cos 2x) + C$.

14. $x\,dy - y\,dx = 4x\,dx$. Dividing by $x\,dx$ we get $\frac{dy}{dx} - \frac{y}{x} = 4$. $I(x,y) = e^{\int \frac{-1}{x}\,dx} = e^{-\ln x} = e^{\ln \frac{1}{x}} = \frac{1}{x}$. Multiplying by this we get $\frac{dy}{dx} \cdot \frac{1}{x} - \frac{y}{x^2} = \frac{4}{x}$. Integration now yields $\frac{y}{x} = 4\ln x + C$.

15. $4x\,dy - y\,dx = 8x\,dx$. Dividing by $4x\,dx$ we get $\frac{dy}{dx} - \frac{y}{4x} = 2$. $I(x,y) = e^{-\int \frac{1}{4x}dx} = e^{-\frac{1}{4}\ln x} = e^{\ln x^{-\frac{1}{4}}} = x^{-\frac{1}{4}}$. Multiplying by this we get $\frac{dy}{dx}x^{-1/4} - \frac{y}{4x}x^{-1/4} = 2x^{-1/4}$. Now integrating we get $yx^{-1/4} = \frac{8}{3}x^{3/4} + C$.

16. $\sec x\,dy = (y-1)\,dx$. Dividing by $\sec x\,dx$ we get $\frac{dy}{dx} = (y-1)\cos x$ or $\frac{dy}{dx} - y\cos x = -\cos x$. The integrating factor is $I(x,y) = e^{-\sin x}$ and multiplying by this we get $\frac{dy}{dx}e - ye^{-\sin x} = -\cos xe^{-\sin x}$. Integration yields $ye^{-\sin x} = e^{-\sin x} + C$.

17. $\frac{dy}{dx} - y\sec^2 x = \sec^2 x$. $I(x,y) = e^{\int -\sec^2 x\,dx} = e^{-\tan x}$ Multiplying we get $\frac{dy}{dx}e^{-\tan x} - y\sec^2 xe^{-\tan x} = \sec^2 xe^{-\tan x}$. Now integration yields $ye^{-\tan x} = -e^{-\tan x} + C$.

18. $x\frac{dy}{dx} = y + (x^2-1)^2$ is equivalent to $\frac{dy}{dx} - \frac{y}{x} = \frac{(x^2+1)^2}{x}$. $I(x,y) = e^{-\int \frac{1}{x}} = e^{-\ln x} = \frac{1}{x}$. Multiplying we get $\frac{dy}{dx}\cdot\frac{1}{x} - \frac{y}{x^2} = \frac{(x^2-1)^2}{x^2} = \frac{x^4 - 2x^2 + 1}{x^2}$. Integrating we get $\frac{y}{x} = \frac{1}{3}x^3 - 2x - \frac{1}{x} + C$.

19. $x\,dy + (1-4x)y\,dx = 4x^2e^{4x}\,dx$. Dividing by $x\,dx$ we obtain $\frac{dy}{dx} + \frac{1-4x}{x}y = 4xe^{4x}$. To get the integratng factor we most find $\int \frac{1-4x}{x}\,dx = \left(\frac{1}{x} - 4\right)\,dx = \ln x - 4x$. Hence $I(x,y) = e^{\ln x - 4x}$. Multiplying by this we obtain the equation $\frac{dy}{dx}e^{\ln x - 4x} + \frac{1-4x}{x}ye^{\ln x - 4x} = 4xe^{4x}e^{\ln x - 4x} = 4xe^{\ln x} = 4x^2$. Now integration yields $ye^{\ln x - 4x} = \frac{4}{3}x^3 + C$. The left side of this equation is $ye^{\ln x}\cdot e^{-4x} = yx\cdot e^{-4x} = \frac{yx}{e^{4x}}$. Hence the answer is $\frac{yx}{e^{4x}} = \frac{4}{3}x^3 + C$.

20. $x\,dy - y\,dx = x^3\sin x^2\,dx$. Dividing by $x\,dx$ we get $\frac{dx}{dx} - \frac{y}{x} = x^2\sin x^2$. $I(x,y) = \frac{1}{x}$ and multiplying by this we get $\frac{dy}{dx}\cdot\frac{1}{x} - \frac{y}{x^2} = x\sin x^2$. Integration yields $\frac{y}{x} = -\frac{1}{2}\cos x^2 + C$.

21. $\frac{dy}{dx} + \frac{2y}{x} = x$. $I(x,y) = e^{\int \frac{2}{x}} = e^{2\ln x} = e^{\ln x^2} = x^2$. Multiplying we obtain $\frac{dy}{dx}x^2 + 2yx = x^3$. Integration yields the general solution $yx^2 = \frac{1}{4}x^4 + C$. Substituting $y = 3$ and $x = 2$ we solve for C. $3\cdot 4 = \frac{1}{4}\cdot 2^4 + C$ or $12 = 4 + C$ or $C = 8$. The particular solution is $yx^2 = \frac{1}{4}x^4 + 8$.

22. $\frac{dy}{dx} + 3y = e^{-2x}$ has the integrating factor e^{3x}. Multiplying we get $\frac{dy}{dx}e^{3x} + 3ye^{3x} = e^x$. Integration yields the general solution $ye^{3x} = e^x + C$. Substituting $y = 2$ and $x = 0$ we solve for C. $2e^0 = e^0 + C$ or $C = 1$. The particular solution is $ye^{3x} = e^x + 1$.

23. $y' - \frac{2y}{x} = x^2\sin 3x$. The integrating factor is $e^{\int \frac{-2}{x}} = e^{-2\ln x} = \frac{1}{x^2}$. Multiplying by $\frac{1}{x^2}$ yields $\frac{dy}{dx}\cdot\frac{1}{x^2} - \frac{2y}{x^3} = \sin 3x$. Integrating we get the general solution $\frac{y}{x^2} = \frac{-1}{3}\cos 3x + C$. Substituting $y = \pi^2$ and $x = \pi$ we solve for C. $1 = \frac{-1}{3}\cos 3\pi + C$. $1 = \frac{1}{3} + C$ or $C = \frac{2}{3}$. The particular solution is $yx^{-2} = -\frac{1}{3}\cos 3x + \frac{2}{3}$.

24. $\sin x\,dy + (y\cos x - 1)\,dx = 0$ is equivalent to $\frac{dy}{dx} + y\cot x = \csc x$. The integrating factor is $e^{\int \cot x\,dx} = e^{\ln|\sin x|} = \sin x$. Multiplying by $\sin x$ we obtain the equation $\frac{dy}{dx}\sin x + y\cos x = 1$. Integration yields the general solution $y\sin x = x + C$. Now substituting $y = \pi$ and $x = \frac{\pi}{6}$ we solve for C and obtain $\pi\sin\frac{\pi}{6} = \frac{\pi}{6} + C$ and so $\frac{1}{2}\pi - \frac{\pi}{6} = C$ or $C = \frac{\pi}{3}$. The particular solution is $y\sin x = x + \frac{\pi}{3}$.

32.5 APPLICATIONS

1. As in example 32.22, $\frac{dN}{dt} - kN = 0$ has general solution $N = Ce^{kt}$. At $t = 0$, $N = 100$ which gives us that $C = 100$. Now when $t = 15$ days, $N = 75$ g and we can solve for k. $75 = 100e^{15k}$, so $\frac{75}{100} = e^{15k}$ and taking natural logarithms we get $15k = \ln(\frac{3}{4})$ or $k = -0.0192$. The equation is $N = 100e^{-0.0192t}$ where N is in grams and t in days.

2. The half-life is when $N = 50$, hence we get $50 = 100e^{-0.192t} \Rightarrow \frac{1}{2} = e^{-0.0192t} \Rightarrow -0.0192t = \ln\frac{1}{2} \Rightarrow t = 36.1413$ days $= 36$ days 3 hr 23 min 29.41 s.

3. If the half-life is 1000 years, we get $\frac{1}{2} = e^{k1000}$ or $k = \frac{\ln\frac{1}{2}}{1000} = -6.93147 \times 10^{-4}$. When 10% has decayed we have $0.9 = e^{-6.93147\times10^{-4}t}$ or $-6.93147 \times 10^{-4}t = \ln(0.9) \Rightarrow t = 152.0031$ yr.

4. As in example 32.22 $\frac{dN}{dt} = kN$ or $N = Ce^{kt}$. Now substituting the particular values we get $2000 = Ce^{k}$ and $6000 = Ce^{4k}$. Dividing the second equation by 2000 or Ce^{k} we get $3 = e^{3k}$ so $k = 0.3662$. Now back substituting into the first equation we get $2000 = Ce^{0.3662}$ or $C = \frac{2000}{e^{0.3662}} = 1386.7$ or 1387. The equation is $N = 1387e^{0.3662t}$, t in hours.

5. When $t = 0, N = 1387$.

6. $e^{0.3662t} = 2 \Rightarrow 0.3662t = \ln 2$ or $t = \frac{\ln 2}{0.3662} = 1.8928$ h.

7. First we must find k in the equation $N = Ce^{kt}$. since the half-life is 5600 years, $\frac{1}{2} = e^{5600k}$ or $k = -1.23776 \times 10^{-4}$. Now we solve for t in the equation $\frac{1}{500} = e^{-1.23776\times10^{-4}t}$, with the result $t = \frac{\ln(\frac{1}{500})}{-1.23776\times10^{-4}} = 50208.4$ years ≈ 50210 years.

8. Since $\frac{dI}{dt} = kI$ we get $I = Ce^{kt}$ where I is the light intersity and t is the thickness. At 3 ft, $I = 25\%$ of the orginal light intersity. Hence $0.25 = e^{3k}$ or $k = \frac{\ln(0.25)}{3} = -0.4621$. At 18 ft, $\frac{I}{C} = e^{-0.4521\times18} = 2.4414 \times 10^{-4} = 0.0244\%$ of the original intensity.

9. (a) As in example 32.23 we have $\frac{dT}{dt} + kT = 30k$. Solving this differential equation we get $Te^{kt} = 30e^{kt} + C$ or $T = 30 + Ce^{kt}$. When $t = 0, T = 10°$, hence $C = -20$ or $T = 30 - 20e^{kt}$. Now when $t = 10, T = 15°$ or $15 = 30 - 20e^{-k10}$ or $\frac{-15}{-20} = e^{10k}$. Hence $k = \frac{\ln(0.75)}{-10} = 0.02877$. The equation is $T = 30 - 20e^{-0.02877t}$ (b) $22 = 30 - 20e^{-0.02877t}$, or $\frac{-8}{-20} = e^{-0.02877t}$, and so $t = \frac{\ln(0.4)}{-0.02877} = 31.85$ min. (c) $t = 30 - 20e^{-0.02877\times60} = 26.44$°C

10. As in exmaple 32.23, $T = 40 + Ce^{-kt}$. Substituting the particular values we get $0 = 40 + Ce^{10k}$ and $10 = 40 + Ce^{-20k}$. These are equivalent to $\frac{-40}{C} = e^{-10k}$ and $\frac{-30}{C} = e^{-20k}$, respectively. Squaring both sides of $\frac{-40}{C} = e^{-10k}$ we get $\frac{1600}{C^2} = e^{-20k} = \frac{-30}{C}$. Hence $1600C = -30C^2$ or $C = \frac{1600}{-30} = -53.33$. This yields the equation $T = 40 - 53.33e^{-kt}$. When $t = 0, T = 40 - 53.33 = -13.33$°F.

11. As in example 32.23, $T = 40 + Ce^{-kt}$. At $t = 0$, $T = 375$°F so $C = 375 - 40 = 335$°F. Now, when $t = 15$ min, $T = 280$°F so $280 = 40 + 335e^{-15k}$ or $\frac{240}{335} = e^{-15k}$, and so $k = 0.02223$. Thus, we get $T = 40 + 335e^{-0.02223t}$. Hence $75 = 40 + 335e^{-0.02223t}$, or $\frac{35}{335} = e^{-0.02223t}$, and, as a result, $t = \frac{\ln\left(\frac{35}{335}\right)}{-0.02223} = 101.6$ min or 1 hr 41.6 min.

12. As in Example 32.23, $T = 20 + Ce^{-kt}$. When $t = 0$, $T = -20°$ so $-20 = 20 + C$ or $C = -40°$, and hence $T = 20 - 40e^{kt}$. At $t = 1$, $T = -16°$ or $-16 = 20 - 40e^{-k}$, or $\frac{36}{40} = e^{-k}$, or $k = 0.1054$, so $T = 20 - 40e^{-0.1054t}$. Now substituting 15 for T we get $15 = 20 - 40e^{-0.1054t}$ or $\frac{1}{8} = e^{-0.1054t}$ or $t = 19.74$ min $= 19$ min 44 s.

13. This is an RL circuit and so $\frac{dI}{dt} + \frac{R}{L}I = \frac{E}{L}$ or $\frac{dI}{dt} + \frac{10}{2}I = \frac{100}{2}$ or $\frac{dI}{dt} + 5I = 50$. The general solution is $I = Ce^{-5t} + 10$. At $t = 0, I = 0$ so $C = -10$. The particular solution is $I = -10e^{-5t} + 10$.

14. In this problems, $\frac{dI}{dt} + 5I = 10\cos 5t$. The integrating factor is e^{5t} and so we get $\left(\frac{dI}{dt} + 5I\right)e^{5t} = 10e^{5t}\cos 5t$. Integrating we get $Ie^{5t} = \frac{10e^{5t}}{50}(5\cos 5t + 5\sin 5t) + C$ or $I = (\cos 5t + \sin 5t) + Ce^{-5t}$. When $t = 0, I = 0$, so $0 = 1 + C$ or $C = -1$. The solution is $I = (\cos 5t + \sin 5t) - e^{-5t}$.

15. $\frac{dI}{dt} + 50I = 5$. The integrating factor is e^{50t} and we get $Ie^{50t} = \frac{1}{10}e^{50t} + C$. Hence $I = 0.1 + Ce^{50t}$. When $t = 0, I = 0$, so $C = -0.1$ and the particular solution is $I = 0.1 - 0.1e^{-50t}$

16. $\frac{dI}{dt} + 20I = 6\sin 2t$. The integrating factor is e^{20t} so we get $\frac{dI}{dt}e^{20t} + 20Ie^{20t} = 6e^{20t}\sin 2t$. Integrating we obtain $Ie^{20t} = \frac{6e^{20t}}{404}(20\sin 2t - 2\cos 2t) + C$ or $I = \frac{3}{101}(10\sin 2t - \cos 2t) + Ce^{-20t}$. When $t = 0$, $I = 6$ so $6 = \frac{3}{101}(-1) + C$, and so $C = \frac{606}{101} + \frac{3}{101} = \frac{609}{101}$. The particular solution is $I = \frac{609}{101}e^{-20t} + \frac{3}{101}(10\sin 2t - \cos 2t)$

32.6 MORE APPLICATIONS

1. $x^2 + y^2 = C$ is a family of circles with center at the origin. Taking the derivative implicitly we get $2x + 2yy' = 0$ or $\frac{dy}{dx} = -\frac{x}{y} = f(x, y)$. The orthogonal trajectories satisfy the equation $\frac{dy}{dx} = \frac{-1}{f(x, y)}$ or $\frac{dy}{dx} = \frac{y}{x}$. Separating variables we get $\frac{dy}{y} = \frac{dx}{x}$. Integration gives $\ln y = \ln x + C_1$ or $y = e^{\ln x + C_1} = e^{C_1}x$. Letting $e^{C_1} = k$ we get the family of curves $y = kx$ or the set of lines through the origin.

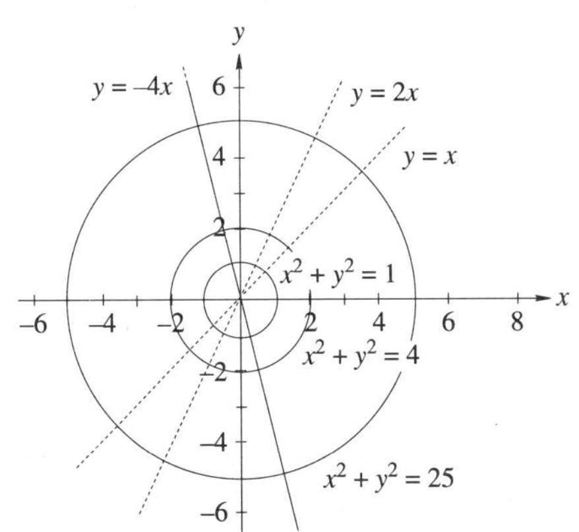

2. $xy = C$ is a family of hyperbolas with the x and y axis as asympototes. Taking the derivative implicitly we obtain $xy' + y = 0$ or $y' = -\frac{y}{x} = f(x, y)$ The family we need satisfies $\frac{dy}{dx} = \frac{-1}{f(x, y)}$ or $\frac{dy}{dx} = \frac{x}{y}$. Separating variables we have $y\,dy = x\,dx$. Integration yields $\frac{1}{2}y^2 = \frac{1}{2}x^2 + C_1$ or equivalently $y^2 - x^2 = k$.

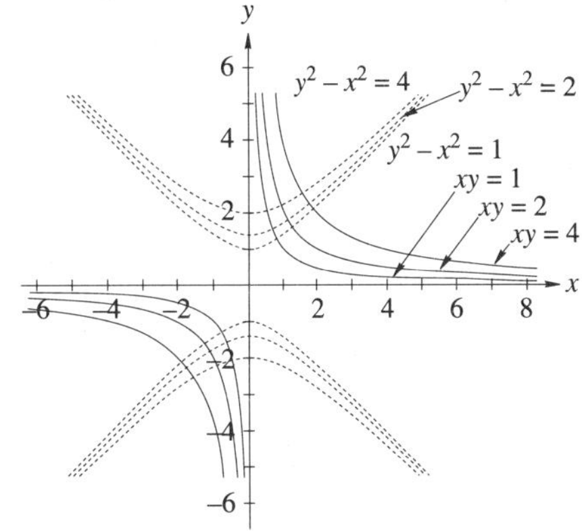

3. $y = Ce^x$ had derivative $\frac{dy}{dx} = Ce^x$. Substituting ye^{-x} for C we obtain $\frac{dy}{dx} = y$. The family of curves we seek satisfies the equation $\frac{dy}{dx} = -\frac{1}{y}$ or $y\,dy = -\,dx$. Integrating we get $\frac{1}{2}y^2 = -x + C$ or $y^2 = -2x + k$ or $y^2 + 2x = k$.

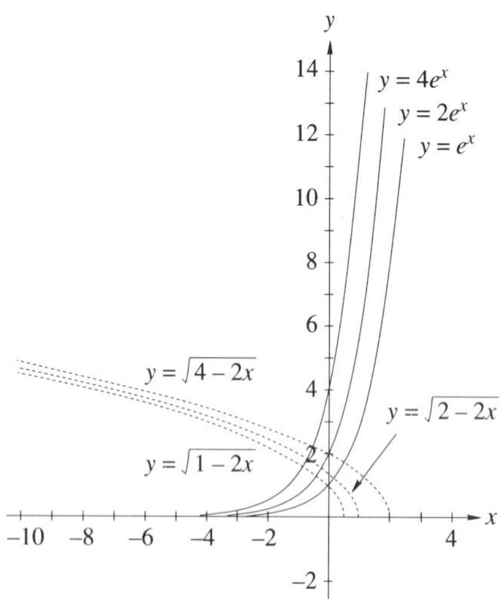

4. $x^2 - y^2 = C$ has implicit derivative $2x - 2yy' - 0$ or $y' = \frac{x}{y}$. The equation we need to solve is $\frac{dy}{dx} = -\frac{y}{x}$ or $\frac{dy}{y} = -\frac{dy}{x}$. This has solution $\ln y = -\ln x + C$ or $y = k \cdot \frac{1}{x}$ or $xy = k$.

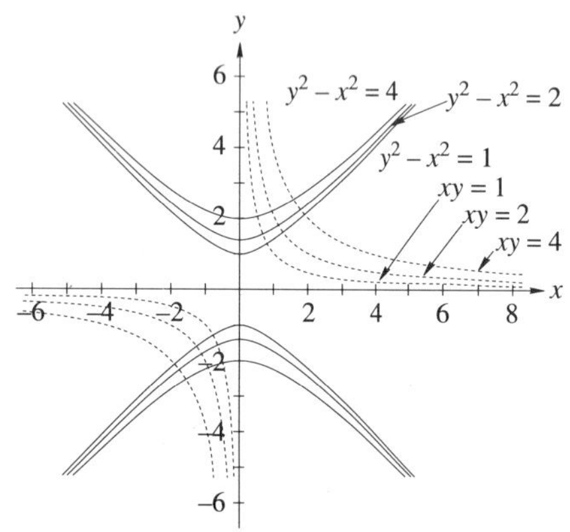

5. $x^2 + \frac{1}{2}y^2 = C^2$ has implicit derivative $2x + yy' = 0$ or $y' = \frac{-2x}{y}$. The orthogonal trajectories satisfy the equation $\frac{dy}{dx} = \frac{y}{2x}$. Separating variables we get $\frac{dy}{y} = \frac{dx}{2x}$. Integration yield $\ln y = \frac{1}{2}\ln x + C_1$ or $\ln y^2 = \ln x + C_2$ or $y^2 = kx$.

6. $x^2+2y^2 = C$ has the implicit derivative $2x+4yy' = 0$ or $y' = \frac{-x}{2y}$. We seek he solution to $\frac{dy}{dx} = \frac{2y}{x}$. Separating variables we obtain $\frac{dy}{y} = 2\frac{dx}{x}$. Integration yields $\ln y = 2\ln x + C_1$ or $\ln y = \ln x^2 + C_1$ or $y = kx^2$, where $k = \ln C_1$.

7. As in example 32.26 we start with the equation $\frac{dv}{dt} + \frac{kv}{m} = g$ or $\frac{dv}{dt} + \frac{0.2v}{10} = 9.8$. This is of linear form and the integrating factor is $e^{0.02t}$. Hence we obtain the equation $\frac{dv}{dt}(ve^{0.02t}) = 9.8e^{0.02t}$. Integration yield $ve^{0.02t} = 490e^{0.02t} + C_1$ or $v = 490 + C_1e^{-0.02t}$. Since the initial velocity is 0, $C_1 = -490$ and the equation becomes $v(t) = 490(1 - e^{-0.02t})$ m/s or $v(t) = 490(1 - e^{-\frac{t}{50}})$ m/s.

8. After 10 seconds $v = 490(1 - e^{-\frac{1}{5}}) = 88.82$ m/s

9. (a) Using the general solution from problem 7 and substituting 1 for v and 0 for t we find $C_1 = 489$ so the particular solution is $v(t) = 490 - 489e^{-\frac{t}{50}}$, (b) $v(10) = 490 - 489e^{-\frac{1}{5}} = 89.64$m/s, (c) $v_L = \frac{10\times 9.8}{0.2} = 490$ m/s. (Note: $\lim_{t\to\infty} v(t) = 490$).

10. (a) Using $v_L = \frac{mg}{k}$ we can solve for k. $k = \frac{mg}{v_L} = \frac{10 \cdot 32}{320} = 1$. This yields the differential equation $\frac{dv}{dt} + \frac{v}{10} = 32$. The integrating factor is $e^{0.1t}$ and we get $\frac{dv}{dt}(ve^{0.1t}) = 32e^{0.1t}$ which has general solution $ve^{0.1t} = 320e^{0.1t} + C_1$. Since $v = 0$ when $t = 0$, $C_1 = -320$ and the particular solution is $v(t) = 320(1 - e^{-0.1t})$ ft/s.
(b) The position with respect to where it was dropped is $\int v(t)\,dt = \int 320(1 - e^{-0.1t})dt = 320(t + 10e^{-0.1t} + C_1)$. Since the position is 0 when $t = 0$, then $C_1 = -10$, or $d(t) = 320(t + 10e^{-0.1t} - 10)$. From the ground, $h(t) = 1000 - d(t) = 1000 - 320(t + 10e^{-0.1t} - 10) = 1000 + 3200 - 320(t - 10e^{-0.1t}) = 4200 - 320(t + 10e^{-0.1t}) = 320(13.125 - t - 10e^{-0.1t})$.

11. As in Example 32.27, $\frac{dQ}{dt} + \frac{2}{50+(2-2)t}Q = 10$ or $\frac{dQ}{Q} = -\frac{dt}{25}$. Separating variables we get $\frac{dQ}{Q} + \frac{dt}{25} = 0$. Integration yields $\ln Q = -\frac{1}{25}t + C$ or $Q = ke^{-t/25}$. When $t = 0$, $Q = 25$ so $k = 25$ and $Q = 25e^{-t/25}$. After 15 minutes $Q = 25e^{-15/25} = 13.72$ lb.

12. First we must find a formula for Q. As in Example 32.27 we get $\frac{dQ}{dt} + \frac{5}{100}Q = 0$ or $\frac{dQ}{dt} = -\frac{1}{20}dt$. Integrating we get $\ln Q = -\frac{1}{20}t + C$ or $Q = ke^{-t/20}$. Since $Q = 40$ when $t = 0$ the particular solution is $Q(t) = 40e^{-t/20}$ (a) $Q(10) = 40e^{-10/20} = 24.26$ lb, (b) $Q(30) = 40e^{-30/20} = 8.93$ lb, (c) $Q(1\text{ hr}) = Q(60) = 1.99$ lb.

13. (a) $\frac{dQ}{dt} + \frac{4}{8}Q = 12$. This equation is linear and the integrating factor is $e^{t/2}$. Multiplying yields $\frac{dQ}{dt}(Qe^{t/2}) = 12e^{t/2}$. Integrating we get $Qe^{t/2} = 24e^{t/2} + C$ or $Q = 24 + Ce^{-t/2}$. Since $Q = 2$ when $t = 0$, $C = -22$ so the particular solution is $Q(t) = 24 - 22e^{-t/2}$ (b) $Q(8) = 24 - 22e^{-4} = 23.597$ lb.

14. $\frac{dQ}{dt} + \frac{2}{60}Q = 6$; $\frac{dQ}{dt}(Qe^{t/30}) = 6e^{t/30}$; $Qe^{t/30} = 180e^{t/30} + C_1$ or $Q = 180 + Ce^{-t/30}$. When $t = 0$, $Q = 60 \times 2 = 120$, so $C = -60$, and we have $Q = 180 - 60e^{-t/30}$. Now, setting $Q = 150$ we solve for t. $150 = 180 - 60e^{-t/30}$, or $\frac{1}{2} = e^{-t/30}$; or $\frac{t}{30} = \ln\frac{1}{2}$, and $t = -30\ln\frac{1}{2} = 20.79$ min.

CHAPTER 32 REVIEW

1. We have $y' = x + y$ or $\frac{dy}{dx} - y = x$. This is linear with integrating factor e^{-x}. Hence we get $\frac{dy}{dx}(ye^{-x}) = xe^{-x}$. Integrating with form 84 yields $ye^{-x} = e^{-x}(-x-1) + C$ or $y = Ce^x - (x+1)$.

2. $x\,dx - y^2\,dy = 0$ or $\frac{1}{2}x^2 - \frac{1}{3}y^3 = C$

3. $y' = y^2x^3$ or $\frac{dy}{y^2} = x^3\,dx$. Integrating produces $-y^{-1} = \frac{1}{4}x^4 + C_1$ or $x^4 + 4y^{-1} = C$.

4. $y' = 8y$ or $\frac{dy}{y} = 8\,dx$ and integrating gives us $\ln y = 8x + C$, or $y = ke^{8x}$.

5. $e^x\,dx - 2y\,dy = 0$ and integtating produces $e^x - y^2 = C$.

6. $(y^2 - y)\,dx + x\,dy = 0$ or $\frac{dx}{x} + \frac{dy}{y^2-y} = 0$. Integrating by Form 48 we get $\ln x - \ln\left|\frac{y}{y-1}\right| = C$ or $\ln\frac{x(y-1)}{y} = C_1$ or $-x + \frac{x}{y} = C$.

7. $(y - xy^3)\,dx + (x - 2x^2y^2)\,dy = 0$ is equivalent to $y\,dx + x\,dy - xy^3\,dx - 2x^2y^2\,dy = 0$. Now we multiply by $\frac{1}{xy}$ to get $\frac{y\,dx + x\,dy}{xy} - (y^2\,dx + 2xy\,dy) = 0$. The first part has integral $\ln|xy|$. The second part has integral $-xy^2$. Putting it together we get $\ln|xy| - xy^2 = C$.

8. $(y+1)\,dx - x\,dy = 0$ is equivalent to $\frac{dx}{x} - \frac{dy}{y+1} = 0$. This has solution $\ln x - \ln(y+1) = \ln C$ or $\frac{x}{y+1} = C$ or $\frac{y+1}{x} = C$.

9. $y\,dx + (2-x)\,dy = 0$ is equivalent to $\frac{dx}{2-x} + \frac{dy}{y} = 0$. This has solution $-\ln|2-x| + \ln y = \ln C$ or $\ln\frac{y}{x-2} = \ln C$ or $\frac{y}{x-2} = C$ or $\frac{x-2}{y} = C$.

10. $(y + x^3y)\,dx + x\,dy = 0$ or $y\,dx + x\,dy = -x^3y\,dx$. Multiplying by the integrating factor $\frac{1}{xy}$, we get

$\dfrac{y\,dx + x\,dy}{xy} = \dfrac{-x^3y}{xy}\,dx = -x^2\,dx$. Integrating, we get $\ln xy = -\frac{1}{3}x^3 + C$

11. $y^2\,dx + xy\,dy = 0$. Multiplying by $\dfrac{1}{xy^2}$ to separate variables we get $\dfrac{dx}{x} + \dfrac{dy}{y} = 0$. Integration yields $\ln|x| + \ln|y| = \ln C$ or $\ln|xy| = C$.

12. $dy - 3y\,dx = 6\,dx$ is linear first-order and has integrating factor e^{3x}. Multiplying we get $e^{-3x}\,dy - 3ye^{-3x}\,dx = 6e^{-3x}\,dx$ and integrating we have $ye^{-3x} = -2e^{-3x} + C$ or $y = -2 + Ce^{3x}$

13. $y' - 4xy = x$ in linear first-order and has integratng factor e^{-2x^2}. Its solution is $ye^{-2x^2} = \int xe^{-2x^2}\,dx$ or $ye^{-2x^2} = \frac{-1}{4}e^{-2x^2} + C$ or $y = \frac{-1}{4} + Ce^{2x^2}$.

14. $y' + \dfrac{12y}{x} = x^{12}$ is linear first-order with integrating factor $e^{12\ln x} = x^{12}$. Multiplying and integrating we get $yx^{12} = \int x^{24}\,dx$ or $yx^{12} = \dfrac{x^{25}}{25} + C$.

15. $y' + y = \sin x$ is linear first-order with integrating factor e^x. Multiplying and integrating we get $ye^x = \int e^x \sin x\,dx$. Using form 87 this becomes $ye^x = \dfrac{e^x}{2}(\sin x - \cos x) + C$ or $y = \frac{1}{2}(\sin x - \cos x) + Ce^{-x}$.

16. $y' - 7y = e^x$ is linear first-order with integrating factor e^{-7x}. Multiplying and integratng we have $ye^{-7x} = \int e^{-6x}\,dx$ or $ye^{-7x} = \frac{1}{6}e^{-6x} + C$ which is equivalent to $y = -\frac{1}{6}e + Ce^{7x}$.

17. $\cos x\,dx + y\,dy = 0$ has general solution $\sin x + \frac{1}{2}y^2 = C$. Substituting $y = 2$ and $x = 0$ we find $C = \sin 0 + \frac{1}{2}\cdot 2^2 = 2$. The particular solution is $\sin x + \frac{1}{2}y^2 = 2$.

18. $(x^2 + y + y^2)\,dx - x\,dy = 0$ is equivalent to $y\,dx - x\,dy + (x^2 + y^2)\,dx = 0$. Using the integrating factor $\dfrac{-1}{x^2 + y^2}$ we get the general solution $x + \arctan\dfrac{y}{x} = C$. Substituting $y = \frac{\pi}{3}$ and $x = \frac{\pi}{3}$ we solve for C. $C = -\frac{\pi}{3} + \arctan 1 = -\frac{\pi}{3} + \frac{\pi}{4} = -\frac{\pi}{12}$. The particular solution is $x + \arctan\dfrac{y}{x} = -\dfrac{\pi}{12}$.

19. $y' + \dfrac{2}{x}y = x$ is of linear first-order form with integrating factor x^2. Multiplying and integrating we get the general solution $yx^2 = \int x^3\,dx$ or $yx^2 = \frac{1}{4}x^4 + C$. Substituting $x = 1$ and $y = 0$ we find that $C = -\frac{1}{4}$. The particular solution is $yx^2 = \frac{1}{4}x^4 - \frac{1}{4}$ or $4yx^2 = x^4 - 1$.

20. $dy + 6xy\,dx = 0$ This is linear first-order with integrating factor e^{3x^2}. Mutiplying and integrating we get $ye^{3x^2} = C$. Substituting $y = 5$ and $x = \pi$ we get $C = 5e^{3\pi^2}$. Hence the particular solution is $ye^{3x^2} = 5e^{3\pi^2} \approx 3.6134 \times 10^{13}$.

21. (a) As in Example 32.22 we get $N = Ce^{kt}$. After 2 years $\dfrac{N}{C} = 0.9$ so $e^{2k} = 0.9$, and so $k = -0.05268$. $N(t) = Ce^{-0.0527t}$ where $C = N(0)$; (b) $e^{-0.0527t} = \frac{1}{2} \Rightarrow -0.0527t = \ln\frac{1}{2}$ or $t = 13.15$ years or 13 years 57 days 12 hours.

22. $y = cx^2$ had derivative $\dfrac{dy}{dx} = 3cx^2$. Substituting $\dfrac{y}{x^3}$ for c we get $\dfrac{dy}{dx} = 3\cdot\dfrac{y}{x^3}\cdot x^2 = \dfrac{3y}{x}$. The family we seek satisfies $\dfrac{dy}{dx} = -\dfrac{x}{3y}$ or $3y\,dy = -x\,dx$ or $\frac{3}{2}y^2 = -\frac{1}{2}x^2 + k$ or $\frac{3}{2}y^2 + \frac{1}{2}x^2 = k$ or $3y^2 + x^2 = C$.

23. As in example 32.33 we get $T = 72 + Ce^{-kt}$. When $t = 0$, $T = 0$ so $C = -72$ and hence $T = 72 - 72e^{-kt}$. When $t = 15$ min, $T = 20°$F, so we get $20 = 72 - 72e^{-15t}$ or $e^{-15t} = \frac{52}{72}$ or $t = \frac{\ln\frac{52}{72}}{-15} = 0.021695$. This gives $T = 72(1 - e^{-0.021695t})$. (a) Substituting 50 for T and solving we get $1 - \frac{50}{72}e^{-0.021695t}$ or $t = \dfrac{\ln\left(\frac{20}{72}\right)}{0.021695} = 59.04$ min. (b) $T(60) = 72(1 - e^{-0.021695\times 60}) = 52.41°$F.

24. Using the formula $\dfrac{dI}{dt} + \dfrac{R}{L}I = \dfrac{E}{L}$ we get $\dfrac{dI}{dt} + 30I = \dfrac{25}{2}$. This equation is linear first-order with integratng factor e^{30t}. Multiplying and integrating we

get $Ie^{30t} = \frac{25}{60}e^{30t} + C$ or $I = \frac{5}{12} + Ce^{-30t}$. The intial current is 0, so $C = -\frac{5}{12}$ and the particular solution is $I = \frac{5}{12}(1 - e^{-30t})$

25. Using the equation $\frac{dq}{dt} + \frac{1}{RC}q = \frac{E}{R}$ we get $\frac{dq}{dt} + 1.25q = 5\cos 2t$. This is linear first-order with integrating factor $e^{1.25t}$. Multiplying and integrating we get $qe^{1.25t} = 5\int e^{1.25t}\cos 2t$. Using form 88 we get $qe^{1.25t} = \frac{5e^{1.25t}}{17.5625}(1.25\cos 2t + 2\sin 2t) + C$ or $q = \frac{1}{3.5125}(1.25\cos 2t + 2\sin 2t + ke^{-1.25t})$. When $t = 0, q = 0$ so $k = -1.25$. Finally, multiplying by $\frac{80}{80}$ to clear up fractions and decimals we get $q = \frac{1}{281}(100\cos 2t + 160\sin 2t - 100e^{-1.25t})$.

26. First we convert 192 lb to slugs: $\frac{192}{32} = 6$ slugs. Since the limiting velocity is 16 ft/s, we get $k = \frac{192}{16} = 12$. Now using the equation $\frac{dv}{dt} + \frac{kv}{m} = g$, we have $\frac{dv}{dt} + \frac{12}{6}v = 32$. This is linear first-order with integrating factor e^{2t}. Multiplying and integrating we get $ve^{2t} = 16e^{2t} + C$ or $v = 16 + Ce^{-2t}$. Since $v_0 = 0, C = -16$ and $v(t) = 16(1 - e^{-2t})$ (a) $v(1) = 13.83$ ft/s. (b) $15 = 16(1 - e^{-2t}) \Rightarrow e^{-2t} = \frac{1}{16}$ or $t = \frac{\ln\left(\frac{1}{16}\right)}{-2} = 1.386$ s.

27. (a) As in exmaple 32.28, $\frac{dQ}{dt} + \frac{100}{2000}Q = 0.035$. This is linear first-order with integrating factor $e^{0.05t}$. Multiplying and integrating we get $Qe^{0.05t} = 0.7e^{0.05t} + C$ or $Q = 0.7 + Ce^{-0.05t}$. The initial quantity of CO_2 is 2000×0.001 so $C = 1.3$. The particular solution is $Q(t) = 0.7 + 1.3e^{-t/20}$ ft^3, (b) $Q(60) = 0.7547$ ft^3, (c) $Q(180) = 0.7002$ ft^3.

28. (a) As in example 32.28, $\frac{dQ}{dt} + \frac{1}{250}Q = 0.0005$. The general solution is $Qe^{t/250} = 0.125e^{t/250} + C$ or $Q = 0.125 + Ce^{-t/250}$. Since $Q_0 = 2.5$, $C = 2.375$ and the solution is $Q(t) = 0.125 + 2.375e^{-t/250}$ ft^3, (b) $Q(60) = 1.8807$ ft^3.

29. (a) $\frac{dQ}{dt} + \frac{25}{100000}Q = 0.0125$ has solution $Qe^{t/4000} = 50e^{-t/4000} + C$ or $Q = 50 + ce^{-t/4000}$. Since $Q_0 = 100{,}000 \times 0.0001 = 10, C = -40$ and $Q(t) = 50 - 40e^{-t/4000}$ m^3, (b) Solving $20 = 50 - 40e^{-t/4000}$ for t we get $\frac{3}{4} = e^{-t/4000}$ or $t = -4000\ln\left(\frac{3}{4}\right) = 1151$ days or 3 years 56 days.

30. (a) $\frac{dQ}{dt} + \frac{500}{10^6}Q = 0.00074 \times 500$ or $\frac{dQ}{dt} + \frac{Q}{20000} = 0.37$ has solution $Qe^{t/20\,000} = \int 0.37e^{t/20\,000} = 7400e^{t/20\,000} + C$ or $Q = 7400 + Ce^{-\frac{t}{20000}}$. Since $Q_0 = 10^3$, $C = -6400$. Hence $Q(t) = 7400 - 6400e^{-t/20\,000}$, (b) Solving $2000 = 7400 - 6400e^{-t/20\,000}$ for t we get $\frac{54}{64} = e^{-t/20\,000}$ or $t = -20000\ln\frac{54}{64} = 3398$ days or 9 years 3.6 months, (c) As t approaches infinity, Q approaches 7400 m^3.

CHAPTER 32 TEST

1. Differentiating $y = 4e^{2x} + 2e^{-3x}$ we get $\frac{dy}{dx} = 8e^{2x} - 6e^{-3x}$ and $\frac{d^2y}{dx^2} = 16e^{2x} + 18e^{-3x}$. Substituting these in the given differential equation, we get $\left(16e^{2x} + 18e^{-3x}\right) + \left(8e^{2x} - 6e^{-3x}\right) - 6\left(4e^{2x} + 2e^{-3x}\right) = (16 + 8 - 24)e^{2x} + (18 - 6 - 12)e^{-3x} = 0$.

2. Here $M = y^2$ and $N = 2xy$. Both M and N are continuous. Since $\frac{\partial M}{\partial y} = 2y = \frac{\partial N}{\partial x}$, the equation is exact.

3. Here $y' = \frac{dy}{dx} = \frac{5x^4}{y^3}$ can be rewritten as $y^3\,dy = 5x^4\,dx$. Integrating, we obtain $\frac{1}{4}y^4 = x^5 + C$.

4. Dividing this equation by x^2, we obtain $\frac{x^2+1}{x^2}dx + y^2\,dy = 0$ or $\left(1 + \frac{1}{x^2}\right)dx + y^2\,dy = 0$. Integrating

produces the solution $x - \frac{1}{x} + \frac{1}{3}y^3 = C$.

5. Rewriting the given equation as $(1+y^2)\,dx + (x^2+1)\,dy = 0$ we see that the equation is separable. If the variables are separated by dividing by $(1+y^2)(x^2+1)$, we obtain $\frac{dx}{x^2+1} + \frac{y}{1+y^2}\,dy = 0$. Integrating we obtain the solution $\arctan x + \frac{1}{2}\ln(1+y^2) = C$.

6. Dividing by $\sin x$ we obtain $\frac{dy}{dx} = \frac{1}{\sin x} - 2y\frac{\cos x}{\sin x} = \csc x - 2y\cot x$, which can be written in the standard form for a linear first-order differential equation as $\frac{dy}{dx} + (2\cot x)y = \csc x$. This has in integrating factor of $e^{\int 2\cot x\,dx} = e^{\ln|\sin x|} = \sin x$ and the solution is $y\sin x = \int(\sin x)(\csc x)\,dx = \int dx = x + C$, and so the solution is $y\sin x = x + C$.

7. The equation contains the combination $x\,dy - y\,dx = -(y\,dx - x\,dy)$. Using integrating factor (4) from Table 32.1, we multiply the given equation by $-\frac{1}{x^2}$ to obtain $\frac{y\,dx - x\,dy}{x^2} + \frac{x^6}{x^2}\,dx = 0$ or $\frac{y\,dx - x\,dy}{x^2} + x^4\,dx = 0$. The left-hand term is the derivative of $-\frac{y}{x}$ so integrating produces the solution $-\frac{y}{x} + \frac{1}{5}x^5 = C$.

8. The given equation contains the combination $x\,dy + y\,dx$. Using integrating factor (1) from Table 32.1, we multiply the given equation by $I(x,y) = \frac{1}{xy}$ obtaining $4\left(\frac{x\,dy + y\,dx}{xy}\right) + \frac{6x^2y}{xy}\,dx = 4\left(\frac{x\,dy + y\,dx}{xy}\right) + 6x\,dx = 0$. Integrating, we get $4\ln xy + 3x^2 = C$. We are given $x = 1$ when $y = e$. Substituting, we obtain $4\ln e + 3 = 7 = C$, so the particular solution is $4\ln xy + 3x^2 = 7$.

9. We have $F(x,y,c) = y - ce^{-2x}$. Differentiating the given equation, we get $y' = -2ce^{-2x}$. Solving the given equation for c, we obtain $c = ye^{2x}$ and so $y' = -2\left(ye^{2x}\right)e^{-2x} = -2y$. The orthogonal trajectories are the solutions of $\frac{dy}{dx} = \frac{-1}{-2y} = \frac{1}{2y}$ or $2y\,dy - dx = 0$. The solution of this differential equation is $y^2 - x = k$.

10. Let N represent the amount of the radioactive substance present at time t. Because the rate of decay is proportional to the amount of substance remaining we have $\frac{dN}{dt} = kN$ or $\frac{dN}{dt} - kN = 0$. Using the integrating factor e^{-kt} and integrating, we obtain $Ne^{kt} = C$ or $N = Ce^{-kt}$. If N_0 represents the amount when $t = 0$, the equation becomes $N = N_0e^{-kt}$. We are told that $N = 0.70$ when $t = 40$ years, and so we get $0.7N_0 = N_0e^{-40k}$ or $0.7 = e^{-40k}$ and so $k = \frac{\ln 0.7}{-40} \approx 0.008\,917$ and the desired equation is $N = N_0e^{-0.008\,917t}$. To find the half-life, we solve $e^{-008\,917t} = \frac{1}{2}$ and obtain $t \approx 77.73$ years.

11. The body satisfies the equation $\frac{dv}{dt} + \frac{0.2v}{5} = 32$ or $\frac{dv}{dt} = 32 - 0.04v$. Separating variables, we obtain $\frac{dv}{32 - 0.04v} = dt$ and integrating produces $-25\ln(32 - 0.04v) = t + C_1$ which can be written as $\ln(32 - 0.04v) = -\frac{t}{25} + C$ where $-25\ln C = C_1$. This is equivalent to $32 - 0.04v = Ce^{-t/25}$. At $t = 0$, we are given $v = 0$ so $C = 32$ which makes the equation $32 - 0.04v = 32e^{-t/25}$ or $v = \frac{3s}{0.04}\left(1 - e^{-t/25}\right) = 800\left(1 - e^{-t/25}\right)$. When $t = 6$ s, $v = 800\left(1 - e^{-6/25}\right) \approx 170.70$ ft/s.

12. If T represent the temperature of the object at some time t minutes and T_m the temperature of the surrounding medium, then according to Newton's law of cooling, we have $\frac{dT}{dt} = -k(T - T_m)$. Here $T_m = 30°\text{C}$ and so $\frac{dT}{dt} + kT = 30k$. The solution of this differential equation is $Te^{kt} = 30e^{kt} + C$ or $T = 30 + Ce^{-kt}$. Since $T = 50°\text{C}$ when $t = 0$, we get $C = 20$ and the equation becomes $T = 30 + 20e^{-kt}$. When $t = 20$, we are given $T = 45$ which gives $45 = 30 + 20e^{-20k}$ or $0.75 = e^{-20k}$. Taking the natural logarithm we get $\ln = 0.75 = -20k$ or $k = \frac{0.75}{-20} \approx 0.0144$. The desired equation is $T = 30 + 20e^{-0.0144t}$.

CHAPTER

33

Higher-Order Differential Equations

33.1 HIGHER-ORDER HOMOGENEOUS EQUATIONS WITH CONSTANT COEFFICIENTS

1. $(D^2 - 7D + 10)y = 0$ has auxiliary equation $m^2 - 7m + 10 = 0$ which factors into $(m-2)(m-5) = 0$ and so has roots $m_1 = 2$ and $m_2 = 5$. The general solution is thus $y = c_1e^{2x} + c_2e^{5x}$.

2. $(D^2 + 7D + 12)y = 0$ has auxiliary equation $m^2 + 7m + 12 = 0$ which has roots -3 and -4. The solution is $y = c_1e^{-3x} + c_2e^{-4x}$.

3. $D^2y - 8Dy + 12y = 0$ has auxiliary equation $m^2 - 8m + 12 = 0$ which has roots 2 and 6. The solution is $y = c_1e^{2x} + c_2e^{6x}$.

4. $6D^2y + Dy = y$ has auxiliary equation $6m^2 + m - 1 = 0$ which factors into $(3m - 1)(2m + 1) = 0$ and has roots $\frac{1}{3}$ and $-\frac{1}{2}$. Hence the solution is $y = c_1e^{\frac{1}{3}x} + c_2e^{-\frac{1}{2}x}$.

5. $2\frac{d^2y}{dx^2} + 9\frac{dy}{dx} - 5y = 0$ has auxiliary equation $2m^2 + 9m - 5 = 0$ which factors into $(2m-1)(m+5) = 0$. The roots are $\frac{1}{2}$ and -5 so the solution is $y = c_1e^{\frac{1}{2}x} + c_2e^{-5x}$.

6. $\frac{d^2y}{dx^2} = 9$ has auxiliary equation $m^2 - 9 = 0$. This has roots -3 and 3 so the solution is $y = c_1e^{-3x} + c_2e^{3x}$.

7. $4D^2y + 7Dy - 2y = 0$ has auxiliary equation $4m^2 - 7m - 2 = 0$. This factors into $(4m+1)(m-2) = 0$ and has roots $-\frac{1}{4}$ and 2. The solution is $y = c_1e^{-\frac{1}{4}x} + c_2e^{2x}$.

8. $4y'' + 11y' - 3y = 0$ has auxiliary equation $4m^2 + 11m - 3 = 0$ which factos into $(4m-1)(m+3) = 0$. The roots are $\frac{1}{4}$ and -3 so the solution is $y = c_1e^{\frac{1}{4}x} + c_2e^{-3x}$.

9. $y'' - 5y' + 4y = 0$ has auxiliary equation $m^2 - 5m + 4 = 0$. This has roots 1 and 4 so the solution is $y = c_1e^{x} + c_2e^{4x}$.

10. $\frac{d^2y}{dx^2} + 4\frac{dy}{dx} + 3y = 0$ has auiliary equation $m^2 + 4m + 3 = 0$ with roots -1 and -3. Hence the solution is $y = c_1e^{-x} + c_2e^{-3x}$.

11. $y'' - y = 0$ has auxiliary equaton $m^2 - 1 = 0$. The roots are 1 and -1 so the solution is $y = c_1e^{x} + c_2e^{-x}$.

12. $D^2y - Dy + 30y = 0$ has auxiliary equation $m^2 - m - 30 = 0$ with roots 6 and -5. The solution is $y = c_1e^{6x} + c_2e^{-5x}$.

13. $y'' = 7y$ has auxiliary equation $m^2 - 7 = 0$. This has roots $\sqrt{7}$ and $-\sqrt{7}$ so the solution is $y = c_1e^{\sqrt{7}x} + c_2e^{-\sqrt{7}x}$.

14. $D^2y + Dy - y = 0$ has auxiliary equation $m^2 + m - 1 = 0$. Using the quadratic formula we get the roots $\frac{-1 \pm \sqrt{5}}{2}$. Hence the solution is $y = c_1e^{\frac{-1+\sqrt{5}}{2}x} + c_2e^{\frac{-1-\sqrt{5}}{2}x}$.

15. $2\frac{d^2y}{dx^2} + \frac{dy}{dx} - y = 0$ has auxiliary equation $2m^2 + m - 1 = 0$. This factors into $(2m-1)(m+1) = 0$ and has roots $\frac{1}{2}$ and -1. Hence the solution is $y = c_1e^{x/2} + c_2e^{-x}$.

16. $D^2y + 2Dy - y = 0$ has auxiliary equation $m^2 + 2m - 1 = 0$. Using the quadratic formula we get roots $\frac{-2\pm\sqrt{8}}{2} = -1 \pm \sqrt{2}$. Hence the solution is $y = c_1e^{(-1+\sqrt{2})x} + c_2e^{(-1-\sqrt{2})x}$.

17. $D^3y - 6D^2y + 11Dy - 6y = 0$ has auxiliary equation $m^3 - 6m^2 + 11m - 6 = 0$. This factors into $(m-1)(m-2)(m-3) = 0$ and has roots 1, 2 and 3. The solution is $y = c_1e^x + c_2e^{2x} + c_3e^{3x}$.

18. $D^3y - D^2y - 4Dy + 4y = 0$ has auxiliary equation $m^3 - m^2 - 4m + 4 = 0$. Factoring by grouping we get $m^2(m-1) - 4(m-1)$ or $m^2 - 4)(m-1)$ or $(m+2)(m-2)(m-1) = 0$. The solution is $y = c_1e^x + c_2e^{2x} + c_3e^{-2x}$

19. $y''' + 6y'' + 11y' + 6y = 0$ has auxiliary equation $m^3 + 6m^2 + 11m + 6 = 0$. This has only negative roots. With synthetic division we can verify that the roots are -1, -2, and -3. Hence the solution is $y =_1 e^{-x} + c_2e^{-2x} + c_3e^{-3x}$.

20. $D^3y - D^2y - 17Dy = 15y$ has auxiliary equation $m^3 - m^2 - 17m - 15 = 0$. Using synthetic division we see 5 is a root. The other two roots are -3 and -1. Hence the solution is $y = c_1e^{-x} + c_2e^{-3x} + c_3e^{5x}$.

21. $D^2y + 2Dy - 15y = 0$ has auxiliary equation $m^2 + 2m - 15 = 0$. This has roots 3 and -5 and so the general solution is $y = c_1e^{3x} + c_2e^{-5x}$. When $x = 0$ we have $y = 2$ so $2 = c_1 + c_2$ or $c_1 = 2 - c_2$ finding $y' = 3c_1, e^{3x} - 5c_2e^{-5x}$ we also have that $y' = 6$ when $x = 0$. Hence $6 = 3c_1 - 5c_2$. Substituting $2 - c_2$ for c_1 we get $6 = 3(2 - c_2) - 5c_2$ or $6 = 6 - 3c_2 - 5c_2$. Hence $c_2 = 0$ and $c_1 = 2$. The particular solution is $y = 2e^{3x}$.

22. $3D^2y - 14Dy + 8y = 0$ has auxiliary equation $3m^2 - 14m + 8 = 0$. This factors into $(3m-2)(m-4) = 0$. Hence the general solution is $y = c_1e^{4x} + c_2e^{2x/3}$. When $x = 0$, $y = 3$ so $3 = c_1 + c_2$ or $c_1 = 3 - c_2$. Also, $y' = 4c_1e^{4x} + \frac{2}{3}c_2e^{2x/3}$ and $y' = 12$ when $x = 0$. This gives $12 = 4c_1 + \frac{2}{3}c_2$. Substituting this for c_1 we have $12 = 4(3 - c_2) + \frac{2}{3}c_2$ or $12 = 12 - 3\frac{2}{3}c_2$. Hence, $c_2 = 0$ and $c_1 = 3$. The particular solution is $y = 3e^{4x}$.

33.2 AUXILIARY EQUATIONS WITH REPEATED OR COMPLEX ROOTS

1. $D^2y + 2Dy + y = 0$ has auxiliary equation $m^2 + 2m + 1 = 0$ This has -1 as a double root. Hence the general solution is $y = c_1e^{-x} + c_2xe^{-x}$ or $y = (c_1 + c_2x)e^{-x}$.

2. $(D^2 - 2D + 1)y = 0$ has auxiliary equation $m^2 - 2m + 1 = 0$. This has 1 as a double root. Hence the general solution is $y = c_1e^x + c_2xe^x$ or $y = (c_1 + c_2x)e^x$.

3. $D^2y - 4Dy + 4y = 0$ has auxiliary equation $m^2 - 4m + 4 = 0$. This has 2 as a double root and hence the solution is $y = (c_1 + c_2x)e^{2x}$.

4. $y'' - 10y' + 25y = 0$ has auxiliary equation $m^2 - 10m + 25 = 0$. 5 is a double root and the solution is $y = (c_1 + c_2x)e^{5x}$.

5. $9y'' - 6y' + y = 0$ has auxiliary equation $9m^2 - 6m + 1 = 0$. $\frac{1}{3}$ is a double roots and the solution is $y = (c_1 + c_2x)e^{x/3}$.

6. $16'' + 8y' + y = 0$ has auxiliary equation $16m^2 + 8m + 1 = 0$. $-\frac{1}{4}$ is a double root and so the solution is $y = (c_1 + xc_2)e^{-x/4}$.

7. $4y'' + 12y' + 9y = 0$ has auxiliary equation $4m^3 + 12m + 9 = 0$. This has $-\frac{3}{2}$ as a double root and hence the solution is $y = (c_1 + c_2x)e^{-3x/2}$

8. $9y'' - 12y' + 4y = 0$ has $9m^2 - 12m + 4 = 0$ as its auxiliary equation. This has $\frac{2}{3}$ as a double root and hence the solution is $y = (c_1 + xc_2)e^{2x/3}$.

9. $y'' + 4y' + 5y = 0$ has $m^2 + 4m + 5 = 0$ as its auxiliary equation. By the quadratic formula, this has

$\frac{-4\pm\sqrt{16-20}}{2} = -2 \pm j$ as its complex roots. Hence the solution is $y = e^{-2x}(c_1 \cos x + c_2 \sin x)$.

10. $y'' + 4y = 0$ has auxiliary equation $m^2 + 4 = 0$. This has complex roots $2j$ and $-2j$ so $a = 0$ and $b = 2$. The solution is $y = c_1 \cos 2x + c_2 \sin 2x$.

11. $D^2y + 9Dy = 0$ has auxiliary equation $m^2 + 9m = 0$. This has 0 and -9 for its roots and hence the solution is $y = c_1 + c_2e^{-9x}$.

12. $D^2y = 0$ has auxiliary equaton $m^3 = 0$. This has 0 as a triple roots so the solution is $y = c_1 + c_2x + c_3x^2$.

13. $D^3y = y$ has auxiliary equation $m^3 - 1 = 0$. 1 is real solution and synthetic division yields $m^2 + m + 1$ as the depressed quotient. The other two roots are $-\frac{1}{2} \pm \frac{\sqrt{3}}{2}j$. Hence the solution is $y = c_1e^x + e^{-x/2}(c_2 \cos \frac{\sqrt{3}}{2}x + c_3 \sin \frac{\sqrt{3}}{2}x)$.

14. $\frac{d^3y}{dx^3} + y = 0$ has auxiliary equation $m^3 + 1 = 0$. We can determine that -1 is a real root and, using synthetic division, we get $m^2 - m + 1$ as the depressed quotient. This has complex roots $\frac{1}{2} \pm \frac{\sqrt{3}}{2}j$. Hence the solutions are $y = c_1e^{-x} + e^{x/2}(c_2 \cos \frac{\sqrt{3}}{2}x + c_3 \sin \frac{\sqrt{3}}{2}x)$.

15. $\frac{d^3y}{dx^3} + 8y = 0$ has auxiliary equation $m^3 + 8 = 0$. We can see that -2 is one root. Using synthetic division we get the depresssed quotient $m^2 - 2m + 4$ which has roots $1 \pm \sqrt{3}j$ Hence the solution is $y = c_1e^{-2x} + e^x(c_2 \cos \sqrt{3}x + c_3 \sin \sqrt{3}x)$.

16. $\frac{d^3y}{dx^3} - 27y = 0$ has auxiliary equation $m^3 - 27 = 0$. This equation factors into $(m-3)(m^2+3m+9) = 0$ so the roots are -3, $\frac{-3}{2} \pm \frac{3\sqrt{3}}{2}j$. The solution is $y = c_1e^{-3x} + e^{-3x/2}(c_2 \cos \frac{3\sqrt{3}}{2}x + c_3 \sin \frac{3\sqrt{3}}{2}x)$.

17. $y''' - 6y'' + 11y' - 6y = 0$ has auxiliary equation $m^3 - 6m^2 + 11m - 6 = 0$. Using the rational root theorem and synthetic division we find the roots are 1, 2 and 3. Hence the solution is $y = c_1e^x + c_2e^{2x} + c_3e^{3x}$.

18. $D^4y - 9D^2y + 20y = 0$ has auxiliary equation $m^4 - 9m^2 + 20 = 0$. This factors as $(m^2-4)(m^2-5) = 0$ and the roots are $2, -2, \sqrt{5}$ and $-\sqrt{5}$. Hence the solution is $y = c_1e^{2x} + c_2e^{-2x} + c_3e^{\sqrt{5}x} + c_4e^{-\sqrt{5}x}$.

19. $D^4y + 8D^3y + 24D^2y + 32Dy + 16y = 0$ has auxiliary equation $m^4 + 8m^3 + 24m^2 + 32m + 16 = 0$. This has -2 as a root four times. Hence the solution is $y = (c_1 + c_2x + c_3x^2 + c_4x^3)e^{-2x}$.

20. $y''' - y'' + y' - y = 0$ has auxiliary equation $m^3 - m^2 + m - 1 = 0$. The real root is 1 and using synthetic division we get the depressed quotient $m^2 + 1$ which as complex roots $\pm j$. Hence the solution is $y = c_1e^x + c_2 \cos x + c_3 \sin x$.

21. $(D-1)^2(D+2)^3y = 0$ has auxiliary equations with 1 as a double root and -2 as a triple root. Hence the solution is $y = (c_1 + c_2x)e^x + (c_3 + c_4x + c_5x^2)e^{-2x}$.

22. $(D-2)^4(D+5)y = 0$ has auxiliary equation with roots 2 of multiplicity four and -5. The solution ia $y = (c_1 + c_2x + c_3x^2 + c_4x^3)e^{2x} + c_5e^{-5x}$.

23. $(D-3)^2(D^2 - 6D - 9)y = 0$ has auxiliary equation with 3 as a double root and irrational roots $\frac{6\pm\sqrt{36+36}}{2} = 3 \pm 3\sqrt{2}$. Hence the solution is $y = (c_1 + c_2x)e^{3x} + c_3e^{(3+3\sqrt{2})x} + c_4e^{(3-3\sqrt{2})x}$.

24. $(D+1)^2(D^2 + 4D + 9)y = 0$ has auxiliary equation with -1 as a double root and complex roots $\frac{-4 \pm \sqrt{16-36}}{2} = -2 \pm \sqrt{5}j$. Hence the solution is given by $y = (c_1 + c_2x)e^{-x} + e^{-2x}(c_3 \cos \sqrt{5}x + c_4 \sin \sqrt{5}x)$.

25. $(3D^2 - 2D + 1)y = 0$ has $3m^2 - 2m + 1 = 0$ as its auxiliary equation. This equation has complex roots $\frac{2 \pm \sqrt{4-12}}{6} = \frac{1}{3} \pm \frac{\sqrt{2}}{3}j$. Hence the solution is $y = e^{x/3}(c_1 \cos \frac{\sqrt{2}}{3}x + c_2 \sin \frac{\sqrt{2}}{3}x)$.

26. $\frac{d^2y}{dx^2} - 2\frac{dy}{dx} + 3y = 0$ has auxiliary equation $m^2 - 2m + 3$. This has complex roots $\frac{2 \pm \sqrt{4-12}}{2} = 1 \pm \sqrt{2}j$. Hence the solution is $y = e^x(c_1 \cos \sqrt{2}x + c_2 \sin \sqrt{2}x)$.

27. $y'' - 2y' + 5y = 0$ has auxiliary equation $m^2 - 2m + 5 = 0$. This has complex roots $1 \pm 2j$. Hence the general solution is $y = e^x(c_1 \cos 2x + c_2 \sin 2x)$. When $x = 0$, y is 4 and so $c_1 = 4$. When $x = 0$, $y' = 7$ so $y' = e^x(c_1 \cos 2x + c_2 \sin 2x) + e^x(-2c_1 \sin 2x + 2c_2 \cos 2x) = e^x[(c_1+2c_2)\cos 2x + (c_2 - 2c_1)\sin 2x]$. Hence $c_1 + 2c_2 = 7$ or $4 + 2c_2 = 7$ or $c_2 = \frac{3}{2}$. The particular solution is $y = e^x(4\cos 2x + \frac{3}{2}\sin 2x)$.

28. $y'' + 8y' + 16y = 0$ has auxiliary equation $m^2 + 8m + 16 = 0$. We see that -4 is a double root and the general solution is $y = (c_1 + c_2x)e^{-4x}$. When $x = 0$, $y = 4$ and so $c_1 = 4$. $y' = c_2e^{-4x} - 4(c_1 + c_2x)e^{-4x}$. When $x = 0$, $y' = c_2 - 4c_1 = -6$ so $c_2 = -6 + 16 = 10$. The particular soluation is $y = (4 + 10x)e^{-4}$.

29. $(4D^2 - 12D + 9)y = 0$ has auxiliary equation with double root $\frac{3}{2}$. Hence the general solution is $y = (c_1 + c_2x)e^{3x/2}$. When $x = 0$, $y = 1$ so $c_1 = 1$ or $y = (1 + c_2x)e^{3x/2}$. When $x = 1$, $y = 0$ so $0 = (1 + c_2)e^{3/2}$, which means that $c_2 = -1$. Hence the particular solution is $y = (1 - x)e^{\frac{3}{2}x}$.

30. $(D^2 - 4D + 5)y = 0$ has auxiliary equation with complex roots $\frac{4 \pm \sqrt{-4}}{2} = 2 \pm j$. Hence the general solution is $y = e^{2x}(c_1 \cos x + c_2 \sin x)$. When $x = 0$, $y = 1$ yields $c_1 = 1$, so $y = e^{2x}(\cos x + c_2 \sin x)$. Now when $x = \frac{\pi}{2}$, $y = 3e^{\pi}$ so $3e^{\pi} = e^{\pi}(\cos\frac{\pi}{2} + c_2 \sin\frac{\pi}{2})$ and so $c_2 = 3$. Hence the particular solution is $y = e^{2x}(\cos x + 3\sin x)$.

33.3 SOLUTIONS OF NONHOMOGENEOUS EQUATIONS

1. $(D^2 - 10D + 25)y = 4$. To find y_c we solve the auxiliary equation $m^2 - 10m + 25 = 0$. This has double root 5. Hence $y_c = (c_1 + xc_2)e^{5x}$. For y_p, we use Case 1 and $y_p = A_0, y_p' = 0$. Substituting into the original equation we get $25A_0 = 4$ or $A_0 = \frac{4}{25}$. Hence the solution is $y = (c_1 + c_2x)e^{5x} + \frac{4}{25}$.

2. $(D^2 - 10D + 25)y = 4x^2$. Using the auxiliary equation $m^2 - 10m + 25 = 0$, we get a double root 5. Hence $y_c = (c_1 + c_2x)e^{5x}$. To find y_p we use Case 1 and set $y_p = A_2x^2 + A_1x + A_0$, $y' = 2A_2x + A_1$, and $y'' = 2A_2$. Substituting into the original equation we get $2A_2 - 10(2A_2 + A_1) + 25(A_2x^2 + A_1x + A_0) = 4x^2$ or $25A_2x^2 + (25A_1 - 20A_2)x + 25A_0 - 10A_1 + 2A_2 = 4x^2$. Hence $A_2 = \frac{4}{25}$, $A_1 = \frac{20A_2}{25} = \frac{80}{625} = \frac{16}{125}$, and $A_0 = \frac{10A_1 - 2A_2}{25} = \frac{\frac{160}{125} - \frac{8}{25}}{25} = \frac{24}{625}$. The solution is $= (c_1 + c_2x)e^{5x} + \frac{4}{25}x^2 + \frac{16}{125}x + \frac{24}{625}$

3. $(D^2 - 10D + 25)y = e^{3x}$ has the same complentary solution as problems 1 and 2, namely $y_c = (c_1 + c_2x)e^{5x}$. To find the particular solution y_p, we use Case 2 with $y_p = A_0e^{3x}$, $y' = 3A_0e^{3x}$, and $y'' = 9A_0e^{3x}$. Substituting into the original equation we get $9A_0e^{3x} - 10 \cdot 3A_0e^{3x} + 25A_0e^{3x} = 3^{3x}$ or $4A_0e^{3x} = e^{3x}$. Hence $A_0 = \frac{1}{4}$. The solution is $y = (c_1 + c_2x)e^{5x} + \frac{1}{4}e^{3x}$.

4. $(D^2 - 10D + 25)y = 2xe^{3x}$. Again $y_c = (c_1 + c_2x)e^{5x}$. To find y_p we use Case 2 with $y_p = e^{3x}(A_1x + A_0)$, $y_p' = 3e^{3x}(A_1x + A_0) + A_1e^{3x} = 3A_1xe^{3x} + (A_1 + 3A_0)e^{3x}$, and $y_p'' = 9A_1xe^{3x} + 3A_1e^{3x} + 3(A_1 + 3A_0)e^{3x}$. Substituting into the original equation we get $9A_1xe^{3x} + (6A_1 + 3A_0)e^{3x} - 30A_1xe^{3x} - (10A_1 + 30A_0)e^{3x} + 25(A_1x + A_0)e^{3x} = 2xe^{3x}$. Hence $9A_1 - 30A_1 + 25A_1 = 2$ and $6A_1 + 9A_0 - 10A_1 - 30A_0 + 25A_0 = 0$. So $A_1 = \frac{1}{2}$ and $4A_0 = 4A_1 = 2$, so $A_0 = \frac{1}{2}$. The solution is $y = (c_1 + c_2x)e^{5x} + (\frac{1}{2}x + \frac{1}{2})e^{3x}$

5. $(D^2 - 10D + 25)y = 10 + e^x$. Again $y_c = (c_1 + c_2x)e^{5x}$. The y_p has two parts. Case 1 $y_{p1} = A_0$ and we get $25A_0 = 10$ or $A_0 = \frac{2}{5}$. Case 2, $y_{p2} = B_0e^x$, $y_{p2}' = B_0e^x$, $y_{p2}'' = B_0e^x$. Substituting we get $16B_0e^x = e^x$ or $B_0 = \frac{1}{16}$. The solution is $y = (c_1 + c_2x)e^{5x} + \frac{2}{5} + \frac{1}{16}e^x$.

6. $(D^2 - 10D + 25)y = 3\sin 2x$. This is Case 3 so we set $y_p = A\sin 2x + B\cos 2x$. $y_p' = 2A\cos 2x - 2B\sin 2x$, $y_p'' = -4A\sin 2x - 4B\cos 2x$. Substituting into the orginial equation and collecting like terms we get $(-4A + 20B + 25A)\sin 2x + (-4B - 20A + 25B)\cos 2x = 3\sin 2x$. Hence $21A + 20B = 3$ and

$-20A + 21B = 0$. Eliminating the B term we get $841A = 63$ or $A = \frac{63}{841}$. Also, $B = \frac{20A}{21} = \frac{60}{841}$. The solution is $y = (C_1 + C_2x)e^{5x} + \frac{63}{841}\sin 2x + \frac{60}{841}\cos 2x$

7. $(D^2 - 10D + 25)y = 29\sin 2x$. Again $y_c = (C_1 + C_2x)e^{5x}$. Using Case 3 we set $y_p = A\sin 2x + B\cos 2x$, and so $y_p' = 2A\cos 2x - 2B\sin 2x$, and $y_p'' = -4A\sin 2x - 4B\cos 2x$. Substituting into the original equation we get $(-4A+20B+25A)\sin 2x + (-4B - 20A + 25B)\cos 2x = 29\sin 2x$. Hence $21A + 20B = 29$ and $-20A + 21B = 0$. Eliminating B we get $(441 + 400)A = 29 \cdot 21$ or $A = \frac{609}{841} = \frac{21}{29}$. Also, $B = \frac{20A}{21} = \frac{20}{21} \cdot \frac{21}{29} = \frac{20}{29}$. The solution is $y = (C_1 + C_2x)e^{5x} + \frac{21}{29}\sin 2x + \frac{20}{29}\cos 2x$.

8. $(D^2 - 9)y = 3\cos x$ has complementary solution $y_c = c_1e^{3x} + c_2e^{-3x}$. To find y_p we use Case 4 with $y_p = A\sin x + B\cos x, y_p' = A\cos x - B\sin x$ and $y_p'' = -A\sin x - B\cos x$. Substituting into the original equation we get $-10A\sin x - 10B\cos x = 3\cos x$. Solving we get $A = 0$ and $B = -\frac{3}{10}$. The solution is $y = c_1e^{3x} + c_2e^{-3x} - \frac{3}{10}\cos x$.

9. $(D^2 - 4)y = x^2e^x - 3x$ has complementary auxiliary equation with roots ± 2. Hence $y_c = c_1e^{2x} + c_2e^{-2x}$. y_p will have 2 parts. x^2e^x is Case 2 so we set $y_{p1} = e^x(A_2x^2 + A_1x + A_0)$, with $y_{p1}' = e^x(A_2x^2 + A_1x + A_0) + e^x(2A_2x + A_1) = A_2e^xx^2 + (A_1 + 2A_2)e^xx + (A_0 + A_1)e^x$ and $y_{p1}'' = A_2e^xx^2 + 2xA_2e^x + (A_1 + 2A_2)e^xx + (A_1 + 2A_2)e^x + (A_0 + A_1)e^x = A_2e^xx^2 + (2A_2 + A_1 + 2A_2)e^xx + (2A_2 + 2A_1 + A_0)e^x$. Substituting into the original equation yields $(A_2 - 4A_2)e^xx^2 + (4A_2 + A_1 - 4A_1)e^xx + (2A_2 + 2A_1 + A_0 - 4A_0)e^x = x^2e^x$. Hence $-3A_2 = 1$ or $A_2 = -\frac{1}{3}$. Next, $-\frac{4}{3} - 3A_1 = 0$ or $A_1 = -\frac{4}{9}$ and finally $3A_0 = 2A_2 + 2A_1$ and so, $A_0 = \dfrac{-\frac{2}{3} - \frac{8}{9}}{3} = \dfrac{-14}{27}$. For the second part of y_p, we have $y_{p2} = B_1x + B_0$, $y_{p2}' = B_1$ and $y_{p2}'' = 0$. Hence $-4B_1x = -3x$ and $B_1 = \frac{3}{4}$. $B_0 = 0$. The solution is $y = c_1e^{2x} + c_2e^{-2x} - \left(\frac{1}{3}x^2 + \frac{4}{9}x + \frac{14}{27}\right)e^x + \frac{3}{4}x$.

10. $(D^2 + 4)y = 6x + 3$. The auxiliary equation for y_c has complex roots $\pm 2j$. Hence $y_c = c_1\cos 2x + c_2\sin 2x$. For y_p we use Case 1 with $y_p = A_1x + A_0$, $y_p' = A_1$, and $y_p'' = 0$. Substituting into the original equaton we get $4(A_1x + A_0) = 6x + 3$. So $4A_1 = 6$ gives $A_1 = \frac{3}{2}$ and $4A_0 = 3$ gives $A_0 = \frac{3}{4}$. The solution is $y = c_1\cos 2x + c_2\sin 2x + \frac{3}{2}x + \frac{3}{4}$.

11. $(D^2 + 9)y = 4\cos x + 2\sin x$. y_c has auxiliary equation with roots $\pm 3j$. Hence $y_c = c_1\cos 3x + c_2\sin 3x$. y_p fits Cases 3 and 4 so we set $y_p = A\cos x + B\sin x, y_p' = -A\sin x + B\cos x$ and $y_p'' = -A\cos x - B\sin x$. Substituting into the original equation we get $(-A\cos x - B\sin x) + 9(A\cos x + B\sin x) = 4\cos x + 2\sin x$ so $8A = 4$ and $8B = 2$ giving $A = \frac{1}{2}$ and $B = \frac{1}{4}$. The solution is $y = c_1\cos 3x + c_2\sin 3x + \frac{1}{2}\cos x + \frac{1}{4}\sin x$.

12. $y'' - 5y' + 6y = 9x + 2e^x$. y_c has auxiliary equation with roots 2 and 3. Hence $y_c = c_1e^{2x} + c_2e^{3x}$. y_p fits Case 1 and 2. We set $y_{p1} = A_1x + A_0$, with $y_{p1}' = A_1$, and $y_{p1}'' = 0$. Substituting we get $-5(A_1) + 6(A_1x + A_0) = 9x$. Hence $6A_1 = 9$, and so $A_1 = \frac{3}{2}$ and $6A_0 - 5A_1 = 0$, which means that $A_0 = \frac{5A_1}{6} = \frac{5}{4}$. Now we set $y_{p2} = Be^x$ and $y_{p2}' = y_{p2}'' = Be^x$. Substituting into the original equation and collection like terms, we obtain $2Be^x = 2e^x$ so $B = 1$. Putting these all together we get $y = c_1e^{2x} + c_2e^{3x} + \frac{3}{2}x + \frac{5}{4} + e^2$.

13. $y'' + y = 2e^{3x}$. y_c has auxililary equation with roots $\pm j$. Hence $y_c = c_1\cos x + c_2\sin x$. y_p fits Case 2 so we set $y_p = Ae^{3x}$, $y_p' = 3Ae^{3x}$, and $y_p'' = 9Ae^{3x}$. Substituting we get $9Ae^{3x} + Ae^{3x} = 2e^{3x}$ or $10A = 2 \Rightarrow A = \frac{1}{5}$. The solution is $y = c_1\cos x + c_2\sin x + \frac{1}{5}e^{3x}$.

14. $(D^2 + 2D + 1)y = 4\sin 2x$. The auxiliary equation $m^2 + 2m + 1 = 0$ has -1 for a double roots. Hence $y_c = (c_1 + c_2x)e^{-x}$. Now, y_p fits Case 3, so we set $y_p = A\sin 2x + B\cos 2x$, with $y_p' = 2A\cos 2x - 2B\sin 2x$, and $y_p'' = -4A\sin 2x - 4B\cos 2x$. Substituting we get $(-4A\sin 2x - 4B\cos 2x) + 2(2A\cos 2x - 2B\sin 2x) + (A\sin 2x + B\cos 2x) = 4\sin 2x$. Hence $-4A - 4B + A = 4$ and $-4B = 4A + B = 0$. These simplify to $-3A - 4B = 4$ and $A - 3B = 0$ with $A = -\frac{12}{25}$ and $B = -\frac{16}{25}$. The solution is therefore $y = (c_1 + c_2x)e^{-x} - \frac{12}{25}\sin 2x - \frac{16}{25}\cos 2x$.

15. $y'' - 4y = 8x^2$. First, $y_c = c_1e^{2x} + c_2e^{-2x}$. Now set $y_p = A_2x^2 + A_1x + A_0$, $y_p' = 2A_2x + A_1$, and $y_p'' = 2A_2$. Substituting we get $2A_2 - 4(A_2x^2 + A_1 + A_0) = 8x^2$.

Hence $-4A_2 = 8$ and $A_2 = -2$. Also, $A_1 = 0$. Finally, $2A_2 - 4A_0 = 0$, so $A_0 = \frac{1}{2}A_2 = -1$. The solution is $y = c_1e^{2x} + c_2e^{-2x} - 2x^2 - 1$.

16. $y'' + 4y' + 5y = e^{-x} + 10x$. The auxiliary equation for y_c is $m^2 + 4m + 5$ and has complex roots $-2 \pm j$. Hence $y_c = e^{-2x}(c_1 \cos x + c_2 \sin x)$. y_p fits Cases 1 and 2. Set $y_{p1} = -Ae^{-x}$, $y'_{p1} = -Ae^{-x}$, and $y''_{p1} = Ae^{-x}$. Substituting we get $Ae^{-x} - 4Ae^{-x} + 5Ae^{-x} = e^{-x}$ or $2A = 1$ and so $A = \frac{1}{2}$. Thus, $y_{p1} = \frac{1}{2}e^{-x}$. Now set $y_{p2} = B_1x + B_0$, with $y'_{p2} = B_1$ and $y''_{p2} = 0$. Substituting we get $4B_1 + 5(B_1x + B_0) = 10x$. Hence $5B_1 = 10 \Rightarrow B_1 = 2$ and $4B_1 + 5B_0 = 0$ so $B_0 = \frac{4}{5}B_1 = -\frac{8}{5}$. Hence $y_{p2} = 2x - \frac{8}{5}$. Putting y_c, y_{p1}, and y_{p2} together we get $y = e^{-2x}(c_1 \cos x + c_2 \sin x) + \frac{1}{2}e^{-x} + 2x - \frac{8}{5}$.

17. $D^2y + y = 4 + \cos 2x$. $y_c = c_1 \cos x + c_2 \sin x$. Set $y_{p1} = A$ and we get $A = 4$ so $y_{p1} = 4$. Now set $y_{p2} = A \sin 2x + B \cos 2x$, $y'_{p2} = 2A \cos 2x - 2B \sin 2x$, and $y''_{p2} = -4A \sin x - 4B \cos 2x$. Hence, $-4A + A = 0$ and $-4B + B = 1$. Thus, we see that $A = 0$ and $B = -\frac{1}{3}$. The solution is $y = c_1 \cos x + c_2 \sin x - \frac{1}{3} \cos 2x + 4$.

18. $3D^2y + 2Dy - y = 4 + 2x + 6x^2$. The auxiliary equation for y_c is $3m^2 + 2m - 1 = 0$ which factors into $(3m - 1)(m + 1) = 0$ so it has roots $\frac{1}{3}$ and -1. Hence $y_c = c_1e^{-x} + c_2e^{x/3}$. Now we set $y_p = Ax^2 + Bx + C$, with $y'_p = 2Ax + B$, and $y''_p = 2A$. Substituting we get $3 \cdot 2A + 2(2Ax + B) - (Ax^2 + Bx + C) = 4 + 2x + 6x^2$. Collecting like terms we get $-Ax^2 + (4A - B)x + (6A + 2B - C) = 4 + 2x + 6x^2$. Hence $-A = 6$, or $A = -6$. We also have $4A - B = 2$ or $B = 4A - 2 = -26$ and $6A + 2B - C = 4$ or $C = 6A + 2B - 4 = -36 - 52 - 4 = -92$. So $y_p = -6x^2 - 26x - 92$. The solution is $y = y_c + y_p = c_1e^{-x} + c - 2e^{x/3} - 6x^2 - 26x - 92$.

19. $3D^2y + 2Dy + y = 4 + 2x + 6x^2$. Here the auxiliary equation is $3m^2 + 2m + 1 = 0$ has complex roots $\frac{-1}{3} \pm \frac{\sqrt{2}}{3}j$ so $y_c = e^{-x/3}\left(c_1 \cos \frac{\sqrt{2}}{3}x + c_2 \sin \frac{\sqrt{2}}{3}x\right)$. As in problem 18 set $y_p = Ax^2 + Bx + C$ with $y'_p = 2Ax + B$ and $y''_p = 2A$. Now substituting, we get $3 \cdot 2A + 2(2Ax + B) + (Ax^2 + Bx + C) = 4 + 2x + 6x^2$. Collecting like terms we obtain $Ax^2 + (4A + B)x + (6A + 2B + C) = 4 + 2x + 6x^2$. So $A = 6$, $4A + B = 2$, and so $B = 2 - 4A = 2 - 24 = -22$. We also obtain $6A + 2B + C = 4$ or $C = 4 - 6A - 2B = 4 - 36 + 44 = 12$. The solution is $y = e^{-x/3}\left(c_1 \cos \frac{\sqrt{2}}{3}x + c_2 \sin \frac{\sqrt{2}}{3}x\right) + 6x^2 - 22x + 12$.

20. $y'' + 9y' - y = x^2 + 6e^{2x}$. The auxiliary equation is $m^2 + 9m - 1 = 0$ and the quadratic formula yields the roots $\frac{-9\pm\sqrt{85}}{2}$ which are real. Hence $y_c = c_1e^{\frac{-9+\sqrt{85}}{2}x} + c_2e^{\frac{-9-\sqrt{85}}{2}x}$. Now set $y_{p1} = Ax^2 + Bx + C$, $y'_{p1} = 2Ax + B$ and $y''_{p2} = 2A$. Substituting we get $2A + 9(2Ax + B) - (Ax^2 + Bx + C) = x^2$. Collecting like terms $-Ax^2 + (18A - B)x + (2A + 9B - C) = x^2$. Hence $-A = 1$ or $A = -1$, $18A - B = 0$, and so $B = -18$ and $2A + 9B - C = 0$, which means that $C = -2 - 162 = -164$. So $y_{p1} = -x^2 - 18x - 164$. Now set $y_{p2} = Ae^{2x}$, $y'_{p2} = 2Ae^{2x}$, $y''_{p2} = 4Ae^{2x}$. Substituting we have $(4A + 9 \cdot 2A - A)e^{2x} = 6x^{2x}$ or $21A = 6$, or $A = \frac{6}{21} = \frac{2}{7}$. So $y_{p2} = \frac{2}{7}e^{2x}$. The answer is $y' = c_1e^{\frac{-9+\sqrt{85}}{2}x} + c_2^{\frac{-9-\sqrt{85}}{2}x} + \frac{2}{7}e^{2x} - x^2 - 18x - 164$.

21. $y'' - 2y' + y = x^2 - 1$. The auxiliary equation $m^2 - 2m + 1 = 0$ has 1 as a double root. Hence $y_c = (c_1 + c_2x)e^x$. Now set $y_p = Ax^2 + Bx + C$, $y'_p = 2Ax + B$, and $y''_p = 2A$. Substituting we get $2A - 2(2Ax + B) + Ax^2 + Bx + C = x^2 - 1$. Collecting like terms we have $Ax^2 + (-4A + B)x + (2A - 2B + C) = x^2 - 1$. Hence, $A = 1$, $-4A + B = 0$, and $2A - 2B + C = -1$. $B = 4A = 4$, and $C = -1 + 2B - 2A = -1 + 8 - 2 = 5$. Thus, $y_p = x^2 + 4x + 5$. Now to find c_1 and c_2 of y_c when $x = 0$, $y = (c_1 + 0)1 + 5 = 7$ and so $c_1 = 2$. $y' = (2 + c_2x)e^x + c_2e^x + 2x + 4$ any when $x = 0$, $y' = 2 + c_2 + 4 = 15$, or $c_2 = 9$. Hence the particular solution is $y = (2 + 9x)e^x + x^2 + 4x + 5$.

22. $y'' - 2y' + y = 10$. Here $y_c = (c_1 + c_2x)e^x$. Also, $y_p = A$, $y'_p = 0$, so $A = 10$. Thus, $y = (c_1 + c_2x)e^x + 10$. When $x = 0$, $y = c_1 + 10 = 20$, so $c_1 = 10$. Also $y' = (10 + c_2x)e^x + c_2e^x$. When $x = 0$, $y' = 10 + c_2 = 5$, and so $c_2 = -5$. The particular solution is $y = (10 - 5x)e^x + 10$.

23. $y'' - y' - 2y = \sin 2x$. $y_c = c_1e^{2x} + c_2e^{-x}$. Now set $y_p = A\sin 2x + B\cos 2x$, with $y_p' = 2A\cos 2x - 2B\sin 2x$ and $y_p'' = -4A\sin 2x - 4B\cos 2x$. Substituting we obtain the equation $(-4A\sin 2x - 4B\cos 2x) - (2A\cos 2x - 2B\sin 2x) - 2(A\sin 2x + B\cos 2x) = \sin 2x$. This yields $-4A + 2B - 2A = 1$ or $-6A + 2B = 1$ and $-4B - 2A - 2B = 0$ or $-2A - 6B = 0$. Solving we get $A = -\frac{3}{20}, B = \frac{1}{20}$. Hence $y = c_1e^{2x} + c_2e^{-x} - \frac{3}{20}\sin 2x + \frac{1}{20}\cos 2x$. When $x = 0$, $y = c_1 + c_2 + \frac{1}{20} = 1$, or $c_1 + c_2 = \frac{19}{20}$. Differentiating, we obtain $y' = 2c_1e^{2x} - c_2e^{-x} - \frac{6}{20}\cos 2x - \frac{1}{10}\sin 2x$, and when $x = 0$, we know that $y' = 2c_1 - c_2 - \frac{6}{20} = \frac{7}{4}$ or $2c_1 - c_2 = \frac{41}{20}$. So, $3c_1 = \frac{60}{20}$ and $c_1 = 1$; also $c_2 = \frac{19}{20} - 1 = -\frac{1}{20}$. The particular solution is $y = e^{2x} + \frac{1}{20}e^{-x} - \frac{3}{20}\sin 2x + \frac{1}{20}\cos 2x$.

24. $y'' - y' - 2y = e^{3x}$, $y_c = c_1e^{2x} + c_2e^{-x}$. Now set $y_p = Ae^{3x}$, with $y_p' = 3Ae^{3x}$, and $y_p'' = 9Ae^{3x}$. Substituting we get $9Ae^{3x} - 3Ae^{3x} - 2\cdot Ae^{3x} = e^{3x}$ so $9A - 3A - 2A = 1$ and $A = \frac{1}{4}$. Hence the general soution is $y = c_1e^{2x} + c_2e^{-x} + \frac{1}{4}e^{3x}$. When $x = 0$, $y = c_1 + c_2 + \frac{1}{4} = 2$ and so, $c_1 + c_2 = \frac{7}{4}$. Differentiating, we obtain $y' = 2c_1e^{2x} - c_2e^{-x} + \frac{3}{4}e^{3x}$. When $x = 0$, $y' = 2c_1 - c_2 + \frac{3}{4} = 11$, so $2c_1 - c_2 = \frac{41}{4}$. This gives $3c_1 = \frac{48}{4} = 12$ or $c_1 = 4$. Thus, $c_2 = \frac{7}{4} - 4 = \frac{-9}{4}$. The particular solution is $y = 4e^{2x} - \frac{9}{4}e^{-x} + \frac{1}{4}e^{3x}$.

33.4 APPLICATIONS

1. Since $F = kx$, we have $k = \dfrac{F}{x} = \dfrac{20 \text{ lb}}{5 \text{ in.}} = 4$ lb/in. $= 48$ lb/ft.

2. Since $F = kx$, we have $k = \dfrac{F}{x} = \dfrac{15 \text{ lb}}{3 \text{ in.}} = 5$ lb/in. $= 60$ lb/ft.

3. Here $F = mg$ and $F = kx$ or $k = \dfrac{F}{x} = \dfrac{mg}{x} = \dfrac{5 \times 98 \text{ N}}{0.2 \text{ m}} = 245$ N/m.

4. $k = \dfrac{mg}{x} = \dfrac{8 \times 98}{0.4} = 176$ N/m

5. We know that $m = \frac{20}{32} = 0.625$ Slugs, $\ell = 2$ and $k = 48$, $2b = \dfrac{\ell}{m} = 3.2$, and $\omega^2 = \dfrac{k}{m} = \dfrac{48}{0.625} = 76.8$. So the auxiliary equation is $m^2 + 3.2m + 76.8 = 0$. This has solutions $-1.6 \pm \sqrt{1.6^2 - 76.8} = -1.6 \pm \sqrt{75.2}j$. Hence the general solution is $x = e^{-1.6t}(c_1\cos\sqrt{75.2}t + c_2\sin\sqrt{75.2}t)$. Since $x(0) = 0$, $c_1 = 0$, and $x'(t) = -1.6c_2e^{-1.6t}\sin\sqrt{75.2}t + c_2\sqrt{75.2}e^{-1.6t}\cos\sqrt{75.2}t$ and since $x'(0) = c_2\sqrt{75.2} = 4$, then $c_2 = \frac{4}{\sqrt{75.2}}$. So, $x = \frac{4}{\sqrt{75.2}}e^{-1.6t}\sin\sqrt{75.2}t \approx 0.4613e^{-1.6t}\sin 8.6178t$.

6. $m = \dfrac{F}{g} = \dfrac{15}{32} = 0.46875$ slugs, $\ell = 1$, $k = 60$, so $2b = \frac{\ell}{m} = \frac{32}{15}$ and $\omega^2 = \dfrac{k}{m} = 128$. The auxiliary equation is $m^2 + \frac{32}{15}m + 128 = 0$ has roots $-\frac{16}{15} \pm \sqrt{\frac{16}{15}^2 - 128} = \frac{-16}{15} \pm \sqrt{\frac{1904}{15}}j$. Hence, the general solution is $x = e^{-16t/15}\left(c_1\cos\sqrt{\frac{1904}{15}}t + c_2\sin\sqrt{\frac{1904}{15}}t\right)$. Since $x(0) = 0$, we find that $c_1 = 0$, and since $x'(t) = -\frac{16}{15}e^{-16t/15}c_2\sin\sqrt{\frac{1904}{15}}t + c_2\sqrt{\frac{1904}{15}}e^{-16/15t}\cos\sqrt{\frac{1904}{15}}t$ and $x'(0) = c_2\sqrt{\frac{1904}{15}} = 2$, we determine that $c_2\frac{2\sqrt{15}}{\sqrt{1904}}$. Thus, we have found that the particular solution is $x = \frac{2\sqrt{15}}{\sqrt{1904}}e^{-16t/15}\sin\sqrt{\frac{1904}{15}}t \approx 0.1775e^{-1.0667t}\sin 11.2665t$.

7. $m = 5$, $\ell = 40$, and $k = 245$, so $2b = 8$ and $\omega^2 = 49$. The auxiliary equation is $m^2 + 8m + 49 = 0$ has solutions $-4 \pm \sqrt{16 - 49} = -4 \pm \sqrt{33}j$ and the general solution is $x = e^{-4t}(c_1\cos\sqrt{33}t + c_2\sin\sqrt{35}t)$. Since $x(0) = 0$, $c_1 = 0$, and we have $x'(t) = -4e^{-4t}c_2\sin\sqrt{33}t + c_2\sqrt{33}e^{-4t}\cos\sqrt{33}t$. Also, $x'(0) = c_2\sqrt{33} = 2$, so $c_2 = \frac{2}{\sqrt{33}}$. The

particular solution is $x = \frac{2}{\sqrt{33}}e^{-4t}\sin\sqrt{33}t \approx 0.3482e^{-4t}\sin 5.7446t$

8. $m = 8$, $\ell = 80$, and $k = 196$, so $2b = 10$ and $\omega^2 = 24.5$. The auxiliary equaiton is $m^2+10m+24.5 = 0$ has solutions $-5 \pm \sqrt{25-24.5} = -5 \pm \sqrt{0.5}$. The general solution is $x = c_1e^{(-5+\sqrt{0.5}t)}+c_2e^{(-5-\sqrt{0.5}t)}$. Since $x(0) = 0$, then $c_1 + c_2 = 0$. Also, $x'(t) = (-5+\sqrt{0.5})c_1e^{(-5+\sqrt{0.5})t}-(5+\sqrt{0.5})c_2e^{(-5-\sqrt{0.5})t}$. It is given that $x'(0) = 3$, and so $(-5+\sqrt{0.5})c_1 - (5+\sqrt{0.5})c_2 = 3$, or, after substituting $-c_1$ for c_2, we have $(-5+\sqrt{0.5})c_1 + (5+\sqrt{0.5})c_1 = 3$. Hence, $2\sqrt{0.5}c_1 = 3$, and so $c_1 = \frac{3}{2\sqrt{0.5}}$, and $c_2 = -\frac{3}{2\sqrt{0.5}}$. The particular solution is $x = \frac{3}{2\sqrt{0.5}}e^{(-5+\sqrt{0.5})t} - \frac{3}{2\sqrt{0.5}}e^{(-5-\sqrt{0.5})t}$ or $x = 2.1213e^{-4.2929t} - 2.1213e^{-5.7071t}$.

9. $E(t) = L\frac{d^2q}{dt^2} + R\frac{dq}{dt} + \frac{q}{c}$ so $50\sin 100t = 0.1\frac{d^2q}{dt^2} + 6\frac{dq}{dt} + 100q$. q_c has auxiliary equation $0.1m^2 + 6m + 100 = 0$ with roots $\frac{-6\pm\sqrt{36-40}}{0.2} = -30 \pm 10j$. The general complementary solution is $q_c = e^{-30t}(c_1\cos 10t + c_2\sin 10t)$. For q_p we set $q_p = A\sin 100t + B\cos 100t$, $q_p' = 100A\cos 100t - 100B\sin 100t$, and $q_p'' = -10^4A\sin 100t - 10^4\cos 100t$. Substituting we get $(-10^3A\sin 100t - 10^3\cos 100t) + 6(100A\cos 100t - 100B\sin 100t) + 100(A\sin 100t + B\cos 100t) = 50\sin 100t$ so $-1000A - 600B + 100A = 50$ and $-1000B + 600A + 100B = 0$; $-900A - 600B = 50$, so $-18A - 12B = 1$. Also, $600A - 900B = 0$ or $2A - 3B = 0$. Hence, $A = -\frac{1}{26}, B = -\frac{1}{39}$. So $q_p = -\frac{1}{26}\sin 100t - \frac{1}{39}\cos 100t$. The general solution is $q = e^{-30t}(c_1\cos 10t + c_2\sin 10t) - \frac{1}{26}\sin 100t - \frac{1}{39}\cos 100t$. Since $q(0) = c_1 - \frac{1}{39} = 0$, then $c_1 = \frac{1}{39}$. Also, $i(t) = q'(t) = -30e^{-30t}(c_1\cos 10t + c_2\sin 10t) + e^{-30t}(-10c_1\sin 10t + 10c_2\cos 10t) - \frac{100}{26}\cos 100t + \frac{100}{39}\sin 100t$. Since $q'(0) = -30c_1 + 10c_2 - \frac{100}{26} = 0$, we have $c_2 = \frac{\frac{30}{39}-\frac{100}{26}}{10} = \frac{3}{39} - \frac{10}{26} = -\frac{24}{78} = -\frac{4}{13}$. The particular solution is $q(t) = e^{-30t}\left(\frac{1}{39}\cos 10t - \frac{4}{13}\sin 10t\right) - \frac{1}{26}\sin 100t - \frac{1}{39}\cos 100t$.

10. $E(t) = L\frac{d^2q}{dt^2} + R\frac{dq}{dt} + \frac{9}{c}$ so we get the equation $\frac{1}{8}\frac{d^2q}{dt^2} + 5\frac{dq}{dt} + 100q = \sin t$. The auxiliary equation is $\frac{1}{8}m^2 + 5m + 100 = 0$ is equivalent to $m^2+40m+800 = 0$ and it has solutions $-20\pm 20j$ and so $q_c = e^{-20t}(c_1\cos 20t + c_2\sin 20t)$. Now we set $q_p = A\sin t + B\cos t$, $q_p' = A\cos t - B\sin t$, and $q_p'' = -A\sin t - B\cos t$. Substituting, we have $\frac{1}{8}(-A\sin t - B\cos t) + 5(A\cos t - B\sin t) + 100(A\sin t + B\cos t) = \sin t$. Hence $-\frac{A}{8} - 5B + 100A = 1$ and $-\frac{B}{8}+5A+100B = 0$ or $\frac{779}{8}A - 5B = 1$ and $5A + \frac{779}{8}B = 0$ or $799A - 40B = 8$ and $40A = 799B = 0$.

Using Cramer's rule $A = \frac{6392}{640001}$ and $B = \frac{-320}{640001}$. The general solution is $q(t) = e^{-20t}(c_1\cos 20t + c_2\sin 20t) + \frac{6392}{640001}\sin t - \frac{320}{640001}\cos t$. Since $q(0) = 0$, $c_1 = \frac{320}{640001}$, $q'(t) = -20e^{-20t}(c_1\cos 20t + c_2\sin 20t) + e^{-20t}(-20c_1\sin 20t + 20c_2\cos 20t) + \frac{6392}{640001}\cos t + \frac{320}{640001}\sin t$. $q'(0) = -20c_1 + 20c_2 + \frac{6392}{640001} = 0$ means that $c_2 = \frac{320}{640001} - \frac{319.6}{640001} = \frac{0.4}{640001}$. The particular solution is $q(t) = e^{-20t}\left(\frac{320}{640001}\cos 20t + \frac{0.4}{640001}\sin 20t\right) + \frac{6392}{640001}\sin t - \frac{320}{640001}\cos t$.

11. (a) $E(t) = L\frac{d^2q}{dt^2} + R\frac{dq}{dt} + \frac{q}{c}$ so we get $\frac{1}{10}\frac{d^2q}{dt^2} + \frac{100}{4}q = 180\cos 60t$. The auxiliary equation is $0.1m^2 + \frac{1000}{4} = 0$, so $m = \pm 50j$ and as a result, $q_c = c_1\cos 50t + c_2\sin 50t$. Now we get $q_p = A\sin 60t + B\cos 60t$, $q_p' = 60A\cos 60t - 60B\sin 60t$, and $q_p'' = -3600A\sin 60t - 3600\cos 60t$. Substituting we get $-360A\sin 60t - 360B\cos 60t + 250A\sin 60t + 250B\cos 60t = 180\cos 60t$. Hence we have the equations $-360A + 250A = 0$ or $A = 0$ and $-360B + 259B = 180$ or $B = \frac{180}{-110} = -\frac{18}{11}$. The general equation is $q(t) = c_1\cos 50t + c_2\sin 50t - \frac{18}{11}\cos 60t$. Since $q(0) = 0$, $c_1 = \frac{18}{11}$. Also, since $q'(t) = -50c_1\sin 50t + 50c_2\cos 50t + \frac{18\cdot 60}{11}\sin 6t$, and $q'(0) = 50c_1 = 0$, we find that $c_2 = 0$. The particular solution is $q(t) = \frac{18}{11}\cos 50t -$

$\frac{18}{11}\cos 60t$ (b) The steady-state current is $i = \frac{dq_p}{dt} = \frac{d}{dt}\left(\frac{-18}{11}\cos 60t\right) = \frac{1080}{11}\sin 60t$.

12. (a) $1\frac{d^2q}{dt} + 10\frac{dq}{dt} + 100q = 50\cos 10t$. The auxiliary equation is $m^2 + 10m + 100 = 0$ which has roots $-5 \pm 5\sqrt{3}j$. Hence $q_c = e^{-5t}(c_1\cos 5\sqrt{3}t + c_2\sin 5\sqrt{3}t)$. To find q_p we set $q_p = A\sin 10t + B\cos 10t$, $q_p' = 10A\cos 10t - 10B\sin 10t$, and $q_p'' = -100A\sin 10t - 100B\cos 10t$. Substituting, we get $(-100A\sin 10t - 100B\cos 10t) + 10(10A\cos 10t - 10B\sin 10t) + 100(A\sin 10t + B\cos 10t) = 50\cos 10t$. This yields the linear equations $-100A - 100B + 100A = 0$ or $B = 0$ and $-100B + 100A + 100B = 50$ or $A = \frac{1}{2}$. The general solution is $q(t) = e^{-5t}(c_1\cos 5\sqrt{3}t + c_2\sin 5\sqrt{3}t) + \frac{1}{2}\sin 10t$. (b) The steady-current is $\frac{dq_p}{dt} = 5\cos 10t$.

CHAPTER 33 REVIEW

1. $D^2y = 0$ has auxiliary equation $m^2 = 0$ with 0 as a double root. Hence $y = (c_1 + c_2x)e^{0x}$ or more simply $y = c_1 + c_2x$.

2. $(D^2 - 9)y = 0$. The auxiliary equation $m^2 - 9 = 0$ has roots 3 and -3. Hence $y = c_1e^{3x} + c_2e^{-3x}$.

3. $(D^2 - 5D + 6)y = 0$. The auxiliary equaiton $m^2 - 5m + 6 = 0$ has roots 2 and 3. Hence $y = c_1e^{2x} + c_2e^{3x}$.

4. $(D^2 - 5D - 14)y = 0$. The auxiliary equation $m^2 - 5m - 14 = 0$ has roots 7 and -2. Hence the solution is $y = c_1e^{7x} + c_2e^{-2x}$.

5. $(D^2 - 6D + 25)y = 0$. The auxiliary equation $m^2 - 6m + 25 = 0$ has complex roots $3 + 4j$. Hence the solution is $y = e^{3x}(c_1\cos 4x + c_2\sin 4x)$

6. $(D^2 + 2D - 2)y = 0$. The auxiliary equation $m^2 + 2m - 2 = 0$ has roots $-1 \pm \sqrt{3}$. Hence $y = c_1e^{(-1+\sqrt{3})x} + c_2^{(-1-\sqrt{3})x}$

7. $(D^3 - 8)y = 0$; $m^3 - 8$ factors as $(m-2)(m^2+2m+4)$ and has roots $2, -1 \pm \sqrt{3}j$. Hence $y = c_1e^{2x} + e^{-x}(c_2\cos\sqrt{3}x + c_3\sin\sqrt{3}x)$

8. $y''' + y'' - y' - y = 0$ has auxiliary equation $m^3 + m^2 - m - 1 = 0$. This factors by grouping into $(m^2 - 1)(m + 1) = 0$ and has roots -1 twice and 1. Hence, the solution is $y = c_1e^x + c_2e^{-x} + c_3xe^{-x}$

9. $(D^2 + 3D - 4)y = 9x^2$. The auxiliary equation is $m^2 + 3m - 4 = 0$ and it has roots -4 and 1. Hence $y_c = c_1e^x + c_2e^{-4x}$. Now we set $y_p = Ax^2 + Bx + C$, $y_p' = 2Ax + B$, and $y_p'' = 2A$. Substituting we get $2A + 3(2Ax + B) - 4(Ax^2 + Bx + C) = 9x^2$ so $-4A = 9$ or $A = -\frac{9}{4}$, and also $6A - 4B = 0$ so $B = \frac{6A}{4} = \frac{-27}{8}$. Now, since $2A + 3B - 4C = 0$, we see that $C = \frac{2A + 3B}{4} = \frac{-\frac{9}{2} - \frac{81}{8}}{4} = \frac{-117}{32}$. So $y_p = -\frac{9}{4}x^2 - \frac{27}{3}x - \frac{117}{32}$ and $y = y_c + y_p = c_1e^x + c_2e^{-4x} - \frac{9}{4}x^2 - \frac{27}{8}x - \frac{117}{32}$.

10. $(D^2 - 4D - 5)y = 3e^{2x}$. The auxiliary equation has roots 5 and -1 so $y_c = c_1e^{5x} + c_2e^{-x}$. Now we set $y_p = Ae^{2x}$, $y_p' = 2Ae^{2x}$ and $y_p'' = 4Ae^{2x}$. Substituting we get $4Ae^{2x} - 4(2Ae^{2x}) - 5(Ae^{2x}) = 3e^{2x}$. Hence $4A - 8A - 5A = 3$ or $-9A = 3$ or $A = -\frac{1}{3}$. The solution is $y = c_1e^{5x} + c - 2e^x - \frac{1}{3}e^{2x}$

11. $(D^2 + 7D + 12)y = \cos 5x$. The auxiliary equation $m^2 + 7m + 12 = 0$, has roots -3 and -4. So $y_c = c_1e^{-3x} + c_2e^{-4x}$. Now we set $y_p = A\sin 5x + B\cos 5x$, $y_p' = 5A\cos 5x - 5B\sin 5x$, and $y_p'' = -25A\sin 5x - 25B\cos 5x$. Substituting we get $(-25A\sin 5x - 25B\cos 5x) + 7(5A\cos 5x - 5B\sin 5x) + 12(A\sin 5x + B\cos 5x) = \cos 5x$. This yields the linear equations $-25A - 35B + 12A = 0$ and $-25B + 35A + 12B = 1$. These simplify to $-13A - 35B = 0$ and $35A - 13B = 1$. Using Cramer's Rule we get $A = \frac{35}{1394}$ and $B = \frac{-13}{1394}$. So $y_p = \frac{35}{1394}\sin 5x - \frac{13}{1394}\cos 5x$. The solution is $y = y_c + y_p = c_1e^{-3x} + c_2e^{-4x} + \frac{35}{1394}\sin 5x - \frac{13}{1394}\cos 5x$.

12. $(D^2-6D+8)y = 6x^2-2$. The auxiliary equation has roots 2 and 4 so $y_c = c_1e^{2x}+c_2e^{4x}$. Now we set $y_p = Ax^2+Bx+C$, $y_p' = 2Ax+B$, $y_p'' = 2A$. Substituting we get $2A-6(2Ax+B)+8(Ax^2+Bx+C) = 6x^2-2$. Hence $8A = 6$ or $A = \frac{3}{4}$ and since, $-12A+8B = 0$, we obtain $B = \frac{12A}{8} = \frac{3}{2}\cdot\frac{3}{4} = \frac{9}{8}$, and since $2A-6B+8C = -2$, we find $C = \dfrac{-2-2A+6B}{8} = \dfrac{-2-\frac{3}{2}+\frac{27}{4}}{8} = \dfrac{13}{32}$. So we get $y_p = \frac{3}{4}x^2+\frac{9}{8}x+\frac{13}{32}$ and $y = y_c+y_p = c_1e^{2x}+c_2e^{4x}+\frac{3}{4}x^2+\frac{9}{8}x+\frac{13}{32}$.

13. $(D^2+4D+4)y = 4x+e^{2x}$. The auxiliary equation has double root at -2, so $y_c = (c_1+c_2x)e^{-2x}$. Now y_p has two parts. Set $y_{p1} = Ax+B$, with $y_{p1}' = A$, and $y_{p1}'' = 0$. Substituting we get $4A+4Ax+4B = 4x$. So $A = 1$ and $B = -1$. Next we set $y_{p2} = Ae^{2x}$ so $y_{p2}' = 2Ae^{2x}$ and $y_{p2}'' = 4Ae^{2x}$. Substituting we have $(4A+8A+4A)e^{2x} = e^{2x}$ or $A = \frac{1}{16}$. $y = y_c+y_{p1}+y_{p2} = (c_1+c_2x)e^{-2x}+x-1+\frac{1}{16}e^{2x}$.

14. $(D^2+4D-9)y = x\sin 2x$. The auxiliary equation $m^2+4m-9 = 0$ has solutions $-2\pm\sqrt{3}$ and so $y_c = c_1e^{(-2+\sqrt{13})x}+c_2e^{(-2-\sqrt{13})x}$. Now we set $y_p = A_1\sin 2x+A_0\sin 2x+B_1x\cos 2x+B_0\cos 2x$, $y_p' = 2A_1x\cos 2x+A_1\sin 2x+2A_0\cos 2x-2B_1x\sin 2x+B_1\cos 2x-2B_0\sin 2x$, and $y_p'' = 2A_1x\cos 2x-4A_1x\sin 2x+2A_1\cos 2x-4A_0x\sin 2x-2B_1\sin 2x-4B_1x\cos 2x-2B_1\sin 2x-4B_0\cos 2x$. Substituting we get $y_p''+4y_p'-9y = x\sin 2x$. Collecting like terms we have the following four linear equations.

$$\begin{aligned}
-4A_1+4(-2B_1)-9A_1 &= 1 && x\sin 2x\\
-4B_1-4A_0+4(-2B_0+A_1)-9A_0 &= 0 && \sin 2x\\
-4B_1+4\cdot 2A_1-9B_1 &= 0 && x\cos 2x\\
-4B_0+4A_1+4(B_1+2A_0)-9B_0 &= 1 && \cos 2x
\end{aligned}$$

These simplify to

$$\begin{aligned}
&-13A_1 && -8B_1 &&= 1\\
&4A_1-13A_0 && -4B_1-8B_0 &&= 0\\
&8A_1 && -13B_1 &&= 0\\
&4A_1+8A_0 && +4B_1-13B_0 &&= 0
\end{aligned}$$

Solving the first and third equations with Cramer's Rule yields $A_1 = \frac{-13}{233}$ and $B_1 = \frac{-8}{233}$. Substituting these results into the 2nd and 4th equations we get

$$13A_0+8B_0 = \frac{20}{233}$$
$$8A_0-13B_0 = \frac{84}{233}$$

Again using Cramer's Rule we have $A_0 = \frac{932}{54289}$ and $B_0 = \frac{932}{54289}$. Putting this all together we have $y = c_1e^{(-2+\sqrt{13})x}+c_2e^{(-2-\sqrt{13})x}-\frac{x}{233}(13\sin 2x+8\cos 2x)+\frac{932}{54289}(\sin 2x+\cos 2x)$.

15. First we find the spring constant $k = \dfrac{mg}{x} = \dfrac{4\cdot 9.8}{0.2} = 196$ N/m. So $m = 4$ kg, $k = 196$ N/m and $\ell = 10$. Hence we get auxiliary equation $4m^2+10m+196 = 0$. This auxiliary equation has roots $\frac{-5}{4}\pm\frac{\sqrt{759}j}{4}$. Hence $x = e^{-5t/4}\left(c_1\cos\frac{\sqrt{759}}{4}t+c_2\sin\frac{\sqrt{759}}{4}t\right)$. Assuming $x(0) = 0$, we obtain $c_1 = 0$. Now, differentiating, we see that $x'(t) = -\frac{5}{4}e^{-5t/4}\left(c_1\sin\frac{\sqrt{759}}{4}t\right)+c_2\frac{\sqrt{759}}{4}e^{-5t/4}\cos\frac{\sqrt{759}}{4}t$. Since $v(0) = x'(0) = c_2\frac{\sqrt{759}}{4} = 1$, we have $c_2 = \frac{4}{\sqrt{759}}$. Hence the solution is $x = \frac{4}{\sqrt{759}}e^{-5t/4}\sin\frac{\sqrt{759}}{4}t$.

16. (a) The given information yields the equation $\dfrac{d^2q}{dt^2}+6\dfrac{dq}{dt}+2500q = 16\cos 10t$. The auxiliary equation $m^2+6m+2500$ has solutions $-3\pm\sqrt{2491}j$. So, we see that $q_c = e^{-3t}(c_1\cos\sqrt{2491}t+c_2\sin\sqrt{2491}t)$. To find q_p we set $q_p = A\sin 10t+B\cos 10t$, $q_p' = 10A\cos 10t-10B\sin 10t$, and $q_p'' = 100A\cos 10t-100B\sin 10t$. Substituting we get $(-100A\sin 10t-100B\cos 10t)+6(10A\cos 10t-10B\sin 10t)+2500(A\sin 10t+B\cos 10t) = 16\cos 10t$. As a result, $-100A-60B+2500A = 0$ and $-100B+60A+2500B = 16$. These simplify to $2400A-60B = 0$ and $60A+2400B = 16$. Using Cramer's Rule we get $A = \frac{960}{5763600}$ and $B = \frac{38400}{5763600}$ which reduce to $A = \frac{4}{24015}$ and $B = \frac{160}{24015}$. Thus, $q_p = \frac{1}{24015}(4\sin 10t+160\cos 10t)$. Hence the general solution is $q(t) = e^{-3t}(c_1\cos\sqrt{2491}t+$

$c_2 \sin \sqrt{2491}t) + \frac{1}{24015}(4 \sin 10t + 160 \cos 10t)$. Since $q(0) = c_1 + \frac{160}{24015} = 0, c_1 = \frac{-160}{24015}$. Also $i(t) = q'(t) = -3e^{-3t}(c_1 \cos \sqrt{2491}t + c_2 \sin \sqrt{2491}t) + e^{-3t}(-\sqrt{2491}c_1 \sin \sqrt{2491}t + \sqrt{2491}c_2 \cos \sqrt{(2491t)} + \frac{40}{24105} \cos 10t - \frac{1600}{24015} \sin 10t$. So $q'(0) = -3c_1 + \sqrt{2491}c_2 + \frac{40}{24015} = 0$ and so $c_2 = \frac{-520}{24015\sqrt{2491}}$. Putting it all together we get $q(t) = \frac{1}{24015}\left[e^{-3t}(-160 \cos \sqrt{2491}t - \frac{520}{\sqrt{2491}} \sin \sqrt{2491}t) + 4 \sin 10t + 160 \cos 10t\right]$.

(b) The steady state current is $\frac{dq_p}{dt} = \frac{1}{24015}[40 \cos 10t - 1600 \sin 10t]$.

CHAPTER 33 TEST

1. The auxiliary equation is formed by replacing the operators, D, with place-holders for the roots, usually m and replacing the variable, y, with 1. In this problem the resulting auxiliary equation is $5m^2 + 2m - 3 = 0$.

2. The auxiliary equation is $m^2 - 25 = 0$. This factors as $m^2 - 25 = (m - 5)(m + 5) = 0$ and has roots $m_1 = 5$ and $m_2 = -5$. So, the general solution of this differential equation is $y = c_1e^{5x} + c_2e^{-5x}$.

3. The auxiliary equation is $m^2 + 8m - 20 = 0$ which factors as $(m + 10)(m - 2) = 0$. The roots of the auxiliary equation are $m_1 = -10$ and $m = 2$. The solution of the differential equation is $y = c_1e^{-10x} + c_2e^{2x}$.

4. The auxiliary equation is $(m - 5)^3(m^2 + 4) = 0$. This has the real root 5 with multiplicity 3 and the complex roots $\pm 2j$. So, the general solution is $y = (c_1 + c_2x + c_3x^2)e^{5x} + c_4 \cos 2x + c_5 \sin 2x$.

5. The auxiliary equation is $m^2 + 6m + 13 = 0$. Using the quadratic formula we get the roots $m = \frac{-6\pm\sqrt{6^2-4(13)}}{2} = \frac{-6\pm\sqrt{-16}}{2} = -3 \pm 2j$. So the general solution to this differential equation is $y = e^{-3x}(c_1 \cos 2x + c_2 \sin 2x)$.

6. The auxiliary equation is $m^2 + 6m - 7 = 0$ which factors as $(m + 7)(m - 1) = 0$. The roots of this auxiliary equation are $m = -7$ and $m = 1$. The general solution is $y = c_1e^{-7x} + c_2e^x$.

7. From Exercise 0 we have the complementary solution $y_c = c_1e^{-7x} + c_2e^x$. The given equation is a nonhomogeneous equation with $f(x) = 6e^{2x}$ and so this is a Case 2 problem using the Method of Undetermined Coefficients with $a = 2$ and $p_n = 6$ so $y_p = A_0e^{2x}$. Differentiating, we obtain $y_p' = 2A_0e^{2x}$ and $y_p'' = 4A_0e^{2x}$. Substituting this in the given differential equation produces $4A_0e^{2x} + 12A_0e^{2x} - 7A_0e^{2x} = 6e^{2x}$. Solving for A_0, we determine that $A_0 = \frac{2}{3}$ and so the general solution is $y = c_1e^{-7x} + c_2e^x + \frac{2}{3}e^{2x}$.

8. Again the complementary solution is $y_c = c_1e^{-7x} + c_2e^x$. This is a nonhomogeneous equation with $f(x) = 3 \cos x$, a Case 4 problem using the Method of Undetermined Coefficients with $a = 0$, $p_n = 3$, and $b = 4$, so $y_p = A_0 \sin 4x + B_0 \cos 4x$. Differentiating, produces $y_p' = 4A_0 \cos 4x - 4B_0 \sin 4x$ and $y_p'' = -16A_0 \sin 4x - 16B_0 \cos 4x$. Substituting in the given differential equation we get $(-16A_0 \sin 4x - 16B_0 \cos 4x) + 6(4A_0 \cos 4x - 4B_0 \sin 4x) - 7(A_0 \sin 4x + B_0 \cos 4x) = 3 \cos 4x$. Multiplying and collecting terms produces $-23(A_0+B_0) \sin 4x + 24(A_0-B_0) \cos 4x = 3 \cos 4x$. Thus, $A_0 + B_0 = 0$ and $A_0 - B_0 = 3$. Solving this two equations we determine that $A_0 = \frac{3}{2}$ and $B_0 = -\frac{3}{2}$ and so the general solution is $y = c_1e^{-7x} + c_2e^x + \frac{3}{2} \sin 4x - \frac{3}{2} \cos 4x$.

9. Again the complementary solution is $y_c = c_1e^{-7x} + c_2e^x$. This is a nonhomogeneous equation with $f(x) = 8x$, a Case 1 problem using the Method of Undetermined Coefficients, so $y_p = A_1x + A_0$. Differentiating, we obtain $y_p' = A_1$ and $y_p'' = 0$. Substituting these in the given differential equation produces $0 + 6(A_1) - 7(A_1x + A_0) = 8x$. Multiplying and collecting terms produces $-A_1x + A_0 = 8x$. Thus, $A_1 = -8$ and $A_0 = 0$ and so the general solution is $y = c_1e^{-7x} + c_2e^x - 8x$.

10. Once again the complementary solution is $y_c = c_1e^{-7x} + c_2e^x$. This is a nonhomogeneous equation with $f(x) = e^{4x}$, a Case 2 problem using the Method of Undetermined Coefficients with $a = 4$, and $p_n = 1$, so $y_p = A_0e^{4x}$. Differentiating, we get $y_p' = 4A_0e^{4x}$ and $y_p'' = 16A_0e^{4x}$. Substituting these in the given differential equation produces $16A_0e^{4x} + 6(4A_0e^{4x}) - 7(A_0e^{4x}) = e^{4x}$. Collecting terms produces $33A_0e^{4x} = e^{4x}$. Thus, $A_0 = \frac{1}{33}$ and the general solution is $y = c_1e^{-7x}+c_2e^x+\frac{1}{33}e^{4x}$. To find the particular solution, we find $y' = -7c_1e^{-7x}+c_2e^x + \frac{4}{33}e^{4x}$. Substituting $x = 0$ and $y = \frac{4}{11}$ in the general solution produces $\frac{4}{11} = c_1+c_2+\frac{1}{33}$ and substituting $x = 0$ and $y' = \frac{5}{11}$ in the derivative gives $\frac{5}{11} = -7c_1+c_2+\frac{4}{33}$. Solving these two equation we find $c_1 = 0$ and $c_2 + \frac{1}{3}$. Thus, the particular solution of the given differential equation is $y = \frac{1}{3}e^x + \frac{1}{33}e^{4x}$.

11. Here $E(t) = 5\sin 10t$ and the desired differential equation is $5\sin 10t = 0.5\frac{d^2q}{dt^2} + 30\frac{dq}{dt} + \frac{q}{2\times 10^{-3}} = 0.5\frac{d^2q}{dt^2} + 30\frac{dq}{dt} + 500q$. The auxiliary equation is $0.5m^2 + 30m + 500 = 0$ or $m^2 + 60m + 1000 = 0$. Using the quadratic equation we get $m = -30 \pm 20j$. These are complex roots and so the complementary solution is $q_c = e^{-30t}(c_1\cos 20t + c_2\sin 20t)$. For the particular solution we have a case 3 situation with $a = 0, b = 10$, and $p_n = 5$, and so $q_p = A_0\sin 10t+B_0\cos 10t$. Differentiating, we get $q_p' = 10A_0\cos 10t-10B_0\sin 10t$ and $q_p'' = -100A_0\sin 10t - 100B_0\cos 10t$. Substituting these in the differential equation, we have $0.5(-100A_0\sin 10t - 100B_0\cos 10t) + 30(10A_0\cos 10t - 10B_0\sin 10t) + 500(A_0\sin 10t + B_0\cos 10t) = 5\sin 10t$. Multiplying and collecting terms, we get $(450A_0 - 300B_0)\sin 10t + (450B_0 + 300A_0)\cos 10t = 5\sin 10t$. Hence, $450A_0 - 300B_0 = 5$ and $450B_0 + 300A_0 = 0$. Solving these two equations simultaneously gives $A_0 = 0.02$ and $B_0 = \frac{1}{75}$. Thus, the general solution is $q(t) = e^{-30t}(c_1\cos 20t + c_2\sin 20t) + 0.02\sin 10t + \frac{1}{75}\cos 10t$. The steady-state solution is $q_p = 0.02\sin 10t + \frac{1}{75}\cos 10t$. Consequently, the steady-state current is $i = \frac{dq_p}{dt} = 0.2\cos 10t - \frac{10}{75}\sin 10t$.

CHAPTER

34

Numerical Methods and LaPlace Transforms

34.1 EULER'S OR THE INCREMENT METHOD

1. See *Computer Programs* in the main text.

2.

x	y		Correct solution
	$h = 0.1$	$h = 0.05$	$y = \frac{1}{2}x^2 + 2x + 1.5$
1.00	4	4	4
1.05	—	4.15	4.15125
1.10	4.3	4.3025	4.305
1.15	—	4.575	4.46125
1.20	4.61	4.615	4.62
1.25	—	4.775	4.78125
1.30	4.93	4.9375	4.95
1.35	—	5.1025	5.11125
1.40	5.26	5.27	5.28
1.45	—	5.44	5.45125
1.50	5.6	5.6125	5.625
1.55	—	5.7875	5.80125
1.60	5.95	5.965	5.98
1.65	—	6.145	6.16125
1.70	6.31	6.3275	6.345
1.75	—	6.5125	6.53125
1.80	6.68	6.7	6.72
1.85	—	6.89	6.91125
1.90	7.06	6.0825	7.105
1.95	—	7.2775	7.30125
2.00	7.45	7.475	7.5

3.

x	y		Correct solution
	$h = 0, 1$	$h = 0.05$	$y = -\frac{1}{2}e^x + \frac{1}{2}e^{3x}$
0	0	0	0
0.5	—	0.5	0.05528
0.10	0.1	0.1101	0.12234
0.15	—	0.1818	0.2032
0.20	0.2405	0.2672	0.3004
0.25	—	0.3683	0.4165
0.30	0.4348	0.4878	0.5549
0.35	—	0.6285	0.7193
0.40	0.7002	0.7937	0.9141
0.45	—	0.9873	1.1446
0.50	1.0595	1.2138	1.4165
0.55	—	1.4784	1.7369
0.60	1.5422	1.7868	2.1138
0.65	—	2.1459	2.5566
0.70	2.1871	2.5636	3.0762
0.75	—	3.0488	3.6854
0.80	3.0046	3.6120	4.3988
0.85	—	4.2650	5.2337
0.90	4.1805	5.0218	6.2101
0.95	—	5.8980	7.3510
1.00	5.6807	6.9120	8.6836

4.

x	y $h = 0.1$	Correct solution $y = -1 + 2e^{x^2}$
0	1	1
0.1	1	1.0201
0.2	1.04	1.0816
0.3	1.1216	1.1883
0.4	1.2489	1.3470
0.5	1.4288	1.5681
0.6	1.6717	1.8667
0.7	1.9923	2.2646
0.8	2.4112	2.7930
0.9	2.9570	3.4958
1.0	3.6693	4.4366

5.

x	y $h = 0.05$	Correct solution $y = \frac{1}{53}(55e^{7x} - 2\cos 2x - 7\sin 2x)$
0	1	1
0.05	1.35	1.4219
0.10	1.8275	2.0265
0.15	2.4770	2.8904
0.20	3.3588	4.1220
0.25	4.5538	5.8753
0.30	6.1717	8.3686
0.35	8.3600	11.9117
0.40	11.3182	16.9442
0.45	15.3154	24.0898
0.50	20.7149	34.2336

6.

x	y
0	0
0.1	0.5
0.2	1.005
0.3	1.5251
0.4	2.0709
0.5	2.6537
0.6	3.2864
0.7	3.9836
0.8	4.7624
0.9	5.6434
1.0	6.6513

7.

x	y
0	1
0.1	1.1
0.2	1.222
0.3	1.3753
0.4	1.5735
0.5	1.8371
0.6	2.1995
0.7	2.7193
0.8	3.5078
0.9	4.8023
1.0	7.1895

8.

x	y
0	1
0.1	1.4
0.2	2.185
0.3	4.0987
0.4	10.8274
0.5	57.7364
0.6	1391.1571
0.7	175,518.399
0.8	2.406×10^{19}
0.9	2.316×10^{22}
1.0	

9.

x	y
0	0
0.1	0.1
0.2	0.2010
0.3	0.3051
0.4	0.4147
0.5	0.5327
0.6	0.6633
0.7	0.8121
0.8	0.9887
0.9	1.2092
1.0	1.5062

10.

x	y
0	1
0.05	1.1
0.1	1.2007
0.15	1.3022
0.20	1.4046
0.25	1.5080
0.30	1.6126
0.35	1.7185
0.40	1.8258
0.45	1.9345
0.50	2.0449

11.

x	y
$\pi/2 \approx 1.57$	1
1.62	1.0270
1.67	1.0507
1.72	1.0711
1.77	1.0882
1.82	1.1022
1.87	1.1130
1.92	1.1209
1.97	1.1259
2.02	1.1281
2.07	1.1277
2.12	1.1249
2.17	1.1197
2.22	1.1123
2.27	1.1028
2.32	1.0913
2.37	1.0779

12.

x	y
$\pi \approx 3.14$	2
3.19	2.05
3.24	2.1026
3.29	2.1579
3.34	2.2164
3.39	2.2782
3.44	2.3437
3.49	2.4131
3.54	2.4868
3.59	2.5651
3.64	2.6483
3.69	2.7368
3.74	2.8311
3.79	2.9313
3.84	3.0381
3.89	3.1518
3.94	3.2728

13. See *Computer Programs* in the main text.

14. (Problem #3)

x	y	
	$h = 0.1$	$h = 0.05$
0	0	0
0.05	—	0.0513
0.1	0.1053	0.1131
0.15	—	0.1871
0.2	0.2547	0.2755
0.25	—	0.3805
0.3	0.4636	0.5048
0.35	—	0.6517
0.4	0.7517	0.8246
0.45	—	1.0279
0.5	1.1455	1.2664
0.55	—	1.5457
0.6	1.6798	1.8722
0.65	—	2.2535
0.7	2.4007	2.6982
0.75	—	3.2163
0.8	3.3689	3.8194
0.85	—	4.5207
0.9	4.6644	5.3358
0.95	—	6.2823
1.00	6.3926	7.3808

(Problem #4)

x	y $h = 0.1$
0	1
0.1	1.021
0.2	1.0837
0.3	1.1911
0.4	1.3492
0.5	1.5674
0.6	1.8592
0.7	2.2440
0.8	2.7485
0.9	3.4105
1.0	4.2826

(Problem #5)

x	y $h = 0.5$
0	1
0.05	1.3612
0.10	1.8571
0.15	2.5356
0.20	3.4624
0.25	4.7263
0.30	6.4479
0.35	8.7913
0.40	11.9793
0.45	16.3143
0.50	22.2077

34.2 SUCCESSIVE APPROXIMATIONS

1. $y' = x + 2, (1,4), x = 2$. Integrating we get $y = \int (x+2)dx = \frac{1}{2}x^2 + 2x + C$. Now substituting 1 for x and 4 for y we obtain $C = 4 - 2\frac{1}{2} = \frac{3}{2}$. Hence $y = \frac{1}{2}x^2 + 2x + \frac{3}{2}, y(2) = 7.5$ the exact value.

2. $y' = -3y + e^x$. Substituting 0 for y and integratng we have $y = \int e^x\,dx = e^x + C_1$. Substituting $(0,0)$ we obtain $C_1 = -1$ so the first approximation is $y = e^x - 1$. Substituting this for y in the original equation we have $y' = 3(e^x - 1) + e^x$ or $y' = 4e^x - 3$. Integrating we have $y = \int (4e^x - 3)\,dx = 4e^x - 3x + C_2$. Again substituting $(0,0)$ we obtain $C_2 = -4$ so $y = 4e^x - 3x - 4$ is the second approximation. Substituting this expression for y in the original equation yields $y' = 3(4e^x - 3x - 4) + e^x = 13e^x - 9x - 12$. Integrating once more $y = 13e^x - \frac{9}{2}x^2 - 12x + C_3$. Substituting $(0,0)$ yields $C_3 = -13$. The third approximation is $y = 13e^x - \frac{9}{2}x^2 - 12x = 13$. $y(1) = 13e - 29.5 \approx 5.8377$. This answer is not very close to the correct solution, 8.6836.

3. $y' = 2xy + 2x$. Substituting 1 for y we have $y' = 4x$ so $y = 2x^2 + C_1$. Substituting $(0,1)$ we obtain $C_1 = 1$. The first approximation is $y = 2x^2 + 1$. Substituting this for y in the orginal equation we have $y' = 2x(2x^2 + 1) + 2x = 4x^3 + 4x$. Integration yields $y = x^4 + 2x^2 + C_2$. Substituting $(0,1)$ again we get $C_2 = 1$ so the second approximation is $y = x^4 + 2x^2 + 1$. Substituting for y once more in the original equation we have $y' = 2x(x^4 + 2x^2 + 1) + 2x = 2x^5 + 4x^3 + 4x$. Integration yields $y = \frac{1}{3}x^6 + x^4 + 2x^2 + C_3$. Again $C_3 = 1$ so the third approximation is $y = \frac{1}{3}x^6 + x^4 + 2x^2 + 1$. $y(1) = 4\frac{1}{3}$ compares favorably with 4.4366, the correct solution.

4. $y' = 7y + \sin 2x$. Substituting 1 for y we have $y' = 7 + \sin 2x$. Integration yields $y = 7x - \frac{1}{2}\cos 2x + C_1$. Substituting $(0,1)$ we obtain $C_1 = \frac{3}{2}$ so the first approximation is $y = 7x - \frac{1}{2}\cos 2x + \frac{3}{2}$. Substituting this into the original equation we have $y' = 49x - \frac{7}{2}\cos 2x + \frac{21}{2} + \sin 2x$. Integration yields $y = \frac{49}{2}x^2 - \frac{7}{4}\sin 2x + \frac{21}{2}x - \frac{1}{2}\cos 2x + C_2$. Substituting $(0,1)$ we obtain $C_2 = \frac{3}{2}$ and the second approximation is $y = \frac{49}{2}x^2 - \frac{7}{4}\sin 2x + \frac{21}{2}x - \frac{1}{2}\cos 2x + \frac{3}{2}$. Again substituting into the original equation we get $y' = \frac{343}{2}x^2 - \frac{49}{4}\sin 2x + \frac{147}{2}x - \frac{7}{2}\cos 2x + \frac{21}{2} + \sin 2x$. Integration yields $y = \frac{343}{6}x^3 + \frac{49}{8}\cos 2x + \frac{147}{4}x^2 - \frac{7}{4}\sin 2x + \frac{21}{2}x - \frac{1}{2}\cos 2x + C_3 = \frac{343}{6}x^3 + \frac{45}{8}\cos 2x + \frac{147}{4}x^2 - \frac{7}{4}\sin 2x + \frac{21}{2}x + C_3$. Substituting $(0,1)$ we obtain $C_3 = -\frac{37}{8}$ and the third approximation is $y = \frac{343}{6}x^3 + \frac{45}{8}\cos 2x + \frac{147}{4}x^2 - \frac{7}{4}\sin 2x + \frac{21}{2}x - \frac{37}{8}$. Thus, $y(0.5) = 18.5250$. This compares poorly with the correct answer 34.2750.

5. $y' = x^2 + y^2$. Substituting 1 for y we have $y' = x^2 + 1$. By integration $y = \frac{1}{3}x^3 + x + C_1$. Substituting $(0,1)$ we get $C_1 = 1$. The first approxiamtion is $y = \frac{1}{3}x^3 + x + 1$. Substituting this for y in the original equation yields $y' = x^2 + \frac{1}{9}x^6 + \frac{2}{3}x^4 + \frac{2}{3}x^3 + x^2 + 2x + 1$. Integration yields $y = \frac{1}{63}x^7 + \frac{2}{15}x^5 + \frac{1}{6}x^4 + \frac{2}{3}x^3 + x^2 + x + C_2$. Again, $C_2 = 1$ so the second approximation is $y = \frac{1}{63}x^7 + \frac{2}{15}x^5 + \frac{1}{6}x^4 + \frac{2}{3}x^3 + x^2 + x + 1$.

6. $y' = ye^x$. Substituting 1 for y we have $y' = e^x$. Integration yields $y = e^x + C_1$. Substituting $(0,1)$ we get $C_1 = 0$, so $y = e^x$. Substituting this into the original equation we have $y' = e^{2x}$. Integrating we get $y = \frac{1}{2}e^{2x} + C_2$. This time $C_2 = \frac{1}{2}$. The second Approximation is $y = \frac{1}{2}e^{2x} + \frac{1}{2}$ or $\frac{1}{2}(e^{2x} + 1)$.

7. $y' = 4 + xy$. Substituting 1 for y and integrating we have $y = 4x + \frac{1}{2}x^2 + C_1$. With the point $(0,1)$ we get $C_1 = 1$ and $y = 4x + \frac{1}{2}x^2 + 1$. Substituting this for y in the original equation yields $y' = 4 + 4x^2 + \frac{1}{2}x^3 + x$. Integrating we get $y = \frac{1}{8}x^4 + \frac{4}{3}x^3 + \frac{1}{2}x^2 + 4x + C_2$. Again $C_2 = 1$ so the second approximation is $y = \frac{1}{8}x^4 + \frac{4}{3}x^3 + \frac{1}{2}x^2 + 4x + 1$.

8. $y' = x^2 + 4y^2, (0,1)$. Substituting 1 for y we have $y' = x^2 + 4$. Integration yields $y = \frac{1}{3}x^3 + 4x + C_1$ and $C_1 = 1$. Substituting into the original equation yields $y' = x^2 + 4\left(\frac{1}{9}x^6 + \frac{8}{3}x^4 + \frac{2}{3}x^3 + 16x^2 + 8x + 1\right)$ or $y' = \frac{4}{9}x^6 + \frac{32}{3}x^4 + \frac{8}{3}x^3 + 65x^2 + 32x + 4$. Integration yields $y = \frac{4}{63}x^7 + \frac{32}{15}x^5 + \frac{8}{12}x^4 + \frac{65}{3}x^3 + 16x^2 + 4x + C_2$.

Again $C_2 = 1$ so the second approximation is $y = \frac{4}{63}x^7 + \frac{32}{15}x^5 + \frac{2}{3}x^4 + \frac{65}{3}x^3 + 16x^2 + 4x + 1$.

9. $y' = \sin 2x + y$; $\left(\frac{\pi}{2}, 1\right)$. Substituting 1 for y and integrating we get $y = -\frac{1}{2}\cos 2x + x + C_1$. Using $\left(\frac{\pi}{2}, 1\right)$ we get $C_1 = 1 - \frac{1}{2} - \frac{\pi}{2} = \frac{1-\pi}{2}$. The first approximation is $y = -\frac{1}{2}\cos 2x + x + \frac{1-\pi}{2}$. Substituting this for y in the original equation we get $y' = \sin 2x - \frac{1}{2}\cos 2x + x + \frac{1-\pi}{2}$. Integration yields $y = -\frac{1}{2}\cos 2x - \frac{1}{4}\sin 2x + \frac{1}{2}x^2 + \frac{1-\pi}{2}x + C_2$. Using $\left(\frac{\pi}{2}, 1\right)$ again we get $C_2 = 1 - \frac{1}{2} - \frac{\pi^2}{8} - \frac{1-\pi}{2}\cdot\frac{\pi}{2} = \frac{4 - 2\pi - \pi^2}{8}$. The second approximation is $y = -\frac{1}{2}\cos 2x - \frac{1}{4}\sin 2x + \frac{1}{2}x^2 + \frac{1-\pi}{2}x + \frac{4 - 2\pi - \pi^2}{8}$.

10. $y' = y + \cos x, (\pi, 2)$. $y' = 2 + \cos x$ so $y = 2x + \sin x + C_1$. Using $(\pi, 2)$ we have $C_1 = 2 - 2\pi$. The first approximation is $y = 2x + \sin x + 2 - 2\pi$. Substituting this into the first equation we get $y' = 2x + \sin x + 2 - 2\pi + \cos x$. Integration yields $y = x^2 - \cos x + (2 - 2\pi)x + \sin x + C_2$. Using $(\pi, 2)$ again we obtain $C_2 = 2 - \pi^2 - 1 - 2\pi + 2\pi^2$ or $C_2 = 1 - 2\pi + \pi^2$. The second approximation is $y = x^2 + (2 - 2\pi)x - \cos x + \sin x + 1 - 2\pi + \pi^2$.

11. The Maclaurin series for e^{-x} is $1 - x + \frac{x^2}{2!} - \frac{x^3}{3!} + \frac{x^4}{4!} - \frac{x^5}{5!} + \cdots$. Substituting into $3e^{-x} + x - 1$ we obtain $3\left(1 - x + \frac{x^2}{2} - \frac{x^3}{6} + \frac{x^4}{24} - \frac{x^5}{120}\right) + x - 1 = 2 - 2x + \frac{3x^2}{2} - \frac{x^3}{2} + \frac{x^4}{8} - \frac{x^5}{40}$. The first four terms are the same.

12. The Maclaurin series for e^{x^2} is $1 + x^2 + \frac{x^4}{2!} + \frac{x^6}{3!} + \frac{x^8}{4!} + \cdots$. Substituting into $2e^{x^2} - 1$ we get $1 + 2x^2 + \frac{x^6}{3} + \frac{x^8}{12}$. The first four terms are the same.

13. $I'(t) = -2I(t) + \sin t$ is like $I' = -2I + \sin t$. Substitute 0 for I and integrate we get $I = -\cos t + C_1$. Using $(0, 0)$ we get $C_1 = 1$. The first approximation is $I = -\cos t + 1$. Substituting into the original equation to get $I' = -2(-\cos t + 1) + \sin t = 2\cos t + \sin t - 2$. Integratng again yields $I = 2\sin t - \cos t - 2t + C_2$. Using $(0, 0)$ again we obtain $C_2 = 1$. The second approximation is $I = 2\sin t - \cos t - 2t + 1$. Substituting once more we have $I' = -3\sin t + 2\cos t + 4t - 2$. Integrating we get $I = 3\cos t + 2\sin t + 2t^2 - 2t + C_3$. Using $(0, 0)$ we obtian $C_3 = -3$. The third approximation is $I(t) = 3\cos t + 2\sin t + 2t^2 - 2t - 3$. $I(0.5) \approx 0.0916$.

14. Since it is dropped frrom rest, when $t = 0$, $v = 0$. Substituting we have $v' = 9.8$ so $v = 9.8t + C_1$. $C_1 = 0$ and the first approximation is $v = 9.8t$. Substituting this into the original we get $v' = 9.8 - 0.4(9.8t)^2 = 9.8 - 38.416t^2$. Integrating we have $v = 9.8t - 12.8053t^3 + C_2$. Again $C_2 = 0$. The second approximation is $v(t) = 9.8t - 12.8053t^3$. Substituting again we have $v' = 9.8 - 0.4(9.8t - 12.8053t^3)^2 = 9.8 - 38.416t^2 + 100.39355t^4 - 65.59028t^6$. Integrating we obtain $v = 9.8t - 12.8053t^3 + 20.0787t^5 - 9.37t^7 + C_3$. Again $C_3 = 0$, $v(0.5) = 3.85359$ m/s.

34.3 LAPLACE TRANSFORMS

1. $f(t) = n,\ L(f) = L(n) = \int_0^\infty e^{-st} n\, dt$

$$= \lim_{b\to\infty} \int_0^b e^{-st} n\, dt = \lim_{b\to\infty} \left[\frac{-n}{s} e^{-st}\right]_0^n$$

$$= \lim_{n\to\infty} \frac{-n}{b} e^{-sb} + \frac{n}{s} = \frac{n}{s}$$

2. $f(t) = e^{-at}$, $L(f) = L(e^{-at}) = \int_0^\infty e^{-st}e^{-at}\,dt$
$= \lim_{b\to\infty} \int_0^b e^{(-s-a)t}\,dt = \lim_{b\to\infty}\left[\frac{-1}{s+a}e^{-(s+a)t}\right]_0^b$
$= \lim_{b\to\infty} \frac{-1}{s+a}e^{-(s+a)b} + \frac{1}{s+a} = \frac{1}{s+a}$

3. $f(t) = \cos at$, $L(f) = L(\cos at) = \int_0^\infty e^{-st}\cos at\,dt = \lim_{b\to\infty}\int_0^b e^{-st}\cos at\,dt$. By form 88, we see that this is $\lim_{b\to\infty}\left[\frac{e^{-st}}{s^2+a^2}(-s\cos at + a\sin at)\right]_0^b = \lim_{b\to\infty}\frac{e^{-sb}}{s^2+a^2}(-s\cos ab + a\sin ab) - \frac{1}{s^2+a^2}(-s) = \frac{s}{s^2+a^2}$.

4. $f(t) = te^{at}, L(f) = L(te^{at}) = \int_0^\infty e^{-st}te^{at}\,dt = \lim_{b\to\infty}\int_0^b te^{(a-s)t}\,dt$. By form 84, we find that this is $\lim_{b\to\infty}\left[\frac{1}{(a-s)^2}((a-s)t-1)e^{(a-s)t}\right]_0^b = \lim_{b\to\infty}\left[\frac{1}{(a-s)^2}((a-s)b-1)e^{(a-s)b}\right] - \left[\frac{1}{(a-s)^2}(-1)\right] = \frac{1}{(a-s)^2}$ or $\frac{1}{(s-a)^2}$.

5. $f(t) = t^3 = 3!\cdot\frac{t^3}{3!}$. $L(f) = 3!L\left(\frac{t^3}{3!}\right) = 3!\frac{1}{s^4} = \frac{3!}{s^4}$. (form 3)

6. $f(t) = e^{2t}, L(f) = \frac{1}{s-2}$. (form 5)

7. $f(t) = \sin 6t, L(f) = \frac{6}{s^2+36}$. (form 8)

8. $f(t) = e^{2t}\sin 6t$, $L(e^{2t}\sin 6t) = \frac{6}{(s-2)+36}$. (form 19)

9. $f(t) = t^3e^{5t}, L(f) = \frac{3!}{(s-5)^4}$. (form 7)

10. $f(t) = e^{-5t}\sin 10t$, $L(f) = \frac{10}{(s+5)^4+100}$. (form 19)

11. $f(t) = \sin 4t + 4t\cos 4t$, $L(f) = \frac{2\cdot 4s^2}{(s^2+4^2)^2} = \frac{8s^2}{(s^2+16)^2}$. (form 14)

12. $f(t) = t\cos 7t$, $L(f) = \frac{s^2-7^2}{(s^2+7^2)^2} = \frac{s^2-49}{(s^2+49)^2}$. (form 15)

13. $L(\sin 5t + \cos 3t) = L(\sin 5t) + L(\cos 3t) = \frac{5}{s^2+25} + \frac{s}{s^2+9}$. (forms 8 and 9)

14. $L(e^{3t}\sin 5t + 4t^8) = L(e^{3t}\sin 5t) + 4L(t^8) = \frac{5}{(s-3)^2+25} + \frac{4.8!}{s^9}$. (forms 3 and 19)

15. $L(3t\sin 5t - \cos 4t + 1) = 3L(t\sin 5t) - L(1 - \cos 4t) = \frac{30s}{(s^2+25)^2} - \frac{16}{s(s^2+16)}$. (forms 10 and 13)

16. $L(t^3e^{4t} + 4e^{3t}) = L(t^3e^{4t}) + 4L(e^{3t}) = \frac{3!}{(s-a)^4} + \frac{4}{s-3}$. (forms 7 and 5)

17. $L(e^{-2t}\cos 3t - 2e^{5t}\sin t) = L(e^{-2t}\cos 3t) - 2L(e^{5t}\sin t) = \frac{s+2}{(s+2)^2+9} - \frac{2}{(s-5)^2+1}$. (forms 19 and 20)

18. $L(t\cos 3t - 2t\sin 2t) = L(t\cos 3t) - 2L(t\sin 2t) = \frac{s^2-9}{(s^2+9)^2} = \frac{8s}{(s^2+4)^2}$

19. $L(4+3t+2e^t) = 4L(1)+3L(t)+2L(e^t) = \frac{4}{s}+\frac{3}{s^2}+\frac{2}{s-1}$. (forms 1, 2 and 5)

20. $L(6-4t+2\sin 10t+e^{5t}) = \frac{6}{s}-\frac{4}{s^2}+\frac{20}{s^2+100}+\frac{1}{s-5}$

34.4 INVERSE LAPLACE TRANSFORMS AND TRANSFORMS OF DERIVATIVES

1. $\frac{4}{s-5}$. $s-a$ appears in the denominator of form 5. $L^{-1}(F) = 4e^{5t}$.

2. $\frac{6}{s^2+4} = 3\frac{2}{s^2+2^2}$. $L^{-1}(F) = 3\sin 2t$ (form 8)

3. $\frac{16}{s^3+16s} = \frac{16}{s(s^2+16)} = \frac{4^2}{s(s^2+4^2)}$ fits form 10. $L^{-1}(F) = 1 - \cos 4t$

4. $\frac{4}{(s+9)^2} = 4\cdot\frac{1}{(s-(-9))^2}$ fits form 6. $L^{-1}(F) = 4te^{-9t}$.

5. $\frac{3s}{s^2+9} = 3\cdot\frac{s}{s^2+3^2}$. $L^{-1}(F) = 3\cos 3t$ (form 9).

6. $\frac{s+6}{(s+16)^2} = \frac{s+16-10}{(s+16)^2} = \frac{1}{s+16} - \frac{10}{(s+16)^2}$. By forms 5 and 6 we get $L^{-1}(F) = e^{-16t} - 10te^{-16t} = e^{-16t}(1-10t)$

7. $\frac{5s}{s^2+9s+14} = \frac{5s}{(s+7)(s+2)}$. Using form 18 with $a=-2$ and $b=-7$ we get $-2e^{-2t}+7e^{-7t}$

8. $\frac{s^2-16}{(s^2+4)^2} = \frac{s^2-4-12}{(s^2+4)^2} = \frac{s^2-4}{(s^2+4)^2} - \frac{12}{(s^2+4)^2} = \frac{s^2-2^2}{(s^2+2^2)^2} - \frac{3}{4}\frac{2\cdot 2^3}{(s^2+2^2)^2}$. By forms 15 and 12, $L^{-1}(F) = t\cos 2t - \frac{3}{4}(\sin 2t - 2t\cos 2t)$.

9. $\frac{2s+6s^2}{(s^2+25)^2} = \frac{1}{5}\cdot\frac{2\cdot 5s}{(s^2+5^2)^2} + \frac{6}{10}\frac{2\cdot 5s^2}{(s^2+5^2)^2}$. By forms 13 and 14, $L^{-1}(F) = \frac{1}{5}t\sin 5t + \frac{3}{5}(\sin 5t + 5t\cos 5t)$.

10. $\frac{s+5}{s^2+10s+26} = \frac{s+5}{s^2+10s+25+1} = \frac{s+5}{(s+5)^2+1}$. By form 20, $L^{-1}(F) = e^{-5t}\cos t$

11. $L(f'')+4L(f') = [s^2L(f)-sf(0)-f'(0)]+4[sL(f)-f(0)] = (s^2+4s)L(f)-s-4 = (s^2+4s)L(f)-(s+4)$

12. $L(f'')-6L(f) = s^2L(f)-sf(0)-f'(0)-6L(f) = (s^2-6)L(f)-3s-2 = (s^2-6)L(f)-(3s+2)$

13. $2L(f'')-3L(f')+L(f) = 2[s^2L(f)-sf(0)-f'(0)] - 3[sL(f)-f(0)]+L(f) = (2s^2-3s+1)L(f)-(2s-3)$

14. $L(f'')-4L(f') = s^2L(f)-sf(0)-f'(0)-4[sL(f)-f(0)] = (s^2-4s)L(f)-2s-1+8 = (s^2-4s)L(f)-(2s-7)$

15. $L(f'')+6L(f')-L(f) = [s^2L(f)-sf(0)-f'(0)] + 6[sL(f)-f(0)]-L(f) = (s^2+6s-1)L(f)-s+1-6 = (s^2+6s-1)L(f)-(s+5)$

16. $L(f'')-3L(f')+4L(f) = s^2L(f)-sf(0)-f'(0) - 3[sL9f)-f(0)]+4L(f) = (s^2-3s+4)L(f)+2s-1-6 = (s^2-3s+4)L(f)+(2s-7)$

34.5 PARTIAL FRACTIONS

1. $\frac{1}{x(x+1)} = \frac{A}{x} + \frac{B}{x+1}$. Multiplying by $x(x+1)$ we get $1 = A(x+1)+Bx$. If $x=0$ we get $A=1$, If $x=-1$, we get $B=-1$. The partial fraction decomposition is $\frac{1}{x} + \frac{-1}{x+1}$.

2. $\frac{4}{x(x-3)} + \frac{A}{x} + \frac{B}{x-3}$. Multiplying by $x(x-3)$ we get $4 = A(x-3)+Bx$. Setting $x=0$ we get $A = -\frac{4}{3}$. Setting $x=3$ we get $B = \frac{4}{3}$. The partial fraction decomposition is $\frac{-4/3}{x} + \frac{4/3}{x-3}$.

3. $\frac{4}{x(x-1)(x+1)} = \frac{A}{x} + \frac{B}{x-1} + \frac{C}{x+1}$. Multiplying by the LCD we get $4 = A(x^2-1)+B(x^2+x)+C(x^2-x) = (A+B+C)x^2+(B-C)x-A$. Hence $A=-4$. Substituting this we have $B+C-4=0$ and $B-C=0$. Solving we get $B=2$ and

$C = 2$. The partial fraction decomposition is $\frac{-4}{x} + \frac{2}{x-1} + \frac{2}{x+1}$.

4. $\frac{5x}{(x-2)(x+3)} = \frac{A}{(x-2)} + \frac{B}{(x+3)}$. Multiplying by the LCD we get $5x = A(x+3)+B(x-2)$. Setting $x = 2$ we find $A = \frac{10}{5} = 2$. Setting $x = -3$ we obtain $B = \frac{-15}{-15} = 3$. The answer is $\frac{2}{x-2} + \frac{3}{x+3}$.

5. $\frac{3x}{x^2-8+15} = \frac{3x}{(x-3)(x-5)} = \frac{A}{x-3} + \frac{B}{x-5}$ Multiplying by the LCD we get $3x = A(x-5) + B(x-3)$. Setting $x = 3$ yields $A = \frac{9}{-2} = -\frac{9}{2}$. Setting $x = 5$ yield $B = 15/2$. Hence the answer is $\frac{-9/2}{x-3} + \frac{15/2}{x-5}$.

6. $\frac{4x}{(x+1)(x+2)(x+3)} = \frac{A}{x+1} + \frac{B}{x+2} + \frac{C}{x+3}$. This is equivalent to $4x = A(x^2+5x+6)+B(x^2+4x+3) + C(x^2+3x+2)$. Collecting like terms we have $(A+B+C)x^2+(5A+4B+3C)x+6A+3B+2C = 4x$. This gives the linear systems $A+B+C = 0$; $5A+4B+3C = 4$; and $6A+3B+2C = 0$. The solution is $A = -2$, $B = 8$ and $C = -6$. The partial fraction decomposition is $\frac{-2}{x+1} + \frac{8}{x+2} + \frac{-6}{x+3}$.

7. $\frac{x-1}{x^2+2x+1} = \frac{x-1}{(x+1)^2} + \frac{A}{x+1} + \frac{B}{(x+1)^2}$. This is equivalent to $x-1 = A(x+1) + B$. Setting $x = 1$ yields $B = -2$. Substituting this we have $x-1 = Ax+A-2$. Using the x-term we see $A = 1$. The answer is $\frac{1}{x+1} + \frac{-2}{(x+1)^2}$.

8. $\frac{x}{(x+1)^2(x-2)} = \frac{A}{x+1} + \frac{B}{(x+1)^2} + \frac{C}{x-2}$ This is equivalent to $x = A(x+1)(x-2) + B(x-2) + C(x+1)^2$. Setting $x = -1$ we obtain $B = \frac{-1}{-3} = \frac{1}{3}$. Setting $x = 2$ we obtain $2 = 9C$ or $C = \frac{2}{9}$. The expansion of the equation (1) yields an x^2-term of $(A+C)x^2 = 0$ so $A = -\frac{2}{9}$. The partial fraction decomposition is $\frac{-2/9}{x+1} + \frac{1/3}{(x+1)^2} + \frac{2/9}{x-2}$.

9. $\frac{x^3+x^2+x+2}{(x^2+1)(x^2+2)} + = \frac{Ax+B}{x^2+1} + \frac{Cx+D}{x^2+2}$ is equivalent to $x^3+x^2+x+2 = (Ax+B)(x^2+2) + (Cx+D)(x^2+1) = (A+C)x^3+(B+D)x^2+(2A+C)x+(2B+D)$. So $A+C = 1$ and $2A+C = 1$ yield $A = 0$ and $C = 1$. Also, $B+D = 1$ and $2B+D = 2$ yield $B = 1$ and $D = 0$. The partial fraction decomposition is $\frac{1}{x^2+1} + \frac{x}{x^2+2}$.

10. $\frac{3x^2+5}{(x^2+1)^2} = \frac{Ax+B}{x^2+1} + \frac{Cx+D}{(x^2+1)^2}$ is equivalent to $3x^2+5 = (Ax+B)(x^2+1)+Cx+D = Ax^3+Bx^2+Ax+B+Cx+D = Ax^3+Bx^2+(A+C)x+(B+D)$. Hence $A = 0$, $A+C = 0 \Rightarrow C = 0$, $B = 3$, and $B+D = 5$ so $D = 5-3 = 2$. The partial fraction decomposition is $\frac{3}{x^2+1} + \frac{2}{(x^2+1)^2}$.

11. $L^{-1}\left[\frac{1}{s(s+1)}\right] = L^{-1}\left[\frac{1}{s}\right] + L^{-1}\left[\frac{-1}{s+1}\right] = 1 - e^{-t}$.

12. $L^{-1}\left[\frac{4}{s(s-3)}\right] = L^{-1}\left[\frac{-4/3}{s}\right] + L^{-1}\left[\frac{4/3}{s-3}\right] = -\frac{4}{3} + \frac{4}{3}e^{3t}$.

13. $L^{-1}\left[\frac{4}{s(s-1)(s+1)}\right] = L^{-1}\left[\frac{-4}{s}\right] + L^{-1}\left[\frac{2}{s-1}\right] + L^{-1}\left[\frac{2}{s+1}\right] = -4 + 2e^{t} + 2e^{-t}$.

14. $L^{-1}\left[\frac{5s}{(s-2)(s+3)}\right] = L^{-1}\left[\frac{2}{s-2}\right] + L^{-1}\left[\frac{3}{s+3}\right] = 2e^{2t} + 3e^{-3t}$.

15. $L^{-1}\left[\frac{3s}{s^2-8s+15}\right] = L^{-1}\left[\frac{-9/2}{s-3}\right] + L^{-1}\left[\frac{15/2}{s-5}\right] = -\frac{9}{2}e^{3t} + \frac{15}{2}e^{5t}$.

16. $L^{-1}\left[\frac{4s}{(s+1)(s+2)(s+3)}\right] = L^{-1}\left[\frac{-2}{s+1}\right] + L^{-1}\left[\frac{8}{s+2}\right] + L^{-1}\left[\frac{-6}{s+3}\right] = -2e^{-t} + 8e^{-2t} - 6e^{-3t}$.

17. $L^{-1}\left[\frac{s-1}{s^2+s+1}\right] =$
$L^{-1}\left[\frac{1}{s+1}\right] + L^{-1}\left[\frac{-2}{(s+1)^2}.\right] = e^{-t} - 2te^{-t}$

18. $L^{-1}\left[\frac{s}{(s+1)^2(s-2)}\right] =$
$L^{-1}\left[\frac{-2/9}{s+1}\right] + L^{-1}\left[\frac{1/3}{(s+1)^2}\right] + L^{-1}\left[\frac{2/9}{s-2}\right] =$
$-\frac{2}{9}e^{-t} + \frac{1}{3}te^{-t} + \frac{2}{9}e^{2t}.$

19. $L^{-1}\left[\frac{s^3+s^2+s+2}{(s^2+1)(s^2+2)}\right] =$
$L^{-1}\left[\frac{1}{s^2+1}\right] + L^{-1}\left[\frac{s}{s^2+2}\right] = \sin t + \cos\sqrt{2}t.$

20. $L^{-1}\left[\frac{3s+5}{(s^2+1)^2}\right] =$
$L^{-1}\left[\frac{3}{s^2+1}\right] + L^{-1}\left[\frac{2}{(s^2+1)^2}\right] =$
$3\sin t + \sin t - t\cos t = 4\sin t - t\cos t.$

21. First we must find the partial fraction decomposition of $\frac{2}{s^2(s-1)}$. We set it equal to $\frac{A}{s} + \frac{B}{s^2} + \frac{C}{s-1}$. Multiplying by the LCD we obtain $2 = As(s-1) + B(s-1) + Cs^2$. Setting $s = 0$ yields $B = -2$. Setting $s = 1$ yields $c = 2$. Expanding the s^2-term is $(A+C)s^2$ so $A = -C = -2$. $L^{-1}\left[\frac{2}{s^2(s-1)}\right] =$ $L^{-1}\left[\frac{-2}{s}\right] + L^{-1}\left[\frac{-2}{s^2}\right] + L^{-1}\left[\frac{2}{s-1}\right] = -2 - 2t + 2e^t$.

22. $\frac{s}{(s+1)(s^2+1)} = \frac{A}{s+1} + \frac{Bs+C}{s^2+1}$ is equivalent to $s = A(s^2+1) + (Bs+C)(s+1) = (A+B)s^2 + (B+C)s + (A+C)$. So $A+B = 0$, $B+C = 1$, and $A+C = 0$. Solving we get $A = -\frac{1}{2}$, $B = \frac{1}{2}$ and $C+\frac{1}{2}$. Hence $L^{-1}\left[\frac{s}{(s+1)(s^2+1)}\right] = L^{-1}\left[\frac{-1/2}{s+1}\right] +$ $L^{-1}\left[\frac{s/2+1/2}{s^2+1}\right] = -\frac{1}{2}e^{-t} + \frac{1}{2}\cos t + \frac{1}{2}\sin t.$

23. $\frac{1}{(s+4)(s^2+9)} = \frac{A}{s+4} + \frac{Bs+C}{s^2+9}$ is equivalent to $1 = A(s^2+9) + (Bs+C)(s+4) = (A+B)s^2 + (4B+C)s + 9A + 4C$. Hence $A+B = 0$, $4B+C = 0$ and $9A+4C = 1$. Solving we find $A = \frac{1}{25}$, $B = -\frac{1}{25}$, and $C = \frac{4}{25}$. $L^{-1}\left[\frac{1}{(s+4)(s^2+9)}\right] =$ $L^{-1}\left[\frac{1/25}{s+4}\right] + L^{-1}\left[\frac{-s/25+4/25}{s^2+9}\right] = \frac{1}{25}e^{-4t} - \frac{1}{25}\cos 3t + \frac{4}{25}\sin 3t.$

24. $\frac{1}{(s^2+1)(s^2-1)} = \frac{1}{(s^2+1)(s+1)(s-1)} =$ $\frac{A}{s+1} + \frac{B}{s-1} + \frac{Cs+D}{s^2+1}$. Multiplying by the LCD we get $1 = A(s^2+1)(s-1) + B(s+1)(s^2+1) + (Cs+D)(s^2-1)$. Setting $s = 1$ we obtain $1 = B(2\cdot 2)$ or $B = \frac{1}{4}$. Setting $s = -1$ we obtain $1 = A(2)(-2)$ or $A = -\frac{1}{4}$. The s^3-term of the expansion is $(A+B+C)s^3$ and this is 0 so $C = -A-B = 0$. The constant term of the expansion is $-A+B-D = 1$ so $D = B-A-1 = -\frac{1}{2}$. $L^{-1}\left[\frac{1}{(s^2+1)(s^2-1)}\right] =$ $L^{-1}\left[\frac{-1/4}{s+1}\right] + L^{-1}\left[\frac{1/4}{s-1}\right] + L^{-1}\left[\frac{-1/2}{s^2+1}\right] =$ $-\frac{1}{4}e^{-t} + \frac{1}{4}e^t - \frac{1}{2}\sin t.$

25. $\frac{1}{(s^2+1)(s-1)^2} = \frac{A}{s-1} + \frac{B}{(s-1)^2} + \frac{Cs+D}{s^2+1}$. Multiplying by the LCD we obtain $1 = A(s-1)(s^2+1) + B(s^2+1) + (Cs+D)(s-1)^2$. Setting $s = 1$ we get $B + \frac{1}{2}$. Expanding the right-hand side of the equation we get $As^3 - As^2 + As - A + Bs^2 + B + Cs^3 - 2Cs^2 + Cs + Ds^2 - 2Ds + D$. Collecting like terms we have $(A+C)s^3 + (-A+B-2C+D)s^2 + (A+C-2D)s + (A+B+D)$. So $A+C = 0$, $(-A+B-2C+D) = 0$, $(A+C-2D) = 0$, and $(-A+B+D) = 1$. Solving we get $A = -\frac{1}{2}$, $B = \frac{1}{2}$, $C = \frac{1}{2}$, and $D = 0$. Hence $L^{-1}\left[\frac{1}{(s^2+1)(s-1)^2}\right] =$ $L^{-1}\left[\frac{-1/2}{s-1}\right] + L^{-1}\left[\frac{1/2}{(s-1)^2}\right] + L^{-1}\left[\frac{s/2}{s^2+1}\right] =$ $-\frac{1}{2}e^t + \frac{1}{2}te^t + \frac{1}{2}\cos t.$

26. $\frac{1}{(s^2+4s+7)(s+1)} = \frac{A}{s+1} + \frac{Bs+C}{s^2+4S+7}$. Multiplying by the LCD we get $1 = A(s^2+4s+7) + (Bs+C)(s+1)$. Expanding we have $1 = As^2 + 4As + 7A + Bs^2 + Bs + Cs + c$. Collecting like terms we have $(A+B)s^2 + (4A+B+C)s + (7A+C) = 1$. Hence $A+B=0$, $4A+B+C=0$, and $7A+C=1$. Solving this system we get $A = \frac{1}{4}$, $B = -\frac{1}{4}$, and $C = -\frac{3}{4}$. Thus, $L^{-1}\left[\frac{1}{(s^2+4s+7)(s+1)}\right] = L^{-1}\left[\frac{1/4}{s+1}\right] + L^{-1}\left[\frac{-s/4-3/4}{s^2+4s+7}\right]$. The first inverse Laplace transform is $\frac{1}{4}e^{-t}$ but the second needs work. So, $\frac{-\frac{1}{4}s-\frac{3}{4}}{s^2+4s+7} = \frac{-\frac{1}{4}s-\frac{3}{4}}{s^2+4s+4+3} = \frac{-\frac{1}{4}s-\frac{3}{4}}{(s+2)^2+\sqrt{3}^2} = -\frac{1}{4}\left(\frac{s+3}{(s+2)^2+\sqrt{3}^2}\right) = -\frac{1}{4}\left(\frac{s+2+1}{(s+2)^2+\sqrt{3}^2}\right) = -\frac{1}{4}\left(\frac{s+2}{(s+2)^2+(\sqrt{3})^2}\right) - \frac{1}{4\sqrt{3}}\left(\frac{\sqrt{3}}{(s+2)^2+(\sqrt{3})^2}\right)$. Hence, $L^{-1}\left[\frac{-\frac{1}{4}s-\frac{3}{4}}{s^2+4s+7}\right] = -\frac{1}{4}L^{-1}\left[\frac{s+2}{(s+2)^2+\sqrt{3}^2}\right] - \frac{1}{4\sqrt{3}}L^{-1}\left[\frac{\sqrt{3}}{(s+2)^2+\sqrt{3}^2}\right] = -\frac{1}{4}e^{-2t}\cos\sqrt{3}t - \frac{1}{4\sqrt{3}}e^{-2t}\sin\sqrt{3}t$. Putting these answers together we get $\frac{1}{4}e^{-t} - \frac{1}{4}e^{-2t}\cos\sqrt{3}t - \frac{1}{4\sqrt{3}}e^{-2t}\sin\sqrt{3}t$.

34.6 USING LAPLACE TRANSFORMS TO SOLVE DIFFERENTIAL EQUATIONS

1. $y' - y = 1$, $y(0) = 0$, $L(y') = sL(y) - y(0) = sL(y)$. Hence the Laplace Transform is $sY - Y = \frac{1}{s}$ or $Y(s-1) = \frac{1}{s}$ or $Y = \frac{1}{s(s-1)}$. The partial fraction decomposition for $\frac{1}{s(s-1)} = \frac{A}{s} + \frac{B}{s-1}$ or $1 = A(s-1) + Bs$. Setting $s = 0$ yields $A = -1$. Setting $s = 1$ yields $B = 1$ so $Y = \frac{-1}{s} + \frac{1}{s-1}$ and $L^{-1}(Y) = L^{-1}\left(\frac{-1}{s}\right) + L^{-1}\left[\frac{1}{s-1}\right]$ or $y = -1 + e^t$.

2. $y' + 2y = t$, $y(0) = -1$. The Laplace Transform is $L(y') + 2L(y) = L(t)$ or $sL(y) - y(0) + 2L(y) = \frac{1}{s^2}$ or $(S+2)Y + 1 = \frac{1}{s^2}$. This yields $(s+2)Y = \frac{1}{s^2} - 1 = \frac{1-s^2}{s^2}$ or $Y = \frac{1-s^2}{s^2(s+2)}$. The partial fraction decomposition is as follows: $\frac{1-s^2}{s^2(s+2)} = \frac{A}{s+2} + \frac{B}{s} + \frac{C}{s^2}$ or $1 - s^2 = As^2 + B(s+2)s + C(s+2)$. Setting $s = -2$ yields $A = -\frac{3}{4}$. Setting $s = 0$ yields $C = \frac{1}{2}$. The s^2-term of the expansion yields $-1 = A + B$ so $B = -1 - A = -\frac{1}{4}$. Thus, $Y = \frac{-\frac{3}{4}}{s+2} + \frac{-\frac{1}{4}}{s} + \frac{\frac{1}{2}}{s^2}$ and $y = -\frac{3}{4}e^{-2t} - \frac{1}{4} + \frac{1}{2}t = -\frac{1}{4} + \frac{1}{2}t - \frac{3}{4}e^{-2t}$.

3. $y' + 6y = e^{-6t}$, $y(0) = 2$. The Laplace Transform is $sY - y(0) + 6Y = \frac{1}{s+6}$ or $(s+6)Y - 2 = \frac{1}{s+6}$. So $(s+6)Y = \frac{1}{s+6} + 2 = \frac{1+2s+12}{s+6} = \frac{2s+13}{s+6}$. Hence $Y = \frac{2s+13}{(s+6)^2} = \frac{A}{s+6} + \frac{B}{(s+6)^2}$. Multiplying by the LCD we have $2s + 13 = A(s+6) + B = As + 6A + B$. Hence $A = 2$ and $6A + B = 13$ or $B = 1$, $y = L^{-1}\left[\frac{2}{s+6}\right] + L^{-1}\left[\frac{1}{(s+6)^2}\right] = 2e^{-6t} + te^{-6t}$.

4. $y' - y = \sin 10t$, $y(0) = 0$. The Laplace Transform is $sY - y(0) - Y = \frac{10}{s^2+100}$ or $(s-1)Y = \frac{10}{s^2+100}$. Hence $Y = \frac{10}{(s-1)(s^2+100)} = \frac{A}{s-1} + \frac{Bs+C}{s^2+100}$. This yields $10 = A(s^2+100) + (Bs+c)(s-1)$.

Letting $s = 1$ we get $A = \frac{10}{101}$. Expanding the s^2-term is $(A + B)s^2$ which is zero so $B = -A = -\frac{10}{101}$. The s-term is $(-B + C)s$ which is also 0 so $C = B = \frac{-10}{101}$. Hence $Y = \frac{10}{101}\left(\frac{1}{s-1} + \frac{-1s-1}{s^2+100}\right)$ and $y = \frac{10}{101}L^{-1}\left[\frac{1}{s-1} - \frac{s+1}{s^2+100}\right] = \frac{10}{101}\left[e^t - \cos 10t - \frac{1}{10}\sin t\right]$.

5. $y' + 5y = te^{-5t}$, $y(0) = 3$. The Laplace Transform is $sY - y(0) + 5Y = \frac{1}{(s+5)^2}$. Hence $(s + 5)Y = \frac{1}{(s+5)^2} + 3 = \frac{1 + 3s^2 + 30s + 75}{(s+5)^2} = \frac{3s^2 + 30s + 76}{(s+5)^2}$. This yields $Y = \frac{3s^2 + 30s + 76}{(s+5)^3} = \frac{A}{s+5} + \frac{B}{(s+5)^2} + \frac{C}{(s+5)^3}$ or $3s^2+30s+76 = A(s+5)^2+B(s+5)+C$. Letting $s = -5$ we get $C = 1$. The s^2-term gives $A = 3$. The s-term yields $10A+B = 30$, so $B = 30-10A = 30 - 30 = 0$. Hence $Y = \frac{3}{s+5} + \frac{1}{(s+5)^3}$ and $y = L^{-1}\left[\frac{3}{s+5} + \frac{1}{(s+5)^3}\right] = 3e^{-5t} + \frac{1}{2}t^2e^{-5t}$.

6. $y' + y = \sin t$, $y(0) = -1$. The Laplace Transform is $sY + 1 + Y = \frac{1}{s^2+1}$ so $(s + 1)Y = \frac{1}{s^2+1} - 1 = \frac{-s^2}{s^2+1}$ and $Y = \frac{-s^2}{(s+1)(s^2+1)} = \frac{A}{s+1} + \frac{Bs+C}{s^2+1}$. Multiplying by the LCD yields $-s^2 = A(s^2 + 1) + (Bs + C)(s + 1)$. Setting $s = -1$ we get $A = -\frac{1}{2}$. Expanding, the s^2-term yields $A + B = -1$, so $B = -\frac{1}{2}$. The constant term is $A + C = 0$ so $C = \frac{1}{2}$. Hence $Y = \frac{-\frac{1}{2}}{s+1} + \frac{-\frac{1}{2}s}{s^2+1} + \frac{\frac{1}{2}}{s^2+1}$ and $y = -\frac{1}{2}L^{-1}\left[\frac{1}{s+1} + \frac{s}{s^2+1} - \frac{1}{s^2+1}\right] = -\frac{1}{2}\left[e^{-t} + \cos t - \sin t\right]$

7. $y'' + 4y = 0$, $y'(0) = 3$, $y(0) = 2$. The Laplace Transform is $s^2Y - sy(0) - y(0) + 4Y = 0$ or $(s^2 + 4)Y - 2s - 3 = 0$. Hence $Y = \frac{2s+3}{s^2+4} = 2\frac{s}{s^2+2^2} + \frac{3}{s^2+2^2}$ and $y = 2\cos 2t + \frac{3}{2}\sin 2t$.

8. $y'' - 5y' + 4y = 0$ has Laplace Transform $s^2Y - sy(0) - y'(0) - 5sY + 5y(0) + 4Y = 0$ or $(s^2 - 5s + 4)Y + 2s - 4 - 10 = 0$. Hence $Y = \frac{-2s+14}{s^2-5s+4} = \frac{-2s+14}{(s-4)(s-1)} = \frac{A}{s-4} + \frac{B}{s-1}$. Multiplying by the LCD yields $-2s+14 = A(s-1)+B(s-4)$. Setting $s = 4$ we get $3A = 6$ or $A = 2$. Setting $s = 1$ we get $-3B = 12$ or $B = -4$. Hence $Y = \frac{2}{s-4} + \frac{-4}{s-1}$. The inverse Laplace transform yields $y = 2e^{4t} - 4e^t$.

9. $y'' + 4y = e^t$, $y(0) = y'(0) = 0$. The Laplace transform is $s^2Y + 4y + \frac{1}{s-1}$ or $Y = \frac{1}{(s-1)(s^2+4)} = \frac{A}{s-1} + \frac{Bs+C}{s^2+4}$. Multiplying by the LCD we get $A(s^2 + 4) + (Bs + C)(s - 1) = 1$. Setting $s = 1$ we get $A = \frac{1}{5}$. Expanding we get $(A + B)s^2 + (-B + C)s + (4A - C) = 1$. Hence $B = -A = -\frac{1}{5}$, and $C = B = -\frac{1}{5}$. So the Laplace becomes $Y = \frac{1}{5}\left(\frac{1}{s-1} + \frac{-s-1}{s^2+4}\right) = \frac{1}{5}\left(\frac{1}{s-1} - \frac{s}{s^2+2^2} - \frac{1}{s^2+2^2}\right)$ and the inverse Laplace transform yields $y = \frac{1}{5}(e^t - \cos 2t - \frac{1}{2}\sin 2t)$.

10. $y'' + 2y' - 3y = 5e^{2t}$, $y(0) = 2$, $y'(0) = 3$. The Laplace transform is $s^2Y - sy(0) - y'(0) + 2sY - 2y(0) - 3Y = \frac{5}{s-2}$ or $(s^2+2s-3)Y - 2s - 3 - 4 = \frac{5}{s-2}$ or $(s^2 + 2s - s)Y = \frac{5}{s-2} + (2s + 7) = \frac{5 + 2s^2 + 3s - 14}{s-2}$ so $Y = \frac{2s^2+3s-9}{(s-2)(s+3)(s-1)} = \frac{A}{s-2} + \frac{B}{s+3} + \frac{C}{s-1}$. Mutiplying by the LCM we get $2s^2+3s-9 = A(s+3)(s-1)+B(s-2)(s-1)+C(s-2)(s+3)$. Setting $s = 2$ we get $A = \frac{5}{5} = 1$. Setting $S = -3, B = \frac{0}{20} = 0$ and setting $s = 1, C =$

$\frac{-4}{-4} = +1$. Hence $Y = \dfrac{1}{s-2} + \dfrac{0}{s+3} + \dfrac{1}{s-1}$ and the inverse Laplace transform yields $y = e^{2t} + e^t$.

11. $y'' - 6y' + 9y = t$, $y(0) = 0$, $y'(0) = 1$. The Laplace Transform is $s^2Y - sy(0) - y'(0) - 6(sY - y(0)) + 9 = \dfrac{1}{s^2}$ or $(s^2 - 6s + 9)Y - 1 = \dfrac{1}{s^2}$ or $(s^2 - 6s + 9)Y = \dfrac{1}{s^2} + 1 = \dfrac{1+s^2}{s^2}$. Hence $Y = \dfrac{s^2+1}{s^2(s-3)^2} = \dfrac{A}{s} + \dfrac{B}{s^2} + \dfrac{C}{s-3} + \dfrac{|Q}{(s-3)^2}$. Multiplying by the LCD we have $s^2 + 1 = As(s-3)^2 + B(s-3)^2 + C(s-3)s^2 + Ds^2$. Setting $s = 0$, we find $B = \frac{1}{9}$, and setting $s = 3$, we get $D = \frac{10}{9}$. The s-term yields $9A - 6B = 0$ so $A = \dfrac{6B}{9} = \dfrac{6}{81} = \dfrac{2}{27}$. The s^3-term yields $A + C = 0$ so $C = -A = -\frac{2}{27}$. $Y = \dfrac{2}{27}\dfrac{1}{s} + \dfrac{1}{9}\dfrac{1}{s^2} - \dfrac{2}{27}\dfrac{1}{s-3} + \dfrac{10}{9}\cdot\dfrac{1}{(s-3)^2}$. The inverse Laplace Transform yields $y = \dfrac{2}{27} + \dfrac{1}{9}t - \dfrac{2}{27}e^{3t} + \dfrac{10}{9}te^{3t}$.

12. $y'' + y = 1$, $y(0) = y'(0) = 1$. The Laplace Transform is $s^2Y - s\cdot 1 - 1 + Y = \frac{1}{s}$ so $(s^2+1)Y = \dfrac{1}{s} + s + 1 = \dfrac{s^2+s+1}{s}$ and $Y = \dfrac{s^2+s+1}{s(s^2+1)} = \dfrac{A}{s} + \dfrac{Bs+C}{s^2+1}$. Multiplying we get $s^2 + s + 1 = A(s^2+1) + (Bs + C)(s) = (A+B)s^2 + Cs + A$. So $A = 1$, $C = 1$, and $A + B = 1 \Rightarrow B = 0$. $Y = \dfrac{1}{s} + \dfrac{1}{s^2+1}$. The inverse Laplace Transform is $y = 1 + \sin t$.

13. $y'' + 6y' + 13y = 0$, $y(0) = 1$, $y''(0) = -4$. The Laplace Transform is $s^2Y - s\cdot 1 + 4 + 6(sY - 1) + 13Y = 0$ which simplifies to $(s^2 + 6s + 13)Y = s + 2$, so $Y = \dfrac{s+2}{s^2+6s+13} = \dfrac{s+2}{s^2+6s+9+4} = \dfrac{s+3-1}{(s+3)^2+2^2} = \dfrac{s+3}{(s+3)^2+2^2} - \dfrac{1}{2}\dfrac{2}{(s+3)^2+2^2}$. The inverse Laplace Transform yields $y = e^{-3t}(\cos 2t - \frac{1}{2}\sin 2t)$.

14. $y'' + 2y' + 5y = 8e^t$, $y(0) = y'(0) = 0$. The Laplace Transform is $s^2Y + 2sY + 5Y = \dfrac{8}{s-1}$, so $Y = \dfrac{8}{(s-1)(s^2+2s+s)} = \dfrac{A}{s-1} + \dfrac{Bs+C}{s^2+2s+5}$, and $8 = A(s^2+2s+5) + (Bs+C)(s-1) = (A+B)s^2 + (2A-B+C)s + 5A - C$. So $A + B = 0$, $2A - B + C = 0$, and $5A - C = 8$. This system yields $A = 1$, $B = -1$, $C = -3$ so $Y = \dfrac{1}{s-1} - \dfrac{s+3}{s^2+2s+5} = \dfrac{1}{s-1} - \dfrac{s+2+1}{s^2+2s+1+4} = \dfrac{1}{s-1} - \dfrac{s+2}{(s+1)^2+2^2} - \dfrac{1}{2}\cdot\dfrac{2}{(s+1)^2+2^2}$. The inverse Laplace Transform yields $y = e^t - e^{-t}\cos 2t - \frac{1}{2}e^{-t}\sin 2t$.

15. $y'' - 4y = 3\cos t$, $y(0) = y'(0) = 0$. The Laplace Transform is $s^2Y - 4Y = \dfrac{3s}{s^2+1}$ so $Y = \dfrac{3s}{(s^2+1)(s^2-1)} = \dfrac{3s}{(s^2+1)(s+2)(s-2)} = \dfrac{A}{s+2} + \dfrac{B}{s-2} + \dfrac{Cs+D}{s^2+1}$. Multiplying by the LCD we get $3s = A(s-2)(s^2+1) + B(s+2)(s^2+1) + (Cs+D)(s^2-4)$. Setting $s = -2$, we get $A = \frac{-6}{(-4)(5)} = \frac{3}{10}$. Setting $s = 2$, we get $B = \frac{6}{4\cdot 5} = \frac{3}{10}$. The s^3-term of the expansion is $A + B + C = 0$ so $C = -\frac{6}{10} = -\frac{3}{5}$. The constant term of the expansion is $-2A + 2B - 4D = 0$ so $D = \dfrac{-2A+2B}{4} = 0$. Hence $Y = \dfrac{3}{10}\cdot\dfrac{1}{s+2} + \dfrac{3}{10}\cdot\dfrac{1}{s-2} - \dfrac{3}{5}\cdot\dfrac{s}{s^2+1}$ and the inverse Laplace Transform is $y = \frac{3}{10}e^{-2t} + \frac{3}{10}e^{2t} - \frac{3}{5}\cos t$.

16. $y'' - y = e^t$, $y(0) = 1$, $y'(0) = 0$. The Laplace Transform is $s^2Y - s - Y = \dfrac{1}{s-1}$. This is equivalent to $(s^2 - 1)Y = \dfrac{1}{s-1} + s = \dfrac{s^2-s+1}{s-1}$ or $Y = \dfrac{s^2-s+1}{(s^2-1)(s-1)} = \dfrac{s^2-s+1}{(s-1)(s-1)^2}$. Finding the partial fraction decomposition we have $\dfrac{s^2-s+1}{(s-1)(s-1)^2} = \dfrac{A}{s+1} + \dfrac{B}{s-1} + \dfrac{C}{(s-1)^2}$. This becomes $s^2 - s + 1 = A(s-1)^2 + B(s^2-1) + C(s+1)$. Setting $s = -1$ we get $A = \frac{3}{4}$, Setting $s = 1$ we get $C = \frac{1}{2}$. The s^2-term yields $A + B = 1$ or

$B = 1 - A = 1 - \frac{3}{4} = \frac{1}{4}$. Hence $Y = \frac{3}{4} \cdot \frac{1}{s+1} + \frac{1}{4} \cdot \frac{1}{s-1} + \frac{1}{2} \cdot \frac{1}{(s-1)^2}$. The inverse Laplace transform yields the answer $y = \frac{3}{4}e^{-t} + \frac{1}{4}e^{t} + \frac{1}{2}te^{t}$.

17. $y'' + 2y' + 5y = 3e^{-2t}, y(0) = y'(0) = 1$. The Laplace transform is $(s^2Y - s - 1) + 2sY - 2 + 5Y = \frac{3}{s+2}$. This is equivalent to $(s^2 + 2s + 5)Y - (s + 3) = \frac{3}{s+2}$. So $(s^2 + 2s + 5)Y = \frac{3}{s+2} + (s + 3) = \frac{3 + (s^2 + 5s + 6)}{s+2} = \frac{s^2 + 5s + 9}{s+2}$. Hence $Y = \frac{(s^2 + 5s + 9)}{(s+2)(s^2 + 2s + 5)} = \frac{A}{s+2} + \frac{Bs + C}{s^2 + 2s + 5}$. This gives $(s^2+5s+9) = A(s^2+2s+5)+(Bs+C)(s+2) = (A+B)s^2+(2A+2B+C)s+5A+2C$. And we have linear equations $A + B = 1$, $2A + 2B + C = 5$ and $5A + 2C = 9$. Solving we obtain $A = \frac{3}{5}$, $B = \frac{2}{5}$, and $C = 3$. So $Y = \frac{3}{5} \cdot \frac{1}{s+2} + \frac{\frac{2}{5}s + 3}{s^2 + 2s + 5}$. The second fraction is $\frac{2}{5} \frac{s + \frac{15}{2}}{s^2 + 2s + 1 + 4} = \frac{2}{5} \frac{s + 1 + \frac{13}{2}}{(s+1)^2 + 2^2} = \frac{2}{5} \frac{s+1}{(s+1)^2 + 2^2} + \frac{13}{10} \frac{2}{(s+1)^2 + 2^2}$. The inverse Laplace transform is $y = \frac{3}{5}e^{-2t} + e^{-t}\left(\frac{2}{5}\cos 2t + \frac{13}{10}\sin 2t\right)$.

18. $y'' - 6y' + 9y = 12t^2e^{3t}$, $y(0) = y'(0) = 0$. The Laplace transform is $s^2Y - 6sY + 9Y = \frac{24}{(s-3)^3}$ or $(s^2 - 6s + 9)Y = \frac{24}{(s-3)^3}$. Hence $Y = \frac{24}{(s-3)^5} = \frac{4!}{(s-3)^5}$. The inverse Laplace Transform yields the answer $y = t^4e^{3t}$.

19. $y''+2y'-3y = te^{2t}, y(0) = 2, y'(0) = 3$. The Laplace transform is $(s^2Y - 2s - 3) + 2sY - 4 - 3Y = \frac{1}{s-2)^2}$. Regrouping we get $(s^2 + 2s - 3)Y - (2s + 7) = \frac{1}{(s-2)^2}$. Hence $(s^2 + 2s - 3)Y = \frac{1}{(s-2)^2} + (2s + 7) = \frac{2s^3 - s^2 - 20s + 29}{(s-2)^2}$ or $Y = \frac{2s^3 - s^2 - 20s + 29}{(s-2)^2(s+3)(s-1)} = \frac{A}{s+3} + \frac{B}{s-1} + \frac{C}{s-2} + \frac{|Q}{(s-2)^2}$. Multiplying by the LCD we get $2s^3 - s^2 - 20s + 29 = A(s-1)(s-2)^2 + B(s+3)(s-2)^2 + C(s+3)(s-1)(s-2) + D(s+3)(s-1)$. Setting $s = -3$ we get $A = \frac{26}{-100} = \frac{-13}{50}$. Setting $s = 1$ we get $B = \frac{10}{4} = \frac{5}{2}$. Setting $s = 2$, we get $D = \frac{1}{5}$. The s^3-term yields $A + B + C = 2$ or $C = 2 - A - B = \frac{6}{25}$. Hence we have $Y = -\frac{13}{50} \cdot \frac{1}{s+3} + \frac{5}{2} \cdot \frac{1}{s-1} - \frac{6}{25} \cdot \frac{1}{s-2} + \frac{1}{5} \cdot \frac{1}{(s-2)^2}$. The inverse Laplace Transform gives the answer $y = -\frac{13}{50}e^{-3t} + \frac{5}{2}e^{t} - \frac{6}{25}e^{2t} + \frac{1}{5}te^{2t}$.

20. $y'' + 2y' - 3y = \sin 2t$, $y(0) = y'(0) = 0$. The Laplace Transform is $s^2Y + 2sY - 3Y = \frac{2}{s^2+4}$. This is equivalent to $(s^2+2s-3)Y = \frac{2}{s^2+4}$ or $Y = \frac{2}{(s^2+4)(s+3)(s-1)} = \frac{A}{s+3} + \frac{B}{s-1} + \frac{Cs + D}{s^2+4}$. Multiplying by the LCD we have $2 = A(s-1)(s^2+4)+B(s+3)(s^2+4)+(Cs+D)(s+3)(s-1)$. Setting $s = -3$, we get $A = \frac{2}{-52} = \frac{-1}{26}$. Setting $s = 1$, we get $B = \frac{2}{20} = \frac{1}{10}$. The s^3-term yields the equation $A+B+C = 0$ or $C = \frac{1}{26} - \frac{1}{10} = \frac{-16}{260} = \frac{-4}{65}$. The constant term yields the equation $-4A + 12B - 3D = 2$ or $D = \frac{4A - 12B + 2}{-3} = \frac{-\frac{2}{13} - \frac{6}{5} + 2}{-3} = \frac{-14}{65}$. Hence $Y = -\frac{1}{26} \cdot \frac{1}{s+3} + \frac{1}{10} \cdot \frac{1}{s-1} - \frac{4}{65} \cdot \frac{s}{s^2+4} - \frac{7}{65} \cdot \frac{2}{s^2+4}$. The inverse Laplace Transform yields the answer $y = -\frac{1}{26}e^{-3t} + \frac{1}{10}e^{6} - \frac{4}{65}\cos 2t - \frac{7}{65}\sin 2t$.

21. $y'' + 2y' + 5y = 10\cos t$, $y(0) = 2$, $y'(0) = 1$. The Laplace transform is $s^2Y - 2s - 1) + (2sY - 4) + 5Y = \frac{10s}{s^2+1}$. This is equivalent to $(s^2+2s+5)Y - (2s+5) = \frac{10s}{s^2+1}$ or $(s^2+2s+5)Y = \frac{10s}{s^2+1} + (2s + 5) = \frac{2s^3 + 5s^2 + 12s + 5}{s^2+1}$ or $Y =$

$\frac{2s^3+5s^2+12s+5}{(s^2+1)(s^2+2s+5)} = \frac{As+B}{s^2+1} + \frac{Cs+D}{s^2+2s+5}$. Multiplying by the LCD we get $2s^3+5s^2+12s+5 = (As+B)(s^2+2s+5)+(Cs+D)(s^2+1) = As^3+(2A+B)s^2+(5A+2B)s+5B+Cs^3+Ds^2+Cs+D$. The yields the linear equations $A+C=2$, $2A+B+D=5$, $5A+2B+C=12$, and $5B+D=5$. Solving this system we get $A=2$, $B=1$, $C=0$, and $D=0$. Hence the partial fraction decomposition is $Y = \frac{2s+1}{s^2+1} = 2\frac{s}{s^2+1} + \frac{1}{s^2+1}$. The inverse Laplace transform yields the answer $y = 2\cos t + \sin t$. (Note: $2s^3+5s^2+12s+5$ factors into $(2s+1)(s^2+2s+5)$ so we could have reduced $\frac{2s^3+5s^2+12s+5}{(s^2+1)(s^2+2s+5)}$ to $\frac{2s+1}{s^2+1}$ and not used partial fractions).

22. $y''+2y'+5y = 10\cos t$, $y(0)=0$, $y'(0)=3$. The Laplace transformation is $(s^2Y-3)+2sY+5Y = \frac{10s}{s^2+1}$ or $(s^2+2s+5)Y-3 = \frac{10s}{s^2+1}$. Thus $(s^2+2s+5)Y = \frac{10s}{s^2+1}+3 = \frac{3s^2+10s+3}{s^2+1}$. This gives $Y = \frac{3s^2+10s+3}{(s^2+1)(s^2+2s+5)} = \frac{As+B}{s^2+1} + \frac{Cs+D}{s^2+2s+5}$. Multiplying we get $3s^2+10s+3 = (As+B)(s^2+2s+5)+(Cs+D)(s^2+1) = As^3+(2A+B)s^2+(5A+2B)s+5B+Cs^3+Ds^2+Cs+D$. The four like terms yield the four linear equations $A+C=0$, $2A+B+D=3$, $5A+2B+C=10$, and $5B+D=3$. Solving this system we get $A=2$, $B=1$, $C=-2$, and $D=-2$. So the partial fraction decomposition of $Y = \frac{2s+1}{s^2+1} - 2\cdot\frac{s+1}{s^2+2s+5} = 2\frac{s}{s^2+1} + \frac{1}{s^2+1} - 2\frac{s+1}{(s+1)^2+2^2}$. The inverse Laplace transform yields the answer $y = 2\cos t + \sin t - 2e^{-t}\cos 2t$.

23. We start with the equations $L\frac{d^2q}{dt^2} + R\frac{dq}{dt} + \frac{1}{C}q = E(t)$ which becomes $q''+20q'+200q = 150$. The Laplace transform is $s^2Q+20sQ+200Q = \frac{150}{s}$. Hence $(s^2+20s+200)Q = \frac{150}{s}$ or $Q = \frac{150}{s(s^2+20s+200)} = \frac{A}{s} + \frac{Bs+C}{s^2+20s+200}$. Multiplying by the LCD we get $150 = A(s^2+20S+200)+(Bs+C)s = (A+B)s^2+(20A+C)s+200A$. The constant term yields $A = \frac{150}{200} = \frac{3}{4}$. The s^2-term yields $A+B=0$ so $B = -\frac{3}{4}$. The s-term yields $20A+C=0$ or $C=-15$. Hence $Q = \frac{3}{4}\cdot\frac{1}{s} + \frac{-\frac{3}{4}s-15}{s^2+20s+200} = \frac{3}{4}\cdot\frac{1}{s} + \frac{-\frac{3}{4}(s+20)}{s^2+20s+100+100} = \frac{3}{4}\cdot\frac{1}{s} - \frac{3}{4}\cdot\frac{s+10+10}{(s+10)^2+10^2} = \frac{3}{4}\frac{1}{s} - \frac{3}{4}\frac{s+10}{(s+10)^2+10^2} - \frac{3}{4}\frac{10}{(s+10)^2+10^2}$. The inverse Laplace transform yields the answe $q(t) = \frac{3}{4} - \frac{3}{4}e^{-10t}(\cos 10t + \sin 10t)$. $i(t) = q'(t) = \frac{15}{2}e^{-10t}(\cos 10t + \sin 10t) - \frac{15}{2}e^{-10t}(-\sin 10t + \cos 10t) = 15e^{-10t}\sin 10t$. The steady state current is 0.

24. We begin with the equation $i'+10i = \sin t$. The Laplace transform is $SI+10I = \frac{1}{s^2+1}$. Hence $I = \frac{1}{(s^2+1)(s+10)} = \frac{A}{s+10} + \frac{Bs+C}{s^2+1}$. Multiplying by the LCD we get $1 = A(s^2+1)+(Bs+C)(s+10) = (A+B)s^2+(10B+C)s+(A+10C)$. This yields the linear equations $A+B=0$, $10B+C=0$, and $A+10C=1$. Solving we get $A = \frac{1}{101}$, $B = \frac{-1}{101}$, and $C = \frac{10}{101}$. So $I = \frac{1}{101}\cdot\frac{1}{s+10} + \frac{1}{101}\cdot\frac{-s+10}{s^2+1} = \frac{1}{101}\left(\frac{1}{s+10} - \frac{s}{s^2+1} + \frac{10}{s^2+1}\right)$. The inverse Laplace transform yields the answer $i(t) = \frac{1}{101}(e^{-10t} - \cos t + 10\sin t)$.

25. We begin with the equation $0.9q''+6q'+50q = 6$. The Laplace transform is $0.1s^2Q+6sQ+50 = \frac{6}{s}$ so $Q = \frac{6}{s(0.1s^2+6s+50)}$ or $Q = \frac{60}{s(s^2+60s+500)}$ or $Q = \frac{60}{s(s+10)(s+50)} = \frac{A}{s} + \frac{B}{s+10} + \frac{C}{s+50}$. Multiplying by the LCD we get $60 = A(s+10)(s+50)+Bs(s+50)+Cs(s+10)$. Setting $s=0$, we get $A = \frac{60}{500} = \frac{3}{25}$. Setting $s=-10$, we get $B = \frac{60}{-400} = -\frac{3}{20}$. Setting

$s = -50$, yields $C = \frac{60}{2000} = \frac{3}{100}$. Hence we get $Q = \frac{3}{25} \cdot \frac{1}{s} - \frac{3}{20} \cdot \frac{1}{s+10} + \frac{3}{100} \cdot \frac{1}{s+50}$. The inverse Laplace transform yields the answer $q(t) = \frac{3}{25} - \frac{3}{20}e^{-10t} + \frac{3}{100}e^{-50t}$. Thus, $i(t) = q'(t) = \frac{3}{2}e^{-10t} - \frac{3}{2}e^{-50t} = \frac{3}{2}(e^{-10t} - e^{-50t})$.

26. We begin with the equation $0.05q'' + 20q' + 10{,}000q = 100\cos 200t$. Multiplying by 20 to eliminate decimals yields $q'' + 400q' + 200{,}000q = 2000\cos 200t$. The Laplace transform is $s^2Q + 400sQ + 200{,}000Q = \frac{2{,}000s}{s^2+(200)^2}$. Solving for Q, we get $Q = \frac{2{,}000s}{(s^2+40000)(s^2+400s+200{,}000)} = \frac{As+B}{s^2+40{,}000} + \frac{Cs+D}{s^2+400s+200{,}000}$. Multiplying by the LCD we get $2000s = (As+B)(s^2+400s+200000) + (Cs+D)(s^2+40{,}000) = (A+C)s^3 + (400A+B+D)s^2 + (200{,}000A+400B+40{,}000C) + (200000B+40000D)$. This yields the linear equations $A+C=0$, $400A+B+D=0$, $200000A+400B+40000C=2000$, and $200000B+40000D=0$. Solving we get $A = \frac{1}{100}$, $B = 1$, $C = \frac{-1}{100}$, and $D = -5$. Hence, $Q = \frac{1}{100}\left(\frac{s+100}{s^2+200^2}\right) - \frac{1}{100}\left(\frac{s+500}{s^2+400s+200{,}00}\right) = \frac{1}{100}\frac{s}{s^2+200^2} + \frac{1}{200}\frac{200}{s^2+200^2} - \frac{1}{100}\left(\frac{s+200+300}{(s+200)^2+160{,}000}\right) = 0.01\frac{s}{s^2+200^2} + 0.005\frac{200}{s^2+200^2} - 0.01 \cdot \frac{s+200}{(s+200)^2+400^2} - 0.0075\frac{400}{(s+200)^2+400^2}$. The inverse Laplace transform yields $q(t) = 0.01\cos 200t + 0.005\sin 200t - 0.01e^{-200t}\cos 400t - 0.0075e^{-200t}\sin 400t = 0.01\cos 200t + 0.005\sin 200t - \frac{1}{400}e^{-200t}(4\cos 400t + 3\sin 400t)$. Thus, $i(t) = q'(t) = -2\sin 200t + \cos 200t + \frac{1}{2}e^{-200t}(4\cos 400t + 3\sin 400t) - e^{-200t}(-4\sin 400t + 3\cos 400t) = -2\sin 200t + \cos 200t + e^{-200t}(\frac{11}{2}\sin 400t - \cos 400t)$. The steady-state current is $-2\sin 200t + \cos 200t$ or $\cos 200t - 2\sin 200t$.

27. First, a 16 pound weight has a mass of $\frac{1}{2}$ slug. The differential equation is $\frac{1}{2}\frac{d^2x}{dt^2} = -4x - 64\frac{dx}{dt}$, we need not bother changing to slugs. The differential equation is $16\frac{d^2x}{dt^2} = -4x - 64\frac{dx}{dt}$ which is equivaltent to $x'' + 128x' + 8x = 0$. This has auxiliary equation $m^2 + 128m + 18 = 0$, which has roots $\frac{-128 \pm \sqrt{128^2 - 32}}{2} = -64 \pm \sqrt{4088} = -64 \pm 2\sqrt{1022}$. Hence the general solution is $x = c_1e^{\left(-64+2\sqrt{1022}\right)t} + c_2e^{\left(-64-2\sqrt{1022}\right)t}$. When $t = 0$, then $x = 2$, so $c_1 + c_2 = 2$. When $t = 0$, $x' = 0$, so $\left(-64+2\sqrt{1022}\right)c_1 + \left(-64-2\sqrt{1022}\right)c_2 = 0$. Substituting $2 - c_2$ for c_1, we get $\left(-64+2\sqrt{1022}\right)(2-c_2) + \left(-64-2\sqrt{1022}\right)c_2 = 0$ or $-128 + 4\sqrt{1022} + 64c_2 - 2\sqrt{1022}c_2 - 64c_2 - 2\sqrt{1022} = 0$, or $-4\sqrt{1022}c_2 = 128 - 4\sqrt{1022}$, and so $c_2 = 1 - \frac{32}{\sqrt{1022}}$. Thus, $c_1 = 2 - c_2 = 2 - \left(1 - \frac{32}{\sqrt{1022}}\right) = 1 + \frac{32}{\sqrt{1022}}$. Thus, the final solution is $x = \left(1 + \frac{32}{\sqrt{1022}}\right)e^{\left(-64+2\sqrt{1022}\right)t} + \left(1 - \frac{32}{\sqrt{1022}}\right)e^{\left(-64-2\sqrt{1022}\right)t}$.

28. As in Exercise #27, we have the differential equation $\frac{1}{2}x'' + 64x' + 4x = \cos 4t$. This has the same complementary solution as in Exercise #27. To complete the particular solution, we assume $x = A\cos 4t + B\sin 4t$. Then, $x' = -4A\sin 4t + 4B\cos 4t$ and $x'' = -16A\cos 4t - 16B\sin 4t$. Substituting, we obtain $-8A\cos 4t - 8B\sin 4t - 256A\sin 4t + 256B\cos 4t + 4A\cos 4t + 4B\sin 4t = \cos 4t$. Collecting like functions, we obtain the equations

$$-8A + 256B + 4A = 1$$

and

$$-8B - 256A + 4B = 0$$

or

$$-4A + 256B = 1$$

and

$$-256A - 4B = 0$$

Solving, the last equation for B, we have $B = -64A$. Substituting, we get $-4A + 256(-64A) = 1$ or $A = \frac{-1}{16{,}388}$. Back substituting, we find $B = \frac{64}{16{,}388}$. Hence, the final solution is $x = \left(1 + \frac{32}{\sqrt{1022}}\right) e^{\left(-64+2\sqrt{1022}\right)t} + \left(1 - \frac{32}{\sqrt{1022}}\right) e^{\left(-64-2\sqrt{1022}\right)t} - \frac{1}{16{,}388} \cos 4t + \frac{64}{16{,}388} \sin 4t$.

CHAPTER 34 REVIEW

1.

x	y	
	$h = 0.1$	$h = 0.05$
0	2	2
0.05	—	1.8
0.10	1.6	1.6299
0.15	—	1.4838
0.20	1.3184	1.3572
0.25	—	1.2467
0.30	1.1098	1.1495
0.35	—	1.0636
0.40	0.9497	0.9873
0.45	—	0.9190
0.50	0.8234	0.8578
0.55	—	0.8026
0.60	0.7217	0.7527
0.65	—	0.7074
0.70	0.6384	0.6661
0.75	—	0.6284
0.80	0.5691	0.5938
0.85	—	0.5621
0.90	0.5108	0.5329
0.95	—	0.5059
1.00	0.4612	0.4809

2.

x	y	
	$h = 0.1$	$h = 0.05$
0	1	1
0.05	—	1.1
0.10	1.2	1.2052
0.15	—	1.3157
0.20	1.4218	1.4313
0.25	—	1.5520
0.30	1.6649	1.6778
0.35	—	1.8088
0.40	1.9288	1.9448
0.45	—	2.0859
0.50	2.2133	2.2321
0.55.	—	2.3833
0.60	2.5183	2.5395
0.65	—	2.7001
0.70	2.8436	2.8671
0.75	—	3.0385
0.80	3.1892	3.2149
0.85	—	3.3963
0.90	3.5549	3.5827
0.95	—	3.7742
1.00	3.9408	3.9706

3.

x	y
	$h = 0.05$
0	60
0.05	59.9
0.10	59.8
0.15	59.6990
0.20	59.5961
0.25	59.4905
0.30	59.3812
0.35	59.2678
0.40	59.1496
0.45	59.0263
0.50	58.8975
0.55	58.7630
0.60	58.6228
0.65	58.4767
0.70	58.3249
0.75	58.1675
0.80	58.0047
0.85	57.8366
0.90	57.6635
0.95	57.4857
1.00	57.3034

4.

x	y $h = 0.01$
0	0
0.01	0.06
0.02	0.1200
0.03	0.1800
0.04	0.2400
0.05	0.2999
0.06	0.3598
0.07	0.4197
0.08	0.4796
0.09	0.5394
0.10	0.5991

5. $y' = xy + 4$, $(0, 1)$. Substituting 1 for y we get $y' = x + 4$. Integrating we have $y = \frac{1}{2}x^2 + 4x + C_1$. Solving for C_1 we get $C_1 = 1$, so the first approximation is $y = \frac{1}{2}x^2 + 4x + 1$. Now we substitute this for y and get $y' = \frac{1}{2}x^3 + 4x^2 + x + 4$. Integration this we get $y = \frac{1}{8}x^4 + \frac{4}{3}x^3 + \frac{1}{2}x^2 + 4x + C_2$. Again $C_2 = 1$ so the second approximation is $y = \frac{1}{8}x^4 + \frac{4}{3}x^3 + \frac{1}{2}x^2 + 4x + 1$.

6. $y' = 2x + y$. Substituting 0 for y we have $y' = 2x$. Integration yields $y = x^2 + C_1$. $C_1 = 0$ so the first approximation is $y = x^2$. Substituting into the original equation we get $y' = 2x + x^2$. Integration we have $y = x^2 + \frac{1}{3}x^3 + C_2$. Again $C_2 = 0$ so the second approximation is $y = x^2 + \frac{1}{3}x^3$.

7. $y' = y\sqrt{x-9}$, $(10, 1)$. Substituting 1 for y we get $y' = \sqrt{x-9} = (x-9)^{1/2}$. Integration yields $y = \frac{2}{3}(x-9)^{\frac{3}{2}} + C_1$. Substituting $(10, 1)$ we get $1 = \frac{2}{3}(1) + C_1$ or $C_1 = \frac{1}{3}$. The first approximation is $y = \frac{2}{3}(x-9)^{3/2} + \frac{1}{3}$. Now substituting this for y in the original equation we obtain $y' = (\frac{2}{3}(x-9)^{3/2} + \frac{1}{3})(x-9)^{1/2} = \frac{2}{3}(x-9)^2 + \frac{1}{3}(x-9)^{1/2}$. Integrating again we get $y = \frac{2}{9}(x-9)^3 + \frac{2}{9}(x-9)^{\frac{3}{2}}$. Solving for C_2 we have $1 - \frac{2}{9} - \frac{2}{9} = \frac{5}{9}$. The second approximation is $y = \frac{2}{9}(x-9)^3 + \frac{2}{9}(x-9)^{\frac{3}{2}} + \frac{5}{9}$.

8. $y' = y + \cos x$, $(\frac{\pi}{2}, 0)$. Substituting 0 for y we have $y' = \cos x$ so $y = \sin x + C_1$. Using $(\frac{\pi}{2}, 0)$ we obtain $C_1 = -1$ so $y = \sin x - 1$. Substituting this in the original equation we get $y' = \sin x + \cos x - 1$. Integration yields $y = -\cos x + \sin x - x + C_2$. $C_2 = \frac{\pi}{2} - 1$ and the second approximation is $y = -\cos x + \sin x - x + \frac{\pi}{2} - 1$.

9. $2y' - y = 4$, $y(0) = 1$ has Laplace transform $2sY - 2 - Y = \frac{4}{s}$ or $(2s-1)Y = \frac{4}{s} + 2 = \frac{4+2s}{s}$. So we get $Y = \frac{4+2s}{s(2s-1)} = \frac{A}{s} + \frac{B}{2s-1}$. Multiplying by the LCD we have $4 + 2s = A(2s-1) + Bs$. Setting $s = 0$, we obtain $A = \frac{4}{-1} = -4$. Setting $s = \frac{1}{2}$, we obtain $B = \frac{5}{1/2} = 10$. So we have $Y = -4\frac{1}{s} + 10 \cdot \frac{1}{2s-1} = -4\frac{1}{s} + 5\frac{1}{s - \frac{1}{2}}$. The inverse Laplace transformation yields the solution $y = -4 + 5e^{t/2}$.

10. $3y' + y = t$, $y(0) = 2$. The Laplace transform is $3sY - 6 + Y = \frac{1}{s^2}$ so $(3s+1)Y = \frac{1}{s^2} + 6 = \frac{1+6s^2}{s^2}$ and $Y = \frac{1+6s^2}{s^2(3s+1)} = \frac{A}{3s+1} + \frac{B}{s} + \frac{C}{s^2}$. Multiplying by the LCD we have $1 + 6s^2 = As^2 + Bs(3s+1) + C(3s+1)$. Setting $s = -\frac{1}{3}$, we obtain $A = \frac{5/3}{1/9} = 15$. Setting $s = 0$, we obtain $C = 1$. The s^2-term yields $6 = A + 3B$ so $B = \frac{6-A}{3} = \frac{-9}{3} = -3$. So the partial fraction decomposition is $Y = \frac{15}{3s+1} - \frac{3}{s} + \frac{1}{s^2} = \frac{5}{s+\frac{1}{3}} - 3 \cdot \frac{1}{s} + \frac{1}{s^2}$. The inverse Laplace transform yields the answer $y = 5e^{-t/3} - 3 + t$.

11. $y' + 2y = e^t$, $y(0) = 1$. The Laplace transform is $sY - 1 + 2Y = \frac{1}{s-1}$ or $(s+2)Y = \frac{1}{s-1} + 1 = \frac{s}{s-1}$. So $Y = \frac{s}{(s-1)(s+2)} = \frac{A}{s-1} + \frac{B}{s+2}$. This gives $s = A(s+2) + B(s-1)$. Setting $s = 1$, we get $A = \frac{1}{3}$. Setting $s = -2$, we get $B = \frac{2}{3}$. Hence $Y = \frac{1}{3} \cdot \frac{1}{s-1} + \frac{2}{3} \cdot \frac{1}{s+2}$. The inverse Laplace transform yields $y = \frac{1}{3}e^t + \frac{2}{3}e^{-2t}$.

12. $y' + 5y = 0$, $y(0) = 1$. The Laplace transform is $sY - 1 + sY = 0$ or $(s+5)Y = 1$ or $Y = \frac{1}{s+5}$. The inverse Laplace transform yields $y = e^{-5t}$.

13. $y'' - y = 0$, $y(0) = y'(0) = 1$. the Laplace transform is $s^2Y - s - 1 - Y = 0$ or $(s^2 - 1)Y = s + 1$. Hence $Y = \frac{s+1}{s^2-1} = \frac{1}{s-1}$. The inverse Laplace transform yields $y = e^t$.

14. $y'' - y = e^t$, $y(0) = y'(0) = 0$. The Laplace transform is $s^2Y - Y = \frac{1}{s-1}$ or $Y = \frac{1}{(s-1)(s^2-1)} = \frac{1}{(s-1)^2(s+1)}$. Getting partial fractions we have $\frac{1}{(s-1)^2(s+1)} = \frac{A}{s-1} + \frac{B}{(s-1)^2} + \frac{C}{s+1}$. Multiplying we get $1 = A(s^2-1) + B(s+10 + C(s-1)^2$. Setting $s = 1$, we obtain $B = \frac{1}{2}$. The s^2-term yields $A + C = 0$. The constant term yields $-A + B + C = 1$ or $-A + C = \frac{1}{2}$. Solving, we get $C = \frac{1}{4}$ and $A = -\frac{1}{4}$. Hence $Y = -\frac{1}{4}\frac{1}{s-1} + \frac{1}{2} \cdot \frac{1}{(s-1)^2} + \frac{1}{4} \cdot \frac{1}{s+1}$. The inverse Laplace transform yields $y = -\frac{1}{4}e^t + \frac{1}{2}te^t + \frac{1}{4}e^{-t}$.

15. $y'' + 2y' + 5y = 0$, $y(0) = 1$, $y'(0) = 0$. The Laplace transform is $s^2Y - s + 2sY - 2 + 5Y = 0$ or $(s^2 + 2s + 5)Y = s + 2$. So $Y = \frac{s+2}{s^2+2s+5} = \frac{s+2}{(s+1)^2+2^2} = \frac{s+1}{(s+1)^2+2^2} + \frac{1}{2}\frac{2}{(s+1)^2+2^2}$. The inverse Laplace transform yields the answer $y = e^{-t}\cos 2t + \frac{1}{2}e^{-t}\sin 2t$.

16. $y'' + y' + y = 0$, $y(0) = 4$, $y'(0) = -2$. The Laplace transform is $(s^2Y - 4s + 2) + (sY - 4) + Y = 0$ or $(s^2 + s + 1)Y = 4s + 2$. Hence $Y = \frac{4s+2}{s^2+s+1} = \frac{4s+2}{s^2+s+\frac{1}{4}+\frac{3}{4}} = 4\frac{s+\frac{1}{2}}{\left(s+\frac{1}{2}\right)^2+\frac{3}{4}}$. The inverse Laplace transform gives $y = 4e^{-t/2}\cos\frac{\sqrt{3}}{2}t$.

17. $y'' + 2y' + 5y = 3e^{-2t}$, $y(0) = y'(0) = 1$. The Laplace transform is $(s^2Y - s - 1) + 2sY - 2 + 5 = \frac{3}{s+2}$ or $(s^2 + 2s + 5)Y - (s + 3) = \frac{3}{s+2}$. Hence $(s^2+2s+5)Y = \frac{3}{s+2} + s + 3 = \frac{s^2+5s+9}{s+2}$ and $Y = \frac{s^2+5s+9}{(s+9)(s^2+2s+5)} = \frac{A}{s+2} + \frac{Bs+C}{s^2+2s+5}$. Multiplying we obtain $s^2+5s+9 = A(s^2+2s+5) + (Bs+C)(s+2) = (A+B)s^2 + (2A+2B+C)s + (5A+2C)$. Hence we get the linear equation $A + B = 1$, $2A + 2B + C = 5$, and $5A + 2C = 9$. Solving we obtain $A = \frac{3}{5}$, $B = \frac{2}{5}$, and $C = 3$. Thus we have $Y = \frac{3}{5} \cdot \frac{1}{s+2} + \frac{2}{5} \cdot \frac{s}{s^2+2s+5} + \frac{3}{s^2+2s+5} = \frac{3}{5} \cdot \frac{1}{s+2} + \frac{2}{5}\frac{s+1}{(s+1)^2+2^2} + \frac{13}{10}\frac{2}{(s+1)^2+2^2}$. The inverse Laplace transform yields the answer $y = \frac{3}{5}e^{-2t} + \frac{2}{5}e^{-t}\cos 2t + \frac{13}{10}e^{-t}\sin 2t$.

18. $y'' + 2y' - 3y = 5e^{2t}$, $y(0) = 2$, $y'(0) = 3$. The Laplace transform is $s^2Y - 2s - 3) + (2sY - 4) - 3Y = \frac{5}{s-2}$ or $(s^2 + 2s - 3)Y - (2s + 7) = \frac{5}{s-2}$ so $(s^2 + 2s - 3)Y = \frac{5}{s-2} + (2s + 7) = \frac{2s^2+3s-9}{s-2}$. This yields $Y = \frac{2s^2+3s-9}{(s-2)(s^2+2s-3)} = \frac{(2s-3)(s+3)}{(s-2)(s+3)(s-1)} = \frac{2s-3}{(s-2)(s+1)} = \frac{A}{s-2} + \frac{B}{s-1}$. Multiplying we have $2s - 3 = A(s-1) + B(s-2)$. Setting $s = 2$ we get $A = 1$. Setting $s = 1$, we get $B = 1$. The partial fraction decomposition of Y is $\frac{s}{s-2} + \frac{1}{s-1}$. The inverse Laplace transform yields the answer $y = e^{2t} + e^t$.

19. $y'' + 4 = \sin t$, $y(0) = y'(0) = 0$. The Laplace tranform is $s^2Y + \frac{4}{s} = \frac{1}{s^2+1}$. So $s^2Y = \frac{1}{s^2+1} - \frac{4}{s}$ and $Y = \frac{1}{(s^2+1)s^2} - \frac{4}{s^3}$. The inverse Laplace transform (by forms 11 and 3) is $y = t - \sin t - 2t^2$.

20. $y'' + 4 = t$, $y(0) = -1$, $y'(0) = 0$. The Laplace transform is $s^2Y + s + \frac{4}{s} = \frac{1}{s^2}$. So $s^2Y = \frac{1}{s^2} - \frac{4}{5} - s$ and

$Y = \frac{1}{s^4} - \frac{4}{s^3} - \frac{1}{s}$. The inverse Laplace transform is $y = \frac{t^3}{6} - 2t^2 - 1$.

21. We begin with the equation $\frac{1}{2}q'' + 10q' + 100q = 10\sin t$. The Laplace transform is $\frac{1}{2}(s^2Q - 10) + 10(sQ) + 100Q = \frac{10}{s^2+1}$ or $\left(\frac{1}{2}s^2 + 10s + 100\right)Q = \frac{10}{s^2+1} + 5 = \frac{5s^2+15}{s^2+1}$. Multiplying by 2 to clear fractions we have $(s^2 + 20s + 200)Q = \frac{10s^2+30}{s^2+1}$. Hence $Q = \frac{10s^2+30}{(s^2+1)(s^2+20s+200)} = \frac{As+B}{s^2+1} + \frac{Cs+D}{s^2+20s+200}$. Multiplying by the LCD we obtain the equation $10s^2 + 30 = (As+B)(s^2+20s+200) + (Cs+D)(s^2+1) = (A+C)s^3 + (20A+B+D)s^2 + (200A+20B+C)s + 200B + D$. Hence we have linear equations $A + C = 0$, $20A + B + D = 10$, $200A + 20B + C = 0$, and $200B + D = 30$. Solving this system, we obtain $A = \frac{-400}{40{,}001}$, $B = \frac{3980}{40{,}001}$, $C = \frac{400}{40{,}001}$, and $D = \frac{404{,}030}{40{,}001}$.

Substituting, we obtain $Q = \frac{10}{40{,}001}\left[\frac{-40s+398}{s^2+1} + \frac{40s+40{,}403}{s^2+20s+200}\right]$. Working each part separately, we have: $\frac{-40s+398}{s^2+1} = -40\frac{s}{s^2+1} + 398\frac{1}{s^2+1}$. The inverse Laplace transform yields $q = -40\cos t + 398\sin t$. For $\frac{40s+40{,}403}{s^2+20s+200}$, we complete the square in the denominator, getting $\frac{40s+40{,}403}{s^2+20s+200} = \frac{40s+40{,}403}{s^2+20s+100+100} = \frac{40s+40{,}403}{(s+10)^2+10^2} = 40\frac{s+10}{(s+10)^2+10^2} + \frac{40{,}003}{(s+10)^2+10^2} = 40\frac{s+10}{(s+10)^2+10^2} + \frac{40{,}003}{10}\cdot\frac{10}{(s+10)^2+10^2}$. The inverse Laplace transform yields $40e^{-10t}\cos 10t + \frac{40{,}003}{10}e^{-10t}\sin 10t$. Hence, we get that

$$q(t) = \frac{10}{40{,}001}\left(-40\cos t + 398\sin t + 40e^{-10t}\cos 10t + \frac{40{,}003}{10}e^{-10t}\sin 10t\right)$$

Thus, we have

$$\begin{aligned} i(t) &= q'(t) \\ &= \frac{10}{40{,}001}(40\sin t + 398\cos t - 400e^{-10t}\cos 10t \\ &\quad - 400e^{-10t}\sin 10t - 40{,}003e^{-10t}\sin 10t \\ &\quad + 40{,}003e^{-10t}\cos 10t) \\ &= \frac{10}{40{,}001}(40\sin t + 398\cos t + 39{,}603e^{-10t}\cos 10t \\ &\quad - 40{,}403e^{-10t}\sin 10t) \end{aligned}$$

The steady-state current is $\frac{10}{40{,}001}(40\sin t + 398\cos t)$.

22. The initial differential equation is $20x'' + 90x' + 700x = 5\sin t$. This is equivalent to $2x'' + 9x' + 70x = \frac{1}{2}\sin t$ with auxiliary equation $2m^2 + 9m + 70 = 0$. This has solutions $\frac{-9 \pm \sqrt{81-560}}{4} = -\frac{9}{4} \pm \frac{\sqrt{479}}{4}j$. Hence, the general complementary solution is $x_c = e^{-9t/4}\left(C_1\cos\frac{\sqrt{479}}{4}t + C_2\sin\frac{\sqrt{479}}{4}t\right)$. For x_p we set $x_p = A\sin t + B\cos t$. Thus, $x_p' = A\cos t - B\sin t$ and $x_p'' = -A\sin t - B\cos t$. Substituting these expressions into the equivalent equation above, we have $2(-A\sin t - B\cos t) + 9(A\cos t - B\sin t) + 70(A\sin t + B\cos t) = \frac{1}{2}\sin t$. Collecting like terms produces $-2A - 9B + 70A = \frac{1}{2}$ and $-2B + 9A + 70B = 0$. These equations are equivalent to $68A - 9B = \frac{1}{2}$ and $9A + 68B = 0$. Solving, we get $A = \frac{68}{9410} = \frac{34}{4705}$ and $B = -\frac{9}{9410}$. This produces $x_p = \frac{1}{9410}(68\sin t - 9\cos t)$. Hence, the general solution is $x = x_c + x_p = e^{-9t/4}\left(c_1\cos\frac{\sqrt{479}}{4}t + c_2\sin\frac{\sqrt{479}}{4}t\right) + \frac{1}{9410}(68\sin t - 9\cos t)$.

Now, $x(0) = 0$ yields $c_1 = \frac{9}{9410}$. We find that $x'(t) = -\frac{9}{4}e^{9t/4}\left(\frac{9}{9410}\cos\frac{\sqrt{479}}{4}t + c_2\sin\frac{\sqrt{479}}{4}t\right) + e^{-9t/4}\left(-\frac{9\sqrt{479}}{9410\cdot 4}\sin\frac{\sqrt{479}}{4}t + \frac{\sqrt{479}}{4}c_2\cos\frac{\sqrt{479}}{4}t\right) + \frac{1}{9410}(68\cos t + 9\sin t)$. Now, using the fact that $v(0) = -1$ or $x'(0) = -1$, (recall, down is positive), we obtain $-1 = -\frac{9}{4}\cdot\frac{9}{9410} + \frac{\sqrt{479}}{4}c_2 + \frac{68}{9410}$ and we obtain $c_2 \approx -0.183692$. The particular solution is

$$x(t) = e^{-9t/4}\left(\frac{9}{9410}\cos\frac{\sqrt{479}}{4}t - 0.1837\sin\frac{\sqrt{479}}{4}t\right) + \frac{1}{9410}(68\sin t - 9\cos t)$$
$$\approx e^{-2.25t}(0.0009564\cos 5.4715t - 0.1837\sin 5.4715t) + 0.007226\sin t - 0.0009564\cos t$$

CHAPTER 34 TEST

1. From Entry 5 in Table 34.1, we see that the Laplace transform of e^{at} is $\frac{1}{s-a}$. Here $a = -3$, so the Laplace transform is $5\left(\frac{1}{s+3}\right) = \frac{5}{s+3}$.

2. From Entry 3, we have the Laplace transform of $2t^3 = 6\left(\frac{t^3}{3}\right)$ as $6\left(\frac{1}{s^4}\right)$ and from Entry 8, the Laplace transform of $\sin 8t$ is $\frac{8}{s^2+8^2}$. Adding these together, we get the Laplace transform of $2t^3 + \sin 8t$ as $\frac{6}{s^4} + \frac{8}{s^2+64}$.

3. From Entry 20, we have $a = -5$ and $b = 4$, so the Laplace transform is $\frac{s-5}{(s-5)^2+4^2}$.

4. From Entry 7, we have $n - 1 = 9$ and $a = 5$, so the Laplace transform is $8\left(\frac{9}{(s-5)^{10}}\right) = \frac{72}{(s-5)^{10}}$.

5. We see that this is of the form in Entry 19 of Table 34.1 with $a = 6$ and $b = \sqrt{7}$. So, $L^{-1}(F) = e^{-6t}\sin\sqrt{7}t$.

6. Completing the square of the denominator we get $F(s) = \frac{s}{(s-6s+9)+4} = \frac{s}{(s-3)^2+2^2}$. This does not satisfy any of the entries in Table 34.1, but it is close to Entry 20. If we rewrite $F(s)$ by adding $0 = -3 + 3$ to the numerator and regrouping, we obtain $F(s) = \frac{s-3+3}{(s-3)^2+2^2} = \frac{s-3}{(s-3)^2+2^2} + \frac{3}{(s-3)^2+2^2} = \frac{s-3}{(s-3)^2+2^2} + \frac{3}{2}\left(\frac{2}{(s-3)^2+2^2}\right)$. Applying Entries 20 and 19 of Table 34.1, respectively, we get $L^{-1}(F) = e^{-3t}\cos 2t + \frac{3}{2}e^{-3t}\sin 2t$.

7. Using partial fraction decomposition we have $\frac{5}{(s+3)(s^2+1)} = \frac{A}{s+3} + \frac{Bs+C}{s^2+1}$. Multiplying by $(s+3)(s^2+1)$ produces $5 = A(s^2+1) + (Bs+C)(s+3) = As^2 + A + Bs^2 + Cs + 3Bs + 3C = (A+B)s^2 + (3B+C)s + (A+3C)$. Solving the simultaneous equations $A + B = 0$, $3B + C = 0$, and $A + 3C = 5$ we obtain $A = 0.5$, $B = -0.5$, and $C = 1.5$. So, we can rewrite $F(s)$ as $\frac{0.5}{s+3} + \frac{-0.5s+1.5}{s^2+1} = \frac{0.5}{s+3} - \frac{0.5s}{s^2+1} + \frac{1.5}{s^2+1}$. From Entries 5, 9, and 8 of Table 34.1, respectively, we obtain $L^{-1}(F) = 0.5e^{-3t} - 0.5\cos t + 1.5\sin t$.

8. Taking Laplace transforms of both sides produces $L(y') + 5L(y) = L(0)$. Since $L(y') = sY - y(0) = sY - 2$, this equation becomes $sY - 2 + 5Y = 0$ or $Y(s+5) = 2$ and so $Y = \frac{2}{s+5}$. Taking the inverse Laplace transform of both sides we obtain $y = L^{-1}\left(\frac{2}{s+5}\right) = 2e^{-5t}$.

9. Taking the Laplace transform of the given equation produces $L(y'') + 2L(y) = L(e^{-2t})$. Since

$L(y'') = s^2Y - sy(0) - y'(0)$, $y(0) = 0$ and $y'(0) = 0$, we can rewrite this as $s^2Y - s + 2Y = \frac{1}{s+2}$ or $(s^2+2)Y = s + \frac{1}{s+2}$ and so $Y = \frac{s}{s^2+2} + \frac{1}{(s^2+2)(s+2)}$. Using partial fractions, we determine that $\frac{1}{(s^2+2)(s+2)} = \frac{1/6}{s+2} + \frac{-s/6+1/3}{s^2+2} = \frac{1}{6}\left(\frac{1}{s+2}\right) - \frac{1}{6}\left(\frac{s}{s^2+2}\right) + \frac{1}{3}\left(\frac{1}{s^2+2}\right)$. Combining this partial fraction with the first $\frac{s}{s^2+2}$, we obtain $Y = \frac{1}{2}\left(\frac{1}{s+2}\right) + \frac{5}{6}\left(\frac{s}{s^2+2}\right) + \frac{1}{3}\left(\frac{1}{s^2+2}\right)$. So, $y = \frac{1}{6}L^{-1}\left(\frac{1}{s+2}\right) + \frac{5}{6}L^{-1}\left(\frac{s}{s^2+2}\right) + \frac{1}{3}L^{-1}\left(\frac{1}{s^2+2}\right) = \frac{1}{6}L^{-1}\left(\frac{1}{s+2}\right) + \frac{5}{6}L^{-1}\left(\frac{s}{s^2+2}\right) + \frac{1}{3\sqrt{2}}L^{-1}\left(\frac{\sqrt{2}}{s^2+2}\right)$. Using Entries 5, 9, and 8 in Table 34.1 we determine that $y = \frac{1}{6}e^{-2t} + \frac{5}{6}\cos\sqrt{2}t + \frac{1}{3\sqrt{2}}\sin\sqrt{2}t$.

10. Taking the Laplace transform of the given equation produces $L(y'') + 2L(y') + L(y) = 0$. Since $L(y') = sY - y(0) = sY^{-1}$ and $L(y'') = s^2Y - sy(0) - y'(0) = s^2Y - s + 1$, we can rewrite this as $(s^2Y - s + 1) + 2sY + Y = 0$ or $(s^2 + 2s + 1)Y = s + 1$ and so $Y = \frac{s+1}{s^2+2s+1} = \frac{1}{(s+1)}$ and $y = L^{-1}(Y) = L^{-1}\left(\frac{1}{(s+1)}\right)$. By form 5, we get $y = e^{-t}$.

11. Using the formula $L\frac{d^2q}{dt^2} + R\frac{dq}{dt} + \frac{q}{C} = E(t)$ we have $0.2q'' + 8q' + \frac{q}{0.0025} = 0.2q'' + 8q' + 400q = 12\sin 20t$. Taking the Laplace transform of this equation, we get $0.2L(q'') + 8L(q') + 400L(q) = L(12\sin 20t)$. Since $L(q') = sQ - q(0) = sQ$, $L(q'') = s^2Q - sq(0) + q'(0) = s^2Q$, and $L(12\sin 20t) = 12\left(\frac{20}{s^2+20^2}\right) = \frac{240}{s^2+400}$, we can rewrite our equation as $0.2s^2Q + 8sQ + 400 = \frac{240}{s^2+400}$ or $(0.2s^2+8s)Q = -400 + \frac{240}{s^2+400}$ and so $Q = \frac{-400}{0.2s^2+8s} + \frac{240}{(0.2s^2+8s)(s^2+20^2)} = \frac{-2000}{s^2+40s} + \frac{1200}{(s^2+40s)(s^2+20^2)}$. Using partial fraction decomposition, we can rewrite this as $Q = \left(\frac{-50}{s} + \frac{50}{s+40}\right) + \left(\frac{0.075}{s} + \frac{-0.015}{s+40} - \frac{0.06s+0.6}{s^2+400}\right) = \frac{-49.925}{s} + \frac{49.985}{s+40} - \frac{0.06s}{s^2+400} - \frac{0.6}{s^2+400}$. Taking the inverse Laplace transform, we obtain $q = L^{-1}(Q) = -49.925 + 49.985e^{-40t} - 0.06\cos 20t - 0.03\sin 20t$.

12. Using Euler's method where $y_{n+1} = y_n + f(x_n, y_n)h$ we have $f(x,y) = 2x + y$. Thus we get the formula $y_{n+1} = y_n + 0.2(2x_n + y_n)$. We are given $x_0 = 0$, $y_0 = 2$, and $h = 0.2$. Using this data and applying the results successively, we obtain the following:

x	y $h = 0.2$
0	2
0.2	2.4
0.4	2.96
0.6	3.712
0.8	4.6944
1.0	5.95328

APPENDIX

A Solutions for Additional Computer Exercises

Technical Mathematics contains 32 exercises which ask students to write a computer program and *Technical Mathematics with Calculus* contains and additional 6 such exercises. This appendix consists of programs written in GWBASIC that serve as answers to each of those exercises. Just as the accompanying textbook increases in sophistication and complexity as you progress from beginning to end, the computer programs in this Program Book also increase in sophistication and complexity.

The name of each program corresponds to its location in the text. Each name consists of the three letters PRG followed immediately by five digits. The first two digits represent the chapter in the text. Thus, Chapter 9 is 09 and chapter 25 is 25. The next (third) digit denotes the section in the chapter; and the last two digits, the exercise in that section where the program was assigned.

Using the above description, the program for Chapter 9, Section 5, Exercise 43, has the name PRG09543. Similarly, the computer program for Chapter 17, Section 3, Exercise 22, is named PRG17322.

* * * * *

You may wish to write your own versions of these program. If you do, then use the following programs as guidelines. In some cases, we have added some additional steps to the programs to make them easier to use and more "user friendly." You may have some better ideas on how this might be done.

However, it is important that anyone programming a computer remember that a computer must be told what to do and how it is done. If the directions are wrong, the computer will give a wrong answer.

There are several special commands, statements, and functions that are used in BASIC. In fact, most of these, with some variation such as SQRT rather than SQR, are used in other computer programming languages. This is not intended as a programming course, and so we assume that you are familiar with BASIC.

Commands, such as LOAD, RUN, LIST, and SAVE are not part of a program, but allow you to tell the computer to do something with a program.

Program *statements* used in these programs include PRINT, PRINT USING, INPUT, REM, GOTO (or GO TO), DEF FN, DIM, FOR ...NEXT, IF ...THEN, GOSUB ...RETURN, CLS, CLEAR, and END.

Functions used in these programs include ABS, ATN, COS, EXP, INT, LEFT$, LOG, SGN, SIN, GQR, TAB, and TAN. You should remember that a computer evaluates trigonometric functions expressed in radians and not degrees. You will need to include conversion factors in any programs where you will to enter, or have the computer print, angles given in degrees.

You may also have noticed that BASIC includes only the sine, cosine, tangent, and arctangent functions. If your programs include any other trigonometric formulas, you will have to define them in that particular program. Naturally, since these are programs involving mathematics, you will need to use the symbols for the mathematical *operations*. The computer uses +, −, ∗, and / for addition, subtraction, multiplication, and division, respectively. For exponentiation, these BASIC programs use ^. (Some computers and some languages use either ∗∗ or ↑.)

* * * * *

Not everyone wishes to take the time to write a computer program. Thus, each of these programs is available on an IBM formatted disk in either $5\frac{1}{4}''$ or $3\frac{1}{2}''$ sizes.

At the end of these program listings is an exercise set of 30 exercises. Some of the exercises ask you to modify a program so that it will do a better job of performing the desired task or so that it can be used in some additional situations. The remaining exercises ask you to use one of the programs with a given set of numbers or equations.

A solution to each of these additional exercises in given in the *Instructor's Solutions Guide*; solutions to the odd numbered exercises are in the *Student's Solutions Guide*. Whenever possible, each solution contains a copy of the exact wording and format used when the computer printed each solution.

1. If you select $x = -1.9$, you get $x = -1.90625$, $y = 1.160156$. If you select $y = 1.16$, you get $x = -1.90625$, $y = 1.1625$.

2. (a) Using 1.414 for $\sqrt{2}$ and 0.3780 for $\dfrac{1}{\sqrt{7}}$ produces an answer of 1.463, (b) If you let $\pi = 3.1416$ and $-\sqrt{19} = -4.3590$, you get the result -13.68424.

3. (a) -12.000000, (b) If you let $\pi = 3.1416$, $\frac{4}{3} = 1.3333$, $-\frac{17}{6} = -2.8333$, $1 + \sqrt{2} = 2.4142$, and $\sqrt{11} = 3.3166$, you get the result 13.879020.

4. In order to avoid any places in PRG05517 that directs the computer to go to lines 440 or 450, change lines 440 and 450 to

```
440 '
450 '
```

Add the following lines to program PRG05517:

```
481 IF ABS(DET) >= .00001
        THEN GOTO 490
482 IF ABS(DX) < .00001 AND ABS(DY)
    < .00001 AND ABS(DZ) < .00001
    THEN PRINT "THIS SYSTEM IS
    DEPENDENT.": GOTO 520
483 PRINT "THIS SYSTEM IS
    INCONSISTENT.": GOTO 520
```

5. (a) `THIS SYSTEM IS DEPENDENT.`

   (b) `THE SOLUTIONS ARE: X = 0, Y = 0, Z = 0.`

   (c) `THIS SYSTEM IS INCONSISTENT.`

6. Replace line 70 and add new lines 72–78, as shown below

```
70 Y = SQR(ABS(D))
72 X1 = -B/(2*A)
74 X2 = Y/(2*A)
76 PRINT:PRINT"THE ROOTS ARE NOT REAL
   NUMBERS. THE ROOTS ARE: ";X1;" + ";
   X2"J AND "X1;" - ";X2"J"
78 GOTO 140
```

7. (a) `THE ROOTS ARE: 1.75 AND .5`

   (b) `THE ROOTS ARE: 3.651635 AND -6.480089`

   (c) `THE ROOTS ARE BOTH THE SAME: 2.3 AND 2.3`

   (d) `THE ROOTS ARE NOT REAL NUMBERS. THE ROOTS ARE: 1.109375 + .6523321J AND 1.109375 - .6523321J`

8. In PRG09236, change line 70 to `FOR X = 0 TO 3.2 STEP .1`. (You may also want to change lines 30 and 50 to read "from 0 to 3.2 …".)

   In PRG09237, change line 70 to `FOR I = 1 TO 180`. (You may also want to change line 30 and 50 to read "from 0 to 180 …".)

9. (a) THE HORIZONTAL VECTOR HAS
MAGNITUDE: 5.17638
AND DIRECTION: 0 RADIANS
OR 0 DEGREES
THE VERTICAL VECTOR HAS
MAGNITUDE: 19.31852
AND DIRECTION 1.570796 RADIANS
OR 90 DEGREES.

(b) THE HORIZONTAL VECTOR HAS
MAGNITUDE: 13.56877
AND DIRECTION: 3.141593 RADIANS
OR 180 DEGREES
THE VERTICAL VECTOR HAS
MAGNITUDE: 8.478708
AND DIRECTION 4.712389 RADIANS
OR 270 DEGREES.

(c) THE HORIZONTAL VECTOR HAS
MAGNITUDE: 4.687785
AND DIRECTION: 0 RADIANS
OR 0 DEGREES
THE VERTICAL VECTOR HAS
MAGNITUDE: 17.79283
AND DIRECTION 4.712389 RADIANS
OR 270 DEGREES.

(d) THE HORIZONTAL VECTOR HAS
MAGNITUDE: 14.32789
AND DIRECTION: 3.141593 RADIANS
OR 180 DEGREES
THE VERTICAL VECTOR HAS
MAGNITUDE: 18.8786
AND DIRECTION 1.570796 RADIANS
OR 90 DEGREES.

10. (a) THE RESULTANT VECTOR HAS
MAGNITUDE = 222.6022
AND DIRECTION = .9193124 RADIANS
OR 52.67272 DEGREES

(b) THE RESULTANT VECTOR HAS
MAGNITUDE = 41.06048
AND DIRECTION = 1.790467 RADIANS
OR 102.5862 DEGREES

(c) THE RESULTANT VECTOR HAS
MAGNITUDE = 58.91126
AND DIRECTION = 1.234034 RADIANS
OR 70.70492 DEGREES

11. (a) −4.96 + 2.25 J,

(b) −5.709999 + .8699999 J

12. (a) THE PRODUCT IS:
-198.1568 + 165.6024 J

(b) THE PRODUCT IS:
.0744 + 2.2719 J

13. (a) THE QUOTIENT IS:
-.0450621 + -.7665063 J

(b) THE QUOTIENT IS:
.1509434 + .5283019 J

14. (a) THE NUMBER IS
-12.4 + 3.7 J
OR
12.94025 CIS 2.851617 RADIANS
12.94025 CIS 163.3856 DEGREES

(b) THE NUMBER IS
-1.641071 + -3.86612 J
OR
4.2 CIS 4.310963 RADIANS
4.2 CIS 247 DEGREES

(c) THE NUMBER IS
4.562753 + -10.28318 J
OR
11.25 CIS 5.13 RADIANS
11.25 CIS 293.9274 DEGREES

15. (a) THE TOTAL IS -198.1568 + 165.6024 J
OR
258.2446 CIS 2.445451 RADIANS
258.2446 CIS 140.114 DEGREES

(b) THE TOTAL IS
18.89179 + -10.47188 J
OR
21.6 CIS
5.77704 RADIANS
21.6 CIS 331 DEGREES

(c) THE TOTAL IS
-17.64838 + -20.43367 J
OR
27 CIS 4 RADIANS
27 CIS 229.1831 DEGREES

16. (a)
```
THE TOTAL IS
-.0450621  $+ -$.7665063 J
OR
.7678297 CIS  1.512075  RADIANS
.7678297  CIS  86.63551  DEGREES
```
(b)
```
THE TOTAL IS
.9270507  $+ -$2.853169 J
OR
3 CIS  5.026549  RADIANS
3  CIS  288  DEGREES
```
(c)
```
THE TOTAL IS
-3.413423  + -2.085316 J
OR
4 CIS  3.69  RADIANS
4  CIS  211.4214  DEGREES
```

17. (a)
```
THE TOTAL IS
-28.31983 + -153.408 J
OR
156.0001  CIS  4.52984  RADIANS
156.0001  CIS  259.5407  DEGREES
```
(b)
```
THE FIRST ROOT IS
1.271921  $+ -$.2530007 J
OR
1.29684  CIS  6.086836  RADIANS
1.29684  CIS  348.75  DEGREES
THE 2ND ROOT IS
.2530007  +  1.271921 J
 OR
1.29684  CIS  1.374447  RADIANS
1.29684  CIS  78.75  DEGREES
THE 3RD ROOT IS
-1.271921  +  .2530008 J
OR
1.29684  CIS  2.945243  RADIANS
1.29684  CIS  168.75  DEGREES
THE  4 TH ROOT IS
-.253001  + -1.271921 J
OR
1.29684  CIS  4.51604  RADIANS
1.29684  CIS  258.75  DEGREES
```

18. (a)
```
THE TOTAL IMPEDANCE FOR A
SERIES CIRCUIT IS
3  + -2 J,
3.605551  CIS -.5880026  RADI-
ANS OR
3.605551 CIS -33.69007  DEGREES.
AND FOR A
PARALLEL CIRCUIT IS
5  +  .9999999 J,
5.09902 CIS  .1973955 RADIANS OR
5.09902 CIS  11.30993  DEGREES.
```
(b)
```
THE TOTAL IMPEDANCE FOR A
SERIES CIRCUIT IS
10  + -5 J,
11.18034  CIS -.4636476  RADI-
ANS OR
11.18034 CIS -26.56505  DEGREES.
AND FOR A
PARALLEL CIRCUIT IS
-5.692307  +  .5384617 J,
5.717718 CIS -9.431399E-02 RADI-
ANS OR
5.717718 CIS -5.403794  DEGREES.
```
(c)
```
THE TOTAL IMPEDANCE FOR A
SERIES CIRCUIT IS
4.069313  +  5.451568 J,
6.80286  CIS  .9295722  RADI-
ANS OR
 6.80286 CIS  53.26057  DEGREES.
AND FOR A
PARALLEL CIRCUIT IS
1.401921  +  .8910406 J,
1.661125 CIS  .5661751 RADIANS OR
1.661125 CIS  32.43944  DEGREES.
```
(d)
```
THE TOTAL IMPEDANCE FOR A
SERIES CIRCUIT IS
2.799219  +  2.23625 J,
3.582798  CIS  .6740596  RADI-
ANS OR
3.582798 CIS  38.62077  DEGREES.
AND FOR A
PARALLEL CIRCUIT IS
.6744572  +  .9550899 J,
1.169226 CIS  .9559405 RADIANS OR
1.169226 CIS  54.77136  DEGREES.
```

19. (a)
```
THE ANSWER IS
 7.7000      -2.0000
-2.9000      10.0000
11.1000       6.0000
```

(b)
```
THE ANSWER IS
2.4000        -1.3000        3.5000
5.7000         8.2000       -2.0000
```

20.
```
THE ANSWER IS
  8.8500        9.0000
-14.7000        7.5000
 13.3000       13.0000
```

21.
```
THE ANSWER IS
 42.8600      94.1100    -206.0900
108.8600     -23.3600     -64.0500
```

22. (a)
```
THE INVERSE MATRIX IS
0.1818   -0.6182
0.9091    -2.4242
```

(b)
```
THIS MATRIX IS SINGULAR. IT DOES
NOT HAVE AN INVERSE.
```

(c)
```
THE INVERSE MATRIX IS
 0.4484  0.3139 -0.3049  0.1928
-1.9058  1.1659 -0.7040 -0.0695
-4.4843  1.8610  0.0493  0.0717
 2.9821 -1.4126  0.3722  0.1323
```

23. (a)
```
THE MATRIX IS SINGULAR. THIS SYS-
TEM HAS NO SOLUTION.
```

(b)
```
THE INVERTED MATRIX IS
-.067         -.333          .267
-.067          .667         -.233
 .333         -.333          .167
THE ANSWER IS
x( 1 ) =  2
x( 2 ) = -3
x( 3 ) =  3
```

24.
```
LEFT ENDPOINT = ? 0
RIGHT ENDPOINT = ? 1
THE ROOT IS APPROXIMATELY .6850733
THE ACTUAL ROOT IS BETWEEN .6850733 AND 1

THE ROOT IS APPROXIMATELY .7827885
THE ACTUAL ROOT IS BETWEEN .7827885 AND 1
THE ROOT IS APPROXIMATELY .7982691
THE ACTUAL ROOT IS BETWEEN .7982691 AND 1

THE ROOT IS APPROXIMATELY .8006438
THE ACTUAL ROOT IS BETWEEN .8006438 AND 1
INTERPOLATION COMPLETED.
THE ROOT IS APPROXIMATELY .8006438
AND F(C) =  8.363128E-04
```

25.
```
THE TOTAL NUMBER IN THE SAMPLE IS:  150
THE MEAN IS:  2.833333
THE VARIANCE IS:
1.938889
AND THE STANDARD DEVIATION IS:  1.39244
THE NUMBER OF SCORES WITHIN ONE STAN-
DARD DEVIATION OF THE MEAN IS:  109  OR
72.66666  PERCENT OF THE TOTAL VALUE.

AND THE NUMBER WITHIN TWO S.D. OF THE
MEAN IS:  144  OR  96 PERCENT.
```

26.
```
THE FOLLOWING ARE THE TOTALS

M =  1.886307
B =  37.4829
R =  .8217226
```

27.
```
ENTER THE LEFT END POINT, A=? 0

NEXT ENTER THE RIGHT POINT,
B =? 3.1416

AND FINALLY, THE NUMBER OF SEGMENTS,
N= ? 50
A              B              N
0              3.1416         50

THE AREA IS  3.915187
```

28.
```
I      X(I)         F(X(I))        F'(X(I))
1    .75          9.592449E-02   -2.268887
2    .7922783     1.709938E-02   -2.282505
3    .7997697     2.546013E-03   -2.283336
4    .8008848     3.644228E-04   -2.283416
5    .8010444     5.185604E-05   -2.283427
6    .8010671     7.331372E-06   -2.283429
7    .8010703     1.013279E-06   -2.283429
8    .8010707     2.384186E-07   -2.283429

   THE ROOT IS APPROXIMATELY  .8010707
```

29.
```
THE AREA UNDER F(X) OVER THE INTERVAL
FROM  0  TO  3.1416  IS
```

```
3.91427 BY SIMPSON'S RULE AND
3.912377 BY THE TRAPEZOIDAL RULE
```

30. (a)
```
CYLINDRICAL COORDINATES OF THE
RECTANGULAR POINT (-2 ,  5 , -
4.5) ARE
R =  5.385165
THETA = 1.951303 = .621119 *PI
Z = -4.5

THE SPHERICAL COORDINATES OF THE
RECTANGULAR POINT (-2 ,  5 , -
4.5 ) ARE
RHO =  7.017834
THETA = 1.951303 = .621119 *PI
PHI = 2.266888 = .7215729  *PI
```

(b)
```
THE RECTANGULAR COORDINATES OF THE
CYLINDRICAL COORDINATE POINT
( 3,  1 * PI/6, -2 ) ARE
X =  2.598076
Y =  1.5
Z = -2
THE SPHERICAL COORDINATES OF THE
CYLINDRICAL COORDINATE POINT
( 3,  1 * PI/6, -2 ) ARE
RHO = 3.605551
THETA = .5235988 = .1666667  * PI
PHI = 2.158799 = .6871671  * PI
```

(c)
```
THE RECTANGULAR COORDINATES OF THE
SPHERICAL COORDINATE POINT
( 5,  1 * PI/6, 1 * PI/4 ) ARE
X=  3.061863
Y=  1.767767
Z=  3.535534
THE CYLINDRICAL COORDINATES OF THE
SPHERICAL COORDINATE POINT
( 5,  1 * PI/6, 1 * PI/4 ) ARE
R =  3.535534
THETA = .5235988 = .1666667  * PI
Z =  3.535534
```

Valdosta Technical Institute Library
PO Box 928/4089 Val Tech Road
Valdosta, Georgia 31603-0928